JPC	J. Phys. Chem.	PIA	Proc. Indian Acad. Sci.
JPR	J. Prakt. Chem.	PIA(A)	Proc. Indian Acad. Sci., Sect. A
JPS	J. Pharm. Sci.	PMH	Phys. Methods Heterocycl. Chem.
JSP	J. Mol. Spectrosc.	PNA	Proc. Natl. Acad. Sci. USA
JST	J. Mol. Struct.	PS	Phosphorus Sulfur
K	Kristallografiya	QR	Q. Rev., Chem. Soc.
KGS	Khim. Geterotsikl. Soedin.	RCR	Russ. Chem. Rev. (Engl. Transl.)
LA	Liebigs Ann. Chem.	RRC	Rev. Roum. Chim.
M	Monatsh. Chem.	RTC	Recl. Trav. Chim. Pays-Bas
MI	Miscellaneous [book or journal]	S	Synthesis
MIP	Miscellaneous Pat.	SA	Spectrochim. Acta
MS	Q. N. Porter and J. Baldas, 'Mass Spectrometry of Heterocyclic Compounds', Wiley, New York, 1971	SA(A)	Spectrochim. Acta, Part A
		SAP	S. Afr. Pat.
		SC	Synth. Commun.
		SH	W. L. F. Armarego, 'Stereochemistry of Heterocyclic Compounds', Wiley, New York, 1977, parts 1 and 2
N	Naturwissenschaften		
NEP	Neth. Pat.		
NJC	Nouv. J. Chim.		
NKK	Nippon Kagaku Kaishi	SST	Org. Compd. Sulphur, Selenium, Tellurium [R. Soc. Chem. series]
NMR	T. J. Batterham, 'NMR Spectra of Simple Heterocycles', Wiley, New York, 1973		
		T	Tetrahedron
		TH	Thesis
OMR	Org. Magn. Reson.	TL	Tetrahedron Lett.
OMS	Org. Mass Spectrom.	UKZ	Ukr. Khim. Zh. (Russ. Ed.)
OPP	Org. Prep. Proced. Int.	UP	Unpublished Results
OR	Org. React.	USP	U.S. Pat.
OS	Org. Synth.	YZ	Yakugaku Zasshi
OSC	Org. Synth., Coll. Vol.	ZC	Z. Chem.
P	Phytochemistry	ZN	Z. Naturforsch.
PAC	Pure Appl. Chem.	ZN(B)	Z. Naturforsch., Teil B
PC	Personal Communication	ZOB	Zh. Obshch. Khim.
PH	'Photochemistry of Heterocyclic Compounds', ed. O. Buchardt, Wiley, New York, 1976	ZOR	Zh. Org. Khim.
		ZPC	Hoppe-Seyler's Z. Physiol. Chem.

COMPREHENSIVE HETEROCYCLIC CHEMISTRY
IN 8 VOLUMES

COMPREHENSIVE HETEROCYCLIC CHEMISTRY

The Structure, Reactions, Synthesis and Uses of Heterocyclic Compounds

Volume 5

Chairman of the Editorial Board
ALAN R. KATRITZKY, FRS
University of Florida

Co-Chairman of the Editorial Board
CHARLES W. REES, FRS
*Imperial College of Science and Technology
University of London*

Part 4A
Five-membered Rings with Two or More Nitrogen Atoms

EDITOR
KEVIN T. POTTS
Rensselaer Polytechnic Institute

PERGAMON PRESS
OXFORD · NEW YORK · TORONTO · SYDNEY · PARIS · FRANKFURT

U.K.	Pergamon Press Ltd., Headington Hill Hall, Oxford OX3 0BW, England
U.S.A.	Pergamon Press Inc., Maxwell House, Fairview Park, Elmsford, New York 10523, U.S.A.
CANADA	Pergamon Press Canada Ltd., Suite 104, 150 Consumers Road, Willowdale, Ontario M2J 1P9, Canada
AUSTRALIA	Pergamon Press (Aust.) Pty. Ltd., P.O. Box 544, Potts Point, N.S.W. 2011, Australia
FRANCE	Pergamon Press SARL, 24 rue des Ecoles, 75240 Paris, Cedex 05, France
FEDERAL REPUBLIC OF GERMANY	Pergamon Press GmbH, Hammerweg 6, D-6242 Kronberg-Taunus, Federal Republic of Germany

Copyright © 1984 Pergamon Press Ltd.

All rights reserved. No part of this publication may be reproduced, stored in a retrieval system or transmitted in any form or by any means: electronic, electrostatic, magnetic tape, mechanical, photocopying, recording or otherwise, without permission in writing from the publishers

First edition 1984

Library of Congress Cataloging in Publication Data

Main entry under title:

Comprehensive heterocyclic chemistry.

Includes indexes.
Contents: v. 1. Introduction, nomenclature, literature, biological aspects, industrial uses, less-common heteroatoms –
v. 2. Six-membered rings with one nitrogen atom – [etc.] –
v. 8. Indexes.
1. Heterocyclic compounds.   I. Katritzky, Alan R. (Alan Roy)
II. Rees, Charles W. (Charles Wayne)
QD400.C65   1984   547'.59   83-4264

British Library Cataloguing in Publication Data

Comprehensive heterocyclic chemistry
1. Heterocyclic compounds.
I. Katritzky, Alan R.   II. Rees, Charles W.
547'.59   QD400
ISBN 0-08-030705-1 (vol. 5)
ISBN 0-08-026200-7 (set)

Typeset by J. W. Arrowsmith Ltd., Bristol
Printed in Great Britain by A. Wheaton & Co. Ltd., Exeter

Contents

	Foreword	vii
	Contributors to Volume 5	ix
	Contents of All Volumes	xi
4.01	Structure of Five-membered Rings with Two or More Heteroatoms A. R. KATRITZKY, *University of Florida*, and J. M. LAGOWSKI, *University of Texas*	1
4.02	Reactivity of Five-membered Rings with Two or More Heteroatoms A. R. KATRITZKY, *University of Florida*, and J. M. LAGOWSKI, *University of Texas at Austin*	39
4.03	Synthesis of Five-membered Rings with Two or More Heteroatoms K. T. POTTS, *Rensselaer Polytechnic Institute, New York*	111
4.04	Pyrazoles and their Benzo Derivatives J. ELGUERO, *Instituto de Quimica Medica, Madrid*	167
4.05	Pyrazoles with Fused Six-membered Heterocyclic Rings J. V. GREENHILL, *University of Bradford*	305
4.06	Imidazoles and their Benzo Derivatives: (i) Structure M. R. GRIMMETT, *University of Otago*	345
4.07	Imidazoles and their Benzo Derivatives: (ii) Reactivity M. R. GRIMMETT, *University of Otago*	373
4.08	Imidazoles and their Benzo Derivatives: (iii) Synthesis and Applications M. R. GRIMMETT, *University of Otago*	457
4.09	Purines G. SHAW, *University of Bradford*	499
4.10	Other Imidazoles with Fused Six-membered Rings J. A. MONTGOMERY and J. A. SECRIST III, *Southern Research Institute, Birmingham, Alabama*	607
4.11	1,2,3-Triazoles and their Benzo Derivatives H. WAMHOFF, *Universität Bonn*	669
4.12	1,2,4-Triazoles J. B. POLYA, *University of Tasmania*	733
4.13	Tetrazoles R. N. BUTLER, *University College, Galway*	791

4.14 Pentazoles   839
      I. UGI, *Technische Universität, München*

4.15 Triazoles and Tetrazoles with Fused Six-membered Rings   847
      S. W. SCHNELLER, *University of South Florida*

      References   905

Foreword

Scope

Heterocyclic compounds are those which have a cyclic structure with two, or more, different kinds of atom in the ring. This work is devoted to organic heterocyclic compounds in which at least one of the ring atoms is carbon, the others being considered the heteroatoms; carbon is still by far the most common ring atom in heterocyclic compounds. As the number and variety of heteroatoms in the ring increase there is a steady transition to the expanding domain of inorganic heterocyclic systems. Since the ring can be of any size, from three-membered upwards, and since the heteroatoms can be drawn in almost any combination from a large number of the elements (though nitrogen, oxygen and sulfur are the most common), the number of possible heterocyclic systems is almost limitless. An enormous number of heterocyclic compounds is known and this number is increasing very rapidly. The literature of the subject is correspondingly vast and of the three major divisions of organic chemistry, aliphatic, carbocyclic and heterocyclic, the last is much the biggest. Over six million compounds are recorded in *Chemical Abstracts* and approximately half of these are heterocyclic.

Significance

Heterocyclic compounds are very widely distributed in Nature and are essential to life; they play a vital role in the metabolism of all living cells. Thus, for example, the following are heterocyclic compounds: the pyrimidine and purine bases of the genetic material DNA; the essential amino acids proline, histidine and tryptophan; the vitamins and coenzyme precursors thiamine, riboflavine, pyridoxine, folic acid and biotin; the B_{12} and E families of vitamin; the photosynthesizing pigment chlorophyll; the oxygen transporting pigment hemoglobin, and its breakdown products the bile pigments; the hormones kinetin, heteroauxin, serotonin and histamine; together with most of the sugars. There are a vast number of pharmacologically active heterocyclic compounds, many of which are in regular clinical use. Some of these are natural products, for example antibiotics such as penicillin and cephalosporin, alkaloids such as vinblastine, ellipticine, morphine and reserpine, and cardiac glycosides such as those of digitalis. However, the large majority are synthetic heterocyclics which have found widespread use, for example as anticancer agents, analeptics, analgesics, hypnotics and vasopressor modifiers, and as pesticides, insecticides, weedkillers and rodenticides.

There is also a large number of synthetic heterocyclic compounds with other important practical applications, as dyestuffs, copolymers, solvents, photographic sensitizers and developers, as antioxidants and vulcanization accelerators in the rubber industry, and many are valuable intermediates in synthesis.

The successful application of heterocyclic compounds in these and many other ways, and their appeal as materials in applied chemistry and in more fundamental and theoretical studies, stems from their very complexity; this ensures a virtually limitless series of structurally novel compounds with a wide range of physical, chemical and biological properties, spanning a broad spectrum of reactivity and stability. Another consequence of their varied chemical reactivity, including the possible destruction of the heterocyclic ring, is their increasing use in the synthesis of specifically functionalized non-heterocyclic structures.

Aims of the Present Work

All of the above aspects of heterocyclic chemistry are mirrored in the contents of the present work. The scale, scope and complexity of the subject, already referred to, with its correspondingly complex system of nomenclature, can make it somewhat daunting initially. One of the main aims of the present work is to minimize this problem by presenting a comprehensive account of fundamental heterocyclic chemistry, with the emphasis on basic principles and, as far as possible, on unifying correlations in the properties, chemistry and synthesis of different heterocyclic systems and the analogous carbocyclic structures. The motivation for this effort was the outstanding biological, practical and theoretical importance of heterocyclic chemistry, and the absence of an appropriate major modern treatise.

At the introductory level there are several good textbooks on heterocyclic chemistry, though the subject is scantily treated in most general textbooks of organic chemistry. At the specialist, research level there are two established ongoing series, 'Advances in Heterocyclic Chemistry' edited by Katritzky and 'The Chemistry of Heterocyclic Compounds' edited by Weissberger and Taylor, devoted to a very detailed consideration of all aspects of heterocyclic compounds, which together comprise some 100 volumes. The present work is designed to fill the gap between these two levels, *i.e.* to give an up-to-date overview of the subject as a whole (particularly in the General Chapters) appropriate to the needs of teachers and students and others with a general interest in the subject and its applications, and to provide enough detailed information (particularly in the Monograph Chapters) to answer specific questions, to demonstrate exactly what is known or not known on a given topic, and to direct attention to more detailed reviews and to the original literature. Mainly because of the extensive practical uses of heterocyclic compounds, a large and valuable review literature on all aspects of the subject has grown up over the last few decades. References to all of these reviews are now immediately available: reviews dealing with a specific ring system are reported in the appropriate monograph chapters; reviews dealing with any aspect of heterocyclic chemistry which spans more than one ring system are collected together in a logical, readily accessible manner in Chapter 1.03.

The approach and treatment throughout this work is as ordered and uniform as possible, based on a carefully prearranged plan. This plan, which contains several novel features, is described in detail in the Introduction (Chapter 1.01).

ALAN R. KATRITZKY
Florida

CHARLES W. REES
London

Contributors to Volume 5

Professor R. N. Butler
Department of Chemistry, University College, Galway, Eire

Professor J. Elguero
Instituto de Quimica Medica, Juan de la Cierva 3, Madrid 6, Spain

Dr J. V. Greenhill
Department of Pharmaceutical Chemistry, University of Bradford, Bradford, West Yorkshire BD7 1DP, UK

Professor M. R. Grimmett
Department of Chemistry, University of Otago, PO Box 56, Dunedin, New Zealand

Professor A. R. Katritzky, FRS
Department of Chemistry, University of Florida, Gainesville, FL 32611, USA

Professor J. M. Lagowski
Department of Zoology, University of Texas, Austin, TX 78712, USA

Dr J. A. Montgomery
Southern Research Institute, 2000 Ninth Avenue South, Birmingham, AL 35255, USA

Dr J. B. Polya
Department of Chemistry, University of Tasmania, Box 252 C, GPO Hobart, Tasmania, Australia

Professor K. T. Potts
Department of Chemistry, Rensselaer Polytechnic Institute, Troy, NY 12181, USA

Dr J. A. Secrist III
Southern Research Institute, 2000 Ninth Avenue South, Birmingham, AL 35255, USA

Professor S. W. Schneller
Department of Chemistry, University of South Florida, Tampa, FL 33620, USA

Professor G. Shaw
School of Studies in Chemistry, University of Bradford, Bradford, West Yorkshire BD7 1DP, UK

Professor Dr I. Ugi
Organische-Chemisches Institut, Technische Universität München, Lichtenberstrasse 4, D-8046 Garching, Federal Republic of Germany

Professor Dr H. Wamhoff
Institut für Organische Chemie und Biochemie, Universität Bonn, Gerhard-Domagk-Strasse 1, D-5300 Bonn 1, Federal Republic of Germany

Contents of All Volumes

Volume 1 (Part 1: Introduction, Nomenclature, Review Literature, Biological Aspects, Industrial Uses, Less-common Heteroatoms)

1.01 Introduction
1.02 Nomenclature of Heterocyclic Compounds
1.03 Review Literature of Heterocycles
1.04 Biosynthesis of Some Heterocyclic Natural Products
1.05 Toxicity of Heterocycles
1.06 Application as Pharmaceuticals
1.07 Use as Agrochemicals
1.08 Use as Veterinary Products
1.09 Metabolism of Heterocycles
1.10 Importance of Heterocycles in Biochemical Pathways
1.11 Heterocyclic Polymers
1.12 Heterocyclic Dyes and Pigments
1.13 Organic Conductors
1.14 Uses in Photographic and Reprographic Techniques
1.15 Heterocyclic Compounds as Additives
1.16 Use in the Synthesis of Non-heterocycles
1.17 Heterocyclic Rings containing Phosphorus
1.18 Heterocyclic Rings containing Arsenic, Antimony or Bismuth
1.19 Heterocyclic Rings containing Halogens
1.20 Heterocyclic Rings containing Silicon, Germanium, Tin or Lead
1.21 Heterocyclic Rings containing Boron
1.22 Heterocyclic Rings containing a Transition Metal

Volume 2 (Part 2A: Six-membered Rings with One Nitrogen Atom)

2.01 Structure of Six-membered Rings
2.02 Reactivity of Six-membered Rings
2.03 Synthesis of Six-membered Rings
2.04 Pyridines and their Benzo Derivatives: (i) Structure
2.05 Pyridines and their Benzo Derivatives: (ii) Reactivity at Ring Atoms
2.06 Pyridines and their Benzo Derivatives: (iii) Reactivity of Substituents
2.07 Pyridines and their Benzo Derivatives: (iv) Reactivity of Non-aromatics
2.08 Pyridines and their Benzo Derivatives: (v) Synthesis
2.09 Pyridines and their Benzo Derivatives: (vi) Applications
2.10 The Quinolizinium Ion and Aza Analogs
2.11 Naphthyridines, Pyridoquinolines, Anthyridines and Similar Compounds

Volume 3 (Part 2B: Six-membered Rings with Oxygen, Sulfur or Two or More Nitrogen Atoms)

2.12 Pyridazines and their Benzo Derivatives
2.13 Pyrimidines and their Benzo Derivatives

2.14 Pyrazines and their Benzo Derivatives
2.15 Pyridodiazines and their Benzo Derivatives
2.16 Pteridines
2.17 Other Diazinodiazines
2.18 1,2,3-Triazines and their Benzo Derivatives
2.19 1,2,4-Triazines and their Benzo Derivatives
2.20 1,3,5-Triazines
2.21 Tetrazines and Pentazines
2.22 Pyrans and Fused Pyrans: (i) Structure
2.23 Pyrans and Fused Pyrans: (ii) Reactivity
2.24 Pyrans and Fused Pyrans: (iii) Synthesis and Applications
2.25 Thiopyrans and Fused Thiopyrans
2.26 Six-membered Rings with More than One Oxygen or Sulfur Atom
2.27 Oxazines, Thiazines and their Benzo Derivatives
2.28 Polyoxa, Polythia and Polyaza Six-membered Ring Systems

Volume 4 (Part 3: Five-membered Rings with One Oxygen, Sulfur or Nitrogen Atom)

3.01 Structure of Five-membered Rings with One Heteroatom
3.02 Reactivity of Five-membered Rings with One Heteroatom
3.03 Synthesis of Five-membered Rings with One Heteroatom
3.04 Pyrroles and their Benzo Derivatives: (i) Structure
3.05 Pyrroles and their Benzo Derivatives: (ii) Reactivity
3.06 Pyrroles and their Benzo Derivatives: (iii) Synthesis and Applications
3.07 Porphyrins, Corrins and Phthalocyanines
3.08 Pyrroles with Fused Six-membered Heterocyclic Rings: (i) *a*-Fused
3.09 Pyrroles with Fused Six-membered Heterocyclic Rings: (ii) *b*- and *c*-Fused
3.10 Furans and their Benzo Derivatives: (i) Structure
3.11 Furans and their Benzo Derivatives: (ii) Reactivity
3.12 Furans and their Benzo Derivatives: (iii) Synthesis and Applications
3.13 Thiophenes and their Benzo Derivatives: (i) Structure
3.14 Thiophenes and their Benzo Derivatives: (ii) Reactivity
3.15 Thiophenes and their Benzo Derivatives: (iii) Synthesis and Applications
3.16 Selenophenes, Tellurophenes and their Benzo Derivatives
3.17 Furans, Thiophenes and Selenophenes with Fused Six-membered Heterocyclic Rings
3.18 Two Fused Five-membered Rings each containing One Heteroatom

Volume 5 (Part 4A: Five-membered Rings with Two or More Nitrogen Atoms)

4.01 Structure of Five-membered Rings with Two or More Heteroatoms
4.02 Reactivity of Five-membered Rings with Two or More Heteroatoms
4.03 Synthesis of Five-membered Rings with Two or More Heteroatoms
4.04 Pyrazoles and their Benzo Derivatives
4.05 Pyrazoles with Fused Six-membered Heterocyclic Rings
4.06 Imidazoles and their Benzo Derivatives: (i) Structure
4.07 Imidazoles and their Benzo Derivatives: (ii) Reactivity
4.08 Imidazoles and their Benzo Derivatives: (iii) Synthesis and Applications
4.09 Purines
4.10 Other Imidazoles with Fused Six-membered Rings
4.11 1,2,3-Triazoles and their Benzo Derivatives
4.12 1,2,4-Triazoles
4.13 Tetrazoles
4.14 Pentazoles
4.15 Triazoles and Tetrazoles with Fused Six-membered Rings

Volume 6 (Part 4B: Five-membered Rings with Two or More Oxygen, Sulfur or Nitrogen Atoms)

4.16 Isoxazoles and their Benzo Derivatives
4.17 Isothiazoles and their Benzo Derivatives
4.18 Oxazoles and their Benzo Derivatives
4.19 Thiazoles and their Benzo Derivatives
4.20 Five-membered Selenium–Nitrogen Heterocycles
4.21 1,2,3- and 1,2,4-Oxadiazoles
4.22 1,2,5-Oxadiazoles and their Benzo Derivatives
4.23 1,3,4-Oxadiazoles
4.24 1,2,3-Thiadiazoles and their Benzo Derivatives
4.25 1,2,4-Thiadiazoles
4.26 1,2,5-Thiadiazoles and their Benzo Derivatives
4.27 1,3,4-Thiadiazoles
4.28 Oxatriazoles and Thiatriazoles
4.29 Five-membered Rings (One Oxygen or Sulfur and at least One Nitrogen Atom) Fused with Six-membered Rings (at least One Nitrogen Atom)
4.30 Dioxoles and Oxathioles
4.31 1,2-Dithioles
4.32 1,3-Dithioles
4.33 Five-membered Rings containing Three Oxygen or Sulfur Atoms
4.34 Dioxazoles, Oxathiazoles and Dithiazoles
4.35 Five-membered Rings containing One Selenium or Tellurium Atom and One Other Group VI Atom and their Benzo Derivatives
4.36 Two Fused Five-membered Heterocyclic Rings: (i) Classical Systems
4.37 Two Fused Five-membered Heterocyclic Rings: (ii) Non-classical Systems
4.38 Two Fused Five-membered Heterocyclic Rings: (iii) 1,6,6aλ^4-Trithiapentalenes and Related Systems

Volume 7 (Part 5: Small and Large Rings)

5.01 Structure of Small and Large Rings
5.02 Reactivity of Small and Large Rings
5.03 Synthesis of Small and Large Rings
5.04 Aziridines, Azirines and Fused-ring Derivatives
5.05 Oxiranes and Oxirenes
5.06 Thiiranes and Thiirenes
5.07 Fused-ring Oxiranes, Oxirenes, Thiiranes and Thiirenes
5.08 Three-membered Rings with Two Heteroatoms and Fused-ring Derivatives
5.09 Azetidines, Azetines and Azetes
5.10 Cephalosporins
5.11 Penicillins
5.12 Other Fused-ring Azetidines, Azetines and Azetes
5.13 Oxetanes and Oxetenes
5.14 Thietanes, Thietes and Fused-ring Derivatives
5.15 Four-membered Rings with Two or More Heteroatoms and Fused-ring Derivatives
5.16 Azepines
5.17 Oxepanes, Oxepins, Thiepanes and Thiepins
5.18 Seven-membered Rings with Two or More Heteroatoms
5.19 Eight-membered Rings
5.20 Larger Rings except Crown Ethers and Heterophanes
5.21 Crown Ethers and Cryptands
5.22 Heterophanes

Volume 8 (Part 6: Indexes)

Subject Index
Author Index
Ring Index
Data Index

4.01

Structure of Five-membered Rings with Two or More Heteroatoms

A. R. KATRITZKY
University of Florida

and

J. M. LAGOWSKI
The University of Texas at Austin

4.01.1 SURVEY OF POSSIBLE STRUCTURES	2
4.01.1.1 Aromatic Systems without Exocyclic Conjugation	2
4.01.1.2 Aromatic Systems with Exocyclic Conjugation	2
4.01.1.3 Non-aromatic Systems	4
4.01.2 THEORETICAL METHODS	5
4.01.2.1 The MO Approximation	5
4.01.2.2 Electron Densities	5
4.01.2.3 Frontier Electron Densities	6
4.01.2.4 Localization Energies	6
4.01.2.5 Semiempirical Methods	7
4.01.2.6 Other Applications of Theory	7
4.01.3 STRUCTURAL METHODS	8
4.01.3.1 X-Ray Diffraction	8
4.01.3.2 Microwave Spectroscopy	8
4.01.3.2.1 Molecular geometry	8
4.01.3.2.2 Partially and fully saturated ring systems	8
4.01.3.3 ^1H NMR Spectroscopy	13
4.01.3.4 ^{13}C NMR Spectroscopy	15
4.01.3.5 Nitrogen NMR Spectroscopy	16
4.01.3.6 UV Spectroscopy	16
4.01.3.6.1 Parent compounds	16
4.01.3.6.2 Benzo derivatives	21
4.01.3.6.3 Effect of substituents	24
4.01.3.7 IR Spectroscopy	24
4.01.3.7.1 Aromatic rings without carbonyl groups	24
4.01.3.7.2 Azole rings containing carbonyl groups	24
4.01.3.7.3 Substituent vibrations	24
4.01.3.8 Mass Spectrometry	30
4.01.3.9 Photoelectron Spectroscopy	30
4.01.4 THERMODYNAMIC ASPECTS	31
4.01.4.1 Intermolecular Forces	31
4.01.4.1.1 Melting and boiling points	31
4.01.4.1.2 Solubility of heterocyclic compounds	31
4.01.4.1.3 Gas–liquid chromatography	32
4.01.4.2 Stability and Stabilization	32
4.01.4.2.1 Thermochemistry and conformation of saturated heterocycles	32
4.01.4.2.2 Aromaticity	32
4.01.4.3 Conformation	34
4.01.5 TAUTOMERISM	35
4.01.5.1 Annular Tautomerism	35
4.01.5.2 Substituent Tautomerism	36
4.01.5.2.1 Azoles with heteroatoms in the 1,2-positions	36
4.01.5.2.2 Azoles with heteroatoms in the 1,3-positions	37

4.01.1 SURVEY OF POSSIBLE STRUCTURES

We classify compounds as aromatic, if there is continuous conjugation around the ring, or non-aromatic. Aromatic compounds are further subdivided into those without exocyclic double bonds and those in which important canonical forms containing exocyclic double bonds contribute.

4.01.1.1 Aromatic Systems without Exocyclic Conjugation

The neutral aromatic azole systems (without exocyclic conjugation) are shown in Scheme 1; throughout, Z is O, S or NR. There are, thus, 24 possible systems; however, for NR = NH, tautomerism renders (**3**) ≡ (**5**), (**4**) ≡ (**6**), and (**7**) ≡ (**8**). Ring-fused derivatives without a bridgehead nitrogen atom are possible for systems (**1**), (**2**), (**3**) and (**5**). Ring-fused derivatives with a bridgehead nitrogen atom can be derived from all except (**5**) and (**8**).

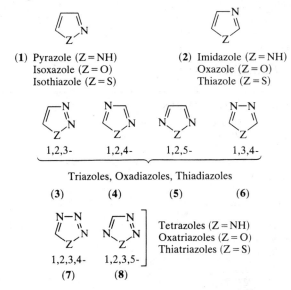

Scheme 1 Neutral aromatic azoles (no exocyclic double bonds) (Z = O, S or NR)

The five possible azole monoanions are shown (one canonical form only) in Scheme 2; all heteroatoms are now nitrogens.

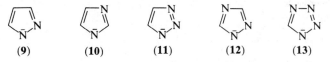

Scheme 2 Monoanionic aromatic azoles

The aromatic azole monocations are given in Scheme 3; here Z and Y are both O, S or NR; there are therefore three mixed sets. If Z = Y, then (**16**) ≡ (**18**), (**20**) ≡ (**21**), (**22**) ≡ (**24**), and (**25**) ≡ (**26**). Hence there are $(3 \times 14) + (3 \times 10) = 72$ possible systems.

4.01.1.2 Aromatic Systems with Exocyclic Conjugation

Each of the aromatic monocationic systems (**14**)–(**27**) can be converted into a neutral system by substitution of an anionic O, S or NR group on to a ring carbon atom. However, (**14**) and (**15**) each give three such systems, (**16**)–(**21**) two each, and (**22**)–(**27**) one each. The resulting 24 systems can be divided into two groups: 12 systems for the azolinones and related compounds (Scheme 4) and 12 systems for the mesoionic (betaine) compounds (Scheme 5).

Of the mesoionic systems, (**40**) and its aza derivatives (**43**), (**44**) and (**49**) have been designated as Class B by Ollis ⟨76AHC(19)1⟩, including compounds with X = CRR'; there

Structure of Five-membered Rings with Two or More Heteroatoms

(14)
Pyrazolium (Z = Y = NH)
Isoxazolium (Z = O, Y = NH)
Isothiazolium (Z = S, Y = NH)
1,2-Dioxolylium (Z = Y = O)
1,2-Oxathiolylium (Z = O, Y = S)
1,2-Dithiolylium (Z = Y = S)

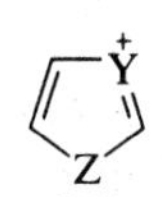

(15)
Imidazolium (Z = Y = NH)
Oxazolium (Z = O, Y = NH)
Thiazolium (Z = S, Y = NH)
1,3-Dioxolylium (Z = Y = O)
1,3-Oxathiolylium (Z = O, Y = S)
1,3-Dithiolylium (Z = Y = S)

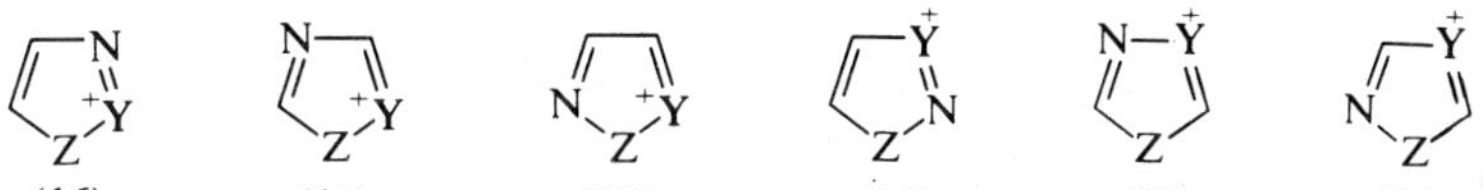

(16) 1,2,3-   (17) 1,2,4-   (18) 1,2,5-   (19) 1,3,2-   (20) 1,3,4-   (21) 1,3,5-

Triazolium, Oxadiazolium, Thiadiazolium, Dioxazolium, Oxathiazolium, Dithiazolium

(22) (23) (24) (25) (26) (27)

Tetrazolium, Oxatriazolium, Thiatriazolium, Dioxadiazolium, Oxathiadiazolium, Dithiadiazolium

Scheme 3 Monocationic azoles

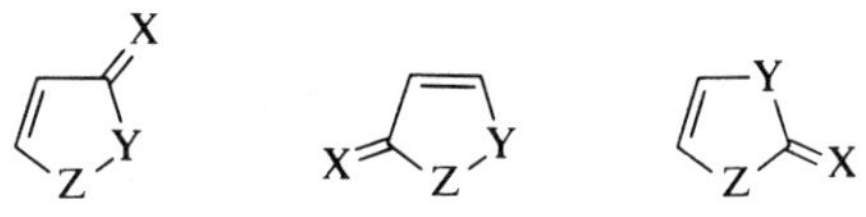

(28) (29) (30)

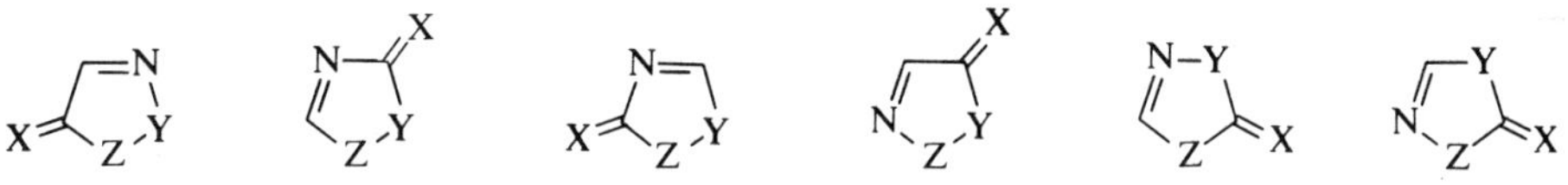

(31) (32) (33) (34) (35) (36)

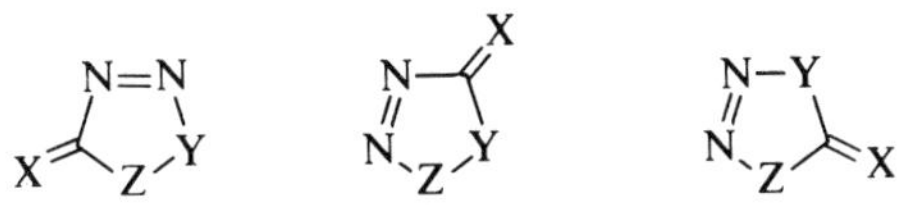

(37) (38) (39)

Scheme 4 Azolinones and related compounds (X = O, azolinones; X = S, azolinethiones; X = NR, azolinimines)

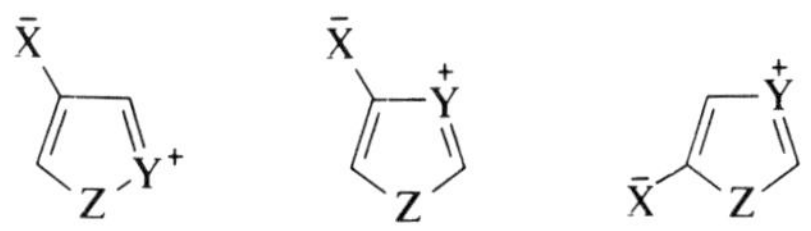

(40) (41) (42)

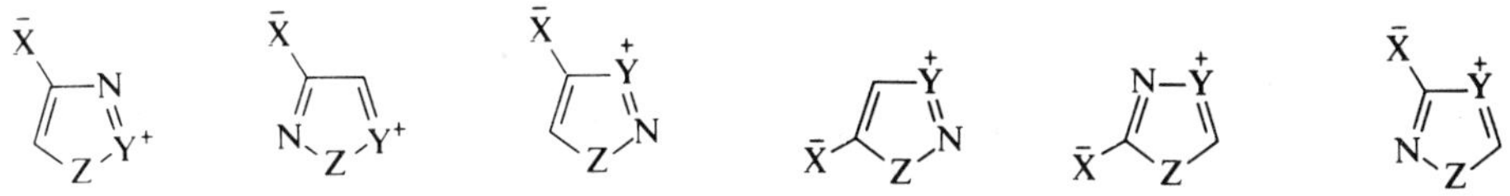

(43) (44) (45) (46) (47) (48)

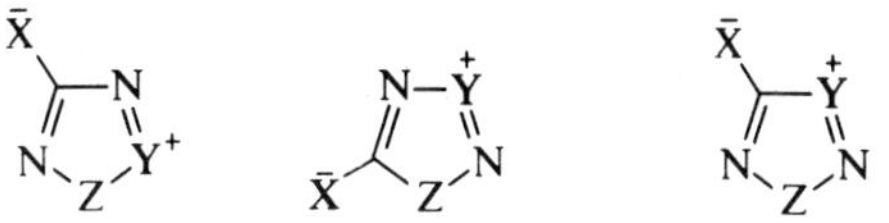

(49) (50) (51)

Scheme 5 Mesoionic compounds (Z = O, S or NR; X = O, S, NR or CR$_2$)

are 88 total systems. Class A mesoionic compounds include (**41**), (**42**) and their aza derivatives (**45**)–(**48**), (**50**) and (**51**), giving a total of 144 systems. Members of the latter group contain 1,3-dipoles, often reflected in their pronounced ability to undergo cycloaddition reactions.

4.01.1.3 Non-aromatic Systems

These are subdivided into: (a) compounds isomeric with aromatic compounds in which the ring contains two double bonds but also an sp^3-hybridized carbon (7 systems; Scheme 6) or a quaternary nitrogen atom (9 systems; Scheme 7).

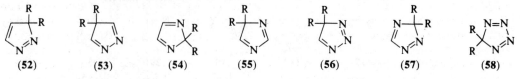

Scheme 6 Isomers of aromatic compounds with an sp^3-hybridized carbon atom

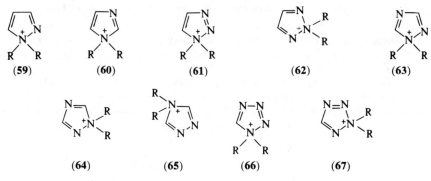

Scheme 7 Isomers of aromatic compounds with a quaternary nitrogen atom

(b) Dihydro compounds in which the ring contains one double bond (66 systems; Scheme 8).

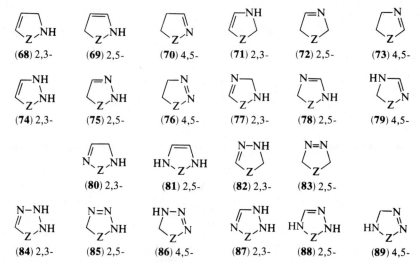

Scheme 8 Dihydroazoles (Z = O, S or NR)

(c) Tetrahydro compounds in which the ring contains no double bonds (24 systems; Scheme 9).

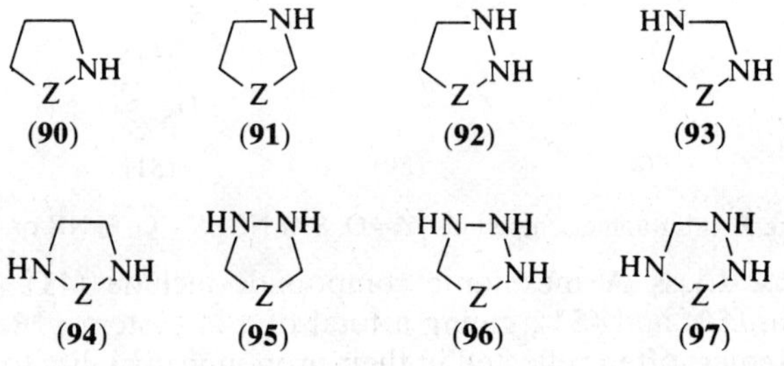

Scheme 9 Tetrahydro compounds (Z = O, S or NR)

4.01.2 THEORETICAL METHODS

4.01.2.1 The MO Approximation

In the simpler MO approximations the π-electrons are assumed to move independently in MOs that can be represented as linear combinations of the atomic p-orbitals. When these MOs are used as the basis for an approximate wave mechanical calculation, the energy of the MOs and the distribution of the π-electrons in each depend on the values of certain integrals. These integrals are of two types, termed Coulomb integrals and resonance integrals; both are negative and have the dimensions of energy. The Coulomb integrals are characteristic of an atomic π-orbital in a given molecular environment and are a measure of the effective electronegativity of that atom toward π-electrons. The resonance integrals are characteristic of a π-bond between two atoms and are a measure of the stability that a localized π-bond would have if formed between them. Neither the Coulomb integrals nor the resonance integrals are usually directly evaluated by integration; the calculations are based on certain simplifying assumptions concerning the relative values of the different Coulomb integrals and the relative values of the different resonance integrals.

For unsubstituted aromatic hydrocarbons all the carbon atoms are assigned the same Coulomb integral (α) and all C—C bonds are assigned the same resonance integral (β).

In heteroaromatic molecules the approximate Coulomb integral for the heteroatom is expressed in terms of α and β, the standard integrals associated with the carbocyclic aromatic hydrocarbons. There has been considerable variation in the Coulomb integral used for the neutral nitrogen atom as typified by that in pyridine; the values used have ranged from $\alpha + 2.0\beta$ to $\alpha + 0.2\beta$, the lower values being the more recent. The Coulomb integral for nitrogen is considered to depend considerably on the chemical environment, and the approximate values are discussed under the compounds concerned. It has been suggested that a different Coulomb integral should be used for the nitrogen atom of pyrrole in simple Hückel-type MO calculations, giving a value of $\alpha + 1.9\beta$. Non-empirical calculations indicate that the π-electrons of the pyrrole anion are almost evenly distributed over all five atoms, so the carbon and nitrogen atoms should be assigned the same Coulomb integrals in the simple MO treatment.

The Coulomb integral (α_C) of the carbon atoms adjacent to the heteroatom (more electronegative than the carbon atoms in benzene because of the inductive effect of the heteroatom as relayed through the σ-bonds) is accommodated by making the appropriate Coulomb integrals more negative than α. This increases the agreement between the simple MO calculations and those based on elaborate, non-empirical treatments. It is convenient to express the value of α_C in terms of α and β, and so the Coulomb integrals used for the heteroatom and for the adjacent carbon atoms may be defined by two parameters h and h' according to equations (1) and (2).

$$\text{Heteroatom} \qquad \alpha_X = \alpha + h\beta \qquad (1)$$

$$\text{Adjacent carbon atoms} \qquad \alpha_C = \alpha + h'\beta \qquad (2)$$

The resonance integral of the π-bond between the heteroatom and carbon is another possible parameter in the treatment of heteroatomic molecules. However, for nitrogen compounds more detailed calculations have suggested that this resonance integral is similar to that for a C—C bond and moreover the relative values of the reactivity indices at different positions are not very sensitive to change in this parameter.

Theoretical reactivity indices of heteroaromatic systems distinguish reactivity toward electrophilic, nucleophilic and homolytic reactions.

4.01.2.2 Electron Densities

The π-electron density refers to the electron density at a given carbon atom obtained by summing the contributions from all the filled molecular orbitals. Electrophilic attack occurs where this density is highest, and nucleophilic attack where it is lowest; π-electron densities are not dominant in determining the orientation of homolytic substitution.

Values for the π-electron densities in imidazole and two related ions are given in Figure 1. The π-electron densities in the conjugate acid of imidazole are greater at the 2-position than at the 4-position and thus bear no relation to the chemical reactivity, as nucleophiles attack at the 2-position and electrophiles at the 4-position. The results for the neutral molecule and the conjugate base are more satisfactory; however, the uncertainty in these values, estimated as 0.02 units, exceeds the difference between the π-electron density at the 2-position and that at the more reactive 4(5)-positions in all three reactive forms of imidazole.

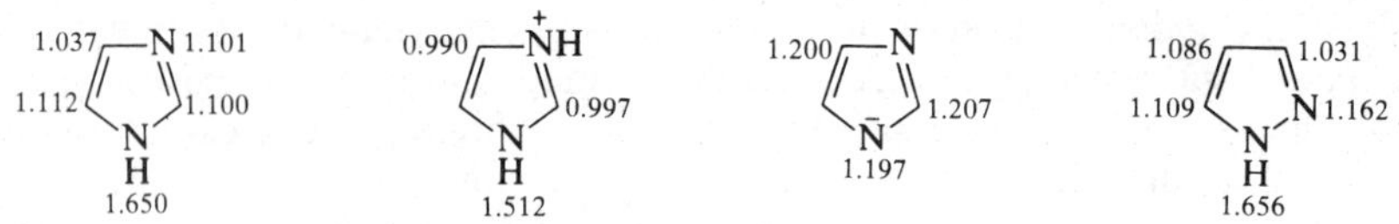

Figure 1 The π-electron densities in imidazole, its conjugate acid and base, and pyrazole

Results for the neutral pyrazole molecule show a considerable spread. The π-electron and total $(\pi+\sigma)$ densities predict electrophilic substitution at the 4-position as found. Results for thiazole also agree with experimentally determined electrophilic and nucleophilic reactivity.

The π-electron distribution in benzimidazole favors substitution at the 4-position in the conjugate acid, at the 4- and 5-positions in the neutral molecule, and at the 2-position in the conjugate base. These results do not explain the apparently exclusive substitution at the 5-position in nitration. There is thus no general agreement between the π-electron distribution and the chemical reactivity.

Values for π-electron density on nitrogen atoms generally indicate the position of electrophilic attack, e.g. at the 3-position of 1,2,3-thiadiazoles.

The net σ-charge on ring hydrogen atoms can also be significant; for thiazole the order of decreasing acidity of the hydrogens is H-2 $\geqslant$ 5 > 4, consistent with experiment.

4.01.2.3 Frontier Electron Densities

An alternative approach is in terms of frontier electron densities. In electrophilic substitution, the frontier electron density is taken as the electron density in the highest filled MO. In nucleophilic substitution the frontier orbital is taken as the lowest vacant MO; the frontier electron density at a carbon atom is then the electron density that would be present in this MO if it were occupied by two electrons. Both electrophilic and nucleophilic substitution thus occur at the carbon atom with the greatest appropriate frontier electron density.

The significance of frontier electron densities is limited to the orientation of substitution for a given aromatic system, but this approach has been developed to give two more complex reactivity indices termed superdelocalizabilities and Z values, which indicate the relative reactivities of different aromatic systems.

4.01.2.4 Localization Energies

Localization energies refer to the difference between the π-electron energy in the isolated molecule and that in the related system where the carbon atom at the point of substitution has been removed from the cyclic conjugation by bonding with the reagent. The Wheland structure, originally considered to be a suitable model of the transition state, generally corresponds to an intermediate. Nevertheless, the relative stabilities of the different transition states in the same molecule appear to be in the same order as the calculated localization energies, and this reactivity index is probably the best guide to the orientation of substitution in aromatic and heteroaromatic systems.

The localization energies for the neutral diazoles were originally calculated on the assumption that the Coulomb integrals for the two nitrogen atoms are equal; for the neutral diazoles these results may therefore require some modification. For the neutral molecule and conjugate acid of pyrazole the orientation of electrophilic attack is correctly assigned to the 4-position. For the neutral molecule and conjugate acid of imidazole the orientation of electrophilic attack is correctly assigned to the 4(5)-positions. If the Coulomb integral of the nitrogen atoms in the conjugate base of imidazole is set equal to that for the carbon atoms, then the localization energies for the 2- and 4,5-positions become equal. In reactions of the conjugate base of imidazole the orientation appears to depend on the reagent, but there is probably little difference between the reactivity of the 2- and 4-positions.

The localization energies for electrophilic substitution in benzimidazole predict that all three reactive forms should undergo substitution in the 4-position. This does not explain the formation of the 5-nitro compound or that of the 2-deutero compound. It is doubtful whether any electrophilic substitution occurs preferentially in the 4-position.

The calculation of localization energies in heteroaromatic systems derived from alternant hydrocarbons has been simplified by Dewar and Maitlis ⟨57JCS2521⟩. This approach has had considerable success; the results provide a somewhat empirical index of reactivity.

4.01.2.5 Semiempirical Methods

The simple, or Hückel based, molecular orbital theory described above frequently provides useful quantitative insights but cannot be used reliably in a quantitative sense. For this purpose it is necessary to use a method which takes account of all the electrons as well as their mutual repulsions. This is usually done within the framework of a formalism developed independently by Roothaan ⟨51MI40100⟩ and Hall ⟨51MI40101⟩. Procedures of this kind are variously known as RH, non-empirical or, most frequently, *ab initio* molecular orbital calculations. Here, once the forms of the atomic orbitals, *i.e.* basis sets, are specified, no further approximations are made. While numerous alternative basis sets are possible, most practical applications of *ab initio* MO theory now utilize one of the carefully optimized expansions of Gaussian functions introduced by Pople and his co-workers ⟨74MI40100⟩. The simplest (and least accurate) is designated STO-3G, while the most complex in common use is designated 6-31G. Fortunately, even the simplest basis sets seem to give excellent descriptions of the electron distribution, especially when expressed in terms of integrated spatial populations ⟨80MI40101⟩. Palmer, in particular, has presented electron distribution data in many heterocyclic systems ⟨78JST(43)33⟩. Unfortunately, apart from those special cases in which species having the same numbers of each type of bond are compared, the quantitative estimation of relative energies in *ab initio* theory usually requires a basis set of at least 3-21G quality. Moreover, for reliable results the geometries of the species being compared must be calculated at the *ab initio* level. The expense inherent in the use of the more complex basis set, as well as that of geometry optimization, has tended to limit the most detailed studies to rather small heterocyclic ring systems ⟨77JA7806, 78JA3674, 83JA309⟩.

The major bottleneck in *ab initio* calculations occurs in the computation and storage of the enormous number of electron-repulsion integrals involved. Early efforts by Pople and co-workers to reduce this problem led to the CNDO, INDO and NDDO approximations ⟨70MI40100⟩. Although once popular, these methods in their original forms are now largely superseded. Dewar has implemented modifications of the two latter approximations which have proven to be very successful. The most recent versions are designated MINDO/3 ⟨75JA1285⟩ and MNDO ⟨77JA4899⟩ respectively. Since their inception these methods have been somewhat controversial since they were parameterized to reproduce not the results of the *ab initio* methods from which they are derived (as was done in CNDO/2 and INDO), but experimental geometries and energies. These semiempirical MO procedures therefore appear to involve several theoretical inconsistencies which have yet to be fully resolved ⟨B-77MI40100⟩. Nevertheless, judged on purely empirical criteria, they seem to work well. Indeed they are frequently comparable in accuracy (relative to experiment) to *ab initio* calculations using moderately sized basis sets and much better than those using minimal basis sets ⟨79JA5558⟩. Semiempirical procedures of this kind have been used to study the reactions of various five-membered heterocyclic systems ⟨77JCS(P2)724⟩.

4.01.2.6 Other Applications of Theory

Applications of MO methods to such diverse problems as aromaticity, tautomeric structure, dipole moments, and UV, NMR and PE spectroscopy are discussed in various monograph chapters.

4.01.3 STRUCTURAL METHODS

4.01.3.1 X-Ray Diffraction

Details of bond lengths and bond angles for all the X-ray structures of heterocyclic compounds through 1970 are listed in 'Physical Methods in Heterocyclic Chemistry', volume 5. This compilation contains many examples for five-membered rings containing two heteroatoms, particularly pyrazoles, imidazoles, isoxazoles, oxazoles, isothiazoles, thiazoles, 1,2-dithioles and 1,3-dithioles. Further examples of more recent measurements on these heterocyclic compounds can be found in the monograph chapters.

For compounds with three or four heteroatoms in the ring the number of measurements is much fewer, and these are summarized in Table 1.

4.01.3.2 Microwave Spectroscopy

Microwave spectra provide a rich source of minute details of molecular structures. They tell us about the molecular geometry because the spectra are primarily analyzed in terms of the accurate average values of the reciprocals of the three moments of inertia. This generally gives at once the general molecular conformation and some precise structural features may emerge. To obtain a complete structure it is necessary to measure the changes in moments of inertia which accompany the isotopic replacements of each atom in turn ⟨74PMH(6)53⟩.

From accurate measurements of the Stark effect when electrostatic fields are applied, information regarding the electron distribution is obtained. Further information on this point is obtained from nuclear quadrupole coupling effects and Zeeman effects ⟨74PMH(6)53⟩.

Microwave studies also provide important information regarding molecular force fields, particularly with reference to low frequency vibrational modes in cyclic structures ⟨74PMH(6)53⟩.

4.01.3.2.1 Molecular geometry

Structural parameters in aromatic five-membered rings are shown in Table 2. All the C—H distances are near 107.5 pm, close to the C—H link in ethylene. With heteroatoms at adjacent ring positions, the C—H groups are displaced from the bisector of the ring angles toward the adjacent heteroatom ⟨74PMH(6)53⟩.

The N—H bond lengths in pyrazole and imidazole (99.8 pm) are a little shorter than those found in dimethylamine. Delocalization in pyrazole, imidazole, 1,2,3-triazole and 1,2,4-triazole is sufficient to bring the hydrogen attached to nitrogen into the plane of the other atoms. The N—H bond in pyrazole does not lie in the bisector of the ring angle (as is required by symmetry in pyrrole), but is displaced by around 5° toward the second nitrogen ⟨74PMH(6)53⟩.

The ring angles at C and N are usually 108±5°, but drop nearer to 90° at sulfur and selenium.

Minor variations in the bond lengths reflect variations in the double bond character and, for example, suggest larger delocalization in 1,3,4-thiadiazole than in 1,3,4-oxadiazole ⟨74PMH(6)53⟩. In thiazole the geometry of the SCN part of the ring resembles the corresponding part of 1,3,4-thiadiazole, while the remaining part of the ring resembles the corresponding thiophene.

Microwave spectroscopy distinguishes readily between possible tautomeric forms of 1,2,3- and 1,2,4-triazole, which are both in the $1H$-form. In tetrazole both the $1H$- and $2H$-forms are detected ⟨74PMH(6)53⟩.

Dipole moments can also be obtained from the microwave spectral data ⟨74PMH(6)53⟩ and available values are given in Table 2.

4.01.3.2.2 Partially and fully saturated ring systems

Relatively few such heterocyclic systems have been studied by microwave spectroscopy; some data are included in Table 2. In 1,3-dioxolane the bent form is more stable than the

Table 1 X-Ray Structures of Compounds with Five-membered Rings and Two, Three or Four Heteroatoms[a]

| Ring | Ring position | | | | | Examples of compounds studied |
	1	2	3	4	5	
C_3N_2	N	N	—	—	—	Pyrazole;[b] substituted pyrazoles; Δ^1- and Δ^2-pyrazolines; pyrazolinones; pyrazolidines; pyrazolidinones
	N	—	N	—	—	Imidazole;[b] 4,5-di-t-butylimidazole; histamine dihydrochloride; 2-thiohydantoin
C_2N_3	N	N	N	—	—	1,3-Dimethyl-4-(1,2,3-triazolyl) sulfide; 3-methyl-2-phenyl-1,2,3-triazol-1-ine-4-thione
	N	N	—	N	—	1,2,4-Triazole[b]
CN_4	N	N	N	N	—	5-Amino-2-methyltetrazole;[b] 5-aminotetrazole monohydrate;[b] sodium tetrazolate monohydrate[b]
C_3NO	O	N	—	—	—	5,5'-Biisoxazole;[b] 3-hydroxy-5-phenylisoxazole;[b] 3,3'-bi-2-isoxazoline;[b] 3-hydroxy-5-phenylisoxazole;[b] 3-phenylisoxazolin-5-one[b]
C_3NS	S	N	—	—	—	Methyl 3-hydroxy-4-phenylisothiazole-5-sulfonate; dehydromethionine
C_3NO	O	—	N	—	—	2,2'-p-Phenylenebis(5-phenyloxazole); 2-(4-pyridyl)oxazole;[b] 2,4-dimethyl-5-(p-nitrophenyl)oxazole;[b] 2-oxazolidinone[b]
C_3NS	S	—	N	—	—	Thiamine hydrochloride monohydrate;[b] rhodanine;[b] 2-imino-5-phenyl-4-thiazolidinone[b]
C_2N_2O	O	N	N	—	—	N-(p-Bromophenyl)sydnone;[b] 4,4'-dichloro-3,3'-ethylenebis(sydnone)
	O	N	—	N	—	3-(2-Aminopyridyl)-5-methyl-1,2,4-oxadiazole
	O	N	—	—	N	3-(p-Bromophenyl)-4-methyl-1,2,5-oxadiazole 2-oxide;[b] 3-(p-bromophenyl)-4-methyl-1,2,5-oxadiazole 5-oxide;[b] 3,4-diphenyl-1,2,5-oxadiazole
	O	—	N	N	—	Monoaryl-1,3,4-oxadiazoles[c]
C_2N_2S	S	N	N	—	—	5-Acylamino-3-methyl-1,2,3-thiadiazole; 5-phenyl-1,2,3-thiadiazole 3-oxide
	S	N	—	N	—	5-Imino-4-phenyl-3-phenylamino-4H-1,2,4-thiadiazoline[d]
	S	N	—	—	N	3,4-Diphenyl-1,2,5-thiadiazole; 1,2,5-thiadiazole-3,4-dicarboxamide
	S	—	N	N	—	1,3,4-Thiadiazole; 2,5-diphenyl-1,3,4-thiadiazole[b]
CN_3O	O	N	N	N	—	Mesoionic 3-phenyl-1,2,3,4-oxatriazole-5-phenylimine; mesoionic 3-phenyl-1,2,3,4-oxatriazol-5-one
CN_3S	S	N	N	N	—	5-Phenyl-1,2,3,4-thiatriazole; 5-amino-1,2,3,4-thiatriazole; 5-phenyl-1,2,3,4-thiatriazole 3-oxide
	S	N	N	—	N	2-Acetyl-5-chloro-2H-1,2,3,5-thiatriazolo[4,5-a]isoquinoline 5-oxide
C_3O_2	O	—	O	—	—	Bis(dioxolane); ethylene carbonate; cis-t-butyl-5-carboxymethyl-1,3-dioxolan-4-one; 2-methyl-1,3-dioxolan-2-ylium perchlorate
C_3OS	O	S	—	—	—	3,3-Diphenyl-1,2-oxathiolane 2,2-dioxide; 5H-1,2-benzoxathiole 2,2-dioxide
	O	—	S	—	—	Cholestan-4-one-3-spiro(2,5-oxathiolane)
C_3S_2	S	S	—	—	—	1,2-Dithiolane-4-carboxylic acid;[b] 3-phenyl-1,2-dithiolylium iodide;[b] 4-methyl-1,2-dithiole-3-thione[b]
	S	—	S	—	—	Bis-1,3-dithiol-2-yl;[b] 4,5-dioxo-2-thioxo-1,3-dithiolane;[b] 1,3-dithiolane-2-thione 5-oxide;[b] tetrathiafulvalene[e]
C_2O_3	O	O	—	O	—	$trans$-5-Anisyl-3-methoxycarbonyl-1,2,4-trioxolane[f]
C_2S_3	S	S	—	S	—	1,2,4-Trithiolane-3,5-dione diphenylhydrazone[g]
C_2O_2S	O	S	O	—	—	1,3,2-Dioxathiolane 2,2-dioxide;[h] 1,3,2-dioxathiole 2,2-dioxide[h]
C_2NS_2	S	S	N	—	—	2,4,6-Tri-t-butyl-7,8,9-dithiazabicyclo[4.3.0]nona-1(9),2,4-triene[i]
C_2NS_2	S	S	—	N	—	5-Amino-1,2,4-dithiazol-3-one ('Rhodan hydrate');[j] 5-amino-1,2,4-dithiazoline-3-thione ('Xanthane hydride');[k] 3,5-diamino-1,2,4-dithiazolium chloride ('Thiuret hydrochloride')[l]
C_2NOS	O	S	N	—	—	2,4-Dioxo-2-(4-methylphenyl)-1,2,3-oxathiazoline;[m] 1,2,3-oxathiazolo[5,4-d][1,2,3]oxathiazole 2,2,5,5-tetraoxide[n]
	O	S	—	N	—	6,10b-Dihydro-3-(2,2,6,6-tetramethylcyclohexyliden)-1,2,4-oxathiazolo[5,4-a]isoquinoline[o]
	O	N	S	—	—	4-Phenyl-1,3,2-oxathiazolin-5-one[p]
	O	—	S	N	—	2-Trichloromethyl-5-phenyl-Δ^4-1,3,4-oxathiazoline[q]

[a] Unless otherwise indicated data are taken from the appropriate chapter of 'Comprehensive Heterocyclic Chemistry'. [b] Data taken from ⟨72PMH(5)1⟩, which gives references to the original literature. [c] X-Ray powder data; cf. Chapter 4.23. [d] ⟨78CC652⟩. [e] ⟨71CC889⟩. [f] ⟨70ACS2137⟩. [g] ⟨71JCS(B)415⟩. [h] ⟨68JA2970⟩. [i] ⟨80AX(B)1466⟩. [j] ⟨66ACS754⟩. [k] ⟨63ACS2575, 63AX1157⟩. [l] ⟨66ACS1907⟩. [m] ⟨71TL4243⟩. [n] ⟨80MI40100⟩. [o] ⟨78AG(E)455⟩. [p] ⟨72G23⟩. [q] ⟨81JCS(P1)2991⟩.

Table 2 Structure Parameters in Five-membered Rings from Microwave Spectra

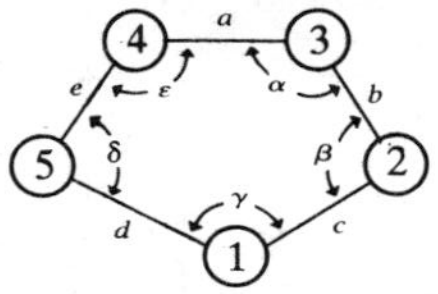

Compound	Bond length (pm)[a,b]					Angle (°)[b]					Dipole moment (10^{-30} C m)[a]
	a	b	c	d	e	α	β	γ	δ	ε	
[c] ⟨N-NH⟩	141.6	133.1	134.9	135.9	137.3	111.9	104.1	113.1	106.4	104.5	7.37
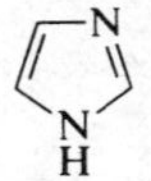 [d] ⟨N NH⟩	(137.8)	(132.6)	(134.9)	(136.9)	(135.8)	(105.4)	(111.3)	(107.2)	(106.3)	(109.8)	12.8
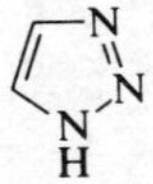 [e] ⟨N=N NH⟩	—	—	—	—	—	—	—	—	—	—	5.97
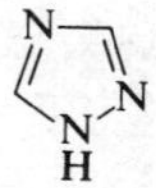 [e] ⟨N N NH⟩	135.9	132.3	135.9	133.1	132.4	114.6	102.1	110.2	110.1	103.0	9.07
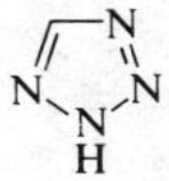 [f] ⟨N=N N NH⟩	135.1	128.4	132.4	133.4	131.0	—	—	—	—	—	7.31
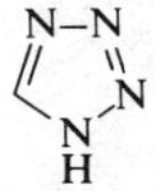 [g] ⟨N-N N NH⟩	(134.5)	(128.3)	(134.7)	(135.1)	(129.0)	(112.2)	(106.9)	(105.3)	—	—	17.05
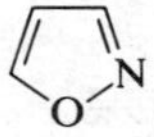 [h] ⟨N O⟩	—	—	—	—	—	—	—	—	—	—	9.21

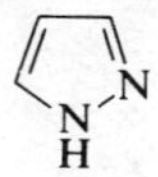

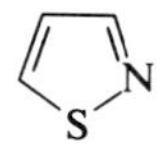

Structure of Five-membered Rings with Two or More Heteroatoms

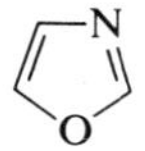

Compound	1	2	3	4	5	6	7	8	9	10	11	12
	—	—	139.5	137.2	(133.4)	(138.0)	142.1	141.7	139.9	136.6	136.6	142.0
	—	—	129.3	130.4	(131.3)	(130.3)	130.0	132.7	129.7	129.0	131.7	132.8
	—	—	135.7	172.4	(138.9)	(141.8)	138.0	163.0	134.8	169.2	164.9	163.1
	—	—	137.0	171.3	—	—	138.0	163.0	134.8	168.9	170.7	163.1
	—	—	135.3	136.7	—	—	130.0	132.7	129.7	136.9	131.3	132.8
	—	—	103.9	110.1	(111.3)	(106.1)	109.0	113.8	105.6	114.0	120.1	113.8
	—	—	115.0	115.2	(103.8)	(103.2)	105.8	106.5	113.4	111.2	107.1	106.4
	—	—	103.9	89.3	(114.4)	(114.2)	110.4	99.4	102.0	92.9	92.8	99.6
	—	—	108.1	109.6	—	—	105.8	106.5	113.4	107.8	112.3	106.4
	—	—	109.1	115.8	—	—	109.0	113.8	105.6	114.2	107.7	113.8
	—	8.67	5.00	5.37	21.12	5.44	11.28	5.27	10.14	11.98	1.5	5.24

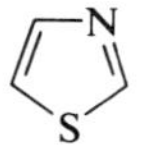

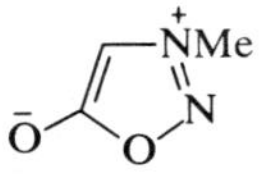

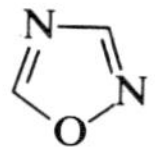

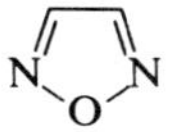

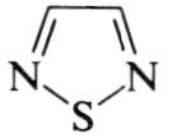

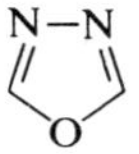

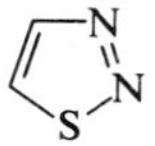

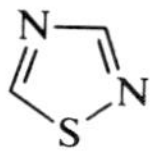

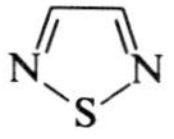

Table 2 (continued)

Compound	Bond length (pm)[a,b]					Angle (°)[b]					Dipole moment $(10^{-30}$ C m$)$[a]
	a	b	c	d	e	α	β	γ	δ	ε	
N—N [c] ring	137.1	130.2	172.1	172.1	130.2	112.2	114.6	86.4	114.6	112.2	10.94
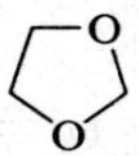 O ring [e]	—	—	—	—	—	—	—	—	—	—	3.97
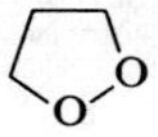 O—O ring [m]	156.3	142.8	—	142.8	156.3	105.1	—	—	105.1	101.7	—
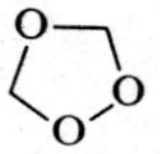 O—O ring [n]	143.6	139.5	147.0	139.5	143.6	106.2	99.2	99.2	106.2	99.2	3.64
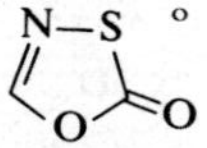 N—S C=O ring [o]	169.0	176.6	140.2	135.6	128.6	93.8	106.3	110.8	121.1	107.9	—

[a] 1 Å = 100 pm; 1 D = 3.336×10^{-30} C m. [b] X-Ray diffraction data are enclosed in parentheses. [c] Data taken from ⟨74PMH(6)53⟩, which see for references to the original literature. [d] Dipole moment is concentration-dependent (cf. Section 4.06.3.2). [e] Data taken from appropriate chapter, 'Comprehensive Heterocyclic Chemistry'. [f] No bond lengths or angles given in ⟨74JSP(49)423⟩; calculated values taken from Chapter 4.13. [g] Bond lengths and angles for 5-bromotetrazole; dipole moment for tetrazole in dioxane. [h] Measured in benzene; cf. Section 4.16.1.4.1(i). [i] Bond lengths and angles from X-ray data for 4,4'-dichloro-3,3'-ethylenebis(sydnone) ⟨67JA5977⟩; calculated dipole moment for 3-methylsydnone ⟨71JST(9)321⟩. [j] Bond lengths and angles from X-ray data for 3-(2-aminopyridyl)-1,2,4-oxadiazole ⟨79AX(B)2256⟩; dipole moment for 3-methyl-5-phenyl-1,2,4-oxadiazole ⟨35G152⟩. [k] Dipole moment from ⟨76MI40100⟩. [l] ⟨74TH40100⟩. [m] Values calculated from experimentally derived rotational constants; cf. Section 4.30.1.3.2. [n] ⟨72JA6337⟩. [o] Values for 1,3,4-oxathiazolin-2-one determined from electron diffraction measurements and refined using rotational parameters from microwave spectra; cf. Section 4.34.2.3.2.

twisted, and pseudorotation occurs. In 1,2,4-trioxocyclopentane the equilibrium conformation is twisted, and there is a barrier of 6.3 kJ mol^{-1} opposing pseudorotation ⟨74PMH(6)53⟩.

4.01.3.3 ^{1}H NMR Spectroscopy

Proton chemical shifts and spin coupling constants for ring CH of fully aromatic neutral azoles are recorded in Tables 3–6. Vicinal CH—CH coupling constants are small; where they have been measured (in rather few cases) they are found to be 1–2 Hz.

Table 3 ^{1}H NMR Spectral Data for Ring Hydrogens of Nitrogenous Azoles: (a) NH Derivatives

Compound	^{1}H Chemical shifts (δ, p.p.m.)				Coupling constants, J (Hz)	Solvent	Ref.
	H-2	H-3	H-4	H-5			
Pyrazole	—	7.61	7.31	7.61	2.1	—	71PMH(4)121, B-73NMR
Imidazole	7.86	—	7.25	7.25	1.0	CDCl$_3$	71PMH(4)121, B-73NMR
1,2,3-Triazole	—	—	7.75	7.75	—	CDCl$_3$	a
1,2,4-Triazole	—	7.92	—	8.85	—	HMPT	71PMH(4)121, B-73NMR
Tetrazole	—	—	—	9.5	—	D$_2$O	

^{a} Data taken from appropriate chapter, 'Comprehensive Heterocyclic Chemistry'.

Table 4 ^{1}H NMR Spectral Data for Ring Hydrogens of Nitrogenous Azoles: (b) N(1)-Methyl Derivatives

Compound	^{1}H Chemical shifts^a (δ, p.p.m.)			
	H-2	H-3	H-4	H-5
Pyrazole^{b,c}	—	7.49	6.22	7.35
Imidazole^b	7.47	—	7.08	6.88
1,2,3-Triazole^b	—	—	7.74	7.59
1,2,5-Triazole^b	—	7.75	7.75	—
1,2,4-Triazole^b	—	7.94	—	8.09
1,3,4-Triazole^b	8.23	—	—	8.23
1,2,3,4-Tetrazole^d	—	—	—	8.98
1,2,3,5-Tetrazole^d	—	—	8.60	—

^{a} Spectra measured in CDCl$_3$.
^{b} Data taken from ⟨B-73NMR⟩ which contains references to the original literature.
^{c} Coupling constants are $J_{3,4} = 2.0$ Hz; $J_{3,5} = 0.7$ Hz; $J_{4,5} = 2.3$ Hz.
^{d} Value for N-methyl derivative; cf. Chapter 4.13.

Table 5 ^{1}H NMR Spectral Data for Ring Hydrogens of Azoles Containing Oxygen

Compound	^{1}H Chemical shifts (δ, p.p.m.)				Solvent
	H-2	H-3	H-4	H-5	
Isoxazole^a	—	8.14	6.28	8.39	CS$_2$
Oxazole^b	7.95	—	7.09	7.69	CCl$_4$
1,2,4-Oxadiazole^c	—	8.2	—	8.7	C$_6$H$_6$
1,2,5-Oxadiazole^d	—	8.19	8.19	—	CHCl$_3$
1,3,4-Oxadiazole^e	8.73	—	—	8.73	CDCl$_3$

^{a} Coupling constants: $J_{3,4} = 1.78$ Hz; $J_{3,5} = 0.27$ Hz; $J_{4,5} = 1.69$ Hz ⟨74CJC833⟩.
^{b} Coupling constants: $J_{2,4} = 0$ Hz; $J_{2,5} = 0.8$ Hz; $J_{4,5} = 0.8$ Hz.
^{c} ⟨76AHC(20)65, 64HCA942⟩.
^{d} Section 4.22.1.3.1.
^{e} Section 4.23.2.2.1.

For the NH azoles (Table 3), the two tautomeric forms are usually rapidly equilibrating on the NMR timescale (except for triazole in HMPT). The N-methyl azoles (Table 4) are 'fixed'; chemical shifts are shifted downfield by adjacent nitrogen atoms, but more by a 'pyridine-like' nitrogen than by a 'pyrrole-like' N-methyl group.

Table 6 ^{1}H NMR Spectral Data for Ring Hydrogens of Azoles Containing Sulfur

Compound	^{1}H Chemical shifts (δ, p.p.m.)				Solvent
	H-2	H-3	H-4	H-5	
Isothiazole[a]	—	8.54	7.26	8.72	CCl$_4$
Thiazole[b,c]	8.88	—	7.98	7.41	CDCl$_3$
1,2,3-Thiadiazole[g]	—	—	(AB multiplet centered at 8.80)		CCl$_4$
1,2,4-Thiadiazole	—	8.66[d]	—	9.90[e]	
1,2,5-Thiadiazole[h]	—	8.70	8.70	—	CCl$_4$
1,3,4-Thiadiazole[f]	7.55	—	—	7.55	CDCl$_3$

[a] Coupling constants: $J_{3,4} = 11.66$ Hz; $J_{3,5} = 0.15$ Hz; $J_{4,5} = 4.66$ Hz.
[b] Coupling constants: $J_{2,4} = 0$ Hz; $J_{2,5} = 1.95$ Hz; $J_{4,5} = 3.15$ Hz.
[c] ⟨79HC(34-1)67, 79HC(34-1)73⟩.
[d] Value given for 5-phenyl derivative ⟨80JOC3750⟩.
[e] Value given for 3-phenyl derivative ⟨74JOC962⟩.
[f] ⟨78BAP291⟩. [g] ⟨78JOC2487⟩. [h] ⟨64DIS2690⟩.

Comparison of the relevant data shows that an adjacent oxygen (Table 5) and especially a sulfur atom (Table 6) induce lower field shifts than either type of nitrogen atom.

The effects of anion and cation formation on ^{1}H chemical shifts can be assessed from data in Tables 7 and 8. Anion formation always results in shifts to higher field; however, the effect is relatively modest except for the 4-position of pyrazole because in all other cases the adjacent nitrogen lone pair partially cancels the shift. Conversely, in the cations (Table 8), the downfield shift is especially large for the CH groups next to nitrogen. The coupling constants appear to be significantly greater in the cations.

Table 7 ^{1}H NMR Spectral Data for Ring Hydrogens of Azole Anions

Compound	^{1}H Chemical shifts (δ, p.p.m.)				Solvent	Ref.
	H-2	H-3	H-4	H-5		
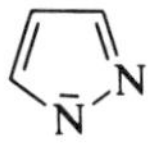 pyrrolide	—	7.35	6.05	7.35	KOD/D$_2$O	68JA4232
 pyrazolide	7.80	—	7.21	7.21	—	71PMH(4)121
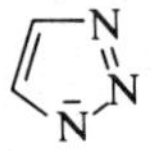 imidazolide	—	—	7.86	7.86	NaOD/D$_2$O	[a]
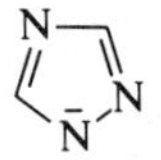 1,2,4-triazolide	—	8.19	—	8.19	NaOD/D$_2$O	71PMH(4)121
 tetrazolide	—	—	—	8.73	—	71PMH(4)121

[a] Data taken from Chapter 4.11.

Relatively few data are available on the ^{1}H NMR spectra of azolinones and related thiones and imines (Table 9).

Some available data on ^{1}H NMR spectra of non-aromatic azoles containing two ring-double bonds are given in Table 10. Here there is no ring current effect and the chemical shifts are consequently more upfield.

Tables 11 and 12 give some available chemical shifts for azolines and azolidines, respectively. Unfortunately data for many of the parent compounds are lacking, often because the compounds themselves are unknown.

Table 8 ^{1}H NMR Spectral Data for Ring Hydrogens of Azole and Related Cations: Two Heteroatoms

Compound	^{1}H Chemical shifts (δ, p.p.m.)				Solvent	Coupling constants, J (Hz)				
	H-2	H-3	H-4	H-5		2,4	2,5	3,4	3,5	4,5
Pyrazolium[a,b]	—	8.57	6.87	8.57	DMSO-d_6	—	—	2.9	—	2.9
Imidazolium[c]	8.6	—	7.5	7.5	H$_2$SO$_4$	1.4	1.4	—	—	2.4
Isoxazolium[i]	—	9.18	7.26	9.01	D$_2$SO$_4$	—	—	2.8	—	1.9
Isothiazolium[a]	—	9.1	7.9	9.6	H$_2$SO$_4$	—	—	2.7	0.6	5.6
Thiazolium[d]	9.55	—	8.23	7.93	—	0.70	1.55	—	—	3.10
1,2-Oxathiolylium[e]	—	—	7.64	—	—	—	—	—	—	—
1,2-Dithiolylium[c]	—	6.71	1.44	−0.26	—	—	—	4.9	—	4.9
1,3-Dioxolylium[f]	10.4	—	—	—	—	—	—	—	—	—
1,3-Oxathiolylium[g]	—	—	8.12	—	—	—	—	—	—	—
1,3-Dithiolylium[h]	11.65	—	9.67	9.67	—	2.0	2.0	—	—	—

[a] Data taken from appropriate chapter, 'Comprehensive Heterocyclic Chemistry'.
[b] Values given for 1,2-dimethylpyrazolium.
[c] Data taken from ⟨71PMH(4)121⟩ and ⟨B-73NMR1⟩ which see for references to the original literature.
[d] ⟨66BSF3524⟩.
[e] Value given for 5-methyl-3-(2-oxo-1-propyl)-1,2-oxathiolylium perchlorate.
[f] Value given for 1,3-benzodioxolylium fluorosulfonate.
[g] Value given for 2,5-diphenyl-1,3-oxathiolylium perchlorate.
[h] ⟨74JOC3608⟩. [i] ⟨83PC40100⟩.

Table 9 ^{1}H NMR Spectral Data for Ring C—H of Azolinones

Compound	^{1}H Chemical shifts (δ, p.p.m.)				Solvent
	H-2	H-3	H-4	H-5	
Pyrazolin-3-one[b]	[a]	—	5.45	7.22	CDCl$_3$
Imidazolin-2-one[c]	—	—	6.50	6.50	—
Pyrazoline-3-thione[b]	—	—	6.23	[d]	—
1,2,4-Triazoline-3-thione	—	—	—	8.20	DMSO
Pyrazolin-3-imine[f]	—	—	5.46	[e]	CDCl$_3$
1,2,4-Triazolin-5-imine	—	2.05	—	—	CDCl$_3$
Δ^2-1,3,4-Oxadiazoline-5-thione[g]	8.88	—	—	—	DMSO-d_6
Isothiazolin-3-one[i]	—	—	6.05	7.98	—
1,3-Thiazolin-2-one[j,k]	—	−1.14	3.21	3.7	DMSO-d_6
Isothiazoline-3-thione[l,m]	—	—	6.90	8.25	—
1,3-Thiazoline-2-thione[k,n]	—	6.68	2.7	3.05	C$_3$D$_6$O
1,3-Thiazolin-2-imine[o,p]	—	—	3.03	3.37	CDCl$_3$

[a] Values given for 1,2-dimethylpyrazolin-3-one; $J_{4,5} = 3.5$ Hz. [b] ⟨76AHC(S1)1⟩. [c] Data taken from ⟨B-73NMR⟩ which contains references to the original literature. [d] Values given for 1,5-dimethyl-2-phenylpyrazoline-3-thione. [e] Values given for 1,5-dimethyl-2-phenylpyrazolin-3-imine. [f] ⟨72BSF2807⟩. [g] Section 4.23.2.2.1. [h] Values given for 2-methylisothiazolin-3-one; $J_{4,5} = 6.0$ Hz. [i] ⟨71JHC571⟩. [j] Coupling constants: $J_{3,4} = 2.5$ Hz; $J_{3,5} = 1.1$ Hz; $J_{4,5} = 5.3$ Hz. [k] ⟨79HC(34-2)385⟩. [l] Values given for 2-methylisothiazoline-3-thione; $J_{4,5} = 6.0$ Hz. [m] ⟨80CPB487⟩. [n] $J_{4,5} = 4.6$ Hz. [o] Values given for 2-ethoxycarbonylimino-3-ethyl-Δ^4-1,3-thiazoline. [p] ⟨79HC(34-2)26⟩.

Proton–proton coupling constants of benzo rings of benzazoles can illuminate the bonding in such compounds. Thus, comparison of the J values for naphthalene with those for benzotriazoles of different types (Table 13) shows evidence of bond fixation, particularly in the 2-methyl derivative (**98**) ⟨71PMH(4)121⟩.

(**98**)

4.01.3.4 ^{13}C NMR Spectroscopy

Chemical shifts for aromatic azoles are recorded in Tables 14–17. As for the proton spectra, fast tautomerism renders two of the chemical shifts equivalent for the NH derivatives (Table 14). However, data for the N-methyl derivatives (Table 15) clearly indicate that the

Table 10 ^{1}H NMR Spectral Data (δ, p.p.m.) for Ring Hydrogens of Non-aromatic Azoles with Two Ring Double Bonds ⟨83UP40100⟩

3*H*-Pyrazoles			
4*H*-Pyrazoles			
2*H*-Imidazoles			
4*H*-Imidazoles			

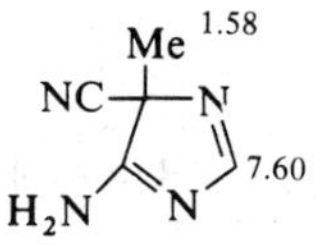

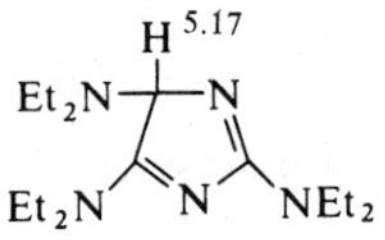

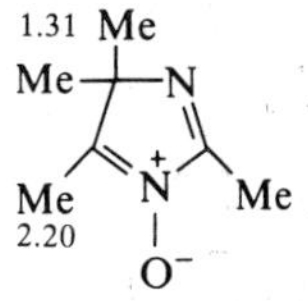

carbon adjacent to a 'pyridine-like' nitrogen shows a chemical shift at lower field than that adjacent to a 'pyrrole-like' *N*-methyl group (in contrast to the H chemical shift behavior). In azoles containing oxygen (Table 16) and sulfur (Table 17), the chemical shifts are generally at lower field than those for the wholly nitrogenous analogues, but the precise positions vary.

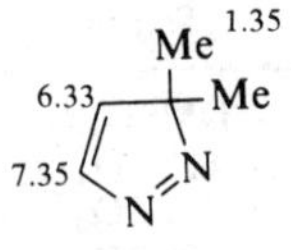

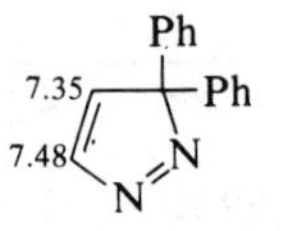

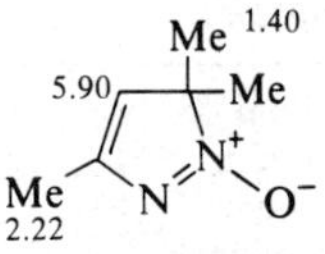

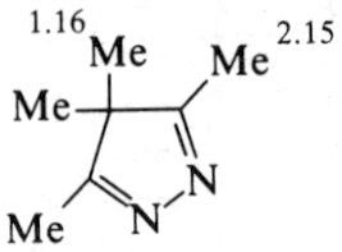

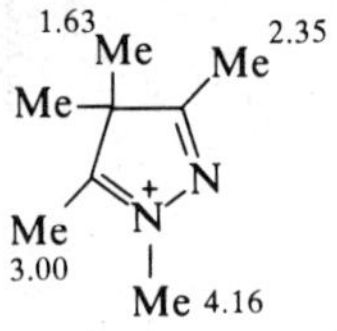

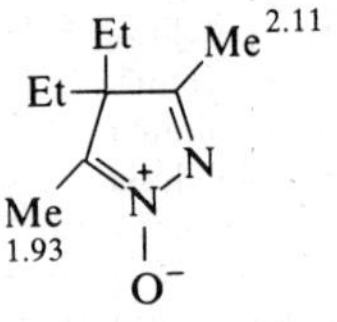

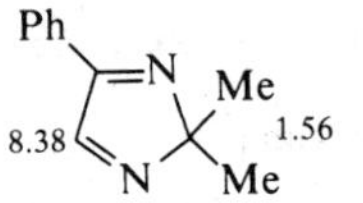

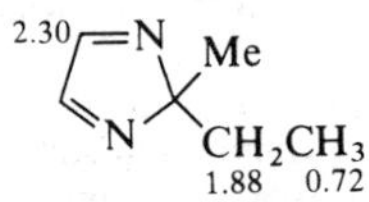

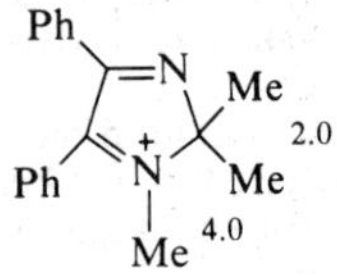

Azolinone derivatives and the corresponding thiones and imines are listed in Table 18; only substituted derivatives have been measured frequently. The ^{13}C chemical shifts of non-aromatic azole derivatives are given in Tables 19–21; relatively few data are available and these are generally for substituted derivatives rather than for the parent compounds.

4.01.3.5 Nitrogen NMR Spectroscopy

Available data are recorded in Table 22. In azoles the chemical shift of a 'pyridine-like' nitrogen atom is around −125 p.p.m. in *N*-methylimidazole or oxazole. Additional nitrogen atoms in the ring cause small upfield shifts if they are non-adjacent, but sizable downfield shifts for adjacent nitrogen atoms. Adjacent ring oxygen gives a large downfield shift, but the effect of ring sulfur is less if adjacent.

A 'pyrrole-like' nitrogen in a ring *N*-methyl group gives a peak around −220 p.p.m. if adjacent to two carbon atoms; one or two neighboring nitrogens shift the peak downfield by about 40 and 90 p.p.m., respectively.

4.01.3.6 UV Spectroscopy

4.01.3.6.1 Parent compounds

In general, aza substitution (replacement of cyclic CH by N) has little effect on UV spectra (Table 23). Typically, aza analogs of pyrrole show λ_{max} at 217 nm or lower with log ε *ca.* 3.5, whereas aza analogs of thiophene show λ_{max} at 230–260 nm with log ε *ca.* 3.7 (Table 23); insufficient data are available for aza analogs of furan to generalize but these compounds appear to have maxima below 220 nm.

Table 11 ^{1}H NMR Spectral Data for Ring Hydrogens of Azolines (Non-aromatic Azoles with One Ring Double Bond)

Compound	Substituents	^{1}H Chemical shifts (δ, p.p.m.)					Solvent	Coupling constants, J (Hz)
		H-1	H-2	H-3	H-4	H-5		
Pyrazole								
2,3-Dihydro-[a]	1,2,3-trimethyl-	2.55	2.60	1.71	4.60	3.65	CDCl$_3$	3,5 = 1.8; 4,5 = 1.8
4,5-Dihydro-3H-[a]	—	—	—	4.27	1.46	4.27	Neat	—
4,5-Dihydro-[a]	—	5.33	—	6.88	2.65	3.31	CDCl$_3$	3,4 = 1.4; 4,5 = 9.8
1,2,3-Triazole								
4,5-Dihydro-[b,d]	—	—	—	—	5.35	4.77	—	4,5 = 2.0
[c,d]	—	—	—	—	5.73	4.60	—	4,5 = 7.5
1,2,4-Oxadiazole								
2,5-Dihydro-[e]	2,5-dimethyl-3-phenyl-	—	—	—	—	6.0	—	—
4,5-Dihydro-[f]	5-ethyl-3-phenyl-	—	—	—	5.38	5.64	—	4,5 = 4.5
1,3,4-Oxadiazole								
2,3-Dihydro-[a]	3-benzoyl-5-phenyl-	—	5.9	—	—	—	—	—
Thiazole								
2,5-Dihydro-[i]	—	—	4.79^g	3.03^g	6.12^g	—	—	2,4 = 2.2; 4,5 = 8.6
4,5-Dihydro-[j]	—	—	2.26	—	5.83	6.84	—	—
1,3,4-Thiadiazole								
2,3-Dihydro-[k]	2,3,5-triphenyl-	—	—	—	—	6.38	—	—
1,2,4-Dioxazole								
2,3-Dihydro-[a]	3-carboxymethyl-5-phenyl-	—	—	6.39	—	—^{l}	CDCl$_3$	—
1,3,4-Dioxazole								
2,3-Dihydro-[a]	2-benzyl-5-phenyl-	—	6.20	—	—	—^{l}	CDCl$_3$	—
1,3,4-Oxathiazole								
2,3-Dihydro-[a]	5-methyl-2-trichloromethyl-	—	6.30	—	—	—^{l}	CDCl$_3$	—
1,2,4-Dithiazole								
2,3-Dihydro-[a]	5-phenyl-	—	—	5.70	—	—^{l}	CCl$_4$	—

[a] Data taken from appropriate chapter, 'Comprehensive Heterocyclic Chemistry'. [b] Values given for trans-5-propyloxy-4-methyl-1-(4-nitrophenyl) derivative. [c] Values given for cis isomer of compound named in footnote b. [d] ⟨65CB1153⟩. [e] ⟨73BSF2996⟩. [f] ⟨77OC1555⟩. [g] Value taken from 5,5-dimethyl-Δ^3-thiazoline. [h] Value taken from 2,2,4-trimethyl-Δ^3-thiazoline. [i] ⟨66BSF3524⟩. [j] ⟨64M140100⟩. [k] ⟨81JCS(P1)360⟩. [l] Substituent in this position.

Table 12 ^{1}H NMR Spectral Data for Ring Hydrogens of Azolidines (Non-aromatic Azoles without Ring Double Bonds)

| Compound | Substituents | ^{1}H Chemical shifts (δ, p.p.m.) | | | | Solvent |
		H-2	H-3	H-4	H-5	
Tetrahydro						
-pyrazole[a]	1,2-dimethyl-3-phenyl-	—	6.51	7.85	6.76	CDCl$_3$
-thiazole[b,c]	—	5.9	8.2	6.8	7.2	—
-1,2,4-oxadiazole[d]	2-t-butyl-3,4-diphenyl-5-thioxo-	—	5.93	—	—	—
-1,3,4-oxadiazole[e]	3,4-dimethyl-	4.27	—	4.27	—	—
-1,2,4-dioxazole[e]	3,5-di-n-propyl-	—	4.63	—	4.63	—
-1,3,2-dioxazole[e]	N-alkyl-	—[f]	—	3.97	3.97	—
-1,3,4-dioxazole[e]	2,5-di-t-butyl-4-phenyl-	4.96	—	—[f]	4.58	—
-1,2,4-dithiazole[e]	3,4-dialkyl-3-phenyl-	—	—[f]	—[f]	4.72	—
-1,3,2-dithiazole[e]	2-methyl-	—[f]	—	3.58	3.58	—

[a] ⟨71T133⟩. [b] $J_{4,5}$ values are $J_{A,B'} = J_{A',B} = 7.61$ Hz; $J_{A,B} = J_{A',B'} = 4.71$ Hz; $J_{5,5} = -13.7$ Hz; $J_{4,4} = -7.2$ Hz.
[c] ⟨74JA1465⟩. [d] ⟨74JOC957⟩. [e] Data taken from appropriate chapter, 'Comprehensive Heterocyclic Chemistry'.
[f] Substituent in this position.

Table 13 Proton–Proton Coupling Constants (Hz) in Benzotriazoles[a]

Compound	$J_{4,5}$	$J_{5,6}$	$J_{4,6}$
Benzotriazole[b]	8.3	6.7	1.4
2-Methylbenzotriazole	9.4	3.6	0.5
Naphthalene	8.6	6.0	1.4

[a] Data taken from ⟨71PMH(4)141⟩ which contains references to the original literature.
[b] Rapid tautomerism between 1- and 3-positions occurs.

Table 14 ^{13}C NMR Chemical Shifts for Nitrogenous Azoles: (a) NH Derivatives

| Compound | ^{13}C Chemical shifts (p.p.m.)[a] | | | | Solvent | Ref. |
	C-2	C-3	C-4	C-5		
Pyrazole	—	134.6	105.8	134.6	CH$_2$Cl$_2$	b
Imidazole	135.9	—	122.0	122.0	—	c
1,2,3-Triazole	—	—	130.4	130.4	Me$_2$CO	b
1,2,4-Triazole	—	147.8	—	147.8	—	c
Tetrazole	—	—	—	144.2	—	c

[a] All chemical shifts expressed in p.p.m. from TMS (original values converted where necessary).
[b] Data taken from appropriate chapter, 'Comprehensive Heterocyclic Chemistry'.
[c] Data taken from ⟨71PMH(4)121⟩, which contains references to the original literature.

Table 15 ^{13}C NMR Chemical Shifts for Nitrogenous Azoles: (b) N(1)-Methyl Derivatives

| Compound | ^{13}C Chemical shifts (p.p.m.)[a] | | | | Solvent | Ref. |
	C-2	C-3	C-4	C-5		
Pyrazole	—	138.7	105.1	129.3	CDCl$_3$	b
Imidazole	138.3	—	129.6	120.3	—	74JOC357
1,2,3-Triazole	—	—	134.3	125.5	DMSO-d_6	b
1,2,3,4-Tetrazole	—	—	—	142.1	DMSO-d_6	74JOC357
1,2,3,5-Tetrazole	—	—	151.9	—	DMSO-d_6	74JOC357

[a] All chemical shifts expressed in p.p.m. from TMS (original values converted where necessary).
[b] Data taken from appropriate chapter, 'Comprehensive Heterocyclic Chemistry'.

Table 16 ^{13}C NMR Chemical Shifts for Azoles Containing Oxygen[a]

Compound	^{13}C Chemical shifts (p.p.m.)[b]				Solvent
	C-2	C-3	C-4	C-5	
Oxazole	150.6	—	125.4	138.1	CDCl$_3$
1,3,4-Oxadiazole[c]	159.5	—	—	166.3	—
1,2,5-Oxadiazoles[d]	—	139–160	—	—	—

[a] Data taken from appropriate chapter, 'Comprehensive Heterocyclic Chemistry'.
[b] All chemical shifts expressed in p.p.m. from TMS (original values converted where necessary).
[c] Values given for 5-methoxy-2-methyl derivative.
[d] 3-Substituted 4-phenyl derivatives.

Table 17 ^{13}C NMR Chemical Shifts for Azoles Containing Sulfur

Compound	^{13}C Chemical shifts (p.p.m.)[a]				Solvent	Ref.
	C-2	C-3	C-4	C-5		
Isothiazole	—	157.0	123.4	147.8	CDCl$_3$	[b]
Thiazole	153.4	—	143.7	119.7	—	79HC(34-1)76
1,2,3-Thiadiazole	—	—	147.3	135.8	CDCl$_3$	[b]
1,2,4-Thiadiazole	—	170.0	—	187.1	—	81JPR279
1,3,4-Thiadiazole 3-amino-5-methylthio-	152.7	—	—	152.7	CDCl$_3$	78BAP291
1,2,3,4-Thiatriazole 5-phenyl-	—	—	—	178.5	CDCl$_3$	[b]

[a] All chemical shifts expressed in p.p.m. from TMS (original values converted where necessary).
[b] Data taken from appropriate chapter, 'Comprehensive Heterocyclic Chemistry'.

Table 18 ^{13}C NMR Chemical Shifts for Azolinones

Compound	^{13}C Chemical shifts (p.p.m.)[a]				Solvent	Ref.
	C-2	C-3	C-4	C-5		
2-Pyrazolin-5-one 3-methyl-1-phenyl-	—	156.2	43.0	170.6	—	[b]
3-Pyrazolin-5-one 2,3-dimethyl-1-phenyl-	—	156.0	98.1	165.7	—	[b]
Imidazolin-2-one	183.8	—	44.4	44.4	DMSO-d_6	80CS(15)193
Imidazoline-2-thione	164.8	—	40.3	40.3	DMSO-d_6	80CS(15)193
Tetrazolin-5-one 1-phenyl-	—	—	—	148.7	—	[b]
Tetrazoline-5-thione	—	—	—	162.9	—	[b]
Δ^4-Thiazoline-2-thione	—	118.4	128.9	114.0	CHCl$_3$	79HC(34-1)388
1,2,3-Oxadiazolin-5-imine 3-methyl-	—	—	97.3	170.4	—	80RCR28
Δ^2-1,3,4-Oxadiazolin-5-one	145.7	—	—	155.7	—	82OMR(18)159
Δ^2-1,3,4-Thiadiazoline-5-thione	157.2	—	—	186.4	—	77JOC3725
1,3,4-Oxathiazolin-2-one 5-methyl-	174.2	—	—	158.7	—	[b]
1,2,4-Dithiazoline-3-thione 5-anilino-	—	209.3	—	179.1	—	[b]

[a] All chemical shifts expressed in p.p.m. from TMS (original values converted where necessary).
[b] Data taken from appropriate chapter, 'Comprehensive Heterocyclic Chemistry'.

Table 19 ^{13}C NMR Chemical Shifts for Non-aromatic Azoles with Two Ring Double Bonds
⟨83UP40100⟩

Compound	^{13}C Chemical shifts (p.p.m.)			
	C-2	C-3	C-4	C-5
4H-Pyrazoles	—	178–182	63–68	178–182
3H-Pyrazoles	—	93–110	125–150	130–170
2H-Imidazoles	101–119	—	158–165	158–165
4H-Imidazoles	163–181	—	66–115	180–190

Table 20 ^{13}C NMR Chemical Shifts for Azolines (Non-aromatic Azoles with One Ring Double Bond)

Compound	^{13}C Chemical shifts (p.p.m.)[a]				Ref.
	C-2	C-3	C-4	C-5	
Δ^2-Pyrazoline	—	142.9	33.2	45.4	80TH40100
Δ^2-Imidazoline					
1-methyl-2-methylthio-	165.3	—	[b]	54.3	76TL3313
Δ^4-Imidazoline-2-thione					
1-methyl-	161.3	—	[b]	119.8	76TL3313
Δ^2-Thiazoline	229.4	—	321.7	354.7	70CR(C)(270)1688
1,3,4-Thiadiazoline					
2-benzylamino-5-phenyl-5-tosyl-	151.5	—	—	149.5	[c]
Δ^2-1,3,4-Oxathiazoline					
2-phenyl-5-trichloromethyl-	95.9	—	—	157.6	[c]
1,2,4-Dithiazoline	—	146.36[d]	—	156.19[e]	[c]

[a] All chemical shifts expressed in p.p.m. from TMS (original values converted where necessary).
[b] No C-4 shift reported.
[c] Data taken from appropriate chapter, 'Comprehensive Heterocyclic Chemistry'.
[d] $^1J_{C,F} = 275.7$ Hz.
[e] $^3J_{C,F} = 15.9$ Hz.

Table 21 ^{13}C NMR Chemical Shifts for Azolidines (Non-aromatic Azoles without Ring Double Bonds)

Compound	^{13}C Chemical shifts (p.p.m.)[a]				Ref.
	C-2	C-3	C-4	C-5	
Imidazoline-2-thione, 1-methyl-	183.2	—	[b]	50.4	76TL3313
Oxazolidine, cis-4-methyl-5-phenyl-	85.4	—	60.3	80.3	79MI40100
Thiazole, tetrahydro-	248.4	—	245.9	226.9	74CR(C)(279)717
1,3,4-Thiadiazolidine, 2,2-dimethyl-4-phenyl-					
5-phenylimino-	79.2	—	—	176.3	80JCS(P1)574
1,3,2-Dioxazole, tetrahydro-N-alkyl-	—	—	67.6	67.6	[c]
1,2,3-Oxathiazole, dihydro-N-phenyl, S-oxide	—	—	45.9	70.9	[c]

[a] All chemical shifts expressed in p.p.m. from TMS (original values converted where necessary).
[b] C-4 shift not reported.
[c] Data taken from appropriate chapter, 'Comprehensive Heterocyclic Chemistry'.

Table 22 ^{15}N NMR Chemical Shifts for Nitrogenous Azoles[a,b]

Nitrogen position(s)	Positions of ring nitrogen	^{15}N Chemical shifts with heteroatom in 1-position (p.p.m.)			
		NH	NMe	O	S
2-Aza	1	−135	−181	—	—
	2	−135	−76	+3	−82
3-Aza	1	−173	−221	—	—
	3	−173	−125	−124	−57
2,3-Diaza	1	−128	−144	—	—
	2	−60	−15	−35[c]	—
	3	−60	−28	−106[d]	—
2,4-Diaza	1	−174	−173	—	−106
	2	—	−84	−20	−186[e]
	4	−132	−130	−140	−70
2,5-Diaza	1	−128	−132	—	—
	2,5	−60	−53	+33	−35
3,4-Diaza	1	—	−222	—	—
	3,4	—	−82	−82	−10
2,3,4-Triaza	1	−106	−159	—	—
	2	−15	−14	—	—
	3	−25	+9	—	—
	4	−106	−54	—	—
2,3,5-Triaza	1	—	−99	—	—
	2	—	−4	—	—
	3	—	−43	—	—
	5	—	−69	—	—

[a] Chemical shifts expressed in p.p.m. referred to internal MeNO$_2$.
[b] Data taken from ⟨B-73MI40100⟩ or ⟨81MI40100⟩, which contain references to the original literature.
[c] Value for 3-methylsydnone ⟨80OMR(13)274⟩.
[d] Value for 5-acetyl-3-methylsydnonimine ⟨80OMR(13)274⟩.
[e] Value for 2-amino-5-methyl-1,3,4-thiadiazole ⟨78H(11)121⟩.

Relatively few data are available for protonated cationic species, but from what there are it appears that protonation has little effect on the position and intensity of the absorption.

4.01.3.6.2 Benzo derivatives

Benzo derivatives show at least two, and up to seven, maxima in the range 200–320 nm (Table 24). The longest wavelength maximum occurs at 275–315 nm and generally at rather longer wavelengths for the sulfur derivatives than for their N or O analogues, and also for the benzo[c] compared with the benzo[b] derivatives.

Table 23 UV Absorption Maxima for Azoles[a]

Additional nitrogen atoms	Z = NH λ_{max} (nm) (log ε)	Z = O λ_{max} (nm) (log ε)	Z = S λ_{max} (nm) (log ε)
2	210 (3.53)	211 (3.60)	244 (3.72)
3	207–208 (3.07) end absorption	240	207.5 (3.41), 233 (3.57)[b]
2,3-di	210 (3.64)	—	211 (3.64), 249 (3.16), 294 (2.29)[c]
2,4-di	216.5 (3.66)	—	229 (3.73)
2,5-di	—	220	250 (3.86), 253 (3.87), 257 (2.83), 260 (3.68)[d]
3,4-di	—	200[e]	220
2,3,4-tri	205	—	280 (4.03)[f]

[a] Unless otherwise indicated data taken from ⟨71PMH(3)67⟩ which contains references to the original literature.
[b] Measured in ethanol.
[c] Measured in cyclohexane.
[d] Measured in isooctane; cf. Section 4.26.2.4.
[e] 2-Methyl-1,3,4-oxadiazole exhibits a maximum at 206 nm (log ε 2.62) in MeOH; cf. Section 4.23.2.2.2.
[f] 5-Phenyl-1,2,3,4-thiatriazole.

Table 24 UV Absorption Maxima for Benzazoles[a]

X	Y	Z = NH λ_{max} (nm) (log ε)	Z = O λ_{max} (nm) (log ε)	Z = S λ_{max} (nm) (log ε)
N	CH	250 (3.65), 284 (3.63), 296 (3.52)	235 (4.00), 243 (3.91), 280 (3.46)	205 (4.20), 222 (4.37), 252 (3.56), 261 (3.39), 297 (3.58), 302 (3.57), 3.08 (3.56)[b]
CH	N	242 (3.72), 265 (3.58), 271 (3.70), 277 (3.69)	231 (3.90), 263 (3.38), 270 (3.53), 276 (3.51)	217 (4.27), 251 (3.74), 285 (3.23), 295 (3.13)
N	N	259 (3.75), 275 (3.71)	[c]	213.5 (4.20), 266 (3.72), 312.5 (3.40)
S+	N	—	—	238 (3.91), 350 (4.35), 425 (3.29)[d]

X	Y	Z = NMe λ_{max} (nm) (log ε)	Z = O λ_{max} (nm) (log ε)	Z = S λ_{max} (nm) (log ε)
N	CH	275 (3.80), 292 (3.79), 295 (3.78)	—	203 (4.16), 221 (4.21), 288s (3.88), 298 (3.46), 315s (3.60)[e]
N	N	274 (3.96), 280 (3.98), 285 (3.97)	310 (3.5), 275 (3.6)	221–222 (4.16), 304 (4.14), 310 (4.14), 330s (3.39)

[a] Unless otherwise indicated data taken from ⟨71PMH(3)67⟩ which contains references to the original literature.
[b] ⟨73JHC267⟩.
[c] Unknown; see Scheme 2 in Section 4.21.1.
[d] In concentrated H_2SO_4 ⟨69ZOR153⟩.
[e] Data taken from Section 4.17.3.6.

Table 25 Effects of Substituents on UV Absorption Maxima (nm)[a,b]

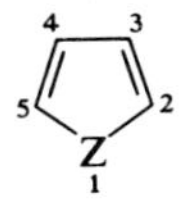

		Z = NH Substituent position				Z = O Substituent position				Z = S Substituent position			
Substituent	Aza positions	2	3	4	5	2	3	4	5	2	3	4	5
Me	2	—	213	220	(≡3)	—	217	221	213	—	246.5	251[c]	243[c]
	3	210	—	215	(≡4)	235	—	—	—	235	—	241	239
	2,3	—	—	216	(≡4)	—	—	—	—	—	—	213	217
	2,4	—	<220	—	(≡3)	—	—	(≡3)	—	—	257[m]	(≡3)	—
	2,5	—	—	—	—	<230	—	—	(≡2)	<230	—	—	(≡2)
	3,4	—	—	—	—	206	—	—	—	—	—	—	—
Ph	2	—	248	250.5	(≡3)	—	239	—	260	—	270,291	266[d]	266[d]
	3	271	—	260	(≡4)	263	—	243	267	287	—	252	275
	2,3	—	—	245	(≡4)	—	310[e]	—	—	—	—	—	—
	2,4	—	241.5	—	(≡3)	—	238[e]	(≡3)	250[e]	—	280[m]	(≡3)	—
	2,5	—	—	—	—	247.5	—	—	—	268	—	—	—
	3,4	—	—	—	241	—	—	—	(≡2)	—	—	—	(≡2)
	2,3,4	—	—	—	—	—	—	—	—	—	—	—	—

Substituent	Pos.	C1	C2	C3	C4	C5	C6	C7	C8	C9	C10	C11	C12
Br (Cl)	2	—	—	221	(≡3)	—	—	—	245	—	(244)ᶠ	256ᶜ	241ᶜ
	3	220*	—	217	(≡4)	—	—	—	—	—	—	246	247
OMe (OH)	2	—	—	—	(≡3)	—	220	—	—	239	255ᵍ	—	—
	3	—	—	—	(≡4)	—	—	—	—	—	—	—	—
	2,3	—	—	234*	(≡4)	—	—	—	—	—	—	—	—
	2,5	—	—	—	(≡3)	—	—	—	—	—	273	(≡3)	—
	2,3,4	—	—	—	213*	—	—	—	—	—	—	—	—
NH₂ (NMe₂)	2	—	228	—	(≡3)	245.5	228.5	—	—	254.4*	278ʰ	286ᶜ	(258)
	3	—	—	230*	(≡4)	—	—	—	—	—	—	—	—
	2,3	—	—	228	(≡4)	—	—	—	—	—	—	—	—
	2,4	<220	—	—	(≡3)	—	—	—	—	—	—	—	—
	3,4	—	—	—	—	—	—	—	—	251	—	—	247ⁱ
	2,3,4	—	—	—	218	(≡2)	—	—	—	—	—	—	(≡2)
CO₂Me (CO₂H, CHO)	2	—	217	—	(≡3)	216ʲ	—	—	—	244	256	(250)ᶠ	263ᶜ
	3	—	—	285	(≡4)	—	—	—	—	—	—	230	—
	2,5	—	—	—	(≡3)	(≡2)	—	243ᵏ	—	—	263ᵐ	(≡3)	—
	3,4	—	—	—	—	(≡2)	—	—	—	(≡2)	—	—	(≡2)
NO₂	2	—	261	275	(≡3)	239	—	—	—	—	272	—	—
	3	—	—	298	(≡4)	—	—	—	—	—	—	—	—
	2,4	—	—	—	(≡3)	—	—	—	—	—	—	—	—
	2,5	325	215, 245ˡ	—	—	(≡3)	—	—	—	—	—	(≡3)	—

ᵃ Unless otherwise indicated, data taken from ⟨71PMH(3)67⟩, which contains references to the original literature. A dash indicates that the value has not been reported. ᵇ Asterisk (*) indicates, exceptionally, acidified or basified solvent; for details, see references quoted in ⟨71PMH(3)67⟩. ᶜ ⟨64JCS446⟩. ᵈ In cyclohexane ⟨66T2119⟩. ᵉ ⟨64HCA942⟩. ᶠ Value based on 'average increment' observed for a large number of compounds; cf. Chapter 4.17. ᵍ ⟨70T2497⟩. ʰ Perchlorate salt in MeOH ⟨71JHC657⟩. ⁱ ⟨54CB57⟩. ʲ 2-Benzyl-5-methyl derivative. ᵏ 2-Benzyl-5-methyl derivative. ˡ In water. ᵐ cf. Section 4.26.2.4.

4.01.3.6.3 Effect of substituents

Although UV spectra have been measured for a large number of substituted azoles, there has been no systematic attempt to explain substituent effects on such spectral maxima. Readily available data are summarized in Table 25, and some major trends are apparent. However, detailed interpretation is hindered by the fact that different solvents have been used and that in aqueous media it is not always clear whether a neutral, cationic or anionic species is being measured. Furthermore, values below 220 nm are of doubtful quantitative significance.

4.01.3.7 IR Spectroscopy ⟨71PMH(4)265, 63PMH(2)161⟩

4.01.3.7.1 Aromatic rings without carbonyl groups

For many of the parent compounds, complete assignments have been made ⟨71PMH(4)265⟩. For substituted derivatives, group frequencies have been derived. In the ring systems under discussion, IR bands may be placed in the following categories. (a) CH stretching modes near 3000 cm^{-1} which are of little diagnostic utility. (b) Ring stretching modes at 1650–1300 cm^{-1} (Table 26), four or five bands generally being found which lie in well-defined regions. The intensities of these bands vary according to the nature and orientation of substituents. (c) CH and ring deformation modes at 1300–1000 cm^{-1} (Table 27) and at 1000–800 cm^{-1} (Table 28). The former are largely in-plane CH and the latter out-of-plane CH and in-plane ring deformation modes. (d) Substituent vibrations: these are discussed later.

4.01.3.7.2 Azole rings containing carbonyl groups

The carbonyl group is found in a wide variety of situations in five-membered rings accommodating diverse heteroatoms. Carbonyl frequencies to be discussed range from about 1800 to below 1600 cm^{-1}; some such frequencies are so low that they often fail to be recognized as carbonyl absorptions at all, e.g. carbonyl joined to a heterocyclic ring via an exocyclic double bond as in structure (**99**). Low frequency ν(C=O) bands are widespread because five-membered heterocyclic rings are inherently π-electron donors. The transfer of electrons into the exocyclic double bond [ν(C=Z) of **99**] is likely to be extensive when both X and Y are electron donor atoms.

(**99**)

In Table 29 the ν(C=O) and other characteristic bands are given for some saturated five-membered heterocycles, and compared with the corresponding absorption frequencies for cyclopentanone. Adjacent NH groups and sulfur atoms have the expected bathochromic effect on ν(C=O), whereas an adjacent oxygen atom acts in the reverse direction. The CH$_2$ vibrations of cyclopentanone are repeated to a considerable extent in the heterocyclic analogs.

Table 30 reports ν(C=O) for a variety of azolinones containing ring double bonds. The hypsochromic effect of an oxygen atom or CR$_2$ group versus the bathochromic effect of NR, S or C=C can readily be traced.

4.01.3.7.3 Substituent vibrations

In general, substituent frequencies in azoles are consistent with those characteristic of the same substituents in other classes of compounds. Some characteristic trends are found, and these have been used to measure electronic effects. Thus, for example, the frequencies of ν(C=O) in 3-, 4- and 5-alkoxycarbonylisoxazoles (cf. **100**) are respectively 9–12, 2–8 and 17–18 cm^{-1} higher than those of the corresponding alkyl benzoates, indicating the

Table 26 Azoles: IR Ring Stretching Modes in the 1650–1300 cm^{-1} Region[a]

Compound	Stretching modes (cm^{-1})				
Isoxazoles	1650–1610	1580–1520	1510–1470	1460–1430	1430–1370
Isothiazoles	—	—	1488	1392	1342
Pyrazoles	—	1600–1570	1540–1510	1490–1470	1380–1370
Oxazoles	1650–1610	1580–1550	1510–1470	1485	1380–1290
Thiazoles	1625–1550	—	1550–1470	1440–1380	1340–1290
Imidazoles	1605	1550–1520	1500–1480	1470–1450	1380–1320
1,2,4-Oxadiazoles	—	1590–1560	—	1470–1430	1390–1360
1,2,5-Oxadiazoles	—	1630–1560	1530–1515	1475–1410	1395–1370
1,3,4-Oxadiazoles	1680–1650	1630–1610	1600–1580	1430–1410	—
1,2,4-Thiadiazoles	—	1590–1560	1540–1490	—	—
1,2,3-Thiadiazoles	1650–1590	—	1560–1420	1350–1325	1260–1180
1,2,5-Thiadiazoles[b]	—	—	—	1461	1350
1,2,3,4-Thiatriazoles	1720–1690	1610–1530	—	—	1300–1260
1,2,3-Triazoles	1650–1615w	—	1530–1485	1440w	1420–1400w
1,2,4-Triazoles	—	—	1545–1535	1470–1460	1365–1335
Tetrazoles	1640–1615	—	1450–1410	1400–1335	1300–1260

[a] Data taken from ⟨71PMH(4)265⟩, which contains references to the original literature; w = weak.
[b] Data taken from Section 4.26.2.5.

Table 27 Azoles: Characteristic IR Bands in the 1300–1000 cm^{-1} Region[a]

Compound	β(CH) modes (cm^{-1})			Ring breathing (cm^{-1})
Isoxazoles	1218m	1155–1130	1088s	1028–1000
Isothiazoles	—	1070m	1060m	980s
Pyrazoles	1310–1130	1160–1090	1090–990	1040–975
Thiazoles	1240–1230	1160–1075	1105–1055	1040
Imidazoles	1285–1260	1140	1100	1060
1,2,4-Oxadiazoles	—	—	1070–1050	—
1,2,5-Oxadiazoles	1360–1175	1190–1150	1160–1150	1035–1000
1,2,4-Thiadiazoles	1270–1215	1185–1170	1160–1080	1050
1,2,3-Thiadiazoles	—	—	1150–950	—
1,2,5-Thiadiazoles[b]	1251–1227	—	1041	—
1,3,4-Thiadiazoles	1230–1165	1190–1120	1075–1045	1040
1,2,3,4-Thiatriazoles	1235–1210	1120–1090	1060–1030	—
1,2,3-Triazoles	1300–1275	1150–1070	1095–1045	1005–970
Tetrazoles	1210–1110	1170–1035	1060	995–900

[a] Data taken from ⟨71PMH(4)265⟩, which contains references to the original literature; s = strong, m = medium, w = weak.
[b] Data taken from Section 4.26.2.5.

Table 28 Azoles: Characteristic IR Bands below 1000 cm^{-1} [a]

Compound	CH modes (?) (cm^{-1})		β-Ring (?) (cm^{-1})	CH modes (?) (cm^{-1})
Isoxazoles	970–920	899	945–845	774
Isothiazoles	915w	—	810s	740vs
Pyrazoles	960–930	860–855	805–790	765–750
Thiazoles	980–880	890–785	800–700	745–715
Imidazoles	970–930	895	840	760
1,2,4-Oxadiazoles	—	—	915–885	750–710
1,2,5-Oxadiazoles	980–900	—	890–825	715–700
1,2,3-Thiadiazoles	—	—	910–890	—
1,2,4-Thiadiazoles	1030–935	890	860–795	750–740
1,2,5-Thiadiazoles	—	—	860–800	780[b]
1,3,4-Thiadiazoles	975–905	905–875	850	775–750
1,2,3,4-Thiatriazoles	960–930	—	910–890	—
1,2,3-Triazoles	—	855–825	970–700	—
1,2,4-Triazoles	—	—	865–855(?)	—
Tetrazoles	960	—	—	810–775

[a] Data taken from ⟨71PMH(4)265⟩ which contains references to the original literature; vs = very strong, s = strong, w = weak.
[b] Data taken from Section 4.26.2.5.

Table 29 IR Absorption Assignments for 2,5-Dihetero Derivatives of Cyclopentanone[a]

C=X: *2-Position:* *5-Position:* *Vibration type*	*C=O* CH_2 *O* (cm^{-1})	*C=O* CH_2 *N* (cm^{-1})	*C=S* CH_2 *N* (cm^{-1})	*C=O* *O* *O* (cm^{-1})	*C=O* *S* *S* (cm^{-1})	*C=S* *S* *S* (cm^{-1})	*C=S* *O* *NH* (cm^{-1})	*C=O* *O* *NH* (cm^{-1})	*C=S* *S* *NH* (cm^{-1})	*C=O* *NH* *NH* (cm^{-1})	*C=S* *NH* *NH* (cm^{-1})
ν(C=X)	1770	1695	1109	1795	1638	1058	1171	1724	1047	1661	1208
CH_2 scissors	1490 1466 1426 1379	1494 1440 1426 1378	1478 1458 1420	1483 1422 1394	1434 1422	1416 1370	1464 1402	1485 1412	1458 1432 1380	1488 1449	1459 1368
CH_2 wag	1284 1242	1270 1230	1309 1215	1226	1275 1254	1275 1243	1319 1230	1333 1230	1250 1203	1200	—
CH_2 twist	1192	1169	1168	1175	1158	1148	1203	—	1160	—	—
CH_2 rock	990	995	970	1005	983	983	969	967	998	988	980
ν ring	1166 1037 890 929	1285 1068 915 887	1290 1059 915 880	1140 1071 971 894	888 826 677 939	882 831 670 946	1290 1035 942 914	1250 1077 1021 918	1294 1080 650 927	1270 1103 1037 933	1273 1042 1003 919
γ ring	800	805	493	773	—	457	—	770	—	768	—

[a] Data taken from ⟨63PMH(2)161⟩, which contains references to the original literature.

Table 30 ν(C=O) Frequencies for Some Azolinones[a]

Ring system substituent(s) R = H or Me	Z = NH (cm^{-1})	Z = O (cm^{-1})	Z = S (cm^{-1})
(structure)	1705–1695[b]	—	—
(structure)	1680–1630	—	—
(structure)	—	—	1660[c]
(structure)	1684–1677[d]	1780–1750	1640[e]
(structure)	—	1770–1740	1780–1700
(structure)	—	ca. 1820	1725[f]

[a] Unless otherwise indicated, data taken from appropriate chapter, 'Comprehensive Heterocyclic Chemistry'.
[b] Z = NR', R' = alkyl, aryl.
[c] R = Me ⟨64TL1477⟩.
[d] Data taken from ⟨63PMH(2)161⟩ which contains references to the original literature.
[e] ⟨79HC(34-2)421⟩.
[f] ⟨79HC(34-2)430⟩.

following order of electron donor power: phenyl > 4- > 3- > 5-position of isoxazole. Similar work has been reported for other ring systems and substituents ⟨63PMH(2)161⟩.

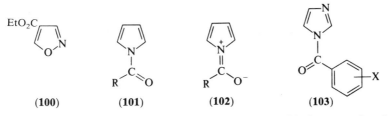

(100)   (101)   (102)   (103)

Otting ⟨56CB1940, 57MI40100⟩ has shown that the ν(C=O) for acetylazoles (cf. **101**) increases with the number of cyclic nitrogen atoms (Table 31). Additional nitrogen atoms in the ring act as powerful electron-withdrawing substituents and decrease the importance of forms such as (**102**). Staab et al. ⟨57MI40100⟩ give data for benzo analogs (Table 31) and

Table 31 Carbonyl Frequencies for N-Acetylazoles[a]

N-Acetylazole	ν(C=O) (cm^{-1})	N-Acetylazole	ν(C=O) (cm^{-1})
Pyrrole	1732	Indole	1711
Imidazole	1747	Benzimidazole	1729
1,2,4-Triazole	1765	Benzotriazole	1735
Tetrazole	1779		

[a] Data taken from ⟨56CB1940⟩ and ⟨57MI40100⟩ which contain references to the original literature.

Table 32 Melting and Boiling Points[a,b,c]

Ring system	H	Me	Et	COMe	CO$_2$H	CO$_2$Et	CONH$_2$	CN	NH$_2$	OH	OMe	SH	SMe	Cl	Br
Benzene	80	111	136	212	122	211	130	190	184	43	154	168	187	131	155
Pyrrole-1	130	114	129	180	95d	180	166	—	—	—	—	—	—	—	—
Pyrrole-2	130	148	181	90	205d	39	174	—	—	83g	—	—	208–14	—	—
Pyrrole-3	130	158	179	115	148	78d	152	147	68	80	110	—	—	78	103
Furan-2	31	64	92	31	133	34	142	—	—	58	—	—	—	80	103
Furan-3	31	65	—	54	122	179	168	196	214	217	156	166	—	128	150
Thiophene-2	84	113	133	214	129	218	180	179	—	—	—	171	—	136	157
Thiophene-3	84	115	135	57	138	208	178	37	—	—	—	—	—	—	—
Pyrazole-1	70	127	137	234	103	213	141	150	40	166	—	—	>370d	40	70
Pyrazole-3	70	205	209	101	214d	160	148	92	81	118	62	—	—	77	97
Pyrazole-4	70	207	244	114	278	79	—	168	—	—	—	—	—	—	—
Isoxazole-3	95	118	139	16	149d	—	134	—	—	—	—	—	—	130	130
Isoxazole-4	95	127	—	52	—	—	174	—	—	—	—	—	—	—	—
Isoxazole-5	95	121	—	102	149	218	—	—	—	—	—	—	—	—	—
Imidazole-1	90	199	226	80	164d	157	—	—	—	250d	—	227	139	165	207
Imidazole-2	90	141	80	—	275d	—	215	—	—	—	—	—	—	—	130
Imidazole-4	90	56	—	—	—	—	—	—	97	—	—	—	—	—	—
Oxazole-2	69	87	—	—	142	48	118	—	92	—	164	—	230	145	147
Oxazole-4	69	—	—	226	102d	48	150	31	83	—	—	79	—	165	190
Thiazole-2	118	128	158	56	196	52	186	60	51	—	176	—	222	140	192
Thiazole-4	118	133	142	—	218	217	—	53	—	—	—	—	—	—	—
Thiazole-5	118	141	238	40–2	—	230	138	—	159	234	oil	216	105	81–2d	136–8d
1,2,3-Triazole-1	206	228	199	—	137	178	312	187	—	—	—	—	—	167	189
1,2,4-Triazole-1	121	20	65–6	—	—	—	—	—	—	—	—	—	—	—	—
1,2,4-Triazole-3	121	95	—	—	—	—	—	—	—	—	—	—	—	—	—

Structure of Five-membered Rings with Two or More Heteroatoms

Compound														
1,2,4-Triazole-4	121	90	oil	—	—	—	—	76-7	—	—	—	—	—	—
Tetrazole-5	**156**	**148**	—	—	—	**234**	**99**	**203**	**260**	**154**	**205**	**151**	**73**	—
Tetrazole-1	**156**	**39**	**265**	—	**86**	—	—	>370	—	**147**	—	—	—	**148**
Tetrazole-2	**156**	**147**	**163**	—	—	—	—	>370	—	—	—	—	—	—
Isothiazole-3	113	134	—	**32**	**135**	**290**	**154**	**33**	**74**	—	—	—	**160**	—
Isothiazole-4	113	146	—	—	**161**	**174**	**192**	**45**	—	—	—	—	**143**	**34**
Isothiazole-5	113	142	—	**250**	**201d**	—	**172**	**112**	—	—	—	—	**149**	150
1,2,3-Oxadiazole (sydnone)	—	—	340	—	—	—	—	—	—	—	—	—	—	—
1,2,4-Oxadiazole-3	87	36	—	—	—	—	—	—	—	—	—	—	—	—
1,2,4-Oxadiazole-5	87	105	—	—	—	—	—	—	—	—	—	—	—	—
1,3,4-Oxadiazole-2	150	104	—	—	—	**255**	—	**156**	**120**	—	**89-91**	—	**170**	—
1,2,3-Thiadiazole-4	160	164	174.5	**140**	**227-8**	**86**	**62-3**	**44-6**	—	—	—	—	—	—
1,2,3-Thiadiazole-5	160	**87-9**	—	—	**104-6**	**222**	—	**145-7**	—	—	—	—	—	—
1,2,4-Thiadiazole-3	121	132	—	—	—	—	—	—	—	—	—	**32**	—	—
1,2,4-Thiadiazole-5	121	—	—	—	—	—	—	**119**	**120**	**44-5d**[e]	**93**	**220**	**122**	—
1,2,3,4-Thiatriazole-5	—	—	—	—	—	**221**	—	**128-30**	—	—	**50-65d**	**34**	expl.[f]	—
1,3,4-Thiadiazole	**43**	**201d**	—	—	—	—	—	**193**	—	—	**143**	—	**33**	**73**
1,3-Dioxole-4	51	76	—	—	—	—	—	—	—	—	—	—	—	—
1,2-Dithiol-3-one-4	>370	>370	>370	—	**214-6**	**223**	—	79.5-80.5	—	—	—	—	**62-3**	—
1,2-Dithiol-3-one-5	>370	—	—	—	—	**>370**	—	—	—	—	—	—	—	—
1,2,4-Trioxolane-3,3-di	116	88	—	—	—	**42**	—	—	—	—	—	—	—	—
1,2,4-Trioxolane-3,5-di	116	90	140	—	—	—	—	—	—	—	—	—	—	—
1,2,4-Trithiolane-3,5-di	78	>200	>230	—	—	—	—	—	—	—	—	—	—	—

[a] Melting points above 30 °C are given in bold; melting points below 30 °C are not included.
[b] Boiling points are given at atmospheric pressure to facilitate comparison; those reported at other than atmospheric pressure were converted using a nomogram ⟨57MI40101⟩.
[c] A dash indicates that the compound is unstable, unknown, or the data are not readily available.
[d] Value given for EtS derivative.
[e] Value given for EtO derivative.
[f] Explodes.
[g] Value given for the monohydrate.

show that $\nu(C=O)$ values for substituted benzoyl-imidazoles (**103**) and -triazoles follow the Hammett equation. For *N*-acylpyrazoles, see reference ⟨59MI40100⟩.

Substituent vibrations in IR spectra have been extensively used to determine tautomeric structure, particularly C=O, NH, OH and SH stretching modes. For example, 3-hydroxy-isoxazoles show broad $\nu(OH)$ at 2700 cm^{-1} (dimers). Boulton and Katritzky developed a technique for determining the structure of potentially tautomeric amino compounds by partial deuterium exchange. If the compound under investigation contains an amino group, the change from NH$_2$ to NHD produces a new single $\nu(NH)$ at a frequency between those of the original doublet for asymmetrical and symmetrical $\nu(NH_2)$. If the compound does not contain an amino group and the two bands are derived from two separate NH groups, there should be no new band between the original two in the partially deuterated derivative. The method was applied to aminoisoxazoles ⟨61T(12)51⟩.

4.01.3.8 Mass Spectrometry ⟨66AHC(7)301, B-71MS⟩

Among the most important fragmentation pathways of the molecular ions of azoles are the following. (a) Loss of RCN or HCN; this occurs particularly readily for systems containing (**104**), *i.e.* imidazoles and thiazoles. It does not occur so readily for oxazoles, but is found again in the 1,2,4-oxadiazoles (**105**) and also for pyrazoles (**106**). (b) Loss of RCO$^+$; this is important for systems (**107**) in oxazoles and in 1,3,4-oxadiazoles. It is also found in isoxazoles, where it probably occurs after skeletal rearrangement of the isoxazole *via* (**108**) into an isomeric oxazole.   (c) Loss of NO$^+$ and/or NO$^{\cdot}$ occurs for furazans (**109**) and sydnones (**110**). (d) Loss of N$_2$ from triazoles and tetrazoles (but *not* pyrazoles). (e) Whenever the structural element (**111**) occurs, a McLafferty rearrangement can take place (**111** → **112**). (f) Isothiazoles are rather stable and show intense molecular ions with some HCN loss, probably *via* rearrangements analogous to (**108**) to give thiazoles.

(**104**) X = S or NR'   (**105**)   (**106**)   (**107**)

(**108**)   (**109**)   (**110**)   (**111**)   (**112**)

There are correlations between mass spectral fragmentations and thermal and photochemical fragmentations and rearrangements; see Sections 4.02.1.2.1 and 4.02.1.2.2.

4.01.3.9 Photoelectron Spectroscopy ⟨74PMH(6)1⟩

In this method, photons of an energy well in excess of the ionization potential are directed onto a molecule. The photoelectron spectrum which results allows assessment of the energies of filled orbitals in the molecule, and thus provides a characterization of a molecule. Comparisons between photoelectron spectra of related compounds give structural information, for example, on the tautomeric structure of a compound by comparison of its spectrum with those of models of each of the fixed forms.

Photoelectron spectra have been discussed and assigned in the following series: pyrazole (Section 4.04.1.4.9), 1,2,3-triazole (Section 4.11.3.2.9), isothiazole (Section 4.17.2.2), 1,3,4-oxadiazole (Section 4.23.2.2.5), 1,2,5-thiadiazole (Section 4.26.2.2), 1,3,4-thiadiazole (Section 4.27.2.3.10), 1,3-dioxolane (Section 4.30.1.4.5), 1,2-dithiole (Section 4.31.1.4), and 1,2,4-trioxolane and 1,2,4-trithiolane (Section 4.33.2.2.5).

4.01.4 THERMODYNAMIC ASPECTS

4.01.4.1 Intermolecular Forces

4.01.4.1.1 Melting and boiling points

In the parent unsubstituted ring systems (*cf.* first column of Table 32) replacement of a —CH=CH— group with a sulfur atom has little effect, and replacement of a —CH=CH— group with an oxygen atom lowers the boiling point by *ca.* 40 °C.

Introduction of nitrogen atoms into the ring is accompanied by less regular changes. Substitution of either a —CH=CH— group by an NH group or of a =CH— group by a nitrogen atom increases the boiling point. When both of these changes are made simultaneously the boiling point is increased by an especially large amount due to the possibilities of association by hydrogen bonding.

The effect of substituents on melting and boiling points can be summarized as follows. (a) Methyl and ethyl groups attached to ring carbon atoms usually increase the boiling point by *ca.* 20–30 and *ca.* 40–60 °C, respectively. However, conversion of an NH group into an NR group results in large decreases in the boiling point (*e.g.* pyrazole to 1-methylpyrazole) because of decreased association. (b) The acids and amides are all solids. Many of the amides melt in the range 130–180 °C; the melting points of the acids vary widely. (c) Compounds containing a hydroxy, mercapto or amino group are usually relatively high-melting solids. For many hydroxy and mercapto compounds this can be attributed to their tautomerism with hydrogen-bonded 'one' and 'thione' forms. However, hydrogen bonding can evidently also occur in amino compounds. (d) Methoxy, methylthio and dimethylamino derivatives are often liquids. (e) Chloro compounds are usually liquids which have boiling points similar to those of the corresponding ethyl compounds. Bromo compounds boil approximately 25 °C higher than their chloro analogs.

4.01.4.1.2 Solubility of heterocyclic compounds ⟨63PMH(1)177⟩

In general, the solubility of heterocyclic compounds in water (Table 33) is enhanced by the possibility of hydrogen bonding. 'Pyridine-like' nitrogen atoms facilitate this (compare benzene and pyridine). In the same way, oxazole is miscible with water, and isoxazole is very soluble, more so than furan.

The effect of amino, hydroxy or mercapto substituents is to increase hydrogen bonding properties. However, if stable hydrogen bonds are found in the crystal, then this can decrease

Table 33 Solubilities of Some Five-membered Heterocycles in Water at 20 °C[a]

Compound	Parts soluble in 1 part of water	Compound	Parts soluble in 1 part of water
Furan	0.03	Pyrrole	0.06
Isoxazole	0.02	Pyrazole	0.40
Oxazole	Misc.[b]	2-amino-4-hydroxy-	0.009
Benzoxazole	0.008	Imidazole	1.8
Sydnone	—	2,5-dihydroxy-	0.02
3-methyl-[c]	Misc.	1-methyl-	Misc.
1,2,4-Oxadiazole[d]	Misc.	1,2,4-Triazole	1
Thiophene	0.001	3-amino-	0.3
Isothiazole	0.03	3-hydroxy-	0.05
Thiazole	—	Tetrazole	1
2-methyl-	Misc.	Benzimidazole	0.002
2-amino-	0.05	Indazole	0.0008
1,2,3-Thiadiazole	0.03	3-hydroxy-	0.002
1,3,4-Thiadiazole	Misc.	Purine	0.5

[a] Unless otherwise indicated, data taken from ⟨63PMH(1)177⟩ or from appropriate chapter, 'Comprehensive Heterocyclic Chemistry', which contain references to the original literature.
[b] Misc. = miscible.
[c] At 40 °C ⟨80JCS(P2)553⟩.
[d] ⟨65MI40100⟩.

their solubility in water ⟨63PMH(1)177⟩, *e.g.* indazole and benzimidazole are less soluble than benzoxazole.

Other solvents can be divided into several classes. In hydrogen bond-breaking solvents (dipolar aprotics), the simple amino, hydroxy and mercapto heterocycles all dissolve. In the hydrophobic solvents, hydrogen bonding substituents greatly decrease the solubility. Ethanol and other alcohols take up a position intermediate between water and the hydrophobic solvents ⟨63PMH(1)177⟩.

4.01.4.1.3 Gas–liquid chromatography ⟨71PMH(3)297⟩

Gas–liquid chromatography has been widely used for the identification of reaction mixtures and for the separation of heterocycles. Some typical conditions are shown in Table 34.

Table 34 Operating Conditions for GLC Separation of Five-membered Heterocycles with More Than One Heteroatom[a]

Compound	Conditions
Pyrazolines	10% Cyanethylated mannite on Celite 545.
Imidazoles	5% OV-17 on Chromosorb W, AW-DMCS (H.P.).[b]
Thiazoles	Carbowax 4000, dioleate on firebrick, 190 °C.
Oxadiazoles	Silicone grease on Chromosorb P.
1-Phenylpyrazoles	Apiezon L on firebrick C-22, 220 °C.
Purines	15% Hallcomid M-18 on firebrick.
Dioxolanes	Carbowax 20M on Gas-Chrom P.
Oxazolines	Ethyleneglycol succinate on Diatoport-S.

[a] Data taken from ⟨71PMH(3)297⟩, which contains references to the original literature.
[b] Simple alkyl- and aryl-imidazoles. *N*-Unsubstituted compounds are *N*-acylated prior to injection.

4.01.4.2 Stability and Stabilization

4.01.4.2.1 Thermochemistry and conformation of saturated heterocycles ⟨74PMH(6)199⟩

Calculation of group increments for oxygen, sulfur and nitrogen compounds has allowed the estimation of conventional ring-strain energies (CRSE) for saturated heterocycles from enthalpies of formation. For 1,3-dioxolane, CRSE is about 20 kJ mol^{-1}. In 2,4-dialkyl-1,3-dioxolanes the *cis* form is always thermodynamically the more stable by approximately 1 kJ mol^{-1}.

For 1,3-dithiolanes the ring is flexible and only small energy differences are observed between the diastereoisomeric 2,4-dialkyl derivatives. The 1,3-oxathiolane ring is less mobile and pseudoaxial 2- or 5-alkyl groups possess conformational energy differences (*cf.* **113** ⇌ **114**); see also the discussion of conformational behavior in Section 4.01.4.3.

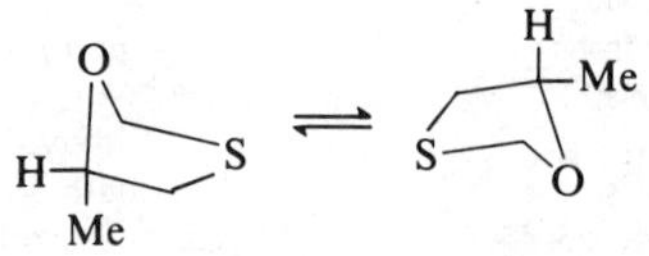

(**113**) 13% (*a*)   (**114**) 87% (*e*)

4.01.4.2.2 Aromaticity

This subject has been dealt with in ⟨74AHC(17)255⟩, which should be consulted for further details and references to the original literature.

(i) *Imidazole, pyrazole, thiazole, isothiazole and oxazole*

ERE (empirical resonance energy) and 'conjugation energy' data (Table 35) suggest that pyrazole is more aromatic than both imidazole and pyrrole. LCAO–SCF calculations on various azoles, however, lead to the conclusion that the stability of azoles decreases on increasing substitution of nitrogen atoms for carbon atoms, with the stabilities of pyrazole and imidazole being comparable.

Table 35 Empirical Resonance Energy Data (kJ mol^{-1}) for Azoles[a,b]

Compound	Method A[c]	Method B[d]	Method C[e]
Pyrazole	173.6	112.1	136.8
Imidazole	134.3	53.1	74.1
1,2,4-Triazole	205.8	83.7	151.5
Tetrazole	—	231.0	264.0
Indazole	309.2	246.9	248.9
Benzimidazole	287.9	203.8	204.2
Benzotriazole	—	346.9	312.5

[a] Adapted from ⟨74AHC(17)255⟩ which contains further details and references to the original literature.
[b] 1×10^{-7} J = 2.3901×10^{-8} cal.
[c] Results obtained using Pauling's bond energy terms.
[d] Using the bond energy terms reported by Cottrell.
[e] Using Coates and Sutton's bond energy terms.

Data relevant to the aromaticity of these compounds have been obtained by a variety of other techniques, many of which are discussed elsewhere in this chapter. Aromaticity implies planarity; X-ray and microwave data support planarity. The NMR method has been extensively applied; thus ^{1}H NMR shifts for imidazoles, pyrazoles and pyrroles are comparable; ^{13}C NMR spectral data for imidazole show appreciable ring current anisotropy. The ring protons of thiazoles and the methyl protons in methylthiazoles are deshielded relative to signals from the corresponding imidazoles. Elvidge has interpreted these NMR data as evidence that the thiazole ring has the same degree of aromatic character as benzene ⟨65PC40100⟩.

The precise geometrical data obtained by microwave spectroscopy allow conclusions regarding bond delocalization and hence aromaticity. For example, the microwave spectrum of thiazole has shown that the structure is very close to the average of the structures of thiophene and 1,3,4-thiadiazole, which indicates a similar trend in aromaticity. However, different methods have frequently given inconsistent results.

^{1}H NMR data for 4-methyloxazole have been compared with those of 4-methylthiazole; the data clearly show that the ring protons in each are shielded. In a comprehensive study of a range of oxazoles, Brown and Ghosh also reported NMR data but based a discussion of resonance stabilization on pK and UV spectral data ⟨69JCS(B)270⟩. The weak basicity of oxazole (pK_a 0.8) relative to 1-methylimidazole (pK_a 7.44) and thiazole (pK_a 2.44) demonstrates that delocalization of the oxygen lone pair, which would have a base-strengthening effect on the nitrogen atom, is not extensive. It must be concluded that not only the experimental measurement but also the very definition of aromaticity in the azole series is as yet poorly quantified. Nevertheless, its importance in the interpretation of reactivity is enormous.

(ii) *Diheterolylium ions and related compounds*

The 1,2-dithiolylium and 1,3-dithiolylium ions (**115**) and (**116**) are iso-π-electronic with the tropylium ion, from which they may be formally derived by replacing double bonds by sulfur atoms. Various calculations and structural data demonstrate that the rings are stabilized by π-electron delocalization.

(**115a**)   (**115b**)   (**116a**)   (**116b**)   (**117**)

The NMR spectrum of the 2-hydroxy-1,3-dioxolylium cation (**117**) ⟨68JA1884⟩ shows a significant ring current. The aromaticity of vinylene carbonate was pointed out by Balaban ⟨59MI40100⟩.

(iii) *Compounds containing three heteroatoms*

The aromatic character is critically dependent upon the position of the heteroatoms in the ring, and oxygenated compounds have marked diene character. Various ERE determinations of 1,2,4-triazole have given values ranging between 83.7 and 205.8 kJ mol^{-1} (Table 35). LCAO–SCF calculations, however, suggest that the ring is substantially less stable than the diazoles but more stable than tetrazole.

Microwave spectra of 1,3,4-thiadiazole (**118**), thiophene (**119**), 1,2,5-thiadiazole (**120**) and 1,2,5-oxadiazole (**121**) yield the following order of decreasing aromaticity based on bond lengths: (**120**), (**119**), (**118**), (**121**), (**122**). 1,2,5-Oxadiazole is planar with a C=N bond length of 130.0 pm, intermediate between that of formaldoxine (127.0 pm) and pyridine (134.0 pm), suggestive of a high degree of 'diene' character. Comparison with cyclopentadiene indicates that 1,2,5-oxadiazole has some small degree of aromatic character.

(**118**) (**119**) (**120**) (**121**) (**122**)

The sydnones may be represented by structures (**123a–d**), of which the zwitterionic structure (**123a**) most clearly implies an aromatic sextet. The diamagnetic susceptibility exaltation for *N*-phenylsydnone of 11.0×10^{-6} cm^3 mol^{-1} is comparable with the corresponding value for pyrrole (10.2×10^{-6}). 3-*p*-Bromophenylsydnone (**123**; R = H, R' = *p*-bromophenyl) is essentially planar; however, the O—N bond and O(1)—C(5) bond lengths are not very different from normal single bond distances.

(**123a**) (**123b**) (**123c**) (**123d**)

(iv) *Benzo derivatives*

In the series 1,3-benzoselenazole (**124**; X = Se), -benzothiazole (**124**; X = S) and -benzoxazole (**124**; X = O) the deshielding effect on the methyl protons on changing the heteroatom X is in the opposite sense to the electronegativity of the atom. This has been explained in terms of decreasing aromaticity in the order (**124**; X = Se), (**124**; X = S), (**124**; X = O).

(**124**) (**125**) (**126**)

In 2,1,3-benzoselenadiazole (**125**; X = Se), 2,1,3-benzothiadiazole (**125**; X = S) and 2,1,3-benzoxadiazole (**125**; X = O) the effect of the heteroatom X on the ring is similar to that in the previous system; thus a decreasing scale of aromatic character (**125**; X = Se), (**125**; X = S), (**125**; X = O) is again suggested. The *ortho* proton–proton coupling constants (5,6), and hence bond fixation and aromatic character, decrease in the order (**126**; X = S), (**126**; X = NMe), (**125**; X = NMe), (**125**; X = S), (**125**; X = O). The ratios of $J_{5,6}:J_{4,5}$ for these compounds are 0.859, 0.824, 0.781, 0.748 and 0.706, respectively.

4.01.4.3 Conformation

Saturated five-membered heterocyclic compounds are non-planar, existing in half-chair or envelope conformations. The far-IR spectra of THF and 1,3-dioxolane (**127**) show both to have barriers of *ca.* 0.42 kJ mol^{-1}.

(**127**) (**128**) (**129**) (**130**)

1,3-Dioxolane also pseudorotates essentially freely in the vapor phase. 2,2'-Bi-1,3-dioxolane (**128**) has been shown by X-ray crystallography to have a conformation midway between the half-chair and envelope forms. The related compound 2-oxo-1,3-dioxolane (**129**) shows a half-chair conformation. This result is confirmed by microwave spectroscopy and by ^{13}C NMR data. Analysis of the AA'BB' NMR spectra of the ring hydrogen atoms in some 1,3-dioxolane derivatives is in agreement with a puckered ring. Some 2-alkoxy-1,3-dioxolanes (**130**) display *anti* and *gauche* forms about the exocyclic C(2)—O bond.

The 1,2,4-trioxolane ring prefers a half-chair conformation (**131**); the C—O—C portion of the ring forms the reference plane, and alkyl substituents prefer the equatorial positions.

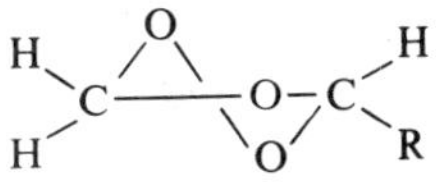

(**131**) 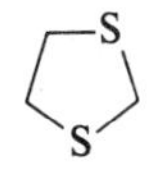   (**132**)

1,3-Dithiolane (**132**) derivatives also possess non-planar skeletons; the most important conformation is probably of symmetry C_2 (half-chair). The dithiolane ring may be quite flexible and a minimum energy conformation is only well defined if there is a bulky substituent at the 2-position.

4.01.5 TAUTOMERISM ⟨76AHC(S1)1⟩

4.01.5.1 Annular Tautomerism

Annular tautomerism (*e.g.* **133** ⇌ **134**) involves the movement of a proton between two annular nitrogen atoms. For unsubstituted imidazole (**133**; R = H) and pyrazole (**135**; R = H) the two tautomers are identical, but this does not apply to substituted derivatives. For triazoles and tetrazoles, even the unsubstituted parent compounds show two distinct tautomers. However, interconversion occurs readily and such tautomers cannot be separated. Sometimes one tautomeric form predominates. Thus the mesomerism of the benzene ring is greater in (**136**) than in (**137**), and UV spectral comparisons show that benzotriazole exists predominantly as (**136**).

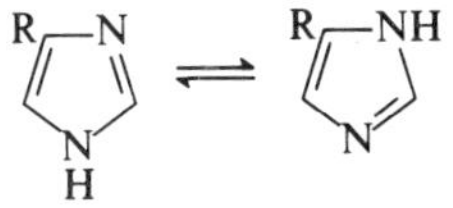

(**133**)   (**134**) 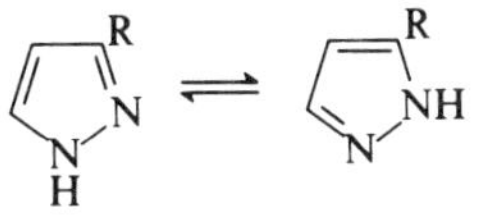   (**135**)

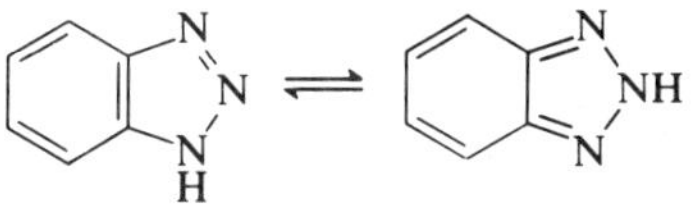

(**136**)   (**137**) 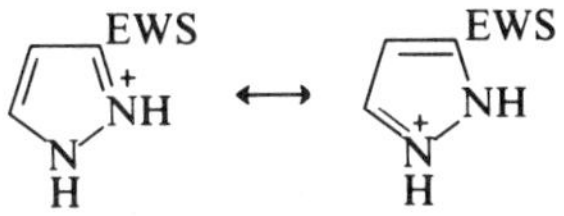   (**138**)

Table 36 summarizes the known annular tautomerism data for azoles. The tautomeric preferences of substituted pyrazoles and imidazoles can be rationalized in terms of the differential substituent effect on the acidity of the two NH groups in the conjugate acid, *e.g.* in (**138**; EWS = electron-withdrawing substituent) the 2-NH is more acidic than 1-NH and hence for the neutral form the 3-substituted pyrazole is the more stable.

Table 36   Annular Tautomerism of Azoles[a]

Azole	Parent	Substituted compounds
Pyrazole	Equivalent	Electron-withdrawing group prefers 3-position
Imidazole	Equivalent	Electron-withdrawing group prefers 4-position
v-Triazole	1,2,5 > 1,2,3	—
s-Triazole	1,2,4 > 1,3,4	—
Tetrazole	1,2,3,4 > 1,2,3,5	—

[a] Data taken from ⟨76AHC(S1)296⟩ which contains references to the original literature.

The situation is more complex for triazoles and tetrazoles where other effects such as lone-pair repulsions intervene; see discussion in ⟨76AHC(S1)296⟩.

4.01.5.2 Substituent Tautomerism

4.01.5.2.1 *Azoles with heteroatoms in the 1,2-positions*

3-Substituted isoxazoles, pyrazoles and isothiazoles can exist in two tautomeric forms (**139, 140**; Z = O, N or S; Table 37). Amino compounds exist as such as expected, and so do the hydroxy compounds under most conditions. The stability of the OH forms of these 3-hydroxy-1,2-azoles is explained by the weakened basicity of the ring nitrogen atom in the 2-position due to the adjacent heteroatom at the 1-position and the oxygen substituent at the 3-position. This concentration of electron-withdrawing groups near the basic nitrogen atom causes these compounds to exist mainly in the OH form.

$$\text{(139)} \quad \rightleftharpoons \quad \text{(140)}$$

Table 37 Tautomerism of 3-Substituted Azoles with Heteroatoms-1,2[a]

Substituent	Ring	Phase(s)	Conclusions
OH	1-Substituted pyrazole	C_6H_{12}, $CHCl_3$, H_2O, xtal.	OH except in H_2O where OH/NH coexist
	Isoxazole	C_6H_{12}, $CHCl_3$, H_2O, xtal.	Mainly OH in all media
	Isothiazole	MeOH, xtal.	OH
NH_2	1-Substituted pyrazole	MeOH, KBr	NH_2
	Isoxazole	$CHCl_3$, H_2O	NH_2
	Isothiazole	CCl_4	NH_2
SH	Isothiazole	CCl_4	SH

[a] For further details and original references see ⟨70C134⟩ and ⟨76AHC(S1)1⟩.

The 4-substituted analogs can exist in two uncharged tautomeric forms (**141**) and (**142**) and, in addition, in the zwitterionic form (**143**), but all the evidence shows that the compounds all exist predominantly in the NH_2 or OH form (**141**).

$$\text{(141)} \quad \rightleftharpoons \quad \text{(142)} \quad \rightleftharpoons \quad \text{(143)}$$

For 5-substituted isoxazoles, pyrazoles and isothiazoles, three uncharged tautomeric forms are possible: (**144**), (**145**) and (**146**). Some conclusions are recorded in Table 38. Again, the amino derivatives exist as such. In the case of the hydroxy compounds, the hydroxy form is of little importance (except in special cases where a suitable substituent in the 4-position can form a hydrogen bond with the 5-hydroxy group). The relative occurrence of the 4*H*-oxo form (**145**) and 2*H*-oxo tautomer (**146**) depends on the substitution pattern and on the solvent. Tautomer (**146**) is considerably more polar than (**145**), with a large contribution from the charge-separated canonical structure (**147**). Hence, it is not unexpected that the 2*H*-oxo tautomer (**146**) is strongly favored by polar media. A substituent at the 4-position also tends to favor form (**146**) over (**145**) because of conjugation or hyperconjugation of the 4-substituent with the 3,4-double bond.

$$\text{(144)} \quad \rightleftharpoons \quad \text{(145)} \quad \rightleftharpoons \quad \text{(146)} \quad \leftrightarrow \quad \text{(147)}$$

Table 38 Tautomerism of 5-Substituted Azoles with Heteroatoms-1,2[a]

Substituent	Ring	Phase(s)	Conclusions
OH	1-Substituted pyrazole	C_6H_{12}, $CHCl_3$, EtOH, H_2O, xtal.	CH in non-polar;
	Isoxazole	C_6H_{12}, $CHCl_3$, EtOH, H_2O, xtal.	Increasing NH in polar; NH favored by 4-substituent
NH_2	1-Substituted pyrazole	CCl_4	NH_2
	Isoxazole	$CHCl_3$, H_2O	NH_2
	Isothiazole	CCl_4	NH_2
SH	Isoxazole	C_6H_{12}, CCl_4, MeOH, xtal.	SH

[a] For further details and original references see ⟨70C134⟩ and ⟨76AHC(S1)1⟩.

Complex tautomerism for azoles with heteroatoms in the 1,2-positions occurs for pyrazoles which are not substituted on nitrogen. Scheme 10 shows the four important tautomeric structures (**148**)–(**151**) for 3-methylpyrazolin-5-one, and (**152**) and (**153**) as examples of other possible structures. A detailed investigation of this system disclosed that in aqueous solution (polar medium) the importance of the tautomers is (**149**) > (**151**) ≫ (**150**) or (**148**), whereas in cyclohexane solution (non-polar medium) (**151**) > (**148**) ≫ (**149**) or (**150**).

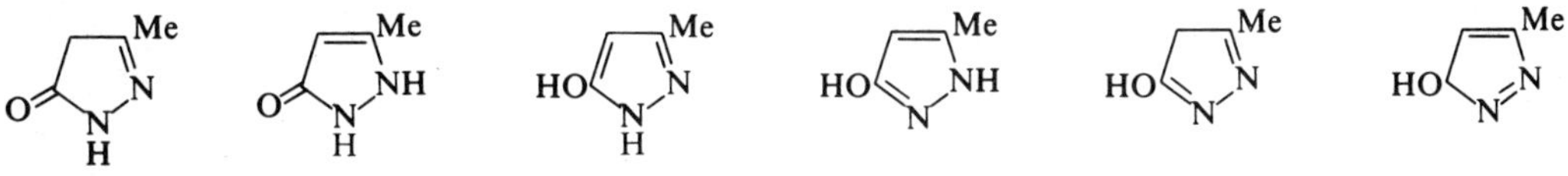

(**148**) (**149**) (**150**) (**151**) (**152**) (**153**)

Scheme 10

4.01.5.2.2 Azoles with heteroatoms in the 1,3-positions

The tautomerism of 2-substituted 1,3-azoles (**154** ⇌ **155**) is summarized in Table 39. Whereas amino compounds occur invariably as such, all the potential hydroxy derivatives exist in the oxo form, and in this series the sulfur compounds resemble their oxygen analogs. There is a close analogy between the tautomerism for all these derivatives with the corresponding 2-substituted pyridines.

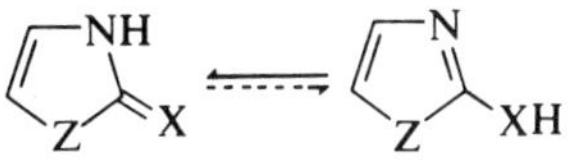

(**154**) (**155**)

Table 39 Tautomerism of 2-Substituted Azoles with Heteroatoms-1,3[a]

Substituent	Ring	Phase(s)	Conclusions
OH	1-Substituted imidazole	KBr disc	NH, C=O
	Oxazole	MeOH, CCl_4	NH, C=O
	Thiazole	Alcohol, CS_2	NH, C=O
NH_2	1-Substituted imidazole	0.1N aq. KCl, EtOH/aq. KCl	NH_2
	Oxazole	MeOH	NH_2
	Thiazole	EtOH, $CDCl_3$	NH_2
SH	1-Substituted imidazole	Liquid film, MeOH, EtOH	NH, C=S
	Oxazole	CCl_4, MeOH	NH, C=S
	Thiazole	CCl_4, EtOH	NH, C=S

[a] For further details and original references see ⟨70C134⟩ and ⟨76AHC(S1)1⟩.

4-Substituted 1,3-azoles exist in two non-charged tautomeric forms (**156**) and (**157**) together with the zwitterionic form (**158**). 5-Substituted 1,3-azoles also exist in forms (**159**) and (**160**) together with the zwitterionic forms (**161**). Some results are summarized in Table

40; for the potential hydroxy forms, the non-aromatic tautomers of types (**157**) and (**160**) clearly can be of importance.

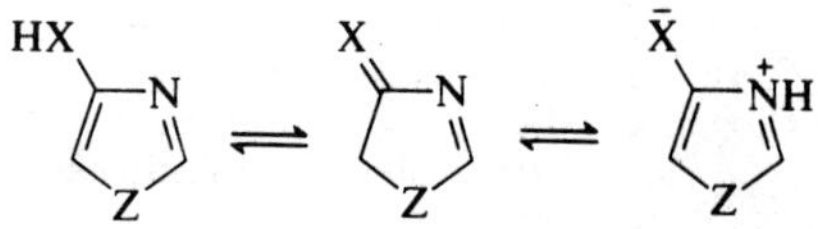

(**156**) ⇌ (**157**) ⇌ (**158**) 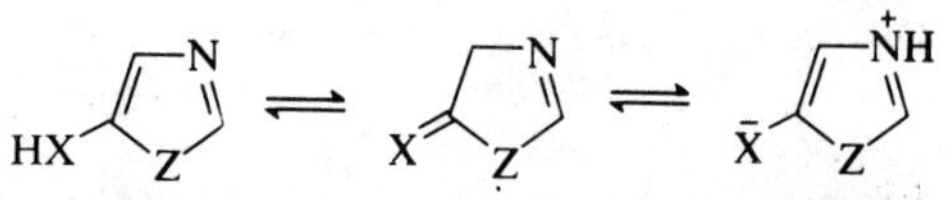   (**159**) ⇌ (**160**) ⇌ (**161**)

Table 40 Tautomerism of 4- and 5-Substituted Azoles with Heteroatoms-1,3[a]

Substituent	Ring	Phase(s)	Conclusions
4-OH	Oxazole	EtOH, xtal., Me$_2$SO	CH
	Thiazole	Me$_2$SO, Me$_2$CO	OH and CH
5-OH	Oxazole	Xtal.	CH
5-NH$_2$	Oxazole	CHCl$_3$, xtal.	NH$_2$

[a] For further details and original references, see ⟨70C134⟩ and ⟨76AHC(S1)1⟩.

4.02

Reactivity of Five-membered Rings with Two or More Heteroatoms

A. R. KATRITZKY
University of Florida

and

J. M. LAGOWSKI
The University of Texas at Austin

4.02.1 REACTIONS AT HETEROAROMATIC RINGS	41
4.02.1.1 General Survey of Reactivity	41
4.02.1.1.1 Reactivity of neutral azoles	41
4.02.1.1.2 Azolium salts	42
4.02.1.1.3 Azole anions	42
4.02.1.1.4 Azolinones, azolinethiones, azolinimines	43
4.02.1.1.5 N-Oxides, N-imides, N-ylides of azoles	43
4.02.1.2 Thermal and Photochemical Reactions Formally Involving No Other Species	44
4.02.1.2.1 Thermal fragmentation	44
4.02.1.2.2 Photochemical fragmentation	45
4.02.1.2.3 Equilibria with open-chain compounds	46
4.02.1.2.4 Rearrangement to other heterocyclic species	46
4.02.1.2.5 Polymerization	47
4.02.1.3 Electrophilic Attack at Nitrogen	47
4.02.1.3.1 Introduction	47
4.02.1.3.2 Reaction sequence	48
4.02.1.3.3 Orientation in azole rings containing three or four heteroatoms	48
4.02.1.3.4 Effect of azole ring structure and of substituents	49
4.02.1.3.5 Proton acids on neutral azoles: basicity of azoles	49
4.02.1.3.6 Proton acids on azole anions: acidity of azoles	50
4.02.1.3.7 Metal ions	51
4.02.1.3.8 Alkyl halides and related compounds: azoles without a free NH group	51
4.02.1.3.9 Alkyl halides and related compounds: compounds with a free NH group	53
4.02.1.3.10 Acyl halides and related compounds	54
4.02.1.3.11 Halogens	54
4.02.1.3.12 Peracids	55
4.02.1.3.13 Aminating agents	55
4.02.1.3.14 Other Lewis acids	55
4.02.1.4 Electrophilic Attack at Carbon	55
4.02.1.4.1 Reactivity and orientation	55
4.02.1.4.2 Nitration	57
4.02.1.4.3 Sulfonation	57
4.02.1.4.4 Acid-catalyzed hydrogen exchange	57
4.02.1.4.5 Halogenation	58
4.02.1.4.6 Acylation and alkylation	58
4.02.1.4.7 Mercuration	59
4.02.1.4.8 Diazo coupling	59
4.02.1.4.9 Nitrosation	59
4.02.1.4.10 Reactions with aldehydes and ketones	59
4.02.1.4.11 Oxidation	60
4.02.1.5 Attack at Sulfur	61
4.02.1.5.1 Electrophilic attack	61
4.02.1.5.2 Nucleophilic attack	61

4.02.1.6 Nucleophilic Attack at Carbon	61
4.02.1.6.1 Hydroxide ion and other O-nucleophiles	62
4.02.1.6.2 Amines and amide ions	64
4.02.1.6.3 S-Nucleophiles	66
4.02.1.6.4 Halide ions	66
4.02.1.6.5 Carbanions	66
4.02.1.6.6 Reduction by complex hydrides	68
4.02.1.6.7 Phosphorus nucleophiles	69
4.02.1.7 Nucleophilic Attack at Hydrogen Attached to Ring Carbon or Ring Nitrogen	69
4.02.1.7.1 Metallation at a ring carbon atom	69
4.02.1.7.2 Hydrogen exchange at ring carbon in neutral azoles	69
4.02.1.7.3 Hydrogen exchange at ring carbon in azolium ions and dimerization	70
4.02.1.7.4 C-Acylation via deprotonation	71
4.02.1.7.5 Ring cleavage via C-deprotonation	71
4.02.1.7.6 Proton loss from a ring nitrogen atom	72
4.02.1.8 Reactions with Radicals and Electron-deficient Species; Reactions at Surfaces	72
4.02.1.8.1 Carbenes and nitrenes	72
4.02.1.8.2 Free radical attack at the ring carbon atoms	72
4.02.1.8.3 Electrochemical reactions and reactions with free electrons	73
4.02.1.8.4 Other reactions at surfaces (heterogeneous catalysis and reduction reactions)	74
4.02.1.9 Reactions with Cyclic Transition States	75
4.02.1.9.1 Diels–Alder reactions and 1,3-dipolar cycloadditions	75
4.02.1.9.2 Photochemical cycloadditions	77
4.02.2 REACTIONS OF NON-AROMATIC COMPOUNDS	**77**
4.02.2.1 Isomers of Aromatic Derivatives	77
4.02.2.1.1 Compounds not in tautomeric equilibrium with aromatic derivatives	77
4.02.2.1.2 Compounds in tautomeric equilibrium with aromatic derivatives	78
4.02.2.2 Dihydro Compounds	78
4.02.2.2.1 Tautomerism	78
4.02.2.2.2 Aromatization	78
4.02.2.2.3 Ring contraction	79
4.02.2.2.4 Other reactions	79
4.02.2.3 Tetrahydro Compounds	80
4.02.2.3.1 Aromatization	80
4.02.2.3.2 Ring fission	80
4.02.2.3.3 Other reactions	80
4.02.2.3.4 Stereochemistry	81
4.02.3 REACTIONS OF SUBSTITUENTS	**81**
4.02.3.1 General Survey of Substituents on Carbon	81
4.02.3.1.1 Substituent environment	81
4.02.3.1.2 The carbonyl analogy	82
4.02.3.1.3 Two heteroatoms in the 1,3-positions	82
4.02.3.1.4 Two heteroatoms in the 1,2-positions	83
4.02.3.1.5 Three heteroatoms	83
4.02.3.1.6 Four heteroatoms	83
4.02.3.1.7 The effect of one substituent on the reactivity of another	83
4.02.3.1.8 Reactions of substituents not directly attached to the heterocyclic ring	84
4.02.3.1.9 Reactions of substituents involving ring transformations	84
4.02.3.2 Fused Benzenoid Rings	85
4.02.3.2.1 Electrophilic substitution	85
4.02.3.2.2 Oxidative degradation	86
4.02.3.2.3 Nucleophilic attack	86
4.02.3.2.4 Rearrangements	86
4.02.3.3 Alkyl Groups	87
4.02.3.3.1 Reactions similar to those of toluene	87
4.02.3.3.2 Alkylazoles: reactions involving essentially complete anion formation	88
4.02.3.3.3 Reactions of alkylazoles involving traces of reactive anions	88
4.02.3.3.4 C-Alkyl-azoliums, -dithiolyliums, etc.	89
4.02.3.4 Other C-Linked Substituents	91
4.02.3.4.1 Aryl groups: electrophilic substitution	91
4.02.3.4.2 Aryl groups: other reactions	91
4.02.3.4.3 Carboxylic acids	92
4.02.3.4.4 Aldehydes and ketones	93
4.02.3.4.5 Vinyl and ethynyl groups	94
4.02.3.4.6 Ring fission	94
4.02.3.5 Aminoazoles	94
4.02.3.5.1 Dimroth rearrangement	94
4.02.3.5.2 Reactions with electrophiles (except nitrous acid)	95
4.02.3.5.3 Reaction with nitrous acid: diazotization	96
4.02.3.5.4 Deprotonation of aminoazoles	97
4.02.3.5.5 Aminoazolium ions/neutral imines	97

4.02.3.6 Other N-Linked Substituents	98
4.02.3.6.1 Nitro groups	98
4.02.3.6.2 Azidoazoles	98
4.02.3.7 O-Linked Substituents	98
4.02.3.7.1 Tautomeric forms: interconversion and modes of reaction	98
4.02.3.7.2 2-Hydroxy, heteroatoms-1,3	99
4.02.3.7.3 3-Hydroxy, heteroatoms-1,2	100
4.02.3.7.4 5-Hydroxy, heteroatoms-1,2	100
4.02.3.7.5 4- and 5-Hydroxy, heteroatoms-1,3 and 4-hydroxy, heteroatoms-1,2	101
4.02.3.7.6 Hydroxy derivatives with three heteroatoms	101
4.02.3.7.7 Alkoxy groups	102
4.02.3.8 S-Linked Substituents	102
4.02.3.8.1 Mercapto compounds: tautomerism	102
4.02.3.8.2 Thiones	102
4.02.3.8.3 Alkylthio groups	103
4.02.3.8.4 Sulfonic acid groups	104
4.02.3.9 Halogen Atoms	104
4.02.3.9.1 Nucleophilic displacements: neutral azoles	104
4.02.3.9.2 Nucleophilic displacements: halogenoazoliums	105
4.02.3.9.3 Other reactions	105
4.02.3.10 Metals and Metalloid-linked Substituents	106
4.02.3.11 Fused Heterocyclic Rings	107
4.02.3.12 Substituents Attached to Ring Nitrogen Atoms	107
4.02.3.12.1 N-Linked azole as a substituent	107
4.02.3.12.2 Aryl groups	107
4.02.3.12.3 Alkyl groups	108
4.02.3.12.4 Acyl groups	109
4.02.3.12.5 N-Amino and N-nitro groups	109
4.02.3.12.6 N-Hydroxy groups and N-oxides	110
4.02.3.12.7 N-Halogeno groups	110

4.02.1 REACTIONS AT HETEROAROMATIC RINGS

4.02.1.1 General Survey of Reactivity

In this initial section the reactivities of the major types of azole aromatic rings are briefly considered in comparison with those which would be expected on the basis of electronic theory, and the reactions of these heteroaromatic systems are compared among themselves and with similar reactions of aliphatic and benzenoid compounds. Later in this chapter all the reactions are reconsidered in more detail. It is postulated that the reactions of azoles can only be rationalized and understood with reference to the complex tautomeric and acid–base equilibria shown by these systems. Tautomeric equilibria are discussed in Chapter 4.01. Acid–base equilibria are considered in Section 4.02.1.3 of the present chapter.

4.02.1.1.1 Reactivity of neutral azoles

Replacing a CH group of benzene with a nitrogen atom gives pyridine (**1**); replacing a CH=CH group of benzene with NH, O or S gives pyrrole, furan or thiophene (**3**), respectively.

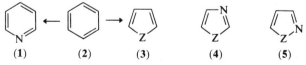

     (**1**)   (**2**)   (**3**)   (**4**)   (**5**)

The azoles (**4**) and (**5**) may be considered to be derived from benzene by two successive steps, one of each of these types. Hence, the chemistry of five-membered aromatic rings with two or more heteroatoms shows similarities to both that of the five- and that of the six-membered aromatic rings containing one heteroatom. Thus, electrophilic reagents attack lone electron pairs on multiply bonded nitrogen atoms of azoles (cf. pyridine) (see Section 4.02.1.3), but they do not commonly attack electron pairs on heterocyclic nitrogen atoms in NR groups or on heterocyclic oxygen or sulfur atoms (cf. pyrrole, furan, thiophene) (for example, see Section 4.02.1.5.1).

The carbon atoms of azole rings can be attacked by nucleophilic (Section 4.02.1.6), electrophilic (Section 4.02.1.4) and free radical reagents (Section 4.02.1.8.2). Some systems, for example the thiazole, imidazole and pyrazole nuclei, show a high degree of aromatic character and usually 'revert to type' if the aromatic sextet is involved in a reaction. Others such as the isoxazole and oxazole nuclei are less aromatic, and hence more prone to addition reactions.

Electron donation from pyrrole-like nitrogen, or to a lesser extent from analogous sulfur or oxygen atoms, helps electrophilic attack at azole carbon atoms, but as the number of heteroatoms in the ring increases, the tendency toward electrophilic attack at both C and N decreases rapidly.

Just as electron displacement toward the nitrogen atom allows nucleophilic reagents to attack pyridines at the α-position, similar displacements toward the nitrogens of the azoles also facilitate nucleophilic attack at carbon. As in similar reactions with pyridine, formation of the initial adduct involves dearomatization of the ring. The subsequent fate of the adduct depends in part on the degree of aromaticity. Those derived from highly aromatic azoles tend to rearomatize, whereas those of lower aromaticity can take alternative reaction paths. For most neutral azoles, nucleophilic attack at a ring carbon atom is only possible with very strong nucleophiles.

Where azoles contain ring NH groups, this group is acidic and nucleophiles can remove a proton. Nucleophilic species can also remove ring-hydrogen atoms, particularly those which are α to a sulfur or oxygen atom, as in base-catalyzed hydrogen exchange and metallation reactions (Section 4.02.1.7).

4.02.1.1.2 Azolium salts

All neutral azoles possess positively charged azolium counterparts. In addition, as discussed in Chapter 4.01, certain 'olylium' species exist which have no neutral counterparts, for example dithiolylium salts.

Azolium systems show much lower reactivity than the corresponding neutral azoles toward electrophiles at ring carbon. Even if an azolium ion contains an additional unquaternized pyridine-like nitrogen, this nitrogen is hardly basic in character. By contrast, azolium cations show a great reactivity toward nucleophiles: at ring carbon atoms, at the hydrogen of ring CH and NH groups, and even at ring sulfur atoms.

In all these azoliums, oxolyliums and thiolyliums the positive charge facilitates attack by nucleophilic reagents at ring carbon atoms α or γ to the charged heteroatom (Section 4.02.1.6). Hydroxide, alkoxide, sulfide, cyanide and borohydride ions, certain carbanions, amines and organometallic compounds react under mild conditions, usually at a position α to the quaternary center as in (**6**), to give initial non-aromatic adducts (**7**) which can be isolated in certain cases but undergo further reaction with alacrity. The most important of these subsequent reactions include: (i) oxidation, *e.g.* the formation of cyanine dyes in the thiazole series (Section 4.02.1.6.5(ii)); (ii) ring opening with subsequent closure, *e.g.* the reaction of dithiolylium salts with amines (Section 4.02.1.6.2); (iii) ring opening without subsequent closure, *e.g.* the reactions of oxazoliums with hydroxide ion (Section 4.02.1.6.5(ii)).

Ring hydrogen atoms can be abstracted from the α-carbon atoms of azolium ions by strong bases, as demonstrated in base-catalyzed hydrogen exchange (Section 4.02.1.7.2).

4.02.1.1.3 Azole anions

Azole anions are derived from imidazoles, pyrazoles, triazoles or tetrazoles by proton loss from a ring NH group. In contrast to the neutral azoles, azole anions show enhanced reactivity toward electrophiles, both at the nitrogen (Section 4.02.1.3.6) and carbon atoms (Section 4.02.1.4.1(i)). They are correspondingly unreactive toward nucleophiles.

4.02.1.1.4 Azolinones, azolinethiones, azolinimines

These compounds are usually written in the unionized form as in (**8**; Z = NH, NR, O, S). Canonical forms of types (**9**) or (**10**) are important, *i.e.* these compounds can also be considered as betaines formally derived from azolium ions. Many compounds of this type are tautomeric and such tautomerism is discussed in Section 4.01.5.2.

(**8**) (**9**) (**10**)

Reactions of these compounds follow logically from the expected electron displacements in the molecules. Their very varied chemical reactivity includes four main possibilities for heterolytic reactions: electrophilic attack at a ring carbon atom β to a ring heteroatom (*e.g.* Section 4.02.1.4.2), or at a carbonyl oxygen atom (Section 4.02.3.7), a thiocarbonyl sulfur atom (Section 4.02.3.8.2) or an imine nitrogen atom (Section 4.02.3.5.5). Nucleophilic attack to remove hydrogen from an NH group (Section 4.02.1.3.6) or at a ring carbon atom α to a ring heteroatom (Section 4.02.3.12.3) also needs to be considered.

The mode of attack of electrophilic reagents (E^+) at ring carbon atoms is β to the heteroatoms as shown, for example, in (**11**) and (**12**); the intermediates usually revert to type by proton loss. Halogenation takes place more readily than it does in benzene (Section 4.02.1.4.5). Nitration and sulfonation also occur; however, in the strongly acidic environment required the compounds are present mainly as less reactive hydroxyazolium ions, *e.g.* (**13**).

(**11**) (**12**) (**13**)

The reactions of electrophilic reagents at a carbonyl oxygen atom, a thiocarbonyl sulfur, and an imino nitrogen atom are considered as reactions of substituents (see Sections 4.02.3.7, 4.02.3.8.2, 4.02.3.5.4 and 4.02.3.5.5).

The removal of a hydrogen atom from a heterocyclic nitrogen atom of azolones by nucleophiles acting as bases, *e.g.* (**14**) → (**15**), gives mesomeric anions, *e.g.* (**15**) ↔ (**16**) ↔ (**17**), which react exceedingly readily with electrophilic reagents, typically: (i) at nitrogen, *e.g.* with alkyl halides (Section 4.02.1.3.9); (ii) at oxygen, *e.g.* with acylating reagents (Section 4.02.3.7); (iii) at the β-carbon atoms, *e.g.* with halogens (Section 4.02.1.4.5), and in the Reimer–Tiemann reaction (Section 4.02.1.4.6).

(**14**) (**15**) (**16**) (**17**)

4.02.1.1.5 N-Oxides, N-imides, N-ylides of azoles

Azole N-oxides, N-imides and N-ylides are formally betaines derived from N-hydroxy-, N-amino- and N-alkyl-azolium compounds. Whereas N-oxides (Section 4.02.3.12.6) are usually stable as such, in most cases the N-imides (Section 4.02.3.12.5) and N-ylides (Section 4.02.3.12.3) are found as salts which deprotonate readily only if the exocyclic nitrogen or carbon atom carries strongly electron-withdrawing groups.

The reactivity of these compounds is somewhat similar to that of the azolonium ions, particularly when the cationic species is involved. However, although the typical reaction is with nucleophiles, the intermediate (**20**) can lose the N-oxide group to give the simple α-substituted azole (**21**). Benzimidazole 3-oxides are readily converted into 2-chlorobenzimidazoles in this way.

(**18**) (**19**) (**20**) (**21**)

4.02.1.2 Thermal and Photochemical Reactions Formally Involving No Other Species

We consider here fragmentations and rearrangements which involve only the azole molecule itself, without the vital involvement of any substituent or other molecule. Many fragmentations of azoles can be summarized by the transformation (22) → (23) + (24), where (23) represents a stable fragment, particularly N_2, but also CO_2, N_2O, COS or HCN.

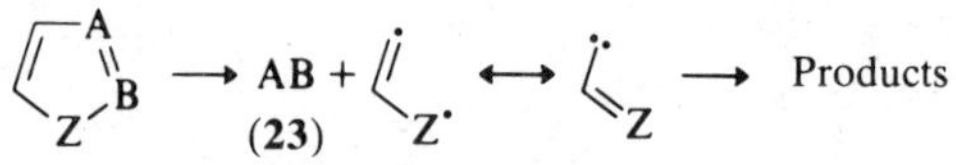

4.02.1.2.1 Thermal fragmentation

Thermal and photochemical fragmentation are often related to the mass spectroscopic breakdown of azole molecules (see Section 4.01.3.8), the latter often providing an indication of the behavior of a given molecule under such stimuli. Such fragmentations are facilitated in the polynitrogenous azoles, and azoles containing several nitrogen atoms undergo ring fission with loss of nitrogen. This is particularly noticeable when two adjacent pyridine-like nitrogen atoms are present. Thermolysis of (25) to give N_2 and a 1,3-dipole, such as $PhC\overset{+}{\equiv}N-\overset{-}{O}$, is a useful and general reaction. Azoles containing only two heteroatoms, such as the pyrazole and thiazole systems, are thermally very stable.

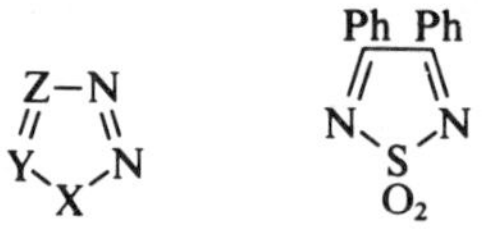

Although unsubstituted 1,2,5-thiadiazole is stable on heating at 220 °C, 3,4-diphenyl-1,2,5-thiadiazole 1,1-dioxide (26) decomposes into benzonitrile and sulfur dioxide at 250 °C ⟨68AHC(9)107⟩.

Thermal reactions of 1,4,2-dioxa-, 1,4,2-oxathia- and 1,4,2-dithia-azoles are summarized in Scheme 1. The reactive intermediates generated in these thermolyses can often be trapped, e.g. the nitrile sulfide dipole with DMAD.

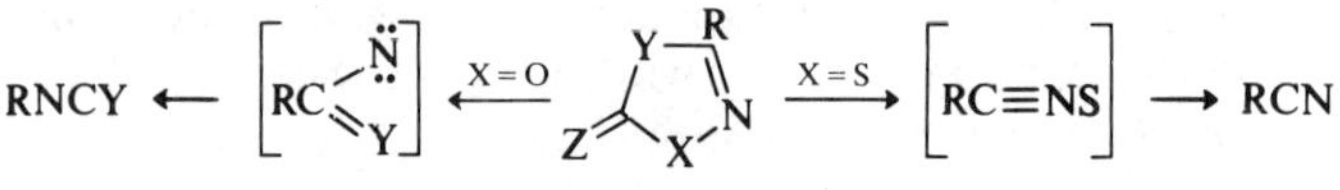

Scheme 1

1,2,3,4-Thiatriazoles readily decompose thermally into nitrogen, sulfur and an organic fragment, usually a cyanide, e.g. (27) → $Bu^iOCN + {}^{15}N{}^{14}N + S$ ⟨76AHC(20)145⟩.

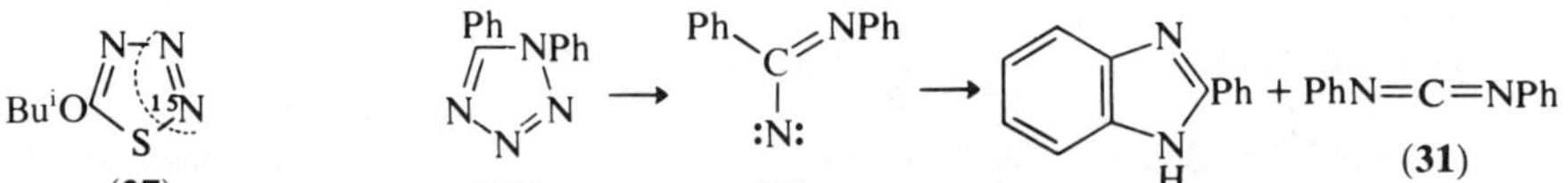

Other classes of heterocycles undergo thermolytic fragmentation to give imidolylnitrenes. As typified by the thermolysis of 1,5-diphenyltetrazole (28), the intermediates (29) can either cyclize on to aromatic rings to form benzimidazoles (30) or undergo a Wolff-type rearrangement to carbodiimides (31) ⟨81AHC(28)231⟩. Compounds (32) and (33) thermolyze to give mainly the carbodiimides (31) ⟨79JA3976⟩. Pentazoles (34) spontaneously form azides, usually below 20 °C.

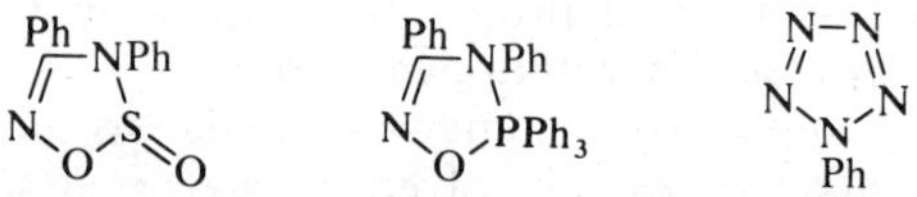

4.02.1.2.2 Photochemical fragmentation

Photolysis of 1,2,3-thiadiazole (**35**) gives thiirene (**36**) which can be trapped by an alkyne ⟨70AHC(11)1⟩. 4,5-Diphenyl-1,2,3-thiadiazole (**37**) is photolyzed at low temperatures to the thiobenzoylphenylcarbene triplet (**38**). Diphenylthioketene (**39**) is formed on warming ⟨81AHC(28)231⟩.

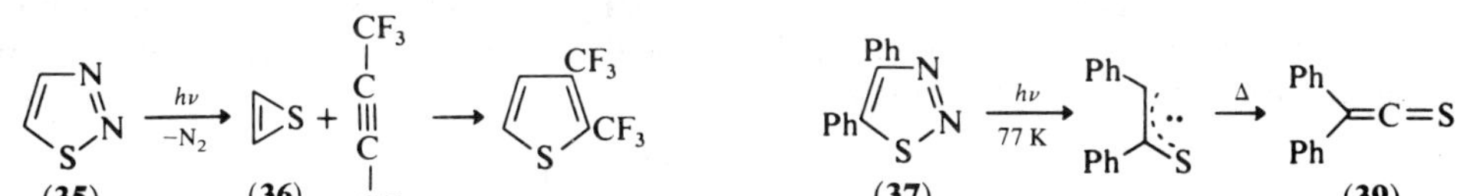

3-Phenylthiazirine (**40**) can be isolated as an intermediate in the photolysis of 5-phenyl-1,2,3,4-thiatriazole and also from other five-membered ring heterocycles capable of losing stable fragments; see Scheme 2 ⟨81AHC(28)231⟩. Photolysis of 5-phenylthiatriazole in the presence of cyclohexene yields cyclohexene episulfide ⟨60CB2353⟩ by trapping the sulfur atom.

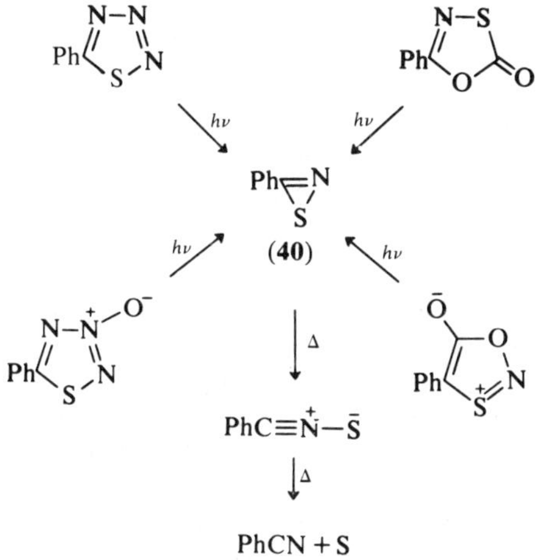

Scheme 2

Diaziridine derivatives (**42**) can be obtained from tetrazoles of type (**41**).

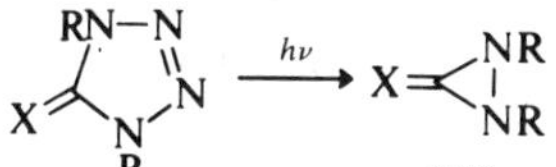

Mesoionic compounds undergo a variety of photochemical fragmentations. Examples are shown in which CO_2 or PhNCO is extruded (Schemes 3 and 4, respectively) ⟨76AHC(19)1⟩.

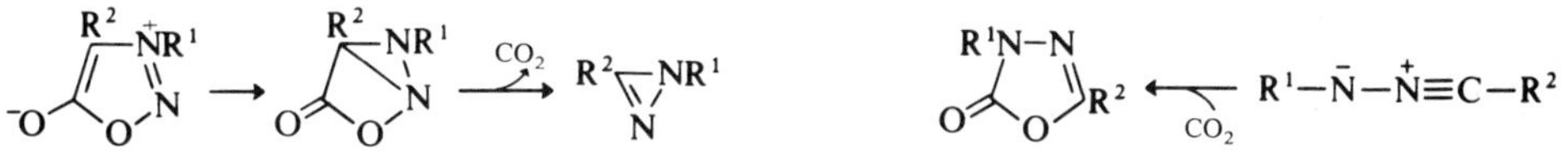

Scheme 3

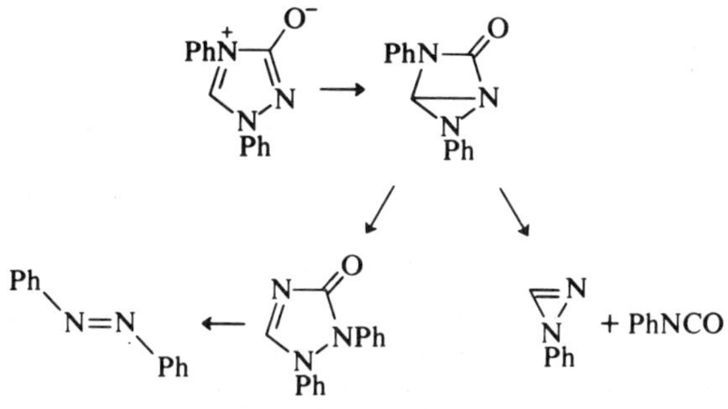

Scheme 4

4.02.1.2.3 Equilibria with open-chain compounds

Azoles of types (**44**) and (**46**) are isomeric with the open-chain compounds (**43**) and (**45**), respectively. Rearrangement between the two pairs is rapid and the thermodynamically stable isomer is encountered. Thus diazoketones (**43**; X = O) exist as such, but diazothioketones (**43**; X = S) spontaneously ring-close to thiadiazoles (**44**). 1,2,3-Triazoles generally exist as such unless the nitrogen carries a strong electron-withdrawing substituent. Thus 1-cyano- and 1-arenesulfonyl-1,2,3-triazoles (**47**) undergo easy, reversible ring-opening to diazo-imine tautomers ⟨74AHC(16)33⟩.

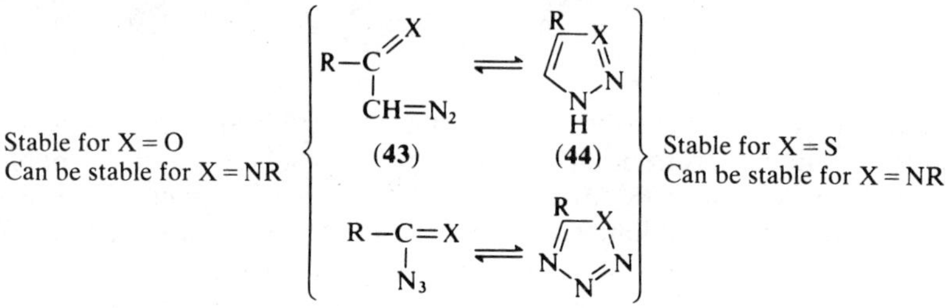

A similar situation exists for molecules containing an azide group bonded to a doubly bound carbon atom as in (**45**). When X is oxygen, the acylazide exists in the acyclic form (**45**), but when X is sulfur the cyclic thiatriazole (**46**) predominates. When X is nitrogen, as in tetrazoles, the imidoylazide (**45**) or the tetrazole (**46**) may predominate, or both may exist in equilibrium. The position of the tetrazole–imidoylazide equilibrium depends on the following factors: (1) electron-withdrawing substituents favor the azide form; (2) higher temperature favors the azide form; and (3) polar solvents tend to favor the tetrazole form, and non-polar solvents the azide form. Ring strain is also important and two fused five-membered rings are in general avoided. For example, in the tetrazolothiadiazole equilibrium (**48**) ⇌ (**49**), the system exists in the bicyclic form in the solid state and in the azide form in carbon tetrachloride solution ⟨77AHC(21)323⟩. Fusion with six-membered rings generally is more favorable to a bicyclic tetrazole form. For example, pyridine fusion gives essentially all tetrazole (**50**) ⟨69TL2595⟩.

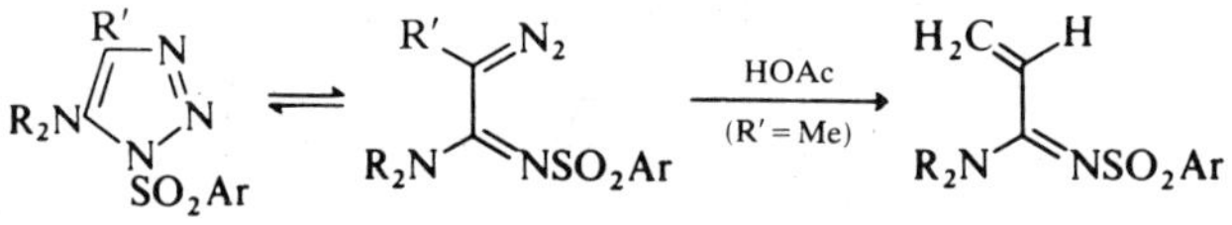

4.02.1.2.4 Rearrangement to other heterocyclic species

Many examples are known of rearrangement of azoles involving scrambling of the ring atoms to give a new isomeric azole molecule. Different mechanisms are involved.

For isoxazoles the first step is the fission of the weak N—O bond to give the diradical (**51**) which is in equilibrium with the vinylnitrene (**52**). Recyclization now gives the substituted 2*H*-azirine (**53**) which *via* the carbonyl-stabilized nitrile ylide (**54**) can give the oxazole (**55**). In some cases the 2*H*-azirine, which is formed both photochemically and thermally, has been isolated; in other cases it is transformed quickly into the oxazole ⟨79AHC(25)147⟩.

The photorearrangement of pyrazoles to imidazoles is probably analogous, proceeding *via* iminoylazirines ⟨82AHC(30)239⟩; indazoles similarly rearrange to benzimidazoles ⟨67HCA2244⟩. 3-Pyrazolin-5-ones (**56**) are photochemically converted into imidazolones (**57**) and open-chain products (**58**) ⟨70AHC(11)1⟩. The 1,2- and 1,4-disubstituted imidazoles are interconverted photochemically.

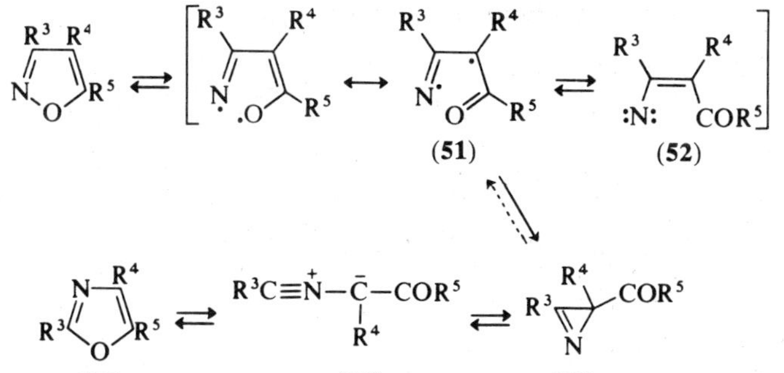

[Structures (51)–(58) depicting isoxazole ring-opening intermediates and photochemistry]

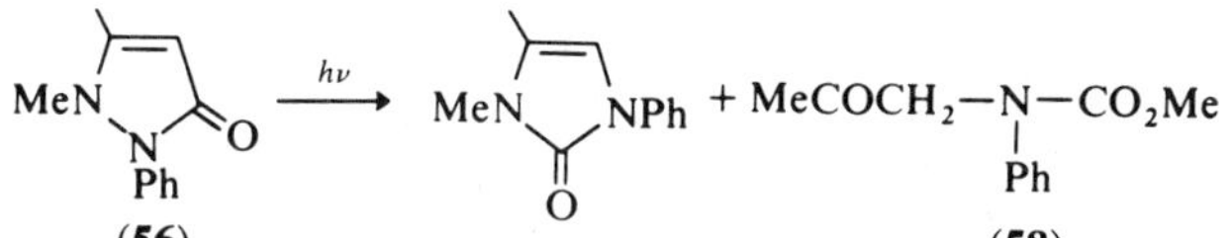

Irradiation of isothiazole gives thiazole in low yield. In phenyl-substituted derivatives an equilibrium is set up between the isothiazole (59) and the thiazole (61) via intermediate (60) ⟨72AHC(14)1⟩.

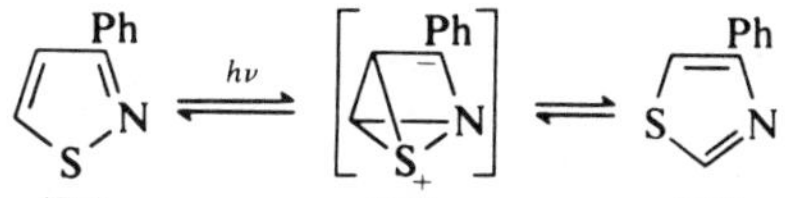

[Structures (59)–(61)]

Iminobenzodithioles (62) and benzisothiazolethiones (63) thermally equilibrate ⟨72AHC(14)43⟩.

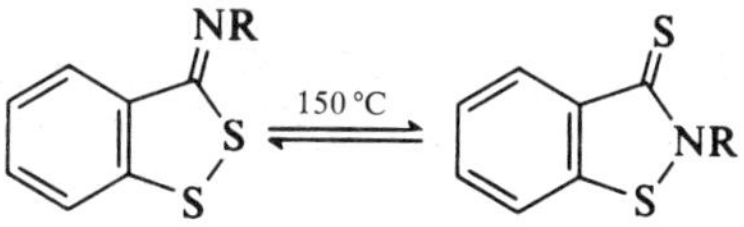

[Structures (62) and (63), equilibrium at 150 °C]

Mesoionic compounds of the type designated ⟨76AHC(19)1⟩ as 'A' are capable of isomerism. In one case in the 1,2,4-triazole series, isomerism of the pair (64) ⇌ (65) has been demonstrated ⟨67TL4261⟩.

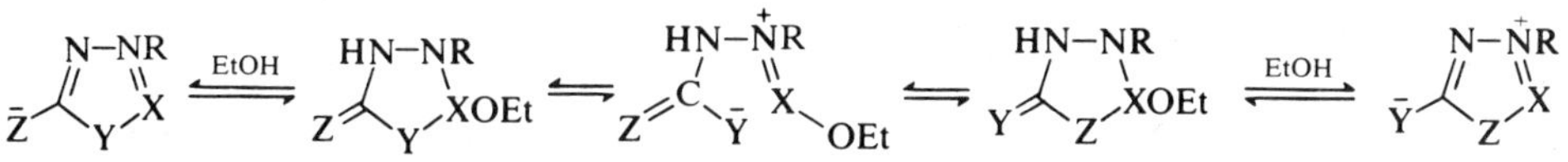

[Structures (64)–(65) with intermediates, EtOH]

4.02.1.2.5 Polymerization

Imidazoles and pyrazoles with free NH groups form hydrogen-bonded dimers and oligomers ⟨66AHC(6)347⟩.

4.02.1.3 Electrophilic Attack at Nitrogen

4.02.1.3.1 Introduction

Reactions of this type can be related to the chemistry of simple tertiary aliphatic amines. Thus the lone pair of electrons on the nitrogen atom in trimethylamine reacts under mild conditions with the following types of electrophilic reagents: (i) proton acids give salts; (ii) Lewis acids give coordination compounds; (iii) transition metal ions give complex formation; (iv) reactive halides give quaternary salts; (v) halogens give adducts; and (vi) certain oxidizing agents give amine oxides. The analogous reactions of pyridines with these electrophilic reagents at the lone pair on the nitrogen atom are well known. All neutral azoles contain a pyridine-like nitrogen atom and therefore similar reactions with electrophiles at this nitrogen would be expected. However, the tendency for such reactions varies considerably; in particular, successive heteroatom substitutions markedly decrease the ease of reaction. One convenient quantitative measure of the tendency for such reactions to occur is found in the basicity of these compounds; this is treated in Section 4.02.1.3.5.

4.02.1.3.2 Reaction sequence

In azoles containing at least two annular nitrogen atoms, one of which is an NH group and the other a multiply-bonded nitrogen atom, electrophilic attack occurs at the latter nitrogen. Such an attack is frequently followed by proton loss from the NH group, e.g. (66) → (67). If the electrophilic reagent is a proton, this reaction sequence simply means tautomer interconversion (see Section 4.01.5), but in other cases leads to the product.

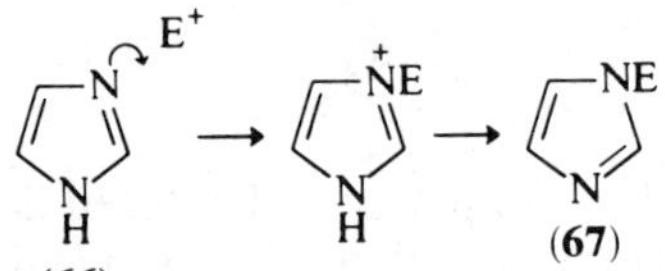

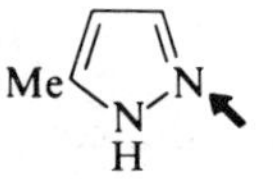

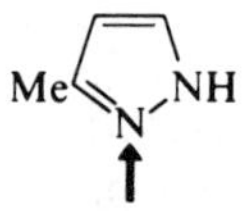

Since the electrophilic reagent attacks the multiply-bonded nitrogen atom, as shown for (68) and (69), the orientation of the reaction product is related to the tautomeric structure of the starting material. However, any conclusion regarding tautomeric equilibria from chemical reactivity can be misleading since a minor component can react preferentially and then be continually replenished by isomerization of the major component.

In addition to reaction sequences of type (66) → (67), electrophilic reagents can attack at either one of the ring nitrogen atoms in the mesomeric anions formed by proton loss (e.g. 70 → 71 or 72; see Section 4.02.1.3.6). Here we have an ambident anion, and for unsymmetrical cases the composition of the reaction product (71) + (72) is dictated by steric and electronic factors.

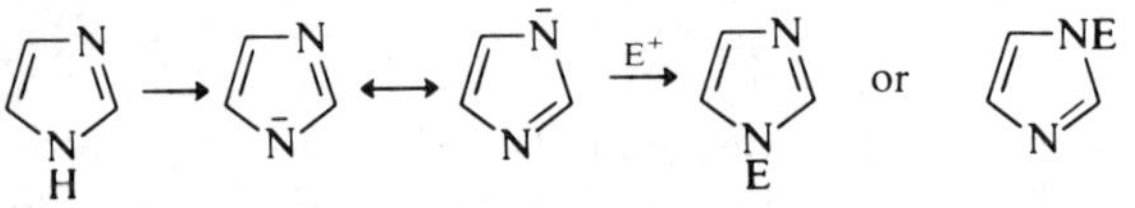

4.02.1.3.3 Orientation in azole rings containing three or four heteroatoms

Such compounds contain two or three 'pyridine-like' heteroatoms. For the symmetrical systems (73) and (74), no ambiguity occurs, but for systems (75)–(78) there are at least two alternative reaction sites. It appears that reaction takes place at the nitrogen atom furthest away from the 'pyrrole-like' heteroatom, as shown in (75)–(77) where evidence is available from reactions with alkylating reagents (Section 4.02.1.3.8).

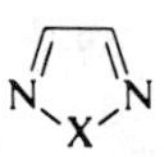

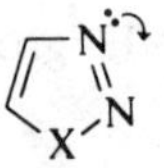

Similar ambiguities arise in the reactions of azole anions. At least as regards alkylation reactions in the 1,2,3-triazole series (79), the product appears to depend on the reagent used. In the 1,2,4-triazole series (80) a single product is formed, whereas tetrazole (81) gives mixtures.

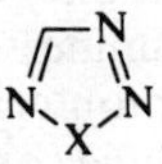

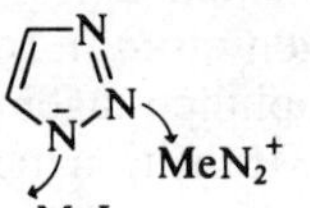

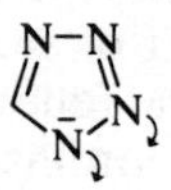

4.02.1.3.4 Effect of azole ring structure and of substituents

The ease of attack by an electrophilic reagent at the nitrogen atom of any azole is proportional to ΔE between the ground state and transition state energies. However, ground state structure largely controls the variation in these differences, which hence depend on the electron density on the basic nitrogen atom and the degree of steric hindrance. The number, orientation and type of heteroatoms are very important in determining electron density. Additional 'pyridine-like' nitrogen atoms always reduce the electron density at another 'pyridine-like' nitrogen (compare the reduced basicities of diazines relative to pyridine). Unshared electron pairs on two 'pyridine-like' nitrogen atoms can interact, but the effect on reactivity appears to be small ⟨78AHC(22)71⟩. In the case of 'pyrrole-like' nitrogen, oxygen and sulfur there are two mutually opposed effects: base-strengthening mesomeric electron donation and base-weakening inductive electron withdrawal. The latter is particularly strong for heteroatoms in the α-position and in fact for oxygen and sulfur always dominates over the base-strengthening effect.

The effects of substituents may be rationalized as follows. (i) Strongly electron-withdrawing substituents (e.g. NO_2, COR, CHO) make these reactions more difficult by decreasing the electron density on the nitrogen atom(s). The effect is largely inductive and therefore is particularly strong from the α-position. (ii) Strongly electron-donating substituents (e.g. NH_2, OR) facilitate electrophilic attack by increasing the electron density on the nitrogen. This is caused by the mesomeric effect and is therefore strongest from the α- and γ-positions. (iii) Fused benzene rings, aryl and alkyl groups, and other groups with relatively weak electronic effects have a relatively small electronic influence.

The foregoing electronic effects are illustrated by the pK_a values given in Section 4.02.1.3.5. Reactions other than proton addition are hindered by all types of α-substituents. However, steric hindrance is less in these five-membered ring heterocycles than that in pyridines because the angle subtended between the nitrogen lone pair and the α-substituent is significantly greater in the five-membered ring compounds and thus the substituent is held further away from the lone pair.

4.02.1.3.5 Proton acids on neutral azoles: basicity of azoles

Mesomeric shifts of the types shown in structures (**82**) and (**83**) increase the electron density on the nitrogen atom and facilitate reaction with electrophilic reagents. However, the heteroatom Z also has an adverse inductive effect; the pK_a of NH_2OH is 6.0 and that of N_2H_4 is 8.0, both considerably lower than that of NH_3 which is 9.5.

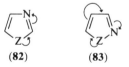

(**82**)   (**83**)

Gas-phase pK_a values for azoles are unavailable except for imidazole and 1-methylimidazole. Imidazole is ca. 75 kJ mol^{-1} more basic than ammonia, i.e. approximately the same as pyridine; 1-methylimidazole is about 29 kJ mol^{-1} more basic than imidazole ⟨82PC40200⟩.

Tetrazoles are preferentially protonated at N-4.

The basicities of the parent azole systems in water are shown in Table 1. When both heteroatoms are nitrogen, the mesomeric effect predominates when the heteroatoms are in the 1,3-positions, whereas the inductive effect predominates when they are in the 1,2-positions. The predominance of the mesomeric effect is illustrated by the pK_a value of imidazole (**82**; Z = NH), which is 7.0, whereas that of pyrazole (**83**; Z = NH) is 2.5 (cf. pyridine, 5.2). An N-methyl group is base-strengthening in imidazole, but base-weakening in pyrazole, probably because of steric hindrance to hydration. When the second heteroatom is oxygen or sulfur the inductive, base-weakening effect increases; the pK_a of thiazole (**82**; Z = S) is 3.5 and that of isoxazole (**83**; Z = O) is 1.3.

Substituents are expected to alter the electron density at the multiply-bonded nitrogen atom, and therefore the basicity, in a manner similar to that found in the pyridine series. The rather limited data available appear to bear out these assumptions. The additional ring nitrogen atoms in triazoles, oxadiazoles, etc. are quite strongly base-weakening; this is as

Table 1 pK_a Values for Proton Addition ⟨63PMH(1)1, 71PMH(3)1, B-76MI40200, B-76MI40201⟩

X	Ring systems				
	 pyrazole	 imidazole	 1,2,3-triazole	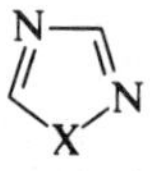 1,2,4-triazole	 1,3,4-triazole
NH	2.52	6.95	1.17	2.45	(1.17)
NMe	2.06	7.33	1.25	3.20	<1
O	−2.97	0.8	—	—	—
S	−0.51	2.53	—	—	−4.9

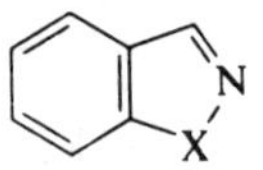

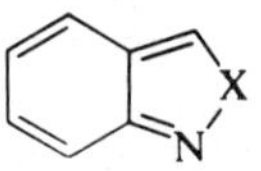

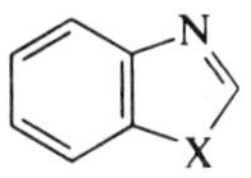

X	indazole	benzisoxazole-type	benzimidazole
NH	1.31	—	5.53
NMe	0.42	2.02	5.57
O	−4.7	−2.20	−0.13
S	—	−0.05	1.2

expected since diazines are weaker bases than pyridine. As regards *C*-substituents, their effects on the pK_a of the parent compounds are as follows. (i) Methyl groups are weakly base-strengthening due to their mesomeric and inductive electron donor effect: thus in the methylthiazoles the base strengths decrease in the order 2>4>5. (ii) Phenyl groups are weak resonance donors, but inductive acceptors. Phenyl groups are therefore expected to reduce the basicity of azoles. (iii) Amino groups are strong resonance electron donors and hence base-strengthening, particularly if directly conjugated with the basic center. (iv) Methoxy groups are resonance donors but inductive acceptors. The inductive effect would be expected to be dominant for azoles. (v) Halogen atoms are inductive acceptors (and weak resonance donors); they are expected to cause a marked decrease in basicity, especially from α-positions. (vi) Fused benzene rings usually have little effect; *cf.* the pK_a values of imidazole and benzimidazole. Substituents on the benzene ring in benzazoles should have little effect on the basicity.

Annular nitrogen atoms can form hydrogen bonds, and if the azole contains an NH group, association occurs. Imidazole (**84**) shows a cryoscopic molecular weight in benzene 20 times that expected. Its boiling point is 256 °C, which is higher than that of 1-methylimidazole (198 °C).

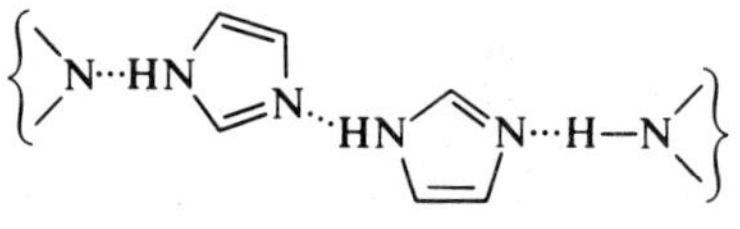

(**84**)

4.02.1.3.6 Proton acids on azole anions: acidity of azoles

The acidities of the five parent compounds are compared with that of pyrrole in Table 2. The acidity of the ring system increases as the number of nitrogens increases, the acidity of pyrrole increasing by approximately 2, 4.5 and 5 pK_a units for each successive addition of a nitrogen atom. 1,2,3-Triazole is slightly more acidic than 1,2,4-triazole, but the effect on NH acidity of nitrogen orientation is much less than the effect of the total number of nitrogens ⟨71PMH(3)1⟩.

Table 2 pK_a Values of Azoles for Proton Loss ⟨B-76MI40200, B-76MI40201⟩

Nitrogen positions	pK_a	Nitrogen positions	pK_a
1	16.5	1,2,3	9.26
1,2	14.21	1,2,4	10.04
1,3	14.44	1,2,3,4	4.89

Ring substituents can have a considerable effect on the acidity of the system. In the 1,2,4-triazole series a 3-amino group decreases the acidity to 11.1, a 3-methyl group to 10.7, whereas a 3-phenyl group increases the acidity to 9.6, and 3,5-dichloro substitution to 5.2 ⟨71PMH(3)1⟩.

4.02.1.3.7 Metal ions

(i) Simple complexes

Many examples are known of complexes between metal cations and both neutral azoles and azole anions. Overlap between the *d*-orbitals of the metal atom and the azole π-orbitals is believed to increase the stability of many of these complexes.

Despite the weak basicity of isoxazoles, complexes of the parent methyl and phenyl derivatives with numerous metal ions such as copper, zinc, cobalt, *etc.* have been described ⟨79AHC(25)147⟩. Many transition metal cations form complexes with imidazoles; the coordination number is four to six ⟨70AHC(12)103⟩. The chemistry of pyrazole complexes has been especially well studied and coordination compounds are known with thiazoles and 1,2,4-triazoles. Tetrazole anions also form good ligands for heavy metals ⟨77AHC(21)323⟩.

Isothiazoles react with hexacarbonyls $M(CO)_6$ to give *N*-coordinate $M(CO)_5$ derivatives.

(ii) Chelate complexes

Chelate rings can be formed by azoles containing α-substituents such as carbonyl or CH=NH groups. An important bidentate chelating agent is histidine (**85**), and many pyrazoles with substituent groups are known which form bis and tris complexes with many metals, *e.g.* (**86**). Similarly, 2- and 4-α-pyridylthiazoles are bidentate chelating agents. Complex formation of this type has analytical applications; thus 1,3,4-thiadiazole-2,5-dithione has been used as a spot test for bismuth and other metals ⟨58MI40200⟩.

4.02.1.3.8 Alkyl halides and related compounds: azoles without a free NH group

Pyrazoles and imidazoles carrying a substituent on nitrogen, as well as oxazoles, thiazoles, *etc.*, are converted by alkyl halides into quaternary salts. This is illustrated by the preparation of thiamine (**89**) from components (**87**) and (**88**).

Azoles having heteroatoms in the 1,3-orientation are more reactive than those in which the arrangement is 1,2. However, the magnitude of the factor varies. Thus oxazole is 68 times more reactive than isoxazole, whereas benzoxazole quaternizes 26 times faster than does 1,2-benzisoxazole ⟨78AHC(22)71⟩.

These reactions are of the S_N2 type and are sensitive to steric effects of substituents in the azole ring. However, these steric effects are significantly less than, for example, in the analogous pyridine derivatives because the angle subtended by the nitrogen lone pair and an α-substituent is about 70° in an azole as opposed to 60° in pyridine. Thus the rate constant for methylation of 2-*t*-butylthiazole by methyl iodide is only 40 times less than that for the corresponding 2-methyl compound. By comparison, in the pyridine series the retardation factor is over 2000 in the same solvent (nitrobenzene). As in six-membered rings, the kinetic consequences of the steric effects of *o*-alkyl groups are related to the E_S parameter. However, buttressing groups can cause special effects as has been investigated extensively in the thiazole series.

Annulation of a five-membered aza ring to a benzo ring generally leads to rate retardation in *N*-quaternization reactions similar in magnitude to that for six-membered rings. Exceptions are known: 2,1-benzisoxazole undergoes *N*-methylation faster than isoxazole, and in 2,1,3-benzoxadiazole and 2,1-benzisothiazole the rates are little changed from the corresponding monocyclic rings; however, here we are dealing with *o*-quinonoid structures. The more usual situation is rate retardation by a moderate amount. This is probably caused not by steric effects, but by electronic effects, as is shown by the corresponding influence on the pK_a values ⟨78AHC(22)71⟩.

Satisfactory Brønsted correlations for α-substituted azoles offer further evidence of the lesser importance of steric effects in the azole series ⟨78AHC(22)71⟩.

For both azole and benzazole rings the introduction of further heteroatoms into the ring affects the ease of quaternization. In series with the same number and orientation of heteroatoms, rate constants increase in the order X = O < S < SMe (cf. Table 3) ⟨78AHC(22)71⟩. The quaternization of triazoles, thiadiazoles and tetrazoles requires stronger reagents and conditions; methyl fluorosulfonate is sometimes used ⟨78AHC(22)71⟩. The 1- or 2-substituted 1,2,3-triazoles are difficult to alkylate, but methyl fluorosulfonate succeeds ⟨71ACS2087⟩.

Table 3 Heteroatom and Benzo-fusion Effects on Relative Rate Constants for *N*-Methylation ⟨78AHC(22)71⟩

Heterocycle	k_{rel}		
	O	S	NMe
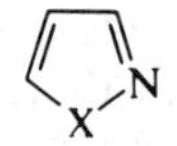 (pyrazole-type)	1	6.9	120
 (isoxazole-type)	1	(15, 1)	(912, 61)
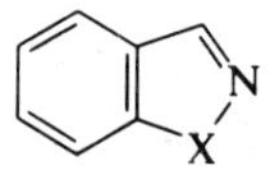 (indazole-type)	1	(20, 1)	(56, 2.8)
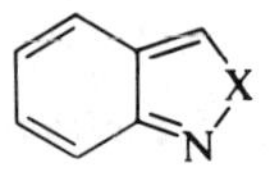 (2,1-benzisoxazole-type)	1	(3.6, 1)	(33, 9.2)
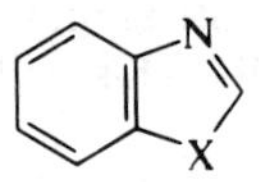 (benzoxazole-type)	1	(9.3, 1)	(708, 76)
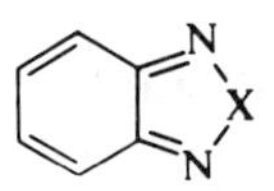 (benzoxadiazole-type)	—	2.8	1

Oxadiazoles are difficult to alkylate, unless the ring contains a strong electron donor group such as an amino substituent.

1,2,3-Thiadiazoles are quaternized to give 3- or mixtures of 2- and 3-alkyl quaternary salts. In 5-amino-1,2,4-thiadiazole, quaternization takes place at the 4-position (**90**) ⟨64AHC(3)1⟩. 1-Substituted 1,2,4-triazoles are quaternized in the 4-position (**91**), and 4-substituted 1,2,4-triazoles are quaternized in the 1- or the 2-position (**92**) ⟨64AHC(3)1⟩.

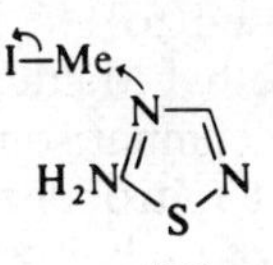

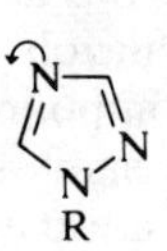

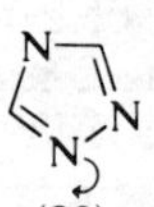

5-Substituted 1,2,3,4-thiatriazoles (**93**) are alkylated only under very forcing conditions with triethyloxonium fluoroborate, but then give the expected products (**94**) ⟨75JOC431⟩.

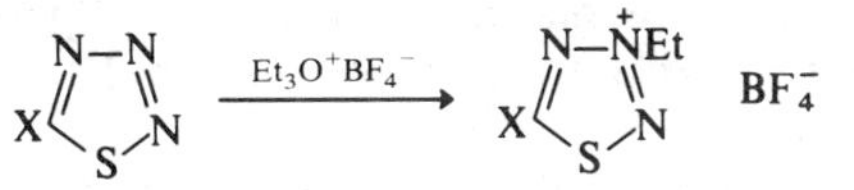

(**93**) X = SEt, SMe, Ph   (**94**) X = SEt, SMe, Ph

1-Alkyl-5-phenyltetrazoles are converted into 2-alkyl isomers on heating with alkyl iodide, presumably by quaternary salt formation followed by elimination of alkyl iodide to give the thermodynamically more stable isomer (Scheme 5) ⟨77AHC(21)323⟩.

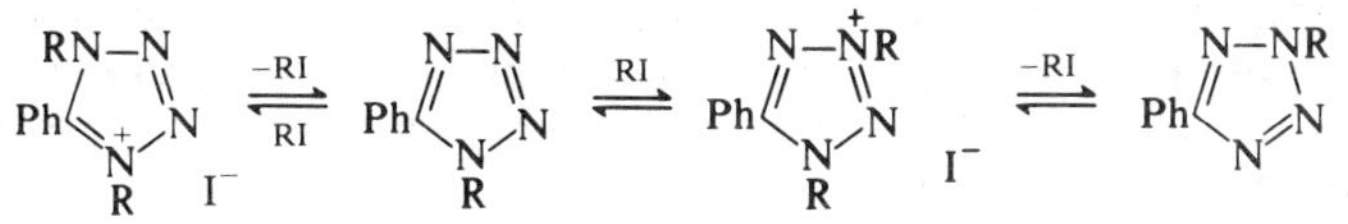

Scheme 5

1,2,4-Thiadiazole with $Me_3O^+BF_4^-$ gives the diquaternary salt (**95**); diquaternary salts are also known in the 1,2,4-triazole series. 1,3-Disubstituted 1,2,4-triazolium salts can be further alkylated to diquaternary derivatives.

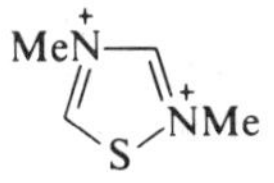

(**95**)

4.02.1.3.9 Alkyl halides and related compounds: compounds with a free NH group

Pyrazoles and imidazoles with free NH groups are readily alkylated, *e.g.* by MeI or Me_2SO_4. A useful procedure is to use the alkyl salt of the azole in liquid ammonia ⟨80AHC(27)241⟩. However, alkylation can also occur under neutral conditions, particularly with imidazoles.

Unsymmetrical imidazoles and pyrazoles usually give a mixture of products, the composition of which may depend on the reaction conditions. Thus the ethoxycarbonylpyrazole (**97**) gives predominantly the isomeric *N*-methyl derivatives (**96**) and (**98**) under the conditions indicated. The difference in orientation can be related to the stabilization of the tautomeric structure (**97**) by hydrogen bonding (possibly intramolecular), which means that alkylation of the free base gives (**98**). The isomer (**96**) is formed *via* the anion. Benzylation of (**97**) gives mainly the analogue of (**98**) ⟨80JHC137⟩.

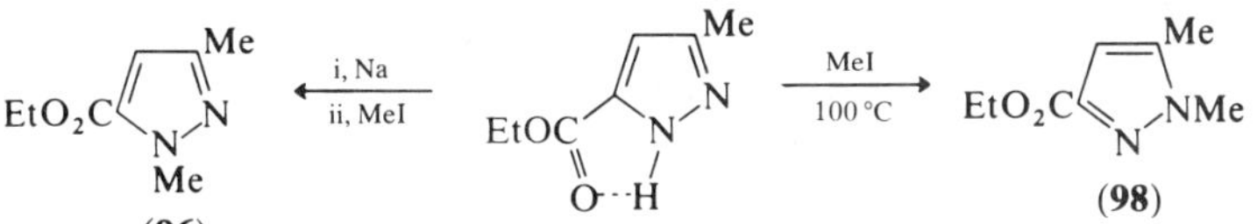

(**96**)   (**97**)   (**98**)

The differential effects of steric hindrance and tautomeric content in the imidazole series are illustrated in Scheme 6 ⟨80AHC(27)241⟩.

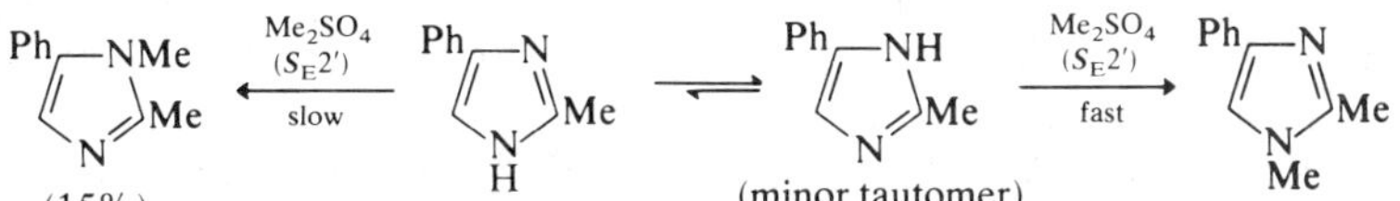

(15%)   (major tautomer)   (minor tautomer)   (85%)

Scheme 6

N-Unsubstituted 1,2,3-triazoles are methylated mainly in the 1-position with methyl iodide and silver or thallium salts, but mainly in the 2-position by diazomethane. There is also some steric control. For example, 4-phenyl-1,2,3-triazole with dimethyl sulfate gives the 2-methyl-4-phenyl (38%) and 1-methyl-4-phenyl isomers (62%), but none of the more hindered 1-methyl-5-phenyltriazole ⟨74AHC(16)33⟩. *N*-Unsubstituted 1,2,4-triazoles are generally alkylated at N-1.

Alkylation of tetrazoles as the anions gives mixtures of 1- and 2-alkyl isomers. In general, electron-donating substituents in the 5-position slightly favor alkylation of the 1-position and electron-withdrawing 5-substituents slightly favor the 2-position.

Azolone anions are readily alkylated at nitrogen, e.g. 2-triazolone with methyl iodide gives the 1-methyl derivative.

N-Arylation of azoles is achieved either by using arynes, activated halobenzenes (*e.g.* dinitro) or under Ullmann conditions. Thus benzyne reacts with imidazoles to give *N*-arylimidazoles ⟨70AHC(12)103⟩, and these compounds have also been prepared under modified Ullmann conditions.

N-Unsubstituted pyrazoles and imidazoles add to unsaturated compounds in Michael reactions, for example acetylenecarboxylic esters and acrylonitrile readily form the expected addition products. Styrene oxide gives rise, for example, to 1-styrylimidazoles ⟨76JCS(P1)545⟩. Benzimidazole reacts with formaldehyde and secondary amines in the Mannich reaction to give 1-aminomethyl products.

4.02.1.3.10 Acyl halides and related compounds

Azoles containing a free NH group react comparatively readily with acyl halides. *N*-Acyl-pyrazoles, -imidazoles, *etc.* can be prepared by reaction sequences of either type **(66)** → **(67)** or type **(70)** → **(71)** or **(72)**. Such reactions have been carried out with benzoyl halides, sulfonyl halides, isocyanates, isothiocyanates and chloroformates. Reactions occur under Schotten–Baumann conditions or in inert solvents. When two isomeric products could result, only the thermodynamically stable one is usually obtained because the acylation reactions are reversible and the products interconvert readily. Thus benzotriazole forms 1-acyl derivatives **(99)** which preserve the 'Kékulé resonance' of the benzene ring and are therefore more stable than the isomeric 2-acyl derivatives. Acylation of pyrazoles also usually gives the more stable isomer as the sole product ⟨66AHC(6)347⟩. The imidazole-catalyzed hydrolysis of esters can be classified as an electrophilic attack on the multiply bonded imidazole nitrogen.

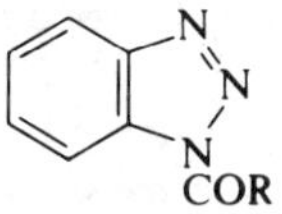

(99)

Since *N*-acylation is a reversible process, it has allowed the regiospecific alkylation of, for example, imidazoles to give the sterically less favored derivative. This principle is illustrated in Scheme 7 ⟨80AHC(27)241⟩.

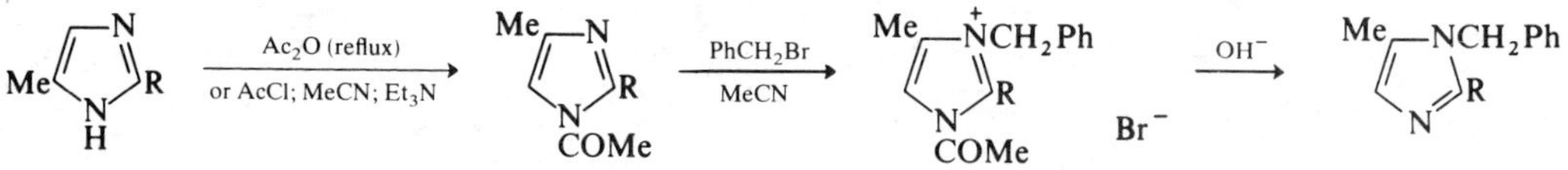

Scheme 7

1,2,3-Triazoles are acylated with acyl halides, usually initially at the 1-position, but the acyl group may migrate to the 2-position on heating or on treatment with base. Thus acetylation with acetyl chloride often gives 1-acetyl derivatives, which rearrange to the 2-isomers above 120 °C ⟨74AHC(16)33⟩.

Whether tetrazoles are acylated in the 1- or 2-position depends on the 5-substituent. 2-Acyltetrazoles are unstable (see Section 4.02.3.12.4) ⟨77AHC(21)323⟩; 1-alkylsulfonyltriazoles are also unstable (see Section 4.02.1.2.3).

4.02.1.3.11 Halogens

At room temperature, *N*-unsubstituted azoles react with halogens and interhalogens (*e.g.* ICl) to give *N*-halogenoazoles, probably *via* unstable adducts. Thus imidazoles with halogens form *N*-halogeno compounds, which easily rearrange to form *C*-halogenoimidazoles ⟨70AHC(12)103⟩. *N*-Halogenopyrazoles are unstable and act as halogenating agents. *N*-

Halogeno-1,2,4-triazoles are more easily isolated, especially when the 3,5-positions are substituted.

4.02.1.3.12 Peracids

Pyridine 1-oxides are formed in good yield on treatment of pyridines with peracids, although pyridine reacts less readily than do aliphatic tertiary amines. Azoles generally react even less readily, and N-oxides of azoles can rarely be formed in this way. However, some examples are known. Thus 1-methylpyrazole gives the 2-oxide, 2,4-dimethylthiazole (**100**) the 3-oxide (**101**) ⟨B-71MI40200⟩ and 1,2,3-thiadiazoles the 3-oxides. 5-Phenylthiatriazole (**102**) gives the 3-oxide (**103**) with peroxytrifluoroacetic acid ⟨75T1783⟩. More frequently, attempted N-oxide formation affords azolones or ring-cleaved derivatives, probably by nucleophilic attack of a peroxide anion on the azolium ion (*cf.* Section 4.02.1.6.1). Thus attempted conversion of oxazoles into N-oxides fails and leads to ring opening ⟨74AHC(17)99⟩, and most thiazoles are extensively degraded.

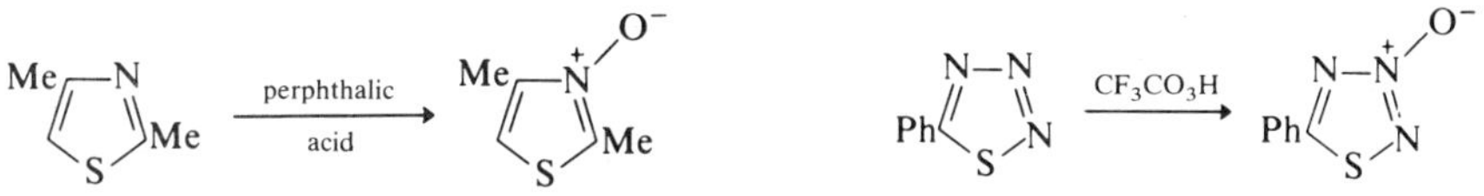

4.02.1.3.13 Aminating agents

Amination at an azole ring nitrogen is known for N-unsubstituted azoles. Thus 4,5-diphenyl-1,2,3-triazole with hydroxylamine-O-sulfonic acid gives approximately equal amounts of the 1- (**104**) and 2-amino derivatives (**105**) ⟨74AHC(16)33⟩. Pyrazole affords (**106**) and indazole gives comparable amounts of the 1- and 2-amino derivatives.

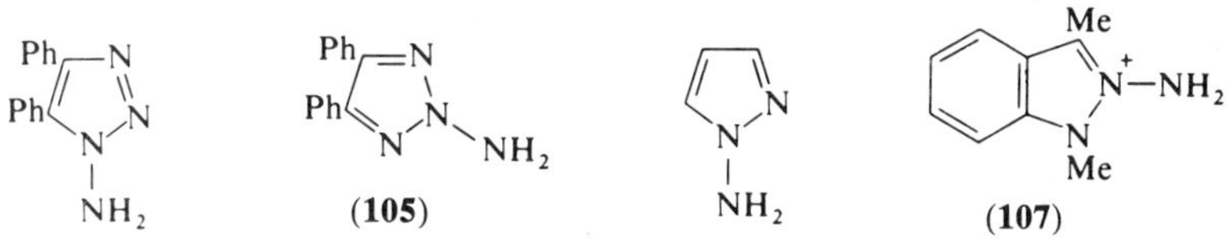

Azoles without a free NH group are also aminated, giving N-aminoazolium salts, *e.g.* (**107**) ⟨81AHC(29)71⟩.

4.02.1.3.14 Other Lewis acids

Azoles can form stable compounds in which metallic and metalloid atoms are linked to nitrogen. For example, pyrazoles and imidazoles N-substituted by B, Si, P and Hg groups are made in this way. Imidazoles with a free NH group can be N-trimethylsilylated and N-cyanated (with cyanogen bromide). Imidazoles of low basicity can be N-nitrated.

4.02.1.4 Electrophilic Attack at Carbon

4.02.1.4.1 Reactivity and orientation

(i) *Ease of reaction*

Replacing a CH=CH group in benzene with a heteroatom (Z) increases the susceptibility of the ring carbon atoms to electrophilic attack noticeably when Z is S, more when Z is O, and very markedly when Z is NH (*cf.* Chapter 3.02). Replacing one CH group in benzene with a nitrogen atom decreases the ease of electrophilic attack at the remaining carbon atoms (cf. Chapter 2.02); replacement of two CH groups with nitrogen atoms decreases it further (cf. Chapter 2.02). Such deactivation is very strong in nitration, sulfonation and

Friedel–Crafts reactions, which proceed in strongly acidic media, *i.e.* under conditions in which the nitrogen atom is largely protonated (or complexed). The effect of a protonated nitrogen atom is considerably greater than, for example, the two nitro groups in *m*-dinitrobenzene. The deactivating effect is less pronounced in reactions conducted under neutral or weakly acidic conditions, where a large proportion of unprotonated free base exists, *i.e.* as in halogenation and mercuration reactions.

In azole chemistry the total effect of the several heteroatoms in one ring approximates the superposition of their separate effects. It is found that pyrazole, imidazole and isoxazole undergo nitration and sulfonation about as readily as nitrobenzene; thiazole and isothiazole react less readily (*ca.* equal to *m*-dinitrobenzene), and oxadiazoles, thiadiazoles, triazoles, *etc.* with great difficulty. In each case, halogenation is easier than the corresponding nitration or sulfonation. Strong electron-donor substituents help the reaction.

Pyrazoles and imidazoles exist partly as anions (*e.g.* **108** and **109**) in neutral and basic solution. Under these conditions they react with electrophilic reagents almost as readily as phenol, undergoing diazo coupling, nitrosation and Mannich reactions (note the increased reactivity of pyrrole anions over the neutral pyrrole species).

(**108**)   (**109**)

(ii) Orientation

A multiply bonded nitrogen atom deactivates carbon atoms α or γ to it toward electrophilic attack; thus initial substitution in 1,2- and 1,3-dihetero compounds should be as shown in structures (**110**) and (**111**). Pyrazoles (**110**; Z = NH), isoxazoles (**110**; Z = O), isothiazoles (**110**; Z = S), imidazoles (**111**; Z = NH, tautomerism can make the 4- and 5-positions equivalent) and thiazoles (**111**; Z = S) do indeed undergo electrophilic substitution as expected. Little is known of the electrophilic substitution reactions of oxazoles (**111**; Z = O) and compounds containing three or more heteroatoms in one ring. Deactivation of the 4-position in 1,3-dihetero compounds (**111**) is less effective because of considerable double bond fixation (cf. Sections 4.01.3.2.1 and 4.02.3.1.7), and if the 5-position of imidazoles or thiazoles is blocked, substitution can occur in the 4-position (**112**).

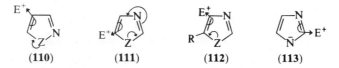

(**110**)   (**111**)   (**112**)   (**113**)

The above considerations do not necessarily apply to reactions of electrophilic reagents with pyrazole and imidazole anions (**108, 109**). The imidazole anion is sometimes substituted in the 2-position (**113**) and the indazole anion in the 3-position (cf. Section 4.02.1.4.5).

(iii) Effect of substituents

Just as in benzene, substituents can strongly activate (*e.g.* NH_2, NMe_2, OMe), strongly deactivate (*e.g.* NO_2, SO_3H, CO_2Et) or have relatively little effect on (*e.g.* Me, Cl) the ring toward further substitution. Further electrophilic substitution generally will not take place on an azole which carries a strongly deactivating substituent unless the azole ring is also carrying a strong electron-donor group or is strongly activated, as it is in the azolone form. However, these considerations can be affected by basicity considerations: thus a strongly deactivating group can also increase the amount of more reactive neutral molecule compared with a less reactive cation.

When the preferred substitution position (cf. **110** and **111**) is occupied, activating substituents can facilitate substitution in other positions (cf. examples in Sections 4.02.1.4.2 and 4.02.1.4.5); *ipso* attack can also occur if the substituent is itself easily displaced.

In benz- and phenyl-azolones, electrophilic substitution often occurs in the benzene ring; such reactions are considered as reactions of substituents (see Sections 4.02.3.2.1 and 4.02.3.4.1).

4.02.1.4.2 Nitration

Nitration of monocyclic compounds is summarized in Table 4. Substitution occurs in the expected positions. The reaction conditions required are more vigorous than those needed for benzene, but less than those for pyridine. Ring nitration of oxazoles is rare, but (114) has been obtained in this way ⟨74AHC(17)99⟩.

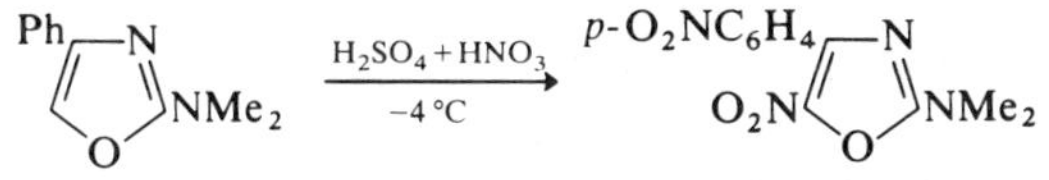

Ph—N $\xrightarrow{H_2SO_4 + HNO_3}_{-4\,°C}$ $p\text{-}O_2NC_6H_4$—N, O_2N, NMe$_2$

(114)

Table 4 Nitration and Sulfonation of Azoles

Heterocycle	Position substituted	Reaction conditions	
		Sulfonation	Nitration
Pyrazole	4	H_2SO_4/SO_3, 100 °C	$HNO_3/H_2SO_4/SO_3$, 100 °C
Imidazole	4 (≡5)	H_2SO_4/SO_3, 160 °C	HNO_3/H_2SO_4, 160 °C
3-Methylisoxazole	4	HSO_3Cl, 100 °C	$HNO_3/H_2SO_4/SO_3$, 70 °C
Isothiazole	4	H_2SO_4/SO_3, 150 °C [a]	HNO_3/H_2SO_4, 230 °C
Thiazole	5	$H_2SO_4/SO_3/Hg$, 250 °C	—
4-Methylthiazole	5	H_2SO_4/SO_3, 200 °C	$HNO_3/H_2SO_4/SO_3$, 160 °C
2,5-Dimethylthiazole	4	H_2SO_4/SO_3, 200 °C	$HNO_3/H_2SO_4/SO_3$, 160 °C

[a] ⟨66GEP1208303⟩.

Substituents are sometimes displaced: thus chloroimidazoles are nitrated normally, but iodo analogues suffer nitro-deiodination.

Pyrazoles can undergo nitration at several positions: 4-bromo-1-methylpyrazole yields the 3,5-dinitro product. 1-Methylpyrazole 2-oxide yields the 5-nitro derivative.

As expected, nitration is facilitated by activating groups such as an amino group; for example, nitration of (115) occurs at about 20 °C (HNO_3/H_2SO_4). Sydnones (116) are nitrated readily. The pyrazolinone (117) is nitrated as indicated, and 1,2,4-triazolinones have also been ring nitrated.

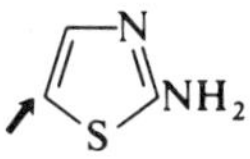

(115) 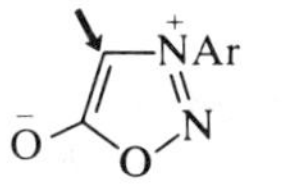   (116) 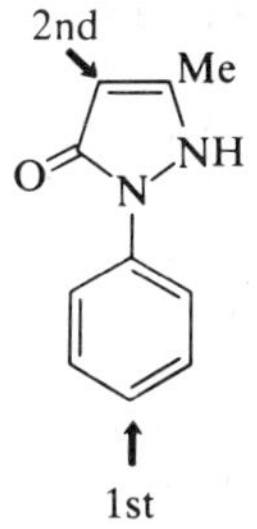   (117)

4.02.1.4.3 Sulfonation

Sulfonation conditions are given in Table 4. The orientation is as expected. Azolinones react as readily as the corresponding azoles; sulfonation of (117) occurs at the positions indicated (H_2SO_4/SO_3 at 100 °C). Imidazoles are readily chlorosulfonated at C-4.

4.02.1.4.4 Acid-catalyzed hydrogen exchange

Acid-catalyzed hydrogen exchange is used as a measure of the comparative reactivity of different aromatic rings (see Table 5). These reactions take place on the neutral molecules or, at high acidities, on the cations. At the preferred positions the neutral isoxazole, isothiazole and pyrazole rings are all considerably more reactive than benzene. Although the 4-position of isothiazole is somewhat less reactive than the 4-position in thiophene, a similar situation does not exist with isoxazole–furan ring systems.

Imidazoles, because of their high basicity, are very unreactive unless electron-withdrawing substituents are present.

Table 5 Reactivities Toward Acid-catalyzed Deuterodeprotonation ⟨79AHC(25)147⟩

Heterocycle	log (partial rate factor)[a] Ring positions			
	2	3	4	5
Isoxazole	—	—	4.3	—
Isothiazole	—	—	3.6	—
Furan	8.2	ca. 4.5	ca. 4.5	8.2
Thiophene	8.6	5.0	5.0	8.6
1-Methylpyrazole	—	5.6	9.8	5.6

[a] Relative to position 1 of benzene = 1.

4.02.1.4.5 Halogenation

Imidazoles and pyrazoles containing an unsubstituted NH group are easily chlorinated (Cl_2/H_2O or N-chlorosuccinimide/$CHCl_3$), brominated ($Br_2/CHCl_3$; $KOBr/H_2O$), and iodinated (I_2/HIO_3). Substitution generally occurs first at the 4-position, but further reaction at other available nuclear positions takes place readily, especially in the imidazole series. When halogenation of the nucleus involves electrophilic attack on anions of type (**109**), the 4-position of imidazole is again initially substituted (earlier work suggested a different orientation). The benzimidazole anion is iodinated at the 2-position; other halogenation generally occurs in the benzene ring.

Isoxazoles can be halogenated in the 4-position ⟨63AHC(2)365⟩. Ring bromination of oxazoles with bromine or NBS occurs preferentially at the 5-position and, if this is occupied, at the 4-position ⟨74AHC(17)99⟩. Aminooxazoles are readily halogenated.

Isothiazoles with electron-releasing substituents such as amino, hydroxy, or alkoxy in the 3- or 5-position are brominated in high yield in the 4-position. Alkylisothiazoles give lower yields, but 3-methylisothiazole-5-carboxylic acid has been brominated in 76% yield ⟨72AHC(14)1⟩. Again, thiazoles with an electron-releasing substituent in the 2- or 4-position are brominated at the 5-position ⟨79HC(34-1)5⟩.

1,2,3-Triazoles are brominated at the 4- or 5-positions, but only if there is no N-substituent ⟨74AHC(16)33⟩. This also applies to 1,2,4-triazoles. N-Halogeno derivatives are frequently isolated as intermediates ⟨81HC(37)289⟩.

3-Amino-1,2,5-thiadiazole is chlorinated or brominated at the 4-position at 20 °C in acetic acid. 3-Methyl-1,2,5-thiadiazole can also be chlorinated in the 4-position ⟨68AHC(9)107⟩. Bromination of 2-amino-1,3,4-thiadiazole succeeds in the 5-position ⟨65ACS2434⟩.

Azolidinones are very easily halogenated, e.g. (**118**) gives (**119**) ⟨79AHC(25)83⟩; similar reactions occur in the isothiazolidinone series.

(**118**) + Br_2 → → (**119**)

4.02.1.4.6 Acylation and alkylation

Although in general azoles do not undergo Friedel–Crafts type alkylation or acylation, several isolated reactions of this general type are known. 3-Phenylsydnone (**120**) undergoes Friedel–Crafts acetylation and Vilsmeier formylation at the 4-position, and the 5-alkylation of thiazoles by carbonium ions is known.

Heating N-substituted pyrazoles with benzoyl chloride at 200 °C gives quite high yields of 4-benzoylpyrazoles, even in the absence of catalysts. Benzylation of N-substituted pyrazoles proceeds similarly in the 4-position ⟨66AHC(6)347⟩.

(**120**)   (**121**)   (**122**)

Heating 2-phenyl-4-benzoyl-1,3,4-thiadiazolium chloride (121) at 200 °C causes the benzoyl group to move to the 2-position as in (122) ⟨68AHC(9)165⟩, probably via deprotonation.

For the C-acylation of imidazoles via deprotonation, see Section 4.02.1.7.4.

4-Aminotriazole is carboxylated at the 5-position by heating with aqueous sodium bicarbonate in a Kolbe-type reaction ⟨71JCS(C)1501⟩. 2-Thiazolinones undergo the Gattermann and Reimer–Tiemann reactions at the 4-position, and 3- and 4-pyrazolinone anions on alkylation give 4-alkyl as well as O- and N-alkyl derivatives.

4.02.1.4.7 Mercuration

While there appears to have been no general study of the mercuration of azoles, the reaction seems to proceed readily in several systems. Thus pyrazoles are 4-chloromercurated by $HgCl_2$. Oxazoles are mercurated in acetic acid: the ring positions react 5 > 4 > 2 ⟨74AHC(17)99⟩. Thiazoles react under the same conditions and show the same order of ring position reactivities. Isoxazoles can be easily mercurated in the 4-position with mercury(II) acetate ⟨63AHC(2)365⟩. Oxadiazoles have been mercurated, e.g. (123) gives (124) ⟨64HCA838⟩. 1-Phenyltetrazole is mercurated in the 5-position, and 3-arylsydnones at the 4-position.

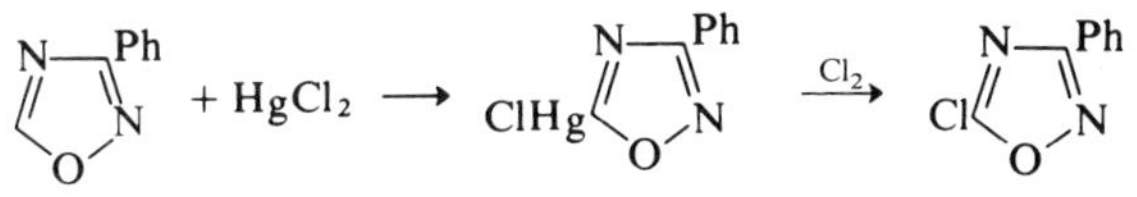

4.02.1.4.8 Diazo coupling

Diazo coupling is expected to occur only with highly reactive systems, and experiment bears this out. Diazonium ions couple with the anions of N-unsubstituted imidazoles at the 2-position (e.g. 125 yields 126) and with indazoles (127) in the 3-position. In general, other azoles react only when they contain an amino, hydroxyl, or potential hydroxyl group, e.g. the 4-hydroxypyrazole (128), the triazolinone (129) and the thiazolidinedione (130) (all these reactions occur on the corresponding anions).

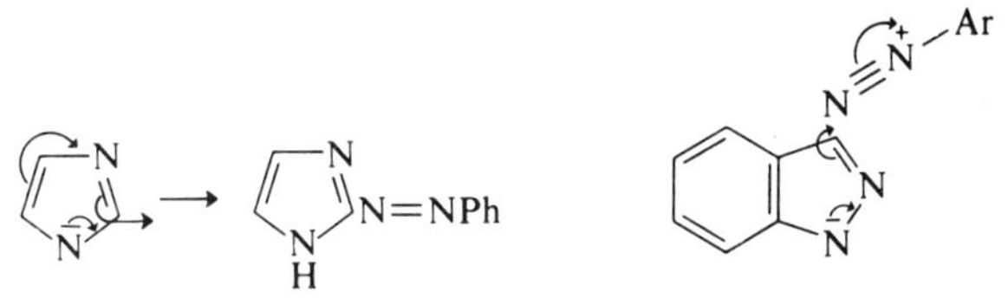

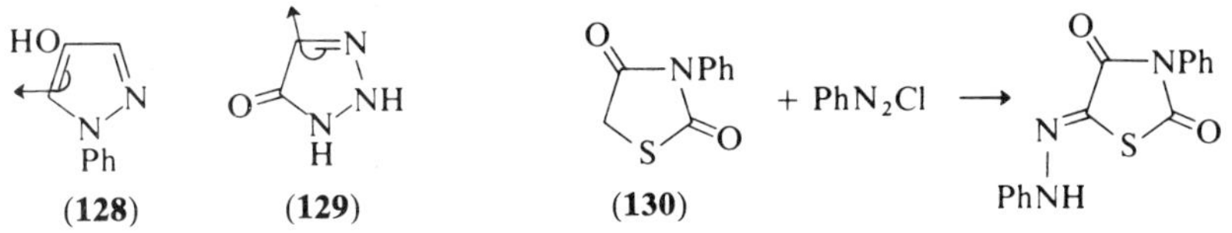

4.02.1.4.9 Nitrosation

Under alkaline conditions, alkyl nitrites nitrosate imidazoles which possess a free NH group in the 4-position ⟨70AHC(12)103⟩. Nitrosation of 3,5-dimethylpyrazole gives the 4-diazonium salt by further reaction of the nitroso compound with more NO^+. 5-Pyrazolinones are often nitrosated readily at the 4-position. 3-Alkyl-5-acetamidoisothiazoles undergo 4-nitrosation.

4.02.1.4.10 Reactions with aldehydes and ketones

Imidazoles are hydroxymethylated by CH_2O at the 4-position; 1-substituted imidazoles react at the 2-position. Isoxazoles can be chloromethylated in the 4-position ⟨63AHC(2)365⟩.

Aldehydes and ketones react with azolinones. The reaction between aldehydes and 2-phenyl-5-oxazolinone (**131**; Y = H), formed *in situ* from PhCONHCH$_2$CO$_2$H and Ac$_2$O, gives azlactones (**131**; Y = RCH). Similar reactions are given by 4-thiazolidinones, *e.g.* (**132**) gives (**133**) ⟨79AHC(25)83⟩, and 4-imidazolinones. In pyrazolin-5-ones the 4-position is sufficiently activated for condensation to occur with ketones in acidic media (Scheme 8) ⟨66AHC(6)347⟩.

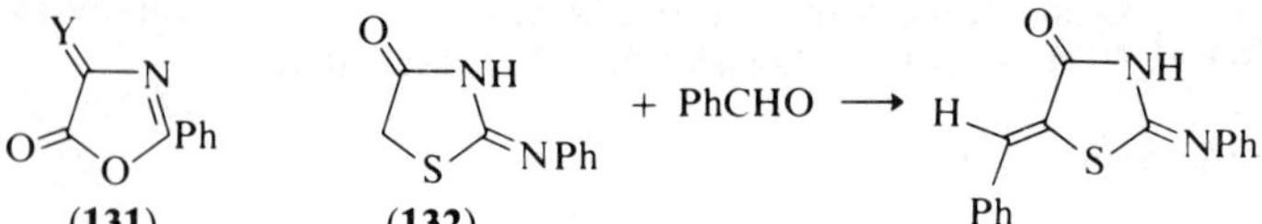

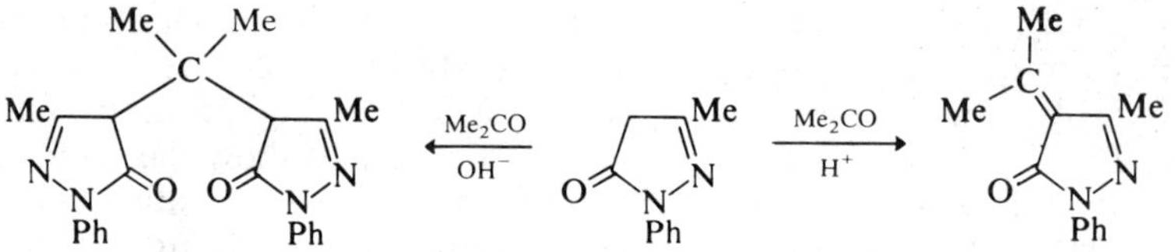

Scheme 8

4.02.1.4.11 Oxidation

The pyrazole ring is generally stable to oxidation and side chains are oxidized to carbonyl groups ⟨66AHC(6)347⟩. 1-Aryl-3-methylpyrazoles (**134**) react with ozone to yield 1,3,4-oxadiazolinones (**135**) ⟨66AHC(7)183⟩.

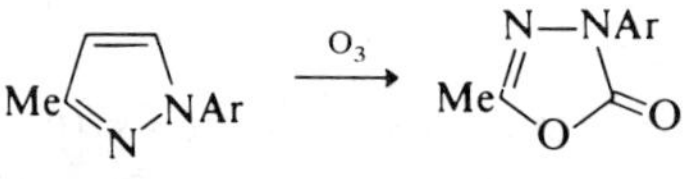

Imidazole rings also survive most oxidation conditions, but photosensitized oxidation of imidazoles can give diarylbenzamidines through a hydroperoxide (**136**) ⟨70AHC(12)103⟩.

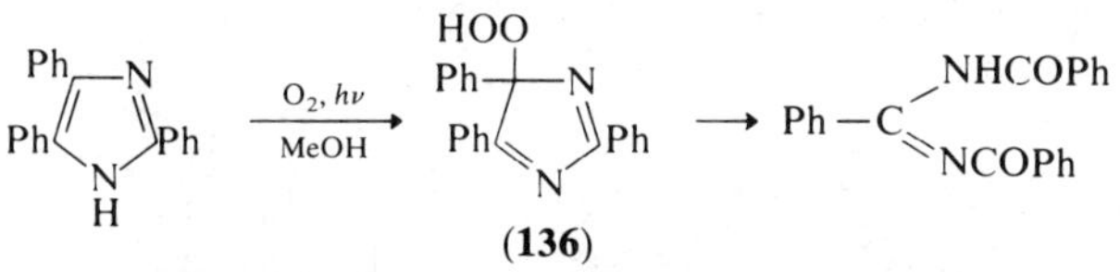

The thiazole, triazole and tetrazole rings are resistant to oxidation (*e.g.* by KMnO$_4$, CrO$_3$). Isoxazoles are more susceptible (*e.g.* **137** → **138**, benzil α-monoxime benzoate). The oxazole ring is relatively readily cleaved by oxidizing agents such as permanganate, chromic acid or hydrogen peroxide to give acids or amides. Oxidation of 4,5-diaryloxazoles with chlorine or bromine gives the corresponding benzils in high yield ⟨74AHC(17)99⟩.

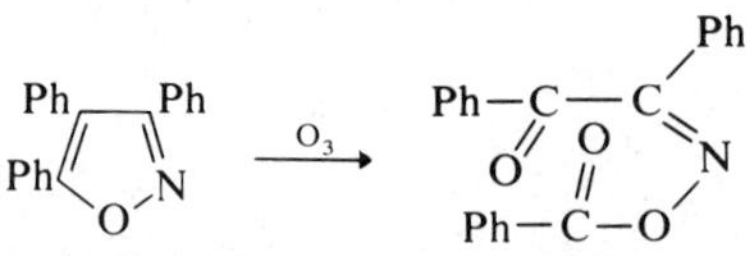

Oxidation of azole anions can give neutral azole radicals which could, in principle, be π (**139**) or σ (**140**) in nature. ESR spectra indicate structure (**141**; hyperfine splittings in G) for imidazolyl radicals, but both π- and σ-character have been observed for pyrazolyl radicals. Tetrazolyl radicals (**142** ↔ **143**) are also well known ⟨79AHC(25)205⟩. Oxidation of 2,4,5-triarylimidazole anions with bromine gives 1,1'-diimidazolyls (**144**) which are in equilibrium with the dissociated free radical (**145**) ⟨70AHC(12)103⟩.

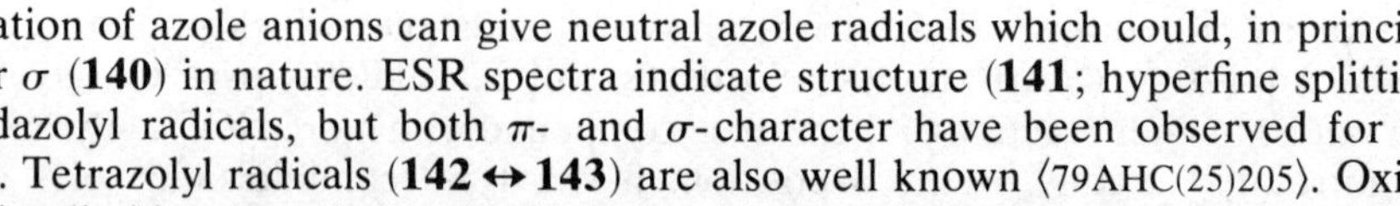

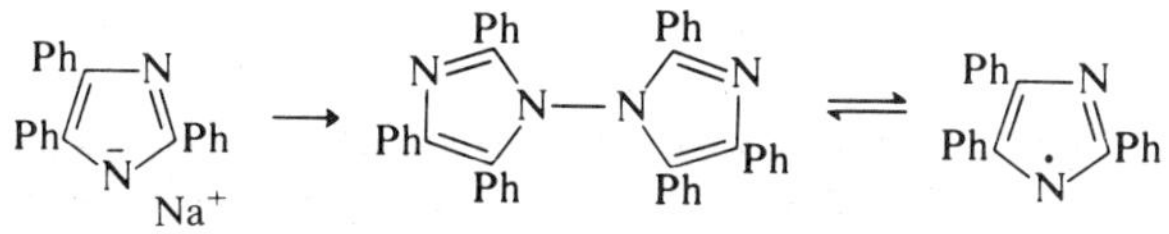

4.02.1.5 Attack at Sulfur

4.02.1.5.1 Electrophilic attack

In contrast to thiazoles, certain isothiazoles and benzisothiazoles have been directly oxidized to sulfoxides and sulfones. 4,5-Diphenyl-1,2,3-thiadiazole is converted by peracid into the trioxide (146). Although 1,2,5-thiadiazole 1,1-dioxides are known, they cannot be prepared in good yield by direct oxidation, which usually gives sulfate ion analogous to the results obtained with 1,2,4- and 1,3,4-thiadiazoles ⟨68AHC(9)107⟩.

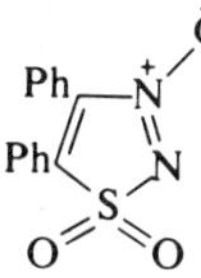

When a hydroxyazole can tautomerize to a non-aromatic structure, oxidation at an annular sulfur atom becomes easy, e.g. as in Scheme 9 ⟨79AHC(25)83⟩.

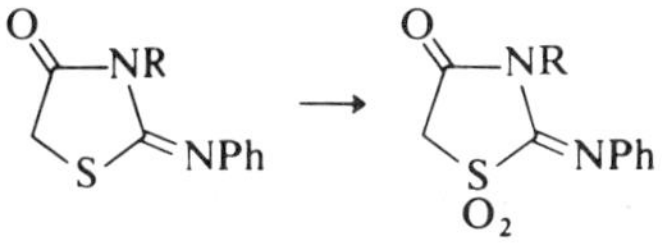

Thiazoles are desulfurized by Raney nickel, a reaction probably initiated by coordination of the sulfur at Ni. The products are generally anions and carbonyl compounds (see Section 4.02.1.8.4).

4.02.1.5.2 Nucleophilic attack

Isothiazoles and isothiazolium cations are attacked by carbanions at sulfur and on recyclization can give thiophenes, illustrated by (147) → (148). 2-Alkyl-3-isothiazolinones (e.g. 149) are also vulnerable to nucleophilic attack at sulfur ⟨72AHC(14)1⟩.

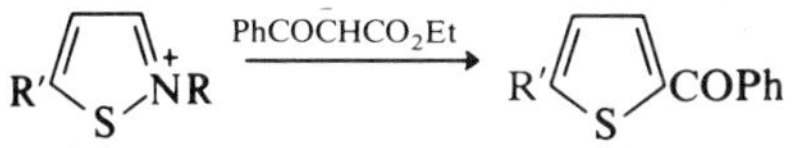

Nucleophilic attack at sulfur is implicated in many reactions of 1,2,4-thiadiazoles; generally, 'soft' electrophiles attack at sulfur, cf. (150) → (151). n-Butyllithium with 4,5-diphenyl-1,2,3-thiadiazole yields PhC≡CPh, probably by initial nucleophilic attack at sulfur.

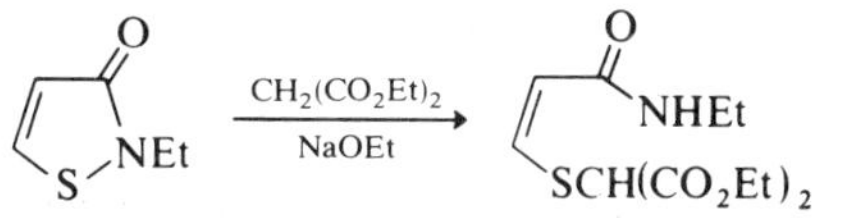

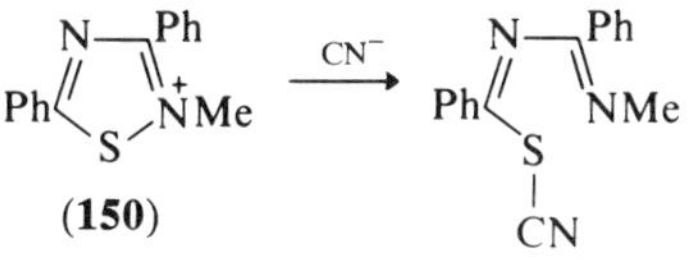

4.02.1.6 Nucleophilic Attack at Carbon

Because of the increased importance of inductive electron withdrawal, nucleophilic attack on uncharged azole rings generally occurs under milder conditions than those required for analogous reactions with pyridines or pyridones. Azolium rings are very easily attacked by nucleophilic reagents; reactions similar to those of pyridinium and pyrylium compounds are known; azolium rings open particularly readily.

Nucleophilic attack on the ring carbon atoms of azoles occurs readily with oxazole and aza analogues. Such reactions are generally facilitated by additional ring heteroatoms and by electron-attracting substituents, and hindered by electron-donating substituents. A fused benzene ring aids nucleophilic attack on azoles, azolium ions and azolones; this may be rationalized because the loss of aromaticity involved in the formation of the initial adduct is less than that in monocyclic compounds. The orientation of attack is generally between two heteroatoms, as in (**152**).

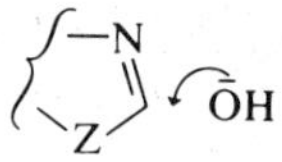

(**152**)

4.02.1.6.1 Hydroxide ion and other O-nucleophiles

(i) *Neutral Azoles*

Uncharged azoles not containing oxygen or sulfur are often resistant to attack by hydroxide ions at temperatures up to 100 °C and above. However, neutral azoles react with hydroxide ions under extreme conditions, e.g. 1-substituted imidazoles such as (**153**) at 300 °C give the corresponding imidazolinone (**154**) ⟨80AHC(27)241⟩.

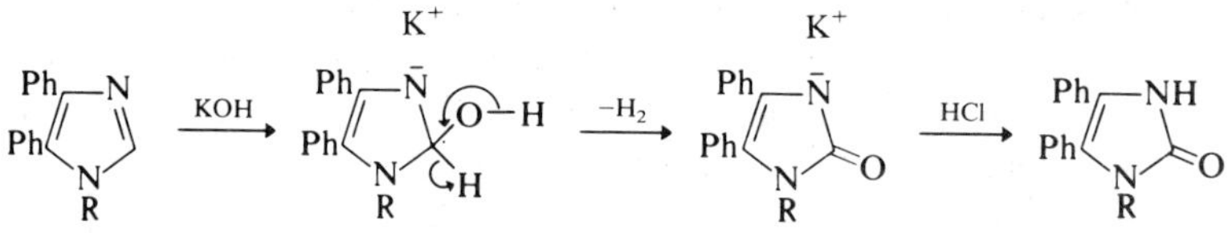

(**153**) → (**154**)

Imidazoles and benzimidazoles (**155**) react with acid chloride and alkali to give compounds of type (**157**), but these are reactions of the cation (**156**). 1,2,4-Triazoles and tetrazoles similarly undergo ring opening.

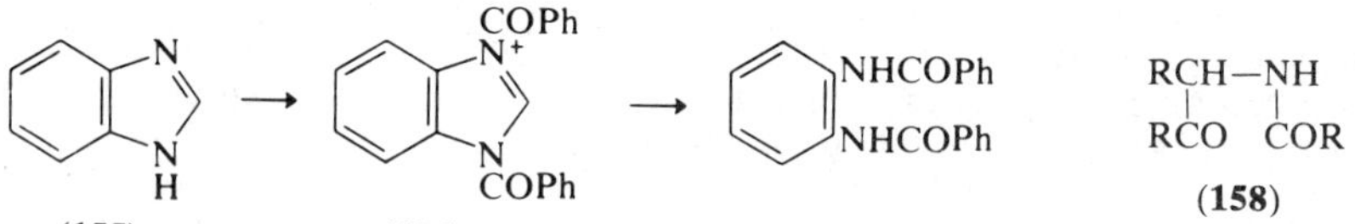

(**155**) (**156**) (**157**) (**158**)

Isoxazoles are also rather stable to nucleophilic attack by OH^- at carbon. For reactions with base at a ring hydrogen atom, leading, for example, to ring opening of isoxazoles, see Section 4.02.1.7.1.

Oxazoles give acylamino ketones (**158**) by acid-catalyzed ring scission, although they are somewhat more stable than furans. The oxazole ring is also moderately stable to alkali ⟨74AHC(17)99⟩; as expected, reaction with hydroxide ions is facilitated by electron-withdrawing substituents and fused benzene rings.

Oxazoles are easily cleaved. 2,5-Dialkyl-1,3,4-oxadiazoles (**159**) in aqueous solution with acid or base give hydrazides (if suitable substituents are present, further reaction can occur; see Section 4.02.3.5.1). 3-Methyl-1,2,4-oxadiazole (**160**) is easily hydrolyzed to acetamidoxime ⟨61CI(L)292⟩.

(**159**) (**160**)

Isothiazoles and thiazoles are rather stable toward nucleophilic attack. Both 5- and 6-nitrobenzothiazole (**161**) add methoxide at C-2; the initial adduct undergoes ring opening to give (**162**) ⟨79AHC(25)1⟩. Unsubstituted 1,2,4-thiadiazole is sensitive to alkali. Substituents stabilize the ring somewhat, but ring-opening reactions are still common, e.g (**163**) → (**164**). 5-Alkylamino-1,2,3,4-thiatriazoles are cleaved by alkali to azide ion and an isothiocyanate; in addition, a Dimroth rearrangement occurs to give a mercaptotetrazole ⟨64AHC(3)263⟩.

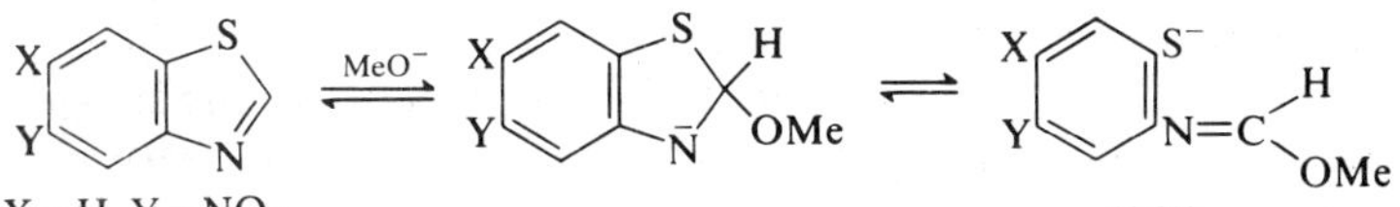

(161) X = H, Y = NO₂; X = NO₂, Y = H

(162)

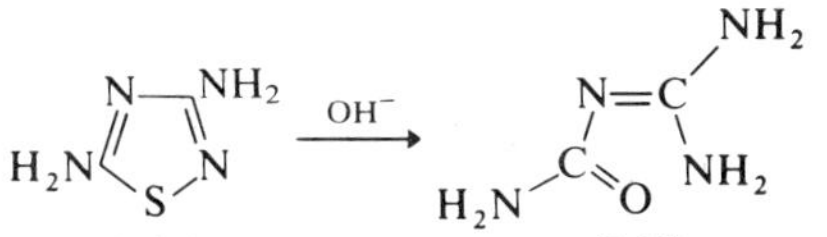

(163)

(164)

(ii) Azolium ions

Azolium ions (**165**) react reversibly with hydroxide ions to form a small proportion of the pseudo base (**166**). The term 'pseudo' is used to designate bases that react with acids measurably slowly, not instantaneously as is normal for acid–base reactions. Fused benzene rings reduce the loss of resonance energy when the heteroring loses its aromaticity, and hence pseudo bases are formed even more readily by benzothiazolium cations than by thiazolium ions. Pseudo bases carrying the hydroxyl group in the α-position are usually formed preferentially. As expected, pseudo base formation for S-containing azoliums is easiest with the dithiolylium and least easy with the thiazolium ⟨79AHC(25)1⟩.

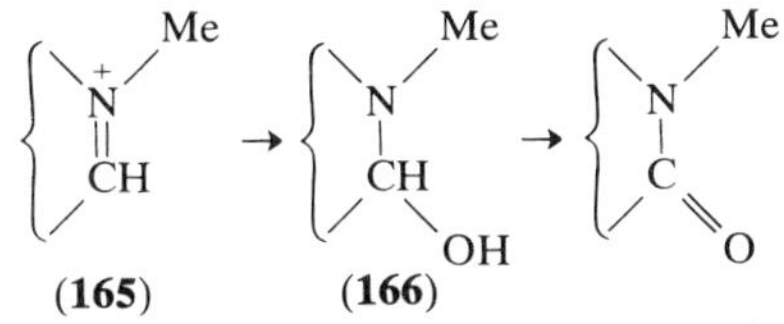

(165) → (166) →

pK_{R^+} values for pseudo base formation are defined by equation (1),

$$K_{R^+} = [H^+][QOH]/[Q^+] \quad (1)$$

where Q^+ and QOH denote the azolium cation and pseudo base, respectively. Some pK_{R^+} values are given in Scheme 10.

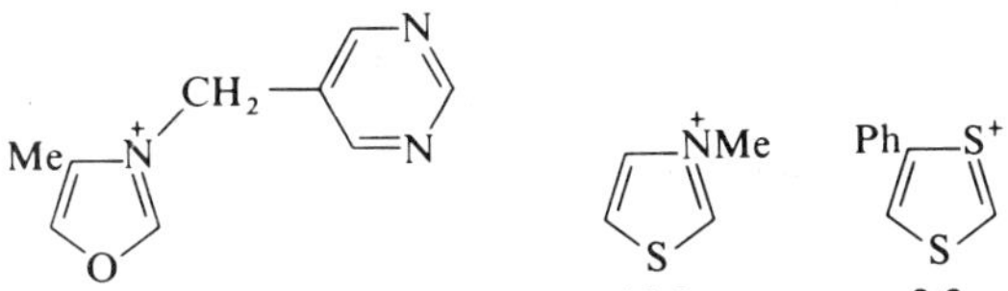

5.8   10.8   2.2

Scheme 10 pK_{R^+} values for azolium cations

Some pseudo bases are stable. 1,3-Dithiolylium adds alkoxide ions at the 2-position to give stable adducts which regenerate the starting salts with acids ⟨80AHC(27)151⟩. Pseudo bases can also lose water to give an ether (*e.g.* **167** → **168**).

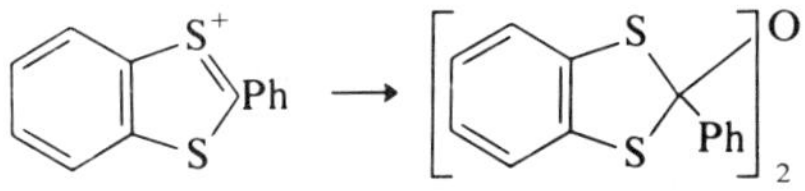

(167)   (168) 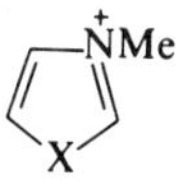   (169)

Oxidation to an azolone is an expected reaction for a pseudo base, but little appears to be known of such reactions. Most commonly, pseudo bases suffer ring fission. Estimated rates of ring-opening of (**169**) are in the ratio $10^9 : 10^{4.5} : 1$ for X = O, S and NMe, respectively ⟨79AHC(25)1⟩.

Thus quaternized thiazoles (**170**) consume two equivalents of OH on titration because the pseudo bases (**171**) ring open to (**172**), which form anions (**173**). Quaternized oxazoles (**174**) are readily attacked by hydroxide to give open-chain products such as (**175**) ⟨74AHC(17)99⟩, and quaternized 1,3,4-oxadiazoles behave similarly. Quaternary isothiazoles (*e.g.* **176**) are cleaved by hydroxide ⟨72AHC(14)1⟩, as are 1,2,4-thiadiazolium salts (**177** → **178**).

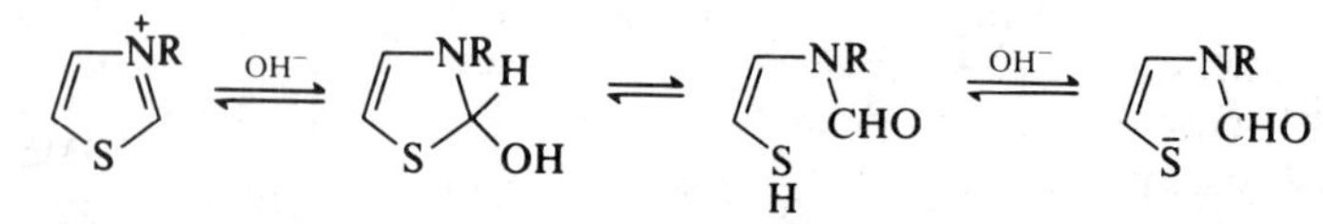

(170) (171) (172) (173)

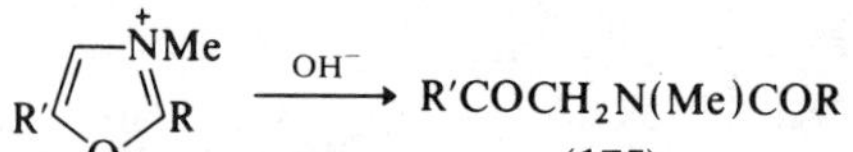

(174) → R'COCH₂N(Me)COR (175)

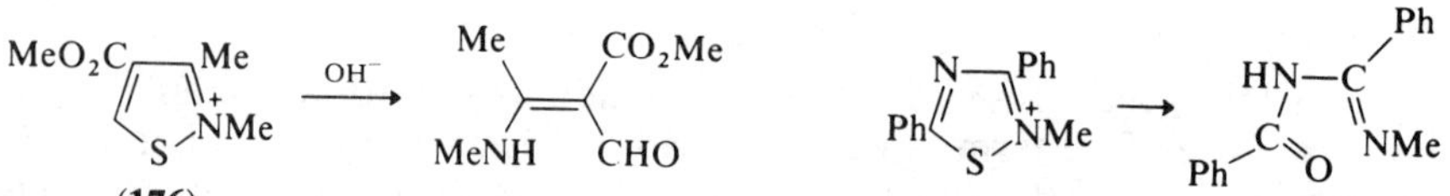

(176) → (177) → (178)

1,2-Dithiolylium ions undergo ring opening and degradation with hydroxide (Scheme 11) ⟨66AHC(7)39, 80AHC(27)151⟩. 1,2-Dimethylpyrazolium is degraded to HCO_2H and MeNHNHMe.

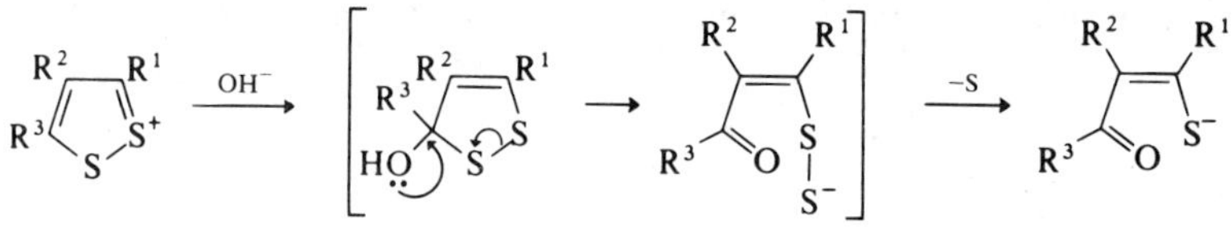

Scheme

Reclosure to form a new heterocyclic or homocyclic ring can occur in azolium ions carrying suitable substituents; these reactions are considered under the appropriate substituents.

Anthranils are readily cleaved by nitrous acid, presumably by attack of water on N-nitroso cations. The first product that can be observed is the nitrosohydroxylamino compound (**179**), which becomes reduced to the diazonium salt (**180**) ⟨67AHC(8)277⟩.

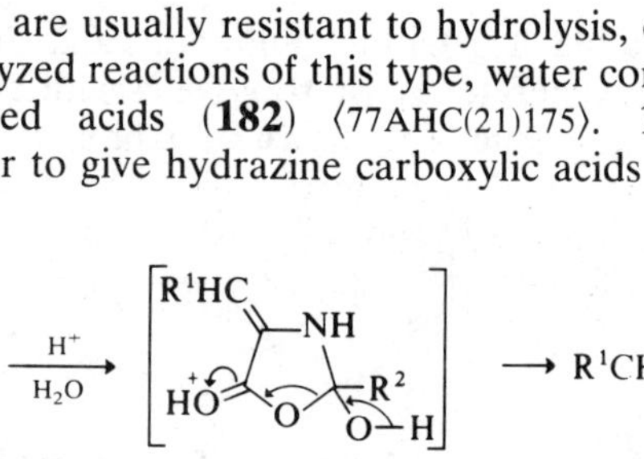

(179) → (180)

(iii) Azolinones

Although imidazolinones are usually resistant to hydrolysis, oxazolinone rings are often easily opened. In acid-catalyzed reactions of this type, water converts azlactones (**181**) into α-acylamino-α,β-unsaturated acids (**182**) ⟨77AHC(21)175⟩. 1,3,4-Oxadiazolinones are readily opened by hot water to give hydrazine carboxylic acids which undergo decarboxylation.

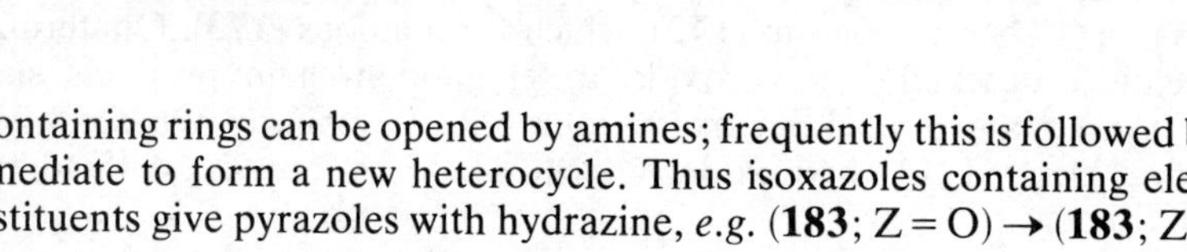

(181) → $R^1CH=C-NHCOR^2$ | CO_2H (182)

4.02.1.6.2 Amines and amide ions

(i) Azoles

Oxygen-containing rings can be opened by amines; frequently this is followed by reclosure of the intermediate to form a new heterocycle. Thus isoxazoles containing electron-withdrawing substituents give pyrazoles with hydrazine, e.g. (**183**; Z = O) → (**183**; Z = NH), and

(184) → (185) ⟨66AHC(6)347⟩. In the benzo series a rather different reaction can occur: 6-nitroanthranil (186) is converted by amines into 2-indazoles (187) ⟨61JOC3714⟩. Oxazoles heated at 180 °C with formamide are transformed into imidazoles ⟨74AHC(17)99⟩. With 2,4-dinitrophenylhydrazine in HCl, oxazoles form hydrazones by ring fission (Scheme 12) ⟨74AHC(17)99⟩. Benzoxazole with hydroxylamine gives 2-aminobenzoxazole by elimination of water from the initial adduct. 1,3,4-Oxadiazoles react readily with ammonia and primary amines to give 1,2,4-triazoles.

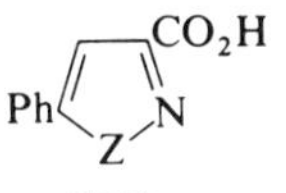

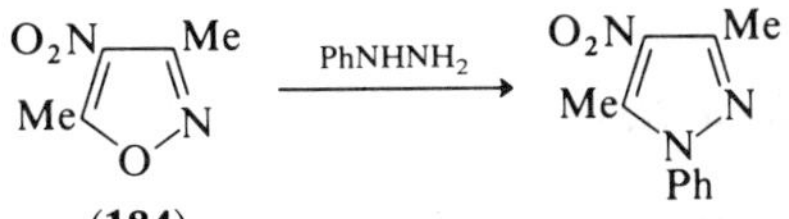

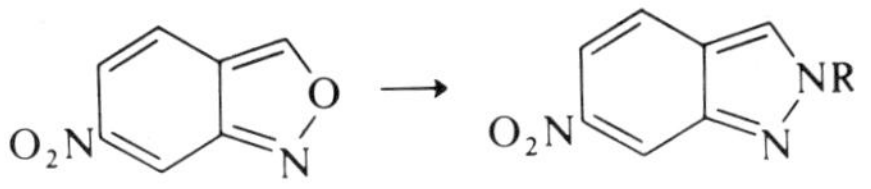

Amines are insufficiently nucleophilic to react with most azoles which do not contain a ring oxygen, and the stronger nucleophile NH_2^- is required. When treated with amide ions, thiazoles can be aminated in the 2-position by $NaNH_2$ at 150 °C. Only N-substituted condensed imidazoles such as 1-alkylbenzimidazole react in such Chichibabin reactions. Imidazoles are aminated by alkaline NH_2OH.

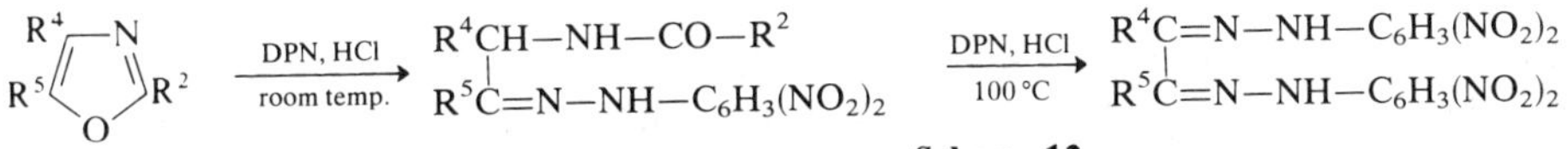

(ii) Azolium ions

Most azolium ions are sufficiently reactive to be attacked by amines. Sometimes the initial adducts are stable: ammonia and primary and secondary amines add to 1,3-dithiolylium salts at the 2-position to give compounds of the types NT_3, RNT_2 and R_2NT, respectively, where T = the 1,3-thiol-2-yl group ⟨80AHC(27)151⟩.

Some azoliums give open-chain products: primary and secondary amines with 1,2-dithiolyliums generally give (188) ⟨80AHC(27)151⟩.

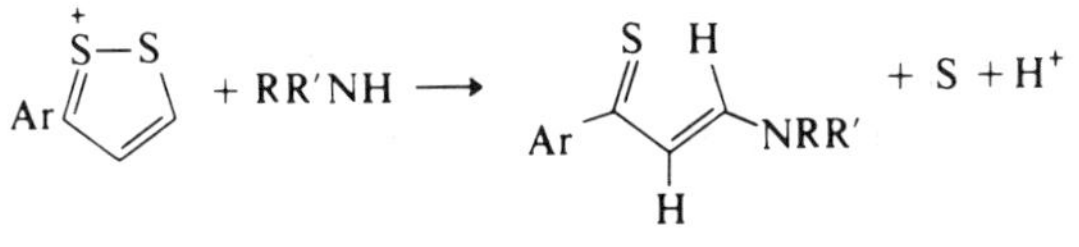

In other cases, reclosure to a new ring occurs: 1,2-dithiolylium ions with ammonia give isothiazoles according to the mechanism shown in Scheme 13 ⟨80AHC(27)151⟩. Treatment of quaternary isothiazoles with hydrazine or phenylhydrazine gives pyrazoles (Scheme 14) ⟨72AHC(14)1⟩, and 1,2,4-thiadiazoliums similarly yield 1,2,4-thiazoles. Oxazolium ions react with ammonium acetate in acetic acid to give the corresponding imidazoles (Scheme 15) ⟨74AHC(17)99⟩.

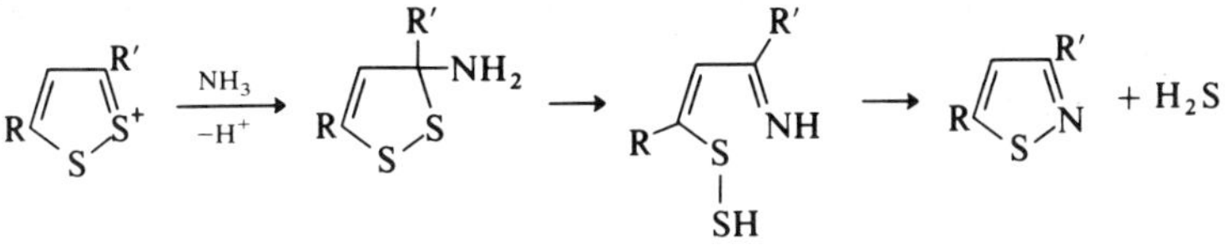

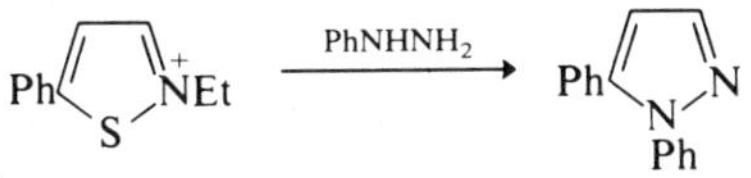

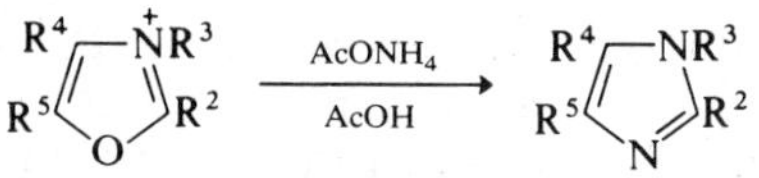

66 Reactivity of Five-membered Rings with Two or More Heteroatoms

Scheme 15

Where an *N*-methoxy group is present, as in 1-methyl-3-methoxybenzimidazole, elimination of MeOH can give an aromatic product.

4.02.1.6.3 S-Nucleophiles

Data on reactions of sulfur nucleophiles with azoles are sparse. Oxazoles are transformed in low yield into the corresponding thiazoles over alumina with H_2S at 350 °C ⟨74AHC(17)99⟩. Sulfur nucleophiles such as SH^- or RS^- add to 1,3-dithiolylium salts at the 2-position ⟨80AHC(27)151⟩.

4.02.1.6.4 Halide ions

Chloride ions are comparatively weak nucleophiles and do not react with azoles. In general, there is also no interaction of halide ions with azolium compounds.

Benzimidazole 3-oxides, *e.g.* (**189**), react with phosphorus oxychloride or sulfuryl chloride to form the corresponding 2-chlorobenzimidazoles. The reaction sequence involves first formation of a nucleophilic complex (**190**), then attack of chloride ions on the complex, followed by rearomatization involving loss of the *N*-oxide oxygen (**191** → **192**).

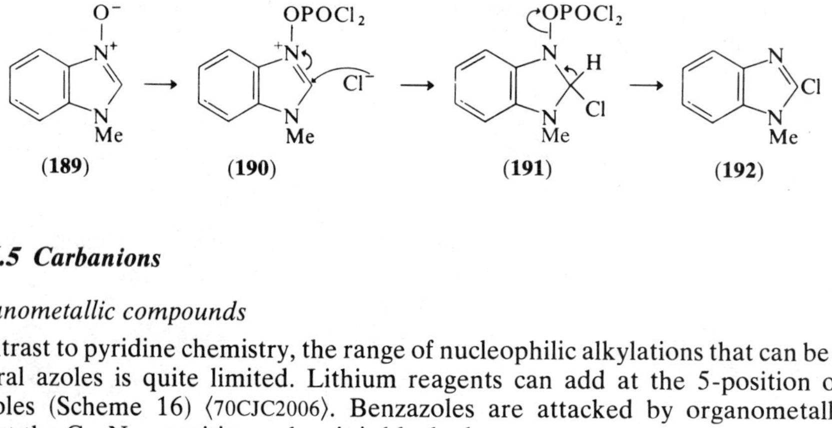

(**189**)   (**190**)   (**191**)   (**192**)

4.02.1.6.5 Carbanions

(i) Organometallic compounds

In contrast to pyridine chemistry, the range of nucleophilic alkylations that can be effected on neutral azoles is quite limited. Lithium reagents can add at the 5-position of 1,2,4-oxadiazoles (Scheme 16) ⟨70CJC2006⟩. Benzazoles are attacked by organometallic compounds at the C=N α-position unless it is blocked.

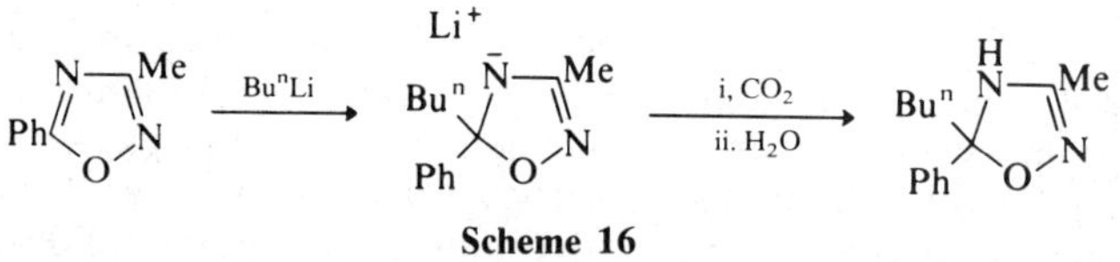

Scheme 16

Azolium rings react readily with organometallic compounds. With a Grignard reagent, conversion (**193**) → (**194**) is known in the benzothiazolium series, and 1,3-benzodithiolyliums give products of type (**195**).

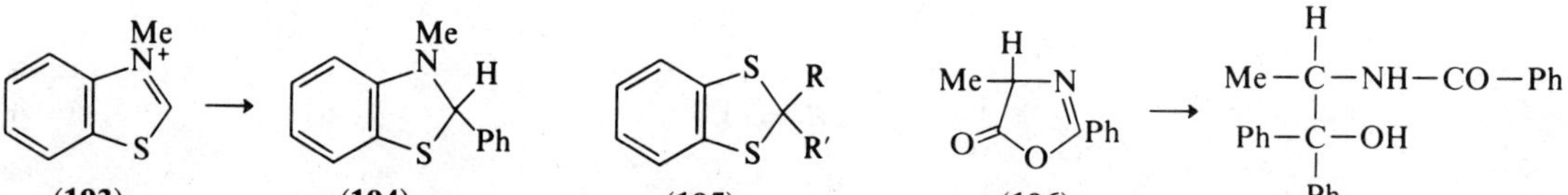

(**193**)   (**194**)   (**195**)   (**196**)   (**197**)

4-Methyl-2-phenyl-5-oxazolinone (**196**) with phenylmagnesium bromide gives (**197**) ⟨65AHC(4)75⟩.

(ii) Activated methyl and methylene carbanions

The mesomeric anions of activated methyl and methylene compounds react with azolium ions. Thus 1,2-dithiolylium ions with a free 3- or 5-position react with various carbon nucleophiles to give products which are oxidized *in situ* to mesomeric anhydro-bases (**198** → **199** → **200**). Dimethylaniline gives an intermediate which is oxidized to a new dithiolylium salt (Scheme 17) ⟨66AHC(7)39⟩. However, if the 3,5-positions are substituted, an alternative ring scission can occur (**199** → **201** → **202**) ⟨80AHC(27)151⟩. In this sequence, (**198**) can be $ArCOCH_2CS_2Me$, $NCCH_2CSNH$ or $NCCH_2CO_2Et$. The conversion in the 1,2,4-thiadiazole series of (**203**) into (**204**) is analogous.

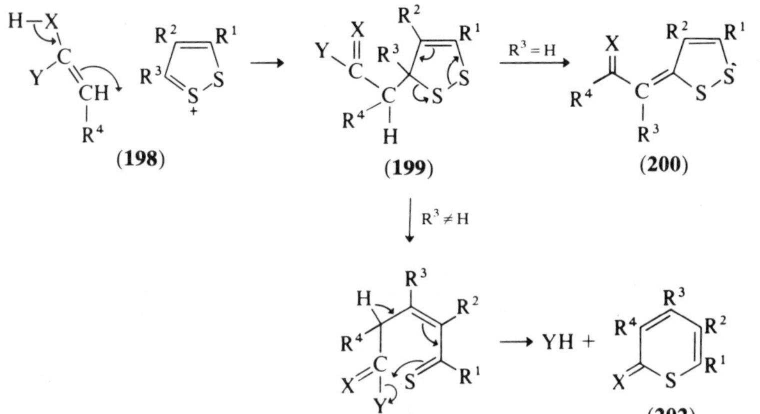

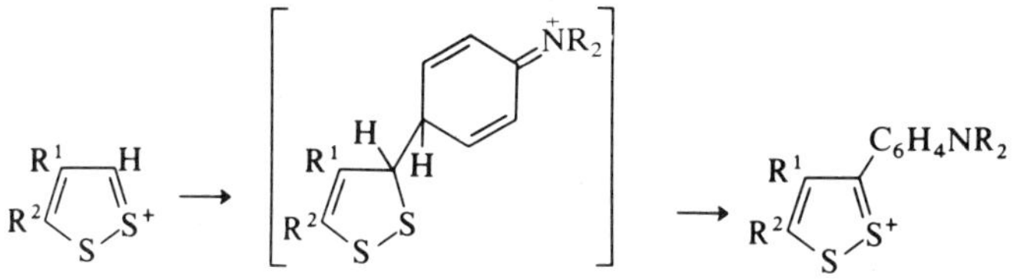

Scheme 17

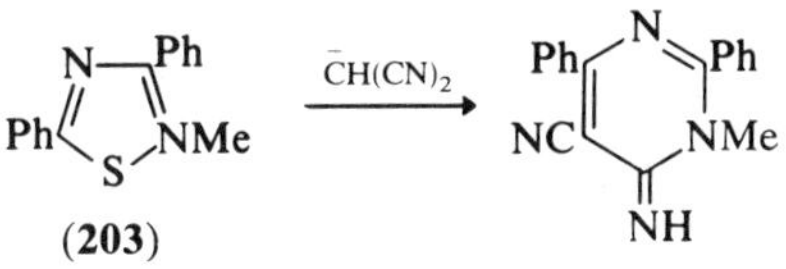

Anhydro bases can attack the α-position, e.g. of thiazolium cations, with the formation of adducts capable of oxidation to cyanine dyes, e.g. Scheme 18 (see Section 4.02.3.3.4).

Active methylene compounds can add to 1,3-dithiolylium ions to give 2-substituted 1,2-dihydro-1,3-dithioles (**206**). Again, addition is often followed by oxidation (to **207**). Alternatively, further addition can occur (to **208**) ⟨80AHC(27)151⟩. In this reaction, (**205**) can be $CH_2(CN)_2$, $CH_2(COMe)_2$ or even MeCOMe. Somewhat similar reactions are shown by 1,3-diarylimidazolium ions.

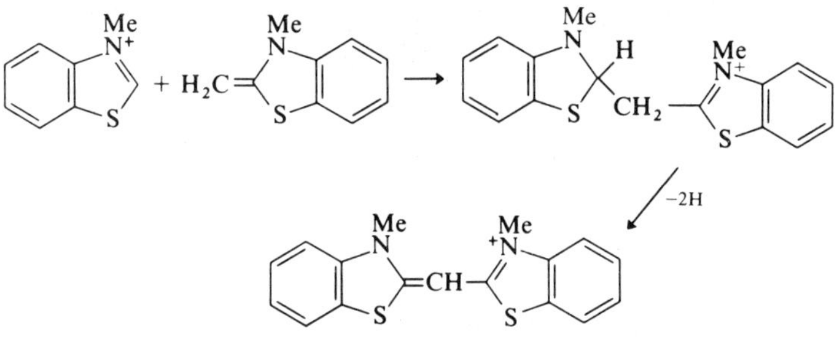

Scheme 18

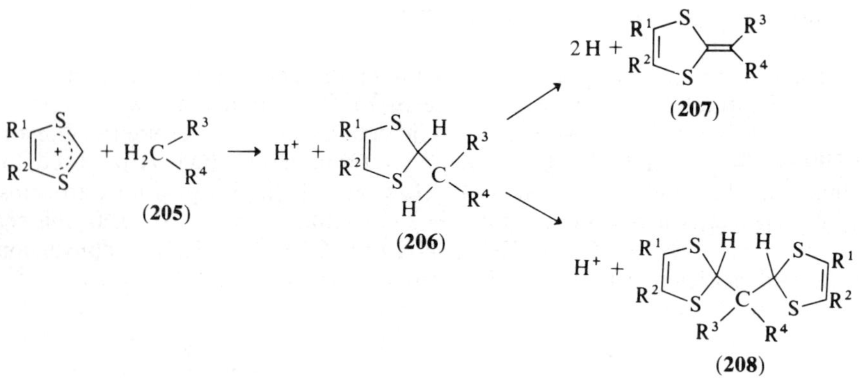

(205) (206) (207) (208)

(iii) Cyanide ions

Cyanopyrazoles are formed by irradiation of pyrazoles in the presence of cyanide ions in photosubstitution reactions.

4.02.1.6.6 Reduction by complex hydrides

Oxygen-containing azoles are readily reduced, usually with ring scission. Only acyclic products have been reported from the reductions with complex metal hydrides of oxazoles (e.g. **209** → **210**), isoxazoles (*e.g.* **211** → **212**), benzoxazoles (*e.g.* **213** → **214**) and benzoxazolinones (*e.g.* **215**, **216** → **214**). Reductions of 1,2,4-oxadiazoles always involve ring scission. Lithium aluminum hydride breaks the C—O bond in the ring (Scheme 19) ⟨76AHC(20)65⟩.

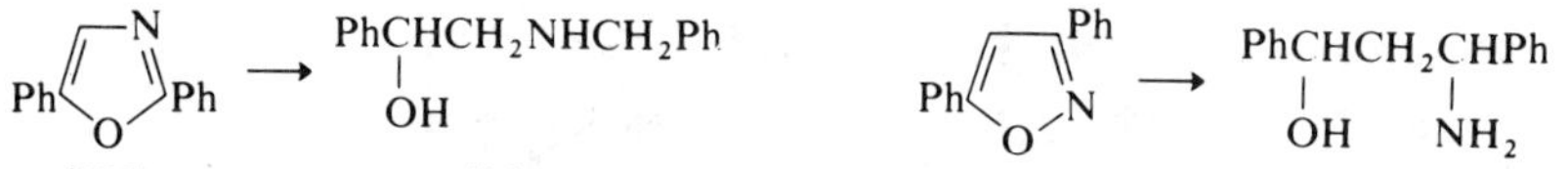

(209) (210) (211) (212)

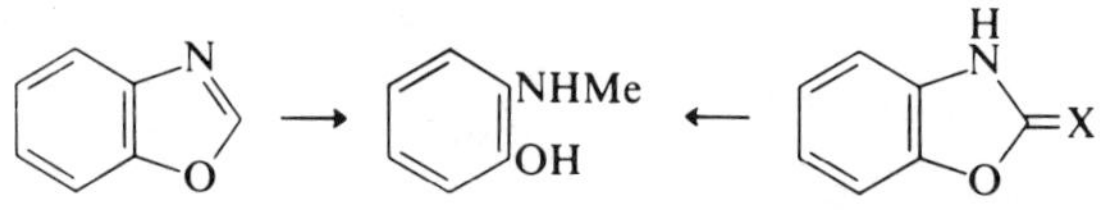

(213) (214) (215) X = O (216) X = S

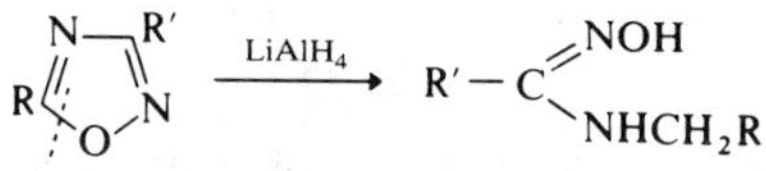

Scheme 19

Nitrogen azoles are less easily reduced: benzimidazole with lithium aluminum hydride gives dihydrobenzimidazole ⟨52CB390⟩.

Cationic rings are readily reduced by complex hydrides under relatively mild conditions. Thus isoxazolium salts with sodium borohydride give the 2,5-dihydro derivatives (**217**) in ethanol, but yield the 2,3-dihydro compound (**218**) in $MeCN/H_2O$ ⟨74CPB70⟩. Pyrazolyl anions are reduced by borohydride to pyrazolines and pyrazolidines. Thiazolyl ions are reduced to 1,2-dihydrothiazoles by lithium aluminum hydride and to tetrahydrothiazoles by sodium borohydride. The tetrahydro compound is probably formed *via* (**219**), which results from proton addition to the dihydro derivative (**220**) containing an enamine function. 1,3-Dithiolylium salts easily add hydride ion from sodium borohydride (Scheme 20) ⟨80AHC(27)151⟩.

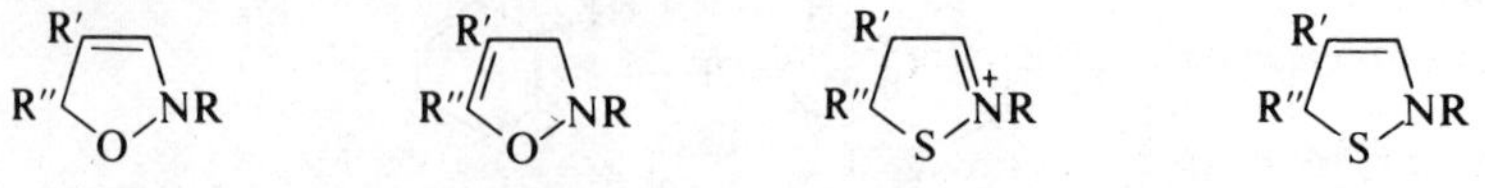

(217) (218) (219) (220)

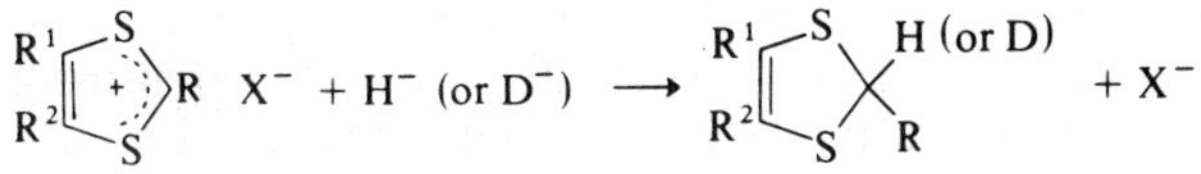

Scheme 20

R = H, SMe, NR$_2$, aryl, cyclohexyl

4.02.1.6.7 Phosphorus nucleophiles

Trialkyl- and triaryl-phosphines react with 1,3-benzodithiolylium ions to give a phosphonium salt which is deprotonated by *n*-butyllithium to give (221) ⟨76TL3695⟩.

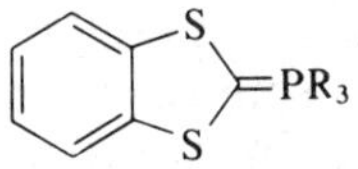

(221) 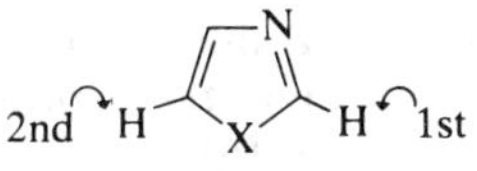   (222) 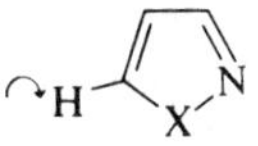   (223)

4.02.1.7 Nucleophilic Attack at Hydrogen Attached to Ring Carbon or Ring Nitrogen

Hydrogens attached to ring carbon atoms of neutral azoles, and especially azolium ions, are acidic and can be removed as protons by bases. Reaction follows the orientations shown in (222) and (223). The anions from neutral azoles can be stabilized as lithium derivatives, except in isoxazoles where ring cleavage occurs. Typically, the anion adds a proton again and hydrogen isotope exchange can result. The zwitterions from azolium rings can react as carbenes.

4.02.1.7.1 Metallation at a ring carbon atom

Neutral azoles are readily *C*-lithiated by *n*-butyllithium provided they do not contain a free NH group (Table 6). Derivatives with two heteroatoms in the 1,3-orientation undergo lithiation preferentially at the 2-position; other compounds are lithiated at the 5-position. Attempted metallation of isoxazoles usually causes ring opening *via* proton loss at the 3- or 5-position (Section 4.02.2.1.7.5); however, if both of these positions are substituted, normal lithiation occurs at the 4-position (Scheme 21).

Table 6 Lithiation of Azoles by *n*-Butyllithium

Heterocycle	Position lithiated	Temperature (°C)	Ref.
3-Methyl-1-phenylpyrazole	5	—	66AHC(6)347, 52JA3242
3,5-Disubstituted isoxazoles	4	−70 to −65	79AHC(25)147, 70CJC1371
3,5-Disubstituted isothiazoles	5	−70	65AHC(4)75, 72AHC(14)1
1-Substituted imidazoles	2	—	—
1-Substituted oxazoles	2	'low'	74AHC(17)99
1-Substituted thiazoles	2	−60	—
2-Substituted thiazoles	5	−100	—
1-Phenyl-1,2,3-triazoles	5	−60 to −20	74AHC(16)33, 71CJC1792
1-Butyltetrazole	5	−60	77AHC(21)323

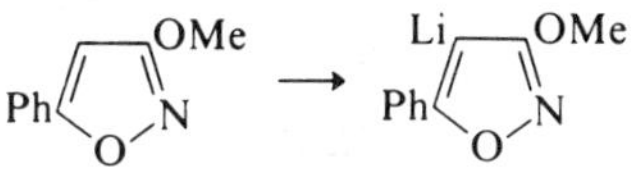

Scheme 21

4.02.1.7.2 Hydrogen exchange at ring carbon in neutral azoles

Base-catalyzed hydrogen exchange has been summarized for five-membered rings ⟨74AHC(16)1⟩. In many reactions of this type the protonated azole is attacked by hydroxide ion to form an ylide in the rate-determining step, *e.g.* for imidazole (Scheme 22) ⟨74AHC(16)1⟩. Deuteration of imidazole is fast at the 2-position and much slower at the 4- and 5-positions. Rates fall off for *N*-unsubstituted imidazoles at high pH values because of the formation of unreactive anions. In the case of 1-methylbenzimidazole the rate of hydrogen exchange in the 2-position is independent of the acidity over a wide range, in agreement with the mechanism shown in Scheme 22 ⟨80AHC(27)241⟩.

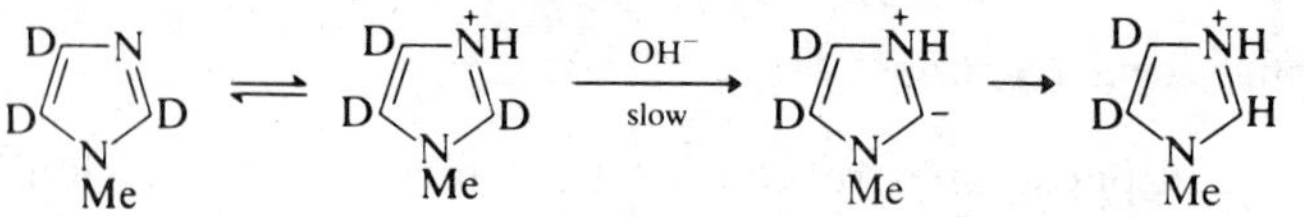

Scheme 22

Under strongly basic conditions oxazoles undergo fast 2-deuteration and slower 5-deuteration ⟨74AHC(17)99⟩. The hydrogen in the 5-position of isothiazoles exchanges rapidly under basic conditions ⟨69JHC199⟩. Neutral thiazoles exchange by two competitive mechanisms: at pD 0–11 the conjugate acid exchanges the 2-H *via* the ylide (**224**), whereas at higher pD exchange is at the 2- and 5-positions *via* the carbanions (**225**) and (**226**). The 1,2,4- (**227**), 1,3,4- (**228**) and 1,2,3-thiadiazoles (**229**) all undergo rapid exchange at the 5-, 2- and 5-positions, respectively ⟨74AHC(16)1⟩.

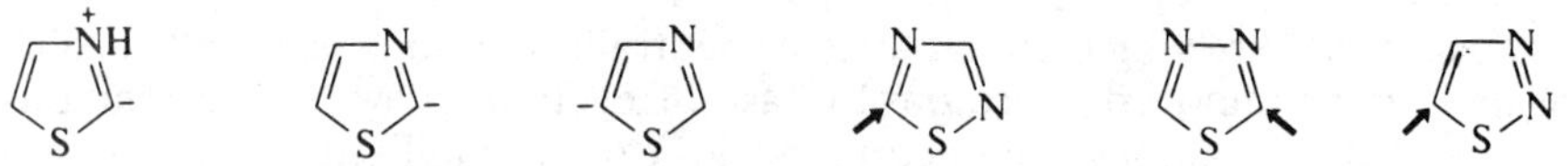

(**224**) (**225**) (**226**) (**227**) (**228**) (**229**)

1-Substituted tetrazoles readily exchange the 5-hydrogen for deuterium in aqueous solution. A major rate-enhancing effect is observed with copper(II) or zinc ions due to σ-complexation with the heterocycle. The rate of base-induced proton–deuterium exchange of 1-methyltetrazole is 10^5 times faster than 2-methyltetrazole ⟨77AHC(21)323⟩.

4.02.1.7.3 Hydrogen exchange at ring carbon in azolium ions and dimerization

Hydrogen atoms in azolium ions can be removed easily as protons (*e.g.* **230** → **232**); exchange with deuterium occurs in heavy water. The intermediate zwitterion (*e.g.* **231**) can also be written as a carbene, and in some cases this carbenoid form can be trapped or isolated as a dimer.

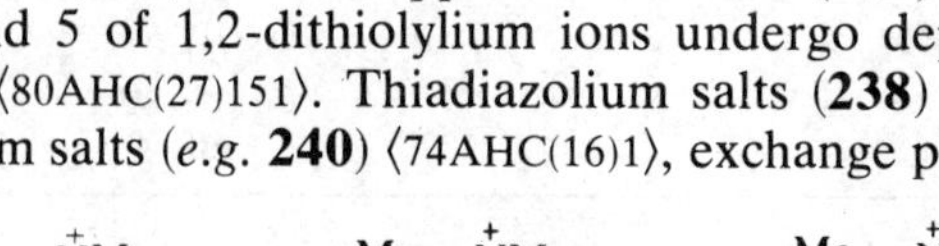

(**230**) (**231**) (**232**)

The relative rates of H-isotope exchange in D_2O/OD^- for oxazolium, thiazolium and imidazolium are shown in formulae (**233**)–(**235**), respectively ⟨71PMH(4)55⟩. The intermediate carbene can form a dimer (**236**) or be trapped with azides (**237**) ⟨74AHC(16)1⟩. Hydrogen atoms in positions 3 and 5 of 1,2-dithiolylium ions undergo deprotonation and can be replaced by deuterium ⟨80AHC(27)151⟩. Thiadiazolium salts (**238**) and (**239**) ⟨74AHC(16)1⟩, and especially tetrazolium salts (*e.g.* **240**) ⟨74AHC(16)1⟩, exchange particularly quickly.

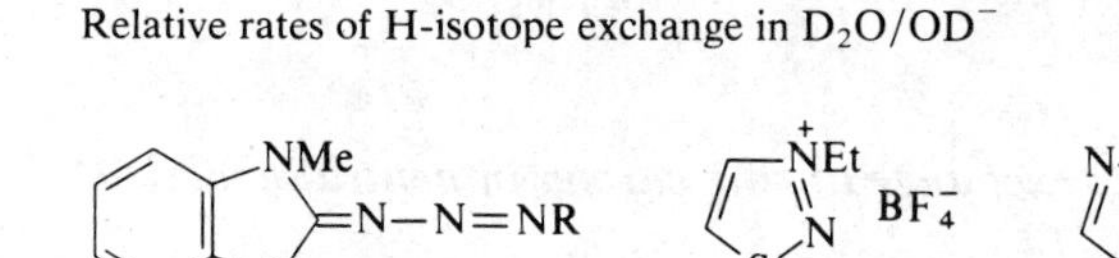

(**233**) (**234**) (**235**)

Relative rates of H-isotope exchange in D_2O/OD^-

(**236**) (**237**) (**238**) (**239**) (**240**)

Base-catalyzed hydrogen exchange occurs at the 3- and 5-positions of 1,2-dimethylpyrazolium salts. 2-Unsubstituted 1,3-dithiolylium salts are easily deprotonated by nucleophilic attack of hydrogen. The intermediate carbene easily undergoes dimerization. Hydrogen exchange can also occur (Scheme 23) ⟨80AHC(27)151⟩.

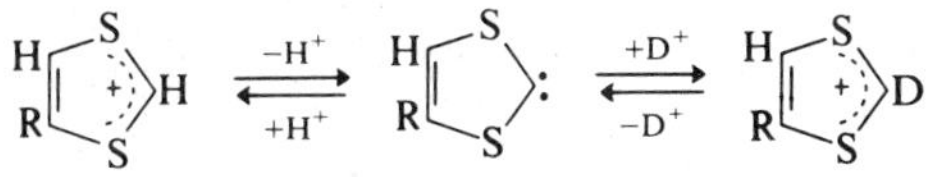

Scheme 23

4.02.1.7.4 C-Acylation via deprotonation

1-Substituted imidazoles can be acylated at the 2-position by acid chlorides in the presence of triethylamine. This reaction proceeds by proton loss on the N-acylated intermediate (241). An analogous reaction with phenyl isocyanate gives (242), probably via a similar mechanism. Benzimidazoles react similarly, but pyrazoles do not ⟨80AHC(27)241⟩ (cf. Section 4.02.1.4.6).

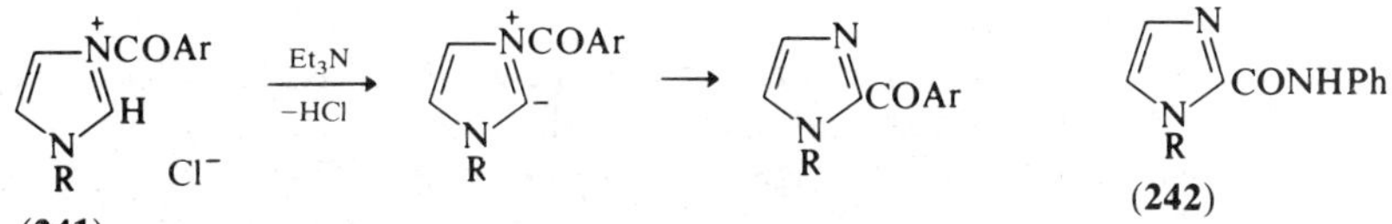

4.02.1.7.5 Ring cleavage via C-deprotonation

Isoxazoles unsubstituted in the 3-position react with hydroxide or ethoxide ions to give β-keto nitriles (243) → (244). This reaction involves nucleophilic attack at the 3-CH group. 1,2-Benzisoxazoles unsubstituted in the 3-position similarly readily give salicylyl nitriles ⟨67AHC(8)277⟩, and 5-phenyl-1,3,4-oxadiazole (245) is rapidly converted in alkaline solution into benzoylcyanamide (246) ⟨61CI(L)292⟩. A similar cleavage is known for 3-unsubstituted pyrazoles and indazoles; the latter yield o-cyanoanilines.

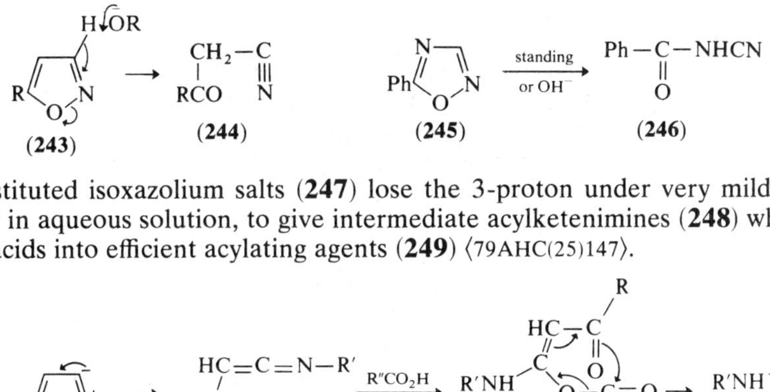

3-Unsubstituted isoxazolium salts (247) lose the 3-proton under very mild conditions, e.g. at pH 7 in aqueous solution, to give intermediate acylketenimines (248) which convert carboxylic acids into efficient acylating agents (249) ⟨79AHC(25)147⟩.

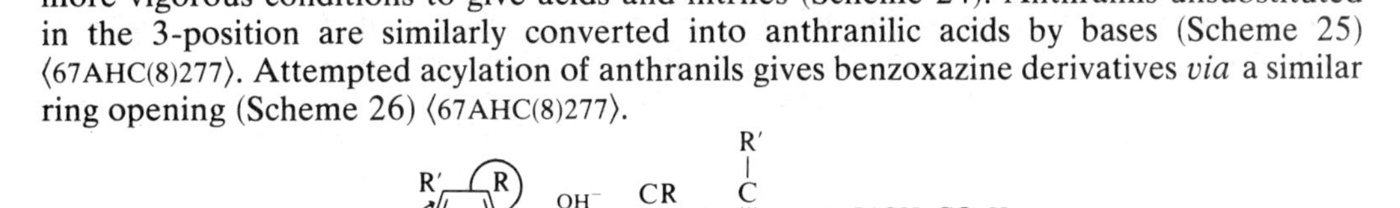

Isoxazoles substituted in the 3-position, but unsubstituted in the 5-position, react under more vigorous conditions to give acids and nitriles (Scheme 24). Anthranils unsubstituted in the 3-position are similarly converted into anthranilic acids by bases (Scheme 25) ⟨67AHC(8)277⟩. Attempted acylation of anthranils gives benzoxazine derivatives via a similar ring opening (Scheme 26) ⟨67AHC(8)277⟩.

Scheme 24

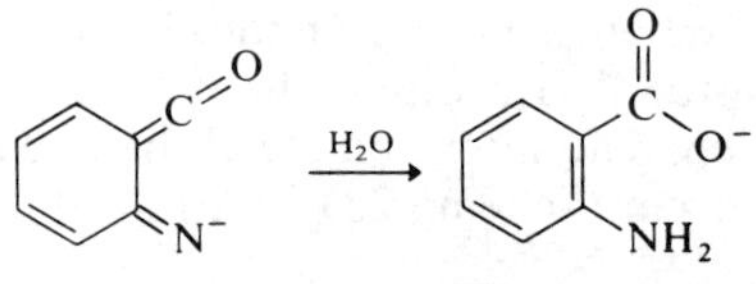

Scheme 25

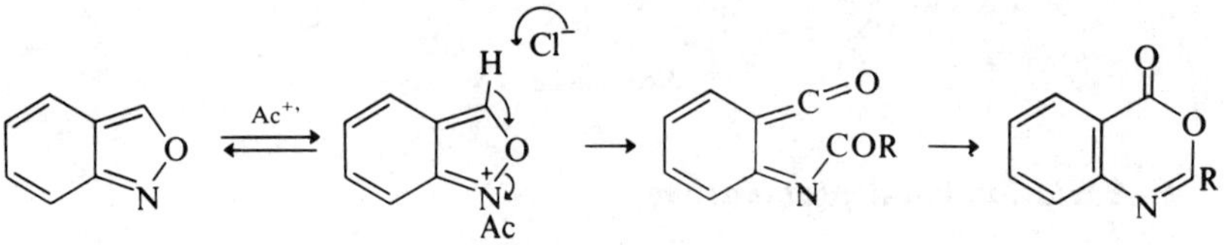

Scheme 26

For ring-opening reactions of *C*-metallated azoles, see Section 4.02.3.8.

4.02.1.7.6 *Proton loss from a ring nitrogen atom*

Pyrazoles, imidazoles, triazoles and tetrazoles are weak acids. They form metallic salts (*e.g.* with NaNH$_2$, RMgBr) which are extensively hydrolyzed by water. The resulting anions react very readily with electrophilic reagents on either ring nitrogen or carbon atoms, as discussed in Sections 4.02.1.3 and 4.02.1.4. For example, proton loss from a ring nitrogen atom gives the highly nucleophilic imidazole anion. This anion can be formed with sodium hydroxide or sodium alkoxide; good results are obtained with sodamide in liquid ammonia ⟨70AHC(12)103⟩.

Azolinones are weak to medium strong acids of pK_a 4–11. They form mesomeric anions which react very readily with electrophilic reagents at the nitrogen, oxygen or carbon atoms, depending on the conditions; see Section 4.02.1.1.4.

4.02.1.8 Reactions with Radicals and Electron-deficient Species; Reactions at Surfaces

4.02.1.8.1 *Carbenes and nitrenes*

Few reactions of azoles with these reagents have been reported. 2-Methylimidazole reacts with :CCl$_2$ to give 5-chloro-2-methylpyrimidine in poor yield. Pyrazoles react with HCCl$_3$ at 550 °C (with :CCl$_2$ formation) to give 2-chloropyrimidines in good yields. 1-Alkyl-1,2,4-triazoles react with nitrenes formed by the irradiation of azides to give *N*-imines (Scheme 27) ⟨74AHC(17)213⟩.

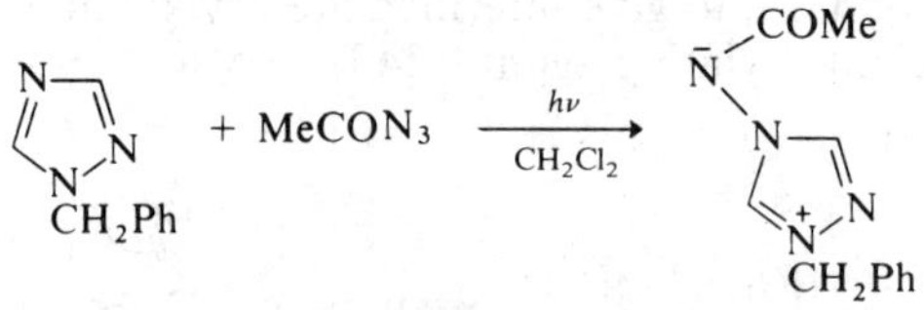

Scheme 27

4.02.1.8.2 *Free radical attack at the ring carbon atoms*

Despite some recent discoveries, free radical reactions are still very much less common in azole chemistry than those involving electrophilic or nucleophilic reagents. In some reactions involving free radicals, substituents have little orienting effect; however, rather selective radical reactions are now known.

(i) Aryl radicals

Phenyl radicals attack azoles unselectively to form a mixture of phenylated products. Relative rates and partial rate factors are given in Table 7. The phenyl radicals may be prepared from the usual precursors: PhN(NO)COMe, Pb(OCOPh)$_4$, (PhCO$_2$)$_2$ or PhI(OCOPh)$_2$. Substituted phenyl radicals react similarly.

Table 7 Relative Rates and Partial Rate Factors for the Homolytic Phenylation[a] of Five-membered Heterocycles ⟨74AHC(16)123⟩

Heterocycle	Relative rates	Partial rate factors			
		2	3	4	5
Thiazole	1.6	6.2	—	1.0	2.8
2-Methylthiazole	0.6	—	—	1.0	2.4
4-Methylthiazole	1.2	2.9	—	—	4.3
5-Methylthiazole	0.8	3.8	—	1.0	—
Isothiazole	0.95	—	2.7	0.5	2.5
1-Methylpyrazole	0.6	—	0.18	0.03	3.4
1-Methylimidazole	1.2	—	2.7	0.5	2.5

[a] Benzoyl peroxide is the source of the phenyl radicals, except for the first entry, where it is nitrosoacetanilide.

(ii) Alkyl radicals

Alkyl radicals produced by oxidative decarboxylation of carboxylic acids are nucleophilic and attack protonated azoles at the most electron-deficient sites. Thus imidazole and 1-alkylimidazoles are alkylated exclusively at the 2-position ⟨80AHC(27)241⟩. Similarly, thiazoles are attacked in acidic media by methyl and propyl radicals to give 2-substituted derivatives in moderate yields, with smaller amounts of 5-substitution. These reactions have been reviewed ⟨74AHC(16)123⟩; the mechanism involves an intermediate σ-complex.

Similar reactions occur with acyl radicals, for example with the ĊONH$_2$ radical from formamide ⟨74AHC(16)123⟩.

4.02.1.8.3 Electrochemical reactions and reactions with free electrons

Neutral rings are reduced by the uptake of an electron to form anion radicals ⟨80AHC(27)31⟩. In isoxazole and oxazole this can be achieved in an argon matrix, but normally ring fission occurs; reduction of 1,2,4-thiadiazoles also usually results in ring cleavage. Thiazoles containing electron-withdrawing groups, 1,3,4-oxadiazoles, 1,2,5-oxadiazoles and 1,2,5-thiadiazoles on electrochemical reduction yield transient anion radicals which can be characterized by ESR, *e.g.* (**250**) and (**251**). Anion radicals from benzazoles can be more stable, *e.g.* (**252**).

Cationic rings are reduced with the uptake of one electron (*e.g.* electrochemically) to give neutral radicals ⟨80AHC(27)31⟩. Examples of radicals which have been detected by ESR are (**253**) and (**254**). Such radicals, *e.g.* those from benzothiazolium ions, can dimerize (to **255**) or undergo further reduction (to **256**). 3-Methylbenzothiazolium (**257**) is reduced in a two-electron wave. The preparative reduction gives a mixture of the dihydro derivative (**258**) and the dimer (**259**) ⟨70AHC(12)213⟩. Benzofurazan is reduced polarographically in a six-electron reaction to *o*-phenylenediamine ⟨59MI40200⟩.

Calculations indicate that 1,2-dithiolyl radicals have large spin densities at the 3,5-positions and that 1,3-dithiolyl radicals have large spin densities at the 2-position. In agreement, radicals of these types unsubstituted at such positions dimerize very readily; when the position is substituted, the radical is more stable (**253** and **254**). Reduction of tetrazolium salts gives tetrazolyl radicals which show appreciable spin density on all four ring nitrogen atoms ⟨77AHC(21)323⟩.

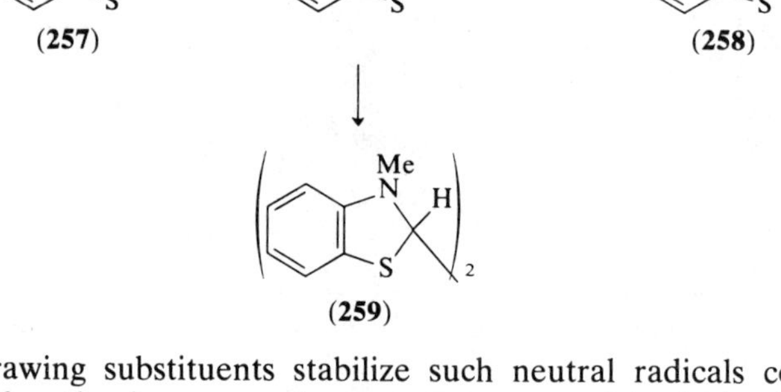

Electron-withdrawing substituents stabilize such neutral radicals considerably. Mero stabilization is found, for example, in the pyrazolyl derivative (260) ↔ (261) ⟨74JCS(P1)1422⟩.

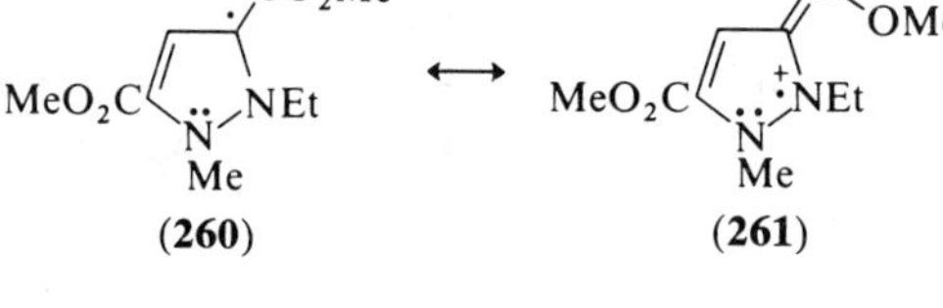

4.02.1.8.4 Other reactions at surfaces (heterogeneous catalysis and reduction reactions)

(i) Catalytic hydrogenation and reduction by dissolving metals

In general, azoles containing a cyclic oxygen atom are readily reduced, those with cyclic sulfur with more difficulty, and wholly nitrogenous azoles not at all.

Pyrazoles are very resistant to catalytic reduction, resisting hydrogenation over nickel at 150 °C and 100 atm ⟨66AHC(6)347⟩. Imidazoles are generally resistant to reduction.

Isoxazoles are readily reduced, usually with concomitant ring fission (*e.g.* **262 → 263**). They behave as masked 1,3-diketones ⟨79AHC(25)147⟩. 1,2-Benzisoxazoles are easily reduced to various products (Scheme 28) ⟨67AHC(8)277⟩. Chemical or catalytic reduction of oxazoles invariably cleaves the heterocyclic ring (Scheme 29) ⟨74AHC(17)99⟩. For similar reactions of thiazoles, see Section 4.02.1.5.1.

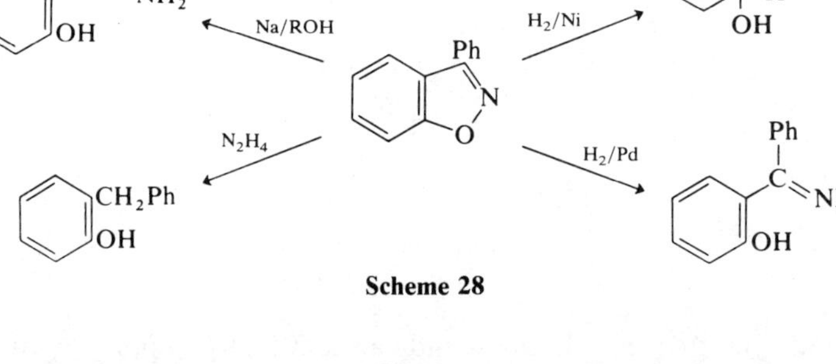

Scheme 28

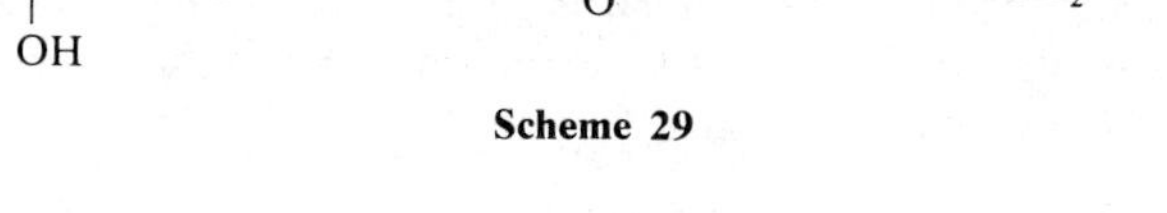

Scheme 29

Catalytic reduction of 1,2,4-oxadiazoles also breaks the N—O bond; *e.g.* (**264**) gives (**265**). Benzofuroxan can be reduced under various conditions to benzofurazan (**266**), the dioxime (**267**) or *o*-phenylenediamine (**268**) ⟨69AHC(10)1⟩. Reduction by copper and hydrochloric acid produced *o*-nitroanilines (Scheme 30) ⟨69AHC(10)1⟩.

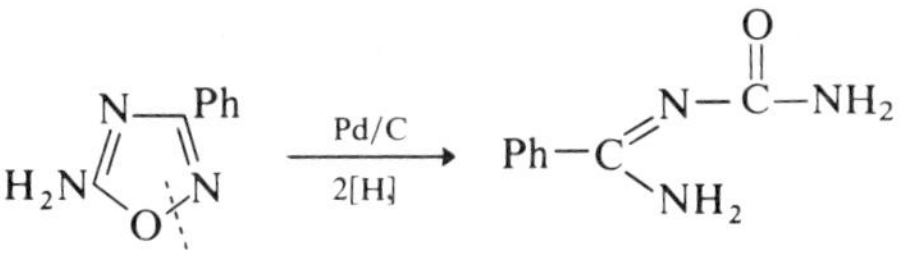

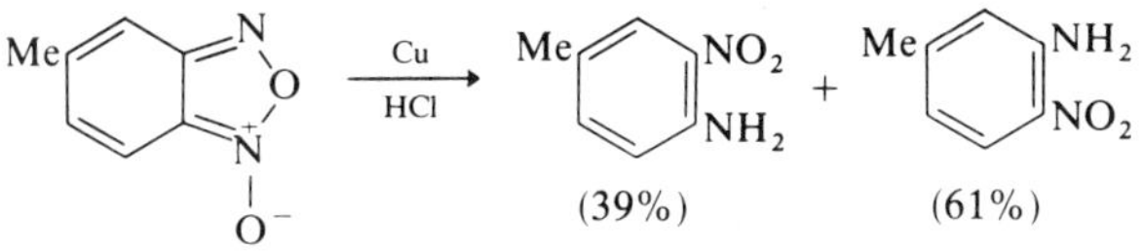

Scheme 30

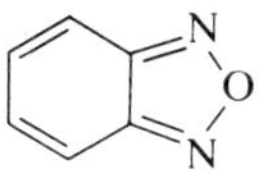

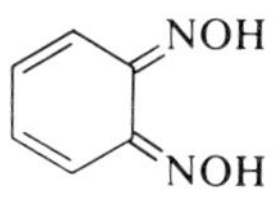

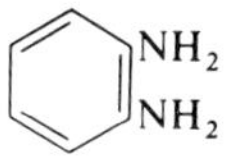

Isothiazoles are reductively desulfurized by Raney nickel, *e.g.* as in Scheme 31 ⟨72AHC(14)1⟩. 1,2,5-Thiadiazoles are subject to reductive cleavage by zinc in acid, sodium in alcohol, or Raney nickel, *e.g.* Scheme 32 ⟨68AHC(9)107⟩.

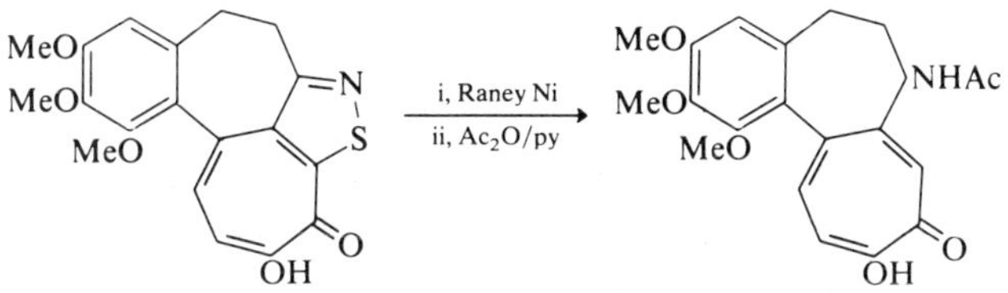

Scheme 31

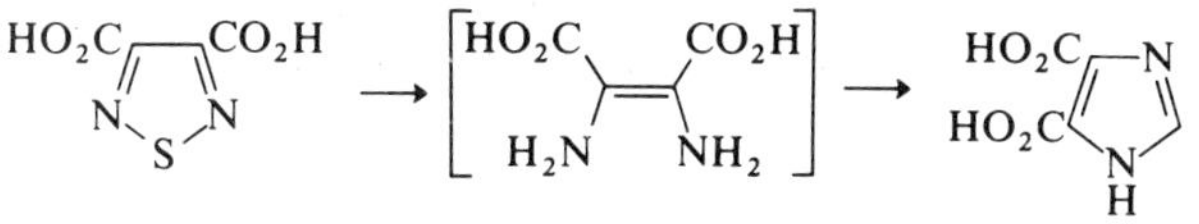

Scheme 32

4.02.1.9 Reactions with Cyclic Transition States

4.02.1.9.1 Diels–Alder reactions and 1,3-dipolar cycloadditions

The distinction between these two classes of reactions is semantic for the five-membered rings: Diels–Alder reaction at the F/B positions in (**269**) (four atom fragment) is equivalent to 1,3-dipolar cycloaddition in (**270**) across the three-atom fragment, both providing the 4π-electron component of the cycloaddition. Oxazoles and isoxazoles and their polyaza analogues show reduced aromatic character and will undergo many cycloadditions, whereas fully nitrogenous azoles such as pyrazoles and imidazoles do not, except in certain isolated cases.

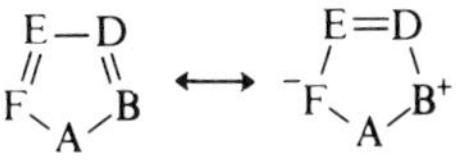

Isoxazolium salts react with enamines to give pyridinium salts (Scheme 33) ⟨69CPB2209⟩.

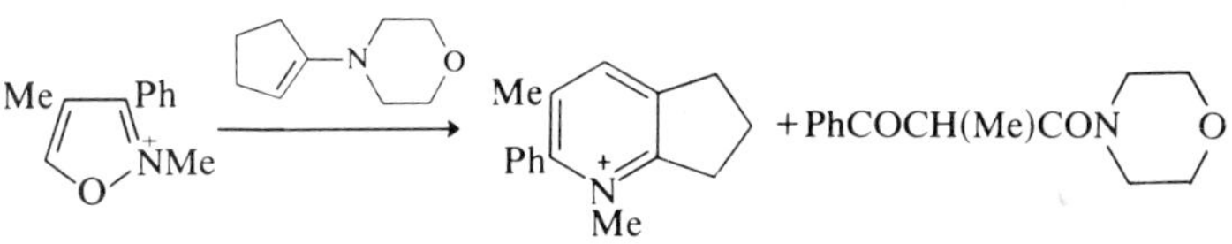

Scheme 33

Diels–Alder reactions of oxazoles afford useful syntheses of pyridines (Scheme 34) ⟨74AHC(17)99⟩. A study of the effect of substituents on the Diels–Alder reactivity of oxazoles has indicated that rates decrease with the following substituents: alkoxy > alkyl > acyl ≫ phenyl. The failure of 2- and 5-phenyl-substituted oxazoles to react with heterodienes is probably due to steric crowding. In certain cases, bicyclic adducts of type (**271**) have been isolated and characterized ⟨74AHC(17)99⟩; they can also decompose to yield furans (Scheme 35). Oxazoles react with singlet oxygen to give bicyclic adducts of type (**272**) which subsequently decompose (*e.g.* to **273**) ⟨74AHC(17)99⟩.

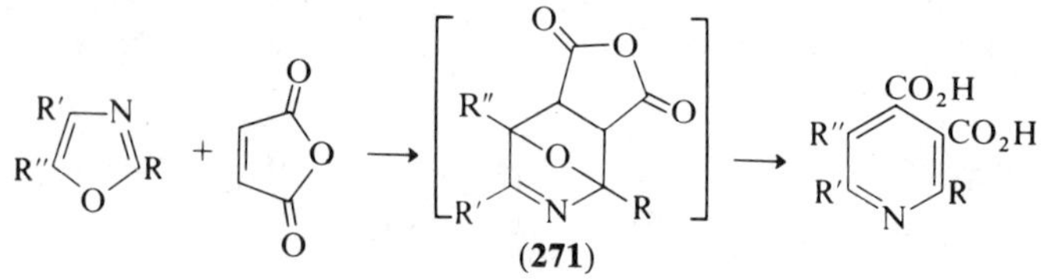

Scheme 34

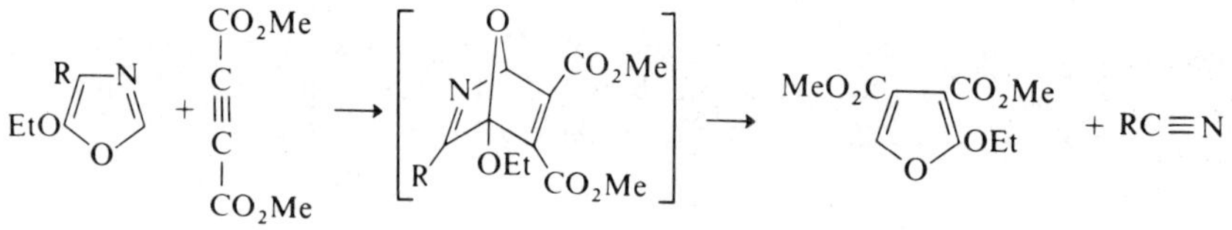

Scheme 35

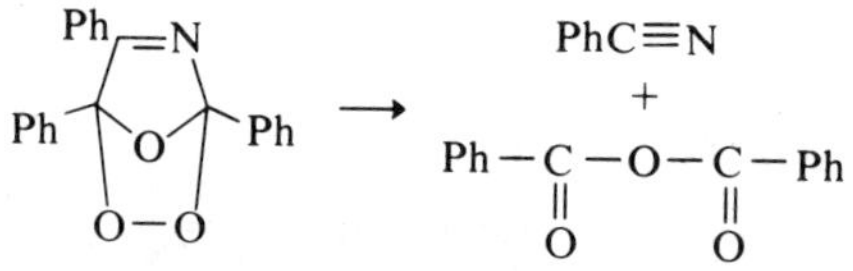

1,2,3,4-Thiatriazolin-5-imines undergo a variety of cycloaddition reactions with the elimination of N_2. 1,3-Dithiolylium-4-olates undergo cycloaddition reactions, *e.g.* as in Scheme 36 ⟨80AHC(27)151⟩. Scheme 37 gives an example of cycloaddition in the oxathiazole series.

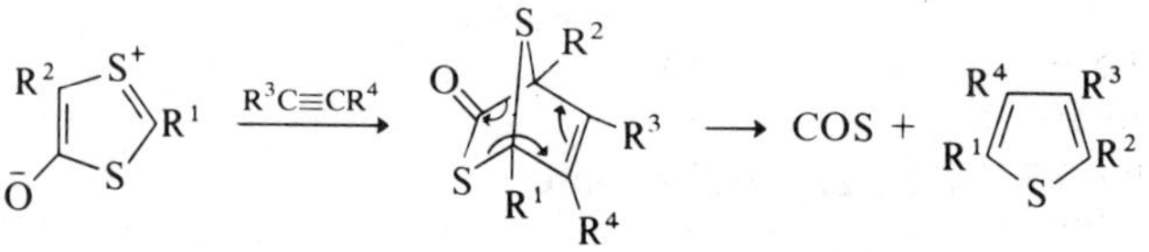

Scheme 36

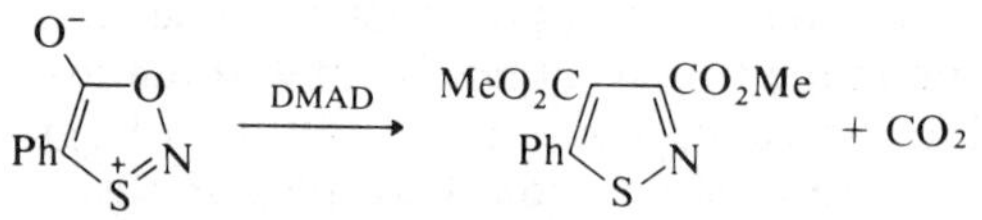

Scheme 37

Enamines and enolate anions react with benzofuroxan to give quinoxaline di-*N*-oxides (Scheme 38) ⟨69AHC(10)1⟩. Sydnones (**274**) with phenyl isocyanate give 1,2,4-triazoles (**275**) ⟨76AHC(19)1⟩, and from (**276**) the intermediate adduct (**277**) can be isolated ⟨73JA8452⟩. This is one of the few instances in which such primary cycloadducts have been isolated in the oxazole series of mesoionic compounds.

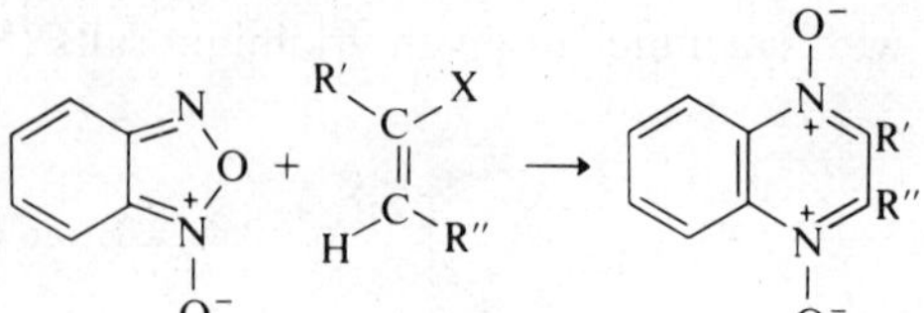

$X = NR_2, O^-$

Scheme 38

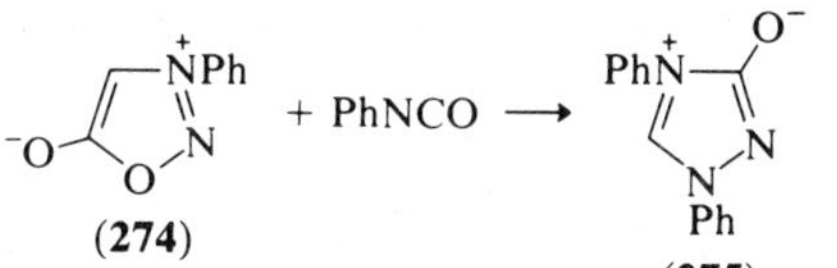

(274) + PhNCO → (275)

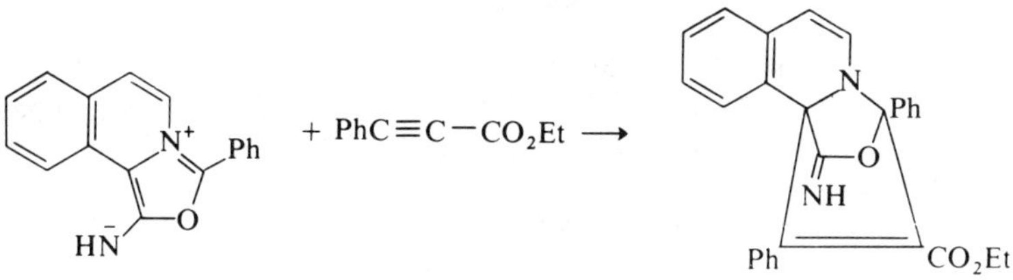

(276) + PhC≡C—CO₂Et → (277)

4.02.1.9.2 Photochemical cycloadditions

Photochemical additions to give four-membered rings are known. Thus the reactions of imidazoles across the 4,5-bond with benzophenone and acrylonitrile are illustrated by (278) → (279) and (280) → (281), respectively ⟨80AHC(27)241⟩. Oxazolin-2-one undergoes acetone-photosensitized photochemical addition to ethylene ⟨80CB1884⟩.

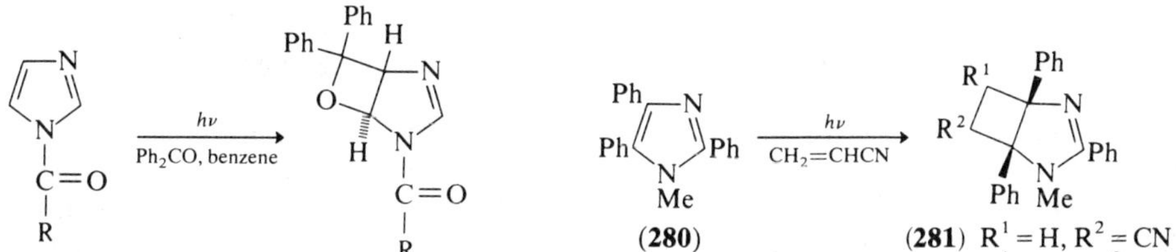

(281) $R^1 = H, R^2 = CN$
$R^1 = CN, R^2 = H$

4.02.2 REACTIONS OF NON-AROMATIC COMPOUNDS

Discussion of these compounds is divided into isomers of aromatic compounds, and dihydro and tetrahydro derivatives. The isomers of aromatic azoles are a relatively little-studied class of compounds. Dihydro and tetrahydro derivatives with two heteroatoms are quite well-studied, but such compounds become more obscure and elusive as the number of heteroatoms increases. Thus dihydrotriazoles are rare; dihydrotetrazoles and tetrahydro-triazoles and -tetrazoles are unknown unless they contain doubly bonded exocyclic substituents.

S-Oxides of sulfur-containing azoles comprise another class of non-aromatic azoles.

4.02.2.1 Isomers of Aromatic Derivatives

4.02.2.1.1 Compounds not in tautomeric equilibrium with aromatic derivatives

The 3*H*- and 4*H*-pyrazoles and 2*H*- and 4*H*-imidazoles ⟨83UP40200⟩ contain two double bonds in the heterocyclic ring, but in each case the conjugation does not include all the ring atoms; hence the compounds are not aromatic.

The quaternization of 5*H*-imidazoles occurs at the 1-position (Scheme 39) ⟨64AHC(3)1⟩. 4*H*-Pyrazoles are also readily monoquaternized.

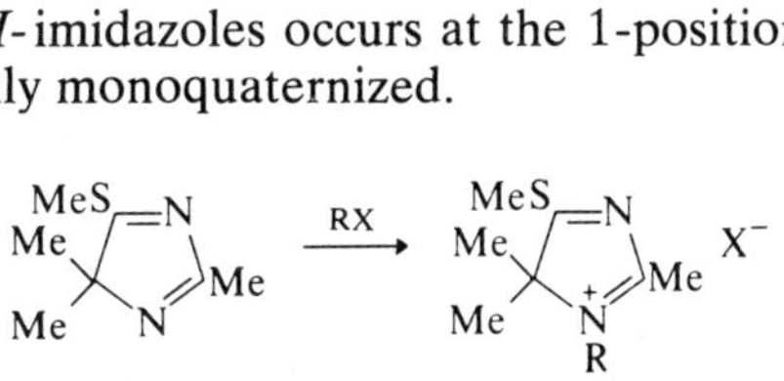

Scheme 39

Dichloropyrazolinones with alkali give alkynic acids (Scheme 40) ⟨58JA599⟩.

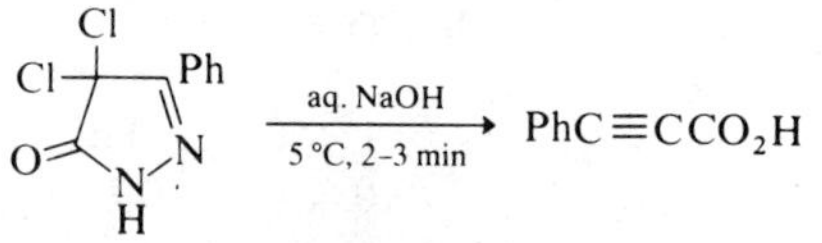

Scheme 40

Migrations of *C*-linked substituents around the ring, on to carbon or nitrogen atoms, are common amongst these compounds. This is the van Alphen–Huttel rearrangement and by it 3*H*-pyrazoles are converted into 1*H*-pyrazoles, and 2*H*-imidazoles are thermally isomerized into 1*H*-imidazoles.

3*H*-Pyrazoles are photochemically converted into cyclopropenes, and 3*H*-indazoles react similarly, *e.g.* (**282**) → (**283**) ⟨70AHC(11)1⟩. If a 3-aryl group is present, an indene can be formed, *e.g.* (**284**) → (**285**) ⟨83UP40200⟩.

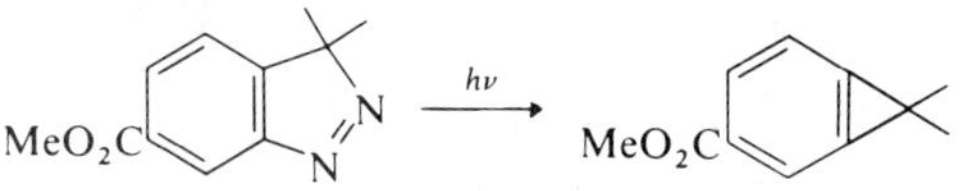

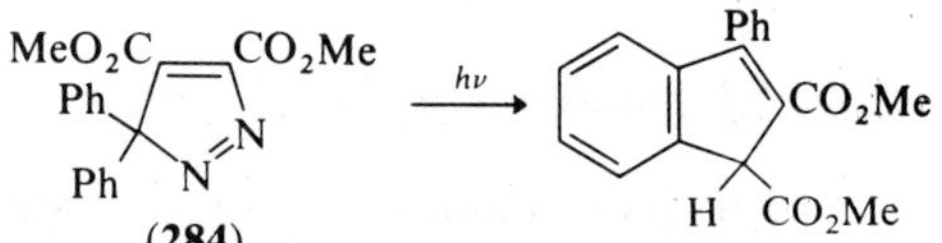

Addition of nucleophiles to C=N bonds is common in these compounds.

4.02.2.1.2 Compounds in tautomeric equilibria with aromatic derivatives

Compounds of types (**286**) and (**287**) are in tautomeric equilibria with 4- or 5-hydroxyazoles. However, the non-aromatic form is sometimes by far the most stable. Thus oxazolinone derivatives of type (**287**) have been obtained as optically active forms; they undergo racemization at measurable rates with nucleophiles ⟨77AHC(21)175⟩. Reactions of these derivatives are considered under the aromatic tautomer.

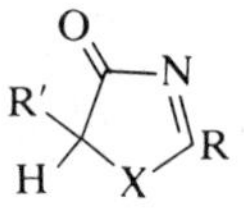

4.02.2.2 Dihydro Compounds

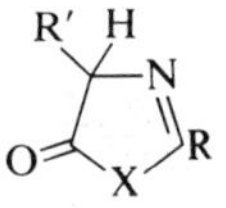

4.02.2.2.1 Tautomerism

Dihydroazoles can exist in at least three forms (*cf.* Section 4.01.1.3), which in the absence of substituents are tautomeric with each other. The forms in which there is no hydrogen on at least one ring nitrogen normally predominate because imines are generally more stable than vinylamines in aliphatic chemistry. Thus for dihydropyrazoles the stability order is Δ^2 (hydrazone) (**288**) > Δ^1 (azo) (**289**) > Δ^3 (enehydrazine) (**290**).

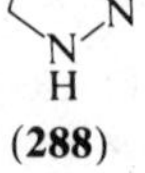

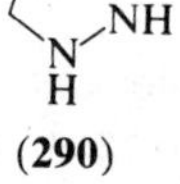

4.02.2.2.2 Aromatization

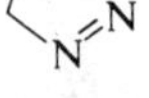

Δ^4-Imidazolines, -oxazolines and -thiazolines (**291**), and their benzo derivatives (**292**), are very easily aromatized (**292** → **293**), and syntheses which might be expected to yield such dihydro compounds often afford the corresponding aromatic products.

Dehydrogenation of Δ^2-imidazolines (**294**; Z = NR) gives imidazoles, but requires quite high temperatures and a catalyst such as nickel or platinum. Alternatively, hydrogen acceptors such as sulfur or selenium can be used ⟨70AHC(12)103⟩.

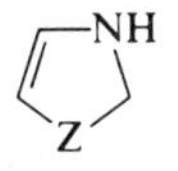

(291)

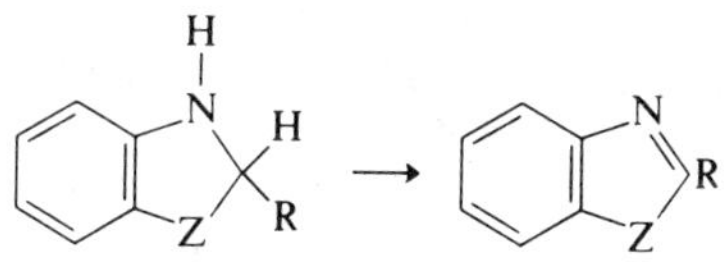

(292) (293)

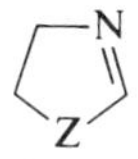

(294)

Δ^2-Pyrazolines are converted into pyrazoles by oxidation with bromine or $Pb(OAc)_4$ and they can also be dehydrogenated with sulfur. 3,5-Diphenylpyrazoline (**295**) on heating with platinum disproportionates to the pyrazole (**296**) and the pyrazolidine (**297**) ⟨66AHC(6)347⟩.

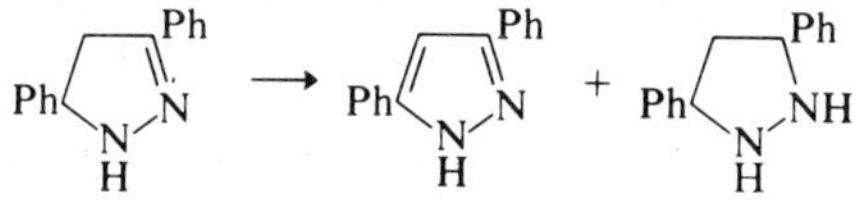

(295) (296) (297)

4.02.2.2.3 Ring contraction

1-Pyrazolines undergo photochemically-induced nitrogen elimination and ring-contraction to cyclopropanes, *e.g.* (**298**) → (**299**). This is particularly useful for the preparation of strained rings, *e.g.* (**300**) → (**301**) ⟨70AHC(11)1⟩. Δ^2-Pyrazolines unsubstituted in the 1-position lose nitrogen on pyrolysis to give cyclopropanes (*e.g.* **302**; Z = NH → **303**), probably *via* Δ^1-pyrazolines.

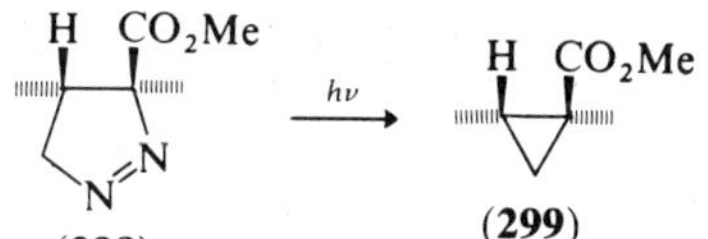

(298) (299)

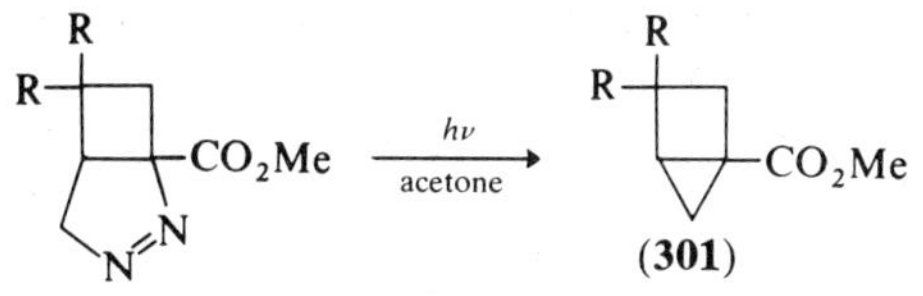

(300) (301)

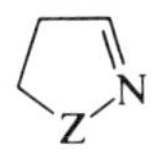

(302)

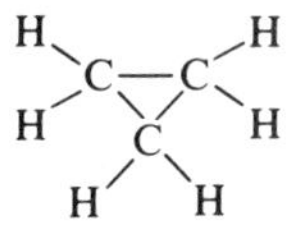

(303)

Photodecomposition of Δ^2-1,2,3-triazolines gives aziridines. In cyclohexane the *cis* derivative (**304**) gives the *cis* product (**305**), whereas photolysis in benzene in the presence of benzophenone as sensitizer gives the same ratio of *cis*- and *trans*-aziridines from both triazolines and is accounted for in terms of a triplet excited state ⟨70AHC(11)1⟩. Δ^2-Tetrazolines are photolyzed to diaziridines.

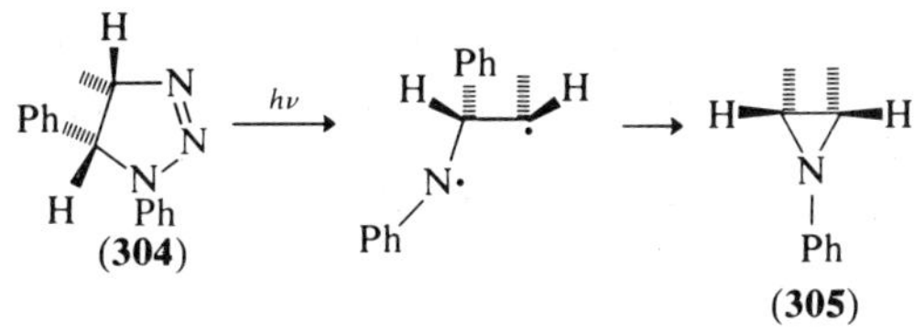

(304) (305)

Fragmentation of Δ^3-1,3,5-thiadiazoline derivatives is summarized in Scheme 41.

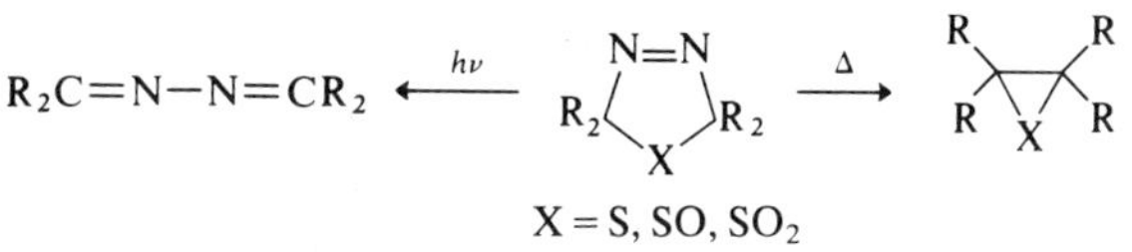

X = S, SO, SO_2

Scheme 41

4.02.2.2.4 Other reactions

Dihydro compounds show reactions which parallel those of their aliphatic analogues provided that the aromatization reactions just discussed do not interfere.

Δ^2-Imidazolines (**294**; Z = NH) are cyclic amidines and exhibit the characteristic resonance stabilization and high basicity. Δ^2-Oxazolines (**294**; Z = O) are cyclic imino ethers, and Δ^2-thiazolines (**294**; Z = S) are imino thioethers; both are consequently easily hydrolyzed by dilute acid.

Δ^2-Pyrazolines and Δ^2-isoxazolines (**302**; Z = NH, O) are cyclic hydrazones and oximes, respectively. 2-Pyrazolines are quaternized at the 2-position (**306** → **307**) ⟨64AHC(3)1⟩. 1,3,4-Oxadiazolines (*e.g.* **308**) are very easily ring-opened ⟨66AHC(7)183⟩.

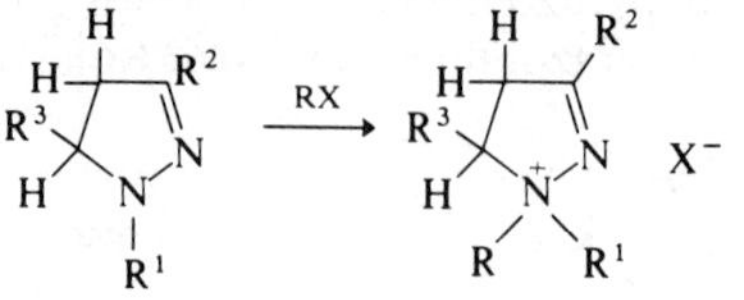

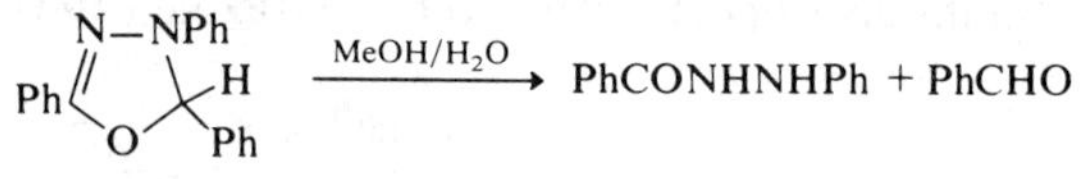

Reduction of dihyro compounds to the tetrahydro derivatives is sometimes possible. For example, thiazolines are reduced to thiazolidines by aluminum amalgam.

4.02.2.3 Tetrahydro Compounds

4.02.2.3.1 Aromatization

Some tetrahydro azoles can be aromatized, but this is more difficult than in the corresponding dihydro series. Thus the conversion of pyrazolidines into pyrazoles is accomplished with chloranil. Imidazolidines are aromatized with great difficulty.

4.02.2.3.2 Ring fission

Cleavage of the heterocyclic ring is usually accomplished using degradative procedures which are also applicable in the aliphatic series. Thus a nitrogen-containing ring can be opened by Hofmann exhaustive methylation (*e.g.* **309** → **310**). Pyrazolidines also undergo reactions of the Fischer indole synthesis type (**311** → **312**). The sulfur-containing ring of thiazolidines can be opened by Raney nickel desulfurization.

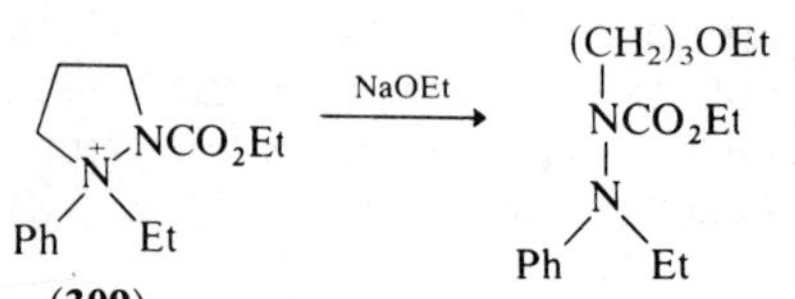

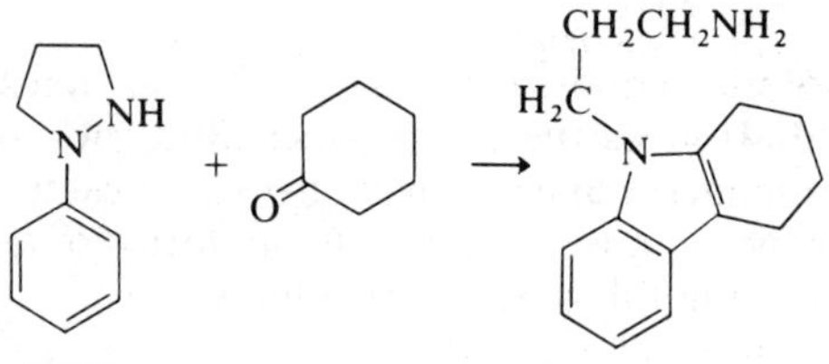

Compounds of types (**313**; R = H) and (**314**; R = H) are in equilibrium with open-chain forms (**315**); such tetrahydro compounds are readily hydrolyzed by dilute acid (R ≠ H).

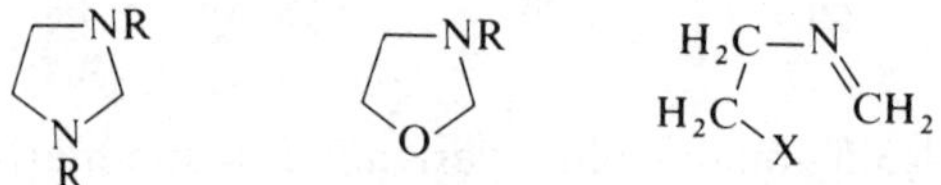

Isoxazolidines sometimes undergo retro 1,3-dipolar cycloaddition to give back alkenes and nitrones ⟨77AHC(21)207⟩.

4.02.2.3.3 Other reactions

These compounds usually show the typical reactions of their aliphatic analogues. 1,3-Dioxolanes (**316**), tetrahydroimidazoles (**313**), tetrahydrooxazoles (**314**) and tetrahydrothiazoles (**317**) are somewhat less easily ring-cleaved than their acyclic analogues (cf. previous section), but their properties are otherwise similar.

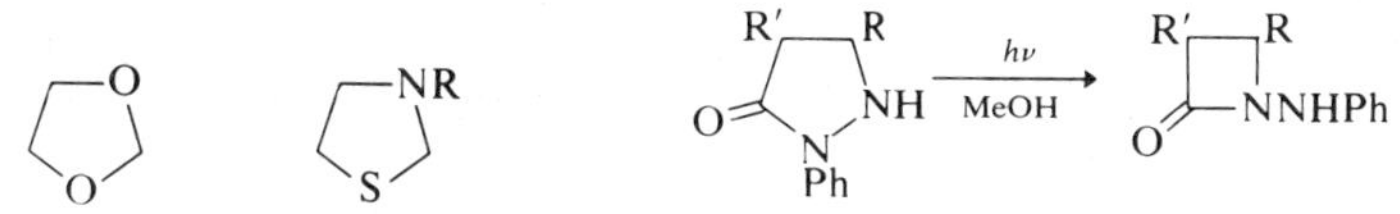

1-Aryl-5-pyrazolidinones (**318**) are photochemically ring-contracted to β-lactams ⟨70AHC(11)1⟩.

4.02.2.3.4 Stereochemistry

Whereas the aromatic systems are planar, fully reduced five-membered rings have non-planar envelope conformations, as is discussed in Section 4.01.4.3.

4.02.3 REACTIONS OF SUBSTITUENTS

Substituents attached to carbon are considered by classes; substituents linked to ring nitrogen are considered separately because of their differing character.

4.02.3.1 General Survey of Substituents on Carbon

If the reactions of the same substituents on heteroaromatic azoles and on benzene rings are compared, the differences in the reactivities are a measure of the heteroatoms' influence. Such influence by the mesomeric effect is smaller when the substituent is β to a heteroatom than when it is α or γ. The influence by the inductive effect is largest when the substituent is α to a heteroatom.

4.02.3.1.1 Substituent environment

The electronic environment of an α-substituent on pyridine (**319**) approaches that of a substituent on the corresponding imino compound (**320**) and is intermediate between those of substituents on benzene and substituents attached to carbonyl groups (**321, 322**) (cf. discussion in Chapter 2.02). Substituents attached to certain positions in azole rings show similar properties to those of α- and γ-substituents on pyridine. However, the azoles also possess one heteroatom which behaves as an electron source and which tends to oppose the effect of other heteroatom(s).

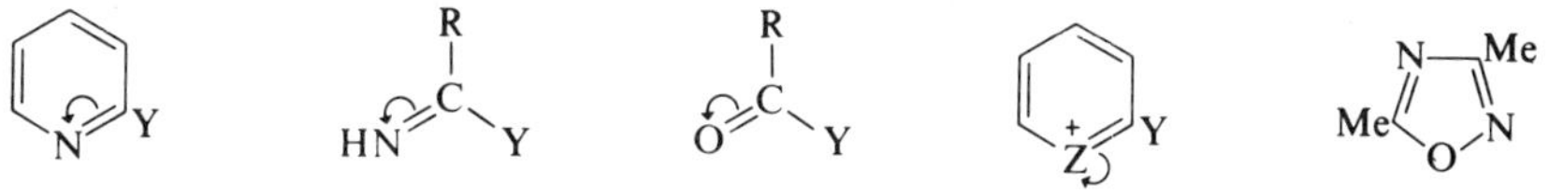

Substituents cannot directly conjugate with β-pyridine-like nitrogen atoms. Azole substituents which are not α or γ to a pyridine-like nitrogen react as they would on a benzene ring. Conjugation with an α-pyridine-like nitrogen is much more effective across a formal double bond; thus the 5-methyl group in 3,5-dimethyl-1,2,4-oxadiazole (**323**) is by far the more reactive.

In azolium cations, the electron-pull of the positively charged heteroatom is strong, and substituents attached α or γ to positive poles in azolium rings show correspondingly enhanced reactivity.

Azolinones and azole N-oxides possess systems which can act either as an electron source or as an electron sink, depending on the requirements of the reaction.

4.02.3.1.2 The carbonyl analogy

In aliphatic compounds, reactions of functional groups are often modified very significantly by an adjacent carbonyl group. As would be expected from the discussion in the preceding section, the reactions of certain substituents α and γ to pyridine-like nitrogen atoms in azole rings are similarly influenced. Such effects on substituents can be classified into six groups. (i) Substituents which can leave as anions are displaced by nucleophilic reagents (**324**). (ii) α′-Hydrogen atoms are easily lost as protons (**325**). (iii) As a consequence of (ii), tautomerism is possible (**326** ⇌ **327**). (iv) Carbon dioxide is readily lost from carboxymethyl (**328**) and carboxyl groups (**329**). (v) These effects are transferred through a vinyl group, and nucleophilic reagents will add to vinyl and ethynyl groups (**330**) (Michael reaction). (vi) Electrons are withdrawn from aryl groups (**331**). Examples of these are listed for both carbonyl and heterocyclic compounds in Table 8.

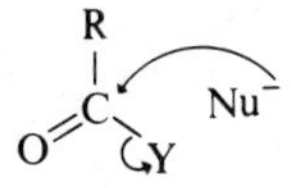

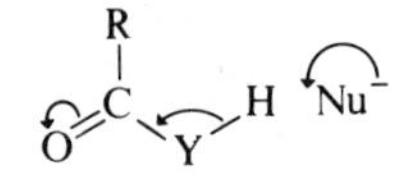

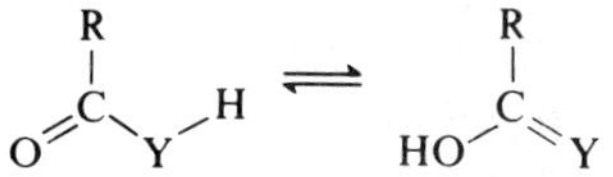

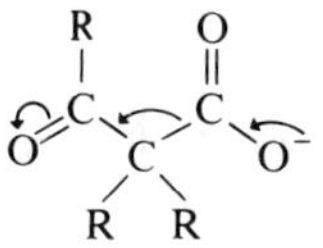

Table 8 Reactivity of Substituents: The Carbonyl Analogy ⟨B-68MI40200⟩

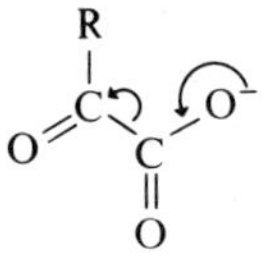

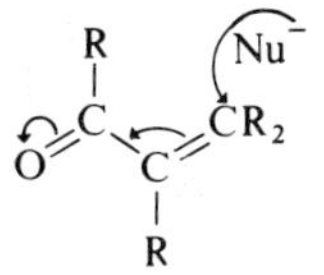

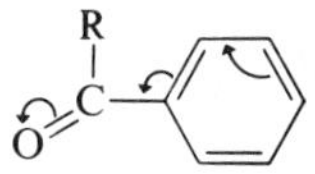

Reaction type	Group	α- or γ-groups	Compare with
Nucleophilic displacement	Nitro	Are displaced readily	—
	Halogen	Are displaced	Acid chloride
	Alkoxyl } Amino }	Are displaced when additionally activated	Ester Amide
Proton loss	Hydroxyl	Are acidic	Carboxylic acid
	Amino	Are less basic	Amide
	Alkyl	Become 'active'	Ketone
Tautomerism	Hydroxyl	Exist largely in the oxo form	Carboxylic acid (two equivalent structures)
	Amino	Exist to a small extent only in the imine form	Amide
	Mercapto	Exist largely in the thione form	Thiocarboxylic acid
Decarboxylation	Carboxyl	Decarboxylate at *ca.* 200 °C	α-Keto acids
	Carboxymethyl	Decarboxylate at *ca.* 50 °C	β-Keto acids
Michael reactions	Vinyl } Ethynyl }	Undergo Michael additions readily	α,β-Unsaturated ketones
	β-Hydroxyethyl	Undergo reverse Michael reaction readily (lose H_2O)	β-Hydroxy ketones
Electrophilic attack on phenyl groups	Phenyl	Undergo electrophilic substitution in the *meta* and *para* positions (*ca.* 1:1)	Phenyl ketones

4.02.3.1.3 Two heteroatoms in the 1,3-positions

The 2-position in imidazoles, thiazoles and oxazoles is electron deficient, and substituents in the 2-position (**332**) generally show the same reactivity as α- or γ-substituents on pyridines. 2-Substituents in azoliums of this type, including 1,3-dithiolyliums, are highly activated.

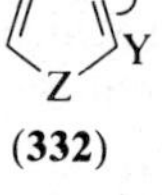

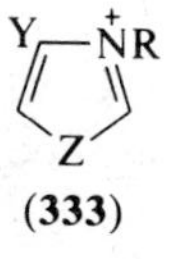

Substituents in the 4-position of these compounds are also α to a multiply-bonded nitrogen atom, but because of bond fixation they are relatively little influenced by this nitrogen atom even when it is quaternized (**333**). This is similar to the situation for 3-substituents in isoquinolines, cf. Chapter 2.02. In general, substituents in the 4- and 5-positions of imidazoles, thiazoles and oxazoles show much the same reactivity of the same substituents on benzeneoid compounds (but see Section 4.02.3.9.1).

4.02.3.1.4 Two heteroatoms in the 1,2-positions

Substituents on pyrazoles and isoxazoles, regardless of their positions, generally show reactivity closer to that of the same substituent on a benzene ring rather than to that of α- or γ-substituents on pyridine. The (electron-releasing) mesomeric effect of the 'pyrrole-type' NH group and 'furan-type' oxygen atom appears to be more important than their (electron-withdrawing) inductive effect in pyrazole and isoxazole (**334**). However, some reactions of these types are known (see *e.g.* Section 4.02.3.3.3) and halogen atoms and methyl groups in the 3- and 5-positions of pyrazoles and isoxazoles (**334**) become 'active' if the ring is quaternized (**335**).

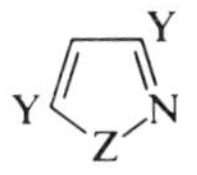

(**334**) 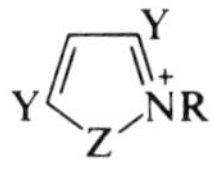   (**335**)

Substituents on the isothiazole ring are a little more reactive, especially in the 5-position. In cationic rings reactivity is much higher, *e.g.* for substituents in 1,2-dithiolylium salts.

4.02.3.1.5 Three heteroatoms

In the 1,2,4-thiadiazole ring the electron density at the 5-position is markedly lower than at the 3-position, and this affects substituent reactions. 5-Halogeno derivatives, for example, approach the reactivity of 4-halogenopyrimidines. The 1,2,4-oxadiazole ring shows a similar difference between the 3- and 5-positions.

Substituents in 1,3,4-thiadiazoles are quite strongly activated, as in the 2-position of pyridine.

In contrast, substituents in 1,2,4-triazoles are usually rather similar in reactivity to those in benzene; although nucleophilic substitution of halogen is somewhat easier, forcing conditions are required.

4.02.3.1.6 Four heteroatoms

Alkyl groups and halogen atoms in tetrazoles are not highly activated unless the ring is quaternized.

4.02.3.1.7 The effect of one substituent on the reactivity of another

The effect of one substituent on the reactivity of another is generally similar to that observed in the corresponding polysubstituted benzenes. However, the partial bond fixation in an azole can lead to differential effects in the mutual interactions of substituents, similar to those found in naphthalene where the benzene ring fusion induces bond fixation. A good example is in the comparison of methyl group reactivity in (**336**); the 5-methyl group condenses with aldehydes easily, the 3-methyl group does not. However, quaternization at nitrogen renders the 3-methyl group reactive.

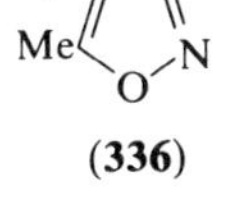

(**336**)

4.02.3.1.8 Reactions of substituents not directly attached to the heterocyclic ring

In general, substituents removed from the ring by two or more saturated carbon atoms undergo normal aliphatic reactions, and substituents attached directly to fused benzene rings or aryl groups undergo the same reactions as do those on normal benzenoid rings.

4.02.3.1.9 Reactions of substituents involving ring transformations

Several classes are known. Dimroth-type rearrangements occur by ring opening and reclosure so that one ring atom changes places with an exocyclic atom. The rearrangement of 5-phenylaminothiatriazole to 1-phenyl-5-mercaptotetrazole in basic solution is reversible (Scheme 42). As the anion it is the tetrazole system which is the stable one, whereas the neutral species is the thiatriazole ⟨76AHC(20)145⟩.

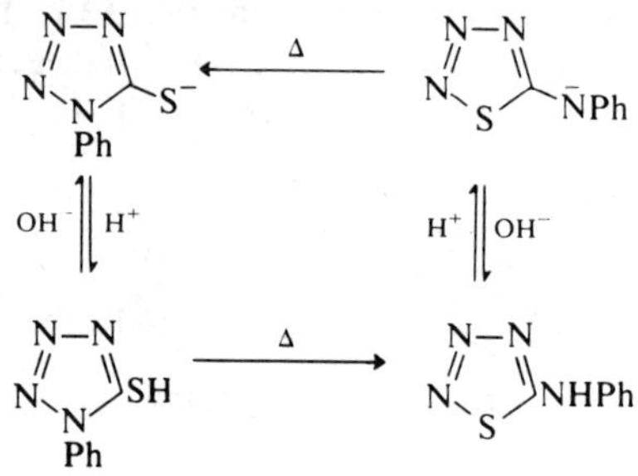

Scheme 42

A different type of rearrangement occurs when suitable side chains are α to a pyridine-like nitrogen atom. In the monocyclic series this can be generalized by Scheme 43. For a given side chain the rate of rearrangement is 1,2,4-oxadiazoles > isoxazoles > 1,2,5-oxadiazoles. Typical side chains include hydrazone, oxime and amidine. Some examples are shown in Table 9 ⟨79AHC(25)147⟩. Similar rearrangements for benzazoles are discussed in Section 4.02.3.2.4.

Table 9 Examples of Rearrangements Involving Three-atom Side Chains of Azoles ⟨79AHC(25)147⟩

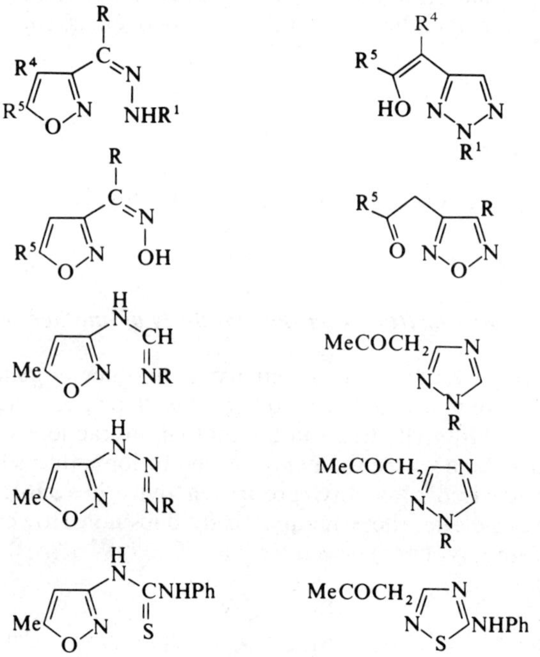

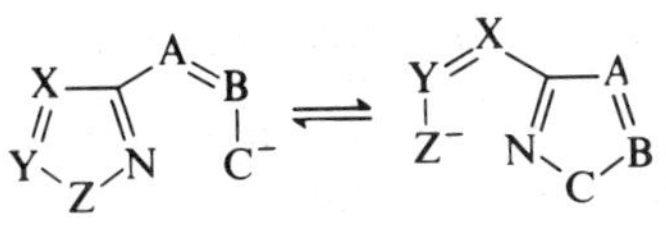

Scheme 43

A somewhat similar type of ring interconversion involving attack on sulfur has been postulated in the 1,2,4-thiadiazole series, *e.g.* (**337**) → (**338**). Such reactions are common in the 1,2,4-dithiazolium series, *e.g.* (**339**) → (**340**).

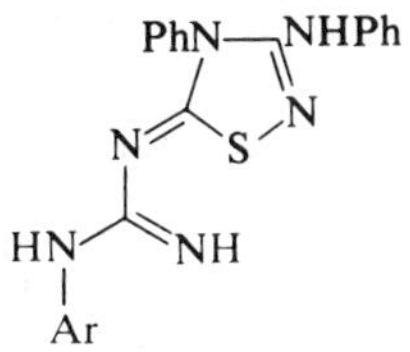

(**337**)

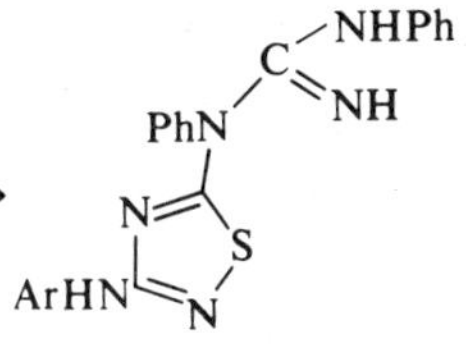

(**338**)

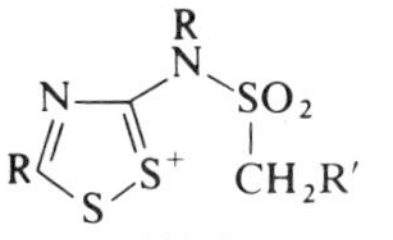

(**339**)

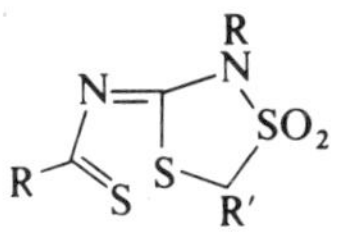

(**340**)

Cycloadditions including a cyclic S atom and an exocyclic C=X bond are known in the dithiazole series, *e.g.* as shown in Scheme 44.

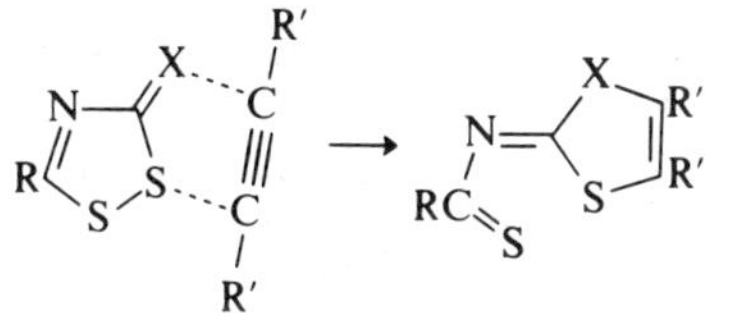

Scheme 44

4.02.3.2 Fused Benzene Rings

4.02.3.2.1 Electrophilic substitution

In compounds with a fused benzene ring, electrophilic substitution on carbon usually occurs in the benzenoid ring in preference to the heterocyclic ring. Frequently the orientation of substitution in these compounds parallels that in naphthalene. Conditions are often similar to those used for benzene itself. The actual position attacked varies; compare formulae (**341**)–(**346**) where the orientation is shown for nitration; sulfonation is usually similar for reasons which are not well understood.

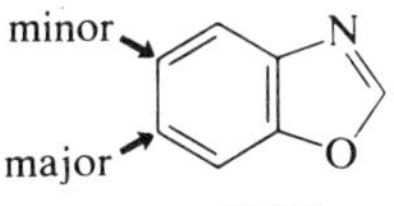

(**341**)

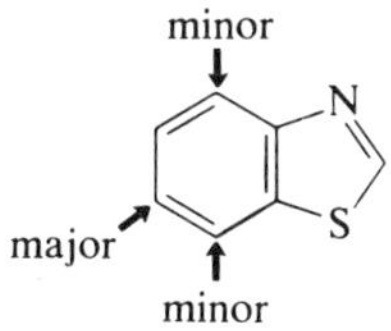

(**342**)

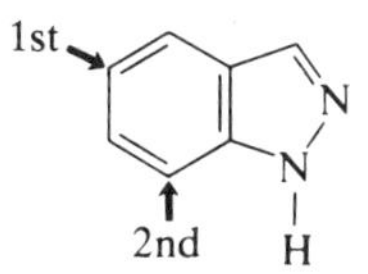

(**343**)

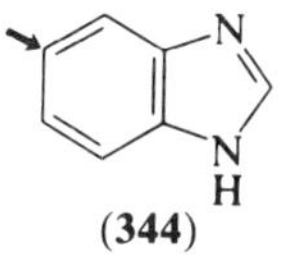

(**344**)

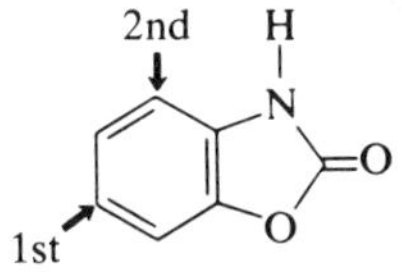

(**345**)

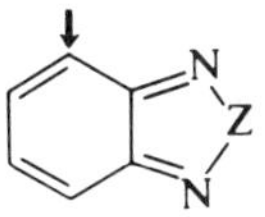

(**346**)

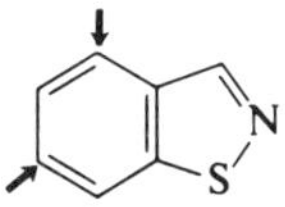

(**347**)

Indazoles show most of the typical benzene electrophilic substitution reactions. Anthranil is halogenated and nitrated in the benzene ring at position 5 ⟨67AHC(8)277⟩. Nitration of 1,2-benzisothiazole gives a mixture of the 5- and 7-nitro derivatives (**347**) ⟨72AHC(14)43⟩. 2,1-Benzisothiazole undergoes electrophilic bromination and nitration in the 5- and 7-positions ⟨72AHC(14)43⟩. Nitration of benzofuroxan gives the 4-nitro and then the 4,6-dinitro compound ⟨69AHC(10)1⟩.

Substituents on the benzene rings exert their usual influence on the orientation and ease of electrophilic substitution reactions. For example, further nitration ($H_2SO_4/SO_3/HNO_3$)

of 4-nitrobenzofuroxan (**348**) gives the 4,6-dinitro derivative, the first nitro group directing *meta* as expected. Strong electron-donating groups enhance electrophilic substitution and direct *ortho/para*. Thus dimethylaminobenzofuroxans can be nitrosated and diazo-coupled ⟨69AHC(10)1⟩; bromination of 4-, 5-, 6- and 7-aminobenzothiazoles occurs *ortho* and *para* to the amino group.

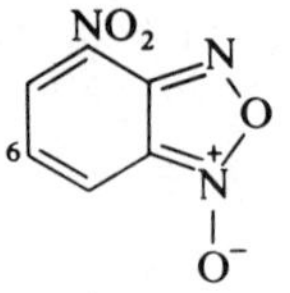

(**348**) 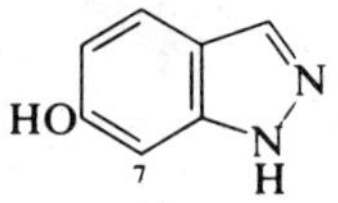   (**349**) 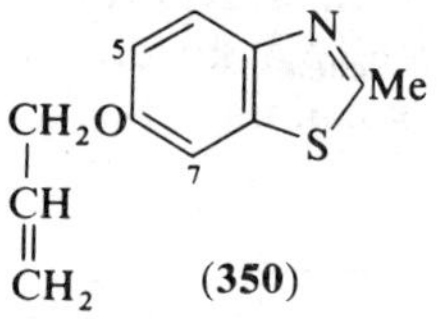   (**350**)

A heterocyclic ring induces partial double-bond fixation in a fused benzene ring. Hence, for example, diazo coupling occurs at the 7-position of 6-hydroxyindazole (**349**), and Claisen rearrangement of 6-allyloxy-2-methylbenzothiazole (**350**) gives the 7- and 5-allyl products in a ratio of 20:1.

4.02.3.2.2 Oxidative degradation

Vigorous oxidation (*e.g.* with $KMnO_4$) usually degrades fused benzene rings in preference to many azole rings, especially under acidic conditions. Thus benzimidazoles are oxidized by chromic acid or 30% hydrogen peroxide to imidazole-4,5-dicarboxylic acid ⟨70AHC(12)103⟩, and 2,1,3-benzothiadiazole is oxidized by ozone or potassium permanganate to the dicarboxylic acid (**351**) ⟨68AHC(9)107⟩.

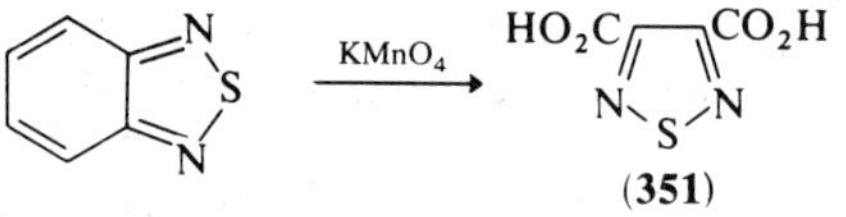

(**351**) 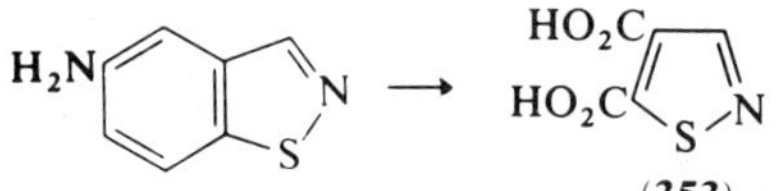   (**352**)   (**353**)

As expected, oxidative degradation of a fused benzene ring is facilitated when it carries electron-donating groups and is hindered by electron-withdrawing substituents. 5-Aminobenzisothiazole (**352**) with potassium permanganate gives the carboxylic acid (**353**) ⟨59JCS3061⟩.

4.02.3.2.3 Nucleophilic attack

Most fused benzene rings are stable toward nucleophilic attack, but exceptions are known for highly electron-deficient benzazoles. Thus aniline and benzofuroxan at 150 °C give the anil (**354**) ⟨45HCA850⟩.

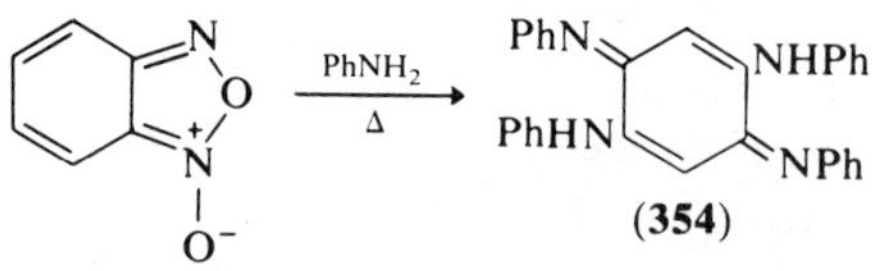

(**354**)

Halogen atoms on benzazole rings can be activated toward nucleophilic displacement by electron-withdrawing groups. Thus azide ion displaces chlorine from 5-chloro-4-nitro- and 4-chloro-7-nitro-benzofuroxan ⟨65JCS5958⟩.

4.02.3.2.4 Rearrangements

In the benzazole series, reactions of the type discussed for monocyclic derivatives in Section 4.02.3.1.9 are generalized by Scheme 45 and examples are given in Table 9.

4-Nitrobenzofuroxan (**355**) undergoes a rearrangement (recognizable as an isomerization in unsymmetrically substituted derivatives) which is an example of this general rearrangement (Scheme 45) ⟨64AG(E)693⟩; see Table 10.

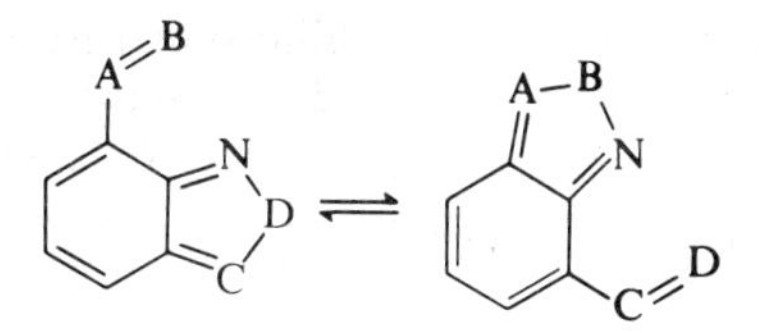

Scheme 45

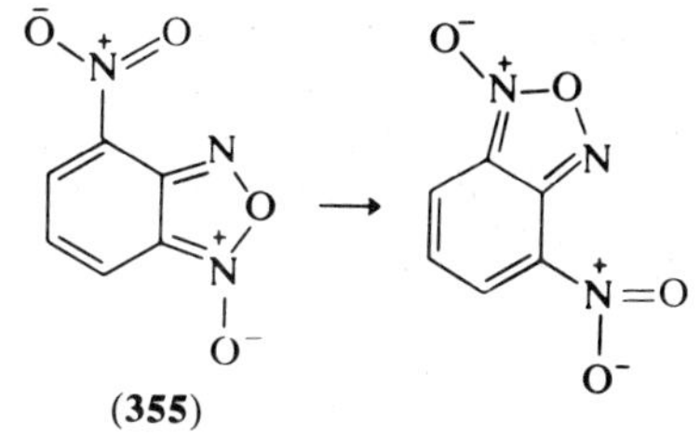

(355)

Table 10 Benzazole Rearrangements ⟨71JCS(C)1193⟩

Examples of involvement of two-atom side chains

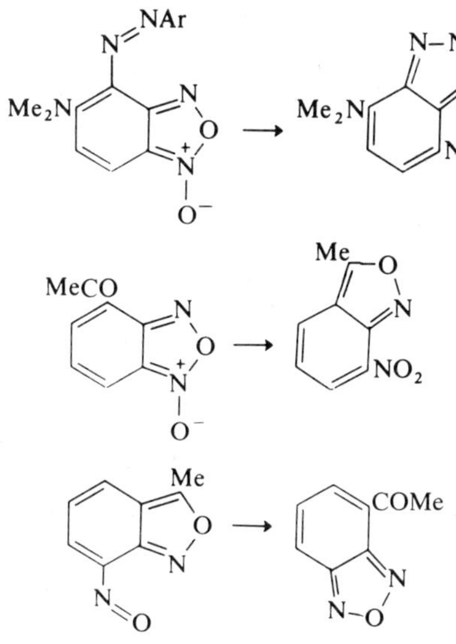

Photolysis of anthranils (**356**) in methanol or amines gives 2-methoxy- or 2-amino-3*H*-azepines (**357**) by ring expansion of intermediate nitrenes ⟨81AHC(28)231⟩. Photolysis of 2-alkylindazoles probably also goes through a nitrene intermediate, which either abstracts hydrogen from the solvent to give (**359**) ⟨81AHC(28)231⟩ or ring expands to yield (**358**).

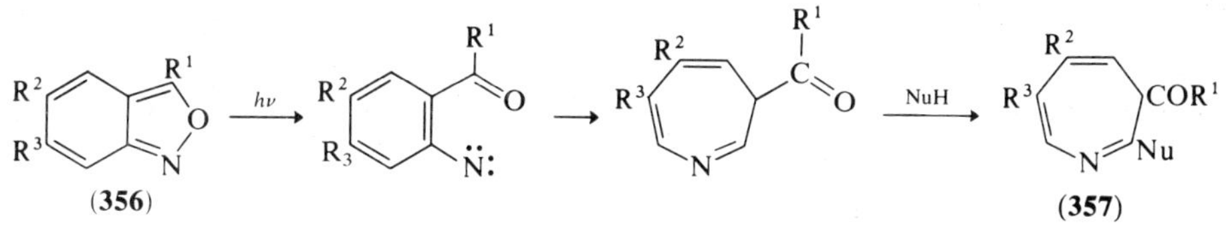

R^1 = Ph or Me; R^2, R^3 = H or Cl; NuH = MeOH, Et_2NH or $PhNH_2$

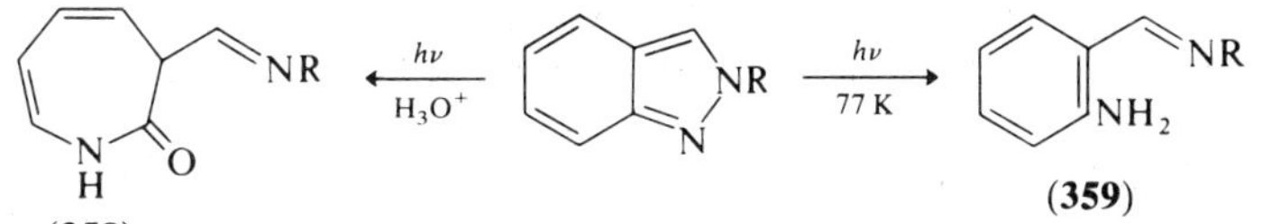

4.02.3.3 Alkyl Groups

4.02.3.3.1 Reactions similar to those of toluene

Alkyl groups attached to heterocyclic systems undergo many of the same reactions as those on benzenoid rings.

(i) Oxidation in solution ($KMnO_4$, CrO_3, *etc.*) gives the corresponding carboxylic acid or ketone; for example, alkyl groups on pyrazoles are oxidized with permanganate to carboxylic acids ⟨66AHC(6)347⟩, 3-methylisothiazoles are converted by chromium trioxide into the 3-carboxylic acids ⟨72AHC(14)1⟩, and methylthiazoles with SeO_2 give thiazole-carbaldehydes.

(ii) Free radical bromination with N-bromosuccinimide often succeeds. Thus 2,5-disubstituted 4-methyloxazoles on bromination give the 4-bromomethyl compounds ⟨74AHC(17)99⟩, and methyl groups in the 4- and 5-positions of isoxazole (360) and (361) have been brominated with NBS ⟨63AHC(2)365⟩.

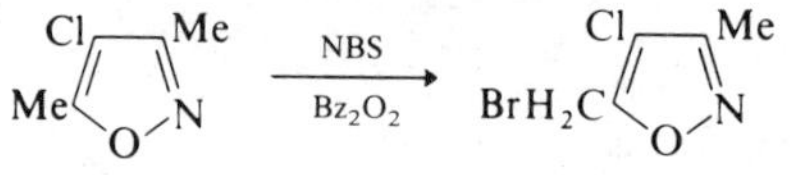

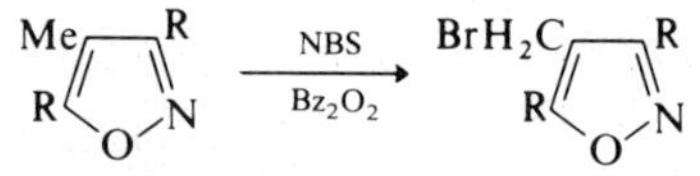

(iii) A fused cyclohexeno ring can be converted into a fused benzene ring, e.g. (362) → (363).

(iv) A trichloromethyl group has been converted by antimony trifluoride into a trifluoromethyl group in the 1,2,4-thiadiazole series (Scheme 46).

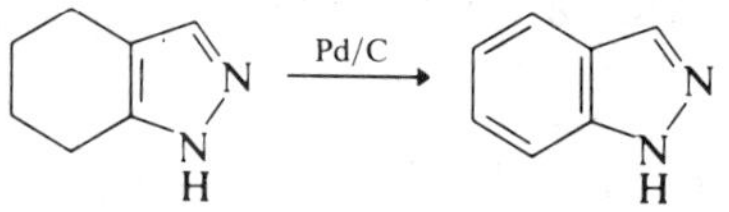

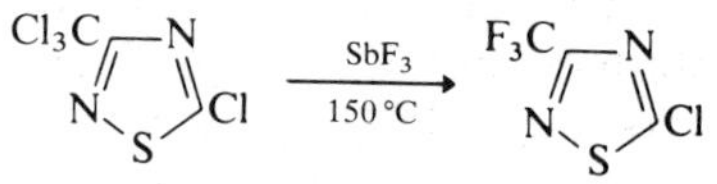

Scheme 46

4.02.3.3.2 Alkylazoles: reactions involving essentially complete anion formation

In addition to the reactions described in the preceding section, alkyl groups in the 2-positions of imidazole, oxazole and thiazole rings show reactions which result from the easy loss of a proton from the carbon atom of the alkyl group which is adjacent to the ring (see Section 4.02.3.1.2).

Additional nitrogen atoms facilitate such reactions, particularly if they are α or γ to the alkyl group, and, if α, act across a formal double bond. Thus, the 5-methyl group in 3,5-dimethyl-1,2,4-oxadiazole is much more reactive than the 3-methyl group in this compound or the methyl groups in 2,5-dimethyl-1,3,4-oxadiazole ⟨76AHC(20)65⟩.

The strongest bases, such as sodamide ($NaNH_2/NH_3$, 40 °C) or organometallic compounds ($BuLi/Et_2O$, 40 °C), convert, for example, 2-methyl-oxazole and -thiazole and 1,2-dimethylimidazole essentially completely into the corresponding anions (e.g. 364), although some ring metallation also occurs (cf. Section 4.02.1.7.1). These anions all react readily even with mild electrophilic reagents; thus the original alkyl groups can be substituted in the following ways. (i) Alkylation, e.g. MeI → CH_2Me for the formation of (365). (ii) Acylation, e.g. the oxadiazole (366) undergoes Claisen condensation with ethyl oxylate ⟨76AHC(20)65⟩. (iii) Carboxylation, e.g. CO_2 → CH_2CO_2H in the tetrazole series. (iv) Reactions with aldehydes, e.g. MeCHO → ·$CH_2CH(OH)Me$ in the 1,2-dimethylimidazole series.

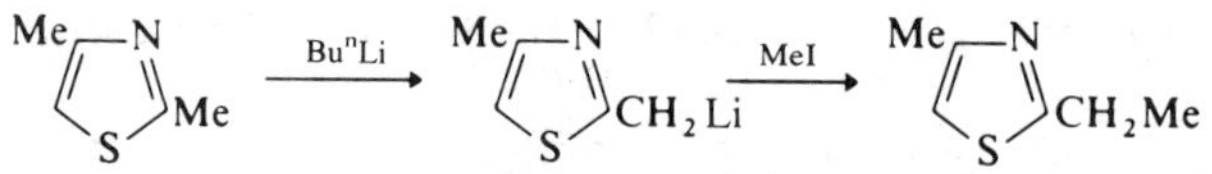

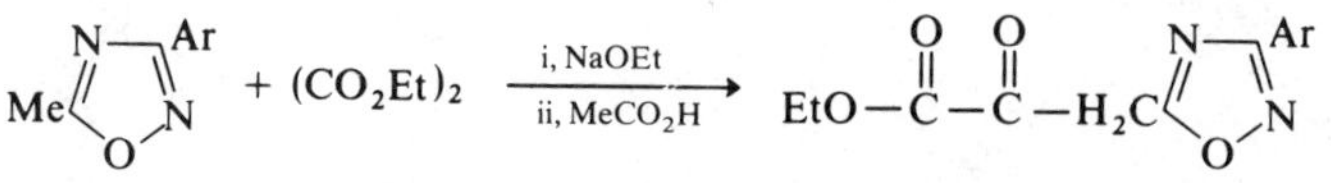

4.02.3.3.3 Reactions of alkylazoles involving traces of reactive anions

In aqueous or alcoholic solution, certain alkylazoles react with bases to give traces of anions of type (367). With suitable electrophilic reagents, these anions undergo reasonably rapid and essentially non-reversible reaction.

(367a) (367b)

(368) (369)

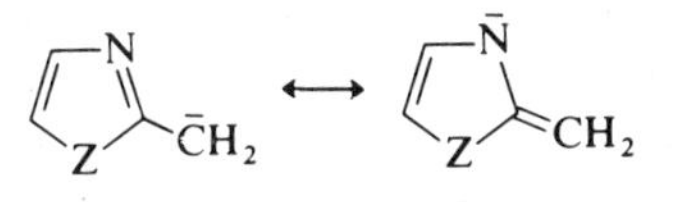

(i) A nitroso group gives an imine, as in the probable mechanism of the conversion of (**368**) into (**369**).

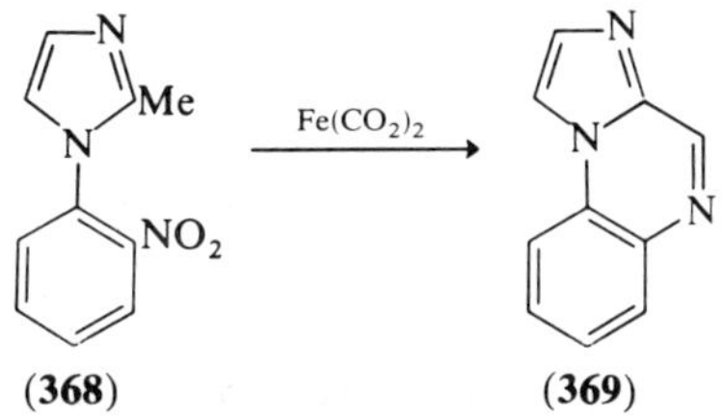

(ii) Aliphatic aldehydes can form monoalcohols, *e.g.* (**370**) gives (**371**) ⟨79HC(34-1)5⟩.

(370) (371)

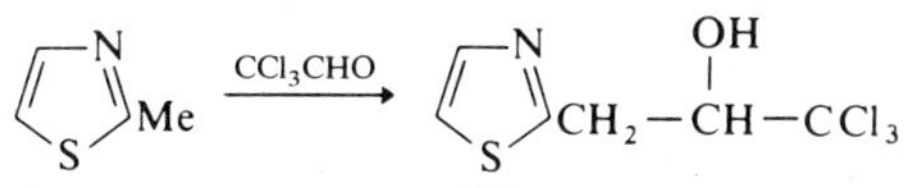

(iii) Aromatic aldehydes give styryl derivatives (*e.g.* **372**) by spontaneous dehydration of the intermediate alcohol (cf. Section 4.02.3.1.2). 5-Methyl-3-phenyl-1,2,4-oxadiazole (**373**) reacts thus with benzaldehyde in the presence of zinc chloride ⟨76AHC(20)65⟩. Benzaldehyde has not been condensed with any of the methylisothiazoles, but 3-nitrobenzaldehyde reacts with the 5-methyl derivative ⟨65AHC(4)75⟩. The 4- and 5-methylthiazoles are unreactive.

(372)

(373)

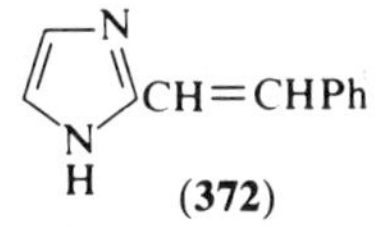

(iv) Halogens displace hydrogen atoms, *e.g.* 3,4,5-trimethylpyrazole (**374**) is converted into (**375**) ⟨56LA(598)186⟩.

(374) (375)

(376) (377)

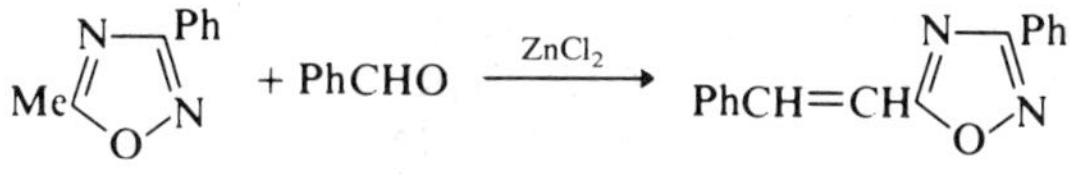

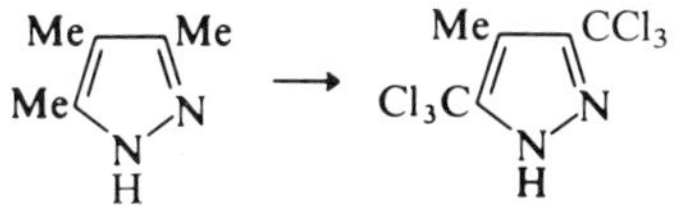

(v) Formamide acetal gives dimethylaminovinyl derivatives, as in (**376**) → (**377**).

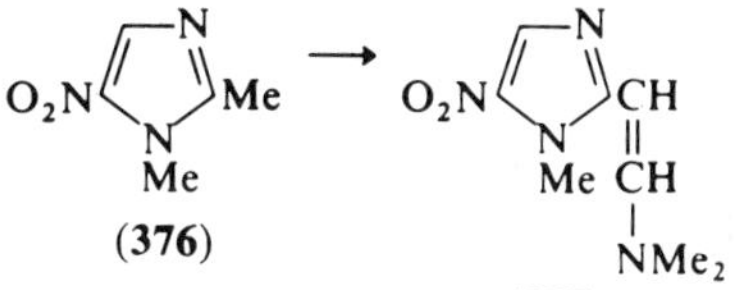

(vi) Pyridine and iodine give pyridinomethyl compounds, *e.g.* (**378**) yields (**379**) ⟨80AHC(27)241⟩.

(378) (379)

(380)

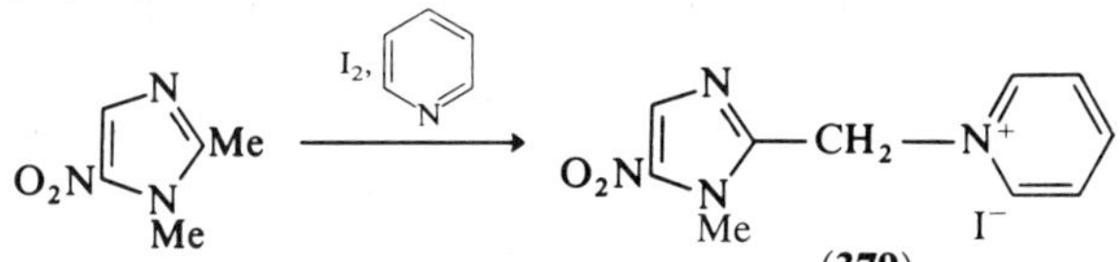

Reactions of types (i)–(vi) can be catalyzed by alkoxide or hydroxide ions, or amines. Alternatively, an acid catalyst forms a complex of type (**380**) from which proton loss is facilitated.

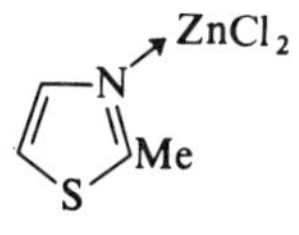

4.02.3.3.4 C-Alkyl-azoliums, -dithiolyliums, etc.

Proton loss from alkyl groups α or γ to a cationic center in an azolium ring is often easy. The resulting neutral anhydro bases or methides (cf. **381**) can sometimes be isolated; they react readily with electrophilic reagents to give products which can often lose another proton to give new resonance-stabilized anhydro bases. Thus the trithione methides are anhydro bases derived from 3-alkyl-1,2-dithiolylium salts (**382** ⇌ **383**) ⟨66AHC(7)39⟩. These methides are stabilized by electron-acceptor substituents such as CN or CO_2R ⟨66AHC(7)39⟩.

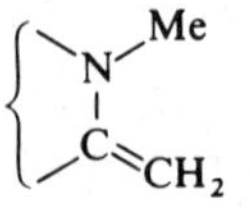

(381) 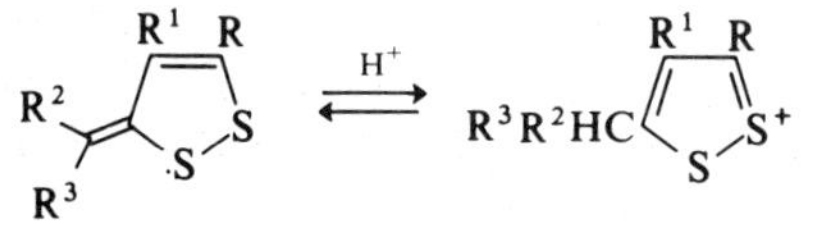   (382)   (383)

Both α- and γ-alkylazolium ions, analogously to the 2- and 4-alkylazoles themselves, can also react with electrophilic reagents without initial complete deprotonation. They undergo the same types of reactions as the alkylazoles but under milder conditions, and these reactions are often catalyzed by piperidine. Thus in quaternized pyrazoles, 5-methyl groups react with benzaldehyde to give styryl derivatives and can be chlorinated ⟨66AHC(6)347⟩. The methyl groups in quaternized isoxazoles are also reactive, and here piperidine is sufficient as catalyst (Scheme 47) ⟨63AHC(2)365⟩.

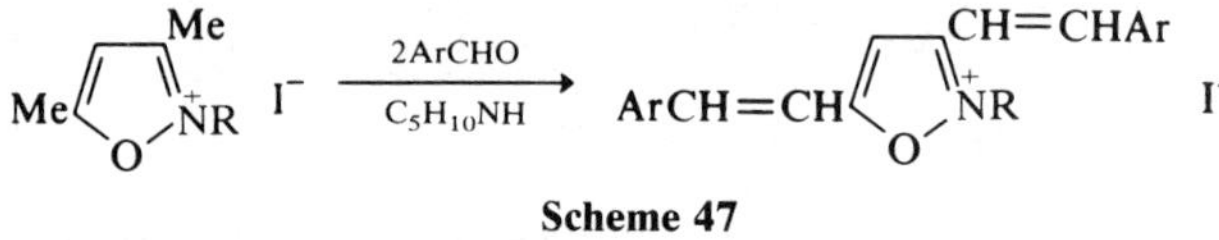

Scheme 47

Some weak electrophilic reagents, which are usually inert toward azoles, also react with quaternized azoles. Diazonium salts yield phenylhydrazones (Scheme 48) in a reaction analogous to the Japp–Klingemann transformation of β-keto esters into phenylhydrazones; in the dithiolylium series illustrated the product has bicyclic character. Cyanine dye preparations fall under this heading (see also Section 4.02.1.6.5). Monomethine cyanines are formed by reaction with an iodo quaternary salt, e.g. Scheme 49. Tri- and penta-methinecarbocyanines (384; $n = 1$ and 2, respectively) are obtained by the reaction of two molecules of a quaternary salt with one molecule of ethyl orthoformate (384; $n = 1$) or β-ethoxyacrolein acetal (384; $n = 2$), respectively.

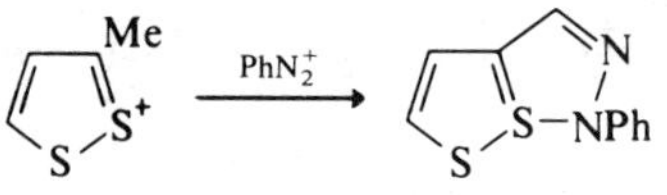

Scheme 48

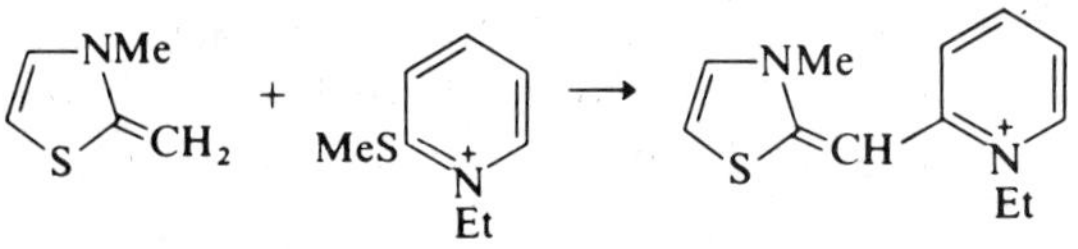

Scheme 49

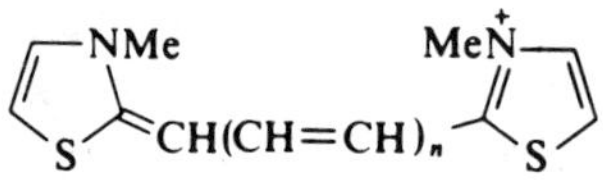

(384) 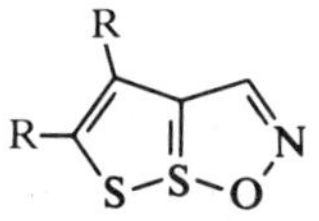   (385)

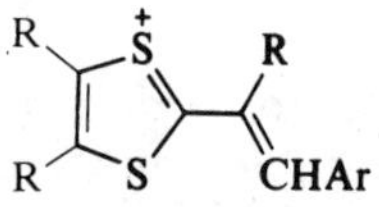

(386) 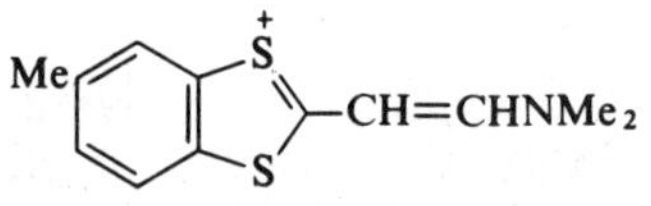   (387)

3-Methyl-1,2-dithiolyliums react with aldehydes to give styryl derivatives, with DMF to give Vilsmeier salts, and on nitrosation form the bicyclic products (385) ⟨80AHC(27)151⟩. 2-Alkyl groups in 1,3-dithiolylium ions also react with aromatic aldehydes to give (386), with DMF to give (387), and with other electrophiles ⟨80AHC(27)151⟩.

In general, methyl groups in the 4- and 5-positions of imidazole, oxazole and thiazole do not undergo such deprotonation-mediated reactions, even when the ring is cationic.

Compounds which can formally be considered as anhydro bases can sometimes react with nucleophiles. Thus unsaturated azlactones with Grignard reagents give saturated azlactones (Scheme 50) ⟨65AHC(4)75⟩.

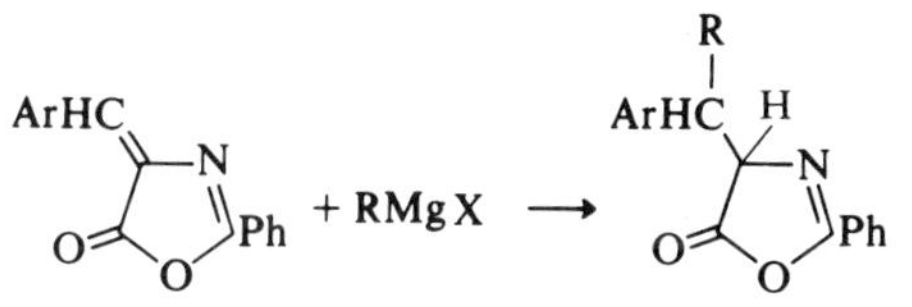

Scheme 50

4.02.3.4 Other C-Linked Substituents

4.02.3.4.1 Aryl groups: electrophilic substitution

Electrophilic substitution occurs readily in C-aryl groups, often predominantly at the *para* position. Thus nitrations of phenyl-thiazoles, -oxazoles and -imidazoles (HNO_3/H_2SO_4 at 100 °C) all yield the corresponding *p*-nitrophenyl derivatives. This is to be contrasted with the situation for α-phenylpyridine, where a mixture of mainly *m*- and *p*-nitrophenyl derivatives is formed. Although in strongly acidic media a C-linked aryl group is generally more readily substituted than the ring, the orientation often changes when C-phenylazole derivatives are nitrated under less acidic conditions. Thus 3- and 5-phenylpyrazoles can give under such conditions the 4-nitro derivatives. Such orientation changes have been demonstrated to result from changes in the species undergoing reaction from the azolium ion to the neutral azole.

3-Methyl-5-phenylisoxazole undergoes nitration as the conjugate acid at the *para* position of the phenyl group. 5-Methyl-3-phenylisoxazole is nitrated as a conjugate acid at the *meta* position but as the free base at the *para* position of the phenyl group ⟨79AHC(25)147⟩. Phenyl groups attached to oxazole rings are nitrated or sulfonated in the *para* position, with relative reactivities of the phenyl groups in the order 5>4>2 ⟨74AHC(17)99⟩.

3-Phenylisothiazole is nitrated predominantly in the *meta* position of the phenyl group, whereas 4-phenylisothiazole is nitrated *ortho* and *para* in the phenyl group ⟨72AHC(14)1⟩. Nitration of 3-phenyl-1,2,4-oxadiazole gives a mixture of *m*- and *p*-nitrophenyl derivatives ⟨63G1196⟩.

In the 1,2-dithiolylium ion system, 3- and 5-phenyl groups on nitration give mixtures of *para* and *meta* orientation, whereas nitration of a 4-phenyl group gives *para* substitution only ⟨61JA2934⟩.

4.02.3.4.2 Aryl groups: other reactions

3-Arylanthranils (**388**) on thermolysis give acridones (**389**) ⟨81AHC(28)231⟩. 3-Phenylanthranils (**390**) also form acridones (**391**) on treatment with nitrous acid ⟨67AHC(8)277⟩. Related rearrangements are found with 3-heteroarylanthranils (*e.g.* **392** → **393**) ⟨81AHC(28)231⟩.

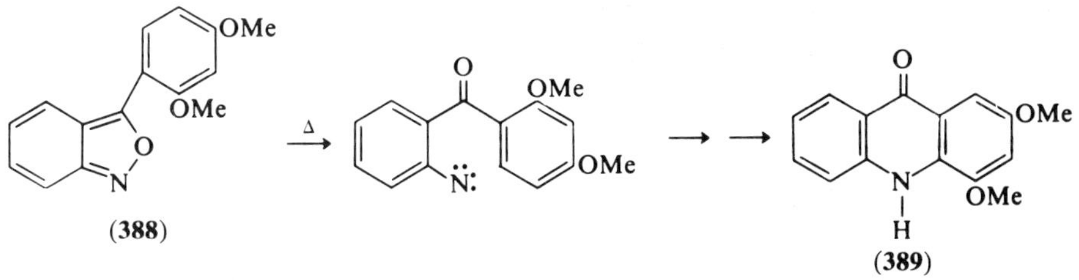

(388)   (389)

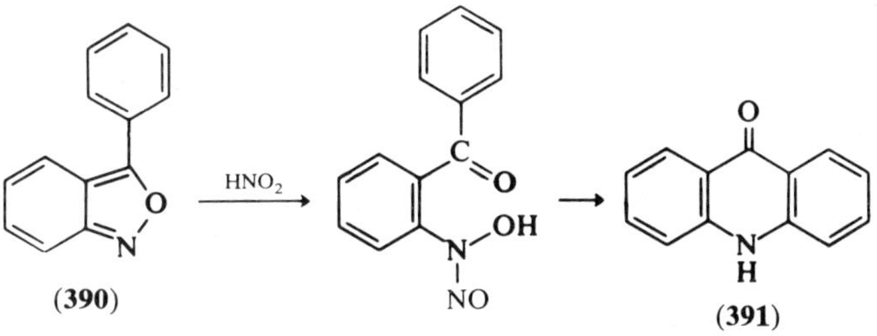

(390)   (391)

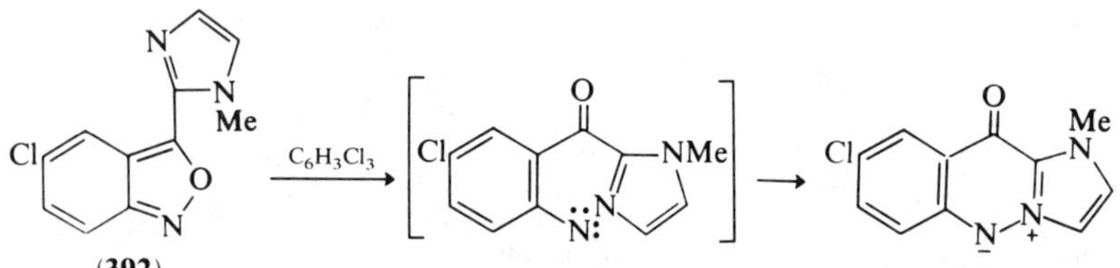

Methyl groups on *C*-linked phenyl attached to oxazoles, isoxazoles and oxadiazoles react with benzylidineaniline to give stilbene derivatives (Scheme 51) ⟨78AHC(23)171⟩.

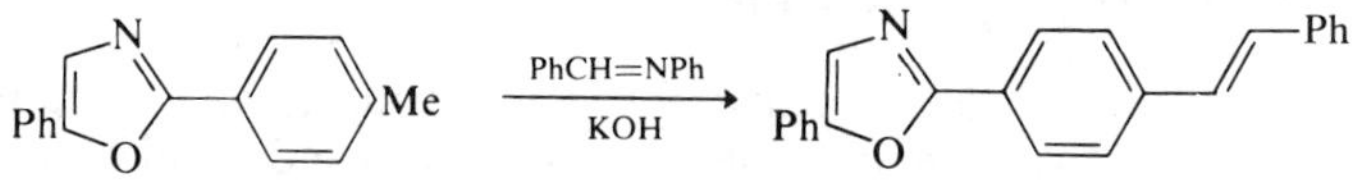

Scheme 51

4.02.3.4.3 Carboxylic acids

Azolecarboxylic acids can be quite strongly acidic. Thus 1,2,5-thiadiazole-3,4-dicarboxylic acid has first and second pK_a values of 1.6 and 4.1, respectively ⟨68AHC(9)107⟩. The acidic strengths of the oxazolecarboxylic acids are in the order $2 > 5 > 4$, in agreement with the electron distribution within the oxazole ring ⟨74AHC(17)99⟩. Azolecarboxylic acids are amino acids and can exist partly in the zwitterionic, or betaine, form (*e.g.* **394**).

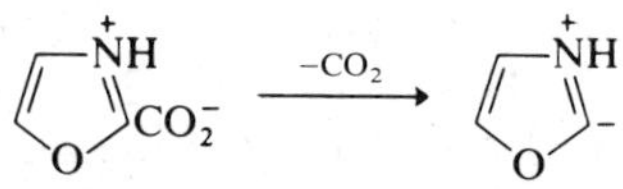

The relatively easy decarboxylation of many azolecarboxylic acids is a result of inductive stabilization of intermediate zwitterions of type (**395**) (cf. Section 4.02.1.7.1). Kinetic studies have shown that oxazole-2- and -5-carboxylic acids both decarboxylate through the zwitterionic tautomers ⟨71JA7045⟩. Thiazole-2-carboxylic acids, and to a lesser extent -5-carboxylic acids, decarboxylate readily; thiazole-4-carboxylic acids are relatively stable. Isothiazole-5-carboxylic acids decarboxylate readily, the 3-isomers less so while the 4-isomers require high temperatures. The 1,2,4-, 1,2,5- and 1,3,4-thiadiazolecarboxylic acids are also easily decarboxylated; their stability is increased by electron-donating substituents ⟨68AHC(9)165⟩. Most 1,2,3-triazolecarboxylic acids lose carbon dioxide when heated above their melting points ⟨74AHC(16)33⟩.

Azoleacetic acids with a carboxymethyl group also decarboxylate readily, *e.g.* all three thiazole isomers, by a mechanism similar to that for the decarboxylation of β-keto acids; cf. Section 4.02.3.1.2. The mechanism has been investigated in the oxazole case, (**396**) → (**397**) → (**398**) ⟨72JCS(P2)1077⟩.

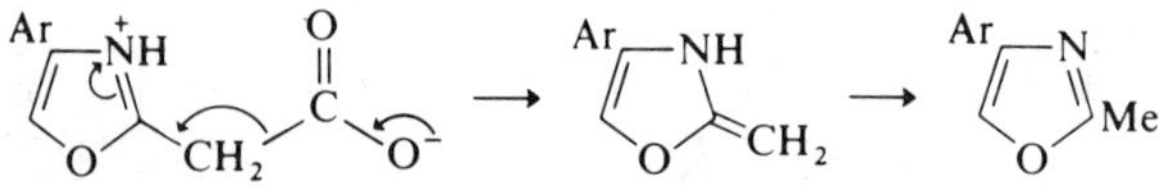

In most other reactions the azolecarboxylic acids and their derivatives behave as expected (cf. Scheme 52) ⟨37CB2309⟩, although some acid chlorides can be obtained only as hydrochlorides. Thus imidazolecarboxylic acids show the normal reactions: they can be converted into hydrazides, acid halides, amides and esters, and reduced by lithium aluminum hydride to alcohols ⟨70AHC(12)103⟩. Again, thiazole- and isothiazole-carboxylic acid derivatives show the normal range of reactions.

However, in some cases carboxylic acid-derived groups can participate in ring fission–reclosure reactions. Thus photolysis of 1,5-disubstituted tetrazole (**399**) gives nitrogen and appears to involve the amino-nitrene intermediate (**400**), which reacts further to give (**401**) ⟨77AHC(21)323⟩.

Scheme 52

(399) **(400)** **(401)**

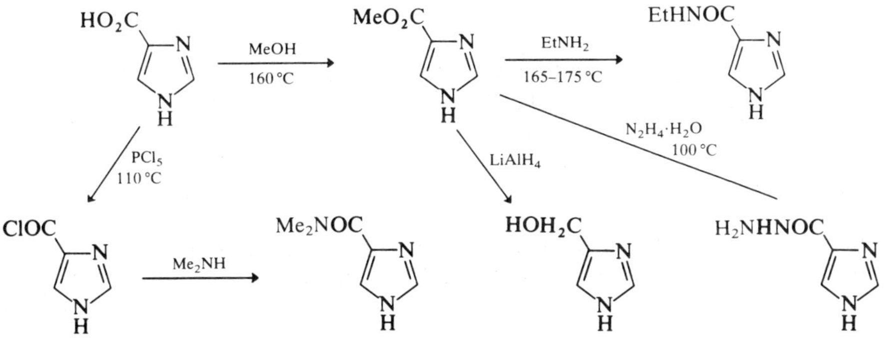

4.02.3.4.4 Aldehydes and ketones

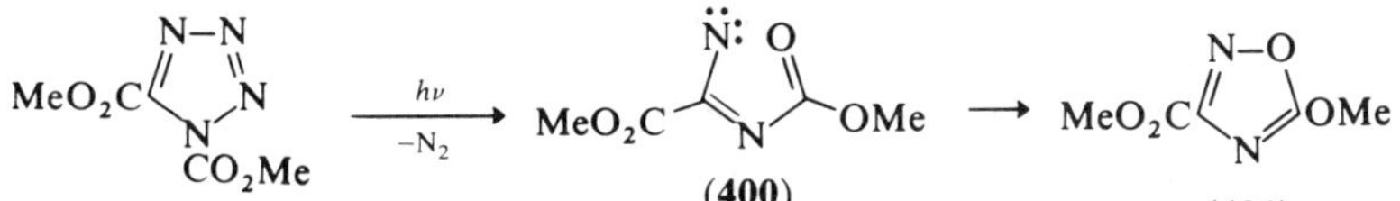

In general, the properties of these compounds and those of their benzenoid analogs are similar. Thus isothiazole aldehydes and ketones behave normally and form the usual derivatives ⟨72AHC(14)1⟩. Imidazole-2-carbaldehyde exists as a hydrate in aqueous solution. 4-Acetyloxazoles are oxidized to the corresponding acids with sodium hypobromite ⟨74AHC(17)99⟩. Thiazole aldehydes undergo the benzoin and Cannizzaro reactions. However, compounds with aldehyde groups α to an NH group sometimes form dimers, *e.g.* as in the 1,2,4-triazole series (**402**) ⟨70TL943⟩.

(402)

(403) **(404)**

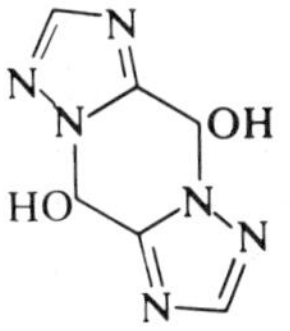

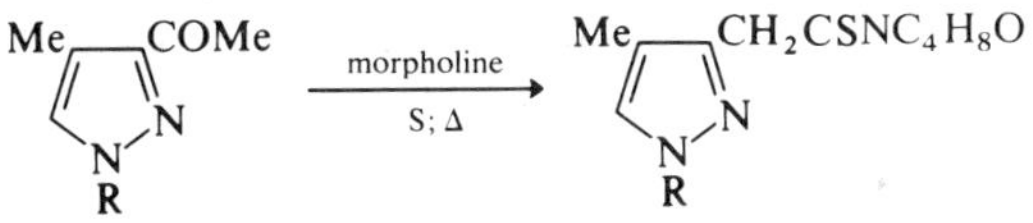

The Willgerodt reaction can proceed normally. Thus the 3-acetylpyrazole (**403**) is converted into the morpholide (**404**) ⟨57JCS2356⟩.

Deacylations are known. *C*-Acyl groups in 1,3,4-thiadiazoles are cleaved by sodium ethoxide in ethanol ⟨68AHC(9)165⟩. Imidazole-2-carbaldehyde behaves similarly, yielding imidazole and ethyl formate; this reaction involves an ylide intermediate. 3-Acylisoxazoles (**405**) are attacked by nucleophiles in a reaction which involves ring opening ⟨79AHC(25)147⟩.

(405)

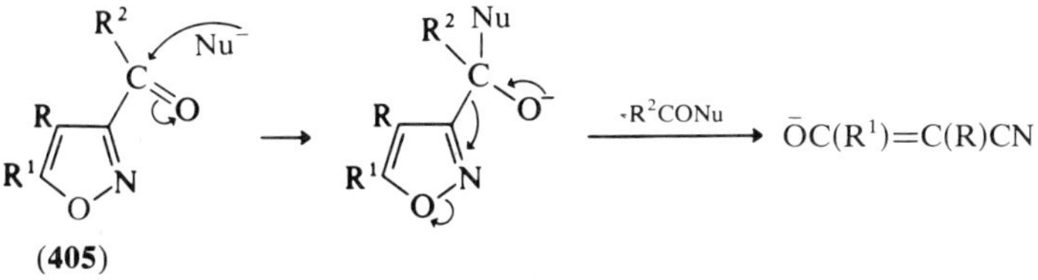

Sometimes ring opening and reclosure can occur with participation of a *C*-acyl group. Thus oxazole derivatives of type (**406**; X = H, Cl or NH_2; Y = OH or OEt) rearrange on heating to 255 °C by ring opening and recyclization ⟨74AHC(17)99⟩. 3-Acylanthranils (**407**) rearrange to benzoxazinones (**408**) on heating ⟨67AHC(8)277⟩.

(**406**) R = alkyl or aryl

(407) **(408)**

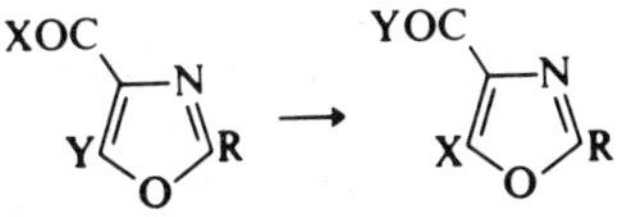

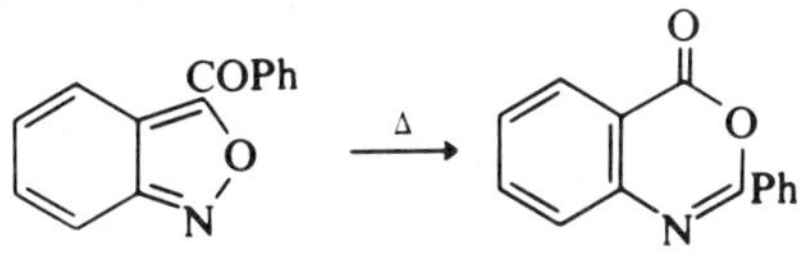

4.02.3.4.5 Vinyl and ethynyl groups

Such groups α to a pyridine-like nitrogen atom are expected to undergo Michael additions. Examples are known in the imidazole series.

4.02.3.4.6 Ring fission

Certain α-substituted alkyltetrazoles on pyrolysis yield nitrogen and an alkyne by the mechanism shown in Scheme 53 ⟨77AHC(21)323⟩.

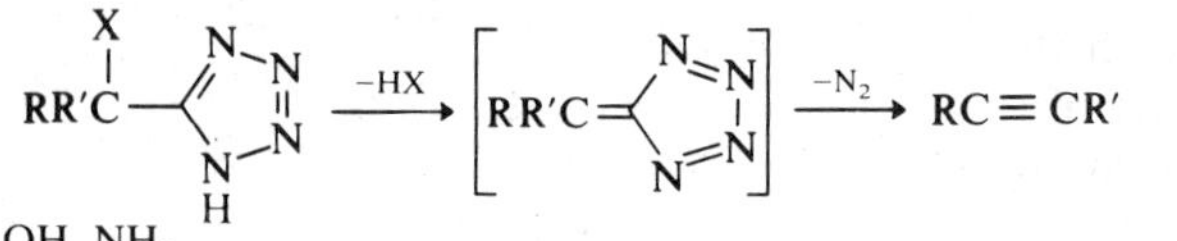

Scheme 53

When an azole carbene is formed, spontaneous ring fission can occur. The prototypes for these reactions are shown: (409) → (410), (411) → (412).

4.02.3.5 Aminoazoles

4.02.3.5.1 Dimroth rearrangement

The thermal acid- or base-catalyzed interconversion of 5-amino-1-phenyltriazoles (413) and 5-anilinotriazoles (415) was discovered by Dimroth. It is an example of a general class of heterocyclic rearrangements (416 ⇌ 417) now known by the name Dimroth rearrangements ⟨74AHC(16)33⟩. The original Dimroth rearrangement probably involves a tautomeric diazoimine intermediate (414) ⟨74AHC(16)33⟩. Electron-attracting and large groups tend to favor the tautomer in which they are on the exocyclic nitrogen. Alkyl groups tend to prefer to reside on the cyclic nitrogen ⟨74AHC(16)33⟩. 5-Aminotetrazoles similarly rearrange via azidoamidines (418).

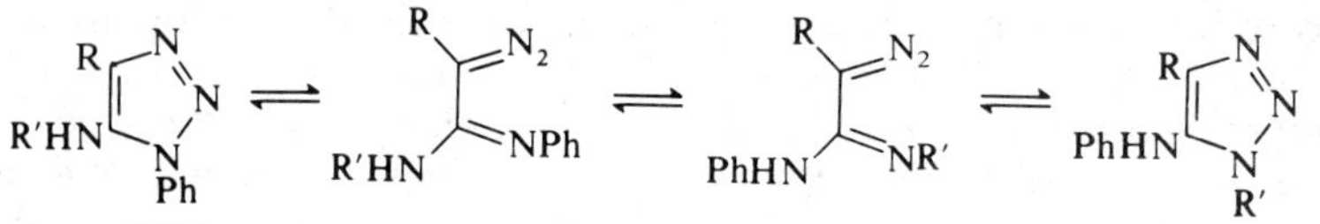

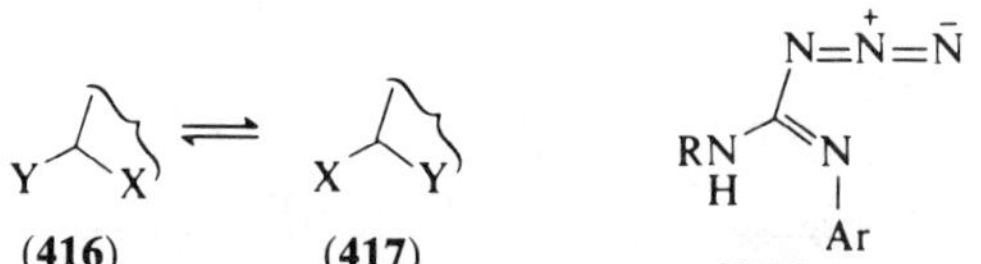

There are many related examples which are now known as the general Dimroth rearrangement. For example, 3-ethylamino-1,2-benzisothiazole (419) is in equilibrium in aqueous solution with the 2-ethyl-3-imino isomer (420) ⟨72AHC(14)43⟩. Dimroth rearrangements are known in the 1,2,4-thiadiazole series (421 → 422), and in the 1,3,4-thiadiazole series as products of reactions of halogeno-1,3,4-thiadiazoles; see Section 4.02.3.9.1 ⟨68AHC(9)165⟩. For a similar example in the 1,2,3,4-thiatriazole series, see Section 4.02.3.1.9.

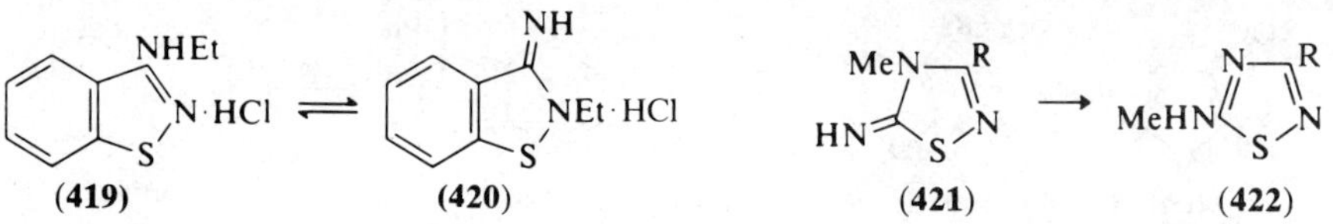

2-Amino-1,3,4-oxadiazoles (**423**) ring-open and the products immediately recycle to triazolinones (**424**) ⟨66AHC(7)183⟩.

4.02.3.5.2 Reactions with electrophiles (except nitrous acid)

In aminoazoles with the amino group α or γ to C=N, canonical forms of type (**425b**) increase the reactivity of the pyridine-like nitrogen atom toward electrophilic reagents, but decrease that of the amino group. Even when the amino group is β to C=N there is still a smaller electron flow in the same sense. Consequently, protons, alkylating agents and metal ions usually react with aminoazoles at the annular nitrogen atom (cf. Section 4.02.1.3). There are exceptions to this generalization, e.g. 4-aminoisothiazole is methylated to the 4-trimethylaminoisothiazole and both 3- and 4-dimethylaminopyrazoles are alkylated at the NMe$_2$ group.

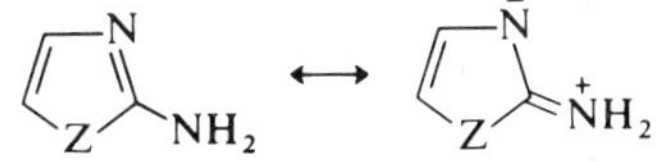

Other electrophilic reagents form products of reaction at the amino group. This occurs when initial attack at the pyridine-like nitrogen atom forms an unstable product which either dissociates to regenerate the reactants or undergoes rearrangement inter- or intramolecularly. In reactions of this type, carboxylic and sulfonic acid chlorides and anhydrides give acylamino- and sulfonamido-azoles, respectively. Thus 3-, 4- and 5-aminothiazoles form acetyl derivatives, sulfonamides and ureas. The 3- and 5-amino-1,2,4-thiadiazoles ⟨65AHC(5)119⟩ can be acylated and sulfonylated; 3-amino-1,2,5- ⟨68AHC(9)107⟩ and 2-amino-1,3,4-thiadiazoles ⟨68AHC(9)165⟩ also behave normally on acylation.

3-Amino-2,1-benzisothiazole (**426**) is acylated both at the cyclic and exocyclic nitrogen atoms to give (**427**) ⟨71AJC2405⟩. 5-Aminotetrazoles with nitric acid give nitramines. Sulfonation of the 5-aminopyrazole (**428**) gives first the expected product, (**429**), then a disulfonyl derivative (**430**), which rearranges on heating to the more stable (**431**). Aminothiazoles react with aldehydes to give Schiff bases.

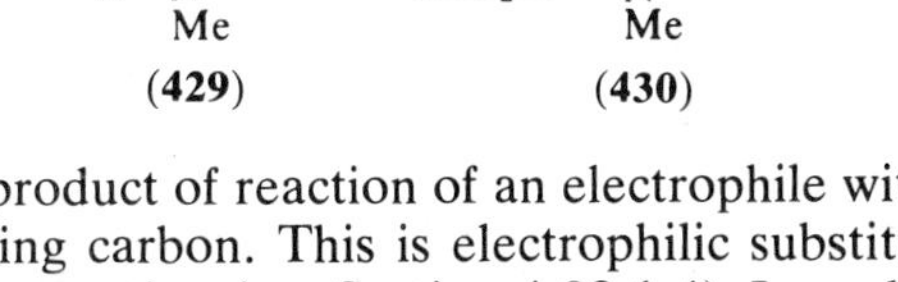

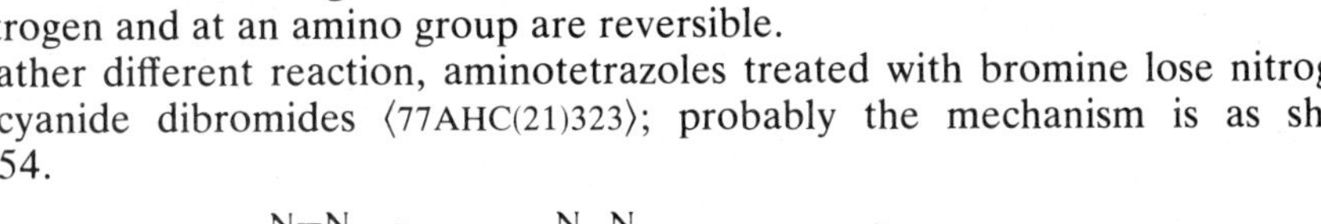

In still other cases, the product of reaction of an electrophile with an aminoazole is from electrophilic attack at a ring carbon. This is electrophilic substitution and is the general result of nitration and halogenation (see Section 4.02.1.4). In such cases, reactions at both cyclic nitrogen and at an amino group are reversible.

In a rather different reaction, aminotetrazoles treated with bromine lose nitrogen and give isocyanide dibromides ⟨77AHC(21)323⟩; probably the mechanism is as shown in Scheme 54.

Scheme 54

4.02.3.5.3 Reaction with nitrous acid: diazotization

Primary amino groups attached to azole rings react normally with nitrous acid to give diazonium compounds *via* primary nitroso compounds. However, the azole series shows two special characteristics: the primary nitroso compounds can be stable enough to be isolated, and diazo anhydrides are formed easily from azoles containing ring NH groups.

(i) Primary nitroso compounds

Attempted diazotization in dilute acid sometimes yields primary nitroso compounds. Reactions of 3- and 5-amino-1,2,4-thiadiazoles with sodium nitrite and acid give primary nitrosamines (*e.g.* **432 → 433**) ⟨65AHC(5)119⟩ which can be related to the secondary nitrosamines (**434**) prepared in the normal way. 1-Substituted 5-aminotetrazoles with nitrous acid give stable primary nitrosamines (**435**). Primary nitrosamines have been isolated in the imidazole series.

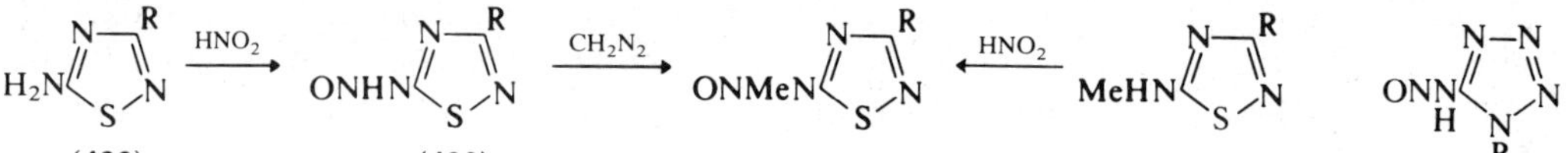

(ii) Diazo anhydrides

Diazotization of aminoazoles with free cyclic NH groups can give diazo anhydrides which show many of the normal reactions of diazoniums ⟨67AHC(8)1⟩. In the pyrazole series these diazo anhydrides are particularly stable.

3-Diazopyrazole (**436**) undergoes gas-phase thermal extrusion to form an azirine, probably by the mechanism shown ⟨81AHC(28)231⟩; 4-diazopyrazoles show normal diazonium-type reactions (Schemes 55 and 56) ⟨67AHC(8)1⟩. Analogous diazoimidazoles and diazopurines are known ⟨67AHC(8)1⟩.

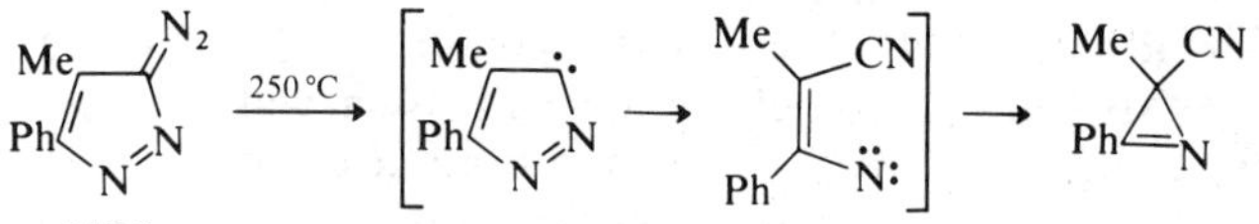

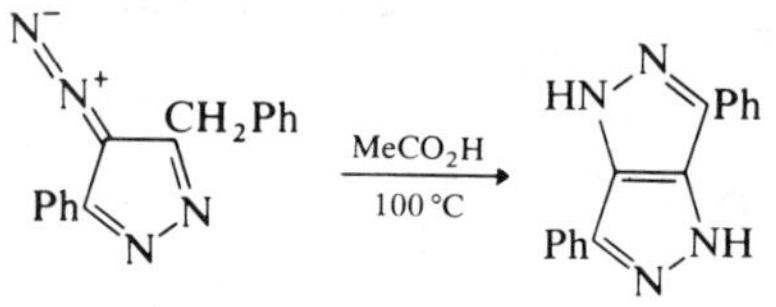

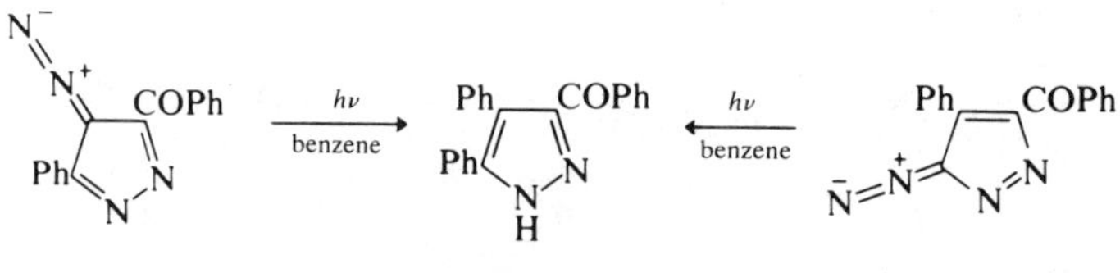

Diazotetrazole (**437**) has been prepared; on pyrolysis it yields carbon atoms and nitrogen ⟨79JA1303⟩.

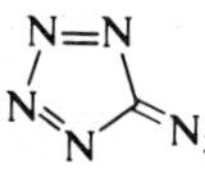

(iii) Diazonium salts

Pyridine-2- and -4-diazonium ions are far less stable than benzenediazonium ions. Azolediazonium salts generally show intermediate stability; provided diazotization is carried out in concentrated acid, many of the usual diazonium reactions succeed. Indeed, azolediazonium salts are often very reactive in coupling reactions.

2-Nitroimidazoles and 2-azidoimidazoles are available *via* the diazonium fluoroborates, and photolytic decomposition of the fluoroborates gives 2-fluoroimidazoles ⟨80AHC(27)241⟩.

3-Amino-2,1-benzisothiazole is readily diazotized to (**438**), which gives coupling products and the cyanide (**439**) ⟨72AHC(14)43⟩. Diazonium salts from 3-, 4- and 5-aminothiazoles undergo Sandmeyer reactions (to give halogenoisothiazoles), reductive deaminations and Gomberg–Hey reactions ⟨72AHC(14)1⟩. 5-Aminooxazoles can be satisfactorily diazotized, but the 2-amino compounds cannot ⟨74AHC(17)99⟩.

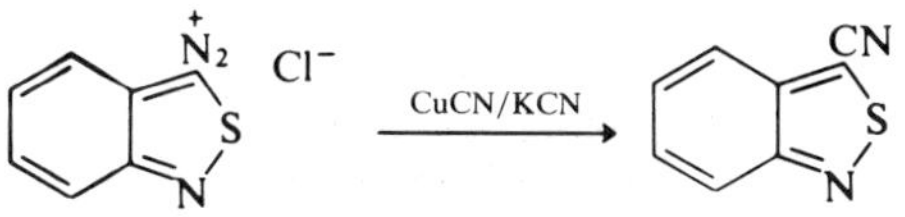

The 4- and 5-amino-1,2,3-triazoles are diazotizable, e.g. the diazonium salt from 4-aminotriazole-5-carboxamide with potassium iodide gives the 4-iodo derivative, and that from 4-amino-1,5-diphenyltriazole gives 1,5-diphenyltriazole in ethanol ⟨74AHC(16)33⟩.

In strong acid the 1,2,4-thiadiazole-3- and -5-diazonium salts have been prepared; the 5-derivatives are very reactive in coupling reactions and undergo Sandmeyer reactions. Diazonium salts from 3-amino-1,2,4-thiadiazoles are less reactive with coupling reagents ⟨65AHC(5)119⟩. Amino-1,3,4-thiadiazoles undergo diazotization smoothly provided the solution is sufficiently acidic. The diazonium salts (**440**) show remarkably strong coupling activity and will even couple with mesitylene ⟨68AHC(9)165⟩. 3-Amino-1,2,5-thiadiazole on attempted diazotization forms only the diazoamino compound (**441**) ⟨68AHC(9)107⟩.

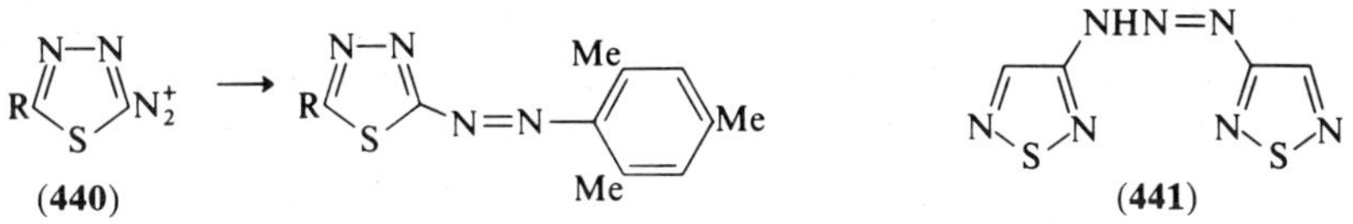

4.02.3.5.4 Deprotonation of aminoazoles

Canonical forms of type (**442b**) facilitate proton loss from the amino groups; the anions formed react easily with electrophilic reagents, usually preferentially at the exocyclic nitrogen atom (e.g. **443** → **444**) ⟨79HC(34-2)9⟩.

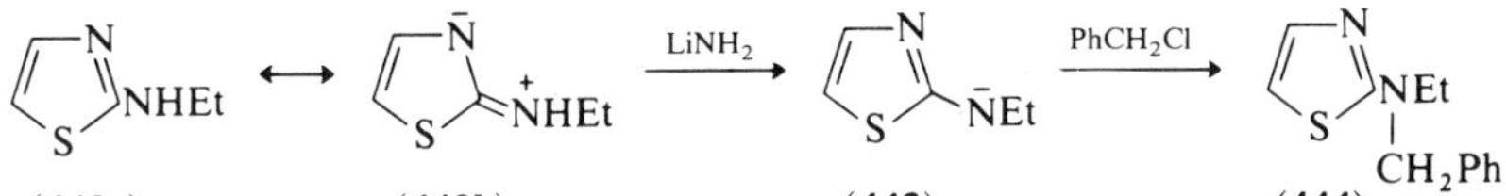

4.02.3.5.5 Aminoazolium ions/neutral imines

Amino groups on azolium rings can lose a proton to form strongly basic azolinimines, e.g. (**445**) yields (**446**). 2-Iminobenzothiazoline with acrylic acid yields (**447**).

In the 1,2-dithiole series such imines are readily isolated; they can be alkylated or protonated, e.g. (**448**) ⇌ (**449**) ⟨66AHC(7)39⟩.

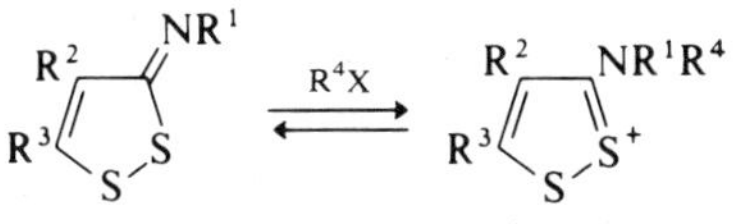

4.02.3.6 Other *N*-Linked Substituents

4.02.3.6.1 Nitro groups

Nitro groups on azole rings are often smoothly displaced by nucleophiles even more readily than are halogen atoms in the corresponding position. Thus 2,4,5-trinitroimidazole (**450**) is converted by HCl successively into (**451**) and (**452**) ⟨80AHC(27)241⟩.

(**450**) (**451**) (**452**)

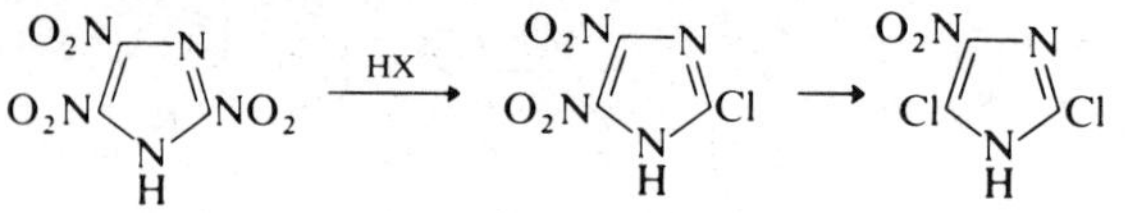

Nitro groups are easily reduced, catalytically or chemically, to give amino compounds, *e.g.* 4-nitroisothiazoles give the corresponding 4-amino derivatives ⟨72AHC(14)1⟩. In the pyrazole series, intermediate nitroso compounds can be isolated. Nitrosoimidazoles are also relatively stable.

4.02.3.6.2 Azidoazoles

The most important chemistry of azidoazoles is the fragmentation of derived nitrenes of which the prototypes are (**453**) → (**454**) and (**455**) → (**456**). Thus 5-azido-1,4-diphenyltriazole (**457**) evolves nitrogen at 50 °C ⟨70JOC2215⟩. 4-Azido-pyrazoles and -1,2,3-triazoles (**458**) undergo fragmentation with formation of unsaturated nitriles ⟨81AHC(28)231⟩.

(**453**) (**454**) (**455**) (**456**)

(**457**) (**458**) X = N or CR

(**459**) (**460**) (**461**) (**462**)

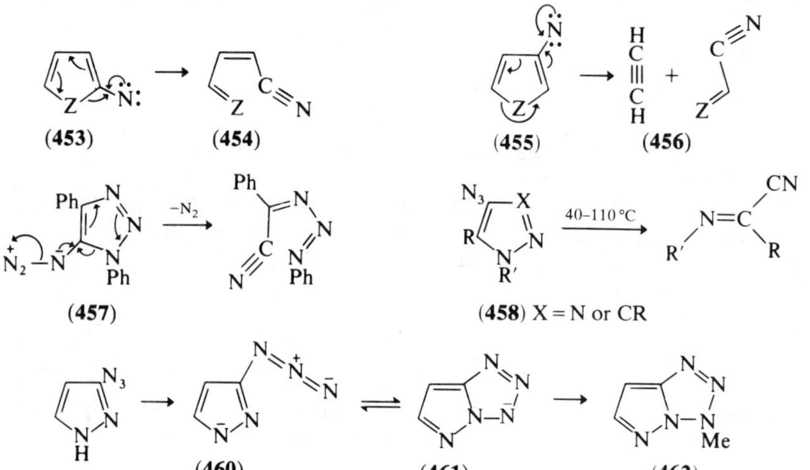

3-Azidopyrazoles exist as such (**459**), but their anions (**460**) are in equilibrium with tetrazole anions (**461**) which can be trapped as (**462**).

4.02.3.7 *O*-Linked Substituents

4.02.3.7.1 Tautomeric forms: interconversion and modes of reaction

As discussed in Section 4.01.5.2, hydroxyl derivatives of azoles (*e.g.* **463**, **465**, **467**) are tautomeric with either or both of (i) aromatic carbonyl forms (*e.g.* **464**, **468**) (as in pyridones), and (ii) alternative non-aromatic carbonyl forms (*e.g.* **466**, **469**). In the hydroxy 'enolic' form (*e.g.* **463**, **465**, **467**) the reactivity of these compounds toward electrophilic reagents is greater than that of the parent heterocycles; these are analogs of phenol.

(**463**) (**464**) (**465**) (**466**)

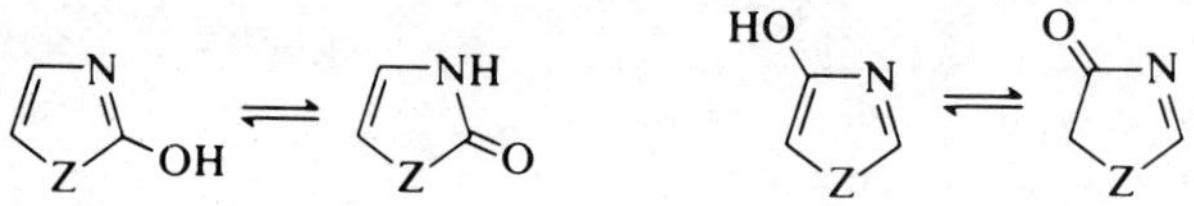

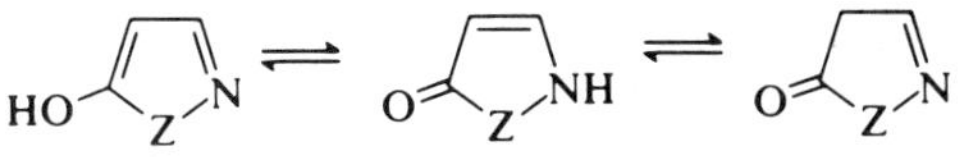

(467) (468) (469)

Interconversion of the hydroxyl and carbonyl forms of these heterocycles proceeds through an anion (as **471**) or a cation (as **472**), just as the enol (**474**) and keto forms (**477**) of acetone are interconverted through the ions (**475**) or (**476**). Reactions of the various species derived from the heterocyclic compounds are analogous to those of the corresponding species from acetone: hydroxyl forms react with electrophilic reagents (**478**) and carbonyl forms with nucleophilic reagents (**479**). In addition, either form can lose a proton (**480**, **481**) to give an anion which reacts very readily with electrophilic reagents on either oxygen (**482**) or carbon (**483**).

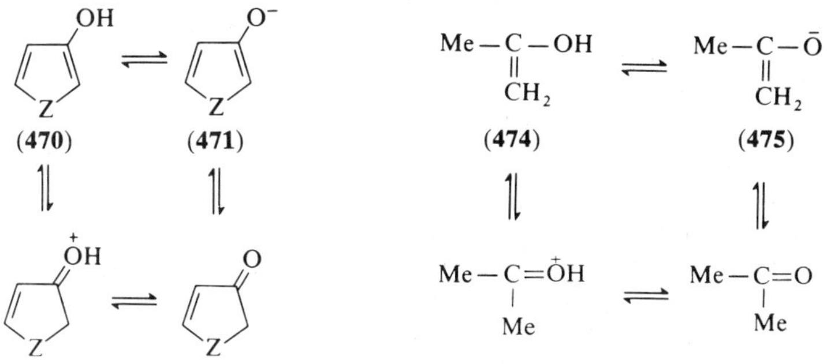

(470) (471) (474) (475)

(472) (473) (476) (477)

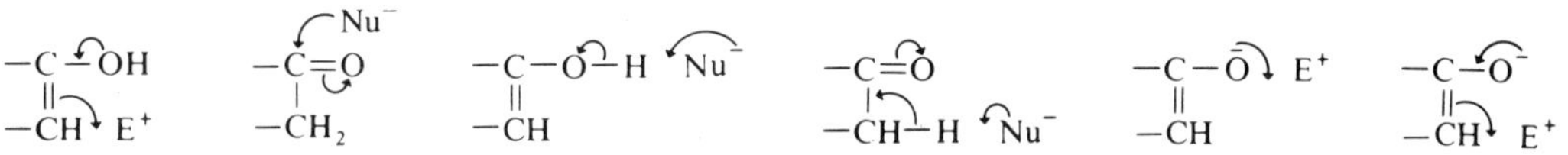

(478) (479) (480) (481) (482) (483)

The completely conjugated carbonyl forms are usually quite stable and highly aromatic in that after reaction they revert to type (Section 4.02.1.1.4). An overall treatment of their reactivity is given in Section 4.02.1.1.4. Electrophilic attack on the oxygen atom of the carbonyl group, and nucleophilic attack at the carbonyl carbon atom, in reactions which lead to substitution rather than ring opening are discussed in this section. Electrophilic attack at ring carbon (Section 4.02.1.4) and ring nitrogen (Section 4.02.1.3) and nucleophilic attack at ring carbon (Section 4.02.1.6) (other than C=O replacement) are discussed in the sections indicated.

4.02.3.7.2 2-Hydroxy, heteroatoms-1,3

2-Hydroxy-imidazoles, -oxazoles and -thiazoles (**484**; Z = NR, O, S) can isomerize to 2-azolinones (**485a**). These compounds all exist predominantly in the azolinone form and show many reactions similar to those of the pyridones. They are mesomeric with zwitterionic and carbonyl canonical forms (*e.g.* **485a** ↔ **485b**; Z = NR, O, S).

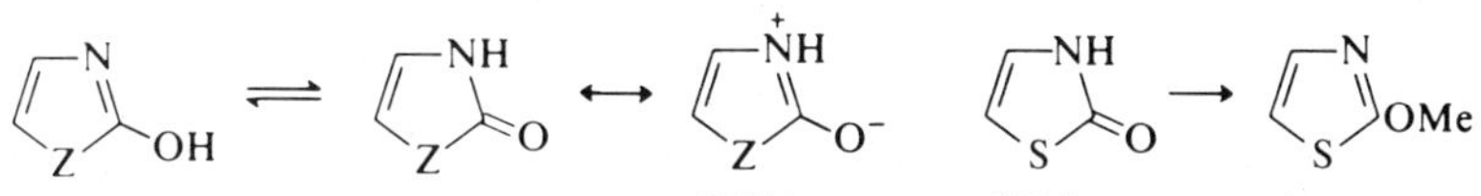

(484) (485a) (485b) (486) (487)

(i) Electrophilic attack on oxygen

2-Azolinones are protonated on oxygen in strongly acidic media. *O*-Alkylation of 2-azolinones can be effected with diazomethane; thiazolinone (**486**) forms (**487**). Frequently *O*- and *N*-alkylation occur together, especially in basic media where proton loss gives an ambident anion.

(ii) Nucleophilic displacements

2-Imidazolinones, 2-oxazolinones and 2-thiazolinones behave as cyclic ureas, thiocarbmates and carbamates, and predictably do not normally react with nucleophilic 'ketonic reagents' such as HCN, RNH$_2$, NaHSO$_3$, NH$_2$OH, N$_2$H$_4$, PhN$_2$H$_3$ or NH$_2$CON$_2$H$_3$. Stronger nucleophilic reagents, *i.e.* those of the type that attack amides, generally also react with azolinones. Thus they can be converted into chloroazoles with POCl$_3$ or PCl$_5$, *e.g.* (**489**) → (**488**). Similarly, bromoazoles may be prepared using PBr$_5$. Alkyl substituents on the azole nitrogen atom are usually lost in reactions of this type. Phosphorus pentasulfide converts carbonyl groups into thiocarbonyl groups (*e.g.* **489** → **490**).

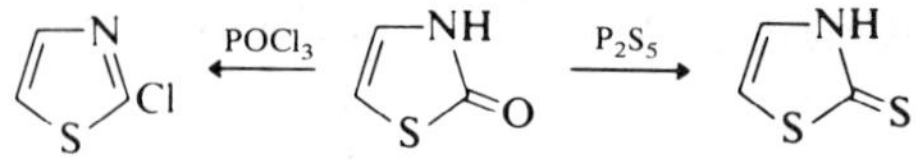

4.02.3.7.3 3-Hydroxy, heteroatoms-1,2

Pyrazoles, isoxazoles and isothiazoles with a hydroxyl group in the 3-position (**491**; Z = NR, O, S) could isomerize to 3-azolinones (**492**). However, these compounds behave as true hydroxy derivatives and show phenolic properties. They give an intense violet color with iron(III) chloride and form a salt (**493**) with sodium hydroxide which can be *O*-alkylated by alkyl halides (to give **494**; R = alkyl) and acylated by acid chlorides (to give **494**; R = acyl).

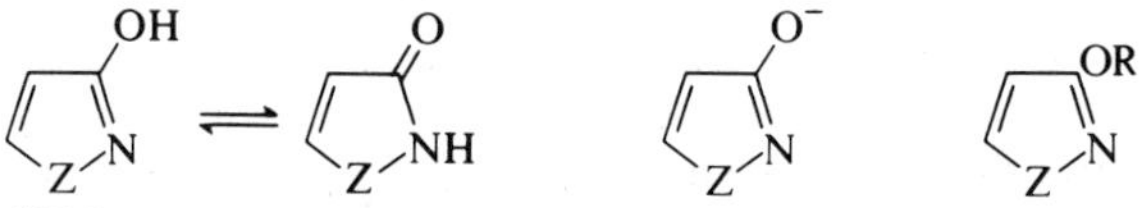

Sometimes compounds which exist predominantly in the hydroxyl form give products of *N*-methylation with diazomethane, for example 3-hydroxy-5-phenylisothiazole ⟨63AHC(2)245⟩; of course, the ambident anion (**493**) is an intermediate. 3-Hydroxypyrazoles, under rather severe conditions, can be converted into 3-chloropyrazoles with POCl$_3$ ⟨66AHC(6)347⟩.

4.02.3.7.4 5-Hydroxy, heteroatoms-1,2

5-Hydroxy-isoxazoles and -pyrazoles can tautomerize in both of the ways discussed in Sections 4.02.3.7.3 and 4.02.3.7.5 (**495** ⇌ **496** ⇌ **497**). The hydroxy form is generally the least stable; the alternative azolinone forms coexist in proportions depending on the substituents and the solvent, with non-polar media favoring the CH form (**497**) and polar media the NH form (**496**). The derived ambident anion can react with electrophiles at N, C or O depending on the reagent and conditions.

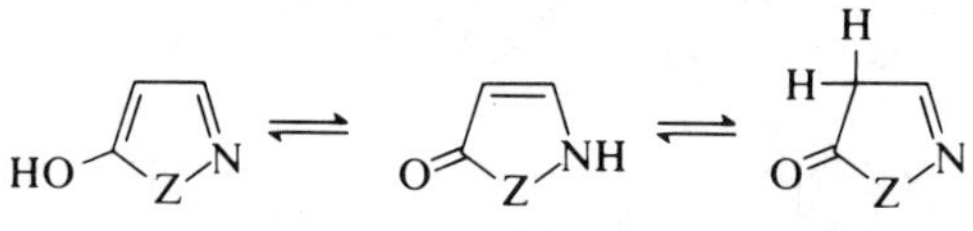

The hydroxyl groups of 5-hydroxypyrazoles (**498**) are readily replaced by halogens by the action of phosphorus halides.

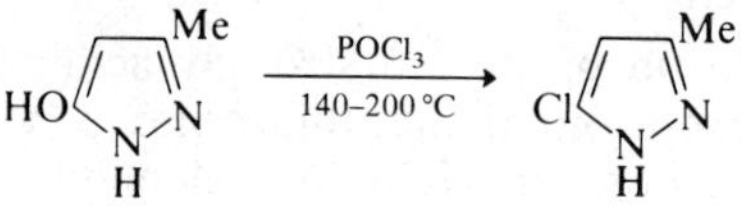

4.02.3.7.5 4- and 5-hydroxy, heteroatoms-1,3 and 4-hydroxy, heteroatoms-1,2

The 4- and 5-hydroxy-imidazoles, -oxazoles and -thiazoles (**499**, **501**) and 4-hydroxy-pyrazoles, -isoxazoles and -isothiazoles (**503**) cannot tautomerize to an aromatic carbonyl form. However, tautomerism similar to that which occurs in hydroxy-furans, -thiophenes and -pyrroles is possible (**499** ⇌ **500**; **503** ⇌ **504**; **501** ⇌ **502**), as well as a zwitterionic NH form (*e.g.* **505**). Most 4- and 5-hydroxy compounds of types (**500**) and (**502**) exist largely in these non-aromatic azolinone forms, although the hydroxyl form can be stabilized by chelation (*e.g.* **506**). The derived ambident anions react with electrophiles at O or C. Replacement of the hydroxyl group is sometimes possible provided electron-withdrawing groups are present as, for example, in 5-substituted 4-hydroxypyrazoles.

(**499**) (**500**) (**501**) (**502**)

(**503**) (**504**) (**505**) (**506**)

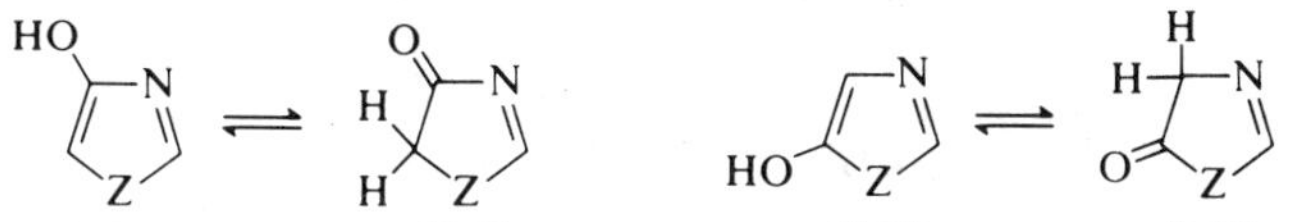

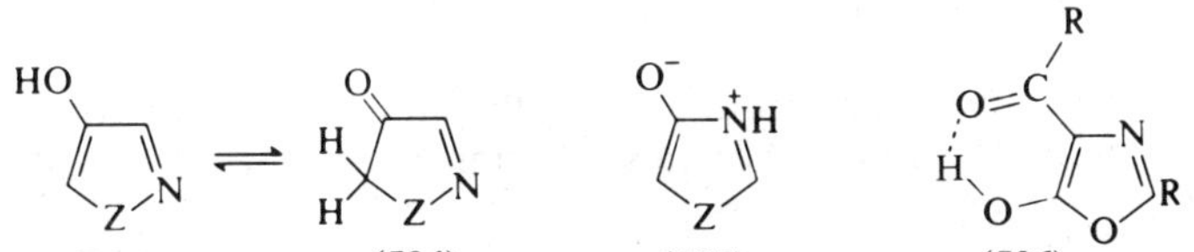

4-Hydroxy derivatives of type (**503**) show more phenolic character; thus 4-hydroxy-isothiazoles are normally *O*-methylated and *O*-acylated ⟨72AHC(14)1⟩.

Ring fission occurs readily in many of these compounds. For example, azlactones, *i.e.* 4*H*-oxazolin-5-ones containing an exocyclic C=C bond at the 4-position (**508**), are hydrolyzed to α-benzamido-α,β-unsaturated acids (**509**), further hydrolysis of which gives α-keto acids (**510**). Reduction and subsequent hydrolysis *in situ* of azlactones is used in the synthesis of α-amino acids (*e.g.* **508** → **507**).

(**507**) (**508**) (**509**) (**510**)

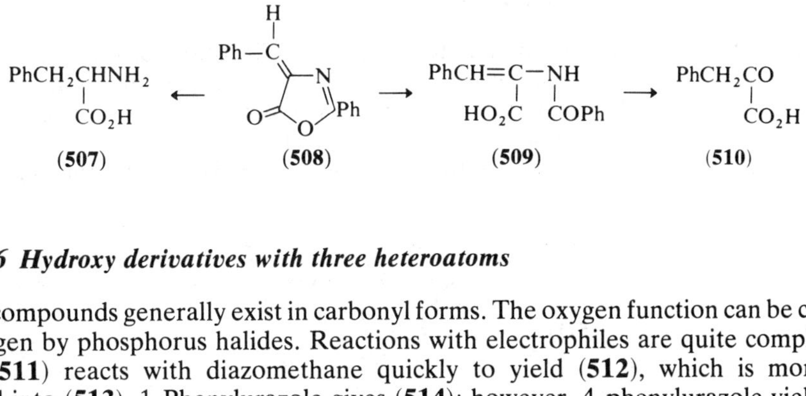

4.02.3.7.6 Hydroxy derivatives with three heteroatoms

These compounds generally exist in carbonyl forms. The oxygen function can be converted into halogen by phosphorus halides. Reactions with electrophiles are quite complex. Thus urazole (**511**) reacts with diazomethane quickly to yield (**512**), which is more slowly converted into (**513**). 1-Phenylurazole gives (**514**); however, 4-phenylurazole yields (**515**). Oxadiazolinones of type (**516**) can be alkylated at both O- and N-atoms.

(**511**) (**512**) (**513**)

(**514**) (**515**) (**516**)

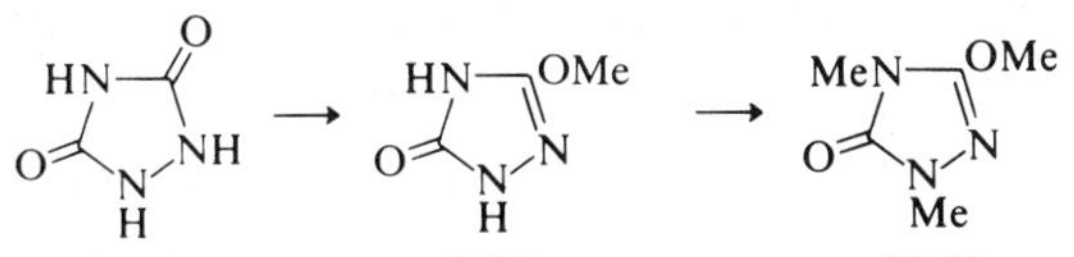

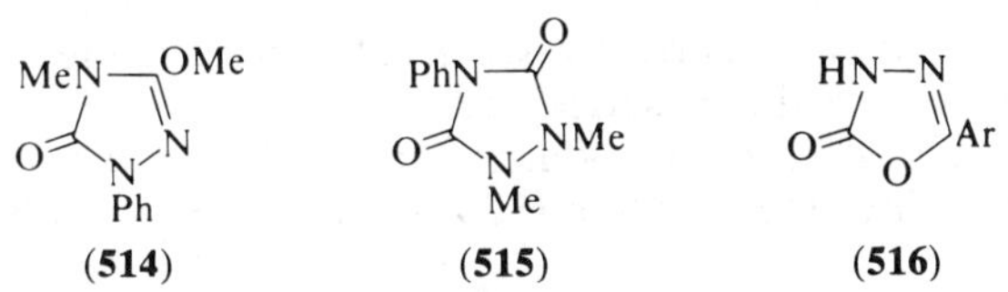

4.02.3.7.7 Alkoxy groups

The alkoxy groups in alkoxyazoles undergo easy dealkylation to the corresponding hydroxyazoleazolinone when several nitrogen atoms are present or when they are additionally activated by another substituent. Thus pyrazolyl ethers are cleaved under vigorous conditions, or more easily if a nitroso group is present. Nucleophilic displacement of alkoxy groups on cationic rings occurs readily, *e.g.* in quaternary 1,2,3-triazole ethers.

Azoles with alkoxy groups α to nitrogen can rearrange to *N*-alkylazolinones on heating; thus 2-alkoxy-1-methylimidazoles give 3-alkylimidazolin-2-ones and 2-methoxythiazoles behave similarly. *O*-Allyl groups rearrange considerably more readily, *e.g.* 2-allyloxybenzimidazole (**517**) gives 1-allyl-2-benzimidazolinone (**518**) at 180 °C ⟨67AHC(8)143⟩. 5-Allyloxypyrazoles undergo Claisen rearrangement of the allyl group to the 4-position.

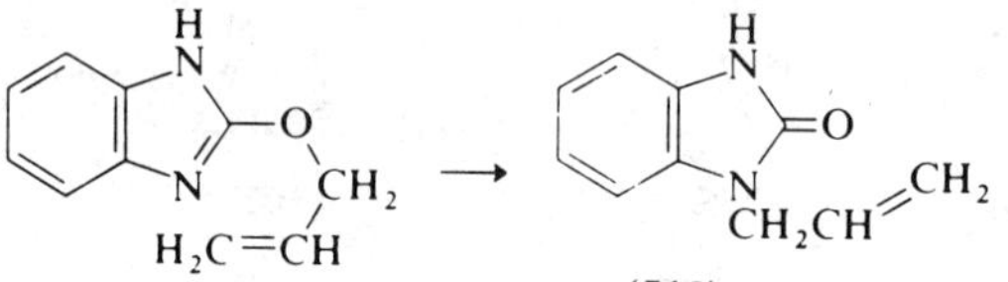

Aryl tetrazolyl ethers (**519**) are reduced by palladium on charcoal to give the arene and the tetrazolinone (**520**) ⟨77AHC(21)323⟩; this reaction is used for the removal of phenolic functionality.

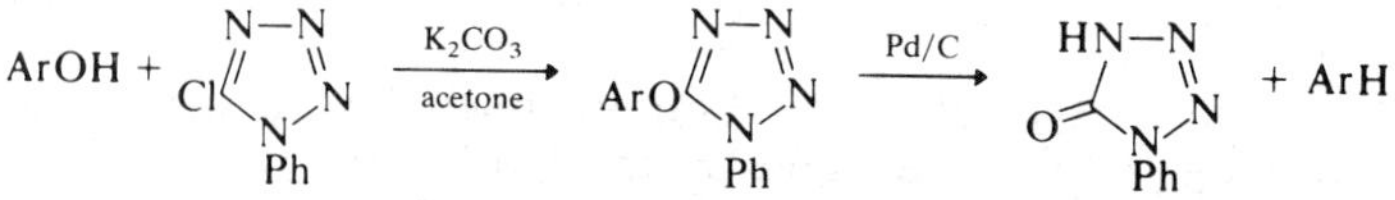

4.02.3.8 *S*-Linked Substituents

4.02.3.8.1 Mercapto compounds: tautomerism

Many mercaptoazoles exist predominantly as thiones. This behavior is analogous to that of the corresponding hydroxyazoles (*cf.* Section 4.02.3.7). Thus oxazoline-, thiazoline- and imidazoline-2-thiones (**521**) all exist as such, as do compounds of type (**522**). However, again analogously to the corresponding hydroxyl derivatives, other mercaptoazoles exist as such. 5-Mercaptothiazoles and 5-mercapto-1,2,3-triazoles (**523**), for example, are true SH compounds.

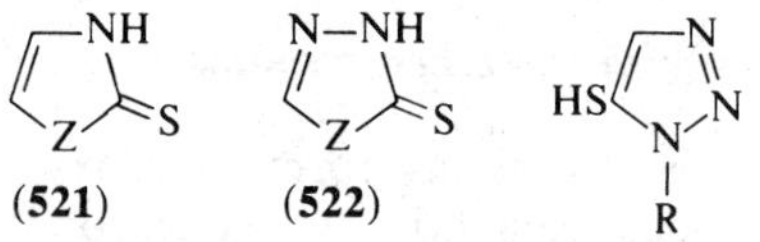

The pattern of reactivity is similar to that discussed for the azolinones in Sections 4.02.1.1.4 and 4.02.3.7.1. A difference is the greater nucleophilicity of sulfur, and thus more reaction of the ambident anion with electrophiles occurs at sulfur.

4.02.3.8.2 Thiones

Many azolinethiones show reactions typical of thioamides; in particular, they react with electrophiles at the sulfur atom.

(i) Alkyl halides give alkylthio derivatives, *e.g.* in the imidazoline-2-thione, thiazoline-2-thione and 1-arylpyrazoline-5-thione series.

(ii) Thiones are oxidized, *e.g.* by iodine, into disulfides. Thus 5-mercapto-1,2,3,4-thiatriazole is converted into the disulfide (**524**) ⟨64AHC(3)263⟩; similar behavior is known in the tetrazole series.

(524)

(525) (526)

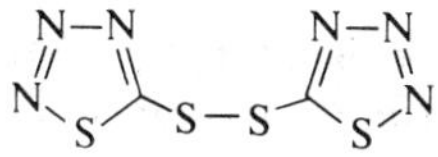

(iii) Thione groups can often be eliminated by oxidation; probably the sulfinic acid is the intermediate. Sometimes the sulfinic acid can be isolated (*e.g.* **525** → **526**), but more usually it spontaneously loses SO_2. In this way, thiazoline-2-thiones give thiazoles, 1,2-dithiole-3-thiones (**527**) are converted into 1,2-dithiolylium salts (**528**) and 1,3-dithiole-2-thiones (**529**) into 1,3-dithiolylium salts (**530**) ⟨66AHC(7)39⟩. In the pyrazole series (**531**) also loses an *N*-methyl group to yield (**532**).

(527) (528)

(529) (530)

(531) (532)

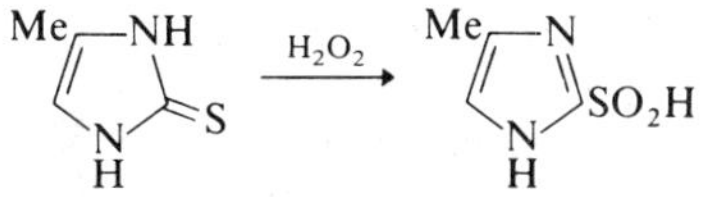

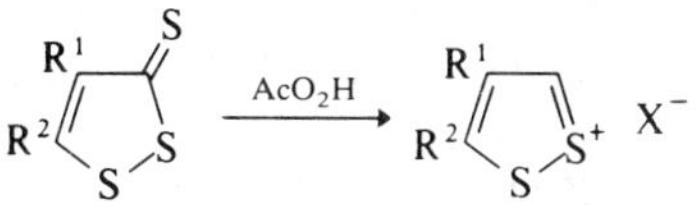

(iv) However, aryl-1,2,4-dithiazoline-3-thiones are oxidized to the 3-ones (**533** → **534**).

(533) (534)

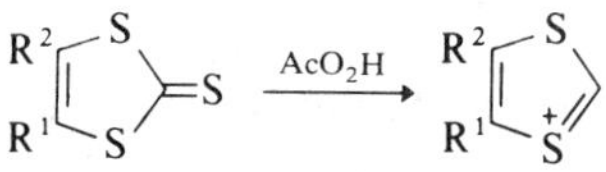

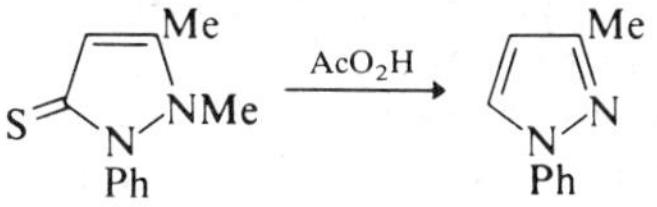

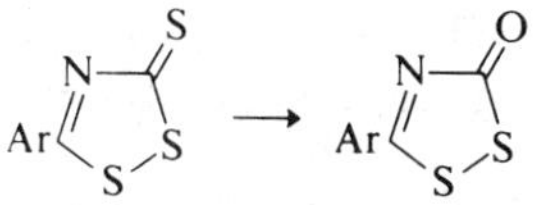

(v) Strong oxidation forms a sulfonic acid or betaine as, for example, in the pyrazole (**535** → **536**), thiazole and tetrazole series.

(535) (536)

(537) (538)

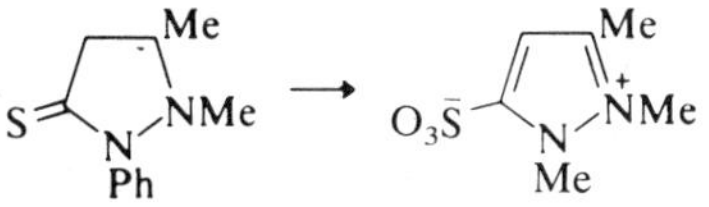

(vi) Cycloaddition across the C=S bond can lead to spiro derivatives, *e.g.* (**537**) → (**538**).

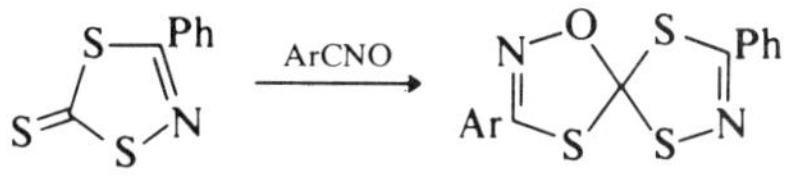

4.02.3.8.3 Alkylthio groups

2-Alkylthiothiazoles rearrange thermally into the 3-alkylthiazoline-2-thiones; in the imidazole series a thermal equilibrium is reached.

Alkylthio groups are oxidized to sulfoxides by H_2O_2 and readily by various oxidizing reagents to sulfones, *e.g.* in the imidazole series. The SR group is replaced by hydrogen with Raney nickel, and dealkylation is possible, *e.g.* of 3-alkylthio-1,2-dithiolyliums to give 1,2-dithiole-3-thiones by various nucleophiles ⟨80AHC(27)151⟩.

Alkylthio groups are replaced in nucleophilic substitutions. Such reactions are easy in cationic derivatives; for example, in the 1,2-dithiolylium series (**539**), substituted cyclopentadienyl ion gives fulvene derivatives (**540**) ⟨66AHC(7)39⟩. 2-Methylthio groups in 1,3-dithiolylium ions are substituted by primary amines or secondary amines ⟨80AHC(27)151⟩, and similar reactions are known for 2-alkylthiothiazoles.

(539) (540)

(541)

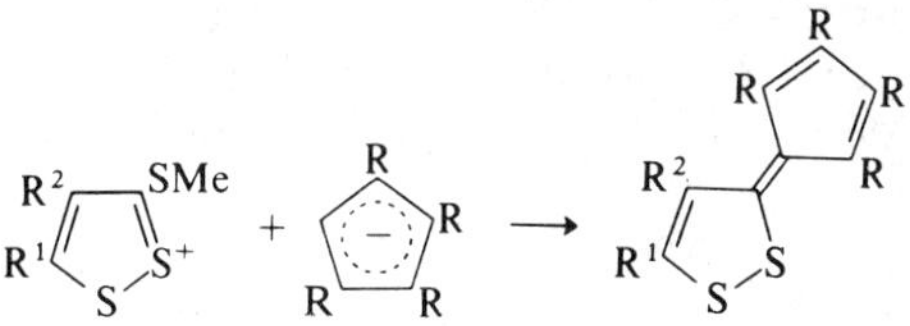

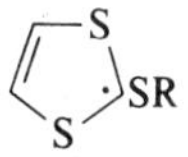

1,3-Dithiole-2-thiones trap radicals to give neutral stabilized radicals (**541**) ⟨80AHC(27)31⟩.

4.02.3.8.4 Sulfonic acid groups

Azolesulfonic acids frequently exist as zwitterions. The usual derivatives are formed, e.g. pyrazole-3-, -4- and -5-sulfonic acids all give sulfonyl chlorides with PCl_5. The sulfonic acid group can be replaced by nucleophiles under more or less vigorous conditions, e.g. by hydroxyl in imidazole-4-sulfonic acids at 170 °C, and by hydroxyl or amino in thiazole-2-sulfonic acids.

4.02.3.9 Halogen Atoms

4.02.3.9.1 Nucleophilic displacements: neutral azoles

As discussed in Section 4.02.3.1, nucleophilic replacements of halogen atoms are facilitated by mesomeric stabilization in the transition state for some halogenoazoles, depending on the number and orientation of the ring heteroatoms and halogen. Additional to this, and just as in benzene chemistry, all types of halogen atoms are activated toward nucleophilic displacement by the presence of other electron-withdrawing substituents. Halogen atoms in the 4- and 5-positions of imidazoles, thiazoles and oxazoles and those in all positions of pyrazoles and isoxazoles are normally rather unreactive, but are labilized by an α or γ electron-withdrawing substituent. Reactions of N-unsubstituted azoles containing a ring NH group are often difficult because of the formation under basic conditions of unreactive anions.

Halogen atoms in the 2-position of imidazoles, thiazoles and oxazoles (**542**) undergo nucleophilic substitution reactions. The conditions required are more vigorous than those used, for example, for α- and γ-halogenopyridines, but much less severe than those required for chlorobenzene. Thus in compounds of type (**542**; X = Cl, Br) the halogen atom can be replaced by the groups NHR, OR, SH and OH (in the last two instances, the products tautomerize; see Sections 4.02.3.7 and 4.02.3.8.1).

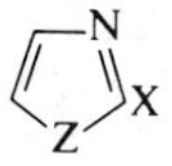

(**542**)

The 4- and 5-halogenoimidazoles and 4- and 5-halogenooxazoles are less reactive toward nucleophilic substitution than the 2-halogeno analogs, but still distinctly more reactive than unactivated phenyl halides. Thus a bromine atom in the 4- or 5-position of 1-methylimidazole requires lithium piperidide to react, whereas the 2-bromo analogue is converted into 2-piperidinoimidazole by piperidine at 200 °C. The chloro group of 5-chloro-4-nitroimidazole can be replaced by an alkylmercapto group ⟨70AHC(12)103⟩. The relative reactivities with respect to nucleophilic displacement increase in the order Cl < Br < I; fluoro compounds have been little studied but 4-fluoroimidazoles are relatively unreactive. 5-Halogenothiazoles react unexpectedly rapidly with methoxide, the 4-halogenothiazoles less readily.

3-Chloro-5-arylisoxazoles undergo nucleophilic displacement with alkoxide ion. Halogen atoms in the 5-position of the isoxazole nucleus are readily displaced if an activating group is present in the 4-position ⟨63AHC(2)365⟩.

Halogen atoms at the various positions of isothiazoles show considerable differences in reactivity. 5-Halogens, particularly when activated by an electron-withdrawing group such as nitro in the 4-position, readily undergo nucleophilic displacement to give hydroxy, alkoxy, alkylthio, amino and cyano derivatives. However, a 3-halogen atom, even when activated, is less reactive than a halogen in the 5-position, and replacement is often accompanied by ring cleavage, e.g. Scheme 57 ⟨72AHC(14)1⟩. 4-Halogenoisothiazoles are still less reactive, but can react with copper(I) cyanide to give the corresponding nitrile ⟨65AHC(4)107⟩.

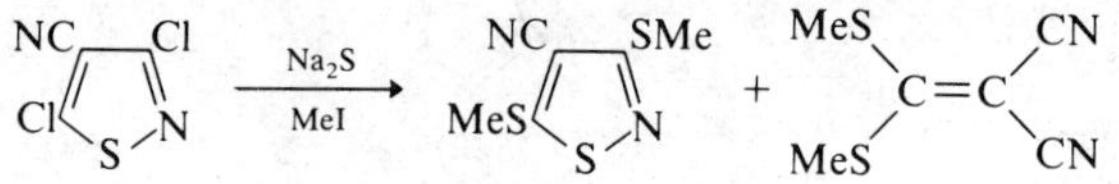

Scheme 57

Halogens attached to the pyrazole nucleus are normally very inert. If there is an electron-withdrawing group in the 4-position, then the halogen atom in the 5-position of a pyrazole ring becomes activated, for example Scheme 58 ⟨66AHC(6)347⟩. However, such an electron-withdrawing group in the 4-position only activates the chlorine atom in the 5-position and not one in the 3-position because of the influence of partial bond fixation ⟨66AHC(6)347⟩ (see discussion in Section 4.02.3.1).

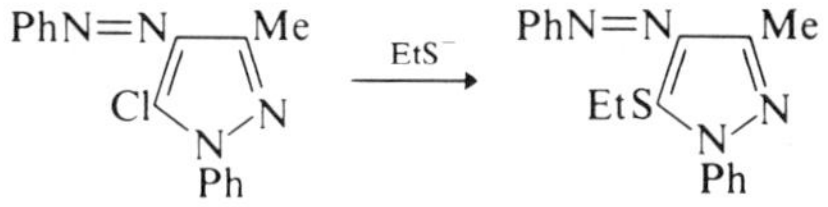

Scheme 58

5-Halogeno-1-methyl-1,2,3-triazoles undergo substitution reactions with amines, but the 4-halogeno analogs do not. 5-Chloro-1,4-diphenyl-1,2,3-triazole with sodium cyanide in DMSO gives the cyano derivative ⟨63JCS2032⟩. 1-Substituted 3-chloro- and 5-chloro-1,2,4-triazoles both react with amines.

5-Chlorine atoms in 1,2,4-oxadiazoles can be replaced by amino, hydroxy or alkoxy groups ⟨76AHC(20)65⟩. 5-Halogeno-1,2,4-thiadiazoles are also quite reactive: silver fluoride gives the fluorides, in concentrated hydrochloric acid a 5-hydroxy group is introduced, and thiourea reacts, as do various amines. Sodium sulfite gives sulfonic acids, and reactive methylene compounds give the expected substitution products ⟨65AHC(5)119⟩. By contrast, halogens in the 3-position of 1,2,4-thiadiazoles are inert toward most nucleophilic reagents: thus 3-chloro-5-phenyl-1,2,4-thiadiazole resists aminolysis and thiourea; however, a 3-alkoxy group is introduced by sodium alkoxide ⟨65AHC(5)119⟩.

Halogens on the 1,2,5-thiadiazole ring are highly reactive and easily converted into ethers by refluxing with alkoxides ⟨68AHC(9)107⟩. 2-Chloro-1,3,4-thiadiazole and benzylamine give a mixture of (**543**) and (**544**) ⟨68AHC(9)165⟩, the latter resulting from a Dimroth rearrangement (see Section 4.02.3.5.1). With hydrazine, (**545**) is similarly formed ⟨68AHC(9)165⟩.

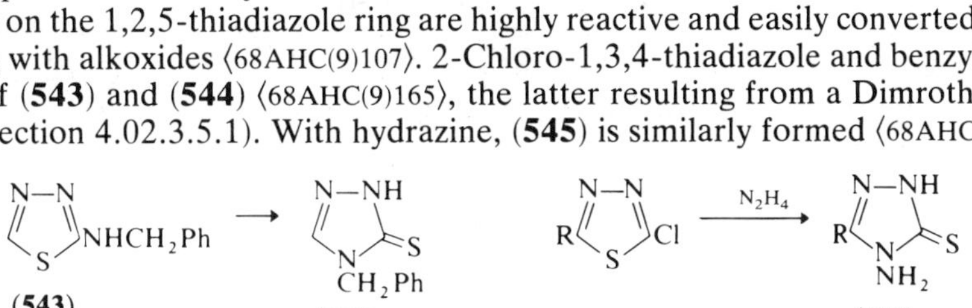

Halogen atoms at the 5-position of tetrazoles are reactive and easily replaced by nucleophiles. 5-Bromo-1-methyltetrazole is significantly more reactive than the 2-methyl isomer ⟨77AHC(21)323⟩.

4.02.3.9.2 Nucleophilic displacements: halogenoazoliums

Halogen atoms in cationic olium rings are very reactive. The halogen atom in the quaternary salts of 3- and 5-halogeno-1-phenylpyrazoles is replaced at 80–100 °C by hydroxyl, alkoxyl, thiol, amino or cyano groups ⟨66AHC(6)347⟩. 3-Halogeno-1,2-dithiolyliums are converted into 1,2-dithiol-3-ones by water and react readily with other nucleophiles ⟨80AHC(27)151⟩. Displacement of bromine from triazolium salts takes place easily, *e.g.* as in Scheme 59 ⟨74AHC(16)33⟩.

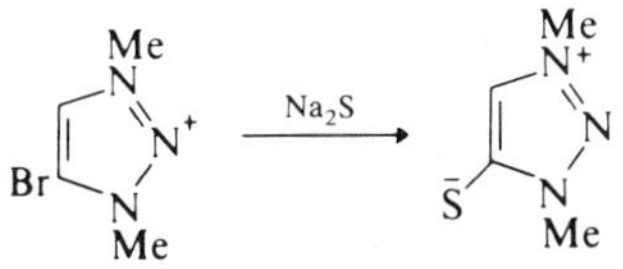

Scheme 59

4.02.3.9.3 Other reactions

Nuclear halogen atoms also show many of the reactions typical of aryl halogens. (i) They can be replaced with hydrogen atoms by catalytic (Pd, Ni, *etc.*) or chemical reduction (HI or Zn/H_2SO_4). For example, halogenopyrazoles with HI and red phosphorus at 150 °C give pyrazoles ⟨66AHC(6)347⟩, and 5-bromo-1,2,4-thiadiazole is reduced by Raney nickel to the parent heterocycle. 2-Bromothiazole can be reduced electrochemically. (ii) They give Grignard reagents; however, in the preparation of these it is sometimes necessary to add ethyl bromide to activate the magnesium ('entrainment method'). Pyrazolyl Grignard reagents have been obtained by the entrainment reaction ⟨66AHC(6)347⟩. 4-Iodoisoxazoles give Grignard reagents ⟨63AHC(2)365⟩. The 4- and 5-halogenooxazoles undergo halogen–metal exchange with *n*-butyllithium to give 4- and 5-lithiooxazoles ⟨74AHC(17)99⟩. Halogenothiazoles give Grignard reagents and lithio derivatives.

4.02.3.10 Metals and Metalloid-linked Substituents

Metalloid azoles frequently show expected properties, especially if not too many heteroatoms are present. Thus Grignard reagents prepared from halogen–azoles (see Section 4.02.3.9.3) show normal reactions, as in Scheme 60. 2-Lithioimidazoles react normally, *e.g.* with acetaldehyde (Scheme 61) ⟨70AHC(12)103⟩; 5-lithioisothiazoles (see Scheme 62) ⟨72AHC(14)1⟩ and 2-lithiothiazoles undergo many of the expected reactions.

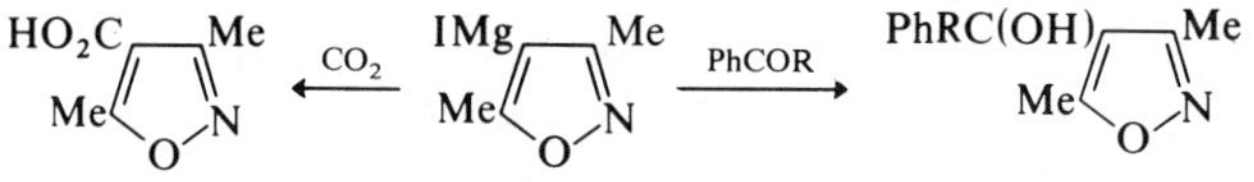

Scheme 60

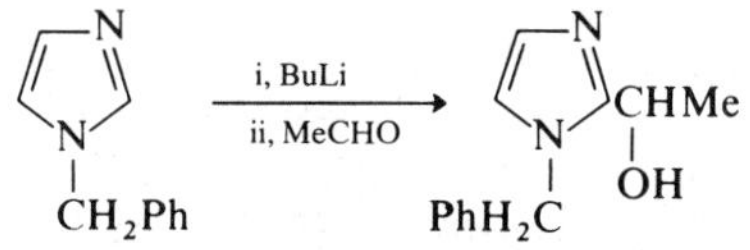

Scheme 61

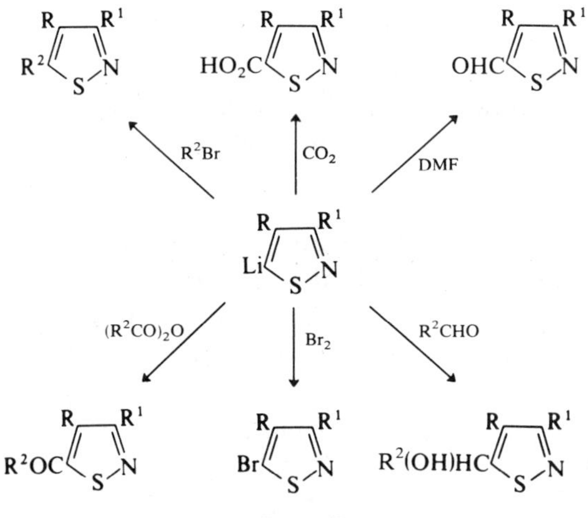

Scheme 62

However, as the number of heteroatoms increases, the stability decreases: the 5-lithio derivatives of 1,2,3-triazoles (**546**) ring-open spontaneously ⟨74AHC(16)33⟩. 1-Methyltetrazol-5-yllithium decomposes to nitrogen and lithium methylcyanamide above −50 °C, although it gives the expected Grignard-like reactions with bromocyanogen, esters, ketones and sulfur at lower temperatures ⟨77AHC(21)323⟩.

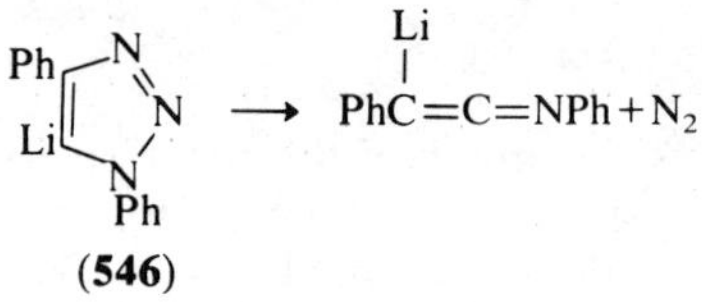

(**546**)

Acetoxymercurioxazoles ⟨74AHC(17)99⟩ and acetoxymercuriothiazoles with halogens give the corresponding halogenooxazoles in good yield. 4-Acetoxymercuriopyrazoles show many of the reactions of phenylmercury(II) acetate: removal by HCl, conversion to Br by bromine, and to SCH$_2$Ph by (SCN)$_2$/PhCH$_2$Cl.

4.02.3.11 Fused Heterocyclic Rings

A wide variety of such derivatives is known; their properties are usually those of the individual ring systems. However, some unique reactions arise from the special juxtaposition of the two rings, e.g. pyridotetrazoles (**547**) on photolysis yield cyanopyrroles (**548**) ⟨81AHC(28)231⟩.

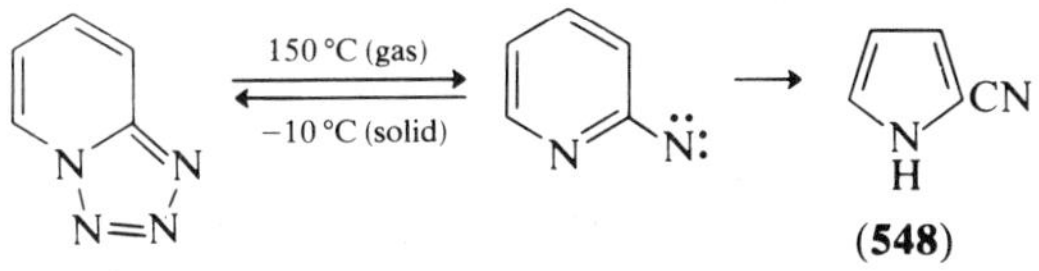

4.02.3.12 Substituents Attached to Ring Nitrogen Atoms

4.02.3.12.1 N-Linked azole as a substituent

It is instructive to consider N-substituted azoles in reverse, i.e. the azole ring as the substituent linked to some other group. Hammett and Taft σ-constant values for azoles as substituents are given in Table 11. The values show that all the azoles are rather weak net resonance donors, imidazole being the strongest. They are all rather strong inductive acceptors, with pyrazole considerably weaker in this respect than imidazole or the triazoles.

Table 11 Hammett and Taft σ-Constants for Azoles as Substituents in a Benzene Ring ⟨81JCR(S)364⟩

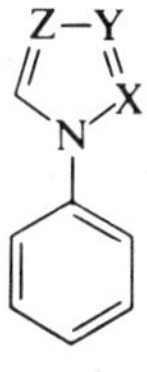

	Azole			σ_I	$\sigma_{R°}$
	X	Y	Z		
3	N	—	—	0.30	−0.06
2	—	N	—	0.51	−0.15
7	N	N	—	0.53	−0.10
5	N	—	N	0.53	−0.12
4	N	N	—	0.66	−0.10

N-Linked azole rings behave as good leaving groups, the more so the more nitrogen atoms contained in the ring (cf. Section 4.02.3.12.4).

4.02.3.12.2 Aryl groups

Electrophilic substitution occurs readily in N-phenyl groups, e.g. 1-phenyl-pyrazoles, -imidazoles and -pyrazolinones are all nitrated and halogenated at the para position. The aryl group is attacked preferentially when the reactions are carried out in strongly acidic media, where the azole ring is protonated.

The azole ring can activate metallation at the ortho position of an N-phenyl group, as in 1-phenylpyrazoles.

If the N-aryl group is strongly activated, then it can be removed in nucleophilic substitution reactions in which the azole anion acts as leaving group. Thus 1-(2,4-dinitrophenyl)pyrazole reacts with N_2H_4 or NaOMe.

On pyrolysis, 1-arylimidazoles rearrange to 2-arylimidazoles. In other systems pyrolysis causes more deep-seated changes. 1-Arylbenzotriazoles (**549**) on pyrolysis or photolysis give carbazoles (**550**) via intermediate nitrenes ⟨81AHC(28)231⟩. 1-Phenyl-1,2,4-triazole (**551**) pyrolyzes to isoindole (**552**) via a carbene intermediate ⟨81AHC(28)231⟩ and another example of participation of N-phenyl groups is found in the formation of benzimidazoles from tetrazoles (see Section 4.02.1.2.1). In the oxadiazolinone series (**553**), a nitrene intermediate (**554**) is also probably formed, which then ring closes ⟨70AHC(11)1⟩.

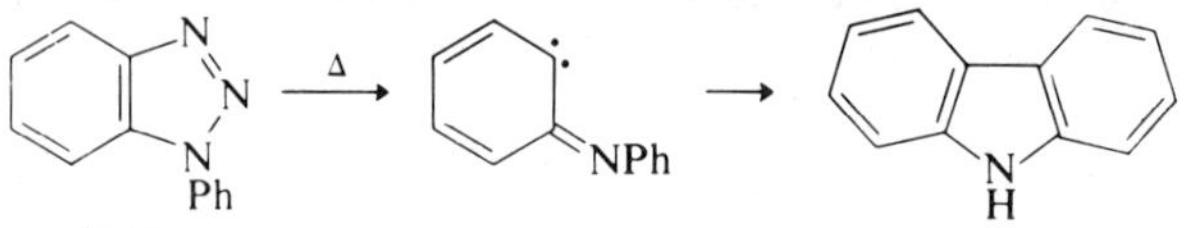

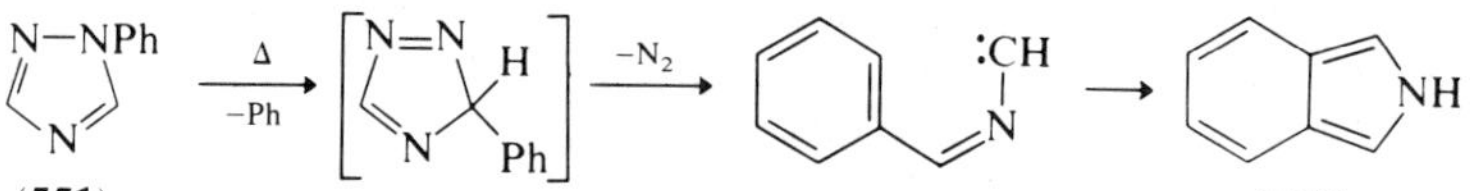

4.02.3.12.3 Alkyl groups

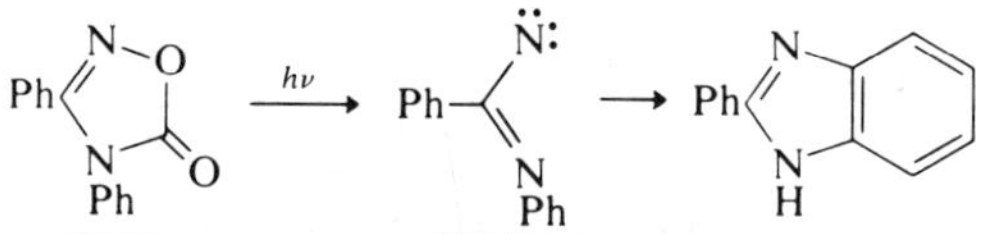

N-Alkyl groups in azolium salts can be removed by nucleophilic S_N2 reactions; soft nucleophiles such as PPh_3 and I^- are effective. Sometimes there is competition; for example, in (**555**) the methyl group is the more readily removed to give mainly N-ethylimidazole (**556**) ⟨80AHC(27)241⟩. This reaction has been studied quite extensively in the imidazolium series. The 1,2- and 1,3-dialkyltriazolium salts undergo nucleophilic displacement on heating ⟨74AHC(16)33⟩, and 2-alkylisothiazolium salts are reconverted into isothiazoles on distillation ⟨72AHC(14)1⟩. Pyrazolium salts similarly give pyrazoles. The benzyl group in 1-benzyl-1,2,3-triazoles is removed by reduction with sodium in liquid ammonia or catalytic reduction.

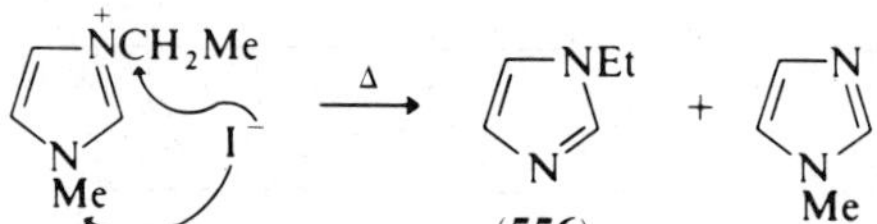

N-Alkyl groups in neutral azoles can rearrange thermally to carbon. For example, 2-alkylimidazoles can be prepared in this way in a reaction which is irreversible, uncatalyzed, intramolecular and does not involve radicals ⟨80AHC(27)241⟩. N-Vinyl and N-alkyl groups in imidazoles also rearrange thermally to the 2- and 4-ring positions.

Deprotonation can occur at the α-CH of azole N-alkyl groups: treatment of 1-methylpyrazole with n-BuLi followed by aldehydes gives products of type (**557**). Such proton loss is facilitated in cationic azido rings, and the ylides so formed sometimes undergo rearrangement. Thus quaternized 1,2-benzisoxazoles (**558**) lose a proton and then rearrange to 1,3-benzoxazines (**559**) ⟨67AHC(8)277⟩. Quaternized derivatives of benzofuroxan formed in situ undergo rearrangement to hydroxybenzimidazole N-oxides (**560**) ⟨69AHC(10)1⟩. Reactions of this type are also known for N-alkylazolinones.

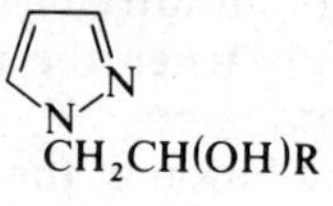

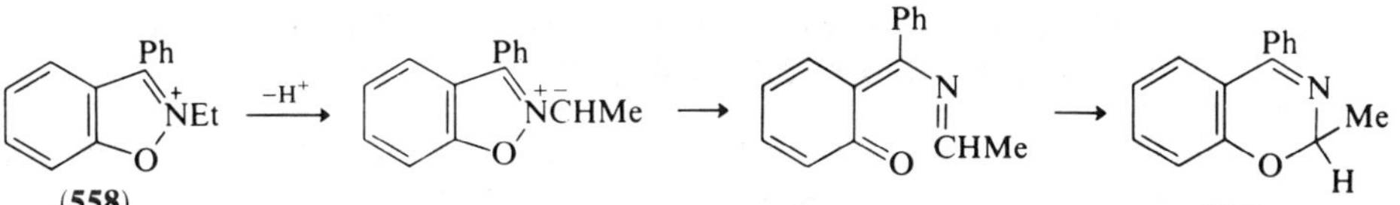

4.02.3.12.4 Acyl groups

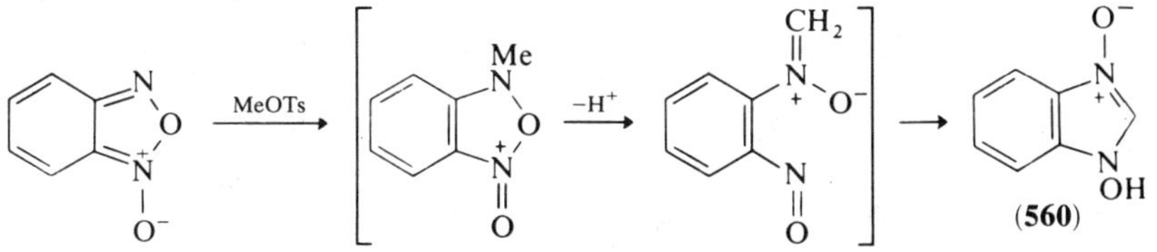

An azole ring is quite a good leaving group, far better than NR_2. Hence N-acylazoles are readily hydrolyzed. Their susceptibility to nucleophilic attack gives rise to the synthetic utility of compounds such as carbonyldiimidazole (**561**) which have been used, for example, in peptide syntheses. N-Acylazoles offer mild and neutral equivalents of acid chlorides. The leaving group ability of the azole ring increases with the number of nitrogen atoms it contains.

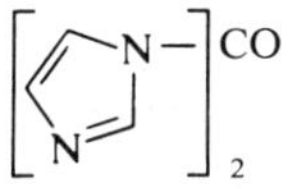

Acyl derivatives of azoles containing two different environments of nitrogen atoms can rearrange. For example, 1-acyl-1,2,3-triazoles are readily isomerized to the $2H$-isomers in the presence of triethylamine or other bases; the reaction is intermolecular and probably involves nucleophilic attack by N-2 of one triazole on the carbonyl group attached to another ⟨74AHC(16)33⟩.

2-Acyltetrazoles lose nitrogen spontaneously to give oxadiazoles, and thiadiazoles can be prepared similarly from 2-thioacyltetrazoles (Scheme 63) ⟨77AHC(21)323⟩.

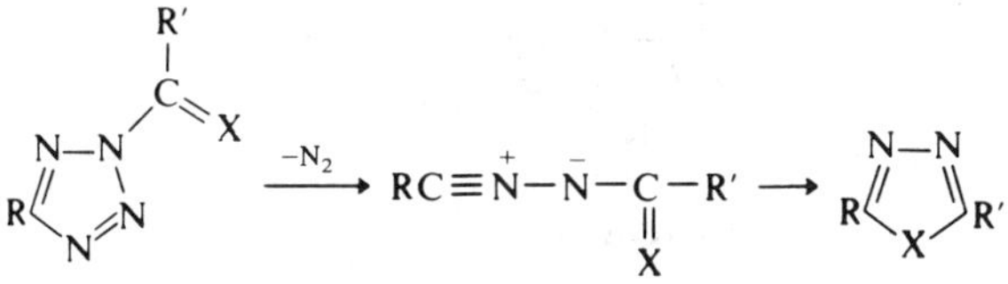

Scheme 63

4.02.3.12.5 N-Amino and N-nitro groups

N-Amino groups are replaced by hydrogen on treatment with nitrous acid (*e.g.* **562** → **563**) ⟨80AHC(27)241⟩ or phosphorus trichloride (1,2,4-triazole-4-acylimines are converted into triazoles ⟨74AHC(17)213⟩).

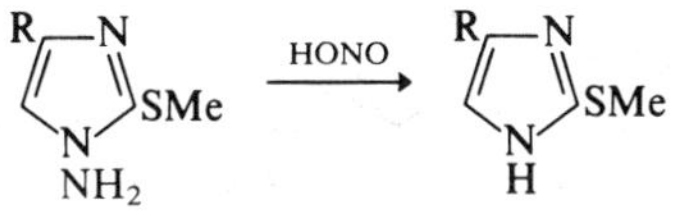

N-Aminoazoles can be oxidized to nitrenes which then fragment or ring expand in various ways. 1-Amino-1,2,3-triazoles (**564**) lose two moles of nitrogen to give alkynes ⟨74AHC(16)33⟩. N-Amino-triazoles (**565**) and -tetrazoles on oxidation with lead tetraacetate fragment to benzonitrile or benzyne ⟨81AHC(28)231⟩; however, the intermediate nitrene can be trapped as an aziridine (**566**). Similarly, the N-aminopyridazinotriazoles (**567**) undergo oxidative fragmentation to give open-chain compounds (**568**) ⟨81AHC(28)231⟩. *N*-Aminopyrazoles can ring expand to 1,2,3-triazines.

(**564**)

(**565**) (**566**)

(**567**) (**568**)

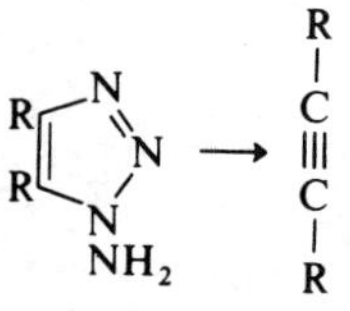

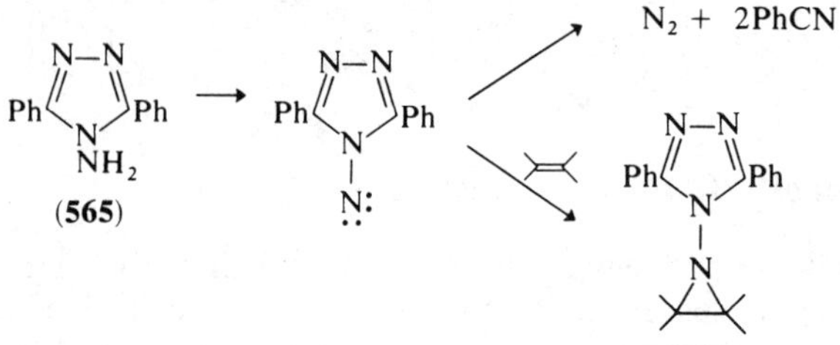

1-Nitropyrazoles rearrange to 4-nitropyrazoles in H_2SO_4 and to 3-nitropyrazoles thermally. Similar rearrangements are known for *N*-nitro-1,2,4-triazoles.

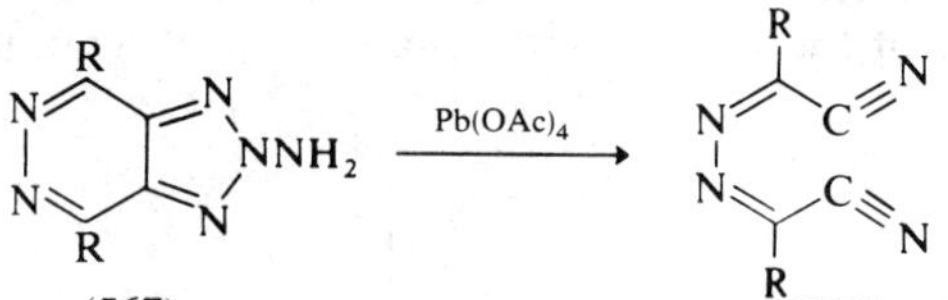

4.02.3.12.6 N-Hydroxy groups and N-oxides

Compounds of this type are tautomeric: in general, the *N*-oxide form (*e.g.* **570**) is favored by polar media, the *N*-hydroxy form (*e.g.* **569**) by non-polar media.

(**569**) (**570**)

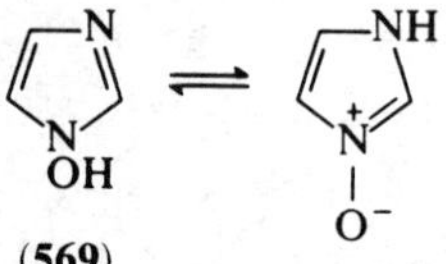

N-Hydroxy groups can be acetylated (Ac_2O) and *O*-alkylated in basic media by methyl iodide. 1-Hydroxypyrazole 2-oxides are quite strong acids.

Azole *N*-oxide groups are readily removed by reduction with Zn/HOAc, HI or PCl_3, *e.g.* in the pyrazole series. 1,2,3-Thiadiazole 3-oxides isomerize on irradiation to the corresponding 2-oxides.

Oxidation of *N*-hydroxyazoles can give cyclic nitroxyls (*e.g.* **571–573**) ⟨79AHC(25)205⟩.

(**571**) (**572**) (**573**)

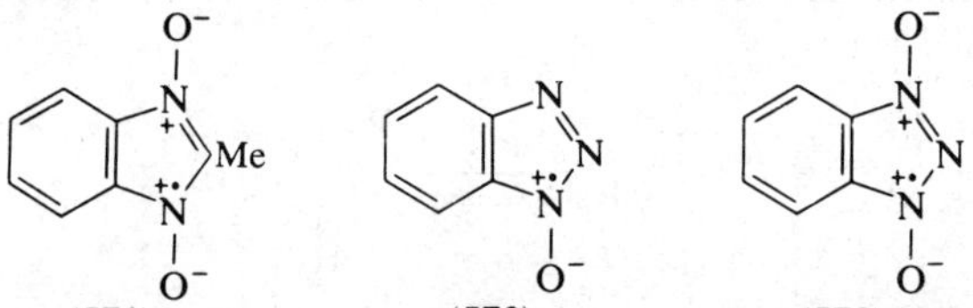

4.02.3.12.7 N-Halogeno groups

Generally these derivatives are rather unstable and behave as oxidizing and halogenating agents. 1-Iodoimidazoles are more stable than other analogs.

4.03

Synthesis of Five-membered Rings with Two or More Heteroatoms

K. T. POTTS

Rensselaer Polytechnic Institute, New York

4.03.1 INTRODUCTION	112
4.03.2 SYNTHESES USING ALDOL-TYPE CONDENSATIONS	112
4.03.2.1 *Intramolecular Condensations*	112
4.03.2.1.1 Ring systems with two heteroatoms	112
4.03.2.1.2 Ring systems with three heteroatoms	115
4.03.2.1.3 Ring-fused systems	116
4.03.2.2 *Intermolecular Condensations*	118
4.03.2.2.1 Ring systems with two heteroatoms	118
4.03.2.2.2 Ring-fused systems	119
4.03.2.2.3 Ring systems with two adjacent heteroatoms	121
4.03.3 SYNTHESES UTILIZING INTERMOLECULAR NUCLEOPHILIC DISPLACEMENTS	122
4.03.3.1 *Classification of Reaction Types*	122
4.03.3.2 *Design of the Appropriate Bielectrophile*	123
4.03.3.2.1 1,1-Bielectrophiles	123
4.03.3.2.2 1,2-Bielectrophiles	123
4.03.3.2.3 1,3-Bielectrophiles	124
4.03.3.2.4 1,4- and higher bielectrophiles	125
4.03.3.3 *Syntheses Using 1,1-Bielectrophiles*	125
4.03.3.3.1 Ring systems with two heteroatoms	125
4.03.3.3.2 Ring systems with three or more heteroatoms	126
4.03.3.3.3 Ring-fused systems	128
4.03.3.4 *Syntheses Using 1,2-Bielectrophiles*	129
4.03.3.4.1 Ring systems with two heteroatoms	129
4.03.3.4.2 Ring systems with three or more heteroatoms	130
4.03.3.4.3 Ring-fused systems	131
4.03.3.5 *Syntheses Using 1,3-Bielectrophiles*	131
4.03.3.5.1 Ring systems with two heteroatoms	131
4.03.3.5.2 Ring systems with three or more heteroatoms	132
4.03.3.5.3 Ring-fused systems	133
4.03.4 INTRAMOLECULAR RING CYCLIZATIONS	133
4.03.4.1 *Oxidative Ring Closure Reactions*	133
4.03.4.1.1 C—N bond formation	133
4.03.4.1.2 N—N bond formation	134
4.03.4.1.3 C—S bond formation	135
4.03.4.1.4 N—S bond formation	135
4.03.4.1.5 O—C bond formation	136
4.03.4.1.6 O—N bond formation	137
4.03.4.1.7 S—S, S—Se and Se—Se bond formation	137
4.03.4.2 *Electrophilic Ring Closures* via *Acylium Ions and Related Intermediates*	138
4.03.4.3 *Ring Closures* via *Intramolecular Alkylations*	139
4.03.4.4 *Ring Closures* via *Nucleophilic Additions*	139
4.03.4.5 *Ring Closures* via *Radical Intermediates*	141
4.03.5 SYNTHESES INVOLVING CONJUGATE ADDITION REACTIONS	142
4.03.6 RING FORMATION *VIA* CYCLOADDITION REACTIONS	143
4.03.6.1 *1,3-Dipolar Cycloadditions*	143
4.03.6.1.1 Monocyclic systems with two or more heteroatoms	146
4.03.6.1.2 Ring-fused systems	147
4.03.6.1.3 Cycloadditions using mesoionic ring systems	149
4.03.6.2 *1,4-Dipolar Cycloadditions*	150
4.03.6.3 *1,5-Dipolar Cycloadditions*	151

4.03.7 HETEROCYCLIC RING INTERCONVERSIONS	153
4.03.7.1 Small Rings as Dienophiles or Dipolarophiles	153
4.03.7.2 Small Rings as a Source of Ylides	153
4.03.7.3 Small Rings as Substrates for Ring-opening Reactions	154
4.03.7.3.1 Three-membered rings	154
4.03.7.3.2 Four-membered rings	155
4.03.7.4 Interconversion of Five-membered Rings	156
4.03.7.5 Interconversion of Six- and Five-membered Rings	157
4.03.7.6 Ring Interconversions Involving Side Chains	158
4.03.8 PHOTOCHEMISTRY IN HETEROCYCLIC SYNTHESIS	159
4.03.8.1 Elimination Processes	159
4.03.8.1.1 Loss of nitrogen	159
4.03.8.1.2 Loss of carbon dioxide	159
4.03.8.2 Photochemically Induced Isomerizations	160
4.03.8.3 Carbon–Carbon Bond Formation	161
4.03.8.3.1 Intramolecular reactions	161
4.03.8.3.2 Intermolecular reactions	162
4.03.9 REACTIVE, ELECTRON-DEFICIENT INTERMEDIATES	162
4.03.9.1 Carbenes	162
4.03.9.2 Nitrenes	163
4.03.10 USE OF YLIDES IN HETEROCYCLIC SYNTHESIS	164
4.03.10.1 Sulfonium Ylides	164
4.03.10.2 Phosphonium Ylides	165
4.03.10.3 Sulfimides	166

4.03.1 INTRODUCTION

Synthetic procedures leading to five-membered rings containing two or more heteroatoms have been reported in the literature for over a century, and the development of these syntheses in many ways reflects the development of organic chemistry. The vast number of five-membered ring systems that incorporate two or more heteroatoms, together with their ring-fused derivatives, presents a considerable challenge for a systematic treatment of their synthesis as the methods utilized embrace the majority of the reaction processes used in organic chemistry.

The synthesis of these five-membered ring systems has classically been discussed in terms of the ring system formed. In recent years many synthetic procedures have been classified in terms of the bonds being formed and the position and nature of the heteroatoms involved. Both methods have their advantages, and also their drawbacks.

In this general chapter on the synthesis of the ring systems described in Volumes 5 and 6 of this work, attention is focused on the reactions involved in forming the bonds involved in making up the heterocycle. This method has many advantages and provides a convenient framework for discussion of not only the heterocycles described in these two volumes, but also of those in the companion volumes in the series. Although the classification of the reaction types is somewhat arbitrary and occasional overlap occurs, this approach emphasizes that heterocyclic synthesis is a vital part of organic synthesis in general, and it is particularly useful as a basis for devising new synthetic approaches to a ring system with a desired substituent pattern. It was initially developed by the author as part of a successful short course on heterocyclic synthesis that was presented at many locations in the United States over the past few years.

4.03.2 SYNTHESES USING ALDOL-TYPE CONDENSATIONS

4.03.2.1 Intramolecular Condensations

4.03.2.1.1 Ring systems with two heteroatoms

Modification of reaction schemes leading to furans, thiophenes and pyrroles by incorporation of a heteroatom into the substrate provides ready access to imidazoles, thiazoles, oxazoles, dithioles, oxathioles, *etc.* Incorporation of two heteroatoms into the substrates enables triazoles, thiadiazoles, oxadiazoles, *etc.* to be obtained. These general reaction schemes are shown in Scheme 1.

Scheme 1 General reaction schemes for the synthesis of five-membered heterocycles containing two or more heteroatoms by aldol-related reactions

Oxazoles may be readily prepared in this manner in good yields and under relatively mild conditions. Thus, 5-ethoxy-4-methyloxazole (**3**) was obtained by treating ethyl 2-formamidopropionate (**2**), itself prepared from alanine (**1**) by formylation and esterification, with phosphorus pentoxide in chloroform at 55 °C ⟨72JCS(P1)909, 914⟩. Known collectively as the Robinson–Gabriel synthesis, these cyclodehydrations can be effected by a variety of acidic ring closure reagents, including sulfuric acid, phosphorus pentachloride, phosgene or anhydrous hydrogen fluoride. α-Acylaminoketones and related systems undergo similar ring closures, and examples illustrating the variety of substituents which may be introduced into the oxazole nucleus are discussed in Chapter 4.18.

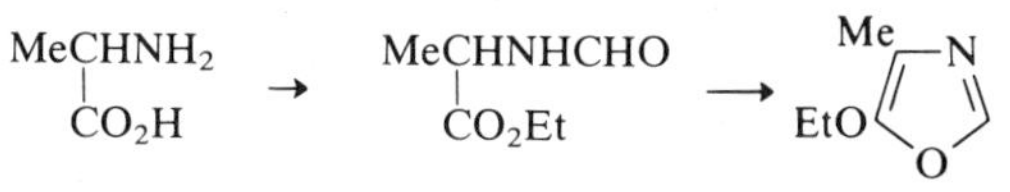

$$\text{MeCHNH}_2 \atop \text{CO}_2\text{H} \quad \rightarrow \quad \text{MeCHNHCHO} \atop \text{CO}_2\text{Et} \quad \rightarrow \quad \text{[Me, EtO, O, N oxazole]}$$

(**1**)   (**2**)   (**3**)

These α-acylaminoketones also provided a convenient synthesis of thiazoles on treatment with phosphorus pentasulfide (Gabriel's method). Although yields range from 45 to 80%, substituents are usually restricted to alkyl, aryl and alkoxy derivatives. Thus, reaction of the α-acylaminoketone (**4**) with P_4S_{10} gave the thiazole (**5**), and thiazole (**7**) itself was prepared in this manner in 62% yield from formylaminoacetal (**6**) ⟨14CB3163⟩. The corresponding 5-ethoxy compound was obtained from the α-formamidoester and phosphorus pentasulfide in an inert solvent.

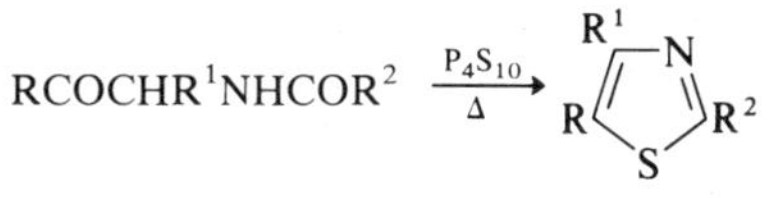

$$\text{RCOCHR}^1\text{NHCOR}^2 \xrightarrow{P_4S_{10}, \Delta} \text{[thiazole with } R^1, R^2, R\text{]} \qquad (\text{EtO})_2\text{CHCH}_2\text{NHCHO} \xrightarrow{P_4S_{10}, \Delta} \text{[thiazole]}$$

(**4**)   (**5**)    (**6**)   (**7**)

Numerous variations of this reaction have been studied, principally those involving a prior inclusion of the nuclear sulfur atom in a thioacylamino compound. Thus, thiobenzamido acetaldehyde diethyl acetal (**8**) underwent ring closure to 2-phenylthiazole (**9**) on gentle heating ⟨57JCS1556⟩. Similarly, *N*-thioacyl α-amino acids also undergo ready ring closure to thiazoles.

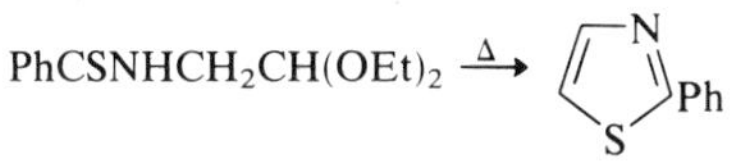

$$\text{PhCSNHCH}_2\text{CH(OEt)}_2 \xrightarrow{\Delta} \text{[2-phenylthiazole]}$$

(**8**)   (**9**)

This reaction may also be used to introduce other substituents into the thiazole ring, and examples are described in Chapter 4.19 and in ⟨69CC818⟩ and ⟨72MI40300⟩. In contrast, this reaction sequence has not been used to prepare selenazole derivatives, these being more readily available by the reactions described in Section 4.03.2.2.2.

α-Acylaminoketones on heating with an ammonia source such as ammonium acetate are converted into imidazoles. 2,4,5-Triarylimidazoles (**11**) were prepared in this way from (**10**) ⟨73CB2415⟩, and the reaction is capable of numerous variations. Formamide on heating at 175 °C with chloroacetaldehyde diethyl acetal in the presence of ammonia resulted in a 60% yield of imidazole.

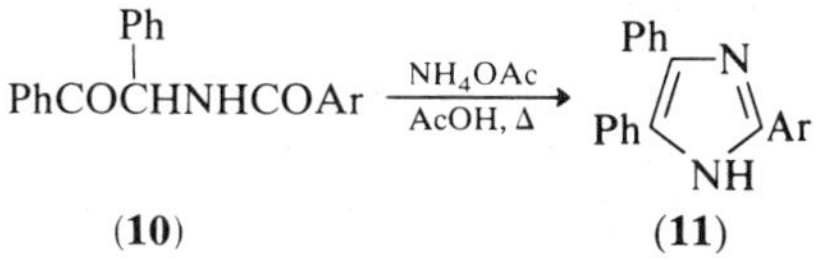

$$\text{PhCOCHNHCOAr} \atop \text{Ph} \quad \xrightarrow{\text{NH}_4\text{OAc} \atop \text{AcOH}, \Delta} \quad \text{[2,4,5-triarylimidazole with Ph, Ph, Ar, NH]}$$

(**10**)   (**11**)

114   Synthesis of Five-membered Rings with Two or More Heteroatoms

An alternative method involves reaction of an α-acylaminoketone (12) with a primary amine and subsequent ring closure of the resultant Schiff's base (13) with phosphoryl chloride. This enables the introduction of a 1-substituent as in (14) to be carried out efficiently, and if the amine were replaced with a monosubstituted hydrazine, the imidazole derivative (15) resulted ⟨78LA1916⟩.

$$\underset{(12)}{\text{RCONHCHCOR}^2}\overset{R^1}{|} \xrightarrow{R^3NH_2} \underset{(13)}{\underset{\overset{\|}{NR^3}}{\text{RCONHCHCR}^2}}\overset{R^1}{|} \xrightarrow{POCl_3/benzene} \underset{(14)}{\underset{R^3}{\overset{R^1}{\text{imidazole}}}} $$

$$R^3 = NHR^4 \rightarrow \underset{(15)}{\underset{NHR^4}{\text{imidazole}}}$$

Both nitrogen atoms of the imidazole ring may be incorporated in the initial 1,4-dicarbonyl or thiocarbonyl system. Thus, diphenylglycine thioamide (16) on heating with aluminum chloride in toluene gave a mixture of the imidazole derivatives (17) and (18) together with the thiazole (19) ⟨80AHC(27)241⟩. Other variations of this approach are discussed in Chapter 4.08 and provide opportunities for diversifying the substituent pattern in the resultant imidazoles.

$$\underset{(16)}{\underset{\text{CSNH}_2}{\text{Ph}_2\text{CNHCOR}}} \xrightarrow[\text{toluene}]{\text{AlCl}_3} \underset{(17)}{\text{imidazole}} + \underset{(19)}{\text{thiazole}} + \underset{(18)}{\text{imidazole}}$$

Analogous open-chain precursors also lead readily to 1,3-dithiolylium salts. S-α-Oxoalkyl thioesters such as (20) on treatment with perchloric acid in glacial acetic acid and H$_2$S undergo ready cyclization to the 1,3-dithiolylium perchlorate (22) ⟨66AHC(7)39⟩. The oxoalkyl dithioesters (21) are probably intermediates in this cyclization as they themselves undergo cyclization with warm 70% perchloric acid or sulfuric acid ⟨80AHC(27)151⟩.

$$\underset{(20)}{\text{RCOCHSCOR}^2}\overset{R^1}{|} \xrightarrow[\text{AcOH, H}_2\text{S}]{\text{HClO}_4} \underset{(21)}{\text{RCOCHSCSR}^2}\overset{R^1}{|} \rightarrow \underset{(22)}{\text{dithiolylium}}\ \text{ClO}_4^-$$

Several variations of this general procedure have been described. The functionalized dithiocarbamate (23) on heating with a mixture of phosphorus pentasulfide and tetrafluoroboric acid gave the 2-amino-substituted 1,3-dithiolylium tetrafluoroborate (24) in moderate to good yield ⟨69CPB1924⟩.

$$\underset{(23)}{\underset{\overset{\|}{S}}{\text{R}_2\text{NCSCH}_2\text{COR}}} \xrightarrow[\text{HBF}_4]{\text{P}_4\text{S}_{10}} \underset{(24)}{\text{dithiolylium}}\ \text{BF}_4^-$$

Use of the β-thiodithiocarbonates (25) and acid results in ring closure to the 1,3-dithiol-2-one (26). Methyl, ethyl and isopropyl groups have been utilized in (25) ⟨76S489⟩, and when R^2 = t-butyl, ring closure occurred in the presence of perchloric acid with extreme ease ⟨74JOC95⟩. Other variations of this synthetic route to 1,3-dithiole derivatives are described in Chapter 4.32.

$$\underset{(25)}{\text{R-C(S)-CHS-C(S)-OR}^2} \rightarrow \underset{(26)}{\text{1,3-dithiol-2-one}}$$

2,5-Diaryl derivatives of the 1,3-oxathiolylium system (29) are prepared by acid-catalyzed cyclization of the β-keto thioesters (28) which are readily prepared from thioacid salts (27) and phenacyl bromides ⟨75H(3)217⟩. The difficulty encountered in preparing representatives of this five-membered heterocyclic system is readily apparent from the examples illustrated in Chapter 4.30.

$$\text{ArCOSK} + \text{PhCOCH}_2\text{Br} \longrightarrow \text{ArCOSCH}_2\text{COPh} \xrightarrow{\text{H}_2\text{SO}_4} \underset{(29)}{\overset{\text{Ph}}{\underset{S_+}{\bigvee\!\!\!\bigvee}}\text{Ar}} \ \text{ClO}_4^-$$

(27) (28) (29)

Application of the above principles to the synthesis of 1,3-diselenolylium ions has been successful. N,N-Dialkyldiselenocarbamates as their oxo esters (30) undergo acid-catalyzed cyclization and dehydration to give the cation (31) in nearly 90% yield ⟨75JOC746, 77CC505, 80CC866, 80CC866⟩.

$$\underset{(30)}{\text{Me}_2\text{N}\overset{R}{\underset{\underset{Se}{\|}}{C}}\text{SeCHCOR}^1} \xrightarrow{H^+} \underset{}{\text{Me}_2\text{N}\overset{R}{\underset{\underset{Se}{\|}}{C}}\text{SeCHCR}^1\underset{^+OH}{}} \longrightarrow \underset{(31)}{\overset{R}{\underset{Se}{\bigvee\!\!\!\bigvee}}\text{NMe}_2}^{Se^+} \ X^-$$

4.03.2.1.2 Ring systems with three heteroatoms

Incorporation of two heteroatoms into the open-chain 1,4-dicarbonyl systems and subsequent ring closure enables a variety of five-membered rings containing three heteroatoms to be prepared in a relatively straightforward manner. 2,5-Disubstituted 1,3,4-oxadiazoles (33) containing alkyl and aryl substituents in the 2,5-positions are most conveniently prepared by the thermal or acid-catalyzed cyclization of 1,2-diacylhydrazines (32). 1-Substituted 1,2-diacylhydrazines (33a) in acetic acid in the presence of a strong acid HX undergo cyclization in a similar fashion. However, in this instance 2,3,5-trisubstituted 1,3,4-oxadiazolium salts (34) were formed ⟨70JCS(C)1397⟩.

$$\text{RCONHNHCOR}^1 \xrightarrow{HX} \underset{(33)}{\overset{N-N}{\underset{O}{R\bigvee\!\!\!\bigvee R^1}}} \qquad \underset{(33a)}{\text{RCONHNCOR}^1}\overset{R^2}{|} \xrightarrow{HX} \underset{(34)}{\overset{N-\overset{+}{N}\overset{R^2}{\diagup}}{\underset{O}{R\bigvee\!\!\!\bigvee R^1}}} \ X^-$$

(32) (33) (33a) (34)

Acylhydrazine derivatives of this general type provide one of the most convenient routes to a variety of 1,3,4-oxadiazoles, and the scope and limitations of this method are discussed in Chapter 4.23.

Both 1,3,4-thiadiazoles and 1,2,4-triazoles may be prepared from derivatives of 1,2-diacylhydrazine by the choice of the appropriate cyclization agent. The majority of 1,3,4-thiadiazole syntheses are based on cyclizations of thiosemicarbazide derivatives or compounds incorporating this basic structural pattern. The simple diacylhydrazines lead to 2,5-disubstituted 1,3,4-thiadiazoles when treated with phosphorus pentasulfide. Alkyl, aryl and heteroaryl groups may be introduced in this way. Cyclization of the acyl thiosemicarbazide (35) with sulfuric acid, polyphosphoric acid or phosphorus halides gave the 5-substituted 2-amino-1,3,4-thiadiazole (36). Methanesulfonic acid has also been used for this cyclization and resulted in high yields of pure products ⟨80JHC607⟩. Numerous variations of this general route to 1,3,4-thiadiazoles have been devised enabling the 2,5-substituents to be varied over a wide range of functional groups. These variations are discussed in Chapter 4.27.

$$\text{RCONHNHCSNHR}^1 \xrightarrow{HX} \underset{(36)}{\overset{N-N}{\underset{S}{R\bigvee\!\!\!\bigvee}}\text{NHR}^1}$$

(35) (36)

1,2-Diacylhydrazines (32) on heating with a primary amine undergo cyclization to the 1,2,4-triazole derivative (38). It is assumed that the amidrazone (37) is an intermediate in the reaction. Most practical syntheses of 1,2,4-triazoles now utilize an amidrazone or a derivative of this alicyclic intermediate containing three nitrogen atoms, and these variations are discussed in Chapter 4.12.

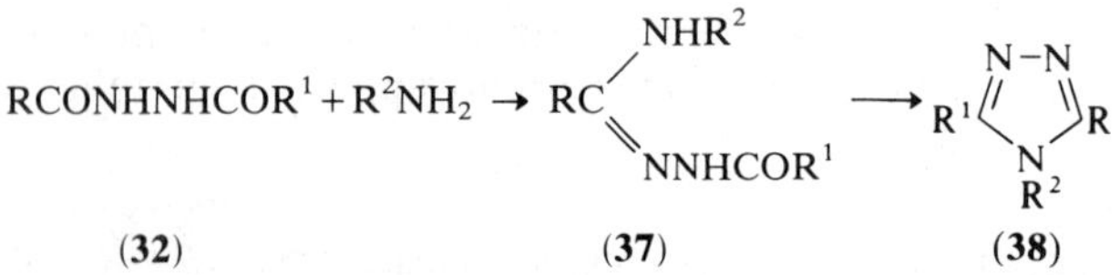

Ring closure of diacylhydrazines (**32**) with phosphorus pentaselenide also provides entry into the 1,3,4-selenadiazoles (**39**). Alkyl and aryl groups may be introduced into the 2,5-positions by this procedure ⟨04JPR(69)509⟩. The corresponding 1,3,4-oxadiazole (**33**) is the major product of this reaction, and as a consequence, attention has focused on variations of this approach using selenium analogs of semicarbazide. Reaction of carboxylic acids with selenosemicarbazide (**40**) yielded the 2-acylamino-1,3,4-selenadiazole (**42**), probably via the intermediate diacyl derivative (**41**). Hydrolysis of the acylamino group in (**42**) occurred readily giving (**43**) ⟨71JHC835, 73JPS839⟩. Other variations are described in Chapter 4.20.

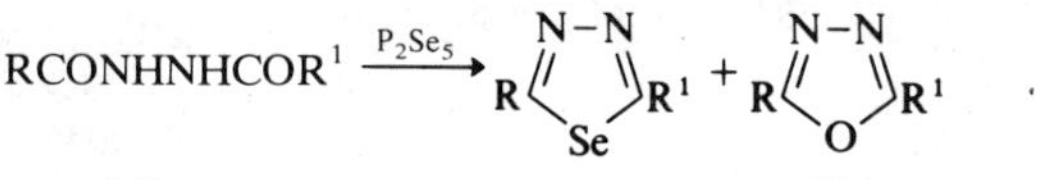

4.03.2.1.3 Ring-fused systems

(i) [5,5] fused systems

Annulation of a second ring to a five-membered heterocycle by an aldol-related reaction requires a 1,4-arrangement of two carbonyl groups or related functional groups in one of the heterocyclic components. This can be achieved in several ways, and syntheses involving vicinal arrangements of the two carbonyl functions are illustrated by the conversion of the 1,3-disubstituted 4,5-diaroylpyrazole (**44**) into the tetrasubstituted thieno[3,4-d]pyrazole (**45**) on heating with phosphorus pentasulfide in an inert solvent ⟨73JOC1769⟩. However, with the diaroyl substituents in the 3,4-positions of the pyrazole nucleus as in (**46**), reaction with phosphorus pentasulfide in boiling pyridine gave the nonclassical heteropentalene thieno[3,4-c]pyrazole (**47**) ⟨74JA4276⟩. Other syntheses of this type are described in Chapter 4.37.

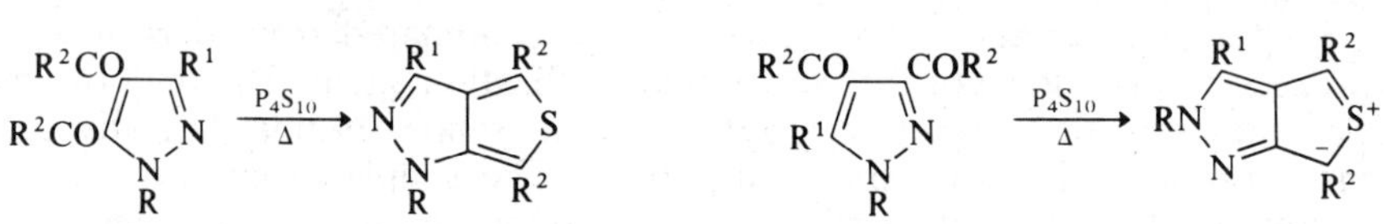

An alternative approach involves an endocyclic–exocyclic arrangement of the 1,4-dicarbonyl functions which also can result in C—C points of ring fusion. Again using phosphorus pentasulfide as the thiation agent, reaction with the α-acylaminocarbonyl system contained in (**48**) led to thieno[2,3-d]thiazoles (**49**) ⟨70CHE1515⟩, pyrazolo[5,4-d]thiazoles, (**50**) ⟨74CHE813⟩ and pyrazolo[4,5-d]thiazoles (**51**) ⟨65CHE165⟩.

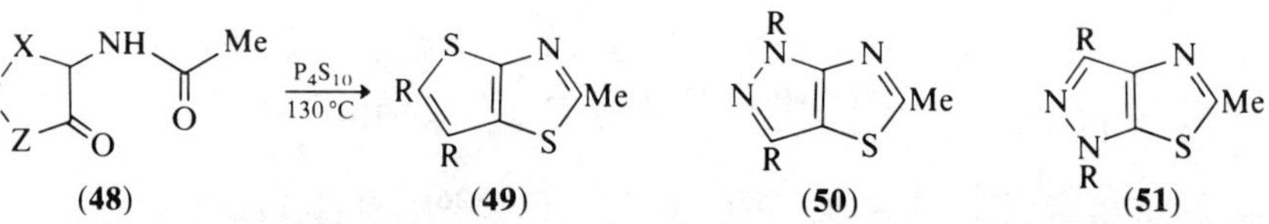

Ring-fused systems involving ring junction nitrogen atoms and additional heteroatoms can also be prepared in this manner. Reaction of an amino-substituted heterocyclic thiol such as (**52**) with an acylating agent gave (**53**), which on heating with phosphorus oxychloride underwent ring closure to give, for example, imidazo[2,1-b][1,3,4]thiadiazoles (**54**) ⟨63LA(663)113⟩.

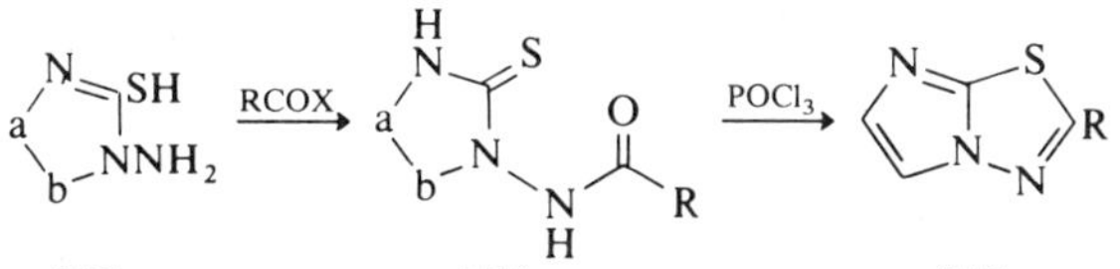

Variations of these approaches to [5,5] ring-fused systems are found in Chapter 4.36.

(ii) [5,6] fused systems

Reaction processes of this type used to form [5,6] ring-fused systems usually involve formation of the five-membered ring. Those in which the six-membered ring is formed have C—C fusion points and are illustrated by the reaction of vicinal dicarbonyl compounds such as (**55**) with hydrazine forming [1,2,3]thiadiazolo[4,5-*d*]pyridazine (**56**) ⟨76JHC301⟩. Other applications of this approach are described in Chapter 4.29, and an interesting variation is the reaction of the vicinal dicarbonyl compound with a primary amine having a methylene group adjacent to the amino group in the presence of DBU. From the 1,2,5-oxadiazole derivative (**57**), the [1,2,5]oxadiazolo[3,4-*c*]pyridine (**58**) was obtained ⟨80S842, 79S687⟩.

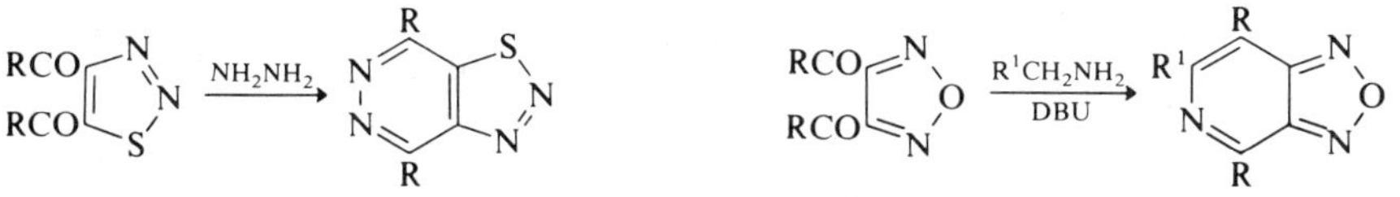

Other C—C fused systems are also available by utilization of 1,4-dicarbonyl-type systems. The substituted pyrimidinethione (**59**) on treatment with polyphosphoric acid readily formed the thiazolo[5,4-*d*]pyrimidine system (**60**) without any of the alternative ring closure product ⟨65JOC1916⟩.

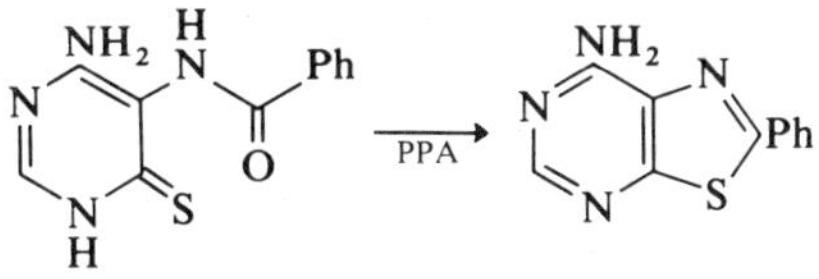

Systems with a nitrogen atom at the ring junction are also readily available by this procedure, a reflection on the case of formation of the requisite 1,4-dicarbonyl compound. The 1-phenacyl-2(1*H*)-pyridinone (**61**) underwent ring closure to the oxazolo[3,2-*a*]pyridinium salt (**62**) on treatment with sulfuric acid/perchloric acid ⟨67JHC66⟩. If the possibility exists for the loss of a proton, then the neutral product is obtained, as in the conversion of (**63**) into the ring-fused system (**64**) ⟨71JOC222⟩. Additional heteroatoms may also be included in the ring-fused systems and examples illustrating these syntheses are discussed in Chapter 4.29.

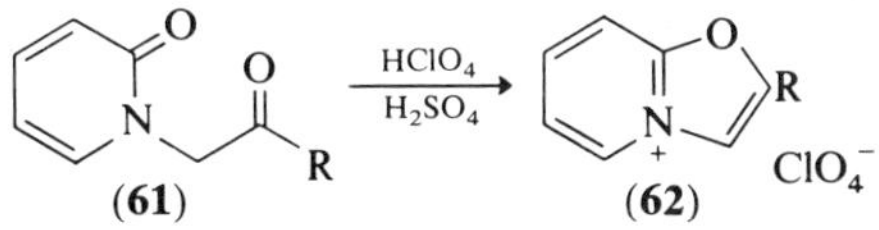

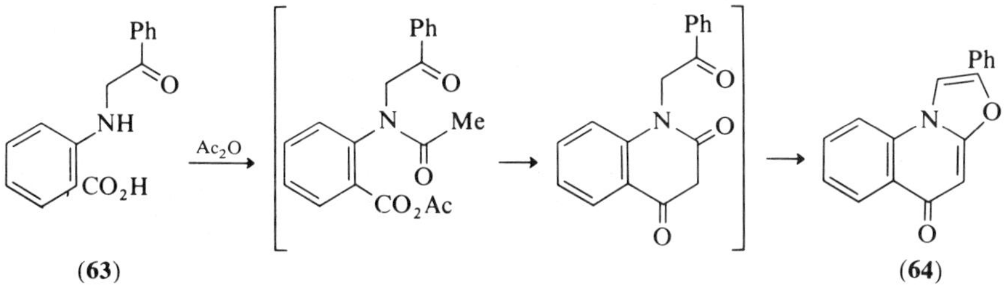

4.03.2.2 Intermolecular Condensations

4.03.2.2.1 Ring systems with two heteroatoms

As shown in Scheme 2, two heteroatom–carbon bonds are constructed in such a way that one component provides both heteroatoms for the resultant heterocycle. By variation of X and Z entry is readily obtained into thiazoles, oxazoles, imidazoles, *etc.* and by the use of the appropriate oxidation level in the carbonyl-containing component, further oxidized derivatives of these ring systems result. These processes are analogous to those utilized in the formation of five-membered heterocycles containing one heteroatom, involving cyclocondensation utilizing enols, enamines, *etc.*

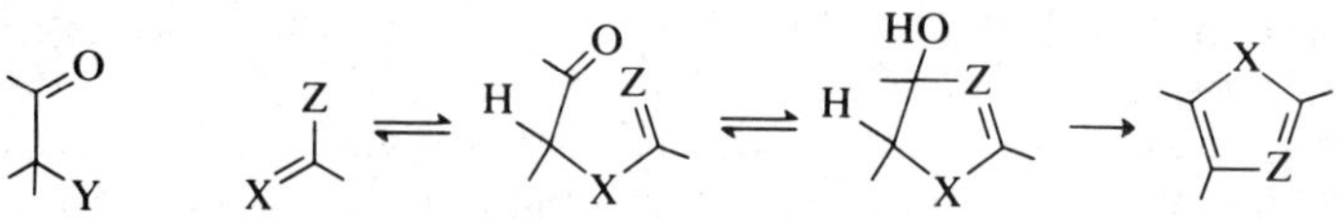

Scheme 2 General reaction scheme for the synthesis of five-membered heterocycles containing two or more heteroatoms by aldol-related reactions

The most widely used method for the synthesis of thiazoles (see Chapter 4.19) is of this type and involves the reaction of α-halo compounds (Y = halogen in Scheme 2) with a reactive component containing an $>$N—C(=S)— structural entity. Reaction of the α-bromoketone (**65**) with the primary thioamide (**66**) in hot benzene gave the intermediate hydroxy compound (**67**), which could be isolated in certain instances but in most cases underwent dehydration to form the thiazole (**68**). The diversity in the substituents capable of being introduced into the resultant thiazole by this procedure is illustrated in Chapter 4.19. Especially noteworthy in this respect is the reaction of the bromopyruvaldehyde oxime (**69**) with thiourea in methanol at room temperature. Neutralization with sodium carbonate resulted in the isolation of the 2-aminothiazole-4-carbaldehyde oxime (**70**) in 39% yield ⟨73JOC806⟩.

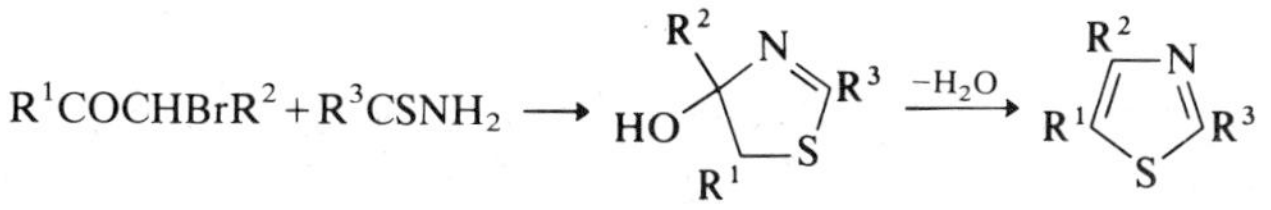

$$R^1COCHBrR^2 + R^3CSNH_2 \rightarrow (67) \xrightarrow{-H_2O} (68)$$

(65)   (66)   (67)   (68)

$$BrCH_2COCH=NOH + NH_2CSNH_2 \rightarrow (70)$$

(69)   (70)

Thiazolines and thiazolidines may also be prepared in this fashion, the structure of the final product determining the substitution pattern to be chosen in the reaction components. Reaction of ethyl bromoacetate with the substituted thioamide (**71**) resulted in formation of the thiazolidin-4-one (**72**) ⟨70KGS1621⟩.

$$BrCH_2CO_2Et + R^1NHCSCHR^2CN \rightarrow (72)$$

(71)   (72)

The comparatively ready accessibility of selenocarboxamides has encouraged the use of this procedure for the synthesis of selenazoles ⟨1889LA(250)294⟩. Reaction of the α-chlorocarbonyl compound (**73**) with the selenocarboxamide (**74**) provided a ready synthesis of a variety of substituted selenazoles (**75**). Useful variations of this general procedure are described in detail in Chapter 4.20, and particularly attractive is the reaction of hydrogen selenide with a mixture of a nitrile and the α-halogenoketone to afford the selenazole ⟨48YZ191, 79S66⟩.

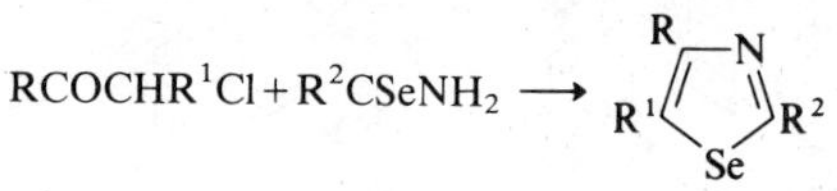

$$RCOCHR^1Cl + R^2CSeNH_2 \rightarrow (75)$$

(73)   (74)   (75)

Amidines and related systems such as guanidines react with α-halogenoketones to form imidazoles. α-Hydroxyketones also take part in this reaction to form imidazoles, and a variety of substituents can be introduced into the imidazole nucleus by these procedures. Reaction of the α-halogenoketone (**73**) with an alkyl- or aryl-substituted carboxamidine (**76**) readily gave the imidazole (**77**) ⟨01CB637, 48JCS1960⟩. Variation of the reaction components that successfully take part in this reaction process is described in Chapter 4.08.

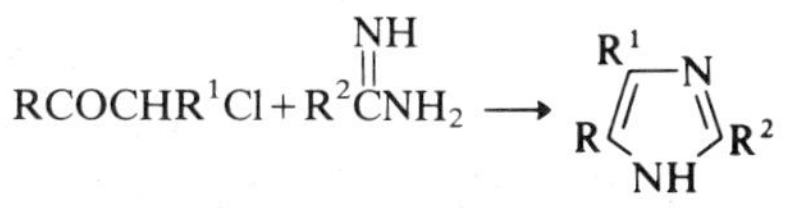

RCOCHR¹Cl + R²C(=NH)NH₂
R¹
N
R
R²
NH

(**73**) (**76**) (**77**)

Oxazoles are also obtained by the reaction of α-halogenoketones (**78**) with primary amides (the Blümlein–Lewy synthesis), and this method is particularly appropriate for oxazoles containing one or more aryl groups as in (**79**). Formamide may also be used in this process, resulting in a free 2-position in the oxazole, and when a urea derivative (**80**) is used, 2-aminooxazoles (**81**) are formed ⟨80ZOR2185, 78IJC(B)1030, 78JIC264⟩. Numerous applications of these procedures are described in Chapter 4.18.

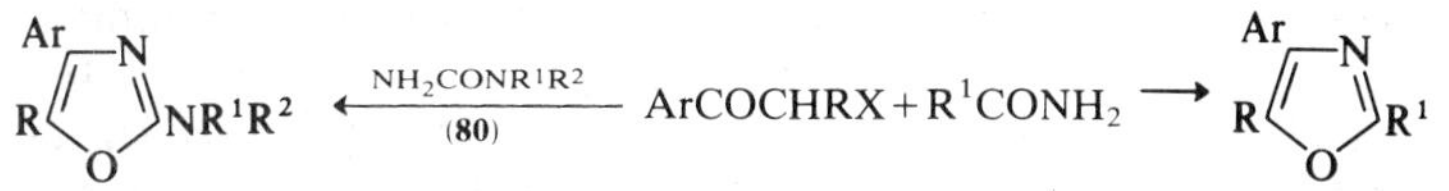

Ar
N
R
NR¹R²
O
NH₂CONR¹R²
(80)
ArCOCHRX + R¹CONH₂
Ar
N
R
R¹
O

(**81**) (**78**) (**79**)

α-Halogenoketones also undergo ready reaction with an excess of thioacids (**82**) or thioesters in the presence of halogen or perchloric acid ⟨71JPR722⟩. 1,3-Dithiolylium salts (**83**) are formed, and replacement of the thioacid with a *gem*-dithiol system as in (**84**) leads readily to 1,3-dithiole derivatives (**85**) ⟨64JHC163, 65JOC732⟩. Other aspects of this reaction type are described in Chapter 4.32.

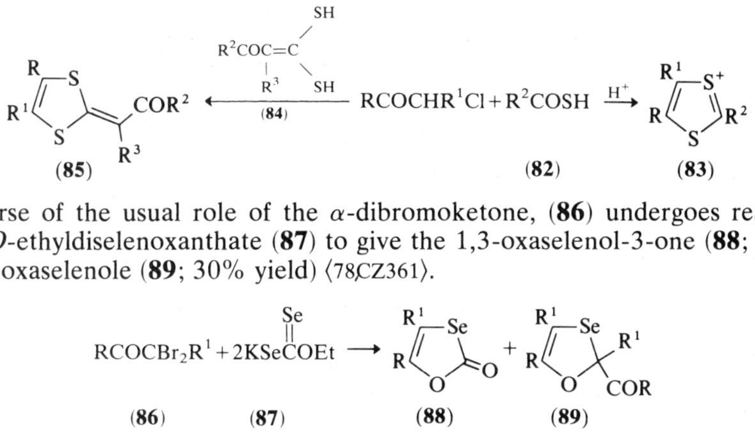

R
S
R¹
COR²
S
R³
SH
R²COC=C
R³
SH
(84)
RCOCHR¹Cl + R²COSH
H⁺
R¹
S⁺
R
R²
S

(**85**) (**82**) (**83**)

In a reverse of the usual role of the α-dibromoketone, (**86**) undergoes reaction with potassium *O*-ethyldiselenoxanthate (**87**) to give the 1,3-oxaselenol-3-one (**88**; 61% yield) and the 1,2-oxaselenole (**89**; 30% yield) ⟨78CZ361⟩.

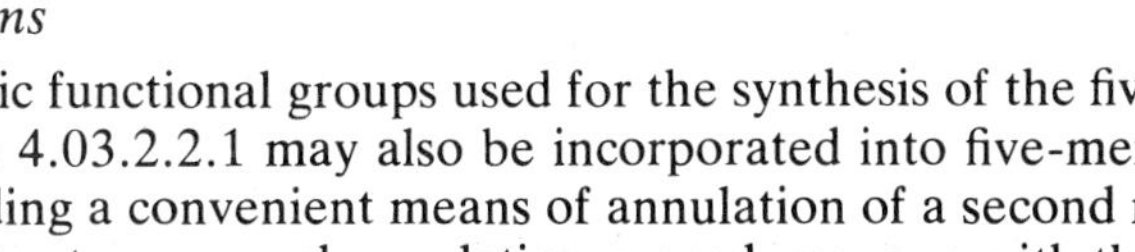

RCOCBr₂R¹ + 2KSeC(=Se)OEt
R¹
Se
R
O
O
R¹
Se
R¹
R
O
COR

(**86**) (**87**) (**88**) (**89**)

4.03.2.2.2 Ring-fused systems

(i) [5,5] fused systems

Many of the acyclic functional groups used for the synthesis of the five-membered system discussed in Section 4.03.2.2.1 may also be incorporated into five-membered heterocyclic systems, thus providing a convenient means of annulation of a second ring.

Scheme 3 illustrates two general annulation procedures, one with the α-halogenoketone moiety incorporated in a cyclic system leading to C—C fused systems. The other, with the heteroatoms already incorporated in a heterocycle, results in a product with a heteroatom at a fusion point.

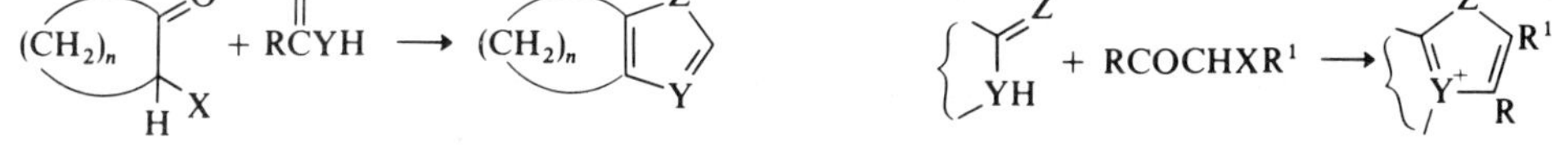

O
(CH₂)ₙ
H
X
Z
+ RCYH
(CH₂)ₙ
Z
Y
Z
YH
+ RCOCHXR¹
Z
R¹
Y⁺
R

Scheme 3 Ring annulations using α-halogenoketones

The latter approach is illustrated by the conversion of the 2-thiazolethione (**90**) into the thiazolo[2,3-b]thiazolylium salt (**92**) by reaction with the α-halogenoketone. The intermediate (**91**) may be isolated, and strong acid was required to effect the final cyclization ⟨77HC(30-1)1⟩. A wide variety of [5,5] fused systems are prepared in this way, and variations of this approach are discussed in Chapter 4.36.

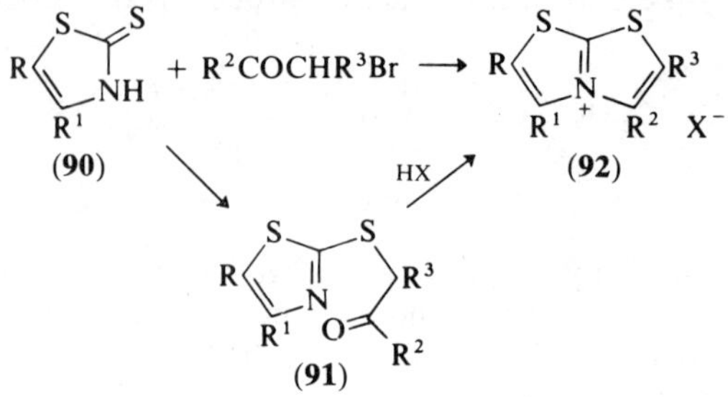

(ii) [5,6] fused systems

Ring-fused systems with C—C fusion points have been prepared by reaction of 2-chlorocyclohexanone (**93**) with an appropriate thioamide (**94**) in boiling chloroform over 24 hours, giving the hexahydrobenzothiazole (**94a**) in modest yield ⟨68BRP1112128⟩. The structural variations in this reaction are the nature of the group R in (**94**), and incorporation of heteroatoms into the carbocyclic ring or increasing the size of the ring. Examples of all variables have been reported. Thus, reaction of 3-bromo-4-piperidone hydrochloride (**95**) with thiourea for three days at 20 °C gave 2-amino-4,5,6,7-tetrahydrothiazolo[5,4-c]pyridine (**96**) ⟨79HC(34-1)230⟩. Increase in ring size such as in the formation of the benzocycloheptathiazoles (**98**) by the reaction of 6-bromo-1-benzosuberone (**97**) with a thioamide in boiling ethanol for four hours (35% yield) was also successful by this route ⟨79HC(34-1)200⟩.

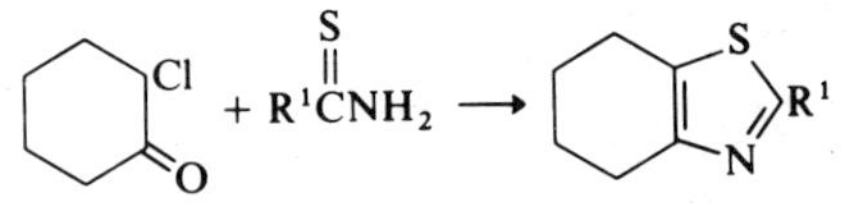

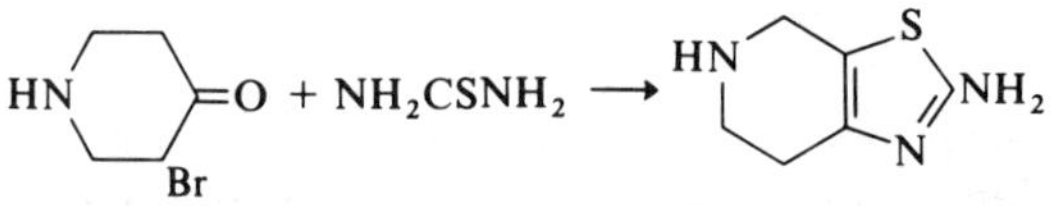

Reaction of α-chlorocyclohexanone (**93**) with selenourea has been reported to yield 2-amino-4,5,6,7-tetrahydrobenzoselenazole (**99**) ⟨43RTC580, 51JA1864⟩.

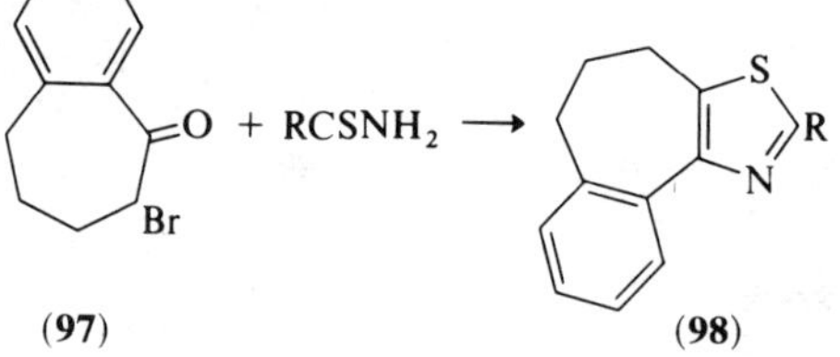

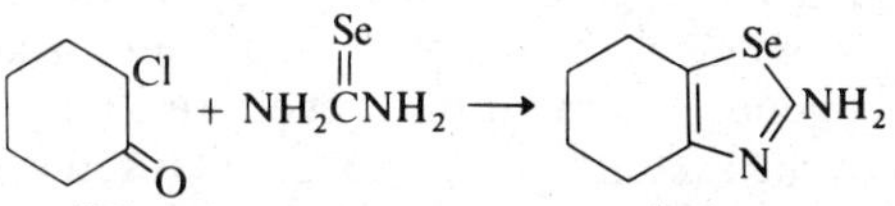

In contrast to the above, the reaction of a heterocyclic thione with an α-halogenoketone (Scheme 3) has been widely exploited as a route to [5,6] ring-fused systems with a heteroatom at a ring fusion point.

2(1H)-Pyridinethione (**100**) reacts readily with an α-halogenoketone to give the 2-β-oxoalkylthiopyridine (**101**) which undergoes ring closure with strong acids such as sulfuric, phosphoric or polyphosphoric to give the thiazolo[3,2-a]pyridinium system (**102**) in good yield. Chloro or nitro substituents in the pyridine ring do not hinder the reaction ⟨66JHC27⟩. The 3-hydroxypyridine analog (103) also cyclized in a similar manner to give (104).

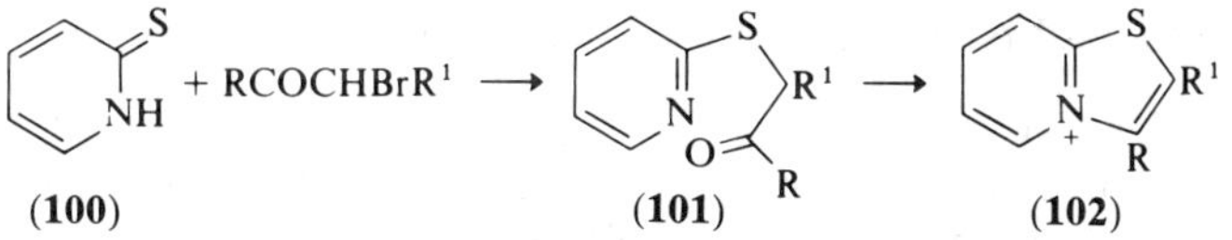

This reaction procedure has also been used for the synthesis of oxazolo[3,2-a]pyridine and imidazo[1,2-a]pyridine derivatives, and the numerous variations studied, allowing changes of the substituent pattern and the number and position of the nitrogen atoms in the six-membered ring, are described in Chapters 4.10 and 4.29.

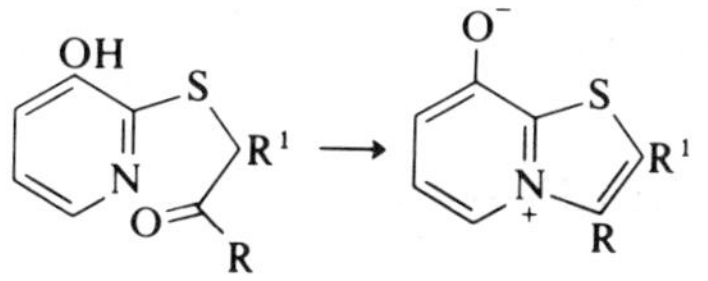

4.03.2.2.3 Ring systems with two adjacent heteroatoms

Scheme 4 shows in a general manner cyclocondensations considered to involve reaction mechanisms in which nucleophilic heteroatoms condense with electrophilic carbonyl groups in a 1,3-relationship to each other. The standard method of preparation of pyrazoles involves such condensations (see Chapter 4.04). With hydrazine itself the question of regiospecificity in the condensation does not occur. However, with a monosubstituted hydrazine such as methylhydrazine and 4,4-dimethoxybutan-2-one (105) two products were obtained: the 1,3-dimethylpyrazole (106) and the 1,5-dimethylpyrazole (107). Although Scheme 4 represents this type of reaction as a relatively straightforward process, it is considerably more complex and an appreciable effort has been expended on its study ⟨77BSF1163⟩. Details of these reactions and the possible variations of the procedure may be found in Chapter 4.04.

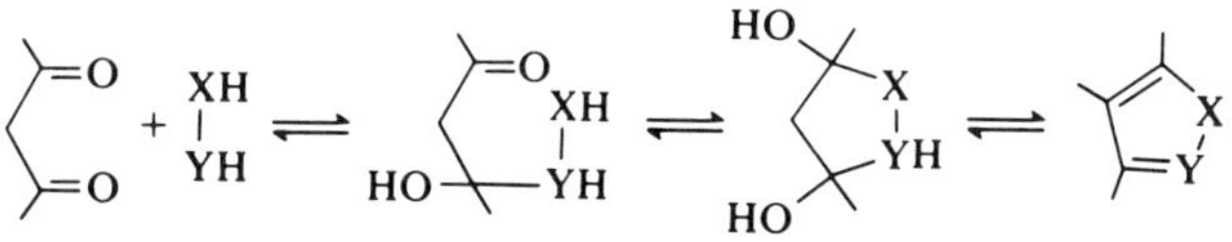

Scheme 4 General reaction scheme for the synthesis of five-membered heterocycles containing two or more heteroatoms by aldol-related reactions

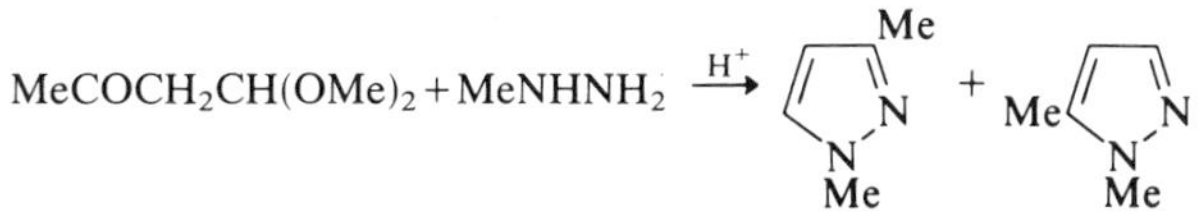

Scheme 4 also represents the classical route to isoxazoles, first studied in 1888 by Claisen and his coworkers ⟨1888CB1149⟩. Reaction of a 1,3-diketone with hydroxylamine gives, via the isolable monoxime (108) and the 4-hydroxyisoxazole (109), the isoxazole (110). Unsymmetrical 1,3-diketones result in both possible isomers (110) and (111), but the ratio of the isomeric products can be controlled by the right combination of the 1,3-dicarbonyl component and the reaction conditions used. These important considerations are described in Chapter 4.16, along with the variations possible in the 1,3-dicarbonyl component designed to yield diverse substituents in the resultant isoxazole.

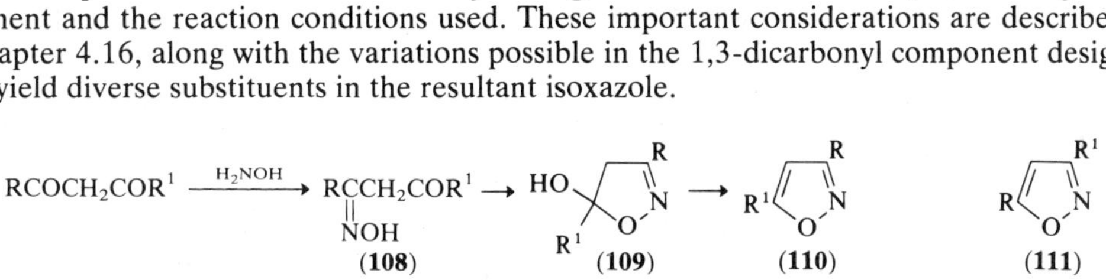

4.03.3 SYNTHESES UTILIZING INTERMOLECULAR NUCLEOPHILIC DISPLACEMENTS

4.03.3.1 Classification of Reaction Types

Reactions considered in this section have, as their guiding principle, bond formation occurring *via* reaction of a binucleophilic component with an electron-deficient bielectrophilic counterpart. As the number of available binucleophiles containing two heteroatoms is comparatively small, the principal emphasis is placed on the bielectrophilic component of the reaction. Reagents under consideration may be arbitrarily classified into three general groups:

Group 1: Binucleophiles (BNu)
Group 2: Bielectrophiles (BE)
Group 3: Reagents containing both a nucleophilic and an electrophilic center (NuE), *e.g.* $HSCH_2CO_2H$

A convenient classification scheme for reactions of this general type expressed below focuses attention on the number of atoms separating the two reactive centers in each component.

BNu + BE → Heterocycle		
Number of atoms separating reactive centers in		Resultant heterocyclic ring system
BNu	BE	
2 +	1	→ 3-membered
2 +	2	→ 4-membered
2 +	3	→ 5-membered
3 +	1	→ 4-membered
3 +	2	→ 5-membered
3 +	3	→ 6-membered
4 +	1	→ 5-membered
4 +	2	→ 6-membered
4 +	3	→ 7-membered
3 +	4	→ 7-membered

Thus, reaction of a 1,2-binucleophile with a 1,3-bielectrophile would lead to a five-membered heterocycle, as would the reaction of a 1,4-binucleophile with a 1,1-bielectrophile. Considering the reactions in this manner has several advantages and provides a convenient framework for considering synthetic approaches to target heterocycles.

Table 1 Some Examples of Commonly Encountered Binucleophiles

1,2-Systems	1,3-Systems	1,4-Systems	1,2-Systems	1,3-Systems	1,4-Systems
H_2NNH_2	$RC(S)NH_2$	$H_2N(CH_2)_2NH_2$	H_2NOH	$RC(NH)NH_2$	$RC(S)NHNH_2$
$RNHNHR$	$RC(S)NHR$	$H_2N(CH_2)_2OH$	$RNHOH$	$RC(NH)NHR$	$RC(Se)NHNH_2$
H_2NNR_2	$RC(Se)NH_2$	$H_2N(CH_2)_2SH$	R_2NOH	$H_2NC(NH)NH_2$	$RC(NH)NHNH_2$
$RNHNR_2$	$RC(Se)NHR$	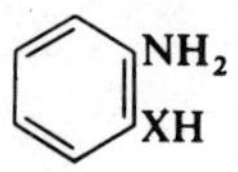 $C_6H_4(NH_2)(XH)$	$RCH=NNH_2$	$H_2NC(S)NH_2$	$RC(NH)NHOH$
			$RCH=NOH$	$H_2NC(Se)NH_2$	$H_2NC(S)NHNH_2$

Synthesis of Five-membered Rings with Two or More Heteroatoms

Table 1 lists some of the common binucleophiles utilized in heterocyclic synthesis, the numerical prefixes referring to the relative positions of the nucleophilic centers to each other. Higher order binucleophiles, *e.g.* 1,5-systems, come readily to mind and the above illustrative examples rapidly increase in scope when the incorporation of these structural elements into heterocyclic systems is considered. This last group offers many opportunities for ring annulations.

4.03.3.2 Design of the Appropriate Bielectrophile

A convenient classification of the bielectrophilic component also relies on the number of atoms separating the reactive centers, resulting in 1,1-bielectrophiles, 1,2-bielectrophiles, 1,3-bielectrophiles and higher order systems. Each group may be further subdivided according to the atoms comprising the bielectrophilic component and, for those involving electron-deficient carbon centers, by the hybridization of the carbon atoms. A further subdivision is that of a bielectrophile equivalent, *i.e.* a reaction component that by virtue of its reaction behaves as a bielectrophile.

4.03.3.2.1 1,1-Bielectrophiles

This group includes many of our common reagents such as the carboxylic acid chlorides, phosgene, thiophosgene, phosgene iminium salts, ortho esters, thionyl chloride, phosphorus oxychloride, isocyanide dichlorides, *etc.* All these reagents have the common feature of at least two electron-withdrawing substituents attached to the same atom, and these substituents also have good leaving group characteristics.

4.03.3.2.2 1,2-Bielectrophiles

By definition, members of this group have a vicinal arrangement of their electron-deficient centers. They may be conveniently considered according to their atom composition and the hybridization state of any carbon atoms involved.

Type 1: Contains two sp^2 hybridized carbon atoms

Type 2: Contains an sp^2 and an sp^3 hybridized carbon atom

Type 3: Contains at least one heteroatom

Type 4: Bielectrophilic equivalents or a 'masked' bielectrophile

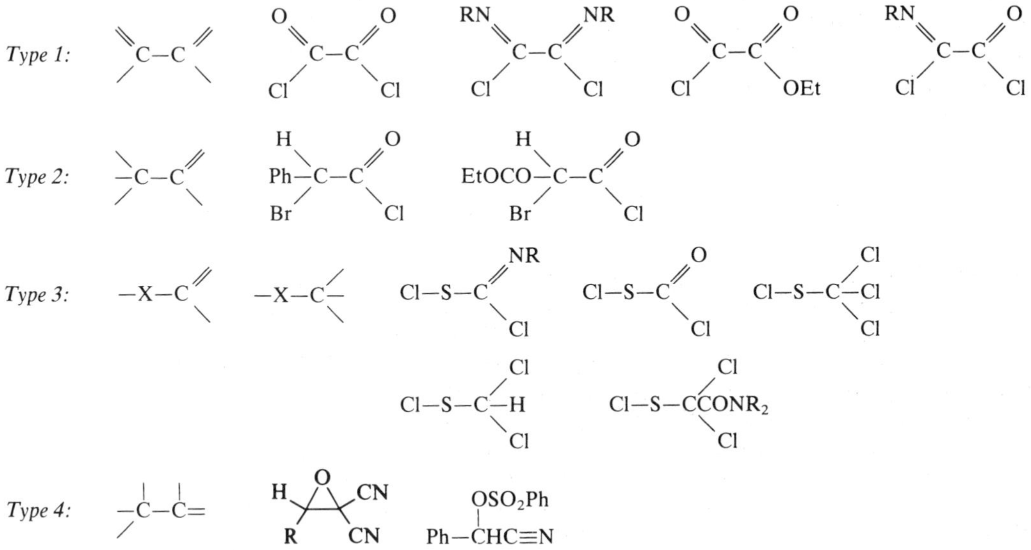

Scheme 5 Illustrative 1,2-bielectrophiles classified by structural type

Representative members of these four groups of 1,2-bielectrophiles are shown in Scheme 5.

4.03.3.2.3 1,3-Bielectrophiles

The possible variations among the three atoms comprising this reactive component make this the largest group and the one with the most potential for further development. They are particularly useful in the synthesis of five-, six- and seven-membered ring systems. They also may be conveniently considered according to their atom composition, and the hybridization state of any carbon atoms involved.

Type 1: Contains three sp^2 hybridized carbon atoms

Type 2: Contains an sp^3 and two sp^2 or sp hybridized carbon atoms

Type 3: Contains at least one heteroatom

Scheme 6 shows representative members of these three groups. Other combinations are possible and the design of an appropriate bielectrophile for use as a versatile synthon presents a considerable synthetic challenge, as by virtue of the structural entities involved they are extremely reactive.

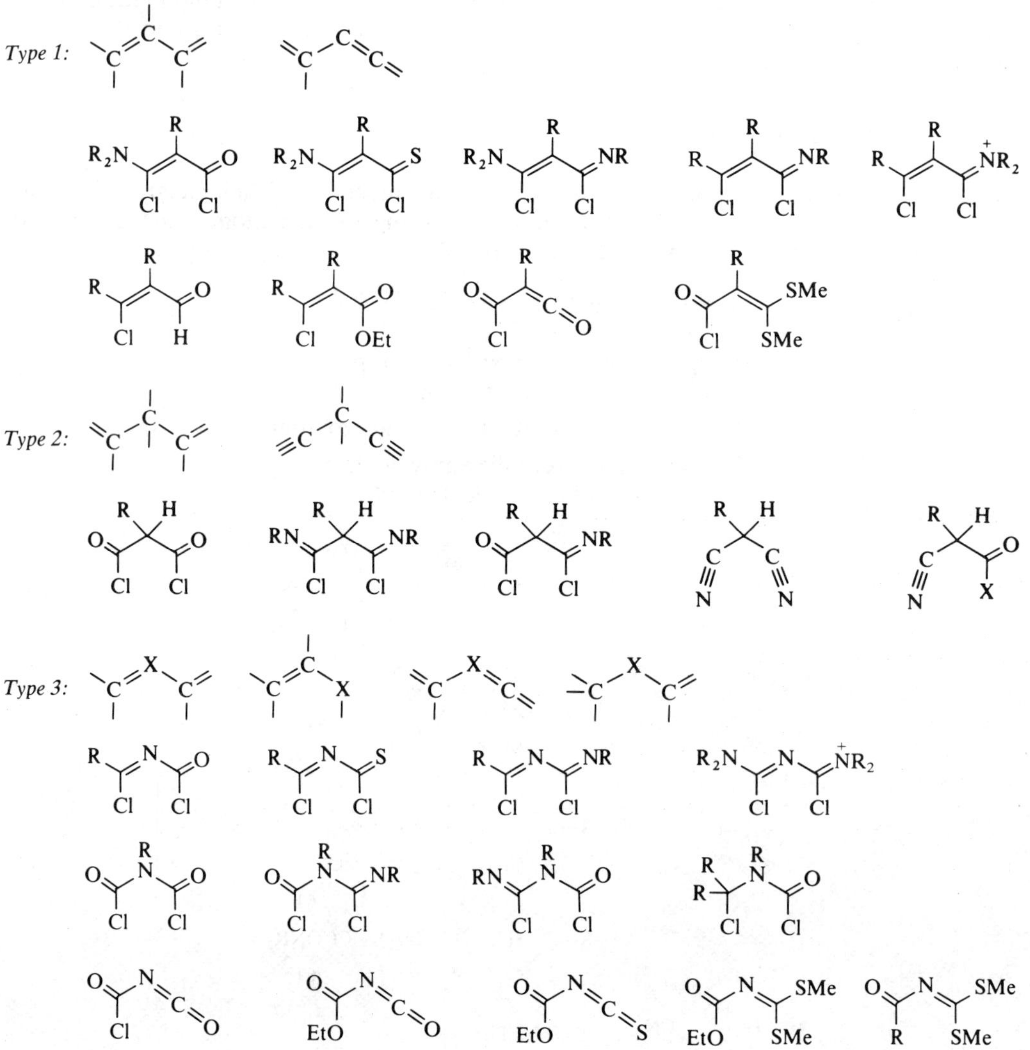

Scheme 6 Illustrative 1,3-bielectrophiles classified by structural type

4.03.3.2.4 1,4- and higher bielectrophiles

Extension of the approach discussed above suggests several structural types which may be classified under this category, the main variable being the nature and hybridization of the electrophilic centers and the nature of the atoms joining these two centers. They may be conveniently divided into two groups:

Type 1: Systems containing all carbon atoms with two sp^2 hybridized carbon atoms, two 'benzylic' carbon atoms or a combination of these types

Type 2: Systems containing two sp^2 hybridized carbon atoms with one or more heteroatoms joining these two centers

Representative examples of these types are shown in Scheme 7. The range of structural types is greatly extended when acyl halides attached to any ring system in a vicinal relationship are considered.

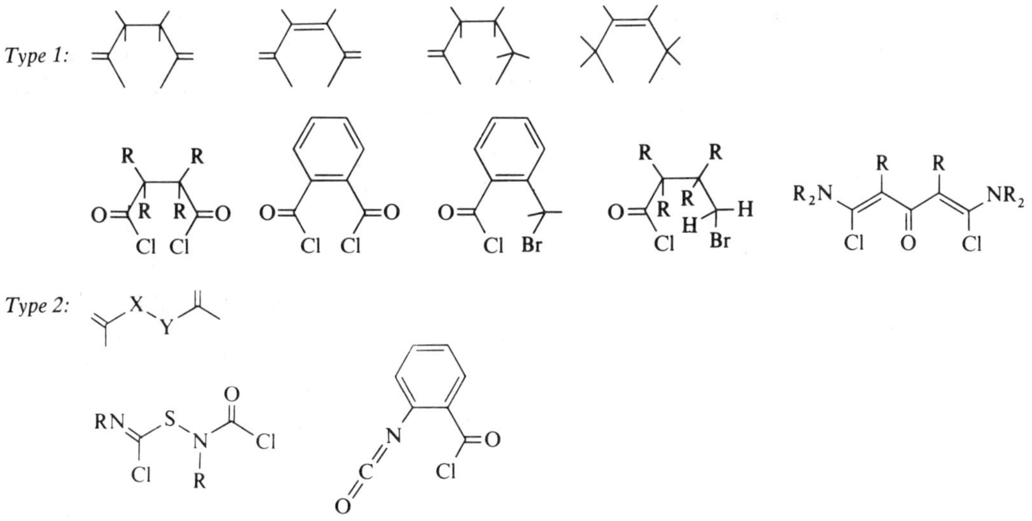

Scheme 7 Illustrative 1,4- and higher order bielectrophiles classified by structural type

4.03.3.3 Syntheses Using 1,1-Bielectrophiles

4.03.3.3.1 Ring systems with two heteroatoms

The role of the 1,1-bielectrophile in ring closures of this type is to provide a one-carbon unit (or heteroatom) to close the cycle. Thus, the synthesis of the four-atom precursor with two nucleophilic centers 1,4 to each other is an appreciable challenge, especially to obtain a heterocycle at the desired oxidation level. The examples below illustrate the way this approach to synthesis may be gainfully utilized.

The methylhydrazone of acetophenone (**112**) underwent ready reaction with *n*-butyllithium giving the dianion (**113**); reaction with acid derivatives such as acid chlorides or esters resulted in pyrazole (**114**) formation whereas with aldehydes, pyrazolines were obtained ⟨76SC5⟩. With dichloromethyleneiminium salts (**115**), 5-dimethylaminopyrazoles

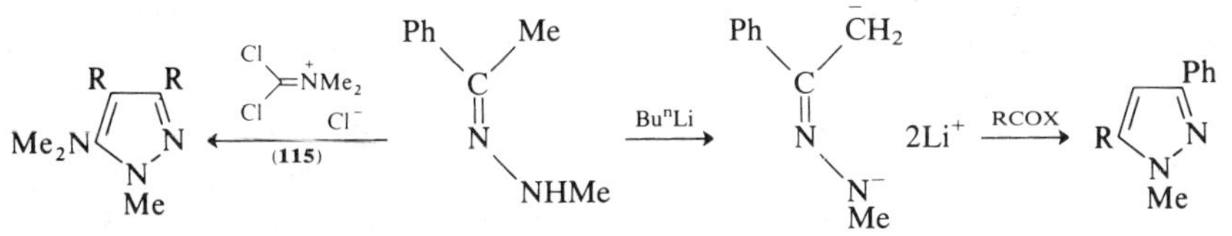

(**116**) resulted ⟨74AG(E)79⟩. Other related ring closures involve the reaction of the Vilsmeier–Haack reagent (POCl₃–DMF) with ketone semicarbazones or ketazines. Two moles of the reagent are used and the resultant pyrazoles contain a 4-formyl substituent after reaction work-up ⟨70JHC25, 70TL4215⟩. Additional reactions of this type are described in Chapter 4.04.

Ethylenediamine derivatives undergo reaction with a wide variety of 1,1-bielectrophiles leading to partially reduced imidazole derivatives (Chapter 4.08). When a dehydrogenation agent is present imidazoles are formed. It is particularly useful for the synthesis of imidazoles with a definitive substitution pattern, illustrated by the preparation of methyl 1-methyl-imidazole-4-carboxylate (**119**) from 2-amino-3-methylaminopropionic acid (**117**) with triethyl orthoformate. The initial imidazoline (**118**), after esterification and subsequent oxidation with activated manganese dioxide, gave (**119**) in very good yield ⟨70AHC(12)103⟩.

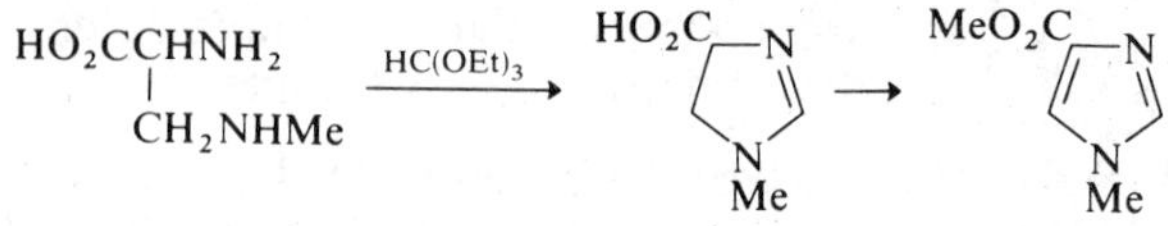

A similar utilization of two heteroatoms is illustrated by the reaction of the ethylenedithiol (**120**) with formic acid in the presence of perchloric acid. 4,5-Diphenyl-1,3-dithiolylium perchlorate (**121**) was formed in 45% yield ⟨69MI40300⟩. Introduction of a 2-oxo or 2-thioxo substituent as in (**123**; X = O, S) is illustrated by the reaction of the disodium salt (**122**) with phosgene and thiophosgene, respectively ⟨76S489⟩.

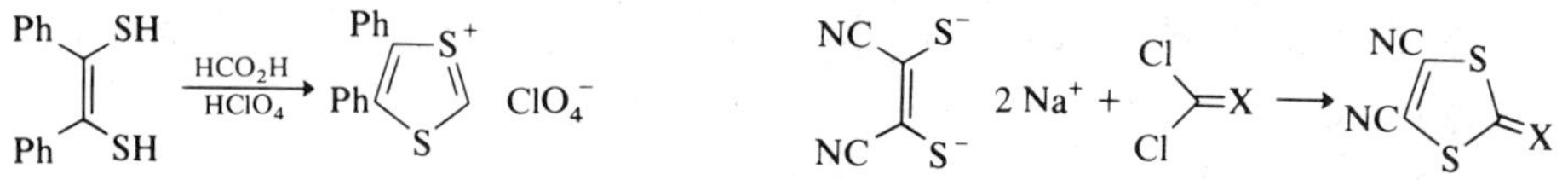

The 1,3-dithiolane ring system may also be prepared in an analogous fashion. Reaction of a carbonyl compound with 1,2-ethanedithiol (**124**) in the presence of BF₃ or p-toluenesulfonic acid readily gave the saturated ring system (**125**) ⟨74S32⟩. Use of dimethylformamide dimethyl acetal provided for the introduction of a 2-dimethylamino substituent as in (**126**), and phosgene in ether at 20 °C gave the corresponding 2-oxo substituent as in (**127**) ⟨41RTC(60)453⟩.

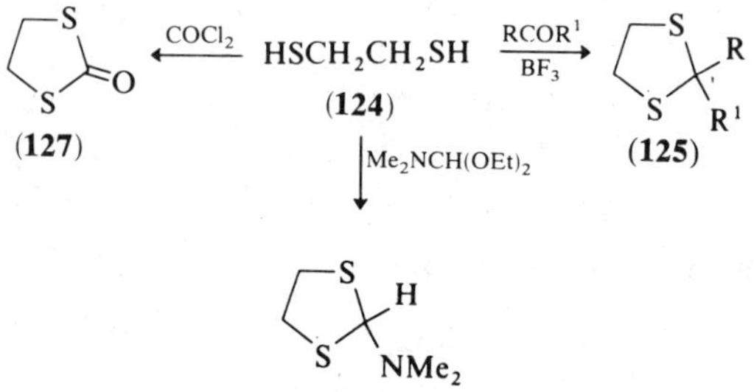

Other applications of this conceptual approach are found throughout the various chapters, and the range of potential 1,1-bielectrophiles, *e.g.* ethyl chloroformate, will be readily apparent.

4.03.3.3.2 *Ring systems with three or more heteroatoms*

Numerous examples of the ring closure of a binucleophilic system with a 1,1-bielectrophile leading to five-membered heterocycles with three or more heteroatoms have been described, the popularity of this route no doubt reflecting the comparative ease with which the penultimate product may be obtained.

Reaction of a hydrazide (**128**) with phosgeneiminium chloride (**115**) led to the 2-dimethylamino-1,3,4-oxadiazole (**129**) in 90% yield ⟨75AG(E)806⟩. The 1,3,4-thiadiazole system was also obtained in an analogous reaction in which the dithioimidate (**130**) underwent reaction with the thiohydrazide (**131**). Depending on the nature of X in (**131**), the 2-substituent in the resultant 1,3,4-thiadiazole (**132**) may be varied ⟨80ZC413⟩. Although (**130**) also has an electrophilic center β to the $C(SMe)_2$ function, reaction of the binucleophile occurs at the latter function to form the five-membered ring (see Section 4.03.3.5).

$$RCONHNH_2 + Me_2\overset{+}{N}{=}\overset{Cl}{\underset{Cl}{C}} \; Cl^- \longrightarrow R\underset{O}{\overset{N-N}{\bigvee}}NMe_2$$

(128)   (115)   (129)

$$MeOCON{=}C\overset{SMe}{\underset{SMe}{\diagdown}} + H_2NNH\overset{S}{\overset{\|}{C}}X \longrightarrow MeOCONH\underset{S}{\overset{N-N}{\bigvee}}X$$

(130)   (131)   (132)

This approach is particularly suited to the formation of nonclassical systems such as mesoionic compounds in the 1,3,4-thiadiazole system. Reaction of the thiohydrazide (133) containing a substituent on N' with phosgene, thiophosgene or an isocyanide dichloride leads to ready ring closure and formation of the mesoionic 1,3,4-thiadiazole (134; X = O, S, NR, respectively) ⟨B-79MI40300⟩. 3,3-Dichloroacrylonitrile (135) and (133) gave the corresponding mesoionic 2-methylene derivative (136) ⟨70TL5083⟩.

(136)   (135)   (133)   (134)

The isocyanide dichlorides are particularly attractive 1,1-bielectrophiles, and the N-sulfonyl derivative (138) underwent reaction with the N-hydroxythioamide (137) to give the 1,3,5-oxathiazole derivative (139) ⟨71AP763⟩. Yields varied from 62% for R = Me and were slightly less for R = Ph (57%) and R = p-MeOC$_6$H$_4$ (50%).

$$PhCSNHOH + \overset{Cl}{\underset{Cl}{\diagup}}{=}NSO_2R \xrightarrow[toluene]{reflux} \underset{O}{\overset{Ph\diagdown S}{\bigvee}}{=}NSO_2R$$

(137)   (138)   (139)

The above examples illustrate reactions at an electron-deficient carbon atom. Other 1,1-bielectrophiles allow the direct introduction of a heteroatom into the resultant heterocycle. The most widely applicable and versatile methods for the synthesis of 1,2,5-thiadiazoles and 1,2,5-selenadiazole rely on this approach.

It has been found that an acyclic NCCN system in which the N—C groups may be sp, sp^2 or sp^3 hybridized reacts with sulfur monochloride or sulfur dichloride to form the appropriately substituted 1,2,5-thiadiazole ⟨68AHC(9)107, 67JOC2823⟩. This reaction is illustrated by the reaction of cis-diaminomaleonitrile (140) with sulfur dichloride, which gave 3,4-dicyano-1,2,5-thiadiazole (141) in 93% yield ⟨72JOC4136⟩. Tables 16 and 17 in Chapter 4.26 list the various substituted systems which may be made in this manner. Selenium monochloride has been used as the cyclization agent in the preparation of 1,2,5-selenadiazoles; however, this series has not been as thoroughly investigated as their sulfur analogs ⟨79S979⟩.

(140)   (141)

4,5-Dicyano-1,2,3-trithiole 2-oxide (143) has been prepared from the silver salt of 2,3-dimercaptomaleonitrile (142) and thionyl chloride ⟨66HC(21-1)67⟩. Similarly, the reaction of ethylene glycol (144) with thionyl chloride gave 1,3,2-dioxathiolane 2-oxide (145), the parent compound of saturated five-membered cyclic sulfites (see Chapter 4.33).

(142)   (143)   (144)   (145)

A number of 1,2,3-oxathiazole S-oxides are prepared from the reaction of thionyl chloride with various ethane derivatives having vicinal oxygen- and nitrogen-containing groups. Reaction of the 2-aminoethanol derivative (146) with SOCl₂ gave (147) (see Chapter 4.34).

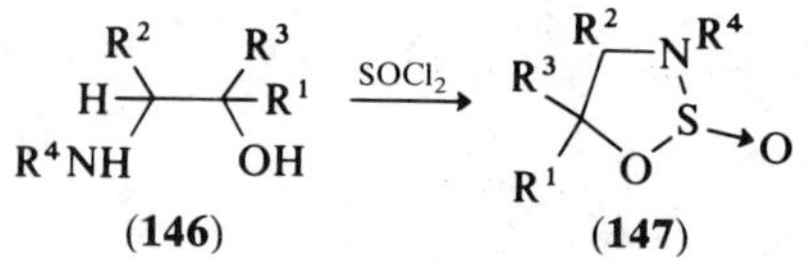

Applications of these ring closure sequences to the synthesis of 1,2,3,5-thiatriazolines are described in Chapter 4.28.

4.03.3.3.3 Ring-fused systems

The syntheses of ring-fused systems with both C—C and C—N fusion points are particularly suited for cyclocondensations of this type. Treatment of the 4,5-diaminopyrazole (148) with thionyl chloride under reflux gave the pyrazolo[3,4-c][1,2,5]thiadiazole S-oxide (149) ⟨68JMC1164⟩, and when (148) was treated with sulfur monochloride, the aromatic system (150) was obtained ⟨81JOC4065⟩. The corresponding 1,2,5-selenadiazole was isolated when (148) was heated with selenium dioxide (52% yield).

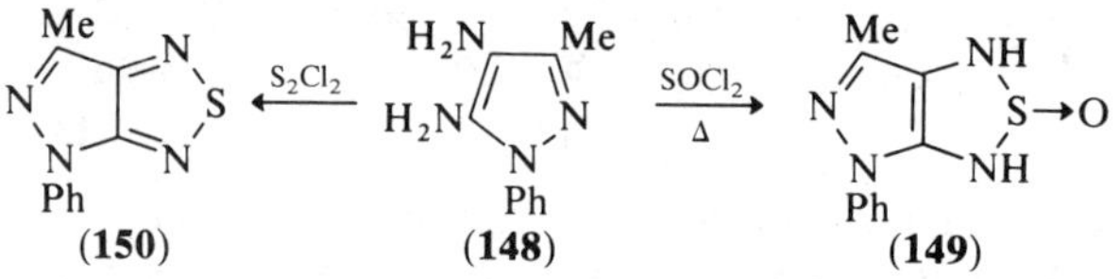

As appropriately substituted o-disubstituted benzene derivatives are readily available, this procedure has found widespread application in the synthesis of benzo-fused five-membered heterocycles. Examples abound in the various chapters in these volumes and the following few examples illustrate the general trend.

o-Substituted anilines (151) react readily with phosgeneiminium chloride (115) to give benzoxazoles, benzothiazoles and benzimidazoles (152) containing a 2-dimethylamino substituent in yields ranging from 60 to 97% ⟨73AG(E)806⟩. Similar vicinally substituted heterocycles also react in a similar way, providing entry into a large number of ring-fused systems. An interesting variation of this reaction is the formation of the imidazole (154) from 1,1-diphenylhydrazine (153). Here a carbon atom of the benzene ring acts as the second nucleophilic component.

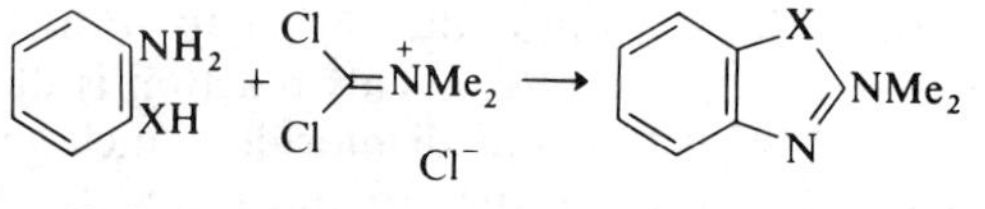

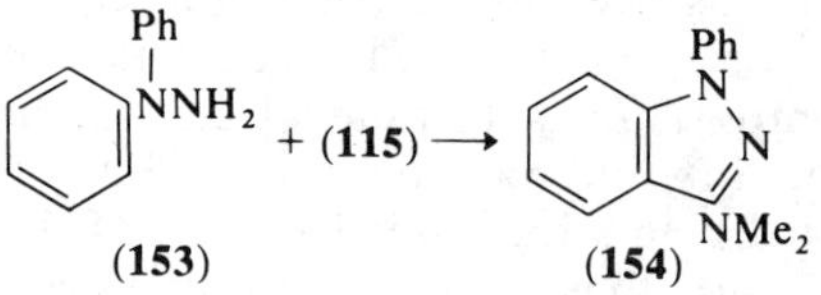

The above procedure is also of major importance for the synthesis of ring-fused systems with a nitrogen atom at the ring junction. Several methods are available for the synthesis of the four-atom precursor. Amination of 2-amino-4-methylthiazole occurs on the ring nitrogen and base treatment leads to the free base (155). Reaction with phosgeneiminium chloride (115) gave the thiazolo[3,2-b][1,2,4]triazole derivative (156) in good yield ⟨73AG(E)806⟩. The isomeric ring system thiazolo[2,3-c][1,2,4]triazole (158) may be prepared from the corresponding hydrazine (157) with a variety of 1,1-bielectrophiles such as carboxylic acids, ortho esters, carbon disulfide and cyanogen bromide ⟨71JOC10⟩.

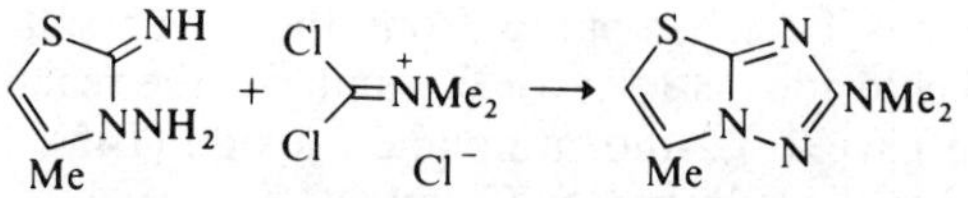

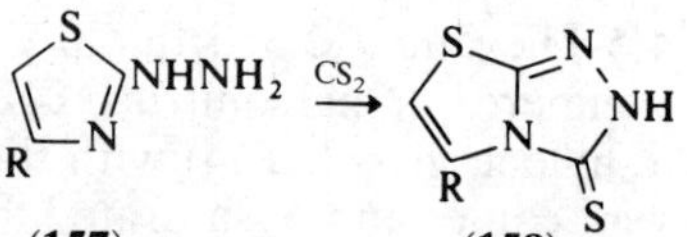

This method of ring closure is the most convenient for the synthesis of ring-fused 1,2,4-triazoles with this arrangement of nitrogen atoms. It has been used extensively to form the analogous [5,6] ring-fused systems, and structural ambiguity is only encountered when the six-membered ring is capable of undergoing ring opening–ring closure in a Dimroth rearrangement (see Chapter 4.15).

4.03.3.4 Syntheses Using 1,2-Bielectrophiles

4.03.3.4.1 Ring systems with two heteroatoms

As the bielectrophile contains two atoms of the five-membered ring, reactions to be considered under this category require a 1,3-binucleophile (see Table 1). An appreciable number of applications have been reported in the literature and illustrative examples are described below. One should also consider reactions discussed in Section 4.03.2.2 above as extensions of this concept.

Because of the structural requirements of the bielectrophile, fully aromatized heterocycles are usually not readily available by this procedure. The dithiocarbamate (**159**) reacted with oxalyl chloride to give the substituted thiazolidine-4,5-dione (**160**) (see Chapter 4.19), and the same reagent reacted with *N*-alkylbenzamidine (**161**) at 100–140 °C to give the 1-alkyl-2-phenylimidazole-4,5-dione (**162**) (see Chapter 4.08). Iminochlorides of oxalic acid also react with *N,N*-disubstituted thioureas; in this case the 2-dialkylaminothiazolidine-2,4-dione bis-imides are obtained. Thiobenzamide generally forms linear adducts, but 2-thiazolines will form under suitable conditions ⟨70TL3781⟩. Phenyliminooxalic acid dichloride, prepared from oxalic acid, phosphorus pentachloride and aniline in benzene, likewise yielded thiazolidine derivatives on reaction with thioureas ⟨71KGS471⟩.

(**159**) (**160**)

(**161**) (**162**)

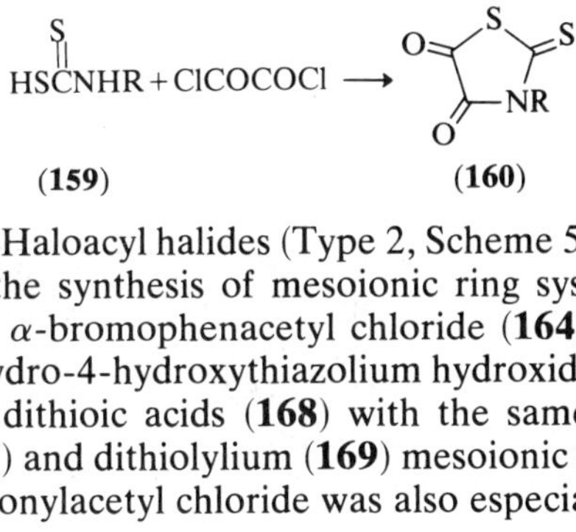

α-Haloacyl halides (Type 2, Scheme 5) are particularly reactive bielectrophiles well suited for the synthesis of mesoionic ring systems. Reaction of the secondary thioamide (**163**) with α-bromophenacetyl chloride (**164**) in the presence of triethylamine readily gave the anhydro-4-hydroxythiazolium hydroxide system (**165**). Similarly substituted amidines (**166**) and dithioic acids (**168**) with the same reagents formed the corresponding imidazolium (**167**) and dithiolylium (**169**) mesoionic systems ⟨77JOC1633, 77JOC1639⟩. 2-Bromo-2-ethoxycarbonylacetyl chloride was also especially useful in reactions of this type.

(**163**) (**164**) (**165**)

(**166**) (**167**)

(**168**) (**169**)

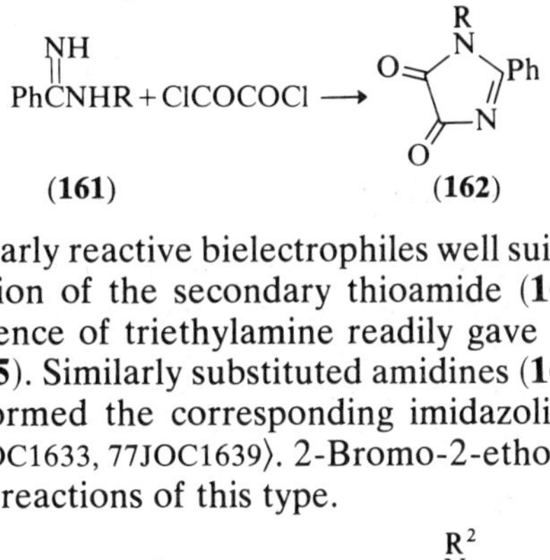

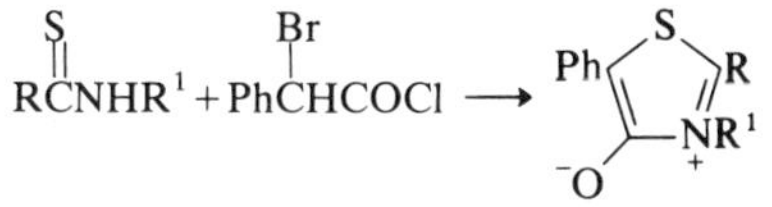

Several reactive species which may be considered 'bielectrophilic equivalents' (Type 4, Scheme 5) have been utilized in cyclocondensations. The aryl-substituted dicyanooxirane (**170**), prepared from aromatic aldehydes, malononitrile and subsequent oxidation, is a particularly interesting application of this idea. It undergoes ready reaction with a 1,3-binucleophile such as a thioamide to give the 4-thiazolinone (**173**). Intermediates such as (**171**) and (**172**) may be involved in this reaction ⟨76S261, 76CC23, 78JOC3732, 73CR(C)(277)1153⟩. Thiourea may be used in the condensation giving the 2-aminothiazolin-4-one, and if a secondary thioamide were used, the mesoionic system (**174**) was obtained.

(**174**) (**170**) R = aryl (**171**) (**172**) (**173**)

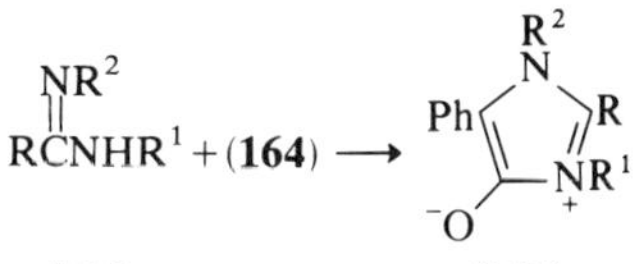

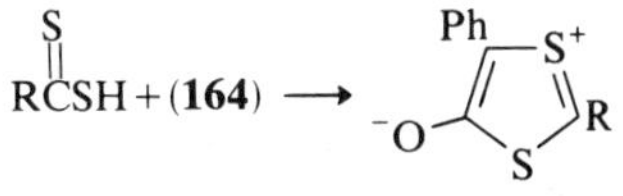

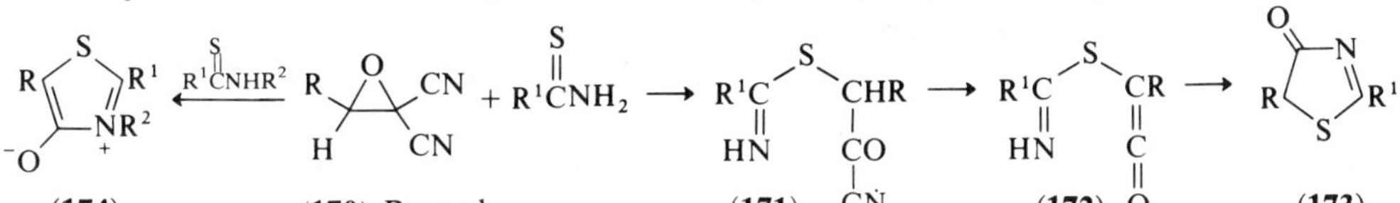

A variation of this idea utilizes the glycidic ester (**175**). This underwent condensation with thioureas and phenyl dithiocarbamate to give (**176**; $R^1 = NR_2$, SPh, respectively) ⟨71BSF4021⟩.

(**175**) (**176**)

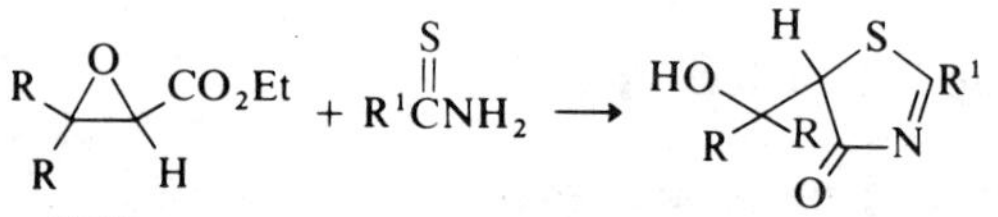

The reactions of the benzenesulfonyl ester of mandelonitrile (**177**) provide another illustration of the 'masked' bielectrophile approach. On reaction with a primary thioamide the 4-aminothiazole (**178**) was obtained and this is a convenient route to these derivatives. With a thiourea, the thiazoline (**179**) was the initial product, and this on treatment with water gave the thiazolidin-4-one (**180**) ⟨61JOC2715⟩.

(**177**) (**178**) (**179**) (**180**)

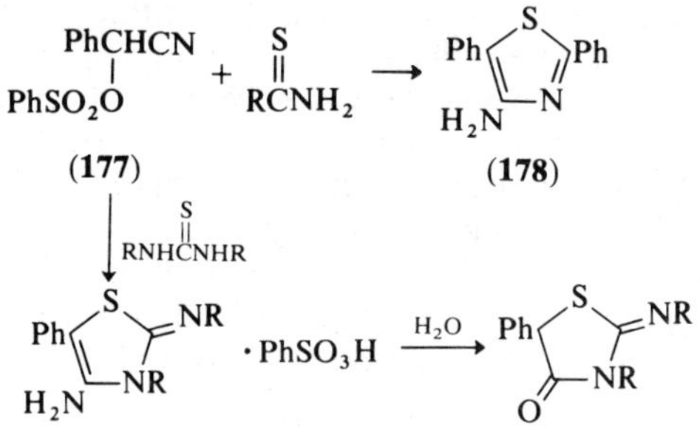

4.03.3.4.2 Ring systems with three or more heteroatoms

Ring systems of this nature are most readily prepared from 1,2-bielectrophiles already containing one of the desired heteroatoms. As sulfur can be readily converted into sulfenyl chlorides, substrates containing this functional group have found considerable use. The reaction of trichloromethanesulfenyl chloride (**181**) with amidines (**182**) in the presence of base under mild conditions is a general method for the preparation of 5-chloro-1,2,4-thiadiazoles (**183**) ⟨65AHC(5)119⟩. Variation of (**181**) to contain one or two chlorine atoms attached to carbon greatly extends the usefulness of these reagents, and these aspects are discussed in Chapter 4.25.

(**182**) (**181**) (**183**)

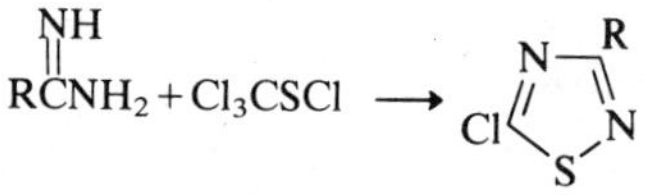

Several related derivatives have also been utilized in this type of synthesis. Iminochloromethanesulfenyl chlorides (**184**), prepared by the controlled addition of chlorine to isothiocyanates, react with amidines (**161**) to give 1,2,4-thiadiazolines (**185**) ⟨71T4117⟩. Chlorocarbonylsulfenyl chloride (**186**), prepared by the hydrolysis of trichloromethanesulfenyl chloride with sulfuric acid, reacted with ureas, thioureas and guanidines to give 1,2,4-thiadiazolidine derivatives (**187**) ⟨70AG(E)54, 73CB3391⟩.

(**184**) (**161**) (**185**) R = aryl, ArCO (**186**) (**187**)

(**188**) (**189**)

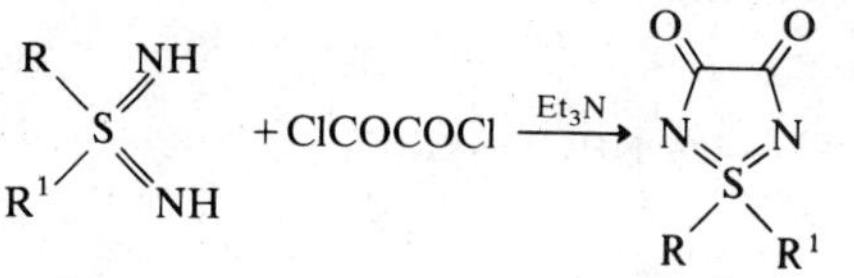

Of particular interest is the reaction of *S,S*-disubstituted sulfur diimides (**188**) with oxalyl chloride in dilute solution in the presence of triethylamine. The 1,2,5-thiadiazole-3,5-dione (**189**) was formed in almost quantitative yield ⟨72LA(759)107⟩.

4.03.3.4.3 Ring-fused systems

Ring-fused systems with C—C points of fusion cannot be prepared by this approach. However, numerous sytems with a nitrogen atom at the fusion point have been synthesized, reflecting incorporation of the above binucleophiles in a heterocyclic ring. Reaction of 2-aminobenzothiazole (**190**) with (phenylimino)oxalic acid dichloride (**191**) gave 3-(phenylimino)-2-oxoimidazo[2,1-*b*]benzothiazole (**192**), and 2-mercaptobenzimidazole (**193**) afforded 2-phenylimino-3-oxothiazolo[3,2-*a*]benzimidazole (**194**) ⟨71KGS471⟩. These authors based their assignment of the direction of ring closure on spectral data.

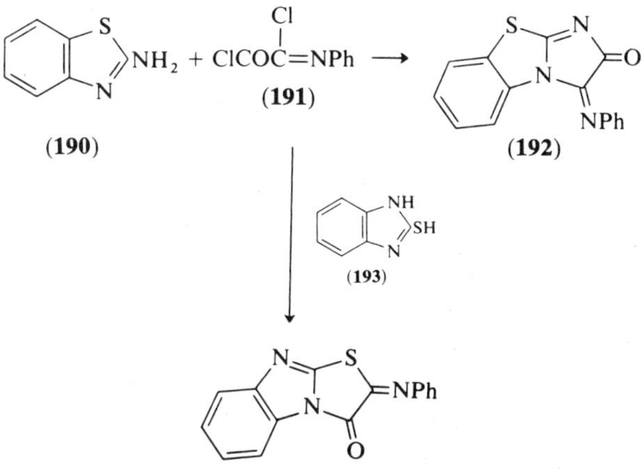

Bielectrophiles containing a heteroatom are equally effective in ring annulations of this type. Both [5,5] and [5,6] ring-fused systems may be readily prepared by reaction of the appropriate 2-amino heterocycle containing an amidine function with trichloromethanesulfenyl chloride (**181**). Reaction of 2-aminopyridine (**195**) with (**181**) in chloroform gave the intermediate sulfenimide (**196**) which with an aromatic amine in the presence of triethylamine underwent ring closure to (**198**), probably *via* the intermediate (**197**). Treatment of (**195**) with the iminochloromethanesulfenyl chloride (**184**) led to (**198**) directly ⟨70JOC1965, 71JOC1846, 73JOC3087, 75JOC2600⟩.

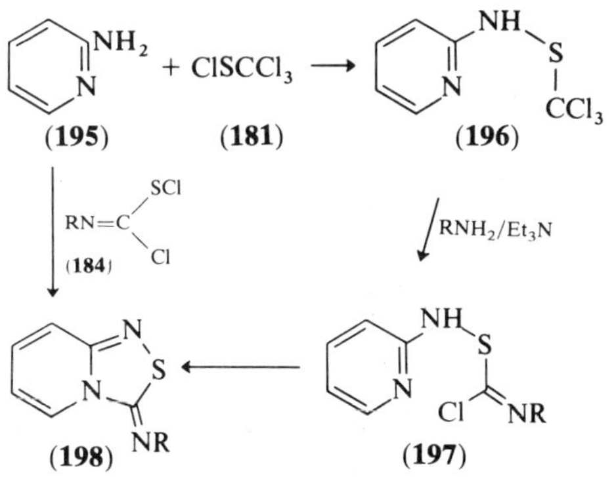

4.03.3.5 Syntheses Using 1,3-Bielectrophiles

4.03.3.5.1 Ring Systems with Two Heteroatoms

Although the wide diversity of structural types illustrated in Scheme 6 suggests that numerous examples of their application to five-membered heterocycles would be known, their utilization has been restricted because of the comparative shortage of 1,2-binucleophiles. However, they are more suited to the synthesis of six-membered systems, and it is in this area that their potential application lies.

The imonium salt (**199**), obtained from ynamines and phosgeneimonium chloride, underwent ready reaction with monosubstituted hydrazines to give the 3,5-bis(dimethylamino)pyrazole (**200**) ⟨68T4217, 69T3453⟩. Similarly, the adduct (**201**), resulting from the addition of phosgene to ynamines, likewise reacted with *sym*-disubstituted hydrazines to give pyrazoles (**202**). With hydroxylamine derivatives the isoxazolinone (**203**) was obtained.

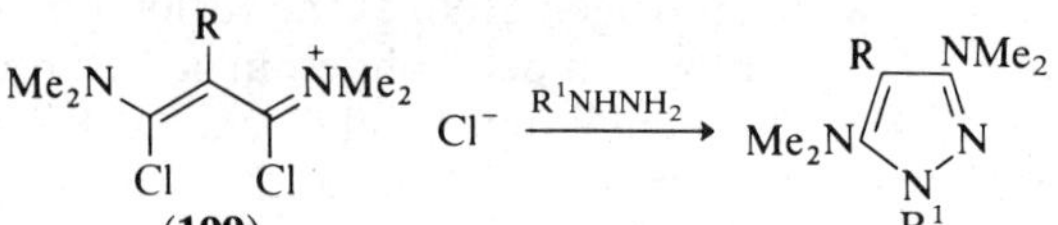

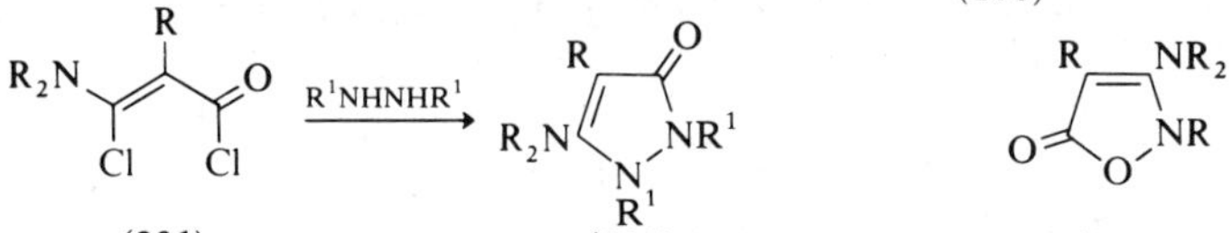

α-Functionalized ketene dithioacetals (Type 3, Scheme 6) and thioacetals are also useful synthons for five-membered rings. Thus, the α-oxoketene dithioacetal (**204**) reacted with a monosubstituted hydrazine on heating in *n*-butanol to give the pyrazole (**205**) in greater than 62% yield ⟨76ZC398⟩. The dithioacetal (**206**), derived from ethyl cyanoacetate, reacted with hydrazine to give products dependent on the reaction conditions. With hydrazine in a 5:1 ratio the pyrazole (**207**) was formed; heating in a 1:1 ratio in ethanol resulted in condensation at the cyano group giving the pyrazole (**208**) in 66% yield ⟨76ZC16⟩.

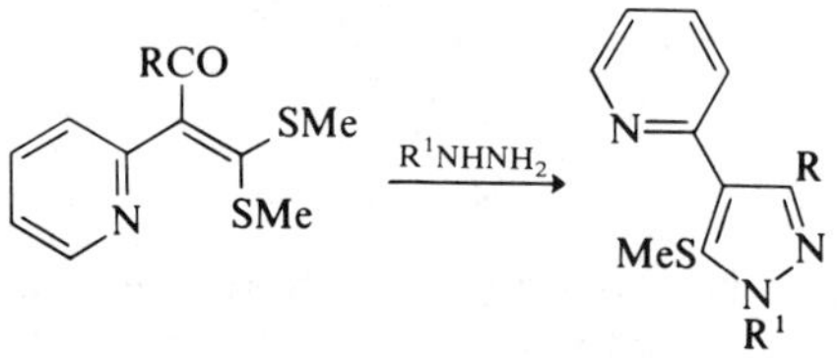

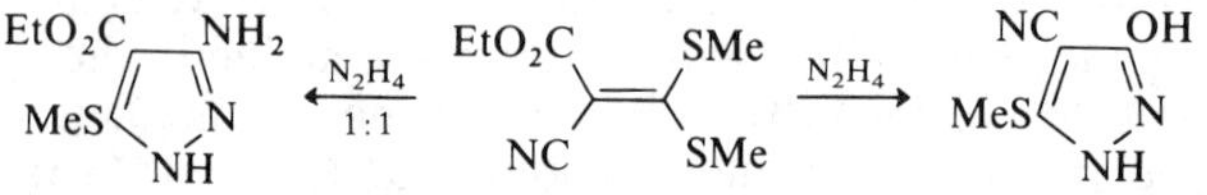

4.03.3.5.2 Ring systems with three or more heteroatoms

To form a five-membered ring in this manner, the 1,3-bielectrophile must contain a heteroatom, and several systems such as the aza analog of (**204**) are known ⟨68T4217, 69T3453⟩. Dimethyl *N*-ethoxycarbonylthiocarbonimidate (**209**) reacted readily with a monosubstituted hydrazine to give the 1,2,4-triazolinone (**210**), and with hydroxylamine the 1,2,4-oxadiazolinone (**211**) was obtained ⟨73JCS(P1)2644⟩.

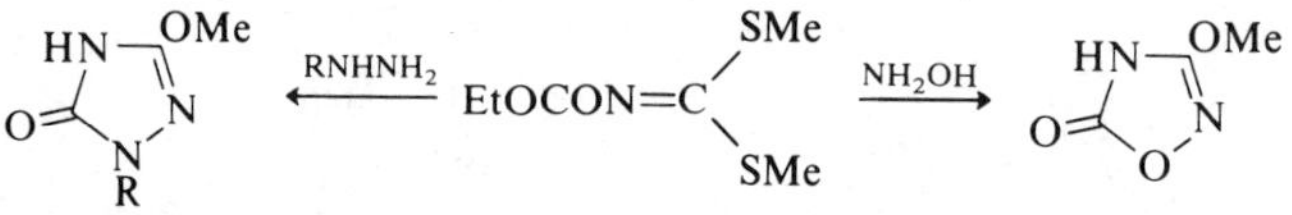

An interesting illustration of a bielectrophile contributing two heteroatoms to the resultant five-membered ring is the 2-alkyl-2-chloro- (or fluoro-) sulfonylcarbamoyl chlorides (**212**). With methylhydrazine initial attack by the more basic nitrogen occurred on the carbamoyl chloride, and this was followed by base-induced cyclization to 1,2,3,5-thiatriazolidine derivative (**213**) ⟨77JCR(S)238, 77JCR(M)2813⟩. Other reactions of this type are discussed in Chapter 4.28.

(212) (213)

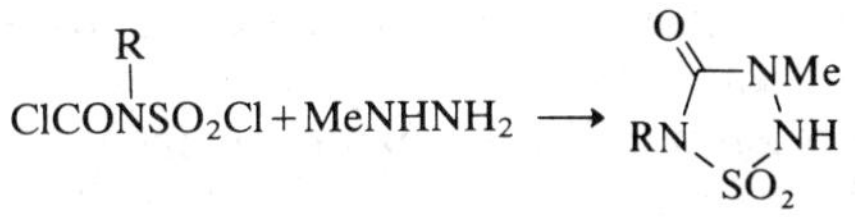

4.03.3.5.3 Ring-fused systems

1,3-Bielectrophiles have found appreciable applications in the synthesis of ring-fused systems, especially those involving [5,6] fused systems. The following serve to illustrate these applications. Reaction of pyrazole with (chlorocarbonyl)phenyl ketene (**214**) (Type 1, Scheme 6) readily formed the zwitterionic pyrazolo[1,2-*a*]pyrazole derivative (**215**) ⟨80JA3971⟩. With 1-methylimidazole-2-thione (**216**), anhydro-2-hydroxy-8-methyl-4-oxo-3-phenyl-4*H*-imidazo[2,1-*b*][1,3]thiazinium hydroxide (**217**) was obtained ⟨80JOC2474⟩.

(215) (214) (217)

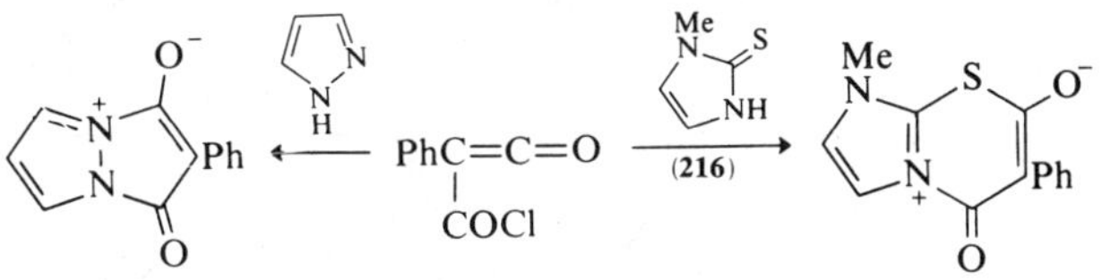

The 3-chloroacrylaldehyde (**218**) (Type 1, Scheme 6) reacted readily with 2-amino-1,3,4-selenadiazole (**219**). The final product was the [1,3,4]selenadiazolo[3,2-*a*]pyridinium salt (**220**) ⟨76ZC337⟩.

(218) (219) (220)

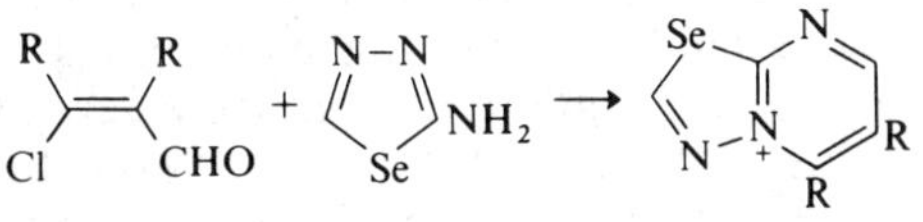

4.03.4 INTRAMOLECULAR RING CYCLIZATIONS

4.03.4.1 Oxidative Ring Closure Reactions

4.03.4.1.1 C—N bond formation

Oxidative procedures have been utilized for the synthesis of both monocyclic five-membered heterocycles and their ring-fused analogs, although the ease of synthesis of the precursors for the latter ring closures results in wider application of this procedure. A variety of oxidizing agents have been used and the conversion of the benzylidene hydrazidines (**221**) into the 4-arylamino-1,2,4-triazole (**222**) was effected with mercury(II) oxide ⟨77BCJ953⟩.

(221) (222)

(225) (223) (224)

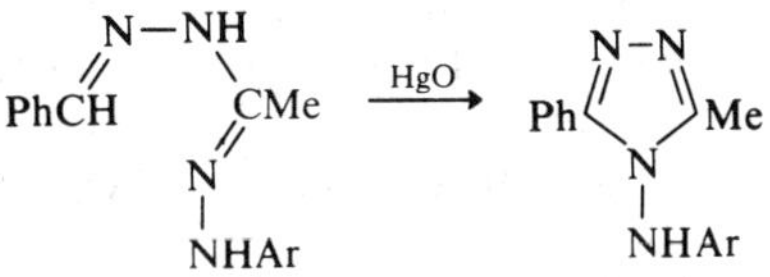

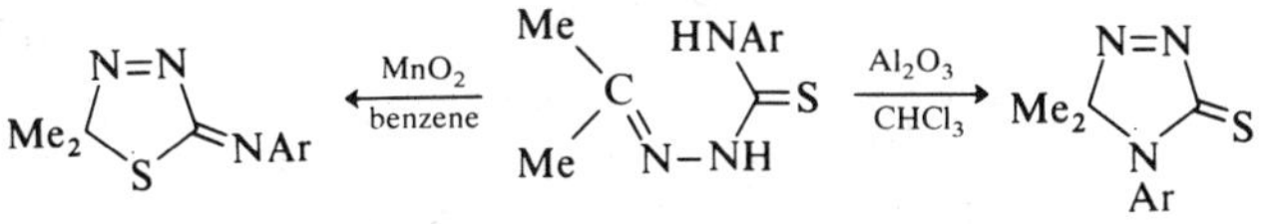

The direction of ring closure can often be influenced by the nature of the dehydrogenation agent. The substituted thiosemicarbazone (**223**) with Al_2O_3 in chloroform formed the

1,2,4-triazoline-3-thione (**224**) but if manganese dioxide in benzene were used, the thiadiazoline (**225**) resulted ⟨70JCS(C)63⟩.

Lead tetraacetate is often used as the oxidizing agent for the conversion of hydrazones into ring-fused systems.

Oxidation of the hydrazone of 2-hydrazinopyrazole (**226**) with $Pb(OAc)_4$ in CH_2Cl_2 is a two-step reaction. The azine (**227**) was formed as an intermediate and this underwent ring closure to the 3*H*-pyrazolo[5,1-*c*][1,2,4]triazole (**228**) ⟨79TL1567⟩. A similar reaction applied to the benzal derivative of 2-hydrazinobenzothiazole (**229**) gave 3-phenyl-[1,2,4]triazolo[3,4-*b*]benzothiazole (**230**) together with a by-product (**231**) ⟨72JCS(P1)1519⟩.

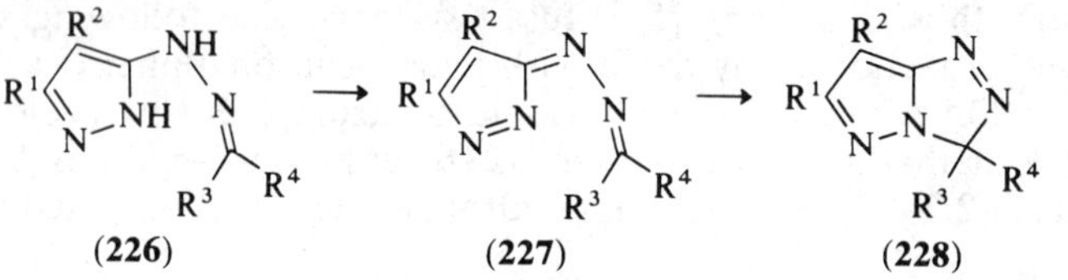

(**226**) (**227**) (**228**)

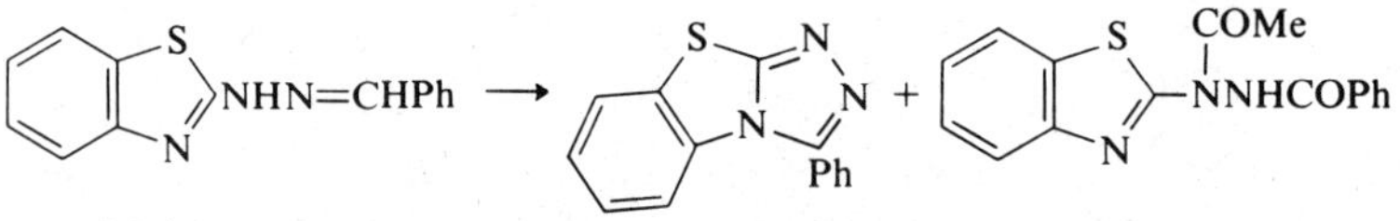

(**229**) (**230**) (**231**)

4.03.4.1.2 N—N bond formation

Oxidative processes leading to N—N bond formation have been utilized to form a variety of five-membered heterocycles and also in ring annulations. The bis-hydrazone (**232**) underwent ready ring closure to the 1-amino-1,2,3-triazole derivative (**233**) with mercury(I) acetate or manganese dioxide ⟨67TL3295, 71JPR882, 71ZC179⟩. In a related oxidative ring closure, the osazone (**234**) with potassium ferricyanide gave the deeply colored zwitterionic 1,2,3-triazole derivative (**235**) ⟨74T445⟩.

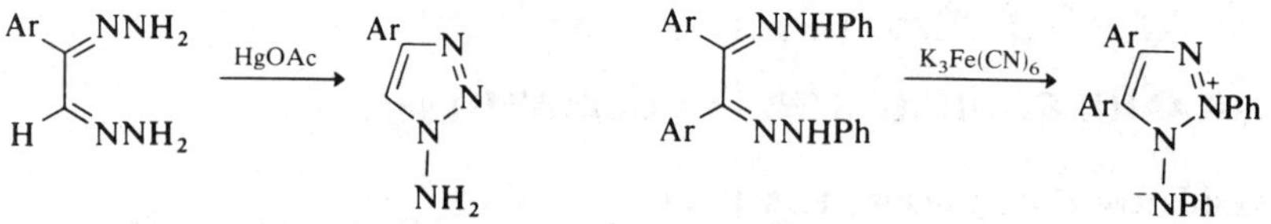

(**232**) (**233**) (**234**) (**235**)

Formation of ring-fused systems *via* N—N bond formation can occur in several ways, depending on the structure of the particular precursor. A variety of oxidizing agents have been employed and the following reactions illustrate the various possibilities. Oxidation of the hydrazone (**236**) with NBS gave *v*-triazolo[5,1-*b*]-thiazolylium and -imidazolylium salts (**237**; X = S and NH, respectively). The imidazolylium salts readily lost HBr in the presence of pyridine to form the free base ⟨67AG(E)261⟩. A precursor amidine (**238**) with $Pb(OAc)_4$ underwent facile ring closure to *s*-triazolo[1,5-*a*]pyridine derivatives (**239**) and a wide variety of amidines will undergo related ring closures (see Chapter 4.15) ⟨66JOC260⟩.

Ring closure to give C—C fusion points in the resultant ring-fused system is also possible. Lead tetraacetate in glacial acetic acid effects closure between the primary amino group and the azo group of (**240**) giving the *v*-triazolo[4,5-*d*]pyrimidine (**241**). It is possible that this closure involves a nitrenoid-type intermediate (see Section 4.03.9.2) ⟨72CPB605⟩.

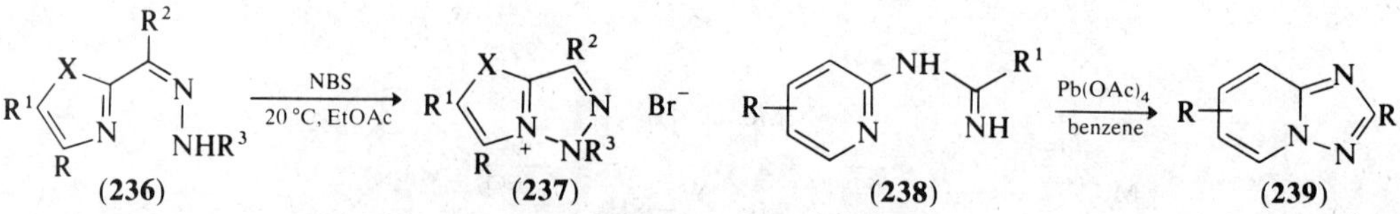

(**236**) (**237**) (**238**) (**239**)

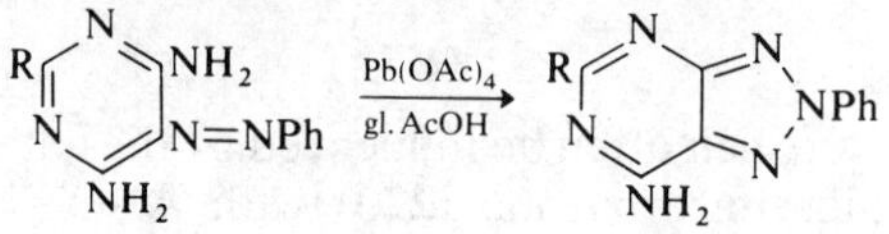

(**240**) (**241**)

4.03.4.1.3 C—S bond formation

This process is not as common as the other oxidative procedures and usually involves ring closure onto an aromatic or heteroaromatic ring. The following examples illustrate the structural types required for this cyclocondensation.

It is particularly appropriate for the synthesis of benzothiazoles (the Jacobson–Hugershoff synthesis) and an example among many is the oxidative ring closure of the thioanilide of diethyl methanetricarboxylate (**242**) with bromine in polar solvents. Sulfenyl bromides (**243**) most likely function as intermediates yielding the benzothiazole derivative (**244**) ⟨73ZC176⟩. A variety of oxidizing agents will effect ring closures of this type, including thionyl chloride ⟨70BCJ2535⟩.

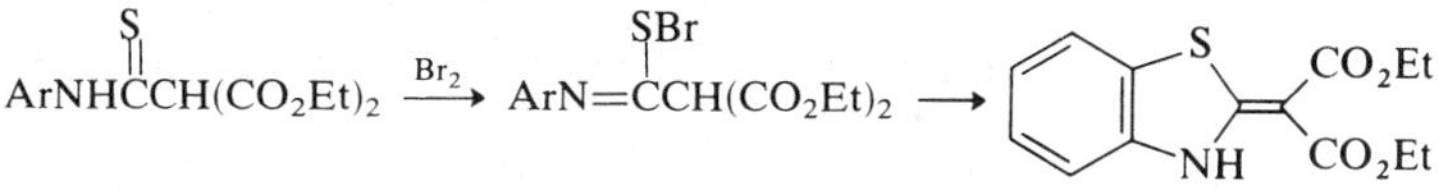

The 2-thienylthiourea (**245**) on oxidation with bromine in acetic acid gave the thieno[3,2-*d*]thiazole (**247**). It has been suggested that the intermediate electrophilic sulfenyl bromide adds to the 2,3-bond of the thiophene ring to form (**246**) when then loses HBr to give (**247**) ⟨71AJC1229, 78JHC81⟩. Pyrazolo(3,4-*d*]thiazoles are formed in a similar fashion ⟨76GEP2429195⟩.

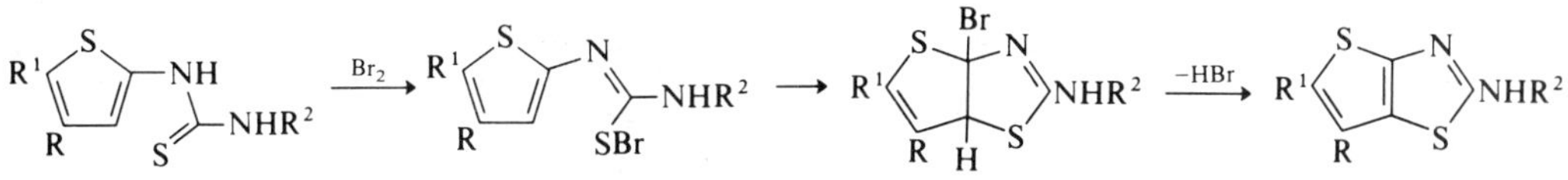

4.03.4.1.4 N—S bond formation

Numerous examples of N—S bond formation using oxidative conditions have been described in the literature. A convenient synthesis of isothiazoles involves the direct oxidation of γ-iminothiols and numerous variations have been studied (see Chapter 4.17). The oxidation of the amidine (**248**) to give the 3-aminoisothiazole (**249**) illustrates the reaction scheme ⟨65AHC(4)107, 72AHC(14)1⟩, which has been extended to the synthetically useful 5-amino-4-cyano-3-methylisothiazole (**251**) obtained by oxidation of (**250**) with hydrogen peroxide ⟨75JHC883⟩.

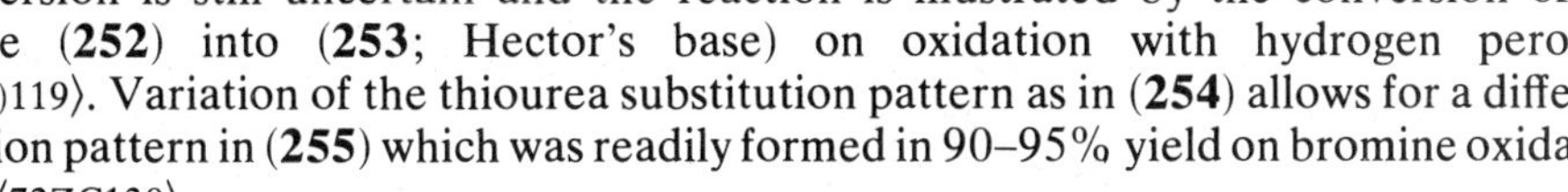

1,2,4-Thiadiazoles are conveniently prepared from thioamides or substrates containing an analogous functional group on treatment with a variety of oxidizing agents including halogens, hydrogen peroxides, sulfur halides, *etc.* (see Chapter 4.25). The mechanism of the conversion is still uncertain and the reaction is illustrated by the conversion of the thioamide (**252**) into (**253**; Hector's base) on oxidation with hydrogen peroxide ⟨65AHC(5)119⟩. Variation of the thiourea substitution pattern as in (**254**) allows for a different substitution pattern in (**255**) which was readily formed in 90–95% yield on bromine oxidation of (**254**) ⟨72ZC130⟩.

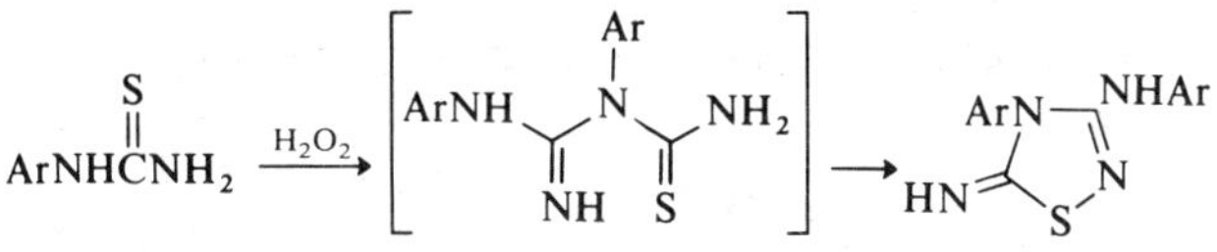

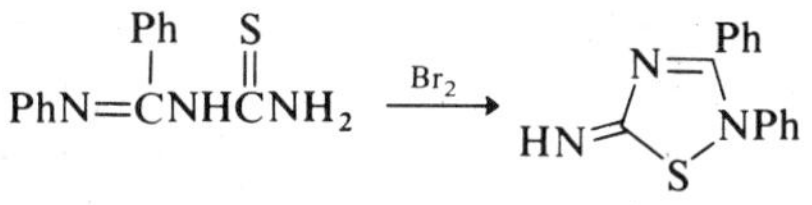

This procedure is equally successful in the formation of ring-fused systems. Hydrogen peroxide oxidation of the vicinal aminothioamide (**256**) readily gave 3-aminoisothiazolo[3,4-*b*]pyridine (**257**) ⟨73CJC1741⟩, and the same oxidant readily converted (**258**) into 3-aminonaphtho[1,2-*c*]isothiazole (**259**) ⟨77JPR65⟩. Oxidative ring closure onto a ring nitrogen atom also occurs readily as in the formation of the [1,2,4]thiadiazolo[2,3-*a*]pyridine (**261**) from the 2-pyridylthiourea (**260**). This ring closure was equally successful with aza analogs of (**260**) ⟨75JHC1191⟩. Bromine oxidation of 2-thiazolylthiourea (**262**) gave the 2-aminothiazolo[3,2-*b*][1,2,4]thiadiazolylium bromide (**263**) ⟨71JPR1148⟩, and the action of the same oxidant on the cyclic thiourea (**264**) resulted in formation of the interesting 2,3,5,6-tetrahydroimidazo[1,2-*d*][1,2,4]thiadiazol-3-ones (**265**) ⟨73JPR539⟩.

(**256**) (**257**) (**258**) (**259**)

(**260**) (**261**) (**262**) (**263**)

(**264**) (**265**)

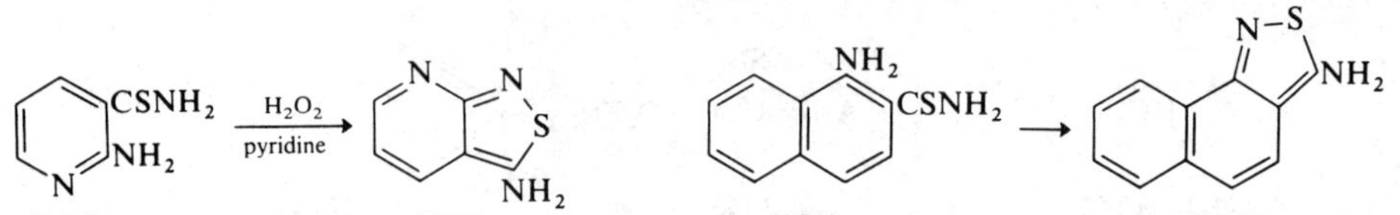

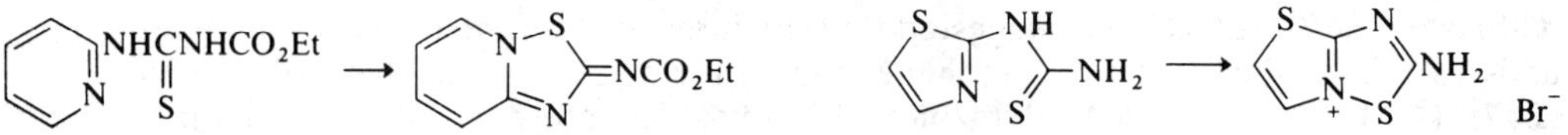

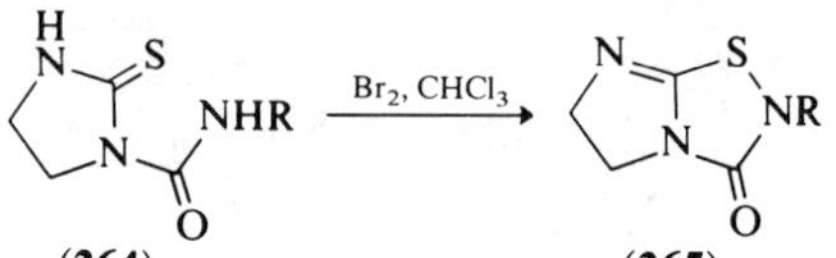

4.03.4.1.5 O—C bond formation

A variety of oxidizing agents can be utilized for O—C bond formation leading to both five- and six-membered rings. Lead tetraacetate has been used for hydrazone cyclization onto a carbonyl oxygen atom as in the conversion of the hydrazone (**266**) into the 1,2,4-oxadiazolyl ether (**267**) ⟨76NKK782⟩. The semicarbazones (**268**) also were converted into the 2-amino-5-benzoyl-1,3,4-oxadiazoles (**269**) ⟨72AC(R)11, 76MI40300⟩. α,β-Unsaturated ketoximes such as (**270**) undergo oxidative ring closure with bis-triphenylphosphinepalladium dichloride to give 3,5-diphenylisoxazole (**271**) in 92% yield ⟨73TL5075⟩ (see also Chapter 4.16).

(**266**) (**267**) (**268**) (**269**)

(**270**) (**271**)

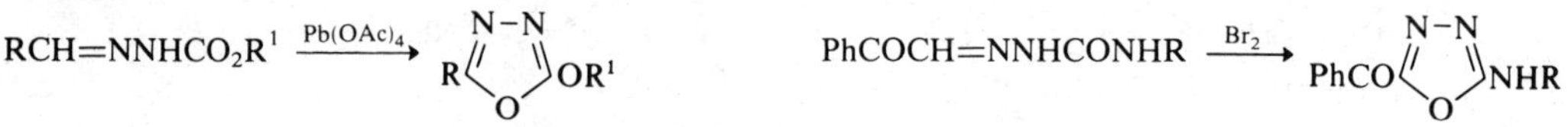

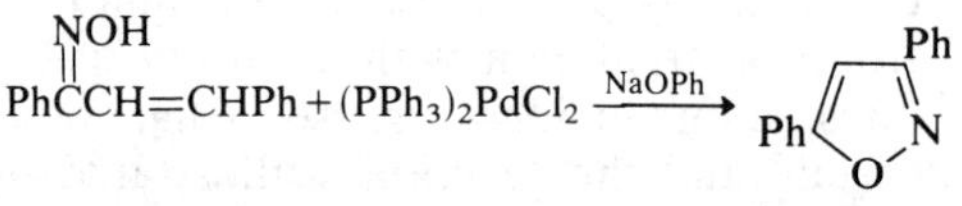

Formation of ring-fused systems by this procedure is not as common. If the vicinally substituted hydroxyamino derivative is available, then Pb(OAc)$_4$ treatment will lead to a ring-fused oxazole as in the conversion of (**272**) into (**273**) ⟨71JCS(C)1482⟩. In a different approach the CH=N— functional group is generated *in situ*. The 2-(1-pyrrolidinyl)ethanol

(**272**) (**273**)

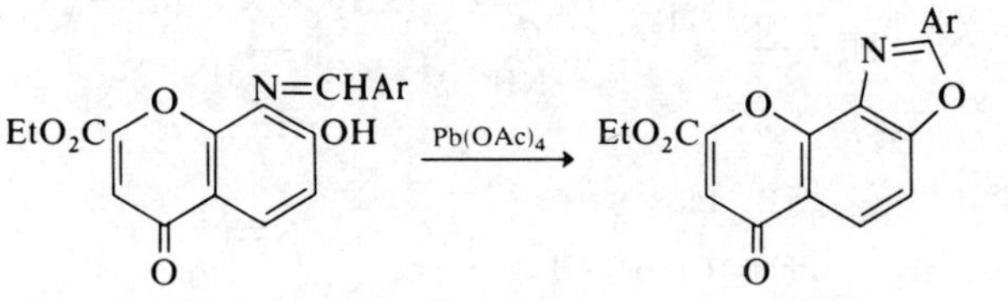

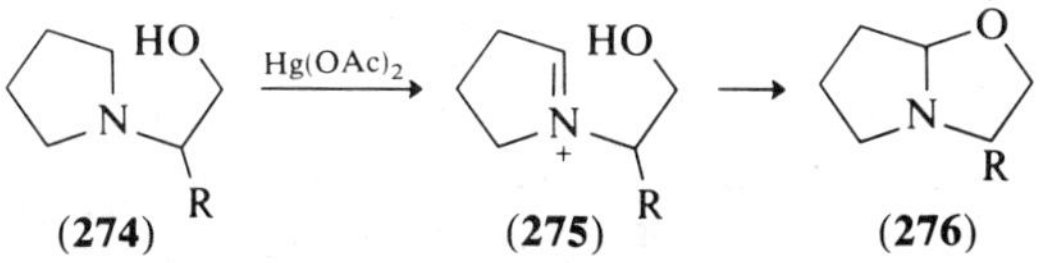

(**274**) on oxidation with mercury(II) acetate is ultimately converted into the pyrrolidino[2,1-b][1,3]oxazolidine derivative (**276**). The intermediate imonium salt (**275**) is considered to undergo nucleophilic attack by the hydroxy group ⟨60JA5148⟩.

4.03.4.1.6 O—N bond formation

In comparison to N—S bond formation, O—N bond formation by essentially oxidative procedures has found few applications in the synthesis of five-membered heterocycles. The 1,2,4-oxadiazole system (**278**) was prepared by the action of sodium hypochlorite on N-acylamidines (**277**) ⟨76S268⟩. The N-benzoylamidino compounds (**279**) were also converted into the 1,2,4-oxadiazoles (**280**) by the action of t-butyl hypochlorite followed by base. In both cyclizations N-chloro compounds are thought to be intermediates ⟨76BCJ3607⟩.

$R = Ph, MeO, C_5H_{10}N$

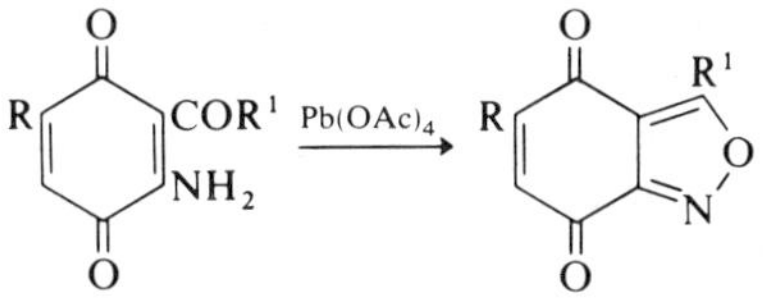

Ring-fused systems have also been prepared using this bond formation approach. Treatment of the β-aminoketone (**281**) with lead tetraacetate gave the isoxazole system (**282**) ⟨74S30⟩.

4.03.4.1.7 S—S, S—Se and Se—Se bond formation

A general route to 1,2-dithiolanes (**284**) involves the direct oxidation of the 1,3-dithiol (**283**) with hydrogen peroxide at 75 °C in acetic acid containing potassium iodide. Iron(III) chloride or lead tetraacetate may be used as the oxidizing agent but the instability of the product makes the H_2O_2 method the one of choice ⟨69JOC36⟩ (see Chapter 4.31). β-Dithio compounds such as (**285**) undergo oxidation in acid medium to yield the 1,2-dithiolylium salt (**286**) ⟨76JCS(D)455⟩.

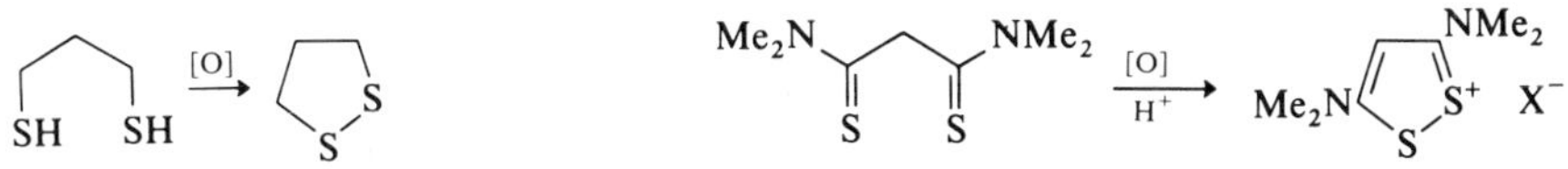

In an analogous fashion the oxidative coupling of 1,3-diselenols is used for the preparation of 1,2-diselenolanes, the precursors being obtained by a variety of methods (see Chapter 4.35). Similar reactions have been used for the preparation of 1,2-thiaselenolanes. Substituted propane-1,3-diselenones (**287**) are oxidized by halogen to the corresponding 3,5-diamino-1,2-diselenolylium chloride (**288**) ⟨67AJC1991⟩.

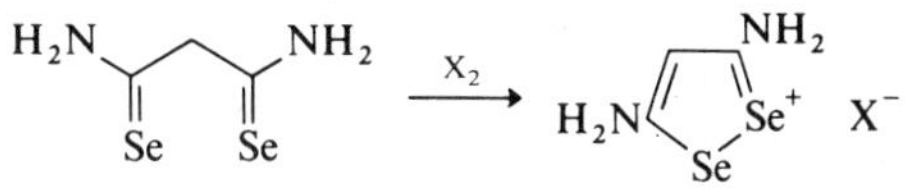

4.03.4.2 Electrophilic Ring Closures *via* Acylium Ions and Related Intermediates

Ring closures considered under this heading are essentially intramolecular cyclodehydrations (or their equivalent) and bear many resemblances to the intermolecular processes discussed in Section 4.03.3. They are more suited to the synthesis of ring-fused systems because of the ease of access to the requisite precursors.

Acyl or thioacyl α-amino acids and their derivatives undergo cyclization to oxazole and thiazole derivatives with a variety of cyclodehydration agents. Reaction of an amino acid with thioacetic acid under controlled conditions gives the *N*-thioacetyl derivative (**289**). If the reaction is carried out at 100 °C over 16 hours, then cyclization occurs to the thiazole (**290**) ⟨69CC818⟩. This process has been developed as a step-wise process for the degradation of polypeptides ⟨69MI40301⟩. Numerous variations of this reaction have been described (see Chapter 4.19), an interesting variation being the use of thiocarbonyl amino acid silyl esters. Treatment of the trimethylsilyl ester (**291**) with phosphorus tribromide gave the thiazolidine-2,5-dione (**292**) in high yield and purity ⟨71CB3146⟩. This procedure offers considerable advantages over previous ones. If trifluoroacetic acid is used as the cyclodehydration agent for (**289**), then 2-thiazolin-5-ones are obtained ⟨69JCS(C)1117⟩.

MeCSNHCHR^{1}CO$_2$H ⟶ (**289**)   (**290**)   ROCSNHCHR^{1}CO$_2$SiMe$_3$ $\xrightarrow{PBr_3}$ (**291**)   (**292**)

Substitution of the nitrogen atom in (**289**) and subsequent ring closure of (**293**) under acid cyclodehydration conditions gave the mesoionic system anhydro-5-hydroxythiazolium hydroxide (**294**). These reactions are analogous to the cyclodehydration of the *N*-nitrosoglycines (**295**) with acetic anhydride to give the sydnones (**296**) (see Chapter 4.21).

RCSNCHR^{2}CO$_2$H ⟶ (**293**)   (**294**)   RNCHR^{1}CO$_2$H | NO ⟶ (**295**)   (**296**)

Various ring annulations are possible using reactions analogous to those described above. 2-Acetamido-5-thiazolylthioglycolic acid (**297**) on heating with phosphorus oxychloride underwent ring closure to 2-acetamidothieno[3,2-*d*]thiazolin-4-one (**298**) ⟨56AC(R)275⟩. Corresponding ring closures in the thiophene series result in isomerization occurring in the final product ⟨62ACS(B)155⟩. Ring closure onto a ring junction heteroatom, as in the ready conversion of (1-methylimidazol-2-yl)thioglycolic acid (**299**) into the ring-fused mesoionic system (**300**), occurred with acetic anhydride ⟨79JOC3803⟩. Acylations of this type are characteristic of reactive mesoionic systems without a substituent adjacent to the carbonyl group.

HO$_2$CCH$_2$S—(**297**)—NHCOMe $\xrightarrow{POCl_3}$ (**298**)

(**299**) $\xrightarrow{Ac_2O}$ (**300**)

The cyano group is also a suitable precursor for an analogous intramolecular cyclization. Reaction of the thioacetonitrile derivative (**301**) with HBr resulted in ready ring closure to form the 3-amino-2-ethoxycarbonylthiazolo[3,2-*a*]pyridinium bromide (**302**) ⟨78JCR(S)407⟩. This is a general reaction and is a successful procedure for the annulation of a 3-aminothiazole function to a variety of five- and six-membered heterocycles containing an analogous relationship of cyano and nitrogen functions. It may also be applied to the synthesis of amino-substituted monocyclic systems, such as 4-aminoimidazoles (see Chapter 4.08).

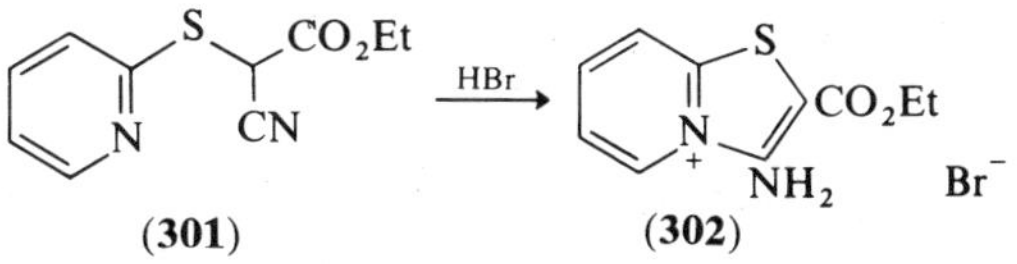

Synthesis of Five-membered Rings with Two or More Heteroatoms 139

4.03.4.3 Ring Closures *via* Intramolecular Alkylations

This synthetic process is more suited to the preparation of ring-fused sytems as the precursors are more readily available than their alicyclic counterparts. The pyrazoline (**303**) with a hydroxy group appropriately situated to form a ring underwent ring closure to the pyrazolinium salt (**304**) on heating with thionyl chloride ⟨68JOC3941⟩. Related ring closures can be effected in alternative ways, *e.g.* the alcohol (**305**) on heating with sulfuric acid or other reagents such as $SOCl_2$, $POCl_3$ or PCl_5 underwent ring closure to the racemate of tetramisole (**306**) ⟨66JMC545, 68JOC1350⟩.

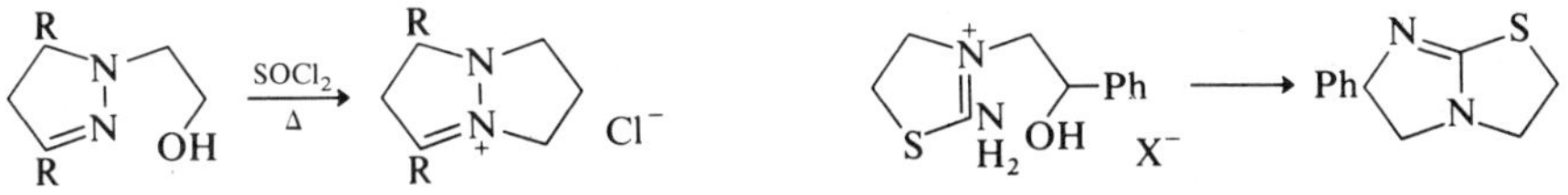

Sulfur may also be alkylated under appropriate conditions. On formation, the *N*-β-chloroethylthiazolidine-2-thione (**307**) underwent spontaneous ring closure to the tetrahydrothiazolo[2,3-*b*]thiazolylium chloride (**308**) ⟨71CHE1534⟩ and a similar reaction occurred in the imidazole series.

2-Chloromethylpyrrolidine (**309**) when treated with CS_2 in DMF in the presence of potassium carbonate readily formed the dithiocarbamate (**310**). In the presence of base, intramolecular alkylation on sulfur occurred to give (**311**) ⟨63JOC981⟩.

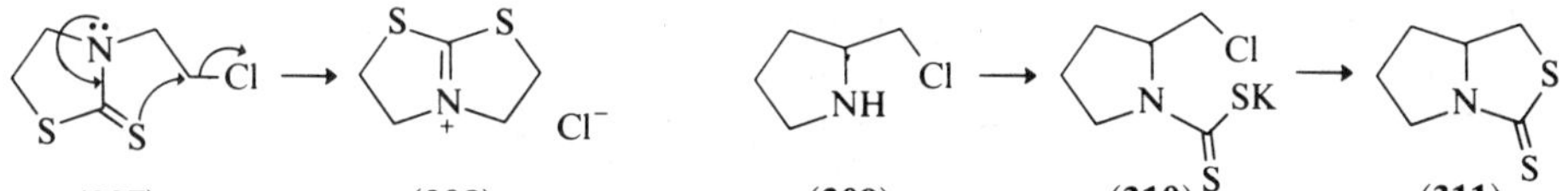

Similar alkylations may be effected on oxygen. 1-(2-Chloroethyl)imidazolidin-2-one (**312**) when treated with potassium hydroxide or sodium hydride underwent ring closure to the tetrahydroimidazo[2,1-*b*]oxazole (**313**) ⟨57JA5276⟩. This approach can be used for the preparation of bicyclic hydantoins and the corresponding dihydro derivatives of (**313**) using the mesylate of (**312**) and NaH ⟨77JHC511, 79JMC1030⟩.

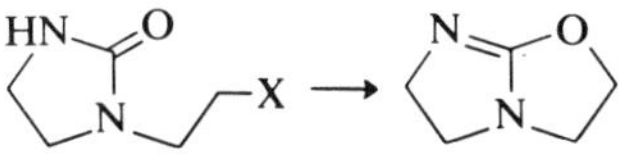

4.03.4.4 Ring Closures *via* Nucleophilic Additions

The addition of nucleophiles to double and triple bond systems is often a convenient way of effecting an intramolecular ring closure. Addition to cyano groups has received considerable attention, as in addition to ring formation it provides a convenient method for the introduction of an amino group. Reaction of methyl *N*-cyanodithiocarbimidate with *N*-methylaminoacetonitrile resulted in displacement of methanethiol and formation of (**314**). Sodium ethoxide treatment in DMF converted (**314**) into a 4-amino-5-cyanoimidazole

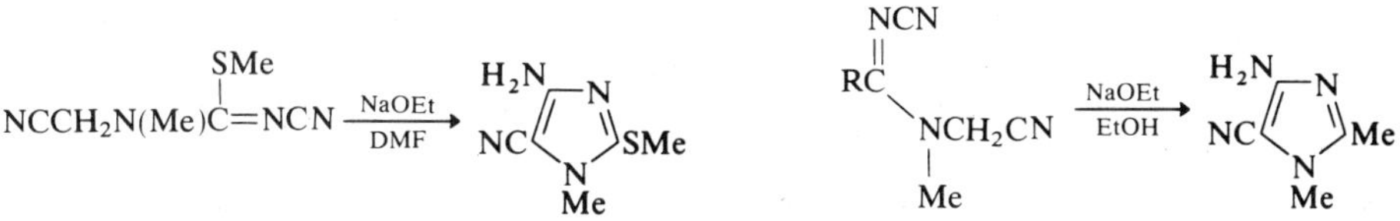

(315) ⟨70AHC(12)103⟩. A similar ring closure involving a carbanion occurred when an N-cyanoamidine such as (316) was treated with sodium ethoxide in ethanol to give (317) in moderate to good yields ⟨75HCA2192⟩.

Five-membered ring formation in this way occurs with an appreciable number of suitably substituted substrates. Reaction of 1,3-oxathiolylium salts with cyanamide gave the ring-opened product (318). When this reaction was carried out in the presence of sodium hydroxide, (318) underwent a facile ring closure to afford the aminobenzoylthiazole (319) in good yield ⟨71ZC421, 73JPR497⟩. Opening of the oxathiolylium nucleus with a variety of carbonyl-stabilized carbanions and subsequent ring closure in this fashion lead to amino-substituted thiophenes.

N-Nucleophiles also add readily to a cyano group, the product depending on the degree of substitution of the nucleophilic amino group. Conversion of an amidine into its N-chloro derivative (320) with sodium hypochlorite and addition of potassium thiocyanate in ethanol at 0 °C resulted in good yields of 5-amino-1,2,4-thiadiazoles (322), the intermediate (321) undergoing ready ring closure under the reaction conditions. This is a mild method for producing 5-amino-1,2,4-thiadiazoles and is discussed further in Chapter 4.25 ⟨65AHC(5)119⟩.

A related ring closure is also successful in the thiazole series. Treatment of a β-amino-α,β-unsaturated ketone with thiocyanogen gave the intermediate thiocyano compound (323) which underwent ring closure to the 2-iminothiazoline derivative (324) ⟨83MI40300⟩. Related reactions are described in Chapter 4.19, and for those involving potassium selenocyanate see Chapter 4.20.

Oxygen nucleophiles may also add to cyano groups, in many instances providing entry into the isoxazole system (Chapter 4.16). These reactions are illustrated by the conversion of the 2-arylhydrazono-3-oxonitrile (325) into a 5-aminoisoxazole (327). Treatment of (325) with hydroxylamine in the presence of sodium ethoxide gave the oximino derivative (326) which then underwent base-catalyzed ring closure to (327) ⟨75JOC2604⟩. If the hydroxylamine addition is carried out in refluxing ethanol, the 3-aminoisoxazole is formed via an intermediate amidoxime.

Examples are known of nitrogen, oxygen and sulfur nucleophiles adding to unsaturated carbon-carbon systems to provide ready syntheses of five-membered ring systems with two heteroatoms. The nucleophile supplies one of the heteroatoms, the other being incorporated into the precursor at some suitable stage in its synthesis. Reaction schemes utilizing this synthetic approach are illustrated by the reaction of the α-aminoacetylene (328) with carbon disulfide. The intermediate dithiocarbamic acid (329) underwent ring closure to the thiazole (330) which with sulfuric acid underwent a prototropic rearrangement to (331) ⟨49JCS786⟩. This addition of CS_2 is, however, sensitive to the degree of substitution at the amino group, 3-t-octylamino-1-butyne not undergoing reaction ⟨63JOC991⟩.

Propargyl alcohol (332) and (328) react readily with isocyanates in the presence of a basic catalyst to give 4-methylene-2-oxazolidinones (334) and 4-methylene-2-imidazolinones (336), respectively ⟨63JOC991⟩. In the absence of sodium methoxide the intermediate methanes (333) and ureas (335) were obtained and on treatment with sodium methoxide underwent ring closure. Moderate to excellent yields were obtained.

(336) (335) (328) (329)

(330) (331)

(332) (333) (334)

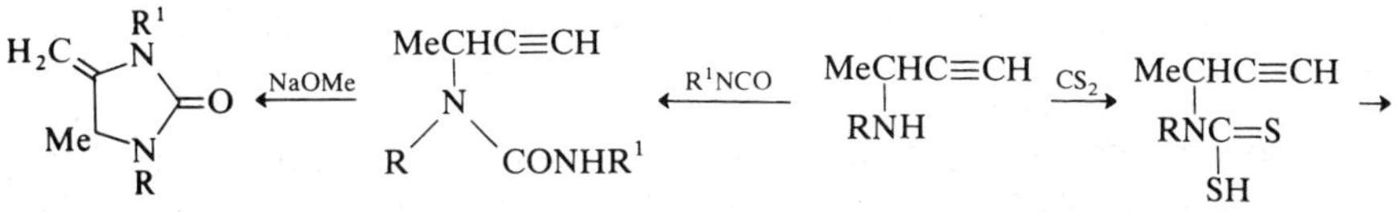

Ring-fused systems may also be prepared by this synthetic approach. Reaction of 2-mercaptobenzimidazole with 3-bromo-1-phenyl-1-propyne gave the thioether (**337**). Addition of the ring nitrogen to the triple bond is controlled by the reaction conditions. In alcohol with sodium ethoxide the 3-benzylthiazolo[3,2-*a*]benzimidazole (**338**) was obtained. On heating (**337**) in acetic acid with mercury(I) acetate and sulfuric acid the 4-phenyl-2*H*-thiazino[3,2-*a*]benzimidazole (**339**) was formed. In contrast the thioether (**340**), obtained from 2-mercaptobenzimidazole and propargyl bromide, under the latter conditions gave 3-methylthiazolo[3,2-*a*]benzimidazole (**341**) ⟨76S189⟩. Other ring closures of this general type are described in Chapter 4.36.

(338) (337) (339)

(340) (341)

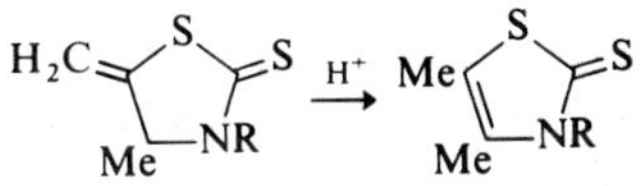

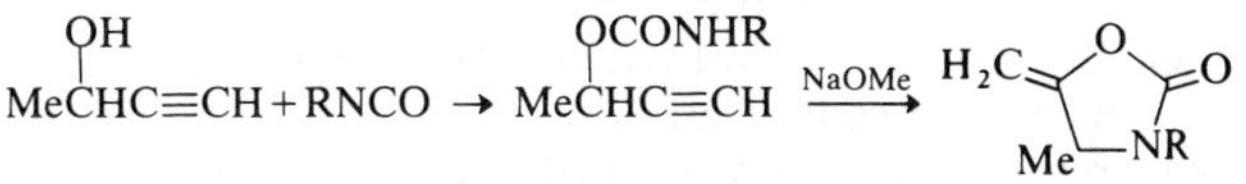

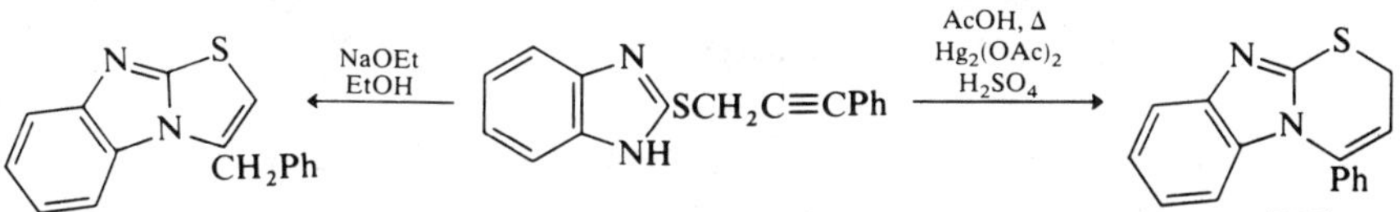

In contrast to the above additions *N*-allyl- and substituted *N*-allyl-amides, -urethanes, -ureas and -thioureas undergo intramolecular cyclization only in 60–96% sulfuric acid to give the corresponding oxazolinium and thiazolinium salts. Treatment of these cations with base yields 2-oxazolines and 2-thiazolines in moderate to good yields. The reaction is illustrated by the conversion of *N*-2-phenylallylacetamide (**342**) into 2,5-dimethyl-5-phenyl-2-oxazoline (**343**) in 70% yield ⟨70JOC3768⟩ (see also Chapter 4.19).

(342) (343)

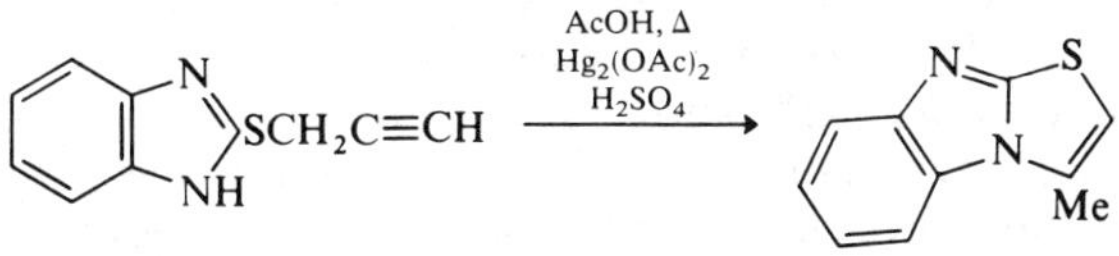

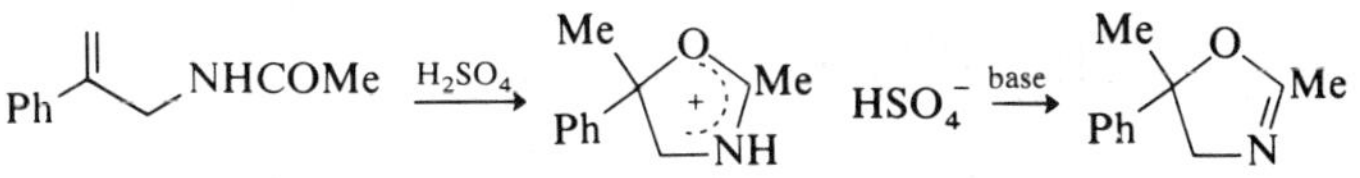

In this group of intramolecular additions the reaction apparently proceeds *via* an intermediate carbocation whose presence was established by NMR data.

4.03.4.5 Ring Closures *via* Radical Intermediates

Although some of the oxidative ring closures described above, *e.g.* reactions with lead tetraacetate (Section 4.03.4.1.2), may actually involve radical intermediates, little use has been made of this reaction type in the synthesis of five-membered rings with two or more heteroatoms. Radical intermediates involved in photochemical transformations are described in Section 4.03.9. Free radical substitutions are described in the various monograph chapters.

4.03.5 SYNTHESES INVOLVING CONJUGATE ADDITION REACTIONS

Conjugate addition reactions leading to five-membered heterocycles with two or more heteroatoms are predominantly intermolecular in nature, and a variety of nucleophiles have been added to alkynes conjugated with electron-withdrawing groups. Dimethyl acetylenedicarboxylate has found extensive use, both for monocyclic ring formation and for ring annulation reactions in which six-membered rings are usually formed.

Hydrazine and its mono- and di-substituted derivatives all undergo addition to acetylene mono- and di-carboxylic acid esters. Hydrazine hydrate and diethyl acetylenedicarboxylate formed ethyl 5-oxopyrazoline-3-carboxylate (**344**) ⟨1893CB1719⟩. Phenylhydrazine and DMAD formed a 1:1 adduct (**345**) that under basic conditions gave the *N*-phenylpyrazolinone (**346**) ⟨1889CB2929⟩. 1,2-Dimethyl- and 1,2-diphenyl-hydrazines gave analogous pyrazolinones. In ether solution 1,2-dimethylhydrazine and ethyl propiolate formed (**347**) but in ligroin a dicondensation product was obtained ⟨65LA(686)134⟩. Reaction conditions also control product formation in the reaction of hydrazobenzene with DMAD. In glacial acetic acid the pyrazolinone (**349**) was obtained, but in xylene methyl indole-2,3-dicarboxylate was produced on heating the intermediate enamine (**348**). In pyridine a quinoline derivative was isolated ⟨34LA(511)168⟩.

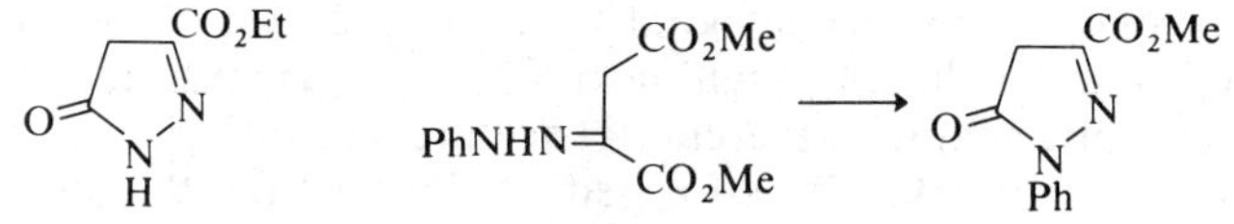

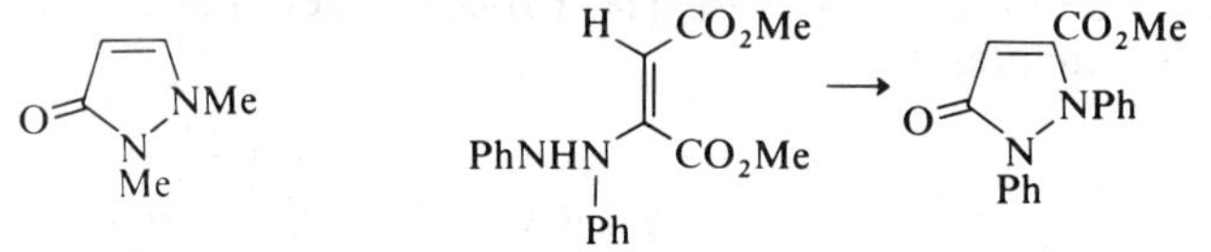

Ring annulations can be effectively carried out using a reaction of this type. As one of the unsaturated groups attached to the triple bond is usually also involved in the reaction, six-membered rings result. Reaction of 3-aminobenzisoxazole (**350**) with methyl propiolate gave a mixture of regioisomers (**351**) and (**352**) ⟨72CB794⟩. With 2-aminoselenazoles (**353**) only one cyclized product, 7*H*-selenazolo[3,2-*a*]pyrimidin-7-one (**354**), was obtained, but it was accompanied by the uncyclized product (**355**) and a 1:2 adduct (**356**) ⟨75JHC675⟩.

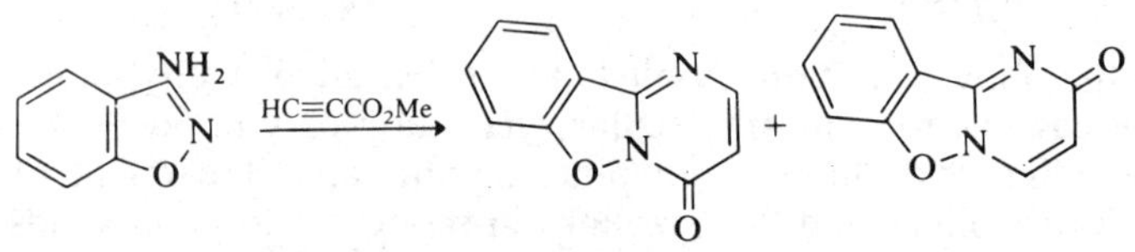

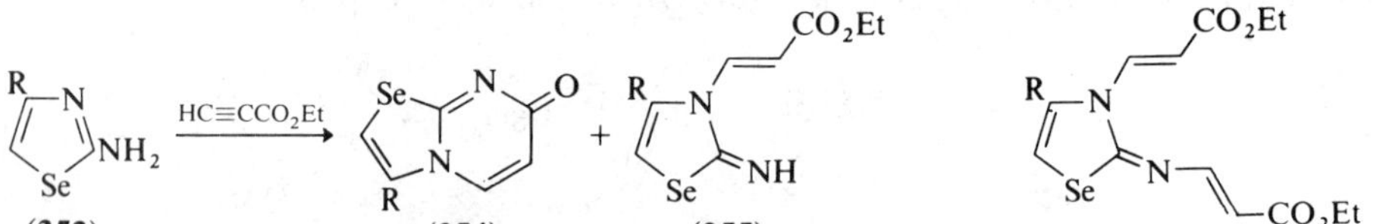

In contrast, 2-amino-5-ethyl-1,3,4-selenadiazole (**357**) and ethyl propiolate gave the ring-closed product 2-ethyl-7*H*-[1,3,4]selenadiazolo[3,2-*a*]pyrimidin-7-one (**358**) together with 2- or 3-ethoxycarbonyl-5*H*-selenazolo[3,2-*a*]pyrimidin-5-one (**359**), apparently formed as the major product in a ring opening–ring closure sequence ⟨75JHC675⟩. With DMAD (**357**) gave (**360**) as the principal product. Amino-1,3,4-oxadiazoles (Chapter 4.23 and 4.29) form analogous products as do the 1,3,4-thiadiazoles (Chapter 4.29).

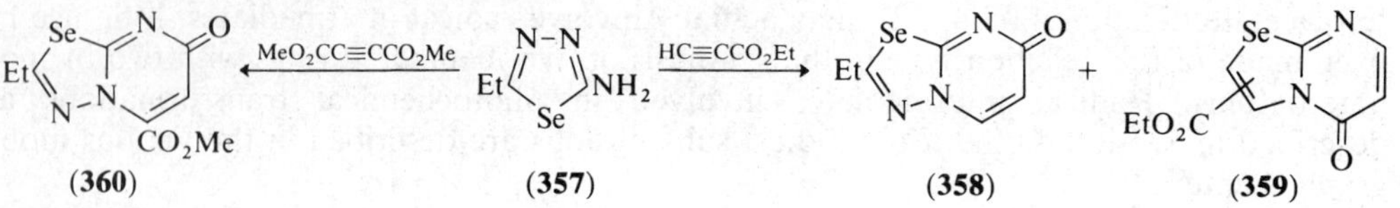

4.03.6 RING FORMATION *VIA* CYCLOADDITION REACTIONS

4.03.6.1 1,3-Dipolar Cycloadditions

A versatile method for the synthesis of a variety of five-membered heterocycles and their ring-fused analogs involves the reaction of a neutral 4π-electron–3-atom system with a 2π-electron system, the dipolarophile, which is usually electron deficient in nature. Available evidence, *e.g.* retention of dipolarophile stereochemistry in the product and solvent polarity exerting only a moderate influence on the reaction, indicates that the cycloaddition proceeds *via* a concerted mechanism ⟨63AG(E)565, 63AG(E)633, 68JOC2291⟩ and may be represented in general terms by the expression in Scheme 8.

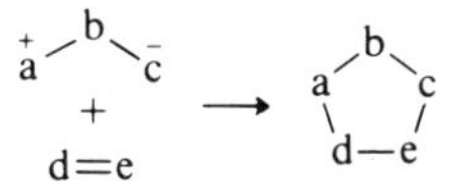

$$\overset{+}{a}\diagup^{b}\diagdown\overset{-}{c} + d{=}e \longrightarrow a\diagup^{b}\diagdown_{c} \atop d{-}e$$

Scheme 8 A 1,3-dipolar cycloaddition reaction

Table 2 illustrates 1,3-dipoles with a double bond and with internal octet stabilization, commonly referred to as the propargyl-allenyl anion type. These are all reactive dipoles and a large number of five-membered heterocycles can be constructed from these readily available dipoles, especially when the dipolarophile is varied to include heterocumulenes, *etc.*

Table 2 1,3-Dipoles with a Double Bond and Internal Octet Stabilization; Propargyl-allenyl Anion Type

$$\begin{array}{c} \overset{+}{a}{=}b{-}\overset{-}{c} \leftrightarrow a{\equiv}\overset{+}{b}{-}\overset{-}{c} \\ \updownarrow \qquad\qquad \updownarrow \\ \overset{-}{a}{=}b{-}\overset{+}{c} \leftrightarrow \overset{-}{a}{=}\overset{+}{b}{=}c \end{array}$$

Sextet structures   Octet structures

Dipole	Structure	Source
Nitrile ylide	$-\overset{+}{C}{=}N{-}\overset{-}{C}\diagdown^{\diagup} \leftrightarrow -C{\equiv}\overset{+}{N}{-}\overset{-}{C}\diagdown^{\diagup}$	*in situ* from $-C{=}N{-}CH\diagdown^{\diagup}$ with Cl
Nitrile imine	$-\overset{+}{C}{=}N{-}\overset{-}{N}{-} \leftrightarrow -C{\equiv}\overset{+}{N}{-}\overset{-}{N}{-}$	*in situ* from $-C{=}N{-}NH{-}$ with Cl
Nitrile oxide	$-\overset{+}{C}{=}N{-}\overset{-}{O} \leftrightarrow -C{\equiv}\overset{+}{N}{-}\overset{-}{O}$	*in situ* from $-C{=}NOH$ with Cl
Nitrile sulfide	$-\overset{+}{C}{=}N{-}\overset{-}{S} \leftrightarrow -C{\equiv}\overset{+}{N}{-}\overset{-}{S}$	Thermal fragmentation of an oxathiazolone
Diazoalkane	$\diagdown_{\diagup}\overset{+}{C}{-}N{=}\overset{-}{N} \leftrightarrow \diagdown_{\diagup}C{=}\overset{+}{N}{=}\overset{-}{N}$	Usually stable
Azide	$-\overset{+}{N}{-}N{=}\overset{-}{N} \leftrightarrow -N{=}\overset{+}{N}{=}\overset{-}{N}$	Usually stable
Nitrous oxide	$-\overset{+}{N}{=}N{-}\overset{-}{O} \leftrightarrow -N{\equiv}\overset{+}{N}{-}\overset{-}{O}$	Stable

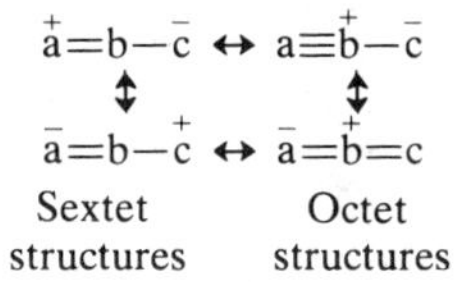

1,3-Dipoles without a double bond but with internal octet stabilization, referred to as the allyl anion type, are shown in Table 3. A third group, 1,3-dipoles without octet stabilization such as vinyl carbenes, iminonitrenes, *etc.*, is known, but these are all highly reactive intermediates with only transient existence. Reference is made to this type where appropriate and in Table 4 (p. 146).

Dipolarophiles utilized in these cycloadditions leading to five-membered heterocycles contain either double or triple bonds between two carbon atoms, a carbon atom and a heteroatom, or two heteroatoms. These are shown in Scheme 9 listed in approximate order of decreasing activity from left to right. Small rings containing a double bond (either C=C or C=N) are also effective dipolarophiles, but these result in six- and seven-membered ring systems.

Table 3 1,3-Dipoles Without a Double Bond but With Internal Octet Stabilization; Allyl Anion Type

	Sextet structures		Octet structures	
	$\overset{+}{a}\diagup\overset{b}{}\diagdown\overset{-}{c}$	↔	$a\diagup\overset{\overset{+}{b}}{=}\diagdown\overset{-}{c}$	
	↕		↕	
	$\overset{-}{a}\diagup\overset{b}{}\diagdown\overset{+}{c}$	↔	$\overset{-}{a}\diagdown=\overset{\overset{+}{b}}{}\diagup c$	

Name	Sextet		Octet	Notes
Azomethine ylide	$-\overset{+}{C}\diagup\overset{\mid}{N}\diagdown\overset{-}{C}-$	↔	$-C\diagup\overset{\overset{\mid}{N^+}}{=}\diagdown\overset{-}{C}-$	*in situ* from $-\overset{\mid}{C}\diagup\overset{\overset{\mid}{N^+}\,X^-}{=}\diagdown\overset{\mid}{C}-$ or aziridines
Azomethine imine	$-\overset{+}{C}\diagup\overset{\mid}{N}\diagdown\overset{-}{N}-$	↔	$-C\diagup\overset{\overset{\mid}{N^+}}{=}\diagdown\overset{-}{N}-$	*in situ* from $-\overset{\mid}{C}\diagup\overset{\overset{\mid}{N^+}\,X^-}{=}\diagdown NH-$
Nitrone	$-\overset{-}{C}\diagup\overset{\overset{\mid}{N^+}}{}\diagdown O$	↔	$-C\diagup\overset{\overset{\mid}{N^+}}{=}\diagdown\overset{-}{O}$	Stable
Azimine	$-\overset{+}{N}\diagup\overset{\mid}{N}\diagdown\overset{-}{N}-$	↔	$-N\diagup\overset{\overset{\mid}{N^+}}{=}\diagdown\overset{-}{N}-$	From heterocycles
Azoxy compound	$-\overset{+}{N}\diagup\overset{\mid}{N}\diagdown\overset{-}{O}$	↔	$-N\diagup\overset{\overset{\mid}{N^+}}{=}\diagdown\overset{-}{O}$	Stable
Nitro compound	$\overset{-}{O}\diagup\overset{\overset{\mid}{N^+}}{}\diagdown\overset{-}{O}$	↔	$O\diagup\overset{\overset{\mid}{N^+}}{=}\diagdown\overset{-}{O}$	Stable
Nitroso imine	$-\overset{+}{N}\diagup\overset{O}{}\diagdown\overset{-}{N}-$	↔	$-N\diagup\overset{\overset{+}{O}}{=}\diagdown\overset{-}{N}-$	
Nitroso oxide	$-\overset{+}{N}\diagup\overset{O}{}\diagdown\overset{-}{O}$	↔	$-N\diagup\overset{\overset{+}{O}}{=}\diagdown\overset{-}{O}$	
Carbonyl ylide	$-\overset{+}{C}\diagup\overset{O}{}\diagdown\overset{-}{C}-$	↔	$-C\diagup\overset{\overset{+}{O}}{=}\diagdown\overset{-}{C}-$	From oxiranes or heterocycles
Carbonyl oxide	$-\overset{+}{C}\diagup\overset{O}{}\diagdown\overset{-}{O}$	↔	$-C\diagup\overset{\overset{+}{O}}{=}\diagdown\overset{-}{O}$	From carbene + O_2
Carbonyl imine	$-\overset{+}{C}\diagup\overset{O}{}\diagdown\overset{-}{N}-$	↔	$-C\diagup\overset{\overset{+}{O}}{=}\diagdown\overset{-}{N}-$	
Ozone	$\overset{+}{O}\diagup\overset{O}{}\diagdown\overset{-}{O}$	↔	$O\diagup\overset{\overset{+}{O}}{=}\diagdown\overset{-}{O}$	Stable
Thiocarbonyl ylide	$-\overset{+}{C}\diagup\overset{S}{}\diagdown\overset{-}{C}-$	↔	$-C\diagup\overset{\overset{+}{S}}{=}\diagdown\overset{-}{C}-$	From heterocycles
Selenocarbonyl ylide	$-\overset{+}{C}\diagup\overset{Se}{}\diagdown\overset{-}{C}-$	↔	$-C\diagup\overset{\overset{+}{Se}}{=}\diagdown\overset{-}{C}-$	From heterocycles

Several five-membered ring systems readily available by 1,3-dipolar cycloadditions are shown in Scheme 10. The dotted line indicates how the system was constructed, the line bisecting the two new bonds being formed in the cycloaddition. The majority of chapters in these volumes make some reference to 1,3-dipolar cycloadditions.

Alkynic dipolarophiles

NCC≡CCN, CF$_3$C≡CCF$_3$, RO$_2$CC≡CCO$_2$R, benzyne, R^{1}COC≡CCOR, HC≡CCO$_2$R, R^{2}C≡CCO$_2$R, R^{2}C≡CR^2, R^{2}C≡CH (R = Me, Et; R^1 = alkyl, aryl; R^2 = aryl, heteroaryl)

Alkenic dipolarophiles

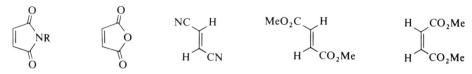

R = Me, Et, Ph

CH$_2$=CHCN, CH$_2$=CHCOMe, CH$_2$=CHCO$_2$Et, CH$_2$=CMeCO$_2$Me, EtO$_2$CN=NCO$_2$Et, PhCH=CHNO$_2$, PhCCl=CHNO$_2$, PhCH=CH$_2$, PhCH=CHPh

Heterocumulenes

R^{1}CONCO, R^{1}CONCS, RNCO, RNCS
R^1 = aryl, CCl$_3$; R = alkyl, aryl

Scheme 9 Frequently used dipolarophiles

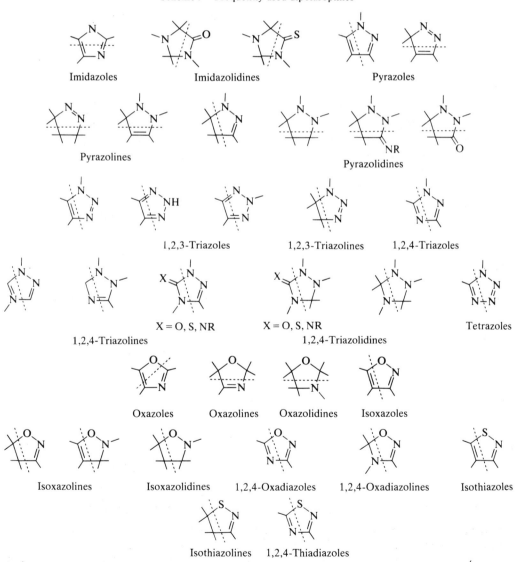

Scheme 10 Some five-membered ring systems available by 1,3-dipolar cycloadditions

Table 4 1,3-Dipoles Without Octet Stabilization

Vinylcarbenes	$-\overset{+}{C}\diagup\overset{\|}{\underset{\|}{C}}\diagdown\overset{-}{C}-$	$\leftrightarrow$ $-\overset{..}{C}\diagup\overset{\|}{\underset{\|}{C}}\diagdown C=$
Iminocarbenes	$-\overset{+}{C}\diagup\overset{\|}{\underset{\|}{C}}\diagdown\overset{-}{N}-$	$\leftrightarrow$ $-\overset{..}{C}\diagup\overset{\|}{\underset{\|}{C}}\diagdown N=$
Ketocarbenes	$-\overset{+}{C}\diagup\overset{\|}{\underset{\|}{C}}\diagdown\overset{-}{O}$	$\leftrightarrow$ $-\overset{..}{C}\diagup\overset{\|}{\underset{\|}{C}}\diagdown O$
Vinylnitrenes	$\overset{+}{N}\diagup\overset{\|}{\underset{\|}{C}}\diagdown\overset{-}{C}-$	$\leftrightarrow$ $\overset{..}{N}\diagup\overset{\|}{\underset{\|}{C}}\diagdown C=$
Iminonitrenes	$\overset{+}{N}\diagup\overset{\|}{\underset{\|}{C}}\diagdown\overset{-}{N}-$	$\leftrightarrow$ $\overset{..}{N}\diagup\overset{\|}{\underset{\|}{C}}\diagdown N=$
Ketonitrenes	$\overset{+}{N}\diagup\overset{\|}{\underset{\|}{C}}\diagdown\overset{-}{O}$	$\leftrightarrow$ $\overset{..}{N}\diagup\overset{\|}{\underset{\|}{C}}\diagdown O$

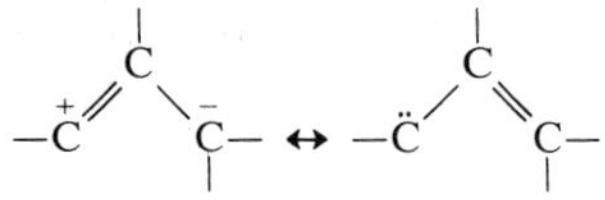

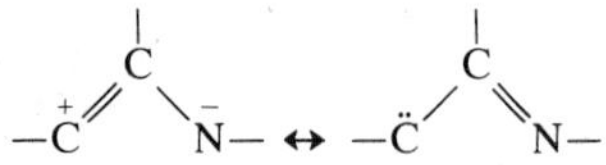

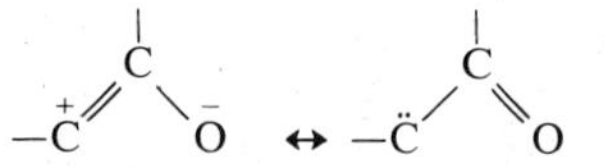

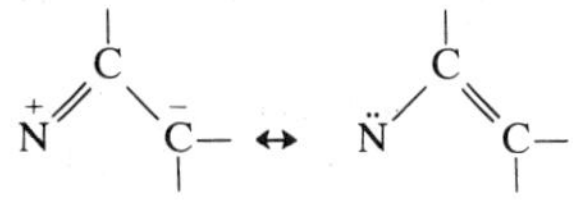

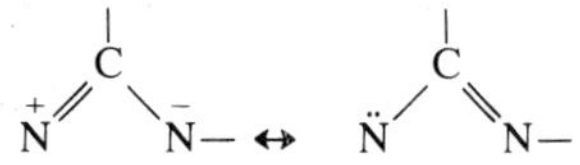

These are all highly reactive intermediates with only transient existence

4.03.6.1.1 Monocyclic systems with two or more heteroatoms

(i) Generation of the 1,3-dipole from alicyclic precursors

Nitrilimines (**361**) are generally generated *in situ* from hydrazonyl halides by dehydrohalogenation with an organic base such as Et_3N. Trapping with alkynic dipolarophiles leads readily to 3,4,5-trisubstituted pyrazoles (**362**) and with activated nitriles 1,3,5-trisubstituted 1,2,4-triazoles (**362a**) result. Applications to the synthesis of pyrazoles and 1,2,4-triazoles are discussed in Chapters 4.04 and 4.12, respectively. A wide variety of examples of this type of reaction utilizing nitrile ylides, nitrile oxides, diazoalkanes, azides, azomethine ylides, azomethine imines and related 1,3-dipoles generated from alicyclic precursors are described in the various monograph chapters.

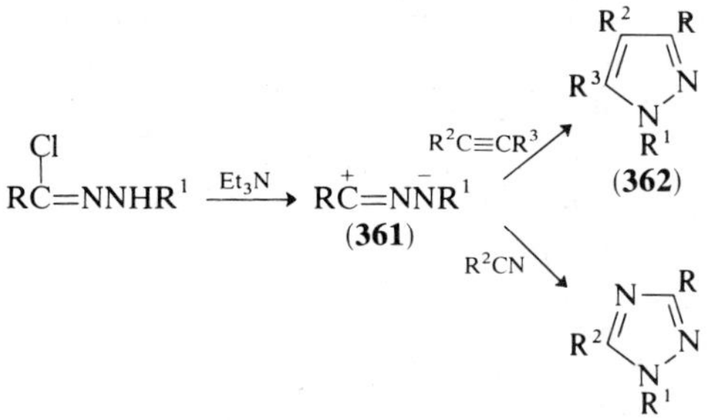

$$R\overset{Cl}{\underset{}{C}}=NNHR^1 \xrightarrow{Et_3N} R\overset{+}{C}=N\overset{-}{N}R^1 \quad \begin{matrix} \xrightarrow{R^2C\equiv CR^3} & (\mathbf{362}) \\ \xrightarrow{R^2CN} & (\mathbf{362a}) \end{matrix}$$
(**361**)

(ii) Generation of the 1,3-dipole by thermal elimination of nitrogen

Generation of a 1,3-dipole in this manner may be expressed in general terms as shown in Scheme 11. Nitrilimines, thiocarbonyl ylides and carbonyl ylides may all be generated in this fashion from the appropriate heterocycle and trapped by the dipolarophile.

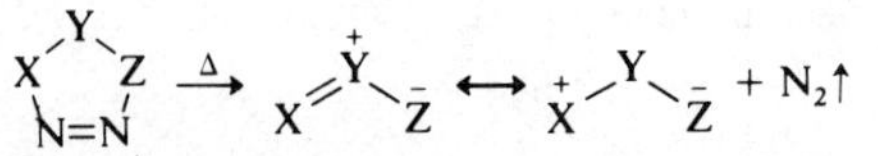

$$\underset{N=N}{X\diagup\overset{Y}{\underset{}{\diagdown}}Z} \xrightarrow{\Delta} X=\overset{+}{Y}\diagdown\overset{-}{Z} \leftrightarrow \overset{+}{X}\diagup\overset{Y}{\diagdown}\overset{-}{Z} + N_2\uparrow$$

Scheme 11 Generation of a 1,3-dipole by thermal elimination of nitrogen

(a) Tetrazoles → nitrilimines ⟨61CB2503, 76T2615⟩. 2,5-Diphenyltetrazole (**363**) on heating at 160 °C gives the diphenylnitrilimine (**364**) which can undergo cycloaddition to alkynes,

alkenes, imines and nitriles. With a nitrile the 1,2,4-triazole (**365**) was obtained. This general reaction adds importance to tetrazoles as synthetic precursors, and it is described fully in Chapter 4.13. Intramolecular reactions of this type leading to ring-fused systems are described in Section 4.03.6.1.2.

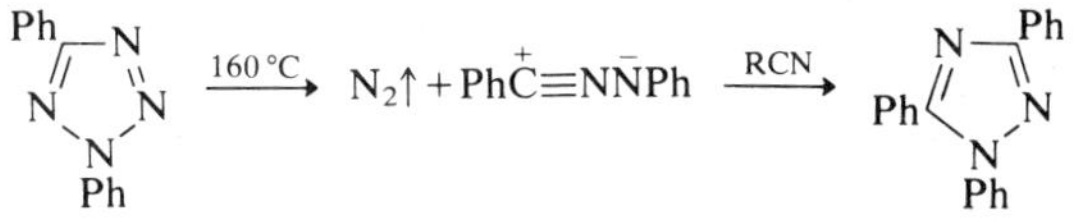

(*b*) Δ^3-*1,3,4-Thiadiazolines* → *thiocarbonyl ylides* ⟨72JOC4045, 73JOC844⟩. Thermolysis of the 2,5-dialkyl-Δ^3-1,3,4-thiadiazoline (**366**) in a hydrocarbon solvent in the presence of diethyl azodicarboxylate (excess) gave the thiadiazolidine (**368**) in good yield. When the pyrolysis was carried out in the presence of diphenylketene (equimolar amount), the *trans*-2,4-disubstituted 5-diphenylmethylene-1,3-oxathiolane (**369**) was obtained indicating that the *trans*-thiocarbonyl ylide (**367**) added suprafacially in a 1,3-manner over the carbonyl function of diphenylketene.

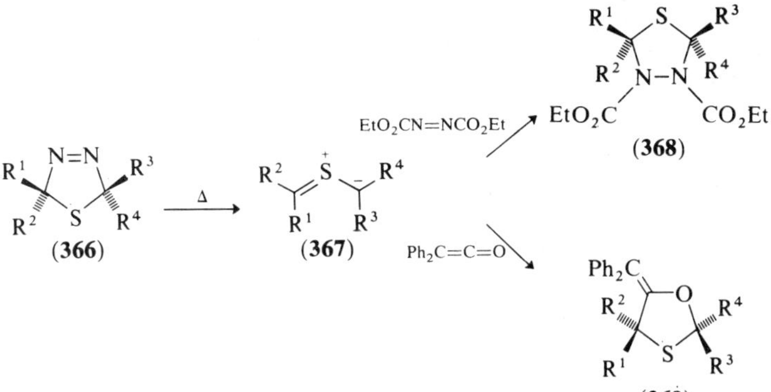

(*iii*) *Generation of the 1,3-dipole by thermal elimination of carbon dioxide* ⟨78JOC3736⟩

5-Phenyl-1,3,4-oxathiazol-2-one (**370**), prepared from primary amides and trichloromethanesulfenyl chloride, undergoes a ready thermal elimination of carbon dioxide with the formation of the nitrile sulfide ylide (**371**). This can be trapped by a wide variety of unsaturated dipolarophiles, and with an alkyne provides a ready route to isothiazoles (**372**) (see Chapter 4.17). Applications to 1,2,4-thiadiazole synthesis are described in Chapter 4.25. Thermolysis of 1,3,4-oxadiazolin-5-ones (500 °C/10^{-2} mmHg) results in the loss of CO_2 and generation of the corresponding nitrilimine ⟨78JOC2037⟩.

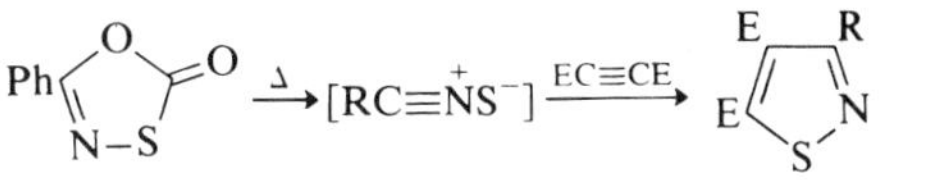

4.03.6.1.2 Ring-fused systems

Cycloadditions resulting in ring-fused systems may be classified in two principal groups: intramolecular and intermolecular cycloadditions. As the synthesis of the requisite precursor for an intramolecular cycloaddition is often more involved, more emphasis has been placed on those of an intermolecular nature.

(*i*) *Intramolecular 1,3-dipolar cycloadditions*

The synthetic strategy in this reaction class is the fabrication of a precursor in which the 1,3-dipole moiety is separated from the unsaturated linkage by a suitable chain of sufficient length and flexibility for the reaction centers to adopt an appropriate conformation for orbital overlap to occur with subsequent bond formation. It is expressed in general terms in Scheme 12. This concept of intramolecular 1,3-dipolar cycloaddition has been extended to include nitrones, azides, nitrile oxides, nitrilimines and azomethine imines, and in many instances the cycloaddition results in products accessible only with difficulty by other means.

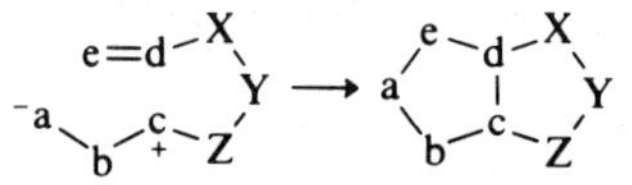

Scheme 12 Intramolecular 1,3-dipolar cycloaddition

The reaction is illustrated by the intramolecular cycloaddition of the nitrilimine (**374**) with the alkenic double bond separated from the dipole by three methylene units. The nitrilimine (**374**) was generated photochemically from the corresponding tetrazole (**373**) and the pyrrolidino[1,2-*b*]pyrazoline (**375**) was obtained in high yield ⟨82JOC4256⟩. Applications of a variety of these reactions will be found in Chapter 4.36. Other aspects of intramolecular 1,3-dipolar cycloadditions leading to complex, fused systems, especially when the 1,3-dipole and the dipolarophile are substituted into a benzene ring in the *ortho* positions, have been described ⟨76AG(E)123⟩.

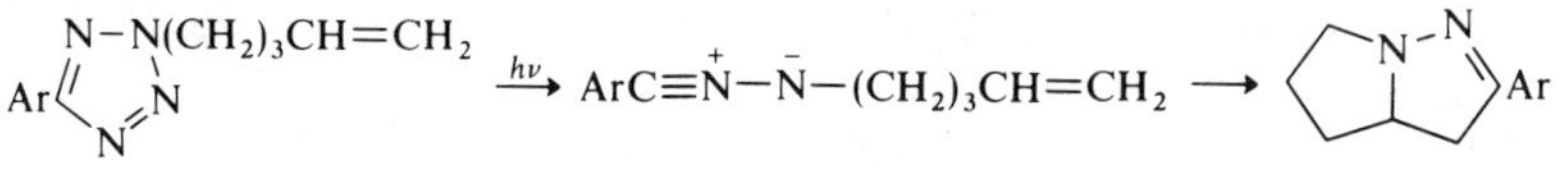

(ii) Intermolecular 1,3-dipolar cycloadditions

Formation of ring-fused systems by reactions of this type can be achieved in two ways: in one the heterocycle to which ring annulation is to occur acts as the dipolarophile; the alternative mode utilizes incorporation of the 1,3-dipole into the heterocyclic ring for reaction with the dipolarophile. Both approaches have been investigated intensively.

2,3-Dihydrofuran (**376**) and 2,5-dihydrofuran (**377**) react with nitrile oxides to give furo[2,3-*d*]isoxazoles (**378**) and furo[3,4-*d*]isoxazoles (**379**), respectively, as cycloadducts. The double bonds of furan, pyrrole and thiophene also react when the nitrile oxide is generated *in situ*. Thus furan and benzonitrile oxide gave (**380**), and with 2-methyl-2-oxazoline the cycloadduct (**381**) was obtained ⟨71AG(E)810⟩. These and related cycloadditions are discussed in Chapter 4.36.

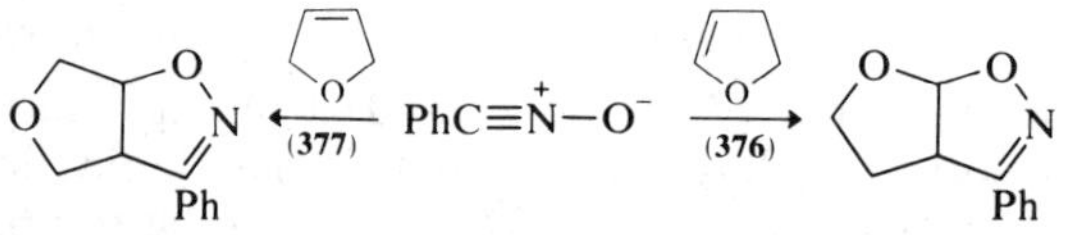

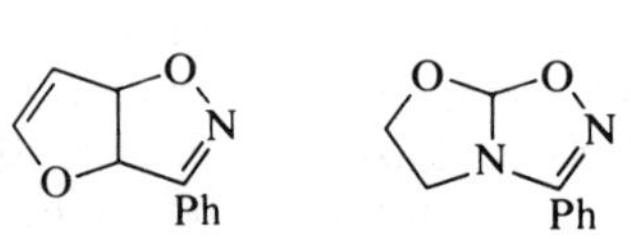

In the alternative approach the 1,3-dipolar system can be constructed in several ways. Treatment of α-chloroacylhydrazones of diaryl ketones and certain aralkyl and dialkyl ketones (**382**) with NaH in anhydrous THF gives 1-(disubstituted methylene)-3-oxo-1,2-diazetidinium inner salts (**383**). Reaction of (**383**) with DMAD in methylene chloride gave (**384**), a 2:1 adduct with loss of CO. Double bond migration in (**384**) occurred on heating to give (**385**). The intermediate in the cycloaddition was found to be (**386**), which on heating lost CO to form a new ylide system which in turn underwent reaction with more DMAD ⟨81JA7743⟩.

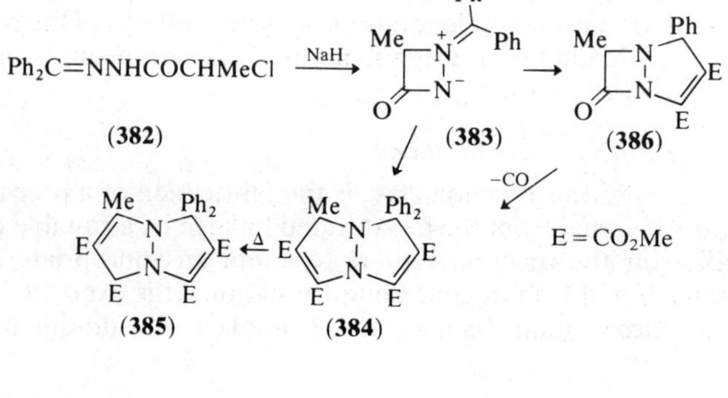

$E = CO_2Me$

As the sp^2 nitrogen atom in many heterocycles can be alkylated and aminated, the construction of an azomethine ylide or azomethine imine dipole is readily attainable as shown in Scheme 13. These ylides are very reactive and undergo cycloaddition with a

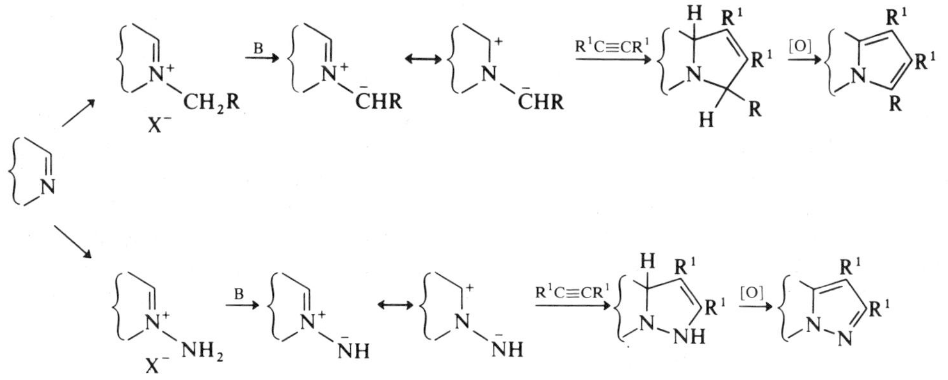

Scheme 13 Azomethine ylides and azomethine imines incorporated in heterocyclic systems

variety of alkynic dipolarophiles. Oxidation of the intermediate dihydro product to the aromatic system usually occurs during the reaction work-up. The reaction is illustrated by the conversion of pyridine into its pyridinium salt (**387**) with hydroxylamine *O*-sulfonic acid. Treatment with base generated the ylide (**388**) which underwent cycloaddition with DMAD to give as the final product the pyrazolo[1,5-*a*]pyridine (**389**). These reactions have been applied to a variety of azines and are described in Chapter 4.15.

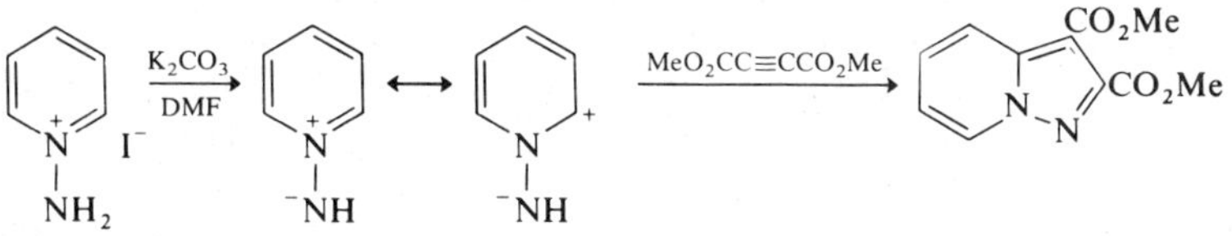

However, depending on the nature of the initial heterocycle, rearrangements are possible. Alkylation of thiazole to form the thiazolium salt (**390**) and generation of the ylide (**391**) with triethylamine in the presence of DMAD gave not (**392**) but the isomeric product (**393**) by the rearrangement indicated ⟨76JOC187⟩. Rearrangements of these types are described in Chapters 4.07 and 4.19.

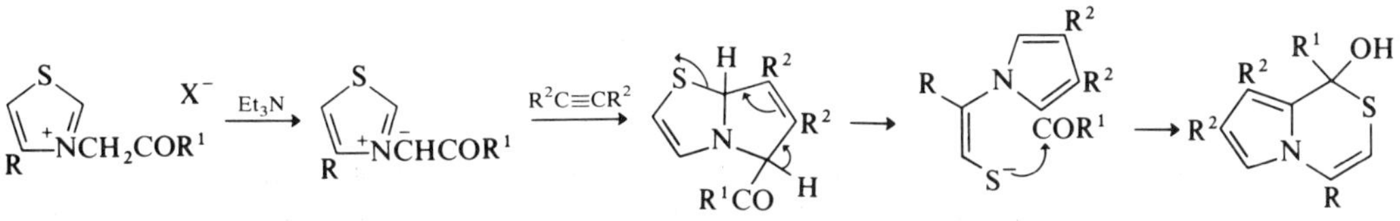

4.03.6.1.3 Cycloadditions using mesoionic ring systems

Use of mesoionic ring systems for the synthesis of five-membered heterocycles with two or more heteroatoms is relatively restricted because of the few readily accessible systems containing two heteroatoms in the 1,3-dipole. They are particularly suited for the unambiguous synthesis of pyrazoles as the azomethine imine is contained as a 'masked' 1,3-dipole in the sydnone system. An attractive feature of their use is that the precursor to the mesoionic system may be used in the presence of the cyclodehydration agent and the dipolarophile, avoiding the necessity for isolating the mesoionic system.

Reaction of the *N*-nitrosoglycine (**394**) with acetic anhydride gave the anhydro-5-hydroxy-1,2,3-oxadiazolium hydroxide (**395**). Reaction with DMAD resulted in formation of the intermediate 1:1 cycloadduct (**396**) which was not isolated and which lost CO_2 under the thermal reaction conditions to give dimethyl 1-phenylpyrazole-3,4-dicarboxylate (**397**) ⟨83MI40300⟩. This reaction is capable of considerable variation in terms of the substituents attached to the 1,2,3-oxadiazole ring and in the nature of the dipolarophile, especially as alkenic dipolarophiles may also be utilized (see Chapter 4.21).

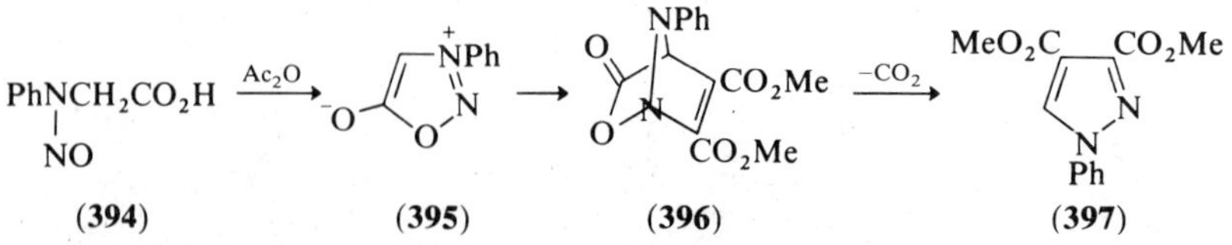

A similar product is obtained from the reaction of anhydro-4(5)-hydroxy-1,2,3-triazolium hydroxide (**398**). In this case reaction with DMAD occurred in 1 hour in boiling benzene. Extrusion of methyl isocyanate from the initial 1:1 cycloadduct (**399**) occurred during the reaction giving (**400**).

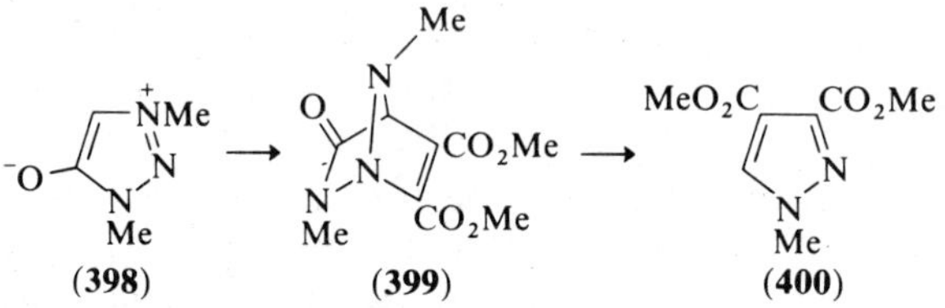

Thiocarbonylimines are present as the 'masked' 1,3-dipole in the anhydro-5-hydroxy-1,3,2-oxathiazolylium hydroxide system (**402**) which may be prepared by the cyclodehydration of S-nitroso-α-phenyl-α-mercaptoacetic acid (**401**). Under thermal conditions reaction with DMAD occurred to give an intermediate 1:1 cycloadduct (**403**) which lost CO_2 under these reaction conditions forming dimethyl 2-phenylisothiazole-3,4-dicarboxylate (**404**) ⟨72CB196⟩. In the cycloaddition of ethyl propiolate the addition was not regiospecific, the 4-ester being obtained in greater amount (56%). However, under photochemical conditions an isomeric product was obtained via the intermediacy of the nitrile sulfide dipole. Irradiation in neat DMAD has been postulated to result in formation of the valence tautomer (**405**) which lost CO_2 to give the thiazirene (**406**) which underwent rearrangement to the nitrile sulfide dipole (**371**). Reaction with DMAD led to dimethyl 3-phenylisothiazole-4,5-dicarboxylate (**372**) ⟨72CB188, 75JA6197⟩ (see also Chapter 4.17).

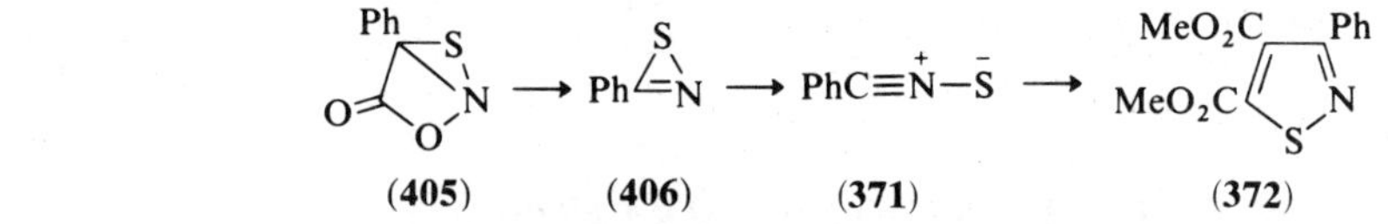

Use of mesoionic systems in the synthesis of [5,5] ring-fused systems has received little attention. However, they are useful for the synthesis of [5,6] ring-fused systems where the six-membered ring is constructed via the dipolar cycloaddition process. Annulation of a pyridone ring to the imidazole, 1,2,4-triazole ⟨79JOC3803⟩, thiazole and 1,3,4-thiadiazole ring systems ⟨79JOC3803⟩ has been achieved via cycloadditions with the appropriate bicyclic mesoionic ring system.

Anhydro-3-hydroxy-2-phenylthiazolo[2,3-b]thiazolylium hydroxide (**407**) underwent ready thermal reaction with alkynic and alkenic dipolarophiles in refluxing toluene. With the former dipolarophile sulfur was lost from the intermediate 1:1 cycloadduct (**408**) to give the substituted 5H-thiazolo[3,2-a]pyridin-5-ones (**409**). With the latter, the intermediate (**410**) lost H_2S, also forming (**409**).

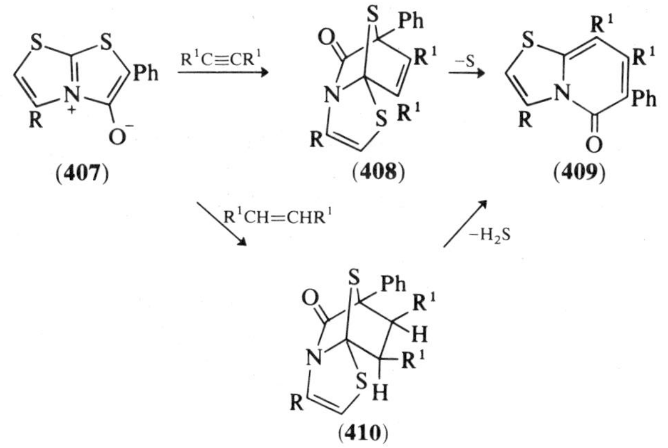

4.03.6.2 1,4-Dipolar Cycloadditions

Just as in the Diels–Alder reaction, 1,4-dipolar cycloadditions lead to six-membered rings. Their principal use in five-membered heterocycles is for ring annulations giving [5,6] ring-fused systems.

Reaction of thiazoles with DMAD illustrates the overall reaction and the rearrangements which may be encountered. Thiazole or 2-methylthiazole (**411**; R=H and Me, respectively) in DMF reacted with DMAD to give an initial 1,4-dipolar species (**412**). Reaction with a second DMAD gave the 1:2 molar adduct, presumably (**413**). Ring opening to (**414**), followed by cyclization in the alternative mode, resulted in the formation of (**415**), the structure of which (R=Me) was established by X-ray analysis ⟨78AHC(23)263⟩ (see also Chapter 4.19).

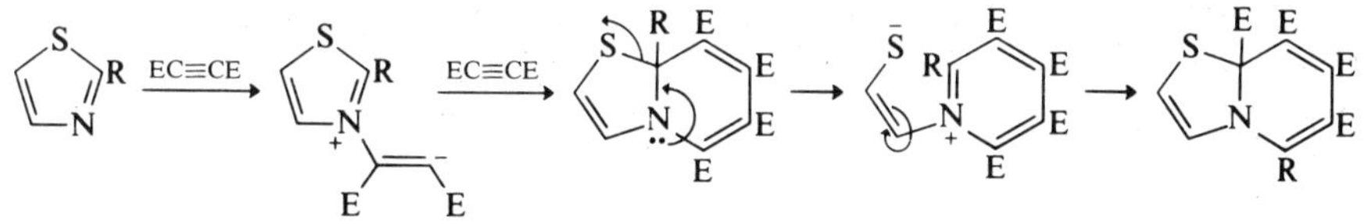

Solvent has an important influence on the course of this cycloaddition, and in the reaction of 2,5-dimethylthiazole with DMAD in DMF the product analogous to (**415**) was obtained. However, in DMSO or acetonitrile a thiazolo[3,2-*a*]azepine was formed in addition to this product, whereas with THF, dichloromethane or nitromethane, only the thiazoloazepine was isolated.

1,4-Dipoles can also be built into heterocyclic systems, and though of limited use, they may also be utilized for the synthesis of [5,6] ring-fused systems. Reaction of 2(3*H*)-benzothiazolethione with (chlorocarbonyl)phenylketene in warm anhydrous benzene gave the heteroaromatic betaine (**416**). On heating with DMAD in boiling toluene the tricyclic pyridinone (**418**) was obtained, presumably by elimination of COS from the intermediate cycloadduct (**417**) ⟨80JOC2474⟩.

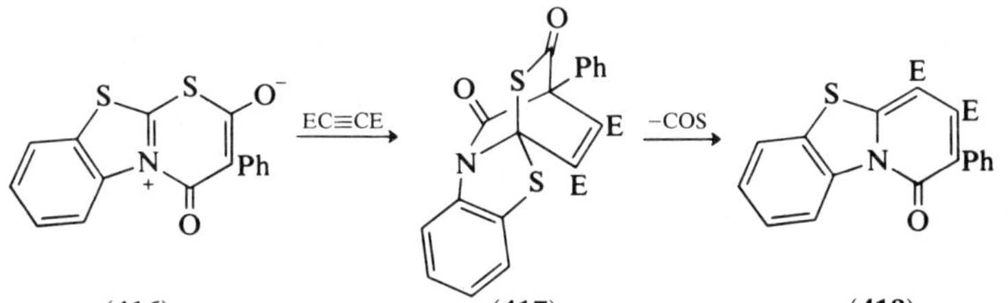

Ethyl propiolate reacted with (**416**) in boiling benzene giving ethyl 1-oxo-2-phenyl-1*H*-pyrido[2,1-*b*]benzothiazole-4-carboxylate in 83% yield, and no trace of the other possible isomer was found.

4.03.6.3 1,5-Dipolar Cycloadditions

The concept of a 1,5-dipolar cyclization gives rise to a general method for the synthesis of an appreciable number of heterocyclic systems. 1,5-Dipoles are derived from 1,3-dipoles by conjugation with different double bond systems, and it is possible to derive 98 theoretically possible 1,5-dipolar systems. The general expression for a 1,5-dipole and some possible combinations of double bond systems are shown in Scheme 14.

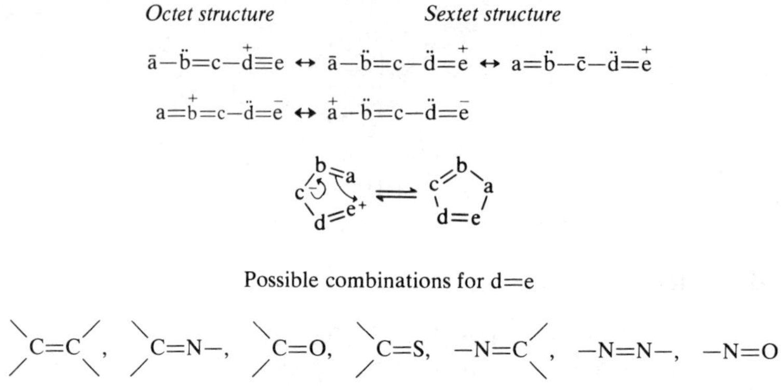

Octet structure   *Sextet structure*

$$\bar{a}-\bar{b}=c-\overset{+}{d}\equiv e \leftrightarrow \bar{a}-\bar{b}=c-\overset{+}{d}=\ddot{e} \leftrightarrow a=\overset{+}{b}-\bar{c}-\ddot{d}=\overset{+}{e}$$

$$a=\overset{+}{b}=c-\ddot{d}=\bar{e} \leftrightarrow \overset{+}{a}-\ddot{b}=c-\ddot{d}=\bar{e}$$

Possible combinations for d=e

$$\rangle C=C\langle, \quad \rangle C=N-, \quad \rangle C=O, \quad \rangle C=S, \quad -N=C\langle, \quad -N=N-, \quad -N=O$$

Scheme 14   1,5-Dipoles

1,5-Dipolar cyclizations leading to monocyclic and ring-fused systems are known. The 2,5-disubstituted tetrazole (**419**; X=O, S, N—) on heating underwent ring opening to the 1,5-dipolar species (**420**) which readily lost N_2 to give the conjugated nitrilimine (**421**). 1,5-Dipolar cyclization of (**421**) through X leads to oxadiazoles, thiadiazoles and triazoles ⟨70CB1918, 65CB2966⟩ (see also Chapter 4.13).

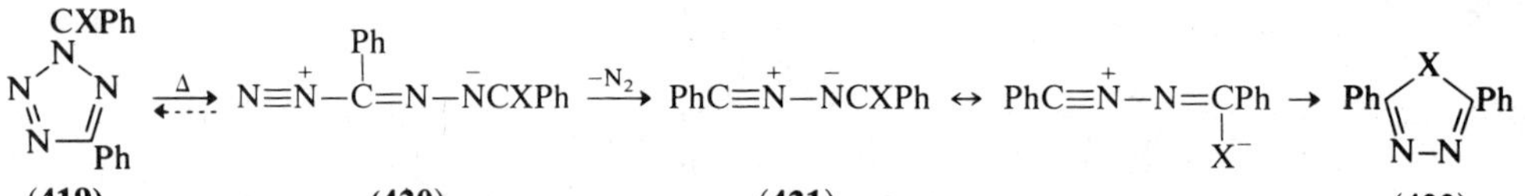

The thermal conversion of 1-substituted 5-aminotetrazole (**423**) into a 5-substituted aminotetrazole (**424**) can be rationalized by a similar 1,5-dipolar cyclization.

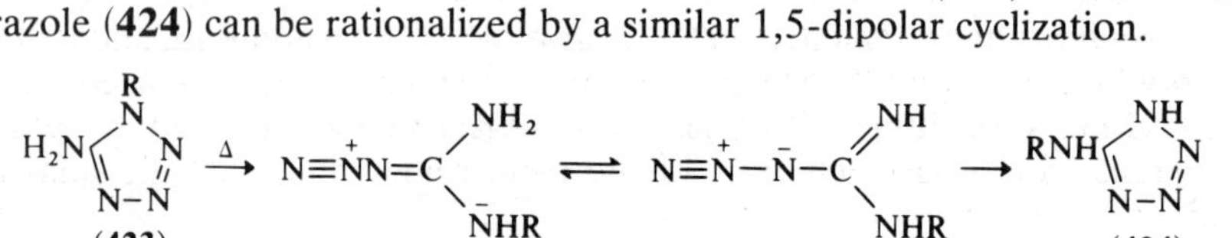

Reaction of 1-chloroisoquinoline with 5-phenyltetrazole gave 3-phenyl[1,2,4]triazolo[3,4-*a*]isoquinoline (**428**). It has been suggested that the initial 2-tetrazolylisoquinoline (**425**) underwent ring opening to (**426**) which readily lost N_2 giving (**427**). This nitrilimine is a 1,5-dipole, and on cyclization the [1,2,4]triazolo[3,4-*a*]isoquinoline (**428**) is formed ⟨70CB1918⟩.

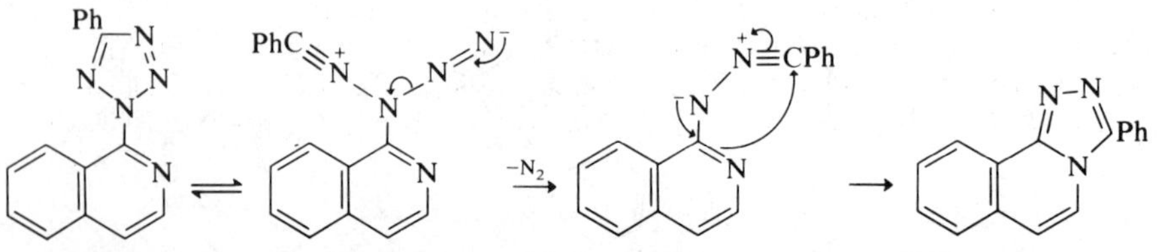

The reaction of 1-aminopyridinium iodide (**429**) with dimethyl chlorofumarate in ethanol/K_2CO_3 to form, ultimately, a pyrazolo[1,5-*a*]pyridine also occurs *via* a 1,5-dipolar mechanism. The initially formed 1:1 adduct (**430**), stabilized by delocalization of the negative charge, underwent disrotatory ring closure as shown to give (**431**) in which the 3 and 3a hydrogens are *cis* to each other. Pd dehydrogenation gave dimethyl pyrazolo[1,5-a]pyridine-2,3-dicarboxylate (**432**) ⟨72JOC3106⟩. This reaction also occurs with the corresponding alkylated pyridinium salt ⟨73JCS(P1)2089⟩.

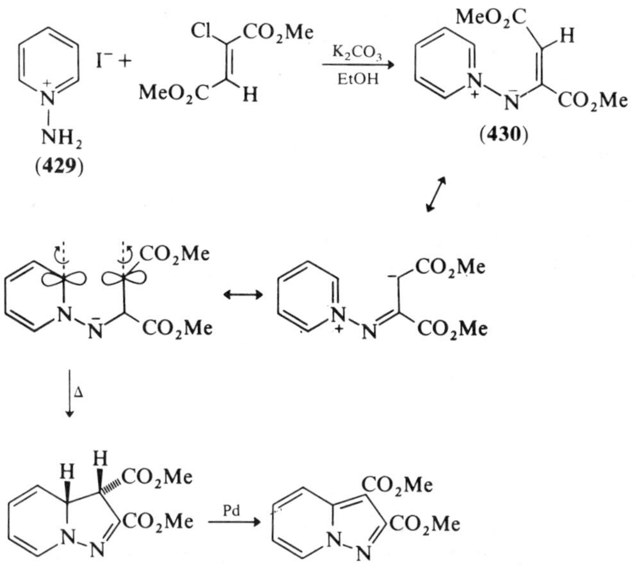

4.03.7 HETEROCYCLIC RING INTERCONVERSIONS

4.03.7.1 Small Rings as Dienophiles or Dipolarophiles

Small unsaturated rings are usually very reactive undergoing ring opening in a number of ways, and this characteristic has been utilized in heterocyclic synthesis. In their role as dienophiles or dipolarophiles, the initial cycloaddition is usually followed by a valence tautomerism resulting in a six-membered or larger ring system. Several examples exist, however, where this does not occur, and these are described below.

The thermal reaction of 2-phenyl-1-azirine (**433**) with carbon disulfide (sealed tube, 100 °C, 3 hours) gave 2-methylthio-5-phenylthiazole (**435**). It has been suggested that this reaction occurred via a regioselective cycloaddition of carbon disulfide to the π-bond of the 1-azirine by either a concerted [$_\pi 2_s + _\pi 2_a$] or stepwise mechanism to give the intermediate (**434**). Ring opening and hydrogen shift would result in (**435**) ⟨75JOC1348⟩. Similar results were obtained with 3-methyl-2-phenylazirine.

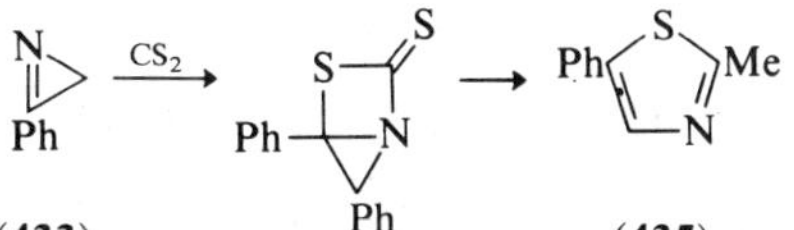

A similar regiospecific [2+2] cycloaddition across a C=S group occurred when benzoyl isothiocyanate (**436**) and 2,3-diphenyl-1-azirine were heated in refluxing benzene for 12 hours. The product obtained was shown to be (**438**) and an intermediate such as (**437**) could also be involved in this cycloaddition ⟨74JOC3763⟩. In contrast, thiobenzoyl isocyanate added in a [4+2] fashion, and after ring expansion gave a thiadiazepine derivative.

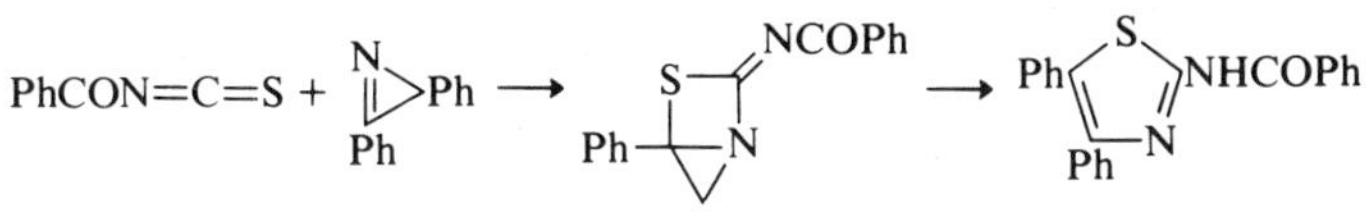

Although in its reactions with several mesoionic systems diphenylthiirene dioxide (**439**) does not lose SO_2 from the cycloadducts, in its reactions with pyridinium, quinolinium and isoquinolinium phenacylides it behaves as an acetylene equivalent. Thus, reaction of (**439**)

with the pyridinium phenacylide (**440**) gave the indolizine (**441**) with elimination of SO_2 and loss of hydrogen ⟨73BCJ667⟩.

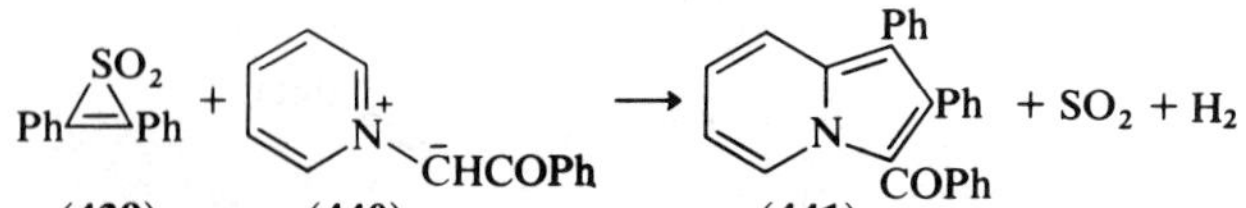

$$\text{(439)} \quad + \quad \text{(440)} \quad \rightarrow \quad \text{(441)} \quad + SO_2 + H_2$$

4.03.7.2 Small Rings as a Source of Ylides

Photolysis of 2,3-diphenyl-Δ^1-azirine (**442**) generates benzonitrile ylide (**443**). Irradiation in the presence of ethyl cyanoformate resulted in a mixture of the oxazoline (**444**) and the imidazole (**445**) by 1,3-dipolar cycloaddition to the carbonyl and nitrile group, respectively ⟨72HCA919⟩.

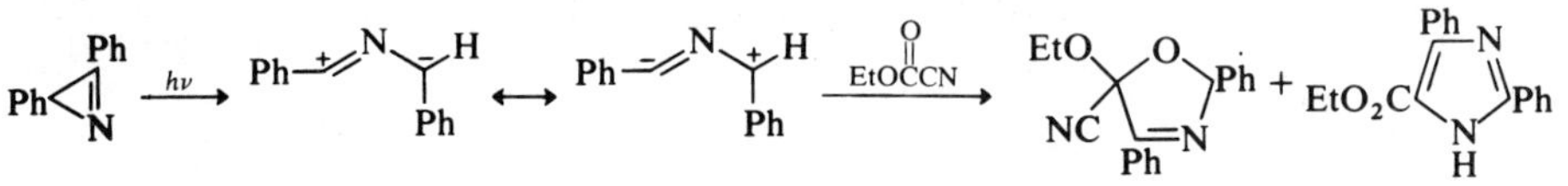

(**442**) → (**443**) → (**444**) + (**445**)

Azomethine ylides (Section 4.03.6.1.1) have been generated from a wide variety of aziridines using both thermal and photochemical methods. With carbon–carbon unsaturated dipolarophiles, pyrrolines or pyrrolidines are obtained. With hetero double bonds, however, ring systems of interest to this discussion result.

Thermolysis of the aziridine (**446**) in the presence of diphenylketene gave a mixture of the pyrrolidone (**447**; minor product) and the oxazolidine (**448**; major product). In this instance the preferential addition to the C=O bond is explained in terms of steric effects ⟨72CC199⟩. Similar addition to diphenylacetaldehyde takes place with the same orientation and the oxazolidine (**448a**) was obtained. When the reaction of the aziridine with the aldehyde was carried out in the presence of hydrogen selenide a selenazolidine was obtained ⟨72BSB295⟩.

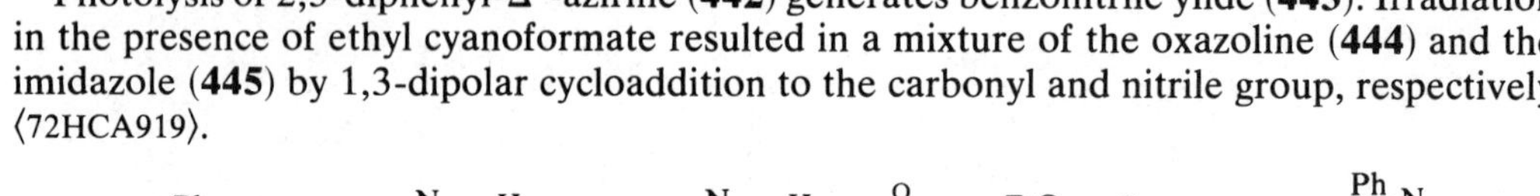

(**446**) —112 °C→ (intermediate) —$Ph_2C=C=O$→ (**447**) + (**448**)

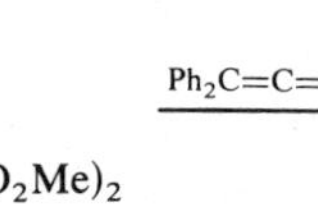

(**448a**)

These thermal or photochemical ring openings of aziridines when applied to ring-fused aziridines generate ylides which may be trapped, leading to structures of considerable interest which are difficult to obtain by other means. As an illustration in five-membered ring chemistry, the ring-fused aziridine (**449**) underwent ring opening in refluxing xylene to the azomethine ylide (**450**). In the presence of diethyl acetylenedicarboxylate the 3H-pyrrolo[1,2-c]imidazole (**451**) was obtained. With maleic anhydride the ylide was equally readily trapped giving the 1:1 cycloadduct (**452**) ⟨68JOC1097⟩. Similar adducts are formed with diethyl fumarate, cis or $trans$-dibenzoylethylene, diethyl azodicarboxylate and N-phenylmaleimide.

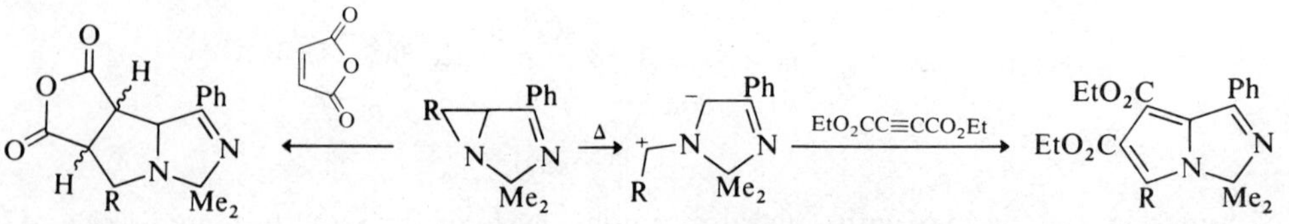

(**452**) ← (**449**) →Δ→ (**450**) —$EtO_2CC\equiv CCO_2Et$→ (**451**)

$R = p\text{-}NO_2C_6H_4$

4.03.7.3 Small Rings As Substrates for Ring-opening Reactions

4.03.7.3.1 Three-membered rings

Three-membered rings, especially oxiranes and aziridines, are susceptible to nucleophilic ring opening and form five-membered rings with a variety of nucleophiles. Chloromethyloxirane (**453**) and a hydrazine derivative at −25 °C readily gave the 4-hydroxypyrazolidine (**454**). A variety of nucleophiles is effective in ring opening and the reaction has been adapted for the synthesis of ring-fused systems. Reaction of the oxazoline (**455**) with the oxirane (**456**) gave the oxazolidino[2,3-*b*]oxazolidine (**457**) ⟨66LA(698)174⟩. Imidazoline and thiazoline formed the analogous ring-fused systems. Oxiranes containing 1,1-dicyano substituents undergo nucleophilic ring opening with ease. Treatment of 1,1-dicyano-2-phenyloxirane with potassium selenocyanate gave 2-*N*-acylimino-1,3-oxaselenoles ⟨78CC447⟩.

(**453**) (**454**) (**455**) (**456**) (**457**)

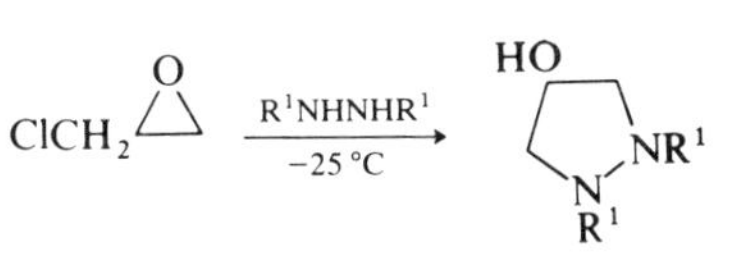

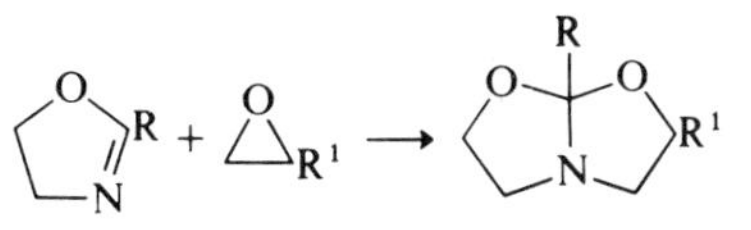

Aziridines represented by the general structure (**458**; X = O, S, NR) undergo a facile ring opening and subsequent closure on heating with sodium iodide in acetone or acetonitrile. For (**458**; X = O) the oxazoline (**460**) was formed, presumably *via* the intermediate (**459**) ⟨66JOC59⟩.

On heating the fused aziridine (**461**) at temperatures greater than 250 °C, the resultant product was the furo[3,4-*d*]oxazole system (**462**).

(**458**) (**459**) (**460**) (**461**) (**462**)

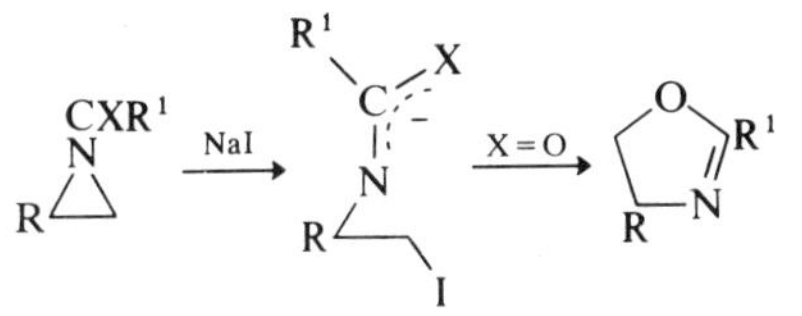

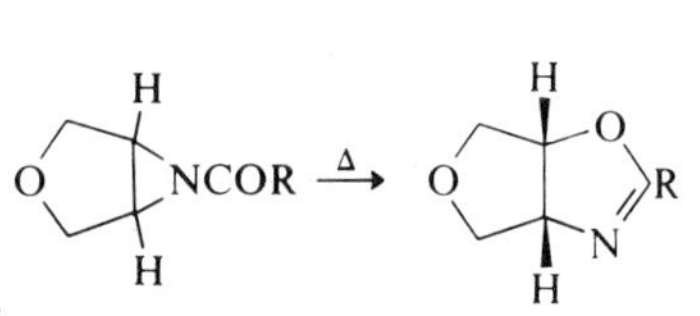

This aziridine ring-opening reaction is a particularly attractive route to imidazolines and ring-fused imidazolines. The 1-substituted aziridine (**463**), prepared from the imino ether and aziridine, on treatment with iodide ion or thiocyanate ion gave (**464**) ⟨62JOC2943, 60JOC461⟩. The 2-(1-aziridinyl)quinoxaline (**465**) under related conditions formed the tricyclic system (**466**), and (**467**) with iodide ion underwent two ring-opening reactions to give initially (**468**) and then (**469**).

(**463**) (**464**) (**465**) (**466**) R = Cl, OMe

(**467**) (**468**) (**469**)

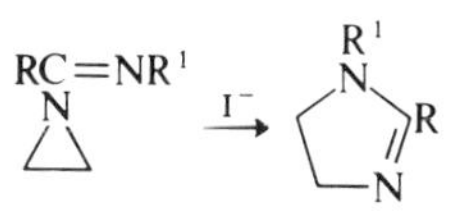

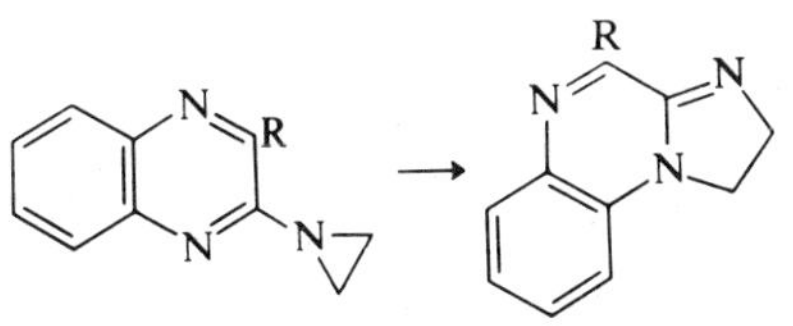

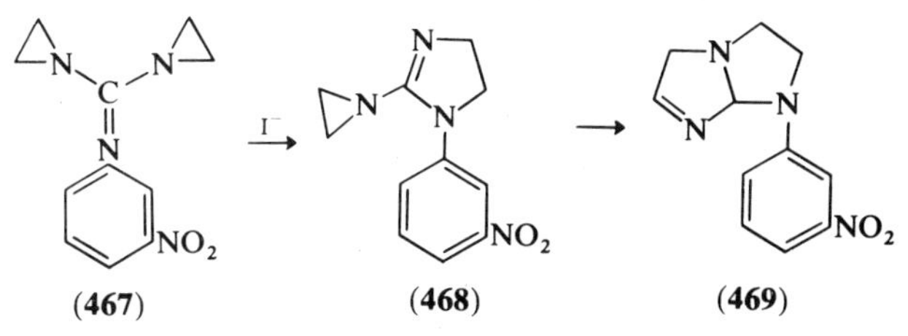

Aziridines also undergo ring enlargement when treatment with thiocyanic acid. *cis*- and *trans*-2,3-Dimethylaziridine (**470**) with thiocyanic acid gave exclusively *trans*- and *cis*-2-amino-4,5-dimethyl-2-thiazoline (**471**) ⟨72JOC4401⟩.

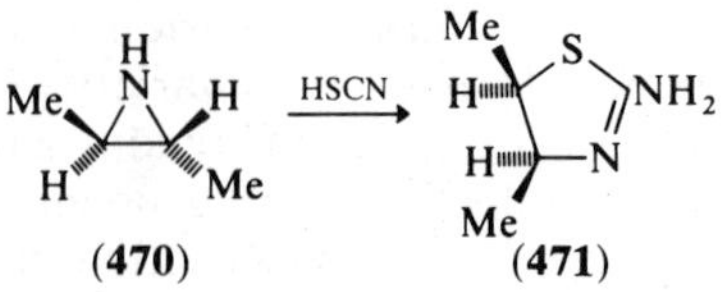

(470)　(471)

4.03.7.3.2 Four-membered rings

Azetidines under analogous reaction conditions to those above result in six-membered ring formation. However, diketene (**472**), an oxetan-2-one, offers considerable promise for five-membered heterocycle formation. With hydroxylamine the 3-methylisoxazolin-5-one (**473**) was formed. Phenylhydrazine gave the corresponding 3-methyl-1-phenylpyrazolin-5-one.

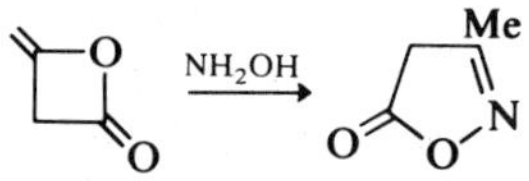

(472)　(473)

4.03.7.4 Interconversion of Five-membered Rings

Interconversion of five-membered heterocycles may be effected in numerous ways, and in general, oxygen in a five-membered ring may be readily replaced by sulfur and by nitrogen. Similarly, sulfur may be replaced by nitrogen with an appropriate reagent. Ring opening with subsequent ring closure is usually involved.

Heating oxazoles with a source of nitrogen leads to imidazoles. Thus, the oxazole (**474**) when heated under reflux with formamide gave the imidazole (**475**) with a variety of alkyl and aryl groups substituted in the nucleus. Using a primary amine with the oxazole-4-carboxylic acid (**476**) in an aqueous environment at 150 °C gave the imidazole (**477**) with accompanying decarboxylation ⟨53CB88, 53CB96, 48JCS1960⟩ (see Chapters 4.07 and 4.18).

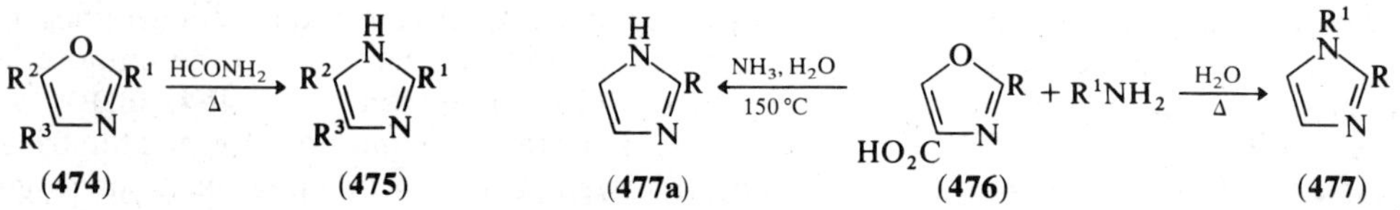

(474)　(475)　(477a)　(476)　(477)

Isoxazoles (**478**) in the presence of base undergo ring opening to α-ketonitriles (**479**). When the reaction was carried out in the presence of hydrazines, 5-aminopyrazoles (**480**) were obtained. The reaction is also a convenient source of imidazoles. For example, when the 1,2-benzisoxazole (**481**) was treated with phenylhydrazine, decarboxylation initially occurred with subsequent ring closure to (**482**) (see Chapter 4.16).

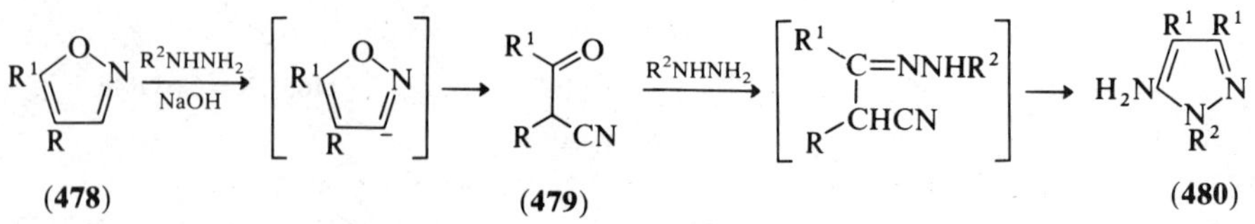

(478)　(479)　(480)

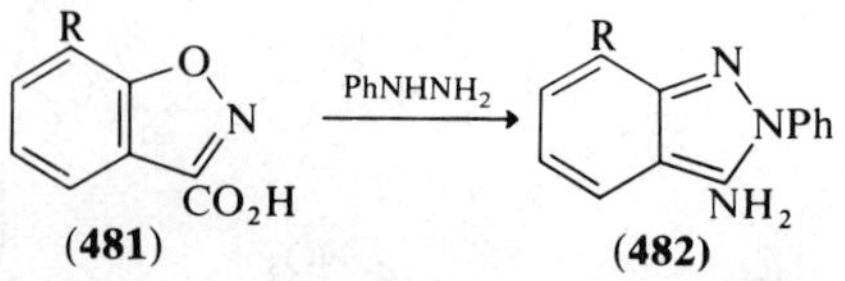

(481)　(482)

1,2-Dithiolylium salts (**483**) may be converted into pyrazoles, pyrazolium salts and isothiazoles depending on the type and degree of substitution of the nitrogen source.

4-Phenyl-1,2-dithiolylium salt (**483**) with hydrazine, methylhydrazine or phenylhydrazine yielded the corresponding pyrazole (**485**) *via* the intermediates (**484a–c**). The ring-fused system (**486**) is a convenient source of the ring-fused pyrazole (**487**) when treated with hydrazine (see Chapter 4.31).

(**483**) (**484a**) (**484b**) (**484c**) (**485**)

(**486**) (**487**)

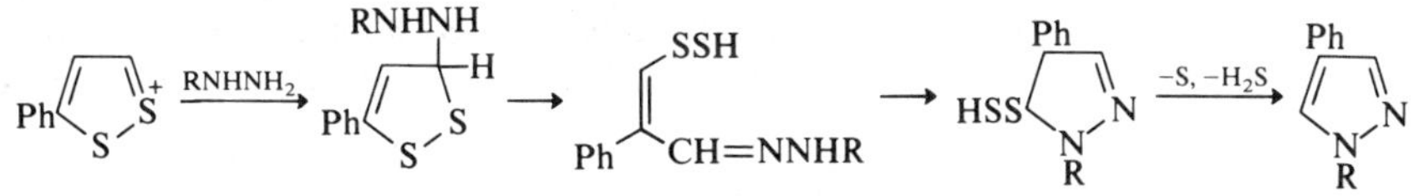

3,4-Dimethyl-1,2-dithiolylium perchlorate (**488**) when treated with ammonia at −5 °C underwent nucleophilic ring opening to (**489**). Subsequent ring closure with loss of H_2S gave 4,5-dimethylisothiazole (**490**) in 85% yield (see Chapter 4.17). This reaction involved an initial attack of the ammonia at the unsubstituted 3-position. A small amount of the 3,4-dimethylisothiazole (**491**; 15%) indicated that some attack of the ammonia also occurred at the 5-position.

(**488**) (**489**) (**490**) (**491**)

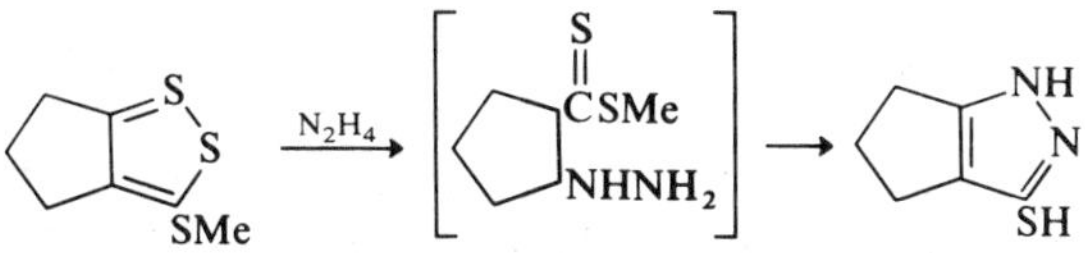

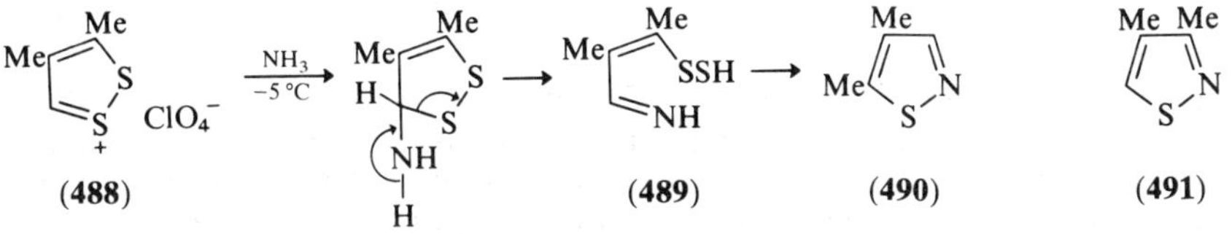

4.03.7.5 Interconversion of Six- and Five-membered Rings

Ring expansion of five- to six-membered rings such as oxazole → pyridine derivatives *via* a Diels–Alder reaction is a well-established procedure. However, the conversion of a six-membered heterocycle into a five-membered ring system has not been exploited to any great extent, and those systems that have been studied usually involve a cationic species.

Pyrylium salts are susceptible to nucleophilic ring opening by attack at the 2-position. When treated with hydroxylamine, the pyrylium salt (**492**) underwent ring opening to an intermediate oxime of a 1,5-enedione (**493**). Conjugate addition of the oxime group to the α,β-unsaturated ketone gave (**494**) which in the presence of acid gave the isoxazole (**495**). A similar reaction with a monosubstituted hydrazine resulted in the pyrazoline (**496**).

(**492**) (**493**) (**494**) (**495**) (**496**)

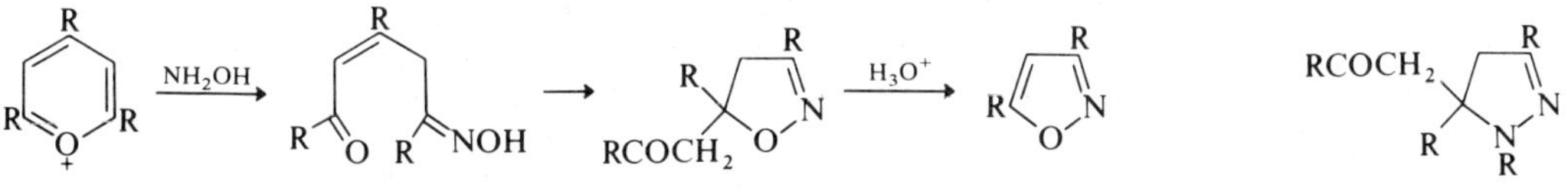

1,3-Oxazinium salts (**497**) and hydrazine undergo an analogous reaction giving pyrazoles (**498**) ⟨69CB269, 65CB334, 62CB937⟩. 1,3,5-Oxadiazinium salts (**499**) reacted with hydroxylamine to give the 1,2,4-oxadiazoles (**500**) and with hydrazines to form the 1,2,4-triazole derivatives (**501**). The substituents in these cationic species are usually aryl, restricting the appeal of these ring interconversions ⟨65CB334, 67CB3736⟩.

(**497**) (**498**) (**500**) (**499**) (**501**)

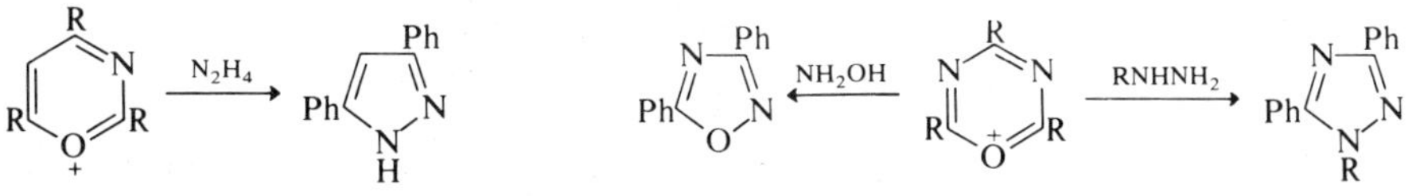

4.03.7.6 Ring Interconversions Involving Side Chains

This type of ring interconversion is represented by the general expression shown in Scheme 15. Analogous rearrangements occur in benzo-fused systems. The known conversions are limited to D = O in the azole system, *i.e.* cleavage of the weak N—O bond occurs readily. Under the reaction conditions, Z needs to be a good nucleophile in its own right or by experimental enhancement (base catalysis, solvent, *etc.*) and Z is usually O, S, N or C.

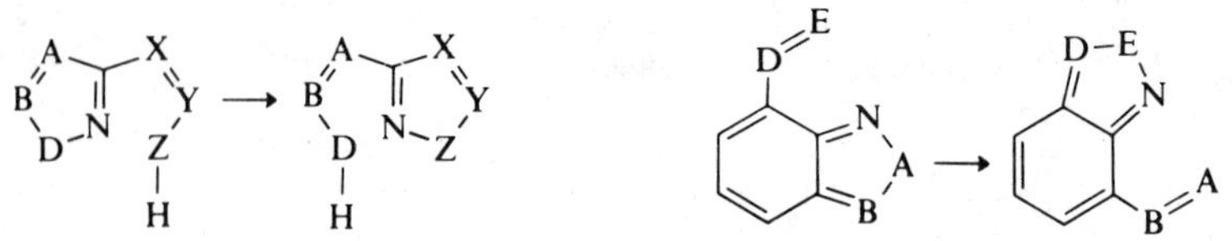

Scheme 15 General expression for mononuclear heterocyclic rearrangements

The reaction is illustrated by the conversion of the 1,2,4-oxadiazole oxime (**504**) into the 3-acylamino-1,2,5-oxadiazole (**505**). This irreversible rearrangement occurred on heating (**504**) in hydrochloric acid ⟨81AHC(29)141⟩. Isoxazoles also undergo this rearrangement and these are discussed in Chapter 4.16.

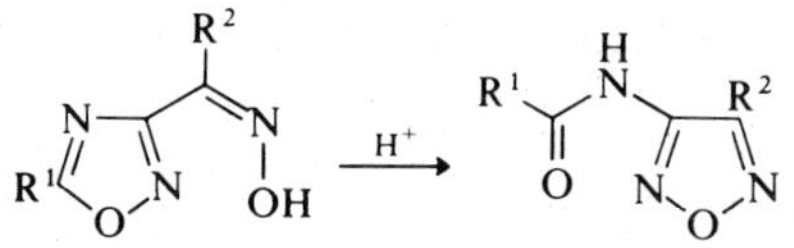

Among ring-fused systems, the 3-(*o*-aminoaryl)benzisoxazole (**506**) underwent rearrangement on refluxing in THF with sodium hydride or lithium aluminum hydride. In this case the 3-(*o*-hydroxyphenyl)benzimidazole (**506a**) was obtained.

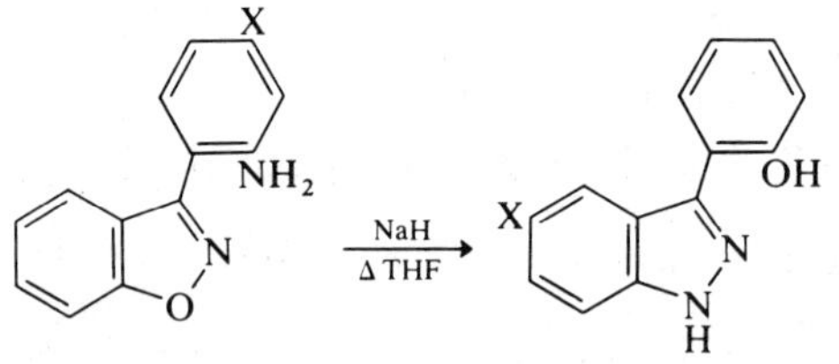

Table 5 Heterocyclic Rearrangements Involving Side Chains (Mononuclear Heterocyclic Rearrangements)

Initial ring system (**502**)	ABD	XYZ	Final ring System (**503**)
1,2,4-Oxadiazole	NCO	CNO	1,2,5-Oxadiazole
Isoxazole	CCO	CNO	1,2,5-Oxadiazole
1,2,5-Oxadiazole	CNO	CNO	1,2,5-Oxadiazole
1,2,4-Oxadiazole	NCO	NCO	1,2,4-Oxadiazole
Isoxazole	CCO	NCO	1,2,4-Oxadiazole
1,2,4-Oxadiazole	NCO	CCO	Isoxazole
1,2,5-Oxadiazole	CNO	CCO	Isoxazole
1,2,4-Oxadiazole	NCO	CNN	1,2,3-Triazole
Isoxazole	CCO	CNN	1,2,3-Triazole
1,2,5-Oxadiazole	CNO	CNN	1,2,3-Triazole
1,2,4-Oxadiazole	NCO	NCN	1,2,4-Triazole
Isoxazole	CCO	NCN	1,2,4-Triazole
1,2,5-Oxadiazole	CNO	NCN	1,2,4-Triazole
1,2,4-Oxadiazole	NCO	NCS	1,2,4-Thiadiazole
Isoxazole	CCO	NCS	1,2,4-Thiadiazole
1,2,5-Oxadiazole	CNO	NCS	1,2,4-Thiadiazole
Isoxazole	CCO	NNN	Tetrazole
1,2,4-Oxadiazole	NCO	NCC	Imidazole
1,2,4-Oxadiazole	NCO	CCN	Pyrazole
Isoxazole	CCO	CCN	Pyrazole

The various combinations of XYZ and ABD in the monocyclic system in Scheme 15 which illustrate the scope of the rearrangement are shown in Table 5.

4.03.8 PHOTOCHEMISTRY IN HETEROCYCLIC SYNTHESIS

4.03.8.1 Elimination Processes

4.03.8.1.1 Loss of nitrogen

A number of five-membered heterocycles undergo ready fragmentation into a reactive segment and a stable molecule, a process usually induced by heat or light. Loss of nitrogen from compounds containing an azo function often occurs on photolysis, generating a 1,3-diradical which undergoes ring closure to a third species. The loss of nitrogen from pyrazolines to form cyclopropanes and from 1,2,3-triazolines to form aziridines are well established examples of this methodology.

To form ring systems of the type under discussion in this volume, two heteroatoms must remain in the reactive fragment. Although 2,5-diphenyltetrazole undergoes photochemical fragmentation to the 1,3-dipole, diphenylnitrilimine (see Section 4.03.6.1.1(ii)), 1,5-diphenyltetrazole (**507**) underwent fragmentation to (**508**), which by an intramolecular trapping process formed (**509**) ⟨67JA5958, 67JA5959⟩.

A similar intramolecular trapping of the intermediate (**511**) from the photolysis of the corresponding methyl tetrazole-1,5-dicarboxylate (**510**) gave methyl 5-methoxy-1,2,4-oxadiazole-3-carboxylate (**512**).

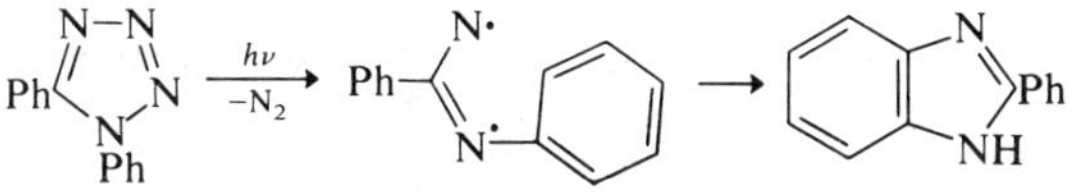

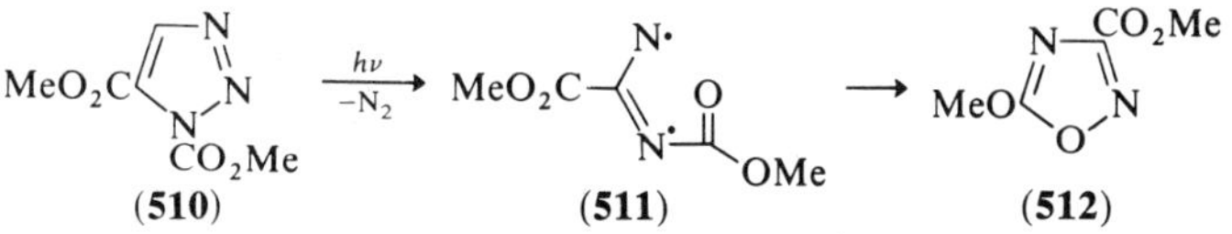

4-Phenyl-1,2,3-thiadiazole (**513**) as well as the 5-phenyl isomer (**514**) both lose N_2 on photolysis. The 1,3-dithiole derivative shown was formed from both thiadiazoles ⟨58LA(614)4⟩.

4.03.8.1.2 Loss of carbon dioxide

Photochemical elimination of carbon dioxide from suitable precursors has given a variety of reactive intermediates at low temperatures where they are often stable and can be studied further. This approach has been utilized in attempts to generate new 1,3-dipolar species, and photolysis of (**515**) gave an azomethine nitrene intermediate (**516**) (see Section 4.03.6)

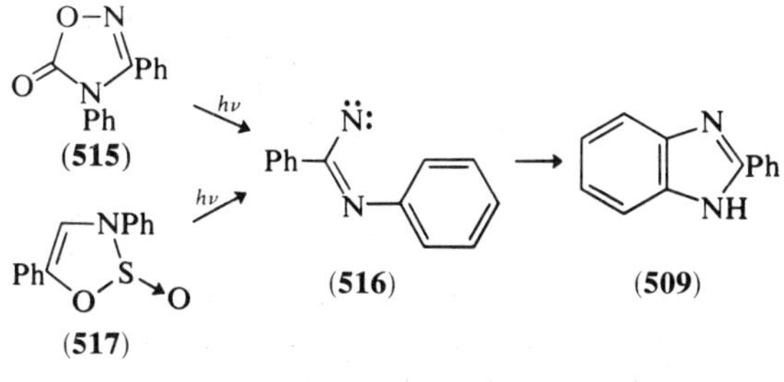

which was trapped intramolecularly to yield (**509**) ⟨68TL325⟩. A small amount of (**509**) was also obtained from photolysis of the oxathiadiazole 2-oxide (**517**) which apparently lost SO_2 to give (**516**) as an intermediate ⟨68TL325⟩.

A number of mesoionic compounds also lose a small, stable fragment on photolysis. N-Phenylsydnone (**518**) on irradiation gave 4-phenyl-1,2,4-oxadiazolin-5-one (**522**). When labelled carbon dioxide was passed through the solution during irradiation, it was incorporated into (**522**) ⟨66TL4043, 71JOC1589⟩. These results were rationalized in terms of an initial valence isomerization to (**519**), and loss of CO_2 to give (**520**) which underwent ring opening to the 1,3-dipole (**521**). This dipole subsequently reacted with carbon dioxide to give (**522**).

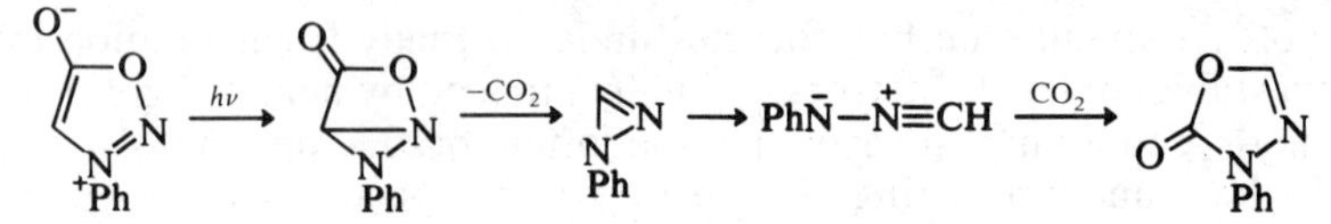

4.03.8.2 Photochemically Induced Isomerizations

The systems discussed here are aromatic systems which undergo a variety of isomerizations on irradiation. Irradiation of imidazoles led to a scrambling of substituents, whereas such scrambling has not been observed in the pyrazoles which undergo photoisomerization to imidazoles.

Irradiation of the substituted pyrazole (**523**) gave the imidazoles (**524**) and (**525**). The amount of each isomer formed is solvent dependent. In ethanol 7% of (**524**) was formed together with 2% of (**525**). In cyclohexane, however, isomerization was more efficient, the percentages of the two isomers being 20% and 10%, respectively.

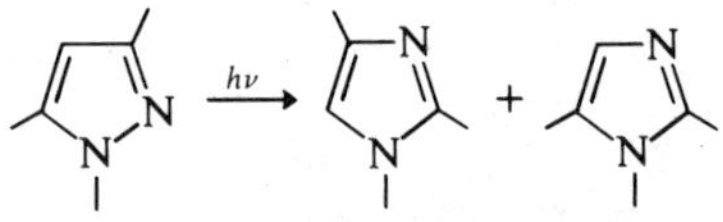

Mechanisms which explain these isomerizations are shown in Scheme 16, and by the use of deuterated 1,4-dimethylimidazole it was concluded that pathway (a) in Scheme 16 was the main one ⟨69T3287⟩ (see Chapter 4.04).

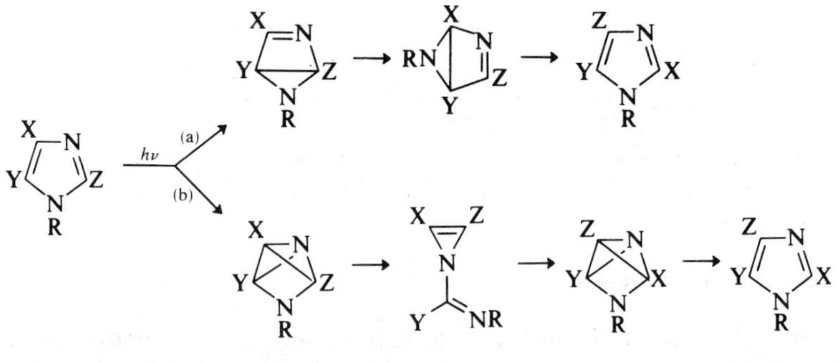

Scheme 16 Mechanistic pathways for the photoisomerization of imidazoles

Indazole → benzimidazole photoisomerization involves a singlet state and has been determined to be monomolecular and monophotonic. The UV spectrum of an intermediate with a lifetime of the order of seconds was recorded. Irradiation of the indazole (**526**) resulted in a 96% yield of the 1-methylbenzimidazole (**528**) probably via the intermediate (**527**) ⟨70PAC(24)495, 69C109⟩.

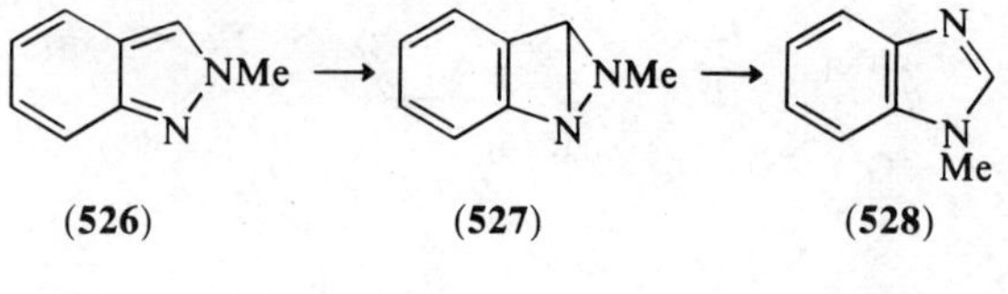

Irradiation of 1H-indazoles under nonacidic conditions resulted in isomerization to benzimidazoles and also ring opening to isomeric benzonitriles. With 1-substituted benzimidazoles and sensitized irradiation, nitriles were formed, but these are only minor products with other substitution patterns ⟨67HCA2244, 64TL2999⟩. Irradiation of benzimidazoles leads to oxidative dimerization.

Isoxazoles largely undergo photochemical isomerization to azirines, which sometimes undergo a further thermal or photochemical reaction. 3,4,5-Triarylisoxazole (**529**) formed the 2,3-diphenyl-3-benzoylazirine (**530**) which underwent further reaction to the oxazole (**531**) ⟨72JA1199⟩. A small amount of the corresponding benzoyl ketenimine was also obtained.

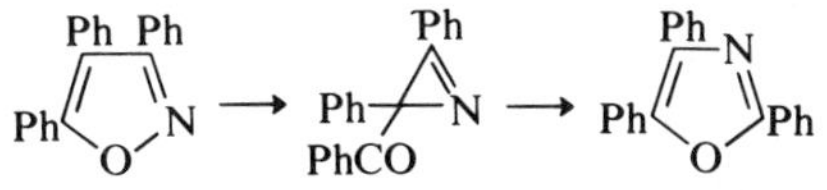

4.03.8.3 Carbon–Carbon Bond Formation

4.03.8.3.1 Intramolecular reactions

With an appropriately substituted precursor, photochemical carbon–carbon bond formation is a convenient method for the synthesis of ring-fused systems of the type under discussion.

N-Bridgehead compounds have been obtained from the photochemical cyclization of cis-1-styrylimidazoles. For example, irradiation of the imidazole (**532**) in methanol in the presence of I_2 resulted in cyclization at the 2-position of the imidazole ring with the formation of an imidazo[2,1-a]isoquinoline (**533**) ⟨76JCS(P1)75⟩. Isomerization of the trans- to the cis-styrylimidazole was followed by photodehydrocyclization.

trans-1-Styrylbenzimidazole (**534**) was isomerized under Pyrex-filtered light in the presence of one molar equivalent of I_2. The resulting cis isomer on irradiation through quartz gave the benzimidazo[2,1-a]isoquinoline (**535**) in 53% yield.

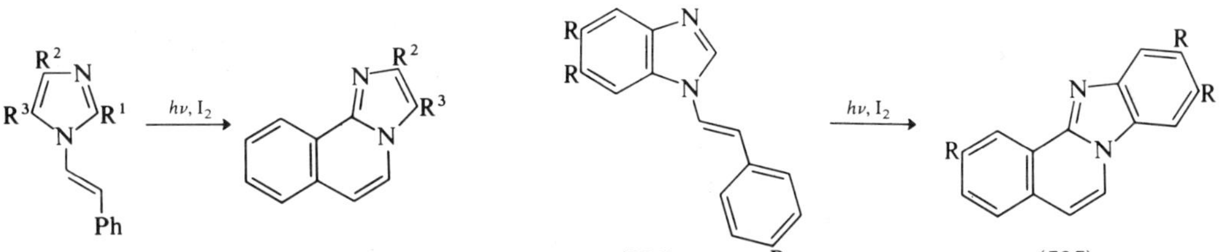

Oxidative cyclization of the fully methylated 6-(benzylidenehydrazino)uracils (**536**) provides a 90% yield of the pyrazolo[3,4-d]pyrimidines (**537**) ⟨75BCJ1484⟩.

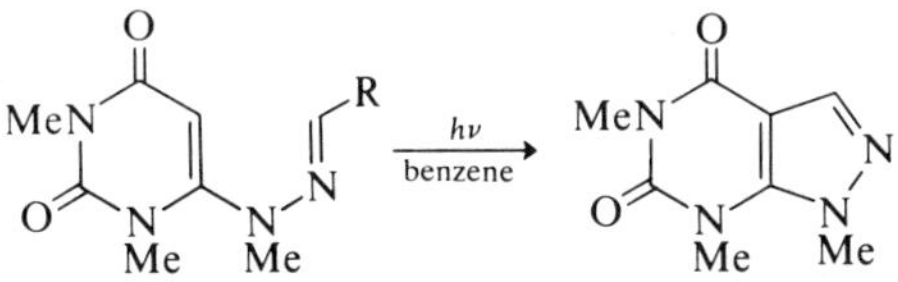

Ring contraction and intramolecular cyclization constitute a convenient route to ring-fused systems that would be difficult to synthesize in other ways. 1H-1,2-Diazepines (**538**) undergo electrocyclic ring closure to the fused pyrazole system (**539**) ⟨71CC1022⟩. Azepines also undergo similar valence bond isomerizations.

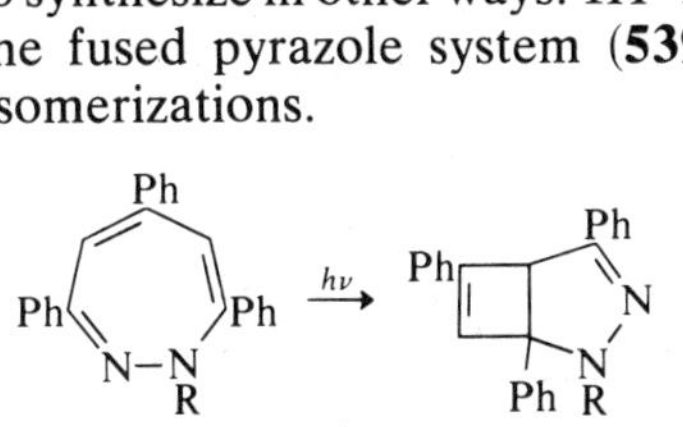

4.03.8.3.2 Intermolecular reactions

Unusual heterocyclic systems can be obtained by photodimerizations and for five-membered heterocycles with two or more heteroatoms such dimerizations need be effected on their ring-fused derivatives. Cyclobutanes are usually obtained as in the photodimerization of the s-triazolo[4,3-a]pyridine (**540**) to the head-to-head dimer (**541**). These thermally labile photodimers were formed by dimerization of the 5,6-double bond in one molecule with the 7,8-double bond in another ⟨77T1247⟩. Irradiation of the bis(1,2,4-triazolo[4,3-a]pyridyl)ethane (**542**) at 300 nm gave the *cisoid*-fused cyclobutane dimer (**543**). At 254 nm the 'cage-like' structure (**544**) was formed ⟨77T1253⟩.

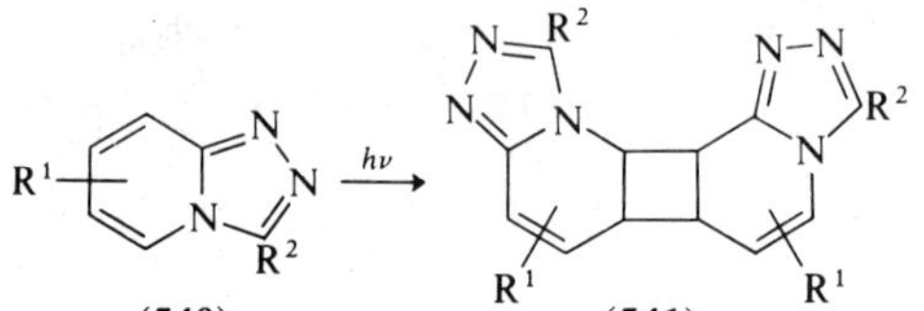

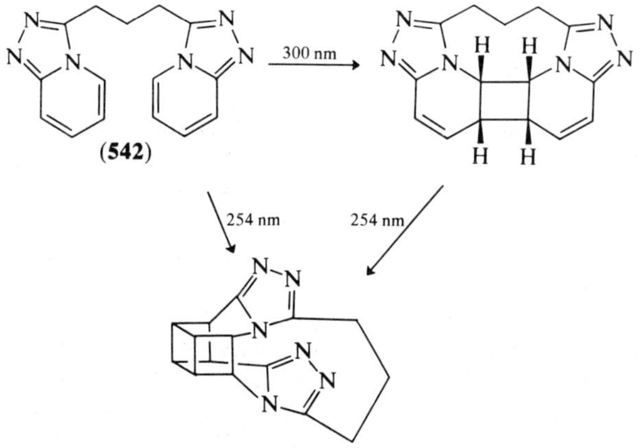

4.03.9 REACTIVE, ELECTRON-DEFICIENT INTERMEDIATES

4.03.9.1 Carbenes

Utilization of carbenes in the synthesis of five-membered heterocycles with two or more heteroatoms has not been featured prominently in the synthetic strategies developed for these ring systems. The following illustrations show their considerable promise.

Thermolysis of the 1,2,3-thiadiazoles (**545**) in the presence of carbon disulfide leads to the thiocarbonyl carbene (**546**) adduct, the ring-fused 1,3-dithiole-2-thione (**547**) ⟨76JOC730⟩.

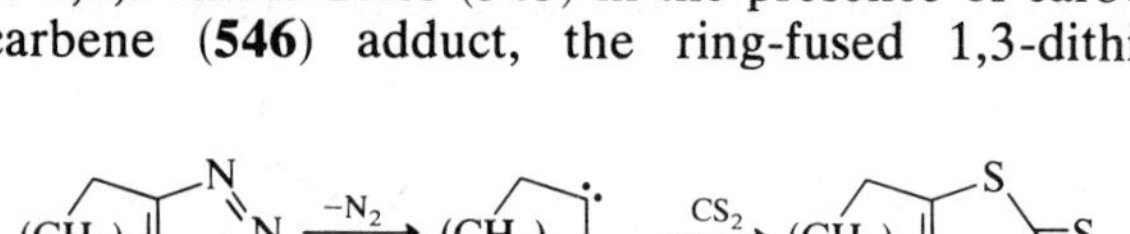

Decomposition of the diazoacetic ester (**548**) to the keto carbene (**549**) is promoted by copper(II) trifluoromethanesulfonate. In the presence of nitriles, 1,3-dipolar addition to the nitrile occurred giving the oxazole (**550**) ⟨75JOM(88)115⟩ (see also Section 4.03.8.1).

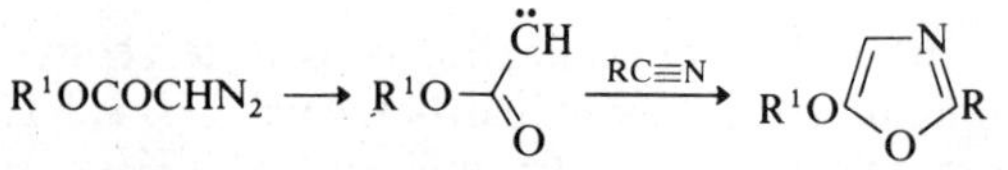

Decomposition of the diazoimide (**551**) by heating in the presence of copper acetylacetonate also generated a ketocarbene (**552**). This undergoes an intramolecular condensation to give the anhydro-4-hydroxy-3-methyl-4-*p*-nitrophenyl-2-phenyloxazolium hydroxide (**553**), which cannot be prepared by more classical means ⟨75CL499⟩.

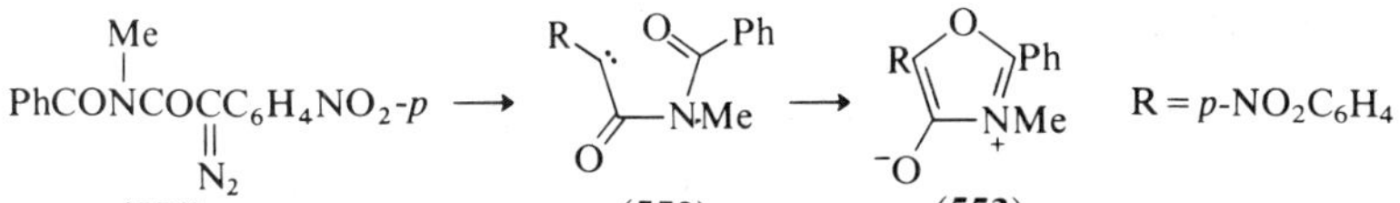

4.03.9.2 Nitrenes

Nitrenes have enjoyed appreciable application in the synthesis of a wide variety of heterocyclic systems, and the majority of the methods used for generating nitrenes have been utilized in these syntheses.

β-Nitrovinylazides (**554**), prepared by nitration of vinyl azides with nitryl fluoroborate, undergo spontaneous cyclization to the furoxan (**555**) with loss of N_2 ⟨75AG(E)775⟩, the Z isomer being implicated in the ring closure onto the nitro group. The E isomer, on heating or photolysis, gives the corresponding nitroazirine (see Chapter 4.22).

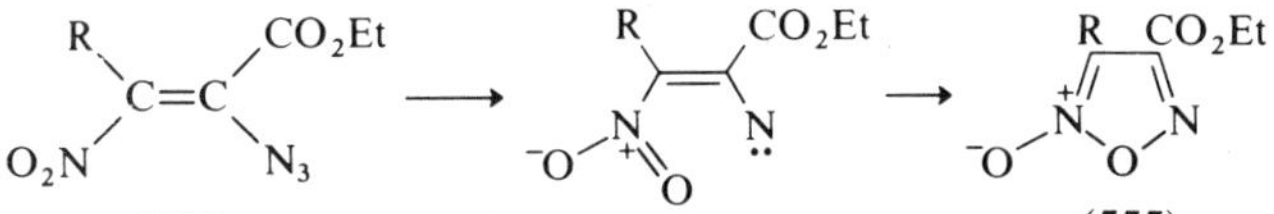

The behaviour of β-oxovinylazides is quite similar to those above. The Z isomer (**556**), formed from the β-halo carbonyl compound and sodium azide, is unstable losing N_2 and forming the isoxazole (**557**) in an anchimerically assisted concerted reaction ⟨75AG(E)775, 78H(9)1207⟩. At moderate temperatures (50–80 °C) the E isomer formed acylazirines which at higher temperatures rearranged to oxazoles and isoxazoles.

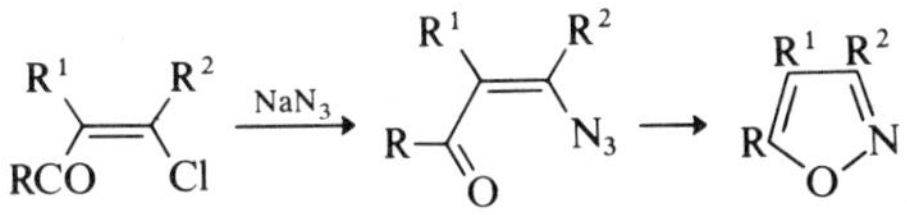

The synthetic potential of nitrenes is more readily apparent in the synthesis of ring-fused systems ⟨81AHC(28)309⟩, which can be accomplished by cyclization onto a heteroatom or onto an adjacent ring, the latter having the possibility of reaction at carbon or at a heteroatom.

Heating of the O,O'-diazidoazobenzene (**558**) at 58 °C resulted in formation of (**559**); raising the temperature to 70 °C resulted in a second ring closure to the tetraazapentalene (**560**) ⟨62JA2453, 67JA2618⟩.

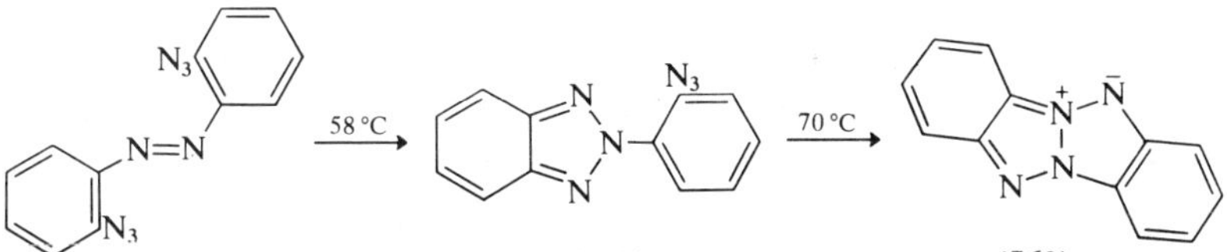

Triethyl phosphite is an effective reagent for the deoxygenation of appropriate nitro (or nitroso) aromatic systems. Free nitrenes or some 'nitrenoid-like' species may be involved, and the use of this reagent is illustrated by the examples below. It has the advantage over the azide approach in that two steps in the synthesis can be avoided.

Treatment of the substituted azobenzene (**561**) with triethyl phosphite gave the triazapentalene (**563**). It is likely that the 2-substituted benzotriazole (**562**) was initially formed which then underwent deoxygenative cyclization to (**563**) ⟨74CL951⟩.

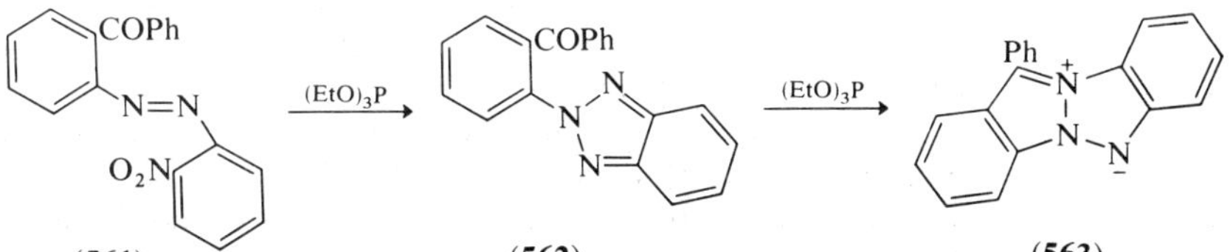

A similar treatment of the Schiff's base derived from 3-nitro-2-thienylcarbaldehyde (564) with triethyl phosphite gave the thieno[3,2-b]pyrazole (565) ⟨78CC453⟩:

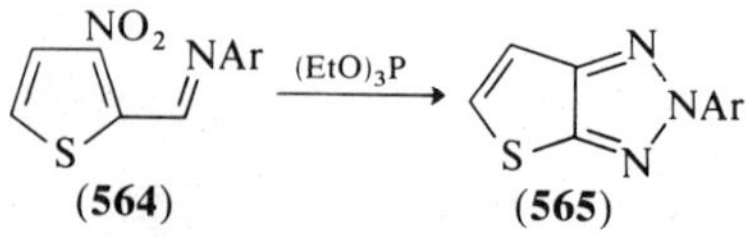

Cyclization onto a heterocyclic ring also readily occurs, as when the 2-substituted pyridine (566) was treated with triethyl phosphite. In this case the pyrrolopyrazole (567) was obtained ⟨79JOC622⟩.

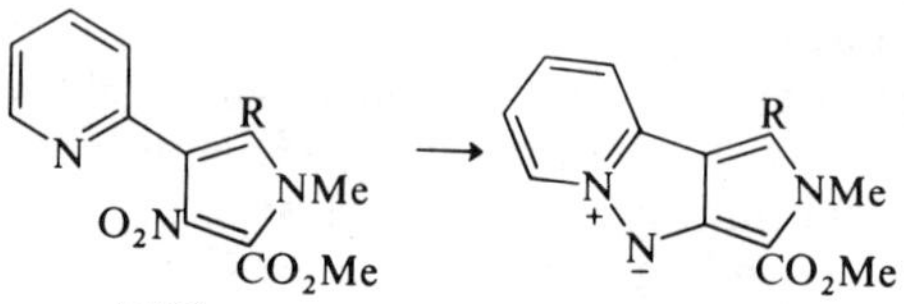

4.03.10 USE OF YLIDES IN HETEROCYCLIC SYNTHESIS

As in organic synthesis in general, ylides have been applied in the synthesis of heterocycles with increasing frequency. In five-membered ring chemistry they are more suited to the synthesis of systems with one heteroatom, but applications to ring systems with two or more heteroatoms are known, as well as to ring-fused systems. Two approaches have evolved: the ylide may be an acyclic synthon used to bring a particular functional group into construction of the heterocycle; or a heterocycle may be modified to incorporate an ylidic function. This latter approach is useful for functional group modification or for ring annulations.

4.03.10.1 Sulfonium Ylides

The dimethylsulfonium ylide (568) added readily to aroyl isocyanates to give the intermediate addition product (569). This on heating underwent ring closure with loss of dimethyl sulfide to form the 4-hydroxyoxazole (570) ⟨73T1983⟩. This normal C-acylation of the sulfonium ylide also leads to thiazoles with thiobenzoyl isocyanate; in this case the initial acylation product was not isolated, the thiazole being obtained directly.

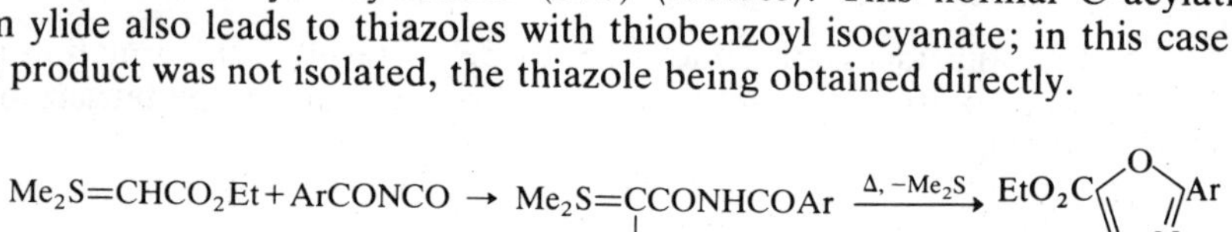

A convenient route to Δ^2-isoxazoline N-oxides has been developed from nitrostyrenes using dimethylsulfoxonium methylide. The addition of the ylide (572) to the nitrostyrene (571) was greatly facilitated by the presence of copper(I) salts, the isoxazoline N-oxide (573) being obtained in excellent yield ⟨76JOC4033⟩.

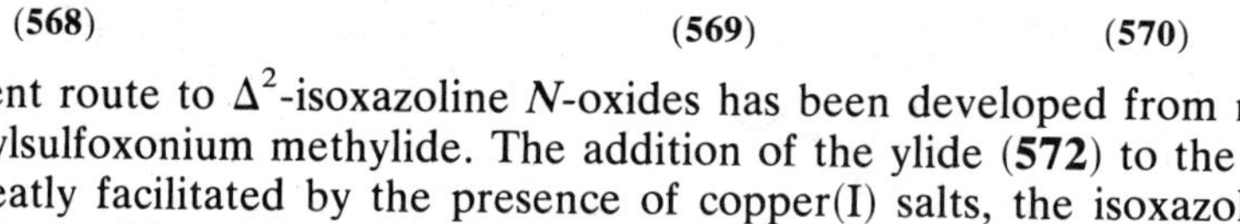

Dimethylsulfonium phenacylide (574) underwent C-alkylation with α-chloronitroso compounds such as (575). The intermediate (576) immediately cyclized to the isoxazoline (577). With a more basic ylide such as dimethylsulfonium methoxycarbonylmethylide the initial alkylation product underwent elimination of the sulfonium group to an alkene rather than its displacement ⟨72T3845⟩.

Modification of functional groups incorporated in a heterocycle is possible via ylide reactions. The 5-methylisoxazole (578) on reaction with n-butyllithium and methanesulfenyl

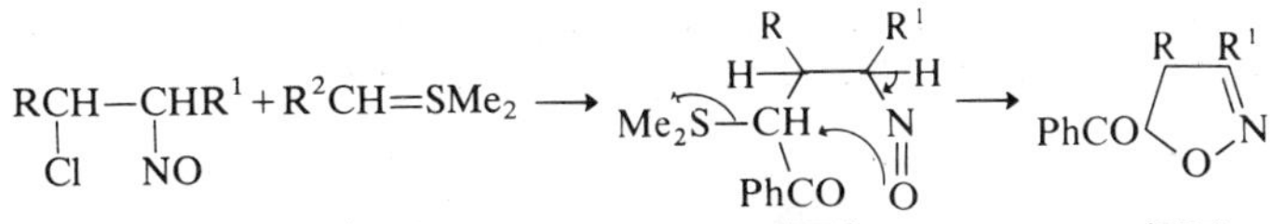

Synthesis of Five-membered Rings with Two or More Heteroatoms 165

$$\text{RCH—CHR}^1 + \text{R}^2\text{CH}=\text{SMe}_2 \longrightarrow \text{(576)} \longrightarrow \text{(577)}$$

(575)   (574)

chloride in THF readily formed the sulfide (**579**). Alkylation followed by treatment of the resultant salt with *n*-butyllithium gave the dimethylsulfonium ylide (**580**). This undergoes the normal reactions of sulfonium ylides: with aldehydes, the oxiranes (**581**) were obtained; with aroyl chlorides, the *C*-acylation products (**582**) were isolated; and with α,β-unsaturated ketones the cyclopropane (**583**) formed with displacement of dimethyl sulfide ⟨77JHC37⟩.

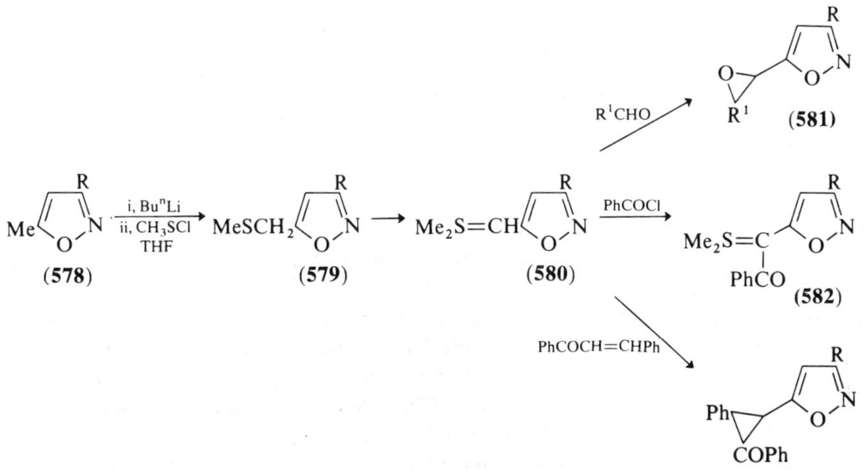

4.03.10.2 Phosphonium Ylides

Although widely applied in functional group modification in a variety of heterocyclic systems, phosphorus ylides have only been employed sparingly in heterocyclic ring construction with two or more heteroatoms in the nucleus. Their potential is shown in the applications illustrated below.

The isoxazoles (**585**) were formed regioselectively from the (dioxoalkyl)phosphonium salts (**584**) with hydroxylamine hydrochloride, the direction of cyclization being different from that of the nonphosphorus-containing 1,3-dioxo compound (see Chapter 4.16). Aqueous sodium hydroxide converted (**585**) into the isoxazole (**586**) and triphenylphosphine oxide. Treatment of (**585**) with *n*-butyllithium and an aldehyde gave the alkene (**587**). With hydrazine or phenylhydrazine analogous pyrazoles were formed ⟨80CB2852⟩.

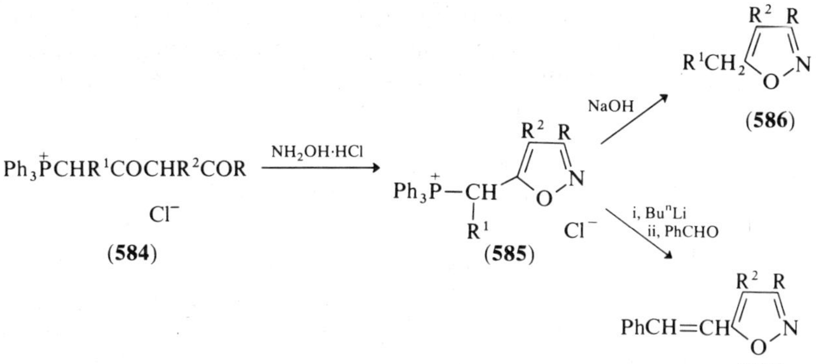

Reaction of the keto ylide (**588**) with cyanazide (**589**) provided a convenient route to the 1,2,3-triazole-1-carbonitrile (**590**). On heating, (**590**) lost N_2 readily to form the α-cyanoimino carbene (**591**) ⟨81AG(E)113⟩.

(**588**) (**589**) (**590**) (**591**)

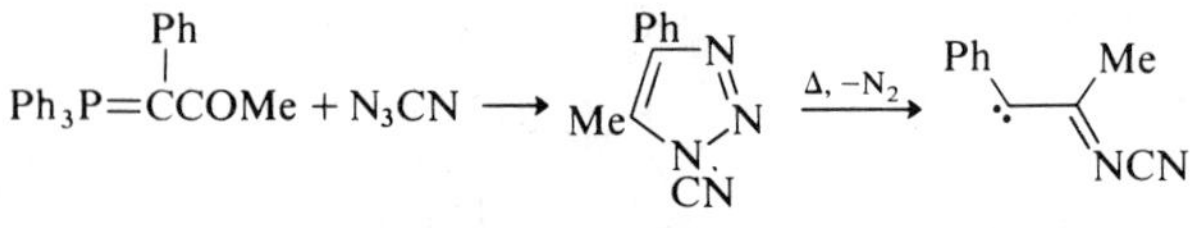

An interesting application of a phosphorus ylide in heterocyclic synthesis is in a ring annulation. The diazopyrazole (**592**) when treated with various phosphorus ylides gave the 3*H*-pyrazolo[5,1-*c*][1,2,4]triazole derivatives (**593**) with elimination of triphenylphosphine ⟨79TL1567⟩.

(**592**) (**593**)

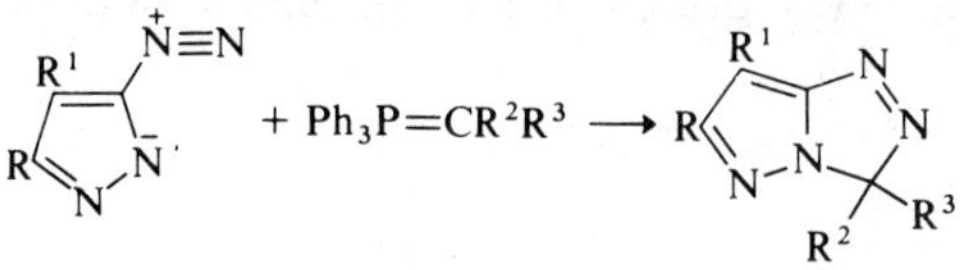

4.03.10.3 Sulfimides

Sulfimides have been utilized extensively for the synthesis of monocyclic ring systems as well as ring annulations.

Formation of the imine (**594**) by reaction of the phenyl thioether with mesitylsulfonylhydroxylamine and treatment with base provided a reactive substrate that could be halogenated with NBS or NCS, acylated with a variety of acylating agents, and was sufficiently reactive to add to benzoylalkynes forming the imine (**595**), which on heating underwent cyclization to isoxazoles (**596**). With the vinyl ether (**597**), substitution occurred giving (**598**), which then cyclized to the isoxazole (**598a**) ⟨73JOC4324⟩.

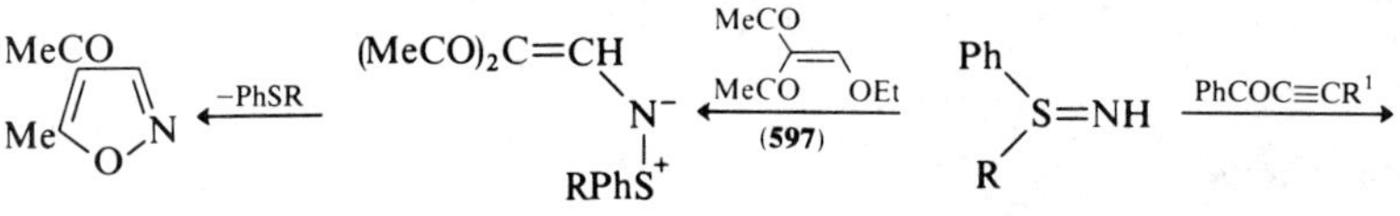

(**598a**) (**598**) (**594**)

(**595**) (**596**)

Sulfimides have been key synthons in a variety of ring annulations, reacting readily with ketenes, nitrile oxides and electron-withdrawing alkenes. These reactions are illustrated by the conversion with diphenylketene of the sulfimide (**599**) derived from 2-aminopyridine by *N*-chlorination, reaction with methyl sulfide and treatment with base. The imidazo[1,2-*a*]pyridin-3-one (**600**) was formed in good yield ⟨77H(8)109⟩. Reaction of (**599**) with a nitrile oxide led to the [1,2,4]triazolo[1,5-*a*]pyridine 3-oxide (**601**), also in good yield. This method provides an unambiguous route to this *N*-oxide and is also applicable to analogous pyrimidines and pyrazines ⟨76JCS(P1)2166, 78BCJ563⟩.

(**601**) (**599**) (**600**)

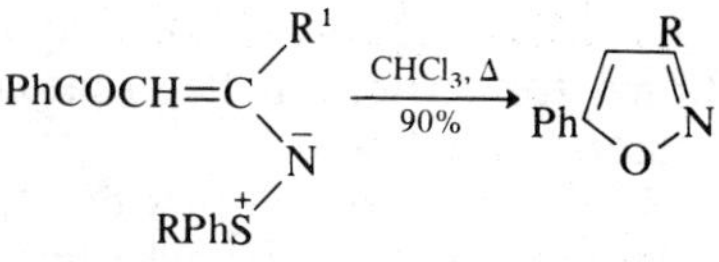

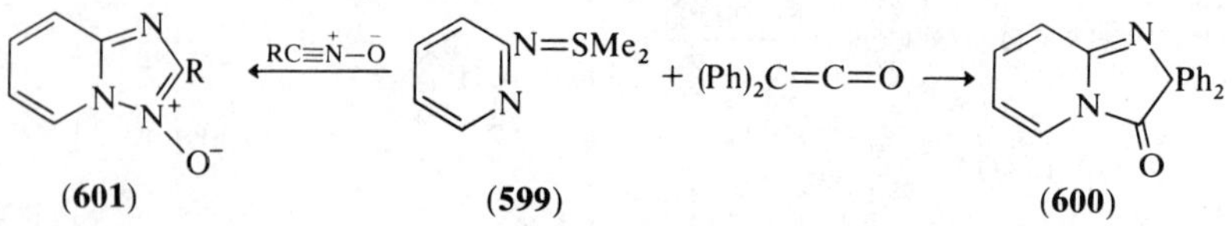

4.04

Pyrazoles and their Benzo Derivatives

J. ELGUERO
Instituto de Química Medica, Madrid

4.04.1 STRUCTURE	169
4.04.1.1 Survey of Possible Structures	169
4.04.1.1.1 Aromatic systems without exocyclic conjugation	170
4.04.1.1.2 Aromatic systems with exocyclic conjugation	170
4.04.1.1.3 Non-aromatic systems	171
4.04.1.2 Theoretical Methods	171
4.04.1.2.1 Structure and reactivity of neutral pyrazole, its anion and its cation	171
4.04.1.2.2 Structure and reactivity of substituted pyrazoles	173
4.04.1.2.3 Structure and reactivity of indazoles	175
4.04.1.2.4 Dipole moments	176
4.04.1.3 Structural Methods	178
4.04.1.3.1 X-Ray diffraction	178
4.04.1.3.2 Microwave spectroscopy	181
4.04.1.3.3 ^1H NMR spectroscopy	182
4.04.1.3.4 ^{13}C NMR spectroscopy	190
4.04.1.3.5 Nitrogen NMR spectroscopy	193
4.04.1.3.6 UV spectroscopy	197
4.04.1.3.7 IR spectroscopy	199
4.04.1.3.8 Mass spectrometry	202
4.04.1.3.9 Photoelectron spectroscopy	205
4.04.1.3.10 Electron spin resonance spectroscopy	206
4.04.1.4 Thermodynamic Aspects	206
4.04.1.4.1 Intermolecular forces	206
4.04.1.4.2 Stability and stabilization	208
4.04.1.4.3 Conformation and configuration	208
4.04.1.5 Tautomerism	210
4.04.1.5.1 Annular tautomerism	211
4.04.1.5.2 Substituent tautomerism	213
4.04.2 REACTIVITY	217
4.04.2.1 Reactions at Aromatic Rings	217
4.04.2.1.1 General survey of reactivity	217
4.04.2.1.2 Thermal and photochemical reactions formally involving no other species	218
4.04.2.1.3 Electrophilic attack at nitrogen	222
4.04.2.1.4 Electrophilic attack at carbon	236
4.04.2.1.5 Nucleophilic attack at carbon	243
4.04.2.1.6 Nucleophilic attack at hydrogen	245
4.04.2.1.7 Reaction with radicals and electron-deficient species; reaction at surfaces	246
4.04.2.1.8 Reactions with cyclic transition states	247
4.04.2.2 Reactions of Non-aromatic Compounds	249
4.04.2.2.1 Isomers of aromatic derivatives	249
4.04.2.2.2 Dihydro derivatives	253
4.04.2.2.3 Tetrahydro compounds	256
4.04.2.3 Reactions of Substituents	257
4.04.2.3.1 General survey of substituents on carbon	257
4.04.2.3.2 Indazoles	258
4.04.2.3.3 Other C-linked substituents	260
4.04.2.3.4 N-Linked substituents	261
4.04.2.3.5 O-Linked substituents	264
4.04.2.3.6 S-Linked substituents	265
4.04.2.3.7 Halogen atoms	266
4.04.2.3.8 Metals and metalloid-linked substituents	267
4.04.2.3.9 Fused heterocyclic rings	267
4.04.2.3.10 Substituents attached to ring nitrogen atoms	268
4.04.2.4 Pyrazoles, Indazoles, and their Derivatives as Starting Materials for the Syntheses of Fused Ring Systems	271

4.04.3 SYNTHESIS	273
4.04.3.1 Ring Synthesis from Non-heterocycles	274
4.04.3.1.1 Formation of one bond	274
4.04.3.1.2 Formation of two bonds	277
4.04.3.1.3 Formation of three bonds	284
4.04.3.2 Ring Synthesis from Heterocycles	285
4.04.3.2.1 Ring opening	285
4.04.3.2.2 Ring transformations	286
4.04.3.3 Best Practical Methods	288
4.04.3.3.1 General methods	288
4.04.3.3.2 Nucleosides	288
4.04.3.3.3 Labelled compounds	289
4.04.4 APPLICATIONS AND IMPORTANT COMPOUNDS	291
4.04.4.1 Synthetic Products	291
4.04.4.1.1 Pharmaceuticals	291
4.04.4.1.2 Agrochemicals	297
4.04.4.1.3 Dyestuffs and analytical applications	298
4.04.4.1.4 Plastics	300
4.04.4.2 Metabolism and Toxicity	301
4.04.4.3 Natural Products	302

Pyrazoles belong to the family of azoles, *i.e.* five-membered rings containing only nitrogen and carbon atoms, ranging from pyrrole to pentazole. According to Albert's classification ⟨B-59MI40400⟩ they are π-excessive N-heteroaromatic derivatives, and according to Kauffmann's arenology principle ⟨71AG(E)743⟩ as a substituted carbon they are analogues of amines (Section 4.04.2.3.1(iii)) and as a substituted nitrogen they are analogues of halogens, *i.e.* pseudohalogens. These two arene analogies will be discussed later in the reactivity sections.

The pyrazole ring is particularly difficult to cleave and, amongst the azoles, pyrazoles together with the 1,2,4-triazoles are the most stable and easiest to work with. This qualitative description of pyrazole stability covers the neutral, anionic and cationic aromatic species. On the other hand, the saturated or partially saturated derivatives can be considered as hydrazine derivatives; their ring opening reactions usually involve cleavage of the N—C bond and seldom cleavage of the N—N bond. It should be noted, however, that upon irradiation or electron impact the N—N bond of pyrazoles can be broken.

In contrast to pyrrole, imidazole, indole and benzimidazole, pyrazole, indazole and their derivatives are of little biological significance. This is probably due to the difficulty for living organisms to construct an N—N bond. On the other hand, the compounds discussed in this chapter have interesting pharmacological properties (Section 4.04.4.1.1) and for this reason they have had application in drug synthesis despite the recent tendency to suppress compounds containing an N—N bond in human medicine.

Several books and their revisions on pyrazoles and their derivatives have appeared, the most recent being that of Grimmett in 'Comprehensive Organic Chemistry' ⟨B-79MI40415⟩. Jacobs' review ⟨B-57MI40400⟩ was the leading reference for many years and a permanent source of research ideas due to its critical presentation. The work of von Auwers, probably the chemist who contributed the most to the development of pyrazole chemistry, is clearly described there. About 10 years later, two of the most prolific workers in this field, Kost and Grandberg, published a chapter on pyrazoles ⟨66AHC(6)347⟩ that it is still useful because it contains an authoritative summary of the Russian school's results. At about the same time a volume on pyrazoles appeared in the Weissberger series ⟨67HC(22)1⟩ and this is still the most important source of information on these compounds, containing a comprehensive chapter on pyrazoles but less detailed and critical chapters on pyrazolines and pyrazolidines, as well as on indazoles and condensed types. In a narrower field, a different point of view and an up-to-date bibliography can be found in the book on hydrazones by Kitaev and Buzykin ⟨B-74MI40407⟩. In the excellent book on azoles ⟨B-76MI40402⟩, pyrazoles are discussed in a comparative fashion with other azoles, and there are numerous very useful tables of physical properties. The book concentrates on pyrazoles with the exclusion of reduced and benzo derivatives.

A number of general reviews related to the topics covered in this chapter have appeared (Table 1) and others will be discussed in the section dealing with synthetic aspects (Section 4.04.3.1). The Sokolov review is particularly noteworthy since it is quite recent and complete, containing 304 references to pyrazoles and isoxazoles.

Table 1 Reviews on Pyrazoles and Related Topics

Main author	Title	Ref.
Huisgen	Cycloaddition reactions of alkenes	B-64MI40400
Busev	Pyrazolones as analytical reagents	65RCR237
Wagner	Pyrazolines as optical brighteners	66AG(E)699
Grandberg	Ring opening reactions of pyrazolines	66RCR9
Elguero	Reaction of hydrazines with 1,3-difunctional compounds	70BSF2717
Trofimenko	Polypyrazolylborates	71ACR17
Lablache-Combier	Photoisomerization of aromatic five-membered rings	71BSF679
Trofimenko	Coordination chemistry of pyrazole ligands	72CRV497
Katritzky	Tautomerism of heteroaromatic compounds	76AHC(S1)1
Elguero	Synthesis and reactivity of pyrazolinium salts	76MI40403
Volkov	Pyrazole chemistry	77RCR374
Bouchet	Fluoroazoles	77BSF171
Baiocchi	Synthesis, properties and reactions of indazolones	78S633
Sokolov	Chemistry of pyrazoles and isoxazoles	79RCR289
Bonati	Pyrazole metallic derivatives	80MI40402
Tišler and Stanovnik	Heterocyclic diazo compounds	80CHE443
Dorn	Pyrazolidones, iminopyrazolidines, amino- and hydroxy-pyrazoles	80CHE1

In addition to the authors quoted above, the names of Boyd, Finar, Habraken, Jacquier, Lynch, Reimlinger, Schmid and Sucrow are all associated as main contributors to pyrazole chemistry over the past three decades.

4.04.1 STRUCTURE

4.04.1.1 Survey of Possible Structures

The most rational way to classify structures related to pyrazole is to take into account the number of double bonds present, either endo- or exo-cyclic. An endocyclic double bond can formally be considered part of a benzene ring in this context.

Three double bonds. The most fully oxidized pyrazoles, the typical non-aromatic representatives of which are the pyrazoline-4,5-diones (**1**) and the pyrazolidine-3,4,5-triones (**2**), should be included here.

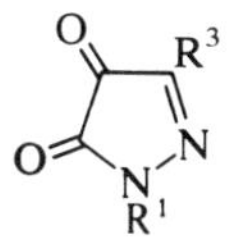

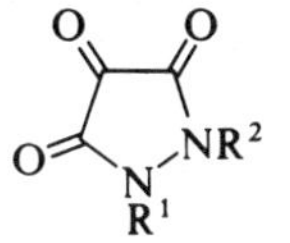

Two double bonds. This is the most important class which includes the aromatic compounds pyrazole (**3**), indazole (**4**) and isoindazole (**5**), their non-aromatic isomers, pyrazolenines (or 3H-pyrazoles; **6**), isopyrazoles (or 4H-pyrazoles; **7**) and 3H-indazoles (**8**), and the carbonyl derivatives of pyrazolines with the endocyclic double bond in positions 1, 2 or 3, *i.e.* (**9**), (**10**) and (**11**), respectively. The indazolones (**12**) and the pyrazolidinediones (**13**) and (**14**) also belong to this group.

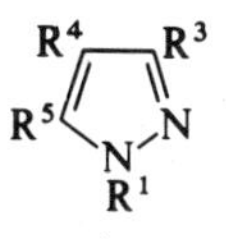

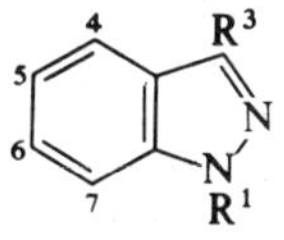

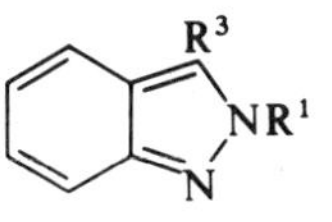

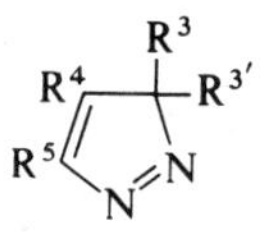

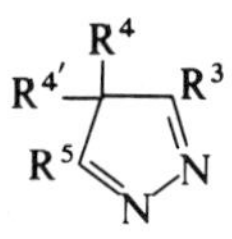

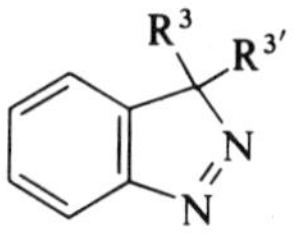

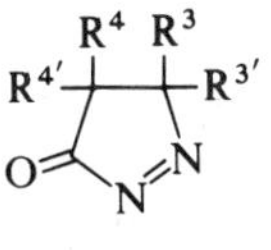

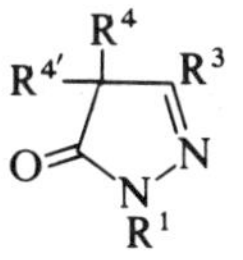

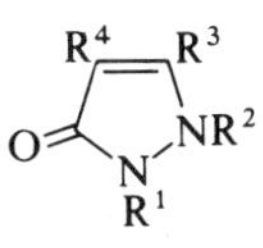

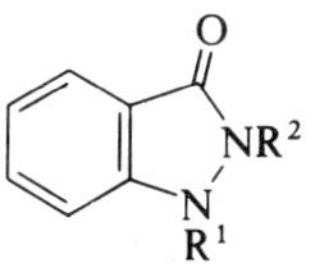

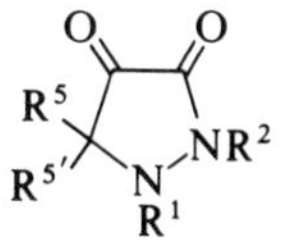

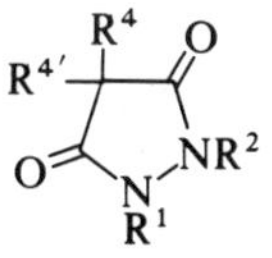

One double bond. The three pyrazolines, Δ¹- (**15**), Δ²- (**16**) and Δ³-pyrazoline (**17**), and the carbonyl derivatives of pyrazolidines like (**18**), are the most representative compounds of this group.

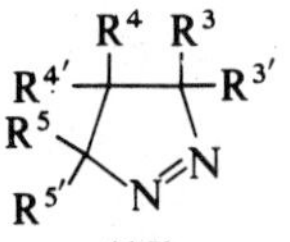

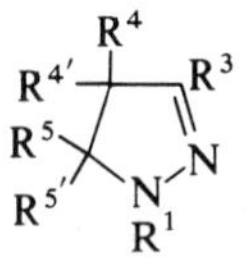

Without double bonds. Pyrazolidines (**19**) are the only representative of this group.

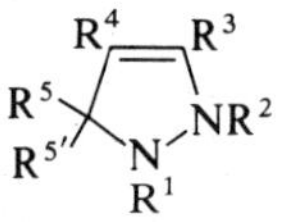

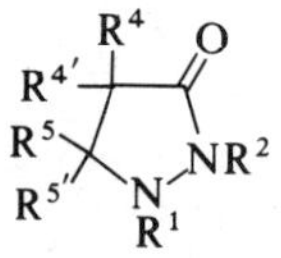

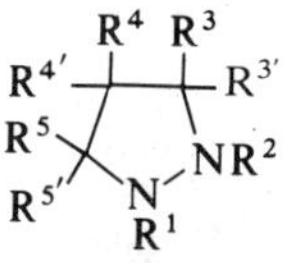

This natural classification, with the number of substituents R increasing regularly from two to eight (the unsubstituted 1-pyrazoline-3,4,5-trione is unknown), is useful for redox studies, but in order to be consistent with other parts of this work the following classification will be used throughout the chapter.

4.04.1.1.1 Aromatic systems without exocyclic conjugation

Only three systems belong to this group: pyrazole (**3**), 1*H*-indazole (**4**) and 2*H*-indazole (isoindazole; **5**). The fused carbon atoms in indazoles are numbered 3a and 7a. When $R^1 = H$, annular tautomerism ⟨76AHC(S1)1⟩ makes the 3- and 5-positions of pyrazoles equivalent and thus the name 3(5)-R-pyrazole means that the compound is a mixture of tautomers with the substituent R in position 3 and in position 5. The same applies to *N*-unsubstituted indazoles; however, the numbering is identical in both tautomers and thus 3-R-indazole means either (**4**) or (**5**) (R^1 or $R^2 = H$). Since the indazole tautomer is largely predominant (Section 4.04.1.5.1), indazoles are usually represented by the formula (**4**).

Only one monoanion and one monocation are possible for each of structures (**20**), (**21**), (**22**) and (**23**). In spite of the localized representation, when $R^1 = R^2$ and $R^3 = R^5$, the anion (**20**) and the cation (**22**) are fully delocalized structures of C_{2v} symmetry. Thus in many publications they are represented as shown in (**20′**)–(**23′**).

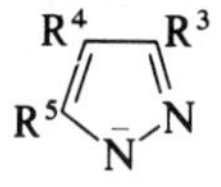

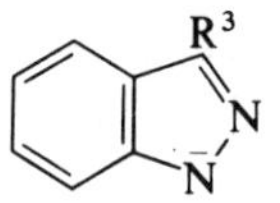

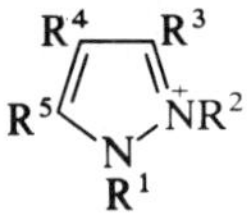

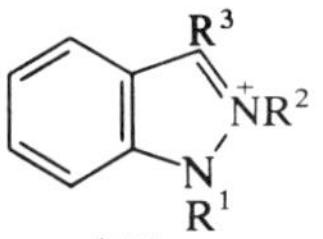

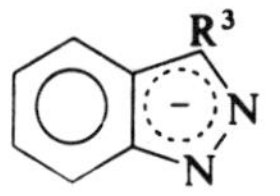

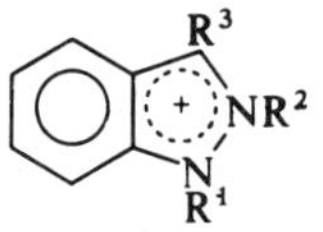

4.04.1.1.2 Aromatic systems with exocyclic conjugation

Formally, these are derived from the monocationic structures (**22**) and (**23**) by replacing one of the CR groups by a negatively charged heteroatom (generally O⁻, S⁻ or RN⁻). Two groups of unequal importance are included here: the pyrazolinones and the betaines.

The pyrazolinones (**11**) (pyrazolinethiones and iminopyrazolines) correspond to the replacement of R^3 or R^5 in (**22**) by O⁻, and are normally represented by their neutral structure, but it is important to bear in mind their dipolar formula (**26**) when discussing their reactivity. Similarly, the indazolinones (**12**) ↔ (**27**) result from the indazolium salts (**23**).

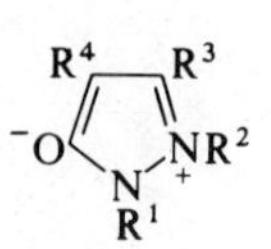

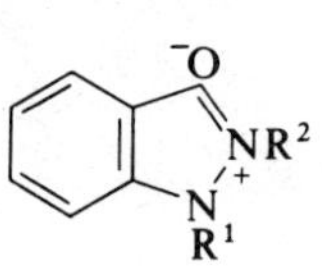

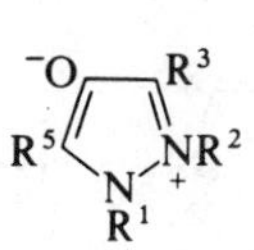

The mesoionic compounds are derived from pyrazolium salts (**22**) when R^4 is replaced by a negatively charged heteroatom, like the anhydro-4-hydroxypyrazolium hydroxide (**28**). According to Ollis and Ramsden ⟨76AHC(19)1⟩ they belong to the mesoionic class B type.

4.04.1.1.3 Non-aromatic systems

Following the classification of Chapter 4.01, three classes will be considered. *(a) Compounds isomeric with aromatic compounds:* (**6**), (**7**) and (**8**). The quaternary, non-aromatic salts (Scheme 7, Chapter 4.01) will be discussed only in connection with protonation studies which lead to the conclusion of their non-existence. The carbonyl derivatives (**9**), (**10**), (**13**) and (**14**) will also be included here because it is possible to write an aromatic tautomer for each one, (**9′**)–(**14′**), even if it is energetically unfavoured. *(b) Dihydro compounds.* In this class not only pyrazolines (**15**), (**16**) and (**17**) but also pyrazolidinones (**18**) and pyrazolinediones like (**1**) are included. *(c) Tetrahydro compounds.* Besides the pyrazolidines (**19**), the pyrazolidinetriones (**2**) are included here.

(**9′**) $R^{3'} = R^{4'} = H$ (**10′**) $R^{4'} = H$ (**13′**) $R^2 = R^{5'} = H$ (**14′**) $R^2 = R^{4'} = H$

4.04.1.2 Theoretical Methods

Theoretical methods ranging from the now obsolete HMO studies to *ab initio* calculations have been used extensively on pyrazoles. Although not emphasized in earlier reviews ⟨66AHC(6)347, 67HC(22)1⟩, the most recent publications ⟨B-76MI40402, 79RCR289⟩ contain several references to theoretical studies. Some publications related to structural studies are to be found in the following sections, especially in connection with ^{13}C NMR spectroscopy (Section 4.04.1.3.4), UV spectroscopy (Section 4.04.1.3.6), PE spectroscopy (Section 4.04.1.3.9) and tautomerism (Section 4.04.1.5).

In this section, reactivity studies will be emphasized while in those devoted to synthesis (Section 4.04.3) theoretical calculations on reactions leading to the formation of pyrazoles (mainly 1,3-dipolar cycloadditions) will be discussed. It should be emphasized that the theoretical treatment of reactivity is a very complicated problem and for this reason, most of the calculations have been carried out on aromatic compounds, as they are the easiest to handle. In general, solvents are not taken into account; thus, at the best, the situation described theoretically corresponds to reactions taking place in the gas phase.

4.04.1.2.1 Structure and reactivity of neutral pyrazole, its anion and its cation

Burton and Finar published a comparative table of calculations on pyrazole (**29**) ⟨70JCS(B)1692⟩. That table, updated with recent references and extended to the pyrazole anion (**30**) and cation (**31**), is the present Table 2.

Concerning the π-framework a considerable spread of results is observed, the most reliable ones being those of Roche and Pujol ⟨69BSF1097⟩ and Olivella and Vilarrasa ⟨81JHC1189⟩ who optimized the geometry and used, respectively, the Julg improved LCAO method and the Dewar MNDO method. Figure 1 contains the optimized geometries used by the last authors; for pyrazole (**29**) there is good agreement with the geometrical structure obtained from microwave spectroscopy (see Section 4.04.1.3.2) ⟨74PMH(6)53⟩, all bond

Table 2 Calculated Electron Densities of Pyrazole, its Anion and its Cation

Molecule	Method	Densities[a]	N-1	N-2	C-3	C-4	C-5	H-1	H-3	H-4	H-5	Ref.
Pyrazole	HMO	π	1.64	1.38	0.93	1.11	0.94	—	—	—	—	51MI40400
Pyrazole	VESCF	π	1.656	1.162	1.031	1.042	1.109	—	—	—	—	60AJC49
Pyrazole	HMO	π	1.776	1.045	0.930	1.137	1.108	—	—	—	—	62T985
Pyrazole	HMO	π	1.700	1.269	0.975	1.107	0.949	—	—	—	—	66CHE413
Pyrazole	EHT	π	1.57	1.54	0.89	1.11	0.90	—	—	—	—	67MI40401
Pyrazole	EHT	σ	3.49	4.28	2.85	3.12	2.89	0.708	0.888	0.890	0.888	67MI40401
Pyrazole	VESCF	π	1.745	1.139	1.015	1.066	1.035	—	—	—	—	67MI40400
Pyrazole	HMO	π	1.453	1.293	1.064	1.150	1.040	—	—	—	—	68JCS(B)725
Pyrazole	ILCAO	{π	1.75	1.18	0.97	1.11	0.99	—	—	—	—	69BSF1097
Pyrazole	SCF	σ}	3.43	4.01	3.06	3.00	3.03	0.81	0.90	0.89	0.88	70BSF273
Pyrazole	CNDO/2	{π	1.596	1.279	0.994	1.105	1.026	—	—	—	—	70JCS(B)1692
Pyrazole	CNDO/2	σ}	3.444	3.867	2.951	2.968	2.911	0.885	0.989	0.992	0.994	70JCS(B)1692
Pyrazole	ab initio	π	1.626	1.093	1.104	1.101	1.077	—	—	—	—	B-70MI40401
Pyrazole	ab initio	σ	3.722	4.038	3.053	3.165	3.037	0.610	0.780	0.806	0.789	B-70MI40401
Pyrazole	MNDO	{π	1.599	1.228	1.003	1.125	1.044	—	—	—	—	81JHC1189
Pyrazole	MNDO	σ}	3.577	3.898	3.053	3.048	2.958	0.772	0.894	0.906	0.896	81JHC1189
Pyrazole anion	VESCF	π	1.210	1.210	1.190	1.199	1.190	—	—	—	—	60AJC49
Pyrazole anion	EHT	{π	—	—	0.898	1.11	0.898	—	—	—	—	67T2513
Pyrazole anion	EHT	σ}	—	—	2.80	3.05	2.80	—	—	—	—	67T2513
Pyrazole anion	ILCAO	π	1.25	1.25	1.12	1.26	1.12	—	—	—	—	69BSF1097
Pyrazole cation	VESCF	π	1.618	1.618	0.913	0.936	0.913	—	—	—	—	60AJC49
Pyrazole cation	EHT	{π	—	—	0.898	1.11	0.898	—	—	—	—	67T2513
Pyrazole cation	EHT	σ}	—	—	2.83	3.04	2.83	—	—	—	—	67T2513
Pyrazole cation	HMO	π	1.512	1.512	0.924	1.128	0.924	—	—	—	—	68JCS(B)725
Pyrazole cation	ILCAO	π	1.74	1.74	0.73	1.06	0.73	—	—	—	—	69BSF1097
Pyrazole cation	CNDO/2	{π	1.617	1.617	0.838	1.090	0.838	—	—	—	—	70JCS(B)1692
Pyrazole cation	CNDO/2	σ}	3.394	3.394	2.973	2.971	2.973	0.788[b]	0.908	0.905	0.908	70JCS(B)1692
Pyrazole cation	CNDO/2	π	1.585	1.585	0.875	1.081	0.875	—	—	—	—	73T3469
Pyrazole cation	CNDO/2	σ	3.399	3.399	2.945	2.975	2.945	0.796[b]	0.916	0.912	0.916	73T3469
Pyrazole cation	MNDO	π	1.604	1.604	0.851	1.092	0.851	—	—	—	—	81JHC1189
Pyrazole cation	MNDO	σ	3.531	3.531	2.973	3.063	2.973	0.709[b]	0.833	0.844	0.833	81JHC1189

[a] Total charges are not given but they can be easily calculated adding the π- and σ-densities. [b] The same value for H-2.

lengths [except N(1)—C(5)] and angles being ±0.02 Å and ±2°, respectively, within experimental values. The N(1)—C(5) bond length (1.398 Å) is somewhat longer than the experimental values obtained by neutron diffraction (1.356 Å) ⟨70ACS3248⟩ and by microwave spectroscopy (1.360 Å) ⟨74JST(22)401⟩. The ILCAO N(1)—C(5) calculated distance is 1.375 Å ⟨69BSF1097⟩. The CNDO force method has been used to calculate fully optimized geometries for five-membered heterocycles including pyrazole ⟨73MI40400⟩.

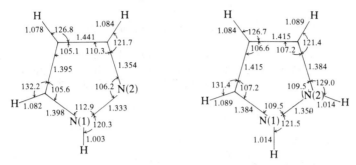

Figure 1 MNDO optimized geometries for pyrazole (**29**) and pyrazole cation (**31**)

HMO calculations ⟨66CHE413⟩ showed a considerable localization of the π-system (bond orders: N(1)N(2), 0.396; N(2)C(3), 0.765; C(3)C(4), 0.581; C(4)C(5), 0.770; C(5)N(1), 0.441). However, a more recent HMO calculation using new parameters gave a more 'aromatic' picture (bond orders: N(1)N(2), 0.5045; N(2)C(3), 0.6712; C(3)C(4), 0.6709; C(4)C(5), 0.6625; C(5)N(1), 0.6434) ⟨70CB3289⟩.

The most complete discussion of the electrophilic substitution in pyrazole, which experimentally always takes place at the 4-position in both the neutral pyrazole and the cation (Section 4.04.2.1.1), is to be found in ⟨70JCS(B)1692⟩. The results reported in Table 2 show that for (**29**), (**30**) and (**31**) both π- and total ($\pi + \sigma$)-electron densities predict electrophilic substitution at the 4-position, with the exception of an older publication that should be considered no further ⟨60AJC49⟩. More elaborate models, within the CNDO approximation, have been used by Burton and Finar ⟨70JCS(B)1692⟩ to study the electrophilic substitution in (**29**) and (**31**). Considering the substrate plus the properties of the attacking species (H^+, Cl^+), they predict the correct orientation only for perpendicular attack on a planar site. For the neutral molecule (the cation is symmetrical) the second most reactive position towards H^+ and Cl^+ is the 5-position. The activation energies (kJ mol^{-1}) relative to the 4-position are: H^+, C-3, 28.3; C-5, 7.13: Cl^+, C-3, 34.4; C-5, 16.9.

The binding energy of pyrazole (0.309 a.u.) is slightly lower than that of imidazole (0.335 a.u.) due to the vicinal position of the nitrogen atoms in the former, a situation that parallels the heat of atomization ⟨B-69MI40401⟩. The relative basicity of these two azoles (Section 4.04.2.1.3(iv)) has been discussed on theoretical grounds ⟨81JHC1189⟩. After application of the partitioning of energy in the MNDO method, the authors concluded that the greater basicity of imidazole in comparison with that of pyrazole can be primarily ascribed to the extra destabilization of the π-bonding in the pyrazole after protonation. The destabilization of the π-bonding for pyrazole is larger than for imidazole because the π lone pair on the N-1 atom of the former compound does not contribute appreciably to the delocalization of the positive charge which appears after protonation at N-2. For a more rigorous treatment of the basicity of pyrazoles, see ⟨83H(20)1717, 83UP40400⟩.

4.04.1.2.2 Structure and reactivity of substituted pyrazoles

Minkin et al. ⟨66CHE413⟩ have used Stretwieser 'optimal' values of Coulomb and resonance integrals to calculate the π-electron densities and bond orders of 1-phenylpyrazole (**32**; Figure 2). From the values of the π-electron densities they concluded that a preferential electrophilic attack takes place at the 4-position, as in pyrazole itself, and that pyrazole as a substituent of the phenyl ring has an *ortho–para* directing effect (pseudohalogen character). The low value of the N(1)—C(1) bond order in (**32**) shows that the interaction energy between the two aromatic rings is small in the ground state.

However, the fundamental contribution to the theoretical study of the structure of substituted pyrazoles has been made by Finar ⟨68JCS(B)725, 70JCS(B)1692⟩. Figure 3 shows

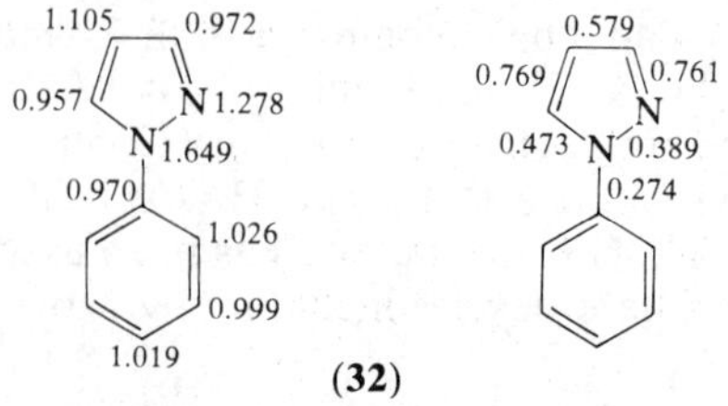

Figure 2 LCAO MO calculations of π-electron densities and bond orders

the π-electron densities and bond orders of 1-phenyl- (**32**) and 1-methyl-pyrazole (**33**) and their corresponding cations (**34**) and (**35**) calculated by the HMO method ⟨68JCS(B)725⟩. Figure 4 shows the CNDO/2 calculations on (**33**) and (**35**) ⟨70JCS(B)1692⟩.

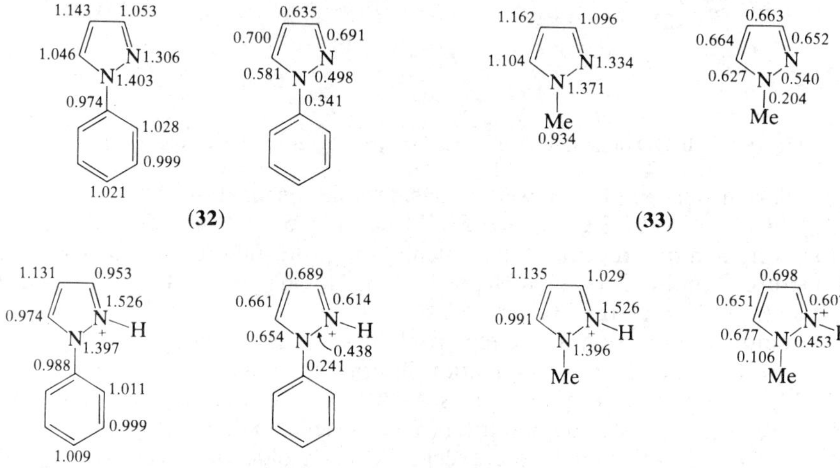

Figure 3 HMO calculations of π-electron densities and bond orders

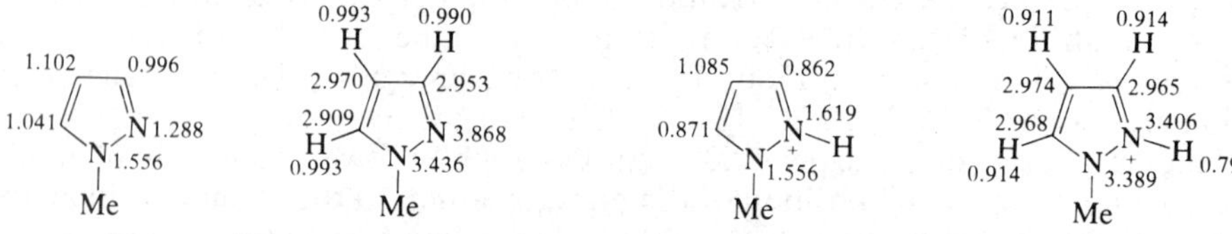

Figure 4 CNDO/2 calculations of π- and σ-electron densities

The results obtained for 1-phenylpyrazole (**32**) and its conjugate acid (**34**) are consistent with those of Minkin. The bond order between the two rings decreases by protonation (from 0.341 to 0.241) and this is in agreement with the expected effect of the N-2 positive charge on the delocalization of the lone pair on N-1 over the phenyl ring.

The reactivities of 1-methylpyrazole (**33**) and pyrazole (**29**) are similar and so are the corresponding charges. The competition between C-4 and C-4' in 1-phenylpyrazole depends on the electrophile and on the experimental conditions (Section 4.04.2.3.10(i)). Thus in an acidic medium the reaction takes place on the conjugate acid (**34**) and considering the calculated charge densities the attack on C-4 would always be favoured.

An alternative method for studying electrophilic substitution is to calculate the energy required to localize a pair of π-electrons at the site of substitution and to assume that this localization energy (E_L) is a measure of the energy of activation. When the attacking reagent is a cation (NO_2^+ or Br^+), then besides E_L (which represents the energy required to reorganize the π-electrons in pyrazole), Finar ⟨68JCS(B)725⟩ takes into account the electrostatic energy required to bring the cation to the site of attack. The calculation of this potential energy (P.E.) is carried out for each case where the unit positive charge was placed at a point 3 Å perpendicular to the site of substitution. Molecular bromine is treated as an uncharged species (P.E. = 0). Another possibility is to calculate the polarization energies (ΔE_π) and to correct them with a P.E. term if the nucleophile carries a positive charge. By considering these different possibilities, Finar has succeeded in explaining the different experimental results.

Free valences and localization energies have been calculated for a series of pyrazoles (neutral molecules and conjugate acids) for homolytic substitution. In all the compounds the site with the lowest localization energy has the highest free valence index. This parallel between the two indices of reactivity is maintained in pyrazole, 1-methylpyrazole and their conjugate acids, but not in 1-phenylpyrazole and its conjugate acid. For the three compounds examined experimentally, (32), (33) and (35) (Section 4.04.2.1.8(ii)), only the predictions for (33) are in agreement with the experimental results.

There is a rather poor agreement between the basicities of pyrazoles (Section 4.04.2.1.3 (iv)) and either the difference in total π-energy between the free base and its conjugate acid (ΔE_π) or the net charge on N-2 of the free base ⟨68JCS(B)725⟩. The inclusion of the σ-electrons and the calculation of the proton affinities (P.A.) does not improve the results ⟨70JCS(B)1692⟩ (see, however, ⟨83H(20)1717, 83UP40400⟩). The lowering effect of N-methylation (Section 4.04.2.1.3(iv)) cannot be explained by indices unable to take solvation phenomena into account. Only gas phase proton affinities can possibly correlate with Burton and Finar's calculations.

The hydrogen–deuterium exchange rates for 1,2-dimethylpyrazolium cation (protons 3 and 5 exchange faster than proton 4; Section 4.04.2.1.7(iii)) have been examined theoretically within the framework of the CNDO/2 approximation ⟨73T3469⟩. The final conclusion is that the relative reactivities of isomeric positions in the pyrazolium series are determined essentially by inductive and hybridization effects.

4.04.1.2.3 Structure and reactivity of indazoles

Indazoles have been subjected to certain theoretical calculations. Kamiya ⟨70BCJ3344⟩ has used the semiempirical Pariser–Parr–Pople method with configuration interaction for calculation of the electronic spectrum, ionization energy, π-electron distribution and total π-energy of indazole (36) and isoindazole (37). The π-densities and bond orders are collected in Figure 5; the molecular diagrams for the lowest $^1(\pi,\pi^*)$ singlet and $^3(\pi,\pi^*)$ triplet states have also been calculated; they show that the isomerization (36) → (37) is easier in the excited state.

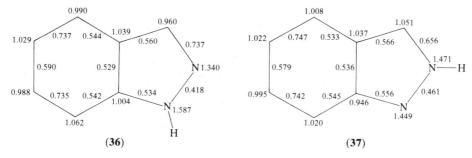

Figure 5 PPP calculations of π-electron densities and bond orders

The positional preference for electrophilic substitution in 1H-indazole (36) and 2H-indazole (37) is predicted to be 7 > 5 and 3 > 5 > 7 > 4, respectively, although it is known that experimentally substitution takes place at C-3 and C-5 (Section 4.04.2.1.4(i)). Since in this study the molecular geometry of indazoles was tentatively approximated as a regular hexagon and pentagon with a bond length of 1.394 Å on account of a lack of experimental data, Faure et al. have used the X-ray structure of benzazoles (indazole, benzimidazole, benzotriazole) as a basis for CNDO/2 and CNDO/S calculations ⟨74T2903⟩. The electron density diagram with the charges (π and σ) and the bond orders is shown in Figure 6. The π-electron densities seem more realistic, with the 3-position rich in electrons (7 > 5 > 3). Nevertheless, no diagram of electron densities, however accurate, would correctly predict the chemistry of a molecule like indazole, with many sites of comparable reactivity.

On the other hand, theoretical methods allow an insight into the structure of 'non-existent' molecules like 2H-indazole (37) or the anion of indazole (38). INDO calculations have been performed by Palmer et al. on the anion of indazole (38) ⟨75JCS(P1)1695⟩. The optimized geometry obtained by them is shown in Figure 7. The N—N bond distance is longer in the

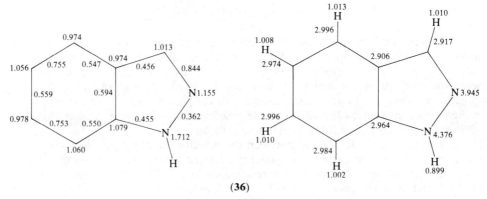

(36)

Figure 6 CNDO/2 calculations of π- and σ-electron densities and bond orders

indazole anion than in the benzotriazole anion as a consequence of the charge repulsion between the negatively charged nitrogen atoms (total net charges: N-1, −0.320; N-2, −0.169). The correlation between N-methylation ratios and charge is less satisfactory; a possible explanation could be that there is more than one mechanism for the reaction between the indazole anion and alkylating agents (Section 4.04.2.1.3(viii)).

(38)

Figure 7 INDO optimized geometry for indazole anion

4.04.1.2.4 Dipole moments

Dipole moments seem to exert a sort of fascination on theoretical chemists, who often check their calculations by comparing their calculated values of the dipole moment with the experimental ones. This is particularly startling since the experimental dipole moments generally arise from dielectric data ⟨63PMH(1)189⟩ and thus only the modulus is known.

Experimental values are collected in the McClellan book ⟨B-63MI40400⟩ and in a review on dipole moments and structure of azoles ⟨71KGS867⟩. Some selected values are reported in Table 3. The old controversy about the dipole moment of pyrazole in solution has been settled by studying its permittivity over a large range of concentrations ⟨75BSF1675⟩. These measurements show that pyrazole forms non-polar cyclic dimers (**39**) when concentration increases and, in consequence, the permittivity value decreases.

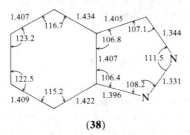

(39)

A selection of calculated dipole moments are shown in Table 4. Except from the aberrant EHT value, the others are quite similar. CNDO/2 calculations using experimental geometries ⟨74T2903, 75BSF1675⟩ provide better results than when ideal geometries are used ⟨70JCS(B)1692⟩. Theoretically optimized geometries have been used in several cases ⟨67MI40400, 69BSF1097, 73MI40400, 81JHC1189⟩. The papers by Palmer et al. ⟨74JCS(P2)420, 78JST(43)203⟩ contain an important discussion on dipole and second moments of five- and six-membered heterocycles based on ab initio calculations and on the linear combination of gaussian orbitals (LCGO) procedure.

Table 3 Experimental Dipole Moments of Pyrazoles and Indazoles at 25 °C

Molecule	Method[a]	Solvent	μ (D)	Ref.
Pyrazole[b]	M	None	2.21	74PMH(6)53
Pyrazole	D	Benzene	1.92	75BSF1675
Pyrazole	D	Dioxane	2.32	75BSF1675
1-Methylpyrazole	D	Benzene	2.25	75BSF1675
3(5)-Methylpyrazole	D	Benzene	1.43	71KGS867
1-Phenylpyrazole	D	Benzene	2.00	71KGS867
3(5)-Phenylpyrazole	D	Benzene	2.26	71KGS867
3,5-Dimethylpyrazole	D	Benzene	2.31	71KGS867
3,4-Dibromopyrazole	D	Benzene	4.66	71KGS867
3,4,5-Tribromopyrazole	D	Benzene	2.48	71KGS867
3(5)-Azidopyrazole	D	Dioxane	3.73	78BSB189
1-Acetylpyrazole	D	Dioxane	1.55	77RRC471
Indazole	D	Benzene	1.60	75BSF1675
1-Methylindazole	D	Benzene	1.50	75BSF1675
2-Methylindazole	D	Benzene	3.40	75BSF1675

[a] Microwave spectroscopy (Stark effect). [b] θ, the probable approximate angle between the direction of μ and the symmetry (or near symmetry) axis of the ring through atom N-1, in the molecular plane, has the value 69°.

Table 4 Calculated Dipole Moments of Pyrazoles and Indazoles

Molecule	Experimental value[a]	Calculated values
Pyrazole	2.21 (G), 2.32 (D)	2.28 (VESCF) ⟨67MI40400⟩; 5.05 (EHT) ⟨67MI40401⟩; 2.50 (ILCAO) ⟨70BSF273⟩; 2.52 (CNDO/2) ⟨70JCS(B)1692⟩; 2.21, $\theta = +40°$ (ab initio) ⟨B-70MI40401⟩; 2.33 (CNDO/2) ⟨73MI40400⟩; 2.85, $\mu_\perp$ 2.23, $\mu_\parallel$ 1.77 (ab initio LCGO) ⟨74JCS(P2)420⟩; 2.30 (CNDO/2) ⟨75BSF1675⟩; 2.11 (MNDO) ⟨81JHC1189⟩
1-Methylpyrazole	2.25 (B)	2.51 (CNDO/2) ⟨70JCS(B)1692⟩; 2.35 (CNDO/2) ⟨75BSF1675⟩
1-Phenylpyrazole	2.00 (B)	2.27 (HMO) ⟨66CHE413⟩
Indazole	1.60 (B)	1.67 (CNDO/2) ⟨74T2903⟩; 1.97 (ab initio LCGO) ⟨78JST(43)203⟩

[a] G, gas phase; D, dioxane; B, benzene.

The comparison of the experimental mean values with the theoretically calculated ones for individual tautomers (Section 4.04.1.5.1) ⟨76AHC(S1)1⟩ or conformers (Section 4.04.1.4.3) has been used in the literature to determine equilibrium constants. Thus, the experimental value for 1,1'-thiocarbonylbis(pyrazole) (**40**) is 3.19 D and the vector sums of the simple group moments after addition of the extra mesomeric moments are shown in Figure 8. From these values Carlsson and Sandström ⟨68ACS1655⟩ concluded that conformation (**40b**) exerts the largest influence.

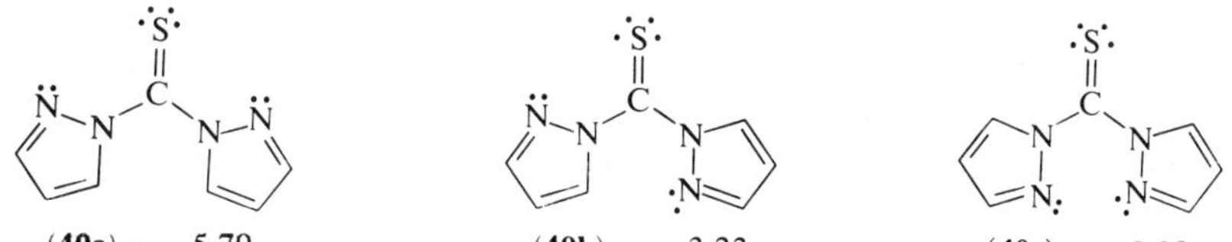

(**40a**) μ_{ZZ} 5.79   (**40b**) μ_{ZE} 3.23   (**40c**) μ_{EE} 0.38

Figure 8 Vectorially calculated dipole moments for the three planar conformations of 1,1'-thiocarbonyl-bis-pyrazole

A similar approach has been used in the case of azolides (N-acylazoles). The experimental dipole moment of 1-acetylpyrazole (1.55 D, Table 3) was compared with the CNDO/2 calculated values for the E and Z conformations, respectively 1.87 and 5.46 D. Fayet et al. concluded that the E conformer (**41**) largely predominates ⟨77RRC471⟩.

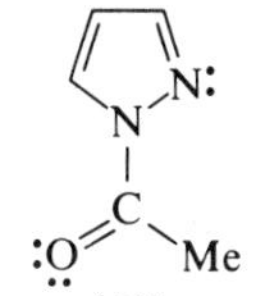

(**41**) 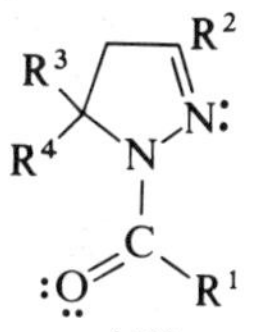   (**42**)

The method has also been applied to partially saturated systems. For instance, the dipole moments of a series of 1-acylpyrazolines (**42**) with $R^1 =$ H, Me, Et and Ph have been measured ⟨72CHE445⟩; they range from 3.46 to 4.81 D. When compared with values computed by the fragmentary calculation method, the conclusion was reached that here also the *E* form predominates. In all these examples the lone pair–lone pair repulsions determine the most stable conformation.

The last example is somewhat more complicated since four isomers (two tautomers and two conformations) are present at equilibrium (Figure 9) ⟨78BSB189⟩. The experimental value (3.73 D, Table 3) establishes the predominance of the 3-azido tautomer but does not allow the determination of the conformational equilibrium; other methods (Section 4.04.2.3.4(v)) are necessary to establish definitely the *Z* conformation (**43b**).

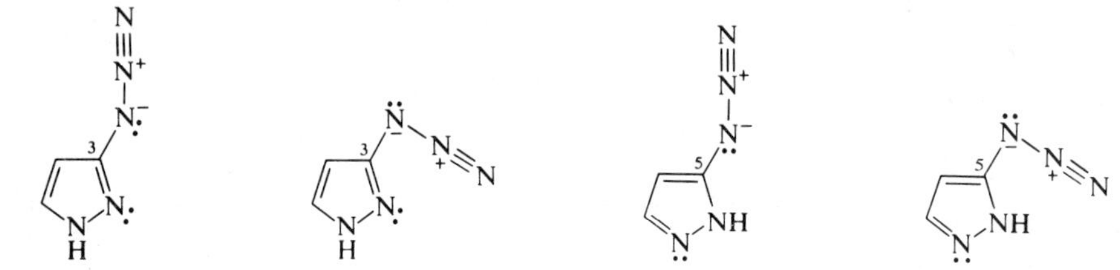

(**43a**) μ_E 3.81   (**43b**) μ_Z 3.56   (**43c**) μ_E 1.29   (**43d**) μ_Z 0.96

Figure 9 Vectorially calculated dipole moments for the four planar isomers of 3(5)-azidopyrazole

4.04.1.3 Structural Methods

4.04.1.3.1 X-Ray diffraction

The fact that only 11 structures were reported in ⟨72PMH(5)1⟩ whereas in Table 5, which is non-exhaustive, 36 more have been added shows the great development of X-ray structural determinations in recent times. Some of these structures deserve brief comment.

Table 5 X-Ray Structures of Pyrazoles and Indazoles

Type of structure	Name[a]	Ref.
Pyrazole	Pyrazole	60AX946, 70AX(B)1880, 73ACS1845
	3-Methyl-5-phenylpyrazole	74JCS(P2)1298
	3,5-Dimethyl-1-phenylpyrazolium nitrate	82CJC97
	4-Bromo-1-(2,4-dinitrophenyl)pyrazole (*101*)	72PMH(5)1
	4-Chloro-1-(2,4-dinitrophenyl)pyrazole	70AX(B)380
	4-Bromo-3-methyl-1-(4-nitrophenyl)pyrazole	72AX(B)791
	1-Acetyl-4-bromopyrazole	72AX(B)3316
	3-Azido-4-phenylpyrazole (**45**)	74CSC713
	3-Amino-4,5-dicyano-1-methylpyrazole	76AX(B)853
	5-Amino-4-cyano-3-dimethylamino-1-phenylpyrazole	77AX(B)413
	3-Methyl-(5-amino-3-methylpyrazol-1-yl)-3-acrylonitrile	78JHC185, 75AX(B)2119
	1-Pyrazol-3-yl-pyrrole-2-carboxylic acid (**46**)	81CPB3214
	4-[4-(*p*-Chlorophenyl)-3-pyrazolyl]-4*H*-1,2,4-triazole (**47**)	77JHC65
	Phenylazo-5-diethylamino-4-methyl-1-phenylpyrazole	74CC339
	1,9-Dihydro[1]benzothiopyrano[4,3-*c*]pyrazole *S,S*-dioxide (**48**)	77JHC387
	4-Ethoxycarbonyl-5-methoxy-1-phenyl-3-(α-phthalimidoethyl)pyrazole (**138**)	80CSC1127
	1,1,1-Trimethylhydrazinium 3-methoxycarbonyl-5-pyrazolecarboxylate	74AX(B)2505
Pyrazolone	4-Bromo-2,3-dimethyl-1-phenyl-5-pyrazolone (*60*)	72PMH(5)1
	3-Methyl-1-phenyl-5-pyrazolone (**49**)	73CSC469
	3-Hydroxy-5-methyl-1-phenylpyrazole	73CSC473
	4-Ethoxycarbonyl-5-hydroxy-1-phenyl-3-(α-phthalimidoethyl)pyrazole	80CSC1121
	2-Bromo-1-methylbenzo[*c*]pyrazolo[1,2-*a*]pyrazole-3,9-dione	73ACS661
	4-(*p*-Chlorobenzal)-1,3-diphenyl-2-pyrazolin-5-one	72CSC253
	3-Methyl-3-pyrazolin-5-one	71AX(B)1227
	Antipyrine (2,3-dimethyl-1-phenyl-5-pyrazolone)	74AX(B)557, 73AX(B)714

Table 5 (*continued*)

Type of structure	Name[a]	Ref.
Betaine	1,1-Dimethyl-3-phenylpyrazolium 5-oxide (**50**)	70JHC895
Pyrazoline Δ²	Δ²-Pyrazoline hydrochloride (*64*)	72PMH(5)1
	3-*p*-Bromophenyl-1-nitroso-2-pyrazoline	71AX(B)986
	1,3-Diphenyl-Δ²-pyrazoline (*89*)	72PMH(5)1
	1-*p*-Iodophenyl-3-phenyl-Δ²-pyrazoline adduct of isocolumbin acetone solvate (*1303*)	72PMH(5)1
	3-Benzoylamino-1-phenyl-Δ²-pyrazoline	81MI40401
	3-Acetamido-1-benzyl-Δ²-pyrazoline	81MI40401
	3a,8b-Dihydro-1,3-diphenylindeno[1,2-*c*]pyrazol-4(1*H*)-one	76AX(B)2314
	Dimers of 3-alkoxycarbonyl-5-formyl-5-methyl-Δ²-pyrazoline	75AX(B)548, 76AX(B)2216
	9-Oxo-2a,3,5,6,7,8,8a,9-octahydroindolizino[1',2':5,4]pyrazolo-[1,5-*a*]indole	81HCA769
	1-(3,5,5-Trimethyl-1-pyrazolinyl) phthalazine hydrobromide	77CPB147
Δ¹	1-(4-Methyl-1-pyrazolin-3-yl)-5,5-bis(trifluoromethyl)-Δ²-1,2,3-triazoline	73CB288
	3-Methoxycarbonyl-*trans*-3,5-dimethyl-Δ¹-pyrazoline hydrobromide (*72*)	72PMH(5)1
Salts	1,2,3,5,5-Pentamethyl-2-pyrazolinium perchlorate (*94*)	72PMH(5)1
	1,2,3,5,5-Pentamethyl-1-phenyl-2-pyrazolinium perchlorate	72CR(C)(274)1192
Pyrazolidine	2-Acetyl-5-(methoxycarbonylmethyl)-1-methyl-3,4,5-pyrazolidinetricarboxylic acid, trimethyl ester	79CB1719
Pyrazolidinone	4'-Methylene-1,2-di-*m*-bromophenyl-1',2',6',7'-tetraphenylspiro{pyrazolidine-4,8'-[8'*H*,4'*H*]-benzo[1,2-*c*;4,5-*c*']dipyrazoline}-3,3',5,5'-tetraone	74AX(B)273
	4-*n*-Butyl-1,2-diphenyl-3,5-dioxopyrazolidine (phenylbutazone) piperazine salt	74AX(B)590
Indazole	Indazole	74AX(B)2009, 74T2903
	3-Diazoindazole (**51**)	78AX(B)293
	Substituted pregn-4-en-20-one[3,2-*c*]-2'-phenylpyrazole	71AX(B)573
Indazolone	2-Acetyl-3-indazolinone (*896*)	72PMH(5)1
Indazoline	2,3,4,7-Tetrahydro-3a,4-bis(methoxycarbonyl)-2,6-dimethyl-5-phenylindazol-7-one	74JOC1007

[a] Numbers in *italic* type are those used in ⟨72PMH(5)1⟩.

The molecular structure of pyrazole was determined for the first time by Ehrlich ⟨60AX946⟩ who noticed that there are two crystallographically independent molecules in the cell, both with strong hydrogen bonds. This structure was later proved to be wrong, but Ehrlich deduced from the bond lengths (C(4)—C(5) longer than C(3)—C(4), Figure 10) that there is a large contribution from the betaine form (**44**) to the resonance hybrid.

(**44**)

This conclusion and the experimental geometry caused considerable confusion among theoretical chemists, who, following a suggestion by Mighell and Reimann ⟨67JPC2375⟩, had to exchange the assignment of the two nitrogen atoms to be consistent with their calculations. This inverted Ehrlich geometry was used explicitly for calculations in several theoretical papers ⟨69BSF1097⟩ but not in others ⟨70CB3289⟩. Mighell and Reimann's proposal was sustained by X-ray structural data from coordinated pyrazole derivatives (Section 4.04.2.1.3 (vi)). In fact the correct structure of pyrazole is not the 'inverted-Ehrlich' structure but one having a more regular geometry compatible with the classical representation of pyrazole ⟨70AX(B)1880⟩. The definitive structure of pyrazole was established by two important publications by Rasmussen *et al.* ⟨70ACS3248, 73ACS1845⟩, the first one dealing with a neutron diffraction study and the second one with an X-ray diffraction study (at 295 and 108 K).

In Figure 10 the five structural determinations of pyrazole (**29**) are represented. They have to be compared with the experimental microwave spectroscopic structure (Section 4.04.1.3.2, Figure 12) and with the theoretically optimized geometries (for instance, Figure 1,

Section 4.04.1.2.1). The spectroscopic and the diffraction results refer to molecules in different vibrational quantum states. In neither case are the distances those of the hypothetical minimum of the potential function (the optimized geometry). Nevertheless, the experimental evidence appears to be strong enough to lead to the conclusion that the electron redistribution, which takes place upon transfer of a molecule from the gas phase to the crystalline phase, results in experimentally observable changes in bond lengths.

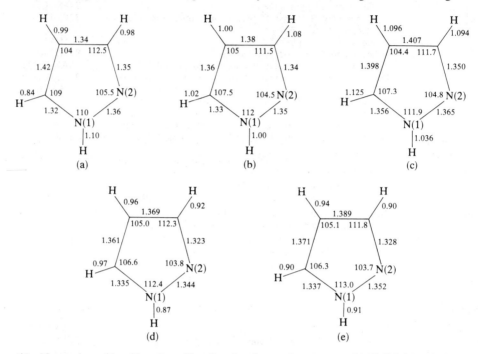

Figure 10 Mean values of bond lengths and bond angles of pyrazole structures: (a) Ehrlich (X-ray); (b) Berthou et al. (X-ray); (c) Rasmussen et al. (neutron, corrected for rigid body motion); (d) Rasmussen et al. (X-ray, 295 K); (e) Rasmussen et al. (X-ray, 108 K)

The indazole molecular structure (Figure 11) shows the tautomeric proton bonded to N-1 (1H-indazole, Section 4.04.1.5.1). A linear correlation between the bond lengths and the bond orders calculated by the CNDO/2 method was observed ⟨74T2903⟩.

Figure 11 Bond lengths and bond angles of indazole X-ray structure

For N-unsubstituted pyrazoles the tautomeric proton was generally located without ambiguity. 3-Substituted tautomers were favoured in the solid state: (**45**), (**46**) and (**48**) (Table 5). For the pyrazolyltriazole (**47**) the authors ⟨77JHC65⟩ concluded that the X-ray analysis indicates that the proton on the pyrazole ring populates either nitrogen atom to

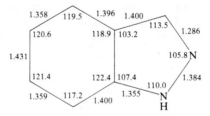

an indistinguishable extent. This conclusion is doubtful, since other similar claims have been proved to be false ⟨81CC1207⟩.

An interesting report on 1-phenyl-3-methyl-5-pyrazolone (**49**) shows that when two tautomers are of comparable energy they could both be present in the crystal cell ⟨73CSC469⟩. In this case NH and OH tautomers, (**49a**) and (**49b**) respectively, coexist in the crystal (Section 4.04.1.5.2).

The structure of the unusual betaine (**50**) has been determined ⟨70JHC895⟩. The bond lengths and angles suggest that a significant contribution to the structure is made by a resonance form (**50b**) in which the N(1)—C(5) bond does not exist ('ketene' form).

3-Diazoindazole (**51**) is one of the few heterocyclic diazo compounds whose structure has been determined ⟨78AX(B)293⟩. The diazo group shows a substantial 'carbanionic' character (**51b**).

4.04.1.3.2 Microwave spectroscopy

Since there is an excellent review by Sheridan ⟨74PMH(6)53⟩ referencing all the relevant publications on this topic, only the microwave geometry of pyrazole determined by the group at the University of Copenhagen ⟨74JST(22)401⟩ will be shown here (Figure 12).

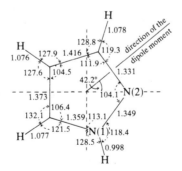

Figure 12 Microwave structure of pyrazole

Quantum chemists refer now in their calculations of optimized geometries to this standard geometry ⟨73MI40400, 81JHC1189⟩. Another conclusion of this study is that the N—N bond is the weakest bond in pyrazoles and is one that is stretched in order to close the ring. This is in accord with electron impact (Section 4.04.1.3.8) and photoinduced fragmentations (Section 4.04.2.1.2(i)). The somewhat longer N—H bond in the crystal structure (0.03 Å, Figure 10) was to be expected, as it had been shown that pyrazole forms strong hydrogen bonds in the crystal state. Structurally, there is no prototropy, at least in the gas phase on the time scale of a microwave experiment. This allows the experimental determination that 3(5)-D-pyrazole is a 50:50 mixture of the 3-D and the 5-D tautomers.

4.04.1.3.3 ^1H NMR spectroscopy

(i) *Aromatic systems*

The excellent book by the late Professor T. J. Batterham contains all the available information (up to 1973) on pyrazoles and their non-aromatic derivatives ⟨B-73NMR165⟩ and on indazoles ⟨B-73NMR263⟩. The bibliography for pyrazoles and pyrazolones has been updated in ⟨B-76MI40402⟩. It should be emphasized that almost all the principal results about the ^{1}H NMR spectra of these heterocycles were published at that time, and thus only a summary of the principal conclusions is needed here.

In the assignment of signals and the identification of isomeric pyrazoles by this technique, one must bear in mind that for 'symmetrical' systems, *i.e.* those for which $R^3 = R^5$, which are N-substituted (**52**) the problem is to assign the signals (protons or methyl groups) of the 3- and 5-positions (those of the 4-position always appear at higher fields (lower frequencies)). When $R^1 = H$, tautomeric equilibrium usually makes the 3- and 5-positions equivalent (Section 4.04.1.5.1). On the other hand, 'asymmetric' pyrazoles, *i.e.* those for which $R^3 \neq R^5$, the problem of distinguishing unambiguously between isomers (**53**) and (**54**) is often a difficult one.

<pre>
      R⁴   R³   R⁴   R³   R⁴   R³
       \\ //   \\ //   \\ //
    R⁵   N   R⁵   N   R⁵   NR
       N   N   N
       R   R

      (52)   (53)   (54)
</pre>

These two problems are closely related and will be discussed together under the title of assignment of signals since it is evident that if the signals are correctly assigned then the isomeric structure of the pyrazole follows immediately. The main methods to assign signals in pyrazoles employ the following:

(i) The fact that $J_{4,5}$ is always larger than $J_{3,4}$; the more the electron-withdrawing ability of R^1 the more the difference between both ^3J constants as a consequence of the localization of the π-system. This fact can be used for compounds having two adjacent protons like 3- or 5-amino- or hydroxy-pyrazoles ⟨74JPR705⟩ or a proton and a methyl group ($R^3 = R^4 = H$; $R^{3(5)} = Me$, $R^4 = H$; $R^{3(5)} = H$, $R^4 = Me$). Caution must be exercised not to extend mechanically these rules to ^1H–^{19}F ⟨77BSF171⟩ or ^{19}F–^{19}F ⟨79TL3179⟩ coupling constants in pyrazoles.

(ii) The fact that the signal of the proton in position 3 is broadened by the nuclear quadrupole relaxation effect of N-2 (the broadening disappears by heteronuclear decoupling). If the H-3 signal is a singlet ($R^4 \neq H$), it is smaller than the H-5 signal; if it is a doublet ($R^4 = H$), it is less well resolved than that of H-5.

(iii) The fact that the signal of the substituent in position 5 is more sensitive to solvent effects than that of the 3-substituent. This is true for protons and methyl groups, the best solvents for these studies being benzene, $CDCl_3$, DMSO and HMPT. A beautiful illustration of this method is provided by the establishment of the structure of the six 1,1'-dimethyl-bipyrazolyl isomers ⟨72JHC1373⟩.

(iv) The appearance of the *C*- or *N*-phenyl signal when a substituent, generally a methyl group, is in a vicinal position. In this case the phenyl group gives rise to a singlet (in $CDCl_3$); if there is no substituent, the signal appears as two multiplets (*ortho* protons at low fields, *meta* and *para* protons at high fields). The difference, $\Delta\delta$, between *ortho* and *meta–para* protons is characteristic of the position of a phenyl group in the azole ring ⟨73CJC2315⟩.

(v) Paramagnetic metals ⟨71OMR(3)595, 79CJC1186⟩ associated with the N-2 lone pair (Section 4.04.2.1.3(vi)). Consequently, the signal of the 3-position moves further than that of the 5-position.

The mean chemical shifts of N-unsubstituted pyrazoles have been used to determine the tautomeric equilibrium constant, but the method often leads to erroneous conclusions ⟨76AHC(S1)1⟩ unless the equilibrium has been slowed down sufficiently to observe the signals of individual tautomers (Section 4.04.1.5.1). When acetone is used as solvent it is necessary to bear in mind the possibility (depending on the acidity of the pyrazole and the temperature) of observing the signals of the 1:1 adduct (**55**) whose formation is thermodynamically favoured by lowering the solution temperature ⟨79MI40407⟩. A similar phenomenon is observed when SO_2 is used as solvent.

The ^{1}H chemical shifts shown in Table 6 for 3-substituted pyrazoles are mean values of 3- and 5-substituted tautomers. Only in the case of the 3-azidopyrazole (**56**) have the

(55) (56) (57)

couplings with the N(1)—H proton been observed, demonstrating the presence of the azido group in position 3 exclusively. In the spectrum of 4-formylpyrazole (57) at low temperature the signals of protons H-3 and H-5 are different. This is not related to the annular tautomerism but is the consequence of a slow rotation about the C—C(4) formyl–pyrazole bond ($\Delta G_C^{\ddagger} = 36.7$ kJ mol^{-1}).

The chemical shifts have been discussed using empirical additive rules ⟨B-73NMR165⟩. Several authors have tried to correlate ^{1}H chemical shifts and charge densities of the carbon atom to which the proton is bound. The success has only been moderate due to the fact that the underlying hypothesis is an oversimplification of the theory of the chemical shift ⟨67MI40402⟩. Much better results are obtained with ^{13}C chemical shifts (Section 4.04.1.3.4). However, some significant papers must be quoted in this respect. An old paper by Lynch and Dou ⟨66TL2627⟩ correlates ^{1}H chemical shifts measured in TFAA with HMO π-electron densities of 1-methyldiazoles. They found an empirical equation: $\delta = 7.98 - 7.90(1 - q_\pi)$. Adam and Grimison ⟨67T2513, 67MI40401⟩ compare the total electron densities calculated by the EHT method (Section 4.04.1.2.1) with the ^{1}H chemical shifts of a series of five-membered rings, including pyrazole. A reasonable correlation was obtained. Finally, all-electron *ab initio* calculations ⟨B-70MI40401⟩ show that proton chemical shifts do not correlate with the π-electron densities but with the σ- and total-electron densities (however H-3(5) of pyrazole clearly deviates). All the studies quoted above suffer from using the mean signals of N-substituted azoles.

The results in Table 7 show that the chemical shift of H-5 is very sensitive to the nature of the R^1 substituent. The ratio $J_{4,5}:J_{3,4}$ also varies with R^1, from 1.0 (R^1 = NH$_2$) to 1.82 (R^1 = NO$_2$). One publication ⟨78MI40403⟩ describes the chemical shifts of twenty N-substituted derivatives of pyrazole (mainly R^1 = alkyl). Table 8 is a selection of pyrazoles substituted both on the nitrogen atom and on the carbon atoms. The small coupling constant between the N-methyl group and H-5 could be used to assign isomeric structures (it disappears in 5-substituted derivatives). When the substituent on the nitrogen atom or the carbon atoms has a lone pair in a position *ortho* to it (as in 2'-pyridyl or 3'-pyridazinyl), a deshielding effect is observed on the opposite proton ⟨70BSF1345, 70BSF1346, 81JHC9⟩. The isomers (58) and (59) clearly illustrate this phenomenon ⟨80JHC137⟩. The N-methyl chemical shifts of 22 pyrazoles have been gathered and compared with those of other azoles; they are characteristic of the azole ring ⟨73CJC2315⟩.

(58) (59)

Work on indazoles has the drawback of having to employ computer procedures to analyze the ABCDX system of C-unsubstituted indazoles. The first five examples in Table 9 correspond to LAOCOON-analyzed spectra and the last two to simpler systems for which the first-order analysis gives acceptable chemical shifts.

The most characteristic coupling constant in indazoles is the 'cross-ring' $^5J_{3,7}$ present both in indazoles and in isoindazoles unsubstituted in positions 3 and 7. 2-Methyl isomers show an additional $^4J_{Me,H}$ coupling which can serve to identify an isoindazole unsubstituted in position 3. In 3-azidoindazole, as in 3-azidopyrazole (56), the prototropic exchange is slowed down sufficiently to allow the measurement of a 'zig-zag' $^5J_{1,4}$ coupling constant. The deshielding effects observed in N-acetyl derivatives, *e.g.* 1-acetyl (60) on H-7 and 2-acetyl (61) on H-3, are related to a preferred E conformation (Section 4.04.1.4.3).

Table 6 ^{1}H NMR Spectral Data for N-Unsubstituted Pyrazoles[a]

Pyrazole	R^3	R^4	R^5	Solvent	H-1	$\delta(^1H)$ (p.p.m.) H-3	H-4	H-5	Coupling constants[j] (Hz) 3,4	4,5	3,5	Others
Pyrazole	H	H	H	CCl_4	12.64	7.61	6.31	7.61	2.0	2.0	—	—
3-Methyl	Me	H	H	$CDCl_3$	—	2.32	6.06	7.48	—	1.7	—	—
4-Methyl	H	Me	H	$CDCl_3$	—	7.36	2.09	7.36	—	—	—	—
3-t-Butyl	Bu^t	H	H	$CDCl_3$	—	1.37	6.06	7.44	—	1.9	—	—
3,5-Dimethyl	Me	H	Me	$CDCl_3$	—	2.21	5.76	2.21	—	—	—	—
3,4-Dimethyl	Me	Me	H	$CDCl_3$	—	2.20	1.98	7.26	—	—	—	—
3,4,5-Trimethyl	Me	Me	Me	$CDCl_3$	—	2.20	1.87	2.20	—	—	—	—
3-Phenyl	Ph	H	H	$CDCl_3$	—	7.77	6.53	7.52	—	—	—	—
4-Phenyl[b]	H	Ph	H	DMSO-d_6	—	8.00	7.3–7.6	8.00	—	2.2	—	—
3-Fluoro[c]	F	H	H	$CDCl_3$	12.40	—	5.83	7.39	(6.0)	2.7	(2.7)	—
4-Fluoro[c]	H	F	H	$CDCl_3$	—	7.49	—	7.49	(4.5)	(4.5)	—	—
3-Chloro	Cl	H	H	$CDCl_3$	—	—	6.24	7.53	—	1.9	—	—
4-Chloro	H	Cl	H	$CDCl_3$	12.75	7.53	—	7.53	—	—	—	—
4-Bromo	H	Br	H	$CDCl_3$	12.23	7.60	—	7.60	—	—	—	—
4-Iodo	H	I	H	$CDCl_3$	—	7.65	—	7.65	—	—	—	—
3-Nitro[d]	NO_2	H	H	DMSO-d_6	7.69	—	7.00	8.02	—	2.6	—	—
4-Nitro[d]	H	NO_2	H	DMSO-d_6	11.20	8.54	—	8.54	—	—	—	—
3-Nitroso[e]	NO	H	H	$CDCl_3$	14.3	—	5.97	7.99	—	2.8	—	—
4-Nitroso[e]	H	NO	H	DMF-d_7	13.1	8.68	—	8.68	—	—	—	—
3-Azido[f]	N_3	H	H	$CDCl_3$	3.34	—	6.01	7.55	—	2.35	—	$J_{1,4}$ 2.10, $J_{1,5}$ 1.45[k]
3-Amino	NH_2	H	H	$(CD_3)_2CO$	—	5.12(b)	5.52	6.30	—	2.2	—	—
4-Ethoxy	H	OEt	H	$CDCl_3$	10.96	7.25	3.92, 1.36	7.25	—	—	—	—
4-Formyl[g]	H	CHO	H	THF (163 K)	—	8.12	9.91	8.62	—	—	—	—
3-Ethoxycarbonyl	CO_2Et	H	H	$CDCl_3$	—	5.51, 1.40	6.86	7.80	—	2.3	—	—
3-Cyano	CN	H	H	$CDCl_3$	12.37	—	6.80	7.81	—	2.5	—	—
3-Trimethylsilyl[i]	Me_3Si	H	H	CCl_4	13.78	0.30	6.37	7.62	—	1.9	—	—
4-Trimethylsilyl[i]	H	$SiMe_3$	H	CCl_4	14.78	7.50	0.22	7.50	—	—	—	—

[a] Unless otherwise indicated, data taken from ⟨B-73NMR165⟩ and ⟨B-76MI40402⟩, for which see references to the original literature. [b] ⟨75MI40403⟩. [c] ⟨75LA470⟩. [d] ⟨79MI40407⟩. [e] ⟨71CB3062⟩. [f] ⟨74JHC921⟩. [g] ⟨75BSB499⟩. [h] ⟨71CS(C)2147⟩. [i] ⟨72CB1759⟩. [j] Values in parentheses are ^{1}H–^{19}F coupling constants. [k] Couplings with NH-1 are observed only in DMSO.

Table 7 ^{1}H NMR Spectral Data for N-Substituted Pyrazoles[a]

Pyrazole	Solvent	$\delta(^1H)$ (p.p.m.)				Coupling constants (Hz)		
		H-1	H-3	H-4	H-5	3,4	4,5	3,5
1-Methyl	CDCl$_3$	3.88	7.49	6.22	7.35	2.0	2.3	0.7
1-Ethyl	CDCl$_3$	4.18, 1.47	7.50	6.23	7.38	2.0	2.3	—
1-Phenyl	CDCl$_3$	7.2–7.6	7.72	6.46	7.87	1.9	2.5	0.7
1-(2',4',6'-Trinitrophenyl)	CDCl$_3$	8.92	7.85	6.60	7.74	1.9	2.6	—
1-Acetyl	CDCl$_3$	2.70	7.71	6.44	8.25	1.6	2.9	0.6
1-Ethoxycarbonyl	CDCl$_3$	4.54, 1.46	7.77	6.44	8.18	—	—	—
1-Carboxamido	CDCl$_3$	—	7.66	6.40	8.26	1.6	2.7	0.7
1-Tosyl	DMSO-d_6	Me: 2.36	7.88	6.58	8.45	1.7	2.8	0.6
1-Nitro[b]	Acetone-d_6	—	7.77	6.66	8.65	1.7	3.1	0.9
1-Amino[b]	CDCl$_3$	6.3(b)	7.30	6.06	7.30	2.2	2.2	—
1-Trimethylsilyl	CCl$_4$	0.41	7.63	6.21	7.53	1.5	2.4	—
1-Trimethylgermyl	CCl$_4$	0.63	7.57	6.19	7.42	1.6	2.2	—
1-P(NMe$_2$)$_2$	CDCl$_3$	2.66	7.70	6.30	7.53	1.9	2.2	—

[a] Unless otherwise indicated, data taken from ⟨B-73NMR165⟩ and ⟨B-76MI40402⟩ which see for references to the original literature. [b] ⟨81UP40400⟩.

(60)   (61)   (62)

The chemical shifts of the pyrazolate anion (62) represented an upfield effect of 0.3 p.p.m. compared with pyrazole itself ⟨68JA4232⟩.

The ^{1}H NMR spectra of pyrazolium and indazolium ions have been widely studied (Table 10), including both the conjugate acids and the quaternary salts. The spectra obtained in sulfuric acid prove unquestionably that the protonation of these species takes place at position 2 (Section 4.04.2.1.3(iv)).

The N-methyl chemical shifts of quaternary pyrazolium and indazolium salts have been discussed in connection with the resonance effect of the heterocycle ⟨74JHC1011⟩.

The spectra of pyrazolinones have been used in tautomeric studies (Section 4.04.1.5.2). The ^{1}H chemical shifts shown in Table 11 are representative of the three classes of fixed derivatives [OR (63) and (64), NR (65) and CR (66)], and many others are discussed in ⟨76AHC(S1)1⟩. Some tautomeric compounds are also included in Table 11.

(63)   (64)   (65)   (66)

(ii) Non-aromatic systems

Sparse data on the pyrazole isomers, pyrazolenines and isopyrazoles, are presented in Table 12. Besides the obvious upfield effect on the chemical shift due to the suppression of the ring current, these compounds behave normally. Data on pyrazolidinones and their salts show the behaviour of cyclic hydrazides ⟨66T2461, 67BSF3502⟩.

Considerable effort has been devoted to the NMR study of pyrazolines. The results shown in Table 13 represent a minute part of the literature values, and more can be found in ⟨B-73NMR165⟩.

NMR studies of Δ^1-pyrazolines, mainly due to Crawford, McGreer and Carrié, cover two main topics: determination of the stereochemistry of carbon atoms C-3, C-4 and C-5 (*cis–trans* pairs) and calculation of the ring puckering. For the latter purpose, chemical shifts and coupling constants *via* the Karplus equation have been used. A reasonable estimation of the θ angle of Δ^1-pyrazolines (67) is 30° ⟨73T1135, 76BSB545⟩. When the substituents on the 4-position are different, the equilibrium constant between the two envelope conformations (67a) and (67b) may be determined by ^{1}H NMR ⟨73OMR(5)453⟩.

Table 8 ¹H NMR Spectral Data for N,C-Substituted Pyrazoles[a]

Pyrazole	R^1	R^3	R^4	R^5	Solvent	$\delta(^1H)$ (p.p.m.) H-1	H-3	H-4	H-5	Coupling constants (Hz) 3,4	4,5	3,5	Others
1,3-Dimethyl	Me	Me	H	H	CDCl₃	3.80	2.23	5.95	7.22	—	2.0	—	^4J(NMe-H(5)) 0.4[c]
1,4-Dimethyl	Me	H	Me	H	CDCl₃	3.80	7.28	2.04	7.12	—	—	—	
1,5-Dimethyl	Me	H	H	Me	CDCl₃	3.73	7.36	5.98	2.22	2.0	—	—	
1,3,5-Trimethyl	Me	Me	H	Me	CDCl₃	3.66	2.17	5.78	2.17	—	—	—	
1,3,4,5-Tetramethyl	Me	Me	Me	Me	CDCl₃	3.68	2.14	1.89	2.12	—	—	—	
1-Methyl-3-phenyl[b]	Me	Ph	H	H	CDCl₃	3.85	7.4–7.8	6.50	7.75	—	2.0	—	^4J(NMe-H(5)) 0.4[c]
1-Methyl-5-phenyl[b]	Me	H	H	Ph	CDCl₃	3.82	7.50	6.30	7.40	1.5	—	—	
1,3-Dimethyl-5-phenyl	Me	Me	H	Ph	CDCl₃	3.80	2.30	6.09	7.43	—	—	—	
1,5-Dimethyl-3-phenyl	Me	Ph	H	Me	CDCl₃	3.80	7.78	6.30	2.30	—	—	—	
3-Methyl-1-phenyl	Ph	Me	H	H	CDCl₃	7.2–7.7	2.35	6.23	7.78	—	2.6	—	
5-Methyl-1-phenyl	Ph	H	H	Me	CDCl₃	7.43	7.56	6.19	2.34	1.5	—	—	

[a] Unless otherwise indicated, data taken from ⟨B-73NMR165⟩ and ⟨B-76MI40402⟩. [b] ⟨69BSF3306⟩. [c] ⟨80M775⟩.

Table 9 ¹H NMR Spectral Data for Indazoles

Indazole	Solvent	NR	$\delta(^1H)$ (p.p.m.) H-3	H-4	H-5	H-6	H-7	Other data	Ref.
Indazole	DMSO-d_6	13.04	8.08	7.77	7.11	7.35	7.55	$J_{3,7}$ 1.11, $J_{4,5}$ 8.12, $J_{4,6}$ 0.96, $J_{4,7}$ 1.05, $J_{5,6}$ 6.89, $J_{5,7}$ 0.78, $J_{6,7}$ 8.46	77OMR(9)235
3-Methyl	CDCl₃	—	2.63	7.64	7.10	7.32	7.40	Coupling constants	75JCS(P1)1695
1-Methyl	CDCl₃	3.95	7.94	7.71	7.10	7.34	7.59	Coupling constants	75JCS(P1)1695
2-Methyl	CDCl₃	3.80	7.67	7.56	7.02	7.26	7.68	J(Me-H(3)) 0.5–0.6	75JCS(P1)1695, 80M775
3-Azido	DMSO-d_6	12.92	—	7.58	7.13	7.41	7.51	$J_{1,4}$ 0.65	77OMR(9)235
1-Acetyl-5-nitro	CDCl₃	2.83	8.30	8.70	—	8.42	8.57	Coupling constants	66BSF2075
2-Acetyl-5-nitro	CDCl₃	2.95	9.05	8.75	—	8.15	7.80	Coupling constants	66BSF2075

Table 10 ^{1}H NMR Spectral Data for Pyrazole and Indazole Cations

Compound	N(1)-R	N(2)-R	Solvent	H-1	H-2	δ(^{1}H) (p.p.m.) H-3	H-4	H-5	Other data	Ref.
Pyrazole	H	H	H$_2$SO$_4$	12.5	12.5	8.18	6.90	8.18	$J_{1,3}=J_{2,3}$ 2.25, $J_{2,4}$ 2.0, $J_{3,4}$ 2.8	81UP40400
1-Phenylpyrazole	Ph	H	H$_2$SO$_4$	—	12.2	7.75	6.50	7.75	$J_{2,3}=J_{2,5}$ 2.0, $J_{2,4}$ 1.6, $J_{3,4}=J_{4,5}$ 2.95	74OMR(6)272
Indazole	H	H	H$_2$SO$_4$	11.35	12.11	8.13	—	—	$J_{1,3}$ 0, $J_{2,3}$ 2.3	74OMR(6)272
1,2-Dimethylpyrazolium	Me	Me	DMSO-d_6	4.20	4.20	8.57	6.87	8.57	$J_{3,4}=J_{4,5}$ 3.0±0.1, $J_{3,5}$ 0.9±0.1	B-73NMR165, 69BSF1687
2-Methyl-1-phenylpyrazolium	Ph	Me	DMSO-d_6	7.82	3.99	8.95	7.18	8.88	$J_{3,4}=J_{4,5}$ 3.0±0.1, $J_{3,5}$ 0.9±0.1	B-73NMR165, 69BSF1687
1,2,3,4,5-Pentamethylpyrazolium	Me	Me	DMSO-d_6	3.96	3.96	2.37	2.01	2.37		B-73NMR165
1,2-Dimethylindazolium	Me	Me	DMSO-d_6	4.34	4.49	9.28	—	—		69BSF1687

Table 11 ^{1}H NMR Spectral Data for Pyrazolinones^a

Compound	Solvent	H-1	δ(^{1}H) (p.p.m.) H-2	H-3	H-4	H-5
3-Methylpyrazolin-5-one	DMSO	10.94	—	2.12	5.31	—
3-Methyl-1-phenylpyrazolin-5-one	CDCl$_3$	7.3–7.8	—	2.12	3.37^b	—
3-Hydroxy-5-methyl-1-phenylpyrazole	CHCl$_3$	7.33	—	11.1	5.54	2.16
3-Methoxy-5-methyl-1-phenylpyrazole	Neat	7.37	—	3.87	5.66	2.10
5-Ethoxy-3-methyl-1-phenylpyrazole	CHCl$_3$	7.0–7.9	—	2.26	5.50	4.14, 1.41
2,3-Dimethyl-1-phenyl-3-pyrazolin-5-one	CHCl$_3$	7.30	3.07	2.20	5.37	—
3,4,4-Trimethyl-1-phenyl-2-pyrazolin-5-one	CDCl$_3$	7.2–7.9	—	1.94	1.18^c	—

^{a} Data taken from ⟨B-73NMR165⟩, which see for references to the original literature. ^{b}CH$_2$. ^{c}CMe$_2$.

Table 12 ^{1}H NMR Spectral Data for Pyrazolenines and Isopyrazoles

Compound	Solvent	H-1	H-3	H-3'	δ(^{1}H) (p.p.m.) H-4	H-4'	H-5	Ref.
5-Methoxycarbonyl-3,3,4-trimethylpyrazolenine	CDCl$_3$	—	1.39	1.39	2.31	—	3.98	69JCS(C)1065
4-Methoxycarbonyl-3,3,5-trimethylpyrazolenine	CDCl$_3$	—	1.52	1.52	3.86	—	2.69	69JCS(C)1065
3,3,5-Trimethylpyrazolenine 1-oxide	CCl$_4$	—	1.38	1.38	7.10	—	2.08	62IOC1309[a]
3,3,5-Trimethylpyrazolenine 2-oxide	CCl$_4$	—	1.40	1.40	5.90	—	2.12	62IOC1309[a]
3,4,4,5-Tetramethylisopyrazole	CDCl$_3$	—	2.16	—	1.17	1.17	2.16	68BSF3866
4-Benzyl-3,4,5-trimethylisopyrazole	CDCl$_3$	—	2.15	—	1.24 (Me)	2.97 (CH$_2$Ph)	2.15	70A6218
3,5-Dimethyl-4-hydroxy-4-phenylisopyrazole	CDCl$_3$	—	2.03	—	7.33 (Ph)	7.73 (OH)	2.03	76IOC2874
1-Ethyl-3,4,4,5-tetramethylisopyrazolium iodide	CDCl$_3$	4.44, 1.63	2.35	—	1.63	1.63	3.00	68BSF3866

[a] See also (B-73NMR165).

Table 13 ^{1}H NMR Spectral Data for Pyrazolines and their Salts

Pyrazoline	Solvent	H-1	H-2	H-3	δ(^{1}H) (p.p.m.) H-3'	H-4	H-4'	H-5	H-5'	Other data	Ref.
Unsubstituted-Δ^1	Neat	—	—	4.27	4.27	1.46	1.46	4.27	4.27	—	65JA3023
3-Acetyl-3,4-dimethyl-Δ^1 (E)	CCl$_4$	—	—	3.60 (Ac)	1.26 (Me)	2.15 (H)	0.87 (Me)	4.57	3.95	Coupling constants	B-73NMR165
3-Acetyl-3,4-dimethyl-Δ^1 (Z)	CCl$_4$	—	—	3.70 (Ac)	1.53 (Me)	0.90 (Me)	1.72 (H)	4.68	3.98	Coupling constants	B-73NMR165
3,5-Dimethyl-Δ^1 (E)	CCl$_4$	—	—	1.29 (Me)	4.57 (H)	1.27	1.27	4.57 (H)	1.29 (Me)	—	B-73NMR165
3,5-Dimethyl-Δ^1 (Z)	CCl$_4$	—	—	4.20 (H)	1.54 (Me)	2.08	0.48	4.20 (H)	1.54 (Me)	—	B-73NMR165
3,3-Dimethoxycarbonyl-4-methyl-Δ^1	CDCl$_3$	—	—	3.81	3.86	—	2.82	4.82	4.27	Coupling constants	73OMR(5)453
Unsubstituted-Δ^2	CDCl$_3$	5.33	—	6.88	—	2.65	2.65	3.31	3.31	Coupling constants[a]	70UP40400
1,3-Dimethyl-Δ^2	CDCl$_3$	2.74	—	1.95	—	2.96	2.96	3.58	3.58	Coupling constants[a]	70UP40400
1-(2,4-dinitrophenyl)-3-methyl-Δ^2	CDCl$_3$	—	—	2.12	—	3.01	3.01	3.79	3.79	Coupling constants[a]	70UP40400
1-Formyl-3-methyl-Δ^2	CDCl$_3$	8.81	—	2.05	—	2.88	2.88	3.90	3.90	Coupling constants[a,b]	70BSF3466
3,5,5-Trimethyl-Δ^2	CDCl$_3$	4.4	—	1.99	—	2.42	2.42	1.25	1.25	Coupling constants	70UP40400
3,4,4,5,5-Pentamethyl-1-phenyl-Δ^2	CDCl$_3$	—	—	1.95	—	0.91	1.03	3.55 (H)	1.02 (Me)	Coupling constants[c]	68BSF3866
1,3-Diphenyl-Δ^2	CDCl$_3$	—	—	—	—	3.12	3.12	3.76	3.76	Coupling constants[d]	B-73NMR165
3,5-Diphenyl-Δ^2	CDCl$_3$	—	—	—	—	2.89	3.35	4.02	4.02	Coupling constants[d]	74MI40401
1,2,3-Trimethyl-Δ^3	CDCl$_3$	2.55	2.60	1.71	—	4.60	—	3.65	3.65	Coupling constants[e]	69BSF3316
1,2,3,5,5-Pentamethyl-Δ^3	CDCl$_3$	2.47	2.56	1.80	—	4.53	—	1.17	1.17	—	70BSF3147
2,3,5-Trimethyl-1-phenyl-Δ^3	CDCl$_3$	7.2	2.70	1.80	—	4.69	—	4.20 (H)	1.37 (Me)	Coupling constants	70BSF1129
1H-Δ^3-Pyrazoline (**70**)	CDCl$_3$	4.96	4.17 (CH)	1.80	—	1.80	—	—	—	^{1}H-^{19}F Coupling constants	79T389
1,1,3,5,5-Pentamethyl-Δ^2-pyrazolinium	DMSO-d$_6$	3.16	—	2.15	—	3.28	3.28	1.48	1.48	—	67BSF3516
1,2,3,5,5-Pentamethyl-Δ^2-pyrazolinium	CDCl$_3$	2.92	3.78	2.57	—	3.23	3.23	1.35	1.35	—	67BSF3516

[a] $J_{3,4} = 1.0$ (67MI40402). [b] $J_{1,5} = 1.0$ (70BSF3466). [c] $J_{4,4}(gem) = -16.0$, $J_{4,5}(cis) = 12.0$, $J_{5,5}(gem) = -10.0$ (B-73NMR165). [d] $J_{4,4}(gem) = -16.32$, $J_{4,5}(trans) = 8.74$, $J_{4,5}(cis) = 10.68$ (74MI40401). [e] $J_{3,4} = J_{3,5} = J_{4,5} = 1.8$ (69BSF3316). [f] endo–exo equilibrium of the double bond (70BSF3147).

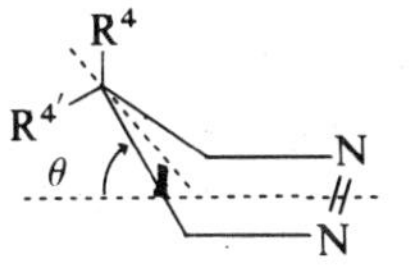

(67a) 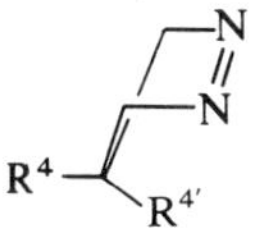 (67b)

The work on Δ²-pyrazolines of the Montpellier group is well summarized in ⟨B-73NMR 165⟩ except for some unpublished results ⟨70UP40400⟩ covering about 100 compounds. The 2-pyrazoline ring is also puckered (68) and a method based on the Karplus equation that uses the ratio of the vicinal coupling constants ⟨74MI40401⟩ has been used for the calculation of the puckering angle (20° < θ < 40°) and the conformation populations. An X-ray study ⟨quoted in 76BSB545⟩ gives a value of θ = 27° for the 5-methoxycarbonyl-4,5-diphenyl-2-pyrazoline.

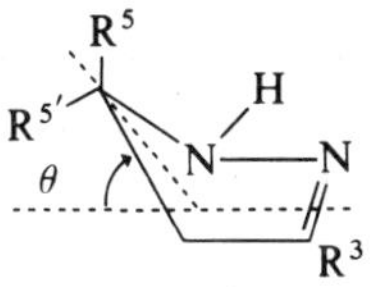

(68)

The Eu(dpm)₃ shift reagent has been used to assign the cis–trans configuration of 3,4-dimethoxycarbonyl-1-phenyl-2-pyrazolines since the value of the $^3J_{4,5}$ coupling constant is almost the same for both compounds ⟨74CB1318⟩. Eu(fod)₃ has also been used in connection with bicyclic pyrazolines ⟨79JOC2513⟩. The results in Table 13 show that for 1,3-diphenyl- and 3,5-diphenyl-2-pyrazolines $J_{cis} > J'_{trans}$, but this criterion could be misleading since it depends on the nature of the substituent on N-1 ⟨74MI40401⟩. Another criterion for the assignment of pairs of cis–trans isomers, when $R^4 = R^5 = Me$, uses the McConnell anisotropy effect of the methyl group on the vicinal proton ⟨68BSF4403, 74CJC2296⟩.

Chemical shifts for some representatives of the less common class of Δ³-pyrazolines are shown in Table 13. These cyclic enehydrazines are also probably puckered (69); however, no X-ray determination nor conformational study has been carried out to solve this problem. Compound (70) is representative of the rare N(1)-unsubstituted Δ³-pyrazolines prepared by Burger et al. ⟨79T389⟩.

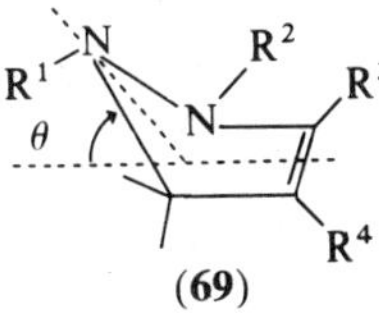

(69) 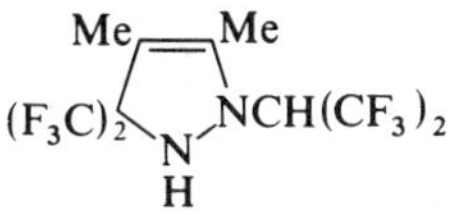   (70)

The salts (71) and (72) are both Δ²-pyrazoline derivatives. However, the first one is obtained by quaternization of a Δ²-pyrazoline and the second one by protonation of a Δ³-pyrazoline (Section 4.04.2.2.2(v)). Several Δ²-pyrazolinium salts (72) and their NMR spectra are described in ⟨69BSF3292, 71T123⟩ and for Δ³-pyrazolinium salts (73) see ⟨69BSF3316⟩.

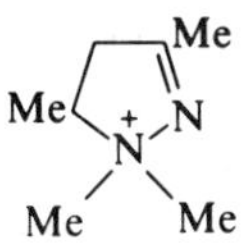

(71) 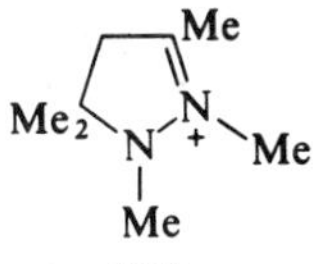   (72) 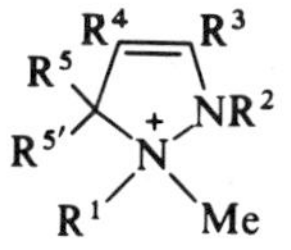   (73) 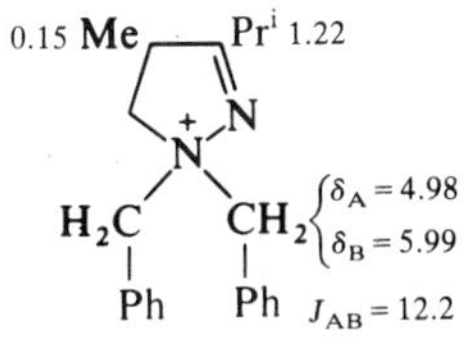   (74)

When the N-1 atom of salts of type (71) bears two prochiral groups an anisochrony of protons analogous to that of diethyl acetals is observed. For instance in compound (74) the benzylic protons show chemical shifts at δ 4.98 and 5.99 p.p.m. with $J_{A,B} = 12.2$ Hz ⟨69OMR(1)249⟩.

The fully saturated pyrazolidines have been utilized as models for the study of the nitrogen inversion of hydrazines. For instance, (75), a 2,3-diazabicyclo[2.2.1]heptene derivative, presents a consecutive inversion process at two nitrogen atoms with an activation barrier of 59 kJ mol⁻¹ determined by proton line-shape measurements ⟨67JA81⟩. Monocyclic pyrazolidines are discussed in ⟨B-73NMR165⟩. It has been shown by ^{1}H NMR data that the tetramethyl derivative (**76**) exists in a conformation such that the steric effects are minimized ⟨69OMR(1)249⟩.

4.04.1.3.4 ^{13}C NMR spectroscopy

In contrast to the previous section, there is little information available on ^{13}C NMR spectra of pyrazoles and indazoles in books or reviews. Consequently, this section will treat the literature results in some detail, giving the principal references and some illustrative examples.

Pyrazole itself (Table 14) has been studied several times, the most remarkable conclusion being that, depending on the solvent and the temperature, one or two signals are observed for carbons C-3 and C-5. This is common for all the N-unsubstituted pyrazoles with not too different tautomer populations. It has become quite common to observe broad signals for these carbons in spectra run in polar aprotic solvents, like DMSO. The factors that influence the prototropic exchange rate have been discussed ⟨79JA545, 79MI40407⟩; in the solid state the proton is firmly bound to the nitrogen atom ⟨81CC1207⟩.

For N-substituted pyrazoles, Table 14 shows a regular trend in the effect of the R^1 substituent; when the electron-withdrawing properties of the substituent increase all the signals shift to lower fields and, simultaneously, the ^1J coupling constants increase. Figure 13 represents the variation of the ^1J coupling constants at the three carbon atoms for 11 R^1 substituents (first-order analyzed spectra). It appears that it could be dangerous to assign C-3 and C-5 signals from the ^1J value, unless R^1 is an electron-withdrawing substituent, and therefore $^1J_{C-5,H-5} > {}^1J_{C-3,H-3} > {}^1J_{C-4,H-4}$.

For 'asymmetric' pyrazoles, ^{13}C NMR data, both chemical shifts and coupling constants, prove superior to ^{1}H NMR data for identifying isomeric structures and even for deciding the major tautomer in solution (Section 4.04.1.5.1). For instance, values in Table 14 for 3(5)-nitropyrazole and its two derived nucleosides clearly show the 3-nitro tautomer preponderance. The C—Me signal is characteristic of its position on the pyrazole ring (3-Me ~ 13.5, 5-Me ~ 11 p.p.m.), and the N—Me signal is sensitive to the presence or absence of a substituent in the 5-position. The difference, $\Delta\delta = \delta_o - \delta_m$, is related to the torsion angle between the two aromatic rings in N-phenylazoles ⟨73ACS3101⟩, and it has been used to calculate a dihedral angle of 25° for 1-phenylpyrazole (Section 4.04.1.4.3) ⟨78T1139⟩. However, $\Delta\delta$ depends not only on steric factors but also on the electronic properties of the azole ⟨81UP40400⟩.

Linear relationships between chemical shifts and the total charge densities in azoles, including pyrazole, have been described several times ⟨67JA6835, 67T2513, 68JA4232, B-70MI40401⟩. However, a careful study with a large set of data on N-methylazoles shows that if a general trend is actually observed ($\delta = -216 q_{\pi+\sigma} + 988$, $R = 0.919$ ⟨78OMR(11)234⟩), the possibility of assigning carbon signals using calculated charge densities is presently excluded.

Table 15 contains the ^{13}C chemical shifts of some selected indazoles. The major difference between indazoles and isoindazoles lies in the chemical shifts of carbons C-3 and C-7a. The substituent chemical shifts (SCS) induced by the substituent in position 3 have been discussed using an empirical model ⟨77OMR(9)716⟩. The model that gives the best results, $\Delta\delta = a\mathscr{F} + b\mathscr{R} + c\mathscr{Q}^*$ ($\mathscr{F}$ and $\mathscr{R}$ are the Swain–Lupton parameters and $\mathscr{Q}^*$ is the Schaefer parameter), has been used with success to correlate SCS ⟨78OMR(11)617⟩. However, the tendency today is to apply the principal component analysis to the data without making any assumption ⟨81JCS(P2)403⟩. The benzo condensation effect, i.e. the effect on the heterocyclic carbon chemical shifts when pyrazole and indazole are compared, has been discussed in connection with other such pairs ⟨78OMR(11)617⟩.

Other pyrazole derivatives have not been the subject of systematic studies. Owing to the variety of structures involved, figures instead of tables are used to describe chemical shift

Table 14 ^{13}C NMR Chemical Shifts for Pyrazoles[p]

Compound	Solvent	C-1	C-2	C-3	C-4	C-5	Other data	Ref.
Pyrazole	CH_2Cl_2	—	—	134.6	105.8	134.6	—	74JOC357
Pyrazole	Acetone	—	—	133.4	104.6	133.4	Coupling constants[a]	76JCS(P2)1736
Pyrazole	HMPT (−17°C)	—	—	138.1	103.9	127.6	—	77JOC659
Pyrazole	DMSO-d_6 (29°C)	—	—	138.6	104.5	128.2	Coupling constants[b]	79JA545
Pyrazole	DMSO-d_6 (34°C)	—	—	133.7(b)	104.8	133.7(b)	—	79OMR(12)587
Pyrazole	Solid state	—	—	138.7	107.0	128.8	—	81CC1207
3(5)-Methylpyrazole	CH_2Cl_2	—	—	144.4 (CMe)	105.3	135.8	CMe 12.6	74JOC357
3-Methylpyrazole[c]	HMPT (−17°C)	—	—	146.05	103.2	128.3	3-Me 13.7	77JOC659
5-Methylpyrazole[c]	HMPT (−17°C)	—	—	137.2	103.2	138.6	5-Me 10.55	77JOC659
3,5-Dimethylpyrazole	CH_2Cl_2	—	—	145.3	104.8	145.3	—	74JOC357
3(5)-Nitropyrazole	Acetone-d_6	—	—	d	102.4	132.6	Coupling constants	79MI40407
4-Nitropyrazole	Acetone-d_6	—	—	132.5	135.6	132.5	Coupling constants	79MI40407
3-Azidopyrazole[e]	Acetone-d_6	—	—	146.8 (CN_3)	95.3	131.1	Coupling constants[f]	79MI40407
3(5)-Aminopyrazole	DMSO-d_6	—	—	154.0 (CNH_2)	91.5	132.0	—	79OMR(12)587
3(5)-Amino-5(3)-Methylpyrazole	DMSO-d_6	—	—	155.1 (CNH_2)	92.9	142.9	5(3)-(Me) 11.9	79OMR(12)587
1-Methylpyrazole	$CDCl_3$	38.4	—	138.7	105.1	129.3	Coupling constants	78JCS(P2)99
1-Ethylpyrazole	$CDCl_3$	15.6 (Me), 46.8 (CH_2)	—	139.0	105.3	128.1	—	79MI40410
1-t-Butylpyrazole	$CDCl_3$	29.9 (Me), 58.1 (C)	—	138.9	104.7	125.3	Coupling constants	81UP40400
1-Allylpyrazole	$CDCl_3$	—	—	139.2	105.7	128.9	g	79MI40410
1-Benzylpyrazole	$CDCl_3$	55.8 (CH_2)	—	139.4	105.7	129.1	g	79MI40410
1-Phenylpyrazole	$CDCl_3$	—	—	140.7	107.3	126.2	Coupling constants[g]	73ACS3101
1-p-Nitrophenylpyrazole	$CDCl_3$	—	—	142.7	109.3	128.8	g	81UP40400
1-(2,4-Dinitrophenyl)pyrazole	$CDCl_3$	—	—	143.8	109.6	131.0	g	81UP40400
1-Acetylpyrazole	$CDCl_3$	21.3 (Me), 169.2 (CO)	—	143.6	109.3	127.8	Coupling constants[h]	78JCS(P2)99
1-Ethoxycarbonylpyrazole	DMSO-d_6	149.1 (CO)	—	144.4	109.3	131.2	g	81UP40400
1-Carboxamidopyrazole	DMSO-d_6	—	—	142.3	108.6	128.8	g	81UP40400
1-Tosylpyrazole	DMSO-d_6	21.1 (Me)	—	145.8	109.6	132.4	g	81UP40400
1-Trifluoromethylsulfonylpyrazole	$CDCl_3$	119.3 (CF_3)	—	148.4	111.9	134.1	—	81UP40400
1-Nitropyrazole	DMSO-d_6	—	—	141.6	109.8	126.8	—	81UP40400
1-Aminopyrazole	DMSO-d_6	—	—	136.9	104.2	129.2	—	80JCR(M)0514

continued overleaf

Table 14 ^{13}C NMR Chemical Shifts for Pyrazoles (continued)

Compound	Solvent	$\delta(^{13}C)$ (p.p.m. from TMS)					Other data	Ref.
		C-1	C-2	C-3	C-4	C-5		
1-Trimethylsilylpyrazole	CDCl$_3$	−0.8 (Me)	—	143.5	106.3	134.0	—	81UP40400
1,3-Dimethylpyrazole	CH$_2$Cl$_2$	39.1	—	148.8	105.6	131.0	3-Me 13.9	74JOC357
1,5-Dimethylpyrazole	CH$_2$Cl$_2$	36.7	—	138.7	105.9	139.1	5-Me 11.9	74JOC357
1,3,5-Trimethylpyrazole	CH$_2$Cl$_2$	36.3	—	147.5	105.4	139.8	3-Me 14.0, 5-Me 11.7	74JOC357
5-Amino-1,3-dimethylpyrazole	DMSO-d_6	33.9	—	146.2	88.9	147.9	3-Me 13.9	79OMR(12)587[i]
3-Amino-1-methylpyrazole	DMSO-d_6	38.0	—	155.7	92.2	131.5	j	79OMR(12)587
5-Amino-3,4-dicyano-1-methylpyrazole	DMSO-d_6	38.0	—	126.5	78.9	155.1	j	75IOC1815
3-Amino-4,5-dicyano-1-methylpyrazole[k]	DMSO-d_6	41.8	—	160.3	84.6	120.6	j	75IOC1815
1,5-Dimethyl-3,4-diphenylpyrazole	CDCl$_3$	36.2	—	—	—	—	5-Me 9.9[l]	79CJC1186
5-Acetyl-4-ethoxycarbonyl-1,3-dimethylpyrazole	CDCl$_3$	38.2	—	149.3	110.6	143.0	3-Me 13.6[j,m]	78IOC2665
3-Bromo-1-phenylpyrazole	CDCl$_3$	—	—	127.9	110.4	128.4	Coupling constants[g]	73ACS3101
4-Bromo-1-phenylpyrazole	CDCl$_3$	—	—	141.3	95.5	126.8	Coupling constants[g]	73ACS3101
5-Bromo-1-phenylpyrazole	CDCl$_3$	—	—	141.1	110.2	112.5	Coupling constants[g]	73ACS3101
3-Ethoxycarbonyl-4-methyl-1-phenylpyrazole	CDCl$_3$	—	—	142.7	122.1	127.7	4-Me 9.9[g]	79OMR(12)205
5-Ethoxycarbonyl-4-methyl-1-phenylpyrazole	CDCl$_3$	—	—	141.1	123.7	130.7	4-Me 10.31[g]	79OMR(12)205
3-Nitro-1-(β-D-ribofuranosyl)pyrazole[n]	—	—	—	156.8	103.4	131.9	j	79MI40402
5-Nitro-1-(β-D-ribofuranosyl)pyrazole[n]	—	—	—	139.6	108.0	146.5	j	79MI40402
Pyrazolate anion (potassium pyrazolate)	H$_2$O	—	—	138.7	103.6	138.7	—	68JA4232
Pyrazolium cation (pyrazole hydrochloride)	H$_2$O	—	—	135.2	109.1	135.2	—	68JA4232
1,2-Dimethylpyrazolium iodide	DMSO-d_6	36.9	36.9	137.6	106.7	137.6	Coupling constants[o]	78OMR(11)234
1,2,3-Trimethylpyrazolium iodide	DMSO-d_6	37.2	34.1	146.2	106.7	136.2	Coupling constants	78OMR(11)234
2,3-Dimethyl-1-phenylpyrazolium iodide	CDCl$_3$	—	35.4	148.4	108.2	137.3	3-Me 12.5[g]	82UP40400
2,5-Dimethyl-1-phenylpyrazolium iodide	DMSO-d_6	—	37.6	138.2	107.6	147.7	5-Me 12.1[g]	82UP40400

[a] At low temperature (−40 °C) the signals of the acetone adduct (**55**) were observed ⟨75T1463, 76JCS(P2)736⟩. [b] At room temperature in acetone C-3 and C-5, ^1J = 185.5, ^2J = ^3J = 6.6, C-4, ^1J = 176.7, ^2J = 9.9 ⟨76JCS(P2)736⟩; in DMSO-d_6 at 10 °C, C-3, ^1J = 185.9, C-5, ^1J = 186.1, C-4, ^1J = 175.1, ^2J = 10.1 ⟨79JA545⟩. [c] Separated signals for each tautomer were observed in HMPT at −17 °C. [d] Not observed. [e] As discussed in Section 4.04.1.3.3(i) only the 3-azido tautomer is present in DMSO-d_6 solution. [f] Coupling of the three carbons with the proton linked to nitrogen is observed. [g] Chemical shifts of the N(1)-substituted compound are also given in the quoted reference. [h] C-3, ^1J = 187.1, ^2J = 9.2, ^3J = 6.1, C-4, ^1J = 178.7, ^2J = 11.8 and 8.8, C-5, ^1J = 190.3, ^2J = 8.8, ^3J = 4.4. [i] See also ⟨79TL2991⟩. [j] Chemical shifts of the carbons at the 3-, 4- and 5-positions are also given in the quoted reference. [k] Isomeric structure established by X-ray crystallography. [l] Its isomer, 1,3-dimethyl-4,5-diphenylpyrazole shows a signal at 12.7 p.p.m. corresponding to 3-Me. [m] Its isomer, 3-acetyl-4-ethoxycarbonyl-1,5-dimethylpyrazole shows a signal at 10.4 p.p.m. corresponding to 5-Me. [n] 2′,3′,5′-tri-O-acetyl derivatives. [o] C-3, C-5, ^1J = 195, C-4, ^1J = 188. [p] For the spectra in the solid state see ⟨83H(20)1713⟩.

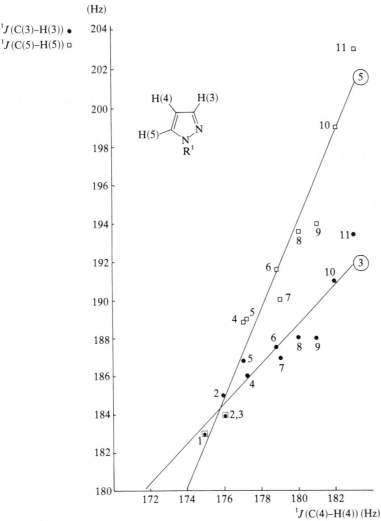

1, SiMe₃; 2, Me; 3, Buᵗ; 4, Ph; 5, NH₂; 6, 4'-NO₂C₆H₄; 7, Ac; 8, 2',4'-(NO₂)₂C₆H₃; 9, CO₂Et; 10, SO₂CF₃; 11, NO₂

Figure 13 Variation of the ^1J coupling constant in pyrazole with the nature of R^1

effects. Some values for pyrazolinones, pyrazolinethiones and pyrazolidinediones are shown in Figure 14. These ^{13}C chemical shifts have been used with success in tautomeric studies (Section 4.04.1.5.2). As in ^{1}H NMR studies, the CH tautomer is easily identified but both NH and OH tautomers give rise to mean signals; the NH/OH exchange rate is still too fast on the NMR time scale. However, even the mean signals are useful for tautomeric studies since the C-5 signal is quite different in the oxo derivatives and very different in the thio derivatives for the NR and OR fixed forms.

A large amount of data on Δ^2-pyrazolines has been collected but not completely published. Some values can be found in Figure 15.

4.04.1.3.5 Nitrogen NMR spectroscopy

The possibility offered by new instruments to obtain ^{15}N NMR spectra using natural abundance samples has made ^{14}N NMR spectroscopy a method which holds no interest for the organic chemist, since the chemical shifts are identical and the signal resolution incomparably better with the ^{15}N nucleus ($I = \frac{1}{2}$) than with ^{14}N ($I = 1$). ^{1}H–^{15}N coupling constants could be obtained from natural abundance samples by ^{15}N NMR and more accurately from ^{15}N-labelled compounds by ^{1}H NMR. Labelled compounds are necessary to measure the ^{13}C–^{15}N and ^{15}N–^{15}N coupling constants.

In the coming years, one can expect a great development in nitrogen NMR spectroscopy for the study of azoles (for recent work see ⟨82JOC5132, 83HCA1537⟩). At this moment most

Table 15 ^{13}C NMR Chemical Shifts for Indazoles

Compound	Solvent	C-3	C-4	C-5	C-6	C-7	C-3a	C-7a	Other data	Ref.
Indazole	DMSO-d_6	133.4	120.4	120.1	125.8	110.0	122.8	139.9	Coupling constants[a]	76CJC1329
3-Methylindazole	DMSO-d_6	140.9	119.8	119.3	125.8	109.9	122.1	140.8	—[b]	77OMR(9)716
3-Fluoroindazole	DMSO-d_6	156.8	118.2	121.0	128.0	111.1	107.7	141.8		77OMR(9)716
3-Azidoindazole	DMSO-d_6	139.6	118.5	120.4	127.2	110.7	114.0	141.6	Coupling constants[a]	77OMR(9)235
1-Methylindazole	CDCl$_3$	132.4	120.8	120.2	125.9	108.6	123.8	139.7	Coupling constants[c]	78JCS(P2)99[d]
2-Methylindazole	CDCl$_3$	123.1	119.6	121.2	125.4	116.8	121.8	148.7	Coupling constants[e]	78JCS(P2)99[d]
1-Acetylindazole	CDCl$_3$	139.4	120.6	124.1	129.2	115.3	126.1	139.4	Coupling constants[f]	78JCS(P2)99
2-Acetylindazole	CDCl$_3$	122.1	124.0	121.4	129.0	118.8	122.1	150.7	Coupling constants[g]	78JCS(P2)99
1,2-Dimethylindazolium iodide	DMSO-d_6	132.6	122.7	124.7	132.6	111.1	118.6	139.9	Coupling constants[h]	78OMR(11)234

[a] For a rigorous determination of the coupling constants see ⟨79MI40411⟩. [b] ^{13}C–^{19}F coupling constants: C-3, $^1J=224$, C-4, $^3J=4$, C-3a, $^2J=25.5$. [c] NMe 35.2. [d] See also ⟨77OMR(9)716⟩. [e] NMe 39.7. [f] NAc 22.7 (Me), 170.7 (CO). [g] NAc 21.8 (Me), 170.8 (Co). [h] 1-Me 33.9, 2-Me 38.5.

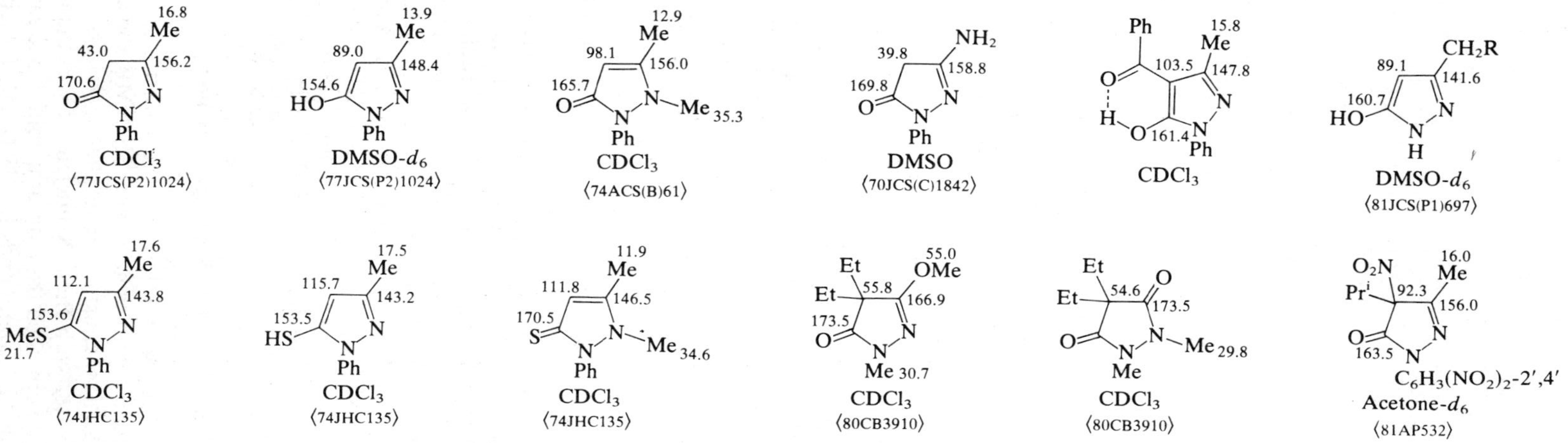

Figure 14 ^{13}C chemical shifts (p.p.m.) of pyrazolinones, pyrazolinethiones and pyrazolidinediones

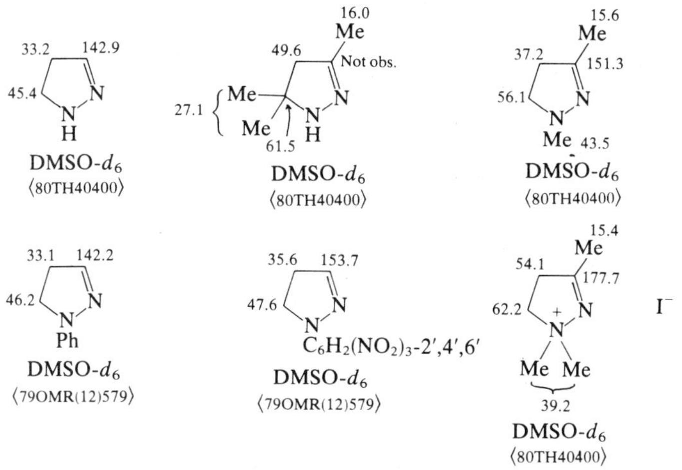

Figure 15 ^{13}C chemical shifts (p.p.m.) of Δ^2-pyrazolines and their salts

of the work on ^{14}N NMR spectra has been done by Witanowski, Stefaniak and Webb ⟨B-73MI40401, 78OMR(11)27, 78OMR(11)385⟩ and will not be discussed here. Study of ^{15}N NMR spectra owes a lot to the pioneering work of Roberts, whose publications on pyrazoles and indazoles ⟨79JOC1765⟩ will be examined in detail. The ^{14}N nuclear quadrupole coupling constants (measured either by pure quadrupole resonance or by microwave spectroscopy) of heterocycles including the pyrazoles are discussed in ⟨71PMH(4)21, 71MI40401, 81ZN(A)34⟩.

All the chemical shifts in this section are from nitromethane as external reference; the values for recalculating data from other external and internal standards are to be found in ⟨77JMR(28)217, 77JMR(28)227⟩. In particular, Roberts' positive values (reference: 1M nitric acid) have been transformed by the following equation: $\delta(\text{MeNO}_2) = -\delta_{\text{Roberts}} - 4.43$ p.p.m. according to the general agreement that negative values denote shifts to higher magnetic field from the reference.

Most available literature results on pyrazoles and indazoles are shown in Tables 16 and 17.

Since nitrogen atoms are directly involved in exchange processes, the best way to study tautomerism, isomerism and protonation is to observe the nitrogen signals. In pyrazoles, protons H-3 and H-5 are separated by 0.15 p.p.m. (Table 7), carbons C-3 and C-5 by 10.5 p.p.m. (Table 14) and nitrogens N-1 and N-2 by 93 p.p.m. (Table 16). Thus two ^{15}N signals are observed for pyrazole in DMSO at room temperature and a broad signal for N-2 of 3-methyl-1-phenylpyrazolone under the same conditions. This broad nitrogen signal corresponds to a slow exchange rate between the OH and NH tautomers ⟨77JCS(P2)1024⟩, whereas the corresponding ^{1}H and ^{13}C NMR spectra show only mean narrow signals. Isomerism is easily determined by ^{15}N NMR; for example, the absolute difference in p.p.m. between the pyridine and pyrrole nitrogen atoms is 146 for the indazole N-methyl structure and 70 for the isoindazole structure. Finally, protonation produces dramatic changes in the chemical shifts, so much so that it is possible to use them to determine accurate pK_a values: pyrazole, 2.63 (thermodynamic value, 2.52 ⟨68BSF5009⟩), N-methylpyrazole, 2.15 (thermodynamic value, 2.09 ⟨68BSF5009⟩).

Nitrogen chemical shifts of nitrogen heterocycles, including pyrazole, N-methylpyrazole and indazole, have been compared with calculated nitrogen screening constants ⟨78OMR(11)27⟩ and a fairly good correlation ($\delta_{\text{exp}} = 1.18\delta_{\text{calc}} - 104.2$, correlation coefficient 0.901) is obtained. The results in Table 16 show the great sensitivity of N-1 chemical shifts to the nature of R^1 ($R^1 = \text{Me, Ph, CMe}_2\text{OH, NO}_2$).

The coupling constants are very rich in structural information. For the sake of comparison, indazole and isoindazole have been numbered like pyrazoles (Figure 16).

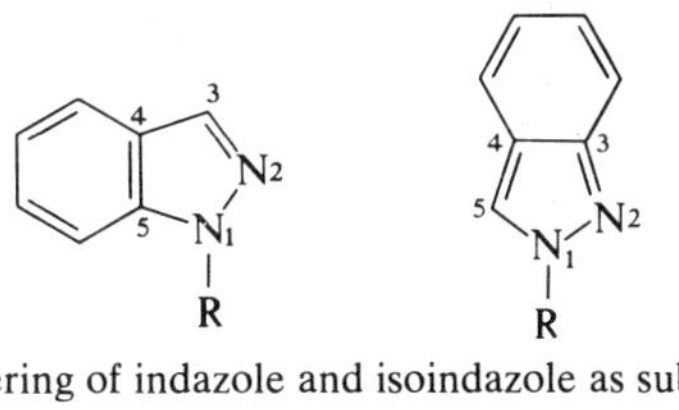

Figure 16 Numbering of indazole and isoindazole as substituted pyrazoles

Table 16 ^{15}N NMR Chemical Shifts for Pyrazoles and Indazoles

Compound	Solvent	$\delta(^{15}N)$ (p.p.m. from MeNO$_2$) N-1	N-2	Other data	Ref.
Pyrazole	CHCl$_3$	−132.9	−132.9		79JOC1765
Pyrazole	DMSO	−171.3	−78.3		79JOC1765
Pyrazole	HCl (pH 0.45)	−183.5	−183.5		79JOC1765
3(5)-Methylpyrazole	CHCl$_3$	−138.2	−132.5	$^3J(N(1)-H(4)) \sim 6$, $^2J(N(1)-H(5)) \sim 14$	79JOC1765
3(5)-Methylpyrazole	HCl (pH 0.45)	−189.0	−185.5	$^3J(N(1)-H(4))$ 4.6, $^2J(N(1)-H(5))$ 6.6, $^3J(N(2)-H(4)) = {}^3J(N(2)-H(5)) = 3.4$	79JOC1765
3,5-Dimethylpyrazole	CHCl$_3$	−138.0	−138.0		79JOC1765
3,5-Dimethylpyrazole	HCl (pH 0.45)	−190.1	−190.1		79JOC1765
1-Methylpyrazole	CHCl$_3$	−179.0	−74.7		79JOC1765
1-Methylpyrazole	HCl (pH 0.06)	−188.1	−174.6		79JOC1765
1,2-Dimethylpyrazolium iodide	H$_2$O	−183.9	−183.9		79JOC1765
1-Phenylpyrazole	CDCl$_3$	−164.2	−78.2	^{15}N–^{15}N, ^{13}C–^{15}N and ^{1}H–^{15}N coupling constants	77JCS(P2)1024
Pyrazole–acetone adduct (55)	Acetone (−50 °C)	−140.4	−80.6	^{15}N–^{15}N, ^{13}C–^{15}N and ^{1}H–^{15}N coupling constants	82UP40401
1-Nitropyrazole	Acetone	−107.8	−82.9	^{15}N–^{15}N, ^{13}C–^{15}N and ^{1}H–^{15}N coupling constants	82UP40401
3-Methyl-1-phenyl-2-pyrazolin-5-one	CDCl$_3$	−186.4	−54.5	^{15}N–^{15}N, ^{13}C–^{15}N and ^{1}H–^{15}N coupling constants	77JCS(P2)1024
Indazole	DMSO-d_6	−194.4	−65.6	^{13}C–^{15}N and ^{1}H–^{15}N coupling constants	82OMR(18)10
1-Methylindazole	DMSO-d_6	−202.8	−56.6	^{13}C–^{15}N and ^{1}H–^{15}N coupling constants	82OMR(18)10
2-Methylindazole	DMSO-d_6	−91.2	−161.0	^{13}C–^{15}N and ^{1}H–^{15}N coupling constants	82OMR(18)10

Table 17 Coupling Constants for Pyrazoles and Indazoles

Compound	Coupling constants (Hz)													Ref.
	N(1)–N(2)	N(1)–C(3)	N(1)–C(5)	N(1)–C(4)	N(2)–C(3)	N(2)–C(5)	N(2)–C(4)	N(1)–H(3)	N(1)–H(5)	N(1)–H(4)	N(2)–H(3)	N(2)–H(5)	N(2)–H(4)	
Pyrazole-1-^{15}N	13.4	—	13.4	—	—	—	—	3.4^a	8.9^a	4.3^a	8.9^a	3.4^a	4.3^a	73MR(10)130
Pyrazole–acetone adduct (55)	—	1.2^c	12.1	6.2	1.2^c	0	2.1	7.5	4.0	5.2	12.9	0	1.1	82UP40401
1-Phenylpyrazole-1,2-^{15}N	12.8	1.2^c	12.1	6.2	1.2^c	0	2.1	7.4	4.4	6.0	14.2	—	1.0	77JCS(P2)1024
1-Nitropyrazole-1,2-^{15}N	14.7	^{d}	11.6	7.9	^{d}	0	2.0	10.6	3.1	7.6	14.4	0.9	0.9	82UP40401
Indazole-1- or -2-^{15}N	—	0	13.6	5.9	0	1.1	1.8	7.4	—	—	—	—	—	82OMR(18)10
1-Methylindazole-1- or -2-^{15}N	—	0	15.0	6.6	0	0.8	1.8	7.6	—	—	12.8	—	—	82OMR(18)10
2-Methylindazole-1- or -2-^{15}N	—	1.3	13.2	5.0	1.8	^{d}	0.7	—	4.3	—	12.8	<0.5	—	82OMR(18)10
1-Phenyl-3,5-dimethylpyrazole-1,2-^{15}N	—	—	—	6.4	—	—	1.9	—	—	—	—	—	—	80OMR(13)197

^{a} Mean values. ^{b} Blocked exchange. ^{c} Not assigned; a coupling of 1.2 Hz is observed on the C-3 resonance of the doubly labelled 1-phenylpyrazole. ^{d} Nonmeasurable on a broadened signal.

The values of Table 17 are remarkably coherent. As Begtrup has pointed out ⟨74CC702⟩ coupling constants are less sensitive to substituent effects than chemical shifts, and for this reason are preferable for tautomeric studies. For instance, the mean value of the ^{1}H–^{15}N coupling constants measured in the acetone adduct (55) is close to the values measured in CCl$_4$ for the monolabelled pyrazole (rapid exchange situation; Figure 17). These last values have been compared with CNDO/2 and INDO calculated spin–spin coupling constants ⟨73JMR(10)130⟩.

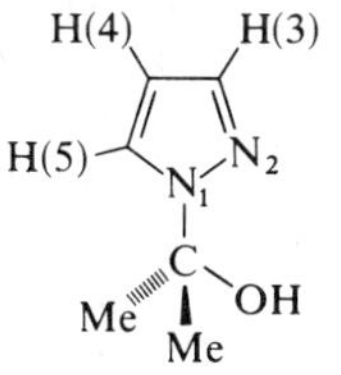

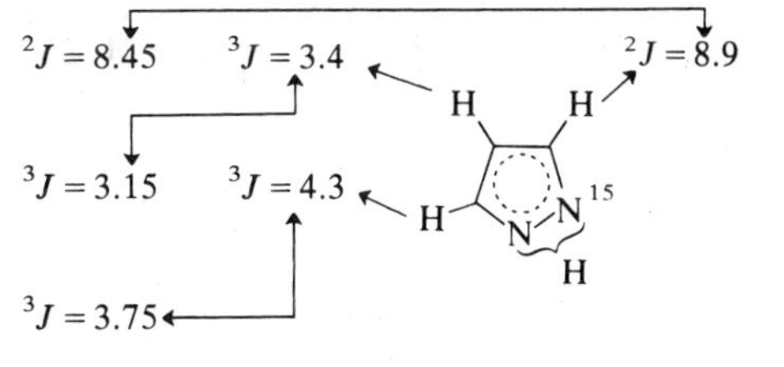

Figure 17 ^{1}H–^{15}N coupling constants in pyrazole

The coupling constants measured in the ^{15}N NMR spectrum of 3(5)-methylpyrazole (Table 16) are probably inaccurate. Since the equilibrium constant is almost 1 (Section 4.04.1.5.1), the expected values are $^3J_{\text{N-1,H-4}} = 3$ to 4 Hz and $^2J_{\text{N-1,H-5}} = 8$ to 9 Hz, instead of 6 and 14 Hz respectively. Significant differences are observed in ^{1}H–^{15}N coupling constants determined from ^{1}H and ^{15}N NMR spectra. For example, the $^1J(^1\text{H}-^{15}\text{N})$ coupling constant in indazole is 106.4 Hz by ^{1}H and 104.8 Hz by ^{15}N NMR ⟨82OMR(18)10⟩. Incidentally, the existence of this coupling proves the 1H structure of indazole (^{15}N-2 is not coupled with H-1).

Axenrod et al. have used ^{1}H–^{15}N ⟨79OMR(12)476⟩ and ^{13}C–^{15}N ⟨80OMR(13)197⟩ coupling constants, measured by ^{1}H and ^{13}C NMR spectroscopy respectively, to assign pyrazole isomeric structures. For a pyrazole of general structure (77), only the isomer (77a) (prepared from PhNH—^{15}NH$_2$) shows a $^3J(^1\text{H}-^{15}\text{N})$ coupling constant of 2.2–3.0 Hz and a $^2J(^{13}\text{C}-^{15}\text{N})$ coupling constant of 8.2–8.8 Hz.

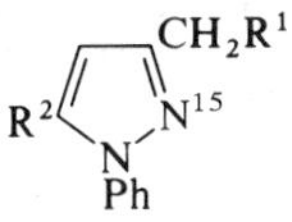

4.04.1.3.6 UV spectroscopy

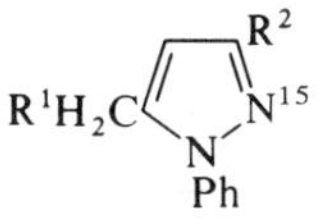

Electronic absorption spectroscopy, UV and visible, has been covered extensively by ⟨63PMH(2)1⟩ and ⟨71PMH(3)67⟩. Table A2 of ⟨B-76MI40402⟩ contains UV absorption data for nearly 70 pyrazoles and pyrazolones.

The use of UV spectroscopy as an identification method is continuously decreasing in relative importance compared to the use of NMR or mass spectrometry. However, due to the general validity of Beer's law, it continues to be an appropriate method for quantitative studies such as the measurement of ionization constants (Section 4.04.2.1.3(iv) and (v)) and the determination of tautomeric equilibrium constants (Section 4.04.4.1.5).

Discussion of the UV spectra of heteroaromatic molecules has changed with time, from the classical way illustrated by the Italian publications ⟨56G797⟩ to the modern quantum mechanical calculations.

(i) Pyrazole

Absorption maxima for pyrazole itself and for monosubstituted pyrazoles are shown in Table 18.

The only observed transition for pyrazole is a $\pi \to \pi^*$ one of the ($^1A_1 \to {}^1B_1$) class; $n \to \pi^*$ bands have not been reported. The $\pi \to \pi^*$ band occurs at 211 nm (47 390 cm^{-1}) in ethanol and at 203.2 nm (49 210 cm^{-1}) in the vapour phase, corresponding to ΔE of 5.88 and 6.10 eV.

Table 18 UV Spectral Maxima (nm) for Pyrazoles in Ethanol

Pyrazole	Substituent at position				λ_{max} (log ε)	Ref.
	1	3	4	5		
Unsubstituted	H	H	H	H	211 (3.49)[a]	66BSF3744[b,c]
3(5)-Methyl	H	Me	H	H	213 (3.57)	66BSF3744[c]
4-Methyl	H	H	Me	H	219 (3.67)	66BSF3744[c]
3(5)-Phenyl	H	Ph	H	H	248 (4.14)	66BSF3744[c]
4-Phenyl	H	H	Ph	H	250.5 (4.18)	66BSF3744[c]
4-Chloro	H	H	Cl	H	220 (3.46)	66BSF3744
4-Bromo	H	H	Br	H	221 (3.47)	66BSF3744
4-Iodo	H	H	I	H	227 (3.38)	66BSF3744
3(5)-Amino	H	NH_2	H	H	228 (3.28)	—
3(5)-Nitro	H	NO_2	H	H	261 (3.79)[d]	73JHC1055
4-Nitro	H	H	NO_2	H	275 (3.91)	71PMH(3)67
3(5)-Methoxycarbonyl	H	CO_2Me	H	H	217 (3.94)	71PMH(3)67
1-Methyl	Me	H	H	H	217 (3.61)	66BSF3744[c]
1-Acetyl	MeCO	H	H	H	239 (4.06)	71PMH(3)67
1-Carboxamido	$CONH_2$	H	H	H	233 (3.98)	66BSF3744
1-Phenyl	Ph	H	H	H	255 (4.15)	68JCS(B)725[c]
1-p-Nitrophenyl	$C_6H_4NO_2$	H	H	H	313 (4.24)	66BSF3744
1-(2,4-Dinitrophenyl)	$C_6H_3(NO_2)_2$	H	H	H	308 (4.07)	66BSF3744
1-(2,4,6-Trinitrophenyl)	$C_6H_2(NO_2)_3$	H	H	H	305 (3.81)	66BSF3744
1-Tosyl	$MeC_6H_4SO_2$	H	H	H	240 (4.25)	66BSF3744
1-Nitro	NO_2	H	H	H	265 (3.69)	—

[a] 203 (vapor phase) ⟨70TH40400⟩. [b] Theoretical calculations ⟨70BCJ3344⟩. [c] Theoretical calculations ⟨68JCS(B)725⟩. [d] In 0.05N HCl.

(ii) Indazoles

The UV spectra of indazole and its two *N*-methyl derivatives can be seen in Table 23 (Section 4.01.3.6). Calculated values for the absorption maxima of indazole agree fairly well with the experimental ones (Table 19).

Table 19 Experimental and Calculated UV Spectral Data for Indazole

Experimental UV maxima			PPP calculation[a] ⟨70BCJ3344⟩			CNDO/S calculation[a] ⟨74T2903⟩		
λ_{max} (nm)	log ε	ΔE (eV)	ΔE (eV)	f (cgs)	α (°)	ΔE (eV)	f (cgs)	θ (°)
296	3.52	4.19 }	4.47	0.18	84	{ 4.32	0.045	23
284	3.63	4.37 }	—	—	—	{ 4.87	$n \to \pi^*$	
250	3.65	4.95	4.88	0.13	−48	5.10	0.124	−76
221	—	5.60	6.11	0.94	7	5.79	0.754	59
—	—	—	6.22	0.56	18	5.81	0.544	−61
—	—	—	—	—	—	5.99	$n \to \pi^*$	
—	—	—	6.53	0.36	−16	6.27	0.427	−41

[a] α = polarization direction measured counterclockwise to the *x*-axis, f = oscillator strength.

(iii) Effect of substituents on aromatic compounds

A publication by Elguero *et al.* ⟨66BSF3744⟩ discusses UV spectra of 170 pyrazoles determined in 95% ethanol. Rigorously, the UV substituent effects must be discussed using wavenumbers, since only wavenumbers (in cm^{-1}) are proportional to transition energies, ΔE. However, in order to have more familiar values, data on 1-(2,4-dinitrophenyl)pyrazoles ⟨66BSF3744⟩ have been transformed from wavenumbers to wavelengths (in nm) (Table 20).

Substituents in positions 3 and 4 produce bathochromic shifts whereas substituents in position 5 produce hypsochromic shifts, which become more pronounced as the bulk of

Table 20 Substituent Effects in the UV Spectra of 1-(2,4-Dinitrophenyl)pyrazoles ⟨68BSF707⟩

Unsubstituted 308, 4-Cl +4, 4-Br +4, 4-Me +14, 3-Cl +4, 3-Br +8, 3-Me +16, 3-Et +16, 3-Pr^i +20, 3-Bu^t +20, 5-Cl −30, 5-Me −16, 5-Et −17, 5-Pr^i −18, 5-Bu^t −39 the 5-substituent increases (steric inhibition of the resonance). When another substituent is introduced into position 4, the steric effect produced by the substituent in position 5 increases (buttressing effect) by about 8 nm (4-Cl/5-Me, 4-Br/5-Me, 4-Br/5-Cl). Braude's law, used with discretion, allows the calculation of a dihedral angle of 55° between pyrazole and phenyl for 1-phenyl-5-methylpyrazole ⟨78T1139⟩.

From UV studies of 4-phenyl-, 4-nitro- and 4-nitroso-pyrazoles, Habraken *et al.* ⟨67RTC1249, 72JHC939⟩ conclude that the 4-pyrazolyl group acts as an electron-donating group. UV spectra of pairs of 1-aryl- and 2-aryl-indazoles and their utility in the determination of isomeric structures are discussed in ⟨67BSF2619⟩; many other UV data on indazole derivatives can be found in ⟨71PMH(3)67⟩.

Simple HMO calculations ⟨68JCS(B)725⟩ satisfactorily account for the UV spectra of a great number of pyrazoles substituted by methyl and phenyl groups. The spectra of pyrazolium and indazolium salts (free bases in 1N HCl) have been compared with calculated transitions (Pariser–Parr–Pople method) ⟨74MI40403⟩.

(iv) Non-aromatic derivatives

Neither ⟨71PMH(3)67⟩ nor ⟨B-76MI40402⟩ contains information on non-aromatic derivatives of pyrazole. Table 21 gives some references to these compounds, including aromatic pyrazolones.

Table 21 Some References to UV Studies of Pyrazole Derivatives

Pyrazole derivatives	Ref.
Pyrazolones	76AHC(S1)1, B-76MI40402, 78JPR521
Pyrazolidones	67BSF3502
Δ^1-Pyrazolines[a]	78JA5122
Δ^2-Pyrazolines[a]	66BSF610, 67BSF3516[b], 80CHE66[c]
Δ^3-Pyrazolines[a]	69BSF3316, 70BSF1129
Pyrazolidines	70CHE744

[a] See also ⟨67HC(22)1⟩. [b] Pyrazolinium salts. [c] Fluorescence.

4.04.1.3.7 IR spectroscopy

IR and Raman studies of heterocycles today cover two different fields. For simple and symmetrical molecules very elaborate experiments (argon matrices, isotopic labelling) and complex calculations lead to the complete assignment of the fundamentals, tones and harmonics. However, the description of modes ought to be only approximate, since in a molecule like pyrazole there are no pure ones. This means that it is not correct to write that the band at 878 cm^{-1} is γ(CH), and the only correct assertion is that the γ(CH) mode contributes to the band. On the other hand, IR spectroscopy is used as an analytical tool for identifying structures, and in this case, bands are assigned to ν(CO) or δ(NH) on the basis of a simple Nujol mull spectrum and conventional tables. Both atttitudes, almost antagonistic to each other, are discussed in this section.

(i) Aromatic rings without carbonyl groups

IR spectroscopy of pyrazoles has been reviewed twice by Katritzky *et al.* ⟨63PMH(2)161, 71PMH(4)265⟩, where much qualitative information is to be found. Recently, several groups of workers have begun an exhaustive study of pyrazole, the two most important studies being those of Tabačik and Pellegrin ⟨78MI40400, 79SA(A)1055⟩ and Troitskaya and Tyulin ⟨74CHE471, 74CHE1215, 74MI40405⟩. The first study involved eight molecules: pyrazole-d_0, pyrazole-1-d_1, pyrazole-4-d_1, pyrazole-1,4-d_2 pyrazole-3,5-d_2, pyrazole-1,3,5-d_3, pyrazole-3,4,5-d_3 and pyrazole-d_4; it employed IR (gas and argon matrix) and Raman (crystal, CCl$_4$ and gas) procedures. The second study utilized pyrazole-d_0, pyrazole-1-d_1, pyrazole-4-d_1, pyrazole-1,4-d_2, pyrazole-3,4,5-d_3, pyrazole-d_4 and ^{15}N$_2$-pyrazole, with IR (gas, solution and crystal) and Raman (crystal, H$_2$O and CCl$_4$) procedures; they also made use of the CdCl$_2$: pyrazole complexes. Both groups have identified the 21 normal modes of pyrazole (point symmetry group C_s; Table 22). Although for the a'' group and the four high-frequency a' there is complete agreement, some discrepancies persist for

Table 22 Pyrazole-d_0: Assignment of Vibrational Spectra (21 Normal Modes) (cm^{-1})

Gas^a	Argon^a matrix	Assignment	Remarks concerning the approximate description of modes^a	Assignment^b	Gas^b
515	525	ν_{21} (a″)	The six vibrations of the a″ class	ν_{21} (a″)	515
—	612	ν_{20} (a″)	include two Γ (ring bending out-of-	ν_{20} (a″)	612
673	677	ν_{19} (a″)	plane), three γ(CH) (CH bending	ν_{19} (a″)	670
744	748	ν_{18} (a″)	out-of-plane) and one γ(NH) (NH	ν_{18} (a″)	746
833	834	ν_{17} (a″)	bending out-of-plane).	ν_{17} (a″)	833
879	878	ν_{16} (a″)	ν_{21} (a″) ≈ γ(NH)	ν_{16} (a″)	880
				ν_{15} (a′)	865
909	909	ν_{15} (a′)	The 11 low-frequency vibrations of the	ν_{14} (a′)	910
1008	1015	ν_{14} (a′)	a′ class include three δ(CH) (CCH	ν_{13} (a′)	1010
			bending in-plane), one δ(NH) (CNH	ν_{12} (a′)	1010
1054	1063	ν_{13} (a′)	bending in-plane), two Δ (ring bending	ν_{11} (a′)	1060
1121	1127	ν_{12} (a′)	in plane), one BR (ring breathing) and	ν_{10} (a′)	1125
(1156)^c	—	ν_{11} (a′)	four ω (ring stretching in plane).	ν_9 (a′)	(1155)^c
1254	1256	ν_{10} (a′)	ν_{11} (a′) ≈ BR (it appears only in the	ν_8 (a′)	1255
1343	1345	ν_9 (a′)	Raman spectrum)^c		
1358	1357	ν_8 (a′)		ν_7 (a′)	1360
1395	1393	ν_7 (a′)		ν_6 (a′)	1400
1449	1449	ν_6 (a′)			
				ν_5 (a′)	1470
1535	1530	ν_5 (a′)			
(3115)^c	3116	ν_4 (a′)	The four high-frequency vibrations of	ν_4 (a′)^f	(3110)^c
3130^c	—	ν_3 (a′)	the a′ class include three ν(CH) (CH	ν_3 (a′)^f	(3135)^d
(3148)^c	—	ν_2 (a′)^e	stretching) and one ν(NH) (NH	ν_2 (a′)^f	(3155)^d
3524	3507	ν_1 (a′)	stretching). ν_1 (a′) ≈ ν(NH)	ν_1 (a′)^f	3525

^{a} ⟨79SA(A)1055⟩. ^{b} ⟨74MI40405⟩. ^{c} Raman. ^{d} CdCl$_2$ complex. ^{e} ν_2 (a′) ≈ ν(CH-4) ⟨78MI40400⟩. ^{f} Respectively, ν(CH-5), ν(CH-3), ν(CH-4) and ν(NH) ⟨74MI40405⟩.

the low frequency a′: the 1345, 1449 and 1530 cm^{-1} bands determined by the French group and the 865, 1010 and 1470 cm^{-1} bands of the Russian group.

Pyrazole, in all possible states, either free or complexed, shows C_s symmetry, with positions 3 and 5 clearly differentiated, even if it is not possible to assign individual bands to these positions, for example ν(CH) or γ(CH). Both research groups carried out normal coordinate calculations using different potentials. In spite of the large amount of data obtained, the problem is not definitively settled and more experimental work and quantum mechanical calculations are to be expected in the next few years.

(ii) Pyrazole rings containing carbonyl groups

In this subsection compounds with a pyrazole C—O bond will be discussed independently of their aromatic character. In solution the tautomers of pyrazolinones, e.g. (**78a**), (**78b**) and (**78c**), are easily identified by their IR spectra (Figure 18) ⟨76AHC(S1)1⟩.

(**78a**)   (**78b**)   (**78c**)

ν(CO) $\begin{cases} 1695 \text{ (R}^1 = \text{alkyl)} \\ 1705 \text{ (R}^1 = \text{aryl)} \end{cases}$ ν(CH(4)) 3150   ν(CO) 1630–1680
ν(ring) 1560   ν(C=C) ~1590

Figure 18 IR characteristics of pyrazolone tautomers

In the solid state the CH tautomer (**78a**) is always readily identified, but the OH and NH tautomers (that, in some cases, coexist in the crystal ⟨73CSC469⟩) are difficult to differentiate due to the strong hydrogen bonds that shift the ν(CO) band to the region of pyrazole vibrations. This source of complications is not present in the 'fixed forms' that can always be identified by their IR spectra, both in solution and in the solid state.

The spectra of pyrazolidinones have often been described ⟨63PMH(2)161, 67BSF3502, 70CHE1568⟩. In the 1-aryl-3-pyrazolidones (**79**), both the ν(CO) and ν(NH) bands changed considerably between the crystal (cyclic dimer) and the solution spectra. Typical values for ν(CO) range from 1700 to 1730 cm^{-1} in solution. For the 1-aryl-5-pyrazolidones (**80**),

(79) (80) (81) (82)

1690 cm^{-1} is often the frequency of the carbonyl band, moving to 1730 cm^{-1} in the conjugate acids and to 1740 cm^{-1} in the quaternary salts, these frequencies being compatible only with an electrophilic attack on N-2, (81) and (82) respectively.

The fundamental vibrations of 1,2-dimethyl-3-pyrazolidinethione (83), a cyclic thiohydrazide, have been tentatively assigned ⟨81ACS(A)767⟩. The X-ray structure of the 5-*p*-chlorophenyl derivative has been determined in order to have an experimental geometry for (83) suitable for normal coordinate analysis. 3,5-Pyrazolidinediones (84) show two carbonyl absorptions at 1715–1725 and 1745–1755 cm^{-1} and these are probably coupled modes. Finally, the mesoionic anhydro-4-hydroxy-1,2-dimethyl-3,5-diphenylpyrazolium hydroxide (85) shows no CO stretching vibration in the usual carbonyl region ⟨73CC402⟩, but a band at 1546 cm^{-1} in DMSO has been assigned to $\nu(CO)$ because it is the band exhibiting the largest shift in changing from aprotic to protic solvent (the UV band also shifted from 447 nm in C_6H_6 to 325 nm in H_2O). In its IR characteristics, compound (85) differs markedly from other mesoionic compounds, like the sydnones ($\nu(CO)$ 1750–1770 cm^{-1}), and this has been explained by the fact that compound (85) belongs to mesoionic heterocycles of type B ⟨76AHC(19)1⟩. On reduction of the hydrochloride of (85) by means of LAH ⟨73TH40400⟩ the 4-pyrazolidone (86) was obtained; this compound showed a carbonyl vibration at 1750 cm^{-1}, similar to those of cyclopentanone (1744 cm^{-1}).

(83) (84) (85) (86)

(iii) Substituent vibrations

Substituent vibrations are discussed in ⟨71PMH(4)265⟩. The C=O frequencies of some methylpyrazoles containing in the 3-, 4- or 5-position a formyl, acetyl or ethoxycarbonyl group have been reported ⟨69JHC545⟩. A comparison of these frequencies shows that with the exception of the ethoxycarbonylpyrazoles, they are 2 to 35 cm^{-1} lower than their corresponding parent benzene derivative (benzaldehyde, acetophenone, ethyl benzoate), indicating that, in general, the pyrazole ring donates electrons more readily than the phenyl ring. The 4-position appears to have the highest electron-releasing capacity ($\Delta \nu = -12$ to -35 cm^{-1}). This conclusion agrees with the observation of a slow rotation about the C—C bond in 4-formylpyrazole (57) (Section 4.04.1.3.3) and shows that resonance structures like (87) are important in these compounds. When both phenyl rings of benzophenone are replaced by 4-pyrazolyl rings (88) the $\nu(CO)$ band shifts from 1665 to 1610 cm^{-1} ($\Delta \nu = -55 \text{ cm}^{-1}$) ⟨78M337⟩.

(87) (88) (89) (90)

The characteristic absorption at 1746 cm^{-1} of 1-acetylpyrazole (89) has been related to the pseudohalogen behaviour of the 1-pyrazolyl substituent ($\sigma_p = 0.23$) by the equation $\sigma_p = 0.0103 \nu(CO) - 17.7$ ⟨74JCS(P2)449⟩. The σ_p value was calculated from the $\nu_{as}(NH)$ band of 1-*p*-aminophenylpyrazole (90) measured in CCl_4 ($\sigma_p = 3.11 \times 10^{-2} \Delta \nu_{as} - 12.6 \times 10^{-2}$).

The two NH_2 stretching vibrations of aminopyrazoles follow the Bellamy–Williams relation $[\nu_s(NH) = 345.5 - 0.876 \nu_{as}(NH)]$ ⟨76AHC(S1)1⟩ establishing the amino tautomeric structure of (91). Azidopyrazoles also show the ν_s and ν_{as} bands of the azido group. The last absorption is very strong and appears at 2110–2190 cm^{-1} depending on the substituent and the sample state. The band at 2130 cm^{-1} of 3-azido-5-methylpyrazole (**92**) in ethanol shifts to 2120 cm^{-1} in basic medium, corresponding to the anion (**93**), and subsequently disappears since in the end the compound is exclusively in the tetrazole form (**94**) ⟨74JHC921⟩. The intensity of the ν_{as} N$_3$ band has been used to measure the rate of the transformation (**93**) → (**94**).

(iv) Non-aromatic derivatives

The three isomeric pyrazolines can be identified by their IR spectra. Δ^1-Pyrazolines show the $\nu(N=N)$ vibration at 1545 ± 5 cm^{-1} (the complete assignment for Δ^1-pyrazoline itself has been published, $\nu(N=N)$ occurring at 1552 cm^{-1}), and in Δ^2-pyrazolines the $\nu(C=N)$ band occurs at 1590–1625 cm^{-1} depending on the substituents in positions 1 and 3 ⟨71PMH(4)265⟩. Δ^3-Pyrazolines substituted in the 3-position by hydrogen or an alkyl group have only the C=C band at 1665–1675 cm^{-1} ⟨69BSF3316⟩. IR spectroscopy has been used in a careful study of the protonation of Δ^2-pyrazolines ⟨71PMH(4)265, 75APO(11)267⟩. The cation (**95**) was fully characterized by the $\nu(C=N)$ band at 1649 cm^{-1} (in the free base it occurred at 1624 cm^{-1}) and the group of bands belonging to the $\overset{+}{N}H_2$ group.

4.04.1.3.8 Mass spectrometry

An introductory comment should be made concerning this section. The structures represented in the following fragmentation schemes are more mnemonic devices than structural formulae in the meaning they have in the ground state. It is necessary to remember that the ions generated in a mass spectrometer have a large excess of energy. In the most recent publications the tendency is to represent the ions by their molecular formulae unless the structure has been established definitively. Often this structure corresponds to an open-chain compound, which is more stable than the cyclic one.

A book ⟨B-71MS⟩ and a review by Nishiwaki ⟨74H(2)473⟩ contain much information about the behaviour of pyrazoles under electron impact. The Nishiwaki review covers mainly the hydrogen scramblings and the skeletal rearrangements which occur. One of the first conclusions reached was that pyrazoles, due to their aromatic character, are extremely stable under electron impact ⟨67ZOR1540⟩. In the dissociative ionization of pyrazole itself, the molecular ion contributes about 45% to the total ion current; thus, the molecular ion is the most intense ion in the spectrum.

A series of papers by van Thuijl *et al.* ⟨70OMS(3)1549, 71JHC311, 71OMS(5)1101, 73OMS(7)1165, 79OMS577⟩ constitutes the fundamental work on the mass spectrometry of pyrazole and its derivatives. For pyrazole itself the fragmentation pattern shown in Scheme 1 was proposed, based on extensive isotopic labelling. The loss of H· from the molecular ion [(**96**) → (**97**)] comes predominantly from H-3 and H-5 (92%); H-4 and H-1 contribute little to this loss. Analogously, elimination of HCN [(**96**) → (**98**)], involves mainly H-3 and H-5 (86%).

Another interesting fact is that hydrogen scrambling, *i.e.* randomization of the ring hydrogens of pyrazole to lose positional identity on electron impact, has not been observed to any significant extent ⟨see however 78OMS575⟩.

For *N*-methylpyrazole (**99**), the molecular ion of which is less intense than pyrazole (a common feature for methyl-substituted pyrazoles ⟨67ZOR1540⟩), the fragmentation pattern involves the methyl group (Scheme 2). These results were established using ^{2}H, ^{13}C and ^{15}N labelling studies.

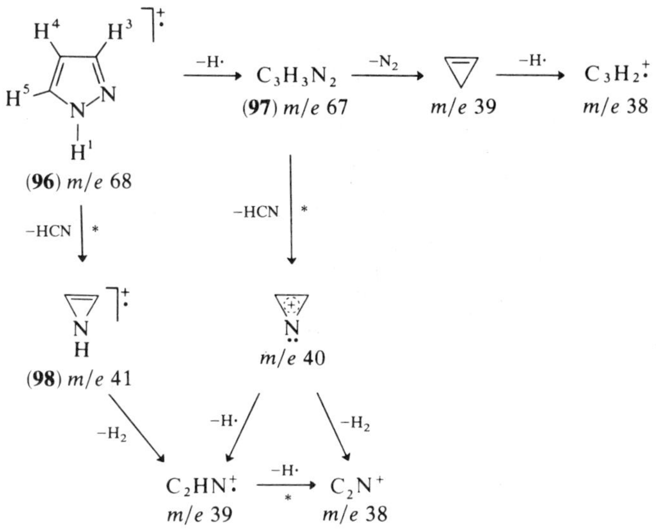

Scheme 1

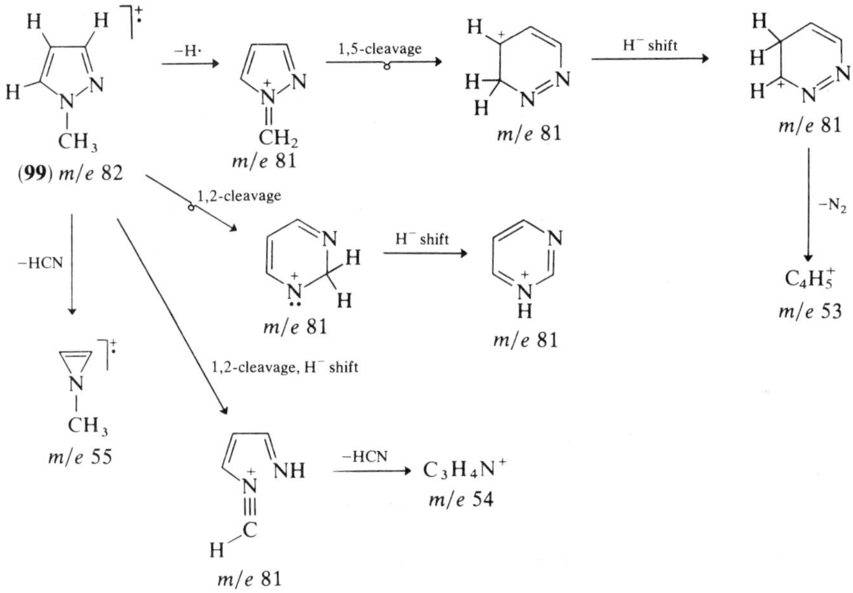

Scheme 2

A mass spectrometric study of indazole and its nitro derivatives has been carried out by the Mons group ⟨75OMS558, 77BSB281, 78OMS518, 79OMS114, 79OMS117, 80OMS144⟩. In indazole itself, the loss of HCN and DCN from indazole-1-d (**100**) and indazole-3-d (**101**) in a near unity ratio leads to the assumption that a molecular ion isomerization occurs prior to the fragmentation (Scheme 3) ⟨75OMS558⟩.

A small preference for the involvement of the N-1 nitrogen is observed in the loss of HCN from the molecule of ^{15}N-labelled indazole ⟨80OMS533⟩. In addition, there is a small isotope effect (1.08).

The structure of the radical cation $[C_6H_5N]^{+}$ (**102**) has been established by mass-analyzed ion kinetic energy spectrometry (MIKE) ⟨78OMS518⟩. From structure (**103**), structures (**104**) and (**105**) are formed (Scheme 4).

Although the conventional mass spectra of the five C-nitro derivatives of indazole are nearly identical, the corresponding metastable peak shapes associated with the loss of NO· can be used to differentiate the five isomers ⟨79OMS114⟩. The protonation and ethylation occurring in a methane chemical ionization source have been studied for a variety of aromatic amines, including indazoles ⟨80OMS144⟩. As in solution (Section 4.04.2.1.3), the N-2 atom is the more basic and the more nucleophilic (Scheme 5).

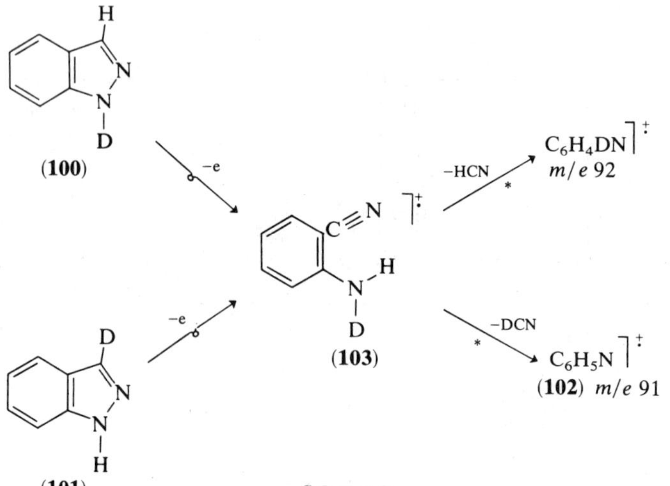

Scheme 3

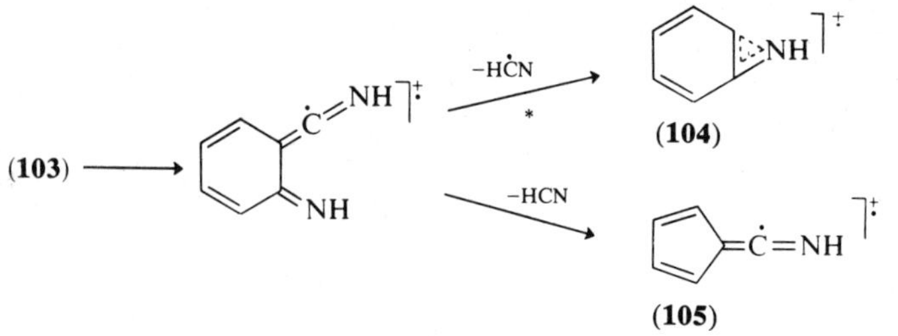

Scheme 4

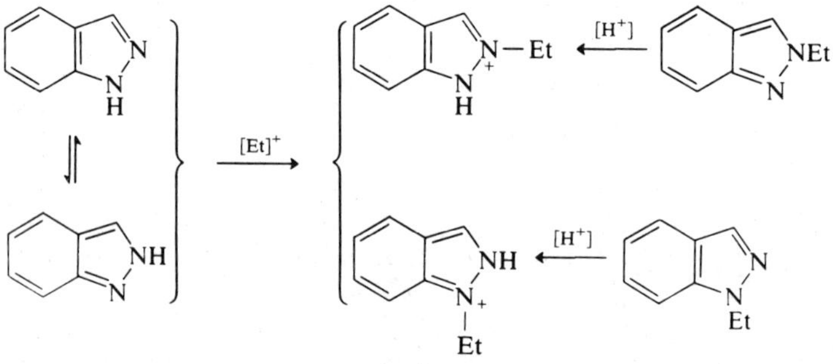

Scheme 5

Substituted pyrazoles fragment following pathways that are strongly dependent on the nature of the substituent. Thus N- and C-phenylpyrazoles lead to N-phenylaziridinium, -cyclopropenium and benzodiazepinium ions ⟨68ZOR689, 78CHE1123⟩, C-diphenylpyrazoles to the fluorenium ion (m/e 165) ⟨69OMS(2)739⟩, and 1-phenylpyrazol-4-yl oximes to the 1-phenylpyrazolium ion (m/e 144) ⟨69JCS(C)2497⟩.

The behaviour under electron impact of N- and C-trimethylsilylpyrazoles (mono-, di- and tri-substituted) has been studied by Birkofer et al. ⟨74OMS(8)347⟩. Loss of a methyl radical followed by loss of HCN is the most common fragmentation feature of these compounds. When more than one trimethylsilyl group is present, a neutral fragment C_3H_8Si is expelled. Mass spectrometry of pyrazolium salts has been studied by Larsen et al. ⟨81OMS377, 83OMS52⟩.

Amongst the non-aromatic derivatives, 3,4,4,5-tetramethylisopyrazole has a molecular ion of low intensity (7%) in contrast to aromatic pyrazoles. It loses one of the 4-methyl groups leading to a pyrazolium ion and after than the fragmentation is similar to that of trimethylpyrazole ⟨67ZOR1540⟩. This tendency towards aromatic structures is also found in the mass spectrometry of 2-pyrazolines ⟨67JCS(B)583, 69OMS(2)729, 73KGS64, 76BSF869⟩. In the last paper, Aubagnac et al. have proposed a retro-dipolar cycloaddition to explain the formation of ethylene from 1,3-disubstituted pyrazolines (Scheme 6), an observation previously noted by Hammerum ⟨72JOC3965⟩.

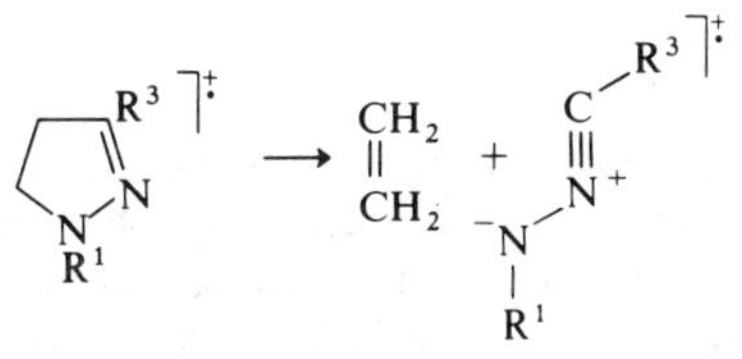

Scheme 6

Finally, Δ^1-pyrazolines lose N_2 ⟨73KGS64⟩ but it is possible that this takes place thermally before ionization, since thermolysis of Δ^1-pyrazolines usually gives cyclopropanes (Section 4.04.2.2.2(iv)).

4.04.1.3.9 Photoelectron spectroscopy

Results on the PE spectroscopy of pyrazoles are few. After a preliminary paper by Baker et al. ⟨70CC286⟩ in which the lowest ionization energies of pyrazole are given (9.5, 10.1 and 10.8 eV) and briefly discussed, the fundamental contribution of Palmer et al. appeared ⟨73T2173⟩. The PE spectrum of pyrazole in the gas phase using a He-I source consists of a series of bands belonging to three regions, A, B and C (Table 23). Only region A shows a resolved vibrational structure, probably involving an in-plane deformation mode (660 cm^{-1}), a 'breathing vibration' (900 cm^{-1}) and a symmetrical skeletal stretching (1050 cm^{-1}). However, the complete assignment of the vibrational spectra of pyrazole (Table 22; Section 4.04.1.3.7) proves that these bands must be assigned to ν_{19} (a″), ν_{15} (a′) and ν_{13} (a′), respectively.

Table 23 Experimental and Calculated Vertical Ionization Energies (eV) in Pyrazole

Regions	Experimental values	Assigned levels	Eigenvalue	Principal character	Centres/bond orbitals
A	9.15	3a″	−11.32	$2p_z$	N(1)N(2)–C(2)C(3)
	9.88	2a″	−13.01	$2p_z$	N(2)C(3)C(4)–N(1)C(5)
	10.7	15a′	−13.84	sp^2	N(2)
B	13.6	{14a′	−17.58	$2p, 1s_H$	C(4)C(5)–C(3)C(4)
		{13a′	−18.59	$2p, 1s_H$	C(4)H–C(5)H
	14.7	1a″	−19.15	$2p_z$	N(1), C
	15.1	12a′	−19.24	$2p, 1s_H$	NH–C(3)H
C	17.5	11a′	−22.79	$2p, 1s_H$	C(5)N(1), C(4)H + C(5)H

The assignment of bands has been carried out using ab initio calculations, unfortunately using the erroneous Ehrlich geometry (Section 4.04.1.3.1). There is a linear relationship between the calculated energy levels (eigenvalues) and the experimental ones: $IE_{exp} = 0.37 + 0.75 IE_{calc}$, (c.c.)$^2 = 0.994$.

The same group has published two articles on indazoles [indazole (**36**), 1-methyl- (**106**) and 2-methyl-indazole (**107**)]. The He-I and He-II spectra have been obtained and satisfactorily interpreted by means of ab initio LCGO calculations ⟨78JST(43)33, 78JST(43)203⟩. The PE spectra support the conclusion (Section 4.04.1.5.1) that the $1H$-tautomer is by far the more stable. X-Ray photoelectron spectra of some pyrazole and pyrazoline derivatives have been reported ⟨83MI40400⟩.

Experimental and calculated (CNDO/S) vertical ionization energies have been measured for pyrazol-3-ine-5-thiones (**108**; R = H, Me). These compounds exhibit an intense low-energy band (7.55–7.60 eV) corresponding to the ionization of both a thiocarbonyl π-electron and the sulfur n electron ⟨78JA1275⟩.

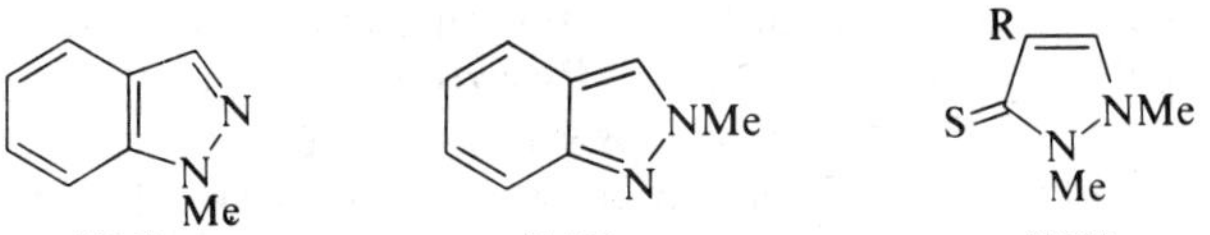

(106)   (107)   (108)

4.04.1.3.10 ESR spectroscopy

ESR spectroscopy of heterocyclic radicals has been reviewed ⟨74PMH(6)95⟩ and two publications are briefly reported in the appendix to this reference. Kasai and McLeod ⟨73JA27⟩ generated radical anions in argon matrices at 4 K by reaction of the parent heterocycles with sodium atoms. Subsequently the matrix was irradiated with 'yellow' light ($\lambda > 5500$ Å). The photoinduced radical anion of pyrazole shows an ESR spectrum with the characteristic triplet-of-doublets feature; this excludes a π-type radical anion possessing the original structure. The authors assigned the spectrum to a non-aromatic tautomeric form (**109**). The triplet feature with a spacing of ~50 G is attributed to the two protons at C-5, and the doublet with a spacing of ~15 G to proton 4. INDO-calculated values for the two isotropic hyperfine coupling constants are 52 and 9 G, respectively. The calculated coupling constant with H-3 is 1 G. The radical formed by the reaction of OH· with pyrazole has the structure (**110**) according to the ESR spectrum (hyperfine constants (G): H-3, H-1, 1.5; H-4, 10; H-5, 31) ⟨73JPC1629⟩.

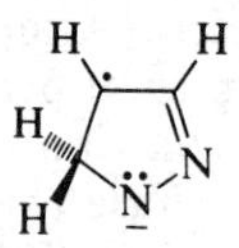

(109) 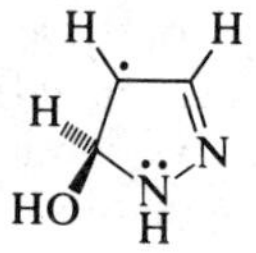   (110)

Janssen and co-workers prepared t-butyl 1-pyrazolecarboxylate (**111**) via 1-pyrazolecarbonyl chloride ⟨75JOC915⟩. The thermolysis of this compound in benzene solution at ~140 °C led to 1-phenylpyrazole. The N-pyrazolyl radical (**112**), which is proposed as an intermediate, proved elusive and neither CIDNP nor ESR signals were observed.

Since the authors did not succeed in obtaining an ESR spectrum, they were unable to decide whether the N-pyrazolyl radical is of the σ (**112a**) or the π (**112b**) type. *Ab initio* calculations indicate that the radical has B_1 (π) symmetry ⟨76T1555⟩. However, the radical is formed from (**111**) as a σ radical and is able to react as such in its lifetime. This is in agreement with the experimental results ⟨75JOC915⟩, no C-phenylated pyrazoles being detected.

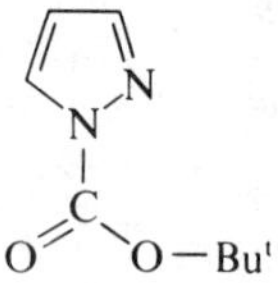

(111) 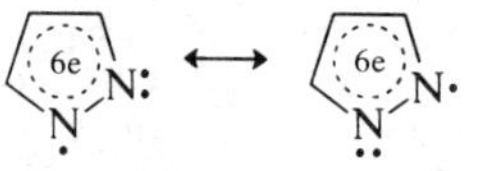   (112a)    (112b)

Solutions of colourless pyrazolidinediones give highly coloured solutions of radicals (**113**) when treated with lead dioxide. The ESR spectra of these radicals have been recorded ⟨78JOC808⟩. They dimerize to tetrazenes (**114**) which appear to be indefinitely stable.

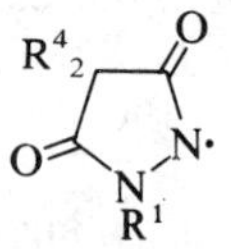

(113) 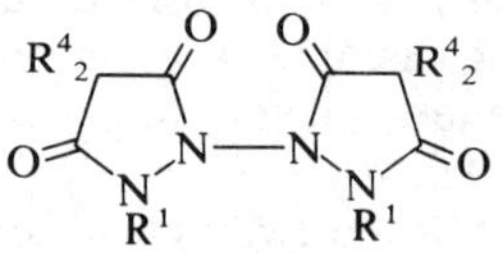   (114)

4.04.1.4 Thermodynamic Aspects

4.04.1.4.1 Intermolecular forces

(i) *Melting and boiling points*

A large collection of melting and boiling points of pyrazoles may be found in ⟨67HC(22)1, B-76MI40402⟩; for pyrazolines, pyrazolidines and indazoles see ⟨67HC(22)1⟩, and for pyrazolones and pyrazolidones see ⟨64HC(20)1⟩. Pyrazole itself melts at 69–70 °C; it is very stable to heat and can be distilled at atmospheric pressure without decomposition (b.p. 186–188 °C at 757.9 mmHg, 96 °C at 16 mmHg). The general conclusions (Section

4.01.4.1.1) still hold for pyrazoles; for instance, when the hydrogen bonds are suppressed (N- or O-methylation) the boiling (melting) point decreases. A point worthy of note is that 3(5)-substituted pyrazoles are mixtures of tautomers and the boiling point may possibly correspond to an azeotrope. Melting and boiling points for some monosubstituted pyrazoles are collected in Table 24.

Table 24 Melting and Boiling Points (°C) of Monosubstituted Pyrazoles[a,b]

	Me	Et	Bu^t	Ph	CHO	COMe	CO_2H	CO_2Me
1-Substituted	127/760	137/760	80/30	246/760	(c?), e	234/760	**103**[d]	35
4-Substituted	207/760	122/16	e	**230**	**83**	**114**	**278**	**137**
3(5)-Substituted	205/760	209/760	**53**	**78**	**150**	**101**	**214**[d]	**141**

	CO_2Et	$CONH_2$	CN	NH_2	N_3	NO	NO_2	OH
1-Substituted	213/760	**141**	37	c	(c?), e	(c?), e	**93**	e
4-Substituted	**79**	e	**92**	**81**	e	**234**	**164**	**118**
3(5)-Substituted	**160**	**148**	**150**	**40**	**56**	**161**	**175**	**166**

	OMe	SEt	SO_3H	F	Cl	Br	I	$SiMe_3$
1-Substituted	e	e	e	(c?), e	(c?), e	(c?), e	e	152/760
4-Substituted	**62**	e	**345**	**38**	**77**	**97**	**109**	**64**
3(5)-Substituted	e	103/0.5	**257**	78/12	**40**	**70**	**73**	**80**

[a] Melting points above 30 °C are given in **bold**; melting points below 30 °C are not included. [b] Boiling points are given in °C/mmHg. [c] Unstable. [d] Decomposes. [e] Unknown.

The melting point of indazole has been reported at various temperatures in the range of 145 to 149 °C. Indazole boils at 269–270 °C at 743 mmHg and 146 °C at 15 mmHg, can be readily sublimed on a water bath and is also steam-volatile. Its two N-methyl derivatives have similar melting points: 1-methyl (**106**), m.p. 61 °C, 2-methyl (**107**), m.p. 56 °C, but they are easily separated by fractional distillation [(**106**), 120/15; (**107**), 141/15].

(ii) Solubility of pyrazoles and indazoles

From the data reported in ⟨63PMH(1)177⟩ it was concluded that hydrophobic substituents reduce the solubility of pyrazole in water (at 20 °C: pyrazole, 1 part in 2.5; 3,5-dimethylpyrazole, 1 part in 52). Another determination gives the following values for the solubilities of pyrazole at 25 °C in water, benzene and cyclohexane (expressed as g/100 g of solvent): 130, 18 and 3 ⟨66AHC(6)347⟩. Indazole is soluble in hot water and most organic solvents, but less so in cold water.

The partition coefficient P, defined as the equilibrium concentration of the compound in n-octanol divided by that in the aqueous phase, has been measured for pyrazole and indazole ⟨B-79MI40416⟩. It was found that $\log P = 0.13$–0.26 for pyrazole and 1.82 for indazole, clearly showing the greater hydrophobicity (lipophilicity) of the indazole ring, due to the benzenoid moiety.

(iii) Chromatography

In addition to the two references, one dealing with pyrazoles and the other with pyrazolines, given in ⟨71PMH(3)297⟩, other references describing the gas chromatography (GC) of pyrazolines, pyrazolidines and pyrazolidones can be found in ⟨66T2461, 67BSF3502, 69BSF3306, 70BSF1129⟩. It has been shown ⟨69BSF3306⟩ that Δ^3-pyrazolines are partially pyrolyzed on the column (an SE 30 silicone) to pyrazoles. The proportion of pyrazoles increases when the column temperature rises and the flow diminishes.

An important publication by Kost *et al.* ⟨63JGU525⟩ on thin-layer chromatography (TLC) of pyrazoles contains a large collection of R_f values for 1:1 mixtures of petroleum ether–chloroform or benzene–chloroform as eluents and alumina as stationary phase. 1,3- and 1,5-disubstituted pyrazoles can be separated and identified by TLC (R_f 1,3 > R_f 1,5). For another publication by the same authors on the chromatographic separation of the aminopyrazoles, see ⟨63JGU2519⟩. N-Unsubstituted pyrazoles move with difficulty and it is necessary to add acetone or methanol to the eluent mixture. Other convenient conditions for NH pyrazoles utilize silica gel and ethyl acetate saturated with water (a pentacyanoamine ferroate ammonium disodium salt solution can be used to visualize the pyrazoles).

4.04.1.4.2 Stability and stabilization

(i) Thermochemistry

Two now classical papers ⟨61MI40400, 62JCS2927⟩ contain almost all the experimental thermochemical data on pyrazole and indazole. Heats of combustion determined by Zimmerman ⟨61MI40400⟩ have been used by Dewar to calculate the heats of atomization (Table 25) ⟨69JA796⟩. Quantum mechanical calculations, carried out by Dewar ⟨69JA796⟩ or Olivella ⟨81JHC1189⟩, gave accurate empirical values.

Table 25 Experimental and Calculated Thermochemical Data of Pyrazole and Indazole (kJ mol^{-1})

Compound	Heat of combustion (crystal)^a	Heat of combustion (gas)^a	−(Heat of atomization) exp.^{b}	−(Heat of atomization) calc.^{b}	Heat of formation exp.^{c}	Heat of formation calc.^{d}
Pyrazole	1858.5	1926.3	3632.4	3638.1	181 ± 9	190.0
Indazole	3763.5	3860.6	6721.6	6744.0	—	—

^{a} ⟨61MI40400⟩. ^{b} ⟨69JA796⟩. ^{c} ⟨62JCS2927⟩. ^{d} ⟨81JHC1189⟩.

(ii) Aromaticity

Since aromaticity is, at best, a relative value, the problem of the aromaticity of pyrazole, compared to other azoles, is to be found in Section 4.01.1.2, in which the authoritative review by Cook et al. ⟨74AHC(17)255⟩ is summarized.

From a qualitative viewpoint there is no doubt that the compounds classified as aromatic in Section 4.04.1.1 (pyrazoles, indazoles, isoindazoles, pyrazolones, indazolones, mesoionic derivatives) indeed have aromatic properties to a greater or lesser extent. Dihydrobenz[cd]indazole (**115**) is also formally an aromatic compound since it has an odd number of electron pairs ($n = 7$). However, due to the presence of two adjacent nitrogen atoms the 'aromatic' tautomer (**115a**) is not stable, and the compound exists as the 1,5- or 1,3-dihydro tautomers (**115b**) or (**115c**) ⟨72JCS(P2)68⟩. The 'antiaromatic' benz[cd]indazole (**116**), although unstable, has been fully characterized by ^{15}N, ^{1}H and ^{13}C NMR spectroscopy ⟨82CC86⟩

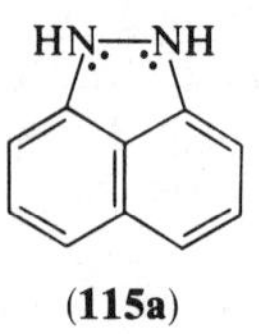

(**115a**) 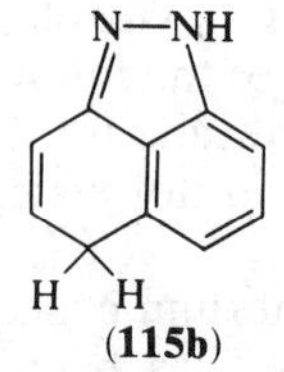 (**115b**) 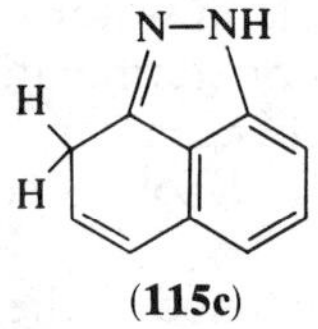 (**115c**) 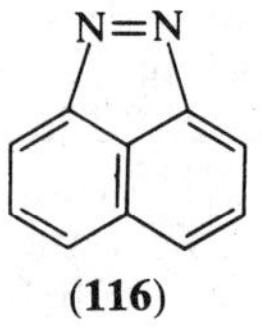 (**116**)

4.04.1.4.3 Conformation and configuration

Considerable information is available on this topic which will be discussed according to the classification shown in Scheme 7.

Aromatic compounds (planar)
- Isomerism about a formal Csp^2—Csp^2 double bond
- Isomerism about a formal sp^2–sp^2 single bond
  - C—C: C-aryl and C-acyl derivatives
  - C—N: N-aryl and N-acyl derivatives
- Conformation about a Csp—Csp^3 single bond

Non-aromatic compounds (non-planar)
- Conformation of the ring
- Nitrogen inversion

Scheme 7

The configuration of pairs of isomeric 4-arylidene-5-pyrazolones, (Z)- and (E)-(**117**), was determined by ^{1}H NMR data ⟨72G491⟩. When R^3 is H, the E configuration is preferred; when it is a methyl or a phenyl group, the Z configuration predominates. The presence of an exocyclic sulfur atom as in (**118**) lowers the interconversion barrier and the products are equilibrium mixtures in solution at room temperature (an X-ray determination shows that (Z)- and (E)-(**118**; R^3 = Ph, R = Bu^t) are present in the crystal in an 87:13 ratio) ⟨76BSB697⟩.

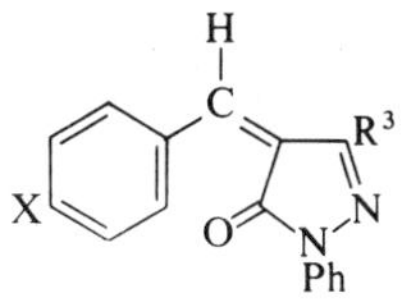

(Z)-(**117**) 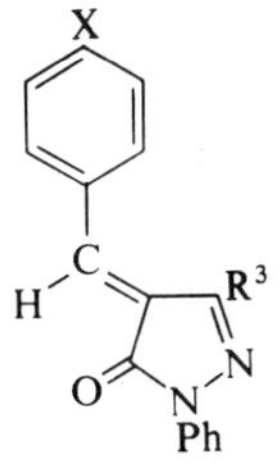   (E)-(**117**) 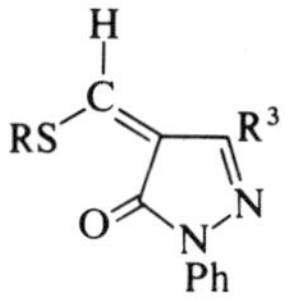   (Z)-(**118**) 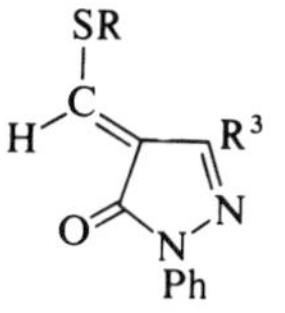   (E)-(**118**)

Another example of the influence of a heteroatom on the barrier to rotation about a formal C—C double bond is provided by Mannschreck ⟨B-75MI40404⟩. In compound (**119**) the rotation about the C—N bond (process 1) is more hindered than the rotation about the C=C bond (process 2), the free energies of activation being 89.1 and 80.3 kJ mol^{-1}, respectively. Other activation energies are given in Figure 19. Compound (**120**) exists in the Z configuration, as it has been represented in Figure 19. The effect of the bulky substituents in (**121b**) on the barrier results in an increase of the dihedral angle between the pyrazole ring and the thiocarbamoyl group, thus reducing their resonance interaction. For compound (**122**) Sandström has calculated the entropy and enthalpy of activation using a very accurate bandshape analysis.

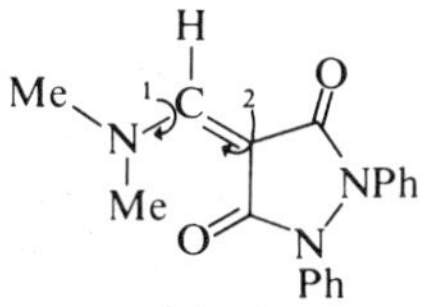

(**119**) $\Delta G_c^{\ddagger}$ = 89.1
⟨B-75MI40404⟩

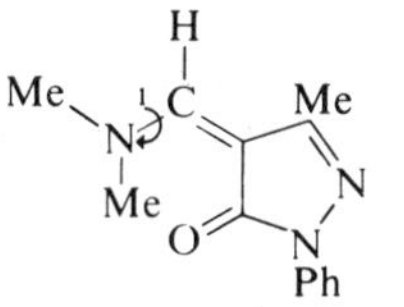

(**120**) $\Delta G_c^{\ddagger}$ = 77.8
⟨77JPR911⟩

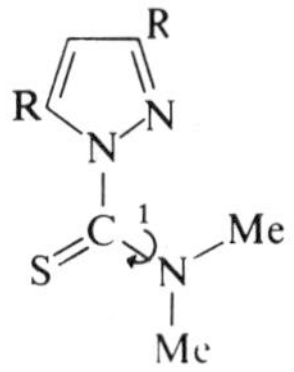

(**121a**) R = H, $\Delta G_c^{\ddagger}$ = 64.0
(**121b**) R = Bu^t, $\Delta G_c^{\ddagger}$ = 88.7
⟨72ACS21⟩

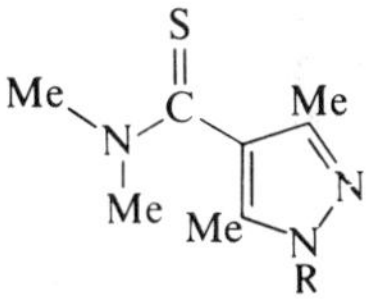

(**122**) R = Me, $\Delta G_c^{\ddagger}$ = 72.0
⟨77OMR(9)1⟩

Figure 19 Barriers to internal rotation (in kJ mol^{-1}) for pyrazole derivatives

In addition to the dynamic study of compound (**120**), Freyer ⟨77JPR895⟩ has used ^{15}N-labelled compounds (^{1}H–^{15}N coupling constants) to assign the configuration of compounds (**123**) and (**124**). For the latter, a $^1J(^1\text{H}-^{15}\text{N})$ coupling constant of 92 Hz establishes the tautomeric structure as (**124**) ⟨78JPR508⟩.

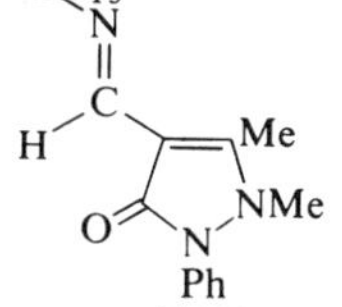

(**123**) 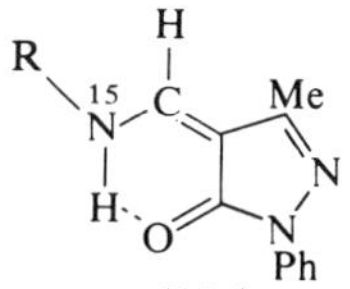   (**124**)

From studies reported in the references in Table 5 (Section 4.04.1.3.1) the dihedral angle between a phenyl and a pyrazole ring in the crystalline state, falls between 4° and 22° when the phenyl group is in the 3- or 4-position. The planar conformation of C-formylpyrazoles (**57**) and the resonance interaction between them (**87**) has already been discussed in connection with ^{1}H NMR (Section 4.04.1.3.3(i)) and IR studies (Section 4.04.1.3.7(iii)).

Considerable effort, with both N-aryl and N-acyl derivatives, has been devoted to studying the rotational isomerism and equilibrium dihedral angles of N-substituted azoles. The results obtained with N-acylazoles and azolides (Sections 4.04.1.3.1 and 4.04.1.3.3(i)) are summarized in ⟨81RCR336⟩. N-Acetyl-pyrazole (**41**), -indazole (**60**) and -isoindazole (**61**) are planar or almost planar compounds with an E configuration. As in other azoles, the predominant structure is that in which the carbonyl oxygen atom and the pyridine-type peripheral nitrogen atom are as far apart as possible. If a pyrrolic (NH) nitrogen atom is present in the α-position, as in compound (**125**) (see Table 5, Section 4.04.1.3.1), then the hydrogen-bonded Z-isomer is found in the crystal ⟨69AX(B)2355⟩.

(125)

The atropoisomerism of N-phenylazoles, *i.e.* an isomerism of the type shown by biphenyl derivatives, has been studied by several authors. Some torsional angles, measured by X-ray crystallography (Section 4.04.1.3.1) are shown in Table 26.

Table 26 Dihedral Angles Between the Pyrazole and the Phenyl Ring in N-Phenylpyrazoles Determined by X-Ray Crystallography (for References see Table 5)

Compound	2-Substituent	5-Substituent	θ (°)
4-Bromo-3-methyl-1-*p*-nitrophenylpyrazole	—	H	20
5-Hydroxy-3-methyl-1-phenylpyrazole (**78b**, R^1 = Ph)	—	OH	38[a]
3-Methyl-1-phenyl-Δ^3-pyrazolin-5-one (**78**, R^1 = Ph)	H	=O	23[a]
3-Hydroxy-5-methyl-1-phenylpyrazole	—	Me	33–44[b]
5-Amino-4-cyano-3-dimethylamino-1-phenylpyrazole	—	NH$_2$	60
3,5-Dimethyl-1-phenylpyrazolium nitrate	H$^+$	Me	48[c]

[a] Calculated torsional angles of minimum energy (method HMO–NBI), 15 and 20°, respectively ⟨75MI40401⟩. [b] Two independent molecules in the unit cell. [c] The calculated dihedral angle corresponding to the conformation of minimum intramolecular van der Waals energy was 60° ⟨82CJC97⟩.

Theoretical calculations on the conformation of 1-phenylpyrazole (**126**) and 5-methyl-1-phenylpyrazole (**127**) (extended Hückel method) have been compared with experimental (including X-ray) values for the dihedral angle of minimum energy ⟨78T1139⟩. The authors conclude that values of θ of 26 and 55°, respectively, are a reasonable estimation for the molecule in solution. Fong has calculated the dihedral angles of a large number of azole derivatives, including many N-phenylpyrazoles ⟨80AJC1763⟩. For the 5-methyl derivative, a value of θ of 53° was found, but the author assumed that compounds (**126**) and (**128**) are planar, a highly improbable hypothesis. The rates of quaternization of N-arylpyrazoles have been discussed on the basis that steric interactions are more important in the products (**128**) than in the reactants (**126**) (phenyl substituted in the 3'- and 4'-positions) and result in twisting of the aryl ring away from coplanarity with the pyrazole nucleus ⟨75AJC1861⟩.

(126)   (127)   (128)

The *syn–anti* conformational problem of α- and β-pyrazofurins (**756**; one of the rare naturally occurring pyrazole compounds, see Section 4.04.4.4.3), which involves a rotation around a pyrazolic sp^2 carbon atom and a sugar sp^3 carbon atom, has been studied theoretically using the PCILO method ⟨81MI40403⟩. In agreement with the experimental observations, the β anomer is energetically more favourable than the α anomer, the preferred conformations being *anti* and *syn*, respectively.

The conformation of heteroethylenic rings, Δ^1- and Δ^2-pyrazolines, and the nitrogen inversion and preferred orientation of the N-methyl groups in pyrazolidines ⟨71T123⟩ have been discussed in connection with their ^{1}H NMR spectra (Section 4.04.1.3.3).

4.04.1.5 Tautomerism

After the publication of a book on the prototropic tautomerism of aromatic heterocycles ⟨76AHC(S1)1⟩ which covered the literature up to 1975, the study of the tautomerism of pyrazoles has not made great strides. In this section the main conclusions of this earlier review will be summarized and comments on a few recent and significant references added.

An account of non-prototropic tautomerism (mainly annular metallotropy) and non-aromatic functional tautomerism will also be included.

From a general point of view, the tautomeric studies can be divided into 12 areas (Figure 20) depending on the migrating entity (proton or other groups, alkyl, acyl, metals...), the physical state of the study (solid, solution or gas phase) and the thermodynamic (equilibrium constants) or the kinetic (isomerization rates) approach.

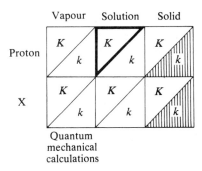

Figure 20 The 12 areas of tautomeric studies

About 90% of all the studies on tautomerism refer to the determination of prototropic equilibrium constants in solution. Most probably in the next years there will be further studies in the remaining areas, some of them almost a desert at the present time.

4.04.1.5.1 Annular tautomerism

All the N-unsubstituted pyrazoles (**129**) in solution (and probably in the gas phase) are mixtures of annular tautomers in different proportions, depending on the nature of the substituents R^1 and R^2. In the majority of cases the difference of free energy between both tautomers is low enough for the chemical reactivity to be unrelated to the equilibrium constant.

$$K_T = [(\mathbf{129a})]/[(\mathbf{129b})]$$

(**129a**)   (**129b**)

When $R^2 = H$, in all the known examples, the 3-substituted tautomer (**129a**) predominates, with the possible exception of 3(5)-methylpyrazole ($R^1 = Me$, $R^2 = H$) in which the 5-methyl tautomer slightly predominates in HMPT solution at $-17\,°C$ (54%) ⟨77JOC659⟩ (Section 4.04.1.3.4). For the general case when $R^2 =$ or $\neq H$, a dependence of the form $\log K_T = a \sum_{3,5} \sigma_I + b \sum_{3,5} \sigma_R$, with $a > 0$, $b < 0$ and $a > |b|$, has been proposed for solutions in dipolar aprotic solvents ⟨79OMR(12)587⟩. The equation predicts that the 5-trimethylsilyl tautomer is more stable than the 3-trimethylsilylpyrazole, since σ_I and σ_R for Me_3Si are -0.12 and 0.07, but more experimental work has to be done to understand the influence of the substituents on the equilibrium constant which is solvent dependent ⟨78T2259⟩. There is no problem with indazole since the $1H$ tautomer is always the more stable ⟨83H(20)1713⟩.

The observation of individual tautomers by ^{13}C and ^{15}N NMR spectroscopy (Sections 4.04.1.3.4 and 4.04.1.3.5) avoids the complication of interpolation methods. As line intensities in ^{15}N NMR spectra are strongly influenced by spin–lattice and spin–spin relaxation times, relaxation mechanism and experimental conditions, special care has to be taken in using ^{15}N spectra for quantitative purposes ⟨80OMR(14)129⟩. However, interpolation methods continue to be used, sometimes without sufficient attention to the implicit assumptions. For example, ⟨75JOC1815⟩ contains a study on the tautomerism of pyrazoles. Using averaged ^{13}C chemical shifts without taking into account any corrective term (effect of the N-methyl group on the chemical shifts), these authors calculated some tautomer predominances that are in contradiction with reliable literature results. From other authors' data ⟨74JOC357⟩ they concluded that the 3-methylpyrazole tautomer (**129a**; $R^1 = Me$, $R^2 = H$) is definitely the preferred one.

The 'magic spinning angle' (MAS) technique gives the possibility of recording high resolution NMR spectra of nuclei other than the proton, thus extending the study of tautomerism in the solid state to non-crystalline solids (Section 4.04.1.3.4). Since the method gives clear results in the case of pyrazole itself, it would be interesting to reexamine those X-ray studies in which the tautomeric proton seems non-localized (Section 4.04.1.3.1), as in the case of 5(3)-methyl-3(5)-phenylpyrazole (**130**) ⟨74JCS(P2)1298⟩.

A large programme utilizing temperature-jump relaxation methods for the study of tautomerism in aqueous solution has led the Dubois group to determine the kinetic and thermodynamic parameters of the equilibrium (**130a**) ⇌ (**130b**) ⟨78T2259⟩. The tautomeric interconversion involves two successive intermolecular proton transfers.

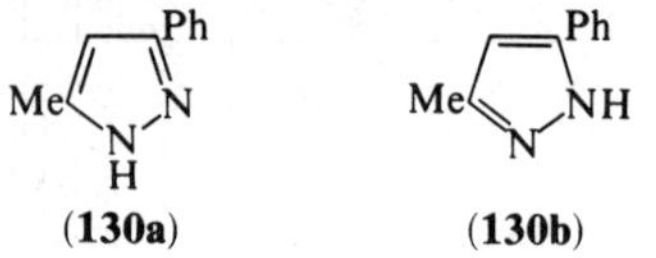

(**130a**)   (**130b**)

A well-known example of non-prototropic tautomerism is that of azolides (acylotropy). The acyl group migrates between the different heteroatoms and the most stable isomer (annular or functional) is obtained after equilibration. In indazoles both isomers are formed, but 2-acyl derivatives readily isomerize to the 1-substituted isomer. The first order kinetics of this isomerization have been studied by NMR spectroscopy ⟨74TL4421⟩. The same publication described an experiment (Scheme 8) that demonstrated the intermolecular character of the process, which has been called a dissociation–recombination process.

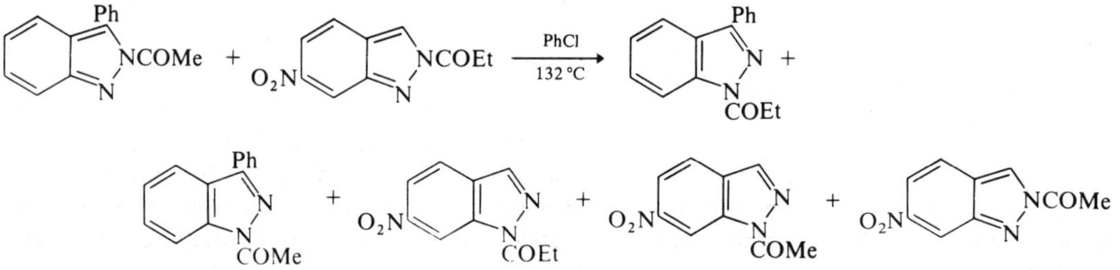

Scheme 8

2-Acylindazoles are good acylating agents, particularly the 2-acetyl-7-nitro derivative (**131**). The acetylation of pyrazole by this compound proceeds quickly in chloroform and more slowly (about 15 min) in DMSO ⟨66BSF3041⟩.

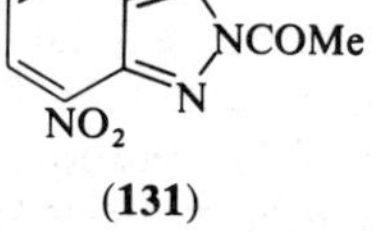

(**131**)

Although the thermodynamic aspects of acylotropy are well documented, there have been few kinetic studies of the process. The activation barrier is much higher than for prototropy and only Castells et al. ⟨72CC709⟩ have succeeded in observing a coalescence phenomenon in ^{1}H NMR spectra. At 215 °C in 1-chloronaphthalene the methyl groups of N-phenyl-3,5-dimethylpyrazole-1-carboxamide coalesce. The mechanism of dissociation–combination explains the reversible evolution of the spectra (Scheme 9).

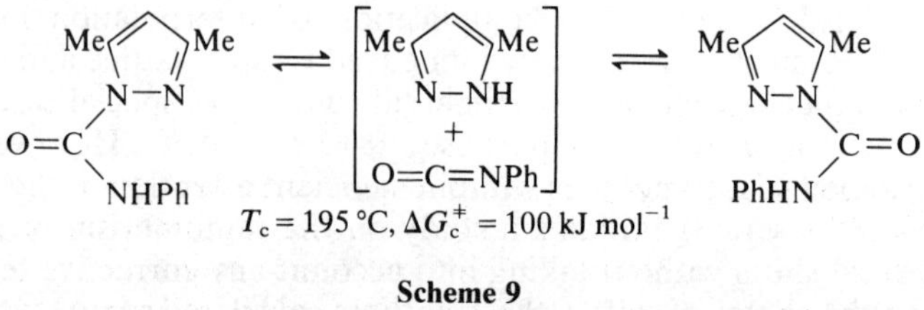

$T_c = 195\,°C$, $\Delta G_c^{\ddagger} = 100\,\text{kJ mol}^{-1}$

Scheme 9

Many other groups migrate with activation energies low enough to determine equilibrium constants. In the absence of strong steric and hydrogen-bond effects the position of the equilibrium is independent of the nature of the N-substituent ⟨78JCS(P2)99⟩. Groups other than protons are sensitive to steric effects; for this reason, the major tautomers of 7-nitroindazole are the 1H and the 2-acetyl (**131**). Table 27 includes several activation energies for pyrazolic annular tautomerism.

Table 27 Kinetic Data for Autotrope Rearrangements in Pyrazoles

NR^1 pyrazoles	R^1	NMR	Coalescence temperature (°C)	Activation energy (kJ mol^{-1})	Ref.
Unsubstituted	H	^{13}C	−100	46[a]	75T1461
Unsubstituted	H	^{13}C	47	63[b]	77JOC659
Unsubstituted	H	^{13}C	64	62[c]	79JA545
3,5-Dimethyl	Me$_3$Ge	^{1}H	95–100	83	72MI40400
3,5-Dimethyl	Me$_3$Ge	^{13}C	—	84	74JOM(70)347
Unsubstituted	Me$_3$Ge	^{1}H	150–160	92	72MI40400
3,5-Dimethyl	Me$_3$Si	^{1}H	104	117	71JOM(27)185
3,5-Dimethyl	Me$_3$Si	^{1}H, ^{13}C	—	96	74JOM(70)347
3,5-Dimethyl	Et$_3$Si	^{1}H, ^{13}C	—	96	74JOM(70)347
3,5-Dimethyl	(EtO)$_3$Si	^{1}H, ^{13}C	—	106	74JOM(70)347
Unsubstituted	Me$_3$Si	^{1}H	165	~134	71JOM(27)185
3,4,5-Trimethyl	Me$_3$Si	^{1}H	90	100	71JOM(27)185
3,5-Bis(trifluoromethyl)	Me$_3$Si	^{1}H	170	—	71JOM(27)185
Unsubstituted	Me$_2$Ga	^{1}H	<30	—	81JOM(215)157
3,5-Dimethyl-4-R	PhHg	^{1}H	~−110	—	71MI40400
Unsubstituted	Bu$_3$Sn	^{1}H, ^{13}C	≪−80	—	77JOM(132)69
Unsubstituted	(CO)$_2$ClRh	^{1}H	low temp.	—	74JOM(65)C51
3,5-Dimethyl	(CO)$_2$ClRh	^{1}H	low temp.	—	74JOM(65)C51
3,5-Dimethyl	(Pr^{i}TPP)(CO)Ru [d]	^{1}H	100	71	72JOM(39)179

[a] In a mixture of ether and tetrahydrofuran. [b] In HMPT the same barrier is obtained for the 3,5-dimethylpyrazole. [c] In DMSO. [d] Pr^{i}TPP = tetrakis(p-isopropylphenyl)porphinate dianion.

Dipolar aprotic solvents raise dramatically the coalescence temperature (other factors are also important; Section 4.04.1.3.4) but the phenomenon is insensitive to steric effects (no differences between pyrazole and 3,5-dimethylpyrazole). The metallotropy activation energies are independent of the solvent and the concentration but are sensitive to steric and to electronic effects of substituents in 3,5-positions. It has been concluded that metallotropy, contrary to prototropy and acylotropy, follows an intramolecular mechanism with a cyclic symmetric transition state (**132**). The migratory aptitude of R^1 in pyrazoles is in the following order: Me < COR < SiR$_3$ < GeR$_3$ < H ≃ HgR < SnR$_3$. For the tin derivatives the transition state is probably pentacoordinated and this aptitude to expand the valency stabilizes the transition state (reactional intermediate?). In the case of the rhodium derivative (**133**), there is a concomitant prototropy and metallotropy of the ligand (**133a**) ⇌ (**133b**). In another fluxional process involving ruthenium instead of rhodium, it has been shown that the rate-controlling step is the complex dissociation and that the ligand exchanges between the two annular nitrogen atoms by an intermolecular process.

4.04.1.5.2 Substituent tautomerism

Together with pyridones, the tautomerism of pyrazolones has been studied most intensely and serves as a model for other work on tautomerism ⟨76AHC(S1)1⟩. 1-Substituted pyrazolin-5-ones (**78**) can exist in three tautomeric forms, classically known as CH (**78a**), OH (**78b**) and NH (**78c**). In the vapour phase the CH tautomer predominates and in the solid state there is a strongly H-bonded 'mixture' of OH and HN tautomers (Section 4.04.1.3.1). However, most studies of the tautomerism of pyrazolones correspond to the determination of equilibrium constants in solution (see Figure 20).

Non-polar solvents resemble the vapour phase and the CH tautomer is favoured. In dipolar aprotic solvents the NH form changes into the OH form; the dielectric constant and the dipole moment of the solvent do not play an important role in this transformation. In protic solvents the amount of the CH form is low, but as the acidity of the solvent increases, the percentage of the NH form present increases to the detriment of that of the OH form. These qualitative conclusions can be converted into quantitative terms by applying the scaled Palm–Koppel equation ($\log K = \log K_0 + yY + pP + eE + bB$, K_0, Y, P, E and B being the equilibrium constant in the vapour phase, the polarity, the polarizability, the acidity and the basicity of the solvent, respectively) to the percentage of the CH tautomer of 1,3-dimethylpyrazolin-5-one in 16 solvents. The following result was obtained: $\log K_0 = 2.2 \pm 0.2$, $e = -1.4 \pm 0.2$, $b = -1.9 \pm 0.1$, y and p not significant. The value of $\log K_0$, 2.2 ($K_0 = 160$), corresponds to the predominance of the CH tautomer in the vapour phase. Only the acidity E and the basicity B of the solvent are important, and as they increase the percentage of the CH form decreases ⟨75BSB1189⟩. A ^{13}C NMR study of 26 4-substituted 3-methyl-1-phenylpyrazolin-5-ones ⟨81JPR188⟩ shows that in non-polar solvents like CDCl$_3$, where the CH form is preferred at room temperature, cooling shifts the tautomeric equilibrium to yield a higher amount of the NH form.

The effect of the substituent in the 1-position on the tautomeric equilibrium has received little attention. However, it is known that the change from the N-Me to the N-Ph series increases the percentage of the CH form to the detriment of the NH form. The effect of the substituent in the 3-position has been thoroughly studied and the percentage of the CH form decreases in the order: $NH_2 > OEt > Me > Ph > H > CO_2Et$. The study of the effect of the 4-substituents shows that the percentage of the CH form decreases as the electronegativity of R^4 increases ⟨81JPR188⟩. The fact that 4-fluoro-5-hydroxypyrazole (**134**) exists as such in DMSO (^{1}H and ^{19}F NMR data) has been attributed to hydrogen bond formation between the hydroxy group and the fluorine atom ⟨81BCJ3221⟩. However, the high electronegativity of the fluorine atom is probably sufficient to justify the observed tautomer.

Hydrogen bonding plays a major role in pyrazolone tautomerism, and the formation of a chelate structure can shift the equilibrium towards the chelated form. Structures (**135**) and (**136**) are two representative examples of such stabilized tautomers. Structure (**137**) is a hypothetical example of stabilization of the NH tautomer.

(**134**)   (**135**)   (**136**)   (**137**)

4-Acyl-, 4-alkoxycarbonyl- and 4-phenylazo-pyrazolin-5-ones present the possibility of a fourth tautomer with an exocyclic double bond and a chelated structure. The molecular structure of (**138**) has been determined by X-ray crystallography (Table 5). It was shown that the hydroxy group participates in an intramolecular hydrogen bond with the carbonyl oxygen atom of the ethoxycarbonyl group at position 4 ⟨80CSC1121⟩. On the other hand, the fourth isomer is the most stable in 4-phenylazopyrazolones (**139**), a chelated phenylhydrazone structure.

(**138**)   (**139**)

The problem of tautomerism is simpler in the case of 1-substituted pyrazolin-3-ones since only two forms, the OH (**140a**) and the NH (**140b**), are possible. The OH form is the more stable and is the only one present in the crystal (Section 4.04.1.3.1). In protic solvents, like water or methanol, the equilibrium position is much more evenly balanced between the OH and NH forms. Finally, 4-hydroxypyrazoles (**141**) exist as such. A CNDO/2 calculation justifies the result that 4-hydroxy tautomers are relatively more stable than

(140a) (140b) (141)

5-hydroxy tautomers and used as an analogy the fact that ketones are easier to enolize than amides ⟨76CJC1752⟩.

The N-unsubstituted pyrazolones constitute a complex case of tautomerism, with no less than eight tautomers being possible. However, there are only four probable tautomers (142a)–(142d). The equilibrium between (142b) and (142c) is a case of annular tautomerism and, generally (Section 4.04.1.5.1), the 3-OH tautomer (142b) is the more stable. The problem thus can be reduced to the study of the equilibria between three forms: the CH (142a), the 3-OH (142b) and the NH (142d). In non-polar solvents there is an equilibrium between the CH and the 3-OH tautomers, in aprotic dipolar solvents the 3-OH tautomer predominates, and in protic solvents there is a mixture of the 3-OH and the NH tautomers.

(142a) (142b) (142c) (142d)

1-Substituted indazolones exist in the OH form (143b) and 2-substituted indazolones exist in the NH form (144a), whereas the structure of N-unsubstituted indazolones varies with the physical state. This difference of behaviour, depending on the position of the N-R substituent, corresponds to the 'aromatic' structure of the indazole derivatives compared with the 'quinonoid' structure of isoindazoles.

(143a) (143b) (144a) (144b)

When the oxygen atom is replaced by a sulfur atom as in pyrazoline-5-thiones, the CH form (145a) never occurs, and the tautomerism implies only the SH (145b) and the NH (145c) forms ⟨77BSB949⟩.

(145a) (145b) (145c)

For amino derivatives, the problem of tautomerism is clear cut; all the derivatives exist in the amino form (3-, 4- and 5-aminopyrazoles, aminoindazoles and 3,5-diaminopyrazoles). IR spectroscopy and especially the Bellamy–Williams relationship (Section 4.04.1.3.7(iii)) are useful for the study of amino tautomerism.

A theoretical, comparative study of the tautomerism of 56 five-membered heterocyclic rings announced in ⟨76AHC(S1)1⟩ has appeared ⟨81MI40402⟩. The stabilities of the three forms for 5-pyrazolones, 5-pyrazolethiones and 5-aminopyrazoles have been calculated by a simple Hückel ω iterative method. The relative energies and the substituent and solvent effects are in agreement with the experimental results.

The tautomerism of bifunctional derivatives is quite complicated. Figure 21 represents the most stable tautomer for some 3,5-disubstituted pyrazoles such as 3-amino-2-pyrazolin-5-ones (146), 5-amino-3-hydroxypyrazoles (147a), 3-hydroxy-2-pyrazolin-5-ones (148) and 1,2-disubstituted 3,5-pyrazolidinediones (149). The structure of (147a) has been questioned ⟨81CHE375⟩; the authors consider the tautomer (147b) to be the more stable on the basis of a band at 1640 cm^{-1} in the IR spectrum, this band previously being assigned to a $\delta(NH_2)$, but no clear proof was given. Recently ⟨82JOC295⟩ it was concluded from IR and UV data that the dihydroxy form (150) is the predominant tautomer when the substituent at position 4 is aromatic.

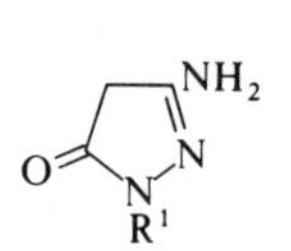

(146)

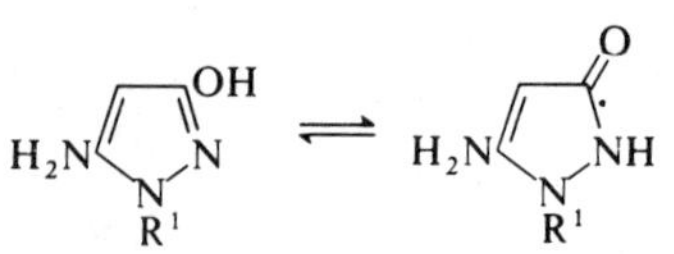

(147a) **(147b)**

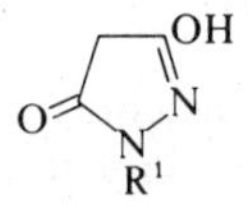

(148)

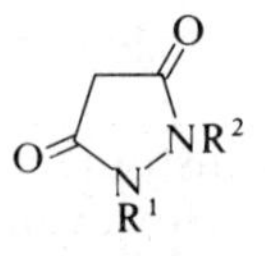

(149)

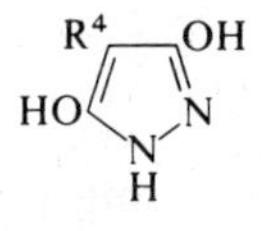

(150)

Figure 21 Structure of the most stable tautomers of difunctional pyrazoles

The 1-hydroxypyrazoles exist as the neutral structure (**151a**) and not in the *N*-oxide form (**151b**), although in water there is a small percentage of the latter present. For tautomerism of 2-hydroxyindazole see ⟨82JOC4323⟩.

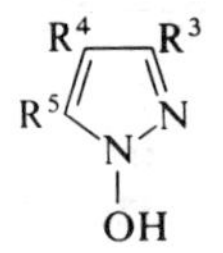

(151a)

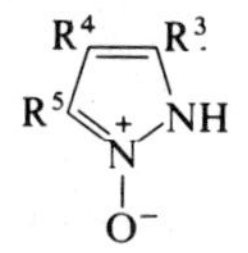

(151b)

An example of non-aromatic tautomerism has already been quoted (Table 13, Section 4.04.1.3.3(ii)); the equilibrium between the two enamines (**152a**) and (**152b**) is solvent and temperature dependent ⟨70BSF3147⟩.

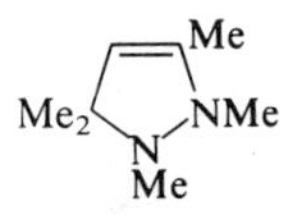

(152a)

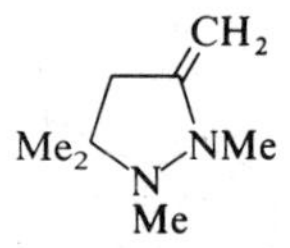

(152b)

Δ^3-Pyrazolines unsubstituted on both nitrogen atoms (**153a**) are unknown. Δ^1-Pyrazolines (**153b**) undergo thermal isomerization to the Δ^2 derivatives (**153c**), and there is only one report of the reverse thermal isomerization (**153c**) → (**153b**) involving the 5,5-dimethyl-Δ^2-pyrazoline ⟨65CHE375⟩ (in the presence of hydrazine hydrochloride the equilibrium is reversed). Thus, the normal order of stability for the three pyrazoline tautomers is: $\Delta^2 > \Delta^1 > \Delta^3$. However, when strong electron-attracting groups are present in both positions 1 and 4, the Δ^3 tautomer (**154a**) is more stable than the Δ^2 tautomer (**154b**) ⟨63YZ373, 65YZ158⟩.

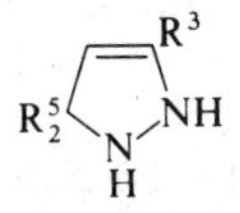

(153a)

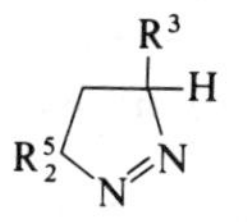

(153b)

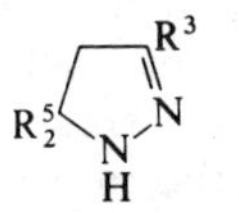

(153c)

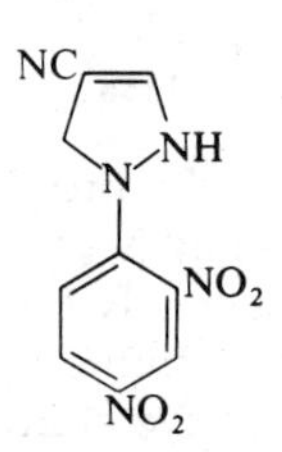

(154a)

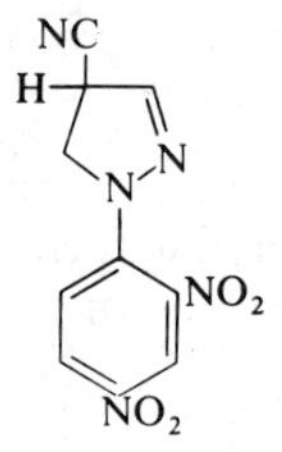

(154b)

A study of the tautomerism of aminopyrazolines ⟨70BSF1571, 70BSF1576, 71TH40400⟩ has led to the conclusion that the imino structure (**155a**) of 5-aminopyrazolines and the amino structure (**155b**) of 3-aminopyrazolines are the predominant tautomers. However, in the last case a broadening of the signals in the ^{1}H and ^{13}C NMR spectra points to a solvent dependent dynamic phenomenon, the origin of which is not yet clearly understood (in $CDCl_3$–TFAA mixtures the broadening is so appreciable that the ^{1}H NMR spectra show no signals, even those of the phenyl group disappearing in the base line).

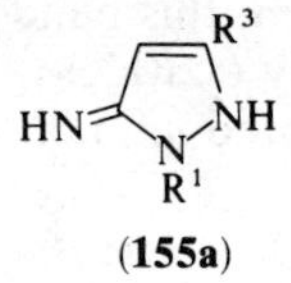

(155a)

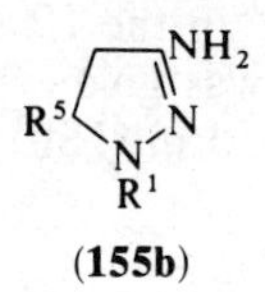

(155b)

4.04.2 REACTIVITY

4.04.2.1 Reactions at Aromatic Rings

4.04.2.1.1 General survey of reactivity

Aromatic pyrazoles and indazoles, in the broad sense defined in Sections 4.04.1.1.1 and 4.04.1.1.2, will be discussed here. Tautomerism has already been discussed (Section 4.04.1.5) and acid–base equilibria will be considered in Section 4.04.2.1.3. These two topics are closely related (Scheme 10) as a common anion (**156a**) or a common cation (**156b**) is generally involved in the mechanism of proton transfer ⟨e.g. 78T2259⟩. For aromatic pyrazoles with exocyclic conjugation there is also a common anion (**157**) for the three tautomeric forms

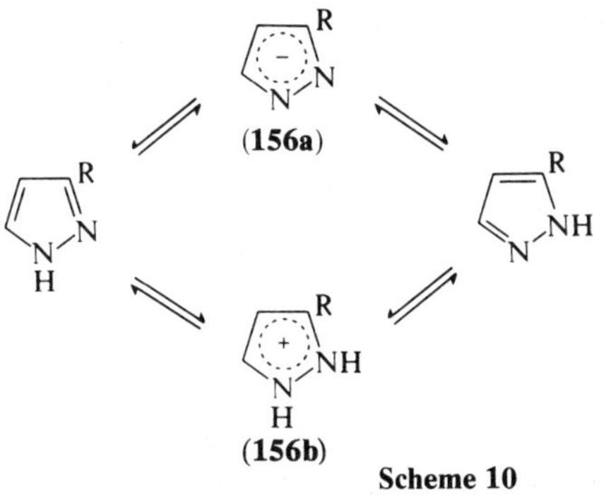

Scheme 10

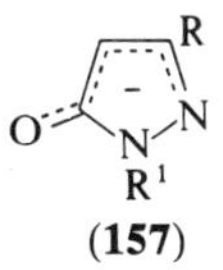

(i) Reactivity of neutral pyrazoles and indazoles

The pyrazole molecule resembles both pyridine (the N(2)—C(3) part) and pyrrole (the N(1)—C(5)—C(4) part) and its reactivity reflects also this duality of behaviour. The 'pyridinic' N-2 atom is susceptible to electrophilic attack (Section 4.04.2.1.3) and the 'pyrrolic' N-1 atom is unreactive, but the N-1 proton can be removed by nucleophiles. However, N-2 is less nucleophilic than the pyridine nitrogen atom and N(1)H more acidic than the corresponding pyrrolic NH group. Electrophilic attack on C-4 is generally preferred, contrary to pyrrole which reacts often on C-2 (α attack). When position 3 is unsubstituted, powerful nucleophiles can abstract the proton with a concomitant ring opening of the anion.

Having its pyrazolic 4-position substituted, electrophilic attack on indazoles takes place in the 3-position and in the 'homocycle' (the 5- and 7-positions). The condensation of a benzene ring results in a decrease of the 'aromaticity' of the pyrazole moiety, as in naphthalene compared to benzene, and therefore basic ring cleavage is easier in indazoles than in pyrazoles (Section 4.04.2.1.7(v)).

Finally, it should be mentioned (Section 4.04.2.1.9) that pyrazole never reacts as an alkene towards a diene or a 1,3-dipole nor as an azadiene towards a dienophile.

(ii) Pyrazolium and indazolium cations

Figure 22 represents the main consequences on pyrazole reactivity when a positive charge is present at the 2-position of the nucleus. A similar situation occurs in the indazolium salts, which thermally decompose into an alkylindazole and an alkyl halide, a reaction sequence described by von Auwers.

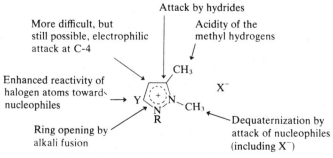

Figure 22 Reactivity of pyrazolium cations

218   *Pyrazoles and their Benzo Derivatives*

(iii) Pyrazole and indazole anions

Pyrazole and indazole anions, in a manner similar to other azole anions, show the expected inversion of reactivity when compared with the cations. They are more reactive towards electrophiles, both at the nitrogen and carbon atoms, and less reactive towards nucleophiles than the corresponding neutral molecules. For practical purposes most of the N-alkylated pyrazoles and indazoles are prepared from the corresponding anions.

(iv) Pyrazolones, pyrazolethiones, pyrazoleimines and their benzo derivatives

The tautomerism of these compounds has already been discussed (Section 4.04.1.5.2). Owing to the extra functionality they have a complex reactivity, and to the reactions of the substituent must be added those of the pyrazole ring itself (Section 4.04.2.3), which are different for each tautomer. In addition to the three neutral tautomers, one has to bear in mind that the NH and OH tautomers of 5-pyrazolones can be written as betaines as in formula (**158**) which is important in understanding their reactivity.

(**158**)   (**159**)

In the case of 4-hydroxypyrazole, tautomer (**159**), although not abundant in the equilibrium mixture (Section 4.04.1.5.2), must be considered when discussing their reactivity.

The mesomeric anion (**157**; Scheme 10) reacts readily with electrophilic reagents such as alkyl or acyl halides at N-2, C-4 and the exocyclic oxygen atom. The percentages of the different products formed are controlled by the HSAB principle. The acylium ion (hard) attacks preferentially at oxygen (hard), whilst the softer alkylating agents attack the nitrogen atom.

(v) N-Oxides, N-imides and N-ylides of pyrazoles and indazoles

The tautomerism of N-hydroxypyrazoles (Section 4.04.1.5.2) is displaced towards the neutral structure (**151a**), but for reactivity studies the N-oxide tautomer (**151b**) must be considered. The reactions at the oxygen atom will be discussed in Section 4.04.2.3.10, and in other aspects they behave like pyrazolium ions. Quaternary salts (**160**) can be obtained by amination of N-alkylpyrazoles and by alkylation of N-aminopyrazoles. However, the corresponding N-imides (**161**), as the N-ylides (**162**), are unknown.

(**160**)   (**161**)   (**162**)

4.04.2.1.2 Thermal and photochemical reactions formally involving no other species

In addition to fragmentation on electron impact (Section 4.04.1.3.8), most of the references in this section involve photochemical studies and an excellent account of these has been given by Lablache-Combier ⟨B-76PH123⟩.

(i) Fragmentation

The photochemistry of antipyrine derivatives (Scheme 11) involves, depending on the substituent in position 4, ring opening fragmentations and rearrangements. In all cases the N—N bond is cleaved. Irradiation of the 4-unsubstituted derivative (**163**; R^4 = H) in alcohols leads to transposition of the ring atoms to yield 1,5-dimethyl-3-phenyl-4-imidazolin-2-one (**164**) plus several ring-opened fragments ⟨69JCS(C)2624⟩. Similar results are obtained when irradiation is carried out in the presence of diethylamine ⟨73AP155⟩.

Irradiation of 4-hydroxy- and 4-alkoxy-3-pyrazolin-5-one derivatives (**163**; R^4 = OH, OR) leads to ring cleavage with the formation of β-diamides (**165**) ⟨69TL271⟩. The methylene blue sensitized rearrangement of the same pyrazolinone (R^4 = H) to the oxindole (**166**) also

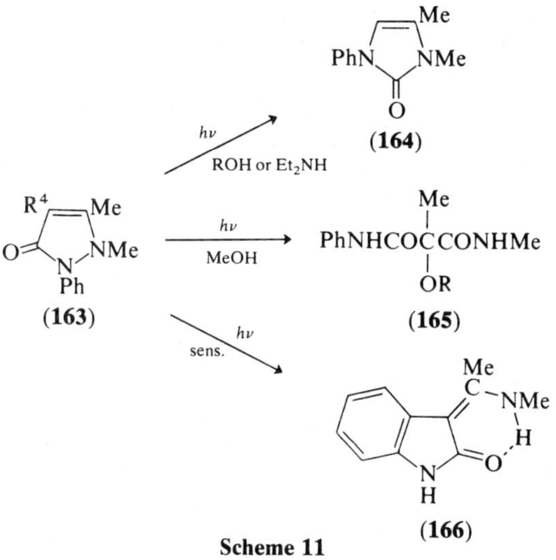

Scheme 11

involves an initial nitrogen–nitrogen bond cleavage ⟨73CB2070⟩. Finally, the 4-amino derivative (163; $R^4 = NH_2$) behaves similarly to the 4-hydroxy derivative giving compound (165; R = H) together with other ring-opened products ⟨74AP211⟩. Flash thermolysis of 3,5-dimethylpyrazole affords 1,3-pentadiene and nitrogen ⟨82JOC2221⟩.

(ii) Rearrangement

The azido/tetrazole rearrangement of pyrazole and indazole derivatives will be discussed in the corresponding section (4.04.2.3.4(v)).

(a) Ring opening without fragmentation. Ring opening to 2-alkylaminobenzonitriles (167) occurs when N-1-alkylated indazoles (168) are irradiated ⟨67HCA2244⟩.

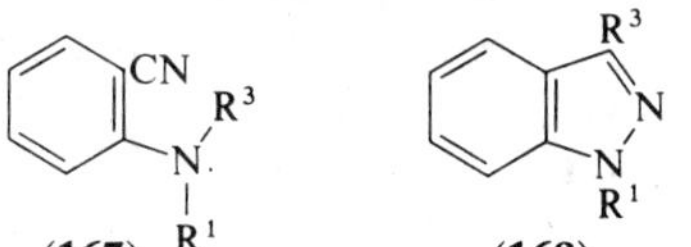

A new type of rearrangement has been reported for certain 1-(o-nitrophenyl)pyrazoles (169), giving *cis*- and *trans*-benzotriazole 1-oxides (170; Scheme 12) ⟨73TL891⟩. The reaction was rationalized in terms of an intermediate azo compound (171), formed in turn either from the diradical species (172) or from the intramolecular 1,4-adduct (173). Subsequently

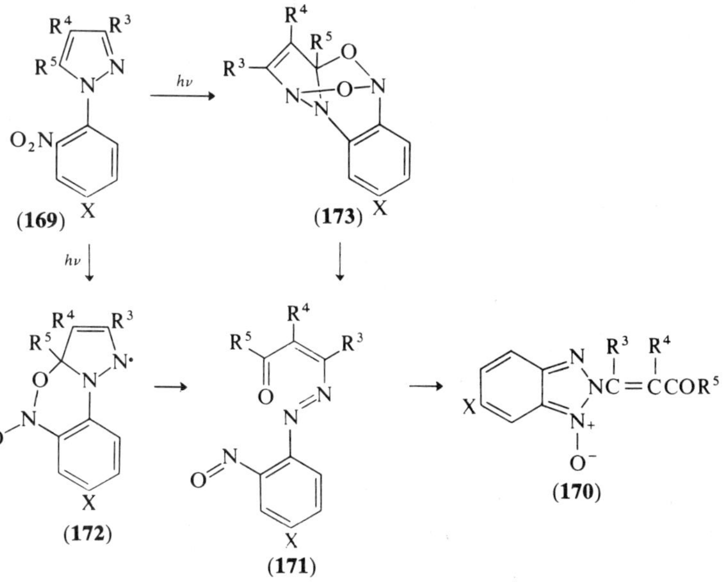

Scheme 12

the radical mechanism was preferred ⟨76BSF184⟩, both for *o*-nitrophenyl and 2,4-dinitrophenyl derivatives. Only nitro group reduction products are obtained from the corresponding 1-(*p*-nitrophenyl)pyrazoles ⟨76BSF195⟩.

Cyclodehydrogenation of 1,4,5-triphenylpyrazole (**174**) to give 1-phenyl-1*H*-phenanthro[9,10-*c*]pyrazole (**175**) has been observed (Scheme 13), but in the presence of benzophenone ring opening to 3-anilino-2,3-diphenylacrylonitrile (**176**) was preferred ⟨77JCS(P1)2096⟩.

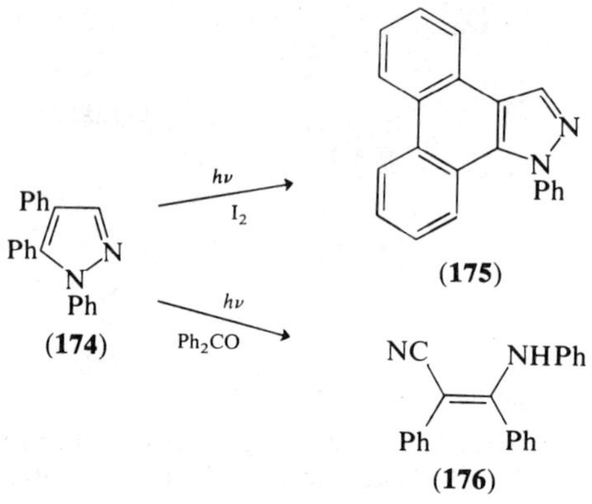

Scheme 13

(*b*) *Ring contraction*. There are no examples of light-induced ring contractions of aromatic pyrazoles (pyrazoles, indazoles and pyrazolones) similar to those described for pyrazolidinones (Section 4.04.2.2.2(iv)).

(*c*) *Ring isomerization*. One of the most important photochemical reactions of pyrazoles is their transformation into imidazoles. Discovered independently by Schmid ⟨67HCA2244⟩, Beak ⟨67TL5315⟩ and Ege ⟨67CC488⟩, this reaction generally involves the exchanges of positions N-2 and C-3 of the pyrazole with positions C-2 and N-3 of the imidazole.

Van Tamelen and Whitesides ⟨71JA6129; see also 71BSF1925⟩ have proposed a general mechanism to account for the rearrangement products of five-membered heteroaromatic systems. It involves initial cleavage of the weakest single bond (in pyrazoles (**177**) the N(1)—N(2) bond) followed by cyclization of the resulting diradical intermediate (**178**) to form the azirine (**179**; Scheme 14). They also propose that species (**179**) is in equilibrium with valence bond isomers (**180**), (**181**) and (**182**), thus accounting for the formation of imidazoles (**183**) and (**184**), the latter being the one most frequently obtained. The mechanism is equally valid for the isomerization of isoindazoles (**177**; R^3—R^4 = C_4H_4) to benzimidazoles (**183**; R^3—R^4 = C_4H_4).

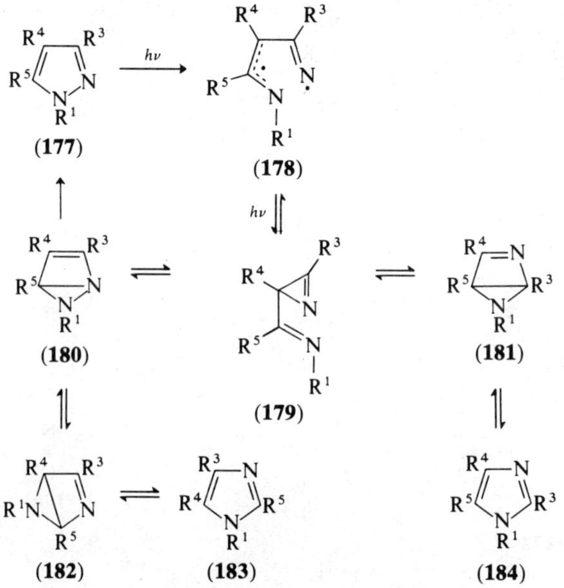

Scheme 14

In this pioneering work Schmid ⟨67HCA2244⟩ studied a large collection of examples of the transformation (**177**) → (**184**), with R^1 = H, Me, Bz and Ph, and R^3, R^4, R^5 = H or Me. The reaction was conveniently carried out in DME in the presence of benzophenone, and the yields were in the range 20–30%. In ethanol solution, 1,3,5-trimethylpyrazole (**177**; $R^1 = R^3 = R^5$ = Me, R^4 = H) yielded (**183**) and (**184**) ⟨69T3287⟩. In contrast with the previously observed sensitized photorearrangements ⟨67HCA2244, 67CC488⟩, this last reaction was not sensitized by acetophenone or benzophenone.

In addition to reduction, ring transposition of the type (**177**) → (**184**) is observed in the conversion of (**185**) into 1,2-bis(1-phenylimidazol-4-yl)ethane-1,2-diol (**186**) by irradiation at 254 nm ⟨74JCS(P1)1871⟩.

(**185**)   (**186**)

The phototransposition reaction in the isoindazole series (**187**) to give benzimidazoles (**188**) has been carefully investigated when R^1 = Me, Et, Pr^i, Bu^t, Bz and Ph ⟨67HCA2244⟩. The reaction is sensitive to steric effects; thus for $R^1 = Bu^t$ and R^4 = Me, the 1-*t*-butyl-7-methylbenzimidazole is not formed.

(**187**)   (**188**)   (**189**)

In the case of 2-alkylindazoles (**187**; $R^3 = R^4$ = H) the reaction has been studied with regard to variations of quantum yield with exciting wavelength, solvent, temperature and concentration ⟨69C109⟩. It has been assumed from theoretical considerations based on the change in bond orders upon excitation that the reaction occurs from the lowest excited singlet state. The most likely structure for the intermediate formed in quantitative yield at −60 °C is the tricyclic compound (**189**), the indazolic counterpart of (**180**). The temperature dependence of the quantum yields of fluorescence, intersystem crossing and photoreactions has been measured for several neutral and protonated 2-alkylindazoles ⟨78HCA234⟩. Based on kinetic and photochemical data, the authors conclude that the very efficient radiationless deactivation process that takes place at 150 K proceeds *via* a hypersurface crossing in the course of the photochemical rearrangement (**187**) → (**189**).

N-Unsubstituted indazoles (R^1 = H) behave like a mixture of N(1)H and N(2)H tautomers giving rise to *o*-aminobenzonitriles (**190**) similar to 1-substituted isomers and to benzimidazoles (**191**) like the isoindazoles ⟨67HCA2244, 74JA2010, 74JA2014⟩.

(**190**)   (**191**)

(*d*) *Ring enlargement*. In dilute sulfuric acid (pH 2–4) rearrangement of isoindazoles into benzimidazoles (**187** → **188**) is suppressed and the dihydroazepinones (**192**) and (**193**) are obtained ⟨76HCA2632⟩, whereas in strongly acid solution the *o*-aminoacetophenones (**194**) are formed. The mechanism of formation of compounds (**194**) has been elucidated ⟨78HCA618⟩. It occurs *via* arylnitrenes formed directly in their triplet ground state from the triplet state of the isoindazoles.

(**192**)   (**193**)   (**194**)

(iii) Polymerization and disproportionation

Pyrazoles with free NH groups form hydrogen-bonded cyclic dimers (**195**) and trimers (**196**) as well as linear polymers, depending on the substituents at positions 3 and 5. For R = H, Me or Et, the oligomers are preferred, but for R = Ph, the cyclic dimer and the linear polymers exist. The cyclic trimer (**196**; R = Ph) is) is not formed because of steric hindrance ⟨B-76MI40402⟩.

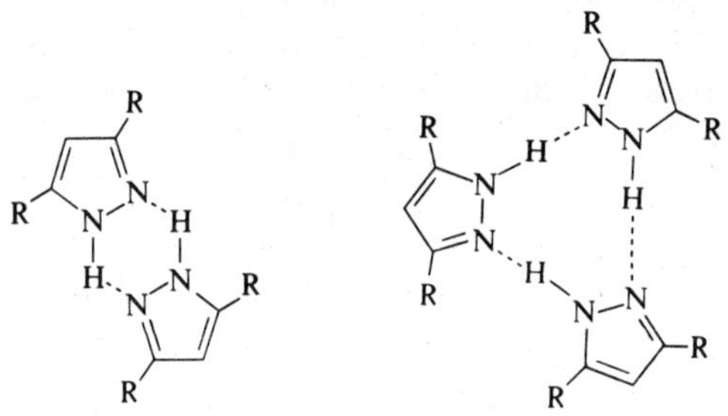

(**195**)   (**196**)

There are no known examples of pyrazole or indazole disproportionation.

All the examples quoted in this section concerning fragmentations or rearrangements involve photochemistry. An interesting thermal reaction has been described ⟨72TL2235⟩ in which the pyrolysis of indazole between 700 and 800 °C leads to a mixture of (**197**) and (**198**; Scheme 15). A mechanism involving the $3H$ tautomer and the carbene seems reasonable.

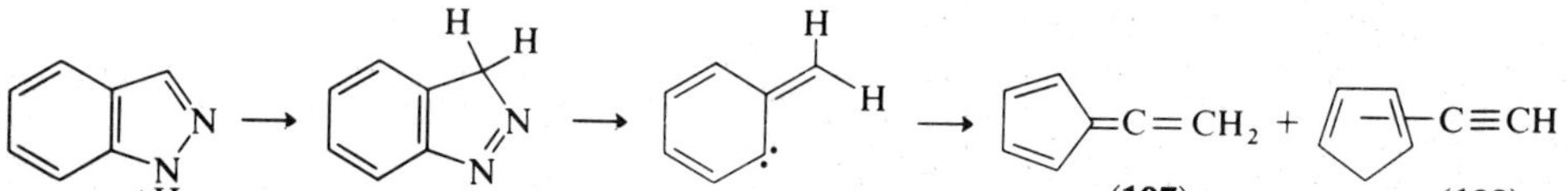

Scheme 15

4.04.2.1.3 Electrophilic attack at nitrogen

(i) Introduction

The reactivity of the pyridine-like N-2 atom of pyrazoles and indazoles follows the general trends indicated in Section 4.02.1.3 for azoles. Since this part belongs to the classical chemistry of pyrazoles, earlier results in ⟨66AHC(6)347, 67HC(22)1, B-76MI40402, 79RCR289⟩ will be summarized and occasionally updated.

(ii) Reaction sequence

In a neutral azole, the apparent rate of formation of an N-substituted derivative depends on the rate of reaction of a given tautomer and on the tautomeric equilibrium constant. For example, with a 3(5)-substituted pyrazole such as (**199**), which exists as a mixture of two tautomers (**199a**) and (**199b**) in equilibrium, the product composition [(**200**)]/[(**201**)] is a function of the rate constants k_A and k_B, as well as of the composition of the tautomeric mixture (Scheme 16) ⟨76AHC(S1)1⟩.

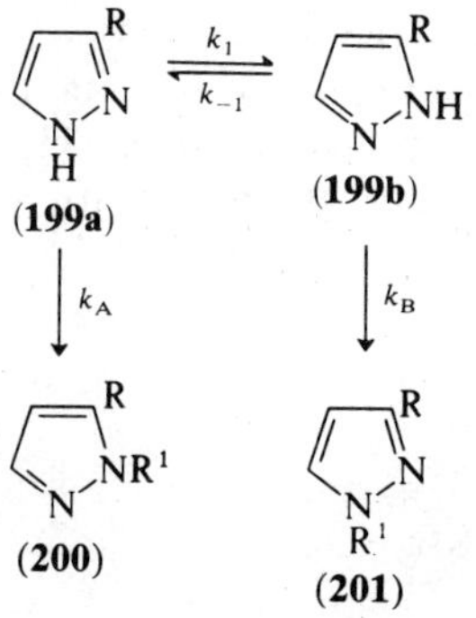

$K_T = [(199a)]/[(199b)]$

$[(200)]/[(201)] = \dfrac{k_A}{k_B} K_T$

Scheme 16

Two points must be stressed: that a 3-R-pyrazole (**199a**) gives a 1,5-disubstituted derivative (**200**) and that often the less abundant tautomeric species (Section 4.04.1.5.1) is the more reactive. It is theoretically possible to use the tautomeric composition to direct the synthesis towards one or another derivative. For instance, if (**199a**) is the more abundant tautomer, the condition to obtain a large proportion of (**200**) is that $k_A \gg k_1$. The factors influencing the tautomeric exchange rates have already been described (Section 4.04.1.3.4) and the reaction rates k_A and k_B will be discussed later (Section 4.04.2.1.3(vii)). Thus it will be possible to direct the synthesis by slowing down the exchange rate (lowering the temperature in a dipolar aprotic solvent or changing the proton by another group) and by using powerful alkylating reagents (like Meerwein's oxonium salts).

Often electrophilic reagents can attack both nitrogen atoms of the mesomeric pyrazolate and indazolate anions. In this case there is no simple relationship between the tautomeric constant and the product composition.

(iii) Effect of substituents

The steric and electronic effects of substituents on the electrophilic attack at the nitrogen atom have been discussed in the general chapter on reactivity (Section 4.02.1.3). All the conclusions are valid for pyrazoles and indazoles. The effect on equilibrium constants will be discussed in detail in the sections dealing with pK_a values (Sections 4.04.2.1.3(iv) and (v)) and the kinetic effects on the rates of quaternization in the corresponding section (4.04.2.1.3(vii)).

(iv) Proton acids on neutral compounds: basicity of pyrazoles and indazoles

The basicity of pyrazole and its relation with imidazole basicity (due both to enthalpy and entropy changes ⟨77MI40403⟩) have been discussed on theoretical grounds (Section 4.04.1.2.1). The pK_a values of 90 pyrazoles have been determined by Gonzalez et al. ⟨68BSF707, 68BSF5009⟩ and it is essentially his work that will be discussed below. A selection of pK_a values are shown in Table 28. The pK_a values for some other pyrazoles have been measured in connection with nitration studies (Section 4.04.2.1.4(ii)) ⟨71JCS(B)2365⟩.

Table 28 pK_a Values for Proton Addition (in Water at 20 °C)

Pyrazole	pK_a (±0.04)
Unsubstituted	2.52
3(5)-Methyl	3.32
3(5)-Ethyl	3.30
3(5)-t-Butyl	3.30
4-Methyl	3.09
3,5-Dimethyl	4.12
3,4,5-Trimethyl	4.63
4-Nitro	−1.96
4-Chloro	0.60
4-Bromo	0.64
4-Iodo	0.82
3(5)-Chloro	−0.48
3(5)-Phenyl	2.13
1-Methyl	2.09
1-Ethyl	2.00
1-t-Butyl	1.95
1-Phenyl	0.44
1-p-Nitrophenyl	−0.65
1-(2,4-Dinitrophenyl)	−1.40
1-(2,4,6-Trinitrophenyl)	−2.0[a]

[a] ±0.1.

The pK_a values are approximately additive and a linear relationship of the type $pK_a^{calc} = pK_a^0 + \sum \Delta \overline{pK_m^n}$ holds for the whole set (pK_a^0 is the pK_a of pyrazole itself and $\Delta \overline{pK_m^n}$ is the effect of a substituent m at position n). Deviation from the additivity is found when two bulky substituents are in contiguous positions. Instead of discussing pK_a values, the authors consider $\Delta \overline{pK_m^n}$ which are mean values and thus more significant since they correspond to several pairs of compounds.

The electronic effect of the substituents in N-unsubstituted pyrazoles roughly follows a Hammett relationship, $-\Delta \overline{pK_m^n} = 6.57\sigma_m$, without taking into account the tautomeric equilibrium. The N-methyl group is base-weakening ($\Delta \overline{pK_a} = -0.45$) and considering the statistical factor ⟨76AHC(S1)1⟩, the real effect is $-0.75\ pK_a$ units ⟨83UP40400⟩. Since in imidazoles the experimental value lies between 0.05 and 0.30, it was concluded that the effect on the pK_a of a methyl group in position 1 is due to a steric hindrance to cation solvation, which is larger in N-ethyl- and N-t-butyl-pyrazole (Table 28). Steric effects of the 3-substituents are discussed in ⟨68BSF707⟩. It is well known that they do not hinder the proton approach but the stabilizing solvation of the conjugate acid.

The basicities of indazole (1.31), 1-methyl (0.42) and 2-methyl (2.02) derivatives and of eight other substituted indazoles have been measured ⟨67BSF2619⟩. The effect of substituents in the 3-position is similar in pyrazoles and indazoles with $\Delta \overline{pK_m^3}$ values as follows: Me, 0.80 and 0.86; Cl, -3.01 and -2.98; Br, -2.85 and -2.82, respectively. A nitro group in the homocycle has an expected base-weakening effect of $-2\ pK_a$ units, whether it is at the 5- or the 6-position.

pK_a values for pyrazolones and alkoxypyrazoles ⟨B-76MI40402⟩ have been used as a tool in tautomeric studies ⟨76AHC(S1)1⟩. Pyrazolone itself has a pK_a of 2.32. The OH form is more basic than the NH form (for N-Ph derivatives the pK_a values are 2.5 and 1.3, respectively) and the CH form is a very weak base ($pK_a \sim -4$). The structures involved in acid–base equilibria in 5-pyrazolones are shown in Scheme 17. The OH (**202b**) and NH (**202c**) forms yield a common cation with a structure of type (**203**) whilst the CH form (**202a**) gives a different cation, probably having structure (**204**). The protonation constants of 4-hydroxy-pyrazoles have been determined in the course of studies of their tautomerism ⟨76AHC(S1)1⟩.

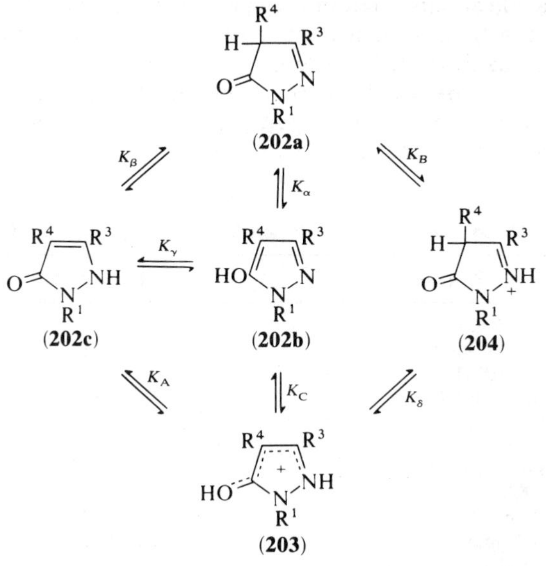

Scheme 17

(v) *Proton acids at anionic compounds: acidity of pyrazoles and indazoles*

Pyrazole and indazole are very weak acids (Table 29) and $-R$, $-I$ groups considerably increase the acidity (3,5-dinitropyrazole has a pK_a of 3.14 ⟨73JHC1055⟩, i.e. it is a stronger acid than formic acid). However, when in the 3- and 4-positions the acids are the same strength. Habraken has pointed out the importance of resonance forms like (**205**), (**206**) and (**207**) in the stabilization of nitro- and nitroso-pyrazole anions. Bulky substituents in the 3- and 5-positions considerably decrease the acidity of these compounds, because they prevent the coplanarity of the nitro or nitroso group at the 4-position ⟨72JHC939, 73JHC1055⟩.

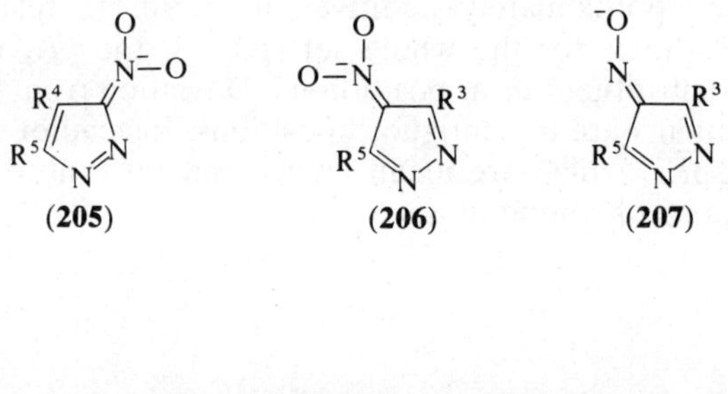

Table 29 pK_a Values for Proton Loss[a]

Compound	pK_a
Pyrazole	14.21
3-Acetylpyrazole	11.85
Pyrazole-3-carboxylic acid	3.74[b]
4-Hydroxy-1-phenylpyrazole	9.05[c]
4-Nitropyrazole	9.67
3,5-Dimethyl-4-nitropyrazole	10.65
3-Nitropyrazole[d]	9.81
3,5-Dimethyl-4-nitrosopyrazole[e]	9.14
Indazole[f]	13.80
Pyrazolone	7.94

[a] From ⟨B-76MI40402⟩, which see for references to the original literature. [b] Involves the CO$_2$H function. [c] Involves the 4-OH function. [d] ⟨73JHC1055⟩. [e] ⟨72JHC939⟩. [f] ⟨67T2855⟩.

A Hammett relationship of the form $\Delta pK_a = 5.8\sigma_m$ has been proposed for 4-substituted pyrazoles ⟨74TL1609⟩ in order to explain the effect of 4-nitro ($\Delta pK_a = 4.5$, $\sigma_m = 0.71$) and 4-diazo groups ($\Delta pK_a = 10.0$, $\sigma_m = 1.76$). The acidity constants of a series of pyrazolidine-3,5-diones have been determined ⟨75AJC1583⟩ and the 4-n-butyl-1,2-diphenyl derivative phenylbutazone has a pK_a of 4.33.

(vi) Metal ions

(a) Ionic compounds of electropositive metals. Pyrazoles and indazoles form sodium, potassium, magnesium and silver salts which are hydrolyzed to a large extent by water. The resulting anions react very readily with electrophiles (Section 4.04.2.3.10(viii)). The sodium salts, prepared from the metal itself, are probably ionic, Pz$^-$ Na$^+$, but the insoluble silver salt AgPz is probably polymeric as in (**208**). The silver derivative of 4-i-C$_3$F$_7$-pyrazole is soluble in acetone where it is tetrameric ⟨72CRV497⟩.

(**208**)

(b) Simple complexes. The neutral and anionic pyrazoles are excellent ligands, and the coordination chemistry of pyrazoles has been extensively studied. Two comprehensive reviews deal with this topic: the one by Trofimenko ⟨72CRV497⟩ stresses the importance of poly(pyrazol-1-yl)borates as ligands, while that of Bonati ⟨80MI40402⟩ covers all the metallic derivatives of pyrazoles. Addition of the name of Reedijk to those of these two important authors completes the list of main contributors to the chemistry of pyrazole complexes.

Bonati has classified the pyrazole complexes into two groups: compounds containing neutral pyrazoles (HPz), called 2-monohaptopyrazoles since it is the N-2 pyridinic nitrogen lone pair which confers on them the ligand properties; and compounds containing pyrazole anions (Pz) which can act as monodentate or, more often, as exobidentate ligands ⟨72CRV497⟩.

Pyrazole and its *C*-methyl derivatives acting as 2-monohaptopyrazoles in a neutral or slightly acidic medium give M(HPz)$_n$X$_m$ complexes where M is a transition metal, X is the counterion and m is the valence of the transition metal, usually 2. The number of pyrazole molecules, n, for a given metal depends on the nature of X and on the steric effects of the pyrazole substituents, especially those at position 3. Complexes of 3(5)-methylpyrazole with salts of a number of divalent metals involve the less hindered tautomer, the 5-methylpyrazole (**209**). With pyrazole and 4- or 5-monosubstituted pyrazoles M(HPz)$_6$X$_2$

(**209**)

complexes are obtained when X is an anion of low coordination potency, such as NO_3^-, BF_4^- or ClO_4^-. When X is a halide or an NCS^- ion, complexes $M(HPz)_4X_2$ are formed in which the anion is coordinated as in (209). 3,5-Disubstituted and 3,4,5-trisubstituted pyrazoles form complexes of the type $M(HPz)_4X_2$ with low coordinating potency anions and of the type $M(HPz)_2X_2$ with halide ions.

Both Bonati and Trofimenko include in the group containing pyrazole anions compounds like (210) and (211), which contain an N—C bond, and compounds like (212), which are reported in Section 4.04.2.1.3(xv).

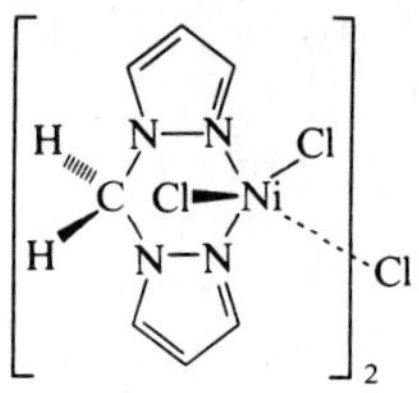

(210) 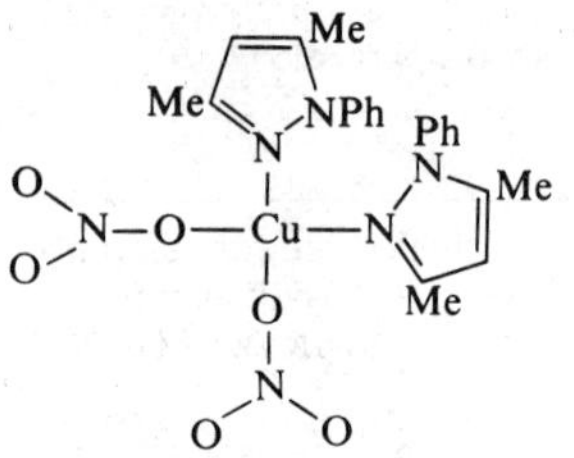   (211) 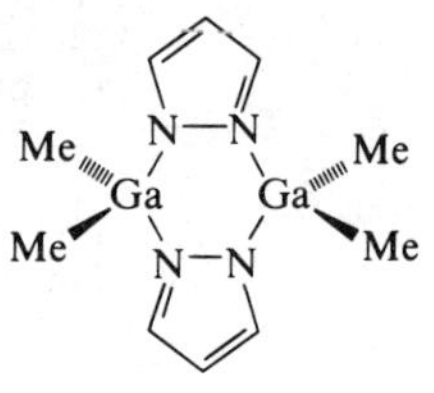   (212)

Bonati considers that due to the spatial disposition of lone pairs in the Pz^- anion, it could be a good exobidentate ligand but not a chelating agent. However it was found recently that the uranium complex (213) has an endobidentate structure which was attributed to the highly ionic character of the U—N bond. There are many examples of pyrazole anion complexes and their study goes beyond the limits of this chapter. Some highly symmetrical complexes, e.g. (214) ⟨79IC658⟩, which resembles the pyrazole N—H···N trimer (196), and (215) of C_{3v} symmetry, are shown.

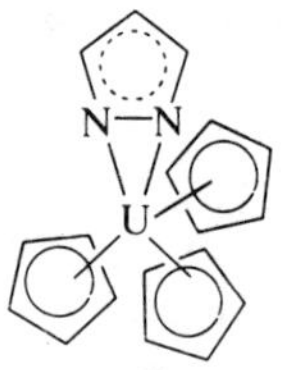

(213) 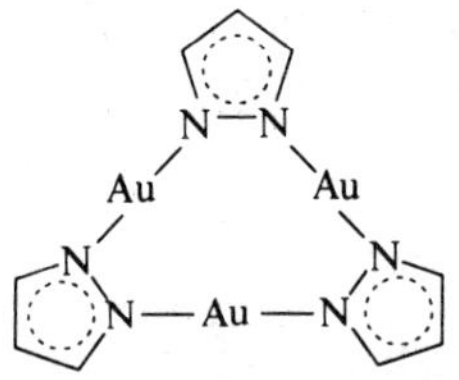   (214) 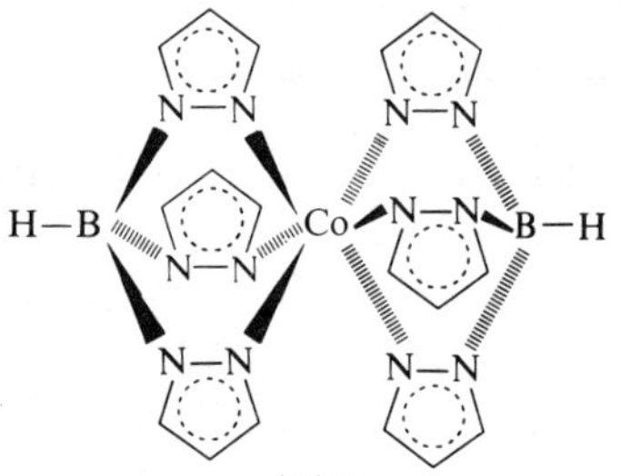   (215)

The only structures definitely established are those obtained from X-ray measurements. Some of these are shown in Table 30, from which a large number of aromatic pyrazole geometries can be found.

Studies on metal–pyrazole complexes in solution are few. The enthalpy and entropy of association of Co(II), Ni(II), Cu(II) and Zn(II) with pyrazole in aqueous solution have been determined by direct calorimetry ⟨81MI40406⟩. The nature of the nitrogen atom, pyridinic or pyrrolic, involved in the coordination with the metal cannot be determined from the available thermodynamic data. However, other experiments in solution (Section 4.04.1.3.3(i)) prove conclusively that only the N-2 atom has coordinating capabilities.

(c) *Chelate complexes.* Apart from the uranium derivative (213), the other chelate complexes involving endobidentate ligands are derived from pyrazoles bearing functional groups containing heteroatoms. Trofimenko ⟨72CRV497⟩ has described several examples of chelates, the endobidentate ligands of which are either *N*-substituted pyrazoles (216) and (217) or *C*-substituted pyrazoles such as (218) and (219).

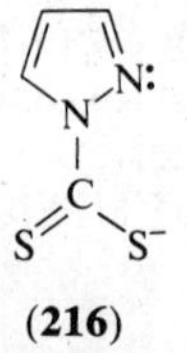

(216) 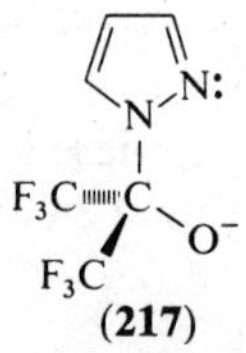   (217) 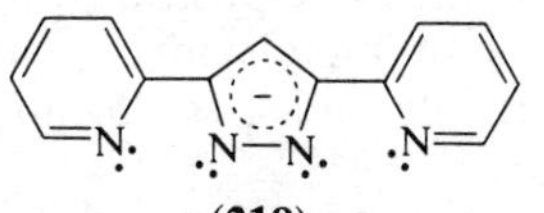   (218) 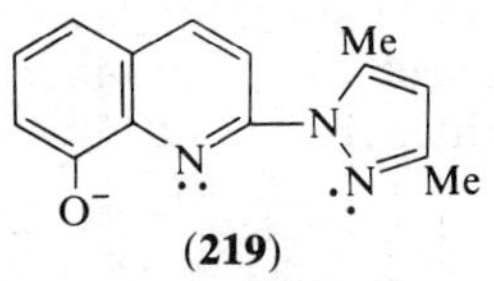   (219)

Copper salt (220; Table 30) involves the nitrogen atom of the carbamoyl group, but in the complex (221) the ligand is coordinated to the divalent metal ions through the nitrogen atom of the pyrazole ring and the oxygen atom of the carbamoyl group ⟨80MI40403⟩.

Table 30 X-Ray Structures of Pyrazole Complexes

Metal	Ligand	Formula	Ref.
Cr(II)	3(5)-Methylpyrazole (HMPz)	[Cr(HMPz)$_4$F$_2$](BF$_4$)	79JCR(S)374
Mn(II)	3(5)-Methylpyrazole (HMPz)	Mn(HMPz)$_4$Br$_2$ (**209**)	71IC2594
Co(II)	3,5-Dimethylpyrazole (HDMPz)	[CoF(HDMPz)$_3$]$_2$(BF$_4$)$_2$	78IC1990
Co(II)	3,5-Dimethyl-1-phenylpyrazole (DMPPz)	Co(NCS)$_2$(DMPPz)$_2$	80AX(B)159
Co(II)	Hydrobis(pyrazol-1-yl)borate [BH$_2$(Pz)$_2$]	Co[BH$_2$(Pz)$_2$]$_2$	73IC508
Co(II)	Hydrotris(pyrazol-1-yl)borate [BH(Pz)$_3$]	Co[BH(Pz)$_3$]$_2$ (**215**)	70IC1597
Ni(II)	Pyrazole (HPz)	Ni(HPz)$_6$(NO$_3$)$_2$	70AX(B)521
Ni(II)	Pyrazole (HPz)	Ni(HPz)$_4$Cl$_2$	67AX135
Ni(II)	Pyrazole (HPz)	Ni(HPz)$_4$Br$_2$	69AX(B)595
Ni(II)	3,5-Diethyl-4-methylpyrazole (HDEMPz)	Ni(HDEMPz)$_4$Cl$_2$	79MI40413
Ni(II)	3,5-Diethyl-1-phenylpyrazole (DEPPz)	Ni(DEPPz)$_2$(NO$_3$)$_2$	79MI40403
Ni(II)	Bis(3,5-dimethylpyrazol-1-yl)methane [CH$_2$(DMPz)$_2$]	{Ni[CH$_2$(DMPz)$_2$]Cl$_2$}$_2$ (**210**)	80IC170
Ni(II)	Dimethylbis(pyrazol-1-yl)gallate [GaMe$_2$(Pz)$_2$]	Ni[GaMe$_2$(Pz)$_2$]$_2$	75JCS(D)176
Ni(II)	Dimethylbis(3,5-dimethylpyrazol-1-yl)gallate [GaMe$_2$(DMPz)$_2$]	Complicated structure[a]	80CJC1091
Ni(II)	3,5-Diethyl-4-methylpyrazole (HDEMPz)	Cu(HDEMPz)$_2$Cl$_2$	79MI40412
Cu(I)	1-Allyl-3,5-dimethylpyrazole (ADMPz)	[Cu(ADMPz)Cl]$_2$	76BCJ143
Cu(II)	3,5-Dimethyl-1-phenylpyrazole (DMPPz)	Cu(DMPPz)$_2$Cl$_2$	80MI40401
Cu(II)	3,5-Dimethyl-1-phenylpyrazole (DMPPz)	Cu(DMPPz)$_2$(NO$_3$)$_2$ (**211**)	79AX(B)1468
Cu(II)	1-Carboxamido-3,5-dimethylpyrazole (CDMPz)	Cu(CDMPz)$_2$ (**220**)	79JCS(D)1867
Cu(II)	Hydrotris(pyrazol-1-yl)borate [BH(Pz)$_3$]	Cu[BH(Pz)$_3$]$_2$	79JCS(D)1646
Cu(I)	Hydrotris(pyrazol-1-yl)borate [BH(Pz)$_3$]	[Cu{BH(Pz)$_3$}]$_2$	76JA711
Cu(II)	Dimethylbis(pyrazol-1-yl)gallate [GaMe$_2$(Pz)$_2$]	Cu[GaMe$_2$(Pz)$_2$]$_2$	75JCS(D)718
Cu(II)	Dimethylbis(3,5-dimethylpyrazol-1-yl)gallate [GaMe$_2$(DMPz)$_2$]	Cu[GaMe$_2$(DMPz)$_2$]$_2$	75JCS(D)718
Cu(II)	Dimethylbis(3-methylpyrazol-1-yl)gallate [GaMe$_2$(MPz)]	{CuO[GaMe$_2$(MPz)$_2$](3MPz)}$_2$	76CJC343
Cu(II)	Antipyrine (APy)	Cu(APy)$_2$(NO$_3$)$_2$, form I[b]	74AX(B)2246
Cu(II)	Antipyrine (APy)	Cu(APy)$_2$(NO$_3$)$_2$, form II[b]	74AX(B)2500
Mo(II)	Dimethylbis(pyrazol-1-yl)gallate [GaMe$_2$(Pz)$_2$]	Complicated structure[a]	80CJC1080
Mo(II)	Dimethylbis(3,5-dimethylpyrazol-1-yl)gallate [GaMe$_2$(DMPz)$_2$]	Complicated structure[a]	80CJC1080
Mo(II)	Methylbis(3,5-dimethylpyrazol-1-yl)hydroxygallate [GaMe(OH)(DMPz)$_2$]	Mo(CO)$_2$(η^3-C$_4$H$_7$)[GaMe(OH)(DMPz)$_2$]	79CJC139
Mo(II)	Hydrobis(3,5-dimethylpyrazol-1-yl)borate [BH$_2$(DMPz)$_2$]	Mo(CO)$_2$(η^3-C$_3$H$_5$)[BH$_2$(DMPz)$_2$]	71AX(B)1859
Mo(V)	Hydrotris(3,5-dimethylpyrazol-1-yl)borate [BH(DMPz)$_3$]	MoOCl$_2$·C$_6$H$_5$Cl[BH(DMPz)$_3$]	82JCR(S)6
Rh(I)	Pyrazole (HPz)	Rh(CO)P(OPh)$_3$(HPz)]$_2$	81JOM(205)247
Pd(II)	3,5-Dimethylpyrazole (HDMPz)	[PdCl(PEt$_3$)$_2$(HDMPz)](BF$_4$)	81IC1545
Ag(I)	3,5-Dimethyl-1-phenylpyrazole (DMPPz)	[Ag(DMPPz)$_3$](NO$_3$)	79AX(B)177
Pt(II)	Pyrazole (HPz)	[PtCl$_2$(C$_2$H$_4$)$_2$(HPz)]Cl	81IC1519
U(IV)	Pyrazole (HPz)	U(Cp)$_3$(HPz)[c] (**213**)	81IC1553

[a] Due to the presence of a second, more complex pyrazole ligand. [b] Cu—O=C(5) bonds. [c] Cp = cyclopentadienyl.

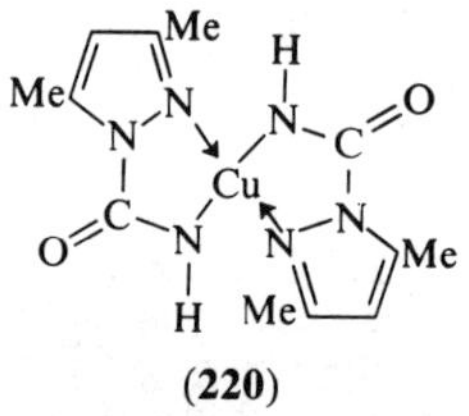

Ligands like (**222**) are useful analytical reagents for Cu(II) ⟨64HC(20)1⟩, whilst ligands (**223**) and (**224**) are used as dyes (Section 4.04.4.1.3). In dyes such as (**223**), which are applied to wool, the usual metal employed is chromium; for cotton dyes like (**224**) it is more common to use copper.

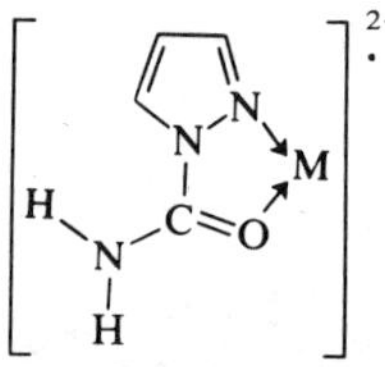

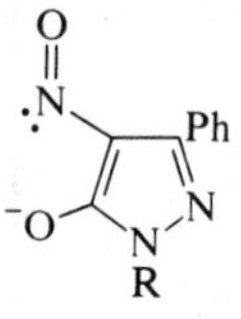

The ^{15}N, ^{13}C and ^{1}H chemical shifts and the corresponding coupling constants have been determined for the chelate (**225**; M = Zn(II)) ⟨81M105⟩.

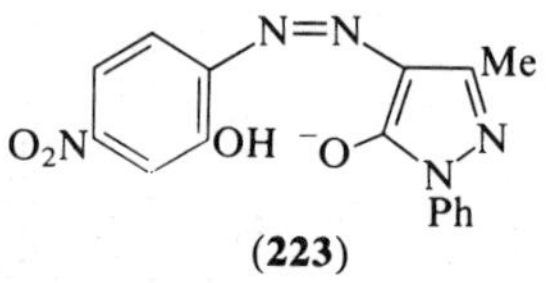

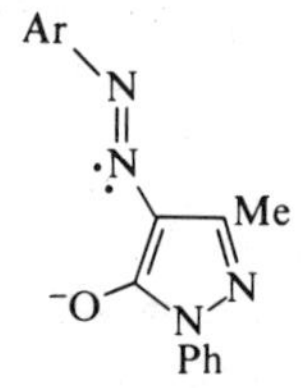

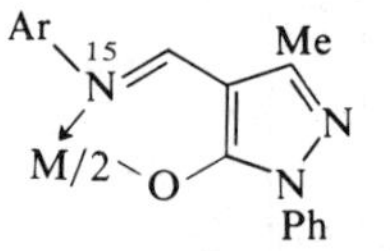

(d) Azocrown ethers and porphyrinogens. In an interesting series of papers, Tarrago ⟨*e.g.* see 81T991⟩ has shown that polypyrazolic macrocycles (**226**), (**227**) and (**228**) are excellent complexing agents for the alkali metal cations due to the electronegative cavity formed by the sp^2 lone pairs. The hexapyrazolic derivative (**227**) is an exceptional complexing agent for Cs$^+$. Compound (**226**) also forms classical complexes with Hg^{2+}, Ag$^+$ and Cd^{2+} salts.

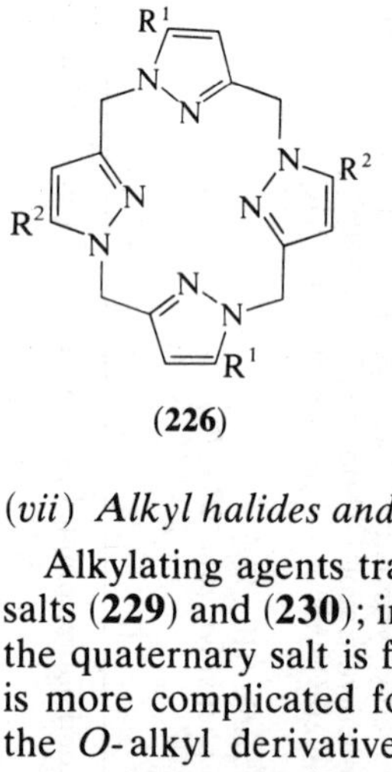

(vii) Alkyl halides and related compounds: compounds without a free NH group

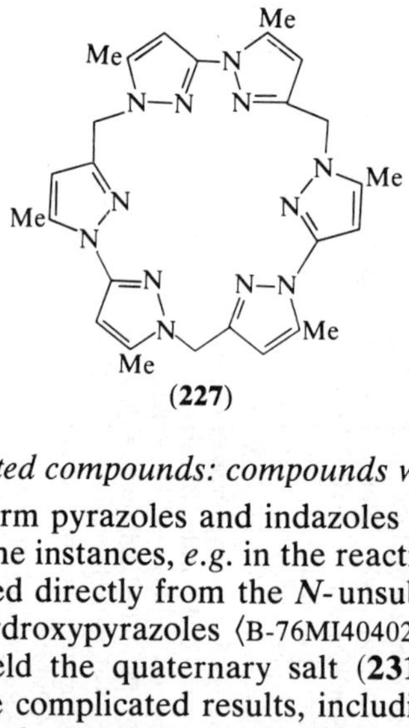

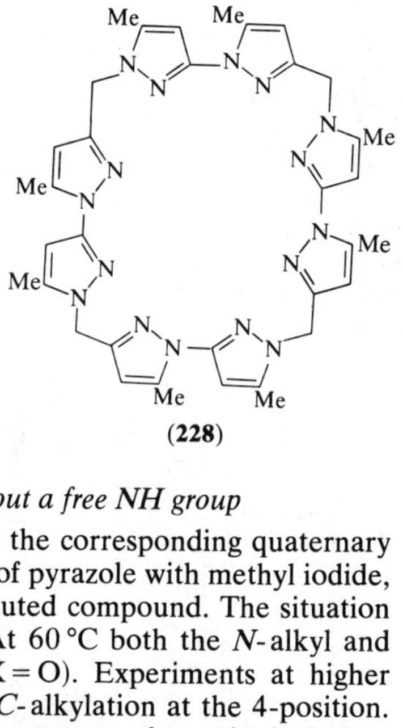

Alkylating agents transform pyrazoles and indazoles into the corresponding quaternary salts (**229**) and (**230**); in some instances, *e.g.* in the reaction of pyrazole with methyl iodide, the quaternary salt is formed directly from the *N*-unsubstituted compound. The situation is more complicated for hydroxypyrazoles ⟨B-76MI40402⟩. At 60 °C both the *N*-alkyl and the *O*-alkyl derivatives yield the quaternary salt (**231**; X = O). Experiments at higher temperatures produce more complicated results, including *C*-alkylation at the 4-position. 4-Hydroxypyrazoles lead to the quaternary salts (**232**), the precursors of mesoionic derivatives (**28**) ⟨B-76MI40402⟩.

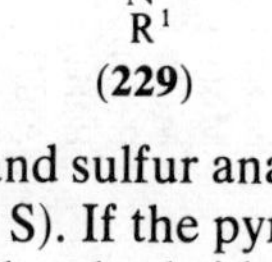

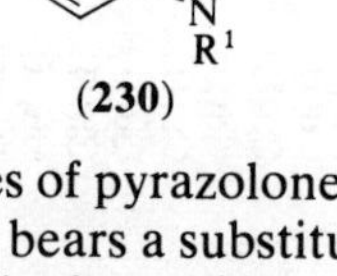

Amino and sulfur analogues of pyrazolones also yield the 'aromatic' quaternary salt (**231**; X = NH or S). If the pyrazole bears a substituent with a second pyridine-like nitrogen atom, an intramolecular bridge can be formed by reaction with a dihalogenoalkane. Thus pyrazol-1'-ylpyridines react with 1,2-dibromoethane to form (**233**) ⟨81JHC9⟩.

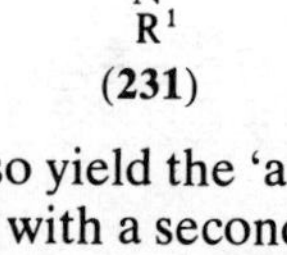

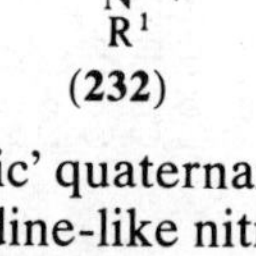

(233)

The kinetic aspects of the S_N2 quaternization (the Menschutkin reaction) have been covered by Zoltewicz and Deady ⟨78AHC(22)72⟩, the latter author having done most of the experimental work related to pyrazoles and indazoles. In C-unsubstituted N-methyl derivatives a Brønsted relationship ($\log k_{rel} = 0.43 pK_a - 2.75$) is followed, and therefore the quaternization rate can be predicted from the pK_a value (Section 4.04.2.1.3(iv)). The point corresponding to 2-methylindazole deviates from the regression line by a factor of 2.7, corresponding to a rate-retarding steric effect. Even if there are no experimental results on the effect of C-substituents on the reaction rate in the case of pyrazoles and indazoles, an equation of the form $\log k/k_H = \delta S^o + \rho_I \sigma_I + \rho_R \sigma_R$ derived for pyridines can be equally valid for *ortho*-substituted pyrazoles (S^o is an *ortho*-steric parameter ⟨80JCS(P2)1350⟩). As commented on in Section 4.04.1.4.3, the rates of N-methylation of N-arylpyrazoles ($\log k/k_H = -1.10\sigma$) show that steric interactions cause the aryl group to rotate out of the plane of the heterocyclic ring in the transition state.

(viii) Alkyl halides and related compounds: compounds with a free NH group

Although N-alkylation is one of the most important and most studied reactions of pyrazoles, its quantitative (orientation ratios) and qualitative (mechanism) aspects are still unclear. All the classical results are reported in ⟨67HC(22)1, B-76MI40402⟩ and can be summarized as follows.

(1) Alkylations have been carried out on the pyrazole ring using alkyl halides (usually iodides and bromides), dialkyl sulfates, arenesulfonates and diazomethane as alkylating agents. Trialkyl phosphates give excellent yields (80–90%) of N-alkylpyrazoles ⟨73JCS(P1)2506⟩. With neutral pyrazoles and alkyl halides the alkylation yields salts of the corresponding acids liberated in the reaction.

(2) In the formation of monoalkyl derivatives from 'unsymmetrical' pyrazoles, not even empirical rules can predict the orientation of the entering alkyl group on the basis of the nature of the existing substituents. For the same reactant, the position taken by an entering alkyl group may be dependent on the nature of the alkylating agent and on experimental conditions. In Scheme 18 some orientation ratios are shown.

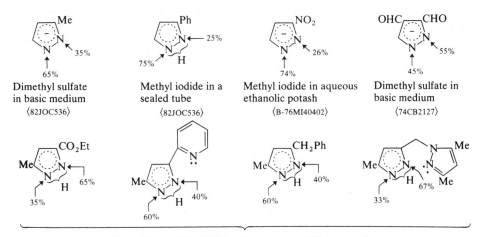

Benzyl bromide in DMF in the presence of KI ⟨80JHC137⟩

Scheme 18

The more recent results of Scheme 18 can be summarized as follows: the less hindered isomer is formed unless the substituent has a lone pair, the electrostatic field of which enhances the nucleophilicity of the adjacent nitrogen atom. This is true not only for pyridine and pyrazole, but probably also for CHO and CO_2Et.

Studies of the alkylation of indazoles ⟨67HC(22)1⟩ have been updated by Nunn ⟨73JCS(P1)2371⟩ and Palmer ⟨75JCS(P1)1695⟩. The ratio of methylation at positions 1 and 2 is relatively sensitive to the steric effect of substituents at positions 3 and 7 as shown by the results obtained in basic medium for unsubstituted indazole (55:45) and its 3-phenyl (74:26) and 7-nitro derivatives (29:71).

Phase transfer catalysis has been used with success to prepare N-substituted pyrazoles ⟨78MI40403, 79MI40408, 70JHC1237, 80JOC3172⟩ and this procedure can be considered the simplest and most efficient way to obtain these compounds. Experimental design methodology has been used to study the influence of the factors on the reaction between pyrazole and n-butyl bromide under phase transfer conditions ⟨79MI40408⟩.

When polyhalogenoalkanes are used as alkylating agents, poly-N-pyrazolylalkanes are obtained. In this way, compounds (**234**), (**235**; R = H) and (**236**) have been prepared from pyrazole anions and methylene iodide ⟨70JA5118⟩, chloroform ⟨B-76MI40402⟩, and carbon tetrachloride ⟨70JA5118⟩, respectively. Phase transfer catalysis is a better way to obtain azolylmethanes ⟨82JHC1141, 83BSF(2)5⟩.

(**234**)   (**235**)   (**236**)

Another possibility is that both nitrogen atoms react with a double alkylating agent. In this way fused pyrazole derivatives (pyrazolo[1,2-a]pyrazoles) like (**237**) can be obtained by reaction of 3,5-dimethylpyrazole with 1,3-dichloropropane or 1-chloro-3-propanol ⟨69BSF2064⟩. More surprising is the reaction with α-chlorocarbonylphenylketene which yields the paraionic compound (**238**) ⟨80JA3971⟩ which can also be obtained from 3,5-dihydroxy-4-phenylpyrazole and β-dicarbonyl compounds ⟨82JOC295⟩.

(**237**)   (**238**)

A parallel exists between the results of protonation and alkylation of pyrazolones since there is an alkyl derivative for each tautomer. The main difference is that the percentage of the different tautomers is thermodynamically controlled whereas that of alkyl derivatives is kinetically controlled. One has to remember that the alkyl derivatives thus obtained are the 'fixed' compounds used in tautomeric studies.

Scheme 19 contains all the reactions observed in different examples, none being so complex ⟨B-76MI40402⟩. Most studies deal with the methylation of 3-methyl-1-phenylpyrazolone, since in this case one of the products obtained is antipyrine (2,3-dimethyl-1-phenyl-3-pyrazolin-5-one), one of the classical antipyretic agents (Section 4.04.4.1.1).

Only the anion (**239**) is susceptible to C-alkylation in the 4-position. The two other anions (**240**) and (**241**) can only give rise to O- and N-substituted derivatives which result in the asymmetry of Scheme 19.

(**239**)   (**240**)   (**241**)

In the alkylation of 1-phenylpyrazole-3,5-diones both carbon and oxygen compete successfully with nitrogen ⟨B-76MI40402⟩. Phenylbutazone (Section 4.04.4.1.1) can be prepared by reaction of 1,2-diphenylpyrazole-3,5-dione with n-butyl halides and alkali, even if the industrial procedure uses n-butyl malonic diesters and hydrazobenzene (Section 4.04.3.1.2(ii)).

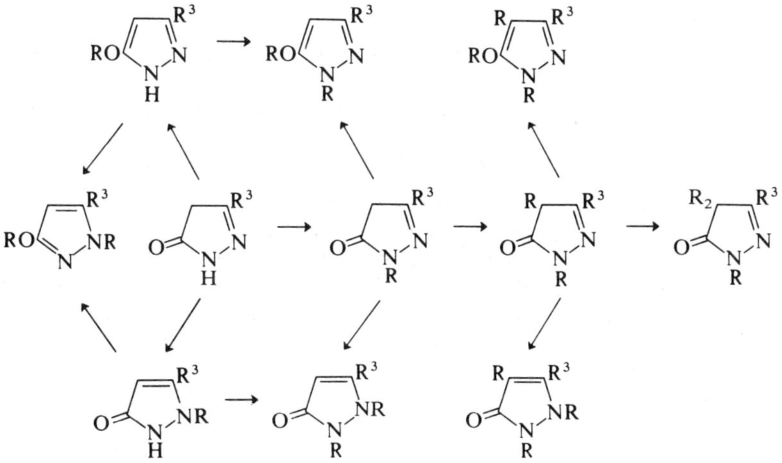

Scheme 19

(ix) Aryl halides and related compounds

The S_NAr reaction of pyrazoles with activated halogenated benzenes has been mainly studied by the Montpellier group (Scheme 20) ⟨66BSF619, 66BSF775, 66BSF2832, 66BSF3727, 67BSF289, 68BSF707, 68BSF5019, 70BSF231⟩.

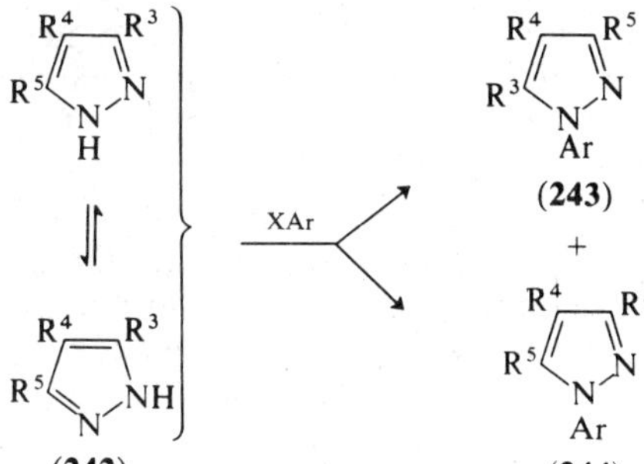

Scheme 20

The principal conclusions from these studies are as follows.

(1) Only *N*-unsubstituted pyrazoles react, thus no 1,2-disubstituted quaternary salts can be obtained in that way.

(2) Only activated halogenated benzenes react (*p*-fluoronitrobenzene, 1-fluoro-2,4-dinitrobenzene, picryl chloride). 3,6-Dichloropyridazine reacts with two moles of pyrazole to yield (**245**), an excellent chelating agent ⟨70BSF1346⟩. With non-activated halogenated benzenes (bromobenzene for instance), the Ullmann condensation should be used ⟨70MI40400⟩. Using this method polymeric pyrazoles were prepared from 1,4-diiodobenzene and 4,4'-bipyrazolyl ⟨69RRC1263⟩.

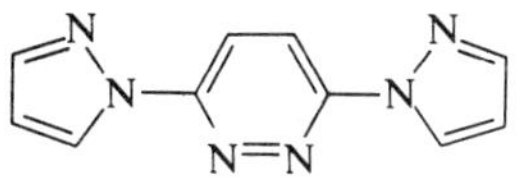

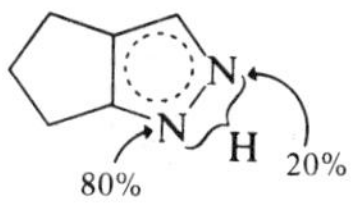

(3) Concerning the orientation of the reaction, *i.e.* the ratio (**243**):(**244**) when asymmetrical pyrazoles are used (**242**; $R^3 \neq R^5$), the tendency of a substituent R^3 to occupy the 3-position (isomer **244**) follows in the order: $CO_2Et > Ph > {>}C{=}C{<} > Et > Me \geqslant Br > Cl > H$. Other experiments prove that $Me > OMe$ and $Bu^t > H$. Finally, the methylene group of a fused saturated ring $>H$, except in the case of the 3(5),4-trimethylenepyrazole (**246**).

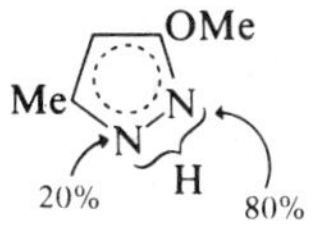

The results obtained with methoxypyrazole (**247**) are attributed to an electronic effect, whereas those obtained with (**246**) can be explained by taking into account the angular strain in the minor isomer. Reaction of 3(5)-nitropyrazole with 1-fluoro-2,4-dinitrobenzene affords exclusively 1-(2,4-dinitrophenyl)-3-nitropyrazole ⟨70JHC1237⟩.

(4) Concerning the reaction rate, a considerable decrease is observed qualitatively when $R^3 = CO_2Et$ or Ph. The presence of two Bu^t groups in the 3- and 5-positions (**242**; $R^3 = R^5 = Bu^t$) completely inhibits the reaction. Structure (**248**) for the transition state has been established from a kinetic study of the reaction between pyrazole and 1-fluoro-2,4-dinitrobenzene ⟨72JCS(P2)1420⟩.

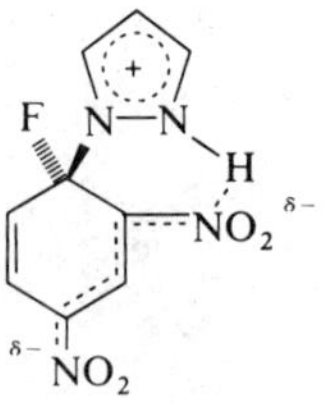

(**248**)

The reaction has been extended to indazole ⟨67BSF2619⟩ from which both isomers have been obtained, and to pyrazolones ⟨68BSF5019⟩. In the latter system N-, O- and C-aryl and even diaryl derivatives have been isolated from the reaction with 1-fluoro-2,4-dinitrobenzene.

(x) Acyl halides and related compounds

Since N-acylpyrazoles have been fully described in ⟨B-76MI40402⟩ only a summary and some new references will be discussed here. Only neutral pyrazolides have been described, but a cationic intermediate (**249**) is involved in the Olofson and Kendall method of obtaining minor N-alkylated isomers ⟨70JOC2246⟩.

N-Acetylated pyrazoles are obtained from N-unsubstituted pyrazoles by treatment with acetyl chloride (alone or in the presence of pyridine) or acetic anhydride. N-Trimethylsilyl- ⟨80JOM(188)141⟩ and N-dimethylgallyl-pyrazoles ⟨81JOM(215)349⟩ (Section 4.04.2.3.10(viii)) are also useful precursors. From an asymmetric pyrazole two acylated derivatives can be obtained, but only one, the less hindered, is often isolated. This happens because the less stable isomer is transformed in the reaction medium into the more stable one (acylotropy; Section 4.04.1.5.1). The same behaviour is observed in indazoles. The unstable 2-acylindazoles can be prepared working at low temperature ⟨76JMC839⟩ or using the indazole silver salt. In the case of pyrazolones the situation is more complicated and N-, O- and C-acyl derivatives are obtained, the first two being subject to acylotropy (no C—C bond breaking has been observed) ⟨64HC(20)1, 74CPB207, 74CPB214⟩. Indazolone (**250**) yields two N-acetyl derivatives (**251**) and (**252**), and a diacetyl derivative (**253**). The unstable isomer (**252**), the structure of which has been determined by X-ray crystallography (Table 5; Section 4.04.1.3.1), is transformed into the stable isomer (**251**) in basic medium ⟨57AG732⟩.

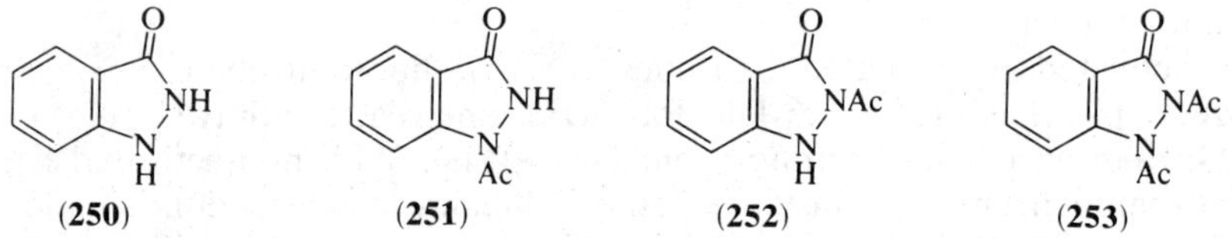

(**250**)   (**251**)   (**252**)   (**253**)

The fact that the isomeric structure of azolides is thermodynamically controlled has been used by Olofson and Kendall to prepare 1-alkylazoles regioselectively ⟨70JOC2246⟩. An asymmetric pyrazole yields two alkylated derivatives (Scheme 21; see Section 4.04.2.1.3 (viii)), but the alkylation with a powerful alkylating agent of the acetylated derivative leads to the less abundant isomer *via* the salt (**249**), which is too unstable to be isolated.

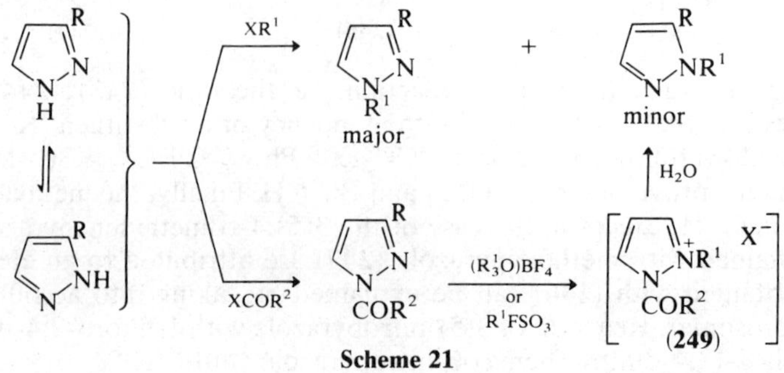

Scheme 21

In addition to the *N*-acylated derivatives described above, *N*-CO$_2$R, *N*-CONR^{1}R^2, *N*-C(=NR^1)R^2 ⟨81JMC1521⟩, *N*-CN ⟨70CB1954⟩, *N*-SO$_2$R^1, *etc.* derivatives have also been prepared from pyrazoles and indazoles. Some of these compounds are biologically active: the *N*-carbamoylindazoles are anthelmintic agents ⟨76JMC839⟩ and the *N*-benzimidoyl-pyrazoles are hypoglycemic agents ⟨81JMC1521⟩.

When pyrazolecarboxylic acids (Section 4.04.2.3.3(iii)) are treated with thionyl chloride instead of the acid chloride, compound (**254**) is isolated. This corresponds to a double *N*-acylation ⟨67HC(22)1⟩.

(**254**)

(xi) Michael addition on double and triple bonds

N-Unsubstituted pyrazoles and indazoles add to compounds containing activated double and triple bonds ⟨67HC(22)1, B-76MI40402⟩. Amongst C—C double and triple bonds, maleic anhydride, acrylic acid esters and nitriles, acetylene-carboxylic and -dicarboxylic esters ⟨78AHC(23)263⟩, quinones, and some α,β-unsaturated ketones have been used with success. Phenylacetylene reacts with pyrazole in the presence of Na/HMPT as catalyst to yield the *Z* isomer of 1-styrylpyrazole in a highly stereoselective reaction ⟨78JHC1543⟩.

With an activated C—C triple bond two successive additions can occur if the intermediate alkene is reactive enough. DMAD and 3,5-dimethylpyrazole give an initial fumarate (**255**) which reacts further at the other end to form regioselectively the succinates (**256**). On the other hand, methyl ethynyl ketone reacts twice at the same carbon atom with pyrazole to form 1,1-pyrazolylbutanone (**258**) ⟨68ZC458⟩. The probable intermediate, a pyrazolide vinylogue (**257**), can be prepared from methyl β-chlorovinyl ketone and pyrazole, in a reaction which is similar to acetylation (Section 4.04.2.1.3(x)).

(**255**)   (**256**)   (**257**)   (**258**)

Indazole adds to acrylonitrile and 4-vinylpyridine giving 1- or 2-substituted derivatives depending on the 7-substituents ⟨69BSF2064⟩. It also reacts with epoxystyrene to form (**259**).

Addition of pyrazole to C—X double bonds is also common. Formaldehyde gives stable adducts (**260**) and (**261**) ⟨69BSF2064⟩, but in the addition to ketones, (**262**) is only observed at low temperatures (Section 4.04.1.3.3(i)). However, hexafluoroacetone forms a stable adduct (**262**; R = CF$_3$) that has been used as a chelating agent (Section 4.04.2.1.3(iv)). Addition of pyrazoles to aryl isocyanates affords (**263**); the addition is also reversible, but it requires high temperatures to dissociate the adduct (Section 4.04.1.5.1).

(**259**)   (**260**)   (**261**)   (**262**)   (**263**)

(xii) Halogens

N-Halogenated pyrazoles are unstable compounds (Cl > Br > I) that are seldom isolated. 1-Bromopyrazoles resemble NBS and may be important in the process of *C*-bromination, not because of an *N* to *C* rearrangement but by acting as a source of the powerfully electrophilic brominium ion (Section 4.04.2.1.4(v)). 4-Substituted pyrazoles can form perbromides which are formulated as (**264**). Such compounds readily lose hydrogen bromide to give 1-bromopyrazoles, the corresponding 1-bromo compound (**265**) from 4-nitropyrazole having been isolated ⟨B-76MI40404⟩.

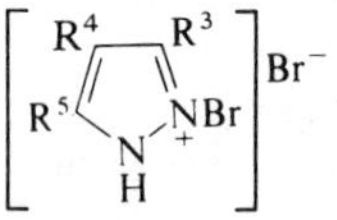

(**264**) 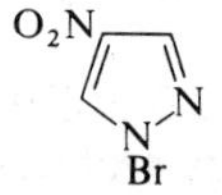 (**265**) 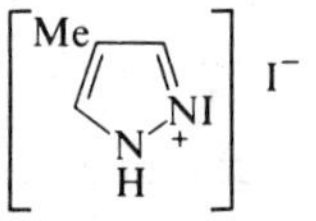 (**266**) 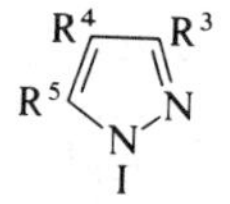 (**267**)

With iodine in carbon tetrachloride, 4-methylpyrazole affords a deep-red oil for which the structure (**266**) has been proposed. Nitric acid, silver nitrate and iodine together convert pyrazole into 1,3,4-triiodopyrazole (**267**; $R^3 = R^4 = I, R^5 = H$). The N-iodopyrazoles are quite stable and well-characterized derivatives, several having been prepared by Hüttel and by Reimlinger ⟨70CB1949⟩ (**267**; $R^3 = R^4 = Me$, $R^5 = H$; $R^3 = R^4 = Me$, $R^5 = I$; $R^3 = R^4 = R^5 = Me$; $R^3 = R^5 = Me$, $R^4 = $ halogen).

(xiii) Peracids

Surprisingly, there are very few examples of successful N-oxidation of pyrazoles. Simple N-alkylpyrazoles generally do not react with peracids ⟨B-76MI40402, 77JCS(P1)672⟩. The only two positive results are the peracetic acid (hydrogen peroxide in acetic acid) transformation of 1-methylpyrazole into 1-methylpyrazole 2-oxide (**268**) in moderate yield and the peroxytrifluoroacetic acid (90% hydrogen peroxide in trifluoroacetic acid) transformation of 5-amino-1-methylpyrazole into 1-methyl-5-nitropyrazole 2-oxide (**269**).

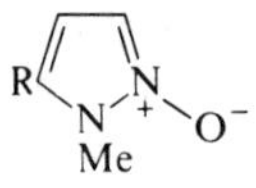

(**268**) 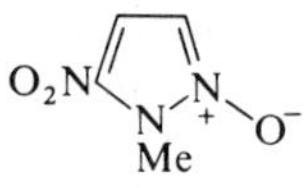 (**269**)

Formation of an N,N'-linked bipyrazole was observed during the oxidation of 5-amino-3-anilino-4-ethoxycarbonylpyrazole by means of dibenzoyl peroxide ⟨82ZC56⟩.

(xiv) Aminating and nitrating agents

The powerful aminating agents hydroxylamine O-sulfonic acid and O-mesitylenesulfonylhydroxylamine have been used to prepare compounds (**270**)–(**274**) ⟨80JCR(M)0514, 76CPB2267⟩.

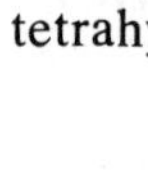

(**270**) 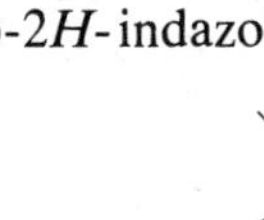 (**271**) 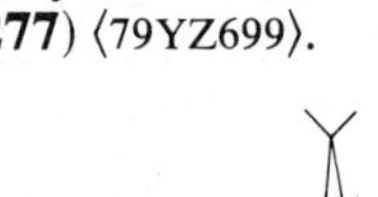 (**272**)  (**273**) 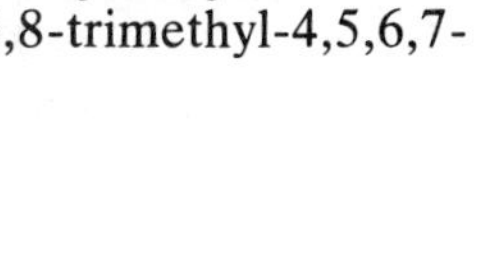 (**274**)

N-Amination of indazole affords a mixture of 60% (**271**) and 40% (**272**), which compares with the 55:45 ratio obtained in methylation (Section 4.04.2.1.3(viii)). A camphopyrazole derivative (a mixture of tautomers **275** and **276**) when treated with hydroxylamine O-sulfonic acid yields exclusively the (4S,7R)-4,7-methano-2-amino-7,8,8-trimethyl-4,5,6,7-tetrahydro-2H-indazole (**277**) ⟨79YZ699⟩.

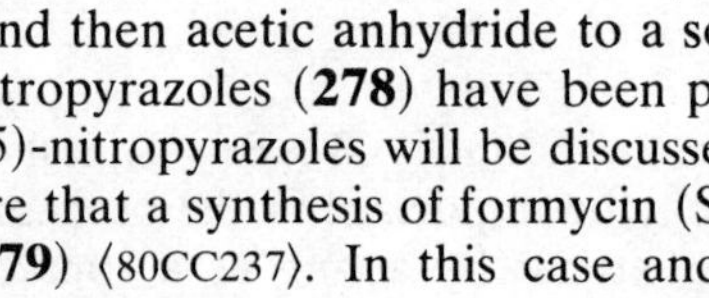

(**275**) (**276**) 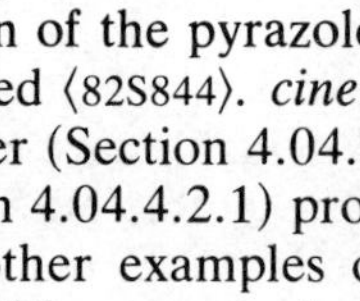 (**277**)

The N-nitration of pyrazoles discovered by Hüttel and extensively studied by Habraken is a reaction of considerable interest ⟨B-76MI40402⟩. Using solutions of acetyl nitrate, or by adding nitric acid and then acetic anhydride to a solution of the pyrazole in acetic acid, a wide range of N-nitropyrazoles (**278**) have been prepared ⟨82S844⟩. *cine*-Substitution and isomerization to 3(5)-nitropyrazoles will be discussed later (Section 4.04.2.3.10(iv)), but it should be noted here that a synthesis of formycin (Section 4.04.4.2.1) proceeds through an N-nitropyrazole (**279**) ⟨80CC237⟩. In this case and in other examples of unsymmetrical pyrazoles, the less hindered isomer is isolated (**278c, d**). With an equivalent of acetyl nitrate

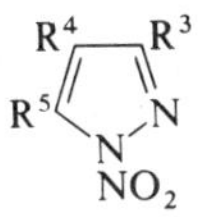

(278)

a; $R^3 = R^5 = H, R^4 = alkyl$
b; $R^3 = R^5 = H, R^4 = NO_2$
c; $R^3 = Bu^t, R^4 = R^5 = H$
d; $R^3 = NO_2, R^4 = R^5 = H$
e; $R^3 = Me, R^4 = R^5 = H$
f; $R^3 = R^4 = H, R^5 = Me$

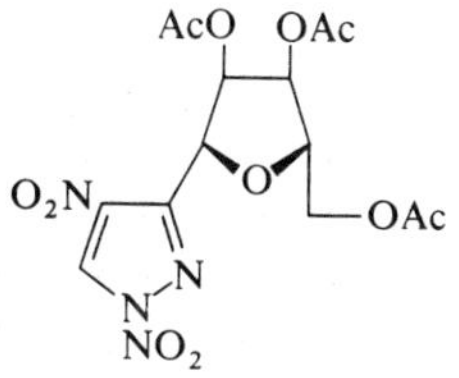

(279)

the 3(5)-methylpyrazole gave a mixture of 3-methyl- (**278e**) and 5-methyl-1-nitropyrazoles (**278f**), but with a large excess of the reagent only the 1,3-compound (**278e**) was obtained.

Depending on the reaction conditions both *N*-nitroindazoles (**280**) and (**281**) have been isolated ⟨71JOC3084, 82OMR(19)225⟩ and their structures were established by ^1H and ^{13}C NMR data. They will be discussed in Section 4.04.2.3.10(vi).

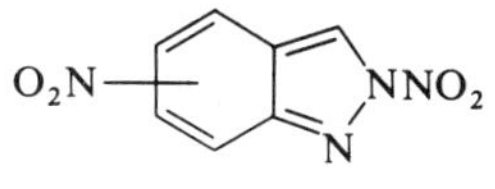

(280)

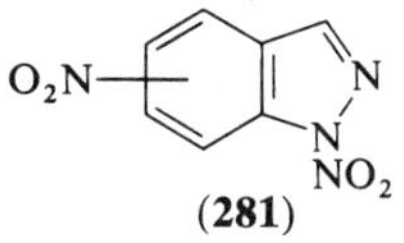

(281)

(*xv*) *Other Lewis acids*

In the preceding parts of Section 4.04.2.1.3 the electrophilic attack on pyrazolic nitrogen with the concomitant formation of different classes of N—R bond has been examined: N—H (iv, v), N—metal (vi), N—C(sp^3) (vii, viii, xi), N—C(sp^2) (ix, x, xi), N—SO_2R (x), N—halogen (xii), N—O (xiii) and N—N (xiv). In this last part the reaction with other Lewis acids leading to the formation of pyrazole N—metalloid bonds will be discussed, and the study of their reactivity will be dealt with in Section 4.04.2.3.10(viii).

Pyrazoles, and some indazoles, substituted on the nitrogen by B, Al, Ga, In, Si, Ge, Sn, P and Hg are known. Poly(pyrazol-1-yl)borates have been studied by Trofimenko ⟨72CRV497⟩ who found that they were excellent ligands (Section 4.04.2.1.3(vi)). The parent ligands (**282**), (**283**) and (**284**) are available by the reaction of an alkali metal borohydride with pyrazole, the extent of substitution depending on the reaction temperature (Scheme 22).

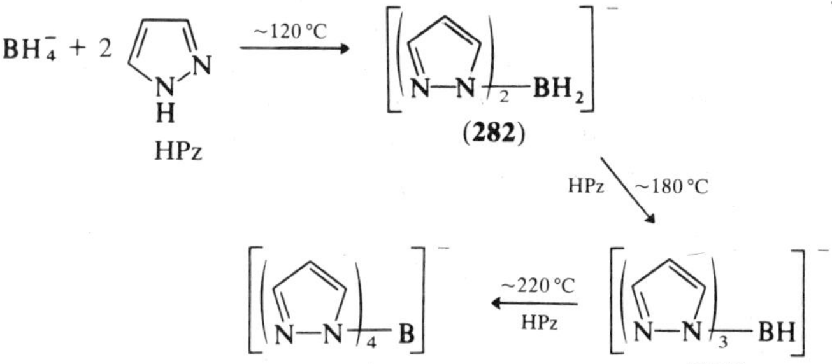

Scheme 22

The three ligands are still widely used to prepare complexes, for instance of gold ⟨80AJC949⟩, palladium ⟨80BCJ961⟩ and platinum ⟨references cited in 81JOM(215)131⟩. Neutral pyrazol-1-ylboranes generally exist in the dimeric pyrazabole structure (**285**) while cations may have the trimeric D_{3h} structure (**286**). Recently monomeric *N*-borylated pyrazole derivatives (**287**) have been synthesized ⟨81JOM(205)147⟩.

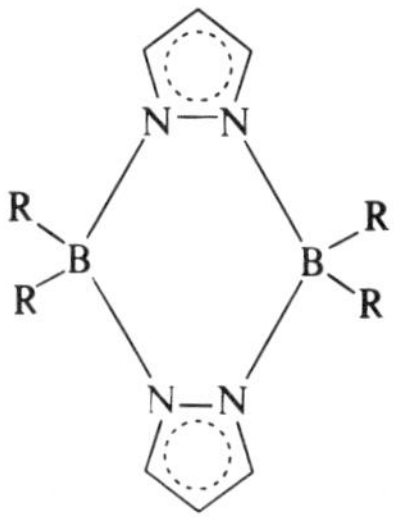

(285)

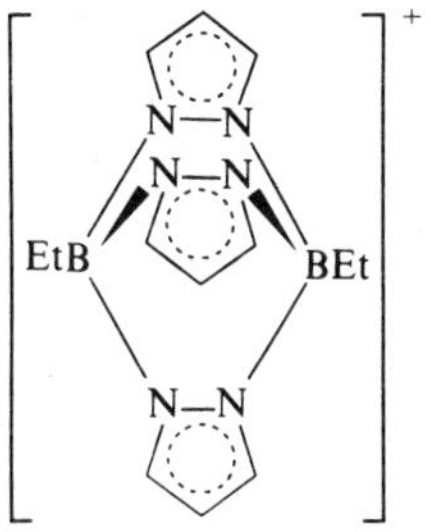

(286)

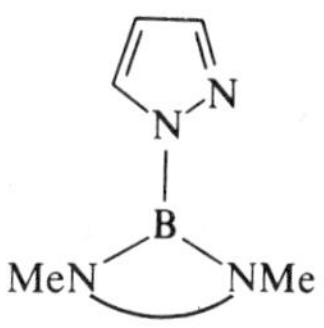

(287)

Analogously, pyrazolyl-aluminate and -indate ligands have been prepared ⟨75JCS(D)749⟩ and their chelating properties evaluated with cobalt, nickel, copper and zinc. Gallyl derivatives of pyrazoles and indazoles have been extensively studied by Storr and Trotter ⟨e.g. 75CJC2944⟩ who determined several X-ray structures of these compounds. These derivatives exist in the solid state as dimers, such as (212) and (288). A ^{1}H NMR study in acetone solution showed the existence of a slow equilibrium between the dimer (212) and two identical tautomers (289) and (290) (Section 4.04.1.5.1) ⟨81JOM(215)157⟩.

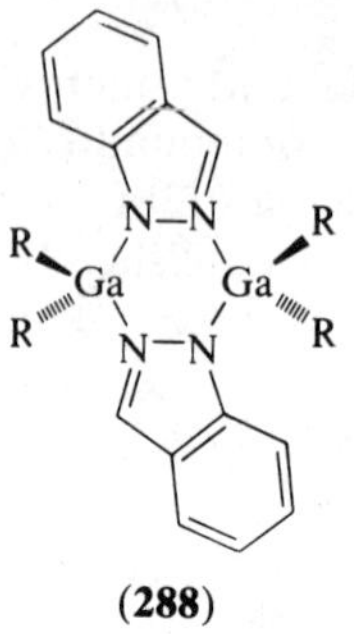

(288) 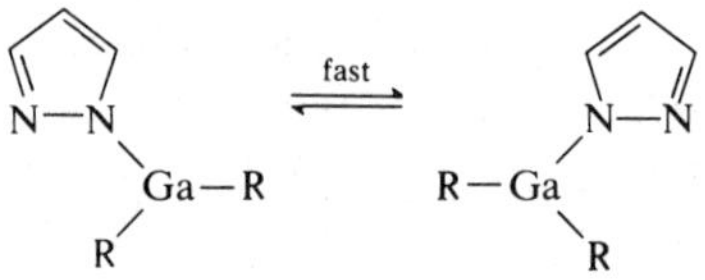   (289)   (290)

1-Silyl-, 1-germyl- and 1-stannyl-pyrazoles and -indazoles are known compounds ⟨80MI40402⟩, the most important being the silylated derivatives ⟨80ZOB875⟩ since they are useful starting materials for other N-substituted derivatives (Section 4.04.2.3.10(viii)). Several reagents can be used to introduce the metalloid on the nitrogen atom including R$_3$MNMe$_2$ (transamination), Me$_3$MCl, Me$_3$M$_2$O ⟨68CR(C)(266)44⟩ and HN(MMe$_3$)$_2$ ⟨80ZOB875⟩ (M = Si, Ge, Sn).

Phosphorus derivatives of different structures have been prepared including pyrazol-1-ylphosphines PPz$_3$, PhPPz$_2$ and Ph$_2$PPz (Pz for pyrazolate anion ⟨72CRV497, 80MI40402⟩). By transamination with tris(dimethylamino)phosphine, pyrazoles and indazole are converted into (291) and (292), respectively ⟨67CR(C)(265)1507⟩. 3,5-Dimethylpyrazole reacts with amidodichlorophosphates to yield triamides (293) whereas 1-substituted pyrazolones yield amidophosphates (294) ⟨71LA(750)39⟩.

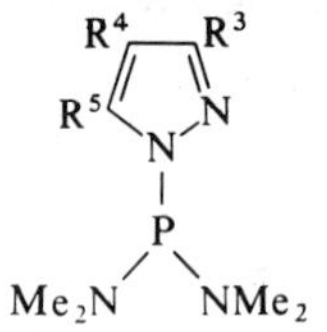

(291) 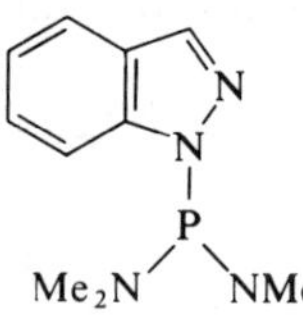   (292)

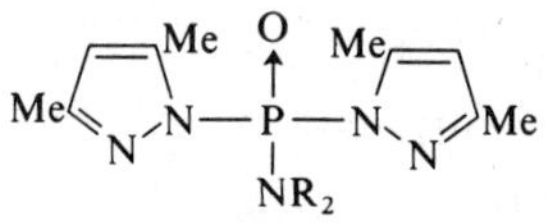

(293) 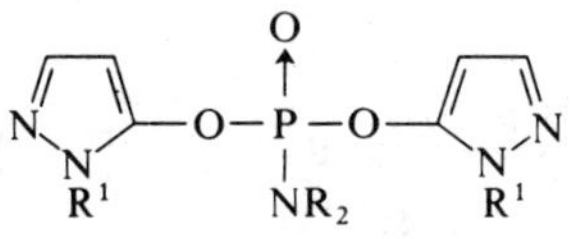   (294) 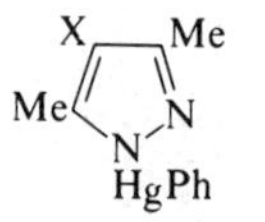   (295) X = Cl, Br or NO$_2$

3,5-Dimethylpyrazole (L) reacts with mercury(II) chloride to give complexes of the structure L$_2$(HgCl$_2$)$_3$. In connection with metallotropy (Section 4.04.1.5.1) the behaviour of compounds (295) has been described. These phenylmercury derivatives were synthesized by the action of phenylmercury hydroxide on the appropriate pyrazole ⟨71MI40400⟩.

4.04.2.1.4 Electrophilic attack at carbon

(i) Reactivity and orientation

The general discussion (Section 4.02.1.4.1) on reactivity and orientation in azoles should be consulted as some of the conclusions reported therein are germane to this discussion. Pyrazole is less reactive towards electrophiles than pyrrole. As a neutral molecule it reacts as readily as benzene and, as an anion, as readily as phenol (diazo coupling, nitrosation, etc.). Pyrazole cations, formed in strong acidic media, show a pronounced deactivation (nitration, sulfonation, Friedel–Crafts reactions, etc.). For the same reasons quaternary pyrazolium salts normally do not react with electrophiles.

As discussed in the theoretical section (4.04.1.2.1), electrophilic attack on pyrazoles takes place at C-4 in accordance with localization energies and π-electron densities. Attack in other positions is extremely rare. This fact, added to the deactivating effect of the substituent introduced in the 4-position, explains why further electrophilic substitution is generally never observed. Indazole reacts at C-3, and reactions taking place on the fused ring will be discussed in Section 4.04.2.3.2(i). Reaction on the phenyl ring of C- and N-phenyl-pyrazoles will be discussed in Sections 4.04.2.3.3(ii) and 4.04.2.3.10(i), respectively. The behaviour of pyrazolones is quite different owing to the existence of a non-aromatic tautomer.

The effect of the substituents on the reactivity of pyrazoles towards electrophiles follows the general pattern discussed in Section 4.02.1.4.1.

(ii) Nitration

Pyrazole is very stable in acid media and even under more vigorous nitration conditions neither ring opening nor ring oxidation was observed (for oxidation to pyrazolones with a mixture of bromine and nitric acid see Section 4.04.2.1.4(v)).

The principal results on the nitration of pyrazoles are shown in Scheme 23. If the substituent is a phenyl group, it can compete with the pyrazole ring and *para*-nitration is often observed (Sections 4.04.2.3.3(ii) and 4.04.2.3.10(i)).

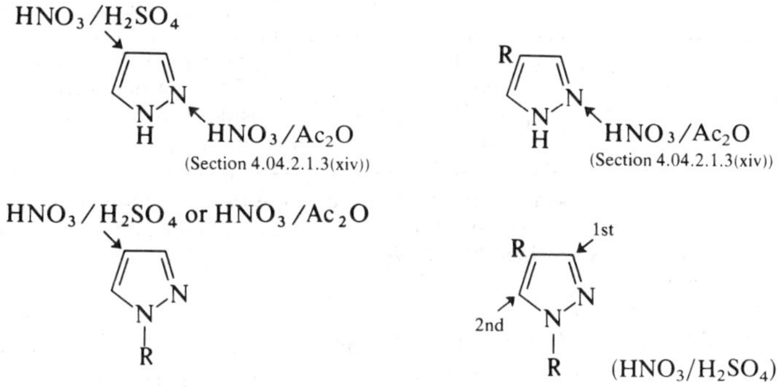

Scheme 23

The mechanism of the reaction is now well known due to a series of kinetic studies by Katritzky *et al.* (Table 31). The nature, free base or conjugate acid, of the substrate depends on the substituents in the pyrazole ring and on the acidity of the nitrating mixture.

Nitration of pyrazolones and alkoxypyrazoles takes place at the 4-position (or in the *para* position of the 1-phenyl substituent) ⟨B-76MI40402⟩. Treatment of propyphenazone (**296**) with fuming nitric acid results in at least 27 products ⟨81AP532⟩, amongst them (**297**) which was identified by ^{1}H and ^{13}C NMR data.

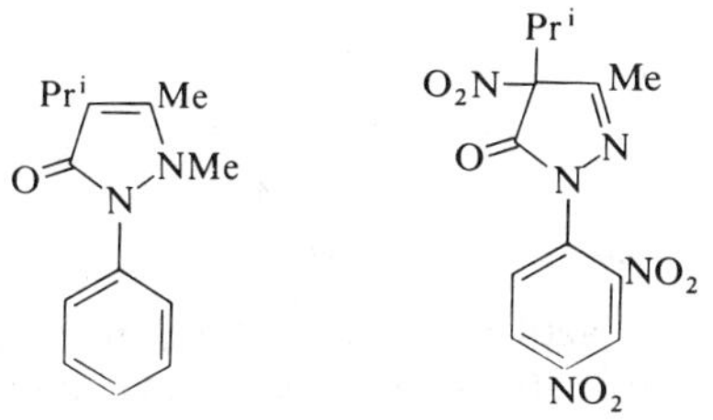

(**296**)   (**297**)

4-Bromopyrazoles (**298**) react with fuming nitric acid in 80% sulfuric acid to give 4-nitropyrazoles (*ipso* nitration; Section 4.04.2.3.7). When R was an alkyl group, nitration took place at C-3 giving (**299**) (a small amount of the dinitro derivative (**300**) was also obtained) ⟨79AJC1727⟩. Nitrodebromination proceeds from the protonated pyrazolium ion whereas 3- and 5-nitration were expected to involve the free base.

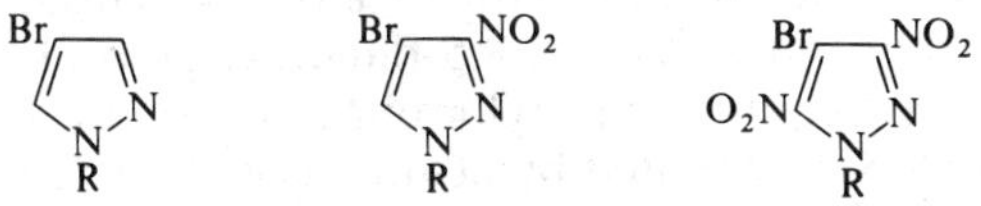

(**298**)   (**299**)   (**300**)

Table 31 Kinetic Nitration Studies on Pyrazoles and Aromatic Pyrazolones (Reagent: HNO_3/H_2SO_4)

Substrate	Conclusions	Ref.
1,3,5-Trimethylpyrazole 1,2,3,5-Trimethylpyrazolium	Nitration occurs on the conjugate acid at the 4-position	71JCS(B)2365
3-Methyl-1-phenylpyrazolone and its N- and O-methyl derivatives	Nitration occurs at the 4- and *para*-positions on the conjugate acid at high acidities. At low acidity some react as free bases	74JCS(P2)382
1-Phenylpyrazoles and their 2-methyl quaternary salts	Nitration occurs at the 4- and *para*-positions; nitro derivatives react as free bases. A methyl group in the 2-position has a high deactivating effect (selective *para* attack)	74JCS(P2)389, 72JCS(P2)1654
Different pyrazoles	Calculation of standard k_0 values at 25 °C and H_0 −6.6. Thermodynamic parameters for pyrazole itself: $\Delta H^{\ddagger} = 69$ kJ mol^{-1}, $\Delta S^{\ddagger} = -109$ J mol^{-1} K^{-1}	75JCS(P2)1600
3-Hydroxy-1-phenylpyrazoles	The 4-nitro derivative nitrates in the *para*-position and the *para*-nitro derivative in the 4-position. At low acidity the reaction occurs on the free base and at high acidity on the conjugate acid	75JCS(P2)1609
Methylpyrazoles	Comparison of standard rates for nitration: benzene, $\log k_0 = 0.45$, 1-methylpyrazole, $\log k_0 = -7.60$, 1,3,5-trimethylpyrazole, $\log k_0 = -4.79$	75JCS(P2)1624
1,4-Dimethylpyrazole 1,4-Dimethyl-3-nitropyrazole	The attack at the 3- and 5-positions occurs on the free base.[a] Standard rates: 1,4-dimethylpyrazole (attack at the 3-position), $\log k_0 = 3.55$, 1,4-dimethyl-3-nitropyrazole (attack at the 5-position), $\log k_0 = -4.73$ (deactivating effect of the nitro group)	75JCS(P2)1632

[a] The difference in the susceptibility of the 3- and 4-positions in the *free-base form* of pyrazole towards nitration is a relatively small factor in favour of the 4-position (*ca.* 1 log unit). Doubtless for the more usual nitration *via* the conjugate acid the difference is considerably greater.

Nitration of 4-(2-thienyl)- (**301**) and 4-(3-thienyl)-pyrazoles (**302**) mainly occurs on the thiophene ring, but when acetyl nitrate is used as the nitration agent small quantities of products nitrated on the pyrazole ring are isolated (position of the nitro group uncertain) ⟨80CS(15)102⟩. Pyrazol-1'-ylpyridines (**303**) undergo electrophilic reactions (bromination, chlorination and nitration) preferentially in the pyrazole ring. Thus, the nitration of (**303**; $R^3 = R^4 = R^5 = H$) either with a mixture of nitric acid and sulfuric acid at 10–15 °C or with acetyl nitrate at 0–5 °C yields the 4'-nitro derivative ⟨81JHC9⟩. When the 4'-position is substituted by bromine, *ipso* nitration and nitration on the pyridine ring are observed.

(**301**)   (**302**)   (**303**)

With mixed acid 1-methylpyrazole 2-oxide (**268**) gives high yields of 1-methyl-5-nitropyrazole 2-oxide (**269**) ⟨B-76MI40402⟩. The form undergoing the reaction is the base, for which first-order kinetics are observed in the H_0 range from −5 to −6.5; a dinitro derivative is also formed under slightly different conditions. The reaction of indazole with fuming nitric acid affords a nearly quantitative yield of 5-nitroindazole (Section 4.04.2.3.2(i)).

(iii) Sulfonation

Direct sulfonation of the pyrazole ring is a rather difficult reaction due to cation formation and takes place only on prolonged heating with 20% oleum. The sulfonic acid group enters at position 4 of pyrazole, 3(5)-methyl- and 3,5-dimethyl-pyrazole ⟨67HC(22)1⟩. Under the vigorous reaction conditions required, phenyl groups, if present, also undergo sulfonation ⟨66AHC(6)347⟩. Indazole can be sulfonated by heating it with 20% oleum for $1\frac{1}{4}$ hours. The product is indazole-7-sulfonic acid (Section 4.04.2.3.2(i)) ⟨67HC(22)1⟩.

Sulfonation of 2- and 3-pyrazolin-5-ones occurs at position 4. Thus 3-methyl-1-phenyl-pyrazolin-5-one with 20% oleum at 10–15 °C yields the corresponding 4-sulfonic acid (304) ⟨64HC(20)1⟩. Higher temperatures cause sulfonation not only of the heterocycle but also of the phenyl group. Antipyrine treated with sulfuric acid and acetic anhydride yields (305).

(iv) Acid-catalyzed hydrogen exchange

Progress in this field is due mainly to the work of Katritzky *et al.* reported in two fundamental papers ⟨73JCS(P2)1675, 75JCS(P2)1624⟩. Qualitatively ⟨B-76MI40402⟩ it was observed that in D_2SO_4 exchange in 1-methylpyrazole occurred initially at C-4 and then simultaneously at C-3 and C-5, and in 1,2-dimethylpyrazolium it occurred only at C-4. Other methylated pyrazoles and pyrazolium ions have been examined, and for all the compounds studied with unsubstituted 4-positions, exchange occurred there preferentially (pyrazoles) or exclusively (pyrazoliums). Rate profiles (70–90 °C) for 4-exchange revealed a change from reaction *via* the free base to reaction *via* the cation at $D_0 = -5$ to -6, a conclusion confirmed by the behaviour of the quaternary salts. At higher temperatures (140–222 °C) exchange occurred at C-3 and C-5 on the free base without mechanistic changeover.

Rates of exchange are collected in Table 32. Methyl groups invariably increase the rate of exchange, but their quantitative effect depends on the position at which reaction is occurring, the position of the methyl group, and the species undergoing reaction. The effect of a methyl group in the 5-position is slightly larger (+1.6 log units) than one in the 3-position (+0.9 log units) upon hydrogen exchange at C-4 in the free base; in the cation the situation is reversed, the effect being +1.0 and +1.5 log units, respectively. The reactivity of protonated pyrazoles is greater than predicted by additivity; this happens when both an electron-donating and an electron-withdrawing group are present in the molecule.

Table 32 Logarithms of the Partial Rate Factors Calculated According to the Standard Conditions (pD = 0, 100 °C)

Pyrazole	Species	3	4	5
1-Methyl	Free base	5.6	9.8	5.6
1-Methyl	Conjugate acid	−2	3.4	−2
1,2-Dimethyl	Pyrazolium cation	a	3.0	a
1,3-Dimethyl	Free base	—	10.7	5.8
1,3-Dimethyl	Conjugate acid	—	4.9	a
1,5-Dimethyl	Free base	5.0	11.4	—
1,5-Dimethyl	Conjugate acid	a	4.4	—
1,2,3-Trimethyl	Pyrazolium cation	—	3.4	a
1,3,5-Trimethyl	Free base	—	13.4	—
1,3,5-Trimethyl	Conjugate acid	—	6.0	—
1,2,3,5-Tetramethyl	Pyrazolium cation	—	5.6	—
1,4-Dimethyl	Free base	6.8	—	6.8
1,2,4-Trimethyl	Pyrazolium cation	a	—	a
1,3,4-Trimethyl	Free base	—	—	7.5
1,4,5-Trimethyl	Free base	7.2	—	—
1,2,3,4-Tetramethyl	Pyrazolium cation	—	—	a

^a Exchange did not occur at any measurable extent.

If the reactivity of the 4-position alone in pyrazolium cations is considered, there is a linear relationship between standard rates for nitration ($k°_2$) (Section 4.04.2.1.4(ii)) and hydrogen exchange ($k°$) ⟨75JCS(P2)1624⟩: $\log k°_2 = 0.89 \log k° - 0.88$, $(\text{c.c.})^2 = 0.98$.

(v) Halogenation

Halogenation is one of the most studied electrophilic substitutions in the pyrazole series ⟨67HC(22)1, B-76MI40402⟩. The results concern chlorination, bromination and iodination since there is no report on direct fluorination of pyrazoles (fluoropyrazoles are prepared by other ways, Sections 4.04.2.3.4(v) and 4.04.3.1.2(ii)). The main results can be summarized as follows: halogenation of neutral and anionic C(4)—H pyrazoles always occurs at C-4; halogenation of anionic C(3)—H indazoles (silver salts) occurs at C-3; polyhalogenation of pyrazoles is observed in N-unsubstituted derivatives and generally occurs via N-halogenation (Section 4.04.2.1.3(xii)); aminopyrazoles, 3-pyrazolin-5-ones (like antipyrine) and 3-hydroxypyrazoles behave normally and are halogenated at C-4. On the other hand, 2-pyrazolin-5-ones yield 4,4-dihalogeno derivatives (306). It was also reported that 4-hydroxypyrazoles are dichlorinated at position 5 to give (307).

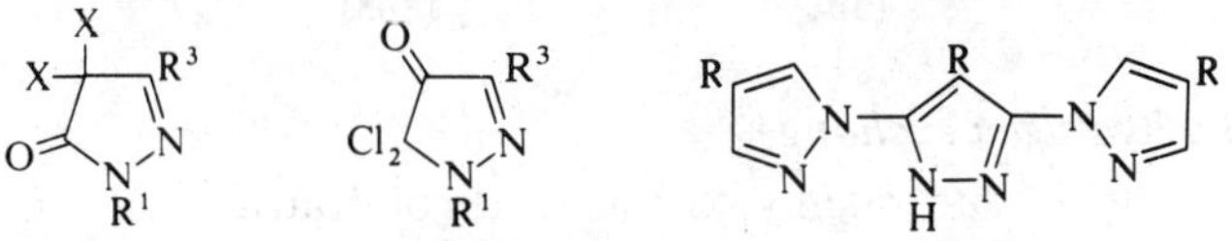

Many reagents are able to chlorinate aromatic pyrazole derivatives: chlorine–water, chlorine in carbon tetrachloride, hypochlorous acid, chlorine in acetic acid (one of the best experimental procedures), hydrochloric acid and hydrogen peroxide in acetic acid, sulfuryl chloride (another useful procedure), etc. N-Unsubstituted pyrazoles are often used as silver salts. When methyl groups are present they are sometimes chlorinated yielding CCl_3 groups. Formation of dimers and trimers (308; R = Cl) has also been observed.

Bromine in chloroform and bromine in acetic acid are the reagents used most often to brominate pyrazole. When nitric acid is used as a solvent, both bromine and chlorine transform pyrazoles into pyrazolones (Scheme 24). Thus 3-methyl-1-(2,4-dinitrophenyl)pyrazole is brominated at the 4-position (309). The product reacts with chlorine and nitric acid to give the pyrazolone (310). The same product results from the action of bromine and nitric acid on (311). The electrophilic attack of halogen at C-4 is followed by the nucleophilic attack of water at C-5 and subsequent oxidation by nitric acid.

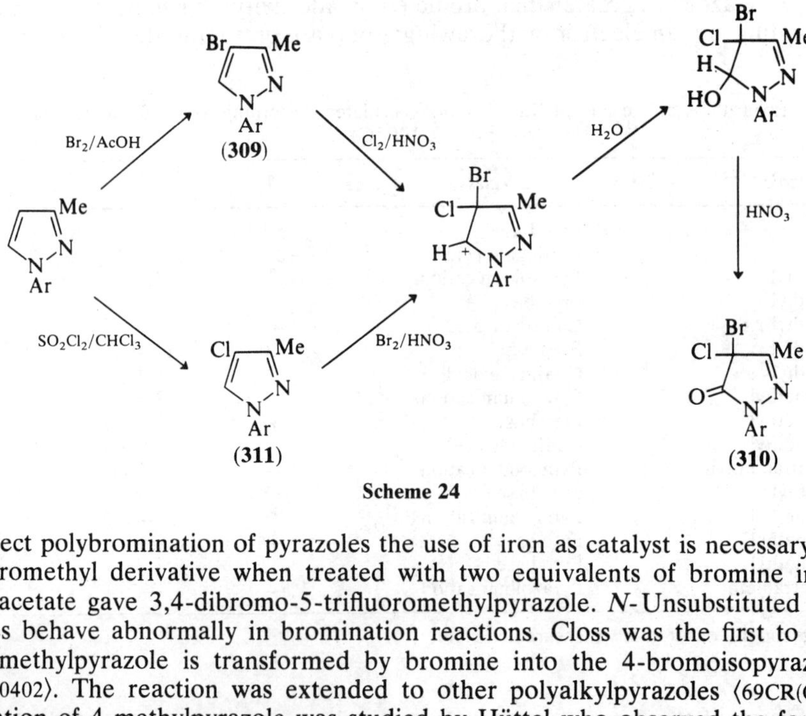

Scheme 24

To effect polybromination of pyrazoles the use of iron as catalyst is necessary; only the 3-trifluoromethyl derivative when treated with two equivalents of bromine in aqueous sodium acetate gave 3,4-dibromo-5-trifluoromethylpyrazole. N-Unsubstituted 4-methylpyrazoles behave abnormally in bromination reactions. Closs was the first to show that 3,4,5-trimethylpyrazole is transformed by bromine into the 4-bromoisopyrazole (312) ⟨B-76MI40402⟩. The reaction was extended to other polyalkylpyrazoles ⟨69CR(C)(269)570⟩. Bromination of 4-methylpyrazole was studied by Hüttel who observed the formation of dimers (313) and trimers (308; R = Me) ⟨69CR(C)(269)570⟩. Reimlinger et al. described the formation of the cyclic dimer (314) from the reaction between bromine and the silver salt of the 3,4,5-trimethylpyrazole ⟨70CB1949⟩. It was later shown ⟨82UP40402⟩ that the correct structure of the product is (315), a structure which accounts for the anisochrony of the methylene protons ('crown' conformation).

Aqueous bromination of indazole shows the relative order of reactivity for various positions to be 5 > 7 ≫ others, 5 > 3 > 7 ≫ others, and 3 ≫ others for reactions of cationic,

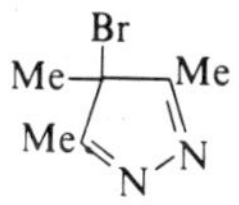

(312) 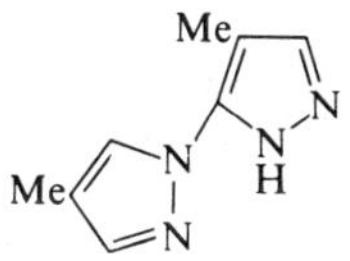   (313) 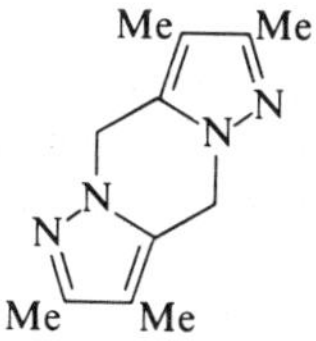   (314) 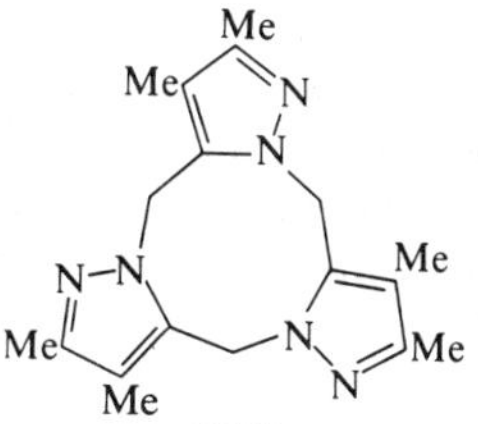   (315)

neutral and anionic species, respectively ⟨74AJC2343⟩. This may account for the variable substitution patterns found in other reactions, viz. 3-bromination in dioxane and 3-azo coupling; 3-, 3,5- and 3,5,7-halogenation in acetic acid; 5- and 5,6-nitration; and 7-substitution (sulfonation in oleum) (Section 4.04.2.3.2(i)).

Pyrazole does not react with iodine although pyrazolylsilver is converted into 4-iodopyrazole. 3-Iodoindazole can be obtained by the reaction of iodine with the silver salt of indazole. Kinetic studies on pyrazole iodination have been carried out by Vaughan et al. ⟨71PMH(4)55, B-76MI40402⟩. Coordination of pyrazole by nickel(II) in aqueous solution increases the rate of iodination by factors of two at pH 6 and eight at pH 7.2 ⟨72JA2460⟩.

(vi) Acylation and alkylation

1-Substituted pyrazoles are formylated (Vilsmeier–Haack reaction) and acetylated (Friedel–Crafts reaction) at C-4 ⟨B-76MI40402⟩. Both hydroxy and amino substituents in positions 3 and 5 facilitate the reaction ⟨80ACH(105)127, 80CHE1⟩, but the heteroatoms compete with the *C*-substitution. For instance, when the amino derivative (**91**; $R^1 = R^3 = Ph$, $R^4 = H$) was submitted to the Vilsmeier–Haack reaction, the 4-formyl-5-dimethylaminomethylene derivative was isolated.

C-Alkylation of pyrazoles was a rather uncommon reaction until Grandberg and Kost found the experimental conditions necessary to obtain high yields of 4-benzylpyrazoles ⟨66AHC(6)347⟩. With *N*-unsubstituted pyrazoles a large excess of aluminum chloride is necessary to accomplish alkylation at C-4.

(vii) Mercuration

Pyrazoles were chloromercurated at C-4 giving (**316**) by reaction with mercury(II) chloride ⟨B-76MI40402, 66AHC(6)347⟩. The mercury group in (**316**) behaves in a variety of reactions just as phenylmercury chloride. 1-Phenylpyrazole reacts with mercury(II) acetate to yield the 4-acetoxymercury derivative (**317**). The same reactions with 3(5)-phenylpyrazole also introduce the acetoxymercury group at position 4 and not on the nitrogen atom.

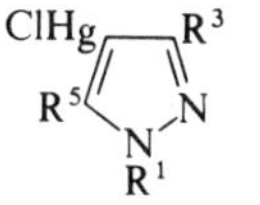

(316) 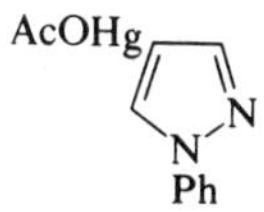   (317)

(viii) Diazo coupling

Generally, pyrazoles do not react with diazonium salts. The literature ⟨B-76MI40402⟩ reports only one such example, the coupling between 3,5-dimethylpyrazole and diazotized 3-aminopyrazole. When an activating group (hydroxy, alkoxy, amino) is present at position 3 or 5, the reaction proceeds easily at position 4. Thus 5-aminopyrazoles give the azo derivatives (**318**) and (**319**) by reaction with the appropriate diazonium salt ⟨78HCA108, 80CHE1160⟩.

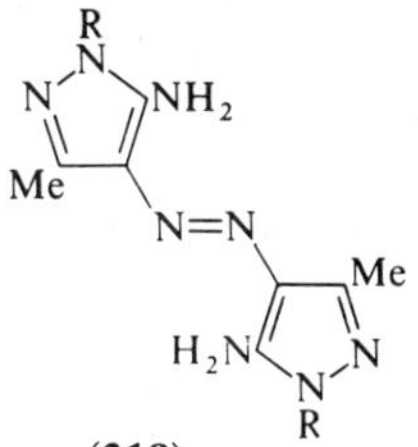

(318) 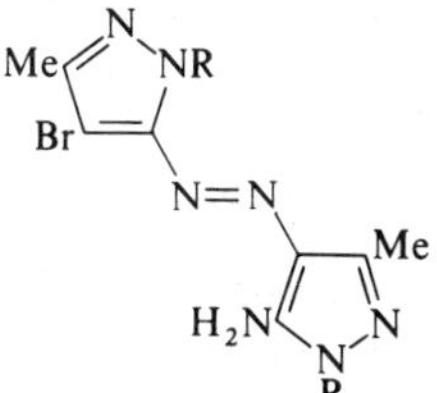   (319)

The reaction is very common in pyrazolone chemistry. Since alkoxypyrazoles and tautomerizable pyrazolones undergo this reaction and 3-pyrazolin-5-ones, like antipyrine, do not, it is assumed that the reaction takes place at C-4 of the OH tautomer. Pyrazolone diazo coupling is an important industrial reaction since the resulting azo derivatives are used as dyestuffs. For instance, tartrazine (Section 4.04.4.1.3) has been prepared this way. 3,5-Pyrazolidinediones react with aryldiazonium salts resulting in the introduction of a 4-arylazo group. As has been described in Section 4.04.2.1.4(v), diazonium salts couple in the 3-position with indazole to give azo compounds.

(ix) Nitrosation

The behaviour of pyrazoles towards nitrosation is similar to their behaviour described above towards diazo coupling, *i.e.* aminopyrazoles and pyrazolones readily react with nitrosation agents, like alkyl nitrites ⟨81FES1019⟩, to afford stable nitroso derivatives. Some simple nitrosopyrazoles have been isolated, for example the blue-green 3,5-dimethyl-4-nitrosopyrazole, and many others have been proposed as reactive intermediates in the direct conversion of pyrazoles into diazonium or diazo derivatives (Scheme 25) ⟨B-76MI40402⟩.

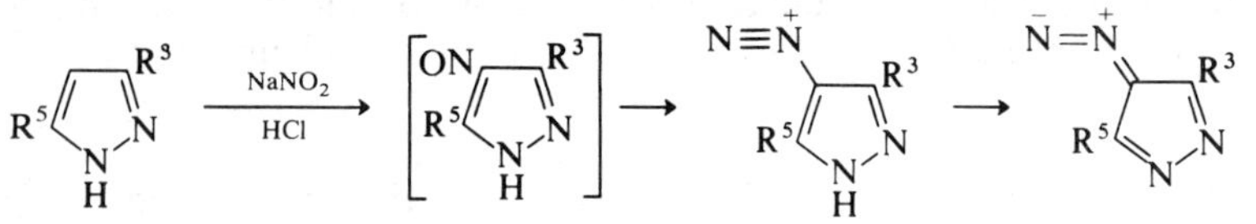

Scheme 25

(x) Reaction with aldehydes and ketones

N-Unsubstituted pyrazoles and indazoles react with aldehydes and ketones to give *N*-hydroxyalkyl compounds (**260**)–(**262**) (Section 4.04.2.1.3(xi)). *N*-Substituted pyrazoles, even those containing a CH$_2$OH group, react with paraformaldehyde in the presence of acids to give a variety of products ⟨B-76MI40402, 68JHC407, 82UP40403⟩ depending on the pyrazole and the experimental conditions. Compounds (**320**) to (**323**) have been isolated and identified.

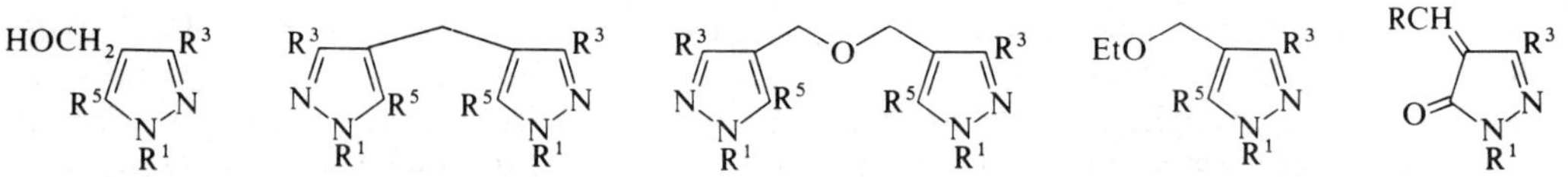

(**320**)   (**321**)   (**322**)   (**323**)   (**324**)

Pyrazolones show a great variety of reactions with carbonyl compounds ⟨B-76MI40402⟩. For instance, antipyrine is 4-hydroxymethylated by formaldehyde and it also undergoes the Mannich reaction. Tautomerizable 2-pyrazolin-5-ones react with aldehydes to yield compound (**324**) and with acetone to form 4-isopropylidene derivatives or dimers (Scheme 8; Section 4.02.1.4.10).

(xi) Oxidation

N-Oxidation with peracids (Section 4.04.2.1.3) and the transformation of pyrazoles into 4,4-dihalogeno-2-pyrazolin-5-ones (Section 4.04.2.1.4(v)) have already been discussed. Transformation of non-aromatic 2-pyrazolin-5-ones into the 4-oxo derivatives will be examined in Section 4.04.2.2.1(ii).

C-Oxidation of pyrazoles and indazoles disrupts the azole ring. Potassium permanganate transforms *C*-methylpyrazoles into pyrazolecarboxylic acids (Section 4.04.2.3.3(i)), and under more drastic conditions indazole is completely destroyed ⟨67HC(22)1⟩. Open-chain compounds (glyoxal, benzoic acid, phenylazocrotonitrile, hydrazine, *etc.*) are isolated by oxidation of pyrazoles with ozone or peracids ⟨B-76MI40402⟩. In some instances ring transformations are observed; 1-aryl-3-methylpyrazoles react with ozone to yield 1,3,4-oxadiazolinones (**325**; Section 4.02.1.4.11), 3,5-diphenylpyrazole reacts with peracetic acid to yield 2,5-diphenyl-1,3,4-oxadiazole (**326**), and 1,4-dihydroxy-3,5-diphenylpyrazole reacts with peracids to give an entry into the 1,3,4-oxadiazin-6-one 4-oxides (**327**). Hydrolysis of (**327**) was accompanied by a facile rearrangement to the corresponding oxadiazole (**326**).

(325)   (326)   (327)

4.04.2.1.5 Nucleophilic attack at carbon

Very little is known about nucleophilic attack on an unsubstituted carbon atom of pyrazoles and their aromatic derivatives (pyrazolones, pyrazolium ions). The $S_N Ar$ reaction of halogenopyrazoles will be discussed in Section 4.04.2.3.7. Sulfur nucleophiles do not attack the ring carbon atoms of pyrazolium salts but instead the substituent carbon linked to nitrogen with concomitant dequaternization (Section 4.04.2.3.10(ii)). The ring opening of pyrazolium salts by hydroxide ion occurs only if carbon C-3 is unsubstituted; the exact mechanism is unknown and perhaps involves an initial attack of OH^- on C-3.

Only two topics are of importance for this section: the reduction of pyrazolium salts and 3-pyrazolin-5-ones by complex hydrides, and the nucleophilic photosubstitution of pyrazoles and indazoles.

(i) Reduction by complex hydrides

Δ^2-Pyrazolines are obtained by the reduction of pyrazoles with sodium and alcohol, by catalytic hydrogenation on palladium or by electrochemical means ⟨B-76MI40402⟩. In some cases the reduction proceeds further yielding pyrazolidines and open-chain compounds.

Δ^3-Pyrazolines and pyrazolidines are isolated from the lithium aluminum hydride reduction of 3-pyrazolin-5-ones and pyrazolium salts. Tautomerizable 2-pyrazolin-5-ones are not reduced (Section 4.04.2.2.1(ii)). A large number of publications by the Montpellier group dealing with this topic are summarized in ⟨B-76MI40402⟩. Three papers ⟨69BSF4466, 69BSF4469, 69BSF4474⟩ describing thiopyrine, pyramidon and thiopyramidon reduction should also be consulted.

A mechanism has been proposed to rationalize the results shown in Figure 23. The relative proportion of the Δ^3-pyrazolines obtained by the reduction of pyrazolium salts depends on steric and electronic effects. When all the substituents are alkyl groups, the hydride ion attacks the less hindered carbon atom; for example when $R^3 = Bu^t$ only C-5 is attacked. The smaller deuterohydride ion is less sensitive to steric effects and consequently the reaction is less selective ⟨73BSF288⟩. Phenyl substituents, both on the nitrogen atom and on the carbon atoms, direct the hydride attack selectively to one carbon atom and the isolated Δ^3-pyrazoline has the C—C double bond conjugated with the phenyl (328; R^1 or R^5 = Ph). Open-chain compounds are always formed during the reduction of pyrazolium salts, becoming predominant in the reduction of amino substituted pyrazoliums.

1-Isopropyl-3-phenylindazole is cleaved when it reacts with phenyllithium, followed by hydrolysis ⟨73KGS138⟩ giving, after acetylation, o-benzoyl-N-isopropylacetanilide.

(ii) Nucleophilic photosubstitution

The research carried out at Leiden by Cornelisse and Havinga and at Montpellier by Bouchet and Lazaro has opened the way for the introduction of different nucleophiles into the pyrazole and indazole ring. Reactions unknown in the ground state are possible in the singlet excited state. Upon irradiation, not only halogens or a nitro group but also the hydrogen atom can be replaced by the nucleophile. Irradiation of pyrazoles (329; R^1 = Me, p-$NO_2C_6H_4$, $R^3 = NO_2$, Br, Me) ⟨78RTC35, 79T1331⟩ in the presence of cyanide ions yields 4-cyanopyrazoles (330); if the 4-position is substituted as in (331), the cyano group enters in the 5-position (332; $R^1 = p$-$NO_2C_6H_4$). Analogously, irradiation in the presence of methanolic sodium acetate or potassium cyanate gives the 1-p-nitrophenylpyrazoles formylated in positions 4 and 5, respectively, in moderate yields ⟨79T1331⟩.

(329)   (330)   (331)   (332)

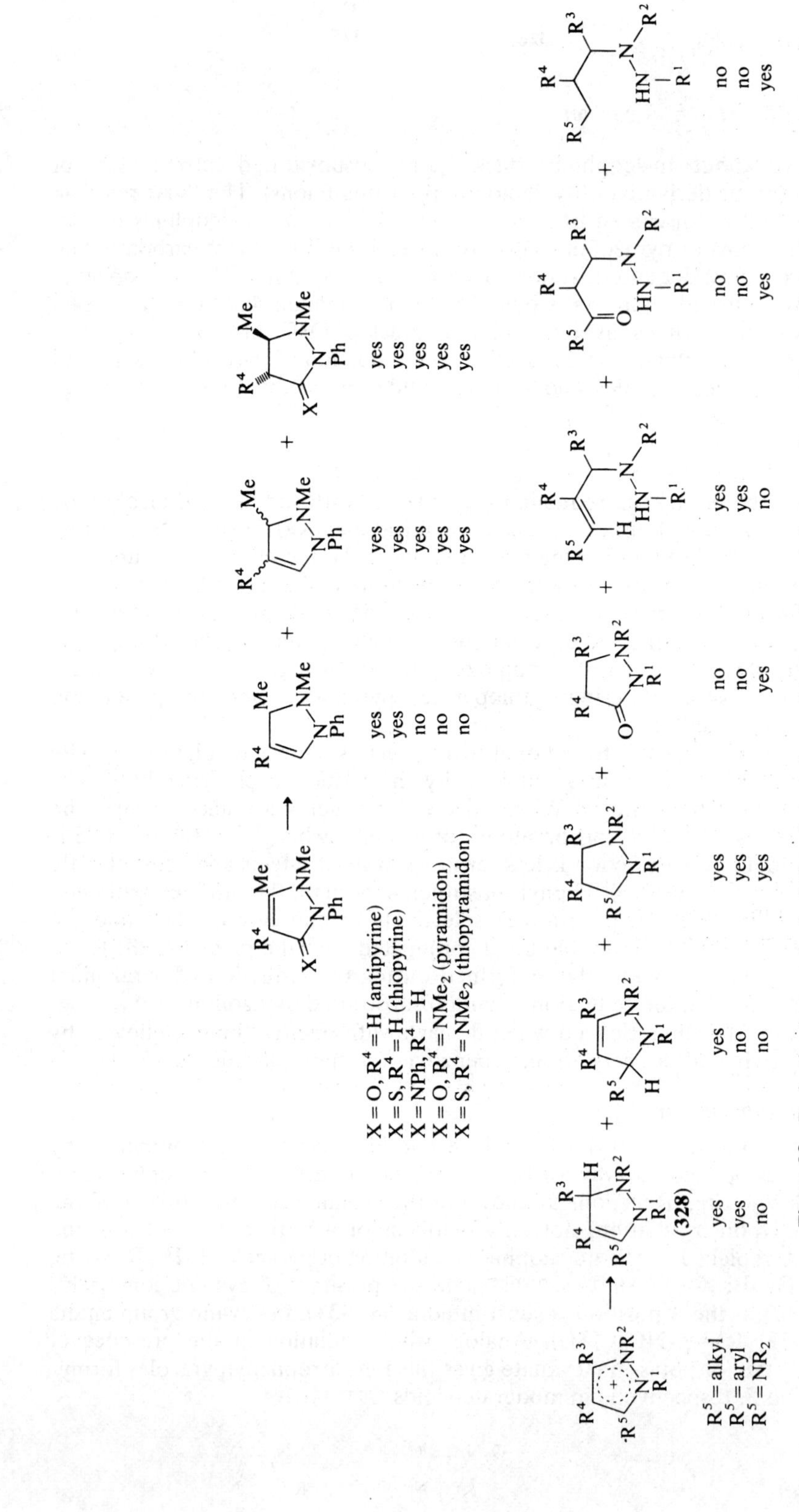

Figure 23 Reduction by complex hydrides of pyrazolones, thiopyrazolones, iminopyrazolines and pyrazolium salts

Similarly, the hydrogen atom at the 3-position of the indazole ring has been replaced by NH_2, NHMe, NMe_2 and NEt_2 upon irradiation in the presence of the appropriate nucleophile ⟨80T3523⟩.

4.04.2.1.6 Nucleophilic attack at hydrogen

(i) Metallation at a ring carbon atom

Lithiation of pyrazoles has been studied as a synthetic route to pyrazolecarboxylic acids, *n*-butyllithium being more effective than phenyllithium. *N*-Substituted pyrazoles are metallated at C-5 ⟨B-76MI40402, 79RCR289⟩ (Table 6 in Section 4.02.1.7.1); with a substituent on C-5, lithiation occurred at the *N*-methyl group (**333**; X = Li). 1-Methylindazole reacted with butyllithium at −15 °C to form 1-lithiomethylindazole (**334**; X = Li) whereas 2-methylindazole gave the corresponding 3-lithio derivative (**335**; X = Li) ⟨70CHE1339⟩. Treatment of these lithio derivatives with carbon dioxide gave the respective acids (**333**)–(**335**) ($X = CO_2H$).

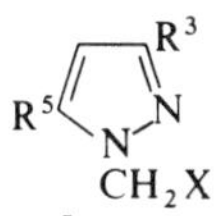

(**333**) R^5 = Me or OMe

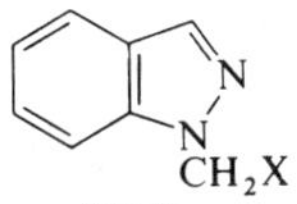

(**334**)

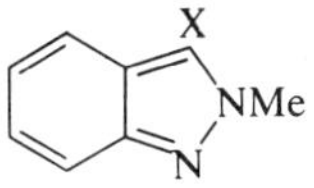

(**335**)

(ii) Hydrogen exchange at ring carbon in neutral pyrazoles

The base-catalyzed hydrogen exchange of pyrazoles has been studied by Vaughan and conveniently summarized in ⟨74AHC(16)1⟩ and ⟨B-76MI40402⟩. The method has been used to prepare the isotopic species necessary to assign the fundamental vibration of pyrazole (Sections 4.04.1.3.7(i) and 4.04.3.3.3).

(iii) Hydrogen exchange at ring carbon in pyrazolium cations

The base-catalyzed deuteration of the 1,2-dimethylpyrazolium ion has been studied by ^1H NMR spectroscopy. Exchange occurs at positions 3 and 5, and position 4 does not exchange. A complementary situation arises under acid-catalyzed (D_2SO_4) conditions where, for the same salt, only H-4 exchanges (Section 4.04.2.1.4(iv)). Other quaternary azolium salts exchange at a faster rate than do pyrazolium salts ⟨B-76MI40402⟩, for example, H-2 of 1,3-dimethylimidazolium exchanges 3×10^4 times faster. The presence of a bromine atom on C-4 significantly increases the acidity of nuclear hydrogen atoms in pyrazolium salts.

(iv) C-Acylation via *deprotonation*

Neither pyrazoles nor pyrazolium salts react by this mechanism which has been described for imidazoles and imidazolium salts (Section 4.01.1.7.4). As exchange rates show (Section 4.04.2.1.7(iii)), it is considerably more difficult to generate an ylide from a pyrazolium salt than from an imidazolium salt (at C-2).

(v) Ring cleavage via *C-deprotonation*

3-Unsubstituted pyrazoles and indazoles are cleaved by strong bases. Indazoles undergo ring opening more readily than pyrazoles since the fused benzene ring enhances the ease of removal of the C-3 proton in the initiating step.

Potassium *t*-butoxide in *t*-butyl alcohol requires powerful electron-attracting substituents at C-4 to effect ring opening of pyrazoles but sodamide does not (Scheme 26) ⟨B-76MI40402⟩. As the key to the transformation is the generation of the anion, similar results were obtained by heating some pyrazole-3-carboxylic acids with quinoline.

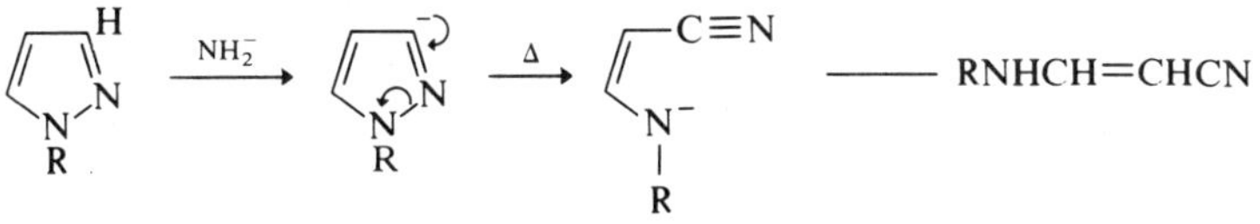

Scheme 26

Similar mechanisms explain the ring opening of 1-alkyl- and 1-aryl-indazoles by sodamide to *o*-cyanoanilines ⟨65CHE531⟩ and the ring opening of 1-arylindazole-3-carboxylic acid during decarboxylation in boiling quinoline (Scheme 27) ⟨73AJC2683⟩. The presence of electron-withdrawing substituents in the 1-aryl group favours the ring opening during the formation of 1-arylindazoles.

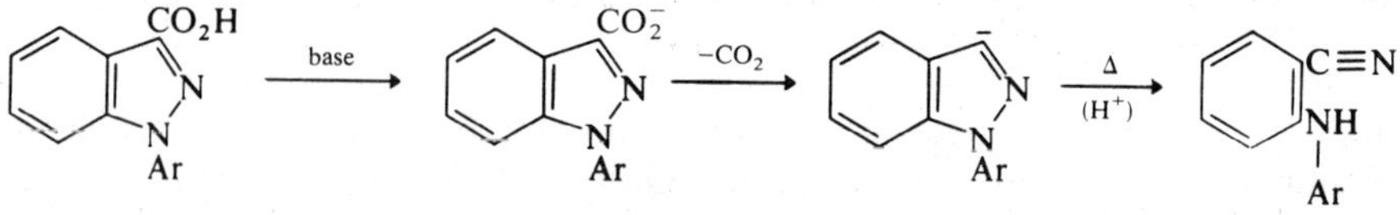

Scheme 27

(vi) *Proton loss from a ring nitrogen atom*

Pyrazoles are weak acids unless they carry powerful electron-withdrawing groups (Section 4.04.2.1.3(v)). They form metallic salts which are readily hydrolyzed by water (Section 4.04.2.1.3(vi)).

4.04.2.1.7 *Reaction with radicals and electron-deficient species; reaction at surfaces*

(i) *Carbenes and nitrenes*

3,4,5-Trimethylpyrazole (336) adds dichlorocarbene generated under basic conditions ($CHCl_3$–EtONa) to give 10% of 4-dichloromethyl-3,4,5-trimethylisopyrazole (337; Scheme 28) (bromine also transforms (336) into an isopyrazole (312); Section 4.04.2.1.4(v)). Treatment with sodium ethoxide results in ring expansion of (337) into an ethoxymethylpyridazine (338) ⟨B-76MI40402⟩.

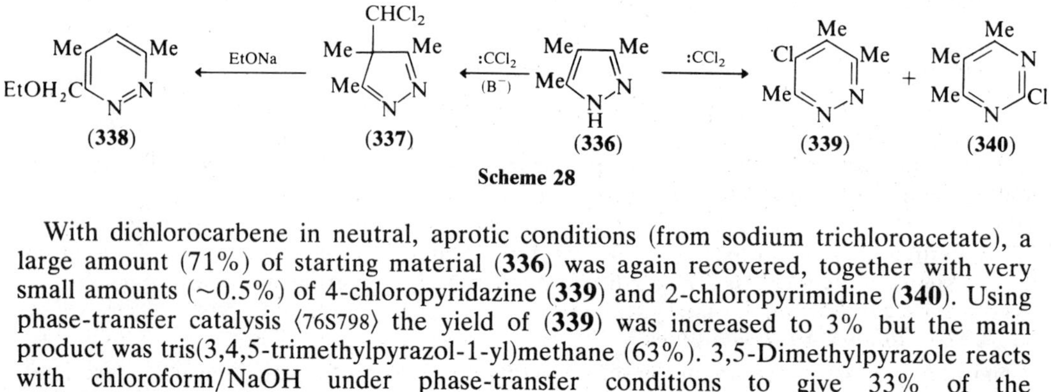

Scheme 28

With dichlorocarbene in neutral, aprotic conditions (from sodium trichloroacetate), a large amount (71%) of starting material (336) was again recovered, together with very small amounts (~0.5%) of 4-chloropyridazine (339) and 2-chloropyrimidine (340). Using phase-transfer catalysis ⟨76S798⟩ the yield of (339) was increased to 3% but the main product was tris(3,4,5-trimethylpyrazol-1-yl)methane (63%). 3,5-Dimethylpyrazole reacts with chloroform/NaOH under phase-transfer conditions to give 33% of the tris(pyrazolyl)methane (235; R = Me) and 1% of 2-chloro-4,6-dimethylpyrimidine ⟨79LA1456⟩.

Better results were obtained when a mixture of chloroform and either pyrazole or *C*-methylpyrazoles was heated at 555 °C in a continuous-flow vapour phase reactor ⟨79JCS(P1)2786⟩. 2-Chloropyrimidines were obtained in high yields (51–89%). Indazole similarly gave only 2-chloroquinazoline (68%).

An interesting example of photolytic nitrene cyclization, (341) → (342), has been described by Tsuge and Samura ⟨71JHC707⟩. This method has often been used to prepare azapentalenes with two ring junction nitrogen atoms ⟨78AHC(22)184, 81AHC(28)232, 81CL331⟩ from 1-arylpyrazoles and indazoles.

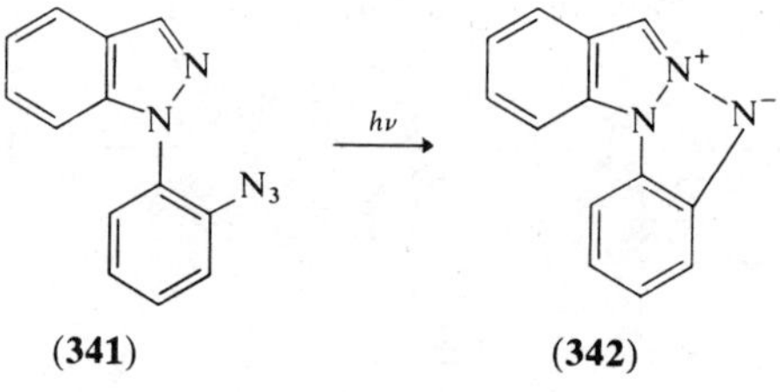

(ii) Free radical attack at the ring carbon atoms

Methyl radicals generated by the decarboxylation of acetic acid with ammonium peroxydisulfate and pyrazole gave about 10% of 3(5)-methylpyrazole ⟨B-76MI40402⟩. Another highly selective alkylation takes place when pyrazole is treated with *t*-butyl radicals ⟨73OPP105⟩, 3(5)-*t*-butylpyrazole being the only product.

Radical phenylation of *N*-substituted pyrazoles has been studied by Lynch *et al.* ⟨B-76MI40402⟩. The most significant results are shown in Figure 24.

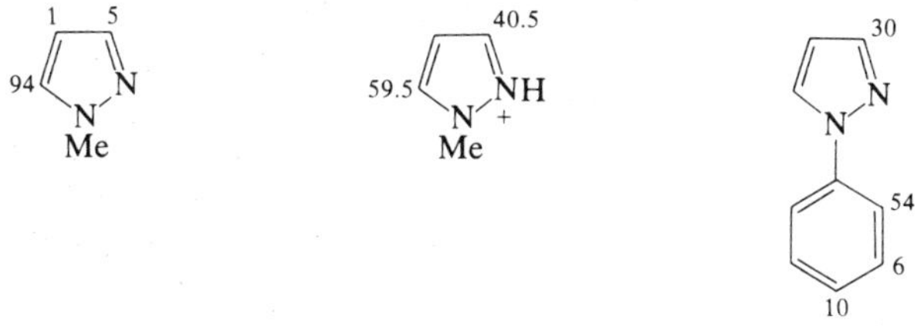

Figure 24 Percentages of radical phenylation attack

Clearly, the proportion of substitution occurring adjacent to the 'pyridinic' nitrogen atom is increased by protonation. Also noteworthy are the high proportion of *ortho* substitution product and the selective attack at C-3 in the *N*-phenyl derivative.

(iii) Electrochemical reactions and reactions with free electrons

Electrochemical oxidation of 3-methyl-1-phenylpyrazole gave the 3-carboxylic acid whereas electrochemical reduction (Section 4.04.2.1.6(i)) of 1,5-diphenyl-3-styrylpyrazole produced the Δ^2-pyrazoline ⟨B-76MI40402⟩ with concomitant reduction of the exocyclic double bond (**343**).

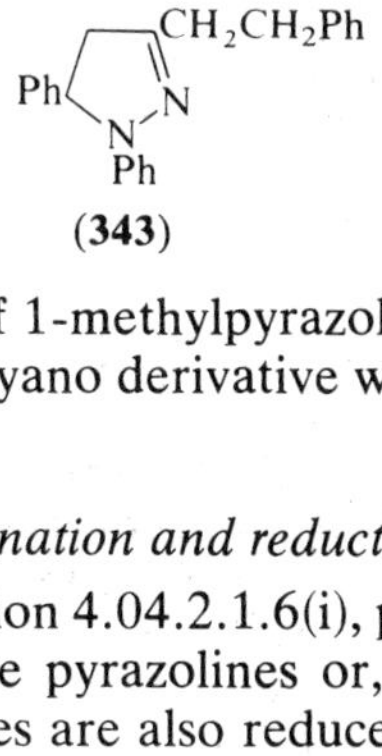

(**343**) 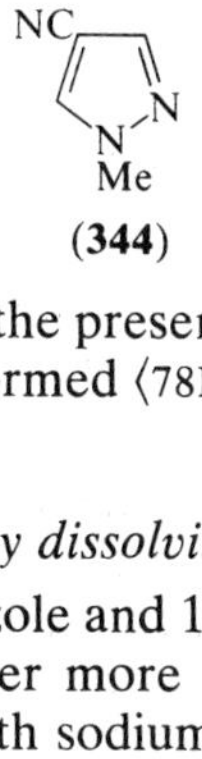   (**344**) 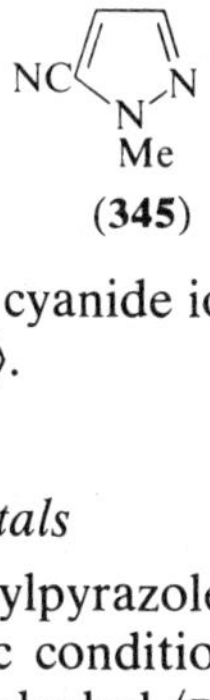   (**345**)

Anodic oxidation of 1-methylpyrazole in the presence of cyanide ions yielded 33% (**344**) and 6% (**345**); no 3-cyano derivative was formed ⟨78RTC35⟩.

(iv) Catalytic hydrogenation and reduction by dissolving metals

As reported in Section 4.04.2.1.6(i), pyrazole and 1-phenylpyrazole under reduction with H₂ palladium give the pyrazolines or, under more drastic conditions, the pyrazolidines. 1-Substituted pyrazoles are also reduced with sodium and alcohol ⟨B-76MI40402⟩.

4.04.2.1.8 Reactions with cyclic transition states

(i) Diels–Alder reactions and 1,3-dipolar cycloadditions

In theory, a pyrazole could react towards dienophiles or dipolarophiles as an azadiene (A) or as a 1,3-dipole of the azomethine imine category (B), both situations being identical with regard to the number of π-electrons involved (Figure 25) (see also Section 4.02.1.9.1). There is also the possibility that it may react as an alkene (C) or as an imine (D) towards dienes or 1,3-dipoles. In the case of ethenylpyrazoles a final possibility of a Diels–Alder reaction involving an exo- and endo-cyclic double bond must be considered.

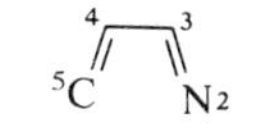

(A) 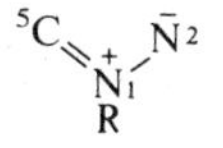   (B) 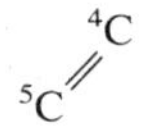   (C) 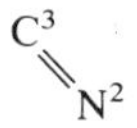   (D)

All these six reactions are theoretically allowed, since they have aromatic transition states; however all of them destroy the 'aromaticity' of pyrazole, and for this reason they are almost unknown at the present time.

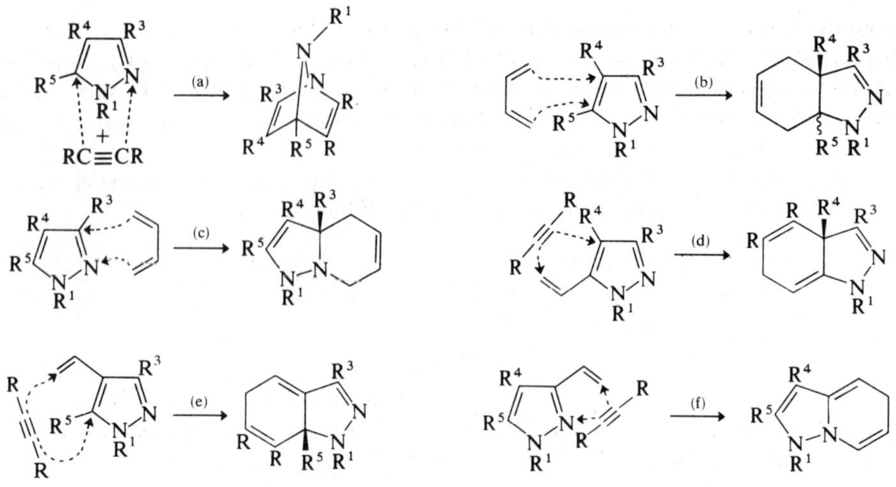

Figure 25 Pericyclic reactions involving the pyrazole nucleus

In 1973 two papers appeared almost simultaneously ⟨73T101, 73CPB2026⟩ describing the formation, as a minor product, of 3,4,5-trimethoxycarbonyl-1-phenylpyrazole (**346**) in the reaction between benzaldehyde phenylhydrazone and DMAD (EC≡CE). To account for the formation of (**346**) George *et al.* ⟨73T101⟩ proposed a tentative mechanism (Scheme 29) involving a Diels–Alder reaction of type (a; Figure 25), followed by a retro-Diels–Alder elimination of methyl phenylpropiolate (**347**).

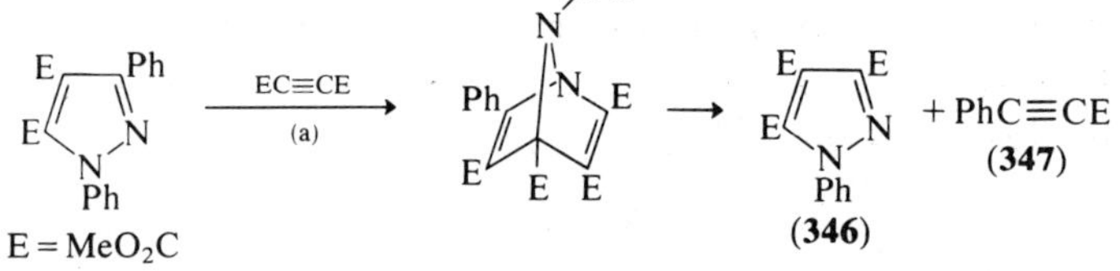

Scheme 29

Further evidence showed this mechanism to be incorrect, especially the fact that it was methyl cinnamate and not (**347**) which was isolated from the reaction ⟨73CPB2026⟩. Also 1-phenylpyrazoles did not react with DMAD under the reaction conditions ⟨74BSF2547⟩. The origin of (**346**) remains obscure, but in no circumstances does it imply a Diels–Alder reaction of a pyrazole. For Ogura *et al.*, it has its origin in an intermediate Δ^3-pyrazoline ⟨73CPB2026⟩.

At about the same time a paper from Abjean appeared ⟨74CR(C)(278)359⟩ describing the sequence of reactions shown in Scheme 30.

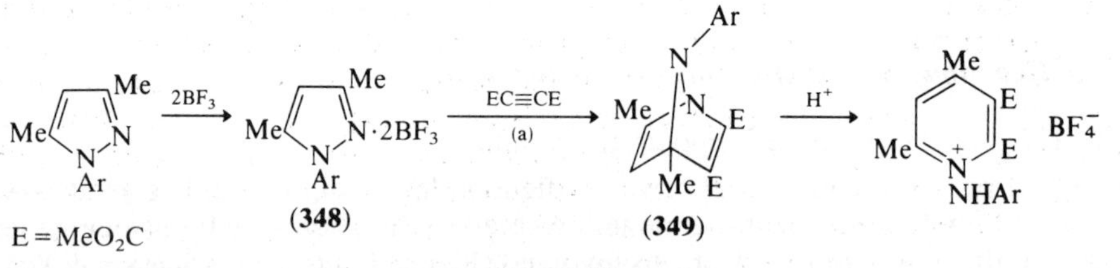

Scheme 30

These interesting results have been quoted by Sokolov ⟨79RCR289⟩ and by Acheson and Elmore ⟨78AHC(23)263⟩. However, they proved to be erroneous, as to both structure (**348**) (a fluoroborate, not a complex) and structure (**349**) (for which (**350**) represents the correct structure established by X-ray crystallography ⟨83T2193⟩).

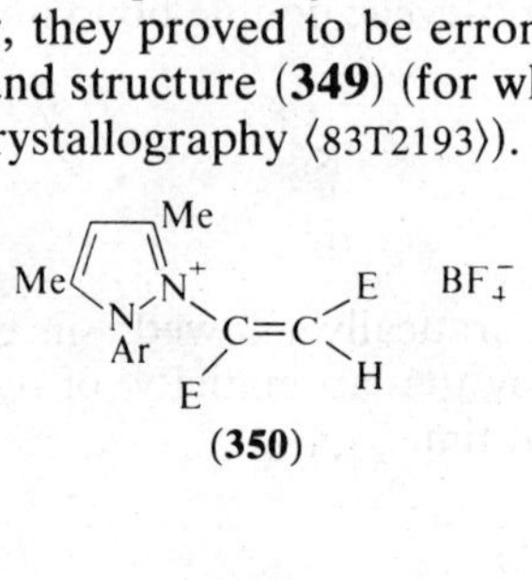

The conclusion is that there are no known examples of behaviour of type (a) for pyrazoles. The same negative conclusion could have been extended to the other reaction pathways of Figure 25 until very recently when an example of type (e) was described (Scheme 31 ⟨82PC40400⟩). Other dienophiles do not take part in this reaction.

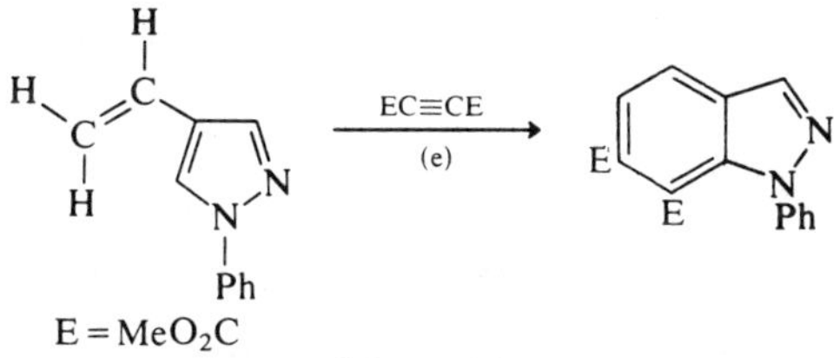

Scheme 31

In the indazole series an example of a type (c) reaction (Figure 25) has been described utilizing a 1,3-dipole instead of a diene (Scheme 32) ⟨76H(4)1655⟩. The cycloadduct (**351**) is transformed into the triazole (**352**) on reaction with hydrochloric acid.

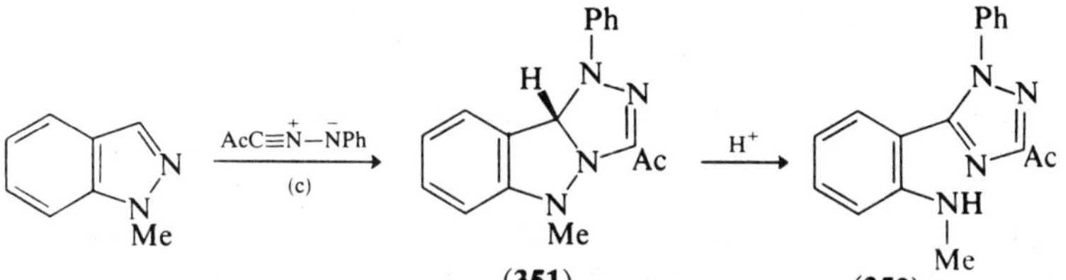

Scheme 32

(ii) Photochemical cycloadditions

[2+2] photochemical additions to give four-membered rings (Section 4.02.1.9.2) are unknown in the pyrazole series.

4.04.2.2 Reactions of Non-aromatic Compounds

4.04.2.2.1 Isomers of aromatic derivatives

(i) Compounds not in tautomeric equilibrium with aromatic derivatives

This important section deals with pyrazolenines, isopyrazoles and 4,4-disubstituted 2-pyrazolin-5-ones whose thermal and photochemical reactions will be discussed in this order.

A thermal rearrangement, the so-called Van Alphen–Hüttel rearrangement (Scheme 33), relates pyrazolenines (**353**), isopyrazoles (**354**) and pyrazoles (**355**) by a series of [1,5]-sigmatropic rearrangements ⟨73TL3781, 79CJC1186⟩. If R^4 is an alkyl or aryl group, the isopyrazole (**354**) obtained from (**353**) is stable; if $R^4 = H$, it undergoes isomerization into an NH-pyrazole; finally, if R^4 is an ester group, under drastic conditions of heating or acidity the isopyrazole loses this substituent from the 4-position affording the NH-pyrazole. When $R^3 \neq R^{3'}$, the migration tendency depends on the experimental conditions (neutral or acidic). In the uncatalyzed rearrangement the relative rates are: Me (1), Et (23), PhCH$_2$ (157), Ph (530), CO$_2$Me (19 000) ⟨80TL1417⟩. Thermal isomerization of tetramethylisopyrazole (**354**) into tetramethylpyrazole (**355**; $R^{3'} = R^3 = R^4 = R^5 = $ Me) occurs at 400 °C ⟨76CRV187⟩; a mechanism involving the tetramethylpyrazoline is possible, although there is evidence that a competing radical pathway is also involved.

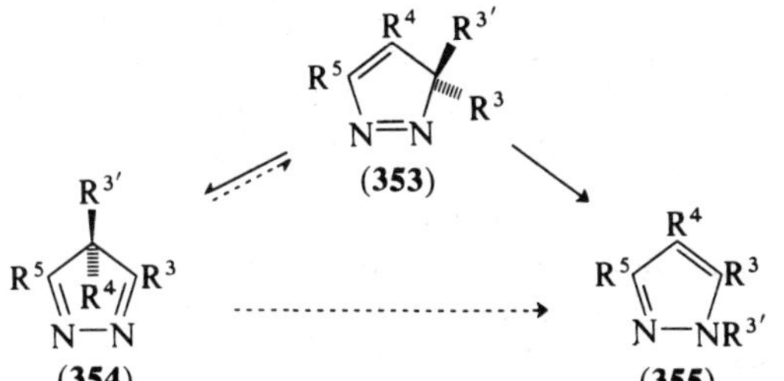

Scheme 33

4-Hydroxyisopyrazoles (**354**; $R^{3'}$ = alkyl or phenyl, $R^{4'}$ = OH) rearrange to 4-hydroxypyrazolenines (**353**; R^4 = OH) which in turn tautomerize to the more stable 2-pyrazolin-4-ones (**356**) ⟨76JOC2874⟩. Finally, some pyrazolenines (**357**) undergo a ring expansion to the 3H-1,2-benzodiazepines (**358**) rather than a thermal Van Alphen rearrangement ⟨75CC128⟩.

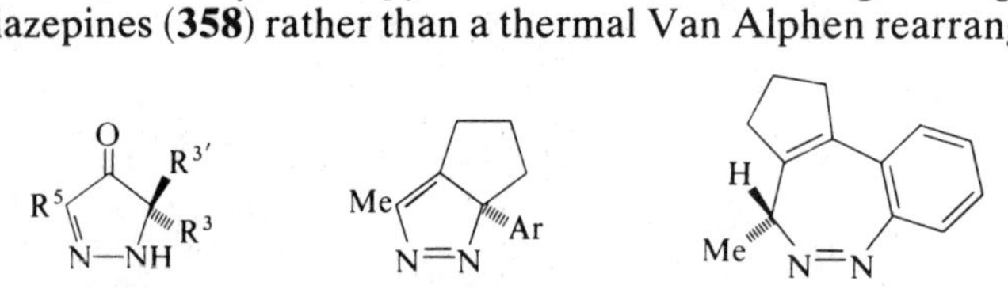

(**356**)   (**357**)   (**358**)

Diazoalkanes and ynamines react with the electrophilic C(4)—C(5) double bond of pyrazolenine (**359**) to afford the cycloadducts (**360**) and (**361**), respectively, whereas diphenylketene yields the bicyclic diazetidinone (**362**) by reaction with the *cis*-azo system of (**359**) ⟨79CC568⟩.

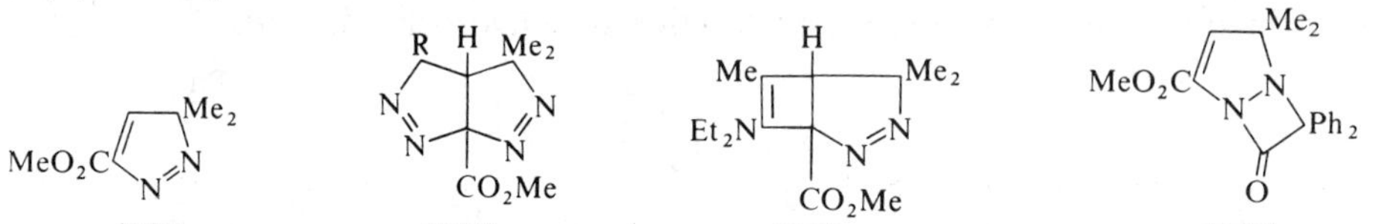

(**359**)   (**360**)   (**361**)   (**362**)

The tetrahydropyrazolo[3,4-c]pyrazole (**360**) is unstable and, by loss of N_2, leads to 3-R-4-isopropyl-5-methoxycarbonylpyrazole.

Isopyrazole quaternary salts (**363**) are key intermediates leading to the highly substituted Δ^2-pyrazolines. Lithium aluminum hydride gives the pentasubstituted derivatives (**364**; R = H) and Grignard reagents provide access to the fully substituted Δ^2-pyrazolines (**364**; R ≠ H) ⟨68BSF3866, 70BSF1121⟩.

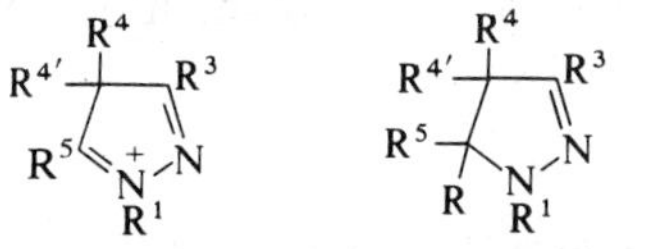

(**363**)   (**364**)

Two reactions of the non-aromatic 4,4-disubstituted pyrazolones are worthy of mention. Carpino discovered that 4,4-dihalogenopyrazolones (**365**) and 4-substituted 4-halogenopyrazolones (**366**) when treated with bases yield α,β-alkynic and -alkenic acids, respectively ⟨66JOC2867⟩. The reaction proceeds through an oxopyrazolenine (2,3-diazacyclopentadienone; **367**) ⟨B-74MI40408⟩. A modification of the experimental procedure transforms (**365**) into bimanes (**368**) ⟨82JOC214⟩, which are formed from (**367**; R = X).

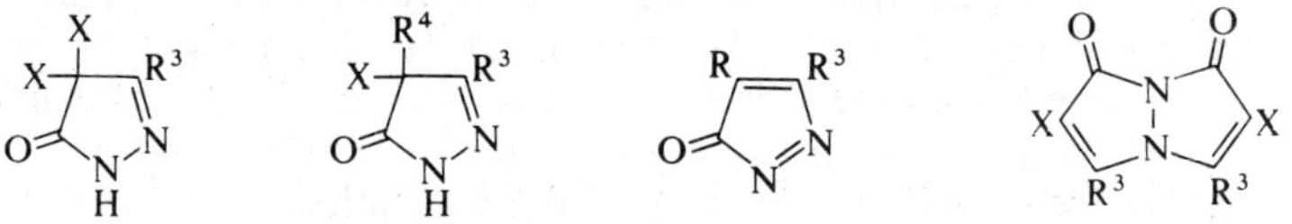

(**365**)   (**366**)   (**367**) R = X or R^4 (**368**)

2-Pyrazolin-5-ones with an exocyclic double bond at the 4-position (**369**; X = CMe_2) react as heterodienes towards alkyl vinyl ethers ⟨77G91⟩. The kinetics of this Diels–Alder reaction giving pyrazolopyrans (**370**) have been studied.

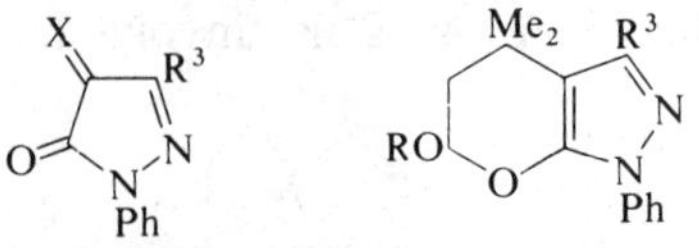

(**369**)   (**370**)

Progress in the photochemistry of pyrazolenines and isopyrazoles is largely due to the studies of Dürr and Franck-Neumann. Since this subject has not previously been treated systematically, a relatively detailed account will be given here.

Considerable effort has been devoted to studying the photolysis of pyrazolenines since it still constitutes the best way to obtain cyclopropenes. An important publication by Closs

⟨68JA173⟩ illustrates this possibility (Scheme 34). Thus 3,3,5-trimethyl-3H-pyrazole (**371**; $R^4 = H$) on irradiation in pentane solution gives 1,3,3-trimethylcyclopropene (**372**; $R^4 = H$); the intermediate diazoalkene (**373**) has been characterized. The tetramethyl derivative (**371**; $R^4 = Me$) when irradiated at −50 °C in methylene chloride leads to a species believed to be a 1,2-diazabicyclo[2.1.0]pent-2-ene (**374**). This isomerization is thermally reversible, the 3H-pyrazole being regenerated at room temperature.

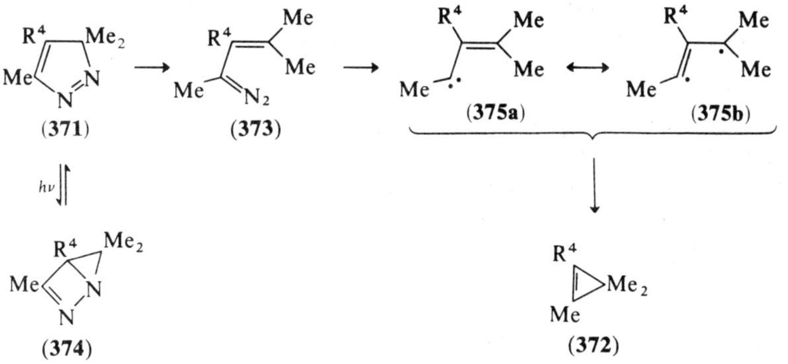

Scheme 34

The mechanism involves the cleavage of the N(2)—C(3) bond of pyrazolenine (**371**) with formation of the isomeric diazoalkene (**373**) (in some cases they are stable for hours at room temperature in benzene solutions ⟨74JA3708⟩). Loss of nitrogen yields a carbene (**375a**) which in turn cyclizes to the cyclopropene (**372**) ⟨B-77MI40402⟩. Acyl- ⟨77T751⟩ and sulfinyl-cyclopropenes ⟨77AG(E)323⟩ have been prepared by photolysis of the corresponding pyrazolenines.

The cyclohexylpyrazole (**376**) and the azirine (**377**) are formed by irradiation of 3-diazo-4-methyl-5-phenylpyrazolenine (**378**) in cyclohexane (Scheme 35) ⟨77JA633⟩. The former is the result of carbene insertion into cyclohexane followed by a [1,5] hydrogen shift, whereas the latter arises by ring cleavage of nitrene (**379**) or by a concerted pathway.

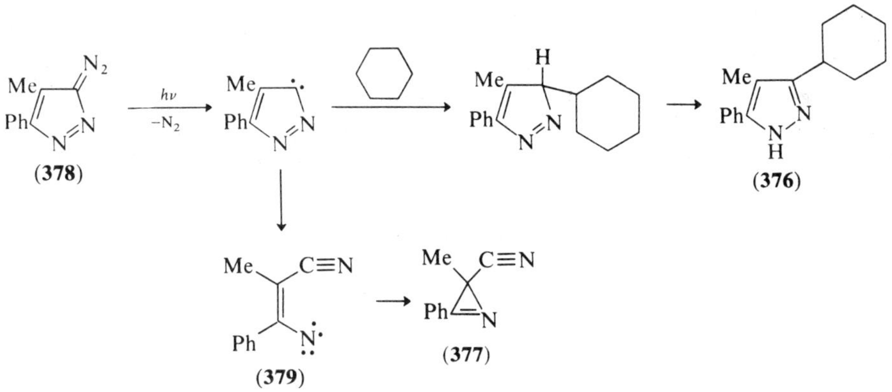

Scheme 35

The prospect of obtaining new ring systems of unusual structure has led to much work on the photolysis of cyclopentadiene-spiropyrazolenines and related compounds. The results obtained have justified the expectations. Figure 26 contains some structures prepared in this way.

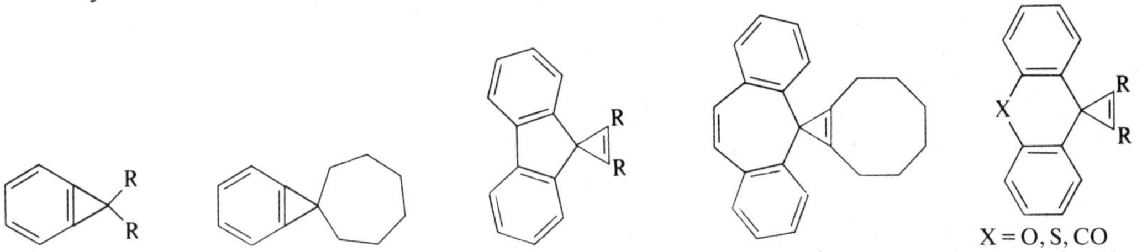

Figure 26 Cyclopropenes obtained by photolysis of spiropyrazolenines

A radical pathway has been proposed to account for the isolation of (**381**)–(**383**) from the irradiation of the 3*H*-indazole (**380**) ⟨73T1833⟩. Results of sensitization and quenching experiments suggest that the reaction proceeds from the excited singlet state.

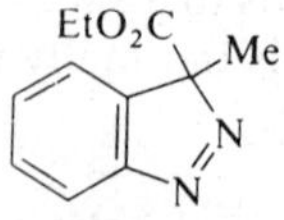

(**380**)    (**381**) 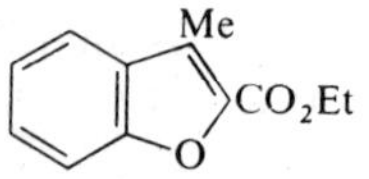   (**382**) 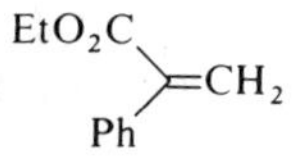   (**383**)

Photolysis of spiro[fluorene-9,3'-indazole] (**384**) to the tribenzopentalene (**385**) has been rationalized in terms of the initial formation of triplet diradical (**386**) ⟨76JOC2120⟩. The spiroindazole (**387**) behaves differently and on irradiation in THF is converted into the dimer (**388**) and the stable *N*-ylide (**389**) ⟨76CB2596⟩.

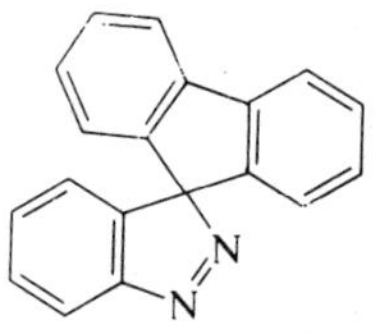

(**384**) 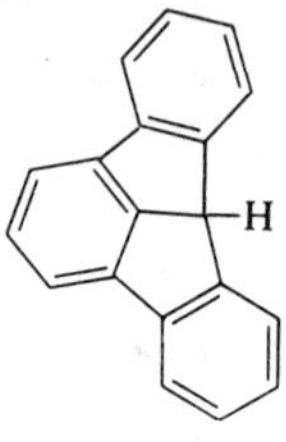   (**385**) 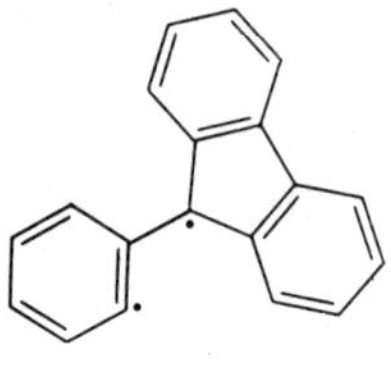   (**386**)

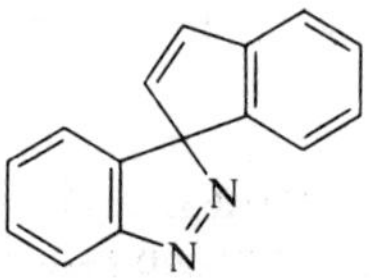

(**387**) 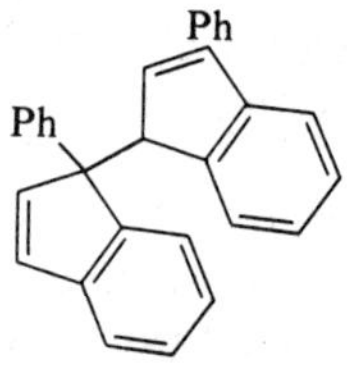   (**388**) 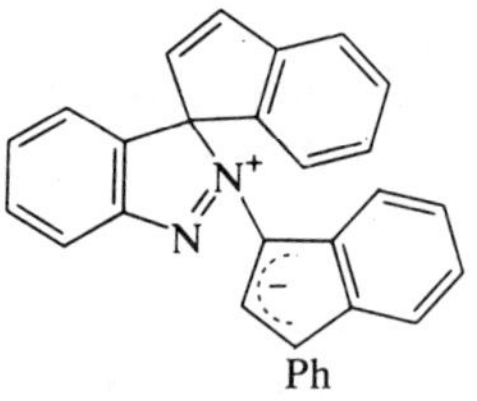   (**389**)

In contrast to pyrazolenines, there are only a few publications on the photochemistry of isopyrazoles and they concern exclusively their *N*-oxides (**390**). Irradiation of (**390**) affords the *N*-oxides of pyrazolenine (**391**) ⟨70CC289⟩. Bicyclic intermediates (**392**) and (**393**; Scheme 36) are believed to be implicated in this reaction ⟨75MI40400⟩. The final step is similar to that reported from studies of the valence bond isomerization of pyrazolenines ⟨68JA173⟩.

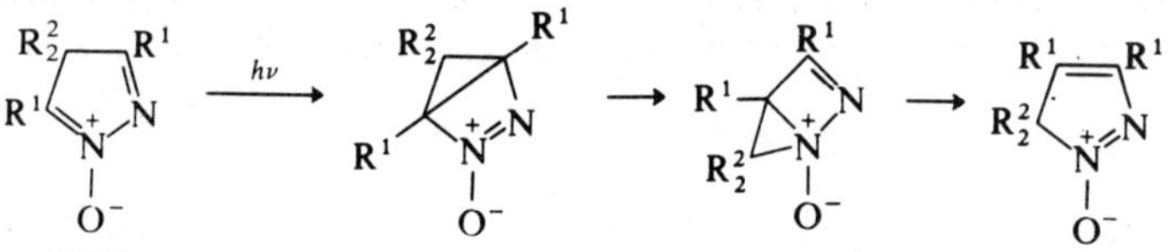

(**390**)   (**392**)   (**393**)   (**391**)

Scheme 36

The photolysis of the diazopyrazolone (**369**; $X = N_2$, $R^3 = Me$) in methanol yields two isomeric forms of methyl 3-phenylazo-2-butenoate (**394**) ⟨80CC1263⟩. The azo esters may arise *via* protonation of the carbene (**395**) with a concurrent opening of the ring by the nucleophilic solvent.

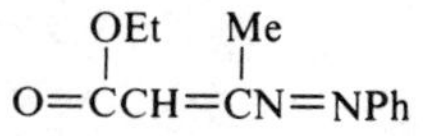

(**394**) 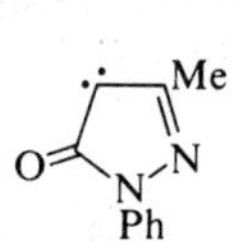   (**395**)

(ii) Compounds in tautomeric equilibrium with aromatic derivatives

The non-aromatic CH tautomers of pyrazolones have a malonic-like reactivity due to the fact that the hydrogen atoms of the CH_2 group at position 4 are acidic because the conjugate anion (**239**) is fully delocalized. Figure 27 illustrates some classical reactions of 3-methyl-1-phenyl-2-pyrazolin-5-one (**396**) ⟨64HC(20)1⟩.

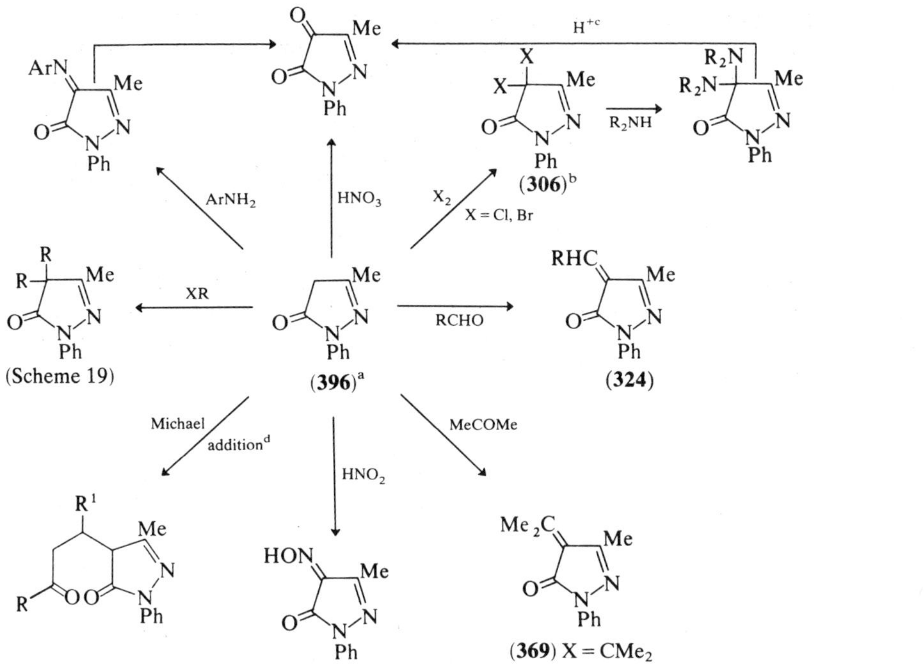

Figure 27 Some reactions of 2-pyrazolin-5-ones

[a] LiAlH$_4$ did not reduce 2-pyrazolin-5-ones ⟨70BSF1974⟩. [b] See also formulae (**365**) and (**366**). [c] ⟨68M2157⟩. [d] ⟨80CHE180⟩.

If the pyrazolone is unsubstituted on position 1, there is still another possibility: lead tetraacetate oxidation of (**397**) yields the oxopyrazolenine (**398**) ⟨73CRV93⟩ which is a very reactive azadienophile ⟨73JCS(P1)221⟩.

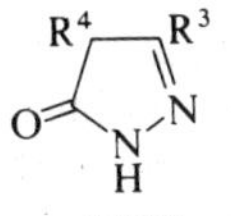

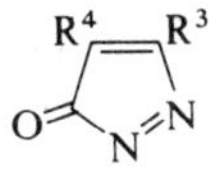

Hydrazinolysis of the ethoxypyrazolone (**399**) leads to the hydrazino derivative (**400**) ⟨70ZC224⟩. A probable mechanism involves intermediates (**401**; R = OEt) and (**401**; R = NHNH$_2$).

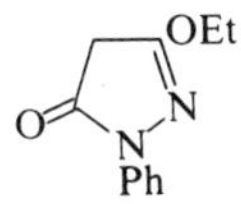

C—N bond homolysis may be the initial step in the conversion of (**396**) into the phenylazopyrazole (**402**), a product which is believed to arise by interaction of ground-state pyrazole with a photochemically generated phenyldiazonium ion ⟨76CC685⟩.

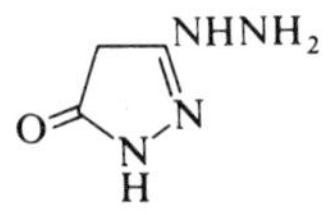

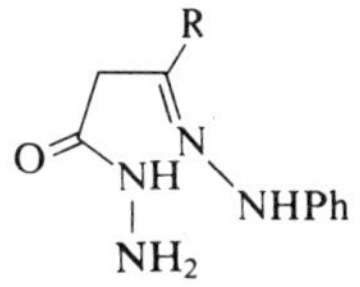

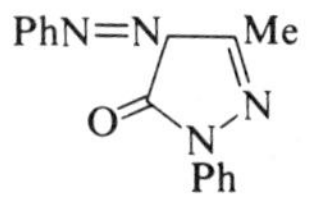

4.04.2.2.2 Dihydro derivatives

According to the classification of Section 4.04.1.1.3 this section deals with Δ^1-(**15**), Δ^2-(**16**) and Δ^3-pyrazolines (**17**), and with pyrazolidones (**18**) and aminopyrazolines. The number of publications dealing with these compounds is so large that only the most relevant results and most important references will be described. The Jarboe review on pyrazolines ⟨67HC(22)1⟩ contains information up to 1963 and fortunately the chemistry of pyrazolidones and aminopyrazolines is covered by an excellent review by Dorn in 1980 ⟨80CHE1⟩.

(i) Tautomerism and isomerization

Tautomerism has been discussed in Section 4.04.1.5.2. It concerns prototropic tautomerism and the decreasing order of stability is Δ^2 (hydrazone) $>\Delta^1$ (azo) $>\Delta^3$ (enehydrazine). The isomerization $\Delta^2 \rightarrow \Delta^3$ occurs via a Δ^1-pyrazoline ⟨65BSF769⟩. Pyrazolidones and amino-Δ^2-pyrazolines exist as such. The only example of non-prototropic tautomerism deals with the isomerization (**403**) → (**404**) ⟨74CJC3474⟩. This intramolecular process is another example (Section 4.04.1.5) of the thermodynamic analogy between prototropy and metallotropy.

(ii) Aromatization

A variety of oxidizing agents transform pyrazolines into pyrazoles: sulfur, bromine, chloranil, potassium permanganate, lead dioxide and mercury(II) acetate are all effective ⟨66AHC(6)347⟩. The use of lead tetraacetate is well documented ⟨73CRV93⟩. Photooxygenation ⟨79CRV447⟩ of 1,3-disubstituted 2-pyrazolines yields pyrazoles and open-chain compounds. If a 2-pyrazoline is disubstituted at position 5, as in (**405**), the oxidation is accompanied by migration of a methyl group and a 3,4,5-trisubstituted pyrazole such as (**406**) is obtained.

2-Pyrazolines substituted at position 4 or 5 with hydroxy or amino groups readily eliminate a molecule of water or amine yielding pyrazoles. The 4-substituted derivatives are relatively more stable than the 5-substituted ones, because for the last group the lone pair at N-1 assists the elimination (**407**) → (**408**) → (**409**). The sulfonyl group at position 1 is also easily eliminated and this property is taken advantage of in Dorn's elegant synthesis of 3-aminopyrazole (Section 4.04.3.3.1).

Δ^3-Pyrazolines such as (**410**) are oxidized by iodine, mercury(II) acetate and trityl chloride to pyrazolium salts (**411**), and compound (**410**) even reduces silver nitrate to Ag^0 ⟨69JOU1480⟩. Electrochemical oxidation of 1,3,5-triaryl-2-pyrazolines has been studied in detail ⟨74BSF768, 79CHE115⟩. They undergo oxidative dimerization and subsequent transformation into the pyrazole derivative (**412**).

Photooxidation has also been used to transform pyrazolines into pyrazoles ⟨74AJC2267⟩ and 3-pyrazolidones into 3-hydroxypyrazoles ⟨76CC685⟩.

(iii) Reduction

All three isomers of pyrazolines have successfully been reduced to pyrazolidines, the Δ^1 catalytically ⟨65JA3768⟩, the Δ^2 polarographically ⟨65JOU95⟩ and the Δ^3 by means of sodium borohydride ⟨68JOU692⟩. 1,2-Disubstituted pyrazolinium salts (**229**) are reduced to pyrazolidines by complex hydrides ⟨69BSF3302⟩ and 3-pyrazolidinones form mixtures of Δ^3-pyrazolines and pyrazolidines ⟨70BSF1936⟩. Normally Δ^2-pyrazolines are not reduced by complex hydrides unless an electron-withdrawing group is present at N-1. Thus, aluminum hydride reduces 1-acetylpyrazolines to 1-ethylpyrazolidines through prior complexing at the N-2 atom of the pyrazoline ring ⟨80CHE936⟩.

(iv) Thermolysis and photolysis

The fundamental subject of this section is the transformation of Δ^1-pyrazolines into cyclopropanes (Buchner–Curtius and Kishner cyclopropane syntheses). The cyclopropane is often accompanied by alkenes ⟨67HC(22)1⟩. When applied to Δ^2-pyrazolines the reaction occurs via the Δ^1 isomers (Scheme 37).

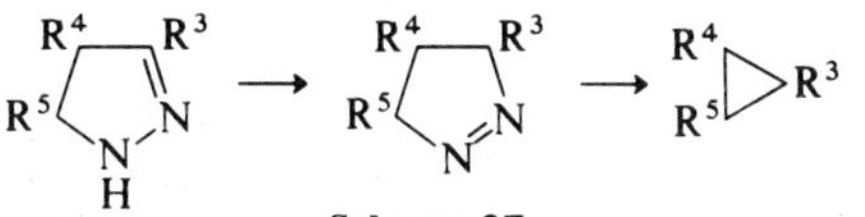

Scheme 37

The thermal and photolytic processes have been extensively studied in connection with the mechanism (concerted or diradical) and the stereochemistry of the decomposition. For a classical paper see ⟨66JA3963⟩ and for more recent studies, Table 33.

Table 33 Thermolysis and Photolysis of 1-Pyrazolines

1-Pyrazoline	Δ	hν	Conclusions	Ref.
Methoxycarbonyl and acetyl derivatives	*		Relative rates of decomposition into cyclopropanes and alkenes	75JCS(P2)1791
Methoxycarbonyl and cyano derivatives	*		One-step process mechanism	75LA449
Cis- and trans-3,5-diphenyl derivatives	*	*	1,3-Diradical mechanism	76T619
3-Ethyl-5-methyl (3R, 5R) and (3R, 5S) derivatives	*		Elucidation of the stereochemistry of the pyrazoline → cyclopropane reaction	77JA2740
Bicyclic derivatives		*	Rate constants for nitrogen extrusion	78JA5122
4-Methylene-1-pyrazoline	*		Secondary deuterium isotope effects on the reaction rate	81CJC2556
Simple derivatives	*	*	Barriers to decomposition	82JA1698

Ring contraction to cyclopropanes (**413**) with the formation of azo side-chains is observed in the photolysis of 5-phenylpyrazolines ⟨71BSF1925⟩ and to azetidinones (**414**) in the photolysis of 5-pyrazolidones ⟨69JCS(C)2624, 83JCS(P2)1111⟩.

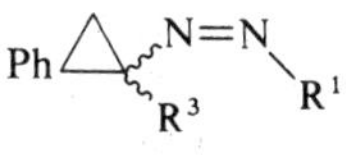

(**413**) 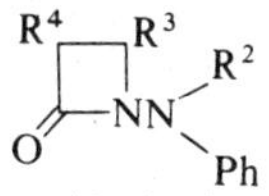   (**414**)

(v) Other reactions

Other significant reactions of Δ^2- and Δ^3-pyrazolines are shown in Figures 28 and 29, respectively.

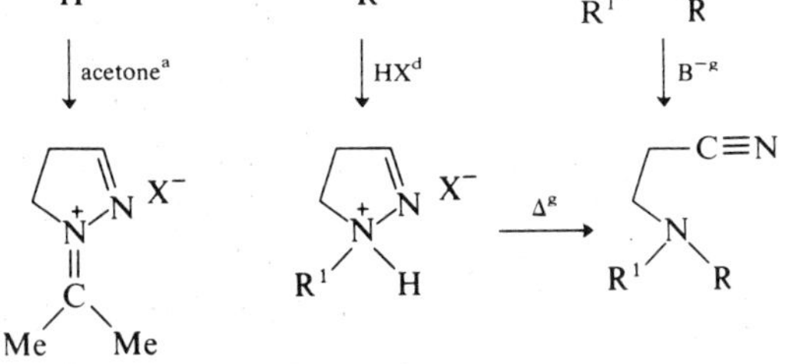

[a] ⟨76MI40403⟩. [b] R^1 = COR or NO ⟨67HC(22)1⟩, R^1 = nitrophenyl ⟨66BSF610⟩. [c] Enamine behaviour (Z = Br, CHO, PhN₂, etc.) ⟨67BSF4716⟩. [d] Protonation ⟨75APO(11)267, 76MI40403⟩; R^1 = H, X-ray structure (Table 5). [e] Quaternization ⟨76MI40403⟩. [f] Dequaternization ⟨76MI40403⟩. [g] Aminonitrile rearrangement ⟨66RCR9, 80ZC167⟩.

Figure 28 Some reactions of Δ^2-pyrazolines

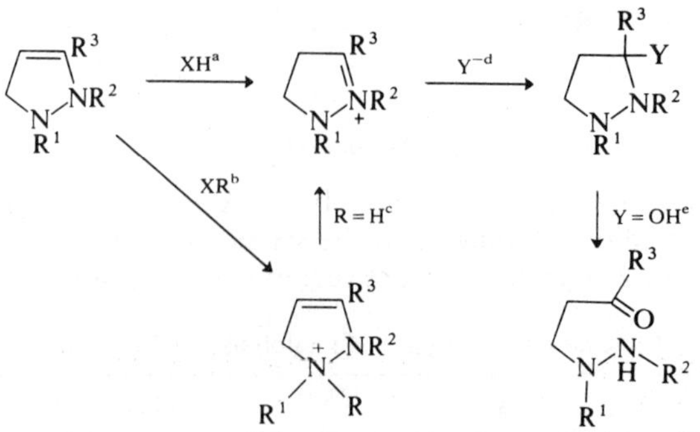

^a Protonation ⟨75APO(11)267, 76MI40403⟩; X-ray structures of the salts (Table 5). ^b Quaternization ⟨76MI40403⟩.
^c ⟨75APO(11)267⟩. ^d Nucleophilic attack (Y = CN, OH, H, R, Ar) ⟨76MI40403⟩. ^e Ring opening ⟨76MI40403⟩.

Figure 29 Some reactions of Δ^3-pyrazolines

For a discussion of the reactivity of pyrazolidones and aminopyrazoles see ⟨80CHE1⟩.

(vi) Stereochemistry

The stereochemistry of pyrazolines and pyrazolidines has already been discussed (Section 4.04.1.4.3). Optically active Δ^1- and Δ^2-pyrazolines have seldom been described ⟨77JA2740, 79CJC360⟩, but *cis-trans* isomeric pairs are common. The C-4 acid-catalyzed epimerization involves the mechanism shown in Scheme 38 ⟨70TL3099⟩, but in spite of some inconclusive arguments the C-5 epimerization has never been established with certainty.

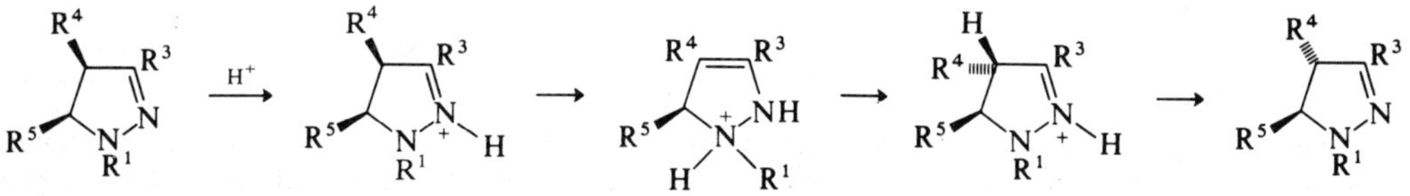

Scheme 38

4.04.2.2.3 Tetrahydro compounds

Pyrazolidines are cyclic hydrazones and their reactivities are comparable, the main difference being found in the oxidation of pyrazolidines to pyrazolines and pyrazoles.

(i) Aromatization

Analogous to the oxidation of hydrazones to azo compounds, *N*-unsubstituted pyrazolidines are oxidized to Δ^1-pyrazolines. For example, the bicyclic pyrazolidine (**415**) when treated with silver oxide yields the pyrazoline (**416**) ⟨65JA3023⟩. Pyrazolidine (**417**) is transformed into the perchlorate of the pyrazolium salt (**411**) by reaction with mercury(II) acetate in ethanol followed by addition of sodium perchlorate ⟨69JOU1480⟩.

The pyrazole (**420**) is formed when the pyrazolidine (**418**) is heated with chloranil and the intermediate Δ^3-pyrazoline (**419**) (one of the rare *N*(1)-unsubstituted derivatives) can be isolated ⟨78TL4503⟩.

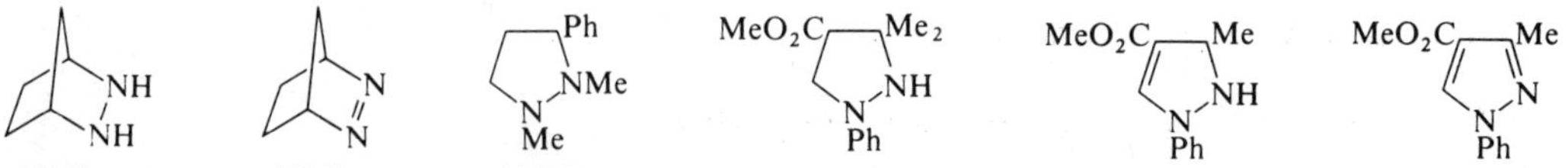

(ii) Ring fission

Formation of a 1,2-disubstituted hydrazine by acid hydrolysis of an appropriately substituted pyrazolidine has been noted ⟨67HC(22)1⟩, but the most interesting ring fission of pyrazolidines involves the N(1)—N(2) bond of 1-phenylpyrazolidines (**421**). If, instead of phenylhydrazone, compound (**421**) is used in the Fischer indole synthesis, *N*-aminopropylindoles are formed ⟨73T4045⟩. Scheme 39 shows the reaction with cyclohexanone.

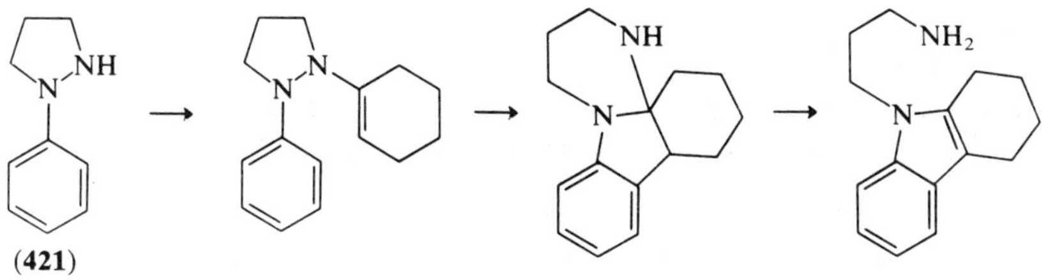

(421)

Scheme 39

Eberle has explored the synthetic possibilities of (**421**); with aldehydes it leads to hexahydropyrimido[1,2-a]indoles ⟨73T4049⟩. Its 5-oxo derivative (1-phenyl-5-pyrazolidinone) has a comparable reactivity ⟨76JOC3775⟩.

Kost *et al.* have studied related reactions of 2-acyl-1-phenylpyrazolidines (**422**) and 1-phenyl-2-thiocarbamoylpyrazolidines (**423**). The former are converted on reaction with phosphorus oxychloride into tetrahydropyrimido[1,2-a]indoles (**424**) ⟨72CHE57⟩ and the latter into tetrahydropyrimido[2,1-b]benzothiazoles (**425**) under the influence of acidic agents ⟨80CHE169⟩.

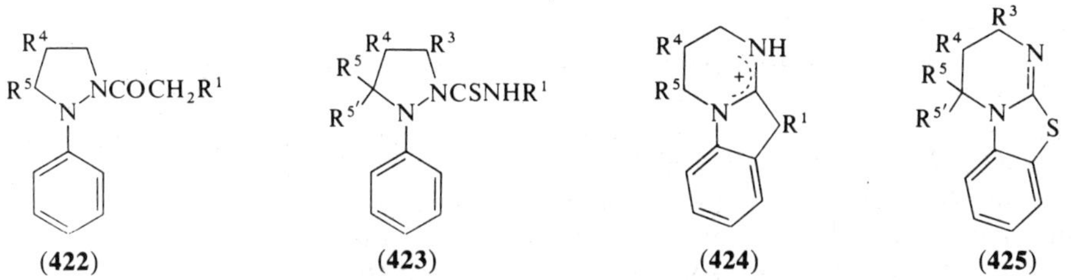

(422)   (423)   (424)   (425)

(iii) Other reactions

Reaction of 2-ethoxycarbonyl-1-phenylpyrazolidine with Meerwein's salt (boron trifluoride etherate) yields the quaternary ammonium salt (**426**) which is not stable and undergoes ring opening to the hydrazine (**427**) with sodium ethoxide ⟨76JOC1244⟩.

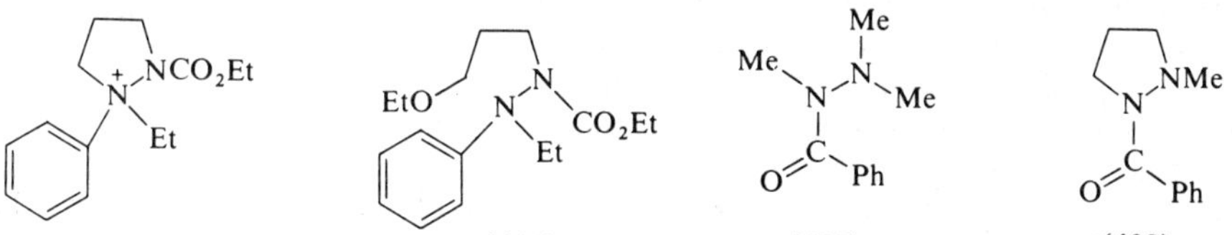

(426)   (427)   (428)   (429)

The reactivities of benzoyltrimethylhydrazine (**428**) and 1-benzoyl-2-methylpyrazoline (**429**) towards methyl iodide have been discussed in connection with the corresponding geometries determined by X-ray crystallography ⟨81JOC2490⟩.

(iv) Stereochemistry

The stereochemistry of pyrazolidines has briefly been discussed in Section 4.04.1.4.3. The pyrazolidine ring in (**429**) in the crystal state is not planar and the configuration about the amide bond is *E* as represented.

4.04.2.3 Reactions of Substituents

4.04.2.3.1 General survey of substituents on carbon

The general discussion in Section 4.02.3.1 is valid for pyrazoles and therefore only minor points will be discussed here.

(i) Substituent environment

In pyrazoles the simplest way to characterize the three ring carbon atoms is to consider that C-3 is similar to the pyridine α-position, C-4 to the pyrrole β-position and C-5 to both the γ-pyridine and the α-pyrrole positions. In indazoles C-3 corresponds to pyrazolic C-3 and in isoindazoles to pyrazolic C-5.

(ii) The carbonyl analogy

The analogy between a substituent linked to a carbonyl group and a substituent in an α or, to a lesser extent, a γ-position to a pyridinic nitrogen has been discussed in Section 4.02.3.1.2). The conclusions hold for pyrazoles and indazoles.

(iii) Pyrazoles and indazoles

As indicated in the introduction to this chapter, Kauffmann ⟨71AG(E)743⟩ remarked on the fact that π-electron-rich arenes, like 1-methylpyrazoles, are comparable with functional groups that have a $+R$ and a $-I$ effect, such as the amino group. As an example of this analogy (the arenology principle) he quotes the fact that the hydrogen atoms of *C*-methylazoles are not acidic just as the hydrogen atoms of *N*-methylamines are not acidic. This analogy is particularly valid for substituents at position 4. As has been discussed before (Sections 4.04.1.3.6(iii) and 4.04.1.3.7(iii)), the pyrazole ring donates electrons to substituents like COR, NO or NO_2 attached to the 4-position.

(iv) The effect of one substituent on the reactivity of another

In pyrazoles the two most important effects are produced by nitro groups and by quaternization (Section 4.02.3.1.7). Both enhance considerably the reactivity of a second substituent, for example, that of a methyl group towards aldehydes or of a chloro substituent towards amines.

(v) Reactions of substituents not directly attached to the heterocyclic ring

According to Section 4.02.3.1.8 substituents removed from the pyrazole ring by two or more saturated carbon atoms and substituents on the benzene ring of indazoles are similar in reactivity to the corresponding aromatic derivatives. For instance, chloromethylpyrazoles are comparable to benzyl chlorides and 5-hydroxyindazoles to β-naphthols in their reactivity.

(vi) Reaction of substituents involving ring transformations

Reaction of substituents of other azoles leading to pyrazoles and indazoles will be discussed in the corresponding preparative section (4.04.3.2.2). Azido–tetrazole isomerism, an interesting property common to all heterocycles having an azido substituent α to a pyridinic nitrogen atom, will be examined in Section 4.04.2.3.4(v). It has been established ⟨78MI40402⟩ that by heating 4-acetyl-3-methyl-1-phenylpyrazol-5-one phenylhydrazone (**430**) in the presence of an acid 3,5-dimethyl-1-phenyl-4-α-phenylhydrazidopyrazole (**431**) is obtained. This constitutes a new example of the Boulton–Katritzky rearrangement, now widely found in oxygen heterocycles (Section (4.04.3.2.2) but still rare in pyrazoles.

(**430**)   (**431**)

4.04.2.3.2 Indazoles

This section deals with reactions taking place at positions 4, 5, 6 and 7 of the indazole ring. The reactivity of position 3 has already been discussed with that of pyrazoles.

(i) Unsubstituted indazoles

The ring enlargement of indazoles to yield dihydroazepinones (**192**) and (**193**) has already been described (Section 4.04.2.1.2(ii)). Another reaction that destroys the aromatic ring is catalytic hydrogenation. Unexpectedly, it is the homocyclic six-membered ring and not the heterocyclic five-membered ring which is slowly reduced by platinum and hydrogen in acetic acid ⟨67HC(22)1⟩. 4,5,6,7-Tetrahydroindazole (3,4-tetramethylenepyrazole) and its two *N*-methyl derivatives have been prepared by this method.

In the section dealing with electrophilic attack at carbon some results on indazole homocyclic reactivity were presented: nitration at position 5 (Section 4.04.2.1.4(ii)), sulfonation at position 7 (Section 4.04.2.1.4(iii)) and bromination at positions 5 and 7 (Section 4.04.2.1.4(v)). The orientation depends on the nature (cationic, neutral or anionic) of the indazole. Protonation, for instance, deactivates the heterocycle and directs the attack towards the fused benzene ring. A careful study of the nitration of indazoles at positions 2, 3, 5 or 7 has been published by Habraken ⟨71JOC3084⟩ who described the synthesis of several dinitroindazoles (5,7; 5,6; 3,5; 3,6; 3,4; 3,7). The kinetics of the nitration of indazole to form the 5-nitro derivative have been determined ⟨72JCS(P2)632⟩. The rate profile at acidities below 90% sulfuric acid shows that the reaction involves the conjugate acid of indazole.

All these results are represented schematically in Figure 30.

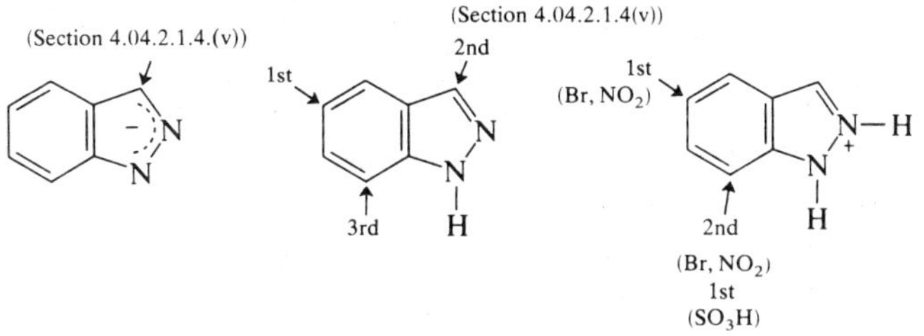

Figure 30 Indazole reactivity towards electrophilic reagents

(ii) Substituted indazoles

Substituents in the indazole ring may direct a given reaction towards another position either by their R and I electronic properties or simply by protecting the most reactive position. Examples of both types are found in sulfonation studies ⟨67HC(22)1⟩. As indicated before (Section 4.04.2.3.2(i)), sulfonation takes place at position 7. However, the presence of an amino group at positions 5 or 7 directs the attack towards the 4-position (Scheme 40). To obtain the indazole-5-sulfonic acid a more complicated procedure has been used but it is still based on the same ideas.

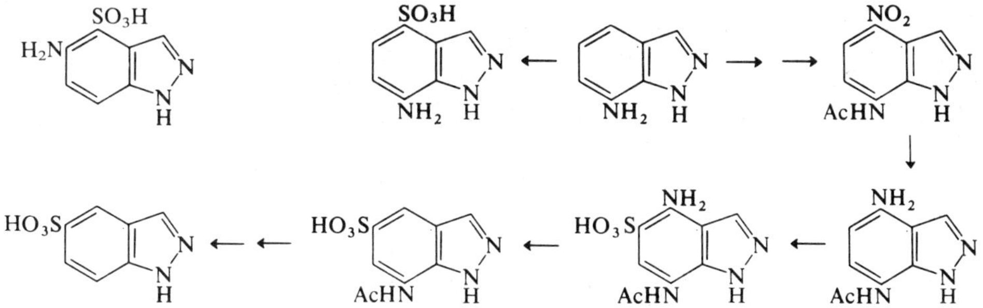

Scheme 40

Indazole substituents can be engaged in a variety of reactions. Those at position 3 react in a similar manner to pyrazolic 3(5)-substituents and will be discussed in the following sections. Substituents in the fused benzene ring have a naphthalene-like reactivity (Section 4.04.2.3.1(v)). Several classical examples are described in ⟨67HC(22)1⟩: $NO_2 \rightarrow NH_2$ (Pd/H_2); $NH_2 \rightarrow$ condensed pyridines (Skraup reaction); $NH_2 \rightarrow N_2^+ \rightarrow$ H, Cl, I, CN,...; $CO_2H \rightarrow Br$ (Hunsdiecker reaction); Br $\rightarrow$ H (Na/Hg); *etc.* As a matter of fact, the whole spectrum of aromatic reactivity can be transferred to indazoles substituted at the fused benzene ring. Compare, for example, the reactions described by Suschitzky *et al.* ⟨68JCS(C)1937⟩ for 2-azidonaphthalene (**432**) and for 5-azidoindazole (**433**). Pyrolysis of these compounds in a mixture of acetic and polyphosphoric acid yields (**434**) and (**435**), respectively.

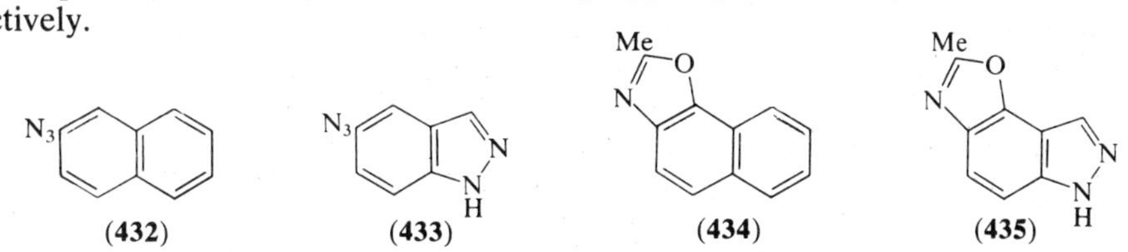

(**432**)　　(**433**)　　(**434**)　　(**435**)

4.04.2.3.3 Other C-linked substituents

C-Linked substituents behave in pyrazoles and indazoles as in other azoles (Section 4.02.3.3). The classical aromatic chemistry of these compounds has given rise to a great number of publications ⟨66AHC(6)347, 67HC(22)1, B-76MI40402⟩, but not to a specific pyrazole chemistry. For this reason, only a brief survey will be given here.

(i) Alkyl groups

We have already noted (Section 4.04.2.1.4(xi)) that alkyl groups on pyrazoles are oxidized with permanganate to carboxylic acids. Silver nitrate and ammonium persulfate transform 4-ethyl-1-methylpyrazole (**436**) into the ketone (**437**) ⟨72JHC1373⟩. The best yield was obtained starting with the alcohol (**438**) and using an acid dichromate solution as oxidizing agent.

The synthesis of indazoles from their 4,5,6,7-tetrahydroderivatives (**439**) by means of sulfur or, better, by catalytic dehydrogenation over palladium on charcoal ⟨67HC(22)1⟩ can also be included here.

The acidic character of the hydrogen atoms of C-methyl groups linked to the pyrazolium ring (Figure 22; Section 4.04.2.1.1(ii)) facilitates a number of reactions difficult to carry out with neutral pyrazoles. Since efficient methods of dealkylation have been described (Section 4.04.2.3.10(ii)), the synthesis via the pyrazolium salt is a useful alternative. The same behaviour is observed for indazolium salts, for example, nucleophilic addition to aromatic aldehydes ⟨78JOC1233⟩.

(ii) Aryl groups

C-Aryl groups carrying hydroxy or amino substituents can be oxidized without affecting the pyrazole nucleus ⟨B-76MI40402⟩. In the discussion of electrophilic reactions the fact that the p-position competes with the pyrazolic 4-position has been mentioned (Sections 4.04.2.1.4(i) and (ii)); for example, nitration occurs on the phenyl ring under conditions in which the pyrazole nucleus is protonated ⟨B-76MI40402⟩. Nitration and bromination of (**301**) and (**302**) (Section 4.04.2.1.4(ii)) have been studied in order to evaluate the directing effect of a 4-pyrazolyl group in electrophilic substitution in the thiophene series ⟨80CS(15)102⟩.

(iii) Other carbon-linked functionality

Pyrazolecarboxylic acids (pK_a ⟨66RTC1195⟩) are readily decarboxylated and this provides a method for introducing a deuterium atom into the pyrazole ring. 3-Deuteroindazole has also been prepared in this way ⟨76CJC1329⟩. Lithium aluminum hydride reduces 3-ethoxycarbonyl-3'(5'),5-dimethyl-5'(3')-pyrazol-1-ylpyrazole (**440**; R = CO$_2$Et) to the corresponding alcohol (**440**; R = CH$_2$OH) ⟨81T987⟩.

Acid chlorides are useful reagents, but when the pyrazole is N-unsubstituted a dimerization occurs and the diketopiperazine (**254**) is isolated (Section 4.04.2.3.3(x)). However, (**254**) reacts with many compounds as an acid chloride would, for example with amines to yield amides ⟨67HC(22)1⟩. The difunctional pyrazole derivative (**441**) affords polymers by reaction with diphenols ⟨69RRC763⟩. Cyanopyrazoles can be hydrolyzed to the corresponding carboxylic acids ⟨68CB829⟩.

Owing to the number of different reactions in which they can be used, pyrazole aldehydes and ketones are the most interesting group of C-linked substituents. Transformations of the CHO group into CO$_2$H, CH(OR)$_2$, CH=CHR, etc., and of the COR group into

C(OH)RR', oximes (Semmler–Wolf and Beckmann rearrangements ⟨76CB1898⟩) and many other functional groups have been mentioned in the literature ⟨67HC(22)1, B-76MI40402⟩. As an example, the aldehyde (**442**) yields the alcohol (**438**) when treated with methylmagnesium iodide, and the ketone (**437**) affords the malonylaldehyde sodium salt (**443**) by formylation ⟨72JHC1373⟩. Compound (**443**) treated with methylhydrazine affords a mixture of 1,1'-dimethyl-3,4'-bipyrazolyl (**444**) and 1,1'-dimethyl-4,4'-bipyrazolyl (**445**).

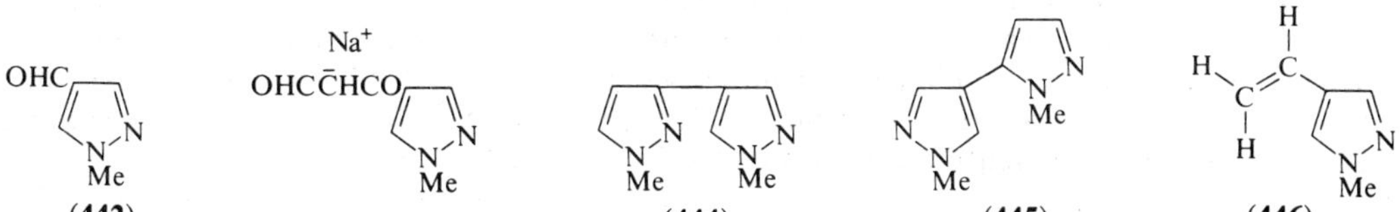

Pyrazolecarbinols can be dehydrated to vinylpyrazoles, (**438**) → (**446**) ⟨72JHC1373⟩, or transformed into chloromethyl derivatives ⟨81T987⟩. Compound (**440**; R = CH$_2$Cl) thus prepared is the starting material for the synthesis of the macrocycles (**226**)–(**228**) (Section 4.04.2.1.2(vi)). Vinyl- and ethynyl-pyrazoles have been extensively studied ⟨B-76MI40402⟩ and many vinylpyrazoles are polymerized by free radical initiators.

4.04.2.3.4 N-Linked substituents

Most of these substituents are interrelated as shown in Scheme 41.

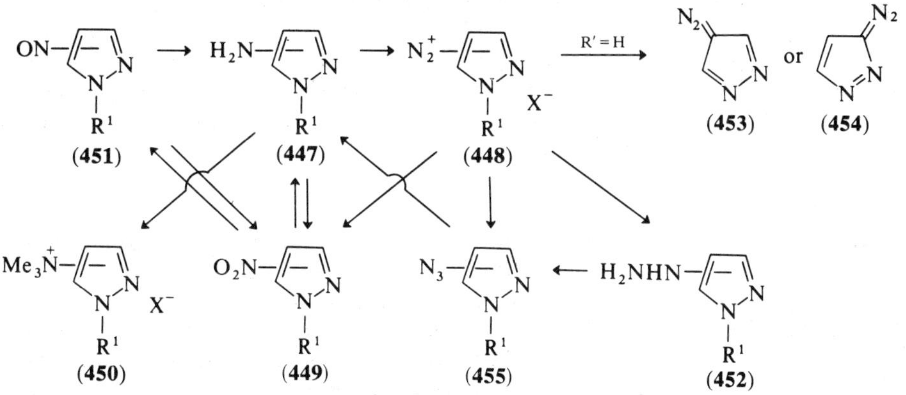

Scheme 41

(i) Amino-pyrazoles and -indazoles

Aminopyrazoles do not undergo the Dimroth rearrangement, since more heteroatoms are necessary for this rearrangement to occur (Section 4.02.3.1.9, 4.02.3.5.1). 3-Aminopyrazole itself (**447**; R^1 = H) can react with electrophiles at four positions: both ring nitrogen atoms, the amino group and the 4-position. Dorn, for example, has studied its reaction with toluene-*p*-sulfonyl chloride ⟨B-76MI40402, 80CHE1⟩ and has identified mono-, di- and tri-substituted derivatives on nuclear and amino nitrogen atoms. A similar situation is found in acylation studies, but in this case the nuclear acyl groups (azolides; Section 4.04.2.3.10(iii)) are very readily hydrolyzed. Other reactions of aminopyrazoles, like diazotization (formation of **448**), peracid oxidation (formation of **449**) and permethylation (formation of **450**) are well documented ⟨B-76MI40402, 80CHE1⟩.

Dimethylaminopyrazoles react with alkylating agents to afford quaternary salts ⟨72BSF2807⟩. The nitrogen atom of the dimethylamino group is the most reactive in the case of 3-dimethylamino- and 4-dimethylamino-pyrazoles (formation of **456** and **457** salts, respectively) whereas 5-dimethylaminopyrazoles yield aminopyrazolium salts (**458**).

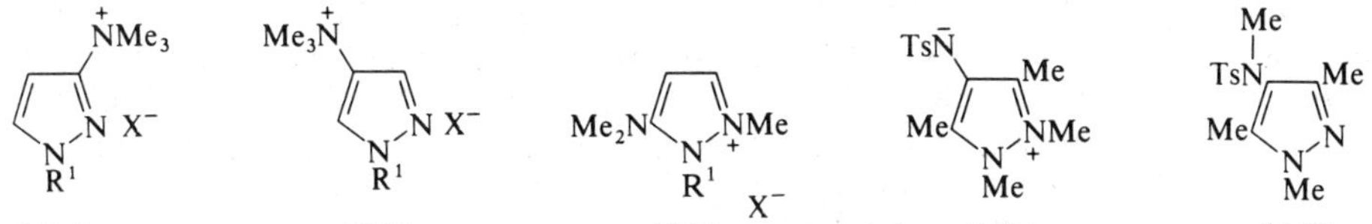

The chemical behaviour of the mesoionic pyrazole (**459**) has been studied by Boyd *et al.* ⟨74JCS(P1)1028⟩. Protonation and alkylation take place on the exocyclic nitrogen atom and a thermal rearrangement of a methyl group is observed when (**459**) is boiled in benzonitrile for several hours giving (**460**).

(ii) Aminopyrazolium ions

Besides the salts (**458**) and (**459**) previously described, aminopyrazolium salts can be obtained from the reaction between amines and chloropyrazolium salts (Section 4.04.2.3.7(ii)) or by quaternization of iminopyrazolines as in (**461**) → (**462**) ⟨72BSF2807⟩. The lithium aluminum hydride reduction of the salt (**462**) affords mixtures of reduced and open-chain pyrazoles (Figure 23; Section 4.04.2.1.6(i)).

(iii) Imines

In addition to (**461**), Dorn has described the imine (**463**) isolated from 5-amino-1-methylpyrazole and arenesulfonyl chloride ⟨80CHE1⟩. Upon heating, or in the presence of triethylamine, it undergoes rearrangement to the more stable 5-bis(arylsulfonamido)pyrazoles (**464**). 5-Iminopyrazolines (**461**) react with acyl chlorides at the exocyclic nitrogen atom to afford amidopyrazolium salts ⟨B-76MI40402⟩.

(iv) Nitro and nitroso groups

4-Nitropyrazoles are prepared by direct nitration of pyrazoles (Section 4.04.2.1.4(ii)) whereas the 3-nitro derivatives are derived from the rearrangement of *N*-nitropyrazoles (Section 4.04.2.3.10(vi)). Partial reduction of 4-nitropyrazoles (**449**) to the corresponding nitroso compounds (**451**) is possible with alkali stannites ⟨B-76MI40402⟩, but the most important reaction of the nitro compound is the complete reduction to amino derivatives (**447**). This reduction may be achieved using zinc and acetic acid (this method has been used in one step of the synthesis of the *C*-nucleoside antibiotic formycin ⟨80CJC2624⟩), tin and hydrochloric acid, sodium hydrosulfite, aluminum amalgam with moist ether or alcohol, catalytically, and with red phosphorus and hydriodic acid ⟨B-76MI40402⟩. The nitroso (**451**) to amino group (**447**) conversion has principally been accomplished using hydrazine and alcohol, but tin and hydrochloric acid and zinc and acetic acid are also effective. This is one of the best methods of synthesis of 4-aminopyrazoles from pyrazoles ⟨67HC(22)1⟩. Oxidation to nitropyrazoles, *e.g.* (**451**) → (**449**), has been carried out with potassium permanganate or nitric acid. 4-Nitrosopyrazoles (**451**) can also react with 3-aminopyrazoles (**447**) to yield azopyrazoles, for example (**465**).

(v) Azo-, diazo- and azido-pyrazoles

4-Arylazo-5-pyrazolones are very common products (**139**; Section 4.04.1.5.2) readily converted into 4-arylazopyrazoles (**466**; $R^5 = H$) *via* the 5-chloro derivatives (**466**; $R^5 = Cl$) ⟨67HC(22)1⟩. Other azopyrazoles (**318**) and (**319**) have been described in Section 4.04.2.1.4(viii). These compounds can be reduced to amines (zinc and acetic acid) or to arylhydrazines (zinc in alkaline medium) ⟨67HC(22)1⟩.

Pyrazole- and indazole-diazonium salts are compounds of exceptional thermal stability. Their chemistry has been studied mainly by Reimlinger, and their application to the synthesis of heterocyclic systems by Tišler and Stanovnik who have reviewed this topic ⟨80CHE443⟩. As in classical aromatic chemistry, the pyrazolediazonium salts (**448**) have been used as starting materials for the synthesis of chloro-, bromo-, cyano- and nitro-pyrazoles ⟨B-76MI40402⟩. The transformation of (**448**) into fluoropyrazoles using the Balz–Schiemann procedure yields only insignificant amounts of 3(5)-fluoropyrazole ⟨77BSF171⟩, but irradiation of the diazonium fluoroborate salt (Kirk and Cohen method) can be used to prepare relatively large quantities of fluoroazoles. 3-Fluoroindazole ⟨77BSF171⟩, mono- ⟨78JHC1447⟩ and di-fluoropyrazoles have been prepared ⟨79TL3179⟩ by this procedure.

Pyrazolediazonium salts (**448**) couple with activated aromatic molecules, like naphthols ⟨79KGS805⟩, and can be reduced to hydrazines (**452**) with tin(II) chloride ⟨74MI40406⟩.

The most interesting property of diazonium salts (**448**; $R^1 = H$) is their conversion into 4- or 5-diazopyrazoles, (**453**) and (**454**). 3-Diazoindazole (**467**) was the first heterocyclic diazo compound to have its structure determined by X-ray diffraction analysis (Section 4.04.1.3.1). In addition to the reactions discussed in Section 4.01.3.4.1, it should be stressed that diazopyrazoles are useful synthons for many condensed heterocyclic systems, such as azapentalenes ⟨78AHC(22)184, 80CHE443⟩. Other reactions of diazopyrazoles are their reduction to pyrazoles with titanium(III) and iron(II) salts ⟨79S194⟩ and their cycloaddition with ylides ⟨79TL1567⟩ and with electron-rich alkenes ⟨81TL1199⟩ to afford 5,5- and 5,6-fused heterocycles, respectively. There is evidence from mass spectrometry that the formation of 3-azidoindazole by the action of azide ion on 3-diazoindazole (**467**) involves a spectacular intermediate, the indazolylpentazole (**468**) ⟨80CHE443⟩.

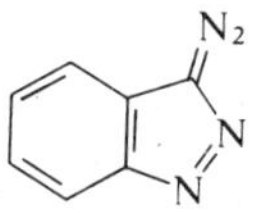

(**467**) 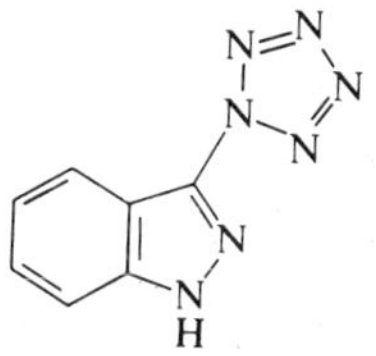   (**468**)

Azidopyrazoles (**455**) exist as such even when the azido group is at the 3-position where an azido–tetrazole equilibrium is possible (**469**) ⇌ (**470**) ⟨77AHC(21)72, 78AHC(22)184⟩. However, when the molecules are converted into their anions, the cyclic pyrazolo[1,5-d]tetrazole (**471**) form predominates. It is possible to 'trap' the cyclized anions such as (**471**) by methylation and to obtain stable azapentalenes (**472**) and (**473**).

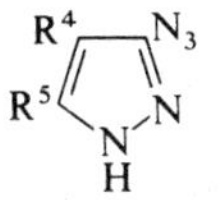

(**469**) 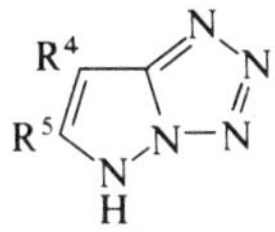   (**470**) 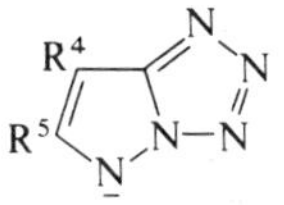   (**471**) 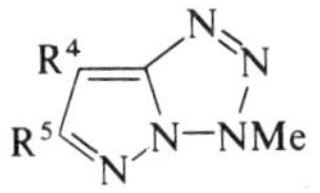   (**472**) 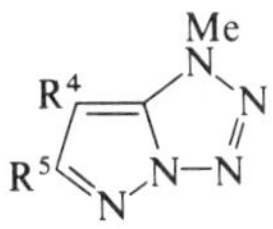   (**473**)

Thermolysis of 4- and 5-azidopyrazoles has been studied by Smith ⟨B-70MI40402, 81AHC(28)232⟩. These compounds undergo fragmentation with formation of unsaturated nitriles via the nitrenes (**474a**) and (**474b**; Scheme 42).

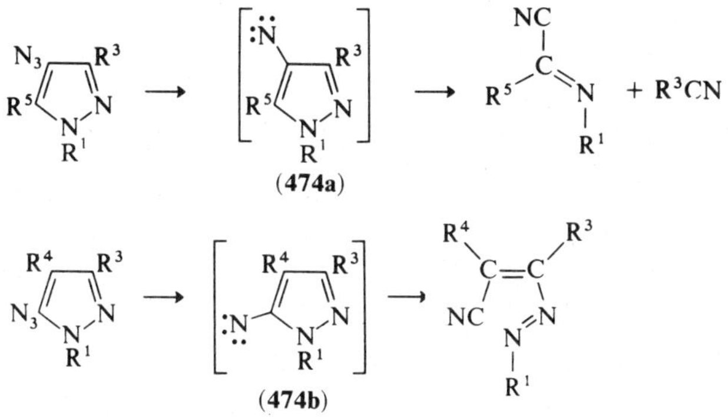

(**474a**)

(**474b**)

Scheme 42

4.04.2.3.5 O-Linked substituents

The fact that an OH or OR group in position 3 or 5 enhances the reactivity of the pyrazolic 4-position towards electrophiles has already been discussed (Section 4.04.2.1.3). Likewise, the non-aromatic reactivity of 2-pyrazolin-5-ones has been examined in Section 4.04.2.2.1(ii).

(i) Interconversion and modes of reaction of tautomeric forms

The tautomerism of pyrazolones has been discussed in Section 4.04.1.5.2 and their multiple reactivity in Sections 4.04.2.1.1(iv) and 4.04.2.1.3(viii). Interconversion of the different tautomers proceeds readily through the anion (**157**; Scheme 10) or the cations (**203**) and (**204**; Scheme 17), thus providing pyrazolones with exceptional possibilities of reaction.

(ii) 3-Hydroxy-pyrazoles and -indazoles

From the results quoted in Section 4.04.2.1.3(x) the stability of acetyl derivatives of indazolone decreases in the order $N(1)$-acetyl $> N(2)$-acetyl $> O(3)$-acetyl. The reactivity of 3-hydroxypyrazoles is covered by Dorn's comprehensive review ⟨80CHE1⟩. Amongst the results reported there are the Claisen rearrangement of allyloxypyrazoles (**475a**) → (**475b**) and a method for transforming 3-hydroxy- into 3-chloro-pyrazoles via the pyrazolone (**476**) and the chloropyrazolium chloride (**477**). Methylation of 3-hydroxypyrazole (**478**; X = OH) affords the pyrazolone (**476**), which in turn is transformed into the salt (**477**) by reaction with phosphorus oxychloride. The final step is the thermolysis of (**477**) that yields the 3-chloropyrazole (**478**; X = Cl)

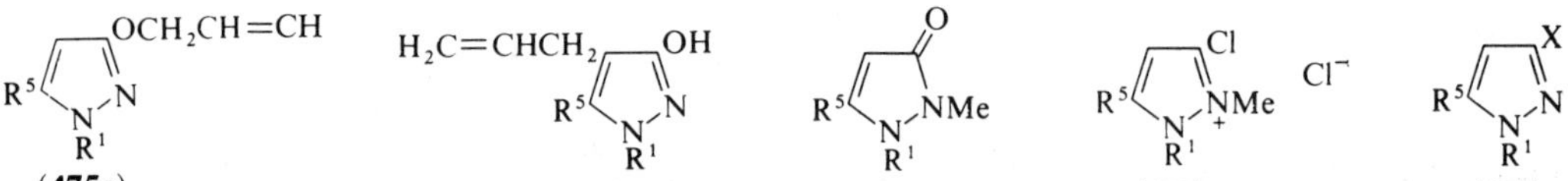

Thermolysis of the betaine (**479**) affords the aromatic 3-methoxypyrazole (**480**) ⟨80CHE1⟩

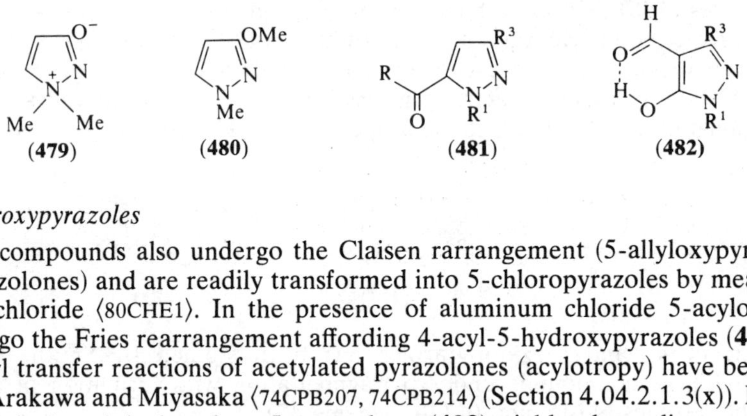

(iii) 5-Hydroxypyrazoles

The title compounds also undergo the Claisen rarrangement (5-allyloxypyrazoles → 4-allyl-5-pyrazolones) and are readily transformed into 5-chloropyrazoles by means of phosphorus oxychloride ⟨80CHE1⟩. In the presence of aluminum chloride 5-acyloxypyrazoles (**481**) undergo the Fries rearrangement affording 4-acyl-5-hydroxypyrazoles (**482**).

The acetyl transfer reactions of acetylated pyrazolones (acylotropy) have been carefully studied by Arakawa and Miyasaka ⟨74CPB207, 74CPB214⟩ (Section 4.04.2.1.3(x)). Methylation of 3-methyl-1-phenyl-4-phenylazo-5-pyrazolone (**402**) yields, depending on the experimental conditions, the N- and the O-methylated derivatives (**483**) and (**484**) ⟨66BSF2990⟩. These derivatives have been used as model compounds in a study of the tautomerism of (**402**) (structure **139**; Section 4.04.1.5.2).

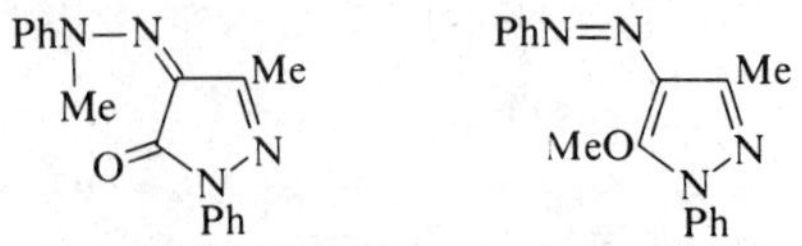

(iv) 4-Hydroxypyrazoles

4-Hydroxypyrazoles are amphoteric compounds which form salts with alkalies and with mineral acids ⟨B-76MI40402⟩. The 4-hydroxy group directs electrophilic substitution towards the 5-position (diazo coupling, chlorination, nitrosation). Alkylation of 1-substituted 4-hydroxypyrazoles (**485**; R = H) takes place on the oxygen atom giving (**485**; R = Me) or on the nitrogen atom giving (**486**), depending on the experimental conditions ⟨B-77MI40402, 79CJC904⟩. The salts (**486**) on deprotonation yield the mesoionic pyrazoles (**28**; Section 4.04.1.1.2). Oxidation of *N*-unsubstituted 4-hydroxypyrazoles (**485**; R = R^1 = H) affords 3,4-diazacyclopentadienone (**487**) which behaves either as a diene or as a dienophile towards a wide variety of cycloaddition reagents ⟨79CJC904⟩.

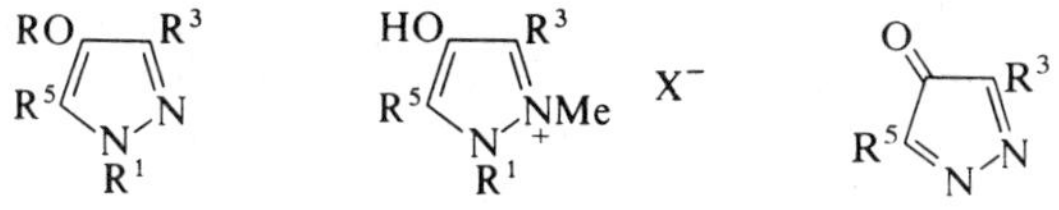

(v) Alkoxy groups

An alkoxy group may be replaced by chlorine upon treatment of a suitably substituted pyrazole with phosphorus oxychloride ⟨67HC(22)1⟩. Heating at 230–300 °C results in a methyl group rearrangement from the oxygen atom in (**488**) to the nitrogen atom at position 2 giving (**489**; R = Me). Quaternary salts like (**490**) lose an alkyl halide molecule upon heating to yield pyrazolones (**489**; R = Et) and not alkoxypyrazoles (**488**) ⟨67HC(22)1⟩.

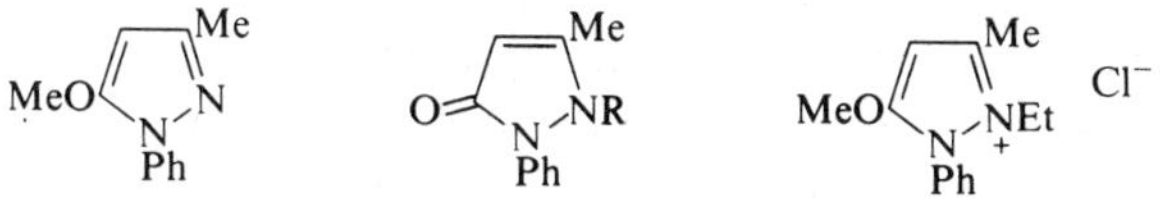

4.04.2.3.6 S-Linked substituents

(i) Mercapto compounds: tautomerism

5-Mercaptopyrazoles exist as a mixture of the SH and NH tautomers (Section 4.04.1.5.2). 1-Substituted 3-mercaptoindazoles exist as such whereas 2-substituted derivatives exist as thiones (NH tautomer) ⟨76AHC(S1)1⟩.

(ii) Thiones

Most of the reactions described in Section 4.01.3.6.2 have their counterpart in thiopyrazolones ⟨67HC(22)1, B-76MI40402⟩: *S*-methylation (**491**; R^5 = Me), *S*-acylation (**491**; R^5 = COR), oxidation to disulfides (**492**) or, with a stronger oxidizing agent, to sulfonic acids (**493**). Oxidative desulfurization is also known and of particular interest is the reaction of thiopyrine (**494**) with hydrogen peroxide which yields 3-methyl-1-phenylpyrazole with loss of sulfur and the 2-methyl group ⟨67HC(22)1⟩.

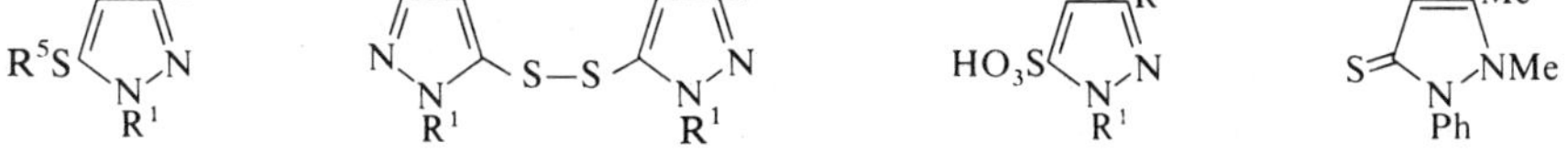

(iii) Alkylthio groups

Derivatives like (**491**: R^5 = Me) can be de-*S*-methylated by Raney nickel in ethanol or concentrated hydrochloric acid. Acid hydrolysis of (**491**; R^5 = acyl) also affords 5-mercaptopyrazoles, whereas alkaline hydrolysis of the pyrazolium salt (**495**) furnishes methanethiol and antipyrine.

Methylation of (**494**) with methyl iodide affords the salt (**495**) which on distillation loses the *N*-methyl group yielding the 5-methylthiopyrazole (**491**; R^1 = Ph, R^3 = R^5 = Me). The direct transformation (**494**) → (**491**) occurs by simple heating ⟨67HC(22)1⟩, although the quaternary salt is probably the catalyst for the *N*- to *S*- rearrangement. This reaction has been thoroughly studied by Maquestiau *et al.* ⟨77BSB961⟩. Alkylthio derivatives (**491**) can be readily oxidized to sulfones (**496**).

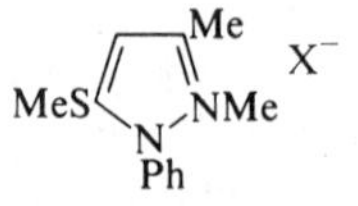

(495), 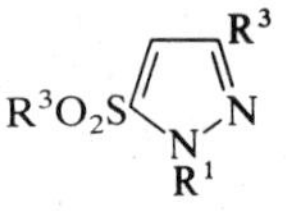 (496), 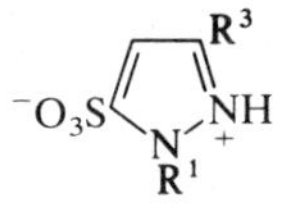 (497)

(iv) Sulfonic acid groups

Pyrazolesulfonic acids, like (**493**), have high melting points (Table 24) and probably exist as the zwitterions (**497**). They are very stable to hydrolysis and only afford pyrazolones at high temperatures. The replacement of the SO₃H group by bromine has also been reported ⟨B-76MI40402⟩. Pyrazole-3-, -4- and -5-sulfonic acids react with phosphorus pentachloride to form sulfonyl chlorides.

4.04.2.3.7 Halogen atoms

(i) Nucleophilic substitution reactions: neutral pyrazoles and indazoles

Halogeno-pyrazoles and -indazoles are quite unreactive compounds. For instance 3-bromo-5-methylpyrazole does not react with hydrazine nor with sodium azide, even in DMSO or HMPT ⟨74MI40406⟩. Neither 3-chloro- nor 3-bromo-indazole reacts with hydrazine ⟨78MI40401⟩ whilst the 3-iodo derivative is reduced to indazole. These compounds need an electron-withdrawing group in the α (pyrazoles) or γ (indazoles) position for the halogen atom to be reactive. Thus 5-chloropyrazoles (**498**; X = Cl) bearing an electron-withdrawing group Y at the 4-position react with secondary amines to afford 5-aminopyrazoles ⟨B-76MI40402⟩. The reactivity depends on the nature of Y and decreases in the order NO₂ > PhN₂ > PhCO ≫ Br ≫ H. 3-Bromo-5-nitroindazole (**499**: X = Br) also reacts with secondary amines to afford the 3-amino derivatives ⟨78MI40401⟩. Nucleophilic photosubstitution of both (**498**; Y = NO₂, X = Cl) and (**499**; X = Br) has been studied ⟨79T1331, 80T3523⟩ and in this way 5-cyanopyrazoles (**498**; Y = NO₂, X = CN) and 3-aminoindazoles (**499**; X = NR¹R²) have been prepared.

Reaction of the pyrazole anion with (**498**; Y = NO₂, X = Cl) in DMSO yields the pyrazolyl-pyrazole (**500**) ⟨81M675⟩.

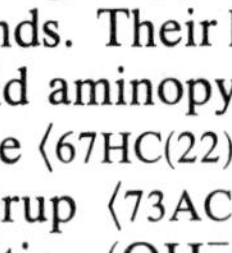

(498), 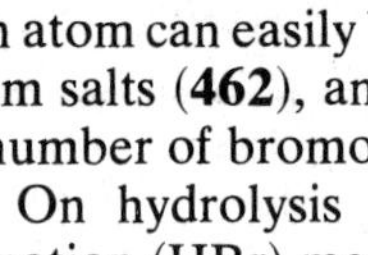 (499), 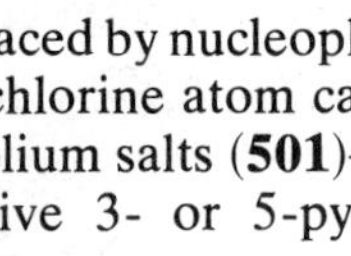 (500)

(iii) Nucleophilic displacements: halogenopyrazolium salts

The 3- or 5-halogenopyrazolium salts such as the chloro or bromo derivatives (**477**) are reactive compounds. Their halogen atom can easily be replaced by nucleophiles, for example by amines to yield aminopyrazolium salts (**462**), and the chlorine atom can be replaced by bromine or iodine ⟨67HC(22)1⟩. A number of bromopyrazolium salts (**501**)–(**503**) have been studied by Begtrup ⟨73ACS2051⟩. On hydrolysis they give 3- or 5-pyrazolones by an 'anomalous' addition (OH⁻)–elimination (HBr) mechanism.

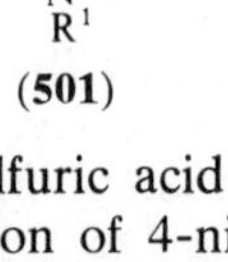

(501), 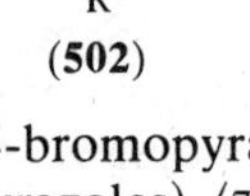 (502), 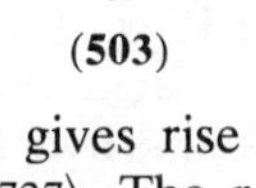 (503)

Nitration in 80% sulfuric acid of 4-bromopyrazoles gives rise to considerable nitrodebromination (formation of 4-nitropyrazoles) ⟨79AJC1727⟩. The reaction takes place on the protonated pyrazolium ion (Section 4.04.2.1.4(ii)).

(iii) Other reactions

Reduction of the halogen substituent has been carried out by different procedures such as catalytic hydrogenation using palladium–carbon or Raney nickel, red phosphorus and hydroiodic acid, and zinc and sulfuric acid ⟨66AHC(6)347⟩. 3-Deuteropyrazole has been prepared by Zn/Cu–AcOD reduction of 3-bromopyrazole ⟨70BSF1974⟩. Grignard reagents are only formed from bromopyrazoles when promoted by active alkyl halides ⟨B-76MI40402⟩.

Two important problems in this section still remain unsolved. The first one concerns the relative leaving group ability of the four halogen atoms. As is well known the experimental order of reactivity of halogenonitrobenzenes in the S_NAr reaction is F ≫ Cl > Br > I. This order has also been observed for 2-halogeno-6-nitrobenzothiazoles ⟨77BSF171⟩. In pyrazoles and indazoles no clear-cut conclusion can be drawn from the scattered literature results. It has been clearly established that in pyrazolium salts chlorine can be replaced by the more nucleophilic bromine and iodine (Section 4.04.2.3.7(ii)) but not by fluorine ⟨77BSF171⟩. On the other hand, fluoropyrazoles (Section 4.04.2.3.4(v)) are quite stable compounds.

The second problem is the relationship between the position of the substituent in the pyrazole nucleus and its mobility. In the 1-phenylpyrazole series in their reactions with Grignard reagents, the bromine reactivity decreases in the order 5-Br > 4-Br > 3-Br ⟨B-76MI40402⟩. When an electron-withdrawing group is present at the 4-position, the 5-chloropyrazole is more reactive than 3-chloropyrazole, but this has been attributed to bond fixation (Section 4.02.3.9). Thus, this problem needs further clarification.

4.04.2.3.8 Metal and metalloid-linked substituents

In different sections of this chapter, pyrazoles and indazoles C-linked to a metal or a metalloid have been described or they will be described in the preparative sections, including lithio derivatives (Section 4.04.2.1.7), pyrazolylmagnesium reagents (Section 4.04.2.3.7(iii)), chloromercury derivatives (Section 4.04.2.1.4(vii)) and silylpyrazoles (Section 4.04.3.1.2(ii)). All these compounds are useful intermediates and some of their most characteristic reactions will be discussed here.

Lithiated pyrazoles prepared from C-H or preferably from C-Br pyrazoles ⟨B-76MI40402⟩ can be transformed into carboxylic acids. In this way pyrazole-4-carboxylic acids (**504**; R^1 = H or Me) have been prepared from the corresponding 4-bromopyrazoles.

Finar and coworkers ⟨B-76MI40402⟩ have made a detailed study of the pyrazolyl Grignard reagent (**505**). In addition to alcohols and ketones, the hydrocarbons (**506**) and (**507**) could be isolated, the first by heating the Grignard reagent (**505**) and the second by reaction with diethyl carbonate.

4-Chloromercurypyrazoles (**316**) react like phenylmercury(II) chloride ⟨B-76MI40402⟩. The group is removed by dilute hydrochloric acid and replaced by bromine in the reaction with bromine in acetic acid. C-Trimethylsilylpyrazoles have been studied by Birkofer and coworkers ⟨72CB1759, B-76MI40402⟩. Sodium hydroxide selectively desilylates the Me₃Si group at the 3-position, while sulfuric acid removes the 4-SiMe₃ substituent. The 3- and 4-nitrosopyrazoles, difficult to obtain by direct nitrosation (Section 4.04.2.1.4(ix)), are readily prepared by nitroso-detrimethylsilylation ⟨B-76MI40402⟩.

4.04.2.3.9 Fused heterocyclic rings

There are many representatives of fused pyrazoles. By far the most intensively studied are the [5.6] bicyclic derivatives which include indazoles and azaindolizines ⟨77HC(30)117, 77HC(30)179⟩ (see also Section 4.04.2.4). Less studied but still well known are the [5.5] bicyclic systems, the azapentalenes ⟨77HC(30)1, 77HC(30)317, 78AHC(22)184⟩. Pyrazoles fused to a seven-membered ring have few representatives. Compound (**508**) when irradiated affords 3-phenylindazole amongst other products ⟨78H(11)293⟩. The cycloheptapyrazoles

(508) **(509)** **(510)** **(511)** **(512)**

(509) and (510) are the major constituents of the reaction mixture obtained in the reaction of diphenylnitrilimine with tropone ⟨77JCS(P1)939⟩.

The 1,2-diazepine ring system is related, thermally and photochemically, to two pyrazole [4.5] bicyclic systems (511) and (512). A large number of publications by Moore, Sharp, Snieckus and Streith deal with these isomerizations ⟨72CC827, 80CC444⟩.

4.04.2.3.10 Substituents attached to ring nitrogen atoms

This is a most original situation in azoles since there is no equivalent in either aromatic or six-membered ring heterocyclic chemistry, at least in neutral molecules. With a substituent on the nitrogen atom, pyrazoles behave as halogens ⟨71AG(E)743⟩; thus N-acetylpyrazole can be regarded as the arenologue of acetyl chloride.

(i) Aryl groups

Spectroscopic measurements of N-arylpyrazoles have been used to calculate the Hammett and Taft parameters of pyrazole and some of its derivatives ⟨74JCS(P2)449, 80AJC1763, 81JCR(S)364⟩. In the last reference a critical examination of the literature values is given. Typical values are as follows: pyrazole, $\sigma_I = 0.300$, $\sigma_R° = -0.061$; indazole, $\sigma_I = 0.361$, $\sigma_R° = -0.154$; isoindazole, $\sigma_I = 0.420$, $\sigma_R° = -0.126$. A typical pseudohalogen substituent, NCS, has $\sigma_I = 0.42$ and $\sigma_R° = -0.07$ ⟨B-79MI40416⟩.

Fong ⟨80AJC1763⟩ assumes that $(\sigma_R°)_0$ (the substituent constant for the conformation where the azole is coplanar with the benzene ring) is the same for all the pyrazoles. From the experimental value of $\sigma_R°$ he calculated the dihedral angle (Section 4.04.1.4.3) assuming that $(\sigma_R°)_\theta = (\sigma_R°)_0 \cos^2 \theta$. If instead of -0.165 we assume that $(\sigma_R°)_0$ is equal to -0.20 then θ for 1-phenylpyrazole and 5-methyl-1-phenylpyrazole is 25° and 57°, respectively.

In the section dealing with electrophilic substitution on the carbon atom (Section 4.04.2.1.4) it has been pointed out that the *para*-position of the 1-phenyl ring competes with the pyrazolic 4-position. For instance, bromination of 1-phenylpyrazole affords 4-bromo-1-*p*-bromophenylpyrazole ⟨B-76MI40402⟩. In a strong acid medium, the 1-phenylpyrazolium cation is nitrated exclusively on the *para*-position ⟨72JCS(P2)1654⟩. In this respect pyrazoles also behave like pseudohalogens (chlorine is an *ortho*–*para* directing substituent, 30–70), but steric hindrance prevents the *ortho* attack. The reactivity of the 1-thienylpyrazoles (513) and (514) has been studied by Gronowitz *et al.* ⟨78CS(13)157⟩. Reactions occur at the thiophene ring and the pyrazole ring behaves as a +R substituent.

(513) **(514)** **(515)** **(516)** **(517)**

Another example of the analogy between pyrazole and chlorine is provided by the alkaline cleavage of 1-(2,4-dinitrophenyl)pyrazoles. As occurs with 1-chloro-2,4-dinitrobenzene, the phenyl substituent bond is broken with concomitant formation of 2,4-dinitrophenol and chlorine or pyrazole anions, respectively ⟨66AHC(6)347⟩. Heterocyclization of N-arylpyrazoles involving a nitrene has already been discussed (Section 4.04.2.1.8(i)). Another example, related to the Pschorr reaction, is the photochemical cyclization of (515) to (516) ⟨80CJC1880⟩. An unusual transfer of chlorine to the side-chain of a pyrazole derivative was observed when the amine (517; X = H, Y = NH_2) was diazotized in hydrochloric acid and subsequently treated with copper powder ⟨72TL3637⟩. The product (517; X = Cl, Y = H) was isolated.

(ii) Alkyl groups

Lithiation on the *N*-methyl group of (**334**) has previously been discussed (Section 4.04.2.1.7(i). *N*-Vinylpyrazoles are polymerized by free radical initiators ⟨B-76MI40402⟩. However, the most important reactions of this section are the dequaternization reactions of pyrazolium and indazolium salts ⟨67HC(22)1, B-76MI40402⟩, which have been extensively studied by von Auwers. Pyrolysis of pyrazole quaternary iodides results in the loss of one of the nitrogen substituents as alkyl iodide. The groups most readily cleaved are benzyl > allyl > methyl > ethyl > propyl > phenyl. The thermal decomposition of indazolium salts produces mixtures of 1- and 2-alkylindazoles, the 1-isomer being generally predominant. These thermal dealkylations require reaction temperatures of 200 °C or higher and are often accompanied by extensive degradation and isomerization. Recently, several less drastic methods of dequaternization have been described involving reagents such as triphenylphosphine, the thiophenolate anion under phase transfer catalytic conditions ⟨78CR(C)(287)439⟩, and piperidine or 3-methylpyridine ⟨80JHC905⟩.

(iii) Acyl groups

The solvolysis of 1-acetyl-pyrazoles and -indazoles has been studied by Staab *et al.* ⟨B-76MI40402⟩ in connection with other azolides, specially 1-acetylimidazole. These compounds, for example (**131**; Section 4.04.1.5.1), are useful acylating agents. *N*-Alkoxycarbonylpyrazoles (**518**) on heating in a polar, aprotic solvent yield *N*-alkylpyrazoles. The mechanism of this alkylative decarboxylation ⟨74JOC1909⟩ involves the pyrazole anion and a nucleophile such as I$^-$ as catalyst. Butyl isocyanate transforms the 2-ethoxycarbonylindazole (**519**) into (**520**) with migration of the CO_2Et group ⟨77S804⟩. 1-Cyanopyrazoles (**521**) are relatively unstable compounds, readily hydrolyzed with formation of the carboxamides (**522**) ⟨70CB1954⟩.

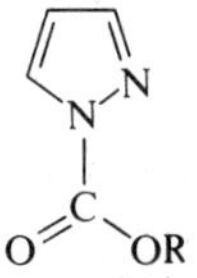

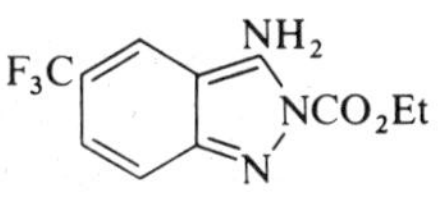

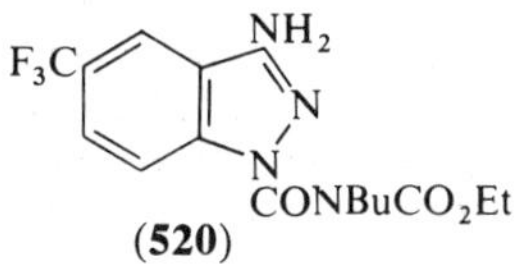

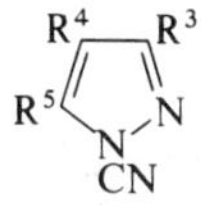

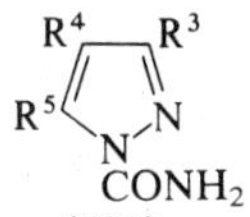

(iv) N-Oxides

The *N*-hydroxypyrazoles (**523**; R = H) and the pyrazole *N*-oxides (**268**; Section 4.04.2.1.3 (xiii)) have been reduced to pyrazoles by means of zinc in acetic acid and catalytic hydrogenation, respectively ⟨75MI40402⟩.

The OH group of (**523**; R = H) can be methylated ⟨73JCS(P2)164⟩, acetylated and tosylated ⟨74JOC2663⟩. The reaction of the 2-hydroxyindazole (**524**) with diazomethane affords a mixture of *O*- and *N*-methylated products, (**525**) and (**526**) ⟨81BSB645⟩. Jones and coworkers have studied the rearrangements of 1-methoxypyrazoles (**523**; R = Me) ⟨73JCS(P1)170⟩.

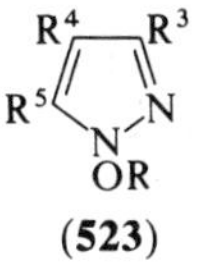

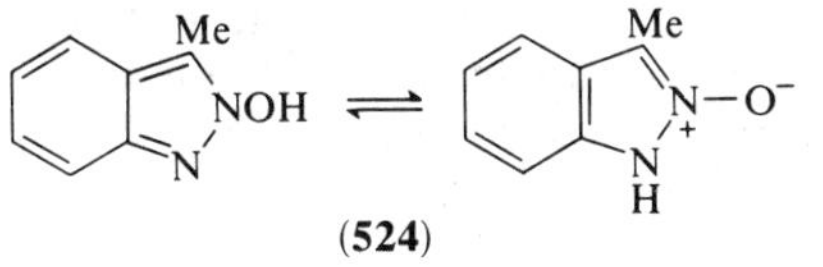

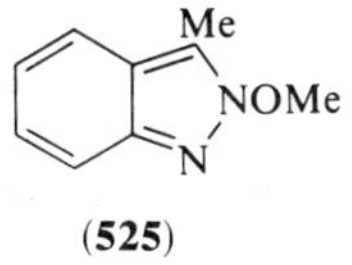

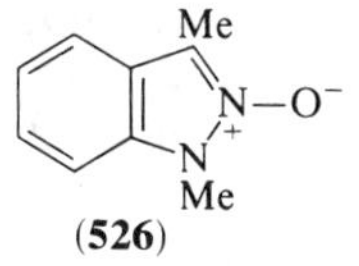

(v) N-Amino groups

N-Aminopyrazoles (Section 4.04.2.1.3(xiv)) have been used for the synthesis of *N,N'*-linked biazoles (**527**; R^5 = H, OH, OMe, Cl) and (**528**) ⟨80JCR(M)0514, 81JHC957⟩.

Vapour phase pyrolysis of sulfoximides (**529**) results in the formation of the nitriles (**530**) ⟨75JCS(P1)41⟩. The tosylate (**273**), when treated with acetic anhydride, rearranges to (**531**)

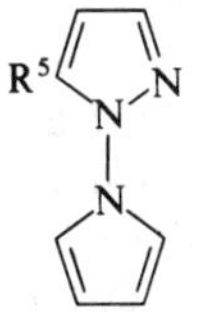

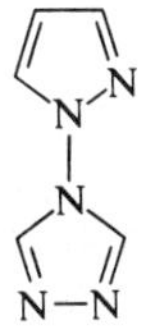

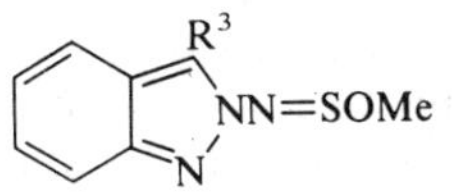

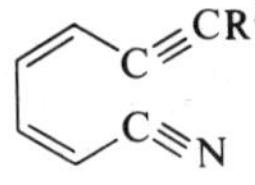

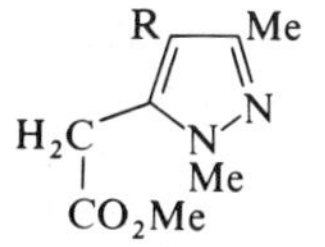

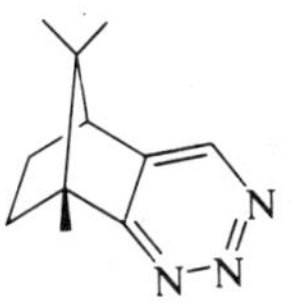

⟨76CPB2267⟩ and the oxidation of (**277**) with lead tetraacetate results in ring expansion to give the 1,2,3-triazine (**532**) ⟨79YZ699⟩.

(vi) N-Nitro groups

The chemistry of *N*-nitro-pyrazoles and -indazoles has been developed by Habraken and coworkers ⟨B-76MI40402⟩. 1-Nitropyrazoles (Section 4.04.2.1.3(xiv)) rearrange to 4-nitropyrazoles in sulfuric acid and thermally to 3(5)-nitropyrazoles, making these last compounds readily accessible. All experimental data ⟨76JOC1758⟩ on the *N*- to *C*-isomerization are compatible with a rate determining [1,5] shift of NO_2 to give a pyrazolenine as an intermediate, which subsequently isomerizes to the 3(5)-nitropyrazole (Scheme 43).

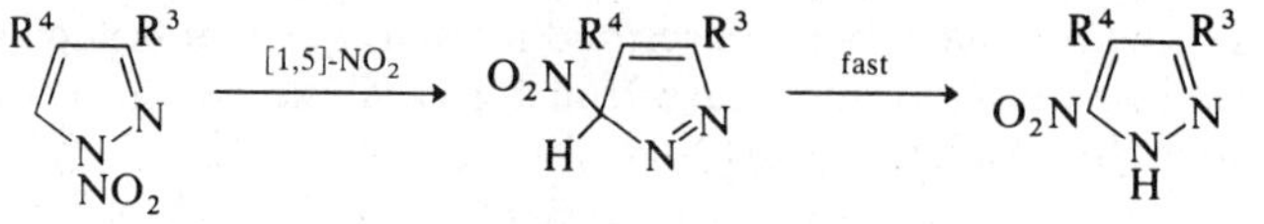

Scheme 43

2-Nitroindazoles rearrange to 3-nitro derivatives on heating ⟨71JOC3084⟩. Another important property of *N*-nitropyrazoles was also discovered by Habraken ⟨77JOC2893⟩ who showed that 1,4-dinitropyrazoles undergo '*cine*' substitution with secondary amines to give aminonitropyrazoles (Scheme 44). Using pyrazoles as nucleophiles this reaction affords 3(5)-(pyrazol-1-yl)pyrazoles ⟨79JOC4156⟩ and tripyrazoles ⟨81JHC559⟩.

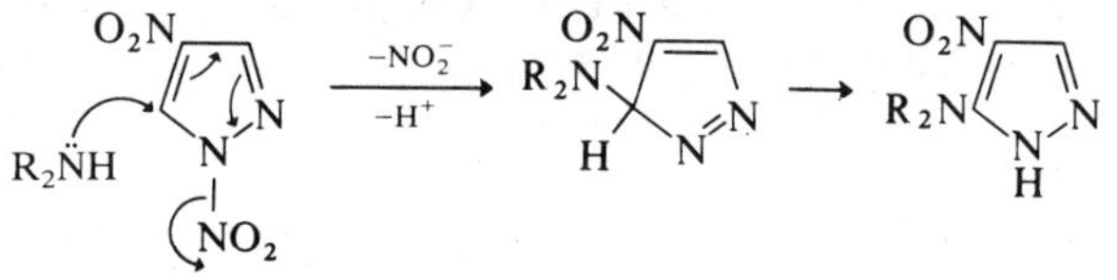

Scheme 44

This reaction has been used as the key step in an original synthesis of formycin ⟨80CJC2624⟩. By a similar mechanism, *via* a 3*H*-indazole, the 2,5-dinitro derivative reacts with secondary amines to afford 3-amino-5-nitroindazoles ⟨80MI40404, 81JOC2706, 82PNA4487⟩.

As Olah *et al.* have reported ⟨81JOC2706⟩, *N*-nitropyrazole in the presence of Lewis or Brønsted acid catalysts is an effective nitrating agent for aromatic substrates. The greater lability of the N—NO_2 bond in *N*-nitropyrazole compared with aliphatic nitramines was discussed on the basis of its molecular structure as determined by X-ray crystallography.

(vii) Halogens

N-Chloro- and *N*-bromo-pyrazoles are unstable compounds, whilst *N*-iodopyrazoles are stable (Section 4.04.2.1.3(xii)). These last compounds readily yield the iodonium ion, which in turn acts as an iodinating agent. Thus, the formation of 3-iodo-4,5-dimethylpyrazole from 1-iodo-4,5-dimethylpyrazole ⟨B-76MI40402⟩ is explained. The different steps in the chlorination of 3,5-diphenylpyrazole have been characterized (Scheme 45). The *N*-chloro derivative (**533**) was fairly stable when stored at low temperatures, but it was converted into the isopyrazole (**534**) on standing for several hours at room temperature ⟨80JOC76, 82S844⟩.

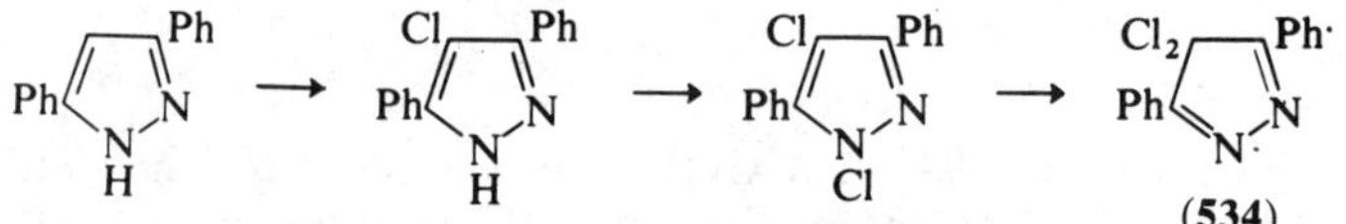

Scheme 45

(viii) N-Metallic derivatives

Sodium and silver pyrazole salts (Section 4.04.2.1.3(vi)) are often used instead of neutral pyrazoles to facilitate electrophilic attack on the ring nitrogen atoms. For example, pyrazolylmethanes (**234**)–(**236**) have been prepared from pyrazole anions (Section 4.04.2.1.3(viii)). Unstable 2-acetylindazoles are obtained from the reactive silver salts of indazoles (Section 4.04.2.1.3(x)). Electrophilic attack on the ring carbon atoms also occurs more readily in the anion than in the neutral molecule. For example, bromination is often carried out on pyrazole and indazole silver salts (Section 4.04.2.1.4(v)).

Pyrazoles bearing metalloid groups on the nitrogen atom (Section 4.04.2.1.3(xv)) such as $SiMe_3$, $GeMe_3$, $SnMe_3$, $GaMe_2$, *etc.*, readily react with alkylating and acylating agents to afford *N*-alkyl- and *N*-acyl-pyrazoles ⟨80ZOB875, 81JOM(215)349⟩, respectively. Aldehydes and ketones yield compounds (**535**) when treated with 1-trimethylsilylpyrazole ⟨81JOM(208)309⟩.

(**535**)　(**536**)　(**537**)

The decomposition of (**536**) with hydrogen sulfide yields pyrazole ⟨76T1909⟩. The 1-phosphorylpyrazoles (**537**) are suitable reagents for the phosphorylation of alcohols, amines, hydrazines and azides ⟨76AG(E)378⟩.

4.04.2.4 Pyrazoles, Indazoles and their Derivatives as Starting Materials for the Syntheses of Fused Ring Systems

Fused ring systems containing a pyrazole unit can be prepared either from the heterocyclic moiety by formation of a pyrazole ring or from the reaction between a pyrazole derivative and a suitably functionalized reagent. The ring systems thus obtained are discussed in detail in other chapters (Chapters 4.05, 4.35, 4.36) but it is of interest to discuss here those methods which start from a pyrazole derivative as the reactions involved can be considered as examples of the reactivity of pyrazoles. The most widely studied fused ring systems are the [5.6] systems and the examples described in this section will be chosen from this group and, occasionally, from [5.5] and [5.7] systems.

The 3- or 5-aminopyrazoles are the synthons used most frequently. The second heterocyclic ring is created between the amino group and the 1-position (if unsubstituted) or between the amino group and the 4-position. Thus 3-substituted 5-aminopyrazoles react with 1,3-difunctional compounds to afford pyrazolo[1,5-*a*]pyrimidine derivatives (**538**) (Table 34). Aminopyrazolinones ($R^3 = OH$) can be used instead of aminopyrazoles. Similarly 3-aminoindazole yields pyrimido[1,2-*b*]indazoles (**539**).

(**538**)　(**539**)

This reactivity of *N*-unsubstituted amino-pyrazoles and -indazoles which can be regarded as 1,3-diamino derivatives has been used to build a great variety of fused six-membered heterocycles such as the 1,2,4-triazine derivatives (**540**) and (**541**), the 1,3,5-triazine derivatives (**542**) and (**543**), and benzothiadiazines (**544**).

(**540**)　(**541**)　(**542**)　(**543**)　(**544**)

When the 1-position is substituted, 3- and 5-aminopyrazoles react at the C-4 carbon atom, the reactivity of which is enhanced by the amino group. Thus pyrazolo[3,4-*b*]pyridines (**545**) are obtained either by the Skraup synthesis or from 1,3-difunctional compounds. Here also aminopyrazolinones have been used instead of aminopyrazoles to prepare (**545**; $R^3 = OH$). If 1,4-ketoesters (succinic acid derivatives) are used instead of β-ketoesters, pyrazolo[3,4-*b*]azepinones (**546**) are obtained.

When the pyrazole ring bears two adjacent functional substituents, it reacts like an *o*-substituted benzene. For example, 4,5-diaminopyrazoles behave similarly to

Table 34 Synthesis of Fused Ring Systems from Pyrazoles and Indazoles

Ring system	Ref.
Pyrazolo[1,5-a]pyrimidine (**538**)	67CB2577, 70BCJ849, 70JHC247, 74JHC423, 77JHC155, 78JHC185, 79FES898, 80JCS(P1)481, 81JHC163, 81KGS1554
2-Hydroxypyrazolo[1,5-a]pyrimidine (**538**; R^3 = OH)	70CB3252, 79JHC773
Pyrimido[1,2-b]indazole (**539**)	76T493
Pyrazolo[1,5-c][1,2,4]triazine (**540**)	76JOC3781
[1,2,4]Triazino[3,2-c]indazole (**541**)	74JOC1833, 76H(4)1115
Pyrazolo[1,5-a][1,3,5]triazine (**542**)	73JHC885, 76JHC1305
[1,3,5]Triazino[1,2-b]indazole (**543**)	76T493
Pyrazolo[1,5-b][1,2,4]benzothiadiazine (**544**)	76JHC395
Pyrazolo[3,4-b]pyridine (**545**)	68AHC(6)347, 67AG981, 68CB3265, 70JHC247, 72JHC235, 75JHC517, 78JHC319, 79JPR881, 80JCS(P1)938
3-Hydroxypyrazolo[3,4-b]pyridine (**545**; R^3 = OH)	79JHC773, 80BSB51
Pyrazolo[3,4-b]azepinone (**546**)	79CJC3034
Pyrazolo[3,4-b]pyrazine (**547**)	67TL2029
Pyrazolo[3,4-b]quinoxaline (**548**)	62JGU1876
Pyrazolo[3,4-b][1,4]diazepine (**549**)	77JHC1013
Pyrazolo[3,4-d][1,2,3]triazole (**550**)	62JGU1876, 75JHC279
Pyrazolo[3,4-d]pyrimidine (**552**)	70TL4611, 73JCS(P1)1903, 75JHC1199, 76JOC3781, 79AP610
Pyrazolo[2,3-e]diazepinone (**553**)	75MI40402
Pyrazolo[4,3-d]pyrimidine (**554**)	68LA(713)149, 69TL289
Pyrazolo[3,4-d]pyridazine (**555**)	69BSF2061, 69JPR1058, 70TL4611, 77JHC375, 77T45, 78JHC813
Pyrazolo[3,4-e][1,2,4]triazine (**556**)	70JCS(C)1313

o-phenylenediamine, yielding pyrazolo[3,4-b]pyrazines (**547**) and the corresponding pyrazolo[3,4-b]quinoxalines (**548**). The last ring system can also be obtained from o-phenylenediamine and 4,5-pyrazolinedione ⟨68M2157⟩. Similarly, pyrazolo[3,4-b][1,4]-diazepines (**549**) have been prepared.

However, there are differences between o-phenylenediamine and 4,5-diaminopyrazoles in the synthesis of [5.5] systems. For example, the formation of pyrazolo[3,4-d]-[1,2,3]triazole (**550**) is similar to that of benzotriazole, but all attempts to extend the synthesis of benzimidazoles to the preparation of the imidazo[4,5-c]pyrazole system (**551**) have failed ⟨78TH40400⟩.

The pyrazole analogues of anthranilic acids or anthranilonitriles are a convenient source of [5.6] fused systems (for a general review see ⟨80T2359⟩). Thus 5-amino-4-cyanopyrazoles (in some examples an ester or a hydrazido group replaced the cyano group) have been transformed into pyrazolo[3,4-d]pyrimidines (**552**) and into pyrazolo[2,3-e]diazepinones (**553**), and 4-amino-5-methoxycarbonylpyrazoles have been converted into pyrazolo[4,3-d]pyrimidines (**554**).

Pyrazolo[3,4-d]pyridazines (**555**) can be prepared readily from hydrazines and pyrazoles substituted in positions 4 and 5 with an acyl and an ester group, or with two ester groups. 4,5-Pyrazolinediones have been used as starting materials for the synthesis of the quinoxaline derivatives (**548**) (see above) and of pyrazolo[3,4-e][1,2,4]triazines (**556**)

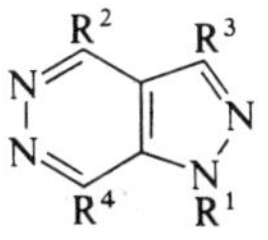

(555) 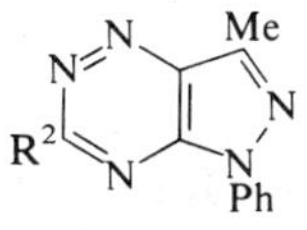 (556)

Pyrazolopyridines isomeric to those described previously have been obtained by other methods. Thus, the derivative (**558**) was formed by Raney nickel reduction of the 4-nitrosopyrazole (**557**) ⟨71JHC1035⟩, and the pyrazolo[3,4-c]pyridine derivative (**560**) was prepared from the azide (**559**) ⟨79CC627⟩.

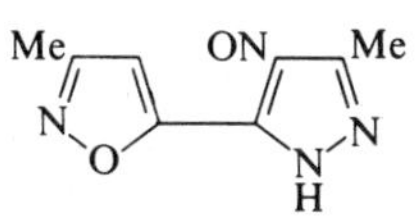

(557) 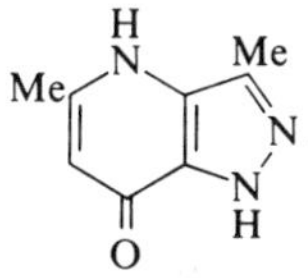 (558) 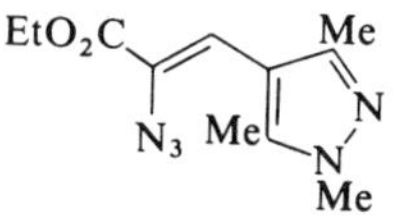 (559) 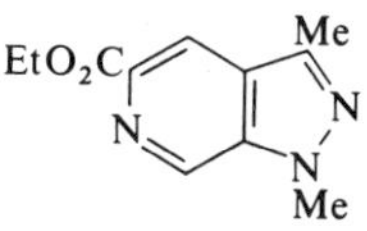 (560)

A heterocyclization involving the N-2 atom of indazole affords [1,2,4]triazino[4,5-b]-indazoles (**561**) from the 3-carbohydrazinoindazole ⟨78JHC1159, 79JHC53⟩. The corresponding pyrazolo[1,5-d][1,2,4]triazine (**562**) has been prepared by the same procedure ⟨81JHC1319⟩.

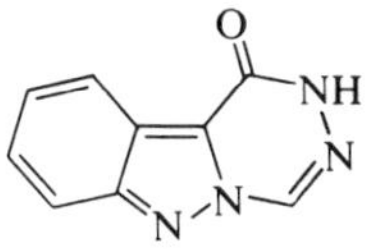

(561) 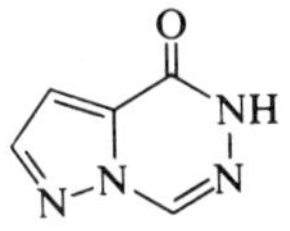 (562) 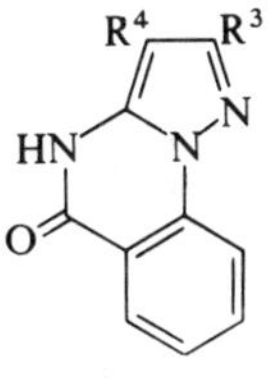 (563) 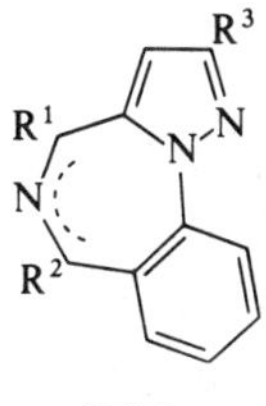 (564)

Some tricyclic systems have been prepared by intramolecular cyclization from N-aryl-pyrazoles carrying substituents both in the pyrazole ring at C-5 and in the phenyl ring at the o-position. Thus pyrazolo[1,5-a]quinazolines (**563**) ⟨69JHC947⟩ and pyrazolo[1,5-a]-[1,4]benzodiazepines (**564**) ⟨77JHC1163, 77JHC1171⟩ can be prepared from suitable precursors.

The Pschorr reaction has been employed successfully in the preparation of a pyrazolo[3,4-c]isoquinoline derivative (**565**) ⟨78JHC1287⟩.

Finally, some results obtained from indazoles substituted in the carbocycle are of interest, even though in these cases the reaction does not involve the heterocyclic moiety (Section 4.04.2.3.2(ii)). For example, pyrazolo[3,4-f]- (**566**) and pyrazolo[4,3-f]-quinolines (**567**) have been obtained from aminoindazoles by the Skraup synthesis ⟨76JHC899⟩. Diethylethoxymethylenemalonate can also be used to give (**566**; $R^1 = CO_2Et$, $R^2 = OH$) ⟨77JHC1175⟩. Pyrazolo-[4,3-f]- and -[4,3-g]-quinazolones (**568**) and (**569**) have been obtained from the reaction of formamide with 5-amino-4-methoxycarbonyl- and 6-amino-5-carboxyindazole, respectively ⟨81CB1624⟩.

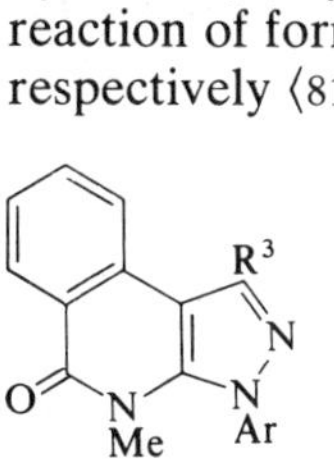

(565) 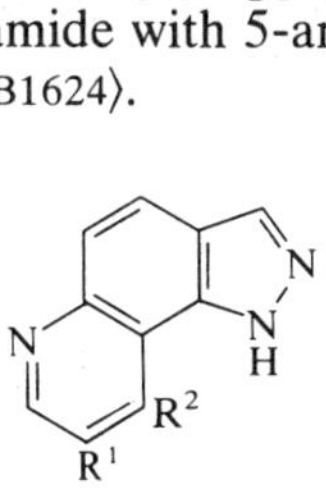 (566) 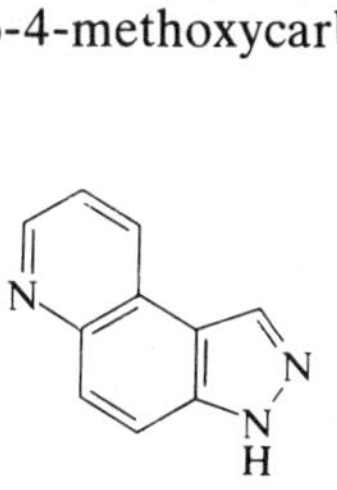 (567) 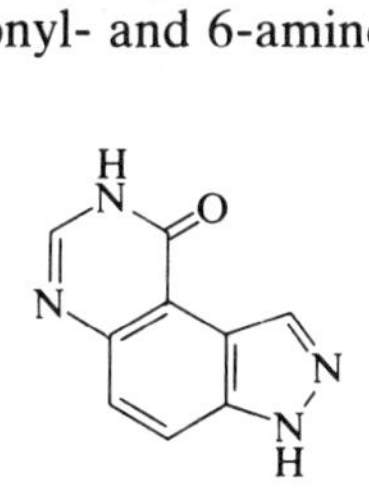 (568) 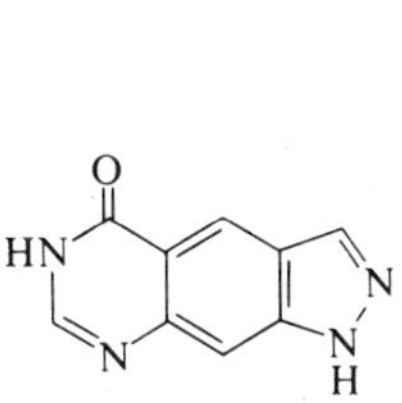 (569)

4.04.3 SYNTHESIS

The synthesis of pyrazoles, indazoles and their derivatives generally follows classical methods, the two most important methods for practical purposes being the reaction between hydrazines and β-difunctional compounds, and 1,3-dipolar cycloadditions (Section 4.04.3.1.2). Both procedures are well documented ⟨64HC(20)1, 66AHC(6)327, 67HC(22)1⟩ and thus the length of the sections in this part of the chapter reflects not only the number of publications dealing with a particular method but also its interest and novelty.

4.04.3.1 Ring Synthesis from Non-heterocycles

The different possibilities for the creation of the pyrazole ring according to the bonds formed are shown in Scheme 46. It should be noted that this customary classification lacks mechanistic significance; actually, only two procedures have mechanistic implications: the formation of one bond, and the simultaneous formation of two bonds in cycloaddition reactions (disregarding the problem of the synchronous *vs.* non-synchronous mechanism).

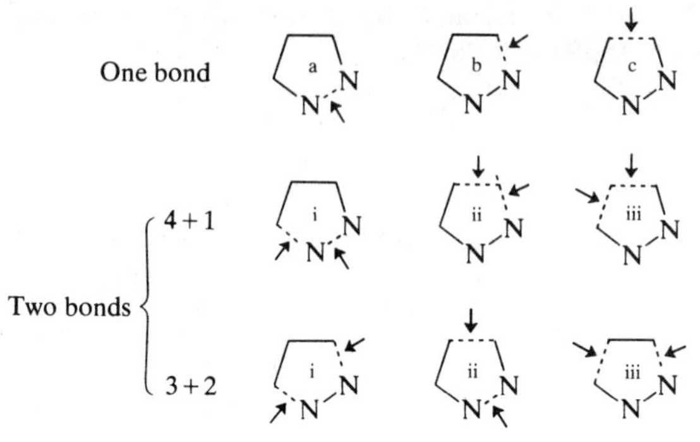

Scheme 46

4.04.3.1.1 Formation of one bond

(*i*) *Between two heteroatoms*

The creation of the N—N bond as the last step of the ring synthesis is common in indazoles and very rare in pyrazoles. In indazoles this method is well known (type B synthesis ⟨67HC(22)1⟩, for example, the dehydration of oximes (**570**) with acetic anhydride yields 1-acetylindazoles (**571**), and in basic medium the indazole 1-oxides (**573**) are formed from the nitro derivatives (**572**).

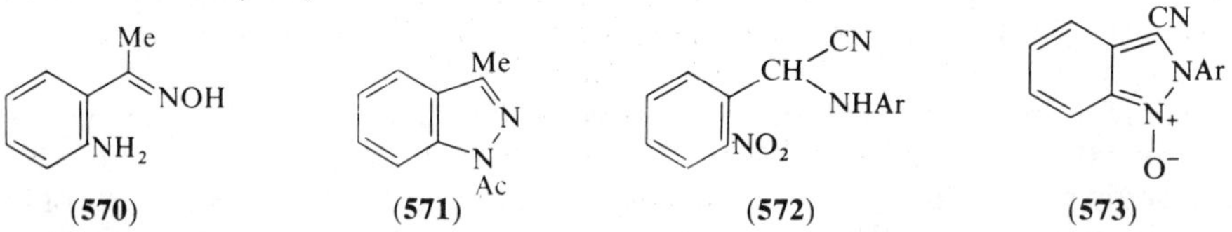

(**570**) (**571**) (**572**) (**573**)

The photochemical cyclization of anthranilonitriles (**190**; Section 4.04.2.1.2) yields indazoles ⟨72CC126⟩. Reduction of *o*-benzonitrile with lithium aluminum hydride yields indazole ⟨75ACS(B)1089⟩. 2-Aminoindazole (**576**) has been prepared in 94% yield from (**574**) *via* the *o*-phthaloyl derivative (**575**; $R^2 = o$-phthaloyl) ⟨72JOC2351⟩. Similarly, treatment of the *o*-azidobenzanilide (**577**) with thionyl chloride affords 2-aryl-3-chloroindazole (**578**) ⟨79S308⟩.

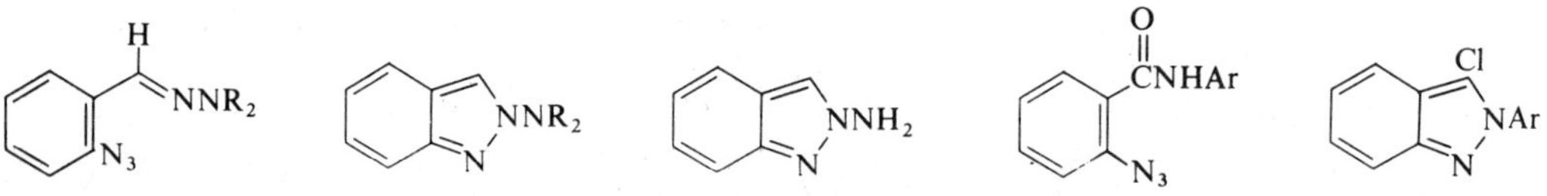

(**574**) (**575**) (**576**) (**577**) (**578**)

The 1,3-dioxime (**579**) with thionyl chloride yields the isopyrazole *N*-oxide (**580**) ⟨66CB618⟩. A general method of pyrazole synthesis involving the creation of an N—N bond has been described by Barluenga and coworkers ⟨81JCS(P1)1891, 83JCS(P1)2273⟩. Rigorously it should be classified in Section 4.04.3.2.2(iii) since it involves a ring transformation. However, the overall reaction goes from the 1-azabutadiene (**581**) to the tetrasubstituted pyrazole *via* the intermediate 1,2,6-thiadiazine *S*-oxide (**582**) which loses sulfur monoxide thermally.

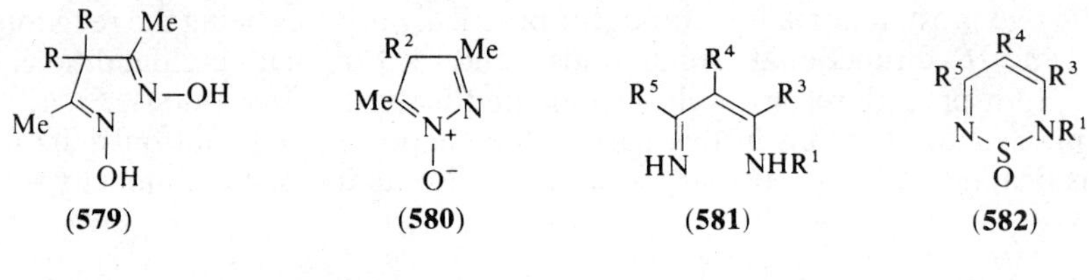

(**579**) (**580**) (**581**) (**582**)

(ii) Adjacent to a heteroatom

This ring closure is the final step of the reaction of hydrazines with 1,3-difunctional compounds (Section 4.04.3.1.2(ii)), and numerous examples in the literature of pyrazoles have been described. In some cases the N—C ring closure occurs by a concerted mechanism, classified by Huisgen ⟨80AG(E)947⟩ as 1,5-electrocyclizations.

Sucrow, studying the chemistry of the enehydrazines ⟨83CB1520⟩, has found several methods for the synthesis of pyrazoles. For example, he has described ⟨79CB1712⟩ the cyclization of monomethylhydrazones of dialkyl oxalacetates to 5-pyrazolones via an enehydrazine (Scheme 47).

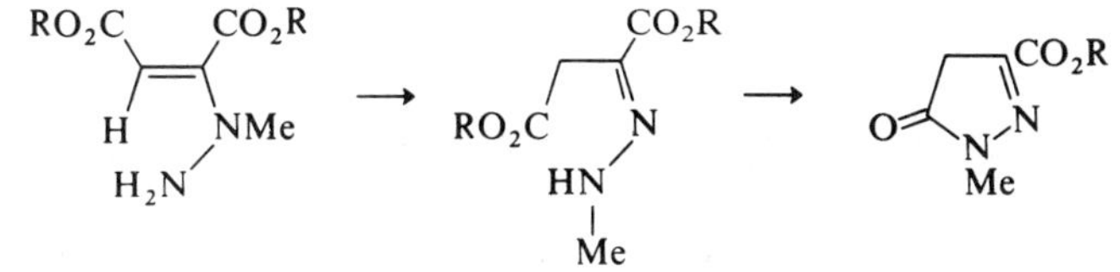

Scheme 47

The azine (**583**) of hexafluoroacetone adds the ynamine (**584**) to give the azetine (**585**). On heating (**585**) in the presence of triethylamine it is converted into the pyrazole (**586**) as indicated in Scheme 48 ⟨75ZN(B)622⟩.

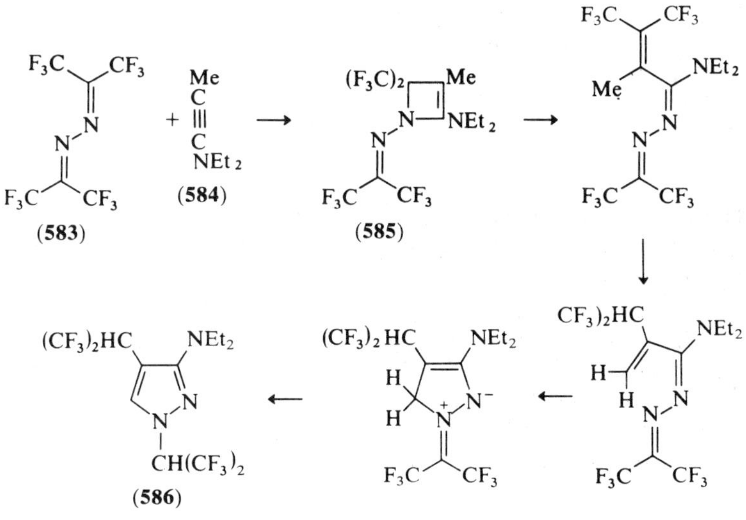

Scheme 48

Hart and Brewbaker have described the cyclization of 1,3-bis(diazopropane) to pyrazole (Scheme 49) by a concerted, intramolecular 1,3-dipolar cycloaddition ⟨69JA711⟩.

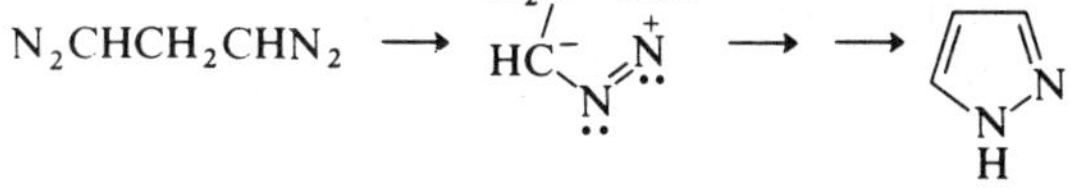

$$N_2CHCH_2CHN_2 \rightarrow \text{...} \rightarrow \text{...}$$

Scheme 49

In indazoles there are two possibilities for ring closure by creation of an N—C bond, depending on whether the bond is the N(2)—C(3) or the N(1)—C(7a) bond. Both are classical methods of indazole synthesis (types A and C ⟨67HC(22)1⟩). An example of each class is shown in Scheme 50 ⟨78S633⟩.

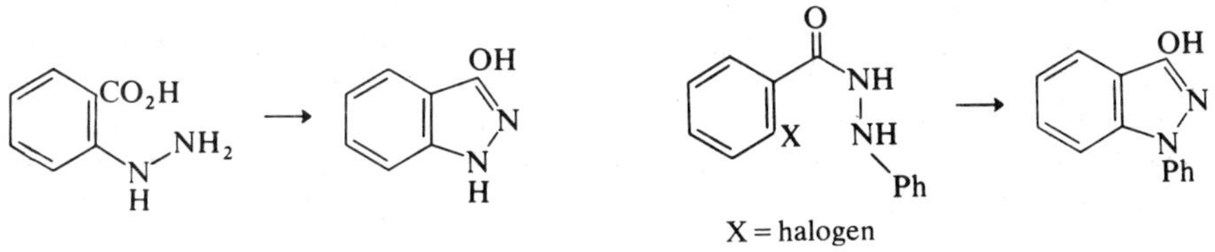

X = halogen

Scheme 50

Recently ⟨82CC450⟩ the conversion of aromatic hydrazides into indazolones in good yield when treated with three equivalents of n-butyllithium has been described. For example, benzoylhydrazine afforded indazolone in an 80% yield (Scheme 51).

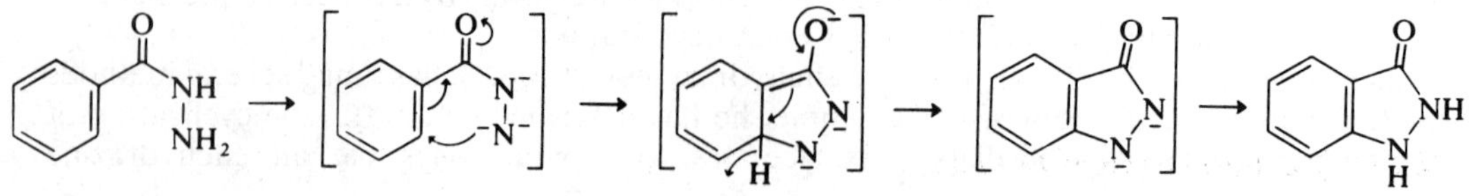

Scheme 51

(iii) β to a heteroatom

The important synthesis of pyrazoles and pyrazolines from aldazines and ketazines belongs to this subsection. Formic acid has often been used to carry out the cyclization ⟨66AHC(6)347⟩ and *N*-formyl-Δ²-pyrazolines are obtained. The proposed mechanism ⟨70BSF4119⟩ involves the electrocyclic ring closure of the intermediate (**587**) to the pyrazoline (**588**; $R^1 = H$) which subsequently partially isomerizes to the more stable *trans* isomer (**589**; $R^1 = H$) (Section 4.04.2.2.2(vi)). Both isomers are formylated in the final step ($R' = CHO$).

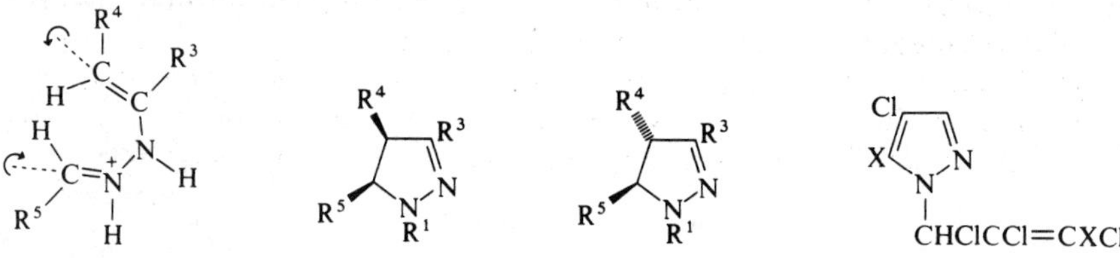

(587) (588) (589) (590)

The synthesis of 1,2-dimethylpyrazolinium salts from dimethylhydrazine and carbonyl compounds is thought to involve a mechanism of the same type ⟨71BSF1925⟩. Pyrazoles (**590**) have been obtained by thermal cyclization of azines derived from di- and tri-halogenoacroleins ⟨77T2069⟩. The dianion of benzyl phenyl ketazine, generated by treating the ketazine with two equivalents of lithium diisopropylamide, rearranged to 3,4,5-triphenylpyrazole ⟨78JOC3370⟩. A new synthesis ⟨71TL1591⟩ of 4-hydroxypyrazole-5-carboxylic acids (**592**) entails cyclization of the hydrazones (**591**).

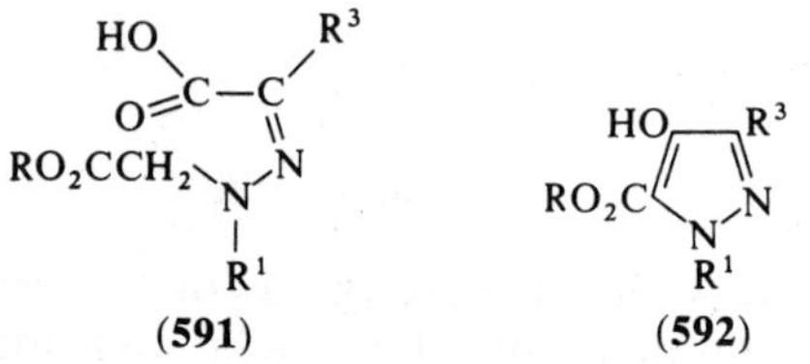

(591) (592)

Although formally it could be classified with the ring transformations (Section 4.04.3.2.2), conversion of 2,4-diphenyl-1,3,4-oxadiazol-2-one (**593**) by flash vacuum pyrolysis at 500 °C into 3-phenylindazole (**595**) involves a C(3)—C(3a) ring closure of the diphenylnitrilimine (**594**) ⟨79AG(E)721⟩.

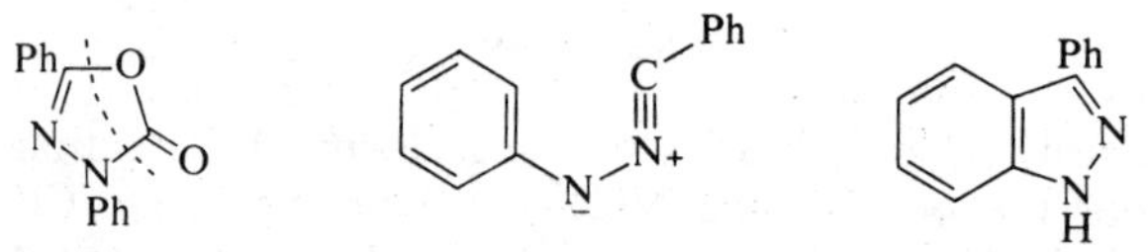

(593) (594) (595)

The Frasca method for obtaining 1-arylindazoles also involves a C(3)—C(3a) ring closure ⟨67CJC697⟩. It consists in the cyclization of *p*-nitrophenylhydrazones of ketones and aldehydes with polyphosphoric acid. The Barone computer-assisted synthetic design program has found several new methods for preparing indazoles ⟨79MI40409⟩. The selected method involves the transformation of *N'*,*N'*-diphenylhydrazides (**596**) into 1-phenylindazoles (**597**) by means of trifluoromethanesulfonic anhydride. The yields vary from 2% ($R^3 = H$) to 50% ($R^3 = Ph$).

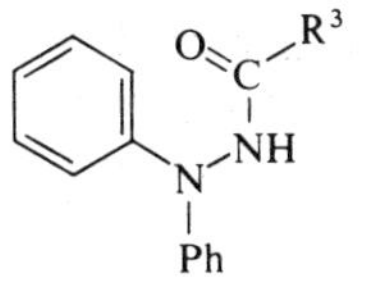

Pyrazoles and their Benzo Derivatives 277

(596) 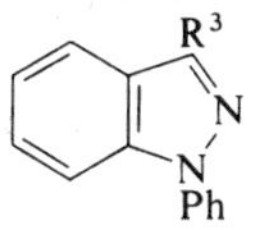 (597)

4.04.3.1.2 Formation of two bonds

(i) From 4+1 atom fragments

The reactions described here are not synchronous and accordingly can be divided into two successive steps, the last step belonging to one of the three classes of Section 4.04.3.1.1. However, several syntheses of pyrazoles result from the reaction of two compounds, one component contributing four atoms to the ring and the other the fifth atom. There are three possibilities for 4+1 reactions shown in Scheme 46, but only those represented by (ii) are described in the literature.

The synthesis of pyrazolines and pyrazoles of the [CCNN+C] type with the creation of two bonds, N(2)—C(3)+C(3)—C(4) (or N(1)—C(5)+C(5)—C(4)), has been studied by several groups. Beam and coworkers have published a series of papers on the synthetic utility of lithiated hydrazones. Thus, the methylhydrazone of acetophenone (598) is converted by butyllithium into the dianion (599), which in turn reacts with methyl benzoate to afford the pyrazole (600) ⟨76SC5⟩. In earlier publications Beam *et al.* have used aldehydes and acyl chlorides to obtain pyrazolines and pyrazoles by the same method.

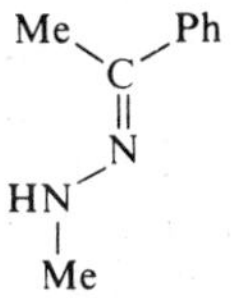

(598) 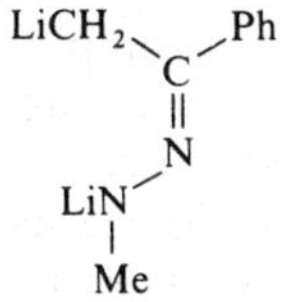 (599) 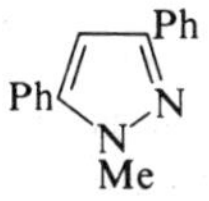 (600)

The stabilized phosphonium ylide (601) reacts with aromatic aldehydes to give *N*-phenacylpyrazoles (602) in good yields ⟨73CC7⟩. Ketone semicarbazones and ketazines react with two moles of phosphorus oxychloride–DMF, the Vilsmeier–Haack reagent, with the formation of 4-formylpyrazoles (603; R^1 = H or PhC=CH$_2$) ⟨70JHC25, 70TL4215⟩.

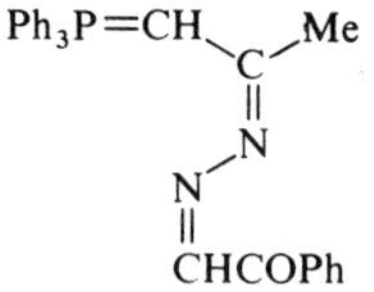

(601) 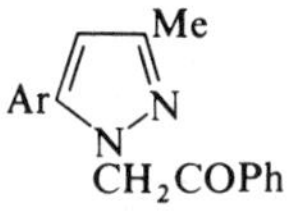 (602) 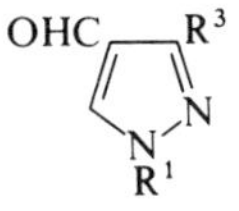 (603)

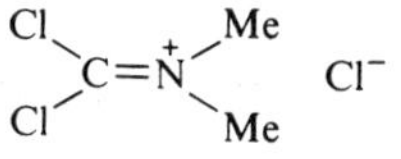

(604) 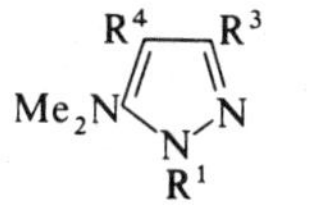 (605) 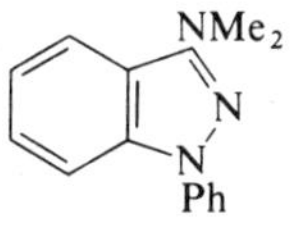 (606)

Viehe and coworkers have used dichloromethyleneammonium salts (604) to obtain 5-dimethylaminopyrazoles (605) from *N*-methylhydrazones ⟨74AG(E)79⟩. The same reagent (604) affords 3-dimethylaminoindazoles (606) by reaction with *N,N*-diphenylhydrazine ⟨73AG(E)405⟩.

(ii) From 3+2 atom fragments

(a) [CCC+NN]. The standard method of synthesis of pyrazoles consists in the condensation of a 1,3-difunctional compound with hydrazine or its derivatives (Scheme 46; [3+2], i). A review has been published on this subject ⟨70BSF2717⟩ and many examples may be found in ⟨64HC(20)1, 66AHC(6)347, 67HC(22)1, 80CHE1⟩. A typical experiment is the formation of 1,3- and 1,5-dimethylpyrazoles from the reaction between methylhydrazine and 4,4-dimethoxy-2-butanone (Scheme 52). The structure of both isomers has aroused much interest in the past ⟨see Appendix 1 of B-76MI40402⟩ and the reaction has been studied using the methodology of design of experiments ⟨77BSF1163⟩ which has allowed the determination of the effect of seven factors on the yield and selectivity of this reaction.

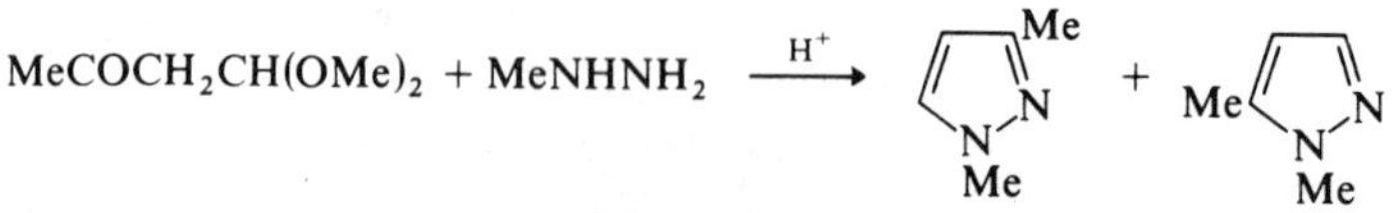

$$MeCOCH_2CH(OMe)_2 + MeNHNH_2 \xrightarrow{H^+}$$ [pyrazole products]

Scheme 52

Although most syntheses of the pyrazole ring use hydrazines and hundreds of publications have been devoted to this method, it must be acknowledged that the mechanism of the reaction is still only partially known and the structural reasons that favour one or the other isomer are not clearly understood. The complexity of the problem appears when considering that a 1,3-dicarbonyl compound, for example, can react in six different ways with a nucleophile (taking into account both carbon atoms of the β-diketo tautomers and the two enols) and that hydrazines have two nucleophilic centers. Thus, there are 12 different possibilities to initiate a reaction which finally ends in only two different compounds. Under neutral or acidic conditions and if the nucleophilicity of the two nitrogen atoms of the hydrazine is very different as in 2,4-dinitrophenyl, tosyl and acetyl derivatives, the most abundant isomer corresponds to the 1,2-addition at the terminal NH_2 group of the hydrazine to the most reactive carbonyl group of the β-dicarbonyl compound ($COCO_2R > COR > COAr$).

Phenylhydrazine reacts in basic medium at the N-1 atom of the conjugate base, $PhN̄NH_2$, with α,β-unsaturated ketones, esters, acids and nitriles. In neutral or acidic medium it reacts at the N-2 atom of the neutral molecule, $PhNHN̄H_2$ ⟨70BSF2717⟩. This proprty has been used to obtain a specific isomer, for example, of pyrazolidones. The mechanism of the Δ^2-pyrazoline ring formation from α,β-unsaturated carbonyl compounds and hydrazines has been examined with regard to the stereochemistry of the 4- and 5-positions in the product ⟨73JCS(P2)936⟩. There is no relationship between the E/Z configuration of the alkene and the stereochemistry of the resulting pyrazoline ⟨73BSF3401⟩, since the configuration at C-4 results from the protonation of a common intermediate, the Δ^3-pyrazoline (**607**; Scheme 53).

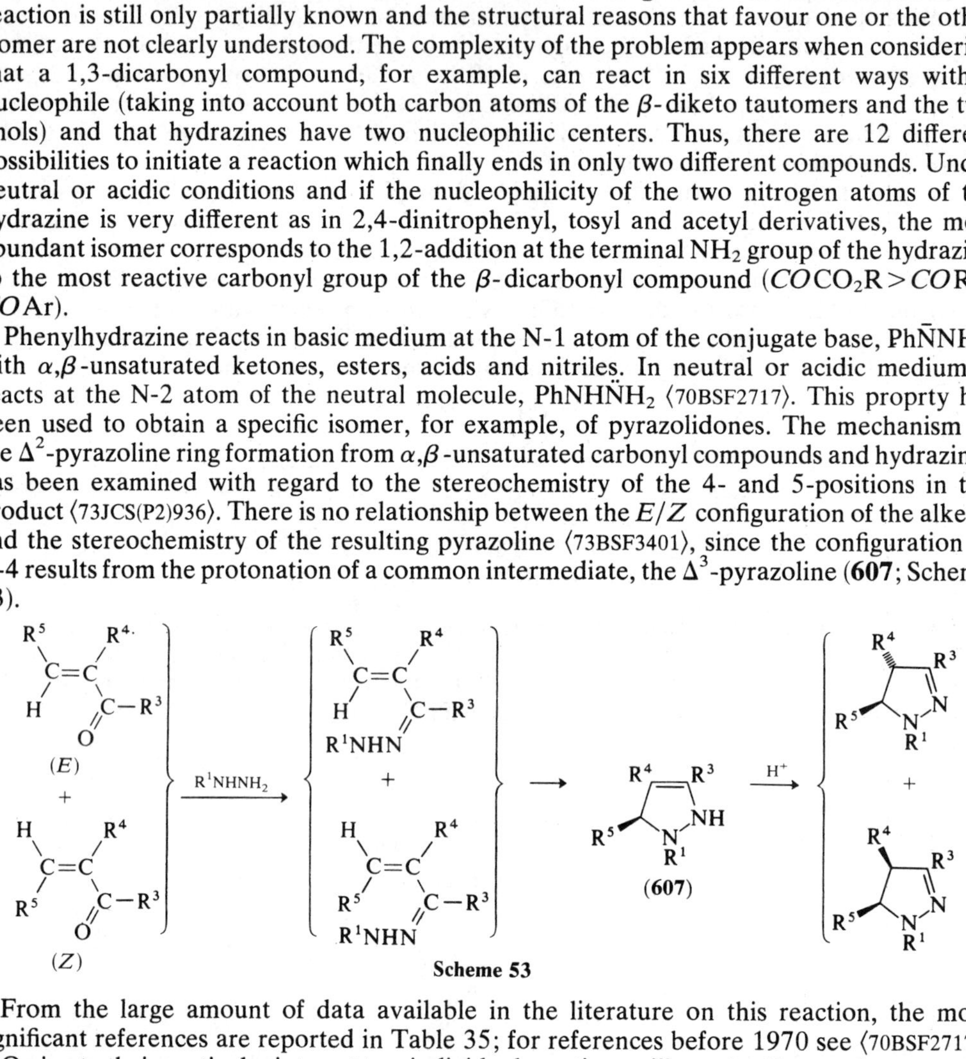

Scheme 53

From the large amount of data available in the literature on this reaction, the most significant references are reported in Table 35; for references before 1970 see ⟨70BSF2717⟩.

Owing to their particular interest two individual reactions will now be discussed separately. The reaction of methoxycarbonylhydrazine and 3-bromo-2,4-pentanedione affords, in addition to the expected pyrazole (**608**), a pyrazolium salt (**609**), the structure of which was established by X-ray crystallography ⟨74TL1987⟩. Aryldiazonium salts have been used instead of arylhydrazines in the synthesis of pyrazolines (**610**) and pyrazoles (**611**) ⟨82JOC81⟩. These compounds are formed by free radical decomposition of diazonium salts by titanium(II) chloride in the presence of α,β-ethylenic ketones.

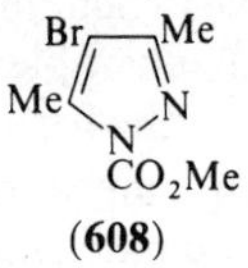

(**608**) 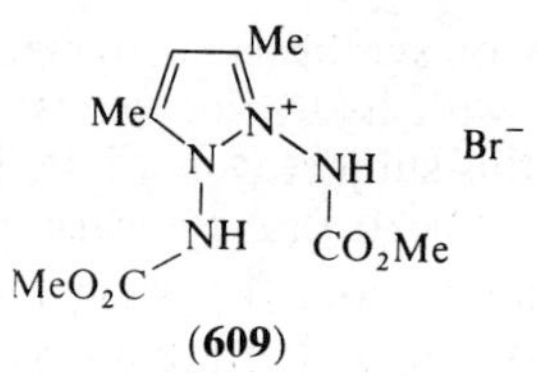 (**609**) 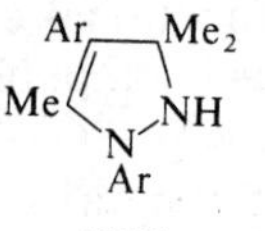 (**610**) 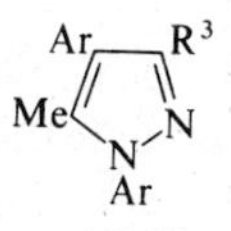 (**611**)

Pyrazoles and their Benzo Derivatives

Table 35 Syntheses of Pyrazoles from Hydrazines

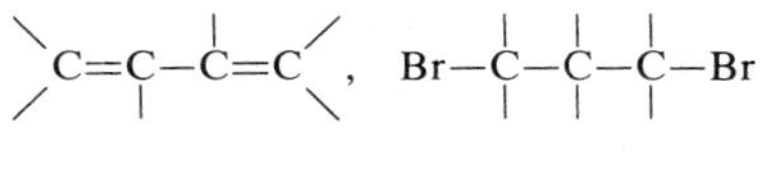

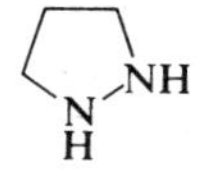

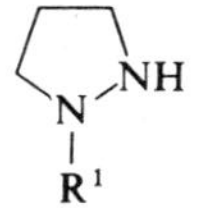

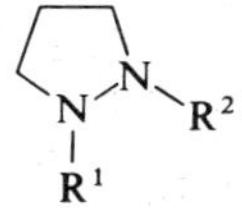

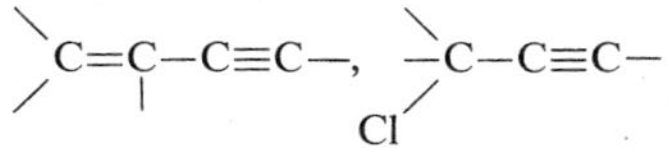

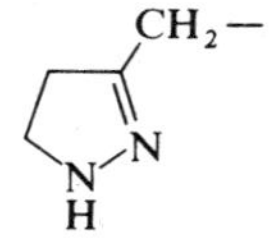

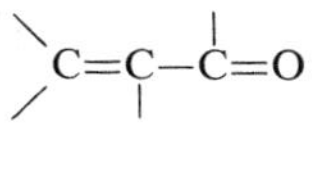

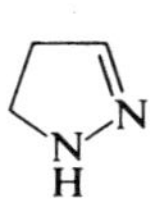

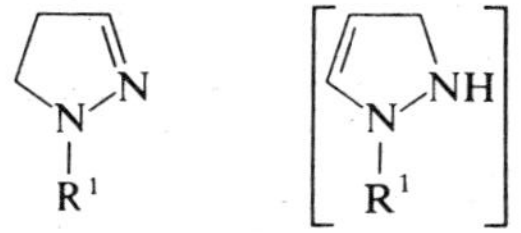

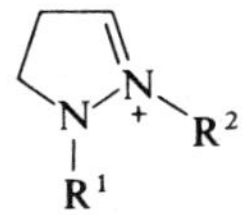

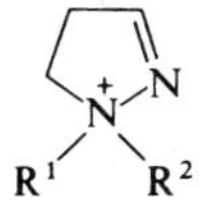

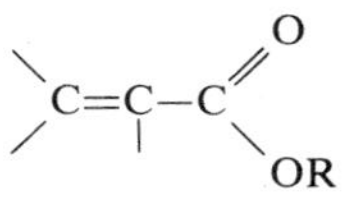

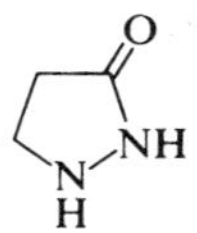

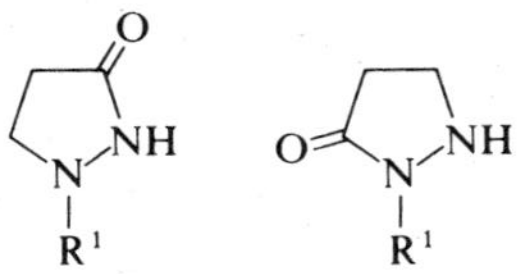

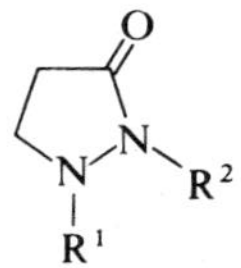

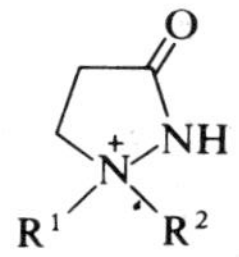

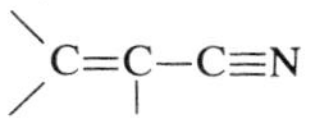

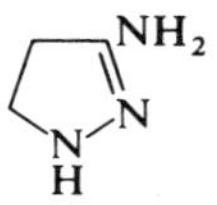

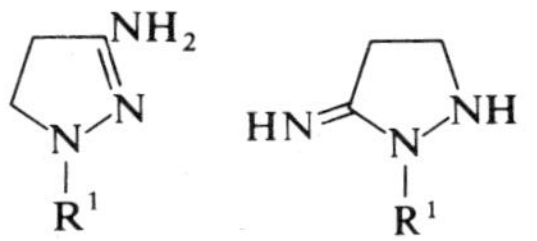

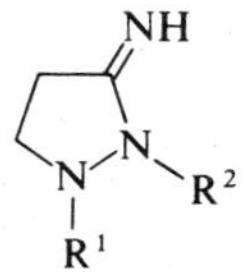

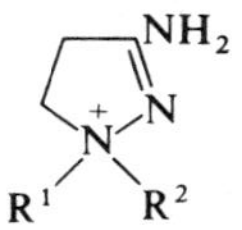

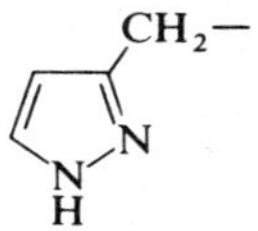

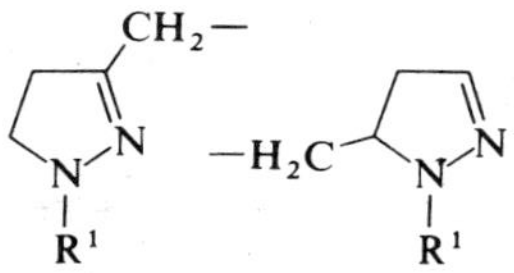

Table 35 (continued)

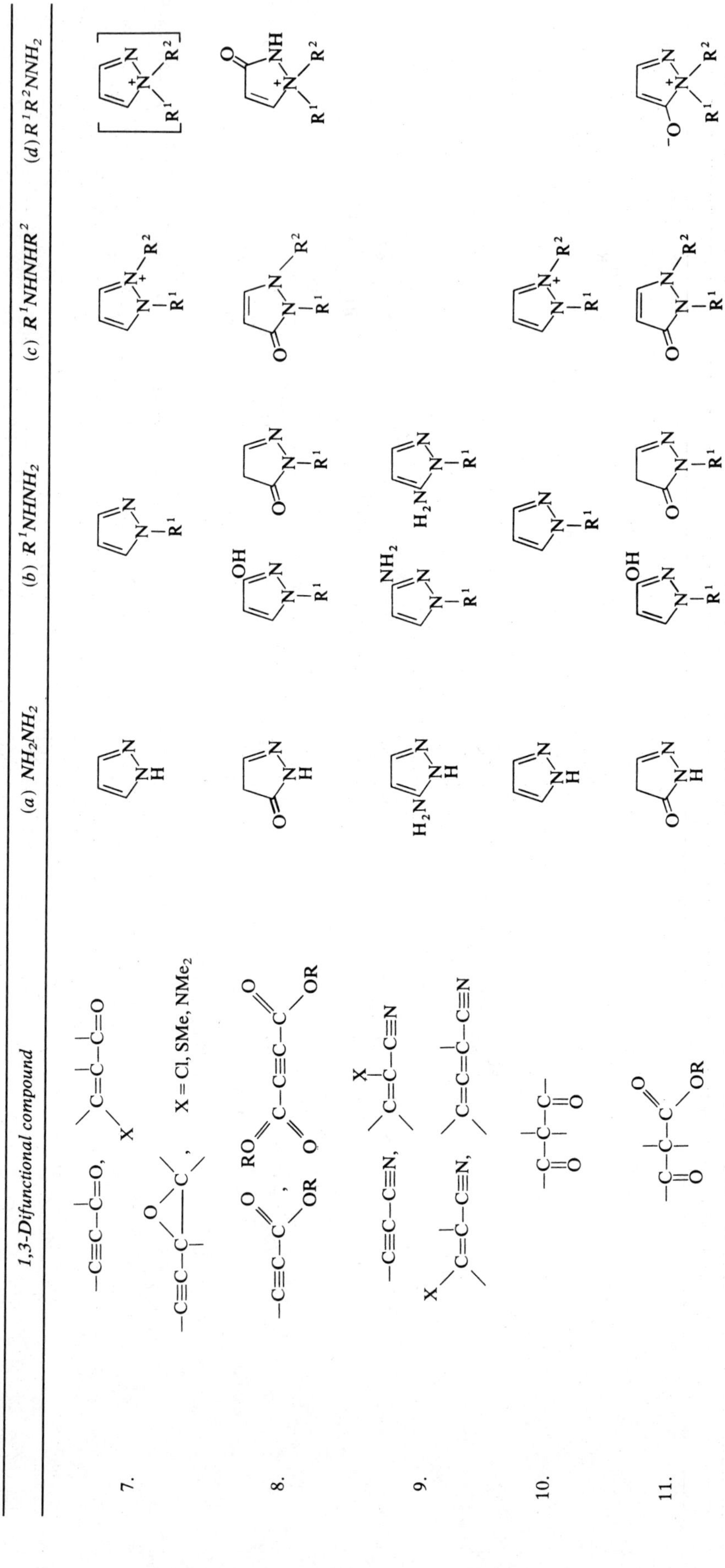

1,3-Difunctional compound	(a) NH_2NH_2	(b) R^1NHNH_2	(c) R^1NHNHR^2	(d) $R^1R^2NNH_2$
7. $-C\equiv C-C=O$, $\overset{\displaystyle }{\underset{X}{>}}C=C-C=O$ X = Cl, SMe, NMe$_2$				
8. $-C\equiv C-\overset{O}{C}-OR$, $RO-\overset{O}{C}-C\equiv C-\overset{O}{C}-OR$				
9. $-C\equiv C-C\equiv N$, $\overset{X}{>}C=C-C\equiv N$, $>C=C-C\equiv N$, $X-\overset{}{\underset{}{C}}=C-C\equiv N$				
10. $-\overset{O}{C}-\overset{}{C}-\overset{O}{C}-$				
11. $\overset{O}{C}-C-\overset{O}{C}-OR$				

Table 35 (continued)

	1,3-Difunctional compound	(a) NH_2NH_2	(b) R^1NHNH_2	(c) R^1NHNHR^2	(d) $R^1R^2NNH_2$
12.	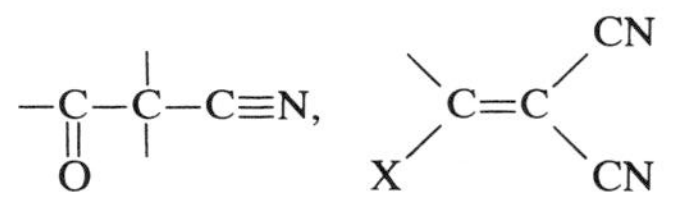 –C(=O)–C(CN)=C(X)CN, X = OEt, NHAr	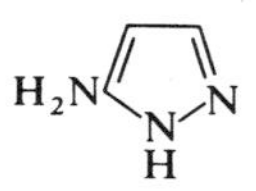 3-amino-pyrazole (NH)	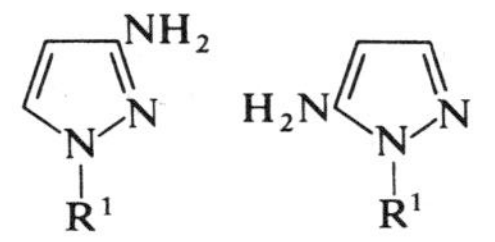 3-amino-1-R¹-pyrazole and 5-amino-1-R¹-pyrazole		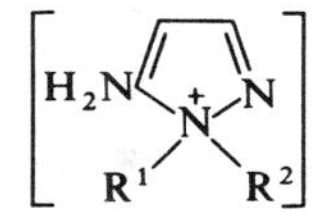 3-amino-1,2-R¹R²-pyrazolium
13.	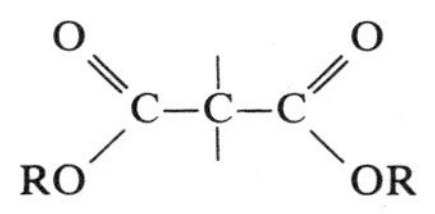 RO(O=)C–CH₂–C(=O)OR	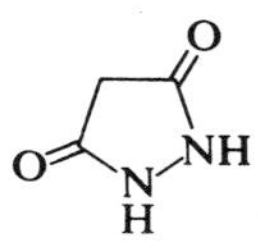 pyrazolidine-3,5-dione (NH)	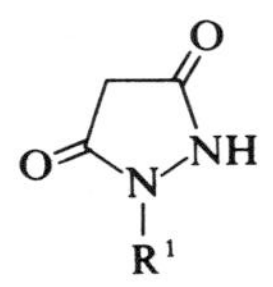 1-R¹-pyrazolidine-3,5-dione	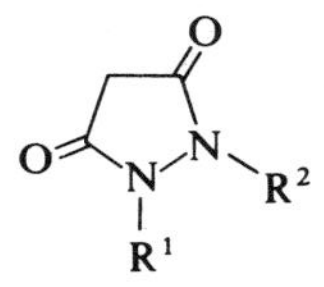 1,2-R¹R²-pyrazolidine-3,5-dione	
14.	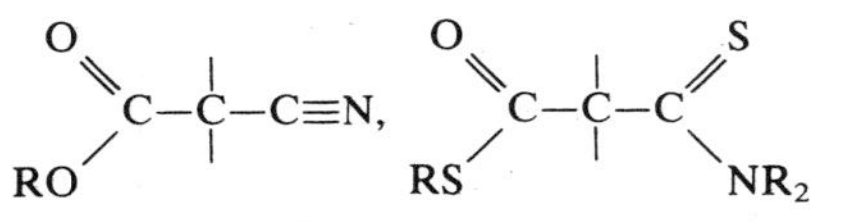 RO(O=)C–CH₂–C(=S)SR / NR₂	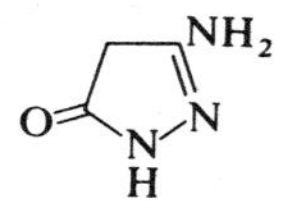 5-amino-pyrazol-3(2H)-one (NH)	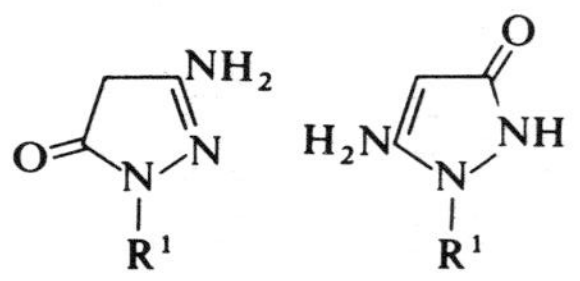 5-amino-1-R¹-pyrazol-3-one	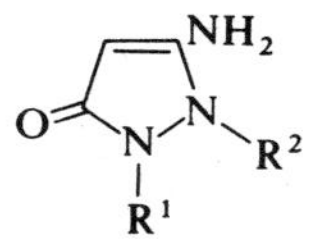 4-fluoro / 4-t-butyl pyrazoles	
15.	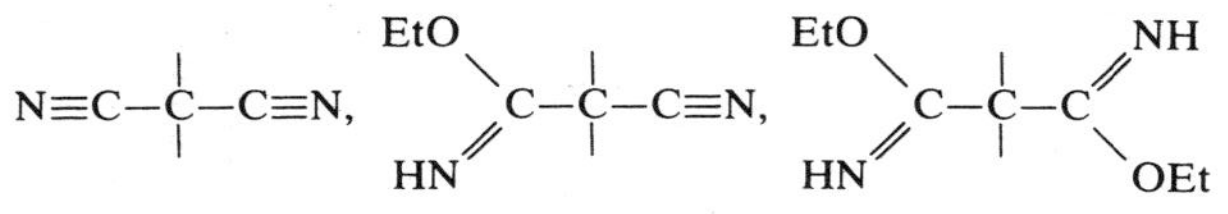 N≡C–C–C≡N (EtO, HN); EtO–C–C–C–OEt (HN, NH)	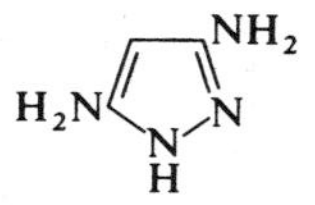 3,5-diamino-pyrazole (NH)	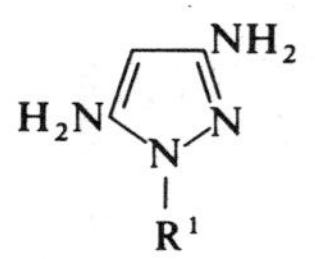 3,5-diamino-1-R¹-pyrazole		

1. Most syntheses start from 1,3-dibromopropanes ⟨67HC(22)1⟩. 1a. The pyrazolidine obtained from dienes is further oxidized. 1c. ⟨75JHC1233⟩. 2a. Δ²-Pyrazolines from chloroalkynic compounds ⟨70ZOR908⟩. 3b. The Δ⁴-pyrazoline is not isolated since it is oxidized to pyrazole ⟨74AJC2041⟩. 3d. ⟨76BSF476⟩. 4. ⟨80CHE1⟩. 4d. ⟨76sSF476, 78BSF(2)602, 80AJC1365⟩. 5. ⟨80CHE1⟩. 5d. ⟨76BSF476⟩. 7a. ⟨75S798, 76JHC455⟩. 7b. ⟨73CB3092, 76JHC257, 76JHC455, 79H(12)657, 79JCR(M)2921⟩. 7d. The non-aromatic pyrazolium salt has never been isolated, since it decomposed to 1-substituted pyrazoles by loss of R²X ⟨70CHE1321⟩ or to open-chain compounds ⟨79JCR(M)2921⟩. 8. ⟨80CHE1⟩. 8b. 3-Methoxycarbonylpyrazolones have been prepared from acetylenedicarboxylic esters ⟨74CS(P1)297⟩. 8d. ⟨75AG(E)560, 83CC1144⟩. 9a. ⟨76552, 79TL2991, 81JCS(P1)2997⟩. 9b. ⟨75JOC1815, 76S52⟩. 10. Reactions of 1,3-cyclohexanedione with hydrazines ⟨73CB745⟩. 10a. For Reichardt's synthesis of 4-fluoro- and 4-t-butyl-pyrazoles from substituted malonic aldehydes, see ⟨75LA470, 75AG(E)86⟩; for synthesis of bis-pyrazoles by this method, see ⟨78M337⟩. 10b. 4-Fluoro- and 4-t-butyl-pyrazoles ⟨74CB3454, 75AG(E)86⟩. 10c. 4-Fluoropyrazolium salts ⟨75LA470, 75AG(E)86⟩; for the reaction of diazetidinone as 1,2-disubstituted hydrazine, see ⟨81A7660⟩. 10d. Formation of mono- and di-substituted hydrazones. 11. ⟨80CHE1⟩. 11d. ⟨70HC923, 82JHC1215⟩. 12. ⟨80CHE1⟩. 12a. The aminopyrazole reacts further with the ketonitrile to afford condensed pyrazoles ⟨78JHC185⟩; for the synthesis of aminopyrazoles from enaminonitriles, see ⟨81M875⟩. 12b. ⟨80JHC1527⟩. 12d. A 1-substituted 5-aminopyrazole has been isolated by dequaternization of the 1,1-disubstituted salt ⟨71CB2053⟩. 13b. ⟨75AJC133⟩. 14. Dialkylaminopyrazolones have been prepared from thioamides ⟨70LA(734)173⟩. 15. Malonitrile itself yields 5-amino-4-cyano-3-cyanomethylpyrazoles ⟨64JOC942, 68JOC2606⟩; 3,5-dimethylaminopyrazoles have been prepared from dichloromalonyl cyanines ⟨74JOC1233⟩.

Indazoles can also be prepared by the [CCC+NN] method. For instance indazolones have been obtained rather simply by the action of hydrazine hydrate on 2-halobenzoic esters ⟨78S633⟩.

(b) *[CCN + NC]*. No definite examples of this way of forming the pyrazole nucleus are described in the literature. The closest example is the transformation of N-substituted anilines (**612**) into 1-substituted indazolones (**613**) via the carbamoyl azides (**614**) or the isatin (**615**) ⟨78S633⟩.

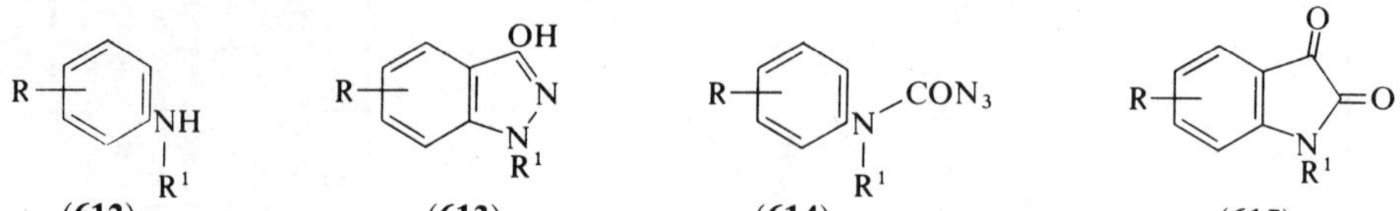

(c) *[CNN + CC]*. This section is second in importance to the use of hydrazines for preparative purposes, but it is the more important one from the theoretical and mechanistic point of view. More diversified than the section dealing with hydrazines, it has the advantage of having been treated in several comprehensive reviews and books (Table 36) to which the reader should refer for the original literature references. Most of these syntheses of pyrazoles and pyrazolines have been known for a long time and many examples are described in ⟨67HC(22)1⟩.

Table 36 General References to Pyrazole Syntheses of the [CNN+CC] Class

Title	Ref.
1,3-Dipolar cycloadditions to alkenes	B-64MI40400
1,3-Dipolar cycloadditions to alkynes	B-78MI40404
Synthetic applications of diazoalkanes	B-78MI40405
Thermal pericyclic reactions	74AG(E)47
1,3-Anionic cycloadditions	74AG(E)627
Reactions of azines and imines (including 'criss-cross' cycloaddition)	76S349
1,3-Dipolar cycloreversions	79AG(E)721
Hydrazoyl halides	80JHC833

Diazo compounds react with alkenes to afford Δ^1-pyrazolines, which in turn izomerize to Δ^2-pyrazolines if there is a hydrogen atom α to the N=N bond (Scheme 54). In those cases where two possible ways of isomerization exist, the more acidic hydrogen migrates preferentially. The alkene configuration is conserved on the Δ^1-pyrazoline (stereospecificity) but the regioselectivity depends on the substituents of both the alkene and the diazo compound.

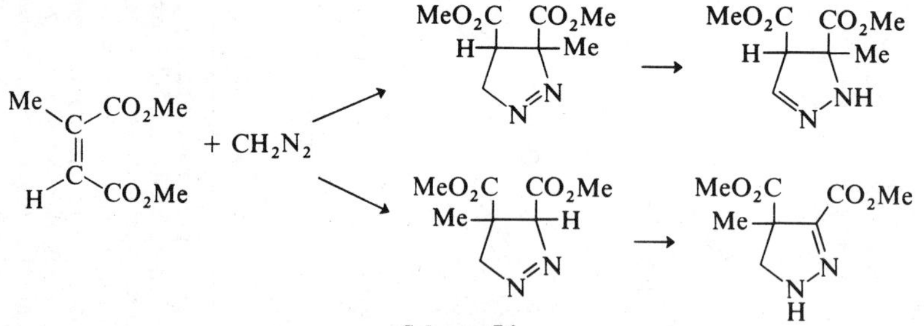

Scheme 54

Pyrazoles are formed when the diazo compounds react with alkynes or with functionalized alkenes, viz. the enols of β-diketones. Pyrazolenines (**353**; Section 4.04.2.2.1) are isolated from disubstituted diazomethanes. Many pyrazoles, difficult to obtain by other methods, have been prepared by this procedure, for example 3-cyanopyrazole (**616**) is obtained from cyanoacetylene and diazomethane ⟨71JCS(C)2147⟩, 3,4,5-tris(trifluoromethyl)pyrazole (**617**) from trifluorodiazoethane and hexafluoro-2-butyne ⟨81AHC(28)1⟩, and 4-phenyl-3-triflylpyrazole (**618**; R^1=H) from phenyltriflylacetylene and diazomethane ⟨82MI40402⟩. An excess of diazomethane causes N-methylation of the pyrazole (**618**; R^1 = H) and the two isomers (**618**; R^1 = Me) and (**619**) are formed in a ratio of 1:1.

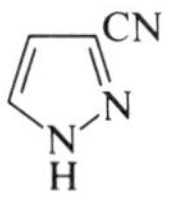

(616) 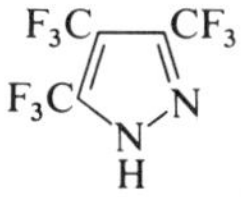 (617) 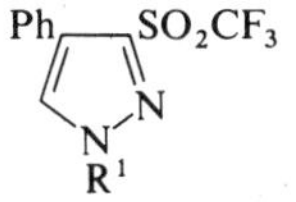 (618) 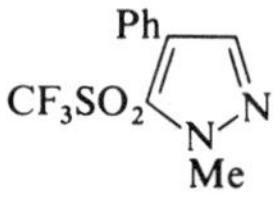 (619) 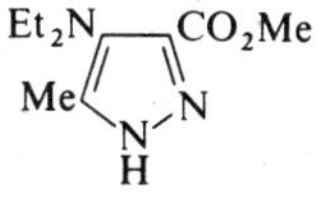 (620)

Since 1,3-dipolar cycloadditions of diazomethane are HOMO (diazomethane)–LUMO (dipolarophile) controlled, enamines and ynamines with their high LUMO energies do not react ⟨79JA3647⟩. However, introduction of carbonyl functions into diazomethane makes the reaction feasible in these cases. Thus methyl diazoacetate and 1-diethylaminopropyne furnished the aminopyrazole (620) in high yield.

The theoretical interest that these reactions aroused was enormous. Presently 1,3-dipolar cycloadditions are one of the cornerstones of theoretical chemistry and many references are to be found in the publications quoted in Table 36. For some recent publications related to this chapter see ⟨78JA5701, 79TL2621, JST(89)147⟩.

Nitrilimines (621) are another class of 1,3-dipoles which provide a useful entry into the pyrazole ring. They are often generated by cycloreversion ⟨79AG(E)721⟩ of tetrazoles, triazolopyridines and oxadiazolones ⟨79JOC2957⟩ (Scheme 55). Pyrazolines of known stereochemistry, pyrazoles and indazoles (Section 4.04.3.1.1(iii)) have all been prepared from nitrilimines.

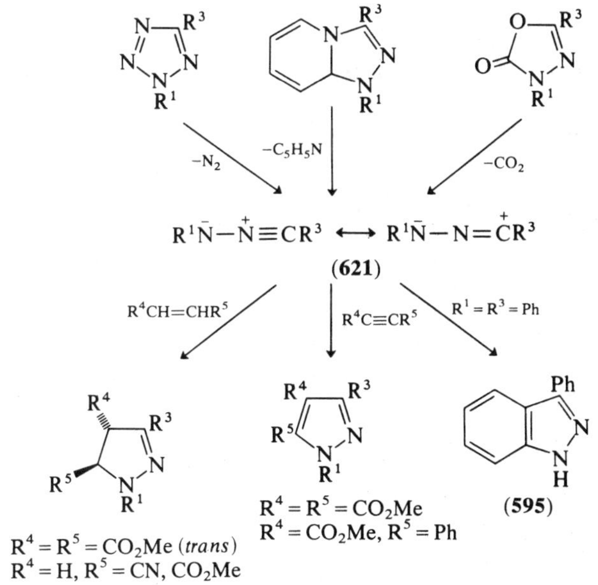

Scheme 55

Azomethines add with great ease and stereospecificity to carbon–carbon double bonds affording pyrazolidines. They also add to carbon–carbon triple bonds with the formation of pyrazolines. Benzimidazolium N-imines, e.g. (622), are a special case of azomethines. Its reaction with DMAD yields 1-(2-methylaminophenyl)pyrazole (624), which is formed by cleavage of the initial adduct (623) ⟨75JHC225⟩.

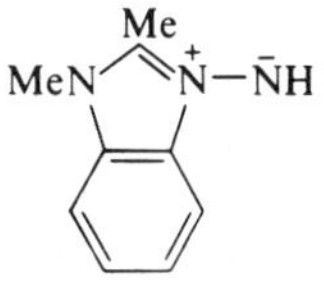

(622) 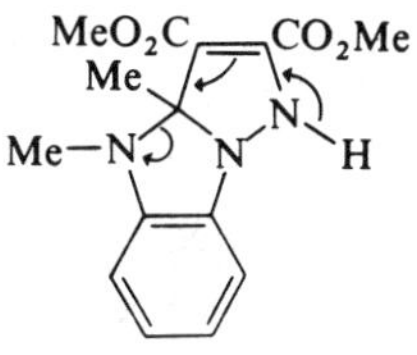 (623) 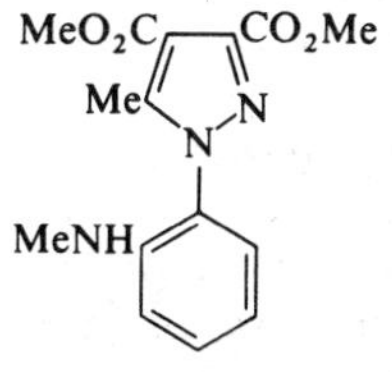 (624)

Burger's 'criss-cross' cycloaddition reaction of hexafluoracetone–azine ⟨76S349⟩ is also a synthetic method of the [CNN+CC] class. In turn, the azomethines thus produced, (625) and (626) ⟨79LA133⟩, can react with alkenes and alkynes to yield azapentalene derivatives (627) and (628), or isomerize to Δ^4-pyrazolines (629) which subsequently lose HCF_3 to afford pyrazoles (630; Scheme 56) ⟨82MI40401⟩.

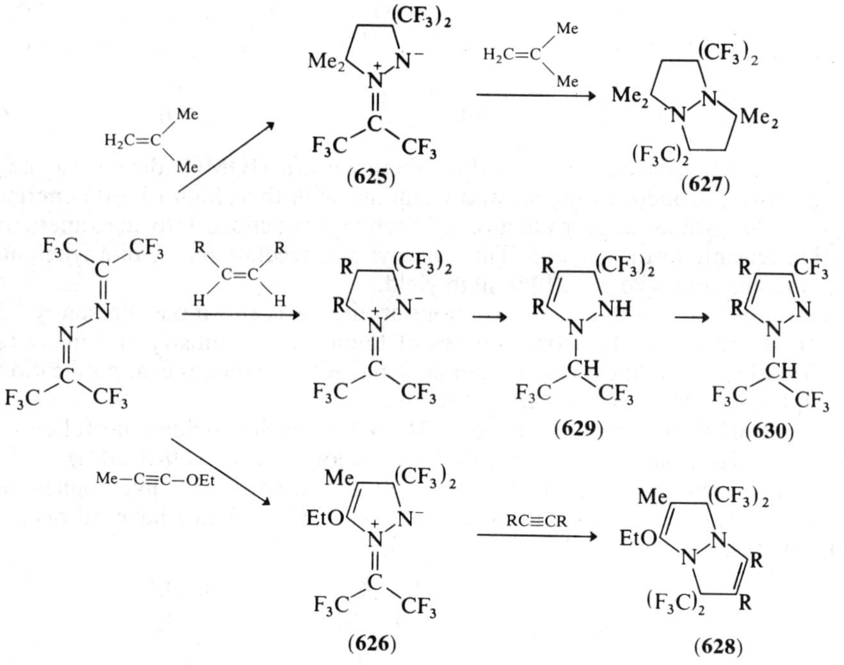

Scheme 56

Hydrazoyl halides are useful reagents for the synthesis of pyrazolines and pyrazoles ⟨80JHC833⟩. The elimination of HX, usually with triethylamine, is now the preferred method for the generation of the nitrilimine (621) *in situ*. Although in some cases it is not clear if the mechanism involves a nitrilimine (621) (as for example in the Fusco method in which sodium salts of β-diketones are used), in other reactions it is the most reasonable possibility. For example, the synthesis of pyrazolobenzoxazine (633) from the hydrazoyl halide (631) probably occurs *via* the nitrilimine (632). Trifluoromethylpyrazoles (634) have been prepared by the reaction of a hydrazoyl halide and an alkynic compound in the presence of triethylamine ⟨82H(19)179⟩.

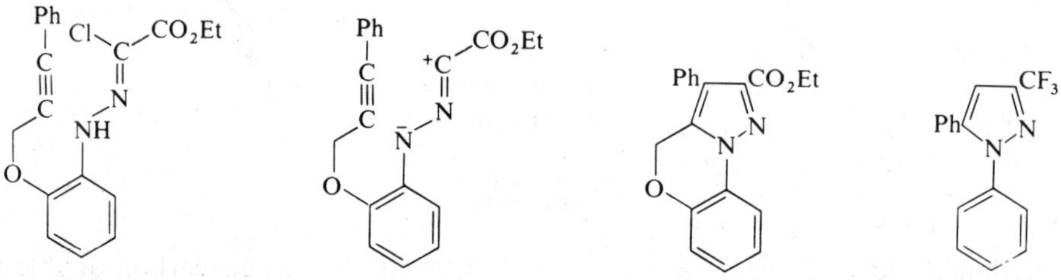

Monosubstituted hydrazones react with alkenes and alkynic compounds to yield pyrazolidines and pyrazolines, respectively ⟨71LA(743)50, 79JOC218⟩. Oxidation often occurs during the reaction and pyrazoles are isolated as the end product.

4.04.3.1.3 Formation of three bonds

Pyrazolines can be prepared from the reaction between a hydrazine and two carbonyl compounds, one of them having at least one hydrogen atom α to the carbonyl group. Formally, these reactions correspond to the [NN+C+CC] class. However, if one considers the different steps in the ring formation, they more properly belong to the [CNN+CC] (Section 4.04.3.1.2(ii)), the [CCNN+C] (Section 4.04.3.1.2(i)), or the formation of one bond (Section 4.04.3.1.1) classes.

The synthesis of Δ^3-pyrazolines by reaction of a 1,2-disubstituted hydrazine, formalin and a carbonyl compound, known as the Hinman synthesis, probably proceeds by the mechanism shown in Scheme 57 ⟨69BSF3300⟩.

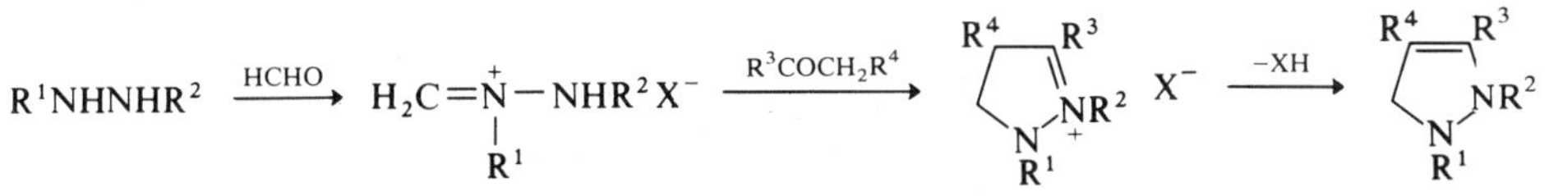

Scheme 57

When two moles of a carbonyl compound are used instead of formalin, the mechanism is different (Scheme 58) ⟨70BSF3147⟩. In one example ⟨80CCC2417⟩ the product of the nucleophilic addition of the hydrazine to the pyrazolinium salt (**635**; $R^1 = R^2 = Ph$, $R^3 = R^4 = H$) has been isolated (**636**). With monosubstituted hydrazines and aldehydes, Δ^2-pyrazolines (**637**) are formed ⟨69CHE805⟩.

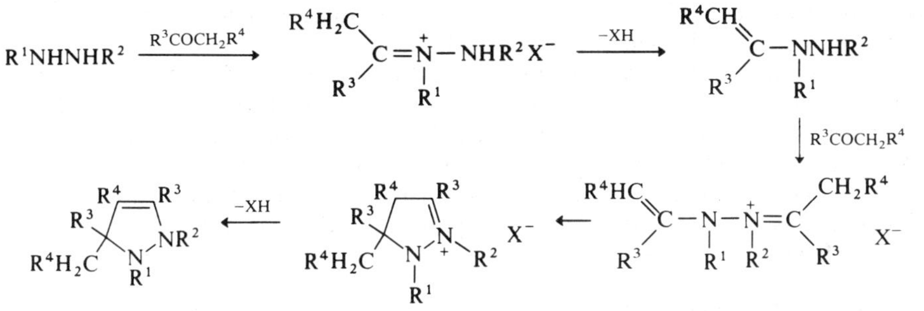

Scheme 58

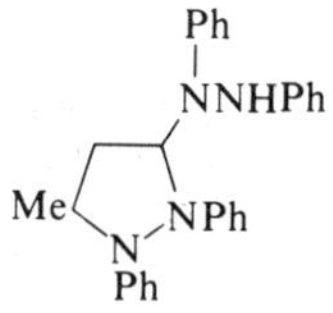

4.04.3.2 Ring Synthesis from Heterocycles

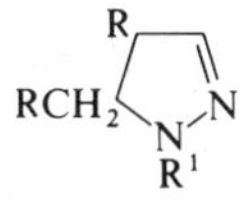

4.04.3.2.1 Ring opening

Pyrazoles can be prepared by ring opening reactions of fused systems already containing the pyrazole nucleus. Thus several [5.5], [5.6] and [5.7] fused heterocycles have been opened to substituted pyrazoles, usually in basic medium. In general, the method has little preparative interest since another pyrazole derivative has usually been used to build the ring-fused system. However, due to the unexpected structures obtained, two publications are worthy of notice. 6H-Cyclopropa[5a,6a]pyrazolo[1,5-a]pyrimidine (**638**) was readily obtained from the corresponding pyrazolopyrimidine by the action of diazomethane at room temperature (Scheme 59) ⟨81H(15)265⟩. When (**638**) was treated with potassium hydroxide, the pyrazole (**640**) was formed, probably via the diazepine (**639**).

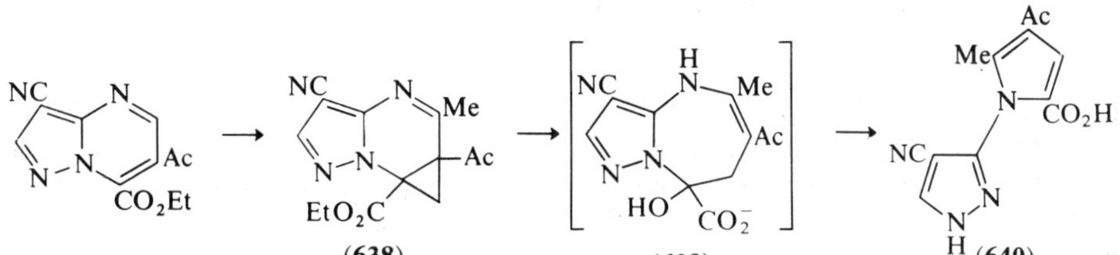

Scheme 59

Other still more unusual structures were obtained in the reaction of the 3a,6a-diazapentalenes with DMAD ⟨78CL1093⟩. Depending on the substituents on the diazapentalene (**641**; $R^1 = R^2 = H$ or $R^1 = COAr$, $R^2 = Ar$), the rearranged 1:2 adducts (**642**; $E = CO_2Me$) and (**643**; $E = CO_2Me$), respectively, were obtained.

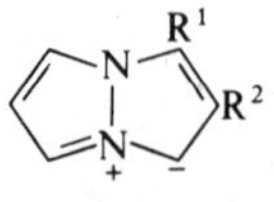

(641) 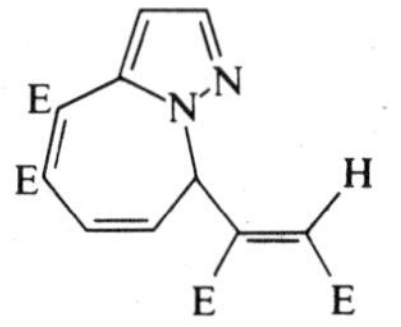   (642) 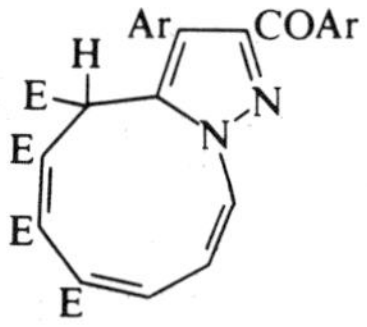   (643)

4.04.3.2.2 Ring transformations

The transformations between pyrazole derivatives of different oxidation levels, *i.e.* between pyrazolones and pyrazolines, will not be discussed here since they have been examined in the reactivity sections (Section 4.04.2).

The richness and complexity of the present section is considerable. Almost any heterocycle conveniently substituted can be transformed into another chosen ring system, and this is shown in the two excellent volumes of van der Plas ('Ring Transformations of Heterocycles') ⟨B-73MI40402⟩. The arrangement of the aforementioned book (from the starting heterocycle point of view) does not suit this section and for the purposes of this chapter an alternative classification has been selected. When no explicit references are given, the material has been taken from ⟨B-73MI40402⟩.

(i) Masked β-difunctional compounds

Pyrazoles have been prepared by the action of hydrazines on heterocyclic derivatives which react as masked functional groups of a 1,3-difunctional derivative (Section 4.04.3.1.2(ii)).

(a) One functional group. A carbonyl group can be replaced by a three-membered ring, usually an oxirane ⟨73BSF3159⟩ or an aziridine ⟨74JHC1093⟩, or by a β-substituted pyrrole or indole derivative ⟨67HC(22)1⟩. In Scheme 60 two classical examples are shown.

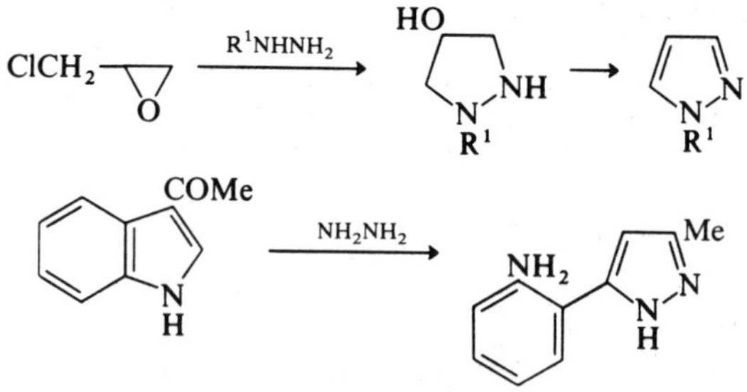

Scheme 60

(b) Both functional groups. Most of the ring interconversions forming pyrazoles involve the action of hydrazine upon heterocycles which function as masked 1,3-difunctional compounds ⟨B-73MI40403⟩. These include 1,2-dithiolium salts, isoxazolinones, oxazoles ⟨77JCR(S)140⟩, furoxans, pyrylium salts, benzopyrylium salts, dehydroacetic acid, 1,3-oxazine-2,4-diones ⟨82JCS(P1)473⟩, pyrimidines ⟨76JCS(P2)315⟩, thymine, cytosine, uracil ⟨81CPB3760⟩ and 1,7- and 1,8-naphthyridines.

(ii) Bimolecular reactions

In this section formation of pyrazoles from the reaction of heterocycles with compounds other than hydrazines will be discussed.

(a) Sydnones. Owing to their preparative and fundamental importance, sydnones have been classified separately. Mesoionic anhydro-5-hydroxy-1,2,3-oxadiazolium hydroxides (sydnones) ⟨76AHC(19)1, 81CB2450⟩ react with alkenes and alkynes to afford pyrazolines and pyrazoles with evolution of carbon dioxide (Scheme 61). With benzyne, 2-substituted indazoles are formed. Isosydnones react similarly ⟨76AHC(19)1⟩.

(b) Others. Thermolysis of 2,5-diphenyltetrazole yields, with nitrogen evolution, diphenylnitrilimine (**621**; $R^1 = R^3 = Ph$), a useful source of pyrazolines and pyrazoles (Section 4.04.3.1.2(ii)). The following interesting reaction belongs to this section. Anthranils react with anilines to yield 2-arylindazoles. Hydrazine behaves as a diamine this reaction, affording a 2,2-diindazolyl derivative (Scheme 62).

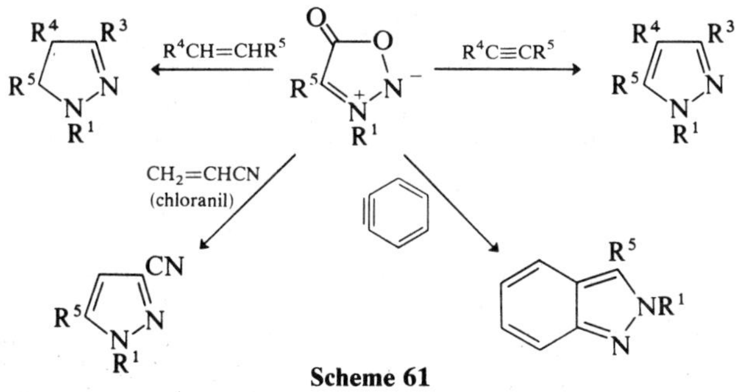

Pyrazoles and their Benzo Derivatives

Scheme 61

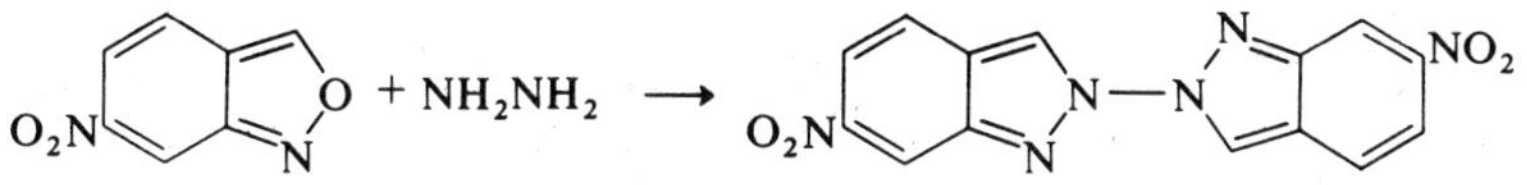

Scheme 62

(iii) Monomolecular rearrangements

This section includes thermal and photochemical rearrangements of heterocycles to pyrazoles without using any other reagent. It is probably the most interesting part but also the most diversified and difficult to summarize. Two examples have already been mentioned: the rearrangement of 1,2-diazepines (Section 4.04.2.3.9) and sulfur monoxide extrusion from thiadiazine S-oxides (Section 4.04.3.1.1(i)). Other examples are shown in Figure 31. They include the isomerization of vinylaziridines (**644**) to pyrazoles (**645**) and of 5-aminoisoxazoles (**646**) to pyrazolones (**647**). The anion (**648**) derived from $6H$-1,3,4-thiadiazine undergoes sulfur extrusion at low temperatures yielding the pyrazole (**649**) ⟨75TL33⟩. Photolysis transforms dipolar imines (**650**) into N-ethoxycarbonylpyrazoles (**651**) ⟨76CC250⟩. 1-Cyanopyrazole (**653**) has been obtained in 62% yield from 2-azidopyrimidine (**652**), via the 2-pyrimidylnitrene ⟨81AHC(28)232⟩.

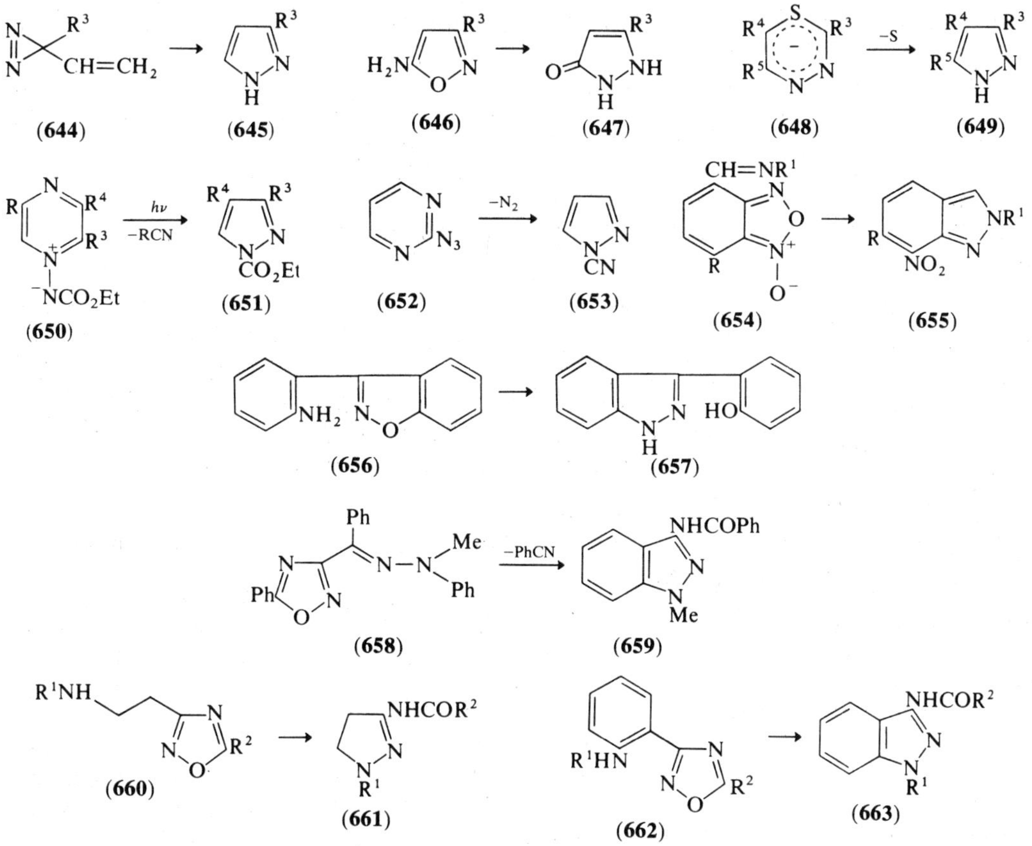

Figure 31 Synthesis of pyrazoles and indazoles by ring transformation

A series of rearrangements involving oxygenated azoles (however, see **430 → 431**, Section 4.04.2.3.1(vi)) related to the Boulton–Katritzky rearrangement yielded pyrazoles and indazoles, for example, the transformation of benzofurazan oxides (**654**) into indazoles (**655**) ⟨80JOC1653⟩. The treatment of 3-(o-aminophenyl)-1,2-benzisoxazole (**656**) with lithium aluminum hydride affords the hydroxyphenylindazole (**657**) ⟨74JHC885⟩. The N-methyl-N-phenylhydrazone of 3-benzoyl-5-phenyl-1,2,4-oxadiazole (**658**) undergoes a rearrangement to (**659**) with loss of benzonitrile ⟨82JCS(P1)165⟩. The transformation (**660**) → (**661**) can be used to prepare 3-acylamino-Δ^2-pyrazolines and the related reaction (**662**) → (**663**) affords 3-acylaminoindazoles ⟨82JCS(P1)759⟩.

4.04.3.3 Best Practical Methods

4.04.3.3.1 General methods

In the synthesis sections (4.04.3.1 and 4.04.3.2) the more useful methods to obtain the heterocycles which this chapter covers have been described. Table 37 contains a list of compounds described in *Organic Syntheses*.

Table 37 Pyrazole and Indazole Derivatives Described in *Organic Syntheses*

Compound	Ref.
3,5-Dimethylpyrazole	51OS(31)43
3-Aminopyrazole	68OS(48)8
3-Amino-1-phenyl-5-pyrazolone	48OS(28)87
Indazole	49OS(29)54, 59OS(39)27, 62OS(42)69
5-Nitroindazole	40OS(20)73
2-Phenylindazole	68OS(48)113

Most of these compounds, for instance pyrazole itself, are today commercially available, so there is only a minor interest in detailing the experimental procedures used. The best way to prepare pyrazole is the Protopopova method (Section 4.04.3.3.2) and a modification using hydrazine hydrate instead of a hydrazine salt has recently been patented ⟨80GEP2922591⟩.

4.04.3.3.2 Nucleosides

Amongst the N-substituted azoles, the nucleosides constitute a special class both for their biological significance ⟨B-81MI40407⟩ and for the particular synthetic methods involved in their preparation. Pyrazoles and indazoles react with suitable sugar derivatives to give the corresponding N-glycosides. These glycosidations have been carried out by conventional methods.

Glycosidation of symmetrical pyrazoles leads to a unique N-substituted pyrazole. However, depending on the glycosidation method used and on the starting materials, one or two isomers (anomers) which differ only in the configuration of the C-1 carbon of the sugar moiety can be obtained. Thus, the reaction of 3,5-dimethylpyrazole and 2,3,5-tri-O-acetyl-D-ribofuranosyl chloride, in the presence of mercury(II) cyanide as a hydrogen halide acceptor, yields the nucleoside (**664**) of β-anomer configuration ⟨80JHC113⟩. However, the fusion reaction of another symmetrical pyrazole, the 4-nitro derivative, with 1,2,3,5-tetra-O-acetyl-D-ribofuranose, in the presence of iodine as a catalyst, gives a mixture of both anomers, represented by (**665**) and (**666**) ⟨74MI40400⟩.

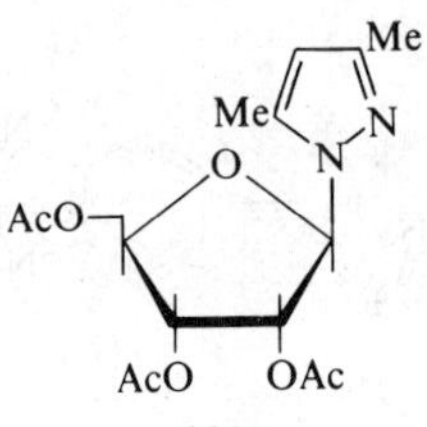

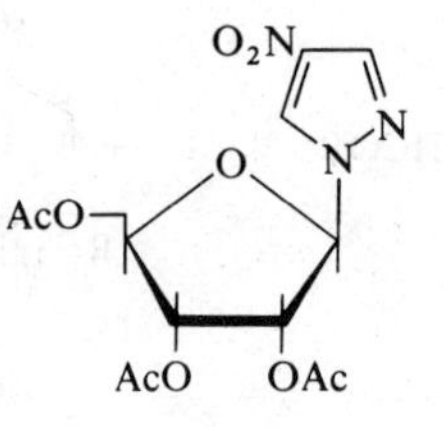

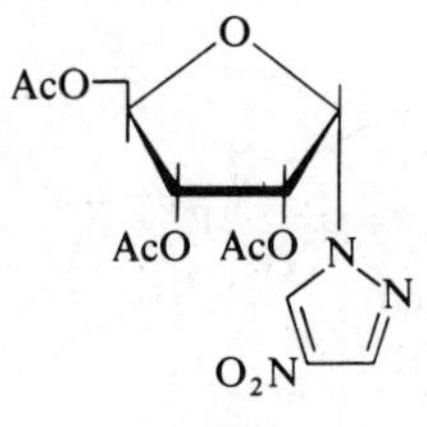

The reaction between asymmetrically substituted pyrazoles ⟨74MI40400, 74MI40404, 74ZOR1787, 79JMC807, 80JMC657⟩ or indazoles ⟨70JHC117, 70JHC1329, 70JHC1435, 73TL4619, 74MI40402, 76BSF1983, 76MI40401⟩ with sugar derivatives leads to glycosyl derivatives on N-1 and N-2, or to mixtures of both derivatives, depending on the glycosidation method, the experimental conditions and the structure of the starting materials.

The fusion reaction between 3(5)-ethoxycarbonyl-5(3)-bromopyrazole and penta-O-acetyl-β-D-glycopyranose in the presence of p-toluenesulfonic acid as a catalyst gives a mixture of the two possible isomers (**667a**) and (**668a**). The same reaction, but with tetra-O-acetyl-β-D-ribofuranose or tetra-O-acetyl-β-D-ribopyranose, yields only the 5-ethoxycarbonyl derivatives (**667b**) and (**667d**) ⟨79JMC807⟩. The fusion of 3(5)-ethoxycarbonylpyrazole with peracylated sugars in the presence of iodine gives preferentially the 3-ethoxycarbonyl derivatives (**669a**), (**669b**) and (**669d**) ⟨74MI40404, 74ZOR1787⟩.

a; R^1 = 2,3,4,6-tetra-O-acetyl-β-D-glucopyranosyl
b; R^1 = 2,3,5-tri-O-acetyl-β-D-ribopyranosyl
c; R^1 = 2,3,4-tri-O-acetyl-β-D-ribopyranosyl
d; R^1 = 2,3,5-tri-O-benzoyl-β-D-ribofuranosyl

In the case of indazoles the reaction of indazole, 5-nitroindazole or 6-nitroindazole with glycosyl halides and mercury(II) cyanide gives exclusively 2-glycosylindazoles (**670**), (**673**) and (**675**) ⟨70JHC1435⟩. Similarly, the reaction of 1-trimethylsilyl derivatives of indazole, 3-cyanoindazole, 4-nitroindazole, 5-nitroindazole and 6-nitroindazole with 2,3,5-tri-O-acetyl-D-ribofuranosyl bromide gives only, or preferentially, the 2-ribofuranosyl derivatives (**670**)–(**674**) ⟨70JHC117, 70JHC1329⟩.

(**670**) $R^1 = R^2 = R^3 = R^4 = H$
(**671**) $R^1 = CN, R^2 = R^3 = R^4 = H$
(**672**) $R^1 = R^3 = R^4 = H, R^2 = NO_2$
(**673**) $R^1 = R^2 = R^4 = H, R^3 = NO_2$
(**674**) $R^1 = R^2 = R^3 = H, R^4 = NO_2$

1-Glycosylindazoles (**675**) are obtained by fusion of peracylated sugars with indazoles in an iodine-catalyzed reaction ⟨73TL4619⟩. However, by changing the catalyst or its amount, mixtures of 1- and 2-(β-D-ribofuranosyl)indazoles are obtained ⟨74MI40402, 76BSF1983⟩. Another way to obtain 1-substituted indazoles is by the nucleophilic addition of an indazole (indazole, 3-bromoindazole, 5-chloroindazole, 5-nitroindazole, 6-nitroindazole) to an α,β-unsaturated cyclic ether, such as 2,3-dihydrofuran or 2,3-dihydro-$4H$-pyran in the presence of an acid as catalyst ⟨76MI40401⟩. Thus the deoxy analogues of nucleosides are obtained. The other general method of obtaining pyrazoles, the reaction of hydrazines and β-dicarbonyl compounds (Section 4.04.3.1.2(ii)), has also been applied to pyrazologluconucleosides ⟨81LA2309⟩.

4.04.3.3.3 Labelled compounds

In the structure sections, labelled compounds have often been used to solve a spectroscopic problem involved in microwave (Section 4.04.1.3.2), nitrogen NMR (Section 4.04.1.3.5), IR (Section 4.04.1.3.7(i)) or mass spectrometry (Section 4.04.1.3.8). The synthesis usually involves non-radioactive compounds (^2H, ^{15}N) by classical methods that must be repeated several times in order to obtain good yields.

The synthesis of the 11 deuterated derivatives of pyrazole (**676**) is described in ⟨70BSF1974⟩. Seven can be obtained by H/D exchanges making use of the fact that the exchange rates decrease in the order 1-position > 4-position > 3(5)-position (Figure 32). The four others have been prepared by reduction of the 3(5)-bromopyrazole (**677**) followed by H/D exchanges.

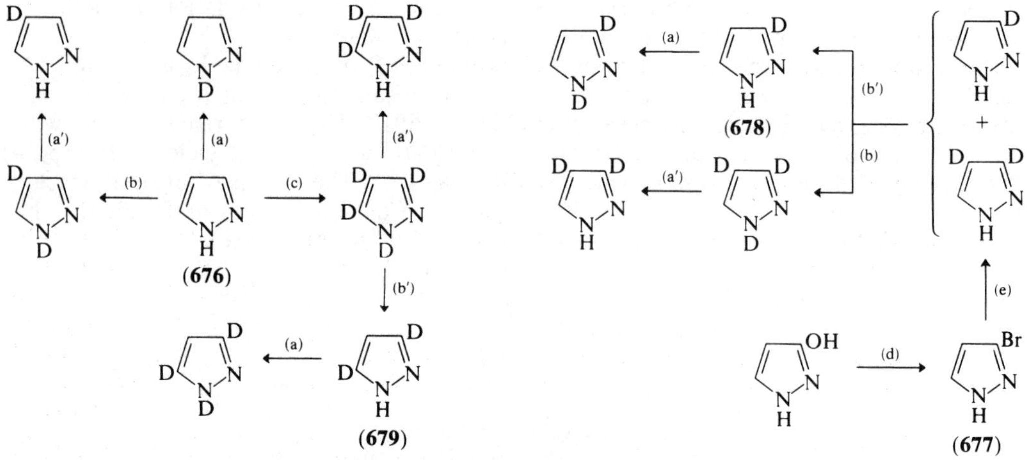

(a) Three 3 h exchanges in D_2O at room temperature
(a') Three 3 h exchanges in H_2O at room temperature
(b) Two 12 h exchanges in an autoclave at 200 °C with D_2O
(b') Two 12 h exchanges in an autoclave at 200 °C with H_2O
(c) Two 12 h exchanges in an autoclave at 200 °C with 1N NaOD
(d) 12 h in an autoclave at 180 °C with $POBr_3$
(e) Reduction by Zn–Cu in AcOD at reflux

Figure 32 Synthesis of deuterium labelled pyrazoles

The isotopic purity on the carbons was about 95%; the experimental conditions, found empirically, are near the theoretical conditions deduced from the exchange rates of Wu and Vaughan ⟨70JOC1146, 79SA(A)1055⟩. Some authors have used similar conditions to prepare deuteropyrazoles ⟨71JHC311, 71JST(9)465⟩ and others, using slightly different procedures such as perdeuterated acetone in a sealed tube at 190 °C ⟨74JST(22)401⟩ or a solution of D_2SO_4 in D_2O at 100 °C ⟨74CHE471⟩ have prepared pyrazole-1,4-d_2. Decarboxylation of 3,5-pyrazoledicarboxylic acid-d_3 has been used to obtain a mixture of (**678**) and (**679**) ⟨74JST(22)401⟩. Decarboxylation of 1-methylpyrazole-5-deuterocarboxylic acid yields 1-methylpyrazole-5-d_1 with a 50–60% labelling with apparently no scrambling ⟨69OMR(1)431⟩, but it is known that this method leads generally to compounds labelled not only on the *ipso* carbon atom but also on the *ortho* one. The five C-monodeuterated indazoles have been prepared ⟨76CJC1329⟩, the 3-d_1 by decarboxylation and the others from aminoindazoles by nitrous acid deamination using deuterated hypophosphorous acid.

^{15}N-Labelled (mono and di) pyrazole has been prepared several times ⟨73JMR(10)130, 74CHE471, 74JST(22)401, 82UP40401⟩, always using the Protopopova method (reaction between a hydrazine salt and tetramethoxypropane) ⟨57ZOB1276⟩. Pellegrin has compared Jones' ⟨49JA3994⟩ and Protopopova's procedures and found that the second one gives better results ⟨82UP40401⟩. In a typical experiment a solution of hydrazine sulfate ($^{15}N_2$) (1 g) in (1.5 cm^3 of water and 0.4 cm^3 of 1N HCl was treated with 2 cm^3 of 1,1,3,3-tetraethoxypropane (malonic aldehyde dimethyl acetal) during 1 h at 50 °C. The volume was reduced by evaporation to 1–1.5 cm^3 and 2.5 cm^3 of 30% NaOH was added. The mixture was extracted with ether (5 times 10 cm). The ethereal solution was dried over sodium sulfate and evaporated. The residue, 500 mg (96% yield), was chromatographically pure pyrazole. The labelling of the hydrazine (95%) is conserved in the pyrazole. ^{15}N-Labelled phenylhydrazine (mono- and di-labelled) has been used to prepare several 1-phenylpyrazoles ⟨77JCS(P2)1024, 79OMR(12)476, 80OMR(13)197⟩. An original synthesis of 1-benzylpyrazole ^{15}N-labelled at N-2 has been reported recently (Scheme 63) ⟨81OPP371⟩.

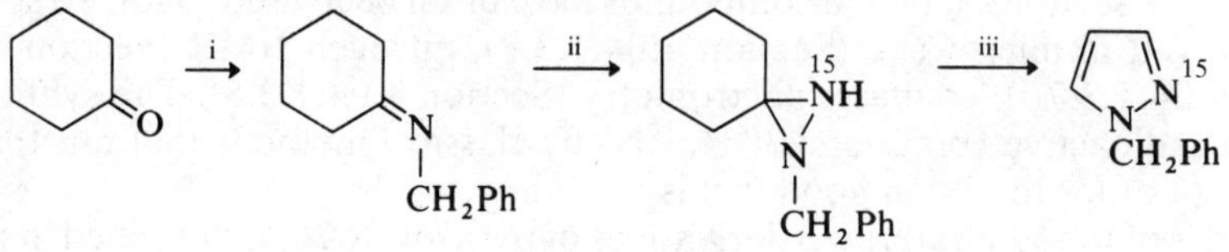

i, $PhCH_2NH_2$; ii, $^{15}NH_2OSO_3H$; iii, $CH_2[CH(OMe)_2]_2/H^+$

Scheme 63

Indazole-1-^{15}N has been prepared from 2-nitrotoluene-^{15}N (Table 37) in 51% yield. Indazole-2-^{15}N has been prepared from 2-methyl-4-nitroaniline and sodium nitrite-^{15}N, through 5-nitroindazole-2-^{15}N, in three steps (overall yield 45%) ⟨81MI40405⟩.

4.04.4 APPLICATIONS AND IMPORTANT COMPOUNDS

4.04.4.1 Synthetic Products

4.04.4.1.1 Pharmaceuticals

Some of the most important aromatic pyrazoles with biological activity are shown in Table 38. Pyrazole itself and several N-unsubstituted pyrazoles are inhibitors and deactivators of liver alcohol dehydrogenase ⟨79JMC356, 79ACS(B)483, B-79MI40414, 82ACS(B)101⟩.

The analgesic activity of difenamizole (**680**) was comparable with those of aminopyrine and aspirin and less than that of mefenamic acid. Its antiinflammatory activity was greater than that of aspirin ⟨B-76MI40404⟩. 1-(4-Fluorophenyl)-4-(4-chlorophenyl)-3-pyrazolylacetic acid (Pirazolac) has recently been reported ⟨82MI40400⟩ to have antiinflammatory activity, and the advantageous relationship between activity and gastric tolerance led to detailed biological investigations ⟨83AP588, 83AP608⟩. Interest in the effect of the CF_3 substituent on the activity of pyrazole derivatives prompted some authors to investigate compounds (**681**) and (**682**). 3-Trifluoromethyl-4,5-trimethylenepyrazole (**681**) was about 0.3% as effective as an amebicide compared with emetine, and 5,5-dimethyl-3-trifluoromethyl-4,5-dihydroindazole (**682**) had 0.5 to 1% the trichomonacidal activity of metronindazole ⟨71JMC997⟩.

(**681**)   (**682**)

Betazole (**683**) has been used as a chemical control substance for pharmacological characterization of histamine receptors ⟨B-80MI40406⟩, and shows a relative selective activity towards the H_2-receptor. Betazole hydrochloride is used to diagnose impairment of the acid-producing cells of the stomach.

Sulfaphenazole (**684**) and sulfazamet (**685**) are both examples of relatively short acting sulfonamides ⟨B-80MI40406⟩ and their antibacterial activity has been tested against *Escherichia coli*, the former being more effective than the latter. Sulfaphenazole also displaces sulfonyl ureas from protein binding sites on human serum albumin and consequently increases the concentration of the free (active) drug and produces a more intense reaction that may result in hypoglycemia.

3,5-Dimethylpyrazole possesses hypoglycemic activity in glucose primed and diabetic rats because of its metabolism to the corresponding 5-carboxylic acid. This metabolism product acts primarily on adipose tissue by inhibiting lipolysis, promoting the conversion of glucose to triglyceride, and stimulating glycogenesis ⟨B-76MI40404⟩. Salts of 5-methylpyrazole-3-carboxylic acid with basically substituted adenine derivatives (**686**) which have significantly more lipolysis inhibitory activity than the free acid itself have been described recently ⟨81AF2096⟩.

(**686**)

A number of 4-pyrazolylpyridinium salts (**687**) have been found to have hypoglycemic activity in several animal models ⟨68JMC981⟩. The possibility that the corresponding 4-pyrazolylpyridine 1-oxides might also exhibit this type of activity was investigated without success ⟨69JMC945⟩.

1-Methyl-4-[3(5)-pyrazolyl]quinolinium iodides (**688**) also failed to depress blood sugar levels significantly ⟨69JMC1124⟩. Neuroleptic-like effects of some β-aminoketones (**689**) containing a pyrazole nucleus have been described in the literature ⟨B-80MI40406⟩. The

Table 38 Some Selected Pyrazole Derivatives with Biological Activity

Generic name	No.	1	3	4	5	Activity	Ref.
Difenamizole	(680)	Ph	Ph	H	NHCOCH(Me)NMe$_2$	Analgesic–antiinflammatory–antipyretic	B-76MI40404
—	(681)	H	CF$_3$	—(CH$_2$)$_3$—		Amebicide	71JMC997
—	(682)	H	CF$_3$	—CH$_2$C(Me)$_2$CH=CH—		Trichomonacidal	71JMC997
Betazole	(683)	H	CH$_2$CH$_2$NH$_2$	H	H	Diagnostic aid (gastric secretion)	B-76MI40404
Sulfaphenazole	(684)	Ph	H	H	NHSO$_2$C$_6$H$_4$NH$_2$-p	Antibacterial–antimicrobial	B-76MI40404
Sulfazamet	(685)	Ph	Me	H	NHSO$_2$C$_6$H$_4$NH$_2$-p	Antibacterial	B-76MI40404
—	—	H	Me	H	Me	Hypoglycemic	B-80MI40406
—	—	H	CO$_2$H	H	Me	Hypoglycemic	B-76MI40404
—	(687)	H	C$_5$H$_4$ṆMe	H	Me	Hypoglycemic	68JMC981
—	(689)	Ph	H	COCH$_2$NC$_4$H$_8$NC$_6$H$_4$Me-o	Me	Antipsychotic	B-80MI40406
EMD-16923	(690)	H	CH=CMe$_2$	H	H	Sedative–hypnotic	79MI40400
—	(691)	H	(CH$_2$)$_2$NC$_4$H$_8$NC$_6$H$_4$Cl-m	H	Me	Sedative–hypnotic	80MI40400

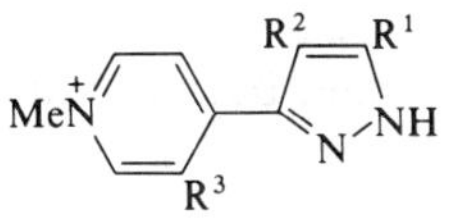

pyrazole (**690**) potentiates hexobarbital-induced sleep time ⟨79MI40400⟩ and EMD-16923 (**691**) was shown to be clinically effective in anxiety, producing an EEG-pattern in rabbits indicative of drowsiness ⟨80MI40400⟩.

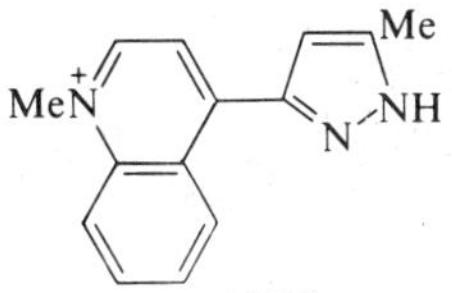

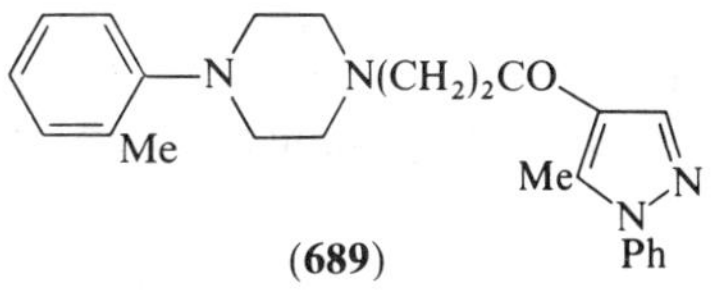

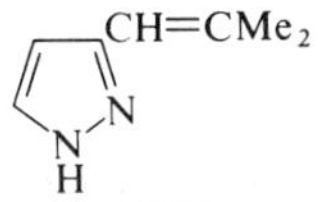

Research for an antidepressant among non-tricyclic compounds with pharmacological effects qualitatively different from those of the conventional tricyclic compounds led to the preparation and testing of a series of indazole derivatives for reserpine-like activity in mice. 1-[3-(Dimethylamino)propyl]-5-methyl-3-phenyl-1H-indazole (FS-32; **692**) antagonizes reserpine-induced effects and potentiates amphetamine-induced self-stimulation and L-Dopa-induced increase in motor activity. FS-32 produces an anticholinergic action mainly on the central nervous system, while the action of imipramine occurs centrally as well as peripherally ⟨79AF511⟩.

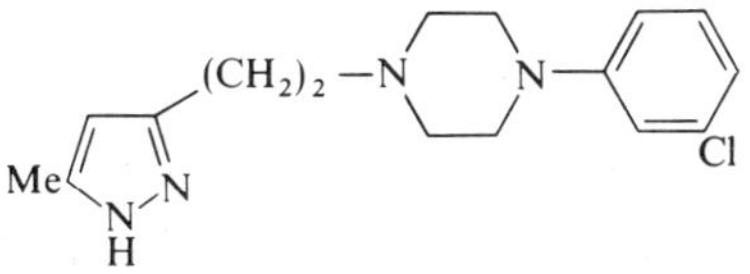

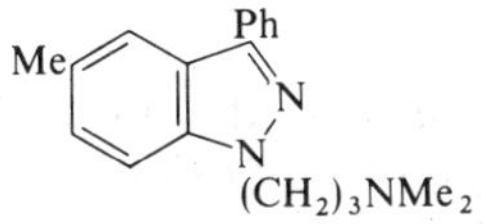

Most of the pharmacologically active indazole derivatives are useful as antiinflammatory drugs, e.g. bendazac or [(1-benzyl-1H-indazol-3-yl)oxy]acetic acid (LD_{50} in mice and rats of 355 and 388 mg kg^{-1} i.p., respectively) and benzydamine or 1-benzyl-3-[3-(dimethylamino)propoxy]-1H-indazole (LD_{50} in mice and rats of 110 and 100 mg kg^{-1} i.p., respectively). The last cited compound also has analgesic and antipyretic properties ⟨B-76MI40404⟩.

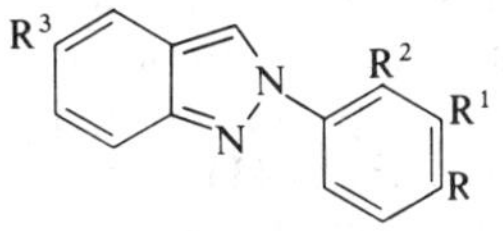

Recently, Picciola et al. ⟨81FES1037⟩ prepared some 2H-indazole derivatives containing a phenylalkanoic acid residue with potential antiinflammatory activity (**693**). M.G. 18755 (R = CHMeCO$_2$H, R^1 = R^2 = R^3 = H) and its lysine salt, M.G. 18334, showed greater activity than ibuprofen, and the homologous butyric acid derivative M.G. 18860 showed good activity as a platelet aggregation inhibitor.

Antiarthritic effects on primary or secondary arthritis have been shown by the ethyl ester of 1-(p-chlorobenzoyl)-5-methyl-1H-indazole-3-carboxylic acid (**694**) ⟨81FES315⟩.

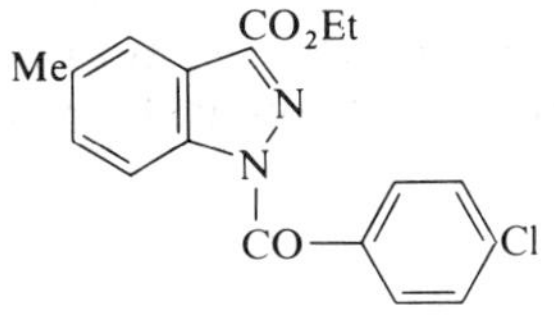

Finally, a new class of antispermatogenic agents containing the same fundamental structure cited above has been described ⟨76JMC778⟩. 1-Halobenzyl-1H-indazole-3-carboxylic acids are potentially useful for birth control, and because they act after a single administration they are of interest for the physiological study of the male reproductive system.

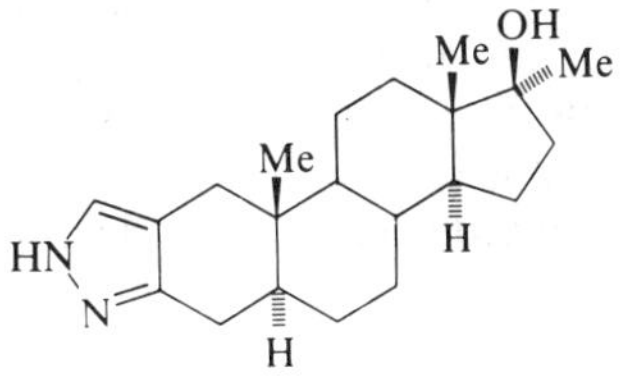

In 1959 Clinton and coworkers reported the first synthesis of some pyrazole fused androstane derivatives and described their biological activity ⟨B-76MI40404⟩. Stanazolol (**695**) or 17-methyl-2H-5α-androst-2-eno[3,2-c]pyrazol-17β-ol was 10 times as active as 17α-methyltestosterone in improving nitrogen retention in rats ⟨B-80MI40406⟩, and its myotrophic activity was only twice that of 17α-methyltestosterone. It is used as an anabolic steroid with no lasting adverse side effects.

11β,17,21-Trihydroxy-6,16α-dimethyl-2'-phenyl-2'H-pregna-2,4,6-trieno[3,2-c]-pyrazol-20-one-21-acetate, commonly named cortivazol (**696**), has been used since 1962 by Merck & Co. as a glucocorticoid ⟨B-76MI40404⟩. Tricyclic pyrazoles (**697**) have been found to possess antiarrhythmic and antiinflammatory activities ⟨76JHC545⟩.

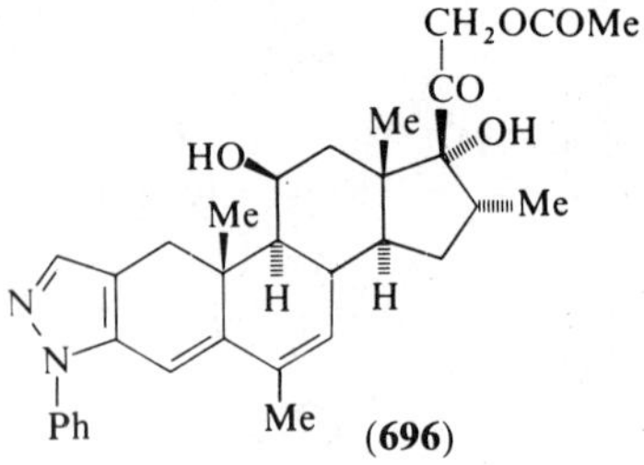

R = Me or $(CH_2)_2NEt_2$
A = CH_2, $(CH_2)_2$, $(CH_2)_3$

(**696**) 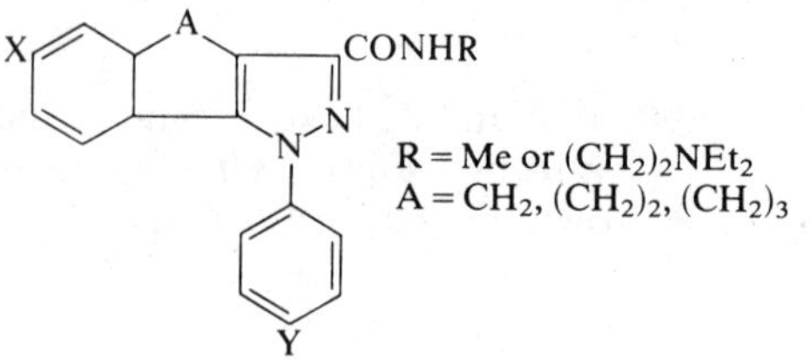   (**697**)

Tricyclic pyrazole derivatives (**698**) are described by Hashem *et al.* as inhibitors of the growth of *Bacillus subtilis*, *Pseudomonas fluorescens*, *Staphylococcus aureus* and KB cells at moderate concentrations ⟨76JMC229⟩.

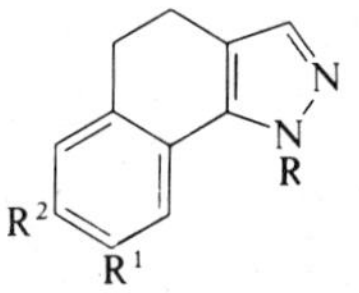

(**698**) 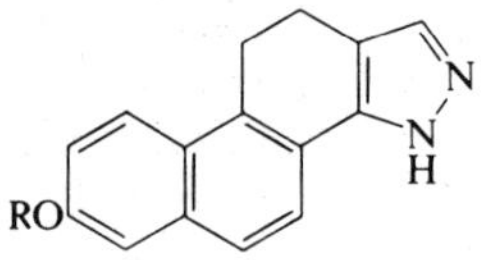   (**699**) 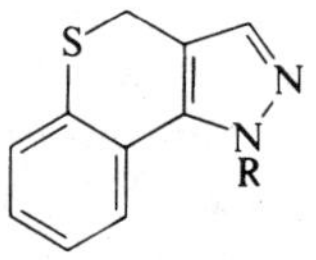   (**700**)

Steroid-type systems (**399**) ⟨77JMC664⟩ and 1,4-dihydro[1]benzothiopyrano[4,3-c]-pyrazole (**700**) also have antimicrobial activity ⟨77JMC847⟩.

In the pharmaceutical literature some pyrazoles fused to the prostaglandin structure are cited. Thus, 9,11-azo-PGE_2 (**701**), a stable endoperoxide analogue that is eight times as potent as PGG_2, is a compound that specifically blocks TXA_2 synthetase ⟨79MI40401⟩.

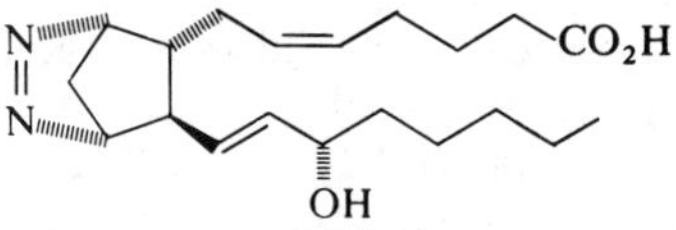

(**701**) 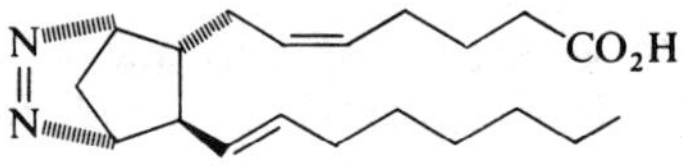   (**702**)

Methoxy-pyrazole and -indazole derivatives constitute a family of interesting pharmaceutical agents. Thus, 4-methoxy-2-(5-methoxy-3-methylpyrazol-1-yl)-6-methyl-pyrimidine (mepirizole), prepared by Naito *et al.* in 1968, is about twice as effective as aminopyrine (see later) by intraperitoneal injection and about as effective as aminopyrine by oral administration in analgesic tests. Its antipyretic activity is low compared with aminopyrine, but its antiinflammatory activity is comparable. This fact suggests that the introduction of a methoxy group into the 5-position of the pyrazole moiety is important in improving activity ⟨B-76MI40404⟩. The search for a drug containing a pyrazole ring with both antiinflammatory, analgesic and antipyretic activities afforded the 3-carbamoyl-1-(*m*-chlorophenyl)-5-methoxypyrazole derivatives (**703**) ⟨76MI40400⟩. The most potent and least toxic derivative of this series was PZ-177 in which the side chain at the 3-position was an *N,N*-disubstituted amide ($R^1 = R^2 = Me$).

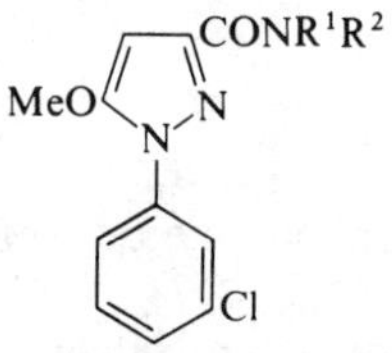

(**703**)

In the search for new structures with antiinflammatory activities some 1-substituted 3-dimethylaminoalkoxy-1*H*-indazoles (**704**) have been synthesized and pharmacologically tested ⟨66JMC38⟩. Doses of 20–40 mg g^{-1} i.p. produced sedation, muscle relaxation and motor incoordination, whereas doses of 80–100 mg kg^{-1} produced depression. Toxicity was fairly constant in all series, varying from 120 to 150 mg kg^{-1} i.p., with the exception of compounds possessing a nitro group or an amino group in the indazole nucleus, which provoked cyanosis.

Some of these indazoles showed analgesic activity greater than that of aspirin and several possessed significant antiinflammatory action. Both properties were the most striking and characteristic ones of the series. In a few cases the authors isolated the corresponding indazolones (**705**) which were less active than (**704**) in terms of analgesic and antiinflammatory properties.

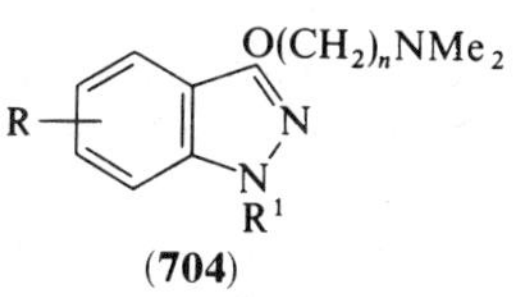

(**704**) 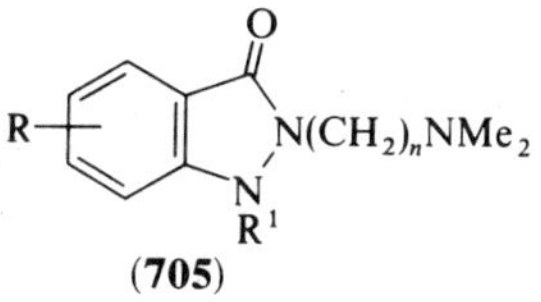   (**705**)

Probably the best known class of pharmacological agents derived from pyrazole are the 5-pyrazolones. They are useful as analgesics, antipyretics and antiinflammatories ⟨B-76MI40404, B-80MI40406, B-80MI40407⟩. Antipyrine, 1-phenyl-2,3-dimethyl-3-pyrazolin-5-one, was one of the first synthetic compounds used in medicine (1884) and its analgesic, antipyretic and antirheumatic activities are similar to those of aspirin and sodium salicylate ⟨B-80MI40407⟩. It is also administered as antipyrine salicylate which is prepared by fusing antipyrine and salicylic acid. Some other 5-pyrazolones ⟨B-76MI40404⟩ are listed in Table 39, but because of their toxicity their use is steadily diminishing. Aminopyrine and dipyrone are no longer official drugs in the United States due to their tendency to cause fatal agranulocytosis, but they are still available in other countries ⟨B-80MI40407⟩. Aminopyrine has been used as a sedative hypnotic combined with chloral hydrate (under the name dichloralphenazone) to avoid the unpleasant taste, odor and gastric irritation of chloral hydrate itself. However, the combination did not achieve much popularity ⟨B-80MI40406⟩.

Table 39 Some Active 3-Pyrazolin-5-ones

Generic name	Substituents			
	1	2	3	4
Antipyrine	Ph	Me	Me	H
Propyphenazone	Ph	Me	Me	Pr^i
Ampyrone	Ph	Me	Me	NH_2
Aminopyrine	Ph	Me	Me	NMe_2
Isopyrine	Ph	Me	Me	$NHPr^i$
Pidylon	Ph	Me	Me	$NHCOCH_2OCOMe$
Nifenazone	Ph	Me	Me	NHCO-(3-pyridyl) 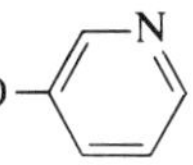
Dipyrone	Ph	Me	Me	$NMeCH_2SO_3Na$
Dibupyrone	Ph	Me	Me	$N(CH_2Pr^i)CH_2SO_3Na$
Piperylon	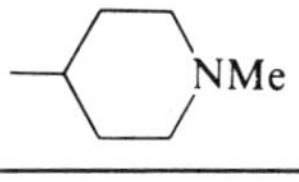 4-(N-methylpiperidyl)	H	Ph	Et

Forbisen, 2,2',3,3'-tetramethyl-1,1'-diphenyl-4,4'-bi-3,3'-pyrazoline-5,5'-dione (**706**) ⟨B-76MI40404⟩, a by-product obtained in the manufacture of antipyrine, has been used in bovine anaplasmosis.

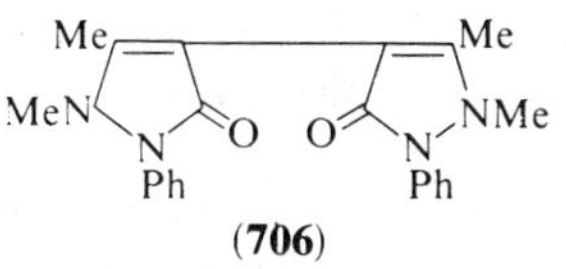

(**706**) 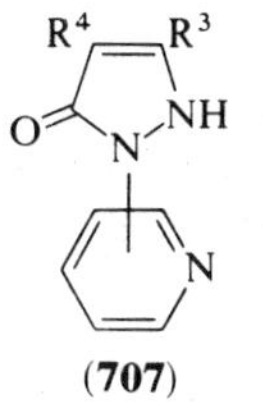   (**707**) 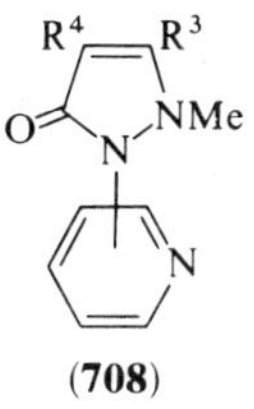   (**708**)

Some active 5-pyrazolone derivatives (**707**) and (**708**) in which the 1-phenyl substituent of antipyrine was replaced by 2'-, 3'- and 4'-pyridyl groups have been prepared ⟨66HCA272⟩. A series of aminoesters substituted at the nitrogen atom of the ester grouping with an antipyryl residue (**709**) were found to possess local anesthetic properties ⟨69MI40400⟩.

(709)

n = 2 or 3
R^1 = H, OMe, OEt, OPr^n, OPr^i, NO$_2$, NH$_2$
R^2 = Me, Et

Finally, benzpiperylon or 1-(1-methyl-4-piperidyl)-3-phenyl-4-benzyl-2- (or 3-)pyrazolin-5-one has been utilized by Sandoz as an investigational agent in connective tissue disorders ⟨B-76MI40404⟩.

Muzolimine (710), a 1-substituted 2-pyrazolin-5-one derivative, is a highly active diuretic, differing from the structures of other diuretics since it contains neither a sulfonamide nor a carboxyl group. It has a saluretic effect similar to furosemide and acts in the proximal tubule and in the medullary portion of the ascending limb of the loop of Henle. Pharmacokinetic studies in dogs, healthy volunteers and in patients with renal insufficiency show that the compound is readily absorbed after oral administration ⟨B-80MI40406⟩.

(710)

The 3,5-pyrazolidinedione derivatives are an important class of antiinflammatory agents. The most widely used are listed in Table 40.

Table 40 Some 3,5-Pyrazolidinedione Derivatives with Biological Activity

Generic name	No.	Substituents 1	4	4'	Ref.
Phenylbutazone	(711)	Ph	Bu^n	H	B-80MI40406
Oxyphenbutazone	(712)	C$_6$H$_4$OH-p	Bu^n	H	B-80MI40406
Sulfinpyrazone	(713)	Ph	CH$_2$CH$_2$SO$_2$Ph	H	B-80MI40406
Mofebutazone	(714)	H	Bu^n	H	B-76MI40404
Kebuzone	(715)	Ph	CH$_2$CH$_2$COMe	H	B-76MI40404
Prenazone	(716)	Ph	CH$_2$CH=CMe$_2$	H	B-76MI40404
Suxibuzone	(717)	Ph	Bu^n	CH$_2$OCOCH$_2$CH$_2$CO$_2$H	B-76MI40404
Pipebuzone	(718)	Ph	Bu^n	CH$_2$N⟨ ⟩NMe	B-76MI40404

Phenylbutazone (711) has been extensively studied. It was synthesized in 1946 in Switzerland and showed analgesic and antipyretic properties but without any advantage over antipyrine and aminopyrine. The antiinflammatory activity in man was not discovered until 1949. It is also a prostaglandin-synthetase inhibitor. Its acidity is comparable to that of carboxylic groups (pK_a = 4.5) and it is possible to prepare the corresponding sodium salt that solubilizes a number of sparingly soluble drugs. The long half-life of phenylbutazone (72 h in man) is due to the slow rate of glucuronide conjugation. Its two phenyl groups provide the lipophilic characteristics for protein binding and tissue distribution. Phenylbutazone competes for the same binding sites on human plasma albumin as coumarin, anticoagulants, sulfonamides, tolbutamide, indomethacin and glucocorticoids.

Oxyphenbutazone (712), γ-hydroxyphenylbutazone and kebuzone (715) are metabolites of phenylbutazone in liver. The first cited is an equally potent antiinflammatory agent but slightly less toxic. Compounds (711) and (712) are rarely used as analgesics and antipyretics because of their toxicities. The first one is used in therapy of rheumatoid disorders characterized by a lack of detectable antiglobulin and antinuclear antibodies in the serum. The γ-hydroxyphenylbutazone has marked uricosuric activity but little antirheumatic effect. Kebuzone (715) is an antiinflammatory agent still widely used in Europe.

Sulfinpyrazone (**713**) has enhanced uricosuric activity but lacks the antiinflammatory and analgesic properties of phenylbutazone. This modification of the biological activity has been ascribed to the increased acidic character of (**713**) ($pK_a = 2.8$) compared to (**711**). In contrast, the half-life of sulfinpyrazone in man is only 2 h. It is known that increasing the acidity of phenylbutazone derivatives decreases the antiinflammatory activity, half-life and sodium retaining potency but increases their uricosuric activity. The uricosuric action of (**713**) is by inhibiting secretion and is additive to that of prebenecid and phenylbutazone but antagonizes that of salicylates. Sulfinpyrazone also displaces organic anions bonded to plasma protein, thus alternating their tissue distribution and renal excretion.

Although some authors propose that an enolizable β-dicarbonyl system is essential for inflammatory activity, two analogues in which this hydrogen atom at carbon 4 has been substituted, suxibuzone (**717**) and pipebuzone (**718**), are used as antiinflammatory agents, and the latter also possesses antipyretic and analgesic properties. However, these compounds are probably not active *per se* and their activity is due to metabolism to phenylbutazone.

From the various modifications that have been made on the phenylbutazone structure in order to increase activity and reduce toxicity it has been found that the activity persists when methyl, chloro, hydroxy or nitro groups are introduced into the *para* position of one or both benzene rings (see oxyphenbutazone (**712**) as an example). Mofebutazone (**714**) has also been used in Europe for several years as an antirheumatic drug.

Some examples of Δ^1-pyrazoline derivatives with biological activity, (**701**) and (**702**), have been described earlier. BW-755C (**719**), a Δ^2-pyrazoline derivative, has been shown to inhibit cyclooxygenase and lipooxygenase enzymes. It also inhibits leucocyte migration into the site of inflammation in rats ⟨81MI40404⟩. Only a very few examples of drugs containing a pyrazolidine nucleus without exocyclic oxo substitution are known. These pyrazole derivatives have found application as new local anesthetics in the form of esters of 1,2-diethyl-3-hydroxymethylpyrazolidine (**720**), which bear structural resemblances to procaine, piperocaine and other synthetic local anesthetics. The structure–activity relationship results obtained parallel findings previously reported for benzoate esters ⟨66JMC493⟩.

(**719**)   (**720**)   (**721**)

The synthesis of 1-phenyl-2-(phenylcarbamoyl)pyrazolidines (**721**) afforded a new series of anticonvulsant agents ⟨79JPS377⟩.

4.04.4.1.2 Agrochemicals

Some pyrazole derivatives have found application in the agrochemical field as insecticides. Two organophosphate insecticides (**722**) and (**723**) containing a pyrazole ring constitute a good example of how molecular modification, based on a sound understanding of the chemistry and pharmacology of the compound, can lead to safer and more useful insecticides and therapeutic agents ⟨B-76MI40404⟩.

(**722**) X = O
(**723**) X = S

O,O-Diethyl *O*-(3-methyl-5-pyrazolyl) phosphate (**722**) and *O,O*-diethyl *O*-(3-methyl-5-pyrazolyl) phosphorothioate (**723**) were prepared in 1956 by Geigy and they act, as do all organophosphates in both insects and mammals, by irreversible inhibition of acetylcholinesterase in the cholinergic synapses. Interaction of acetylcholine with the postsynaptic receptor is therefore greatly potentiated. *O*-Ethyl-*S*-*n*-propyl-*O*-(1-substituted pyrazol-4-yl)(thiono)thiolphosphoric acid esters have been patented as pesticides ⟨82USP4315008⟩.

Dimetilan (dimethylcarbamic acid 1-[(dimethylamino)carbonyl]-5-methyl-1*H*-pyrazol-3-yl ester; **724**) and isolan (dimethylcarbamic acid 3-methyl-1-isopropyl-1*H*-pyrazol-5-yl ester; **725**) also have useful insecticidal properties, exerting their toxic action on insects by reversibly inhibiting cholinesterase. The carbamoylated enzyme reacts with water much more slowly than does the acetylated enzyme ⟨B-76MI40404⟩.

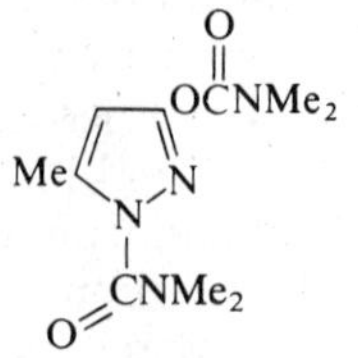

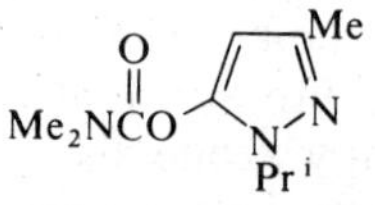

1-Phenylcarbamoyl-2-pyrazolines (**726**) were synthesized and studies on their biological activities demonstrated a high insectidal action for this class of compound. 3,4-Diphenyl substitution in the heterocyclic ring increased the potency of this new class of insecticides compared with 3-phenyl or 3,5-diphenyl substitution by a factor of 3 to 100. The phenomena observed after poisoning are similar in all cases, these derivatives being active against larvae and adult insects ⟨79MI40405⟩.

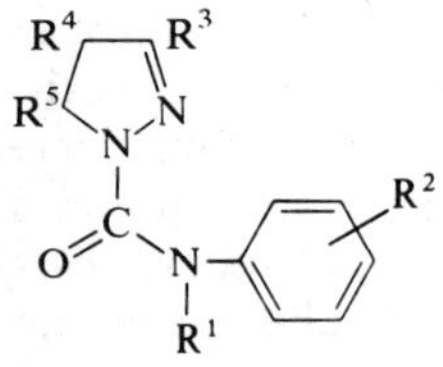

4.04.4.1.3 Dyestuffs and analytical applications

5-Pyrazolone derivatives have found many applications as cotton azo dyes because, even if they were more expensive intermediates, they improved qualities such as brightness and light fastness. Rosanthrene Orange (**727**) and Pyrazol Orange (**728**) are two representative examples.

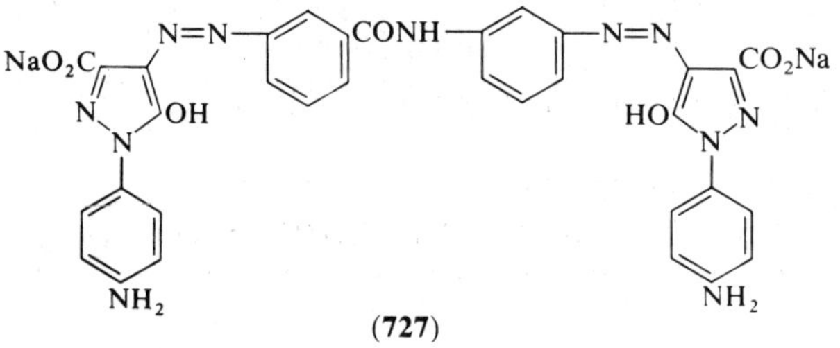

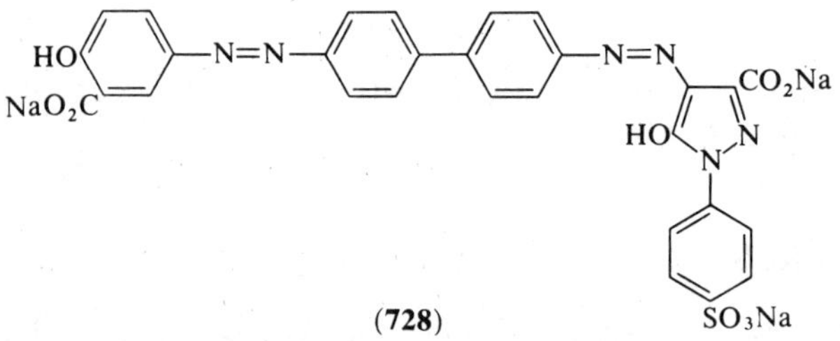

Tartrazine, 4,5-dihydro-5-oxo-1-(4-sulfophenyl)-4-[(4-sulfophenyl)azo]-1H-pyrazole-3-carboxylic acid trisodium salt was discovered by Ziegler in 1884 and is used as a dye for wool and silk. It is used as a colour additive in foods, drugs and cosmetics, and is an adsorption–elution indicator for chloride estimations in biochemistry ⟨B-76MI40404⟩.

Other azopyrazolone acid dyes covering a range of bright yellows have been prepared and among the most important ones are Fast Light Yellow 36 (**729**) ⟨B-70MI40403⟩, Xylene Light Yellow (**730**) ⟨B-70MI40403⟩ and Polar Yellow (**731**) ⟨B-76MI40404⟩.

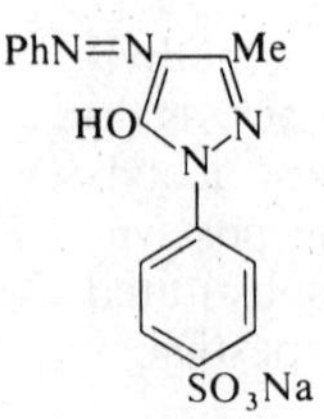

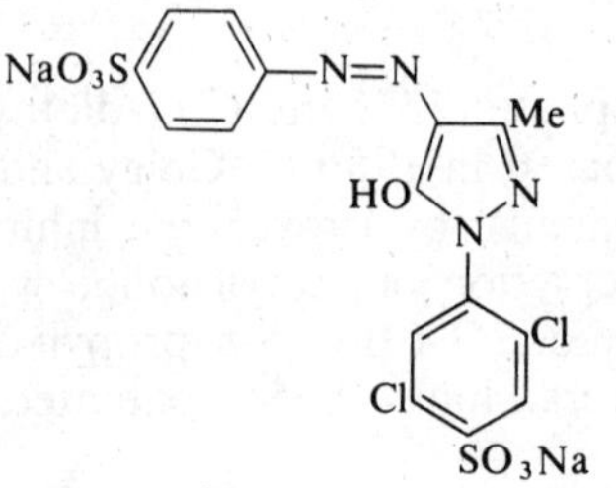

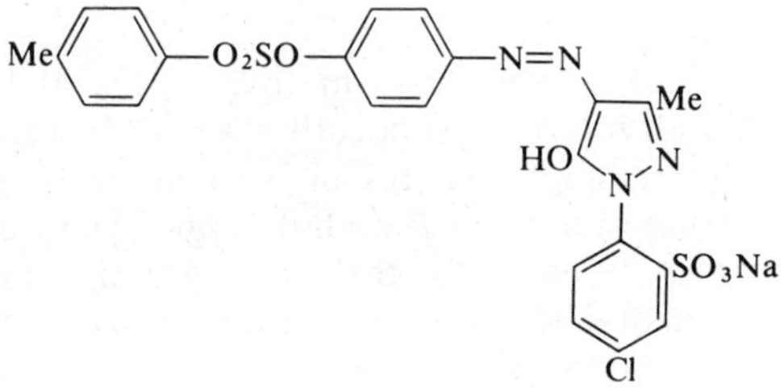

Hagenbach used 5-hydroxy-3-methyl-1-phenylpyrazole as a coupling component and made Eriochrome Red B (**732**), an important chrome dye ⟨B-70MI40403⟩.

(**732**)

(**733**)

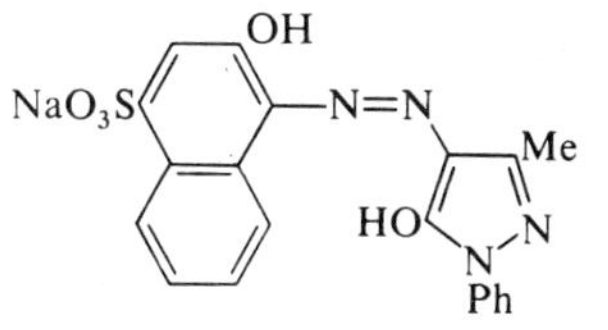

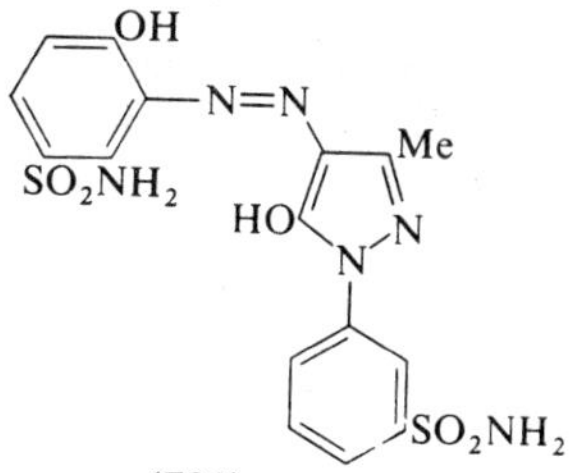

The use of prechromed complexes instead of chroming the dye on the fiber is advantageous; one of the more frequently used ones is Neolan Orange R (**733**) ⟨B-70MI40403⟩.

Pigments derived from pyrazolones include Hansa Yellow R, 1-phenyl-3-methyl-4-[(2,5-dichlorophenyl)azo]-5-hydroxypyrazole and Pigment Chrome Yellow, 1-phenyl-3-methyl-4-[(*o*-tolylazo]-5-hydroxypyrazole, as examples of monoazo compounds.

Vulcan Fast Red GF (**734**), a bisazopyrazolone pigment, has been developed primarily for the coloration of rubber ⟨B-70MI40403⟩.

(**734**)

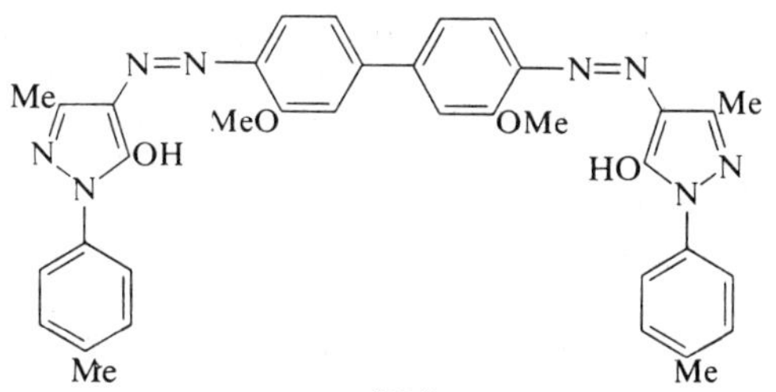

Anthra[1,9]pyrazol-6(2*H*)-one (**735**) dyes have also been described in the literature ⟨B-70MI40403⟩. These compounds result by acid ring closure of 1-anthraquinonyl hydrazine or its *N*-sulfonic acid.

3-Methyl-1-phenyl-3-pyrazolin-5-one gives a green-black dye (**736**) with 4-nitroso- or 4-amino-dimethylaniline and silver chlorides in the presence of light, a process of great importance in colour photography ⟨B-76MI40403⟩.

Other pyrazole derivatives (**737**) suitable as coupling components for diazotype printing have been prepared by Pett *et al.* ⟨79MI40406⟩.

(**735**)

(**736**)

$R = SO_3H, SO_3^-, H_2NMe_2^+$

(**737**)

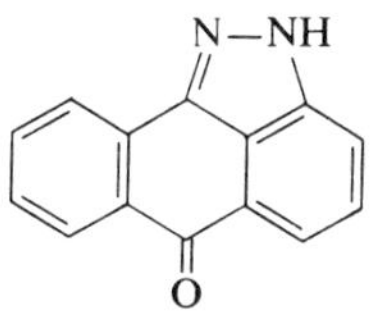

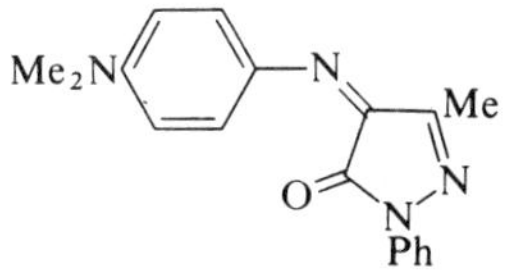

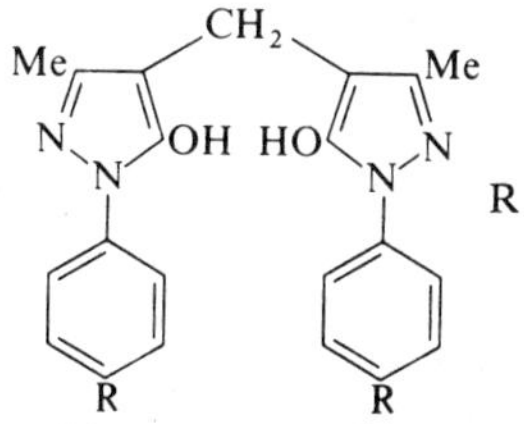

Phenidone, 1-phenyl-3-pyrazolidone (**738**) ⟨B-76MI40404⟩, has been used as a non-staining, high contrast photographic developer. New optical brighteners containing 2-pyrazolines (**739**) and pyrazoles (**740**) have been synthesized recently and their properties and applications reviewed ⟨75AG(E)665⟩.

(**738**)

(**739**)

(**740**)

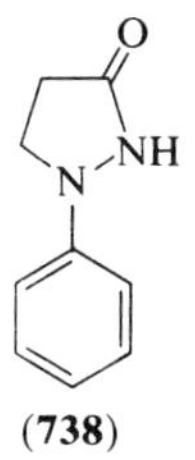

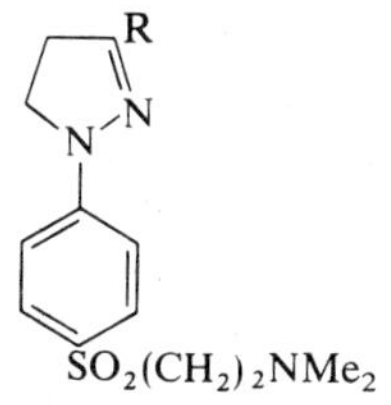

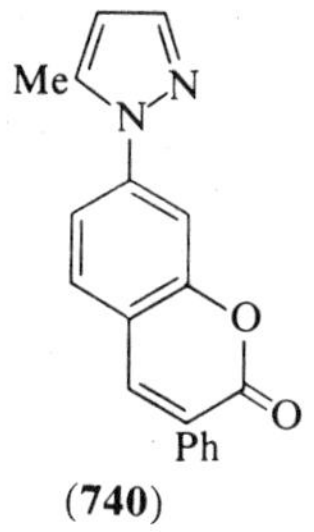

Optical brighteners are colourless fluorescent dyestuffs that absorb in the near UV (350–390 nm) and fluoresce in the violet to blue region of the spectrum (425–445 nm).

Analytical applications of pyrazolones have been reviewed by Busev *et al.* ⟨65RCR237⟩. Organic bases are easily characterized by formation of highly crystalline salts with picrolonic acid (1-(4-nitrophenyl-3-methyl-4-nitro-5-hydroxypyrazole). The last-named compound is used as a reagent for alkaloids, tryptophan, phenylalanine and for the detection and estimation of calcium ⟨B-76MI40404⟩.

4.04.4.1.4 Plastics

Polymers with a backbone of five-membered heterocyclic rings have been developed in the new area of thermally stable materials during the last 10 years ⟨B-80MI40408⟩. The simple polypyrazole (**741**) is prepared by condensation of polydiethynylbenzene with hydrazine in pyridine with yields of 60–97%.

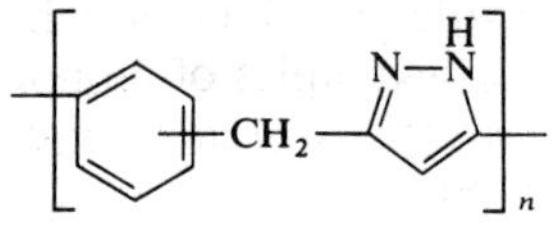
H
N—N
CH2
n

(**741**)

N-Substituted polypyrazoles can also be obtained by using *N*-alkylhydrazines, and it should be noted that these polymers consist of a random mixture of head-to-head and head-to-tail structures. Other syntheses of polypyrazoles have been described in the literature. Thus polyphenylene pyrazoles (**742**) and (**743**) occurred when *m*- or *p*-diethynylbenzene (DEB) reacted with 1,3-dipoles such as sydnones or bis(nitrilimines) (Scheme 64).

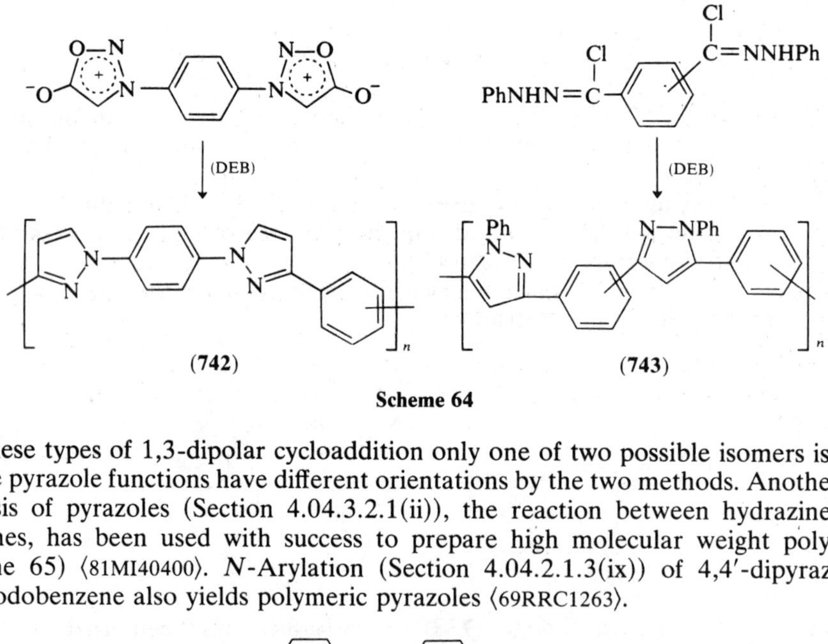
O—N
N—O
(DEB)
Cl
Cl
C=NNHPh
PhNHN=C
(DEB)
Ph
N—NPh
n
(742)
(743)

Scheme 64

In these types of 1,3-dipolar cycloaddition only one of two possible isomers is obtained and the pyrazole functions have different orientations by the two methods. Another classical synthesis of pyrazoles (Section 4.04.3.2.1(ii)), the reaction between hydrazines and β-diketones, has been used with success to prepare high molecular weight polypyrazoles (Scheme 65) ⟨81MI40400⟩. *N*-Arylation (Section 4.04.2.1.3(ix)) of 4,4′-dipyrazolyl with 1,4-diiodobenzene also yields polymeric pyrazoles ⟨69RRC1263⟩.

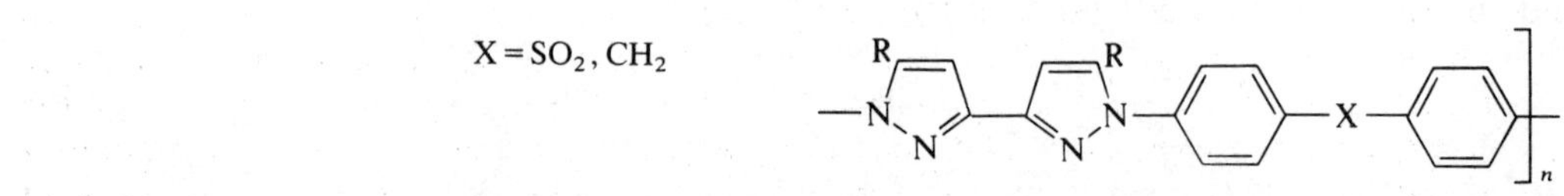
RCOCH2COCOCH2COR + H2NNH
X
NHNH2
X = SO2, CH2
R
R
X
n

Scheme 65

Tetrazoles (**744**), as bis(nitrilimine) (**745**) generators (Section 4.04.3.1.2(ii)), afford polypyrazoles when reacted with diynes. Benzoquinone has also been condensed with bis-sydnones to incorporate a fused pyrazole nucleus (**746**).

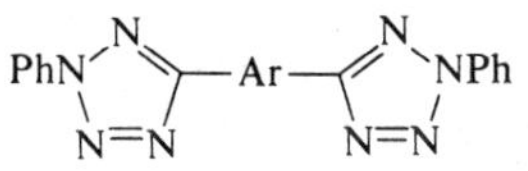

(744)    (745) 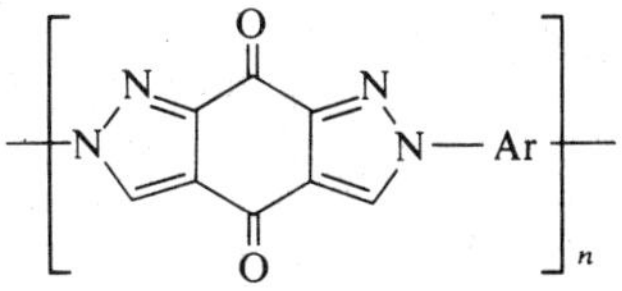   (746)

Finally, using divinyl compounds instead of diynes, surprisingly stable polypyrazolines (**747**) can be obtained.

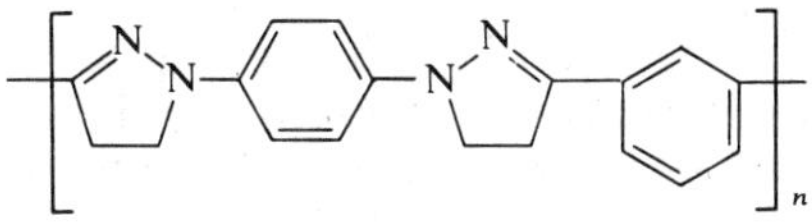

(747)

Properties of ferrocene-containing polymers have been improved by inclusion of pyrazole systems in the backbone. The synthesis of (**748**) was achieved by condensation of bis(β-diketoferrocenes) with aromatic dihydrazines to give polyhydrazones that were later cyclodehydrated ⟨B-80MI40408⟩.

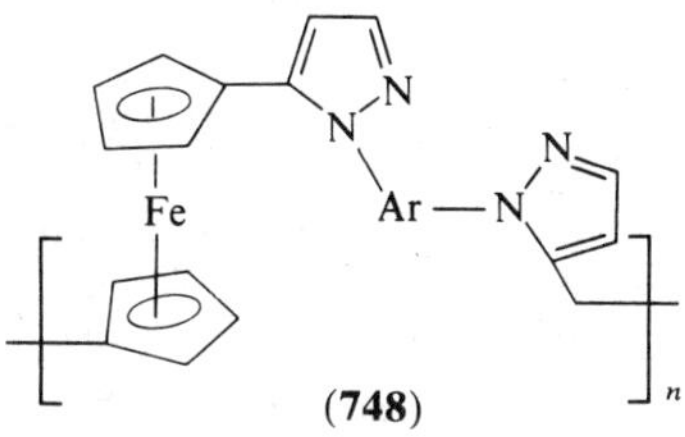

(748)

4.04.4.2 Metabolism and Toxicity

Drugs and other chemicals such as food additives or insecticides foreign to the body undergo enzymatic transformations that result in loss of pharmacological activity (*detoxification*), or lead to the formation of metabolites with therapeutic or toxic effects (*bioactivation*).

One example of bioactivation is found in active 3,5-pyrazolidinedione derivatives (Section 4.04.4.1.1). Thus, phenylbutazone is metabolized almost completely by liver microsomal enzymes to oxyphenbutazone and γ-hydroxyphenylbutazone, which have approximately the same activity ⟨B-80MI40406⟩. Detoxification of phenylbutazone and sulfinpyrazone (Section 4.04.4.1.1) in man occurs by formation of the corresponding *C*-glucuronide metabolites (**749**) and (**750**). The glucuronyl moiety with its ionized carboxyl group and polar hydroxyl groups imparts the water solubility required for ready excretion, and the glucuronides are pharmacologically inactive ⟨B-80MI40406⟩.

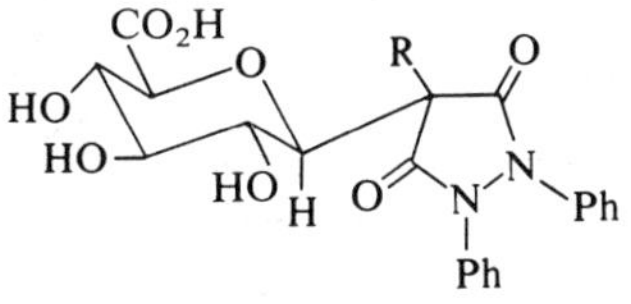

(**749**) R = Bu^n
(**750**) R = CH$_2$CH$_2$SOPh

Several other studies on metabolism of pyrazole derivatives with biological activity have been reported in the literature. 2-Norantipyrine, 4-hydroxyantipyrine and 3-hydroxymethylantipyrine were identified as metabolites of antipyrine in the rat by action of three hepatic monooxygenases ⟨80MI40405⟩. The major metabolite of pyrazole in male rats was found to be 4-hydroxypyrazole ⟨77MI40401⟩. In another study ⟨77MI40400⟩ seven metabolites were identified, including hydroxylated and conjugated derivatives. Two metabolites were conjugated with a pentose sugar, perhaps indicating that pyrazole serves as a substrate in the salvage pathway of purines and pyrimidines, forming pyrazole ribosides.

The metabolic fate of 1-(4-methoxy-6-methyl-2-pyrimidinyl)-3-methyl-5-methoxypyrazole, mepirizole, has been investigated in rats, rabbits and man. Three metabolites, (**751**), (**752**) and (**753**), were identified in human urine ⟨76CPB804⟩.

(751)   (752)   (753)

In man, the metabolic pathways of mepirizole were distinct from those in experimental animals, since hydroxylation on each of the aromatic rings did not occur in man. Compound (752) was obtained by oxidation of the 3-methyl group to the carboxylic acid (a similar process occurs with 5-methylpyrazole-3-carboxylic acid, an active metabolite of 3,5-dimethylpyrazole). However, the carboxylic acid metabolite of mepirizole had no analgesic activity and did not decrease blood glucose.

Many azo dyes, such as tartrazine (Section 4.04.4.1.3), are susceptible to reduction by bacterial reductases in the intestinal flora. Azo reduction is believed to proceed through a hydrazo intermediate that undergoes subsequent reductive cleavage of the nitrogen–nitrogen bond to yield the arylamine derivatives ⟨B-80MI40406⟩.

Organophosphates and carbamates containing a pyrazole ring, useful as insecticides as discussed earlier (Section 4.04.4.1.2), are metabolized mainly through hydrolysis of the ester function ⟨B-80MI40406⟩.

As regards toxicity, pyrazole itself induced hyperplasia of the thyroid, hepatomegaly, atrophy of the testis, anemia and bone marrow depression in rats and mice ⟨72E1198⟩. The 4-methyl derivative is well tolerated and may be more useful than pyrazole for pharmacological and metabolic studies of inhibition of ethanol metabolism. It has been shown ⟨79MI40404⟩ that administration of pyrazole or ethanol to rats had only moderate effects on the liver, but combined treatment resulted in severe hepatotoxic effects with liver necrosis. The fact that pyrazole strongly intensified the toxic effects of ethanol is due to inhibition of the enzymes involved in alcohol oxidation (Section 4.04.4.1.1).

Several comments about the toxicity of pharmaceutical, agrochemical and azo dye pyrazoles have been made previously in Section 4.04.4.1. Some LD_{50} values from the literature are shown in Table 41 ⟨B-76MI40404⟩.

Table 41 50% Lethal Dose of Pyrazole Derivatives

Generic name	LD_{50} (mg kg^{-1})	Section
Difenamizole	103 i.v. in mice	4.04.4.1.1
Sulfaphenazole	5800 orally in mice	4.04.4.1.1
Mepirizole	970–1020 orally in mice	4.04.4.1.1
Antipyrine	1800 orally in rats	4.04.4.1.1
Aminopyrine	1700 orally in rats	4.04.4.1.1
Phenylbutazone	525 orally in mice	4.04.4.1.1
Kebuzone	650 orally in mice	4.04.4.1.1
Prenazone	1065 orally in mice	4.04.4.1.1
Bendazac	1105 orally in mice	4.04.4.1.1
Benzydamine	515 orally in mice	4.04.4.1.1
Phosphate[a]	4 orally in mice	4.04.4.1.2
Dimetilan	25 orally in rats	4.04.4.1.2

[a] O,O-Diethyl-O-(3-methyl-5-pyrazolyl) phosphate.

4.04.4.3 Natural Products

There are very few examples of naturally occurring pyrazoles. As indicated in the introduction to this chapter, compounds containing the N—N bond are rare in higher plants and the biosynthesis and metabolism of N—N bonds is still unknown. Withasomnine, 4-phenyl-1,5-trimethylenepyrazole (754), was isolated from the roots of Indian medicinal plants, *Withania somnifera Dun*, and its structure established by physical methods and total synthesis ⟨68TL5707, 82H(19)1223⟩.

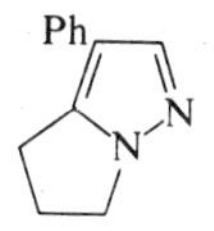

(754) 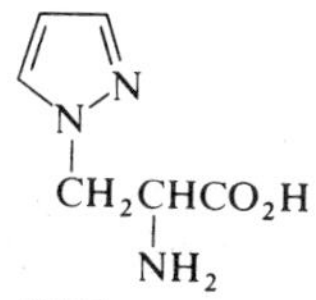 (755) 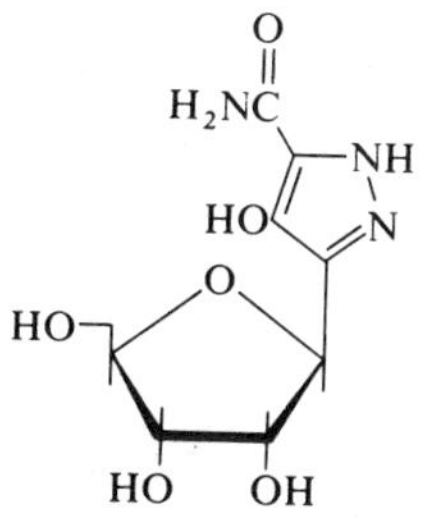 (756)

L-β-Pyrazolylalanine (**755**) and γ-L-glutamyl-β-pyrazol-L-alanine are found in the seeds of many species of *Cucurbitaceae* ⟨65P933⟩. It was subsequently demonstrated that free pyrazole occurs in watermelon seeds in concentrations of 280 to 410 μg g^{-1}, depending on the variety ⟨75P2512⟩. Biosynthetic studies implicate 1,3-diaminopropane as a precursor of the pyrazole moiety of (**755**) ⟨82P863⟩.

An azole *C*-nucleoside, 4-hydro-3-β-D-ribofuranosylpyrazolo-5-carboxamide, pyrazofurin or pyrazomycin (**756**), was isolated in 1969 from the fermentation broth of *Streptomyces candidus*. This antibiotic has broad spectrum antiviral activity and high cytotoxicity. It is now being evaluated as an antitumor agent and has been the object of various syntheses ⟨81JCS(P1)2374⟩. On standing in aqueous solutions the inactive α-anomer is formed. The antiviral activity of pyrazofurin was reversed by uridine and uridine 5′-monophosphate, and pyrazofurin 5′-monophosphate, an apparent metabolite, inhibits orotidylic acid decarboxylase. Activity against Friend leukemia virus infection in mice was shown by acyl derivatives of pyrazofurin ⟨B-80MI40406⟩. Syntheses of pyrazofurin analogues have been reported in the literature ⟨70HCA648⟩.

The influence of pyrazole derivatives on 'Prebiotic Condensation Reactions of α-Amino Acids' induced by polyphosphates in aqueous solution has been studied in comparison with other azoles ⟨81C59⟩. In the case of triglycine formation the presence of pyrazole did not increase the yield significantly.

4.05

Pyrazoles with Fused Six-membered Heterocyclic Rings

J. V. GREENHILL
University of Bradford

4.05.1 INTRODUCTION	305
4.05.2 STRUCTURE	306
4.05.2.1 Theoretical Methods	306
4.05.2.2 NMR Spectroscopy	308
4.05.2.3 UV Spectroscopy	308
4.05.2.4 Tautomerism	308
4.05.3 REACTIVITY	309
4.05.3.1 Electrophilic Attack at Nitrogen	309
4.05.3.2 Electrophilic Attack at Carbon	311
4.05.3.3 Nucleophilic Attack at Carbon	312
4.05.3.4 Reduction	313
4.05.3.5 Reactions of Substituents	315
4.05.4 SYNTHESIS	316
4.05.4.1 From Non-heterocyclic Compounds	316
4.05.4.2 Six-membered Ring Synthesis by Formation of One Bond	318
4.05.4.2.1 α to a heteroatom	318
4.05.4.2.2 γ to a heteroatom	321
4.05.4.3 Six-membered Ring Synthesis by Formation of Two Bonds	322
4.05.4.3.1 From [5+1] atom fragments	322
4.05.4.3.2 From [4+2] atom fragments	326
4.05.4.3.3 From [3+3] atom fragments	331
4.05.4.4 Five-membered Ring Synthesis by Formation of One Bond	334
4.05.4.4.1 Between the heteroatoms	334
4.05.4.4.2 α to a heteroatom	336
4.05.4.4.3 β to a heteroatom	337
4.05.4.5 Five-membered Ring Synthesis by Formation of Two Bonds	338
4.05.4.5.1 From [4+1] atom fragments	338
4.05.4.5.2 From [3+2] atom fragments	339
4.05.4.6 By Transformation of Existing Heterocycles	342
4.05.5 IMPORTANT COMPOUNDS	343

4.05.1 INTRODUCTION

There appears to be a fundamental law of heterocyclic chemistry that the greater the ratio of heteroatoms to carbon atoms the more likely the final compound in the synthetic scheme is to be formed by ring closure. When the references for this chapter were finally assembled, more than 80% dealt only with synthesis by cyclization methods. Much of this work has been stimulated by the discovery of potent antibiotic and antineoplastic compounds (Section 4.05.5).

Most of the compounds reported contain only nitrogen heteroatoms. Under modern rules of nomenclature they are properly called pyrazolopyridines, pyrazolopyrimidines, pyrazolotriazines, *etc.* The literature abounds with other, often more elegant terms; a few examples are given here. Pyrazolo[1,5-*a*]pyridine (**1**) may be called 1,7a-diazaindene,

3,4-diazaindene or 3-azaindolizine. Pyrazolo[1,5-*a*]pyrimidine (**2**) may be 3,8-diazaindolizine. Pyrazolo[3,4-*b*]pyridine (**3**) may be 1,2,7-triazaindene or 7-azaindazole. Pyrazolo[5,1-*c*][1,2,4]triazine (**4**) has been called 3,4,7,8-tetraazaindene or even pyrazolo[3,2-*c*]-*as*-triazine. The older literature uses pyrrocoline for indolizine.

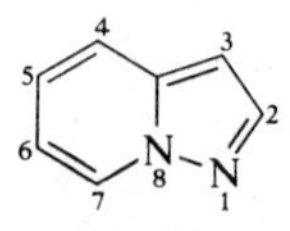

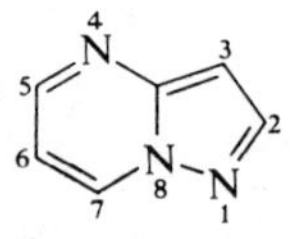

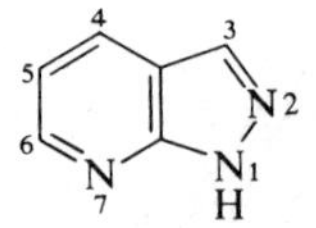

4.05.2 STRUCTURE

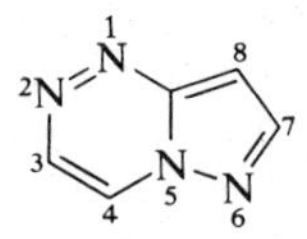

4.05.2.1 Theoretical Methods

The results of various electron density calculations for pyrazolo[1,5-*a*]pyridine are collected in Table 1. All successfully predict the observed electrophilic attack at C(3). The CNDO/2 values ([b]) lead to a theoretical dipole moment of 2.6 D and those of ([c]) to 2.5 D. These compare favourably with the experimental value (2.18 D). The values are also consistent with delocalization of the bridgehead nitrogen lone pair over the 10π-electron system.

Table 1 Electron Density Calculations for Pyrazolo[1,5-*a*]pyridines

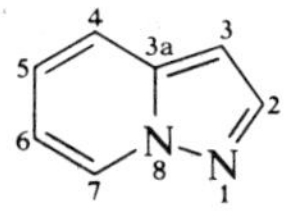

	N(1)	C(2)	C(3)	C(3a)	C(4)	C(5)	C(6)	C(7)	N(8)	Ref.
(i)	1.426	0.939	1.189	0.988	1.040	0.980	1.052	1.012	1.372	71JA1887[a]
(ii)	5.166	3.939	4.067	3.896	4.026	3.966	4.031	3.896	4.894	72CB388[b]
(iii)	5.191	3.853	4.162	3.856	4.059	3.933	4.040	3.902	5.034	74JHC223[b]
(iv)	5.043	3.824	4.124	3.803	4.046	3.898	4.016	3.877	4.928	74JHC223[c]
(v)	1.42	1.01	1.14	0.98	1.01	0.98	1.02	0.95	1.49	65JHC410[d]
(vi)	0.61	0.02	0.46	0.04	0.26	0.18	0.10	0.33	0.01	65JHC410[e]

[a] π-Electron densities, CNDO/2. [b] Total electron densities, CNDO/2. [c] Total electron densities, CNDO/2 for $\overset{+}{N}$(1)—H. [d] π-Electron densities, frontier electron density calculation. [e] Frontier electron densities.

Other available electron densities are shown in Scheme 1. Compound (**2**) shows CNDO/2 calculations ⟨75CJC119⟩, compounds (**3**) and (**5a**) show HMO calculations ⟨69CJC1129⟩, and (**5b**) and (**6**) show the results of LCAO calculations ⟨58JCS2973⟩.

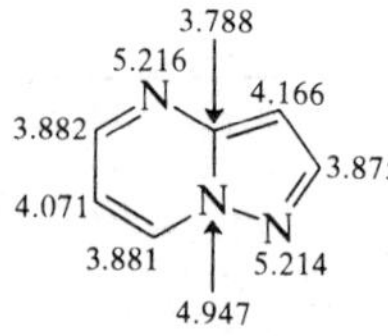

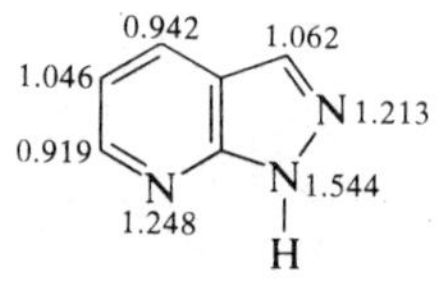

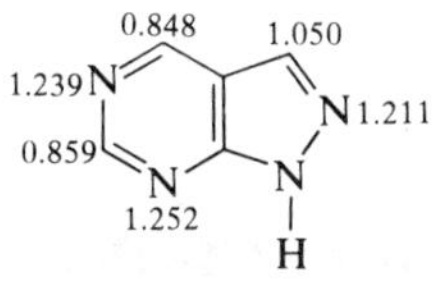

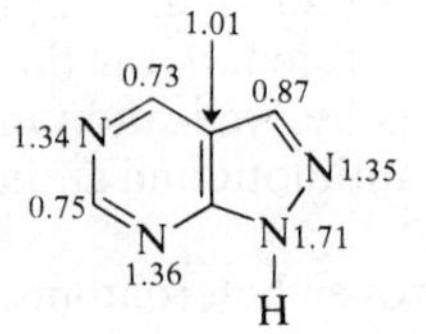

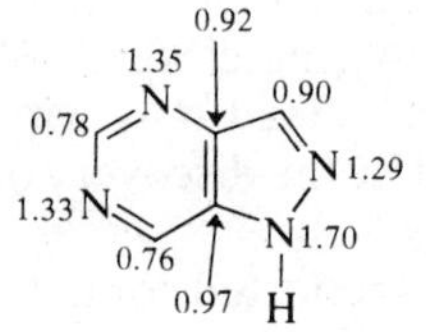

Scheme 1 Electron density calculations for other fused pyrazoles

Table 2 ^{1}H NMR Spectral Data for Fused-ring Pyrazole Derivatives

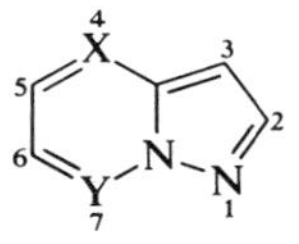

Ring atoms		Solvent	Chemical shifts (p.p.m.)						Coupling constants (Hz)								Ref.
X	Y		$\delta(2)$	$\delta(3)$	$\delta(4)$	$\delta(5)$	$\delta(6)$	$\delta(7)$	$J_{2,3}$	$J_{2,5}$	$J_{3,7}$	$J_{4,5}$	$J_{4,6}$	$J_{4,7}$	$J_{5,6}$	$J_{5,7}$	$J_{6,7}$	
CH	CH	CCl$_4$	7.80	6.38	7.44	6.97	6.62	8.39	2.18	0.5	0.9	8.94	1.17	1.02	6.79	1.00	6.93	64AJC1128
N	CH	CDCl$_3$	8.13	6.72	—	8.69	6.80	8.47	2.0	—	—	—	—	—	5.0	2.0	7.0	75CJC119
N	CH	TFA-d	8.75	7.32	—	9.88	7.72	9.20	3.0	—	—	—	—	—	5.0	2.0	7.0	75CJC119
CH	N	CDCl$_3$	8.02	6.62	7.95	6.95	8.25	—	3.0	—	—	10.0	2.0	—	4.8	—	—	74CC941

Table 3 UV Spectral Data for Fused-ring Pyrazole Derivatives

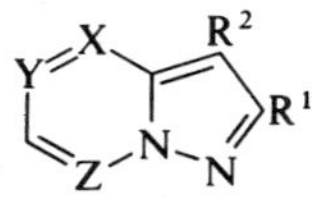

Ring atoms			Substituents		Solvent	λ_{max} (nm) ($\log \varepsilon$)	Ref.
X	Y	Z	R^1	R^2			
CH	CH	CH	H	H	H$_2$O	219 (4.59) 286 (3.65)	64JCS4226
CH	CH	CH	H	H	aq. H$_2$SO$_4$	214 (4.51) 292 (3.77)	64JCS4226
CH	CH	CH	Me	Me	C$_6$H$_{12}$	218 (4.42) 223 (4.56) 289 (3.52) 310 (3.25)	57JCS4510
CH	CH	CH	Me	COMe	MeOH	227 (4.54) 227 (4.68) 287 (3.80) 298 (3.83)	68JOC3766
N	CH	CH	H	H	MeOH	234 (4.65) 223 (4.63) 252 (4.08) 260 (4.10)	68JOC3766
CH	CH	CMe	H	H	EtOH	226 (4.66) 280 (3.30) 320 (3.20)	62CPB620
CH	CH	N	H	H	a	224 (4.57) 276 (3.08) 317 (3.15)	74CC941
N	N	CMe	H	H	a	226 (4.50) 230 (4.60) 287 (3.28) 350 (3.41)	66JCS(C)1127

^{a} Solvent not given.

4.05.2.2 NMR Spectroscopy

Many of the published ^{1}H NMR spectra have not been unequivocally assigned. Some that have are given in Table 2. An interesting comparison between a 1-alkyl- and a 2-alkyl-pyrazolo[4,3-c]pyridine is shown in Scheme 2 ⟨73T441⟩. The ^{13}C chemical shifts for pyrazolo[1,5-a]pyridine are shown in Scheme 3 ⟨81JHC1149, 74HCA873⟩ and ^{13}C chemical shifts (DMSO) for two pyrazolopyrimidinones (7) and (8) are noted in their formulae ⟨73JHC431⟩.

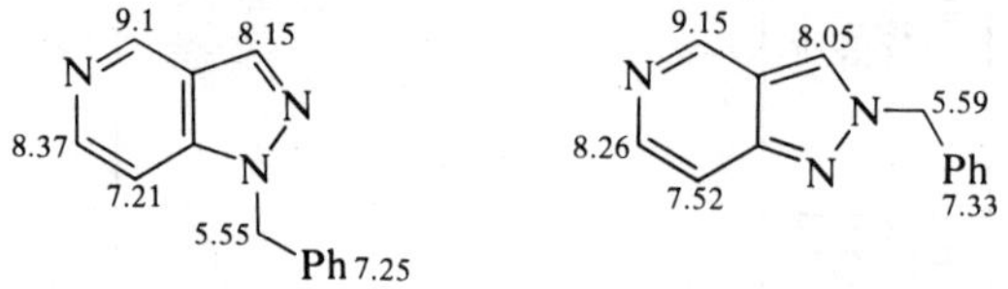

Scheme 2 ^{1}H NMR chemical shifts (δ, p.p.m., CDCl$_3$) for 1-alkyl- and 2-alkyl-pyrazolo[1,5-a]pyridines

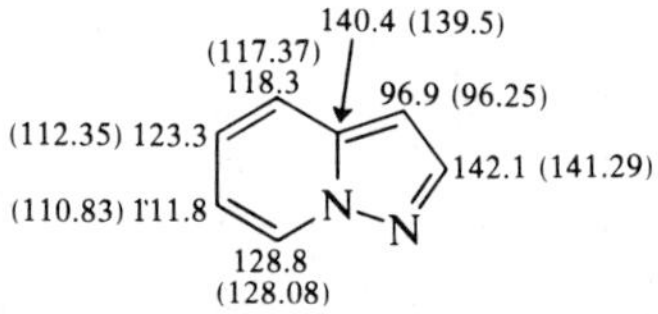

Scheme 3 ^{13}C NMR chemical shifts (p.p.m.) for pyrazolo[1,5-a]pyridine (1) in CDCl$_3$ (neat liquid)

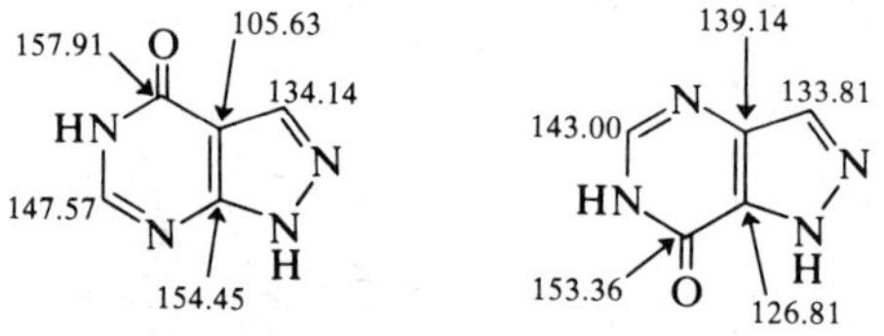

4.05.2.3 UV Spectroscopy

UV spectra in this series show typical aromatic, multipeak, envelopes. Some absorption maxima for simple b-fused pyrazoles are shown in Table 3.

4.05.2.4 Tautomerism

Available examples of compounds carrying oxo substituents in the six-membered ring all have the carbonyl group α and γ to a pyridine-type nitrogen and, as expected, strongly favour the keto form. Scheme 4 gives some typical spectroscopic results in support of this generalization; none of the compounds shows a hydroxy band in the IR spectrum ⟨70CB3252⟩ (but see Scheme 45, Section 4.05.4.2.2 for a special case).

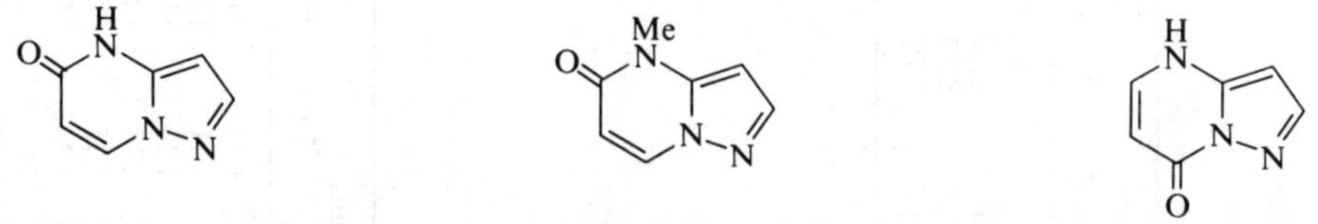

λ_{max} (MeOH) 230 nm (log ε 4.30), 268 (3.81)
IR (KBr) 1738, 1672, 1578 cm^{-1}

λ_{max} (MeOH) 210 nm (log ε 4.37), 215 (4.27), 250 (3.57), 271 (3.64)
IR (KBr) 1678 cm^{-1}

λ_{max} (MeOH) 211 nm (log ε 4.39), 257 (3.88), 297 (3.79)
IR (KBr) 1682, 1624, 1583 cm^{-1}

Scheme 4 UV and IR absorptions for pyrazolo[1,5-a]pyrimidines

Allopurinol (9; X = O) has an IR absorption (solid) at 1695 cm^{-1} (C=O) and thiopurinol (9; X = S) at 1195 cm^{-1} (C=S), both being strong bands with no OH or SH absorption. Similarly (10; X = O) shows 1680 and 1210 cm^{-1} bands, and when X = S, absorptions occur at 1210 and 1200 cm^{-1} ⟨69CR(C)(268)1798⟩.

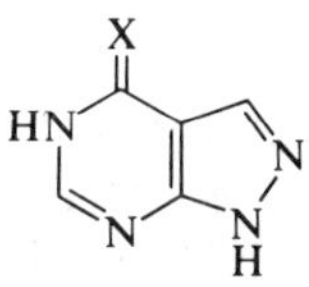

Strong carbonyl bands are shown by the pyrazolopyridazines (**11**) and (**12**) at 1670 and 1680 cm^{-1}, respectively. The IR spectra are little changed by methylation at N(5) ⟨73T435⟩.

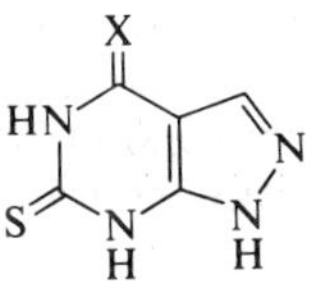

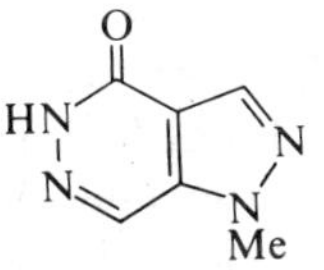

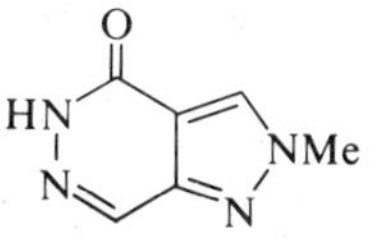

2-Hydroxypyrazolo[1,5-*a*]pyridine (**13**) has been examined carefully and shows no evidence for a carbonyl tautomer. The IR spectrum (CHCl$_3$) has an OH band at 3000 cm^{-1} and a band assigned to C=N at 1635 cm^{-1}. The UV spectrum with λ_{max} (MeOH) at 232 (log ε 4.59), 281 (3.09) and 310 nm (3.10) is very similar to that of the *O*-methyl derivative (**14**) which shows λ_{max} at 234 (log ε 4.57), 282 (3.30) and 307 nm (3.31). The *N*-methyl derivative (**15**) is clearly different, with an IR band at 1660 cm^{-1} (C=O), and λ_{max} (MeOH) at 242 (log ε 4.47), 281 (3.89), 288 (3.88) and 341 nm (3.32) (Scheme 5) ⟨76BCJ1980⟩.

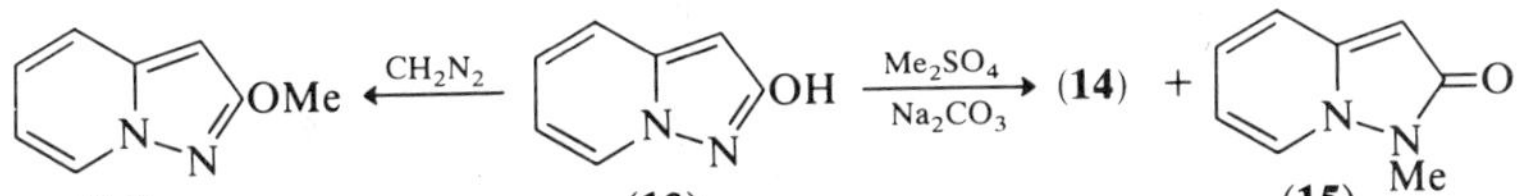

Scheme 5

3-Hydroxy derivatives of pyrazolo *c*-fused compounds give conflicting results. For example, there is evidence for the carbonyl form in 3-hydroxypyrazolo[4,3-*c*]pyridine but the ν(C=O) absorption seems to be absent in 1-substituted 3-hydroxypyrazolo-[3,4-*b*]pyridines and in pyrazolo[3,4-*d*]pyrimidines ⟨B-76MI40500⟩.

4.05.3 REACTIVITY

Electrophilic attack at carbon always occurs at C(3) of the pyrazole ring but if a bridgehead nitrogen is present, a second attack on the six-membered ring may follow almost as rapidly. A bridgehead nitrogen apparently donates its lone pair to complete a 10π-aromatic system so it is not surprising that both rings are activated towards electrophiles.

The conditions used for nucleophilic attack on pyridine and the diazines are probably too severe and would cause decomposition of the fused systems considered here. Further, the *b*-fused pyrazoles would clearly be deactivated to nucleophilic attack and in the *c*-fused pyrazoles, electron withdrawal by the pyrazole ring would make attack on the six-membered ring difficult. However, displacement of a suitably positioned halide or other good leaving group is often possible.

4.05.3.1 Electrophilic Attack at Nitrogen

The 10π-electron system (**1**) is very weakly basic (pK_a 1.43), but in strong acid it protonates at N(1) to give the 10$\pi \leftrightarrow 6\pi$ resonance system (Scheme 6) ⟨71JOC3087⟩. However, the UV spectrum is little changed so presumably the 10π-system makes the major contribution ⟨64JCS4226⟩.

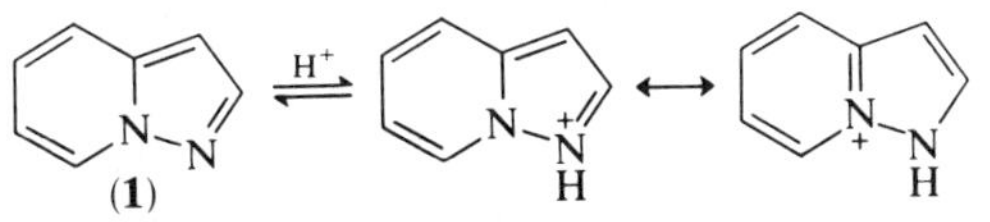

Scheme 6

The pyrazolopyrimidine (**16**; R=H) has pK_a 2.84 and its 7-amino derivative (**16**; R=NH$_2$) has pK_a 5.00 ⟨58JCS2973⟩.

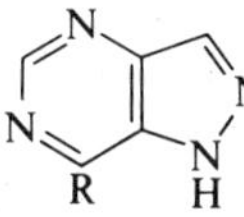

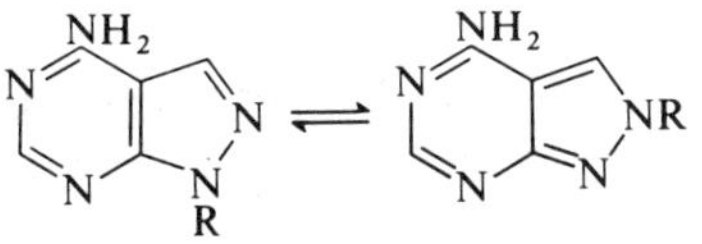

4-Aminopyrazolo[3,4-d]pyrimidine is a tautomeric system with a calculated (from temperature jump ^{13}C NMR relaxation studies) ΔH(tautomerism) of 3.8 kJ mol^{-1} at 10 °C giving K (18; R = H)/(17; R = H) of 0.1. The interconversion is catalyzed by acid (via the common cation) or base (via the anion). The parent system has pK_a 4.61 (for anion formation, pK_a 11.03) and the derivatives (17; R = Me) pK_a 4.30 and (18; R = Me) pK_a 5.11. The sites of protonation have been investigated for the isopropyl derivatives by ^{13}C NMR; compound (17; R = Pr^i) protonates at N(5) but (18; R = Pr^i) protonates equally at N(5) and N(7) ⟨77JA7257⟩. 4-Aminopyrimidine has pK_a 5.7, showing that the pyrazole ring of (17) and (18) exerts, as expected, a net electron-withdrawing effect.

The pyrazolopyrimidinone (19) gives the product of N(4)-methylation with dimethyl sulfate and sodium hydroxide, but diazomethane produces a mixture of N(1)- and O-methyl derivatives (Scheme 7) ⟨78FES14⟩.

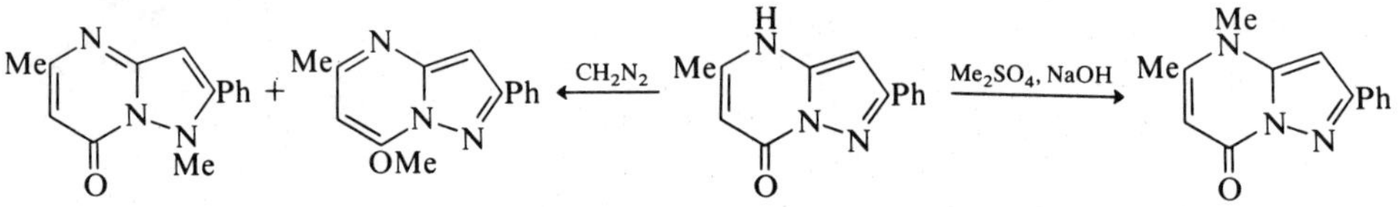

Scheme 7

The 3-nitropyrazolopyridine (20; R = H) is alkylated at N(1) by methyl iodide to give (20; R = Me) ⟨73JCS(P1)2901⟩, but the 7-aminopyrazolopyrimidine (21) alkylates in the six-membered ring while acylation followed by reduction provides a route to the secondary amine (22; Scheme 8) ⟨65CPB142⟩. It is interesting that one of the three entering methyl groups when (23) is treated with excess dimethyl sulfate goes to N(2) rather than N(1) (Scheme 9) ⟨58HCA1052⟩. With the pyrazolo[4,5-d]pyrimidine (24; R = H), dimethyl sulfate gives a mixture of products (25) and (26), but for (24; R = Me) only compound (25) is produced (Scheme 10) ⟨65JOC199⟩ (see Section 4.05.4.3.1 and Scheme 5, Section 4.05.2.4).

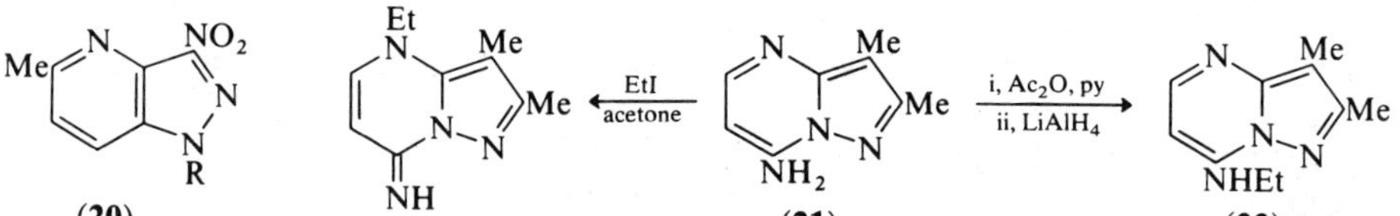

Scheme 8

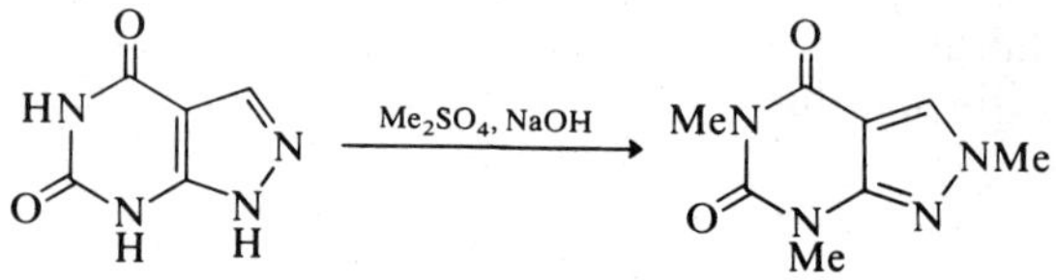

Scheme 9

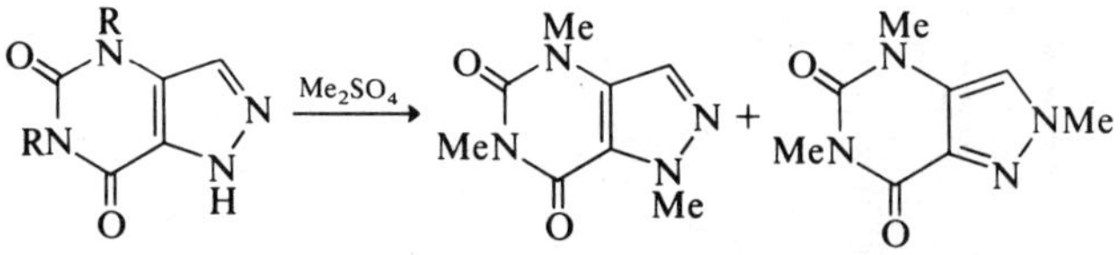

Scheme 10

The useful intermediate (29) is formed when the pyrazolopyrimidine (27) is treated with the hydroxylamine (28) at room temperature (see Section 4.05.3.5) ⟨81JCS(P1)2387⟩.

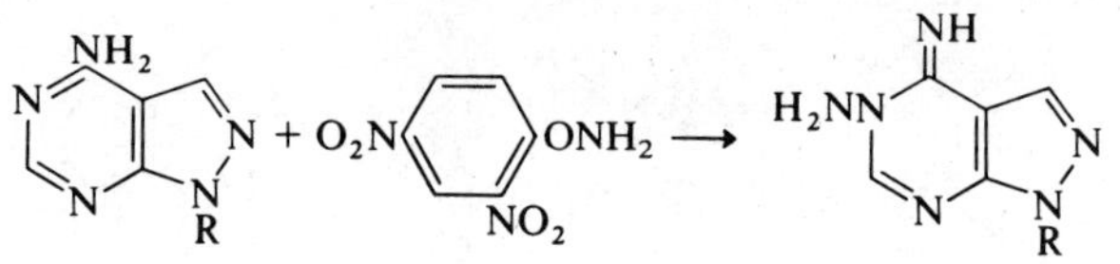

(27) R = protected ribosyl   (28)   (29)

Acylation of a pyrazolopyridine has been shown to be temperature dependent. In the cold, benzoyl chloride and pyridine give the 2-acyl derivative (30), but at reflux the same mixture gives the 1-acyl compound (31; Scheme 11) ⟨73JCS(P1)2901⟩.

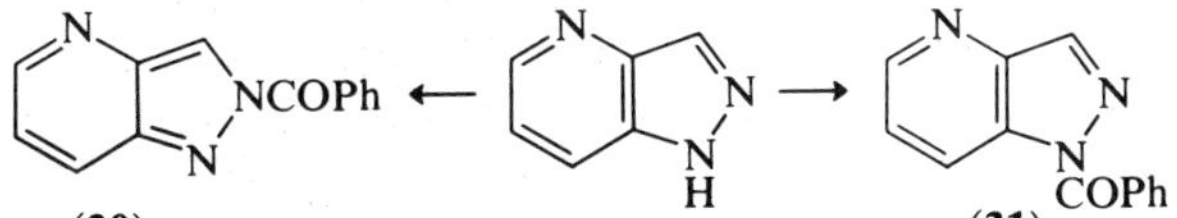

Scheme 11

4.05.3.2 Electrophilic Attack at Carbon

Generally, C(3) of the five-membered ring is the position of electrophilic attack, but the rate of reaction and the possibility of a second substitution in the six-membered ring depend on the fusion system. The *b*-fused pyrazolopyridine (1) is readily nitrated at C(3) at ice bath temperature, but a dinitro derivative (33) is produced at room temperature. Use of fuming nitric acid in acetic anhydride gives a mixture of two C(3) derivatives (Scheme 12). Alternatively, nitrous acid converts (1) to the 3-nitroso compound which may be oxidized to (32) with peracetic acid ⟨74JHC223⟩. Derivatives of (1) carrying electron-donating groups at C(2) substitute very readily at C(3). For example, (34; R = Me) gives the 3-acetyl derivative with acetyl chloride and pyridine at room temperature ⟨68JOC3766⟩. Compound (34; R = OH) gives 3-nitroso (with HNO_2) and 3-nitro (with fuming HNO_3 at 0 °C) derivatives and, with 1 mole of bromine (in acetic acid–sodium acetate), the 3-bromo compound. With a second mole of bromine the unusual 1,3-dibromo derivative (35) is obtained. The hydroxy group of (34; R = OH) shows pyrid-2-one character in being replaced by chlorine with phosphorus oxychloride to give (34; R = Cl) ⟨76BCJ1980⟩.

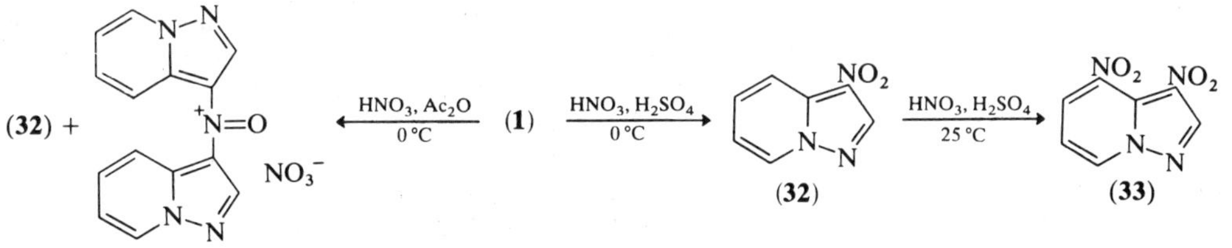

Scheme 12

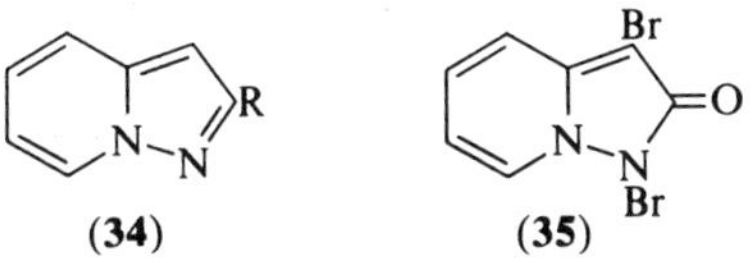

Pyrazolo[1,5-*a*]pyrimidine (2) gives the 3-bromo, followed by the 3,6-dibromo, derivatives. Several attempts failed to find conditions to give the 6-monobromo compound. 'Normal' nitration conditions gave the 3-nitro derivative (36) but nitric acid in acetic anhydride gave exclusively the 6-nitro product (38). It has been suggested that this results from rapid formation of (37), followed by loss of acetic acid on work-up (Scheme 13) ⟨75CJC119⟩. The 5,7-dimethyl derivative (39) of (12) gives exclusively 3-substitution with electrophiles. A nitration mixture at 10 °C gives (40; E = NO_2). *N*-Bromosuccinimide and

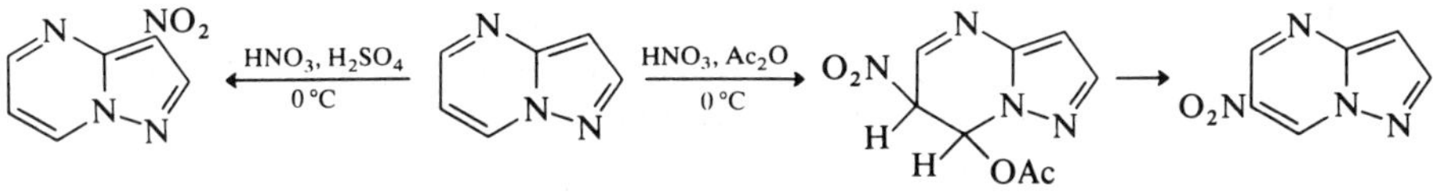

Scheme 13

N-chlorosuccinimide give (**40**; E = Br, Cl) and iodine monochloride gives (**40**; E = I). Friedel–Crafts acylation (acetyl chloride–tin(IV) chloride–dichloromethane) gives (**40**; E = Ac) and the Mannich reaction gives (**40**; E = CH$_2$NMe$_2$) in acetic acid at room temperature ⟨74JMC645⟩. The pyrazolo[1,5-*c*]pyrimidine (**41**) gives first the 3-bromo then the 3,4-dibromo derivatives ⟨72CB388⟩. The pyrazolotriazine (**42**) brominates in the five-membered ring at C(8) on brief boiling with *N*-bromosuccinimide in chloroform ⟨74JHC199⟩.

4.05.3.3 Nucleophilic Attack at Carbon

Despite the significant amount of work that has been done on pyrazolopyridines and pyrazolopyrimidines, there seems to be no report of direct nucleophilic substitution on these systems. The pyrazolotriazine (**43**) reacts with 0.5M sodium hydroxide solution at room temperature. One carbon is removed to give a useful intermediate (**44**; Scheme 14) ⟨74JHC199⟩.

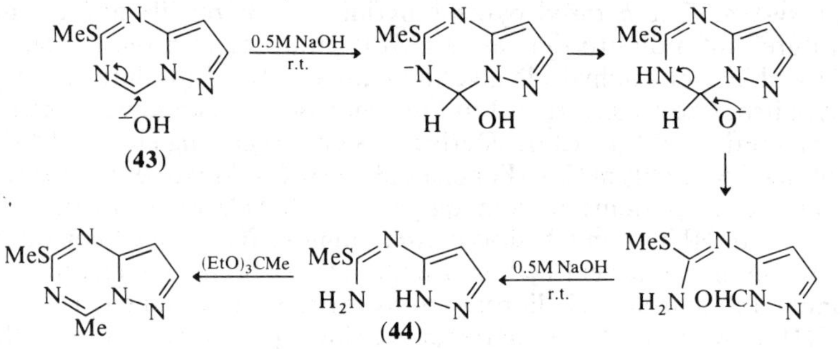

Scheme 14

When the six-membered ring is strongly activated, *e.g.* by a 6-nitro substituent, nucleophilic attack by alcoholic alkali may lead to rearrangement as in Scheme 15 ⟨81T3423⟩. These reactions are analogous to those occurring in the corresponding 1,2,4-triazolo-pyridine and -pyrimidine systems.

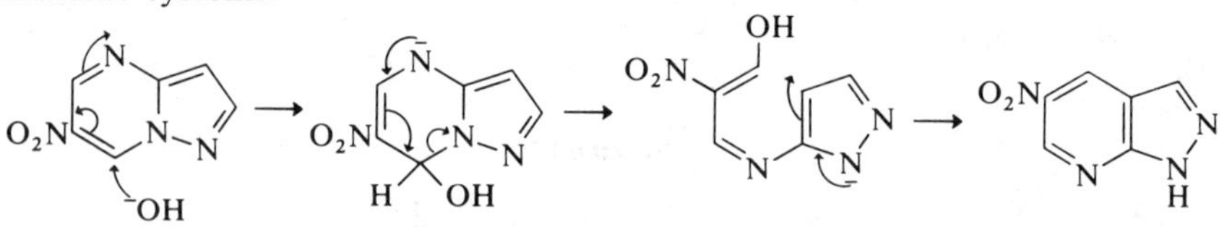

Scheme 15

Many reports of nucleophilic displacement of halogens and methylthio groups have appeared. A few examples will serve to summarize this work.

Scheme 16 illustrates the ease of displacement of chlorine from the C(4) position of (**45**). In the pyrazolopyrimidine derivative, disubstitution can generally be achieved by the use of higher temperatures and longer reaction times. Only dimethylamine has been found to disubstitute too rapidly for the monosubstitution product to be isolated. For the products (**46**; R = OMe or SMe), aqueous methylamine at 80 °C gives (**46**; R = NHMe) without replacing the chlorine atom, but with (**46**; R = SH) the chlorine rather than the thiol group is replaced. Similarly for (**46**; R = OH) the chlorine atom is replaced by ethanolic ammonia at 200 °C. In contrast to the dichloro compound (**45**), the methoxymethylthio derivative (**47**) or the dimethylthio compound gives the 4-dimethylamino derivative (**48**) without displacing the 6-methylthio group. These nucleophilic reactions probably take place on the anions of the pyrazolopyrimidines. If the 1-methyl derivative of (**45**) is employed, the C(6)

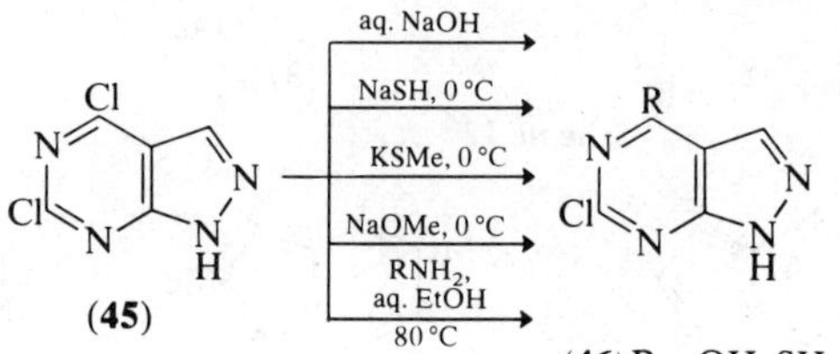

(**46**) R = OH, SH, SMe, OMe, NHR

Scheme 16

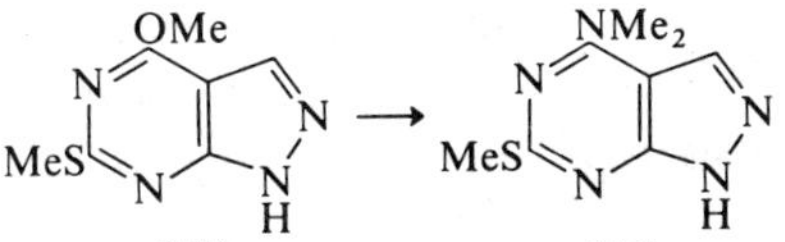

chlorine atom is much more readily displaced ⟨57JA6407⟩. Further examples of nucleophilic attack at C(4) are shown in Scheme 17 ⟨81JCS(P1)2387⟩.

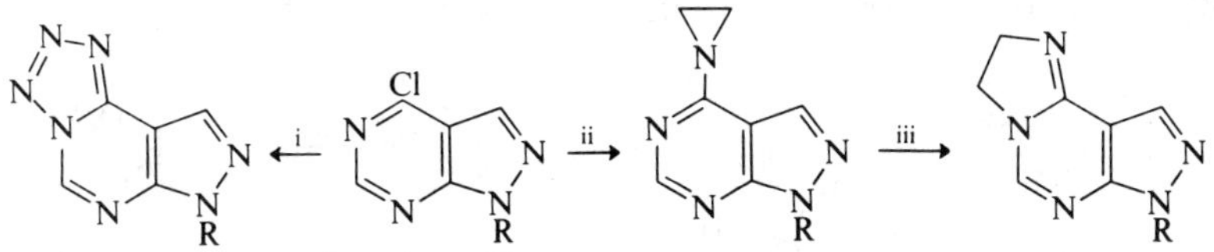

i, LiN₃, acetone, 25 °C; ii, ethyleneimine; iii, NaI, acetone, reflux

Scheme 17

Difficulties encountered in nucleophilic displacement at the C(6) position of pyrazolo-[3,4-d]pyrimidines are illustrated in Scheme 18. Compound (**49**; R = H) shows no carbonyl absorption in its IR spectrum. Electron release from its hydroxy group deactivates the pyrimidine ring so that the methylthio group is not displaced by any of the common nucleophiles. Oxidation with chlorine water gives the sulfone, a much better leaving group, which is displaced by ammonia to give (**51**). Conversion to the 3-chloro compound (**50**) allows attack at C(6). The 3-chloro substituent slightly activates the pyrimidine ring. For (**50**; R = Me) the methylthio group can be replaced by OH (NaOH), OEt (NaOEt) or NH₂ (NaNH₂). Position C(3) is not attacked; the chlorine atom remains in position even after 24 h at 200 °C with ethanolic ammonia. Strong base converts (**50**; R = H) into its anion and no nucleophilic attack occurs ⟨61JOC451⟩.

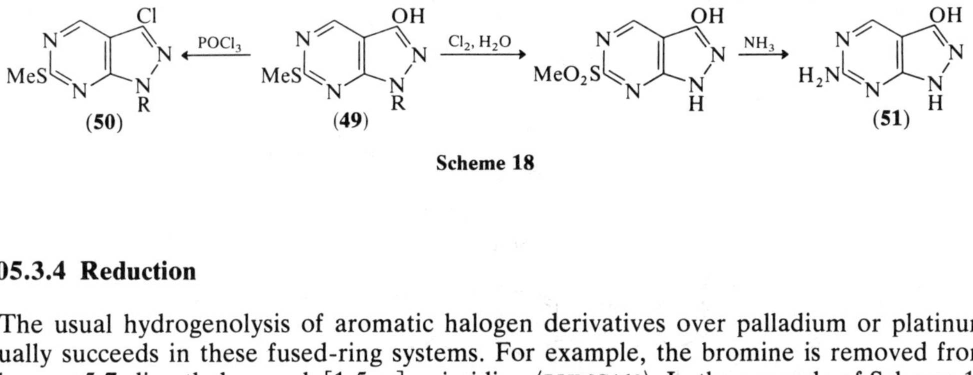

Scheme 18

4.05.3.4 Reduction

The usual hydrogenolysis of aromatic halogen derivatives over palladium or platinum usually succeeds in these fused-ring systems. For example, the bromine is removed from 3-bromo-5,7-dimethylpyrazolo[1,5-a]pyrimidine ⟨75JMC460⟩. In the example of Scheme 19 the halogen is lost after five minutes at room temperature. Within 2 h the six-membered ring is reduced as well ⟨65CPB1207⟩. Partial reduction of (**52**) with sodium borohydride gives (**53**) ⟨79FES751⟩. Reduction of a pyrazolopyridine is accompanied by decarboxylation (Scheme 20) ⟨75JHC517⟩.

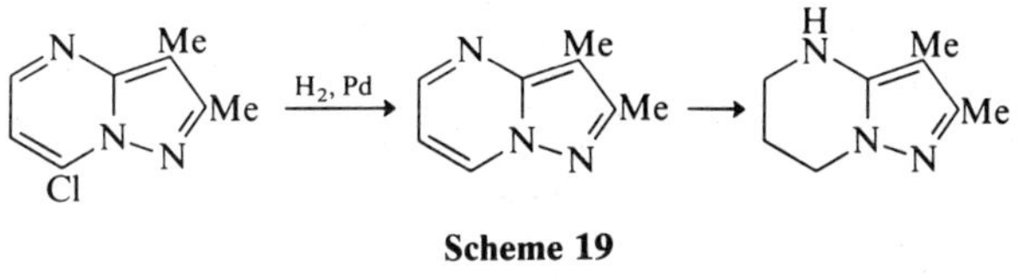

Scheme 19

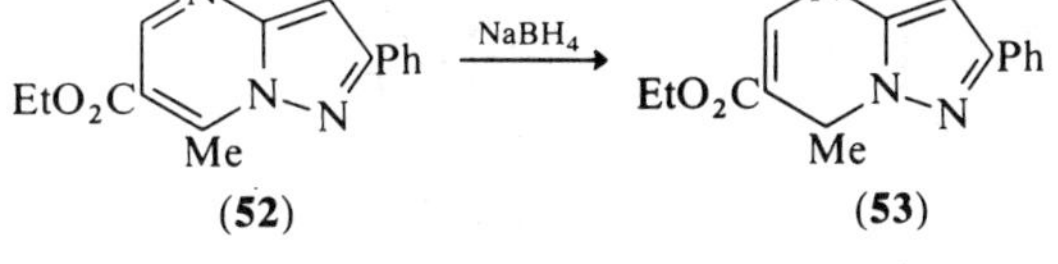

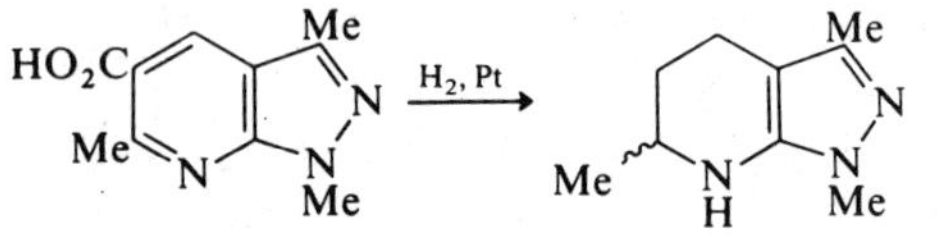

Scheme 20

The 3-hydroxypyrazolopyrazine (**54**) undergoes reductive ring-opening with Raney nickel. A lower reaction temperature suffices for its 1-methyl derivative (Scheme 21) ⟨58JA421⟩. This reaction often succeeds with cyclic hydrazides of this type.

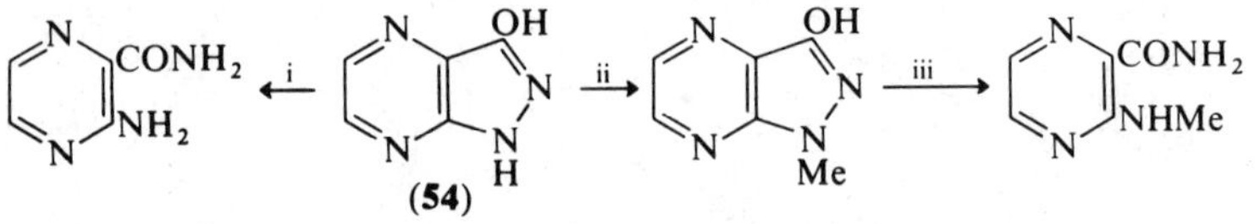

i, Raney Ni, NH$_2$CHO, 115–120 °C; ii, MeI, NaOH; iii, Raney Ni, EtOH, reflux

Scheme 21

The pyrazolo[1,2-*a*]pyridazine (**55**) reacts with sodium methoxide in boiling methanol, presumably *via* preliminary deprotonation at benzylic carbon. After 8 h a mixture of (**56**; 72%) and the Hofmann-type elimination product (**57**; 12%) is obtained, but after 25 h only (**57**; 75%) is isolated. The imino bond (ν 1640 cm^{-1}) of (**57**) may be saturated with sodium borohydride. With potassium *t*-butoxide in *t*-butyl alcohol at reflux, the rearranged isoindole (**58**) is obtained (Scheme 22) ⟨74CPB2142⟩. Scheme 23 shows a rearrangement similar to (**55**) → (**56**) for which a radical mechanism has been suggested. However, here, both radicals from N—N bond homolysis are resonance stabilized. Ring closure is to the thermodynamically more stable system (**59**) ⟨72CPB69⟩.

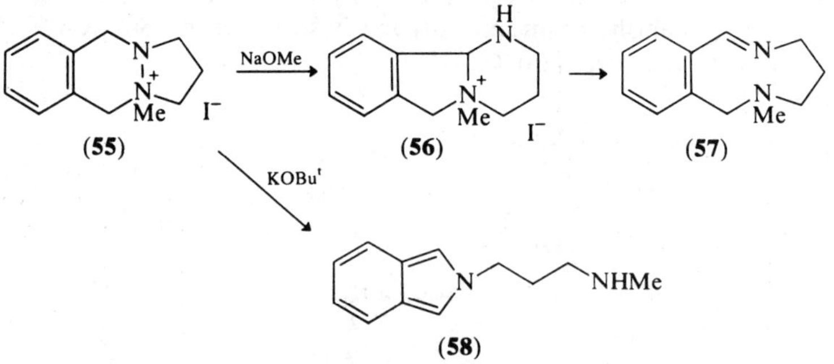

Scheme 22

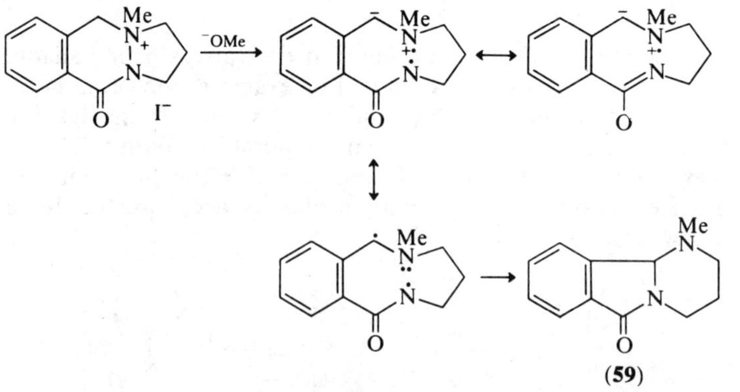

Scheme 23

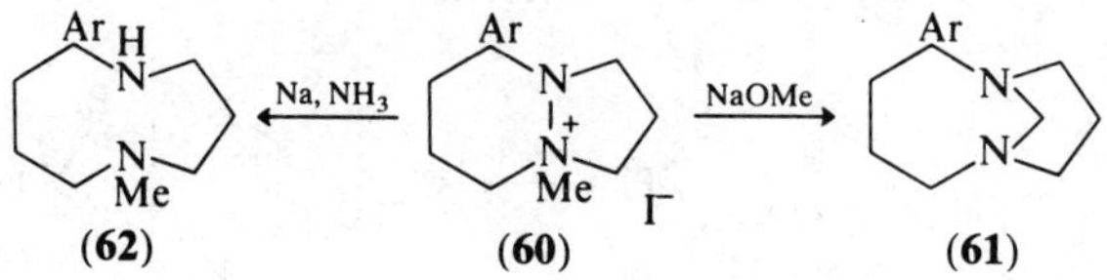

Scheme 24

Compound (**60**), which lacks only the benzo fusion of (**56**), follows a different reaction path with boiling sodium methoxide in methanol (Scheme 24). Deprotonation of the methyl group has been suggested as the first step in the production of the diazabicyclodecane (**61**). Reduction of (**60**) splits the N—N bond to give the diazacyclononane (**62**) ⟨69JOC2720⟩.

4.05.3.5 Reactions of Substituents

The hydrazino compound (**63**) (from nucleophilic displacement of chlorine with hydrazine) reacts with triethyl orthoformate under mild conditions to give the fused triazole (**64**). Heating in a neutral solvent gives the thermodynamically more stable, isomeric fused triazole (**65**) which can also be obtained from (**63**) directly. The 5-amino derivative (**29**; see Section 4.05.3.1) is readily converted into (**65**; Scheme 25) ⟨81JCS(P1)2387⟩.

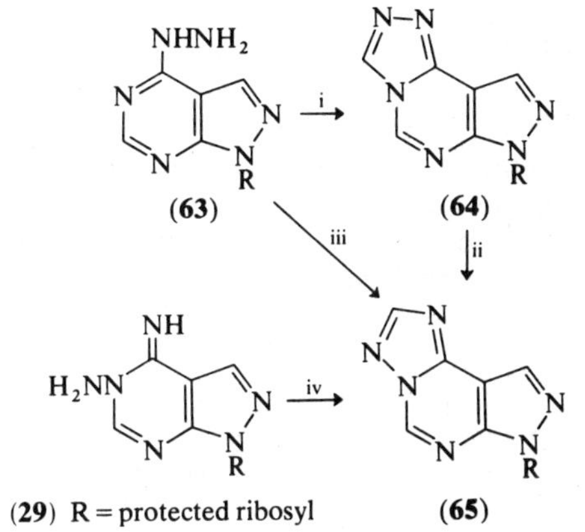

i, (EtO)$_3$CH, 70 °C, 45 min; ii, MeO(CH$_2$)$_2$OH, reflux; iii, (EtO)$_3$CH, reflux, 12 h or TsOH, 25 °C; iv, MeCO$_2$CH(OEt)$_2$

Scheme 25

Catalytic hydrogenation of 3-nitro or 3-nitroso ring-fused pyrazoles (see Section 4.05.3.2) gives the 3-amino derivatives which on diazotization often give stable diazo compounds. For example, compound (**66**) heated with hydrobromic acid gives the 3-bromopyrazolopyridine as expected (Sandmeyer reaction) ⟨73JCS(P1)2901⟩. However, photolysis in benzene produces the phenyl derivative (**67**), or treatment with 2,5-dimethylfuran in methanol gives, after rearrangement, the 3-pyrazolyl derivative (**68**). A salt of (**66**) rapidly undergoes a Diels–Alder reaction to give the 3-pyridazinyl salt (**69**; Scheme 26) ⟨78JHC1175⟩.

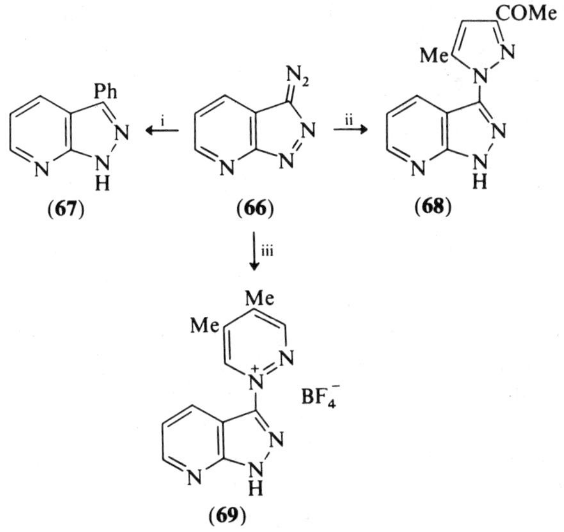

i, C$_6$H$_6$, $h\nu$; ii, 2,5-dimethylfuran, MeOH, 0 °C; iii, for BF$_4^-$ salt, 2,3-dimethylbutadiene, 0 °C

Scheme 26

The diazonium salt (**70**) is prepared from the amine with sodium nitrite in fluoroboric acid. Photolysis of the resulting solution gives the 3-fluoro derivative (**71**) ⟨77JMC386⟩.

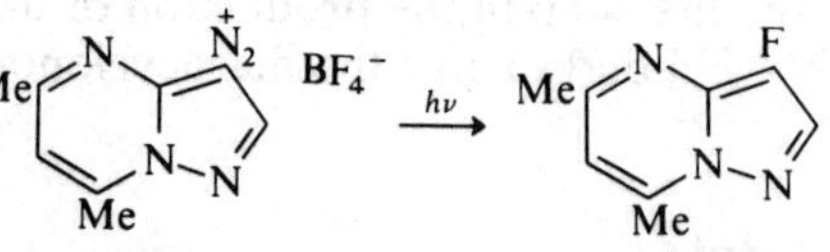

4.05.4 SYNTHESIS

4.05.4.1 From Non-heterocyclic Compounds

In neutral solution, cyanoacetohydrazine (**72**) reacts with acetylacetone to give the enaminone derivative (**73**) which, under carefully controlled alkaline conditions, cyclizes to the pyrazolopyrimidine (**74**). The hydrazine (**72**) also reacts with diketene to give (**75**; Scheme 27) ⟨61LA(647)116⟩. The aldehydo nitrile (**76**) and hydrazine hydrate in refluxing toluene give (**77**; 30–54% depending on conditions) along with a simple pyrazole ⟨71AP121⟩. The enol ether (**78**) and hydrazine give the hydrogen bonded bis-enaminone (**79**) which on heating collapses to the fused heterocycle (**80**). The reactive substituents are removed by standard methods to give the dimethylpyrazolopyrimidine (**81**; Scheme 28) ⟨70BCJ849⟩.

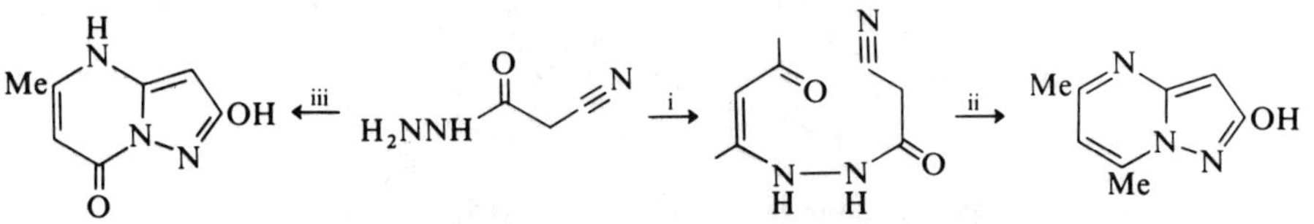

i, pentane-2,4-dione; ii, NaOH (2.4 eq.); iii, diketene

Scheme 27

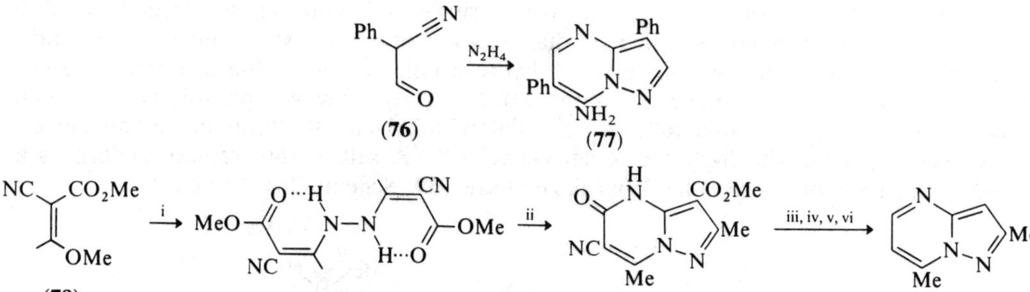

i, N_2H_4, EtOH, C_5H_5N, 25 °C; ii, EtOH, reflux; iii, aq. NaOH; iv, Δ; v, $POCl_3$; vi, H_2, catalyst

Scheme 28

Important syntheses of allopurinol (**85**; $R^2 = H$) and oxyallopurinol (**87**; $R^2 = H$), both used in the treatment of gout (see Section 4.05.5), are shown in Scheme 29. Cyanoacetic acid and formamide react in hot acetic anhydride to give (**82**; $R^1 = H$) which is treated, without isolation, with triethyl orthoformate to give the enol ether (**83**). This is converted into the hydrazine (**84**). Heating (**84**; $R^1 = H$, OEt; $R^2 = H$) for a few minutes gave the pyrazolopyrimidines (**85**) and (**87**) directly, but for (**84**; $R^2 = Me$, Ph) gentle heating during recrystallization from nitromethane gave the isolatable pyrazoles (**86**) which closed the second ring on stronger heating. A further elaboration of this general approach is shown in Scheme 30. The dithiocarbamate (**88**) can be converted directly into the pyrazolopyrimidine (**89**) or taken *via* the substituted pyrazole as before. The product (**89**) is hydrolyzed to (**87**; $R^2 = Ph$) by warm aqueous chloroacetic acid ⟨71JCS(C)1610⟩.

2,3,5-Tri-*O*-benzyl-D-ribose condenses with hydrazine in methanol at room temperature to give the derivative (**90**). This reacts as shown to give the protected nucleosides (**91**; Scheme 31) ⟨81S925⟩.

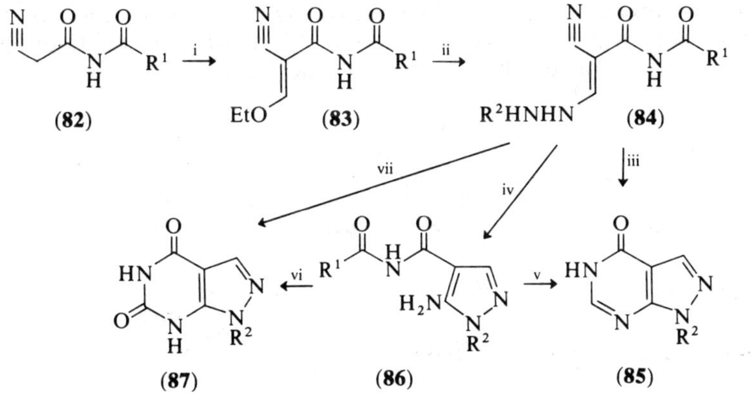

Pyrazoles with Fused Six-membered Heterocyclic Rings 317

i, (EtO)$_3$CH, Ac$_2$O; ii, R^{2}NHNH$_2$; iii, for R^1 = R^2 = H, 150 °C, 15 min; iv, for R^1 = H, R^2 = Ph; or R^1 = OEt, R^2 = Me, Ph, warm; v, for R^1 = H, R^2 = Ph, Δ; vi, for R^1 = OEt, R^2 = Me, Ph, Δ; vii, for R^1 = OEt, R^2 = H, 140 °C

Scheme 29

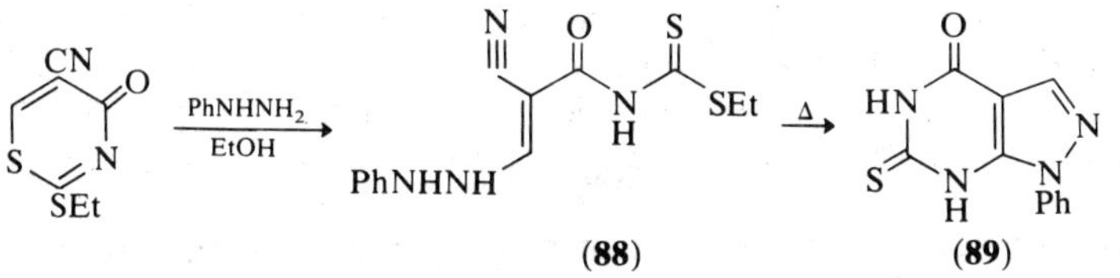

Scheme 30

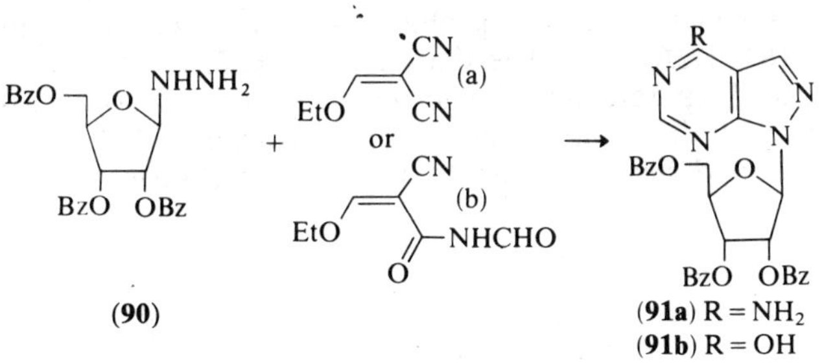

Scheme 31

The pyrazoloquinazolines (92) and (93) are available in high yield as shown in Scheme 32. The reactants are boiled with dilute hydrochloric acid ⟨69JHC947⟩.

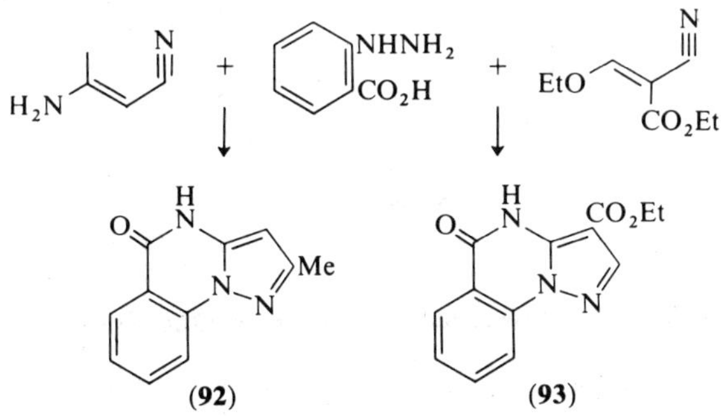

Scheme 32

Nitrilimines (94), which are reactive 1,3-dipoles, undergo [3+2] cycloadditions with alkenic and alkynic dipolarophiles. Their use in the preparation of benzopyrazolooxazines and benzopyranopyrazoles, often in high yields, is illustrated in Scheme 33. The reactive intermediates, *e.g.* (96), are generated by elimination of hydrogen chloride from suitable substrates (95) and (97) ⟨74TL269⟩, or by photolysis of tetrazoles (98) ⟨78JOC1664⟩. Pyrazoles

$$R^1-\bar{C}=\overset{+}{N}=N-R^2 \longleftrightarrow R^1-C\equiv\overset{+}{N}-\bar{N}-R^2 \longleftrightarrow R^1-\overset{+}{C}=N-\bar{N}-R^2$$

(94)

of type (**99**) can be prepared by mild oxidation (DDQ or chloranil ⟨77HCA3035⟩) or directly from the appropriate alkyne. These intramolecular 1,3-dipolar cycloadditions provide entry into fused-ring systems not readily available by other routes.

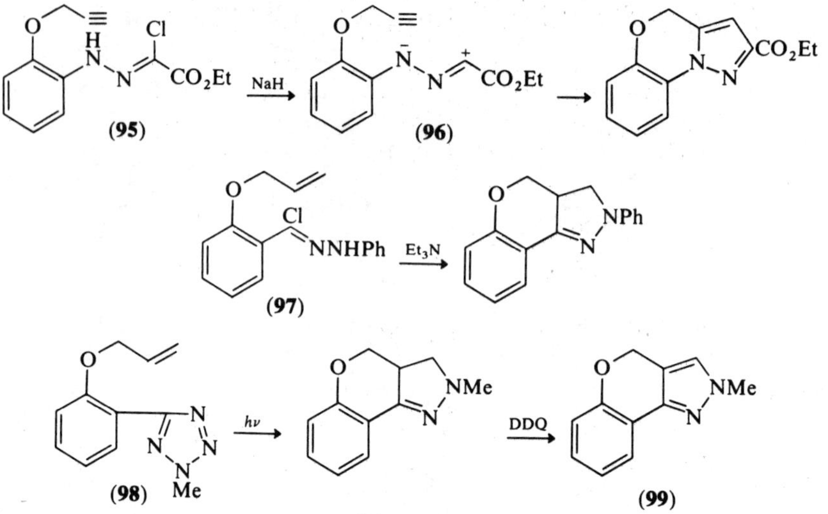

Scheme 33

A general method for the preparation of fused pyrazoles with a bridgehead nitrogen atom is the [2+3] cycloaddition of an alkyne to a diazoketone. A spontaneous 1,5 sigmatropic rearrangement completes the reaction. For example, diazocyclopentanone and dimethyl acetylenedicarboxylate give the unstable adduct (**100**) which rapidly changes into the pyrazolopyridine (**101**) ⟨73JOC825⟩. The diazoketone (**102**) reacts with benzyne (from diazotized anthranilic acid) in refluxing dichloromethane ⟨73TL1417⟩, and the diazobenzopyrazole (**103**) reacts with the ynamine by a similar mechanism (Scheme 34) ⟨78CB2258⟩. Even diazanorbornanone and diazacamphor react in this way ⟨73AG(E)240⟩.

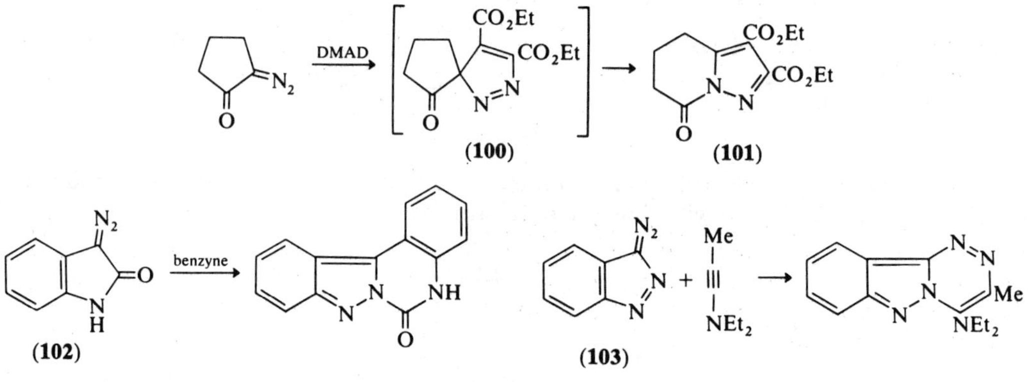

Scheme 34

4.05.4.2 Six-membered Ring Synthesis by Formation of One Bond

4.05.4.2.1 α to a heteroatom

An elegant synthesis of pyrazolo[1,5-*a*]pyridine (**1**) in 47% yield is shown in Scheme 35. Sodium methoxide in diglyme at 180 °C catalyzes the rearrangement of the bis-hydrazone (**104**) and evidence is available for the indicated mechanism. The same technique also allows the preparation of the fused cyclopropane derivative (**105**) ⟨74HCA1259⟩.

Hot aqueous sodium carbonate is sufficiently basic to catalyze the ring closure of (**106**) to (**107**), where the amino group attacks the nitrile carbon atom ⟨60JCS2536⟩. Dinitriles like (**108**) are ring closed under acid catalysis. The reaction usually occurs in one step, but the intermediate acid amide can be isolated if necessary (Scheme 36). Protonation of the carbonyl oxygen atom of the acid group allows attack by the amide nitrogen ⟨59JA2452⟩.

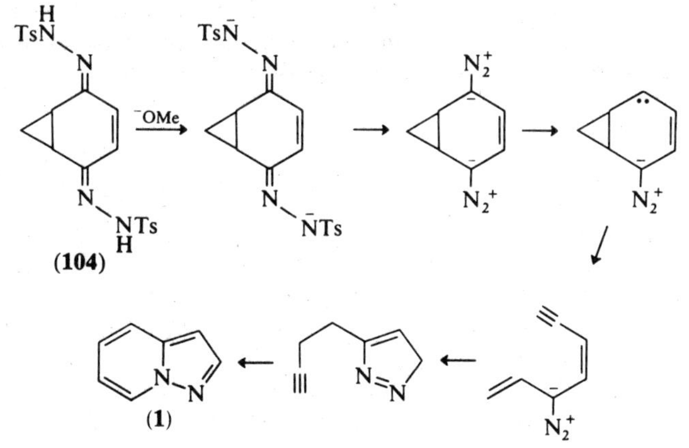

Scheme 35

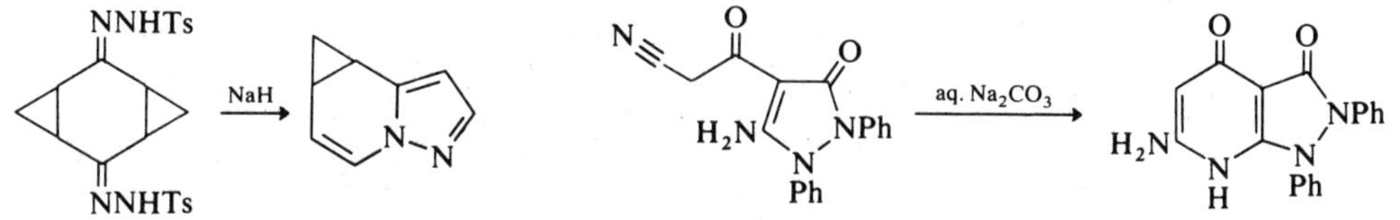

(105) **(106)** **(107)**

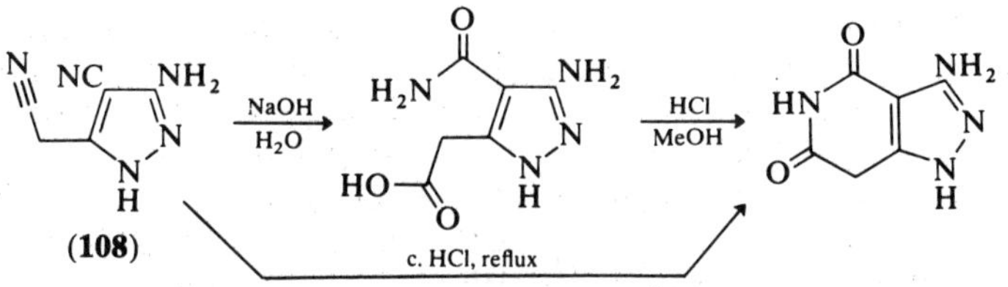

Scheme 36

The appropriate aldehyde treated with nitromethane (NaOEt) and reduced (H_2/Ni) gives the amino alcohol (**109**). Triethylamine catalyzes lactam formation and further elaboration gives the fully aromatic pyrazolopyridine (**110**; Scheme 37) ⟨73T441⟩.

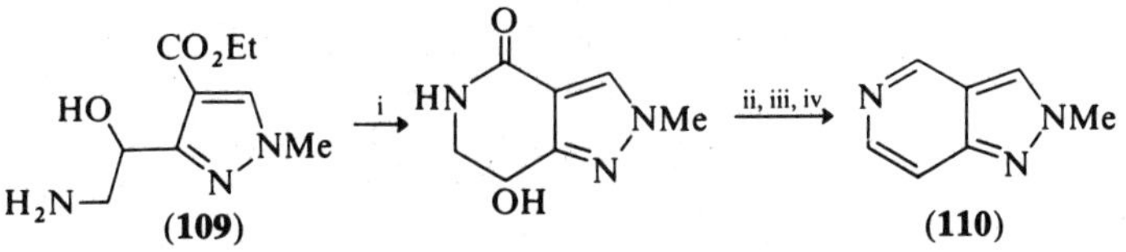

i, Et_3N; ii, $KHSO_4$, melt; iii, $POCl_3$; iv, H_2, Pd

Scheme 37

Sodium hydrosulfide hydrolyzes the nitrile group of (**111**) to the thioamide which spontaneously undergoes ring closure to thiopurinol (**112**) in 97% yield ⟨66AG(E)308⟩. Alkaline peroxide hydrolyzes the nitrile (**113**) and allows attack on the neighbouring acetyl group to give (**114**). This is a general method for the preparation of fused pyrimidines ⟨58JOC191⟩.

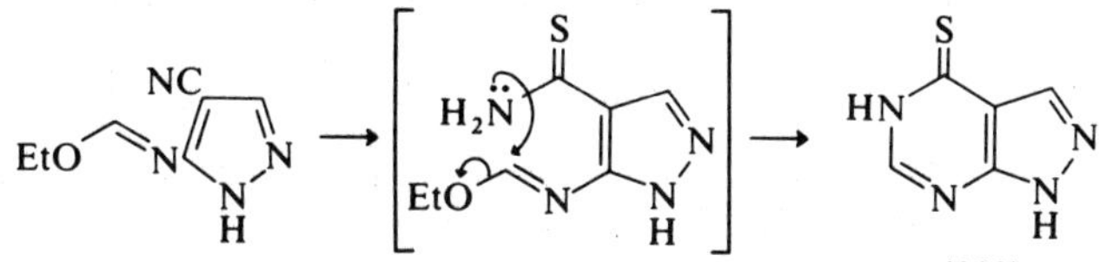

(111) **(112)**

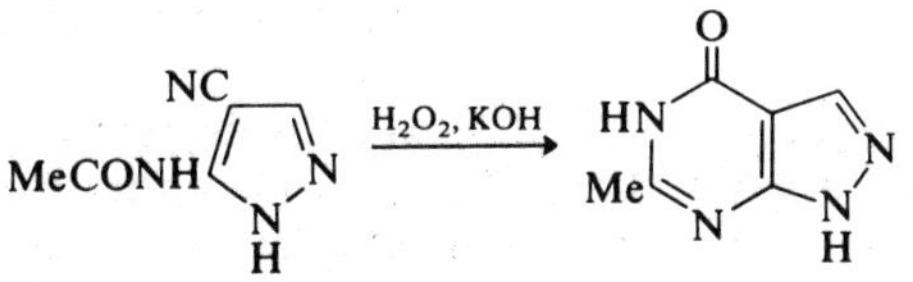

(113) **(114)**

In Scheme 38 an isoxazole is used to generate the vinylogous urea which ring closes with alcoholic hydrogen chloride ⟨72JHC951⟩. In the same way the isoxazolyl carboxamide (**115**) is changed, *via* the substituted enaminone, into (**116**) ⟨74JHC623⟩.

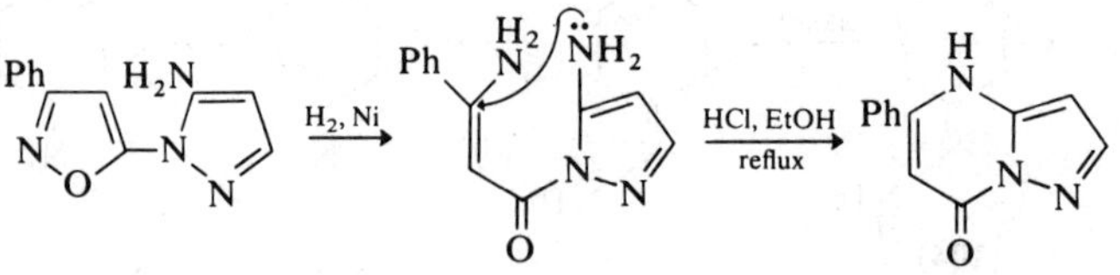

Scheme 38

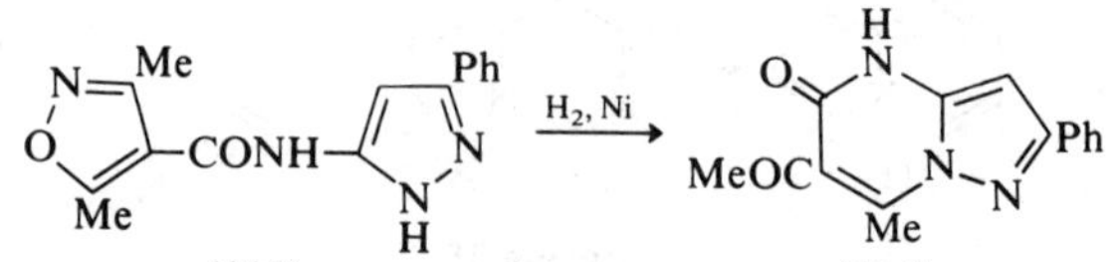

(**115**) (**116**)

The conditions for the general method of formation of pyrazolo[1,5-*a*]pyrimidines shown in Scheme 39 seem to be critical. In one report ⟨77ZN(B)307⟩, prolonged reflux of (**117a**) in glacial acetic acid gave only (**118a**, 75%). In another ⟨70BCJ849⟩, the use of alcoholic hydrogen chloride produced a mixture (**117a** to **118a**, 30% + **118b**, 63%). Clearly there is little difference between the activation energies for attack at nitrile and ester. For (**117b**) only the keto group reacts to give (**118c**) ⟨81JHC163⟩.

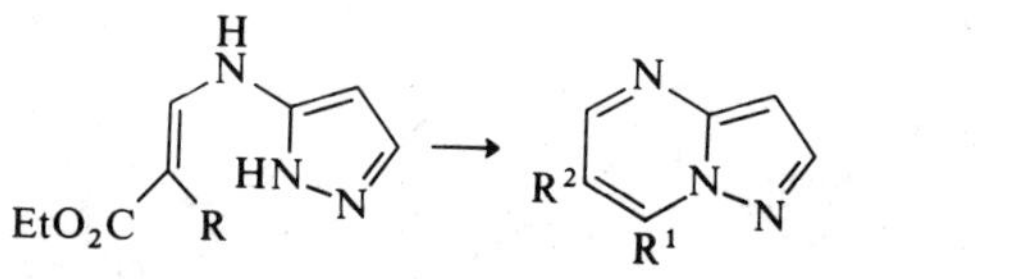

(**117a**) R = CN (**118a**) $R^1 = NH_2$; $R^2 = CO_2Et$
(**117b**) R = $COCO_2Et$ (**118b**) $R^1 = OH$; $R^2 = CN$
  (**118c**) $R^1 = R^2 = CO_2Et$

Scheme 39

3-Amino-5-phenylpyrazole reacts with aqueous nitrous acid to give the stable (m.p. 168 °C) diazonium salt (**119**). With active methylene compounds, this gives the derivatives (**120**; R = CN, CO_2Et) which are isolated and dissolved in concentrated sulfuric acid to produce the pyrazolotriazines (**121**; R = NH_2, OH) ⟨76JOC3781⟩. Two further examples illustrate this widely used procedure. The diazonium salt (**122**) gives the derivatives (**123**; R = Me, OEt) with acetylacetone and ethyl acetoacetate, respectively, and compound (**124**) with 2-naphthol ⟨81M245⟩. The stable diazo compound (**125**), prepared with isopentyl nitrite and fluoroboric acid followed by sodium carbonate, reacts with cyclohexane-1,3-dione to give first the hydrazone and, on acid treatment, the tetracycle (**126**) which is aromatized to the corresponding phenol by strong heating (Scheme 40) ⟨78JHC1175⟩.

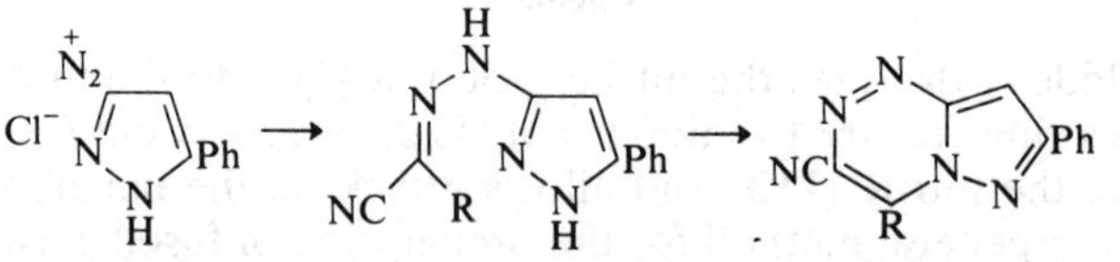

(**119**) (**120**) R = CN, CO_2Et (**121**) R = NH_2, OH

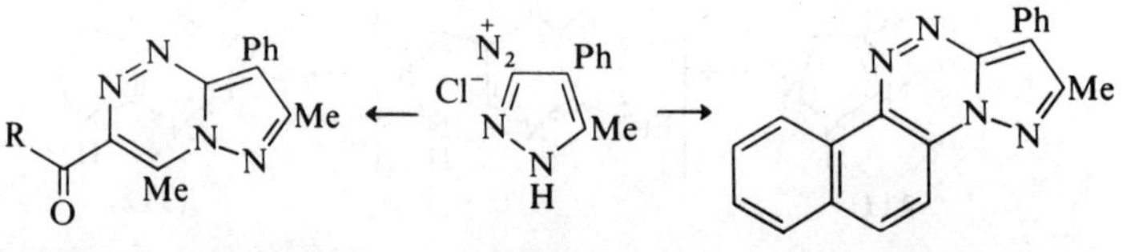

(**123**) R = Me, OEt (**122**) (**124**)

Base-catalyzed ring closure of the isothiocyanate derivative (**127**) gives the pyrazolo-[1,5-*a*][1,3,5]triazine (**128**; Scheme 41) ⟨74JHC199⟩. The sulfur may be removed with Raney nickel in ammonium hydroxide. Another example of this versatile route is the formation of (**130**) in Scheme 42. However, the starting material (**129**) also has a hydrazone group

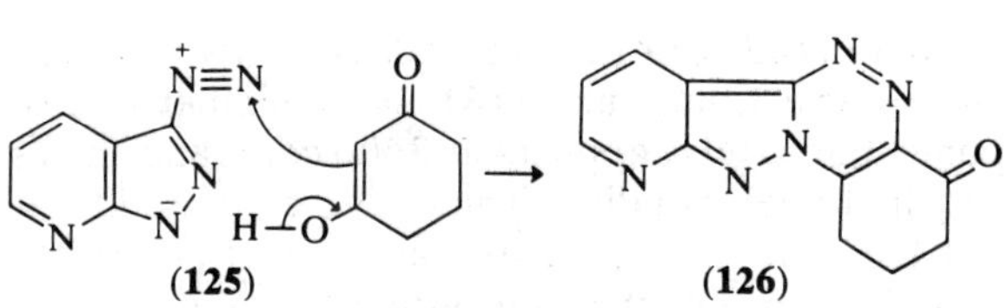

(125) (126)

Scheme 40

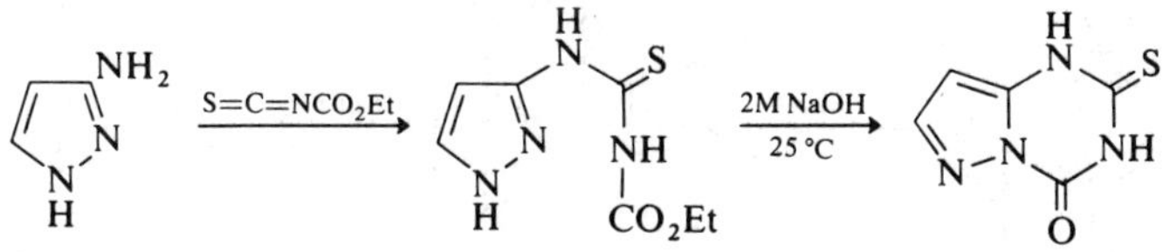

(127) (128)

Scheme 41

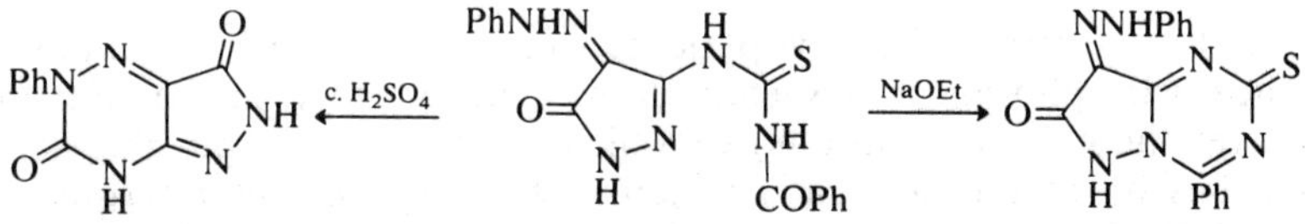

(131) (129) (130)

Scheme 42

which, in acid, takes part in the ring closure. Elimination of benzamide occurs to give the pyrazolo[3,4-*e*][1,2,4]triazine (**131**) ⟨77ZN(B)430⟩.

The ketone (**132**) gives the pyranopyrazole (**133**) on heating with strong acid, but under reducing conditions the tetrahydro derivative (**134**) forms (Scheme 43) ⟨81KGS403⟩. The δ-keto acid (**135**) gives the α-pyranone (**136**) with thionyl chloride. Treatment with hydrazine reopens the ring and the resultant hydrazone/hydrazide (**137**) may be reclosed either to a pyrazolodiazepine (**138**) or to a pyrazolopyridine (**139**) according to the reaction conditions (Scheme 44) ⟨81JHC271⟩.

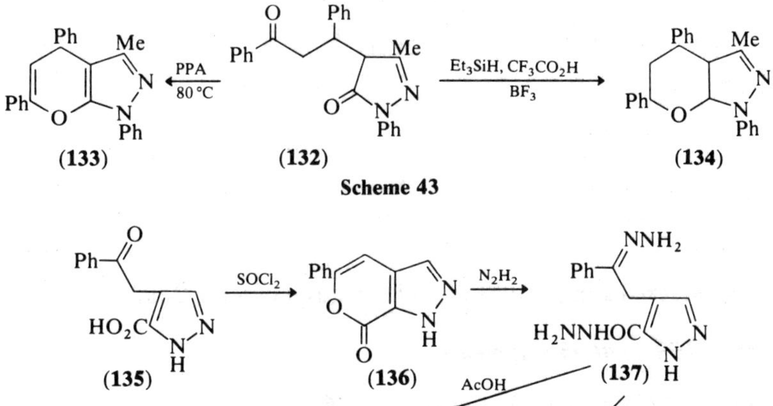

(133) (132) (134)

Scheme 43

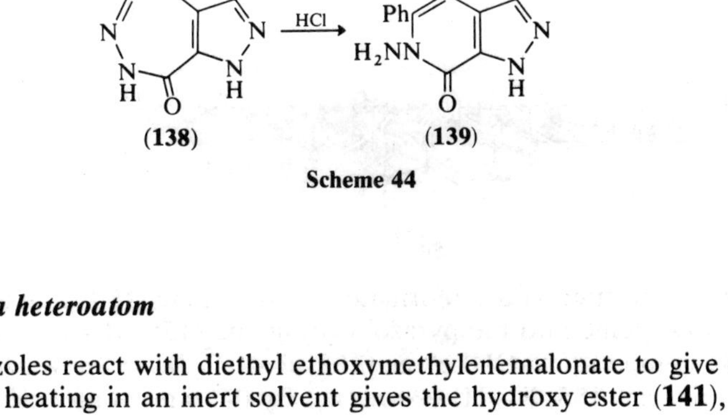

(135) (136) (137) (138) (139)

Scheme 44

4.05.4.2.2 γ to a heteroatom

3-Aminopyrazoles react with diethyl ethoxymethylenemalonate to give derivatives such as (**140**). Strong heating in an inert solvent gives the hydroxy ester (**141**), from which the valuable chloro compound (**142**) can be obtained. Alternatively, the diester (**140**) will go directly to (**142**; Scheme 45). Compound (**141**) is a member of a series which shows no pyridone carbonyl bands but ν(OH) appears at 3400 cm^{-1} and the ester ν(CO) is lowered to 1660–1680 cm^{-1} by hydrogen bonding. The chloro compound (**142**) has a normal ester carbonyl absorption at 1730 cm^{-1}. The hydroxy tautomer (**141**) is stabilized by the hydrogen bond and by the aromatic nature of the ring system ⟨72JHC235⟩.

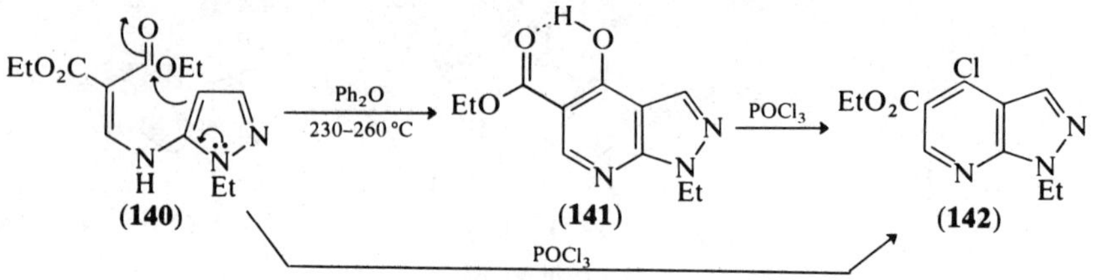

Scheme 45

The standard preparation of a new pyridine ring by treatment of a suitably substituted anthranilic acid with phosphorus oxychloride has examples which are relevant here. The acid (**143**) gives the chloro-substituted tricyclic derivative (**144**), but the isomeric system (**145**) ring closes on to nitrogen rather than carbon and the pyrazoloquinazoline (**146**) results ⟨76JMC262⟩. Examples of the same ring closure from aminopyrazolecarboxylic acids are also available ⟨76JHC155⟩. The triarylpyrazole (**147**) on photolysis cyclizes with loss of hydrogen chloride to (**148**) in 100% yield (Scheme 46) ⟨80CJC1880⟩.

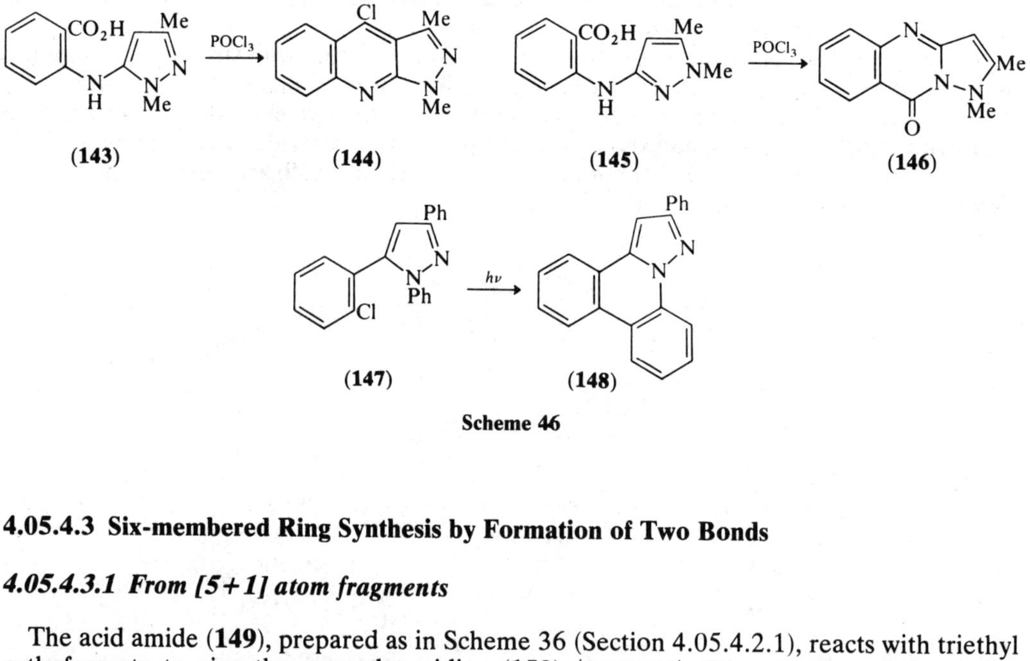

Scheme 46

4.05.4.3 Six-membered Ring Synthesis by Formation of Two Bonds

4.05.4.3.1 From [5+1] atom fragments

The acid amide (**149**), prepared as in Scheme 36 (Section 4.05.4.2.1), reacts with triethyl orthoformate to give the pyrazolopyridine (**150**) ⟨59JA2452⟩. Diazotization of the amine (**151**) leads to immediate insertion into the methyl ketone to give the pyrazolopyridazine (**152**) ⟨50HCA1183⟩.

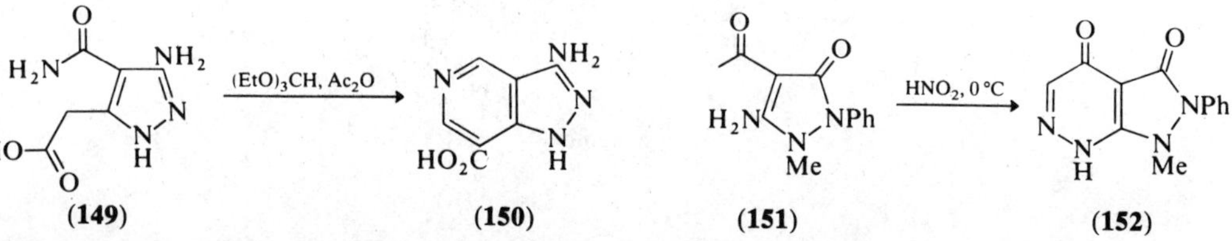

The amine (**153**) and triethyl orthoformate give the imine (**154**) which with ammonia is converted via the amidine into the pyrazolopyrimidine (**155**; R = CN). Hydrolysis of the remaining nitrile group (aq. NaOH) gives (**155**; R = CO$_2$H) which readily decarboxylates when sublimed to form (**155**; R = H) ⟨66JOC342⟩. By the same route, methylamine and the appropriate imine give (**156**). On warming in water, however, this readily undergoes Dimroth rearrangement as shown in Scheme 47 ⟨60JA3147⟩.

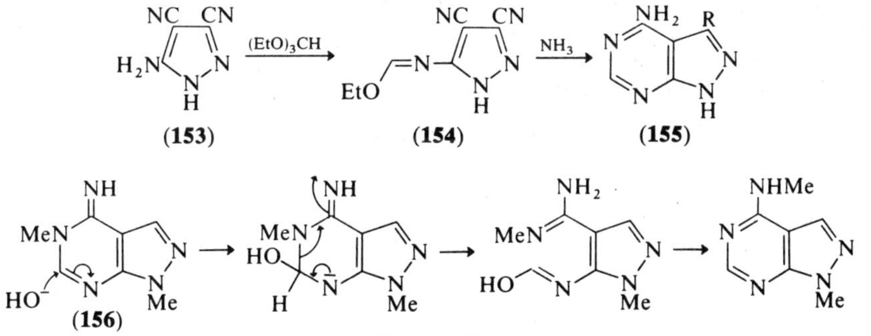

Scheme 47

Acid-catalyzed addition of thioacetamide to the aminonitrile (**157**) in ethanol gives an intermediate pyrazolothiazine (**158**), but basic treatment during work-up catalyzes the Dimroth rearrangement to the pyrazolopyrimidine (**159**). Other thioamides react similarly. A by-product (**160**) from this reaction, believed to arise by an exchange mechanism, provides a route to the 6-unsubstituted derivative (**162**). Hydrogen chloride in DMF gives (**161**) which ring closes under the reaction conditions (Scheme 48) ⟨61JA248⟩.

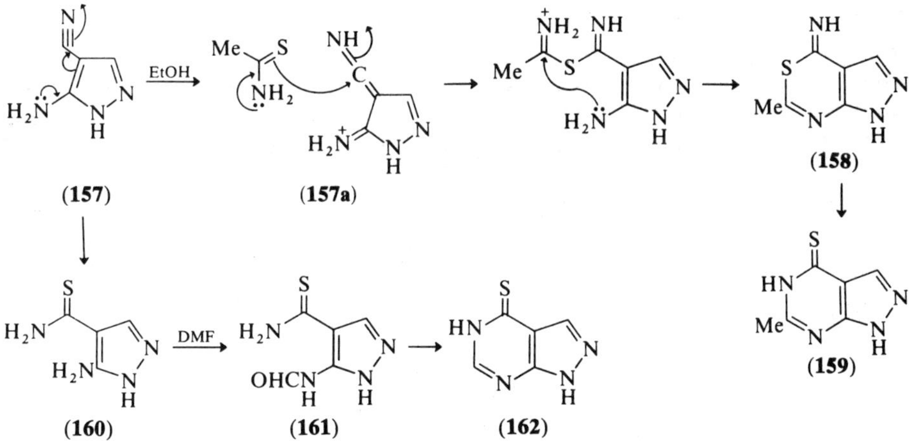

Scheme 48

The last step in Scheme 48 provides a useful way to form a pyrimidine ring by the reaction of a formamide with a vicinal aminoamide. The amine is formylated and ring closure occurs under the conditions of the reaction. For example, the preparation of pyrazolo[3,4-d]-pyrimidine (**5**; Scheme 49) ⟨56JA784⟩ and of the adenine analogue (**165**; Scheme 50) ⟨56JA2418⟩ illustrate this approach. Recent work has provided evidence that the reaction goes *via* an amidine (**167**) which is attacked by the neighbouring amide nitrogen atom. The substituted amides (**166**) gave (**168**) in good yield as the sole product when boiled with formamide (Scheme 51) ⟨81S727⟩.

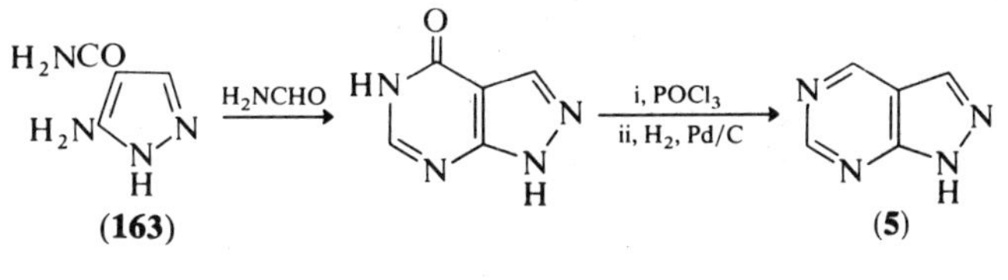

Scheme 49

Less commonly, formic acid or a homologue is used. Presumably the conversion of the amine to an amide is the first step. The aminoamide (**169**) gives (**170**; R = H) in refluxing formic acid or (**170**; R = Me) with acetic acid at 185 °C ⟨76USP3939161⟩. With methyl isocyanate, (**169**) gives (**171**) ⟨70JHC863⟩.

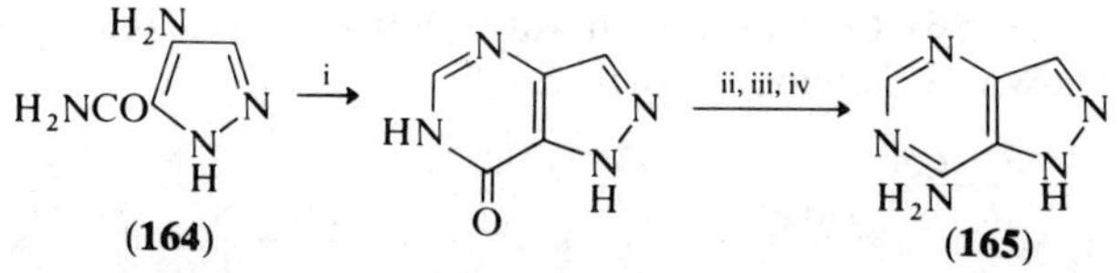

(164) (165)

i, H_2NCHO; ii, P_2S_5, C_5H_5N; iii, MeI, NaOH; iv, NH_3, 200 °C

Scheme 50

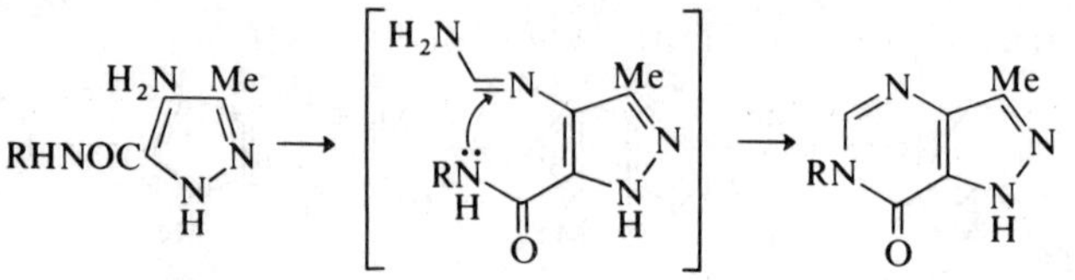

(166) (167) (168)

R = Pr, Bu, $C_5H_{11}{}^i$, Bz, $PhCH_2CH_2$

Scheme 51

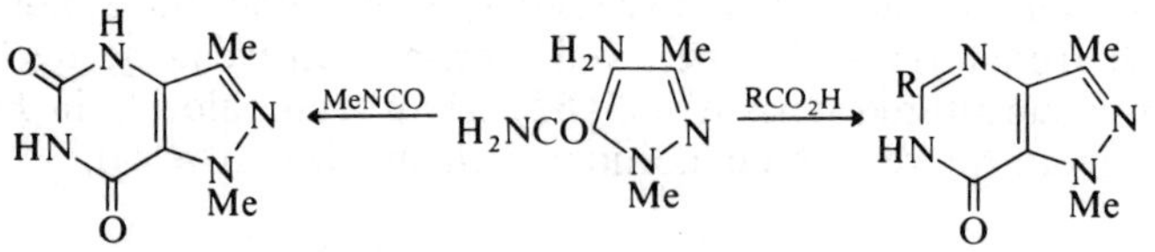

(171) (169) (170)

Fusion of aminoamides with urea or thiourea leads to pyrazolopyrimidinediones or the corresponding one-thiones (Scheme 52) ⟨56JA2418⟩. This method was employed ⟨65JOC199⟩ to establish the structure of the trimethylated compound (**26**; Scheme 53) (Section 4.05.3.1).

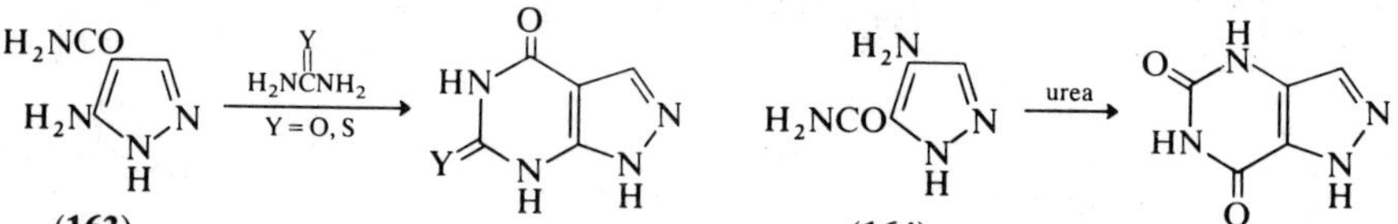

(163) (164)

Scheme 52

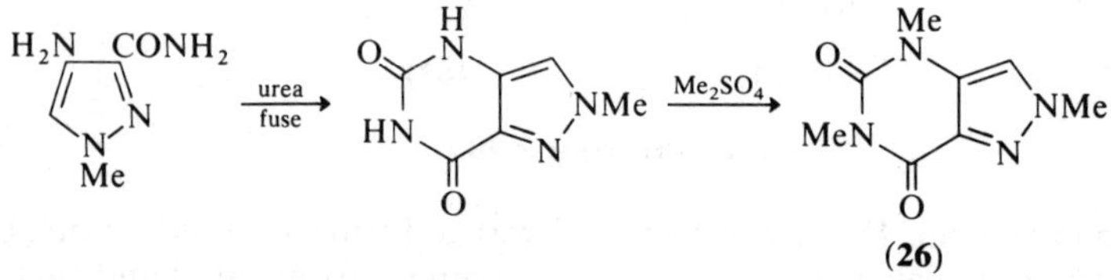

(26)

Scheme 53

The nitrile group of (**172**) reacts with hydroxylamine to give the hydroxyamidine (**173**). Reduction of the nitro to an amino group and then treatment with trimethyl orthoformate gives the aminopyrazolopyrimidine *N*-oxide (**174**) which, with care, can be hydrolyzed to the cyclic hydroxamic acid (**175**; Scheme 54) ⟨70JHC863⟩.

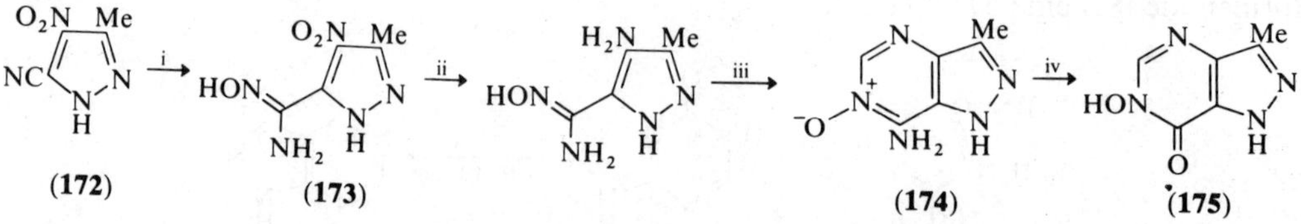

(172) (173) (174) (175)

i, NH_2OH, EtOH, reflux; ii, H_2, Pd; iii, $(MeO)_3CH$, reflux; iv, H_2O, AcOH, DMF, 55 °C

Scheme 54

The pyrazolylaniline (**176**) condenses with benzaldehyde to give the pyrazolo[1,5-*c*]-quinazoline (**177**). It also reacts in low yield with formaldehyde but other aliphatic aldehydes fail to react. With formic acid or acid chlorides, it gives the fully aromatic derivatives (**178**; Scheme 55) ⟨63JOC1336⟩.

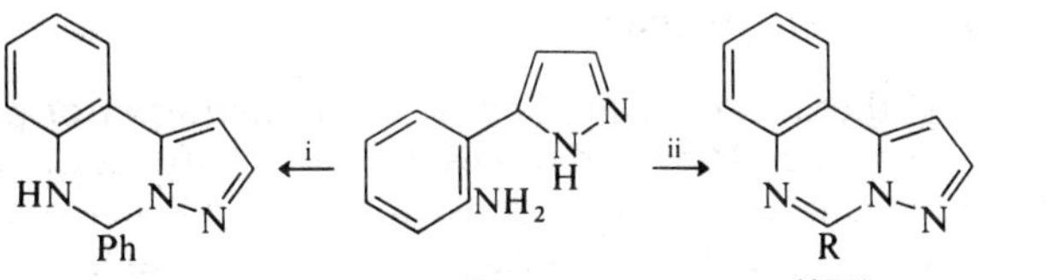

(177) **(176)** **(178)**

i, PhCHO; ii, HCO_2H (R = H) or RCOCl, NaOH, H_2O (R = Me, Et, Ph, *etc.*)

Scheme 55

3-Amino-5-phenylpyrazole and ethyl acetimidate give the pyrazolylamidine (**179**). This reacts with different reagents to give derivatives of pyrazolo[1,5-*a*][1,3,5]triazine (Scheme 56) ⟨74JHC691⟩ (*cf.* Section 4.05.3.2).

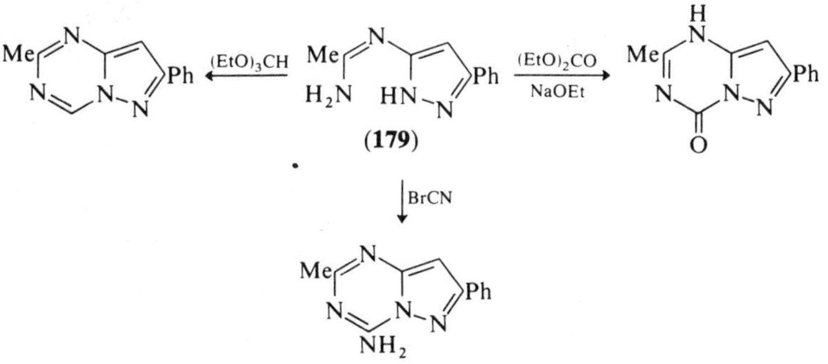

Scheme 56

The thiourea (**180**) rapidly cyclizes with diethoxymethyl acetate to the thione (**181**) from which derivatives can be made *via* nucleophilic displacement of methanethiol (*e.g.* **182**; Scheme 57) ⟨74JHC991⟩.

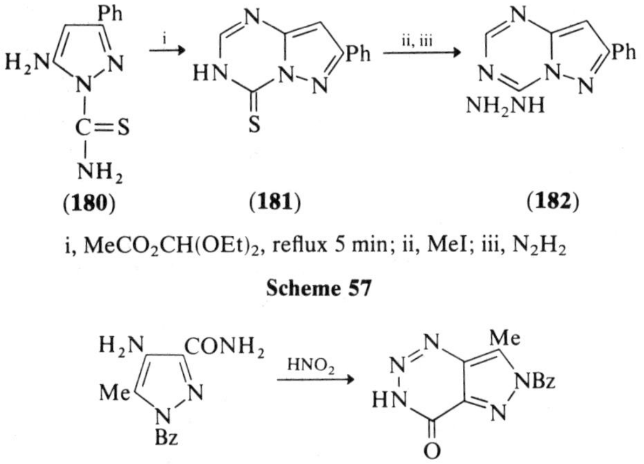

(180) **(181)** **(182)**

i, $MeCO_2CH(OEt)_2$, reflux 5 min; ii, MeI; iii, N_2H_2

Scheme 57

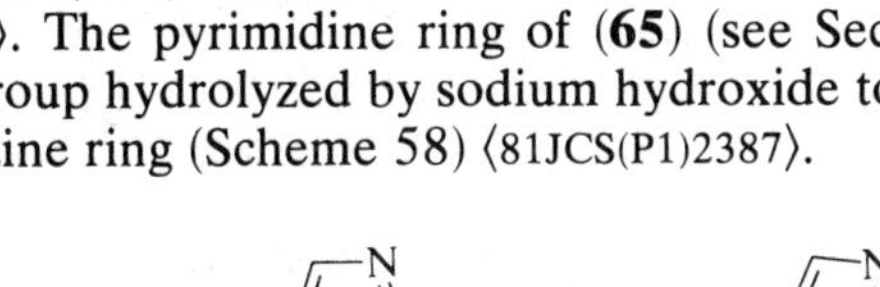

(183) **(184)**

Diazotization of the amine (**183**) is followed by ring closure on to the amide nitrogen to give the pyrazolotriazinone (**184**) which can be extracted from its salt with potassium bicarbonate ⟨81JCS(P1)2374⟩. The pyrimidine ring of (**65**) (see Section 4.05.3.5) is opened and the resulting formyl group hydrolyzed by sodium hydroxide to give (**185**). Nitrous acid produces a new 1,2,3-triazine ring (Scheme 58) ⟨81JCS(P1)2387⟩.

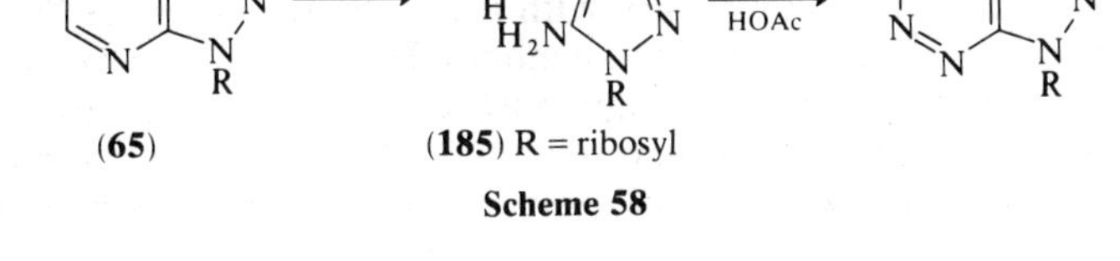

(**65**) (**185**) R = ribosyl

Scheme 58

4.05.4.3.2 From [4+2] atom fragments

The Friedlander synthesis can be applied to the preparation of pyrazolopyridines. For example, the aminopyrazole (**186**) undergoes Friedel–Crafts acylation to give the amino ketone (**187**), two molecules of which condense together to give the pyrazolopyridine (**188**) ⟨78T1175⟩. The Vilsmeier reaction converts (**186**) into the amino aldehyde (**189**) which reacts with ethyl acetoacetate in pyridine to give (**191**), presumably *via* (**190**). At high temperature (**192**) is produced, possibly *via* (**190**), but more likely through intermediate amide formation (Scheme 59) ⟨75JHC517⟩.

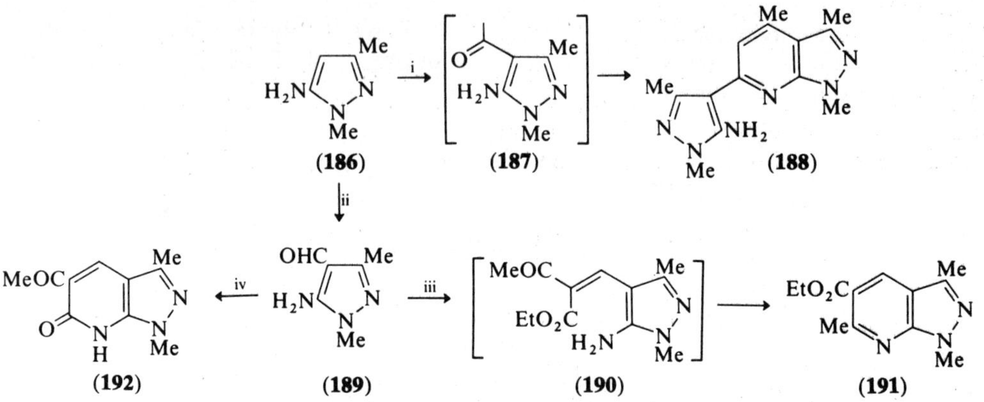

i, Ac_2O, H_2SO_4; ii, $POCl_3$, DMF; iii, EAA, py, reflux; iv, EAA, 160–170 °C

Scheme 59

The diketone (**193**) condenses with ethylglycine hydrochloride to give a mixture of the pyrazolopyridines (**194**; 8%) and (**195**; 37%), which are separated by column chromatography (Scheme 60) ⟨81JHC1073⟩.

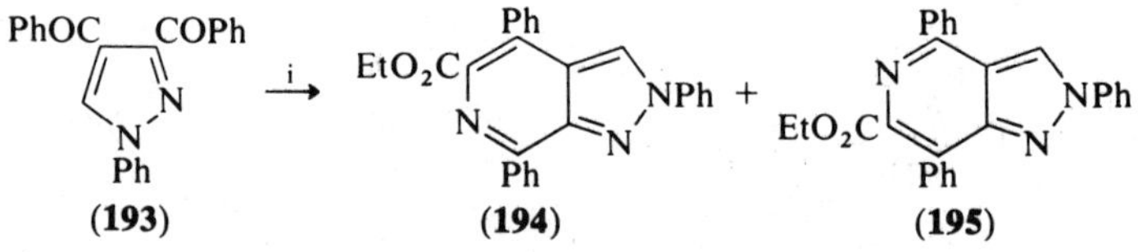

i, $H_2NCH_2CO_2Et \cdot HCl$, butanol, reflux

Scheme 60

Vicinal dicarbonylpyrazoles react with hydrazines to give pyrazolopyridazines. For example, the diketone (**196**) gives (**197**) ⟨77JHC375⟩ and the aldehydo esters (**198**) and (**199**) both give pyrazolopyridazones (Scheme 61) ⟨73T435⟩.

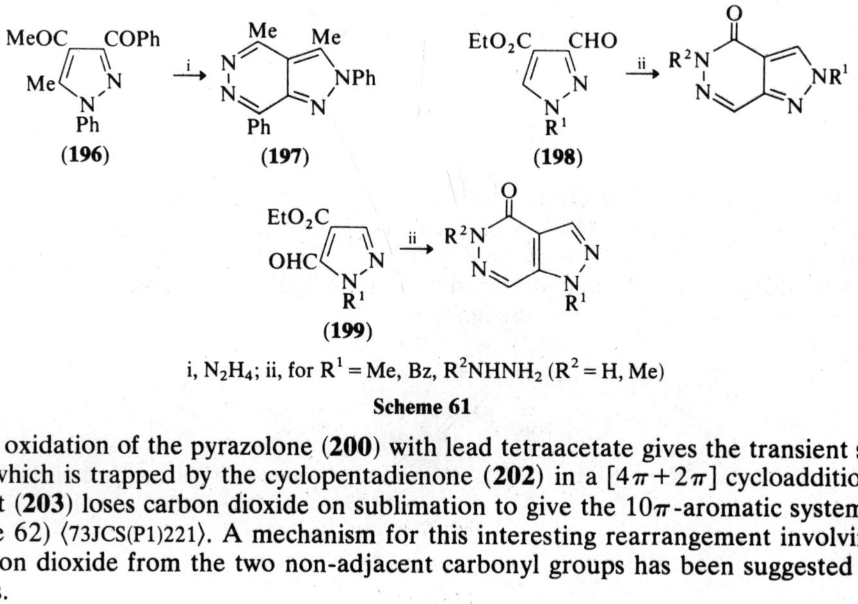

i, N_2H_4; ii, for R^1 = Me, Bz, R^2NHNH_2 (R^2 = H, Me)

Scheme 61

Mild oxidation of the pyrazolone (**200**) with lead tetraacetate gives the transient species (**201**) which is trapped by the cyclopentadienone (**202**) in a $[4\pi + 2\pi]$ cycloaddition. The product (**203**) loses carbon dioxide on sublimation to give the 10π-aromatic system (**204**; Scheme 62) ⟨73JCS(P1)221⟩. A mechanism for this interesting rearrangement involving loss of carbon dioxide from the two non-adjacent carbonyl groups has been suggested by the authors.

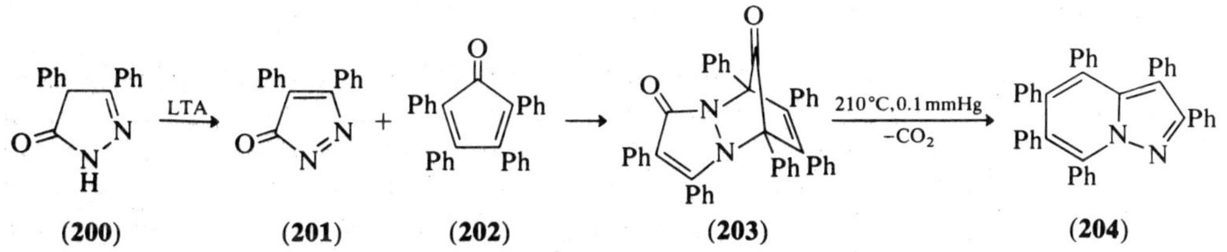

Scheme 62

The bis-diazo ketone (205) refluxed in 2,3-dimethylbuta-1,3-diene gives the pyrazolopyridine (207), presumably via (206) ⟨71CC537⟩. A Diels–Alder reaction gives the pyrazolo[1,2-a]pyridazine (208) which can be further elaborated to the aromatic salt (209; Scheme 63) ⟨71JHC25⟩. An alternative preparation of (201), carried out in the presence of buta-1,3-diene, gives (210), which is also obtained when the tetrahydropyridazine (211) is heated with the β-keto ester (212; Scheme 64) ⟨66JOC2867⟩.

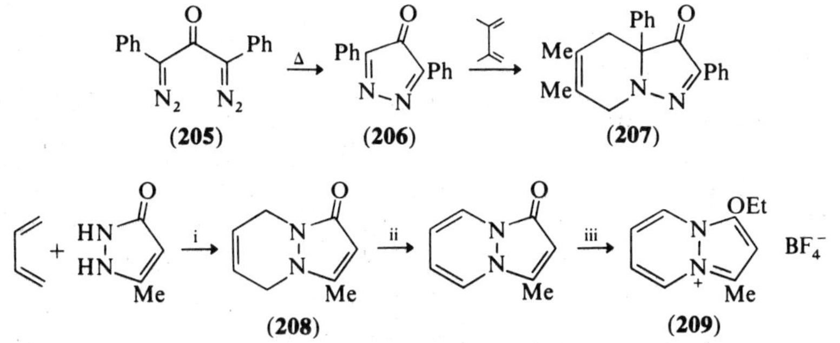

i, LTA; ii, Pd/C, PhCl, boil; iii, Et$_3$O$^+$ BF$_4^-$

Scheme 63

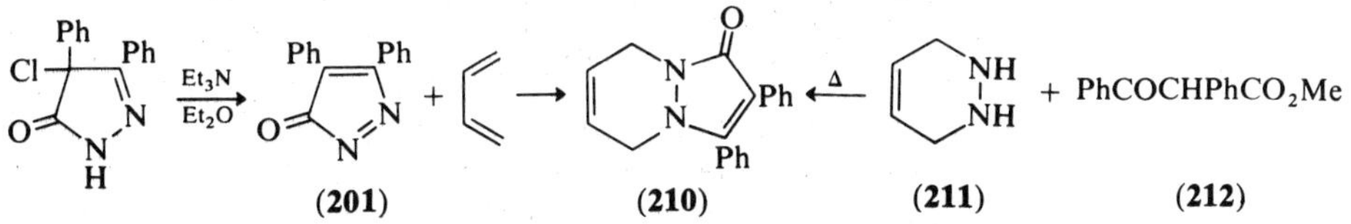

Scheme 64

The Diels–Alder 'ene' (213) is also prepared in situ by mild oxidation (lead tetraacetate or t-butyl hypochlorite). It forms many adducts, two of which are shown in Scheme 65.

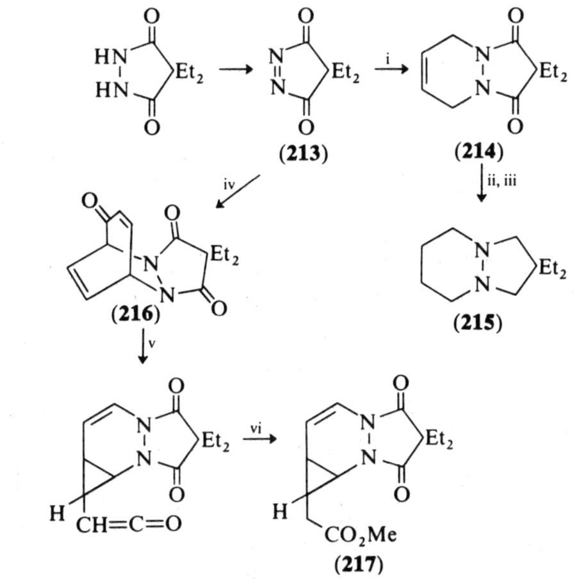

i, buta-1,3-diene; ii, H$_2$, Pd; iii, LAH; iv, tropone; v, hν; vi, MeOH

Scheme 65

With butadiene it gives (**214**), which can be reduced to (**215**) ⟨72JHC41⟩. Tropone gives the bridged ring system (**216**) which on photolysis in methanol is converted into (**217**). In refluxing ethanol this reaction goes thermally to give the ethyl ester equivalent to (**217**; Scheme 65) ⟨72JCS(P1)783⟩.

Diesters, diacid chlorides and dihaloalkanes react with suitable cyclic hydrazines to give saturated or partially unsaturated pyrazolo[1,2-*a*]pyridazines ⟨62HCA37⟩. For example, tetrahydropyrazole and diethyl succinate give the dione (**218**) which under standard amide reduction conditions gives (**219**; Scheme 66). As with most saturated pyrazolo[1,2-*a*]-pyridazines, catalytic hydrogenation splits the N—N bond of (**219**) ⟨72JHC41⟩.

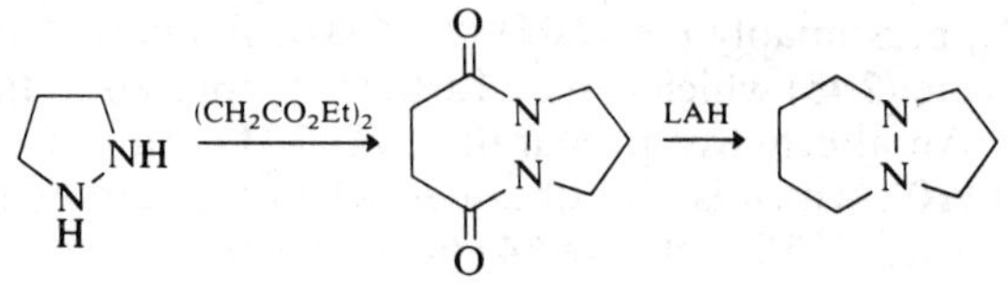

Scheme 66

Aminopyrazolonitriles react with simple nitriles to give aminopyrazolopyrimidines, *e.g.* as in Scheme 67 ⟨61JOC4967⟩. The amino ester (**220**) reacts with phenylacetonitrile *via* nucleophilic attack of the deprotonated amine on the nitrile carbon atom to give (**221**). A minor by-product (**222**) arises due to the benzyl carbanion attacking the ester group (Scheme 68) ⟨62HCA1620⟩.

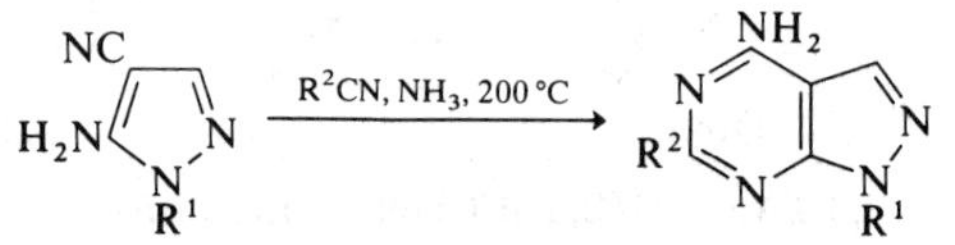

R^1 = H, Me, Ph; R^2 = Me, Bz, Ph, subst. Ph

Scheme 67

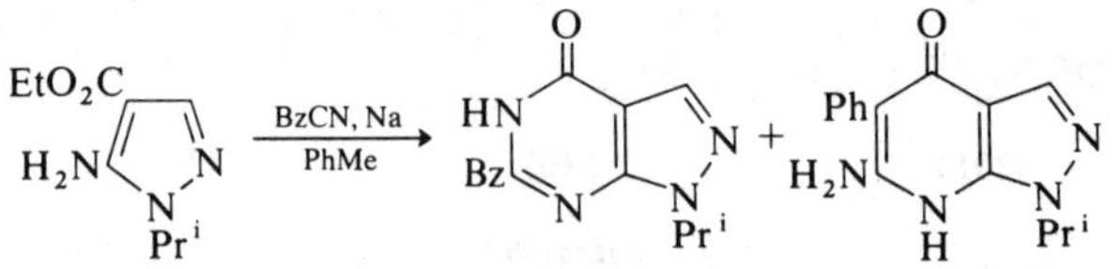

Scheme 68

Formamidine acetate is a useful reagent because it can provide a C—N unit to synthesize a pyrimidine ring, *e.g.* as in Scheme 69 ⟨74JOC2023⟩. Another important example is seen in a synthesis of the antibiotic formycin (**223**; Scheme 70) ⟨80CJC2624⟩.

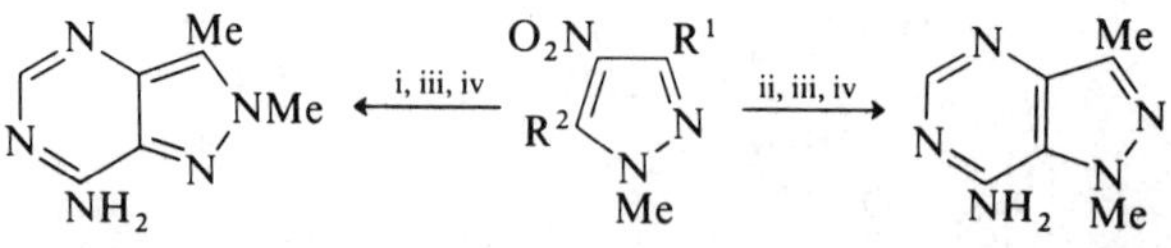

i, R^1 = CN, R^2 = Me; ii, R^1 = Me, R^2 = CN; iii, $Na_2S_2O_4$; iv, $NH_2CH\overset{+}{N}H_2\ \overset{-}{O}Ac$

Scheme 69

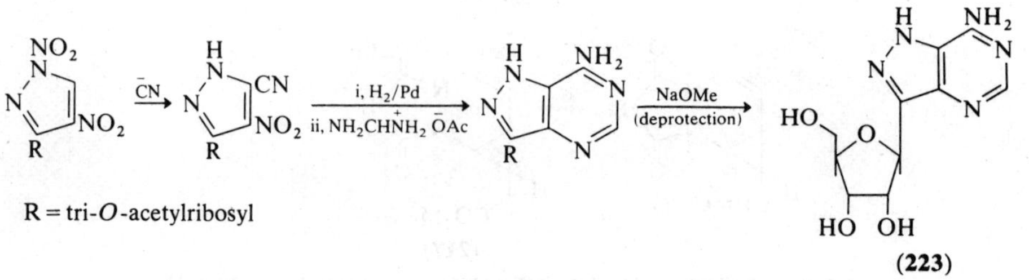

R = tri-*O*-acetylribosyl

(**223**)

Scheme 70

Guanidine or guanidine carbonate can also provide a C—N unit, but the resulting pyrimidine carries an additional amino group (Scheme 71) ⟨61JCS2589⟩.

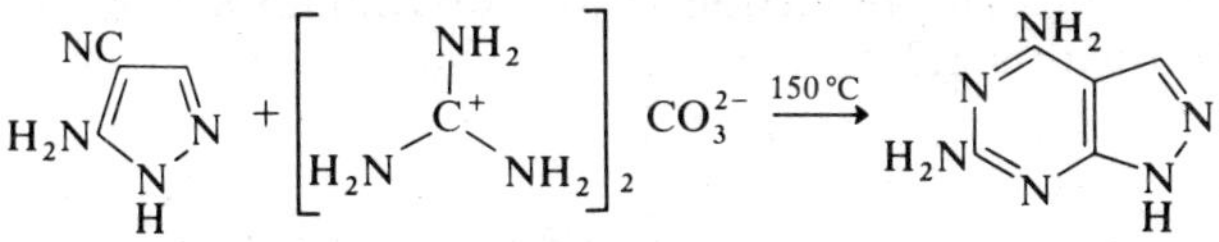

Scheme 71

With aminonitriles, formamides and ureas provide two-atom units to give pyrimidines (Scheme 72). With the derived aminoamides these reagents provide one-atom units (see Section 4.05.4.3.1) ⟨56JA784⟩.

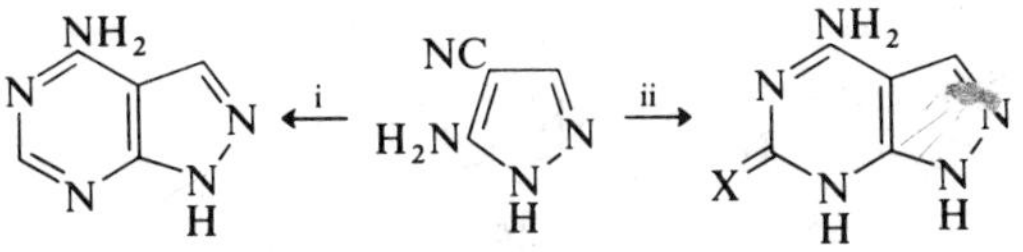

i, $HCONH_2$, boil; ii, NH_2CXNH_2, fuse (X = O, S)

Scheme 72

The pyrazole (**224**) is readily available from tetracyanoethylene and methylhydrazine. It reacts with refluxing *N*-alkylformamides to give the pyrazolopyrimidines (*e.g.* **225**, 62%) after spontaneous Dimroth rearrangement ⟨73JCS(P1)1903⟩. Formamides with amino esters such as (**226**) give pyrazolopyrimidones (**227**) ⟨56HCA986⟩.

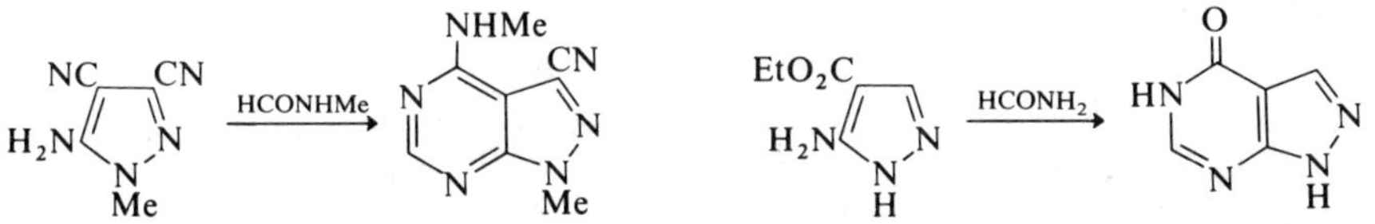

(**224**) (**225**) (**226**) (**227**)

Fusion of amino esters such as (**228**) with urea (200 °C) or thiourea (180 °C) gives pyrazolopyrimidines (**229**), and with alkyl isocyanates, *N*-alkylated derivatives (**230**) are obtained (Scheme 73) ⟨59HCA349⟩.

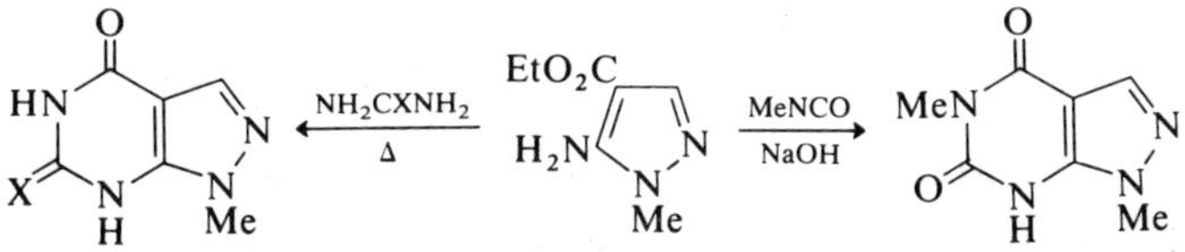

(**229**) X = O, S (**228**) (**230**)

Scheme 73

Carbon disulfide reacts with the amine (**231**) and cyclizes on to the neighbouring nitrile group to give the isolable pyrazolothiazine (**232**). This readily rearranges in alkaline solution (Scheme 74) ⟨66AG(E)308⟩.

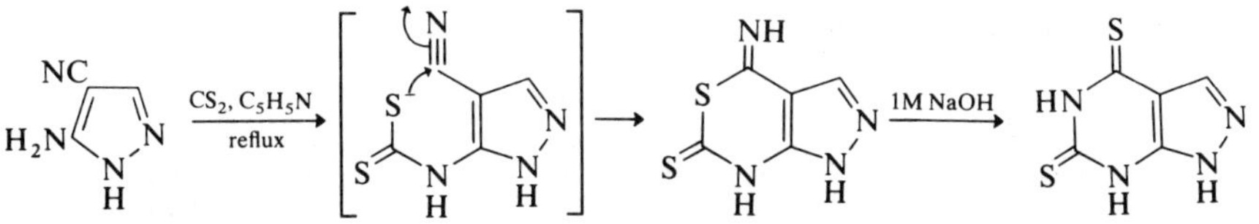

(**231**) (**232**)

Scheme 74

Glyoxal (**233**) and the diaminopyrazole (**234**) react to give 3-hydroxypyrazolo-[3,4-*b*]pyrazine (**235**) ⟨58JA421⟩. 1,2-Diketones also react to give 5,6-disubstituted derivatives of (**235**) ⟨61AG15⟩.

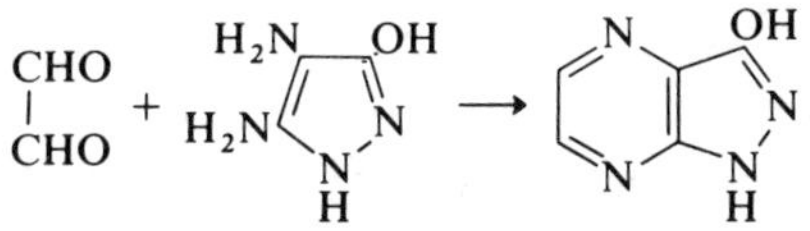

(**233**) (**234**) (**235**)

3-Aminopyrazolediazonium chloride (**236**) reacts with ethyl acetoacetate to give the intermediate (**237**) which ring closes to the pyrazolotriazine (**238**). Other 1,3-diones also react in a similar manner. The reduction of the diazonium salt to the hydrazine (**239**) followed by treatment with a 1,2-diketone gives a similar fused heterocycle (**240**; Scheme 75) ⟨66JCS(C)1127⟩.

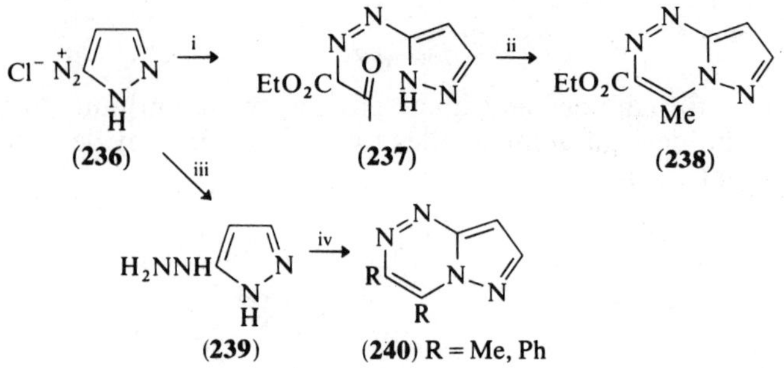

i, EAA, H$_2$O, NaAc, 25 °C; ii, MeOH, reflux; iii, SO$_2$, EtOH; iv, RCOCOR

Scheme 75

Another diazonium salt (**241**) reacts with 3-chloropentane-2,4-dione, or 3-chloroacetoacetic ester, with loss of two carbon atoms to give (**242**) which hydrolyzes and cyclizes under carefully controlled conditions to the hydroxypyrazolotriazine (**243**). The free diazo compound (**244**) gives the adduct (**245**) with dimethyl acetylenedicarboxylate (Scheme 76) ⟨77JHC227⟩.

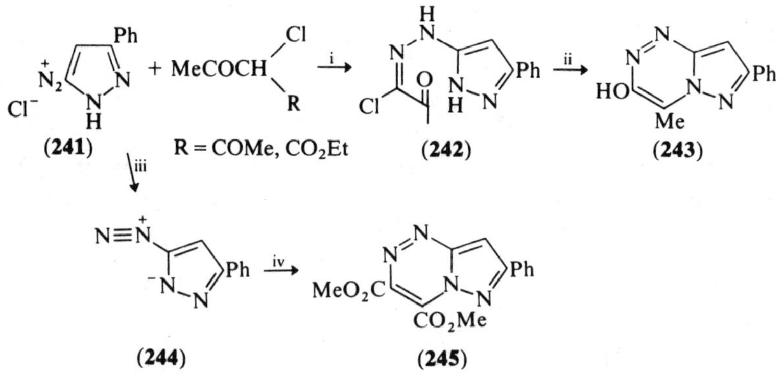

i, NaAc, H$_2$O, EtOH, 25 °C; ii, KCN, H$_2$O, reflux; iii, base; iv, DMAD

Scheme 76

Michael addition of the anion from malononitrile to the conjugated system (**246**) results in the pyranopyrazole (**247**) ⟨74AP444⟩. For (**246**; R^1 = Ph) adducts form with vinyl ethers at 70–80 °C. With extra activation (**246**; R^1 = PhCO or MeCO) the adducts form at room temperature. The reaction is regiospecific but gives a mixture of *cis* (**248**; *ca.* 80%) and *trans* (**249**; *ca.* 20%) isomers (Scheme 77) ⟨80JPR711⟩.

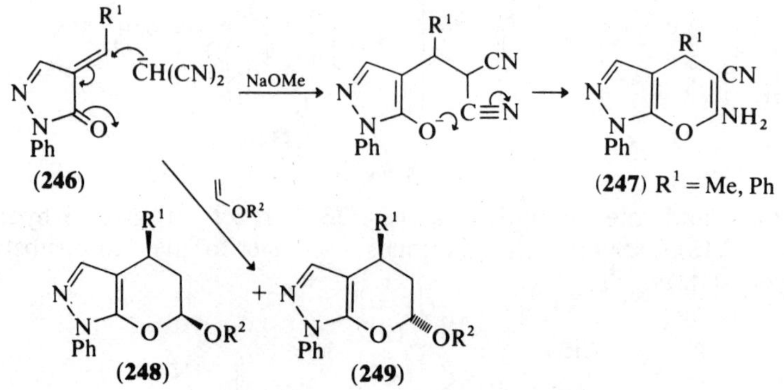

R^1 = Ph, MeCO, PhCO; R^2 = Me, Et, Pr^i, Bu^t

Scheme 77

4.05.4.3.3 From [3+3] atom fragments

The 4-aminopyrazole (**250**) reacts under Skraup conditions to give a useful synthesis of the pyrazolo[4,3-*b*]pyridine (**251**; 19%) ⟨58JCS3259⟩. 1,3-Diketones (**252**) react with 3-aminopyrazoles (*e.g.* **253**) to give simple pyrazolo[1,5-*a*]pyrimidines (**254**). If the two R groups are different, mixtures result, but these can be treated with *N*-bromosuccinimide to give the 3-bromo derivatives, purified, and the hydrocarbons regenerated by reduction ⟨75JMC460⟩.

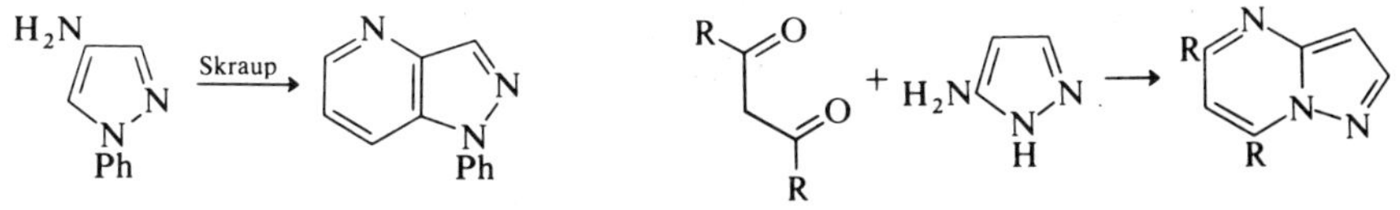

(**250**) (**251**) (**252**) (**253**) (**254**)

The malonaldehyde acetal (**255**) reacts with aminopyrazoles (**256**; R = H, CN, CO_2H) to give the derivatives (**257**) unsubstituted in the pyrimidine ring. The acid (**257**; R = CO_2H) decarboxylates in high yield on pyrolysis, but, surprisingly, a mixture results. Only 14% yield of the expected pyrazolo[1,5-*a*]pyrimidine (**258**) is obtained along with 71% of the rearranged pyrazolo[3,4-*b*]pyridine (**259**). Hydrolysis and decarboxylation of the nitrile (**257**; R = CN) by prolonged boiling in 85% sulfuric acid gives only the rearranged product (**259**; 80%; Scheme 78) ⟨70JHC247⟩.

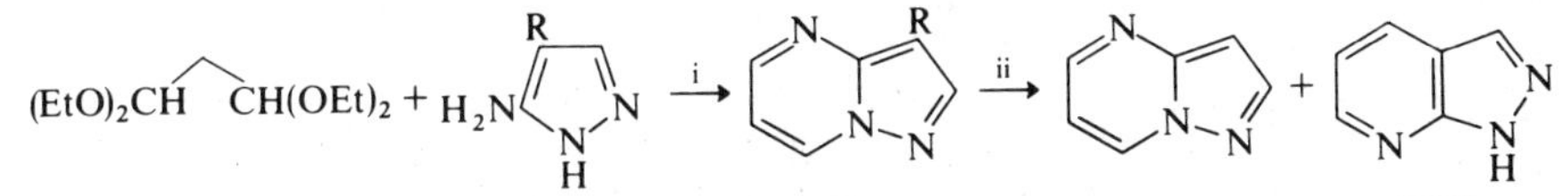

(**255**) (**256**) (**257**) (**258**) (**259**)

i, $ZnCl_2$, EtOH, reflux; ii, R = CO_2H, 290–300 °C; R = CN, 85% H_2SO_4, reflux

Scheme 78

The reactions of 3-aminopyrazole (**253**) with various esters and nitriles are summarized in Scheme 79 ⟨62CPB620, 70CB3252, 74AP177, 81JMC610⟩. In most cases it is clear that initial alkylation of the amine nitrogen is followed by ring closure on to N(1) or in a strongly protic solvent for (**261**) on to C(4) ⟨70CB3252⟩. The mechanisms of formation of compounds (**260**) and (**262**) are uncertain ⟨70CB3252, 62CPB620⟩. 3-Aminopyrazoles carrying substituents on C(4) and/or C(5) react similarly ⟨70BCJ849, 79FES478⟩ (see Section 4.05.4.2.1).

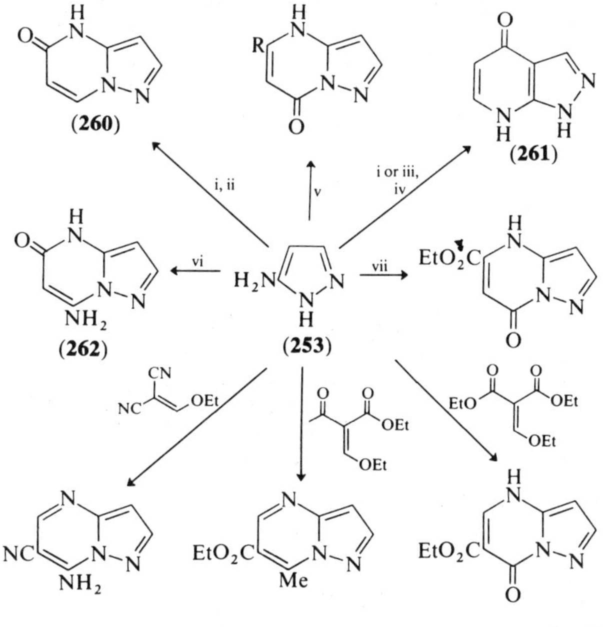

i, HC≡CCO_2Me; ii, THF, 25 °C; iii, EtOCH=$CHCO_2Et$; iv, HOAc, reflux; v, for R = H, NaOCH=$CHCO_2Et$; for R = Me, EAA; vi, $NCCH_2CO_2Et$; vii, DEAD

Scheme 79

3-Amino-5-phenylpyrazole (**263**) forms an amide (**265**) with ethyl benzoylacetate at 160 °C. The ring closes under acid catalysis to (**266**) but on strong heating it rearranges to give (**264**; Scheme 80) ⟨72JHC951⟩.

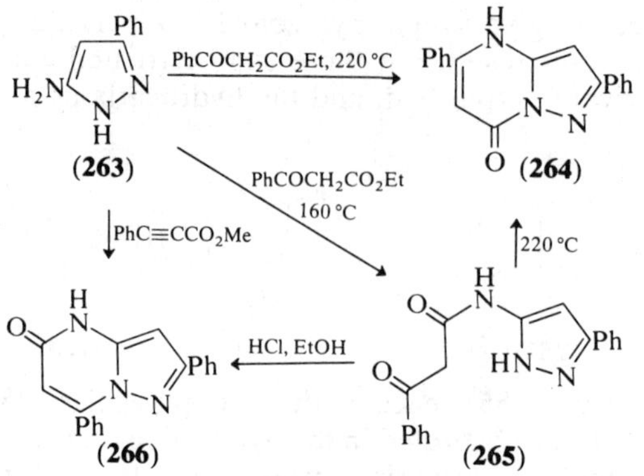

Scheme 80

The aminopyrazolinone (**267**; R = Me) reacts with pentane-2,4-dione to give (**268**; R = Me) ⟨79JHC773⟩, but for (**267**; R = H) alkaline conditions are needed for the reaction to take this course, giving (**268**; R = H). The pyrazolone carbanion ensures that ring closure is to C(4) and not N(1). In acid (aqueous HCl) the pyrazolopyrimidine (**269**) is obtained. Compound (**267**; R = H) reacts with ethyl acetoacetate and diethyl malonate to give pyrazolopyridines ⟨60CB1106⟩, but with diketene a pyrimidine ring forms (Scheme 81) ⟨61LA(647)116⟩. The aminopyrazolone (**270**) reacts with 1,3-diketones to give pyrazolopyridines (**271**) ⟨79JHC773⟩.

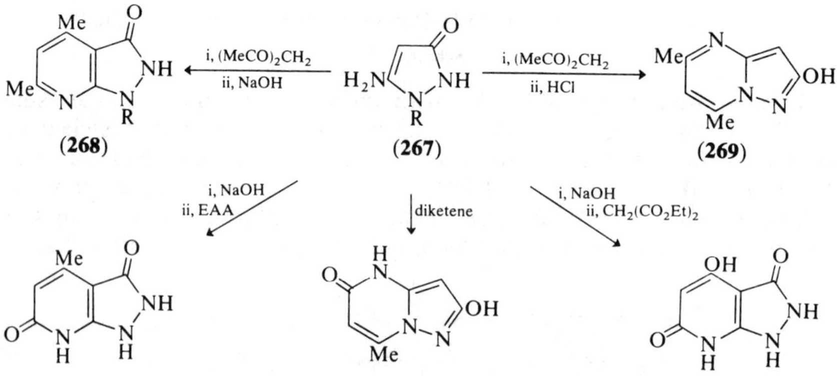

Scheme 81

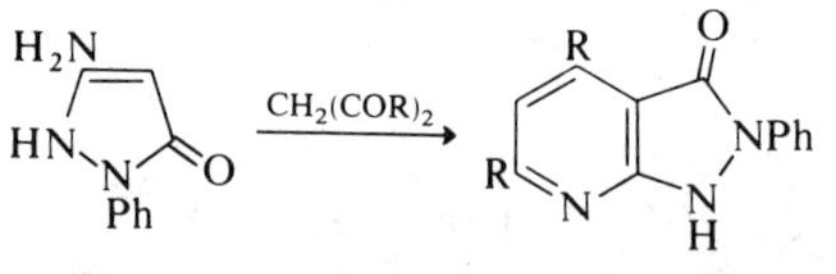

(**271**) R = H, Me, Ph

Ethyl acetoacetate converts the amine (**272**) into the isolable enaminone (**273**). On strong heating in an inert solvent this cyclizes to (**274**) which is deoxygenated to (**275**) by standard means. If the enaminone is heated in glacial acetic acid, it rearranges and ring closes to the alternative pyrazolopyridone (**276**). Deoxygenation gives the trimethylpyrazolopyridine (**277**), which is also available by independent syntheses (Scheme 82). Both (**275**) and (**277**) are converted into tetrahydro derivatives by hydrogen over platinum, only the pyridine ring being reduced. The ^{13}C chemical shifts for the carbonyl groups of (**274**) and (**276**), 178.5 and 163.2 p.p.m. respectively, provide a sensitive method for the assignment of similar structures ⟨75JHC517⟩.

When the aminopyrazole (**272**) reacts with another 3-keto ester (**278**) an unexpected dehydrogenation occurs to give (**279**) ⟨75JHC523⟩. The extra double bond gives an enaminone system, for which the mesomer has aromatic stabilization.

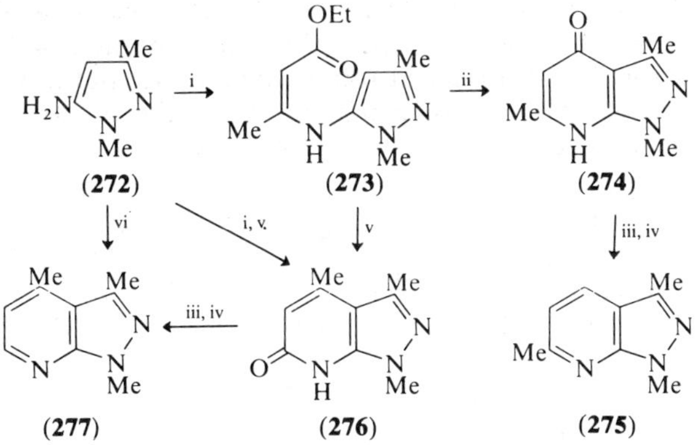

i, EAA; ii, Dowtherm, reflux; iii, POCl₃; iv, H₂, Pd; v, HOAc, reflux; vi, MeCH=CHCHO or MeCOCH=CH₂
Scheme 82

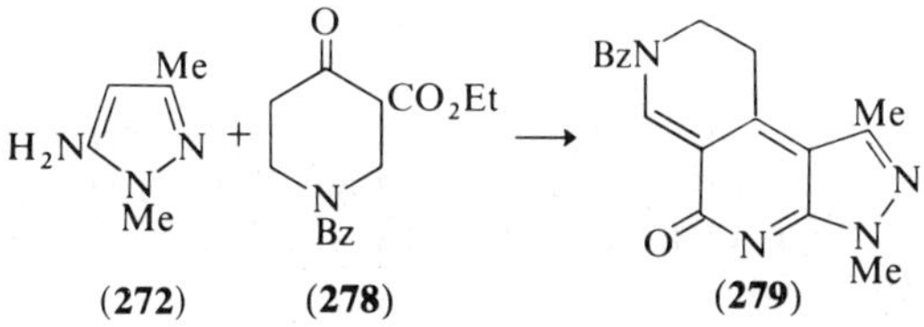

Several enaminones have been shown to react like (**280**) with the aminopyrazolone (**281**) to give pyrazolopyridines (**282**; Scheme 83). For one enaminone (**283**), a mixture of the normal product (**284**; 80%; ν(CO) 1610 cm^{-1}) and the pyrazolopyrimidine (**285**; 18%; ν(CO) 1620 cm^{-1}) was obtained (Scheme 84) ⟨69M1250⟩.

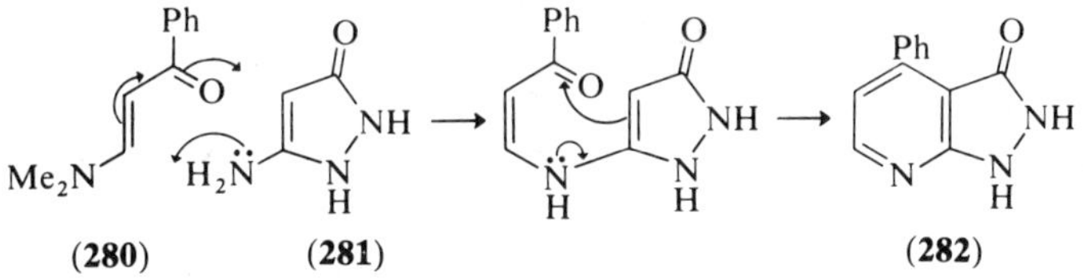

Scheme 83

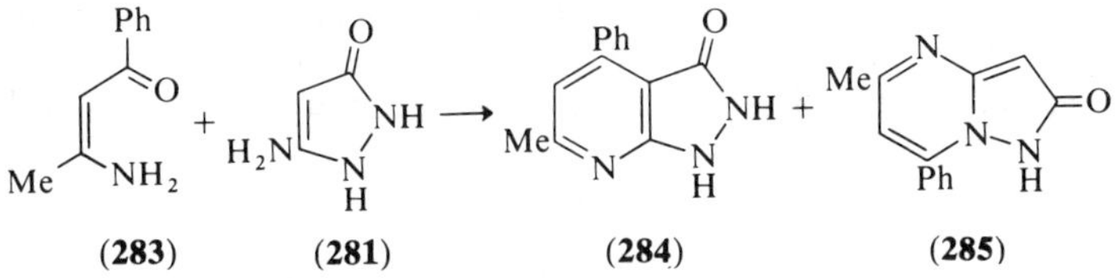

Scheme 84

Hydrazine and 3-aminocrotononitrile (**286**) give the pyrazole (**287**), which can be reacted with a second molecule of starting material or with acetoacetonitrile to give the pyrazolopyrimidine (**288**; Scheme 85). A similar reaction sequence is shown in Scheme 86 ⟨74JHC423⟩.

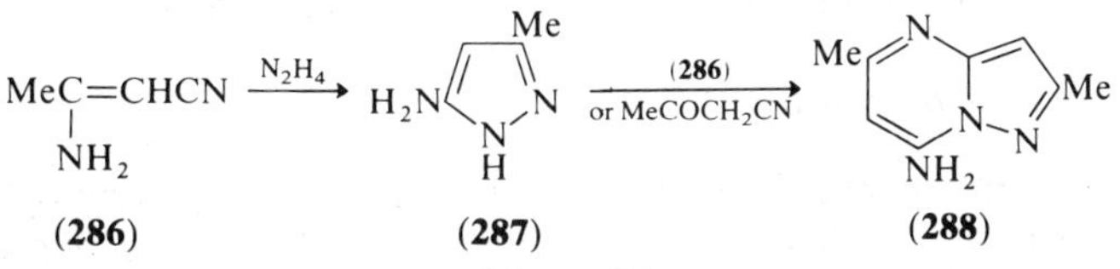

Scheme 85

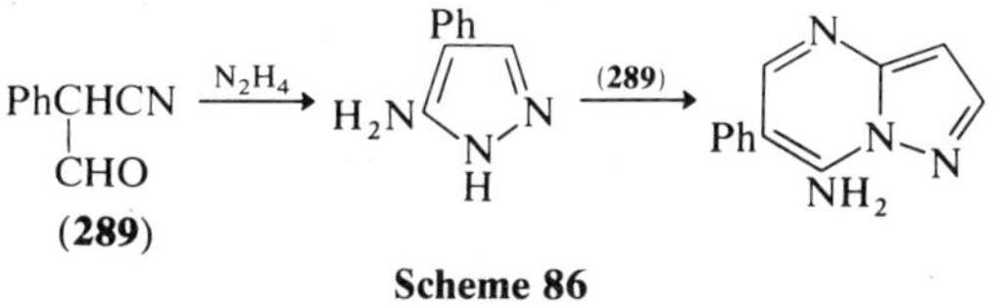

Scheme 86

Scheme 87 illustrates two potentially useful routes to the pyrazolo[3,4-d]pyrimidine system. In (**290**) the usual higher reactivity of the halogen at C(4) is seen (see Section 4.05.3.3). Ethylamine monosubstitutes at room temperature, but disubstitution requires a temperature of 100 °C ⟨75S645⟩. Cyanoguanidine converts the aminopyrazole (**263**) into the pyrazolotriazine (**291**) on heating at 160 °C ⟨57G597⟩.

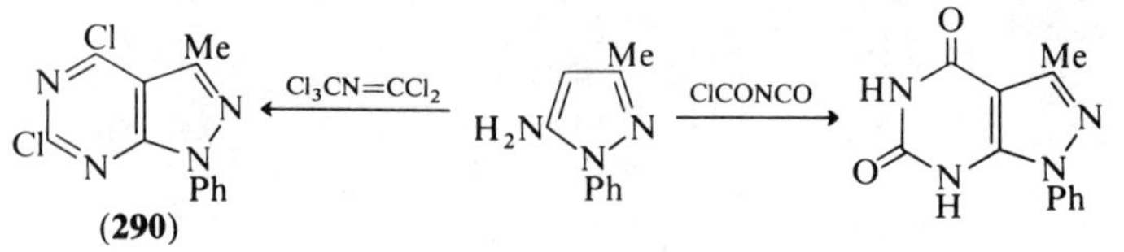

Scheme 87

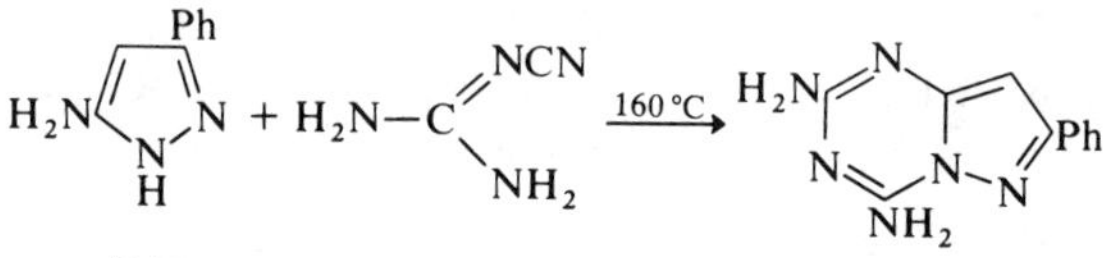

Aminopyrazoles (**292**) carrying ribosyl groups are used to prepare potential antineoplastic compounds (Scheme 88). The aminopyrazolotriazine (**293**) is readily hydrolyzed (see Section 4.05.3.3) to (**294**). After deprotection of the sugar residue, both derivatives showed activity against leukemic cells ⟨B-78MI40500⟩.

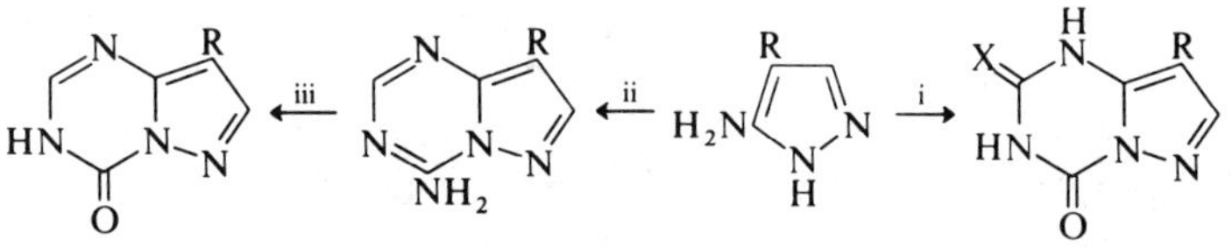

R = protected ribosyl

i, X = O, PhOCONCO, X = S, SCNCO$_2$Et; ii, EtOCH=NCN; iii, K$_2$CO$_3$

Scheme 88

Benzoylhydrazine reacts with ethyl acetoacetate at 130 °C in the absence of solvent to give first the pyrazolone (**295**; R = PhCO). Water from the reaction hydrolyzes the benzoyl group and a second molecule of the ester reacts to give the pyranopyrazolone (**296**) ⟨63BSF2742⟩. The aminopyrazolones (**297**) and carbon disulfide give the zwitterionic pyrazolothiazines (**298**) which S,S-dialkylate with diazomethane or alkyl iodides ⟨78FES799⟩.

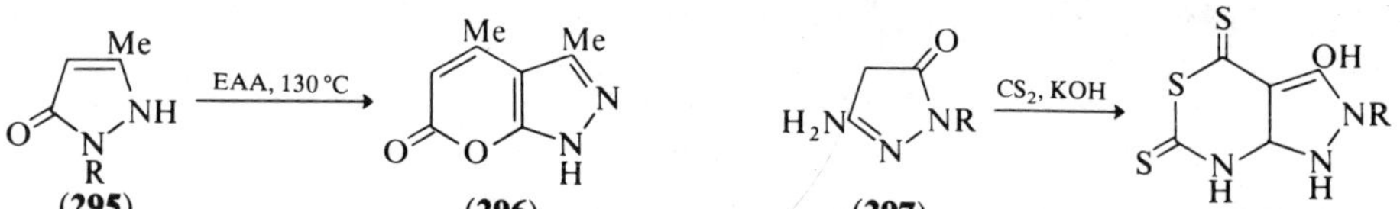

R = H, Me, Ph

4.05.4.4 Five-membered Ring Synthesis by Formation of One Bond

4.05.4.4.1 Between the heteroatoms

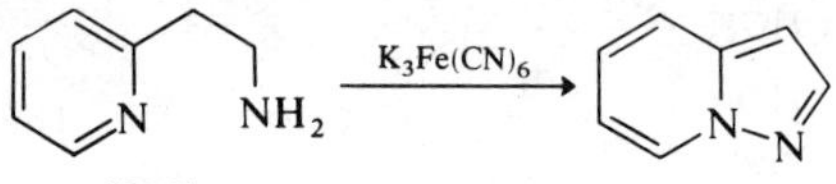

Mild oxidation converts the pyridylethylamine (**299**) into pyrazolo[1,5-a]pyridine (**1**) ⟨57JCS4510⟩. The useful ring closure of the amine on to the nitro group of (**300**) proceeds as shown in Scheme 89. Similar procedures provide the pyridopyrazolobenzimidazole (**301**) and the pyrazolopyridines of Scheme 90 ⟨73JCS(P1)319⟩. However, under different conditions the nitroamines (**302**) give the hitherto difficult to obtain *N*-oxides (**303**). The *N*-oxide groups can be reduced (H_2/Pd/C) ⟨77CC556⟩.

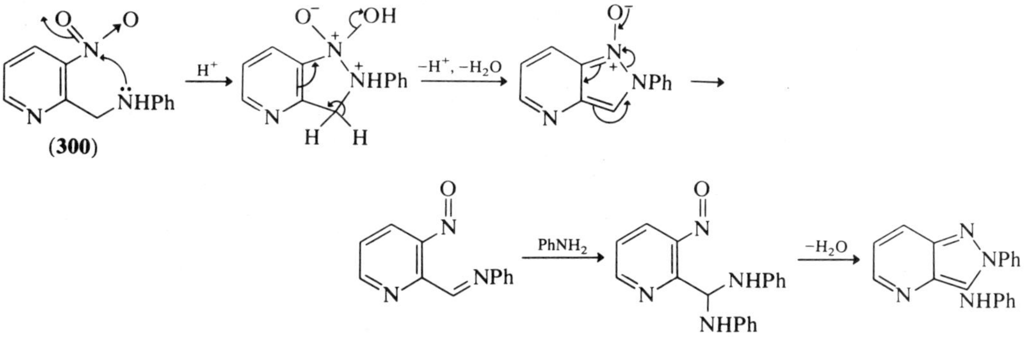

Reagents: TsOH, $PhNH_2$, EtOH, reflux
Scheme 89

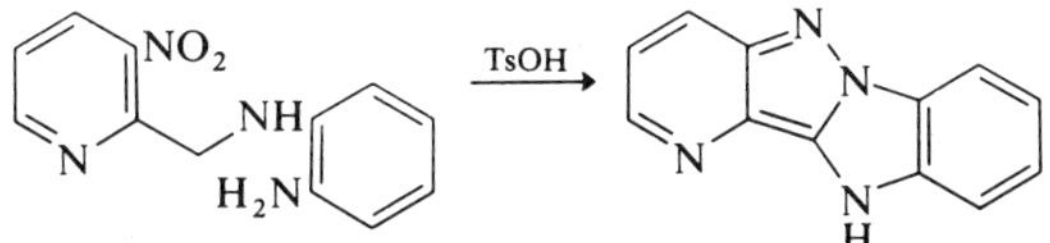

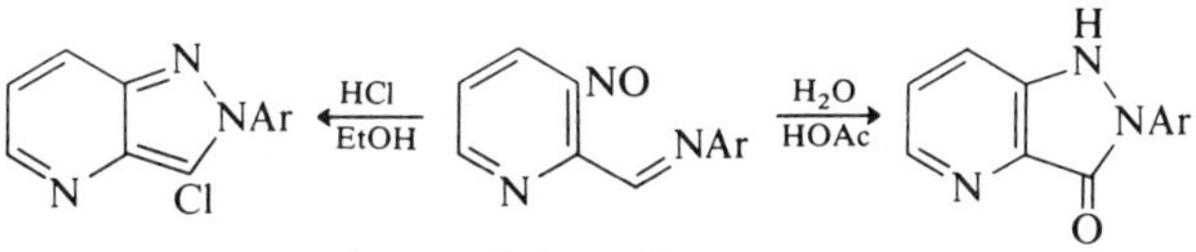

Scheme 90

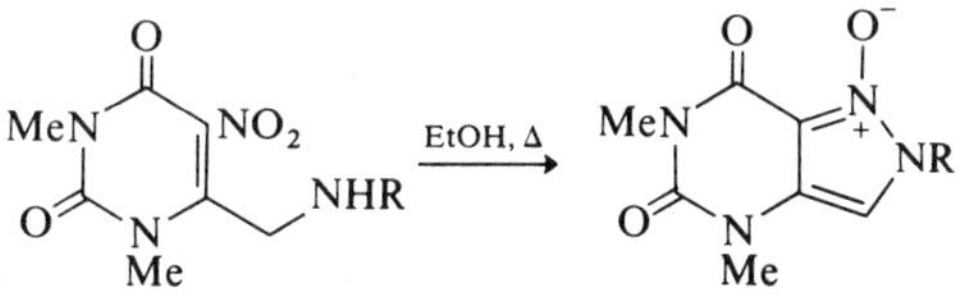

(**302**) R = Me, Pr^i, Bz, $PhCH_2CH_2$ (**303**)

The azide (**304**; R = N_3) ring closes in boiling decalin to give the pyrido[1,2-*b*]indazole (**305**; 60%) as the only product. Surprisingly, (**304**; R = NO_2) gives the same product (**305**) on treatment with iron(II) oxalate at 300 °C, probably also *via* the nitrene intermediate. This reaction succeeds even if the pyridine ring of (**304**) is quaternized or converted into its *N*-oxide, when the extra group is lost before ring closure ⟨61CJC2516⟩. In contrast, the nitrene from (**306**) gives a mixture of the pyrrolopyridine (**307**) and pyrazolopyridine (**308**) in approximately equal amounts ⟨73JOC3995⟩.

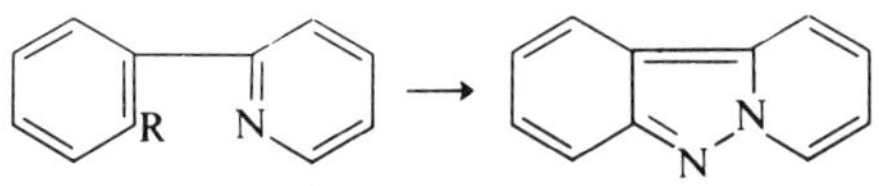

(**304**) R = N_3, NO_2 (**305**)

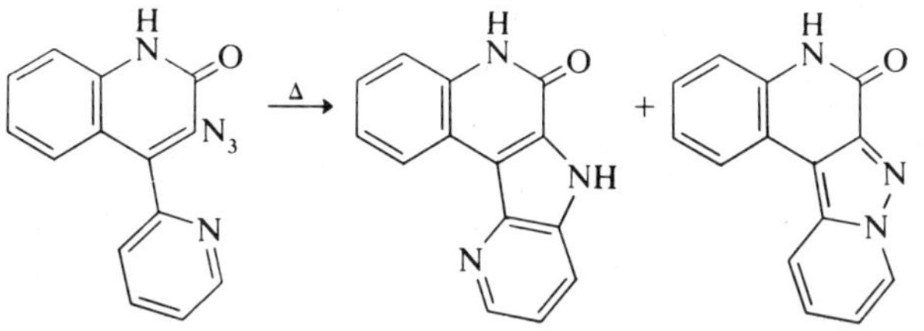

(**306**)   (**307**)   (**308**)

Nitrene generation from nitroso and nitro compounds with triethyl phosphite can also be used. For example, (**304**; R = NO) and TEP give (**305**) in 98% yield ⟨63JCS42⟩. The pyrrole (**309**) and TEP, refluxed in xylene (7 d), give the mesomerically stabilized heteropentalene derivative (**310**) which undergoes cycloaddition reactions (Scheme 91) ⟨79JOC622⟩.

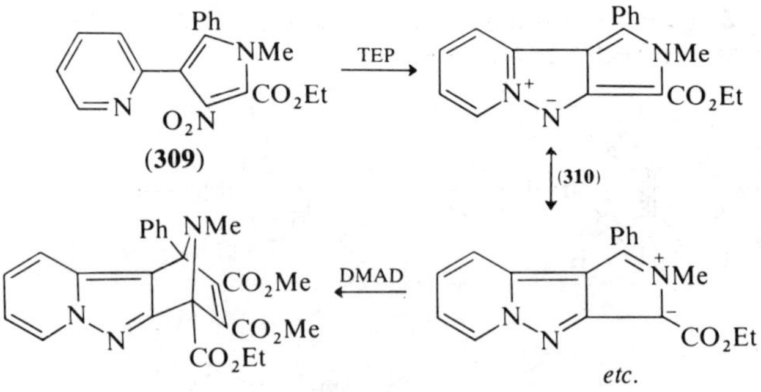

Scheme 91

The hydrazine (**311**; R = NHNH$_2$) heated at 300 °C without solvent gives the triazinoindazole (**312**), but the azide (**311**; R = N$_3$) gives the same product at 200 °C ⟨70JCS(C)2298⟩.

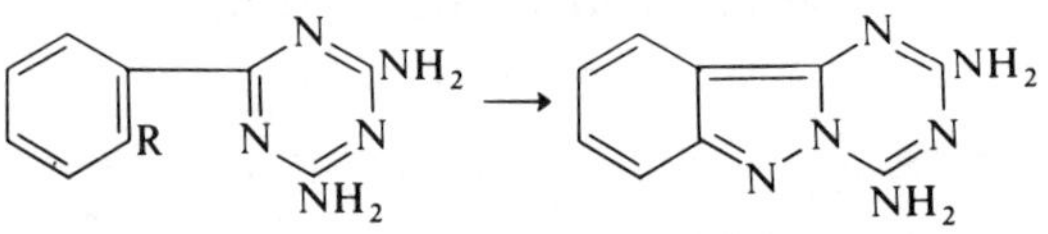

4.05.4.4.2 α to a heteroatom

4-Hydrazinonicotinic acid or its *N*-oxide cyclizes to pyrazolo[4,3-*c*]pyridin-3-one in refluxing hydrochloric acid ⟨65AJC379⟩. The phenylhydrazone (**313**) and cyanoacetamide react in the presence of base to give a mixture of the substituted pyridines (**314**) and (**315**). After separation these can be cyclized to the pyrazolopyridines (**316**) and (**317**). The protonated 'benzylic' methoxy group is displaced in a nucleophilic attack by the hydrazone nitrogen (Scheme 92) ⟨62LA(657)156⟩.

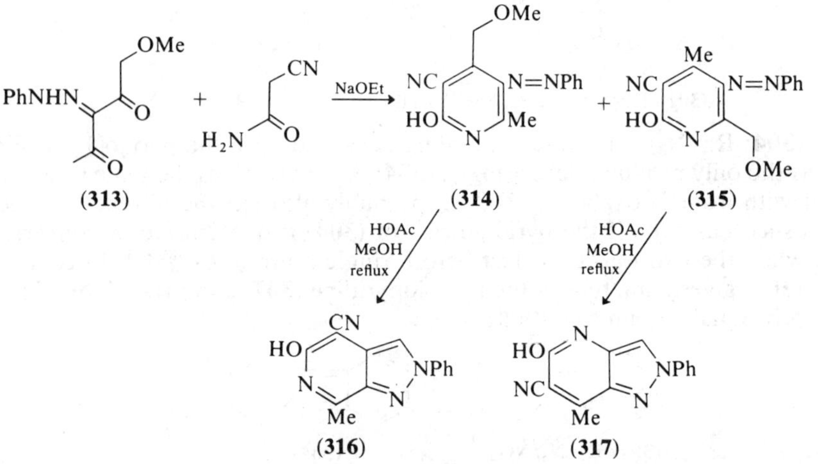

Scheme 92

Several hydrazones like (**318**) are converted by aldehydes in refluxing DMF into pyrazolopyrimidines (**320**). By using ethanol as solvent, the intermediate (**319**) can be isolated (Scheme 93) ⟨77JCS(P1)765⟩.

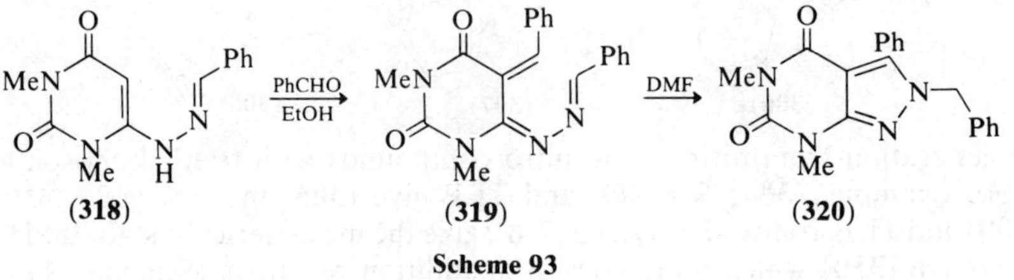

Scheme 93

The phenylhydrazone (**321**) cyclizes in hot hydrochloric acid, presumably *via* nucleophilic attack on the protonated quinoxaline, to (**322**) ⟨71MI40500⟩.

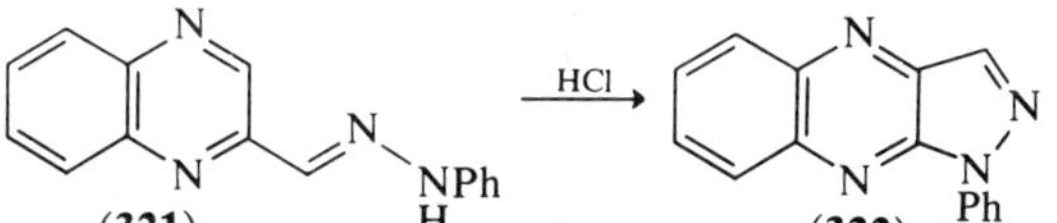

(**321**) (**322**)

4.05.4.4.3 β to a heteroatom

The *N*-iminopyridine (**323**) ring closes and decarboxylates in refluxing xylene to give (**324**) ⟨73JCS(P1)2580⟩. The imine (**325**) similarly cyclizes in boiling toluene to give the ketone (**326**; ν(CO) 1665, ν(C=C) 1630 cm^{-1}) ⟨72T21⟩.

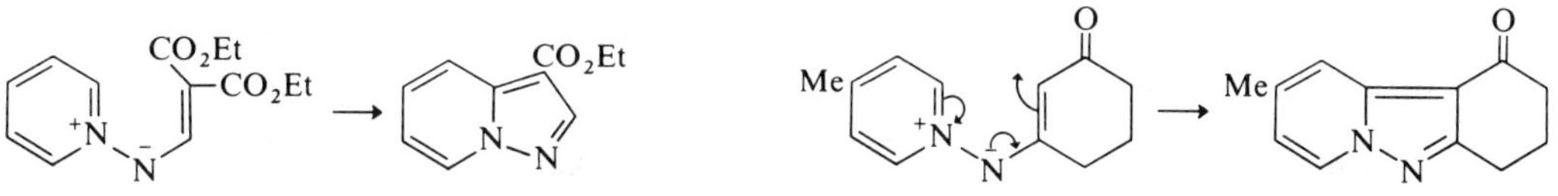

(**323**) (**324**) (**325**) (**326**)

N-Iminopyridines (**327**) react with diketene to give pyrazolo[1,5-*a*]pyridines (**328**) and (**329**) as shown in Scheme 94. Methyl iodide and sodium ethoxide convert (**328**) into a mixture of its *N*- (**329**) and *O*-methyl (**330**) derivatives ⟨75CPB452⟩. Similar cyclizations where ester and nitrile groups stabilize the intermediate carbanion are known ⟨78BCJ251⟩.

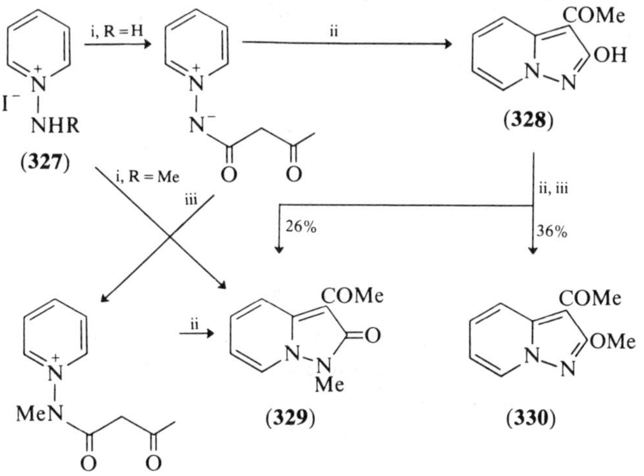

i, K_2CO_3, diketene; ii, NaOEt; iii, MeI

Scheme 94

The hydrazone (**331**; R = H) is converted into (**332**) on heating above its melting point for a few minutes ⟨73S300⟩. The derivative (**331**; R = Me) is not cyclized in this way but refluxing in thionyl chloride ⟨77H(6)945⟩ or photolysis ⟨75BCJ1484⟩ gives (**332**; R = Me). In

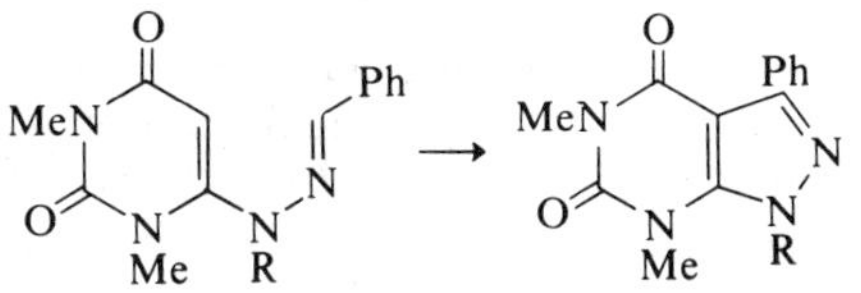

(**331**) (**332**)

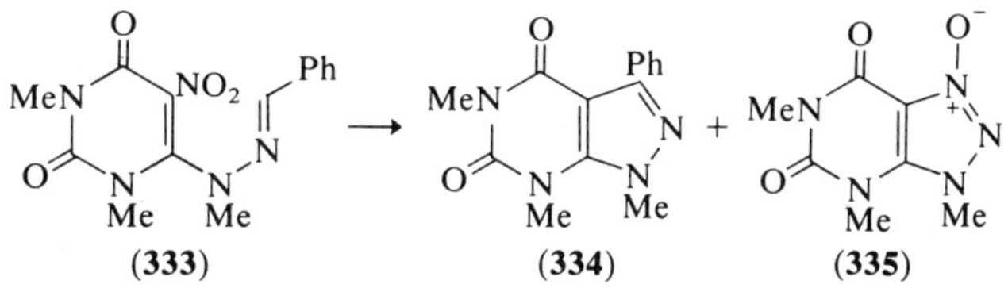

(**333**) (**334**) (**335**)

each case the reaction is accompanied by an oxidation, presumably by air. The nitrohydrazone (333) is ring closed either thermally or photolytically to give a mixture of the pyrazolopyrimidine (334; 40%) with loss of the nitro group and the triazolopyrimidine (335; 35%) with loss of benzaldehyde ⟨74CPB1269⟩.

4.05.4.5 Five-membered Ring Synthesis by Formation of Two Bonds

4.05.4.5.1 From [4+1] atom fragments

Pyrazole rings are formed from 2-alkylpyridine-N-imines and acid chlorides under basic conditions. For example, the 1-aminopyridinium salt (336) gives first (337) but this is very reactive and is acetylated under the reaction conditions to (338; Scheme 95) ⟨68JOC3766⟩. The 2-hydroxymethylpyridine (339) shows similar reactions (Scheme 96) ⟨73CPB2146⟩.

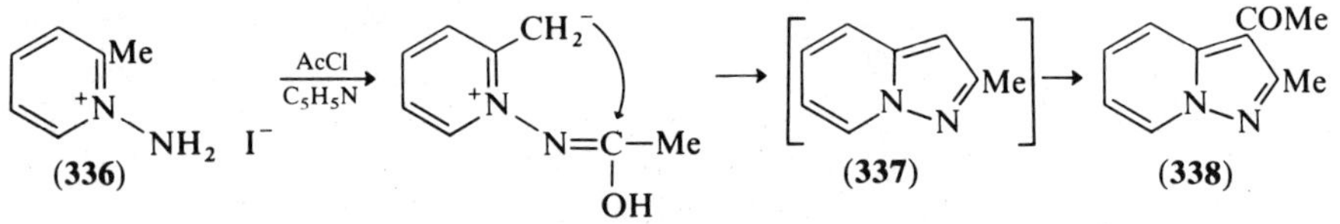

Scheme 95

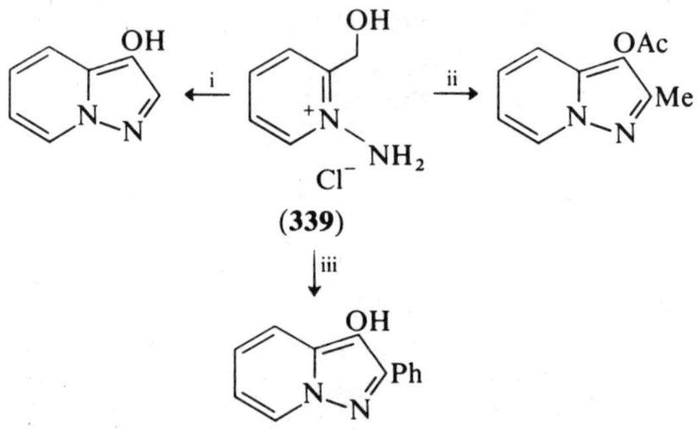

i, (EtO)$_3$CH, HOAc, NaOAc; ii, Ac$_2$O, NaOAc; iii, PhCOCl, K$_2$CO$_3$, H$_2$O

Scheme 96

The hydrazine (340) and acetic anhydride give the pyrazolopyrimidine (341; R = Me). With mixed formic–acetic anhydride the unsubstituted derivative (341; R = H) is obtained ⟨58LA(615)42⟩. The pyridylacetic ester (342) reacts with hydroxylamine-O-sulfonic acid in water at room temperature to give the phenolic compound (13) (see Section 4.05.2.4) ⟨76BCJ1980⟩.

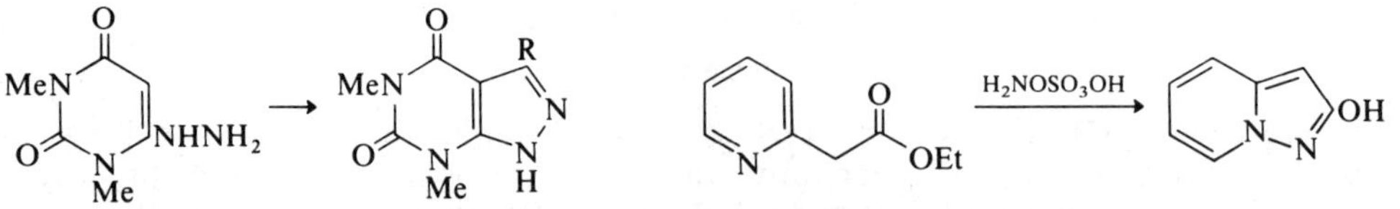

Pyrimidines with alkyl and primary amine groups on neighbouring carbon atoms can be converted into pyrazolopyrimidines with nitrous acid. Compound (343) treated with nitrous acid at 20 °C followed by warming gives (344). Remarkably, the 2-amino group does not seem to be affected by these conditions ⟨52JCS3448⟩. The amine (345) gives a trace of pyrimidotriazine (346), which can be filtered off, and the diazonium salt (347) which only cyclizes in 20% sodium hydroxide (Scheme 97) ⟨65JOC199⟩ (cf. Section 4.05.4.6).

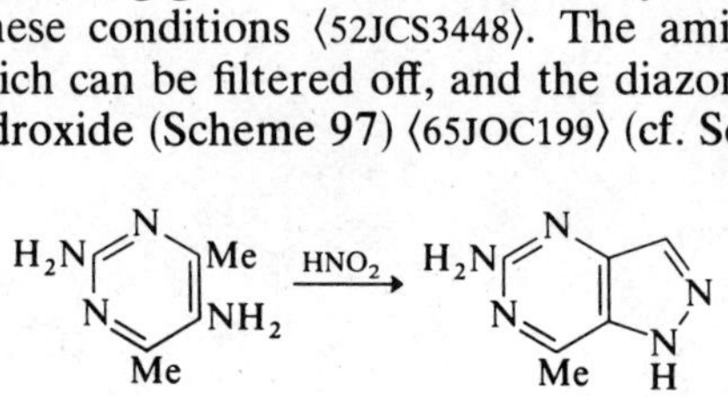

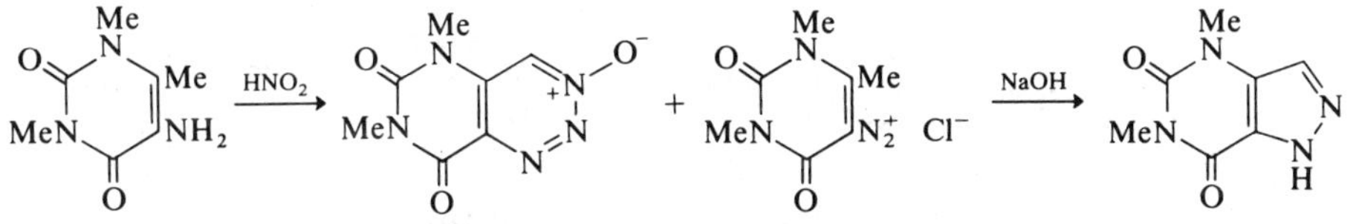

Scheme 97

The amide of 3-amino-2-methylpyridine (**348**) gives the nitroso compound (**349**) which dehydrates in boiling benzene without purification to give the 1-acetylpyrazolopyridine (**350**). Hydrolysis gives the base (**351**) which under mild conditions is reacetylated at N(2) to form (**352**). It is noteworthy that the carbonyl group of (**350**) has $\nu(CO)$ 1710 cm^{-1}, but that of (**352**) has $\nu(CO)$ 1740 cm^{-1}. This high figure presumably reflects the contribution of the dipolar mesomer for which the pyridine ring is a 6π-system (Scheme 98) ⟨73JCS(P1)2901⟩.

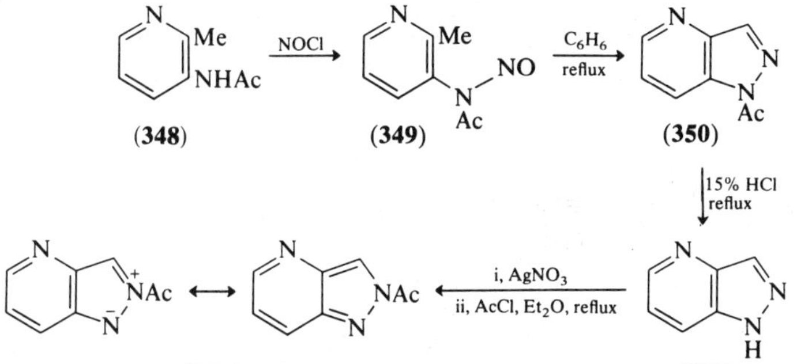

Scheme 98

4.05.4.5.2 From [3+2] atom fragments

Activated alkynes add to N-iminopyridines to provide a simple, general route to pyrazolo[1,5-a]pyridines. Oxidation is usually spontaneous under the reaction conditions and decarboxylation occurs at moderate temperatures (Scheme 99) ⟨68JOC2062⟩. Substituents at position 3 of the pyridine ring generally lead to mixtures of products, but in high total yields ⟨75JCS(P1)406⟩. N-Iminopyridine also reacts with dimethyl 2-chloro-fumarate or -maleate to give in both cases the *trans* product (**353**) which is a stable solid. In solution at room temperature it smoothly ring closes to (**354**), which can be oxidized by palladium at room temperature, by heating in benzene, or photochemically (Scheme 100) ⟨72JOC3106⟩. Ethyl acetoacetate and acetylacetone also undergo useful [2+3] cycloadditions with N-iminopyridines ⟨73JHC821⟩. N-Iminoquinoline and N-iminoisoquinoline react similarly, the latter to give only pyrazolo[3,2-a]isoquinoline ⟨62TL387⟩. Pyrazolo[1,5-a]quinoxalines and pyrazolonaphthyridines are also available ⟨75JHC119⟩. N-Iminopyridazine can be used for the preparation of pyrazolo[1,5-b]pyridazine and derivatives (Scheme 101) ⟨74CC941⟩.

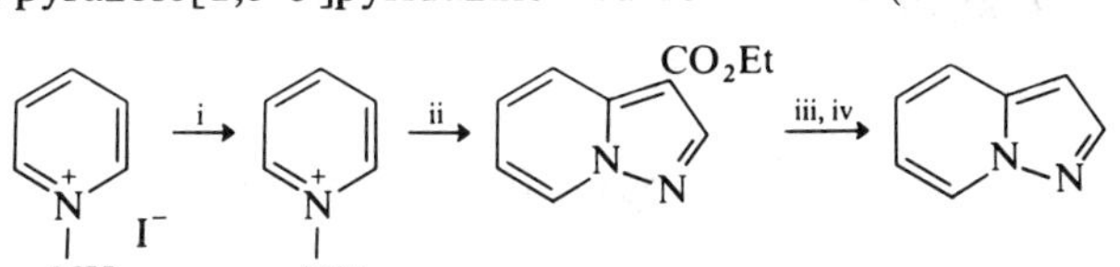

i, K$_2$CO$_3$, DMF, 25 °C; ii, ethyl propiolate; iii, KOH, MeOH, 25 °C; iv, distil

Scheme 99

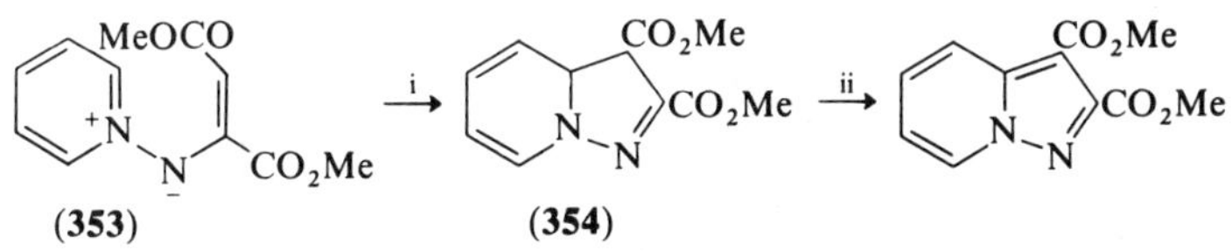

i, CH$_2$Cl$_2$, 25 °C; ii, Pd, C$_6$H$_6$, 25 °C; or C$_6$H$_6$, boil; or $h\nu$

Scheme 100

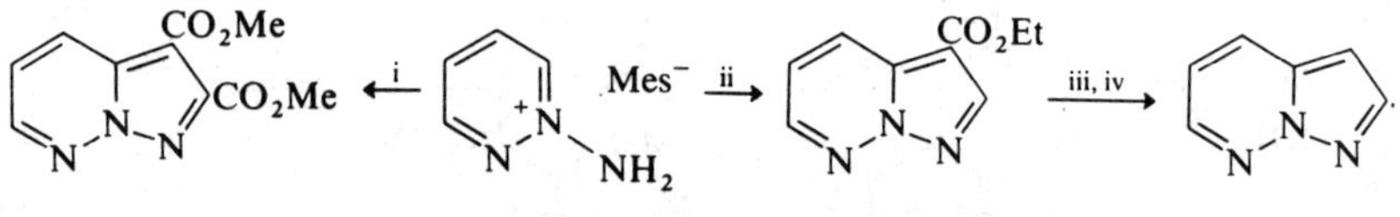

i, DMAD, py; ii, ethyl propiolate; iii, KOH, MeOH; iv, 57% HI

Scheme 101

Some interesting cycloadditions to 3-pyridyne (**355**) are illustrated in Scheme 102. *N*-Phenylsydnone (**356**) gives (**357**), which spontaneously loses carbon dioxide to give the pyrazolopyridine (**358**). Benzaldehyde phenylhydrazone is oxidized by lead tetraacetate to the nitrilimine dipole (**359**), which also reacts with 3-pyridyne to give a pyrazolopyridine (**360**) ⟨71BCJ858⟩.

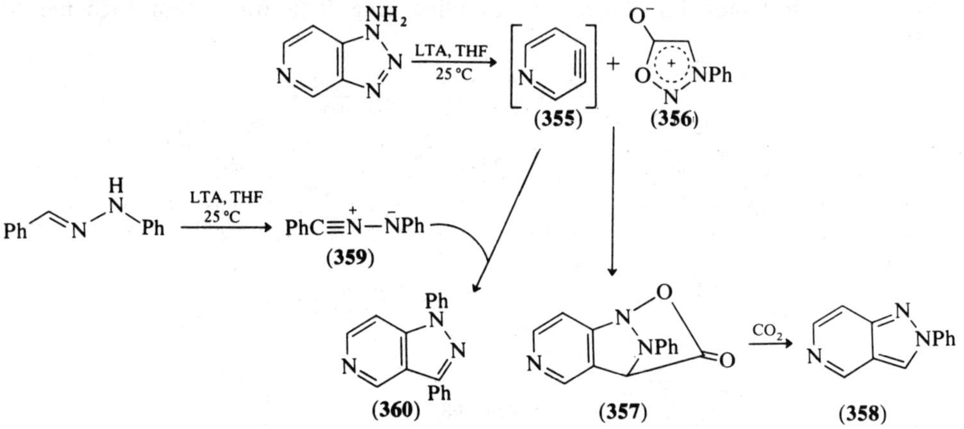

Scheme 102

The ester (**361**) is the first product formed when the corresponding acid is treated with diazomethane. However, it reacts further to give, as the major product of a mixture, the pyrazolopyridone (**362**; 54%), probably by the mechanism shown in Scheme 103 ⟨76CPB1870⟩.

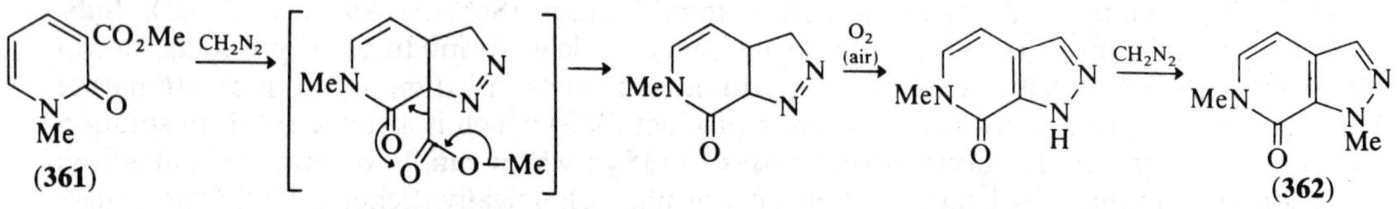

Scheme 103

Hydrazine and its derivatives react with readily available chloro aldehydes, ketones, esters and nitriles to give a variety of fused systems. A few typical examples are shown in Schemes 104 ⟨77H(6)727⟩, 105 (a preparation of allopurinol in 74% yield) ⟨71GEP1950076⟩, 106 ⟨64CB3349⟩ and 107 ⟨81JHC333⟩.

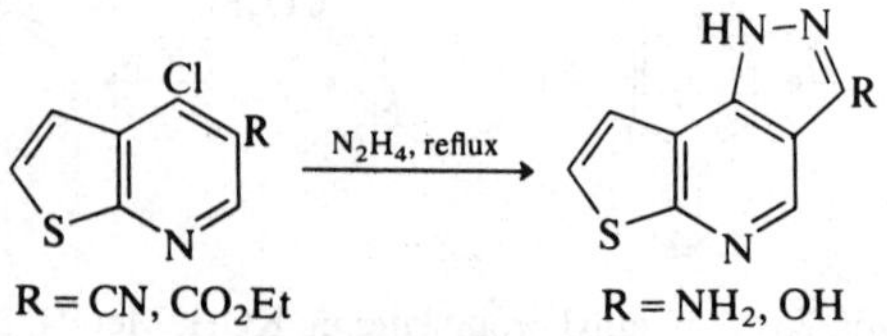

Scheme 104

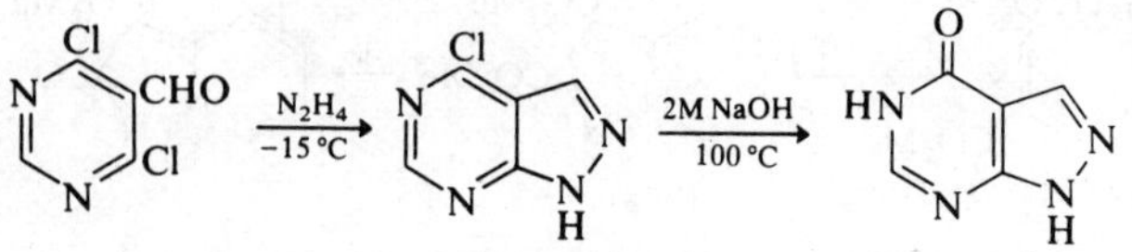

Scheme 105

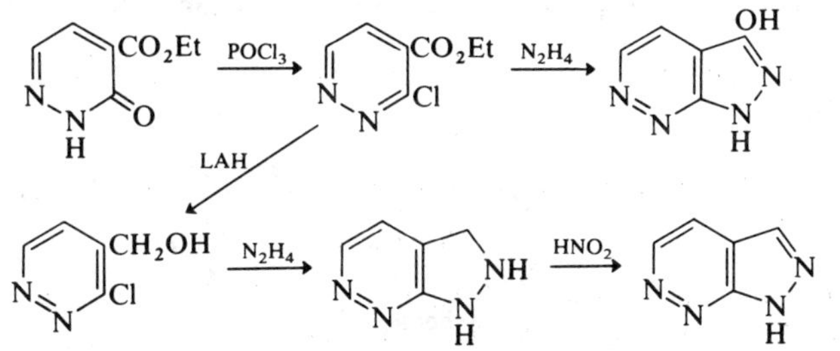

Scheme 106

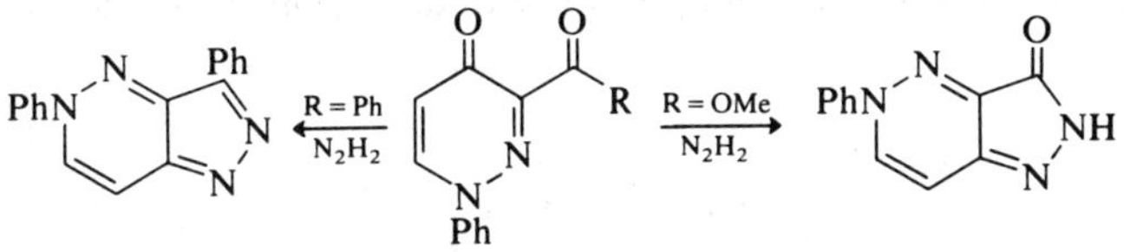

Scheme 107

Pyridazine and its fused derivatives react readily with diphenylcyclopropenethione to give pyrazolo[1,2-a]pyridazines. The derivatives of diphenylcyclopropenone are sometimes also available (Scheme 108). The proximity of the sulfur or oxygen atom to one of the aromatic protons causes a considerable downfield shift, as shown in Scheme 108 ⟨71CJC3119⟩.

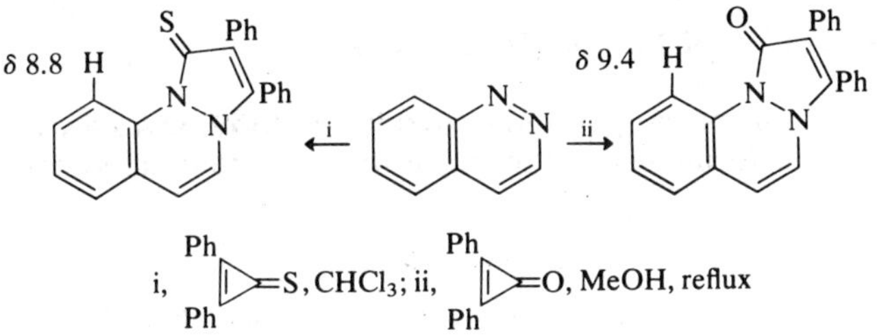

δ 8.8 H δ 9.4 H

i, Ph-C=S, CHCl$_3$; ii, Ph-C=O, MeOH, reflux

Scheme 108

A standard pyrazolone preparation applied to piperidone esters is shown in Scheme 109 ⟨34JA700⟩. Standard procedures for the preparation of saturated and partially unsaturated pyrazolo[1,2-a]pyridazines are shown in Schemes 110 ⟨68BSF4222⟩, 111 ⟨70CPB1526⟩ and 112 ⟨65CR(261)1872⟩.

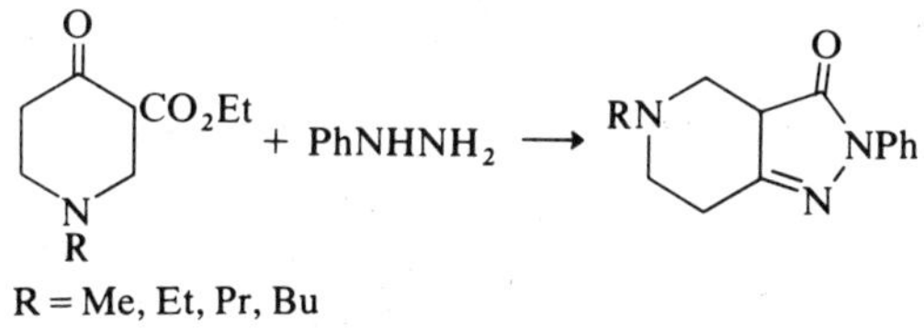

R = Me, Et, Pr, Bu

Scheme 109

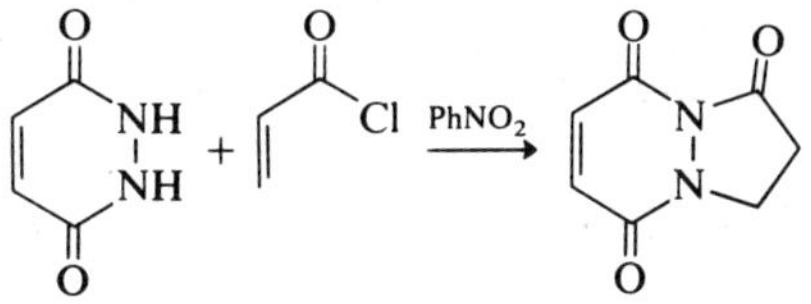

Scheme 110

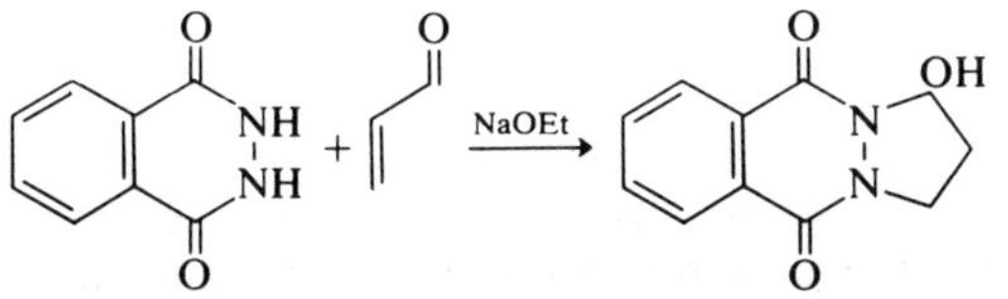

Scheme 111 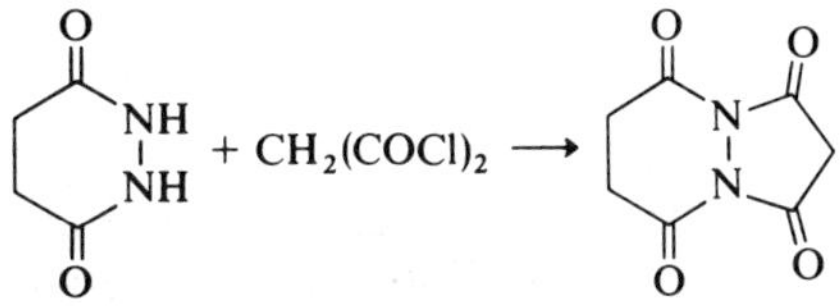   **Scheme 112**

The benzopyran (**363**) and diazomethane give the benzopyranopyrazole (**364**). The chlorohydrazone (**365**) loses HCl in the presence of a weak base to give an azomethine imine which undergoes a 1,3-dipolar cycloaddition with (**363**) to give (**366**; Scheme 113) ⟨81BCJ217⟩.

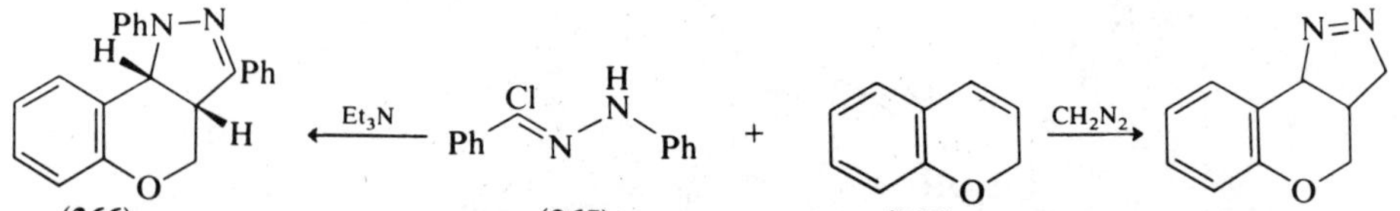

(**366**) (**365**) (**363**) (**364**)

Scheme 113

Scheme 114 shows the preparation of benzothiopyranopyrazolones (**367**) and their further oxidation to (**368**). Both series contain potent immunosuppressants ⟨81JMC830⟩.

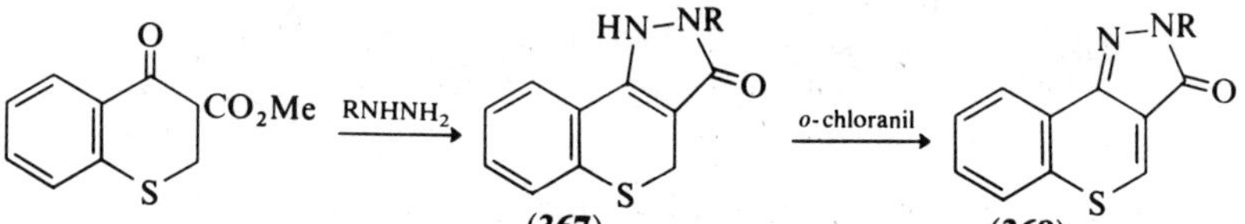

(**367**) (**368**)

R = H, Me, Ph, substituted Ph

Scheme 114

4.05.4.6 By Transformation of Existing Heterocycles

Oxidative rearrangement of the pyrimidinopyridazine (**369**) gives a pyrazolopyrimidine (**370**) which can be transformed into the unsubstituted aromatic system (Scheme 115) ⟨74TL3893⟩.

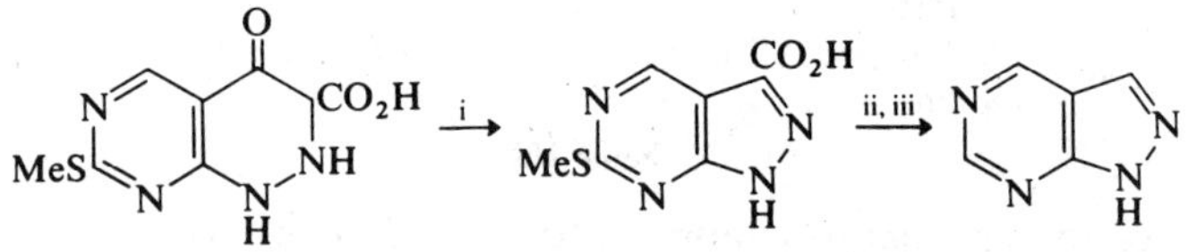

(**369**) (**370**)

i, KOH, $KMnO_4$; ii, 300 °C ($-CO_2$); iii, Raney Ni

Scheme 115

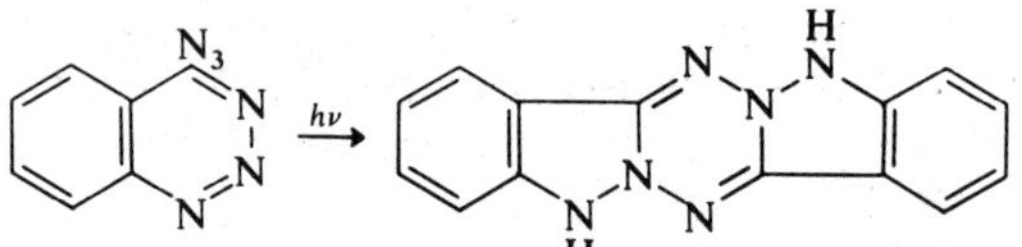

(**371**) (**372**)

The azide (**371**) dimerizes on photolysis to the pentacycle (**372**; 85%) ⟨71JHC785⟩. The *N*-oxide (**373**), available in good yield in this case (*cf.* Section 4.05.4.5.1), is transformed into a pyrazolopyrimidine (**374**) under reductive conditions (Scheme 116) ⟨65JOC199⟩.

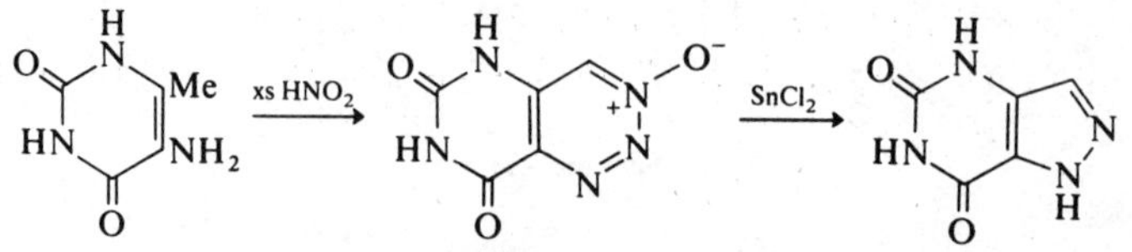

(**373**) (**374**)

Scheme 116

N-Substituted isatoic anhydrides react with anions from pyrazolones to give pyrazoloquinazolinones. A large number of fused pyrazoles has been prepared by this route, *e.g.* as in Scheme 117 ⟨81JHC117, 81JMC735⟩.

Dehydroacetic acid (**375**) reacts with some ketone reagents to give the pyrazolopyrimidines in Scheme 118. A study of the UV spectra of alkyl derivatives showed that the predominant tautomers are as indicated ⟨72CB388⟩.

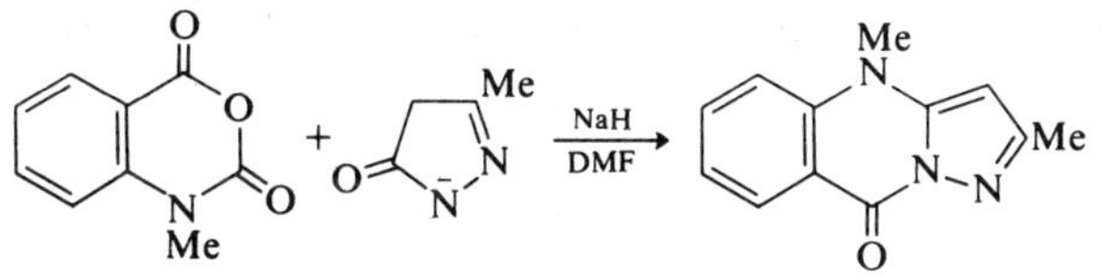

Scheme 117

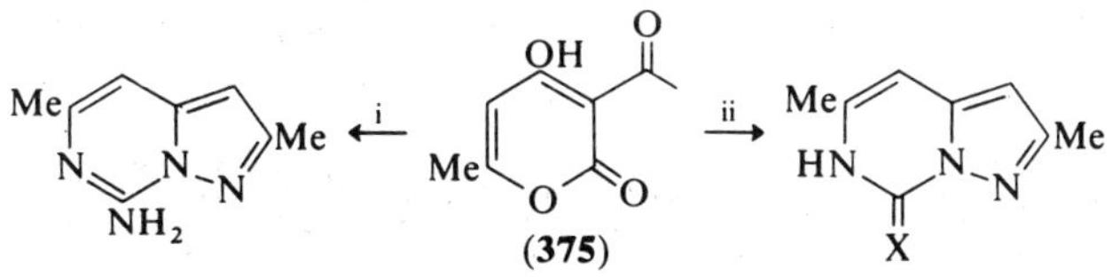

Scheme 118

i, *N*-aminoguanidinium chloride; ii, NH$_2$CONHNH$_2$ or NH$_2$CSNHNH$_2$

Contraction of the pyrazolothiazepine (**376**) on treatment with sodium hydride and methyl iodide in DMF or DMSO probably follows the mechanism shown in Scheme 119 ⟨75JHC1137⟩.

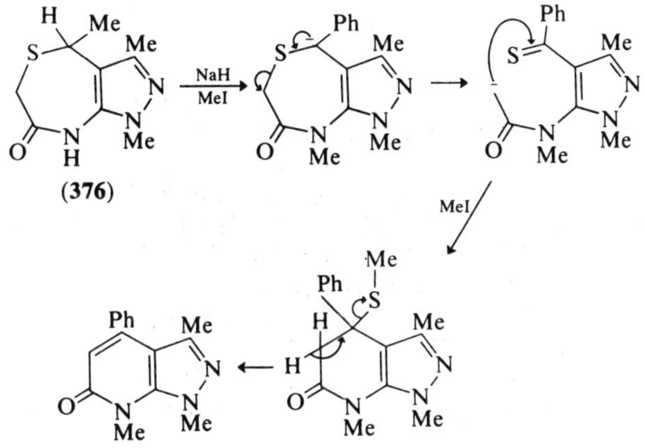

Scheme 119

4.05.5 IMPORTANT COMPOUNDS

Most of the work discussed in this chapter has been stimulated by the discovery of a few biologically active compounds. As is usual in medicinal chemistry, only a few out of many thousands of compounds synthesized have found their way on to the market.

Perhaps the most widely used at present is allopurinol (**85**; R^2 = H) in the treatment of gout. This and its congeners oxyallopurinol (**87**; R^2 = H) and thiopurinol (**112**) inhibit xanthine oxidase and so inhibit the oxidation of xanthine and hypoxanthine to uric acid, the causative agent of gout. This group of compounds also shows some antineoplastic activity ⟨71JCS(C)1610⟩.

Much effort has been expended on the search for antitumor and antiviral compounds based on the antibiotic formycin (**223**). No derivative seems to have reached the market yet, but exciting research results continue to appear ⟨B-70MI40500, 81JCS(P1)2387⟩.

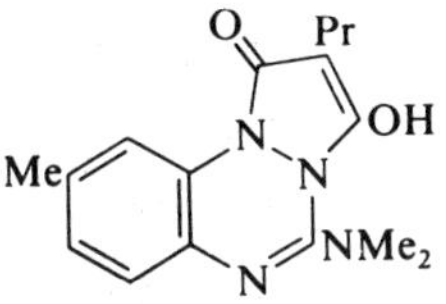

Azapropazone (**377**) has found some use in the treatment of rheumatic disease and other musculoskeletal disorders ⟨71AF1532⟩.

4.06

Imidazoles and their Benzo Derivatives: (i) Structure

M. R. GRIMMETT
University of Otago

4.06.1 INTRODUCTION		345
4.06.2 THEORETICAL METHODS		346
4.06.3 STRUCTURAL METHODS		350
4.06.3.1 X-Ray Diffraction		*350*
4.06.3.2 Dipole Moments		*351*
4.06.3.3 ^1H NMR Spectroscopy		*352*
4.06.3.3.1 Aromatic systems		*352*
4.06.3.3.2 Non-aromatic systems		*353*
4.06.3.4 ^{13}C NMR Spectroscopy		*354*
4.06.3.4.1 Aromatic systems		*354*
4.06.3.4.2 Non-aromatic systems		*355*
4.06.3.5 NMR Involving Other Nuclei		*355*
4.06.3.6 Ultraviolet Spectroscopy		*355*
4.06.3.6.1 Aromatic systems		*355*
4.06.3.6.2 Non-aromatic systems		*357*
4.06.3.7 Infrared Spectroscopy		*357*
4.06.3.7.1 Aromatic systems		*357*
4.06.3.7.2 Non-aromatic systems		*358*
4.06.3.8 Mass Spectrometry		*358*
4.06.3.8.1 Aromatic systems		*358*
4.06.3.8.2 Non-aromatic systems		*360*
4.06.3.9 Electron Spin Resonance Spectra		*361*
4.06.3.10 Photoelectron Spectra		*361*
4.06.4 THERMODYNAMIC ASPECTS		362
4.06.4.1 Intermolecular Forces		*362*
4.06.4.1.1 Melting and boiling points		*362*
4.06.4.1.2 Solubilities		*362*
4.06.4.1.3 Gas–liquid chromatography		*362*
4.06.4.2 Thermochemistry; Aromatic Stability		*362*
4.06.5 TAUTOMERISM		363
4.06.5.1 Annular Tautomerism		*363*
4.06.5.2 Ring-substituent Tautomerism		*365*
4.06.5.3 Ring–Chain Tautomerism		*371*
4.06.5.4 Betaines		*372*

4.06.1 INTRODUCTION

Imidazole (**1**) is a planar, five-membered ring system with three carbons and two nitrogen atoms in the 1- and 3-positions. The systematic name for the compound then is 1,3-diazole. One of the annular nitrogens bears a hydrogen atom and can be regarded as a pyrrole-type nitrogen; the other resembles the nitrogen in pyridine. For this reason it is possible to look upon imidazole as a molecule which has overlapping properties of both pyrrole and pyridine. Contributions of one π-electron from each carbon and the 'pyridine' nitrogen and two from the 'pyrrole' nitrogen make up an aromatic sextet.

Benzimidazole (**2**) has a benzene ring fused in the 4,5-positions of an imidazole ring. The molecules (**1**) and (**2**) are numbered as depicted on the structures.

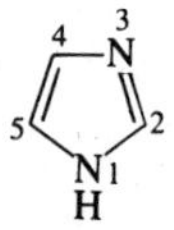

The partially and fully reduced derivatives of imidazole are variously designated as 2-imidazoline (**3**), 3-imidazoline (**4**), 4-imidazoline (**5**) and imidazolidine (**6**). Such compounds are no longer planar and lack the aromatic character of imidazole and benzimidazole. The great bulk of the published material relates to the fully unsaturated compounds.

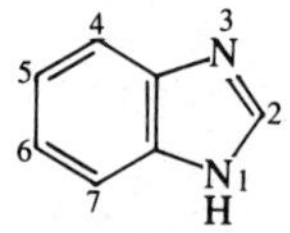

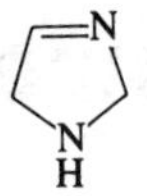

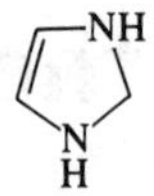

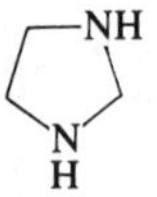

4.06.2 THEORETICAL METHODS

As has been mentioned earlier (Section 4.01.2), many of the theoretical studies applied to the azoles have endeavoured to achieve comparisons within the series. Thus calculations of electron densities, dipole moments, and energies of formation give values which reflect the decrease in azole stability as the numbers of nitrogen atoms increase. Imidazole, therefore, stands near the head of a decreasing stability series.

A survey of quantum mechanical studies shows that the HMO method (which is not of great current interest) has been used mainly in connection with dipole moments, while the VESCF method has been employed likewise in such calculations ⟨B-76MI40600, 74JCS(P2)420, 72CHE1131, 78JCS(P2)865⟩.

Values calculated for π-electron densities at various positions in the imidazole molecule are, at times, contradictory ⟨66RCR122, 70AHC(12)103, 80AHC(27)241⟩, but this merely reflects the gradual development of the computational technique by the MO LCAO method. When the LCAO–SCF method was used, taking into account the differences between N-1 and N-3, it showed that in the imidazole neutral molecule the electron density at C-4 (or C-5) is greater than at C-2 between the two nitrogens; in the corresponding anion and cation the electron density is higher at C-2 (Section 4.01.2.2). Earlier calculations, which did not take into account the differences between the two nitrogen atoms, had shown qualitatively the same trends. Table 1 gathers together the results of a number of these theoretical studies.

More recent calculations of σ-complex energies and total electron densities for the imidazole cation, anion and neutral molecule demonstrate that, where electrophilic attack is believed to involve a σ-complex, good agreement can be obtained with the experimentally observed site of substitution. SCF calculations of π-electron distributions for the ground

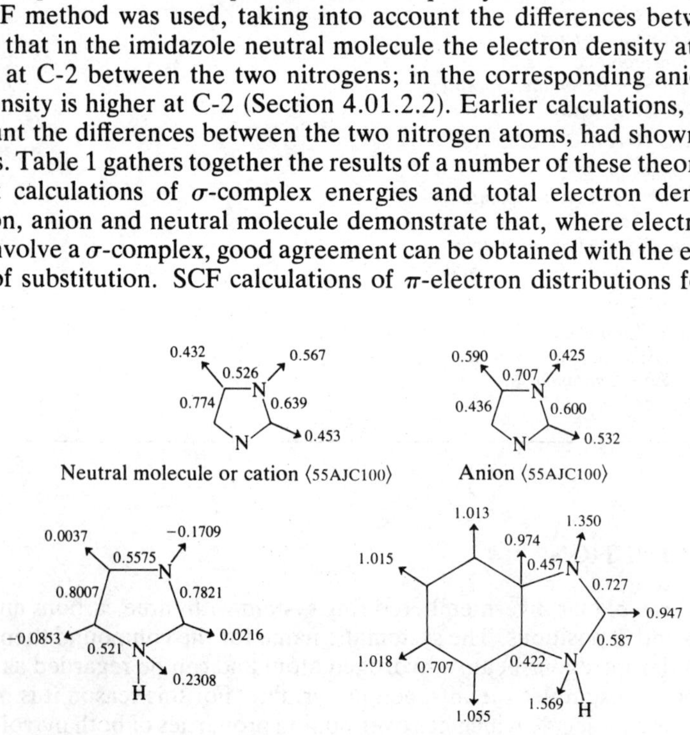

Figure 1 Bond localization energies, π-densities and/or free valencies for imidazole and benzimidazole

Table 1 π-Electron Densities in Imidazole

Neutral molecule					Anion			Cation			Ref.
N-1	C-2	N-3	C-4	C-5	N-1(3)	C-2	C-4(5)	N-1(3)	C-2	C-4(5)	
1.502	0.884	1.502	1.056	1.056	0.903	1.429	1.323	—	—	—	54JCS2701
1.479	0.907	1.479	1.067	1.067	0.835	1.458	1.436	—	—	—	55AJC100
1.650	1.100	1.101	1.037	1.112	1.197	1.207	1.200	1.512	0.997	0.990	59AJC543
0.2308	0.0216	−0.1709	0.0370	−0.0853	—	—	—	—	—	—	62T985
1.347	0.991	1.286	1.104	1.072	—	—	—	—	—	—	B-63MI40600
—	0.833	—	1.06	1.06	—	0.833	1.06	—	0.833	1.06	66T835, 67T2513
1.525	1.012	1.315	1.006	1.142	—	—	—	—	—	—	70BCJ3344
0.225	−0.139	−0.150	−0.066	−0.350	—	—	—	—	—	—	78JCS(P2)865
—	0.031	—	—	—	—	−0.458	—	—	0.289	—	73CHE88
0.298	0.094	−0.287	−0.068	−0.037	—	—	—	—	—	—	79CHE939

state of imidazole have predicted the order of electrophilic substitution as 5>2>4, but these calculations did not take into account the tautomeric equivalence of C-4 and C-5 (see Section 4.06.5.1.1) ⟨70BCJ3344, 78JCS(P2)865⟩. The results of calculations ⟨55AJC100, 62T985⟩ of bond orders and free valences for imidazole are collected in Figure 1.

While the π-densities predict greater reactivity to electrophilic attack at C-2 in the imidazole anion, localization energies suggest an order of reactivity of C-4>C-2. Use of the CNDO/2 method predicts the order of electrophilic substitution in benzimidazole to be 5≈7>6>4>2, whereas the observed order (for aqueous bromination) is 5>7>6,4,2 ⟨78JCS(P2)865⟩. Predictions of preferential nucleophilic attack at C-2 are confirmed by experiment. Table 2 lists π-electron densities in benzimidazole.

Semi-empirical LCAO calculations for all azoles, introducing σ-electrons, indicate that charges are weak except those on NH nitrogen atoms, and that the σ-dipolar moments are close to those of lone pairs. It is therefore inappropriate to take σ-polarity into account in the approximations used in π-calculations for these heterocycles. A number of other quantum mechanical calculations have been applied to reactions of imidazoles ⟨80AHC(27)241⟩, while the nucleophilic substitution reactions at C-2 of benzimidazoles, and diazo coupling at C-2 of uncondensed imidazoles have been discussed from theoretical points of view.

Several workers have examined the electronic spectra of the azoles using Pariser–Parr–Pople type calculations, while even more interest has centred on NMR spectroscopy. Using EHT calculations it has proved possible to obtain good correlation between total electron densities and ^{1}H and ^{13}C chemical shifts, but the method does not work for spin–spin coupling constants ⟨B-76MI40600⟩. The generalized free electron MO method (G-FEMO) gives a good description of the ground state properties of imidazole, and yields results equivalent to Hückel MO calculations. Calculations (LCAO–SCF) of the π-energies of protonation of imidazole free bases indicate that methyl substituents exert a purely inductive electronic effect. The electronic natures of both imidazole and benzimidazole have been computed taking into account both σ- and π-electrons. The energies of bonds between hydrogen atoms and the hetero ring provide information which has been used to explain changes in stability which had been noticed in pyrolysis experiments. Correlation of thermal stabilities with resonance energies and, in particular, with the energies of HOMO was possible.

Further applications of MO treatments to studies of protonation, hydrogen bonding, tautomerism and conformation have been discussed earlier ⟨80AHC(27)241, B-76MI40600⟩; *ab initio* SCF calculations have, for example, shown that imidazole is a stronger proton-donor to water than pyrrole, and as proton-acceptors the two are quite different. Thus, while imidazole forms a strong hydrogen bond with water through the N pair of σ-electrons, pyrrole forms a relatively weak and flexible π-complex. Good agreement with experiment is found in theoretical studies of isomerism and conformation of azolides.

The valence bond approach allows resonance structures to be drawn for imidazole (**7–12**; Scheme 1), among which are a number of dipolar structures. Certainly such dipolar contributors will be much more important than in the case of benzene. They correctly compute the amphoteric nature of the molecule and its susceptibility to electrophilic attack at N-3, C-4 and C-2. When tautomerism is taken into account, electrophilic attack at C-5 could also be expected. There is also an indication in structure (**11**) that nucleophiles will be attracted to C-2. A structure such as (**12**) can have real meaning only in the presence of very strong bases.

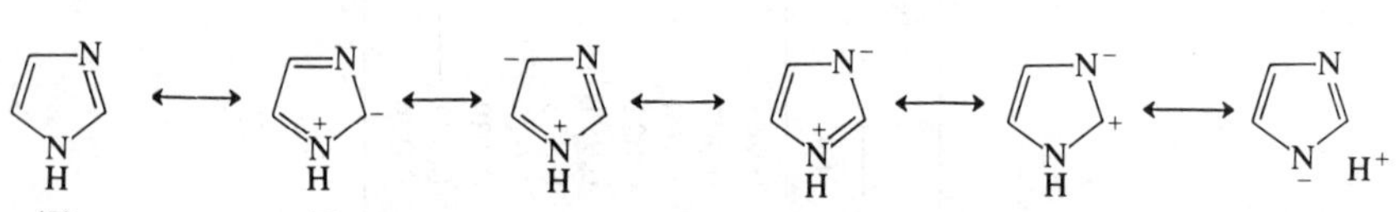

Scheme 1

A set of resonance structures drawn for benzimidazole show similar features as far as amphoteric nature is concerned, but imply that electrophilic attack will be either at N-3 or in the benzene ring; nucleophilic attack at C-2 is predicted (see Scheme 2). Qualitatively the same trends can be noted as are obtained by applications of MO methods.

Table 2 π-Electron Densities in Benzimidazole

	Neutral molecule							Anion			Cation			Ref.
N-1	C-2	N-3	C-4	C-5	C-6	C-7		C-2	C-4	C-5	C-2	C-4	C-5	
0.228	0	−0.121	−0.005	−0.016	−0.011	−0.030		—	—	—	—	—	—	78JCS(P2)865[a]
−0.021	0.174	−0.225	−0.015	−0.024	−0.006	−0.035		—	—	—	—	—	—	78JCS(P2)865[a]
−0.013	0.170	−0.237	−0.010	−0.015	−0.012	−0.015		—	—	—	—	—	—	78JCS(P2)865[a]
—	0.168	—	—	—	—	—		0.391	—	—	0.380	—	—	73CHE88
1.40	0.96	—	1.03	1.04	—	—		1.39–1.73	0.99–1.02	1.14–1.35	0.81	1.03	1.02	56JCS4288
1.569	0.947	1.350	1.013	1.015	1.018	1.055		—	—	—	—	—	—	70BCJ3344
0.256	0.170	−0.312	−0.024	−0.024	−0.021	−0.029		—	—	—	—	—	—	79CHE939

[a] Values quoted from top to bottom used VESCF/BJ, CNDO/2 and CNDO/S respectively.

There has been some success in the application of Hammett treatments to reactions and reactivity of imidazoles ⟨80AHC(27)241, B-76MI40600⟩, but the complications of tautomerism and the different natures of the ring nitrogen atoms have created difficulties. Nevertheless, the use of σ_m and σ_o values of -0.34 and 0.42, respectively, for the ring NH group (replacing a CH=CH group) in conjunction with the Hammett relationship for pyridines has permitted the prediction of reasonably accurate pK_a values for imidazoles. Fairly successful correlation of pK_a values for 1-, 2- and 4-substituted imidazoles is obtained using σ_m and σ_I values; meta- and para-substituted 1-phenylimidazoles have also proved amenable to Hammett treatment. Calculated σ-constants for the imidazole ring in benzimidazole suggest that it is electron releasing, or conversely that the fused benzene ring is electron withdrawing. This is borne out by experimental observations, and indeed follows the predictions of the resonance structures (Scheme 2). Calculations of Taft σ^*-values for alkyl groups have been made from rate constants for N-acylimidazole hydrolysis and imidazole-catalyzed N-acylimidazole hydrolysis in water. The steric effect was found to be the same in both acid- and base-catalyzed reactions, and the range of σ^*_{rel} values ($\sigma^*_{rel} = \sigma^*_{alkyl}/\sigma^*_{ethyl}$) for primary, secondary and tertiary alkyl groups demonstrates no dependence on the degree of branching. The corrected (or 'true') σ^*_{alkyl} values were calculated from the rate constants for C-substituted ester hydrolysis, and show that, in this instance, the steric effects in acid- and base-catalyzed hydrolysis are not the same. This means that the Taft σ^*-values probably give a measure of steric effects. Other applications of Hammett treatments to specific problems such as tautomerism and quaternization studies will be discussed in the appropriate section.

4.06.3 STRUCTURAL METHODS

4.06.3.1 X-Ray Diffraction

The results of a low-temperature study (Figure 2) show imidazole to be a planar molecule with considerable double-bond character in all bonds. The dimensions for 4,5-di-t-butyl-imidazole are essentially similar, but the 4,5-bond is somewhat stretched as might be expected. In imidazole the presence of NH···N bonds with the exceptionally short length of 286 pm can be demonstrated. Such chains of molecules give the crystals a fibrous appearance. A similar X-ray study of a supercooled melt of 4-methylimidazole revealed that in this case the associated molecules in the polymer structure have linear NH···N bonds with a bond length of about 300 pm ⟨B-76MI40600, 80AHC(27)241⟩.

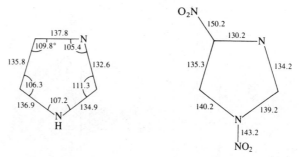

Figure 2 X-ray structures for imidazole and 1,4-dinitroimidazole (bond lengths in pm)

More recently there has been some interest in the structures of the N-nitration products of 4-nitroimidazoles. Both 4-nitro- ⟨81UP40600⟩ and 2-isopropyl-4-nitro-imidazole ⟨72CJC3472⟩ have been shown to N-nitrate with such an orientation that the 1,4-dinitro compound results (see Figure 2).

While imidazolidine-2-thione has been shown to be a fairly planar molecule with only slight distortions caused by crystal forces ⟨53AX369⟩, there is evidence for folded (envelope-type) structures in the 3-aryl-5-benzylhydantoins. Such deviations from the plane may be a consequence of non-bonded aromatic–amide attraction ⟨79JCS(P2)653⟩.

4.06.3.2 Dipole Moments

There has been considerable interest in the measurement of the dipole moments of imidazole and its derivatives since the values obtained should provide valuable information about the structures of the molecules ⟨70AHC(12)103, 80AHC(27)241, 66RCR122⟩.

The high values (16–20 × 10^{-30} C m) obtained by early workers for imidazole resulted in the postulation of a dipolar ion for its structure; even ionic salt-like structures were suggested for imidazole (**13**) and benzimidazole in aqueous solution. Subsequent experiment, however, has demonstrated that these high values were probably obtained as a consequence of their being made in relatively concentrated solutions in which intermolecular association would be significant. The present accepted value for the dipole moment of imidazole is nearer to 13×10^{-30} C m, a value which reflects considerable polarization of the ring, but insufficient to yield an ionic structure or even a dipolar structure. The dipole moment is concentration dependent, a property typical of compounds containing an NH group and forming intermolecular hydrogen bonds of the type N—H···N. At high concentrations in non-aqueous media the association coefficient reaches values between 5 and 20 (**14**), and not 2 as would be expected for an ionogenic structure (**13**).

(**13**)   (**14**)

Dipole moment data indicate that the line of action of the moment in imidazole makes an angle of ~13° with the A-axis (where the proposed A-axis makes an angle of ~28° with the N—H bond and intersects the N(1)—C(2) bond). The molecule is planar, as is benzimidazole, but the application of dipole moment studies to conformational studies has shown that N- and C-aryl, and N- and C-furyl, compounds are not planar, even though planar bicyclic fragments do exist in polyarylimidazoles. In 1-arylimidazoles the dipole is towards the aryl substituent, a finding which is in agreement with the observed reactivity towards electrophilic reagents (Section 4.07.3.2.2). Substituent effects on dipole moments are as expected; non-polar groups and condensed carbocyclic rings have little influence, but polar groups may change matters to a major extent. A selection of dipole moments is listed in Table 3.

Table 3 Dipole Moments for Imidazoles and Benzimidazoles ⟨71CHE809⟩

Compound	p (C m × 10^{30})	Compound	p (C m × 10^{30})
Imidazole	12.8	Benzimidazole	13.3
1-Phenyl-	10.5	1-Methyl-	13.5
1-Propyl-	13.7	1-Benzyl-	11.6
1-Methyl-	12.1	1-Phenyl-	11.2
4,5-Diphenyl-	14.5	5-Nitro-	19.9
2-Formyl-1-methyl-	12.6	1-Ethyl-4,5-dimethyl-	13.7
1,2-Dimethyl-	12.5	2-Methyl-5-nitro-	21.9
1,2-Dimethyl-5-bromo-	11.2	2-Formyl-1-methyl-	11.2
2-Formyl-1-phenyl-	11.8	2-Acetyl-	6.4
2,4,5-Triphenyl-	13.1		

4.06.3.3 ¹H NMR Spectroscopy

4.06.3.3.1 Aromatic systems

The proton magnetic resonance spectrum of imidazole shows a one-proton triplet at δ 7.64 p.p.m. (assigned to the 2-H) and a two proton doublet at δ 7.01 p.p.m. (assigned to the 4- and 5-H atoms). The coupling constants are of the order of 1.0 Hz. Because of the rapid interchange of the N—H proton (see Section 4.06.5.1) there are no spin–spin splittings observed due to interaction with the 1-H. In fact, this NH proton gives rise to a broad signal only in concentrated solutions in aprotic solvents such as benzene, chloroform or acetone; the positions and half-widths of these signals depend very much on the concentration. As the concentration increases the signal shifts to lower field with a relatively large shift (>10 p.p.m.) compared with that observed for other such hydrogen bonded compounds.

One of the major difficulties encountered in attempting to obtain NMR spectra of imidazoles unsubstituted on a ring nitrogen is a consequence of their poor solubility in non-polar solvents (Section 4.06.4.1.2). Spectra in D_2O are frequently obtained, but where there are exchangeable hydrogens present (e.g. as in hydroxyalkyl- or aminoalkyl-imidazoles), incomplete or poorly-defined spectra may often be obtained. In the case of imidazole quaternary salts the 2-proton is frequently labile enough to exchange with deuterium. Some workers have recommended conversion of imidazoles into their N-acetyl derivatives, which are readily soluble in $CDCl_3$ or $(CD_3)_2SO/CDCl_3$ mixtures. Imidazoles substituted on ring nitrogen (and benzimidazoles) are, of course, usually quite soluble in the standard NMR solvents.

In concentrated acid solution there is a reduction in the exchange frequency of the N—H proton to the extent that CH—NH spin splitting may become apparent, and the 4- and 5-protons may no longer be equivalent. The imidazolium ion can be shown by its proton NMR spectrum to have the symmetrical structure (15).

(15)

A study has been made of the effects of solvent on the chemical shifts in the neutral molecule and cationic species of 1-methylimidazole. The shifts induced by protonation, obtained by direct comparison of spectra in $CDCl_3$ and CF_3CO_2H, are consistent with stabilization of the cations by an amidinium-type resonance such as that shown in (15). When N-3 is protonated the signal for H-2 shifts downfield by 1.26 p.p.m., whereas those shifts for H-4 and H-5 move only 0.52 and 0.64 p.p.m. respectively. Such a result is only to be expected since H-4 and H-5 are adjacent to only part of the positive charge; H-2 is much more fully involved. Simple quaternary salts of 1-methylimidazole resemble the protonated species. The process of quaternization deshields the ring protons (H-4, $\Delta\delta$ = +0.95 p.p.m.; H-2, $\Delta\delta$ = +1.45 p.p.m.) and may result in the 2-H signal being broadened. Often it is, as mentioned above, so decidedly acidic that its signal may be absent from a spectrum taken in D_2O. In the case of 1-methylimidazole methiodide it has proved possible to split the methyl signal into a doublet with a coupling constant, $J(N\text{-}CH_3, H\text{-}2) = 0.45$ Hz, while the magnetically equivalent 4- and 5-protons are coupled to H-2 with $J(H\text{-}4,5, H\text{-}2) = 1.8$ Hz. The spectra of 4(5)-bromoimidazole in concentrated acid solutions have permitted inferences to be made as to the tautomeric nature of the compound (Section 4.06.5.1).

Table 4 ¹H NMR Solvent Shifts ($\Delta\delta$) for 1-Methylimidazole ⟨78JOC3565⟩

Position	Solvent			$\Delta\delta$ (p.p.m.)	
	$CDCl_3$	$Me_2SO\text{-}d_6$	D_2O	Δ_1	Δ_2
H-2	7.41	7.55	7.57	−0.14	−0.16
H-4	7.03	6.88	6.99	+0.15	+0.04
H-5	6.87	7.08	7.07	−0.21	−0.20

ᵃ $\Delta_1 = \delta_{Me_2SO\text{-}d_6} - \delta_{CDCl_3}$; $\Delta_2 = \delta_{D_2O} - \delta_{CDCl_3}$.

In 1-alkylimidazoles the ring protons adjacent to an N-methyl group can be differentiated from other ring protons by a characteristic shift in δ with variation in solvent. In non-polar solvents H-5 appears at higher field than H-4, with the order being reversed in polar solvents (see Table 4).

The examination of coupling constants has permitted differentiation of 1,4- and 1,5-disubstituted imidazoles and, when electronegative substituents are present (*e.g.* as in 4-nitroimidazoles), the H-2 signal may not necessarily appear at lowest field. It would appear that a number of assignments in the literature may require revision in the light of these observations. A selection of chemical shift data for imidazoles is listed in Table 5.

Table 5 A Selection of ^{1}H NMR Data for Imidazoles[a,b]

Compound	Ring protons (p.p.m.)			Methyl protons (p.p.m.)			
	H-2	H-4	H-5	Me-1	Me-2	Me-4	Me-5
Imidazole	7.73	7.14	7.14	—	—	—	—
1-Me	7.47	7.08	6.88	3.70	—	—	—
2-Me	—	6.96	6.96	—	2.36	—	—
4-Me	7.47	—	6.81	—	—	2.23	—
1,2-Me$_2$	—	6.79	6.73	3.52	2.30	—	—
1,4-Me$_2$	7.20	—	6.53	3.49	—	2.15	—
1,5-Me$_2$	7.27	6.68	—	3.42	—	—	2.10
2,4-Me$_2$	—	—	6.54	—	2.21	2.04	—
4,5-Me$_2$	7.45	—	—	—	—	2.19	2.19
2,4,5-Me$_3$	—	—	—	—	2.23	1.98	1.98
2-Br-1-Me	—	7.04	7.04	3.64	—	—	—
5-Br-1-Me	7.59	7.07	—	3.63	—	—	—
1-Me-4-NO$_2$	7.54	—	7.87	3.90	—	—	—
1-Me-5-NO$_2$	7.64	8.09	—	4.05	—	—	—
1,2-Me$_2$-5-NO$_2$	—	7.76	—	3.90	2.42	—	—
2-Br-1-Me-4-NO$_2$	—	—	7.91	3.80	—	—	—
5-Br-1-Me-4-NO$_2$	7.68	—	—	3.79	—	—	—

[a] Data largely taken from ⟨70AHC(12)103, B-76MI40600⟩.
[b] Solvents are CDCl$_3$ or CCl$_4$ for N-substituted imidazoles: D$_2$O for others.

In addition to its utility for structure elucidation, and in tautomerism studies, ^{1}H NMR has found applications for analysis in kinetic studies, in conformational studies (*e.g.* 1-arylimidazoles and reduced imidazoles), for the determination of product ratios in isomerization reactions, and in reaction mechanism studies. Thus the technique was able to show that deuterium exchange at the 2-position of imidazoles can occur either in the presence or absence of added base (Section 4.07.1.6.2).

Considerable use has also been made of the technique in the study of benzimidazoles ⟨77JHC1321, 74JCS(P1)903, 72CHE1533, 74CRV279⟩ and quaternary salts ⟨73LA380, 74CRV279⟩. Although an interpretation of the A_2B_2 portion of the spectrum of benzimidazole has been reported, as yet there has been no detailed, specific analysis of the chemical shifts of protons in unsymmetrically substituted derivatives. As with imidazole, rapid N-1–N-3 proton exchange takes place in neutral solutions of benzimidazoles, and in dilute acids fast exchange between solvent protons and the NH is possible. This is severely retarded in concentrated sulfuric acid, when the compound is protonated at N-3. The H-2 appears as a triplet with $J_{1,2} = J_{2,3} = 2.5$ Hz under these conditions demonstrating the position of protonation unequivocally. The use of NMR has led to the conclusion that the transmission of substituent effects from C-2 to C-5(6) is about 20% less effective than in the opposite direction ⟨81OMR(15)219⟩. Lanthanide shift reagents, which complex at N-3, cause a large paramagnetic shift of the C-4 proton in 1-methylbenzimidazoles.

4.06.3.3.2 *Non-aromatic systems*

Although the amount of systematic published material is not great, there have been a number of ^{1}H NMR studies relating to conformations and configurations of imidazolines and imidazolidines, along with their oxo and thione derivatives ⟨B-77SH(1)113⟩. Most interest has been concentrated on the 2-imidazolines and imidazolinones. In many instances, nearby bulky substituents prefer to take up a *trans* relationship. Thus when 1,2-diaminoalkanes react with aldehydes the products are frequently the *cis*-2,4-disubstituted imidazolidines, but when phenyl groups are present on the nitrogen atoms the *trans* isomers begin to predominate. In 4,5-di(substituted amino)-1,2-disubstituted 2-imidazolines the amino groups are also *trans* to each other ⟨81JHC921⟩. With the 2,4-disubstituted 1,3-dibenzylimidazolidines, however, the 2- and 4-substituents adopt the *cis* configuration ⟨71T2453, 79JCS(P1)2289⟩. Proton magnetic resonance studies have allowed differentiation of such geometrical isomers as in the 5-methyl-1,2-diphenylimidazolidines ⟨73BSF2676⟩, and have also provided evidence (in accord with dipole moment studies) for the envelope conformations of some imidazolidines ⟨B-77SH(1)113, 73BSF2676⟩. Such studies have been of value in determining the configurations of compounds with chiral centres. Evidence for the symmetrical cation of imidazolinium comes from ^{1}H NMR studies of 2-imidazolines in trifluoroacetic acid ⟨76CHE917⟩.

The 4- and 5-methylene protons of 2-imidazolines resonate in the region δ 3–4, depending on the substituents present; in imidazolidines the corresponding resonances are usually around 3 p.p.m. with the 2-protons at lower field (~4 p.p.m.). The proximity of an oxo or thione function shifts the signals to even lower field, *e.g.* in 1,3-dimethylimidazolin-2-one, the methylene protons give a singlet at 3.27 p.p.m. in CDCl$_3$ ⟨81JCS(P2)317⟩, while in 2-methyl-2-imidazolin-5-one the H-4 resonance in DMSO-d_6 appears at δ 3.88 ⟨71BSF1040⟩.

4.06.3.4 ^{13}C NMR Spectroscopy

4.06.3.4.1 Aromatic systems

The increasing interest in ^{13}C NMR spectroscopy has resulted in considerable use of the technique, although no detailed study of the whole range of imidazoles and benzimidazoles has been reported. Chemical shifts for a few simple compounds of this type are listed in Table 6. As expected, in the absence of electron-withdrawing substituents, C-2 resonates at lowest field because of its situation between the two annular nitrogen atoms. As long as the neutral molecule is the substrate, the ^{13}C chemical shifts might be expected to provide an indication of the likely sites for electrophilic (and nucleophilic) attack in the imidazole ring. Thus C-5 would appear to be the site of choice for electrophilic attack in 1-methylimidazole, while in the benzimidazoles C-2 should be prone to nucleophilic attack; electrophiles are more likely to favor the fused benzene ring. Studies of ^{13}C NMR have been only of limited value in ascertaining positions of tautomeric equilibrium for rapidly equilibrating tautomers, but there have been some useful conformational studies made using the technique. In particular, the relationships between ^{13}C chemical shift and dihedral angle in phenyl substituted imidazoles have been noted ⟨73ACS3101⟩. In the relatively unhindered 1-phenylimidazole (where there is considerable interannular conjugation), there is a relatively large difference ($\delta_{C-3'} - \delta_{C-2'} = 9.4$ p.p.m.) between the *ortho-* and *meta-*carbon chemical shifts. The *C-*phenyl compounds have the *ortho-*carbons (C-2') more shielded than in the *N-*phenyl isomers, and in the more hindered 1-methyl-2-phenylimidazole the signal for the *ortho*-carbon is shifted appreciably downfield compared with the corresponding

Table 6 ^{13}C NMR Chemical Shifts (p.p.m.) for Imidazoles and Benzimidazoles

Compound	C-2	C-4	C-5	C=O	Me	C-1'	C-2'	C-3'	C-4'	Ref.
Imidazole	136.2	122.3	122.3	—	—	—	—	—	—	76CJC617, 73JMR(12)225
1-Ac	136.2	130.5	115.9	166.3	22.3	—	—	—	—	78JCS(P2)99
1-Me	138.7	130.2	121.0	—	34.2	—	—	—	—	74JOC357
1-Ph	135.0	129.9	117.8	—	—	136.7	121.0	129.4	128.0	73ACS3101
2-Ph	146.7	122.7	122.7	—	—	130.0	125.2	128.4	128.3	73ACS3101
1-Me-2-Ph	147.2	127.8	121.9	—	34.4	130.1	128.0	128.0	128.0	73ACS3101
1-Me-4-Ph	141.7	137.4	122.2	—	—	133.6	124.3	128.1	126.2	73ACS3101

	C-2	C-4	C-5	C=O	Me	C-6	C-7	Ref.
Benzimidazole	144.4	110.1	122.7	—	—	121.9	119.4	79OMR(12)95
1-Me	143.1	119.7	121.5	—	30.4	122.4	108.9	78JCS(P2)99
1-Ac	141.2	120.2	124.7	166.9	23.3	125.5	115.1	78JCS(P2)99

carbon atom in 1-methyl-4-phenylimidazole. Changes in solvent affect the chemical shifts, and lanthanide shift reagents have found application ⟨74JOC357⟩.

4.06.3.4.2 Non-aromatic systems

^{13}C Chemical shifts for imidazoline- and imidazolidine-thiones have been compared ⟨76TL3313⟩. Additionally, it has been shown that in 1-methyl-3-phenyl-4-imidazolin-2-ones there is extensive interannular conjugation, but the larger sulfur atom in the analogous 2-thiones pushes the phenyl group out of the plane of conjugation ⟨74ACS(B)61⟩. Related studies have been applied to 1-aryl-3-phenyl-2-thioxoimidazolidin-4-ones ⟨82OMR(19)27⟩, and to a series of 4-mono- and 4,5-di-substituted imidazolidine-2-thiones with a variety of alkyl substituents on nitrogen. In the latter group it was found that the increase in steric interactions between N-methyl and N-ethyl compounds caused deformations of the ring and variation in the valence angle of the substituted nitrogen atom ⟨80CS(15)193⟩.

On the basis of ^{13}C satellite observations and iterative computer calculations it has proved possible to make tentative assignments of conformation for 1,3-dimethylimidazolidines. At about −150 °C it was possible to detect both cis (**15a**) and trans (**15b**) isomers; the free energy barrier to rotation is approximately 6.2 kJ mol^{-1}. In such compounds typical shifts are 41.8 (CH$_3$), 55.0 (CH$_2$—CH$_2$) and 80.2 (NCH$_2$N) p.p.m. ⟨79OMR(12)362, 71T2453⟩.

Table 6a lists some typical ^{13}C NMR shifts for imidazolinones and imidazolidinones.

Table 6a ^{13}C NMR Chemical Shifts (p.p.m.) for Some Non-aromatic Derivatives of Imidazole

Compound	Solvent	CH$_3$	C-2	C-4,5	Ref.
2-Imidazolidinone	H$_2$O	—	169.4	43.4	81JCS(P2)317
1,3-Me$_2$-2-imidazolidinone	CDCl$_3$	29.5	160.1	43.3	81JCS(P2)317
4,5-Me$_2$-4-imidazolin-2-one	CF$_3$CO$_2$H	8.7	151.2	119.5	81JCS(P2)310

4.06.3.5 NMR Involving Other Nuclei

Studies involving ^{14}N, ^{15}N and ^{19}F nuclei have appeared ⟨80AHC(27)241⟩. The ^{15}N shifts for imidazole and 1-methylimidazole in dichloromethane relative to those in aqueous solution are primarily due to strong hydrogen bonding between ^{15}N-3 and water protons. Both hydrogen bonding and protonation result in upfield shifts, but the magnitudes of the protonation shifts far exceed those associated with hydrogen bonding. In non-aqueous media the imidazolium ions show less than normal shifts, suggesting that there is ion-pair formation resulting from interior hydrogen-bonding. In solvents of higher dielectric constant (and greater solvating power for salts) these effects diminish.

4.06.3.6 Ultraviolet Spectroscopy

4.06.3.6.1 Aromatic systems

Imidazole and its simple alkyl derivatives have little absorption in the near-UV, and this can be attributed to their high aromatic stability. Alkyl groups in any ring position produce a small bathochromic shift of the absorption band at 207–208 nm. When, however, there is an aryl group attached to a ring carbon or nitrogen, new intense bands appear between 250–300 nm as a consequence of conjugation between the aromatic and heteroaromatic systems. Bands of this type are also present in the spectra of 2-arylbenzimidazoles

⟨70AHC(12)103, 66RCR122⟩. In instances where steric hindrance to coplanarity is present, then these absorption bands may be appreciably shifted.

Other groups conjugated with the imidazole ring may profoundly affect the spectrum. Thus imidazole-2-carbaldehyde has λ_{max} 285 nm (log ε 4.10) and imidazole-4-carbaldehyde (as the neutral molecule) has λ_{max} 256 nm (log ε 4.07). For the latter isomer the corresponding anion has λ_{max} 280 nm (log ε 4.17); the cation absorbs at λ_{max} 237 nm (log ε 3.85) ⟨B-76MI40600⟩. The spectra of nitroimidazoles display features which are altered characteristically by substituents and by pH and which, in consequence, are useful for the purposes of orientation. Some examples are listed in Table 7. A method of simultaneous analysis of N-substituted and N-unsubstituted nitroimidazoles depends largely on their spectroscopic properties. The latter form yellow nitroimidazole anions in alkaline medium, and difficulty of reduction of these anions together with the shift in the absorption maximum to longer wavelengths makes the analysis possible.

Table 7 Ultraviolet Absorption Spectral Data for Nitroimidazoles[a]

Imidazole	H_2O λ_{max} (nm) ($\log_{10}$ ε)	Acid medium λ_{max} (nm) ($\log_{10}$ ε)	Basic medium λ_{max} (nm) ($\log_{10}$ ε)
2-NO_2	325 (3.95)	298 (3.91)	372 (4.13)
4-NO_2	298 (3.80)	267 (3.85)	227 (3.52), 352 (4.01)
2,4-$(NO_2)_2$	304 (4.05)	—	354 (4.09)
4,5-$(NO_2)_2$	284, 221	—	255, 370
1-Me-2-NO_2	325 (3.93)	300 (3.89)	—
1-Me-4-NO_2	225 (3.55), 301 (3.83)	268 (3.85)	—
1-Me-5-NO_2	227 (3.53), 305 (3.91)	367 (3.80)	—
2-Br-1-Me-4-NO_2	309 (3.86)	220 (3.76), 280 (3.88)	—
2-Br-1-Me-5-NO_2	316 (3.96)	223 (3.73), 282 (3.83)	—
5-Br-1-Me-4-NO_2	221 (3.68), 314 (3.85)	221 (3.69), 288 (3.87)	—
5-I-4-NO_2	240 (3.66), 325 (3.83)	307 (3.80)	273 (3.53), 363 (3.97)
5-OMe-4-NO_2	237 (3.57), 330 (4.12)	225 (3.54), 307 (3.90)	237 (3.60), 285 (3.53), 366 (4.15)

[a] Most of these data are from ⟨B-76MI40600⟩.

The absorption spectrum of benzimidazole is much more complicated than that of imidazole, with absorption maxima appearing at 244 (log ε 3.74), 248 (3.73), 266 (3.69), 272 (3.71) and 279 (3.73) nm. The short wavelength band (244–248 nm) is attributed to electron transitions localized in the imidazole ring, while the longer wavelength absorption is regarded as a displaced B-band of benzene (255 nm). As seems characteristic of the spectra of other benzazoles, there appears to be no band which corresponds with an interaction between the benzene and azole rings. When more than one ring is fused to imidazole (as in the naphthimidazoles) the spectra become even more complex. There are marked bathochromic shifts in the series imidazole, benzimidazole, angular naphthimidazole, linear naphthimidazole.

UV spectroscopy has been applied to studies of 2,4,5-triarylimidazole radicals, imidazolones and various thioimidazoles. In addition, the rates of hydrolysis, alcoholysis and aminolysis of imidazolides such as 1-acetylimidazole can be determined readily since such azolides show characteristic intense absorptions at wavelengths longer than those of the imidazole products. In the C-acyl- and C-aroyl-imidazoles, those which have the carbonyl group attached at C-2 exhibit a bathochromic shift of about 20 nm, while in m- and p-substituted 2-aroylimidazoles intense absorptions ($\log_{10}$ ε 3.5–4.3) appear in the 295–309 nm region, with slightly less intense peaks at about 233–266 nm ⟨74CRV279⟩.

Applications of UV studies to the determination of the tautomeric natures of imidazoles will be discussed later (Section 4.06.5).

In recent years there has been intensive research into the luminescence and photochemistry of imidazoles. While aspects of this will be discussed in Sections 4.07.1.1 and 4.07.1.4, the spectral characteristics of the compounds will be treated at this stage. The types of compound involved are polyphenyl- or polyaryl-substituted imidazoles such as 1,2,5- or 2,4,5-triphenylimidazole (lophine). It has been established that the introduction of substituents into the 1-phenyl group of 1,2,4-triphenylimidazole has little effect either on the absorption spectrum (λ_{max} 290 nm) or the spectral-luminescence characteristics (λ_{max} 370 nm, $\psi = 0.4$–0.5). The effects of substituents in other phenyl rings lead to a bathochromic effect from electron acceptors in the 2-position or from electron donors in the

5-position. The luminescence, however, is not affected to any extent except in special cases (*e.g.* Br or NO_2 substitutent) in which there is a substantial decrease in the quantum yield due to intensification of the intercombination conversion.

Benzimidazole derivatives absorb at 300–400 nm and fluoresce intensely in solution at 320–450 nm. In 2-substituted benzimidazoles the substituent group causes substantial changes in the spectral characteristics of the absorption and fluorescence. Aryl substituents with acceptor groups cause hypsochromic shifts whereas donor groups shift the spectra bathochromically. Such a substituent effect suggests that the nature of the long-wave electron transition is associated with redistribution (during excitation) of the electron density in a direction away from the C-aryl ring.

Chemiluminescence, which is a low-intensity luminescence which develops as a result of conversion of a portion of the chemical energy into energy of electronic excitation of molecules and subsequent deactivation by the emission of light, was first detected in 1877 when lophine was oxidized by molecular oxygen in 1M ethanolic KOH. In fact this chemiluminescence (broad band at 465–660 nm, λ_{max} 545 nm) actually develops during deactivation of excited peroxide molecules formed in the reaction of lophine free radicals with molecular oxygen ⟨77CHE1160⟩.

4.06.3.6.2 Non-aromatic systems

Only in the presence of such chromophores as C=O and C=S do the non-aromatic derivatives show any marked UV absorption. The π–π* transitions of the N—C(=S)—N chromophore in imidazolidine-2-thiones give rise to absorptions in methanol solution at about 240 nm (log ε ~4.2) ⟨79T511⟩.

4.06.3.7 Infrared Spectroscopy

4.06.3.7.1 Aromatic systems

Assignments of the absorption bands of imidazoles fall into three main regions: (i) ring-skeletal (~1550, 1490, 1325 cm^{-1}), (ii) ring-breathing and C—H in-plane deformations (~1260, 1140, 1060 cm^{-1}), and (iii) C—H out-of-plane deformations (~940, 840, 760 cm^{-1}). When the imidazoles are unsubstituted on nitrogen, the N—H bond stretching frequencies appear in the region 2200–3600 cm^{-1}; association generally makes this band a very broad one. Many of the fundamental frequencies associated with C—H bonds have been assigned through studies of deuterated imidazoles, and similar assignments have been made for 1-methylimidazole in the solid, liquid and gaseous states and in solution. A comparison of such spectra in different states suggests that there is some interaction between ring C—H groups and the 'pyridine-type' nitrogen of another imidazole nucleus. Perhaps this may account in part for the rather high boiling points of compounds such as 1-methylimidazole.

The very strong hydrogen-bonding between imidazoles with free NH groups leads to linear associates of molecules, absorption at frequencies lower than usual, and concentration-dependent spectra. Similar absorption characteristics are evident in the IR spectra of benzimidazoles, and a wide range of substituted imidazoles and condensed imidazoles.

N-Acylazoles have interesting IR characteristics in that their carbonyl stretching bands appear at frequencies which are usually slightly higher than those normally associated with aliphatic ketones (1765–1725 cm^{-1}), and much higher than those for disubstituted amides (1650 cm^{-1}). Thus 1-acetylimidazole has a carbonyl stretching frequency of 1747 cm^{-1}, illustrating the minor importance in these compounds of conjugation between the heteroaromatic ring nitrogen and the acyl group (see structure 16). This results in a carbonyl carbon which is very susceptible to nucleophilic attack, and hence to the extremely facile hydrolysis of such compounds. They are, in fact, powerful acylating agents (Section 4.07.3.8.4). It should be noted, though, that a shift in the carbonyl band towards higher wavelengths (*i.e.* increase in the C—O force constant) can, however, also be attributed to increased electron attraction by the heterocyclic ring.

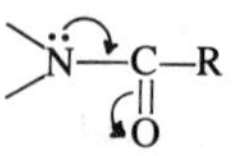

(16) 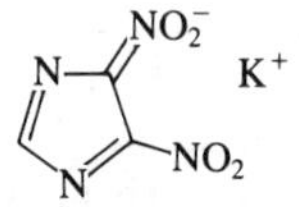   (17)

In general, IR spectroscopy has neither been a particularly valuable structural tool in imidazole chemistry, nor has it found particular application in tautomerism studies. The IR spectra of nitroimidazoles and their salts have been claimed to indicate that the salt formation involves only the nitro function, with the formation of an isoimidazole ring (17). In polynitroimidazoles only one of the nitro functions appears to be involved in such salt formation ⟨70AHC(12)103, 74CRV279, 80AHC(27)241⟩.

4.06.3.7.2 Non-aromatic systems

The ring stretching frequencies for 4H-imidazoles appear in the regions 1625–1650 and 1567–1575 cm^{-1} ⟨75JHC471⟩. In 2-imidazolines there is a strong band for C—N between 1585 and 1625 cm^{-1} ⟨76CHE917, 77LA1183⟩, shifted to 1670 cm^{-1} in 2-amino derivatives ⟨77JHC607⟩. The 2-imidazolones show a strong carbonyl absorption at ~1680 cm^{-1} with the C=C stretch appearing near 1650 cm^{-1}, while 1-hydroxyimidazolin-5-ones show their principal bands at 3250 (OH) and 1690–1700 (C=O) cm^{-1} ⟨70JHC439⟩.

4.06.3.8 Mass Spectrometry

4.06.3.8.1 Aromatic systems

The high stabilities of imidazoles give rise to pronounced molecular ions in their mass spectra which are also notable for their characteristic fragmentation patterns (but few skeletal rearrangements). The major fragmentations in many of the compounds are those due to successive losses of HCN and hydrogen atoms by competing pathways. Thus the mass spectrum of imidazole shows a strong ion at m/e 40 m.u. assigned to an azirinium cation (18; see Scheme 3). Deuterium labelling experiments indicate that the loss of hydrogen atoms is specifically from the 4(5)-positions, but the HCN loss is less specific, originating in order of preference from positions 2 > 4(5) > 1. The manner in which HCN is lost can be profoundly affected by the substitution pattern; it can come from the 2,3-positions (1- and 4-methylimidazole), or from the 1,5- or 3,4-positions (2-methylimidazole). The observation that both 1,2-dimethylimidazole and 2-mercapto-1-methylimidazole show very little loss of HCN from the molecular ions is an indication that such a fragmentation does not readily involve the 3,4-positions ⟨70AHC(12)103⟩. At energies above the range 17–26 eV the azirinium cation (18) loses a hydrogen molecule to give C$_2$N$^+$ (Scheme 3).

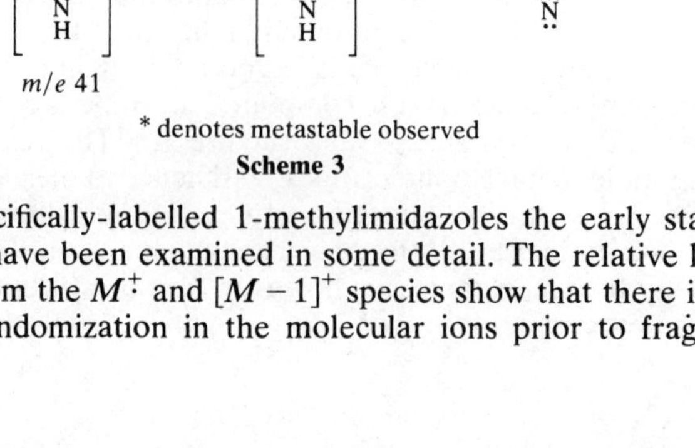

* denotes metastable observed

Scheme 3

For a number of specifically-labelled 1-methylimidazoles the early stages of the mass spectral fragmentation have been examined in some detail. The relative losses of labelled and unlabelled HCN from the $M^{+\cdot}$ and $[M-1]^+$ species show that there is an insignificant amount of hydrogen-randomization in the molecular ions prior to fragmentation. Subsequent elimination of HCN, as mentioned above, involves C-2 and N-3. The molecular ions eject hydrogen radicals largely from the methyl group (Scheme 4), but there is some minor contribution to this process from H-5. The resultant $[M-1]^+$ ions exhibit extensive, but incomplete, hydrogen-randomization, and subsequent loss of HCN from these ions is consistent with the intermediate formation of ring-expanded ions.

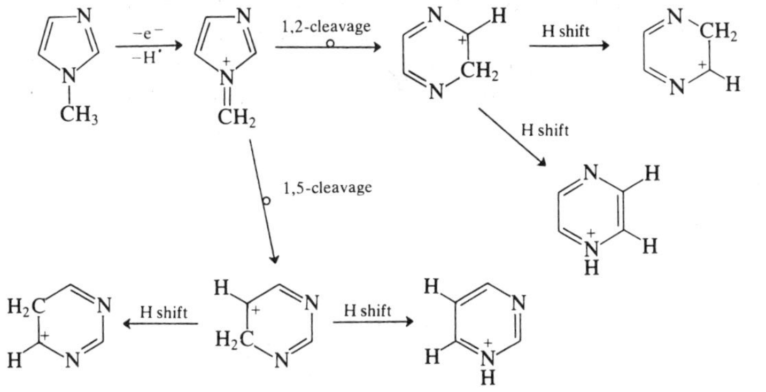

Scheme 4

The mass spectral fragmentation pattern of benzimidazole parallels that of imidazole. Again, the molecular ion is frequently the most intense ion which loses HCN to form an azirinium radical cation by a non-specific process. Further consecutive losses of HCN and a hydrogen atom lead to an ion, $C_5H_3^+$ (Scheme 5). In 2-methylbenzimidazole the loss of MeCN is not a major process; rather, the initial fragmentation is loss of a hydrogen atom (probably from the methyl group) with concomitant ring expansion to give a stable quinoxalinium cation (**19**). Subsequent loss of consecutive molecules of HCN gives rise to the phenyl cation (Scheme 6) ⟨74CRV279, 68T1875, 78M379, 77JHC1321⟩. In 1-methylbenzimidazoles, competitive losses of $CH_3\cdot$, CH_3NC and HCN are possible from the molecular ion.

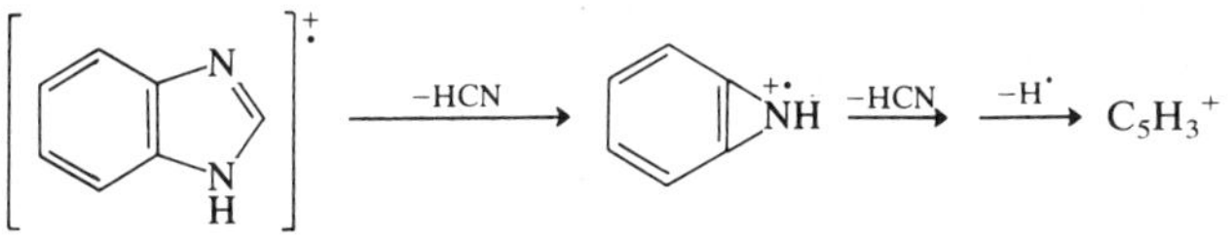

Scheme 5

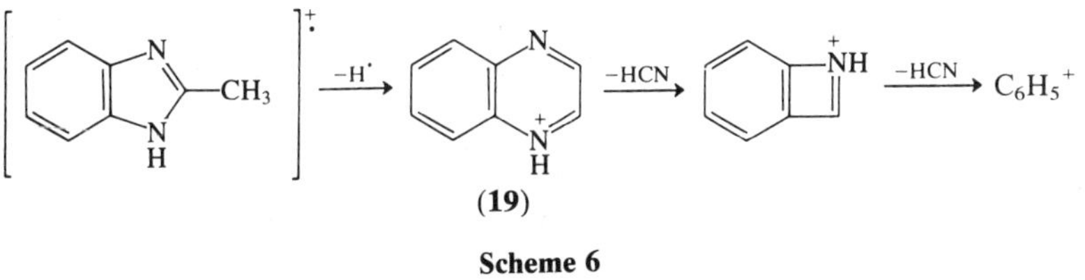

Scheme 6

On electron impact the 4-methyl-5-nitroimidazole molecular ion loses H_2O. Such a loss seems to require an NH group adjacent to the nitro group, *i.e.* loss of H_2O comes from the least dominant tautomer ⟨80AHC(27)241⟩. However, an adjacent CH group may be implicated. Loss of $\cdot$OH by the normal 'ortho' process also takes place (Scheme 7). In 1-alkyl-5-nitroimidazoles there is a similar hydrogen transfer which is followed by an unusual rearrangement, resulting ultimately in the loss of an aldehyde or ketone fragment from the molecular ion (Scheme 8). In 1-(*o*-nitrophenyl)imidazoles the 2-H proton is involved in the loss of $\cdot$OH, a related 'ortho effect' ⟨78OMS(13)575⟩. The mass spectrum of 1-methyl-4-nitroimidazole-5-carboxamide shows elimination of water by electron impact as distinct from the normal thermal process. Most nitro-imidazoles and -benzimidazoles, though, have mass spectra which display the usual characteristic losses of $NO_2\cdot$, NO and $O\cdot$ fragments. Thus fragments arising from the losses of $O\cdot$, NO, $NO_2\cdot$ and $\cdot$OH appear in the spectrum of 1,2,5-trimethyl-4-nitroimidazole. In addition, there is evidence of loss of CHO, a fragmentation which appears to be uncommon in azoles but which has been observed with some other nitroaromatics.

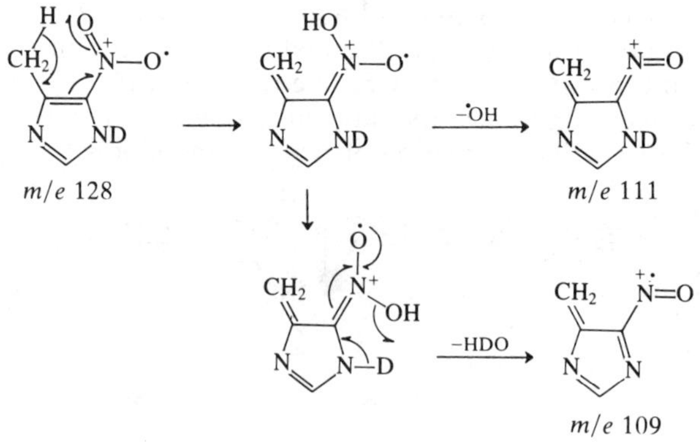

Scheme 7

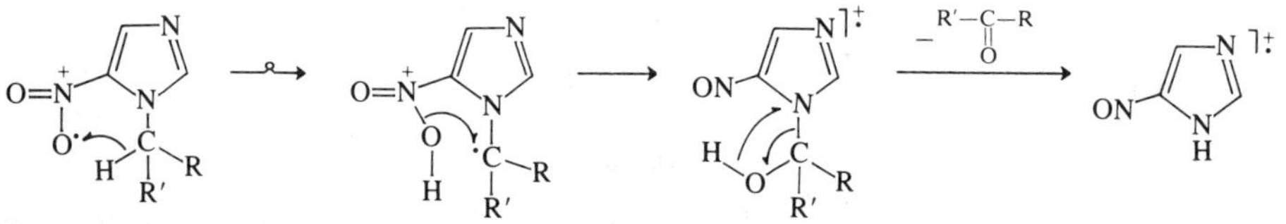

Scheme 8

The mass spectra of 1-benzylimidazoles are dominated by the tropylium ($C_7H_7^+$) ion.

Azolides tend to eliminate ketene, giving rise to 'molecular ions' of the corresponding unsubstituted azoles. Other fragmentation pathways, though, may compete with this. When 1-acetylimidazoles lose CO the resultant ions do not have N- or C-methylimidazole structures. In 2-aroyl-imidazoles and -benzimidazoles a hydrogen atom is commonly eliminated from the molecular ion, while CO loss from either the M^+ or $[M-1]^+$ species is also characteristic; p-methoxybenzoyl derivatives lose either formaldehyde or consecutive CH_3· and CO fragments from the $[M-29]^+$ ion ⟨80AHC(27)241⟩. 2-Formylimidazole, 2-formyl-1,4,5-trimethylimidazole 3-oxide and 2-formyl-1,5-dimethylimidazole 3-oxide all lose CO from the molecular ions. The two principal fragmentation sequences, though, are loss of O· followed by the formyl function and the alternative order of fragmentation. Whereas loss of a hydrogen atom is observed from 2-formyl-1-methylimidazole, this is not the situation with the N-oxides presumably because the hydrogen atom is lost in the ·OH fragment. Loss of ·O and ·OH fragments from most imidazole N-oxides is common. In 2-formyl-1,4,5-trimethylimidazole 3-oxide, HCN is eliminated from the m/e 109 m.u. species, the molecular ion having first lost ·OH and CO in this particular instance; in 2-formyl-1,5-dimethylimidazole the process differs in that the HCN loss from the m/e 95 m.u. fragment follows prior loss of O· and CO from the molecular ion. In the trimethyl compound, not only can a hydrogen radical be removed from the formyl group but it can also originate from the 4-methyl group. In the corresponding 2-carboxylic acid the most striking loss is CO_2, but losses of O·, ·OH and HCN are also present as with the formyl-substituted N-oxides ⟨B-76MI40600⟩. While 2-arylbenzimidazole 3-oxides display the characteristic losses of 16 and 17 m.u., the most intense ion is invariably an RCO^+ ion which must result from some rearrangement ⟨78JCR(S)366⟩.

In the mass spectra of a number of methylated histamine derivatives the base ions arise from β-cleavage, usually with concomitant hydrogen rearrangement. This results in ions of the general form $[M-CHR=NR]^+$, where R = Me, H.

A variety of fragmentations has been noted in the mass spectra of benzimidazoles substituted by ketone functions at C-2. Some of these involve rearrangements such as those illustrated in Scheme 9.

4.06.3.8.2 Non-aromatic systems

The most significant ions in the mass spectra of 4,5-dihydroimidazoles arise from (i) elimination of the C-4 substituent, (ii) cleavage of the 1,2 N—C and 4,5 C—C bonds with, or without, hydrogen migration, and (iii) cleavage of the 1,5 and 2,3 N—C bonds with charge retention on the N-1—C-2 species ⟨78OMS(13)319⟩. In the mass spectrum of 4-

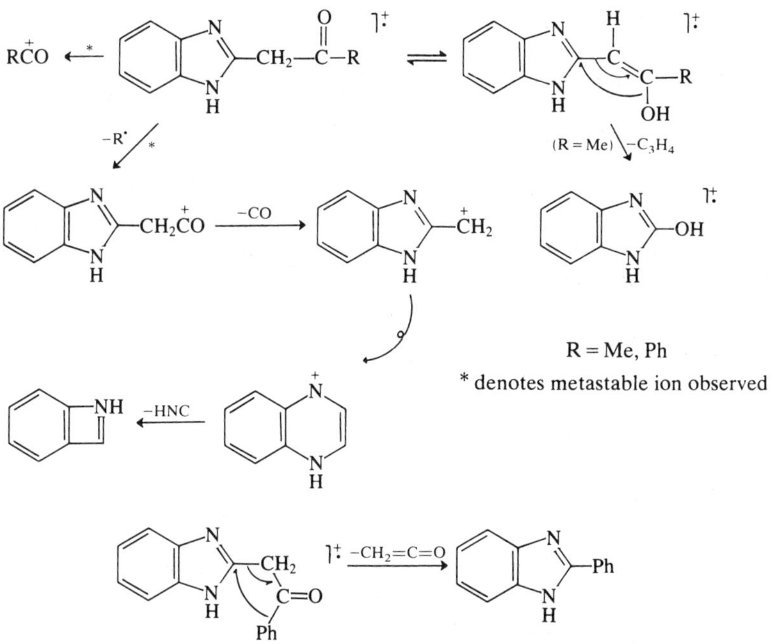

Scheme 9

substituted 2-imidazolones the base peak is the molecular ion, which loses such fragments as CO and RNCO (where R is the N-substituent or hydrogen) ⟨79JHC983⟩. Mass spectrometry has been used in the determination of the *trans* configuration of 4,5-dimorpholino-2-imidazoline ⟨81JHC921⟩.

4.06.3.9 Electron Spin Resonance Spectra

The free radicals produced at room temperature by X-irradiation of crystalline imidazole, benzimidazole and their 2-methyl derivatives are shown to be almost identical (1:2:1 triplet ($a = 45$ G) with each line displaying further, poorly-resolved, hyperfine splitting). The radical produced from imidazole, therefore, must have coupling with two identical protons as in structure (**20**). Examination of the ESR spectra of 2,4,5-triphenylimidazolyl radicals using partially and completely deuterated species has been supplemented by the application of simple MO calculations to demonstrate that the radicals are not planar. There is, nevertheless, some ambiguity since ESR shows the radicals to be of the π-type; INDO calculations point to a σ-radical. Oxidation of 1-hydroxybenzimidazole 3-oxides with silver oxide or lead tetraacetate gives symmetrical nitroxide radicals, and imidazole anion radicals are believed to exist in the tautomeric α-pyrrolenine form (**21**) rather than (**22**). A CNDO study has criticized this conclusion ⟨80AHC(27)241, 74CRV279⟩.

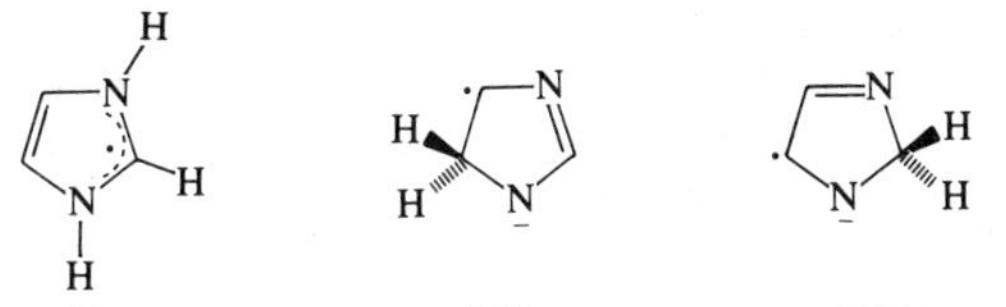

(**20**)   (**21**)   (**22**)

4.06.3.10 Photoelectron Spectra

Comparisons have been made among benzimidazole, benzoxazole and benzoselenazole using MO calculations and PE spectra. The theoretical calculations agree with the observed relative band intensities for the He(I) and He(II) spectra ⟨78MI40600⟩. Application to imidazoles of He(I) PE spectroscopy demonstrates that the lone pair levels are heavily localized on nitrogen, while the other MOs of the valency shell are strongly delocalized ⟨73T2173⟩.

4.06.4 THERMODYNAMIC ASPECTS

4.06.4.1 Intermolecular Forces

4.06.4.1.1 Melting and boiling points

As a general rule the boiling points of imidazoles are relatively high, probably because of the high degree of intermolecular association. When there is a substituent other than hydrogen on an annular nitrogen, there is no longer any chance of such association. As a result there is a considerable lowering of boiling point (*e.g.* imidazole, 256 °C; 4-methylimidazole, 264 °C; 1-methylimidazole, 198 °C; 1,4-dimethylimidazole, 200 °C). There is a similar pattern associated with the melting points (*e.g.* imidazole, 90 °C; 4-methylimidazole, 56 °C; 1-methylimidazole, −6 °C; benzimidazole, 171 °C; 1-methylbenzimidazole, 66 °C).

4.06.4.1.2 Solubilities

The solubilities of imidazoles, too, are affected in a striking fashion by substituents on an annular nitrogen. Those with a free NH group are usually highly soluble in polar, protic solvents such as water and have poor solubility in non-polar solvents. These solubility characteristics are reversed in 1-substituted imidazoles. There are occasional exceptions to this general rule of thumb. The compound 2-acetylimidazole is quite soluble in chloroform and only poorly soluble in water. It appears that intramolecular hydrogen bonding (between the NH and carbonyl oxygen) competes with the normal intermolecular hydrogen bonding common to the series. Such quantities as association energy, dielectric relaxation time, ionization energy and dissociation constant in aqueous medium have been determined for imidazole. An IR examination has been made of the association of 4-methylimidazole in carbon tetrachloride and in 1,1,2,2-tetrachloroethane ⟨64MI40600⟩. The concentration dependence of the maximum extinction coefficient of the free NH valency bond allowed determination of the monomer content, the mean viscosity and the equilibrium constants K_{12} and K_{13}. From the temperature dependence of these values the mean heat of addition (34 ± 2 kJ mol^{-1}) and the heats of formation of the dimers (42.7 kJ mol^{-1}) and trimers (33.9 kJ mol^{-1}) were obtained. Although this mean heat of addition was rather higher than had been obtained by an earlier worker, both results were not contradictory to the assumption that imidazoles form chain-like associations with angled structures.

When one endeavours to apply such separation techniques as solvent extraction and the various forms of chromatography to imidazoles, one must be aware of the significant influence that the substituents (and their positions) have on molecules which have a well-developed polar nature and a tendency to form associates.

4.06.4.1.3 Gas–liquid chromatography

The polar natures of imidazoles with a free NH group have made their separation by gas chromatography difficult ⟨70AHC(12)103⟩. The 1-substituted compounds are much more susceptible to the separation technique, and it is often advantageous to convert *N*-unsubstituted compounds into their *N*-acetyl derivatives before separation. A set of conditions which has proved effective for simple alkyl- and aryl-substituted imidazoles comprised a glass column packed with Chromosorb W, AW-DMCS (H.P.), 80–100 mesh coated with 5% OV-17, and using nitrogen as carrier gas ⟨72MI40600⟩.

4.06.4.2 Thermochemistry; Aromatic Stability

That imidazoles and benzimidazoles have high stability has been known for many years. Resistance to acids, bases, heat and oxidation or reduction are common traits of these compounds, which display considerable aromatic character. Thus treatment of benzimidazole with permanganate leads to imidazole-4,5-dicarboxylic acid; imidazoles, in general, are not easily oxidized or reduced (Sections 4.07.1.4.11, 4.07.1.5.6, 4.07.1.7.4). Thermal stability too is evidenced by the resistance of the imidazole nucleus to ring fission during thermal rearrangements of 1-alkyl- and 1-aryl-imidazoles at temperatures above 600 °C (Section 4.07.1.2.2). The planar natures of the molecules (as shown by X-ray studies), the lack of ring strain, the typical, low-field chemical shifts of the ring protons in the ^{1}H NMR spectra of the molecules, and the appreciable ring current anisotropy in the ^{13}C NMR spectra are all diagnostic of aromatic character. The importance, compared with benzene, of dipolar resonance contributors to the hybrids has been commented on earlier (Section 4.06.2).

From the enthalpy of formation of crystalline imidazole (61 ± 3 kJ mol^{-1}), the heat of sublimation (67 ± 4 kJ mol^{-1}) and the heat of formation of gaseous imidazole (128 ± 8 kJ mol^{-1}), a resonance energy value of 59 kJ mol^{-1} can be calculated. While this is considerably lower than that of benzene, it still indicates considerable stabilization.

4.06.5 TAUTOMERISM

4.06.5.1 Annular Tautomerism

Although it is not possible to separate 4- and 5-methylimidazole isomers, methylation gives an approximately 50:50 mixture of the 1,4- and 1,5-dimethylimidazoles. In addition, the ^{1}H NMR spectrum of imidazole shows but a single peak for the 4- and 5-hydrogen atoms, indicating that they must be magnetically equivalent. Such observations can be explained in terms of a rapid proton exchange between the —NH— and the =N— nitrogen atoms. In neutral, organic solvents this can be an intermolecular process involving two or more imidazole molecules (Scheme 10), while in protic solvents such as water the solvent itself is involved. The NH proton of imidazole can be shown to exchange with deuterium very rapidly in D$_2$O. The rate equation for the tautomerism of imidazole has been shown to contain kinetic terms which correspond to catalysis by H$^+$, OH$^-$ and imidazolium cations. There is also an uncatalyzed term which seems to be unexpectedly large and may involve the bifunctional participation of water (Scheme 10) ⟨76JA3737⟩. Certainly, the large associates of imidazoles, unsubstituted on the NH nitrogen, in organic solvents are hydrogen-bonded in such a way that fast proton exchange is possible. It is this rapid exchange which is responsible for the absence of spin–spin splittings with the NH proton in the ^{1}H NMR spectrum, and for the observation that this proton gives rise to a broad, distinct signal only in concentrated solutions in such solvents as benzene, chloroform and acetone. Thus this prototropy makes the 4- and 5-positions magnetically and chemically equivalent.

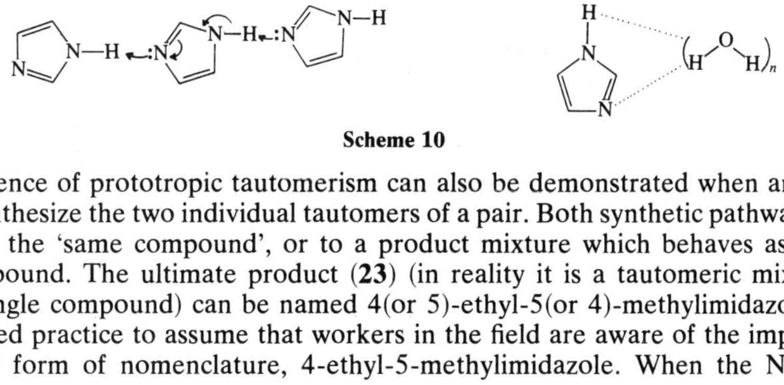

Scheme 10

The presence of prototropic tautomerism can also be demonstrated when an attempt is made to synthesize the two individual tautomers of a pair. Both synthetic pathways (Scheme 11) lead to the 'same compound', or to a product mixture which behaves as if it was a single compound. The ultimate product (**23**) (in reality it is a tautomeric mixture which acts as a single compound) can be named 4(or 5)-ethyl-5(or 4)-methylimidazole, but it is now accepted practice to assume that workers in the field are aware of the implications of the simpler form of nomenclature, 4-ethyl-5-methylimidazole. When the NH group is

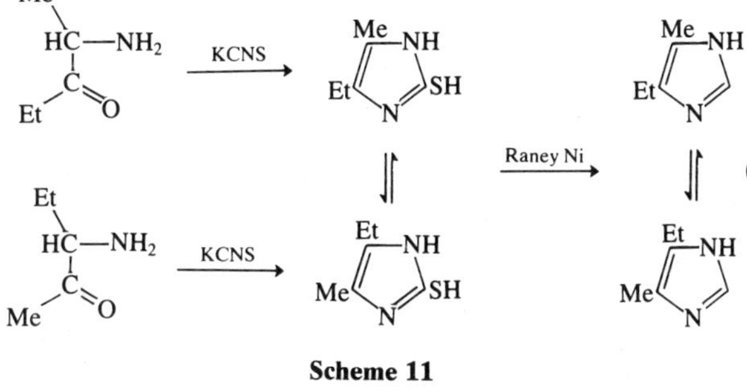

Scheme 11

substituted by, for example, an alkyl group, then prototropic tautomerism is no longer possible. Thus methylation of 4-methylimidazole gives a mixture of two quite distinct compounds, 1,4- and 1,5-dimethylimidazole, in about equal quantities.

To this stage we have only considered imidazoles which have simple alkyl groups at C-4 or C-5. It would not be unreasonable to expect that other groups might favour one tautomeric form at the expense of the other. A case in point is 4(or 5)-bromoimidazole (**24, 25**; Scheme 12), which exists mainly as the 4-bromo form (**24**). Evidence for this comes from examination of the NMR spectra of deuterated derivatives in concentrated acidic medium. Such a medium converts imidazole into the cation which is able to undergo proton exchange only very slowly; coupling with the NH proton can be observed in consequence. When 2-deuterio-4(5)-bromoimidazole (**26**) was treated with concentrated D_2SO_4, the product had a coupling constant of 2.7 Hz. 1,2-Dideuterio-4(5)-bromoimidazole (**27**) in concentrated H_2SO_4 showed a smaller coupling constant of 1.9 Hz. On the basis of the assumption that a larger coupling constant is due to adjacent protons, and the smaller one is due to distant protons (cross-ring coupling), it could be concluded that the compound exists as 4-bromoimidazole ⟨70AHC(12)103⟩. The results of this study could also be interpreted as indicating that the 4-bromo tautomer protonates faster, but in view of the known electronic characteristics of the bromo substituent this is an unlikely state of affairs. The observation that the coupling constants for 4(5)-fluoroimidazole correspond closely with those of 4-fluoro-1-methylimidazole suggests that in solvents of low polarity the 4-fluoro form is the dominant tautomer ⟨78JHC1227⟩.

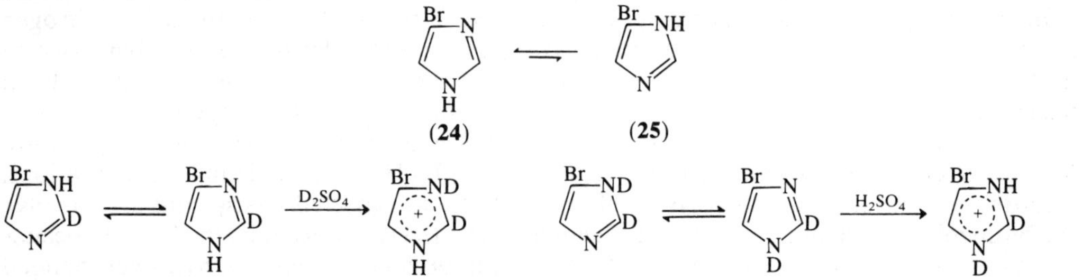

Scheme 12

Another way to study tautomerism is through pK_a measurements. A comparison of the basic pK_a values for 4(5)-nitroimidazole with those of 1-methyl-4- (**29**) and 1-methyl-5-nitroimidazole (**30**) leads one to the conclusion that the 4-nitro tautomer (**28**) predominates (Scheme 13); in fact the 4-nitro:5-nitro ratio has been computed as about 400:1. In this experiment the methylated compounds serve as 'fixed models' in which the nitro substituent is fixed in each of the two possible orientations. The influence of the methyl group can be shown to be small by comparison of the pK_a values for imidazole (~7.0) and 1-methylimidazole (~7.1), and hence 1-methyl-4-nitroimidazole resembles the major tautomer.

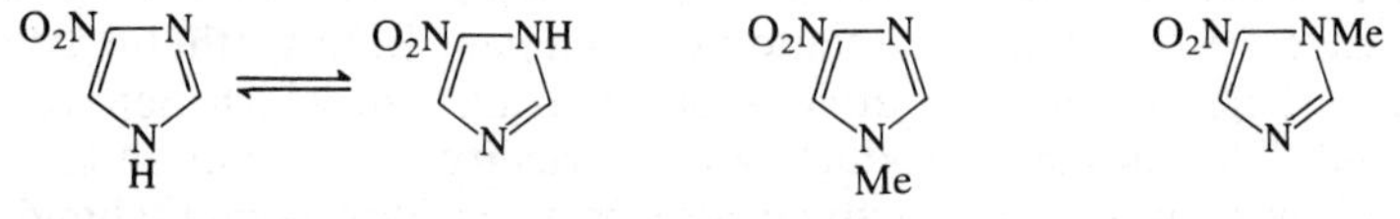

(**28**) $pK_a = -0.05$ (**29**) $pK_a = -0.53$ (**30**) $pK_a = 2.13$

Scheme 13

The K_T values have been summarized for a variety of 4(5)-substituted imidazoles (Table 8) ⟨76AHC(S1)280⟩, and these, along with Charton's application of the Hammett equation to heteroaromatic tautomerism ⟨65JOC3346, 69JCS(B)1240⟩, have suggested qualitatively that electron-donating groups ($\sigma_m < 0$) will favour the 5-substituted tautomer, while electron-attracting groups ($\sigma_m > 0$) will favour the 4-isomer. It is possible to predict the position of tautomeric equilibrium for any unsymmetrically substituted imidazole (provided that $\Delta\rho$ is known) using the relationship $\log K_T = pK_B - pK_A = (\rho_B - \rho_A)\sigma_m$ (where A and B refer to the 4- and 5-tautomers). A relatively recent attempt ⟨70JHC227⟩ to the correlate pK_a values of imidazoles with the sum of the Hammett coefficients, $\Sigma\sigma^*$, in which the σ-values of the substituents were expressed as $\frac{1}{2}(\sigma_o - \sigma_m)$, made the inadmissible assumption that $K_T = 1$ in each case. Although the available experimental evidence tends to support the view that electron-withdrawing groups favour the 4-substituted tautomer and the converse is true for electron donors, as yet the range of substituents on which this is based is very limited. In fact, calculated heats of combustion show that in the 4-aminoimidazole ⇌ 5-aminoimidazole equilibrium there is a slight predominance of the 4-amino form, a conclusion at variance with the foregoing since $\sigma_m(\mathrm{NH_2}) = -0.161$.

Table 8 K_T Values for Imidazole Annular Tautomerism ⟨76AHC(S1)280⟩

Imidazole	K_T	Imidazole	K_T
Unsubstituted	1	4-Ph	10–37
2-Br	1	5-Cl-4-NO$_2$	118–186
4-Br	63–152	2-Br-4-NO$_2$	210
4-Cl	447	2,4-(NO$_2$)$_2$	1.1–3.1
4-NO$_2$	320–500		

^{a} $K_T = [\text{4-NO}_2]/[\text{5-NO}_2]$.

It should be emphasized that none of the tautomerism studies discussed here relies on chemical reactivity, a notoriously misleading measure of tautomeric structure.

X-Ray crystallography has been employed to demonstrate that, in the solid state, histamine is 5-(2-aminoethyl)imidazole, a result which contrasts with the structure determined for its cation. However, from determinations of pK_a values of histamine and its two possible N-methyl derivatives, a K_T value of 4 (4-:5-substituted isomer) has been found. In fact, in aqueous solution histamine exists largely as the protonated form (**31**) which reverts to (**32**) on neutralization. Electron-withdrawing groups in the imidazole ring of histamine cause the 4-isomer to predominate, but methyl groups have little effect ⟨80AHC(27)241⟩. The 4-substituted zwitterionic form for histidine is preferred in basic medium; this form is maintained in a number of histidine derivatives (*e.g.* the polypeptide bacitracin).

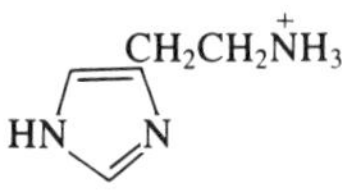

(**31**) 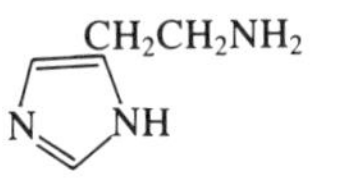   (**32**) 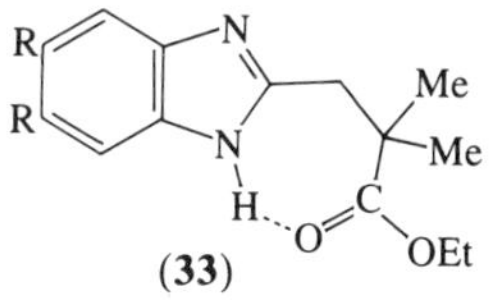   (**33**)

Although, as mentioned above, the prototropy between N-1 and N-3 of imidazole is normally very rapid, as it is with benzimidazole, it is possible to prepare benzimidazoles (of general structure **33**) in which the rate of prototropy has decreased to the extent that the two forms can be demonstrated. The retarding effect described in this instance can be ascribed to intramolecular hydrogen bonding between the NH and the carbonyl oxygen. Ring–chain tautomerism can be eliminated as an explanation because of the low energy barrier to rotation, and the strong carbonyl band in the IR spectrum ⟨75TL4085⟩. Of course, when there is a benzene ring fused in the 4,5-positions of imidazole the prototropic tautomerism makes the 4- and 5-positions equivalent to the 7- and 6-positions, respectively (Scheme 14). K_T is near unity for most benzimidazoles.

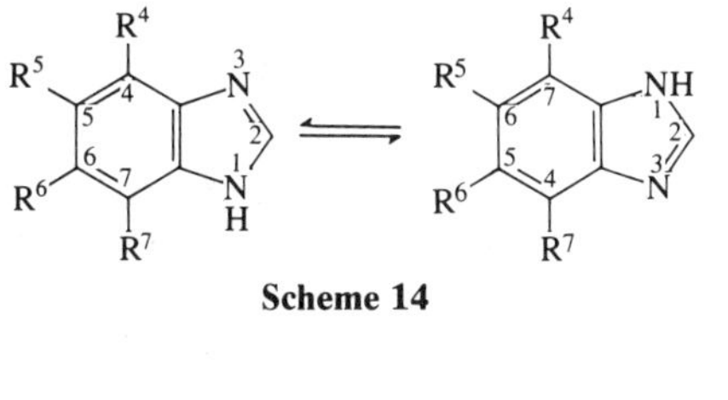

Scheme 14

4.06.5.2 Ring-substituent Tautomerism

As discussed in the General Chapter (Section 4.01.5), imidazoles with potential hydroxy, thiol and amino substituents can exist in a variety of tautomeric forms. In contrast to the 'hydroxypyrazoles' which have been studied in detail, comparatively little is known about the corresponding 1,3-diazoles. Many of the imidazoles are not easily accessible synthetically, and they may not be particularly stable (*e.g.* 4-hydroxyimidazoles). Amino derivatives have also received little attention for similar reasons. The most valuable source material relating to ring-substituent tautomerism is to be found in ⟨76AHC(S1)266⟩. The 1-substituted imidazolin-5-ones can exist in four possible forms (**34–37**; X = O) in tautomeric equilibrium (Scheme 15), and the same could be said for the corresponding sulfur (X = S) and amino (X = NH) compounds ⟨76AHC(S1)280⟩. Certainly both 1,2,4-trimethyl- and 1,4-dimethyl-2-phenyl-imidazolin-5-one exist in the form corresponding to (**34**; X = O) under a wide variety of conditions. Evidence came from UV studies in methanol, IR studies in the solid state, NMR data in a variety of solvents, and comparison with 4,4-dimethyl model compounds. However, the 4-phenyl analogues do not exist in the form (**34**; X = O), but the experimental data available do not adequately define the true structure; the authors suggest that the enolic form (**35**; X = O) predominates.

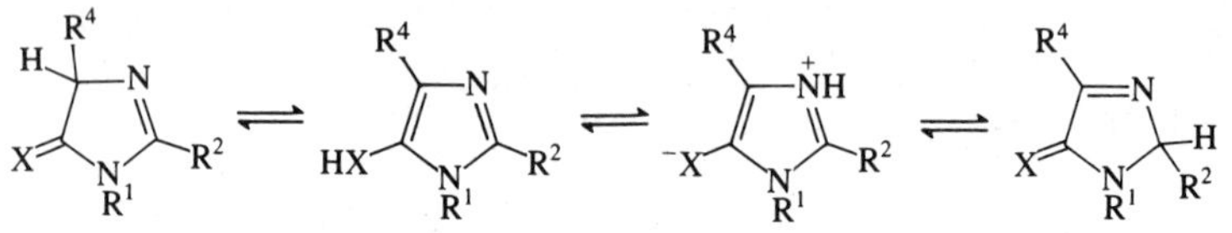

(**34**) X = O, S, NH   (**35**)   (**36**)   (**37**)

Scheme 15

1-Methylimidazolin-4-ones can exist in three possible forms (**38–40**; X = O) (Scheme 16). In the solid state, 1,2-dimethyl-5-phenylimidazolin-4-one exists in the OH form (**39**; X = O), but insolubility precluded studies in solution. Compounds stated to be 1-acyl-imidazolin-4-ones have been formulated as the 4-oxo forms (analogous to **38**; X = O) on IR evidence. Where possible, it seems that a non-conjugated carbonyl form (*e.g.* **34, 42**; X = O) is preferred to the conjugated version (*e.g.* **37, 41**; X = O), but although substituents may have a marked effect, the nature of the solvent is not of particular importance in determining the position of tautomeric equilibrium for these compounds.

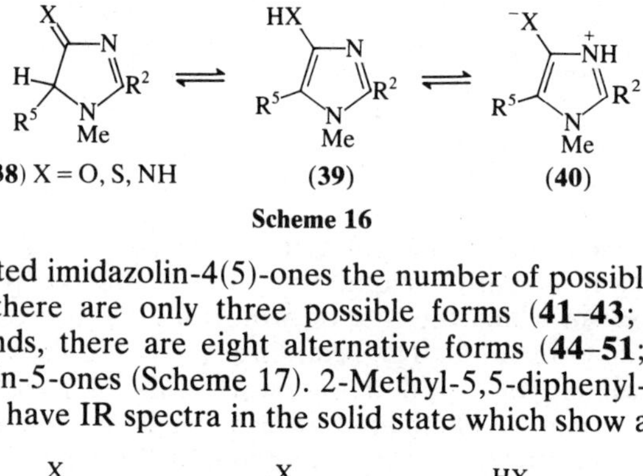

(**38**) X = O, S, NH   (**39**)   (**40**)

Scheme 16

In the *N*-unsubstituted imidazolin-4(5)-ones the number of possible structures is greatly increased. Although there are only three possible forms (**41–43**; X = O) for the 5,5-disubstituted compounds, there are eight alternative forms (**44–51**; X = O) for the 2,4-disubstituted imidazolin-5-ones (Scheme 17). 2-Methyl-5,5-diphenyl- and 2,5,5-triphenyl-imidazolin-4-one both have IR spectra in the solid state which show absorptions belonging

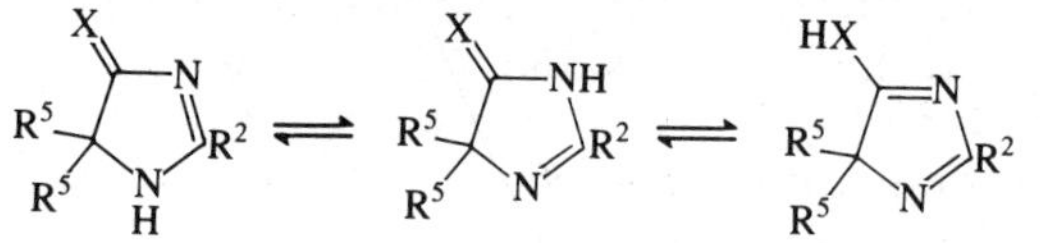

(**41**) X = O, S, NH   (**42**)   (**43**)

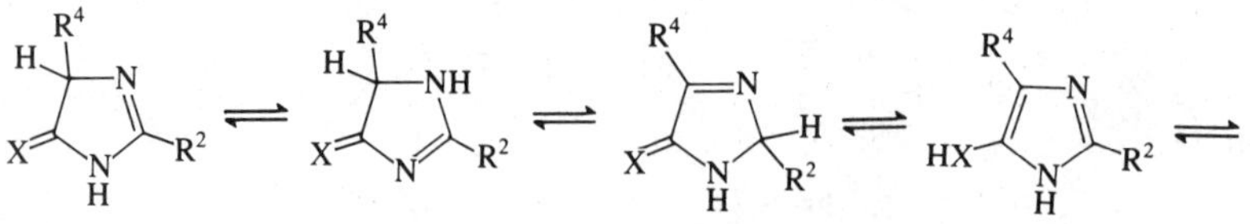

(**44**) X = O, S, NH   (**45**)   (**46**)   (**47**)

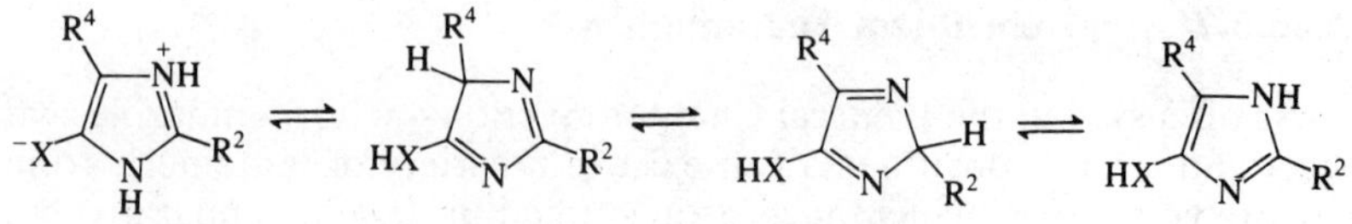

(**48**)   (**49**)   (**50**)   (**51**)

Scheme 17

to individual tautomers (**41**) and (**42**) and it is even possible to separate both 'desmotropic' forms as crystalline solids. However, in solution each form contributes to the same equilibrium mixture with identical spectra. In such compounds the hydroxy form (**43**; X = O) does not seem to be a significant contributor. In non-polar solvents the 5,5-diphenyl and 5-spirocyclohexyl derivatives of imidazolin-4-one exist as the less polar, less conjugated tautomer resembling (**42**), but as the solvent polarity increases the more polar tautomer of type (**41**) increases in importance ⟨72JHC363⟩, particularly when electron-donating groups are present at C-2.

In a recent study ⟨79BSB289⟩ of the prototropy of imidazolin-5(4)-ones, two methods were used: (i) an energetic analysis of the molecular entities with consideration of the solvation term, and (ii) examination of the protonation sites of the common anion. In agreement with the experimental results it was shown that in the vapour phase, as well as in solution, the major tautomeric form is the non-conjugated carbonyl form (**42**; $R^5 = H$, X = O); methylation at positions 1, 2 and 5 leads to the predominant form (**38**; $R^2 = R^5 = Me$, X = O). Until recently, work on the 2,4-disubstituted imidazolin-5-ones was not particularly definitive, but the use of spectroscopic studies allied with CNDO/2 calculations has proved of great value. The summarized conclusions ⟨76AHC(S1)280, p. 374⟩ are (i) in the absence of 4- or 5-substituents, the CH form (**44**; X = O) predominates; (ii) if there is a 4-phenyl group present, one or both hydroxy forms (**47, 51**; X = O) predominates in the solid state or in solution; in CF_3CO_2H the CH form (**44**; X = O) is also present; (iii) of the CH forms (**44, 45**; X = O) the former is always favored; (iv) 2-substituents have less influence than 4-substituents on the position of equilibrium; and (v) solvents do not greatly affect the equilibrium.

If one excludes zwitterionic structures which appear to have less stability than the other forms, there are four possible structures for the imidazolin-2-ones (**52–55**; X = O) (Scheme 18). IR evidence points to a carbonyl form (**53**; X = O) in the solid state, while the characteristic absorption in the UV region supports their designation in solution as the carbonyl forms. If their melting points are compared with imidazoles (*e.g.* imidazole, 90 °C; imidazolin-2-one, 250 °C), the high values may point to the presence in the solid state of salt-like structures. Benzimidazolinones, and other fused-ring imidazolin-2-ones, appear to exist and react as the ketone forms ⟨66RCR122⟩.

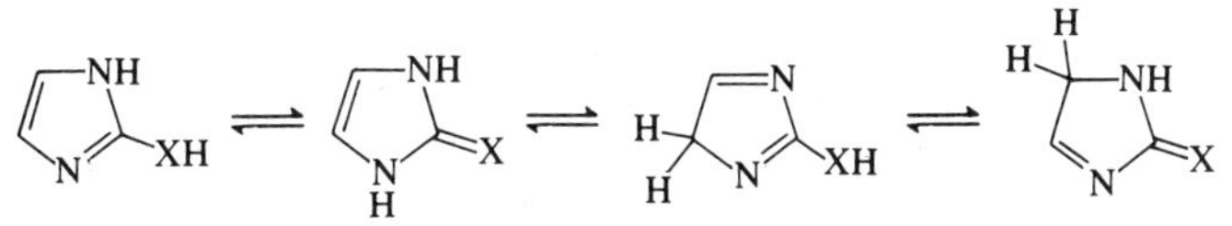

(**52**) X = O, S, NH (**53**) (**54**) (**55**)

Scheme 18

Critical summaries of what is known about the tautomeric natures of imidazolinones have appeared earlier ⟨B-76MI40600, 76AHC(S1)280, 70ZC338, 80AHC(27)241⟩.

The best known of the potential mercaptoimidazoles are the imidazoline-2- and benzimidazoline-2-thiones, which resemble imidazolin-2-ones in that the tautomeric form (**53**; X = S) is the preferred form. The crystal structure and the ^1H NMR spectrum of 1,3-dimethyl-3H-imidazoline-2-thione have been interpreted as showing partial double bond character in the N—C—N system, but no aromaticity ⟨70CC56⟩. However, the preference for a betaine structure (**56**) rather than (**57**) or (**58**) should be accepted with caution since it is really only a resonance structure similar to others which undoubtedly contribute to the overall structures of oxo-, thio- and amino-imidazoles. Measurement of the pK_a values for a series of imidazoline-2-thiones substituted variously on C-4, C-5, N-1 and N-3 by hydrogen, phenyl or methyl shows that all of the values are similar. Approximate K_T values calculated show that these compounds exist even more in the thione forms (**53**, X = S; **58**) than do the corresponding thiazoline-2-thiones and oxazoline-2-thiones. The UV spectra in aqueous solution support thione structures, as do dipole moment and X-ray studies ⟨76AHC(S1)280, p. 400⟩.

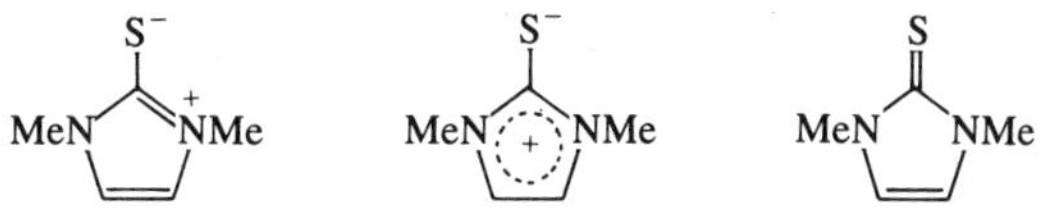

(**56**) (**57**) (**58**)

Imidazolidine-2-thiones (or 2-ones) have been shown by ^{13}C NMR studies to exist entirely in these forms rather than as the possible 2-mercapto-(or 2-hydroxy-)2-imidazolines ⟨80CS(15)193⟩. The thione structure for benzimidazolinethiones is supported by UV, IR and dipole moment measurements. In addition, an X-ray study of 1-(β-D-ribofuranosyl)benzimidazoline-2-thione confirms its thione (**59**) rather than the thiol (**60**) structure (Scheme 19).

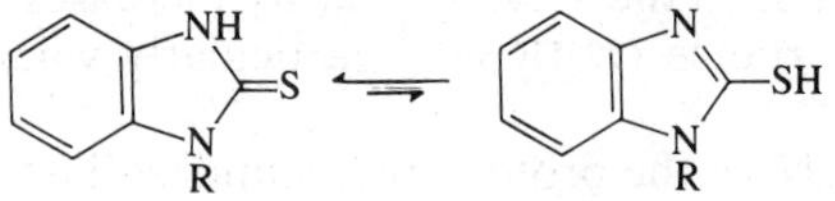

(**59**) R = β-D-ribofuranosyl   (**60**)

Scheme 19

Compounds with a potential thiol group at C-4 or C-5 are not at all well known, but might be expected to display behavior analogous to the related oxygen compounds. Seemingly only the 4,4-diphenylimidazoline-5-thiones (together with their *N*- and *S*-methyl derivatives) have been described, but no study of their tautomerism has been reported.

Aminoimidazoles might be expected to prove more complicated than the analogous 'hydroxy' and 'mercapto' derivatives because of the possibility that an R group on NHR could have a marked effect on the position of tautomeric equilibrium, particularly if R is an acyl or sulfonyl function. However, amines are much weaker acids than hydroxy or mercapto compounds and so it should not be necessary to consider zwitterionic structures to any extent; an NH$^-$ group is not at all likely.

The bulk of the evidence available points to these compounds existing mainly in the amino (NH$_2$) rather than the imino (=NH) forms ⟨71CHE752⟩. With 2-amino-1-methyl-4,5-diphenylimidazole the UV (in methanol), IR (in the solid state) and pK_a values of the two corresponding fixed derivatives (**63**) and (**64**) demonstrate clearly that the compound exists largely as the amino tautomer (**61**) with $K_T = (\mathbf{61})/(\mathbf{62}) = 3 \times 10^4$ (Scheme 20). An amino structure in which 2,4,6-trinitrophenyl groups replace the methyl functions in (**64**) has been incorrectly assigned to the reaction product of 2-aminoimidazole and picryl fluoride ⟨70JHC1391⟩.

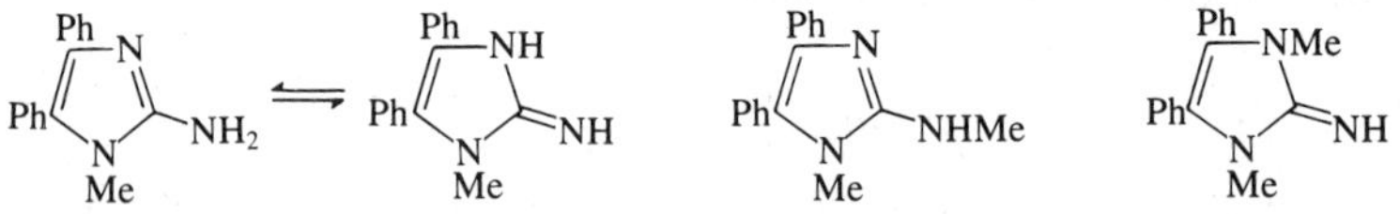

(**61**)   (**62**)   (**63**)   (**64**)

Scheme 20

4-Aminoimidazoles are rather unstable and have not been studied extensively, but despite the absence of a specific study of their tautomerism it is possible to gather some evidence in favour of the amino tautomer of types (**35**; X = NH) (**39**; X = NH) or (**47**; X = NH). Imino structures (*e.g.* **34**, **38**, **44**; X = NH) require the appropriate CH signals in the NMR spectrum: these are absent, and the IR spectra display characteristic bands for NH$_2$. The X-ray structure for 4-acetamido-2-bromo-5-isopropyl-1-methylimidazole confirms the structure (**65**) rather than the tautomer (**66**) (Scheme 21). On the basis of ^{13}C NMR evidence, 2-arylamino-2-imidazolines prefer to exist in the arylimino forms ⟨B-77MI40600⟩.

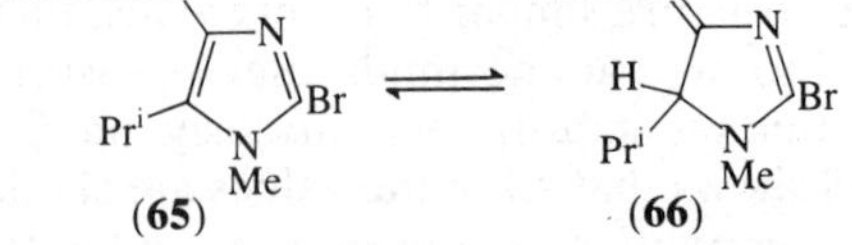

(**65**)   (**66**)

Scheme 21

Like the uncondensed compounds, 2-aminobenzimidazoles exist in the amino forms. Evidence comes from the NH$_2$ bands in the IR (solid state) and comparison of the UV spectra with those of fixed models. A K_T value of 160 in 50% aqueous ethanol has been calculated for 2-amino-1-methylbenzimidazole. Calculations of the π-energies of each tautomer also favour the amino form for this compound and for linear and angular 2-aminonaphthimidazoles. Even in the structure (**67**), where hydrogen bonding might be expected to favour an imino tautomer (**68**), the amino form persists (Scheme 22).

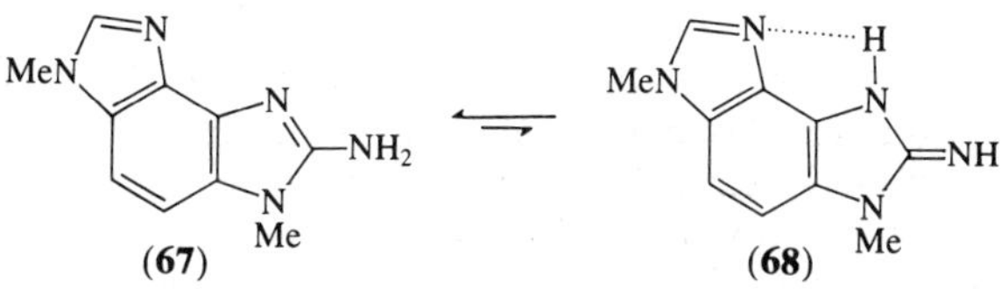

[Scheme 22 structures: (67) and (68)]

Scheme 22

There is a tautomeric equilibrium between N-oxides and N-hydroxyazoles and their benzo derivatives. The equilibrium shown in Scheme 23 has a K_T (**70/69**) value of about 3, but the UV spectrum bears most resemblance to the hydroxy model (**71**). The authors admit, though, that steric hindrance between the N-methyl and adjacent phenyl groups in the oxide model (**72**) renders it less than perfect and the best conclusion (albeit imprecise) may be that in aqueous solution K_T is about 1. In less polar solvents (ethanol, cyclohexane, chloroform, dioxane) UV comparisons still favor the OH form but the foregoing limitations of (**72**) as a model still apply ⟨71JCS(B)2350⟩. In general, such a study of the UV spectra using 'fixed models' when allied with pK_a determinations (which allow one to know with certainly whether an anion, cation or neutral molecule is being studied) provides an excellent example for tautomerism studies. Both hydroxy and oxide forms of 1-hydroxy-4,5-dimethylimidazole are present in chloroform, dioxane and acetonitrile solutions. In the solid state, mass spectrometry favors a polymorphism of a single tautomer. However, mass spectral data on gaseous molecules might not be expected to provide unequivocal evidence for the existence of structures in the solid state, in which only an X-ray crystallographic study can be expected to be completely definitive.

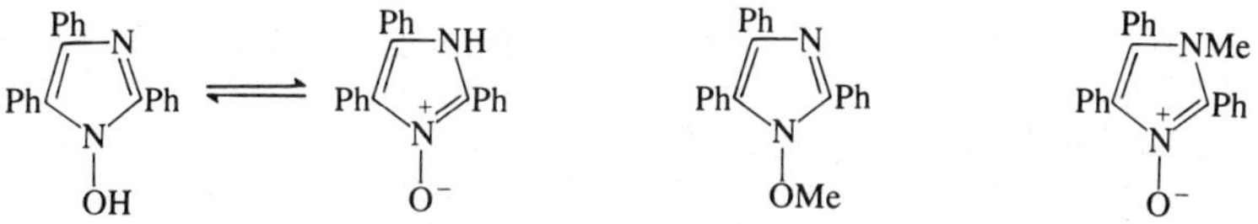

[Scheme 23 structures: (69), (70), (71), (72)]

Scheme 23

One of the common properties of imidazole N-oxides is their strong association even in dilute solution, and in the solid state such association can mask distinction in the IR between NH and OH bands. Thus IR studies of such compounds in the solid state may give rise to confusing results. Careful UV studies of 1-hydroxybenzimidazoles show that with a variety of substituents (*e.g.* 6-nitro, 2-alkyl or 2-aryl) the hydroxy form (**73**) is favoured in organic solvents, and the more polar oxide form (**74**) in water (there is a mixture of both in aqueous ethanol). With 1-hydroxybenzimidazole the N-methyl model (**75**) is not sterically hindered, and the value of K_T (**74/73**) = ~12 deduced from the pK_a values of the models (**75/76**) in water should be much more reliable than that for the hydroxyimidazoles (Scheme 24). In other solvents (benzene, chloroform, acetonitrile, dioxane), UV results show that the OH form (**73**) predominates; in 75% aqueous ethanol $K_T = 1$.

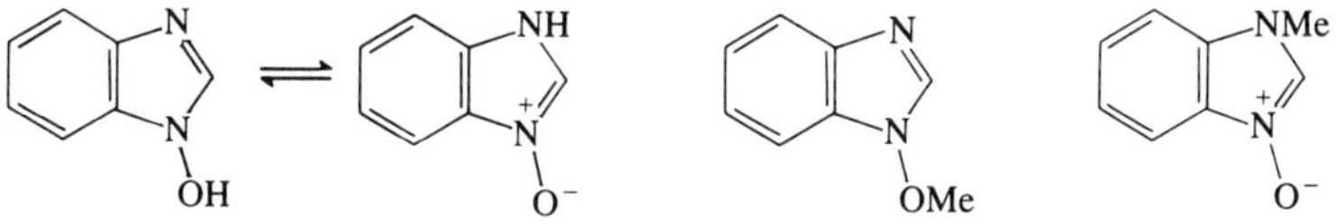

[Scheme 24 structures: (73), (74), (75), (76)]

Scheme 24

On the basis of UV measurements it has been shown that 1-hydroxy-2-imidazolines exist in equilibrium with 2-imidazoline 3-oxides ⟨76CHE1280⟩.

In the 1-hydroxyimidazole 3-oxides a single signal is observed in the NMR spectrum for the 4- and 5-methyl groups of (**77**). This requires an autotropic form of rearrangement (Scheme 25).

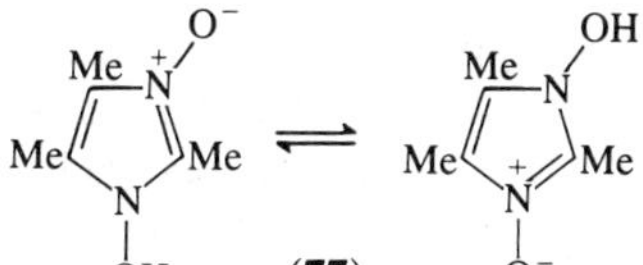

[Scheme 25 structures showing (77)]

Scheme 25

Cases in which more than one tautomerizable substituent are present in the imidazole ring are common, but tautomerism studies of such compounds are less so. Work on the hydantoins (imidazolidine-2,4-diones) is not at all definitive; UV and IR studies suggest the dioxo form (**78**), but evidence for the existence of other tautomeric structures (**79, 80**) is not convincing (Scheme 26). Imidazolidine-4,5-diones also exist as the dioxo forms. NMR studies show that the 2-benzyl derivatives must exist in solution as the NH form (**81**; R = H, Me), and the methoxy derivatives, similarly, are NH (**82**) rather than OH (**83**) derivatives (Scheme 26).

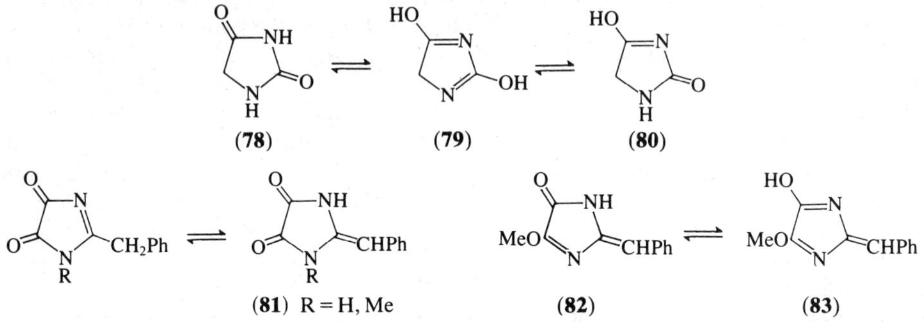

Scheme 26

In contrast to the 2,4-dioxoimidazoles (and such compounds as 5-phenyl-2-thiohydantoin), there is convincing evidence (UV, IR, NMR data) that imidazolidine-2,4-dithiones exist mainly in the thione-thiol forms (**84**; Scheme 27). The two tautomers (**85, 86**) have been separated in 5,5-diphenylimidazolidine-2,4-dithione; in KBr the C=N bands appear at 1495 and 1575 cm^{-1}, respectively. In the 5,5-spiropentane analogues the structure related to (**85**) increases in importance as the solvent polarity increases. As mentioned above, the oxo-thione structures for imidazolin-4-one-thiones is supported by IR spectra in the solid state, and by UV spectra in ethanol. The partially fixed derivatives (**87, 88**; Scheme 28) can be crystallized separately, but give identical spectra in solution. Both X-ray crystal structures and solid state IR spectra confirm the existence of two structures (**87**; γ(C=O) 1690, γ(C=N) 1510–1490 cm^{-1}) and (**88**; γ(C=O) 1720, γ(C=N) 1590–1575 cm^{-1}). As with the dithiones, the importance of the cross-conjugated form (**88**) increases with decrease in solvent polarity.

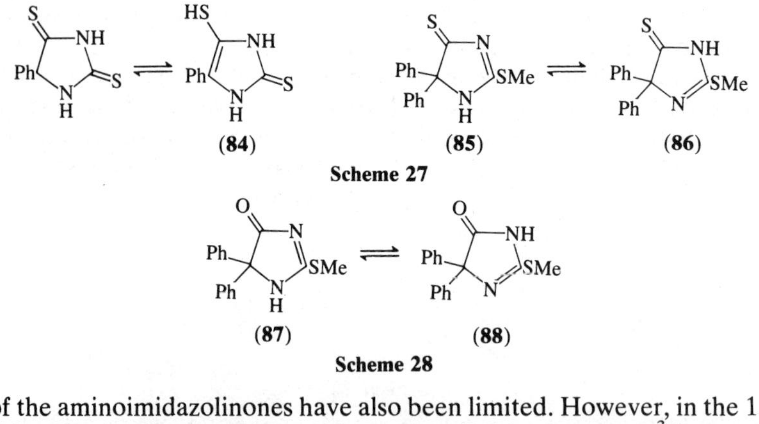

Scheme 27

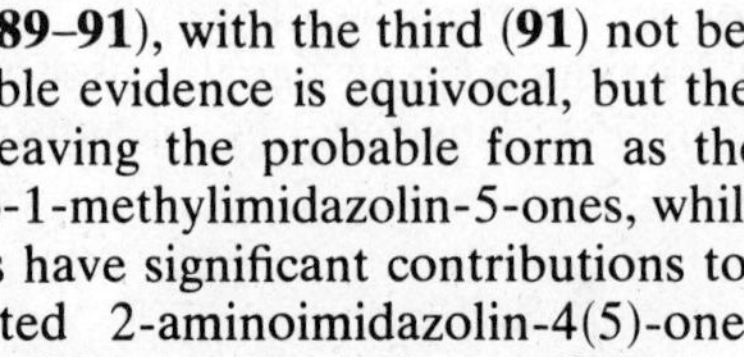

Scheme 28

Studies of the aminoimidazolinones have also been limited. However, in the 1-substituted 2-aminoimidazolin-4-ones the NMR evidence that C-5 is close to sp^3 in nature limits the possible tautomers to three (**89–91**), with the third (**91**) not being at all likely (Scheme 29). Of the other two, the available evidence is equivocal, but the pK_a of 4.55–4.80 rules out (**90**) (expected p$K_a \sim 8$–9) leaving the probable form as the amino-oxo structure (**89**). Similar evidence for 2-amino-1-methylimidazolin-5-ones, while not definitive, suggests that both amino and imino forms have significant contributions to the equilibrium. The added complication in unsubstituted 2-aminoimidazolin-4(5)-ones of annular tautomerism increases the number of possible tautomeric forms (Scheme 30). The UV absorption in ethanol at 213–225 nm, pK_a of ~4.5–4.8 and NMR spectrum showing an sp^3-type C-5 still leave five possible structures, and the problem is as yet unresolved. The distinction between 4-amino and 4-imino forms in 4-aminoimidazolin-2-ones has also not been made.

4.07
Imidazoles and their Benzo Derivatives: (ii) Reactivity

M. R. GRIMMETT
University of Otago

4.07.1	REACTIONS AT AROMATIC RINGS	374
4.07.1.1	*General Survey of Reactivity*	374
4.07.1.1.1	Reactivity of neutral imidazoles	374
4.07.1.1.2	Imidazolium ions	375
4.07.1.1.3	Imidazole anions	375
4.07.1.1.4	Imidazolinones, imidazolinethiones and imidazolinimines	375
4.07.1.1.5	N-Oxides	376
4.07.1.2	*Thermal and Photochemical Reactions Formally Involving No Other Species*	376
4.07.1.2.1	Fragmentation	376
4.07.1.2.2	Rearrangement	376
4.07.1.2.3	Polymerization and disproportionation	380
4.07.1.3	*Electrophilic Attack at Nitrogen*	382
4.07.1.3.1	Introduction	382
4.07.1.3.2	Reaction sequence	382
4.07.1.3.3	Effects of substituents	383
4.07.1.3.4	Proton acids at neutral azoles: basicity of imidazoles and benzimidazoles	383
4.07.1.3.5	Proton acids at imidazole anions: acidity of imidazoles	385
4.07.1.3.6	Metal ions	386
4.07.1.3.7	Alkyl halides and related compounds: compounds without a free NH group	386
4.07.1.3.8	Alkyl halides and related compounds: compounds with a free NH group	387
4.07.1.3.9	Acyl halides and related compounds	390
4.07.1.3.10	Halogens	393
4.07.1.3.11	Other electrophiles	393
4.07.1.4	*Electrophilic Attack at Carbon*	394
4.07.1.4.1	Reactivity and orientation	394
4.07.1.4.2	Nitration	395
4.07.1.4.3	Sulfonation	397
4.07.1.4.4	Acid-catalyzed hydrogen exchange	397
4.07.1.4.5	Halogenation	398
4.07.1.4.6	Acylation	402
4.07.1.4.7	Mercuration	403
4.07.1.4.8	Diazo coupling	403
4.07.1.4.9	Nitrosation	404
4.07.1.4.10	Reactions with aldehydes and ketones	404
4.07.1.4.11	Oxidation	405
4.07.1.5	*Nucleophilic Attack at Carbon*	406
4.07.1.5.1	Hydroxide ion and other O-nucleophiles	407
4.07.1.5.2	Amines and amide ions	409
4.07.1.5.3	S-Nucleophiles	413
4.07.1.5.4	Halide ions	413
4.07.1.5.5	Carbanions	414
4.07.1.5.6	Reduction by complex hydrides	415
4.07.1.5.7	Azide ions and nitrite ions	415
4.07.1.6	*Nucleophilic Attack at Hydrogen*	415
4.07.1.6.1	Metalation at a ring carbon atom	415
4.07.1.6.2	Hydrogen exchange at ring carbon in neutral azoles	416
4.07.1.6.3	Hydrogen exchange at ring carbon in imidazolium ions	417
4.07.1.6.4	C-Acylation *via* deprotonation	417
4.07.1.6.5	Proton loss from a ring nitrogen atom	417
4.07.1.7	*Reactions With Radicals and Electron-deficient Species*	418
4.07.1.7.1	Carbenes	418
4.07.1.7.2	Free radical attack at the ring carbon atoms	418

4.07.1.7.3 Electrochemical reactions and reactions with free electrons	419
4.07.1.7.4 Catalytic hydrogenation and reduction by dissolving metals	419
4.07.1.8 Reactions With Cyclic Transition States	419
4.07.1.8.1 Diels–Alder reactions and 1,3-dipolar cycloadditions	419
4.07.1.8.2 Photochemical cycloadditions	420
4.07.2 REACTIONS OF NON-AROMATIC COMPOUNDS	421
4.07.2.1 Isomers of Aromatic Derivatives	421
4.07.2.1.1 Compounds not in tautomeric equilibrium with aromatic derivatives	421
4.07.2.1.2 Compounds in tautomeric equilibrium with aromatic derivatives	424
4.07.2.2 Didhydro Compounds	424
4.07.2.2.1 Tautomerism	424
4.07.2.2.2 Aromatization	425
4.07.2.2.3 Other reactions	425
4.07.2.3 Tetrahydro Compounds	427
4.07.2.3.1 Aromatization	427
4.07.2.3.2 Ring fission	427
4.07.2.3.3 Other reactions	427
4.07.3 REACTIONS OF SUBSTITUENTS	428
4.07.3.1 Benzenoid Rings	428
4.07.3.1.1 Fused benzene rings: unsubstituted	428
4.07.3.1.2 Fused benzene rings: substituted	429
4.07.3.2 Other C-Linked Substituents	430
4.07.3.2.1 Saturated alkyl groups	430
4.07.3.2.2 Aryl groups	432
4.07.3.2.3 Other C-linked functionality	434
4.07.3.3 N-Linked Substituents	438
4.07.3.3.1 Amino compounds	438
4.07.3.3.2 Aminoazolium ions	441
4.07.3.3.3 Imines	441
4.07.3.3.4 Nitro groups	441
4.07.3.3.5 Nitroso groups	441
4.07.3.3.6 Azidoazoles	442
4.07.3.4 O-Linked Substituents	442
4.07.3.4.1 Hydroxy compounds	442
4.07.3.4.2 Alkoxy and aryloxy compounds	443
4.07.3.5 S-Linked Substituents	443
4.07.3.5.1 Thiones	443
4.07.3.5.2 Disulfides	445
4.07.3.5.3 Thioethers	446
4.07.3.5.4 Sulfinic and sulfonic acids and derivatives	447
4.07.3.6 Halogen Atoms	447
4.07.3.7 Metals and Metalloid-linked Substituents	448
4.07.3.8 Substituents Attached to Ring Nitrogen Atoms	448
4.07.3.8.1 Aryl groups	448
4.07.3.8.2 Alkyl groups	449
4.07.3.8.3 Alkenyl and alkynyl groups	450
4.07.3.8.4 Acyl and aroyl groups	451
4.07.3.8.5 Nitrogen functions	454
4.07.3.8.6 Halogen atoms	454
4.07.3.8.7 Silicon, phosphorus, sulfur and related groups	454
4.07.3.8.8 N-Oxides and N-hydroxy compounds	455
4.07.3.8.9 N-Ylides	456

4.07.1 REACTIONS AT AROMATIC RINGS

4.07.1.1 General Survey of Reactivity

4.07.1.1.1 *Reactivity of neutral imidazoles*

Imidazole (**1**) can be considered as having properties similar to both pyrrole and pyridine. In consequence, one would expect that electrophilic reagents would attack the unshared electron pair on N-3, but not that on the 'pyrrole' nitrogen since it is part of the aromatic sextet. The ring carbons, too, are prone to attack by electrophilic, radical and nucleophilic species to a greater or lesser extent; substitution reactions which do not destroy the aromatic character are predominant. While the imidazole ring is rather susceptible to electrophilic attack on an annular carbon, it is much less likely to become involved in nucleophilic substitution reactions unless there is a strongly electron-withdrawing substituent elsewhere in the ring. In the absence of such activation the position most prone to nucleophilic attack is C-2. Although such reactions do not take place with any great facility with imidazoles, the fused benzene ring in benzimidazoles provides sufficient electron withdrawal to allow a variety of nucleophilic substitution reactions at C-2.

When the imidazole ring is considered to be something resembling a pyrrole–pyridine combination (1) it would appear that any electrophilic attack should take place preferably at C-5 (pyrrole-α, pyridine-β). Such a model, though, fails to take account of the tautomeric equivalence of C-4 and C-5 (Section 4.06.5.1). The overall reactivities of imidazole and benzimidazole can be inferred from sets of resonance structures in which the dipolar contributors have finite importance (Section 4.06.2) or by mesomeric structures such as (2). These predict electrophilic attack in imidazole at N-3 or any ring carbon atom, nucleophilic attack at C-2 or C-1, and also the amphoteric nature of the molecule. In benzimidazole the acidic and basic properties, the preference for nucleophilic attack at C-2 and the tendency for electrophiles to react at the fused benzene ring can be readily rationalized.

4.07.1.1.2 Imidazolium ions

When the multiply bonded nitrogen becomes quaternized either by a proton or some other group it loses its basic properties. The ring now becomes much less susceptible to electrophilic attack, and conversely shows an increasing inclination to react with nucleophiles. The C-2 hydrogen atom between the two quaternary nitrogens may become appreciably acidic, and nucleophilic attack at C-2 becomes more facile, often resulting in ring opening. Since many of the common electrophilic substitution reactions occur in acid medium, they often involve the imidazolium species as substrate. Thus, such reactions as nitration and sulfonation occur with difficulty, and Friedel–Crafts reactions are virtually absent. In the benzimidazolium ion too the reactivity of C-2 with nucleophiles is enhanced compared with the neutral molecule.

4.07.1.1.3 Imidazole anions

In contrast to the neutral molecule of imidazole, the conjugate base is highly susceptible to electrophilic attack, particularly on one or other of the annular nitrogen atoms. Thus, alkylation occurs very rapidly in basic medium. There are also some electrophilic substitution reactions at ring carbon atoms which involve the heterocyclic anion, *e.g.* diazo coupling and iodination.

4.07.1.1.4 Imidazolinones, imidazolinethiones and imidazolinimines

The tautomerism (Section 4.06.5.2) and general reaction types (Section 4.02.1.1.4) of these compounds have been discussed earlier. Although the major tautomers are usually written in the unionized forms, there is undoubtedly polar character, *i.e.* the molecules could be considered as betaines of general structure (3) or (4). With the exceptions of imidazolin-2-ones and -thiones (and the corresponding benzo derivatives) compounds of these classes are not particularly well known. 2-Aminobenzimidazoles have been studied but amino, hydroxy and thiol groups at position 4 or 5 of imidazole often seem to give rise to unstable compounds. In fact, many of the reactions typical of azolinones, azolinethiones and azolinimines are uncommon among the imidazole analogues. The 2-imidazolinones are weakly basic and acidic compounds in line with their tautomeric character, while nucleophilic displacements occur readily at positions adjacent to an oxo function. 2-Phenyl-2-imidazolin-5-ones have an active methylene group at C-4 capable of undergoing condensations with aldehydes ⟨76TL571⟩. Among the hydantoins (imidazolidine-2,4-diones; **5**) the 5-methylene group is acidic enough to form benzylidene derivatives very readily. Such hydantoins are fairly acidic with much of this acidity deriving from the NH group flanked by the carbonyl functions. Parabanic acid (**6**) is a fairly acidic molecule with only negligible basic character. Both 2-imidazolidinones and the corresponding thiones readily form the anions by loss of an NH proton when treated with such reagents as sodium hydride.

(**3**) Z = NH, O, S   (**4**) Z = NH, O, S   (**5**)   (**6**)

Both 4,4-diaryl- and 5,5-diaryl-2-imidazoline-5- and -4-thiones are rearranged by aluminum chloride in benzene to 4,5-diarylimidazoles; 5,5-diphenyl-4-imidazolidine-thiones, though, only give benzo[*b*]thiophenes under these conditions ⟨76T2421⟩.

4.07.1.1.5 N-Oxides

Since the direct *N*-oxidation of imidazoles has not yet been achieved, the synthetic methods available start with acyclic precursors. In consequence the range of compounds is not extensive, and their reactivity is not well known as yet. Since most of the published chemistry relates to the reactivity of the oxide function, it has been considered preferable to discuss all of the reactions of imidazole *N*-oxides in Section 4.07.3.9.8.

4.07.1.2 Thermal and Photochemical Reactions Formally Involving No Other Species

4.07.1.2.1 Fragmentation

Imidazoles in general are very stable to heat. The parent molecule is said to decompose at 590 °C by an unknown process which may be similar to the mass spectral fragmentation (Section 4.06.3.8) since HCN has been identified among the reaction products. The reported stabilities of the thermal rearrangement products of 1-substituted imidazoles at temperatures as high as 650 °C cast some doubt on the above decomposition temperature, but contact time is doubtless important (Section 4.07.1.2.2).

When imidazole 3-oxides are photolyzed the products are unsymmetrical benzil diimines ⟨78CC857⟩. This reaction is believed to proceed *via* a fused oxaziridine intermediate (**7**; Scheme 1). In an analogous process 1-hydroxy-2-methylbenzimidazole 3-oxide (**8**; R = Me) is transformed photolytically into *o*-nitrosoacetanilide (**9**; R = Me). When there is no substituent at C-2 (**8**; R = H) the product isolated is *o*-nitroformanilide, presumably an oxidation product of (**9**; R = H) ⟨78TL4581⟩. The UV irradiation of 2,2-dimethyl-2*H*-benzimidazole 1,3-dioxide (**10**) gives rise to (**11**) by an analogous process. In view of the foregoing results it seems likely that the photosensitized oxygenation of imidazoles to benzamidines may follow a similar process, although a hydroperoxide intermediate has been implicated ⟨68TL3277⟩. Thus, 2,4,5-triphenylimidazole reacts with singlet oxygen to give (**12**; R = H), while a 97% yield of *N,N'*-dibenzoyl-*N*-phenylbenzamidine (**12**; R = Ph) is derived from 1,2,4,5-tetraphenylimidazole (Scheme 2).

4.07.1.2.2 Rearrangement

The equilibrium between azidoimidazoles and the bicyclic tetrazole tautomers has been discussed earlier (Section 4.06.5.3).

Although the reverse reaction is well known, the photochemical conversion of imidazoles into pyrazoles and benzimidazoles into indazoles does not take place. However, 1,4- and 1,2-disubstituted imidazoles are interconverted in *t*-butanol, while 1,4,5-trimethylimidazole gives the 1,2,5-isomer (40%) in ethanol, *t*-butanol or cyclohexane. In the last-named solvent, 1,2,5-trimethylimidazole gives less than 5% of the 1,4,5-isomer, being largely destroyed. When 1,2,4-trimethylimidazole is photolyzed in *t*-butanol it is hardly affected, and though it undergoes a photoreaction in cyclohexane no isomeric imidazoles are formed ⟨69T3287⟩. The above interconversions may involve a disrotatory valence bond isomerization, followed by a [1,3]-sigmatropic shift and another disrotatory isomerization (Scheme 3). Evidence for such a pathway is derived from the observation that 2-deutero-1,4-dimethylimidazole undergoes photoconversion into 4-deutero-1,2-dimethylimidazole. When 2,4,5,5-tetraphenyl-4*H*-imidazole is heated it rearranges to 1,2,4,5-tetraphenylimidazole.

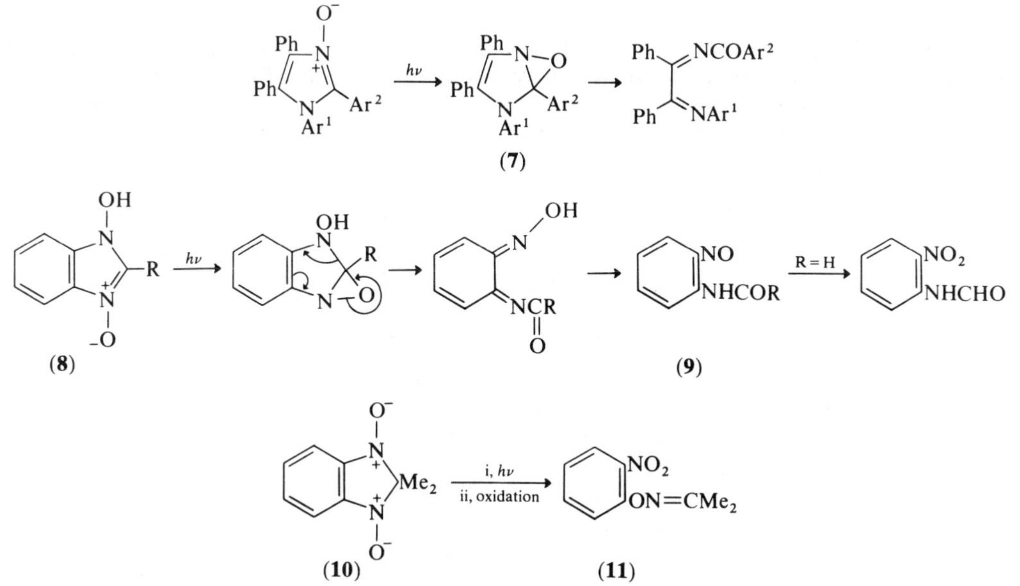

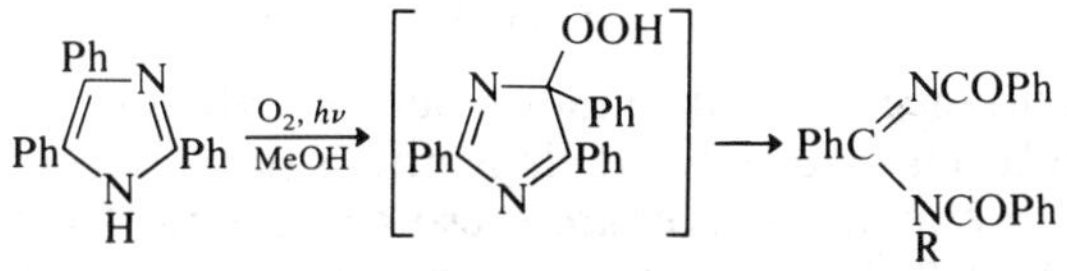

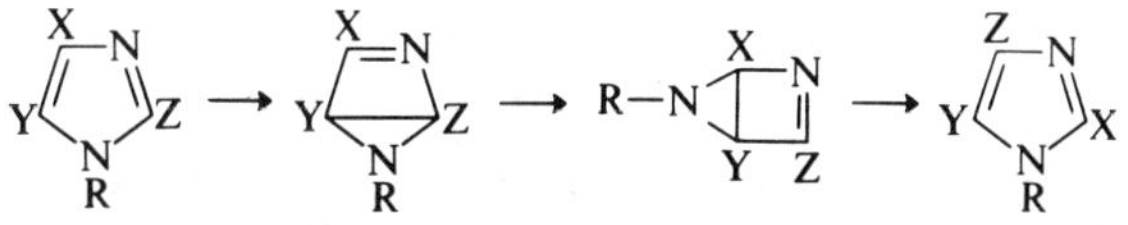

X = Me, H; Y = H, Me;
Z = Me, H, D; R = Me

Scheme 3

In 1883 Wallach discovered that when 1-methylimidazole was heated he obtained 2-methylimidazole and some HCN. The reaction now appears to be much more general ⟨73AJC2435⟩ and of some synthetic importance since it can give rise to a fairly wide variety of 2-substituted imidazoles. The reaction can be shown to be irreversible, uncatalyzed and intramolecular, with no radical involvement. The implication of [1,5]-sigmatropic shifts (Scheme 4) seems likely. Although the 2-substituted product (**13**) predominates, small quantities of the 4- or 5-isomer (**14**) are also formed. Previous reports of related reactions included the isomerization on melting of 1-trityl- to 2- and 4-trityl-imidazoles, but more recent results ⟨81H(15)943⟩ report that at 230 °C 1-tritylimidazoles rearrange in a rather different fashion; no migration to the 2- or 4-positions was noted and the product proved to be (**15**; Scheme 5). Groups such as isopropyl, *n*-propyl and *p*-tolyl at 600–650 °C give products which show no rearrangement of the 'migrating group', but some cracking of the alkyl groups can lead to decreased yields and mixtures of products. That a 1-allyl group migrates with equal facility to both the 2- and 5-positions suggests that a Cope-type rearrangement may also be involved in the case of this group ⟨73AJC2435⟩. In the presence of γ-alumina 1,2,4- and 1,2,5-trimethylimidazoles are thermally rearranged to 2,4,5-trimethylimidazole, while the thermal transformation of 1- to 2-alkylbenzimidazoles appears to proceed normally.

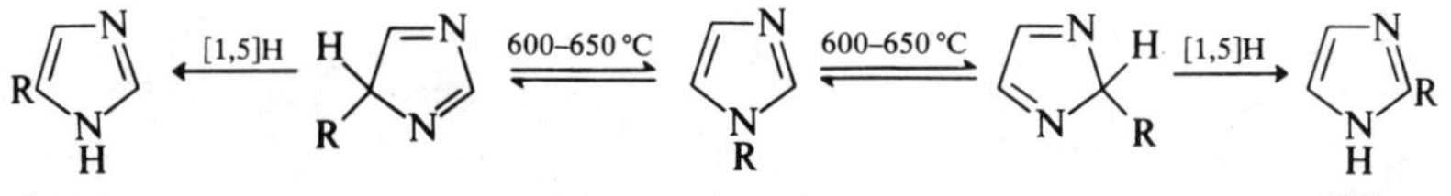

Scheme 4

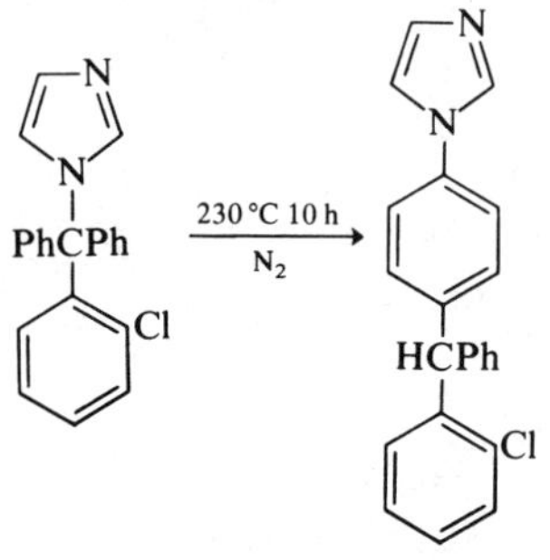

Scheme 5

In view of the importance of 2-nitroimidazoles as antibiotics (*e.g.* 2-nitroimidazole; 'azomycin') and the problems involved in their synthesis (usually *via* the 2-aminoimidazole) it seemed worthwhile to attempt to prepare these compounds by thermal rearrangement of the *N*-nitro compounds (by analogy with the well-known reactions of 1-nitro-pyrazoles and -indazoles) (Chapter 4.04). Thus, when 1,4-dinitroimidazole (**16**) is heated at 140 °C in chlorobenzene, benzonitrile or anisole there is conversion into 2,4- and 4,5-dinitroimidazoles, but considerable denitration can also occur, giving 4-nitroimidazole (Scheme 6). 2-Methyl-1,4-dinitroimidazole rearranges smoothly to 2-methyl-4,5-dinitroimidazole, but 5-methyl-1,4-dinitroimidazole appears to denitrate in preference to any other reaction. Further study of these reactions is indicated.

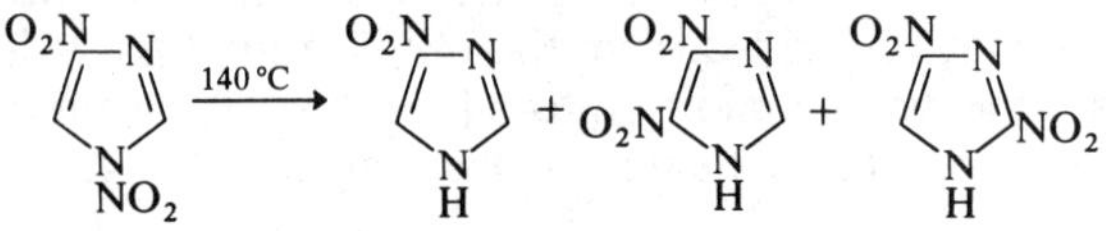

Scheme 6

Under irradiation at 300 nm, *trans*-1-styrylimidazoles are transformed into the *cis* isomers (**17**) which subsequently undergo photocyclization to (**18**; Scheme 7). That this is probably a radical cyclization onto the imidazole ring gains credence both from the position of attack at C-2 and from an apparently analogous reaction in which the diazonium salt of (**19**) reacts in the presence of H_3PO_2 (Scheme 7) ⟨80AHC(27)241⟩.

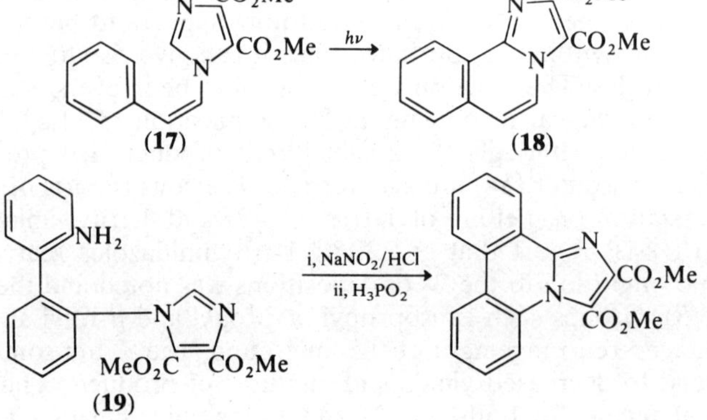

Scheme 7

(20) (21)

Scheme 8

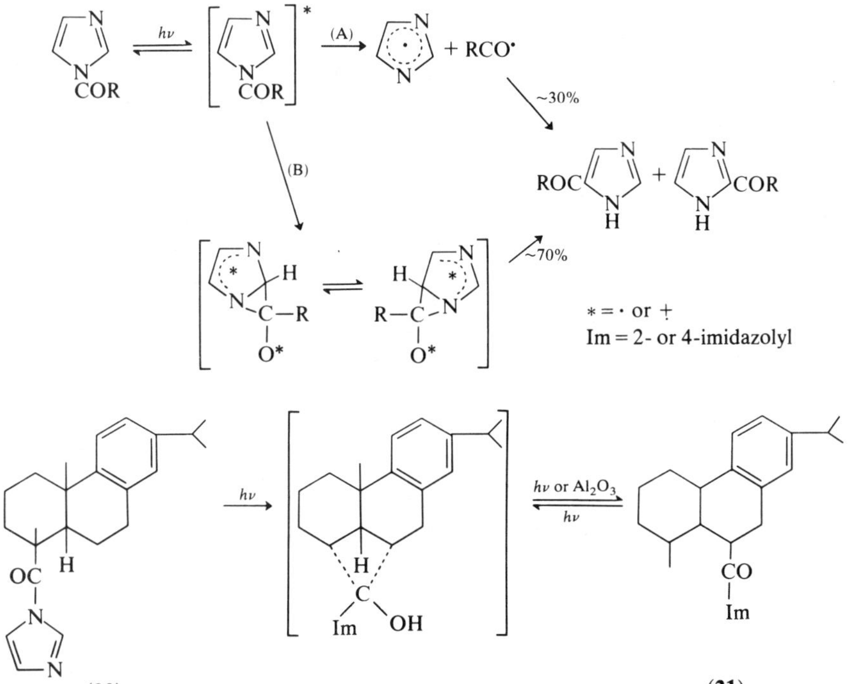

The photo-Fries rearrangement of *N*-substituted imidazoles, where the substituent on nitrogen is an acyl group, to give 2- and 4-substituted isomers could involve either a dissociative path (A) or an intramolecular process (B). The excited species could be a radical or a radical cation (Scheme 8). More complex acyl groups, *e.g.* stearoyl, tend to undergo cleavage in the side-chain, but the 1-acylimidazole (**20**), derived from dehydroabietic acid appears to be subject only to migration, perhaps *via* a cyclobutanol intermediate, to give 2- and 4-acyl derivatives (**21**; Scheme 8) ⟨80AHC(27)241⟩. The photo-Fries rearrangement of *N*-acyl- or -aroyl-benzimidazolinones (**23**) in benzene or alcohol solution gives rise to products of deacylation in addition to so-called *ortho*- and *para*-transposition products (**24**

(22) (23) (24) (25)

(26) (30) (27) (28) (29)

Scheme 9

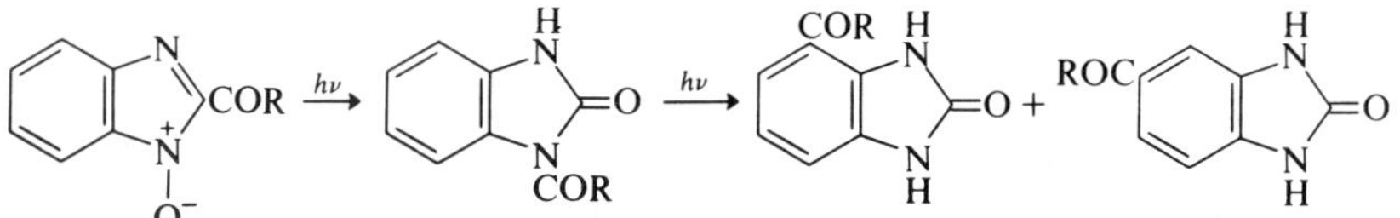

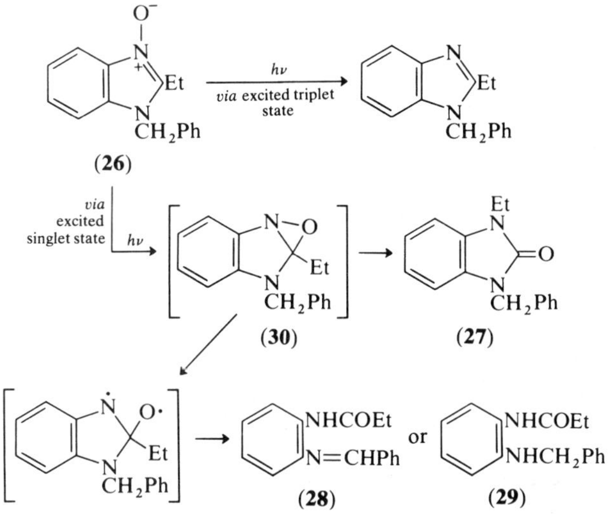

and **25**; Scheme 9). The initial *N*-acyl compounds (**23**) can be obtained by photolysis (or thermolysis) of 2-acylbenzimidazole 1-oxides (**22**) ⟨78JHC625⟩. The initial photorearrangement to the benzimidazolinone occurs also with imidazole *N*-oxides and often involves displacement of an alkyl group from C-2 to a ring nitrogen. For instance, 1-benzyl-2-ethylbenzimidazole 3-oxide (**26**) gives 1-benzyl-3-ethylbenzimidazolinone (**27**) as the major product. There is evidence of some deoxygenation of (**26**) (particularly in the presence of aceto- or benzo-phenone) and, depending on the solvent, products of fragmentation (**28**, **29**) are found (Scheme 9). The products can be rationalized in terms of two different pathways. Deoxygenation of an excited triplet state is enhanced by acetophenone or benzophenone, or the excited singlet state can form an oxaziridine intermediate (**30**) which can either undergo ionic rearrangement in a protic solvent to form (**27**) or *via* a diradical yield the fragmentation products ⟨70CPB964⟩. It does not seem entirely necessary to postulate the diradical intermediate since an ionic mechanism such as that described in Section 4.07.1.2.1 could equally well account for the products. These foregoing reactions bear striking similarities to the effects of an acyl halide (particularly tosyl chloride) in the presence of hydroxide ions on 2,3-polymethylenebenzimidazole 1-oxides (**31**). The benzimidazolinone products (**33**) could arise either from a 1-tosylate intermediate (**32**) (which could follow the familiar process of ring opening of benzimidazoles in the presence of hydroxide ion and acyl halides) or by way of an oxaziridine intermediate (**34**; Scheme 10) ⟨73JCS(P1)705⟩.

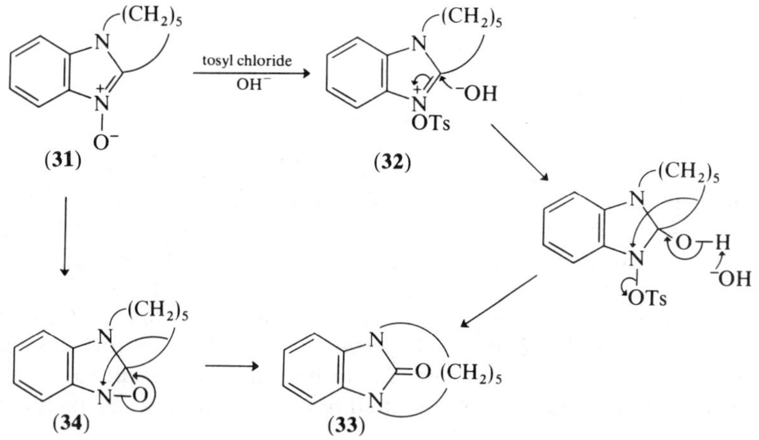

Scheme 10

In the remarkable formation of 2-benzoyl-6-nitrobenzimidazole (**40**) on irradiation of 1-(2,4-dinitrophenyl)-4-phenylimidazole (**35**), either a dipolar cycloaddition or a radical process could lead to the proposed *ortho*-nitrosoimine intermediates (**36, 37**). Certainly the *ortho*-nitro group could take part in a 1,3-cycloaddition at C-2 and C-5, and subsequent rupture of the 1,2- or 1,5-bonds could lead to these intermediates. Recyclization would give rise to (**38**) and (**39**); after hydrolysis and deoxygenation the former would form (**40**), while (**39**) would be expected to produce a 2-aminobenzimidazole product (Scheme 11) ⟨76BSF192⟩.

4.07.1.2.3 Polymerization and disproportionation

When imidazoles and benzimidazoles have free N*H* groups, intermolecular hydrogen bonding gives rise to linear associates of molecules in the crystals and in non-protic solvents. Early determinations of molar masses and dipole moments gave anomalous results because of this phenomenon, particularly when concentrated solutions of the azoles were used. In fact, linear associates of as many as 20 molecules of imidazole are possible at high concentrations in solvents such as benzene. This intermolecular hydrogen bonding gives rise to broad N*H* signals in NMR spectra, and in solvents capable of exchange, *e.g.* D_2O, no signal at all for N*H*. The high melting and boiling points of imidazoles have been discussed (Section 4.06.4.1.1). *N*-Oxides of the series which also contain hydroxy groups, *e.g.* 1-hydroxy-2,4,5-trimethylimidazole 3-oxide, are also capable of forming association complexes; even at high dilutions these complexes exist as trimers.

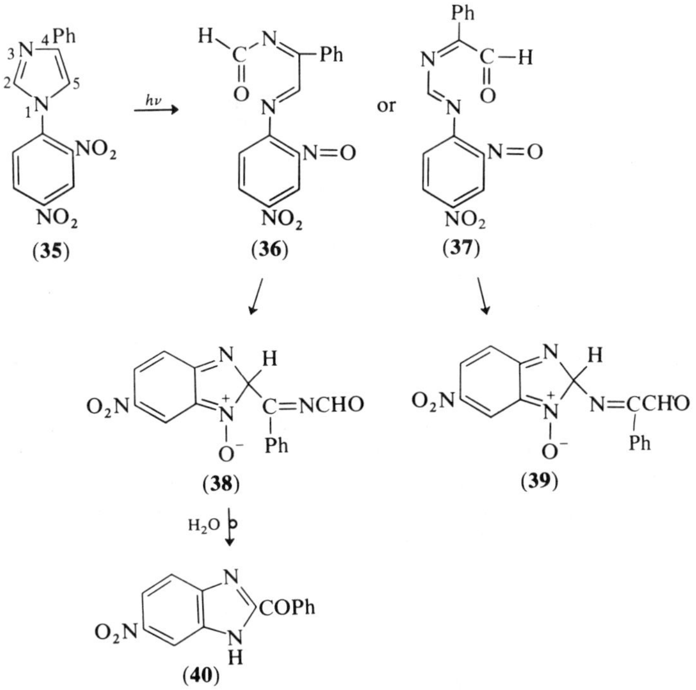

Scheme 11

In the absence of hydrogen sources the thermolysis of 2,4,4-triaryl-5-methylthio-4H-imidazoles (**41**) produces 1,1'-biimidazolyls (**42**), probably by way of a free radical pathway (Scheme 12).

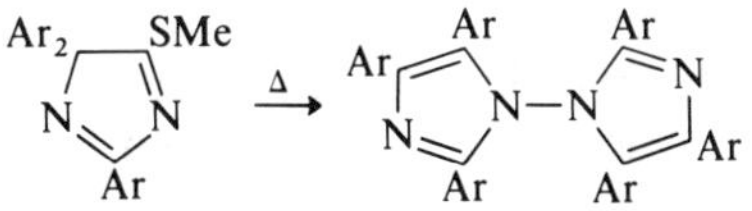

Scheme 12

A number of photochemical processes give rise to dimeric products in which hydrogen is lost, sometimes by concurrent oxidation. Thus, benzimidazole photodehydrodimerizes in 1% ethanol to give a mixture of isomers (**43** and **44**). This reaction probably proceeds through the intermediacy of a 2-benzimidazolyl radical which substitutes the benzene moiety of another benzimidazole molecule ⟨78AJC2675⟩. About 50% of the substrate remains unchanged (Scheme 13). In contrast, 2-alkylbenzimidazoles do not give dimers; rather, mixtures of 1-, 4- and 5-acetylbenzimidazoles are produced through oxidation of transient benzimidazoline intermediates.

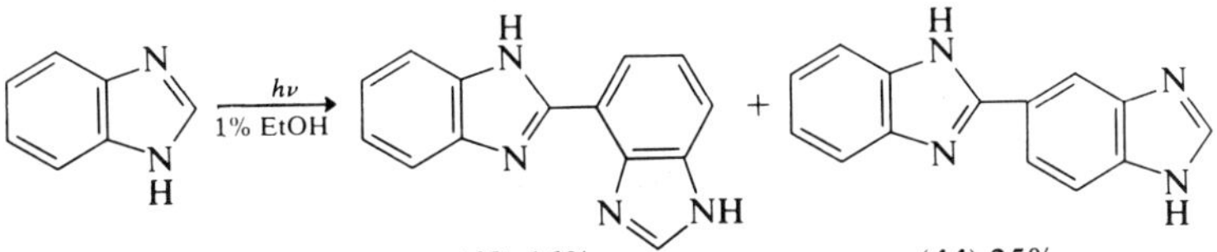

Scheme 13

Great interest has centered on the involvement of 2,4,5-triphenylimidazole (lophine) and similar compounds in the phenomenon of chemiluminescence ⟨77CHE1160⟩. The phenomenon was first observed in 1877 by Radziszewski who noted the emission of light when lophine was decomposed with alcoholic potash in the presence of light. Some aspects of this are discussed elsewhere (Sections 4.07.1.2.1 and 4.07.1.4.11), but some of the dimerization reactions will be treated under this present heading. In the presence of potassium ferricyanide lophine gives rise to a piezochromic dimer (**45**) which gives an intensely violet solution of the radical (**46**) in organic solvents. Other dimeric species (**47**,

48) have also been reported ⟨77CHE1160, B-76MI40701⟩. In a number of instances the photooxidation of the arylimidazoles leads to products of fragmentation, *e.g.* dibenzoylbenzamidine.

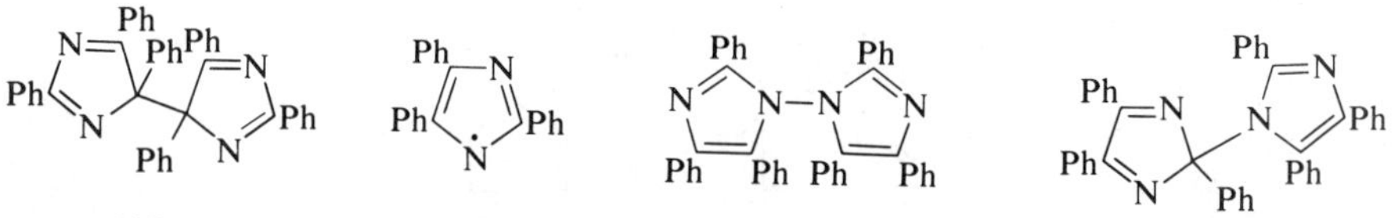

(45)   (46)   (47)   (48)

4.07.1.3 Electrophilic Attack at Nitrogen

4.07.1.3.1 Introduction

The range of reactions which are possible at the ring nitrogens of imidazole and benzimidazole has been outlined (Section 4.02.1.3). As they are the most basic of the azoles, imidazoles are particularly prone to such reactions as protonation, alkylation and acylation, and they readily become involved in complex formation with transition metals. Benzimidazoles, too, are reactive, but less so than the non-condensed species.

4.07.1.3.2 Reaction sequence

In theory, electrophilic attack on imidazole at a ring nitrogen can involve the neutral species, the conjugate base or the conjugate acid. This would lead to four possible transition states (**49**)–(**52**) for the reaction. While either nitrogen of the neutral molecule could be attacked by the electrophile, reaction is normally confined to the multiply bonded nitrogen which has an unshared electron pair orthogonal to the ring. Reaction with the NH nitrogen would require the use of two electrons from the 6π system to form a bond, disrupting the aromaticity. For this reason the transition state (**49**) would be energetically more favorable than (**50**) and, in consequence, reactions with the imidazole neutral molecule follow the sequence shown in Scheme 14, an S_E2' process. When the imidazole already has a group other than hydrogen at N-1 a quaternized product results. Such a reaction sequence is typical of the classical method of alkylation of imidazoles which involved heating the azole with excess alkylating agent, perhaps in a sealed tube. Yields are not high and side reactions complicate the process. For example, either the HI produced or excess MeI can react with imidazole or 1-methylimidazole (Scheme 14). The resulting salts formed are impurities and

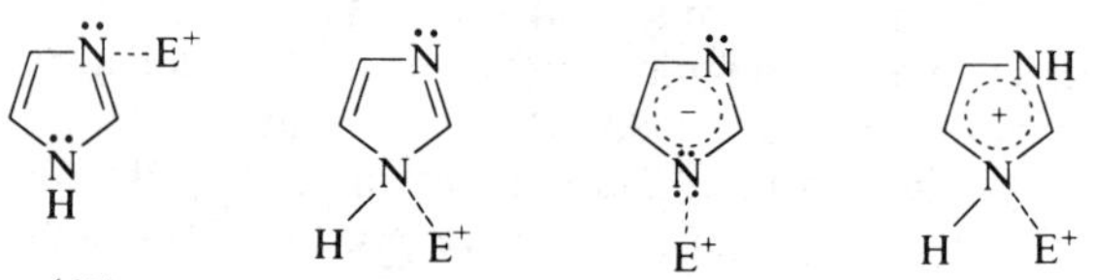

(49)   (50)   (51)   (52)

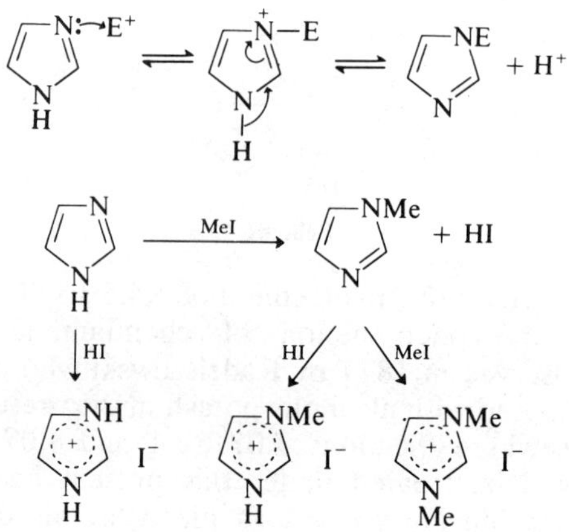

Scheme 14

are also resistant to methylation. In spite of these limitations the reaction has found considerable utility in the synthesis of 1-substituted imidazoles. Attack at either nitrogen of the imidazole anion (**51**) will be a highly favored reaction. Such a reaction sequence applies when imidazoles are alkylated in basic medium, conditions which lead usually to good yields of 1-alkylimidazoles with none of the problems of quaternization which can accompany alkylation of the neutral molecule. Any reaction involving a transition state of type (**52**) would not be favored and such electrophilic substitutions of imidazolium appear to be absent.

4.07.1.3.3 Effects of substituents

When the imidazole ring already has a substituent at C-4 or C-5 then there will be a directional effect imposed on electrophilic attack at annular nitrogen. As outlined earlier (Section 4.02.1.3) this orientation may be related to the tautomeric nature of the substrate. In spite of the inherent pitfalls in equating tautomeric nature with chemical reactivity there are examples which seem only to be explicable in terms of a major tautomer reacting, e.g. methylation of 4-nitroimidazole. Substituents at C-2 can only affect the rate of a reaction at ring nitrogen.

The usual substituent effects apply in the series: electron-withdrawing groups decrease the rate of substitution and, if possible, direct it to the more remote nitrogen atom. This is particularly evident in reactions of the imidazole anion which cannot be subject to complications engendered by tautomerism. Such electron withdrawal is largely inductive. Mesomeric electron-releasing groups will usually be expected to direct electrophilic attack to a nitrogen α or γ to the substituent. In the case of such a 4- or 5-substituted imidazole, however, both annular nitrogens can only be in α or γ positions. Thus inductive and steric effects (or the tautomeric nature) are more likely to determine the orientation of substitution. Since imidazoles with OH, OR, NH_2 or SH groups at the 4- and 5-positions are not well known (in some instances they appear to be unstable), it is not possible to make firm generalizations based on experimental results. Certainly steric effects can be important; electrophilic substitutions leading to N-acylimidazoles are reversible reactions which, in consequence, are subject to thermodynamic control and give the sterically least-hindered isomer (Section 4.07.1.3.9). Steric effects, however, would have less importance in the protonation of imidazoles and benzimidazoles in which the pK_a values of the parent molecules are only modified by the nature and position of the substituent—electronic effects here predominate.

4.07.1.3.4 Proton acids at neutral azoles: basicity of imidazoles and benzimidazoles

Imidazole is the most basic of the azoles and forms salts with a wide variety of acids, both organic and inorganic. Thus, hydrochloride and nitrate salts are well defined, although they may be hygroscopic. Salts of organic acids, e.g. oxalates and picrates, form readily and their relatively low solubilities in aqueous medium have made them extremely useful for the isolation and purification of imidazoles. Picrates, for example, can be decomposed by sulfuric acid and the resulting imidazolium hydrogen sulfate yields the imidazole after sequential neutralization with $BaCO_3$, filtration of the $BaSO_4$ and ether extraction of the aqueous filtrate.

The stability of an imidazole salt is a function of the symmetrical imidazolium cation (**53**) which is resonance stabilized.

$$HN\stackrel{+}{\diagup}NH \longleftrightarrow H\stackrel{+}{N}\diagup NH$$
(**53**)

One major consequence of the basicity of imidazole is that at physiological pH (~7.4) substantial quantities of both the free base and protonated imidazole species are present in the histidine units of a protein. This allows histidine to act as either a proton acceptor or a proton donor according to the demands of its immediate environment. It seems likely that histidine units in a number of enzymes such as ribonuclease, aldolase and some proteases may have this role. The buffering action of histidine in the hemoglobin–oxyhemoglobin

Table 1 Basic pK_a Values of Imidazoles ⟨B-76MI40701, 70AHC(12)103, 80AHC(27)241⟩

Compound	pK_a	Compound	pK_a
Alkylimidazoles		*Arylimidazoles*	
Imidazole	7.00, 7.25	1-Ph	5.10
1-Me	7.30, 7.33	2-Ph	6.40
2-Me	7.85, 7.56	4-Ph	6.10
4-Me	7.56	2,4-Ph$_2$	5.64
1,2-Me$_2$	7.85	4,5-Ph$_2$	5.90
1,4-Me$_2$	7.20	1-Me-4-Ph	5.78
1,5-Me$_2$	7.70	1-(*p*-HOC$_6$H$_4$)	5.35
2,4-Me$_2$	~8.5	1-(*m*-HOC$_6$H$_4$)	5.23
2,4,5-Me$_3$	8.92	1-(*p*-BrC$_6$H$_4$)	4.91
1-Et	7.30	1-(*p*-COMeC$_6$H$_4$)	4.54
2-Et	8.00	1-(*p*-NO$_2$C$_6$H$_4$)	3.96
1-CH$_2$Ph	6.10		
Halogenoimidazoles		*Halogenonitroimidazoles*	
2-F	2.40	4-Cl-5-NO$_2$	−3.62
4-F	2.44	4-Cl-1-Me-5-NO$_2$	−1.42
2-F-1-Me	2.30	5-Cl-1-Me-4-NO$_2$	−3.49
4-F-1-Me	1.90	2-Br-1-Me-4-NO$_2$	−3.07
5-F-1-Me	3.85	2-Br-1-Me-5-NO$_2$	−0.75
4-Cl-1-Me	3.10	5-Br-2-Me-4-NO$_2$	−0.55
5-Cl-1-Me	4.75	5-Br-1-Me-4-NO$_2$	−1.77
2-Br	3.85	2-I-4-NO$_2$	−0.85
4-Br	3.88, 3.60	2-I-1-Me-4-NO$_2$	−1.70
2-Br-1-Me	3.88	2-I-1-Me-5-NO$_2$	−0.14
5-Br-1-Me	5.26		
Nitroimidazoles		*Aminoimidazoles*	
2-NO$_2$	−0.81	2-NH$_2$	8.46
4-NO$_2$	−0.05, −0.16	2-NH$_2$-1-Me	8.65
1-Me-2-NO$_2$	−0.48	2-NH$_2$-4,5-Me$_2$	9.21
1-Me-4-NO$_2$	−0.53	2-NH$_2$-4,5-Ph$_2$	7.04
1-Me-5-NO$_2$	2.13	4-NH$_2$-1,2-Me$_2$-5-NO$_2$	2.50
2-Me-4-NO$_2$	0.50	5-NH$_2$-1,2-Me$_2$-4-NO$_2$	0.33
2,4-(NO$_2$)$_2$	−7.33		
1-Me-2,4-(NO$_2$)$_2$	−7.47		
Acylimidazoles		*Alkoxyimidazoles*	
4-CHO	2.90	2-OMe-5-NO$_2$	−0.90
1-COMe	3.6	2-OMe-1-Me-4-NO$_2$	−0.44
4-CO$_2$H	6.08	2-OMe-1-Me-5-NO$_2$	−1.03
4-CO$_2$Et	3.66	4-OMe-1,2-Me$_2$-5-NO$_2$	2.65
4,5-(CO$_2$H)$_2$	−1.55		
2-Me-4,5-(CO$_2$H)$_2$	4.25	*Miscellaneous imidazoles*	
2-Ph-4,5-(CO$_2$H)$_2$	3.00	4-CH$_2$OH	6.54
4,5-(CN)$_2$	5.2	4-CH$_2$CH$_2$OH	7.26
		4-CH$_2$OAc	6.20
Imidazole 1-oxides		4-CH$_2$NH$_2$	9.68, 5.88
2,4,5-Ph$_3$	3.28	4-(CH$_2$)$_2$NH$_2$	9.8, 6.0
1-Me-2,4,5-Ph$_3$	3.32	4-CF$_3$	2.28
1-OMe-2,4,5-Ph$_3$	3.78	2-(2-imidazolyl)	4.53
		4-(2-pyridyl)	5.42

system is a further consequence. The comment has been made ⟨76MI40700⟩ that the imidazole groups of histidine units in a polypeptide are the strongest bases present in any quantity at physiological pH. Furthermore, the imidazolium cations are the strongest acids present in substantial amount, with variations in pK_a induced by local environment.

The effects of substituents on the basic nature of imidazole are summarized in Section 4.02.1.3.4. Tables 1 and 2 list the basic pK_a values of some representative imidazoles and benzimidazoles. From these it can be seen that methyl and other alkyl substituents exert a weak, base-strengthening inductive effect which is additive. Aromatic substituents decrease basic strength, while groups attached to these aryl rings exert their normal behavior. In fact, Hammett studies have permitted prediction with reasonable accuracy and correlation of pK_a values for a variety of imidazoles. In a study of *meta-* and *para-*substituted 1-phenylimidazoles Pozharskii ⟨70CHE194⟩ has shown not only that 1-phenylimidazole is a

Table 2 Basic pK_a Values of Benzimidazoles ⟨53HC(6-1), 65JCS5590⟩

Compound	pK_a (H_2O)[a]	Compound	pK_a (50% EtOH)
Benzimidazole	5.53 (4.98)	2-NH_2	3.37
1-Et-6-Me	5.07	5-Cl	3.92
1-Me	5.57 (4.88)	5,6-Cl_2	3.26
2-Me	6.19 (5.77)	4-OMe	4.98
4-Me	5.67 (5.16)	5-OMe	5.07
5-Me	5.81 (5.32)	2-OMe-1-Me	4.07
5-NH_2	6.11	2-OEt-1-Me	4.36
4-OMe-1-Me	5.25	2-OMe-1-Me-5-NO_2	2.10
5-OMe-1-Me	5.78	2-Ph	4.51
1-CH_2Ph-5-OMe	5.38	4-NO_2	3.33
6-OMe-1-Me	5.65	5-NO_2	2.67
5-OH-1-Me	5.94	2-Me-5-NO_2	4.37
5-NH_2-1-Me	6.21	2-Me-5,6-$(NO_2)_2$	~0.7

[a] Values in parentheses are in 50% EtOH.

weaker base than imidazole, but also that the phenyl substituents have their usual effects. The ρ value for correlation between these substituents on the benzene ring and the ionization constants is +0.753. This is an indication that the electron transfer effects from the aryl substituent to the imidazole basic nitrogen are weak. It was observed that strongly electron-attracting groups (viz. p-NO_2, p-COMe) behave as though their σ values are greater than normal (i.e. +1.25 and +0.65, respectively). These values are, in fact, close to the σ^- constants (+1.27, +0.87). Such a result implies that there is significant polar conjugation between the π-systems of the heterocycle and the substituent.

The fused benzene ring of benzimidazole decreases the basicity of imidazole by about the same amount as a 1-phenyl or a 5-chloro or -bromo substituent, and the groups attached to that benzene ring (e.g. Cl, NO_2) can enhance this base-weakening effect. The marked difference in basicity between 1- and 2-methylbenzimidazoles may be a consequence of the more symmetrical cation (**54**) formed by the latter (Scheme 15).

Scheme 15

There is commonly a sharp contrast between the pK_a values of 4- and 5-substituted imidazoles. This contrast can only be evident when a ring nitrogen is substituted with consequent definition of the position of the 4- or 5-substituent. Thus 1-methyl-5-nitroimidazole is a much stronger base than the 1-methyl-4-nitro isomer. Certainly an electron-withdrawing substituent α to the basic nitrogen atom has a much greater effect than a β substituent. This observation, which appears to be general, has been utilized in some tautomerism studies (Section 4.06.5.1), and the resulting generalization that electron-withdrawing substituents favor existence mainly as the 4-isomer, and vice versa, is valid.

4.07.1.3.5 Proton acids at imidazole anions: acidity of imidazoles

The NH proton in simple imidazoles and benzimidazoles is weakly acidic. Thus, the compounds are able to form salts with a number of metals (Section 4.07.1.6.5). The anion which forms on loss of the proton is again symmetrical and is highly susceptible to attack by electrophiles. Use is made of this property when imidazoles are alkylated in basic medium (Section 4.07.1.3.8). That substituents can modify considerably the weakly acidic nature of imidazole is evident from the values quoted in Table 3.

A detailed study of fluorinated imidazoles (related to relative reactivities in nucleophilic dehalogenation reactions) has shown that whereas 2-fluoroimidazole dissociates to the anion with p$K = 10.45$, the 4-fluoroisomer shows no sign of such dissociation up to pH 11.7 (p$K = 11.92$). Such results parallel those for the corresponding bromo analogues; the greater acidities of the 2-substituted isomers can be attributed to the symmetrical anions that they

Table 3 Acidic pK_a Values for Imidazoles and Benzimidazoles ⟨80AHC(27)241, 70AHC(12)103⟩

Compound	pK_a	Compound	pK_a	Compound	pK_a
Imidazole	14.90	2-Br	11.03	2,4,5-Ph$_3$ 1-oxide	8.4
2-Me	15.10	4-Br	12.32	Benzimidazole	13.2, 11.26
4-Me	15.10	2-F	10.50	2-Me	11.65
2-Ph	13.32	2-F-4-Me	10.70	2-Ph	11.41
4-Ph	13.42	4-F	11.92	2-(2-Pyridyl)	10.17
2,4-Ph$_2$	12.53	4-NO$_2$	9.30	2-(4-Pyridyl)	10.27
4,5-Ph$_2$	13.43	4-CHO	10.7	Naphth[1,2-d]imidazole	12.52

form. That the fluoroimidazoles are stronger acids (and weaker bases) than their bromo counterparts is an indication that the inductive effect of halogen overshadows its resonance effect.

4.07.1.3.6 Metal ions

There are many examples of complexes which form between transition metal ions and imidazoles; the coordination number is four or six ⟨70AHC(12)103⟩. Thus, coordination between the pyridine-type nitrogen and metal ions such as Cd(II), Co(II), Zn(II), Pt(II), Mn(II), Cu(II) or Cu(I) has been reported on many occasions, giving rise to complexes such as CoCl$_2$·2L and CoCl$_2$·4L (where L = imidazolyl ligand). It is also likely that a number of old reports in the literature of 'metal salts of imidazoles' really refer to complexes. One very important example of such coordination occurs in hemoglobin where Fe(II) coordinates with a histidine unit of the protein. Methylmercury(II) complexes with imidazole in solution usually have a coordination number of two. Under equilibrium conditions tetraamminerutheium(II) complexes bind to imidazoles mainly at N-3, but there are one or two instances in which binding may occur at C-2 (e.g. in 4-methylimidazole and benzimidazole). Equilibrium constants for the formation of six-coordinate complexes of general structure *trans*-Ru(NH$_3$)$_4$(SO$_3$)L (L = 1-methylimidazole, imidazole and histidine) are $>4.6 \times 10^3$, 11.8×10^3 and 0.76×10^3, respectively ⟨78JA2767⟩.

4.07.1.3.7 Alkyl halides and related compounds: compounds without a free NH group

The quaternization of 1-substituted imidazoles is a facile reaction which leads to a stable quaternary salt *via* an S_N2 reaction (Scheme 16) which may be affected by steric factors. It has been shown ⟨78AHC(22)71⟩ that the effects of N-aryl substituents on rate constants for the quaternization of imidazoles can be correlated using a Hammett equation. The value of ρ (−0.45) for the ethylation of imidazoles shows little sensitivity to substituent effects, but this is to be expected since the basic pK_a values of the N-aryl compounds do not vary widely, and the rate of alkylation must depend on the basicity of the nitrogen being quaternized. Certainly the expected sequence of reactivities, 1-methyl > 1-benzyl > 1-phenyl, is observed in the ethylation of 1-substituted imidazoles with iodoethane in ethanol or acetone, and although only qualitative observations are available, 1-methyl-4- and -5-chloroimidazoles react less readily than 1-methylimidazole. Considerable experimental difficulty is experienced in quaternizing nitro-substituted imidazoles.

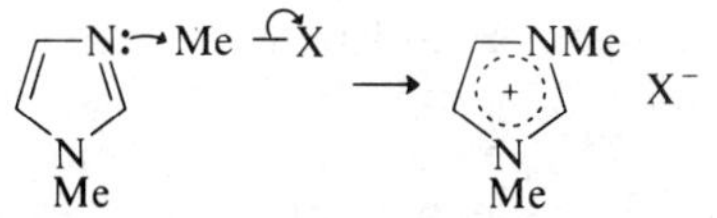

Scheme 16

There are numerous examples of quaternizing alkylations of imidazoles using such diverse reagents as alkyl, alkenyl or aralkyl halides, ethyl chloroacetate, phenacyl bromide or dimethyl sulfate. Since water is frequently held very tenaciously by imidazole quaternary salts the compounds are often best prepared in anhydrous conditions (*e.g.* dry benzene solvent in a dry nitrogen atmosphere) even though the reaction is commonly slower in non-polar solvents.

Reports of quaternization at the already substituted nitrogen should be viewed with suspicion, and one of the products of the reaction of imidazole with 2,2-dichlorodiethyl sulfide (mustard gas) designated as (55) is more likely to be (56) or (57). The observations of Pinner and Schwarz in 1902 that the quaternary salt obtained from 1-methylimidazole and 1-bromopentane could be decomposed by alkali to give both aminomethane and 1-aminopentane was the first piece of evidence (more recently confirmed by NMR studies) to support the accepted view that quaternization takes place at the unsubstituted ring nitrogen.

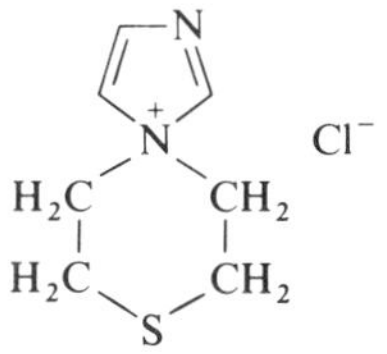

4.07.1.3.8 Alkyl halides and related compounds: compounds with a free NH group

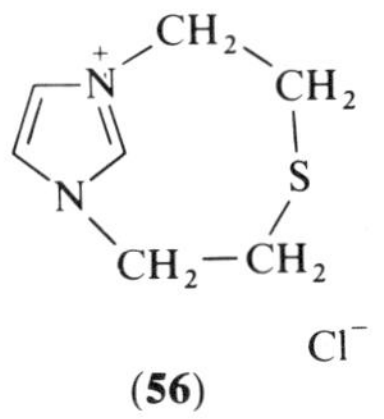

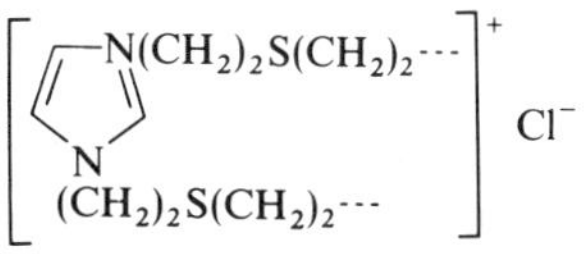

As mentioned above (Section 4.07.1.3.2) imidazoles are readily alkylated in neutral or basic medium following either an S_E2' or an S_E2cB mechanism. The method of choice now commonly involves alkylation of the heterocycle in basic medium (Scheme 17) since these conditions do not suffer from side reactions which produce the imidazolium species. Thus, the reaction of an imidazole with an alkyl halide (or related compound) is carried out in the presence of the oxide or hydroxide of an alkali metal or alkaline earth element, with sodium ethoxide or sodamide, in solvents such as ethanol, dioxane, acetone or liquid ammonia. The use of sodium in liquid ammonia gives excellent yields (>70%) in many instances. Other workers have obtained high yields of N-alkylimidazoles by heating the potassium imidazolate with an alkyl halide either in a sealed tube or under reflux in xylene, benzene or toluene.

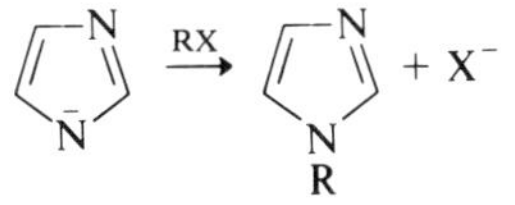

Scheme 17

Related reactions which utilize the imidazole anion include alkylations of the silver salts or an imidazolyl Grignard reagent. The former reaction has been employed specifically for the introduction of trityl and carbohydrate residues. Condensation of either the mercury complex or silver salt of 4-aminoimidazole-5-carboxamide with protected carbohydrates provides a route to imidazole-1-glycosides ⟨80AHC(27)241⟩. The first example of such a synthesis was the preparation of 1-glucopyranosyl-5-methylimidazole from the silver salt of 4-methylimidazole and α-acetobromoglucose. Many alkylations involving unsaturated compounds, too, probably proceed via the imidazole anion. 1-Allylimidazole is accessible by reaction of the heterocycle with 3-chloro-1-propene in the presence of sodium hydroxide in acetonitrile, while the reaction of 4,5-dicyanoimidazole with ethylene oxide in sodium hydroxide solution gives 4,5-dicyano-1-vinylimidazole. Vinyl halides give the same products, and even alkynes in the presence of bases can alkylate imidazoles.

Alkaline medium, however, is not always suitable since branched-chain alkyl halides are subject to elimination, and there are striking changes in orientation of substitution with unsymmetrically substituted imidazoles. Only low yields of 1-(1-imidazolyl)-2-methylpropane are obtained in the reaction of imidazole with 1-iodo-2-methylpropane in sodium–liquid ammonia.

The use of phase-transfer catalysis in the alkylation of imidazoles and benzimidazoles has already proved its value; in ether, 18-crown-6 catalysis gives 64–86% yields of N-methyl products with basic aqueous solutions of the heterocycles and alkyl halides ⟨79TL4709, 76BSF1861, 80JOC3172⟩.

There are a multitude of examples which involve alkylation of the imidazole neutral molecule in spite of the problems of side products which commonly accompany the procedure. The process is of the S_E2' variety (Scheme 14) and involves reagents as diverse as triphenylmethyl chloride or fluoroborate, alkyl esters of phosphonic or phosphinic acids, various carbonyl compounds, oxiranes, unsaturated compounds and diazomethane. The N-alkylation reactivity of trialkyl phosphates decreases with increase in size of the alkyl groups. High yields of N-alkylimidazoles result on treatment of imidazole at 180 °C with dimethyl oxalate, while the sealed-tube reaction of imidazole and formaldehyde at 120–130 °C gave a liquid (described as 1-hydroxymethylimidazole) which formed imidazole when boiled with water. Dimethyl sulfoxide reacts at elevated temperatures with 1-trimethylsilylimidazole in a Pummerer reaction to produce 1-(methylthio)methylimidazoles. The vapor phase methylation of imidazole with methanol in the presence of alumina is accompanied by catalytic rearrangement of an N-methyl group to C-4 or C-5 followed by remethylation at N-1. Such a reaction seems unlikely to involve an electrophilic substitution mechanism and may indeed be a radical process ⟨70AHC(12)103⟩. In the alkylation of 2-amino-1-methylimidazole with α-bromocarbonyl reagents, an N,N-dialkylated iminoimidazole (**58**) is formed at low temperatures; when the temperature is raised above 66 °C, a bicyclic dehydration product (**59**) results (Scheme 18). The use of trimethyl orthoformate to give 1-dialkoxymethyl products, which are readily hydrolyzed in acidic or neutral medium at room temperature, has been recommended for protection of imidazoles during formation of the 2-lithiated derivatives (Section 4.07.1.6.1) ⟨80JOC4038⟩.

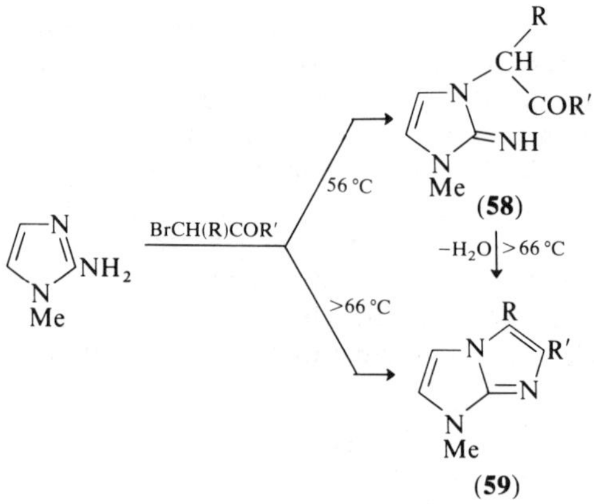

Scheme 18

As intimated earlier, the alkylation products of 'unsymmetrical' imidazoles may vary according to reaction conditions. The product orientation depends not only on the mechanism, but also on such factors as substituent and steric effects, and even the tautomeric nature of the azole cannot be neglected. In basic medium the mechanism (S_E2cB) involves the anion, and the two possible products (**60**, **61**; Scheme 19) are obtained in a ratio which reflects the electronic and steric effects of the substituent group R and those of the reagent. A large R group tends to direct substitution to the more remote nitrogen. An electron-withdrawing group R reduces the basic nature of the adjacent nitrogen more than that of the remote nitrogen and gives the same orientation (**61**); conversely, an electron-releasing group leads to a preponderance of (**60**). Thus, the alkaline benzylation of 2,4-dialkylated imidazoles gives the least sterically-hindered product (1,2,4), and also in basic medium the methylation of 4-nitroimidazole gives a 9:1 ratio of 1-methyl-4-nitro:1-methyl-5-nitro products (electronic effect). Under similar conditions dimethyl sulfate converts 2-chloro-4-nitroimidazole into mainly 2-chloro-1-methyl-4-nitroimidazole.

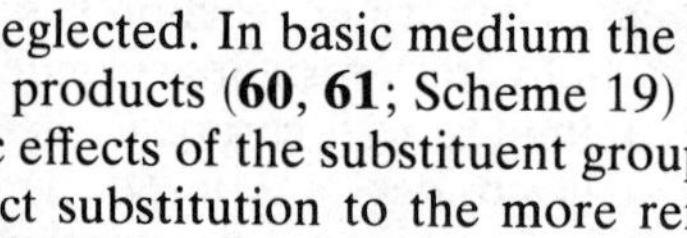

Scheme 19

Matters become more complex still in neutral conditions because now tautomeric effects can be superimposed upon steric and electronic effects. In an S_E2' process 2-methyl-4-phenylimidazole is methylated by dimethyl sulfate to give a mixture of the 4-phenyl and

5-phenyl isomers in the ratio 5.7:1 (Section 4.02.1.3.8). The phenyl group has some steric effect, besides being electron withdrawing; both of these combine to result in a preference for the formation of the 1,4-product in spite of the fact that this necessarily involves methylation of the minor tautomer. There are, however, a number of instances in which the products can only be accounted for in terms of alkylation of a major tautomer. 4-Nitroimidazole and 4-bromoimidazole both give mainly the 1,5-substituted compounds when methylated with dimethyl sulfate, while 2-methyl-4-nitroimidazole can be fused with alkyl tosylates to give mainly the 1,2,5-substituted products. Table 4 lists alkylation results from the literature, but it would be valuable to check and amplify the data using methods of analysis currently available since the total yields quoted are often less than quantitative.

Table 4 Ratios[a] of Isomers Formed in Alkylation of Imidazoles ⟨B-76MI40701, 79H(12)186⟩

Imidazole	MeI	Me_2SO_4	Reagents CH_2N_2	Me_2SO_4/OH^-	Ag^+ salt/MeI
4-Br	—	1:34	—	—	—
4-CN	—	—	—	2.9:1	—
4-CHO	—	b	—	—	—
4-CO_2Me	mixture	b	b	—	—
4-Me	2:1	—	—	2.2:1	—
4-NO_2	—	1:350	1:45	9:1	c
4-Ph	—	5:1	1:2	—	—

[a] 1,4-Isomer:1,5-isomer. [b] 1,5-Isomer only. [c] 1,4-Isomer only.

Ridd and his co-workers ⟨60JCS1352, 60JCS1363⟩ have provided a sound basis for the interpretation of the results of methylation of 4-nitroimidazole, the results of which have been discussed elsewhere ⟨B-76MI40701⟩. The kinetic results are summarized in Figure 1. It is evident that the conjugate base reacts about 10^3 times faster than the neutral molecule, and that in imidazoles with electron-withdrawing groups the changeover from the S_E2cB to S_E2' mechanisms should occur somewhere in the pH range 6–11.

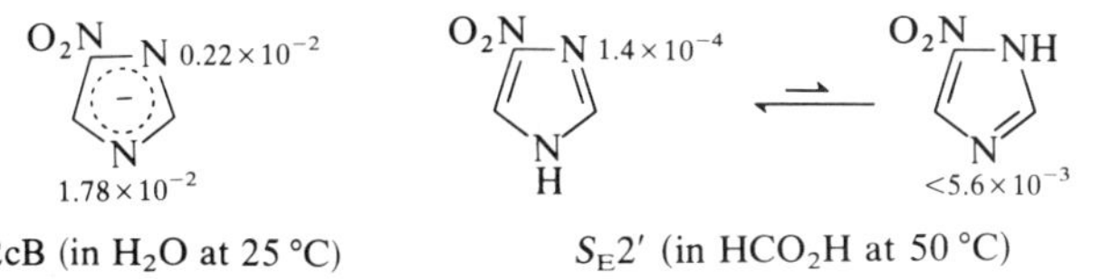

Figure 1 Rate coefficients ($dm^3\,mol^{-1}\,s^{-1}$) for methylation of 4-nitroimidazole with dimethyl sulfate

The orientation of methylation using diazomethane is not easy to explain since the mechanism is in doubt. One interpretation of the preferential formation of 5-substituted 1-methylimidazoles (with a 'highly electron-dense' potential 5-substituent) is via the initial formation of an ion pair [$Im^- MeN_2^+$] in which the cationic portion is situated near to the nitrogen adjacent to the substituent. This may be true, but the results obtained with unsymmetrical pyrazoles ⟨79AJC2203⟩ cannot be explained entirely in this way.

Benzimidazoles, too, are readily alkylated either in neutral or basic medium. As with the uncondensed compounds, the former conditions are complicated by salt formation, but to a lesser extent since benzimidazoles are less basic. It is often valuable to vary the initial amounts of alkali and alkyl halide in order to improve yields. The best yields of 1-alkylbenzimidazole (76–83% for primary and 50–60% for secondary alkyl and aralkyl bromides) result with two moles of bromide and 1.5 moles of alkali per mole of benzimidazole ⟨66RCR122⟩. There is severe steric hindrance to the alkylation of 2-arylbenzimidazoles in alkaline medium, but reactions with the silver salts seem more successful. The rapid and almost quantitative reaction between a benzimidazole and dimethylphenylalkylammonium chlorides in aqueous sodium hydroxide provides a very convenient method of introducing a primary aralkyl group (benzyl, α-naphthylmethyl). There has been little systematic study of the alkylation of unsymmetrical benzimidazoles, though much the same criteria should apply as in the uncondensed compounds. In this respect the observation that 1-bromopropane reacts with the anion of 2,6-dimethyl-4-nitrobenzimidazole to give

2,6-dimethyl-4-nitro-1-propylbenzimidazole is not unexpected. With diethyl sulfate, alkylation of 5-nitrobenzimidazole gives a higher proportion of the 1,6-isomer, but with dimethyl sulfate in basic medium the ratio of 1,5:1,6 is close to 1:1. Alkyl halides, however, *e.g.* methyl iodide and benzyl chloride, in basic conditions give mainly the 1,5-isomer ⟨74CRV279⟩. 1,3-Polymethylenebenzimidazoles are formed when suitable dibromoalkanes react with benzimidazolinone in strongly basic medium (Scheme 20). When $n = 10-12$ yields are high, but reduction of chain length ($n = 5-8$) gives rise to dimers and trimers, and with $n = 3$ and 4 $N-O$-bridged products appear ⟨76TL79⟩. Other reactions in which there is competition for alkylation between an annular nitrogen and an exocyclic group such as amino, hydroxy or mercapto will be discussed with the appropriate functional group.

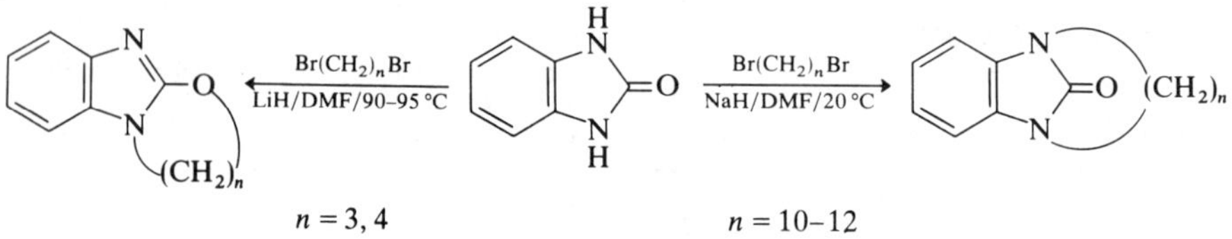

Scheme 20

There are a few examples of intramolecular alkylation and related processes involving N-1. Base-catalyzed cyclization of the appropriate 2-substituted thioacid derivatives have given (**62**) and (**63**). In addition cyclization of a series of ketones (**64**) gives the tertiary alcohols (**65**) which are subject to dehydration (Scheme 21) ⟨74CRV279⟩.

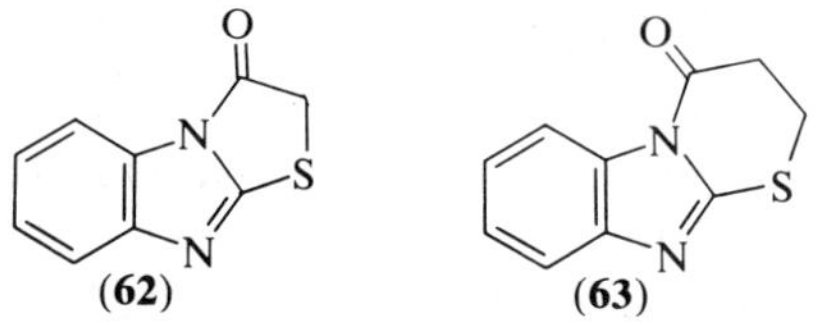

Scheme 21

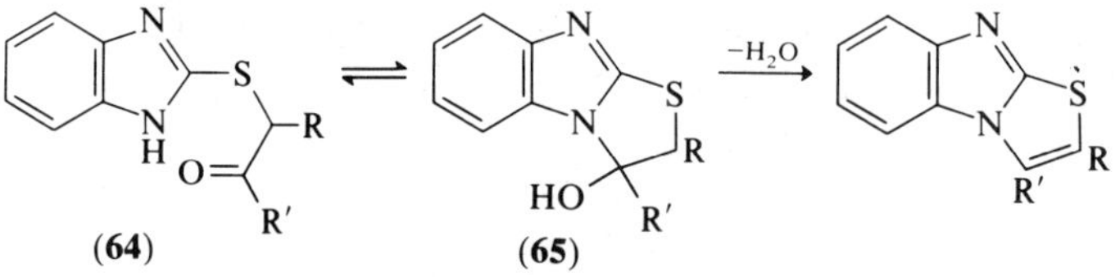

Benzimidazoles, benzimidazolinones and benzimidazolinethiones react with formaldehyde and secondary amines in a Mannich reaction to form N-aminomethyl derivatives. The conversion of benzimidazoles into 1-(2-hydroxyethyl) derivatives with ethylene oxide is well known.

Arylation at a ring nitrogen is not a particularly simple procedure for aryl halides are not usually susceptible to nucleophilic displacement of the halogen group. Nevertheless, a modified Ullmann reaction using the aryl halide in nitrobenzene in the presence of potassium carbonate and copper(I) bromide catalyst gives reasonable yields of 1-arylimidazoles ⟨66KGS143⟩. When there are strongly electron-withdrawing groups associated with the aryl halide less vigorous conditions are needed; 1-fluoro-2,4-dinitrobenzene reacts with imidazoles in benzene containing a little triethylamine to give 77–92% yields of the N-aryl derivatives. The kinetics of the reactions of imidazole with either the above aryl fluoride or picryl chloride are third order overall, and second order with respect to imidazole at low imidazole concentrations. Some leveling off of rates has been interpreted as indicating an addition–elimination mechanism with the decomposition into products being fast for the chloro and slow for the fluoro compound ⟨80AHC(27)241⟩. The use of benzyne leads to the direct N-arylation of imidazoles.

4.07.1.3.9 Acyl halides and related compounds

Reactions leading to N-acylimidazoles have excited considerable interest in recent years in view of the importance of N-acylimidazoles in certain biological processes. In addition they are among the more useful of the 'azolides' which have found synthetic utility in a variety of acylation reactions (Section 4.07.3.8.4). When one adds to these facets the probable intermediacy of N-aroylimidazoles in 2-aroylimidazole formation, and the value of 1-acylimidazoles in the regiospecific alkylation of the heterocycles there is ample reason for the upsurge of interest in these compounds.

As long as there is only a hydrogen on the ring nitrogen of imidazole the N-acyl compounds can be prepared using such diverse reagents as acyl halides and anhydrides, alkyl chlorocarbonates, isocyanates (at elevated temperatures), cyanates, ketene, isopropenyl acetate, acetyladenosine monophosphate, phosphoryl halides and chlorides of sulfonic acids. For simple N-acyl compounds the most common methods of synthesis are, however, reaction of a 1:2 molar mixture of acid chloride and imidazole in an inert solvent at room temperature, or reaction of free carboxylic acids with 1,1'-carbonyldiimidazole at room temperature. With phosgene the latter reagent forms the 1-carbonyl chloride (66) which is in turn converted into the amide (67) by piperidine (Scheme 22) ⟨79LA1756⟩. Note though that 1-acetyl derivatives of imidazole-2-thiols (or -2-ones) must be prepared by heating with acetic anhydride in pyridine; in a reaction medium of ethanol the S-acyl product is formed preferentially. Acid chlorides also react with 1-trimethylsilylimidazoles to give azolides. Trifluoromethylsulfonation forms an imidazolide which is a convenient reagent for the introduction of a 'triflate' group. In the cases of highly basic 1-substituted imidazoles, acid halides can give rise to quaternized products, and in the presence of antimony chloride 1-acetylimidazole is even quaternized by acetyl chloride ⟨70AHC(12)103, 80AHC(27)241⟩.

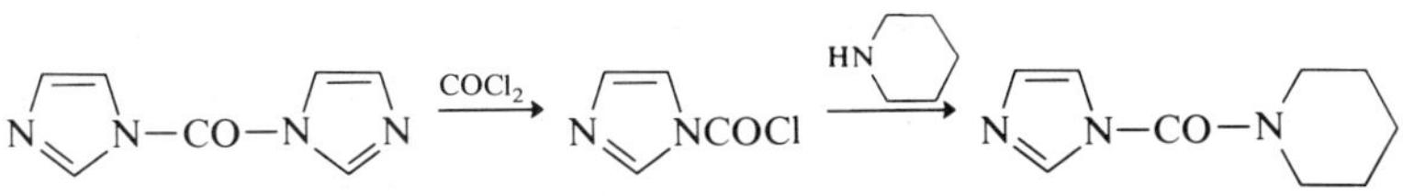

Scheme 22

Benzimidazoles, too, are readily acylated and may also be converted by acylating agents into quaternary salts. In one such reaction the benzimidazole is eventually partially dearomatized to (68) on reaction with bis(1-chloroformyl)diphenylamine (Scheme 23) ⟨76CC48⟩. Under similar conditions imidazole is only acylated.

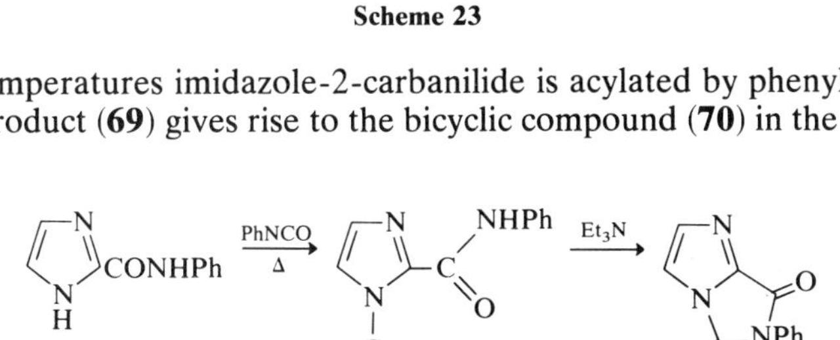

Scheme 23

At elevated temperatures imidazole-2-carbanilide is acylated by phenyl isocyanate. The initial reaction product (69) gives rise to the bicyclic compound (70) in the presence of base (Scheme 24).

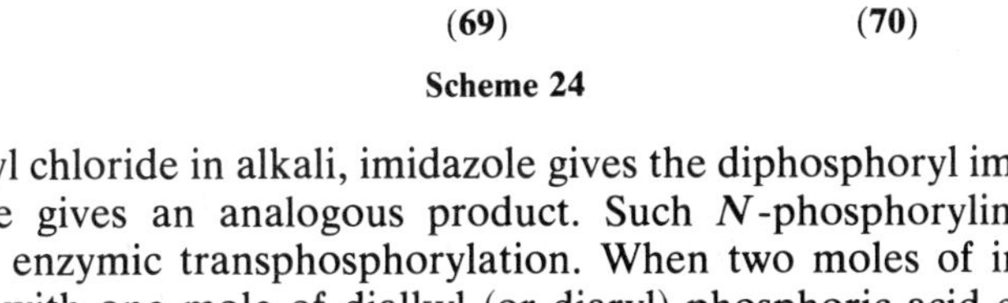

Scheme 24

With phosphoryl chloride in alkali, imidazole gives the diphosphoryl imidazole (71) while 1-methylimidazole gives an analogous product. Such N-phosphorylimidazoles play an important role in enzymic transphosphorylation. When two moles of imidazole (or benzimidazole) react with one mole of dialkyl (or diaryl) phosphoric acid chloride, the compounds (72) are formed, and di- and tri-imidazolides of phosphoric acid (e.g. 73) can be obtained similarly from phosphoric acid dichloride or from phosphoryl chloride (Scheme

25) ⟨66RCR122, 78ZN(B)1033⟩. The derivatives of 1-phosphorylimidazole in which the phosphoryl residue is combined with an organic radical are relatively stable and readily phosphorylate amines, alcohols and carboxylic acids at room temperature.

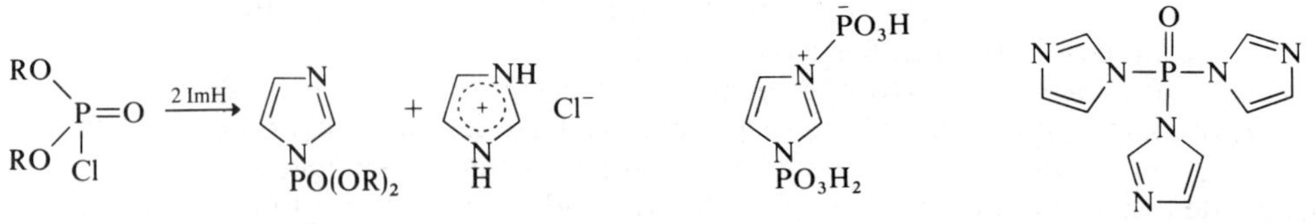

Scheme 25

Kinetic studies of the reactions of imidazoles with benzoyl and sulfonyl halides have been carried out. In acetonitrile the reaction of benzoyl fluoride with imidazole is a mixed third- and fourth-order process, which means that the rate-determining step contains the acid fluoride and two or three molecules of imidazole. With benzoyl chloride in benzene a second-order equation, similar to the aroylation of aniline, is followed. The reactions are subject to steric hindrance, and even a methyl group on an adjacent carbon makes the N-acylation more difficult. Because of this, and because of the reversible nature of the reaction, it has been possible to devise a method of regiospecific alkylation of imidazoles to give the sterically least-favored product ⟨80AHC(27)241⟩. A typical reaction scheme is shown in Section 4.02.1.3.9. When ethyl chloroformate is used as the acylating reagent, high yields of N-alkoxycarbonyl compounds result. For example, 2-ethyl-4-methylimidazole gave greater than 95% of 1-ethoxycarbonyl-2-ethyl-4-methylimidazole with little or none of the 5-methyl isomer. These 1-acyl derivatives can lose carbon dioxide to give the 1-alkylimidazoles, but the decarboxylation process is unfortunately accompanied by some isomerization. The above product gave a 3:1 mixture of 1,2-diethyl-4- and -5-methyl-imidazoles ⟨80AHC(27)241⟩.

Acylation reactions which lead ultimately to 2-acylimidazoles will be discussed as electrophilic substitutions at ring carbon atoms (Section 4.07.1.4.6) in spite of the almost certain intermediacy of the N-acyl compounds.

Considerable effort has been applied to studies of ester hydrolysis catalyzed by imidazoles ⟨76MI40700, 80AHC(27)241⟩. Certainly, 1-acetylimidazole can be made enzymically, probably by the sequence: acetyl phosphate + coenzyme A ⇌ acetylcoenzyme A + phosphate, acetylcoenzyme A + imidazole ⇌ 1-acetylimidazole + coenzyme A. In addition, the imidazolyl group of histidine appears to be implicated in the mode of action of such hydrolytic enzymes as trypsin and chymotrypsin, thereby engendering further interest in the process of imidazole catalysis. The two pathways which have been found to be involved are *general base catalysis* and *nucleophilic catalysis*. In the former (Scheme 26) a basic imidazole molecule can activate a water molecule to attack the ester at the carbonyl carbon, this being followed by the usual sequence of steps as in simple hydroxide ion hydrolysis. At high imidazole concentrations the imidazole molecules may be involved directly.

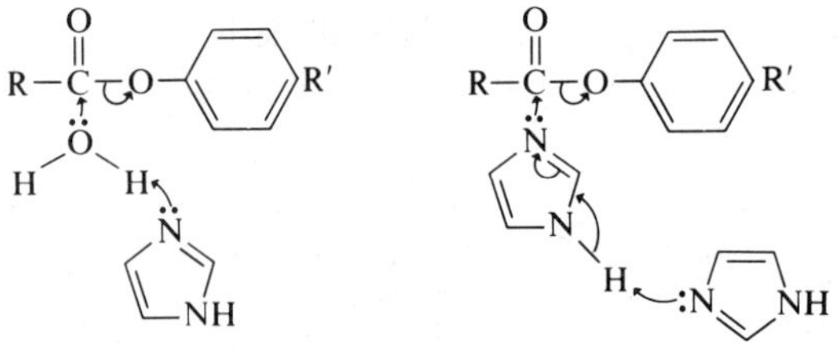

Scheme 26

In nucleophilic catalysis the catalytic properties are a result of the intermediate formation of a 1-acylimidazole (Scheme 27). When the ester has a good leaving group, *e.g.* p-nitrophenyl acetate, the effective catalyst is the imidazole neutral molecule which increases in effectiveness as the basic pK_a of the heterocycle increases. Where, however, the ester has a poor leaving group, *e.g.* p-cresol acetate, the imidazole anion becomes involved and general base catalysis predominates. Thus, for imidazoles with $pK_a \leq 4$ catalysis by the anion is the main reaction. Imidazole is a much more effective nucleophile than other amines in this type of reaction since it is a tertiary amine with little steric hindrance, and it is able to delocalize the positive charge which results from the nucleophilic addition to the ester carbonyl group. Additionally, 1-acylimidazoles are much more reactive in aqueous solution than other substituted amides, a facet which will be discussed with the reactions of the azolides (Section 4.07.3.9.4). The imidazole-catalyzed isomerization of penicillin into penicillenic acids probably involves an initial nucleophilic attack of the heterocycle on the pencillin ⟨B-76MI40701⟩.

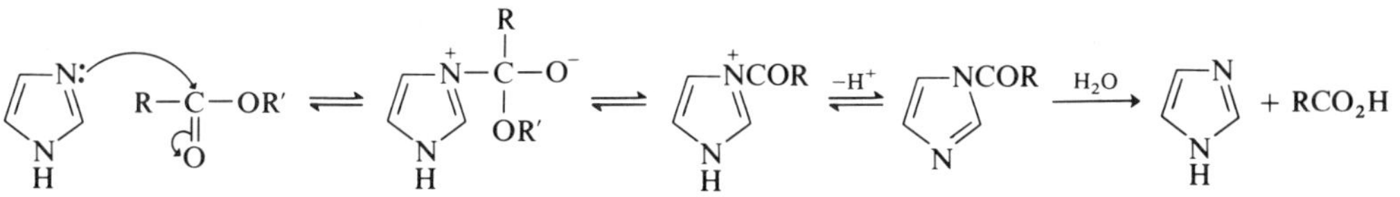

Scheme 27

4.07.1.3.10 Halogens

Examples exist in which some azoles are able to react reversibly with halogens and interhalogen compounds to give unstable adducts which are probably n,σ charge transfer complexes ⟨77JOU1872⟩. A few N-halogenated azoles have been isolated — N-iodoimidazoles would appear to have reasonable stability in some instances. During the C-halogenation of imidazoles there is probably a transhalogenation process occurring since there is evidence for the initial formation of the N-halogen isomers. In a situation where subsequent rearrangement of the N-halogen atom cannot occur, e.g. in the reaction between the sodium salts of 2,4,5-triarylimidazoles and bromine in anhydrous ether, the operation of a radical process is indicated by the formation of 1,1'-biimidazolyls.

4.07.1.3.11 Other electrophiles

There are no known examples of direct N-oxidation of imidazoles or benzimidazoles by peracids. With hydrogen peroxide imidazole is cleaved to give oxamide, while perbenzoic acid destroys the ring giving ammonia and urea.

Trimethylsilylation of N-unsubstituted compounds is common and the derivatives formed have synthetic and analytical applications.

Direct N-cyanation is possible when cyanogen bromide reacts with an imidazole or benzimidazole having a free NH group (Scheme 28) ⟨80H(14)1963⟩.

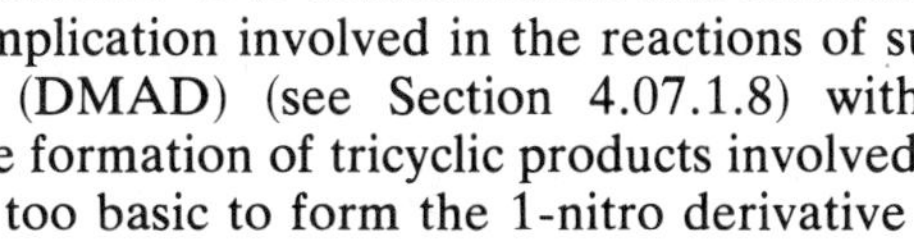

Scheme 28

Reactions of alkynes with benzimidazole are also electrophilic processes. Thus, 1-vinylbenzimidazole and 1,3-divinylbenzimidazolinone are formed by the addition of acetylene in aqueous dioxane at 160 and 200 °C to benzimidazole and benzimidazolinone, respectively. There is rather more complication involved in the reactions of such alkynes as dimethylacetylene dicarboxylate (DMAD) (see Section 4.07.1.8) with not only products of N-alkylation, but also the formation of tricyclic products involved.

Although imidazole is too basic to form the 1-nitro derivative when treated with nitric acid–acetic anhydride (indeed, the nitrate salt forms in preference), less basic imidazoles can be N-nitrated under these conditions. Thus, 4-nitroimidazole gives 1,4-dinitroimidazole ⟨81UP40700, 80AHC(27)241⟩, and the corresponding 2- and 5-methyl-4-nitroimidazoles react in the same way. Of interest is the orientation of N-nitration which parallels that of methylation of the anion. Whether the anion reacts in the nitronium acetate medium or whether steric factors control the site of attack is not known.

With tetracarbonyliron, mercury tetracarbonylferrate(II), ferrocene or dicarbonylcyclopentadienyliron, imidazole forms iron(II) complexes of the general form $Fe(C_3H_3N_2)_2$ or $Fe(C_3H_3N_2)_2 \cdot 0.5(C_3H_4N_2)$. Iron(III) complexes have also been described.

4.07.1.4 Electrophilic Attack at Carbon

4.07.1.4.1 Reactivity and orientation

(i) Ease of reaction

Imidazole varies considerably in its reactivity with electrophiles depending on the type of reagent and the reaction conditions. Reactions which can involve the imidazole neutral molecule or the anion occur with some facility, *e.g.* halogenation and diazo coupling, but when highly acidic conditions are employed the azolium ion is the substrate. In consequence it is considerably deactivated. Such reactions as sulfonation and nitration occur with difficulty. Imidazole is nitrated in mixed acids 10^9 times slower than benzene, but 10^{10} times faster than pyridine (which also reacts as the protonated species). Friedel–Crafts reactions will only take place if there is strong activation from a substituent. The measurements and MO calculations of π-electron densities for imidazole and benzimidazole predict a reactivity with electrophiles similar to that of benzene, and much greater reactivity in the case of the imidazole anion, but the carbon atoms are less prone to attack than the multiply bonded nitrogen atom. A number of reactions which result in ultimate C-substitution may proceed *via* an initial N-substitution, *e.g.* halogenation and acylation, and others may involve an addition–elimination process, *e.g.* bromination of 1-methylimidazole with cyanogen bromide.

(ii) Orientation

In reactions of the imidazole neutral molecule or cation, attack by electrophiles at C-5 is preferred. However, when there is a free NH in the ring, tautomerism makes the 4- and 5-positions equivalent. Conditions in which the imidazole anion exists are reported to be characterized by electrophilic substitution at C-2. There would appear to be exceptions to this generalization, *e.g.* halogenation (Section 4.07.1.4.5). When the preferred positions are blocked it is usually possible to induce electrophiles to enter alternative ring positions and multiple substitution is commonly possible. In fact it is difficult to prepare monobromoimidazole, and imidazole can be dinitrated. More than one benzeneazo group can couple when imidazole reacts with diazonium salts.

Attempts to correlate reaction mechanisms, electron density calculations and experimental results have met with only limited success. As mentioned in the previous chapter (Section 4.06.2), the predicted orders of electrophilic substitution for imidazole (C-5 > -2 > -4) and benzimidazole (C-7 > -6 > -5 > -4 ≫ -2) do not take into account the tautomeric equivalence of the 4- and 5-positions of imidazole and the 4- and 7-, 5- and 6-positions of benzimidazole. When this is taken into account the predictions are in accord with the observed orientations of attack in imidazole. Much the same predictions can be made by considering the imidazole molecule to be a combination of pyrrole and pyridine (**74**) — the most likely site for electrophilic attack is C-5. Furthermore, while sets of resonance structures for the imidazole and benzimidazole neutral molecules (Schemes 1 and 2, Section 4.06.2) suggest that all ring carbons have some susceptibility to electrophilic attack, consideration of the stabilities of the expected σ-intermediates (Scheme 29) supports the commonly observed preference for 5- (or 4-) substitution. In benzimidazole attack usually occurs first at C-5 and a second substituent enters at C-6 unless other substituent effects intervene.

$$\text{deactivated} \begin{cases} \text{pyridine } \alpha \\ \text{pyrrole } \beta \end{cases} \quad \text{activated} \begin{cases} \text{pyridine } \beta \\ \text{pyrrole } \alpha \end{cases} \quad \begin{cases} \text{pyridine } \alpha \\ \text{pyrrole } \alpha \end{cases} \text{deactivated}$$

(**74**)

(iii) Effects of substituents

Normal substituent effects operate in imidazoles and benzimidazoles. Thus, if a nitro group is already present in the ring it is extremely difficult to introduce a second or a third one. Furthermore, when there is a phenyl ring either fused or attached to imidazole, electrophiles may prefer to attack this benzene moiety, particularly electrophiles generated in acid medium. Although methyl groups have little activating effect they may exert an apparent directing effect if they are responsible for a change in reaction mechanism, *e.g.* halogenation of 1-methylimidazole (Section 4.07.1.4.5).

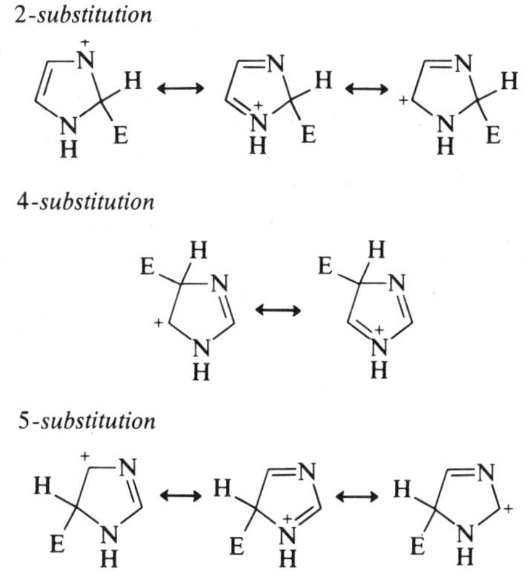

2-substitution

4-substitution

5-substitution

Scheme 29

In an endeavor to study the transmission of substituent effects in imidazoles Noyce ⟨73JOC3762⟩ examined the solvolysis rates of a series of 1-(1-methylimidazolyl)ethyl *p*-nitrobenzoates (**75–77**; OPNB = *p*-nitrobenzoate). The relative rates (**75**:**76**:**77** = 1:13:15) parallel the relative electron densities in 1-methylimidazole as deduced from chemical shift data. By comparison with other heteroarylethyl *p*-nitrobenzoates the effective replacement constants, σ^+_{Ar}, were determined as $\sigma^+_{2\text{-Im}} = -0.82$, $\sigma^+_{4\text{-Im}} = -1.01$ and $\sigma^+_{5\text{-Im}} = -1.02$. The effects on the 2-substituted compound of alkyl, aryl and halogen substituents at the 4- and 5-positions were examined, but though the rates for the 5-substituents could be represented satisfactorily by σ^+_p, σ^+_m failed to account properly for the observed reactivities of the 4-substituted compounds. It is not surprising that the distorting effects of annular heteroatoms make it difficult to superimpose the substitution behavior of benzenoid compounds into this series.

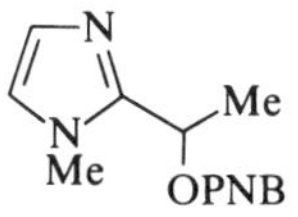

(**75**) 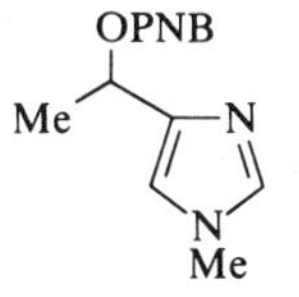 (**76**) 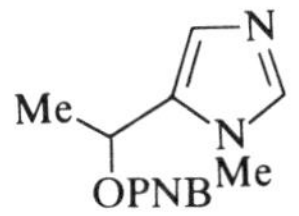 (**77**)

OPNB = *p*-nitrobenzoate

In the benzimidazoles a powerful electron-releasing group at C-5 will direct subsequent substitution to C-4; electron-withdrawing substituents lead to subsequent attack at C-4 or C-6.

4.07.1.4.2 Nitration

When imidazole is treated with a mixture of concentrated nitric and sulfuric acids the 4-nitroimidazole is formed by nitration of the imidazole conjugate acid. There is no substitution at C-2 and so the important antibiotic 2-nitroimidazole (azomycin) cannot be prepared by direct nitration. Variations in reaction conditions such as heating imidazole nitrate with sulfuric acid or addition of sodium or potassium nitrates to imidazole give the same 4-nitro product. Exhaustive nitration in 'mixed acids' gives successively 4-nitro- and 4,5-dinitroimidazoles, but not 2,4,5-trinitroimidazole. The last-named compound can, however, be prepared by nitration of 2,4-dinitroimidazole ⟨73CHE465⟩. A chloro substituent is no hindrance to the nitration process: both 1-methyl- 4- and 1-methyl-5-chloroimidazoles are readily nitrated in sulfuric acid. While nitric acid in acetic anhydride (nitronium acetate) seems to form only the nitrate salt in many instances, dinitrogen tetroxide in acetonitrile is reported to convert 4-substituted imidazoles bearing electron-withdrawing substituents into a mixture of 5- and 2-nitro derivatives. Such nitrations at C-2 are unexpected. With nitric–sulfuric acid mixtures 1-methyl- and 1,2-dimethyl-imidazoles give mixtures of the 4- and 5-nitro isomers in the approximate ratios of 5:2 and 2:1, respectively.

The kinetics of nitration of imidazole have been studied in sulfuric acid medium. The yields of 4-nitroimidazole are dependent on the acidity, e.g. %H$_2$SO$_4$, yield (%): 83.7, 46; 89.6, 26; 93.8, 19; 98.8, 19; 1% SO$_3$, 90. Ring opening accompanies the process giving rise to ammonia and, presumably, oxidation products. The kinetic results which could be separated indicate that the species being nitrated is the imidazole cation, but side reactions complicate matters. The suggested mechanism is outlined in Scheme 30 ⟨B-76MI40701⟩.

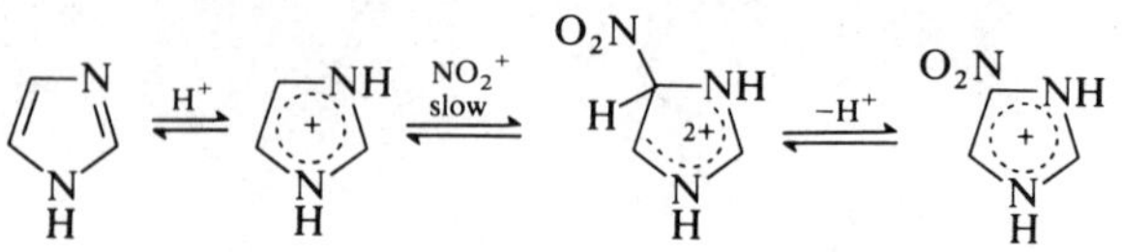

Scheme 30

Both benzimidazoles and aryl-substituted imidazoles are commonly substituted by nitro in the aryl rings, and consequently such reactions will be treated as reactions of substituents (Sections 4.07.3.1 and 4.07.3.2). On occasions, though, reaction can occur in the heterocyclic nucleus. Thus, when 2-(p-fluorophenyl)imidazole is nitrated under mild conditions an 80% yield of the 4-nitro product (**78**) results. Yet a further nitro group can be introduced by the use of nitric and acetic acids at 95 °C. More acidic nitrating agents tend to attack the benzene ring or may result in oxidation. The observation that nitric acid in 20% oleum at −10 °C still nitrates (presumably) the deactivated imidazolium species in preference to the electron-deficient fluorophenyl group must reflect the relative stabilities of the Wheland intermediates (Scheme 31) ⟨79JHC1153⟩.

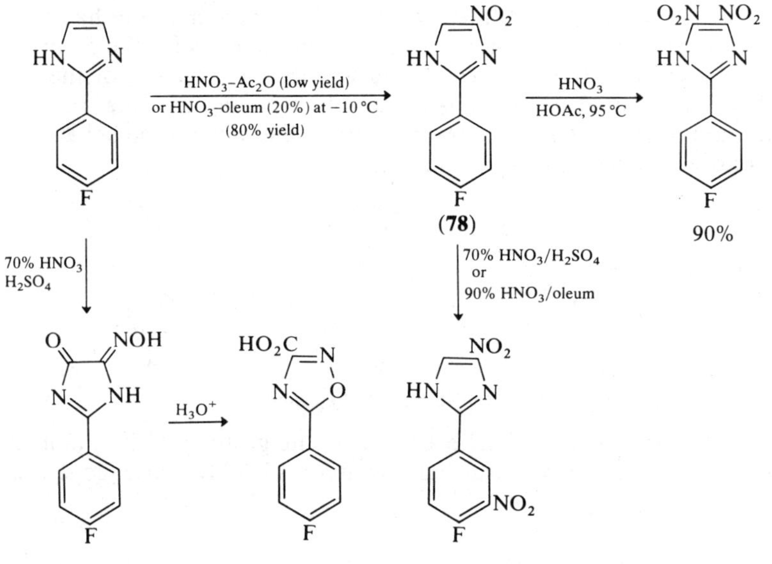

Scheme 31

Attention is also drawn to the conditions under which N-nitration (Section 4.07.1.3.11) and NNO$_2 \rightarrow C$NO$_2$ migrations (Section 4.07.1.2.2) take place.

'ipso'-Nitration of iodoimidazoles has been reported ⟨80AHC(27)241⟩, but recent studies ⟨79CC523, 81JOC1781⟩ have demonstrated that some compounds, reported to be 2-iodo derivatives, are in fact 4-iodoimidazoles. Accordingly, when the results reported earlier are amended in the light of these findings it can be seen that iodo groups in the 4- and 5-positions of imidazole may be replaced by the nitro group (Scheme 32). It had been reported earlier that 4,5-dibromoimidazole could not be nitrated, but that 4-chloro- and 4-bromo-imidazoles undergo nitration at the vacant C-5 (or -4) position. The formation of 2,4,5-trinitroimidazole (**80**) from 2,4,5-triiodoimidazole (**79**) is unexpected as it involves nitration at C-2. An analogous nitration (in 50% HNO$_3$) of 4,5-diiodoimidazole to give 5-iodo-2,4-dinitroimidazole must be viewed with suspicion since the authors thought their starting material was 2,4-diodoimidazole. Furthermore, the suggestion ⟨76JCS(P2)1089⟩ that the facility of nitrodehalogenation decreases with an increase in acid strength (because water is needed to remove the halogen cation from the Wheland intermediate) is not in accordance with the use of 100% nitric acid for the conversion of (**79**) into (**80**). Repetition of this sequence of reactions will be required to clarify the anomalies.

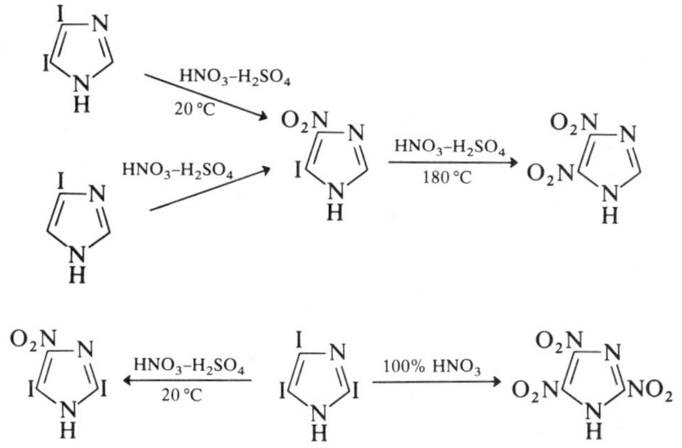

Scheme 32

4.07.1.4.3 Sulfonation

The sulfonation of imidazole occurs at the 4-position when the heterocycle is treated with 50–60% oleum at 160 °C. As with nitration, attack is on the imidazolium species. The 2-methyl, 4-methyl, 4-bromo and 4-bromo-2-methyl compounds are all similarly sulfonated at C-4 (or C-5), and under severe conditions 2-methylimidazole gives some of the 4,5-disulfonic acid. Sulfur trioxide in the presence of air also appears capable of sulfonating imidazole.

Chlorosulfonation of 4-bromo-, 4-chloro-1-methyl-, 5-chloro-1-methyl- and a series of 2-aryl-imidazoles occurs also at C-4 (or C-5), but only decomposition products were isolated from the reaction with 4-acetamidoimidazole.

4.07.1.4.4 Acid-catalyzed hydrogen exchange

While most hydrogen exchange reactions of imidazoles appear to involve a base-catalyzed process (Section 4.07.1.6.2), the relative rates of hydrogen–deuterium exchange at the 2-, 4- and 5-positions of 1-methylimidazole (Path 1, Scheme 33) are quoted as 1 : 73 : 120. This reaction involves an S_EAr pathway through the conjugate acid species. In such an acid-catalyzed process involving a Wheland intermediate the relative reactivities of the various species arising from the heterocycle are conjugate base > neutral molecule > conjugate acid ⟨80AHC(27)241, B-76MI40701⟩. Acid-catalyzed ring proton exchange in 1-methylimidazole occurring by electrophilic attack on the imidazolium ion has significant reaction rates only at high temperatures and in moderately strong acid media, e.g. ~160 °C in 1M DCl in D_2O. If the imidazole has a low enough pK value, then an alternative mechanism can apply involving electrophilic attack at C-4 or C-5 in the neutral molecule. Substituents must be present which are capable of stabilizing a carbonium ion. Under these circumstances the exchange is much more rapid. For example, in 0.1–3M DCl at 50 °C, $t_{1/2} \approx 9$ h for 4-fluoro-2-methyl-, 12 h for 4-fluoro-1-methyl-, 16 h for 2-fluoro-4-methyl- and 54 h for 4-fluoro-imidazole. No exchange at C-2 was noted. In 1M DCl at 50 °C exchange of H-4 in 1-methylimidazole was estimated to occur with $t_{1/2}$ approaching 84 years. To date, a fluorine substituent appears to be the only one capable of providing the necessary combination of electronegativity and resonance overlap to permit facile exchange by this alternative pathway (Path 2, Scheme 33). As a corollary of this, 2-fluoroimidazole resists nitration at 100 °C, but 2-fluoro-4-methylimidazole can be nitrated even at −60 °C. This remarkable behavior is consistent with the observed effect of the 4-methyl group in promoting isotope exchange in 2-fluoro-4-methylimidazole: the carbonium ion (**81**) is stabilized ⟨79JOC4240⟩.

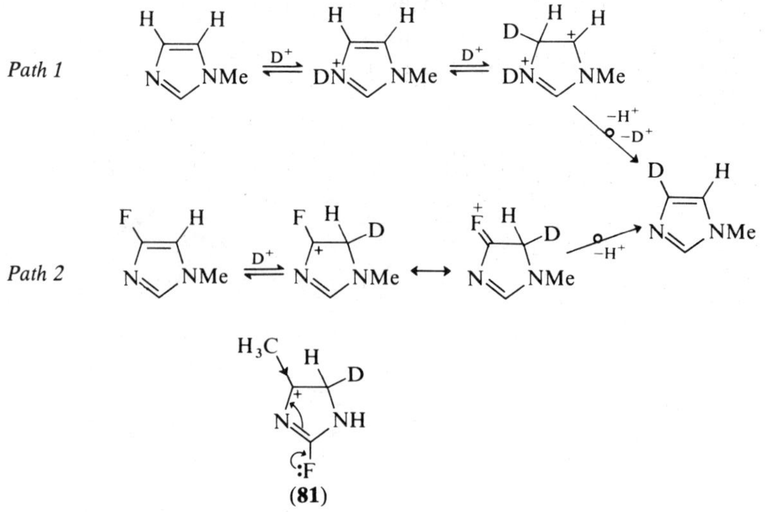

Scheme 33

9.07.1.4.5 Halogenation

The results of halogenations of imidazole and its derivatives have provided a host of complicated but mechanistically interesting results.

(i) Chlorination

It is only in the past 15 years that reports of imidazole chlorination have become at all common. With chlorine, imidazole gives 'undesired products containing carbonyl groups' or 'poor yields of undescribed chloro compounds'. It has been suggested ⟨76CB1625⟩ that the major product is 2,2,4,5-tetrachloro-2H-imidazole. Sulfuryl chloride, too, appears to be incapable of forming simple chloroimidazoles. When, however, sodium hypochlorite is employed under carefully defined conditions the product is 4,5-dichloroimidazole. The behaviour is similar with 2-ethyl-, 2-methyl-imidazole and imidazole-2-carboxylic acid, while 4-bromoimidazole gave the 4-bromo-5-chloro product. There is some evidence that small quantities of 2,4,5-trichloroimidazole are formed in the further chlorination of 4,5-dichloroimidazole. In boiling chloroform N-chlorosuccinimide or N-chlorophthalimide converts imidazole into the 4-chloro- (13%) and 4,5-dichloro- (25%) imidazoles. 2-Methylimidazole behaves in a similar fashion, and 2,4-dimethylimidazole gives a 60% yield of 5-chloro-2,4-dimethylimidazole ⟨B-76MI40701⟩. In such halogenations, though it must be admitted that the initial reaction probably involves radical species; once the reaction has been initiated electrophilic chlorination can take over. Imidazole anions are monochlorinated by hexachloroethane. When subjected to chlorination at 100–200 °C the 1-methyl-2,4,5-trichloroimidazole ring opens forming perchloro-2,5-diazahexa-1,5-diene. When the 2-lithio derivative of 1-triphenylmethylimidazole is treated with chlorine a 39% yield of the 2-chloro derivative is obtained ⟨78JOC4381⟩.

(ii) Bromination

By far the greatest amount of study has been applied to the bromination and iodination of imidazoles and the benzologues. It is, however, becoming apparent that some of the definitive results may have been based on earlier incorrect assignment of structure, particularly in the case of 2-iodoimidazole. Subsequent discussion here will endeavour to examine some of these earlier results in the light of the presently available structural information.

Imidazole brominates very readily; with bromine in aqueous solution or organic solvents 2,4,5-tribomoimidazole is formed. With 1-alkylimidazoles, too, it is difficult to prevent bromination of all vacant ring positions. A feature of these reactions is the ease with which a bromine will enter the 2-position, in sharp contrast to the situation in nitration. With bromine in chloroform at −10 °C 4-methylimidazole gives the dibromo derivative along with 34% of 4-bromo-5-methylimidazole, but even under these mild conditions imidazole gives the tribromo product along with traces of 4,5-dibromoimidazole. Complete C-bromination occurs with a variety of substituted imidazoles: 2-methyl-, 4-methyl-5-nitro-, 2- and 4-ethoxycarbonyl- and 4-ethoxycarbonyl-5-methyl-. In aprotic solvents imidazole is also tribrominated by NBS. It has proved possible to achieve direct monobromination of imidazole and 1-methylimidazole using the reagent 2,4,4,6-tetrabromocyclohexa-2,5-dienone; the bromine enters at C-4. A similar result is achieved when the anion of imidazole reacts with carbon tetrabromide. Using the former reagent the reaction products (Scheme 34) also include some 4,5-dibromo- and 2,4,5-tribromo-imidazoles. That these polybrominated species are formed when only one molar proportion of the tetrabromodienone is used is a consequence of the equilibrium set up between imidazole and 4-bromoimidazole, producing a conjugate base species susceptible to further bromination. Such an explanation accounts for the comparative resistance of 1-methylimidazole to polybromination. With 2,4,4,6-tetrabromocyclohexa-2,5-dienone (1 mole) a 60% yield of 5-bromo-1-methylimidazole is obtained; with 2 moles of brominating agent 65% of 4,5-dibromo-1-methylimidazole forms, and only traces of other products can be detected ⟨72JCS(P1)2567⟩. 4-Arylimidazoles are brominated at C-5 and dibrominated at C-2 and C-5; 4,5-diphenylimidazole gives the 2-bromo derivative.

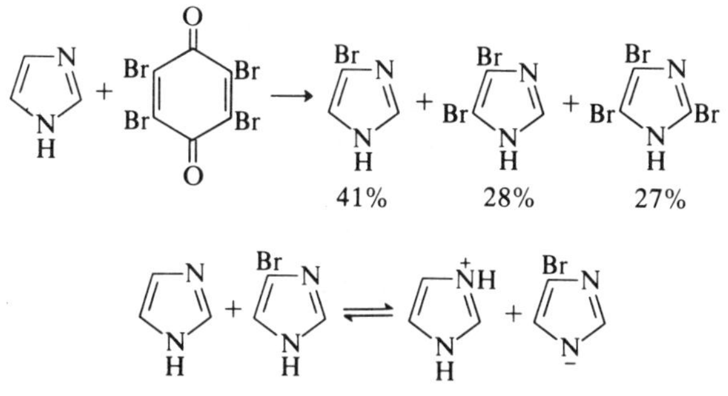

Scheme 34

With bromine in chloroform imidazole forms a complex, $(C_3H_4N_2)_2Br^+ Br_3^-$, in which two imidazoles are joined to a bromonium cation through the N-3 atoms; 1-methylimidazole behaves in the same way. During the bromination of imidazoles there can be considerable ring degradation which is enhanced by an increase in acidity of the medium. Such reactions, which are initiated by electrophilic bromination, will be further discussed as oxidations (Section 4.07.1.4.8)

As with the unsubstituted compounds, 1-alkylimidazoles are so readily brominated that it is difficult not to obtain tribromoimidazoles. Again the 2-position brominates with remarkable facility, although 1,5-dimethylimidazole reacts less readily than the 1,4-isomer. When 1,4,5-trimethylimidazole is warmed with bromine in carbon disulfide a 22% yield of the 2-bromo product is isolated. It may be that the reaction proceeds via an initial N-bromination (or an addition–elimination reaction) as shown in Scheme 35. Such a hypothesis is supported by the reaction of imidazole with cyanogen bromide in which 2-bromoimidazole and HCN are formed (but see Section 4.07.1.3.11). The involvement of the ylide in this process is contraindicated since the second stage of the pathway would have to be attack by Br^+, a species which could not originate from BrCN. Whilst the bromination of 4,5-diarylimidazolin-2-ones leads ultimately to ring opening, one of the

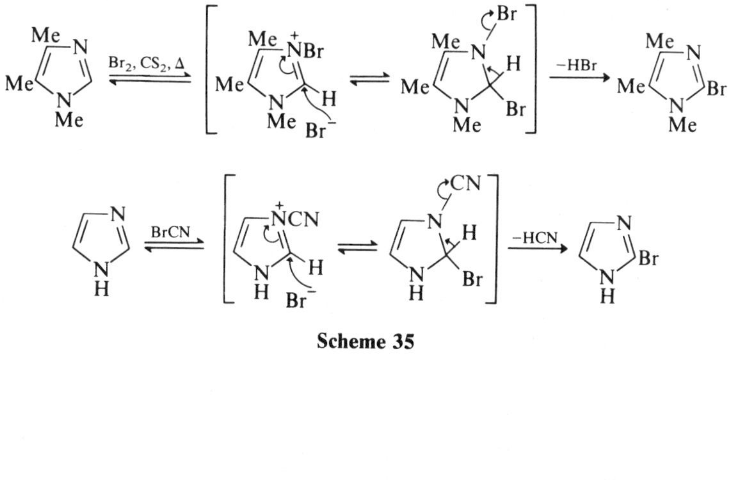

Scheme 35

initial processes is bromination of the aryl rings. In the case of 4,5-di(p-bromophenyl)imidazolin-2-one the first product formed appears to be (**82**; Scheme 36) ⟨B-76MI40701, 70AHC(12)103, 80AHC(27)241⟩.

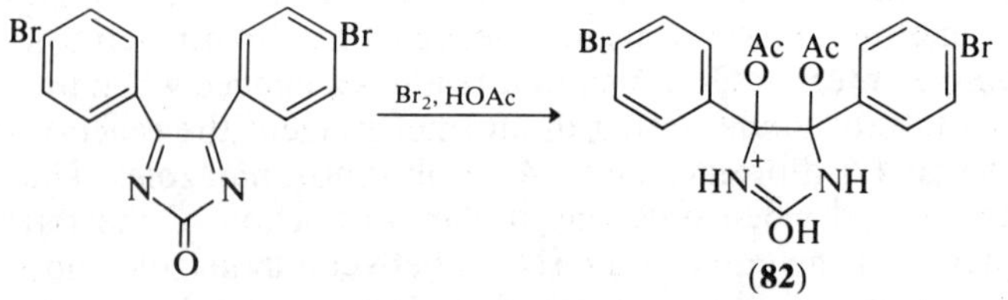

Scheme 36

Determination of the rate laws and isotope effects provides evidence that imidazole is brominated *via* a rate-determining bimolecular process which leads to the formation of a Wheland intermediate ⟨78JCS(P2)865⟩ between the neutral imidazole molecule and molecular bromine. The studies carried out by Coller's group employed halogenation by instalments with couloupotentiometric monitoring. In imidazoles all three hydrogens were found to be replaced with the order of reactivity 5 > 4 > 2 (in agreement with MO calculations). There is still unexpectedly high reactivity at C-2, and it may be that an addition–elimination process is occurring. Rate constants, calculated using literature values of K_a, are listed in Table 5. It appears that the relative rates of successive brominations are pH dependent and it may be possible to design conditions to obtain optimum yields of partially brominated imidazoles. The limited effect of N-methylation (only fourfold across the 1,2- or 1,5-bond) contrasts sharply with the effect of a 2-methyl group (180-fold between C-2 and C-5). Such differences emphasize the importance of bond fixation effects in heterocyclic compounds. Again, the greater reactivity of C-5 towards electrophiles is evident from the relative reactivities of the 4- and 5-bromo-1-methylimidazoles.

Table 5 Bimolecular Rate Constants for Neutral Monobromination of Imidazoles and Benzimidazoles ⟨78JCS(P2)865, 74AJC2331⟩

Substrate	$k_{bi}^{o} \times 10^{-4}$ (dm^3 mol^{-1} s^{-1})			
	C-2	C-4	C-5	C-7
Imidazole	20	40	50	—
1-Methylimidazole	80	170	230	—
2-Methylimidazole	—	9000	9000	—
5-Bromo-1-methylimidazole	4	36	—	—
4-Bromo-1-methylimidazole	<10	—	460	—
2-Bromo-1-methylimidazole	—	~240	~240	—
2,4(5)-Dibromo-1-methylimidazole	—	5	5	—
4,5-Dibromo-1-methylimidazole	0.16	—	—	—
Benzimidazole	<0.0004	—	0.0935	0.004
1-Methylbenzimidazole	—	—	0.0879	—
2-Methylbenzimidazole	—	—	0.63	—
5-Bromobenzimidazole	—	—	—	0.004

The fused benzene ring in benzimidazole deactivates C-2 to electrophilic bromination by a factor of about 5000. However, most bromination occurs in the fused benzene ring with the preferential order in multiple brominations being 5 > 7 > 6, 4 > 2, an order which closely parallels that predicted by MO calculations. These reactions will be discussed in more detail as substituent reactions (Section 4.07.3.1).

(iii) Iodination

In aqueous alkali iodine reacts with imidazole to give 2,4,5-triiodoimidazole and some of the 4,5-diiodo compound (the latter compound had been incorrectly designated as 2,4-diiodoimidazole). The products isolated by Pauly on iodination of 4-methylimidazole are almost certainly 4-iodo-5-methyl- and 2,4-diiodo-5-methyl-imidazole. Imidazole-4-carboxylic acid gives 2,4,5-triiodoimidazole, but 1,4- (or 1,5- ?)-dimethylimidazole is not susceptible to iodination. In alkaline medium a number of imidazoles and benzimidazoles are subject to iodination on a ring nitrogen, giving rise to 1-iodo products, and it is these which may be precursors to the ultimate *C*-iodo products. In addition, it is possible that the complexes formed between molecular iodine and azoles could have some involvement in the process. The overall reaction mechanism still has some regions of uncertainty, but it seems clear that the rate-limiting step is nucleophilic attack by a base on the hydrogen being abstracted from the σ-complex (83). Such acceleration follows the nucleophilicity of the base rather than its basicity. It seems likely that the initial step involves molecular iodine (or perhaps I_3^-) reacting with the imidazole anion (Scheme 37). The involvement of imidazole anions even at pH = 7 accounts for the failure of 1-substituted imidazoles to react. Iodination of 2,4,5-trideuteroimidazole showed a large kinetic isotope effect (k_H/k_D = 4.5), present also for 4,5-dideutero- but not for 2-deuteroimidazole. The relative reaction rates for the above three deuterated substrates proved that the initial substitution occurs at C-4. Even in the presence of a large excess of imidazole the major product is still a diiodoimidazole (almost certainly 4,5-diiodo-) ⟨B-76MI40701, 80AHC(27)241⟩. In fact, the monoiodination of imidazole has proved to be a difficult process. With iodine and iodic acid, 5-chloro-1-methylimidazole was converted into a mixture of 27.5% of the 4-iodo and 12.7% of the 2,4-diiodo compounds. More recently the previously undescribed 2-iodoimidazole (85) (Pauly's '2-iodoimidazole' has been shown to be the 4-isomer) was prepared in low yield (∼5%) *via* the lithiation of 1-benzenesulfonylimidazole (84; R = SO$_2$Ph) and in higher yield (40%) using the 1-trityl analogue (84; R = CPh$_3$; Scheme 38) ⟨78JOC4381⟩.

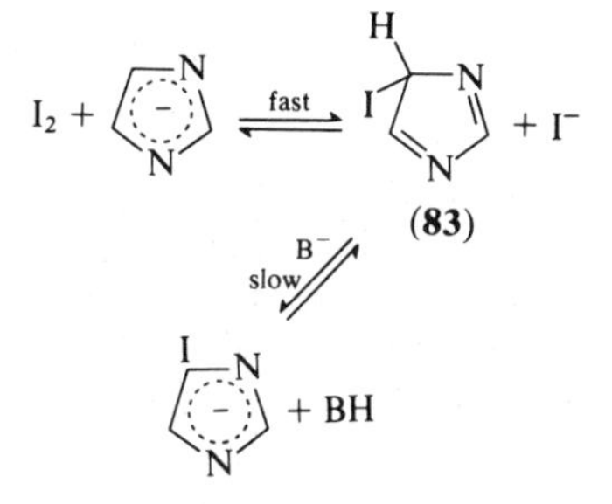

Scheme 37

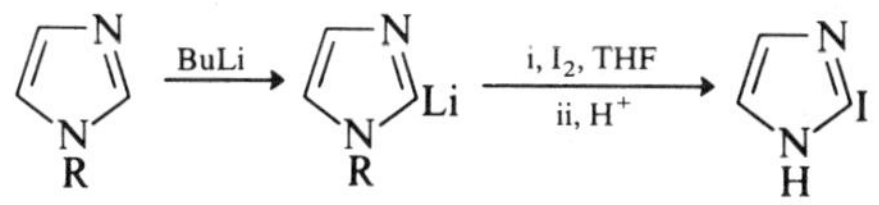

(84) R = SO$_2$Ph, CPh$_3$ (85)

Scheme 38

A detailed study of the kinetics of iodination of 2- and 4-methylimidazole ⟨80JOC3108⟩ showed that, like the iodination of imidazole, the reactions are base catalyzed, but whereas 2-methylimidazole (and imidazole) also undergoes uncatalyzed iodination, 4-methylimidazole does not. The relative rates for uncatalyzed iodination (imidazole:2-methylimidazole:4-methylimidazole) were 1:28:0; those for base-catalyzed iodination were 1:34:667. Further conclusions about the process of 2-iodination drawn from this study have been rendered dubious in the light of the reassignment of 2,4- as 4,5-diiodoimidazole ⟨81JOC1781⟩.

When histidine is iodinated, attack is again by iodine on the anionic species, and it occurs at C-4; proton removal is again the rate-determining step. With histidine and some of its derivatives there is the added complication of hydrogen bonding being possible between the side-chain and the nucleus, hence deactivating the anion. A number of bases, including imidazole, have been found to catalyze the iodination of imidazoles, such catalysis depending on the nucleophilicity of the bases in question. In the iodination of the Ni(II) complex of imidazole there is a negligible catalytic term since the base catalysis is eliminated by Ni—N coordination. Rather, the reaction involves molecular iodine reacting with Ni(C$_3$H$_4$N$_2$)$^{2+}$ and Ni(C$_3$H$_3$N$_2$)$^+$ species ⟨B-76MI40701⟩.

In alkaline solution benzimidazole iodinates at C-2. 1-Alkoxy-2-iodobenzimidazole is formed when the corresponding Grignard reagent reacts with iodine.

4.07.1.4.6 Acylation

It is a general rule that imidazoles and benzimidazoles are resistant to Friedel–Crafts reactions. This is not surprising since such basic compounds must be markedly deactivated in the presence of Lewis acids. Imidazolin-2-ones appear to be an exception and apparently possess sufficient activation to react. Reactions between imidazoles and N-methylformanilide and phosphoryl chloride are also unproductive. With 4,5-diphenylimidazole, phenyl isocyanate at 80 °C gives products of both N- and C-substitution, but in boiling nitrobenzene only the latter (**86**) is formed. 2-Methyl-4-phenylimidazole gives (**87**) under the same conditions, and 1,3-diphenylimidazolium perchlorate is transformed by potassium t-butoxide into a ylide which reacts at C-2 with phenyl isothiocyanate. Sufficient activation is present in 1-methyl-2-phenyl-4-phenylaminoimidazole for it to react by substitution at C-5 with acetic anhydride ⟨71JOC3368⟩.

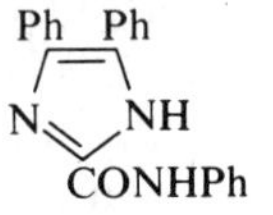

Perhaps the most interesting examples of C-acylation or C-aroylation are those in which a 1-substituted imidazole reacts with an acyl or aroyl halide in acetonitrile in the presence of triethylamine ⟨80AHC(27)241⟩. The reaction may follow the sequence of Scheme 39. Certainly the mechanism has been shown to involve an initial N-acylation or -aroylation, and electron-withdrawing groups in the acid halide assist the reaction. The process by which the N-acyl intermediate rearranges appears to be intramolecular. Benzimidazoles react similarly, as does 1-benzyl-5-methylimidazole, but an exocyclic C-aroyl product is obtained when there is a methyl group blocking C-2. When imidazole itself is subjected to the same reaction conditions the product is a triacylimidazolinylimidazole (**88**). In order to prepare 2-acyl- or 2-aroyl-imidazoles with no ring nitrogen substituent, the acylation can be carried out either on 1-methoxymethylimidazole (here the N-substituent can later be removed under conditions which do not affect the entering ketone function) or by a modification of Regel's original conditions. Imidazole reacts readily with two equivalents of aroyl chloride in pyridine containing two equivalents of triethylamine to give 2-aroylimidazoles in 60–80% yields ⟨78S675⟩. Since 1-benzoylimidazole (**89**; R = H) gives 2-benzoylimidazole under the same conditions it seems likely that the powerful benzoyl chloride–pyridine adduct is the acylating agent in this instance (Scheme 40). Apart from a large number of aroyl halides other acyl compounds such as ethyl chloroformate, phosgene and trifluoroacetyl chloride (but not acetyl chloride) take part in the reaction. The presence of strongly electron-withdrawing groups at C-4 of imidazole prevents the reaction, probably by failing to assist the initial N-acylation. Reactions of imidazoles at elevated temperatures with isocyanates show some resemblance to the foregoing. Imidazolium ions, deprotonated by sodium hydride, react with acetic anhydride to give 2-acyl products.

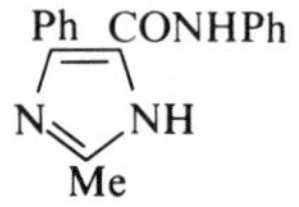

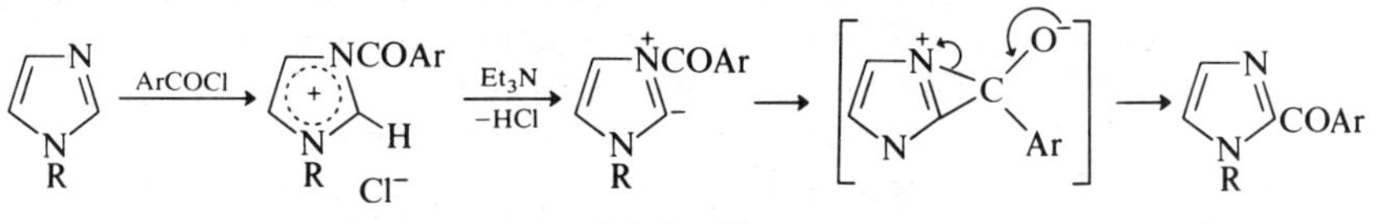

Scheme 39

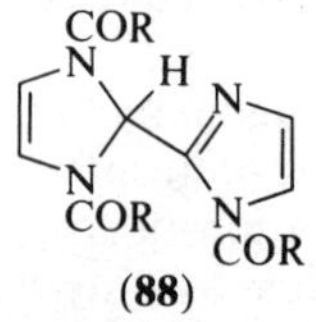

There has been scant attention paid to the carboxylation of imidazoles. Imidazole-4,5-dicarboxylic acids can be prepared by the reaction of carbon dioxide with imidazole (or 2-alkylimidazoles) at 260 °C under pressure in the presence of potassium carbonate and cadmium fluoride. 5-Aminoimidazole, and some of its 1-substituted derivatives, and aqueous bicarbonate are in equilibrium with the 4-carboxylate (**90**). In acid medium decarboxylation occurs, but in the presence of excess bicarbonate carboxylation takes place (Scheme 41).

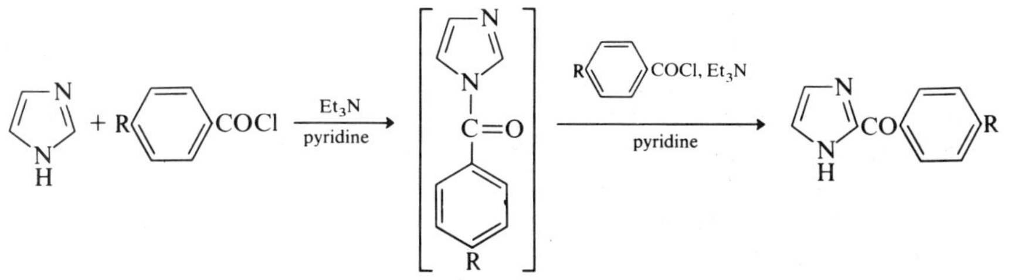

Scheme 40

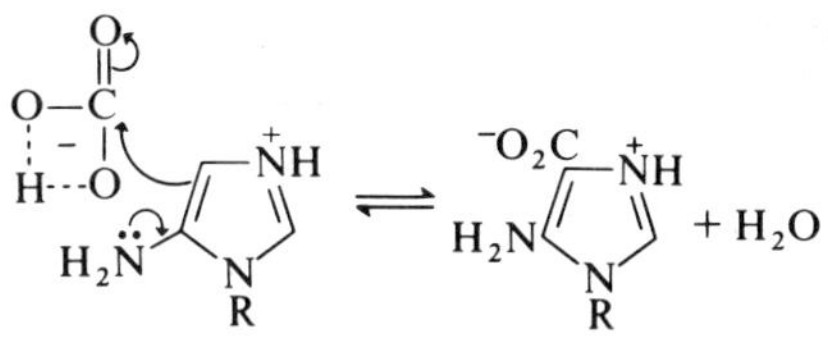

Scheme 41

Both carbon dioxide and potassium carbonate are ineffective in promoting the reaction ⟨B-76MI40701⟩.

4.07.1.4.7 Mercuration

References to mercuration of imidazole are notable for their scarcity. The ylide obtained from 1,3-diphenylimidazolum perchlorate is said to react at C-2 with mercury(II) chloride.

4.07.1.4.8 Diazo coupling

Diazo coupling is an electrophilic substitution reaction which occurs readily only with activated substrates. Thus, N-unsubstituted imidazoles react *via* the conjugate base species, N-substituted imidazoles fail to react, and imidazoles with electron-withdrawing substituents present in the ring are resistant to a greater or lesser extent. The interesting facet of this reaction is that coupling, which takes place in alkaline medium, occurs preferentially at C-2, although other ring positions are also reactive, *e.g.* 4-methylimidazole gives approximately equal amounts of the 2- (**91**) and 4- (**92**) isomers and some 2,4-bis compound (**93**) (Scheme 42). With imidazole itself some tris coupling has been observed. The azo products of the reactions are red, orange or yellow dyes which are of value for the detection of N-unsubstituted imidazoles on chromatographs, and for their spectrophotometric analysis ⟨70AHC(12)103⟩. Diazotized sulfanilic acid is the coupling agent of choice in what has come to be referred to as the Pauly reaction. Imidazole-4-carboxylic acid couples readily, but derived esters and anilides do not; nitroimidazoles will not couple, and imidazole-4,5-dicarboxylic acids couple with the loss of one carboxyl group. Imidazolin-4-ones couple at C-5.

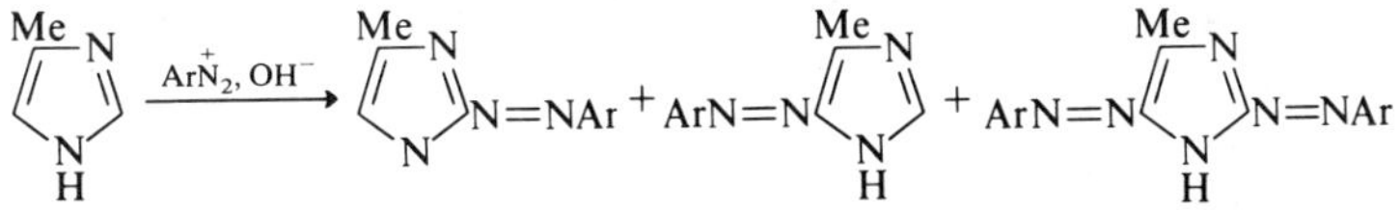

Scheme 42

In the pH range 7–11 the coupling of imidazole with diazotized sulfanilic acid is a bimolecular reaction involving the imidazole anion. For the corresponding reaction with 2,4,5-trideuteroimidazole there is a negligible isotope effect which shows that proton removal is not involved in the rate-limiting step. This is one noteworthy difference from the iodination reaction, while the observation that initial substitution is at C-2 constitutes another. A possible explanation is that the charge density at C-2 in the anion is greater than that at C-4; the localization energies, however, predict that electrophiles should attack C-4 (or C-5). In iodination the rate is determined by proton loss from the σ-complex, and this follows the predictions of localization energies. Diazo coupling occurs preferentially at the site defined by the electron density. The reactions are, however, complicated and the above rationalization may be an oversimplification ⟨B-76MI40701⟩.

The rate of coupling of imidazole has been reported to be even greater in DMSO than in water, but since the reactions involve the imidazole anion results are complicated by the different acidities of the solvents.

Intramolecular diazo coupling occurs with o-(imidazol-1-yl)phenyl diazonium salts (94) which substitute at C-5 at pH 5–6 to give the imidazo[5,1-c][1,2,4]benzotriazine (95; Scheme 43). If the 5-position is blocked, coupling can occur (with greater difficulty) at C-2. In these reactions the coupling does not take place at pH values of less than 4, and consequently the unprotonated imidazole ring participates. The driving force for the reaction could be the tendency for ring closure to produce the imidazotriazine system ⟨70AHC(12)103⟩.

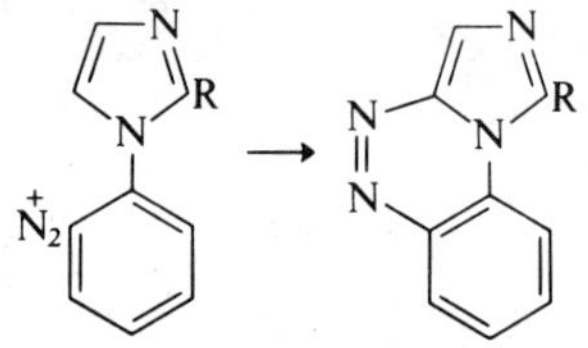

(94)   (95)

Scheme 43

4.07.1.4.9 Nitrosation

To date, there has been no report of the nitrosation of imidazole itself. There are, however, a number of examples of nitrosation of substituted imidazoles in basic medium. The observation that such reactions do not occur with N-substituted imidazoles implicates the imidazole anion as the likely substrate. Substitution generally occurs at C-4 or C-5 and may be accompanied by some nitration. In the presence of sodium ethoxide, pentyl nitrite nitrosates 4-phenyl-, 2-methyl-5-phenyl- and 2,5-diphenyl-imidazoles at the vacant 4- or 5-positions. The reaction apparently failed with 5-methyl-2-phenyl- and 4,5-diphenyl-imidazoles ⟨58G463⟩.

4.07.1.4.10 Reactions with aldehydes and ketones

When imidazoles substituted on nitrogen are treated with formalin, either neat or in a solvent such as DMSO, hydroxymethylation takes place at C-4 (or C-5); 1-substituted imidazoles react at C-2. Thus, 4-methyl- and 4-bromo-imidazoles give the 5-hydroxymethyl derivatives (96): 1-methyl-, 5-halo-1-methyl-, 1-benzyl-, 1-aryl- and 1,5-dimethyl-imidazoles give the 2-hydroxymethyl products (97; Scheme 44). 1,4-Dimethylimidazole, however, is converted into the 5-hydroxymethyl derivative, but a number of imidazoles with electron-withdrawing substituents, e.g. 1-methyl-2- and -4-nitro-, fail to react. Nevertheless, 1-methyl-5-nitroimidazole undergoes hydroxymethylation at C-2 ⟨79JHC871⟩. When heated in a sealed tube imidazole is hydroxymethylated by formaldehyde not only at C-2, but also (in low yield) at the other vacant ring positions (Scheme 44). 1,2,4-Trialkylimidazoles react at the 5-position. A change in reaction conditions to weakly acid medium (e.g. formate buffer) with 1,2-dimethylimidazole gives a 25–50% yield of a mixture of the 5- and 4,5-dihydroxymethylimidazoles; sealed tube reaction gives a low yield of the former product. The reaction mechanism could involve either direct attack by formaldehyde at the ring carbon or N-hydroxymethylation followed by rearrangement. The intermediacy of a quaternary salt would have to be conceived for 1-substituted imidazoles (and perhaps also for N-unsubstituted compounds), but this is not too dissimilar to the acylation reactions in acetonitrile with triethylamine (Section 4.07.1.4.6). Thermal condensation of N-substituted imidazoles with other aldehydes (except acetaldehyde) or reaction of aldehydes with 2-lithio- or 2-trimethylsilyl-imidazoles also results in the production of 2-hydroxymethylimidazoles ⟨80AHC(27)241⟩.

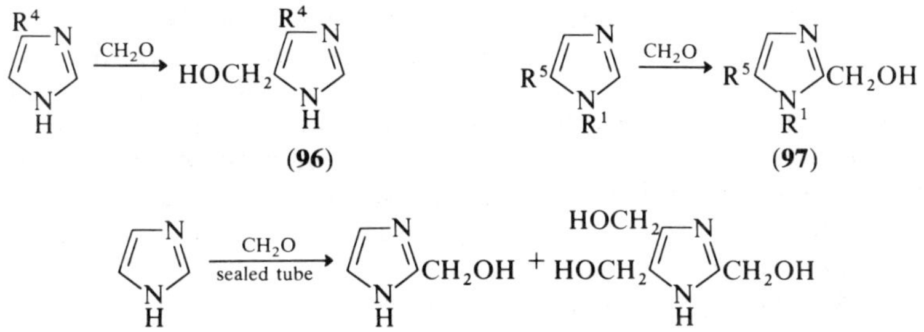

Scheme 44

The related Mannich reaction is not common. Under the usual acidic reaction conditions N-substitution occurs, but this is a reversible reaction in the presence of base. Therefore, in basic medium, C-substituted products accumulate, and all positions can be substituted. 2-Methylimidazole gives 1,4,5-tri, 4,5-di- and 4-mono-substituted products. The observation that with formaldehyde and hydrochloric acid histamine gave (**98**) is at variance with the apparent requirement for basic medium. Since 1-substituted imidazoles do not react it is likely that the imidazole conjugate base is the reactive species. Unless imidazoles contain activating substituents they are not very susceptible to reaction with aldehydes (except HCHO) and ketones. An exception appears to be the product (**99**) of interaction between imidazole and hexane-2,4-dione. An activated compound such as 4-methylimidazoline-2-thione gives the 5-dimethylamino compound (**100**); imidazoline-2-thione gave only the N-hydroxymethyl product under the same reaction conditions. Imidazolin-4-ones with a free 5-position readily form benzylidene derivatives ⟨B-76MI40701⟩.

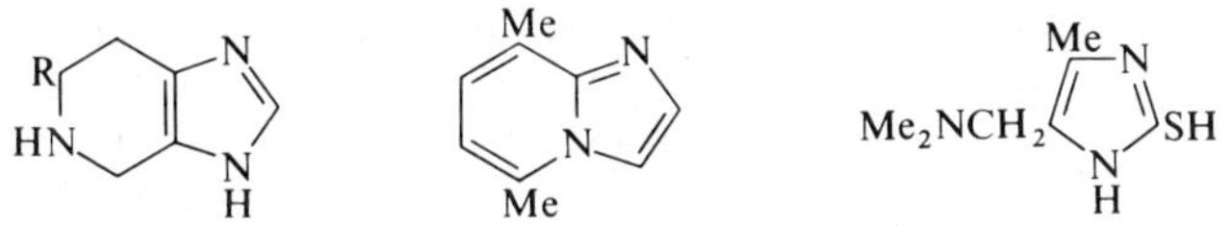

(**98**) R = H, CO$_2$H (**99**) (**100**)

Both benzimidazole and 1-acetylbenzimidazole react with diketene to form products of N-3 and C-2 acetylation.

4.07.1.4.11 Oxidation

The imidazole ring is not at all susceptible to oxidation, although the parent compound can be degraded by permanganate to formic acid and carbon dioxide, as can some derivatives. An exception to the usual resistance to chromic acid oxidation is the conversion of 4,5-diphenylimidazoline-2-thione into dibenzoylurea; nitric acid converts the same compound into benzil. While hydrogen peroxide has been reported to oxidize some imidazoles to oxamide, under alkaline conditions it transforms 4-ethyl-2-phenylimidazole into 5-ethyl-3-phenyl-1,2,4-oxadiazole along with some carbonate, benzoate and propanoate. Although this appears to be a general reaction for a number of 2,4-disubstituted imidazoles, 4,5-di- and 2,4,5-triphenylimidazoles are unaffected, and both 2- and 4-phenylimidazoles formed only benzoic acid, perhaps via the oxadiazoles. Perbenzoic acid transforms imidazole into ammonia and urea; direct N-oxidation of imidazoles does not occur and the synthesis of imidazole N-oxides can be a difficult procedure.

Benzimidazole (but not 1-methylbenzimidazole) is oxidized by permanganate, dichromate or hydrogen peroxide to imidazole-4,5-dicarboxylic acid, while napth-[1,2-d]- and -[2,3-d]-imidazoles also form products in which the heterocyclic ring remains intact, hence demonstrating its stability to these conditions. With lead peroxide benzimidazole is subject to an unusual oxidation as it forms (**101**), also the reaction product of lead dioxide and 2,2'-bibenzimidazolyl. In dioxane, selenium dioxide oxidizes 2-methylbenzimidazole to o-hydroxyacetanilide ⟨66RCR122⟩.

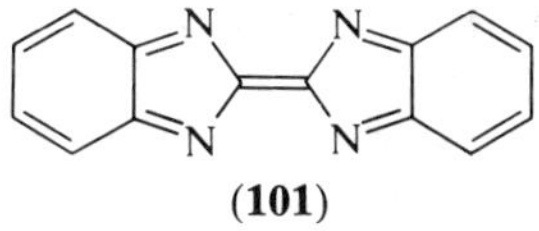

(**101**)

Oxidative degradation of the imidazole ring can accompany certain electrophilic substitution reactions such as bromination, and perhaps nitration. When NBS reacts in aqueous medium with imidazole and its 4-substituted derivatives, ammonia, glyoxal (or the appropriate 1,2-dicarbonyl compound) and formamide can be detected among the reaction products. The aqueous conditions suggest that this ring degradation may be initiated by nucleophilic attack. In boiling acetic acid 4,5-di(p-bromophenyl)imidazolin-2-one reacts with bromine to give (**82**; Scheme 36). Hydrolysis of this product forms urea and 4,4'-dibromobenzil, while reaction of the urea with 4,5-diacetoxy- or 4,5-dihydroxy-4,5-di(p-bromophenyl)-2-imidazolidinone from (**82**) gives rise to glycouril (**102**). Histidine is degraded to iodoform and oxalate during iodination.

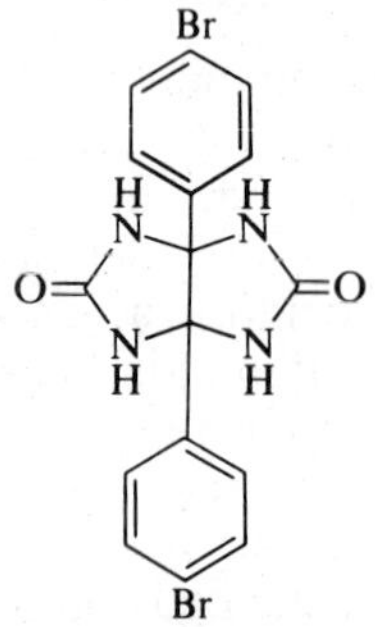

(**102**)

The oxidation reactions of 2,4,5-triphenylimidazole (lophine) have received considerable attention. With chromic acid it gives benzamide and benzanilide, but even more interest has centred on its involvement in the phenomenon of chemiluminescence. Some of this material has been discussed earlier (Sections 4.06.3.6, 4.07.1.2.1 and 4.07.1.2.3). The oxidative decomposition of lophine in the presence of air is accompanied by the emission of light, and it is the excited singlet state of the diaroylarylamidine (**12**; Scheme 2) which is the light emitter. The radical (**46**) derived from oxidation of lophine with aqueous ferricyanide and ethanolic KOH forms a hydroperoxide with hydrogen peroxide with consequent luminescence. When 2,4,5-tri- and 1,2,4,5-tetra-phenylimidazoles are oxidized in dilute methanol solution in the presence of methylene blue, the dibenzoylbenzamidine is also formed under circumstances in which hydroperoxides cannot be intermediates ⟨B-76MI40701⟩.

When phenylimidazoles react with singlet oxygen, although there is some evidence that 2,5-endoperoxides or zwitterionic intermediates may be formed initially, the products which are isolated (Scheme 45) appear to arise from the decomposition of hydroperoxides or dioxetanes ⟨81T(S9)191⟩.

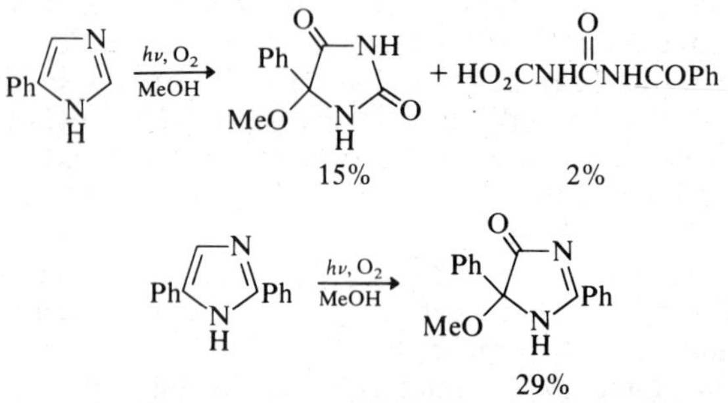

Scheme 45

Reactions of the sodium salts of 2,4,5-triarylimidazoles with bromine and oxidations of N-hydroxybenzimidazoles to cyclic nitroxyls have been discussed elsewhere (Section 4.02.1.4.11).

4.07.1.5 Nucleophilic Attack at Carbon

Nucleophilic reactions on imidazoles are not at all facile unless the ring is activated to such attack by an electron-withdrawing group or in some other way such as in the reactions of the quaternized compounds. Furthermore, a fused benzene ring (as in benzimidazole, benzimidazolium and benzimidazolinone) aids such reactions since any loss in aromaticity during the reaction is minimal compared with the monocyclic analogues.

MO calculations predict that in unsubstituted imidazoles and benzimidazoles the 2-position should be the most susceptible to nucleophilic attack and quaternization of the heterocycle should enhance this orientation. The presence of electron-withdrawing groups elsewhere in the molecule can modify this preferred site of attack.

Some of the reactions described below lead to products of nucleophilic displacement; others result in ring opening.

4.07.1.5.1 Hydroxide ion and other O-nucleophiles

(i) Neutral imidazoles

As mentioned earlier (Section 4.02.1.6.1) 1-substituted 4,5-diphenylimidazoles react at 300 °C with hydroxide ion to give the imidazolin-2-one. The reaction of a neutral imidazole with hydroxide ion, however, rarely leads to hydroxylation. Both 1,2-dimethyl-4- and -5-nitroimidazoles show signs of reacting with hydroxide ion, and ammonia is produced on prolonged heating. Bis(1-benzimidazolyl)alkanes (103) are hydroxylated to (104) by potassium hydroxide as long as the linking hydrocarbon chain is at least three carbon atoms long (Scheme 46). Compounds with $n = 1$ or 2 decompose under the influence of KOH. The reason is probably not steric but rather a function of the greater pK_a values of the compounds in which $n = 3-5$ allowing easier reaction with the K^+ ion ⟨72CHE1280⟩. At high temperatures (~250 °C) 2-alkyl and 2-aryl groups of benzimidazoles can be replaced by hydroxy groups (Scheme 46) ⟨76CHE437⟩. Whereas bulky groups at N-1 do not significantly hinder amination (Section 4.07.1.5.2), they exert a much greater influence on hydroxylation. Thus, whilst 1-isopropylbenzimidazole gives a 33% yield of the benzimidazolone at 250–290 °C with potassium hydroxide, the 1-*t*-butyl compound resinifies. Groups such as nonyl and decyl with no branching provide little steric hindrance to the reaction.

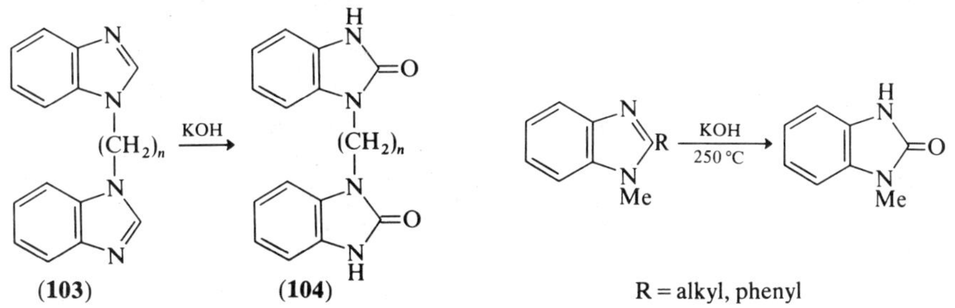

Scheme 46

There are a few examples of displacement of groups other than hydrogen by hydroxy. Halogen atoms, particularly at C-2, are replaced by hydroxyl, as are sulfo groups. However, hydroxide ion (or thiolate anion) often causes decomposition. The cleavage of a 5-nitroimidazole derivative by hydroxide has been ascribed to nucleophilic attack at C-4. Recently, this instability of 1-alkylnitroimidazoles has been examined and it was found that the rate of decomposition increases sharply with the base concentration and temperature ⟨80AHC(27)241⟩. Under comparable conditions it can be demonstrated that 1-methyl-4-nitro- and 1-methyl-2-nitro-imidazoles are 50–150 times more stable than the 1-methyl-5-nitro isomer. In view of the postulate that the first step in the breakdown of the 4- and 5-nitro compounds is β-addition of hydroxide ion to the 4,5-bond leading to (105) and (106), respectively, the greater stability of the 4-nitro compound may be a function of the fact that (105) cannot form so readily—it is subject to an adjacent lone pair effect not present in (106) ⟨78JOC3570⟩. However, this argument may have less validity if the negative charge resides on the nitro group (Section 4.06.3.7).

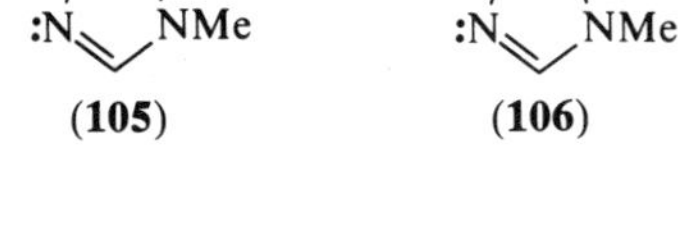

When the mild fluorinating agent, xenon difluoride, reacts with 2,4,5-tribromo-1-methylimidazole the product is 1-methyl-2,4,5-imidazolidinetrione, formed by successive fluorination and hydrolysis steps ⟨79JGU1251⟩. There are frequent references to the replacement of a halogen atom by alkoxy, hydroxy or aryloxy groups, particularly in imidazoles which already carry a nitro (or adjacent formyl) group. The development of blue-green colours during the reaction of alkoxide with 4-bromo-1,2-dimethyl-5-nitroimidazole has been attributed to the formation of the intermediate complex. However, the corresponding 5-bromo-4-nitro compound failed to react under the same conditions.

The formation of 1-benzyl-4,5-dimethylimidazolin-2-one (**107**) when acetic anhydride reacts with 1-benzyl-4,5-dimethylimidazole 3-oxide parallels similar reactions in the pyridine series (Scheme 47).

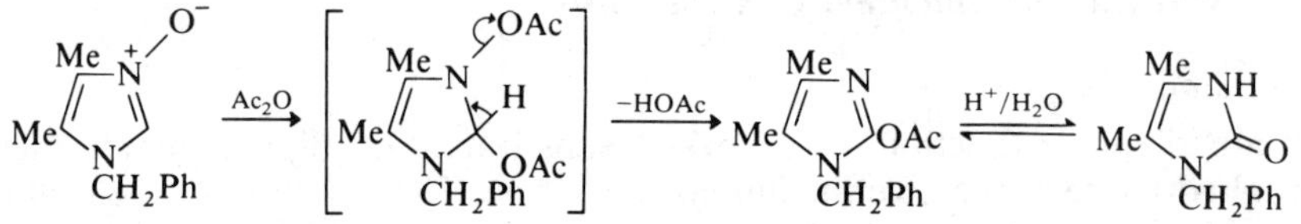

Scheme 47

One important point regarding relative reactivities of *N*-unsubstituted and *N*-substituted 2-chloroimidazoles with powerful nucleophiles such as alkoxide is that in the former there is competition between proton abstraction from N-1 (which severely hampers any nucleophilic attack at C-2) and the nucleophilic substitution at the 2-position. Thus, chloride is not displaced from 2-chlorobenzimidazole by alkoxide ions, whereas 2-chloro-1-methylbenzimidazole reacts readily. That steric effects can also be important is evidenced by the lack of reactivity of 2-chloro-1-methylbenzimidazole with *t*-butoxide, and similar resistance to nucleophilic attack of 2-chloro-1-isopropylbenzimidazole. Alkoxide readily displaces chloride from 2-chloro-1-isopropenylbenzimidazole, and the *N*-substituent may be removed subsequently by oxidative cleavage ⟨74CRV279⟩.

There appear to be no examples of displacement of amino or diazonium by hydroxy groups.

(ii) Azolium ions

The formation of pseudo bases when *N*-methylazolium ions react with hydroxide has been discussed earlier (Section 4.02.1.6.1). Alkalies convert imidazolium and benzimidazolium species (*via* these pseudo bases, *e.g.* **108**) into alkenes or *o*-disubstituted benzene derivatives (Scheme 48) by attack at the 2-position. Such a process is responsible for the Bamberger degradation reaction in which benzoyl chloride in alkaline medium cleaves the imidazole ring to form a 1,2-dibenzamidoethylene (**109**; Scheme 48). The intermediate formation of a 1,3-dibenzoyl quaternary salt, readily attacked by hydroxide ion, is assumed. It is this reaction which prevents the employment of Schotten–Baumann conditions for the acylation of imidazoles. The reaction is also general for benzimidazoles which are converted into amides of *o*-phenylenediamine. When a non-hydroxylic base is used, 2-acylimidazoles can be prepared (Section 4.07.1.4.6). A kinetic study of the Bamberger cleavage of imidazole and histidine derivatives by diethyl pyrocarbonate in aqueous solution has indicated that at low concentrations of the acylating agent, the carboxyethylation

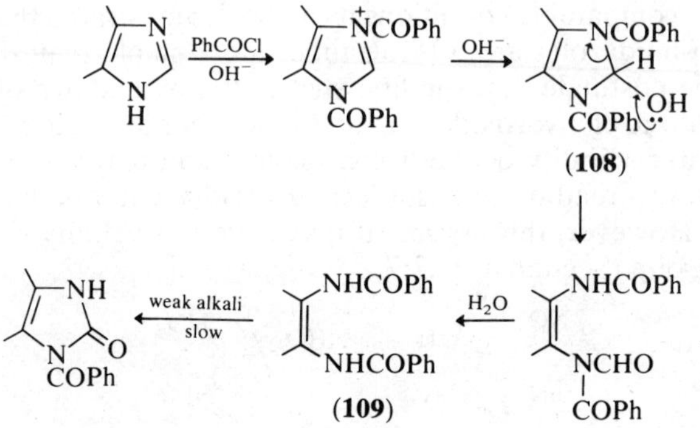

Scheme 48

is rate determining; at higher concentrations the breakdown of an intermediate resembling (**108**) becomes the slow step. If the solution is only weakly alkaline the diacyl- or diaroyl-alkene (**109**) can recycle slowly to give a 2-imidazolinone ⟨81TL2431⟩.

There have been a number of studies of the behaviour of benzimidazolium species with nucleophiles, and ring opening is especially easy when there is a 1-(2,4-dinitrophenyl) substituent. Even weak bases such as aniline or pyridine can effect the transformations. Under the influence of benzoyl chloride in aqueous alkali, 1,2-disubstituted benzimidazoles are cleaved in a rather complex reaction. Provided that a methylene or methyl group is present at C-2, compounds of type (**110**) result (Scheme 49) ⟨74CRV279⟩.

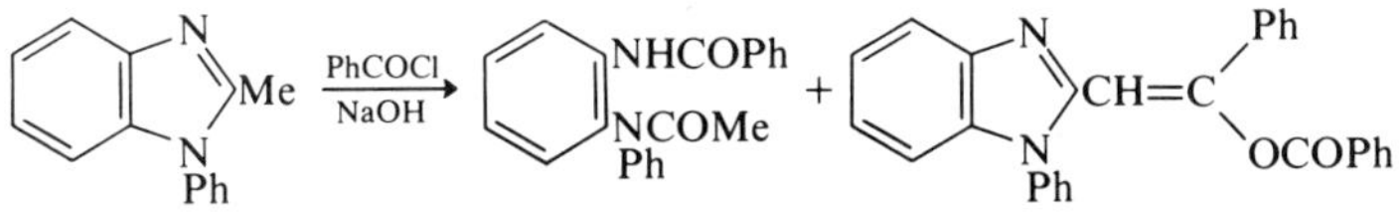

Scheme 49

In what appears to be a related reaction, benzimidazole N-oxides give rise to benzimidazolinones as by-products when treated with an acyl halide in the presence of alkali. With tosyl chloride the sole product is the benzimidazolinone (**111**) which may be formed as in Scheme 50. The C → N rearrangement could also take place via the oxaziridine (**112**) ⟨73JCS(P1)705⟩.

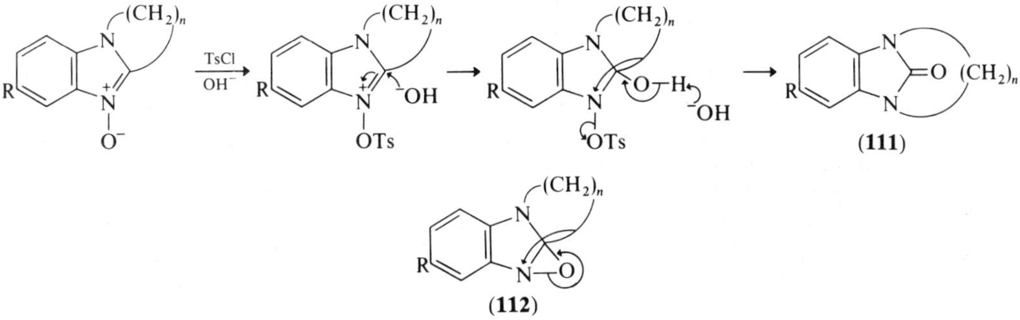

Scheme 50

In addition to products of ring opening, the alkaline hydrolysis of 1,3-dimethyl-4-nitrobenzimidazolium salts gives the benzimidazolinone and 4,4'-azoxybis(1,3-dimethylbenzimidazolin-2-one). The Wheland intermediates (**113**) formed in the amination of imidazolium and benzimidazolium species are also subject to hydrolysis reactions which ultimately lead to the formation of pseudo bases (Scheme 51).

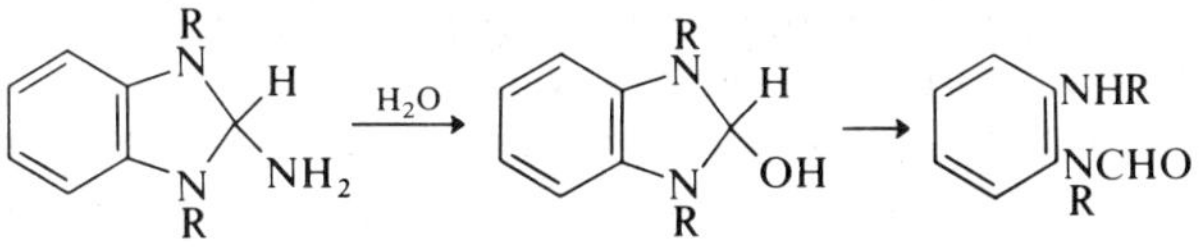

Scheme 51

4.07.1.5.2 Amines and amide ions

(i) Azoles

Usually amines are too weakly nucleophilic to react with imidazoles and benzimidazoles unless there are activating groups present elsewhere in the molecule. For this reason, most successful aminations have utilized the NHR⁻ anionic species. Imidazole itself cannot be directly aminated at C-2, nor do 1-alkyl-4,5-diphenylimidazoles react with sodamide. In reaction with alkaline hydroxylamine, 1,2-dimethyl-4- and 1,2-dimethyl-5-nitroimidazole give respectively the 5- and 4-amino derivatives, with replacement of H-4 seemingly more facile than H-5. It is only condensed imidazoles which will enter the Chichibabin reaction ⟨74CRV279⟩. Thus, the reaction of sodamide in xylene with 5-methoxy-1-methylbenzimidazole gives a good yield of the 2-amino product (**114**; Scheme 52), a reaction which can be extended to a wide variety of N-substituted benzimidazoles and naphthimidazoles. Benzimidazoles react more readily than pyridine as long as the ring nitrogen is substituted. When it is not, the anion formed initially from the heterocycle cannot be subject to nucleophilic attack, and so the reaction fails. The comparison with pyridine was deduced from the reactions of (**115**) and (**116**) with sodamide. On the basis of NMR data for the two bases, it was concluded that the reason for the greater reactivity of benzimidazole is the greater polarizability of the C=N bond in benzimidazole during coordination with the sodamide ⟨72CHE1131⟩.

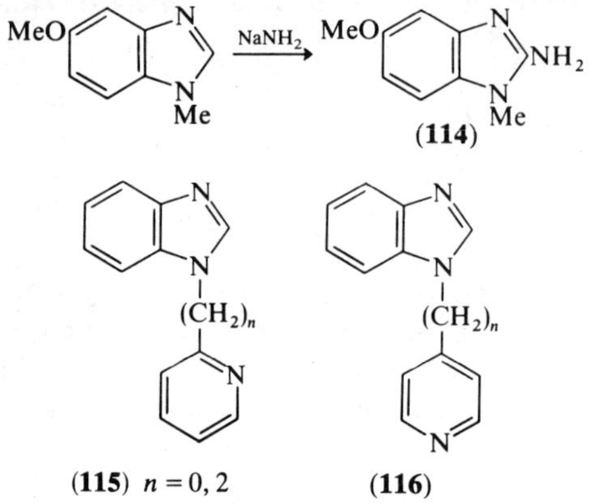

Scheme 52

When 1-arylbenzimidazoles take part in the reaction there is often an accompanying ring opening of the imidazole moiety, and reductive debenzylation of a 1-benzylimidazole or 1-benzimidazole can also occur with sodamide. The amination of N-alkylbenzimidazoles is affected by a substituent at C-5; thus, while 5-alkyl and 5-alkoxy derivatives are aminated, those with halogen, hydroxy, nitro and carboxy groups at C-5 are resistant ⟨66RCR122⟩. When there are two methoxy groups in the fused benzene ring the situation becomes more complex. Compounds with two methoxy groups *ortho* to each other (*viz.* 1-benzyl-5,5-, -6,7- and -4,5-dimethoxybenzimidazole) fail to react with sodamide with the exception that all experience some demethylation, and the last named, surprisingly, gives a 75% yield of benzimidazole. The other dimethoxy compounds (substituted 4,7, 4,6 and 5,7) substitute normally to give the 2-amino product ⟨71CHE1036⟩. The process of amination is also affected to some extent by the substituent on N-1. If there is some branching at the α-carbon of this substituent (*e.g.* isopropyl, α-phenylethyl) the retarding effect is only minimal and good yields are obtained as long as excess sodamide is used and the reaction time is lengthened. Reaction, however, is much more difficult when there is a 1-*t*-butyl group present. Although at first glance steric factors may appear to be responsible there is some evidence that the basicity of the heterocycle may be the controlling factor. Amination of compounds (**103**; Scheme 46) parallels the hydroxylation behaviour.

Under vigorous conditions (NaNH₂, 250 °C) a 2-alkyl or 2-aryl group in 1-methylbenzimidazole can be replaced by amino, although other products are also formed (Scheme 53) ⟨76CHE437⟩.

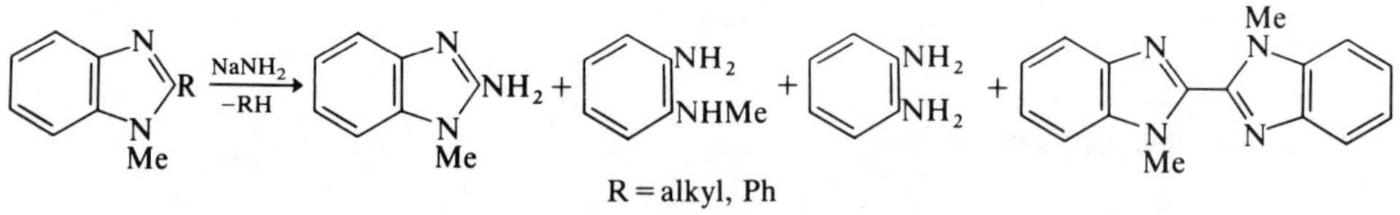

Scheme 53

In contrast to the limited range of examples of replacement of a hydrogen atom by amino, the replacement of halogen is much more common. In the non-condensed imidazoles difficulty may still be experienced, but electron-withdrawing groups on adjacent ring carbons usually ensure the successful outcome of such reactions. Thus 1-alkyl-4-halogeno-5-nitro- and 1-alkyl-5-halogeno-4-nitro-imidazoles react with ammonia and amines, 2-bromo-1-methyl-5-nitroimidazole reacts with boiling piperidine and 1-alkyl-2-bromo-4,5-diphenylimidazole reacts by displacement of bromine with ammonia and a variety of alkyl- and aryl-amines in an autoclave or in DMF. In a sealed tube at 80 °C 4-chloro-1-methyl-5-

nitroimidazole reacts with cyclic secondary amines to give some 4,5-diamino product. The behaviour of 5-bromo-4-sulfonamidoimidazole is interesting; ammonia replaces the bromine by amino, while ethanolic ammonia at 180 °C (or diethylamine) also displaces the sulfonamido group. Nitro groups are particularly efficient in promoting such nucleophilic substitutions as the above, in fact more effective than an extra annular nitrogen. When treated with piperidine both 2-bromo-1-methyl-5-nitro- and 5-bromo-1-methyl-4-nitro-imidazoles are much more reactive than 5-bromo-1-methyl-1,2,4-triazole. Although 2-bromo-1-methylimidazole is fairly resistant to reaction with piperidine (1-methyl-2-piperidinoimidazole is formed after 60 hours at 200 °C), the 5-chloro and 5-bromo isomers (**117**) are even less reactive and require the much more nucleophilic lithium piperidide for any substitution to be achieved ⟨B-76MI40701⟩. The 4-halo isomers are so unreactive that they give no trace of either the 2- or 4-piperidino compounds. The following reaction products were reported from reaction of 5-chloro- and 5-bromo-1-methylimidazole (**117**) with lithium piperidide in ether ⟨71RTC594⟩: from (**117**; X = Cl), 5-chloro-1-methylimidazole (70%), 4-chloro-1-methylimidazole (trace), 1-methyl-5-piperidinoimidazole (15–20%), 1-methyl-2-piperidinoimidazole (**118**) (3–5%); the yields of the corresponding four compounds from (**117**; X = Br) were 49, 30, 25 and 16%, respectively. The possibility that a hetaryne-type mechanism might be operating is negated by the absence of any 4-piperidino product, while the 4-halo products are probably a result of normal base-induced isomerization of (**117**). The formation of (**118**) may be accounted for in terms of the pathway shown in Scheme 54.

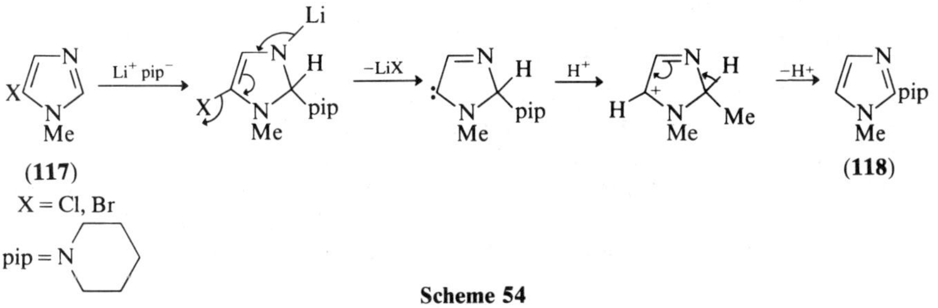

Scheme 54

The reaction of 5-bromo-1-methylimidazole with potassium amide in liquid ammonia does not result in nucleophilic substitution, but rather in migration and elimination of halogen producing a mixture of 1-methyl- and 4-bromo-1-methyl-imidazole. Usually, a 2-chloroimidazole will be quite unreactive in the presence of an amine, and even the corresponding 2-bromo derivatives need very forcing conditions to effect substitution. 2-Chlorobenzimidazoles, too, only react sluggishly. A study of the effects of benzo-annulation on the amination of compounds (**119**)–(**122**) gave relative rates of 383:12:1:0 for displacement of chlorine. It is noteworthy that in the Chichibabin reaction, which involves replacement of the much more tightly bonded hydrogen, the order of reactivity is quite different with ⟨(**121**) > (**120**) > (**119**); compound (**122**) does not react ⟨79CHE218⟩. However, in the presence of copper salts the reactions are assisted to the extent that 2-chlorobenzimidazoles (**121**) and 2-chloroimidazoles (**122**), albeit less readily, undergo nucleophilic substitution. The reaction may be accompanied by some dehalogenation (Scheme 55).

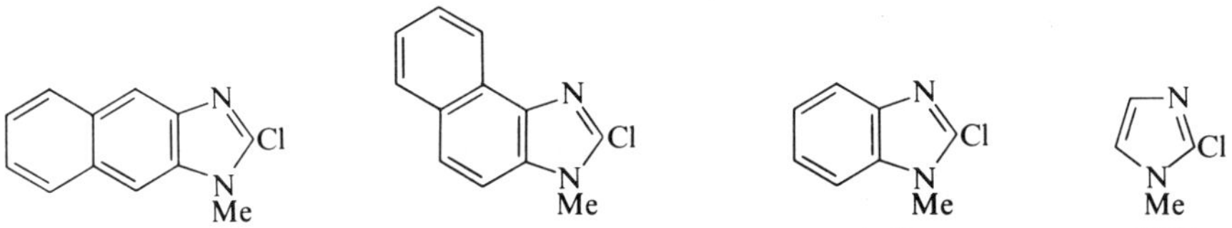

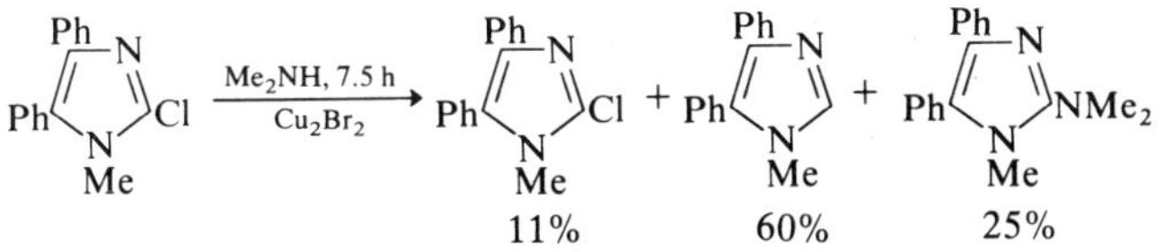

Scheme 55

The 2-halo-substituted benzimidazoles, then, more or less readily react with ammonia, amines and hydrazines ⟨80AHC(27)241, 70AHC(12)103, 74CRV279⟩. It is likely that the formation of the 2-pyridinium compound (**123**) when cycloheptimidazolin-2(1*H*)-one reacts with phosphoryl chloride in pyridine occurs *via* the 2-chloro intermediate (Scheme 56). The sulfur function of benzimidazole-2-sulfonic acid can be displaced by amino or alkylamino groups.

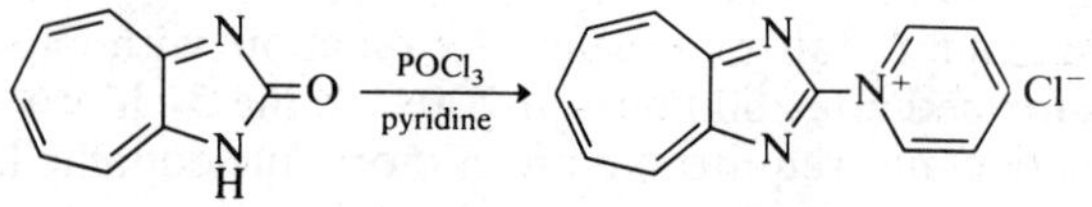

Scheme 56

(ii) Imidazolium and benzimidazolium ions

In the cationic forms the compounds are more readily attacked by amines, particularly the benzimidazoles. Thus, benz- and naphth-[1,2-*d*]imidazolium salts substituted at N-1 by a 2,4-dinitrophenyl group substitute at C-2 even with bases as weak as aniline and pyridine. Less activated salts such as 4,5-diphenylimidazolium require more severe conditions. In the reactions of potassium amide with 1,3-dimethyl-imidazolium and -benzimidazolium salts the difficulties of aromatization of the adducts (**113**; Scheme 51) mean that they tend to undergo a variety of alternative transformations, the chief of which is hydrolysis to pseudo bases ⟨76CHE304⟩. However, the fundamental possibility of amination of such cations by KNH_2 is demonstrated in the case of 1-methoxybenzimidazolium (**124**; Scheme 57). The general scope of the reaction of (**124**; R = Me) with nucleophiles has been studied. Yields are generally very high, and the process is useful in the preparation of 2-alkylaminobenzimidazoles. It is likely, though, that the reaction mechanism cannot be described in terms of a simple addition–elimination process. Rather, it may involve carbene intermediates. When treated with triethylamine in acetonitrile (**124**; R = Me) gave a 64% yield of the bibenzimidazolyl 3-oxide (Scheme 57) ⟨74CRV279⟩.

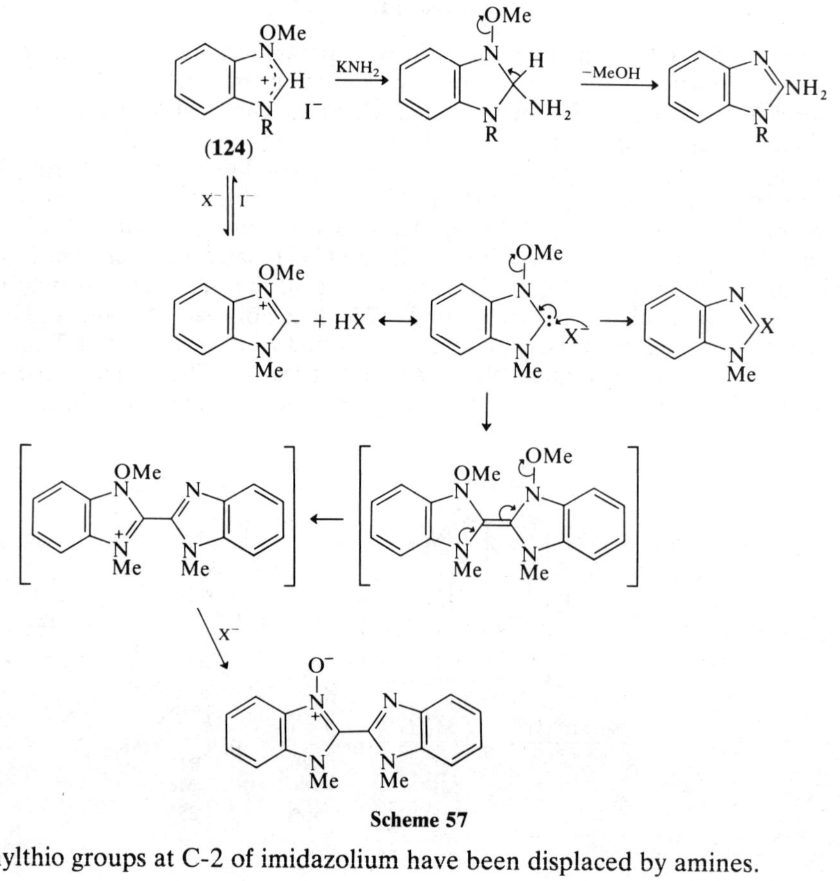

Scheme 57

Methylthio groups at C-2 of imidazolium have been displaced by amines.

4.07.1.5.3 S-Nucleophiles

5-Halogeno-4-nitro- (and -4-formyl-) imidazoles, whether or not substituted at N-1 by methyl, react with ammonium sulfite or thiols to give the corresponding sulfur derivatives. Other successful reagents have been sodium or ammonium sulfide and thiourea. Conversion of halogen-substituted imidazoles into the sulfonic acids by treatment with sulfite succeeds only when extra activation is present in the ring; otherwise the sulfur compound merely reduces the halogen compound. For this reason sodium sulfite produces only a small quantity of 5-bromoimidazole-4-sulfonic acid when it reacts with 2,4,5-tribromoimidazole. When there is a nitro group present as in 5-bromo-4-nitro- or 5-chloro-1-methyl-4-nitro-imidazole (but not 2-bromo-1,5-dimethyl-4-nitroimidazole) the sulfonic acids can be prepared by halogen displacement. A carboxyl group is much less effective in assisting these reactions.

Nucleophilic displacement of halogen by thiophenol occurs in 1-substituted 2-halogenobenzimidazoles and in 1,3-dialkyl-2-chlorobenzimidazolium tetrafluoroborates. The reactions of 4-haloimidazolium salts with sulfide or methanethiolate ions are accompanied by some replacement of halogen, but reduction and ring cleavage to give N-methyl-thioamides can also take place.

While the fluoro group in 4-fluoroimidazoles is stable in the presence of nucleophiles, a 2-fluoro group is labile under fairly mild conditions. The ease of displacement by thiol is doubtless due to the intermediate formation of the symmetrical 2-fluoroimidazolium cation (125) which is subject to addition–elimination processes (Scheme 58) ⟨80AHC(27)241, B-76MI40701⟩.

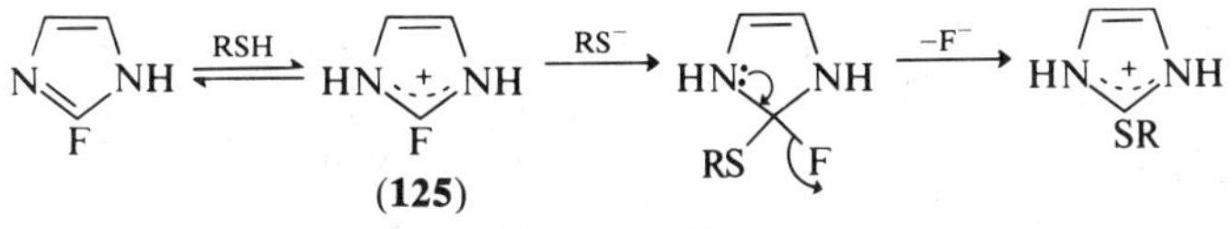

Scheme 58

4.07.1.5.4 Halide ions

While nucleophilic replacement of hydrogen by halogen is unknown in these compounds there are a number of examples of other groups which are displaced by halide and related nucleophiles.

The most common is the conversion of imidazolinones and benzimidazolinones into the halo compounds. Phosphoryl chloride transforms 4,5-diphenylimidazolin-2-one (126) into the 2-chloro compound, and the reaction is even more rapid with benzimidazolinones ⟨66RCR122⟩. The same reagent will convert imidazole N-oxides into 2-chloroimidazoles (Scheme 59) (see Section 4.02.1.6.4).

Scheme 59

Halogen replacement by another halogen is also quite common. At 150 °C 2,5-dibromoimidazole-4-carboxylic acid reacts with hydrochloric acid to give a mixture of chloro compounds, but 2,4,5-tribromo- and 2-bromo-4,5-dichloro-imidazoles in boiling HCl gave 2,4,5-trichloroimidazole. With potassium iodide in DMF 1-alkyl-4-chloro- (or bromo-) 5-nitroimidazole gives the 4-iodo analogue. Replacement of bromine by fluorine has been reported; xenon hexafluoride converts 2,4,5-tribromoimidazole into the trifluoro compound ⟨79JGU1251⟩, but 4-fluoroimidazoles could not be prepared by halogen exchange (using KF, CsF, AgF) on 4-bromo-5-nitro- or 4-bromo-5-ethoxycarbonyl-imidazoles. Nor can a chlorine atom be displaced in this way ⟨80AHC(27)241⟩.

Sulfo and hydrazine functions at C-2 of benzimidazoles can be replaced by chloro. The decomposition of imidazolediazonium fluoroborates might be expected to lead to fluoroimidazoles, but such diazonium salts are remarkably stable to heat. However, UV irradiation of the compounds leads to 30–40% yields of the corresponding fluoroimidazoles. The reaction works well for 2- and 4-fluoro- and 2-chloro-imidazoles, but fails for the bromo- and iodo-azoles ⟨78JOC4381⟩. The formation of stable diazo compounds may also account for the reluctance of the Sandmeyer reaction to operate when ethyl 5-aminoimidazole-4-carboxylate is diazotized in the presence of copper(I) chloride; only low yields of the 5-chloro products are obtained. However, 1-substituted 5-aminoimidazoles give diazonium salts which are much more reactive giving good yields of the 5-chloroimidazoles ⟨80JCS(P1)2310⟩.

When there is more than one nitro group present in an imidazole ring, halogenodenitrations are possible with the displacement order being 2-NO_2 > 5-NO_2 > 4-NO_2 (Scheme 60). Phosphoryl chloride in pyridine or DMF is successful for 4,5-dinitro compounds, but only the latter solvent is suitable for 2,4-dinitroimidazoles ⟨80AHC(27)241⟩.

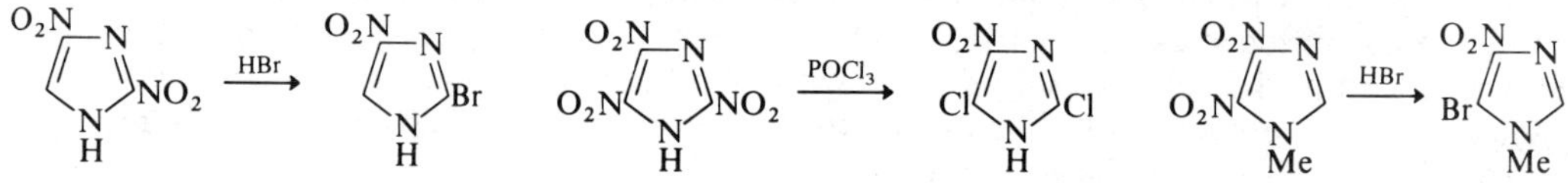

Scheme 60

4.07.1.5.5 Carbanions

(i) Activated methyl and methylene carbanions

Few examples of this type of reaction exist, but the chlorine atom of 5-chloro-1-methyl-4-nitroimidazole is replaced in reaction with sodiomalonic ester. The reactions of 1,3-diacylimidazolium at C-3 of a 1,2-disubstituted indole provides a further example of this type; such reactions are used in the synthesis of aromatic aldehydes. The reactive aromatic compound (pyrrole, indole, thiophene, anisole) reacts with 1,3-diacyl-imidazolium or -benzimidazolium to give the 2-substituted 1,3-diacyl-4-imidazoline. Alkaline hydrolysis of this gives the aldehydes (Scheme 61) ⟨80T2505⟩.

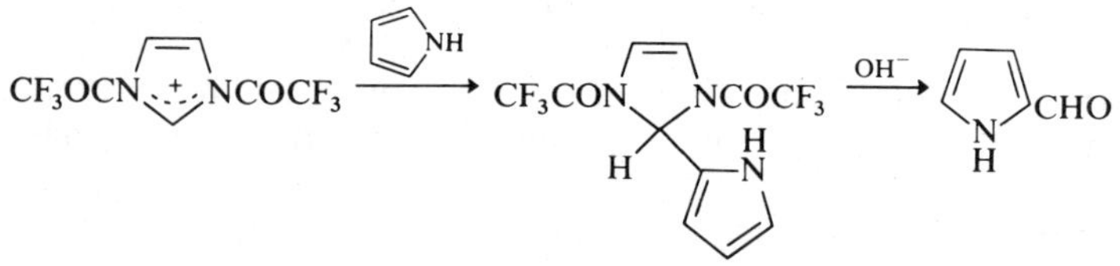

Scheme 61

(ii) Cyanide ions

In 1-methylimidazoles, nitro and methoxy groups, especially at C-5, can be replaced by cyanide in nucleophilic photosubstitution reactions. The following products and yields have been reported: 1-methyl-5-nitroimidazole gives 5-cyano-1-methylimidazole (65%); 1-methyl-4-nitroimidazole gives 4-cyano-1-methyl-(3%) and 5-cyano-1-methyl-4-nitroimidazole (1.5%); 1-methyl-2-nitroimidazole gives no stable products; 1-methyl-5-methoxyimidazole gives 2-cyano-5-methoxy-1-methylimidazole (13%); 1-methylimidazole gives 5-cyano-1-methylimidazole (3%). The final two listed reactions show that the replacement of hydrogen by cyanide is not impossible in the series, and indeed the reaction has also been noted with imidazole quaternary salts such as (**127**), in which instance the reaction is probably of the S_N1 type (Scheme 62). In the corresponding N-oxide the Reissert reaction failed to introduce a cyanide group at C-2. With potassium cyanide in ethanol 1,2-dimethyl-4-nitroimidazole gives (**128**); the conversion of a nitro function into azoxy is not peculiar to this series.

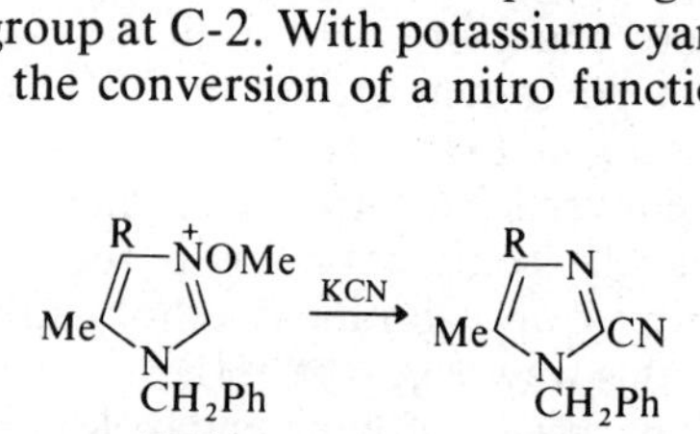

(**127**)

Scheme 62

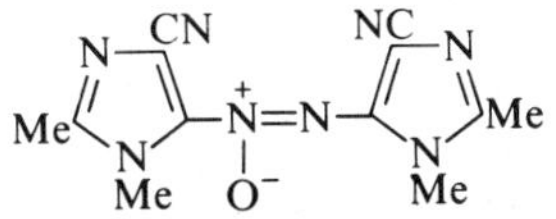

(128) 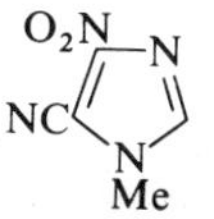   (129)

Displacement of halogen by cyanide is much more common, occurring in preference at C-5. Thus, while 1-alkyl-5-chloro- (or bromo-) 4-nitroimidazoles react with potassium cyanide in ethanol to give the 5-cyano product, the 1-alkyl-4-chloro- (or bromo-) 5-nitro isomers would not react. Under much more forcing conditions though (20 h, 120–130 °C with KCN and KI in DMF) the earlier unreactive isomers gave (129). This surprising transformation does not take place in the absence of iodide and may involve transmethylation giving at first 5-cyano-1,3-dimethyl-4-nitroimidazolium which would decompose preferentially to (129) ⟨B-76MI40701⟩.

4.07.1.5.6 Reduction by complex hydrides

Sodium borohydride converts 4H-imidazole N-oxides into 1-hydroxy-imidazoline or -imidazolidine derivatives. Under the same conditions 1,3-dioxides form 1,3-dihydroxy-imidazolidines ⟨76CHE1280⟩.

Unless the NH of imidazole or benzimidazole is substituted, the preferential reaction with a complex hydride will be salt formation which leaves a negative charge on the ring nitrogen. In consequence the species becomes resistant to reduction. Thus, at room temperature or below, benzimidazole forms only a salt with lithium aluminum hydride, but in benzene–ether as solvent the benzimidazoline (130) results. Benzimidazoline-2-thione does not react, but hydantoins (imidazolidine-2,4-diones) give a variety of products depending on the reaction conditions. With lithium aluminum hydride at room temperature 1-methyl-4-phenylimidazolidine-2,5-dione gave 1-methyl-4-phenylimidazolin-2-one (131) which could be reduced further to (132) and (133) ⟨66AHC(6)45⟩.

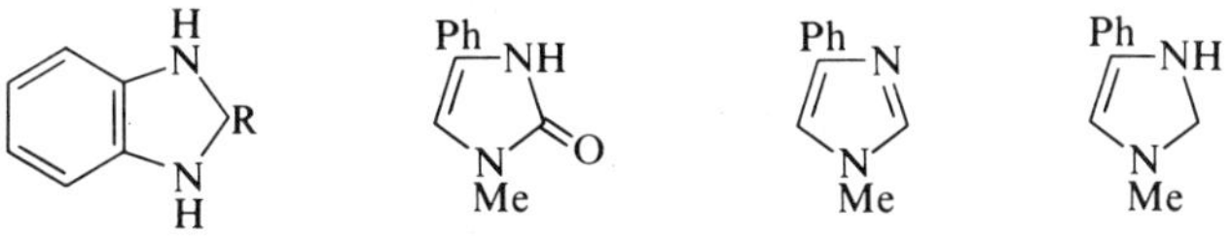

(130)   (131)   (132)   (133)

2,4,5-Tris(diethylamino)imidazolium chloride is reduced by sodium borohydride in ethanol to the 1,3-diazacyclopentadiene (134; Scheme 63) ⟨81TL2973⟩.

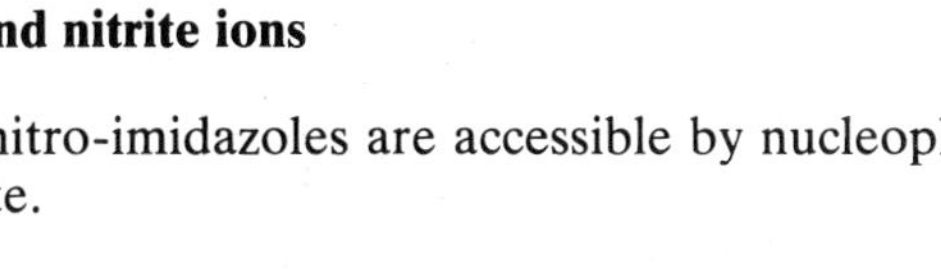

Scheme 63

4.07.1.5.7 Azide ions and nitrite ions

Both 2-azido- and 2-nitro-imidazoles are accessible by nucleophilic displacement of the 2-diazonium fluoroborate.

4.07.1.6. Nucleophilic Attack at Hydrogen

4.07.1.6.1 Metalation at a ring carbon atom

N-Alkyl-imidazoles and -benzimidazoles react with lithium or butyllithium at low temperatures to give the 2-lithio derivatives. It has been reported, though, that in the metalation of 1-methylimidazole a small amount of the 5-substituted compound is also formed. Reagents such as phenyllithium or phenylsodium are also effective, but 5-bromo-1-ethylbenzimidazole would not react with phenyl- or pentyl-sodium.

These metalation reactions require an *N*-protected imidazole (otherwise the reagent would preferentially deprotonate the NH to give an unreactive anion), and the protecting group should be capable of subsequent removal. Benzyl can be removed reductively or by oxidation, but it does suffer from the deficiency that deprotonation can occur at the benzyl methylene group rather than in the heterocyclic ring. A trityl group is removed by strong acid hydrolysis as is benzenesulfonyl. The most useful *N*-protecting group would appear to be alkoxymethyl (Section 4.07.1.3.8) which is easily hydrolyzed under acidic or neutral conditions. The 1-ethoxymethylimidazoles lithiate readily at −40 °C in THF. While earlier reports suggested that C-5 metalation is difficult, 1-ethoxymethyl-2-phenylthioimidazole (135) is lithiated at C-5 in 83% yield by butyllithium in THF at −78 °C. The 5-methylthio analogue of (135) failed to metalate at C-4 with butyllithium, but potassium diisopropylamide–lithium *t*-butoxide at −78 °C was successful (Scheme 64). It has proved possible to replace successively all bromine atoms by lithium in 1-ethoxymethyl-2,4,5-tribromoimidazole ⟨81CC1095, 80JOC4038⟩.

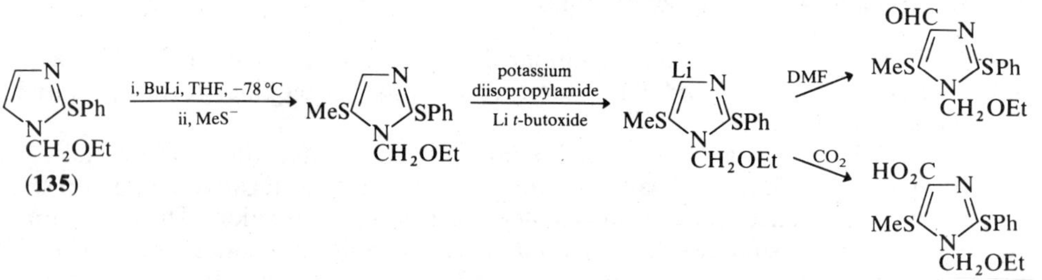

Scheme 64

Thus, when the 2-position of imidazole is blocked it is possible for alternative metalation to take place at C-5. The process does not occur particularly readily, especially at low temperatures, and an alternative lateral metalation of the 2-methyl group may take place.

Butyllithium reacts with 1-methylbenzimidazole at room temperature to form the 2,2'-dibenzimidazolyl (136) *via* its dihydro derivative (Scheme 65).

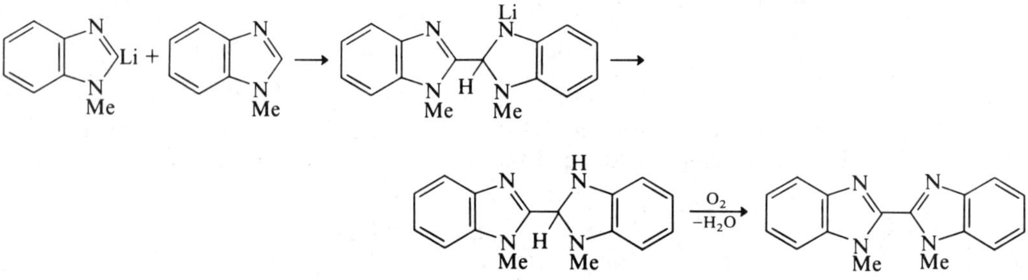

Scheme 65

4.07.1.6.2 Hydrogen exchange at ring carbon in neutral azoles

The basic features of hydrogen and deuterium exchange have been summarized in the general chapter (Section 4.02.1.7.2). In the parent imidazole molecule the proton attached to nitrogen exchanges in D_2O, while H-2 is exchanged at 37 °C with $t_{1/2}$ ~700 minutes. At 150 °C the reaction is virtually complete in two hours with or without added base; in acid medium exchange does not take place. Thus exchange at C-2 did not take place over a period of days at 37 °C in D_2SO_4 or $MeCO_2D$, but complete exchange of all imidazole protons occured at 250 °C in D_2O.

Two parallel processes appear to be occurring: rate-determining proton abstraction from the imidazolium ion by D_2O and by OD^- to give the ylide at C-2 (137) followed by deuteration there (Scheme 66). The pD profile for 4-substitution can be accounted for by an additional path involving proton abstraction from the imidazole neutral molecule. In strongly alkaline medium imidazoles with no nitrogen substituent exchange more readily at C-4 or C-5 by a carbanion process which involves deprotonation of the free base. There is a further mechanistic possibility involving σ-intermediates at high pD values. In the ylide mechanism shown here the conjugate acid is naturally more reactive than the neutral molecule, while the conjugate base is unreactive. In any process via a Wheland intermediate the opposite reactivity sequence applies, viz. conjugate base > neutral molecule > conjugate acid.

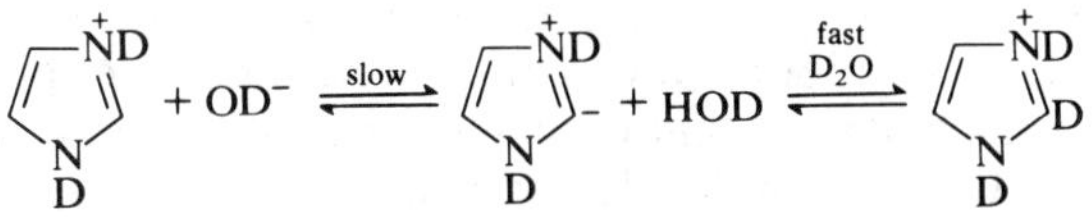

Scheme 66

1-Methylimidazole is deuterated at C-2 by D_2O with the rate of the reverse reaction being virtually independent of pH, as long as conditions do not become too acidic when the rate rapidly falls to zero. Again an ylide mechanism applies (Scheme 9; Section 4.01.1.7.2). The relative rates of exchange at the 2-, 4- and 5-positions of 1-methylimidazole are $54.5 \times 10^3 : 1.6 : 1$ (base-catalyzed exchange) for the ylide mechanism, and $1 : 73 : 120$ (acid-catalyzed exchange) for an S_EAr process involving a conjugate acid species. Even in basic conditions, then, 1-methylimidazole acts as the cation which loses a proton to form an ylide in the rate-determining step ⟨80AHC(27)241, B-76MI40701⟩. The resistance of H-4 to exchange in these compounds has been attributed to the adjacent lone-pair effect, i.e. electrostatic repulsion between the lone pairs of the sp^2 orbitals at N-3 and C-4.

Substituted imidazoles, too, can be deuterated: 4-bromoimidazole with D_2O for 7.5 hours at 100 °C gave the 1,2-dideutero compound, and 4-nitroimidazole at 100 °C for 13.5 hours gave the 2- and 5-deutero compounds in the ratio 5 : 3. When sodium deuteroxide is present the exchange of all ring hydrogens is complete in 12 hours at room temperature. 4-Methylimidazole gave the 2,5-dideuterated species after two exchanges of 4–5 hours at 100 °C in D_2O–NaOD, but exchange was slow at 37 °C, and in acid only H-5 exchanged. Under identical conditions 1-methyl-4- and -5-nitroimidazoles behave quite differently. In the former, after 12.5 hours at 100 °C in D_2O exchange at C-5 was 90% complete, and that at C-2 was 50% complete. With the latter, exchange at C-2 was 100% complete, but only 10% complete at C-4. If the reaction is assumed to depend on a rate-limiting proton abstraction at each site, then these results are in line with the ^1H NMR shifts (p.p.m.) quoted for the molecules, viz. 1-methyl-4-nitroimidazole: H-5 8.30, H-2 7.54 ($CDCl_3$), H-5 8.94, H-2 8.38 (CF_3CO_2D); 1-methyl-5-nitroimidazole: H-4 7.64, H-2 8.09 ($CDCl_3$), H-4 8.54, H-2 9.03 (CF_3CO_2D) ⟨80AHC(27)241, B-76MI40701⟩.

Hydrogen exchange at C-2 in 1-methylbenzimidazole has been discussed earlier (Section 4.02.1.7.2).

4.07.1.6.3 Hydrogen exchange at ring carbon in imidazolium cations

The observation that ^1H NMR spectra of imidazole quaternary salts in D_2O show relatively rapid loss of the H-2 signal is evidence for facile exchange at that position.

The base-catalyzed deuteration at C-2 of imidazole quaternary salts proceeds via an ylide. Exchange at C-2 in 1,3-dimethylimidazolium is 3×10^4 times as fast as 3- (or 5-) deuteration in 1,2-dimethylpyrazolium.

4.07.1.6.4 C-Acylation via deprotonation

The acylation reactions at C-2 of imidazoles and benzimidazoles are discussed elsewhere (Section 4.07.1.4.6).

4.07.1.6.5 Proton loss from a ring nitrogen atom

The salts formed when imidazoles and benzimidazoles react with metallic reagents such as alkali metal amides and Grignard reagents are very easily hydrolyzed. The resulting anions, though, are very strong nucleophiles, a property which is utilized in such reactions as alkylation. The weakly acidic nature of the compounds has been discussed in Section 4.7.1.3.5, and pK_a values are listed in Table 3.

The sodium salts of imidazoles are so readily hydrolyzed that they are only stable in such media as ethanol or liquid ammonia. Sparingly soluble salts, *e.g.* silver salts, are stable in aqueous medium, and have been utilized for isolation of imidazoles and in the formation of some 1-substituted derivatives, *e.g.* 1-trityl, 1-glycosyl (Section 4.07.1.3.8). Mercury salts of imidazole are believed to adopt a linear polymeric structure, and results of studies of thallium and tin derivatives suggest that they, too, are polymeric, at least in the solid state. Stable organotin derivatives of imidazoles are non-explosive and often stable in boiling water.

4.07.1.7 Reactions with Radicals and Electron-deficient Species

4.07.1.7.1 Carbenes

When imidazole, or 1- and 2-substituted imidazoles react with chloroform at 550 °C the products include 5-chloro- (**138**) and 4-chloro-pyrimidines, and 2-chloropyrazines. Carbene insertion from the chloroform into either a C—C or C—N bond accounts for the products among which (**138**) predominates (Scheme 67). With hexachloroacetone at room temperature 2-methylimidazole gives a 5–7% yield of (**138**; R = Me) ⟨80JCS(P1)1427⟩.

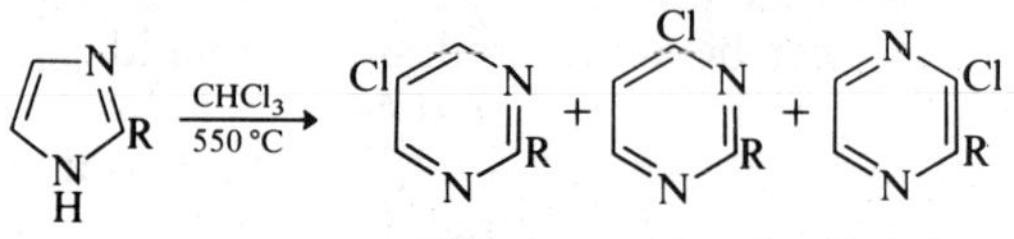

(**138**)   R = Me; 11 : 1 : 8
R = H; 9 : 0 : 1

Scheme 67

4.07.1.7.2 Free radical attack at the ring carbon atoms

(i) Aryl radicals

A study of the phenylation of 1-methylimidazole using *N*-nitrosoacetanilide or diazoaminobenzene as the sources of phenyl radicals gave about 10–20% conversion to phenylimidazoles with a relative rate compared with benzene of 1.2. Most substitution was at C-2 (radicals follow the substitution patterns of nucleophiles largely) with partial rate factors relative to benzene: $f_2 = 4.8$, $f_4 = 0.14$, $f_5 = 2.2$. Later experiments using benzene diazonium fluoroborate with 1-methylimidazole gave phenylated products with isomeric proportions similar to the foregoing, but which depended on the concentration of the salt. There is a suggestion that some of this radical substitution might be occurring with the more reactive quaternized substrate (**139**). When phenylation, using the decomposition of benzoyl peroxide at 118 °C, was studied under two sets of conditions, one in which 1-methylimidazole itself acted as the solvent, and the other in acetic acid, there was no change in the overall yield. In acetic acid, however, only 1-methyl-2-phenylimidazole was formed; with excess heterocycle 1-methyl-2-phenyl- and -5-phenyl-imidazoles were isolated in the ratio 29:21. The values of the relative rates (9.0 in acetic acid, 5.7 in 1-methylimidazole) are much higher than those quoted above, but this may be a function of the reagent, since the quaternized species (**140**) may be responsible for the values obtained.

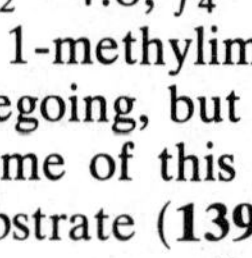

(**139**) 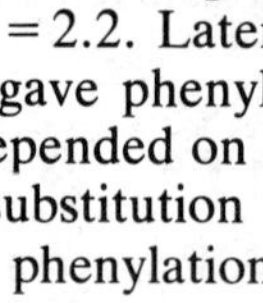   (**140**)

(ii) Alkyl and other carbon radicals

The nucleophilic radicals formed by the silver-catalyzed, oxidative decarboxylation of carboxylic acids by peroxydisulfate ions attack protonated imidazoles mainly at the 2-

position, the most electron-deficient site. The catalytic action of the silver salt follows the reaction sequence shown.

$$2Ag^+ + S_2O_8^{2-} \rightarrow 2Ag^{2+} + 2SO_4^{2-}$$

$$RCO_2H + Ag^{2+} \rightarrow R\cdot + CO_2 + H^+ + Ag^+$$

The medium has no specific oxidative action towards the alkyl radicals, and in the presence of protonated base the major reaction is a substitution. Thus, imidazoles and 1-substituted imidazoles are alkylated exclusively at C-2, albeit in rather low yields. The use of isopropyl and *t*-butyl radicals gives improved yields (80–90%) but benzyl and allyl radicals tend to dimerize in preference. Benzimidazoles are also alkylated at C-2, and with isopropyl and *t*-butyl radicals yields of 50–80% can be achieved ⟨80AHC(27)241, 74AHC(16)123⟩.

Under UV irradiation 1-(*p*-nitrophenyl)-4-phenylimidazole reacts with cyanide ion to give the 2-cyano substitution product (80%); the corresponding 2-methyl analogue gave a mixture of 4-cyano (30%) and 4-hydroxymethyl (20%) products ⟨79T1331⟩.

4.07.1.7.3 *Electrochemical reactions and reactions with free electrons*

In the electrochemical reduction of 5-bromo-1-methyl-4-nitroimidazole some cleavage of the C—Br bond is evident giving rise to the debrominated product ⟨79JGU1877⟩. Imidazole itself is not reducible cathodically in aqueous media, but electrons have been attached to imidazole and histidine in aqueous solution; the rate of oxidation depends on pH. Protonated or quaternized imidazoles form the neutral conjugate acids of the true anion radical, and a number of anion radicals have been made from nitroimidazoles under various radiolytic conditions ⟨79AHC(25)205⟩. The electrochemical reduction of 2-cyanobenzimidazole 3-oxide gives sequentially 2-cyanobenzimidazole and 2-aminomethylbenzimidazole ⟨80ZC263⟩.

4.07.1.7.4 *Catalytic hydrogenation and reduction by dissolving metals*

Imidazoles and benzimidazoles are generally quite resistant to such reduction reactions. Catalytic reduction of aryl- and furyl-substituted imidazoles affects only the aryl and furyl rings, but there has been an isolated report that 2-methyl-4,5-diphenylimidazole can be reduced over palladium to give the corresponding imidazolidine ⟨70AHC(12)103⟩. In the presence of palladium, platinum and rhodium catalysts in acid media, benzimidazoles were found to be reduced only in the fused benzene ring ⟨73MI40700⟩.

4.07.1.8 Reactions with Cyclic Transition States

4.07.1.8.1 *Diels–Alder reactions and 1,3-dipolar cycloadditions*

Reactions of imidazoles with dienophiles such as DMAD usually lead not to normal Diels–Alder adducts, but to products of *N*-alkylation (Scheme 68; see also Section 4.07.1.3.11) ⟨78AHC(23)265, 79JCS(P1)1239⟩.

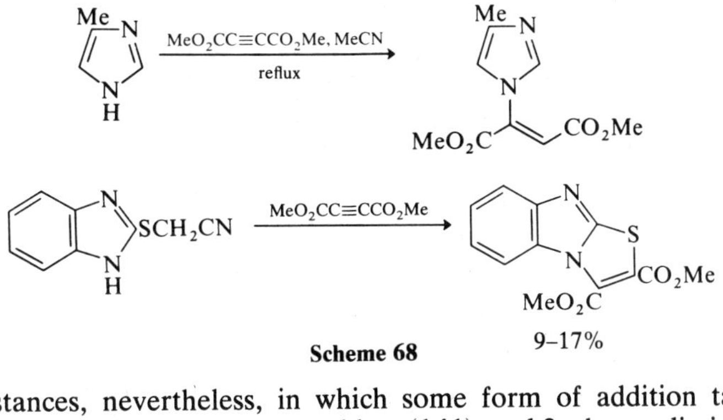

Scheme 68

There are instances, nevertheless, in which some form of addition takes place. For example, 1,2-dimethylimidazole gives the adduct (**141**), and 2-phenacylimidazole forms the imidazopyridine (**144**), presumably *via* (**142**) which cyclizes to (**143**) before undergoing a Stobbe-type ring opening under the influence of solvent (Scheme 69). The mesoionic imidazole (**145**) gives the pyrrole (**146**) almost quantitatively in what appears to be a typical 1,3-dipolar cycloaddition reaction (Scheme 69).

(**141**)

(**142**) (**143**) (**144**)

(**145**) (**146**)

Scheme 69

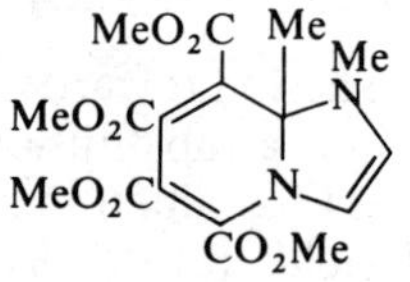

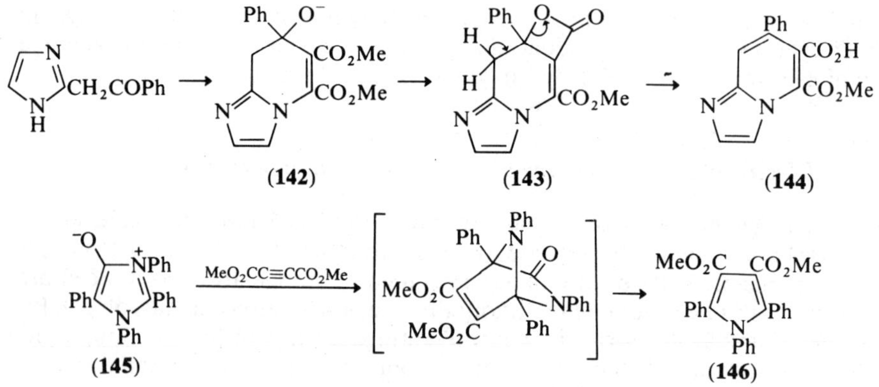

Most of the reactions with benzimidazoles have been summarized ⟨78AHC(23)265, 74CRV279⟩. Benzimidazolium ylides (**147**) react with 1,3-dipolarophiles to give the tricyclic products (**148**) in low yield (Scheme 70).

(**147**) (**148**) 4%

Scheme 70

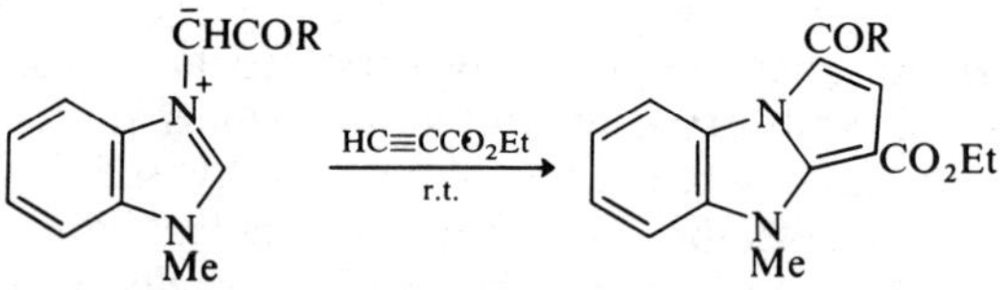

Imidazole reacts very slowly with singlet oxygen to form the imidazolidinone (**150**) through an elimination reaction which involves proton loss and cleavage of the oxygen–oxygen bond of the cyclic peroxide (**149**). Under similar conditions 4-phenylimidazole forms the hydantoin derivative (**151**) and *N*-benzoyl-*N′*-methoxycarbonylurea (Scheme 71). Apparently more than one mole of oxygen per mole of substrate is taken up in this latter instance, which is a parallel with the oxygenation of histidine during the photooxidative inactivation of some enzymes ⟨70AHC(12)103⟩. Other photooxidation reactions have been discussed earlier (Section 4.07.1.2).

(**149**) (**150**) (**151**)

Scheme 71

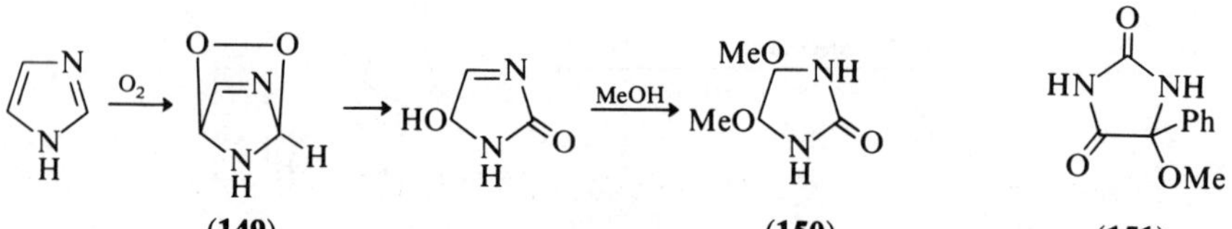

4.07.1.8.2. *Photochemical cycloadditions*

Many of these reactions give rise to four-membered rings. Additions across the 4,5-bond with benzophenone and acrylonitrile have already been referred to briefly (Section 4.02.1.9).

In the photoaddition of acetone and other ketones to 1-, 2- and 1,2-di-methylimidazoles the products are α-hydroxyalkylimidazoles (**153**) which are derived from the selective attack of excited carbonyl oxygen at C-5. In the case of 2-methylimidazole the products are the 4-mono- (8%) and 4,5-di- (14.5%) substituted compounds, but imidazole itself does not react. The suggestion that it is not a sufficiently electron-rich substrate is not particularly convincing. The reaction mechanism (Scheme 72) may reflect the greater radical reactivity at C-5, and the comparative stabilities of the radical intermediates derived from carbonyl attack at this position. Hückel calculations of radical reactivity indices show that, indeed, C-5 is more reactive, and the radical intermediate at C-5 is more stable than that at C-4, but a concerted cycloaddition could also give rise to the oxetane (**152**). Such an oxetane can be isolated in the photochemical addition of benzophenone to 1-acetylimidazole.

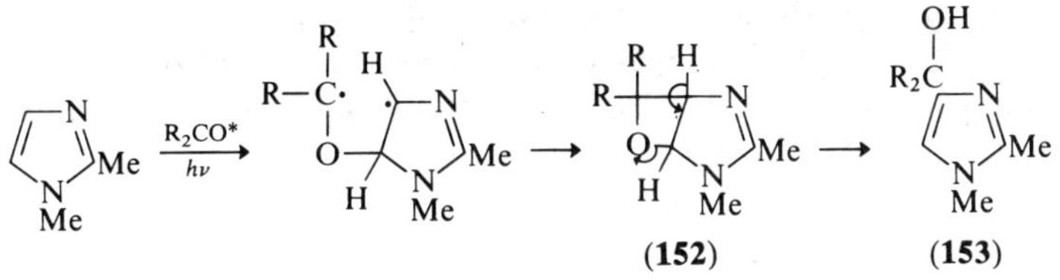

Scheme 72

Benzophenone reacts differently with 1,2-dimethylimidazole. While acetone gives (**153**), the diarylketone adds at the 2-methyl group, and with 1-benzylimidazole reaction is at the exocyclic methylene (Scheme 73) ⟨80AHC(27)241, 80JHC1777⟩.

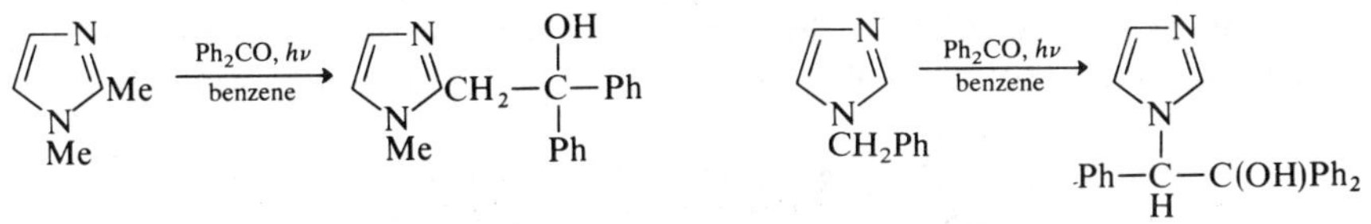

Scheme 73

When acrylonitrile adds across the 4,5-bond of 1-methyl-2,4,5-triphenylimidazole a four-membered ring is also formed, but the reaction is very sensitive to substituents and to solvents. While 2,4,5-triphenylimidazole reacts in both ethanol and acetonitrile to give (**154**), the 1-methyl analogue gives the [2+2]-cycloaddition products (**155**) in ethanol, but (**156**) when the solvent is acetonitrile.

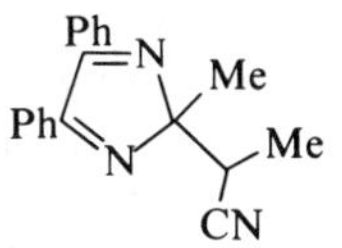

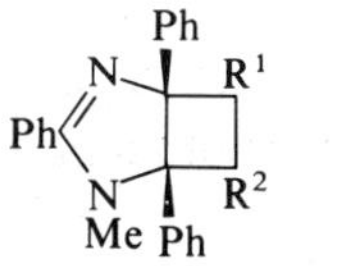

(**155**) $R^1 = H, R^2 = CN$
$R^1 = CN, R^2 = H$

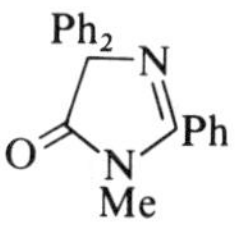

It has been concluded that encounter complexes (or exciplexes) and ion-pairs are the key intermediates in these reactions. Addition of acrylonitrile to N-unsubstituted imidazoles gives the N-substitution product, while the N-substituted compounds undergo cycloaddition ⟨80AHC(27)241⟩.

4.07.2 REACTIONS OF NON-AROMATIC COMPOUNDS

4.07.2.1 Isomers of Aromatic Derivatives

4.07.2.1.1 Compounds not in tautomeric equilibrium with aromatic derivatives

These compounds include the 2H- (**157**) and 4H- (**158**) imidazoles ('isoimidazoles'), the isoimidazolones (**159, 160**) and such derivatives as 5,5-disubstituted hydantoins (**161**). There will also be a number of other possibilities in which the exocyclic oxygens of (**159**)–(**161**) are replaced by S, NH or CR_2.

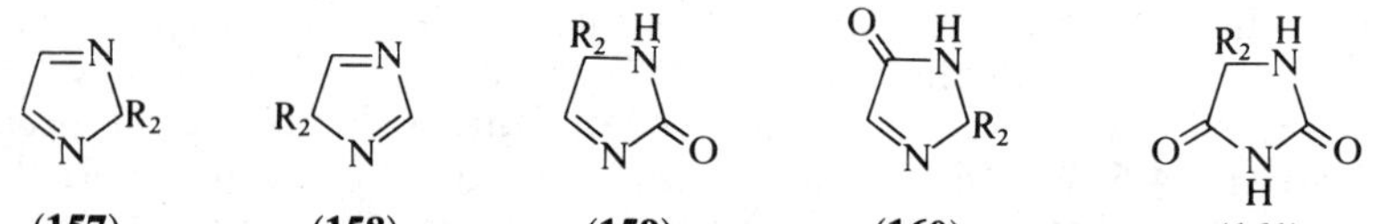

(157) (158) (159) (160) (161)

Some of the properties such as salt formation and alkylation to quaternary salts have already been mentioned for the 2*H*- and 4*H*-imidazoles (Section 4.02.2.1.1), *e.g.* 5*H*-imidazoles quaternize at N-1. If, however, the 4-substituent is thiol, then methylation with methyl iodide in basic medium gives the *S*-methyl compound before subsequent methylation leads to quaternization. There are a number of examples of such exocyclic alkylation and acylation ⟨69T4265, 75M1461⟩. The spiro 1-substituted 2-imidazolin-5-ones (162) alkylate readily to form the quaternary salts ⟨61JOC4480⟩, while similar methylation in basic medium of 2-substituted 4,4-diphenyl-4*H*-imidazolin-5-one (or its other isomer) gives the *N*-methyl derivative (163; Scheme 74) ⟨69T4265⟩. Hydantoins unsubstituted on nitrogen can be alkylated there by a variety of reagents.

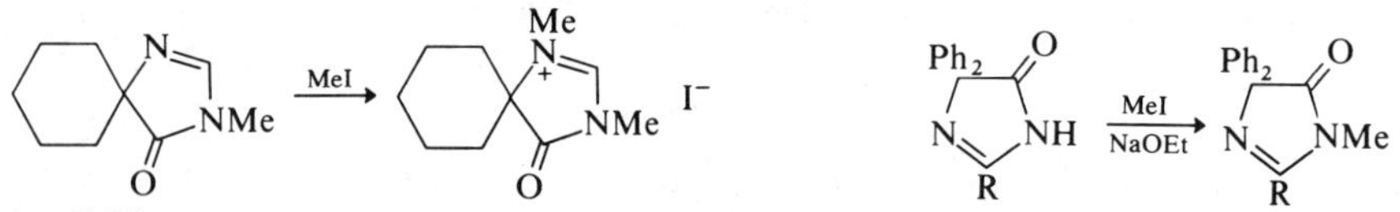

Scheme 74

The acylation of 3-imidazoline-4-thiones gives 1-acylimidazoline-5-thiones which rearrange on heating to the 'isoimidazoles' (164; Scheme 75) ⟨75M1461⟩.

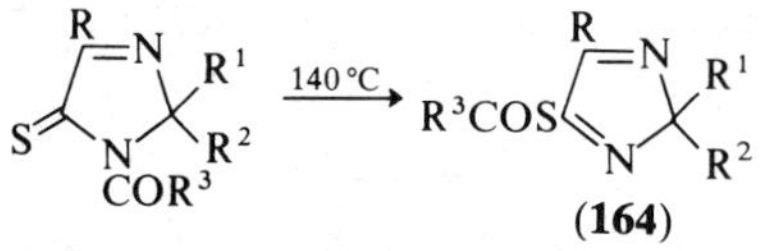

Scheme 75

There are a number of thermal reactions which transform 'isoimidazoles' into the fully aromatic compounds, *e.g.* the rearrangement of 2,2-dialkyl-2*H*-imidazoles (165) to the 1,2-dialkyl isomers proceeds by a concerted process. In the absence of hydrogen sources thermolysis of 2,4,4-triaryl-5-methylthio-4*H*-imidazoles gives 1,1'-biimidazolyl products (166) by a radical process ⟨80AHC(27)241⟩. Catalytic hydrogenation of the 4*H*-imidazole (167) leads to aromatization, and the compound is indeed unstable enough for treatment with concentrated hydrochloric acid at room temperature also to give an imidazole product (Scheme 76) ⟨75JHC471⟩.

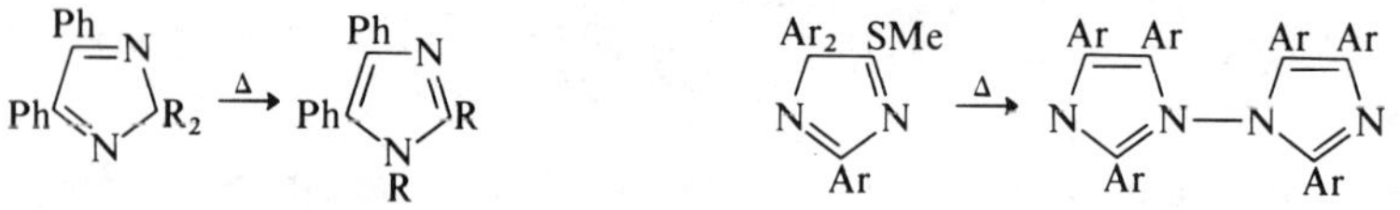

(165) (166)

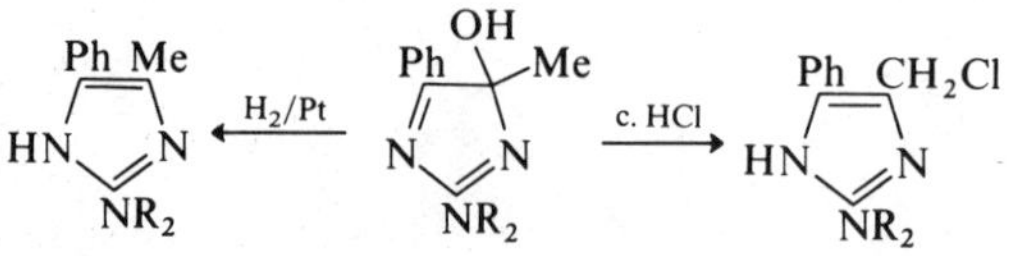

(167)

Scheme 76

Some imidazolinones can be hydrated when heated under reflux with water which adds across the 2,3-double bond. Catalytic hydrogenation also operates at the same site indicating high double-bond character at that site (Scheme 77) ⟨61JOC4480⟩. On the other hand, the 4*H*-imidazole (168) is transformed into a 3-acyl-5-amino-1,2,4-oxadiazole when heated in aqueous solution (Scheme 77).

When heated under reflux with anhydrous AlCl₃ in an aromatic hydrocarbon solvent, 5,5-diphenyl-2-imidazoline-4-thiones are both aromatized and desulfurized ⟨69T4265⟩.

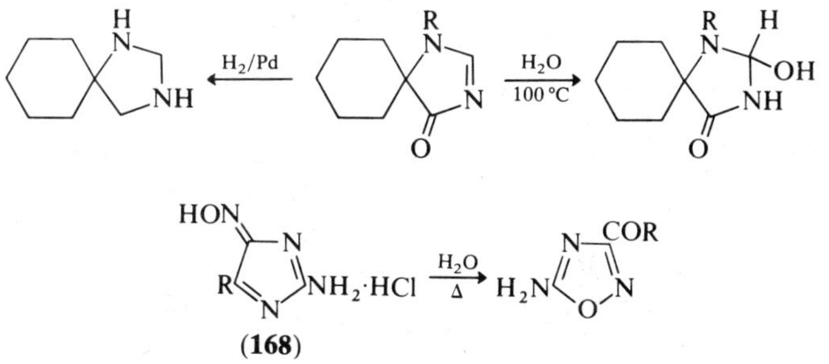

[Scheme 77 structures]

(168)

Scheme 77

The isoimidazole species can sometimes be reduced to imidazoline, e.g. 2,2-pentamethylene-2H-benzimidazole (169) is reduced by sulfinic acids under mild conditions at the same time as being substituted at C-5 (Scheme 78) ⟨77JOU1817⟩.

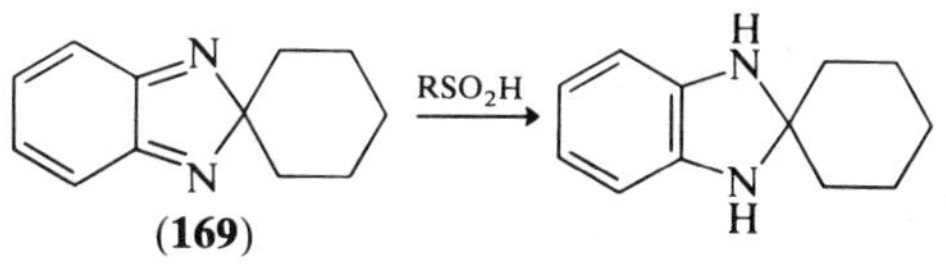

(169)

Scheme 78

Nucleophiles attack 2,4,5-trichloro-2H-imidazoles or 2,4-dichloro-4H-imidazoles with displacement of halogen (Scheme 79) ⟨76CB1638⟩.

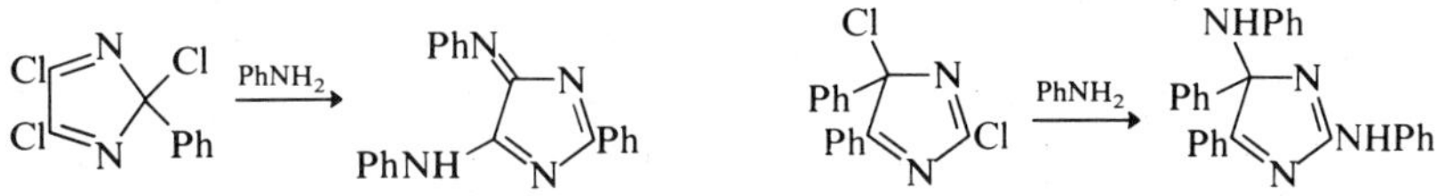

Scheme 79

5,5-Disubstituted thiohydantoins can be oxidized by permanganate to 5H-2-imidazolin-4-ones.

The N-oxide derivatives of compounds of this class are subject to a variety of typical reactions. As with the fully aromatic imidazoles the oxide function can be removed: hexachlorodisilane induces deoxygenation of 2,2-dimethyl-4-phenyl-2H-imidazole 3-oxide ⟨78JOC2289⟩. Such compounds act as nitrones with reagents like LiAlH$_4$, Grignard reagents and DMAD. Thus the compounds can be reduced to the imidazolines by the hydrides (Scheme 80) ⟨80CHE628⟩. Depending on the position of the N-oxide group similar NaBH$_4$ reduction of 4H-imidazole N-oxides can give 1-hydroxyimidazolines or imidazolidine derivatives, and under the same conditions 4H-imidazole 1,3-dioxides give 1,3-dihydroxyimidazolidines ⟨76CHE1280⟩. The N-oxide functions can be alkylated and acetylated. Sometimes the ring is cleaved by nucleophiles such as Grignard reagents or alkali (Scheme 80).

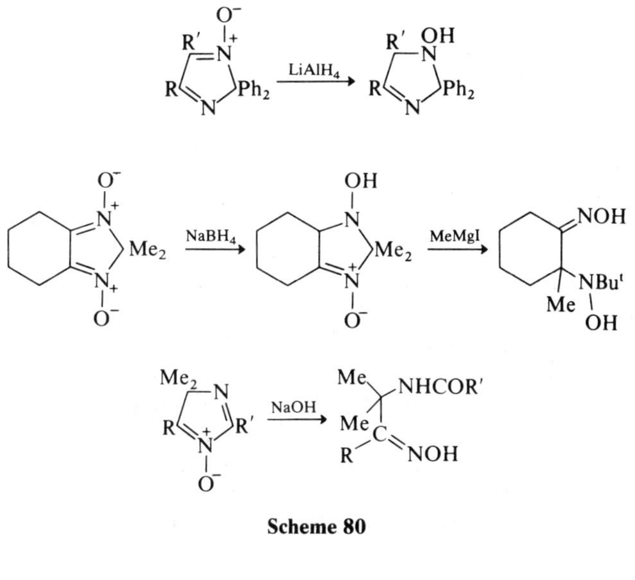

Scheme 80

4.07.2.1.2 Compounds in tautomeric equilibria with aromatic derivatives

Compounds of this type include some of the imidazolinones or imidazolinethiones which can be in equilibrium with the fully aromatic hydroxy or thiol forms (Section 4.06.5.1). The non-aromatic forms may be quite stable in some instances, but may also react in the aromatic form. Thus, 5H-imidazolin-4-ones (**170**) are acylated under Schotten–Baumann conditions on the exocyclic oxygen. When the reaction conditions are altered it is possible to obtain products of N-acylation and N,O-diacylation ⟨71CHE746⟩. With the same compound a Mannich reaction takes place at the 5-position, and phosphoryl chloride gives products of O- and N-phosphorylation rather than nucleophilic chlorination (Scheme 81).

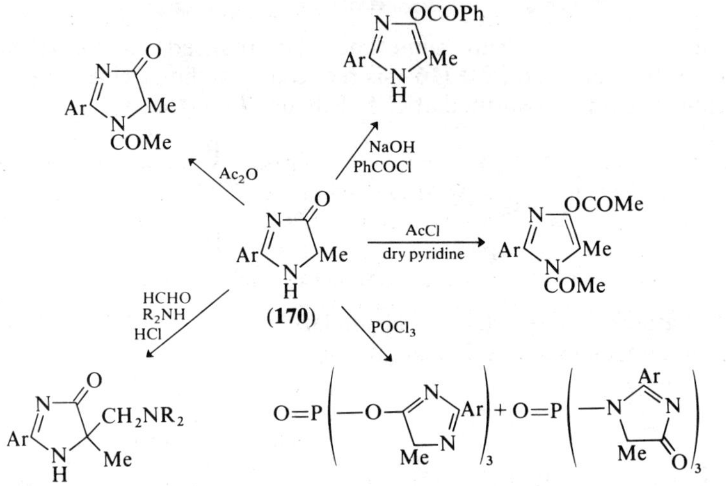

Scheme 81

Imidazolin-2-one (see Scheme 18, Chapter 4.06) has enamine reactivity and dimerizes in dilute acid, rather like a pyrrole, by protonation at C-4 (Scheme 82). Such imidazolinones are weakly acidic compounds as well as being basic by virtue of their tautomeric character. With phosphoryl chlorides they give 2-chloroimidazoles, and the compounds may also be dialkylated. Bromine adds across the 4,5-bond, while treatment with permanganate yields the glycol. They can also undergo Friedel–Crafts reactions with acyl halides and Lewis acids to form 4-acyl derivatives. The compounds are therefore characterized by a mixture of aromatic and non-aromatic behaviour. The same is true for the 4-imidazolinones which can couple with diazonium salts as well as the active methylene adjacent to the carbonyl function being involved in the formation of benzylidene derivatives ⟨B-57MI40700⟩.

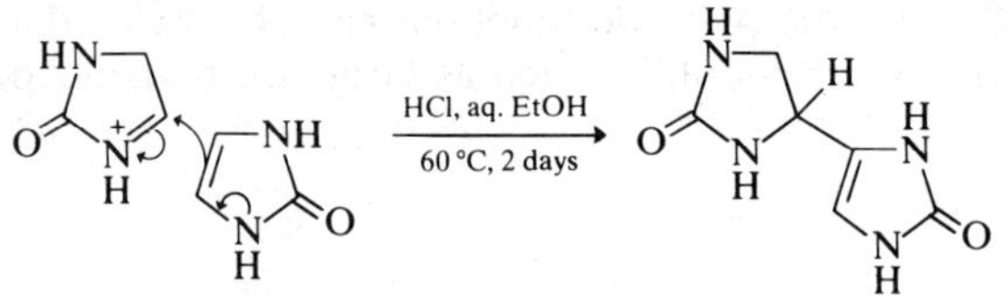

Scheme 82

4.07.2.2 Dihydro Compounds

4.07.2.2.1 Tautomerism

Imidazolines can exist as three possible structures: 2-imidazolines (**171**), 3-imidazolines (**172**) or 4-imidazolines (**173**) (Scheme 83). The first-named can exist as a pair of tautomers, but any proton shift in (**172**) will give (**173**) by rearrangement. In fact the hydrolysis of N-unsubstituted 3-imidazolines to α-aminoketones occurs presumably via the 4-imidazoline. 2-Imidazolines are cyclic amidines, and as such exhibit the characteristic resonance stabilization and strongly basic natures of these compounds. Protonation occurs on the unsubstituted nitrogen to give a resonance-stabilized imidazolinium ion. Examples of pK_a

values, which demonstrate that the 2-imidazolines are slightly weaker bases than the corresponding amidines, are: 2-methyl 11.09, 2-ethyl 11.05, 2-phenyl 9.88, 1,2-diphenyl 9.26 ⟨69BSF4075, 73JCS(P2)1371⟩.

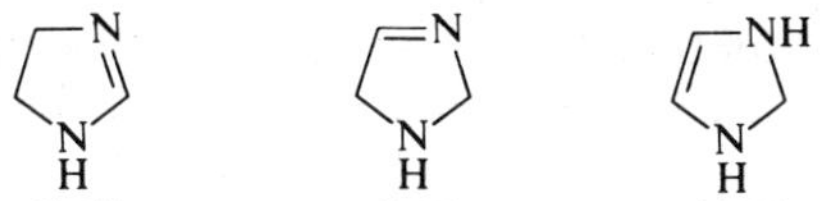

Scheme 83

Positions 4 and 5 are equivalent as in the fully aromatic imidazoles, but when these positions are occupied by the same substituent groups *meso* and racemic forms are possible.

4.07.2.2.2 Aromatization

Both 3- and 4-imidazolines (and benzimidazoline) are converted very readily into the fully aromatic compounds (Section 4.02.2.2.2). The 2-imidazolines, though, are much more resistant to dehydrogenation. Among the reagents and methods which have proved effective are Raney nickel at 170–200 °C, metals such as Pt or Pd at high temperature and hydrogen-acceptors such as S, Se, CuO or cyclohexanone. Even mild oxidation with active manganese dioxide has given high yields of the aromatic compounds from substituted 2-imidazolines ⟨70AHC(12)103, 80AHC(27)241⟩.

4.07.2.2.3 Other reactions

The strongly basic 2-imidazolines dissolve in water to give basic solutions and form salts with acids. Although the hydrochlorides tend to be hydroscopic the picrates are well-developed crystalline compounds. 3-Imidazolines, too, form hydrochloride salts readily in anhydrous medium, but these are hydrolyzed with some facility to ring-opened products. The picrates are more stable.

Alkylation can lead to several products. When heated in a variety of solvents with alkyl halides, 2-imidazolines give low yields (30–40%) of 1-alkyl-2-imidazolines, accompanied by some 1,3-dialkyl quaternary salts. If water is present these salts are readily hydrolyzed to aliphatic secondary diamines. In the presence of reducing agent, *e.g.* H_2/Ni, 1-benzyl-2-imidazoline is methylated at N-1 by formaldehyde. When 2,2,4-triethyl-5-methyl-3-imidazoline (**174**) is treated with dimethyl sulfate in aqueous alkali the 1-methyl product is formed (Scheme 84).

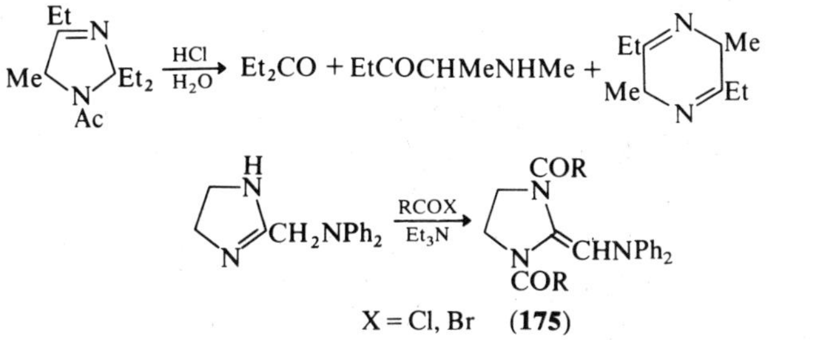

Scheme 84

Acylation of imidazolines also occurs at a ring nitrogen ⟨72JOC2158, 59M402, B-73MI4071⟩. As with imidazoles, acylation under Schotten–Baumann conditions brings about ring cleavage to aliphatic diamides. Under very mild conditions, though, it is possible to form N-acyl compounds in which the heterocyclic ring is intact, e.g. acylation of (**174**) with acetic anhydride, and formation of the diacyl product (**175**), probably via the mono-N-acyl derivative (Scheme 84).

Imidazolines are rather susceptible to hydrolytic reagents which cause fission of the ring. Aqueous acids and alkalies react with 2-imidazolines to give N-acylated derivatives of ethylenediamine (Scheme 85). The acid hydrolysis appears to involve a rate-limiting attack by water on the heterocyclic dication ⟨B-73MI40701⟩. Even heating in water can bring about this transformation with the 2-aryl derivatives being more resistant than the 2-alkyl compounds ⟨77LA1183⟩. Hydrolysis of acylated 2-imidazolines takes place with cleavage of the 1,2-bond to give a diamide (Scheme 85), while acylated 3-imidazolines under similar conditions give an aldehyde (or ketone) and an α-aminoketone (Scheme 84) ⟨72JOC2158⟩.

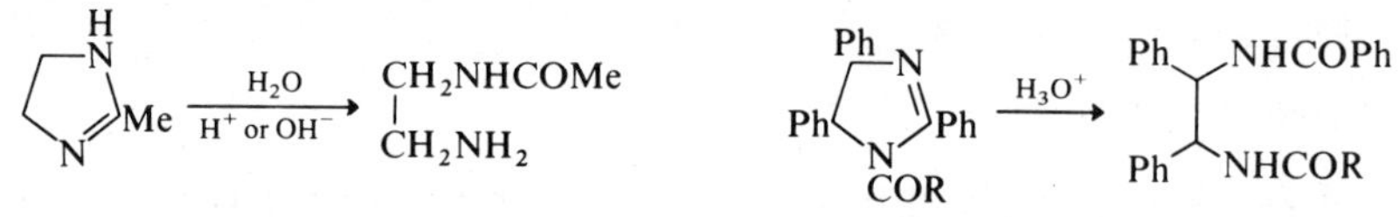

Scheme 85

Exposure of 2-methyl-2-imidazoline to a mixture of nitric and acetic acids leads to both oxidation and nitration, giving 1,3-dinitroimidazolidin-2-one (**179**). When 2-nitramino-2-imidazolines are further nitrated they give the 1-nitro products (**176**) which are dinitroguanidines. Addition of amines and alcohols leads to (**177**) and (**178**), respectively, while allylic rearrangement followed by further nitration gives (**179**; Scheme 86).

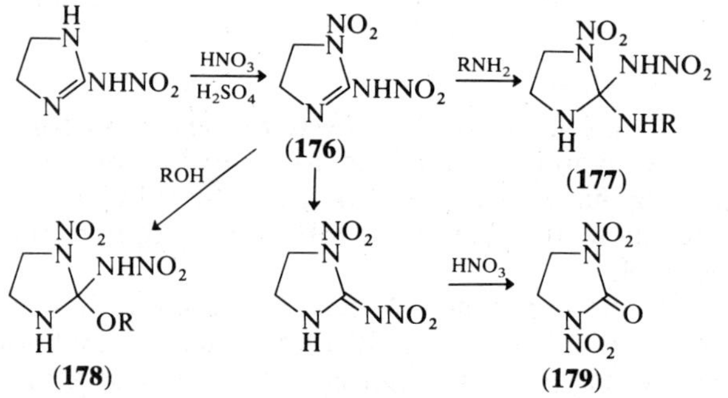

Scheme 86

It is not possible to form imidazolidines by hydrogenation of imidazolines. Hydroxylamine converts 2-methylthio-2-imidazoline into 2-hydroxyiminoimidazolidine. When heated at 170 °C with sodium metal cis-2,4,5-triphenyl-2-imidazoline rearranges to the trans form ⟨72JOC2158⟩.

There are a few examples in which imidazolines act as dienophiles in Diels–Alder reactions ⟨81TL2063⟩, while some related compounds can also react with monoalkenes in 'ene' reactions. Nucleophiles may also add across the C=N bond (Schemes 86 and 87) ⟨70JOC3097⟩.

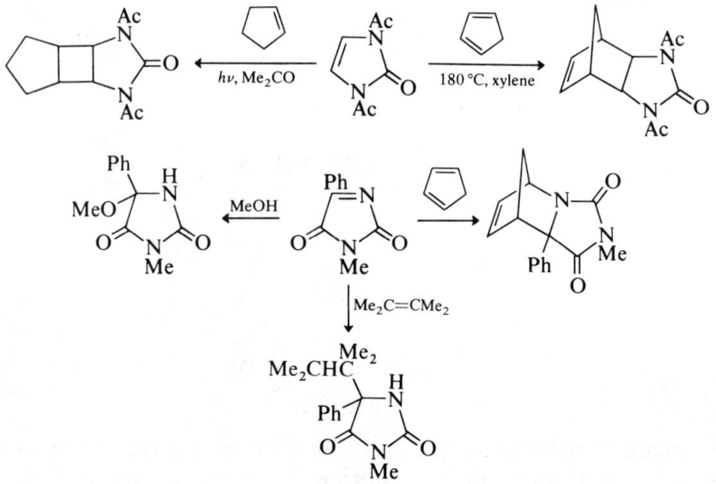

Scheme 87

4.07.2.3 Tetrahydro Compounds

4.07.2.3.1 Aromatization

Imidazolidines are not particularly well known; they may be considered as aldehyde–ammonias, and can be converted into the aromatic imidazoles only with extreme difficulty if at all. Quinones and azo compounds dehydrogenate 1,3-diphenylimidazolidine, but only as far as the imidazolinium ion ⟨B-77MI40700⟩.

4.07.2.3.2 Ring fission

General methods of ring opening have been summarized earlier (Section 4.02.2.3.2). A variety of reagents are capable of cleaving the ring, *e.g.* acid chlorides and HCN, borane in THF, and even cold dilute acid, which brings about rapid hydrolysis of imidazolidines to N,N'-disubstituted ethylenediamine salts and an aldehyde. The compounds are, however, resistant to cold dilute alkali ⟨B-77MI40700, B-74MI40700⟩.

Ring–chain tautomerism of the type shown in Scheme 88 is well known, but the ring form is generally dominant ⟨80H(14)1313⟩.

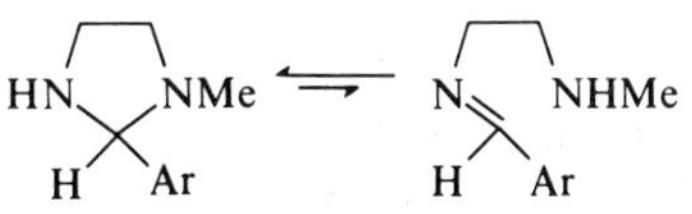

Scheme 88

Aqueous permanganate converts 1,3-dibenzyl-2-phenylimidazolidine into a mixture of benzamide and benzoic acid. When 1-alkyl-3-phenylimidazolidines are subjected to N-oxidation the unstable N-oxides rearrange to give tetrahydro-1,2,5-oxadiazines (**180**) *via* the ring-opened product (Scheme 89) ⟨77LA956⟩.

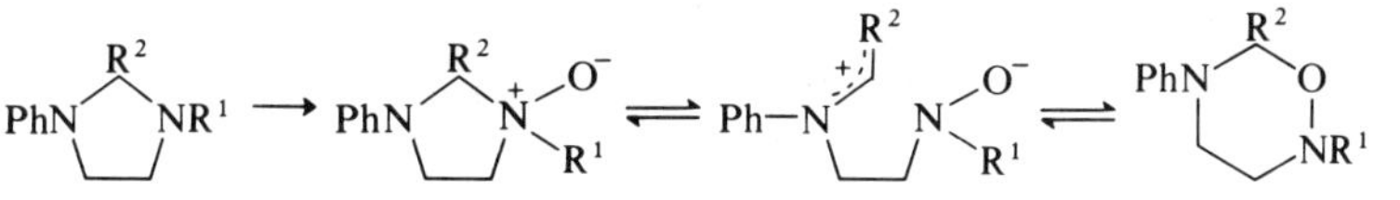

Scheme 89

4.07.2.3.3 Other reactions

Many of the reactions discussed here will be those of oxo and thio derivatives of imidazolidine, although such reactions might also be classified as those of imidazolines or properties of oxygen- and sulfur-linked functional groups.

Alkylation and acylation on ring nitrogen should occur readily with simple imidazolidines; alkylation in particular occurs with some facility in the presence of strong bases.

When 1-substituted 2-iminoimidazolidines are nitrated in acetic anhydride the initial product is the nitrate salt (**181**). This can be further nitrated under the influence of chlorine catalysis (Scheme 90).

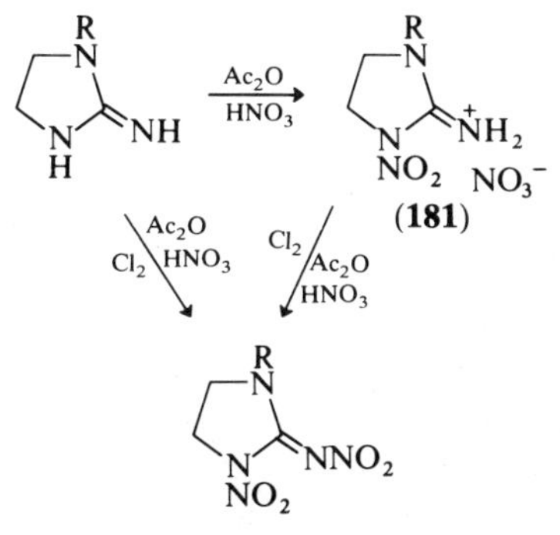

Scheme 90

Treatment of 2-imidazolidinone with a carboxylic acid (and heat) gives a 2-alkyl-2-imidazoline. The corresponding 2-thione (**182**) undergoes nucleophilic displacement by primary amines to form the 2-alkylamino-2-imidazoline; the same result is obtained using a 2-methylthio-2-imidazoline as substrate. In the presence of strong bases (**182**) can be acylated on a ring nitrogen, but methylation occurs on the exocyclic sulfur atom (Scheme 91) ⟨81JCS(P1)2499⟩.

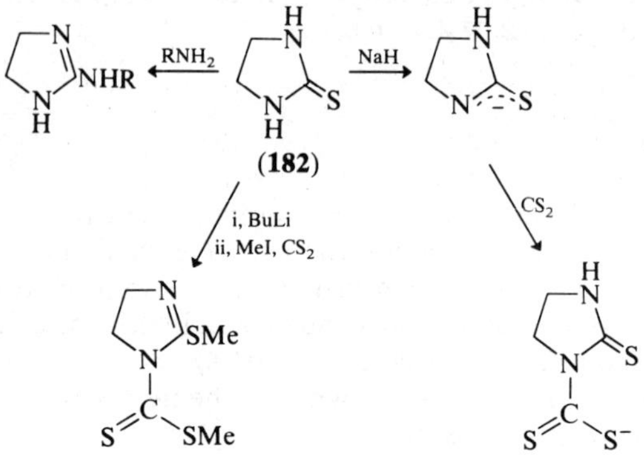

Scheme 91

The hydantoins (2,4-dioxoimidazolidines) when unsubstituted at N-3 are fairly acidic ($K_a = 7.6 \times 10^{-10}$ for the parent compound). The hydrogen atoms at C-5 are also acidic enough to be able to form benzylidene compounds in alkaline solution. This lability of the hydrogens at C-5, while present, is negligible when compared with that of the NH protons. In basic solution oxidation can take place at the 5-position, and the ring is cleaved by initial hydroxide ion attack at C-4. Thus barium hydroxide hydrolysis gives hydantoic acids, with the ease of hydrolysis dependent on substituents at C-5 (Scheme 92). Acetylation and nitration of hydantoin give the 1-substituted products, but halogenation occurs at the 5-position to give products susceptible to facile nucleophilic displacement.

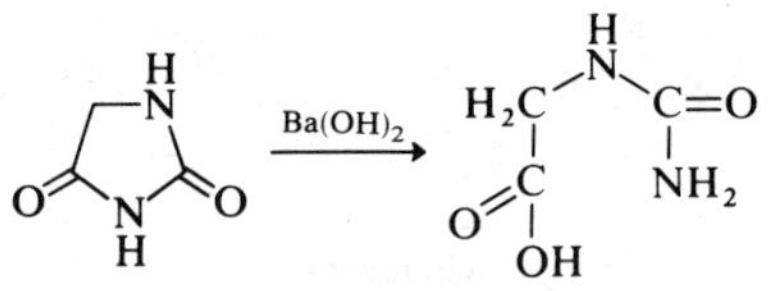

Scheme 92

Parabanic acid (2,4,5-trioxoimidazolidine) has NH groups which are fairly acidic ($K_a = 3.7 \times 10^{-6}$) and have very little basic character. The compound acts as a dibasic acid which forms readily alkylated salts. There are a variety of analogues of parabanic acid in which one or more of the ketone functions is replaced by =S or =NH. Such compounds revert to the oxo compound when treated with peracetic acid. Even hydrolysis with aqueous ethanolic mineral acid serves to convert =NH into =O ⟨79JOC3858⟩.

4.07.3 REACTIONS OF SUBSTITUENTS

Attention is drawn to the general survey of reactivities and reaction types in Chapter 4.02.3.1.

4.07.3.1 Benzenoid Rings

4.07.3.1.1 Fused benzene rings: unsubstituted

Electrophilic reagents preferentially attack benzimidazoles in the fused benzene ring, while nucleophiles react at C-2 which has enhanced nucleophilic activity because of the electron-withdrawal effect of the benzene moiety. The fused aryl ring appears to exhibit less aromatic stability than the heteroring as evidenced by the ready oxidation of benzimidazole to imidazole-4,5-dicarboxylic acid, and by its catalytic reduction over platinum oxide to 4,5,6,7-tetrahydrobenzimidazole. Benzimidazoles (but not 1-methylbenzimidazoles) are also oxidized by chromic acid or by 30% hydrogen peroxide.

The SCF calculations (Section 4.06.2) predict the sequence of electrophilic attack in benzimidazole as 7 > 6 > 5 > 4, although positions 4 and 7, and 5 and 6 are tautomerically equivalent in N-unsubstituted benzimidazoles. Substitution appears to occur with greatest facility in the 5- (or 6-) positions. Bromination in acetic acid of 2-amino-1-methylbenzimidazole occurs at C-6. It can be seen from Table 5 (Section 4.07.1.4.5) that aqueous bromination of benzimidazole itself occurs most readily at C-5, followed by C-7, as it does in the 1- and 2-methyl analogues ⟨78JCS(P2)865⟩. Some of the experimental results are summarized in Scheme 93. Chlorination of both benzimidazole and 2-methylbenzimidazole give the 4,5,6-trichloro product. Nitration and sulfonation, too, occur most readily at position 5 with either sulfuric or chlorsulfonic acids being effective in the latter reaction.

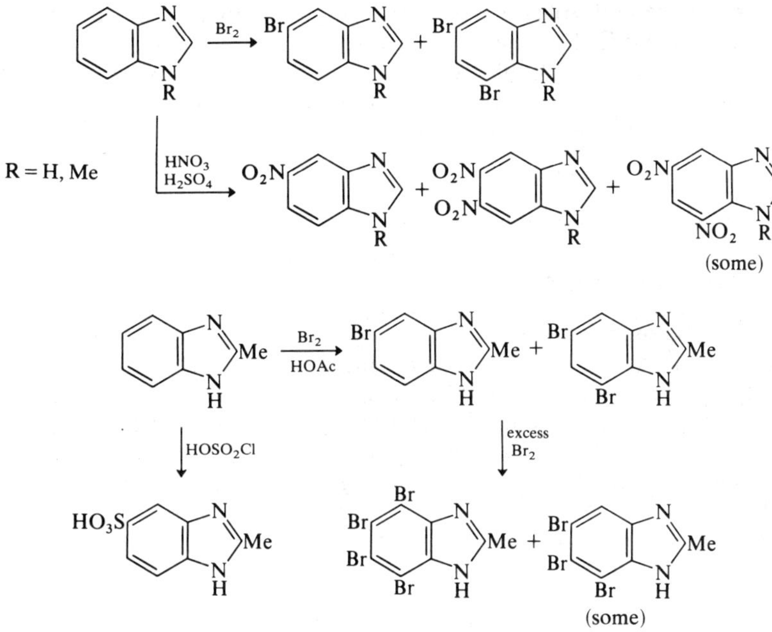

Scheme 93

In acetic anhydride at 30 °C nitration of benzimidazolinone gives 5-nitrobenzimidazolinone; at 70 °C the 5,6-dinitro product is obtained, while fuming nitric acid can give 4,5,6-trinitrobenzimidazolinone ⟨76M1307⟩. In a Friedel–Crafts reaction benzimidazolinone (and its 6-methyl and 1,3-dimethyl derivatives) reacted with phthalic anhydride and $AlCl_3$ to form the 5-(2-carboxybenzoyl) derivatives ⟨74CRV279⟩.

4.07.3.1.2 Fused benzene rings: substituted

Substituents on the benzene ring can play some part in subsequent electrophilic substitution reactions, but the situation is not a simple one. As can be seen in Scheme 93 the normal *meta* directing effect of a nitro substituent can be offset by the directional effect of the fused imidazole ring, as can the usual *ortho–para* directing effect of bromo.

Under the influence of nitrosylsulfuric acid, 2-amino-5- and -6-bromo-1-methylbenzimidazoles (**183**) are converted, as a result of self-coupling, into the 2′,5- (**184**) and 2′,6- (**185**) azobenzimidazoles. Even the 5,6-dibromo analogues can react in a similar way by loss of one of the bromine atoms (Scheme 94) ⟨72CHE1533⟩. Coupling with diazonium salts of 4-amino- and 2-methyl-4-amino-benzimidazole occurs at C-7, while a protected 4-amino group (acetylated) directs nitration into both the 5- and 7-positions. A 4-fluoro substituent likewise leads to 7-nitration ⟨74CRV279⟩.

Oxidation (with such reagents as iron(III) chloride, potassium dichromate, silver oxide or nitrous acid) of 4,7-, 6,7-, and 5,6-dihydroxy-, and 5,6-dimethoxy-benzimidazoles gives the corresponding quinones. A 5-methyl group is oxidized by permanganate to carboxy ⟨74CRV279⟩. In the presence of copper(II)–piperidine or –dimethylamine complexes oxygen converts 5-hydroxybenzimidazoles in 70% yield into 4,5-dioxo products which have also been aminated at C-8. No C-2-amination was noted ⟨78CHE1366⟩.

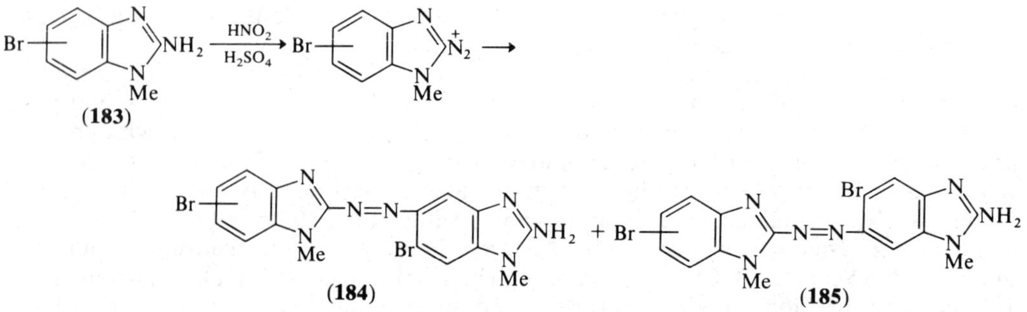

Attempts to carry out nucleophilic substitution of bromine by arylamino failed for 5-bromo-, 5-bromo-6-nitro- or 6-bromo-5-nitro-1-ethylbenzimidazole, as it did likewise for the corresponding benzimidazolium salts ⟨70CHE628⟩.

4.07.3.2 Other C-Linked Substituents

4.07.3.2.1 Saturated alkyl groups

Alkyl groups can be oxidized under a variety of conditions either to the aldehydes or to the carboxylic acids. Reagents such as permanganate or dichromate yield carboxylic acids, but milder conditions can lead to the aldehyde. A 4-hydroxymethyl group can be converted into an aldehyde using selenium dioxide, lead tetraacetate, or active manganese dioxide, with the last named reagent being the one of choice. A 70% yield of imidazole-2-aldehyde was obtained from 2-hydroxymethylimidazole using a sixfold excess of activated manganese dioxide in anhydrous ether, acetone or carbon tetrachloride ⟨70AHC(12)103⟩. Periodate will cleave polyhydroxyalkyl chains to aldehyde in the normal way.

Hydroxyalkyl groups can be converted by thionyl chloride into haloalkyl, or reduced to alkyl.

Methyl groups in the 2-position of imidazole are subject to sufficient electron withdrawal by the flanking nitrogen atoms to act as if they are 'active'. Proton loss in the presence of strongly basic reagents is possible giving rise to the anions. Even in the cationic form of the heterocycle methyl groups at C-4 or C-5 are unaffected; 2-alkyl groups, though, are very prone to proton loss under these conditions, e.g. formation of stable anhydro bases such as the diazafulvenes (**186**; Scheme 95). When there are strongly electron-withdrawing groups at the 4- or 5- positions a C-4 (or C-5) methyl group becomes reactive, but in these circumstances a 2-methyl group reportedly failed to condense with benzaldehyde.

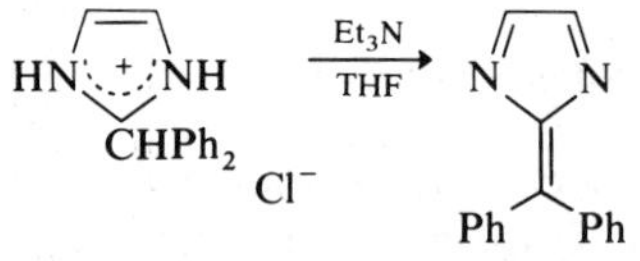

Butyllithium reacts with 1,2-dimethylimidazole at −80 °C to lithiate the 2-methyl group, and at higher temperatures some 5-metalation also occurs ⟨74JOC2301⟩. Reaction of 2-methyl-, 5-chloro-2-methyl- and 2-benzyl-benzimidazoles with two molar equivalents of butyllithium in THF–hexane at 0 °C results in dilithiation, although subsequent reactions at the side-chain carbanion centre also take place (Scheme 96) ⟨73JOC4379⟩.

In the presence of weaker bases than butyllithium or sodamide active methyl groups are still able to take part in a variety of condensation reactions with carbonyl compounds. In fact, only catalytic amounts of the proton acceptor may be needed. Thus, 2-methylimidazole reacts with benzaldehyde to form the 2-styryl compound ((**187**), and 1,2-dimethyl-5-nitroimidazole condenses with dimethylformamide dicyclohexyl acetal to give an 86% yield

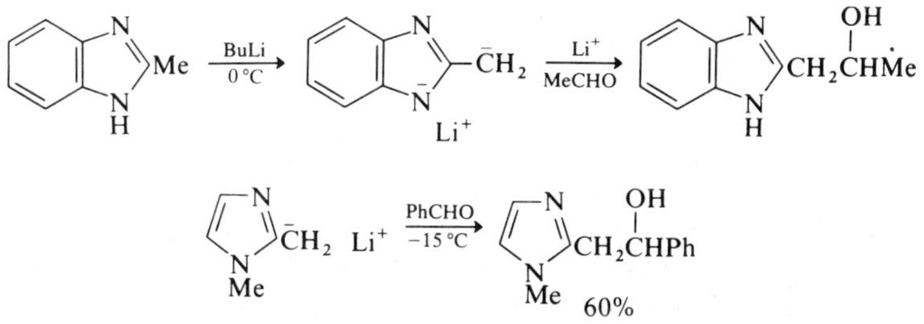

Scheme 96

of (**188**); with benzoyl chloride in the presence of a tertiary amine base the product is the 2-phenacyl derivative (**189**; Scheme 97) ⟨80AHC(27)241⟩. An analogous reaction to the last occurs with 1-benzyl-2-methylimidazole, but with 1-benzyl-5-methylimidazole the product is 2-benzoyl-1-benzyl-5-methylimidazole as a result of substitution at C-2 ⟨75JOC252⟩ (Section 4.07.1.4.6).

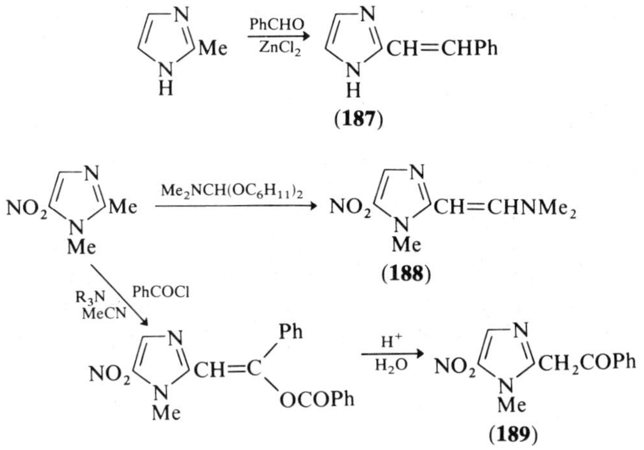

Scheme 97

A recent study of the kinetics of base-catalyzed exchange of *C*-methyl protons in *C,N*-dimethylnitroimidazoles has shown that the most reactive isomer is 1,5-dimethyl-4-nitroimidazole, and the 'least acidic' compounds are 1,2-dimethyl-4-nitro- and 1,4-dimethyl-2-nitro-imidazoles. These results parallel the reactivities in aldol condensations, demonstrating low activity in compounds with nitro and methyl groups '*meta*' to each other (this orientation hinders resonance stabilization of the anion). There is a suggestion that in 1-substituted imidazoles there may be significant localization of the 4,5-bond ⟨81JOC4717⟩.

Related reactions include the formation of the 2-cyano compounds (**190**) when 1,2-dimethyl-5-nitroimidazole is heated with nitrosyl chloride or an *N*-oxide, and when 2-methyl-1-(*o*-nitrophenyl)imidazoles (**191**) cyclize under the influence of iron(II) oxalate (Scheme 98) ⟨74JCS(P1)1970⟩. The last reaction product is contaminated by a large amount of amine reduction product (~64%) but there is also some cyclization with the 4-methyl isomer of (**191**). In the presence of trimethylamine, 2-cyanomethylbenzimidazole condenses with acetone to give the unsaturated derivative (**192**; Scheme 99) ⟨77CPB3087⟩. Neither 2-methylimidazole nor 2-methylbenzimidazole reacts with formamide in the presence of phosphoryl chloride.

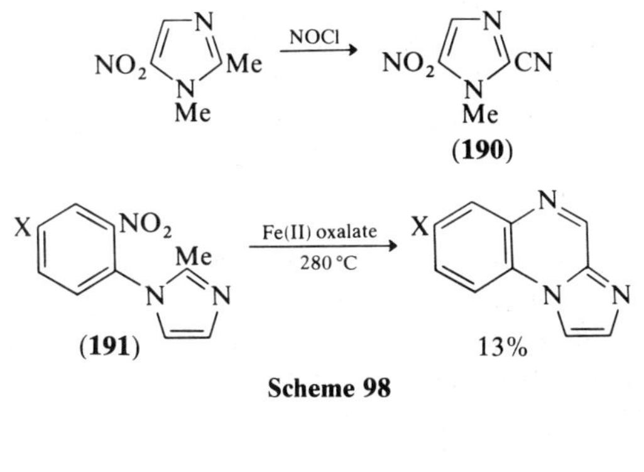

Scheme 98

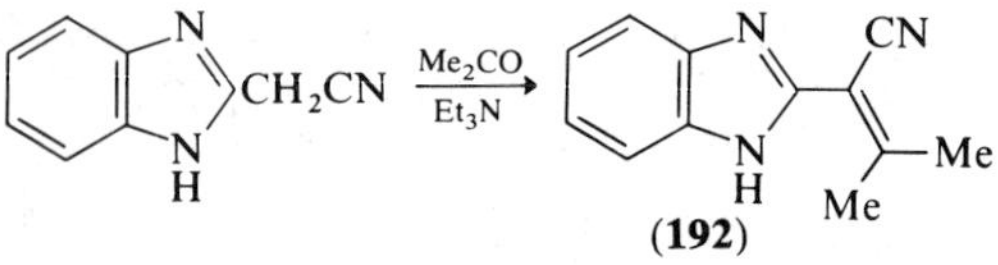

Scheme 99

Examples of side-chain halogenation of alkyl groups are uncommon, but 1,2-dimethyl-5-nitroimidazole reacts with iodine and pyridine to form 30% of the pyridinium iodide (**193**), presumably *via* an initial lateral iodination of the Me group at the 2-position (Scheme 100) ⟨80AHC(27)241⟩.

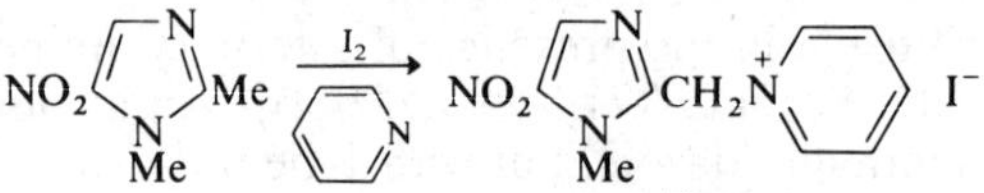

Scheme 100

Substituted alkyl groups attached to an imidazole ring can take part in a wide variety of typical reactions. The halomethyl compounds are subject to nucleophilic substitution reactions forming the corresponding alcohols, amines, cyanides, *etc*. Investigations of the mode of action of 2-trichloromethylbenzimidazoles with nucleophiles have been initiated since these may serve as model compounds in studies related to the possible involvement of benzimidazoles in purine antimetabolite behaviour. The benzimidazoles may act biologically by competitive inhibition of nucleic acid synthesis. Thus, with aqueous ammonia, 2-trichloromethylbenzimidazole forms a mixture of 2-cyanobenzimidazole (50%) and the pyrazine diimine (**194**) (30%). With anhydrous ammonia, only the cyano compound (86%) is obtained (Scheme 101). Primary and secondary amines also react to give amidines and anilides, respectively, but 1-methyl-2-trichloromethylbenzimidazole is much less reactive in such substitution reactions, perhaps implying involvement of the benzimidazole anion ⟨74CRV279⟩.

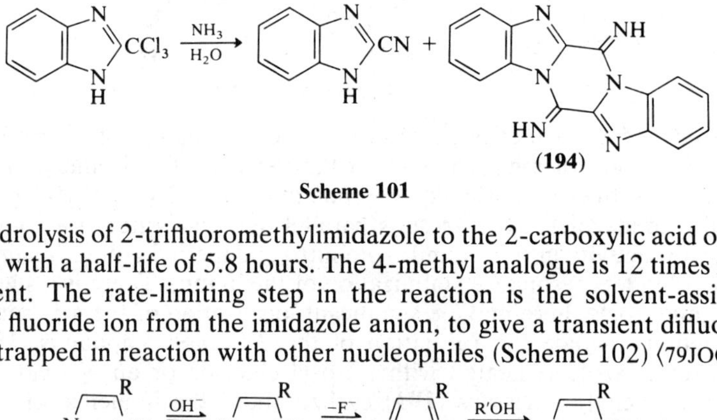

Scheme 101

Alkaline hydrolysis of 2-trifluoromethylimidazole to the 2-carboxylic acid occurs at 30 °C in 0.1M KOH with a half-life of 5.8 hours. The 4-methyl analogue is 12 times more reactive than the parent. The rate-limiting step in the reaction is the solvent-assisted, internal elimination of fluoride ion from the imidazole anion, to give a transient difluoroazafulvene which can be trapped in reaction with other nucleophiles (Scheme 102) ⟨79JOC2902⟩.

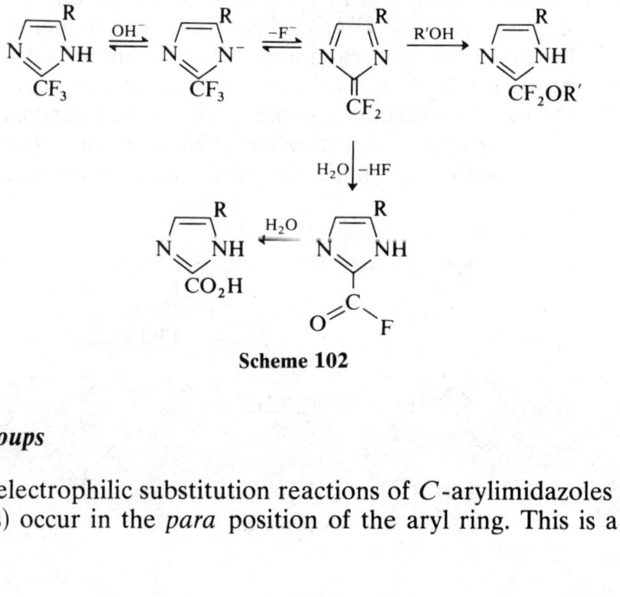

Scheme 102

4.07.3.2.2 Aryl groups

The majority of electrophilic substitution reactions of *C*-arylimidazoles (involving acidic reaction conditions) occur in the *para* position of the aryl ring. This is a consequence of

Table 6 Nitration of C-Arylimidazoles ⟨B-76MI40701⟩

Imidazole	Products	Conditions[a]
2-Ph	50% 2-(p-$NO_2C_6H_4$), 1.5% 2-(o-$NO_2C_6H_4$), 0.2% 2-(m-$NO_2C_6H_4$)	A (100 °C)
1-Me-2-Ph	~43% 1-Me-2-(p-$NO_2C_6H_4$)	A (100 °C)
4-Me-2-Ph	4-Me-5-NO_2-2-(p-$NO_2C_6H_4$)	B
4-CO_2H-2-Ph	52% 4-CO_2H-2-(p-$NO_2C_6H_4$), 19% 4-CO_2H-2-(m-$NO_2C_6H_4$)	A
4,5-(CO_2H)$_2$-2-Ph	52% 4,5-(CO_2H)$_2$-2-(m-$NO_2C_6H_4$), 19% 4,5-(CO_2H)$_2$-2-(p-$NO_2C_6H_4$)	C
4-Br-2-Ph	31% 4-Br-2-(p-$NO_2C_6H_4$)	D
4,5-Br$_2$-2-Ph	63% 4,5-Br$_2$-2-(p-$NO_2C_6H_4$), 1.8% unidentified isomer	D
4-Ph	69% 4-(p-$NO_2C_6H_4$), 25% 4-(o-$NO_2C_6H_4$)	A
4-Ph	4-(p-$NO_2C_6H_4$)	B or C
1-Me-4-Ph	1-Me-4-(p-$NO_2C_6H_4$)	A
1-Me-5-Ph	1-Me-5-(p-$NO_2C_6H_4$)	A
2,4,5-Ph$_3$	2,4,5-(p-$NO_2C_6H_4$)$_3$	B

[a] Conditions: A, nitrate of base added to H_2SO_4; B, fuming HNO_3; C, HNO_3–H_2SO_4; D, solution of base in H_2SO_4 added to KNO_3.

the substrate being the imidazolium species which is rather resistant to electrophilic attack, but which is electron releasing to the aryl substituent. Table 6 lists results of nitration of some aryl-substituted imidazoles. In contrast to the arylpyrazoles the preference for *para* nitration is not nearly so clear-cut. Undoubtedly the nitration proceeds *via* the cationic species in each case, and hence imidazolium acts largely as an *ortho,para* director. The introduction of electron-withdrawing groups into the imidazole ring can modify this directional effect to the extent that *meta* nitration becomes predominant (Scheme 103) ⟨B-76MI40701⟩.

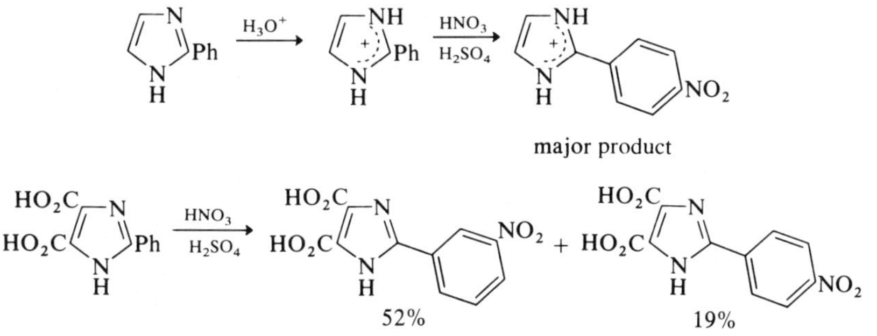

Scheme 103

Substituents in the aryl ring, too, can exert their own directional effects: a *p*-nitro group deactivates the benzene ring and directs further attack into the heterocyclic ring; in 4-(3,4-dichlorophenyl)imidazole nitration is mainly at C-5; in 4-(*p*-methoxyphenyl)imidazole (**195**) subsequent nitration takes place adjacent to the methoxy function and then at C-5 (Scheme 104) ⟨77CHE1110⟩. The nitration of 2-(*p*-fluorophenyl)imidazole has been discussed earlier (Section 4.07.1.4.2).

Oxidation of aryl-imidazoles and -benzimidazoles can give carboxylic acids, in some instances with breakdown of the imidazole ring. Thus, (**196**) is transformed by alkaline permanganate into 4-methoxy-3-nitrobenzoic acid. The effects of oxidizing agents in alkaline medium on triarylimidazoles have been covered earlier in relation to luminescence phenomena (Sections 4.07.1.2.1, 4.07.1.2.3 and 4.07.1.4.8). The polarographic oxidation of such compounds is facilitated by electron donor groups in the aryl rings.

Rearrangement reactions of 3-(imidazol-2-yl)anthranils have been described (Section 4.02.3.4.2).

Irradiation at 313 nm converts 4,5-diphenyl- and 2,4,5-triphenyl-imidazoles into the phenanthroimidazoles (**197**) by a process which may involve hydrogen abstraction by the solvent (Scheme 105) ⟨72JPC3362⟩.

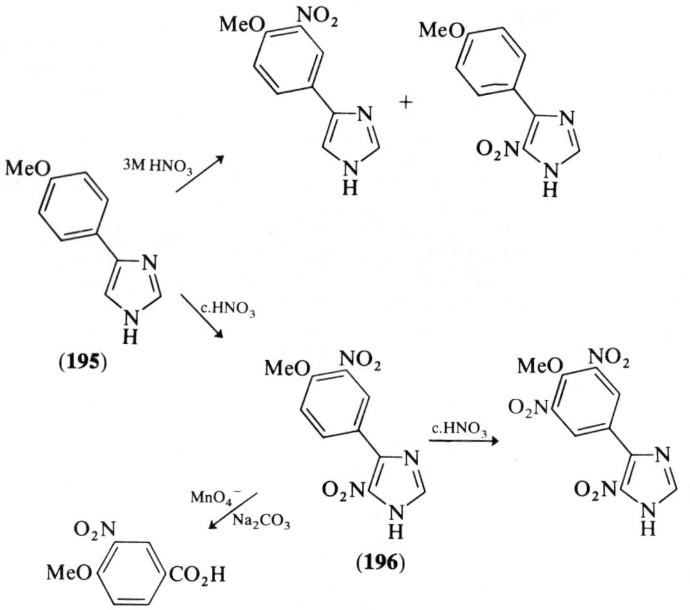

Substituents attached to aryl groups on imidazole or its benzo derivatives have their normal individual reactivities, e.g. reduction of nitro to amino, diazotization of amino, etc., although there may be some modification of these properties because of the particular environment.

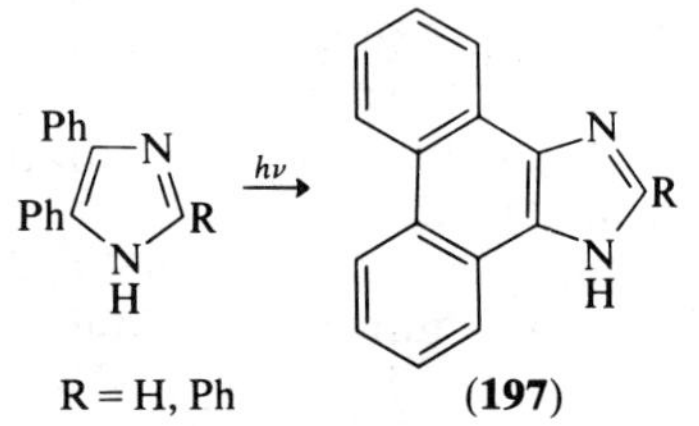

4.07.3.2.3 Other C-linked functionality

(i) Carboxylic acids and derivatives

Imidazolecarboxylic acids are stable, crystalline compounds which form salts with metals, and which may exist in zwitterionic forms such as (**198**; Scheme 106). Such zwitterionic forms have been implicated in the decarboxylation mechanisms of the compounds. The carboxyl functions exert a pronounced base-weakening effect on the parent molecules, but it is surprising that a 5-bromo substituent in imidazole-4-carboxylic acids is not subject to ready nucleophilic displacement by CN^- or SO_3^{2-}.

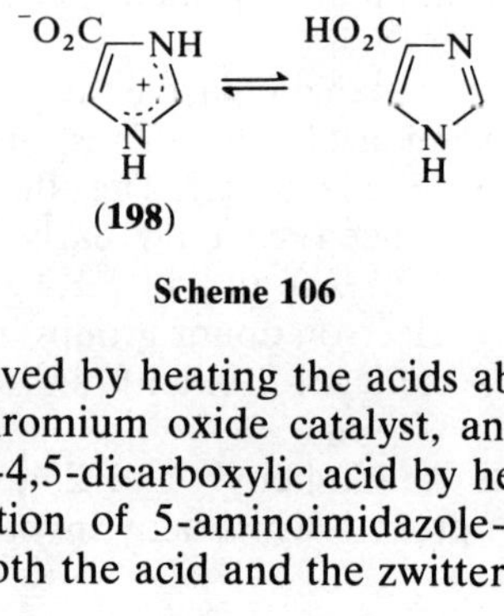

Decarboxylation can be achieved by heating the acids above their melting points, often in the presence of a copper–chromium oxide catalyst, and it is possible to remove one group at a time from imidazole-4,5-dicarboxylic acid by heating the monoanilide. Studies of the kinetics of decarboxylation of 5-aminoimidazole-4-carboxylic acids implicate a first-order decarboxylation of both the acid and the zwitterion; the anion is stable and not decarboxylated. Some evidence exists for catalysis by hydrogen ions at low pH values, and of general base or nucleophilic catalysis. The rate of loss of CO_2 is decreased by complex formation with transition metal ions. When imidazolium-2- and -5-carboxylates are decarboxylated ylides are formed, presumably again through the dipolar tautomers. When imidazole-4-carboxylic acid is treated with iodine in alkaline solution the product is 2,4,5-triiodoimidazole ⟨B-76MI40701⟩. Benzimidazole-2-carboxylic acids decarboxylate readily ⟨75S703⟩. When 1-methylimidazole-4,5-dicarboxylic acid is treated with PCl_5 in thionyl chloride there is some decarboxylation and conversion of carboxyl functions into trichloromethyl ⟨73NKK996⟩.

The usual range of carboxylic acid derivatives can be prepared and interconverted. Both carboxylic acid and ester functions are capable of reduction by lithium aluminum hydride to alcohols, or by controlled potential reduction to aldehydes. Attempts to form the anhydride from imidazole-4,5-dicarboxylic acid by heating with acetic anhydride failed. Instead, compound (**199**) is formed. This product forms the monoester (**200**) when heated with methanol and the hydrazide (**201**) when treated similarly with hydrazine (Scheme 107) ⟨75S162⟩. The corresponding 1-methyl-4,5-dicarboxylic acid loses the 4-carboxyl group when heated with acetic anhydride, but in boiling aniline it is transformed into the 1-methyl-4-carboxanilide ⟨79H(12)186⟩.

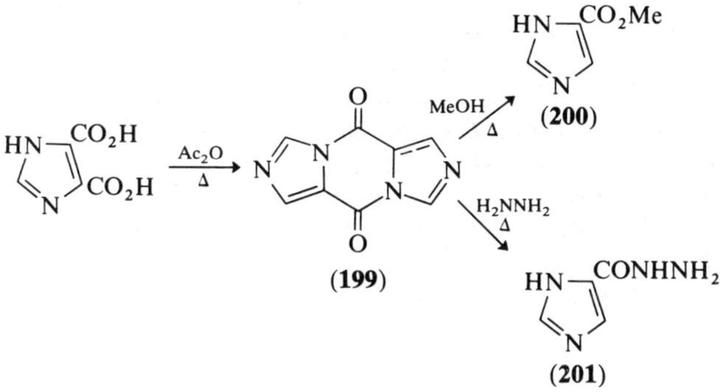

Scheme 107

Imidazole-4-carboxamides do not take part in the Hofmann reaction. 5-Aminoimidazole-4-carboxamide forms a fairly stable diazocarboxamide when treated with nitrous acid; under acid, basic or neutral aqueous conditions this product cyclizes to a triazine. Amide functions can be converted into thioamide which is then susceptible to methylation on the sulfur atom.

Cyanoimidazoles have been of interest because of their utility as starting materials for the synthesis of purines, *e.g.* 4-cyanoimidazole-5-carboxamide. The function reacts in most normal cyanide transformations, *e.g.* with aqueous alkali it forms the amide or the acid, and it can be reduced with $LiAlH_4$, sodium in alcohol or by hydrogenation over Raney Ni. In 1-substituted 4,5-dicyanoimidazoles (**202**) the two groups exhibit different reactivities. Even with 4,5-dicyanoimidazole itself, partial hydrolysis to the monoamide requires 1M NaOH, and stronger base at higher temperatures converts the remaining cyano group

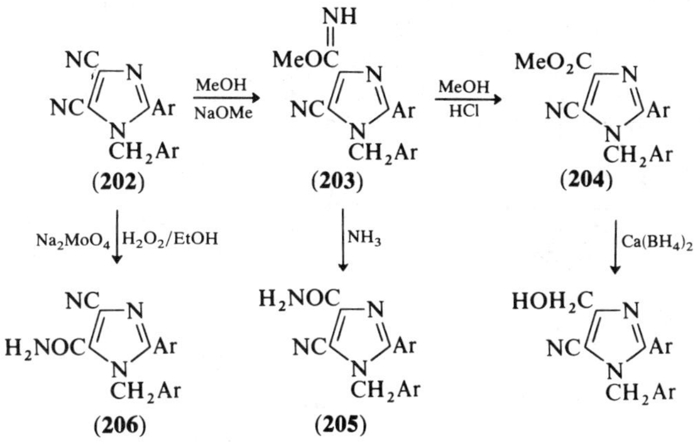

Scheme 108

⟨B-76MI40701, 80AHC(27)241⟩. Regiospecific conversion of the cyano groups in (**202**) may afford purines of unequivocal structure. Thus, methanol in sodium methoxide selectively adds at the 4-cyano group to give the methyl imidate (**203**) which can be converted into either the ester (**204**) or the amide (**205**); selective reduction of (**204**) to the hydroxymethyl compound has been accomplished. On the other hand (**202**) can be directly hydrolyzed to the 5-carboxamide (**206**), thus making the two different positional isomers of cyanoimidazolecarboxamide (**205** and **206**) available for purine synthesis (Scheme 108).

The zwitterion, or carbene, derived from 4,5-dicyano-2-diazoimidazole reacts with acetic acid to give 4,5-dicyanoimidazolin-2-one ⟨79JOC1717⟩. Photolysis of 4-amino-5-cyanoimidazole gives 1,6-dihydroimidazo[4,5-d]imidazole (**207**; Scheme 109) ⟨74JA2010⟩.

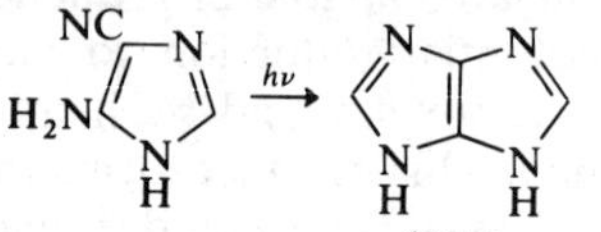

Scheme 109

(ii) Aldehydes and ketones

Acyl-substituted imidazoles have distinctive UV and IR spectra, and can exist as hydrates in solution, *e.g.* imidazole-2-carbaldehyde. The aldehyde group is sufficiently electron withdrawing to assist nucleophilic displacement of an adjacent halogen atom. The normal aldehyde derivatives such as oximes, acetals and hydrazones can be formed, and the kinetics of oxime formation with the 4-carbaldehyde have been studied. The tautomeric ratio (zwitterion:uncharged aldehyde form) of 1-substituted imidazole- and benzimidazole-2-aldoximes rises in parallel with an increase in basicity of the parent molecule ⟨73CHE1074⟩. These compounds have been shown by NMR studies to adopt the *syn* configuration. With hydrazine hydrate, imidazole-4,5-dicarbaldehydes give imidazo[4,5-d]pyridazines (**208**; Scheme 110).

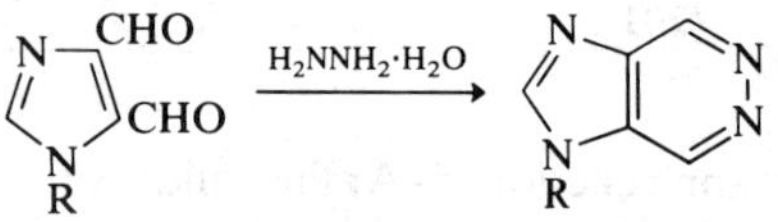

Scheme 110   **Scheme 111**

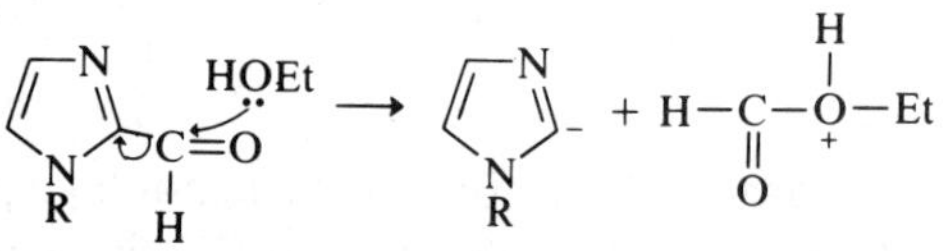

When the 2-aldehyde of imidazole (**209**) is treated with ethanol some decarbonylation takes place by way of a nucleophilic attack of the alcohol on the carbonyl carbon (Scheme 111). The products isolated are imidazole and ethyl formate. One would normally expect that acetal formation would result under these conditions, but it appears that in the acid-catalyzed formation of acetals from (**209**) the yield of acetal depends on the amount of catalyst. When the ratio of H^+:imidazole is >1, then the diethyl acetal is produced in high yield. When this ratio falls to less than unity there is some decarbonylation, and in the absence of acid only decarbonylation occurs. Electron-withdrawing groups in the heterocyclic nucleus (or quaternization) assist this reaction which does not appear to occur with 4- or 5-aldehyde functions ⟨80AHC(27)241⟩.

Both imidazole-2- and -4-carbaldehydes resist the Cannizzaro, Perkin and benzoin reactions, but the 1-substituted compounds react normally. 1-Substituted 5-nitroimidazole-2-aldehydes condense readily with alkyl aryl ketones. Examples of successful condensations

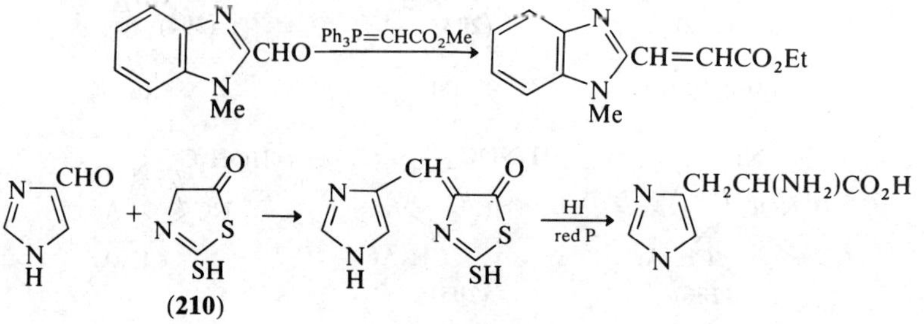

Scheme 112

with carbanions, active methylene and similar compounds include reaction with nitroalkanes, with Grignard reagents ⟨73JOC3495⟩, with phenyllithium in THF, and a number of Wittig reactions (Scheme 112) ⟨74CHE358⟩. In the synthesis of histidine use is made of the reaction of imidazole-4-carbaldehyde at the active methylene of 2-mercapto-5-thiazolone (**210**; Scheme 112).

Hydroxylamine-*O*-sulfonic acid converts an imidazole-2-aldehyde into a mixture of the carboximidamide and the 2-cyanoimidazole ⟨79JHC871⟩.

Most usual oxidizing agents act normally with imidazole aldehydes and ketones but 1-benzylimidazole-2-carbaldehyde is reportedly somewhat resistant to selenium dioxide oxidation. Reduction of ketone functions under Clemmensen and Wolff–Kischner conditions is usually successful. Zinc dust and acetic acid reduce acetyl groups to a mixture of secondary alcohol and ethyl; borohydride gives the alcohol exclusively ⟨B-76MI40701⟩.

(iii) Unsaturated alkyl groups

Studies of alkenyl- and alkynyl-imidazoles in recent years have increased because of the interest in the polymerization of these compounds under the influence of free radical initiators.

Oxidation of the substituents usually gives carboxylic acids, or sometimes the aldehydes. Thus, ozonolysis of 1-methyl-5-nitro-2-styrylimidazole gives the corresponding 2-aldehyde ⟨75LA1465⟩.

While addition reactions occur normally with some unsaturated derivatives, there are instances in which this is not the case. 2-Vinylimidazole resists epoxidation and hydroboration, and this resistance is even more marked in the *N*-substituted compounds (Section 4.07.3.8.3).

The vinyl- and ethynyl-substituted compounds are able to participate in Michael addition reactions with a variety of nucleophiles, particularly if the group is attached to C-2. When the sodium salt of 2-ethynyl-1-methylbenzimidazole is treated in liquid ammonia with an aliphatic alcohol, the vinyl ether (**211**) is produced, but KOH in ethanol surprisingly gives 1,2-dimethylbenzimidazole (Scheme 113) ⟨75JOU1130⟩.

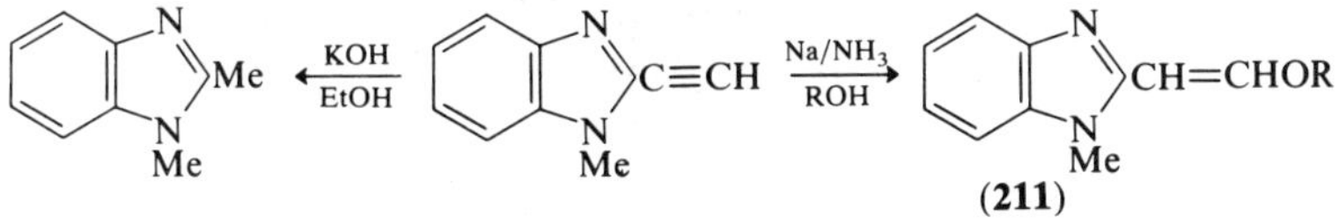

Scheme 113

A number of the condensation reactions of 2-ethynyl-1-methylbenzimidazole are summarized in Scheme 114. The oxidative self-condensation of the compound can take two courses depending on the reaction solvent ⟨74CHE1491⟩. Under Favorskii reaction conditions (KOH in ether) the expected product (**212**) with acetone is not formed. Instead, further

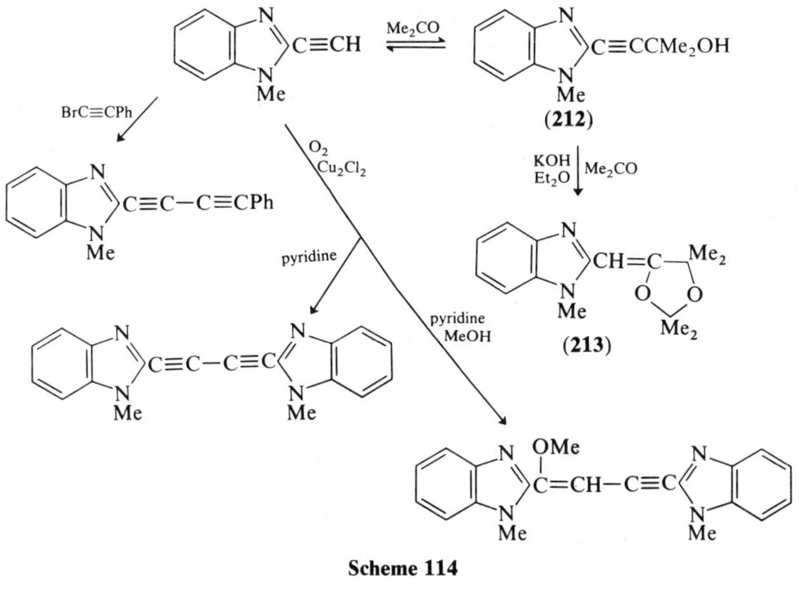

Scheme 114

rapid condensation with another molecule of acetone gives the dioxolane (**213**) ⟨75CHE123⟩. Some of these reactions are a function of the acidic nature of the ethynyl group, which is a weak electron acceptor when attached to azoles and which is more acidic than acetylene itself in these situations ⟨75CHE718⟩.

Both acrylic and alkynic acids of imidazoles decarboxylate normally on heating, and bromoacrylic acids in the series are dehydrobrominated under the influence of ethanolic alkali ⟨71CHE132⟩.

4.07.3.3 *N*-Linked Substituents

4.07.3.3.1 *Amino compounds*

Aminoimidazoles are not particularly well-described compounds (Section 4.07.1.1.4). There are synthetic difficulties involved, and 4-amino and 1-substituted 5-amino compounds are often unstable; they are certainly much more labile than the 2-aminoimidazoles ⟨81JOC4717⟩.

(i) Dimroth rearrangement

Very few examples of this reaction exist among the imidazoles. Nevertheless, treatment of 4-amino-1-methyl-2,3-diphenylimidazolium chloride (**214**) with warm, dilute potassium hydroxide solution gives 4-anilino-1-methyl-2-phenylimidazole (**215**; Scheme 115) ⟨71JOC3368⟩. The thermolysis of a 2-amino-1-aroyl-3-methylbenzimidazolium chloride, which gives a 2-aroylamino-1-methylbenzimidazole, provides another example ⟨74CHE1225⟩. Heating in concentrated aqueous ammonia converts ethyl 5-methylaminoimidazole-4-carboxylate into 5-amino-1-methylimidazole-4-carboxamide, again via a Dimroth product ⟨79JCS(P1)3107⟩.

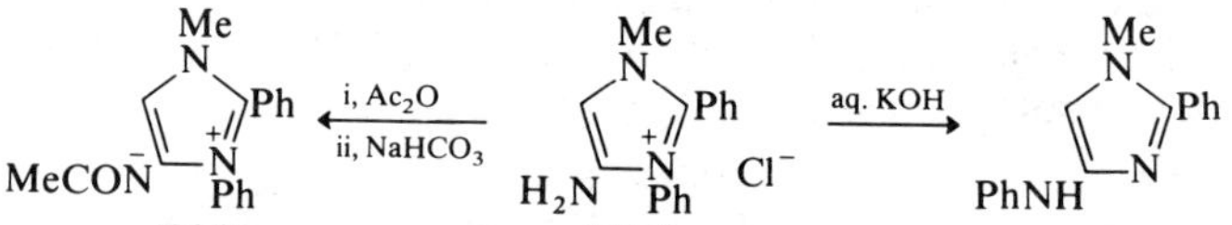

Scheme 115

(ii) Reactions with electrophiles

Whereas 2-aminoimidazole is a monoacidic base which protonates on the annular nitrogen, 4-aminoimidazole can form dihydrochlorides and dipicrates. In this respect 4-aminoimidazoles resemble true aromatic amines. Even the salts of these compounds, however, are relatively unstable in aqueous solution, and 4-aminoimidazole itself is deaminated and undergoes ring fission under the usual Van Slyke amino–nitrogen determination conditions. Usually alkylation and acylation occur on a ring nitrogen atom, but if initial reaction at this site gives rise to an unstable product the ultimate product may turn out to be the exocyclically substituted one. Consecutive substitution at ring and exocyclic nitrogens is possible. Thus, the imidazole quaternary salt (**214**) acetylates on the amino function and subsequent treatment of the product with bicarbonate gives the mesoionic system (**216**; Scheme 115). 2-Aminoimidazole and some of its derivatives give products of acetylation and aroylation which are almost certainly 2-acylaminoimidazoles; 4-aminoimidazoles can give 4-acylamino compounds and sometimes diacylated products ⟨B-76MI40701⟩.

The acylation of 2-amino-1-methylbenzimidazole with unsaturated acid chlorides gives the 3-acyl quaternary salt (**217**). With triethylamine in acetonitrile at 10–15 °C this dehydrochlorinates to give the 2-imino derivative (**218**), and both (**217**) and (**218**) can rearrange to give the exocyclic amide (**219**) which is capable of cyclization (Scheme 116) ⟨78CHE73⟩. Other similar examples exist ⟨74CHE1225⟩ in which an increase in temperature is generally sufficient to promote the acyl rearrangement — treatment with base gives the imino product.

In the presence of sodamide the anionic form of 2-amino-1-ethylbenzimidazole is substituted by alkyl halides on both the annular and exocyclic nitrogens. With butyl and isopropyl iodides the proportion of dialkylated product is increased. The synthetic utility of such nucleophilic reactions of 2-aminobenzimidazole is exemplified by reactions with ethyl cyanoacetate, acetoacetic ester and ethyl benzoylacetate, when subsequent cyclization of the initial products also gives pyrimidobenzimidazole derivatives (Scheme 117).

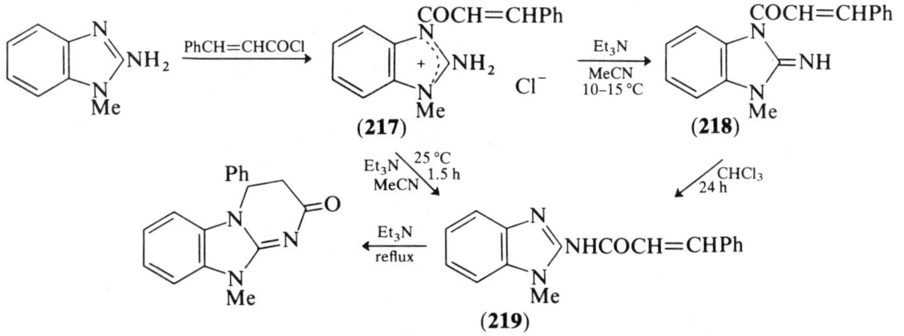

Scheme 116

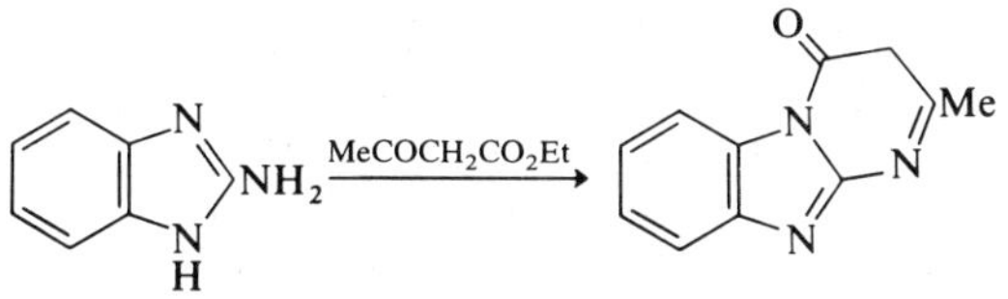

Scheme 117

(iii) Diazotization

Diazonium salts of both 2- and 4-aminoimidazoles have been prepared, with a strongly acidic medium being preferred for the reactions. On occasion it is possible to isolate the intermediate nitrosoamines which are resistant to the action of dilute acid on storage, but which are subject to denitrosation when heated in alcohol solutions. The nitrosoamines can be converted into the diazonium salts. Diazonium fluoroborates have proved synthetically valuable for they can be transformed into the 2-nitro-, 2-azido- and 2-fluoro-imidazoles ⟨80AHC(27)241⟩. The derived diazonium salts are also capable of coupling reactions (Scheme 118) ⟨72CHE1533⟩.

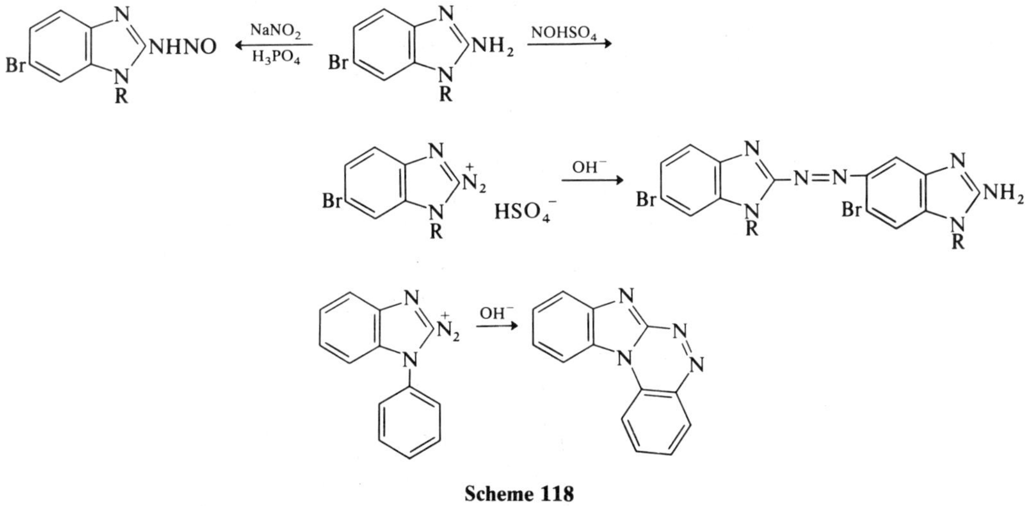

Scheme 118

(iv) Condensation reactions with carbonyl compounds

It has been reported that 2-aminoimidazoles do not react readily with aromatic aldehydes to form Schiff's bases, and even though the 4-amino compounds have more aromatic amine character, their instability has meant that there have been few studies of these reactions. Triethyl orthoformate reacts with 5-amino-4-cyano-1-methylimidazole to give (**220**) which is cyclized by ammoniacal ethanol to 9-methyladenine ⟨78JCS(P1)1381⟩. The cyclization of 2-amino-1-methylimidazole with an α-bromoketone provides another example. There are also a number of examples of lactam formation as when polyphosphoric acid reacts with 5-amino-1-imidazolylacetic acid, or when 2-aminobenzimidazole and propiolic esters interact (Scheme 119) ⟨73JCS(P1)1588⟩.

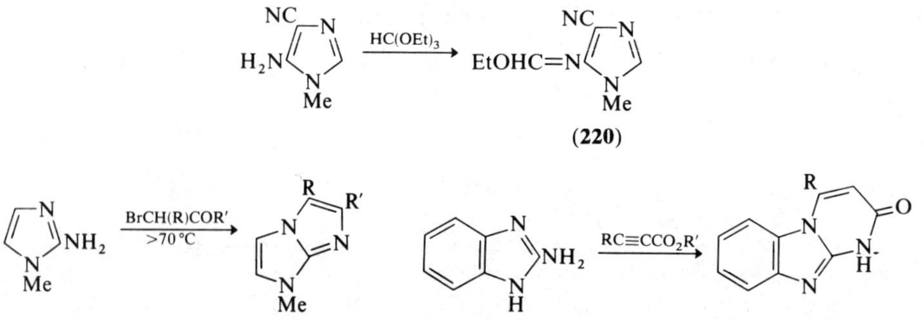

440   Imidazoles and their Benzo Derivatives: (ii) Reactivity

(220)

Scheme 119

(v) Oxidation

In the presence of atmospheric oxygen 2-amino-1-benzylbenzimidazole reacts with two equivalents of sodium metal to give 2-aminobenzimidazole and 2,2′-azodibenzimidazole (**221**; Scheme 120). When four equivalents of sodium are present the products are (**221**) (55%) and 2-nitrobenzimidazole (45%) ⟨70CHE1183⟩. The proposed mechanism for such oxidations, based on such circumstantial evidence as the unreactivity of 1-alkyl analogues, the requirement for the presence of atmospheric oxygen in the process, and the apparent requirement for a large excess of 1-benzylbenzimidazole for the oxidation of 2-aminobenzimidazole, is shown in Scheme 120. The last requirement suggests that benzylsodium plays a significant role in the process.

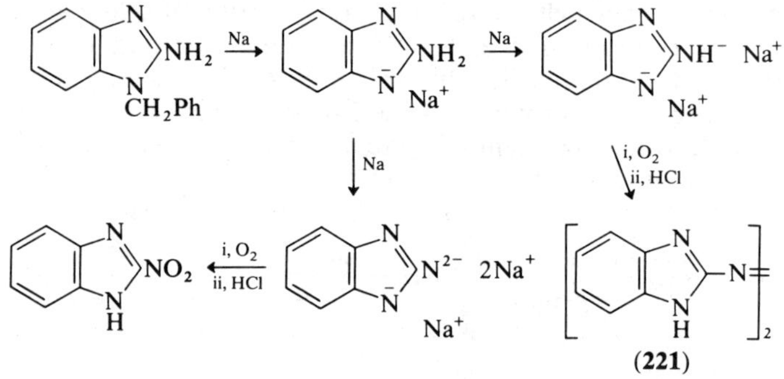

Scheme 120

1,2-Diamino-4-phenylimidazole is oxidized by manganese dioxide to a mixture of 4-phenyl-1,2,3-triazole and 3-amino-5-phenyl-1,2,4-triazine ⟨80AHC(27)241⟩. A *C*-nitrene intermediate (**222**) is probably involved, although an *N*-nitrene intermediate cannot be ruled out (Scheme 121).

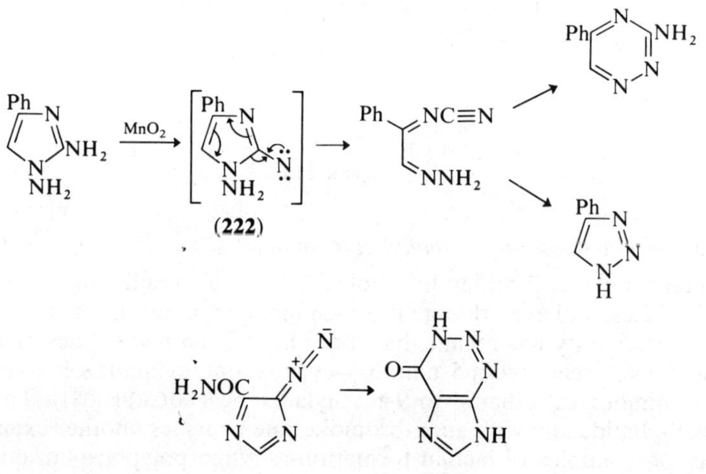

Scheme 121

4.07.3.3.2 Aminoazolium ions

5-Diazoimidazole-4-carboxamide is an explosive compound which cyclizes to 2-azahypoxanthine (**222a**; Scheme 121) ⟨67AHC(8)1⟩.

4.07.3.3.3 Imines

Proton loss from cationic species of type (**217**; Scheme 116) can give rise to relatively stable imines which can be alkylated or acylated with some facility. However, imines are much more common in compounds which are not in tautomeric equilibrium with fully aromatic imidazoles, and among the imidazolidines.

4.07.3.3.4 Nitro groups

IR studies have shown that strongly electropositive metals form salts with nitroimidazoles in which the negative charge is largely associated with the nitro function (Section 4.06.3.7). Nitroimidazoles dissolve in aqueous solutions of alkali metal hydroxides, carbonates or ammonia to form yellow solutions containing the nitroimidazole anion ⟨B-76MI40701⟩.

While there are a number of examples of nucleophilic displacement of nitro groups attached to imidazole rings (Section 4.07.1.5), the nitro groups themselves facilitate the nucleophilic displacement of other functional groups, and also have a profound effect on the orientation of such reactions as alkylation (Section 4.07.1.3). The strongly base-weakening effects are well known and may even prevent the formation of salts with acids.

Reduction of nitro groups can be accomplished using either chemical or catalytic means. Such reagents as zinc and acetic acid, tin and HCl, or hydrogen with Raney nickel, Adams catalyst or palladium–carbon, in addition to hydrazine with the latter catalyst, usually reduce a nitro group smoothly to the amino group. Problems which arise are often a function of the instability of the reduction products. Thus, when 2-nitroimidazole is reduced with Adams catalyst an unstable reduction product is formed. Nitro groups attached to the benzene ring of benzimidazole have been reduced with hydrazine and Raney nickel.

When ethylene oxide is used to alkylate 2,4-dinitroimidazole, 60% of the N-alkylated product is formed, but 25% of the product is a bicyclic nitroimidazo[2,1-b]oxazole (**223**) which must be formed as a result of intramolecular nucleophilic displacement of the 2-nitro group (Scheme 122) ⟨79JHC1499⟩.

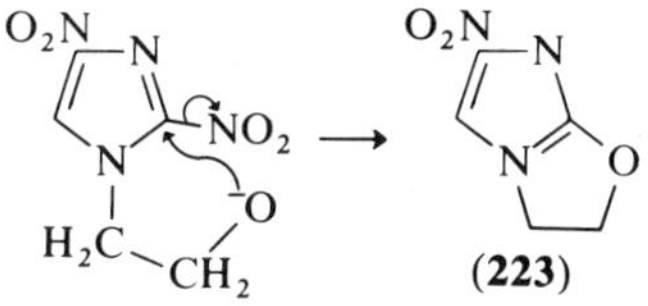

Scheme 122

4.07.3.3.5 Nitroso groups

Although not particularly well known, nitrosoimidazoles appear to be quite stable compounds. The nitroso function can be reduced to amino, or oxidized to nitro. When 5-nitroso-2,4-diphenylimidazole is subjected to dropwise treatment with phenylhydrazine there is some reduction, but ring modification with the formation of the oxadiazole (**224**) accompanies this reaction (Scheme 123) ⟨60G831⟩. The nitroso function is able to take part in condensation reactions with compounds which possess active methylene groups, and related species ⟨B-76MI40701⟩.

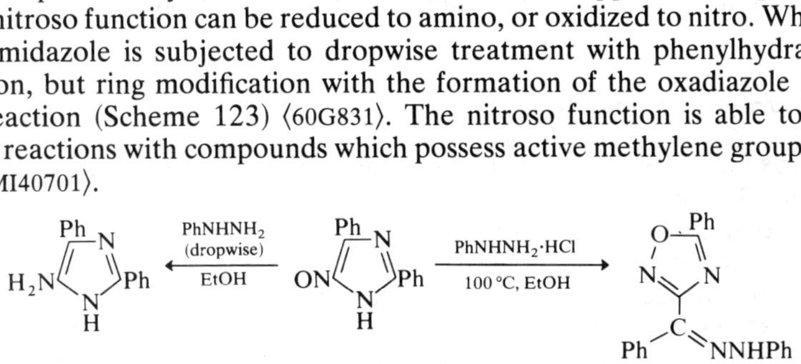

Scheme 123

4.07.3.3.6 Azidoazoles

Azido–tetrazole tautomerism has been discussed earlier (Section 4.06.5.3).

Many of the compounds which are known are crystalline, but they can decompose (often explosively) on melting. They exhibit an IR absorption band in the 2300 cm^{-1} region.

Reaction of sodium metal in liquid ammonia with 2-azidoimidazoles (or 2-azidobenzimidazole) gives the N-dianions (**225**) which methylate very readily (Scheme 124) ⟨71CHE1156⟩. Other reactions of 2-azido-1-methylbenzimidazole are also summarized in this scheme.

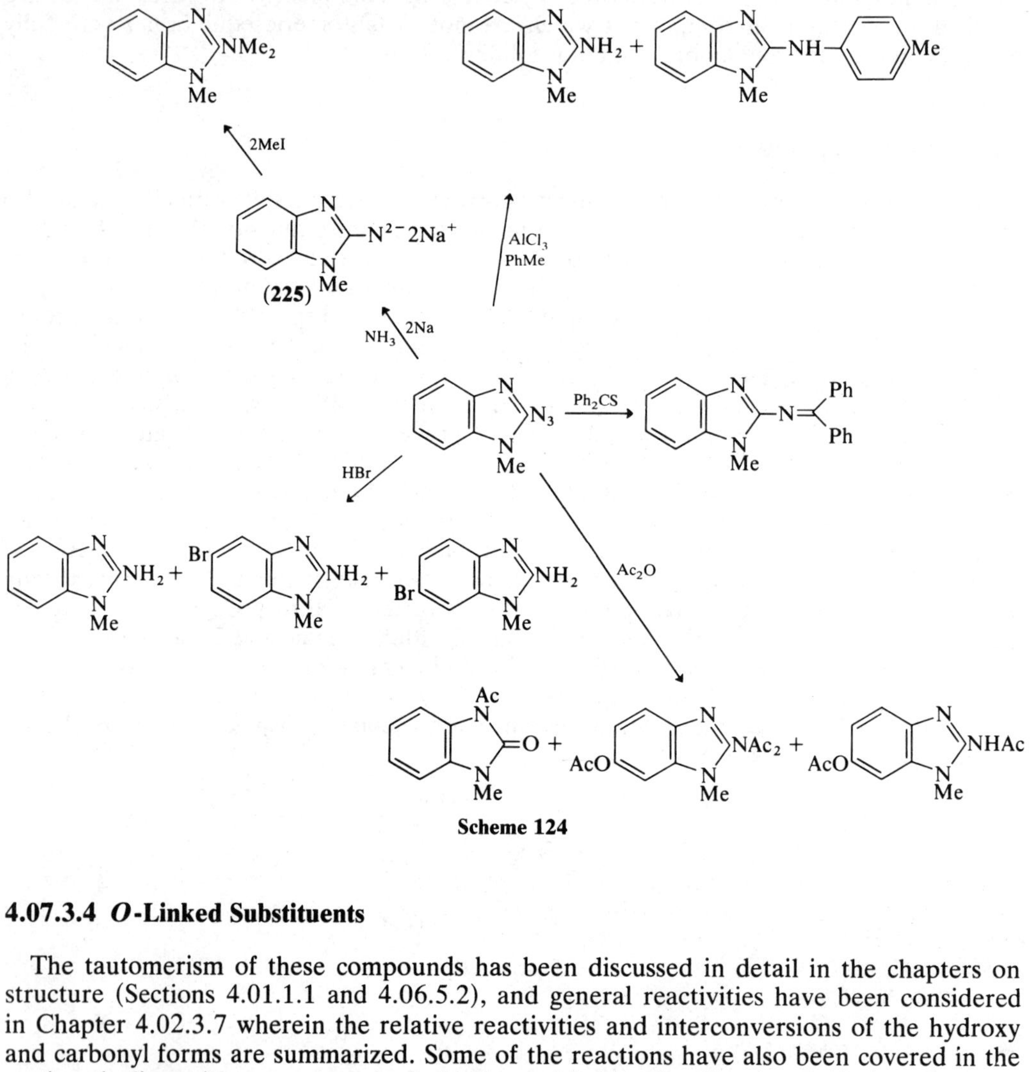

Scheme 124

4.07.3.4 O-Linked Substituents

The tautomerism of these compounds has been discussed in detail in the chapters on structure (Sections 4.01.1.1 and 4.06.5.2), and general reactivities have been considered in Chapter 4.02.3.7 wherein the relative reactivities and interconversions of the hydroxy and carbonyl forms are summarized. Some of the reactions have also been covered in the section dealing with non-aromatic derivatives of imidazoles (Section 4.07.2). Discussion here will be limited to reactions which do not lead to ring fission.

4.07.3.4.1 Hydroxyl compounds

The 2-hydroxyimidazoles are acidic compounds which have lost the normal imidazole basic properties. In dilute alkaline solution they form salts, and they are commonly stable, crystalline compounds with high melting points. Stability to acid hydrolysis is typical and there are few indications of phenolic properties, the compounds best being regarded as imidazolinones (or benzimidazolinones). The imidazolin-4 (or 5) -ones, though, have both amidine and keto–enol systems. Accordingly they are amphoteric compounds forming, for example, both picrates and silver salts, and they are rather less stable than the 2-isomers. Again the keto forms are preferred.

Resistance to reduction processes seems to be a general characteristic as most catalytic methods (as well as sodium in ethanol) reduce only the ring. However hydantoin can be reduced by diisobutylaluminum hydride to imidazolin-2-one ⟨81TL2063⟩, and imidazoline-2-thiones can be prepared from 2-thiohydantoins ⟨70AHC(12)103⟩. Oxidative procedures often result in ring opening ⟨B-76MI40701⟩.

The oxygen function is strongly activating for electrophilic substitution. In consequence, imidazolin-2-ones are readily substituted at C-4 and C-5, while imidazolin-4-ones couple with diazonium salts. Nucleophilic displacement of the oxygen functions has been discussed elsewhere (Section 4.07.1.5).

Both acylation and alkylation occur most readily on an annular nitrogen, but there are examples of O-acylation which have been reported (albeit tentatively). The acetyl derivative of 4-methyl-2-phenylimidazolin-5-one has been formulated as 4-acetoxy-5-methyl-2-phenylimidazole because of its basic properties, and derivatives of 1-substituted imidazolin-5-ones are also probably the O-acetyl compounds ⟨70JHC961, 71CHE746⟩. From the limited results available it appears that alkylation of imidazolin-2-ones is confined to the annular nitrogens; nucleophilic displacement of halogen is the most usual approach to the synthesis of alkoxy-substituted imidazoles. 1-Acyl-2-imidazolidinones, however, are alkylated on the exocyclic oxygen by oxonium fluoroborates ⟨77JOC941⟩. One of the most notable differences in reactivity between the oxygen and sulfur derivatives of imidazoles is in their reactivity with alkylating and acylating reagents.

4.07.3.4.2 Alkoxy and aryloxy compounds

When 2-alkoxy-1-methylimidazoles are heated they rearrange to alkylimidazolin-2-ones. The reaction occurs also with 2-alkoxybenzimidazoles. The 2-allyloxy compounds undergo a similar Claisen rearrangement but 15–20 times more rapidly (Section 4.02.3.7.7). At 150 °C in the absence of light (in nitrobenzene with AlCl₃) 2-aryloxybenzimidazoles give no Fries-rearranged products, but photorearrangement leads to 2-(2- and 4-hydroxyphenyl)benzimidazoles (**226, 227**; Scheme 125) ⟨73BCJ2600⟩.

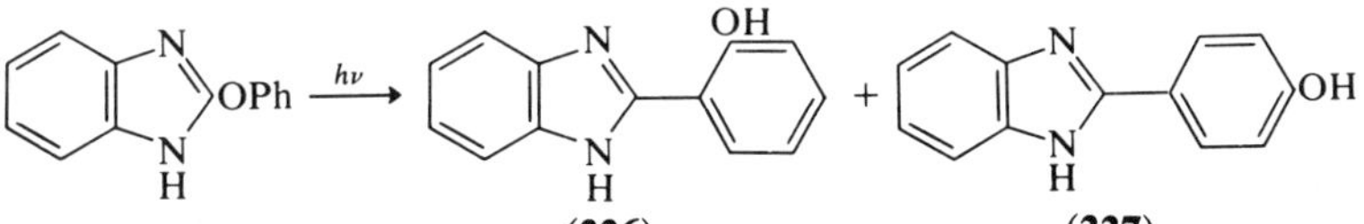

Scheme 125

The kinetics and mechanism of the transalkylation reaction between 2-alkoxy-1-methylbenzimidazoles and thiophenol have been studied. The results demonstrate a rapid protonation of the azole followed by an S_N2 reaction which forms the benzimidazolinone. The basicity of the benzimidazole and the pH both affect the reaction rate (Scheme 126). The corresponding sulfur analogue, 1-methyl-2-methylthiobenzimidazole, does not react with thiophenol, probably because of the poor electron-donating ability of the sulfur atom of thiophenate ⟨71JCS(B)2299⟩.

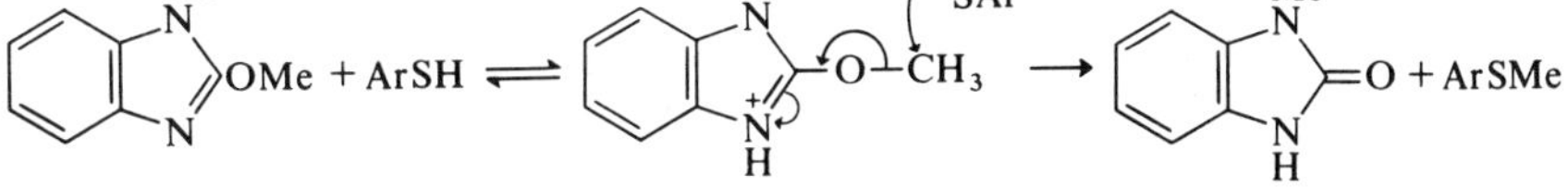

Scheme 126

4.07.3.5 S-Linked Substituents

4.07.3.5.1 Thiones

Compounds with a potential SH group exist predominantly in the thione form (Sections 4.02.3.8 and 4.06.5.2). The UV maxima for the thiones lie in the range 260–270 nm, shifted to shorter wavelengths in the quaternary salts. Imidazoline-2-thiones are high-melting solids which form stable salts in aqueous alkali, but they are too weakly basic to form stable picrate or hydrochloride salts. The compounds can indeed be quite strong acids, e.g. 5-mercapto-1-phenylimidazole-4-carbaldehyde, $pK_a = 4.63$ (admittedly the formyl group is here exhibiting a strong effect).

In contrast to the oxy analogues, imidazolinethiones are much more subject to alkylation on the exocyclic atom. Heating directly with an alkyl halide, or heating the sodium salt of a thione with an alkylating agent both give S-alkyl products. Dialkylation giving N,S-dialkylimidazoles can occur. Thus, methyl bromoacetate alkylates 1,3-dimethyl-imidazoline-2-thione to give the exocyclic quaternized product (228), which forms the quasi-Wittig reagent (229) when treated with base (Scheme 127) ⟨78JA6538⟩. Under phase-transfer conditions imidazolidine-2-thione is alkylated both on nitrogen and sulfur by 1,2-dibromoethane to give a bicyclic product. With 3-bromopropyne, benzimidazoline-2-thione gives the S-propargyl product (230) ⟨74TL2643⟩, while S-alkenyl compounds can be made by heating imidazoline- or benzimidazoline-2-thiones with alkynes, either under pressure or in the presence of heavy metal salts ⟨74CHE1217, 76CHE433⟩. Both N- and S-products are produced. The addition of alkynic esters to benzimidazoline-2-thiones provides yet another example of S-alkylation (Scheme 127) ⟨79TL53⟩.

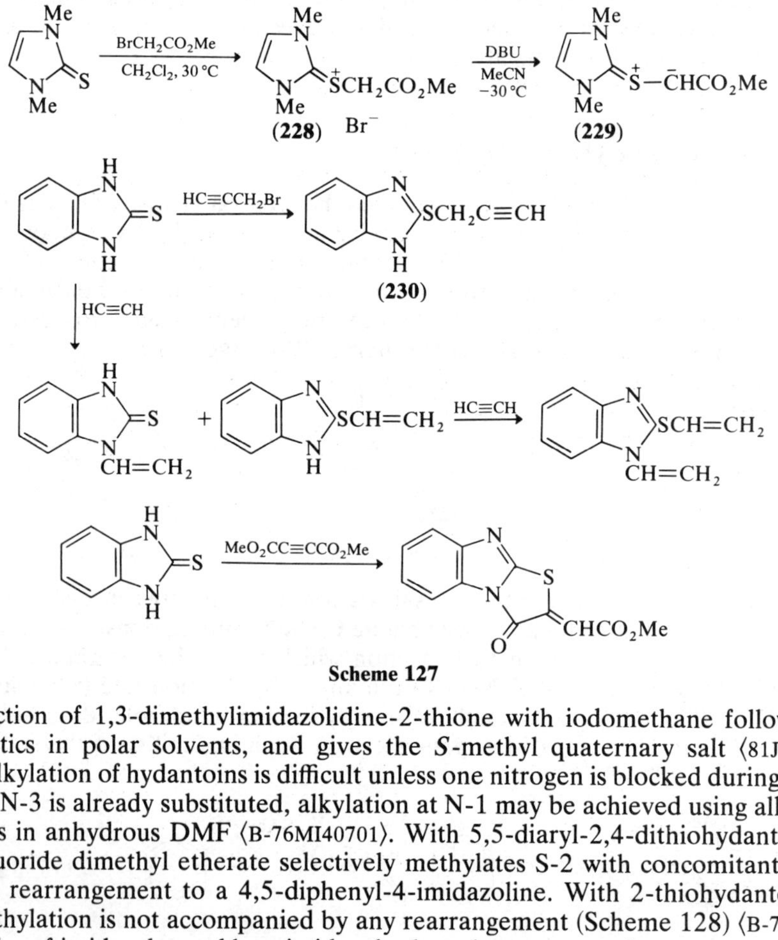

Scheme 127

The reaction of 1,3-dimethylimidazolidine-2-thione with iodomethane follows second order kinetics in polar solvents, and gives the S-methyl quaternary salt ⟨81JCS(P2)414⟩. Selective alkylation of hydantoins is difficult unless one nitrogen is blocked during synthesis. As long as N-3 is already substituted, alkylation at N-1 may be achieved using alkyl halides or tosylates in anhydrous DMF ⟨B-76MI40701⟩. With 5,5-diaryl-2,4-dithiohydantoins (231) boron trifluoride dimethyl etherate selectively methylates S-2 with concomitant extrusion of S-4 and rearrangement to a 4,5-diphenyl-4-imidazoline. With 2-thiohydantoins (232) such S-methylation is not accompanied by any rearrangement (Scheme 128) ⟨B-73MI40701⟩. S-Glycosides of imidazoles and benzimidazoles have been prepared.

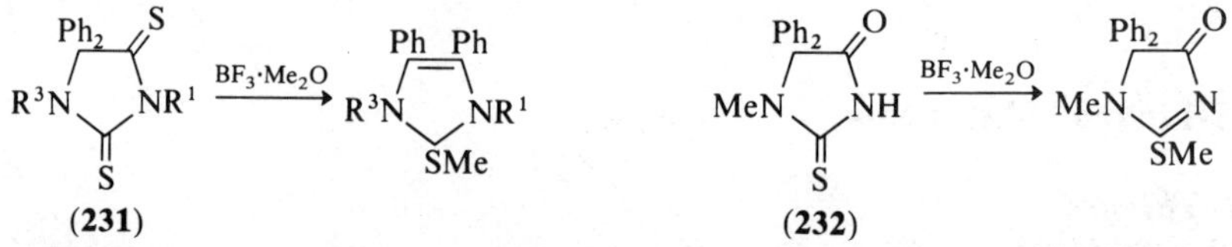

When benzimidazoline-2-thiones react with a mixture of DMSO and acetyl chloride at 50–60 °C a profusion of products are formed (Scheme 129). Rationalization of these is in terms of replacement reactions by benzimidazoline-2-thione on an intermediate sulfonium acetate. At lower temperatures the products can also include the ylide (233) in 20% yield ⟨74CRV279⟩.

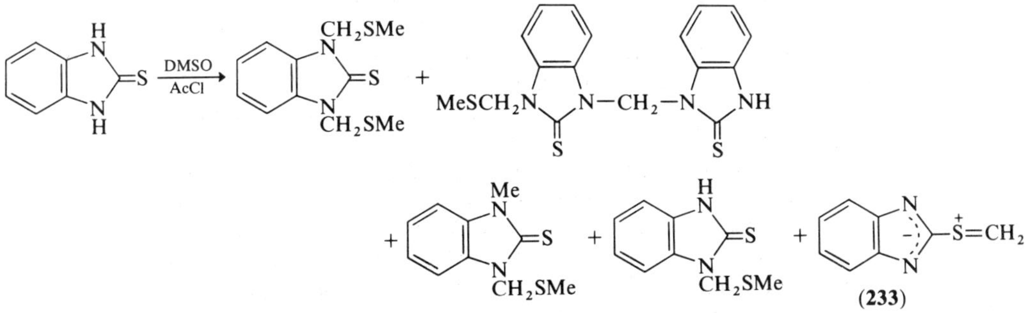

Scheme 129

Removal of the mercapto (or thione) function is most commonly achieved using Raney nickel, although concentrated HCl at high temperatures or nickel boride have been employed. Thiohydantoins are transformed into imidazolidin-4-ones by sodium in amyl alcohol ⟨70JHC439⟩. Oxidative desulfurization procedures too are often used; the reactions probably proceed through the unstable sulfinic acid species which very readily eliminates sulfur dioxide. The high resistance of sulfonic acids to acid hydrolysis renders unlikely the possibility that they are reaction intermediates.

When thiones are exposed to oxidizing agents the reaction products can vary considerably depending on conditions. Mild oxidation (e.g. air, iodine, hydrogen peroxide) converts them into disulfides (234); more vigorous conditions occasionally allow the detection of a sulfinic acid (235), e.g. carefully controlled oxidation with hydrogen peroxide of 4-methyl-imidazoline-2-thione. The most common oxidation products (using hydrogen peroxide, chlorine, bromine, chromic or nitric acids) are the sulfonic acids (236), while yet more vigorous conditions can lead to cleavage of the C—S bond and desulfurization (Scheme 130). Chromic acid oxidation of 4,5-diphenylimidazoline-2-thione gives N,N'-dibenzoylurea ⟨B-76MI40701⟩. With chlorine in the presence of HCl, 4-mercaptoimidazole-5-carboxamide is oxidized to the sulfonyl chloride derivative ⟨76CHE924⟩.

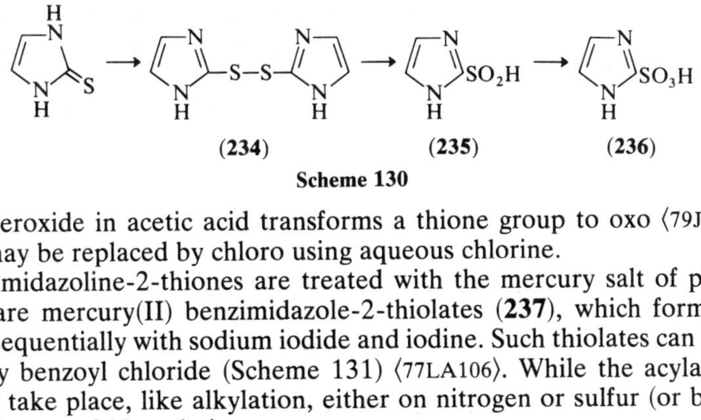

Scheme 130

Hydrogen peroxide in acetic acid transforms a thione group to oxo ⟨79JOC3858⟩, and a sulfur group may be replaced by chloro using aqueous chlorine.

When benzimidazoline-2-thiones are treated with the mercury salt of phenylacetylene the products are mercury(II) benzimidazole-2-thiolates (237), which form the disulfides when treated sequentially with sodium iodide and iodine. Such thiolates can be benzoylated on nitrogen by benzoyl chloride (Scheme 131) ⟨77LA106⟩. While the acylation of thiones in general can take place, like alkylation, either on nitrogen or sulfur (or both), there is a preference in favour of N-acylation.

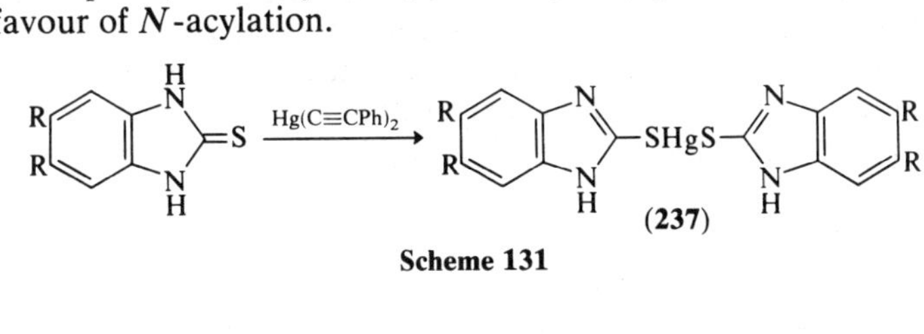

Scheme 131

4.07.3.5.2 Disulfides

Besides their oxidation behaviour (Scheme 130) disulfides can be cleaved by hydrogen sulfide and may be desulfurized by photochemical means ⟨80AHC(27)241⟩. With dibenzoylmethane the disulfide (**238**), derived from benzimidazoline-2-thione, is alkylated via (**239**) to (**240**), a product resulting from a benzoyl migration. Acetylacetone gives only the *S*-alkylation product (**241**), and benzoylacetone gives a product which ultimately rearranges to the *N*-acetyl derivative (**242**; Scheme 132) ⟨74JA1957⟩.

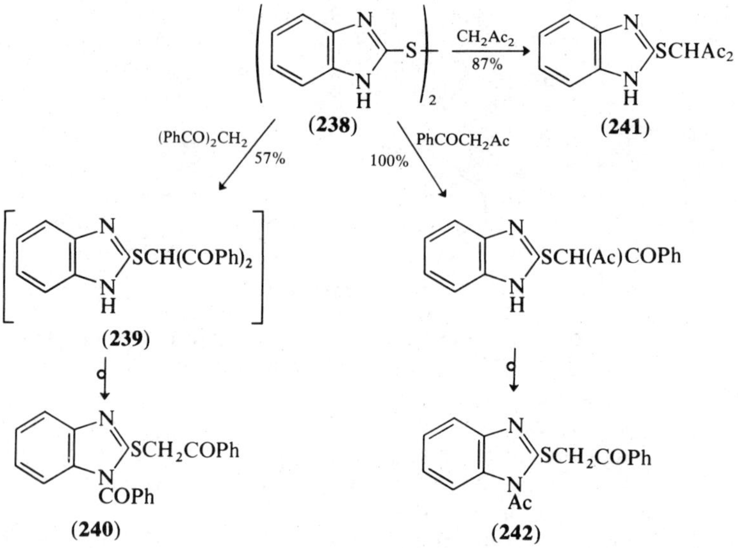

Scheme 132

4.07.3.5.3 Thioethers

Alkyl- and aryl-mercaptoimidazoles are basic compounds which can be oxidized to sulfones with a range of oxidizing agents, and to sulfoxides by careful treatment with peroxide ⟨77JHC889⟩ or periodate ⟨75JHC597⟩. Removal of the sulfur function can be accomplished with Raney nickel or by acids at elevated temperatures. The alkylthio group is subject to cleavage by hydriodic acid, sodium in liquid ammonia, or aluminum bromide in benzene, but not by butyllithium. An equilibrium is set up between *SR*- and *NR*-substituted imidazoles (Scheme 133). Kinetic studies show that the process is an autocatalytic rearrangement in which there is reaction between the thioether (**243**) and the quaternary salt (**244**) ⟨79CJC813⟩.

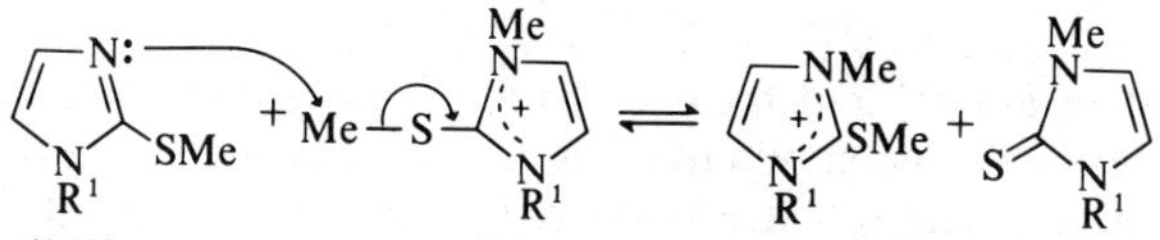

Scheme 133

The vinyl group of 2-vinylthiobenzimidazole can be reduced to ethyl by hydrogen in the presence of a nickel catalyst ⟨74CHE1217⟩, while addition of an alkanethiol occurs under radical conditions in an anti-Markownikov fashion ⟨76CHE1278⟩. 2-Propargylthiobenzimidazoles (**245**) are subject to Claisen rearrangement in a similar way to *S*-alkyl compounds (Scheme 134; see Section 4.01.3.5.8) ⟨74TL2643⟩.

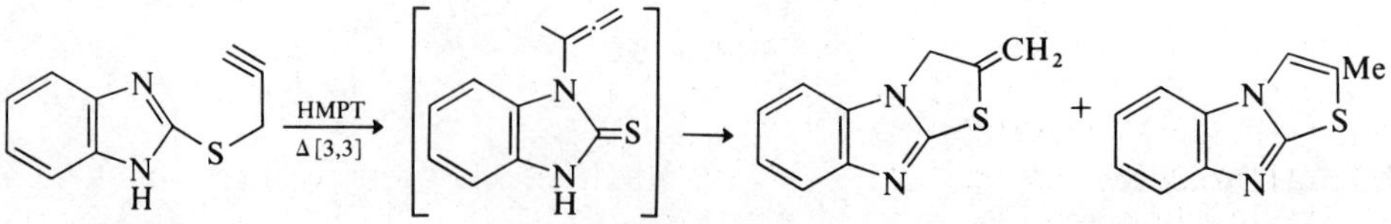

Scheme 134

4.07.3.5.4 Sulfinic and sulfonic acids and derivatives

Sulfinic acids and sulfoxides are not particularly common, being readily oxidized to the sulfonic acids and sulfones, respectively. Sulfonic acids have high melting points and probably exist as zwitterions. They are amphoteric, but mainly display the characteristics of weak acids. The sulfonic acid group activates an adjacent halogen to nucleophilic displacement, and may itself be displaced, *e.g.* reaction of alkylamines with benzimidazole-2-sulfonic acid. Imidazolesulfonic acids resist esterification and acid chloride formation, and are only hydrolyzed by concentrated hydrochloric acid at 170 °C; the 2-isomers are more resistant than the 4- or 5-isomers. Aqueous alkali converts the free acids into hydroxy derivatives. Sulfonyl chlorides are accessible *via* the thiols (Section 4.07.3.6.1) which react with ammonia to form sulfonamides, or are reduced by tin(II) chloride to thiols ⟨77JHC889⟩.

4.07.3.6 Halogen Atoms

Reactions in which halogens are displaced by nucleophiles have been discussed earlier (Sections 4.02.3.9 and 4.07.1.5) as have electrophilic nitrodehalogenations (Section 4.07.1.4.2).

Halogen substituents are electron withdrawing and decrease the basic strength of imidazole. Conversely the acid strength is augmented by these atoms (Tables 1 and 3). Salts formed with concentrated mineral acids often hydrolyze on dilution. The fluoroimidazoles have become more well known in recent years and are weaker bases and stronger acids than the corresponding bromo compounds. The inductive effect of the halogen atom obviously overshadows any resonance effect (Section 4.07.1.3.5) ⟨75JCS(P2)928⟩. The effects of a fluorine substituent on ring-hydrogen exchange have been outlined (Section 4.07.1.4.4).

Among the catalytic and other methods which have been used to replace a halogen atom by hydrogen, reagents such as Pd–C and Raney Ni are generally successful, while red phosphorus and HI, sodium and alcohols, and a variety of metal–acid combinations have been employed with varying success ⟨B-76MI40701⟩ (see also Section 4.07.1.7.3). It is frequently necessary to adjust the dehalogenation conditions in order to avoid modifying other groups. For this reason hydrogenation of 1-benzyl-5-chloro-2-phenylimidazole over Raney Ni was carried out at low temperature in order to prevent debenzylation and conversion of phenyl to cyclohexyl ⟨72CHE742⟩.

Reactions of sulfite with haloimidazoles can take two courses. If sufficient activation is present nucleophilic replacement will occur; if not the sulfite will merely reduce the halogen function. For this reason the major reaction of 2,4,5-tribromoimidazole with sulfite is reduction, although a little 4-bromoimidazole-5-sulfonic acid appears among the products (Scheme 135), and even 2-bromo-1,5-dimethyl-4-nitroimidazole gave no sulfonic acid product. The 2-sulfonic acid was, however, formed from the 2-bromo-1,4-dimethyl-5-nitro isomer. Apparently the position of the activating group is important. When there is little or no activation for nucleophilic substitution, then in N-unsubstituted imidazoles 2-bromo and (with less facility) 4-bromo groups are easily replaced by hydrogen. In 1-methylimidazoles even a 2-bromo group is only slowly reduced. An earlier observation ⟨B-76MI40701⟩ that in iodoimidazoles it is the 2-iodo substituent that is more resistant to sulfite reduction should now be modified, since the conclusion was based on Pauly's incorrect assignment of structure to 4,5-diiodoimidazole (Section 4.07.1.4.5). It would be of value to attempt sulfite reduction of authentic 2,4-diiodoimidazole.

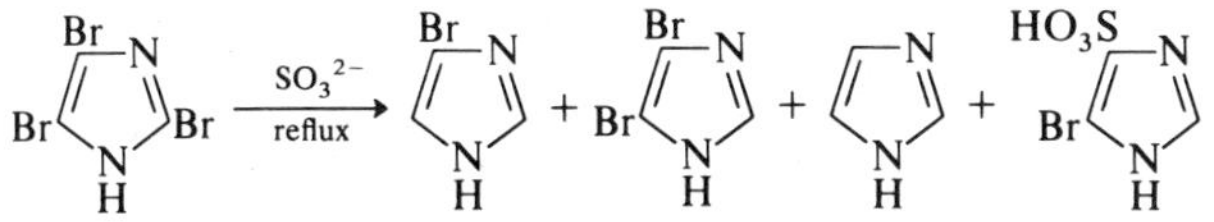

Scheme 135

The catalytic replacement of halogen atoms by alkynic groups is reported to take place more readily with 2- than with 4-iodoimidazoles ⟨B-76MI40701⟩. Again, this observation must be treated with suspicion.

1-Methyl-5-chloroimidazole reacts with naphthyllithium or naphthylsodium more quickly than with butyllithium to give the 5-lithio (or -sodio)-1-methylimidazole along with the 2-metalated isomer. Bromine and iodine atoms are replaced with much more facility than chlorine ⟨75CHE343⟩.

4.07.3.7 Metals and Metalloid-linked Substituents

The reactions of 2-lithio- and 2-sodio-imidazoles and -benzimidazoles are not particularly novel. The compounds do, however, prove a means of introducing a variety of functional groups into the 2-position of the heterocyclic ring. Such metalation reactions at C-2 can only occur readily when there is no alternative site for the metal. Therefore, only N-substituted imidazoles are of synthetic utility, and it may be necessary to select an N-substituent which can be removed later. For this reason, benzyl (removed by reductive or oxidative methods), benzenesulfonyl (removed by ammoniacal ethanol), trityl (hydrolyzed by mild acid treatment) and alkoxymethyl (easily hydrolyzed in acid or basic medium) groups have proved useful in this context. A typical reaction sequence is shown in Scheme 136 ⟨78JOC4381, 77JHC517⟩. In addition, reactions with aldehydes and ketones (to form alcohols), with ethyl formate (to form the alcohol) and with carbon dioxide (to form carboxylic acids) have found application ⟨B-76MI40701⟩.

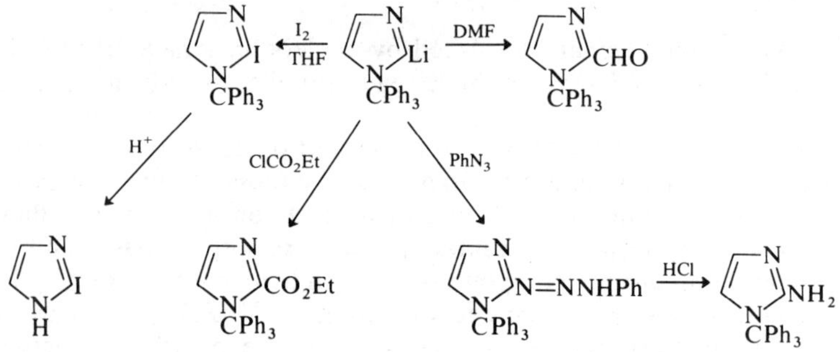

Scheme 136

While 1-alkylbenzimidazoles readily form the 2-sodio and 2-lithio derivatives, 5-bromo-1-ethylbenzimidazole does not react with phenylsodium or amylsodium. In addition, the metalation of 1-arylbenzimidazoles is appreciably complicated by side reactions. Some of these problems can be overcome by the use of o-anisylsodium ⟨71CHE1163⟩.

Lateral metalation can be a complication, and this occurs in compounds such as 1,2-dimethyl- and 1-benzyl-imidazoles, and with 2-methyl- and 2-benzyl-benzimidazoles. Subsequent reactions of these compounds take place at the side-chain carbanion centre ⟨73JOC4379, 74JOC2301⟩.

4.07.3.8 Substituents Attached to Ring Nitrogen Atoms

4.07.3.8.1 -Aryl groups

N-Aryl substituents usually depress the basic properties of the imidazole or benzimidazole nucleus. The effects of substituents on the aryl group are evident from the data listed in Table 7 which also lists rate constants for quaternization with iodoethane, a reaction which is dependent upon the nucleophilic character of the pyridine-type nitrogen ⟨70CHE194⟩. A phenyl group withdraws electrons from the imidazole ring, but does so rather weakly. The ρ value of +0.753 for the protonation process is in accord with weak electron transfer from the phenyl substituent to the basic nitrogen.

Table 7 Basic pK_a Values and Rate Constants for Ethylation of 1-Arylimidazoles in Acetone ⟨70CHE194⟩

1-Aryl substituent	pK_a	$k \times 10^6$ (dm^3 mol^{-1} s^{-1})	1-Aryl substituent	pK_a	$k \times 10^6$ (dm^3 mol^{-1} s^{-1})
None	5.10	15.7	m-Me	5.24	16.9
p-OH	5.35	22.6	p-Br	4.91	11.0
m-OH	5.23	19.6	p-Ac	4.54	10.8
p-Me	5.24	19.0	p-NO$_2$	3.96	7.8

In many instances the aryl and azole rings may not be completely coplanar. This state of affairs becomes more common when there are other aryl groups present and where there are substituents (even methyl) on adjacent carbon atoms. Both dipole moment and spectroscopic studies have demonstrated this. In the ^{1}H NMR spectra, the *ortho* protons of phenyl are deshielded, and in instances in which there is considerable loss of coplanarity the phenyl resonance often appears as a singlet.

Electrophilic substitution reactions of 1-phenylimidazoles frequently involve substitution on the benzene ring. When 1-phenylimidazole is nitrated with a mixed nitric–sulfuric acid reagent the major product is 1-(*p*-nitrophenyl)imidazole. Such nitration has been shown to involve the imidazolium species as substrate. Nitration in acetic anhydride yields only the nitrate salt.

Oxidation and reduction procedures have little effect on 1-aryl substituents which are also very difficult to remove. When, however, there are strongly electron-withdrawing groups present in the benzene ring, nucleophiles are effective in promoting dearylation. Thus, a 2,4-dinitrophenyl group at N-1 of histidine is cleaved by alkaline hydrolysis, aminolysis or hydrazinolysis. On the other hand, 1-(2-pyridyl)imidazole is cleaved neither by 2M sodium hydroxide nor by 2M hydrochloric acid.

The thermal rearrangement of *N*- to *C*-arylimidazoles has been discussed earlier (Section 4.07.1.2.2).

4.07.3.8.2 Alkyl groups

The *N*-alkyl compounds have lower boiling and melting points than their *C*-alkyl isomers, as well as being much more soluble in non-polar solvents (Section 4.06.4.1).

While removal of an *N*-alkyl substituent is not always a feasible process benzyl groups can be removed by reduction with sodamide or by catalytic hydrogenolysis. If such reductive methods fail, an oxidation with chromium trioxide in acid may be successful. Other groups are not so readily displaced, but a procedure involving transalkylation with benzyl chloride followed by debenzylation can be employed to convert 1-methylimidazole into imidazole (Scheme 137). The reaction is capable of extension and operates because the quaternary salt is in equilibrium with both 1-alkylimidazoles and the alkyl halides. Under conditions in which the more volatile alkyl halide can escape from the system the 1-benzylimidazole builds up.

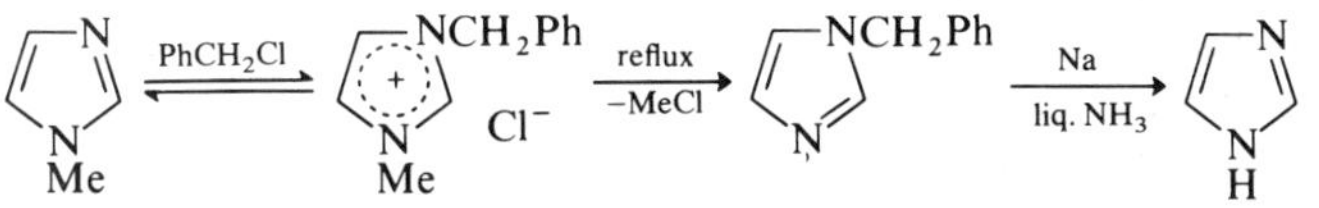

Scheme 137

The dequaternization of imidazole quaternary salts is a process which has been known for many years, but which has only recently been studied systematically ⟨77AJC2005⟩. Pyrolysis of imidazolium halides disubstituted on nitrogen by alkyl (or aryl) groups gives 1-substituted imidazoles. Usually the reaction is carried out at about 240 °C *in vacuo* so that the 1-substituted imidazoles can be distilled from the reaction vessel as they are formed. It is likely that the process is of the S_N2 type — this is evident from the behaviour of differing alkyl groups, and from a study of the effects of different anions on the reaction. The facility with which groups are cleaved from imidazolium diminishes in the order: allyl (2.8) > methyl (1.0) > benzyl (0.6) > ethyl (0.14) > butyl, propyl (~0.04–0.09) > isopropyl (~0.01–0.03) ≫ phenyl, vinyl (0). The values in parentheses refer to relative cleavage rates compared with methyl. The reaction conditions are not typical of an S_N1 reaction, and the above reactivity sequence does not correspond with carbonium ion stability. In addition, non-nucleophilic and very bulky anions (*e.g.* perchlorate, tetraphenylborate) will not promote the reaction. In the decomposition of (**246**; R = Me, R' = Et, X = I, Br, Cl) the proportion of 1-methylimidazole increases as the nucleophile becomes smaller (in the ratio I$^-$:Br$^-$:Cl$^-$ = 1:1.5:2.3). The process can then be envisaged as competing S_N2 attacks at R and R' with the least sterically hindered alkyl group being converted most readily into the alkyl halide (Scheme 138). An S_N2' process could account for the high reactivity of allyl, although some S_N1 character in this instance cannot be ruled out. Electron-withdrawing groups at C-4 of imidazole favour cleavage of the adjacent N—alkyl bond. Conjugated groups such as vinyl and phenyl are not cleaved.

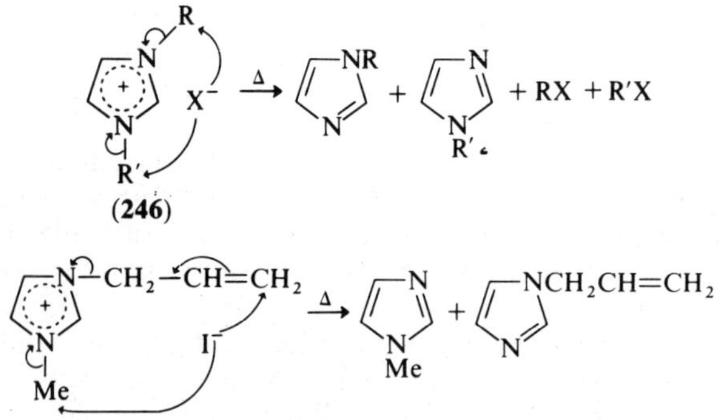

Scheme 138

Methyl and methylene groups attached to nitrogen are not normally regarded as 'reactive', but some metalation experiments using 1-benzylimidazole as substrate indicate reaction at the exocyclic methylene ⟨B-76MI40701⟩.

4.07.3.8.3 Alkenyl and alkynyl groups

UV studies suggest that in 1-vinyl-imidazoles and -benzimidazoles there is conjugation with the heteroaromatic ring. Certainly the reactivities of such compounds would indicate that they are not normal alkenes, nor are the alkynyl compounds typical of other alkynes. They are very prone to polymerization processes, and electrophilic addition reactions are often quite difficult to accomplish. When 1-styrylimidazoles are prepared, the *trans* compound is much more likely to eventuate as the more stable isomer ⟨73CJC3765, 76JCS(P1)545⟩.

1-Vinylimidazole does not react at low temperatures with hydrogen halides, but at 25 °C it forms a hydrochloride when treated with HCl gas in carbon tetrachloride. Halogenation, too, is resisted and can be achieved only partially. The reaction is believed to proceed through initial complex formation followed by addition at the exocyclic double bond. The vinyl function can, however, be reduced by catalytic hydrogenation over Raney nickel, but not as readily as an exocyclic *S*-vinyl group ⟨74CHE1217, 80AHC(27)241⟩.

Oxidation sometimes gives the carboxylic acid, but the use of hydrogen peroxide (as with other free radical initiators) causes polymerization. Under the influence of radical initiators it is possible to induce thiols to add in an anti-Markownikov manner across an *N*-vinyl group (Scheme 139), while ionic catalysts promote normal electrophilic addition ⟨75CHE1416⟩.

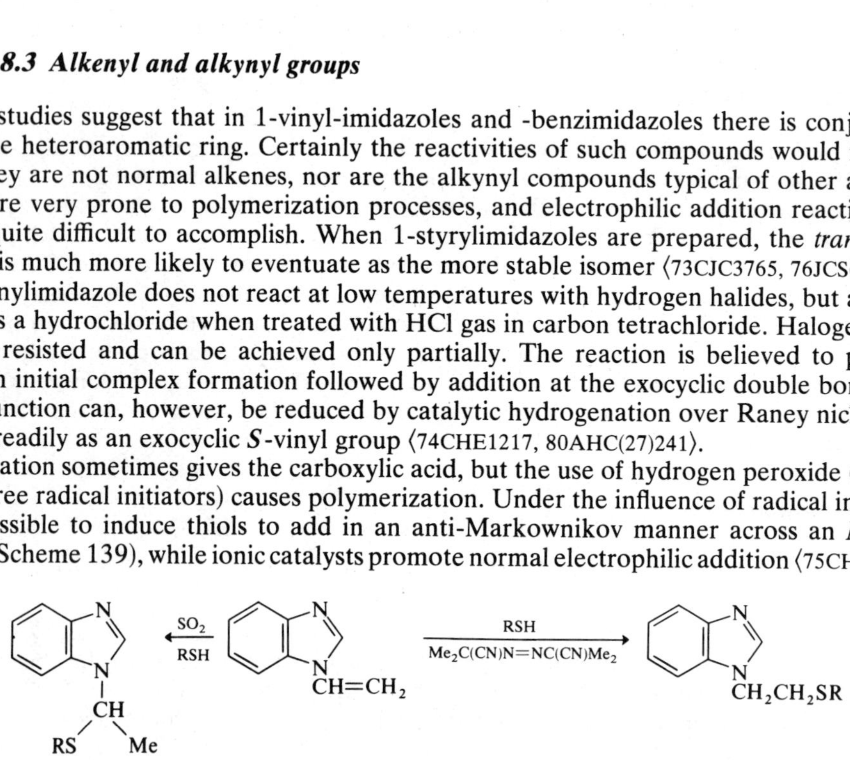

Scheme 139

Thermal rearrangement of 1-vinylimidazole gives a 7:1 ratio of 2- and 4-vinyl isomers. Similar treatment of 1-allylimidazole gives about equal quantities of the 2- and 4-allyl compounds. Irradiation at 300 nm transforms *trans*-1-styrylimidazole into the *cis* isomer (**247**) which then photocyclizes to (**248**; Scheme 140) ⟨76JCS(P1)75⟩.

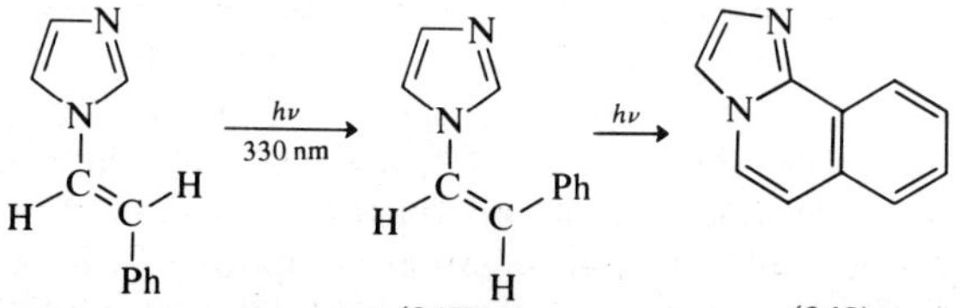

Scheme 140

With potassium hydroxide in THF at 0 °C 3-(2-amino-1-benzimidazolyl)propyne (**249**) isomerizes to the allene (**250**; Scheme 141) ⟨75CHE461⟩.

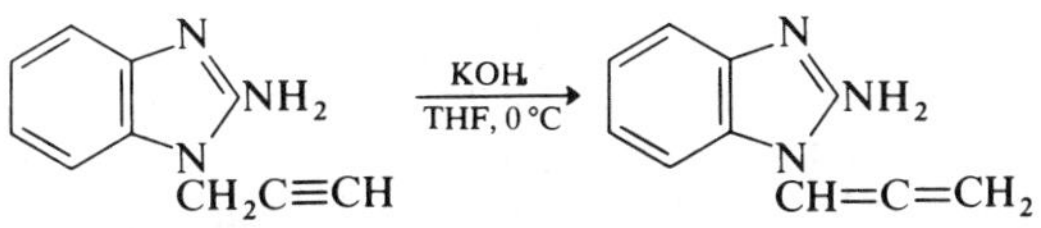

Scheme 141

4.07.3.8.4 Acyl and aroyl groups

Probably the most important property of these compounds is the propensity of *N*-acylimidazoles and -benzimidazoles (as well as other azoles) to become involved in reactions which result in acylation of an attacking nucleophile. The compounds are unlike other tertiary amides in that there is little or no contribution from resonance structures of type (**251**) to the hybrid (Scheme 142); hence the positive nature of the carbonyl carbon is undiminished. The electron pair on the annular nitrogen is part of the aromatic sextet. The compounds are known as 'azolides' generally, and more specifically as 'imidazolides'. Because the annular nitrogens are not directly adjacent imidazolides are more reactive than the corresponding pyrazolides.

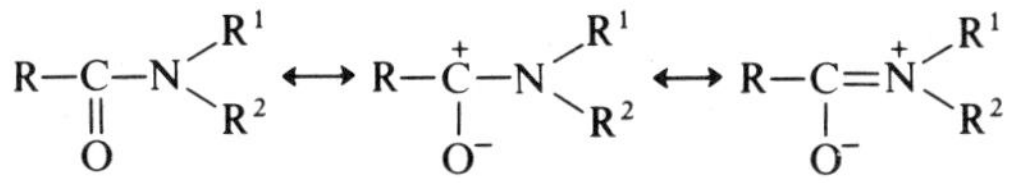

Scheme 142

The azolides are able to participate in a wide variety of nucleophilic 'olysis' reactions which form aldehydes, ketones, carboxylic acids, esters, amides, thiol esters, hydrazides and anhydrides (Scheme 143). In addition, 1-trifluoroacetylimidazole (**252**) is a convenient reagent for the conversion of aldoximes into nitriles (Scheme 144) ⟨81BCJ1579⟩.

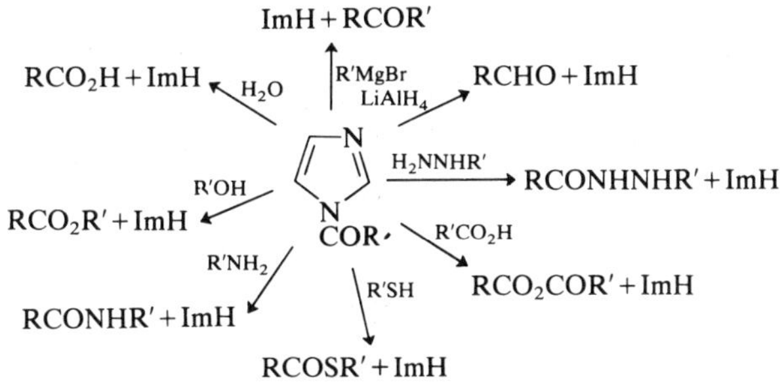

Scheme 143

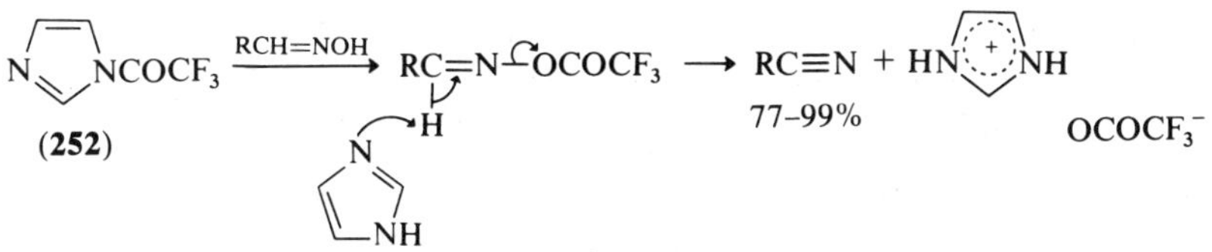

Scheme 144

Among the most useful of these azolides is 1,1'-carbonyldiimidazole (**253**). This compound is extremely reactive towards nucleophilic reagents because the carbonyl group is subject to electron withdrawal from both sides. Also, although it is very rapidly hydrolyzed by water at room temperature with vigorous carbon dioxide evolution, the compound is crystalline, and much more easily handled than phosgene which has similar reactivity. In the formation of 1-acylimidazoles compound (**253**) reacts in equimolar proportions with a carboxylic acid in an inert solvent to give practically quantitative yields. This reaction comprises a two-step mechanism (Scheme 145) in which the carboxylic acid reacts initially with the diazolide, and then an intermolecular transacylation, in which imidazole is released, gives the 1-acylimidazole. Such transacylations render 1,1′-carbonyldiimidazole a valuable synthetic reagent with applications in the synthesis of esters, amides (and peptides), hydroxamine acids, hydrazides, anhydrides, acyl halides, diacyl peroxides, *C*-acyl triphenylphosphine alkylenes and peracid esters. In peptide synthesis, the mild conditions of the reaction minimize racemization (Scheme 145).

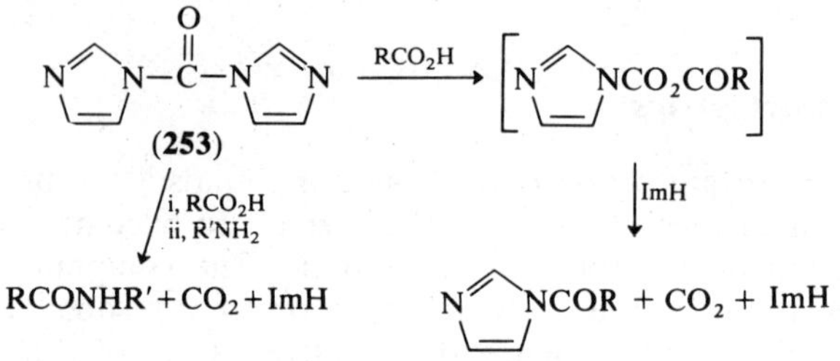

Scheme 145

When (253) reacts with phosgene the 1-acyl chloride product (254) can react with amines to give amides ⟨79LA1756⟩, while in a further transfer reaction with ketones the compounds (255) and (256) are produced (Scheme 146) ⟨80H(14)97⟩. Acylation of aromatic hydrocarbons using 1-acylimidazoles in the presence of trifluoracetic acid gives high yields provided that the aryl compounds are electron rich, *e.g.* *p*-dimethoxybenzene, thiophene, anisole ⟨80BCJ1638⟩.

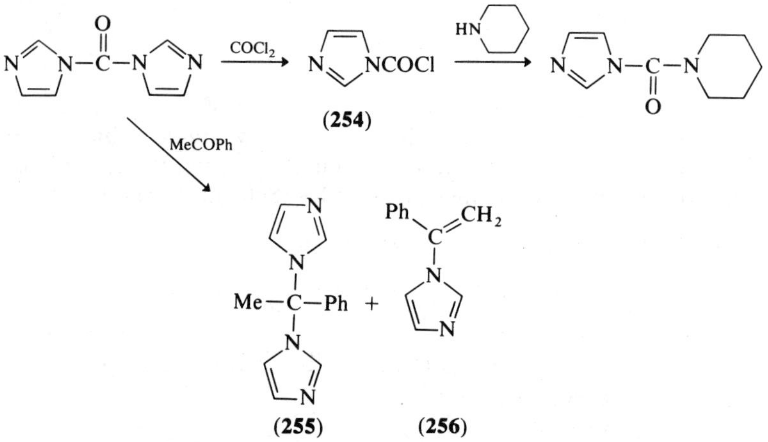

Scheme 146

In addition to transacylations, imidazolides possess a number of other useful synthetic applications. 1-Formylimidazole is an effective reagent for introducing an aldehyde group; it decomposes above its melting point with evolution of carbon monoxide to reform imidazole. The formation of aldehydes and ketones (in 50–95% yields) when lithium aluminum hydride and Grignard reagents, respectively, react with imidazolides has been commented on above. A further source of aldehydes is the reaction in which 1-acetylimidazoles are treated by the slow addition of a reactive molecule such as indole in acetic anhydride at 125 °C. The major product (257), formed by electrophilic attack at C-3 of indole by 1,3-diacetylimidazolium, can be hydrolyzed to the aldehyde (258; Scheme 147). Further reactions analogous to those of phosgene are shared by 1,1′-carbonyldiimidazole. In inert solvents ureas are formed with primary amines; secondary amines yield imidazole-1-carboxamides. If the reaction conditions are controlled it is possible to achieve a 1:1 reaction between the primary amine and (253) resulting in the formation of 1-carboxamides which dissociate at low temperatures to isocyanates. With alcohols and phenols (253) forms diesters of carbonic acid. Monoesters of phosphoric acid, and the acid itself, react with (253) at room temperature with elimination of carbon dioxide and formation of the imidazolium salts of imidazolides, $(ImH_2)^+$ $(ROPO_2Im)^-$, which are effective phosphorylating agents in solution. Furthermore, the reaction of (253) with water to produce carbon dioxide has been recommended in preference to the use of PCl_5 for the microanalytical estimation of carbon and hydrogen in organic compounds.

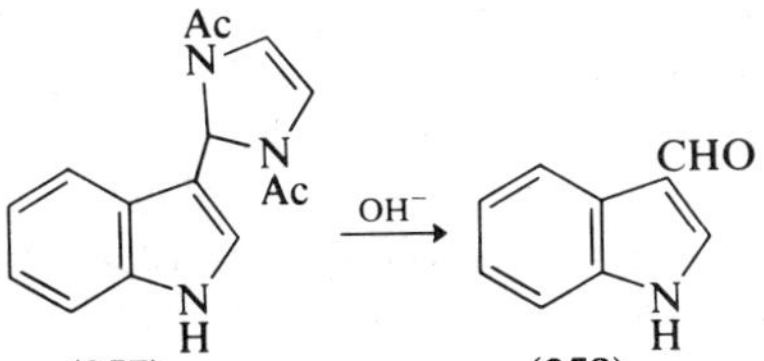

Scheme 147

Kinetic studies of the nucleophilic reactions of azolides have demonstrated that the aminolyses and alcoholyses proceed via a bimolecular addition–elimination reaction mechanism, as does the neutral hydrolysis of azolides of aromatic carboxylic acids. Aliphatic carboxylic acid azolides which are subject to steric hindrance can be hydrolyzed in aqueous medium by an S_N1 process. There have been many studies of these reactions, and evidence supporting both S_N1 and S_N2 processes leaves the impression that there are features of individual 'olysis' reactions which favour either an initial ionization or a bimolecular process involving a tetrahedral intermediate ⟨80AHC(27)241, B-76MI40701⟩.

The imidazole-catalyzed hydrolysis of esters has been described earlier (Section 4.07.1.3.9). In the uncatalyzed aminolysis of 1-acetylimidazole, strong bases react with the azolide in a second order reaction (which is faster than general base catalysis of hydrolysis), while weak bases simply act as general base catalysts of hydrolysis. The rate-determining step is the dissociation of the ion-pair, $R\overset{+}{N}H_2Ac\cdot Im^-$ (Scheme 148). Usually the expulsion of an amine anion from an amide in aqueous solution is very difficult without general acid–base catalysis, and the fact that such reactions of 1-acylimidazoles with nucleophiles occur at all is very largely a consequence of the much greater stability of an imidazole anion compared with anions of other amines.

$$RNH_2 + AcIm \underset{k_{-1}}{\overset{k_1}{\rightleftharpoons}} R\overset{+}{N}H_2Ac\cdot Im^- \underset{k_{-2}}{\overset{k_2}{\rightleftharpoons}} R\overset{+}{N}H_2Ac + Im^-$$

$$\Big\updownarrow \pm H^+ \quad \Big\updownarrow \pm H^+$$

$$RNHAc \quad ImH$$

Scheme 148

In a study of the kinetics of transfer of an acetyl group from 1-acetylimidazolium and 1-acetylimidazole to phenols the rate constants were found to be in close agreement with those obtained for similar acetylation of thiols. As well as this, it was discovered that the acetyl transfer reactions of the non-dissociating model, 1-acetyl-3-methylimidazolium chloride, compared very closely with those of 1-acetylimidazolium. This provided evidence that many reactions of acetylimidazolium with reagents of type HY are merely acid-catalyzed reactions of the anion Y^- ⟨B-76MI40701⟩.

Diimidazolides analogous to 1,1'-carbonyldiimidazole, e.g. 1,1'-thiocarbonyldiimidazole, are similar to the oxy compounds but are rather less reactive; even so they are powerful thioacylating agents.

Deacylation of azolides can be accomplished in weakly acidic or basic solutions, with boiling water (or even a moist atmosphere) and by heating. Electron-withdrawing groups elsewhere in the imidazole ring assist the process. A corollary of the decarbonylation by heat of 1-acetylimidazole is the thermal decarboxylation of 1-alkoxycarbonylimidazoles (**259**) to 1-alkylimidazoles (Scheme 149) ⟨80AHC(27)241⟩. When, however, the methyl ester of 2-aminobenzimidazole-1-carboxylate is heated above its melting point in boiling anisole, the ester function migrates to the exocyclic nitrogen. A cyano group at N-1 of imidazole (1-cyanoimidazole can be prepared by the action of phosphoryl chloride on imidazole-1-carboxamide) can be removed in aqueous solution ⟨80H(14)1963⟩.

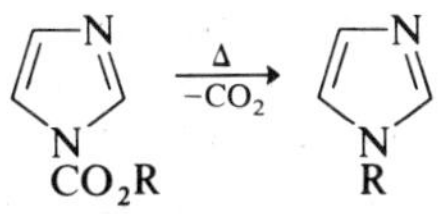

Scheme 149

4.07.3.8.5 Nitrogen functions

(i) Nitro

The few 1-nitroimidazoles which are known are relatively stable, crystalline compounds with characteristic UV and IR spectra (N—NO antisymmetric vibration band in the region $\sim 1640\,\text{cm}^{-1}$). The thermolysis of these compounds has been mentioned earlier (Section 4.07.1.2.2).

(ii) Amino

1-Aminoimidazoles can only be prepared by ring-synthetic methods. They are stable, crystalline compounds with melting points usually in excess of 100 °C. Removal of the amino group occurs on treatment with nitrous acid, but oxidation causes extensive modification. Manganese dioxide converts 1,2-diamino-4-phenylimidazole into a mixture of 4-phenyl-1,2,3-triazole and 3-amino-5-phenyl-1,2,4-triazine, while 1-aminobenzimidazole similarly gives 3-amino-1,2,4-benzotriazines ⟨80AHC(27)241, 77JOC542⟩ (see also Section 4.07.1.7). Oxidation of 1-aminoimidazole quaternary salts gives azoimidazolium salts.

When β-diketones and related carbonyl compounds react with 1,2- and 1,5-diaminoimidazoles, bicyclic products form after initial attack at the C-amino function ⟨80AHC(27)241⟩.

4.07.3.8.6 Halogen atoms

Most N-haloimidazoles and their benzo analogues are rather labile compounds, but the N-iodo compounds are stable probably because the iodonium cation is more stable than other halogen cations. 1-Iodoimidazoles do not dissolve in acids or alkalies and decompose on heating with liberation of iodine. The rearrangement of N- to C-halopyrazoles has been observed and has also been postulated to occur with imidazoles; N-iodoimidazoles may be intermediates in the C-iodination process.

During the nitration of 1,2,4,5-tetraiodoimidazole the N-iodo function is lost ⟨B-76MI40701⟩.

4.07.3.8.7 Silicon, phosphorus, sulfur and related groups

1-Trialkylsilyl-imidazoles and -benzimidazoles are prepared by reaction of the bases with trialkylsilyl chlorides or hexaalkyldisilazanes. The derivatives are thermally stable but can be cleaved by hydrolysis or alcoholysis. The compounds also react smoothly in acylation reactions. When treated at 180 °C with DMSO, 1-trimethylsilylimidazole gives the N-(methylthio)methylimidazole (**260**) via a Pummerer reaction. The corresponding derivatives of 2-methylimidazole and benzimidazole react likewise (Scheme 150) ⟨79JHC415⟩. Oxidation of 1,1'-bis(trimethylsilyl)-2,2'-bi(2-imidazolinyl) with manganese dioxide gives 2,2'-biimidazolyl ⟨74S815⟩.

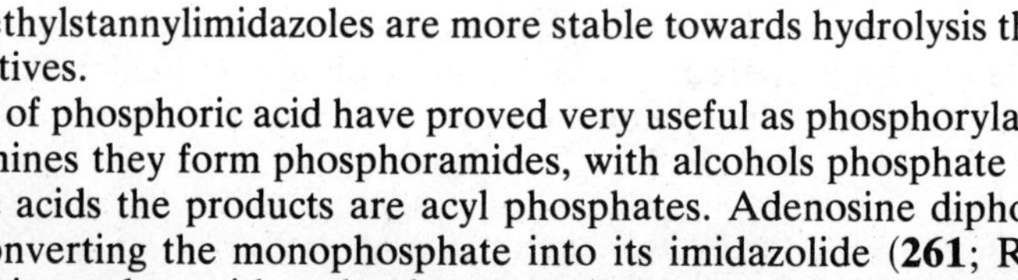

$$2\,\text{N}\!\frown\!\text{NSiMe}_3 \xrightarrow[180\,°C,\,6\,h]{\text{Me}_2\text{SO}} \text{N}\!\frown\!\text{NCH}_2\text{SMe} + (\text{Me}_3\text{Si})_2\text{O} + \text{ImH}$$

(**260**)

Scheme 150

Most 1-trimethylstannylimidazoles are more stable towards hydrolysis than the foregoing silylated derivatives.

Imidazolides of phosphoric acid have proved very useful as phosphorylating agents. With ammonia or amines they form phosphoramides, with alcohols phosphate esters result, and with carboxylic acids the products are acyl phosphates. Adenosine diphosphate has been prepared by converting the monophosphate into its imidazolide (**261**; R = adenosyl) and then treating this product with orthophosphate (Scheme 151). The imidazolides of phosphoric acid are hydrolyzed very rapidly by water, the di- and tri-imidazolides being particularly susceptible to moisture. Greater than 80% conversions of aldoximes into nitriles are possible using imidazolides of phosphorus ⟨78ZN(B)1033⟩. Sulfur oxidizes 1-diphenylphosphinoimidazole to the phosphinato analogue ⟨74CJC3981⟩. When 1-dialkylphosphorylimidazoles with short alkyl chains decompose below 80 °C to give the 1-alkylimidazole, trialkyl phosphate and 1,3-dialkylimidazolium polyphosphate, the reaction is considered to result from nucleophilic attack of the pyridine nitrogen of one 1-dialkylphosphorylimidazole on the α-alkyl carbon of another ⟨78JOC4853⟩.

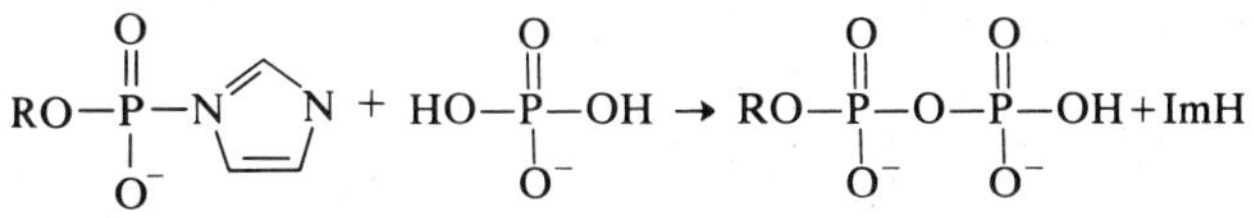

Scheme 151

4.07.3.8.8 N-Oxides and N-hydroxy compounds

The photolysis of imidazole and benzimidazole N-oxides has been described earlier (Section 4.07.1.2) as has their tautomerism (Section 4.06.5.2).

The compounds are mostly crystalline, although hygroscopic, and are often light sensitive. This hygroscopic nature may lead to their isolation as hydrates. Deoxygenation can occur on heating, while some compounds of this type decompose above their melting points. When there are both N-hydroxy and N-oxide functions in the same molecule inter- and intra-molecular hydrogen bonds can form. Thus, 1-hydroxy-2,4,5-triphenylimidazole 3-oxide is associated in chloroform solution and even exists as trimers at high dilutions.

The N-oxides and N-hydroxy compounds have acidic properties, but basic properties are not suppressed to a great extent since salts with picric and hydrochloric acids can still be formed.

Removal of an N-oxide function utilizes the usual reagents such as zinc and acetic or hydrochloric acids, sodium borohydride, hydriodic acid, phosphoryl chloride, sodium dithionite, phosphorus trichloride or hydrogenation over Raney nickel. When 1-hydroxy-2-methyl-5-phenylimidazole (262) reacts with butyllithium and hexachlorodisilane this also induces dehydroxylation (Scheme 152) ⟨80AHC(27)241⟩. At times, as with other heterocyclic N-oxides, deoxygenation with phosphorus halides can introduce a halogen atom at C-2 of imidazole ⟨75JCS(P1)275⟩.

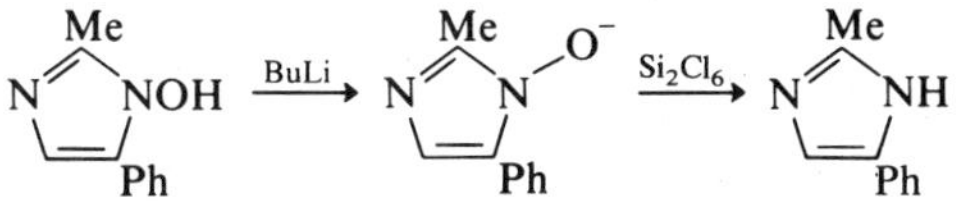

Scheme 152

When 1-hydroxy-3-imidazoline 3-oxides (263) are treated with HCl they dehydrate forming imidazole 3-oxides (Scheme 153) ⟨73CHE1175⟩. Reduction of 4H-imidazole N-oxides with borohydride leads to either 1-hydroxy-imidazolines or -imidazolidines, depending on the position of the oxide function. Under the same conditions 4H-imidazole 1,3-dioxides give 1,3-dihydroxyimidazolidines ⟨76CHE1280⟩. The 4H-imidazole N-oxides are prepared by heating 5,5-disubstituted 1-acyloxy-3-imidazoline 3-oxides *in vacuo*.

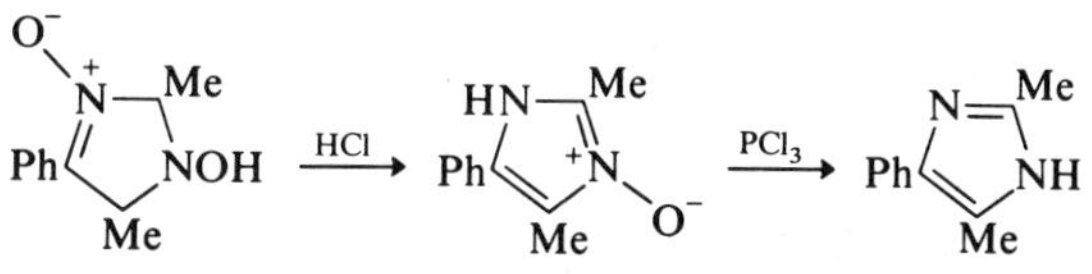

Scheme 153

Dehydrogenation of 1-hydroxyimidazoles and their 3-oxides with lead dioxide gives N-oxides and N,N'-dioxides of imidazolyls. These short-lived radicals are also formed when alkali salts of 1-hydroxyimidazole 3-oxides react with halogen in polar organic solvents. Hydroquinone converts the N,N'-dioxides back into 1-hydroxy 3-oxides.

Both the N-hydroxy compounds and the N-oxides which are in tautomeric equilibrium with hydroxy compounds form O-acyl and O-aroyl compounds, and such esters can act as acyl transfer reagents. Methylation, either with iodomethane in alkaline medium or with diazomethane, converts imidazole N-oxides into the N-methoxy derivatives ⟨B-76MI40701⟩.

Alkaline cleavage of 2,4,4-trimethyl-5-phenyl-4H-imidazole 1-oxide gives oximes of α-acylaminoketones as products ⟨80AHC(27)241⟩.

One of the difficulties in studying directional effects of an N-oxide group in imidazoles has been the synthesis of examples with free ring-carbon positions; few of these have been described ⟨74JHC615, 77JCS(P1)672⟩. Indeed, the 3-oxide function of 1,4,5-trimethylimidazole 3-oxide (**264**; R = Me) exerts a similar effect to that in pyridine N-oxide: nitration is directed to C-2 via the free base species (Scheme 154) ⟨77JCS(P1)672⟩. Although the Reissert reaction failed to introduce a cyanide group into the ring, when the oxides are converted into quaternary salts and treated with potassium cyanide, the 2-cyanoimidazoles result. As had been found earlier with benzimidazole N-oxides, the imidazole 3-oxides can undergo 1,3-dipolar additions. With phenyl isocyanate (**264**; R = CH$_2$Ph) gives N-(1-benzyl-4,5-dimethylimidazol-2-yl)-N,N'-diphenylurea (**266**) via the cycloaddition product (**265**) which is subject to ring cleavage, decarboxylation and reaction with further phenyl isocyanate. Similarly, with DMAD (**267**) is formed (Scheme 154) ⟨75JCS(P1)275⟩.

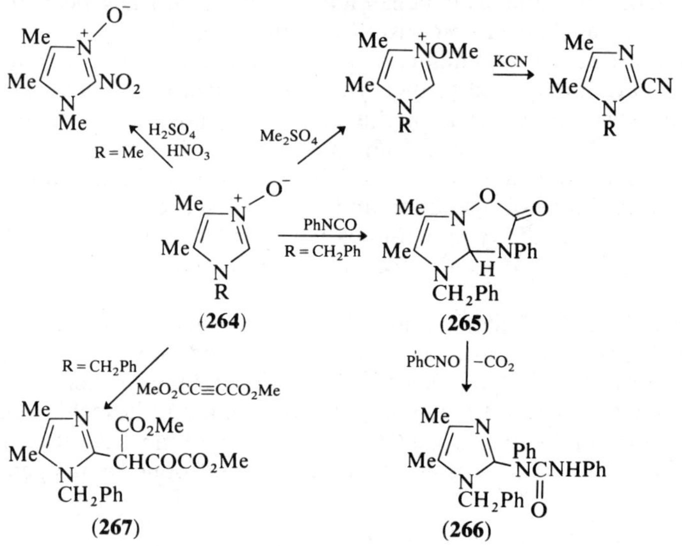

Scheme 154

2,2-Disubstituted 2H-benzimidazole N,N'-dioxides (**268**), when treated with sodium borohydride, are reduced in a stepwise manner to the mono-N-oxides and then to the isobenzimidazoles (Scheme 155).

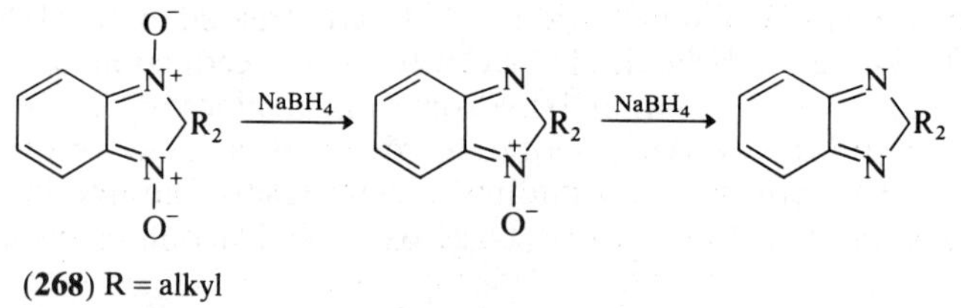

(**268**) R = alkyl

Scheme 155

4.07.3.8.9 N-Ylides

A number of the reactions of imidazoles and benzimidazoles with dienophiles such as DMAD appear to be reactions of the azole-N-ylides (see Section 4.07.1.8.1).

4.08
Imidazoles and their Benzo Derivatives: (iii) Synthesis and Applications

M. R. GRIMMETT
University of Otago

4.08.1	RING SYNTHESIS FROM NON-HETEROCYCLIC COMPOUNDS	457
4.08.1.1	Formation of One Bond	457
4.08.1.1.1	Formation of the 1,2- (or 2,3-) bond	457
4.08.1.1.2	Formation of the 1,5- (or 3,4-) bond	464
4.08.1.1.3	Formation of the 4,5-bond	466
4.08.1.2	Formation of Two Bonds	467
4.08.1.2.1	From [4 + 1] carbon fragments	467
4.08.1.2.2	From [3 + 2] carbon fragments	473
4.08.1.3	Formation of Three or Four Bonds	482
4.08.1.3.1	From reactions of ammonia, amines or formamide with functionalized carbonyl compounds	482
4.08.2	SYNTHESIS BY TRANSFORMATION OF OTHER HETEROCYCLES	487
4.08.2.1	Ring Expansions	487
4.08.2.2	Transformations of Other Five-membered Rings	489
4.08.2.2.1	Pyrazoles and indazoles	489
4.08.2.2.2	Formation of benzimidazoles from imidazoles	489
4.08.2.2.3	2-Imidazolines from cyclic ureas	490
4.08.2.2.4	From other five-membered heterocycles containing nitrogen only	490
4.08.2.2.5	From oxazoles and isoxazoles	491
4.08.2.2.6	From oxadiazoles	492
4.08.2.2.7	From five-membered rings containing sulfur	493
4.08.2.2.8	Imidazoles from condensed heterocycles	493
4.08.2.3	Ring Contractions	493
4.08.2.3.1	From pyrimidines	494
4.08.2.3.2	From pyrazines and quinoxalines	495
4.08.2.3.3	From other six-membered heterocycles	496
4.08.2.3.4	From seven-membered heterocycles	496
4.08.3	APPLICATIONS AND IMPORTANCE	497

In contrast to the pyrazoles, there is no single method of wide application for the synthesis of imidazoles, although the use of *o*-phenylenediamine and related compounds in benzimidazole synthesis might merit such a description. The common synthetic methods are divided into those which build up the imidazole ring from largely acyclic precursors, and transformations of other heterocyclic species. Ring-synthetic procedures can entail the formation of between one and four bonds.

4.08.1 RING SYNTHESIS FROM NON-HETEROCYCLIC COMPOUNDS

4.08.1.1 Formation of One Bond

4.08.1.1.1 Formation of the 1,2- (or 2,3-) bond

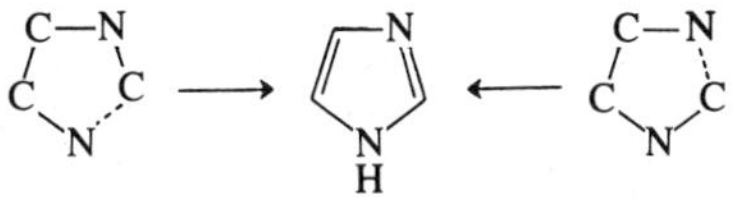

The earliest example of this type of synthesis was probably the Wallach synthesis ⟨70AHC(12)103⟩ in which an N,N'-disubstituted oxamide underwent ring closure with phosphorus pentachloride. The chloro-containing intermediate which formed was reduced by hydriodic acid to give a 1-substituted imidazole. An adaptation of this reaction to the synthesis of 5-chloro-1-methylimidazole is shown in Scheme 1.

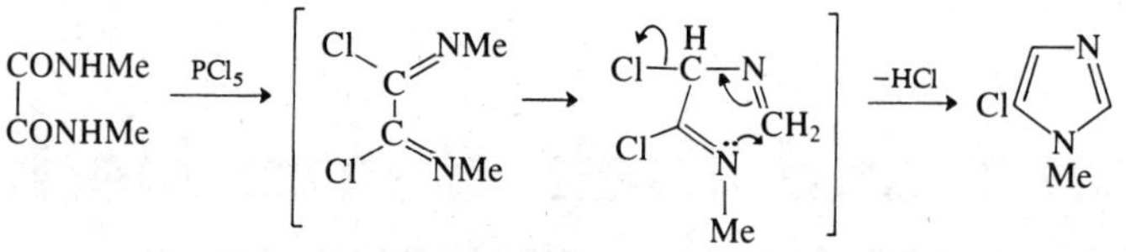

Scheme 1

There are a number of syntheses which use an alkylene or arylene 1,2-diamino compound in which one of the amino groups is acylated or thioacylated. Functionalization of the starting material must also be such that cyclization can give an aromatic product. Perhaps the best-known example of this type is the formation of 2-substituted benzimidazoles by the cyclodehydration of o-aminobenzanilides (**1**) in mineral acid (Scheme 2). This reaction can be adapted to the preparation of 2-aminobenzimidazoles if an alkyl halide is utilized to induce cyclodesulfurization of an N-(2-aminophenyl)thiourea (**2**) ⟨B-73MI40800, 75MI40800, 74S41⟩. Such thioureas are available by condensation of phenyl isothiocyanate with o-phenylenediamines. Depending on the nature of the substitution, these thioureas can be decomposed thermally to give benzimidazoline-2-thiones (**3**); in some instances the energy input required is only minimal. For example, when the o-phenylenediamine is substituted on nitrogen by a β-dimethylamino function, the reaction with phenyl isothiocyanate leads directly to a 1-substituted benzimidazoline-2-thione. There is, of course, some classification difficulty with these reactions since they could be considered to involve formation of both the 1,2- and 2,3-bonds (Section 4.08.1.2.1). N,N'-Diacylated o-arylenediamines (**4**) cyclize in the melt to give 2-methyl- and 2-phenyl-benzimidazoles in a 3:1 ratio (Scheme 2) ⟨74JOU1542⟩.

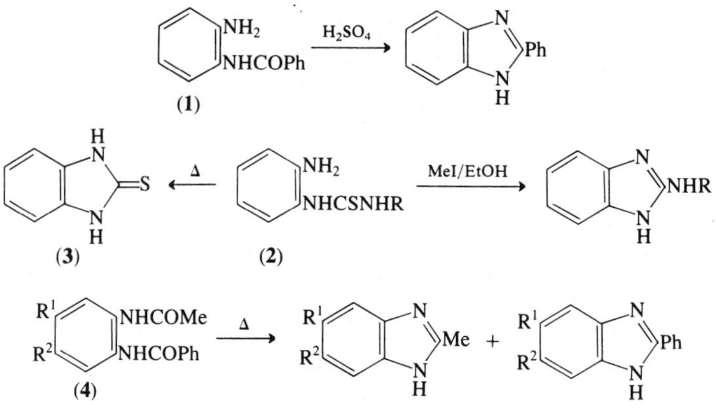

Scheme 2

The reductive cyclization of o-benzoquinonedibenzimide (**5**) with triphenylphosphine gives 1-benzoyl-2-phenylbenzimidazole, perhaps by the mechanism outlined in Scheme 3 ⟨74CRV279⟩.

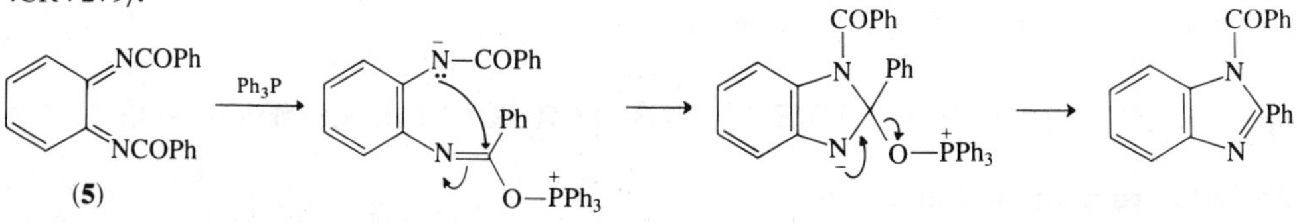

Scheme 3

In a study of the formation of benzimidazoles by the peroxy acid-catalyzed cyclization of o-acylamino-N,N-dialkylanilines, it has been demonstrated that the reaction proceeds via an N-oxide intermediate, since it has proved possible to synthesize this N-oxide and cyclize it under the reaction conditions ⟨B-73MI40800⟩. Both ortho-N-acylated and -N-

aroylated arylamines and nitrobenzenes are able to be cyclized under suitable conditions to give benzimidazoles. The tricyclic derivative (6) involves a skeletal rearrangement in its formation (Scheme 4). In a reaction of the same class, sunlight converts quinone dibenzenesulfonamides (7) into a number of products, among which are benzimidazoles (8; Scheme 4).

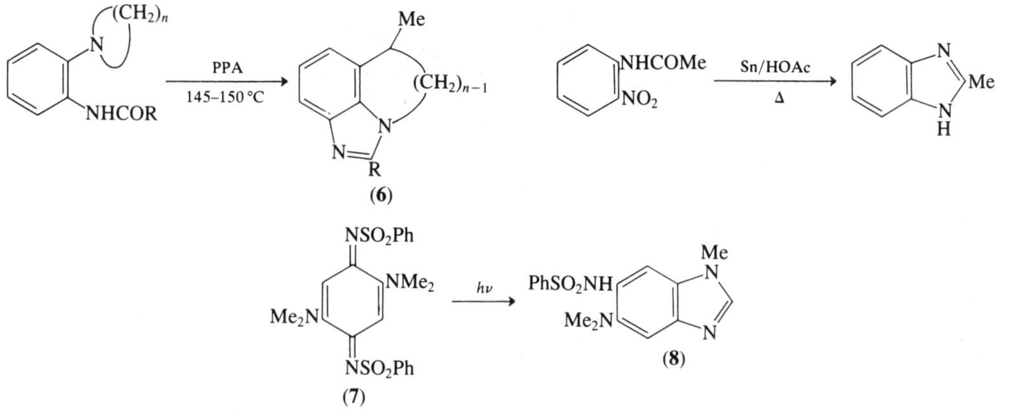

Scheme 4

One of the amino functions in *o*-phenylenediamine can be functionalized in other ways. Thus high yields of polychlorobenzimidazoles (9) are formed when sulfuryl chloride reacts with *o*-amino-*N*,*N*-dialkylanilines, while the Schiff base (10) eliminates methanol when it cyclizes to the 2-substituted benzimidazole (Scheme 5) ⟨75MI40800⟩.

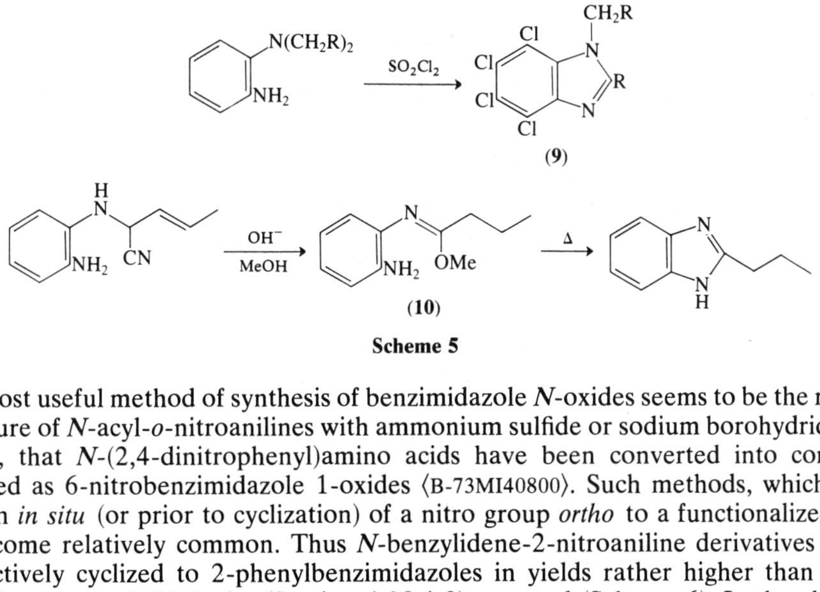

Scheme 5

The most useful method of synthesis of benzimidazole *N*-oxides seems to be the reductive ring closure of *N*-acyl-*o*-nitroanilines with ammonium sulfide or sodium borohydride. Note, however, that *N*-(2,4-dinitrophenyl)amino acids have been converted into compounds designated as 6-nitrobenzimidazole 1-oxides ⟨B-73MI40800⟩. Such methods, which involve reduction *in situ* (or prior to cyclization) of a nitro group *ortho* to a functionalized amino, have become relatively common. Thus *N*-benzylidene-2-nitroaniline derivatives (11) can be reductively cyclized to 2-phenylbenzimidazoles in yields rather higher than when *o*-arylenediamines and aldehydes (Section 4.08.1.2) are used (Scheme 6). In the absence of oxygen, acids catalyze the formation of 2-substituted benzimidazoles from the aldehyde *o*-aminophenylhydrazones, again available by reduction of the nitro compounds (12). Although this method is complicated it gives yields of 67–86% and provides a useful alternative for benzimidazoles with sensitive groups at C-2. The reaction may proceed through intermediate 1,2,3,4-tetrahydro-1,2,4-benzotriazines (13) which are formed by addition of the amino group to the hydrazone double bond (Scheme 6) ⟨79JHC1005⟩.

Under reductive conditions an *o*-nitro-*N*,*N*-dimethylaniline (14; R = Me) is converted into a benzimidazolinone in a reaction which may involve transition metal complexes. When, however, the *N*,*N*-diethylaniline (14; R = Et) is subjected to the same reaction conditions, the product is the benzimidazole (15; Scheme 7).

In the formation of the 1,1'-bibenzimidazole (16) the second benzimidazole ring is formed by cyclization of a monoacylated *o*-phenylenediamine ⟨81JCS(P1)403⟩, while thermolysis of the anils (17) derived from 2-azidoaniline also gives good yields of 2-aryl-substituted

(11)

R=H, 47%
R=Me, 33%

(12) **(13)**

Scheme 6

(15) **(14)**

Scheme 7

benzimidazoles. Yields are in the range 48–96% (Scheme 8) ⟨74CRV279⟩. Remarkably high yields of benzimidazolinones are obtained when aroyl azides (**18**) are heated in acetic anhydride or xylene. The reaction is of unknown mechanism, but that some of these cyclizations which form benzimidazoles involve nitrene intermediates is evident. The participation of such intermediates in, for example, the thermolysis of *o*-azidoanils and the deoxygenation of *o*-nitroanils of aromatic aldehydes has been reviewed recently ⟨79CHE467⟩.

(16)

(17)

(18) R=H, Me, COPh

R′=Ac if R=H
R′=Me if R=Me

Scheme 8

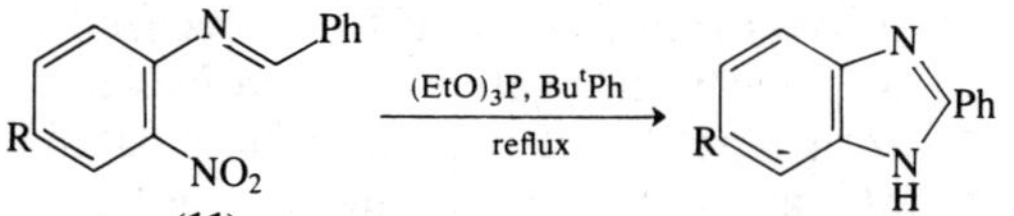

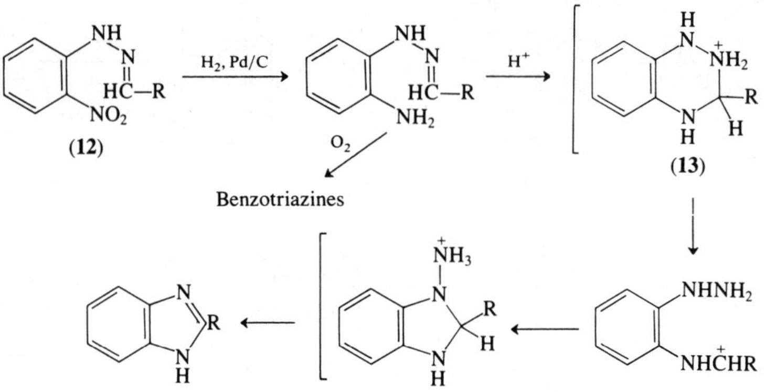

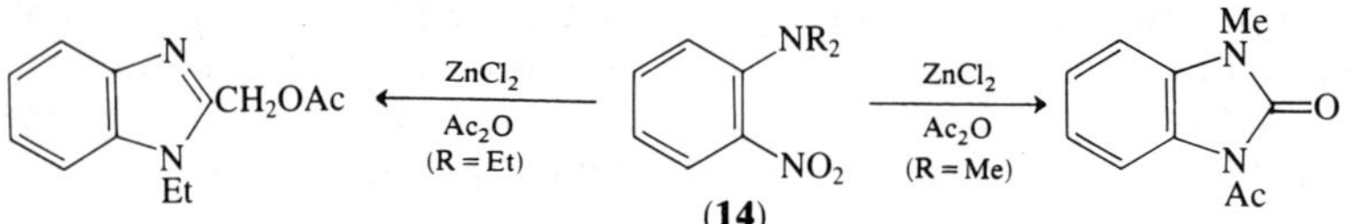

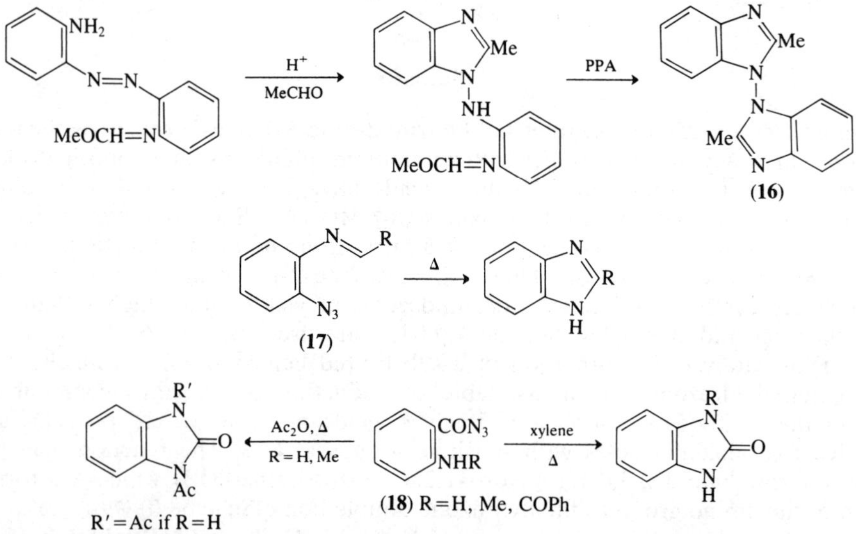

When an *o*-arylenediamine is treated with phosgene a polymer is formed which can be degraded thermally to produce the *o*-diisocyanate (**19**), a source of a range of benzimidazolinones; the corresponding diisothiocyanates produce benzimidazolinethiones when heated with alcohols (yields about 60%) (Scheme 9) ⟨80JCS(P1)2608⟩.

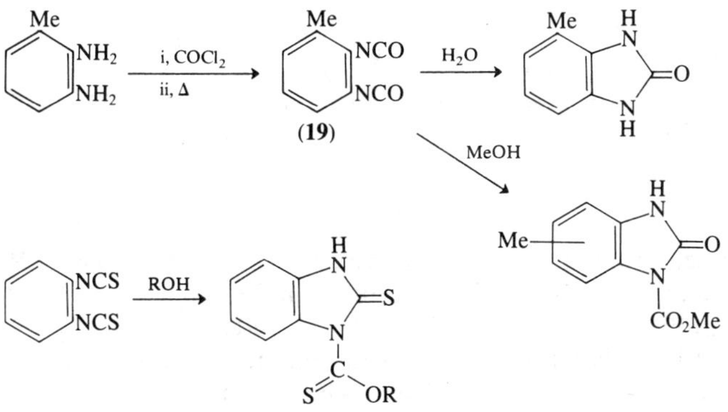

Scheme 9

Although synthesis of the desired acyclic precursors may not always be a simple procedure, there is a host of reactions of suitably functionalized 1,2-diaminoalkanes which give uncondensed imidazole products. Many of these are related to some of the foregoing preparations of benzimidazoles. For example, when the β-glucopyranosylamine (20) is treated in sulfuric acid with an acetylating mixture, the imidazole product (21) is formed in good yield (Scheme 10).

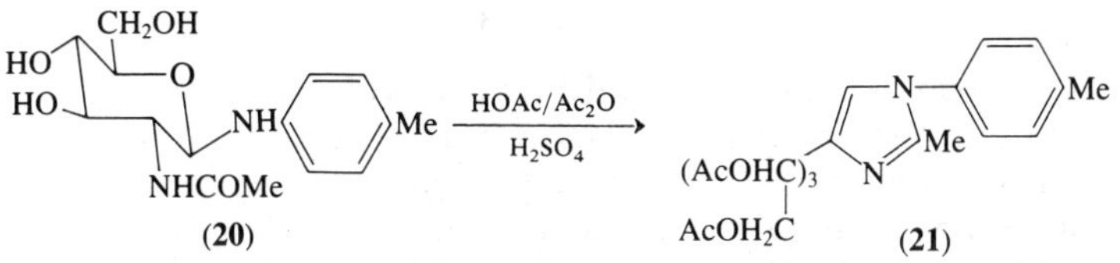

Scheme 10

Oxidation of the Schiff bases (22) of diaminomaleonitrile monoamide gives 4-cyanoimidazole-5-carboxamides with a variety of 2-substituents. When diaminomaleonitrile reacts with formamidine, the condensation product (23) can form imidazoles in two ways. Direct loss of ammonia gives 4,5-dicyanoimidazole, but isomerization, followed by cyclization in which HCN is eliminated, gives 4-amino-5-cyanoimidazole. With further formamidine this latter product is converted into adenine (Scheme 11) ⟨79JOC4532⟩.

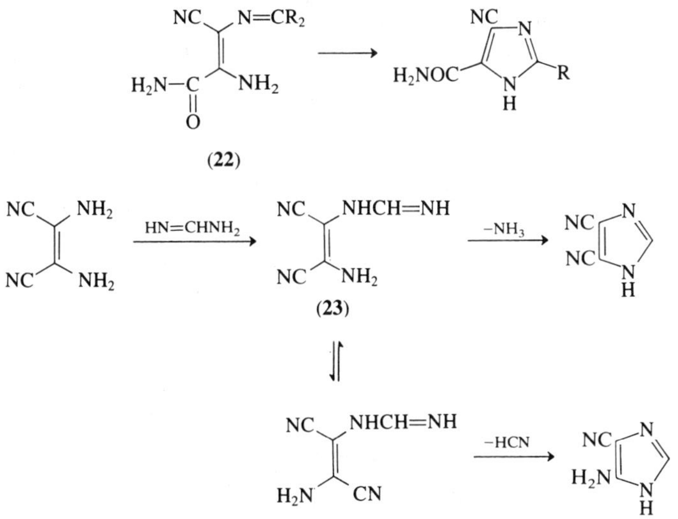

Scheme 11

Diphenylimidazoles, along with 4H-imidazoline-5-thiones (24) and the iminothiazolines (25), are formed when diphenylglycine thioamides are cyclized in boiling toluene by aluminum chloride (Scheme 12). Use of hydrogen chloride in hot dioxane gave only (24). Rather similar treatment of N-formylsarcosine-N-methylthioamide with acid leads to 1,3-dimethylimidazolium-4-thiols ⟨80AHC(27)241⟩.

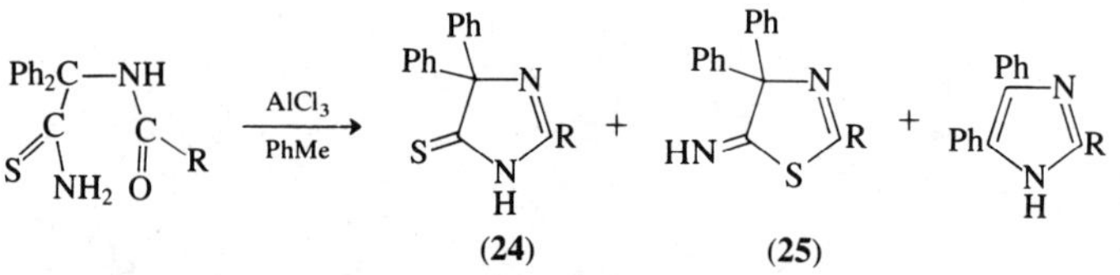

Scheme 12

Dry hydrogen chloride induces ring closure of PhCH=NC(CN)=C(R)SMe to chloroimidazoles, and N-(1-cyanoalkyl)alkylideneamine N-oxides (**26**) can be converted by way of nucleophilic attack by thiophenol into imidazoles substituted at C-5 by thiophenoxy (Scheme 13). This latter reaction is accelerated by small amounts of piperidine, but inhibited by temperatures above the melting point of the N-oxide. The action of cyanide ion on similar nitrones gives cyanoimines which cyclize to imidazoles in 40–90% yield ⟨80JCS(P1)244⟩.

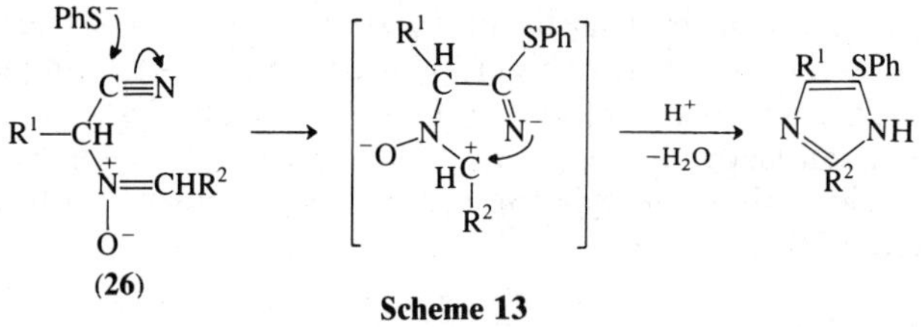

Scheme 13

The Schiff bases, which can be formed by the reactions of primary amines with α-acylamino ketones (**27**), cyclize in the presence of phosphoryl chloride or phosphorus pentachloride to give 1-substituted imidazoles. If a substituted hydrazine is used instead of a primary amine the reaction gives 1-aminoimidazoles, and the versatility of this reaction is such that it can also be adapted to the synthesis of bicyclic products such as the imidazopyridine (**28**; Scheme 14). The starting α-acylamino ketones can be prepared quite readily from α-amino acids ⟨78LA1916⟩. Ring closure of formylglycine amidines (**29**; $R^1 = R^2 = R^4 = H$; $R^5 =$ NH-alkyl), either induced thermally or by the action of phosphoryl chloride, gives 1-alkyl-5-aminoimidazoles ⟨70AHC(12)103⟩.

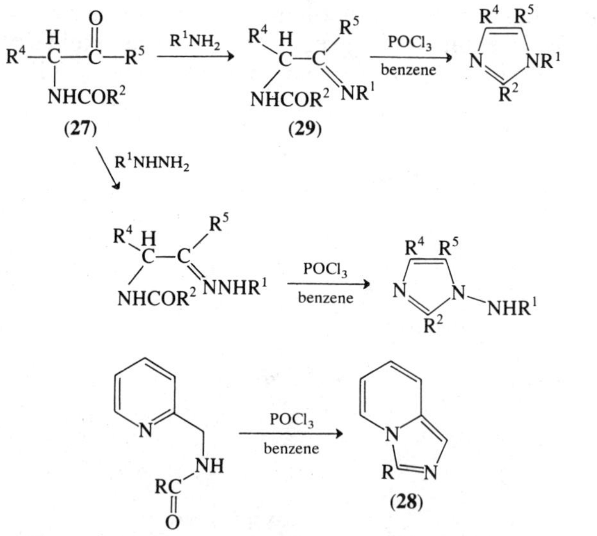

Scheme 14

Adaptation of similar functionality to the synthesis of imidazole N-oxides is illustrated by the fluoride ion-promoted cyclization of the trimethylsilyl ether of the oxime (**30**) derived from 1-(dichloroacetylphenyl)aminopropanone (Scheme 15) ⟨80AHC(27)241⟩.

The reaction of dimethylformamide diethyl acetal with the carbanion which forms when the isocyanide (**31**) is metalated gives an alkene (**32**) which reacts with methyl iodide to give an imidazole product ⟨79LA1444⟩. In the presence of base, 2-isocyanoalkane nitriles (**33**) react with alcohols, thiols or hydrogen sulfide to form imidazoles which have an oxygen or sulfur substituent at C-5 (Scheme 16) ⟨79LA1602⟩.

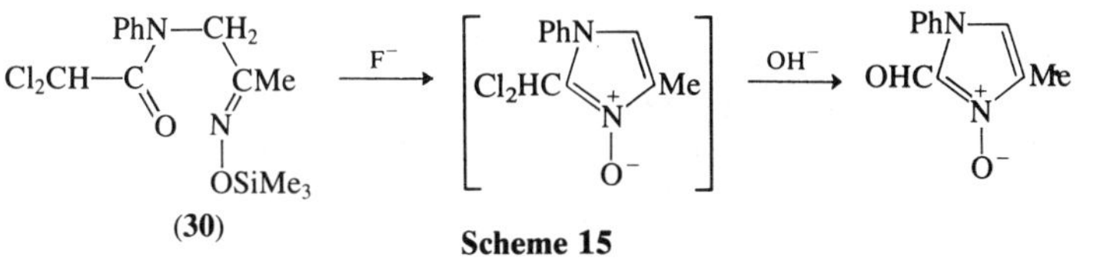

Scheme 15

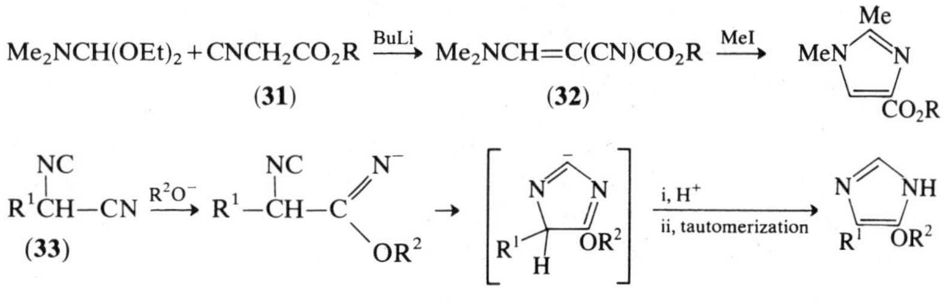

Scheme 16

In addition to the reactions of diaminomaleonitrile shown in Scheme 11 there are also some examples of photochemical transformations which lead to imidazole products. The initial reaction isomerizes the *cis*-dinitrile to the *trans* form which then forms a 5-aminoimidazole-4-carbonitrile *via* the iminoazetine (**34**). There are a number of related reactions, although the photochemical isomerization of enaminonitriles (**35**) probably involves an azirine intermediate (Scheme 17) ⟨81JOC2872⟩.

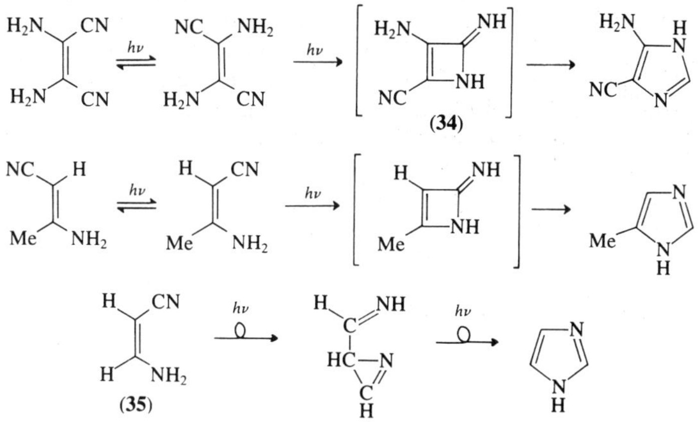

Scheme 17

Some of the above synthetic procedures can be adapted for the synthesis of imidazolines. If a 1,2-diamide is heated alone, or in the presence of aluminum or magnesium, ring closure brings about the formation of a 2-alkyl- or -aryl-imidazoline. Better results are obtained when the diamide and the corresponding diamine hydrochloride are heated at about 300 °C, and good yields (85–95%) of 1,2-diaryl-2-imidazolines result when polyphosphoric acid is used to cyclize *N*-aryl-*N'*-benzoylethylenediamines (Scheme 18) ⟨70JHC791⟩. When monoacylated 1,2-diamines carry some substituent other than hydrogen on the amino function, it is normal to obtain imidazolines rather than imidazoles, and a further example is provided by the cyclization of the oximinoisocyanate to an imidazoline *N*-oxide (Scheme 18).

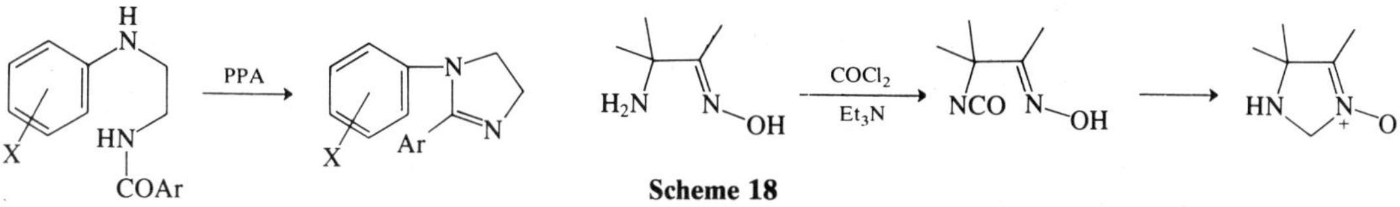

Scheme 18

Usually the 2- or 4-imidazolines that are formed in the cyclization of 3- or 2-aza-1,3-butadienylnitrenes are dehydrogenated under the reaction conditions to imidazoles. There are, however, some cases in which the hydrogenated diazoles can be isolated in moderate yields ⟨79CHE467⟩.

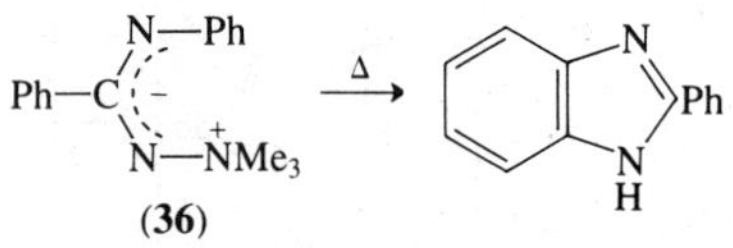

Scheme 19

Above 170 °C the amidrazone ylide (**36**) decomposes with loss of triethylamine and concurrent cyclization to give an 85% yield of 2-phenylbenzimidazole (Scheme 19) ⟨B-75MI40800⟩. Poorer yields (~40%) are obtained when *N*-benzyl-*o*-nitroaniline is pyrolyzed in the presence of iron oxalate. No doubt this last reaction is similar in many respects to the reactions shown in Scheme 2. Both 2-phenyl-imidazoles and -benzimidazoles (as well as other 2-substituted analogues) can be obtained as a result of thermal rearrangement of the 1-substituted isomers (Section 4.07.1.2.2), by radical substitution methods (Section 4.07.1.7) or *via* the 2-lithio derivatives (Sections 4.07.1.6, 4.07.3.7).

4.08.1.1.2 Formation of the 1,5- (or 3,4-) bond

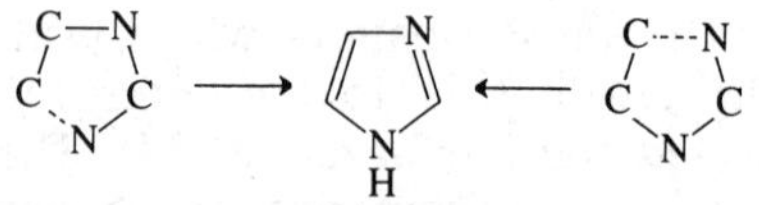

Compounds related to amidines are frequently involved in syntheses of this class. Thus methyl 3-hydroxypropionimidate condenses with the acetal of an aminoaldehyde to give an amidine salt (**37**) which ring closes in acid solution to the imidazole (Scheme 20). This reaction has been applied to the synthesis of 2-phenylimidazole and a number of its 4-substituted derivatives, and to 'isohistamine' [2-(2-aminoethyl)imidazole ⟨70AHC(12)103⟩]. Ring closure in the same way of a cyanide carbon on to an imidic nitrogen allows access to a series of mesoionic imidazoles (**40**) *via* the 4-aminoimidazolium salts (**38, 39**) ⟨80AHC(27)241⟩. Should stronger base treatment (*e.g.* aqueous KOH) be applied to (**38**), it is likely to undergo a Dimroth rearrangement (Section 4.07.3.3.9), and concentrated alkali causes ring fission (Scheme 21).

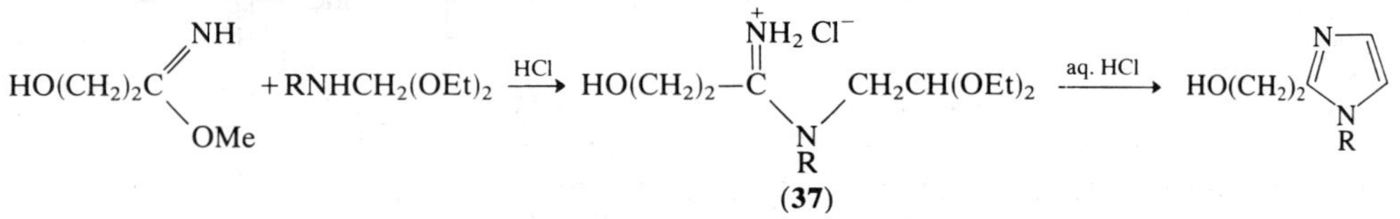

Scheme 20

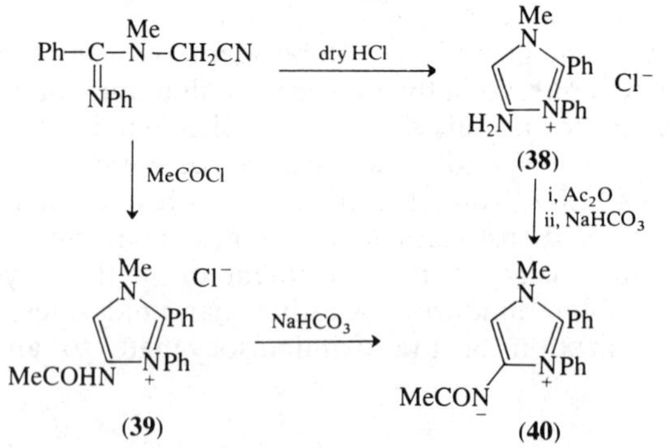

Scheme 21

An electron-deficient trifluoromethyl carbon becomes part of an imidazole ring when (**41**) is heated with tin(II) chloride (Scheme 22).

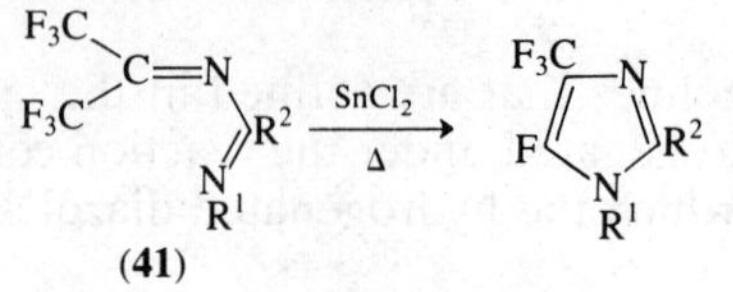

Scheme 22

There is one example in which part of the amidine system is a C—N bond in a heterocyclic ring. The enamino ketone condensation products (**42**) of 3-amino-1,2,4-oxadiazoles and 1,3-dicarbonyl compounds cyclize in basic medium to form 60–80% yields of imidazoles. The driving force for this reaction is provided by the well-established, general attack of a nucleophilic centre in the side-chain at N-2 of the heterocyclic ring, but it is unusual in that a carbon nucleophile (rather than an oxygen or nitrogen species) is implicated (Scheme 23).

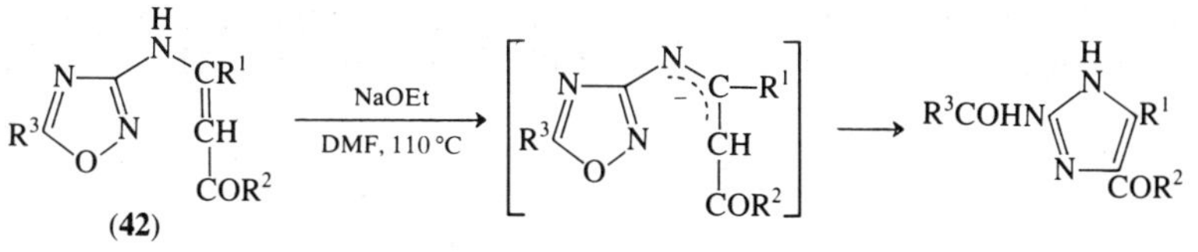

Scheme 23

Claisen rearrangement of the adduct (**43**) formed when a propiolate ester and an aryl aldoxime combine leads to imidazole-4-carboxylates in 61–72% yields. This reaction also proceeds by ring closure of an amidine (**44**; Scheme 24) ⟨80AHC(27)241⟩.

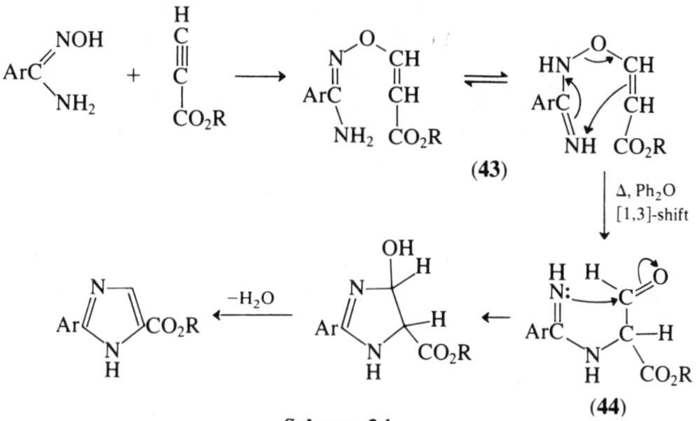

Scheme 24

The formation of benzimidazoles by cyclization of aryl amidines can be accomplished in a number of ways, but is usually carried out under anhydrous conditions in the presence of a base. Thus cyclizations of (**45**; R = OH) induced by benzenesulfonyl chloride in pyridine or triethylamine generally give yields of greater than 60%. The more direct process of oxidizing the parent compounds (**45**; R = H) with sodium hypochlorite under basic conditions gives even higher yields (70–98%). Intermediates in these reactions are the N-haloamidines (**45**; R = Cl); a nitrene intermediate may be involved or merely dehydrochlorination with concomitant cyclization (Scheme 25). Alternative cyclization procedures include oxidation of the parent amidines (**45**; R = H) by manganese dioxide or lead tetraacetate and thermal

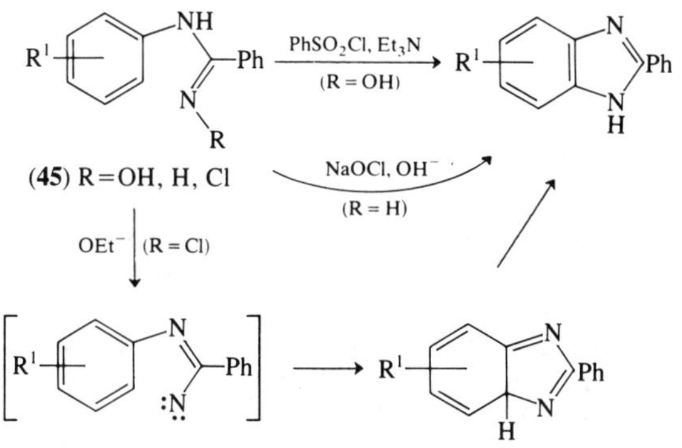

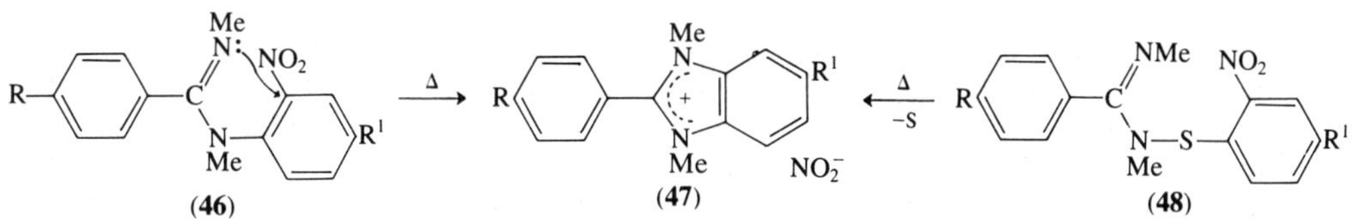

Scheme 25

and photochemical methods. The substituted benzamidine (46), which has a nitro group in the *ortho*-position of the *N*-phenyl ring, undergoes thermal cyclization *via* a nucleophilic substitution reaction, giving the benzimidazolium nitrite (47). The sulfenamide (48) in which the nitrophenyl moiety is linked through sulfur gives the same product with extrusion of the sulfur atom (Scheme 25) ⟨77TL3453, 80AHC(27)241⟩.

Reactions of imidoylaziridines (49) (amidine-type compounds), which ring open and cyclize in the presence of such reagents as oxalyl chloride or ethanolic HCl, provide further examples of 1,5-bond formation (Scheme 26). The products, though, are reduced imidazoles, being triketoimidazolidines (50) and 2-imidazolines (51), respectively ⟨77JOC847⟩.

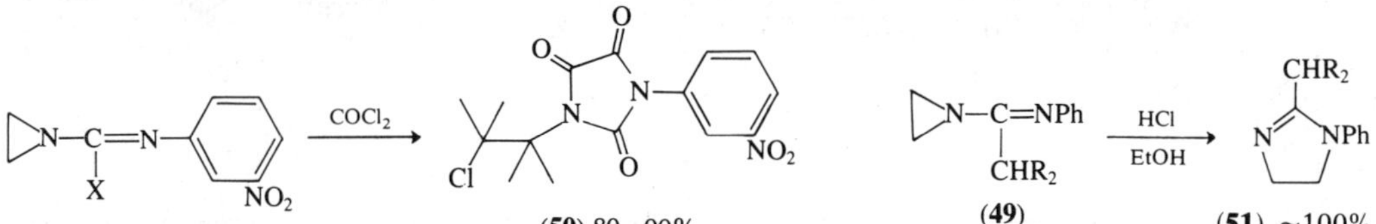

Scheme 26

3-Butynylthiourea and 3-butynylurea are cyclized in sulfuric acid to 4,5-dimethyl-imidazoline-2-thione and the corresponding imidazolin-2-one, respectively (Scheme 27), while the imidazolin-2-ones are also accessible when suitable urea derivatives are cyclized thermally ⟨79JHC983⟩. The use of urea derivatives for the synthesis of 2-imidazolidinones and hydantoins is relatively common (Scheme 27) ⟨79JHC607, B-75MI40800⟩.

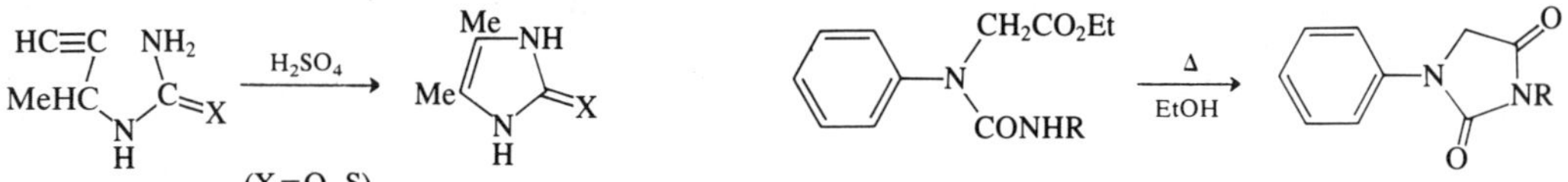

Scheme 27

Thermal ring-closure of enecarbamoylazides (52) gives rise to 4-imidazolin-2-ones (Scheme 28) ⟨71JHC557⟩.

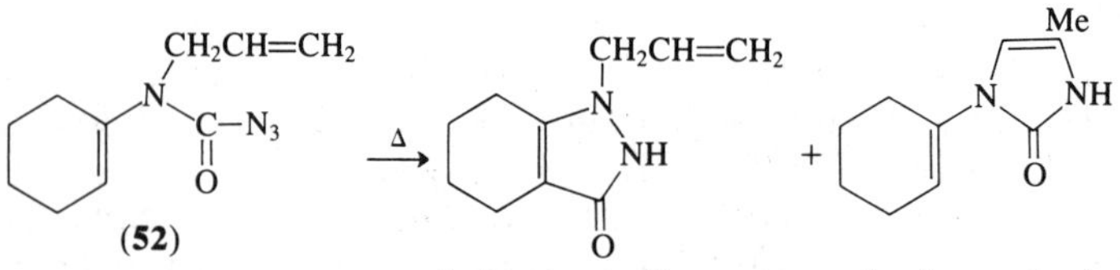

Scheme 28

4.08.1.1.3 Formation of the 4,5-bond

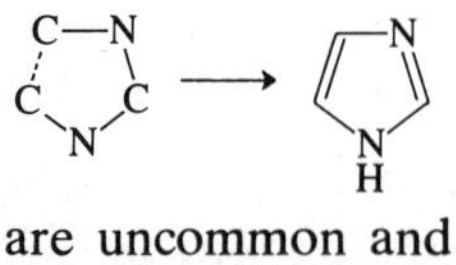

Synthetic methods of this class are uncommon and are limited largely to examples in which an active methylene group, in basic medium, cyclizes on to the carbon of a nitrile function. The acyclic precursors required for the reactions can be made by alkylation of arylaminomethylene cyanamides (53), or cyanoimidothiocarbamates, with α-halogen-substituted carbonyl compounds ⟨76MI1413⟩, or by the condensation in the presence of triethylamine of 2-aminonitriles (or glycine esters) with *N*-cyanoacetimido esters (54) ⟨75HCA2192⟩. Subsequent base-catalyzed cyclization gives 4-aminoimidazoles in yields between 40 and 90% (Scheme 29). Such 4-aminoimidazoles substituted at C-5 by nitrile or carbonyl functions are useful intermediates for purine synthesis.

A further example of a similar sequence of reactions is provided by the reactions of *N*-cyaniminodithiocarbonic esters (55) with α-cyanoammonium salts. The initial condensation product (56) cyclizes when heated with sodium ethoxide in DMF, again giving a 4-amino-5-cyanoimidazole. Related is the reaction of (55) with the hydrochloride salt of ethyl methylaminoacetate, which gives imidazoles directly; perhaps this might be classified

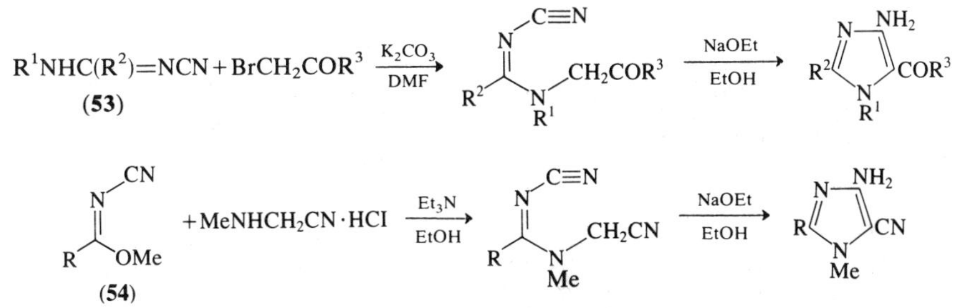

Scheme 29

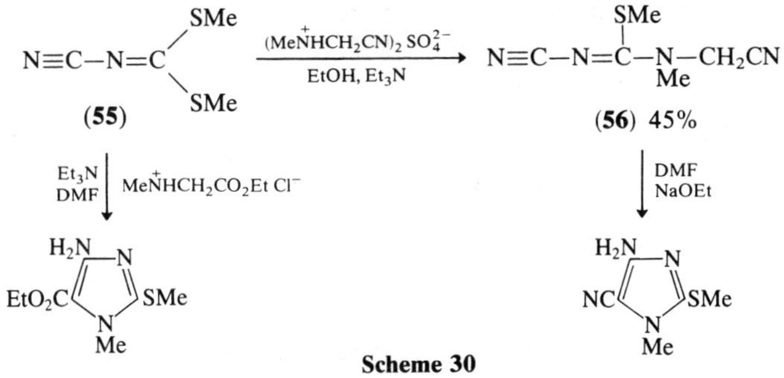

Scheme 30

more correctly as a cycloaddition leading to the formation of the 1,2- and 1,5-bonds (Scheme 30) ⟨70AHC(12)103⟩.

In the presence of a catalytic amount of a metal carbonyl, benzylamine reacts with carbon tetrachloride to form a mixture of 2,4,5-triphenylimidazole and 2,4,5-triphenyl-2-imidazoline. The reaction is probably of the free radical type involving, ultimately, cyclization of a species such as (**57**). When hydrobenzamide (**58**) is treated with an acid chloride in the presence of triethylamine the product is an *N*-acyl-3-imidazoline (**59**). When, however, (**58**) is heated and subsequently reacts with an acid chloride in the presence of triethylamine, the isomeric 2-imidazoline is formed (Scheme 31) ⟨72JOC2158⟩.

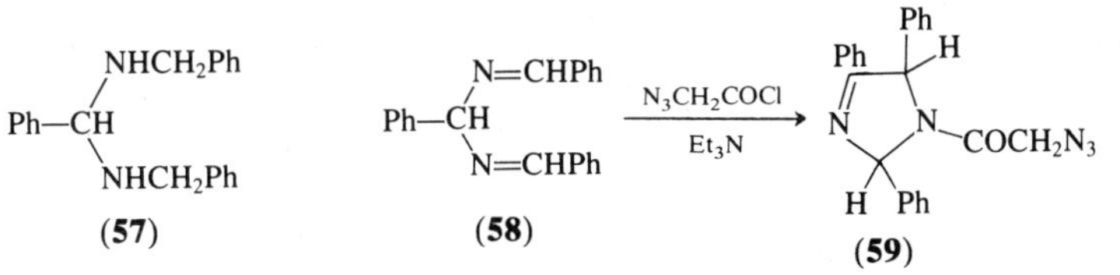

Scheme 31

4.08.1.2 Formation of Two Bonds

4.08.1.2.1 From [4+1] carbon fragments

(i) Formation of the 1,5- and 1,2-bonds

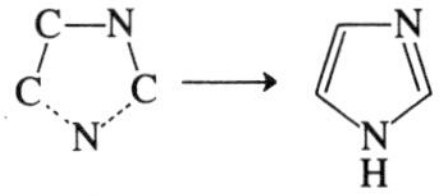

Alkyl *N*-cyanoalkylimidates (**60**) react with primary amines to form 1-substituted 5-aminoimidazoles. This reaction has been applied to the synthesis of 5-aminoimidazole nucleosides (Scheme 32). The reaction appears to be quite general for a wide variety of such imidates with not only primary alkyl- and aryl-amines, but also hydrazines (not 1,2-disubstituted hydrazines) and semicarbazides. Hydrazines lead to 1-aminoimidazole products: thus 1,5-diaminoimidazole (**61**) can be prepared from hydrazine and *N*-cyanomethylformimidate ⟨80AHC(27)241, 66RCR122, 80JCS(P1)2310⟩.

In the presence of phosphorus trichloride, phenacylbenzamides react with primary arylamines with the formation of 1,2,5-triarylimidazoles (**62**), albeit in less than 10% yields

(61)

Scheme 32

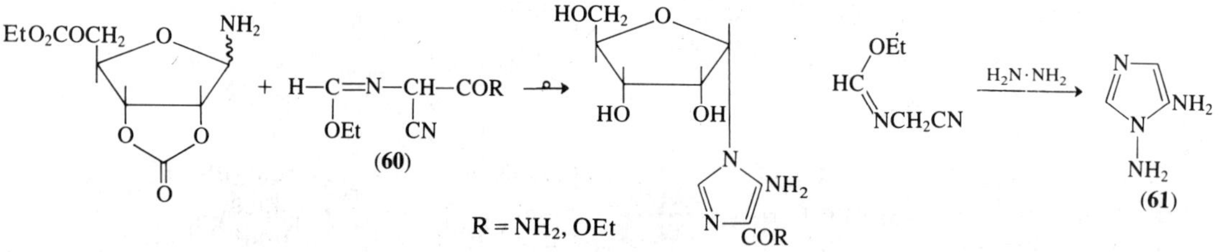

(Scheme 33) ⟨72CHE1142⟩. Both 1,2,4,5-tetraarylimidazoles and the pentaarylimidazolium salts are formed when compounds of type (**63**) react with either ammonia or primary aromatic amines; in the latter instance a follow-up treatment with thionyl chloride is required to produce the quaternary compound (Scheme 33) ⟨70AHC(12)103⟩.

Scheme 33

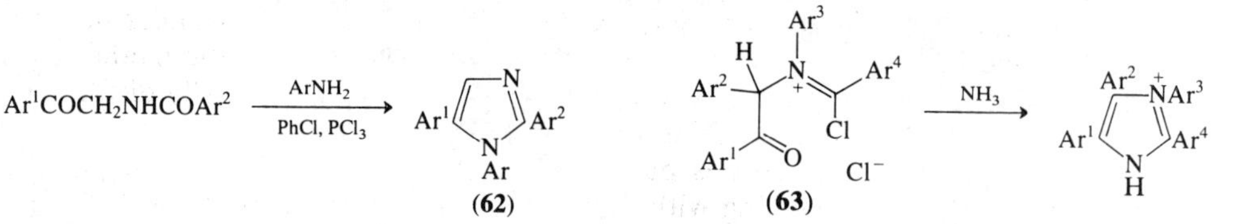

When 1-substituted *N*-formylglycine esters take part in a carbanion condensation reaction with formate esters in the presence of base, the resulting *N*-alkenylformamides (**64**) can be cyclized with ammonia to give 1-substituted imidazole-4,5-dicarboxylates (strictly speaking, it is the 2,3- and 3,4-bonds which are being formed in this instance). The reaction products, after reduction to the dialdehydes, can be converted by reaction with hydrazine into imidazo[4,5-*d*]pyridazines ⟨80AHC(27)241⟩. In a further reaction which shows some similarities, 4-acylimidazoles (**65**) can be prepared from 1,3-dicarbonyl compounds which are nitrated in their enolic forms. Reduction of the nitroalkene gives an intermediate *N*-alkenylformamide which is cyclized by formamide in formic acid (Scheme 34) ⟨79HCA497⟩.

(65)

Scheme 34

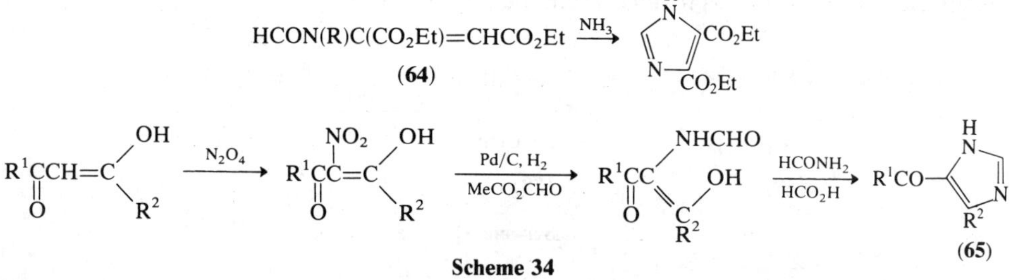

From the interaction at low temperatures in basic medium of an aldehyde or ketone and tosylmethyl isocyanide ('tosmic'), *N*-(1-tosyl-1-alkenyl)formamides are formed. When these are dehydrated they produce 1-isocyano-1-tosylalkenes (**66**) which react with primary amines to form imidazoles in high yields (Scheme 35) ⟨79RTC258⟩.

Scheme 35

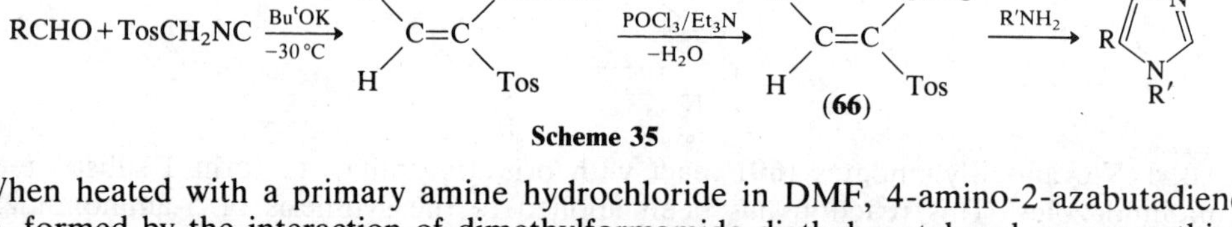

When heated with a primary amine hydrochloride in DMF, 4-amino-2-azabutadienes (**67**), formed by the interaction of dimethylformamide diethyl acetal and an azomethine, cyclize to imidazoles (Scheme 36) ⟨81AG(E)296⟩. Imidazolidin-4-ones are formed when α-(*N*-acylhydroxyamino) esters react with ammonia ⟨B-77MI40800⟩.

Scheme 36

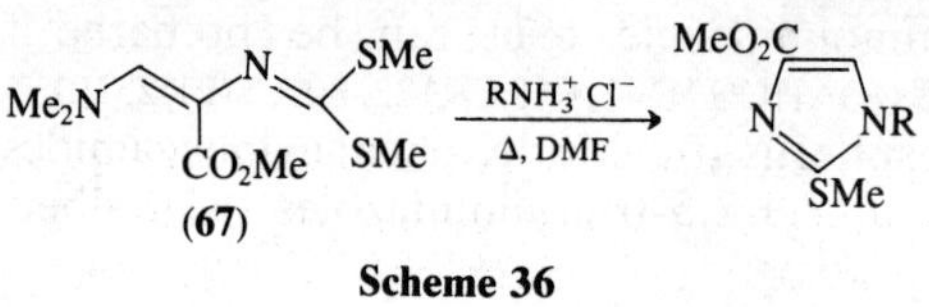

(ii) Formation of the 1,2- and 2,3-bonds

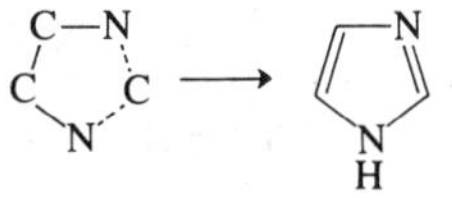

Reactions which involve variations on the cyclization of a 1,2-diaminoalkene are among the most common methods leading to the formation of the imidazole nucleus, and by far the most common approach to the syntheses of benzimidazoles, benzimidazolinones and 2-imidazolines. The cyclization of such compounds in which one of the amino groups is already acylated has been discussed earlier (Section 4.08.1.1).

High yields of 2-alkylimidazoles (**68**) are obtained when a 1,2-diaminoalkene is treated with an alcohol, aldehyde or carboxylic acid at high temperatures, and in the presence of a dehydrogenating agent such as Pt/Al_2O_3 (Scheme 37). In the absence of dehydrogenating agents the products are imidazolines, which are also available from the interaction of the diamines with imino ether hydrochlorides, amidine hydrochlorides, thioamides or nitriles ⟨B-57MI40800⟩. One of the problems associated with this general synthetic method is the purification of the reaction products. The compounds can be azeotropically distilled with an alkylaromatic hydrocarbon possessing a boiling point about 10–40 °C lower than the imidazoles. Thus 2-methylimidazole is purified by distillation with 1- or 2-methylnaphthalene and then isolated by washing with toluene or pentane, solvents in which the azole is only very sparingly soluble.

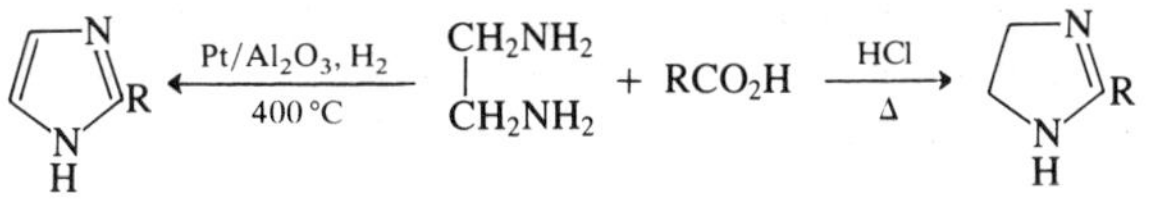

Scheme 37

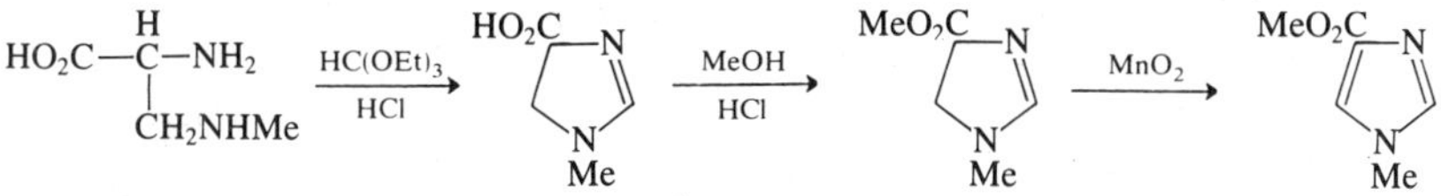

The specific synthesis of 1,4- and 1,5-disubstituted imidazoles in 70% yields has been achieved by cyclizing 2-amino-3-methylaminopropionic acid with triethyl orthoformate. The products are isolated after dehydrogenation with active manganese dioxide of the 2-imidazoline (**69**; Scheme 37) ⟨70AHC(12)103⟩.

When an *o*-diaminobenzene is substituted for ethylenediamine this provides the most important synthetic method for benzimidazoles. In fact, an 83% yield of benzimidazole is produced when *o*-phenylenediamine and formic acid are maintained at room temperature for five days. At 100–110 °C only two hours are required to achieve this yield, testifying to the ease of the reaction (Scheme 38), which is applicable to a wide spectrum of 2-substituted benzimidazoles ⟨70AHC(12)103, 66RCR122⟩. Usually the reaction is carried out in the presence of hydrochloric acid, and the yields can depend not only on the natures of the diamine and carboxylic acid but also on the mineral acid concentration. Almost quantitative yields of 2-benzyl- and 2-(2-phenylethyl)-benzimidazoles can be obtained under the optimum conditions. The 2-arylbenzimidazoles can also be formed in 75–95% yields from *o*-phenylenediamine and benzoic acid or its derivatives (*e.g.* esters, nitriles, amides, imidates, selenoamides) (see Scheme 40). Much more vigorous reaction conditions, however, are required than for the 2-alkyl derivatives. Thus a 95% yield of 2-phenylimidazole can be obtained if the diaminobenzene and benzoic acid are heated for 40 minutes in a sealed tube at 180 °C in the presence of 25% hydrochloric acid. Much poorer yields are obtained in the absence of the mineral acid and with halogen-substituted benzoic acids. In place of the carboxylic acids it is quite common to employ aldehydes in the presence of oxidizing agents. For the synthesis of benzimidazoles using aryl carboxylic acids it is preferable to use polyphosphoric acid in place of hydrochloric acid as the reaction medium. The diamine often competes successfully for the proton of the acid catalyst and this can inhibit attack at the carbonyl carbon. This problem can be overcome by replacing the carbonyl oxygen with the much more basic imino group; thus imino ethers (imidates) have proved very successful in the reaction (Scheme 38) ⟨74CRV279⟩.

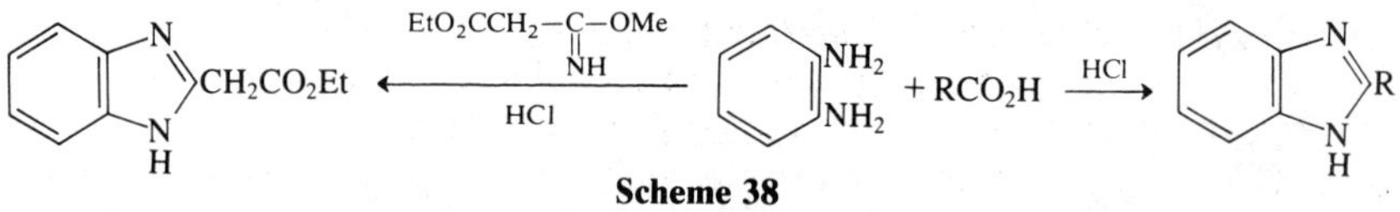

470   *Imidazoles and their Benzo Derivatives: (iii) Synthesis and Applications*

Scheme 38

With dicarboxylic acids, bis(2-benzimidazolyl)alkenes are obtained, and benzimidazoles substituted at C-2 by heterocyclic groups use the heterocyclic carboxylic acid in the presence of a cation exchange resin. As can be seen in Scheme 39, the general reaction can be modified to prepare 1-substituted benzimidazoles by use of an *N*-substituted *o*-diamine; the use of two moles of formaldehyde in boiling ethanolic HCl also gives moderate yields of 1-methylbenzimidazoles ⟨74JCS(P1)903⟩. When, however, an *N,N*-disubstituted *o*-phenylenediamine reacts with chloral hydrate and hydroxylamine, semicarbazide or phenylhydrazine, benzimidazoles substituted in the 2-position by oxime, semicarbazone and hydrazone groups are formed in about 60% yields. A reaction sequence similar to that shown in Scheme 39 may apply, proceeding *via* the chlorimine intermediate (**70**) ⟨74CRV279⟩.

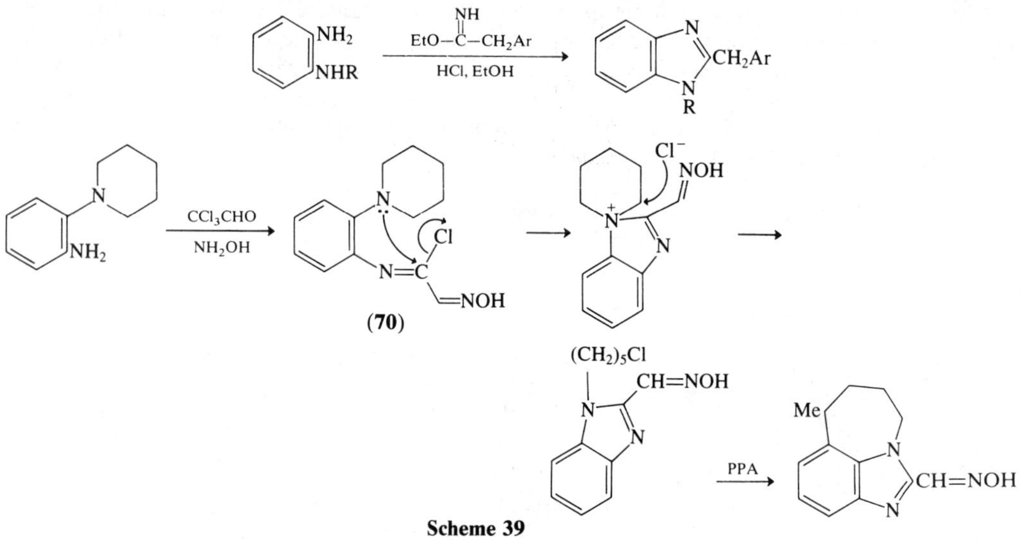

Scheme 39

The great diversity of reagents which can react with *o*-arylenediamines to give 2-substituted benzimidazoles is exemplified in Scheme 40 ⟨66RCR122⟩. It should be noted that in addition to these ring-synthetic methods, 2-substituted benzimidazoles can also be obtained from benzimidazoles by nucleophilic substitution reactions (Section 4.07.1.5), from the 2-lithio derivatives (Section 4.07.1.6) or by thermal rearrangement of 1-substituted benzimidazoles (Section 4.07.1.2).

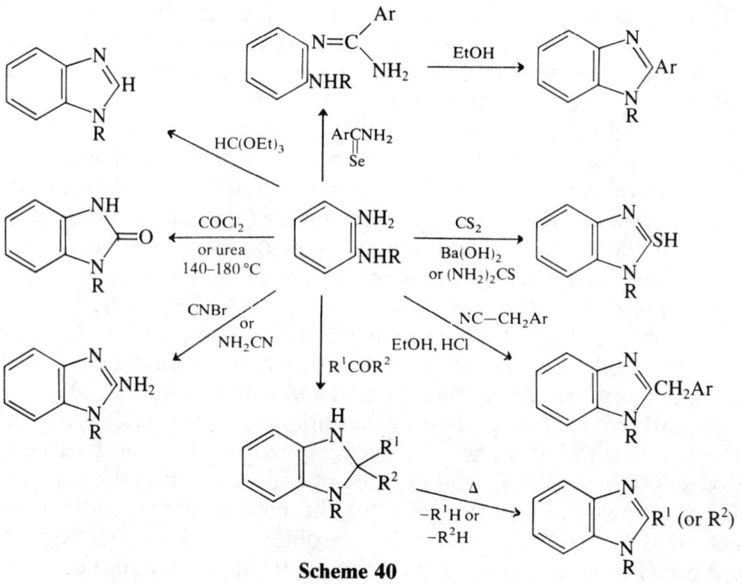

Scheme 40

When there is no dehydrogenating agent present, the reactions of 1,2-diamines and derivatives with carbonyl compounds will lead to reduced imidazoles, either imidazolines or imidazolidines (Scheme 41). Conventional preparation of 2-imidazolines normally requires nitriles or imino ethers, and only in selected cases can carboxylic acid esters be used directly with 1,2-diamines (similar limitations apply to benzimidazole formation). However, when coupled with trimethylaluminum, reagents (**71**) are formed which react quite readily with esters to give high yields of 2-imidazolines ⟨81JOC2824⟩. Imidazolidin-2-ones (**72**) are readily obtained from diamines and urea (or diethyl carbonate) heated together. If the diamine is treated with nitroguanidine, the product is the 2-nitraminoimidazoline (**73**; Scheme 41) ⟨80CB2823, 81JCS(P2)317, 80JHC1789, 79JCS(P1)2289, 81JOC2824⟩.

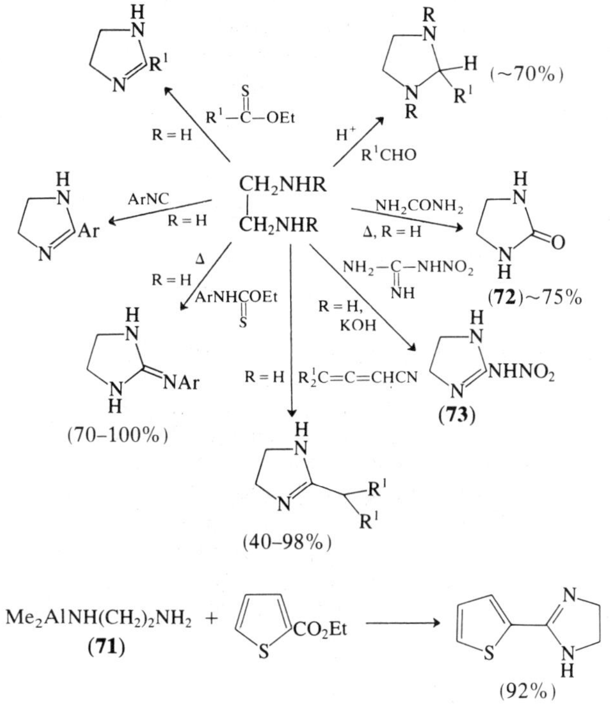

Scheme 41

The reaction between o-phenylenediamine and an equimolar amount of an aromatic or heterocyclic aldehyde has been shown to proceed by initial formation of a monoanil (**74**). In the presence of oxidizing agents (e.g. nitrobenzene, which also acts as the solvent) this can form the 2-substituted benzimidazole. With two moles of aldehyde the bis-anil (**75**) forms, giving rise to a 1,2-disubstituted benzimidazole (Scheme 42). This aldehyde route to benzimidazoles is particularly suited to the synthesis of compounds with a heterocyclic group (e.g. 2-thienyl-, 2-pyridyl-) at C-2. Reaction of 2,2,4,4-tetrakis(trifluoromethyl)-1,3-dithietane with o-phenylenediamine gives 2,2-bis(trifluoromethyl)benzimidazoline ⟨74BCJ785⟩.

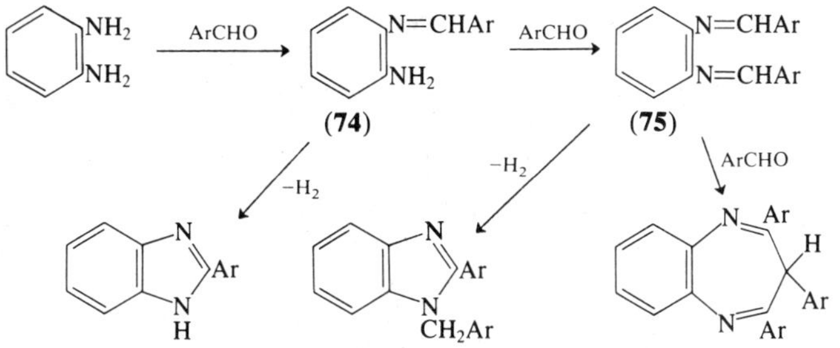

Scheme 42

β-Keto esters react with arylenediamines when heated in neutral solvents to give benzimidazolinones in a reaction which is capable of extension to the β-keto esters derived from cyclic ketones. Under acidic conditions, o-phenylenediamine gives a 78% yield of

2-methylbenzimidazole when treated with acetoacetic ester, but the benzimidazolinone (**76**) predominates in neutral medium (Scheme 43). The compounds can be obtained even more efficiently by heating the nitroamines in sand at 240 °C ⟨74CRV279⟩.

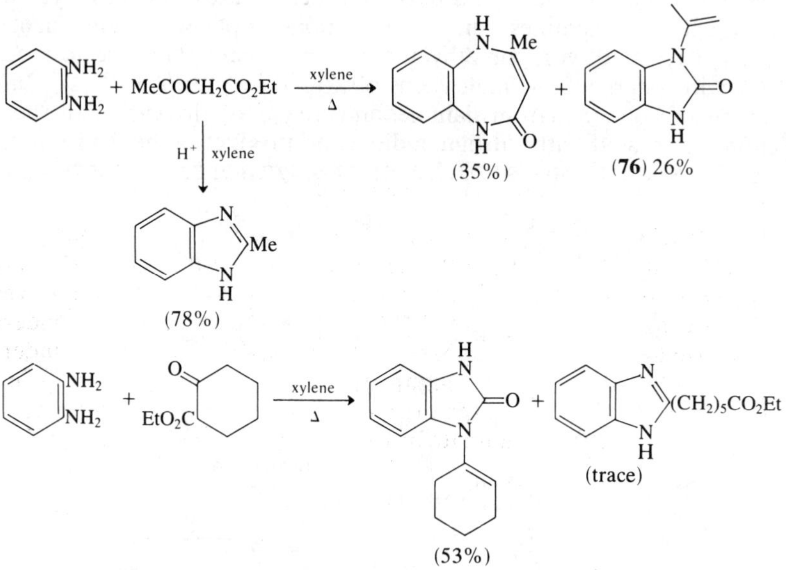

Scheme 43

Such use of *o*-nitroanilines in place of *o*-phenylenediamines has found increasing favour in the field, the cyclization and reduction processes usually being combined into one. Thus in the presence of a suitable reducing agent (*e.g.* metal oxalates, iron pentacarbonyl, triethyl phosphite, titanium(III) chloride, hydrazine hydrate in ethanol with Raney nickel, or hydrogenation using palladium and carbon) the compounds can be cyclized to give high yields of benzimidazoles. In connection with the interest in vitamin B_{12}, much work has been done on the preparation of 5,6-dimethylbenzimidazole and its 1-glycosyl derivative since such compounds have high physiological activity. The acetylated glycosyl compound (**77**) can be prepared by the sequence of reactions in Scheme 44 ⟨66RCR122⟩. The mechanism of benzimidazole formation, which involves the electrochemical reduction of *o*-nitroanilines which have an alkyl substituent on the amino group, involves a nitroso compound. This is formed in a redox reaction and takes part in the cyclization process ⟨74BSF673⟩.

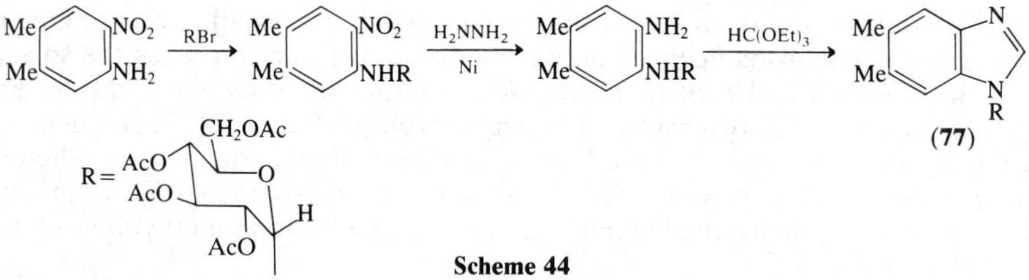

Scheme 44

Although the direct oxidation of the parent compounds with organic peracids has not proved fruitful, benzimidazole *N*-oxides can be made by the action of boiling hydrochloric acid on *o*-nitro-*N*,*N*-dialkylanilines, although prolonged treatment results in the formation of a chloro-substituted benzimidazole ⟨B-73MI40800⟩.

Some of the reactions of diaminomaleonitrile have already been outlined (Schemes 11 and 17). Reactions of this versatile compound which lead to the formation of the 1,2- and 2,3-bonds of imidazole have also been described. When heated in alcoholic solution with an alkyl orthoformate, 4,5-dicyanoimidazole is formed; with formic acid in refluxing xylene, 4-cyanoimidazole-5-carboxamide results; with cyanogen chloride the product is 4,5-dicyanoimidazolin-2-one. With catalytic amounts of acid present, diaminomaleonitrile reacts with acetone to give an 80% yield of 4,5-dicyano-2,2-dimethyl-2*H*-imidazole. Diiminosuccinonitrile (**78**) reacts with acetone and isopropylamine to form the 3-imidazolin-5-one (**79**), while with phosgene the imidazolidinone (**80**) results. The trimethylsilylimine (**81**) gives 4,5-diphenyl-2*H*-imidazolinone (Scheme 45) ⟨72JOC4136⟩.

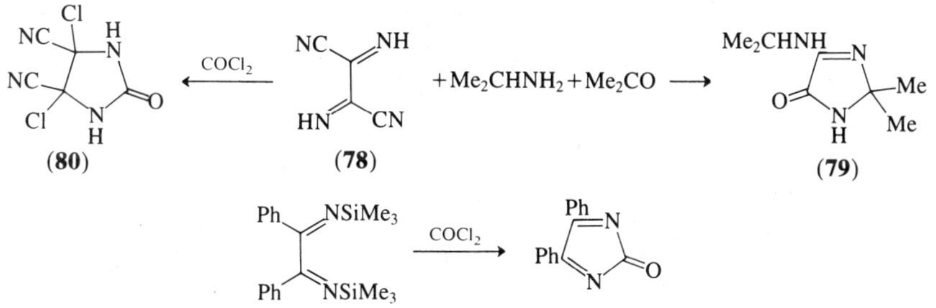

Imidazoles and their Benzo Derivatives: (iii) Synthesis and Applications

Scheme 45

A reaction restricted to the synthesis of 2,4-disubstitued imidazoles is that between 2-aminonitriles and aldehydes ⟨71CB1562⟩, and 4,5-dichloroimidazoles, previously not accessible by general synthetic methods, can be obtained in 36–72% yields when aldehydes react with cyanogen in the presence of HCl. Aldehydes which can enolize undergo aldol-type side reactions, while basic heteroaromatic aldehydes merely form salts under the reaction conditions ⟨80AG(E)130⟩. The condensation of 2-azidoanilines with aromatic aldehydes in the presence of fluoroboric acid gives phenylogous azido-iminium salts (**82**) which cyclize to N-diazonium ions and then eliminate nitrogen to give 2-arylbenzimidazoles (Scheme 46) ⟨80H(14)1725⟩.

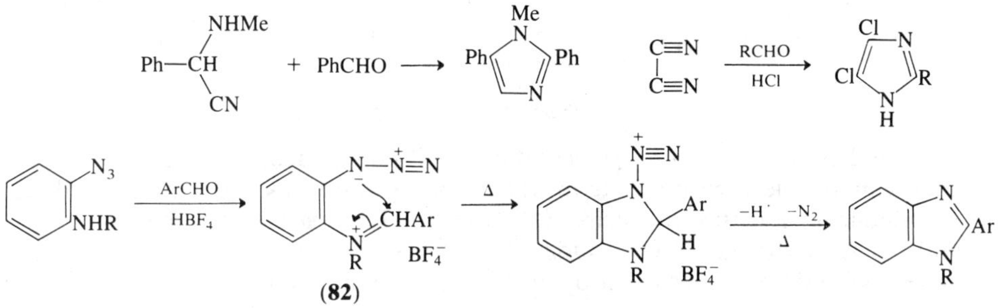

Scheme 46

At 180 °C acetic anhydride reacts with 1,2-dibenzoylaminoethylene to give 2-methylimidazole. Although some other anhydrides can take part in this reaction it is not successful for the synthesis of a wide range of 2-substituted imidazoles ⟨70AHC(12)103⟩.

The reactions of α-aminooximes and α-dioximes with aldehydes to form imidazole N-oxides are discussed with related synthetic methods in Section 4.08.1.2.2.

4.08.1.2.2 From [3+2] carbon fragments

(i) Formation of the 1,5- and 2,3-bonds

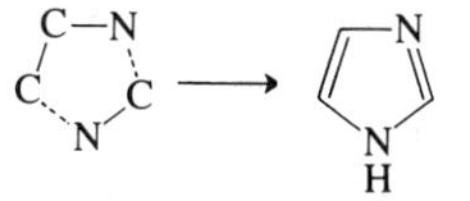

The earliest method of this type, developed by Marckwald, employed the reaction of α-aminocarbonyl compounds (or their acetals) with cyanates, thiocyanates or isothiocyanates to give 3H-imidazoline-2-thiones. These compounds can be converted readily into imidazoles by oxidation or dehydrogenation. The major limitations of this synthetic procedure are the difficulty of synthesis of a wide variety of the α-aminocarbonyl compounds, and the limited range of 2-substituents which are introduced. The reduction of α-amino acids with aluminum amalgam provides one source of starting materials. The method has been applied to the preparation of 4,5-trimethyleneimidazole (**83**) from 2-bromocyclopentanone ⟨70AHC(12)103⟩, and to the synthesis of pilocarpine (**84**; Scheme 47) ⟨80AHC(27)241⟩. If esters of α-amino acids react with cyanates or thiocyanates, the products are hydantoins and 2-thiohydantoins, respectively.

If cyanamide is used in place of isocyanate or thiocyanate, the cycloaddition reaction with an α-aminocarbonyl compound gives a 2-aminoimidazole ⟨80AHC(27)241⟩. Alkylation of the sulfur atom of ethyl 2-thiooxamate gives products (**85**) which contain nucleophilic

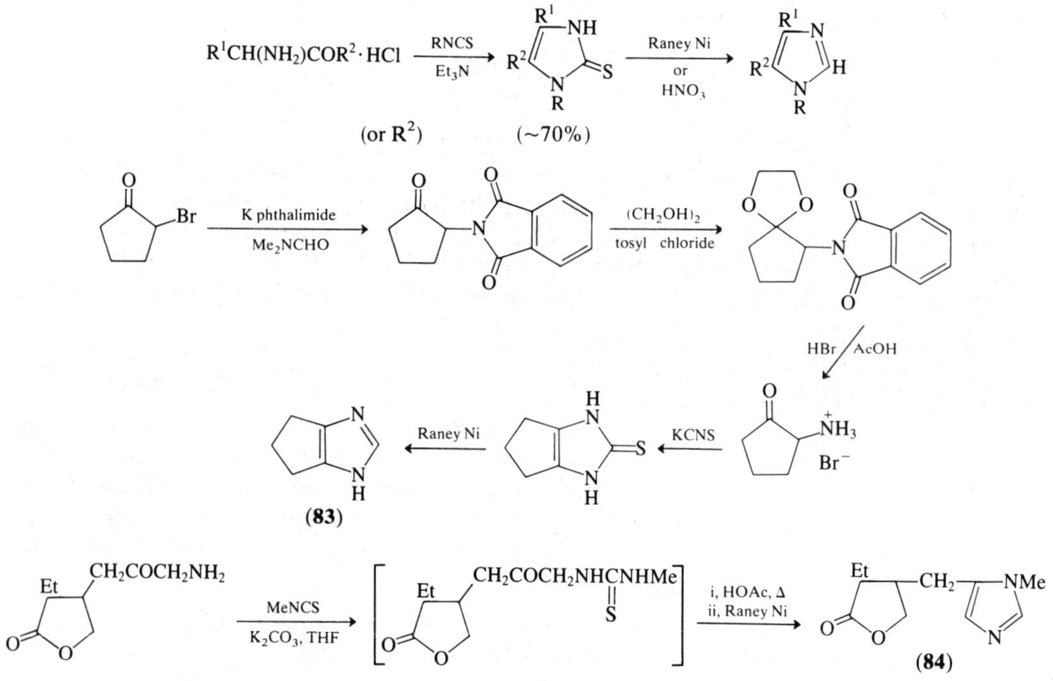

Scheme 47

nitrogen atoms and good leaving groups. Treatment with aminoacetone gives good yields of ethyl 4-methylimidazole-2-carboxylate in what appears to be a useful, general synthetic approach to imidazole-2-carboxylic acids ⟨73JOC1437⟩. The exothermic reaction of α-oxothionamides with aldimines gives imidazole-4-thiols. Cyclization of the isonitrosochromanone (86) with imines gives a 4-aroylimidazole (Scheme 48) ⟨76TL4333⟩.

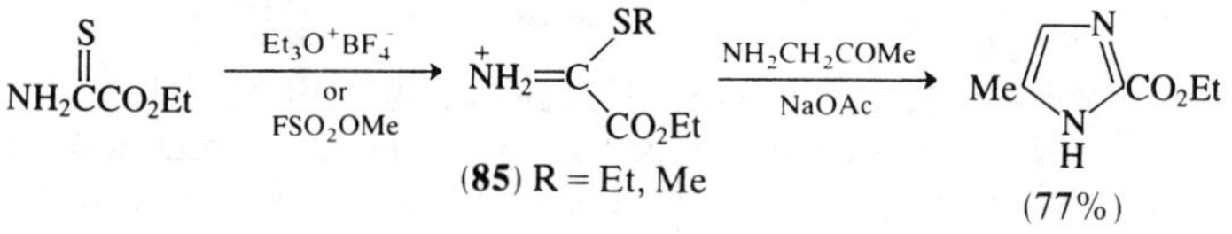

Scheme 48

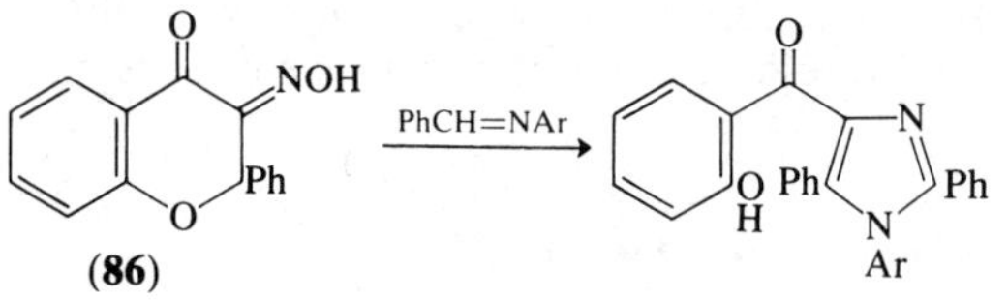

Usually such an oximino ketone would react with an aldimine to form either an N-hydroxyimidazole or an imidazole N-oxide. In aqueous ammonia, butane-2,3-dione monoxime reacts with aldehydes to give 1-hydroxyimidazoles by the pathway shown (Scheme 49). The tautomeric equilibrium between N-hydroxyimidazoles and imidazole N-oxides has been discussed earlier (Section 4.06.5.2). Since direct N-oxidation of the imidazole ring has not been successful, ring-synthetic methods have, of necessity, become important. Most of these will be discussed at this stage even though some do not conform strictly to the classification implied by the heading. The interactions of 1,2-dioximes with aldehydes, or oximino ketones with aldimines, aldehydes and ammonia or amines, or aldoximes give 1-hydroxyimidazoles, imidazole N-oxides and 1-hydroxyimidazole 3-oxides. In addition, α-hydroxyaminooximes give 1-hydroxy-3-imidazoline 3-oxides which can be dehydrated readily to imidazole N-oxides ⟨80AHC(27)241, 73CHE1175⟩.

Cyclohexanedione- and cycloheptanedione-1,2-dioximes react with ketones in acid medium to give 2H-imidazole 1,3-dioxides; α-aminooximes give 3-imidazoline 3-oxides under similar conditions (Scheme 50) ⟨80CHE628⟩.

The cyclization of anti-(E)-α-aminooximes (87) to 3-imidazolin-2-one 3-oxides (89; Scheme 51) can be induced by phosgene (or by the hydrolysis of the urethanes 88). In instances in which (87) has an α-hydrogen atom (R' = H), then the product formed is the

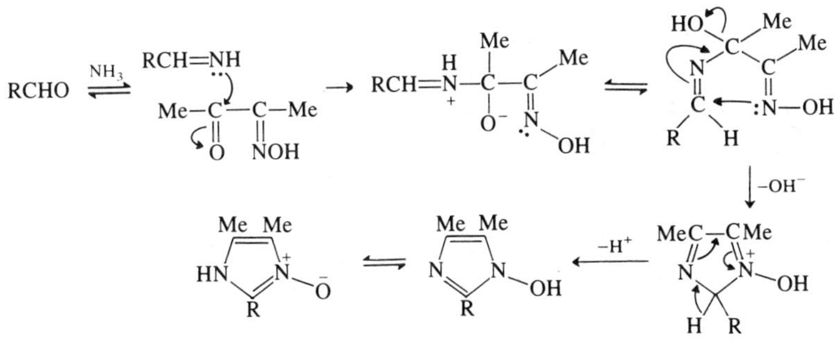

Scheme 49

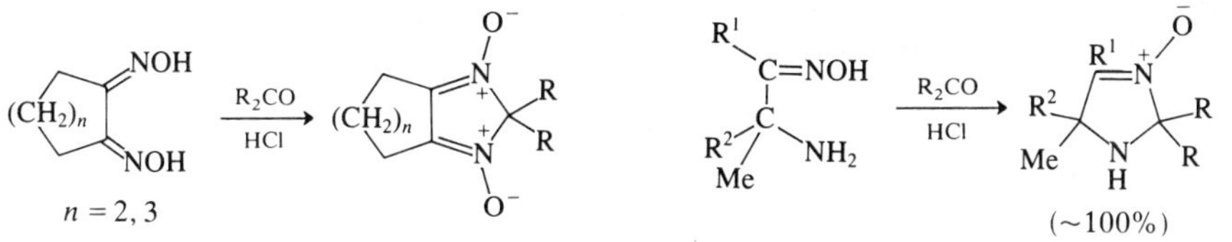

Scheme 50

1-hydroxy-4-imidazolin-2-one (**90**). It is of interest that while both *anti-* and *syn-α*-aminooximes react with acetimidic ester to generate 2-methylimidazoles, orthoacetic ester only reacts with the *anti* isomer to form the dimeric 4*H*-imidazole *N*-oxide; the *syn*-aminooximes are converted instead into pyrazine *N*-oxides. Although the majority of reactions of α-ketooximes lead to imidazoles with *N*-oxygen substituents, there are exceptions such as with (**86**), and the formation of 4-acetyl-5-methyl-2-phenylimidazole from benzylamine and 3-oximinopentane-2,4-dione ⟨80AHC(27)241⟩. In fact, yields of 2-arylimidazoles of 30–80% result from such interaction of benzylamines with α-oximinoketones in solvents such as DMSO, toluene and acetonitrile ⟨80JHC1723⟩.

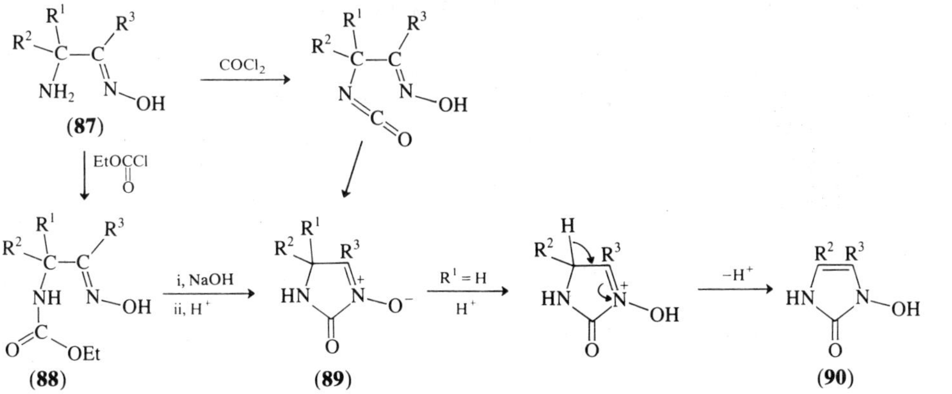

Scheme 51

A 40% yield of 5-hydroxy-1-isopropyl-2-methyl-4-propylimidazole (**91**) has been obtained from the interaction of norvaline methyl ester and *N*-isopropylacetonitrilium tetrachloroferrate (Scheme 52).

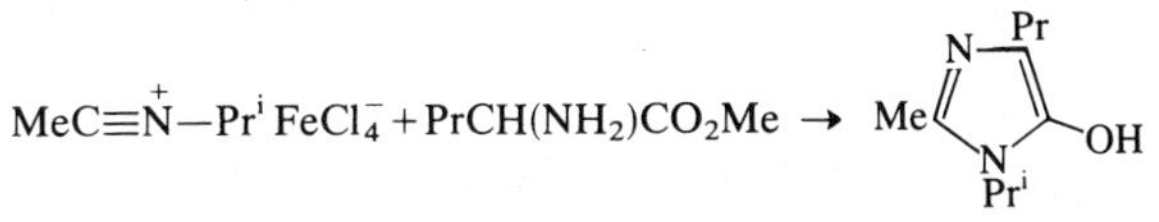

$$MeC\equiv\overset{+}{N}-Pr^i\ FeCl_4^- + PrCH(NH_2)CO_2Me \rightarrow$$ (**91**)

Scheme 52

Dimerization of ethyl 2-amino-2-cyanoacetate gives diethyl 5-aminoimidazole-2,4-dicarboxylate (**92**) as in Scheme 53. At reflux temperatures in propanol, 1,1-dimethyl-1-phenacylhydrazinium bromide is converted into 2-benzyl-4-phenylimidazole (**93**). Again the cyclization is a dimerization process (Scheme 53) ⟨80AHC(27)241⟩.

Among other reactions which fall under this heading and lead to imidazoles are the synthesis of 4-formylimidazoles from 5-aminopyrimidine and *N*-alkyl (or -aryl) imidoyl chlorides in the presence of phosphoryl chloride. These reactions proceed by cyclization of amidines (**94**) on to the pyrimidine ring and subsequent opening of the six-membered ring (Scheme 54) ⟨81JOC608⟩.

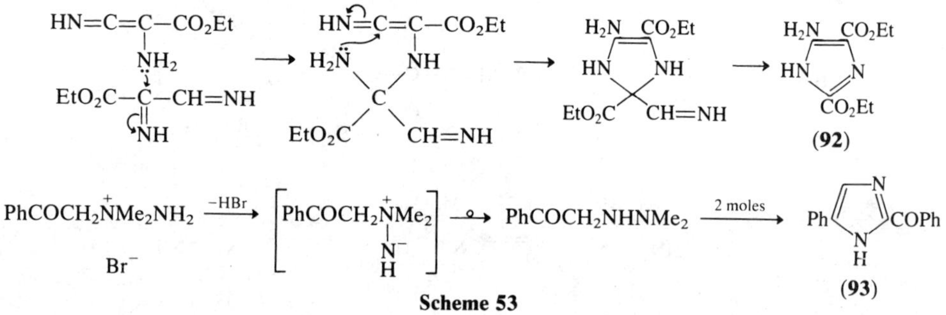

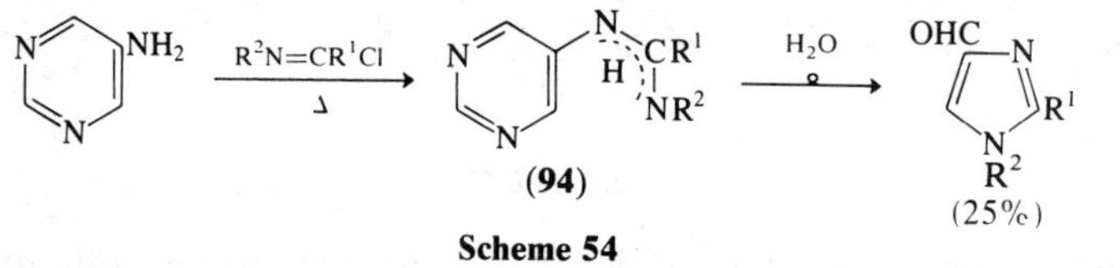

Scheme 54

When 3-aminobut-1-yne is refluxed for 5 hours with acetamide, 2,4,5-trimethylimidazole is formed in a reaction which is capable of application to the synthesis of other *C*-alkylimidazoles ⟨70AHC(12)103⟩. The reaction of Schiff bases with diazenedicarboxylic esters leads, in the presence of phenyllithium or sodium hydride, to 1,3-disubstituted 2-imidazolinones (**95**) in 58–88% yields ⟨81S563⟩. The most general method for preparing 2-imidazolin-4-ones (**96**) consists of reaction between imidates and α-amino acid esters ⟨B-57MI40800⟩. In the presence of potassium thiocyanate in methanol, α-chloroaldimines give 20–90% yields of 2-imidazolidinethiones (**97**), while 1,3-diarylimidazolidinones (**98**) are similarly formed from aryl isocyanates and 2-chloroamines (Scheme 55).

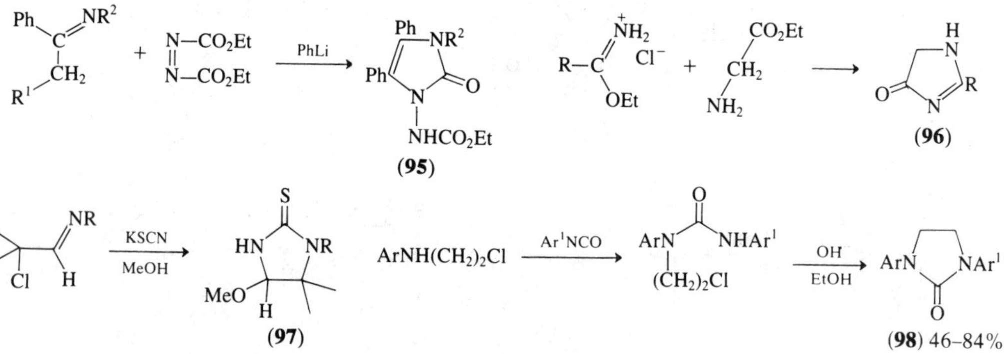

Scheme 55

(ii) Formation of the 1,2- and 4,5-bonds

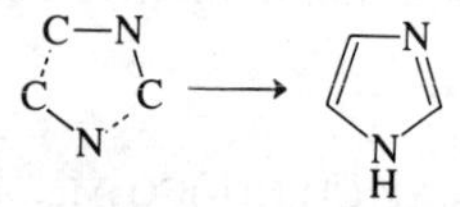

One of the more useful developments in imidazole synthesis in recent years has been the use of the reagent 'tosmic', or toluene-*p*-sulfonylmethyl isocyanide (**99**), which enters into base-induced cycloaddition reactions with aldimines in protic media. Toluene-*p*-sulfinic acid is eliminated to give 1,5-disubstituted imidazoles (**100**; $R^4 = H$). In analogous reactions with imidoyl chlorides, hydrogen chloride is eliminated giving, in these cases, toluene-*p*-sulfonyl derivatives (**100**; $R^4 = p$-tosyl) (Scheme 56). In a similar way, α-tosylbenzyl and α-tosylethyl isocyanides react with aldimines to give 1,4,5-trisubstituted imidazoles. The natures of the substituent groups on the aldimine play an important part and, under aprotic conditions, the primary cycloadducts are imidazolines (which ultimately aromatize to imidazoles). The use of sodium hydride in anhydrous DMSO converts 'tosmic' into its sodium

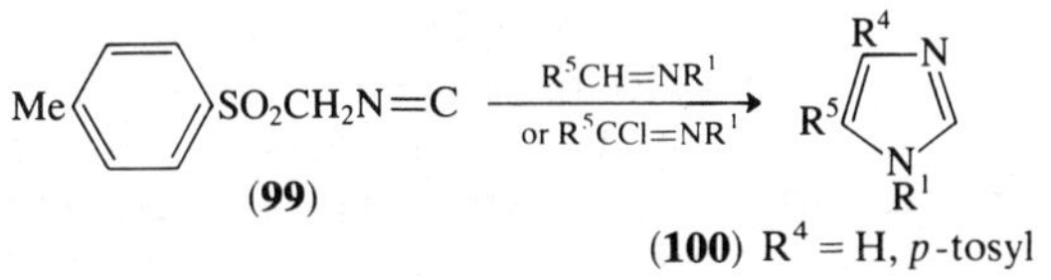

Scheme 56

salt which is now a suitable nucleophile for reaction with the imidoyl compounds. Groups such as *t*-butyl and cyclohexyl on the aldimine render it very prone to hydrolysis and in consequence, even if very dry conditions are employed, only the amide (*e.g.* Bu^{t}CONH-cyclohexyl) is obtained. 'Tosmic' can also be used as a source of *N*-tosylmethylimidic esters or thioesters (**101**), which react with Schiff bases to give 1,2,5-trisubstituted imidazoles (Scheme 57), while the use of nitriles in place of aldimines with 'tosmic' itself leads to the formation of high yields of 4-substituted imidazoles with no nitrogen substituent. This reaction is an example of a $[4\pi + 2\pi]$ cycloaddition, and both reactions in Scheme 57 are capable of extension to other isocyanides (*e.g.* thiomethyl) which can be α-metalated ⟨80AHC(27)241⟩.

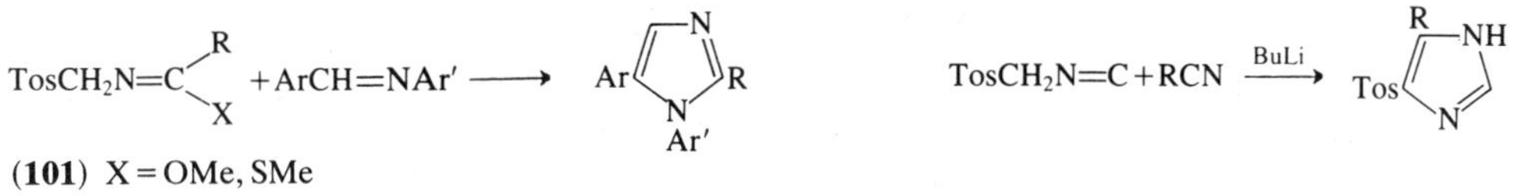

Scheme 57

Should the aldimine have two substituents (other than hydrogen) on the carbon atom then the reaction product, which cannot aromatize, is a 2-imidazoline (**102**; Scheme 58) ⟨77LA1183⟩. A further extension of this type of reaction utilizes dilithiotosylmethyl isocyanide (*i.e.* the dicarbanion of 'tosmic') (**103**), which reacts with carbon–nitrogen multiple bonds with much more facility than does the monoanion. Even such weakly reactive aldimines as isoquinoline will react (albeit in only 44% yield) (Scheme 58) ⟨80TL3723⟩.

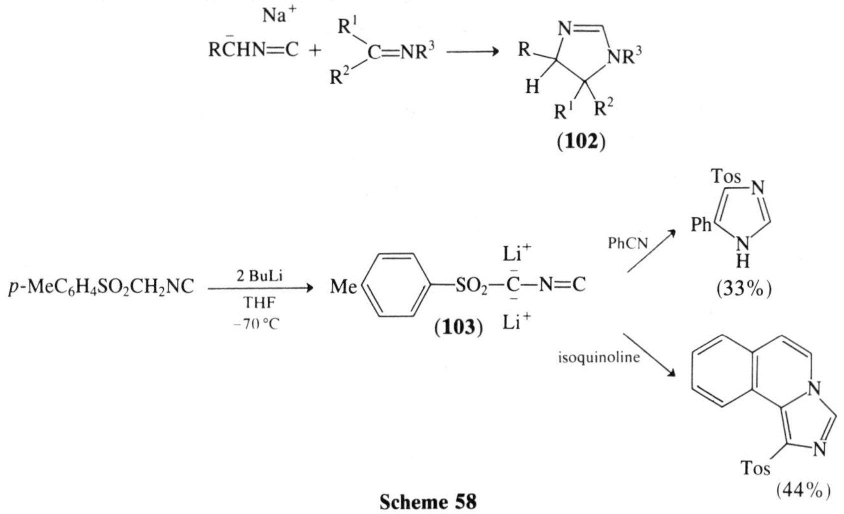

Scheme 58

Similar to the above reactions is the isocyanide cycloaddition which leads to the regiospecific synthesis of ethyl 5-aminoimidazole-4-carboxylates (**104, 105**; Scheme 59).

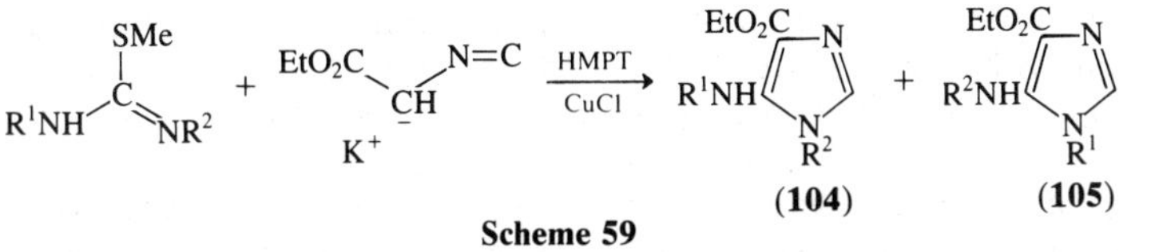

Scheme 59

Treatment of the imidoyl chloride (**106**) with triethylamine yields the nitrile ylide (**107**), which undergoes cycloaddition with a variety of reagents giving imidazole products. With ethyl cyanoformate a mixture of esters is obtained; with 2,4-xylyl cyanate the 4-aryloxy

478   *Imidazoles and their Benzo Derivatives*: (iii) *Synthesis and Applications*

product (**108**) results; with benzylidenemethylamine the initially formed 2-imidazoline dehydrogenates to the fully aromatic tetrasubstituted imidazole (**109**; Scheme 60) ⟨80AHC(27)241⟩. The mesoionic oxazole (**110**) undergoes a 1,3-dipolar cycloaddition with an electron-deficient nitrile to form, after decarboxylation of the intermediate 1:1 cycloadduct (**111**), a compound of type (**109**) ⟨71CB1562⟩. The electron-deficient nitrile can be alkyl- or aryl-substituted (Scheme 60).

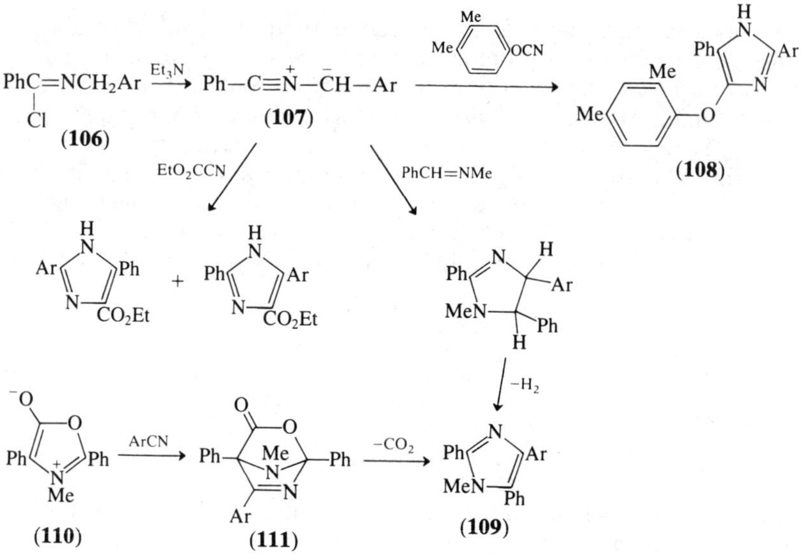

Scheme 60

In the presence of potassium cyanide the *N*-methyl-*C*-phenylnitrone (**112**) takes part in yet a further cycloaddition reaction to give 1-methyl-4,5-diphenylimidazole. With lithium metal the aldimine (**113**) forms a delocalized carbanion which will cyclize with aryl cyanides (Scheme 61) ⟨80AHC(27)241⟩. See also Section 4.08.1.1.3.

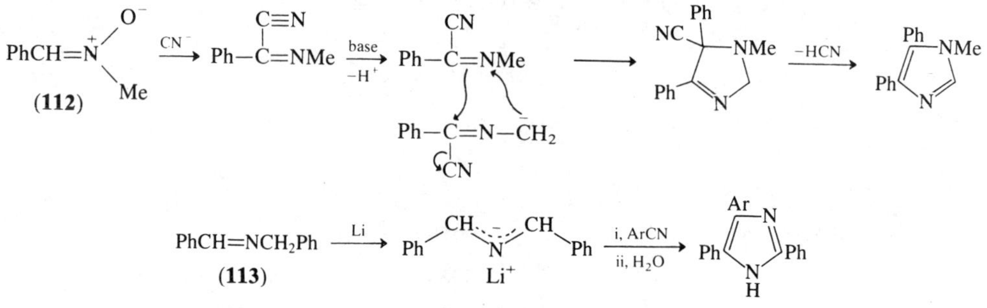

Scheme 61

(iii) Formation of the 1,5- and 3,4-bonds

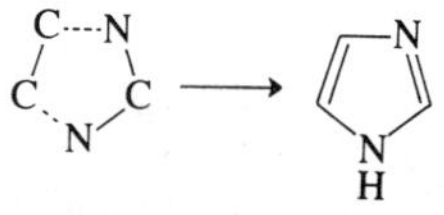

The reactions of amidines or guanidines with α-functionalized carbonyl compounds is a common method of imidazole synthesis. Formamidine will react with α-hydroxy or α-halogeno ketones to form mixtures of oxazoles and imidazoles. Usually, the formamidine is liberated from its hydrochloride by the addition of sodium butoxide in butanol. When α-hydroxy ketones become involved in this synthetic procedure the aliphatic members give mainly imidazoles (35–70% yields) while benzoins give oxazoles (67–80%) in preference. The competing reaction pathways are shown in Scheme 62. Unlike formamidine, both acetamidine and benzamidine react with aromatic and aliphatic α-hydroxy ketones to give imidazoles exclusively. It has been suggested that aryl groups favor the enolic form (**115**) of the tautomeric mixture, resulting in the formation of oxazoles as the major reaction products; aliphatic groups favor the keto form (**114**) from which imidazoles are derived.

That amidines more complex than formamidine give imidazoles exclusively may be a consequence of steric hindrance to the reaction of the enolic oxygen with the amidine carbon atom. The reaction has been widely used to prepare such compounds as 4,5-dipropylimidazole [from tris(formylamino)methane and 5-hydroxyoctan-4-one], and a variety of imidazolin-2-ones and 2-aminoimidazoles.

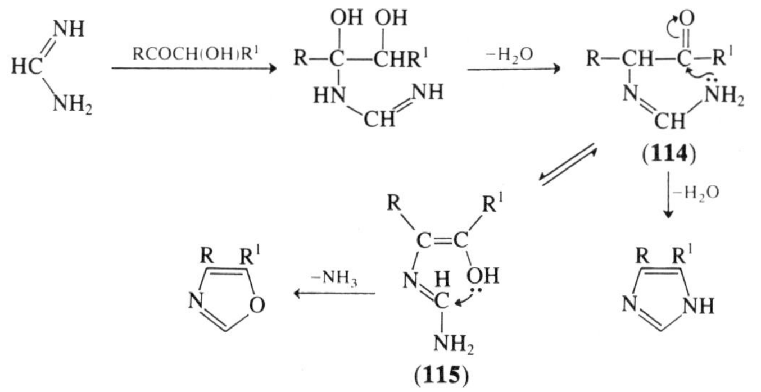

Scheme 62

Reaction of an *N*-alkylbenzamidine with oxalyl chloride gives a 1-alkyl-2-phenyl-imidazole-4,5-dione (**116**), while benzamidine with 3-bromobenzo-4-pyrones can yield both imidazoles and pyrimidines; the former are the favored products in non-polar solvents (Scheme 63). Ethoxyformamidine reacts with 2-methoxycycloheptatrienone to give the imidazole (**117**), which hydrolyzes in acid medium to give the cycloheptimidazolin-2(1*H*)-one. With formamidine acetate, aminomalonitrile gives 4-amino-5-cyanoimidazole. Di-*N*-substituted amidines and guanidines cannot, of course, give fully aromatic products, and therefore they provide a route to imidazolidines. At reflux temperatures aromatic aldehydes condense with benzamidine hydrochloride and ethyl chloroacetate in the presence of bicarbonate to give 2-phenyl-2-imidazolin-5-ones (**118**) ⟨76TL571⟩. Amidines also react with haloacetonitriles or α-bromoacetyl chlorides to give salts of mesoionic 4-amino- or 4-hydroxy-imidazoles (**119**), while *N*-acetylamidrazones (**120**) give 1-aminoimidazoles on reaction with phenacyl chloride (Scheme 64).

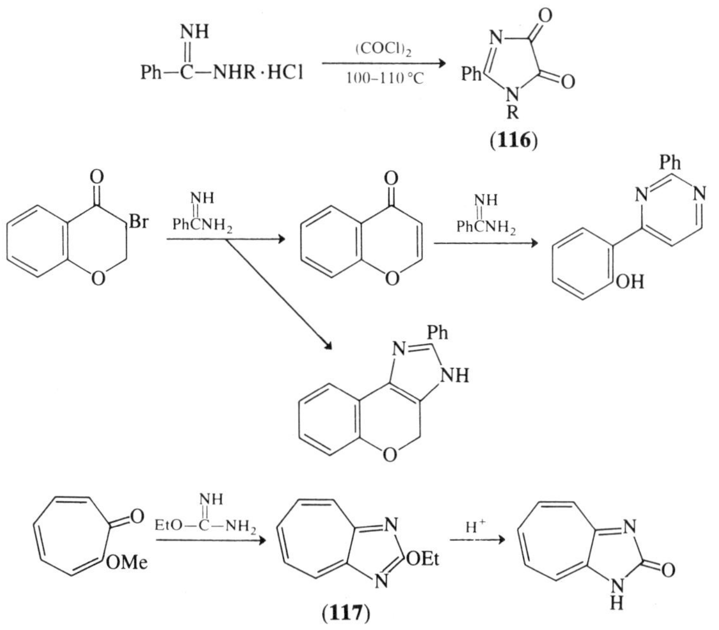

Scheme 63

An interesting corollary of these reactions is the formation of 2-aminoimidazoles from guanidines and suitably substituted ketones. At −10 °C in methanol, 1-phenylpropane-1,2-dione reacts with 1,1-disubstituted guanidines to give 2-(disubstituted amino)-4-hydroxy-4-methyl-4*H*-imidazoles (**121**) which, on catalytic hydrogenation, give excellent yields of

1,3-dimethylurea and imidazolidin-2-one fail to give simple products ⟨81JCS(P2)310⟩. Cyanourea reacts in a similar way with α-halo ketones to give 1-cyanoimidazolin-2-ones which can be hydrolyzed to the corresponding 1-carbamoyl compounds or to the 1-unsubstituted imidazolin-2-ones (Scheme 69).

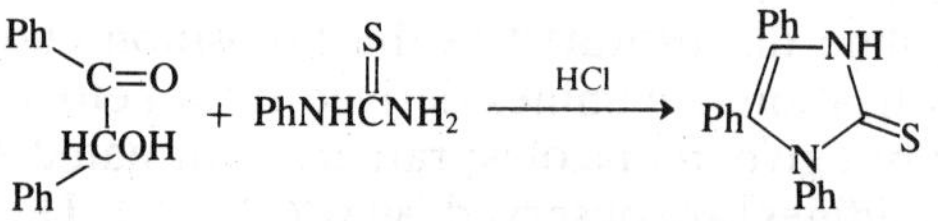

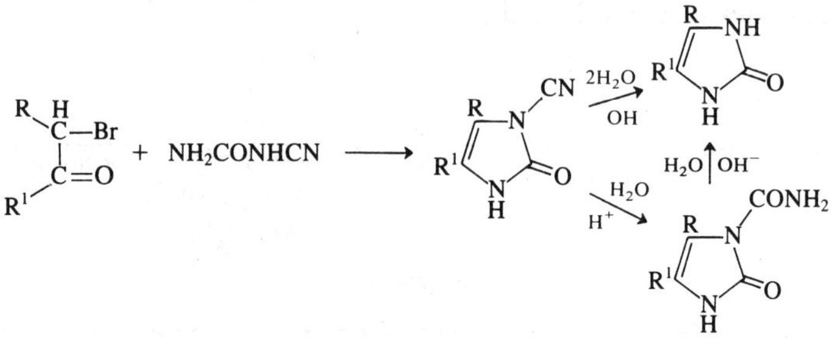

Scheme 69

4.08.1.3 Formation of Three or Four Bonds

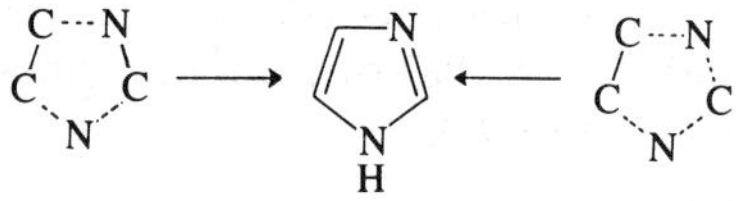

4.08.1.3.1 From reactions of ammonia, amines or formamide with functionalized carbonyl compounds

The earliest synthesis of imidazole was achieved by Debus from glyoxal, formaldehyde and ammonia (Scheme 70), and many of the classical methods of imidazole synthesis were based on this general type of reaction. Initially, most syntheses utilized α-diketones, but in the 1930s it was shown that α-hydroxy ketones could serve equally well provided that some oxidizing agent (e.g. ammoniacal copper(II) acetate, citrate or sulfate) was incorporated in the reaction mixture. Further improvement used ammonium acetate in acetic acid as the nitrogen source.

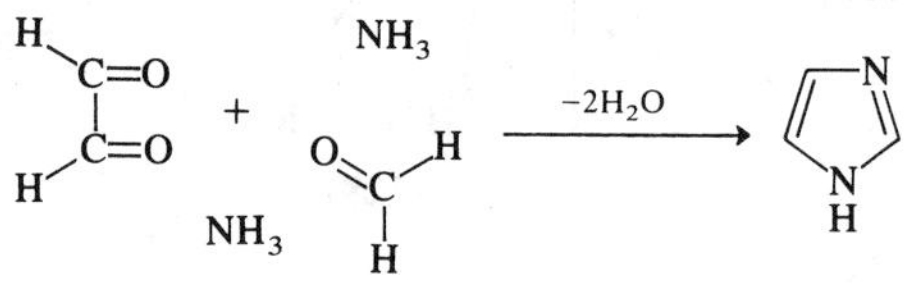

Scheme 70

Although all of these methods suffered from deficiencies (difficulties of synthesis of starting materials, low yields, and more often than not the formation of mixtures of products requiring tedious separation procedures), they still find ample application for the preparation of many C-substituted imidazoles (e.g. 4-alkyl, 4,5-dialkyl, 2,4,5-trialkyl). The old Debus (or Radziszewski) method is still useful for preparing such compounds as 4-methyl- (**132**) and 2,4-dimethyl-imidazoles (**133**) using pyruvaldehyde (Scheme 71). However, alkaline fission of the pyruvaldehyde can result in a mixture of products. When pyruvaldehyde is treated alone with aqueous ammonia there are three main products: (**132**), (**133**) and 2-acetyl-4-methylimidazole. Reversed aldol condensations cause degradation of the pyruvaldehyde, and subsequent cyclization of the fragments as in Schemes 71 and 72 accounts for the products.

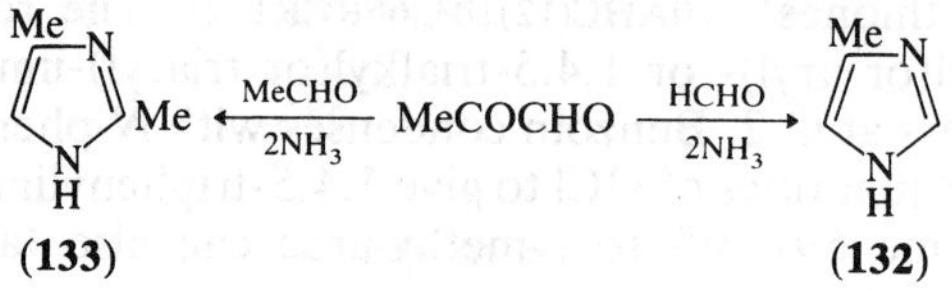

Scheme 71

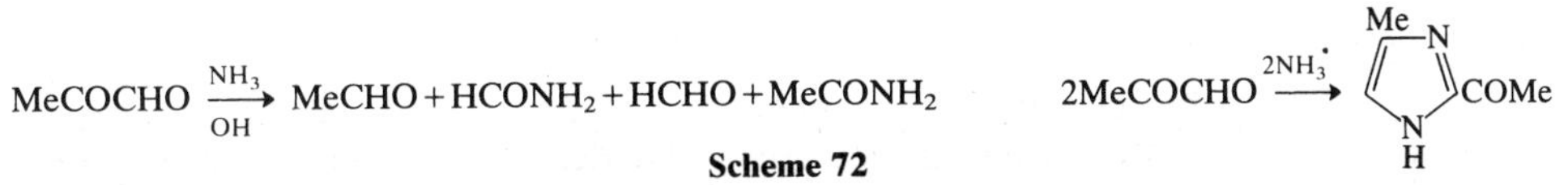

Some of the problems associated with the synthesis of α-dicarbonyl starting materials have been alleviated by the use of propane-1,3-dithiol, which reacts with aldehydes to give cyclic thioacetals. With butyllithium the resulting stable dithiane anions (**134**) can be transformed into α-diketones or α-hydroxy ketones (Scheme 73). A further approach to such compounds is found in the reaction of α-ketonitrate esters with sodium acetate (Scheme 73), while aryl α-diketones are also available from α-ketoanils (prepared from the cyanide-ion-catalyzed transformation of aromatic aldimines) ⟨70AHC(12)103⟩.

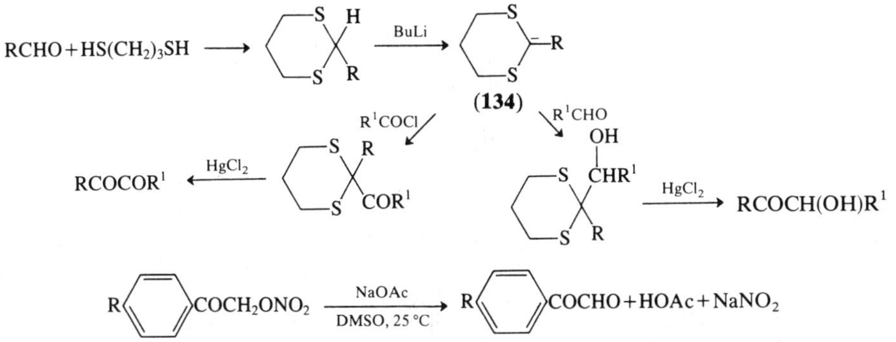

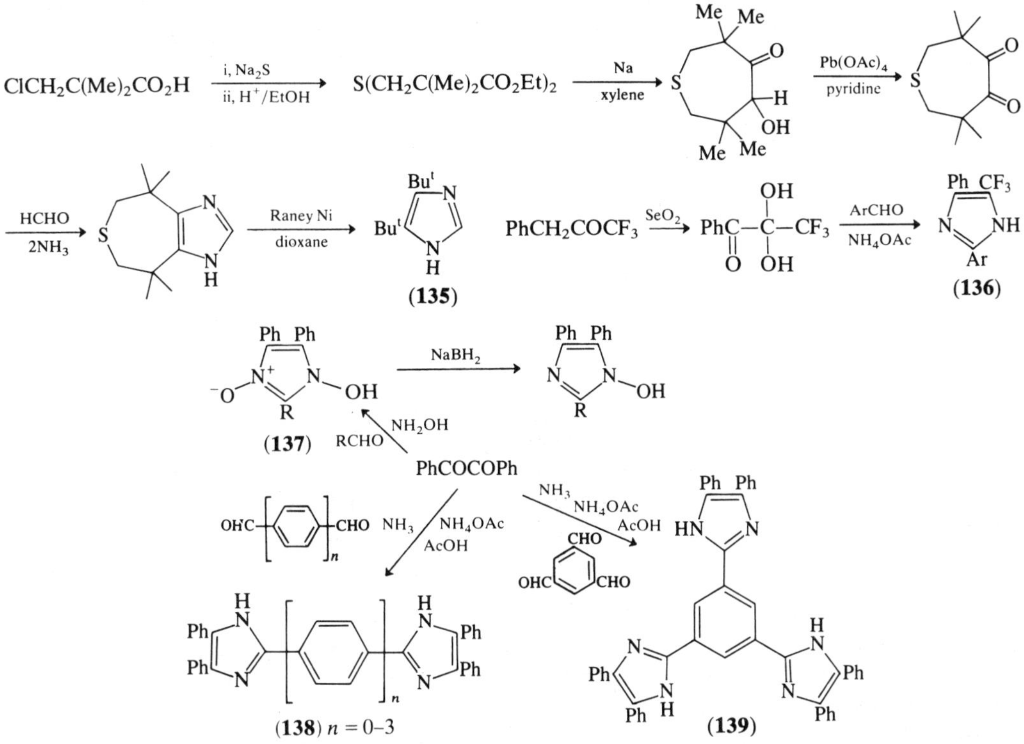

Examples of reactions which have used 1,2-diketones and ammonia with or without added aldehyde include the isolation of good yields of lophine (2,4,5-triphenylimidazole) from benzil and ammonia, a rather elegant synthesis of 4,5-di-*t*-butylimidazole (**135**), and the preparation of a series of 2-aryl-5-trifluoromethyl-4-phenylimidazoles (**136**) from 3,3,3-trifluoro-1-phenylpropane-1,2-dione monohydrate (Scheme 74). Additionally, the formation of 1-hydroxyimidazole 3-oxides (**137**) from α-diketones, hydroxylamine and aldehydes ⟨74CI(L)38⟩, and the preparations of 2,2'-bis- (**138**) and 2,2',2"-tris-imidazolyls (**139**) from benzil, ammonia and polyformyl aromatics provide further examples of this versatile reaction (Scheme 74). There is yet some doubt about the pathway or pathways involved in these reactions; current ideas as to the mechanism have been discussed in detail elsewhere ⟨70AHC(12)103⟩.

Before the turn of the century, Maquenne had prepared imidazole-4,5-dicarboxylates from tartaric ester dinitrates with either an aliphatic aldehyde or precursor in the presence of ammonium ions at pH values 3.5–6.5. Hydrolysis, especially with bromoacetic acid, of the resulting dicarboxylate esters gives imidazole-4,5-dicarboxylic acid (Scheme 75) ⟨70AHC(12)103⟩.

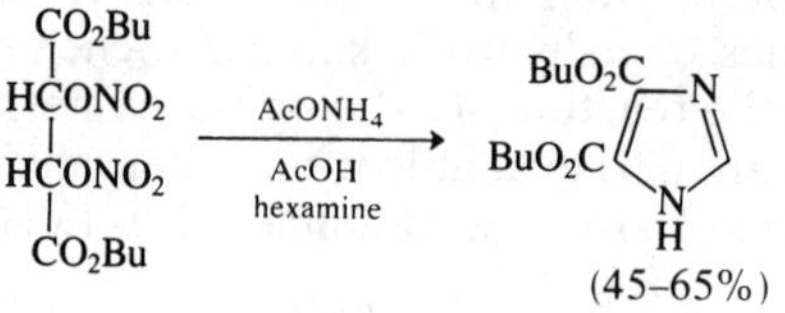

Scheme 75

When Windaus discovered that α-hydroxycarbonyl compounds (acyloins) could be oxidized *in situ* in the presence of ammonia and aldehydes to give imidazoles, he opened up a wide variety of imidazole precursors, including reducing sugars, alicyclic and aromatic ketols, and ketol acetates. Provided that an excess of the acyloin is used in an alcoholic solution of copper(II) acetate and ammonia, the ring closure can give 2-imidazolyl ketones. Thus benzoylmethanol gives a 32% yield of 2-benzoyl-4-phenylimidazole (**140**), and the method has also been employed in the synthesis of histamine analogues and potential histidine decarboxylase inhibitors. Even in the absence of copper(II) salts, acyloins still seem able to react with ammonia to give imidazoles. 1,4-Dihydroxybutan-2-one gives 1-hydroxy-2-(4-imidazolyl)ethane (**141**; R = OH), while 1-hydroxybutan-2-one gives 4-ethylimidazole (**141**; R = H) in low yield (Scheme 76).

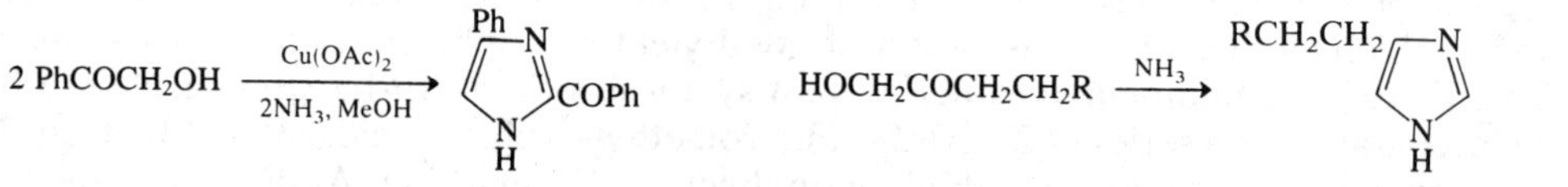

Scheme 76

It is usual for the reaction of α-acyloxy ketones with ammonium acetate to give mixtures of oxazoles and imidazoles, with the result that the method may not be the one of choice if the product mixture complexity can become too great (Scheme 77) ⟨70AHC(12)103⟩. Excess ammonia may help depress oxazole formation, although the oxazoles can be converted into imidazoles (Section 4.08.2.2).

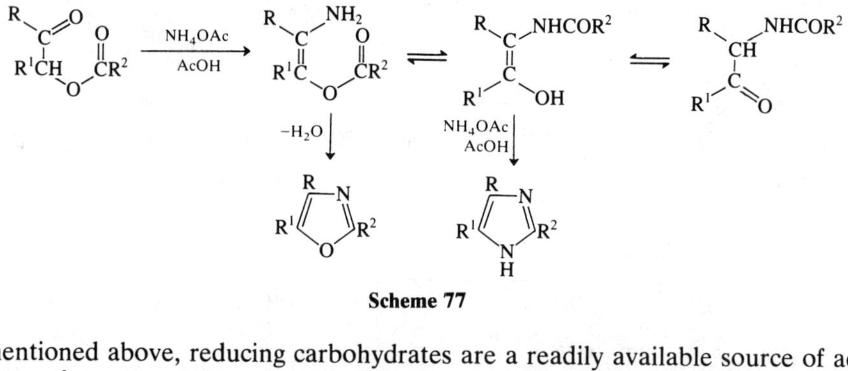

Scheme 77

As mentioned above, reducing carbohydrates are a readily available source of acyloins, and glucose, fructose, *etc.* have been used to prepare both 4-methyl- and 4-hydroxymethylimidazoles. The reactions, though, are very complex, giving low yields along with many other products including pyrazines. Besides the two imidazoles mentioned above (best prepared with added formaldehyde), glucose and ammonia have yielded small quantities of 2-hydroxymethyl-4-methyl and other 4-substituted imidazoles with hydroxyethyl, dihydroxypropyl and trihydroxybutyl side chains. From rhamnose and ammonia, 4-ethyl- and a 4-(dihydroxybutyl)-imidazole have been isolated. When lactose, cellobiose, sucrose and a number of starches react with ammonia at high temperatures and pressures, conditions which cause extensive degradation of any polyhydroxyalkyl side chain, a number of simple alkylimidazoles (and pyrazines) are isolated. Other carbohydrate-type sources which have been used include glycolaldehyde, glyceraldehyde, hydroxypyruvaldehyde and even wood. In only a few instances, however, have the results merited classification as useful organic syntheses. Identification of the imidazoles formed from the interaction of ammonia and reducing oligosaccharides or periodate-oxidized polysaccharides has led to methods of linkage determination in the sugars ⟨70AHC(12)103, 80AHC(27)241⟩, and the field has been reviewed ⟨65MI40800⟩. The reaction of periodate-oxidized glucosamine with ammonium acetate and formaldehyde in the presence of a copper(II) salt provides a route to L-*erythro*-β-hydroxyhistidine ⟨79JA3982⟩.

Probably the most useful extension of the foregoing synthetic methods was developed by Bredereck and Theilig in 1953. Their 'formamide synthesis' reacts α-diketones, α-hydroxy, α-halogeno or α-amino ketones, α-ketol esters or (under reducing condtions) α-oximino ketones with formamide. Yields of imidazoles can approach 90% and are seldom less than 40%. Although the reaction has been reviewed earlier ⟨59AG753, 70AHC(12)103, B-64MI40800⟩ the major features will be summarized here. The general reaction sequence can be formulated as in Scheme 78.

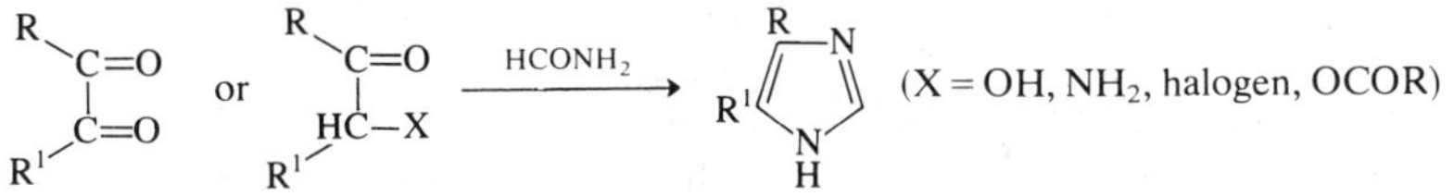

Scheme 78

Applicability of the reaction sequence to the preparation of 2-substituted imidazoles rests on the use of more complex amides than formamide. This constitutes a severe limitation to the process since it has only proved possible to substitute acetamide in a few cases, with the formation of 2-methylimidazoles. Nevertheless, the method still has many applications in the preparation of 4- and 5-substituted imidazoles. Best results are obtained if a large excess of formamide is used at 180–200 °C, or passage of a stream of ammonia at 150–175 °C for 4–6 hours is maintained. The use of sulfuric acid as a condensing agent is also of value. Should these conditions not be adhered to then oxazoles may be formed preferentially.

One of the simplest methods of carrying out the reaction is to brominate the ketone in the presence of formamide; where unsymmetrical ketones are involved, careful control of solvent may allow selective orientation of such bromination. Imidazole itself can be prepared in 60% yield by reaction of bromoethanal (as the glycol acetal), formamide and ammonia at 180 °C. Apparently, the initial step in the reaction is acyloin formation followed by attack of formamide at the carbonyl carbon (Scheme 79). The α-formamido ketone (**142**) is an isolable intermediate which reacts with formamide to give the imidazole (**143**). The alternative possibility that the imidazoles might be formed by the action of formamide on oxazoles formed from (**142**) has been dicounted since it can be shown that α-formamido ketones give oxazoles only to a minor extent if heated in the absence of a condensing agent, and oxazoles are not converted into imidazoles at temperatures as low as that of the reaction.

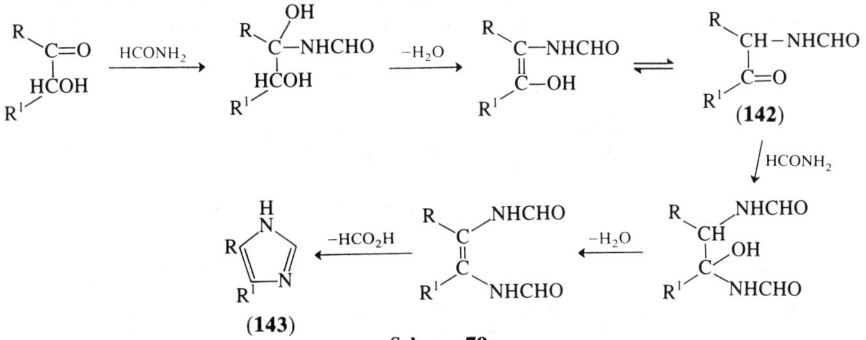

Scheme 79

When α-diketones react with a mixture of formamide and formaldehyde at 180–200 °C it is not possible to detect α-hydroxy ketones during the reaction. It seems, therefore, that the formaldehyde cannot be acting as a reducing agent, and that imidazole formation must be a consequence of generation of ammonia from formamide, and subsequent reaction between the diketone, ammonia and formaldehyde. The major advantage here over earlier methods lies in the reduced decomposition of the α-diketone, which normally results in reduction in yields and mixtures of products.

486 Imidazoles and their Benzo Derivatives: (iii) Synthesis and Applications

Whereas α-amino ketones readily form imidazoles with formamide, they are often not easy to prepare. Accordingly, they can be replaced by precursors, α-oximino ketones, which can be reduced either by dithionite or using catalytic mehods in formamide at 70–100 °C. Ring closure can then be achieved by raising the temperature (Scheme 80). When α-ketol esters are used it appears that the imidazole formation may in this instance proceed by way of the oxazole. A further special case is the formation of 4,5-disubstituted imidazoles from 1-chloro-1,2-epoxides and formamide. One recent example of an application of Bredereck's method is the synthesis of the imidazolepropanol (**144**) from 3-bromo-2-methoxytetrahydropyran (Scheme 81) ⟨80AHC(27)241⟩.

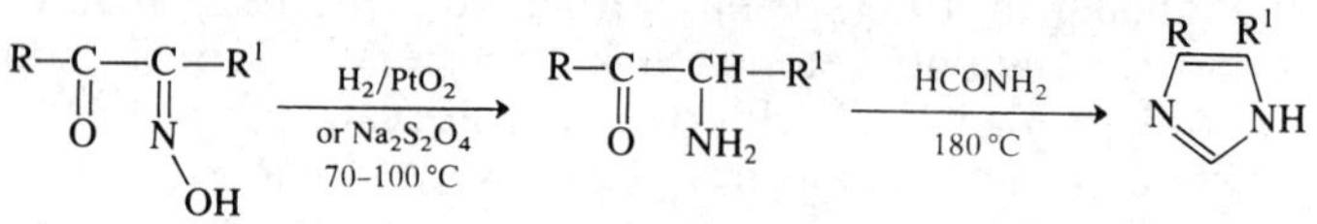

Scheme 80

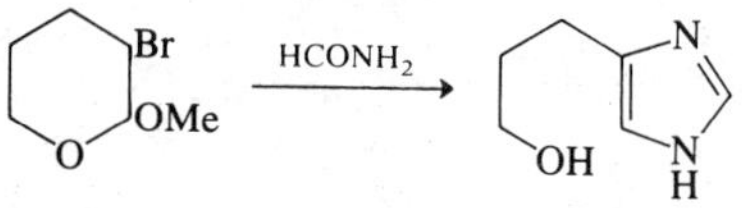

Scheme 81

A reaction which is applicable to the synthesis of imidazoles substituted at C-4 by sulfur substituents is the interaction of α-chloro-α-phenyl thioketones (prepared from the corresponding diazoketones) with ammonia and carboxylic acids. Although the detailed reaction course is yet uncertain, it bears a close resemblance to the reactions of α-chloro ketones with amides. The method has been used to prepare 2-ethyl-4-methyl-5-phenylthioimidazole (**145**) using ammonia, propanoic acid and 1-chloro-1-phenylthiopropanone (Scheme 82).

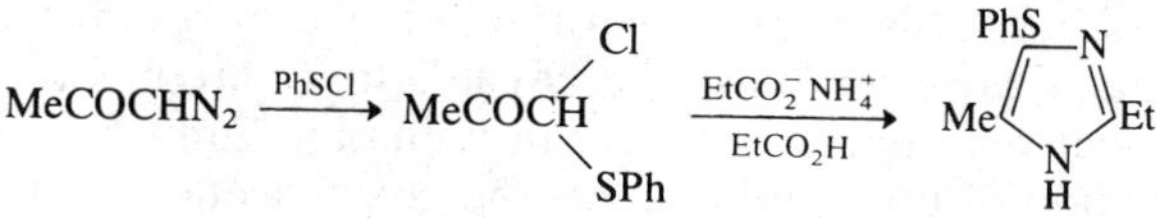

Scheme 82   (**145**) 32%

Yet a further reaction which forms three bonds is the use of imidic esters with suitably functionalized ketones (*e.g.* α-hydroxy, α-bromo). This method is exemplified by the synthesis of a new imidazole alkaloid (**146**), which had been isolated from a member of the *Urticaceae* family ⟨80AHC(27)241⟩, and by the more general synthesis of 4-alkyl-5-methoxymethylimidazole (**147**) ⟨79AP107⟩. The cyclization process, which takes place in liquid ammonia, may be complicated by 2-alkylbutanamide by-products which result from Favorskii rearrangement of the bromomethoxy ketones (Scheme 83).

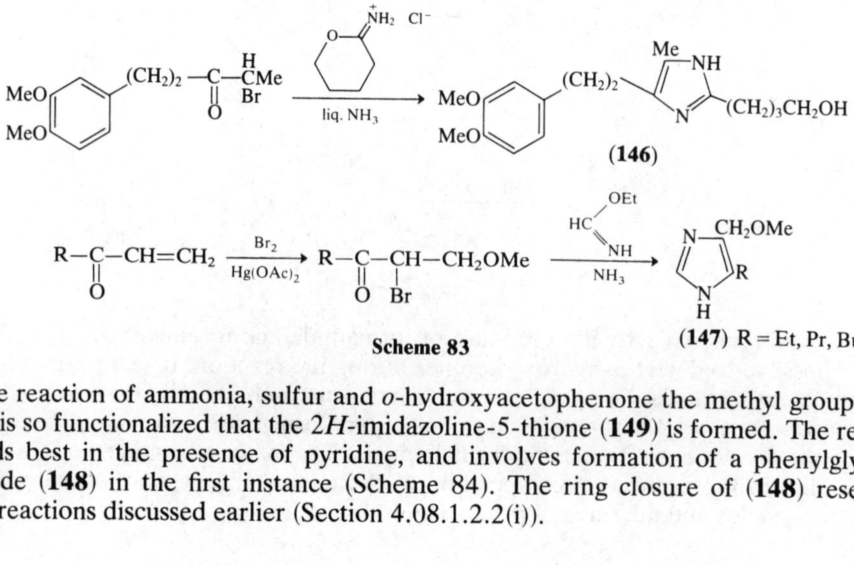

Scheme 83   (**147**) R = Et, Pr, Bu

In the reaction of ammonia, sulfur and *o*-hydroxyacetophenone the methyl group of the ketone is so functionalized that the 2*H*-imidazoline-5-thione (**149**) is formed. The reaction proceeds best in the presence of pyridine, and involves formation of a phenylglyoxylic thioamide (**148**) in the first instance (Scheme 84). The ring closure of (**148**) resembles similar reactions discussed earlier (Section 4.08.1.2.2(i)).

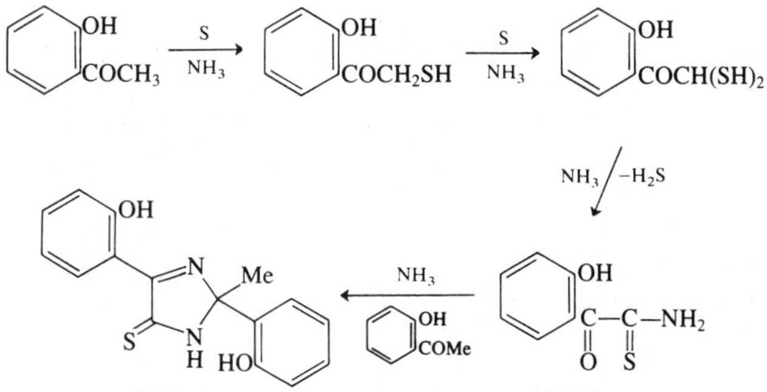

Scheme 84

Under UV irradiation, aqueous solutions of formaldehyde and ammonium salts produce imidazoles (Scheme 85). It is conceivable that the irradiation has catalyzed a 'formose' reaction, yielding trioses and glycolaldehyde, and that reaction of these species with ammonia could account for the products isolated. Such a reaction might be classed as a 'primitive-life' synthesis, as might the observation that hydrogen cyanide and liquid ammonia react to form 4,5-dicyanoimidazole and adenine ⟨70AHC(12)103⟩.

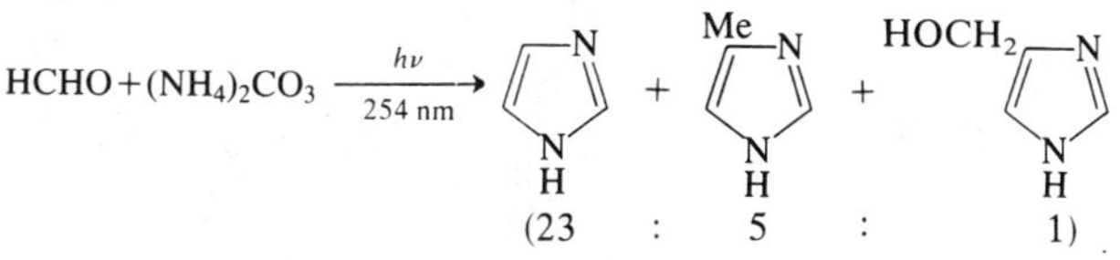

Scheme 85

There have been reports of a number of other processes which lead to imidazoles, but which may have only limited synthetic utility. For example, the preparation of 2,4,5-triarylimidazoles by the rhodium ion-catalyzed combination of alkenes with carbon monoxide and ammonia; 50–60% yields are obtained. Benzylamine and derivatives react with carbon tetrachloride in the presence of catalytic quantities of metal carbonyls to give 2,4,5-triarylimidazoles and some of the corresponding imidazolines through a process in which radical species which coordinate with the metal carbonyl are implicated. Catalytic conversion of acetylene, ammonia and ethylenimine at 45–50 °C produces acetonitrile, 2-methylimidazole and 2-methylimidazoline. The heating of monoamide resins has been reported to give low yields of imidazoles ⟨80AHC(27)241⟩.

4.08.2 SYNTHESIS BY TRANSFORMATION OF OTHER HETEROCYCLES

Syntheses of imidazoles which involve dehydrogenation of imidazolines have been discussed in Section 4.07.2.2.2, as have oxidations of benzimidazoles (Section 4.07.3.1.1).

4.08.2.1 Ring Expansions

There are a few examples of imidazole formation from azirines and aziridines. 2-Phenylimidazole has been obtained in 25% yield from the reaction in ethanol of ammonium carbonate with 1-benzoyl-2-cyanoaziridine (**150**; Scheme 86). The reactions of 2H-azirines (**151**) or their salts with nitriles likewise lead to imidazoles, a process which is promoted by perchloric acid or boron trifluoride, and which is a 1,3-dipolar cycloaddition to the azaallyl cationic intermediates (**152**). When the azirine is disubstituted at C-3, then perchloric acid and acetonitrile convert it into an imidazolium salt. Thus 3,3-dimethyl-2-phenyl-1-azirine (**153**) gives a 77% yield of the 4-hydroxy-2,5,5-trimethyl-4-phenylimidazolidinium perchlorate (**154**). Results from labelling the nitrile nitrogen are consistent with the possible mechanisms included in Scheme 87. It is likely that there is 1,3-bond cleavage and ring enlargement at the cleavage site, since there were no products which would have been obtained as a result of nucleophilic attack at the 1,2-bond ⟨70AHC(12)103⟩.

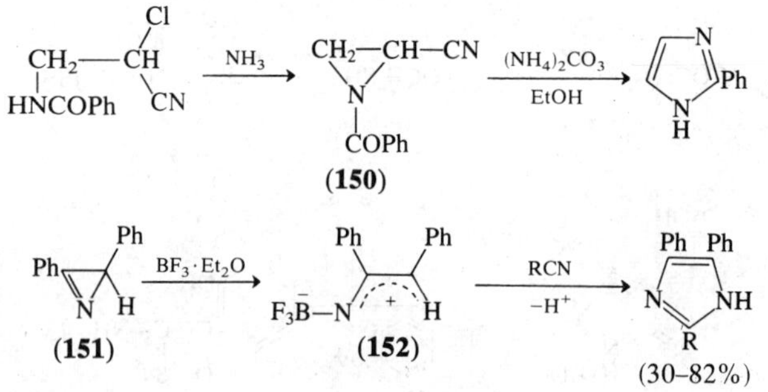

Photolysis, too, of such azirines (**155**) generates benzonitrile ylides (**156**), and in the presence of ethyl cyanoformate a mixture of oxazoline (**158**) and imidazole (**159**) is the consequence of 1,3-dipolar cycloaddition to the carbonyl and nitrile groups respectively. When there is no dipolarophile present, the ylide adds to the ground state azirine, giving firstly the geometrically isomeric 1,3-diazabicyclo[3,1,0]hex-3-enes (**157**), and then a mixture of 2-methyl-4,5-diphenyl-1-vinylimidazole and 2,4-dimethyl-5,6-diphenyl-pyrimidine (Scheme 88). The photolysis of a formylazirine (which rearranges thermally to a pyrazole) gives a 1-phenylimidazole ⟨80AHC(27)241⟩.

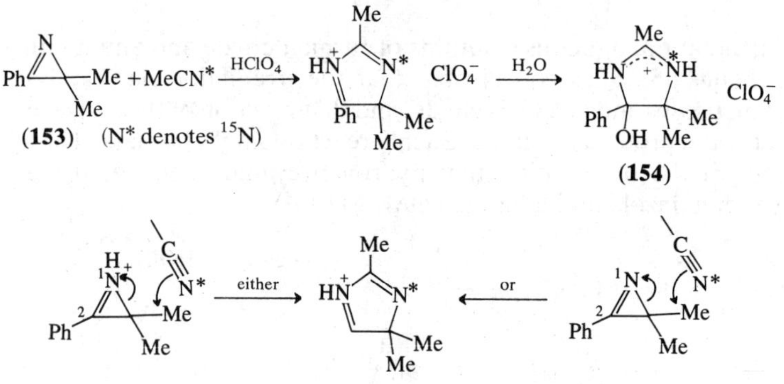

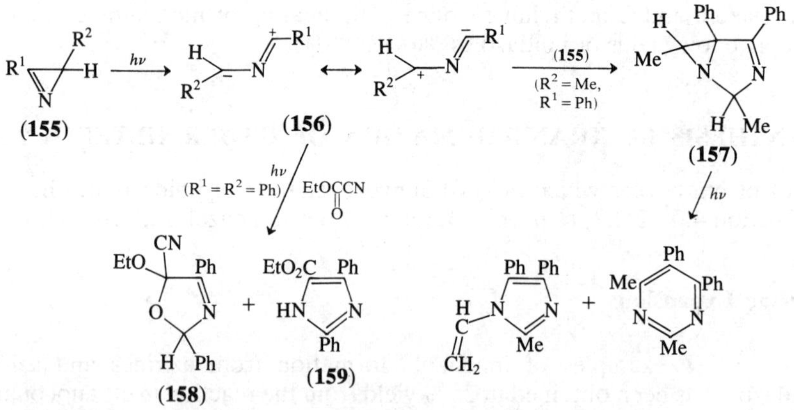

Scheme 88

There have also been a few examples of ring expansions involving azetines. One such instance, which follows a pericyclic mechanism, is the thermolysis of 4-cyano-1-*t*-octyl-3-*t*-octylamino-2-*t*-octyliminoazetine (**160**), producing the 4-amino-5-cyanoimidazole (Scheme 89). Such azetine species have been implicated as intermediates in the photolysis of enaminonitriles to imidazoles (Section 4.08.1.1.1; Scheme 17). In strongly basic medium the azetidinone (**161**), which possesses a lactam group, is ring expanded to the 4*H*-imidazolinone (**163**), probably *via* the anionic acylic species (**162**; Scheme 89) ⟨80AHC(27)241⟩.

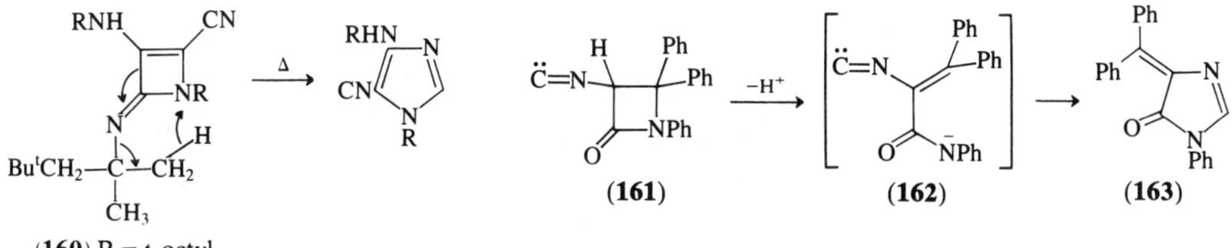

4.08.2.2 Transformations of Other Five-membered Rings

4.08.2.2.1 Pyrazoles and indazoles

Photolysis of *N*- and *C*-substituted pyrazoles and indazoles gives imidazoles and benzimidazoles, respectively, in moderate to high yields (20–90%). Thus 3-methylpyrazole gives a mixture of 2- and 4-methylimidazoles, and 4-aroyl-1-arylpyrazoles give the biimidazol-4-yl derivatives (164). The reaction is potentially of synthetic importance in view of the often greater ease of synthesis of pyrazole derivatives. Reactions are carried out in 1,2-dimethoxyethane in the presence of benzophenone; although the sensitizer is not absolutely necessary it improves the yield. Electron-attracting groups inhibit the reaction, which goes through a ring opening to form an azirine (Scheme 90). The consequence, allowing for tautomerism in the case of *N*-unsubstituted pyrazoles, is interchange of N-2 and C-3 in the pyrazole. 2-Methylindazole similarly gives 1-methylbenzimidazole when irradiated, probably *via* the intermediate (165; Scheme 90). The courses of these reactions, which can give yields as high as 96%, are markedly dependent on the nature and position of substitution in the heterocyclic ring. In the absence of a substituent, benzimidazole (27%) is formed, and whilst 2-alkylindazoles give 1-alkylbenzimidazoles, 1-alkylindazoles only give 2-alkylaminobenzonitriles ⟨B-76MI40800, 80AHC(27)241, 76HCA1512⟩. Two pathways have been suggested for the phototransposition of cyanopyrazoles to cyanoimidazoles, a reaction which gives about 30% yields of isomeric imidazoles ⟨81CC604⟩.

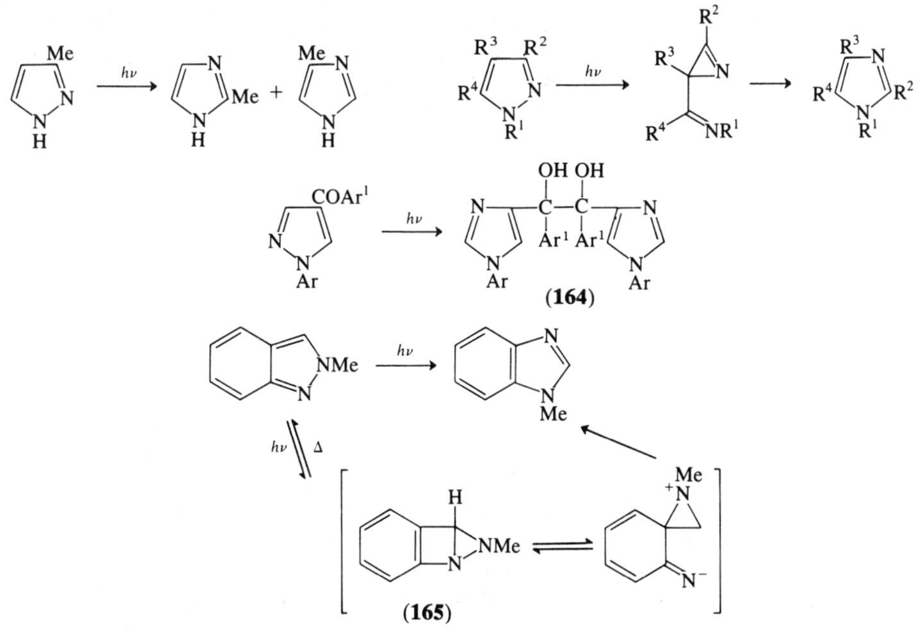

4.08.2.2.2 Formation of benzimidazoles from imidazoles

An unusual reaction which consists of building the fused benzene ring on to a preformed imidazole nucleus utilizes the reaction of a 5-formylimidazole with a Grignard reagent, then cyclization of the resulting alcohol (Scheme 91) ⟨73JOC3495⟩.

Scheme 91

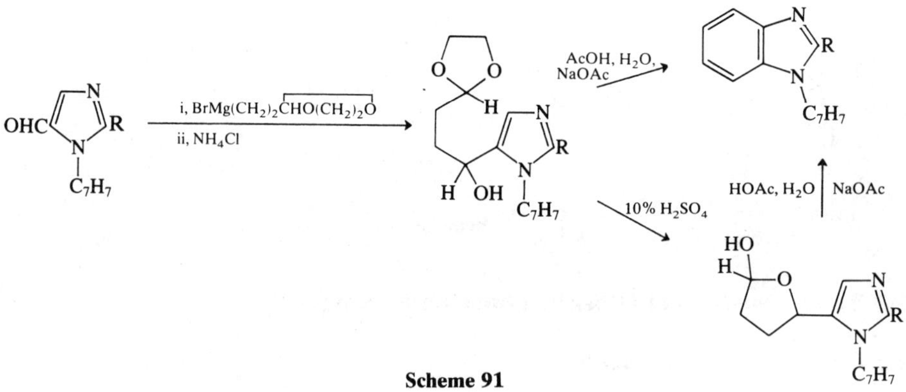

4.08.2.2.3 2-Imidazolines from cyclic ureas

A convenient method for the synthesis of 2-imidazolines with long-chain alkyl or aralkyl substituents at C-2 involves heating a carboxylic acid with a suitable cyclic urea (or thiourea) (Scheme 92) ⟨B-57MI40800⟩.

Scheme 92

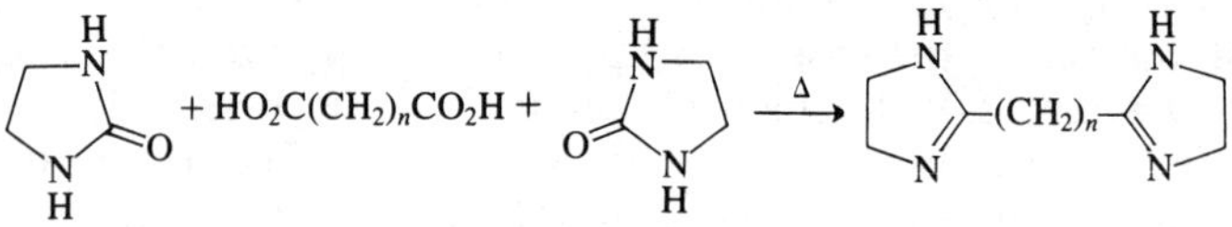

4.08.2.2.4 From other five-membered heterocycles containing nitrogen only

Thermolysis of β-substituted α-(1-tetrazolyl)acrylamides (**166**) in the presence of copper gives 2,4-disubstituted imidazole-5-carboxamides. Since, however, the side chain is incorporated in the new ring, this might be better classified as a reaction in which the 1,5- or 3,4-bond is formed (Scheme 93) ⟨80AHC(27)241⟩.

(166)

Scheme 93

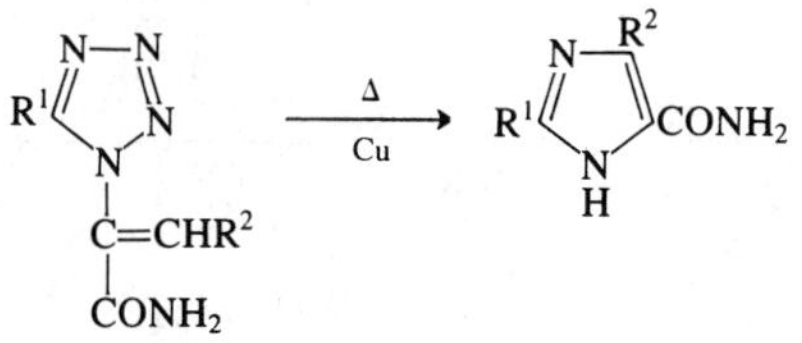

Thermal decomposition of 1,5-diaryltetrazoles gives low yields (8–20%) of benzimidazoles ⟨66RCR122, 74CRV279⟩. Photolysis, too, of tetrazoles can give rise to benzimidazoles in a reaction which may have some synthetic utility. A 42% yield of 2-phenylbenzimidazole was obtained from 1,5-diphenyltetrazole (**167**; R = Ph), while 5-phenoxy-1-phenyltetrazole (**167**; R = OPh) gave 2-phenoxybenzimidazole (36%). Other products, however, were also formed. These reactions can be accounted for in terms of an initial loss of nitrogen, followed by partial secondary decomposition of 2-phenoxybenzimidazole (Scheme 94) ⟨74CRV279⟩.

(167) R = Ph, OPh

Scheme 94

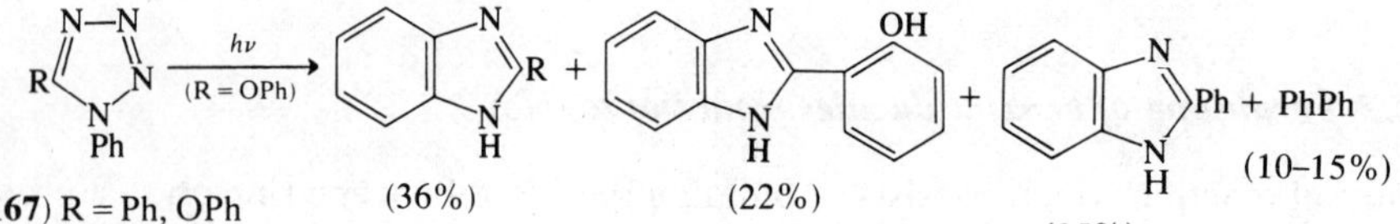

4.08.2.2.5 From oxazoles and isoxazoles

Oxazoles have long provided sources of imidazoles through oxygen–nitrogen exchange. The oxazole can be heated with ammonia, formamide, hydrazine (gives 1-aminoimidazoles) or amines, often in the presence of a Brønsted acid. The use of formamide in place of ammonia has been recommended to improve yields, which generally lie in the range 50–90%. The reaction has been reported to fail with 2,4,5-triethyloxazole and with benzoxazole; thus the employment of this transformation in benzimidazole synthesis seems to be precluded. Aliphatic groups at C-2 and C-5 of oxazole may cause steric hindrance to the reaction since such compounds react only with difficulty with formamide, but even such recalcitrant oxazoles can be converted into imidazoles using formamide and ammonia in an autoclave at 200–210 °C. Aromatic and halogenoalkyl substituents, particularly at C-4 of oxazole, facilitate the reaction, presumably by lowering the electron density at the annular oxygen. The proposed reaction course is outlined in Scheme 95 ⟨66RCR122, 70AHC(12)103⟩.

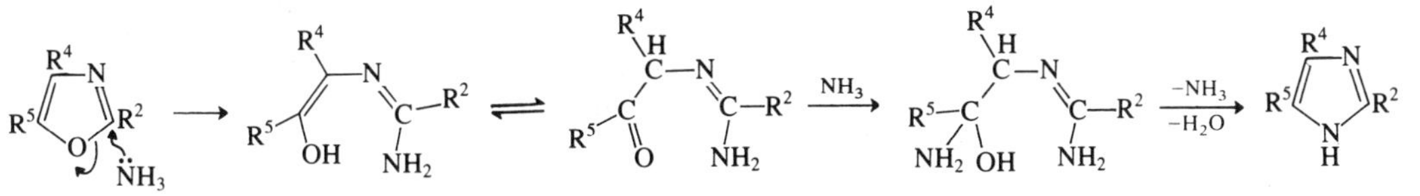

Scheme 95

A novel oxazole to imidazole transformation occurs when the oxazole (**168**) reacts with the imidoyl chloride (**169**) in the presence of phosphoryl chloride. The process probably involves an initial quaternization of the oxazole with subsequent ring opening of the oxazolium salt to form a ketimine (**170**). Intramolecular nucleophilic attack then leads to the imidazole product (Scheme 96).

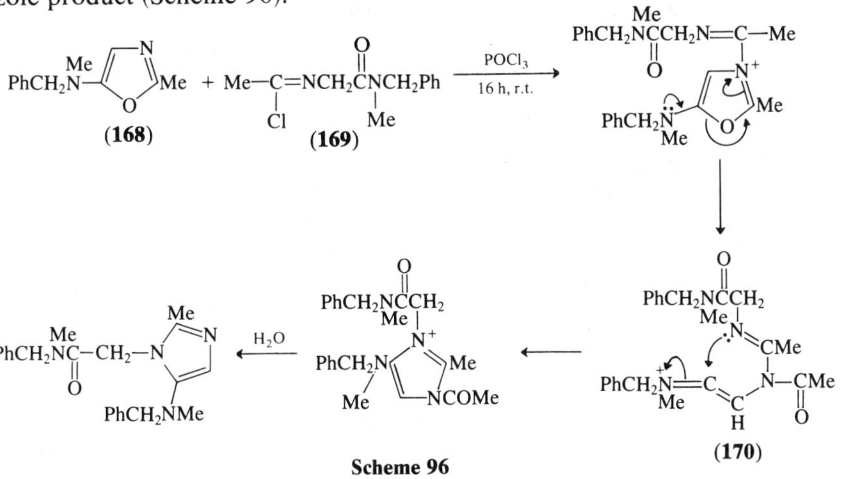

Scheme 96

Oxazolium salts can be converted into imidazoles under much the same conditions as oxazoles are, and in the transformations of 3-phenyloxazolium salts into 1-phenylimidazoles the remarkable stabilities of the non-aromatic intermediates, and the forcing conditions required to dehydrate them, have been remarked upon. Related reactions include conversions of 2-oxo-4-oxazoline-3-carboxamides with strong acid into 4-imidazolin-2-ones, and of isoxazolin-5-imines into related compounds. 2-Substituted 4-methyloxazole N-oxides are attacked at C-2 by phenyl isocyanate with cleavage of the ring which then reforms, yielding 1,2,5-trisubstituted 5-hydroxy-4-methylene-2-imidazolines (not imidazole N-oxides as reported earlier). With ethyl α-amino-α-cyanoacetate, a 2-isoxazoline has been converted into a derivative of 1-(D-xylofuranosyl)imidazole (**171**; Scheme 97) ⟨80AHC(27)241⟩. Thermolysis, photolysis and hydrogenation reactions have also proved of value in effecting ring transformations of isoxazoles into imidazoles. Thus heating 5-amino-3,4-dialkylisoxazoles at 180–190 °C gives 4,5-dialkylimidazolin-2-ones in 40–65% yields (Scheme 98). Extensions of this work have revealed that 4-methyl-5-propyl-, 4-butyl-5-propyl- and 5-benzyl-4-methyl-imidazoles can be prepared from 3,4-disubstituted 5-aminoisoxazoles only if urea is present and the reaction is carried out in the condensed phase. Arylamines such as aniline are also effective in promoting this transformation ⟨75S20⟩.

1,5-Diphenyl-4-aryl(or alkyl)-4-imidazolin-2-ones are formed in yields greater than 70% by the photolysis of 5-imino-2,3-diphenyl-4-aryl(or alkyl)-3-isoxazolines in ethanol or benzene with a high-pressure mercury lamp.

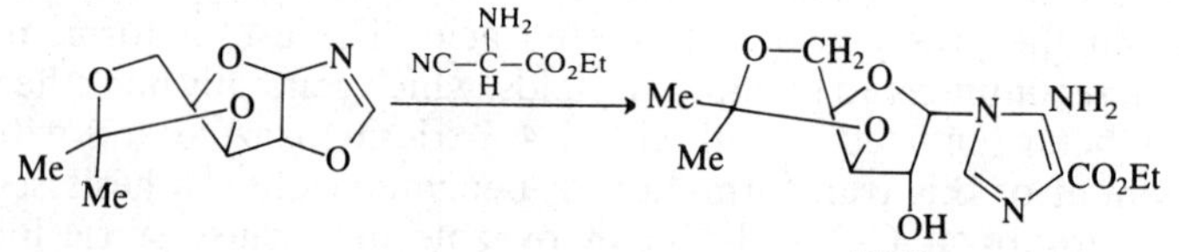

Scheme 97

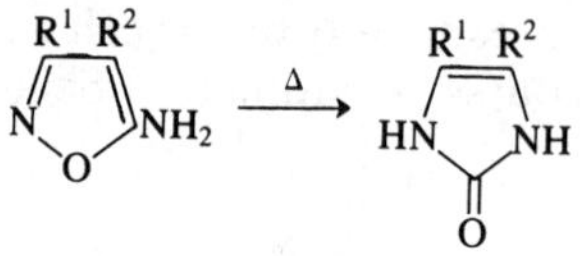

Scheme 98

4.08.2.2.6 From oxadiazoles

In the presence of ammonia, amines or heterocyclic bases, some oxadiazolium salts form 1,2-diaminoimidazoles (Scheme 99) (see also Section 4.08.1.1.2) ⟨67CB3418⟩.

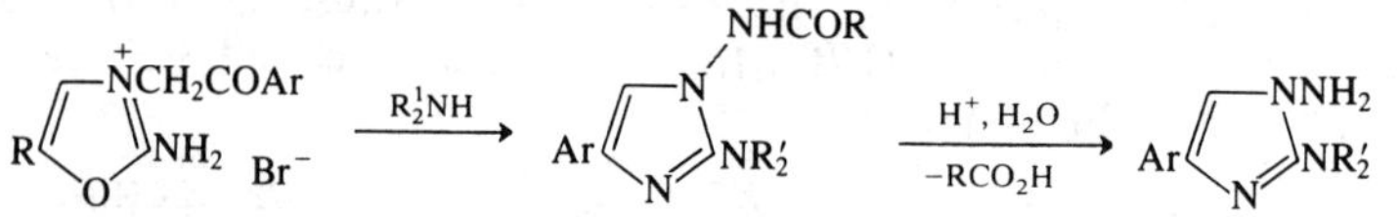

Scheme 99

1-Hydroxybenzimidazoles and their 3-oxides can be prepared from 2,1,3-benzoxadiazole 1-oxides (172). Nitroalkane carbanions can accomplish the transformation by nucleophilic attack at a ring nitrogen, while barbituric acid and 2,4,6-trialkylhexahydro-1,3,5-triazines can effect similar transformations (Scheme 100) ⟨77JCS(P1)470, 75S703⟩.

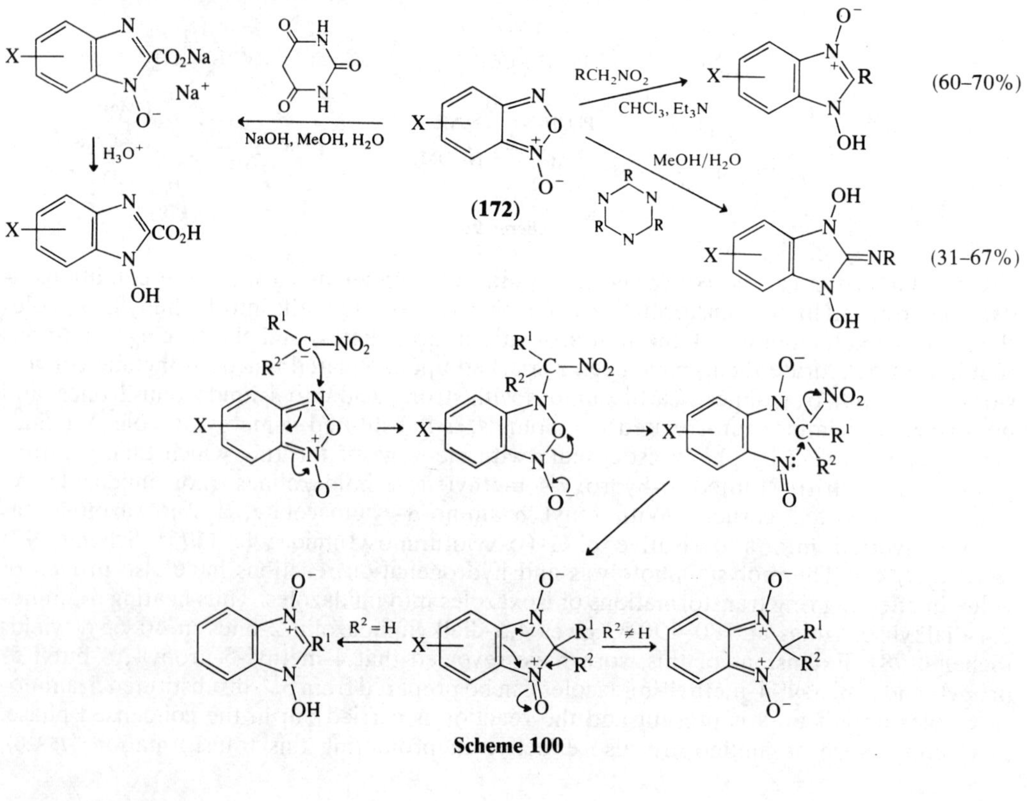

Scheme 100

The thermal decomposition of 2,4-diaryl-1,2,4-oxadiazolin-5-ones gives benzimidazoles in high enough yields to make this a worthwhile synthetic method. The starting materials are easy to prepare (Scheme 101). Modification of this thermolytic procedure by using peroxide initiators, or by refluxing in dioxane, has also proved of value. For example, when the analogous oxadiazolinethiones are heated, 2-(4-thiazolyl)benzimidazoles (**197**; Scheme 113) are formed ⟨74CRV279⟩.

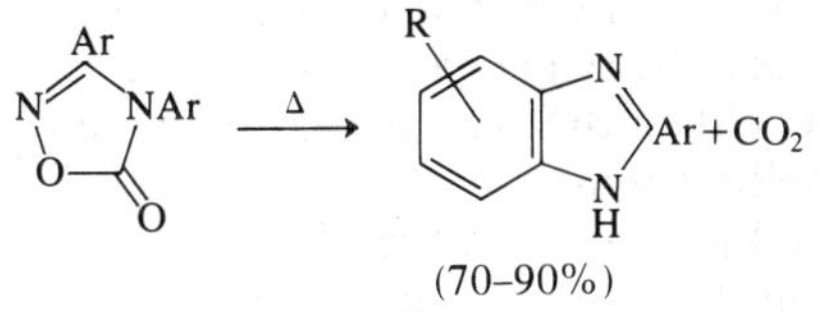

(70–90%)

Scheme 101

4.08.2.2.7 From five-membered rings containing sulfur

When 2,5-diamino-1,3,4-thiadiazole is treated with an α-halo ketone in ethanol, the product is a 2-imino-3-(2-oxoalkyl)-5-amino-1,3,4-thiadiazoline hydrohalide (**173**). Direct treatment with excess hydrazine hydrate, or such treatment subsequent to reaction with aqueous ammonia, gives 1-aminoimidazoline-2-thiones (Scheme 102) ⟨70AHC(12)103⟩.

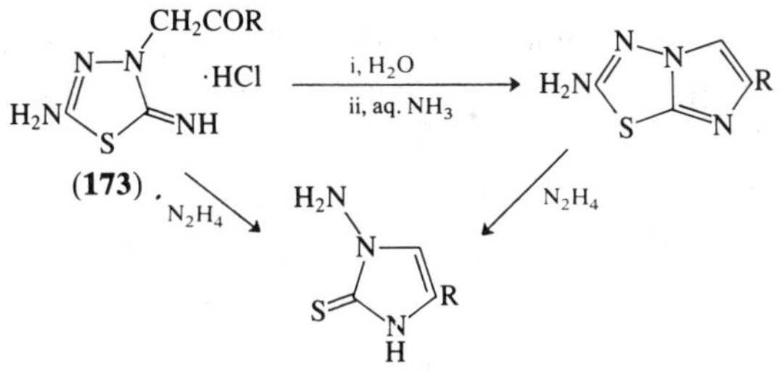

Scheme 102

The dithiazole (**174**) is converted into the imidazodithiazole (**175**) and the imidazolinethione (**176**) by a photochemical process of unknown mechanism (Scheme 103) ⟨80AHC(27)241⟩. However, the similarity of this last scheme to Scheme 103 suggests that (**176**) may well be a breakdown product of (**175**). Treatment of 2,5-diaminothiazoles with alkali gives imidazoline-2-thiones.

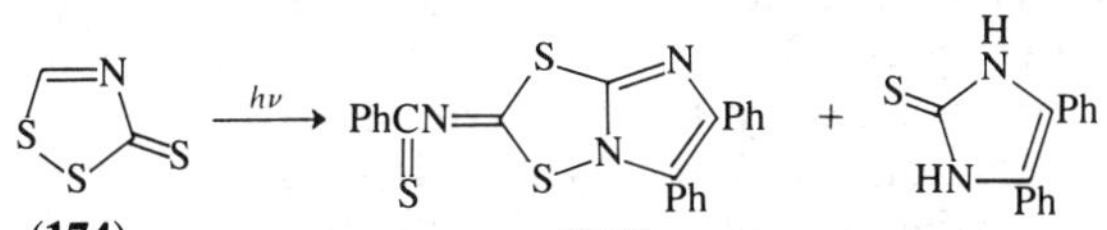

Scheme 103

4.08.2.2.8 Imidazoles from condensed heterocycles

Examples exist of ring fission of purines to give 4,5-disubstituted imidazoles (Scheme 104) ⟨78TL5007⟩. The conversion of benzimidazoles into imidazoles has been discussed as an example of oxidation (Section 4.07.3.1).

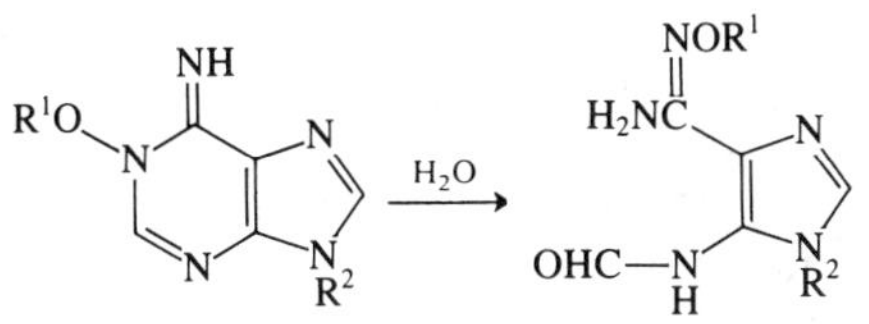

Scheme 104

4.08.2.3 Ring Contractions

There has been considerable literature pertaining to imidazole formation from pyrimidines, pyrazines and triazines ⟨70AHC(12)103, 80AHC(27)241, B-73MI40801⟩.

4.08.2.3.1 From pyrimidines

When 5-amino-4-chloro-2-phenylpyrimidine (**177**) reacts with potassium amide in liquid ammonia, the product is 4-cyano-2-phenylimidazole. Under the same conditions, (**178**) gives 4-ethynyl-2-phenylimidazole (Scheme 105). A pyridine analogous to (**177**) (3-amino-2-bromopyridine) rearranges when subjected to the same conditions, giving 3-cyanopyrrole, but 2-bromo-3-methylpyridine fails to give a pyrrole product. Whereas in the pyridines only the 2,3-bond cleavage can account for the reaction products, with the pyrimidines (**177**) and (**178**) either the 4,5- or 5,6-bonds could be broken. The situation has been clarified by ^{14}C-labelling, which demonstrates unequivocally that it is the 5,6-bond of the pyrimidine which is subject to cleavage. This information permits the writing of a feasible mechanism to account for the products (Scheme 106). There is still uncertainty as to whether ring closure proceeds through a carbenoid species (**180**), or directly from (**179**) ⟨80AHC(27)241⟩.

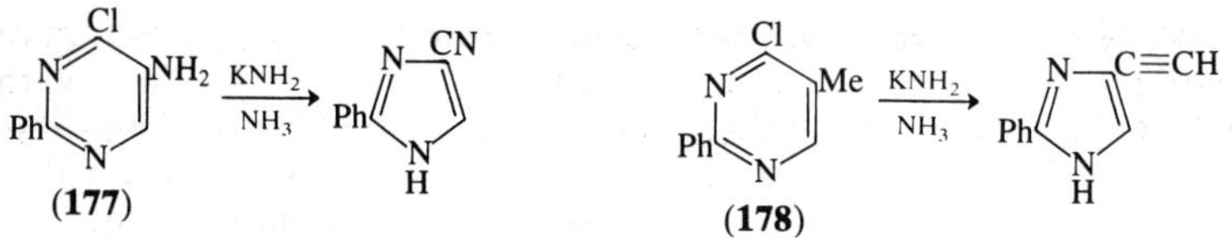

Scheme 105

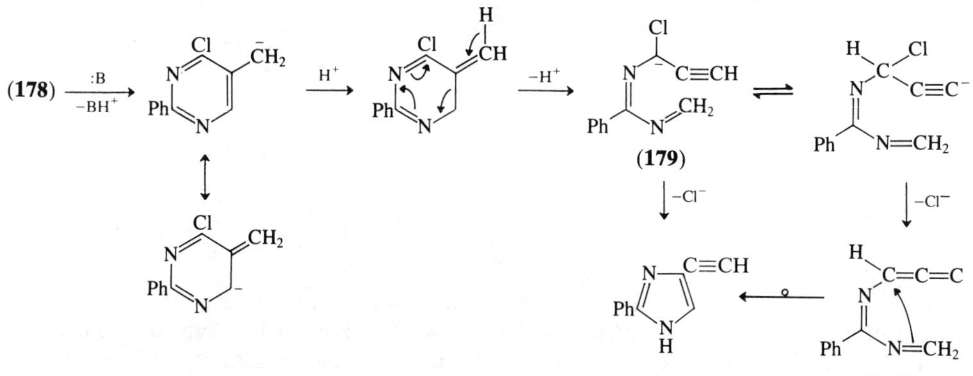

Scheme 106

1-Cyanoimidazoles (**185**) are formed in high yield (80–90%) when 4- or 6-azidopyrimidines are pyrolyzed. The azido compounds are readily prepared from the corresponding chloro compounds by reaction with potassium azide, and probably exist largely as the tetrazolo tautomers (**181**). When (**181**) is labelled as in Scheme 107 (by reaction of the 4-chloropyrimidine with K^{15}NH$_2$), the subsequent pyrolysis gives a product (**185**) in which only a ring nitrogen carries the label. Such a result is explicable only in terms of the formation of a nitrene intermediate (**182**), which rearranges *via* the triazepine (**183**) irreversibly into the pyrazinyl nitrene (**184**). This then undergoes ring contraction ⟨80AHC(27)241⟩. Certainly, thermal (or photochemical) nitrene generation from the tetrazolopyrazine (**186**) also leads to 1-cyanoimidazoles (Section 4.08.2.3.2). Similar thermolysis in benzene at 125 °C of 4-azido-6-methyl-2-methylthiopyrimidine gives a 50% yield of 1-cyano-4-methyl-2-methylthioimidazole.

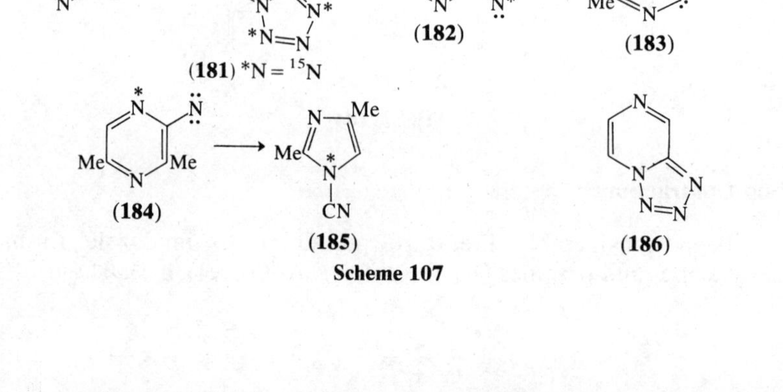

Scheme 107

There are examples of photochemically induced transformations of pyrimidines into imidazoles. In benzene or methanol solution, substituted pyrimidine 1-oxides give low yields of imidazoles (Scheme 108). Such reactions have been explained by invoking oxaziridines, 1,2,4-oxadiazepines and zwitterionic species as intermediates.

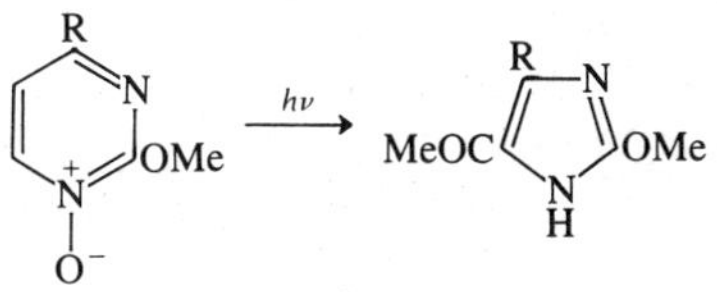

Scheme 108

4.08.2.3.2 From pyrazines and quinoxalines

Both pyrazine N-oxides and 2,3-dihydropyrazines rearrange photochemically to imidazoles. Irradiation of the 2,5-disubstituted oxides (**187**) gives mixtures of the imidazole products (**188, 189**), probably through the intermediacy of oxaziridine intermediates (Scheme 109).

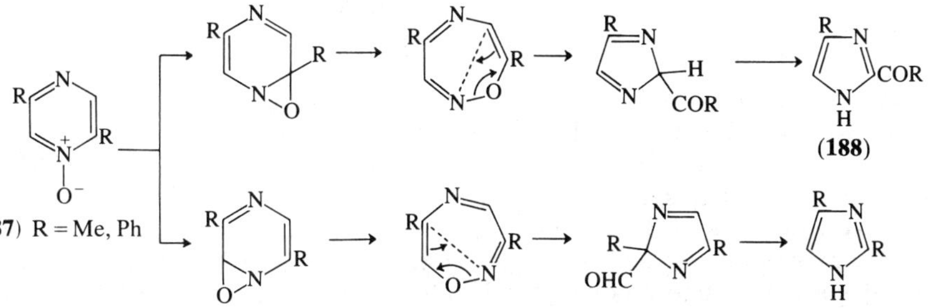

Scheme 109

In a transformation which parallels those of azidopyrimidines (Section 4.08.2.3.1) 2-azidopyrazine 1-oxide (**190**) on thermolysis forms 2-cyano-1-hydroxyimidazole; 2-azidopyrazines at 220 °C give 1-cyanoimidazoles *via* the nitrene intermediate (Scheme 110) ⟨80H(14)1963⟩.

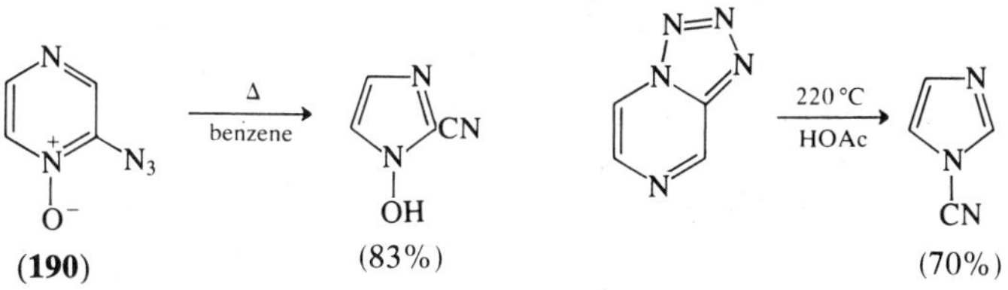

Scheme 110

Certain quinoxaline N-oxides (**191**), too, are converted into 1-acetyl-3-aroylbenzimidazolinones (**193**) when they are heated in acetic anhydride. The reaction proceeds by an initial acetylation at N-3, and this is succeeded by formation and rearrangement of an intermediate oxaziridine (**192**; Scheme 111). Apart from their involvement in other reactions discussed earlier, such intermediates have also been invoked to explain the formation of 1,3-dibenzoylbenzimidazolinone (70%) when quinoxaline 1,3-dioxide is exposed to sunlight ⟨74CRV279⟩.

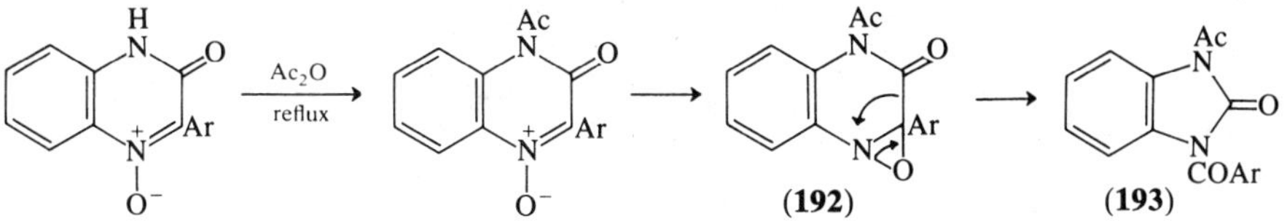

Scheme 111

When treated with potassium amide in liquid ammonia, chloropyrazines undergo ring contraction to imidazoles. The reaction has synthetic importance and resembles similar reactions of pyridines and pyrimidines (Section 4.08.2.3.1). As before, labelling experiments have been utilized to establish the origins of the nitrogen atoms in the products, and complex mechanisms have been proposed (Scheme 112). One factor which stands out is that the products seldom arise from a simple nucleophilic attack at the carbon atom bearing the halogen. The reaction products obtained from 2-chloropyrazine are 2-aminopyrazine (13–15%), imidazole (14–15%) and 2-cyanoimidazole (30–35%). Benzimidazoles are the major products of similar reaction sequences with 2-chloroquinoxalines ⟨80AHC(27)241, 72RTC850⟩.

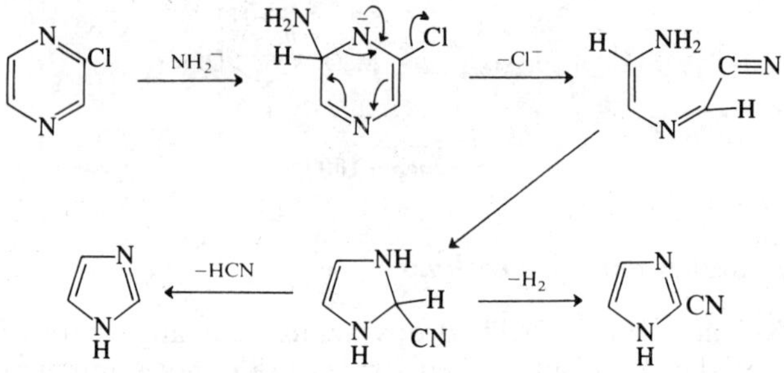

Scheme 112

4.08.2.3.3 From other six-membered heterocycles

When the dihydro-1,2,4-triazinone (**194**) is heated at 180 °C under reducing conditions there is rupture of the 1,2-bond followed by ring closure, producing 4,5-diphenyl-imidazolinone (**195**). Such reductive ring closures of benzotriazine 1-oxides (**196**) can give benzimidazoles, in particular 2-(4-thiazolyl)benzimidazoles (**197**) ('thiabendazole'). Similar reactions take place with 1,3,5-triazines. In fact, primary amines cleave the compounds completely with evolution of ammonia and formation of N,N'-disubstituted formamidines. With suitable primary amines, though, this reaction can be designed to produce imidazolines or benzimidazoles (Scheme 113).

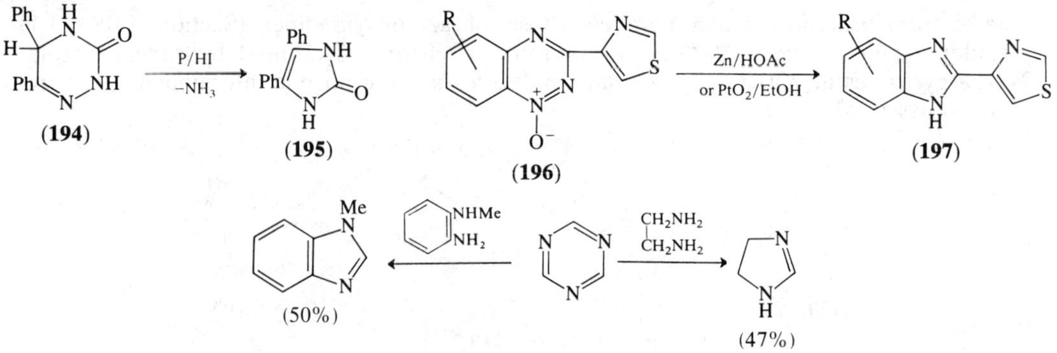

Scheme 113

Treatment of 4H-1,3,5-thiadiazines (**198**) with aliphatic amines at room temperature gives 2,4,5-triarylimidazoles. These may be formed through the 8π 1,3,5-thiadiazine anion (**199**), which could lose sulfur via a thia-α-homoimidazole (**200**; Scheme 114).

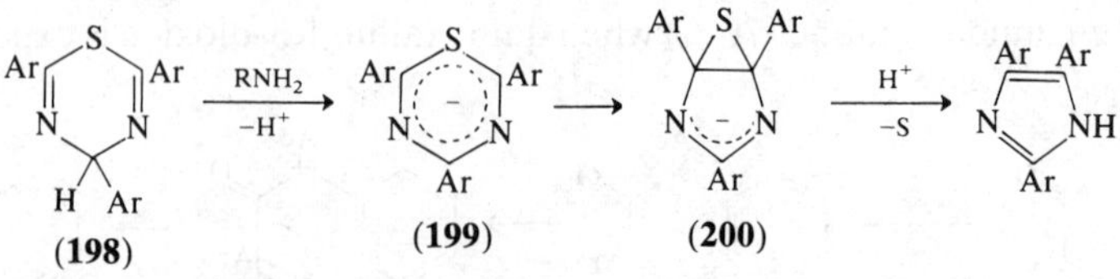

Scheme 114

4.08.2.3.4 From seven-membered heterocycles

Few examples are known in which heterocyclic rings containing more than six atoms contract to form imidazoles. However, the acid-catalyzed contraction of 1,5-benzodiazepines (**201**) into benzimidazoles is well documented. When heated in water, (**201**) form benzimidazoles and ketones (Scheme 115). Both the salts and free bases decompose on standing in aqueous solution at ambient temperatures, while dry distillation of the benzodiazepinium salts has the same result.

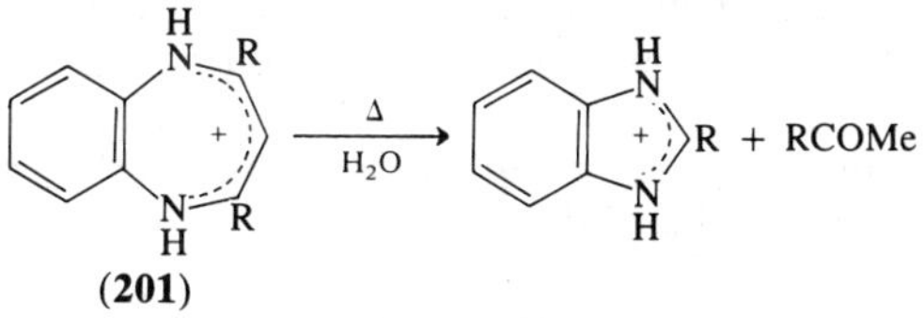

Scheme 115

If 5-benzyl-3,7-diaryl-4,6-dihydro-1,2,5-triazepines (**202**) are treated with bromine or *N*-bromosuccinimide, a process involving consecutive halogenation and dehalogenation has been proposed. The reaction sequence (Scheme 116) has been postulated to pass through bicyclic intermediates ⟨80AHC(27)241⟩.

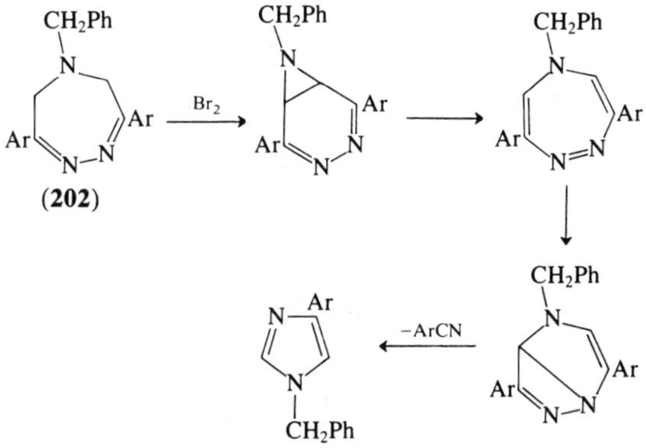

Scheme 116

4.08.3 APPLICATIONS AND IMPORTANCE

The imidazole nucleus appears in a number of naturally occurring products, among which the most important are the amino acid histidine (**203**) and the purines, which comprise many of the important bases in nucleic acids. These purines, of which adenosine (**204**) is an example, are fused imidazopyrimidines. Their chemistry is discussed in Chapter 4.09.

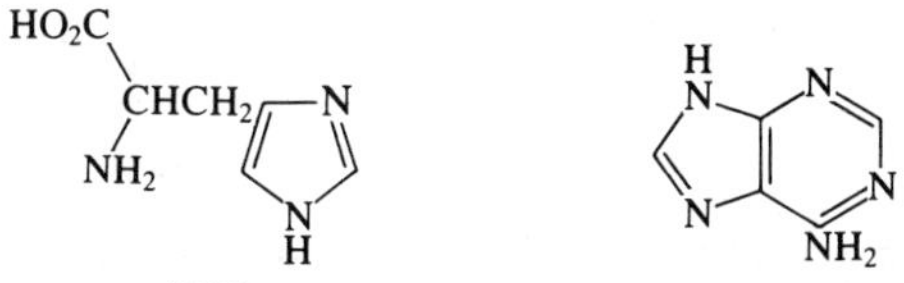

Histidine can be prepared from imidazole as in Scheme 117. It is a basic amino acid, the biosynthesis of which appears to involve adenosine 5'-phosphate, ribose phosphate and glutamine (which supplies a nitrogen for the imidazole ring) (Scheme 118). Histidine is one of the most important amino acids and has involvement in such biological processes as ester hydrolysis, acylation and (by virtue of its complexing power with iron) oxygen

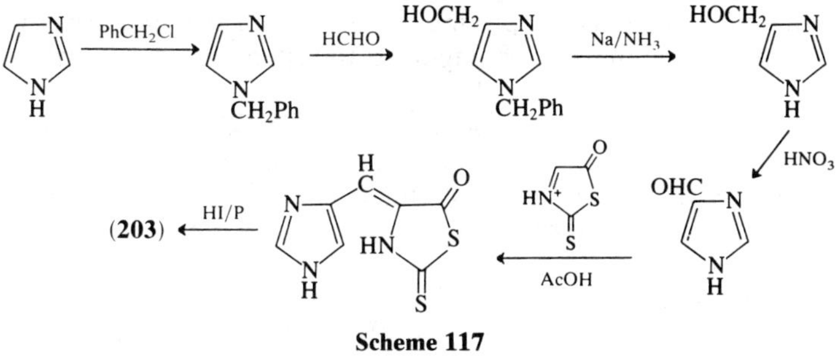

Scheme 117

transport (in haemoglobin). Several mammalian degradation routes of histidine are known, of which decarboxylation to histamine (**205**) is well known. Histamine is an amine which has powerful biological properties, including stimulation of glands and smooth muscle and dilation of capillaries. Release of histamine in the body is believed to be responsible for certain allergic manifestations such as asthma and urticaria. This connection with allergies has stimulated the synthesis of a large number of 'antihistamines' to act as antagonists to the effects of histamine in biological systems. Related (and degradation) products are urocanic acid (**206**) and ergothionine (**207**).

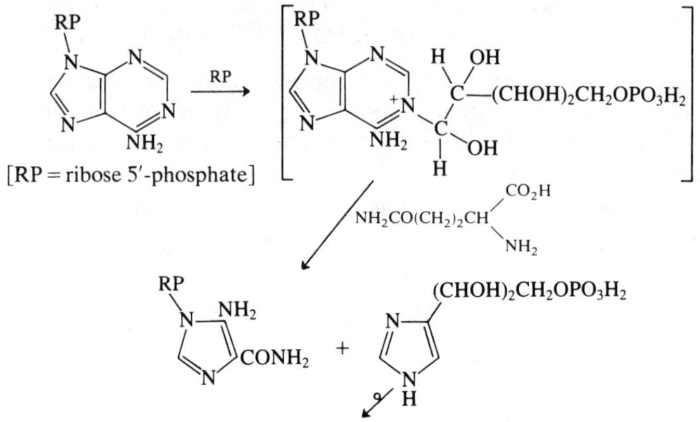

Scheme 118

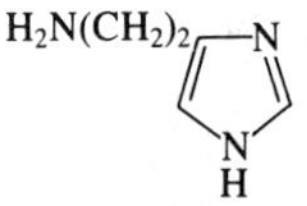

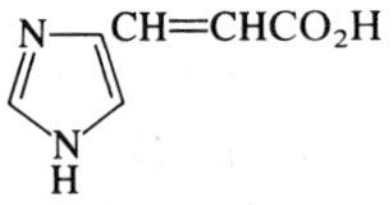

Among other naturally occurring compounds with an imidazole or reduced imidazole nucleus are biotin (**208**), the essential growth factor, the hydantoin (**209**) which occurs in beet sap, and allantoin (**210**) which is related to uric acid.

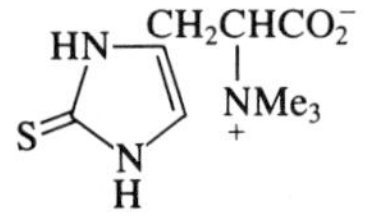

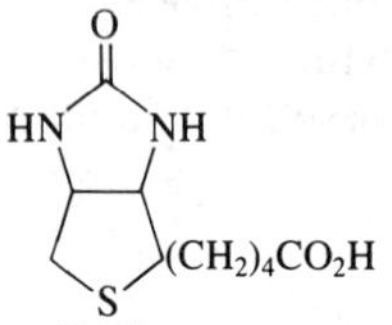

In view of the obvious importance of naturally occurring imidazoles in biological systems, it is not surprising to find that a vast number of synthetic imidazoles have been prepared as potential pharmacological reagents. Of these, a number of nitroimidazoles appear to be finding considerable application. 2-Nitroimidazole (azomycin) and 1-(2-hydroxyethyl)-2-methyl-5-nitroimidazole (metronidazole) are antibacterial agents with particular applications as trichomonacides. Along with metronidazole, other nitroimidazoles (misonidazole, metrazole and clotrimazole) are important anticancer drugs. They are promising radiosensitizers which sensitize hypoxic tumour cells to the lethal effects of ionizing radiation.

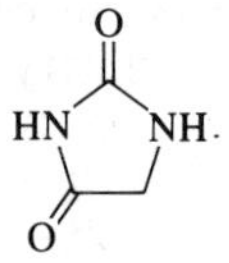

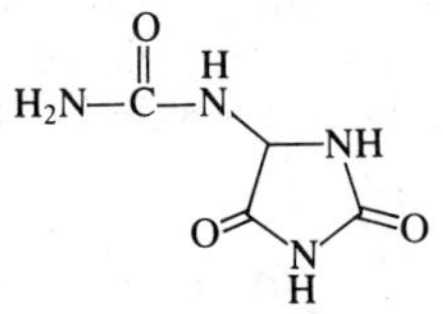

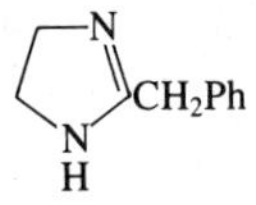

Two imidazolines, priscol (**211**) and privine, are valuable vasodilating and vasoconstricting drugs, while there are also a number of antagonists to histamine, such as antisine, methiamide and cimethidine, which are also imidazolines.

Benzimidazoles have been applied rather more as herbicides, and in veterinary problems. There are, for example, many benzimidazole fungicides and herbicides, and the antiheliminthics thiabendazole (**197**; R = H) and cambendazole (**197**; R = NHCO$_2$Pr^i) are of proven efficiency both for human and veterinary use. There are also some fungicides known among the 2-aminoimidazolines.

A host of other applications of imidazoles and benzimidazoles is listed in the literature with functions as widely divergent as dyestuffs, catalysts, polymerizing agents, drugs, herbicides and fungicides.

4.09
Purines

G. SHAW
University of Bradford

4.09.1 INTRODUCTION	501
4.09.2 NOMENCLATURE OF PURINES	502
4.09.2.1 Nomenclature of Purine Bases	502
4.09.2.2 Nomenclature of Purine Nucleosides and Nucleotides	503
4.09.3 BIOSYNTHESIS OF PURINES	505
4.09.4 STRUCTURE	505
4.09.4.1 Theoretical Methods	505
4.09.4.2 Structural Methods	506
4.09.4.2.1 X-ray diffraction	506
4.09.4.3 Molecular Spectra	510
4.09.4.3.1 ^1H NMR spectra	511
4.09.4.3.2 ^{13}C NMR spectra	514
4.09.4.3.3 ^{15}N NMR spectra	515
4.09.4.3.4 UV spectra	516
4.09.4.3.5 UV photoelectron spectra	517
4.09.4.3.6 IR spectra	518
4.09.4.3.7 Mass spectra	519
4.09.4.3.8 ESR spectra	520
4.09.4.4 Tautomerism in Purines	520
4.09.4.4.1 Introduction	520
4.09.4.4.2 Dipole moments and purine bases	521
4.09.4.4.3 Dipole moments and purine nucleosides	522
4.09.4.4.4 Magnetic circular dichroism	523
4.09.4.4.5 Ionization constants	523
4.09.5 REACTIVITY AT THE RING ATOMS	525
4.09.5.1 Introduction	525
4.09.5.2 Reactions with Electrophiles	526
4.09.5.2.1 ^1H, ^2H and ^3H exchange reactions	526
4.09.5.2.2 N-Alkylation	528
4.09.5.2.3 N-Glycosylation	536
4.09.5.2.4 C-Alkylation and -arylation	536
4.09.5.2.5 Nitration	538
4.09.5.2.6 Reaction with diazonium ions	538
4.09.5.2.7 Oxidation of ring atoms	539
4.09.5.2.8 Halogenation	540
4.09.5.3 Reactions with Nucleophiles	540
4.09.5.3.1 Reduction	540
4.09.5.3.2 Amination	541
4.09.5.4 Reactions with Free Radicals	543
4.09.5.4.1 Alcohols	545
4.09.5.4.2 Amines	545
4.09.5.4.3 Ethers	545
4.09.5.4.4 Acyl radicals	546
4.09.5.4.5 Aryl radicals	546
4.09.6 REACTIVITY OF SUBSTITUENTS	547
4.09.6.1 Introduction	547
4.09.6.1.1 Nucleophilic displacement reactions	547
4.09.6.1.2 Substituent modification	547
4.09.6.2 N- and C-Alkyl Derivatives	547
4.09.6.3 Purine Aldehydes	547
4.09.6.4 Purine Ketones	548
4.09.6.5 Cyanopurines	548

4.09.6.6 Purinecarboxylic Acids	549
4.09.6.7 Purinecarboxylic Acid Esters	550
4.09.6.8 Purinecarboxamides	550
4.09.6.9 Nitropurines	550
4.09.6.10 Aminopurines	551
4.09.6.10.1 Introduction	551
4.09.6.10.2 Acylation	551
4.09.6.10.3 Alkylation	551
4.09.6.10.4 Displacement reactions	552
4.09.6.11 Hydrazinopurines	553
4.09.6.12 Purine N-Oxides	554
4.09.6.13 Hydroxyaminopurines	555
4.09.6.14 Oxopurines	556
4.09.6.14.1 Introduction	556
4.09.6.14.2 Oxo- to halogeno-purines	556
4.09.6.14.3 Thiation	557
4.09.6.14.4 Miscellaneous displacement reactions	557
4.09.6.15 Alkoxypurines	557
4.09.6.16 Thioxopurines	558
4.09.6.16.1 Dethiation	558
4.09.6.16.2 Alkylation and acylation	559
4.09.6.16.3 Halogenation	559
4.09.6.16.4 Amination	560
4.09.6.16.5 Hydrolysis	560
4.09.6.16.6 Oxidation	560
4.09.6.17 Halogenopurines	561
4.09.6.18 Purines with a Fused Heterocyclic Ring System	564
4.09.6.18.1 Introduction	564
4.09.6.18.2 Five-membered fused rings	564
4.09.6.18.3 Six-membered fused rings	566
4.09.6.19 Benzopurines	567
4.09.7 SYNTHESIS	567
4.09.7.1 Introduction	567
4.09.7.2 Synthesis from Acyclic Precursors	567
4.09.7.3 Synthesis from Diaminopyrimidines	570
4.09.7.3.1 Introduction	570
4.09.7.3.2 C(8)-Unsubstituted purines	571
4.09.7.3.3 C(8)-Substituted purines	573
4.09.7.4 Synthesis from Other 4,5-Disubstituted Pyrimidines	579
4.09.7.4.1 Synthesis from 5-amino-4-oxopyrimidines	582
4.09.7.4.2 Synthesis from 4,5-dioxopyrimidines	582
4.09.7.4.3 Synthesis from 4-aminopyrimidine-5-carboxamides	582
4.09.7.5 Purine Synthesis from Imidazoles	583
4.09.7.5.1 Introduction	583
4.09.7.5.2 C(2)-Unsubstituted purines	583
4.09.7.5.3 2-Alkyl-, aryl- and glycosyl-purines	587
4.09.7.5.4 2-Oxopurines	587
4.09.7.5.5 2-Thioxopurines	589
4.09.7.5.6 2-Aminopurines	590
4.09.7.6 Synthesis from Ring-fused Pyrimidine Heterocycles	591
4.09.7.7 Best Practical Methods for Specific Purine Synthesis	592
4.09.7.7.1 Introduction	592
4.09.7.7.2 Purine	592
4.09.7.7.3 C-Linked purines	593
4.09.7.7.4 N-Linked purines	593
4.09.7.7.5 O-Linked purines	596
4.09.7.7.6 S-Linked purines	596
4.09.7.7.7 Se-Linked purines	597
4.09.7.7.8 Halogenopurines	597
4.09.8 NATURALLY OCCURRING PURINES—BIOCHEMISTRY AND APPLICATIONS	598
4.09.8.1 Biosynthesis	598
4.09.8.1.1 De novo biosynthetic pathways	598
4.09.8.1.2 Salvage pathways	598
4.09.8.2 Naturally Occurring Purines	598
4.09.8.2.1 Methylated purines	598
4.09.8.2.2 Cytokinins	601
4.09.8.3 Nucleoside Antibiotics and Related Compounds	602
4.09.8.4 Miscellaneous	604

4.09.1 INTRODUCTION

The purine ring system is undoubtedly among the most ubiquitous of all the heterocyclic compounds. This arises not only from the universal occurrence of adenine and guanine in DNA and RNA and of additional modified derivatives in the various tRNAs but also from the subsidiary uses of the ring system in very many biochemical systems. Indeed across the whole spectrum of biochemical reactions in living systems there is hardly a reaction sequence which does not involve in some way a purine derivative such as the adenosine or guanosine mono-, di- and tri-phosphates, associated cyclic phosphates and nucleotide coenzymes.

In addition to the nucleic acid bases and the long-known methylated purines such as caffeine, in recent years modified purine structures both of natural and synthetic origin have been a rich source of a wide variety of biologically active materials. Such compounds generally include structural modifications in the carbohydrate moiety of ribose or deoxyribose derivatives as with arabinosides, simple changes in known purines including cytokinins, which are 6-N-alkylated adenines, and more deep seated changes in the purine skeleton as with deaza and aza purines which involve carbon to nitrogen skeletal changes.

Table 1 Purines and Derivatives: Bibliography of Important Publications

Subject	Publication type	Author(s)	Ref.
Major chemical review	Book	Lister	71HC(24-2)1
Chemistry and biology	Book	Wolstenholme and O'Connor (eds.)	B-57MI40900
Early review	Chapter	Lister	66AHC(6)1
Early general review	Chapter	Robins	B-67MI40900
Syntheses	Book	Zorbach and Tipson (eds.)	B-68MI40901
Physicochemical aspects	Book	Zorbach and Tipson (eds.)	B-73MI40902
Physicochemical aspects	Chapter	Lister	79AHC(24)215
Purine N-oxides and cancer	Paper	Brown	68MI40903
Purine N-oxides	Book	Katritzky and Lagowski	B-71MI40902
Chemistry of guanine	Review	Shapiro	68MI40902
Uric acid: chemistry and synthesis	Book	Hitchings	B-78MI40900
Nucleophilic and electrophillic reactions	Chapter	Bergman, Lichtenberg, Reichman and Neiman	B-74MI40901
Tautomerism and prototropy	Chapter	Elguero, Marzin, Katritzky and Linda	76AHC(S1)502
Electronic aspects and structure	Chapter	B. Pullman and A. Pullman	71AHC(13)77
Literature July 1978–June 1979	Chapter	Ellis and Smalley	B-80MI40903
6-Mercaptopurine review	Paper	Benezra and Foss	78MI40906
Chemistry of purines and analogs	Chapter	Lunt	B-79MI40905
Recent chemical developments	Paper	Yoneda	75MI40901
General chemistry	Chapter	Shaw	B-80MI40901
Ring opening and closing of adenine	Paper	Fujii, Itaya and Saito	77H(6)1627
X-Ray data	Chapter	Ringertz	B-72MI40902
X-Ray data	Book	Katritzky (ed.)	72PMH(5)1
X-Ray crystallography	Chapter	Rao and Sundaralingham	B-73MI40902, p. 399
X-Ray data	Paper	Taylor and Kennard	82JST(78)1
X-Ray data	Paper	Taylor and Kennard	82JA3209
Gas-phase analysis	Chapter	Pierce	B-73MI40902, p. 125
Dipole moments	Chapter	Bergmann and Weiler-Feilchenfeld	B-73MI40901
IR spectra	Chapter	Tsuboi and Kyogoku	B-73MI40902, p. 215
EPR of irradiated single crystals	Chapter	Herak	B-73MI40901, p. 127
Metal complexes	Chapter	Kistenmacher and Marzilli	B-77MI40901, p. 7
Chromatography	Chapter	Radrazil	B-73MI40902, p. 533
Charge transfer interactions	Chapter	Slifkin	B-73MI40901, p. 67
Mass spectra	Chapter	DeJongh	B-73MI40902, p. 145
UV spectra	Chapter	Albert	B-73MI40902, p. 47
Ionization constants	Chapter	Albert	B-73MI40902, p. 1
H-exchange	Chapter	Jones and Taylor	81MI40900, p. 329

The subsidiary uses of purines, together of course with associated pyrimidines, can hardly be accidental and serve to illustrate the economy shown by nature in its use of available and appropriate feedstock chemicals, and also points to some deeper significance in the original choice of purines for the specific biological functions they possess. In either case the purines assume a philosophical importance possessed by few other heterocyclic compounds in considerations related to the organization of organic matter and the origin of living systems. It is hardly surprising therefore that the purine ring system has been the object of intense study by every relevant physicochemical technique, and syntheses and reactions of purines have been examined in great detail. Some important general reviews of purine chemistry and physical chemistry are collected in Table 1.

4.09.2 NOMENCLATURE OF PURINES

The system was first named by Fischer ⟨1899CB435⟩ but the correct ring structure was proposed much earlier ⟨1875LA(175)243⟩ in a formula for uric acid. In the general literature purine nomenclature has been to a large extent affected by the abundance of trivial names for compounds which were isolated before their structures were known and by variation in description of tautomeric forms.

4.09.2.1 Nomenclature of Purine Bases

The naturally occurring purines fall into 4 main groups. (1) Simple substituted derivatives of purine (**1**) such as adenine (**2**) and various 6-N-substituted derivatives. (2) Monoxodihydropurines such as hypoxanthine (**3**), guanine (**4**), and isoguanine (**5**). (3) Dioxotetrahydropurines such as xanthine (**6**) and methylated derivatives including the 3,7-dimethyl derivative theobromine (**7**), 1,3-dimethylxanthine or theophylline (**8**), and 1,3,7-trimethylxanthine or caffeine (**9**). (4) Trioxohexahydropurines such as uric acid (**10**).

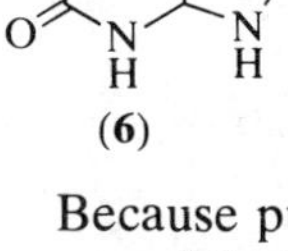

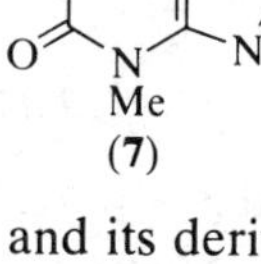

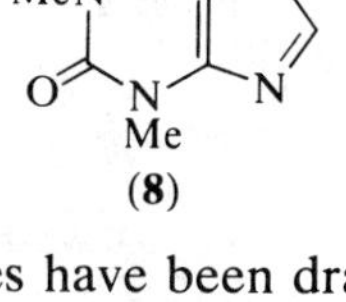

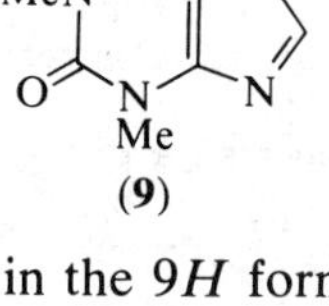

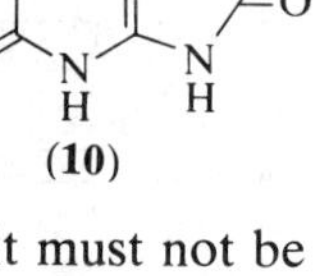

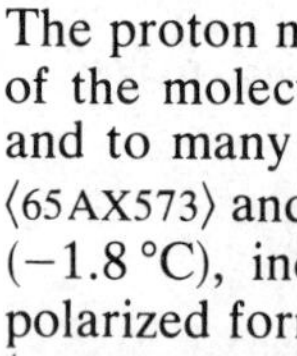

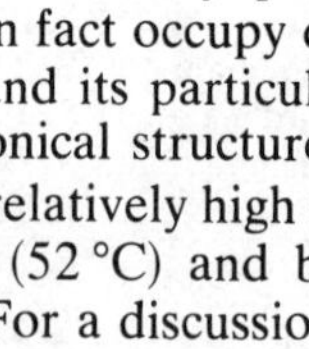

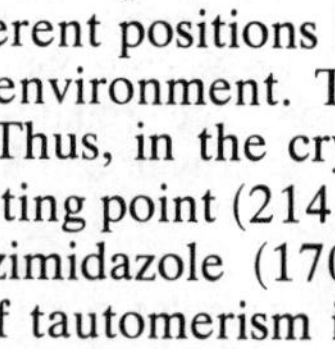

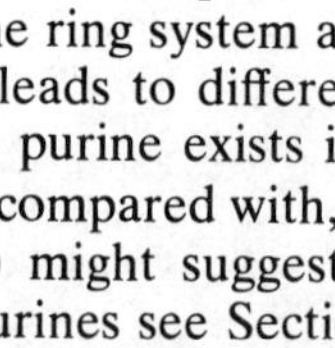

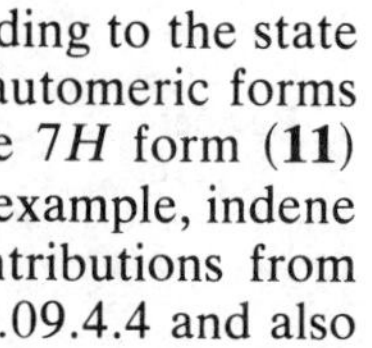

Because purine and its derivatives have been drawn in the 9H form (**1**), it must not be assumed that this necessarily specifies the precise location of that proton in all circumstances. The proton may in fact occupy different positions in the ring system according to the state of the molecule and its particular environment. This leads to different tautomeric forms and to many canonical structures. Thus, in the crystal purine exists in the 7H form (**11**) ⟨65AX573⟩ and its relatively high melting point (214 °C) compared with, for example, indene (−1.8 °C), indole (52 °C) and benzimidazole (170 °C) might suggest contributions from polarized forms. For a discussion of tautomerism in purines see Section 4.09.4.4 and also ⟨76AHC(S1)502⟩. The numbering system adopted is as shown (**12**) and is in accordance with IUPAC rule B-2.11.

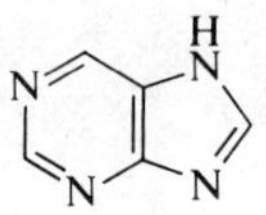

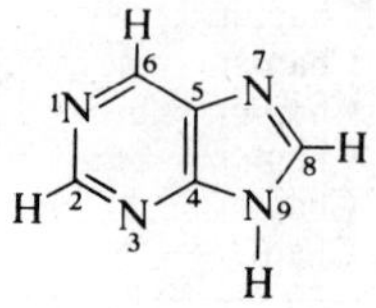

It will be appreciated that it is also possible to write oxo-*nH*-purines in different ways, *i.e.* in the lactam (oxo-*nH*) form as with (**13**) and in a lactim (hydroxy form) as with the alternate structure for hypoxanthine (**14**). However there is little doubt (see later) that both in the solid state and in solution, except at high pH, these particular derivatives exist almost exclusively in the oxo-*nH* forms.

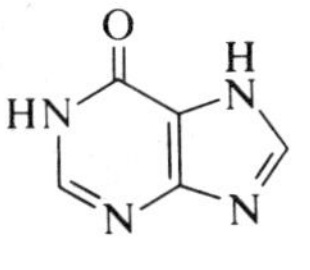

(13) 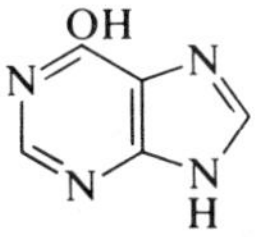   (14)

4.09.2.2 Nomenclature of Purine Nucleosides and Nucleotides

In addition to the free bases outlined in Section 4.09.2.1 many purines also occur as *N*-glycosides or nucleosides. Of special importance are the 9-β-D-ribofuranosyl and 2′-deoxy-9-β-D-ribofuranosyl derivatives of adenine and guanine, adenosine (**15**), guanosine (**16**), inosine (**17**), xanthosine (**18**), 2′-deoxyadenosine (**19**) and 2′-deoxyguanosine (**20**). The nucleosides also occur in various phosphorylated forms or nucleotides which include 5′-phosphates such as adenosine 5′-phosphate (AMP, **21**), guanosine 5′-phosphate (GMP, **22**) inosine 5′-phosphate (IMP, **23**) and corresponding pyro- (di-) and tri-phosphates including ADP (**24**) and ATP (**25**). These last compounds may be regarded as the monomer units of the nucleic acids which can thus be regarded as polynucleotides. The numbering system adopted is outlined in structure (**26**). It will be noted that nucleoside structures are presented in the normally correct *anti* form. See however Section 4.09.4.4.3.

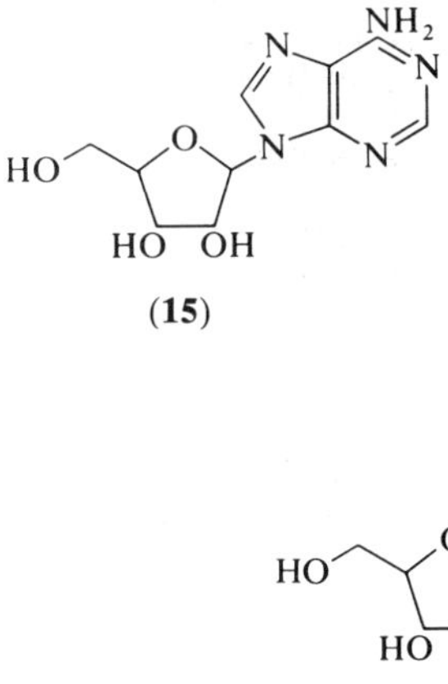

(15) 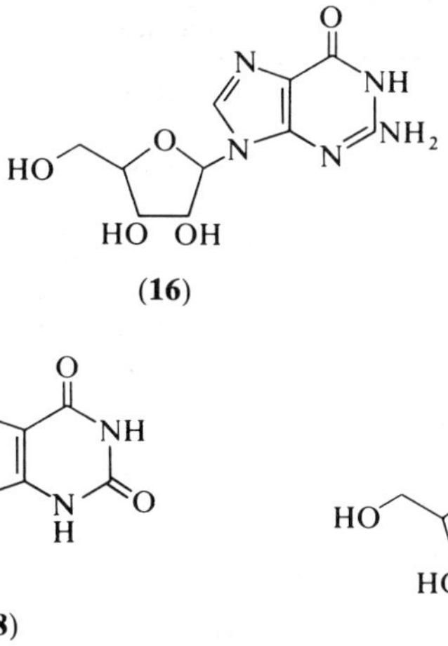   (16) 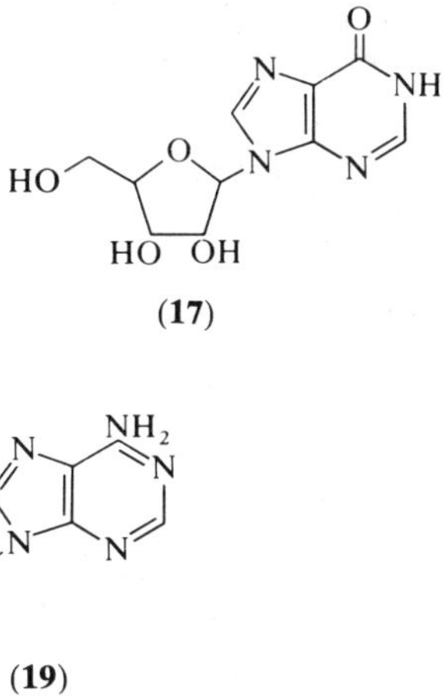   (17)

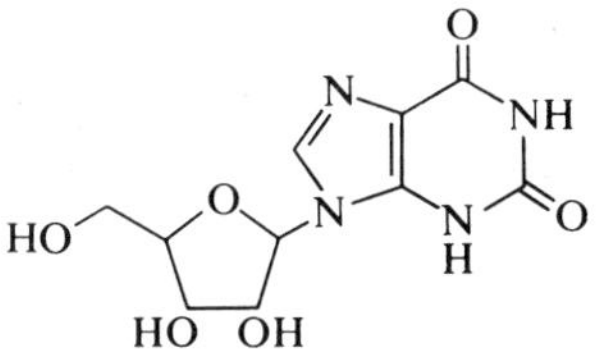

(18) 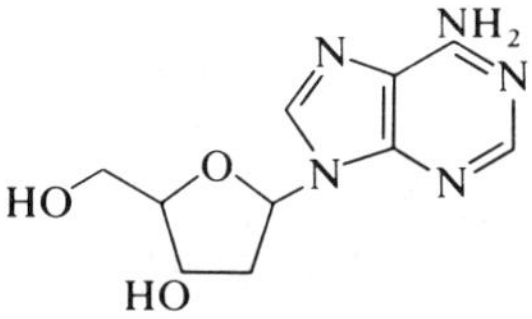   (19)

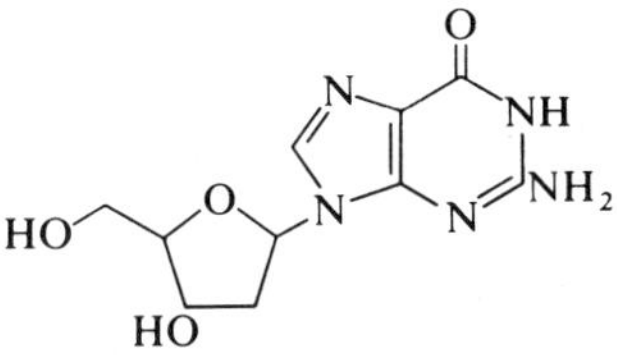

(20) 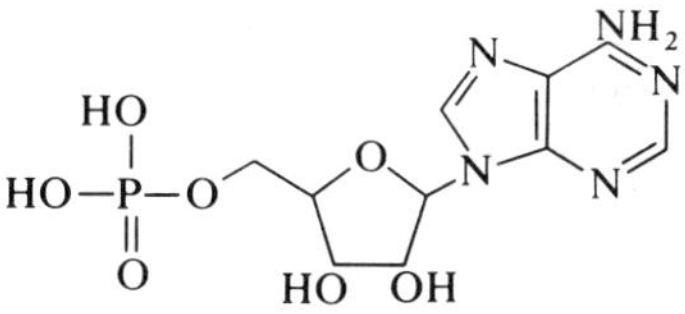   (21)

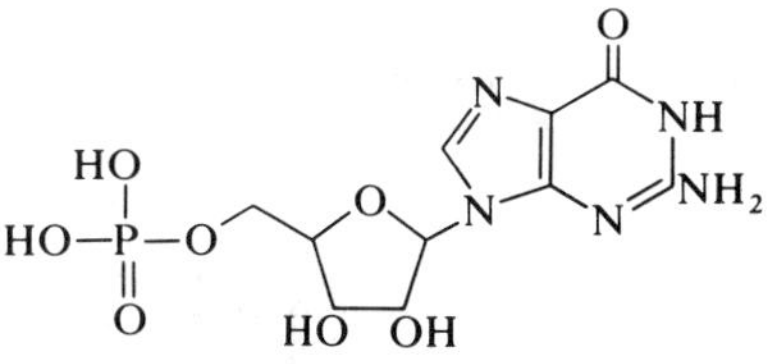

(22) 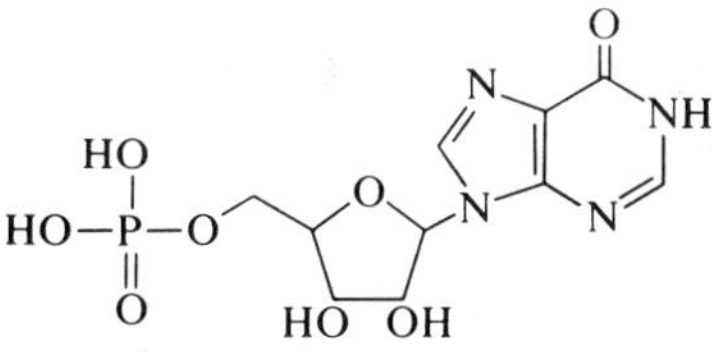   (23)

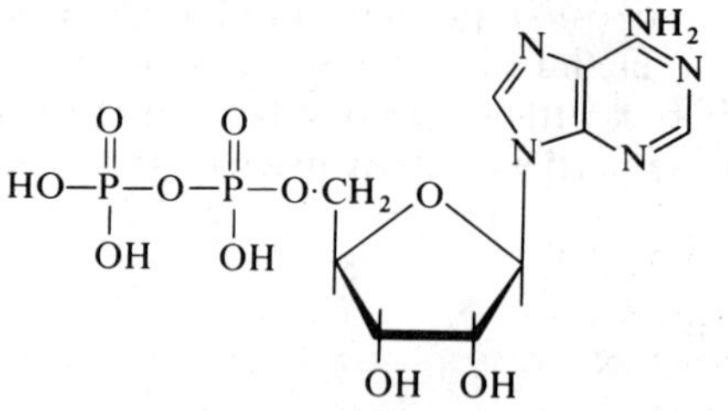

(24) 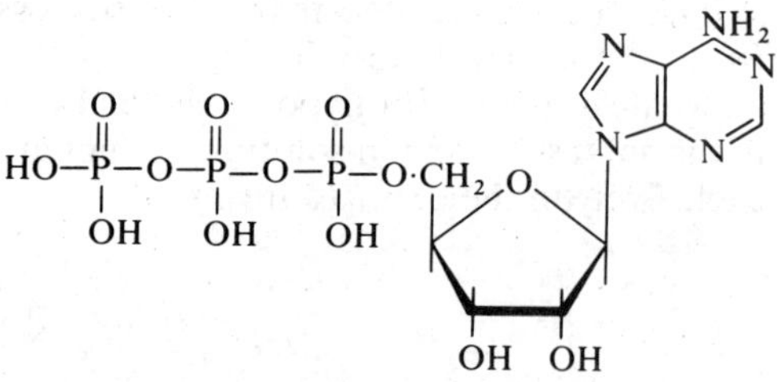 (25)

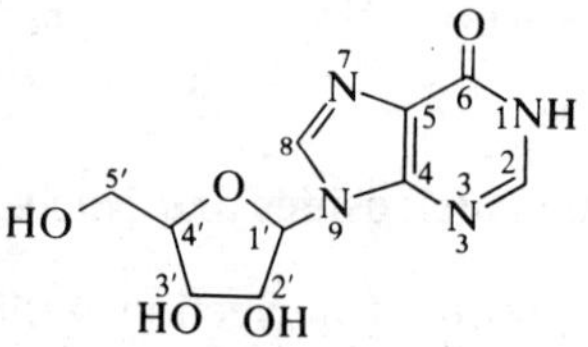

(26)

The importance to science of the nucleic acids has currently reached awesome proportions and the literature is now very substantial indeed. It is therefore beyond the scope of this article to give more than the briefest mention of purine nucleoside, nucleotide and nucleic acid chemistry. However, Table 2 lists some useful reviews on the subject.

Table 2 Purine Nucleosides and Nucleotides: Bibliography of Important Publications

Subject	Publication Type	Author(s)	Ref.
General	Book	Robins and Townsend (eds.)	B-79MI40902
Preparation of nucleosides and nucleotides	Book	Zorbach and Tipson (eds.)	B-68MI40901
Syntheses	Books	Townsend and Tipson (eds.)	B-78MI40901, B-78MI40903
General	Book	Harmon, Robins and Townsend (eds.)	B-78MI40908
Purine nucleosides from imidazoles	Paper	Yamazaki and Okutsu	78JHC353
Nucleosides	Paper	Galankiewicz	78KGS723
Nucleosides and nucleotides	Chapter	Davies	76MI40900
Nucleosides and nucleotides	Chapter	Shaw	B-79MI40903
Nucleosides and nucleotides	Chapter	Shaw	80MI40902
Nucleosides, nucleotides and nucleic acids	Chapter	Jones	B-80MI40901, p. 117
Nucleoside analogues	Book	Walker, De Clercq and Eckstein (eds.)	B-79MI40901
Photochemistry	Review	Rufalska and Wenska	80MI40907
tRNA-modified nucleosides	Paper	McCloskey and Nishimura	77ACR403
C-Nucleosides	Paper	Sato and Noyori	78MI40907
Nucleoside antibiotics	Book	Suhadolnik	B-70MI40900
Nucleoside antibiotics	Chapter	Bloch	B-78MI40903, p. 962
Cyclic nucleotides	Book	George and Ignarro (eds.)	B-78MI40905
Cyclic nucleotides	Book	Brooker, Greengard and Robinson	B-79MI40907
Cyclic nucleotides (mainly biological)	Book	Cramer and Schultz (eds.)	B-77MI40900
Biosynthesis of purine nucleotides	Chapter	Flaks and Lukens	B-63MI40902
Purine nucleotide biosynthesis	Chapter	Buchanan and Hartman	59MI40900, 59MI40901
Purine arabinosides–physiological action and enzyme synthesis	Paper	Yamanaka and Utagawa	80MI40904
Arabinofuranosyl purines	Review	Kaneko and Shimizu	79MI40904
Antitumour nucleosides	Chapter	LePage	B-75MI40900
ORD	Chapter	Ulbricht	B-73MI40902, p. 177

4.09.3 BIOSYNTHESIS OF PURINES

The oxo-nH-purines are frequently end products of adenine or guanine catabolism but are also concerned in the biosynthesis of the purine skeleton. Biosynthetic routes to purines fall into two main categories. (1) *De novo* biosynthesis: this operates at the nucleotide level and involves the elaboration of new adenine and guanine ring systems around the primary amino group in 5-phosphoribosylamine in an enzyme-controlled sequence of reactions involving eleven steps (for further details see references in Tables 2 and Section 4.09.8.1). (2) Salvage pathways: these do not involve formation of new ring systems but can control the ratios of existing nucleotides in the biochemical pool (see Section 4.09.8.1).

4.09.4 STRUCTURE

4.09.4.1 Theoretical Methods

The evolution of molecular biology has stimulated interest in the quantum theory of purines which were the first molecules of specific biological importance to be investigated quantitatively by the methods of quantum chemistry. Indeed, it has been suggested that these particular studies inaugurated the advent of quantum biochemistry ⟨B-72MI40902⟩. The major stages in the development of the calculations on purines and base pairs and appropriate references are briefly set out in Table 3.

Table 3 Stages in the Development of Theoretical Calculations on Purines

Year	Method	Ref.
1956–63	π-Electron calculations. Hückel-type molecular orbital (HMO) method:	
	Bases—first calculations	56CR(243)380
	Base pairs—first calculations	59BBA(36)343
1961–6	π-Electron calculations. Self-consistent field MO without (SCF MO) and with (SCF CI MO) configuration interaction:	
	First calculations	63MI40901
1965	σ-Electron calculations. Del Re's Hückel-type method:	
	First calculations	65MI40905
1968	All valence electron (AVE) calculations (simultaneous treatment of σ- and π-electrons)	
	Three treatments:	
	(i) EHT (extended Hückel theory)	68MI40908
	(ii) IEHT (iterative extended Hückel theory)	68MI40907
	(iii) CNDO treatment (an SCF MO type)	69MI40907
1969–81	All-electron calculations	
	Ab initio calculations using a Gaussian atomic basis set:	69MI40906
	First calculations	69MI40901

The theoretical calculations have been of undoubted value to the chemist in successfully predicting several physicochemical properties of purine molecules such as dipole moments and ionization potentials (see Tables 4 and 5) and providing information about the various tautomeric forms.

Unfortunately, the results of the mathematical treatment normally only reflect ground-state values of the molecules ⟨75IJC668⟩ since it is difficult to find formulae which include all the various intrinsic environmental and extraneous factors such as solvent effects. Thus in the case of purine, CNDO calculations suggest that the $7H$ tautomer should have a slightly lower energy value than the $9H$ form. In the solid state this appears to be the case ⟨65AX573⟩ but in solution there is ample chemical evidence of the equivalence of the $7H$ and $9H$ forms. Thus alkylation of purine with vinyl acetate leads to equal amounts of the 7- and 9-vinylpurines (**27**) and (**28**), respectively, but with other purines the reaction is kinetically rather than thermodynamically controlled and normally only one alkylated purine is produced ⟨75JOC3296⟩ (see also Section 4.09.5.2.2).

Table 4 Calculated Dipole Moments of some Purine Nucleosides[a] ⟨B-73MI40901⟩

Compound	Pucker endo	anti Glycoside torsion angle (°)	μ (D)	syn Glycoside torsion angle (°)	μ (D)
Adenosine	C-3'	20	5.6	240	1.9
				280	2.8
Adenosine	C-2'	70	8.2	260	4.4
Guanosine	C-3'	35	4.6	280	10.0
2'-Deoxyguanosine	C-2'	60	5.9	230	6.3

[a] Orientation at C(4')—C(5') gauche–gauche.

Table 5 Experimental and Calculated Ionization Potentials (eV) of some Purines

Compound	Experimental	Calculated SCF	CNDO	Non-empirical
Purine	9.68 ± 0.1	—	—	—
Xanthine	9.3 ± 0.2	8.82	—	—
Hypoxanthine	9.17 ± 0.1	8.0	—	—
Adenine	8.91 ± 0.1	7.92	10.08	9.41[a], 9.99[b]
Tetramethyluric acid	7.87	≈7	—	—
Guanine	—	7.59	9.06	9.09[b]

[a] ⟨69MI40906⟩. [b] ⟨69MI40901⟩.

(27)   (28)

Electronic structures of purines in their ground and lower excited singlet ($\pi\pi^*$) and triplet ($n\pi^*$) states carried out using the AVE approximation CNDO/2 method have allowed both identification of most photoreactive sites and consideration of the dependence of site locations on the ionic states of the molecules to be made ⟨79MI40906⟩. Net electron charges in the major tautomers of purine, adenine, guanine and hypoxanthine are outlined in Figure 1 ⟨71AHC(13)77⟩. Information about other tautomers is given in the same publication.

4.09.4.2 Structural Methods

4.09.4.2.1 X-ray diffraction

X-Ray crystallographic studies of purine and its various substituted derivatives have been of special value in providing (a) fine structural details of the ring system of simple molecules in their various neutral and protonated, and to a lesser extent, deprotonated forms; (b) a source of molecular geometries for theoretical calculations and related purposes; and (c) information about the precise arrangement of purine (and pyrimidine) bases in the various nucleic acids, and the way in which interaction of such bases with extraneous materials including intercalated or absorbed compounds occurs.

The studies have necessarily been limited by the availability of appropriate crystalline derivatives, and the sparsity of crystalline anionic species is of special note since information about the latter compounds would be of special value to the synthetic organic chemist. The localization of protons in a given compound is to a large extent governed by the stacking and interaction of units in the crystal lattice and hence the structure which eventually emerges may have little relationship to that which may dominate in a reactive environment as in solutions of varying polarities. Consequently, crystal structure data must be accepted with care if the aim of the investigation concerns the prediction of reaction sites under the normal conditions of homogeneous or heterogeneous chemical reactions.

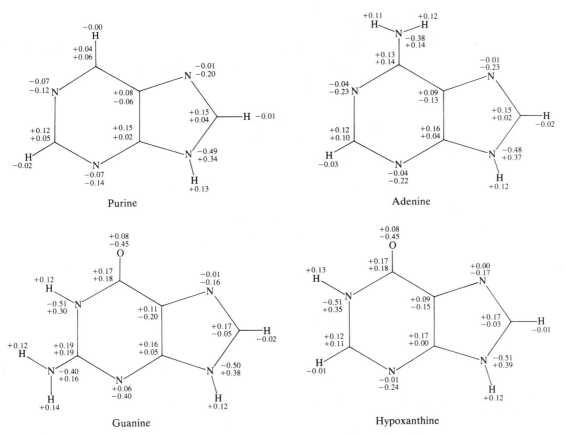

Figure 1 Net electronic charges in some purine N(9)H tautomers evaluated by the CNDO/2 method. Upper numbers: σ-charges; lower numbers π-charges

(i) Purine bases

The first recorded X-ray crystal structures of purines were those of adenine (1948) and guanine (1951) derivatives. Data for these and related purines are recorded in Table 5.1 and additional examples are collected in a publication ⟨72PMH(5)1⟩. Recently, results derived from precisely-determined X-ray crystal structures of 24 adenine and 10 guanine derivatives in both neutral and protonated forms have been published ⟨82JST(78)1, 82JA3209⟩.

The average standard deviations for bond lengths and angles assigned to this latter set of data are 0.011 Å and 0.67° in adenine (neutral form), 0.011 Å and 0.81° in adenine (protonated form), 0.78 Å and 0.48° in guanine (neutral form), and 0.015 Å and 0.83° for guanine (protonated form). See also ⟨B-72MI40902, p. 61⟩ and Table 1 for further publications devoted to bond lengths and bond angles in various purine derivatives.

Oddly enough, it is difficult to find accurate X-ray crystal structure determination of pure anhydrous adenine or guanine. The amount of experimental data available has clearly been limited by the ability of workers to produce appropriate crystalline derivatives. This in general means that published material almost exclusively is concerned with substituted derivatives, frequently at N-9, including nucleosides and nucleotides or protonated forms and hydrates which are found to produce good crystals. Unfortunately, literature abstracts and titles of papers generally fail to make these distinctions, and reviews often quote a structure for say, adenine, close examination of which reveals it to be for perhaps the hydrochloride or other derivative. Also structures, especially tautomeric structures, have frequently been assigned by assuming the existence of close relationships between bond lengths and angles of, say, an aglycone and corresponding nucleoside derivative. Such assumptions are probably reasonable, but are nevertheless assumptions and to some extent must cloud the accuracy of the reported assignments.

In contrast to adenine and guanine the structure of purine in the crystal is well characterized as the N(7)H tautomer ⟨65AX573⟩. However even this is to some extent unexpected since calculations indicate that the N(7)H and N(9)H forms have similar stabilities.

Table 5.1 X-Ray Bond Lengths (Å); and Bond Angles (°) of Some Purines

Bond length (angle)	Purine[a]	Adenine hydrochloride hemihydrate[b]	Guanine hydrochloride dihydrate[c]	Hypoxanthine hydrochloride monohydrate[d]
4—5 (4—5—6)	1.404 (118.2)	1.37 (118)	1.377 (119.9)	1.368 (120.9)
5—6 (5—6—1)	1.389 (119.0)	1.40 (114)	1.414 (110.8)	1.415 (110.2)
1—6 (2—1—6)	1.330 (118.0)	1.38 (123)	1.390 (125.6)	1.382 (124.9)
1—2 (1—2—3)	1.349 (128.2)	1.37 (124.5)	1.374 (123.4)	1.356 (125.4)
2—3 (2—3—4)	1.331 (113.2)	1.30 (112.5)	1.318 (112.8)	1.297 (112.1)
3—4 (3—4—5)	1.337 (123.5)	1.36 (128)	1.345 (127.6)	1.352 (126.5)
5—4 (5—4—9)	1.404 (109.6)	1.37 (107)	1.377 (106.2)	1.368 (106.3)
4—9 (4—9—8)	1.374 (104.2)	1.36 (105)	1.375 (108.6)	1.375 (108.3)
8—9 (7—8—9)	1.312 (114.6)	1.33 (115)	1.335 (109.6)	1.333 (110.0)
7—8 (5—7—8)	1.332 (106.4)	1.35 (102)	1.322 (108.2)	1.318 (107.8)
5—7 (4—5—7)	1.374 (105.3)	1.37 (111)	1.378 (107.4)	1.376 (107.7)

Bond length (angle)	9-Methyladenine dihydrobromide[e]	Sodium xanthinate tetrahydrate[f]	Uric acid[g]	Caffeine monohydrate[h]	Theophylline monohydrate[i]
4—5 (4—5—6)	1.40 (120)	1.37 (118.2)	1.360 (121.2)	1.32 (119.9)	1.37 (122.6)
5—6 (5—6—1)	1.38 (111)	1.41 (113.0)	1.411 (111.3)	1.44 (115.8)	1.41 (112.5)
1—6 (2—1—6)	1.39 (124)	1.37 (125.7)	1.397 (128.8)	1.36 (127.6)	1.38 (126.0)
1—2 (1—2—3)	1.36 (127)	1.38 (120.5)	1.367 (116.0)	1.42 (112.9)	1.40 (117.8)
2—3 (2—3—4)	1.35 (107)	1.33 (114.7)	1.382 (118.4)	1.35 (122.8)	1.35 (119.4)
3—4 (3—4—5)	1.37 (129)	1.36 (127.7)	1.356 (124.1)	1.42 (121.4)	1.37 (121.8)
5—4 (5—4—9)	1.40 (100)	1.37 (105.5)	1.360 (107.5)	1.32 (110.9)	1.37 (112.7)
4—9 (4—9—8)	1.37 (116)	1.41 (106.0)	1.360 (109.2)	1.31 (105.7)	1.33 (102.0)
8—9 (7—8—9)	1.35 (104)	1.37 (115.0)	1.376 (107.1)	1.34 (112.3)	1.31 (114.3)
7—8 (5—7—8)	1.33 (109)	1.30 (101.9)	1.359 (108.1)	1.32 (103.4)	1.31 (106.7)
5—7 (4—5—7)	1.37 (111)	1.41 (111.7)	1.387 (108.0)	1.41 (107.2)	1.34 (104.2)

[a] ⟨65AX573⟩. [b] ⟨51AX81⟩. [c] ⟨65MI40903⟩. [d] ⟨69AX(B)1608⟩. [e] ⟨62AX1179⟩. [f] ⟨69BCJ3099⟩. [g] ⟨66AX397⟩. [h] ⟨58AX453⟩. [i] ⟨58AX83⟩.

An explanation for the existence of the $N(7)H$ tautomer may reside in the fact that the calculated dipole moment of the $N(7)H$ form is much greater than that of the $N(9)H$ form and the stabilization energy for interaction (as in the crystal lattice) of two $N(7)H$ forms is hence greater than that for two $N(9)H$ tautomers. In the gas phase however, the $N(9)H$ tautomer is the dominant form (see Section 4.09.4.3.5).

A summary with references of published X-ray structural data in the solid state of purines other than adenine and guanine has been compiled ⟨B-72MI40902, p. 61⟩.

In almost all cases the imidazole ring system is found to be planar whereas the pyrimidine ring has out-of-plane features. Bond lengths vary with the ionic state. Thus studies of various adenine salts which contain the adeninium (protonated adenine) ion indicate that protonation occurs firstly at N-1 in 9-substituted derivatives, which have been most studied (see 71CRV439 for a discussion of protonation sites), followed by a second protonation at N-7 ⟨62AX1179⟩. Differences in the imidazole ring bond angles between the mono- (29) ⟨74AX(B)166⟩ and the di-cation (30) ⟨74AX(B)1528, 74CL409⟩ of adenine similarly confirm that in the latter, protonation occurs initially at N-7.

A study of 9-methyladenine at 126 K ⟨81AX(B)1584⟩, which reduces atomic vibrations, has shown that the charge density on nitrogen is in the order N-1 > N-7 > N-3 with N-1 and N-7 accordingly the preferred sites for protonation. In unsubstituted adenine the first protonation most likely occurs at N-1 ⟨78AX(B)2229⟩ but a proportion of the protonated imine form (**31**) is also probably present (see Section 4.09.4.3.3). The adenine anion however is assigned the amino structure (**32**) ⟨73AX(B)1974⟩.

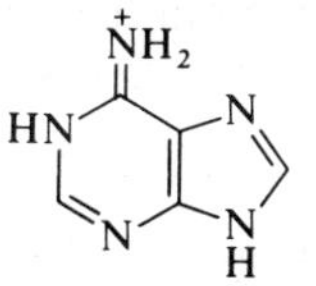

(31) 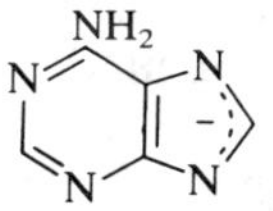 (32)

Crystal structures of several 9-substituted adenines including 9-methyladenine ⟨63AX907, 77AX(B)253⟩, adenosine ⟨72AX(B)1982⟩, and 2′-deoxyadenosine ⟨65AX111⟩ confirm the 6-amino structure. Additional references are given in a review ⟨76AHC(S1)502⟩.

The crystal structure of a zinc chloride:adenine adduct namely (adenine)$_2$: $ZnCl_4 \cdot H_2O$ contains adenine cations which exist in the unusual $N(7)H$ form ⟨81AX(A)C63⟩ analogous to the structure found in purine crystals ⟨65AX573⟩.

With diprotonated 7-methyladenine (*e.g.* the dihydrochloride) the protonation sites are at N-3 and N-9 ⟨75AX(B)211⟩. Hypoxanthine and the analogous 6-mercaptopurine ⟨69AX(B)1608⟩ both exist in the 7*H*-lactam tautomeric forms ⟨69AX(B)1338, 69AX(B)1330⟩ and the lactam structure of inosine has been confirmed in the solid state ⟨69MI40903⟩. Similarly, monoprotonated hypoxanthine exists in the $N(1)H$, $N(7)H$ form with the proton at the charged N-9 position ⟨69AX(B)1608⟩.

Crystal structure determinations of guanine hydrochloride hydrate also indicate that it too exists with protonation assumed to occur at $N(7)H$ as in (**33**) ⟨69AX(B)1608⟩ and, on the assumption that bond lengths and angles in guanine are analogous to derived structures of guanosine and inosine which are very similar, the $N(9)H$ tautomer is also favoured ⟨68BBR(33)436⟩. In contrast the $N(7)H$ tautomer dominates in 6-thioguanine ⟨70JA7441⟩ and crystal structure analysis of 3-methylguanine (**34**) also indicates that it exists as the $N(7)H$ tautomer ⟨76TL3483⟩.

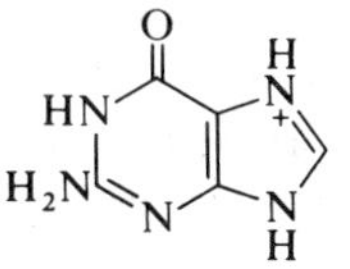

(33) 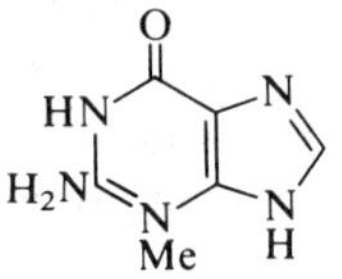 (34)

Puckering or distortion of the pyrimidine ring is especially noteworthy in the case of 1,3,9-trimethylxanthine with the C(2)—O bond displaced out-of-plane ⟨73ACS2757⟩, and similarly the C—S bond in 6-thioxo-(1*H*)-purine is 1.5° out of plane with a dihedral angle of 1.16° between the rings.

In the anionic form of xanthine (as a sodium salt) no hydrogen appears at N-3 or N-7 and the data has been interpreted by assignment of the canonical structures (Scheme 1) ⟨69BCJ3099⟩. It is interesting to note that in this case all the forms are assigned as 9*H* tautomers, the imidazole proton in solution usually being assigned to N-7.

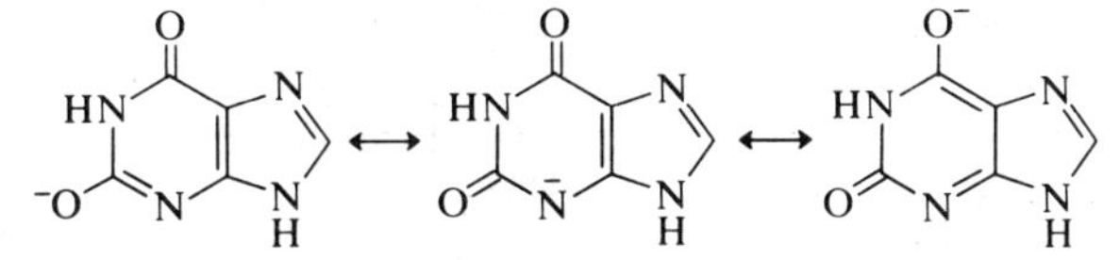

Scheme 1

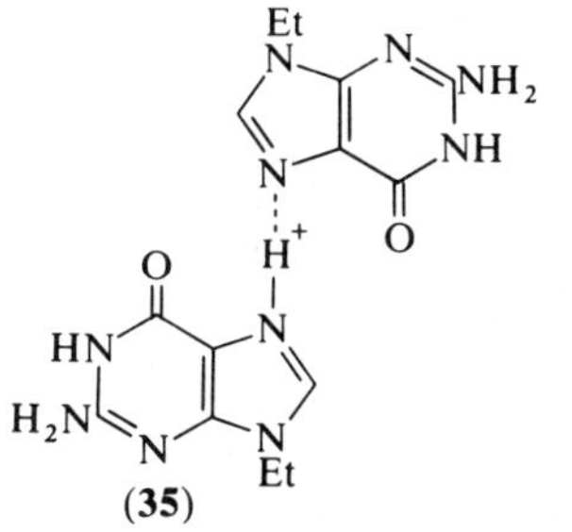

(35) 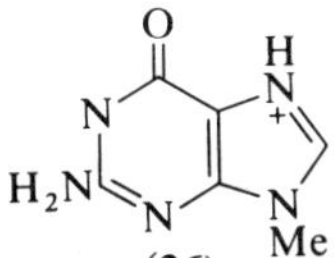 (36)

The effect of close stacking in a crystal structure is illustrated by the hemihydrochloride structure of 9-ethylguanine (**35**) in which one form is protonated and the other hydrogen bonded between the N-7 positions by a bond (2.637 Å) which is the shortest so far observed ⟨75AX(B)2862⟩. The hydrobromide of 9-methylguanine analogously has the $7H^+$ structure (**36**) ⟨64AX126⟩. There can be no certainty, of course, that such a structure would have any relevance in solutions of the compound.

Generally, however, the specific stacking patterns found in crystals of purine derivatives are surprisingly consistent, especially when the influences of additional factors such as hydrogen bonding, van der Waals interactions and steric factors are taken into account.

(ii) Purine nucleosides and nucleotides

The purine nucleosides and nucleotides have similar stacking patterns to the aglycones ⟨B-72MI40902, p. 178⟩ and the purine stacking in double helical polynucleotides is similar to that in simple purine derivatives. The stacking patterns are ascribed to the polar regions of one molecule interacting with the polarizable ring system of an adjacent molecule, the most common type involving the C—N bond of one base interacting with the polarizable ring of the adjacent base. The high dipole moments of the purines ensure strong interactions. The stacking patterns fall into two general types.

In the first type the imidazole rings point roughly in the same direction as in purine and guanine. The second type has the imidazole rings of adjacent bases stacked in opposite directions as in guanosine dihydrate and in interstrand stacking between purines of complementary polynucleotide strands.

A crystal structure determination of guanosine hydrobromide ⟨B-72MI40902, p. 217⟩ has interesting features in that the bromide ions appear to reside over the pyrimidine ring portion of the compound implying a charge delocalization on the bases, in contrast to guanine hydrobromide ⟨64AX126⟩ where there is close approach of the N-7 (presumably protonated in both nucleoside and aglycone) and the bromide ion.

4.09.4.3 Molecular Spectra

Molecular spectra have always played an important part in investigations of purine structure and reactions, and recent improvements, especially in the development of ^{1}H,

Table 6 NMR Spectra of Purines: Bibliography of Important Publications

Subject	Publication Type	Author(s)	Ref.
^{15}N NMR spectra of nucleosides	Paper	Grenner and Schmidt	77CB373
General ^{13}C NMR spectra	Chapter	Stothers	B-72MI40901
General ^{13}C NMR spectra	Book	Levy and Nelson	B-72MI40900
^{13}C NMR spectra (purine and 26 substituted purines)	Paper	Thorpe, Coburn and Montgomery	74JMR(15)98
^{13}C NMR spectrum of purine	Paper	Pugmire, Grant, Robins and Rhodes	65JA2225
^{1}H and ^{13}C NMR spectra	Paper	Cheng, Kan, P.O.P. Tso, Giessner-Prettre and B. Pullman	80JA525
NMR spectra in molecular biology	Book	Davies	B-78MI40902, p. 509
^{15}N NMR spectra of nucleosides and nucleotides	Paper	Markowski, Sullivan and Roberts	77JA714
General NMR spectra	Book	Batterham	B-73NMR1
^{1}H NMR spectra of purines and nucleosides	Chapter	Townsend	B-73MI40902, p. 313
^{15}N NMR spectra—general	Book	Levy and Lichter	B-79MI40900
^{15}N–^{1}H coupling in ^{15}N-labelled adenines	Paper	Leonard and Henderson	75JA4990
^{15}N NMR spectra of nucleotides	Paper	Buchner *et al.*	78JMR(29)45
General review—mainly nucleosides and nucleotides	Book	Pullman (ed.)	B-78MI40902, p. 509

^{13}C and ^{15}N NMR spectral techniques, have resulted in major advances being made in our understanding of these molecules.

There have been many publications representing a large amount of experimental work on the purine system. Table 6 lists some major reviews and reference articles.

4.09.4.3.1 ^{1}H NMR spectra

The preparation of specific, deuterated derivatives has made it possible to assign correctly chemical shifts for the C-2, C-6 and C-8 protons of purine ⟨63TL1499, 65JCS623⟩. It was shown that in neutral aqueous solution purine has three sharp singlets corresponding to H-8 at highest field, followed downfield by H-2 and H-6, and the order is maintained in non-aqueous solvents although accompanied by changes in chemical shift to lower field ⟨65JOC1110, 64JOC1988, 68B3721⟩.

The spectrum of purine and, for that matter, its derivatives is very dependent on pH, solvent, temperature and added chemicals. Thus aniline in aqueous solution causes the proton signals for purine to be shifted upfield ⟨66ZPC(51)297⟩, presumably due to some weak association complex formation. The purine anion (32) has a similar spectrum to that of the neutral molecule but that of the cation (29) is shifted downfield. Measurements of pK_a suggest that purine is largely protonated at N-1 ⟨B-57MI40900, p. 11⟩ but the large shift of H-8 in the cation suggests delocalization of the charge with contributions from the N(3)- and N(7)-protonated forms ⟨65JOC1110⟩. At pH < 2.4 spin–spin coupling between $C(2)H$ and $C(6)H$ occurs suggesting $N(1)H$ protonation ⟨B-73MI40902, p. 313⟩. No such coupling occurs at pH > 2.4.

An alternative (to deuteration) method involves measurements on purines in which each C-proton is systematically replaced by methyl groups, the C-methyl having little effect on the chemical shifts of the remaining protons. Spin–spin coupling also occurs as would be expected in the neutral form of 1-methylpurine where the N-1 site is methylated rather than protonated but not with 3-, 7- or 9-methylpurine where the $C(2)H$, $C(6)H$ and $C(8)H$ protons appear as sharp singlets with no evidence for spin–spin coupling ⟨B-73MI40902, p. 313⟩.

Data for other C-substituted purines are included in Table 7.

Table 7 ^{1}H NMR Spectral Data of Purine and Some Simple Derivatives

| Compound | Solvent | Chemical shift (δ, p.p.m.)[a] | | | Ref. |
		H-2	H-6	H-8	
Purine	DMSO	8.99	9.19	8.68	65JOC1110
1-Methylpurine	DMSO	9.07	9.2	8.67	B-73MI40902, p. 313
3-Methylpurine	DMSO	9.1	9.0	8.45	66JHC241
6-Methylpurine	DMSO	8.80	—	8.55	78JOC2587
6-Aminopurine	DMSO	8.2	—	8.17	78JOC2587
6-Methylaminopurine	DMSO	8.2	—	8.33	B-73MI40902, p. 313
6-Dimethylaminopurine	DMSO	8.19	—	8.07	78JOC2587
2-Azidopurine	DMSO	—	8.03	8.58	66JOC935, 66JOC2210
	TFA	—	8.55	9.35	66JOC935, 66JOC2210
	AcOH	—	8.10	8.58	66JOC935, 66JOC2210
6-Azidopurine	TFA	9.13	—	9.13	66JOC935, 66JOC2210
	AcOH	8.74	—	8.59	66JOC935, 66JOC2210
6-Chloropurine	DMSO	8.80	—	8.73	78JOC2587
6-Iodopurine	DMSO	8.63	—	8.60	78JOC2587
2,6-Dichloropurine	DMSO	—	—	8.77	65JOC1110
6-Methoxypurine	DMSO	8.57	—	8.40	78JOC2587
6-Methylthiopurine	DMSO	8.75	—	8.47	78JOC2587
6-Phenoxypurine	DMSO	8.60	—	8.66	78JOC2587
6-Trimethylammonium chloride	DMSO	9.28	—	9.03	78JOC2587
6-Cyanopurine	DMSO	9.12	—	8.93	78JOC2587
6-Chloro-9-ethylpurine	CDCl$_3$	8.17	—	8.76	80JOC3969
6-Bromo-9-ethylpurine	CDCl$_3$	8.3	—	8.70	80JOC3969
6-Iodo-9-ethylpurine	CDCl$_3$	8.24	—	8.63	80JOC3969
Hypoxanthine	DMSO	8.12	—	7.97	66JOC1417, 66JOC2202
6-Mercaptopurine	D$_2$O	—	—	8.2	B-72MI40902, p. 264

[a] TMS used as internal standard.

It is noteworthy that the chemical shift of C(8)H is affected by substituents at C-2 and C-6. Electron donating groups (*e.g.* NMe$_2$) have a shielding effect and an upfield shift to smaller δ values is observed, whereas electron withdrawing groups (*e.g.* CF$_3$) result in a deshielding effect with concomitant larger δ values. An equation incorporating an electrophilic substitution constant which enables predictions to be made of the chemical shift of C(8)H in purines has been of value ⟨58JA4979⟩. In the case of nucleosides, the ribofuranosyl group exerts a deshielding effect on the immediately adjacent O—H protons.

Adenine has two sharp singlets (δ 8.11 and 8.14 p.p.m.) assigned to H-2 and H-8, respectively ⟨65JOC1114⟩, whereas adenosine also has two singlets δ 8.27 (H-2) and 8.45 p.p.m. (H-8), a downfield shift occurring with both protons but mostly with the adjacent H-8 ⟨B-73MI40902, p. 313⟩.

Information on tautomerism, and sites of protonation of purines has been derived from NMR spectral studies. Thus protonation of adenine is first suggested to occur at N-1 with no evidence for protonation of the amino group ⟨65JOC1110⟩ although protonation of the substituted amino group does occur in 6-dimethylamino-3-methylpurine ⟨68CC1002⟩.

An MO study of the protonation of adenine and guanine using the Hartree–Fock–Roothaan SCF method has suggested, however, that for guanine, and to a lesser extent adenine, several protonated forms may coexist in acid solution ⟨80MI40906⟩. Other naturally occurring methylated purines including caffeine, theophylline and theobromine have similarly been examined and assignments of the methyl protons made ⟨64JPS962, 65T3435, 64JA4721⟩. Table 8 lists some ^{1}H NMR spectral data for oxo- and thioxo-purines and Table 9 for charged purine species.

Table 8 ^{1}H NMR Spectral Data of some Oxo and Thioxopurines ⟨B-72MI40902, p. 264, 65T3435⟩

Compound	Solvent	Chemical shifts (δ, p.p.m.)^a			
		H-Me(1)	H-Me(3)	H-Me(7)	H-8
Xanthine	DMSO (9):D$_2$O (1)	—	—	—	2.15
Theophylline (1,3-dimethylxanthine)	D$_2$O	—	—	—	2.05
	TFA	3.65	3.86	—	—
	PhNO$_2$	3.27	3.45	—	—
Theobromine (3,7-dimethylxanthine)	D$_2$O	—	—	—	2.12
	TFA	—	3.75	4.28	—
	PhNO$_2$	—	3.35	3.81	—
Caffeine(1,3,7-trimethylxanthine)	D$_2$O	—	—	—	2.12
	TFA	3.59	3.8	4.33	—
	PhNO$_2$	3.23	3.38	3.82	—
2-Oxo-(3H)-purine	D$_2$O	—	—	—	1.61
6-Thioxanthine	—	—	—	—	2.05

^{a} TMS used as internal standard.

NMR spectral studies of purine nucleosides have been much concerned with assignment of anomeric structure. The dihedral angle between vicinal protons in the *cis* H(1')–H(2') configuration, as with α-D-ribofuranosyl derivatives, and the *trans* H(1')–H(2') configuration, as with β-D-ribofuranosyl derivatives, may vary from 0–45° and 75–165°, respectively ⟨62JA62⟩.

Coupling constants, $J(1'-2')$, vary from 3.5–8.0 Hz for *cis* protons in the α-D compounds and from 0.0–8.0 Hz for *trans* protons in β-D-ribofuranose derivatives. Hence the β-D-configuration can be assigned when the $J(1'-2')$ values fall within the range 0–3.5 Hz. However subsequent work has indicated that the coupling constants must be ≤1 Hz before acceptance of the *trans* configuration can be made with confidence. Sometimes however a suitable derivative must be examined before assignment is possible. Thus adenosine has $J(1'-2') = 5.2$ Hz ⟨63JA2026⟩ which is diminished to 2.9 Hz in the 2',3'-O-isopropylidene derivative (**37**) and this is sufficiently small to indicate the β-configuration. Further reduction of the J value is observed in the rigid N(3)—C(5') anhydro derivative (**38**) where the anomeric proton now occurs as a singlet (δ 6.8 p.p.m.). It has also been shown that the anomeric proton signal of aldofuranosyl nucleosides with *cis* H(1')–H(2') configurations normally appears at a lower field than the signal for the shielded (by the [C(2')—OH])

Table 9 ^{1}H NMR Spectral Data of some Charged Purine Species

Species	Solvent	Chemical shift (δ, p.p.m.)[a]			Ref.
		H-2	H-6	H-8	
Purine (neutral molecule)	D$_2$O	8.82	8.96	8.60	66JCS(B)433
Purine cation	1N DCl	9.39	9.56	9.12	66JCS(B)433
Purine anion	1N NaOD	8.74	8.93	8.53	63TL1499
Purine anion	NH$_3$/KNH$_2$	8.58	8.79	8.12	79JOC3140
[6-^{2}H]Purine anion	NH$_3$/KNH$_2$	8.62	—	8.15	79JOC3140
6-Methylpurine anion	NH$_3$/KNH$_2$	8.48	—	8.06	79JOC3140
[8-^{2}H]6-Methylpurine anion	NH$_3$/KNH$_2$	8.48	—	—	79JOC3140
2-Methylpurine anion	NH$_3$/KNH$_2$	—	8.68	8.05	79JOC3140
[6-^{2}H]8-Methylpurine anion	NH$_3$/KNH$_2$	8.4	—	—	79JOC3140
6,8-Di-t-butylpurine (neutral molecule)	CDCl$_3$	9.0	—	—	79JOC3140
Hypoxanthine					
(neutral)	DMSO	8.01	—	8.15	79JOC1450
(cation)	D$_2$O/CF$_3$CO$_2$D [2:1]	8.61	—	9.44	79JOC1450
3-Methylhypoxanthine					
(neutral)	DMSO	8.31	—	8.21	79JOC1450
(cation)	D$_2$O/CF$_3$CO$_2$D [2:1]	9.32	—	8.60	79JOC1450

[a] TMS used as internal standard.

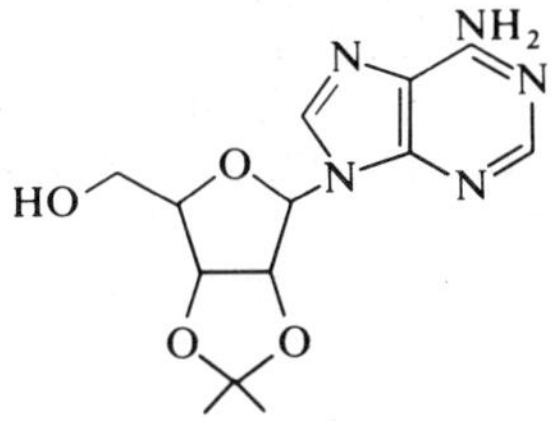

(37) 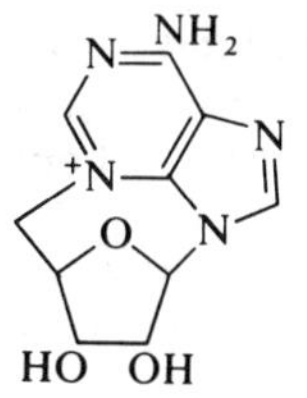   (38)

anomeric proton of the corresponding *trans* H(1')–H(2') anomer, and this has proved a valuable empirical rule for assignment of anomeric configurations. Another useful rule involves examination of 2',3'-*O*-isopropylidene-D-ribofuranosyl derivatives. This (Imbach's) rule proposes ⟨73JHC417, 74TL129, 75ANY(255)177⟩ that the difference in chemical shifts $\Delta\delta$ for the *endo* and *exo* methyl group is >0.15 p.p.m. for β-anomers and <0.15 p.p.m. for α-anomers (Figure 2).

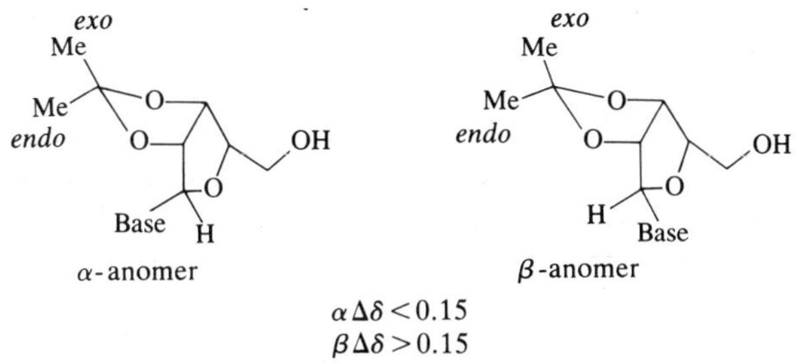

$\alpha\Delta\delta < 0.15$
$\beta\Delta\delta > 0.15$

Figure 2 Imbach's rule

For a more detailed discussion of notation see ⟨B-78MI40902, p. 509⟩.

The variation of spectra in different solvents and under differing conditions can make assignments difficult. However tautomer compositions can be used to explain the mechanism of line broadening in spectra. Thus for guanine, guanine nucleosides and nucleotides, the H-8 line widths were found to be strongly pH and temperature dependent ⟨B-72MI40902, p. 277⟩. However, sharp H-8 resonances were observed with 1-methylguanosine (**39**) and 6-methoxypurine riboside over a wide range of temperatures and pH values. In the former compound the guanine base can only exist in the lactam tautomeric form and in the latter the purine exists solely in the enol (lactim) form. The line broadening is ascribed to chemical exchange between tautomeric lactam–lactim structures for the guanine base. The alternative amino–imino tautomerism is considered to be unlikely since H-8 broadening was still observed in N^2-dimethyl-1-methylguanosine (**40**) where this type of tautomerism is not possible. The ratio of the normal (lactam) to abnormal (lactim) tautomers was approximately 17:3 at room temperature in neutral aqueous solution.

(**39**) R = H
(**40**) R = Me

4.09.4.3.2 ^{13}C NMR spectra

Reviews and major papers dealing with the ^{13}C NMR spectra of purines and derivatives are collected in Table 6.

The ^{13}C chemical shifts are normally reported relative to the peak in carbon disulfide, benzene, DMSO or similar compounds. The first results on purine ⟨65JA2225⟩ indicated that the relative positions of the peaks for the carbon atoms were in the sequence C-4 > C-2 > C-8 > C-6 > C-5; the assignment of C-5 to the highest field was based on theoretical predictions that it would have the highest π-electron density. It is noteworthy that ^1H NMR spectra indicate a different order namely 6 > 2 > 8 for the corresponding hydrogen atoms, suggesting that in the use of chemical shifts for predicting charge densities care should be exercised ⟨B-73MI40902, p. 313⟩.

Comparison of purine in the neutral, anionic and cationic forms indicated that the cation is protonated at N-1 but in the anion, N-7 and N-9 could not be distinguished as a site of protonation ⟨71JA1880⟩.

However, use of 7- and 9-methylpurine as reference compounds and extrapolation of the results to the 7H- and 9H-purines by application of α- and β-substituent parameter corrections indicated 40% of the 7H tautomer in DMSO ⟨75JA4627⟩ and 58% in water ⟨75JA4636⟩. Data for adenine in DMSO indicates about 15% of the 7H form ⟨75JA4627⟩ and this is in fair agreement with a figure of 22% for the 7H form derived from a temperature-jump relaxation technique in aqueous solution ⟨75JA2369⟩. Use of ^1H and ^{13}C NMR spectra in DMSO-d_6 has indicated that in 3-substituted adenine derivatives protonation occurs at N-7 ⟨81CPB2403⟩. The assignments for adenine in the direction of increasing field were found to be in the order of C-6 < C-2 > C-4 > C-8 > C-5 ⟨70JA4079⟩ and this was confirmed in later work ⟨74JMR(15)98⟩.

Table 10 ^{13}C NMR Spectral Data of some Purines

Compound	Solvent	Chemical shifts (δ, p.p.m.)[a]					Ref.
		C-2	C-4	C-5	C-6	C-8	
Purine	DMSO	152.1	154.77	130.46	145.5	146.09	74JMR(15)98
2-Aminopurine	DMSO	160.59	155.11	125.52	147.69	141.02	74JMR(15)98
6-Aminopurine (adenine)	DMSO	152.37	151.30	117.61	155.30	139.29	74JMR(15)98
6-Methylpurine	DMSO	151.76	153.87	129.59	155.67	144.51	74JMR(15)98
9-Ethylpurine	CDCl$_3$	152.5	149.9	134.2	148.6	144.8	80JOC3969
6-Chloro-9-ethylpurine	CDCl$_3$	151.9	151.0	131.7	149.2	144.7	80JOC3969
6-Bromo-9-ethylpurine	CDCl$_3$	151.7	150.5	134.2	144.8	143.0	80JOC3969
6-Iodo-9-ethylpurine	CDCl$_3$	151.8	148.0	122.1	144.2	138.7	80JOC3969
6-Chloropurine	DMSO	151.50	154.16	129.23	147.78	146.18	74JMR(15)98
6-Bromopurine	DMSO	151.52	152.98	131.97	140.12	145.94	74JMR(15)98
6-Iodopurine	DMSO	151.72	150.02	120.18	136.53	145.02	74JMR(15)98
2-Ethyl-6-chloropurine	DMSO	165.01	155.11	127.70	147.59	145.99	74JMR(15)98
9-β-D-Ribofuranosylpurine (nebularine)	D$_2$O	152.6	151.2	134.4	148.7	146.6	80JOC3969

[a] TMS used as internal standard.

The last authors also record the ^{13}C NMR spectra of purine and 26 substituted purines (three 2-substituted, twelve 6-substituted and eleven 2,6-disubstituted) and in most cases proton-coupled in addition to decoupled spectra were obtained. Data for some of these compounds are presented in Table 10 and for some purine species in Table 11. ^{13}C spin lattice relaxation data have also been used to assign a major structural form to the marine sponge nucleoside 1-methylisoguanosine ⟨80CC339⟩ (see Section 4.09.8.2.1). See also ⟨74T3941⟩ for a study of ^{13}C chemical shifts of 6-substituted purines.

Recently the assignments, concentration dependence of chemical shifts in aqueous solution and solvent effects of all eight ^{1}H and ^{13}C resonances of purines have been determined ⟨80JA525⟩.

Table 11 ^{13}C NMR Spectral Data of some Purine Species ⟨79JOC3140⟩

Compound	Solvent	Chemical shifts (δ, p.p.m.)[a]				
		C-2	C-4	C-5	C-6	C-8
Purine (neutral)	CDCl$_3$	152.2	154.4	130.5	145.6	146.1
Purine (anion)	NH$_3$/KNH$_2$	149.5	160.7	134.5	143.7	156.9
2-Methylpurine (neutral)	CDCl$_3$	161.0	155.0	128.8	145.3	145.3
2-Methylpurine (anion)	NH$_3$/KNH$_2$	156.7	164.5	134.5	143.3	158.4
8-Methylpurine (neutral)	CDCl$_3$	151.5	155.9	131.2	143.3	156.2
8-Methylpurine (anion)	NH$_3$/KNH$_2$	148.4	165.2	138.5	140.7	168.2
6,8-Di-t-butylpurine (neutral)	CDCl$_3$	149.7	153.8	132.5	163.0	168.5

[a] TMS used as internal standard.

Table 12 shows the values of C–H coupling constants (Hz) of 1.0M purine at 25 °C.

Natural abundance ^{13}C NMR spectra of purine nucleosides show a clean separation between peaks for the aglycone (−36 to +33 p.p.m.) and the carbohydrate (+37 to +89 p.p.m.) moieties ⟨70PNA(65)27⟩ with a benzene reference.

Table 12 Carbon–Proton Coupling Constants (Hz) of 1.0M Purine ⟨80JA525⟩[a]

Carbon	H-2	H-6	H-8
C-2	$^1J = 206.5$	$^3J = 10.5$	—
C-4	$^3J = 10.8$	$^3J = 4.9$	$^3J = 9.3$
C-5	$^4J = 1.2$	$^2J = 5.9$	$^3J = 8.4$
C-6	$^3J = 10.3$	$^1J = 187.3$	—
C-8	—	$^4J = 0$	$^1J = 212.9$

[a] In D$_2$O at pH 7, 25 °C.

4.09.4.3.3 ^{15}N NMR spectra

N–H coupling has been used to investigate the rearrangement of 3-benzyladenine (**41**) to N^6-benzyladenine (**42**; Scheme 2), the data indicating that the mechanism involves ring opening of both the imidazole and pyrimidine rings ⟨75JA4990⟩ since the ^{15}N label at the 6-NH$_2$ position of 3-benzyladenine appears at N-3 and N-9 in the product.

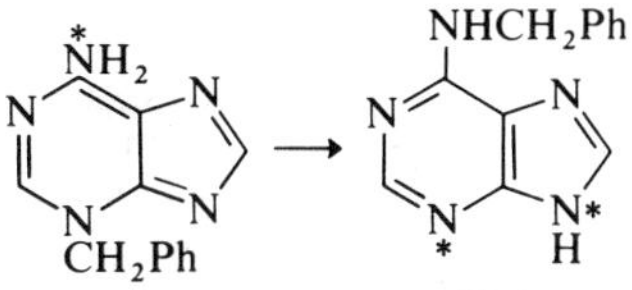

Scheme 2

The mechanism of adenine formation under presumed abiotic conditions (see Section 4.09.7.2) from hydrogen cyanide and formamide has also been investigated in an elegant series of experiments by detection of ^{15}N–^{13}C coupling contants ⟨78JA4617⟩. Thus when doubly enriched HCN was heated with normal formamide at >160 °C the adenine produced had a ^{13}C NMR spectrum in which only the resonances of C-4, C-5 and C-6 were split by ^{15}N nuclei. By comparison, using normal HCN and doubly enriched formamide, in the ^{13}C NMR spectrum of the adenine produced the C-2 and C-8 atoms alone are labelled and there is no ^{15}N–^{13}C coupling. The derivation of the adenine atoms could accordingly be assigned as in Scheme 3. The absence of C–N coupling using enriched formamide was explained in a separate experiment ⟨78CC485⟩ when it was shown that the C—N bond is cleaved and then recombined as shown in Scheme 4.

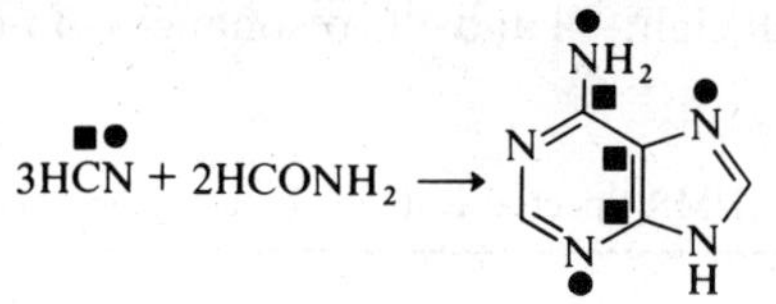

3HCN + 2HCONH$_2$ →

Scheme 3

$H^{12}CO^{14}NH_2 + H^{13}CO^{15}NH_2 \rightarrow H^{12}CO^{15}NH_2 + H^{13}CO^{14}NH_2$

Scheme 4

^{15}N NMR spectra have also been useful in predicting the presence or absence of complexing between nucleoside bases and metal atoms. Thus ATP shows a very broad peak assigned to the 6-NH$_2$ nitrogen, N-9 is a singlet and N-7 a doublet (coupling to H-8) at lowest field. N-1 and N-3 produce doublets by coupling to H-2 with N-1 at highest field. Addition of Mg^{2+} causes no change in chemical shifts, suggesting a lack of complex formation between Mg and N, whereas Zn^{2+} gives evidence which suggests complex interaction at N-7 and the 6-amino group ⟨66JA2077, B-73MI40902, p. 353⟩.

High resolution ^{15}N NMR spectra of six nucleosides and six nucleotides including adenine, guanine and hypoxanthine derivatives has enabled all nitrogen resonances to be assigned by comparison with related compounds ⟨77JA714⟩. The assignments are confirmed in a related study with adenosine and guanosine 3′-phosphates (3′ AP and 3′ GP) ⟨78JMR(29)45, B-78MI40906⟩. On protonation a drastic upfield shift of the N-1 resonance of 3′ AP is interpreted as conversion of a pyridine type nitrogen to a pyrrole type. The other five resonances of 3′ AP shift downfield on protonation. This suggests that the aromatic character of the adenine ring increases, consequently inducing a deshielding of all the nitrogen atoms except N-1. There is also a strong downfield shift of the 6-NH$_2$ nitrogen atom suggesting that its partial double bond character (C=NH) increases during protonation. With 3′ GP the ^{15}N(3)–^{15}NH$_2$ coupling constant is 6.0 Hz suggesting that the lone pair is on N-3 and no change was observed over the pH range 2.2–10.

Table 13 contains ^{15}N NMR spectral data of purine and some common nucleosides and nucleotides.

Table 13 ^{15}N NMR Spectral Data of Purine and Some Purine Nucleosides and Nucleotides ⟨77JA714⟩

Compound	Solvent	Chemical shifts (p.p.m)[a]				
		N-1	N-3	N-7	N-9	NH$_2$
Adenosine	DMSO	139.6	152.7	134.7	205.6	293.8
Guanosine	DMSO	228.0	209.5	128.5	205.3	302.0
Inosine	DMSO	200.7	161.2	126.7	200.7	—
2′-Deoxyadenosine	DMSO	139.1	152.1	134.3	202.1	293.7
Purine	DMSO	94.5	113.6	187 (broad)	187 (broad)	—
Adenosine 5′-phosphate	H$_2$O	152.1	160.4	144.8	207.0	297.0
ATP	H$_2$O	151.7	160.3	145.0	207.2	297.0
ATP (pH 2.5)	H$_2$O	214.3	154.4	139.5	200.1	288.8
GMP	H$_2$O	229.1	211.3	141.2	207.2	303.5

[a] Upfield from D^{15}NO$_3$.

4.09.4.3.4 UV spectra

The characteristic UV absorption spectra of purine derivatives have made the technique widely used for identification, structure determination and tautomerism studies and have resulted in a very large literature. Accordingly it is not possible to give more than a brief account of the subject here. Full details and lists of absorption maxima are given in various publications including ⟨B-73MI40902, 71HC(24-2)1, 72PMH(5)1, 77MI40902⟩.

The absorption spectra of purines fall generally into the following groups. (1) $n \to \pi^* \geq$ 300 nm. These tend to be weak bands (ε 10^3) which arise from 'forbidden transitions' and show bathochromic shifts in passing from polar to non-polar solvents ⟨63JOC1305⟩. (2) $\pi \to \pi^*$ transitions: (a) 230–300 nm. Strong bands with high extinction coefficients (ε 2000 to 20 000) which show hypsochromic shifts in passing from polar to non-polar solvents. (b) >230 nm with high extinction coefficients ($\varepsilon > 20\,000$). The absorption maxima may lie below the limit of most UV spectrometers in normal use (*ca.* 210 nm).

In purine the first band (294 nm) is out-of-plane polarized ($n \to \pi^*$ transition) probably localized at N-3 ⟨69JCP(51)1862, 77JA3934⟩.

The second band (263 nm) is in-plane polarized at 48° from the C(4)—C(5) axis (towards C-6).

The weak band (250 nm) is out-of-plane polarized and assigned to a second $n \to \pi^*$ transition probably located at N-1.

A strong absorption band at 200 nm is polarized parallel to the 263 nm band and the identity of an out-of-plane band (190 nm) is not established.

$n \to \pi^*$ transitions in purines have also been calculated by the CNDO–CI method ⟨74MI40903, 77MI40902⟩.

Since many purines exist in lactam and lactim forms (*cf.* Section 4.09.2.1) the position and intensity of the peaks are often affected by pH and this can be used for pK_a determinations. In addition the position of substituents affects the spectra by increasing the intensity and causing bathochromic shifts usually in the order 6 < 8 < 2 and of increasing hypsochromic effect 2 > 6 > 8 for the longer wavelength band. The effects of the substituents in polysubstituted derivatives are not, however, additive.

Differentiation of tautomers by the use of UV spectra has not been too rewarding. Thus whereas hypoxanthine almost certainly exists as the lactam form (**13**) as indicated by IR and other spectral studies, its UV absorption spectrum (λ_{max} 249 nm, pH 5.6) is similar to that of 6-methoxypurine (λ_{max} 252 nm, pH 5.2), a derivative of the lactim form of hypoxanthine, and 1,7-dimethyl-6-oxo-1(*H*),6(*H*)-purine (**43**) ⟨B-73MI40902⟩, a definite lactam derivative. On the other hand, the UV spectrum of 8-oxo-7(*H*),8(*H*)-purine (**44**) is closer to that of its 7- and 9-monomethyl and 7,9-dimethyl (oxo or lactam) derivatives than to 8-methoxypurine (**45**) ⟨B-73MI40902⟩ implying the oxo structure. Even in basic solutions where the lactim structure is expected, the results may be difficult to interpret. Thus the spectrum of the anion of hypoxanthine is similar to that of the neutral species of adenine, (λ_{max} 258 and 260 nm, respectively) whereas this is not true of the 2- and 8-oxopurines when compared with the corresponding aminopurines.

(43)   (44)   (45)

On the other hand, there is a close correlation in the spectra of the monoaminopurines with those of the corresponding *N*-dimethyl derivatives. This at least does not conflict with the primary amino (as distinct from an imino) structure assigned to the aminopurines which is well characterized by, for example, X-ray crystallography and magnetic resonance studies. However, full comparison with the various imino derivatives would be required before the result could be considered meaningful.

4.09.4.3.5 UV photoelectron spectra

UV photoelectron spectroscopy has recently been used to study the electronic structure and tautomerism in the gas phase of purine, 6-, 7- and 9-methylpurines, adenine, 7- and 9-methyladenines, and 6-*N*-dimethyl- and N^6,9-dimethyladenines ⟨80JA4627 and references therein⟩.

Assignment of the photoelectron spectral data was aided by results from HAM/3 MO calculations ⟨77TL4627⟩. The spectral data indicate that in purine (adenine) the three highest occupied lone-pair orbitals have ionization potentials ~9.6 (~9.6), ~10.6 (~10.5) and ~11.7 (~11.39) eV for the two bases, respectively, confirming the similarity indicated by the calculations. The four highest π-orbitals in both bases were also compared and shown to differ significantly as a result of the planar exocyclic amino group of adenine affecting the π-structure of the molecule.

Comparison of the photoelectron spectra of purine, adenine and their 7- and 9-methyl derivatives in the gas phase revealed that the spectra of purine and adenine are both much more like the spectra of their 9-methyl rather than their 7-methyl derivatives. This suggests that in an isolated environment the $N(9)H$ tautomers of both purine and adenine are more stable than the $N(7)H$ forms.

4.09.4.3.6 IR spectra

The insolubility of many of the naturally occurring purines has restricted the general application of IR spectral studies as a useful routine procedure. Nevertheless there is a substantial literature including reviews of the subject which contain lists of spectral data (see Table 1).

(i) CH stretching bands

C(2)—H, C(6)—H and C(8)—N stretching frequencies for several purines including purine, adenine, hypoxanthine and guanine are 3023, 3060 and 3098 cm^{-1}, respectively ⟨B-73MI40902⟩, and weak bands may be revealed in aminopurines by conversion of NH to ND.

(ii) OH stretching bands

In the solid form the oxohydropurines show no strong absorption in the OH (3650–3590 cm^{-1}) or hydrogen bonded OH (3600–3200 cm^{-1}) regions, implying that they exist as the oxo and not the hydroxy forms.

(iii) NH stretching bands

The NH stretching vibration of purine is broad (3000–2500 cm^{-1}) and similar broad bands occur in the oxopurines centred at 3100 cm^{-1}.

(iv) NH_2 stretching bands

The aminopurines show bands in the 3400–3100 cm^{-1} region with a strong characteristic band at 3300 cm^{-1}.

(v) C=O, C=N and C=C stretching bands

The C=O absorption band in the oxohydropurines is in the region 1700–1620 cm^{-1} and the C=N and C=C bands at about 1600 cm^{-1}.

Adenine in the solid state, for example, has a band at 1672 cm^{-1} assigned to the primary amino group ⟨61JCS504, 71HC(24-2)1, p. 71⟩ and 9-ethyladenine in chloroform has two bands separated by 100 cm^{-1} which is characteristic of primary amines.

Adenine will dissolve in compounds such as antimony trichloride and IR and Raman spectral data in this solvent are also consistent with the primary amino structure as are data on polycrystalline adenine ⟨74JCP(71)415⟩. Similarly, IR spectra of guanine in the solid state support the oxoamino structure, and related work with thioxopurines shows a strong band at 1323 cm^{-1} and absence of bands at 2500 cm^{-1} in the solid state which favor the thioxo structures ⟨55JA2569⟩.

(vi) CH deformation bands

Purine has three sharp bands at 1100, 804 and 790 cm^{-1} which after deuteration at C(6) are replaced by two bands at 770 and 760 cm^{-1}. The first three bands therefore are assigned to C(6)—H deformation motions and the latter pair to C(6)—D deformations.

In the case of adenine a sharp band at 800 cm^{-1} splits into two when deuteration at C-2 occurs and a band in hypoxanthine (600 cm^{-1}) only appears after C-8 deuteration ⟨B-73MI40902, p. 215⟩.

With purine nucleosides the absorptions of the bases are in slightly different positions and of slightly different intensity to those in the aglycone spectra. Thus inosine has bands at 1690(s) and 1600(m) and 1560(m) cm^{-1} compared to hypoxanthine with two bands in this region. The carbohydrate portion of the molecules of course gives strong broad OH stretching bands at 3400 cm^{-1} normally distinguishable from NH$_2$ stretching bands.

4.09.4.3.7 Mass spectra

Simple purines at least are relatively stable compounds so that the highest intensity ion in the mass spectrum is due to the molecular ion. Subsequent fragmentation depends on the nature of the substituents. Purine liberates two molecules of hydrogen cyanide derived from C(2)—N(3) and C(6)—N(1) and similarly 2-, 6- and 8-methylpurines produce hydrogen cyanide and methyl cyanide ⟨66NKK71⟩ whereas adenine liberates three molecules of hydrogen cyanide ⟨67JA2719⟩ with slight involvement of the amino group, in contrast to the substituted adenine, zeatin [**46a**, *trans*-6-(4-hydroxy-3-methylbut-2-enyl)aminopurine]. This shows initial fragmentation at the substituted amino group ⟨64CC230, 66MI40904⟩ in a manner analogous to that of 6-methylaminopurine which eliminates firstly CH$_2$N or CH$_3$N followed by elimination of three hydrogen cyanide units ⟨67JA2719, 66MI40904⟩. 6-Dimethylaminopurine also eliminates CH$_3$N ⟨67JOC3856⟩.

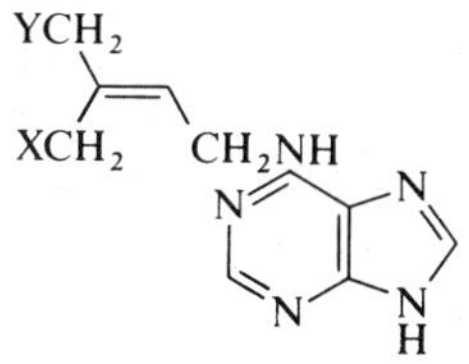

(**46a**) X = H; Y = OH
(**46b**) X = OH; Y = H

The most intense ion in the mass spectrum of guanine is the molecular ion which fragments to cyanamide derived from N(1)—C(2) NH$_2$ ⟨67JA2719⟩ and an ion at *m/e* 109 which loses CO and hydrogen cyanide.

Fragmentation of oxohydropurines such as hypoxanthine, xanthine and uric acid is similar to that of guanine, the main route involving elimination of hydrogen cyanide from N(1)—C(2), followed by ejection of CO and HCN. Xanthine and uric acid eject HCNO followed by decarbonylation ⟨67JA2719, 62M632⟩. Also see ⟨B-72MI40902, p. 402⟩ for a mass spectral study of *N*- and *S*-methylpurines.

Mass spectra of purine nucleosides give valuable information about both the base and the attached carbohydrate. Thus adenosine produces a molecular ion which fragments to produce base plus H, and base plus 2H units, the latter being more prominent in ribosides than deoxyribosides ⟨B-62MI40900, 62JA2005, B-73MI40901⟩. A complete high resolution spectrum of adenosine has also appeared ⟨67C226⟩, and mass spectroscopy has been used to determine the site and extent of benzylation of eleven monobenzylated nucleosides ⟨79JHC585⟩. A major advantage of the mass spectral techniques is the small amount of material required. The relative non-volatility of purine nucleosides, especially oxo derivatives, and nucleotides can be improved by the use of trimethysilyl derivatives as with silylated nucleosides (**47**) and (**48**) and the nucleotide (**49**).

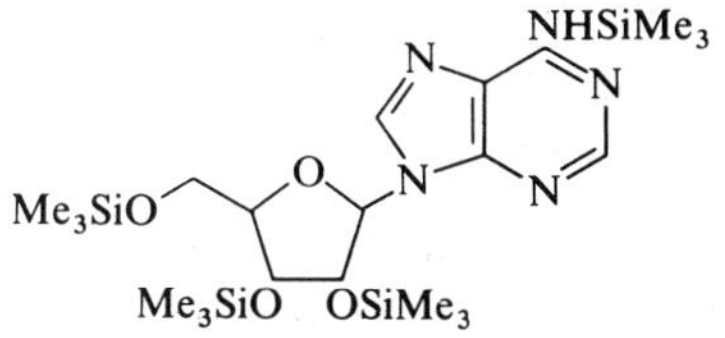

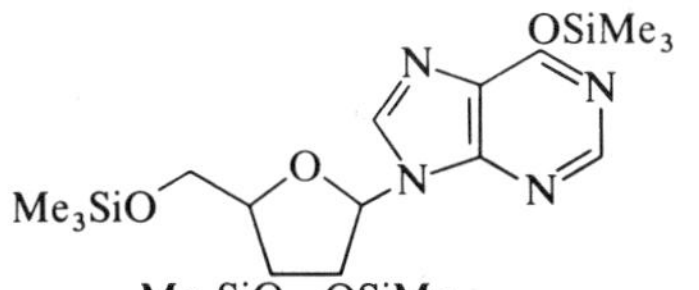

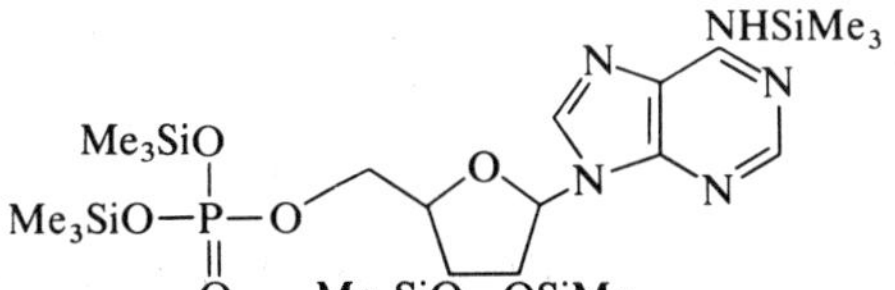

4.09.4.3.8 ESR spectra

In recent years there has been much interest in the chemistry and biochemistry of irradiated nucleic acids and their derivatives ⟨B-76MI40901⟩. ESR spectroscopy is concerned with the structure of radical forms produced by high energy radiation and has been of value in understanding the complex reactions which occur in the nucleic acids and their simpler derivatives. A discussion of these last reactions is beyond the scope of this chapter and the reader is referred to the bibliography in Table 2.

Guanine hydrochloride dihydrate was first examined for three different orientations of the single crystal form in the magnetic field ⟨67PNA(58)1279⟩. The protonated radical observed (**50**) is formed by hydrogen atom addition at C-8 and the structure required hydrogen at N-7 which has also been observed in the undamaged molecule by X-ray crystallographic studies ⟨65MI40903⟩.

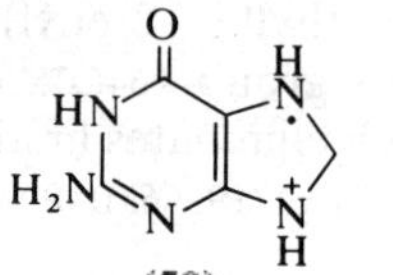

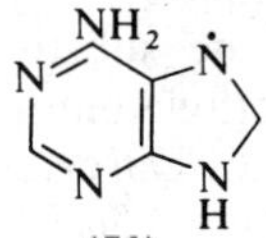

In the case of adenine, analysis of hydrogen addition radicals was achieved with 2'-deoxyadenosine monohydrate crystals ⟨68PNA(60)450⟩ and this suggested hydrogen addition to C-2. However more recent experiments involving selective deuteration at C-8 have shown that in all adenine derivatives hydrogen addition to C-8 occurs as with (**51**) ⟨71MI40908, B-73MI40901, p. 214⟩.

4.09.4.4 Tautomerisms in Purines

4.09.4.4.1 Introduction

There are various forms of tautomerism which operate in the different purine species. (1) Prototropy which involves attachment of the proton to any one of the four ring nitrogen atoms (Scheme 5). Corresponding CH tautomers, for example (**52**), seem to be of little significance. (2) Amine–imine tautomerism which operates in the aminopurines such as adenine (Scheme 6). (3) Lactam–lactim tautomerism as in the hydroxypurines such as hypoxanthine (Scheme 7) and the related thioxo–thiol tautomerism (**53**) and (**54**) in the biologically important mercaptopurines (Scheme 8). The subject has recently been discussed in some detail ⟨76AHC(S1)502⟩.

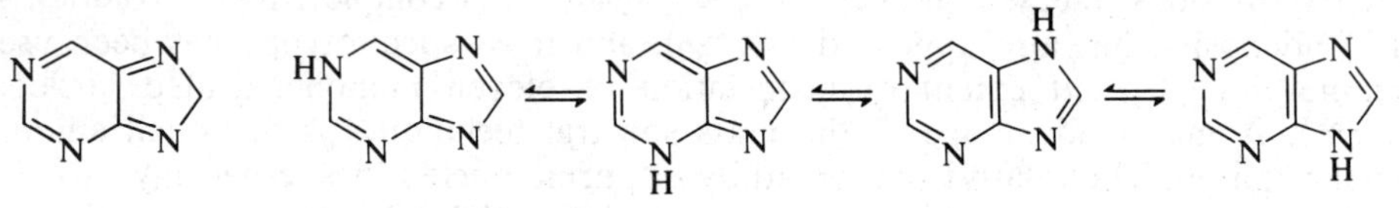

Scheme 5

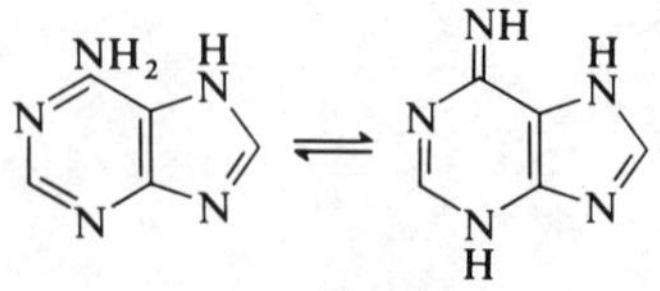

Scheme 6

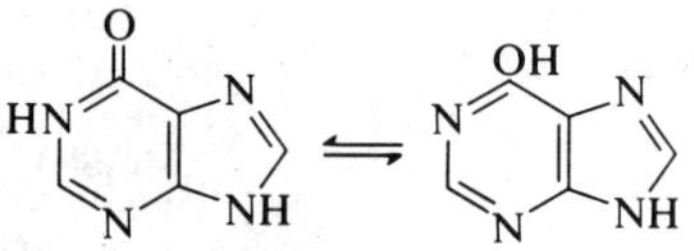

Scheme 7

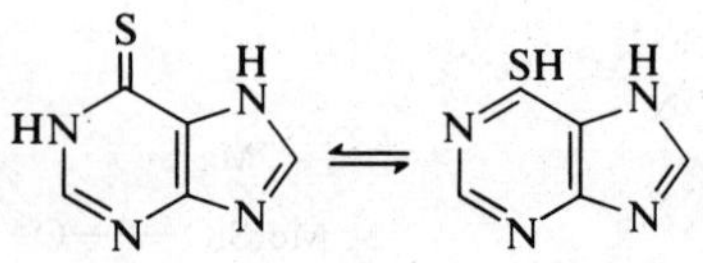

Scheme 8

Information about the precise nature of the tautomeric forms of the various purines is of special and practical importance.

Thus the transfer of genetic information during DNA replication and RNA transcription for protein synthesis depends on the accuracy of base-pairing of the complementary nucleic acid bases. Imperfect pairing of minor base tautomers could lead to genetic mutation. Thus the lactim form of guanine is electronically superimposable with normal adenine and could possibly be misread and paired with thymine to produce an abnormal G–T base pair instead of the normal Watson–Crick G–C pair which involves the lactam form of guanine (Figure 3).

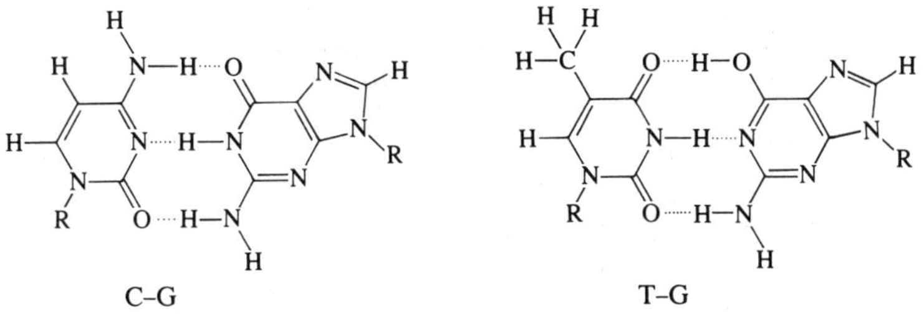

Figure 3 Normal C–G base pair and abnormal T–G base pairs

The comment by Pullman that investigations of the chemical and physicochemical properties of purine tautomers are still in a primitive form ⟨71AHC(13)77⟩ may still be a fair comment on the current state of the subject.

The distribution of negativity in the purine ring system was originally considered ⟨B-68MI40900⟩ to be a function of the annulation of a π-electron excessive imidazole ring with a π-electron deficient pyrimidine ring. Of the four possible NH tautomers, calculated heats of formation (MINDO method) show exothermic values for the $7H$ (**11**) and $9H$ (**1**) tautomers and endothermic values for the $1H$ (**55**) and $3H$ (**56**) forms (Scheme 5). Together with charge distribution calculations (CNDO type), this suggests that in the $7H$ and $9H$ forms a remarkable flow of negative charge has occurred so that the net result is that the pyrimidine ring is π-electron excessive and the imidazole ring is π-electron deficient, and that the original concept only applies to the $1H$ and $3H$ tautomers ⟨75E996⟩.

CNDO calculations also indicate that the $9H$ and the $7H$ tautomer pairs should be more stable (*ca.* 170 kJ mol^{-1}) than the $1H$ and $3H$ forms ⟨71AHC(13)77⟩ and that orders of stability are $9H < 7H < 3H < 1H$ ⟨73CPB1470, 72JCS(P2)585⟩.

Calculations also predict that the $1H$ and $3H$ tautomers should be bathochromic by about 10 nm compared with the $7H$ and $9H$ forms.

4.09.4.4.2 Dipole moments and purine bases

Both the CNDO and PPPCI procedures predict that the dipole moments of $1H$ and $7H$ tautomers should be higher (order of 5–7 D) than the tautomers $9H$ and $3H$ (about 3–4 D), and measurements of three series of derivatives of the four purine protomers, namely 9-methyl-8-phenyl-, *N*-methyl-6-thiomethyl- and *N*-methyl-6-thiomethyl-8-phenyl-purines, satisfactorily confirm the theoretical predictions ⟨68CR(C)(267)1461, B-70MI40902⟩. However, such differences have not always been noted as with the dipole moments of 9-benzyl- and 7-benzyl-6-chloropurines ⟨69JOC973⟩. The differences in the values and also the direction of the dipole moments (Figure 4) of the purine protomers reflects their different charge distributions ⟨71AHC(13)77⟩.

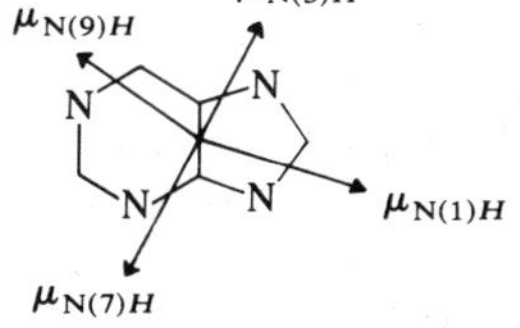

Figure 4 Relative magnitudes and the directions of the dipole moments in the four purine NH tautomers (PPPCI method)

Dipole moments may also be used to some extent to differentiate between tautomeric forms of the monooxopurines ⟨69JOC1205⟩. The tautomers predicted by both the CNDO and the SCF CI methods to be the most stable in the 6-oxo (**57**) and 8-oxo (**58**) series have low moments and the most stable form in the 2-oxo series (**59**) has a high moment.

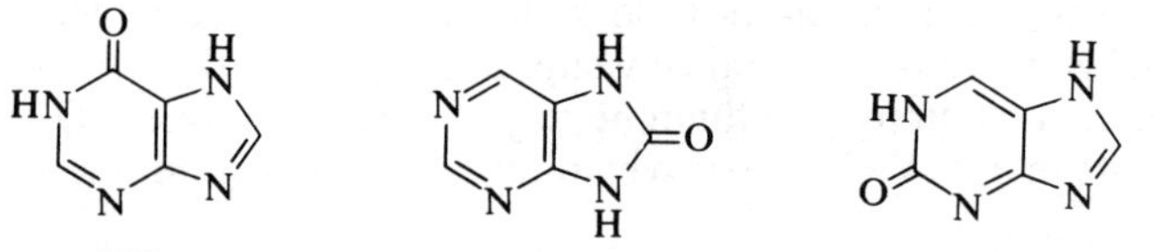

(**57**) (**58**) (**59**)

In the case of xanthine (**4**) the $9H$ tautomer is predicted to have a higher moment than the $7H$ form. In the case of adenine the amine form is similar to the purine series with the calculation indicating that the moment for the $7H$ tautomer is greater than that for the $9H$ form and similarly $\mu\ 1H > \mu\ 3H$.

Table 14 lists experimentally-determined dipole moments of some purines. The compounds examined are limited by their solubilities, hence the absence of the highly polar compounds. Solubility in non-polar solvents is, of course, increased by alkylation or arylation.

Table 14 Dipole Moments of Some Purines

Compound	μ (D)	Ref.
Purine	4.32	B-72MI40902
9-n-Butylpurine	4.3	67BSF1833
6-Chloropurine	5.34	68MI40905
3-Methyladenine	4.98	B-72MI40902
9-Methyladenine	3.25	70JCS(B)596
Theophylline	3.4, 3.94	68MI40905
Caffeine	3.70, 4.6	68MI40905
2′,3′-O-Isopropylideneinosine	5.43	B-72MI40902, p. 72
2′,3′-O-Isopropylideneadenosine	3.04	B-72MI40902, p. 72
5′-Trityl-2′,3′-O-isopropylidineguanosine	6.47	B-72MI40902, p. 72

The experimental value for purine (4.32 D) lies between the calculated values for the $9H$ form (3.6 D) and the $7H$ form (5.5 D). It might therefore be assumed that a solution of purine contains a mixture of the $7H$ and $9H$ tautomers in stark comparison with the $7H$ form found by X-ray crystal structure determinations (Section 4.09.4.2) and the $9H$ form favoured in the gas phase (Section 4.09.4.3.5).

4.09.4.4.3 Dipole moments and purine nucleosides

Whereas simple *N*-alkylation or -arylation of purine does not markedly affect the dipole moment, *e.g.* 9-ethyl-, 9-phenyl- and 9-(2-phenylethyl)-6-chloropurines all have the same dipole moment ⟨67JCS(B)410⟩, products of glycosidation as with purine nucleosides are more complex because of the uncertainties surrounding the precise conformation of the sugar relative to the base, the extremes being termed the *syn* (**60**) and *anti* form (**61**), respectively ⟨72BBR(46)125⟩. The *anti* form is the most common, the *syn* form dominating when there is a large substituent at C-8 as in the example given of 8-bromoadenosine. In the case of guanosine however, whereas the *anti* form is preferred at neutral pH, at pH 1 conversion to the *syn* form occurs ⟨B-72MI40902, p. 297⟩. Details of bond angles, bond lengths and the

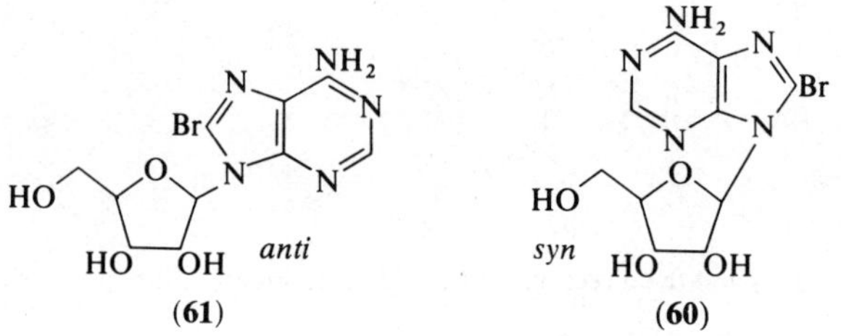

(**61**) anti   syn (**60**)

various tautomeric forms of purine nucleosides have been recorded ⟨73AG(E)591⟩. The five ring atoms in both the riboside and 2-deoxyribose moieties are in a half-chair conformation. NMR data and X-ray crystal analysis ⟨69MI40902, 70MI40901, 60JA222, 70JA4088⟩ indicate that one of the five atoms is displaced from the plane formed by the remaining four. The plane is produced by C-1', C-4', O-1' and either C-2' or C-3'. The out-of-plane atom is termed *endo* if it is on the same side of the plane as the C-5' atom and *exo* if on the other side. The most common forms are C-2' and C-3' *endo* (Figure 5).

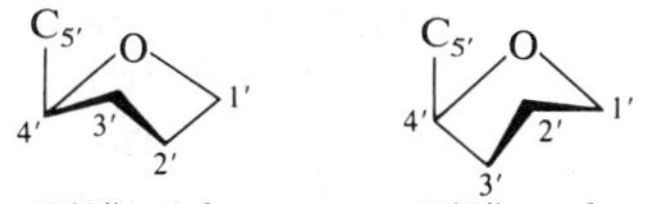

Figure 5 *Endo* configurations of ribofuranosyl derivatives

These various conformations influence the dipole moment so that theoretical values may vary over a wide range and experimental values may agree with more than one conformation ⟨B-73MI40901⟩.

Calculated values for the extreme *syn* and *anti* forms of a few purine nucleosides are given in Table 14 and are derived from the PCILO [Perturbation Configurational Interaction (using) Localized Orbitals] method. The data predict that with the *gauche–gauche* conformation for the C(4')—C(5') bond, the ribosides of purine and xanthine (the $9H$ tautomer) should have higher dipole moments in the *anti* form, the reverse being the case for the hypoxanthine $1H$–$9H$ form ⟨B-72MI40902, p. 21⟩.

4.09.4.4.4 Magnetic circular dichroism

Differences between the two sets of pairs of the four isomeric purine types also arise from magnetic circular dichroism measurements of the four *N*-methylpurines ⟨73JHC419⟩. Thus the MCD bands for the 1-Me and 3-Me forms are of lower intensity and occur at shorter wavelengths than the related bands for the 7-Me and 9-Me isomers. Also protonation results in wavelength shifts and changes in peak intensities and these variations could be used to characterize a particular isomer. Thus peak heights of the bands for the 1-Me and 3-Me isomers after protonation *increase* by 0.25 and three times whereas those for the 7-Me and 9-Me isomers *decrease* by 0.25 and three times, respectively.

4.09.4.4.5 Ionization constants

Imidazole (basic pK_a 7, *i.e.* $N \rightarrow NH^+$) is much more basic a molecule than pyrimidine (basic pK_a 1.31) and the combination of the rings in purine results in a basic pK_a of 2.39. Since the basicity of purine is depressed by electron-withdrawing groups such as trifluoromethyl and since the decrease in basicity is greatest with the 6-substituted purine,

Table 15 pK_a Values of Some *C*-Substituted Purines with Electron-donating Substituents[a–c]

Purine derivative	pK_a(basic) $N \rightarrow NH^+$	pK_a(acidic) $NH \rightarrow N^-$
7-H	2.39	8.93
2-NH$_2$	3.80	9.93
6-NH$_2$	4.25	9.83
8-NH$_2$	4.68	9.36
2,6-Di-NH$_2$	5.09	10.77
2,6,8-Tri-NH$_2$	6.23	10.79
2-NMe$_2$	4.02	10.22
6-NMe$_2$	3.87	10.5
8-NMe$_2$	4.80	9.73
2-NHMe	4.01	10.32
6-NHMe	4.18	9.99
8-NHMe	4.78	9.56

[a] ⟨54JCS2060⟩. [b] ⟨65JCS3770⟩. [c] ⟨60BJ(76)621⟩.

it was suggested many years ago that the most basic centre in purine resides in the pyrimidine ring especially at N-1 ⟨B-57MI40900, p. 3⟩. This has generally been confirmed by other measurements although NMR data (cf. Section 4.09.4.3.1) suggests that protonation also occurs at N-7 and N-9 to a modest degree.

Electron-donating groups increase the basic pK_a values markedly (Table 15), electron-withdrawing groups decrease the pK_a values especially when present in the pyrimidine ring (Table 16), and N-substitution has relatively little effect (Table 17). The hydroxy and thiol purines exist largely in the oxo- and thioxo-hydro forms which results in weakening of acidic properties, and this was deduced by comparing pK_a values, and UV and IR spectra of O- and N-methyl derivatives (Table 18). In all cases the thiol(thioxo) derivatives proved to be stronger acids than the corresponding oxygen derivatives (Table 19).

Linear free energy (LEF) relations for monosubstituted purines relating pK_a to the Hammett constants σ_m and σ_p have made calculations of pK_a values possible ⟨81JPS425⟩. Tables of calculated and observed values are quoted for mono-, di- and some tri-substituted purines. Generally good agreement between calculated and experimental values are obtained. Many additional values have also been recorded ⟨B-73MI40902, p. 1⟩.

Table 20 records pK_a values for some nucleosides and nucleotides.

Table 16 pK_a Values of Some C-Substituted Purines with Electron-withdrawing Substituents

Purine derivative	pK_a (basic) $N \rightarrow NH^+$	pK_a (acidic) $NH \rightarrow N^-$	Ref.
2-Cl	0.69	8.21	65JCS3017
6-Cl	0.45	7.88	65JCS3017
8-Cl	1.77	6.02	65JCS3017
2,6-Di-Cl	−1.16	7.06	65JCS3017
2,6,8-Tri-Cl	−3.1	3.96	65JCS3017
6-CF$_3$	<0	7.35	58JA3932, 58JA5744, 59JA2515
8-CF$_3$	~1.0	5.12	58JA3932, 58JA5744, 59JA2515
6-CN	~0.3	6.88	58JA3932, 58JA5744, 59JA2515

Table 17 pK_a Values of Some N-Substituted Purines

Purine derivative	pK_a (basic) $N \rightarrow NH^+$	Ref.
7-Me	2.29	54JA6073
9-Me	2.36	54JCS2060
2-Cl-9-Me	0.65	65JCS3017
6-Cl-9-Me	0.20	65JCS3017
8-Cl-9-Me	2.00	65JCS3017
2-OEt-9-Me	2.53	67JCS(B)954
6-OEt-9-Me	1.90	67JCS(B)954
8-OEt-9-Me	3.45	67JCS(B)954

Table 18 pK_a Values of Some Oxo(H)purines and Derivatives

Purine derivative	pK_a (basic) $N \rightarrow NH^+$	pK_a (acidic) $NH \rightarrow N^-$	Ref.
6-OH(oxo) (lactam)	1.98	8.94, 12.1	54JCS2060
6-OMe (lactim)	2.21	9.16	54JCS2060
2-OH	1.69	8.43, 11.90	54JCS2060
2-OMe	2.44	9.2	54JCS2060
8-OH	2.58	8.24, >12	54JCS2060
8-OMe	3.14	7.73	57JCS682
2-OH(oxo)-9-Me	<1.5	9.19	57JCS682
6-OH-9-Me	1.86	9.32	57JCS682
8-OH-9-Me	2.80	9.05	57JCS682
2-NH$_2$-6-OH (guanine)	3.0	9.32, 12.6	61LA(647)161
2,6-Di-OH (xanthine)	—	7.7, 11.94	61LA(647)155, 61LA(647)161
2,6,8-Tri-OH (uric acid)	—	5.4, 10.6	52BJ(51)133

Table 19 pK_a Values of Some Thioxo(H)purines and Derivatives

Purine derivative	pK_a (basic) $N \rightarrow NH^+$	pK_a (acidic) $NH \rightarrow N^-$	Ref.
2-SH	~0.5	7.15, 10.4	54JCS2060
2-SMe	1.91	8.91	54JCS2060
6-SH	<0	7.77, 10.84	54JCS2060
6-SMe	1.63	8.74	54JCS2060
8-SH	<0	6.64, 11.16	54JCS2060
8-SMe	2.95	7.67	54JCS2060
8-SH-9-Me	<2.5	7.48	57JCS682
8-SMe-9-Me	2.98	—	57JCS682

Table 20 pK_a Values of Some Purine Nucleosides and Nucleotides in Water at 20–25 °C

Compound	pK_a (basic) $N \rightarrow NH^+$	pK_a (acidic) $NH \rightarrow N^-$	Sugar OH	PO_4	Ref.
Adenosine	3.63	—	12.35	—	51JBC(193)425, 60JA2641
AMP	3.74	—	13.06	6.05	51JBC(193)425, 60JA2641
ADP	4.2	—	—	7.0	62JPC(66)359
ATP	4.0	—	—	6.48	51JBC(193)425
dAMP	~4.4	—	—	6.4	53JBC(204)847
Guanosine	~1.6	9.33	~12.3	—	53BJ(54)646
GMP	~2.4	~9.4	—	~6.1	56MI40900
GDP	~2.9	~9.6	—	~6.3	56MI40900
GTP	~3.3	~9.3	—	~6.5	56MI40900
Inosine	~1.5	8.82	—	—	53BJ(54)646
Xanthosine	—	5.67	—	—	53BJ(54)646

4.09.5 REACTIVITY AT THE RING ATOMS

4.09.5.1 Introduction

The purine ring system is capable of undergoing both electrophilic and nucleophilic reactions at the ring atoms, but the products will often depend on the state of polarization of the molecule and whether the reaction products examined result from kinetic or thermodynamic control. Thus the anionic form of purine (**62**) is readily attacked by electrophiles such as alkylating or glycosylating agents to produce normally, and usually exclusively, the N-9 substituted derivatives. In the neutral form the products may vary. Thus kinetically-controlled vinylation of purine (Section 4.09.4.1) produced a mixture of N(7)- and N(9)-vinyl derivatives whereas normally (thermodynamic control) only one (9-substituted) product is obtained. Once the ring proton has been replaced, further alkylation can occur to produce quaternary salts at either the imidazole or pyrimidine ring nitrogen atoms. In the case of oxo-, thioxo- and amino-purines the additional tautomeric forms available enlarge the possibility for attack by electrophiles at different ring nitrogen atoms. Thus alkylation of adenine with alkyl bromide in dimethylacetamide produced 3-alkyl (55–66%), 9-alkyl (10–14%), and 1-alkyl (7–13%) derivatives ⟨63JA3719⟩.

(**62**)

In addition to the ring nitrogen atoms the adjacent carbon atoms also show degrees of electrophilic or nucleophilic character, the extent of which will depend on the state of ionization of the molecule and the degree and type of substitution.

The large movement of negativity in purines from the π-electron excessive imidazole ring to the π-electron deficient pyrimidine ring system (Section 4.09.4.4.1) results in the C-8 atom becoming the most electron deficient in the unionized purine molecule and substituents in this position are normally displaced first by nucleophiles. The order for displacement is C-8 > C-6 > C-2. On the other hand, purine anion formation appears to cause a return of negativity to the imidazole half of the system with nucleophilic attack now predominating at C-6 in the pyrimidine ring, the order for replacement of appropriate substituents being C-6 > C-2 > C-8.

In addition to its reaction with nucleophiles the C-8 atom in purines may also achieve sufficient electronegativity to permit attack by electrophiles. Such negativity, however, is normally only achieved by the presence in the ring system of one or more strongly electron-donating groups. Thus, whereas purine only forms an adduct with bromine which readily reverts to the free base, adenine produces the 8-bromo derivative. This type of substitution is analogous to reactions at C-2 by electrophiles in imidazoles and is of special interest to studies of the mechanism of oncogenesis in which 8-arylation, especially of guanine moieties, may occur in DNA bases ⟨80CC82, 70MI40904⟩ by aryl cations. Other reactions at C-8 by electrophiles include exchange reactions with $^2H^+$, $^3H^+$, reactions with alkyl radicals, cations and diazonium ions.

4.09.5.2 Reactions with Electrophiles

Much of the earlier literature of purines was concerned with attack by nucleophiles but more recently there has been a marked increase in studies of electrophile attack. This is no doubt encouraged in part by work concerned with the mechanism of oncogenesis which involve arylation or alkylation of purines usually at C-8.

4.09.5.2.1 ^1H, ^2H and ^3H exchange reactions

Most purines readily undergo exchange of H-8 in the neutral or monoprotonated molecule but not in the anionic form. Measurement of the rates of exchange can be followed by using an 8-^2H labelled purine and noting the increase of the peak height of the NMR signal due to the appearance of 8-^1H in the presence of the exchanging medium which is normally water ⟨73JCS(P2)1889, 73JCS(P2)2138, 74JCS(P2)174⟩. Alternatively, using a ^3H-labelled purine the exchange reaction may be monitored by measuring the gain in radioactivity in the exchange medium which again is normally water ⟨73JCS(P2)1889⟩. In all these reactions some contamination can occur due to exchange at other carbon atoms but this is usually small. Thus in the case of adenine, exchange of the 2-^3H derivative is 2×10^3 times slower than that for 8-^3H-adenine ⟨71CC394⟩.

The rate of hydrogen exchange is increased by factors which increase the electron deficiency of the imidazole ring. Thus adenosine with a weakly electron-attracting ribofuranosyl group in the 9-position exchanges at about twice the rate of adenine ⟨71CC394⟩. Similarly, if a positive charge is introduced into the imidazole ring by quaternization of one of the nitrogen atoms, e.g. (63), then a very rapid increase in exchange rate occurs and the signal for 8-H moves strongly downfield ⟨75JHC171, 69JA5625, 73JHC849⟩. The same effect can also be observed in a graduated fashion by reaction of inosine with methylmercury(II) chloride. When more than one molecular equivalent of the mercury derivative has been

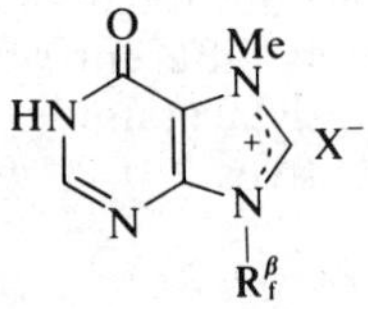

(63) $R_f^\beta = \beta$-D-ribofuranosyl

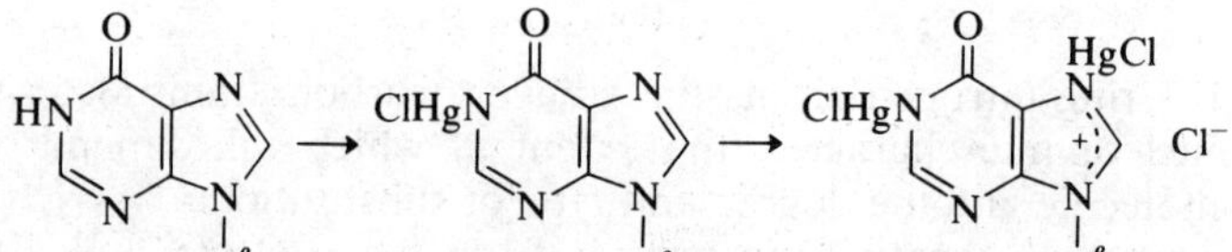

(64)

(65)

Scheme 9

added, 8-H exchange readily occurs. This is explained by initial formation of the 1-mercuriochloride derivative (**64**) which does not affect the rate, followed by quaternization at N-7 to produce (**65**) with the second molecule (Scheme 9) ⟨74CC957⟩. Inosine quaternized at N-7 as in (**65**) has a rate exchange which is too rapid to be measured.

Substituents at C-2 and C-6, whether electron withdrawing or donating, have little effect on the rate of exchange but *N*-alkylation affects the hydrogen exchange to a marked degree. Thus 1-methylhypoxanthine (**66**) has a slightly enhanced rate of exchange at H-8 whereas the 3-methyl derivative (**67**) shows a considerably reduced exchange rate ⟨75CC125⟩. Indeed the decrease in exchange at H-8 is so large that exchange at C-2 takes preference.

(**66**)   (**67**)

The exchange at C-2 rather than C-8 is explained by invoking dipolar forms for the 1-methyl (**68**) and 3-methyl (**69**) derivatives, respectively. These aspects have been reviewed recently ⟨81MI40900⟩.

(**68**)   (**69**)

(i) Effect of pH

Purines not substituted in the imidazole ring have similar exchange rate curves with variation in pH. In particular little exchange occurs at low (<2) and at high (>12) pH values. This suggests that the neutral form and the monocation exchange at roughly similar rates whereas the anionic species and the dication do not exchange. In the case of the anion the excess negative charge is localized in the five membered ring as in (**70**) and exchange does not occur. However, when the ring proton is replaced as in 9-isopropylpurine the exchange rate increases dramatically at high pH ⟨73JCS(P2)1889⟩. In this form of course, a normal anion cannot be formed.

(**70**)   (**71**)

Almost identical bell-shaped rate–pH curves are produced by [8-³H]theophylline (1,3-dimethylxanthine; **8**), xanthine ⟨77T803⟩ and 1-methylxanthine ⟨79T663⟩ whereas the characteristic curve shown by 9-isopropylpurine is also reproduced by [8-³H]caffeine (1,3,7-trimethylxanthine) ⟨77T803⟩, [8-³H]-theobromine (3,7-dimethylxanthine), 7-methylxanthine and xanthosine ⟨79T663⟩ in which substituents also occur in the imidazole ring. 3-Methylxanthine (**71**) on the other hand shows a rate–pH profile which is exceptional.

The explanations for these various phenomena are by no means satisfactory. They have to account for hydroxyl ion catalysis, and hydroxyl ion concentration being a rate determining factor. The explanations which have been proferred generally fall into two separate mechanistic categories: (a) hydrogen abstraction by hydroxyl ion; and (b) hydrogen ion attack at a negatively charged C-8. The conclusions may be summarized as follows. (1) At low pH prior protonation of the imidazole ring nitrogen occurs followed by abstraction of hydrogen at C-8 by hydroxyl ions (Scheme 10) ⟨72B1235⟩. However, simple purines are well known to undergo the first protonation in the pyrimidine ring and this casts some doubt on the scheme. (2) As the pH increases most compounds may participate in hydrogen exchange in the neutral form. In the case of neutral species it is difficult to distinguish between hydrogen abstraction or hydrogen ion attack, since rate–pH plots are the same for both mechanisms. (3) In strongly alkaline solution compounds may participate as mono- or di-anions. The likely mechanism here, of course, involves hydrogen ion attack (Scheme 11). (4) The increase in rates shown by purines substituted in the imidazole ring system which cannot therefore form an anion in that ring is ascribed to hydrogen abstraction. This is not very satisfactory since it is difficult to see why the rate should increase so dramatically at a given pH. The implication is that a new species is being formed by hydroxyl ion attack and that this is the rate determining step. A species such as the azomethine ylide (**72**; Scheme 12) might be an attractive possibility although such a structure would only be expected in very small amounts. A similar mechanism could be used, incidentally, to explain exchange in the (correctly) protonated form (**73**). At the same time a methine intermediate could also possibly explain the unusual rate–pH curve shown by 3-methylxanthine with protonation at N-7 as in (**74**) followed by hydrogen abstractions as in (1) above.

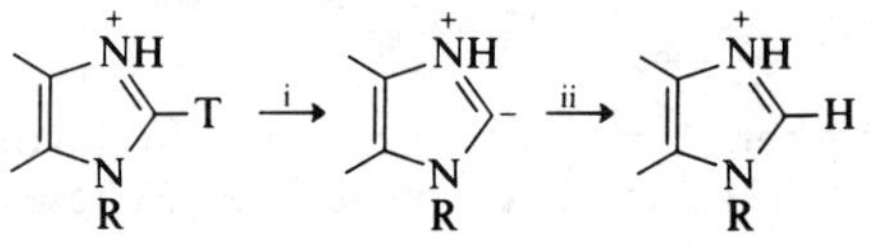

i, OH⁻, slow, rate determining; ii, H₂O, fast

Scheme 10

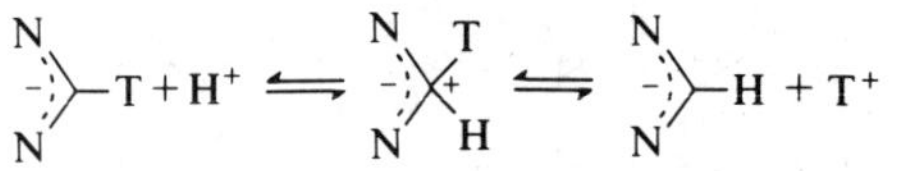

Scheme 11

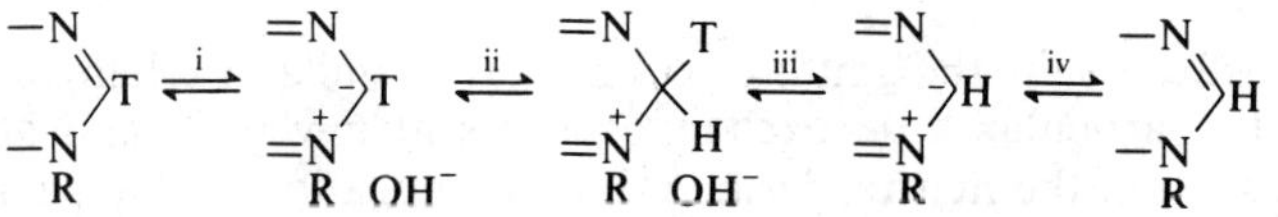

i, OH⁻, slow; ii, H₂O fast; iii, TOH; iv rearrange

Scheme 12

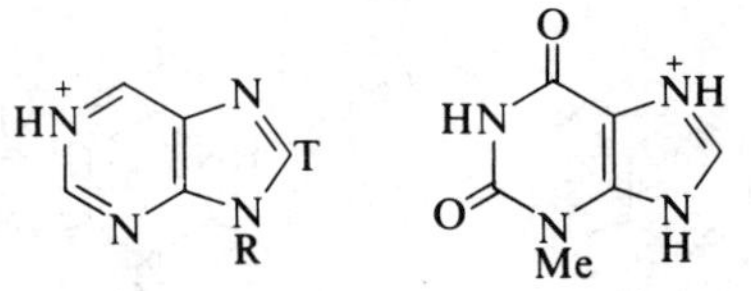

4.09.5.2.2 N-Alkylation

(i) Purine

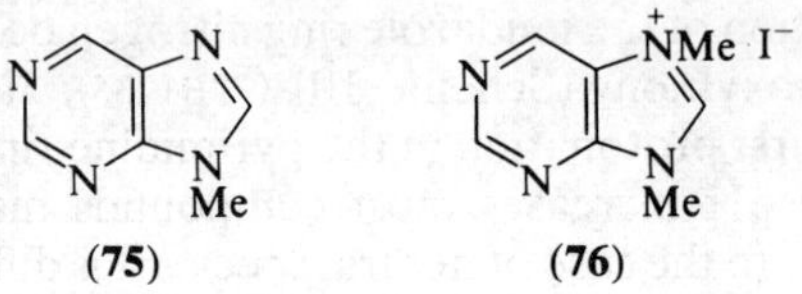

Whereas reaction of purine with dimethyl sulfate in water, or with diazomethane, produces only the 9-methyl derivative (**75**), vinyl acetate and purine give a mixture of the 7- and 9-vinylpurines ⟨75JOC3297⟩ (Section 4.09.4.1). Similarly, methylation of the thallium salt of purine with methyl iodide in DMF at room temperature produces 9-methylpurine together with the 7,9-dimethylpurinium iodide (**76**). The latter is also formed in better yield by reaction of purine with an excess of methyl iodide in methanol ⟨69JOC1170⟩ or from 7-methylpurine and methyl iodide ⟨1898CB2550⟩.

(ii) Halogenopurines

With electron-donating groups in the pyrimidine ring system, however, some alkylation may also occur there. Thus methylation of 6-methylpurine with dimethyl sulfate in methanolic potassium hydroxide produced a mixture of 6,9-dimethylpurine (**77**; 50%) and 3,6-dimethylpurine (**78**; 15%) ⟨66MI40905⟩. In contrast, the presence of electron-withdrawing groups in the pyrimidine ring system enhances alkylation in the imidazole moiety. Thus glycosylation of 6-chloropurine with 2,3,5-tri-*O*-acetyl-β-D-ribofuranosyl chloride (**79**) gave the 9-substituted purine (**80**) ⟨53JBC(204)1019⟩. Generally however, alkylation of mono-, di- or tri-chloropurines gives rise to a mixture of the 7- and 9-alkylpurines, the latter normally being produced in the highest amount although the use of thallium salts results in exclusive 9-alkylation ⟨69JOC1170⟩. Table 21 gives further information with some selected purine derivatives.

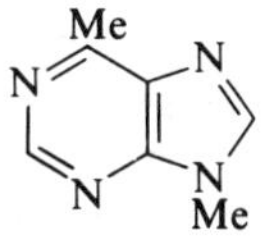

(**77**) 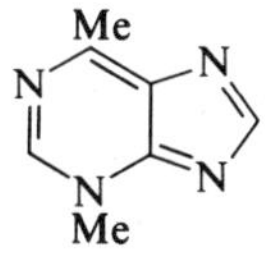 (**78**) 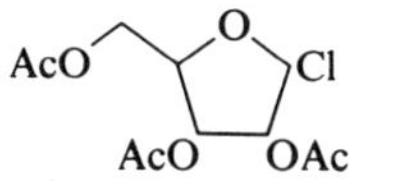 (**79**) 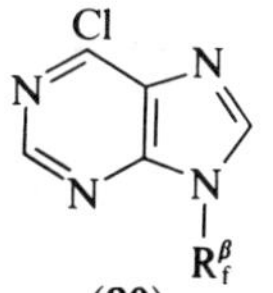 (**80**)

Table 21 Alkylation of Purines

Compound	Alkylating agent	Solvent	Product	Ref.
Theophylline	EtI	NaOEt/EtOH	7-Et	07AP(245)312
Theophylline	Epichlorhydrin	K salt/125 °C	7-Dihydroxypropyl	56JCS3975
Theophylline	Formaldehyde, HCl	H_2O/base	7-CH_2Cl	59JOC562
Theobromine	PrI	KOEt/EtOH	1-Pr	1897AP(235)469
6-Chloropurine	Me_2SO_4	1N NaOH	7- and 9-Me	62JOC2478
6-Chloropurine	EtI	DMSO/K_2CO_3	7-Me (5%) 9-Me (50%)	61JA630
6-Chloropurine	BzCl	DMSO/K_2CO_3	7-Bz (15%) 9-Bz (38%)	61JA630
8-Chloropurine	MeI	DMSO/K_2CO_3	7-Me (4%) 9-Me (8%)	63JOC2310
2-Chloropurine	MeI	DMSO/K_2CO_3	7-Me (9%) 9-Me (85%)	63JOC2310
2,6-Dichloropurine	CH_2N_2	Et_2O	7-Me (11%) 9-Me (20%)	63JOC2310
6,8-Dichloropurine	BzCl	DMF/K_2CO_3	7-Bz (trace) 9-Bz (57%)	62JMC15
2,6,8-Trichloropurine	MeI	DMSO/K_2CO_3	9-Me (50%)	63JOC1662
2-Amino-6-chloropurine	BzCl	DMSO/K_2CO_3	7-Bz (24%) 9-Bz (40%)	62JMC25

6-Fluoropurine behaves similarly and with methyl iodide produces 7-methyl-6-fluoropurine (**81**; 40%) and 9-methyl-6-fluoropurine (**82**; 60%). Base-catalyzed Michael addition reactions leading to alkylated purines are also possible. Thus 6-chloropurine and propenal in DMSO containing potassium carbonate readily gives the 9-(2-cyanoethyl)purine (**83**) ⟨65JOC2857⟩ in 60% yield. Acid-catalyzed alkylations with appropriate cyclic ethers occur, usually in ethyl acetate using toluene-4-sulfonic acid catalysts. Thus 6-chloropurine

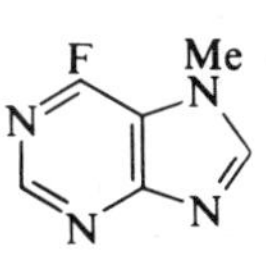

(**81**) 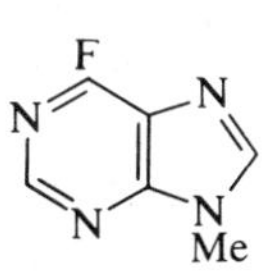 (**82**) 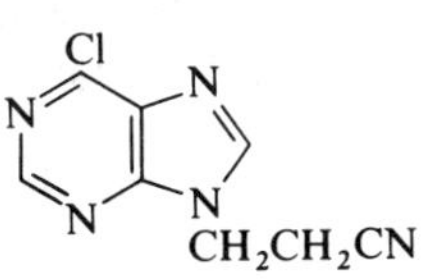 (**83**) 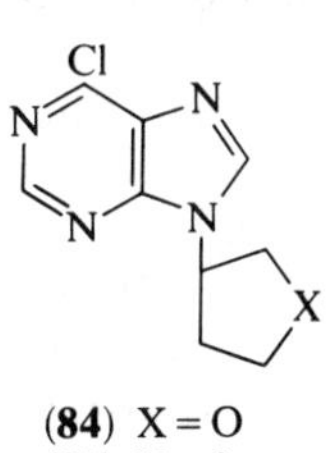 (**84**) X = O (**85**) X = S

530   Purines

and 2,3-dihydrofuran at 70 °C, or 2,3-dihydrothiophene at higher temperatures, produce 6-chloro-9-(tetrahydrofuran-2-yl)purine (**84**) or the analogous 6-chloro-9-(tetrahydrothiene-2-yl)purine (**85**), respectively ⟨61JOC3837⟩. Analogous 9-(tetrahydropyran-2-yl)-purines were also produced from 2,6-dichloropurine ⟨63JMC471⟩ and 2,6,8-trichloropurine ⟨63JOC1662⟩. Similar compounds are furnished by 6-bromo- and 6-iodo-purines ⟨61JA2574⟩.

(iii) Aminopurines

Introduction of amino groups into the pyrimidine ring system can result in substitution by direct alkylation occurring in that ring. The course of alkylation of adenine, for example, depends on several factors including the nature of the solvent and the presence or absence of base or acid. Thus in acid media no alkylation occurs. In neutral aqueous solution or in DMF or dimethylacetamide with dimethyl sulfate, the major product is the 3-alkyladenine plus smaller amounts of the 1- and 9-alkyl derivatives ⟨62B558⟩. Thus alkylation with alkyl bromides in dimethylacetamide produced alkyladenines in yields of the degree, 3-alkyl- (55–66%), 9-alkyl- (10–14%) and 1-alkyl- (7–13%) ⟨63JA3719⟩. Trimethyl phosphate ⟨75JOC385⟩ or epichlorhydrin ⟨75JHC1045⟩ give similar products without use of solvents. By contrast, in alkaline aqueous solutions or in aprotic media with bases including potassium carbonate, sodium hydride and sodium alkoxides, alkylation of adenine gives the 9-alkyladenine as the major product ⟨64JHC115, 65JOC3235⟩. An exception, however, includes the reaction of the sodium salt of adenine in ethanol with 3-methylbut-2-enyl bromide to produce the 3-alkyl derivative (**86**) (triacanthine) with only a small amount of the 9-substituted isomer ⟨62JA2148⟩ being formed.

(**86**)   (**87**)   (**88**)

Acylation of the exocyclic amino group appears to have little effect on the position taken by the entering alkyl group. Thus benzylation of 6-benzamidopurine in DMF ⟨65TL2059⟩ or dimethylacetamide ⟨64JHC115⟩ gave 3-benzyl-N^6-benzoyladenine (**87**), whereas at higher temperatures increasing amounts of the corresponding 9-benzyl derivative (**88**) are produced ⟨65TL2059⟩ in addition to the 3-benzyl derivative.

Further alkylation of an already alkylated aminopurine can be achieved usually by more forcing reaction conditions or by use of a base. Thus 3,7-dimethyladenine (**89**) may be produced by methylation of 3-methyladenine with methyl iodide in methanolic potassium hydroxide or from 7-methyladenine in DMF at higher temperatures ⟨64B494⟩. Similarly benzylation of 3- or 7-benzyladenines using benzyl bromide (not chloride) gave 3,7-dibenzyladenine (**90**) but benzylation in the presence of sodium carbonate produced N^6,3,7-tribenzyladenine (**91**) ⟨64JHC115⟩ which is also formed by benzylating the N^6-benzyl compound. The tribenzyl derivative was probably produced by a Dimroth rearrangement of an intermediate quaternary salt involving a ring opening and closing mechanism.

(**89**)   (**90**)   (**91**)

It is noteworthy however that methylation of N^6-benzyladenine with dimethyloxosulfonium methylide afforded 9-methyl-N^6-benzyladenine ⟨70JOC3981⟩. When a 9- or 1-alkyladenine is further alkylated the 1,9-dialkylpurine is produced ⟨64JHC115⟩. Conditions used may vary.

Thus in dimethylacetamide, methylation of 1-methyladenine with methyl *p*-toluenesulfonate to give 1,9-dimethyladenine required 125 °C/2 h whereas benzylation of 1-benzyladenine occurred at room temperature over 70 h ⟨65JHC291⟩.

Since benzyl groups may be readily removed from *N*-benzyladenines, for example by hydrogenolysis, it is possible to prepare what might otherwise be difficultly available 1- or 7-alkyladenines by hydrogenolysis of an appropriate 1-alkyl-9-benzyl- or 7-alkyl-3-benzyladenine. An alternative to the benzyl group in these reactions is 2-cyanoethyl which may be removed by heating with methanolic potassium *t*-butoxide ⟨68JHC863⟩.

Imine formation in adenines is presumably a factor of some importance to reactivity towards electrophiles since in N^6-dialkylated derivatives where imine formation is not possible, further alkylation becomes more difficult. Thus 6-diethylamino-9-methylpurine required prolonged heating with methyl iodide to produce the quaternary salt (**92**) ⟨65JOC3597⟩ and similarly 6-dimethylamino-9-benzylpurine required 3 days boiling with benzyl bromide in acetonitrile to produce 3,9-dibenzyl-6-dimethylaminopurinium bromide (**93**) ⟨66JOC2202⟩. Under these circumstances it is clearly difficult to control the alkylation step and quaternization is the normal rule.

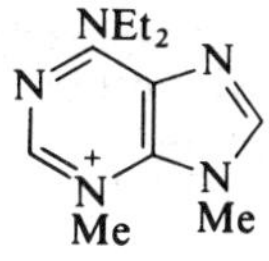

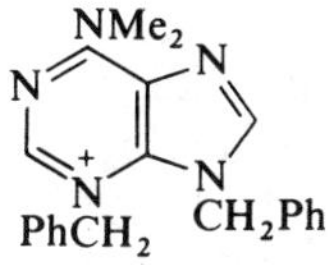

Intramolecular alkylation can also occur. Thus the imidazolopurine (**94**) is produced from N^6-ethyl-N^6-hydroxyethyladenine (**95**) and thionyl chloride (Scheme 13) ⟨62JOC973⟩ (see Section 4.09.6.18). Although alkylation of adenines does not normally occur in acid media, when N^6-formylmethyladenine (**96**) is heated with dilute aqueous acid it rapidly produces the bisimidazole (**97**) presumably *via* prior intramolecular alkylation at N-1 (Scheme 14) ⟨70JCS(C)2206⟩.

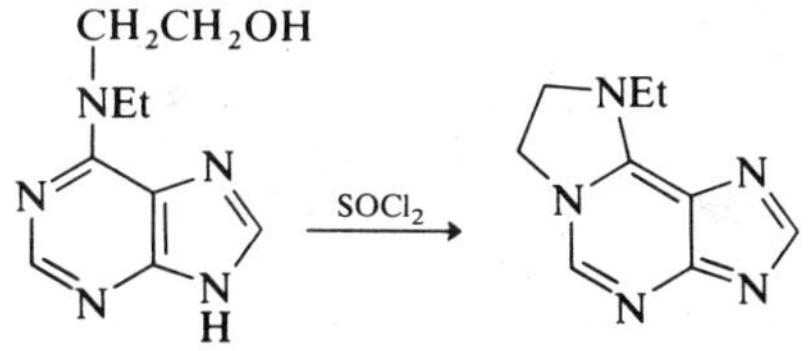

Scheme 13

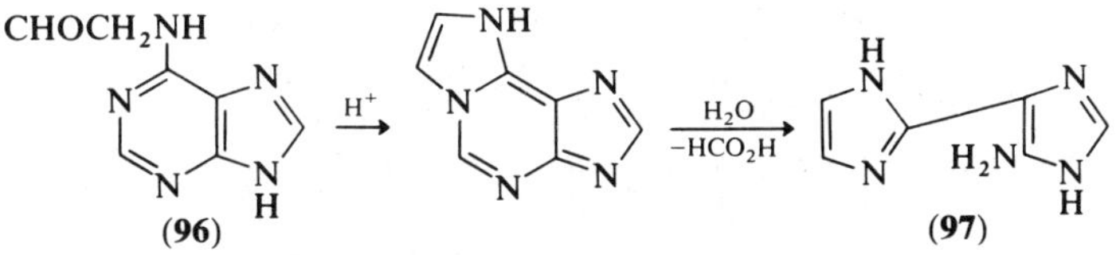

Scheme 14

Alkylation of guanine in alkaline solution leads normally to the 9-alkyl derivative. Thus in sodium hydroxide solution with methyl chloride the products formed included the 9-methyl- (33%), 7-methyl- (18%), 3-methyl- (11%) and 1-methyl-guanine (trace), together with some N^2-methylguanine ⟨66B3007⟩. Higher yields of 9-substituted guanines may also be obtained by fusion of the base with a tetraalkylammonium hydroxide and the same method also furnishes 9-alkyladenines ⟨63JOC2087⟩.

Alkylation of N(9)-blocked guanine derivatives such as nucleosides and nucleic acids, results in $N(7)$-alkylation ⟨71BJ841⟩, and N(9)-blocking can also be achieved in xanthines with cobalt complexes also leading to $N(7)$-alkylation ⟨75JA3351⟩. Several 7-substituted guanine derivatives have been prepared as potential antiprotozoal agents ⟨80JMC357⟩. Further alkylation of guanine or the 7- or 9-alkyl derivatives with alkylating agents in aprotic solvents such as dimethylacetamide furnishes 7,9-dialkylguanines. Thus methylation of 7- or 9-methylguanine with dimethyl sulfate ⟨62JA1914⟩ or methyl *p*-toluenesulfonate ⟨60CB1206⟩ gave the 7,9-dimethylguanine salt (**98**) convertible by ammonia into the betaine (**99**; Scheme 15).

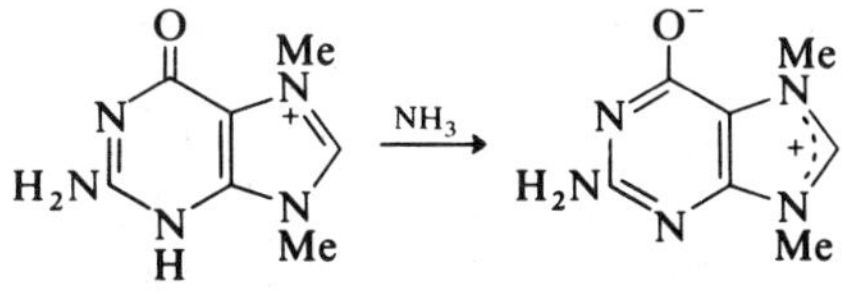

Scheme 15

(iv) Oxopurines

The increase in protonation sites in the oxopurines with increasing oxygen functions extends the possibilities for N-alkylation sites. Normally in the case of monooxo- and dioxo-purines, with the exception of reactions with diazoalkanes, many of which are probably free radical in nature, alkylating agents even in alkaline solutions produce only N-alkyl derivatives. Trioxopurines such as uric acid derivatives are an exception. The nature of the alkylation and the site(s) of attack nevertheless still vary according to the solvent, pH and temperature.

(v) Monooxopurines

The most common monooxopurine is hypoxanthine (**3**). Methylation with methyl iodide in alcoholic sodium ethoxide solution produces 1,7-dimethylhypoxanthine (**100**) as a complex with sodium iodide. 2-Methylhypoxanthine similarly affords 1,2,7-trimethylhypoxanthine (**101**) ⟨1897CB2231⟩. 7-Methylhypoxanthine may be obtained by methylation of N(1)-blocked derivatives containing removable blocking groups such as 1-oxy-2-picolyl. However, in contrast, hypoxanthine or its 7- or 9-methyl derivatives with dimethyl sulfate or methyl p-toluenesulfonate in dimethylacetamide produces a 7,9-methylhypoxanthinium salt which with ammonia gives the betaine (**102**; Scheme 16) ⟨62JA1914, 66CB944⟩. Also methylation of 9-methylhypoxanthine with dimethyl sulfate in alkaline solution affords 1,9-dimethylhypoxanthine (**103**) ⟨62JOC2478⟩.

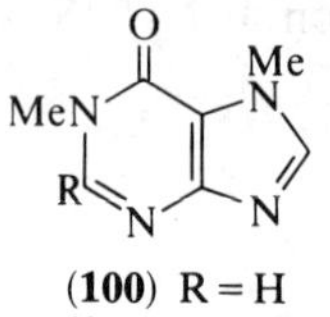

(**100**) R = H
(**101**) R = Me
(**103**) 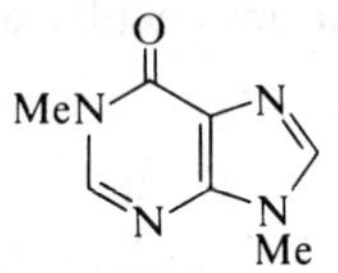
(**102**) 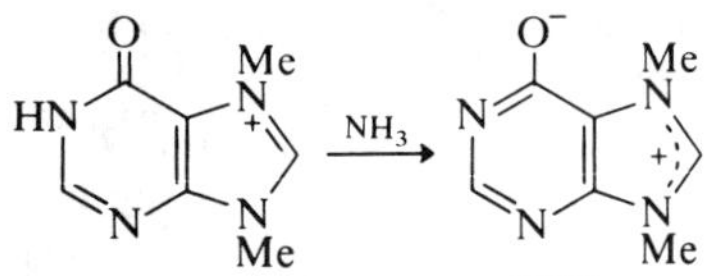

Scheme 16

However benzylation of 1-benzylhypoxanthine in DMF or dimethylacetamide over 22 h at 95 °C afforded 1,7-dibenzyl- (42%) and 1,9-dibenzyl- (21%) hypoxanthines ⟨66JOC2202⟩. On the other hand an improved yield (57%) of 1,7-dibenzylhypoxanthine was produced by benzylation of the 7-benzyl derivative with benzyl bromide. Similarly benzylation of 3-benzylhypoxanthine gave as the main product (74%) 3,7-dibenzylhypoxanthine.

Benzylation of 3-benzylhypoxanthine with benzyl bromide in acetonitrile at 80 °C over 16 h gave 1,3-dibenzylhypoxanthinium bromide but at higher temperatures (110 °C) the major products were 1,7- and 1,9-dibenzyl derivatives obtained in roughly equal amounts and produced by migration from N-3. Evidence for the migration comes from heating the 1,3-dibenzyl salt (**104**) with adenine in dimethylacetamide at 120 °C for 4 h when some 3-benzyladenine is produced (Scheme 17) ⟨66JOC2202⟩.

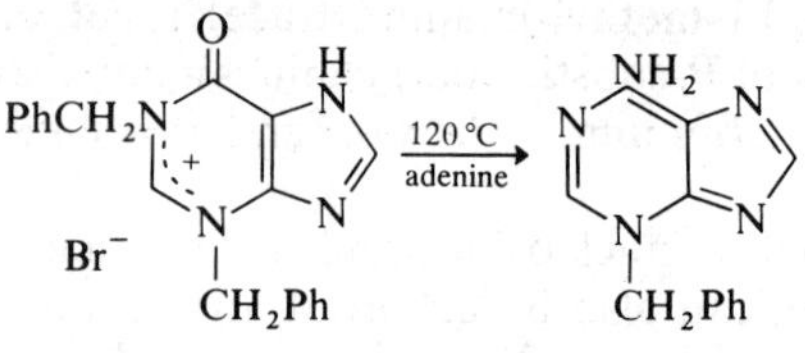

(**104**)

Scheme 17

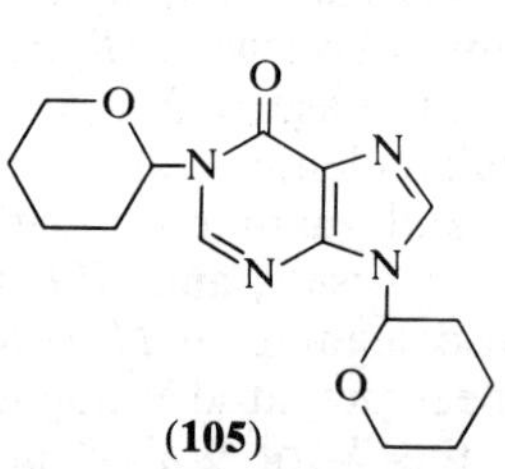

(**105**)

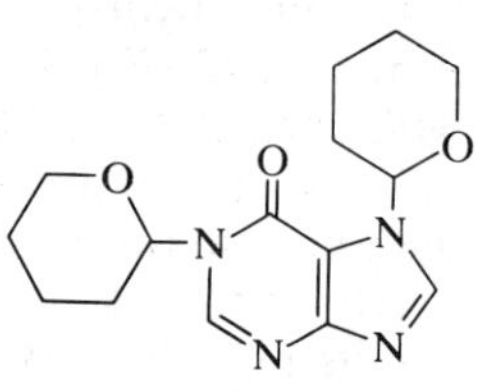

(**106**)

Tetrahydropyranylation of hypoxanthine in DMSO occurs with acidic catalysis and 2,3-dihydropyran, to produce 1,9-bis(tetrahydropyran-2-yl)hypoxanthine (**105**) and the corresponding 1,7 isomer (**106**) ⟨66JOC2685⟩. Alkylation studies of 2-oxodihydropurine have been limited and no data are available for the 8-oxo isomer. Methylation of the 2-oxo derivative with methyl iodide in alkaline solution at 80 °C leads to the 7-methyl, then the 3,7-dimethyl derivatives.

(vi) Dioxopurines

No doubt the commercial importance of the various methylated xanthines such as theobromine, theophylline and caffeine has contributed to the substantial literature on alkylation of xanthine and its derivatives much of which was carried out in the early part of this century. It is not possible to cover all the reagents and conditions used and more extensive data have been collected elsewhere ⟨28JPR(118)198, 71HC(24-2)1⟩.

In alkaline solutions, xanthine alkylates on nitrogen in the order N-3, N-7, N-1, *i.e.* in decreasing order of acidity, and substitution is easier when the first position has been alkylated so to this extent dialkyl derivatives tend to be most easily produced. Thus 3,7-dimethylxanthine (theobromine) is produced by methylation of xanthine or 3-methylxanthine with methyl iodide and barium carbonate ⟨33JCS662⟩ or dimethyl sulfate in potassium hydroxide at 60 °C ⟨50CB201⟩. However when the oxygen functions are blocked by trimethylsilylation, alkylation occurs at N-7 ⟨64CB934⟩.

Further methylation of theobromine readily produces caffeine ⟨31MI40900⟩. Similarly, 1-methylxanthine is methylated by dimethyl sulfate in alkaline solution to furnish 1,3-dimethylxanthine (theophylline) which is further converted into caffeine ⟨31MI40900⟩. Scheme 18 outlines the major stages of xanthine alkylation.

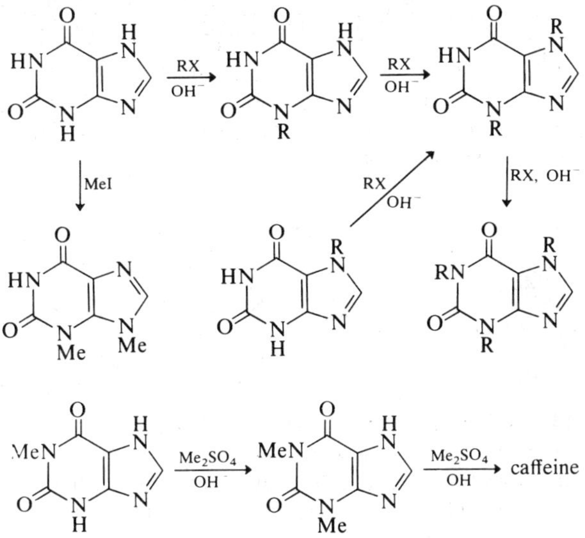

Scheme 18

Under acidic or neutral conditions methylation of xanthine proceeds at N-7 or N-9. Thus with methyl iodide the 7,9-dimethylxanthine betaine (**107**) results after heating in a sealed tube at 150 °C for 5 h. In solvents such as DMSO or dimethylacetamide using dimethyl sulfate or methyl *p*-toluenesulfonate, better yields may be obtained ⟨62CB1812⟩. The same betaine is also produced by similar methylation of 9-methylxanthine ⟨62JA1914, 61LA(647)161⟩.

Further methylation of the betaine (**107**) produces 1,7,9-trimethylxanthine (**108**; Scheme 19) ⟨64LA(673)78⟩ and, analogously, 3,7-dimethylxanthine affords 3,7,9-trimethylxanthine (**109**; Scheme 20) ⟨62CB1812⟩. When the fully methylated derivative caffeine is further methylated with dimethyl sulfate in hot nitrobenzene solution, the quaternary salt (**110**) is produced ⟨21LA(423)200⟩.

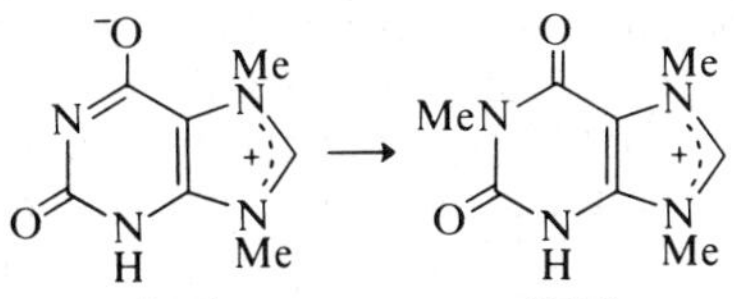

(**107**)   (**108**)

Scheme 19

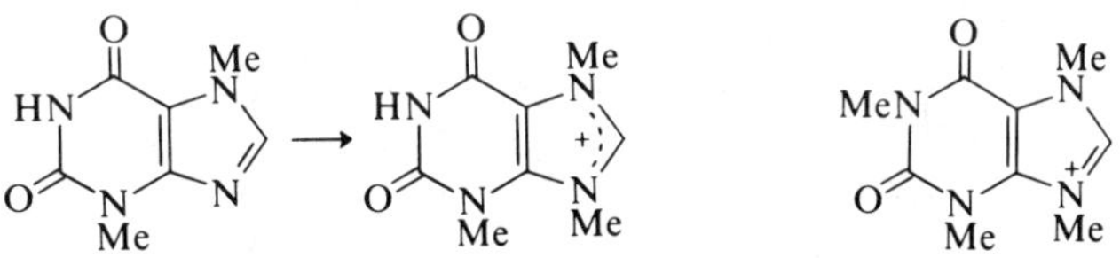

(**109**)   (**110**)

Scheme 20

The tendency to form 7-substituted derivatives in acid media is exemplified by treatment of 9-benzylxanthine with either hot benzyl bromide or hydrogen bromide when a rearrangement to 7-benzylxanthine occurs ⟨79AJC387⟩. Similarly 9-benzyltheophylline undergoes acid-catalyzed rearrangement to 7-benzyltheophylline ⟨77H(6)383⟩. In somewhat analogous fashion it might be noted that tetrahydropyranylation of xanthine in DMSO with 2,3-dihydropyran and acid catalysis with hydrogen chloride produces the 7-tetrahydropyran-2-ylxanthine directly ⟨66JOC2685⟩.

In contrast to xanthine the isomeric 2,8- and 6,8-dioxo compounds have been little studied. However the presence of an oxo function in the imidazole ring ensures that alkylation will very readily and rapidly occur in that ring, followed generally by alkylation elsewhere. Thus in alkaline solution dimethyl sulfate converts 2,8-dioxo-1,2,7,8-tetrahydropurine into a mixture of the 1,7- and the 1,9-dimethyl derivatives (**111**) and (**112**), respectively, in 10 minutes at room temperature, the former compound predominating ⟨14JBC(17)1⟩. Further alkylation then affords the 1,7,9-trimethylpurine (**113**). Similar products are formed with methyl iodide under more forcing conditions (100 °C, 1 h) and the latter also produces the 3,7,9-trimethyl derivative (**114**) from the 3-methyl-2,8-dioxo derivatives ⟨1899CB2721⟩.

In a similar manner 6,8-dioxo-1,6,7,8-tetrahydropurine with methyl iodide and alkali affords the 1,7,9-trimethyl derivative (**115**) and the 3,7-dimethyl-6,8-dioxo derivative also produces the 3,7,9-trimethyl derivative (**116**) ⟨1897CB1846⟩.

A fuller account of the complex alkylation of 6,8-dioxopurines has appeared ⟨74JCS(P1)2229⟩.

(vii) Trioxopurines

The major trioxopurine is uric acid (**10**). As with xanthines *O*-alkylation is rare, alkylation normally occurring on nitrogen. From a study of the ionization constants of uric acid it was shown ⟨21CB1676⟩ that the protons are most easily removed in the order N-3, N-9, N-1, N-7. Accordingly, with methyl iodide in alkaline solution 3-methyluric acid is first formed ⟨1899CB435⟩, then further converted into 1,3,7-trimethyluric acid ⟨1899CB2721⟩ which may be fully methylated to 1,3,7,9-tetramethyluric acid under more forcing conditions (100 °C) ⟨1897CB559⟩. Much of the early literature also records the alkylation of various metal salts of uric acid under dry conditions. The pattern of products formed closely follows that produced in alkaline solution with prior formation of the 3-alkyl derivative, followed by 9-alkylation, and the degree of substitution could be modulated by the degree of metal content.

Occasionally *O*-alkylation products may be isolated. Thus ethylation at 100 °C (with ethyl iodide) of the silver salt of 1,3,7-trimethyluric acid initially gives 8-ethoxy-1,3,7-trimethyluric acid (**117**) which subsequently rearranges to produce 1,3,7-trimethyl-9-ethyluric acid ⟨1882LA(215)316⟩.

(**117**) R = Et
(**118**) R = Me

(**119**)

Alkylation of uric acid with diazoalkanes has been reviewed ⟨32JPR(134)310⟩. With uric acid 1,3,7,9-tetramethyluric acid is produced, presumably *via* an intermediate alkoxy derivative ⟨21ZPC(117)23⟩ since 1,3,7-trimethyluric acid and 1,7,9-trimethyluric acid with diazomethane at room temperature furnished 8-methoxy-1,3,7-trimethyl and 2-methoxy-1,7,9-trimethyl derivatives (**118**) and (**119**), respectively ⟨20CB2327⟩.

Alkylation of the triethylsilylated uric acid (**120**) with methyl iodide is unusual in producing 1-methyluric acid but at 150 °C with dimethyl sulfate the major products isolated were 1,7,9-trimethyl- (**121**) and 1,3,7,9-tetramethyl-uric acids (**122**; Scheme 21) ⟨64CB934⟩.

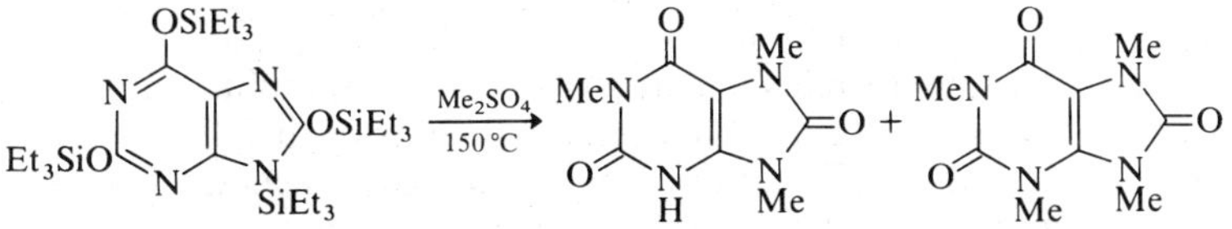

Scheme 21

(viii) Thioxopurines

Alkylation of the thioxopurines may lead to both *S*- and *N*-alkylation but in many examples the *S*-alkyl derivative is probably first formed under mild conditions and then rearranges to the *N*-substituted derivative under more forcing conditions. An exception appears to be 6-thioxo-1,6-dihydropurine (**123**) which with dimethyl sulfate in alkaline solution produces the relatively stable 6-methylthiopurine (**124**) ⟨22LA(426)306⟩. Similar methylation of 6-thiotheobromine in alkali with methyl iodide at room temperature furnishes the methylthio derivative (**125**) whereas, in contrast, the 6-thiocaffeine derivative (**126**) is produced with dimethyl sulfate and alkali at 40 °C ⟨62JCS1863⟩ and corresponding *S*- and *N*-methyl derivatives are produced from the 6-thio analogue of theophylline.

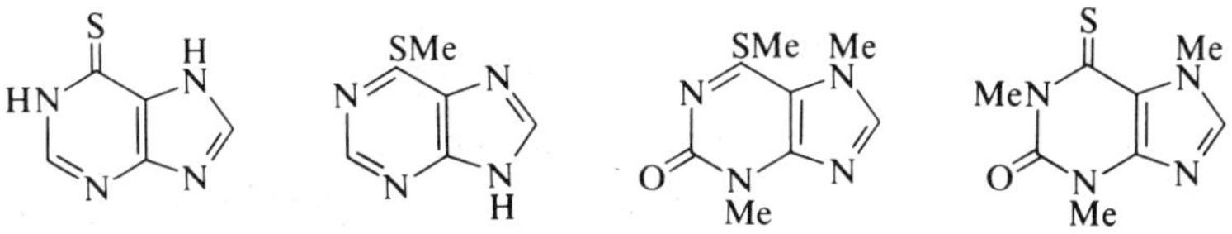

The site of alkylation may be modulated by solvent, temperature and the presence of bases. Thus in DMF solution 6-thioxo-1,6-dihydropurine (**123**) with methyl iodide produces some of the 3-methyl-6-methylthiopurine (**127**) ⟨63MI40904⟩ which in turn with methyl bromide gave 3,7-dimethyl-6-methylthiopurinium bromide (**128**; Scheme 22).

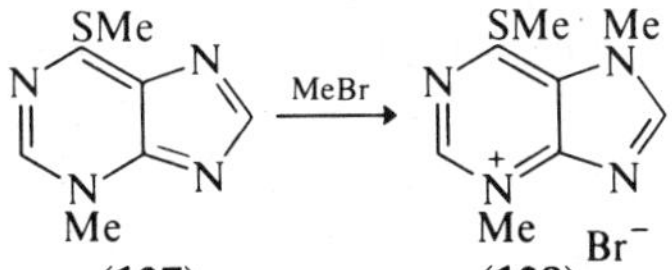

Scheme 22

When the 1-position is blocked so that *S*-alkylation is not possible, alkylation usually occurs at the 7- or 9-positions. Thus benzylation of 1-benzyl-6-thioxo-1,6-dihydropurine with or without a base, afforded a mixture of the 1,7- and 1,9-dibenzyl derivatives ⟨59JOC320⟩.

Methylation of the 2-thio analogue of xanthine (**129**) at 170 °C gave a quaternary salt which with ammonia afforded the betaine (**130**) in which both *S*- and *N*-methylation has occurred. The analogous compound (**131**) is similarly produced as a toluene-4-sulfonate salt from 2-amino-6-thioxo-1,6-dihydropurine and methyl toluene-4-sulfonate ⟨62JA1914⟩.

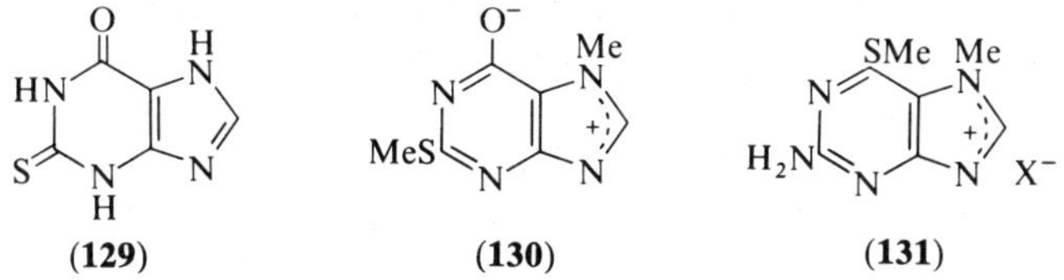

Alkylthiopurines may also undergo Mannich and Michael addition reactions. Thus 6-methylthiopurine with formaldehyde and base gave 9-hydroxymethyl-6-methylthiopurine ⟨67JMC104⟩, and dihydrofuranylation of 2-acetamido-6-benzylthiopurine occurs in the presence of toluene-4-sulfonic acid and 2,3-dihydrofuran to give 2-acetamido-6-benzylthio-9-(tetrahydrofur-2-yl)purine ⟨63JMC471⟩.

4.09.5.2.3 N-Glycosylation

The N-glycosylpurines or purine nucleosides include many naturally occurring compounds, especially 9-β-D-ribofuranosyl and 9-β-D-2-deoxyribofuranosyl derivatives of adenine and guanine which are of course major constituents of the various ribo (RNA) and deoxyribo (DNA) nucleic acids. The literature on purine nucleoside chemistry is very substantial and in this chapter it will only be possible to give a brief survey. However several reviews exist (see Table 2).

Direct condensation of 2,3,5-tri-O-benzoyl-β-D-ribofuranosyl bromide with adenine in acetonitrile at 50 °C leads after deblocking to the 3-β-D-ribofuranosyladenine (**132**) ⟨B-68MI40901⟩. Metal derivatives of purines tend to produce 7- or 9-glycosyl derivatives; thus the silver salt of 2,8-dichloroadenine with 2,3,5-tri-O-acetylribofuranosyl chloride produces the 9-β-D-ribofuranosyl derivative (**133**) ⟨48JCS967, 48JCS1685⟩. However similar reactions with theobromine produce O-glycosides and with theophylline N-7 derivatives ⟨34JCS1639⟩.

(**132**)   (**133**)   (**134**)

With ribofuranosyl halides the major anomer produced is the β-anomer which derives from the neighbouring group (adjacent to the halogeno group) effect of the 2'-O-acyl group which directs the incoming nucleophilic purine *trans* to the latter group. In contrast, condensation of 3,5-di-O-acyl-β-D-2-deoxyribofuranosyl halides with N^6-benzoyladenine in the presence of a molecular sieve produces a mixture of α- and β-anomers of the 9-glycosides ⟨B-68MI40901, p. 183⟩. The same mixture of anomeric 2'-deoxyadenosines may also be produced by glycosylating the mercuriochloride salt of adenine with 5-O-benzoyl-2-deoxy-D-*erythro*-pentose diisopropyl dithioacetal (**134**) ⟨B-68MI40901, p. 132⟩.

If a *cis*-glycofuranosylpurine derivative is required, an appropriate sugar derivative containing a non-participating neighbouring group may be used. Thus, condensation of N^6-benzoyladenine with 2,3,5-tri-O-benzyl-D-arabinofuranosyl chloride gave the corresponding 9-(β-D-arabinofuranosyl)adenine (**135**) after deblocking ⟨B-68MI40901, p. 126⟩.

(**135**) AR_f^β = β-D-arabinofuranosyl   (**136**)

Silyl derivatives have proved valuable intermediates in purine glycosylation reactions. Thus the trimethylsilyl derivative of hypoxanthine (**136**) with 2,3,4,6-tetra-O-acetyl-α-D-glucosyl chloride produced, after deblocking, an anomeric mixture of the two 9-glucosylhypoxanthines, and similar reactions with trimethylsilylated N^6-benzoyladenine led to a mixture of 9-substituted anomers ⟨B-68MI40901, p. 135⟩.

4.09.5.2.4 C-Alkylation and -arylation

Until recently only a few examples of C-alkylation or -arylation by ionic species had been recorded, alkylation normally resulting in N-alkyl derivatives. However in perhaps the first recorded example, theophylline in aqueous alkali with crotyl bromide at room temperature produced 8-(but-2-enyl)theophylline (**137**) and no 7-alkyl derivative was detected ⟨59CB1500, 76AJC891⟩. The latter authors also confirmed an earlier ⟨55FES733⟩ claim to have isolated 8-benzyltheophylline from a benzylation of theophylline under analogous conditions. In this case however the 7-benzyl derivative was also produced but it should be noted that complete exclusion of radicals could not be achieved in this reaction.

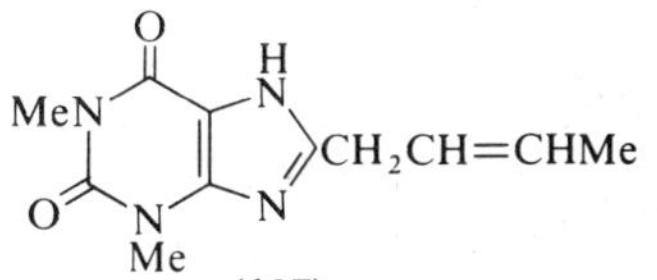

(**137**) 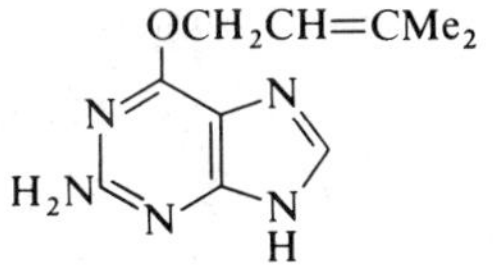   (**138**) 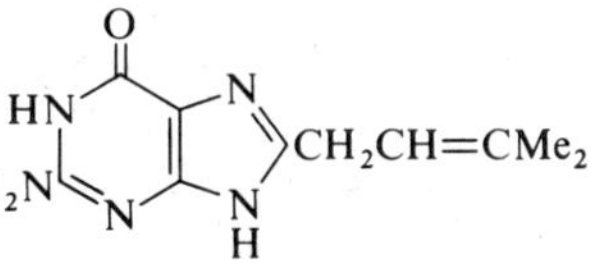   (**139**)

Intramolecular rearrangement of 6-*O*-allyl-, 6-*O*-crotyl- and 6-*O*-2-methylbut-2-enyloxy-2-aminopurine (**138**) occurs in good yield to produce the corresponding 8-alkenylpurines (**139**), especially when the anion is used ⟨73JA7174, 74JA5894⟩. The mechanism proposed involves two anionic 3,3-sigmatropic shifts (Scheme 23), the first forming a C(5)-linked intermediate (**140**) and the second the substituted anion (**141**). When the 8-position is blocked the 3-[3,3] or 7-[3,2] alkenyl derivatives are produced ⟨76JOC568⟩.

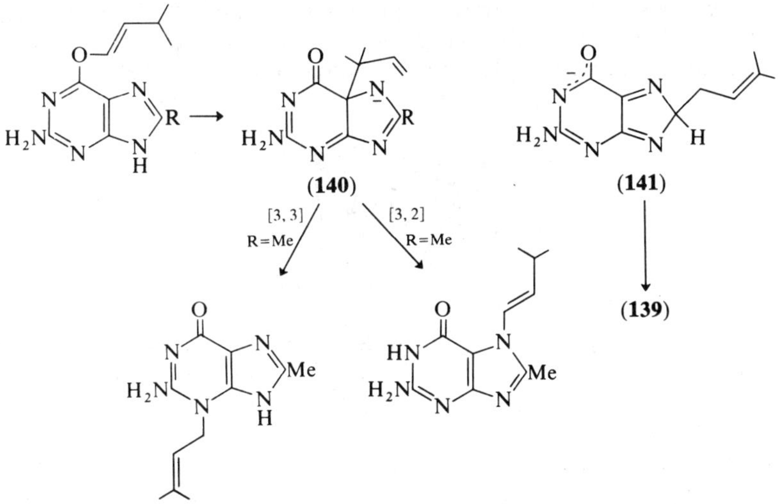

Scheme 23

Recently, treatment of 2-methylthiopurine with diazomethane has been shown to give 9-methyl-2-methylthiopurine as the main product accompanied by the 7-methyl derivative, and also by small but significant amounts of the 7,8- and 8,9-dimethyl-2-methylthiopurines (**142**) and (**143**), respectively. Experiments showed that the 8-methyl group was introduced prior to *N*-methylation ⟨79AJC2771⟩. With diazomethane, however, it is again not possible to eliminate a free radical mechanism.

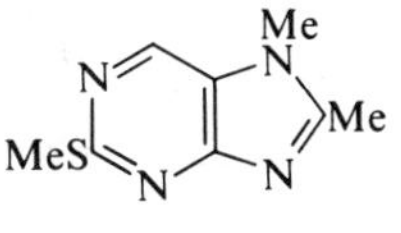

(**142**) 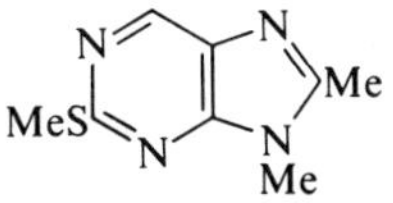   (**143**)

Ionic alkylation of lithiated purine nucleosides has recently proved to be a successful route to 8-substituted derivatives. Thus N^6-dimethyl-5'-*O*-methyl-2',3',*O*-isopropylideneadenosine (**144**) with butyllithium followed by methyl iodide readily gave the 8-methyl derivative (**145**) in good yield (88%) ⟨79TL279⟩. In a more extensive series of reactions several 8-lithiopurine trimethylsilylated nucleosides, *e.g.* the inosine derivative (**146**), were reacted with a variety of electrophilic reagents to produce eventually, after desilylation, good yields of the corresponding 8-substituted derivatives (**147**) ⟨79TL2385⟩. The lithio derivatives were made by reacting several 8-bromo trimethylsilylated purine nucleosides with butyllithium and the nucleosides used included inosine, adenosine, guanosine and xanthosine. The electrophiles used included alkyl halides, ethyl chloroformate, carbon dioxide, benzaldehyde, benzonitrile and acetone to produce a wide variety of appropriate

8-substituted derivatives. In the case of adenosine and guanosine derivatives, N^6-alkylation also occurred.

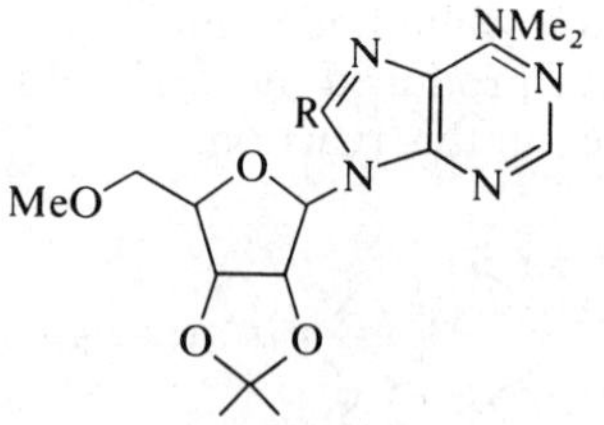

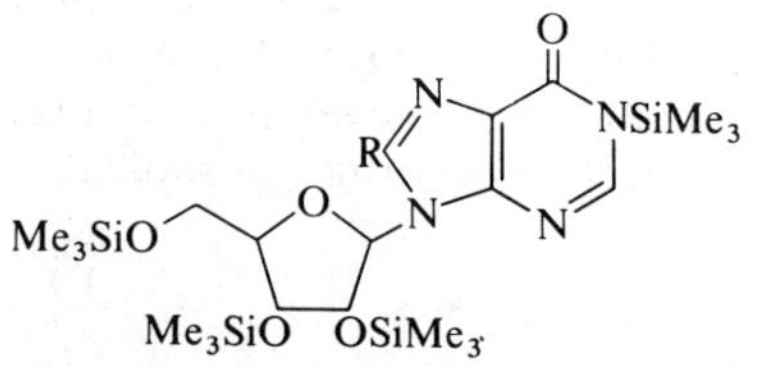

(144) R = H
(145) R = Me
(146) R = Li
(147) R = Me, CO_2H, CO_2Et, PhCHOH, PhCO, Me_2CHOH, $(CH_2)_3CO_2Et$

Silylated 8-bromoadenosine has similarly been converted into various 8-aryladenosines by reaction with arylmagnesium halides in the presence of dichlorobis(triphenylphosphine)palladium ⟨79TL3159⟩.

Similarly, adenosine, inosine and 2'-deoxyinosine have been converted into their 8-trifluoromethyl derivatives by reaction with the copper complex formed from trifluoromethyl iodide and copper in hexamethylphosphorotriamide ⟨80JCS(P1)2755⟩. N^6-Trifluoromethylpurine riboside was also produced from 6-chloropurine riboside. Recently, guanosine has been shown to give the C(8)-substituted derivative (148) by reaction with a benz[a]anthracene 5,6-dioxide at pH 9.5 over 4 days at 37 °C ⟨80CC82⟩.

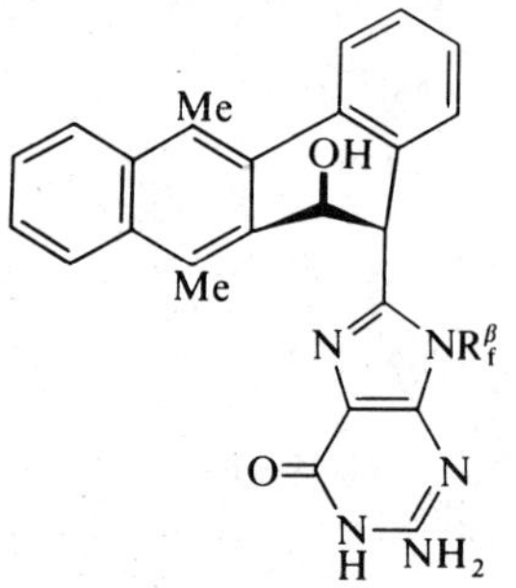

(148)

4.09.5.2.5 Nitration

Electron-releasing groups are required in the purine ring system before nitration can occur. In this way 8-nitro-caffeine (149), -theophylline (150), and -theobromine (151) have been prepared by nitration of the appropriate purine with nitric acid, neat or in acetic acid ⟨71HC(24-2)1, p. 402⟩. Similarly, 9-methylxanthine, but not xanthine, undergoes nitration to 9-methyl-8-nitroxanthine ⟨60JA3773⟩.

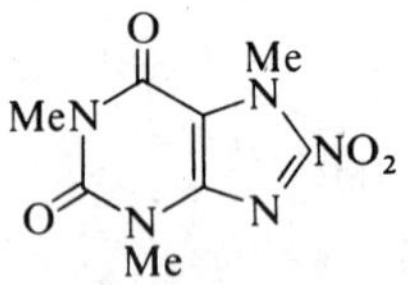

(149) 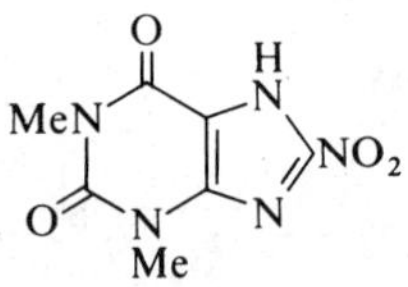   (150) 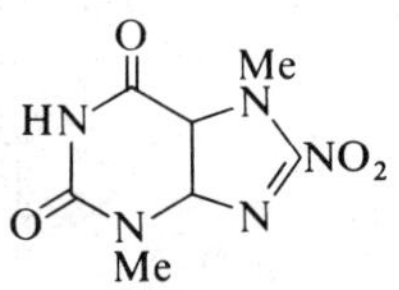   (151)

4.09.5.2.6 Reaction with diazonium ions

Disubstituted oxo- or amino-purines will usually couple with diazonium ions to produce 8-arylazo derivatives. Thus xanthine with diazotized p-chloraniline gave the derivative (152) ⟨60JA3773⟩. Theophylline similarly condenses with diazotized p-aminobenzenesulfonic acid to furnish (153) ⟨07ZPC(51)425⟩, but when alkyl groups are present in the imidazole moiety as with theobromine and caffeine, coupling does not occur ⟨07ZPC(51)425⟩ possibly because anion formation is not possible in these species ⟨63JOC1662⟩. Guanine and isoguanine

⟨60JA2587⟩ both couple with diazotized 2,4-dichloroaniline and several other arylamines to produce the 8-arylazo derivative, and 2,6-diaminopurine also furnishes the 8-arylazopurine (**154**) with diazotized 2,4-dichloroaniline.

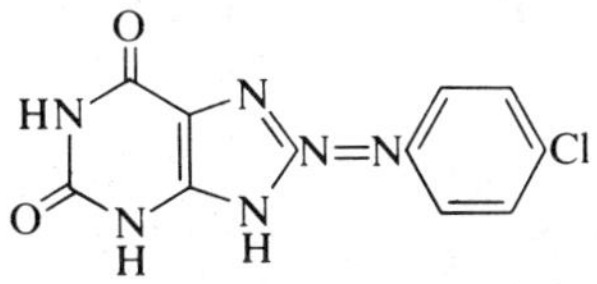

(**152**) 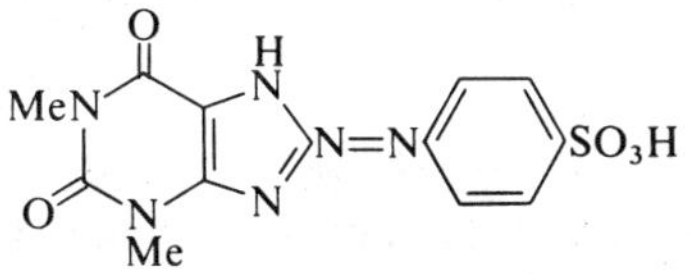   (**153**) 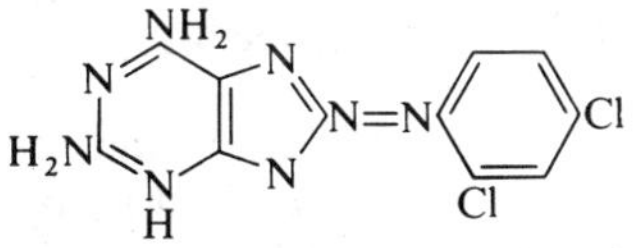   (**154**)

Similar reactions of adenine with arenediazonium chlorides however afforded 8-aryladenines but *via* an internal free radical process from an aryltriazene (see Section 4.09.5.4.5).

4.09.5.2.7 Oxidation of ring atoms

Ionic as distinct from free radical oxidation (see Section 4.09.5.4) of purines leads to *N*-oxides. A major review of the chemistry of heterocyclic *N*-oxides including purine derivatives has been published (see Table 1) and should be consulted for more detailed information. Reagents most commonly used for direct oxidation of purines include hydrogen peroxide and various peracids but especially peracetic acid.

Purine with perbenzoic acid slowly (two weeks) afforded the 1-oxide (**155**) ⟨62JOC567⟩, whereas peracetic acid was used to convert adenine or adenosine ⟨58JA2755⟩ into their 1-oxides, these reactions taking a few days. Similar compounds are produced from other nucleosides ⟨67JMC130⟩ and from adenosine mono-, di- and tri-phosphates ⟨58AG571⟩. Hypoxanthine, however, did not readily yield the 1-oxide ⟨58JA2755⟩ which is best obtained from the adenine derivative by deamination.

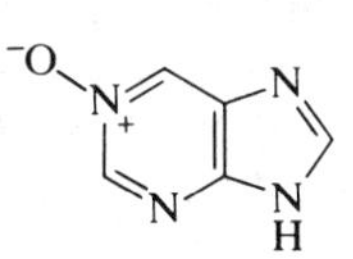

(**155**) 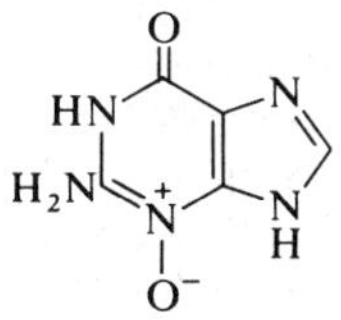   (**156**)

Oxidation of guanine overnight with trifluoroperacetic acid produced the 3-oxide (**156**) in 62% yield ⟨69JOC978⟩ and this may be hydrolyzed to xanthine 3-oxide. Similar 3-oxides are produced from various 6-substituted purines including 6-chloropurine and 6-alkoxy-purines with peracetic acid in ethanol ⟨69BCJ750⟩. The 1-, 7-, 8- and 9-monomethylguanines and 1,7-dimethylguanine with trifluoroperacetic acid also gave corresponding 3-oxides ⟨71JOC2635⟩. Purine furnished the 3-oxide with *m*-chloroperbenzoic acid but 6-methylpurine gave a mixture of the 1- and 3-oxides.

There appear to be no confirmed examples of the direct conversion of purines into 7- or 9-oxides which have been prepared, however, by other methods from imidazole precursors (Section 4.09.7.5).

Methylation of adenine 1-oxide produces the 9-methyl 1-oxide (**157**) *via* an intermediate *O*-methyl derivative (Scheme 24) ⟨71CPB1611⟩.

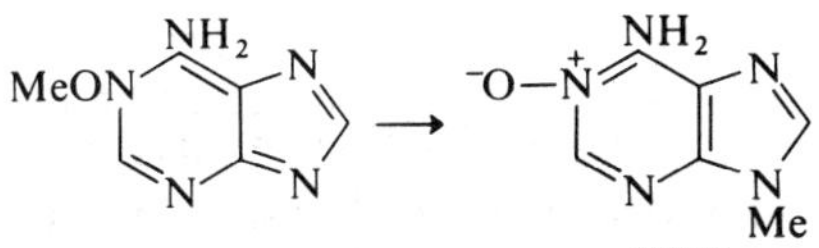

(**157**)

Scheme 24

The purine 3-oxides ⟨69JBC(244)4072, 70MI40905⟩ and 7-oxides ⟨76JOC1889⟩ have potent oncogenic properties.

Thiation of purine, 6-methyl- or 2-amino-purines with hot sulfur leads to the 8-thioxo derivatives ⟨64JOC3209⟩. The reaction failed with guanine, hypoxanthine and 6-chloropurine but was successful with 1,9-dimethyl-, 7,9-dibenzyl- and 7,9-dimethyl-hypoxanthines

⟨68CC200⟩. Thiation of the 2-position also occurred with 1,3,7-trimethylhypoxanthinium nitrate ⟨69CB3000⟩.

4.09.5.2.8 Halogenation

Most halogenopurines are normally prepared by replacement of hydroxy or thio groups by halogen but direct halogenation at an unoccupied 8-position is possible, especially when the purine contains one or more electron-releasing, normally hydroxy or similar, groups. The course of the halogenation is considerably affected by the nature of the solvent used. In non-polar solvents the xanthines furnish corresponding 8-halogenoxanthines; thus theobromine with chlorine in chloroform gave 8-chlorotheobromine ⟨1882LA(215)316⟩ but in acetic acid it produced 5-chloro-3,7-dimethylisouric acid (**158**) derived by prior addition to the C(4)—C(5) bond ⟨14LA(406)22⟩.

(**158**)

Theophylline and caffeine ⟨17LA(413)155, 10CB3553⟩ and thio analogues of these purines chlorinate in a similar manner ⟨28CB1409⟩. 1,9-Dimethylxanthines on the other hand are slowly converted into 8-chloropurines by sulfuryl chloride ⟨60ZOB3339, 64ZOB1137⟩.

Bromination, in contrast, readily occurs even with monosubstituted purines. Thus hypoxanthine and adenine ⟨1890CB225⟩ both produced the corresponding 8-bromo derivatives by direct bromination, and disubstituted derivatives such as xanthine ⟨63JOC2310⟩ and guanine ⟨1883LA(221)336⟩ very readily afforded the corresponding 8-bromo derivatives. Purine nucleosides also may be directly halogenated. Thus 2′,3′,5′-tri-O-acetyladenosine or 3′,5′-di-O-acetyl-2′-deoxyadenosine with N-bromoacetamide gave, after deblocking, 8-bromoadenosine and 8-bromo-2′-deoxyadenosine, respectively ⟨64JA1242⟩, and 8-bromoguanosine is similarly obtained from its tri-O-acetyl derivatives by bromination with bromine in acetic acid followed by deblocking. Recently 8-chloropurine nucleosides have been prepared by reaction of nucleoside with m-chloroperbenzoic acid and hydrogen chloride in aprotic solvents ⟨81JOC2819⟩.

4.09.5.3 Reaction with Nucleophiles

Nucleophilic attack on purines, by replacement reactions involving halogens, alkoxy, alkylthio and similar groups, is very common and offers a valuable route to many purine derivatives and is discussed in Section 4.09.6. Direct attack on ring atoms by nucleophiles on the other hand is relatively rare, but in recent years increasing numbers of examples of these reactions have appeared.

4.09.5.3.1 Reduction

Since many purines may be reduced by hydride-producing agents it is convenient to gather together chemical reduction of purines under the nucleophilic heading. A general review of this topic has appeared ⟨71HC(24-2)1⟩. Reduced purines of the oxo-nH type, however, are perhaps best regarded as special cases since they can at least formally be written in aromatic (hydroxypurine or betaine) forms and are dealt with elsewhere.

Reduction of purine in acetic or hydrochloric acid at a mercury cathode produced the labile 1,6-dihydropurine which has been isolated as a complex with sodium tetraphenylborate ⟨62JA1412⟩. Purine also afforded the 1,6-dihydro derivative by catalytic hydrogenation using palladium or platinum catalysts in acid solution and the unstable compound may be isolated as a 1,9-diacetyl derivative by carrying out the hydrogenation in acetic anhydride ⟨69LA(729)73⟩. The same author claims the formation of 1,3,7,9-triacetyloctahydropurine by continued hydrogenation under similar conditions. In a similar manner, N-acetyl-1,6-

dihydro derivatives of 2-chloro- and 2-chloro-9-methylpurine may be produced without hydrogenolysis of the halogen atom, in contrast to 2,6,8-trichloropurine in which all the halogen atoms are reported to be removed ⟨59JA3789⟩.

Chemical reduction of purine with various metal–acid combinations generally results in ring opening to imidazoles or pyrimidines.

Electrochemical reduction of oxo- and amino-purines tends to be more complex. With a lead cathode in dilute sulfuric acid, caffeine was found to lose the 6-oxo function and afford the 1,6-dihydro derivative (**159**) in 70% yield ⟨1899CB68⟩; see also ⟨B-71MI40903, p. 428⟩ for additional confirmatory comment.

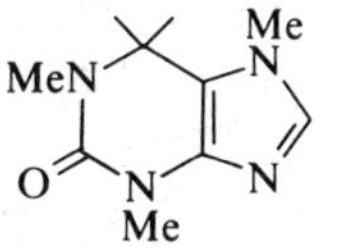

(**159**) 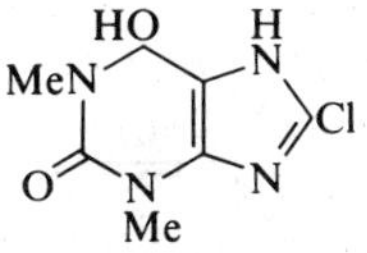   (**160**)

Xanthine ⟨01CB1165⟩ and its 1-, 3- and 7-methyl derivatives similarly produced 6-deoxy derivatives using a lead cathode in 75% sulfuric acid solution, and in 30% sulfuric acid theophylline afforded the 6-deoxypurine ⟨07CB3752⟩. An intermediate product (**160**) has been detected by polarography in an electrolytic reduction of 8-chlorotheophylline ⟨54MI40901, 54MI40902⟩.

Under similar conditions hypoxanthine undergoes ring opening, and uric acid at low temperatures (<8 °C) produced the labile substance 'purone' with the possible structure (**161**), where the C(4)–C(5) bond has been reduced, and this undergoes ready ring opening to give pyrimidines ⟨01CB258⟩. In contrast, methylated uric acids afford the normal 6-deoxy derivatives, as does guanine which gave 2-amino-1,6-dihydropurine (**162**) in 75% yield ⟨01CB1170⟩.

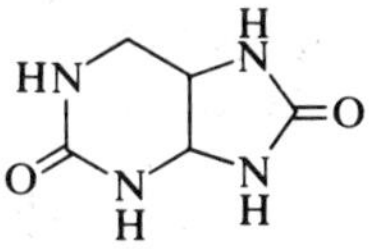

(**161**) 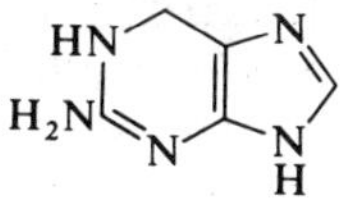   (**162**)

Further reduction of 6-deoxytheophylline in acetic acid and acetic anhydride over palladium afforded the 7-acetylhexahydro derivative (**163**) whilst with a platinum catalyst in acetic anhydride the 7,9-diacetyloctahydropurine (**164**) is formed ⟨69LA(729)73⟩. In contrast, the same author was unable to reduce adenine, hypoxanthine, xanthine and 6-chloropurine in this manner.

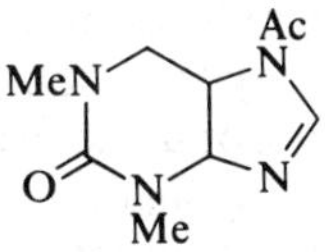

(**163**) 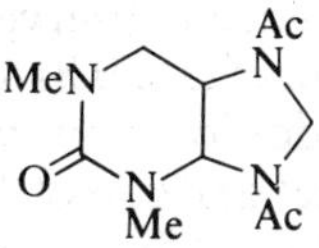   (**164**)

N-Alkylated purines, especially compounds such as quaternized caffeine with a positive charge in the imidazole ring, are readily reduced by sodium borohydride to produce dihydro derivatives, reduction occurring in the charged imidazole ring ⟨76TL1199, 76JOC2303⟩.

Compounds which can be regarded as analogous to reduction products are 1,6-covalent adducts produced by various weak acids including potassium hydrogen sulfite and purine (**165**) and the 2-oxopurine (**166**) ⟨75JCS(P1)2240⟩.

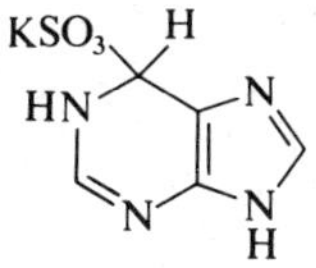

(**165**) 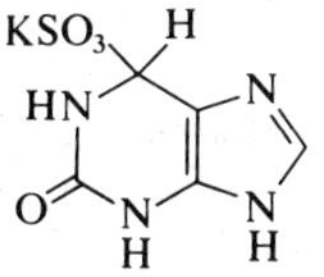   (**166**)

4.09.5.3.2 Amination

It has recently been found that purine in its anionic form undergoes a Chichibabin reaction with potassium amide in liquid ammonia with formation of adenine as the sole product.

The reaction is slow and after 20 h about 30% of purine could be recovered, but after 70 h the reaction conversion was quantitative ⟨79JOC3140⟩. Moreover when the reaction was repeated using ^{15}N-labelled potassium amide, the label appeared exclusively in the exocyclic amino group with no ring labelling. Evidence for the involvement of the purine anion (**167**) and hence the intermediate adduct anion (**168**) in the reaction came from a study of the ^{1}H and ^{13}C NMR spectra of reaction mixtures. The mechanism proposed (Scheme 25) involves addition of the amide ion to the purine anion in which the pyrimidine ring has apparently been so depleted of electronegativity as to make it a readily achievable target for anion attack. Finally a hydride ion is liberated.

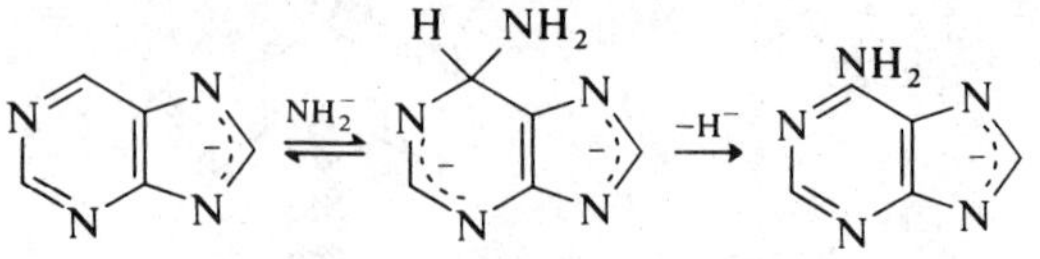

Scheme 25

A similar conversion of 2-methylpurine into the corresponding adenine was also achieved although in lower yield (20%), and evidence for direct amination with no ring opening was again obtained. With 6-methylpurine, however, no adenine derivative was obtained and this is readily ascribed to dianion formation with excess negativity in both rings. ^{1}H NMR spectroscopic evidence was obtained (Scheme 26) for the dianion (**169**). However with 6,8-di-*t*-butylpurine which can only form a monoanion (**170**), not surprisingly there was no reaction with potassium amide.

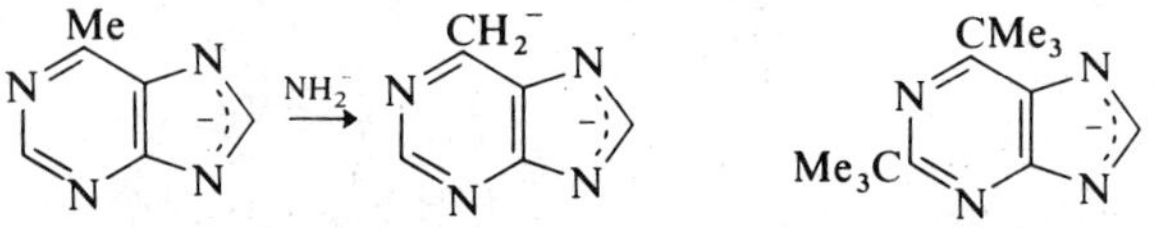

Scheme 26

In the case of 8-methylpurine, a modest yield of 8-methyladenine was produced in spite of the fact that the dianion could be detected in solution. The implication is that the reaction solution contained insufficient dianion to inhibit the reaction of the potassium amide with the monoanion which was also present.

Similar reactions have also been observed with 8-chloro- and 8-methylthio-purines which produce 8-chloro- (10% yield after 20 h) and 8-methylthio-adenines (80% yield after 20 h), respectively ⟨80RTC267⟩. In addition, however, 8-chloropurine produced adenine (30% yield) as the major reaction product and use of ^{15}N-labelling again indicated that no ring opening had occurred. The proposed mechanism is a novel *tele*-amination involving prior amination of the chloropurine to form the adduct (**171**) which then undergoes a 1,4-*tele*-elimination to produce adenine directly or *via* the purine adduct (**172**; Scheme 27).

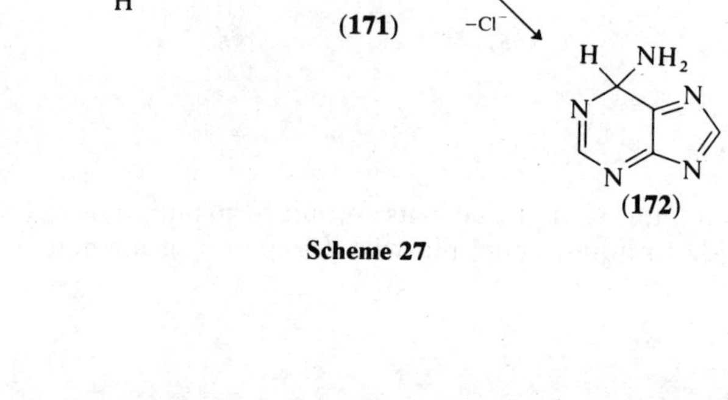

Scheme 27

4.09.5.4 Reactions with Free Radicals

Calculations suggest that free radical attack on purines should marginally favour the 6-position followed closely by C-8 ⟨71MI40903⟩. In support of this, reaction of purine with hydroxyl radicals generated by Fenton's reagent or by ionizing radiations in aqueous solutions produced hypoxanthine, 2-aminopurine afforded guanine, and adenine gave the 8-oxo derivative ⟨61CR(253)687⟩. The nature of the hydrogen addition radicals generated by radiation has been considered elsewhere (Section 4.09.4.3.8).

Free radical alkylation procedures have proved a useful route to alkylpurines which are not readily available by other methods. Thus 6-substituted purines including adenine and hypoxanthine may be converted into 8-methyl derivatives with t-butyl hydroperoxide in the presence of iron(II) ions and acid ⟨74T2677⟩, although small amounts of 2-methyl and 2,8-dimethyl derivatives were formed simultaneously. Adenosine and guanosine similarly furnished the corresponding 8-methyl derivatives with diacetyl peroxide (as a source of methyl radicals) and iron(II) ions ⟨76T337⟩.

Replacement of iron(II) ion as a catalyst by radiation and use of peracid esters leads to more efficient production of radicals. Thus with a 1200 W source and t-butyl peracetate, methylation of caffeine is 28 times faster than using t-butyl hydroxide ⟨77JA5096⟩. An alkaline solution of theophylline containing benzyl bromide similarly gave with UV radiation both the 8- and the 7-benzyl derivatives ⟨76AJC891⟩. Formation of alkyl radicals in these various reactions are summarized in Scheme 28.

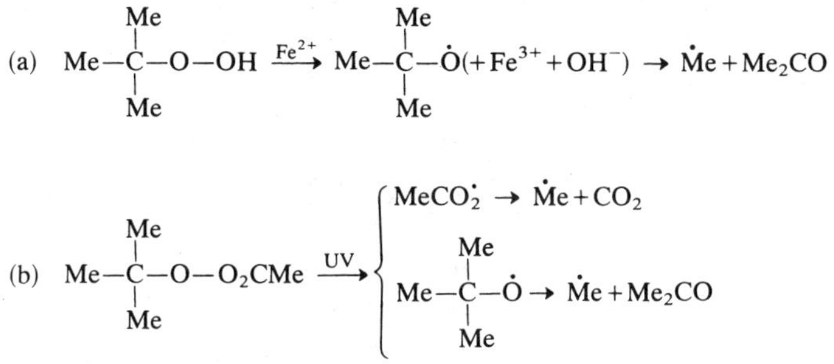

(a) $\text{Me}-\underset{\underset{\text{Me}}{|}}{\overset{\overset{\text{Me}}{|}}{\text{C}}}-\text{O}-\text{OH} \xrightarrow{\text{Fe}^{2+}} \text{Me}-\underset{\underset{\text{Me}}{|}}{\overset{\overset{\text{Me}}{|}}{\text{C}}}-\dot{\text{O}}(+\text{Fe}^{3+}+\text{OH}^-) \rightarrow \dot{\text{Me}} + \text{Me}_2\text{CO}$

(b) $\text{Me}-\underset{\underset{\text{Me}}{|}}{\overset{\overset{\text{Me}}{|}}{\text{C}}}-\text{O}-\text{O}_2\text{CMe} \xrightarrow{\text{UV}} \begin{cases} \text{MeCO}_2^{\cdot} \rightarrow \dot{\text{Me}} + \text{CO}_2 \\ \text{Me}-\underset{\underset{\text{Me}}{|}}{\overset{\overset{\text{Me}}{|}}{\text{C}}}-\dot{\text{O}} \rightarrow \dot{\text{Me}} + \text{Me}_2\text{CO} \end{cases}$

Scheme 28

Irradiation of purines in different environments can, of course, lead to a variety of radical reaction products. Recently the regioselective nature of radical methylation has been observed in several purine derivatives.

Thus using a t-butyl peracetate and a 450 W mercury lamp as a source of methyl radicals it was shown that in acid solution hypoxanthine, 6-methoxy- and 6-methylthio-purines gave mainly the C-8 methyl derivative whereas 3-methylhypoxanthine, 3-methyl-6-methoxy-purine and 3-methyl-6-methylthiopurine gave mainly the C-2 methyl derivatives. The results tend to be reversed at higher pH ⟨79JOC1450⟩.

The selectivity has been correlated with the site of cation formation, methylation occurring fastest on the carbon where the proton resonated at lowest field. Table 22 gives details of

Table 22 Free Radical Methylation of some Purines ⟨79JOC1450⟩[a]

		Yield (%)		
Compound	pH	C-2	C-8	Ratio C-8:C-2 (%)
Hypoxanthine	1.1	<1	85.2	>85
	7.0	9.9	24.6	2.5 ± 0.3
3-Me-hypoxanthine	1.1	47.4	8.8	0.2 ± 0.1
	7.0	7.8	34.9	5.1 ± 0.3
6-MeO-purine	1.1	2.3	67.4	29.2 ± 0.8
	7.0	11.6	22.8	2.0 ± 0.2
3-Me-6-MeO-purine	1.1	25.7	17.5	0.7 ± 0.1
	7.0	<1	24.2	>24
6-MeS-purine	1.1	3.1	53.6	16.8 ± 1.6
	7.0	5.7	34.6	5.9 ± 0.7
3-Me-6-MeS-purine	1.1	21.7	13.5	0.6 ± 0.0
	7.0	16.8	25.5	1.5 ± 0.2

[a] t-Butyl peracetate: irradiation with a 450 W Hg lamp at room temperature.

reaction products and product ratios at different pH values and Table 23 gives rate constants observed in these reactions. The proposed methylation routes for hypoxanthine and 3-methylhypoxanthine are outlined in Schemes 29 and 30, respectively.

Table 23 Rate Constants for Free Radical Methylation of Purines ⟨79JOC1450⟩[a]

Compound	$k_{C\text{-}2} \times 10^5$ (s^{-1})	$k_{C\text{-}8} \times 10^5$ (s^{-1})	$k_{C\text{-}8} : k_{C\text{-}2}$
Hypoxanthine	0.05 (±0.01)	0.88 (±0.01)	17.6
3-Me-hypoxanthine	0.20 (±0.04)	0.06 (±0.01)	0.3
6-MeO-purine	0.07 (±0.01)	0.61 (±0.02)	8.7
3-Me-6-MeO-purine	1.33 (±0.01)	0.27 (±0.06)	0.2
6-SMe-purine	—	1.04 (±0.13)	—
3-Me-6-SMe-purine	1.94 (±0.19)	0.54 (±0.06)	0.28

[a] 1.7M in D$_2$O:CF$_3$CO$_2$D (2:1) and t-butyl peracetate (9 fold excess); irradiation with a 450 W mercury lamp at room temperature.

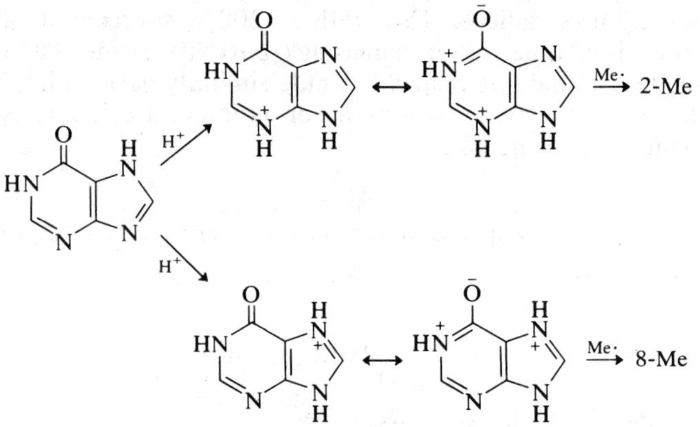

Scheme 29

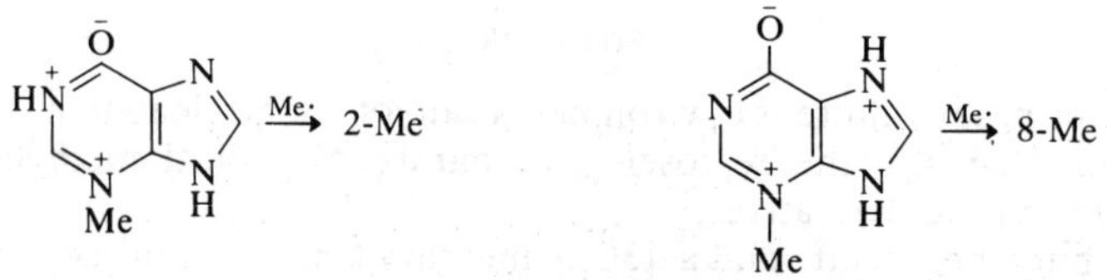

Scheme 30

The methylation of caffeine at C-8 using t-butyl peracetate in acid solution has also been studied in detail ⟨77JA5096⟩ and rate constants were determined (Table 24) over a wide range of conditions including thermal and irradiation production of methyl radicals. The suggested pathway involves radical attack on either the neutral or $N(7)$-protonated forms of caffeine with eventual production in each case of a radical carbocation σ-complex which then suffers loss of hydrogen (Scheme 31). In neutral solution the rate determining step is protonation which follows methylation, and in acid solution methylation of the protonated purine is the controlling factor.

Table 24 Rate Constants for Thermal and Photoinduced Radical Methylation of Caffeine ⟨77JA5096⟩[a]

Reaction conditions	Temp. (°C) (Hg lamp)	Rate constant (5×10^5 s^{-1})
Thermal	58	No reaction after 21 h
Thermal	65	0.40 (±0.01)
Thermal	80	1.67 (±0.18)
Thermal	95	7.61 (±0.37)
Light	(450 W)	1.39 (±0.02)
Light	(1200 W)	6.67 (±0.36)

[a] 1.0M caffeine and 3M t-butyl peracetate in D$_2$O:CF$_3$CO$_2$D (2:1).

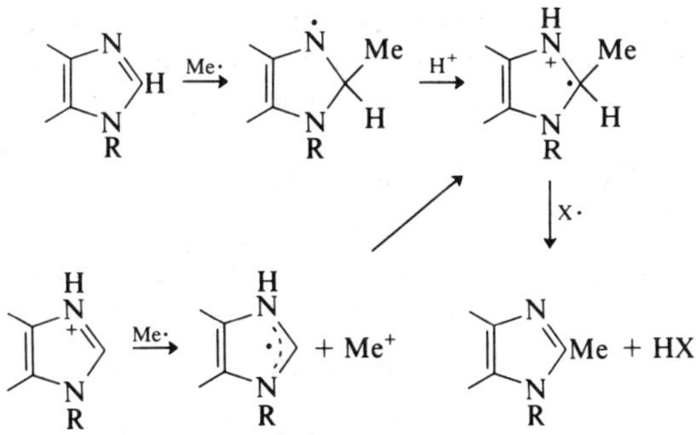

Scheme 31

The same authors have examined the free radical methylation of several nucleosides in the presence of *t*-butyl peracetate and shown that *C*-methylation occurs at pH 1–4 but at pH 4–10 there is increasing *N*-methylation ⟨80JOC2373⟩ as the electron availability on these atoms builds up. *C*-Methylation is proposed to proceed *via* an addition mechanism and *N*-methylation *via* a radical extraction of hydrogen from NH followed by reaction with a methyl radical. Implications of these reactions in radical oncogenesis are discussed.

4.09.5.4.1 Alcohols

UV irradiation of purine ⟨68JA2979⟩, nebularine ⟨70JA4751⟩ and 2-aminopurine ⟨71JOC3594⟩ in methanol and ethanol gave 1,6-adducts. The ethanol adduct (**173**) with purine was isolated in two diasteroisomeric forms.

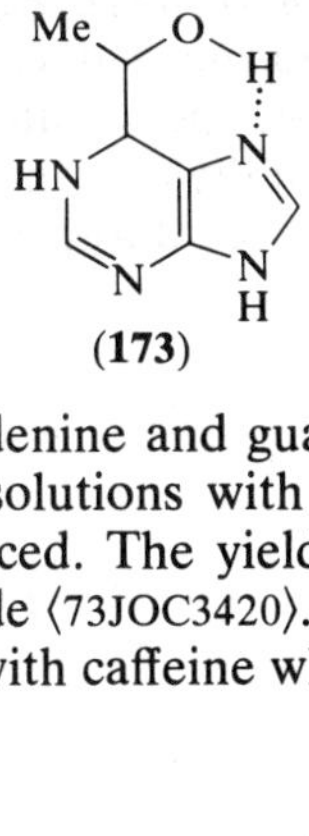

Not surprisingly, adenine and guanine have been a major focus of many studies in this field and in alcohol solutions with irradiation at wavelengths >290 nm, 8-hydroxyalkyl derivatives are produced. The yields are improved by use of sensitizers such as acetone and di-*t*-butyl peroxide ⟨73JOC3420⟩. With substituted purines, attack may also occur at the substituent group as with caffeine which in 2-propanol produced (**174**).

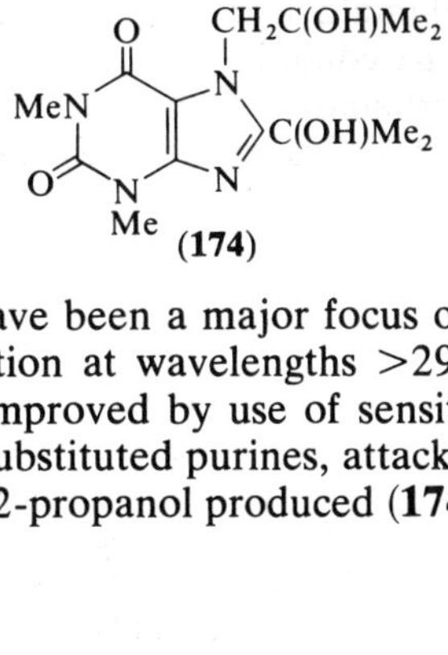

4.09.5.4.2 Amines

Free radical reactions of purines with amines gave similar products to those produced in alcohol solution although deamination may also occur, probably at the post- rather than the pre-adduct stage. Whereas purine and *n*-propylamine afforded 6-*n*-propylpurine ⟨71MI40907⟩, adenine and caffeine produced both the 8-aminoalkyl and corresponding 9-alkyl derivatives ⟨74MI40904⟩. Also irradiation of 8-aminoalkylpurines in methanol furnished the 8-alkyl derivatives. Amino acids as an amine source are of special biochemical interest. They also tend to produce 8-alkylpurines by concomitant deamination and decarboxylation ⟨69CC905⟩.

4.09.5.4.3 Ethers

With more complex substances such as ethers the nature of free radical attack becomes more difficult to forecast. Thus in the presence of an acetophenone sensitizer, caffeine and diethyl ether gave a 3:1 mixture of the 8-(1-ethoxyethyl) and 8-ethyl derivatives (Scheme 32) ⟨70CJC1716⟩. The first formed ether radical, presumably (**175**), is apparently too unstable to produce a reaction product prior to rearrangement (Scheme 33). With unsymmetrical ethers, mixtures of products are inevitably produced. Thus, tetrahydrofuryl alcohol and caffeine gave a mixture of the two 8-substituted derivatives (**176**) and (**177**) in a ratio of 2.5:1.

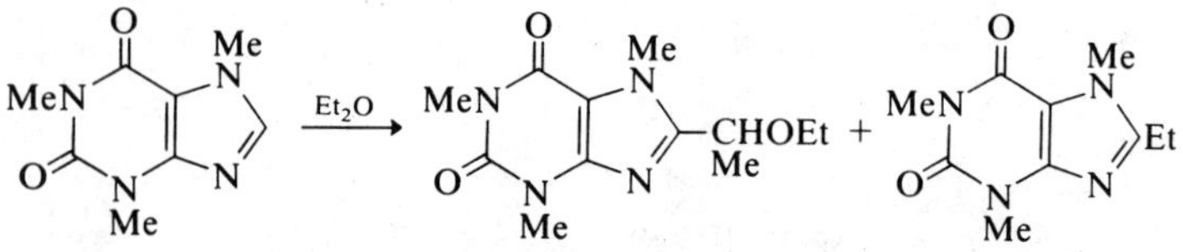

Scheme 32

$$MeCH_2OCH_2CH_2\cdot \rightarrow MeCH_2O\dot{C}HMe$$
(**175**)

Scheme 33

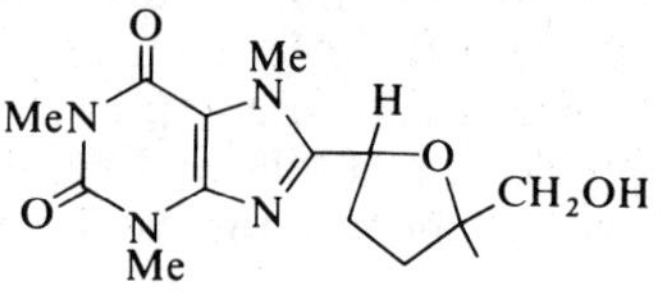

(**176**) 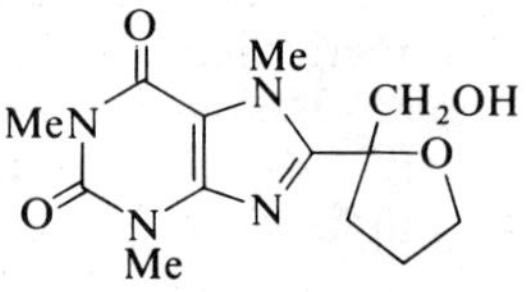   (**177**)

4.09.5.4.4 Acyl radicals

In the presence of substances capable of producing acyl radicals, purines gave 8-acyl derivatives. Thus guanosine and ethanol give 8-acetylguanosine (**178a**), and similar products were produced from other aldehyde sources and from inosine ⟨75B1490⟩. However adenosine was unreactive under the conditions used, namely iron(II) persulfate initiation. Guanosine also produced the 8-carbamoyl (**178b**) and 8-(*N,N*-dimethylcarbamoyl) (**178c**) derivatives with formamide and DMF, respectively, as sources of the appropriate carbamoyl radicals ⟨74T2677⟩.

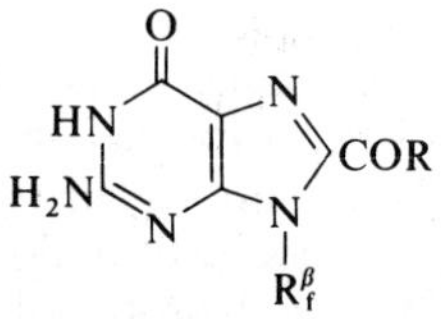

(**178a**) R = Me
(**178b**) R = NH$_2$
(**178c**) R = NMe$_2$

4.09.5.4.5 Aryl radicals

8-Aryladenines have recently been produced in an intermolecular free radical process by heating coupling products (**179**) derived from benzenediazonium chloride at 60–90 °C (Scheme 34) ⟨81JOC2203⟩. The reaction is seen to proceed by formation of a phenyl radical which attacks C-8. Evidence for this includes (a) the corresponding *p*-bromophenyltriazene decomposed in the presence of 2,5-diaminopurine to give a mixture of 8-aryladenine and 8-aryl-2,6-diaminopurine; (b) when adenine is added to a solution of the phenyltriazene

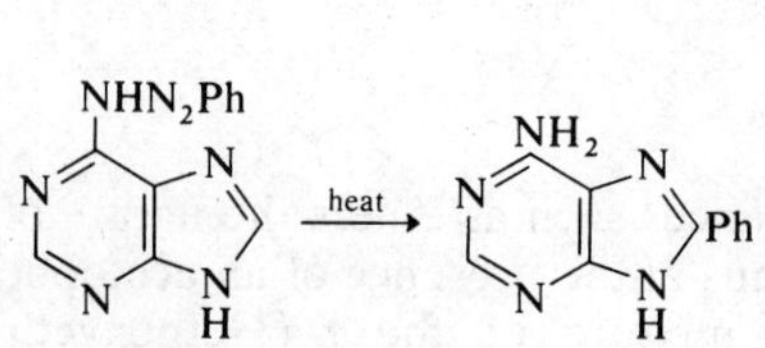

(**179**)

Scheme 34

4.09.6 REACTIVITY OF SUBSTITUENTS

4.09.6.1 Introduction

4.09.6.1.1 Nucleophilic displacement reactions

Much of the early published chemistry of purines was concerned with nucleophilic displacement reactions of halogen, alkylthio and similar substituted purine derivatives. The ease of displacement varies according to the substitution pattern in the ring system and the particular neutral, cationic or anionic form of the purine. Thus in both the neutral and protonated forms ensuing movement of electronegativity from the imidazole to the pyrimidine ring system tends to favour nucleophilic displacement of groups at the C-8 position. On the other hand when anion formation is possible, electronegativity is drawn into the imidazole ring system resulting in preferential nucleophilic attack at C-2 or C-6 in the pyrimidine half of the molecule.

4.09.6.1.2 Substituent modification

Modification of substituents in purines generally follows patterns which are similar to substituent modification in typical aromatic systems. These include interconversion of hydroxyl, thiol, halogeno, amino and similar groups, and Sandmeyer type reactions involving primary amines although isolation of stable diazonium salts is rarely possible. In addition, transformations and reactions of carboxylic acid derivatives follow standard patterns.

4.09.6.2 N- and C-Alkyl Derivatives

Although all the simple alkyl, especially methyl, purines have been prepared there has not been a great deal of interest in their chemistry. 6-Methylpurine shows weak acidic properties and reacts with chloral to produce the adduct (**180**). The compound may also be oxidized to 6-formylpurine (**181**) and to purine-6-carboxylic acid in various ways, and it undergoes successive halogenation with either halogens, especially chlorine, or with *N*-halogenosuccinimides or sulfuryl chloride. 8-Methylpurines may also be halogenated and stable compounds isolated, especially when the nitrogen atoms are blocked as with the various methylated xanthines, although 8-methyl theophylline is unreactive. 6-Alkylpurine 1-oxides undergo exceptional reactions *via* O-acyl rearrangements. Thus 6-methylpurine 1-oxide and methane sulfonyl chloride produced 6-chloromethylpurine (**182**) *via* an *O*-mesyl derivative. These reactions are examined in more detail in Section 4.09.6.12.

There are only a few recorded reactions of *N*-alkylpurines which result in preservation of the purine ring system, since ring opening generally occurs and pyrimidines or imidazoles are produced. Halogenation may, however, lead to *N*-chloromethyl derivatives as with 2,7-dimethylpurine which can be chlorinated to form 7-chloromethyl-2-methylpurine.

Details of reaction products derived from some alkylpurines are collected in Table 25.

4.09.6.3 Purine Aldehydes

A few 6- and 8-formylpurines have been prepared and their reactions investigated. They tend to behave like typical aromatic aldehydes and produce stable anils with aromatic

Table 25 Reactions of Alkylpurines

Purine	Product	Conditions	Ref.
6-Me	6-CHO	SeO_2/DMF	68CB611
6-Me	6-CH=NOH	$NaNO_2$/HOAc	68CB611
6-Me	6-CO_2H (8%)	$KMnO_4$/OH^- (80 °C)	68CB611
6-CH_2OH	6-CO_2H (51%)	$KMnO_4$ (cold)	68CB611
6-CH_2Cl	6-CO_2H (77%)	via Pyridinium salt/$KMnO_4$	59JA2515
6-CH_2Cl	6-CO_2H	$KMnO_4$	63ZOB3342
6-CH_2Cl	6-CHO	Pyridinium salt + $NOC_6H_4NMe_2$ + acid	59JA2515
6-Me	6-CCl_3 (80%)	SO_2Cl_2/TFA	62JOC3545
2-NH_2-6-Me	2-NH_2-6-CCl_3	SO_2Cl_2/TFA	64MI40900
8-CF_3-6-Me	8-CF_3-6-CCl_3	SO_2Cl_2/TFA	64MI40900
2-MeNH-6,8-Me_2	2-MeNH-6-CCl_3-9-Me	SO_2Cl_2/TFA	64MI40900
6-Me	6-$CHBr_2$	N-Bromosuccinimide	62JOC3545
1,7-Et_2-8-Me-xanthine	8-CCl_3-1,7-Et_2-xanthine	Cl_2/$POCl_3$	45JCS751
8-Me-theobromine	8-CCl_3-theobromine	$AlCl_3$/CH_2Cl_2 (60 °C) (8%)	58MI40901
8-Me-caffeine	8-CCl_3-caffeine	Cl_2/PhCl (<7 °C)	58MI40900
8-Me-caffeine	7-$ClCH_2$-8-CCl_3-caffeine	Cl_2/PhCl (>7 °C)	58MI40900
8-Me-caffeine	8-CH_2Br-caffeine	N-Bromosuccinimide $(PhCO_2)_2$, UV	66AF541
8-Me-isocaffeine	8-CCl_3-isocaffeine	N-Chlorosuccinimide	60ZOB3332
8-Me-isocaffeine	8-CCl_3-3-$ClCH_2$-isocaffeine	N-Chlorosuccinimide	60ZOB3332
6-Me	6-$CH_2CHOHCCl_3$	Chloral/AcOH	61LA(649)131
8-Me-caffeine	8-$CHCl_2$-caffeine 8-CH_2Cl-caffeine	SO_2Cl_2/$CHCl_3$	62ZOB2015
6-Me 1-oxide	6-CH_2Cl	$MeSO_2$Cl	62JOC3545
6-Me 1-oxide	6-CH_2OAc	Ac_2O	62JOC567
2,7-Me_2	2-Me-7-CH_2Cl	Cl_2/Cl_3CMe	71HC(24-2)1, p. 131

amines. They undergo Perkin type reactions with malonic acid and will condense similarly with other active methylene derivatives and also undergo benzoin condensation. In addition they give the usual aldehyde derivatives including oximes, hydrazones, semicarbazides, reaction with Wittig reagents, and undergo oxidation to the carboxylic acids and reduction to alcohols. Table 26 lists some of the more important reaction products, most of which, it will be noted, are concerned with highly alkylated compounds. The paucity of examples of highly polarized purine derivatives undoubtedly results from their insolubility in common solvents.

4.09.6.4 Purine Ketones

Even fewer purine ketones have been prepared than aldehydes. They have been reduced to alcohols either by hydrogenation or use of Meerwein–Pondorff methods. Table 27 outlines information on the few compounds examined.

4.09.6.5 Cyanopurines

A few 6- and 8-cyanopurines have been prepared and undergo characteristic nitrile addition reactions rather readily. Thus, alkaline hydrolysis produces carboxamides, then carboxylic acids, alcoholysis leads to imidates, ammonolysis to amidines, hydrazinolysis to amidhydrazines, hydroxylamine to amidoximes, and hydrogen sulfide to thioamides. Acid hydrolysis tends to give the decarboxylated acid derivative. Reduction either by sodium–ethanol or, preferably, by catalytic hydrogenation affords aminoalkylpurines and addition of Grignard reagents produces, in the first place, acylpurines. As with aldehydes, most of the compounds examined have been relatively non-polar derivatives. Table 28 lists some reactions and relevant literature.

Table 26 Reactions of Purine Aldehydes

Purine	Product	Conditions	Ref.
6-CHO	6-CH=NOH	NH_2OH	66JOC4239
6-CHO	6-CH=NNH_2	NH_2NH_2	66JOC4239
6-CHO	6-Me	NH_2NH_2/heat	64JOC3209
6-CHO	6-CH=$NNHCONH_2$	$NH_2CONHNH_2$	59JA2515
6-CHO	6-CO_2H	$KMnO_4$(cold) (82% yield)	59JA2515
6-CHO	6-CH_2OH	Pt/H_2	68CB611
8-CHO-theophylline	8-CO_2H-theophylline	Aqueous $KMnO_4$	62CB414
8-CHO-theobromine	8-CO_2H-theobromine	Aqueous $KMnO_4$	62CB414
8-CHO-caffeine	8-CO_2H-caffeine	Aqueous $KMnO_4$	48ZOB2129
8-CHO-theobromine	8-CH=NOH-theobromine	NH_2OH	66JOC4239
8-CHO-theophylline	8-CH=NOH-theophylline	NH_2OH	66JOC4239
8-CHO-caffeine	8-CH=NOH-caffeine	NH_2OH	48ZOB2129
8-CHO-6-NEt_2	8-CH=NOH-6-NEt_2	NH_2OH	58JCS4069
2-Me-6-CHO	2-Me-6-anil	$PhNH_2$	60ZOB1878
8-CHO-theophylline	8-CH=NPh	$PhNH_2$	62CB403
8-CHO-caffeine	8-CH=NPh	$PhNH_2$	62CB403
8-CHO-isocaffeine	8-CH=NPh	$PhNH_2$	62ZOB2015
8-CHO-theophylline	8-CH=$CHCO_2H$	$CH_2(CO_2H)_2$/pyridine	62CB414
8-CHO-theophylline	8-CH=C$(CO_2Et)_2$	$CH_2(CO_2Et)_2$/pyridine	62CB414
8-CHO-caffeine	8-CH=C$(CO_2Et)_2$	$CH_2(CO_2Et)_2$/pyridine	52ZOB2220
8-CHO-theophylline	8-CH=C$(COR^1R^2)_2$ $R^1 = R^2 = Me$ $R^1 = Me, R^2 = Ph$	$CH_2(COR^1R^2)_2$ $R^1 = R^2 = Me$ $R^1 = Me, R^2 = Ph$ Pyridine/piperidine	62CB414
8-CHO-theobromine	2(8-CH=CH)PyMe$^+$I$^-$	2-Methylpyridine methiodide	63ZOB3205
8-CHO-caffeine	2(8-CH=CH)PyMe$^+$I$^-$	2-Methylpyridine methiodide	62ZOB452
8-CHO-isocaffeine	2(8-CH=CH)PyMe$^+$I$^-$	2-methylpyridine methiodide	63ZOB2942
8-CHO-theobromine	8-CH(OH)PO(OR)$_2$	Pyridine methiodide (RO)$_2$HPO, R = Me, Et, Pr, Bu	61ZOB3406
8-CHO-caffeine	8-CH(OH)PO(OR)$_2$	(RO)$_2$PO, R = Me, Et, Pr, Bu	60ZOB2427
8-CHO-theophylline	8-CH=$CHCO_2Et$	Ph_3P=$CHCO_2Et$ (hot dioxane)	62CB414
8-CHO-isocaffeine	8-$CHCl_2$	$SOCl_2/CHCl_3$	62ZOB2015
8-CHO-caffeine	8-COCHOH-8	KCN/DMF/EtOH	62CB414
8-CHO-theophylline	8-COCHOH-8	KCN/DMF/EtOH	62CB414
8-CHO-theobromine	8-COCHOH-8	KCN/DMF/EtOH	62CB414

Table 27 Reactions of some Purine Ketones ⟨56AP453, 63JOC1379⟩

Purine	Product	Conditions
6-(p-OHC$_6$H$_4$CO)	6-(p-HOC$_6$H$_4$CHOH)	Pd/H$_2$/charcoal
6-(p-OHC$_6$H$_4$CO)	6-(p-HOC$_6$H$_4$C=NNH$_2$)	NH$_2$NH$_2$
6-(p-OHC$_6$H$_4$CO)	6-CO$_2$H	H$_2$O$_2$/OH$^-$
6-(p-OHC$_6$H$_4$CO)	6-OH	H$_2$O$_2$/H$^+$
8-COMe-caffeine	8-CHOHMe-caffeine	Al(OPr^i)$_3$ · HOPr^i
8-COMe-caffeine	8-C(OH)(Ph)Me-caffeine	PhMgBr
8-COCH$_2$Br-caffeine	8-CHOHCH$_2$Br-caffeine	Al(OPr^i)$_3$H · OPr^i
8-COMe-caffeine	8-COCH$_2$Br-caffeine	Dioxane/Br$_2$

4.09.6.6 Purinecarboxylic Acids

Several 6- and 8-carboxypurines have been prepared. They readily undergo decarboxylation when heated, the 8-carboxyl group being liberated the most easily. Thus caffeine-8-carboxylic acid loses carbon dioxide in hot water. The acids also readily form esters normally using the Fischer–Speiers method and a few acid chlorides have been prepared using phosphorus pentachloride or, preferably, thionyl chloride. Table 29 lists several typical reactions of the acids.

Table 28 Reactions of some Cyanopurines

Purine	Product	Conditions	Ref.
6-CN	6-CONH$_2$	2N NaOH 1 h	56JA3511
6-CN	6-CO$_2$H	2N NaOH (hot)	56JA3511
6-CN	6-CSNH$_2$	H$_2$S/NH$_3$/EtOH	56JA3511
6-CN	6-C(=NH)NH$_2$	NH$_3$/EtOH/pressure	56JA3511
6-CN	6-C(=NH)OEt	EtOH	56JA3511
6-CN	6-C(=NH)NHNH$_2$	NH$_2$NH$_2$	59JA2515
6-CN	6-C(=NH)NHNHPh	PhNHNH$_2$	59JA2515
6-CN	6-C(=NH)NOH	NH$_2$OH	59JA2515
6-CN	6-OH	2N H$_2$SO$_4$	59JA2515
6-CN	6-CH=NNHCONH$_2$ (low yield)	H$_2$/Ni/NH$_2$NHCONH$_2$	59JA2515
6-CN	6-CH$_2$NH$_2$	H$_2$/Pd/charcoal	58JA3932
2-NH$_2$-6-CN	2-NH$_2$-6-CONH$_2$	NaOH/H$_2$O$_2$	63JMC289
2-NH$_2$-6-CN	2-NH$_2$-6-CSNH$_2$	H$_2$S/NH$_3$/EtOH	64USP3128274
8-CN-caffeine	8-CH$_2$NH$_2$-caffeine	Na/EtOH	1895MI40900
8-CN-isocaffeine	8-CH$_2$NH$_2$-isocaffeine	H$_2$/Ni	63ZOB1650
8-CN-caffeine	8-COMe-caffeine	MeMgI	56AP453
8-CN-caffeine	8-COEt-caffeine	EtMgI	56AP453

Table 29 Reactions of some Purinecarboxylic Acids

Purine	Product	Conditions	Ref.
6-CO$_2$H	Purine	Heat	56JA3511
8-CO$_2$H	Purine	Heat	60JCS4705
8-CO$_2$H-caffeine	Caffeine	Hot water	48ZOB2129
3-Me-8-CO$_2$H-xanthine	3-Me-8-CO$_2$Et-xanthine	HCl/EtOH	23LA(432)266
8-CO$_2$H-caffeine	8-CO$_2$Me-caffeine	Ag salt/MeI	1895MI40900
6-CO$_2$H	6-COCl	PCl$_5$	58JA3932
8-CO$_2$H-theophylline	8-COCl-theophylline	SOCl$_2$	62CB414

4.09.6.7 Purinecarboxylic Acid Esters

Several purine esters have been prepared, especially 6- and 8-derivatives. They readily aminate to amides and may be reduced with lithium aluminum hydride to the corresponding alcohols. Table 30 lists several reactions of typical esters.

Table 30 Reactions of some Purine Esters

Purine	Product	Conditions	Ref.
9-Me-8-CO$_2$Et-xanthine	9-Me-8-CONH$_2$-xanthine	NH$_3$	64ZOB1137
6-(CH$_2$)$_2$CO$_2$Et	6-(CH$_2$)$_3$OH	LiAlH$_4$	61LA(649)131
8-CO$_2$Et-theobromine	8-CH$_2$OH-theobromine	LiAlH$_4$	66ZOB816
8-COSBu-caffeine	8-CH$_2$OH-caffeine	Ni/H$_2$	52ZOB2225
9-Me-6-CCl(CO$_2$Et)$_2$	6-CH$_2$Cl-9-Me	HCl/EtOH	63ZOB3342
8-CCl(CO$_2$Et)$_2$-caffeine	8-CH$_2$OH-caffeine	HCl/H$_2$O	61ZOB2645

4.09.6.8 Purinecarboxamides

A few carboxamides have been prepared. They may be hydrolyzed to carboxylic acids, dehydrated to nitriles, and homoamides also undergo the Hoffman reaction with alkaline sodium hypobromite to produce aminoalkylpurines. Table 31 lists reaction products of some typical purinecarboxamides.

4.09.6.9 Nitropurines

The nitropurines may be reduced in various ways to provide corresponding aminopurines. Table 32 records some reaction conditions for this process. In addition, nitro groups may

Table 31 Reaction of Some Purinecarboxamides

Purine	Product	Conditions	Ref.
6-CONH$_2$	6-CO$_2$H	2N NaOH	56JA3511
8-CONH$_2$-caffeine	8-CO$_2$H-caffeine	N$_2$O$_3$, 50% H$_2$SO$_4$	1895MI40900
8-CONH$_2$-caffeine	8-CN-caffeine	P$_2$O$_5$	1895MI40900
8-CONH$_2$-caffeine	8-CN-caffeine	POCl$_3$	56AP453
6-CH$_2$CONH$_2$	6-CH$_2$NH$_2$	NaOBr/OH$^-$	61LA(649)131
6-CONH$_2$	6-C(=NH)NH$_2$	NH$_2$NH$_2$	58JA3932
9-Me-6-CONH$_2$	9-Me-6-C(=NH)NH$_2$	NH$_2$NH$_2$	63JMC289

Table 32 Reactions of Some Nitropurines

Purine	Product	Conditions	Ref.
2-Nitroinosine	Guanosine	NaHSO$_3$	64JA2948
8-Nitrotheophylline	8-Aminotheophylline	SnCl$_2$	57AC(R)362
8-Nitrotheophylline	8-Aminotheophylline	NaSH	60JA3773
8-Nitrotheobromine	8-Aminotheobromine	NH$_4$SH	1897CB2584
8-Nitrohypoxanthine	8-Aminohypoxanthine	NaHSO$_3$	60JA3773

occasionally undergo replacement by nucleophiles. Thus 8-nitrotheophylline gave the 8-chloro and 8-bromo derivatives with hydrochloric or hydrobromic acids, respectively ⟨57AC(R)366⟩.

4.09.6.10 Aminopurines

4.09.6.10.1 Introduction

The biological importance of adenine and guanine and their derivatives has ensured that there is a very substantial literature concerned with the aminopurines, and it will only be possible to give a brief account of reactions undergone by the amino group in such compounds. Table 33 gives a representative list of typical reaction products but it is of course far from complete. For more complete details see ⟨71HC(24-2)1⟩. The amino group in aminopurines in the main undergoes most of the reactions of typical aromatic amines.

4.09.6.10.2 Acylation

Acetylation of adenine with acetic anhydride produced the N^6-acetyl and the N^6, N^9-diacetyl derivatives, the N-9 group being readily removed by alcoholysis or by alkali. Other aminopurines similarly may be acylated both on the exocyclic amino group and on ring atoms, the ring acyl groups being labile to both alcoholysis and alkaline hydrolysis. In contrast, adenine with p-nitrobenzenesulfonyl chloride gave a ring sulfonylated derivative, probably the N-9 derivative ⟨66JMC373⟩ which is usual in such acylations. On the other hand, 2-amino-9-methylpurine provides an exceptional case in which sulfonylation with p-acetamidobenzenesulfonyl chloride in pyridine produced the 2-sulfonamido derivative ⟨66JMC373⟩.

4.09.6.10.3 Alkylation

Alkylation of aminopurines normally leads almost exclusively to ring-substituted products (see Section 4.09.5.2.2(iii)). However, alkylation of adenine with formaldehyde probably produces the N^6-hydroxymethyl derivative (**183**) ⟨64JCS792⟩, and the crystalline diadeninyl derivative (**184**) is also obtained from a mixture of equimolar amounts of adenine and formaldehyde in water, especially in acid solution ⟨62MI40902⟩.

Table 33 Reactions of Some Aminopurines

Purine	Product	Conditions	Ref.
Adenine	N^6-Formyladenine	HCO_2H/Ac_2O	80JCS(P1)2728
Adenine	N^6-Acetyladenine	Ac_2O	51JA1650
Adenine	N^6-Benzoyladenine	PhCOCl	57JOC568
N^6-Acetyladenine	N^6-Ethyladenine	$LiAlH_4$	56JOC1276
N^6-Benzoyladenine	N^6-Benzyladenine	$LiAlH_4$	65BSF2617
Adenine	9-Allyladenine	$CH_2=CHCH_2Cl/K_2CO_3/DMA$	65JOC3235
Adenine	9-Benzyladenine	$PhCH_2Cl/K_2CO_3$	64JHC115
Adenine	3-, 9- and 1-Methyladenine	Me_2SO_4/DMF	62B558
Adenine	3- and 9-Alkyladenine	Alkylation in neutral/alkaline solution	64JA5320
Sodioadenine	3-(3-Methylbut-2-enyl)adenine (triacanthine) plus trace of 1-isomer	3-Methylbut-2-enyl bromide in EtOH	62JA148
1-Methyladenine	1,9-Dimethyladenine	Methyl p-toluenesulfonate/DMA	64B494
3-Methyladenine	3,7-Dimethyladenine	MeI/KOH	64B494
N^6-Benzoyladenine	3- and 9-Benzyl-N^6-benzoyladenine	$PhCH_2Br/DMF/80\,°C$	65TL2059
Adenine	8-Bromoadenine	Br_2	1892ZPC329
Adenine	Hypoxanthine	HNO_2	1886ZPC(10)258
Sodioadenine	9-Chloromercurioadenine	$HgCl_2$	66JOC1413
Adenine	5(4)-Aminoimidazole-4(5)-carboxamide	6N HCl/150 °C	49JA3973
2-Aminopurine	2-Fluoropurine	HNO_2/HCl	60JA463
2-Aminopurine	2-Chloropurine	HNO_2/HCl	66JOC3258
Adenosine	6-Fluoroadenosine	HNO_2/HCl	69JOC1396
8-Aminocaffeine	8-Diacetylaminocaffeine	Ac_2O	59JOC1813
2,6-Diaminopurine	$N^2,N^6,9$-Triacetylpurine	Ac_2O	57JOC959
Guanine	7,9-Dimethylguanine	$Me_2SO_4/90\,°C/DMA$	62JA1914
Guanine	9-, 7- and 3-Methylguanine	MeCl/NaOH/70 °C	66B3007
Guanine	9-Alkylguanines	Tetraalkyl ammonium hydroxide/fusion	63JOC2087
1-Alkylguanine	1,7-Dialkylguanines	—	45JCS751
N^2-Acetylguanine	7- and 9-Benzyl-N^2-acetylguanines	$PhCH_2Br/90\,°C/7\,h/DMF$	67CPB1066
Guanine	8-Bromoguanine	Br_2	1883LA(221)336
Guanine	8-Arylazoguanines	Aryldiazonium salts	39JA351
Guanine	Xanthine	HNO_2	1861LA(118)151
Isoguanine	Xanthine	25% HCl	48JA3109
Guanine	Guanidine	$HCl/KClO_4$	58LA(129)141
Guanine	5(4)-Aminoimidazole-4(5)carboxamide	3.4N HCl/160 °C	36MI40900, 36MI40901

Alkylation at the 6-NH_2 group similarly may also occur by Mannich reactions with an amine and formaldehyde. Thus morpholine and adenine afforded the diaminomethylated derivative (**185**) ⟨67AP1000⟩.

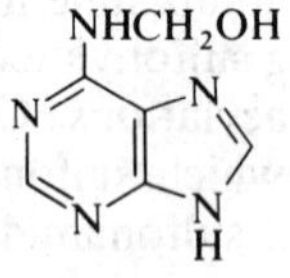

(**183**) 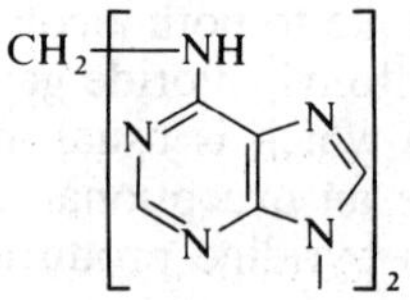   (**184**) 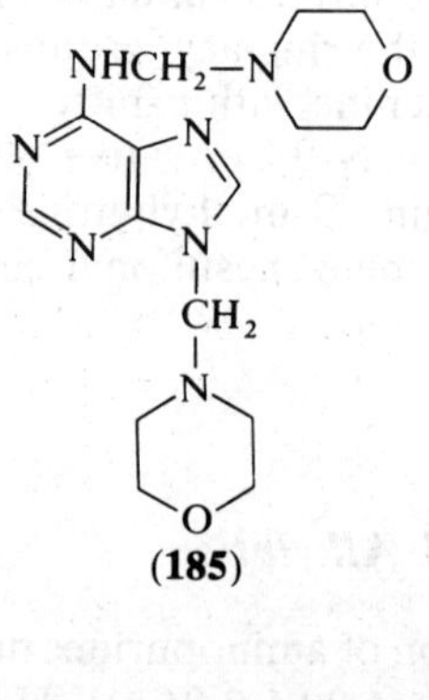   (**185**)

4.09.6.10.4 Displacement reactions

The amino group in adenine and substituted adenines is readily transformed into hydroxyl by nitrous acid to afford hypoxanthine and its derivatives, and guanine similarly produces xanthine, whereas isoguanine does not react but can be hydrolyzed to xanthine by long heating with 25% hydrochloric acid. Amino derivatives resistant to nitrous acid may often be deaminated with nitrosyl chloride and pyridine in DMF ⟨72JOC3985⟩.

Replacement of the amino group in aminopurines by halogens has been achieved in several ways. Thus 2,6-diamino-9-[(2,3,5-tri-O-benzyl)-D-arabinofuranosyl]purine with fluoroboric acid and sodium nitrite in THF at −10 °C gave a 48% yield of the 2-fluoroadenosine derivative (**186**; R = 2,3,5-tri-O-benzyl-β-D-arabinofuranosyl) ⟨79JHC157⟩, and similar use of 2,6-diamino-9-β-D-ribofuranosylpurine in hydrochloric acid solution gave the 2,6-dichloropurine nucleoside (**187**) ⟨60JA463⟩.

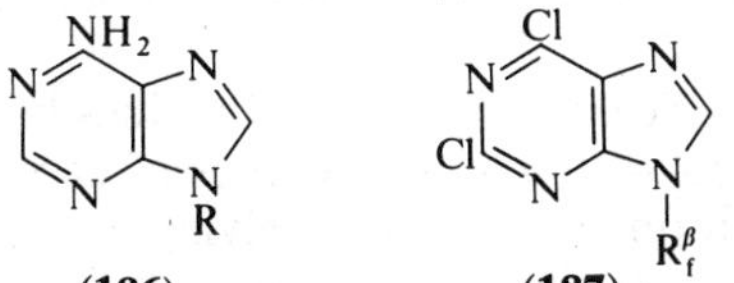

(**186**) (**187**)

Recently several 9-substituted adenines including 2′,3′,5′-tri-O-acetyladenosine, with pentyl nitrite in non-aqueous hydrogen atom donating solvents (especially THF) and photolysis with a 200 W light source, were found to produce good yields (40–70%) of 9-substituted 6H-purines. The reaction is considered to proceed via homolysis of an intermediate diazonium ion with formation of a purinyl radical which then extracts an appropriate atom from the solvent (Scheme 35). In halogen-donating solvents, especially carbon tetrachloride, bromoform or methylene iodide, 6-chloro-, 6-bromo- and 6-iodo-purines, respectively, are similarly produced ⟨80JOC3969⟩.

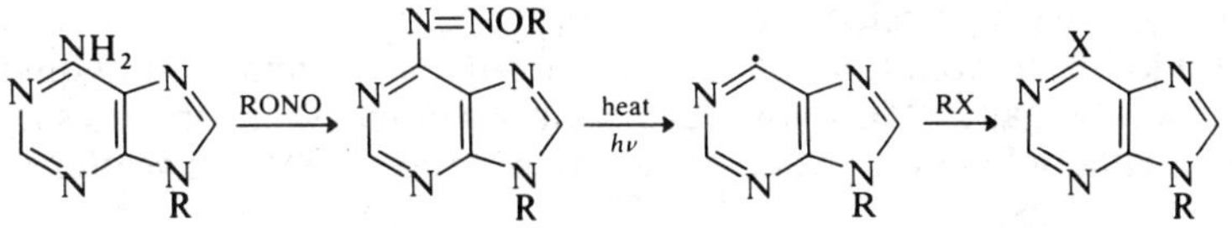

X = Cl, Br, I, H

Scheme 35

In contrast to 6-aminopurines, several 8-aminopurines form stable 8-diazopurines with nitrous acid, especially when dipolar forms are possible. Thus 8-aminoxanthine and nitrous acid at 5–10 °C gave the 8-diazo derivative (**188**) ⟨60JA3773⟩ and similar products are produced from hypoxanthine, isoguanine and theophylline. In contrast 8-amino-9-methylxanthine failed to produce a stable diazo derivative and this tends to favour the particular ionic structure (**188**) which is further confirmed by the absence of imidazole imino groups in the IR spectra of these compounds.

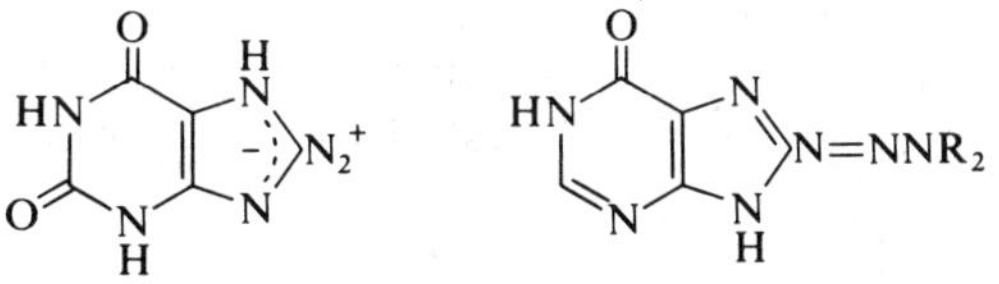

(**188**) (**189**)

The diazo group may be replaced by hydroxyl (water) and thiol (hot potassium hydrosulfide) groups ⟨60JA3773⟩ and also undergoes coupling reactions with amines to produce triazenes including (**189**) from 8-diazohypoxanthine and dialkylamines ⟨61JA1113⟩. Adenine and arenediazonium chlorides also produce triazenes which when heated at 60–90 °C produce 8-aryladenines by an intermediate free (aryl) radical process (see Section 4.09.5.4.3).

4.09.6.11 Hydrazinopurines

Several 2-, 6- and 8-hydrazinopurines have been prepared. The hydrazino group may be replaced by hydrogen by aerial oxidation in alkaline solution or with iron(III) salts. On the other hand, the small yield (6%) of 6-chloropurine first recorded from 6-hydrazinopurine and iron(III) chloride ⟨58JA3932⟩ has been markedly increased (67%) ⟨B-78MI40901, p. 19⟩ (see

Section 4.09.7.7.8), and 8-chlorocaffeine is similarly obtained from 8-hydrazinocaffeine ⟨55CB1932⟩. 6-Hydrazinopurine is also converted into 6-thioxo-1,6-dihydropurine in 60% yield when heated with thiolacetic acid. The hydrazinopurines give azidopurines with nitrous acid and undergo other reactions typical of arylhydrazines, including formation of exocyclic pyrazole and triazole rings.

4.09.6.12 Purine N-Oxides

The chemistry of the heterocyclic N-oxides, including purine N-oxides, has been reviewed extensively ⟨B-71MI40902⟩.

The N-oxides are readily converted into parent purines by reduction using platinum(IV) oxide or Raney nickel catalysts with varying degrees of ease ⟨58JA2755, 69JMC717, 69BCJ750, 69JOC978⟩, or by deoxygenation processes involving heat, as in the conversion of 8-phenyltheophylline 7-oxide into 8-phenyltheophylline by heating in DMF ⟨64JA4721⟩, irradiation ⟨64B880⟩ or by the action of phosphorus halides or pentasulfide ⟨65JOC408⟩ or amines as in the conversion of 6-chloropurine 3-oxide into adenine derivatives ⟨69JOC2157⟩. Enzymic reduction using xanthine oxidase and an electron donor has also been used to reduce purine N-oxides ⟨69JBC2498⟩.

Alkylation of many N-oxides affords corresponding N-alkoxypurines although xanthine 1-oxide produced the 1-hydroxybetaine, and the 3-oxide behaves in a similar manner. Alkylation often proceeds better in the presence of hydrogen peroxide ⟨74CPB2466⟩ as in the conversion of adenine 1-oxide into 1-methoxy-9-methyladenine with methyl iodide. Similarly alkylation of adenosine 1-oxide and its 2',3'-O-isopropylidene derivative afforded the corresponding 1-alkoxy nucleosides ⟨73CPB1676⟩.

1-Alkoxy-9-alkyl-6-iminopurines undergo Dimroth rearrangements (involving ring opening) in aqueous solution at 100 °C to produce 6-alkoxyamino-9-alkylpurines ⟨71T2415⟩. Also 7-alkoxy-8-methyltheophyllines when heated produced 8-alkoxycaffeine derivatives (Scheme 36) ⟨66LA(693)233⟩.

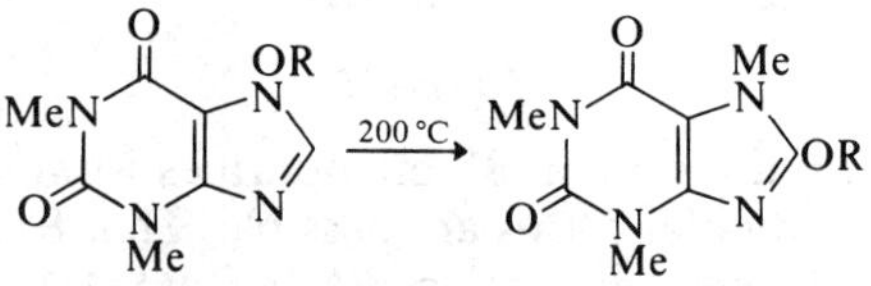

Scheme 36

Acylation of N-oxides usually produces a reactive N,O-acyl derivative. Adenine 1-oxide, however, with acetic anhydride undergoes ring opening to afford an oxadiazole (Scheme 37) ⟨B-72MI40901, p. 550⟩.

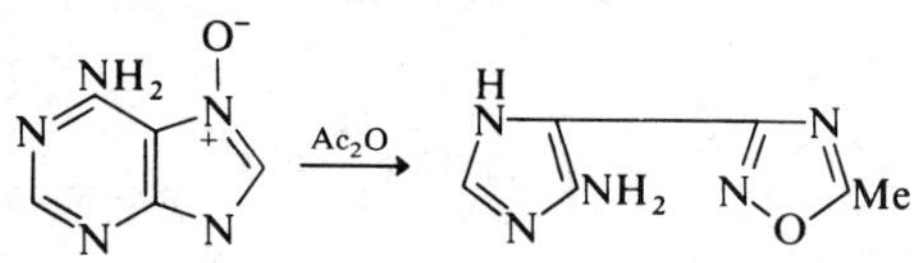

Scheme 37

In addition, migration of methyl groups in unsubstituted methylated N-oxides such as 1-methoxyadenine has been observed (Scheme 38) ⟨71CPB1731⟩ and kinetic studies have been carried out ⟨72CPB958, 75CPB2643⟩.

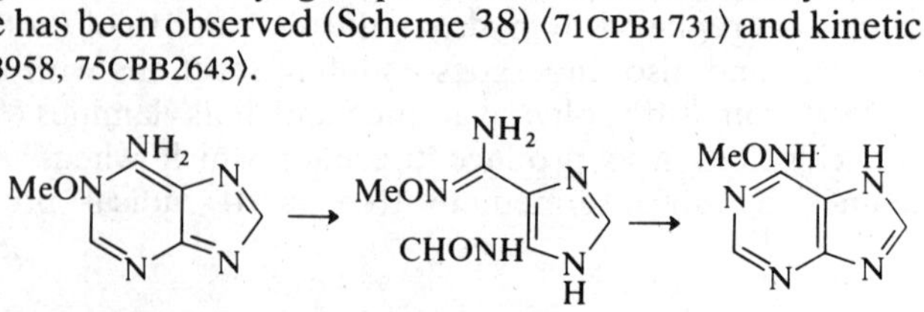

Scheme 38

Xanthine 3-oxide on the other hand is readily converted into 3-acetoxyxanthine (**190**) which reacts under very mild conditions with a wide variety of inorganic or organic nucleophiles to produce 8-substituted xanthines (Scheme 39) in high yields ⟨B-72MI40901, p. 550⟩. The mechanism has been extensively investigated by the same workers and it is proposed that intermediate nitrenium (**191**) and carbo cations (**192**) are involved (Scheme 40).

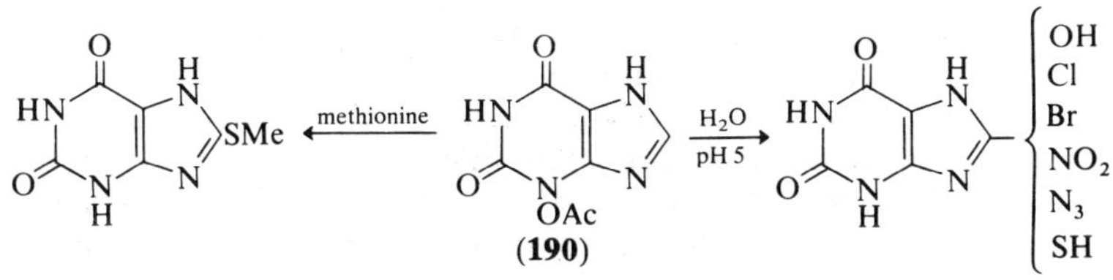

Scheme 39

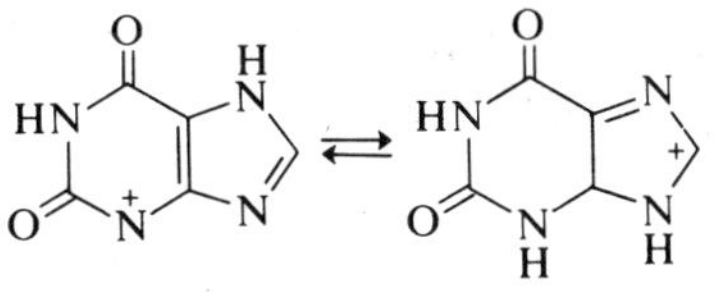

Scheme 40

Table 34 records some reactions of purine *N*-oxides.

Table 34 Reactions of Some Purine *N*-Oxides and Derivatives

Purine	Product	Conditions	
Adenine 1-oxide	Adenine (94%)	Ni/H$_2$/aq. NH$_3$	58JA2755
Adenine 1-oxide	Adenine + isoguanine	UV light	64B880
Adenosine 1-oxide	Adenosine + isoguanosine	UV light	64B880
Adenine 1-oxide	1-Methoxyadenine	MeI/dimethylacetamide	65MI40904
Adenine 1-oxide	1-Methoxy-9-methyladenine	MeI/H$_2$O$_2$	73CPB1676
2-Methyladenine 1-oxide	1-Acetoxy-2-methyladenine	Ac$_2$O	60JA1148
6-Methylpurine 1-oxide	6-Acetoxymethylpurine	Ac$_2$O	62JOC567
Adenine 3-oxide	Isoguanine	Ac$_2$O/(CF$_3$CO)$_2$O	69JOC981
Guanine 3-oxide	6-Amino-2,8-dioxo-2,3,7,8-tetrahydropurine	Ac$_2$O	69JOC981
Hypoxanthine 3-oxide	Xanthine	Ac$_2$O/(CF$_3$CO)$_2$O	69BCJ750
Hypoxanthine 3-oxide	Guanine	Pyridine/AcOH (several hours) + aq. NaOH	69BCJ750
Hypoxanthine 3-oxide	Hypoxanthine	Ni/H$_2$/aq. NH$_3$(22 h)	69BCJ750
Hypoxanthine 3-oxide	2,6-Dichloropurine + a little 6,8-dichloropurine	POCl$_3$/heat	69BCJ750
6-Bromo-, 6-iodo-, 6-methoxy-, 6-sulfo-purines, 3-oxides	6-Bromo-, 6-iodo-, 6-methoxy, 6-sulfo-purines	Ni/H$_2$/aq. NH$_3$/1 h	69JOC2157
6-Chloropurine 3-oxide	6-Chloropurine	Ni/H$_2$/72 h	69JOC2157
6-Methoxypurine 3-oxide	6-Methyl-2-oxo-2,3-dihydropurine	Ac$_2$O/(CF$_3$CO)$_2$O	69JOC981
6-Methoxypurine 3-oxide	2-Thiohypoxanthine	MeCOSH	69BCJ750
Xanthine 3-oxide	Uric acid	Ac$_2$O	69JOC981
Xanthine 3-oxide	8-Thiohypoxanthine (88%)	MeCOSH (4 h reflux)	69BCJ750
Xanthine 3-oxide	8-Aminoxanthine	Heat/aq. NaOH	69TL785
7,9-Dimethylxanthine 1- and 3-oxides	7,9-Dimethylxanthine	Ni/H$_2$	69JOC978
Xanthine 7-oxide[a]	Xanthine	Ni/H$_2$/NaOH (slow)	65MI40904
Theophylline 7-oxide	8-Acetoxytheophylline	Ac$_2$O	66LA(693)233
Theophylline 7-oxide	1,3-Dimethyluric acid	Heat	66LA(691)142
8-Phenyltheophylline 7-oxide	8-Phenyltheophylline	Heat in DMF	64JA4721
8-Methyltheophylline 7-oxide	8-Methyltheophylline	Ni/H$_2$/EtOH	65MI40904
3-Acetoxyxanthine	8-Substituted xanthines	Nucleophiles/H$_2$O pH ~5	B-72MI40902, p. 550
8-Oxo-7,8-dihydropurine	2,8-Dioxo + 6,8-dioxo-dihydropurines	Acetic anhydride	62JOC567

[a] The 1-oxides are very difficult to reduce.

4.09.6.13 Hydroxyaminopurines

Several 2-, 6- and 8-hydroxyaminopurines have been prepared and their chemistry reviewed by the most active workers in this field ⟨B-72MI40901, p. 478⟩. Hydrogenation of hydroxyaminopurines affords aminopurines. Thus 6-hydroxyaminopurine with a palladium catalyst gave adenine ⟨58JA3932⟩ and 2,6-dihydroxyaminopurine afforded 2,6-diaminopurine ⟨68JMC52⟩. Acid hydrolysis of hydroxyaminopurines tends to produce oxopurines, an example includes 6-hydroxyaminopurine which with nitric acid gives hypoxanthine ⟨58JA3932⟩ whereas in TFA containing hydrogen peroxide 2-amino-6-hydroxyamino- and 6-hydroxyamino-2-oxo-2,3-dihydropurines furnish guanine (68%) and xanthine (91%), respectively ⟨66B3057⟩. In aerated alkaline solution however, 6-hydroxyaminopurine produced 6,6'-azoxypurine ⟨B-57MI40900, p. 3⟩. Similarly 2- and 8-hydroxyaminopurines gave 2,2'- and 8,8'-azoxypurines and a 6-hydroxyaminopurine N-oxide was instantly converted by alkali into 6,6'-azoxypurine 3,3'-dioxide ⟨69JMC717⟩ or by hydrogenation with a Raney nickel catalyst into the 6,6'-bis(adenine) (**193**).

(**193**)

With thioacetic acid 6-hydroxyaminopurine afforded 6-thioxo-1,6-dihydropurine ⟨64JOC3209⟩ and oxidation of 6-hydroxyaminopurine with manganese dioxide afforded 6-nitrosopurine ⟨66MI40901⟩.

Table 35 collects data on some of the reaction products of various substituted hydroxyaminopurines.

Table 35 Reactions of Some Hydroxyaminopurines

Hydroxyamino-purine	Product	Conditions	Ref.
6-NHOH	Adenine	Pd/H$_2$	58JA3932
2,6-Di-NHOH	2,6-Di-NH$_2$	Pd/H$_2$	68JMC52
6-NHOH	6-N(NO)OH	HNO$_2$	70JHC75
6-NHOH	6-SH	MeCOSH	64JOC3209
6-NHOH	6,6'-Azoxy	aq. NaOH	B-72MI40902, p. 478
2-NHOH	2,2'-Azoxy	aq. NaOH	B-72MI40902, p. 478
8-NHOH	8,8'-Azoxy	aq. NaOH	B-72MI40902, p. 478
2-F-6-NHOH	2-F-adenine	Ni/H$_2$	B-72MI40902, p. 478
2-F-6-NHOH-9-β-D-ribofuranosyl	2-F-adenosine	Ni/H$_2$	B-72MI40902, p. 478
2-NHOH	2-NH$_2$	Ni/H$_2$	B-72MI40902, p. 478
2-NHOH-6-NH$_2$	2,6-Di-NH$_2$	Ni/H$_2$	B-72MI40902, p. 478
2-F-6-NHOMe	2-F-adenine	Ni/H$_2$	B-72MI40902, p. 478

4.09.6.14 Oxopurines

4.09.6.14.1 Introduction

The main reactions associated with the oxopurines, especially the enolizable derivatives, involve their conversion into halogeno-, especially chloro-, and thioxo-purines.

4.09.6.14.2 Oxo- to halogeno-purines

The main agent used for the conversion of oxopurine into chloropurines is phosphoryl chloride, either alone or with added water (which may produce pyrophosphoryl chloride), phosphorus pentachloride, or with bases, usually tertiary amines and especially N,N-dimethyl- or diethyl-aniline. The use of bases, especially aliphatic amines, may however lead to introduction of dialkyamino groups into the purine or formation of other byproducts ⟨74RCR1089⟩. Other reagents which have been used include (a) thionyl chloride in DMF (Vilsmeier–Haak reagents) which converts 2',3',5'-tri-O-benzoylinosine into the 6-chloropurine nucleoside ⟨64CPB639⟩; and (b) chloromethylenedimethylammonium chloride in chloroform.

Alkylated oxopurines in which no enolization is possible may also react but usually with elimination of one or more alkyl groups; simultaneous halogenation at an unsubstituted

8-position may also occur. Thus theobromine with phosphoryl chloride and dimethylaniline furnished 7-methyl-2,6-dichloropurine ⟨60ZOB327⟩ whereas when a mixture of phosphoryl chloride and phosphorus pentachloride was used, 7-methyl-2,6,8-trichloropurine was produced ⟨1897CB2400, 1895CB2480⟩. Similarly caffeine with the last mixture at 180 °C gave the same trichloropurine. Evidence for the intermediate formation of a dichlorodihydropurine in these reactions has been obtained. For example, 1,9-dimethylxanthine after treatment with phosphoryl chloride was condensed with ammonia to afford the imine (194) ⟨64ZOB2472⟩.

(194)

Aminooxopurines may also produce corresponding halogeno derivatives although guanine is resistant to chlorination. Thus 2-amino-6,8-dioxo-1,6,7,8-tetrahydropurine with phosphoryl chloride and phosphorus pentachloride afforded 2-amino-6-chloro-8-oxo-7,8-dihydropurine ⟨1898CB2619⟩.

Bromopurines have also been prepared from oxopurines and phosphoryl bromide. In this way 6-bromopurine was produced in 40% yield from hypoxanthine ⟨56JA3508⟩ but xanthine only furnished a low (10%) yield of 2,6-dibromopurine.

4.09.6.14.3 *Thiation*

Oxopurines, especially 6-oxo derivatives, are converted into corresponding thioxo derivatives with phosphorus pentasulfide in hot pyridine over a few hours, the order of displacement following the normal order of nucleophilic substitution. Thus hypoxanthine gave 6-thioxo-1,6-dihydropurine in 73% yield ⟨61JA4038⟩. The 2- and 8-oxopurines are less reactive than the 6-oxo derivatives. Xanthine and the isomeric 6,8-dioxo-1,6,7,8-tetrahydropurine with phosphorus pentasulfide give 2- or 8-oxo-6-thioxo-1,6,7,8-tetrahydropurines, respectively ⟨54JA5633, 61JOC1660⟩ and even after prolonged heating with phosphorus pentasulfide in pyridine, uric acid could only produce 6-thiouric acid ⟨59JA3039⟩. Thiation to produce polythioxopurines is possible if the starting materials have thioxo groups in the difficult positions (prepared by other means—see Section 4.09.6.17). Thus 2-thioxo-2,3-dihydrohypoxanthine produced the corresponding 2,6-dithioxopurine ⟨54JA5633⟩, and 6,8-dioxo-2-thioxo-1,2,3,6,7,8-hexahydropurine afforded trithiouric acid ⟨59JA5997⟩. Aminooxopurines may also be thiated. Guanine for example furnished 2-amino-6-thioxo-1,6-dihydropurine ⟨60JA2633⟩ after long reflux in pyridine.

It is not essential to have enolizable oxy groups for thiation to succeed. Thus theophylline and theobromine ⟨62JCS1863⟩ furnished the corresponding 6-thioxo derivatives after 8 and 6 h heating in pyridine, respectively. Caffeine failed to react under these conditions, but produced 6-thiocaffeine with phosphorus trisulfide in kerosine at 150 °C ⟨48ZOB2116⟩ and at higher temperatures (210 °C) 2,6-dithiocaffeine was obtained.

Purine nucleosides also readily undergo thiation. Thus 2′,3′,5′-tri-O-benzoylguanosine gave the 6-thioguanosine derivative ⟨58JA1669⟩.

4.09.6.14.4 *Miscellaneous displacement reactions*

Examples of amination of oxopurines to aminopurines *via* silyl derivatives include conversion of inosine into adenosine with catalysis by mercury ions ⟨76LA745⟩, and hypoxanthines have been converted directly into adenines with phosphoramides at 225–230 °C ⟨69IZV655⟩.

4.09.6.15 Alkoxypurines

Unlike their thio analogs, the alkoxypurines have had relatively little application in purine synthesis perhaps because their preparation is essentially limited to alkoxide displacement of halogens in halogenopurines.

They are relatively stable to alkalis but undergo ready hydrolysis with aqueous acids. Thus 6-methoxy-3-methylpurine furnished 3-methylhypoxanthine when set aside with 5N hydrochloric acid over 12 h at 20 °C ⟨66JCS(C)10⟩. More vigorous conditions (hydrochloric acid at 100 °C for 10 min) converted 8-ethoxycaffeine into 2,3,7-trimethyluric acid ⟨1882CB253⟩, but under these conditions ring opening may also occur to give imidazoles. Thus acid hydrolysis of 6-methoxy-9-methylpurine gave a mixture of products including 5-amino-1-methylimidazole-4-carboxamide (**196**; Scheme 41) ⟨74JCS(P1)1284⟩.

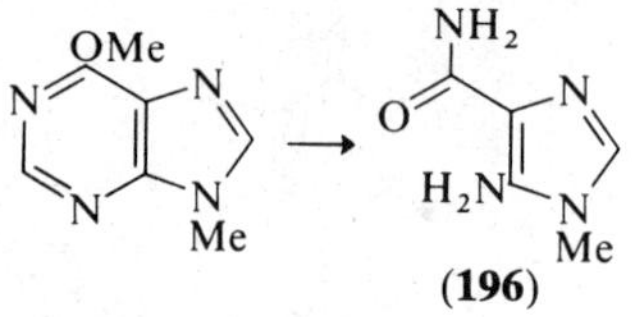

Scheme 41

It has sometimes proved convenient to fix oxopurines in their lactim forms by O-benzylation. The benzyl group in such derivatives may be removed by hydrogenolysis with palladium catalysts. In this way 6-amino-2-benzyloxy-9-methylpurine furnished 9-methylisoguanine with palladium and hydrogen in dilute acid ⟨49JCS2490⟩, and similarly in ethanol solution 6-benzyloxy-2-fluoropurine produced 2-fluorohypoxanthine ⟨67JOC3258⟩. The technique has been applied to nucleosides, including 9-(β-D-ribofuranosyl)-6-benzyloxy-2-dimethylaminoinosine ⟨65JA3752⟩.

Rearrangement of alkoxypurines to their N-alkyl isomers readily occurs on heating, as with 8-methoxytheobromine which gave 2,7,9-trimethyluric acid when heated to 290 °C ⟨30CB2876⟩.

Many other examples exist ⟨71HC(24-2)1⟩ including analogous rearrangements of dialkoxypurines such as 2,6-dimethoxy-7-methylpurine which when heated at 210 °C for 15 min afforded caffeine ⟨35JCS955⟩. The same authors were unable to isomerize a 2,6-dipropoxypurine.

Like their alkylthio analogs the alkoxypurines react with ammonia and amines to produce aminopurines but few examples have been reported. However 2-chloro-6-methoxy-9-β-D-ribofuranosylpurine gave 2-chloroadenosine with methanolic ammonia at 80 °C ⟨58JA3738⟩.

N-Alkoxypurines have also been prepared and their reactions are recorded in Section 4.09.6.1.2 under N-oxides.

4.09.6.16 Thioxopurines

4.09.6.16.1 Dethiation

The thioxopurines undergo a wider variety of useful reactions than the corresponding oxo derivatives. A major reaction is dethiation or desulfurization which enables replacement of SH by hydrogen to be carried out. The dethiation step is commonly achieved by reaction with Raney nickel. In this manner 6-thioxo-1,6-dihydropurine and 2,6-dithioxo-1,2,3,6-tetrahydropurine may be converted into purine with nickel in aqueous ammonia ⟨54JA5633⟩. Table 36 outlines some typical applications of the dethiation process. Dethiation may also be achieved by the use of nitrous acid especially with 2- or 8-thio derivatives ⟨21LA(423)200⟩, or by use of various oxidizing agents such as hydrogen peroxide, when the sulfinic acid is an intermediate ⟨76MI40902⟩. Even dilute nitric acid will sometimes succeed, as in the conversion of 2-thioxanthine into hypoxanthine ⟨06AP(244)11⟩. Similarly acidified hydrogen peroxide converted 2-thioadenine into adenine ⟨06AP(244)11⟩. When enolization of the thioxo group is not possible because of ring substitution, dethiation may give dihydro derivatives, as with 6-thiotheophylline which with nickel gave the 1,3-dimethylpurine (**195**).

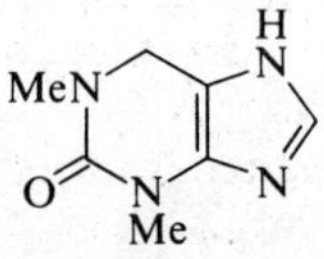

(**195**)

The Eschenmoser sulfur extrusion process has been applied to purines. Dethiation by sulfur extrusion of 6-phenacylthiopurine nucleosides using triphenylphosphine and

Table 36 Reactions of Some Thioxopurines and Derivatives

Purine	Product	Conditions	Ref.
6-Thioxo-1,6-dihydropurine	Purine	Raney Ni/aq. NH$_3$	54JA5633
1-Methyl-2-thioxo-2,3-dihydropurine	6-Methylpurine	Heat, 2 h/Raney Ni/aq. NH$_3$	64JOC3209
1-Methyl-6-thioxo-1,6-dihydropurine	1-Methylpurine	Hot H$_2$O/Ni	62JOC990
6-Thiotheophylline	1,3-Dimethyl-2-oxo-1,2,3,6-tetrahydropurine	Raney Ni, 1 h/heat	B-65MI40902, p. 285
6-Thiocaffeine	1,3,7-Trimethyl-2-oxo-1,2,3,4-tetrahydropurine	Raney Ni, 1 h/heat	66BSF3934
2-Thioxo-2,3-dihydrohypoxanthine	Hypoxanthine	aq. HNO$_3$	06AP(244)11
2-Thioxo-2,3-dihydroadenine	Adenine	H$_2$O$_2$ 20%/H$_2$SO$_4$	06AP(244)11

potassium *t*-butoxide, has led to 6-phenacylpurine nucleosides ⟨76AG724⟩ and similarly 6-phenacylthioinosine produced 6-phenacylpurine riboside ⟨80CPB157⟩.

4.09.6.16.2 Alkylation and acylation

Alkylation of thiopurines may produce both *S*- and *N*-alkyl derivatives in contrast to oxopurines where *O*-alkylation is rare (Section 4.09.5.2.2). The *S*-alkyl derivatives are favoured by carrying out the reaction below 40 °C in aqueous or aqueous alcoholic solutions with equimolar equivalents of sodium hydroxide and either an alkyl halide or sulfate. At higher temperatures the *N*-alkylpurine tends to be produced normally by *S–N* rearrangement of the first formed *S*-alkyl derivative ⟨65MI40906⟩.

Interesting results which illustrate *S–N* migration arise when thioxopurines are acylated with alkyl chloroformates. Thus 6-thioxo-1,6-dihydropurine or 8-thioxo-7,8-dihydropurine with alkyl chloroformates produced 9-alkoxycarbonylpurines. On the other hand when the imidazole ring is substituted as with 9-methyl-6-thioxo-1,6-dihydropurine, then *S*-alkylation occurred with ethyl chloroformate to produce the *S*-ethoxycarbonyl derivative ⟨64JMC10⟩. Evidence supporting the suggestion that the *N*-ethoxycarbonyl derivatives are produced by rearrangement of a first formed *S*-acyl derivative comes from formation of 9-acetyl-6-thioxo-1,6-dihydropurine (**196**) from 6-chloropurine and potassium thiolacetate in DMF (Scheme 42) ⟨69JOC973⟩, and successive treatment of 6-thiocyanatopurine (**197**) with sodium ethoxide followed by acid (Scheme 43) gave the 9-ethoxycarbonyl derivative (**198**).

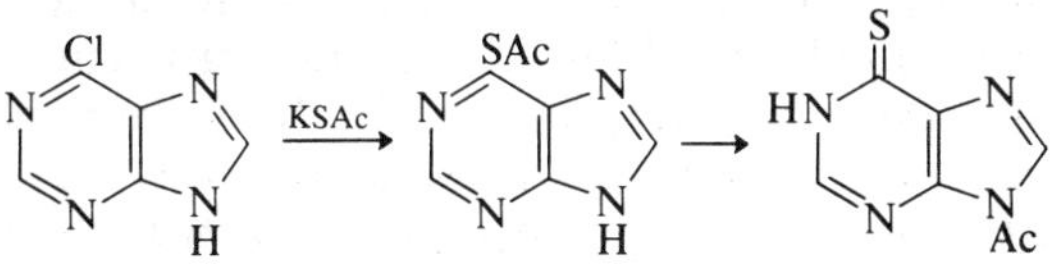

Scheme 42   **Scheme 43**

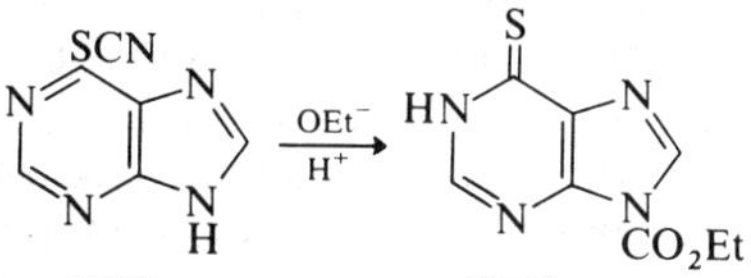

Acetylation of thioxopurines with cyanogen bromides similarly produces thiocyanatopurines. Thus 2-, 6- and 8-thiocyanatopurines may be prepared from the corresponding thioxopurines in sodium hydroxide solution at 0 °C with cyanogen bromide ⟨67CPB909⟩.

4.09.6.16.3 Halogenation

Thioxopurines may often be converted into halogenopurines by reaction with halogens. For example 6-thioxo-1,6-dihydropurine with chlorine in cold ethanol afforded 6-chloropurine. The use of chlorine in methanol or ethanol, especially on a large scale, has been shown to be hazardous and acetonitrile or hydrochloric acid are proposed as alternative and safer solvents ⟨74JHC77⟩. Methylthiopurines are often more suitable in these reactions, and bromopurines may be prepared analogously and with advantage ⟨62JOC986⟩. 8-Iodopurines have occasionally been obtained in a similar manner from 8-thioxopurines and iodine.

The method is especially valuable since it permits preparation of 2- and 8-chloropurines which are not readily obtainable by chlorination of the coresponding oxopurines ⟨B-71MI40901, p. 141⟩. The mildness of the procedure makes it especially useful for analogous syntheses of chloropurine nucleosides. Thus 6-thioxo- or 6-methylthio-purine-9-β-D-ribofuranosides with chlorine in methanol at $-10\,°C$ produced the valuable 6-chloropurine-9-β-D-ribofuranoside in 82% yield ⟨60JA2654⟩.

4.09.6.16.4 Amination

Amination of thioxopurines to aminopurines requires forcing conditions which may bring about ring opening reactions and Dimroth rearrangements. However alkylthiopurines are more reactive and many reactions with amines have been described. The order of replacement is $8>6>2$ as predicted by the Fukui FMO superdelocalizability theory ⟨B-74MI40901⟩. With weak amines such as aniline, mercury(II) chloride is recommended as a catalyst ⟨70IZV420⟩. However alkylthio groups at C-2 normally fail to react ⟨56JA217⟩ although the interesting conversion by aqueous ammonia of 1-methyl-2-methylthio-6-thioxo-1,6-dihydropurine (**199**) into 6-amino-2-methylaminopurine (**200**) is noteworthy ⟨62JOC2478⟩ but probably involves ring opening prior to elimination of methanethiol (Scheme 44).

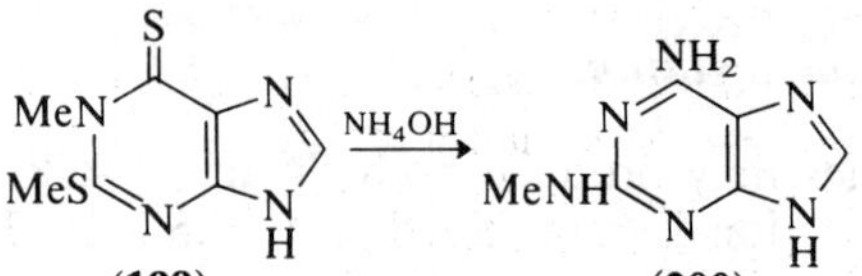

Scheme 44

4.09.6.16.5 Hydrolysis

Thioxo- and S-alkylthio-purines are readily converted into corresponding oxopurines with various reagents including oxidizing agents such as nitric acid (25–50%) which converts 6-thioxo-1,6-dihydropurine into hypoxanthine ⟨65MI40906⟩. More readily hydrolyzed sulfinic acids are likely intermediates in these reactions ⟨74MI40902⟩.

Hydrolysis of alkylthiopurines with acid readily occurs to give oxopurines, although several hours heating may be required, especially with 2-methylthiopurines ⟨62JA3008⟩ which may prove very difficult to hydrolyze when electron donors are present in the ring. 6-Amino-2-methylthiopurine, for example, failed to hydrolyze to any significant extent with 6N hydrochloric acid ⟨49JCS2490⟩. On the other hand 2-methylthiohypoxanthine was readily converted into xanthine ⟨13JBC(14)381⟩ by acid hydrolysis. If an oxidizing agent is included in the acid medium, then the reaction frequently readily proceeds to the oxopurine. Thus 6-amino-2-methylthiopurine in the presence of hydrogen peroxide readily gave isoguanine ⟨59JA2442⟩, again presumably *via* sulfinic acid intermediate. S-Carboxymethyl-purines which are conveniently prepared from thioxopurines in aqueous alkaline solution with chloroacetic acid, have frequently been used as intermediates in the conversion of thioxo- into oxo-purines. In this manner hypoxanthine and 8-oxopurine derivatives may be prepared from 6- or 8-carboxymethylthiopurines, respectively, by acid hydrolysis ⟨58JA6265, 31CB752⟩. However not all such derivatives, *e.g.* 2-carboxymethylthioadenine, undergo ready hydrolysis ⟨48JA3109⟩.

4.09.6.16.6 Oxidation

With appropriate, *i.e.* enolizable, thioxopurines disulfides may be produced, often by treatment with iodine in sodium iodide solution. 6-Thioxo-1,6-dihydropurine is oxidized in this way ⟨61JOC3401⟩. Other reagents include hydrogen peroxide in DMF ⟨66LA(695)77⟩ or butyl nitrite in aqueous methanol ⟨B-71MI40901, p. 291⟩. Mixed purinyl disulfides including *n*-butyl 6-purinyl disulfide (**201**) have been prepared using the reagent (**202**) ⟨70JAP7039709⟩.

Benzyl purinyl disulfide ⟨79CPB2518⟩ has been ribosylated with ribofuranosyl tetra-*O*-acetate and tin(IV) chloride.

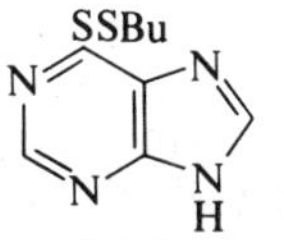

(201)    (202)

Further aerial oxidation of dipurinyl disulfide in alkaline solution gave an excellent (90%) yield of purine-6-sulfinic acid ⟨61JOC3401⟩ and the 2-amino derivative is produced in the same way. More vigorous oxidation of thioxopurines or intermediate sulfinic acids with potassium permanganate in alkaline solution afforded sulfonic acids ⟨61JOC3401⟩. The purinesulfinic acids are readily hydrolyzed to oxopurines including 6-sulfinic acid which immediately gave hypoxanthine with 0.1N hydrochloric acid ⟨61JOC3401⟩, whereas the corresponding sulfonic acid required brief heating to effect hydrolysis. Similar oxidation of methylthiopurines produced methanesulfonyl derivatives which are reported to react with nucleophiles as well as or even better than corresponding chloropurines ⟨76JCS(P2)1176⟩. The methanesulfonyl groups in 2,8-dimethanesulfonyladenosine, for example, with cyanide were replaced to produce 2- and 8-cyanoadenosines ⟨79CPB183⟩. Additionally 6-methanesulfonyl-2′,3′,5′-tri-O-benzoyl-β-D-ribofuranosylpurine reacted smoothly with ethyl acetoacetate and sodium hydride in THF to afford 6-ethoxycarbonylmethylpurine nucleoside. Similar reactions occurred with carbanions derived from diethyl malonate, ethyl cyanoacetate, malononitrile and nitromethane ⟨80CPB150⟩.

The methanesulfonyl derivatives are more usually produced by reaction of appropriate methylthiopurines with chlorine in aqueous or aqueous alcoholic solution at low temperatures, or with N-chlorosuccinimide in methanol–DMF ⟨62CPB1075⟩. Gentle oxidation of methylthiopurines, e.g. 6-methylthiopurine, with restricted amounts of hydrogen peroxide can lead to intermediate methylsulfinylpurines. Thus 2-methylthioadenine gave 6-amino-2-methylsulfinylpurine (203) ⟨63JOC2560⟩.

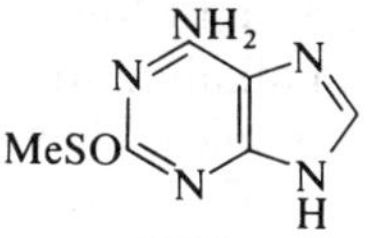

(203)

4.09.6.17 Halogenopurines

Nucleophilic displacement of halogen atoms in the halogenopurines, especially the readily available chloropurines, has provided the major route to a wide variety of substituted purines including nucleosides ⟨B-71MI40901⟩. The reaction of chloropurines with nucleophiles has been found to be second order and bimolecular ⟨65JCS3017, 67JCS(B)954⟩. The ease of replacement of halogen atoms from the 2-, 6- and 8-positions varies according to the state of N-substitution. Thus with unsubstituted 2,6,8-trichloropurine replacement by nucleophiles decreases in the order 6 > 2 > 8 and also in the various monohalogenopurines a similar sequence is imposed. Thus 2-chloropurine is less reactive towards amines than 6-chloropurine ⟨57JA2185⟩. However when the pyrimidine ring is alkylated, nucleophilic displacement occurs in the decreasing order 8 > 6 > 2 ⟨63JOC1662⟩. These differences in reactivity between N-substituted and unsubstituted halogenopurines does not accord with the order predicted for N-substituted and unsubstituted derivatives by the Fukui FMO superdelocalizability theory ⟨B-74MI40901⟩.

The results have been explained by the 'anion theory'. Thus in unsubstituted chloropurines, anion (204) formation is possible with the negative charge located in the imidazole ring so inhibiting attack at C-8 by a nucleophile (Scheme 45). No such anion is possible when the ring is alkylated and attack then tends to occur at the most polarized

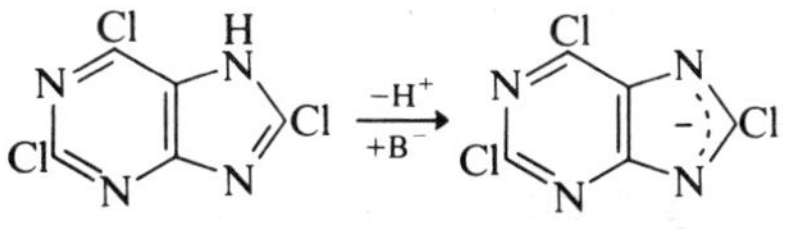

(204)

Scheme 45

carbon–halogen bond, generally C-8. However, apparent exceptions include 2,8-dichloropurine which with bases produced displacement at C-8 rather than C-2 ⟨63CJC1807⟩. It has been suggested ⟨B-71MI40901, p. 153⟩ that this may be due to partial fixation of the C(4)—C(5) double bond, leading to an increased electron density at C-2, possibly as in Scheme 46, involving a mesomeric chlorine atom.

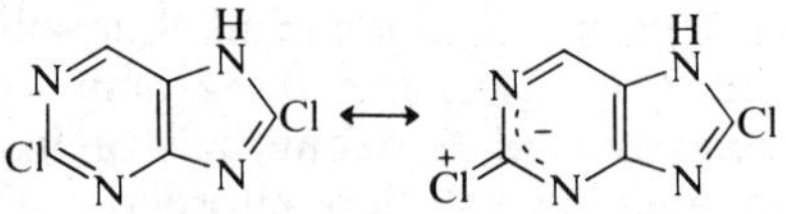

Scheme 46

Conversion of the halogenopurines into protonated forms should lead to a reversal of the substitution order and this is generally but not always the case. Thus reaction of 6,8-dichloropurine ⟨58JA6671⟩ or 2,6,8-trichloropurine ⟨1897CB2220⟩ with hot acid gave 6-chloro- and 2,6-dichloro-8-oxodihydropurines, respectively. These variations in reactivity of the anionic and cationic forms of halogenopurines have frequently been used to direct nucleophiles into specific positions in the purine ring. Acid treatment however may also lead to rearrangements occurring.

The ease of replacement of a halogen atom at a particular position in the ring system also varies, not unexpectedly, according to the degree and type of substitution. Thus whereas 8-bromocaffeine, where no anion is possible, is rapidly hydrolyzed by aqueous alkali, 8-bromoxanthine which can form an anion is stable at 100 °C ⟨1898CB3272⟩. Similarly, N-alkylation increases reactivity towards ammonia and amines. Thus, whereas 6-chloropurine requires a forcing high pressure reaction with ammonia to produce adenine ⟨54JA6073⟩, 6-fluoro-9-methylpurine reacts with aqueous ammonia under relatively mild conditions ⟨62JMC1067⟩ to give 9-methyladenine, and similarly 9-substituted 2,6-dichloropurines react readily with alcoholic ammonia to produce 9-substituted 2-chloroadenines ⟨63JMC471⟩. However, N-substituted, especially N^1-alkylpurines are prone to undergo Dimroth rearrangements in alkaline solutions. In contrast, aminochloropurines tend to be relatively inert towards nucleophiles. Thus 2-amino-6-chloropurine barely reacted with hot 0.1N sodium hydroxide ⟨60JOC1573⟩, presumably due to anion formation. On the other hand, 2,6-dichloropurine reacted readily enough with aqueous ammonia to produce 2,6-diaminopurine ⟨58JOC125⟩ and in this case apparently the base is not strong enough to achieve significant anion formation. However amination of 2,6,8-trichloropurine required aqueous ammonia at 100 °C over 6 h to form 6-amino-2,8-dichloropurine ⟨1897CB2226⟩, the extra halogen presumably deactivating the ring system substantially. Fluorine tends to be more reactive than chlorine, as with 6-chloro-2-fluoropurine riboside which produced 2-amino-6-chloropurine riboside with ammonia at room temperature ⟨66JOC3258⟩.

Similar reactions of halogenopurines occur with primary and secondary amines and large numbers of compounds have been made in this way, including several naturally occurring substances many of which are 6-alkylaminopurines including the plant cytokinin zeatin ⟨66JCS(C)921⟩ (Section 4.09.8.2). For more information see ⟨B-71MI40901⟩.

Halogenopurines will also react with tertiary amines. Thus 2,6,8-trichloropurine with pyridine in ethanol gave, after neutralization, the betaine (**205**) ⟨60AG708⟩, and similarly 7-methyl-2-chlorohypoxanthine gave the betaine (**206**). 6-Chloropurine and trimethylamine afforded a quaternary salt ⟨B-68MI40901, p. 39⟩ and the corresponding 9-β-D-riboside has also been reported ⟨63JOC945⟩. Some reactions and rearrangements of these quaternary derivatives have been recorded ⟨B-71MI40902, p. 504⟩.

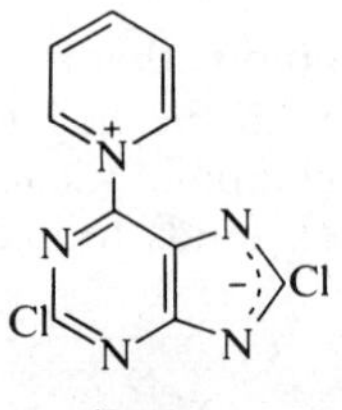

(**205**) 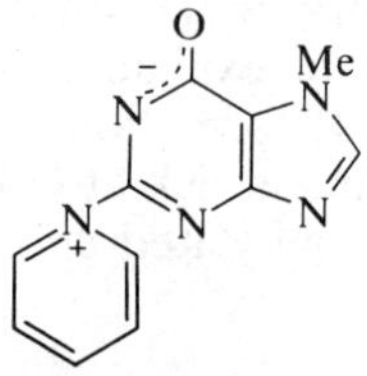   (**206**)

Halogen atoms in the halogenopurines may be replaced by hydrogen, the ease of replacement generally being 6 > 8 > 2. The preferred method is usually hydrogenolysis with palladium catalysts in the presence of magnesium oxide, or use of hydriodic acid alone or with phosphonium iodide, but many other reducing agents have been used. Use of $^2H_2$ or $^3H_2$ provides a useful route to ^2H- or ^3H-purines ⟨71MI40905⟩, complementing hydrogen exchange reactions which are normally restricted to C-8 (see Section 4.09.5.2.1).

Hydrogenolysis is especially useful in that in polyhalogenopurines it may be possible to control the reaction so as to remove only one halogen. Thus hydrogenolysis of 6-amino-2,8-dichloropurine can be adjusted to produce either 6-amino-2-chloropurine ⟨61LA(649)114⟩ or adenine ⟨59AG524⟩. Hydrogenolysis is also possible in alkaline solution which makes it an especially useful technique for reactions involving the oxopurines which are generally insoluble in water but soluble in alkaline solution. Thus hydrogenolysis of 2,8-dichlorohypoxanthine in alkaline solution over palladium may be controlled to produce either 2-chlorohypoxanthine ⟨61LA(649)114⟩ or hypoxanthine ⟨63CI(L)953⟩.

Replacement of halogen atoms by alkoxy groups may be hindered in $N(7)$- and $N(9)$-unsubstituted purines because of anion formation although this has been overcome to some extent by use of antimony chloride followed by the alcohol ⟨71MI40904⟩. Aryloxypurines have similarly been prepared from halogenopurines and hot alkaline phenol solutions. Since alkylation of oxopurines tends to lead almost exclusively to N-alkyl derivatives (see Section 4.09.5.2.2(iv)), direct replacement of halogen atoms assumes special importance as a route to the alkoxypurine derivatives.

Similarly, replacement of halogens by alkylthio or arylthio groups may be achieved using appropriate metal sulfides in aqueous or alcoholic solution, generally under mild conditions. This provides a useful route to those compounds which are, however, unlike their oxo analogues, also available by alkylation (and occasionally arylation) of thioxopurines (Section 4.09.5.2.2(viii)). The biological activity of compounds such as 6-mercaptopurine and azathiopurine (Section 4.09.8.4) has ensured that large numbers of thioxopurines and derivatives have been prepared.

Replacement of halogens by thiol groups may also be achieved using alkali metal hydrosulfides, thioacetates ⟨64JOC3209⟩, thiourea ⟨54JA6073⟩ or thioacetamide ⟨65JMC667⟩ in ethanol solution. In these last reactions, intermediate thiouronium salts are formed which readily hydrolyze or alcoholyze to the thioxopurine. Corresponding selenium derivatives have been similarly prepared ⟨71JHC379⟩.

In these last reaction types, selective replacement of halogens may be possible as with 2,8-dichloropurine where only the 8-chloro group is replaced to produce 2-chloro-8-thioxo-7,8-dihydropurine (**207**), whereas with 6,8-dichloropurine thiourea may be used to furnish either the 8-chloro-6-thioxo derivative (**208**) or the 6,8-dithioxopurine (**209**) according to the amount of thiourea used ⟨58JA6671, 59JOC320⟩.

(**207**)   (**208**)   (**209**)

The halogenopurines react with many other nucleophiles including hydrazines, hydroxylamine and alkoxyamines, sulfonic acids and thiocyanates, and compounds with active methylene groups such as diethyl malonate furnish appropriate purine derivatives which offer valuable routes to alkyl and substituted alkyl derivatives. Also aryl groups have been substituted for halogen using aryllithiums ⟨63N224⟩.

Table 37 gives details of some of these miscellaneous reactions.

Displacement of chlorine atoms in selected chloropurines by other halogens, especially fluorine and iodine, has been of value in the preparation of these derivatives. In particular 6-chloropurine and trimethylamine followed by potassium hydrogen fluoride produced 6-fluoropurine ⟨69CC381⟩, and several fluoropurines have been made by heating the corresponding chloropurine with silver fluoride in high boiling solvents such as toluene or (for 2-chloropurines) xylene ⟨63JOC2310⟩. Relatively few iodopurines have been prepared and then mostly by displacement of chlorine. Thus 6-chloropurine and hydriodic acid (47%) at 0 °C gave 6-iodopurine ⟨56JA3508⟩. The C-2 chlorine in 2,6- and 2,8-dichloropurine is resistant to reaction, however, and with hydriodic acid, 2-chloro-6-iodo- and 2-chloro-8-iodo-purine ⟨63CJC1807⟩ respectively, were formed. In an analogous fashion, 2,6,8-trichloropurine produced 2-chloro-6,8-diiodopurine ⟨61LA(649)114⟩. However when phosphonium iodide was used in conjunction with hydriodic acid, 2,6,8-trichloropurine underwent partial reduction with formation of 2,6-diiodopurine ⟨57ZOB1712⟩.

Table 37 Miscellaneous Reactions of Some Halogenopurines

Purine	Product (purine)	Conditions	Ref.
2-Cl	2-NHNH$_2$	NH$_2$NH$_2$	57JA2185
2-Cl-6-NH$_2$	2-NHNH$_2$-6-NH$_2$	NH$_2$NH$_2$	57JA2185
2-Cl-6-oxo	2-NHNH$_2$-6-oxo	NH$_2$NH$_2$ (91%)	57JA2185
6-Cl	6-NHNH$_2$	NH$_2$NH$_2$ (80%)	57JA2185
6-Cl-7-Me	6-NHNH$_2$-7-Me	NH$_2$NH$_2$	57JA6401
6-Cl-9-Me	6-NHNH$_2$-9-Me	NH$_2$NH$_2$	57JA490
2,6-Cl$_2$	2-Cl-6-NHNH$_2$	NH$_2$NH$_2$, room temp.	57JA2185
2,6-Cl$_2$	2,6-(NHNH$_2$)$_2$	NH$_2$NH$_2$ (80%)	57JA2185
6-Cl	6-NHOH	NH$_2$OH	58JA3932
6-Cl-9-Me	6-NHOH-9-Me	NH$_2$OH	57JA490
6-Cl-2-F	2,6-(NHOH)$_2$	NH$_2$OH	68JMC52
6-Cl	6-NHOMe	NH$_2$OMe	66MI40901
2-Cl-hypoxanthine	2-N$_3$-hypoxanthine	NaN$_3$	66JOC2210
2,6-Cl$_2$	2,6-(N$_3$)$_2$	NaN$_3$	66JOC2210
2,6,8-Cl$_3$	2,6,8-(N$_3$)$_3$	NaN$_3$	64MI40901
8-Br-adenosine	8-N$_3$-adenosine	NaN$_3$ (75%)	65JA1772
6-Cl	6-SO$_3$H	Na$_2$SO$_3$ (3%)	61JOC3401
2-NH$_2$-6-Cl	2-NH$_2$-6-SO$_3$H	Na$_2$SO$_3$	61JOC3401
6-Cl	6-SCN	KCNS	59JA1898
8-Br-caffeine	8-CH(CO$_2$Et)$_2$-caffeine	NaCH(CO$_2$Et)$_2$	35MI40900
6-Cl	6-CH(CO$_2$Et)$_2$	NaCH(CO$_2$Et)$_2$	63N224
6-Cl-9-Me	6-CH(CO$_2$Et)$_2$-9-Me	NaCH(CO$_2$Et)$_2$	63ZOB3342
2-Cl	2-CH(CO$_2$Et)$_2$	NaCH(CO$_2$Et)$_2$	64ZOB3254
2-Cl-7-Me	2,7-Me$_2$	MeMgI	49JAP49177355
6-Cl	6-Ph	PhLi	63N224

4.09.6.18 Purines with a Fused Heterocyclic Ring System

4.09.6.18.1 Introduction

Many examples of compounds in which a heterocyclic ring is fused to a purine have been recorded. These include the various types of cyclonucleosides such as (**210**). However in this chapter a discussion of these compounds is not practical and for further details the reader should consult works devoted to nucleoside chemistry. Appropriate references are collected in Table 2.

(**210**)

The fused heterocyclic rings to be discussed here have normally been produced by an intramolecular cyclization of an $N(6)$-, $N(2)$- or S-substituent onto a ring nitrogen atom in the purine nucleus and are mainly five- or six-membered systems; cyclonucleosides such as (**210**) are of course notable exceptions and a few miscellaneous examples exist.

4.09.6.18.2 Five-membered fused rings

The imidazopurine (**211**) or 1,N^6-ethenoadenine, obtained by cyclization of N^6-formylmethyladenine (**212**), rapidly produced the bisimidazole (**213**) when heated for a short time at 100 °C with hydrochloric acid (Scheme 47) ⟨70JCS(C)2206⟩. Analogous alkyl-substituted adenines and adenosines have also been prepared and similarly were found to produce bisimidazoles by acid treatment ⟨80JA770⟩. The corresponding dihydroimidazopurines (**214**) have also been prepared ⟨62JOC973⟩.

The triazolopurine (**215**) is more resistant to ring opening but with strong hydrochloric acid over 70 hours it produced the imidazolyltriazole (**216**) ⟨61JOC3446⟩. The triazolopurine (**215**) also isomerized to (**217**) when heated in formamide (Scheme 48) ⟨65JOC3601⟩. A

(212) (211) (213) (214)

Scheme 47

similar isomerization appears to occur with tetrazolopurines (**218**) and (**219**). These are formed in equilibrium both in the solid state or in solution with 2-azidopurine (**220**), and the latter azido form is favoured in acid solution, the tetrazolopurines being favored in neutral solution (Scheme 49) ⟨66JOC935⟩.

(217) (215) (216)

Scheme 48

(218) (220) (219)

Scheme 49

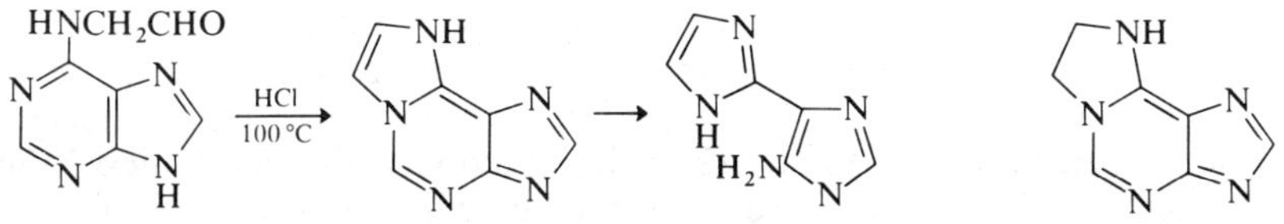

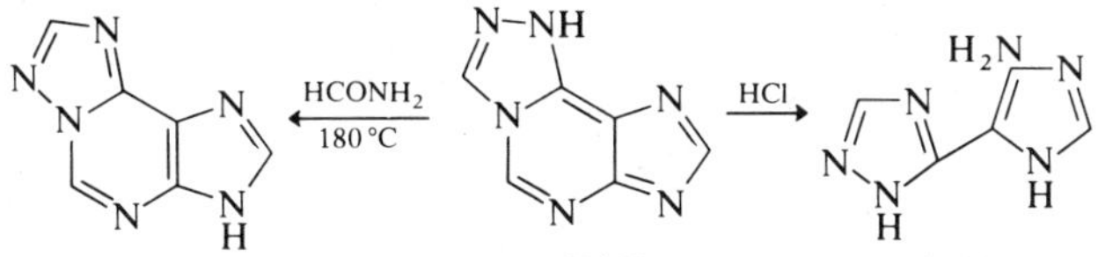

The oxoimidazopurine (**221**), prepared from N^6-chloroacetylpurine (**222**), has also been implicated as an intermediate in the hydrolysis of N^6-glycylpurine (**223**). With hot water the imidazo derivative rapidly hydrolyzed to the carboxymethyladenine (**224**) *via* the intermediate (**225**) ⟨66B2082⟩ and a Dimroth rearrangement (Scheme 50).

Scheme 50

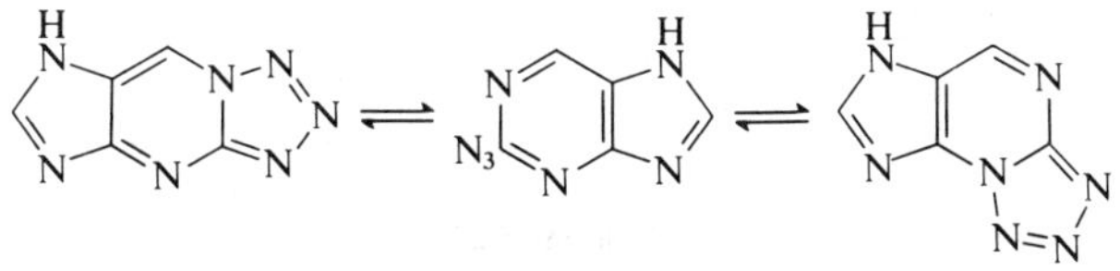

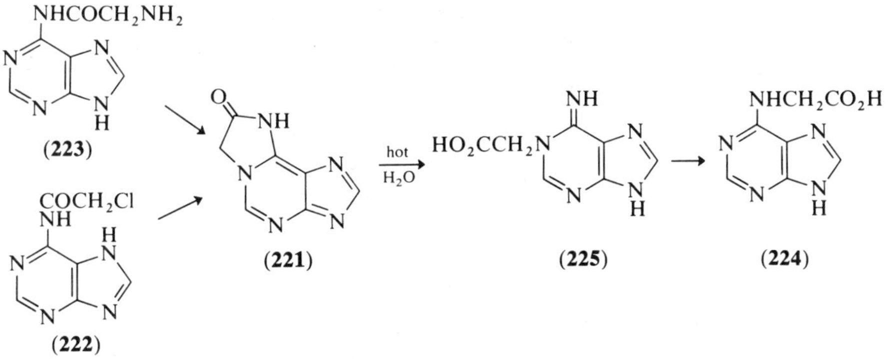

A somewhat different imidazopurine occurs in the naturally occurring nucleoside wyosine (**226**) which is a highly fluorescent compound found in the next position to the 3′-end of the anticodon of yeast $tRNA^{Phe}$ and is very readily and characteristically (of 3-alkylguanosines) hydrolyzed to the aglycone ⟨78TL2579⟩ (see Section 4.09.8.3). Similar compounds, namely wybutosine (**227**) and wybutoxosine (**228**) have been found in rat liver and *Lupinus luteus*, respectively.

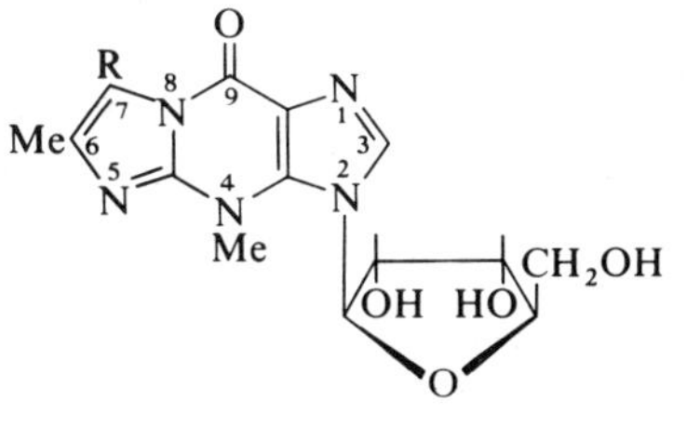

(**226**) R = H

(**227**) R = $CH_2CH_2CH(CO_2Me)NHCO_2Me$

(**228**) R = $CH_2CH(CO_2H)CH(CO_2Me)NHCO_2Me$

Several thiazolopurines have been prepared (Section 4.09.5.2.2(viii)) by simple cyclization of thioxopurines. Such compounds, by dethiation, have proved to be a useful route to alkylpurines. Thus 1-, 3-, 7- and 9-ethylpurines, respectively, are produced from the thiazolopurines (**229**), (**230**), (**231**) and (**232**) by reaction with Raney nickel (Scheme 51) ⟨61JOC3446⟩. Compound (**229**) also reacted readily with hydrogen sulfide to give the thioxopurine (**233**). When the ring proton is substituted the thiazolopurines tend to exist as quaternary forms, *e.g.* (**234**) ⟨62JOC195⟩. The unusual thiazolopurines (**235**) and (**236**) are produced in 13% and 18% yields, respectively, from the corresponding 6-chloropurine and thiourea (the 6-thioxo-1,6-dihydropurine is the main product) and when heated in acid or alkaline solution they rearrange to produce the 8-amino-6-thioxo-1,6-dihydropurines (**237**) and (**238**), respectively (Scheme 52) ⟨66JOC1417⟩.

(**229**) (**230**) (**231**) (**232**)

Scheme 51

(**233**) (**234**)

(**235**) R = H (**237**) R = H
(**236**) R = Et (**238**) R = Et

Scheme 52

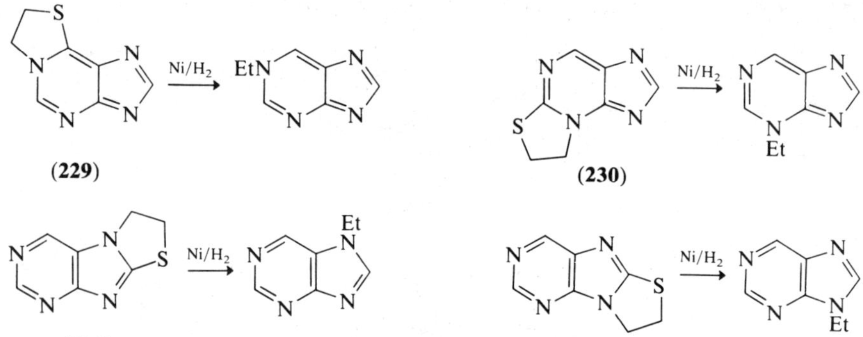

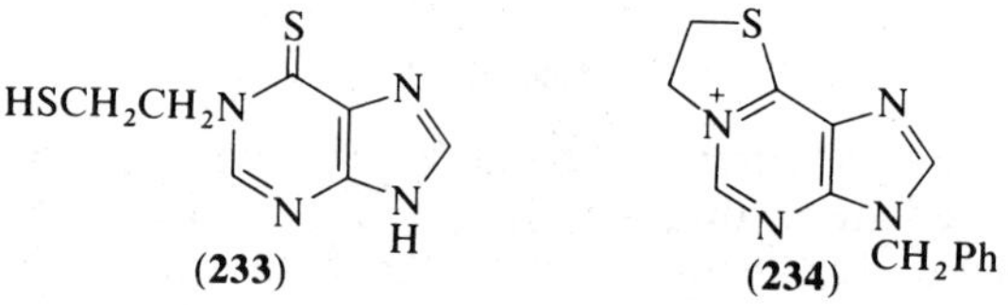

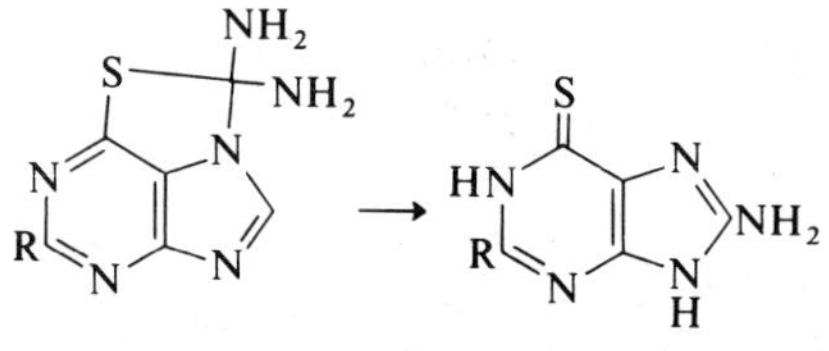

4.09.6.18.3 Six-membered fused rings

The pyrimidopurine (**239**), prepared from N^6-isopentenyladenine (see Section 4.09.5.2.2(iii)) and TFA or fluoboric acid, rearranged when heated at 85 °C for 36 hours to produce the isomeric pyrimidopurine (**240**; Scheme 53) ⟨68JCS(C)1731⟩. Recently methyl 7-oxo-7*H*-pyrido[2,1-*i*]purine-9-carboxylate has been prepared ⟨80RTC20⟩ in addition to a series of 7,8-purine derivatives with ring sizes up to seven atoms ⟨79LA1872⟩ and purine[9,8-*a*]quinolines and some related sulfur derivatives ⟨70MI40903⟩.

(**239**) (**240**)

Scheme 53

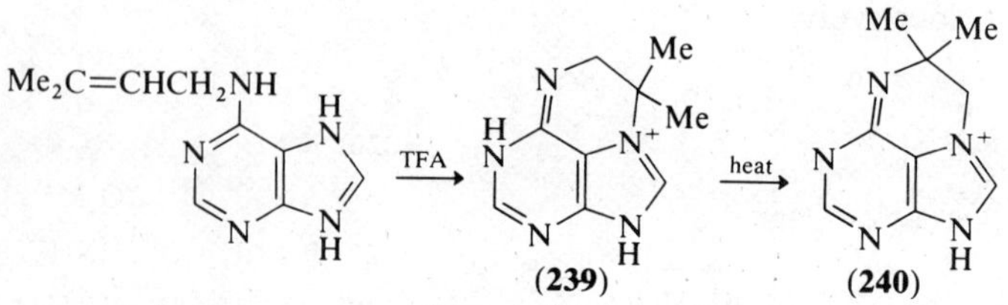

Triazinopurines have also been prepared including (241) and (242) ⟨53CR(236)2519, 56BSF888⟩.

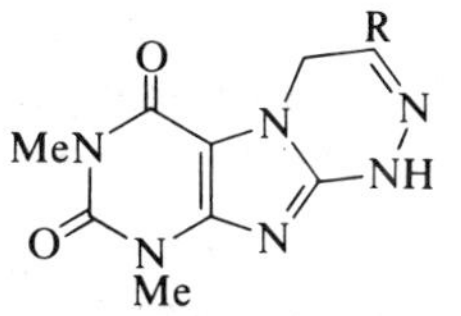

(241) R = Me
(242) R = Ph

4.09.6.19 Benzopurines

An interesting series of so-called stretched purines has been prepared in which a benzene ring is interpolated between the imidazole and pyrimidine moieties of various purines. This leads to three isomeric structures which have been named *lin*-benzoadenine (243a), *prox*-benzoadenine (244a) and *dist*-benzoadenine (245a) ⟨75JOC356, 75JOC363⟩.

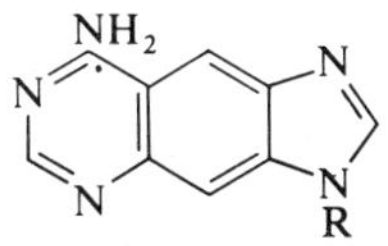

(243a) R = H
(243b) R = β-D-ribofuranosyl
(244) 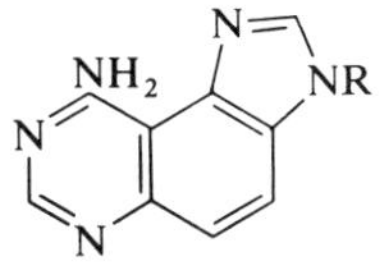
(245) 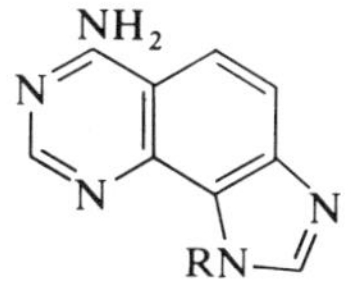

Corresponding ribofuranosyl derivatives (243b–245b) have also been prepared by standard methods ⟨76JA3987⟩ and these converted into 5′-phosphates. The *lin*-benzoadenosine was found to hydrolyze to *lin*-benzoinosine in the presence of adenosine deaminase at a rate comparable to that of the adenosine to inosine conversion. However *lin*-benzoadenine unexpectedly was also converted by the same enzyme into *lin*-hypoxanthine but the *prox* and *dist* isomers were refractory. Similarly *lin*-benzohypoxanthine with xanthine oxidase produced *lin*-benzoxanthine, and *lin*-uric acid and *lin*-benzoinosine gave *lin*-xanthosine with this enzyme, the *prox* and *dist* isomers being less active in this system. A recent publication ⟨79JA1564⟩ discusses some inter- and intra-molecular reactions of the *lin*-benzoadenine ribonucleotides and should be consulted for a more detailed reference list of earlier work.

4.09.7 SYNTHESIS

4.09.7.1 Introduction

Methods for the synthesis of purines may be conveniently separated into three groups. (1) Synthesis from acyclic precursors: These normally consist of one step reactions although they are often known to proceed *via* intermediate imidazoles or pyrimidines. This group also includes the so-called abiotic or abiogenic syntheses but the distinction of these from the first type is blurred and to a large extent based on philosophical grounds. Perhaps the main criteria for an abiotic synthesis is that it proceeds from the simplest molecules and perhaps more importantly that it should occur under reaction conditions which might reasonably have been expected to have existed on the primeval earth near the time of emergence of organic molecules and living systems. Abiotic syntheses have provided a basis for the so-called chemical evolutionary theories which have been used to account for the first formation of life on earth from non-living precursors ⟨B-73MI40900⟩. (2) Synthesis from pyrimidines by completion of an imidazole ring. (3) Synthesis from imidazoles by completion of a pyrimidine ring.

4.09.7.2 Synthesis from Acyclic Precursors

There are relatively few examples of one-step syntheses of purines from acyclic precursors and most of them in any case probably involve intermediate imidazoles or pyrimidines which, however, have not necessarily been isolated in such reactions or their formation confirmed.

Perhaps the earliest recorded example is the formation of uric acid by heating glycine ⟨1882M(3)796⟩ or later by heating together trichlorolactic acid amide and urea (Scheme 54) ⟨1887M(8)201, 25LA(441)215⟩.

$$Cl_3CCH(OH)CONH_2 \xrightarrow{urea} uric\ acid$$

Scheme 54

Purine has also been obtained by heating formamide alone ⟨72CPB623⟩ and detected in reactions involving hot formamide ⟨77CPB1923⟩. Interpretation of mechanisms of reactions carried out in formamide solution must clearly be made with care. However the distribution of ^{13}C and ^{15}N labels in purine derived from [^{13}C, ^{15}N]HCN and formamide has suggested the mechanism outlined (Scheme 55) ⟨78TL4039⟩.

$$2HCONH_2 \rightleftarrows NH_2CH(OH)NHCHO \rightleftarrows NH_2CH=NCHO$$

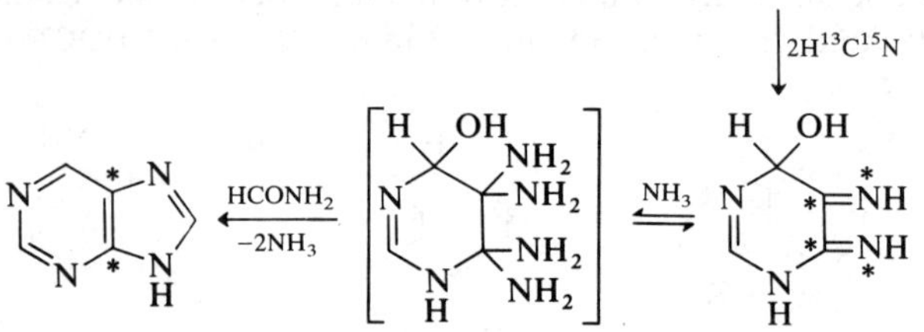

Scheme 55

Purine has also been produced in 40% yield by heating mixtures of N-cyanomethylphthalimide with either trisformamidomethane or formamidine acetate in formamide or n-butanol (Scheme 56) ⟨61AG163⟩.

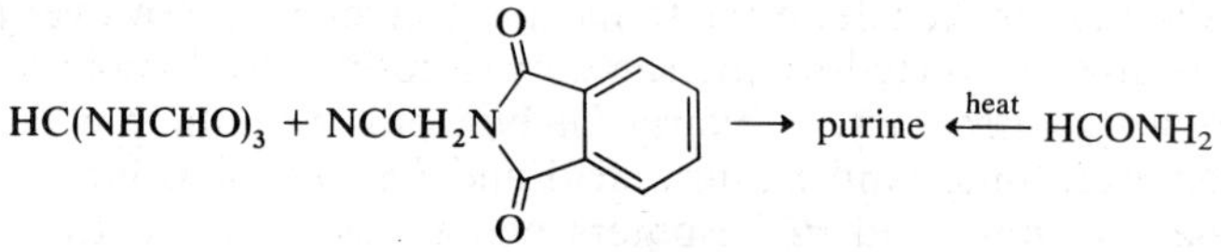

Scheme 56

Several useful preparations of purine derivatives have been recorded from cyanoacetamide and malononitrile derivatives. Thus hypoxanthine was obtained by heating formamidomalonamidinecarboxamide with formamide (Scheme 57) ⟨50JBC(185)439⟩. In a similar manner a 75% yield of hypoxanthine was produced by heating ethyl acetamidocyanoacetate, ethanolic ammonia, ammonium acetate and triethyl orthoformate ⟨59TL9⟩, and in like manner phenylazocyanoacetylurea gave xanthine and guanine whereas the N'-methylurea produced 3-methylxanthine ⟨76CPB1331⟩.

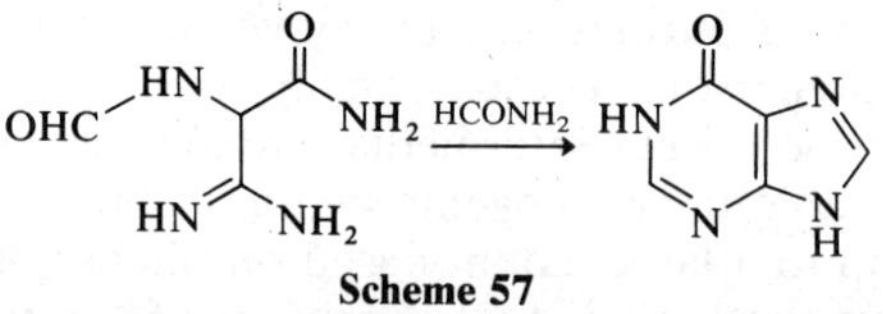

Scheme 57

Analogous reactions of aminomalonamidinecarboxamide with ortho esters in DMF or acetonitrile slowly afforded various 2,8-dialkylpurines (Scheme 58). Thus heating together triethyl orthoformate and aminomalonamidinecarboxamide gave hypoxanthine after 5 minutes whereas triethyl orthoacetate required 10 hours to produce reasonable amounts of 2,8-dimethylhypoxanthine, and triethyl orthopropionate 60 hours to give 2,8-diethylhypoxanthine ⟨60JA3144⟩. In a similar manner aminomalondiamidine and ortho esters furnished

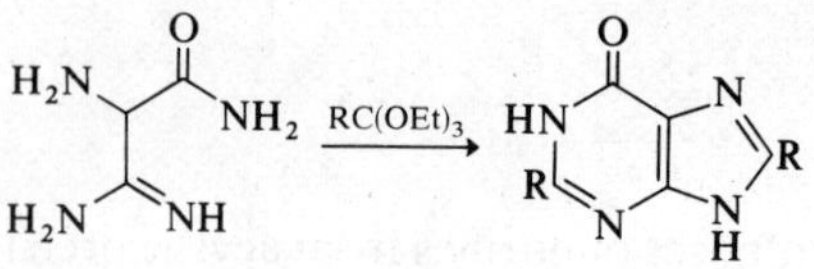

Scheme 58 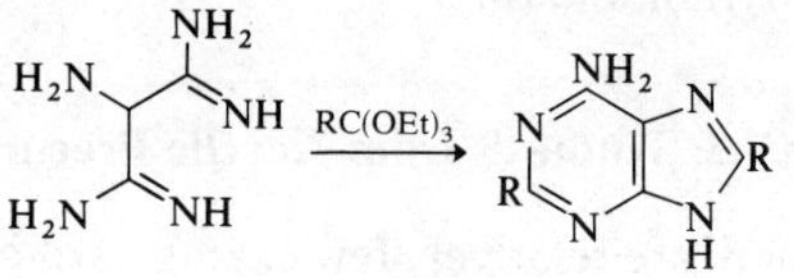   **Scheme 59**

2,8-dialkyladenines (Scheme 59). Purine syntheses from aminomalonamidine derivatives appear to proceed by prior formation of an aminoimidazole followed by cyclization with a second molecule of ortho ester.

In constrast, high pressure hydrogenation of hydroxyiminomalononitrile or related arylazomalononitriles, amides or esters, using palladium or nickel catalysts in the presence of ammonia and formamide, afforded purine derivatives including adenine, but in this case intermediate nitrosopyrimidines are first produced ⟨75CPB2401, 76CPB1331⟩ where cyclization has preceded reduction (Scheme 60). Guanidine carbonate, potassium hydroxyiminomalononitrile and sodium dithionite in formamide also afforded 2,6-diaminopurine (246) in good yield ⟨57JA1518⟩, the reaction again proceeding through an intermediate nitrosopyrimidine (Scheme 61). The use of arylamidines in this reaction has led to syntheses of 2-phenyl- and 2-β-pyridyl-adenines ⟨59JA2442⟩.

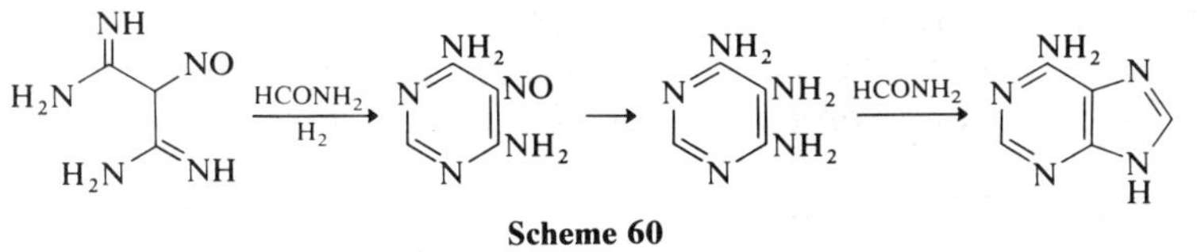

Scheme 60

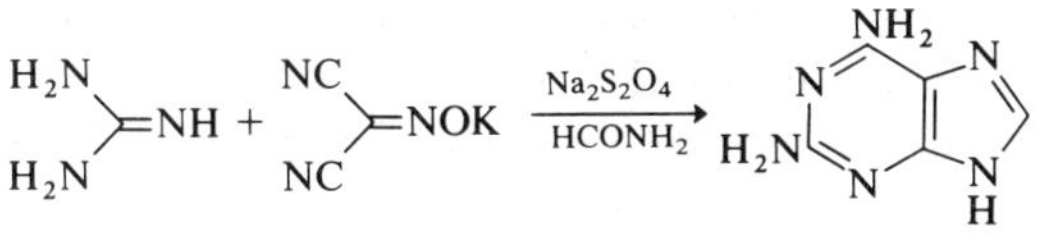

Scheme 61

Many of the so-called abiotic or abiogenic syntheses of purines have had as a basis, pioneer experiments ⟨53MI40902, 55JA2351⟩ derived from theories which required the primeval earth to have a reducing (hydrogen containing) atmosphere ⟨B-52MI40900, B-53MI40900⟩. In these experiments mixtures of hydrogen, methane, ammonia and water vapour were subjected to a continuous spark discharge for several days. Many types of organic compounds, especially amino acids, are produced but originally no purines were detected. However later ⟨63PNA(49)737⟩ a 0.01% yield of adenine was obtained by irradiation of gaseous mixtures of the above type with a 4.5 MeV electron beam, and purines were also produced when similar mixtures were exposed to a high frequency electric discharge ⟨75CR(C)685⟩. Adenine had also been detected at an earlier date ⟨60BBR(2)207⟩ in heated, concentrated solutions of ammonium cyanide. It was accompanied by hypoxanthine (but no guanine) and at least fifty other UV absorbing substances ⟨63MI40905⟩. When 0.1% solutions of ammonium cyanide were heated for 1 day at 90 °C or 70 °C for 25 days, 0.1% yields of adenine could be obtained ⟨61MI40900⟩. In these experiments 5-aminoimidazole precursors such as the 4-carbonitrile and 4-carboxamide were isolated in addition to formamidine ⟨62MI40905⟩ and on the basis of these observations the route to adenine outlined in Scheme 62 was proposed.

(a) $NH_3 + HCN \rightarrow H_2NCH=NH$

(b) $2HCN \rightarrow NH=CHCN \xrightarrow{HCN} H_2NCH(CN)_2 \xrightarrow{NH_3} H_2NCH[C(:NH)NH_2]_2 \xrightarrow{H_2NCH=NH}$

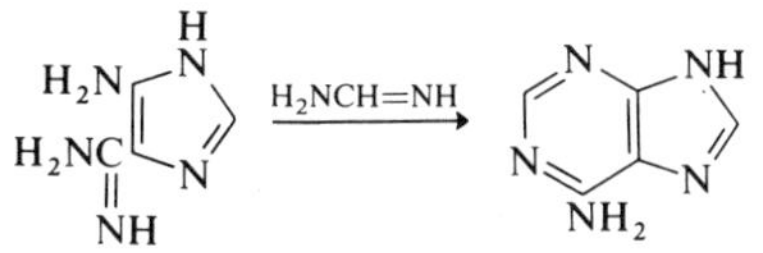

Scheme 62

In addition to adenine, guanine was also produced in yields of 1% and 0.5%, respectively, by UV radiation of dilute aqueous solutions of hydrogen cyanide over 7 days at room temperature ⟨B-65MI40901⟩ and a biradical scheme has been proposed for this reaction (Scheme 63) ⟨62PNA(48)1300⟩. When liquid ammonia was heated with liquid hydrogen cyanide at 120 °C in an autoclave over 20 hours, a 10-15% yield of adenine was obtained ⟨66JAP6610114⟩ by extraction of the residual solid with water.

An intermediate in this reaction using anhydrous hydrogen cyanide and ammonia was diaminomaleonitrile ⟨67MI40901⟩ and the same compound was also obtained from aminomalononitrile and hydrogen cyanide ⟨66JA3829⟩. UV radiation of the maleonitrile

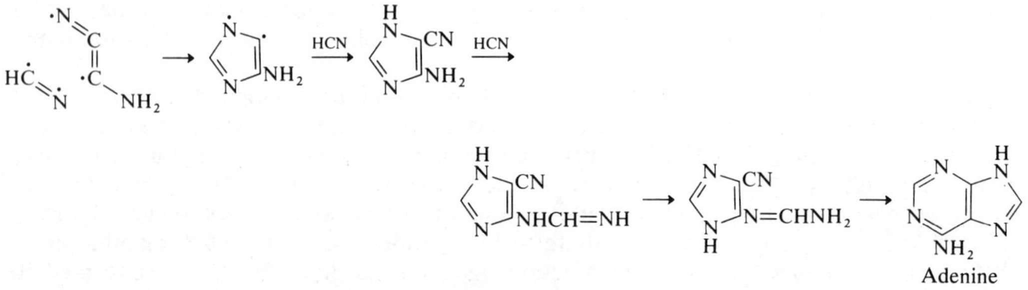

Scheme 63

induced an isomerization *via* the *trans*-nitrile to give 5-aminoimidazole-4-carbonitrile in 80% yield ⟨66JA1074, 69MI40904⟩. The aminoimidazolenitrile with ammonium cyanide solutions further reacted at 30–100 °C to produce adenine, and extrapolation of the results to −20 °C allowed the authors to propose that molar solutions of adenine and similar heterocyclic compounds could have been formed after 10^6 years in a primeval lake which alternately thawed in summer and froze in winter. In addition the aminoimidazole nitrile was readily hydrolyzed to the corresponding carboxamide which with cyanogen or cyanate at 100 °C gave up to 20% yield of guanine (Scheme 64). If acetamide is used in place of hydrogen cyanide then 2-methyl-, 8-methyl- and 2,8-dimethyl-adenines are produced, and liquid methylamine and hydrogen cyanide similarly furnished 7- and 9-methyl-6-methyl-aminopurines ⟨68JOC642⟩.

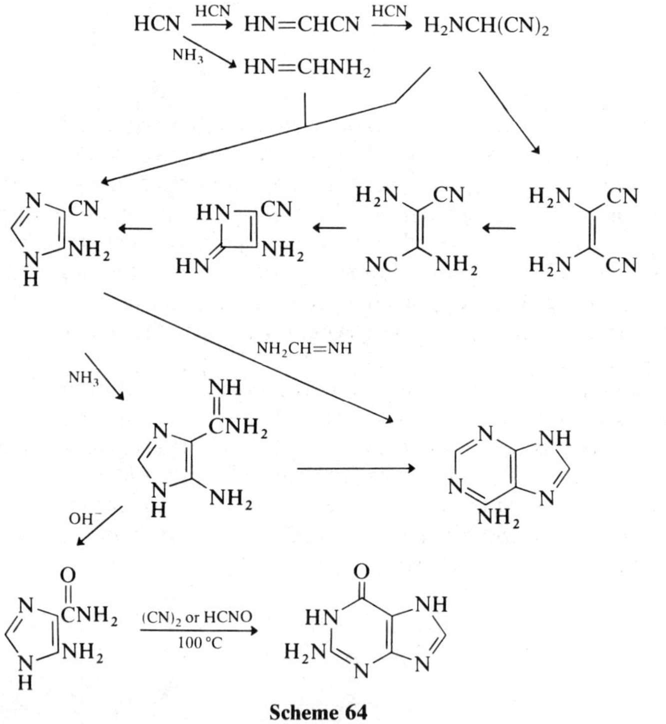

Scheme 64

4.09.7.3 Synthesis from Diaminopyrimidines

4.09.7.3.1 Introduction

The most widely used route to purines has undoubtedly involved the addition of an imidazole ring to an appropriate pyrimidine. This particular route has probably dominated the early literature because of the relative ease of synthesis of the requisite pyrimidines compared with the lack, until recently, of useful routes to imidazole precursors. The synthesis of pyrimidines has been exhaustively reviewed ⟨62HC(16)1⟩ and full details have been provided of the synthesis of the specfic pyrimidines used in purine synthesis. The pyrimidine moiety in such syntheses will of course include substituents in what will become the 1-, 2-, 3- and 6-positions in the resulting purine and the added imidazole ring will incorporate C(8)-substituents whereas substitution at N-7 or N-9 could come from either moiety, but usually from the pyrimidine, according to the precise intermediates used (Scheme 65).

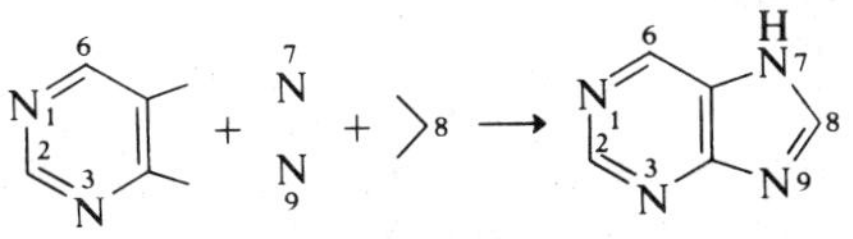

Scheme 65

It is therefore convenient to consider and classify variation in C-8 and N-7 or N-9 substitution in a synthesized purine resulting from different cyclization procedures when applied to a given pyrimidine.

4.09.7.3.2 C(8)-Unsubstituted purines

The simplest example of the synthesis of a $C(8)$-unsubstituted purine from a pyrimidine is illustrated by the first general application, namely the formation of guanine by heating 6-oxo-2,4,5-triamino-1,6-dihydropyrimidine with formic acid (Scheme 66) ⟨00CB1371⟩. This and related reactions, frequently called Traube syntheses, proceed by formation of intermediate N-acyl or N-formyl derivatives (**247**) which can usually be isolated and cyclized in other ways, e.g. by fusion or warming with aqueous alkali. The intermediate N-formyl derivatives may be produced more readily by the use of formic–acetic anhydride, as in the formylation of 4,5-diamino-2-methoxypyrimidine (**248**) which is reluctant to react with formic acid alone (Scheme 67) ⟨54JCS2060⟩. That the site of acylation is usually at position 5 in the pyrimidine as in (**247**), can be shown by comparison with a known 5-acylaminopyrimidine made by an unambiguous synthesis ⟨48JCS1157⟩, or alternatively, by reduction with lithium aluminum hydride to the alkylaminopyrimidine, cyclization of which, with formic acid, then produces a known 7-alkylpurine (Scheme 68) ⟨54MI40900⟩.

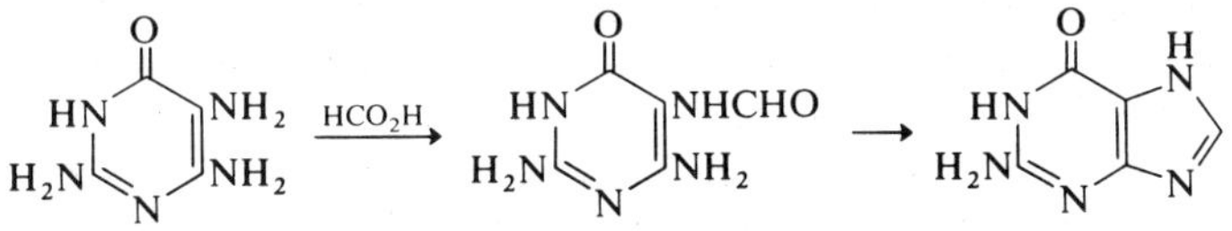

(**247**)

Scheme 66

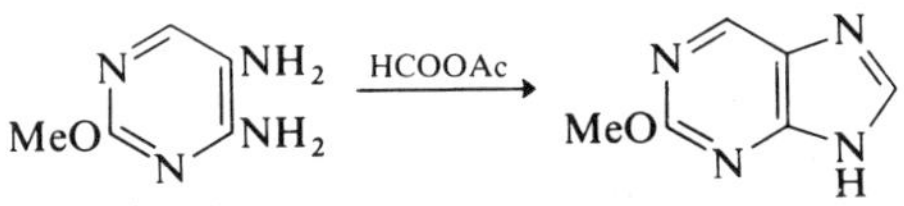

(**248**)

Scheme 67

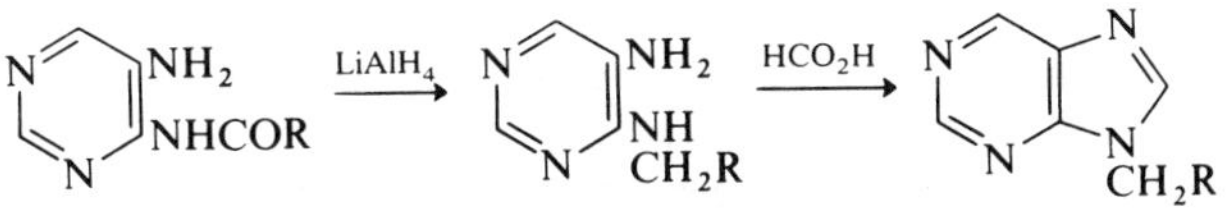

Scheme 68

The ease of cyclization of the 4-amino-5-formamidopyrimidines has been associated with their acidity. Thus the weakly acidic methylated pyrimidine derivative (**249**) cyclized to caffeine spontaneously (Scheme 69) ⟨00CB1371, 00CB3035⟩, whereas the much more strongly

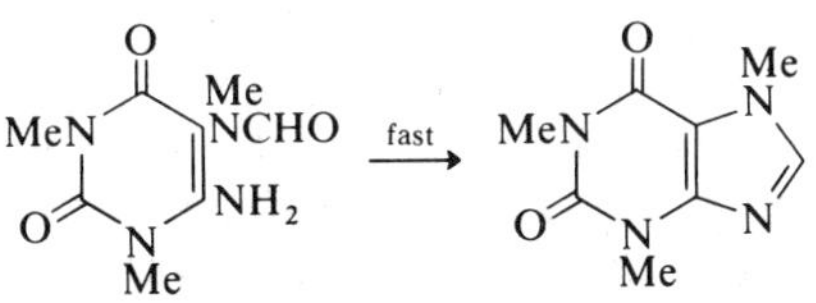

(**249**)

Scheme 69

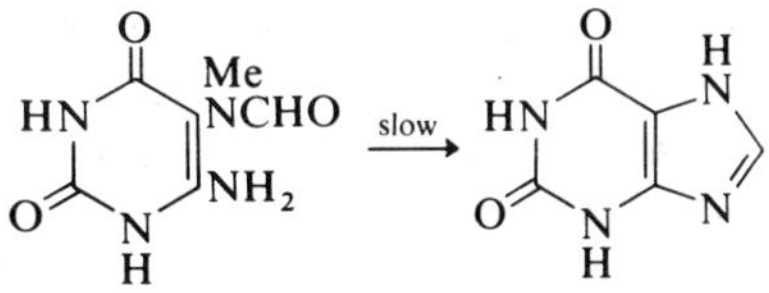

(**250**)

Scheme 70

acidic 4-amino-5-formamido-2,6-dioxo-1,6-dihydropyrimidine (**250**) was only converted into xanthine by heating the sodium salt to 220 °C (Scheme 70).

In early preparations, formamide was commonly used as a solvent with formic acid, but later it was shown that formamide alone is a useful formylating agent superior to formic acid and its effectiveness is improved by the presence of an acid catalyst. It tends to give better yields than formic acid as in a synthesis of adenine (92%) ⟨62ZOB1655⟩ compared with a 62% yield using formic acid. However, hot formamide tends to decompose slightly to produce, among other things, ammonia (and to some extent adenine — see Section 4.09.7.2) and this restricts its use to pyrimidines with non-reactive substituents. Formamide, like formic acid, with 4,5,6-triaminopyrimidines produces largely (75%) the 9-substituted purine accompanied by some of the N^6-substituted purine derivative ⟨58JCS4107⟩. However, the high temperature normally associated with reactions in boiling formamide (160–180 °C) may suffice to cause side reactions and also to displace any acyl groups which may be attached to the aminopyrimidines. Thus the acetyl derivative (**251**) and formamide gave xanthine (Scheme 71), not 8-methylxanthine as might have been expected ⟨51JA4609, 58JOC2010⟩, whilst 2,4,6-triamino-5-(*p*-chlorobenzamido)pyrimidine and formamide gave a mixture of 2,6-diaminopurine and the 8-*p*-chlorophenyl derivative ⟨51JA5235⟩.

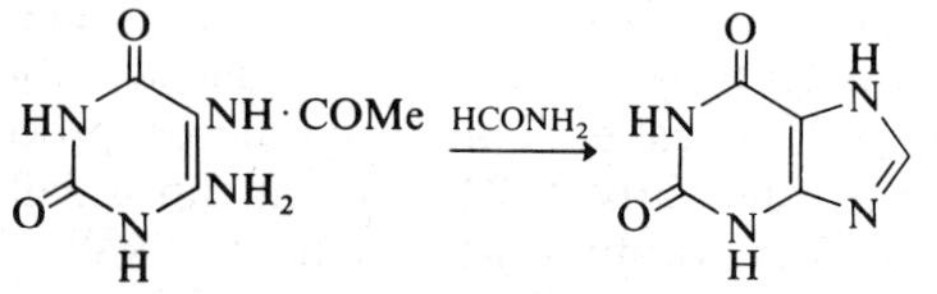

Scheme 71

Thioxopurines may similarly be produced by cyclization of appropriate thioxopyrimidines. Thus 4,5-diamino-2-(or 6-)thioxodihydropyrimidines and formic acid gave 2-thioxo-2,3-dihydropurine ⟨54JCS2060⟩ and 6-thioxo-1,6-dihydropurine ⟨52JA411⟩, respectively. However, in the latter case especially, an alternative ring closure to produce a thiazolopyrimidine (**252**) can also occur (see Section 4.09.6.1.8) ⟨56JA2858⟩ but this is less likely using formamide if only because intermediate thiazolopyrimidines rearrange to the corresponding thioxopurines in the hot solvent (Scheme 72). Thioxopurine formation is improved by the use of high pH and low temperatures. An alternative route to 6-thioxopurines which avoids thiazolopyrimidine formation involves treatment of an aminooxoformamidopyrimidine with phosphorus pentasulfide in a hot base such as pyrimidine. In this way thioguanine is produced from 2,4-diamino-5-formamido-6-oxo-1,6-dihydropyrimidine ⟨60JA2633⟩.

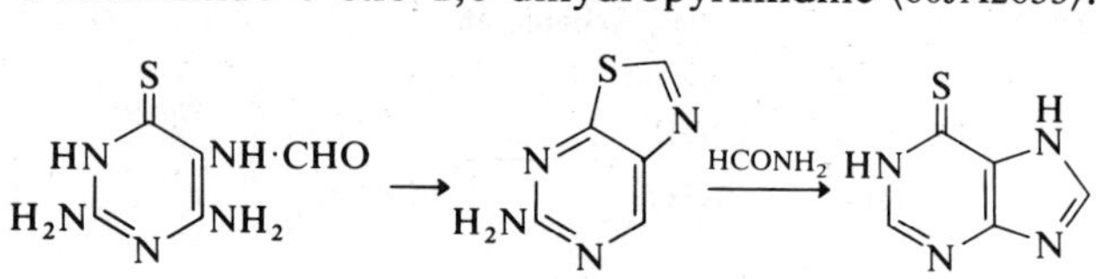

Scheme 72

The use of formic acid, as with formamide, as a cyclizing agent may be unsatisfactory when the pyrimidine contains labile, easily hydrolyzed, or acid sensitive groups such as halogens or sugars. Thus 4,5-diamino-6-chloropyrimidine with formic acid furnished hypoxanthine ⟨54JA6073⟩, and a 2-methylthiopyrimidine and formamide ultimately gave the 2-aminopurine derivative ⟨62JOC2478⟩. In such cases 1-carbon transfer agents have been developed which will react with diaminopyrimidines under milder conditions than formic acid or formamide. These include dithioformic acid which was introduced during early investigations into the synthesis of purine nucleosides. Thioformamidopyrimidines are produced in aqueous solution with this reagent and may be cyclized by heating in solvents such as pyridine ⟨43JCS383⟩, as in the formation of 2-methyladenine (Scheme 73). The method has also been used to prepare 9-glycosylpurines although, if aqueous or alcoholic bases are used for the final cyclization step, the glycosyl derivative is always in the pyranose form even if the original (blocked usually by acylation) sugar was in the furanose form. However, 9-glycofuranosylpurines have been obtained in these reactions by heating the intermediate *N*-glycosyl-*N*-formamidopyrimidine in an aprotic solvent such as acetonitrile with added potassium acetate ⟨46JCS852⟩. However in the case of adenine precursors, mixtures of 9- and N^6-glycosyl derivatives may be obtained ⟨45JCS556⟩. The various problems associated with these nucleoside preparations, however, make them of little value today.

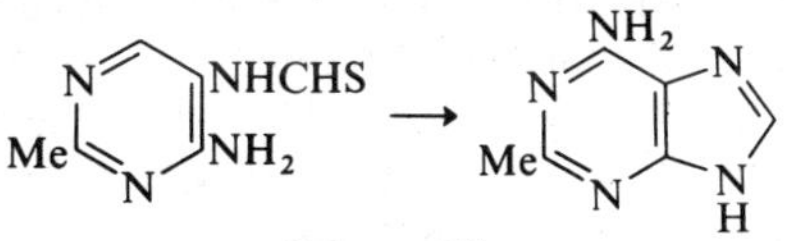

Scheme 73

Other cyclizing agents include the Vilsmeier–Haack reagent prepared from DMF and phosphoryl chloride ⟨27CB119⟩ which permits the formylation and cyclization of 4,5-diaminopyrimidines to be carried out at low temperatures ⟨61JCS5048⟩ and is especially valuable for the preparation of chloropurines from chloropyrimidines. A modified Vilsmeier–Haack procedure which involves cyclization of an intermediate azomethine (**253**) at low temperatures *in situ* has also been applied to chloropurine synthesis ⟨61JCS5048⟩.

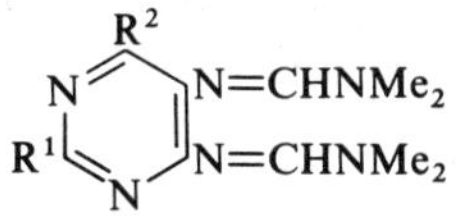

(**253**)

Triethyl orthoformate has proved to be a useful cyclizing agent, especially when mixed with acetic anhydride which produces diethoxymethyl acetate ⟨56JA1928, 56JOC599, B-78MI40903, p. 995⟩.

Whereas triethyl orthoformate will convert 4,5-diamino-2,6-dichloropyrimidine into 2,6-dichloropurine (Scheme 74), many other 4,5-diaminopyrimidines including 4,5-diaminopyrimidine and 2- and 6-chloro-4,5-diaminopyrimidines fail to cyclize and instead produce intermediate ethoxymethylene derivatives (Scheme 75) ⟨60JOC395⟩.

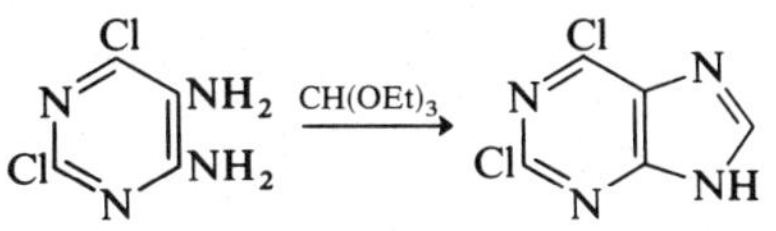

Scheme 74 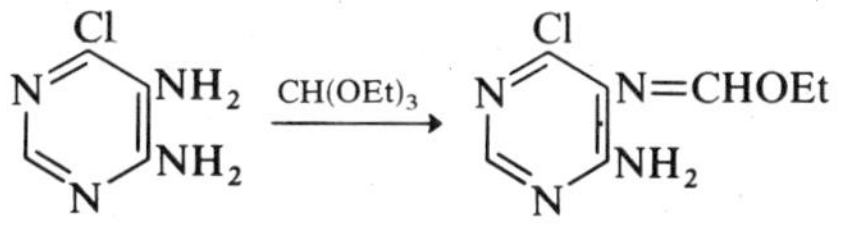   **Scheme 75**

Diethoxymethyl acetate, on the other hand, permits ready synthesis of chloropurines under mild conditions and even 6-fluoropurines, although not 6-fluoropurine itself, have been prepared by this method ⟨62JMC1067⟩. The reagent will also convert 4,5-diaminopyrimidine into purine ⟨60JOC395⟩. The method however is not readily adapted to nucleoside synthesis since under the reaction conditions the sugar is usually lost. However a 1-(β-D-ribofuranosyl)-substituted purine has been obtained using this reagent ⟨61JOC526⟩.

4.09.7.3.3 C(8)-Substituted purines

(i) 8-Alkylpurines

8-Alkylpurines may in general be prepared by methods which closely parallel those used for the unsubstituted derivatives although closure of the imidazole ring is more difficult to achieve, and prolonged heating in acid solution is necessary. Normally, however, the intermediate *N*-acyl derivative may be isolated and cyclized using aqueous alkali, fusion, or similar heat treatment. Thus acetic acid and several homologues with 4,5-diamino-1,3-dimethylpyrimidine gave intermediate 5-acylamidopyrimidines which could then be cyclized to 8-alkyltheophylline derivatives (Scheme 76) ⟨53JA114⟩. Use of hydroxy acids in these

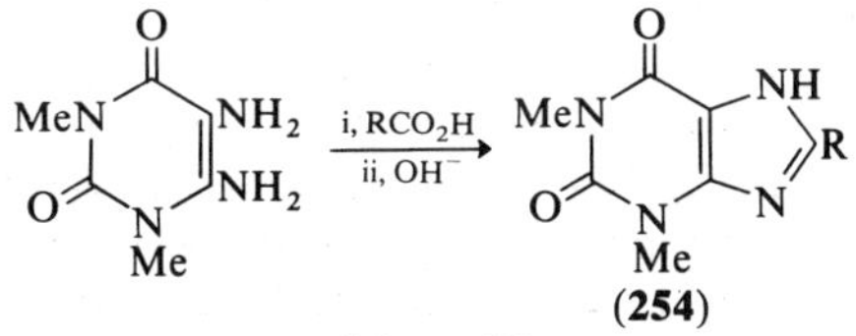

(**254**)

Scheme 76

reactions is also possible; thus lactic acid produced 8-α-hydroxyethylpurine (**254**; R = CHOHMe) ⟨56AP453⟩.

Alkanoic acid anhydrides and to a lesser extent acid chlorides have also been used and intermediate 9-acyl derivatives are then more readily produced. Using basic pyrimidines such as those which lead to adenines or alkylated purines, direct cyclization may be achieved under these conditions. Thus acetic anhydride and 4,5-diamino-1,3-dimethyl-6-oxo-1,2,3,6-tetrahydropyrimidine alone produced 2,3,8-trimethylxanthine ⟨23LA(432)266⟩ but the yield is improved in the presence of pyridine ⟨53CB333⟩. The intermediate N-acetyl derivative (**255**), however, when fused produced the betaine (**256**; Scheme 77) ⟨59CB566⟩. Trifluoroacetic acid or anhydride have similarly been used to prepare 8-trifluoromethylpurines including 8-trifluoromethylpurine itself ⟨58JA5744⟩ and 8-trifluoromethylguanine ⟨72CB1497⟩.

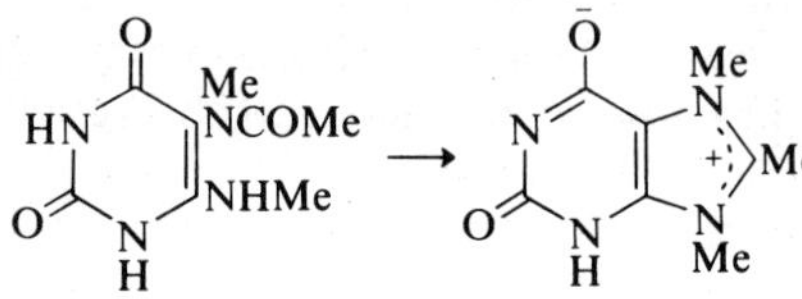

Scheme 77

With thioxopurines, thiazolopyrimidines are usually obtained. Thus 4,5-diamino-6-thioxo-1,6-dihydropyrimidine with acetic anhydride mainly afforded the thiazolopyrimidine (**257**; Scheme 78) ⟨59JA193⟩ which unlike the unsubstituted derivative was stable to alkali treatment and hence failed to produce the 6-thioxopurine by this route. However 5-acetylamino-6-aminopyrimidine-2-thione with phosphorus pentasulfide in pyridine gave a mixture of 6-oxo-2-thioxo- and 2,6-dithioxo-purines (Scheme 79).

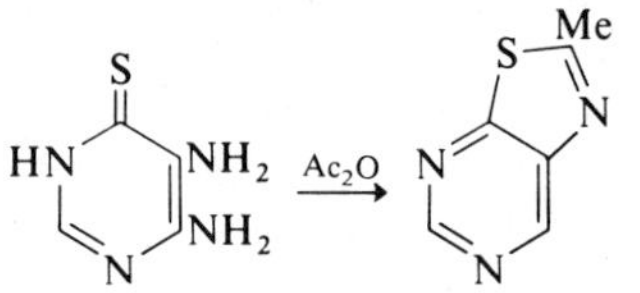

Scheme 78   Scheme 79

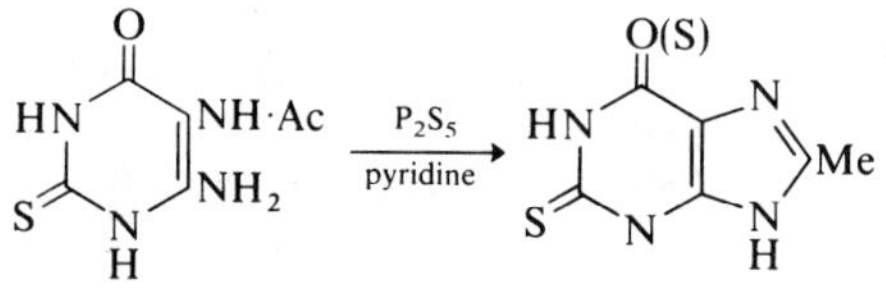

Acid chlorides have proved to be mainly of value for the preparation of di(purin-8-yl)alkanes. Thus succinyl chloride with 4,5-diamino-6-oxo-1,6-dihydropyrimidine furnished a diamide which after cyclization gave a bispurine (Scheme 80) ⟨61JOC1914⟩.

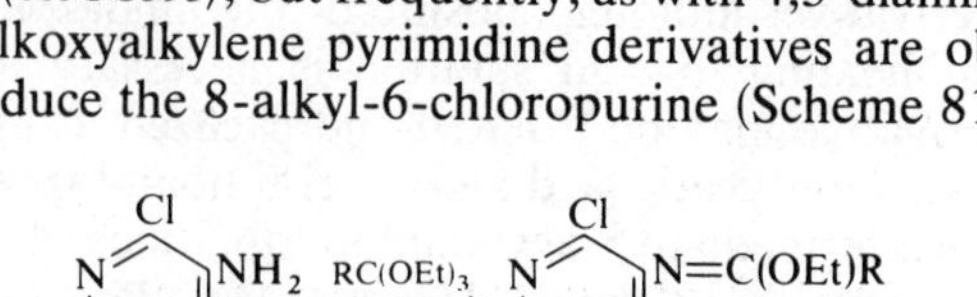

Scheme 80

Ortho esters have also been used to prepare 8-alkylpurines, especially triethyl orthoacetate or orthopropionate ⟨60JOC395⟩, but frequently, as with 4,5-diamino-6-chloropyrimidine, only the intermediate alkoxyalkylene pyrimidine derivatives are obtained. These require further treatment to produce the 8-alkyl-6-chloropurine (Scheme 81).

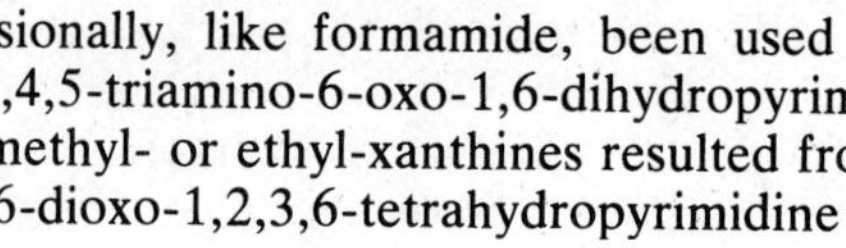

Scheme 81

Acid amides have occasionally, like formamide, been used to produce 8-substituted purines. Acetamide and 2,4,5-triamino-6-oxo-1,6-dihydropyrimidine afforded 8-methylguanine ⟨60JA2633⟩ and 8-methyl- or ethyl-xanthines resulted from acetamide or propionamide and 4,5-diamino-2,6-dioxo-1,2,3,6-tetrahydropyrimidine ⟨53CB333⟩. Diamides such as malonamide and succinamide similarly afforded bistheophylline derivatives (**258**; $n = 1$ or 2), respectively, when heated with the appropriate diaminopyrimidine.

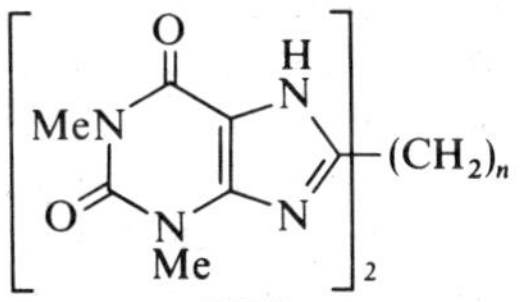

(**258**)

Acetamidine similarly has been used to prepare 8-methylpurines, usually by fusing the hydrochloride with the diaminopyrimidine in the presence of sodium acetate ⟨61JCS4468⟩. The method has also been used for making 8-methylthiopurines and 7-alkyl-8-methylpurines ⟨67CB2280⟩. A few 8-methylpurines have been prepared using an aldehyde, especially as the bisulfite adduct, and an appropriate oxidizing agent ⟨65JHC453⟩. Thus the dioxopyrimidine (**259**) and the adduct of ethanal with sodium hydrogen sulfite gave 1,3,8-trimethylxanthine (**260**) under mild conditions (Scheme 82). In this example the dihydropurine intermediate (**261**) is assumed to be oxidized by the hydrogen sulfite anion. Acetylacetone and (**259**) similarly gave the 8-methyl derivative (**260**) via the azomethine (**262**).

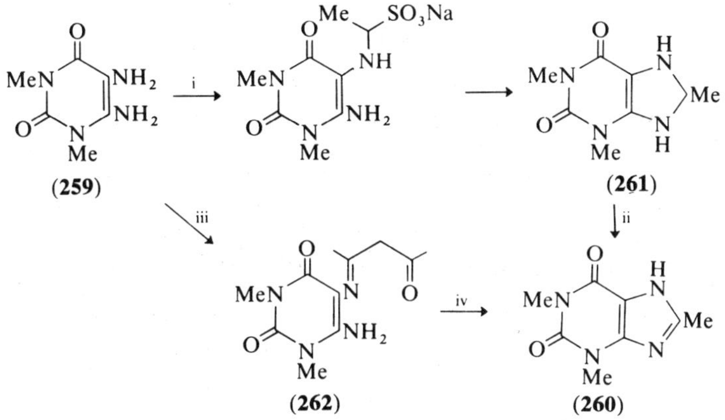

i, MeCH(OH)SO$_3$Na; ii, HSO$_3^-$; iii, MeCOCH$_2$COMe; iv, heat

Scheme 82

(ii) 8-Arylpurines

Several 8-arylpurines have been prepared using the Traube synthesis although little use has been made of free aromatic acids as cyclization agents. However 8-phenylguanine has been prepared from 4,5-diamino-6-oxo-1,6-dihydropyrimidine with benzoic acid in polyphosphoric acid ⟨67JMC109⟩, and phosphoryl chloride has also been used as a cyclization agent ⟨65MI40900⟩. With oxopyrimidines, use of phosphoryl chloride may lead to chloropurines and several 2-amino-6-chloro(bromo)-8-(substituted phenyl)purines have been prepared in this way ⟨52JA4897, 51JA5235⟩. It has been shown that halogenation precedes cyclization in this reaction process.

In contrast to the sparse use of acids, aroyl chlorides, no doubt because of their availability and greater reactivity, have been widely used to provide 8-arylpurines via intermediate aroylamidopyrimidines. These can usually be prepared by reaction of the appropriate diaminopyrimidine with the aroyl chloride in aqueous alkali. The intermediate amides may be cyclized by heating the free pyrimidine or, on occasions, its sodium salt, and in some cases, especially with methylated pyrimidines, spontaneous cyclization can occur during the acylation process.

Examples include 4-amino-5-benzamidopyrimidine which gave 8-phenylpurine when heated briefly at 200 °C ⟨54JCS2060⟩ and 2-amino-, 6-amino- or 2,6-diamino-8-arylpurines were made in a similar manner ⟨52JA4897, 51JA5235⟩. On the other hand, 8-phenyltheophylline was readily formed from the corresponding benzamidopyrimidine and aqueous alkali ⟨65MI40900⟩.

Aromatic amides have been rarely used for Traube cyclizations but examples do occur including 2,4-diamino-5-benzamido-6-oxo-1,6-dihydropyrimidine which furnished 8-phenylguanine when heated with benzamide ⟨51JA5235⟩. Benzamidine has also been used to produce 8-phenylpurines ⟨66JCS(C)10⟩, 8-phenyloxopurines ⟨61JCS4468⟩, N-phenyl-N-alkylpurines ⟨67CB2280⟩ and some 8-(2- and 3-pyridyl)purines were obtained similarly from corresponding nicotinamidines ⟨67JCS(C)1254⟩.

The use of an aldehyde and an oxidizing agent has been more generally applied to arylpurine than to alkylpurine synthesis. In this manner, 8-phenyltheophylline is readily obtained from the diaminouracil, benzaldehyde and oxygen plus a palladium catalyst (Scheme 83) ⟨54LA(590)232⟩. Intermediate 7,8-dihydropurines are probably involved in these reactions. When an excess of aldehyde is used, an 8-aryl-7-benzylpurine results ⟨06CB227⟩. Keto esters have also been used to produce 8-phenyltheophylline from (263) *via* an intermediate azomethine (265) ⟨59CB2902⟩ analogous to (264) obtained from benzaldehyde.

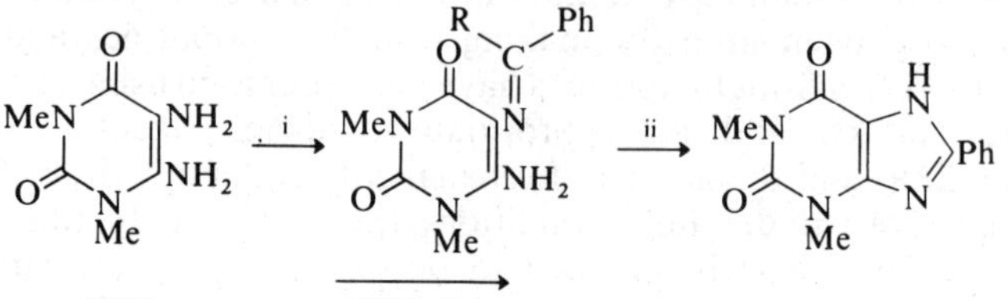

i, (a) PhCHO, (b) PhCOCH$_2$CO$_2$Et; ii O$_2$/Pd, heat 230 °C

Scheme 83

Alternative dehydrogenating agents have been used in these reactions. Thus 8-arylxanthines and 9-substituted 8-aryltheophyllines have been obtained by the hydrogenative cyclization of 6-amino-5-benzylideneaminouracils using diethyl azodicarboxylate ⟨78CPB2905⟩ and thionyl chloride ⟨78CPB3240⟩. 8-Arylaminotheophyllines have also resulted from 6-amino-8-arylazouracils ⟨78JHC64⟩, and 9-aryltheophyllines were obtained in a similar manner from 6-arylamino-5-arylazouracil ⟨78H(9)1437⟩.

(iii) 8-Oxopurines

Synthesis of an 8-oxopurine by the Traube route requires the condensation of a 4,5-diaminopyrimidine with a carbonic acid derivative. Such derivatives include carbon dioxide (rarely), alkyl chloroformates, carbonyl chloride, urea or substituted ureas and cyanic acid derivatives.

Carbon dioxide has been used under pressure and at high temperatures to convert the silylated pyrimidine (266) into uric acid in good yield ⟨60CB2810⟩ (Scheme 84) but otherwise it has found little application.

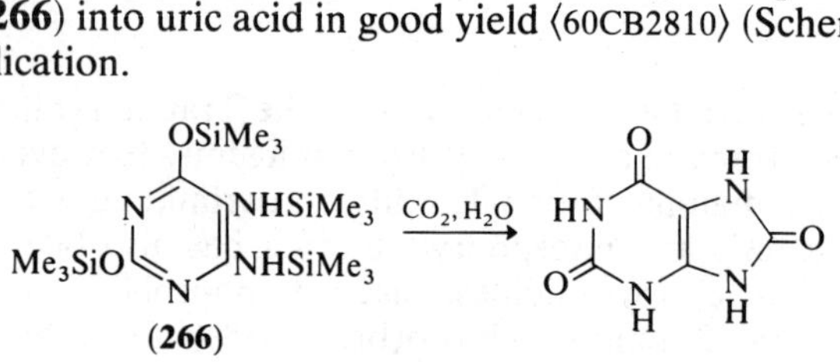

Scheme 84

In contrast alkyl chloroformates have been widely utilized and numerous examples have been described especially in the early literature. Intermediate urethanes are usually obtained as in the conversion of (267) with benzylchloroformate in aqueous sodium hydrogen carbonate into the 8-oxopurine (268) directly (Scheme 85) ⟨61JOC4961, 62MI40901⟩.

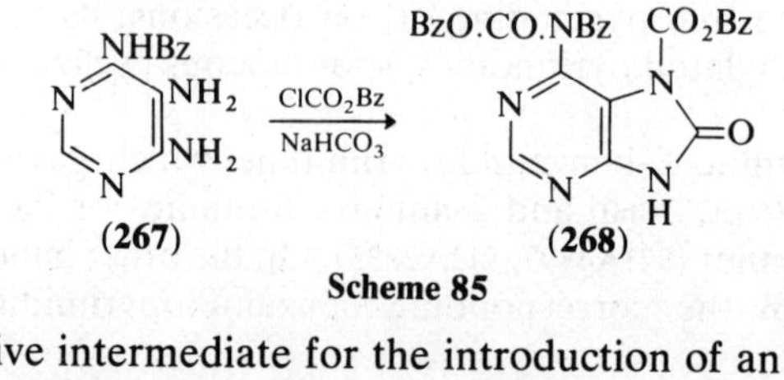

Scheme 85

An even more reactive intermediate for the introduction of an 8-oxo group is phosgene which will frequently produce the desired compound by reaction of a diaminopyrimidine in aqueous sodium hydroxide solution at room temperature. In this way 4,5,6-triaminopyrimidine gave the 8-oxopurine (269; Scheme 86) ⟨50JA2587⟩. Chloropurines including 6-chloro-8-oxo-7,8-dihydropurine have also been made by this method ⟨63JOC2677, 68JOC530⟩. On the other hand, 4,5-diaminouracil with phosgene failed to give uric acid ⟨60CB2810⟩ but trimethylsilylated uracil with phosgene in toluene containing triethylamine furnished uric acid after subsequent deblocking of the silylated purine.

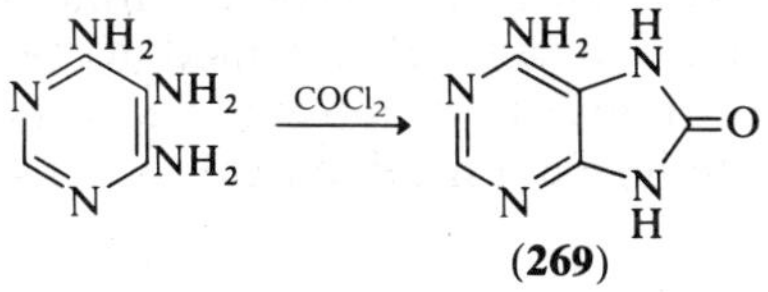

Scheme 86

Many 8-oxopurines have been produced by fusion of a 4,5-diaminopyrimidine with an excess of urea. In this manner 4,5-diaminouracil afforded uric acid (Scheme 87). It has been established by ^{15}N-labelling that the liberated ammonia in this reaction is supplied wholly by the urea nitrogen atoms ⟨48JA1240, 53HCA1324⟩. Similarly 8-oxo-7,8-dihydropurine was produced from 4,5-diaminopyrimidine and N,N'-dimethylurea (Scheme 88). When reactive substituents, e.g. chlorine atoms, are present, the method is usually unsatisfactory; however 2-chloro-8-oxo-7,8-dihydropurine ⟨60BBA(45)49⟩ and 9-benzyl-6-chloro-8-oxo-7,8-dihydropurine ⟨68JHC679⟩ have been prepared by the urea fusion method. Intermediate ureido derivatives are probably produced in all these reactions but not normally isolated under the vigorous reaction conditions employed. However potassium cyanate and 4,5-diaminouracil afforded a 5-ureido derivative which could be cyclized to uric acid by fusion ⟨23LA(432)266⟩ or by heating with hydrochloric acid ⟨16JBC(25)607⟩.

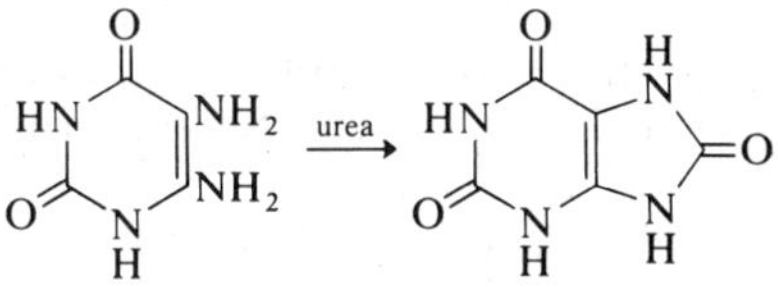

Scheme 87 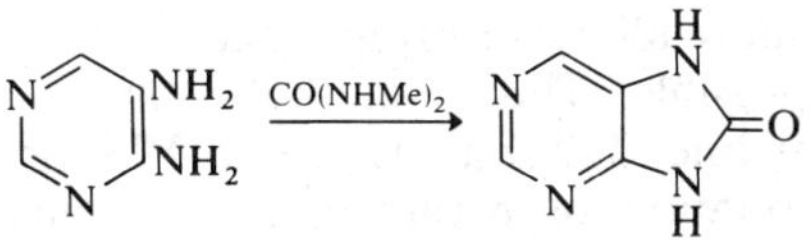   **Scheme 88**

Use of alkyl (or aryl) isocyanates in the same way offers a useful route to 9-alkyl (or aryl)-8-oxopurines. 9-Phenyluric acid is produced in this way by the reaction of 4,5-diaminouracil with phenyl isocyanate and cyclization of the intermediate phenyl-ureidopyrimidine with hydrochloric acid (Scheme 89) ⟨23LA(432)266⟩. It is noteworthy that the cyclization results in loss of ammonia and not aniline. In contrast to this reaction, fusion of the N-methylphenylureidopyrimidine (270) resulted in loss of aniline rather than ammonia (Scheme 90) ⟨66JCS(C)10⟩.

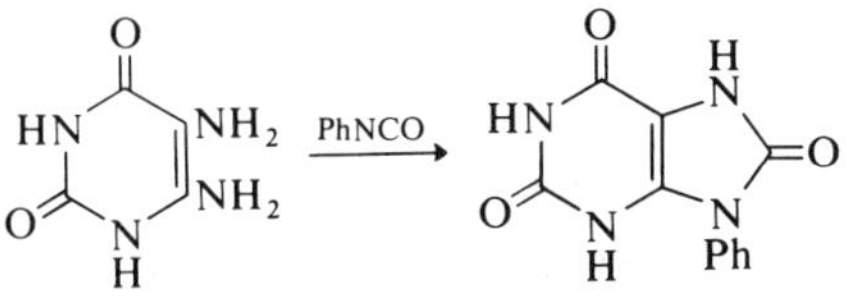

Scheme 89   **Scheme 90**

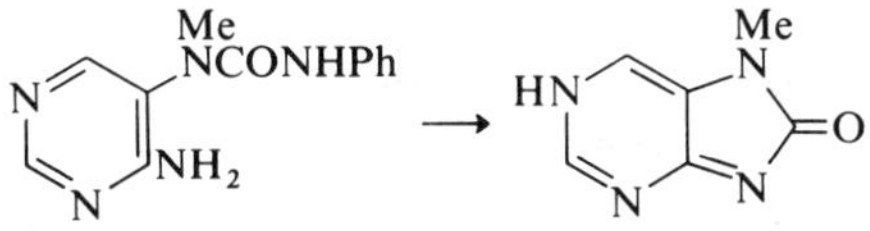

Tetraethyl carbonate has also been used to give 8-ethoxypurines including 8-ethoxy-7-methylpurine ⟨74JCS(P1)349⟩.

(iv) 8-Thioxopurines

8-Thioxopurines have generally been prepared by Traube methods from a diaminopyrimidine and either carbon disulfide, thiourea, thiophosgene or thiocyanic acid derivatives. Carbon disulfide, in contrast to carbon dioxide, has been widely used in Traube syntheses normally in conjunction with a base but it will also react directly. 4,5-Diaminouracil and carbon disulfide under pressure gave 8-thiouric acid ⟨02GEP142468⟩, but in pyridine reflux conditions were sufficient for reaction to occur ⟨49JCS3001⟩. Various solvent combinations have been used according to the solubility of the intermediate diaminopyrimidine and these have included aqueous alkali ⟨66JMC160⟩, but reactions in DMF are reported to give better yields ⟨61JOC3386⟩. The method is relatively mild and in solvents such as pyridine hydrolysis of labile substituents can often be avoided.

Thus 4,5-diamino-2-chloropyrimidine produced 2-chloro-8-thioxo-7,8-dihydropurine ⟨63CJC1807⟩. On the other hand 6-chloropyrimidines furnished thiazolopyrimidines, including 4,5-diamino-6-chloropyrimidine which gave the thiazolopyrimidine (**271**; Scheme 91) ⟨61JOC3386⟩. Such compounds may usually be isomerized to 8-thioxopurines with aqueous alkali. When other leaving groups such as thiol or alkylthio are present in the 6-position of the pyrimidine molecule, similar results are obtained ⟨62JOC986⟩. In contrast 4-amino-5-methylaminopyrimidine failed to give an 8-thioxopurine with carbon disulfide, inhibition of intermediate dithiocarbamate formation being held responsible ⟨60JCS4347⟩.

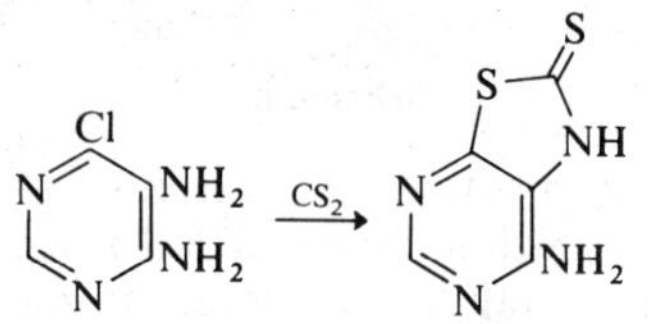

Scheme 91

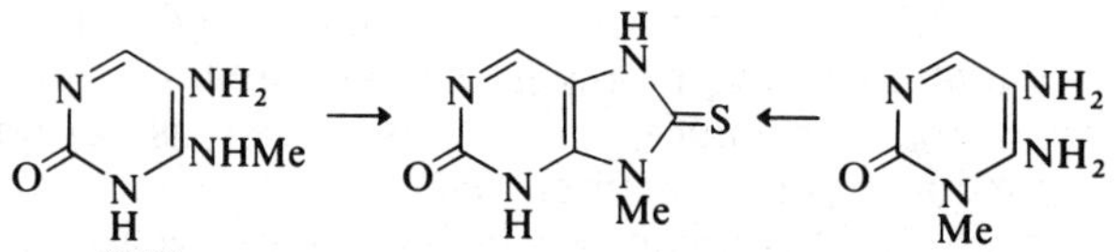

Scheme 92

With readily hydrolyzed (generally fully alkylated) pyrimidines Dimroth rearrangements are possible. Thus 9-methyl-8-thioxo-7,8-dihydropurine results, as expected, from the pyrimidine (**272**) but also from (**273**; Scheme 92) ⟨63JCS1276⟩.

Formation of 8-thioxopurines using thiourea parallels similar reactions with urea. However, more drastic reaction conditions are required and there is more scope for production of by-products. Thus the 2-thioxopyrimidine (**274**) gave the 2,8-dithioxopurine (**275**) ⟨13JBC(14)299⟩, guanidine thiocyanate being considered an intermediate in such fusion reactions (Scheme 93). This suggestion would account for the formation of the 8-aminopurine (**277**) from the 2-alkylthiopyrimidine (**276**; Scheme 94), and the same compound can be produced using guanidine thiocyanate directly ⟨13JBC(14)381⟩.

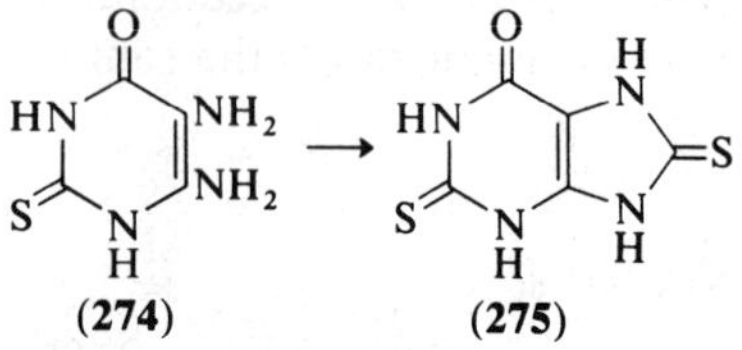

Scheme 93   **Scheme 94**

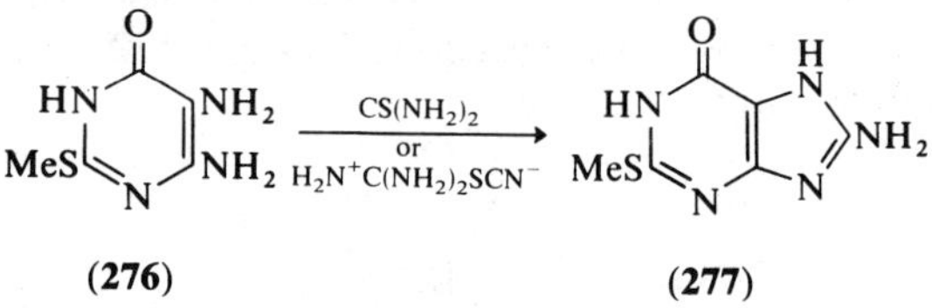

N,N'-Dimethylthiourea also produced 8-thioxopurines at lower temperatures than thiourea but the reaction may be complicated by S-methylation especially at higher temperatures as in the formation of 8-methylthiotheophylline (Scheme 95) ⟨65AF10⟩.

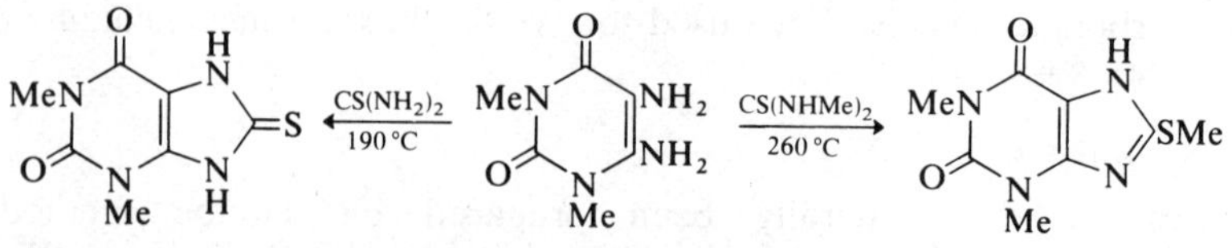

Scheme 95

Both alkyl and aryl isothiocyanates have been used to produce 9-substituted 8-thioxopurines *via* intermediate thioureidopyrimidines. Thus the pyrimidine (**278**) and methyl isothiocyanate gave the 9-methyl-8-thioxopurine (**279**) by loss of ammonia (Scheme 96) ⟨56JA5857⟩, and phenyl isothiocyanate similarly afforded the corresponding 9-phenyl-8-thioxopurine. However in the latter type of compound loss of aniline rather than ammonia may sometimes occur as in the formation of the 6-oxo-8-thioxopurine (**280**) from the phenylthioureidopyrimidine (**281**) (Scheme 97) ⟨47JCS943⟩. Also, in certain circumstances, cyclization involving loss of hydrogen sulfide may occur to afford an 8-aminopurine as in the conversion of (282) into (283) in the presence of mercury(II) oxide (Scheme 98) ⟨57CPB240⟩.

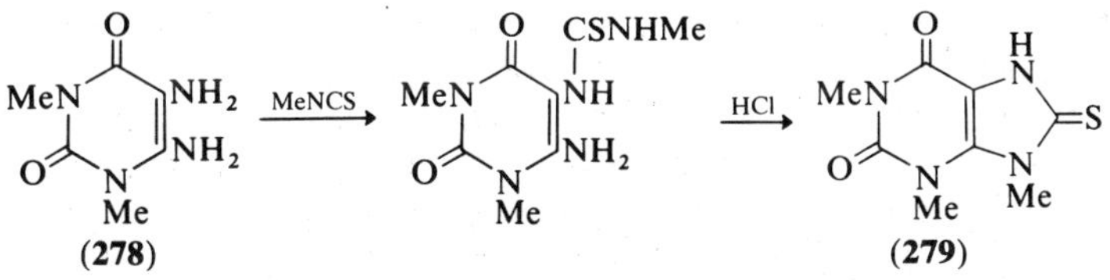

Scheme 96

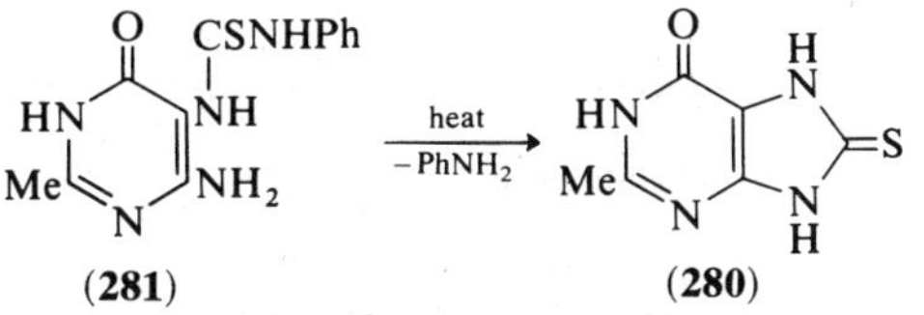

Scheme 97

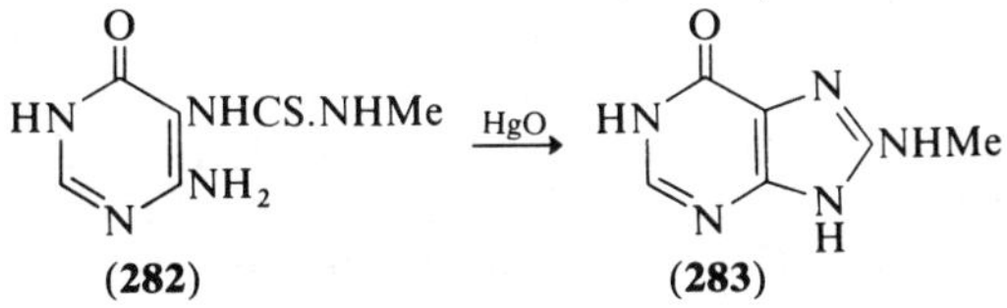

Scheme 98

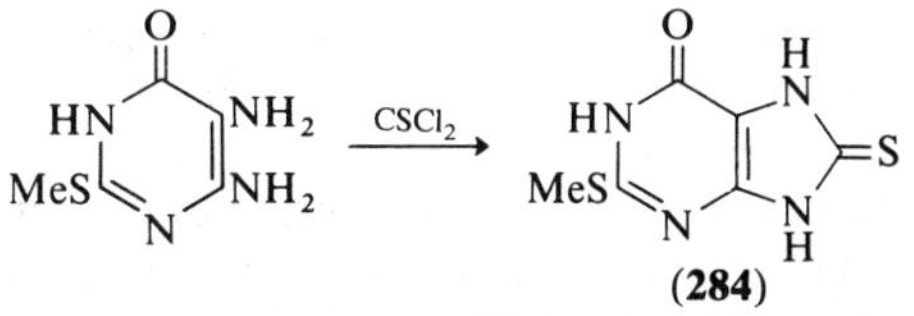

Scheme 99

The use of thiophosgene in Traube reactions has probably been limited by its relative inaccessibility but applications have been made under conditions analogous to those used with phosgene as in the preparation of the 2-methylthio-6-oxo-8-thioxopurine (284; Scheme 99) ⟨13JBC(15)515⟩.

(v) 8-Aminopurines

Guanidine does not react too readily with diaminopyrimidines although a few examples leading to 8-aminopurines have been recorded, including the 2- and 6-oxo and 6-thioxo (but not 2-thioxo) derivatives prepared from appropriate diaminopyrimidines ⟨63CJC1807⟩ (see also Section 4.09.5.2.2). Cyanogen bromide and 4,5-diaminopyrimidine gave a low yield of 8-aminopurine ⟨54JCS2060⟩ whereas 4,5-diamino-6-thioxo-1,6-dihydropyrimidine with cyanogen bromide gave a 35% yield of 8-amino-6-thioxo-1,6-dihydropurine (Scheme 100) ⟨66JOC1417⟩.

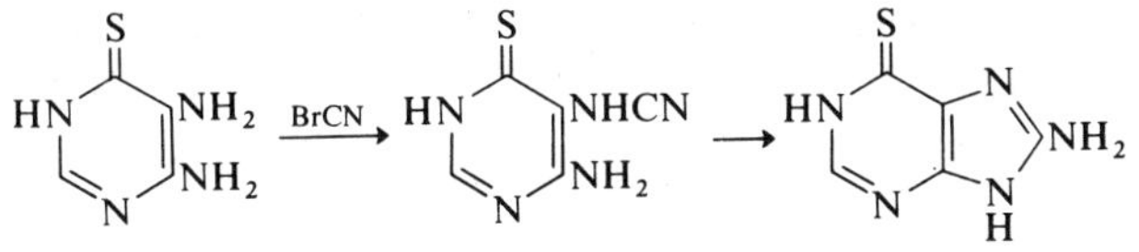

Scheme 100

Intermediate arylguanidinopyrimidines (285) produced from diphenylcarbodiimide and 5-alkyl(or aryl)amino-4-aminopyrimidines furnished 8-anilino-7-alkyl(or aryl)purines (286) in hot DMF (Scheme 101).

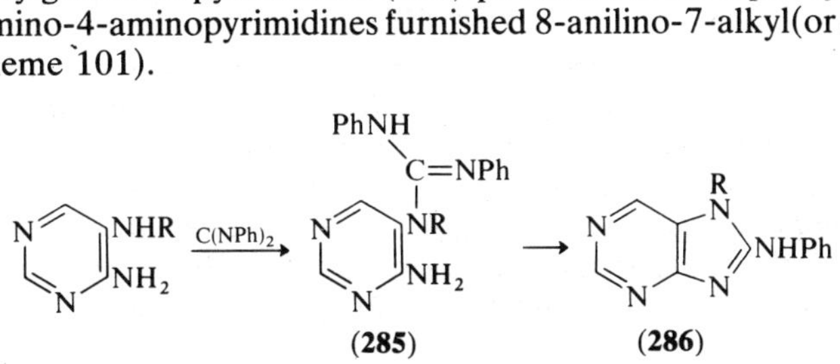

Scheme 101

4.09.7.4 Synthesis from Other 4,5-Disubstituted Pyrimidines

A minor extension of the Traube synthesis involves the reduction of a 4,5-diaminopyrimidine precursor in the presence of an acylating agent such as formic acid.

Appropriate precursors which have been used include 4-amino-5-nitroso-, 4-amino-5-nitro- and 4-amino-5-phenylazo-pyrimidines and 4-acylamino derivatives of these compounds. Various reducing agents and conditions have been used and these are described in Table 38.

Table 38 Synthesis of Purines from Nitroso- or Nitro-pyrimidines by Reductive Acylation

Pyrimidine	Purine formed	Conditions	Ref.
4-NH_2-5-NO	Purine	90% HCO_2H/Zn	62MI40903
5-NH_2-6-NO-uracil	Xanthine	Ni/H_2/NaOAc	62MI40903
5-NH_2-6-NO-uracil	Xanthine	H_2S/$HCONH_2$	54MI40903
5-NH_2-6-NO-uracil	Xanthine	$HCONH_2$/$Na_2S_2O_4$	55CB1306, 57JA1518
2-Ph-5-PhN_2-4,6-di-NH_2	2-Phenyladenine	H_2S/DMF/$(EtO)_3CH$/heat, 6 h	65JA1976
4-NH_2-5-NO_2	Adenine	Ni/H_2	65MI40907
4-Acetamido-6-p-acetamido-styryl-2-diethyl-5-nitro	2-Diethylamino-8-methyl-6-p-acetamidostyryl	Ni/H_2	60JCS1113, 67AG(E)258

In addition to the simple extension of the Traube method attention has been turned in recent years to the direct cyclization of purines from 4-alkylamino-5-nitrosopyrimidines. Thus the benzylaminonitrosopyrimidine (**287**) (which may be prepared directly from the benzylaminopyrimidine and pentyl nitrite and used *in situ*) cyclized in hot xylene or *n*-butanol to afford 8-phenyltheophylline (Scheme 102) ⟨66LA(691)142⟩, and the related methylaminopyrimidine similarly cyclized to theophylline ⟨64N137⟩. The vigorous conditions required for cyclization may be much reduced by employing *O*-acyl derivatives such as (**288**) which cyclized in hot ethanol to give theophylline ⟨67AG(E)258⟩. Analogously, *N*-methylation of the nitrosopyrimidine gave, *via* the intermediate *N*-oxide (**289**), smooth cyclization to caffeine ⟨66LA(698)145⟩.

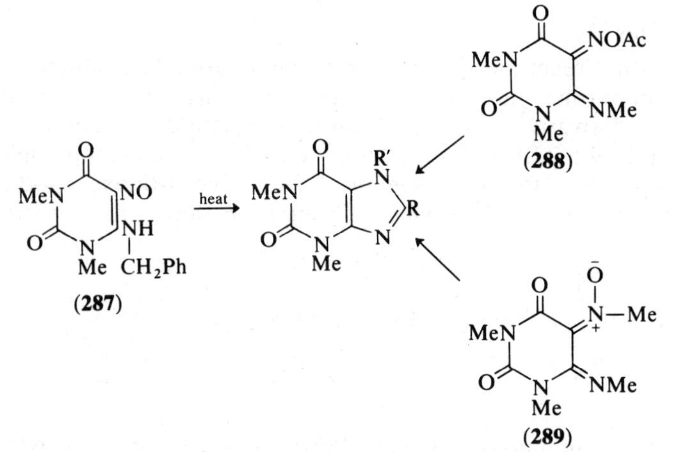

Scheme 102

The use of 4-*s*-alkylamino-5-nitrosopyrimidines can still produce purines as in the case of the 4-isopropylamino-5-nitrosopyrimidine (**290**) which afforded the 8,8-dimethylpurine (**291**). This readily isomerized to 8-methylcaffeine when melted (Scheme 103) ⟨64ZC454⟩. The method may be modified in several ways; thus the 4-amino-5-nitrosopyrimidine with methylamine or benzylamine gave theophylline or 8-phenyltheophylline, respectively ⟨66LA(691)142⟩. Similarly the Vilsmeier–Haack reagent (DMF plus phosphoryl chloride) has also been used to produce 8-dimethylaminotheophylline (Scheme 104) ⟨73BCJ1836⟩.

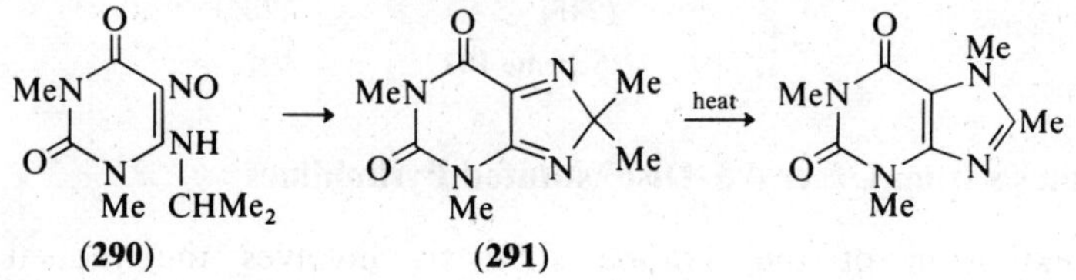

Scheme 103

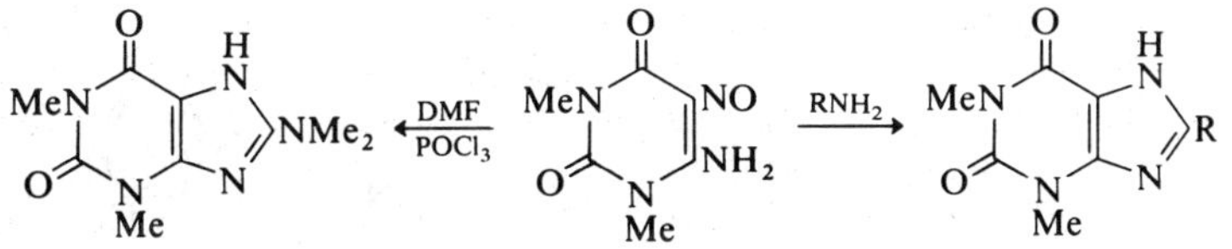

Scheme 104

A variation of these reactions includes introduction of the second nitrogen into a monoalkylaminopyrimidine as in the reaction of the pyrimidine (**292**) with nitrosobenzene which in acetic anhydride gave 7-phenyltheophylline directly (Scheme 105) ⟨72JOC4464⟩. Similarly 6-benzylaminouracil and diethyl azodicarboxylate produced the purine (**293**) *via* the intermediate (**294**; Scheme 106) ⟨75CC146⟩, and the related urazolylpyrimidine (**295**) gave 8-arylpurines with aromatic aldehydes (Scheme 107) ⟨74CC551⟩. In the same vein the 4-benzylaminopyrimidine (**296**) with pentyl nitrite followed by ethanolic hydrogen chloride afforded 8-phenyltheophylline (Scheme 108) ⟨66LA(691)142⟩.

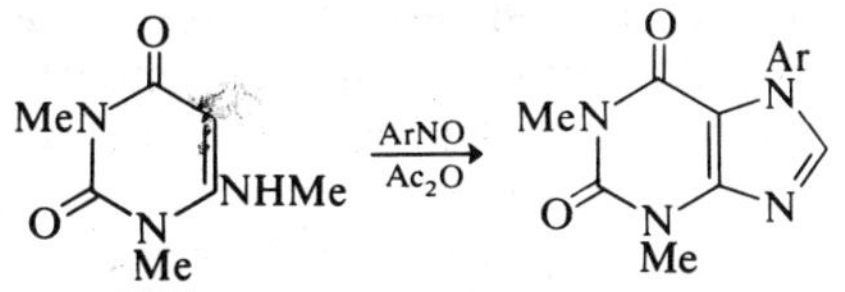

Scheme 105

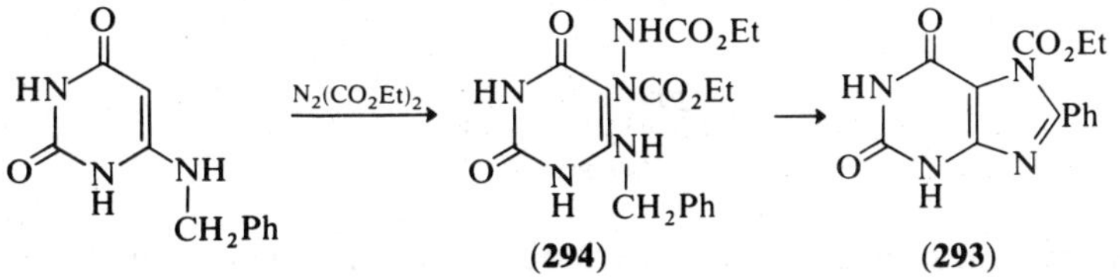

Scheme 106

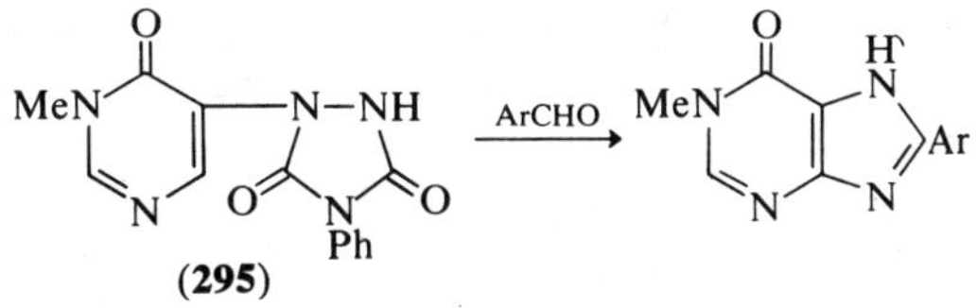

Scheme 107

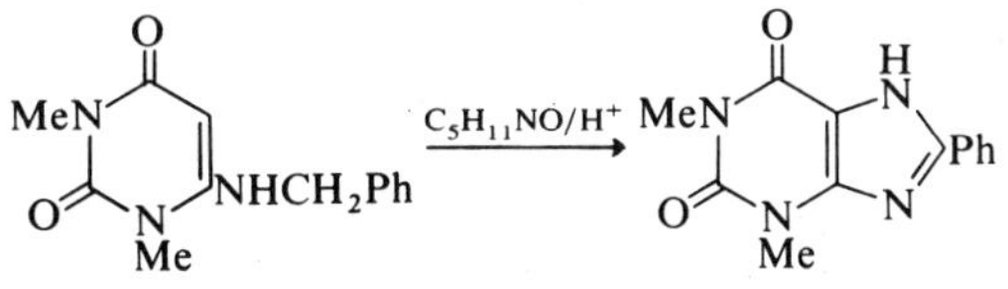

Scheme 108

Benzaldehyde will also react with nitrosoaminopyrimidines but mixtures of the 8-aryl-purines and purine 7-oxides are obtained. The latter can be avoided and good yields of 8-arylpurines obtained by the use of the aldehyde hydrazone ⟨70CC1068⟩ or *N,N*-dimethylhydrazone ⟨76JCS(P1)1547⟩. The Wittig reagent derived from benzyl bromide and triphenylphosphine has also been used to produce 8-arylpurines directly (Scheme 109) ⟨76CC155⟩. Phenylazopyrimidines behave like their nitroso analogues. Thus heating the derivative (**297**) produced theophylline (Scheme 110).

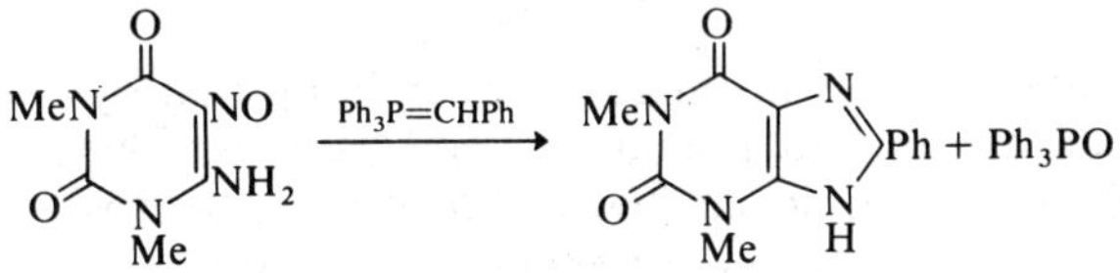

Scheme 109

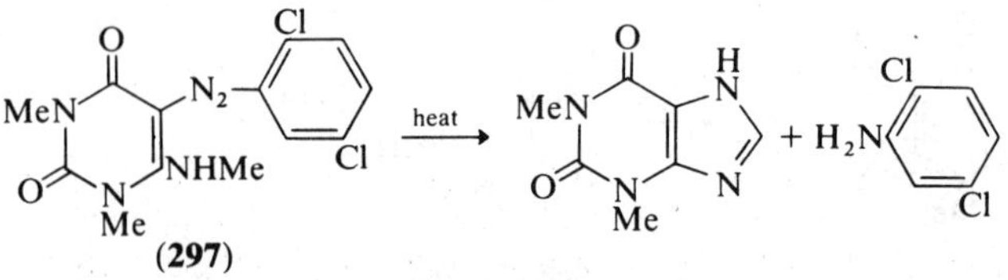

Scheme 110

4.09.7.4.1 Synthesis from 5-amino-4-oxopyrimidines

The method has been largely confined to the synthesis of uric acid and its 7- and 9-alkyl derivatives by reaction of the appropriate 5-amino-4-oxopyrimidine with urea or alkylureas, cyanates or isocyanates. Thus 5-aminobarbituric acid when heated with urea gave uric acid ⟨48JA1240⟩. Potassium cyanate can replace urea and 9-substituted uric acids (**298**) resulted from reaction with an alkyl or aryl isocyanate *via* the intermediate ureidopyrimidine (**299**; Scheme 111) ⟨21LA(423)242⟩. 5-Alkylaminopyrimidines similarly furnished 7-alkyluric acids with urea or potassium cyanate ⟨65B650⟩.

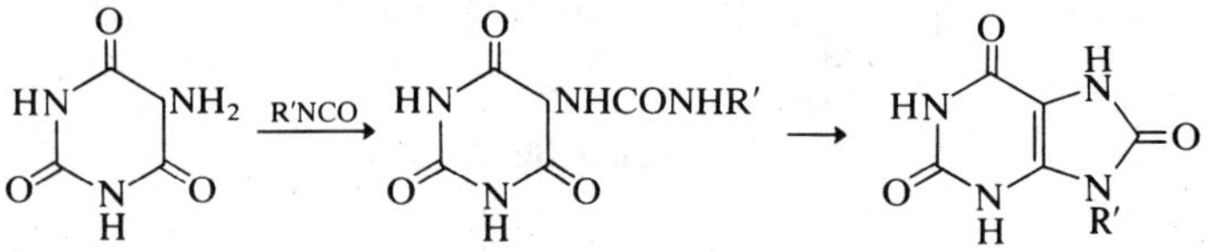

Scheme 111

Alkyl isothiocyanates also afford alkylthioureidopyrimidines, usually by reaction in aqueous alkali, and these may be cyclized by hydrochloric acid to produce 9-alkyl-8-thioxopurines. In this manner 5-aminobarbituric acid furnished 9-alkyl-8-thiouric acids ⟨21LA(423)242⟩. 8-Thiouric acid (**300**) was also obtained by formation of a thioureidopyrimidine *in situ* from the cyanamidopyrimidine (**301**; Scheme 112) ⟨02CB2563⟩.

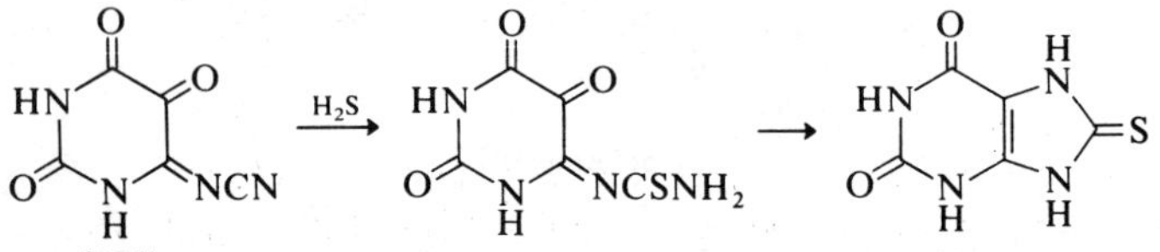

Scheme 112

4.09.7.4.2 Synthesis from 4,5-dioxopyrimidines

Only a few examples have been recorded using ureas as the imidazole nitrogen source. Thus isodialuric acid and urea afforded uric acid when heated in dilute acid and this is one of the earliest recorded preparations of this compound ⟨1889LA(251)235⟩. *N*-Methyl and *N,N*-dimethylureas similarly gave low (10%) yields of 7-methyl- and 7,9-dimethyl-uric acid, respectively ⟨25LA(441)203⟩.

4.09.7.4.3 Synthesis from 4-aminopyrimidine-5-carboxamides

Using a carboxamide group as a source of an amino group by applying Hoffman type reactions, it has been possible to prepare several purines. Thus 4-aminopyrimidine-5-carboxamide and alkaline hypochlorite gave a 60% yield of 8-oxo-7,8-dihydropurine (**302**) ⟨62CB956⟩, and the same authors achieved a Curtius reaction with a 4-aminopyrimidine-5-carbohydrazide and nitrous acid (Scheme 113).

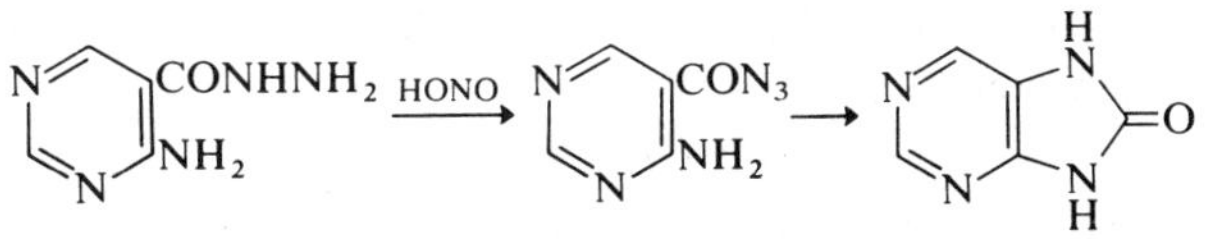

Scheme 113

4.09.7.5 Purine Synthesis from Imidazoles

4.09.7.5.1 Introduction

The preparation of purines *via* an appropriate pyrimidine dominated the synthetic chemistry of purine especially in the early pre Second World War literature. The reason for this is undoubtedly connected with the difficulty of obtaining suitable imidazole, compared with pyrimidine, precursors and additionally by the tendency of imidazole precursors to be rather labile and prone to aerial oxidation. To some extent these disadvantages have been overcome in recent years and this particular route to purines including nucleosides and nucleotides has been used increasingly. The method of course is of special significance in that it is the route adopted in living systems for the *de novo* biosynthesis of purine nucleotides, and interestingly it also appears to be the route favoured in the so-called abiotic syntheses from simple acyclic precursors (see Section 4.09.8.1).

Almost all recorded purine syntheses from imidazoles involve the cyclization of 5(4)-aminoimidazole-4(5)-carboxylic acid derivatives especially the carboxamides, thiocarboxamides, carboxamidines, carboxamidoximes, nitriles and esters. The intermediates used for completion of the purine ring are much the same as have been used for Traube cyclization of diaminopyrimidines (Section 4.09.7.3), especially formic and carbonic acid derivatives, and cyclization generally occurs under much milder conditions. This feature has been of special value in the synthesis of purine nucleosides from imidazole nucleoside precursors. The resultant purine will have variable substituents at C-2 and C-6 and it is convenient to discuss and classify the various preparations largely in terms of the introduced 2-substituents. The C-6 substituents largely reflect the type of carboxylic acid moiety used and do not vary very much between amino, oxo and thioxo.

4.09.7.5.2 C(2)-Unsubstituted purines

Formic–acetic anhydride has been widely used with aminoimidazoles to furnish an *N*-formyl derivative which will cyclize by prolonged heating or better by treatment with a weak base. In this manner 5-aminoimidazole-4-carboxamide produced hypoxanthine ⟨50JBC(185)439⟩ *via* the *N*-formyl derivative (**303**) which readily cyclized when warmed with aqueous sodium hydrogen carbonate ⟨65JHC253⟩.

Similar methods have been used to prepare nucleosides including inosine ⟨62JCS2937⟩ which results from the 1-β-D-ribofuranosylimidazole (**304**; Scheme 114) and in a similar manner, 7-β-D-ribofuranosylhypoxanthine (**306**) was obtained from the isomeric aminoimidazole nucleoside (**307**; Scheme 115) ⟨59JCS2893⟩. 1-Benzylinosine has also been prepared by cyclization of the aminoimidazolebenzylamide (**305**) with cold aqueous alkali ⟨58JA3899⟩. In like manner, 9-methyl- and 8,9-dimethyl-6-thioxo-1,6-dihydropurine (**308**)

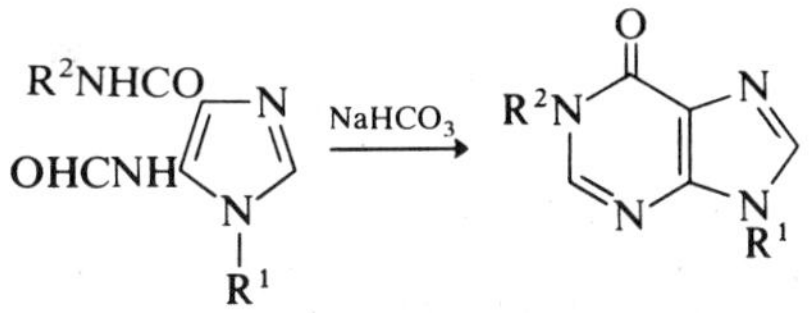

(**303**) $R^1 = R^2 = H$
(**304**) $R^1 = R_f^\beta, R^2 = H$
(**305**) $R^1 = H, R^2 = PhCH_2$

Scheme 114

and (**309**), respectively, were obtained by cyclization of the corresponding formamidoimidazolethioamides (Scheme 116) ⟨59JCS4040⟩.

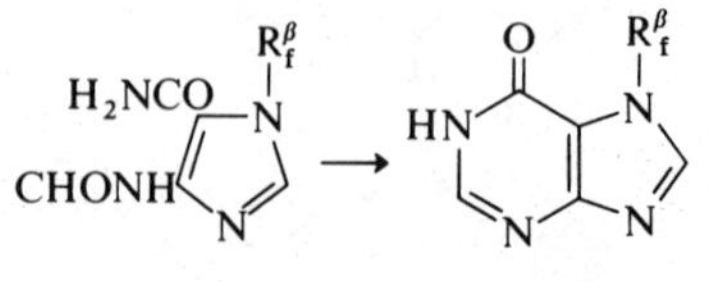

Scheme 115

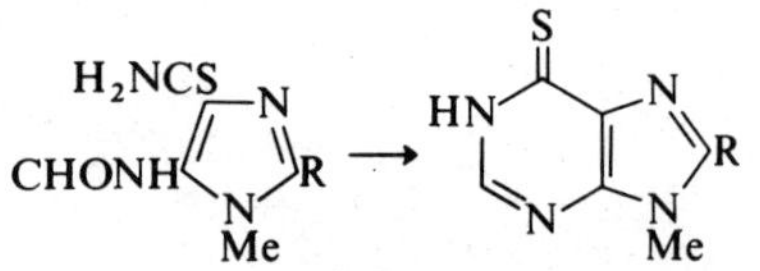

Scheme 116

In contrast to its application in Traube syntheses (Section 4.09.7.3.2) formamide has been little used as a one-carbon cyclizing agent with aminoimidazoles. However 7-methylhypoxanthine (**310**) was obtained by heating 4-amino-1-methylimidazole-5-carboxamide in formamide (Scheme 117) ⟨57JA6401⟩. 9-Benzylinosine was produced from (**311**) and formamide (Scheme 118) ⟨58JA3899⟩, thioinosine was similarly obtained from the corresponding aminoimidazolethioamide (**312**; Scheme 119) ⟨68CPB2172⟩ and in an analogous reaction, adenine was obtained from the aminoimidazolenitrile (**313**) and formamide (Scheme 120) ⟨57JA6401⟩.

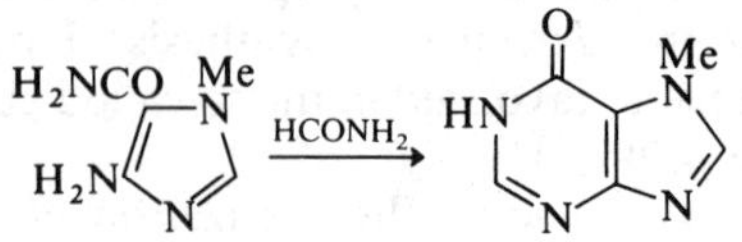

Scheme 117

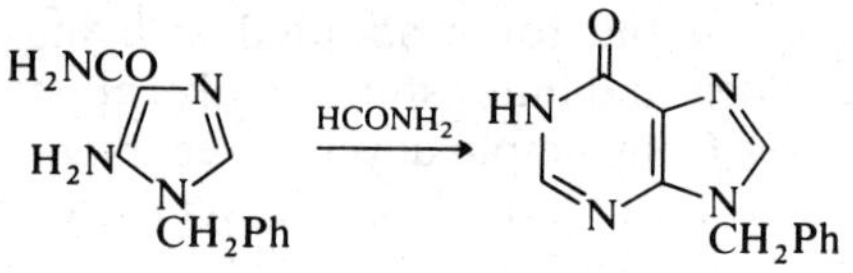

Scheme 118

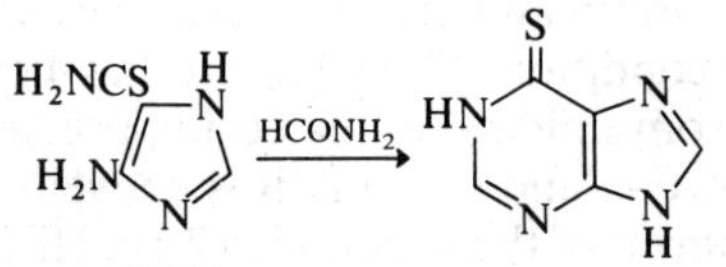

Scheme 119

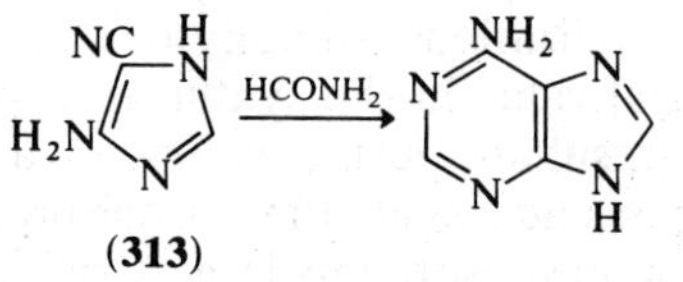

Scheme 120

An excellent yield of adenine was also obtained by heating the aminoimidazolenitrile (**313**) with formamidine acetate ⟨65JA4976, 66JA3829⟩ and 3-methylhypoxanthine was similarly produced from the corresponding aminoimidazole ester and formamidine acetate ⟨79JCS(P1)3107⟩.

Ethyl formate has also been used occasionally as a cyclizing agent as in a preparation of hypoxanthine from 5-aminoimidazole-4-carboxamide in hot ethanolic sodium ethoxide solution ⟨67JOC3258⟩ but triethyl orthoformate has been more widely used. Ring closure may occur directly with this reagent as in the preparation of 8-methylhypoxanthine from the 2-methylimidazole (**314**) using triethyl orthoformate in DMF (Scheme 121) ⟨60JA3144⟩. Hypoxanthine 1-oxide was similarly produced from 4-amino-5-(*N*-hydroxy)carbamoylimidazole (**315**; Scheme 122) ⟨59JOC2019⟩.

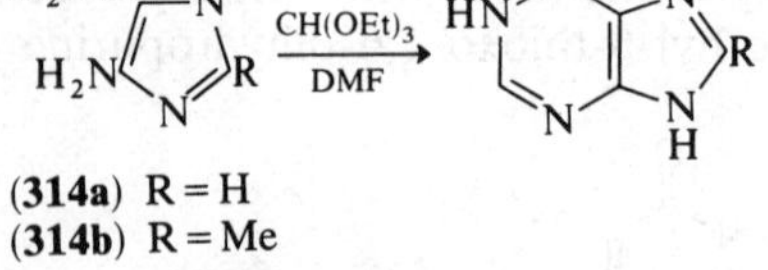

(**314a**) R = H
(**314b**) R = Me

Scheme 121

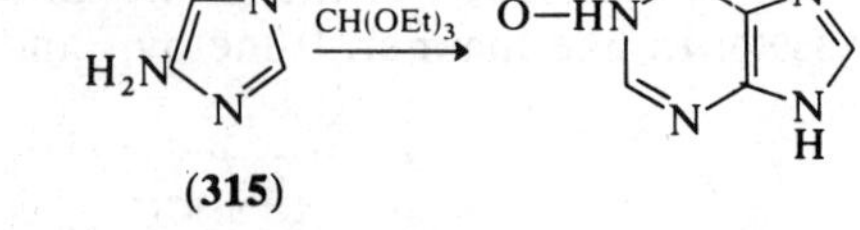

Scheme 122

Glycosyl *N*-oxides have also been prepared from imidazole precursors. Thus the ethoxymethylidene derivatives (**316**) and hydroxylamine gave, presumably *via* the hydroxyaminomethylene derivatives, isopropylideneadenosine (and 8-methyladenosine) 1-oxides (**317**) and (**318**), respectively, in 35% overall yield (Scheme 123) ⟨69CPB1128⟩. Similarly (**316**) and methoxyamine furnished 9-(2,3-*O*-isopropylidene-β-D-ribofuranosyl)-6-methoxyaminopurine (**319**).

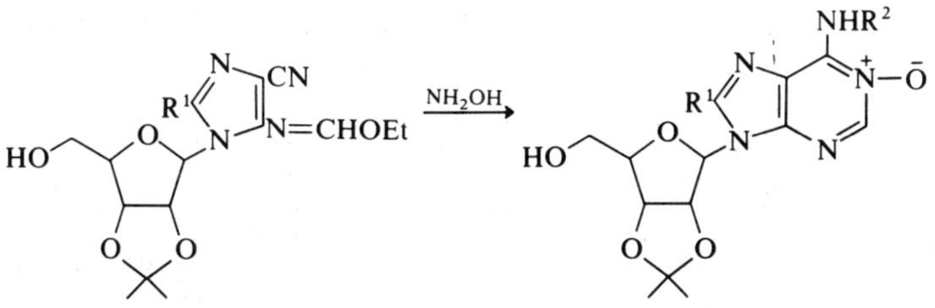

(316)

(317) R¹ = R² = H
(318) R¹ = Me, R² = H
(319) R¹ = H, R² = OMe

Scheme 123

Similar cyclizations occur using mixtures of triethyl orthoformate and acetic anhydride or diethoxyethyl acetate ⟨61JCS4845⟩, but in the absence of the anhydride the intermediate ethoxymethyleneamino derivatives may sometimes be isolated. Thus 5-amino-4-cyano-1-methylimidazole and triethyl orthoformate furnished the ethoxymethylene derivative (320). Compounds of this type have proved to be useful intermediates for purine synthesis. Thus (320) with ethanolic ammonia readily furnished 9-methyladenine (Scheme 124) ⟨59JCS4040⟩. 9-Cyclohexyladenine was similarly prepared ⟨80JCS(P1)2728⟩ using cold alcoholic ammonia. The reaction has been extended to include a glucopyranosylimidazole which produced 9-glucopyranosyladenine ⟨75CC964⟩.

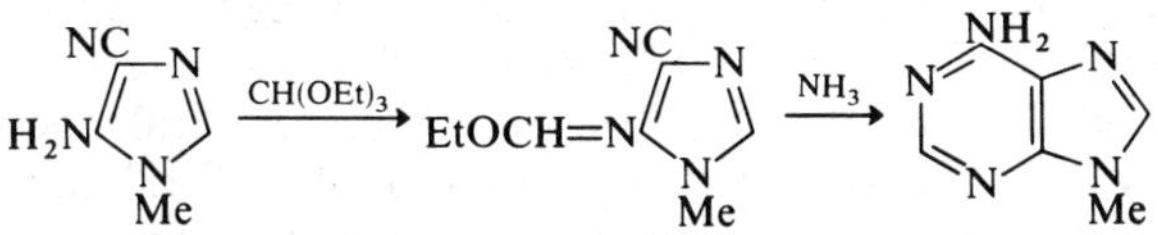

Scheme 124

Several 8-glycosylpurines have similarly been prepared from appropriate imidazoles. Thus 5-amino-4-cyano-2-(2,3,5-tri-*O*-benzoyl-D-arabinofuranosyl)imidazole (321) with triethyl orthoformate followed by ammonia gave, after deblocking, 8-arabinofuranosyladenine. The 8-ribosyl derivative (322) underwent reaction in the same manner (Scheme 125) ⟨76MI40902, 72MI40903, 72CR(C)(274)2192⟩. An alternative successful route to 8-ribofuranosyladenine involved reaction of 5-amino-4-cyanoimidazole and the ribosyl imidate (323; Scheme 126) ⟨76TL45⟩. Analogous 8-(deoxyribosyl)purines have also been prepared. Thus the 4-amino-2-(2-deoxyribosyl)imidazole-5-carbonitrile (324a) with formamidine acetate followed by deacylation produced the 8-(2-deoxy-β-D-ribosyl)adenine (325c), and the related amide (324b) gave the 8-(2-deoxy-β-D-ribosyl)hypoxanthine (326c) after reaction with diethoxymethyl acetate and subsequent deblocking (Scheme 127) ⟨75T2914⟩. 7-Methyladenine was prepared similarly from the isomeric 3-methylethoxymethyleneimidazole ⟨60JA3147⟩ which with methylamine also gave 1,9-dimethyladenine, and with hydrazine 9-methyl-1-aminoadenine. Recently inosine was obtained in low yield (15%) by reaction of the aminoimidazole riboside (327) with chloroform and sodium methoxide, the proposed mechanism involving an intermediate carbene (Scheme 128) ⟨76MI40904⟩.

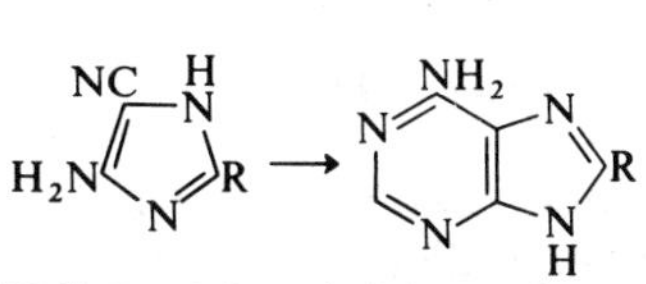

(321) R = 2,3,5-tri-*O*-benzoyl-D-arabinosyl
(322) R = 2,3,5-tri-*O*-benzoyl-D-ribosyl

Scheme 125

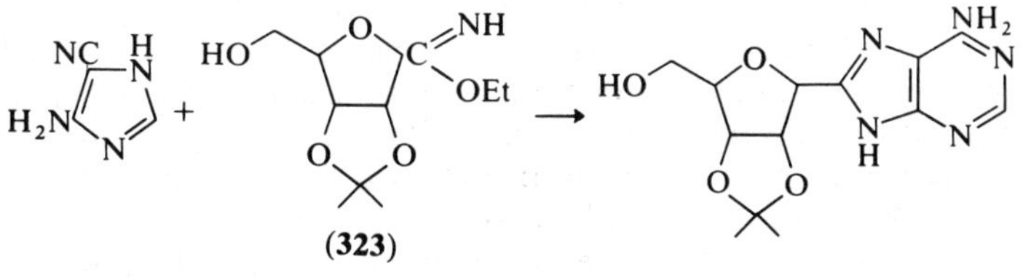

(323)

Scheme 126

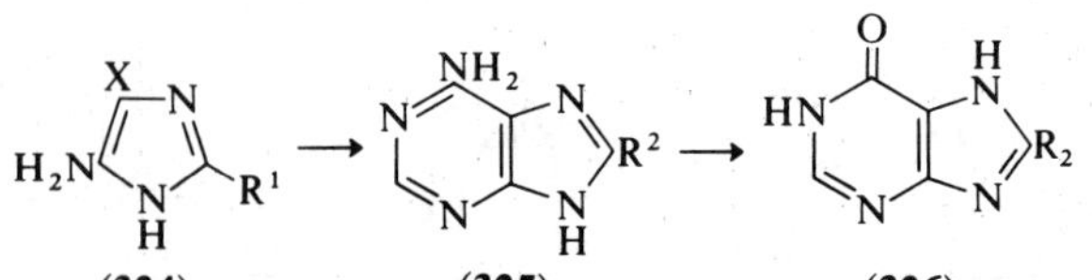

(a) $R^1 = 3,5$-di-O-p-toluoyl-2-deoxyribosyl, $X = CN$; (b) $R^1 = 3,5$-di-O-p-toluoyl-2-deoxyribosyl, $X = CONH_2$; (c) $R^2 = 2$-deoxyribofuranosyl

Scheme 127

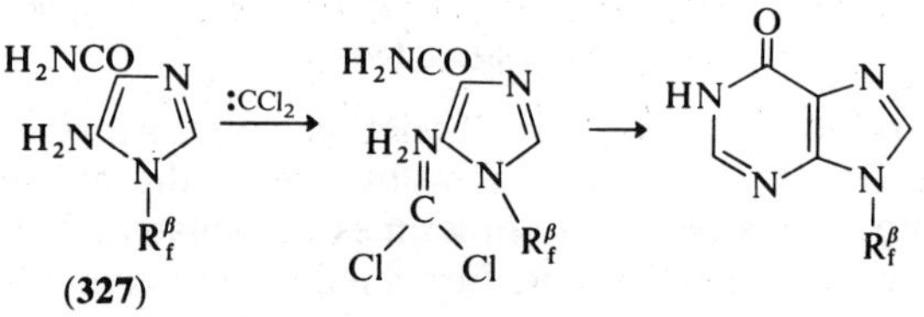

Scheme 128

Adenine was also obtained by formylation of the aminoimidazolecarboxamidine (**328**) and cyclization of an intermediate which was postulated to be the formylamidine derivative (Scheme 129) ⟨50JBC(185)439⟩. Recently this last compound has been shown to be N^6-formyladenine (**329**) and not the formylamidine indicating that cyclization of the intermediate occurs very quickly ⟨80JCS(P1)2728⟩. C(2)-Labelled adenine has been prepared by this method using labelled formic acid ⟨53JA5753⟩. The antiviral agent (**331**) was similarly produced directly by formylation of 5-amino-1-(2,3-dihydroxypropyl)imidazole-4-carboxamidine (**330**) with formic/acetic anhydride and sodium formate and treatment of the resulting N^6-formyladenine derivative with sodium hydrogen carbonate (Scheme 130) ⟨80JCS(P1)2728⟩. Various 9-substituted adenines have also been obtained from 1-substituted aminoimidazolecarboxamidines by similar cyclizing processes ⟨78CPB1929⟩.

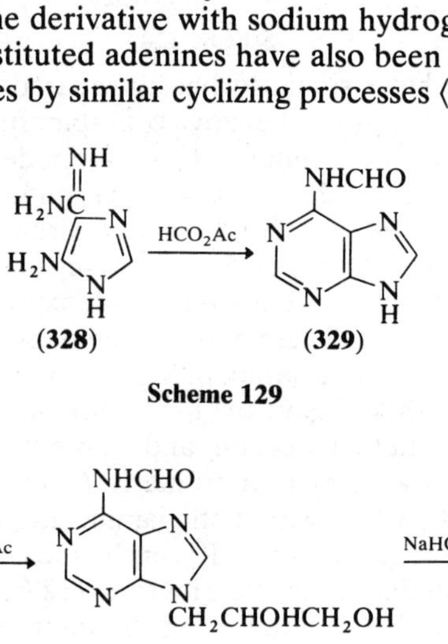

Scheme 129

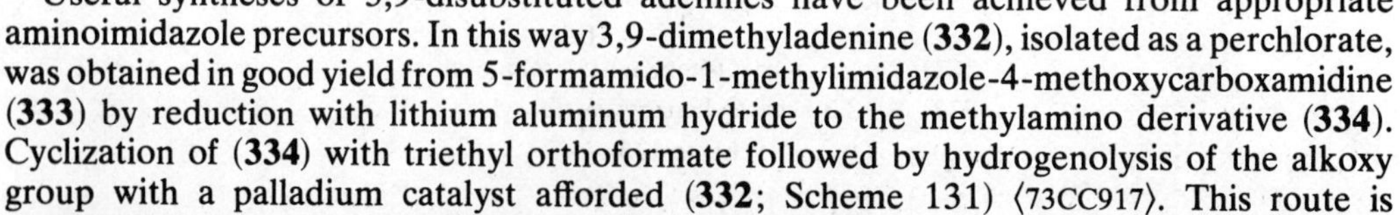

Scheme 130

Useful syntheses of 3,9-disubstituted adenines have been achieved from appropriate aminoimidazole precursors. In this way 3,9-dimethyladenine (**332**), isolated as a perchlorate, was obtained in good yield from 5-formamido-1-methylimidazole-4-methoxycarboxamidine (**333**) by reduction with lithium aluminum hydride to the methylamino derivative (**334**). Cyclization of (**334**) with triethyl orthoformate followed by hydrogenolysis of the alkoxy group with a palladium catalyst afforded (**332**; Scheme 131) ⟨73CC917⟩. This route is necessarily restricted to 3-methylpurine derivatives but it has been extended by alkylation of the formamido derivative (**333**) with sodium hydride and an alkyl halide to give the alkylformamido derivative (**335**) which may be cyclized directly to the alkoxyadenine (**336**) ⟨78TL5007⟩. The alkylation has recently been found to proceed without added base ⟨81H(16)909⟩. The method has been applied to the synthesis of 9-(β-D-ribofuranosyl)-3-methyladenine (**337**). Hydrolysis of this very labile nucleoside with 0.1M hydrochloric acid proved to be a convenient route to 3-methyladenine in 92% yield (Scheme 132) ⟨79CC135⟩.

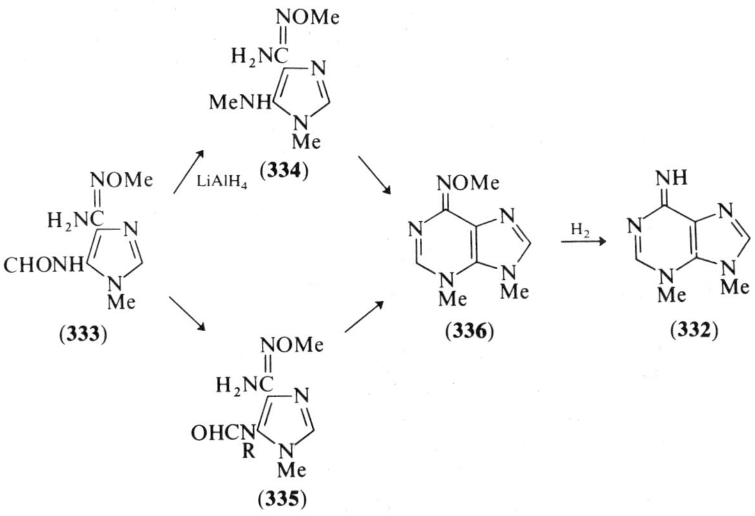

Scheme 131

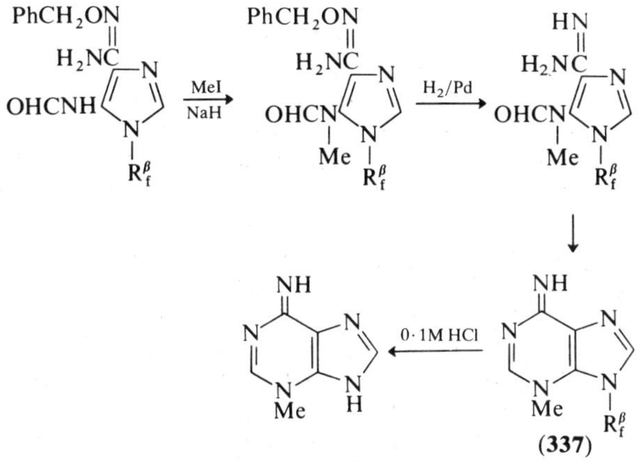

Scheme 132

4.09.7.5.3 2-Alkyl-, aryl- and glycosyl-purines

Ethyl acetate and ethyl propionate in ethanolic sodium ethoxide with 5-aminoimidazole-4-carboxamide and its riboside have provided a route to 2-methyl- and 2-ethyl-hypoxanthine and corresponding 2-alkylinosines, respectively ⟨67JOC3258, 68JA2661⟩. 2-Methylhypoxanthine has also been prepared from the same imidazole and ethyl orthoacetate via an intermediate ethoxymethylene derivative which cyclized to the purine when heated ⟨60JA3144⟩. The same method has been applied to the synthesis of 2,8-dimethyl- and 2,8-diphenyl-hypoxanthine from appropriate imidazole precursors.

Amides appear to have had little application as sources of C-2 purine substituents but trifluoroacetamide when heated with the aminoimidazolecarboxamide or the nitrile afforded 2-trifluoromethylhypoxanthine or adenine, respectively ⟨58JA5744⟩. The ribosyl-aminoimidazoleamidine cyclic phosphate similarly gave 2-trifluoromethyl cAMP with trifluoroacetamide ⟨74JA4962⟩. The same authors also converted the imidazoleamidine into 2-alkyl(aryl) derivatives of cAMP in yields of 19–81% by reaction with an aldehyde and a dehydrogenating agent which included palladium, chloranil in DMF, or aerial oxygen in aqueous ethanol.

4.09.7.5.4 2-Oxopurines

The first purine prepared from an imidazole was a 2-oxopurine, namely 7-methylxanthine (**338**), which was obtained from 4-amino-1-methylimidazole-5-carboxamide and diethyl carbonate in a sealed tube (Scheme 133) ⟨24HCA713⟩.

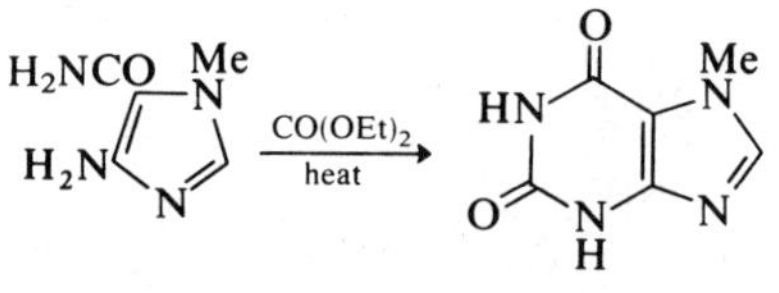

Scheme 133

Xanthine and xanthosine (74% yield), however, are more conveniently prepared from diethyl carbonate and 5-aminoimidazole-4-carboxamide or its riboside, respectively, in refluxing ethanolic sodium ethoxide ⟨67JOC3258⟩. 6-Thioxanthosine was similarly obtained from 5-aminoimidazole-4-thiocarboxamide riboside ⟨68CPB2172, 73CPB692⟩. However, fusion of an aminoimidazole with urea has been more commonly used to prepare xanthine and its derivatives including the parent ⟨50JBC(185)439⟩. 6-Thioxanthine was similarly produced from the aminothioamide (**339**) by urea fusion (Scheme 134) ⟨68CPB2172⟩ and fusion of the related aminoimidazoleamidine with urea furnished isoguanine (Scheme 135) ⟨50JBC(185)439⟩. The amino ester (**340**) in hot pyridine with urea also gave xanthine ⟨50JCS1884⟩, and 8-phenyl-, 8-benzyl- and 7-methyl-8-phenyl-xanthines were prepared in the same way from the appropriately substituted imidazole esters.

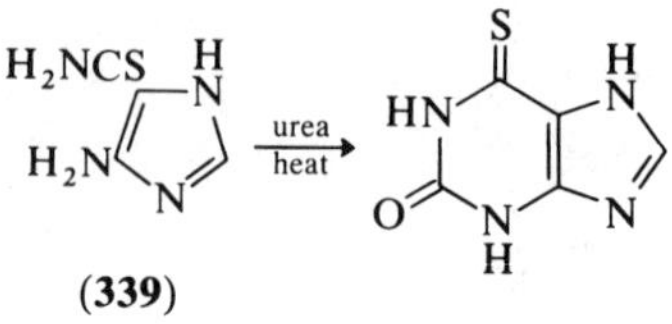

Scheme 134   Scheme 135

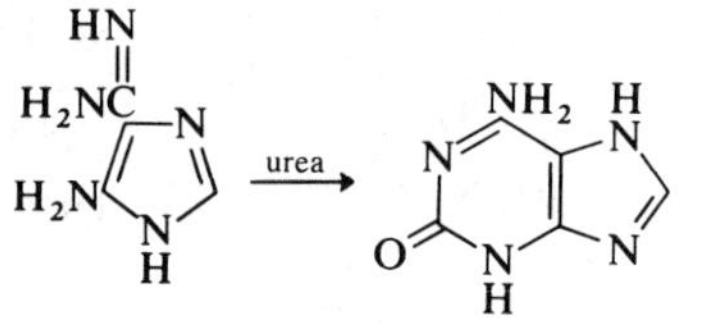

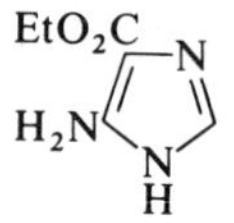

Intermediate ureido derivatives may be isolated in many of these reactions by working at lower temperatures as with 1-methyl-4-methylaminoimidazole-5-*N*-methylcarboxamide (**341**) which with urea at 140 °C in the presence of hydrogen chloride gave the ureidoimidazole (**342**). This then cyclized to theobromine with hot dilute acid (Scheme 136) ⟨28CB1409⟩. Potassium cyanate can sometimes replace urea, and with this reagent (**341**) was readily converted into caffeine (Scheme 136) ⟨28CB1409⟩, but on the other hand 5-amino-2-methylthio-1-methylimidazole-4-carboxamide failed to react.

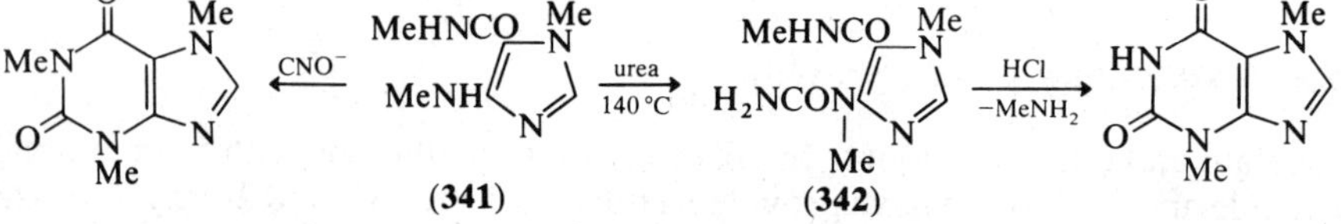

Scheme 136

The more reactive alkyl or aryl chloroformates can be used to produce 2-oxopurines under milder conditions than those obtained in urea fusions. Xanthine for example resulted from the aminoimidazolecarboxamide and ethyl chloroformate at 0 °C followed by cyclization of the intermediate urethane (**343**), either by fusion or by heating with aqueous ammonia (Scheme 137) ⟨50JBC(185)439⟩.

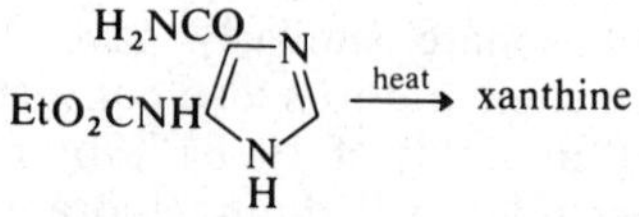

Scheme 137

An alternative route to xanthine involves prior reaction with phenylthiochloroformate as in a synthesis of 9-methyl-8-methylthioxanthine (**344**; Scheme 138) ⟨49JCS2329⟩. Several alkylated xanthines have been prepared from ethyl chloroformate and an appropriate aminoimidazolecarboxamide in the presence of potassium carbonate, including 1,7-dialkyl- ⟨45JCS751⟩, 7,8-dialkyl- ⟨66RTC1254⟩ and 1,3,8,9-tetramethyl-xanthines ⟨63ZOB1650⟩.

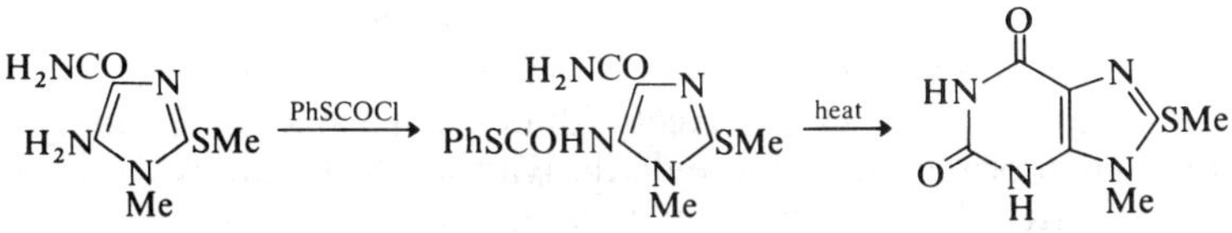

Scheme 138

In specific cases, cyclization to the 2-oxopurine may occur very readily as with the trimethylimidazole (**345**; caffeidine) which with ethyl chloroformate in aqueous sodium hydrogen carbonate readily furnished caffeine (Scheme 139) ⟨28CB1409⟩. In a similar manner, phosgene and the aminoimidazolecarboxamidines (**346**) and (**347**) in aqueous alkali readily afforded isoguanine and isoguanosine, respectively (Scheme 140) ⟨49JA3973⟩. Analogously 1,1′-carbonyldiimidazole has been used to convert the aminoimidazolecarboxamidine (**348**) into 2-oxo cAMP (Scheme 140) ⟨74JA4962⟩, and also to give a 2-oxo-1,N^6-etheno-cAMP (**349**) from (**350**; Scheme 141) ⟨76CPB1561⟩.

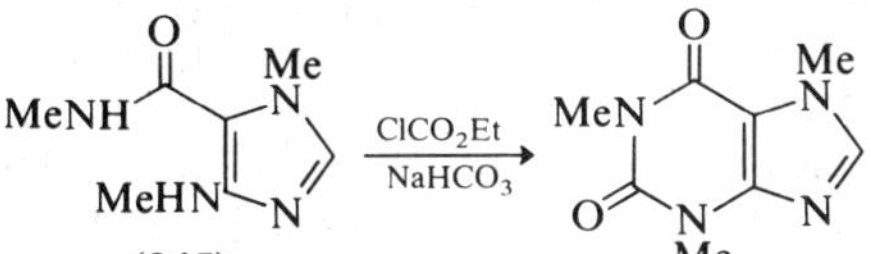

Scheme 139

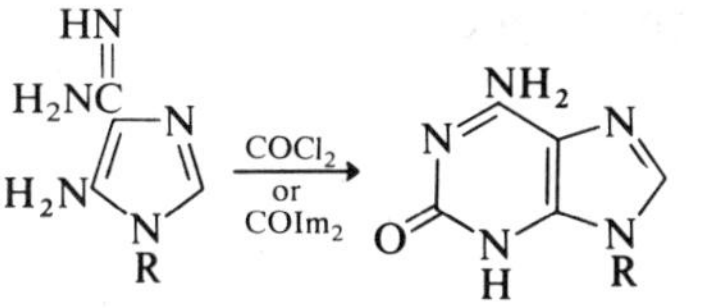

(**346**) R = H
(**347**) R = R_f^β
(**348**) R = R_{cp}

Scheme 140   **Scheme 141**

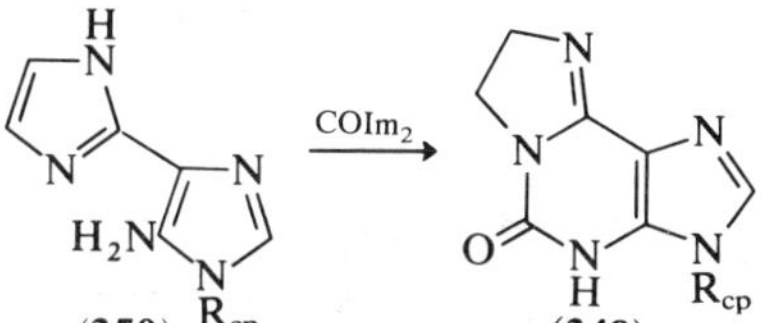

Aminoimidazole esters have also been used as a source of 2-oxopurines, the additional nitrogen being derived from urea, cyanates and similar compounds. In this manner the amino ester (**340**) and potassium cyanate furnished a ureidoimidazole which smoothly cyclized to xanthine with dilute alkali (Scheme 142) ⟨42JCS232⟩. Methyl isocyanate similarly gave 1-methylxanthine and related derivatives have also been prepared ⟨49JCS1071, 50JCS1884⟩.

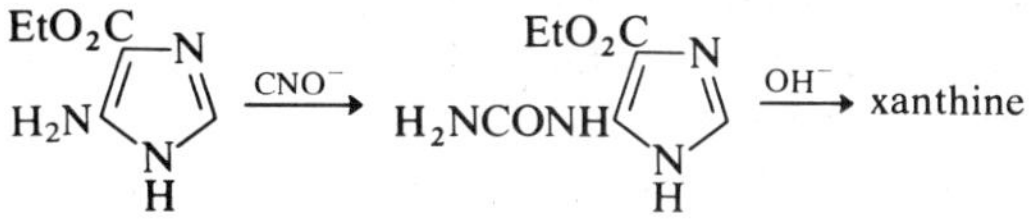

Scheme 142

Uric acid and 1,7-dimethyluric acid were similarly produced from 2-hydroxyimidazoles (imidazolin-2-ones) by reaction with potassium cyanate (Scheme 143) ⟨62JOC4633⟩.

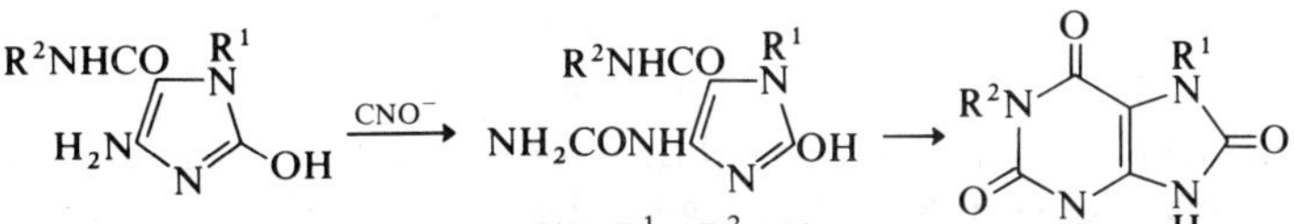

(a) $R^1 = R^2 = H$
(b) $R^1 = R^2 = Me$

Scheme 143

4.09.7.5.5 2-Thioxopurines

Alkyl (or acyl) isothiocyanates have proved to be useful intermediates for thioxopurine synthesis and can also be used to produce guanine derivatives. Methyl isothiocyanate and

5-amino-1-methylimidazole-4-carboxamide or the 2-methylthio derivative, for example in hot pyridine, gave 9-methyl-2-thioxanthine and 9-methyl-8-methylthio-2-thioxanthine, respectively (Scheme 144) ⟨49JCS3001⟩. Acetyl thiocyanate and the 2-methylthioimidazole similarly afforded an intermediate thioureido derivative which readily cyclized to 8-methylthio-2-thioxanthine after short treatment with aqueous sodium hydroxide ⟨49JCS3001⟩.

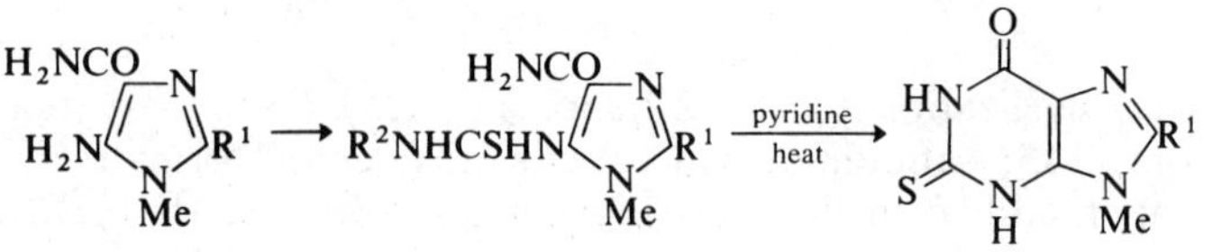

Scheme 144

Thiourea has also been used occasionally for 2-thioxopurine synthesis. 7-Methyl-2-thioxanthine is formed by fusion of 4-amino-1-methylimidazole-5-carboxamide and thiourea ⟨57JA6401⟩. Other cyclization agents include carbon disulfide which converts the aminoimidazole amidoxime (351) into the 2-thioxopurine N-oxide (352) after 2 days in pyridine at room temperature (Scheme 145) ⟨63JOC2560⟩, and sodium methylxanthate with which 5-aminoimidazole-4-carboxamide riboside gave 2-thioxanthosine in almost quantitative yield ⟨67JOC3032, 68MI40903⟩. 9-Methyl-2-thioxanthine and 9-methyl-2,8-dithiouric acid have similarly been obtained using carbon disulfide and appropriate aminoimidazoles ⟨49JCS2329⟩. Aminoimidazolenitriles furnished 2,6-thioxopurines including the 9-ribofuranoside, probably *via* the intermediate (353) under similar conditions ⟨75CPB759⟩. The 7-ribofuranosyl-2,6-dithioxopurine was obtained similarly ⟨68JOC2828⟩, and 1,1'-thiocarbonyldiimidazole and the aminoamidine (346) have been used to produce 2-thiocyclic *iso*-GMP ⟨74JA4962⟩.

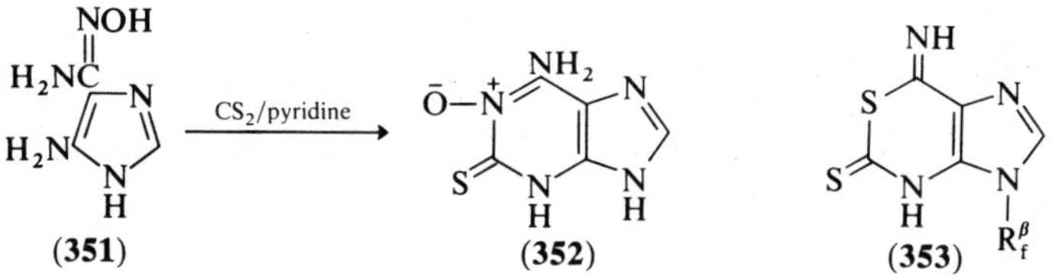

Scheme 145

4.09.7.5.6 2-Aminopurines

Cyclization of thioureidoimidazoles in the presence of sulfur-extracting agents such as mercury salts may favour 2-aminopurine formation. Thus 5-aminoimidazole-4-(N-methyl)carboxamide (354) and some 2-substituted derivatives with phenyl isothiocyanate afforded the phenylthioureidoimidazole which when heated with mercury(II) salts produced 1,N^2-dimethylisoguanine and 8-substituted derivatives (355), respectively (Scheme 146) ⟨50JCS1888⟩.

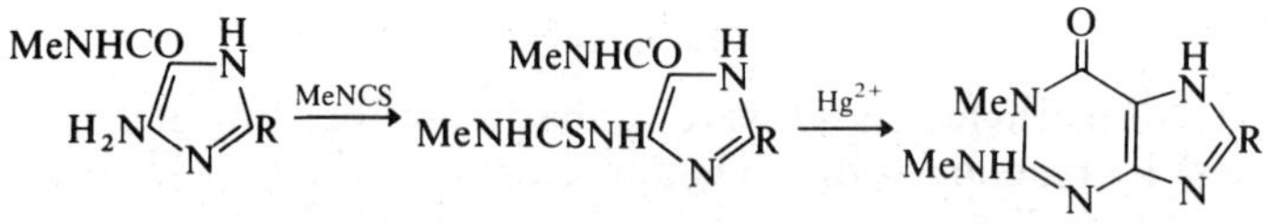

R = H, Ph, CH₂Ph

Scheme 146

The use of benzoyl isothiocyanate has led to a useful route to guanine and its derivatives. Thus 5-aminoimidazole-4-carboxamide and benzoyl isothiocyanate gave a thioureidoimidazole which after S-methylation and amination of the S-methyl derivative readily cyclized in alkaline solution to afford guanine (Scheme 147) ⟨67JOC1825, 67JOC3032⟩. 9-Hydroxy-8-methylguanine has been made by the same route ⟨72JOC1867⟩. The methylaminoimidazole ester (356) was similarly converted into 3-methylguanine ⟨79JCS(P1)3107⟩, and 1-methyl-5-methylaminoimidazole-4-carboxamide produced 3,9-dimethylguanine ⟨78TL1447⟩ which was used in a synthesis of a wyosine base, 3-methylwye (see also Section 4.09.8.2). Similar reactions have led to syntheses of the potent coronary vasodilator, 2-phenylaminoadenosine from aminoimidazole nucleoside precursors ⟨81CPB1870⟩.

Scheme 147

(356)

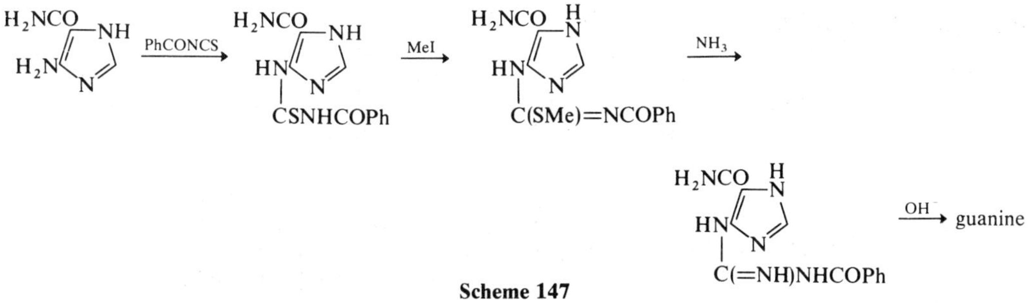

4.09.7.6 Synthesis from Ring-fused Pyrimidine Heterocycles

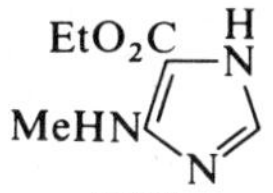

Several compounds exist which contain a pyrimidine ring fused to a heterocyclic system and they are able to produce a purine, either by rearrangement or by reaction with an appropriate reagent. Examples of the first type include the aminothiazolopyrimidines which produce thioxopurines by heating or on treatment with alkali (see Section 4.09.6.16), and the analogous 7-aminooxazolo[5,4-*d*]pyrimidine (**357**) which rearranged to hypoxanthine or its 9-substituted derivatives with hot alkali (Scheme 148) ⟨70BCJ3909⟩. The related ethoxy derivative (**358**) similarly afforded 3-substituted xanthines (Scheme 149) ⟨73BCJ506⟩, and aminopurines are also produced by reaction with ammonia or amines. Thus the oxazolo derivative (**359**) with methylamine gave a 9-methyladenine derivative ⟨52JA4897⟩ (Scheme 150).

(357)

Scheme 148

(358)

Scheme 149

(359)

Scheme 150

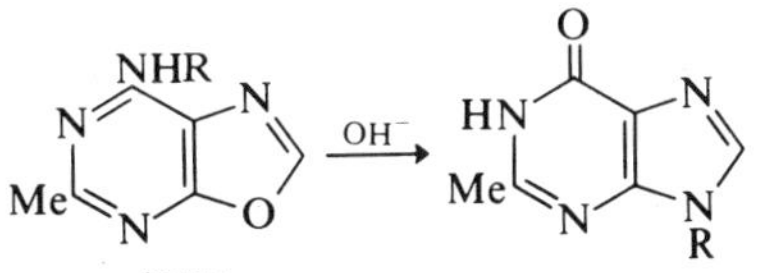

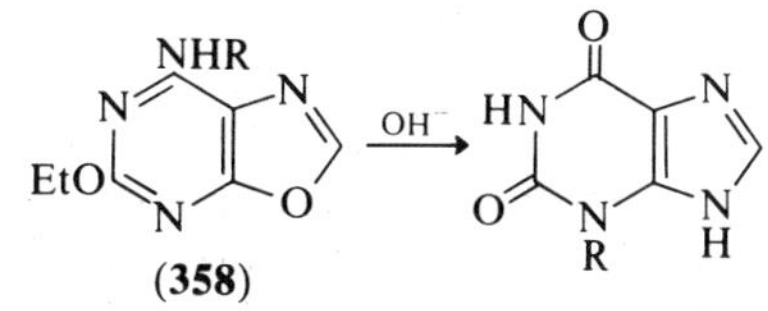

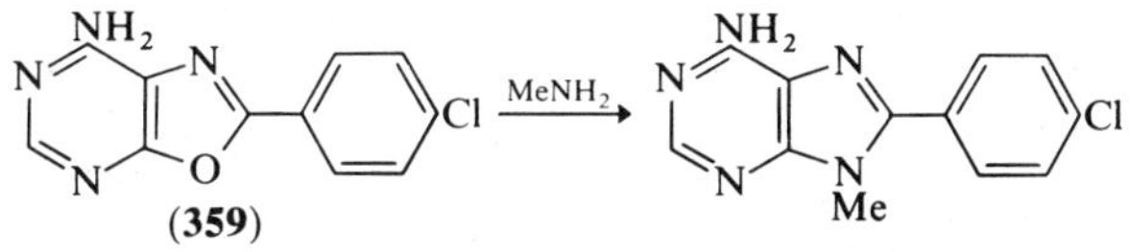

The somewhat analogous furazanopyrimidine (**360**) after reductive cleavage in the presence of an amine produced adenine ⟨71JOC3211⟩ and 9-substituted derivatives (Scheme 151).

The pyrimidino[5,4-*c*][1,2,5]oxadiazine (**361**) similarly when heated to 200 °C lost formaldehyde with formation of 8-methyltheophylline (Scheme 152) ⟨66LA(692)134⟩.

(360)

Scheme 151

(361)

Scheme 152

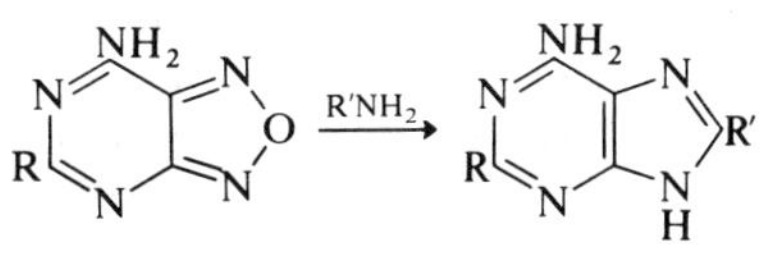

Tetrazolopyrimidines can also act as a source of a diaminopyrimidine and hence of purines. Thus 8-amino-7-chlorotetrazolo[1,5-*c*]pyrimidine (**362**) when heated with diethoxymethyl acetate furnished 6-chloro-8-ethoxypurine (Scheme 153) ⟨62JOC1671⟩ *via* a nitrene intermediate and concomitant loss of nitrogen.

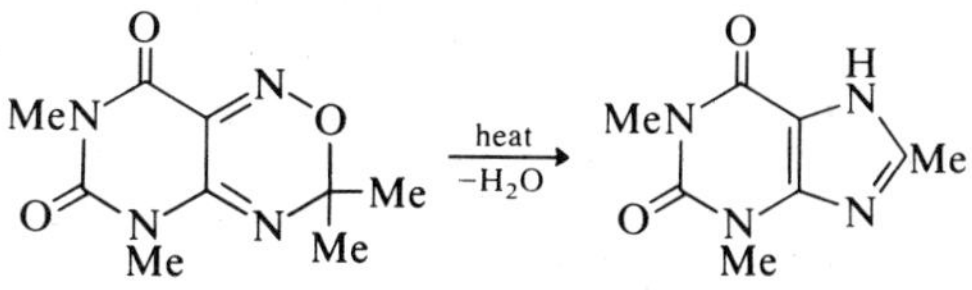

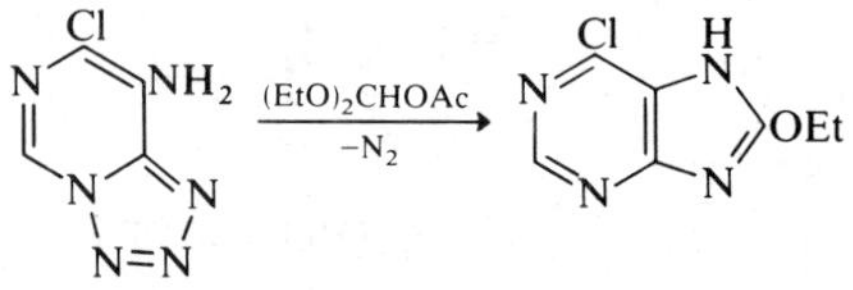

Purines have also been obtained by transformations of appropriate pteridines ⟨75CPB2678, 73CB3203⟩. Also the 6-azapteridine (363) on reduction produced 8-substituted theophylline (Scheme 154) ⟨69MI40900⟩. Similarly the 7-azapteridine (364) with hot formamide afforded 8-phenyltheophylline (Scheme 154) ⟨76H(4)749⟩, and the diazepinone (365) with alkoxide also produced the 9-(2-propenyl)purine (366) by what is ultimately a Traube synthesis (Scheme 155) ⟨70JHC1029⟩.

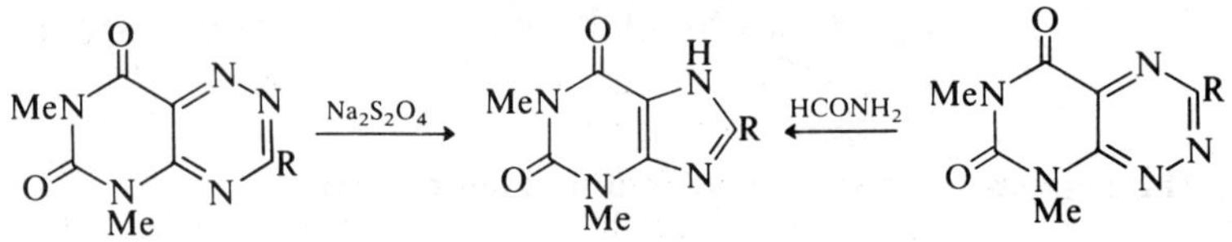

4.09.7.7 Best Practical Methods for Specific Purine Synthesis

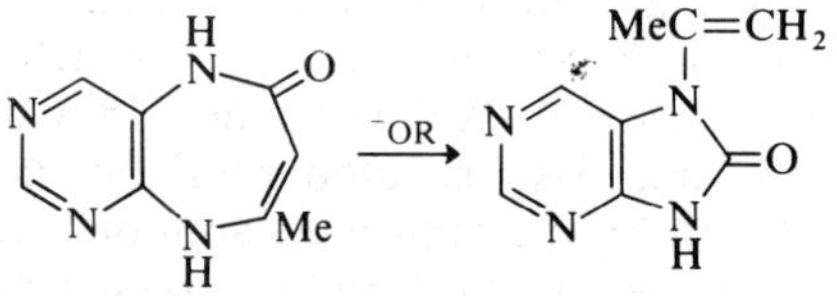

4.09.7.7.1 Introduction

Detailed preparations with full experimental details and additional references of a wide variety of purines and purine nucleosides and nucleotides have been recorded in the publication ⟨B-68MI40901⟩ and the more recent successors to this series ⟨B-78MI40901, B-78MI40903⟩.

The purines are exceptional when compared to the majority of heterocyclic compounds in that the most useful starting materials for their preparation are often of natural occurrence. This reflects of course the widespread occurrence of adenine and guanine and their nucleosides as components of nucleic acids. Thus adenine and guanine may be obtained by acid hydrolysis of nucleic acid ⟨38CB718⟩ and hypoxanthine is readily produced by deamination of adenine with nitrous acid ⟨38CB718⟩ or by adenosine deaminase.

Caffeine occurs in both tea (3.5%) and coffee and is normally manufactured from coffee and tea dust and similarly theobromine is the main purine of the cocoa bean from which it may be prepared. Caffeine has also been manufactured from guanine by hydrolysis to xanthine followed by methylation with methyl acetate or dimethyl sulfate ⟨46USP2509084⟩.

4.09.7.7.2 Purine

Purine is probably best prepared by the Traube route from 4,5-diaminopyrimidine and formic acid in an atmosphere of carbon dioxide (83% yield) ⟨54JA6073⟩, or, better, using diethoxymethyl acetate as the cyclizing agent (82% yield) ⟨60JOC395⟩.

4.09.7.7.3 C-Linked purines

(i) C-Alkylpurines

C-Alkylpurines are usually best obtained by Traube syntheses as in the preparation of 6-methylpurine ⟨65JCS623⟩ and 8-methylpurine ⟨54JCS2060⟩, or by dehydrohalogenation of a halogenoalkylpurine as in the preparation of 2-methylpurine from 2-methyl-6-chloropurine ⟨59JA193⟩.

Recently a valuable route to 8-alkyl- or acyl-purine nucleosides has involved the reaction of an unsubstituted or 8-halogenopurine with butyllithium and an appropriate alkyl or acyl halide, including ethyl chloroformate, when excellent yields of the 8-substituted purine nucleosides are produced ⟨79TL2385⟩.

(ii) Purinecarboxylic acids and derivatives

Purine-6-carboxylic acids are probably best prepared by hydrolysis, using 2N sodium hydroxide, of 6-cyanopurine, itself prepared by heating 6-iodopurine (not chloropurine) and copper(I) cyanide in pyridine over 2 hours ⟨56JA3511⟩. The nitrile may be used also as a convenient source of purine-6-carboxamide and -thiocarboxamide. Homologous 8-, but not 2- or 6-, purinecarboxylic acids are best prepared from 8-chloropurines and the sodium derivative of diethyl malonate and its alkylated derivatives followed by hydrolysis as in the preparation of 8-(α-alkyl)carboxymethylcaffeines from 8-bromocaffeine ⟨53MI40901⟩, and 8-carboxymethylisocaffeine from 8-chloroisocaffeine ⟨60ZOB3628⟩. 8-Carboxy derivatives of adenosine and guanosine may be conveniently prepared in good yield by reaction of 8-lithiopurine trimethylsilyl derivatives with ethyl chloroformate ⟨79TL2385⟩.

Purine aldehydes have only been prepared in recent years. The best route appears to be the hydrolysis of a dichloromethylpurine which may be made from the methyl derivative ⟨62ZOB2015⟩. Thus hydrolysis over several hours of 8-dichloromethylcaffeine produced 8-formylcaffeine ⟨48ZOB2129⟩. The corresponding 6-formylpurine may be prepared as an oxime by reaction of 6-methylpurine with cold nitrous acid and this is converted by hydrazine into the hydrazone ⟨63N224⟩. The hydrazone may be converted by nitrous acid into the aldehyde ⟨59JA2515⟩. Dehydration of the oxime with hot acetic anhydride over 5 hours also afforded an alternative route to 6-cyanopurine ⟨68CB611⟩.

A few purine ketones have been prepared including 8-acetyl- or 8-propionyl-caffeine from 8-cyanocaffeine and the appropriate Grignard reagent ⟨56AP453⟩.

(iii) C-Aminopurines

2-Aminopurine is probably best obtained by a Traube synthesis from 2,4,5-triaminopyrimidine and formic acid in boiling 4-formylmorpholine ⟨54JCS2060⟩. 2-Hydrazinopurine on the other hand is best prepared from 2-chloropurine and hydrazine ⟨57JA2185⟩. Adenine is best prepared in quantity by a modified Traube synthesis ⟨53JA263, 66MI40903⟩. 2,6-Diaminopurine is also produced by the Traube method from 2,4,5,6-tetraaminopyrimidine, and material labelled with ^{15}N in positions 1 and 3 has been prepared in this way ⟨50JBC(185)423, 50JBC(185)435⟩.

8-Aminopurines have usually been prepared from 8-chloropurines as in the amination of 8-chlorocaffeine to 8-aminocaffeine ⟨1882LA(215)265⟩, or alternatively from 8-methylthiopurine as in the synthesis of 8-amino-, 8-methylamino- or 8-dimethylamino-purine from 8-methylthiopurine with ammonia, methylamine or dimethylamine, respectively ⟨54JCS2060⟩. A good yield of 8-amino-6-oxo-1,6-dihydropurine was obtained in a Traube synthesis from 4,5-diamino-6-oxo-1,6-dihydropyrimidine and guanidine at 200 °C ⟨58JA6671⟩.

4.09.7.7.4 N-Linked purines

(i) N-Alkylpurines, nucleosides and nucleotides

The preparation of N-alkylpurine derivatives poses certain practical problems. In the case of the purine anion the normal site of alkylation is N-7 or N-9. Thus benzyladenine may be conveniently prepared from sodium adenine and benzyl chloride ⟨B-68MI40901, p. 3⟩. 9-Ethylpurine is readily obtained in 68% yield from 9-methyladenine and pentyl nitrite with illumination from a 200 W lamp ⟨80JOC3969⟩.

9-Substituted purine nucleosides which include most of the naturally occurring compounds or their analogs are also produced by glycosylation of an appropriate purine anion derivative.

In this way 9-(β-D-xylofuranosyl)adenine is obtained by reaction of 2,3,5-tri-O-acetyl-D-xylofuranosyl bromide with the chloromercurio derivative of N^6-benzoyladenine, followed by deacylation with methanolic sodium methoxide ⟨B-68MI40901, p. 139⟩.

Silyl derivatives of purines are commonly used in the same way as in a synthesis of 9-(α,β-D-glucopyranosyl)adenine from N^6-benzoyl-9-bis(trimethylsilyl)adenine and 2,3,4,6-tetra-O-acetyl-α-D-glucosyl chloride at 150–160 °C over 4 hours to produce, after deblocking, a mixture of anomeric nucleosides ⟨B-68MI40901, p. 135⟩. The hypoxanthine nucleosides may be obtained in a similar fashion or by deamination of the adenine derivatives.

2′,3′,5′-Tri-O-acetyladenosine is also usefully converted into 2′,3′,5′-tri-O-acetylnebularine with pentyl nitrite and 200 W illumination in 45.8% yield ⟨80JOC3969⟩.

In a similar manner purine nucleotides including AMP and its α-anomer may be produced directly by reaction of the N^6-benzoyl-9-bis(trimethylsilyl)adenine with 2,3-di-O-benzoyl-5-O-(diphenylphosphono)-D-ribofuranosyl bromide and deblocking the intermediate with lithium hydroxide and a phosphodiesterase from *Trimeresurus flavoviridis* ⟨B-78MI40903, p. 821⟩. This method however gives a low yield and can hardly be recommended.

Purine nucleotides are probably best made by phosphorylation of the corresponding nucleosides. In recent years phosphoryl chloride has proved most convenient for this purpose and may be used with an unprotected nucleoside to give yields of about 50% ⟨B-78MI40903, p. 827⟩. Yields of purine nucleotides up to 60% may also be obtained using a phosphotransferase and a suitable phosphate donor with the unprotected nucleoside. Typical preparations of this type using an enzyme from wheat shoots and *p*-nitrophenyl phosphate as a phosphate donor have been described ⟨B-78MI40903, p. 955⟩.

The alternative fusion method for nucleoside synthesis involves the reaction of a purine with an acylated sugar in the presence of a suitable acid such as *p*-toluenesulfonic acid or a compound such as tin(IV) chloride or antimony chloride. One of many examples includes the preparation (48% yield) of 2,6-dibromo-9-(β-D-ribofuranosyl)purine by fusion at 130 °C over 30 minutes of 1,2,3,5-tetra-O-acetyl-β-D-ribose with 2,6-dibromopurine in the presence of *p*-toluenesulfonic acid ⟨B-68MI40901, p. 180⟩.

The use of 2-deoxyribose derivatives tends to produce anomeric mixtures of nucleosides. Thus 3,5-di-O-*p*-nitrobenzoyl-2-deoxy-D-ribofuranosyl chloride and N^6-benzoyladenine and molecular sieves gave, after deblocking with methanolic barium methoxide, a mixture of the anomeric 2′-deoxyadenosines ⟨B-68MI40901, p. 183⟩. A variation to this technique includes the condensation of N^6-benzoylchloromercurioadenine with 5-O-benzoyl-2-deoxy-D-*erythro*-pentose diisopropyl dithioacetal in acetonitrile under reflux over 4 hours to produce the same anomeric mixture of 2′-deoxyadenosines ⟨B-68MI40901, p. 132⟩.

1-Methylpurine is conveniently prepared by dethiation of 1-methyl-6-thioxo-(1*H*)-purine with Raney nickel ⟨62JOC990⟩. An alternative route to 1-substituted purines involves protection of the 9-position with a removable group followed by alkylation. Thus 9-propenylhypoxanthine, prepared by a Traube synthesis, is readily methylated by methyl *p*-toluenesulfonate to produce 1-methyl-9-propenylhypoxanthine (**367**) from which 1-methylhypoxanthine may be obtained by oxidative removal of the propenyl group with alkaline potassium permanganate ⟨B-68MI40901, p. 22⟩. 1-Methyladenine has been prepared similarly ⟨65JOC3235⟩. The benzyl group may also be used for protecting the 9-position in such compounds and can subsequently be removed by hydrogenolysis.

(**367**)

In the case of nucleosides which contain a 9-glycosyl group, alkylation occurs at the 1-position. In this way, 1-isopentenyladenosine is obtained by direct alkylation of adenosine with isopentenyl bromide ⟨B-68MI40901, p. 212⟩.

In contrast to 9-substituted purines, direct alkylation of adenine with benzyl chloride in DMA at 115 °C for 18 hours produced 3-benzyladenine (60% yield) which, when deaminated with nitrosyl chloride, afforded 3-benzylhypoxanthine (61% yield) ⟨B-68MI40901, p. 28⟩. Triacanthine (3-isopentenyladenine) was prepared similarly from adenine and isopentenyl bromide ⟨B-68MI40901, p. 13⟩.

In a similar fashion methylation of 6-methylthiopurine with methyl *p*-toluenesulfonate in DMA at 95 °C over 1 hour afforded 3-methyl-6-methylthiopurine which could either be desulfurized with Raney nickel to give 3-methylpurine or reacted with primary amines in hot methanol over 12 hours to produce 3-methyladenine derivatives ⟨B-68MI40901, p. 36⟩. Glycosylation of adenine with 2,3,5-tri-*O*-benzoyl-β-D-ribofuranosyl bromide in acetonitrile at 50 °C over 36 hours also produced, after deblocking with methanolic ammonia, 3-(β-D-ribofuranosyl)adenine (25% overall yield) ⟨B-68MI40901, p. 160⟩. In contrast 3-methylguanine is probably best prepared by a Traube synthesis from 2,5,6-triamino-1-methyl-4-oxo-3,4-dihydropyrimidine in refluxing formamide over 1.5 hours (80% yield), and 3-methyluric acid may be usefully obtained by a similar fusion reaction with urea at 150 °C, followed by alkaline hydrolysis of the intermediate 2-amino-3-methyl-6,8-dioxo-1,9-dihydropurine. 3-Methylxanthine has also been prepared by alkaline hydrolysis of 3-methylguanine (84% yield) ⟨B-68MI40901, p. 18⟩.

7-Substituted purines may arise in mixtures with 9-substituted derivatives from direct alkylation of purine anions. They are best prepared however by Traube syntheses. Thus formylation of 4,5-diamino-6-benzylthiopyrimidine and ethylation of the formyl derivatives gave a formamidopyrimidine which readily cyclized to 6-benzylthio-7-ethylpurine in the presence of potassium carbonate ⟨B-68MI40901, p. 31⟩. The derivative is clearly a ready source of 7-substituted adenines by reaction with ammonia or amines, or of 7-substituted purines by dethiation with Raney nickel. 7-Methylguanine has also been obtained from 7-methylguanosine, sodium borohydride and aniline at pH 4.5 ⟨B-78MI40903, p. 615⟩.

By reaction with appropriate nucleophiles, 6-chloropurine has been the major intermediate for the preparation of 6-substituted purines. It has in particular been widely used to prepare large numbers of 6-aminopurine derivatives. Thus with 3-methylbut-2-enylamine in 2-methoxyethanol, 6-chloropurine produces the cytokinin and tRNA constituent N^6-(3-methylbut-2-enyl)adenine ⟨B-68MI40901, p. 11⟩. This method uses an excess of amine to remove hydrogen chloride. Triethylamine may be used for this purpose to conserve the primary amino compound, as in a synthesis of the cytokinin zeatin from 6-chloropurine and *trans*-4-hydroxy-3-methylbut-2-enylamine and triethylamine in hot butanol ⟨66JCS(C)921⟩. 6-Chloropurine riboside has been used similarly to prepare zeatin riboside ⟨66JCS(C)921⟩. The corresponding 6-chloro-9-(β-D-ribofuranosyl)purine 5′-phosphate has also been prepared by an improved process ⟨68JCS(C)1516⟩ and used to prepare zeatin ribotide and other N^6-substituted AMP derivatives by reactions with appropriate amines.

(ii) N-Aminopurines

N-Aminopurines have been prepared from either pyrimidine or imidazole precursors. Thus 9-aminopurine may be obtained from 5-amino-4-hydrazinopyrimidine and formic acid ⟨60JA4592⟩. On the other hand 9-amino-8-methylhypoxanthine resulted from cyclization of 1,5-diamino-4-carbamoyl-2-methylimidazole with triethyl orthoformate and acetic anhydride ⟨61JCS4845⟩.

1-Aminoadenine is probably best obtained by amination of adenine with mesitylenesulfonylhydroxylamine ⟨74JOC3438⟩ and the related 1-aminoinosine and guanosine were prepared from the nucleosides and hydroxylamine-*O*-sulfonic acid in aqueous sodium hydroxide ⟨69JOC1025⟩. The same reagent is used in the preparation of 6,7-diamino-8-oxo-7,8-dihydro-9-(β-D-ribofuranosyl)purine from 8-oxo-7,8-dihydroadenosine ⟨69JOC2157⟩.

Substituted 3-aminopurines have also been best prepared by Traube syntheses from 3-(substituted amino)-4,5-diaminopyrimidines and triethyl orthoformate ⟨63JOC1162⟩.

(iii) N-Oxides

Examples of all the four purine *N*-oxides are known. Purine 1-oxide may be prepared from purine and perbenzoic acid over two weeks ⟨62JOC567⟩, and adenine 1-oxide is similarly prepared from adenine over 2 days ⟨58JA2759⟩. On the other hand hypoxanthine 1-oxide is best obtained by deamination of adenine 1-oxide ⟨66JOC966⟩. Reaction of 5-amino-4-hydroxyformamidinoimidazole with carbon disulfide may be used to make 2-thioadenine 1-oxide which with alkaline peroxide provides a useful route to isoguanine 1-oxide ⟨67JOC1151⟩.

Purine 3-oxide has been prepared in good yield (64%) by hydrolysis of purine-6-sulfonic acid 3-oxide with 90% formic acid ⟨B-78MI40901, p. 33⟩. Guanine 3-oxide may be obtained directly from guanine and hydrogen peroxide in trifluoroacetic acid and when boiled with 6N hydrochloric acid it readily produced xanthine 3-oxide ⟨69JOC978⟩. 6-Chloropurine

3-oxide may be prepared similarly from 6-chloropurine and peroxyphthalic or peracetic acid in ethanol ⟨69BCJ750, 69JOC2153⟩. Hydrolysis of the latter oxide with sodium hydroxide is a useful route to hypoxanthine 3-oxide and to 6-methoxy, 6-bromo-, 6-iodo- and 6-sulfo-purine 3-oxides by reaction with appropriate nucleophiles ⟨69BCJ750, 69JOC2157⟩.

Purine 7-oxides are generally most conveniently made by Traube syntheses from nitroso- or nitro-pyrimidines as with 7-hydroxytheophylline ⟨66LA(691)142, 66LA(693)233, 66LA(698)145, 66LA(699)145, 72JOC1871⟩.

Purine 9-oxides on the other hand are best made from imidazole precursors. Thus 9-hydroxy-8-methylxanthine (**368**) and hypoxanthine (**369**) have been most conveniently obtained by cyclization of 5-amino-1-benzyloxy-2-methylimidazole-4-carboxamide (**370**) with carbonate or formate esters, respectively, and debenzylation of the intermediate benzyloxy derivatives with hydrogen bromide in acetic acid (Scheme 156) ⟨72JOC1867⟩.

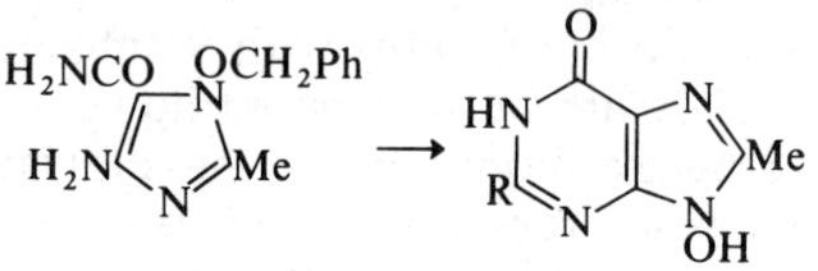

Scheme 156

4.09.7.7.5 O-Linked purines

Hypoxanthine is readily available from a Traube synthesis using triethyl orthoformate and acetic anhydride in 95% yield ⟨60JOC148⟩ and an alternative one-step preparation from aminomalonamidine, triethyl orthoformate in DMF gives an 85% yield. Similarly, xanthine is probably best prepared by acid hydrolysis or nitrous acid deamination of guanine ⟨1861LA(118)151⟩ or by Traube cyclization procedures. Guanine labelled at N-3 and N-1 has also been prepared by the Traube method using ^{15}N-labelled guanidine ⟨44JBC(153)203⟩.

Hypoxanthine is an important primary feedstock intermediate for the synthesis of many 6-substituted purines especially the 6-thioxopurines and adenine derivatives.

Most of the other oxopurines have been prepared by Traube syntheses. Thus 2-oxo-2,3-dihydropurine may be made from 4,5-diamino-2-oxo-2,3-dihydropyrimidine and formic acid ⟨12JBC(11)67⟩, and 8-oxo-7,8-dihydropurine is best prepared from 4,5-diaminopyrimidine and urea ⟨54JCS2060⟩.

2,8-Dioxo-2,3,7,8-tetrahydropurine and 6,8-dioxo-1,6,7,8-tetrahydropurine are similarly prepared from 4,5-diamino-2-oxo-2,3-dihydropyrimidine and 4,5-diamino-6-oxo-1,6-dihydropyrimidine and urea, respectively ⟨54JCS2060⟩.

Uric acid is probably best prepared from 4,5,6-triamino-2-oxo-2,3-dihydropyrimidine and urea, and material labelled at positions 1,3 and 9 by ^{15}N has been made in excellent yield by this procedure ⟨48JA1240⟩.

Alkoxypurines, especially 6-alkoxypurines, are normally best prepared by reaction of a chloropurine with a sodium alkoxide as with 6-methoxypurine from 6-chloropurine and sodium methoxide ⟨57CB698⟩. However Traube syntheses have also been used in the preparation of 2-methoxypurine from 4,5-diamino-2-methoxypyrimidine and formic acid ⟨54JCS2060⟩ or 2,6-dimethoxypurine from the corresponding dimethoxypyrimidine ⟨54JA5087⟩. 8-Ethoxypurine has also been usefully prepared from 4,5-diaminopyrimidine and tetraethoxymethane ⟨74JCS(P1)349⟩.

4.09.7.7.6 S-Linked purines

A major route to thioxopurines involves thiation of the corresponding oxopurine. Thus a good yield of 6-thioxo-1,6-dihydropurine is obtained from hypoxanthine and phosphorus pentasulfide ⟨61JA4038⟩. Thiation of an appropriate diaminopyrimidine and concomitant cyclization to a thioxopurine is sometimes possible as in the preparation of 2-amino-1-methyl-6-thioxo-1,6-dihydropurine from 2,6-diamino-5-formamido-3-methyl-4-oxo-3,4-dihydropyrimidine and phosphorus pentasulfide in refluxing pyridine over 30 hours

⟨B-68MI40901, p. 44⟩. Alternatively, Traube cyclizations with thiourea have provided a valuable route to 8-thioxopurines as in the reaction of 2,5,6-triamino-3-methyl-4-oxo-3,4-dihydropyrimidine and thiourea at 200 °C followed by 1N potassium hydroxide to give 8-mercapto-1-methylguanine ⟨B-68MI40901, p. 45⟩. When the appropriate chloropurine is readily available, the related thioxopurine may be obtained by reaction with various sulfur derivatives including thiourea fusion as in the preparation of 9-benzyl-6-thioxo-1,6-dihydropurine (94% yield) from 9-benzyl-6-chloropurine and thiourea in hot propanol over 2 hours ⟨B-68MI40901, p. 47⟩. Sodium hydrosulfide is also used as in the conversion of 2-acetamido-6-chloro-9-(2-deoxy-3,5-di-O-p-toluoyl)purine into the 6-thioxo derivative ⟨B-68MI40901, p. 272⟩. S-Alkylation of thioxopurines including nucleosides is readily achieved (in yields of 52–78%) by reaction with alkyl halides in DMF in the presence of potassium carbonate and several examples may be found in the publication ⟨B-68MI40901, p. 258⟩.

The synthesis of purinesulfonic acids (yields 87–96%) may be successfully achieved by oxidation of the corresponding thioxo derivatives with sulfite ions in the presence of oxygen ⟨74CB2284, B-78MI40903, p. 677⟩. 6-Methylsulfonylpurines have, on the other hand, been prepared from the methylthiopurine and chlorine, as in a synthesis of 6-methylsulfonyl-9-(β-D-ribofuranosyl)purine (50–75% yield) from chlorine in 90% ethanol solution at 0 °C over 10 minutes ⟨B-78MI40903, p. 651⟩.

Purine-6-sulfinic acid 3-oxide has been prepared in excellent yield (92%) from 6-thioxo-1,6-dihydropurine 3-oxide and active manganese dioxide in aqueous sodium hydrogen carbonate ⟨B-78MI40901, p. 33⟩.

4.09.7.7.7 Se-Linked purines

Related selenoxopurines have been prepared by replacement of the 6-amino group in selected purines. Thus 9-(β-D-ribofuranosyl)-6-selenoxo-1,6-dihydropurine (**371**) was conveniently prepared (56% yield) from adenosine in aqueous pyridine with hydrogen selenide over 5 days at 65 °C. In a similar manner 2-aminoadenosine furnished 6-selenoguanosine (21%) ⟨B-78MI40903, p. 673, 75CC319, 75JHC493⟩. The latter compound may also be obtained (58% yield) from 2-amino-6-chloro-9-(2,3,5-tri-O-acetyl-β-D-ribofuranosyl)purine and sodium hydrogen selenide. The method adopted for the preparation of selenoxo derivatives depends on the relative availability of the appropriate chloro or aminopurine.

(**371**)

4.09.7.7.8 Halogenopurines

2-Chloropurine is best made by the Traube process from 4,5-diamino-2-chloropyrimidine and triethyl orthoformate ⟨56JA1928⟩ whereas 2-fluoropurine is better prepared (in 41% yield) from 2-aminopurine with sodium nitrite and 48% fluoroboric acid. Similarly 2,6-diaminopurine afforded 6-amino-2-fluoropurine in 22% yield with sodium nitrite in liquid hydrogen fluoride ⟨69JOC747⟩.

The valuable 6-chloropurine has been prepared from hypoxanthine by reaction with phosphoryl chloride in the presence of dimethylaniline ⟨54JA6073⟩, or better, N-ethylpiperidine ⟨64ZC430⟩. 6-Bromopurine is prepared in the same way using phosphoryl bromide ⟨56JA3508⟩. An alternative synthesis of 6-chloropurine (67% yield) proceeds from 6-hydrazinopurine and iron(III) chloride ⟨B-78MI40901, p. 19⟩. The hydrazinopurine may be prepared from 6-methylthiopurine and hydrazine ⟨58JA404⟩. The (albeit hazardous) chlorination of thiohypoxanthine has also been recommended ⟨62MI40904⟩ as a route to 6-chloropurine. 6-Iodopurine is best obtained from 6-chloropurine and hydrogen iodide ⟨56JA3508⟩.

6-Chloro-9-(β-D-ribofuranosyl)purine has similarly been prepared from inosine ⟨63JOC945⟩ by chlorination of 6-thioinosine ⟨63MI40903⟩ or, perhaps most successfully, by reaction of 2′,3′,5′-tri-O-acetylinosine with (chloromethylene)dimethylammonium chloride when yields of 70% are produced ⟨B-78MI40903, p. 611⟩. The latter reagent is obtained from phosgene and DMF in chloroform solution ⟨B-78MI40903, p. 989⟩.

6-Bromo- or 6-iodo-9-(β-D-2,3,5-tri-O-acetylribofuranosyl)purines have been prepared in good yields (72.7% and 68.6%) from 2′,3′,5′-tri-O-acetyladenosine with pentyl nitrite in bromoform or diiodomethane (as halogen sources), respectively, under 200 W illumination ⟨80JOC3969⟩.

6-Fluoropurine is probably best prepared from 6-chloropurine. This was used to quaternize trimethylamine and the resultant quaternary salt on treatment with potassium hydrogen fluoride in ethanol at 50 °C afforded 6-fluoropurine ⟨69CC381⟩ in good yield.

2,6-Dichloropurine may be prepared from xanthine with either phosphoryl chloride and a trace of water ⟨56JA3508⟩ or with phosphoryl chloride and phosphorus pentachloride ⟨58JA2751⟩, and similarly 2,6,8-trichloropurine resulted in 90% yield from heating uric acid with an excess of phosphoryl chloride ⟨55MI40900⟩.

8-Halogenopurines are generally most usefully prepared by direct halogenation of the purine ⟨62JOC986⟩ with halogens or halogeno amides.

4.09.8 NATURALLY OCCURRING PURINES—BIOCHEMISTRY AND APPLICATIONS

4.09.8.1 Biosynthesis

4.09.8.1.1 De novo biosynthetic pathways

New purine bases are produced, at the nucleotide level, by the *de novo* biosynthetic route which commences with 5-phospho-α-D-ribosyl pyrophosphate and proceeds *via* 5-phospho-β-D-ribosylamine to produce IMP in a sequence of enzyme-controlled reactions involving aminoimidazole intermediates. IMP is further converted into AMP and GMP by separate pathways (Scheme 157).

4.09.8.1.2 Salvage pathways

The salvage pathway does not involve the formation of new heterocyclic bases but permits variation according to demand of the state of the base (B), *i.e.* whether at the nucleoside (N), or nucleoside mono- (NMP), di- (NDP) or tri- (NTP) phosphate level. The major enzymes and routes available (Scheme 158) all operate with either ribose or 2-deoxyribose derivatives except for the phosphoribosyl transferases. Several enzymes involved in the biosynthesis of purine nucleotides or in interconversion reactions, *e.g.* adenosine deaminase, have been assayed using a method which is based on the formation of hydrogen peroxide with xanthine oxidase as a coupling enzyme ⟨81CPB426⟩.

4.09.8.2 Naturally Occurring Purines

4.09.8.2.1 Methylated purines

Methylation is an important biochemical event and all the common purines and their nucleosides occur naturally in mono- or poly-methylated forms. Some common naturally occurring purines are collected in Table 39.

1-Methylxanthine is the major purine constituent of human urine (3.1 g in 1000 l) ⟨1898ZPC(24)364⟩. 3- and 7-Methylpurines are also minor constituents of urine, especially following large doses of caffeine or other methylated xanthines. 1,3-Dimethylxanthine (theophylline) occurs with caffeine in tea leaves and is a powerful diuretic and has been used clinically for this purpose (generally as an adduct with salts of organic acids) and also in the treatment of asthma. 1,7-Dimethylxanthine (paraxanthine) is also an efficient diuretic and, in addition, possesses antithyroid properties ⟨45JCS751⟩. The main purine constituent

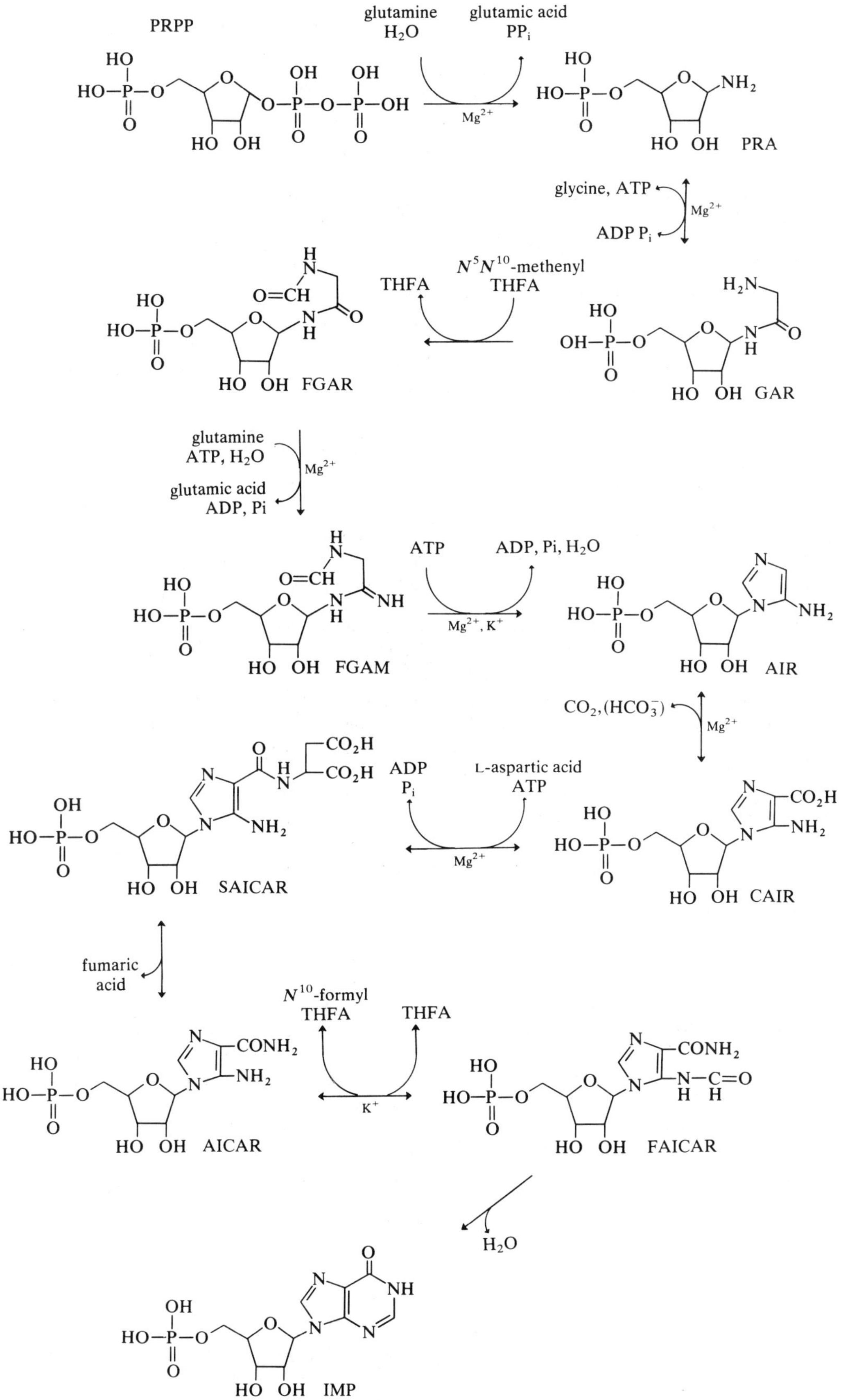
PRPP
glutamine
H_2O
glutamic acid
PP_i
Mg^{2+}
PRA
glycine, ATP
ADP P_i
N^5N^{10}-methenyl
THFA
THFA
FGAR
GAR
glutamine
ATP, H_2O
glutamic acid
ADP, Pi
ATP
ADP, Pi, H_2O
Mg^{2+}, K^+
FGAM
AIR
CO_2, (HCO_3^-)
ADP
P_i
L-aspartic acid
ATP
SAICAR
CAIR
fumaric
acid
N^{10}-formyl
THFA
THFA
K^+
AICAR
FAICAR
H_2O
IMP

Scheme 157

(1) Phosphoribosyl transferases (ribose only)
$$B + R_fTP \rightarrow NMP + (P_2)_i$$
(2) Nucleoside phosphorylases
$$B + 1 - R_fMP \leftrightarrow N + P_i$$
(3) Nucleoside kinases
$$ATP + N \rightarrow NMP + ADP$$
(4) 5'-Nucleotidases
$$NMP + H_2O \rightarrow N + P_i$$
(5) Nucleoside monophosphate kinases
$$N_1MP + N_2TP \leftrightarrow N_1DP + N_2DP$$
(6) Nucleoside diphosphate kinases
$$N_1DP + N_2TP \rightarrow N_1TP + N_2DP$$

Scheme 158

Table 39 Some Naturally Occurring Purines

Trivial name	\multicolumn{7}{c}{Substitution at position}						
	1	2	3	6	7	8	9
Adenine	—	H	—	NH_2	—	H	H
Guanine	—	NH_2	—	OH	—	H	H
Isoguanine	—	OH	—	NH_2	—	H	H
Hypoxanthine	—	H	—	OH	—	H	H
Xanthine	—	OH	—	OH	—	H	H
Uric acid	—	OH	—	OH	—	OH	H
Theophylline	Me	OH	Me	OH	—	H	H
Theobromine	—	OH	Me	OH	Me	H	H
Caffeine	Me	OH	Me	OH	Me	H	H
Isocaffeine	Me	OH	Me	OH	—	H	Me
Triacanthine	—	H	R^1	NH_2	—	H	H
Eritadenine	—	H	—	NH_2	—	H	R^2
Kinetin	—	H	—	NHR^3	—	H	H
6-Methylaminopurine	—	H	—	NHMe	—	H	H
6-Dimethylaminopurine	—	H	—	NMe_2	—	H	H
6-Succinoaminopurine	—	H	—	NHR^4	—	H	H
Discadenine	—	H	R^5	NHR^1	—	H	H
Herbipoline	NH_2	—	O^-	—	—	H	$\overset{+}{Me_2}$

$R^1 = CH_2CH=C(Me)_2$; $R^2 = CH_2(CHOH)_2CO_2H$; R^3 = furfuryl; $R^4 = CH(CO_2H)CH_2CO_2H$;
$R^5 = CH_2CH_2CH(NH_2)CO_2H$.

of the cocoa bean is 3,7-dimethylxanthine (theobromine) ⟨51AG511⟩ and smaller amounts of theophylline are also present ⟨68N299⟩. Theobromine is also strongly diuretic and like theophylline it forms soluble complexes with salts of various organic acids which aids oral use. 1,2,7-Trimethylxanthine (caffeine) is the major purine of the coffee bean and also occurs in tea leaves (*ca.* 3.5%). It is a cardiac and respiratory stimulant and is frequently used in headache powders. It has recently been shown that 7-methylxanthosine (**372**) is an intermediate in caffeine biosynthesis ⟨78P2075⟩. Methylated adenines including 6-methylaminopurine ⟨58JBC(230)717⟩ and 6-dimethylaminopurine ⟨58MI40902⟩ are of common occurrence as constituents of yeast and other tRNAs.

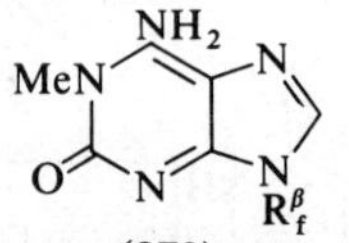

(**372**)

2'-*O*-Methyladenosine (A^m), 2'-*O*-methylguanosine (G^m) and 7-methylguanosine (m^7G) are also involved in capped 5'-termini of messenger RNA (mRNA) produced by vaccinia virus ⟨75PNA318, 75MI40903⟩. Two types of termini are produced, namely $m^7G(5')$-pppG^m and $m^7G(5')$-pppA^m. Similar caps occur in virus-specific mRNA isolated from vaccinia-infected HeLa cells where the termini also carry additional methyl groups on the N-6 atom of the penultimate adenosine and the 2'-hydroxy group of the third nucleoside ⟨77MI40904⟩, and the presence of such methylated 5'-ends (caps) appears to be a feature of eukaryotic mRNA, heterogeneous nuclear RNA and low molecular weight RNAs. Exceptions include polio and other viruses. When the 7-methylguanosine cap is removed from the mRNA of vaccinia virus, significant loss of ability to bind to ribosomes and to stimulate protein synthesis *in vitro* occurs but this is recovered when the 7-methylguanosine is replaced ⟨78JBC(253)1710⟩. In contrast, elimination of the penultimate 2'-O-methyl group appeared to have only a minor effect upon ribosome binding *in vitro*. Inhibition of the capping mechanism may be important as an explanation of antiviral activity with many compounds.

7-Methylguanine or epiguanine ⟨1898CB3272⟩ also occurs in urine (3.4 g per 10^4 l) and 1-methylisoguanosine (doridosine) has been isolated from the marine sponges *Tedania digitata* ⟨80MI40900⟩ and *Anisodoxis nobilis* ⟨80MI40905⟩. The latter compound has powerful muscle relaxant and cardiovascular activity when administered orally in animals and since it is resistant to adenosine deaminase it is thought to act as a long acting adenosine. Adenosine of course is rapidly deaminated to inosine *in vivo*. The enzyme which catalyzes the deamination, namely adenosine deaminase (aminohydrolase E.C. 3.5.4.4), is strongly inhibited by transition state analogs ⟨72ACR10⟩, *i.e.* compounds with enzyme binding properties which resemble those of the highly reactive substrate intermediates approaching the transition state (**373**). The naturally occurring compounds pentostatin (**374**) ⟨74JHC641, 77MI40903⟩ and the ribose analog coformycin (**375**) ⟨B-70MI40900⟩ are two such compounds which possess a tetrahedral carbon at what would be the analogous C-6 position in adenosine and the transition state in the latter may be regarded as the tetrahedral state (**373**). Isocoformycin (**376**) is a related inhibitor of adenosine deaminase ⟨79MI40908⟩. These compounds have been widely used to increase the efficacy of the antitumour drug arabinofuranosyladenine by prolonging its action *in vivo*. Simple transition state analogs including ethanol adducts presumably at C-6 have also shown enzyme-inhibiting activity ⟨70JA4751⟩.

4.09.8.2.2 Cytokinins

The cytokinins are plant growth substances which promote cell division ⟨B-71MI40900, 65MI40908⟩. They are adenine derivatives, the first of which to be isolated from a natural source (*Zea mays*) was zeatin, *trans*-6-(4-hydroxy-3-methylbut-2-enyl)aminopurine (**377a**) ⟨63MI40900⟩, 0.7 mg of which was isolated from 60 kg of corn. The structure ⟨64CC230, 67T479⟩ was confirmed by synthesis ⟨64CC231, 66JCS(C)921⟩ from 6-chloropurine and *trans*-4-hydroxy-3-methylbut-2-enylamine, the latter compound being obtained from methyl (*E*)-2-methylcrotonate. The *cis* isomer of zeatin also occurs naturally but was found to be 50 times less active than the *trans* isomer in a standard tobacco callus bioassay ⟨71JA3056⟩. The *cis* form ⟨72S618⟩ has also been synthesized and occurs as a 9-β-ribonucleoside in *Nicotinia tabacum* ⟨78MI40910⟩.

Other naturally occurring cytokinins include (−)dihydrozeatin (**377b**) from immature seeds of *Lupinus luteus* ⟨67TL1317⟩. N^6-Isopentenyladenine (**377c**) occurs as the free base in *Corynebacterium fasciens* ⟨66PNA(56)52, 66PNA(56)60⟩, as a 9-β-D-ribofuranosyl derivative in tRNA of yeast and calf liver ⟨66JA2614, 66AG392⟩ and in spinach and garden peas ⟨67MI40902⟩. Both zeatin ⟨65PNA(54)1052⟩ and the *cis* isomer ⟨68MI40906, 70B3701⟩ have also been isolated as 9-β-D-ribofuranosides. Various adenine derivatives related to zeatin have also been isolated from *Zea mays*, including 6-(2,3,4-trihydroxy-3-methylbutylamino)purine (**377d**) ⟨73P2445⟩, and O-glucosylzeatin occurs in *Populus* species ⟨79P819⟩.

(377a) $R^1 = CH_2CH=CMeCH_2OH(trans)$; $R^2 = R^3 = H$
(377b) $R^1 = CH_2CH_2CHMeCH_2OH$; $R^2 = R^3 = H$
(377c) $R^1 = CH_2CH=CMe_2$; $R^2 = R^3 = H$
(377d) $R^1 = CH_2CH(OH)C(OH)MeCH_2OH$; $R^2 = R^3 = H$
(377e) $R^1 = $ 2-furfuryl; $R^2 = R^3 = H$
(377f) $R^1 = PhCH_2$; $R^2 = R^3 = H$
(377g) $R^1 = o$-hydroxybenzyl; $R^2 = $ glucofuranosyl; $R^3 = $ SMe

Perhaps the first cytokinin to be examined in detail was kinetin or 6-furfurylaminopurine (377e), originally thought to be of natural occurrence but later found to be an artefact produced in and isolated from autoclaved solutions of DNA ⟨56JA1375⟩. N^6-Benzyladenine (377f) has strong cytokinin activity and in minute amounts has been used commercially (verdan) to keep green vegetables such as lettuces fresh for extended periods. Substituted benzyladenines have also been isolated from natural sources including the N^6-o-hydroxybenzyl-8-methylthio-9-glucofuranosylpurine (377g) from *Zantedeschia aethiopica* (a cuckoo pint) ⟨80TL4387⟩. A recent review of structure–activity relationships of 161 cytokinins including synthetic compounds has appeared ⟨80P2239⟩.

4.09.8.3 Nucleoside Antibiotics and Related Compounds

Several glycosylpurines have been isolated from microorganisms (Table 40) and proved to have substantial biological activity ⟨B-70MI40900⟩. They include substances such as 9-(β-D-arabinofuranosyl)adenine (Ara A) which is a powerful antiviral and antitumour agent and

Table 40 Some Purine Nucleoside Antibiotics and Related Compounds

Purine nucleoside	Trivial name (Origin)	Ref.
6-Dimethylamino-9-[(3-p-methoxy-β-L-phenylalanylamino)-3-deoxy-β-D-ribofuranosyl]purine	Puromycin (*S. alboniger*)	B-70MI40900, p. 13
3'-Deoxyadenosine	Cordycepin (*Cordyceps militaris* and *Aspergillus nidulans*)	B-70MI40900, p. 50
3'-Amino-3'-deoxyadenosine (plus 3'-N-acetylhomocitrullyl and lysyl)	(*Helminthosporium, Cordyceps militaris* and *Aspergillus nidulans*)	B-70MI40900, p. 176
6-Amino-9-(β-D-psicofuranosyl)purine	Psicofuranine (Angustmycin C) (*S. hygroscopicus* var. *decoyicus*)	B-70MI40900, p. 98
9-β-D-(5,6-Psicofuranosenyl)-6-aminopurine	Decoyinine (Angustmycin A) (*S. hygroscopicus*)	B-70MI40900, p. 115
9-(β-D-Ribofuranosyl)-2-methoxyadenine	Spongosine (*C. trypta*)	B-70MI40900, p. 126
9-(β-D-Arabinofuranosyl)adenine	Spongoadenosine (*S. antibioticus*)	B-70MI40900, p. 128
9-[β-(2'α, 3'α-dihydroxy-4'β-(hydroxymethyl)cyclopentyl]adenine	Aristeromycin (*S. citricolor*)	B-70MI40900, p. 236
4',5'-Dehydroaristeromycin	Neplanocin A (*Actinoplanacea ampullariella*)	81CPB597
9-(Fluoro-5-O-sulfamoylpentofuranosyl)adenine	Nucleocidin (*S. clavus*)	B-70MI40900, p. 246
N^6-[Isopalmitoylglycyl-4-amino-4-deoxy-L-glycero-L-glycoheptosyl]purine	Septacidin (*S. fimbriatus*)	B-70MI40900, p. 256
9-(β-D-Ribofuranosyl)purine	Nebularine (*Agaricus* (*clitocybe*) *nebularis* and *S. yokosukanensis*)	B-70MI40900, p. 261
2-Hydroxy-9-(β-D-ribofuranosyl)adenine	Crotonoside (*Croton tiglium*)	B-70MI40900, p. 267
9-(β-D-Ribofuranosyl)hypoxanthine	(*S. antibioticus*)	80JHC461
N^3-(β-D-Ribofuranosyl)-4,6-dimethylimidazo[1,2-a]purine	Wyosine (Nucleoside Yt) (Yeast tRNA^{Phe})	78TL2579
1-Methylisoguanosine	Doridosine (*Tedania digitata* and *Anisodoris nobilis*)	80TL567 80MI40905
2-Methylthio-N-carbonyl-[N-threonyl]adenosine	(Mammalian tRNA)	79JA2224
Adenosine substituted on 5'-O by a (−)-*chiro*-inositol L-gulosamine and L-serine unit	Adenomycin (*S. griseoflavus*)	80TL3203

is used clinically for these purposes ⟨76MI40905, 76MI40903, 76MI40906, 76MI40907⟩. Activity of the compound *in vivo* is severely reduced by its rapid deamination with the enzyme adenosine deaminase. However the activity may be prolonged by using the drug in the presence of an adenosine deaminase inhibitor such as pentostatin (2'-deoxycoformycin) ⟨77MI40903⟩ or by using a precursor (prodrug) of the compound such as the 2'-deoxy-2'-fluoro derivative which produces Ara-A by slow hydrolysis ⟨79JHC157⟩.

Puromycin (**378**) has been extensively studied as an inhibitor of protein biosynthesis in both bacterial and mammalian cells. It is structurally similar to the aminoacyl end of aminoacyl tRNA and functions by producing a codon-independent analog of aminoacyl tRNA and so promotes the release of incomplete peptide chains from the ribosome synthesizing complex. It blocks peptide chain extension by reacting with the growing peptide chain on the peptidyl-tRNA site on the ribosome to produce a peptidylpuromycin derivative. It is highly toxic to mammals and this restricts its use as an antibiotic. The analogous homocitrullyaminoadenosine also inhibits protein synthesis apparently by inhibiting the incorporation of leucine-tRNA ⟨63BBA(72)62⟩.

(**378**) Puromycin; R = *p*-methoxy-L-phenylalanyl

(**379**) Nucleocidin

Nucleocidin (**379**) which contains the first fluorosugar to be obtained from a natural source is a more potent inhibitor of protein synthesis *in vivo* than puromycin but has similar activity *in vitro* ⟨66JBC(24)1091⟩.

The ketosylpurines, psicofuranine and decoyinine, both have antimicrobial activity and activity against adenocarcinoma in rats, and appear to operate by inhibiting XMP aminase ⟨60MI40900, 60BBR(3)596⟩.

Septacidin and nebularine has some activity against tumour cells ⟨B-70MI40900⟩ and nebularine in addition has some antibacterial action. It is, however, very toxic to mice ⟨53JBC(204)1019, 66MI40902⟩.

Carbocyclic nucleoside analogs such as Aristeromycin (**380**) have also been isolated from natural sources (Table 40) and have antimicrobial activity. The 4',5'-dehydro derivative of the latter, namely Neplanocin A, also has antitumour activity and has been synthesized ⟨81CPB597⟩.

(**380**)

Both naturally occurring and synthetic nucleoside derivatives of several purines or purine analogs have a broad spectrum of biological activity ⟨B-74MI40900⟩. Generally most of the compounds only become active after intracellular phosphorylation, usually catalyzed by adenosine kinase to produce nucleotides. It has been concluded ⟨75MI40902⟩ that necessary but not sufficient conditions for substrate activity with this enzyme are (a) the nucleoside should have a 2'-hydroxy group which should be *trans* to the purine unit, (b) there should be a considerable degree of freedom of rotation about the C(1')—N(9) bond. Given these conditions the 3'-hydroxy and 4'-hydroxymethyl group could be either *cis* or *trans* to the purine ring.

In recent years exceptional antiviral activity has been obtained with compounds which show structural analogy to the common purine nucleosides. In particular the substance acycloguanine (**381**) ⟨78MI40909⟩ and 9-(*S*-2,3-dihydroxypropyl)adenine (**382**) have high antiviral activity and are undergoing clinical trials. The latter compound is active in the early stages of viral replication but is not antibacterial since it does not penetrate the bacterial cell wall. It may also be seen to be related to the naturally occurring compound eritadenine (**383**) which was isolated from *Lentinus edodes* ⟨74JMC846⟩ and synthesized ⟨73JOC2887, 73JOC2891⟩. This compound has high hypolipidemic activity which is, however, surpassed by the synthetic analog 9-(4-chlorophenoxy-2-hydroxypropyl)adenine ⟨78JMC1073⟩. Related 2,3-dihydroxypropylguanine derivatives have also been prepared ⟨78CCC3103⟩. Details of the synthesis and properties of various modified nucleosides found as minor components in DNA and RNAs have been recorded ⟨B-71MI40901⟩.

(381)   (382)   (383)

4.09.8.4 Miscellaneous

6-Mercaptopurine (6-thioxo-1,6-dihydropurine) was one of the first purine derivatives to find application in the treatment of leukaemia, especially in children. The drug inhibits the biosynthesis of purine nucleosides, including the conversion of IMP to AMP ⟨54MI40905, B-66MI40900⟩. The *S*-imidazolyl derivative (**384**), azathiopurine or Imvran, has also been of value as a suppressor of the immune response.

(384)

(385a) R = H
(358b) R = α-OH
(385c) R = β-OH

Saxitoxin (**385a**) is an unusual derivative of tetrahydrodiaminopurine. It is produced by various marine dinoflagellates such as *Gonyaulax catanella* which form the red tides off the coasts of Canada and the USA. The compound tends to accumulate in various edible shell fish such as the Alaska butter clam (*Saxidomus giganteus*), toxic mussels (*Mytilus californianus*), and scallops. Saxitoxin is one of the most toxic, non-protein, compounds known and its presence in shell fish has created serious health and economic problems along the North Atlantic coast. The structure of the compound has been determined ⟨75JA1238, 75JA6008⟩ and it has been shown to exist as a hydrated cation (**385a**). Proton relaxation rates and scalar coupling constants have been used to define the total absolute stereochemistry of saxitoxin in solution and to give details of the overall and internal rotation of the molecule ⟨80JA1513⟩. The pK_a values of the compound have been determined by potentiometric titration and observation of the change in the ^{13}C NMR chemical shifts with pH ⟨80JA7335⟩.

A stereospecific total synthesis of (±)-saxitoxin has been achieved ⟨77JA2818⟩. Related hydroxysaxitoxins Gonyaulax II (**385b**) and III (**385c**) have also been isolated from *Gonyaulax tamarensis* ⟨76JA5414, 82TL321⟩ and *N*-sulfo derivatives of these have been isolated from *Protogonyaulax* clone P107 ⟨81JA6977⟩. The latter compounds, reported as the first *N*-sulfocarbamoyl derivatives to be found in a natural source, are relatively non-toxic but readily hydrolyzed to the toxic materials. Related diaminopurines such as 2,6-diaminopurine are also toxic and the latter compound also has antileukaemic activity.

Saxitoxin can be regarded as a 3-substituted purine. The compound discadenine (**386**), isolated from the cellular slime mold *Dictyostelium discoideum*, was the first recorded naturally occurring purine to contain an amino acid residue in the 3-position. The compound has pronounced spore germination inhibitory activity and in addition has cytokinin activity, showing two thirds the activity of kinetin at 10^{-7} mol in the standard tobacco pith test ⟨76TL3807⟩. Discadenine minus the isopentenyl group has the same spore germination inhibitory activity as the parent ⟨80CB2043⟩.

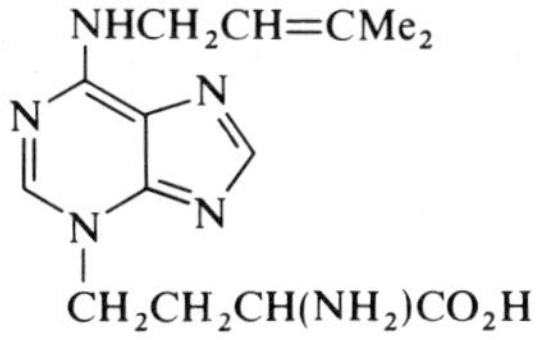

NHCH$_2$CH=CMe$_2$

CH$_2$CH$_2$CH(NH$_2$)CO$_2$H

(**386**)

One of the more unusual bases found in tRNAs (next position of the 3'-end of the anticodon of yeast tRNA^{Phe}) is wyosine (**226**) (see Section 4.09.6.18) or nucleoside Yt, a 3-methylguanine derivative ⟨71TL2725, 76B898⟩ which is characterized by its strong fluorescence. In addition, the nucleoside bond is very readily hydrolyzed in weak acid (pH 4) solution. Similar compounds, wybutosine (**227**) and wybutoxosine (**228**) have been also found in baker's yeast tRNA^{Phe} ⟨67PNA(57)751, 70JA7617⟩, rat liver ⟨71MI40906, 73B188⟩ and *Lupinus luteus* ⟨74JA7797⟩. The lability of the sugar made assignment of its position difficult. However the structure has been confirmed by synthesis ⟨78TL2579⟩.

4.10

Other Imidazoles with Fused Six-membered Rings

J. A. MONTGOMERY and J. A. SECRIST III
Southern Research Institute, Birmingham, Alabama

4.10.1	INTRODUCTION	608
4.10.2	STRUCTURE	610
	4.10.2.1 Imidazo[1,2-c]pyrimidine	610
4.10.3	REACTIVITY	610
	4.10.3.1 Imidazo[1,2-a]pyridine	610
	4.10.3.2 Imidazo[1,5-a]pyridine	614
	4.10.3.3 Imidazo[4,5-b]pyridine	615
	4.10.3.4 Imidazo[4,5-c]pyridine	619
	4.10.3.5 Imidazo[1,2-a]pyrazine	624
	4.10.3.6 Imidazo[1,5-a]pyrazine	625
	4.10.3.7 Imidazo[4,5-b]pyrazine	626
	4.10.3.8 Imidazo[1,2-a]pyrimidine	627
	4.10.3.9 Imidazo[1,2-c]pyrimidine	627
	4.10.3.10 Imidazo[1,5-a]pyrimidine	628
	4.10.3.11 Imidazo[1,2-b]pyridazine	628
	4.10.3.12 Imidazo[4,5-c]pyridazine	628
	4.10.3.13 Imidazo[4,5-d]pyridazine	629
	4.10.3.14 Imidazo[4,5-d][1,2,3]triazine	629
	4.10.3.15 Imidazo[1,2-b][1,2,4]triazine	630
	4.10.3.16 Imidazo[1,5-b][1,2,4]triazine	630
	4.10.3.17 Imidazo[1,5-d][1,2,4]triazine	630
	4.10.3.18 Imidazo[2,1-c][1,2,4]triazine	630
	4.10.3.19 Imidazo[5,1-f][1,2,4]triazine	630
	4.10.3.20 Imidazo[1,2-a][1,3,5]triazine	631
4.10.4	SYNTHESIS	631
	4.10.4.1 Imidazo[1,2-a]pyridine	631
	4.10.4.2 Imidazo[1,5-a]pyridine	634
	4.10.4.3 Imidazo[4,5-b]pyridine	635
	4.10.4.4 Imidazo[4,5-c]pyridine	639
	4.10.4.5 Imidazo[1,2-a]pyrazine	642
	4.10.4.6 Imidazo[1,5-a]pyrazine	644
	4.10.4.7 Imidazo[4,5-b]pyrazine	645
	4.10.4.8 Imidazo[1,2-a]pyrimidine	647
	4.10.4.9 Imidazo[1,2-c]pyrimidine	648
	4.10.4.10 Imidazo[1,5-a]pyrimidine	649
	4.10.4.11 Imidazo[1,5-b]pyrimidine	649
	4.10.4.12 Imidazo[1,2-b]pyridazine	650
	4.10.4.13 Imidazo[1,5-b]pyridazine	651
	4.10.4.14 Imidazo[4,5-c]pyridazine	651
	4.10.4.15 Imidazo[4,5-d]pyridazine	651
	4.10.4.16 Imidazo[4,5-d][1,2,3]triazine	652
	4.10.4.17 Imidazo[1,2-b][1,2,4]triazine	653
	4.10.4.18 Imidazo[1,5-b][1,2,4]triazine	653
	4.10.4.19 Imidazo[1,5-d][1,2,4]triazine and Imidazo[1,2-d][1,2,4]triazine	654
	4.10.4.20 Imidazo[2,1-c][1,2,4]triazine	654
	4.10.4.21 Imidazo[4,5-e][1,2,4]triazine	655
	4.10.4.22 Imidazo[5,1-c][1,2,4]triazine	656
	4.10.4.23 Imidazo[5,1-f][1,2,4]triazine	656
	4.10.4.24 Imidazo[1,2-a][1,3,5]triazine	657
	4.10.4.25 Imidazo[1,5-a][1,3,5]triazine	659

4.10.4.26 Imidazo[1,2-b][1,2,4,5]tetrazine	659
4.10.4.27 Imidazooxazines	660
4.10.4.27.1 Imidazo[1,2-c][1,3]oxazine	660
4.10.4.27.2 Imidazo[1,5-c][1,3]oxazine	660
4.10.4.27.3 Imidazo[2,1-b][1,3]oxazine	660
4.10.4.27.4 Imidazo[4,5-d][1,3]oxazine	660
4.10.4.27.5 Imidazo[5,1-b][1,3]oxazine	661
4.10.4.27.6 Imidazo[2,1-c][1,4]oxazine	661
4.10.4.27.7 Imidazo[5,1-c][1,4]oxazine	661
4.10.4.28 Imidazooxadiazines	662
4.10.4.28.1 Imidazo[1,5-b][1,2,5]oxadiazine	662
4.10.4.28.2 Imidazo[2,1-b][1,3,4]oxadiazine	662
4.10.4.28.3 Imidazo[2,1-c][1,2,4]oxadiazine	662
4.10.4.29 Imidazothiazines	663
4.10.4.29.1 Imidazo[2,1-b][1,3]thiazine	663
4.10.4.29.2 Imidazo[1,2-c][1,3]thiazine	664
4.10.4.29.3 Imidazo[4,5-d][1,3]thiazine	664
4.10.4.29.4 Imidazo[4,5-e][1,3]thiazine	664
4.10.4.29.5 Imidazo[2,1-c][1,4]thiazine	665
4.10.4.29.6 Imidazo[5,1-c][1,4]thiazine	665
4.10.4.30 Imidazothiadiazines	665
4.10.4.30.1 Imidazo[4,5-c][1,2,6]thiadiazine	665
4.10.4.30.2 Imidazo[4,5-e][1,2,4]thiadiazine	666
4.10.4.30.3 Imidazo[2,1-b][1,3,4]thiadiazine	666
4.10.4.30.4 Imidazo[2,1-b][1,3,5]thiadiazine	667
4.10.4.30.5 Imidazo[4,5-d][1,2,3]thiadiazine	667
4.10.4.31 Imidazo[4,5-c][1,5,2]diazaphosphorine and Imidazo[4,5-d][1,3,2]diazaphosphorine	667
4.10.4.32 Imidazo[4,5-d][1,3,2]diazaborine	668

4.10.1 INTRODUCTION

Fifty-two ring systems composed of an imidazole fused with six-membered heterocyclic rings have been described in the literature, and are pictured below. The best known of these is the imidazo[4,5-d]pyrimidine (purine) ring system described in the preceding chapter. The other 51, the topic of this chapter, include the various azaindenes and rings containing oxygen, sulfur, phosphorus, boron and combinations of those heteroatoms in rings with nitrogen. The major interest in most of these ring systems results from their resemblance to the biologically important purine; some of them have been shown to function as antimetabolites with potentially useful activities. In addition, some are of interest as dyes or because of other properties such as the bioluminescence of luciferins, the biological activity of the streptothricin antibiotics, and the fluorescence of certain of the imidazo[1,2-c]pyrimidines.

The literature on these rings systems varies from fairly extensive in the case of the imidazopyridines (deazapurines) to as little as a single journal article in the case of the imidazodiazaphosphorines. In recent years many studies of these systems have included rather complete characterization by means of UV, IR and NMR spectroscopy and mass spectrometry, and these studies have provided much useful information. Owing to space limitations, we have spent very little time on the physical properties of the compounds unless some feature warranted special note. Reviews are available on imidazo[1,2-a]pyridines ⟨61HC(15-1)460, 77HC(30-2)117⟩, imidazo[1,5-a]pyridines ⟨61HC(15-1)643, 77HC(30-2)117⟩, imidazo[1,2-c]pyrazines ⟨61HC(15-2)774, 79HC(35)365⟩, imidazo[1,5-a]pyrazines ⟨61HC(15-2)835, 79HC(35)370⟩, imidazo[4,5-b]pyrazines ⟨79HC(35)376⟩, imidazo[1,2-a]pyrimidines ⟨61HC(15-2)802⟩, imidazo[1,2-c]pyrimidines ⟨61HC(15-2)764, 84MI41000⟩, imidazo[1,2-b]pyridazines ⟨73HC(27)801⟩, imidazo[4,5-c]pyridazines ⟨73HC(27)818⟩, imidazo[4,5-d]pyridazines ⟨73HC(27)807⟩, imidazo[4,5-d][1,2,3]triazines ⟨82MI41000, 75ANY(255)292, 78HC(33)118⟩, imidazo[1,2-b][1,2,4]triazines ⟨61HC(15-2)903⟩, imidazo[2,1-c]-[1,2,4]triazines ⟨78HC(33)882⟩, imidazo[5,1-f][1,2,4]triazines ⟨78HC(33)885⟩, imidazo[1,2-a]-[1,3,5]triazines ⟨59HC(13)432, 61HC(15-2)906⟩, imidazo[1,5-a][1,3,5]triazines ⟨61HC(15-2)916⟩ and imidazo[2,1-b][1,3]thiazines ⟨61HC(15-2)801⟩.

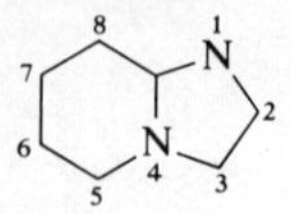

Imidazo[1,2-a]pyridine 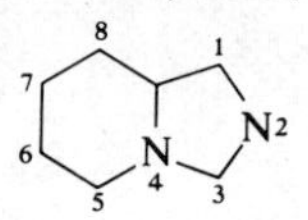   Imidazo[1,5-a]pyridine

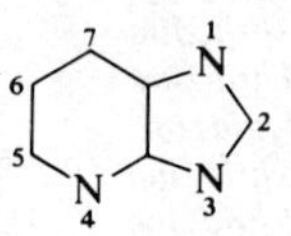

Imidazo[4,5-b]pyridine 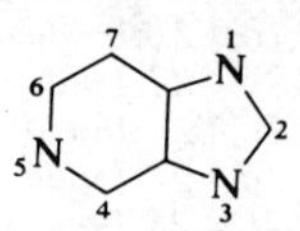   Imidazo[4,5-c]pyridine

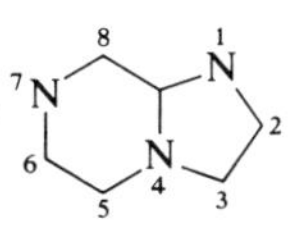

Other Imidazoles with Fused Six-membered Rings

Imidazo[1,2-a]pyrazine 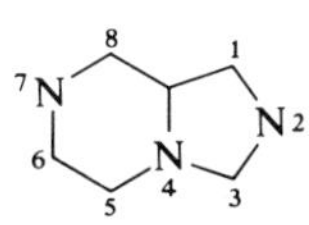 Imidazo[1,5-a]pyrazine 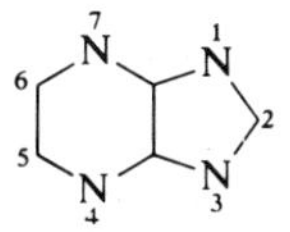 Imidazo[4,5-b]pyrazine 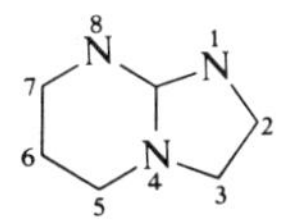 Imidazo[1,2-a]pyrimidine

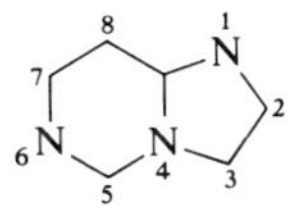

Imidazo[1,2-c]pyrimidine 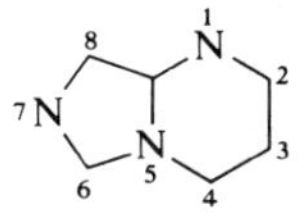 Imidazo[1,5-a]pyrimidine 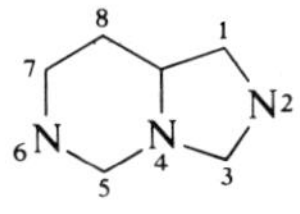 Imidazo[1,5-c]pyrimidine 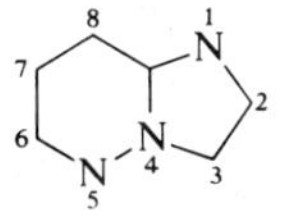 Imidazo[1,2-b]pyridazine

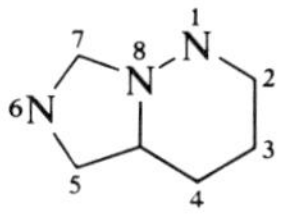

Imidazo[1,5-b]pyridazine 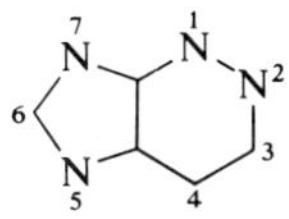 Imidazo[4,5-c]pyridazine 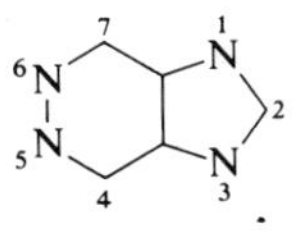 Imidazo[4,5-d]pyridazine 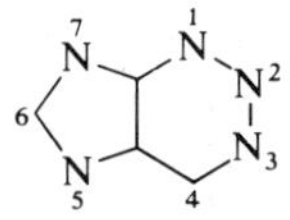 Imidazo[4,5-d][1,2,3]triazine

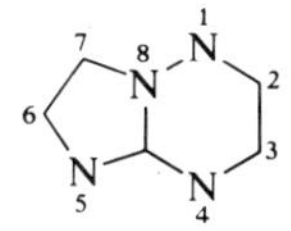

Imidazo[1,2-b][1,2,4]triazine 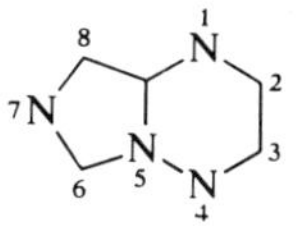 Imidazo[1,5-b][1,2,4]triazine 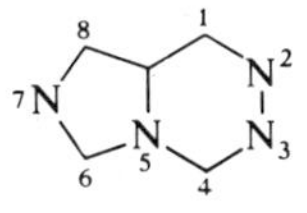 Imidazo[1,5-d][1,2,4]triazine

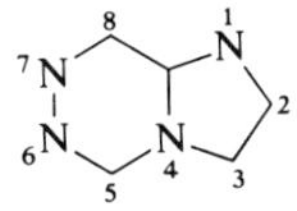

Imidazo[1,2-d][1,2,4]triazine 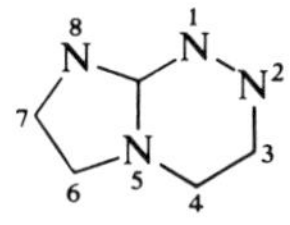 Imidazo[2,1-c][1,2,4]triazine 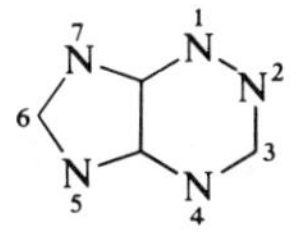 Imidazo[4,5-e][1,2,4]triazine

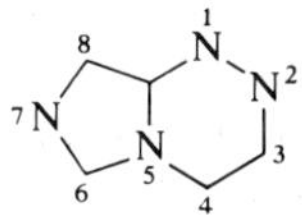

Imidazo[5,1-c][1,2,4]triazine 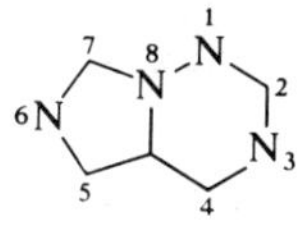 Imidazo[5,1-f][1,2,4]triazine 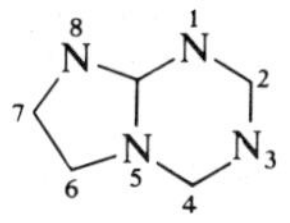 Imidazo[1,2-a][1,3,5]triazine

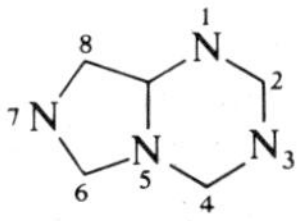

Imidazo[1,5-a][1,3,5]triazine 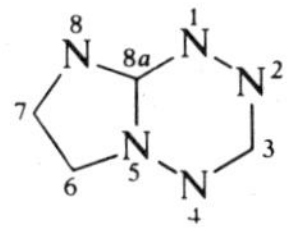 Imidazo[1,2-b][1,2,4,5]tetrazine 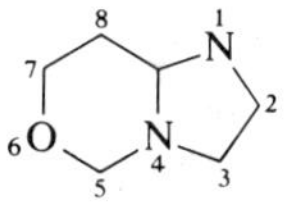 Imidazo[1,2-c][1,3]oxazine

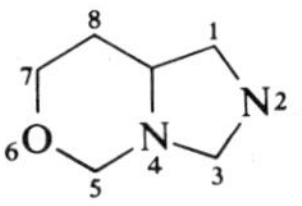

Imidazo[1,5-c][1,3]oxazine 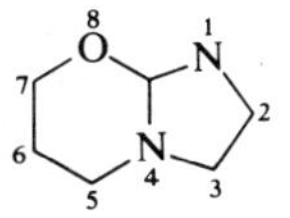 Imidazo[2,1-b][1,3]oxazine 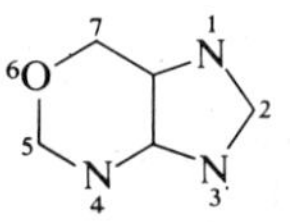 Imidazo[4,5-d][1,3]oxazine

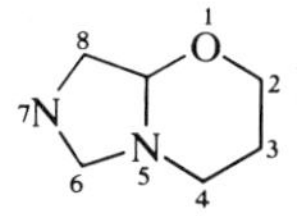

Imidazo[5,1-b][1,3]oxazine 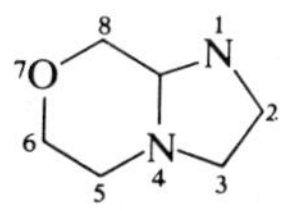 Imidazo[2,1-c][1,4]oxazine 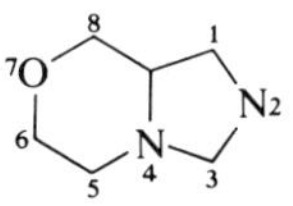 Imidazo[5,1-c][1,4]oxazine

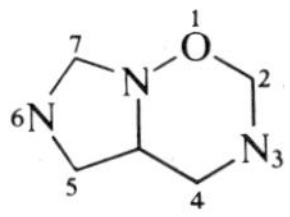

Imidazo[1,5-b][1,2,5]oxadiazine 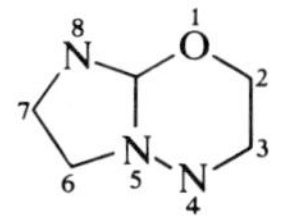 Imidazo[2,1-b][1,3,4]oxadiazine 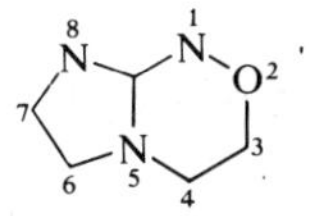 Imidazo[2,1-c][1,2,4]oxadiazine

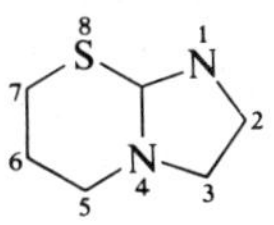

Imidazo[2,1-b][1,3]thiazine 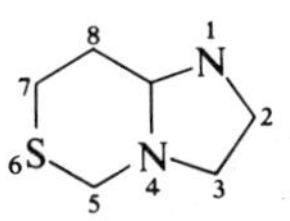 Imidazo[1,2-c][1,3]thiazine 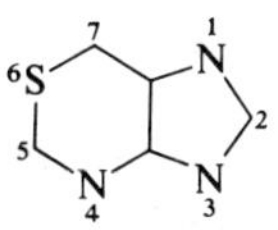 Imidazo[4,5-d][1,3]thiazine

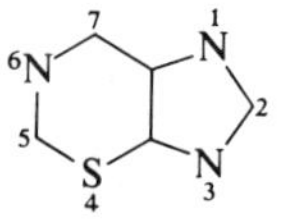

Imidazo[4,5-e][1,3]thiazine 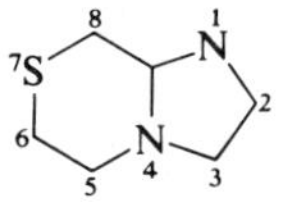 Imidazo[2,1-c][1,4]thiazine 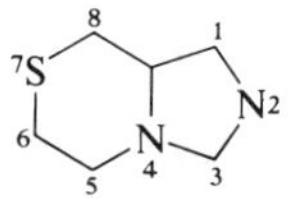 Imidazo[5,1-c][1,4]thiazine

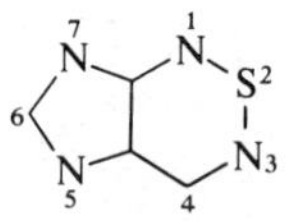

Imidazo[4,5-c][1,2,6]thiadiazine 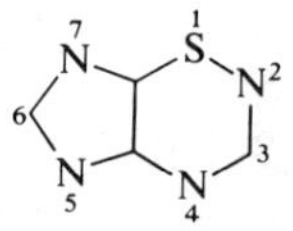 Imidazo[4,5-e][1,2,4]thiadiazine 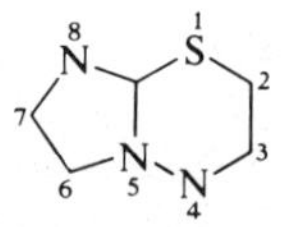 Imidazo[2,1-b][1,3,4]thiadiazine

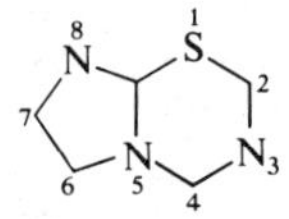

Imidazo[2,1-b][1,3,5]thiadiazine 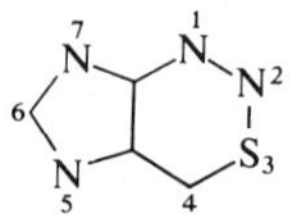 Imidazo[4,5-d][1,2,3]thiadiazine 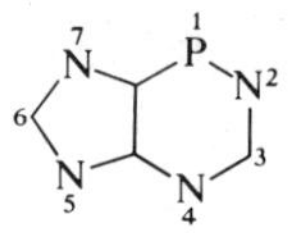 Imidazo[4,5-c][1,5,2]diazaphosphorine

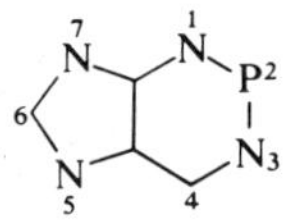

Imidazo[4,5-d][1,3,2]diazaphosphorine 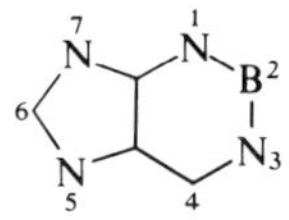 Imidazo[4,5-d][1,3,2]diazaborine

4.10.2 STRUCTURE

Data on the structures of other ring systems covered in this chapter are available primarily from IR, UV and NMR spectroscopy, but nothing deserves special comment. Much of this material is discussed in the reviews cited in Section 4.10.1.

4.10.2.1 Imidazo[1,2-c]pyrimidine

The main imidazo[1,2-c]pyrimidines whose structural characteristics have drawn interest are the compounds prepared from cytidine, the ε-cytidine derivatives such as (1a). X-Ray data on ε-Cyd·HCl (1a) indicate a slight deviation of the nitrogen heterocycle from planarity, an *anti* conformation about the glycosidic linkage, and no base stacking within the crystal ⟨76JA7401⟩. Compound (1a) has a fluorescence emission maximum at 340 nm, with a quantum yield of <0.01 and a fluorescence lifetime of 30 ± 5 ps ⟨72BBR(46)597, 76JA7408⟩. The quantum yield and fluorescence lifetime can both be enhanced by the preparation of substituted derivatives, such as (1b) ⟨76JA7408⟩. Although the protonated forms such as (1a) are fluorescent, the neutral species are not.

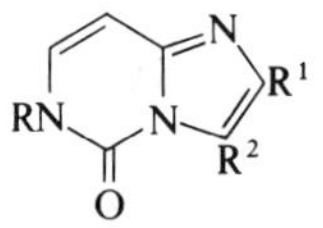

a; R = β-D-ribofuranosyl, R^1 = H, R^2 = H, HCl
b; R = β-D-ribofuranosyl, R^1 = NHCOMe, R^2 = H
c; R = β-D-ribofuranosyl, R^1 = H, R^2 = HgOAc

(1)

4.10.3 REACTIVITY

4.10.3.1 Imidazo[1,2-a]pyridine

The position of methylation of 2-methylimidazo[1,2-a]pyridine (2) has been established as N-1 by an ambiguous synthesis of 1,2-dimethylimidazo[1,2-a]pyridine (3) from (4) ⟨65JHC331⟩. This work was extended by others to protonation and to methylation of other imidazo[1,2-a]pyridines ⟨66JOC1295⟩, and it verified earlier conjectures concerning the position of quaternatization ⟨55CB1093, 26CB2048, 64JCS4226⟩.

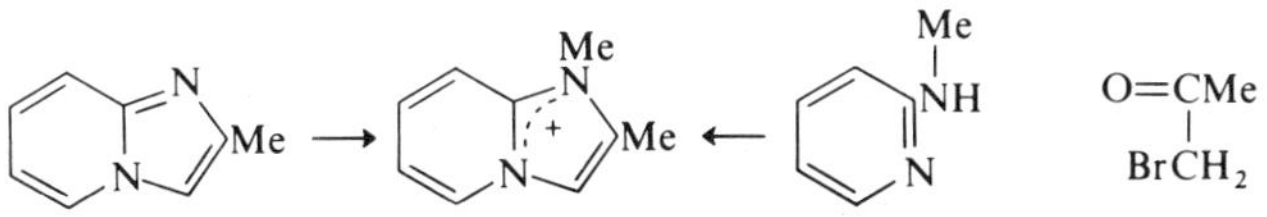

Imidazo[1,2-a]pyridines also react at N-2 with haloketones, especially phenacyl bromide, to give quaternary salts, with the ease of reaction depending upon the basicity of N-1 and steric factors as determined by ring substituents ⟨26CB2048, 55CB1093⟩. These salts undergo a variety of reactions with bases including cyclization with a methyl group at C-2.

Acid-catalyzed exchange of deuterium for hydrogen occurs at C-3 of imidazo[1,2-a]pyridine (**8**) and a number of its methyl derivatives ⟨68JOC1087⟩. The exchange rate of the corresponding 1-methylimidazo[1,2-a]pyridinium halides is considerably slower, suggesting hydrogen–deuterium exchange on the free base rather than the protonated species. Base-catalyzed exchange also occurs readily, but in this case at both C-3 and C-5 to give (**7**), as predicted by resonance considerations. The 1-methyl compounds (**10**) undergo catalyzed exchange at C-2 as well as C-3 and C-5 to give (**11**), which is also predictable.

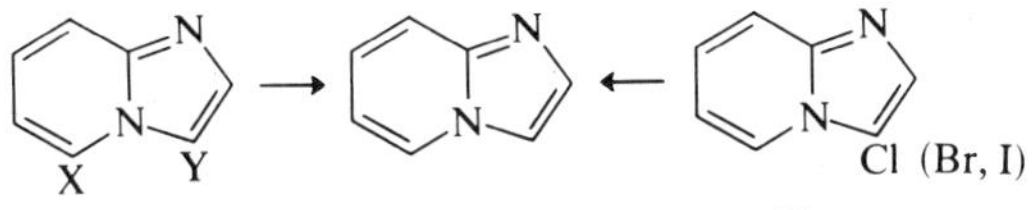

Detailed studies have been carried out on the halogenation of imidazo[1,2-a]pyridine (**8**) ⟨65JOC4085, 65JOC4081, 77MI41000⟩. In all cases where C-3 is unsubstituted, using bromine water, N-bromosuccinimide, N-chlorosuccinimide or iodine, the 3-halo compound (**9**) is obtained. Perchlorimidazo[1,2-a]pyridine has been obtained in high yield by treatment of imidazo[1,2-a]pyridine with phosphorus pentachloride at 275 °C in a sealed tube ⟨71JHC37⟩.

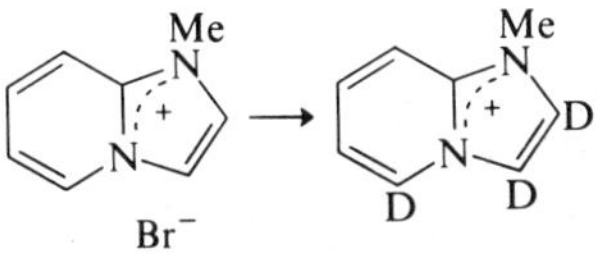

The reaction of 3-methylimidazo[1,2-a]pyridine (**12**) with NBS gave products formed by apparent nucleophilic substitution at C-2 (**13** and **14**) ⟨80JOC3738⟩. A variety of electrophilic substitution reactions have been shown to occur at C-3 ⟨74ZOR2467, 77MI41000, 75JOC2916⟩. Bromination of 2-(2-furyl)imidazo[1,2-a]pyridine (**15**) occurs either at C-3 of the imidazopyridine (**16**) or at C-5 of the furan ring (**17**) depending on the reaction conditions, the reagents used or the position of substituents ⟨72KGS691⟩. On heating in DMF, the 3-bromo compound (**16**) disproportionates to give a mixture of 5'-bromo (**17**), 3,5'-dibromo (**18**) and unbrominated material (**15**) ⟨77ZOR2626⟩. 2-Phenylimidazo[1,2-a]pyridine 1-oxide (**19**) also undergoes electrophilic substitution at C-3 giving (**20**) ⟨78JOC658⟩.

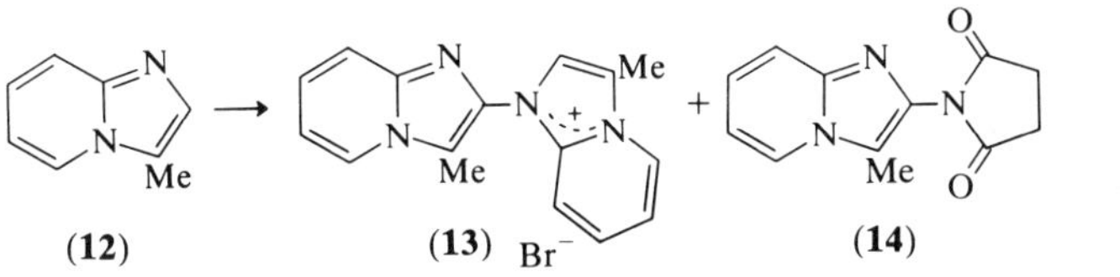

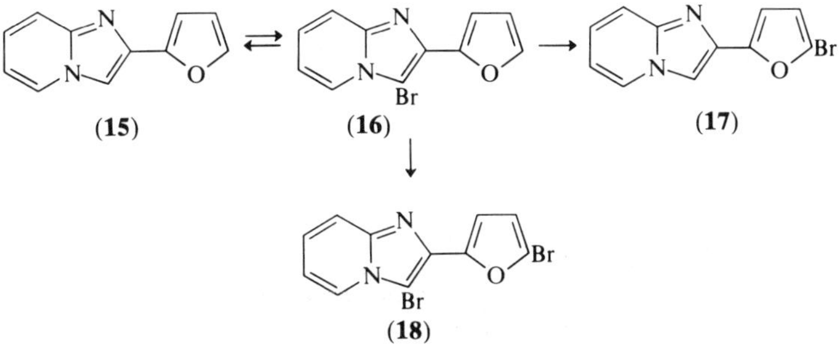

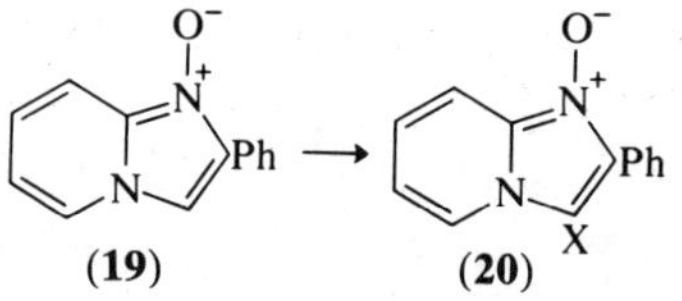

(19) (20)

Imidazo[1,2,-*a*]pyridine (**8**) and several of its derivatives (**21**) were found to react smoothly with nitrous acid to produce high yields of 3-nitroso products (**22a**) ⟨25CB393, 77HC(30-2)117, 76JMC1352, 61HC(15-1)481⟩. Nitration of imidazo[1,2-*a*]pyridine with cold nitric and sulfuric acids gave a high yield of the 3-nitro derivative as did a number of monomethylimidazo[1,2-*a*]pyridines (**22b**) ⟨77HC(30-2)117⟩. Other imidazo[1,2-*a*]pyridines containing a variety of substituents at C-2 and in the pyridine ring also give 3-nitro derivatives ⟨78MI41003, 77KGS556, 79ZOR(15)2534⟩. One study showed that nitration occurs primarily at C-3 of (**23**) but to a lesser extent at C-5, giving in one instance a 3,5-dinitro derivative (**25**). Nitration of 5-ethoxyimidazo[1,2-*a*]pyridine (**26**) gave both the 3,6- (**27**) and the 3,8-dinitro (**28**) derivatives ⟨79BSF(2)529⟩. These results were generally in agreement with the ^{13}C NMR data and estimated CNDO π-charge densities.

(**21**) (**22**) **a**; X = NO
b; $X = NO_2$

(**23**) (**24**) 80% (**25**) 20%

(**26**) (**27**) (**28**)

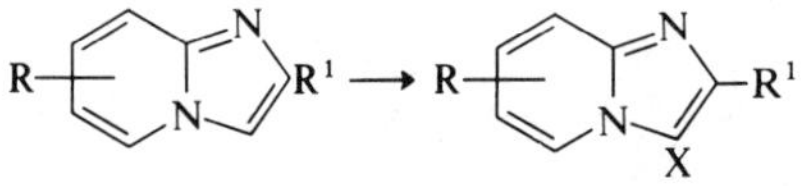

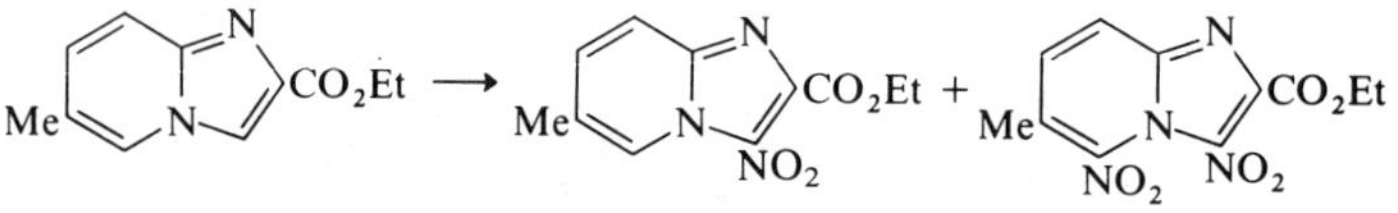

1,2-Dimethylimidazo[1,2-*a*]pyridinium perchlorate (**29**) underwent Friedel–Crafts acylation at C-3 with acetyl and benzoyl chloride to give (**30a** and **b**) ⟨74ZOR2467⟩. Imidazo[1,2-*a*]pyridin-2(3*H*)-one (**31**) undergoes acylation at C-3 forming (**32**) as well as on oxygen giving (**33**) ⟨24CB1381, 24CB2092, 25CB393⟩.

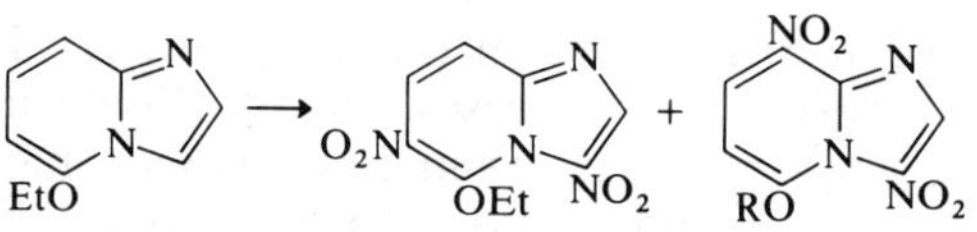

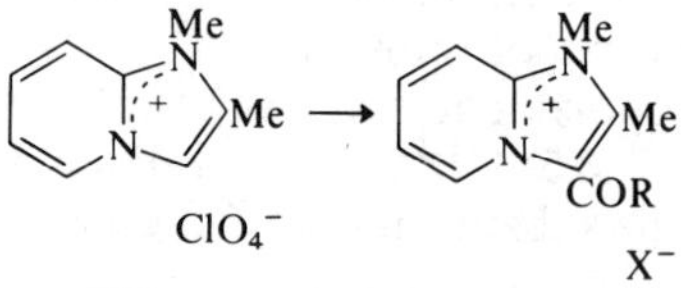

(**29**) (**30**) **a**; R = Me
b; R = Ph

(**31**) (**32**) R = Me, Ph (**33**)

Imidazo[1,2-*a*]pyridine (**8**) and its derivatives (**34**) undergo the Mannich reaction ⟨65JOC2403⟩ as well as reaction with acetaldehyde and chloral to give a variety of products (**35–41**) ⟨77JOC3377, 80KGS528⟩. Aromatic aldehydes (but not acetaldehyde or phenylacetaldehyde) condense with imidazo[1,2-*a*]pyridin-2(3*H*)-one (**31**) at C-3 to give (**42**), which disproportionates to (**43**) ⟨26CB2926, 49USP2481953⟩. The anion of imidazo[1,2-*a*]pyridine or its 2-, 6-, 7- or 8-methyl derivatives (**45**), prepared from (**44**) by means of phenyllithium, adds across the double bond of cylohexanone to give (**46**). 5-Methylimidazo[1,2-*a*]pyridine (**47**) reacts at the methyl group affording (**48**) rather than at the ring ⟨68JOC1638⟩. Reaction of 3-bromoimidazo[1,2-*a*]pyridine (**50**) with strong bases leads to metal–halogen exchange.

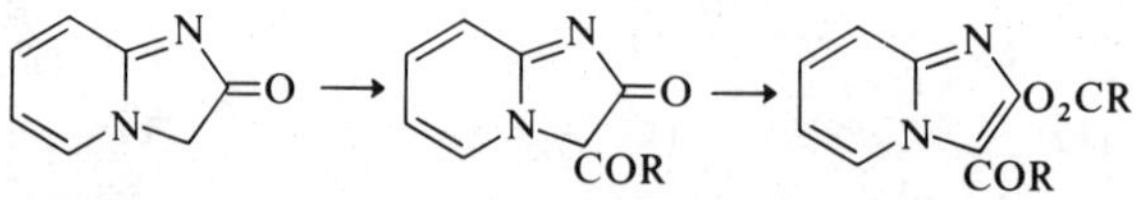

With methyllithium coupling at C-3 to give (**49**) occurs, but with metal amides debromination, coupling at C-5 to give (**51**) and *tele*-substitution at all positions of the pyridine ring (**52**) take place ⟨78JOC2900⟩.

(**34**) (**35**)

(**36**) (**37**) (**38**) (**39**)

(**40**) (**41**)

(**42**) (**43**)

(**44**) (**45**) (**46**)

(**47**) (**48**) (**49**) (**50**) (**51**) (**52**)

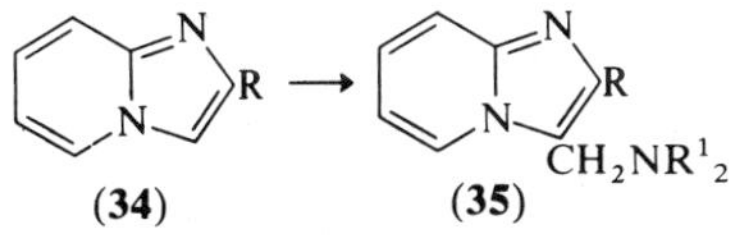

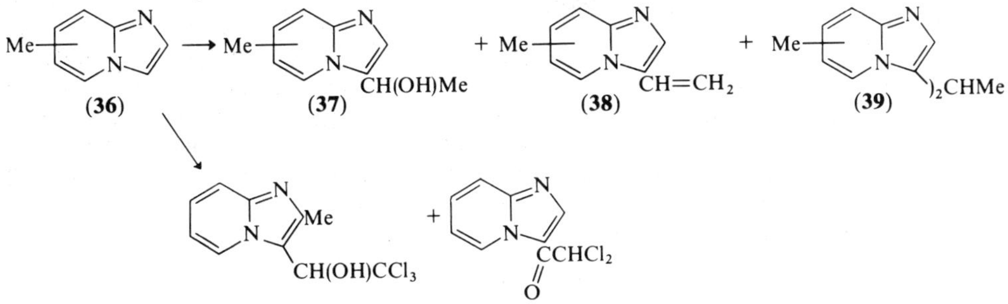

The 5-chloro group is easily displaced from the imidazo[1,2-*a*]pyridine ring with alkoxides, but neither the 2-chloro nor the 7-chloro group is ⟨65JHC53, 77HC(30-2)117⟩. Sodium methoxide displaces only the 5-chloro group of perchloroimidazo[1,2-*a*]pyridine (**53**) giving (**54a**), but piperidine replaces two chlorines, presumably the 5- and 7-chloro groups (**54b**) ⟨71JHC37⟩. The 2-chloro group can be displaced (**56**), however, when activated by a nitro group in the 3-position (**55**) ⟨65JOC4085, 77AF82⟩.

(**53**) (**54**) **a**; R = MeO, R^1 = Cl (**55**) (**56**)

b; R = R^1 = $C_5H_{10}N$

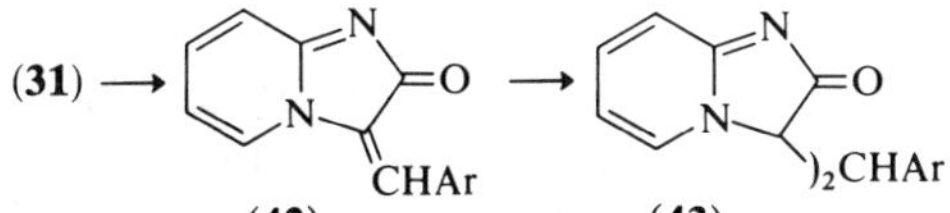

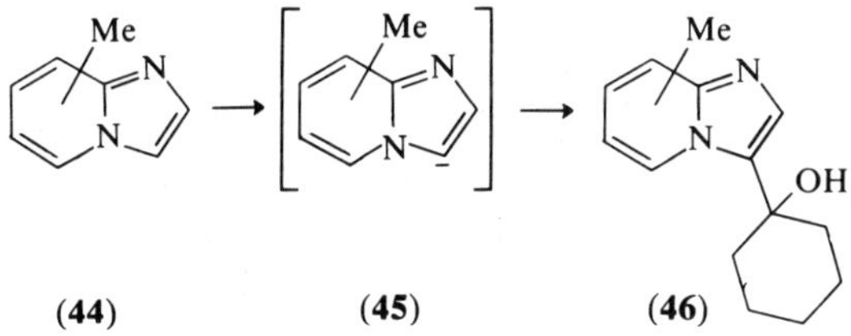

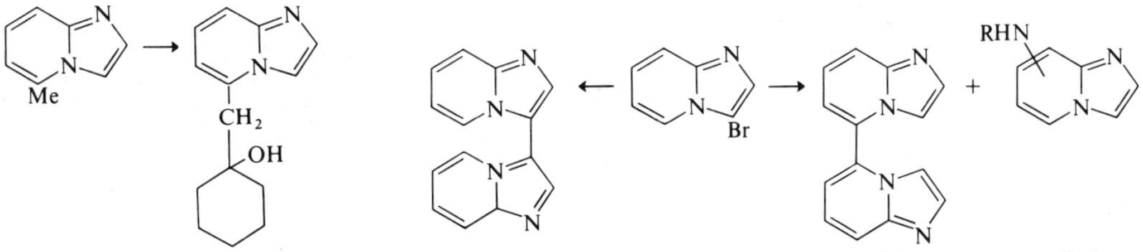

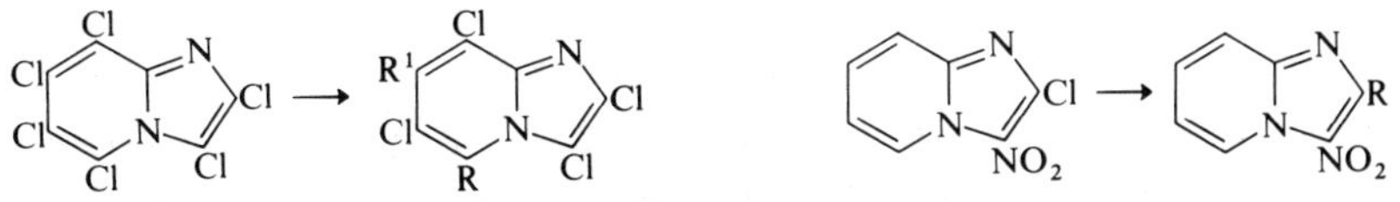

Oxidation of imidazo[1,2-*a*]pyridin-2(3*H*)-one (**31**) results in the formation of a red dye ⟨24CB2092, 23MI41000⟩, which probably has the structure (**57**) ⟨26CB2921⟩.

Imidazo[1,2-*a*]pyridines (**58**) activated by a nitro group at C-6 or C-8 undergo the Dimroth rearrangement to (**60**) in aqueous base. Intermediates of the rearrangement (**59**) were detected and their structures investigated spectroscopically ⟨73JHC755⟩.

(57)

(58) (59) (60)

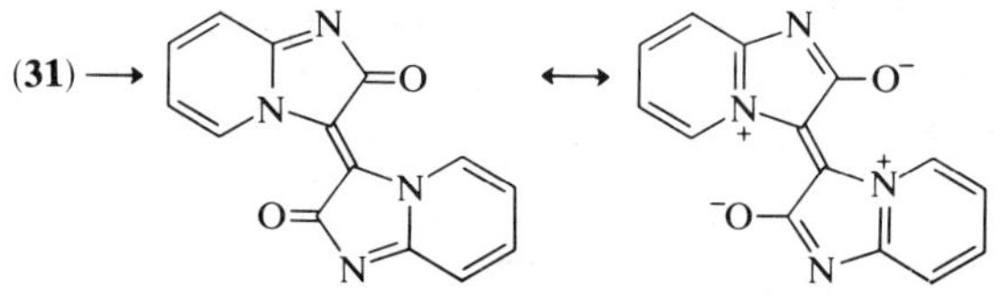

Dipyrido[1,2-*a*:3′,3′-*d*]imidazoles (**62**) have been prepared by the reaction of the corresponding imidazo[1,2-*a*]pyridines (**61**) with α-chloroacrylonitrile in nitrobenzene containing aluminum chloride ⟨78CPB2924⟩.

(61) (62)

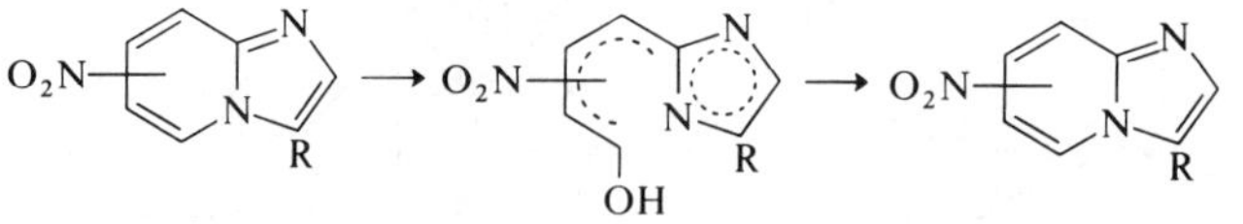

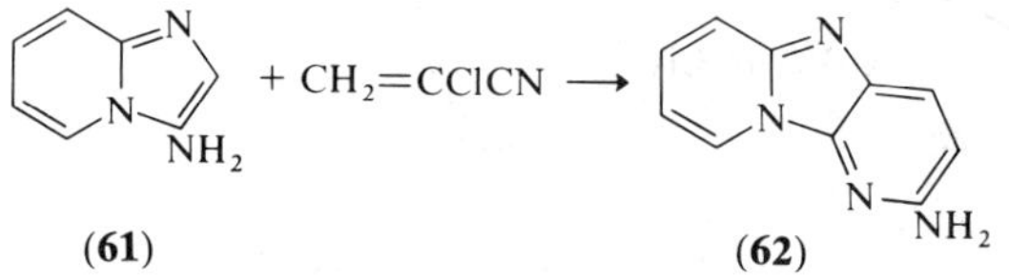

There is little published work on the catalytic hydrogenation of the imidazo[1,2-*a*]pyridines (**63**), but what there is indicates that reduction of the pyridine ring takes place preferentially to give the 5,6,7,8-tetrahydro derivatives (**64**) ⟨77HC(30-2)117, 73JMC1272⟩.

(63) (64)

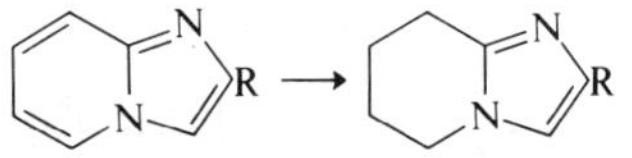

The reactions of groups attached to the imidazo[1,2-*a*]pyridine ring are unremarkable ⟨61HC(15-1)460, 77HC(30-2)117, 76JMC1352, 76JOC3549, 76FES209⟩.

4.10.3.2 Imidazo[1,5-*a*]pyridine

The imidazo[1,5-*a*]pyridine ring system has been successfully functionalized with various electrophilic reagents, metalated, and reduced in one case. No nucleophilic displacements of leaving groups have been reported on appropriate derivatives.

Acetylation ⟨55JCS2834⟩, formylation ⟨75JHC379⟩ and nitration ⟨80JCS(P1)959⟩ all occur mainly at C-1, producing compounds (**65a–c**), respectively, although formylation also produces some C-3 substitution. When C-1 is blocked, electrophilic attack occurs at C-3 ⟨55JCS2834, 80JCS(P1)959⟩. Thus, nitration of 1-phenylimidazo[1,5-*a*]pyridine (**65d**) produces mainly the 3-nitro compound (**65e**), although some dinitration (at C-3 and in the phenyl ring) also occurs. Nitration of the 1-nitro compound (**65c**) gives the 1,3-dinitro compound (**65f**) ⟨80JCS(P1)959⟩. Nitrosation also occurs at C-1, but the product tautomerizes, the ring opens, and recyclization produces (**66**) ⟨67JOC2430⟩. Diazo coupling has also been carried out on imidazo[1,5-*a*]pyridines ⟨78BRP1531752, 78BRP1528801⟩. Imidazo[1,5-*a*]pyridin-3-one (**67a**) can be acylated, alkylated and aminomethylated at N-2 under standard conditions ⟨61AP404⟩. The imidazo[1,5-*a*]pyridin-3-one (**67b**) can be reduced with hydrogen and palladium on charcoal to the tetrahydro derivative (**68**) ⟨75FES197⟩.

(65)

a; $R^1 = COMe, R^2 = H$
b; $R^1 = CHO, R^2 = H$
c; $R^1 = NO_2, R^2 = H$
d; $R^1 = Ph, R^2 = H$
e; $R^1 = Ph, R^2 = NO_2$
f; $R^1 = R^2 = NO_2$

(66)

(67) **a**; R = H
b; $R = CO_2Et$

(68)

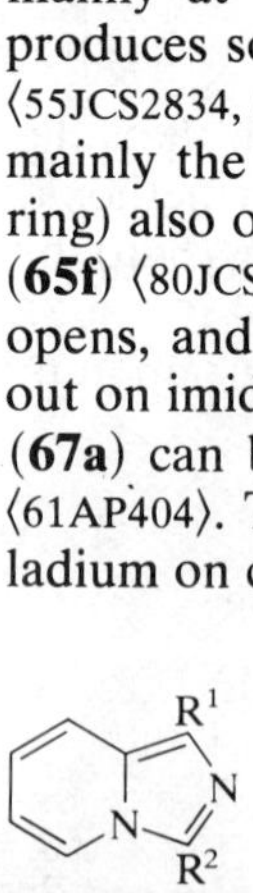

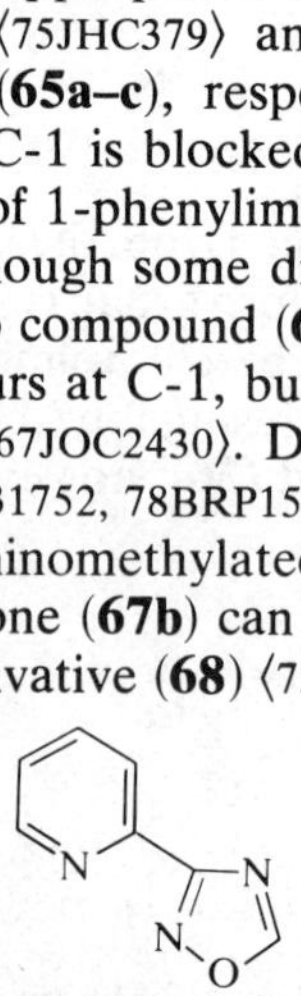

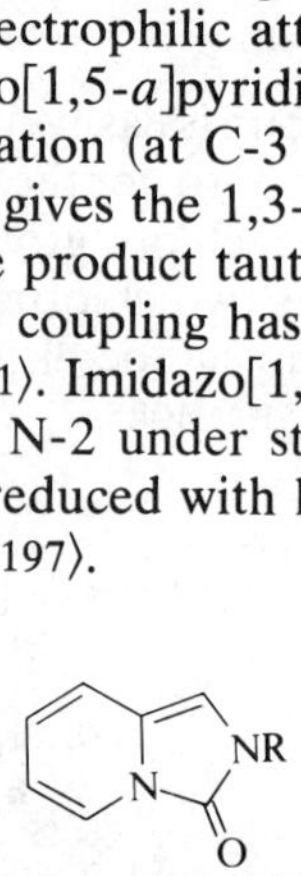

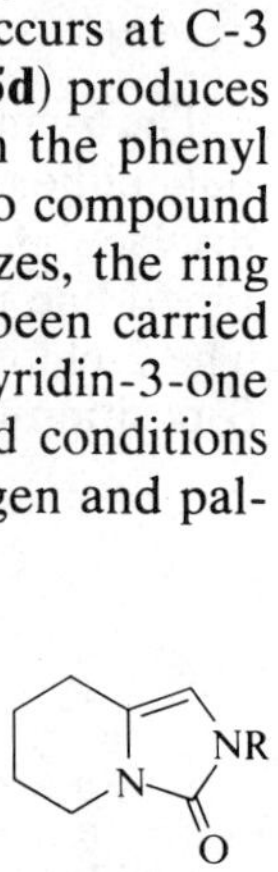

Protonation ⟨65JCS2778, 64JCS4226⟩ and *N*-methylation of the parent heterocycle ⟨77HC(30-2)117⟩ both occur at N-2. *N*-Amination of imidazo[1,5-*a*]pyridine and 1-substituted derivatives with either *O*-mesitylsulfonylhydroxylamine or *O*-*p*-toluenesulfonylhydroxylamine produces (**69**). Oxidation of (**69b**) with saturated aqueous bromine resulted in imidazole ring opening and eventually produced (**70**) rather than a triazine, as seen with the corresponding imidazo[1,2-*a*]pyridine ⟨76JCS(P1)1722⟩. Oxidation with nitric acid results in either imidazole ring opening or C-1 nitration, depending upon the substitution pattern of the heterocycle and the nitration reagent employed ⟨79JCS(P1)1833⟩.

(**69**) a; R = H
b; R = Ph

(**70**) X⁻

Metalation of the parent heterocycle with phenyllithium produced (**71a**) or (**71b**) after quenching with cyclohexane or DMF ⟨72JHC1157, 75JHC379⟩, respectively, indicating metalation at C-3. Treatment of (**71c**) with *n*-butyllithium followed by methyl iodide produced the C-4 substituted compound (**71d**). The anion was also quenched with various carbonyl compounds ⟨80TL2195⟩. Treatment with Raney nickel afforded 4-methylimidazo[1,5-*a*]pyridine (**71e**).

(**71**)
a; R¹ = (cyclohexanol), R² = H
b; R¹ = CHO, R² = H
c; R¹ = SEt, R² = H
d; R¹ = SEt, R² = Me
e; R¹ = H, R² = Me

4.10.3.3 Imidazo[4,5-*b*]pyridine

Treatment of imidazo[4,5-*b*]pyridine or a number of its 5-, 6- and 7-mono- and 5,7-disubstituted derivatives (**72**) with mercury(II) chloride and aqueous sodium hydroxide gave chloromercury derivatives of undefined structure that react with acylglycosyl halides to give 3-glycosyl derivatives (**73**) ⟨60MI41000, 63CPB265, 63IJC30, 63JOC1837, 66IJC403⟩. In some cases the 1-glycosyl derivatives (**74**) were also reported ⟨65JOC4066⟩ in considerably reduced amount, the ratio varying with the substitution on the imidazo[4,5-*b*]pyridine. Fusion of a variety of 7- and 5,7-substituted imidazo[4,5-*b*]pyridines with tetra-*O*-acetyl-β-D-ribofuranose and an acid catalyst gave only substitution at N-3 (**73**), but frequently a small amount of the α-anomer with the predominant β form ⟨66JMC354, 68IJC616, 71RTC654, 72JHC465, 75RTC153⟩. The mercury(II) cyanide method of glycosylation has also been used ⟨75RTC153, 77H(8)433, 78JHC839⟩, but in one case gave predominantly 7-amino-4-(tri-*O*-benzoyl-β-D-ribofuranosyl)imidazo[4,5-*b*]pyridine rather than the 3-isomer ⟨72JHC465⟩. Reaction of imidazo[4,5-*b*]pyridine with tetra-*O*-acetyl-β-D-ribofuranose in the presence of an equivalent of tin(IV) chloride for five hours gave 3-(tri-*O*-acetyl-β-D-ribofuranosyl)imidazo[4,5-*b*]pyridine, but its trimethylsilyl derivative (**75**) gave a 60–70% yield of the 4-isomer (**76**), which on treatment with tin(IV) chloride gave a 40% yield of the 3-isomer (**77**) and a 25% yield of the 1-isomer (**78**). If the initial ribosylation is run for 30 hours, the predominant isomer is still the 4-, but significant amounts of the 1- and 3-isomers are formed, whereas in the presence of 3.6 equivalents of tin(IV) chloride, the major product is the 3-isomer, followed by the 1-isomer and a smaller amount of the 4-isomer. These results indicate that the kinetically controlled product is the 4-isomer and that the 3-isomer is the thermodynamically most stable, followed by the 1-isomer ⟨76H(5)285⟩. The trimethylsilyl derivative of 6-nitroimidazo[4,5-*b*]pyridine produced only the 1- (11%) and 3-isomers (35%) indicating the decreased basicity of N-4 resulting from the nitro group ⟨77H(8)433⟩. The trimethylsilyl derivative (**79**) of imidazo[4,5-*b*]pyridine 4-oxide yielded only 1-(tri-*O*-acetyl-β-D-ribofuranosyl)imidazo[4,5-*b*]pyridine 4-oxide (**80**), perhaps due to a steric effect ⟨82MI41001⟩. Reaction of 5-acetamido-7-chloroimidazo[4,5-*b*]pyridine with tri-*O*-acetyl-D-ribofuranosyl chloride in the presence of molecular sieves gave substitution at

N-3 ⟨78JMC112⟩. Treatment of 7-chloro-1-(tri-*O*-acetyl-β-D-ribofuranosyl)imidazo[4,5-*b*]pyridine with tetra-*O*-acetyl-D-ribofuranose and $HgBr_2$ produced the 3-isomer in 75.9% yield with the indication that transglycosylation proceeds through the 1,3-bis(glycosyl) intermediate to the thermodynamically more stable 3-isomer ⟨82MI41001⟩, confirming the relative stabilites mentioned above.

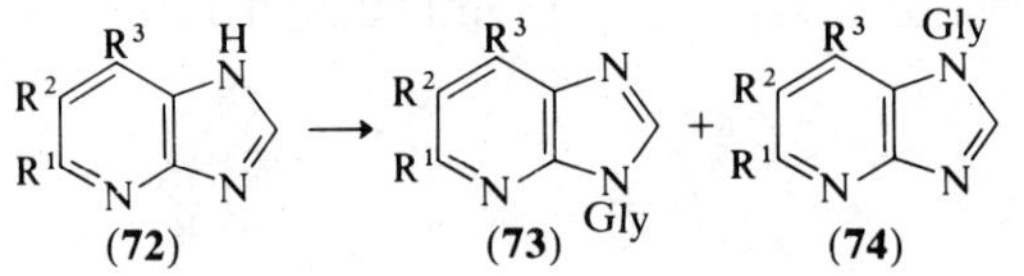

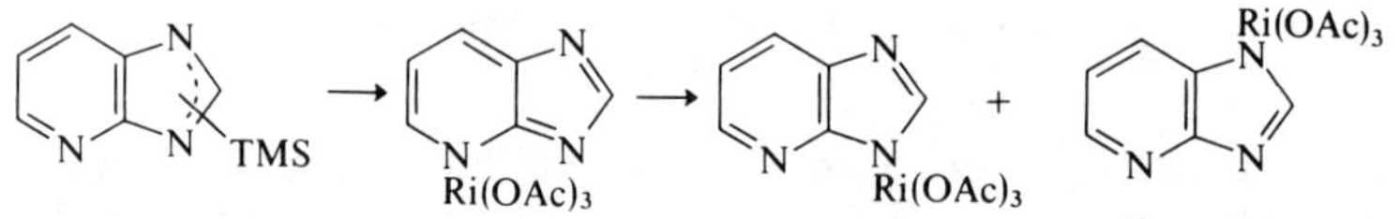

Alkylation of 6-haloimidazo[4,5-*b*]pyridines (**81**) has been reported to occur at N-3 (**82**) ⟨60CPB539⟩ and N-1 (**83**) ⟨63MI41001⟩. Aminomethylation with formaldehyde and piperidine or morpholine has also been reported but the position of attachment was not established ⟨60CPB539⟩. Heating 3-(2,3-*O*-isopropylidine-5-*p*-toluenesulfonyl-β-D-ribofuranosyl)imidazo[4,5-*b*]pyridine (**84**) in boiling acetone caused intramolecular alkylation to occur at N-4 to give the anhydronucleoside (**85**) ⟨63JOC3329⟩. Reaction of primary amines and excess formaldehyde with 2-methoxycarbonylaminoimidazo[4,5-*b*]pyridine (**86**) occurred at N-3 and N of the urethane to give hexahydrotriazinoimidazopyridines (**87**) ⟨77ZC262⟩.

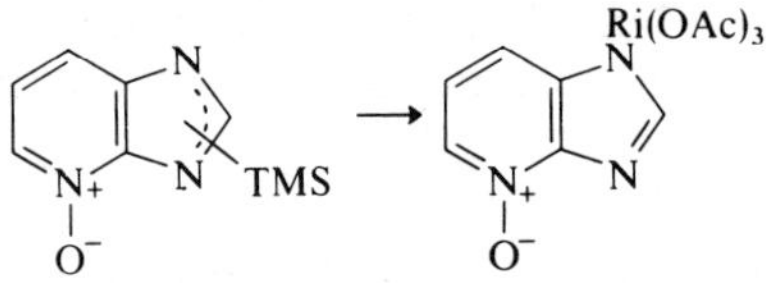

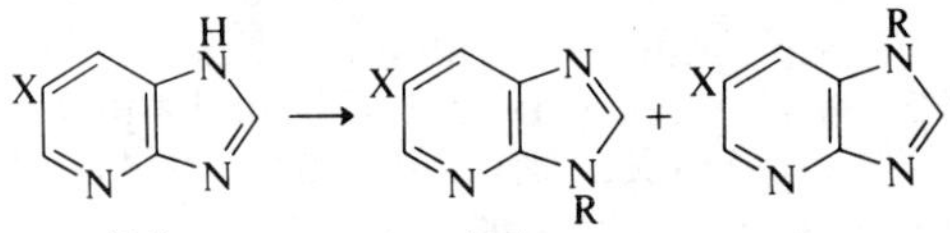

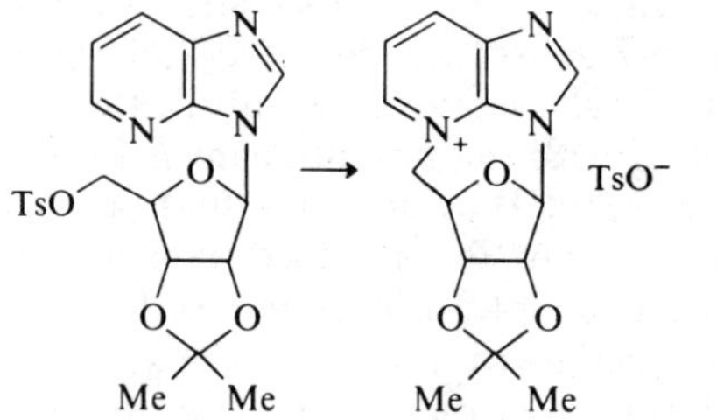

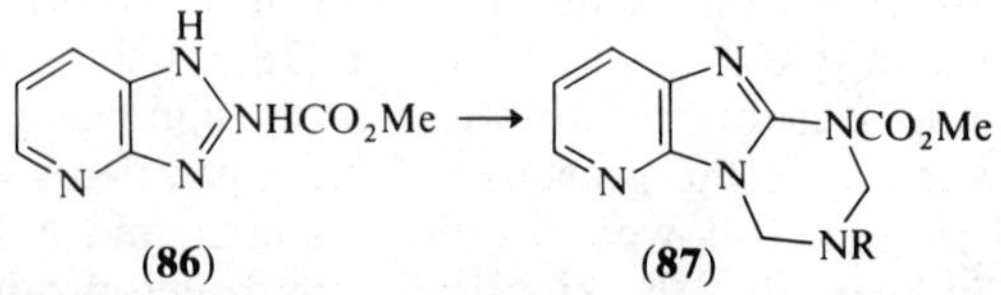

Alkylation of 3-methylimidazo[4,5-*b*]pyridine (**88**) with alkyl halides occurred at N-1 to give the quaternary salts (**89**) ⟨68KGS954, 73KGS570, 73KGS1686⟩, whereas methylation of the 1-methyl isomer (**90**) gave the 1,4-dimethylimidazo[4,5-*b*]pyridinium chloride (**91**) or iodide (**91**) ⟨73KGS570, 73KGS1686⟩. Alkylation of 3-substituted imidazo[4,5-*b*]pyridin-2(1*H*)-ones (**92**) with alkyl halides occurred at N-1 (**93**) ⟨73JHC201, 78JMC965⟩. Reaction of 3-amino (or anilino) -7-methyl-2-phenylimidazo[4,5-*b*]pyridin-5(4*H*)-one (**94**) with diazomethane or methyl iodide took place at both N-4 and the carbonyl oxygen to given almost equal amounts of 3-amino (or anilino) -4,7-dimethyl-2-phenylimidazo[4,5-b]pyridin-5(4H)-one (**95**) and 3-amino (or anilino) -5-methoxy-7-methyl-2-phenylimidazo[4,5-b]pyridine (**96**) ⟨78JHC937⟩.

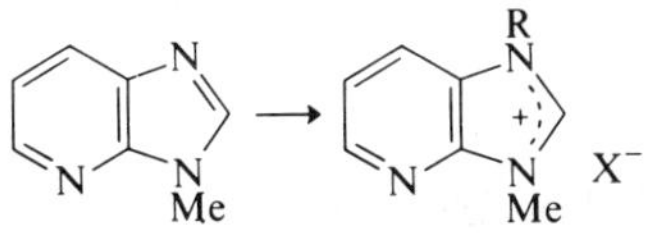

Reaction of imidazo[4,5-b]pyridine (**97**), but not its 5-amino derivative, with perphthalic acid ⟨49JA1885⟩ or hydrogen peroxide in acetic acid ⟨66IJC403⟩ gave the N^4-oxide (**98**). Peracetic acid was used to prepare the N^4-oxides of 6-bromo-, 6-chloro, 6-bromo-5- and -7-methyl-, and 5,7-dimethyl-imidazo[4,5-b]pyridine (**98**) ⟨59JOC1455⟩. The N-oxides can be catalytically reduced to (**97**) ⟨80JHC1757⟩.

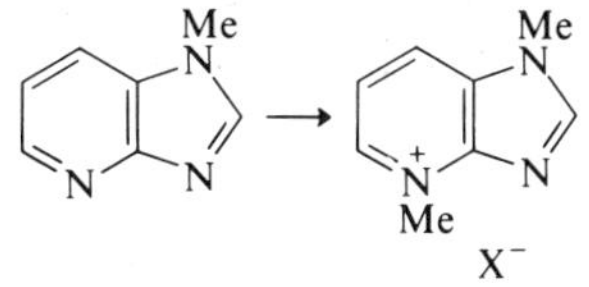

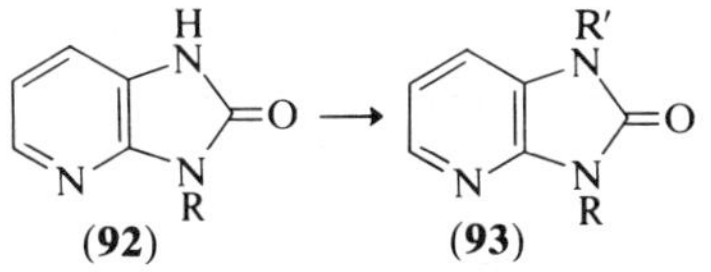

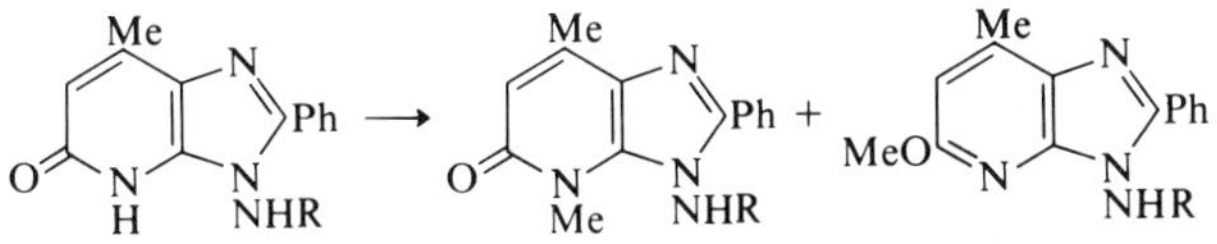

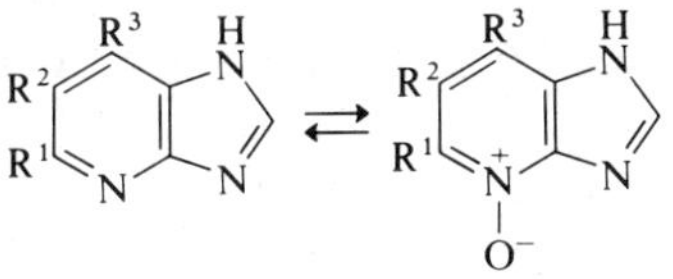

A number of imidazo[4,5-b]pyridines have been acetylated on a nitrogen of the ring as well as exocyclic groups ⟨59JOC1455, 59JCS3157, 66IJC146, 78JMC112⟩, but in only one instance does the position of acylation appear to be established ⟨78JHC937⟩. These N-acetyl groups are labile and can be removed by mild treatment in aqueous media. The N^3-ethoxycarbonyl group can be removed preferentially by treatment with aqueous ammonia ⟨78MI41007⟩. The N^3-isopropyl group is acid labile ⟨69MI41003⟩, and the N-amino group can be removed by diazotization ⟨77KGS411, 78JHC937⟩.

Reaction of 1-(tri-O-acetyl-β-D-ribofuranosyl)imidazo[4,5-b]pyridine 4-oxide (**99**) with phosphorus oxychloride or the Vilsmeier reagent gave 7-chloro-1-(tri-O-acetyl-β-D-ribofuranosyl)imidazo[4,5-b]pyridine (**100**) ⟨82MI41001⟩.

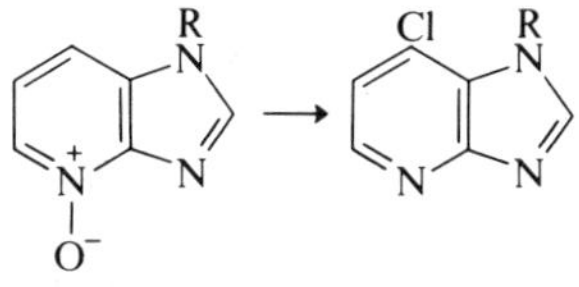

R = tri-O-acetyl-β-D-ribofuranose

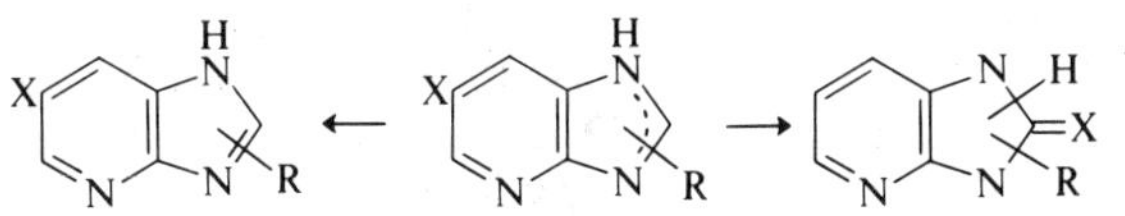

Nitration or bromination of 1- or 3-methylimidazo[4,5-b]pyridines (**102**) occurs at C-6 giving (**101**) ⟨68KGS953, 71KGS279⟩, but nitration of imidazo[4,5-b]pyridine 4-oxide (**104**) takes place at C-7 to give (**105**) ⟨66IJC403⟩, indicating the effect of the N-oxide on the electron distribution of the ring. On the other hand, hydroxylation or thiolation occurs at C-2 ⟨71KGS428, 76KGS1252⟩. Oxidation at C-2 has also been used to prepare 1,3-disubstituted imidazo[4,5-b]pyridine-2(3H)-one (103) ⟨68KGS954⟩.

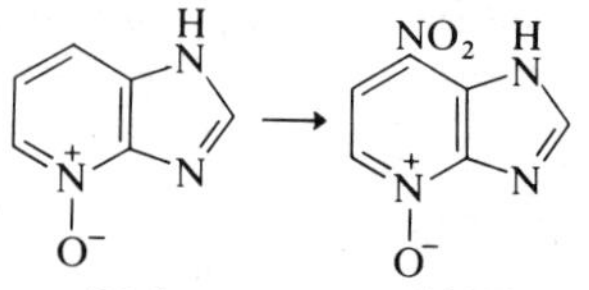

Radical methylation of 3-methylimidazo[4,5-b]pyridine (106) gave in 71% yield a mixture of 2,3,5,7-tetramethylimidazo[4,5-b]pyridine, 2,3,7-trimethylimidazo[4,5-b]pyridine and 2,3-dimethylimidazo[4,5-b]pyridine (107) in a ratio of 1:9.3:14.7. Similar methylation of the dimethyl compound gave a 57% yield of the tetra- and tri-methyl compounds in a 1:5 ratio. 1,2-Dimethylimidazo[4,5-b]pyridine (108) was methylated simultaneously in the 5- and 7-positions (109), with the 7-position being more reactive than the 5-positions ⟨78MI41004⟩.

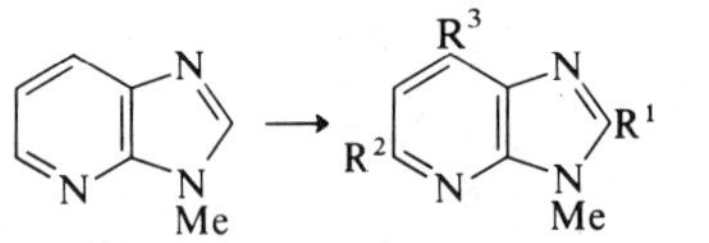

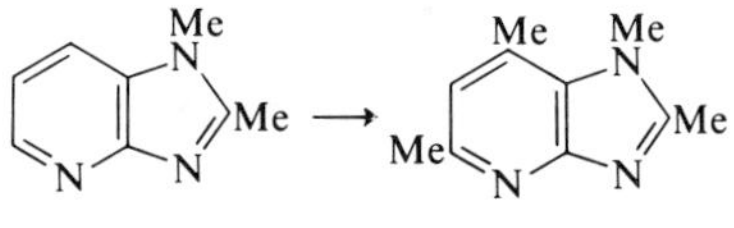

O-Alkylation of imidazo[4,5-b]pyridin-2-ones (110) sometimes occurs on treatment with alkyl halides or diazomethane ⟨78H(10)241, 78JHC937⟩, and one instance of O- as well as N-acetylation as in (111) has been reported (the position of the N-acetyl group was not firmly established) ⟨59JOC1455⟩.

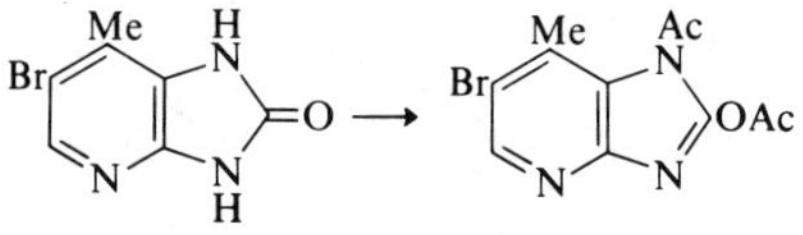

Chlorination of imidazo[4,5-b]pyridinones (112) is normally accomplished by means of phosphorus oxychloride ⟨49RTC1013, 57AP20, 58AP368⟩; in one case phosphorus trichloride was used additionally ⟨70KGS1146⟩ but that seems unnecessary. Chlorine at positions 5, 6, and 7 and bromine at position 6 have been removed by catalytic hydrodehalogenation using palladium catalysts ⟨48RTC29, 49RTC1013, 49JA1885, 61RTC545, 71RTC654, 78JHC839, 57JA6421⟩. Phosphonium iodide and fuming hydriodic acid have been used to remove chlorine at C-2, (113) → (114) ⟨57AP20, 58AP368⟩.

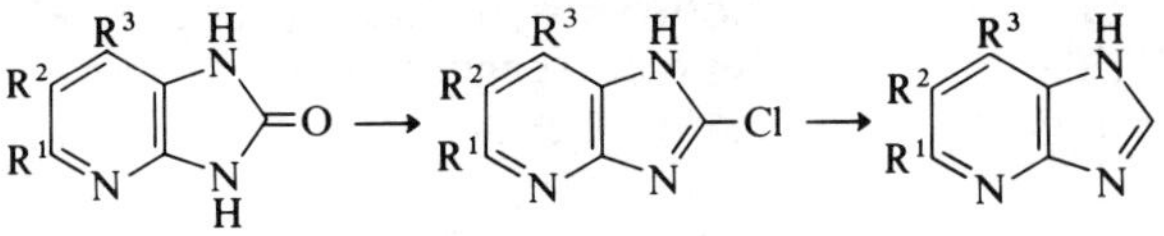

The usual nucleophilic displacement reactions of chlorine have been carried out with ammonia and amines ⟨49RTC1013, 52CB1012, 61RTC545, 70KGS1146, 71RTC654, 71RTC1166, 75MI41001⟩, hydrazine ⟨72RTC650, 78JHC839, 82MI41001⟩, alkoxides ⟨77JHC813, 78JHC839⟩, thiourea, sodium hydrosulfide and sodium methylmercaptide (115) → (116) ⟨65JMC708, 66JMC354, 78JMC112⟩; but, as in the case of the imidazo[4,5-c]pyridines, these displacements are more difficult than those in the purine series because of the lesser electron-withdrawing power of one versus two ring nitrogens.

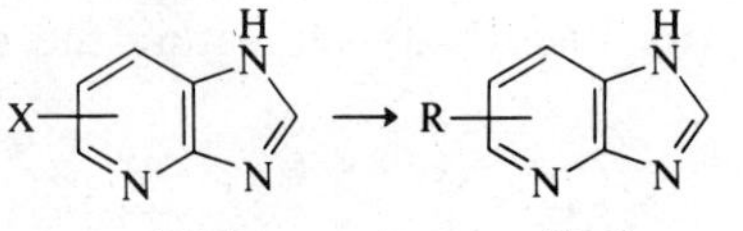

R = H, NH_2, N_2H_3, NR'R", OR', =O, =S, SMe

Diazotization of 5- and 7-aminoimidazo[4,5-b]pyridines (**118**) in concentrated hydrochloric acid with or without copper(I) chloride gave the corresponding chloro compounds (**117**), whereas in sulfuric acid the imidazopyridinones (**119**) and (**120**) were formed ⟨48RTC29, 49RTC1013, 57JA6421, 61RTC545, 65JMC708, 72RTC650, 78JHC839⟩.

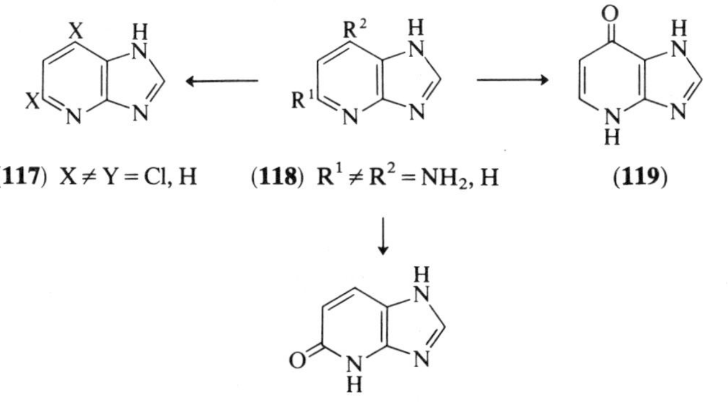

(**117**) X ≠ Y = Cl, H   (**118**) R^1 ≠ R^2 = NH$_2$, H   (**119**)

(**120**)

Aminoimidazo[4,5-b]pyridines (**121**) undergo normal acylation reactions ⟨66IJC146, 68IJC616, 74RTC160, 75P2539⟩, and under forcing conditions with sodium benzylate, 5-acetamido-7-chloro-3-β-D-ribofuranosylimidazo[4,5-b]pyridine (**121b**) undergoes N-benzylation along with displacement of the chloro group to give (**122b**) ⟨78JHC839⟩. The benzyloxy group of (**122a–c**) has been catalytically hydrogenolyzed to give the imidazo[4,5-b]pyridin-7-ones (**123**) ⟨56JA4130, 74RTC160, 75RTC153, 78JHC839⟩. Methyl or benzyl groups on the imidazo[4,5-b]pyridine ring undergo typical reactions such as oxidation ⟨59JOC1455, 75M1059, 75KGS1389⟩ or reaction with aryl aldehydes ⟨79IJC(B)428⟩. The resulting acids, aldehydes or ketones are also typical.

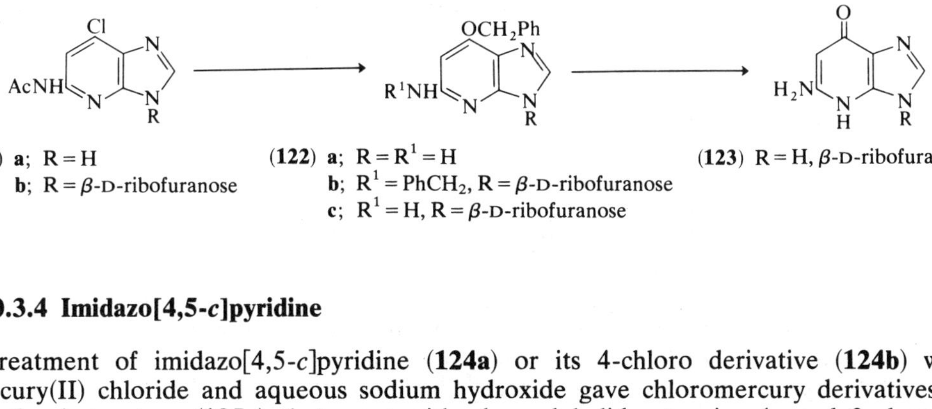

(**121**) a; R = H
b; R = β-D-ribofuranose

(**122**) a; R = R^1 = H
b; R^1 = PhCH$_2$, R = β-D-ribofuranose
c; R^1 = H, R = β-D-ribofuranose

(**123**) R = H, β-D-ribofuranose

4.10.3.4 Imidazo[4,5-c]pyridine

Treatment of imidazo[4,5-c]pyridine (**124a**) or its 4-chloro derivative (**124b**) with mercury(II) chloride and aqueous sodium hydroxide gave chloromercury derivatives of undefined structure (**125a**) that react with glycosyl halides to give 1- and 3-glycosylimidazo[4,5-c]pyridines (**126**) and (**127**) with the 1-isomer (**126**) usually predominating ⟨63JOC1837, 64JOC2611, 68CPB2011⟩, although imidazo[4,5-c]pyridine itself is an exception with equal amounts of the two isomers being formed ⟨65JOC4066⟩. It is likely that, at least in some cases, the 3-isomer (**127**) is formed by kinetic control whereas the 1-isomer (**126**) is thermodynamically more stable. In keeping with this interpretation is the fact that the fusion of (**124b**) or (**124c**) with tetra-O-acetyl-D-ribofuranose at elevated temperatures gave only traces of the 3-isomer ⟨66JMC105, 66B756, 77JHC195⟩. Reaction of trimethylsilyl derivatives of the imidazo[4,5-c]pyrimidines with glycosyl halides ⟨75JCS(P1)125, 75MI41000⟩,

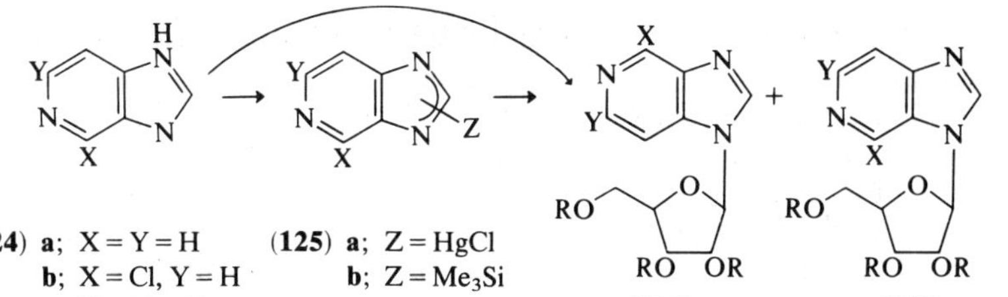

(**124**) a; X = Y = H
b; X = Cl, Y = H
c; X = Y = Cl

(**125**) a; Z = HgCl
b; Z = Me$_3$Si

(**126**)   (**127**)

however, gave a significant amount of the 3-isomer, as well as α- and β-anomers ⟨77JHC195⟩, as did reaction of the base itself with 2,3,5-tri-O-benzyl-β-D-arabinofuranosyl chloride in a nonpolar solvent in the presence of molecular sieves ⟨82JMC96⟩. Methylation of 6-chloroimidazo[4,5-c]pyridine with diazomethane gave a mixture of the 1- and 3-methyl derivatives ⟨66JCS(B)285⟩. Treatment of 1- or 3-methylimidazo[4,5-c]pyridine (**128** or **130**) with alkyl halides gave 64–90% yields of quaternary salts (**129**) and (**131**), respectively.

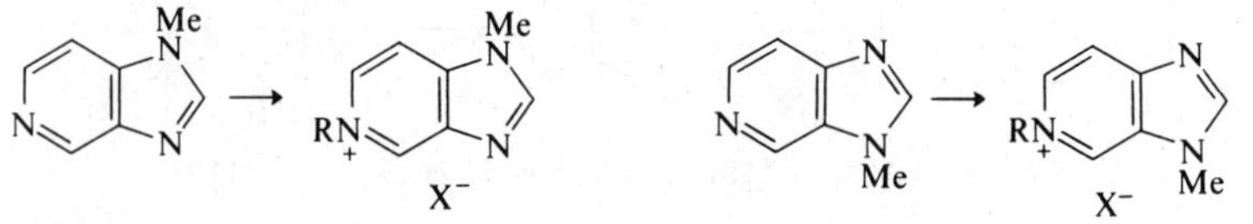

Preparation of 1,3-dimethylimidazo[4,5-c]pyrimidinium iodide (**133**) was accomplished by the lithium aluminum hydride reduction of 1,3-dimethylimidazo[4,5-c]pyridin-2(3H)-one (**132**), prepared by the methylation of imidazo[4,5-c]pyridin-2(3H)-one with dimethyl sulfate ⟨74KGS854⟩, rather than by methylation of (**128**) or (**130**) ⟨79MI41001⟩.

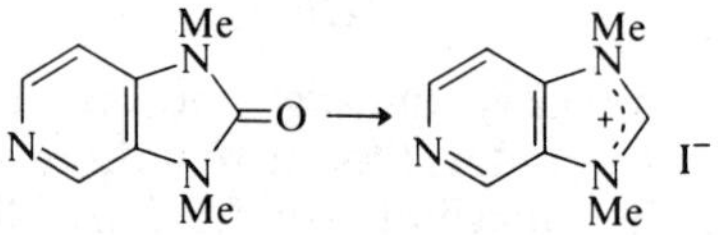

Michael addition of acrylonitrile to imidazo[4,5-c]pyridine-2(3H)-thione (**134**) gave 1,3-bis(cyanoethyl)imidazo[4,5-c]pyridine-3(3H)-thione (**135**) ⟨79MI41000⟩.

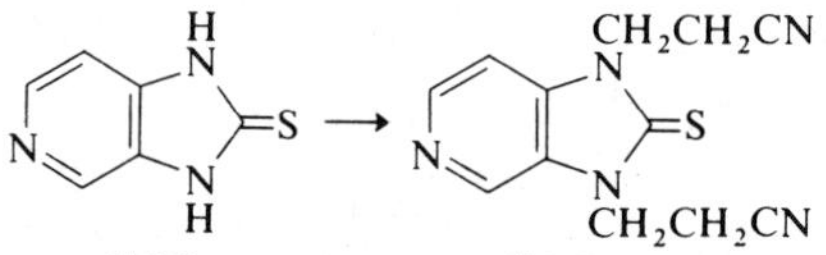

Diazotization of 2-(2-aminophenyl)imidazo[4,5-c]pyridine (**136**) resulted in attack by the diazo group of (**137**) on both N-1 and N-3 giving a mixture of (**138**) and (**139**) ⟨80OPP234⟩.

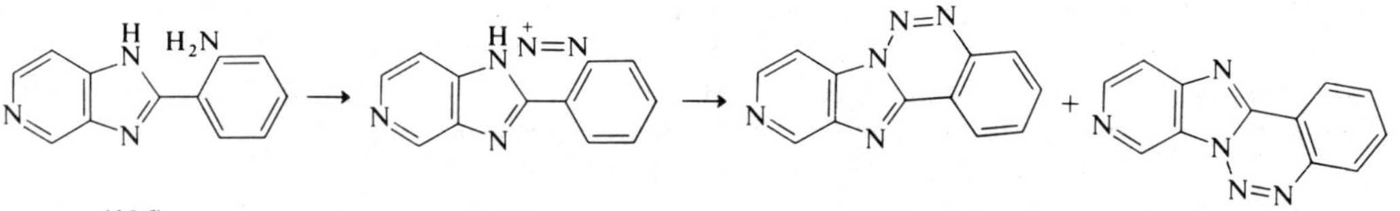

Oxidation at N-4 (**141**) of the imidazo[4,5-c]pyridine ring system (**140**) has been accomplished with hydrogen peroxide in acetic acid and with perphthalic acid ⟨64JOC2611, 80JHC1757, 64CPB866⟩.

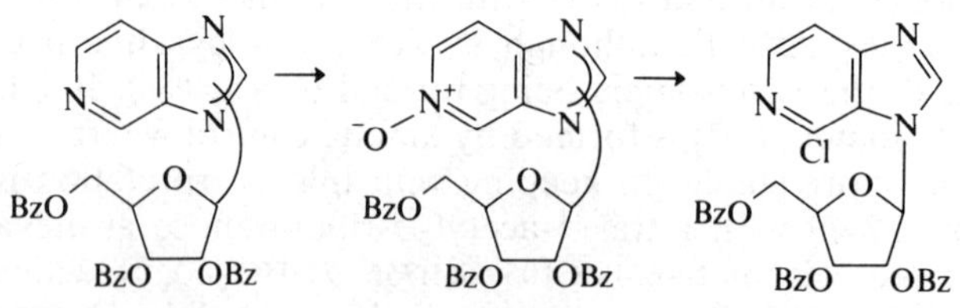

Chlorination of 3-β-D-ribofuranosylimidazo[4,5-c]pyridine 4-oxide (**141**) resulted in the expected reductive chlorination at C-4 to give 4-chloro-3-β-D-ribofuranosylimidazo-[4,5-c]pyridine (**142**), but the reaction failed with the 1-ribosyl derivative ⟨64JOC2611⟩. 4-Chloroimidazo[4,5-c]pyridine was prepared in the same manner from imidazo[4,5-c]pyridine 4-oxide ⟨64CPB866⟩.

Nitration of imidazo[4,5-c]pyridin-2(3H)-one and its methyl derivatives (**143**) to give the 4-nitro compound (**144**) has been reported ⟨73KGS138, 78MI41005⟩ but not confirmed ⟨80JHC1757⟩.

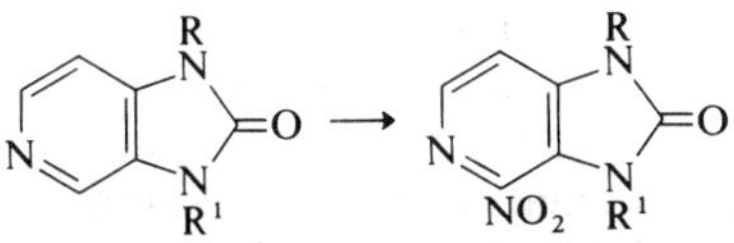

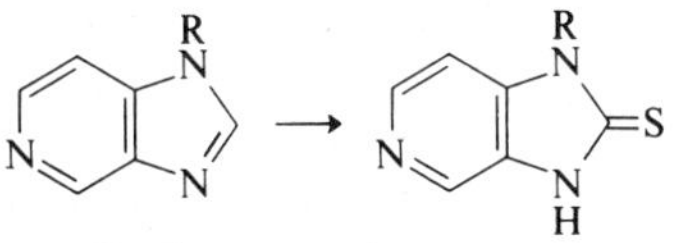

(143) → (144)   (145) → (146)

The thiolation of imidazo[4,5-c]pyridine (145) by heating with sulfur at 230–260 °C for 15–60 minutes gave imidazo[4,5-c]pyridine-2(3H)-thiones (146), which can be desulfurized by heating at 50–90 °C with dilute nitric acid ⟨71KGS428⟩.

Radical alkylation of 1-methyl-, 3-methyl- and 2,3-dimethyl-imidazo[4,5-c]pyridine (147) takes place at C-4 giving (148) with some dimethylation of monomethyl compounds occurring, the second methylation being at C-2 ⟨77KGS933, 78MI41006⟩.

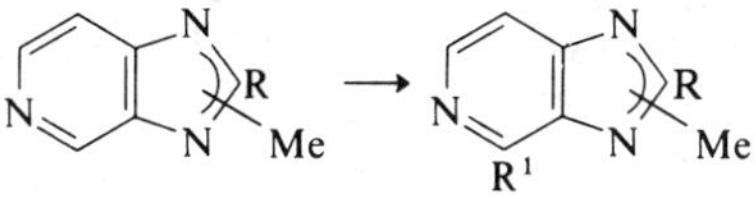

(147) → (148)

Imidazo[4,5-c]pyridin-4(5H)-one (150) has been chlorinated with phosphorus oxychloride and thiated with phosphorus pentasulfide to give 4-chloroimidazo[4,5-c]pyridine (149) and imidazo[4,5-c]pyridine-4(5H)-thione (151), respectively ⟨49RTC1013, 65JMC708⟩.

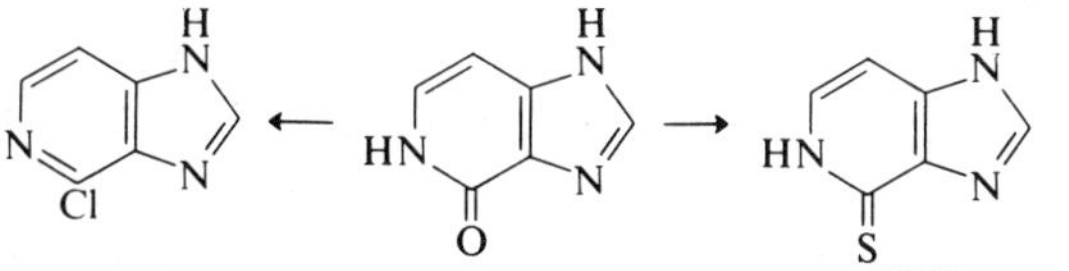

(149) ← (150) → (151)

As expected, the chlorine of (149) is less susceptible to nucleophilic displacement reactions than that of the corresponding purine, but nevertheless has been successfully displaced by ammonia (under drastic conditions) ⟨49RTC1013⟩, hydrazine hydrate ⟨64MI41001⟩, sodium hydrosulfide ⟨65JHC196⟩ and thiourea ⟨65JMC708⟩. The chlorine of the 1-ribosyl derivative (152; R = H) of (149) is somewhat more reactive as is also the case in the purine series, presumably because there is no ionizable NH. It has been displaced by sodium hydrosulfide ⟨66JMC105⟩, anhydrous hydrazine ⟨66B756⟩, water to give the oxo compound, and by amines giving (153) or (154) ⟨68CPB2011, 75MI41001⟩. Decomposition occurred with ammonia and copper sulfate at elevated temperature and sodium azide failed to react ⟨68CPB2011⟩. Oddly enough the reaction of (152; R = H) with aqueous hydrazine resulted in dechlorination to give (153; R = R² = H) ⟨68CPB2011⟩. Again, as in the purine series, a chlorine at C-6 activates one at C-4 so that, in contrast to (152; R = H), (152; R = Cl) reacts with ammonia to give 4-amino-6-chloro-1-β-D-ribofuranosylimidazo[4,5-c]pyridine (152; R = Cl, R² = NH₂), which can be hydrogenolyzed to 4-amino-1-β-D-ribofuranosylimidazo[4,5-c]pyridine (153; R = H, R² = NH₂, 3-deazaadenosine) ⟨75JCS(P1)125⟩. Other cases of hydrodechlorination have been reported ⟨66JCS(B)285⟩. The bromine of 6-amino-4-bromoimidazo[4,5-c]pyridine (155) has been displaced by thiourea and sodium hydroselenide, as has the bromine of the corresponding 1-ribosyl derivative by sodium hydrosulfide, to give (156; X = S, Se) ⟨74JHC233, 77JHC195⟩.

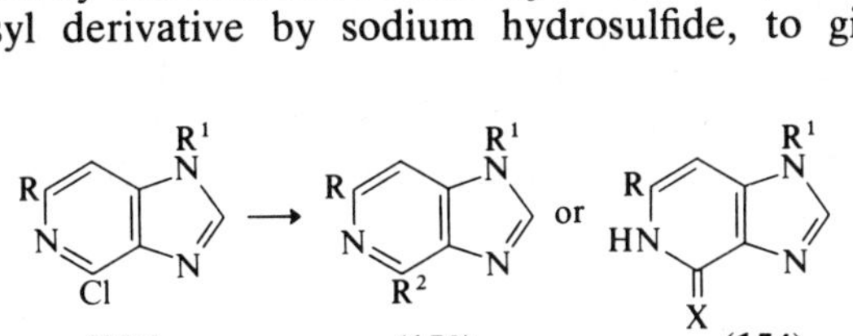

(152) → (153) or (154)

R¹ = (tri-O-acetyl)-β-D-ribofuranose

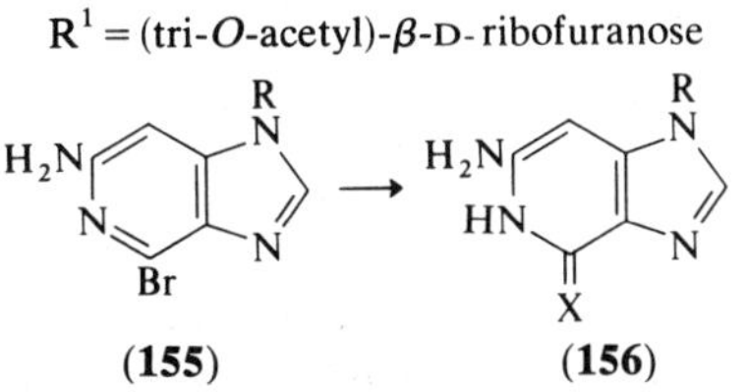

(155) → (156)

R = H or β-D-ribofuranose

A nitro group at C-4 of methylated imidazo[4,5-c]pyridin-2(3H)-ones (**157**) can also be displaced by nucleophilic attack, by halide ions ⟨74KGS854⟩ or alkoxide ions giving (**158**), as well as being reduced to the amine ⟨78MI41005⟩.

Diazotization of the 4-amino group of (**159**) in concentrated hydrochloric acid resulted in the normal displacement reaction to give the 4-chloro compound (**160**) ⟨65JMC708⟩.

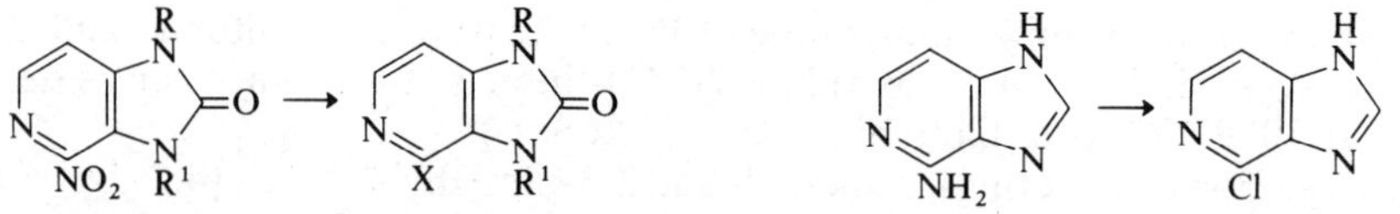

(**157**) (**158**) (**159**) (**160**)

Alkyl and substituted alkyl groups attached to carbon or nitrogen of the imidazo[4,5-c]pyridine ring or through sulfur to carbon undergo the expected reactions. For example, methyl groups (**161**; R = H) react with benzaldehyde to form styryl derivatives (**162**) that can be oxidized to the corresponding acids (**163**) which decarboxylate on heating to (**164**) ⟨62JPR(288)199⟩. A methyl (**161**; R = H) or benzyl group (**161**; R = Ph) can be oxidized with selenium dioxide to the aldehyde (**165**; R = H) or phenyl ketone (**165**; R = Ph), which may be reduced with sodium borohydride to the alcohol (**166**) ⟨72MIP41000, 73JMC1296⟩. Treatment of 2-(hydroxymethyl)imidazo[4,5-c]pyridine (**166**; R = H) with thionyl chloride gave the chloromethyl compound (**167**) ⟨62JPR(288)199⟩.

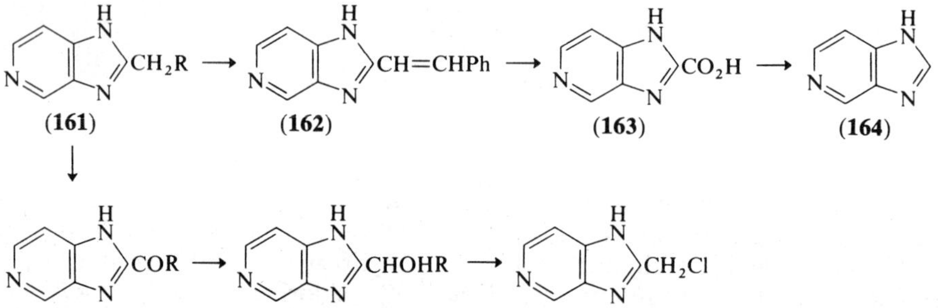

(**161**) (**162**) (**163**) (**164**)

(**165**) (**166**) (**167**)

Thiones such as (**168**) or (**170**) are readily alkylated to (**169**) ⟨66JMC105, 66JCS(B)285, 79MI41000⟩, and the resulting derviatives such as (**171**; R^1 = CH$_2$CO$_2$H or CH$_2$CN, R = X = H) undergo normal reactions ⟨79MI41000⟩. Treatment of the 2-methylthio compound (**171**; R = H, R^1 = Me, X = Cl) with chlorine gas in methanolic HCl causes displacement of the methylthio group by chloride giving (**172**) ⟨66JCS(B)285⟩.

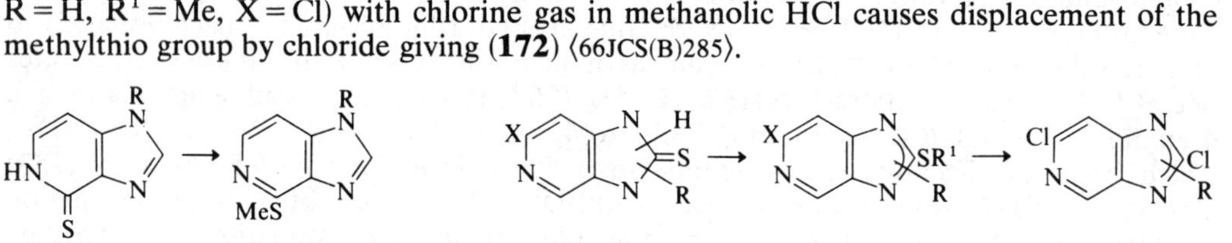

(**168**) (**169**) (**170**) (**171**) (**172**)

Reduction of 1-β-D-ribofuranosyl-4-hydrazinoimidazo[4,5-c]pyridine (**173**) with Raney nickel cleaved the N—N bond in typical fashion to give the adenosine analog 4-amino-1-β-D-ribofuranosylimidazo[4,5-c]pyridine (**174**) ⟨66B756⟩, whereas 1-aminoimidazo[4,5-c]pyridin-2(3H)-one (**175**) was characterized by reaction with veratraldehyde to give the hydrazone (**176**) ⟨76JHC601⟩.

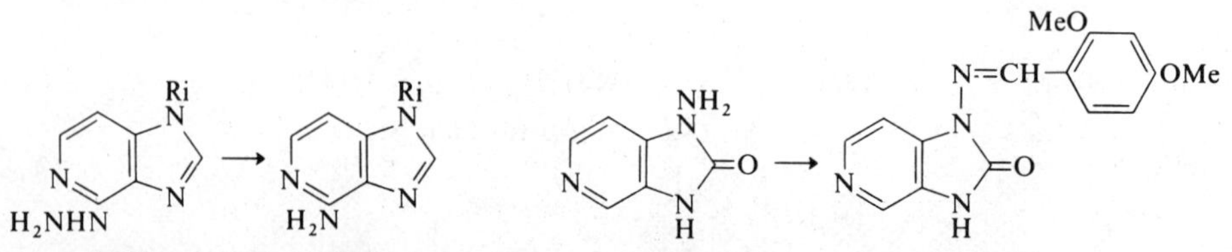

(**173**) (**174**) (**175**) (**176**)
Ri = β-D-ribofuranosyl

Reaction of 1-methylimidazo[4,5-c]pyridine-2(3H)-thione (**177**) with nitric acid, or that of 1-methylimidazo[4,5-c]pyridine and its 2-methyl derivative with a mixture of nitrous and nitric acid, gave *v*-triazolo[4,5-*c*]pyridine (**180**). A mechanism involving hydration of the imidazole ring to give (**178**), cleavage to an *N*-nitroso compound (**179**) and recyclization was suggested ⟨80KGS121⟩. In another ring opening and reclosure, treatment of 7-nitroimidazo[4,5-*c*]pyridin-4(5*H*)-one (**181**) with hydrazine hydrate gave 7-methylimidazo[4,5-*d*]pyridazin-4(5*H*)-one (**182**) ⟨82KGS705⟩.

(**177**) (**178**) (**179**) (**180**)

(**181**) (**182**)

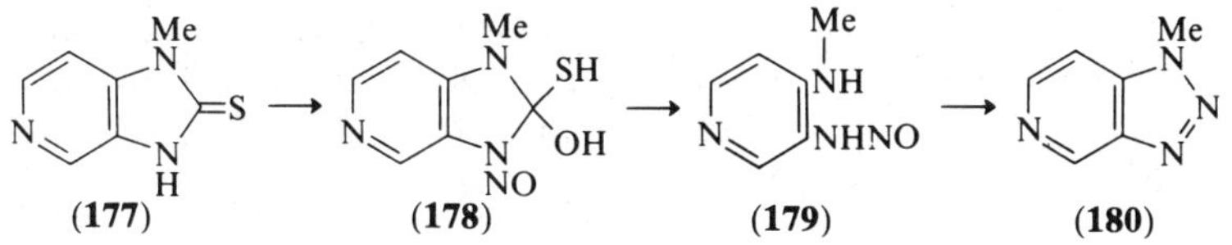

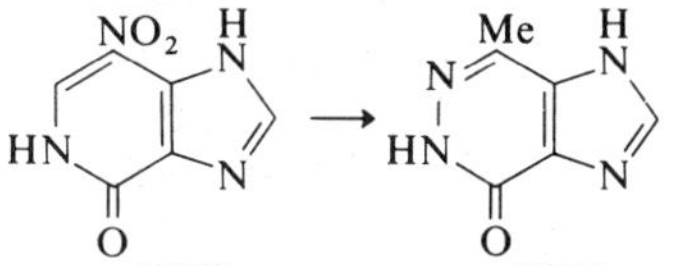

Partially saturated compounds. Only imidazo[4,5-*c*]pyridines saturated in the pyridine ring have been studied. While reaction of histidine with equimolar amounts of formaldehyde gave 4,5,6,7-tetrahydroimidazo[4,5-*c*]pyridine-6-carboxylic acid (**183**), an excess gave an insoluble methylol derivative but the position of the methylol group was not determined (**184**) ⟨44BJ(38)309⟩. When heated at 290 °C (**183**) decarboxylated to give 4,5,6,7-tetrahydroimidazo[4,5-*c*]pyridine (**187**; R^1–R^3 = H), whereas treatment of the silver salt of (**183**) with ethyl iodide gave the ethyl ester (**186**) ⟨13MI41000⟩.

(**183**) (**184**) (**185**) $R^1 = R^2$ = alkyl, R^3 = H
a; R^4 = Me, X = 2H
b; $R^4 = PhCH_2$, X = O
c; $R^4 = R^5NH$, X = O, S, NH
(**186**) (**187**)
(**188**) **a**; R = $CHMe_2$
b; R = EtCO
(**189**) (**190**)

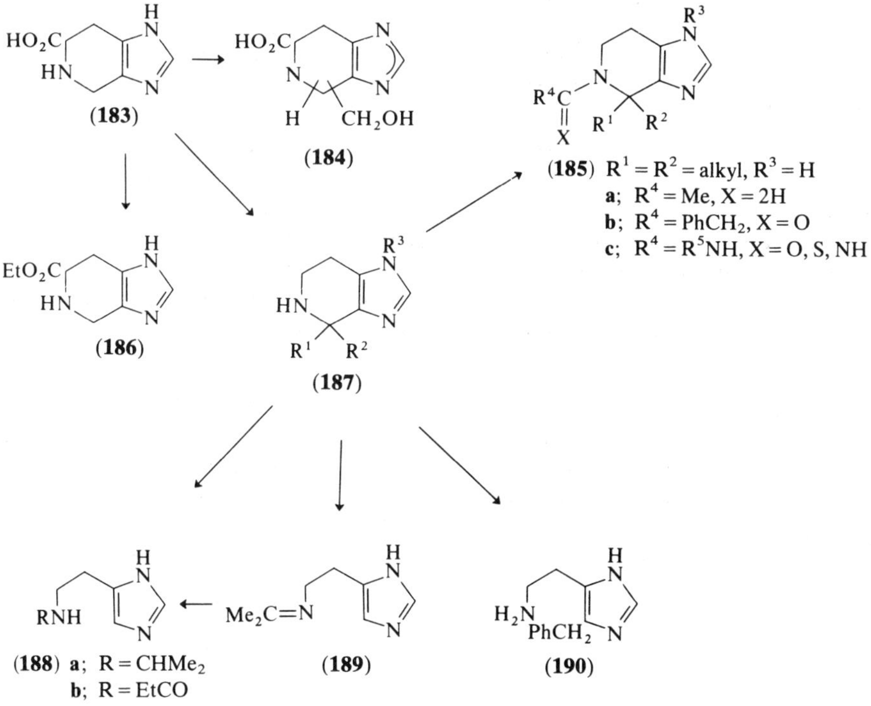

The nitrogen of the saturated pyridine ring of (**187**) behaves like a normal aliphatic amine since it reacts readily with ethyl bromide, benzyl chloroformate, isocyanates, isothiocyanates and *S*-methyl thiouronium salts to give (**185a–c**) ⟨77GEP2700012, 79USP4141899, 79BEP871985⟩. It is likely that this nitrogen is the point of attachment of the methylol group of (**184**).

Catalytic reduction of (**187**; R^1 = Ph, $R^2 = R^3$ = H) with palladium–charcoal resulted in cleavage of the pyridine ring to give the 4-benzylimidazole (**190**). The reaction is probably limited to R^1 = aryl since the dimethyl compound (**187**; $R^1 = R^2$ = Me, R^3 = H) on heating at 140 °C cleaved to (**189**), which was catalytically reduced to 4-(2-isopropylaminoethyl)imidazole (**188a**), and boiling a solution of (**187**; $R^1 = R^2$ = Me, R^3 = H) in propionic acid gave *N*-(2-imidazol-4-ylethyl)propionamide (**188b**) ⟨82JMC1168⟩.

4.10.3.5 Imidazo[1,2-a]pyrazine

Treatment of imidazo[1,2-a]pyrazine (**196**) with methyl iodide resulted in quaternization at N-7 (**192**) and N-1 (**191**) in a ratio of 1:1.6 indicating the relative basicities of these nitrogen atoms ⟨77JOC4197⟩. No reports of N-oxidation have appeared. Bromination with NBS occurred at C-3 to give (**195**), and with bromine in glacial acetic acid reaction occurred at both C-3 and C-5 (**197a**). No deuterium exchange took place in neutral or acid media, but with sodium deuteroxide in DMSO exchange occurred at C-3 and C-5 (**197b**). The pentachloroimidazo[1,2-a]pyridine (**198**) was prepared by treatment of the parent with phosphorus trichloride at 265 °C ⟨75JHC861⟩. Methoxide displaced the chloro groups from C-5 and C-8 to give 2,3,6-trichloro-5,8-dimethoxyimidazo[1,2-a]pyrazine (**199**) which was converted by catalytic hydrogenolysis to 5,8-dimethoxyimidazo[1,2-a]pyrazine (**200**). Treatment of the 5,8-dibromo compound with methoxide gave 5-bromo-8-methoxyimidazo[1,2-a]pyrazine, which was hydrogenolyzed to the 8-methoxy compound. Hydrazine hydrate reacts with imidazo[1,2-a]pyrazine (**196**) by nucleophilic addition and subsequent cleavage and reduction of the Wolff–Kishner type. Attack at C-5 gave 2-methylimidazole (**193**) and at C-8 1-ethyl-2-methylimidazole (**194**) ⟨76H(4)943⟩. Oxidation of 2,8-dimethyl-6-phenyl-imidazo[1,2-a]pyrazin-3(7H)-one (**201**) in the presence of potassium t-butoxide in DMSO caused cleavage of the imidazole ring with loss of CO_2 to give (**202**) ⟨68TL3873⟩.

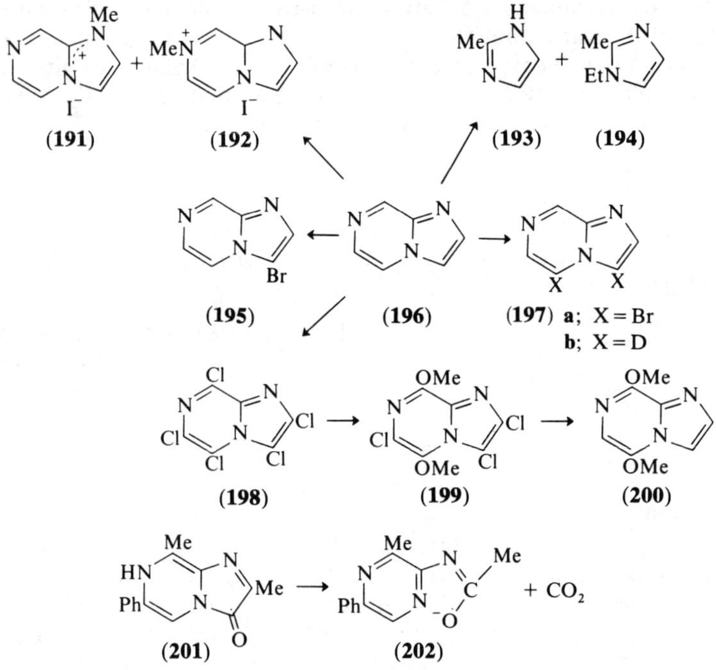

Renilla luciferin was converted to luciferyl sulfate (**204**) by treatment of the diacetate (**203**) with sulfur trioxide in pyridine followed by removal of the acetyl group with ammonia ⟨77TL2685⟩.

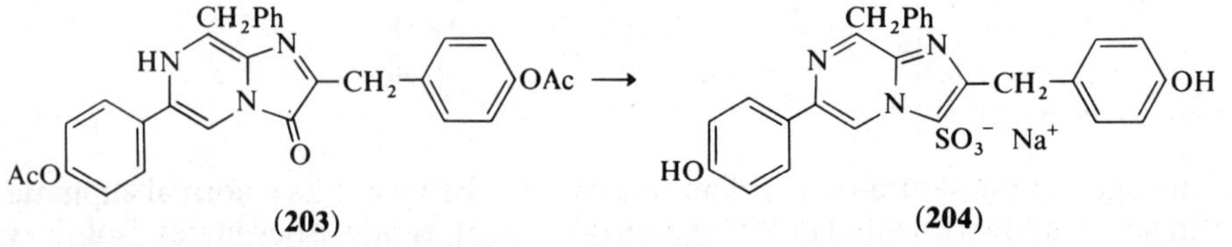

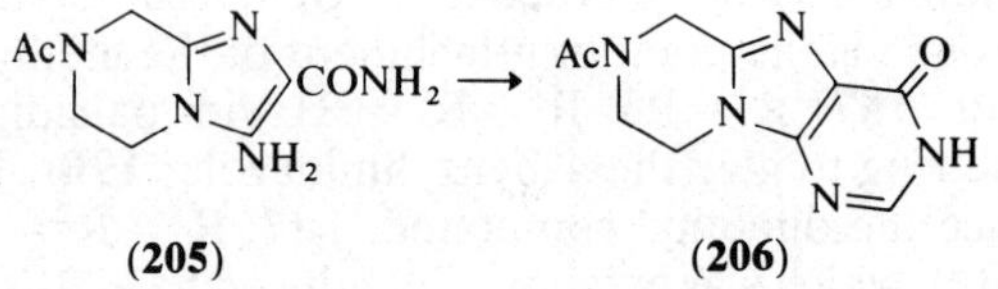

Treatment of (**205**) with triethyl orthoformate gave an 8,9-cyclic derivative of hypoxanthine (**206**) ⟨67MI41000, 67MIP41000⟩.

4.10.3.6 Imidazo[1,5-*a*]pyrazine

Oxidation of imidazo[1,5-*a*]pyrazine and its 3-methyl derivative (**207**) with MCPBA occurred at N-7 giving (**208**) ⟨73JOC2049⟩. Reaction of these heterocycles with acid chlorides and anhydrides resulted in addition rather than substitution, with the addition occurring at the N-7,C-8 double bond to give (**209**), in a manner analogous to the Reissert reaction ⟨75JOC3376⟩. Oxidation of the addition product followed by alcoholysis gave the corresponding imidazo[1,5-*a*]pyrazin-8(7*H*)-ones (**211**). Curiously, reaction of 3-(methylthio)imidazo[1,5-*a*]pyrazine (**212**) with *p*-nitrobenzoyl chloride followed by quenching with water gave a product (**213**) resulting from addition across the N-2,C-3 double bond ⟨75JOC3379⟩.

(**207**) (**208**) (**209**) (**210**) (**211**)

(**212**) (**213**)

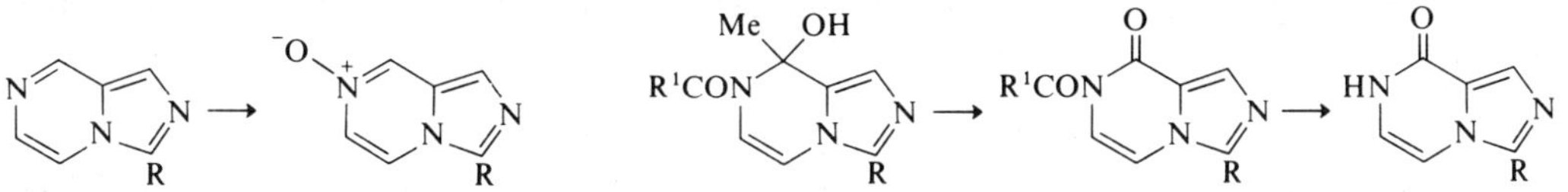

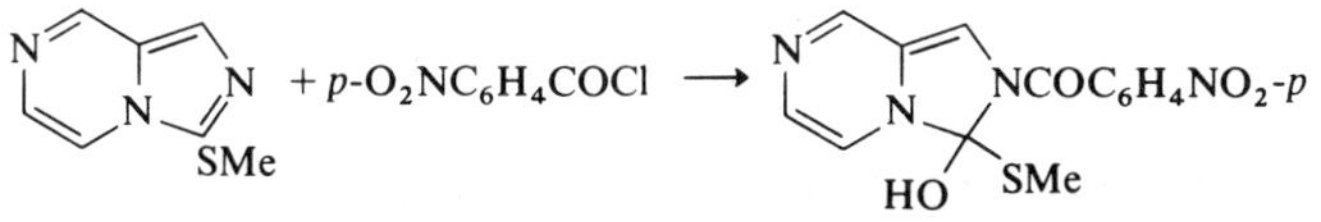

Aromatic substitution reactions on the imidazo[1,5-*a*]pyrazine ring system have been studied using a variety of reagents ⟨75JOC3373, 75JHC207⟩. C-1, C-3 and C-5 are amenable to attack. Chlorination of the parent ring system (**214**) furnished the 1-chloro-, 3-chloro-, 1,3-dichloro- and 1,3,5-trichloroimidazo[1,5-*a*]pyrazines (**215a–d**). Other reactions studied include bromination, reaction with selenium dioxide, and the Mannich reaction; in the latter case, substitution of 3-methylimidazo[1,5-*a*]pyrazine occurred at C-1 and C-5. Reaction of nitrosonium fluoroborate with (**216**) provided 3-(methylthio)-1-nitroimidazo[1,5-*a*]pyrazine (**217**) whereas aqueous nitrous acid resulted in pyrazine ring cleavage ⟨75JOC3379⟩.

(**214**) (**215**)

a; $R^1 = Cl, R^2 = R^3 = H$
b; $R^1 = R^3 = H, R^2 = Cl$
c; $R^1 = R^2 = Cl, R^3 = H$
d; $R^1 = R^2 = R^3 = Cl$

(**216**) (**217**)

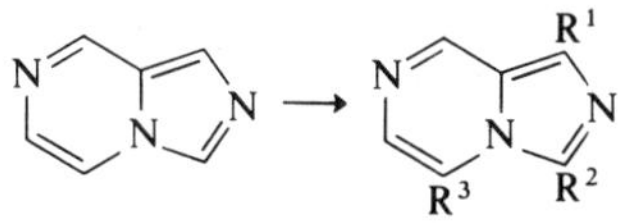

The bromine of 1-bromo-3-methylimidazo[1,5-*a*]pyrazine could not be displaced by a variety of nucleophiles. This contrasts with the reactivity of the 8-chloro substituent of (**219**; R = H or Me), prepared by the usual chlorination of the imidazo[1,5-*a*]pyrazin-8(7*H*)-ones (**218**), which was readily displaced by thiourea to give the thiones (**220**).

(**218**) (**219**) (**220**)

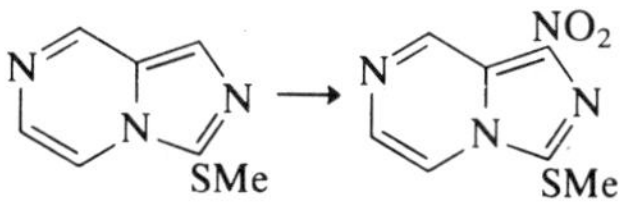

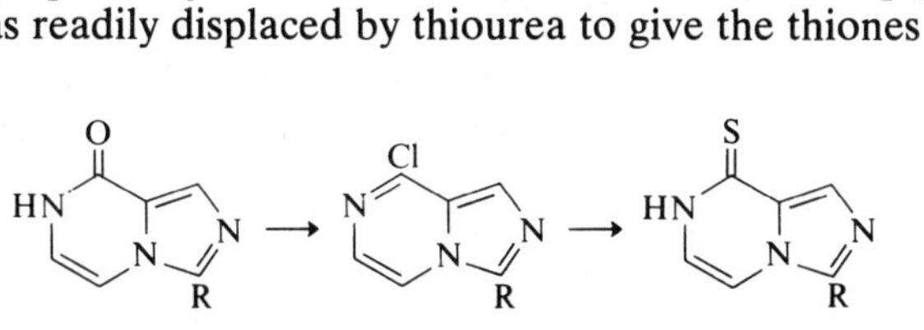

Alkaline hydrolysis of (**221**) resulted in ring opening to give caffeidine (**222**) ⟨61AC(R)1409⟩. Treatment of 1-bromo-3-methylimidazo[1,5-*a*]pyrazine (**223**) with bromine in water followed by neutralization with excess dilute sodium hydroxide also resulted in destruction of the pyrazine ring to give 4-bromo-3-methylimidazole-4-carbaldehyde (**224**) ⟨75JOC3373⟩.

(**221**) (**222**) (**223**) (**224**)

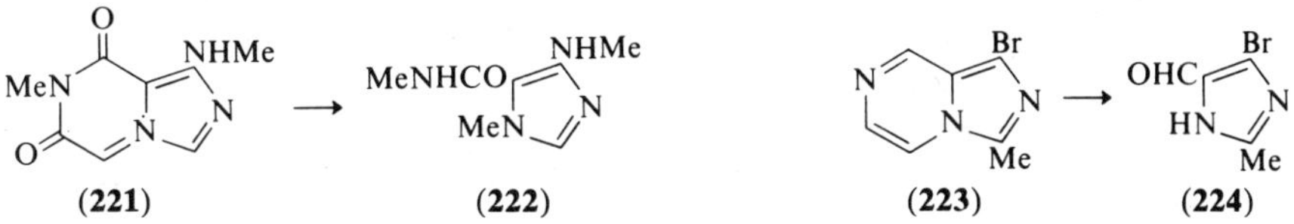

The few studies that have been carried out indicate that substituents attached to the imidazo[1,5-*a*]pyrazine nucleus react normally ⟨61AC(R)1409, 70JCS(C)1540, 75JOC3379⟩.

4.10.3.7 Imidazo[4,5-*b*]pyrazine

Treatment of imidazo[4,5-*b*]pyrazine (**225**; R = H) with hydrogen peroxide in glacial acetic acid gave the 4-oxide (**226**; R = H) as the major product (50%) and the 4,7-dioxide (**227**; R = H) as the minor product. A 4-oxide was also obtained from the 1-methyl- and the 1-(2,3,5-tri-*O*-acetyl-*β*-D-ribofuranosyl)-imidazo[4,5-*b*]pyrazine, in the latter case by means of MCPBA ⟨73JMC643⟩. Other 4-oxides and 4,7-dioxides have been prepared ⟨75KGS690⟩.

(**225**) (**226**) (**227**)

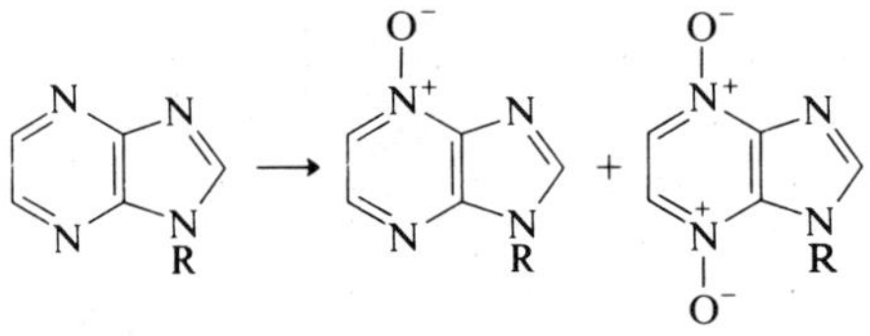

Various imidazo[4,5-*b*]pyrazin-2-ones (**228**) have been subjected to a number of different reactions. Alkylation, acylation and the Mannich reaction all give 1,3-disubstituted derivatives (**229**). Halogenation of the 2-oxo compound (**230**) with bromine or chlorine in acetic acid or with sulfuryl chloride provided monohalo derivatives (**231**) ⟨69FRP1578366, 71BRP1248146⟩. The chlorine atom of (**231**) can be removed by catalytic hydrogenation.

(**228**) (**229**) (**230**) (**231**)

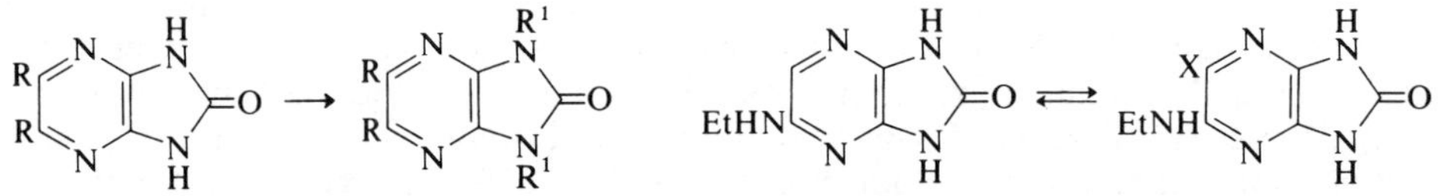

Nitration of 5-chloroimidazo[4,5-*b*]pyrazin-2-one (**232**) provides the 6-nitro derivative (**233**), which can be reduced to the amino compound (**234**; R = H) or the substituted amino compound (**234**; R = alkyl) if the reduction is carried out in the presence of an aldehyde ⟨71BRP1238105⟩.

(**232**) (**233**) (**234**)

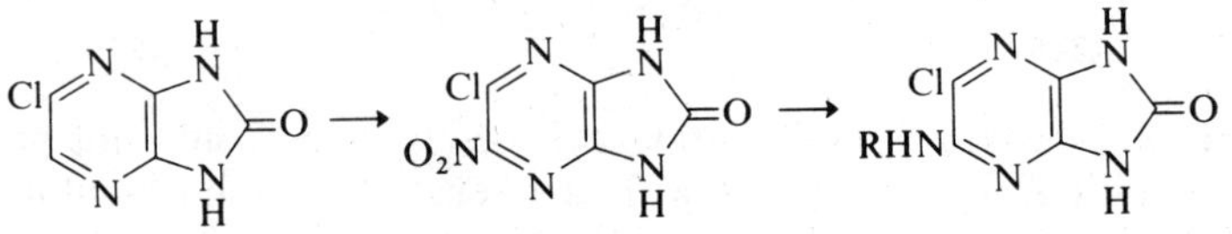

1,5,6-Trimethylimidazo[4,5-*b*]pyrazin-2(3*H*)-one (**235**) has been chlorinated to (**236**) with phosphorus oxychloride and the resultant chlorine at C-2 displaced by ethoxides, amines and diethyl malonate to give (**237**) ⟨75MI41002, 76MI41001⟩. The product of the diethyl malonate reaction was chlorinated and decarboxylated to the 2-chloromethyl compound (**237**; R = CH_2Cl) which was aminated with a variety of amines ⟨76MI41001⟩.

(**235**) (**236**) (**237**)

(**238**) (**239**) (**240**) R = CO_2H, $CONH_2$, CN

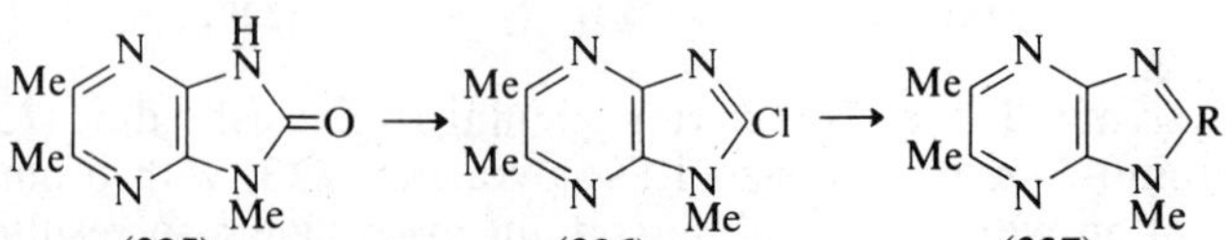

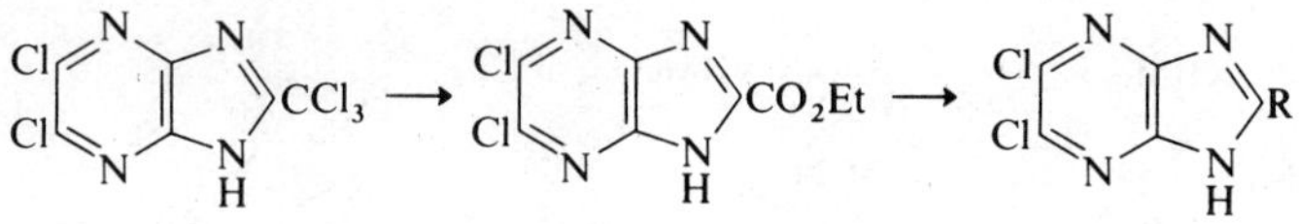

5,6-Dichloro-2-(trichloromethyl)imidazo[4,5-*b*]pyrazine (**238**) was converted into the 2-carboxylic acid ethyl ester (**239**) by treatment with silver nitrate in ethanol ⟨80JHC381⟩. The ester was converted into the acid, amide and nitrile (**240**).

4.10.3.8 Imidazo[1,2-*a*]pyrimidine

As predicted by theoretical calculations ⟨74JHC1013⟩, the initial site of electrophilic attack on imidazo[1,2-*a*]pyrimidines is C-3. Bromination of the parent compound (**241a**) with NBS in chloroform produced (**241b**) ⟨66JOC809⟩. As judged by NMR data, the most probable site of protonation of (**241a**) is N-1 ⟨75OMR(7)455⟩. As expected, the 2-substituted imidazo[1,2-*a*]pyrimidines have a lower reactivity toward electrophiles than the corresponding imidazo[1,2-*a*]pyridines. Nitration of 2-phenylimidazo[1,2-*a*]pyrimidine (**241c**) occurs initially in the phenyl ring, and then at C-3 ⟨70G110, 70G1106⟩. Nitrosation of (**241c**) directly produces (**241d**) ⟨77H(6)929, 66LA(699)112⟩. A properly substituted imidazo[1,2-*a*]pyrimidine can undergo the Dimroth rearrangement ⟨71JHC643⟩, as exemplified later.

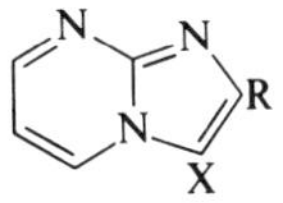

(**241**)
a; R = X = H
b; R = H, X = Br
c; R = Ph, X = H
d; R = Ph, X = NO

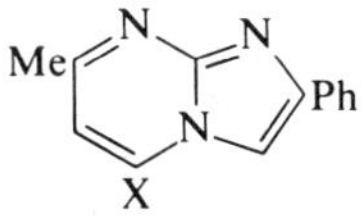

(**242**)
a; X = Cl
b; X = NMe$_2$

The 5-chlorine in (**242a**) is subject to displacement by various nucleophiles, including amines and alkoxides. Treatment of (**242a**) with dimethylamine produces (**242b**). A chlorine at C-5 can be introduced in standard fashion from an oxo group with POCl$_3$ ⟨66LA(699)127⟩.

Certain imidazo[1,2-*a*]pyrimidines undergo amination at N-1 with *O*-*p*-toluenesulfonylhydroxylamine ⟨77JCS(P1)78⟩.

4.10.3.9 Imidazo[1,2-*c*]pyrimidine

The parent imidazo[1,2-*c*]pyrimidine (**243a**) has been found to protonate on N-1 ⟨65JCS2778⟩. Compound (**1a**), as the neutral compound, protonates and alkylates on N-1 ⟨76JA7408⟩.

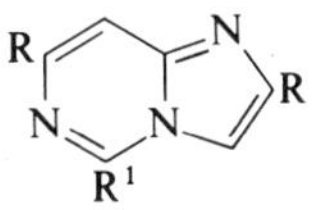

(**243**) a; R = R^1 = H
b; R = R^1 = Cl
c; R = Cl, R^1 = NHMe

Mercuration of (**1a**) with mercury(II) acetate in water affords the mercury derivative (**1c**) ⟨74BBA(361)231⟩.

Treatment of 2,5,7-trichloroimidazo[1,2-*c*]pyrimidine (**243b**) with methylamine results in preferential displacement of the 5-chloro group to produce (**243c**). Even a methylthio group at C-5 is displaced preferentially to the chlorine at C-2 and C-7 ⟨70JHC715⟩.

Nucleosides such as (**246**) have been prepared by the coupling of the trimethylsilyl derivative of (**244**) with 2,3,5-tri-*O*-acetyl-β-D-ribofuranosyl bromide (**245**) followed by deacetylation ⟨75JOC3708⟩. Arabinose derivatives substituted at N-1 have also been prepared ⟨76JMC814⟩.

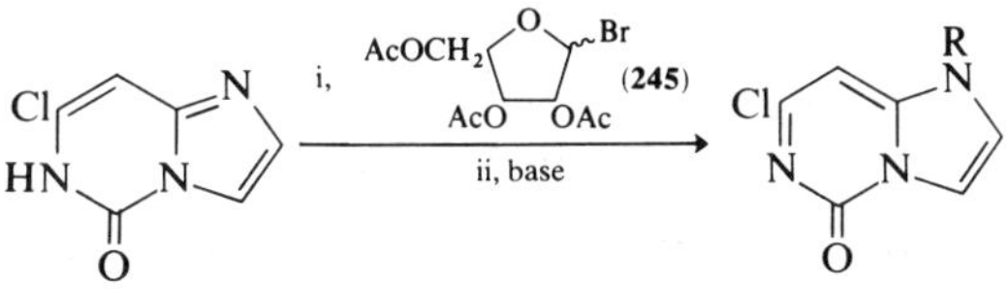

(**244**) (**246**) R = β-D-ribofuranosyl

A Dimroth rearrangement in one particular system is mentioned in the synthesis section (Section 4.10.4.9).

4.10.3.10 Imidazo[1,5-a]pyrimidine

As judged by NMR data, the parent heterocycle (**247a**) is protonated and quaternized with methyl iodide at N-7. Bromination of (**247a**) with either NBS or bromine produces the 6,8-dibromo compound, in agreement with theoretical calculations ⟨72BSF2481⟩.

a; R = H
b; R = Br

(247)

4.10.3.11 Imidazo[1,2-b]pyridazine

Electrophilic substitutions on imidazo[1,2-b]pyridazines produced exclusively 3-substituted products ⟨67T387, 68T239⟩. Treatment of imidazo[1,2-b]pyridazine (**248a**) with either NBS or bromine in acetic acid produced 3-bromoimidazo[1,2-b]pyridazine (**248b**). Nitration and sulfonation have also been examined. Nitration of 2-phenylimidazo[1,2-b]pyridazine (**248c**) resulted in initial nitration on the imidazole ring (3-position) followed by para nitration of the phenyl group if excess reagent were present ⟨68MI41000⟩, producing (**248d**). The parent ring system has a pK_a of ca. 4.5 and undergoes protonation or quaternization at N-1 ⟨73HC(27)801⟩.

a; R = R^1 = H
b; R = Br, R^1 = H
c; R = H, R^1 = Ph
d; R = NO$_2$, R^1 = p-NO$_2$C$_6$H$_4$

(248)

Nucleophilic displacement of a chlorine atom in the 6-position is possible with hydroxide ⟨64CPB1351⟩, alkoxides ⟨64CPB1351, 67T2739, 65JAP6522265⟩, thiophenolate ⟨67T387⟩, hydrazine ⟨67T387, 67T2739⟩, amines ⟨64CPB1351, 65JHC287⟩ and azide ⟨67T387, 68JHC351⟩, but not with ammonia or KSH. Nucleophilic displacements at other positions have not been investigated.

Various other reactions have also been explored. Homolytic methylation of imidazo[1,2-b]pyridazine produced a mixture of 7- and 8-methyl and 7,8-dimethyl derivatives, with the latter predominating ⟨68T2623⟩, while homolytic phenylation produced 8-phenyl and to a lesser extent 3-phenyl derivatives ⟨74M834⟩. N-Oxidation (hydrogen peroxide in PPA) produces the 5-oxide ⟨70JOC2478⟩. A 2-methyl group is susceptible to oxidation and condensation reactions typical of a methyl group at a position adjacent to nitrogen in azines ⟨68JHC35, 67T2739⟩.

4.10.3.12 Imidazo[4,5-c]pyridazine

The adenosine analog containing an imidazo[4,5-c]pyridazine (**252**) has been prepared by the condensation of (**249**) with 2,3,5-tri-O-benzoyl-D-ribofuranosyl bromide (**250**) to produce (**251**). Removal of the benzoyl groups is readily achieved by methanolic ammonia at room temperature, and then the chlorine is displaced with ethanolic ammonia in an autoclave at high temperatures to afford (**252**) ⟨74JHC39⟩.

(249)   (250)   (251)   (252)

R = 2,3,5-tri-O-benzoyl-β-D-ribofuranosyl   R = β-D-ribofuranosyl

Displacement or removal of chlorine is possible at both the 3- and 4-positions, and desulfurization or alkylation on sulfur of compounds such as (**253**) has also been carried out ⟨67JHC555⟩. Ring opening with base occurred at C-6 ⟨67JHC555⟩. Treatment of (**254a**) with dihydropyran under acidic conditions produced only (**254b**). Compared to chlorine displacement in (**254a**), it was found that the THP group in (**254b**) somewhat facilitated displacement of the chlorine with amines, producing, for example, (254c). The THP group was readily removed with aqueous acid ⟨68JHC13⟩.

(253)

(254) a; R = Cl, R¹ = H
b; R = Cl, R¹ = THP
c; R = NH(CH₂)₃NEt₂, R¹ = THP

4.10.3.13 Imidazo[4,5-d]pyridazine

Nucleophilic substitutions have been carried out most commonly with either chloro or methylthio substituents. The order of reactivity is 4 > 7 > 2, with the 4- and 7-positions being distinguished by a group attached at N-1. The 4,7-dichloro derivatives are typically prepared from the 4,7-dioxo compounds with phosphorus oxychloride ⟨68JHC13, 58JA6083⟩. Selective displacement of one chlorine substituent is readily possible. For example, 1-benzyl-4,7-dichloroimidazo[4,5-d]pyridazine (255a) when treated with ammonia or certain mono- or di-alkylamines produced (255b) selectively (site of displacement inferred from steric considerations) ⟨58JA6083⟩. To replace both chlorines, quite vigorous conditions are required. The bis-hydrazino compound (255d) is prepared by heating (255c) with 96% hydrazine ⟨64JHC182⟩. Conversion of oxo or chloro groups to thioxo compounds is readily achieved with phosphorus pentasulfide ⟨64JHC182, 70CPB1685⟩ or sodium hydrogen sulfide ⟨58JA6083⟩. The thioxo groups can be alkylated under standard conditions or removed with Raney nickel ⟨69JHC93, 64JHC182, 65JHC247⟩.

(255)
a; R = CH₂Ph, R¹ = R² = Cl, R³ = H
b; R = CH₂Ph, R¹ = NZ₂ (Z = H or alkyl), R² = Cl, R³ = H
c; R = H, R¹ = R² = SMe, R³ = Ph
d; R = H, R¹ = R² = NHNH₂, R³ = Ph
e; R = β-D-ribofuranosyl, R¹ = NH₂, R² = R³ = H
f; R = β-D-ribofuranosyl, R¹ = H, R² = NH₂, R³ = H
g; R = H, R¹ = R² = NH₂, R³ = H
h; R = R¹ = R² = R³ = H

(256)

Nucleosides relating to adenosine have been made by condensation of an imidazo[4,5-d]pyridazine with a blocked carbohydrate. The chloromercury derivative of 4(7)-benzoylaminoimidazo[4,5-d]pyridazine (256) was condensed with 2,3,5-tri-O-benzoyl-D-ribofuranosyl chloride to give a mixture of nucleosides. After removal of the blocking groups and separation, the products were identified as (255e) and (255f) ⟨57JOC954⟩.

4.10.3.14 Imidazo[4,5-d][1,2,3]triazine

The 4-thiones can be readily S-methylated (257a → 258a) ⟨75JOU902⟩ or chlorinated to the 4-chloro derivative, e.g. (257a → 258b) ⟨76CHE465, 74CHE376⟩. The chlorine can then be displaced with various amines ⟨76CHE465⟩. Nucleophilic displacement of a 6-bromo substituent is also possible ⟨74JOC3651⟩. 7-(β-D-Ribofuranosyl)imidazo[4,5-d]-[1,2,3]triazine nucleosides such as (261) can be prepared by direct condensation of the nitrogen heterocycle (259) with an activated ribose derivative (245) in the presence of mercury(II) cyanide ⟨79CHE685⟩. Treatment of (258c) with hydrogen peroxide in acetic acid affords several N-oxides, the major one being the 3-oxide ⟨60JA3189⟩.

(257) a; X = S, R = H
b; X = O, R = β-D-ribofuranosyl
c; X = O, R = H

(258) a; X = SMe, R = H
b; X = Cl, R = H
c; X = NH₂, R = H
d; X = NH₂, R = β-D-ribofuranosyl

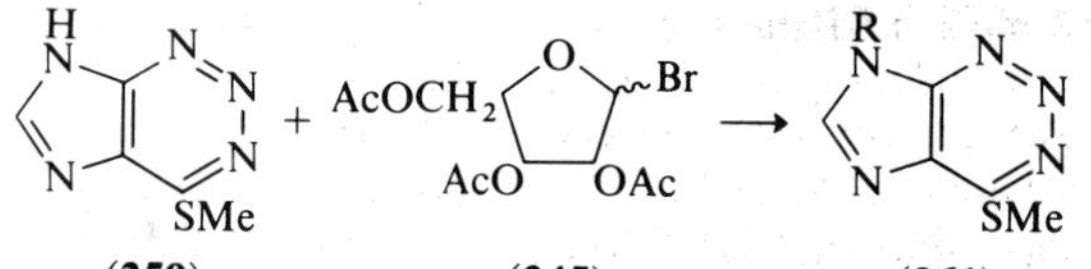

(259) **(245)** **(261)**

R = 2,3,5-tri-*O*-acetyl-*β*-D-ribofuranosyl

4.10.3.15 Imidazo[1,2-*b*][1,2,4]triazine

Several papers deal with the reactivity of this ring system. Bromination of 2,6-diphenylimidazo[1,2-*b*][1,2,4]triazine (**262a**) produced (**262b**). Treatment of the parent ring system with phenyllithium produced the 3-phenyl derivative (**263**) after air oxidation. Greater stabilization of the intermediate anion leading to (**263**) is suggested as the reason for the singular direction of addition ⟨72JHC1157⟩. Alkylation occurred at N-1 ⟨64MI41000⟩.

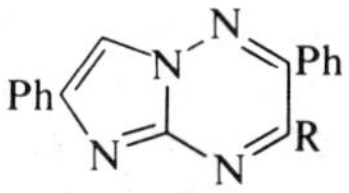

(**262**) **a**; R = H
b; R = Br

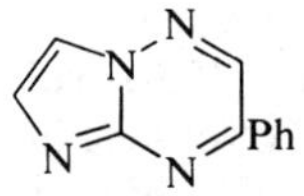

(**263**)

4.10.3.16 Imidazo[1,5-*b*][1,2,4]triazine

Bromination of (**578a**) with NBS afforded the 8-bromo derivative (**578b**) ⟨74BSF1453⟩.

4.10.3.17 Imidazo[1,5-*d*][1,2,4]triazine

Compound (**582a**) can be *S*-methylated and the methylthio group then displaced with various secondary amines ⟨79JHC277⟩.

4.10.3.18 Imidazo[2,1-*c*][1,2,4]triazine

Oxidation of (**264**) with NBS or bromine produced the fully unsaturated compound (**265**) ⟨69C303⟩.

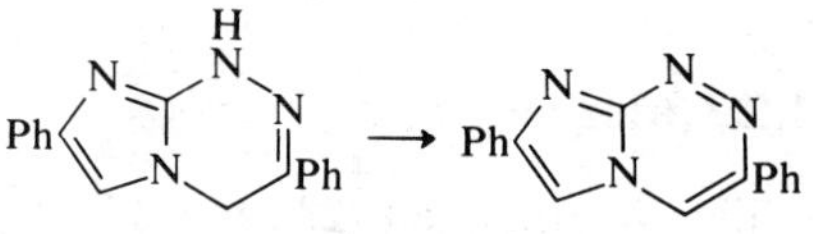

(**264**) (**265**)

4.10.3.19 Imidazo[5,1-*f*][1,2,4]triazine

Protonation of (**266a**) occurred at N-6 ⟨79JCS(P2)1327⟩, although acylation occurred at the exocyclic amino group ⟨79JCS(P1)1120⟩. Reduction of (**266b**) with lithium aluminum hydride produced (**267a**), which could be dehydrogenated to (**268**) with palladium–charcoal. Addition of methylmagnesium iodide to (**268**) occurred across the 3,4-double bond to produce (**267b**) ⟨80JCS(P1)1139⟩.

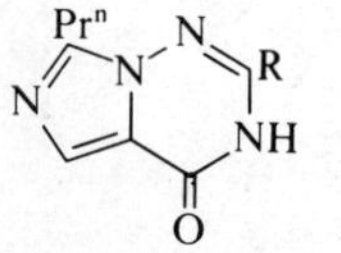

(**266**) **a**; R = NH_2
b; R = Me

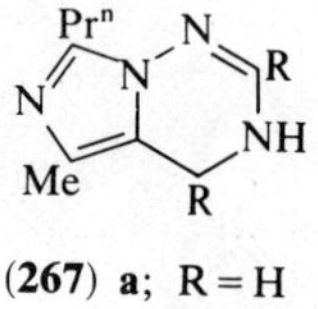

(**267**) **a**; R = H
b; R = Me

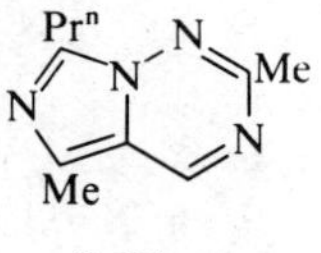

(**268**)

4.10.3.20 Imidazo[1,2-a][1,3,5]triazine

Bromination of (**269a**) (as the free base) with NBS produced (**269b**) in the only example of an electrophilic substitution on a fully aromatic imidazo[1,2-a][1,3,5]triazine ⟨70M724⟩. Ribosidation of (**270**) (as the bis-trimethylsilyl derivative) with 1-*O*-acetyl-2,3,5-tri-*O*-benzoyl-β-D-ribofuranose in 1,2-dichloroethane containing $SnCl_4$ produced the 8-substituted isomer (**271a**), which was readily debenzoylated to (**271b**) ⟨78JMC883⟩.

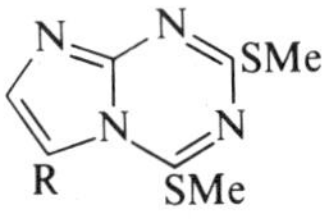

(**269**) a; R = H
b; R = Br

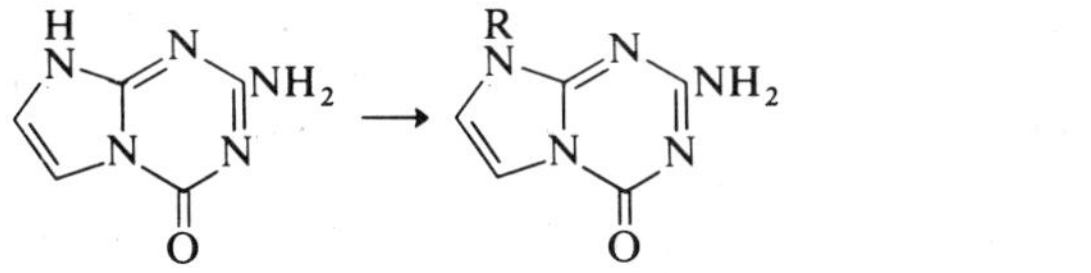

(**270**)

(**271**) a; R = 2,3,5-tri-*O*-benzoyl-β-D-ribofuranosyl
b; R = β-D-ribofuranosyl

4.10.4 SYNTHESIS

4.10.4.1 Imidazo[1,2-a]pyridine

(i) Closure of the imidazole ring

The majority of imidazo[1,2-a]pyridines (**275**) have been prepared by the reaction of a 2-aminopyridine (**272**) with a α-halocarbonyl compound (**273**). 2-Aminopyridine and chloro- or bromo-acetaldehyde gave imidazo[1,2-a]pyridine itself ⟨25CB1704⟩, and this reaction can be carried out under mild conditions ⟨61HC(15-1)460⟩. A variety of 2-aminopyridines and α-halocarbonyl compounds have been used ⟨26CB2048, 51JCS2411, 54JA4470, 53JA746, 55CB1117, 49YZ496, 79FES417, 79KGS643, 65JHC331⟩, but failures have occurred, for example, with 2-amino-6-pyridinone, probably due to the decreased basicity of the ring nitrogen ⟨65JHC53⟩. In some cases the intermediate onium bromides (**274**) cyclize slowly in DMSO at room temperature with concomitant bromination at C-3 ⟨78KGS258⟩. 2,3-Diaminopyridine reacts selectively with bromoacetaldehyde to give only 8-aminoimidazo[1,2-a]pyridine ⟨78JOC2900⟩. 1-Alkyl-2-hydroxy-2,3-dihydroimidazo[1,2-a]pyridinium salts (**278**) were identified as isolatable intermediates in the synthesis of 1-alkylimidazo[1,2-a]pyridinium salts (**279**) from 2-alkylaminopyridines (**276**) ⟨78JHC1149⟩. Methyl ketones in the presence of bromine or iodine give the corresponding 3-bromo- or 3-iodo-imidazo[1,2-a]pyridines ⟨76KGS1396⟩.

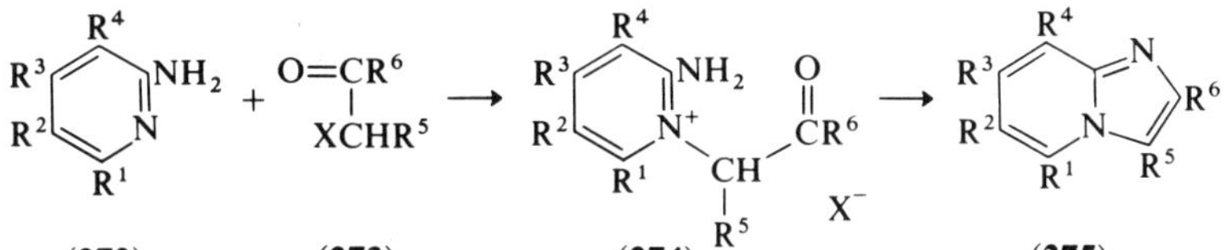

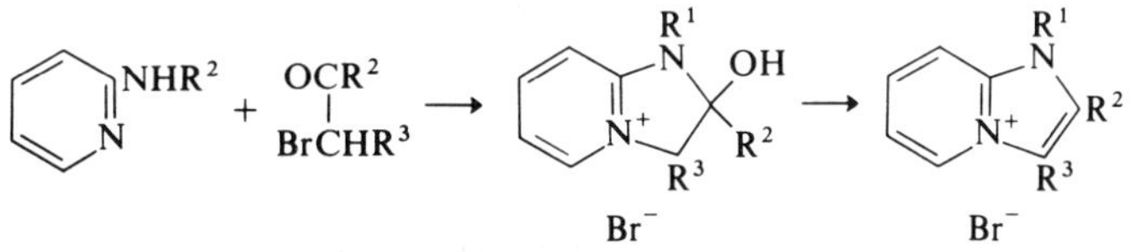

Condensation of 2-amino-*N*-phenacylpyridinium bromide (**280**) with *p*-dimethylaminobenzaldehyde (**281**) ⟨76IJC(B)551⟩ and that of 2-acetoacetylaminopyridine (**283**) with 4-nitroso-*N,N*-dimethylaniline (**284**) gave the corresponding 2,3-disubstituted imidazo[1,2-a]pyridines (**282**) and (**285**), respectively ⟨44JA1805⟩. 2-Aminopyridines react with α-chloro-α-isonitrosoacetone to give 2-methyl-3-nitrosoimidazo[1,2-a]pyridines ⟨78KGS1525⟩ whereas α-ketohydrazidoyl bromides give 3-(substituted phenyldiazo)imidiazo[1,2-a]pyridines ⟨80JHC877⟩. 3-Hydroxyimidazo[1,2-a]pyridines can be prepared by the reaction of 2-aminopyridine 1-oxides with α-halomethyl ketones ⟨79JHC187⟩. 1-Benzyl-2-aminopyridines (**286**) cyclized with difficulty in acid anhydrides to the corresponding

3-arylimidazo[1,2-*a*]pyridines (**287**) ⟨55CB1093⟩. Other carbonyl compounds such as dypnone ⟨26CB1360⟩, diethyl oxalate ⟨78CB2813⟩, benzoin ⟨78HCA129⟩ and isonitrosoflavanone ⟨80JCS(P1)354⟩ have been substituted for the α-halo compounds.

(**280**) (**281**) (**282**)

(**283**) (**284**) (**285**)

(**286**) (**287**)

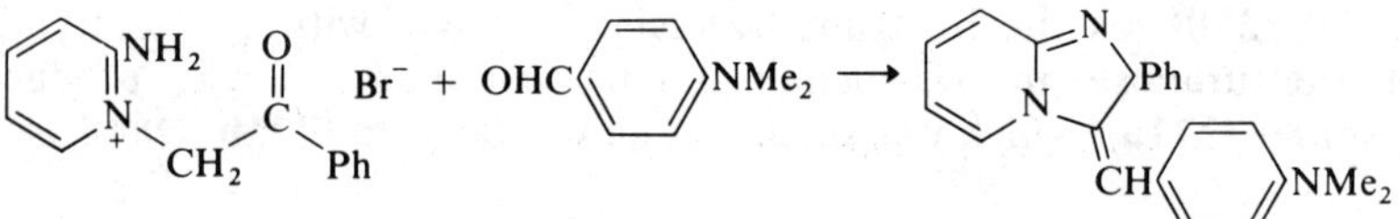

Diphenyl ketone reacts with *N*-pyridin-2-yl-*S*,*S*-dimethylsulfilimine to give 3,3-diphenylimidazo[1,2-*a*]pyridin-2(3*H*)-one ⟨79CPB2116⟩, whereas reaction of the 2-methylpyridin-2-yl-*S*,*S*-dimethylsulfilimine with a variety of α-halocarbonyl compounds gave 3-substituted 5-methyl-2-methylthioimidazo[1,2-*a*]pyridines ⟨78YZ631⟩.

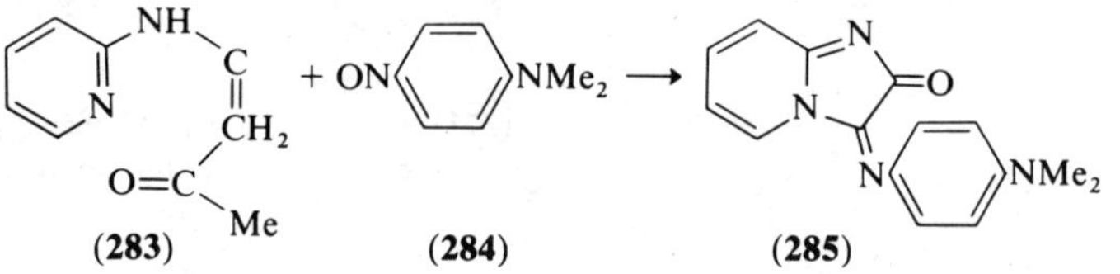

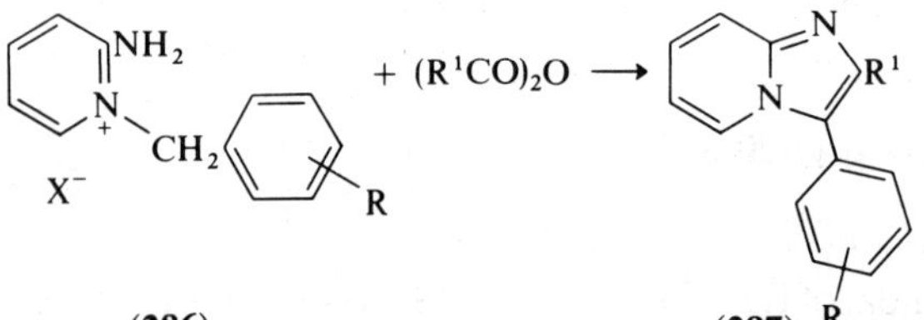

In a slight variation of the above synthetic approach, 2-phenylimidazo[1,2-*a*]pyridine (**290**) was prepared by the reaction of 1-phenacyl-2-chloropyridinium bromide (**288**) with ammonia (**289**) ⟨55CB1117⟩. The use of primary amines (**291**) resulted in 1-alkylimidazo[1,2-*a*]pyridinium halides (**292**) ⟨69JOC2129⟩, whereas hydroxylamine (**293**) gave imidazo[1,2-*a*]pyridine 1-oxides (**294**) ⟨78JOC658⟩.

(**294**) (**288**) (**290**) (**292**)

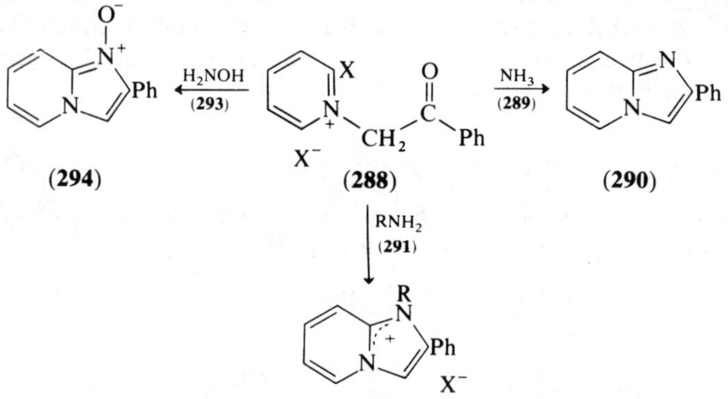

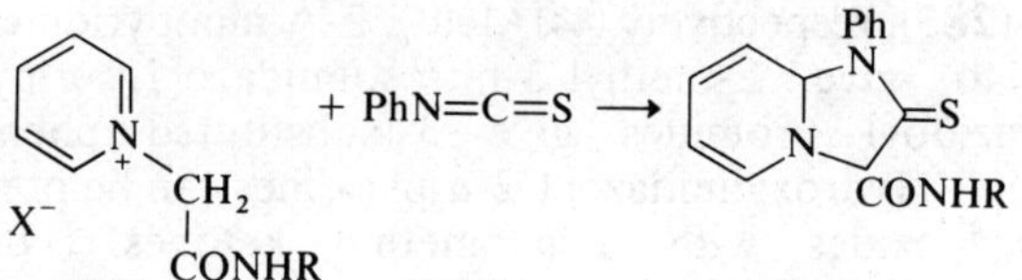

Cycloaddition of phenyl isothiocyanate (**296**) to pyridinium salts (**295**) produced imidazo[1,2-*a*]pyridine-2(3*H*)-thiones (**297**) ⟨77MI41001⟩. Reaction of 1-aryl-2-chloropyridin-6-ones with mono-*N*-substituted ethylenediamines provided dihydroimidazo[1,2-*a*]pyridin-5-ones ⟨79CPB1207⟩. Another procedure involves the polyphosphoric acid cyclization of 2-(2-hydroxyethyl)aminopyridines ⟨76AJC1039⟩. Treatment of certain β-aminoacrylic esters (**298**) with benzenediazonium fluoroborate (**299**) resulted in

(**295**) (**296**) (**297**)

ring closure with loss of aniline to provide tetrahydroimidazo[1,2-*a*]pyridines (**300**) ⟨78H(9)757⟩. The cyclodehydration of cyclic tertiary amines (**301**) containing aliphatic neighboring groups capable of undergoing intramolecular Mannich reactions to (**302**) has been accomplished with the mercury(II) EDTA complex ⟨73AP209⟩.

(**298**) (**299**) (**300**)

(**301**) (**302**)

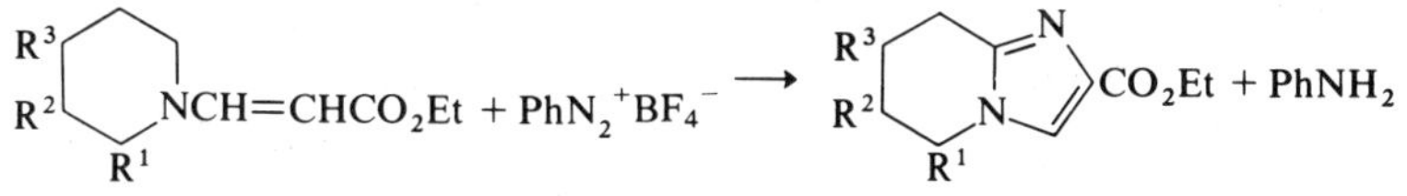

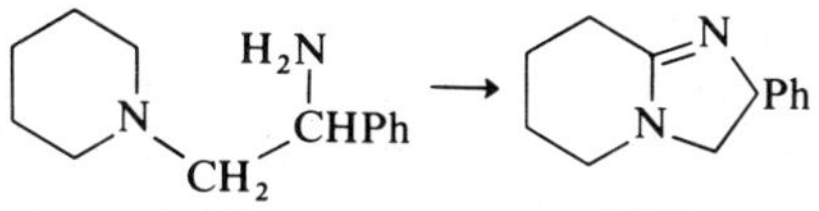

(*ii*) *Closure of the pyridine ring*

The Horner–Wittig reaction of 1-substituted imidazole-2-aldehydes (**304**) gave intermediates (**305**) that cyclized to imidazo[1,2-*a*]pyridin-5-ones (**306**) ⟨80LA542⟩.

(**303**) (**304**) (**305**) (**306**)

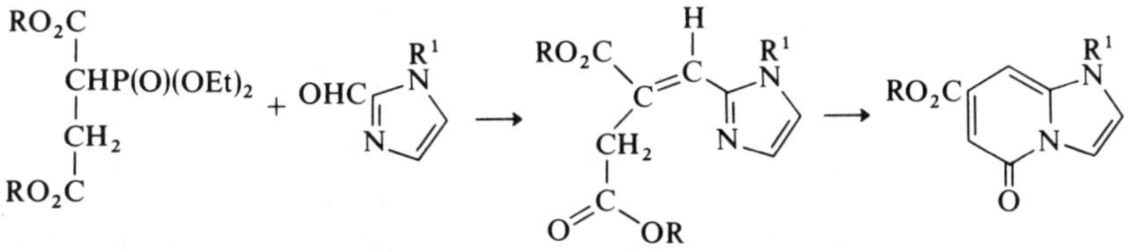

1-Methylimidazole-2(3*H*)-thione (**308**) and α-bromophenacetyl chloride (**307**) in the presence of triethylamine gave anhydro-3-hydroxy-7-methyl-2-phenylimidazo[2,1-*b*]thiazolylium hydroxide (**309**), which with a variety of alkynic and alkenic dipolarophiles (**310**) in refluxing benzene formed 1-methyl-6-phenyl-5*H*-imidazo[1,2-*a*]pyridin-5-ones (**312**) ⟨79JOC3803⟩.

(**307**) (**308**) (**309**) (**311**) (**312**)

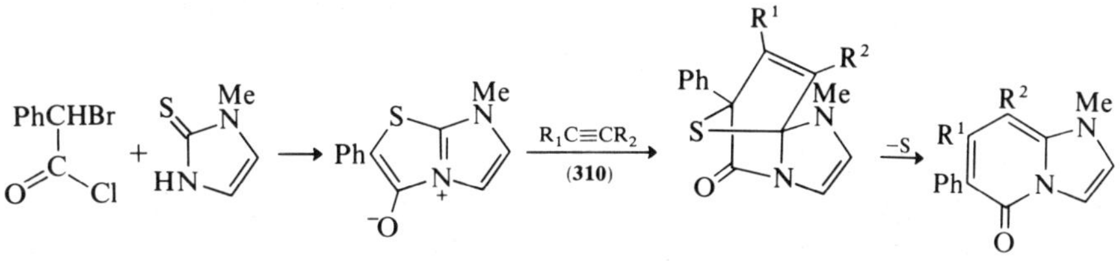

Alkylcobaloximes (**313**) with acetic anhydride in pyridine gave an imidazo[1,2-*a*]pyridine (**314**). A mechanism for the degradation of (**313**) involving formation of (**314**) prior to cleavage of the cobalt complex has been proposed ⟨75JCS(P1)386⟩.

(**313**) (**314**)

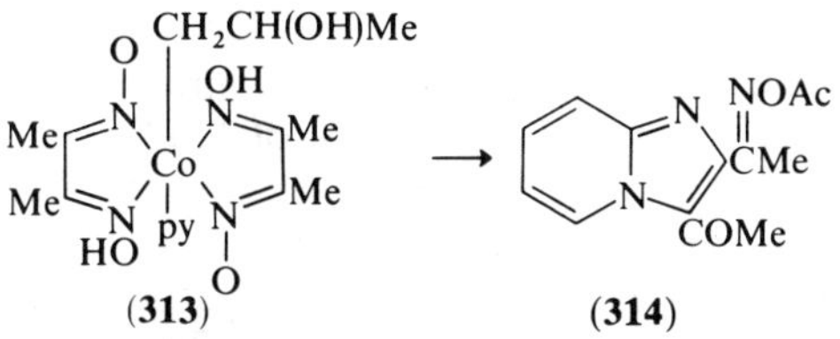

Condensation of 1,2-dimethylimidazole ⟨32LA(498)1⟩ or 1-methyl-2-(methylthio)imidazole (**316**) ⟨74HCA750⟩ gave a dihydroimidazo[1,2-*a*]pyridine (**317**), which in the case of the methylthio compound rearranged to (**318**), and the latter compound could be desulfurized.

(**315**) R = Me or MeS, E = CO_2Me (**316**) (**317**) (**318**) (**319**)

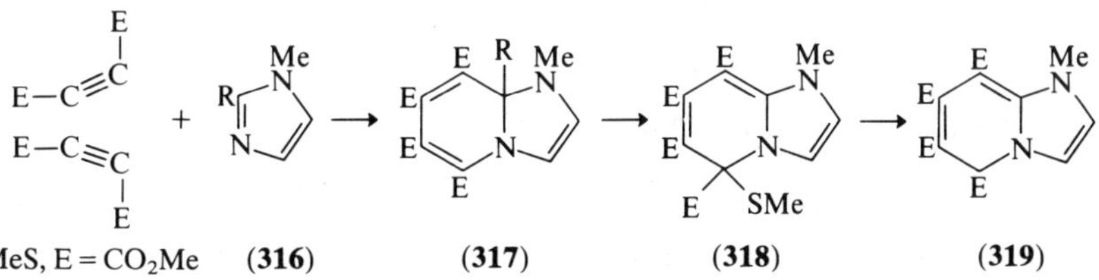

Derivatives of 8-phenyl-2,3-dihydroimidazo[1,2-*a*]pyridin-7-one (**322**) were prepared by the condensation of a series of β-keto esters (**320**) with 2-benzylimidazoline (**321**) ⟨78JHC1021⟩.

(**320**) (**321**) (**322**)

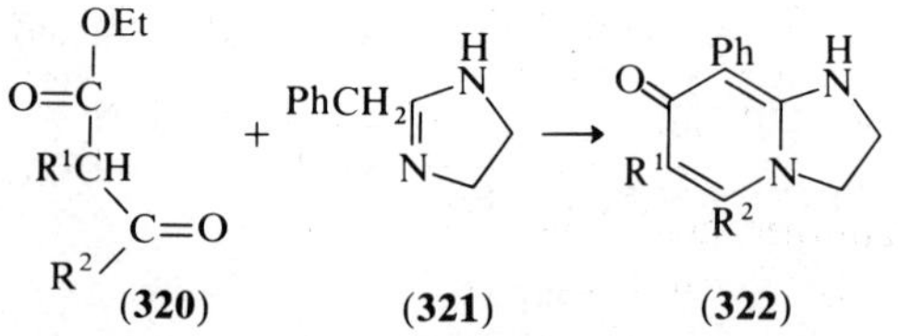

In a related reaction, condensation of 1,3-dicarbonyl compounds (**323**) with methylene bis-imidazolidine (**324**) gave substituted imidazo[1,2-*a*]pyridines (**325**) ⟨78LA1491⟩. The reactions of aldehydes (**326**) with ethyl 2-imidazolidinyleneacetate (**327**), or those of aldehydes (**326**) and β-dicarbonyl compounds with imidazolidines (**330**) gave similar structures (**328**) and (**331**) ⟨78LA1476, 73GEP2210633⟩.

(**323**) (**324**) (**325**)

(**326**) (**327**) (**328**)

(**326**) (**329**) (**330**) (**331**)

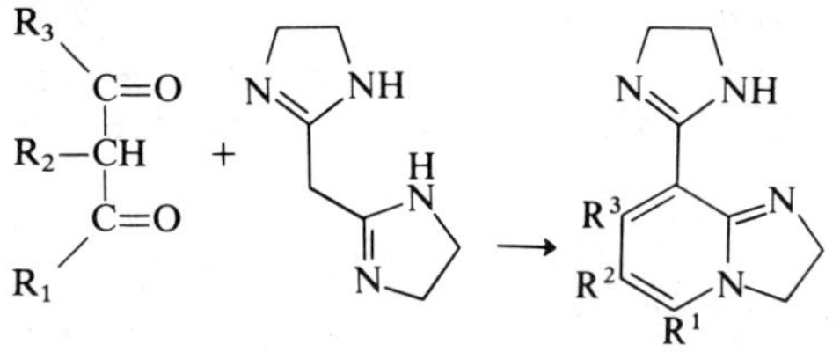

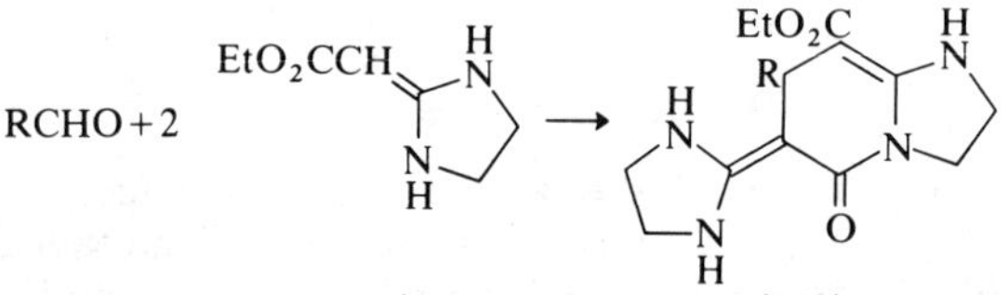

Thermal rearrangement of 2,2-pentamethylene-4,5-diphenyl-2*H*-imidazole (**332**) gave 5,6,7,8-tetrahydro-2,3-diphenylimidazo[1,2-*a*]pyridine (**333**) in 66% yield ⟨75JOC2562⟩.

(**332**) (**333**)

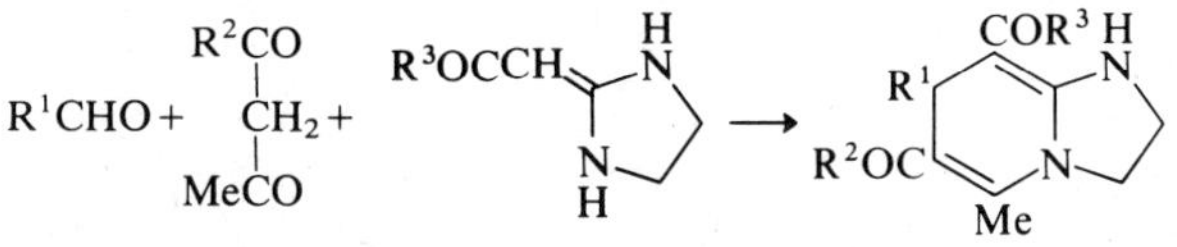

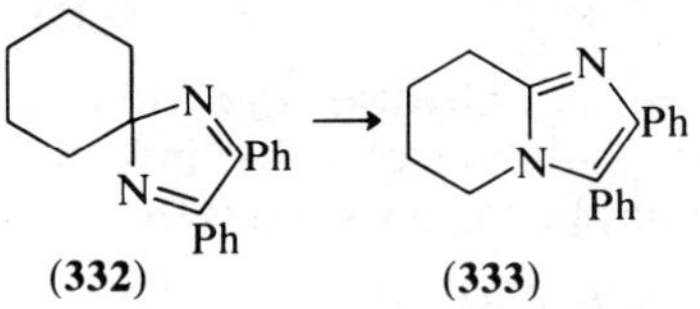

There has been one report of the synthesis of the imidazo[1,2-*a*]pyridine ring system (**336**) by the closure of both rings by the reaction of (**334**) and (**335**), but closure of the pyridine ring probably occurs first ⟨73ZOR2006⟩.

(**334**) (**335**) (**336**)

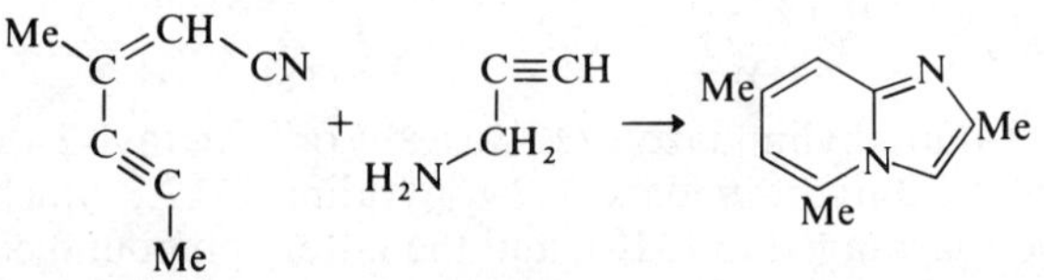

4.10.4.2 Imidazo[1,5-*a*]pyridine

(i) Closure of the imidazole ring

Synthetic procedures exist for both saturated and unsaturated imidazo[1,5-*a*]pyridines. All procedures form the five-membered ring by formation of one bond adjacent to the six-membered ring nitrogen in the last step of a sequence starting from a preformed pyridine or piperidine ring.

All of the synthetic methods of this type are quite similar. A 2-substituted pyridine with an α-amino substituent is treated with an appropriate reagent to generate a trigonal carbon atom next to it, which is then closed in various ways to the imidazo[1,5-*a*]pyridine. As an example, formylation of α-2-pyridylbenzylamine produces (**337**), which cyclizes to 1-phenylimidazo[1,5-*a*]pyridine (**65d**) upon heating in benzene with phosphorus oxychloride ⟨73JCS(P1)2595⟩. Intermediates suitable for cyclization also have been prepared from α-2-pyridylamines and isothiocyanates ⟨80JHC555⟩ and carboxylic acids ⟨63AG1101⟩. Direct closure has been achieved with cyanogen bromide ⟨72JAP(K)7225190⟩ and phosgene ⟨75FES197⟩. In one example 2-pyridylacetohydrazide was treated with one equivalent of nitrous acid, presumably generating after Curtius rearrangement an α-2-pyridylisocyanate which closed immediately to the imidazo[1,5-*a*]pyridine ⟨77JHC993, 79JHC689⟩. With aliphatic compounds, cyclization has occurred by treating pipecolic acid (or an ester) with isothiocyanates ⟨52JA4960, 60AP203, 63MI41000, 68RTC11⟩, isocyanates ⟨60JOC2108, 70JHC355⟩ and by treatment of pipe-colylamines with aldehydes or ketones ⟨58N492⟩.

4.10.4.3 Imidazo[4,5-*b*]pyridine

(i) Closure of the imidazole ring

Almost all of the literature procedures for the preparation of the imidazo[4,5-*b*]pyridine ring system, like the [4,5-*c*] system, involve closure to the imidazole by means of a carboxylic acid or some derivative thereof. The first reported synthesis involved boiling 2,3-diaminopyridine or a chloro derivative (**338**) with acetic anhydride to give 2-methyl-imidazo[4,5-*b*]pyridine and a chloro derivative (**340**) ⟨27CB766⟩. The diacetamidopyridines (**339**; R = Me) have been isolated by milder treatment with acetic anhydride and cyclized to the imidiazopyridines ⟨48JCS1389, 63MI41001⟩, and other anhydrides have been used ⟨63MI41001⟩. Cyclization in 98% formic acid to give imidazopyridines unsubstituted at C-2 (**340**; R = H) has been employed extensively ⟨48JCS1389, 49JA1885, 49RTC1013, 56JA4130, 57JA670, 57JA6421, 59CPB602, 59JOC1455, 60MI41000, 63JOC1837, 65JMC296, 72RTC650⟩. Isolation of the intermediate formamido compounds followed by distillation from magnesium has also been used ⟨52CB1012⟩. A variety of other acids have been used ⟨58CPB443, 67MI41001, 77JMC1189⟩, and polyphosphoric acid facilitates this ring closure ⟨64JOC3403, 64M242, 78IJC(B)531, 79IJC(B)428, 80JHC1757⟩. The aldehyde-oxidation procedure has also been applied to this ring system ⟨48RTC29, 61RTC545, 62JOC2163, 65IJC138, 66IJC403, 79IJC(B)428⟩. Application of this and other methods to 2,3,4-triaminopyridine results in ring closure between the 3-amino group and both the 2- and the 4-amino groups to give a mixture of the imidazo[4,5-*c*]pyridine and the imidazo[4,5-*b*]pyridine ⟨65JMC708, 69RTC1263⟩. Nitrobenzene, instead of copper(II) acetate, has served as the oxidant and the intermediate Schiff's base (**341**) has been isolated and cyclized ⟨79IJC(B)428⟩. Formamidine, phenyl acetamidates and the ethyl ester of phenylacetamidic acid have also been used to close the imidazole ring but seem to offer no particular advantage ⟨63JOC1837, 75M1059, 65IJC138⟩. Ortho esters, primarily triethyl orthoformate, represent another variant ⟨57JA6421, 77JHC813⟩, and this cyclization is facilitated by the addition of a slight excess of concentrated hydrochloric acid (relative to the diamine) ⟨73JOC613, 73JMC292, 78MI41007⟩.

Acetic anhydride has also been employed to enhance the ortho ester reaction ⟨73UKZ277, 74RTC160⟩ as has diethoxymethyl acetate ⟨65JMC708, 66JMC354, 76RTC127⟩, the actual cyclization reagent resulting from the reaction of acetic anhydride and triethyl orthoformate. Reaction of 2,3-diaminopyridine with tetraethoxymethane and with tetrapropoxymethane gave 2-ethoxy- and 2-propoxyimidazo[4,5-c]pyridine ⟨74JCS(P1)349⟩.

Other variations include formamide ⟨57JA6421⟩, sodium dithioformate or phosphorus oxychloride–DMF ⟨69RTC1263⟩, a thioamide ⟨78MI41000⟩ and acetylacetone ⟨77JHC813⟩. Reduction of 2,6-dibenzamido-3-(2-butoxypyrid-5-ylazo)pyridine with iron and hydrochloric acid resulted in ring closure to 5-benzamido-2-phenylimidazo[4,5-b]pyridine ⟨60AC(R)125⟩.

Ring closure to imidazo[4,5-b]pyridin-2(3H)-ones (**343**; X = O) has been accomplished by fusion of 2,3-diaminopyridines (**342**) with urea ⟨48JCS1389, 59JOC1455, 65JMC296, 80JHC1757, 78JMC965⟩, by reaction with phosgene ⟨49JA1885, 78JMC965⟩, or in one instance by means of ethyl pyrocarbonate ⟨76KGS1277⟩. The imidazo[4,5-b]pyridine-2(3H)-thiones (**343**; X = S) are most commonly prepared by closure with carbon disulfide ⟨48JCS1389, 65JMC296, 70MI41001, 77KGS411, 79AJC2713⟩, but also by means of thiophosgene or thiourea ⟨49JA1885, 80JHC1757⟩. 2,3-Diamino-5-nitropyridines, however, failed to react with urea or carbon disulfide ⟨65JMC296⟩. 3-Amino-2-hydrazinopyridine (**345**) closes with carbon disulfide and triethylamine to give 3-aminoimidazo[4,5-b]pyridine-2(3H)-thione (**346**), but under neutral conditions closure occurs to give s-triazolo[4,3-a]pyridine-3(2H)-thione (**344**) ⟨77KGS411⟩. Heating ethyl 1-aminopyridine-3-carbamate (**347**) at 160 °C for 15 minutes gave imidazo[4,5-c]pyridin-2(3H)-one (**348**), but the N-methyl carbamate required 200 °C ⟨57JCS442⟩. An isomeric 3-aminopyridine-2-carbamate cyclized on reflux in n-propanol ⟨66JOC1890⟩. Quaternary salts of 3-pyridylurea (**349**) cyclized on addition of alkoxides to 1,4-dihydroimidazo[4,5-b]pyridine-2(3H)-ones (**350**) ⟨79ZN(B)1473⟩.

Derivatives of 2-aminonicotinamide (**351**) underwent an intramolecular Hofmann rearrangement in which the intermediate isocyanate (**352**) reacted with the 2-amino group to give imidazo[4,5-b]pyridin-2(3H)-ones (**353**) ⟨57AP20⟩. Similarly pyridine-2- or -3-carboxylic acid hydrazides undergo the Curtius reaction with cyclization on the adjacent 3- or 2-amino groups ⟨58AP368, 59JCS3157⟩. Treatment of 2-amino-N-hydroxynicotinamide with tosyl chloride and aqueous base also resulted in cyclization to imidazo[4,5-c]pyridin-2(3H)-one.

The photocyclization of 3-nitro-2-pyridyl-DL-leucine (**354**) to 2-isobutylimidazo[4,5-b]pyridine (**355**) has been studied spectrophotometrically and the quantum yield of the reaction determined as a function of pH ⟨72JCS(P2)2218⟩. 3-N-Methanesulfonamidopyridine 1-oxide (**356**) is acylaminated at C-2 by phenylbenzimidoyl chloride (**357**) and the intermediate 2-acylaminated product (**358**) cyclized to 2,3-diphenylimidazo[4,5-b]pyridine (**359**) ⟨74JOC1802⟩.

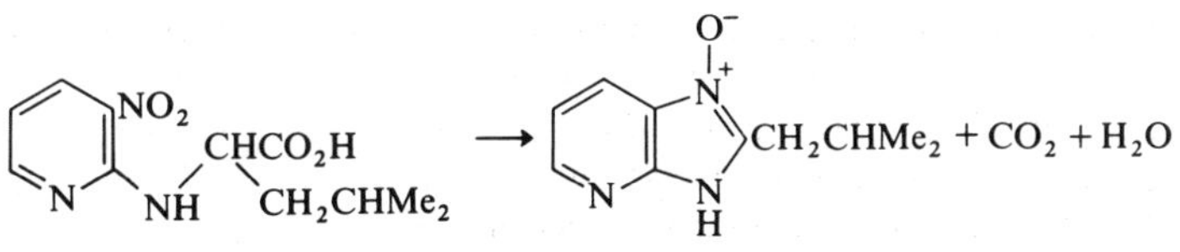

Fusion of 2,3-diaminopyridine (**360**) with 2-(cyanoamino)-4-hydroxy-6-methylpyrimidine (**361**) provided 2-[4-hydroxy-6-methylpyrimidin-2-ylamino]imidazo[4,5-b]pyridine (**362**) ⟨73JHC363⟩. Reaction of tetra-O-acetyl glucopyranosyl isothiocyanate with 2,3-diaminopyridine gave 1-(2-aminopyridin-3-yl)-3-glucopyranosylthiourea (**363**), which was S-methylated to (**364**) which cyclized on heating to 2-(tetra-O-acetylglucopyranosylamino)imidazo[4,5-b]pyridine (**365**) ⟨79CPB1153⟩.

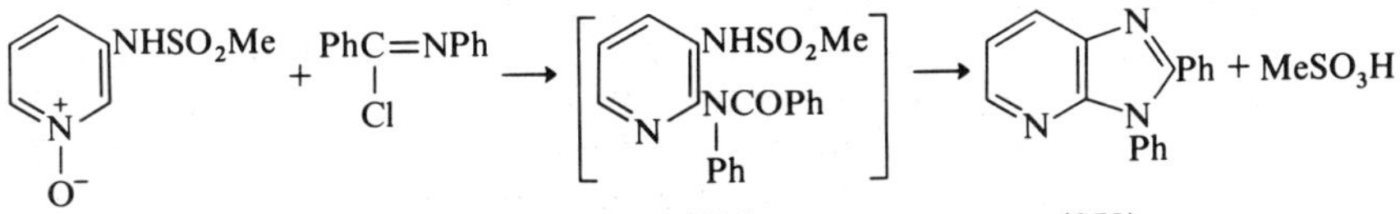

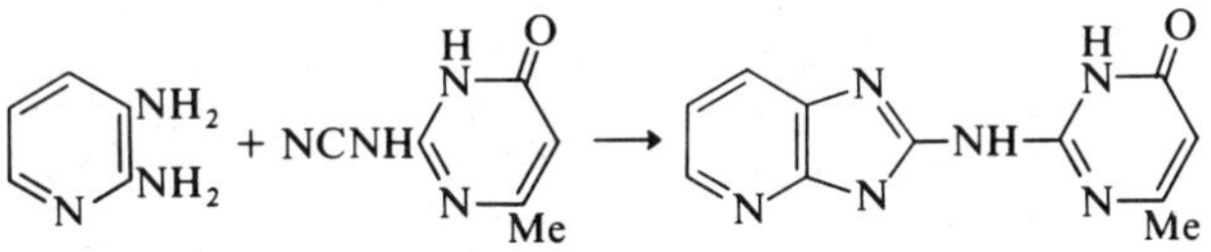

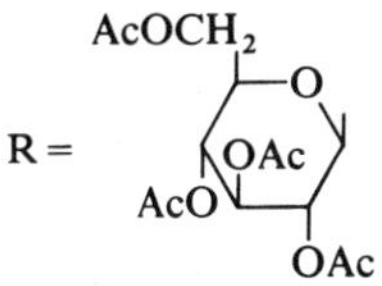

The reactions of β-keto esters with 2,3-diaminopyridine (**360**) and derivatives thereof are complex and dependent upon the substituents on both substrates and on the reaction conditions. Thus, acetoacetic ester (**366**) reacted with 2,3-diaminopyridine (**360**) in xylene to give the diazepinone (**369**) and 1-isopropenylimidazo[4,5-b]pyridin-2(3H)-one (**370**); without solvent the diazepinone (**367**), and 3-isopropenylimidazo[4,5-b]pyridin-2(3H)-one (**368**) were obtained. Heating both these diazepinones caused rearrangement to the corresponding N-isopropenylimidazo[4,5-b]pyridin-2-ones. Reaction of 2,3-diaminopyridine with acetoacetic esters substituted on the α-carbon in xylene gave (**373**). The isomeric compound could be prepared in this case by reaction neat to give (**374**) which cyclized to the diazepinone (**375**) on treatment with sodium ethoxide in ethanol. The diazepinone rearranged on heating to the imidazo[4,5-b]pyridin-2-one (**376**) ⟨73JHC201⟩. On heating with acetoacetic ester in xylene, 2,3-diaminopyridine substituted on the 2-amino group by a methyl or phenyl group (**377**) gave 2-acetonyl-3-phenylimidazo[4,5-b]pyridine (**381**) without the intervention of the diazepinone. Reaction with the phenyl compound (**377**; R = Ph) neat at 70 °C gave ethyl 3-(2-anilino-3-pyridyl)aminocrotonate (**378**; R = Ph), which on treatment with sodium ethoxide in ethanol underwent a reverse Claisen reaction followed by cyclization to give 2-methyl-3-phenylimidazo[4,5-b]pyridine (**379**; R = Ph) ⟨68TL4811⟩. 6,6-Diallyl-5,7-dioxo-1,4,5,7-tetrahydropyrido[2,3-b][1,4]diazepine (**382**), prepared by the reaction of 2,3-diaminopyridine (**360**) with diethyl diallylmalonate, was hydrated to give α-2-(pyrido[2,3-b]imidazolyl)-α-allyl-γ-valerolactone (**383**) ⟨72MI41000⟩.

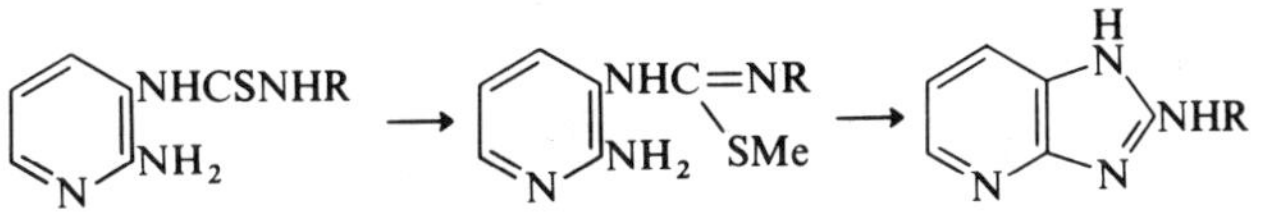

Distillation of 1,2,3,4-tetrahydropyrido[3,2-d]pyrimidine-2,4-dione (**384**) with zinc dust caused a reductive rearrangement to imidazo[4,5-b]pyridine (**385**) ⟨56JCS1045⟩. Reaction of 6-chloropyrido[2,3-b]pyrazines (**386**) with potassium amide in liquid ammonia resulted in attack at C-2 giving (**387**), with ring contraction to (**388**) involving loss of chlorine to give imidazo[4,5-b]pyridine or 2-substituted derivatives thereof (**389**) ⟨79JHC305⟩.

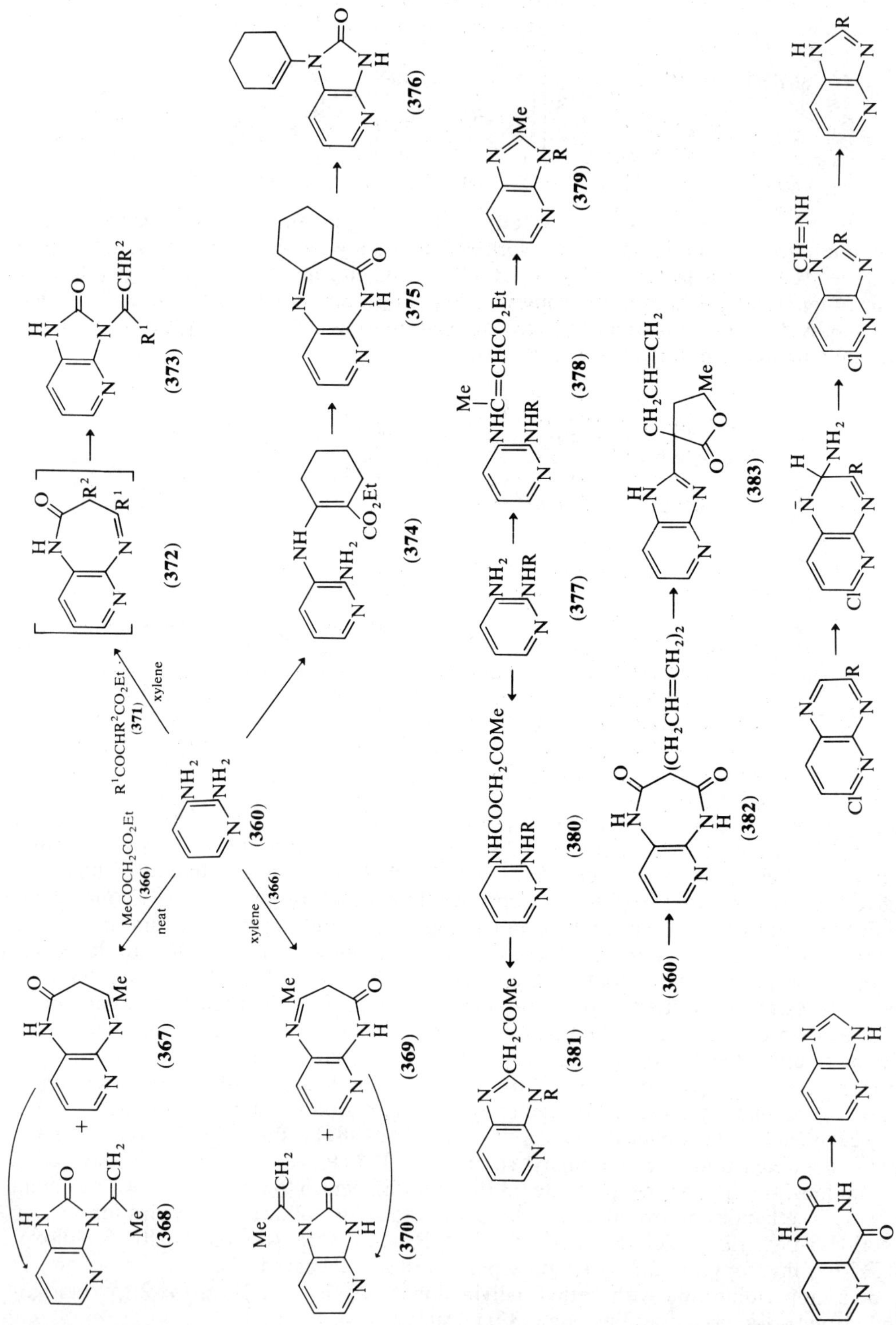

638   *Other Imidazoles with Fused Six-membered Rings*

(ii) Closure of the pyridine ring

Heating equimolar amounts of 5-amino-1-methylimidazole-2(3H)-thione (**391**) with diethyl ethoxymethylenemalonate (**390**) gave the diethyl ester of [(5-amino-2,3-dihydro-1-methyl-2-thioxo-1H-imidazol-4-yl)methylene]propanedioic acid (**392**), which cyclized in 10% aqueous sodium hydroxide to 6-ethoxycarbonyl-3-methyl-2(1H)-thioxoimidazo[4,5-b]pyridin-5(4H)-one (**393**) ⟨78H(10)241⟩.

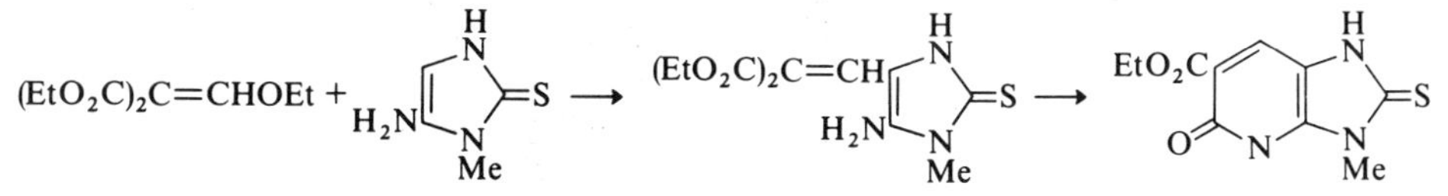

Treatment of 1,5-diaminoimidazoles (**395**) with ethyl acetoacetate or propionate (**394**) yielded imidazo[4,5-b]pyridin-5(4H)-ones (**397**) and imidazo[1,5-b][1,2,4]triazepinones (**398**). The relative yields of these isomeric compounds depend on the substituents on the reactants and on the reaction conditions. Conversion of (**398**) into (**397**) occurred on prolonged heating in water ⟨78JHC937⟩.

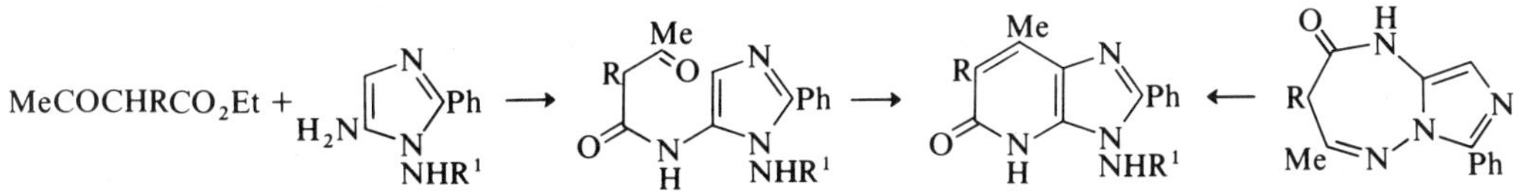

The reaction of β-aminocrotonamide and ethyl N-acylglycinates gave 2-[(acylamino)methyl]-6-methylpyrimidin-4(3H)-ones (**399**), which cyclized to 5-methylimidazo[4,5-b]pyridin-7(4H)-ones (**402**) when heated to 180–190 °C in polyphosphoric acid. At 100–110 °C 2-methylimidazo[1,5-a]pyrimidin-4(1H)-ones (**400**) are formed, which on heating to 180–190 °C (in polyphosphoric acid) rearrange via (**401**) to the imidazo[4,5-b]pyridin-7(4H)-ones (**402**) ⟨82JOC167⟩. Attack by malononitrile on purine (**403**) occurs at C-6 with ring opening (**404**) and reclosure (**405**) to give 5-amino-6-cyanoimidazo[4,5-b]pyridine (**406**) ⟨73JCS(P1)1794⟩.

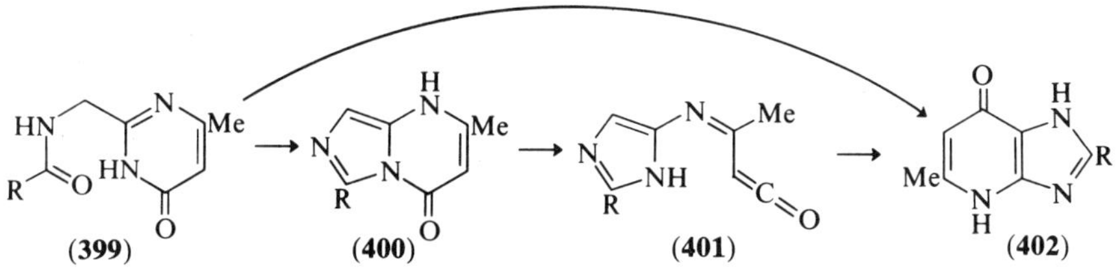

4.10.4.4 Imidazo[4,5-c]pyridine

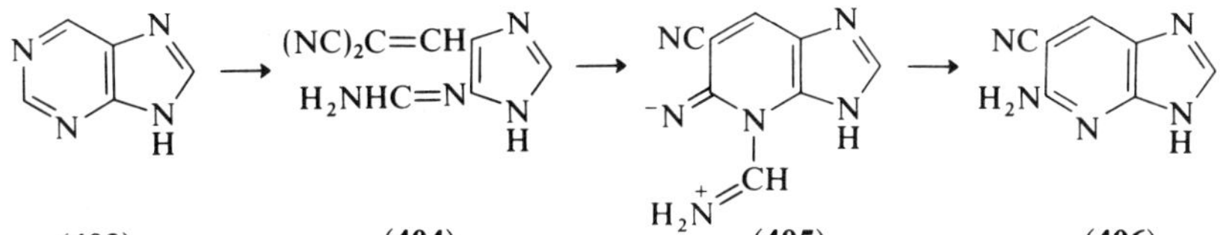

(i) Closure of the imidazole ring

Imidazo[4,5-c]pyridine (**408**; R = R¹ = H) and a number of its 2-substituted and 1,2-disubstituted derivatives have been prepared in moderate to good yields by the reaction of the 3,4-diaminopyridines (**407**) with aliphatic, aromatic and heteroaromatic aldehydes in aqueous alcohol solution in the presence of excess copper(II) acetate in a sealed tube at 130–150 °C ⟨38CB2347, 42CB1936⟩. The Schiff bases (**409**) are formed primarily in this synthesis, and in many cases it is preferable to isolate the base (**409**) and effect the subsequent oxidation in a separate operation in alcohol solution. Ring closure of 5-substituted, 6-substituted and 2,6-disubstituted 3,4-diaminopyridines (**410**) to 7-substituted, 6-substituted and 4,6-disubstituted imidazo[4,5-*c*]pyridines (**411**) has also been accomplished with sodium dithioformate (not satisfactory), 98% formic acid ⟨48RTC29, 49RTC1013⟩, triethyl orthoformate and acetic anhydride ⟨65JHC196, 66JCS(B)285, 73UKZ350⟩, diethoxymethyl acetate ⟨65JMC708, 66JMC105⟩ and triethyl orthoformate and concentrated hydrochloric acid in dimethylacetamide ⟨82JMC626⟩. The use of triethyl orthoacetate and acetic anhydride ⟨73UKZ703⟩, iminoesters ⟨65IJC138, 73JMC1296⟩, urea and carbon disulfide ⟨65JMC296, 66JCS(B)285⟩ provided 2-methyl-, 2-benzyl-, 2-oxo- and 2-thioimidazo-[4,5-*c*]pyridines with substituents in the pyridine ring (**413–416**).

(**407**) (**408**) (**409**)

(**410**) (**411**)

(**416**) (**413**) (**412**) (**415**) (**414**)

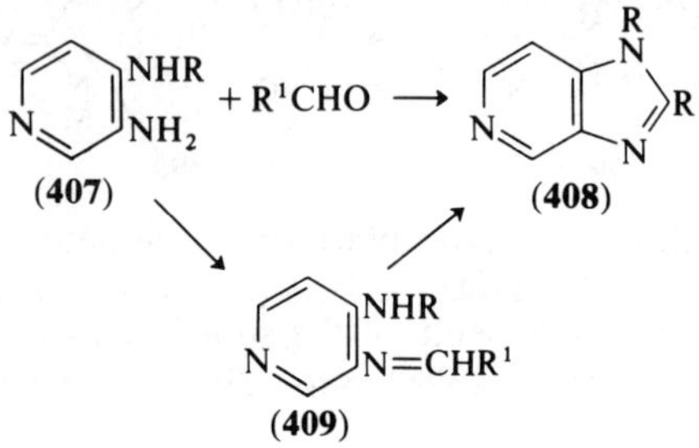

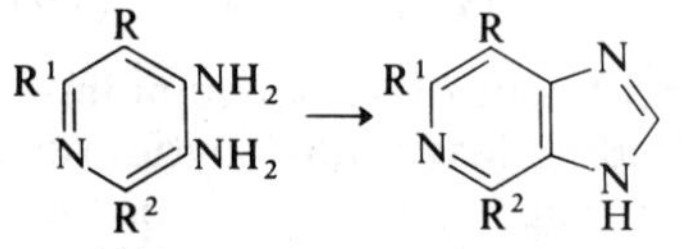

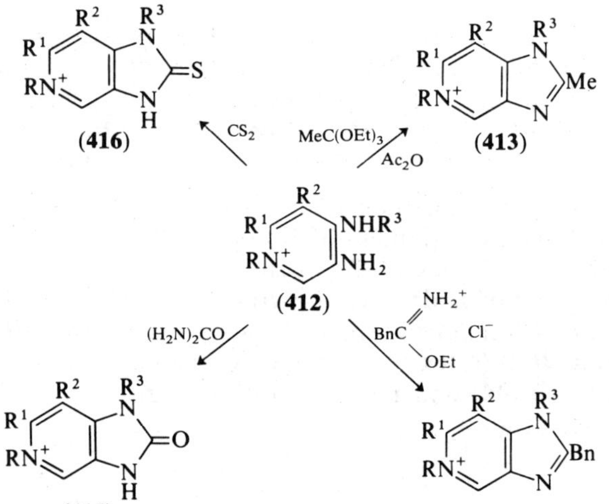

Cyanogen bromide ring closure of (**417**) was used to prepare streptolidine lactam (**418**), a guanidine-containing amino acid lactam moiety of the streptothricin antibiotic group ⟨77BCJ2375⟩, that moiety being a 4,5,6,7-tetrahydroimidazo[4,5-*c*]pyridine.

(**417**) (**418**)

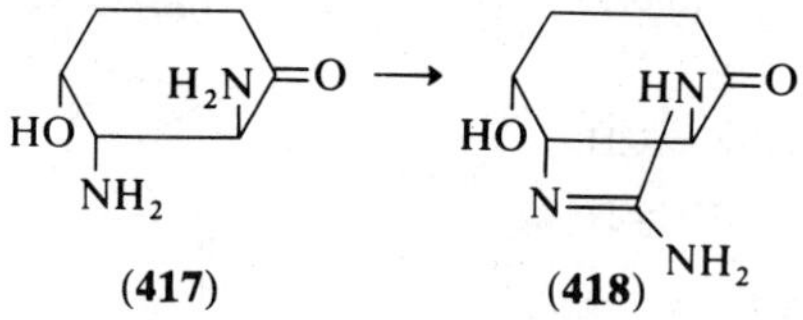

Refluxing ethyl 2-(3-aminopyridin-4-yl)hydrazinecarboxylate (**419**) in 36% methanolic hydrogen chloride for 8 hours gave 1-aminoimidazo[4,5-*c*]pyridin-2(3*H*)one hydrochloride (**423**), rather than the expected pyridotriazinone (**420**), thought to be an intermediate in the formation of (**423**) *via* (**421**) and (**422**) ⟨76JHC601⟩.

(**419**) (**420**) (**421**) (**422**) (**423**)

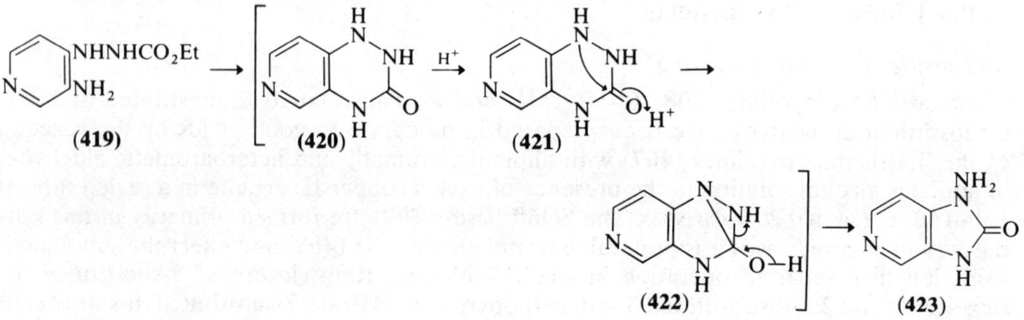

The pyrolysis of 3-azido-4-dialkylaminopyridines (**424**) and (**427**) by heating in nitrobenzene to generate nitrene intermediates (**425**) has also been used to form the imidazole ring of imidazo[4,5-*c*]pyridines (**426**) and (**428**) ⟨66JCS(C)80⟩. In one instance the intermediate dihydro compound (**429**) was trapped by carrying out the reaction in acetic anhydride ⟨63JCS1666⟩.

(**424**) (**425**) (**426**)

(**427**) (**428**) (**429**)

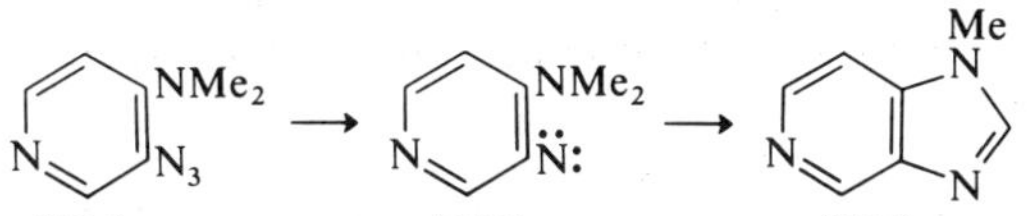

(ii) Closure of the pyridine ring

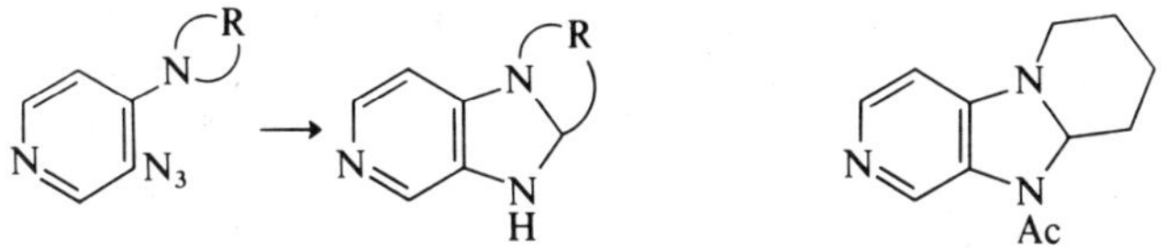

Treatment of both histidine (**430**; R = CO_2H) and histamine (**430**; R = H) with diethoxymethane in hydrochloric acid solution forms a CH_2 bridge between the NH_2 group and C-5 of the imidazole ring to give 4,5,6,7-tetrahydroimidazo[4,5-*c*]pyridine-6-carboxylic acid (spinacine; **432**; R = CO_2H, $R^1 = R^2$ = H) ⟨13MI41000, 44BJ(38)309⟩ and 4,5,6,7-tetrahydroimidazo[4,5-*c*]pyridine (spinaceamine; **432**; R = $R^1 = R^2$ = H) ⟨20MI41000⟩. This reaction has been extended to other aldehydes ⟨66JOC2380⟩, including pyridoxal ⟨48JA3429, 3669⟩, and ketones ⟨65FES634, 67FES821, 76H(5)127, 77IJC(B)629⟩. The cyclization has been carried out in basic and neutral aqueous solution, in organic solvents, and neat. In one investigation the corresponding Schiff bases (**431**) were formed under mild conditions and are presumably intermediates in the cyclization ⟨66JOC2380⟩.

(**430**) (**431**) (**432**)

(**433**) (**434**)

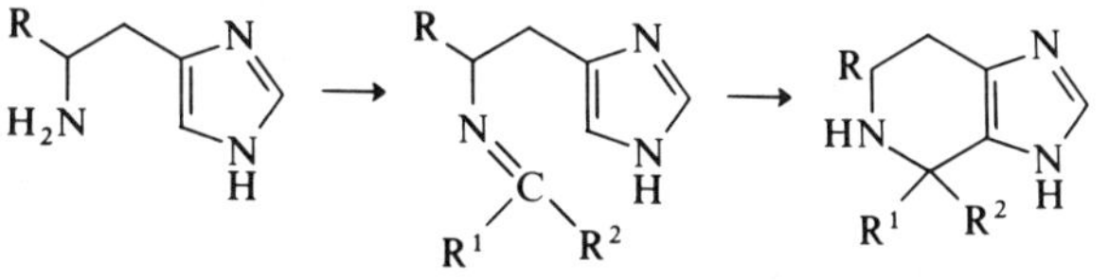

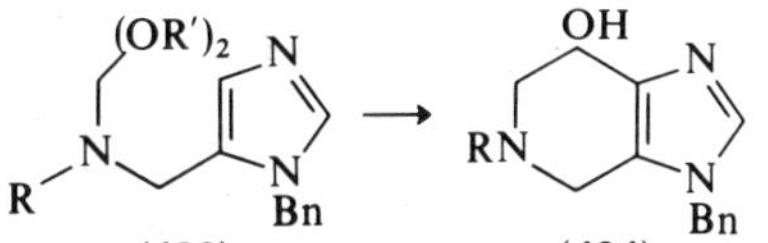

R = Me, Bn
R′ = Me, Et

Intramolecular cyclization of the imidazoles (**433**) also gave rise to 4,5,6,7-tetrahydroimidazo[4,5-*c*]pyridines (**434**) ⟨75JHC1039⟩.

4,6-Dihydroxyimidazo[4,5-*c*]pyridine (3-deazaxanthine; **436**; R = H) and related 2-substituted derivatives (**436**; R = Cl) and (**438**) were prepared from the requisite imidazole 4-acetamide-5-carboxylic acid esters (**435**) and (**437**) by base-catalyzed cyclization to the imide structure ⟨63JOC3041⟩.

(**435**) (**436**) (**437**) X = O, S (**438**)

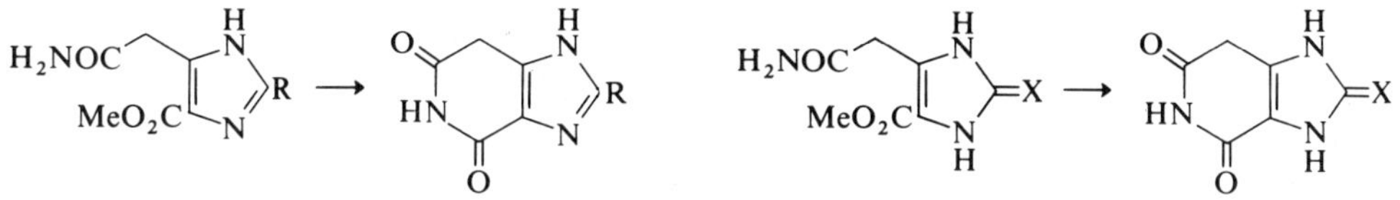

Treatment of 5-cyanoimidazole-4-acetonitrile (**439**; R = H) with anhydrous HBr gave 6-amino-4-bromoimidazo[4,5-*c*]pyridine (**440**; R = H, R^1 = Br) ⟨74JHC233⟩. Likewise, the 1-(tri-*O*-benzoyl-*β*-D-ribofuranose) derivative of (**439**) gave (**440**; R = tri-*O*-benzoyl-*β*-D-ribofuranose, R^1 = Br), whereas treatment with liquid ammonia and with hydrogen sulfide and triethylamine gave the corresponding 4-amino and 4-thio compounds (**440**; $R^1 = NH_2$, SH) ⟨78JOC289⟩.

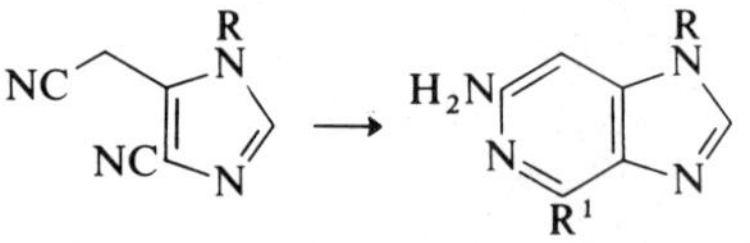

(439)   (440)

Similarly, reaction with liquid ammonia of 4-cyanomethylimidazole-5-carboxylic acid methyl ester and a number of 1- and 3-substituted derivatives thereof (**441**) at 100 °C for 8 days gave 6-aminoimidazo[4,5-c]pyridin-4(5H)-one (3-deazaguanine; **442**; R = H) and 1- and 3-substituted derivatives (**442**; R = tri-O-acyl-β-D-ribofuranose, di-O-acetyl-5-deoxy-β-D-ribofuranose, tetrahydropyranyl and tri-O-benzyl-β-D-arabinofuranose) ⟨76JA1492, 78JMC1212, 79JMC958⟩. Derivatives of 6-aminoimidazo[4,5-c]pyridin-4(5H)-one substituted on the amino group have been prepared by ring closure of the corresponding 3,4-diaminopyridines, but attempts to convert them into (**442**; R = H) itself failed ⟨74RTC3⟩. In a variation of the ring closure described above, the cyano group of (**441**; R = H) was allowed to react with hydroxylamine to give the carboxamide-oxime (**443**), which cyclized on reduction with hydrogen and Raney nickel via (**444**; R = H) to (**442**; R = H) ⟨79JHC1063⟩.

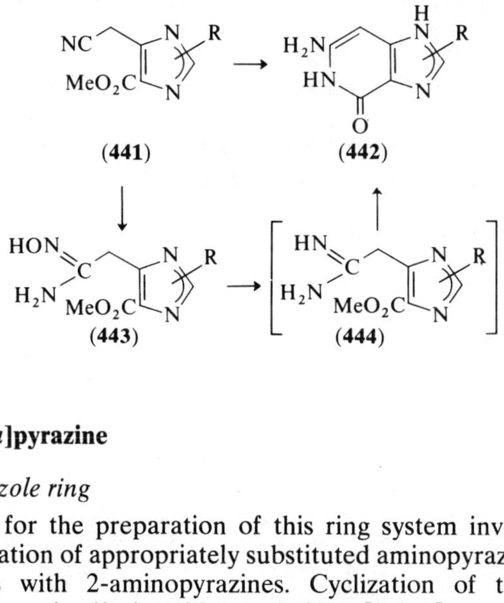

(441)   (442)

(443)   (444)

4.10.4.5 Imidazo[1,2-a]pyrazine

(i) Closure of the imidazole ring

The major methods for the preparation of this ring system involve formation of the imidazole ring by cyclization of appropriately substituted aminopyrazines or by the reaction of bifunctional agents with 2-aminopyrazines. Cyclization of the pyrazin-2-ylamino acetal (**445**) in concentrated sulfuric acid gave imidazo[1,2-a]pyrazine (**196**) in low yield ⟨65JCS2778⟩. Treatment of pyrazin-2-ylaminoethanols (**446**) with thionyl chloride gave tars and low yields of the 2,3-dihydroimidazo[1,2-a]pyrazines (**447**) ⟨57BSB136, 65JCS2778⟩. Cyclization of pyrazin-2-ylaminoacetic acids (**448**) by means of sodium alkoxides affords imidazo[1,2-a]pyrazin-3(2H)-ones (**449**). Basic conditions are superior to acid for this cyclization ⟨67CC1011⟩ (in one instance dicyclohexylcarbodiimide was employed) and esters and amides have also been used ⟨68TL3873, 70YZ423⟩. A number of imidazo[1,2-a]pyrazin-3(2H)-ones (**450**) including the luciferins have been prepared by reaction of 2-aminopyrazines (**451**) with α-keto acids, usually followed by reduction of (**452**) to (**453**),

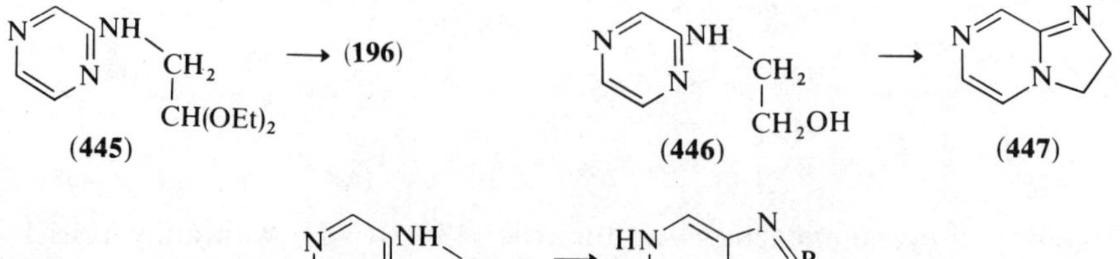

(445)   (446)   (447)

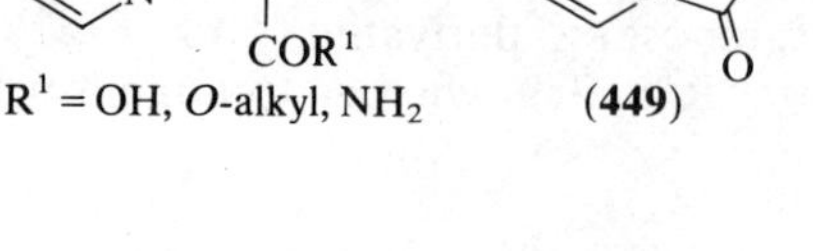

(448) R^1 = OH, O-alkyl, NH_2 (449)

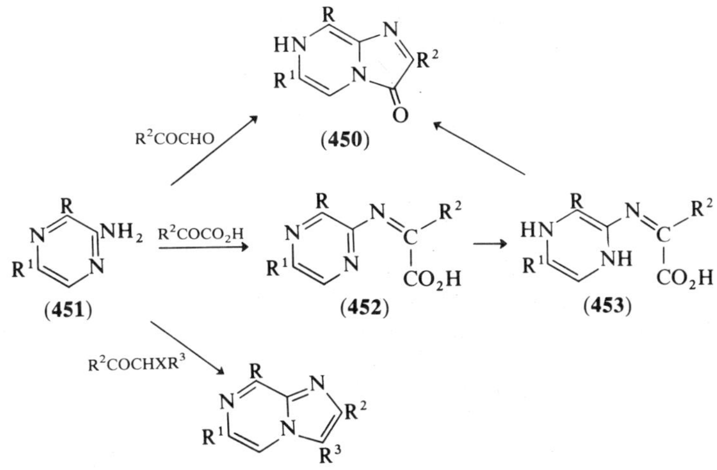

and treatment of (**453**) with dicyclohexylcarbodiimide or acetic anhydride ⟨66TL3445, 69YZ1657, 70YZ441, 70YZ707, 80CL299⟩.

Generally, better yields are obtained by the reaction of 2-aminopyrazines (**451**) with α-keto aldehydes to give (**450**) ⟨70YZ441, 70YZ707⟩ or α-halocarbonyl compounds to give (**454**), the latter being the most widely used reactants ⟨65JHC287, 71JHC643, 75JHC861, 76H(4)943, 77JOC4197, 80EUP13914⟩.

3-Aminoimidazo[1,2-a]pyrazine (**456**) has been prepared by the reaction of 2-aminopyrazine (**455**) with sodium cyanide and the bisulfite addition compound from formaldehyde ⟨68TL3873⟩. Another 3-aminoimidazo[1,2-a]pyrazine (**458**) was made by the reaction of the tetrahydropyrazine (**457**) with α-amino-α-cyanoacetamide ⟨67MI41000, 67MIP41000⟩.

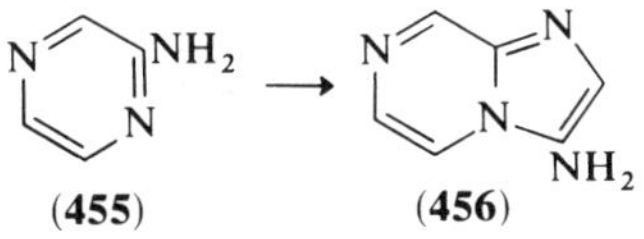

(ii) Closure of the pyrazine ring

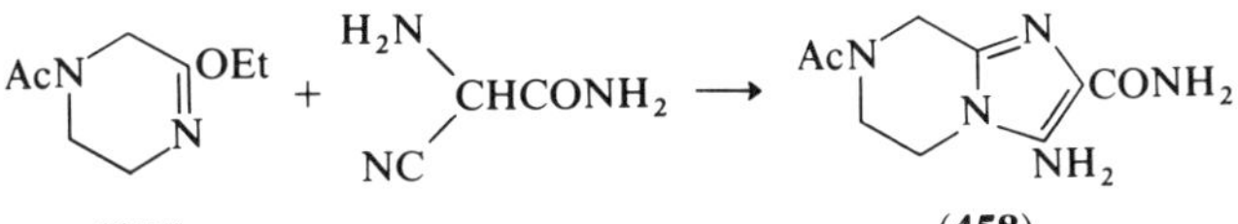

The sodium salt of 2-benzoylimidazole (**459**) can be alkylated with bromoacetal in DMF and the product (**460**) cyclized to 8-phenylimidazo[1,2-a]pyrazine (**461**) by refluxing in acetic acid containing ammonium acetate ⟨69MI41002⟩.

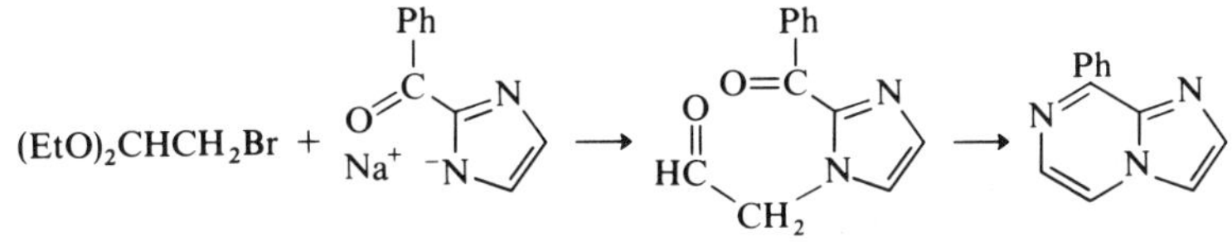

Various 1-alkylimidazo[1,2-a]pyrazine 7-oxide quaternary salts have also been synthesized from imidazoles ⟨71JCS(C)2748⟩. Two approaches were used. Quaternization of the oximes (**462**) with α-bromocarbonyl compounds gave salts (**463**) which cyclized in strong acid to the 7-oxides (**464**), and quaternization of the acetals (**465**) with bromoacetaldehyde oxime afforded salts (**466**) which were treated with concentrated sulfuric acid.

Studies on the tri-, tetra- and penta-peptides of α-methylalanine ⟨65T3209, 73ACS1509⟩ have shown that cyclization to the imidazo[1,2-a]pyrazine ring system occurs and a sequence for its formation from the tripeptide suggested (**467**) → (**468**) → (**469**). Another tripeptide cyclizes to a piperazine and then to the bicyclic system ⟨78TL1009⟩.

(462) (463) (464) (465) (466)

(467) (468) (469)

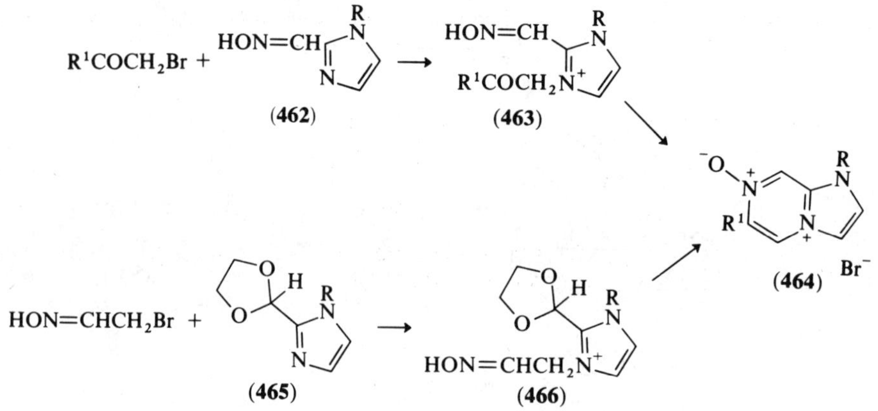

4.10.4.6 Imidazo[1,5-*a*]pyrazine

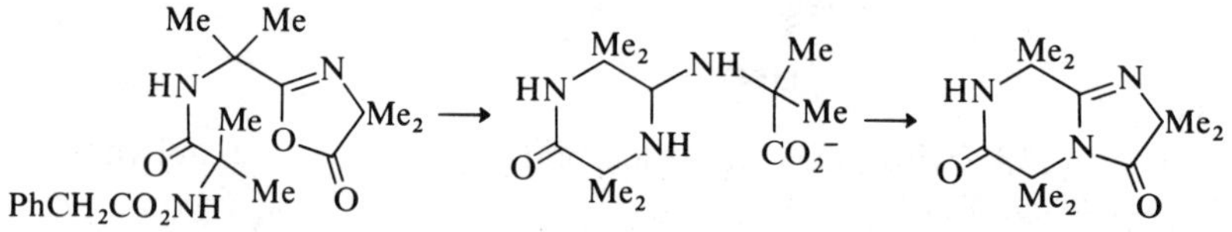

(i) Closure of the imidazole ring

The most useful synthesis of this ring system involves conversion of 2-aminomethylpyrazines (**470**) by means of carboxylic acids or anhydrides into the amides (**471**), which are cyclized to 3-substituted imidazo[1,5-*a*]pyrazines (**472**) ⟨73JOC2049, 75JHC207, 75JHC211, 75JOC3379⟩.

Reaction of (**470**; R = H, NH_2) with carbon disulfide gives 2,3-dihydro-3-thioxoimidazo[1,5-*a*]pyrazine (**473**) and its 8-amino derivative instead of the 3,4-dihydropteridines ⟨70JCS(C)1540, 75JOC3379⟩.

(473) (470) (471) (472)

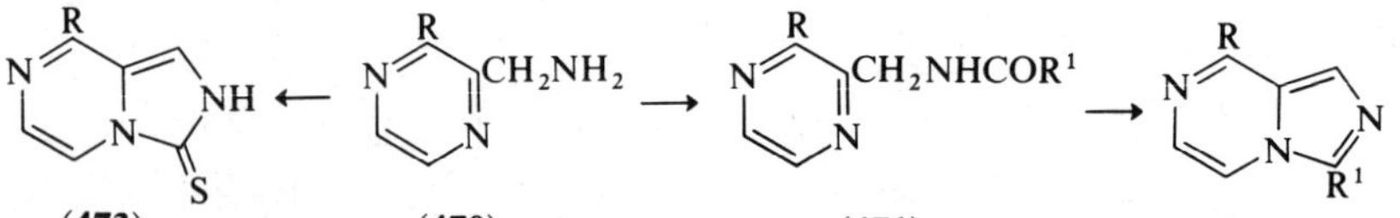

3-(Pyrazin-2-yl)imidazo[1,2-*a*]pyrazine (**475**) was obtained as a by-product (10%) in the ammonium chloride-catalyzed preparation of pyrazine carbaldehyde dimethyl acetal from the aldehyde (**474**) and trimethyl orthoformate ⟨71TL1441⟩.

(474) (475)

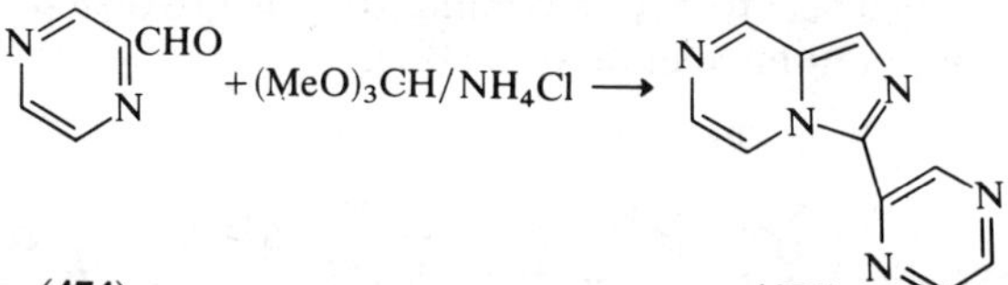

Reaction of 1,2-dihydropyrazines (**476**) and (**478**) with isocyanates gave the imidazo[1,5-*a*]pyrazines (**477**) and (**479**) ⟨79JOC4871⟩.

(476) (477) (478) (479)

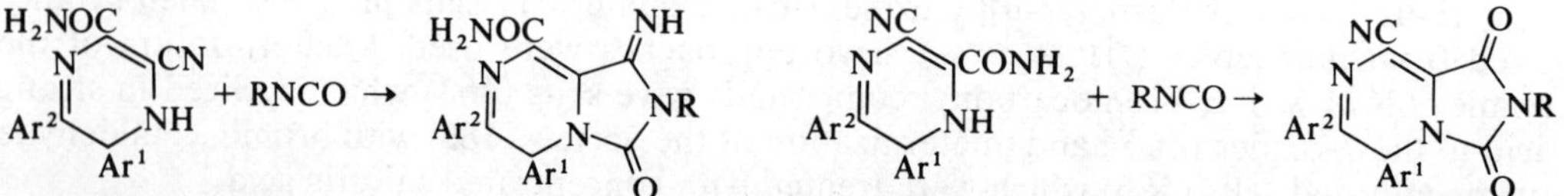

Hexahydroimidazo[1,5-*a*]pyrazines (**481**) have been prepared by closure of substituted aminomethylhexahydropyrazines (**480**) with carbonyldiimidazole ⟨75FES650⟩. Other closures have been performed using ethyl chlorocarbonate with sodium ethoxide and thiophosgene

⟨75IJC468, 78IJC(B)1015⟩. Treatment of the piperazine ester (**482**) with phenyl isocyanate followed by refluxing hydrochloric acid gave the dioxo derivative (**483**) ⟨62HCA2383⟩.

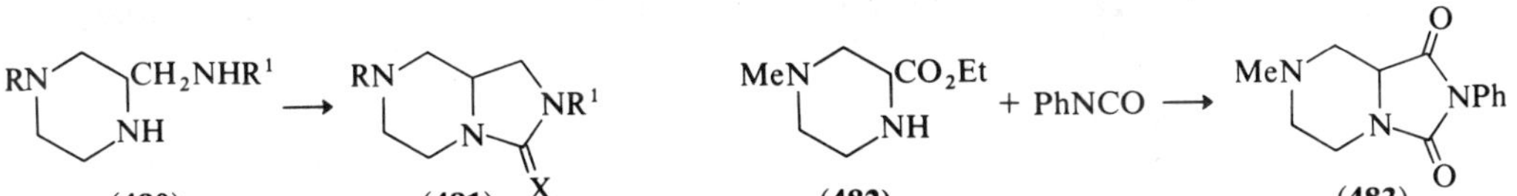

Other hexahydro derivatives (**485**) have been obtained by cyclization of anilinomethyl-piperazines (**484**) with aldehydes ⟨66JMC868, 70JMC77⟩.

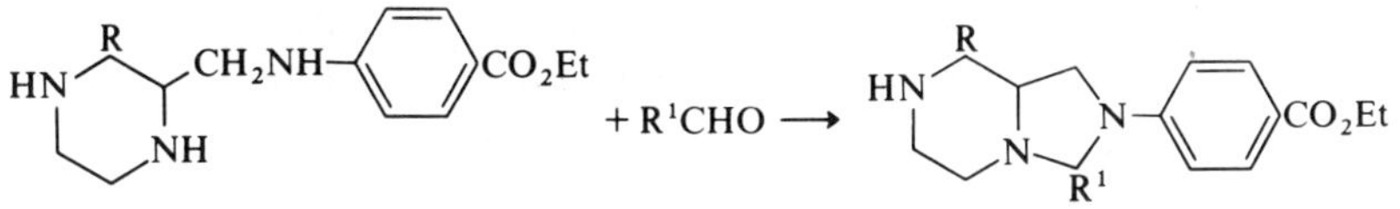

(ii) Closure of the pyrazine ring

Prolonged heating of theophyllin-7-acetic acid (**486**) with barium hydroxide ⟨55FES616⟩ or treatment with sodium hydroxide at room temperature ⟨61AC(R)1409⟩ resulted in fission of the pyrimidine ring and reclosure to the imidazo[1,5-*a*]pyrazine (**487**).

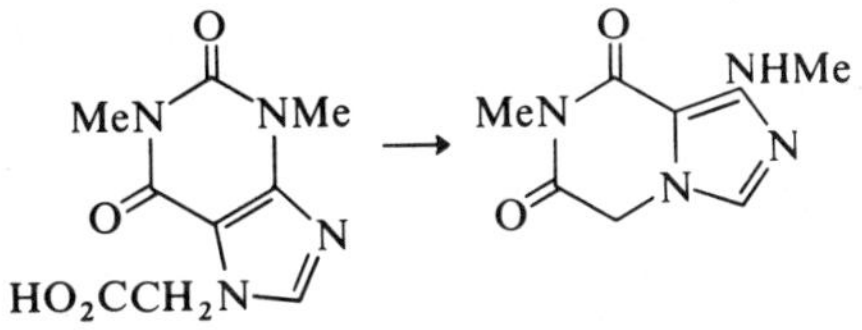

4.10.4.7 Imidazo[4,5-*b*]pyrazine

(i) Closure of the imidazole ring

In general, derivatives of this ring system have been synthesized by the formation of the imidazole ring from 2,3-diaminopyrazines ⟨52JA350, 56JA242⟩, but alternative approaches have been examined particularly for 2-oxo derivatives. The reaction of 2,3-diaminopyrazine or its 5,6-disubstituted derivatives (**488**) with acetic anhydride gave the corresponding 2-methylimidazo[4,5-*b*]pyrazines (**489**; R^1 = Me) ⟨56JA242, 57BSB136, 66FES811⟩. Acid chlorides and carboxylic acids also provide 2-substituted derivatives (**489**) ⟨52JA350, 56JA242, 75JHC1127⟩. Triethyl orthoformate has been used to prepare imidazo[4,5-*b*]pyrazines unsubstituted at C-2 ⟨57BSB136, 52JA350, 56JA242⟩. A pyrazine *N*-oxide (**490**) ⟨73JMC643⟩ and other pyrazines with substituted amino groups such as (**492**) have also been used to form imidazo[4,5-*b*]pyrazines (**491**) and (**493**) ⟨73JMC643, 71JAP7133957, 71JAP7133956⟩.

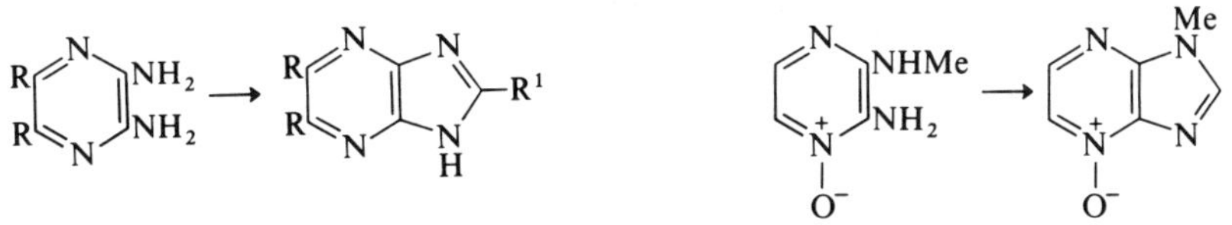

Treatment of 2,3-diaminopyrazine (**488**) with a wide variety of reagents has been shown to give 2-oxoimidazo[4,5-*b*]pyrazines (**496**): typical reagents that have been used are urea, phosgene, diethyl carbonate, potassium cyanate, carbamoyl chlorides and methyl chloroformate ⟨52JA350, 56JA242, 70USP3541093⟩. 2-Aminopyrazine-3-hydroxamic acids (**494**) on treatment with benzenesulfonyl chloride rearrange to give 2-oxo derivatives (**496**) ⟨68JOC2543,

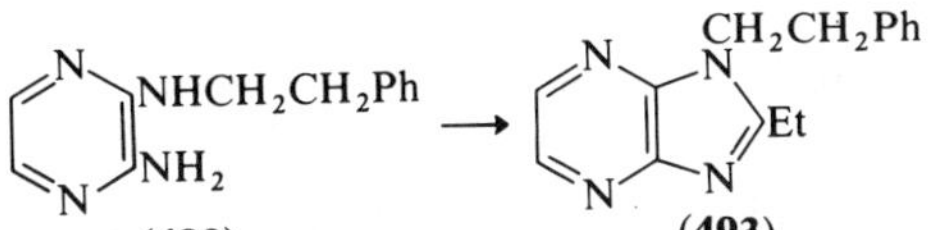

70USP3549633⟩. This rearrangement takes place under basic conditions and it is possible to isolate the benzenesulfonates (**495**). Other acid chlorides have also been used. Fusion of the hydroxamic acid (**494**; R = Cl, R^1 = EtNH) resulted in ring closure ⟨70USP3549633⟩.

(**494**) (**495**) (**496**) (**497**) (**498**)

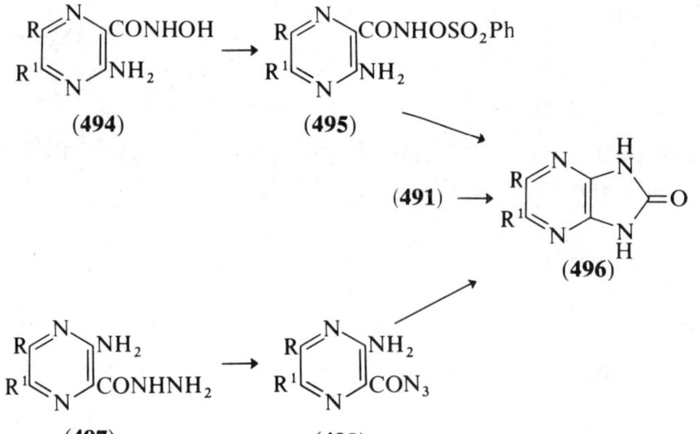

The Curtius rearrangement of acyl azides (**498**) prepared by diazotization of the hydrazides (**497**) also provides imidazo[4,5-*b*]pyrazine-2-ones (**496**) in good yield ⟨73JMC537⟩. Various modifications of this approach have been described ⟨69USP3461123, 69FRP1578366, 71BRP1248146, 71BRP1238105⟩. Thermal cyclization of structures like (**499**) and (**501**) has been utilized to prepare this type of structure (**500**) ⟨70GEP1957711, 71BRP1232758⟩.

(**499**) (**500**) (**501**)

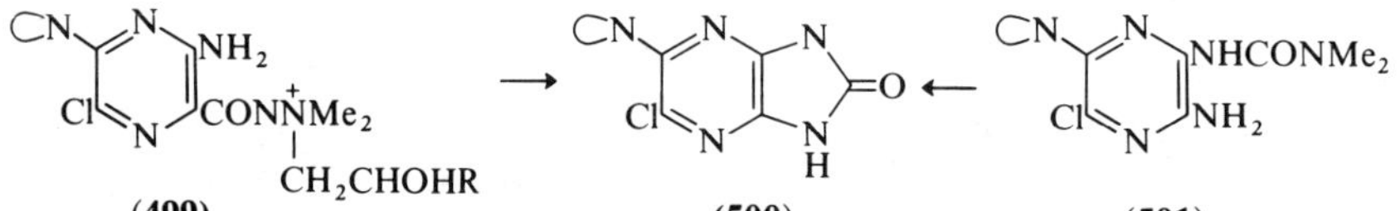

(ii) Closure of the pyrazine ring

One of the few examples of formation of the pyrazine ring as the last step in the synthesis of imidazo[4,5-*b*]pyrazines (**504**) involves the reaction of biacetyl with the diamine (**503**) generated *in situ* from the 4-nitro-5-amino compound (**502**) ⟨70TL1013⟩. Another synthesis involving the use of an imidazole is the condensation of ethylenediamine (**505**) with (**506**) to give the perhydro derivative (**507**).

(**502**) (**503**) (**504**)

(**505**) (**506**) (**507**)

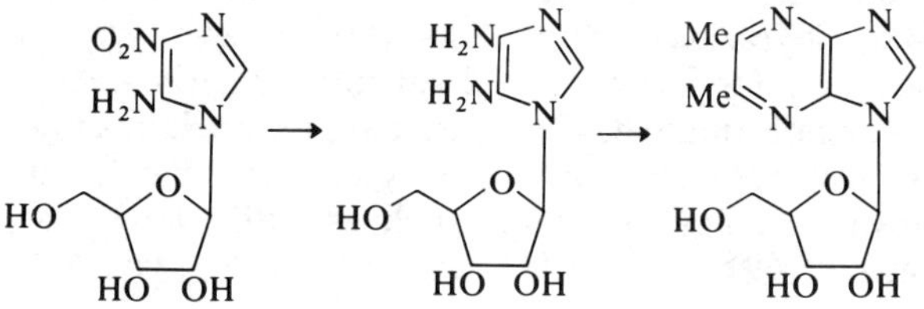

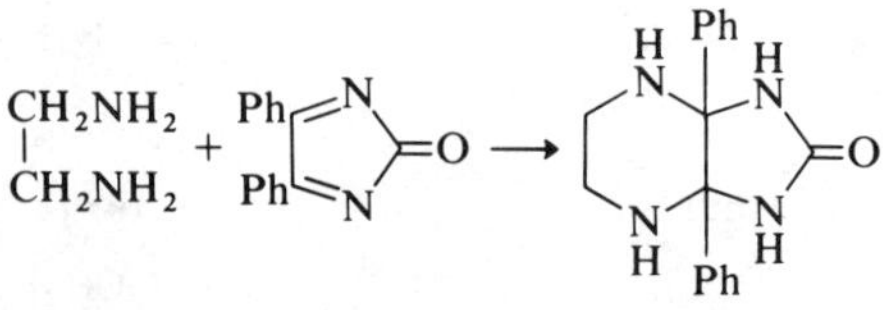

Another synthesis of the fully reduced ring system (**510**) results from the alkylation of the nitramine salt of ethylenediamine (**508**) with 4,5-dichloro-1,3-diacetylimidazolidine (**509**) ⟨79IZV1618⟩.

(**508**) (**509**) (**510**)

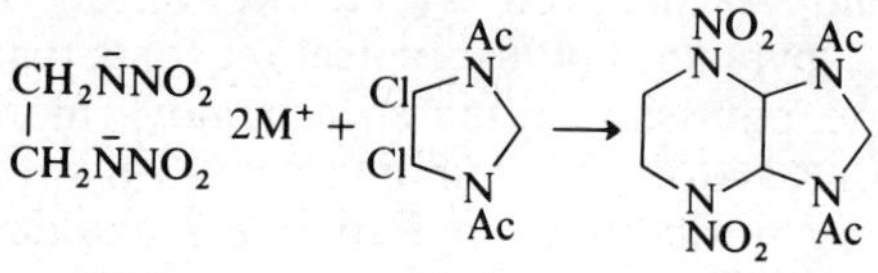

4.10.4.8 Imidazo[1,2-*a*]pyrimidine

Although it has not been synthesized, one natural product containing an imidazo[1,2-*a*]pyrimidine system, alcocherneine, has been characterized ⟨72T5207⟩.

(i) Closure of the imidazole ring

By far the most common synthetic method for the preparation of imidazo[1,2-*a*]pyrimidines involves the condensation of a 2-aminopyrimidine with an α-halocarbonyl compound ⟨39YZ97, 51YZ760, 56CR(243)2090, 59CR(248)1832, 60JA1469, 63LA(663)108, 63BSF1000, 66BSF2529, 66JMC29, 71JPS74, 74CHE992⟩. As an example, treatment of 2-amino-5-methoxypyrimidine (**511a**) with phenacyl bromide (**512**) produces only (**513**) ⟨66JMC29⟩. In all examples thus far the carbonyl carbon has become joined to the exocyclic amino group of the 2-aminopyrimidine. When the pyrimidine moiety is unsymmetrical, two isomers are possible, though apparently only one is seen in many cases. A condensation of a 2-aminopyrimidine with glyoxal and glyoxalic acid has also been carried out ⟨74GEP2513864⟩.

(**511**) **a**; R = OMe (**512**) (**513**)
b; R = H

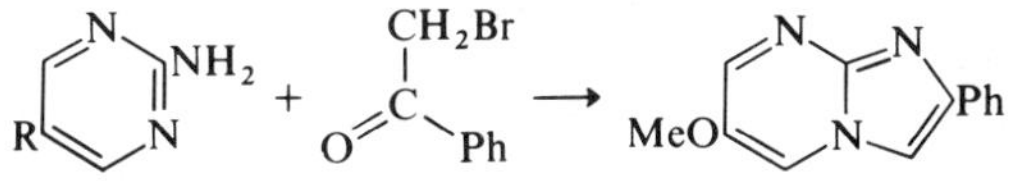

Several other synthetic routes starting with 2-aminopyrimidines are also known. The reaction of 2-aminopyrimidine (**511b**) with (β-acylvinyl)triphenylphosphonium salts such as (**514**) under kinetic conditions produces (**515a**), while under thermodynamic conditions the same reactants produce (**515b**). A Dimroth rearrangement of (**515a**) produces a mixture of (**515a**) and (**515b**) ⟨75LA1934⟩.

(**514**) (**515**) **a**; $R^1 = CH_2P^+Ph_3Br^-$, $R^2 = Me$
b; $R^1 = Me$, $R^2 = CH_2P^+Ph_3Br^-$

(**516**) (**517**)

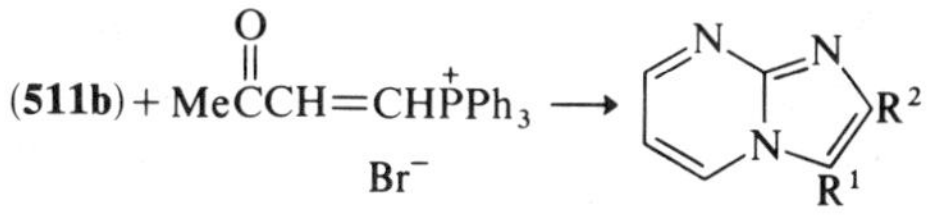

Treatment of (**511a**) with α-bromoglyoxal phenylhydrazone (**516**) in perchloric acid affords (**517**) ⟨76JOU929⟩.

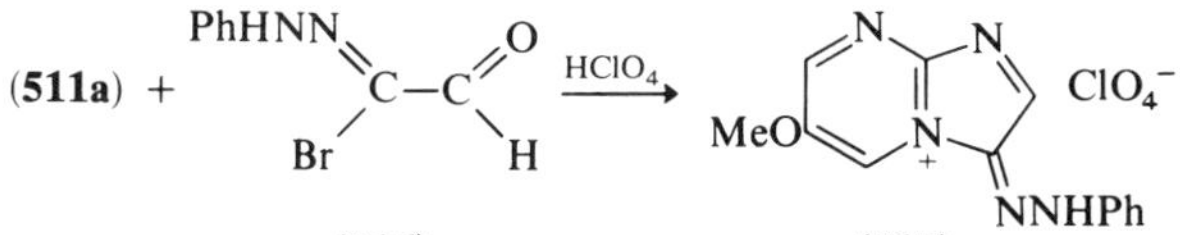

Displacement of the ethylthio group in (**518**) with 2-aminoethanol followed by treatment with thionyl chloride and then cyclization with base and heat afforded a mixture of (**519**) and (**520**) ⟨70TL1411⟩. Other similar sequences have been carried out at this oxidation level or by closing cyclic guanidine acids ⟨08CB176, 71JCS(C)679, 76CHE696, 56JA1618⟩.

(**518**) (**519**) (**520**)

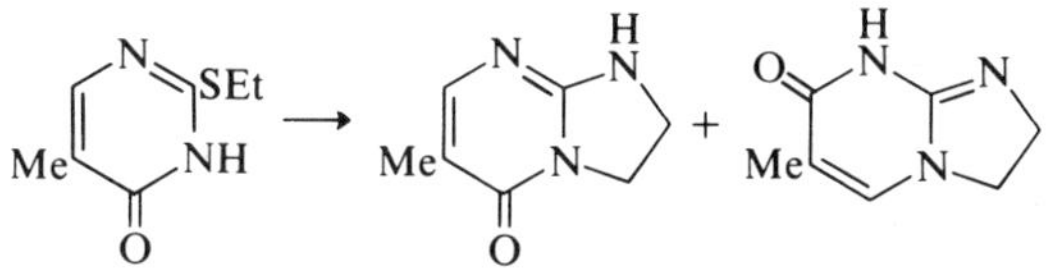

(ii) Closure of the pyrimidine ring

The other common synthetic method for imidazo[1,2-*a*]pyrimidines involves the condensation of a 2-aminoimidazole (or imidazoline) with a β-dicarbonyl compound ⟨50BSB573, 52DOK(87)783, 54JGU1045, 55JGU939, 72BSF3503, 73JHC1021, 73LA103, 76HCA1203, 76JMC512, 78GEP2806199, 80JMC1188, 80JHC337⟩. For example, treatment of 2-aminoimidazole (**521**) with diethyl methylmalonate (**522**) and ethoxide ion produces (**523**). Depending upon the β-dicarbonyl compound employed, a single isomer may be obtained, or a mixture of the two possible isomers in an unsymmetrical case. As might be expected, α,β-unsaturated esters and acid chlorides also react with 2-aminoimidazoline to give imidazo[1,2-*a*]pyrimidines ⟨73GEP2133998, 78AJC179⟩.

(521) + EtO$_2$CCHCO$_2$Et → (523)
       Me
       (522)

4.10.4.9 Imidazo[1,2-c]pyrimidine

Considerable recent interest has arisen in the synthesis of imidazo[1,2-c]pyrimidines due to the ready conversion of cytosine- and cytidine-related compounds into this ring system, and due to the attendant chemical, physical and biological properties of these derivatives ⟨84MI41000⟩.

(i) Closure of the five-membered ring

The most common and also the most versatile synthetic method for imidazo[1,2-c]pyrimidines is the condensation of a 4-aminopyrimidine with an α-halocarbonyl compound. Treatment of 1-methylcytosine (524a) with chloroacetaldehyde (525) produces (526a) ⟨71TL1993⟩. Cytidine (524b) reacts similarly to produce (526b) ⟨72BBR(36)597⟩. A substituted α-halocarbonyl compound such as phenacyl bromide (512) produces exlusively the isomer represented by (526c) ⟨78MI41010⟩. The mechanism of this type of transformation has been well studied ⟨78MI41011, 78MI41008, 77DOK(234)343, 75IZV2766⟩. Recently bromoacetaldehyde has been recommended in place of chloroacetaldehyde since it reacts faster and is apparently less mutagenic ⟨80CPB932⟩. Examples abound with other 4-aminopyrimidines and α-halocarbonyl compounds ⟨76MI41002, 76JCS(P1)1991, 71JHC643, 78JHC119, 80B3773, 64JHC34, 75CHE81⟩. It is also possible to use α-haloacetals directly ⟨75JOC3708⟩. Displacement of a 4-halogen with aminoacetaldehyde diethyl acetal results in an intermediate which can be cyclized with acid ⟨78JHC119, 65JCS2778⟩. It is important to note that not all 4-aminopyrimidines react with α-halocarbonyl compounds to produce imidazo[1,2-c]pyrimidines. With some compounds when C-5 is unsubstituted pyrrolo[2,3-d]pyrimidines are formed, and some of the earlier literature has been found to be in error in this regard ⟨76JCS(P1)1991⟩.

(524) a; R = Me   (525)   (526) a; R = Me, R^1 = H
     b; R = β-D-ribofuranosyl   b; R = β-D-ribofuranosyl, R^1 = H
     c; R = H   c; R = Me, R^1 = Ph

In a related synthesis, cytidine (524b) and some derivatives form imidazo[1,2-c]pyrimidines (528) when treated with β-acetylvinyltriphenylphosphonium bromide (527) ⟨72TL439, 73LA278, 75LA1934, 75M417⟩. Compound (528a) undergoes the Dimroth rearrangement to produce (528b) ⟨75LA1934⟩.

(524b) + MeCCH=CNPPh$_3$ Br$^-$ →

(527)   (528) a; R = β-D-ribofuranosyl, R^1 = Me, R^2 = CH$_2$P$^+$Ph$_3$ Br$^-$
                  b; R = β-D-ribofuranosyl, R^1 = CH$_2$P$^+$Ph$_3$ Br$^-$, R$_2$ = Me

Another common procedure involves treating 4-halopyrimidine with 2-chloroethylamine to produce a 4-(2-chloroethyl)amino derivative which can be cyclized readily. Treatment of 2,4-dichloro-5-nitropyrimidine (529) with 2-chloroethylamine affords (530). Reduction with Raney nickel followed by cyclization gave the imidazo[1,2-c]pyrimidine (531) ⟨52JCS4410⟩. Other similar cases are known ⟨55JCS896, 58JCS2821, 57T(1)75, 71JCS(C)675, 71JCS(C)679⟩ and the corresponding (2-hydroxyethyl)amino derivatives can be prepared and cyclized using thionyl chloride ⟨62JOC4080, 67T1297, 76CHE696⟩.

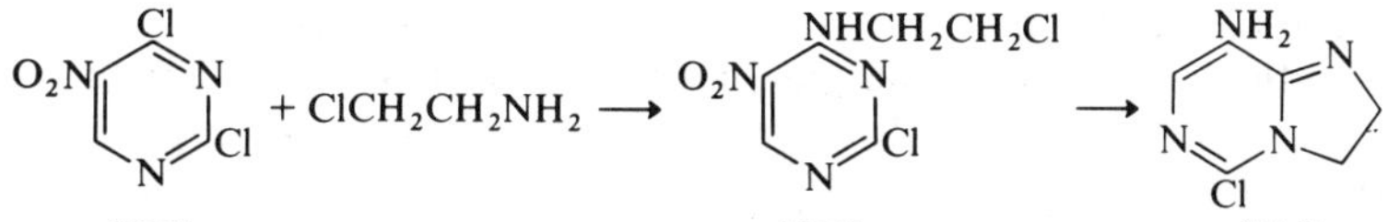

(529) (530) (531)

(ii) Closure of both rings

Condensation of the isothiocyanate (**532**) with ethylenediamine affords (**533**) ⟨76M171⟩.
Treatment of α-(5-bromouracil-1-yl)-α-alanine (**534**) with an amine such as piperidine or morpholine results in the formation of (**535**) by an addition–elimination mechanism ⟨80CHE418⟩.

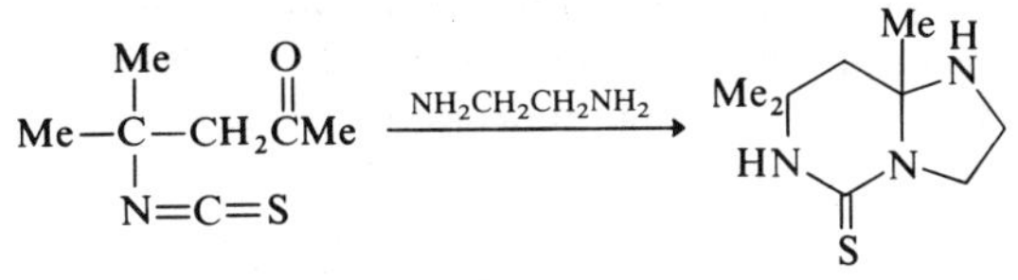

(532) (533)

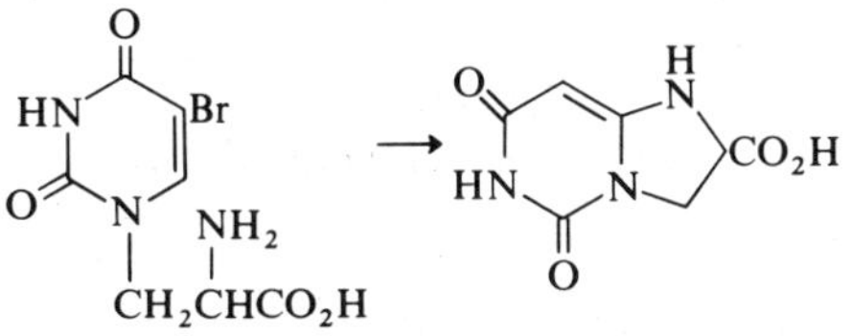

(534) (535)

4.10.4.10 Imidazo[1,5-*a*]pyrimidine

The imidazo[1,5-*a*]pyrimidine ring system has been formed in all examples thus far by annulation of a 4-aminoimidazole with a three-carbon unit. In the first synthesis, a condensation of 5-amino-4-methylimidazole (**536**) with 2,4-pentanedione (**537**) produced (**538**) ⟨39YZ185⟩. Several other similar condensations have also been carried out ⟨65M741, 71BSF1031, 74JHC873⟩. The 1-carboxyethylimidazole (**539**) has been cyclized to (**540**) by heating in polyphosphoric acid ⟨70JHC211⟩.

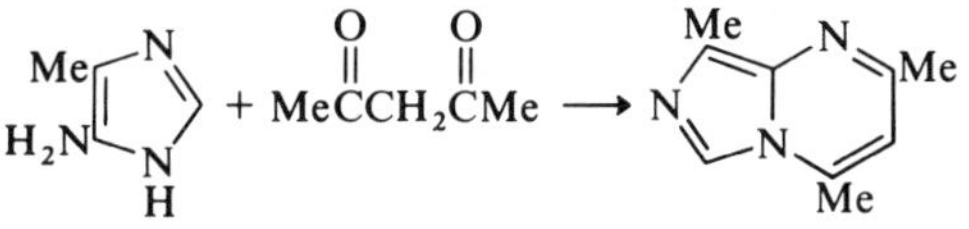

(536) (537) (538)

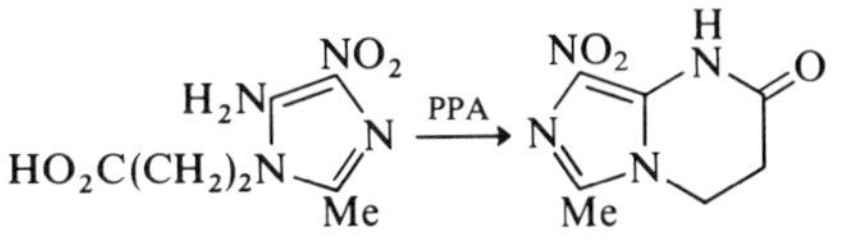

(539) (540)

4.10.4.11 Imidazo[1,5-*b*]pyrimidine

(i) Closure of the pyrimidine ring

Treatment of the methyl ester of L-histidine (**541a**) with phenylthiocarbonyl chloride (**542**) results in the formation of (**544**) ⟨56M679⟩. The same ring closure can be accomplished with carbonyldiimidazole (**543**) and histidine or histamine (**541b**) ⟨77BBA(491)253, 67T239⟩. Ring opening at the C-5 carbonyl group of (**544**) is readily achieved with amines ⟨56M679⟩. Conversion of histidine and histamine derivatives into (**544**) and various other substituted derivatives of (**544**) has been used to achieve specific methylation both on the side-chain nitrogen ⟨78MI41002⟩ and on a ring nitrogen ⟨78RTC293⟩. Compound (**544c**) can be readily polymerized to oligohistidines in an aqueous buffer at 30 °C ⟨77BBA(491)253⟩.

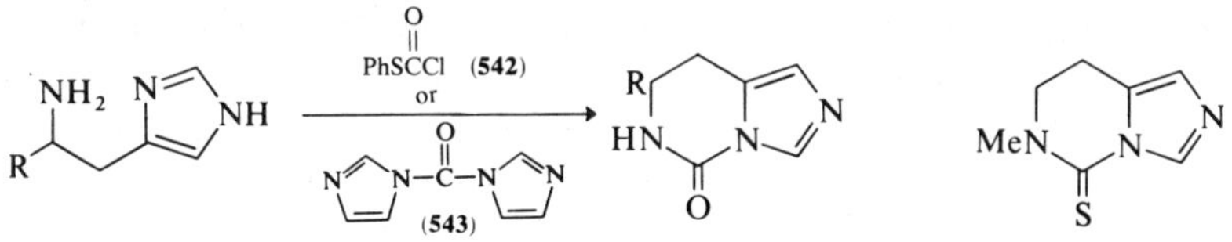

(**541**) **a**; R = CO_2Me
b; R = H

(**544**) **a**; R = CO_2Me
b; R = H
c; R = CO_2H

(**545**)

The sulfur-containing alkaloid zapotidine has been found to have the structure (545) ⟨61JA2022⟩. A synthesis of (545) has been achieved from (544b) by lithium aluminum hydride reduction to N^α-methylhistamine followed by ring closure to (545) with thiocarbonyldiimidazole ⟨67T239⟩.

4.10.4.12 Imidazo[1,2-b]pyridazine

(i) Closure of the imidazole ring

Direct syntheses of the bicyclic system have mainly employed the standard Tschitschibabin cyclization between a 3-aminopyridazine and an α-halocarbonyl compound, or a closely related sequence. For example, treatment of 3-amino-6-chloropyridazine (546) with α-bromoacetophenone (512) in boiling ethanol produces (547a) ⟨64CPB1351, 65JHC287⟩. Typical yields for this general type of cyclization are 30–60%. The parent ring system is best prepared by condensation of (546) with bromoacetaldehyde to produce (547b), from which the chlorine can be readily removed by catalytic hydrogenolysis ⟨67T387⟩. Substituents on the imidazole ring can be varied just by selecting the appropriate α-halocarbonyl compound. For example, condensation of (546) with ethyl 2-bromoacetoacetate produces (547c) in 37% yield ⟨67T2739⟩. Substituents on the pyridazine ring in imidazo[1,2-b]pyridazines are either present on the starting 3-aminopyridazine or derivable from groups present on the starting pyridazine.

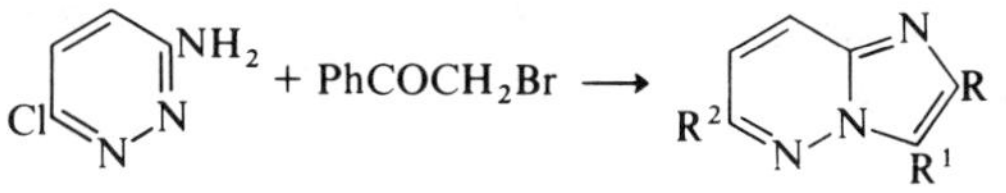

(547) a; R = Ph, R^1 = H, R^2 = Cl
b; R = H, R^1 = H, R^2 = Cl
c; R = Me, R^1 = CO$_2$Et, R^2 = Cl
d; R = Ph, R^1 = R^2 = H
e; R = R^1 = H, R^2 = N$_3$

The mechanism of the ring formation appears to involve initial alkylation at N-2 followed by cyclization to the exocyclic amino group. The evidence for this pathway is the isolation of the salt (548) in the reaction of 3-aminopyridazine with (512) under mild conditions. Heating of (548) in aqueous solution results in the formation of (547d) ⟨64CPB1351⟩. All compounds thus far isolated have substitution patterns consistent with this two-step mechanism in which the intermediate is generally not isolated.

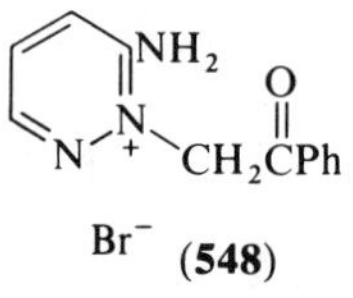

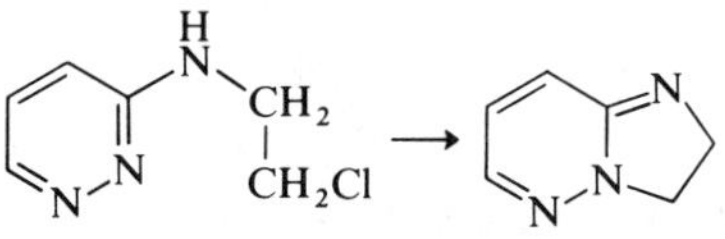

Formation of 2,3-dihydro compounds is possible either by generating a 3-(2-chloroethylamino)pyridazine such as (549) and cyclizing it to (550) ⟨65JCS2778⟩ or by treating 3-aminopyridazine directly with dibromoethane ⟨69MI41001⟩. The fully aromatized material is then available by oxidation.

(ii) From a different ring system

An indirect preparation of the imidazo[1,2-b]pyridazine ring system has also been reported by valence isomerization of a tetrazolo[1,2-b]pyridazine ⟨70JOC1138⟩. Treatment of 6-(2,2-dimethoxyethylamino)tetrazolo[1,5-b]pyridazine (551) with polyphosphoric acid produces a 45% yield of 6-azidoimidazo[1,2-b]pyridazine (547e).

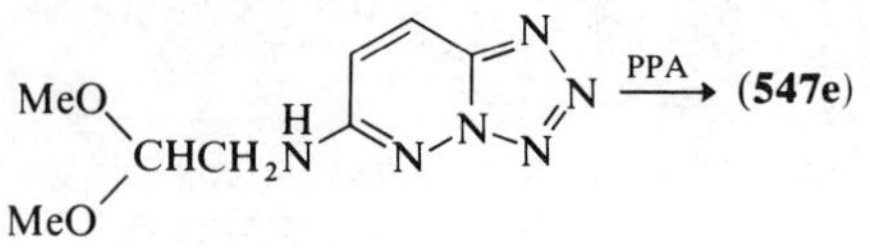

4.10.4.13 Imidazo[1,5-b]pyridazine

Two reports exist on the synthesis of this ring system. A double acid-catalyzed cyclization of the semicarbazone (**552**) produced 2-phenyl-4-acetylimidazo[1,5-b]pyridazine-5,7(3H,6H)-dione (**553a**) in 70% yield ⟨79JHC1105⟩. Several other examples are reported. Methylation of (**553a**) with diazomethane afforded the 6-methyl compound (**553b**). The other synthesis begins with the preparation of pyridazine (**554**) in several steps. Cyclization of (**554**) with polyphosphoric acid followed by treatment with POCl₃ produced the imidazo[1,5-b]pyridazine (**555**) ⟨78GEP2804909⟩.

4.10.4.14 Imidazo[4,5-c]pyridazine

Unless the bonds are fixed by attachment of a group at N-5 or N-7, this ring system has a mobile hydrogen which can reside at either site. For clarity, derivatives with this capability are referred to only as (7H)-imidazo[4,5-c]pyridazines.

Syntheses of this ring system have been carried out exclusively by forming the imidazole ring from a 3,4-diaminopyridazine. Substituents are incorporated by means of groups attached to the pyridazine or by varying the precursor to C-6. Ring closure has been accomplished with either triethyl orthoformate or formic acid to produce 6-unsubstituted compounds. For example, when (**556a**) was treated with formic acid, (**557a**) was obtained in 68% yield. Treatment of (**556a**) with cyanogen bromide produced the 6-amino compound (**557b**), while treatment with carbon disulfide afforded the corresponding 6-thione ⟨67JHC555⟩. Urea has also been used as a condensing agent ⟨81AJC1361⟩. The adenine analog (**557c**) has been prepared by condensing (**556b**) with triethyl orthoformate to give (**557d**), which was converted into (**557c**) by ethanolic ammonia in an autoclave at high temperature ⟨64JHC42⟩. The parent heterocycle has been obtained by several routes, including the catalytic dehalogenation of (**557d**) ⟨67JHC555⟩.

(**556**) a; R¹ = R² = H
b; R¹ = H, R² = Cl

(**557**) a; R¹ = R² = R³ = H
b; R¹ = R² = H, R³ = NH₂
c; R¹ = H, R² = NH₂, R³ = H
d; R¹ = R³ = H, R² = Cl

Derivatives substituted at N-5 or N-7 with methyl groups are available simply by ring closure of the appropriate methylamino-substituted pyridazine ⟨67JHC555⟩.

4.10.4.15 Imidazo[4,5-d]pyridazine

The imidazo[4,5-d]pyridazine ring system has a mobile hydrogen which could give rise to two isomers in an unsymmetrically substituted molecule.

Chemical synthesis of this ring system has been achieved by two basic methods. First, imidazoles with groups at the 4- and 5-positions capable of undergoing condensation to form the pyridazine ring have been employed. Second, 4,5-diaminopyridazines can be condensed with an appropriate one-carbon synthon to produce imidazo[4,5-d]pyridazines.

(i) Closure of the pyridazine ring

As an example of the first method, treatment of dimethyl imidazole-4,5-dicarboxylate (**558**) with hydrazine at elevated temperatures results in direct formation of (**559**) (which can be written in several tautomeric forms). Carrying out the reaction at lower temperatures results in formation of the dihydrazide, which can be further converted into (**559**) upon heating ⟨56JA159⟩. This process has also been accomplished with a nucleoside. Treatment of (**560a**) with 85% hydrazine hydrate in boiling ethanol produced only (**560b**), which could be converted into (**561**) by boiling in 97% hydrazine ⟨76MI41000⟩. Other examples similar to the two described are known ⟨78JHC1, 73JPS1011, 69JMC43, 68JHC13, 65JHC247, 64JHC182, 62JOC2500, 58JA6083, 58JOC1534⟩. Recently 4,5-dicyanoimidazole has been converted with hydrazine directly to the 4,7-diamino compound (**255g**) ⟨78JHC1451⟩.

(**558**)   (**559**)   (**560**) a; R = COMe, R^1 = OMe   (**561**)
              b; R = H, R^1 = NHNH$_2$

(ii) Closure of the imidazole ring

Many different 4,5-diaminopyridazines have been condensed with one-carbon fragments to form imidazo[4,5-*d*]pyridazines. Treatment of (**562**) with formic acid at reflux produces (**255h**) in 78% yield ⟨70CPB1685⟩. Other condensing agents have included triethyl orthoformate ⟨81JOC2467, 60CPB999, 63AF878, 70CPB1685⟩, triethyl orthoacetate and acetic anhydride ⟨60CPB999⟩, formamide ⟨70CPB1685⟩ and carbon disulfide ⟨60CPB999, 63AF878, 70CPB1685⟩.

(**562**)

4.10.4.6 Imidazo[4,5-*d*][1,2,3]triazine

The impetus for research on this ring system comes from its close resemblance to the purine ring. Various derivatives have been prepared as 2-aza analogs of purines and purine nucleosides and evaluated for biological activity ⟨82MI41000, 75ANY(255)292⟩.

Syntheses of imidazo[4,5-*d*][1,2,3]triazines have been carried out exclusively by building the triazine ring onto an existing, properly substituted imidazole ring. Treatment of 5-amino-4-carboxamido-1-(β-D-ribofuranosyl)imidazole (**563a**) with sodium nitrite in aqueous hydrochloric acid, resulting in diazotization of the 5-amino group and then cyclization to the 4-amide, produced (**257b**; 2-azainosine) ⟨72JHC623, 72JMC841⟩. Other imidazo[4,5-*d*]-[1,2,3]triazin-4-ones have been prepared in similar fashion ⟨74JOC3651, 73JHC417, 51JBC(189)401, 60JA3189, 61JOC2396, 61JCS4845⟩. Similar diazotization–cyclization on 5-amino-4-carboxamidinoimidazoles such as (**563b**) produced (**258d**) ⟨69CC458, 72JMC182⟩. The 4-thiones, such as (**257a**), are prepared from the corresponding thiocarboxamide (**563c**) apparently by initial formation of imidazothiadiazines, which readily rearrange in ammonium hydroxide ⟨75CHE995, 73CHE1173⟩. Exposure of 4-(di-*n*-butyltriazeno)-5-carboxamido-imidazole (**564**) to light results in the formation of imidazo[4,5-*d*][1,2,3]triazin-4-one (2-azahypoxanthine; **257c**) ⟨62JOC2150⟩. The starting imidazoles can be prepared by direct synthesis or by one of several methods for the ring opening of a purine.

(**563**) a; X = O, R = β-D-ribofuranosyl   (**564**)
         b; X = NH, R = β-D-ribofuranosyl
         c; X = S, R = H

4.10.4.17 Imidazo[1,2-*b*][1,2,4]triazine

(i) Closure of the imidazole ring

The most versatile method for the preparation of this ring system is the condensation of an α-halocarbonyl compound with a 3-amino-1,2,4-triazine. For example, treatment of 3-amino-5,6-dimethyl-1,2,4-triazine (**565**) with phenacyl bromide (**512**) produced only (**566**) ⟨65JHC287⟩. In addition to α-haloketones, α-haloaldehydes ⟨72JHC1157, 55MI41000⟩, α-haloacetals ⟨80MI41000⟩ and α-haloesters ⟨79CHE1255⟩ have been used. The direction of condensation is always as shown for the preparation of (**566**).

(**565**) (**566**)

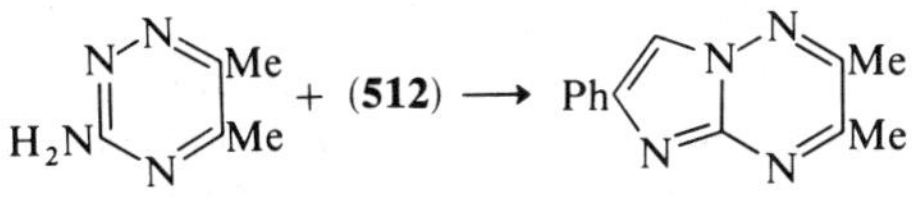

The 6,7-dihydro compounds can be prepared by a standard sequence such as the treatment of (**567**) with 2-aminoethanol to produce (**568a**), which is converted into the chloro compound (**568b**) with $SOCl_2$ and cyclized with potassium *t*-butoxide to (**569**). Full aromatization to (**570**) was carried out by heating with 10% palladium–charcoal ⟨76JHC807⟩.

(**567**) (**568**) **a**; X = OH **b**; X = Cl (**569**) (**570**)

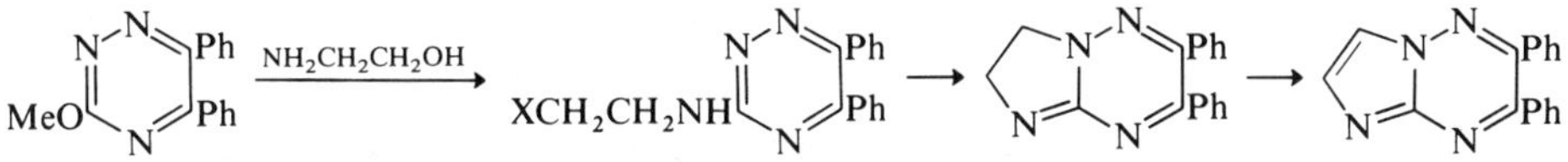

(ii) Closure of the triazine ring

Another useful synthetic method is the condensation of 1,2-diaminoimidazoles with 1,2-dicarbonyl compounds. Treatment of 1-acetylamino-2-amino-4-phenylimidazole (**571**) with biacetyl (**572a**) under acid catalysis produced (**573a**) ⟨70CB3533⟩. The use of phenylglyoxal (**572b**) hydrate gave both (**573b**) and (**573c**), with (**573c**) predominating. The corresponding aldoxime only resulted in (**573b**). Cyclization is also possible with α-ketoacids ⟨74JHC327⟩.

(**571**) (**572**) **a**; $R = R^1 = Me$ **b**; $R = Ph, R^1 = H$ (**573**) **a**; $R = R^1 = Me$ **b**; $R = Ph, R^1 = H$ **c**; $R = H, R^1 = Ph$

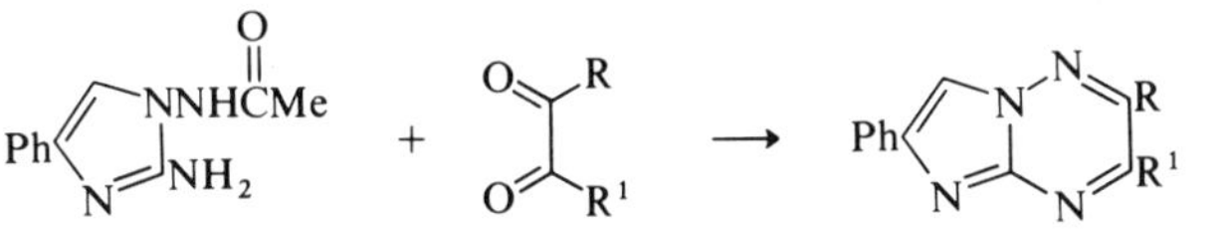

Other, less general syntheses have also been reported. Reaction of phenacyl bromide (**512**) with aminoguanidine produced dihydro compounds (**574**), which are readily oxidizable to the fully aromatic compound (**262**) ⟨68TL789⟩. Cyclization of (**575**) with acid produced the 3,6-dione (**576**) ⟨69MI41000⟩. When 3-amino-1,2,4-triazine was heated in concentrated hydrochloric acid, a small amount of the parent compound imidazo[1,2-*b*][1,2,4]triazine was isolated, presumably by degradation of the triazine to a two-carbon fragment which condensed with the remaining triazine ⟨76RTC74⟩.

(**574**) **a**; R = H **b**; R = Br

(**575**) (**576**)

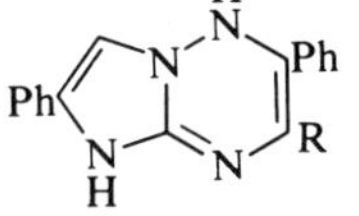

4.10.4.18 Imidazo[1,5-*b*][1,2,4]triazine

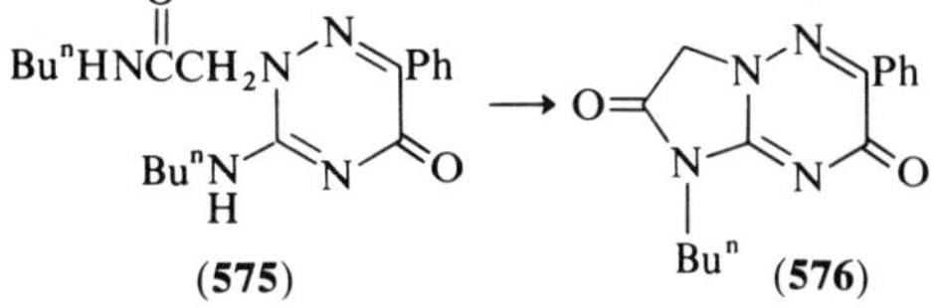

One synthesis of this ring system has been reported, the condensation of 1,5-diaminoimidazoles with glyoxal or biacetyl. For example, treatment of (**577**) with biacetyl produced (**578a**) ⟨74BSF1453⟩.

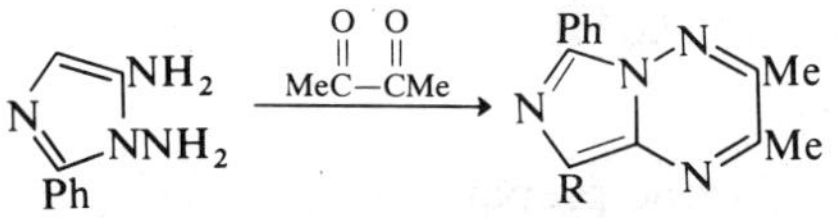

(577) + MeC(O)-C(O)Me → (578) a; R = H; b; R = Br

4.10.4.19 Imidazo[1,5-d][1,2,4]triazine and Imidazo[1,2-d][1,2,4]triazine

These new ring systems are both prepared by annulation of the triazine ring onto an imidazole precursor. Treatment of imidazole-4-carbaldehyde (**579**) with either methyl dithiocarbazinate (**580a**) or ethyl carbazate (**580b**) produced (**581a**) or (**581b**), respectively. Heating of either (**581**) derivative produced the corresponding bicyclic products (**582**) ⟨79JHC277⟩.

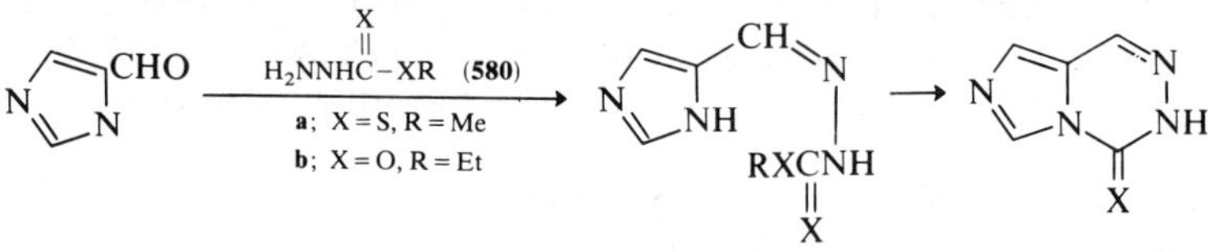

(579) + H₂NNHC(X)−XR (580) a; X = S, R = Me; b; X = O, R = Et → (581) a; X = S, R = Me; b; X = O, R = Et → (582) a; X = S; b; X = O

Similar treatment of imidazole-2-carbaldehyde (**583**) produced the imidazotriazines (**584**) ⟨79JHC277⟩.

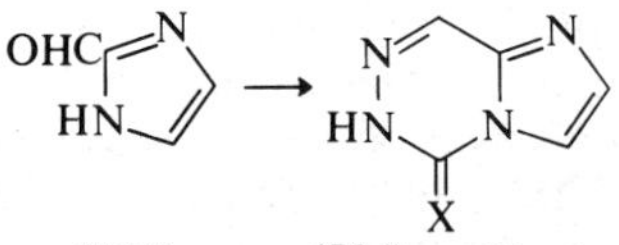

(583) → (584) a; X = S; b; X = O

4.10.4.20 Imidazo[2,1-c][1,2,4]triazine

The more general syntheses of this ring system have all involved annulation of the triazine ring onto a preformed imidazole ring. A series of variations within this theme has been developed.

(i) Closure of the triazine ring

Treatment of 2-methoxyimidazoline (**585a**) with hydrazine produced the cyclic aminoguanidine (**586**), which reacted with α-ketoesters (**587**) to give imidazo[2,1-c]-[1,2,4]triazin-4(8H)-ones such as (**588**) ⟨72LA(764)112⟩

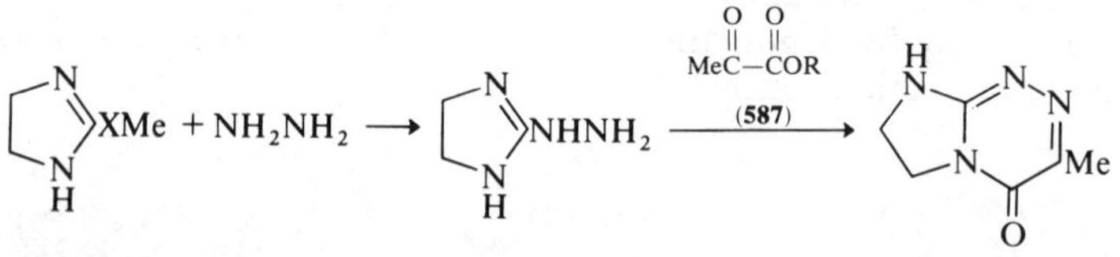

(585) a; X = O; b; X = S + NH₂NH₂ → (586) —MeC(O)−COR (587)→ (588)

Alkylation of (**585b**) with a phenacyl bromide produced (**589**), which can be cyclized with hydrazine to (**590**) ⟨77JHC59⟩. Similar cyclizations have been carried out on a 1-phenacyl-2-bromoimidazole derivative ⟨74UKZ99⟩, a 1-phenacyl-2-methanesulfonyl-imidazole derivative ⟨74CHE1492⟩ and 1-(2-chloroethyl)-2-bromo-4,5-diphenylimidazole ⟨75CHE371⟩ to afford imidazo[2,1-c][1,2,4]triazines.

Treatment of (**586**) with DMAD (**591**) in methanol in the presence of triethylamine produced (**593**), presumably *via* (**592**), which can be isolated under slightly different conditions ⟨74JCS(P1)297⟩.

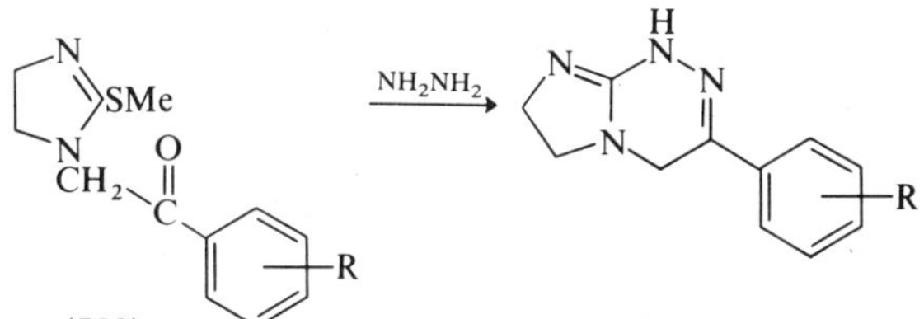

Other Imidazoles with Fused Six-membered Rings 655

(589) → (590)

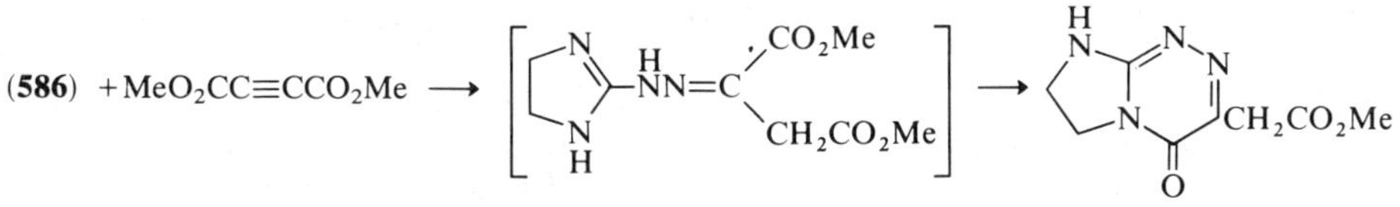

(586) + MeO₂CC≡CCO₂Me → [(591)] → (592) → (593)

The diazonium salt of 2-aminoimidazole (**594**) when treated with ethyl cyanoacetate produced the azo compound (**595**). Cyclization of (**595**) to (**596**) occurred on heating in acetic acid ⟨76JMC517⟩.

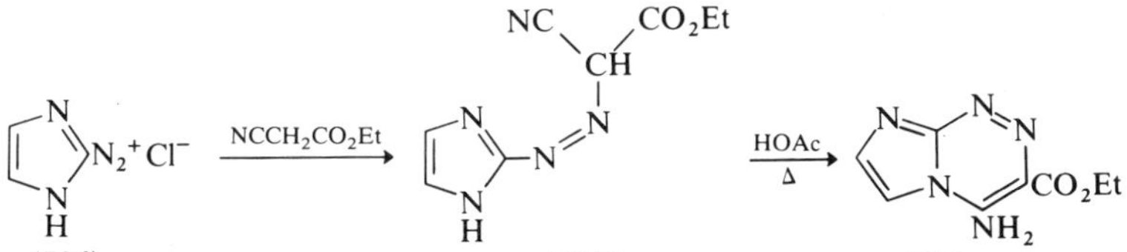

(594) → (595) → (596)

(ii) Closure of the imidazole ring

The synthesis of (**588**) has also been reported by the cyclization of (**597**) ⟨69MI41000⟩. This is the only reported synthesis of an imidazo[2,1-c][1,2,4]triazine starting from a triazine derivative.

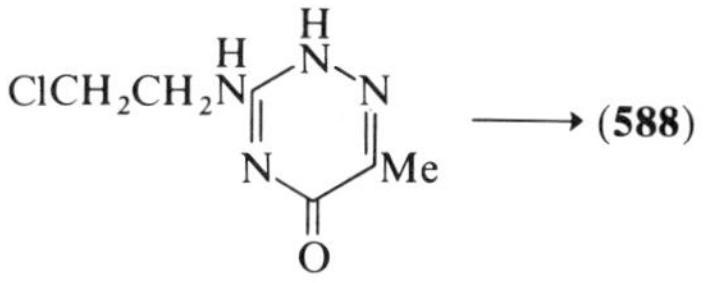

(597) → (588)

(iii) From other ring systems

Several other syntheses of this ring system were accomplished by annulation of the triazine ring onto a five-membered ring, either oxazole ⟨69C303⟩ or thiazole ⟨71JCS(C)1615⟩, with concurrent ring opening and reclosure to form the imidazo[2,1-c][1,2,4]triazine system. For example, treatment of 2-amino-3-phenacyl-5-phenyloxazolium bromide (**598**) with phenylhydrazine produced (**599**).

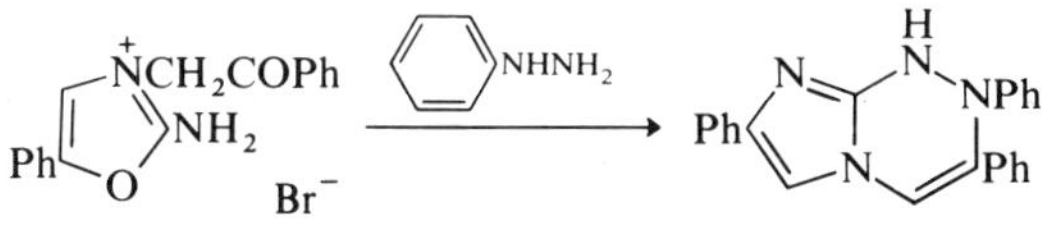

(598) → (599)

4.10.4.21 Imidazo[4,5-e][1,2,4]triazine

Two basic synthetic approaches to this ring system have been developed recently. In one, the imidazole ring is appended onto a 5,6-diaminotriazine precursor. For example, treatment of 2-benzyl-5,6-diamino-1,2,4-triazin-3(2H)-one (**600**) with triethyl orthoformate in nitrobenzene at reflux produced (**601a**) in 72% yield. Similarly, benzaldehyde afforded (**601b**). The tautomeric structure of the imidazole ring is unknown ⟨76CPB2274⟩.

A second synthetic route involves the ring contraction of certain 7-azapteridinones. Treatment of 6-benzylidenehydrazino-3-methyluracils, for example (**602a**), with sodium

(600) (601) a; R = H; b; R = Ph

nitrite in acetic acid produced the 5-nitroso compounds (**602b**), which underwent dehydrative cyclization to the azapteridine (**603**) with acetic anhydride. Treatment of (**603**) with alcoholic sodium hydroxide followed by acidification resulted in the formation of 3-substituted 5-methyl-5*H*-imidazo[4,5-*e*][1,2,4]triazin-6(7*H*)-ones such as (**604**) ⟨78CPB3154⟩. 7-Azapteridine 5-oxides undergo a similar conversion with either acetic anhydride or alcoholic sodium hydroxide ⟨76CC658, 76H(4)1503⟩. Both reactions may occur by a benzilic acid-type rearrangement with initial attack of hydroxide, for example at the C-5 carbonyl group of (**603**).

(602) a; X = H; b; X = NO (603) (604)

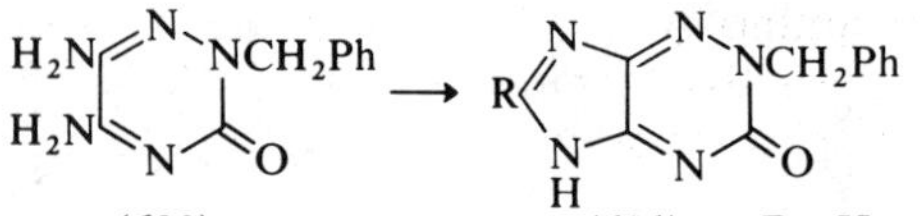

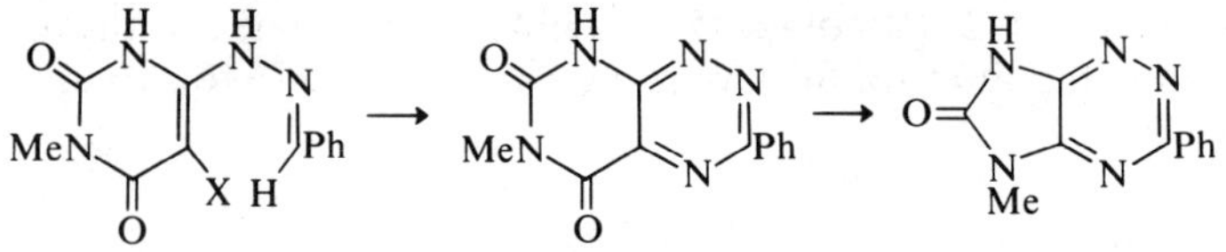

4.10.4.22 Imidazo[5,1-*c*][1,2,4]triazine

This ring system has been prepared by the annulation of the triazine ring onto a substituted imidazole. In the first reported synthesis, treatment of 4-hydrazino-2,5-di-*t*-butyl-2-methyl-(2*H*)-imidazole (**605**) with 3-bromo-4-methyl-2-pentanone (**606**) in the presence of base produced a 44% yield of (**607**) ⟨74M38⟩. In a conceptually similar sequence, alkylation of 2-methyl-4-nitro-5-bromoimidazole (**608a**) with ethyl bromoacetate gave (**608b**), which cyclized upon heating with 85% hydrazine hydrate in boiling DMF to (**609a**) in 78% yield ⟨79CHE1237⟩. The 1-(2-bromoethyl) compound (**608c**) also cyclized upon heating with hydrazine hydrate to form (**609b**). Formation of imidazo[5,1-*c*][1,2,4]triazines has also been accomplished from (**608a**) through initial alkylation with α-bromoketones ⟨75CHE751, 81CHE622⟩.

(605) (606) (607)

(608) a; R = H; b; R = CH_2CO_2Et; c; R = CH_2CH_2Br (609) a; R = O; b; R = H_2

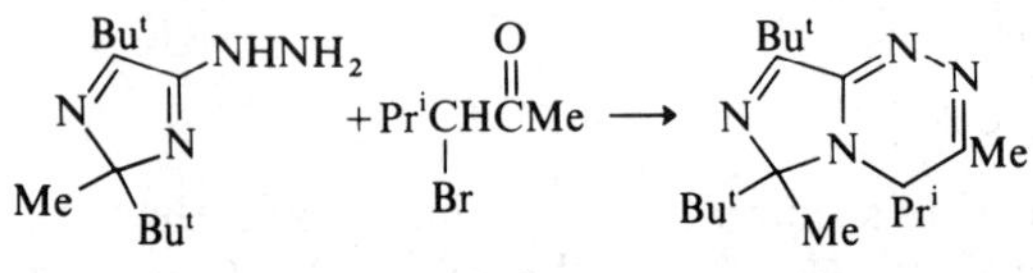

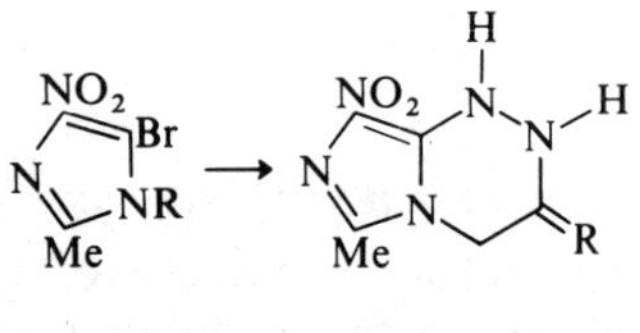

In a different approach, treatment of the diazonium salt (**610**) with 4,6-nonanedione produced azo compound (**611**), which could be cyclized to (**612**) upon heating in acetic acid ⟨76JMC517⟩. Other active methylene compounds have also been employed ⟨81JCS(P1)1424⟩.

(610) (611) (612)

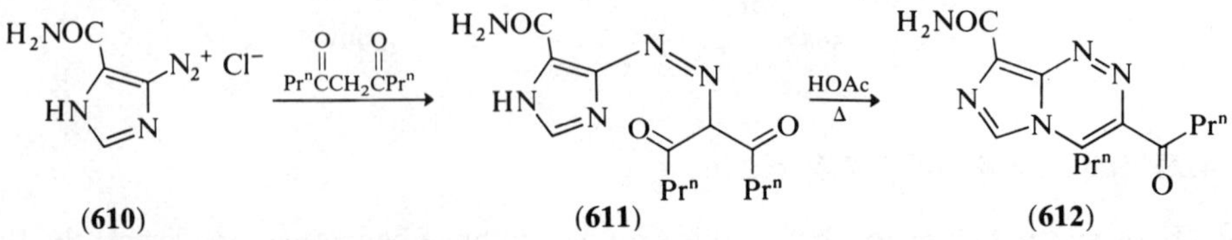

4.10.4.23 Imidazo[5,1-*f*][1,2,4]triazine

(i) Closure of both rings

Treatment of azodicarboxamidine (**613**) with 1,3-dicarbonyl compounds such as (**614**) resulted in the formation of (**615**) directly ⟨73T1413, 73AP561, 73AP801⟩. Acylation of (**615**) occurred at both exocyclic amino groups ⟨73AP730⟩. The mechanism of formation of (**615**) involves initial addition of the active methylene group to the azo linkage followed by closure of the triazine ring and then closure of the imidazole ring ⟨73AP697⟩.

(**613**) (**614**) (**615**)

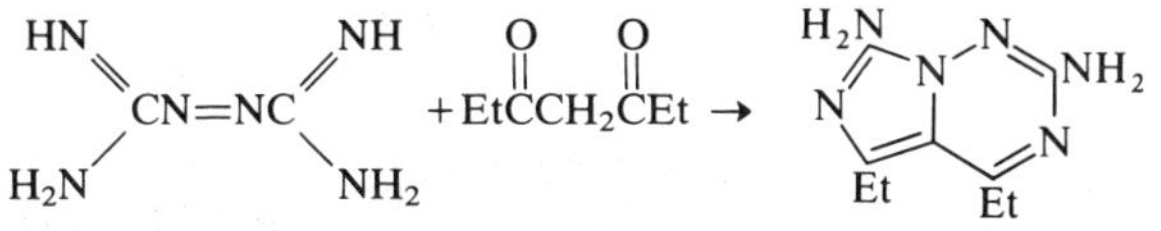

(ii) Closure of the imidazole ring

A different synthetic route to imidazo[5,1-*f*][1,2,4]triazines involves the condensation of acylamino-α-ketoesters (**616**), available from *N*-acylamino acids, with aminoguanidine (**617a**) or amidrazones, for example (**617b**), to produce triazinones (**618**) ⟨80JCS(P1)1139⟩. Cyclization of (**618**) to (**266**) occurred with polyphosphoric acid or phosphorus oxychloride ⟨80JCS(P1)1139, 79JCS(P1)1120⟩. For example, condensation of (**616**) with (**617a**) gave (**618a**), which ring-closed to (**266a**), and likewise (**617b**) led to (**618b**) and then (**266b**).

(**616**) (**617**) **a**; R = NH_2 **b**; R = Me (**618**) **a**; R = NH_2 **b**; R = Me (**266**) **a**; R = NH_2 **b**; R = Me

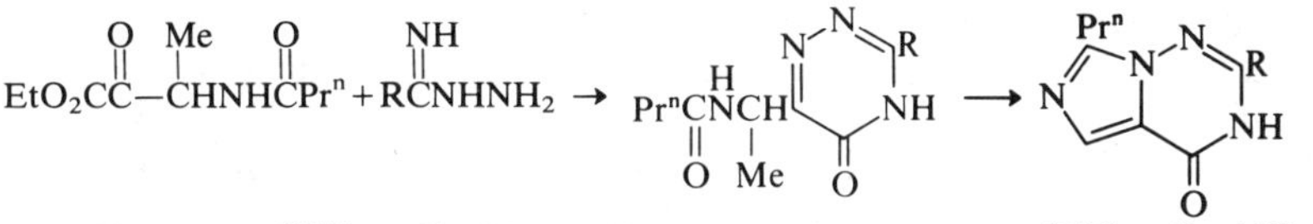

4.10.4.24 Imidazo[1,2-*a*][1,3,5]triazine

Imidazo[1,2-*a*][1,3,5]triazines have been prepared by annulation of the imidazole onto an existing triazine as well as by annulation of the triazine onto an imidazole precursor. Depending upon the desired substituent pattern, both routes allow effective preparation of variously substituted heterocycles. Research on this ring system has been minimal until recently, when several papers have focused on the synthesis of nucleoside derivatives, which can be thought of as 5-aza-7-deazapurine nucleosides.

(i) Closure of the imidazole ring

An early synthesis involved the rearrangement of aziridinyltriazines. For example, heating (**619**) with triethylamine hydrochloride in acetonitrile produced (**620**). The same compound was prepared by the condensation of potassium dicyanoguanidine with 2-chloroethylamine hydrochloride ⟨55JA5922⟩. In a a similar vein, rearrangement of (**621**) occurred upon heating to produce (**622**) ⟨77CHE210, 80CHE1190, 81CHE838⟩.

(**619**) (**620**) (**621**) (**622**)

(**623**)

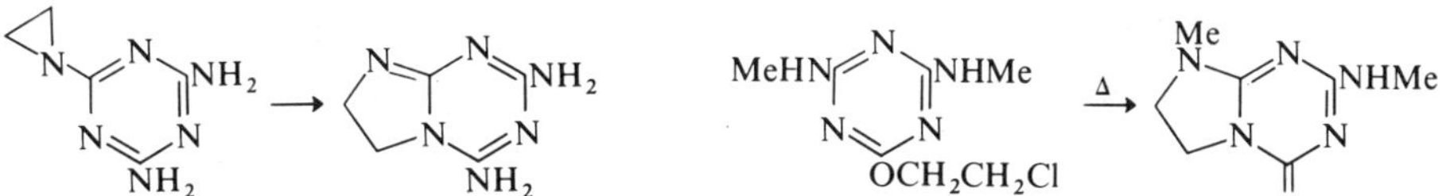

Treatment of 2-amino-4,6-bis(methylthio)-1,3,5-triazine (**623**) with bromoacetaldehyde afforded (**269a**) as the hydrobromide in 55% yield.

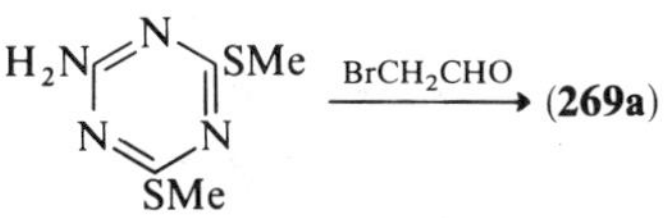

(ii) Closure of the triazine ring

Recently, it has been found that the thermal rearrangement of (**624**) gave (**625**) and (**626**). Treatment of (**625**) with base isomerized it completely to (**626**) ⟨80TL4731⟩.

(**624**) (**625**) (**626**)

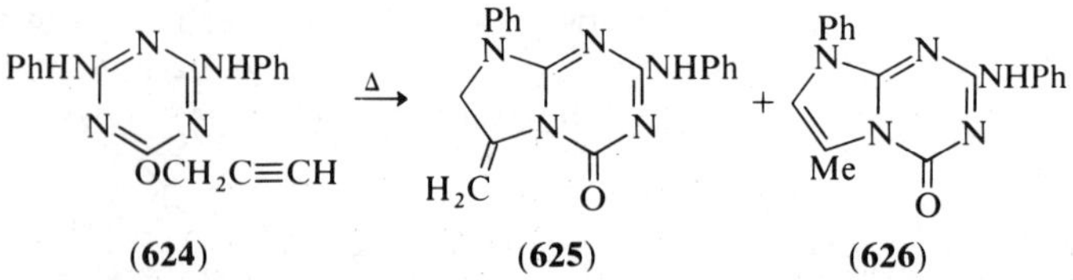

The guanine-related heterocycle (**270**) has been prepared by treatment of (**627**) with aminoacetaldehyde dimethyl acetal to produce (**628**), followed by deblocking and acid-catalyzed cyclization ⟨78JMC883⟩.

(**627**) (**628**)

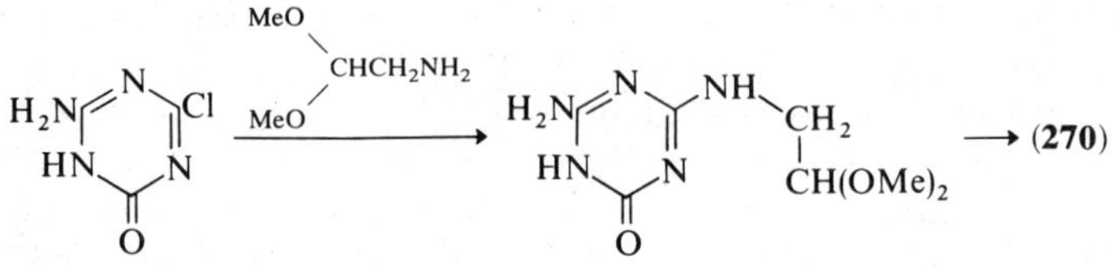

Appending a triazine ring to an imidazole ring to produce imidazo[1,2-*a*]-*s*-triazines has been carried out in two ways. Treatment of 1,2-dimethylimidazoline (**629**) with methyl isothiocyanate gave (**630**) ⟨71G569⟩. Treatment of 2-(2,6-dichloroanilino)-2-imidazoline (**631**) with acid chloride (**632**) in the presence of base afforded (**633**) in 52% yield ⟨74GEP2314488⟩. A much more general route has been developed in connection with the synthesis of certain nucleosides. Treatment of substituted 2-aminoimidazoles such as (**634**) and (**636**) with various aryloxycarbonyl isocyanates and their sulfur analogs produces compounds (**635a**) and (**637a**). Removal of the benzoyl groups affords the nucleosides (**635b**) and (**637b**). When X or Y is sulfur, amino groups can be incorporated by alkylation at sulfur followed by displacement of the methylthio group by ammonia or an amine ⟨78JOC4774, 78JOC4784⟩.

(**629**) (**630**)

(**631**) R = 2,6,-dichlorophenyl (**632**) (**633**) R = 2,6-dichlorophenyl

(**634**) (**635**) **a**; R = 2,3,5-tri-*O*-benzoyl-β-D-ribofuranosyl
b; R = β-D-ribofuranosyl

(**636**) (**637**) **a**; R = 2,3,5-tri-*O*-benzoyl-β-D-ribofuranosyl
b; R = β-D-ribofuranosyl

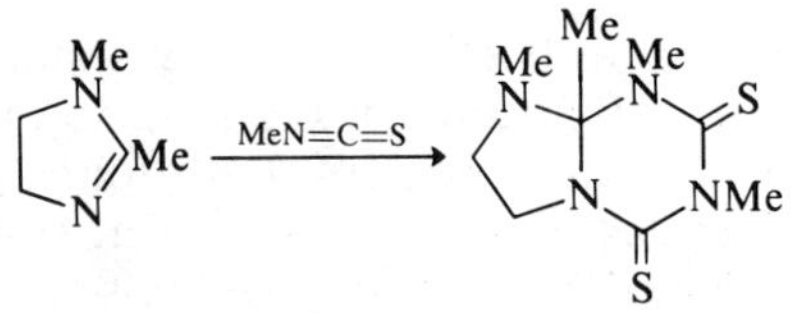

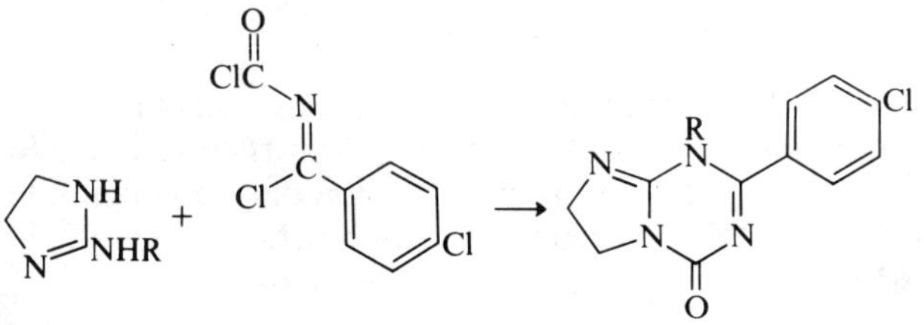

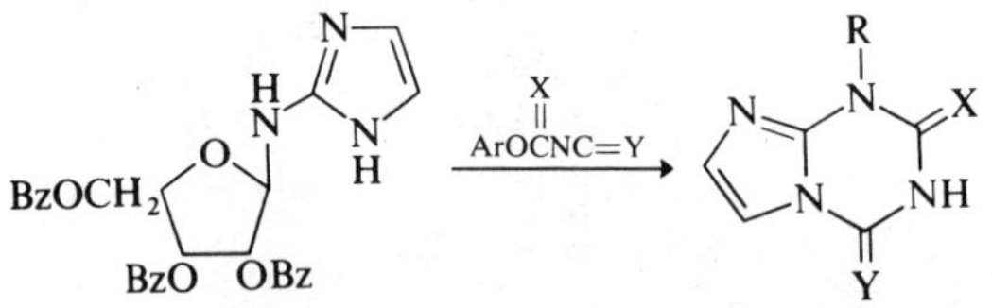

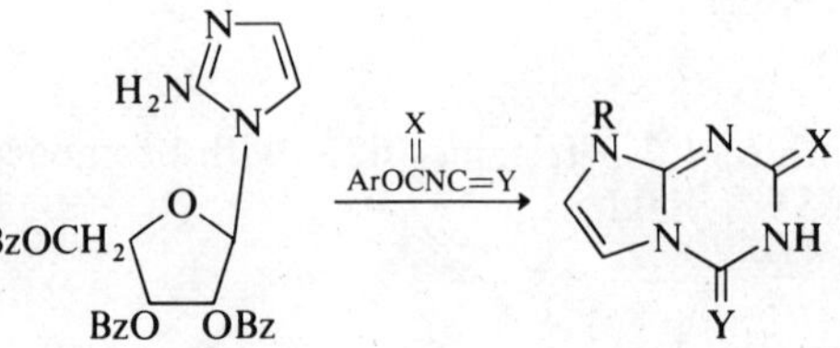

4.10.4.25 Imidazo[1,5-*a*][1,3,5]triazine

Although a compound containing this ring system was prepared in 1934 during the exploration of the chemistry of theobromine ⟨61HC(15-2)916⟩, research has really only begun recently.

(i) Closure of the imidazole ring

Treatment of 4,5-dihydro-5-methylpyrimidin-4-ones such as (**638a**) with chlorotrimethylsilane and hexamethyldisilazane in pyridine produced the imidazo[1,5-*a*][1,3,5]triazinone (**639a**) ⟨79JOC3835, 79JOC1740⟩. 5-Allyl-substituted pyrimidinones, exemplified by (**638b**), undergo the same reaction to afford (**639b**) ⟨81JOC3681⟩. Although no mechanism has been established, the suggested mechanism involves cyclization of a trimethylsilylated (**638a**) to an intermediate (**640**) which might undergo carbon–carbon bond cleavage either heterolytically or electrocyclically to an intermediate such as (**641**), which could reclose on the adjacent imidazole nitrogen to produce (**639a**).

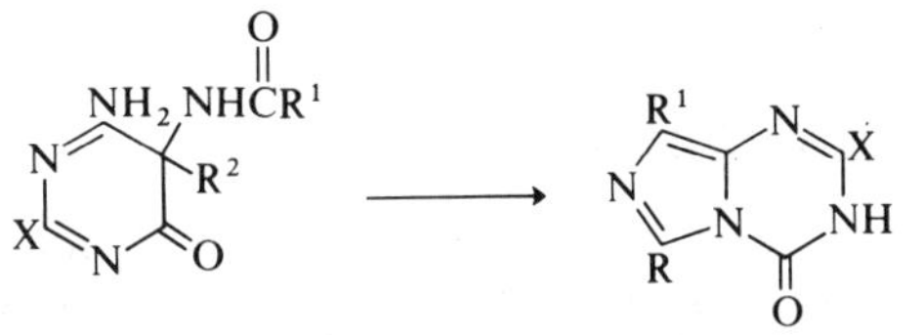

(**638**) a; X = SH, R^1 = H, R^2 = Me
b; X = NH$_2$, R^1 = Me, R^2 = CH$_2$CH=CH$_2$

(**639**) a; X = SH, R = H, R^1 = Me
b; X = NH$_2$, R = Me, R^1 = CH$_2$CH=CH$_2$

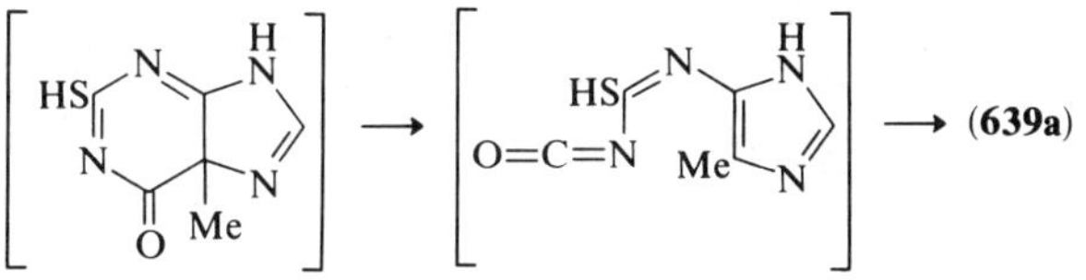

(**640**)   (**641**)

(ii) Closure of the triazine ring

Treatment of 5-amino-4-cyano-4-methyl-4*H*-imidazole (**642**) with formamidine under mild conditions resulted in the formation of (**644**), with 5-methyl-5*H*-adenine (**643**) postulated as a transient intermediate. Rearrangement of (**643**) as suggested above would produce (**644**). An independent synthesis of (**644**) was carried out by treatment of (**645**) with methyl *N*-cyanomethanimidate (**646**) ⟨82JA235, 83JOC3⟩.

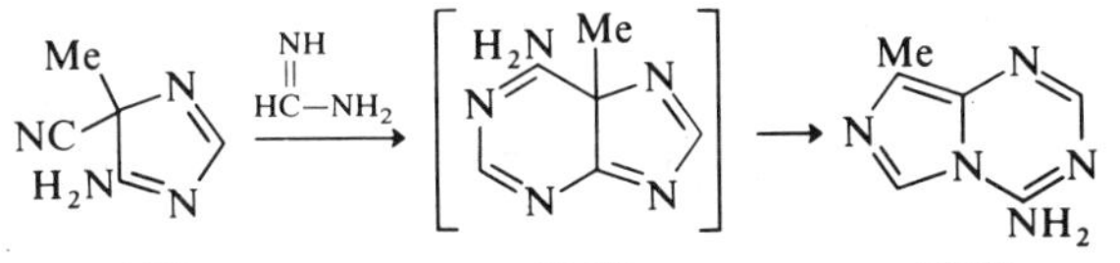

(**642**)   (**643**)   (**644**)

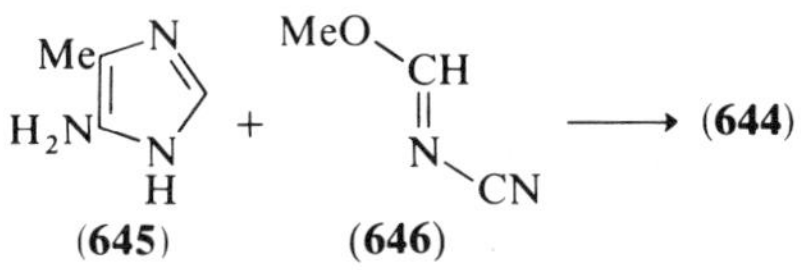

(**645**)   (**646**)   (**644**)

4.10.4.26 Imidazo[1,2-*b*][1,2,4,5]tetrazine

Treatment of 3,6-dibenzyl-1,2,4,5-tetrazine (**647**) with methanolic potassium hydroxide resulted in the formation of (**648**) by a complex mechanism. Other alcohols did not give the corresponding products ⟨79JCS(P1)333⟩.

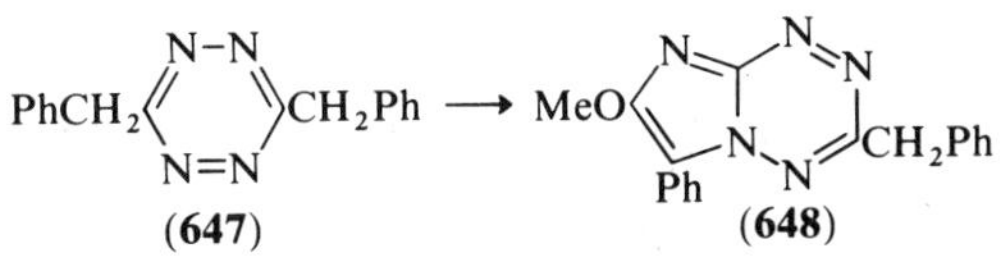

(**647**)   (**648**)

4.10.4.27 Imidazooxazines

The seven imidazooxazine systems pictured in the introduction have been synthesized. With a few exceptions the reaction schemes have little generality.

4.10.4.27.1 Imidazo[1,2-c][1,3]oxazine

Treatment of imidazolylacetophenones such as (**649**) with phenyl isocyanate or isothiocyanate afforded (**650**), which is stable with C-5 tetrahedral ⟨75JOC252⟩.

(**649**) (**650**)

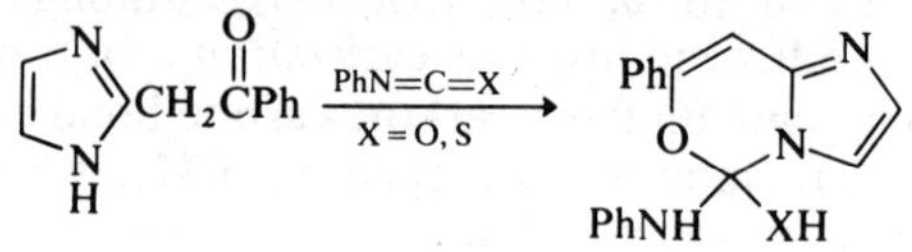

4.10.4.27.2 Imidazo[1,5-c][1,3]oxazine

A similar type of product (**652**) was formed when the imidazole derivative (**651**) was treated with acetyl chloride in pyridine under a specific set of conditions. The imidazo[1,5-*c*][1,3]oxazine was isolated as a stable solid ⟨74JA2481, 73JA4452⟩.

(**651**) (**652**)

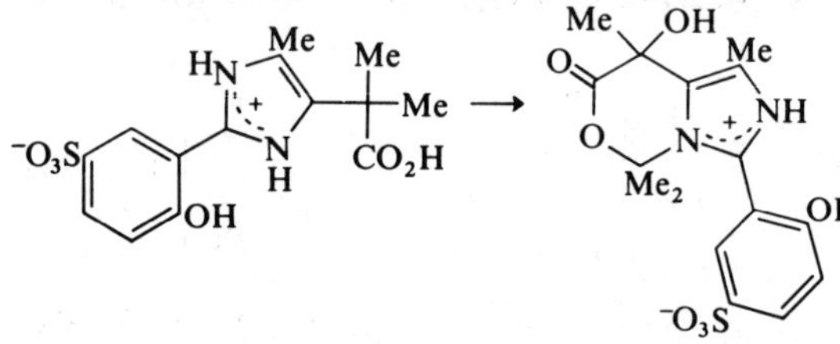

4.10.4.27.3 Imidazo[2,1-b][1,3]oxazine

Treatment of 2-methylthioimidazoline (**653**) with β-trichloromethyl-β-propiolactone (**654**) produced (**655**), presumably by initial *N*-acylation and then cyclization to form the oxazine ring ⟨69JOU1790⟩.

(**653**) (**654**) (**655**)

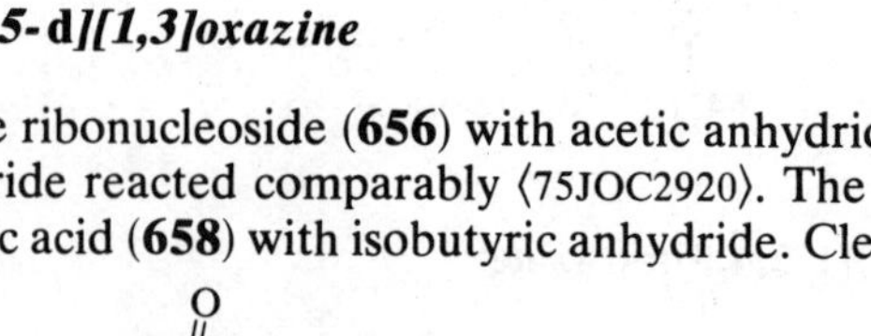

4.10.4.27.4 Imidazo[4,5-d][1,3]oxazine

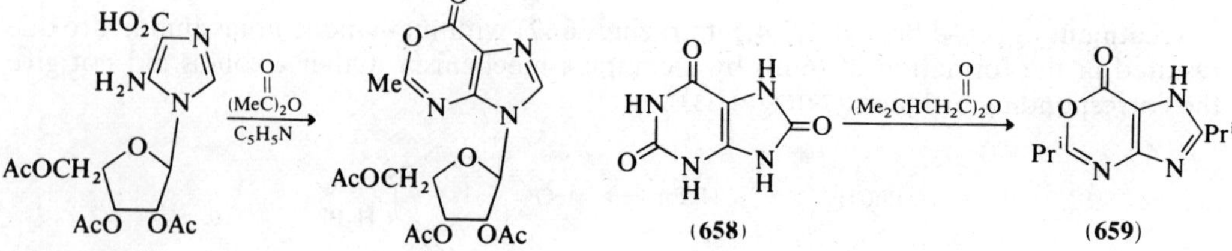

Heating the imidazole ribonucleoside (**656**) with acetic anhydride and pyridine produced (**657**). Propionic anhydride reacted comparably ⟨75JOC2920⟩. The same ring system can be prepared by treating uric acid (**658**) with isobutyric anhydride. Cleavage and rearrangement

(**656**) (**657**) (**658**) (**659**)

in both the imidazole and pyrimidine rings resulted in the formation of (**659**) ⟨77JOC3132⟩. With acetic or propionic anhydride, cleavage and rearrangement occurred only in the imidazole ring.

*4.10.4.27.5 Imidazo[5,1-*b*][1,3]oxazine*

Intramolecular ring closure of (**660**) with dicyclohexylcarbodiimide produced the imidazo[5,1-*b*][1,3]oxazine (**661**) ⟨75CB372⟩.

(**660**) (**661**)

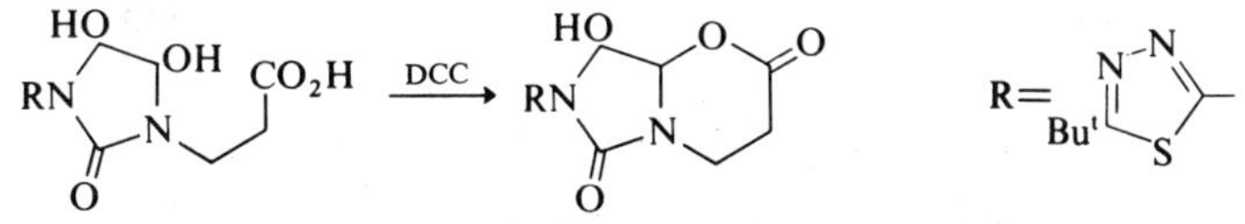

*4.10.4.27.6 Imidazo[2,1-*c*][1,4]oxazine*

The synthesis of imidazo[2,1-*c*][1,4]oxazines has been accomplished by several methods. Several groups have closed the oxazine ring from an imidazole derivative with an *N*-hydroxyethyl group ⟨79JHC871, 67USP3341549, 71USP3565892⟩. For example, treatment of (**662**) with hydrochloric acid and ethanol afforded (**663**) ⟨79JHC871⟩.

(**662**) (**663**)

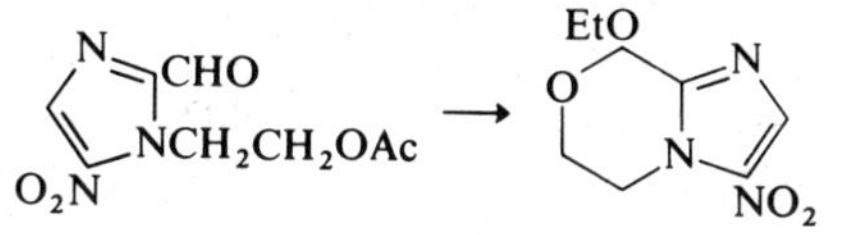

In one report the imidazole ring has been closed from a preexisting 1,4-oxazine derivative. Treatment of (**664**) with 2-amino-2-cyanoacetamide (**665**) under acidic conditions gave (**666**) ⟨66CHE134⟩.

(**664**) (**665**) (**666**)

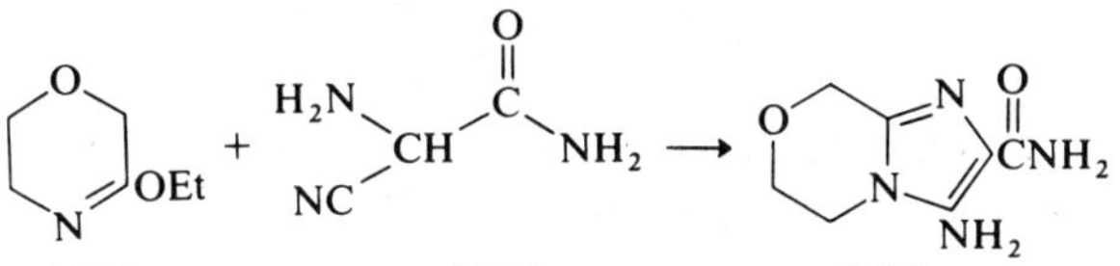

*4.10.4.27.7 Imidazo[5,1-*c*][1,4]oxazine*

Several different routes have been found which lead to imidazo[5,1-*c*][1,4]oxazines. Treatment of the imidazoledicarboxylic acid (**667**) with acetic anhydride afforded (**668**). The initial steps of this transformation presumably involve a modified Dakin–West reaction ⟨66JOC806⟩.

(**667**) (**668**)

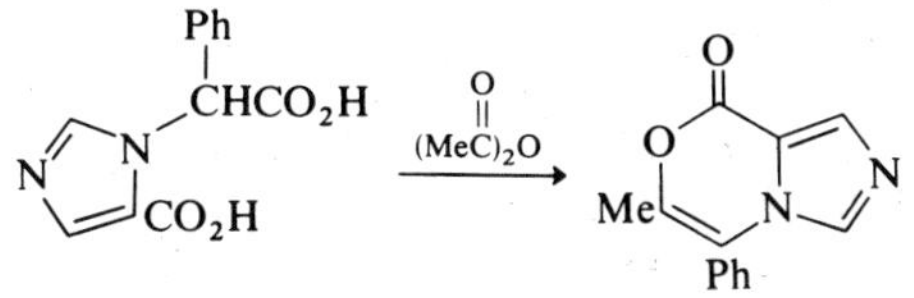

Cyclization of the 2-chloroethyl ester (**669**) with base gave (**670**) ⟨64JOC3707⟩.

(**669**) (**670**)

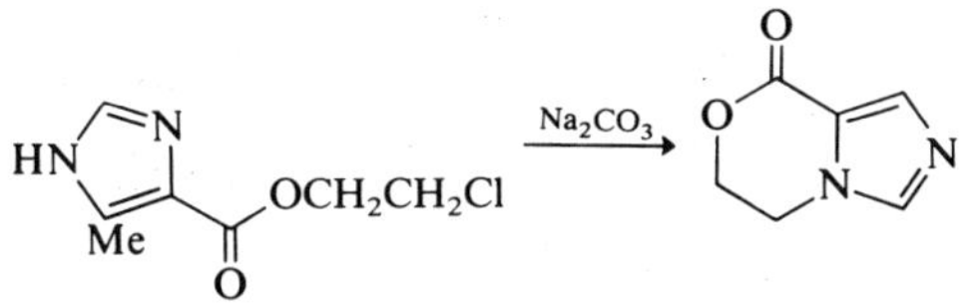

Treatment of the diester (**671**) with styrene oxide and 0.1 mole equivalent of potassium resulted in the formation of (**672**). Larger quantities of potassium gave the imidazole derivative (**673**) *via* an imidazo[5,1-*c*][1,4]oxazine intermediate ⟨75JCS(P1)798⟩.

(**671**) (**672**) (**673**)

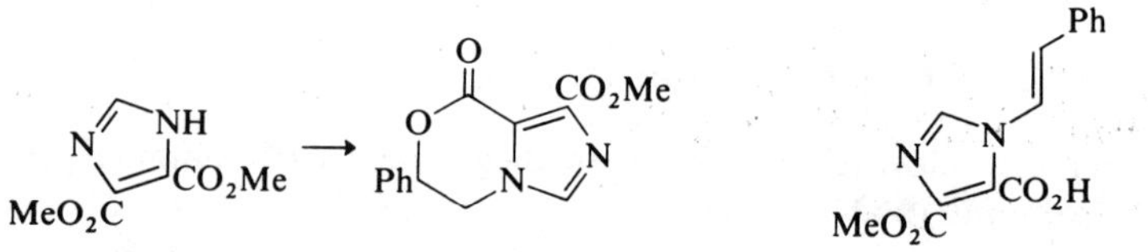

4.10.4.28 Imidazooxadiazines

The three imidazooxadiazines pictured in the introduction have been prepared, the first two by annulation of the oxadiazine ring onto an imidazole-related precursor, and the latter by forming both rings in the same reaction.

*4.10.4.28.1 Imidazo[1,5-*b*][1,2,5]oxadiazine*

Treatment of 3-hydroxyimino-2,4-pentanedione (**674**) with 4-methoxybenzaldehyde in ammonia and ethanol afforded (**675**) ⟨74LA1399⟩.

(**674**) (**675**)

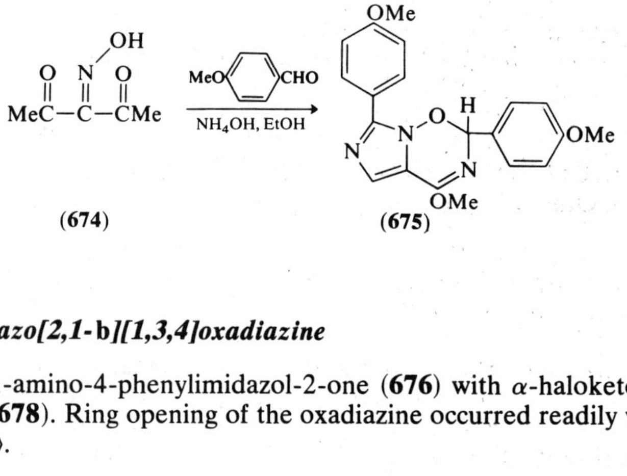

*4.10.4.28.2 Imidazo[2,1-*b*][1,3,4]oxadiazine*

Treatment of 1-amino-4-phenylimidazol-2-one (**676**) with α-haloketones, for example (**677**), produced (**678**). Ring opening of the oxadiazine occurred readily with aqueous HBr ⟨70CB272, 62ZC153⟩.

(**676**) (**677**) (**678**)

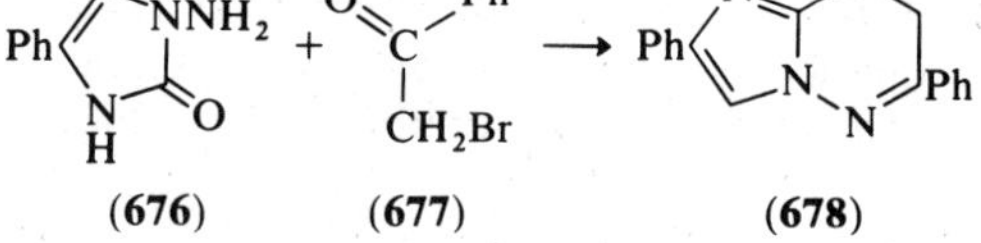

*4.10.4.28.3 Imidazo[2,1-*c*][1,2,4]oxadiazine*

Displacement of the methylthio group from (**679**) with aminooxyacetic acid gave (**680**), which cyclized to (**681**) upon attempted esterification ⟨70TL1879⟩.

(**679**) (**680**) (**681**)

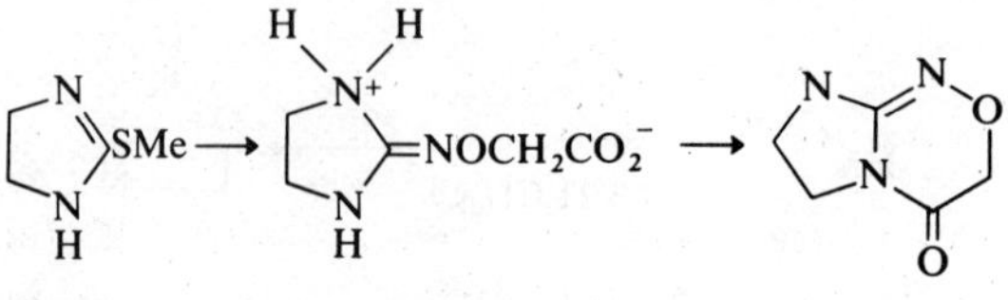

4.10.4.29 Imidazothiazines

Of the six imidazothiazine systems pictured in the introduction, by far the most common is the imidazo[2,1-*b*][1,3]thiazine.

4.10.4.29.1 *Imidazo[2,1-b][1,3]thiazine*

(i) Closure of the thiazine ring

A number of methods employ an imidazole-2-thione or related compound with a bifunctional reagent to build the thiazine ring in one or two steps. Treatment of (**682**) with acrylyl chloride directly produced (**683**) ⟨64JOC1720⟩. Substituting acrylic acid under catalysis allows isolation of the *S*-alkylated intermediate, which cyclized to (**683**) upon heating ⟨64JOC1715⟩.

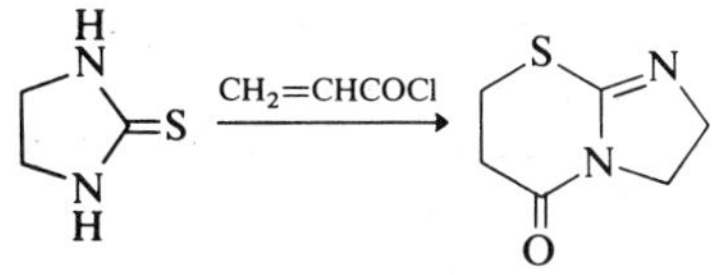

In a similar sequence, treatment of (**684**) with 3-bromopropionic acid produced (**685**), which cyclized to imidazo[2,1-*b*][1,3]thiazine (**686**) upon heating with acetic anhydride ⟨74JPR147⟩.

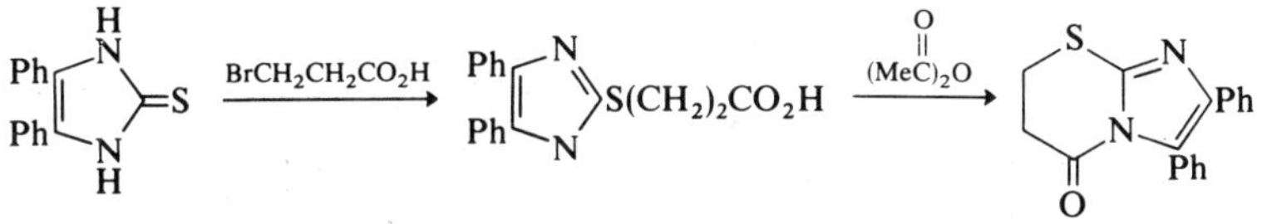

Cyclization occurred readily to (**688**) when (**682**) was treated with methyl methylpropiolate (**687**) ⟨68JCS(C)2510⟩. Other alkynic esters also have been employed ⟨68JCS(C)2510, 67CJC953⟩.

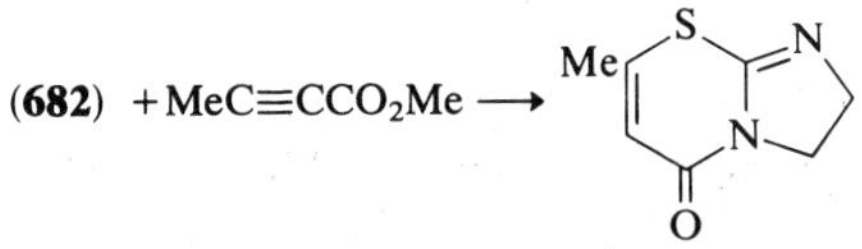

Alkylation of (**684**) with chloroketone (**689**) afforded (**690a**), which was converted into the chloro compound (**690b**) in two steps. Cyclization of (**690b**) to (**691**) occurred upon heating in basic solution ⟨70MI41000⟩. A chloroacetal has also been employed ⟨79JMC237⟩.

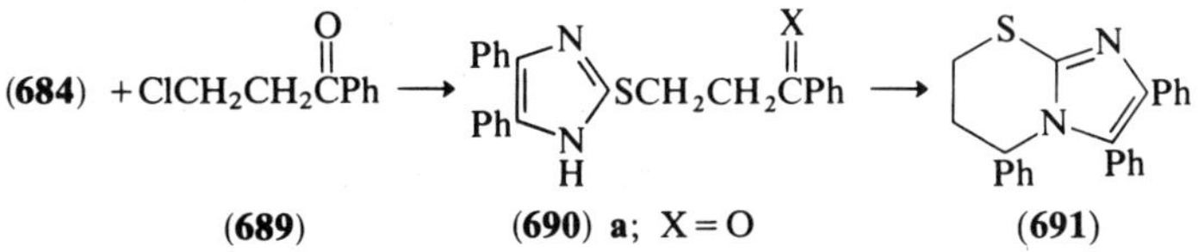

Treatment of (**684**) with 1,3-dibromopropane under the conditions of phase-transfer catalysis (PTC) afforded a high yield of (**692**) ⟨80JHC393⟩. Cyclization with 1,3-dibromopropane and epibromohydrin has been carried out under various reaction conditions ⟨73IJC747, 72JOC1464, 80T1079, 65JCS3456⟩.

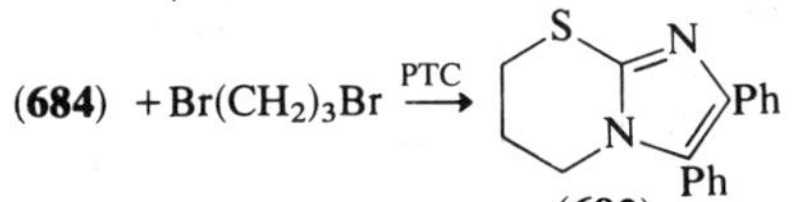

When 2(3*H*)-imidazolethione (**693**) was treated with (chlorocarbonyl)phenylketene (**694**), the imidazo[2,1-*b*][1,3]thiazinedione (**695**) was produced directly. Employment of the *N*-methyl derivative of (**693**) affords a heteroaromatic betaine ⟨80JOC2474⟩.

(693) (694) (695)

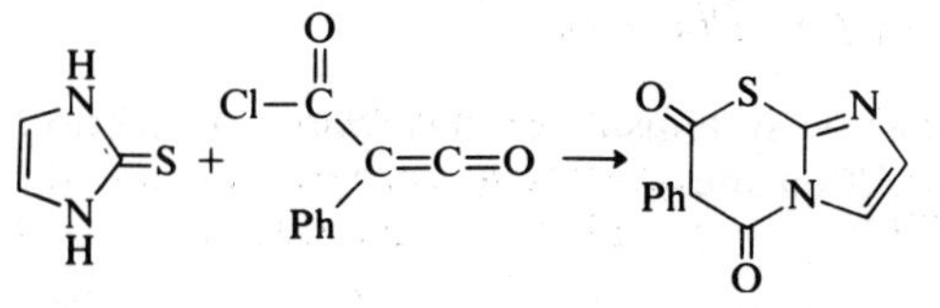

(ii) Closure of the imidazole ring

The synthesis of the imidazo[2,1-*b*][1,3]thiazine system also has been approached starting with a thiazine. Treatment of (**696**) with an α-bromoketone such as phenacyl bromide (**512**) produced an *N*-alkylated material, which could then be cyclized by heating to (**697**) ⟨70LA(742)85⟩. An α-chloroketone has also been used in similar fashion ⟨69ZC27⟩.

(696) (697)

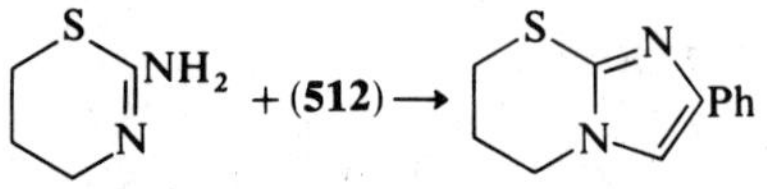

4.10.4.29.2 *Imidazo[1,2-*c*][1,3]thiazine*

Ring expansion of the bicyclic isothiazole (**698**) occurred upon treatment with cyanide ion to give the imidazo[1,2-*c*][1,3]thiazine (**699**). The mechanism involves initial ring opening by attack on sulfur, and then ring closure. The anion of methyl propiolate also causes the same type of rearrangement ⟨79TL1281⟩.

(698) (699)

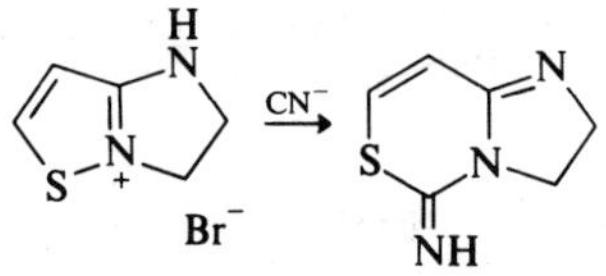

4.10.4.29.3 *Imidazo[4,5-*d*][1,3]thiazine*

Formation of imidazo[4,5-*d*][1,3]thiazines has been carried out in all cases from appropriately substituted imidazoles. Treatment of a 5-amino-4-cyanoimidazole with either an isothiocyanate or carbon disulfide directly affords this ring system ⟨81CPB1870, 81JMC947, 75CPB759⟩. For example, treatment of (**700**) with phenyl isothiocyanate produced (**701**), which could be readily rearranged with base to the purine (**702**) ⟨81CPB1870⟩. Treatment of a 4-amino-5-carboxamidoimidazole with thiophosgene also resulted in an imidazo[4,5-*d*]-[1,3]thiazine after rearrangement ⟨69CB3000⟩.

(700) (701) (702) R = β-D-ribofuranosyl

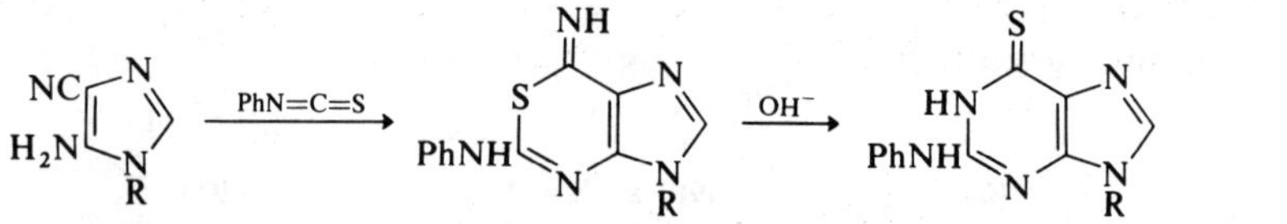

4.10.4.29.4 *Imidazo[4,5-*e*][1,3]thiazine*

In possibly the first synthesis of an imidazothiazine, treatment of (**703**) with potassium isothiocyanate afforded (**704**), which upon heating in pyridine cyclized to 6-amino-2-bromo-1,5-dihydroimidazo[4,5-*e*][1,3]thiazine (**705**) ⟨34JIC867⟩.

(703) (704) (705)

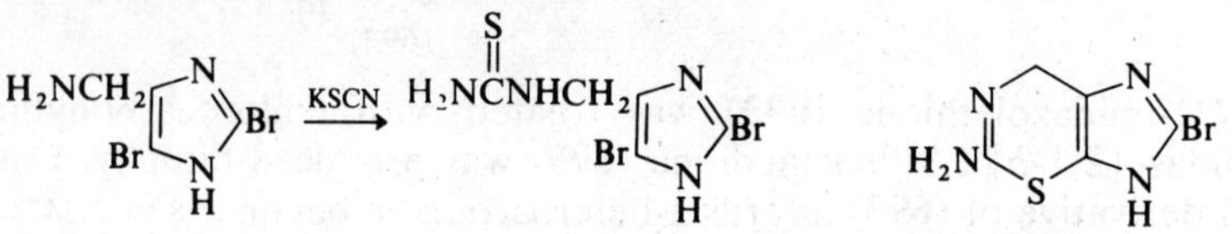

4.10.4.29.5 Imidazo[2,1-c][1,4]thiazine

Synthesis of imidazo[2,1-c][1,4]thiazines has been carried out from the lactam ether (**706**) by treatment with an appropriate two-carbon fragment ⟨68CHE322, 79JMC237⟩. For example, treatment of (**706**) with 2-amino-2-cyanoacetamide under acidic conditions gave (**707**) ⟨68CHE322⟩.

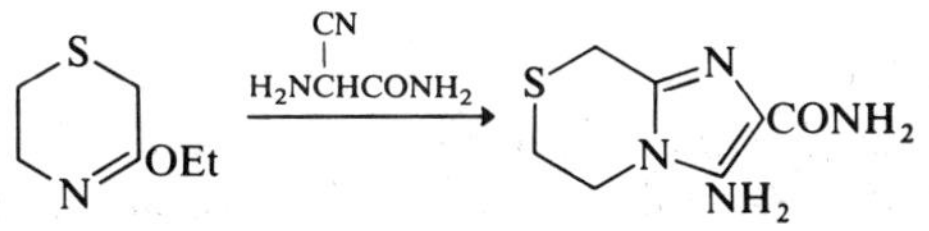

4.10.4.29.6 Imidazo[5,1-c][1,4]thiazine

When the thiazine ester (**708**) was treated with an isocyanate such as (**709**), an imidazo[5,1-c][1,4]thiazine (**710**) was produced ⟨81JAP8161384⟩. Treatment of certain penicillin sulfoxide esters (**711**) with ethoxycarbonyl isocyanate results in a ring expansion of both rings to afford an imidazo[5,1-c][1,4]thiazine (**712**) ⟨81JOC3026⟩.

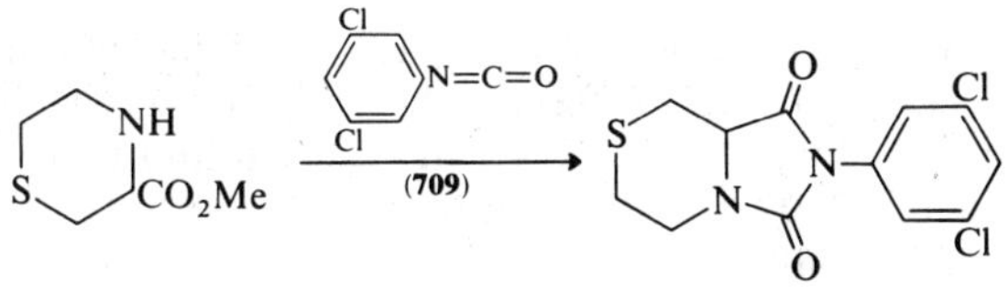

$R = PhOCH_2\overset{O}{\underset{\|}{C}}-, \quad R^1 = CH_2\text{-}C_6H_4\text{-}NO_2$

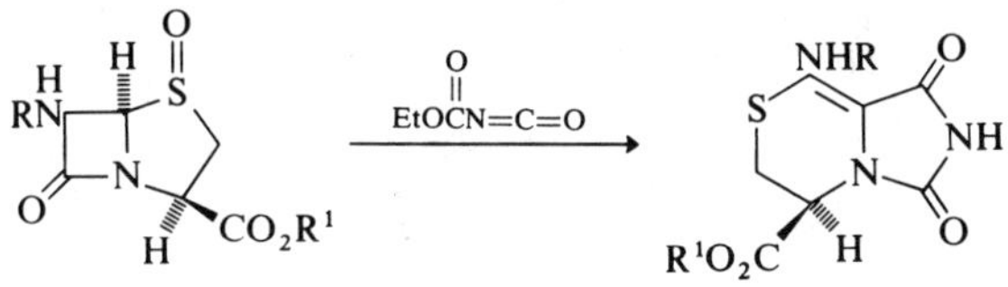

4.10.4.30 Imidazothiadiazines

Derivatives of the five imidazothiadiazines pictured in the introduction are known. The most research has been carried out on 4,5-fused systems, where the impetus, at least partially, has been their resemblance to the purine ring system.

4.10.4.30.1 Imidazo[4,5-c][1,2,6]thiadiazine

The imidazo[4,5-c][1,2,6]thiadiazine system has been synthesized from both imidazole and thiadiazine precursors. Treatment of the thiadiazine dioxide (**713**) with potassium dithioformate under mild conditions allowed the isolation of (**714**), which was readily cyclized to (**715a**) upon heating. Direct conversion of (**713**) to (**715a**) also occurs, with an improvement in yield over the two separate steps ⟨76JHC793⟩. This type of ring closure has

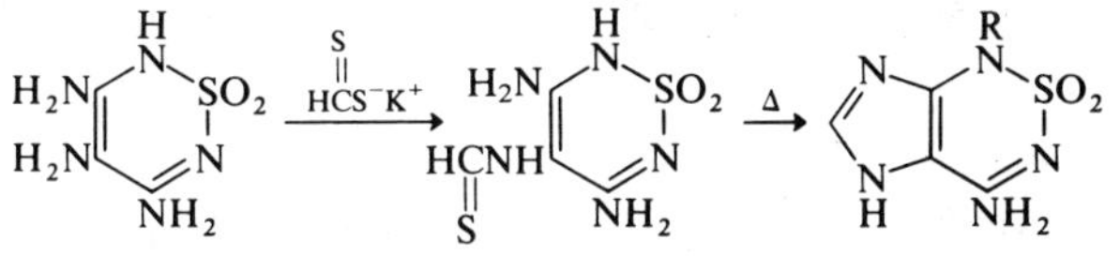

(**715**) a; R = H
b; R = 2,3,4,6-tetra-O-acetyl-β-D-glucopyranosyl

also been used for other derivatives ⟨78JHC221⟩. Compound (**715a**) has pK_a values of 0.28, 4.28 and 12.21, reflecting its unique structure ⟨76JHC793⟩.

Treatment of an appropriately substituted imidazole with sulfamoyl chloride results in closure of the thiadiazine ring ⟨77HCA521, 79JMC944⟩. For example, silylation of the imidazole (**716**) with hexamethyldisilazane followed by treatment with sulfamoyl chloride and then base to cleave silyl groups produced (**715a**). The cyclization was greatly facilitated by amine activation through silylation ⟨79JMC944⟩.

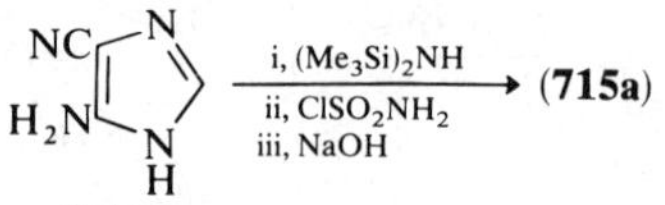

Treatment of the silylated derivative of (**715a**) with 2,3,4,6-tetra-*O*-acetyl-β-D-glucopyranosyl bromide in the presence of mercury(II) cyanide afforded (**715b**). Synthesis of (**715b**) was also possible by attaching the sugar to (**713**), and then closing the ring as previously described ⟨78MI41009⟩.

4.10.4.30.2 Imidazo[4,5-e][1,2,4]thiadiazine

All of the synthetic routes to imidazo[4,5-*e*][1,2,4]thiadiazines have used imidazole precursors. Ring closure of (**717a**) to (**718a**) can be carried out with formic acid, an ortho ester, phosgene or thiophosgene ⟨61JOC1861⟩. It has been found that (**717b**) is unstable on isolation, but it can be generated without isolation and converted directly into (**718b**) with triethyl orthoformate ⟨79JOC4046⟩. Ribosylation of (**718b**) under a set of standard conditions produced (**718c**).

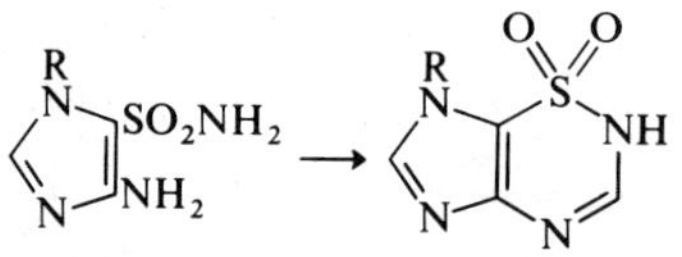

(**717**) a; R = Me   (**718**) a; R = Me
b; R = H   b; R = H
                             c; R = β-D-ribofuranosyl

Heating the acyl azide (**719**) in an inert solvent induced rearrangement to the isocyanate with loss of nitrogen followed by ring closure to (**720**) ⟨76CHE924⟩.

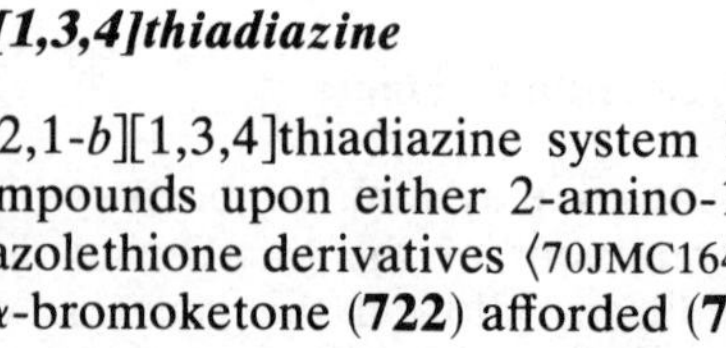

4.10.4.30.3 Imidazo[2,1-b][1,3,4]thiadiazine

Synthesis of the imidazo[2,1-*b*][1,3,4]thiadiazine system has been carried out by the action of α-halocarbonyl compounds upon either 2-amino-1,3,4-thiadiazines ⟨72JHC1385, 75ZC482⟩ or 1-amino-2-imidazolethione derivatives ⟨70JMC164, 75ZC482⟩. Treatment of the readily available (**721**) with α-bromoketone (**722**) afforded (**723**) ⟨72JHC1385⟩.

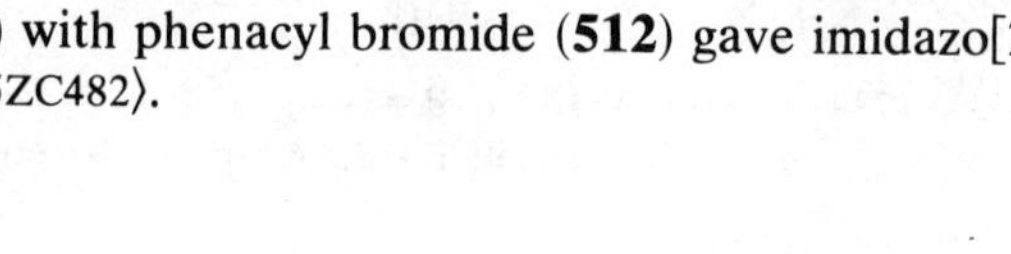

Cyclization of (**724**) with phenacyl bromide (**512**) gave imidazo[2,1-*b*][1,3,4]thiadiazine (**725**) in good yield ⟨75ZC482⟩.

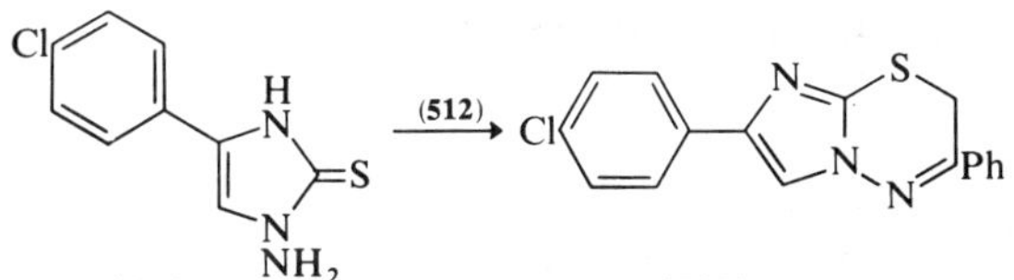

Other Imidazoles with Fused Six-membered Rings 667

4.10.4.30.4 *Imidazo[2,1-b][1,3,5]thiadiazine*

Two quite specific methods have been employed for the synthesis of imidazo[2,1-*b*]-[1,3,5]thiadiazines. In the condensation of 1,2-diphenylimidazoline-4,5-dione (**726**) with a thiocarbonyl isocyanate such as (**727**), the isocyanate added to (**726**) in the manner of a Diels–Alder diene to produce (**728**) ⟨67CB2064⟩.

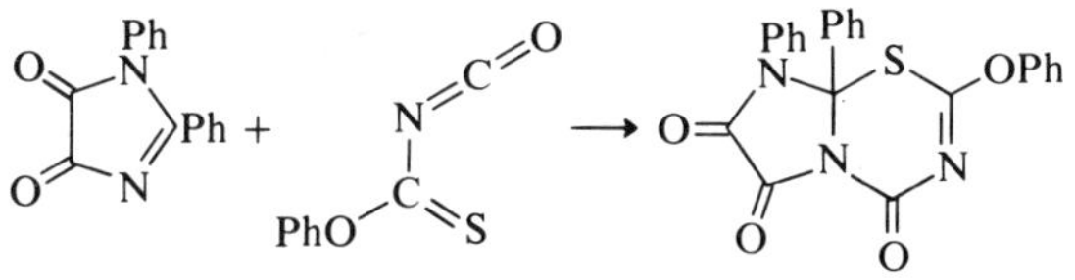

Treatment of imidazolidine-2-thione (**729**) with isothiocyanate (**730**) afforded (**731**) in high yield ⟨75S675⟩.

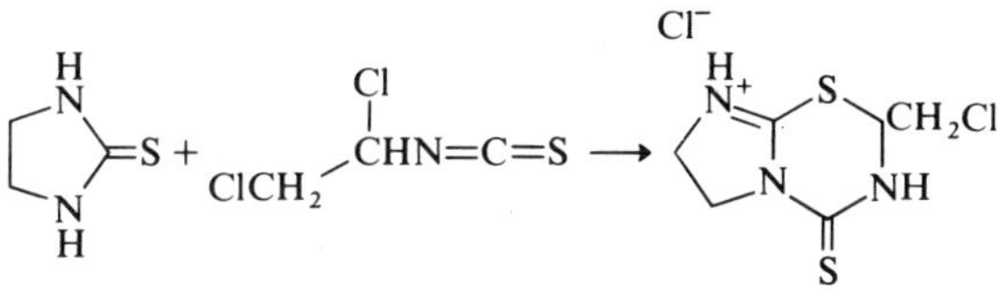

4.10.4.30.5 *Imidazo[4,5-d][1,2,3]thiadiazine*

Formation of the imidazo[4,5-*d*][1,2,3]thiadiazine system has been accomplished by treating the thiocarbamoyl diazonium compound (**732**) with base, producing (**733**). Further base treatment rearranges (**733**) into (**734**) ⟨75CHE995, 73CHE1173⟩.

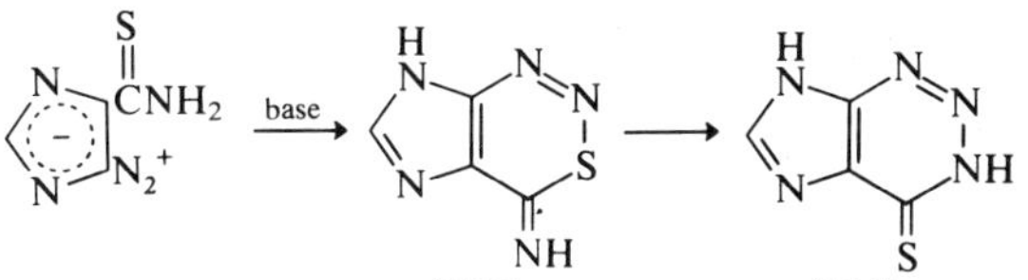

4.10.4.31 Imidazo[4,5-c][1,5,2]diazaphosphorine and Imidazo[4,5-d][1,3,2]diazaphosphorine

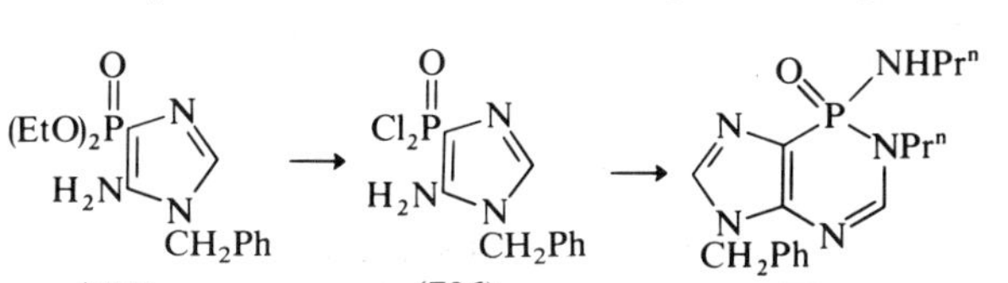

Syntheses of the imidazo[4,5-*c*][1,5,2]diazaphosphorine system, oxidized at phosphorus, was undertaken to prepare transition state analogs for adenosine aminohydrolase. Conversion of imidazole derivative (**735**) to (**736**) required three steps. Treatment of (**736**) with *n*-propylamine followed by either acidic or basic ring closure gave (**737**) ⟨78MI41001⟩. In this same paper an imidazo[4,5-d][1,3,2]diazaphosphorine was also prepared in extremely low yield.

4.10.4.32 Imidazo[4,5-d][1,3,2]diazaborine

The imidazo[4,5-d][1,3,2]diazaborine system was prepared as a boron-containing purine analog. Treatment of (**738**) in hot ethanolic solution with phenylboronic anhydride gave a high yield of (**739**) ⟨61JA2708, 59JA6329⟩.

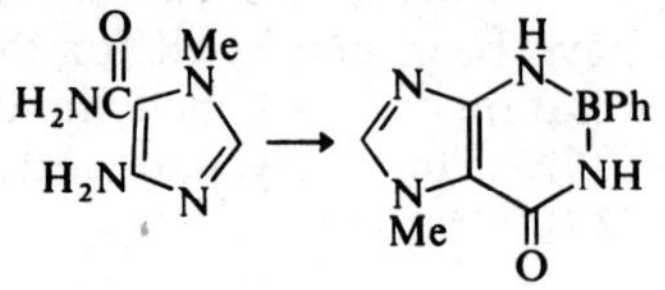

4.11
1,2,3-Triazoles and their Benzo Derivatives

H. WAMHOFF
Universität Bonn

4.11.1 INTRODUCTION AND NOMENCLATURE	670
4.11.2 HISTORY AND REVIEWS	670
4.11.3 STRUCTURE	671
4.11.3.1 Theoretical Methods	671
4.11.3.1.1 HMO methods	672
4.11.3.1.2 Electronic structure; π-electron distribution; dipole moments	672
4.11.3.2 Structural Methods	673
4.11.3.2.1 X-Ray diffraction	673
4.11.3.2.2 Microwave spectroscopy and dipole moment measurements	676
4.11.3.2.3 ^1H NMR spectroscopy	678
4.11.3.2.4 ^{13}C NMR spectroscopy	680
4.11.3.2.5 ^{14}N and ^{15}N NMR spectroscopy	682
4.11.3.2.6 UV spectroscopy	684
4.11.3.2.7 IR spectroscopy	685
4.11.3.2.8 Mass spectrometry	686
4.11.3.2.9 Photoelectron spectroscopy	688
4.11.3.2.10 Polarography	689
4.11.4 REACTIVITY	690
4.11.4.1 Reactions at Aromatic Rings	690
4.11.4.1.1 General survey of reactivity	690
4.11.4.1.2 Thermal and photochemical reactions formally involving no other species	691
4.11.4.1.3 Electrophilic substitution and nucleophilic attack	697
4.11.4.1.4 Reactions with Cyclic Transition States	701
4.11.4.2 Dihydro Compounds (Δ^2-1,2,3-Triazolines)	702
4.11.4.2.1 Stereochemical aspects	702
4.11.4.2.2 Aromatization reactions (dehydrogenation of Δ^2-1,2,3-triazolines)	703
4.11.4.2.3 Ring fission of Δ^2-1,2,3-triazolines	704
4.11.5 SYNTHESIS	705
4.11.5.1 General Remarks	705
4.11.5.2 Triazoles from Non-heterocyclic Compounds	706
4.11.5.2.1 Formation of one bond	706
4.11.5.2.2 Formation of two bonds	708
4.11.5.3 Triazoles by Transformation of Other Heterocycles	718
4.11.5.4 Other Methods of Synthesis of 1,2,3-Triazoles	720
4.11.5.5 1H- and 2H-Benzotriazoles	722
4.11.6 APPLICATIONS OF 1,2,3-TRIAZOLES AND BENZOTRIAZOLES	723
4.11.6.1 Access to Compounds of Preparative Importance	724
4.11.6.2 Triazoles Utilized in Organic Synthesis	726
4.11.6.3 Triazoles in Medicinal Chemistry	728
4.11.6.4 Industrial Applications	730
4.11.6.4.1 Dyestuffs, fluorescent whiteners	730
4.11.6.4.2 Corrosion inhibition	730
4.11.6.4.3 Photostabilizers	731
4.11.6.4.4 Agrochemicals	731
4.11.6.5 Important Natural Products	732

4.11.1 INTRODUCTION AND NOMENCLATURE

1,2,3-Triazoles and their benzo derivatives have attracted considerable attention because of their theoretical interest, and synthetic value, as well as for their numerous applications in industry and agriculture due to their extensive biological activities and their successful application as fluorescent whiteners, light stabilizers and optical brightening agents.

The 1,2,3-triazole system can be formally derived from pyrrole or indole by replacing two carbons by two nitrogen atoms. The resulting parent compounds may be classified in terms of the constitutional formulae (1)–(3). In the literature this ring system is still named 'v-triazole', in order to distinguish it from 's-triazole'. While (1) and (2) can be considered to depict heteroaromatic systems, (3) is a non-aromatic compound and is rarely found in the literature. In the older literature the term 'osotriazole' belongs to derivatives of (2), particularly if these are derived from osazones.

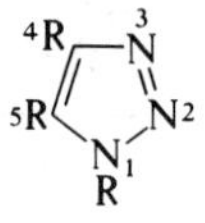

(1) 1H-1,2,3-Triazoles

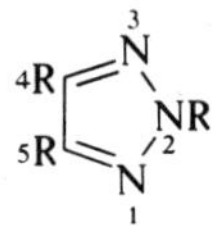

(2) 2H-1,2,3-Triazoles
('osotriazole', 1,2,5-triazole, 2,1,3-triazole)

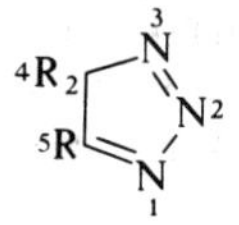

(3) 4H-1,2,3-Triazoles
(1,2,3-isotriazole)

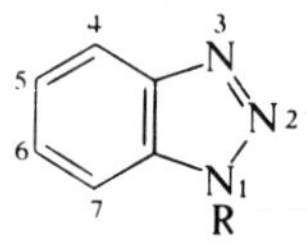

(4) 1H-Benzo[d][1,2,3]triazoles
(azimidobenzene, benzene azoimine, benzisotriazole, 1,2,3-benzotriazole)

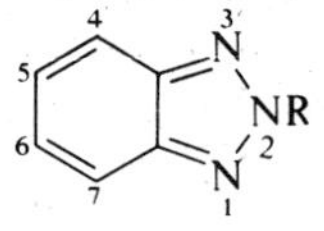

(5) 2H-Benzo[d][1,2,3]triazoles
(pseudoazimidobenzene, 2,1,3-benzotriazole)

The nomenclature of the potential benzotriazole derivatives (4) and (5) was chosen analogously to that of the monocyclic systems.

4.11.2 HISTORY AND REVIEWS

The class of 1,2,3-triazoles and their benzo derivatives represents an overwhelming, rapidly developing field in modern heterocyclic chemistry. This is impressively illustrated by the large number of excellent reviews which have been specially devoted to this class of heterocycle and which are listed below.

As early as 1888 von Pechmann, assistant to von Baeyer, prepared the 1,2,3-triazole compound (6) as well as some derivatives, and correctly characterized his newly invented azole heterocycles as 1,2,3-triazoles ⟨1888CB2751⟩. In a 1950 review it was stated that more than 1400 1,2,3-triazoles had been described in the literature, and in the period 1973–1980 *Chemical Abstracts* reported more than 9000 studies involving 1,2,3-triazoles.

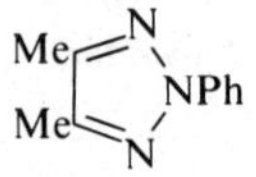

(6) 'Osotriazone' (v. Pechmann)

The following are the most important reviews on 1,2,3-triazoles and their benzo derivatives: ⟨50CRV(46)1, B-57MI41100, B-61MI41100, B-71MI41104, 74AHC(16)33, B-76MI41103, B-79MI41105, 80HC(39)1, B-80MI41102⟩. Further reviews (including syntheses, reactivities, *etc.*) may be found in the following references: ⟨67AG786, B-67MI41101, B-70MI41101, 69CRV345, B-74MI41101⟩.

4.11.3 STRUCTURE

4.11.3.1 Theoretical Methods

Several types of molecular orbital calculation have been carried out on 1H- and 2H-1,2,3-triazole structures, often in combination with predictions of the most reactive sites, and studies of dipole moments, ^{1}H, ^{13}C, ^{14}N and ^{15}N NMR chemical shifts, proton exchange rates, gas-phase tautomerism and PE spectroscopy.

4.11.3.1.1 HMO methods

Using CNDO/2 calculations, relative base strengths for 1H- and 2H-1,2,3-triazole, as well as for 1H,2H- and 1H,3H-triazolium ions, have been obtained and compared with pK_a values. From the E_T values (*i.e.* total molecular energy), it was initially concluded that 1H-1,2,3-triazole is more stable than 2H-1,2,3-triazole; (**7**) → (**8**) is the most likely protonation step ⟨68TL3727, 66JCP(44)759⟩.

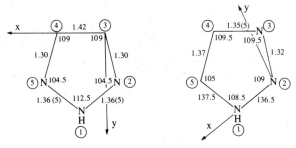

Furthermore, LCAO calculations point towards significant differences between dipole moments and primary $\pi \rightarrow \pi^*$ transitions of 1H- and 2H-triazoles ⟨69BSF1097⟩. Calculated geometries of 1H- and 2H-1,2,3-triazoles are shown in Figure 1 ⟨69BSF1097⟩.

Figure 1 Calculated geometries of 1H- and 2H-1,2,3-triazoles

In turn, from the total energy E_T it now follows that 2H-1,2,3-triazoles are more stable than the 1H-tautomers, where $E_T = E_\sigma + E_\pi + E_{\sigma\pi} + E_{\text{nucleus}}$. This last calculation was confirmed recently by PES spectra (see Section 4.11.3.2.10) and dipole moment measurements (see Section 4.11.3.2.2), comparing them with the corresponding N-methyl derivatives. A refined and optimized molecular structure has also been given for both the 1H- and 2H-tautomers (Figure 2) ⟨81ZN(A)1246⟩.

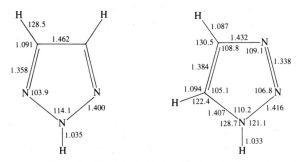

Figure 2 Final optimized molecular structures

In studying the relation between structure and reactivity in azole derivatives, CNDO/2 calculations have been carried out on heteroaromatic compounds including 1,2,3-triazoles

⟨73T3463, 73T3469⟩. This report deals with azolium ions representative of those studied experimentally and with the zwitterions resulting from the deprotonation of these ions.

The correlation between calculated deprotonation energies and observed deprotonation rates was examined, and the calculated charge distributions have been studied. The nature of the important factors (solvent effects, delocalization of the added negative charge into the σ-framework, coulombic effects) affecting deprotonation rates has been considered. Some experimental results (H^+ exchange rates) are in reasonable agreement with these calculations ⟨71ACS(B)249, 71ACS(B)2087⟩.

HMO calculations were carried out on 5-chloro-1-methoxy-1,2,3-benzotriazole (**9**), and the bond orders and π-electron charge densities were found to be as shown ⟨78JHC1043⟩. A comparison of (**9**) with the unsubstituted 1-methoxy-1,2,3-benzotriazole showed a small effect of the 5-chlorine on the bond orders ⟨76JHC509⟩.

(**9**)

4.11.3.1.2 Electronic structure; π-electron distribution; dipole moments

All-electron calculations for azoles have shown that there are good correlations between σ or total electron densities and ^1H and ^{13}C chemical shifts ⟨69MI41100⟩. A good relationship (correlation coefficient: 0.974) is found between π-electron densities calculated for 1,2,3-triazole (and other azoles) by the simple MO method ⟨B-61MI41101⟩ and ^{13}C chemical shifts ⟨68JA3543⟩. Furthermore, the proton shifts show linear correlations with the MO π-electron densities ⟨68CC1337⟩.

Table 1 NMR and π-Electron Density Data ⟨68CC1337⟩

Position on nucleus of 1,2,3-triazole	$\delta(^{13}C)$ (p.p.m.)[a]	$\delta(^1H)$ (p.p.m.)[b]	MO π-electron density
4	62.4	7.90	0.980
		In comparison with other azoles	

[a] Upfield from CS_2. [b] Downfield from TMS; CF_3CO_2H solutions.

LCAO calculations of the π-electron structures of 1H- and 2H-1,2,3-triazoles ⟨cf. 69BSF1097⟩ are found in ⟨70BSF273⟩; consideration is also given to the σ-skeleton with extended Hückel iterations. π-Electron charge distribution diagrams have been given for 1,2,3-triazole, other azoles, their benzo homologs and the 1-vinyl derivatives ⟨73MI41100⟩.

^{14}N chemical shifts (cf. Section 4.11.3.2.6) of 1H-, 2H-, 1-methyl- and 2-methyl-benzotriazoles correlate reasonably well with the SCF-MO-π-charge densities calculated by the Pariser–Parr–Pople approximation using special parameters for the methyl groups. It is rather striking that a fairly linear correlation exists between such calculations of π-charge densities (q_N^π) and the experimental ^{14}N chemical shifts ⟨72T637⟩.

Using *ab initio* wave functions and the full operators, dipole and second moments (cf. Section 4.11.3.2.2) as well as electronic charge distributions have been calculated for 1,2,3-triazoles and a number of other five- and six-membered heterocycles using the linear combination of Gaussian orbitals (LCGO) procedure (Table 2) ⟨74JCS(P2)420⟩. Population analyses for the compounds are found to be additive in bond contributions, and this assists in the interpretation of the σ- and π-dipole moments. Some further calculations of the dipole moments of 1H- and 2H-1,2,3-triazoles have been discussed using the vector additive system for the calculations ⟨75ZOB1821, 75DOK(224)847⟩.

Table 2 Dipole Moments (D) and Vector Components in 1,2,3-Triazoles ⟨74JCS(P2)420⟩

Compound	μ_{total}	$\mu_\perp$	$\mu_\parallel$	μ_σ	μ_π
1H-1,2,3-Triazole	4.50	3.26	3.10	2.79 (+81.7°)	2.87 (+83.0°)
2H-1,2,3-Triazole	−3.24	−3.26	0.0		

In the course of an N-methylation study and the analysis of the ^{1}H NMR spectra of indazoles, INDO calculations of the charge distribution in the benzotriazole anion have been determined (Figure 3). As the molecular geometry was not known, the total energy in the INDO semiempirical method was calculated for various geometric arrangements and allowed to achieve an optimal value by a sequence of one-dimensional parabolic minimizations ⟨75JCS(P2)1695⟩. The large angle at N-2 (for comparison: indazole 111.5°) is at first surprising but is reasonable since much of the negative charge in the anion is localized at N-1 and N-3. MO calculations carried out more recently by Servé et al. ⟨78JHC127⟩ give different values but show that N-1 is a more nucleophilic site than N-2.

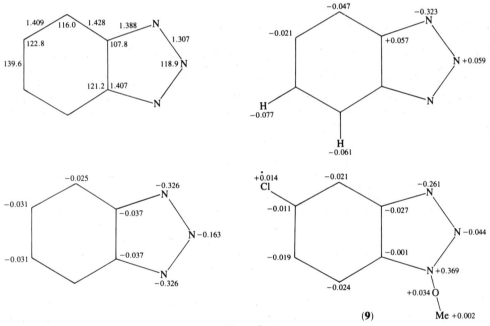

Figure 3

However, the use of HMO calculations to elucidate π-electron charge densities of 5-chloro-1-alkoxy-1,2,3-benzotriazole (9) led, as expected, to somewhat different results (Figure 3) ⟨78JHC1043⟩. Following these results, positions 4, 6 and 7 are the most likely sites of electrophilic attack carrying comparable π-charges.

4.11.3.2 Structural Methods

4.11.3.2.1 X-Ray diffraction

No measurements have been made to date on the parent compounds 1,2,3-triazole and -benzotriazole. However, since 1975 several 1,2,3-triazoliosulfide derivatives have had their crystal structures determined.

The mesoionic 1,3-disubstituted 1,2,3-triazolio-4-sulfides can be regarded as resonance hybrids of several forms (**A**)–(**E**); ^{13}C NMR data indicate a localization of charge on N-1 ⟨74ACS(B)61⟩.

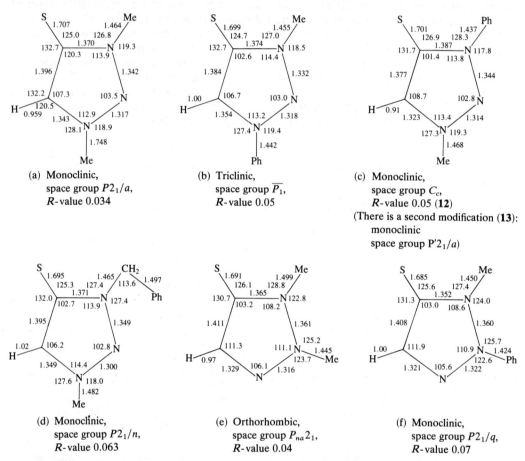

Some examples of differently substituted mesoionic 1,2,3-triazolio-4-sulfides are shown in Figure 4, including interatomic distances (Å), bond angles (°), crystal system and space group. Analysis of the triazole rings in (**10**)–(**14**) shows that the heterocyclic ring is almost planar. However, in (**14**) the methyl group, S atom and benzylic CH$_2$ group are significantly out of plane.

Figure 4 Some parameters of structure determinations of: (a) 1,3-dimethyl-4-(1,2,3-triazolio)sulfide (**10**) ⟨75ACS(A)647⟩; (b) 1-phenyl-3-methyl-4-(1,2,3-triazolio)sulfide (**11**) ⟨79ACS(A)687⟩; (c) 1-methyl-3-phenyl-4-(1,2,3-triazolio)sulfide (**12, 13**) ⟨79ACS(A)687⟩; (d) 1-methyl-3-benzyl-4-(1,2,3-triazolio)sulfide (**14**) ⟨79ACS(A)693⟩; (e) 2,3-dimethyl-1,2,3-triazol-1-ine-4-thione (**15**) ⟨79ACS(A)697⟩; (f) 2-phenyl-3-methyl-1,2,3-triazol-1-ine-4-thione (**16**) ⟨79ACS(A)697⟩

A comparison of the geometry of the triazole nucleus in the above mentioned triazoliosulfides was carried out. The resulting bond lengths and angles for the average triazole nucleus are shown in (**17**). Apparently methyl and phenyl substituents can be interchanged with negligible effect on the geometry of the triazole ring. The almost identical values for the angles at N-1 and N-3 show that these are almost unaffected by the asymmetry induced by the carbon substituents ⟨79ACS(A)693⟩. Calculation of the N(1)—C(5), N(1)—N(2) and N(2)—N(3) bond characters was carried out in order to distinguish between the canonical forms (**A**)–(**E**). This led to the conclusion that the resonance hybrid is made up predominantly of form (**A**), and almost equal amounts of forms (**B**) and (**C**) ⟨75ACS(A)647⟩.

1,2,3-Triazoles and their Benzo Derivatives

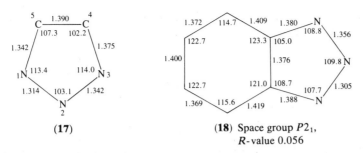

Although the parent compound 1,2,3-triazole has not been studied, the crystal structure of benzotriazole (**18**) has been determined ⟨74AX(B)1490⟩. Benzotriazole crystallizes in the space group $P2_1$. In association with its remarkable anticorrosion activity (Section 4.11.6.4.2) benzotriazole and its conjugate base form stable transition metal complexes whose crystal structures have, in part, been studied. Some examples are presented in Figures 5–7, with special regard to the molecular dimensions of the benzotriazole moiety.

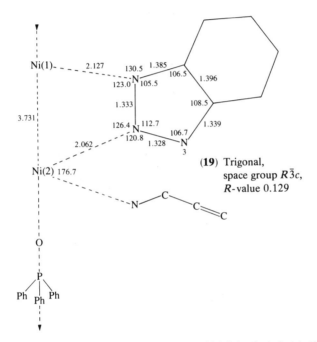

Figure 5 Partial structural details for hexakis(benzotriazolyl)hexakis(allylamine)trisnickel(II) triphenylphosphine oxide complex (**19**; Ni[(BT)$_3$Ni(AA)$_3$]$_2$ · 2Ph$_3$PO) ⟨76AX(B)714⟩

Structure and bonding in transition metal triazenido complexes have attracted much interest in recent years. The triazenido ligand (**20**), *i.e.* the conjugate base of benzotriazole, represents a versatile group which may function as a bridging group between two metal centers, a bidentate three-electron donor or a monodentate one-electron donor. Partial structural details for an organometalic benzotriazenide complex are given in Figure 6. The structures of other triazenido complexes, such as *trans*-bis(triphenylphosphane)carbonyl(1,3-di-*p*-tolyltriazenido)hydridoruthenium(II) and *cis*-bis(triphenylphosphane)-bis(1,3-diphenyltriazenido)platinum(II), have been evaluated ⟨76IC2788, 76IC2794⟩.

(**20**)

Additional benzotriazole transition metal complexes have been described recently, *e.g.* di-μ-chloro-bis[bis(benzotriazole)chlorocopper(II)]monohydrate (**22**; Figure 7) ⟨81ACS(A)733⟩. Coordination to copper has only a small effect on the geometry of the (nearly planar) benzotriazole groups as shown by comparison with benzotriazole itself ⟨74AX(B)1490⟩.

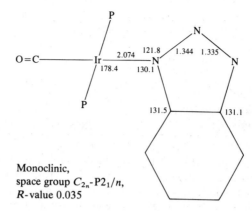

Figure 6 Partial structural details for bis(triphenylphosphane)carbonyl(triazenido)iridium(I) (**21**; Ir(BTA)(CO)-(PPh$_3$)$_2$) ⟨78IC3026⟩

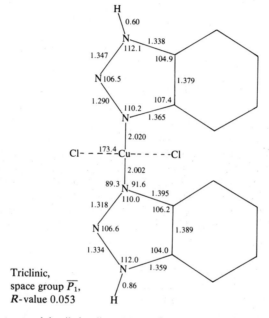

Figure 7 Partial structural details for di-μ-chlorobis[bis(benzotriazole)chlorocopper(II)] monohydrate (**22**)

The structure consists of layers of Cu and Cl atoms separated by layers of almost parallel benzotriazole ligands.

Furthermore, the crystal structures of benzotriazolium tetrachlorocobaltate(II) [(BTAH$_2$)$_2$(CoCl$_4$)], bis(benzotriazole)dichlorozinc(II) [ZnCl$_2$(BTAH)$_2$] and polymeric tetrakis(benzotriazolato)dizinc(II) {[Zn$_2$(BTA)$_4$]$_n$} have been investigated by X-ray diffraction techniques (BTAH = 1H-1,2,3-benzotriazole; BTAH$_2$ = protonated form) ⟨81ACS(A)739⟩. A summary of bond lengths (Å) and angles (°) for the (BTAH$_2$)$_2$ ligand is shown in Table 3. The BTAH$_2$ ion is nearly planar and has nearly C_{2v} symmetry, the H atoms being bonded to N-1 and N-3.

The crystal structure of a trinuclear nickel(II) benzotriazolate amine complex, [Ni$_3$(BTA)$_6$(NH$_3$)$_6$]·2Me$_2$CO·H$_2$O has also been determined ⟨81ACS(A)747⟩.

4.11.3.2.2 *Microwave spectroscopy and dipole moment measurements*

For an introduction to and general literature on, this topic see Section 4.01.3.2 as well as ⟨74PMH(6)53⟩.

Magnetic susceptibility anisotropy (MSA) (measured from the Zeeman splitting of microwave absorption spectra) is defined for planar systems in terms of the in-plane and

Table 3 Some Bond Distances and Bond Angles for $(BTAH_2)_2$ in $(BTAH_2)_2[CoCl_4]$ [a] ⟨81ACS(A)739⟩

Bond distance (Å)		Bond angle (°)	
N(1)—N(2)	1.311	C(5)—N(1)—N(2)	113.4
N(2)—N(3)	1.313	N(1)—N(2)—N(3)	104.6
N(3)—C(4)	1.361	N(2)—N(3)—C(4)	112.7
C(4)—C(5)	1.390	N(3)—C(4)—C(5)	105.1
C(5)—N(1)	1.359	C(4)—C(5)—N(1)	104.1

[a] Monoclinic; space group $C2/c$; $R = 0.025$.

out-of-plane total susceptibilities:

$$\text{MSA} = \chi_{oop}^T - (\chi_{ip}^T)_{av}$$

The diamagnetic susceptibility anisotropy (DSA) is linearly related to the binding energy (BE), the correlation being

$$\chi_{oop}^d - (\chi_{ip}^d)_{av} = 0.048 Be + 109$$

for azoles like 1,2,3-triazoles. DSA seems not to be related to aromaticity ⟨74TL253⟩. Furthermore, values of the diamagnetic susceptibility (χ_{aa}^d) for $1H$-1,2,3-triazole were calculated to be $-\chi_{oop} = 288.8$ and $(-\chi_{ip})_{av} = 174.2$ ⟨74TL253⟩.

The hyperfine structure in the microwave spectra of 1,2,3-triazole and N-deutero-1,2,3-triazole was analyzed. The coupling constants derived from the analysis of each isotopic species were combined and led to the principal nuclear quadrupole coupling constants at the sites of the three non-equivalent ^{14}N nuclei ⟨77JSP(65)313⟩.

Experimental dipole moments of azoles are collected in ⟨B-63MI41100⟩; values for three compounds are as follows: 1,2,3-triazole 1.79 (benzene, 25 °C), benzotriazole 4.10 (dioxane, 25 °C), 1-methylbenzotriazole 4.1 D (benzene, 25 °C). More recent measurements combined with CNDO/2 calculations have been carried out by Mauret et al. ⟨75BSF1675⟩; the results are shown in Table 4. The measurement of molecular polarizations represents not only a valuable method for investigating autoassociation effects of azoles, but can also be widely used for the determination of tautomerism in the azole field ⟨75BSF1675, 76AHC(S1)32, 76AHC(S1)281⟩.

Table 4 Experimental Dipole Moments of Some 1,2,3-Triazoles in Benzene at 25 °C ⟨75BSF1675⟩

Compound	α	β	$P_{2\infty}$	RM_D	μ (D)
1,2,3-Triazole	5.45	−0.38	86.56	16.70	1.85
1,2,3-Triazole (45 °C)	6.34	−0.48	100.02	16.70	2.08
1-Methyl-1H-1,2,3-triazole	26.24	−0.38	429.08	21.54	4.46
2-Methyl-2H-1,2,3-triazole	0.19	−0.27	24.43	21.54	0.37
Benzotriazole	16.23	−0.45	388.42	35.55	4.15
Benzotriazole (45 °C)	14.68	−0.67	325.93	35.55	4.12
1-Methyl-1H-benzotriazole	13.08	−0.32	359.90	40.39	3.95
2-Methyl-2H-benzotriazole	0.41	−0.26	45.41	40.39	0.49

By comparing the dipole moments of 1,2,3-triazole with its two 1-methyl-1H (**23**) and 2-methyl-2H derivatives (**24**) in benzene at 25 °C, it was found that 83% of 1,2,3-triazole existed as the symmetrical 2H-form (**2**). From a second measurement at 45 °C the thermodynamic parameters of the equilibrium were calculated ⟨73MI41101⟩. This result was confirmed by another group ⟨78MI41100⟩. Further, dipole measurements show similarly a large predominance of the 'asymmetric' 1H-benzotriazole tautomer in benzene at 25 and 45 °C established by comparison with its 2-N-methyl derivatives ⟨74MI41100⟩.

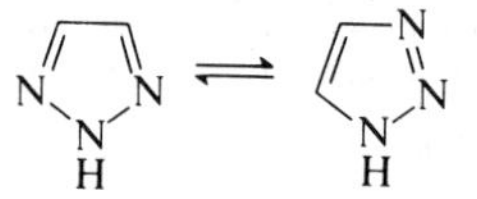

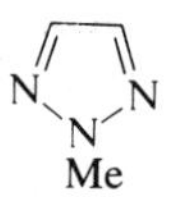

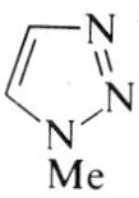

(2) 83%   (1) 17%   (23)   (24)
2H   1H

Dipole moment measurements have been used for structure determination and for discussion of the probable conformation of two possible isomeric 1-(α-aroyloxyarylideneamino)-1,2,3-triazoles (25) formed by oxidation of bis-aroylhydrazones of α-dicarbonyl compounds ⟨77JCS(P2)1779, 79JHC571⟩. The probable conformation of these compounds was determined after theoretical calculation of the dipole moments by vector addition of the moments of all polar groups, the results being very similar to those of X-ray measurements.

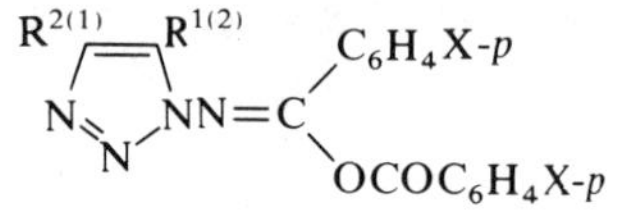

(25)

4.11.3.2.3 ^1H NMR spectroscopy

The first systematic studies of ^{1}H NMR spectra of 1,2,3-triazoles and derivatives have been reported by two groups ⟨67JCS(B)516, 67BSF2998⟩, and a review has appeared ⟨B-73NMR219⟩; see also ⟨71PMH(4)121⟩.

Spectral parameters for a number of simple derivatives are listed in Table 5. The ^{1}H NMR spectrum of the parent compound was first reported by Gold ⟨65LA(688)205⟩, who assigned the signals at δ 7.9 p.p.m. to H-4 and H-5, and that at 15.9 p.p.m. to the NH proton. Furthermore, he showed that the symmetrically substituted N^1-(2-cyanoethyl)-1,2,3-triazole gave a doublet at $\delta \sim 7.93$ p.p.m., while the symmetrical N^2-(2-cyanoethyl)-1,2,3-triazole gave a singlet at δ 8.15 p.p.m.

Table 5 ^{1}H NMR Spectral Data for Triazoles

Substituents	Species	Solvent	$\delta(^1H)$ (p.p.m.) H-4	H-5	NR	Ref.
—		Neat	7.86	7.86	12	67BSF2998
		CDCl$_3$	7.75	7.75	12.05	67BSF2998
		DMSO	7.91	7.91	13.50	67BSF2998
		C$_6$D$_6$	7.30	7.30	13.40	67BSF2998
	Cation	TFA	8.60	8.60		67JCS(B)516
	Anion	NaOD	7.86	7.86		67JCS(B)516
		D$_2$O	8.00	8.00		67JCS(B)516
		MeCN	7.80	7.80		67JCS(B)516
1-Me		Neat	7.89	8.12	4.20	67BSF2998
		CDCl$_3$	7.74	7.59	4.10	67BSF2998
		DMSO	7.72	8.08	4.09	67BSF2998
		C$_6$D$_6$	7.40	6.90	3.37	67BSF2998
		HMPT	7.63	8.37	4.15	67BSF2998
	Cation	TFA	8.44	8.49	4.48	67JCS(B)516
1-Me, 5-Br		CDCl$_3$	7.88		3.98	67JCS(B)516
	Cation	TFA	8.48		4.42	67JCS(B)516
2-Me		CDCl$_3$	7.57	7.57	4.18	67BSF2998
		DMSO	7.77	7.77	4.18	67BSF2998
		C$_6$D$_6$	7.32	7.32	3.65	67BSF2998
		HMPT	7.75	7.75	4.20	67BSF2998
2-Me, 4-Br		CDCl$_3$		7.55	4.14	67JCS(B)516
	Cation	TFA		7.80	4.40	67JCS(B)516
1-Me, 4-CO$_2$H		CDCl$_3$		8.25	4.18	67JCS(B)516
	Cation	TFA		8.81	4.50	67JCS(B)516
1-Tosyl, 5-Me		CDCl$_3$	7.20	2.30	7.50	60JA5007
1-Ph, 5-Me		CDCl$_3$	7.30	2.21	7.30	60JA5007

In the same way, in 1-methyl-1,2,3-triazole (**26**) in $CDCl_3$ H-4 and H-5 are not equivalent and the assignment was made by comparison with 1-methyl-4-bromo-1,2,3-triazole (**27**) ⟨67BSF2998⟩.

(**26**)

(**27**)

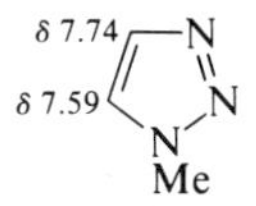

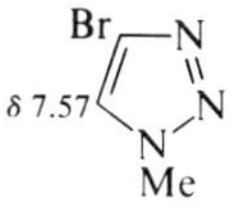

In their study of the stereochemistry of the 1,3-dipolar cycloaddition of azides to enol ethers (*cf.* Section 4.11.5.2.2), Huisgen and Szeimies ⟨65CB1153⟩ have reported ^{1}H NMR spectral data for some Δ^2-1,2,3-triazolines. On the basis of the chemical shifts and coupling constants of H-4 and H-5, structural assignment of the stereoisomeric triazolines (**28**) and (**29**) was made as shown.

(**28**) $^3J_{4,5}$ 2.0 Hz

(**29**) $^3J_{4,5}$ 7.5 Hz

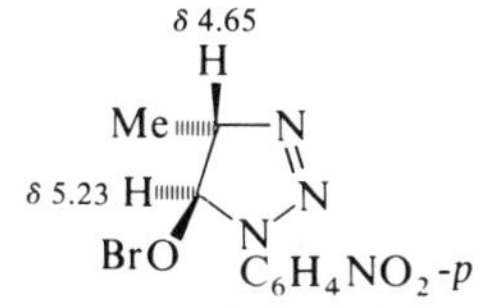

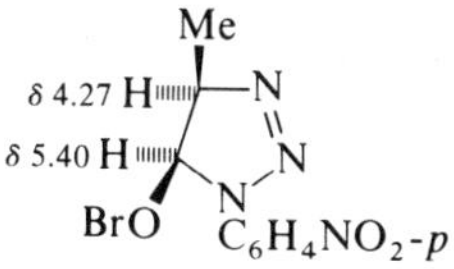

N^2-Silylated 1,2,3-triazole (**30**) reacts with acetyl chloride to afford two isomeric acetyl derivatives (**31a**) and (**31b**) ⟨67CB3485⟩, the assignment being made according to their ^{1}H NMR spectra. Furthermore, at elevated temperatures (>120 °C) 1-acetyl-1,2,3-triazoles like (**32**) are capable of a reversible acyl group migration to (**33**). This equilibrium was analyzed by means of ^{1}H NMR data ⟨67CB3485⟩, H-4 and H-5 resulting in one signal in (**33**).

Using temperature-dependent ^{1}H NMR (in acetone) thermodynamic and kinetic phenomena of the tautomerism in the azole series were studied. For 1,2,3-triazole at room temperature a predominance of the 2*H*-tautomer (singlet at δ 7.83 p.p.m. for H-4 and H-5) was found, and at −90 °C a 50/50 mixture of the 1*H*- and 2*H*-tautomers (two additional doublets at δ 8.32 and 7.83 p.p.m.) was evident ⟨69TL495⟩.

(**30**) (**31a**) 75% (**31b**) 25%

(**32**) $J_{4,5}$ 1.3 Hz (**33**)

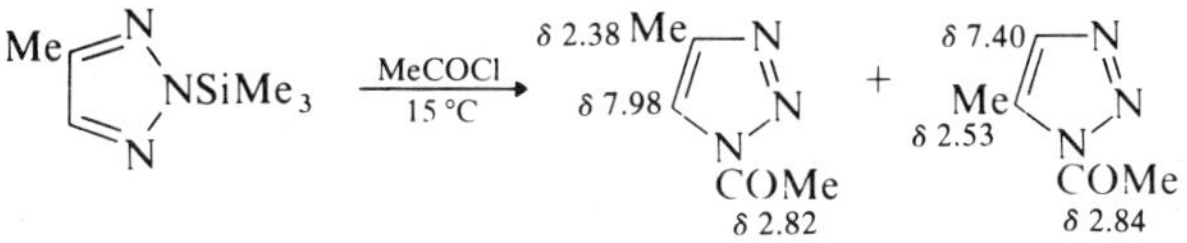

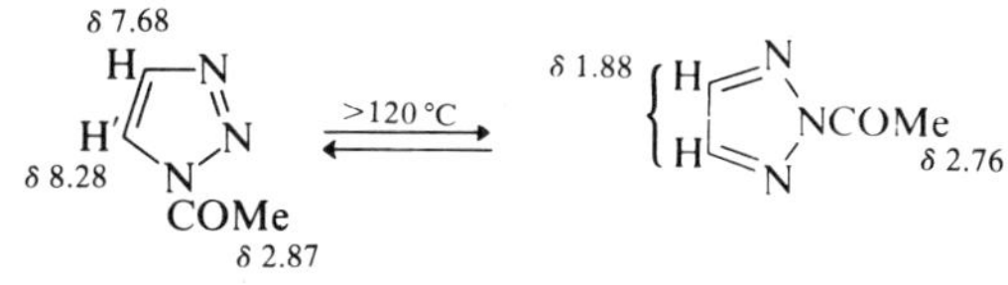

In substituted triazoles (obtained, for example, from phosphorus ylides and carbonyl-azides) ^{1}H NMR is a valuable tool for determining the constituents and the progress of a reaction, as in the acyl group migration (**34**)→(**35**) ⟨70TL5225⟩. In the case of the isomeric and regioisomeric triazoles (**36**, **37**) with adjacent aroyl or ethoxycarbonyl substituents ^{1}H NMR data can be successfully applied in structural determination ⟨71T5623⟩. ^{1}H NMR data have also been used to assign the position of acyl groups in 4-substituted 5-amino-1,2,3-triazoles ⟨71JCS(C)706⟩.

(**34**) (**35**)

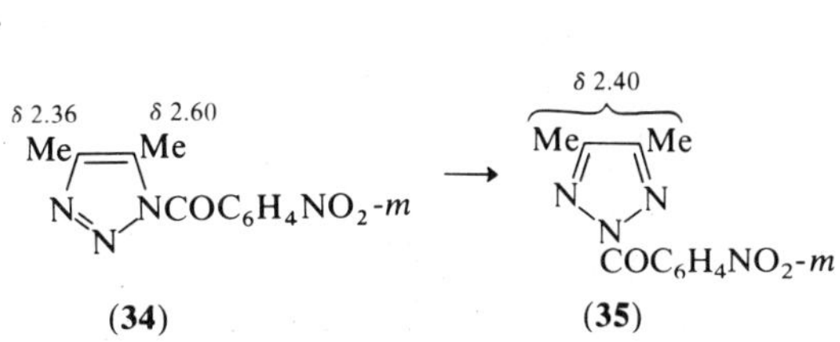

$$PhC\equiv CH + EtO_2CN_3 \longrightarrow$$

δ 8.0–7.3(m) δ 8.36   δ 7.67 δ 7.44(s)

Structures (36) and (37): 1-ethoxycarbonyl-4-phenyl and 1-ethoxycarbonyl-5-phenyl 1,2,3-triazoles.

The NMR spectra of benzotriazole and its 1-methyl and 2-methyl derivatives have been described. The 1-methyl derivative shows strong coupling between the N-methyl group and the benzene ring hydrogen. The coupling constants suggest also a quinone-type structure for the 2-methyl compound (cf. compound (97a) and Table 12a, Section 4.01.3.3) ⟨63JCS5556, 68JSP(26)139⟩.

4.11.3.2.4 ^{13}C NMR spectroscopy

The ^{13}C NMR spectrum of 1,2,3-triazole was first reported by Weigert and Roberts ⟨68JA3543⟩, the solvent being acetone; $\delta(^{13}C)$ (p.p.m.): C-4 = 130.4, C-5 = 130.4; $J_{C,H-4}$ = 205, $J_{C,H-5}$ = 13.4 Hz. This equivalence of both C-4 and C-5 points once again to the symmetrical $2H$-form.

The ^{13}C shieldings for 1,2,3-triazole can be predicted by a four-parameter empirical equation. Relative to the cyclopentadienyl anion, the carbon shifts are given by the equation $\delta = N_\alpha C_\alpha + N_\beta C_\beta + N_{\alpha\beta} C_{\alpha\beta} + N_{\beta\beta} C_{\beta\beta}$ (N_i = number of nitrogen atoms in the ith position; C_i = shielding parameter associated with those nitrogens) ⟨68JA3543⟩.

Empirical values: $C_\alpha = 15$, $C_\beta = 4$, $C_{\alpha\beta} = 10$, $C_{\beta\beta} = -7$ p.p.m.

For 1,2,3-triazole: $\delta = C_\alpha + C_\beta + C_{\alpha\beta} + C_{\beta\beta} = 26$ p.p.m.

$\delta(^{13}C)_{triazole} = 103^* + 26 = 129$ p.p.m.

observed = 130.4 p.p.m.

(* $\delta(^{13}C)$ of a cyclopentadienyl anion)

In a systematic study, the effects of substitution, lanthanide shift reagents, solvent changes and tautomerism on the ^{13}C NMR spectra of azoles were investigated ⟨74JOC357⟩. Some results are given in Table 6.

Table 6 ^{13}C Chemical Shifts of 1,2,3-Triazoles ⟨74JOC357⟩

Compound	Solvent	$\delta(^{13}C)$ (p.p.m.) C-4	C-5	NMe	Predominant tautomeric form
1,2,3-Triazole	Dioxane	131.6	131.6		$2H$
1,2,3-Triazole	DMSO	130.3	130.3		$2H$
2-Methyl-1,2,3-triazole	Dioxane	134.7	134.7	41.5	$2H$
2-Methyl-1,2,3-triazole	DMSO	133.2	133.2		$2H$
1-Methyl-1,2,3-triazole	DMSO	134.3	125.5	35.7	$1H$
1-Methyl-1,2,3-triazole	DMSO	132.6	124.8	35.2	$1H$
1-Methyl-1,2,3-triazole + Eu(fod)$_3$	CDCl$_3$ (1 M)	135.0	125.6	37.4	$1H$
Eu(fod)$_3$ shift		−6.6	+7.4	−5.7	

As complexation of the shift reagent takes place predominantly on N-3, the carbon signal of C-5 is accordingly shifted upfield. From the calculated charge densities ⟨72T637⟩, a reduced affinity of N-3 for Eu(fod)$_3$ and, thus, smaller shift effects can be expected. Furthermore, simple chemical shifts are of limited value in ascertaining the positions of tautomeric equilibria for rapidly interconverting azole tautomers. However, for this purpose $^1J(^{13}C-H)$ values, which are only slightly influenced by N-methylation, can be useful ⟨74CC702⟩. Thus, in 1,2,3-triazole the $^1J(^{13}C-H)$ value of 194.3 Hz for C-4 (i.e. C-5) falls within the range of that for C-4 in 2-methyl-$2H$-1,2,3-triazole ($^1J(^{13}C-H)$ 192.5 Hz). The average value for the C-4 (194.3 Hz) and C-5 (196.6 Hz) methine groups in 1-methyl-$1H$-1,2,3-triazole implies that in CDCl$_3$ solution at 32 °C, 1,2,3-triazole exists as a 2:3 equilibrium mixture of the $2H$- and $1H$-tautomers. The intermediacy of (38) or a cyclic dimer in the interconversion of the tautomers seems likely. The calculated $^1J(^{13}C-H)$ value for the latter species

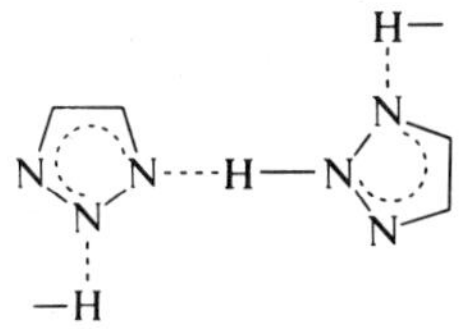

(38)

[(196.5 + 192.5 + 194.3 + 192.5)/4 = 194.0 Hz] is in good agreement with the observed value ⟨74CC702⟩. This is confirmed also by other physical characteristics. All the parent azoles possess similar dipole moments (1.7–2.7 D) in the diluted gas phase according to microwave spectroscopy; in solution however, pyrazole and 1,2,3-triazole exhibit low dipole moments (1.6–1.8 D), whereas imidazole, 1,2,4-triazole and tetrazole have higher values (3.2–5.1 D) ⟨43MI41100, 76JCS(P2)736⟩.

Table 7 ^{13}C–H Coupling Constants of 1,2,3-Triazoles ⟨76JCS(P2)736⟩

Compound	Solvent	$\begin{Bmatrix} ^1J(C-H) \\ ^2J(C-H) \end{Bmatrix}$(Hz)	
		C-4	C-5
1,2,3-Triazole	CDCl$_3$	194.3	194.3
		13.2	13.2
1,2,3-Triazole	Acetone-d_6	193.9	193.9
		13.6	13.6
1-Methyl-1,2,3-triazole	CDCl$_3$	194.3	196.6
		11.7	15.6
2-Methyl-1,2,3-triazole	CDCl$_3$	192.5	192.5
		12.8	12.8
4-Methyl-1,2,3-triazole	CDCl$_3$		191.8
	Acetone-d_6 ^{a}		191.2
1,4-Dimethyl-1,2,3-triazole	CDCl$_3$		192.7
1,5-Dimethyl-1,2,3-triazole	CDCl$_3$	191.3	
2,4-Dimethyl-1,2,3-triazole	CDCl$_3$	189.5	

^{a} ^3J(C–H) = 3.9 Hz.

Table 8 Conformer Distribution in N-Acetyl-1,2,3-triazole (39) and -benzotriazole (40) in CDCl$_3$ ⟨78JCS(P2)99⟩

	%Z at 34 °C	Shift difference calculated for position
(39)	0	C-5
(40)	100	C-7a

In Table 7 are shown some ^{13}C–H coupling constants for a variety of substituted 1,2,3-triazoles. Furthermore, ^{13}C NMR spectral data of N-acetyl-azoles (39) and -benzazoles (40) have been compared with data for the corresponding N-methyl compounds in order to obtain information about the conformation of the acetyl compounds ⟨78JCS(P2)99⟩. These results are given in Table 8.

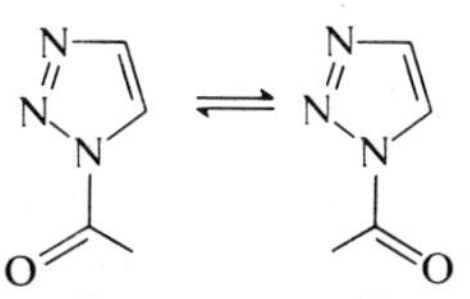

(39) Z ⇌ E   (40) Z ⇌ E

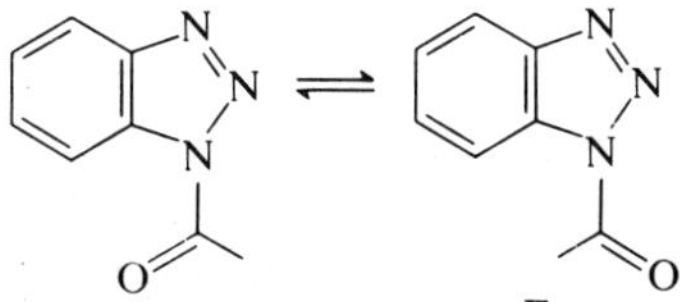

A ^{13}C NMR study of poly[4(5)-vinyl-1,2,3-triazole] in DMSO revealed that the spectrum at room temperature displays three sets of two triazole carbon atoms corresponding to the three tautomers: 1H (15–17%), 2H (66–70%) and 3H (15–17%), according to (41) ⟨76MI41101⟩. While the monomer 4(5)-vinyl-1,2,3-triazole (42) showed only one set of four signals in the same solvent, the two triazole carbon signals being slightly broadened, a ^{13}C NMR study at −55 °C allowed direct observation of the three separate tautomers, as shown

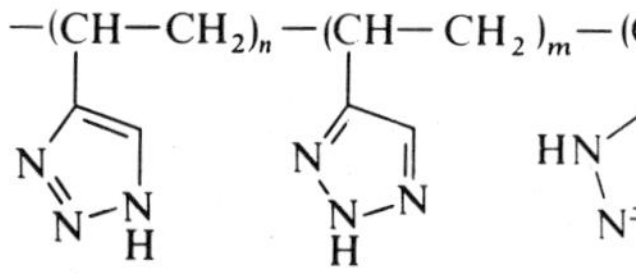

−(CH−CH$_2$)$_n$−(CH−CH$_2$)$_m$−(CH−CH$_2$)$_p$−

(41)   1H   2H   3H

Table 9 ^{13}C NMR Chemical Shifts of (42) (1.6M in DMF) ⟨78OMR(11)578⟩

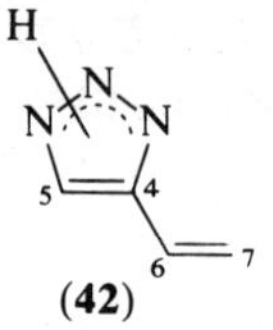

(42)

T (°C)	C-4	C-5	C-6	C-7	% Tautomer
40	142–144	126–129	124.6	115	Averaged
−55	144	119.6	125.4	113.5	1H (20%)
	145.1	130.2	125.4	115.7	2H (70%)
	135	130.2	120.5	116.5	3H (10%)

in Table 9 ⟨78OMR(11)578⟩. The assignment of the absorptions is based on comparison with N-substituted derivatives (43)–(46).

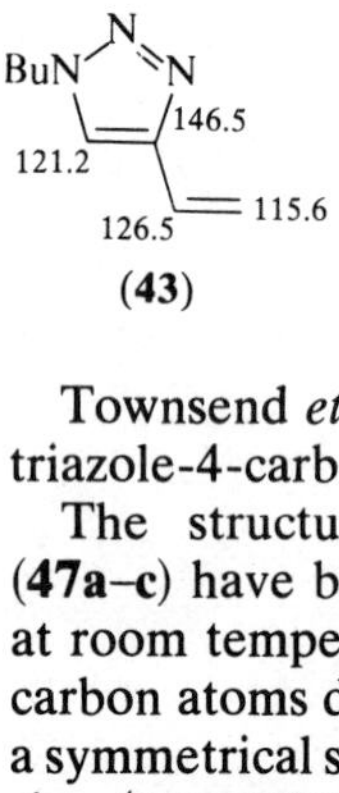

(43) 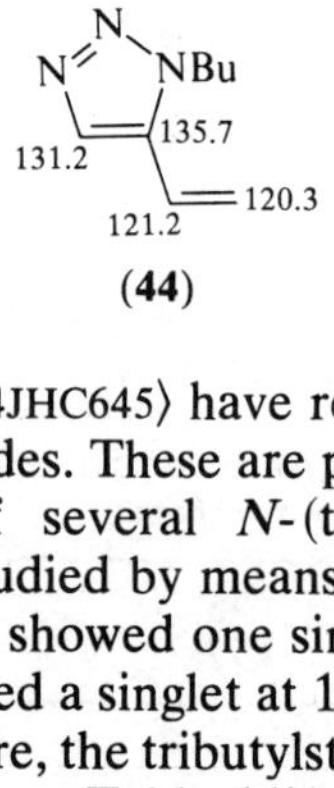   (44) 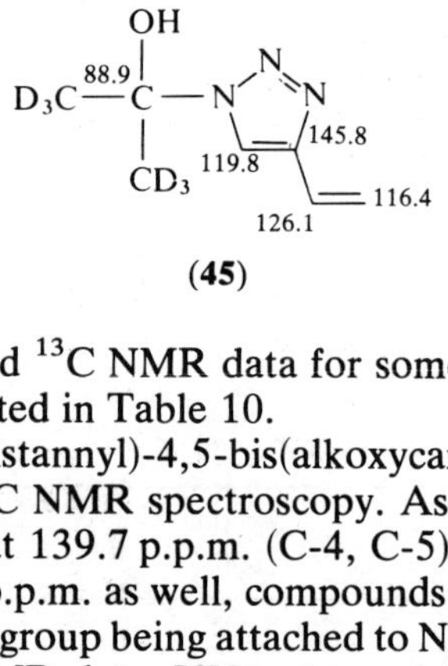   (45) 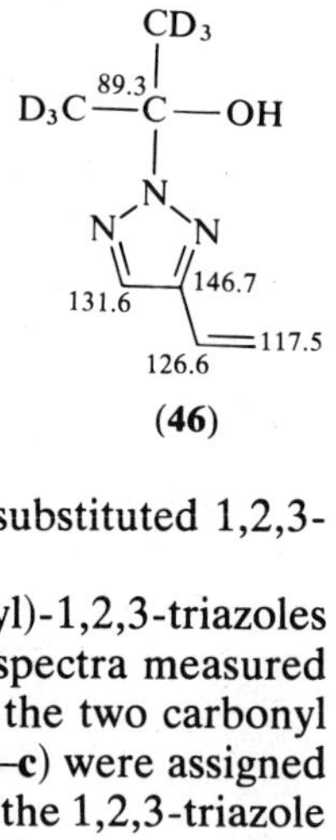   (46)

Townsend et al. ⟨74JHC645⟩ have reported ^{13}C NMR data for some N-substituted 1,2,3-triazole-4-carboxamides. These are presented in Table 10.

The structures of several N-(trialkylstannyl)-4,5-bis(alkoxycarbonyl)-1,2,3-triazoles (47a–c) have been studied by means of ^{13}C NMR spectroscopy. As the spectra measured at room temperature showed one singlet at 139.7 p.p.m. (C-4, C-5) and the two carbonyl carbon atoms displayed a singlet at 161.9 p.p.m. as well, compounds (47a–c) were assigned a symmetrical structure, the tributylstannyl group being attached to N-2 of the 1,2,3-triazole ring ⟨77JOM(127)273⟩ (see Table 11). ^{1}H NMR data, UV spectra and dipole moments also confirm that the trialkylstannyl group is located at N-2 ⟨72JOM(44)117⟩.

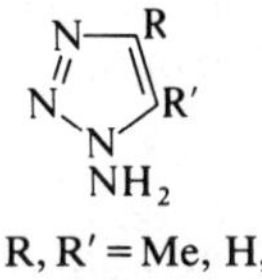

(48) R, R' = Me, H, aryl 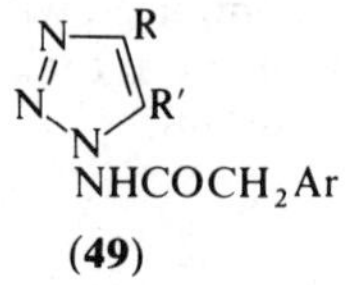   (49)

Recently 4,5-unsymmetrically substituted 1-amino- (48) and 1-(N-arylacetylamino)-1,2,3-triazoles (49) have been identified by their ^{13}C NMR spectra ⟨80JHC1127⟩. A signal at 11 ± 0.6 p.p.m. indicates a 4-methyltriazole derivative, whereas a signal at 7.9 ± 1 p.p.m. indicates a 5-methyltriazole. A signal at 120 ± 5 p.p.m. (C-5) indicates a hydrogen atom in the 5-position. Some values are given in Table 12.

4.11.3.2.5 ^{14}N and ^{15}N NMR spectroscopy

^{14}N and ^{15}N NMR spectroscopy have been applied only recently to heterocyclic compounds, and especially azoles. However, one must take into consideration the fact that the

Table 10 ^{13}C NMR Chemical Shifts for Certain 1,2,3-Triazole-4-carboxamides (in DMSO-d_6) ⟨74JHC645⟩

Structure	CONH$_2$	δ(^{13}C) (p.p.m.) C-4	C-5	ΔC-4	ΔC-5
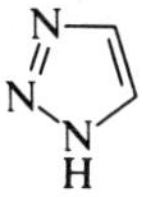 1H-1,2,3-triazole		130.4	130.4		
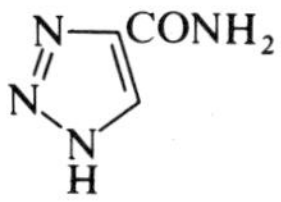 1H-1,2,3-triazole-4-carboxamide	161.65	141.78	129.93		
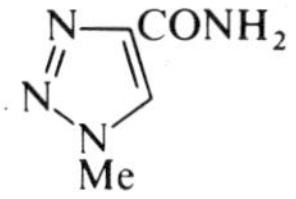 1-Me-1,2,3-triazole-4-carboxamide	161.64	142.96	127.28	−1.18	+2.65
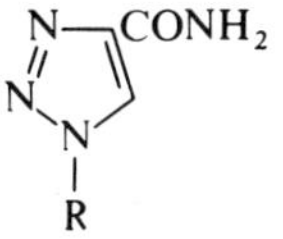 1-R-1,2,3-triazole-4-carboxamide	161.53	143.34	125.01	−1.56	+4.92
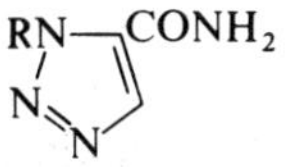 2-R-1,2,3-triazole-4-carboxamide	158.97	131.64	134.63	+10.14	−4.70

R = β-D-ribofuranosyl

Table 11 Some Selected ^{13}C NMR Data for 1,2,3-Triazoles (**47a–c**) ⟨77JOM(127)273⟩

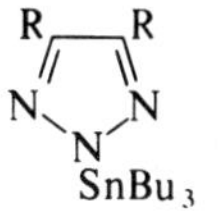

Compound	R	T (°C)	C-4/C-5	δ(^{13}C) (p.p.m.) C=O	CH$_3$	CH$_2$
(**47a**)	CO$_2$Me	r.t.	139.7	161.8	52.5	—
		−50	137.7	161.5	52.8	—
(**47b**)	CO$_2$Et	r.t.	139.5	161.7	14.1	61.6
		−50	139.3	161.8	14.2	61.7
(**47c**)	Et	r.t.	146.8	—	14.2	18.3
		−50	145.8	—	14.7	18.1

Table 12 ^{13}C NMR Chemical Shifts of Various Substituted Triazoles ⟨80JHC1127⟩

Compound	R	R'	δ(^{13}C) (p.p.m.) 4-Me	5-Me
(**48**)	Me	Me	10.6	7.4
(**48**)	Ph	Me	—	8.9
(**48**)	Me	Ph	11.5	—
(**49**)	Me	Me	10.4	6.9
(**49**)	Ph	Me	—	8.3
(**49**)	Me	Ph	11.1	—

^{14}N nucleus is inherently difficult to study, not only on account of its low relative sensitivity but also because it has an electric quadrupole moment. Very often quadrupolar interactions can dominate the information available from a study of the ^{14}N nucleus and other nuclei spin-coupled to it. Two monographs have appeared in this field ⟨B-73MI41103, B-79MI41103⟩.

The first systematic ^{14}N NMR investigations of azoles (and 1,2,3-triazoles and benzotriazoles) was reported by Witanowski *et al.* ⟨72T637⟩. The ^{14}N chemical shifts are referred either to MeNO$_2$ or Me$_4$N$^+$X$^-$; some results are given in Table 13. Furthermore, it was found that triazoles and benzotriazoles show a linear relationship between chemical shifts and the SCF–PPP–MO π-charge densities. In the case of unsubstituted 1,2,3-triazole, the ^{14}N spectra show that it exists 70–100% in the symmetrical 2*H*-form. However, due to the appreciable quadrupole broadening of the signals, the results from ^{14}N NMR spectroscopy are not always reliable, and in some cases are difficult to reproduce.

Table 13 ^{14}N Chemical Shifts of 1,2,3-Triazoles and Benzotriazoles ⟨72T637⟩

Compound	Solvent	$\delta(^{14}N)$ (p.p.m.) MeNO$_2$ [a]	Me$_4$N$^+$ X$^-$ [a]
1,2,3-Triazole	Neat	60 ± 8 (N, N)	272
		132 ± 4 (NH)	200
	MeOH (1:1)	60 ± 8 (N, N)	272
		128 ± 6 (NH)	204
1-Methyl-1,2,3-triazole	Neat	22 ± 1 (N, N)	310
		143 ± 1 (NMe)	188
	MeOH (1:1)	28 ± 1 (N, N)	304
		144 ± 1 (NMe)	188
2-Methyl-1,2,3-triazole	Neat	51 ± 1 (N, N)	282
		130 ± 1 (NMe)	202
	MeOH	53 ± 2 (N, N)	280
		132 ± 2 (NMe)	200
Benzotriazole	Dioxane (satd.)	81 ± 7	252
	MeOH	89 ± 7	244
1-Methylbenzotriazole	Acetone (satd.)	40 ± 6 (N, N)	292
		148 ± 5 (NMe)	184
2-Methylbenzotriazole	Neat	50 ± 8 (N, N)	282
		118 ± 2 (NMe)	214

[a] Internal reference.

Accordingly, nitrogen NMR spectroscopy turns more to the ^{15}N isotope using liquid NH$_3$ at 25 °C in a sealed tube as internal reference (δ 0 p.p.m.); NH$_4$NO$_3$ (in aqueous HNO$_3$, δ 21.6 p.p.m.), MeNO$_2$ (δ 379.60 p.p.m.) and NaNO$_3$ (δ 376.53 p.p.m. in saturated aqueous solution) may also be used. The future technique with a signal enhancement factor of 10 is INEPT (Intensity Nuclei Enhancement by Polarization Transfer); *cf.* ⟨80JA429⟩, using sensitivity enhancement/polarization transfer with INEPT pulse sequence. Up to now no ^{15}N NMR of spectra 1,2,3-triazoles have been published.

4.11.3.2.6 UV spectroscopy

The first UV spectra of 1,2,3-triazoles were reported in 1954 ⟨54JA667⟩, with the parent compound showing a maximum at 210 nm (log ε 3.6), being very similar to the spectrum of pyrrole. Substitution of an alkyl group in the 4-position causes a slight bathochromic shift to 215–216 nm.

More detailed studies on the UV spectra of 1,2,3-triazoles (and benzotriazoles) followed in 1958 ⟨58G977, 58G1035⟩, although benzotriazoles had been investigated much earlier ⟨42CB1338⟩. There exists a rich collection of the UV data for 1,2,3-triazoles and benzotriazoles known today ⟨71PMH(3)77, 103⟩. Some additional and more recent data are presented in Table 14.

Comparing the UV spectrum of benzotriazole in different solvents with that of benzene and *o*-benzoquinone leads to the assumption that the principal contributing forms of the

Table 14 Some UV Spectral Data of 1,2,3-Triazoles in Ethanol (E) and 95% Ethanol (E*)

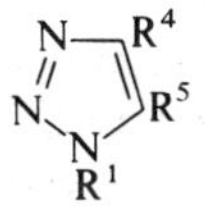

R^1	R^4	R^5	λ_{max} (nm)	log ε	Solvent	Ref.
Ph	H	H	209, 244	4.24, 4.26	E	68JCS(C)1329
p-ClC$_6$H$_4$	H	H	201, 232	4.20, 3.99	E	68JCS(C)1329
p-NO$_2$C$_6$H$_4$	H	H	295	4.47	E	68JCS(C)1329
Ph	H	OH	235	4.03	E*	58CJC1441
Ph	H	Cl	228	4.03	E*	58CJC1441
Ph	Br	H	253	3.96	E	55JA6532
Ph	CO$_2$H	H	247	4.07	E	55JA6532
Ph	CO$_2$H	Cl	231	4.03	E*	58CJC1441
Ph	CO$_2$H	Me	226	4.07	E	68JCS(C)1329
			214	4.22	E	58G977
Ph	H	CO$_2$H	222	4.00	E	68JCS(C)1329
Ph	H	Me	207, 230	3.67, 3.84	E	68JCS(C)1329
p-NO$_2$C$_6$H$_4$	H	Me	215, 272	4.14, 4.42	E	68JCS(C)1329

molecule in neutral, mildly acidic, acidic and basic solutions are, respectively, the neutral species (**A**), the cation (**B**), and the anion (**C**) ⟨73JA4360⟩. UV spectroscopy has also been employed for structure elucidation of 1-benzotriazolyl- and triazolyl-carbonyl moieties ⟨54JA4933⟩.

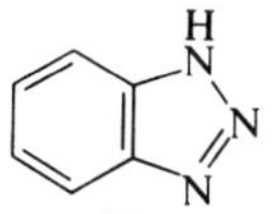

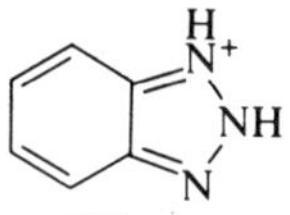

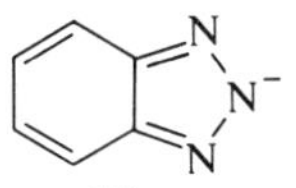

4.11.3.2.7 IR spectroscopy

Complete assignments for the parent compounds have been reported ⟨54JA667⟩; see also Section 4.01.3.7. From symmetry considerations the IR spectrum of 1,2,3-triazole in the vapour and liquid states has been interpreted in terms of an asymmetric 1H-structure, which is contrary to the results obtained with other spectroscopic methods ⟨69JCS(B)307⟩. This interpretation has been disputed ⟨64AC(R)735⟩.

For 4-phenyltriazole the NH-stretching band was measured in all phases (cm^{-1}): 3522 (vap), 3470 (CCl$_4$), 2400–3300 (solid) ⟨65CB4014⟩.

Both 4- and 5-unsubstituted triazoles show C—H stretching at 3100–3410 cm^{-1} ⟨70JHC361⟩. Examination of the IR spectra of several 1,2,3-triazoles yields the following pattern of bands appearing consistently in the ranges given (cm^{-1}): 1290–1150 (CH in-plane deformation), 1130–1105 (CH in-plane deformation), 1090–1010 and 1000–900 (ring breathing), 850–700 (CH out-of-plane deformation) ⟨64CJC43⟩. Furthermore, four different triazoles studied showed the following approximate ring-skeletal vibrations: ~1520, ~1450, ~1410 cm^{-1} ⟨64CJC43⟩.

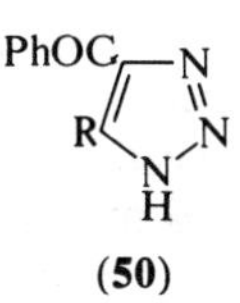

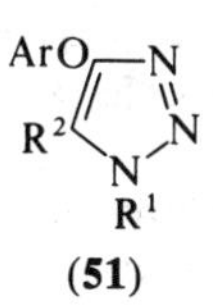

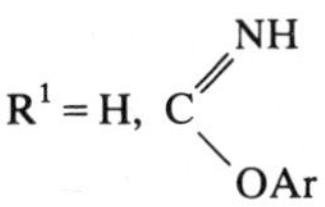

It has been observed that among 16 azoles possessing the common feature of a benzoyl group attached to a C atom of the heteroaromatic nucleus, the 1,2,3-triazoles (**50**) exhibit a strong band in the region 895–920 cm^{-1} with an intensity of the same order as that of the benzoyl stretching band; for example, in (**50**; R = NHCOPh, COPh) there is a characteristic band at 925 and 920 cm^{-1}, respectively, in addition to ν(C=O) at 1645 cm^{-1}. The nature of this band has been established. The ratio of the apparent extinction coefficients of the benzoyl band and that in the 920–895 cm^{-1} region seems to be useful as a diagnostic aid in structural determination ⟨62JOC2209⟩.

In the course of cycloadditions of diazo compounds to cyanates the IR spectra of different substituted 4-aryloxy-1,2,3-triazoles (**51**) have been measured. The absorptions between 957 and 990 cm^{-1} have been attributed to the triazole nucleus while the γ(CH) vibration of the triazole-H (R^2 = H) is found in the range 1030–1045 cm^{-1} and what is presumably an N=N valence band occurs at 1720 cm^{-1} ⟨66CB317⟩.

Raman spectra of several benzotriazoles were reported as early as 1940 ⟨40CB162⟩.

4.11.3.2.8 Mass spectrometry

No mass spectra of 1,2,3-triazoles were described prior to 1968, when a series of highly substituted triazoles, the oxidation products of 1,2-dicarbonyl-bis-benzoylhydrazones, were investigated. The previously assigned structure of a dihydro-1,2,3,4-tetrazine could be excluded in favor of the 1,2,3-triazole-isoimide (**52**), which showed a low intensity molecular ion $M^{\ddot{+}}$, but a prominent $[M-28]$ ion corresponding to loss of N_2 giving (**53**) ⟨68TL231⟩. The $[M-28]$ ion underwent further fragmentations as shown in Scheme 1.

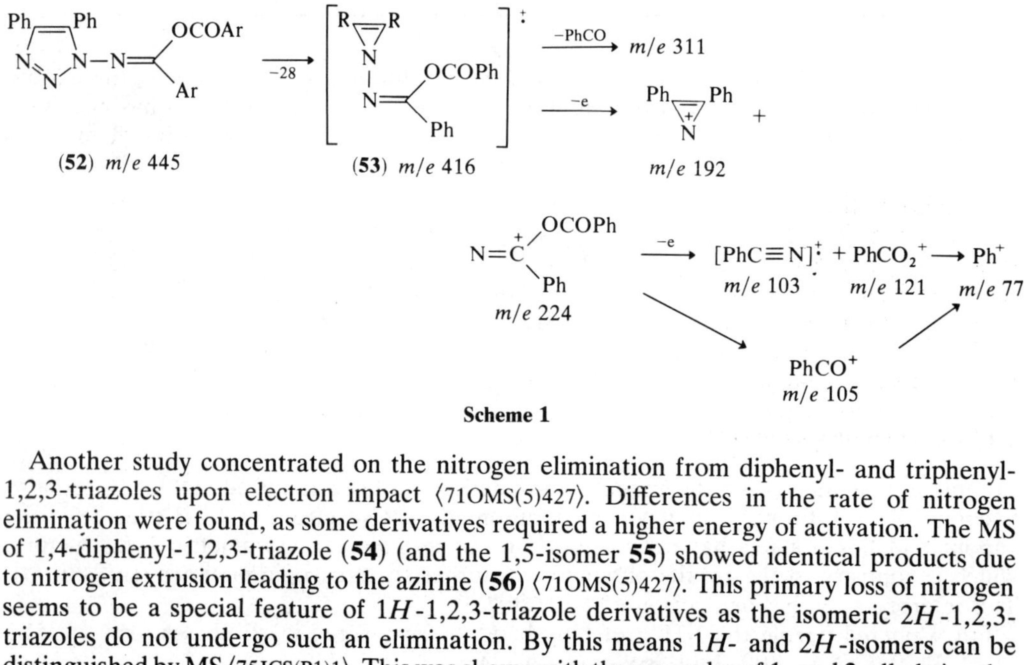

Scheme 1

Another study concentrated on the nitrogen elimination from diphenyl- and triphenyl-1,2,3-triazoles upon electron impact ⟨71OMS(5)427⟩. Differences in the rate of nitrogen elimination were found, as some derivatives required a higher energy of activation. The MS of 1,4-diphenyl-1,2,3-triazole (**54**) (and the 1,5-isomer **55**) showed identical products due to nitrogen extrusion leading to the azirine (**56**) ⟨71OMS(5)427⟩. This primary loss of nitrogen seems to be a special feature of 1H-1,2,3-triazole derivatives as the isomeric 2H-1,2,3-triazoles do not undergo such an elimination. By this means 1H- and 2H-isomers can be distinguished by MS ⟨75JCS(P1)1⟩. This was shown with the examples of 1- and 2-alkyltriazoles (**57**) and (**58**). While the 1-alkyl derivatives (**57**) all show an ion $[M^+ - 28]$, this is absent from the spectra of the 2-alkyl isomers (**58**) ⟨75JCS(P1)1⟩.

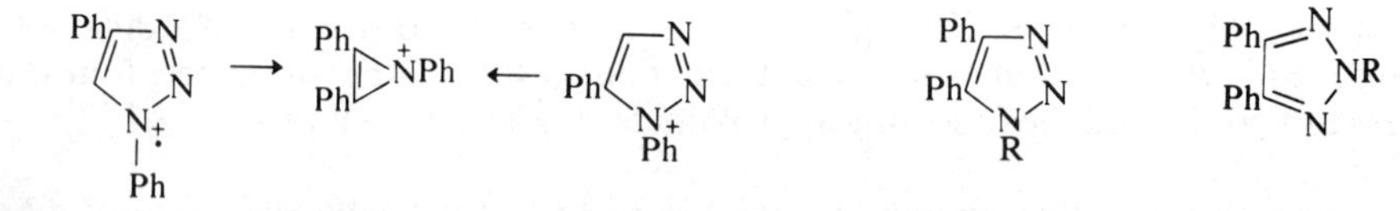

However, unsubstituted 1,2,3-triazole eliminates nitrogen from both the 1H- and 2H-tautomers (Scheme 2) ⟨73OMS(7)271⟩. In addition, 1,2,4- and 1,2,3-triazoles behave similarly upon electron impact, which is ascribed to the formation of common ionic structures ⟨73OMS(7)271⟩.

In the gas phase the MS of 1,2,3-triazole is interpreted as being derived from a mixture of both tautomers; the 1H-triazole thereby eliminates one molecule of nitrogen and, on the other hand, the 2H-form undergoes two other competitive reactions: $M^+ - HCN$ and $M^+ - N_2$, the latter one being considered in terms of a previous isomerization of the fragment ion (Scheme 3) ⟨73OMS(7)271⟩.

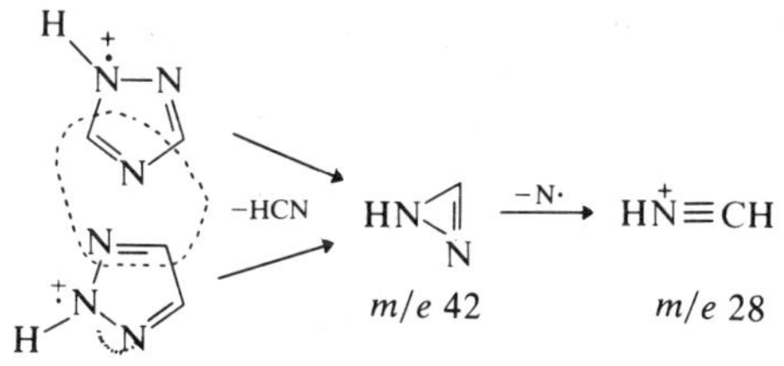

Scheme 2

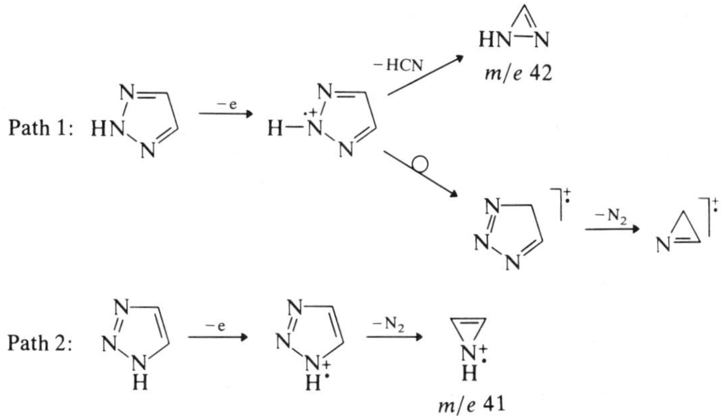

Scheme 3

In addition, the MS of a wide variety of substituted 1,2,3-triazoles have been measured. The molecular ion M^{+} is usually strong and almost half the compounds in the group show an $[M^{+}+1]$ ion. Subsequent fragmentation patterns are strongly dependent on the substitutents R and R′. The cleavage of the triazole ring may begin with one or more accessible cyclic (1H, 2H or 3H) or acyclic (α-diazo-β-imino) molecular ions ⟨79JCS(P1)15⟩.

The tautomerism of the 4- and 5-hydroxy-1,2,3-triazoles (**59**)–(**61**) in the gas phase has been studied by comparison of their MS with those of the fixed dimethyl derivatives, e.g. (**62**). These data suggest that (**59**) exists in the hydroxy form, and (**60**) and (**61**) exist as tautomeric mixtures ⟨76BSB795⟩.

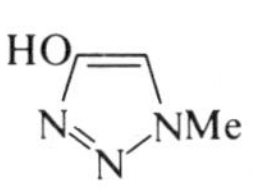

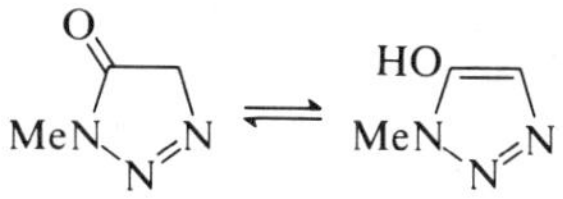

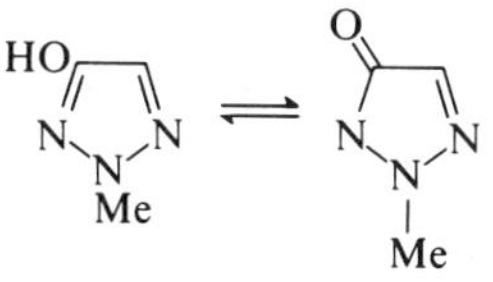

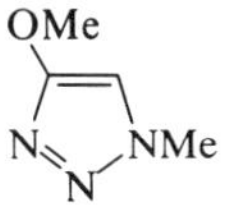

A detailed mechanistic MS study of 1-phenyl-1,2,3-triazole and its D- and ^{13}C-labelled derivatives with regard to competing decomposition pathways has been reported ⟨78OMS571⟩. Similarly, the mass spectra of some alkyl-, aryl- and acyl-benzotriazoles have been reported and discussed on the basis of their thermolytic and photolytic fragmentation products ⟨67TL4379, 68CC1026⟩. In the case of benzotriazole the molecular ion loses nitrogen, followed by loss of HCN, to give a $[C_5H_4]^{+}$ ion, the radical cation (**A**) being more stable by 1.7 eV (Hückel MO calculations) than the azirine ion (**D**) ⟨70OMS(4)203⟩.

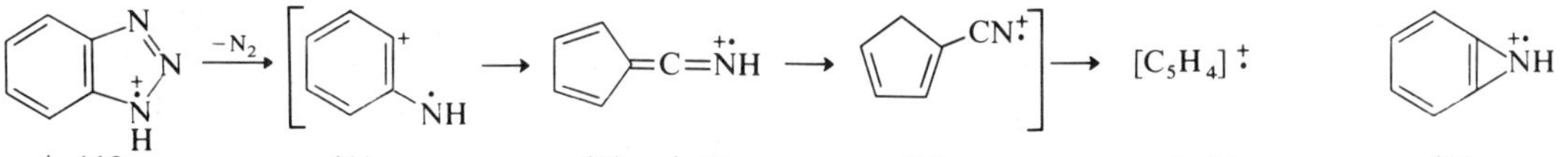

The most intense ion in the MS of 1-phenylbenzotriazole (**63**) at m/e 167 is due to the loss of nitrogen from the molecular ion and the subsequent fragmentations are quite similar to that of carbazole (**65**). Thus, after loss of nitrogen the radical cation (**E**) is formed, which cyclizes to give the molecular ion of a 4a*H*-carbazole (**64**). Isomerization gives rise to the molecular ion of carbazole (**65**) ⟨70OMS(4)203⟩.

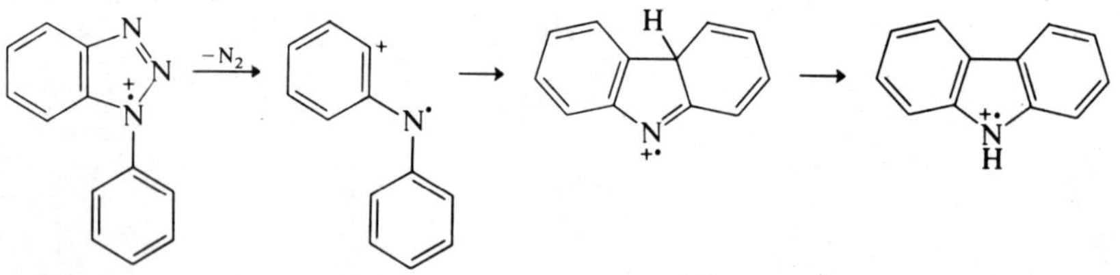

(**63**)   (**E**)   (**64**) *m/e* 167   (**65**) *m/e* 167

More detailed studies of the behavior of benzotriazoles upon electron impact provided evidence for the exclusive presence of the 1*H*-benzotriazole tautomer in the gas phase. The M^{+} of 2*H*-benzotriazole is generated from 2-ethylbenzotriazole (**66**) by ethylene elimination; isomerization to the 1*H*-tautomer precedes the fragmentation ⟨73OMS(7)1267⟩.

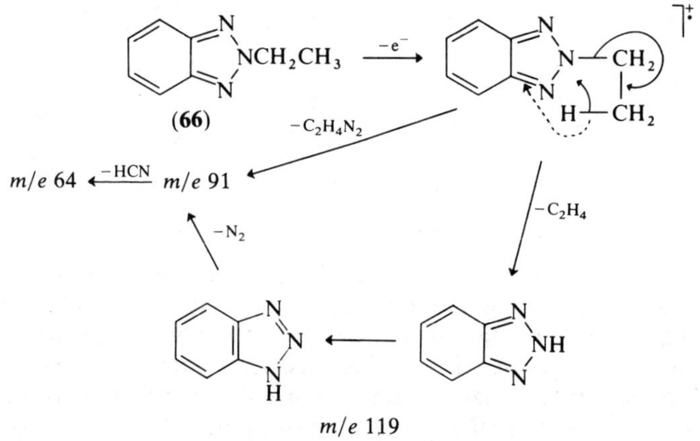

4.11.3.2.9 Photoelectron spectroscopy

The molecular energy levels of azoles and 1,2,3-triazole have been studied by gas phase He(I) PE spectroscopy (Figure 8) and calculated by *ab initio* MO calculations. Thereby sets of energy levels have been determined that correlate well with one another in the upper valence shell region ⟨73T2173⟩.

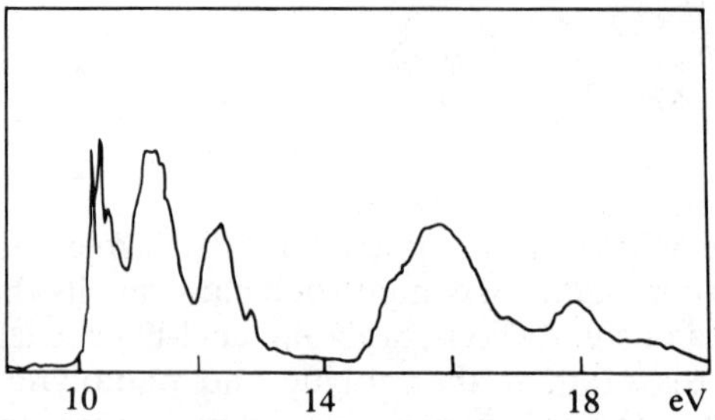

Figure 8 He(I) photoelectron spectrum of 1,2,3-triazole ⟨73T2173⟩

Eigenvalues, principal character and centers/bond orbitals are presented in Table 15. The calculations provide in addition to eigenvalues (energy levels) a set of eigenvectors, permitting analysis of the bonding characteristics of the levels and trends. The spectra relate to a tautomeric mixture of the 1*H*- and 2*H*-forms. In a subsequent reconsideration of this topic a comparison was made of the PE spectra of the tautomeric 1,2,3-triazole with its corresponding 1- and 2-methyl derivatives. 2*H*-1,2,3-Triazole has been identified as the dominant tautomer in the gas phase ⟨81ZN(A)1246⟩.

Table 15 Eigenvalues, Principal Character and Centers/Bond Orbitals of 1,2,3-Triazole
⟨73T2173⟩

	Eigenvalue (eV)	Principal character	Centers/bond orbitals
a'	−429.4	1s	N^1
	−428.2	1s	N^2
	−427.0	1s	N^3
	−312.5	1s	C^5
	−311.5	1s	C^4
	−42.03	2s ('A')	$N^1+N^2+N^3$
	−34.69	2s	N^1-N^3
	−31.98	2s	$N^2-(C^4+C^5)$
	−25.18	2s 2p $1s_H$	NH, N^3C^4, C^4C^5
	−24.09	2s 2p $1s_H$	NN+CN, C^4H-C^5H
	−22.85	2p $1s_H$	CN, C^4H+C^5H
	−18.85	2p $1s_H$	C^4C^5, NH$-C^5H$
	−18.47	2p $1s_H$	C^4H, CN
	−15.46	sp^2	N^3+N^2
	−12.81	sp^2	N^3-N^2
a"	−19.90	$2p_z$	$N^1+N^2+N^3+C^4+C^5$
	−13.71	$2p_z$	$N^3C^4-N^1C^5$
	−12.35	$2p_z$	$(N^1+N^2+N^3)-(C^4+C^5)$

Additionally, the final equilibrium geometries for the tautomeric pair have been computed and are shown below. In agreement with the microwave data (*cf.* Section 4.11.3.2.2) both minimal and double zeta bases predict the 2*H*-tautomer to be more stable than the 1*H*-form ⟨81ZN(A)1246⟩.

The He(I) PE spectra of 4-(1,2,3-triazolio)sulfides have been recorded and assignments made. An unusual stabilization effect obtained by methylation is discussed in terms of perturbation theory ⟨78JA1275⟩.

4.11.3.2.10 Polarography

Substituent effects on the polarographic reduction of 1-amino-4-aryl-1,2,3-triazoles and 1*H*-4-aryl-1,2,3-triazoles of the general formula (**67**) have been reported. Cathodic half-wave potentials ($E_{1/2}$) determined in acetonitrile at 20 °C were between −2.1 and −2.6 V. Linear relations were found between $E_{1/2}$ and the π-electron density at C-5, as well as Hammett σ-values ⟨76ZC280⟩.

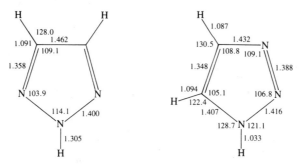

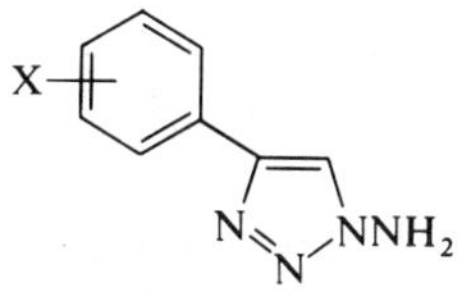

(**67**) X = *p*-F, *p*-Cl, *p*-Br, *p*-Me

Benzotriazoles can take up four electrons in an irreversible reaction with subsequent ring cleavage, whereas in basic medium benzotriazoles are normally not reduced. However, 2-substituted benzotriazoles exhibit a reversible two-electron step yielding dihydro derivatives ⟨68MI41100⟩.

4.11.4 REACTIVITY

General aspects of the reactivity of azoles and 1,2,3-triazoles are discussed in Chapter 4.02. A more detailed discussion of these and other features of the chemistry of 1,2,3-triazoles and benzotriazoles follows.

4.11.4.1 Reactions at Aromatic Rings

4.11.4.1.1 General survey of reactivity

In this section, the reactivity of the major types of aromatic 1,2,3-triazoles and benzotriazoles is considered and compared with that which would be expected on the basis of electronic theory and spectroscopic data presented in Section 4.11.3. Tautomeric equilibria and ring–chain tautomerisms are discussed in Section 4.11.4.1.2(i), Dimroth rearrangements in Section 4.11.4.1.2(iii) and acid–base considerations are discussed in the section immediately following.

(i) Acid–base considerations; ionization constants

1,2,3-Triazole is both a weak base and a weak acid of comparable strength to phenol. The amphoteric properties may be deduced from examination of the mesomeric or resonance structures ⟨63PMH(1)98, 68JA6588⟩. Some representative pK_a values are listed in Table 16.

Table 16 pK_a Values for Triazoles

Compound	Basic pK_a	Acid pK_a	Ref.
1,2,3-Triazole	1.17	9.4	63PMH(1)98
1-Methyl-1,2,3-triazole	1.25	—	63PMH(1)98
2-Methyl-1,2,3-triazole	<1	—	63PMH(1)98
Benzotriazole	—	8.2	51JA4360
5-Amino-1-phenyltriazole	2.27	—	69JCS(C)2379
4-Phenyl-5-(p-nitrophenyl)amino-1,2,3-triazole	—	6.60	58JOC1916
4-Phenyl-5-(p-methoxyphenyl)amino-1,2,3-triazole	—	7.91	58JOC1916
4-Phenyl-1,2,3-triazole	—	7.68	16G308
4,5-Dicyano-1,2,3-triazole	—	2.53	42MI41100
4,5-Dibromo-1,2,3-triazole	—	5.37	68JA6588
5-Chlorobenzotriazole	—	7.7	51JA4360
4,5,6,7-Tetrachlorobenzotriazole	—	5.48	55JA5105

1-Methyl-1,2,3-triazole shows a comparable basicity, but 2-methyl-1,2,3-triazole is a much weaker base. That the N-alkyl groups show no base-strengthening effect can be explained in terms of the formation of imidazolium-type cations (**68**) (1-methyl) and (**69**), whilst 2-methyl-2H-1,2,3-triazole must achieve a pyrazolium-type cation (**70**).

In the case of benzimidazole the fused benzene ring is base-weakening and acid-strengthening; substituents on the aromatic ring show predictable effects. 4,5-Dibromotriazole is a much stronger acid than the parent compound ⟨51JA4360⟩.

Most derivatives with a ring NH are weak acids ⟨58JOC1916⟩, an exception being 4,5-dicyano-1,2,3-triazole ⟨42MI41100⟩ in which resonance stabilization of the corresponding anion plays a certain role. Benzotriazole shows an increased acidity after chlorine is substituted in the benzene ring ⟨51JA4360, 55JA5105⟩.

(ii) Metalloid-linked substituents

4-Benzoyl- and 4-butyryl-triazoles form iron complexes when treated with iron(II) chloride in THF under nitrogen. A Mössbauer spectrum was measured with this complex ⟨66DOK(170)354⟩. Similarly, $Fe_2(CO)_9$ reacts instantly with the above 4-aroyltriazoles to give red solutions of $(PrCO-C_2H_2N_3)_2Fe$ and $(PhCO-C_2H_2N_3)_2Fe$ which have been investigated spectroscopically ⟨68DOK(181)1397⟩.

On oxidation of 1,2-cyclooctanedione dihydrazone with HgO, cyclooctyne and Hg derivatives of triazoles have been identified. The major product is $C_{32}H_{42}N_6Hg$, which was assigned the structure shown ⟨68LA(716)11⟩. Cyclometallations of 2-aryl-4,5-dimethyl-1,2,3-triazoles occur with Pd(II), Pt(II), Rh(II) and Ir(III) chlorides ⟨78MI41105⟩. The triazole acts as an N—C-chelate ligand.

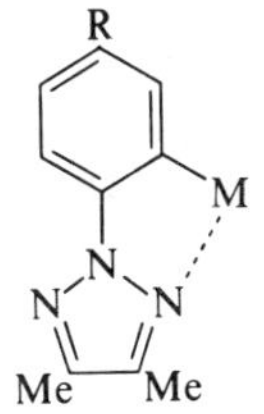

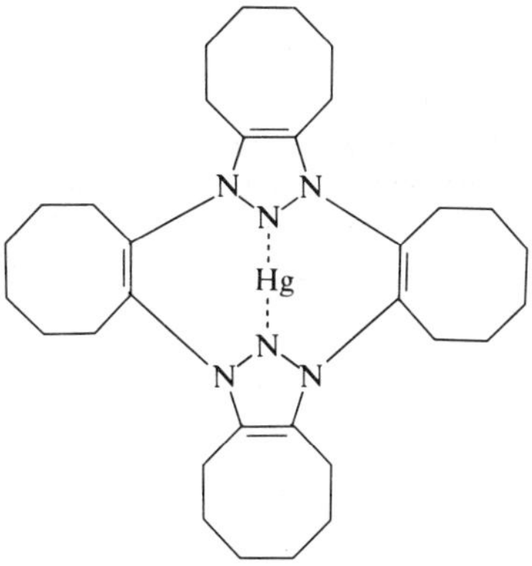

4.11.4.1.2 Thermal and photochemical reactions formally involving no other species

(i) Tautomerism: ring–chain tautomerism

As has been pointed out often in the previous sections 1,2,3-triazole can exist in both the 1H- and 2H-forms. Benzotriazole behaves in a similar fashion (Scheme 4).

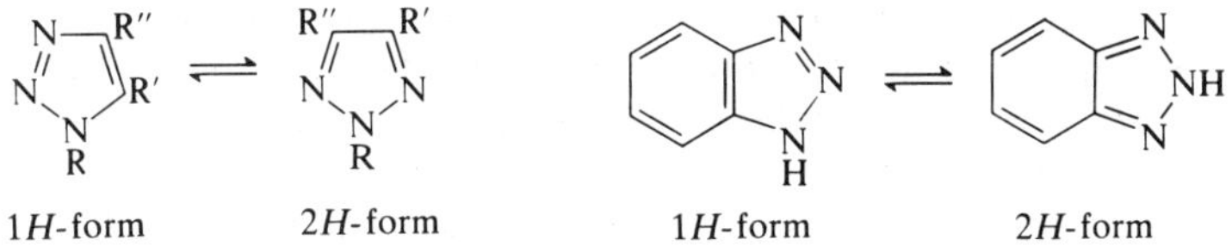

Scheme 4

Considering the spectroscopic data known today leads to the conclusion that in the case of 1,2,3-triazole the 'symmetrical' 2H-form predominates. More evidence is given in ⟨76AHC(S1)281⟩. Similarly, spectroscopic data indicate for benzotriazole that the 1H-form is predominant (for more details see ⟨76AHC(S1)295⟩).

Some features of the tautomerism of both 1,2,3-triazole and benzotriazole are summarized in Table 17 ⟨76AHC(S1)296⟩, where the tautomer with the highest dipole moment (and hence the most likely to be associated) and that with the lowest basicity are listed.

Table 17 Annular Tautomerism of 1,2,3-Triazole and Benzotriazole ⟨76AHC(S1)296⟩

Azole	Definition of K_T	Conclusion	Tautomer with highest μ	Less basic
1,2,3-Triazole (1,2,3-T)	$\dfrac{1H\text{-}1,2,3\text{-}T}{2H\text{-}1,2,3\text{-}T}$	2H-1,2,3-T in solution; mixture of 1H- and 2H-1,2,3-T in the gas phase	1H-1,2,3-T	2H-1,2,3-T
Benzotriazole (BT)	$\dfrac{1H\text{-BT}}{2H\text{-BT}}$	1H; $K_T \gg 1$ in solution and gas phase	1H-BT	2H-BT

Besides the annular tautomerism and the Dimroth rearrangement (Section 4.02.3.5) ring–chain tautomerism of certain 1,2,3-triazoles have been discussed and, in some cases, shown by means of spectroscopic methods. Thus, cyanogen azide reacts with acetylene to form a 1:1 adduct, which was proved by means of IR, UV and temperature-dependent

^{1}H NMR data to exist as a ring–chain tautomeric mixture of 1-cyano-1,2,3-triazole (**71a**) and α-diazo-N-cyanoethylideneimine (**71b**). The equilibrium is sensitive to temperature and solvent ⟨67JA4760⟩.

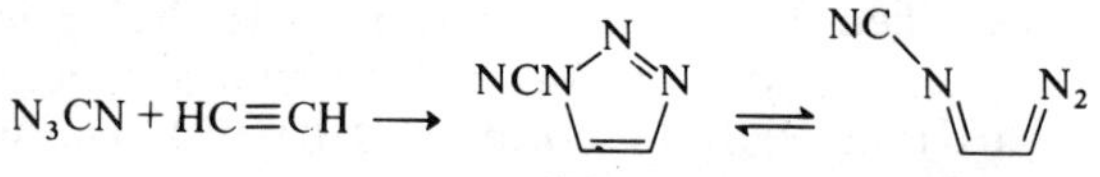

$$N_3CN + HC\equiv CH \longrightarrow (71a) \rightleftharpoons (71b)$$

Similarly, 1,3-dipolar cycloadditions of benzenesulfonyl azides to N,N-diethylaminoprop-1-yne led to 1,2,3-triazoles (**72a**) and α-diazoamidines (**72b**). With the aid of IR and ^{1}H NMR data these were shown to exist in a tautomeric equilibrium ⟨70JOC3444⟩. With IR and ^{1}H NMR data a 5-aminotriazole–diazoamidine equilibrium (**73a**) ⇌ (**73b**) was established ⟨70TL2823, 72CB2963, 72CB2975⟩. However, 4-amino-1,2,3-triazole (**74a**) proved to be stable and no ring–chain tautomerism could be identified ⟨73TL1137⟩.

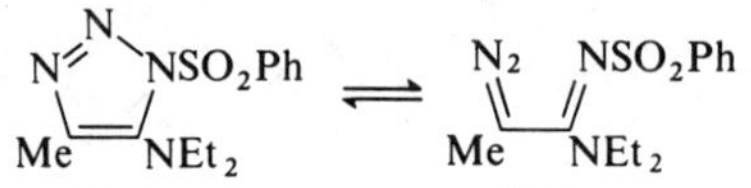

(ii) Thermal and photochemical reactions

Although the ring system of monocyclic 1,2,3-triazoles shows a remarkable stability towards oxidation, reduction and hydrolysis, under forced conditions thermal extrusion of nitrogen can be achieved at elevated temperatures. Some such reactions are reviewed in ⟨B-76MI41103⟩. Similarly, photochemical extrusion of nitrogen can be effected, and this is consistent with their behavior upon electron impact (*cf.* Section 4.11.3.2.8).

After extrusion of nitrogen the intermediate can be formulated in terms of three mesomeric forms: an iminocarbene (**A**), a zwitterion (**B**) or a diradical (**C**). The potential constitutions of (**A**), (**B**) and (**C**) follow from their reaction products involving, for example, a photo-Wolff rearrangement ⟨68JA1923⟩, and a subsequent ring closure (with R = Ph, *cf.* ⟨70CB3811⟩) to indoles, while (**C**) undergoes cyclization to give a 1H-azirine which rearranges in turn to the corresponding 2H-isomer, the latter process being more favorable than the Wolff rearrangement (Scheme 5) ⟨71CC1519, 73JCS(P1)555⟩.

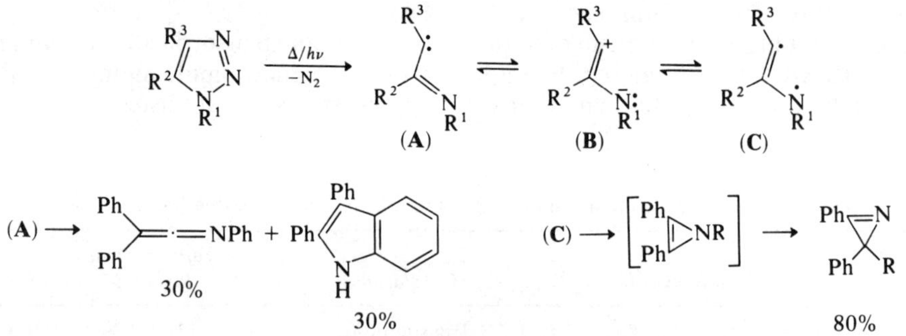

Scheme 5

Vapor phase pyrolysis of two unsymmetrically substituted 1,2,3-triazole isomers, like (**75**) and (**76**), involves the antiaromatic 1H-azirine derivative (**77**) which rearranges to two isomeric 2H-azirines (**78**) and (**79**) ⟨71CC1519, 73JCS(P1)555⟩ and other products. Iminocarbenes (**80**) are most likely involved as intermediates. Furthermore, nitrogen has been extruded from several 1H-1,2,3-triazoles by flash vacuum pyrolysis. 1-Alkyl-1,2,3-triazoles (**81**) give nitriles (**82**) and (**83**) (*via* Wolff rearrangement) and (hydroxy)isoquinolines (**84**) and (**85**) (by [1,4-H] transfer in the iminocarbene) ⟨75JCS(P1)1⟩.

Both 4- and 5-phenyl-1-(1-phenylvinyl)-1,2,3-triazoles are pyrolyzed at 575 °C to afford mixtures of 2,4- and 2,5-diphenylpyrrole; similarly, 1-phenyltriazoles (**86**) react to give 2-

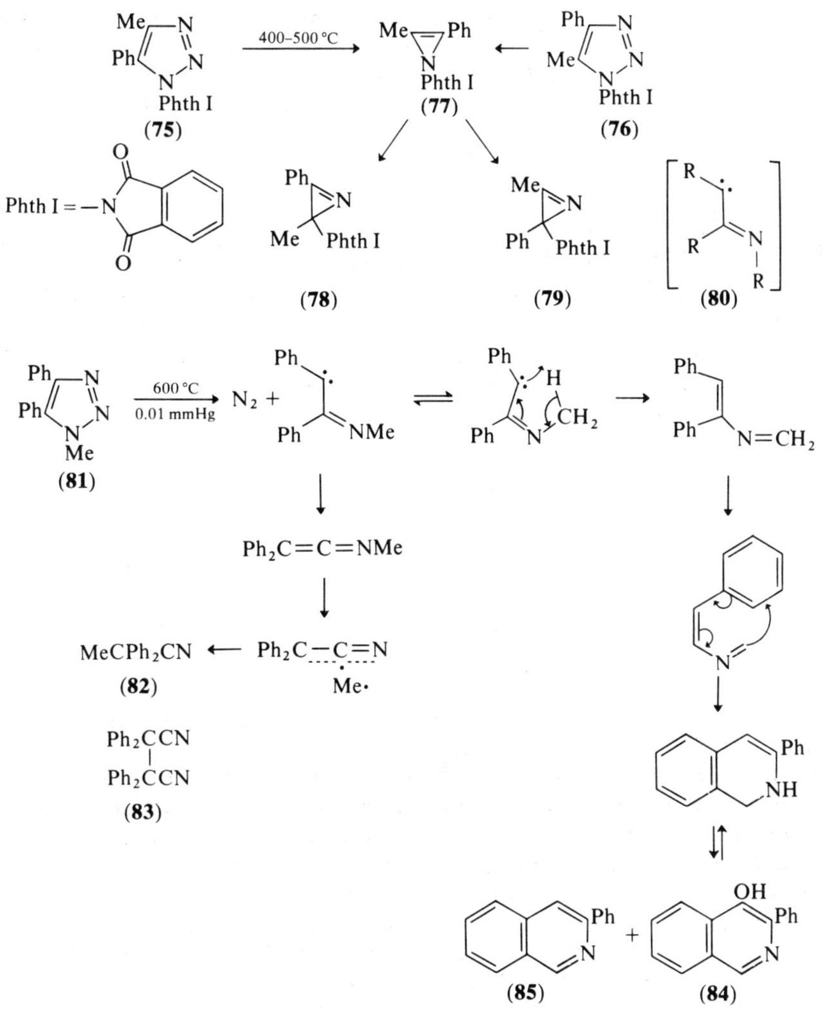

and 3-phenylindoles (**87**), (**88**) ⟨75JCS(P1)1⟩. Photolysis and pyrolysis of 4-phenyltriazole lead mainly to phenylacetonitrile, and pyrolysis of 4,5-diphenyltriazole (**89**) in solution gives 2,3,5,6-tetraphenylpyrazine (**90**) ⟨72JHC87⟩.

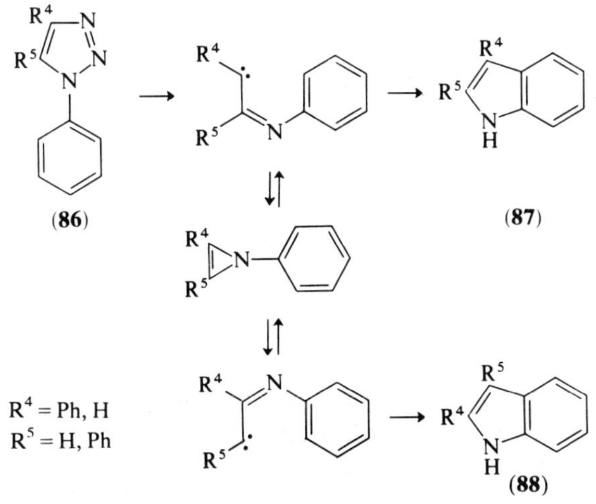

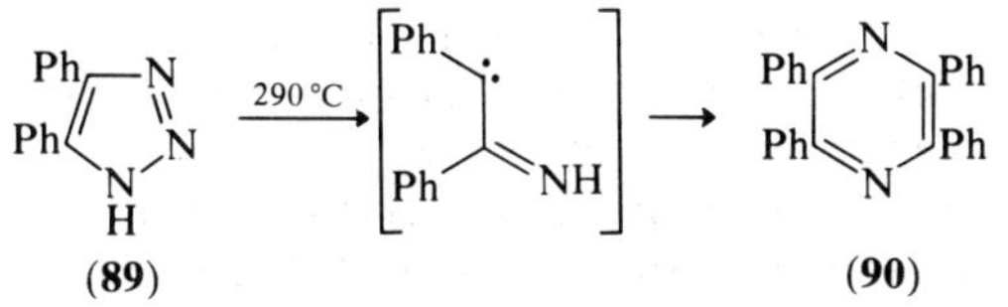

Gas phase pyrolysis of the unsubstituted 1H- and 2H-1,2,3-triazole mixture resulted in the formation of acetonitrile as the main product, and HCN and NH$_3$ occur as by-products. At about 850 °C, an unexpectedly high pyrolysis temperature, 1,2,3-triazole was decomposed quantitatively. This is explained by the predominance of the 2H-form in the gas phase ⟨78JCR(S)298⟩.

Irradiation of 2H-1,2,3-triazoles (**91**) in ether gives 22% MeCN, and in the presence of excess cyclopentene it affords the adduct (**92**) ⟨68CC977⟩. 4H-Triazolines (**93**) bearing methoxycarbonyl and alkyl groups at the β-position give, on thermolysis, 2H-1,2,3-triazoles (**94**) and/or 3-amino-2H-azirines; on photolysis, only azirines are formed ⟨80CC940⟩.

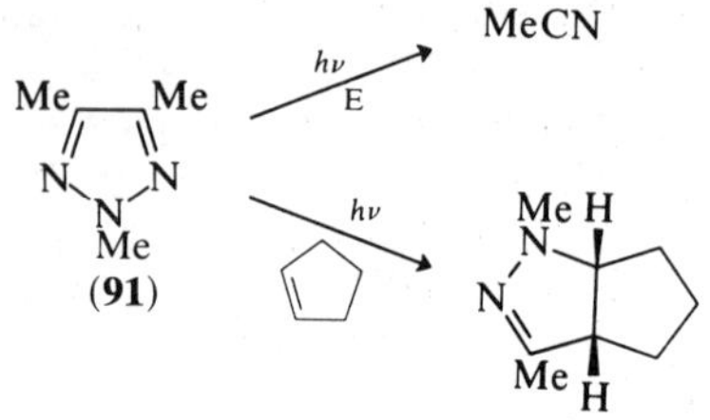

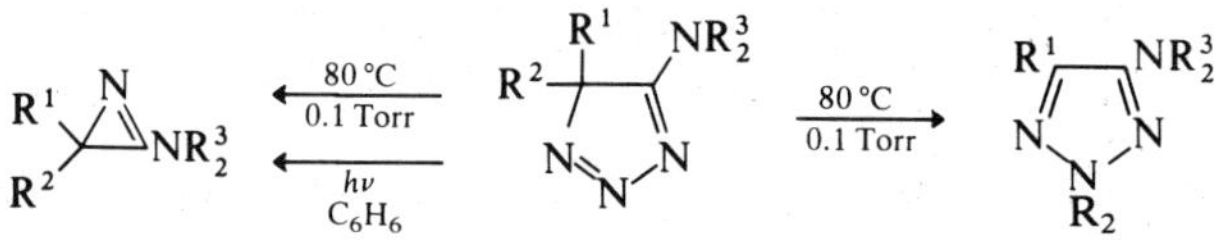

On irradiation, 1-phenylbenzotriazole (**96**) loses nitrogen to give carbazole (**98**) in nearly quantitative yield (*cf.* also the MS fragmentation in Section 4.11.3.2.8) ⟨68JA1923⟩. The nature of the intermediate (**97**) was established as a diradical by photolysis of 5-chloro-1-(4-chlorophenyl)benzotriazole (**99**) which gave (**100**) as the sole product ⟨70CC1089⟩. The antiaromatic azirine (**101**) could be excluded as a potential intermediate ⟨70CC1089⟩. The photochemical reactions of benzotriazoles have been the subject of a review ⟨75MI41100⟩.

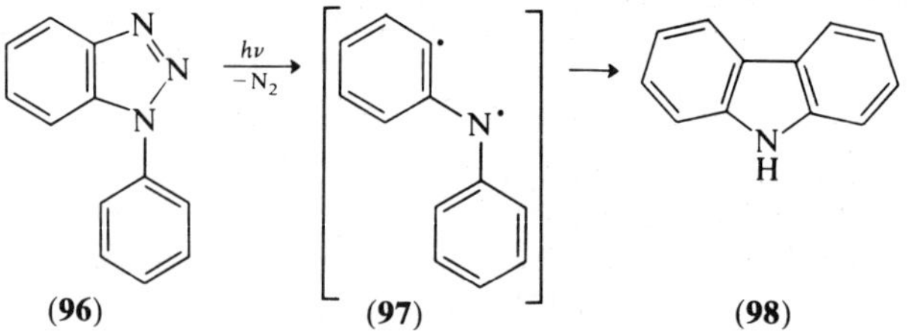

Upon photolysis at 254 nm 1-methoxy- and 1-cycloalkoxy-1,2,3-benzotriazoles (**102**) show rather different photolytic pathways: homolytic cleavage of the N—O—R bond occurs, while a second path proceeds *via* nitrogen extrusion followed by an intramolecular H-abstraction leading to phenylnitrene (**103**). The latter dimerizes to give an azobenzene (**104**) ⟨74JOC3788, 80JHC825⟩. 1-Benzyloxy-1,2,3-benzotriazole shows a similar behavior ⟨80JHC1309⟩.

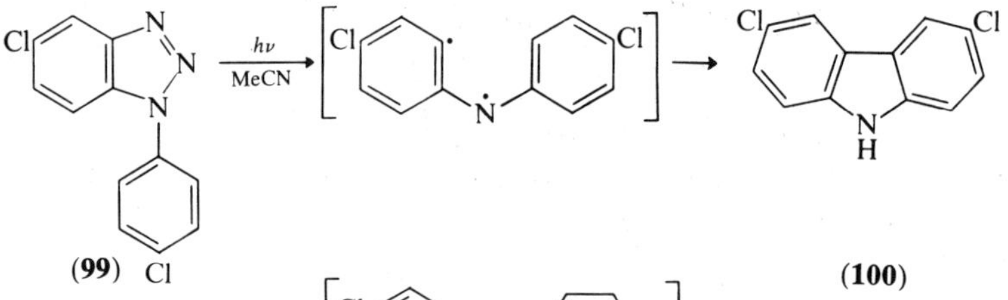

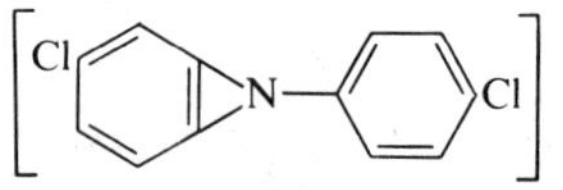

(iii) Dimroth rearrangements

The Dimroth rearrangement is frequently encountered in heteroaromatic chemistry and was first recognized in the triazole series. General aspects of the rearrangement are described

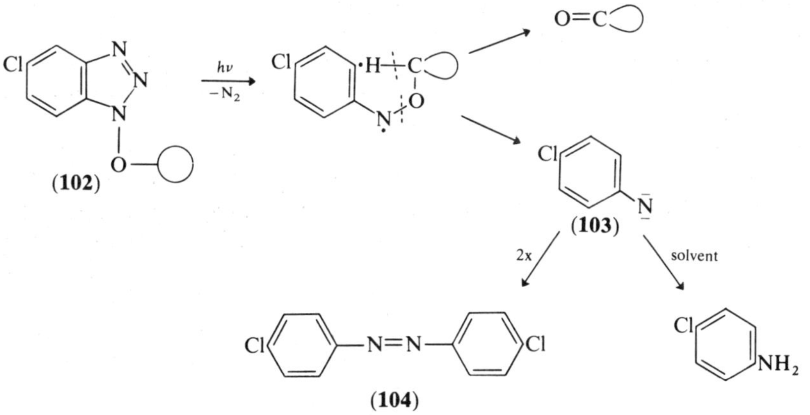

in Section 4.02.3.5 and several reviews have appeared on this subject ⟨B-61MI41100, B-68MI41101, 69ZC241, 74AHC(16)33⟩.

The term Dimroth rearrangement was coined in 1963 as a convenient way of referring to an isomerization proceeding by ring fission and subsequent recyclization, whereby a ring nitrogen and its attached substitutent exchanged place with an imino (or potential imino) group in the position α to it ⟨63JCS1276⟩.

The general reaction scheme for 1,2,3-triazoles is shown in Section 4.02.3.5 by formulae (**240**)–(**244**). If the equilibrium reaction is carried out under thermal conditions, the position of the equilibrium can be influenced by the solvent basicity. In general terms, the more basic the solvent, the greater the proportion of the more acidic NH-triazole (**106**) which exists in the equilibrium mixture (**105**) ⇄ (**106**). Furthermore, electron-attracting groups and bulky, rigid substituents tend to favor the tautomeric form in which they occur on the exocyclic nitrogen; alkyl groups tend to favor the cyclic nitrogen atom. 5-Methylamino-1,2,3-triazole-4-carboxamide (**107**) rearranges to 5-amino-1-methyl-1,2,3-triazole-4-carboxamide (**108**) when heated at 160 °C for one hour. Triazole (**108**) remains unchanged even when heated with ethanolic ammonia at 180 °C ⟨69JCS(C)152⟩.

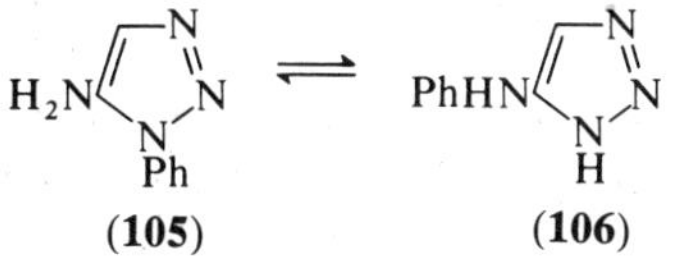

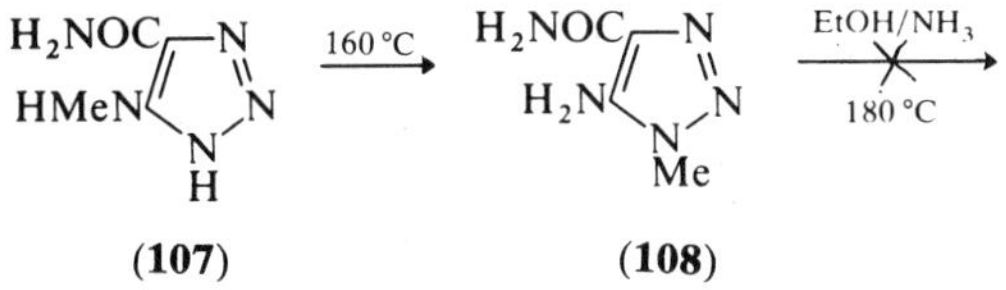

However, the 1-benzyl derivative is rapidly isomerized in base, though the equilibrium favors the ring-substituted isomers under neutral conditions ⟨70JCS(C)230, 72JCS(P1)461⟩. These reaction tendencies are generalized in Scheme 6.

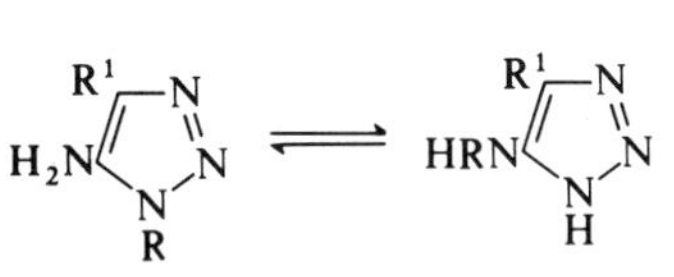

Favored in neutral and basic solution (R = Me), and in neutral solution (R = CH_2Ph)

Favored in basic solution (R = Ar, CH_2Ph)

Scheme 6

Dimroth rearrangement also occurs between the 5-mercapto-1,2,3-triazoles (**109**) and the 5-amino-1,2,3-thiadiazoles (**110**). If the thiadiazoles (**110**) are heated in basic solvents they are converted into the triazoles, whereas the reverse reaction is observed in acidic media ⟨67LA(710)118, 69CB417, 72ACS1243⟩. Thermal equilibria of thiadiazoles and triazoles, with both isomers present, have been reported ⟨62ACS(B)1800, 66CB1618⟩.

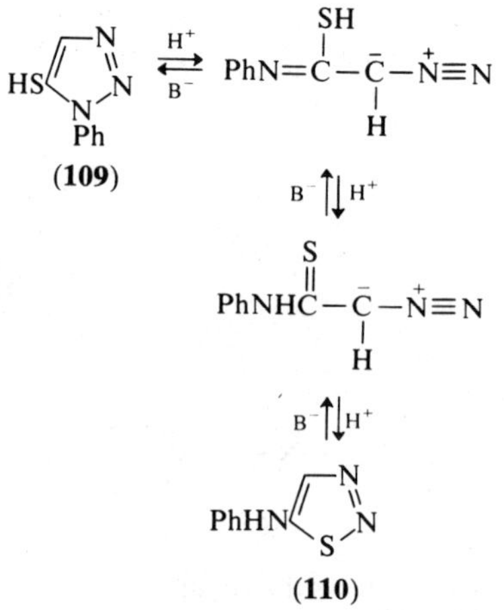

A detailed examination of the acylation of 1-substituted 1,2,3-triazoles ⟨09LA(364)183, 60CB2001, 68JCS(C)344, 69JCS(C)152⟩ revealed that heating of 4-substituted 5-amino-1-phenyl-1,2,3-triazoles in acetic anhydride afford acetyl derivatives of the isomeric 4-substituted 5-anilino-1,2,3-triazoles; in turn these anilino isomers and their acetyl derivatives retrogress to acetyl derivatives of the amino compounds on prolonged refluxing in acetic anhydride. Acylation of 5-amino- or 5-anilino-1,2,3-triazoles at room temperature in the presence of sulfuric acid yields unrearranged products ⟨71JCS(C)706⟩. Some yields of the various products formed are shown in Table 18. The appropriate triazole derivatives (111) or (115) (5 mmol) were refluxed with acetic anhydride (10–20 cm^3).

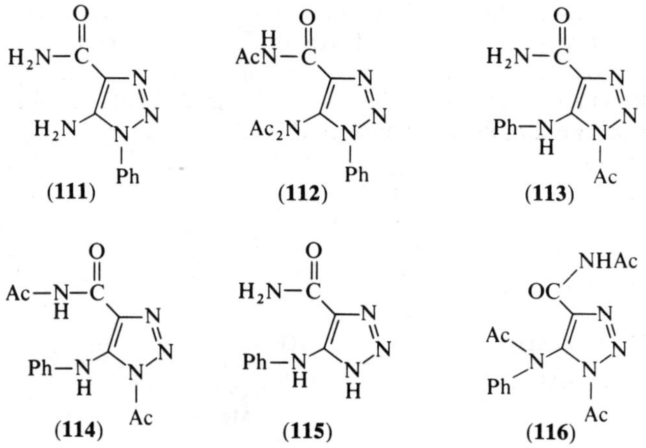

Table 18 Acetylation of Amino-1,2,3-triazoles ⟨71JCS(C)706⟩

Substrate	Time of reflux	Products	Yield (%)
(111)	12 h	(112)	56
(111)	20 min	(113)	70
		(114)	15
(111)	3 or 6 h	(113)	8
		(114)	86
(115)	20 min	(113)	59
		(114)	5
(115)	12 h	(116)	63

Cyclization of 1-aryl-3-(cyanomethyl)triazenes (117) in protic media involves a Dimroth rearrangement to afford 5-(arylamino)-1,2,3-triazoles (118) ⟨81JOC856⟩. As found accidentally, ethyl 5-amino-1-phenyl-1,2,3-triazole-4-carboxylate (119) undergoes a Dimroth rearrangement to (120) by simply refluxing (119) in carbon disulfide ⟨77UP41100⟩. Furthermore (119) proved to be stable under the conditions of direct irradiation. However, in the presence of triphenylene as sensitizer, (119) rearranges to (120), this photorearrangement being the first example of a photoinduced Dimroth rearrangement ⟨76ACH(19)265⟩. Some years later, this photo-Dimroth rearrangement was also observed with 1,4-diphenyl-5-amino-1,2,3-triazoles ⟨77BCJ2505⟩.

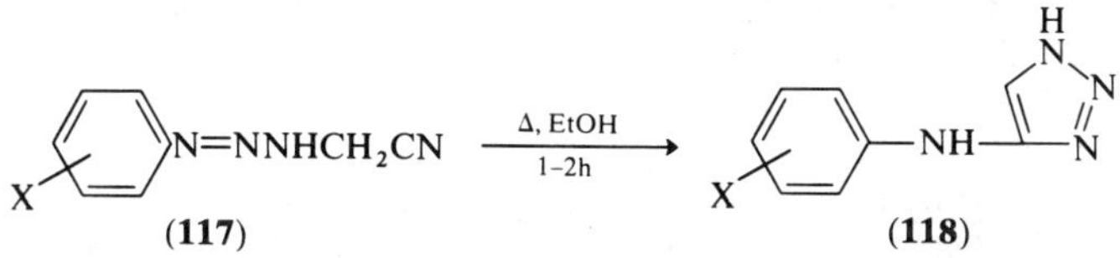

(iv) Ring cleavage reactions: fragmentations

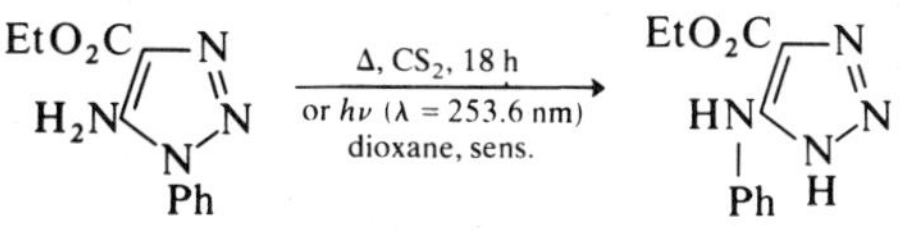

Besides ring–chain tautomerism (Section 4.11.4.1.2(i)) and the Dimroth rearrangement (Section 4.11.4.1.2(iii)) cleavage reactions of some amino-, azido and diazo-methyltriazoles have been reported. Thus, 5-azido-1,4-diphenyl-1,2,3-triazole (**121**) proved to be unusually labile above 50 °C. Nitrogen is evolved with the formation of a conjugated nitrile (**122**) ⟨64JA2025, 70JOC2215⟩. Similarly the diazomethyltriazole (**123**) decomposes to (**124**). Both reactions can be explained in terms of concerted processes.

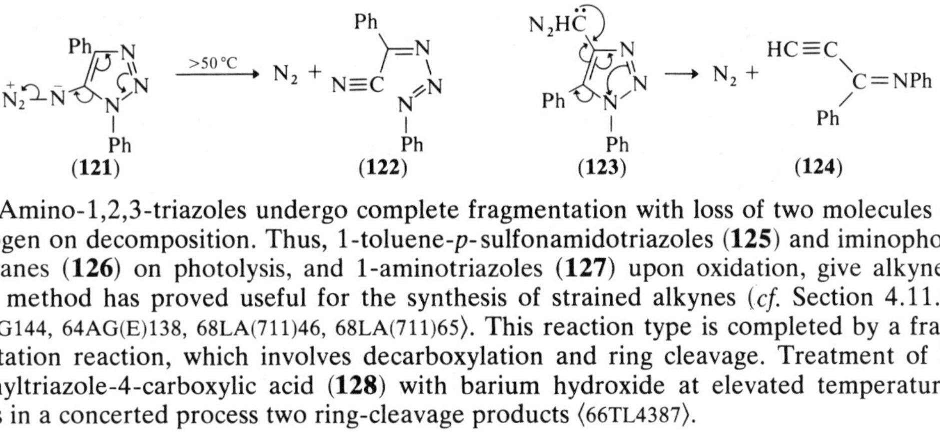

1-Amino-1,2,3-triazoles undergo complete fragmentation with loss of two molecules of nitrogen on decomposition. Thus, 1-toluene-*p*-sulfonamidotriazoles (**125**) and iminophosphoranes (**126**) on photolysis, and 1-aminotriazoles (**127**) upon oxidation, give alkynes. This method has proved useful for the synthesis of strained alkynes (*cf.* Section 4.11.6) ⟨64AG144, 64AG(E)138, 68LA(711)46, 68LA(711)65⟩. This reaction type is completed by a fragmentation reaction, which involves decarboxylation and ring cleavage. Treatment of 2-phenyltriazole-4-carboxylic acid (**128**) with barium hydroxide at elevated temperatures gives in a concerted process two ring-cleavage products ⟨66TL4387⟩.

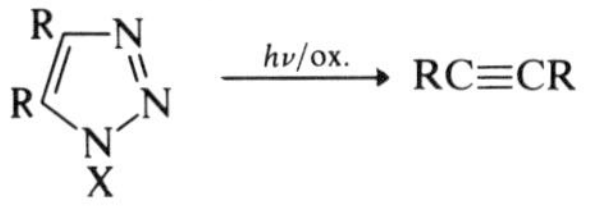

(**125**) X = NTs
(**126**) X = N=PPh$_3$
(**127**) X = NH$_2$

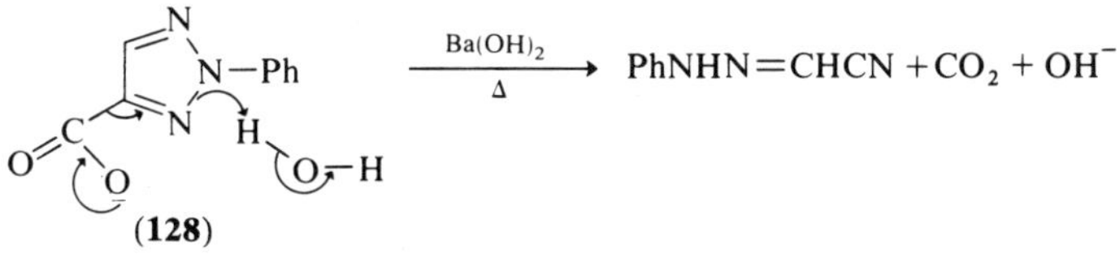

4.11.4.1.3 *Electrophilic substitution and nucleophilic attack*

In general, electrophilic substitution can take place at a ring nitrogen or at a ring carbon atom. Triazole offers three, and benzotriazole five H atoms which can be theoretically replaced by other groups. Additionally, due to the tautomerism in unsymmetrical 1,2,3-triazoles three different positions of nitrogen substitution can be involved whereas symmetrical 1,2,3-triazoles can result in two isomeric *N*-derivatives.

Alkylation on nitrogen can be readily carried out with 1,2,3-triazoles (**129**) by means of alkyl halides, dimethyl sulfate, diazoalkanes or methyl *p*-toluenesulfonate, or by the Mannich reaction. Some of these electrophilic substitution reactions have been described ⟨B-61MI41100, 74AHC(16)69, B-79MI41105⟩. Using methyl iodide and the silver or thallium salts of 1,2,3-triazole, or using an alkaline solution, unsubstituted 1,2,3-triazole is in most cases preferentially attacked on N-1 to give (**132**) ⟨55LA(593)207, 65LA(688)205⟩. However, the action of diazomethane results in a high proportion of 2-methyl-2*H*-1,2,3-triazole (**130**) ⟨75JCS(P1)1⟩. As soon as 1*H*-1,2,3-triazoles are N-1 substituted, they tend to form the 1,3-disubstituted triazolium salts (**133**) ⟨55JA1703⟩. The corresponding N-2-substituted 2*H*-1,2,3-triazoles can be methylated only with powerful methylating agents, *e.g.* 'magic methyl', to give 1,2-disubstituted triazolium salts (**131**) ⟨71ACS(B)2087⟩.

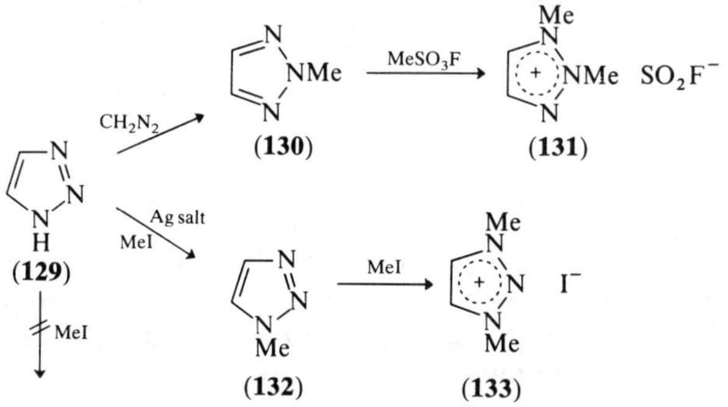

Furthermore, the orientation of alkylation can be directed by the nature of the substituents. 4-Phenyl-1*H*-1,2,3-triazole (**134**) reacts with dimethyl sulfate to give the 1-methyl (**135**) and 2-methyl isomers (**136**) in 62% and 38% yield, respectively. The hindered 1-methyl-5-phenyl derivative was not formed ⟨67LA(707)147⟩. *C*-Methylation of 1-phenyl-1*H*-1,2,3-triazole can be carried out with butyllithium and methyl iodide at low temperatures ⟨71CJC1792⟩.

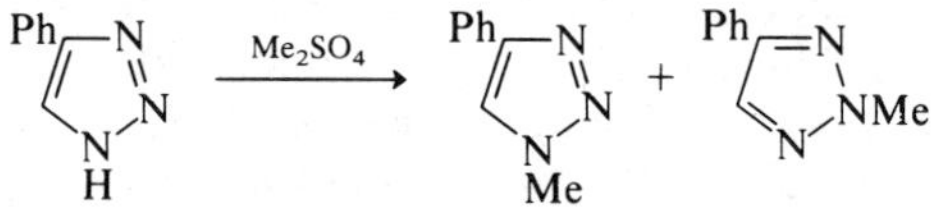

Some special examples follow. Recently phase-transfer-catalyzed alkylation has been carried out on benzotriazole in benzene/aqueous NaOH in the presence of [PhCH$_2$NEt$_3$]$^+$ Cl$^-$. Both possible derivatives (**137**) and (**138**) were obtained ⟨78MI41101⟩.

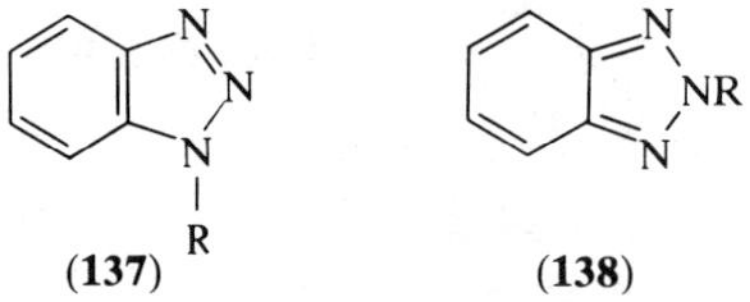

R = Me, Et, Pr, Bu, PhCH$_2$

1,2,3-Triazoles can be acylated in the usual manner, *e.g.* with acyl halides and anhydrides. Isocyanates react with the formation of mixed ureas (**139**), which show characteristic short-wave carbonyl absorptions (1745, 1788 cm^{-1}) ⟨61LA(648)72⟩. Treatment of 1,2,3-triazole with thiobenzoyl chloride in the presence of triethylamine yields a mixture of 1-thiobenzoyl-1*H*- and 2-thiobenzoyl-2*H*-1,2,3-triazoles ⟨73LA636⟩.

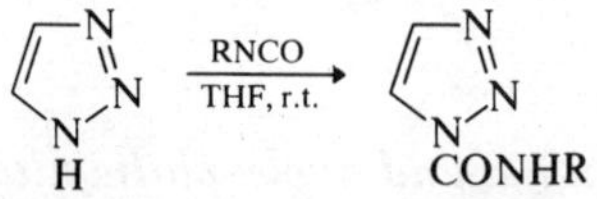

In some cases acyl group migration from the 1-position to the 2-position can be observed when the primary 1-acyl derivative is heated above 120 °C ⟨*cf.* 65AG414, 65AG(E)417, 66CB2512⟩.

2-Trimethylsilyl-1,2,3-triazole (**140**), easily obtained by silylation of 1,2,3-triazole ⟨67CB3485⟩, can be smoothly acylated in the 1-position to give (**141**) by means of acetic anhydride or acetyl chloride ⟨67CB3485⟩. This reaction has been studied intensively using p-substituted monophenylacetylenes and TMS azide; contrary to normal observations, an electron-releasing substitutent induces a higher yield ⟨73T3271, 80MI41101⟩.

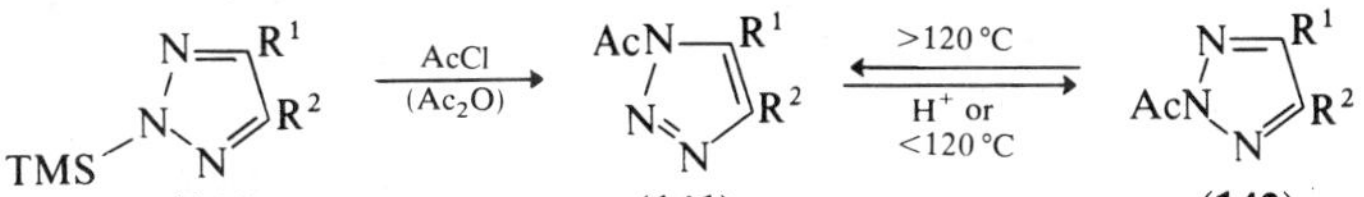

Acylation of benzotriazoles occurs in most instances on the 1- (or 3-)nitrogen atom. From 5-methylbenzotriazole and acetyl chloride or acetic anhydride, a mixture of 1-acetyl-4- and -5-methylbenzotriazoles is obtained ⟨13JCS1391⟩. With acetic anhydride, benzoyl chloride or benzenesulfonyl chloride and 5,6-dimethylbenzotriazole, acylation on N-1 occurs ⟨52JA4917⟩ and no N-2 substitution product could be observed. Analogously, benzotriazole or 1-trimethylsilylbenzotriazole reacts with thiobenzoyl chloride to afford 1-thiobenzoyl-benzotriazole ⟨73LA636⟩. Similarly, with phenyl isocyanate 1-phenylcarbamoylbenzotriazole (143) is formed ⟨49JA2297⟩.

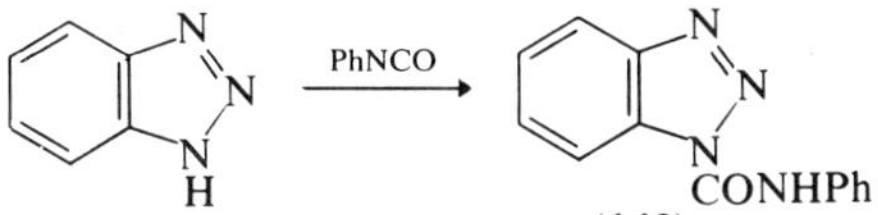

1,2,3-Triazole reacts with bromine to afford the 4,5-dibromo derivative and an excess of hypobromite leads to 1,4,5-tribromo-1,2,3-triazole. 2-Methyl-1,2,3-triazole shows in comparison a decreased readiness towards halogenation. But in the presence of iron filings as catalyst, the bromination affords 2-methyl-4,5-dibromotriazole (144) and no monobromination product could be isolated ⟨55LA(593)207⟩.

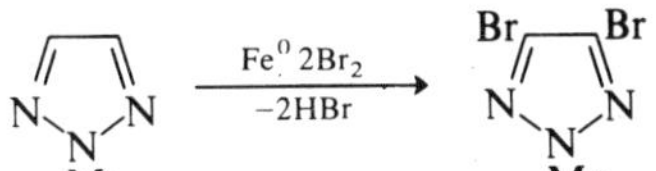

1-Iodo-1,2,3-triazoles have been reported to rearrange into the 4-isomers upon heating ⟨70ZC220⟩. In contrast, bromination of 2-phenyl-1H-1,2,3-triazole in the presence of silver sulfate affords the 2-p-bromophenyl derivative with a typical ^{1}H NMR pattern (A_2B_2 of the aromatic ring; singlet at δ 7.75 p.p.m.: two equivalent triazole-H) ⟨63CJC2380⟩. Similar observations have been reported for the nitration of the same 2H-triazole: nitration occurs firstly in the aromatic ring and then at C-4 of the triazole ring to give (145) and (146) ⟨48JOC815, 63CJC274⟩.

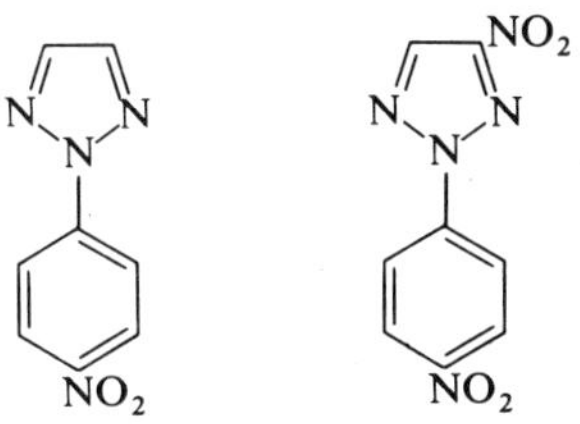

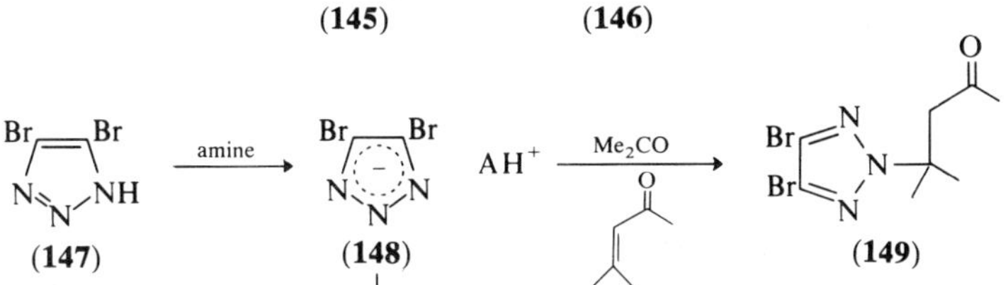

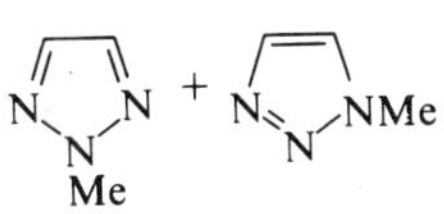

The H-1 atom of 4,5-dibromo-1,2,3-triazole (**147**) ⟨67G109⟩ is very acidic and forms instantly the triazolide salt (**148**) when treated with dimethylamine ⟨67G109⟩. The salt is stable even when heated to 260 °C for 140 hours and reacts smoothly in a conjugate addition reaction to give (**149**) ⟨54JA4933⟩. Methylation of (**148**) gives both the 1-methyl-1*H* and 2-methyl-2*H* derivatives (**150**) and (**151**) ⟨70JHC961⟩.

Similar results have been obtained with 4,5-dimethoxycarbonyl- and 4-phenyl-1,2,3-triazolides, which displace chlorine from ethyl chloroacetate or β-chloropropionate to give both 1-*N*- and 2-*N*-alkylated products. The highest 2-*N* to 1-*N* selectivity was *ca.* 5:1 with the base triethylamine in DMF. Furthermore, triazolides add to alkynes to give Michael adducts at the 2-position exclusively ⟨73T3285, 73JOC2708⟩.

1,3-Dimethyl-1,2,3-triazolylium *p*-toluenesulfonate is brominated in the 4-position (to give **153**) by means of bromine and NBS. When the 4-bromo salt (**153**) is dissolved in base an equilibrium mixture of (**152**)–(**154**) is obtained. Evidently, (**153**) is converted into an ylide which is further brominated by (**153**). Subsequently, (**153**) and the dibromo compound (**154**) undergo very much slower nucleophilic substitution to give 1,3-dimethyl-1,2,3-triazole 4-oxides (**154a**) ⟨69ACS(B)2733, 71ACS(B)249⟩. A similar mixture is obtained from the 5-bromo derivative (**155**).

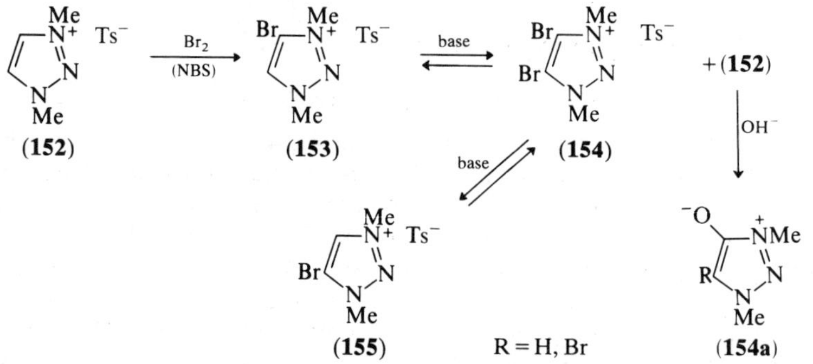

Chlorination of benzotriazole with aqua regia gives 4,5,6,7-tetrachlorobenzotriazole in 87% yield. The 1-methyl and 2-methyl derivatives have been obtained by alkylation and chlorination of 1- and 2-methylbenzotriazole, respectively ⟨55JA5105⟩.

In contrast to various syntheses of 1,2,3-triazole nucleosides by cycloaddition of glycosyl azides to substituted alkynes (see Section 4.11.5.2.2 and *e.g.* ⟨70JHC1269⟩), the acid-catalyzed fusion of methyl 1,2,3-triazole-4-carboxylate, 4-cyano-1,2,3-triazole and 4-nitro-1,2,3-triazole (**156**) with an acylated ribofuranose provides the corresponding 4-substituted 2β-D-ribofuranosyl-1,2,3-triazoles (**157**) along with the 1-β-D-isomers (**158**) ⟨72JHC1195⟩.

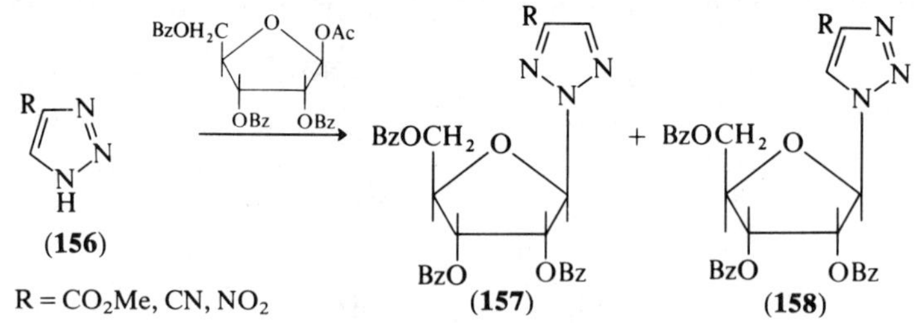

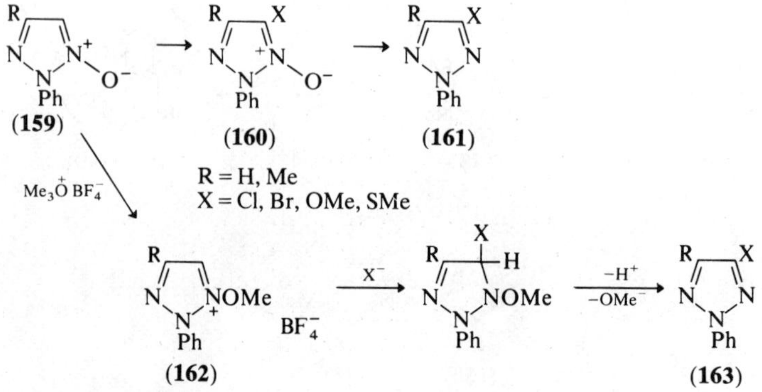

2-Aryltriazole 1-oxides (159), readily availabe by various procedures (cf. Section 4.11.5), undergo both electrophilic and nucleophilic substitution, being activated at C-5. The N-oxides can be selectively halogenated and, in turn, the halogen can be replaced by strong nucleophiles giving (160). On subsequent deoxygenation, the N-oxides yield halogeno-, methylthio- and methoxy-triazoles (161) ⟨81JCS(P1)503⟩. Methylation of the N-oxides with trimethyloxonium tetrafluoroborate produces N-methoxytriazolium salts (162) in which H-5 can be replaced by weak nucleophiles, e.g. fluoride ions and numerous other substitutents (to give 163) ⟨81JCS(P1)503⟩.

Diazonium salts (165), derived from 4- or 5-aminotriazoles (164), undergo ready displacement of the nitrogen like other aromatic diazonium salts. Thus, 5-diazo-1,2,3-triazole-4-carboxamide is rapidly converted into the corresponding iodide (166) ⟨66JMC733⟩.

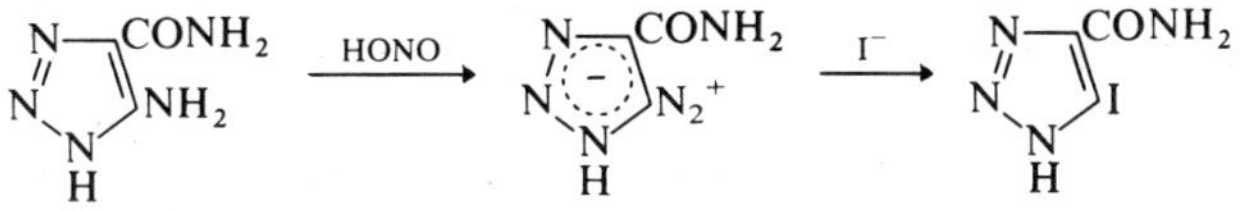

4.11.4.1.4 Reactions with cyclic transition states

(i) Cycloaddition reactions of heterocyclic ylides and mesoionic compounds containing a 1,2,3-triazole nucleus

By means of ^{1}H NMR studies Petersen and Heitzer have shown that upon oxidation of 2,3-butanedionebis(benzoylhydrazone) (168) the tetrazine (167) was not formed ⟨00CB644, 09CB659⟩. Instead, an ylidic anhydro-2-benzoyl-1-benzoylimino-4,5-dimethyl-1,2,3-triazolium salt (169) was obtained ⟨70AG81, 70AG(E)67⟩. The structure of (169) is in agreement with its chemical properties. Some time earlier, anhydro-1,3-dimethyl-4-hydroxy-1,2,3-triazolium hydroxide was obtained by methylation of 1-methyl-1,2,3-triazolin-5-one ⟨66ACS(B)1555, 69ACS(B)2733⟩.

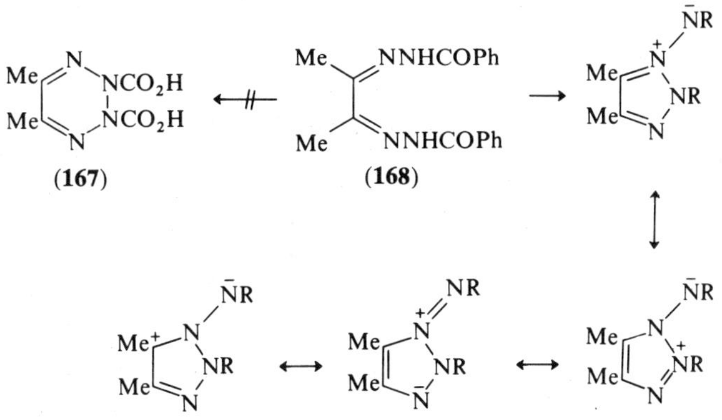

Cyclization ⟨64CRV129⟩ of ethyl N-methyl-N-phenylazoaminoacetate (171) with thionyl chloride–pyridine readily gives anhydro-4-hydroxy-3-phenyl-1-methyl-1,2,3-triazolium hydroxide (172) ⟨70JOC3451⟩ which reacts smoothly with DMAD, leading to the formation of the pyrazole (174) via the trisazanorbornene (173). The intermediate cycloadduct (173) easily aromatizes by loss of an aromatic isocyanate under the thermal reaction conditions ⟨70JOC3451⟩. The polyaza adduct (175) obtained with ethyl azodicarboxylate is stable as aromatization is impossible here. Both isocyanate and phenyl isothiocyanate gave 1:1 cycloadducts with (172), namely 1,2,6,7-tetraazabicyclo[2.2.1]heptanes (176) and (177) ⟨72JOC2049⟩.

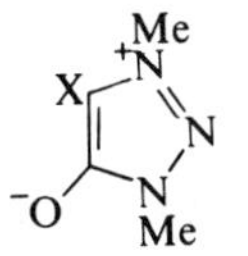

(171) (172)

(172) (173) (174) (175)

(176) (177)

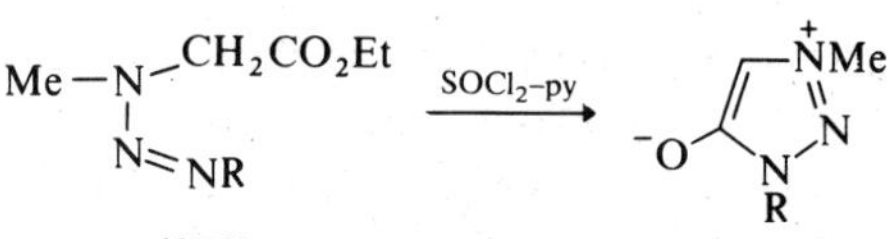

Anhydro-1-phenylimino-2,4,5-triphenyl-1,2,3-triazolium hydroxide (**178**), obtained by oxidation of benzil bis-phenylhydrazone, affords 1:1 cycloadducts with a wide variety of alkynic and alkenic dipolarophiles, yielding pyrazolino[2,3-*c*]- and dihydropyrazolino[2,3-*c*]-[1,2,3]triazoles (**179**). The initial 1:1 cycloadduct (**180**) obtained from (**178**) and carbon disulfide undergoes thermal fragmentation to give 2,4,5-triphenyl-1,2,3-triazole (**181**) ⟨72T3987⟩.

(178) (179) (180) (181)

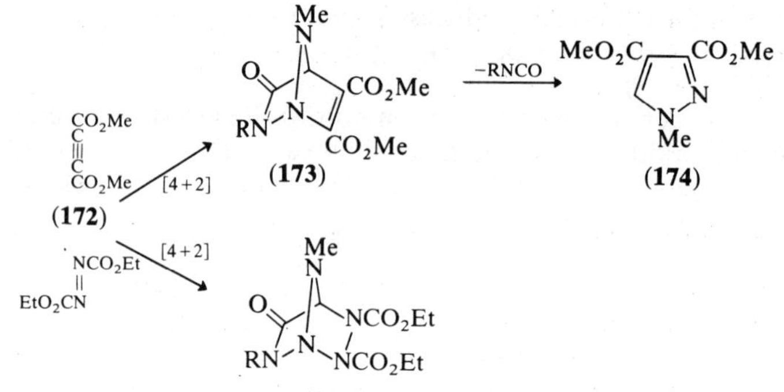

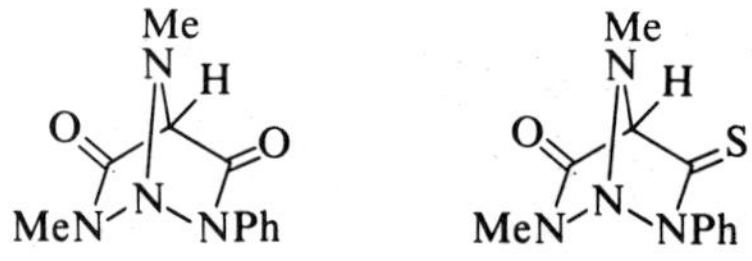

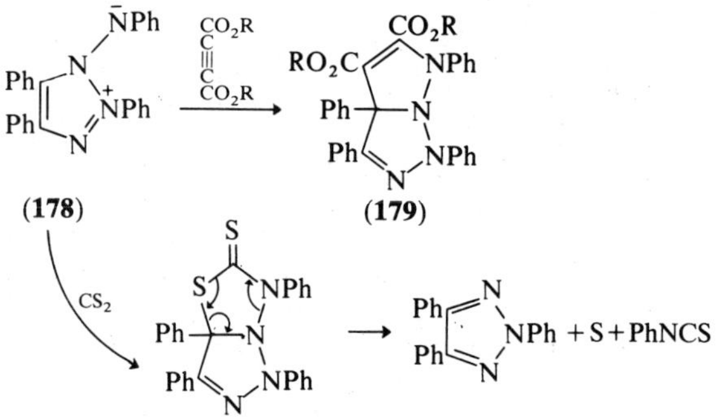

4.11.4.2 Dihydro Compounds (Δ^2-1,2,3-Triazolines)

4.11.4.2.1 Stereochemical aspects

Stereochemical interest has naturally concentrated on the Δ^2-1,2,3-triazolines ⟨review: B-77SH(1)117⟩, especially on the potential asymmetry at C-4 and C-5 and the orientation of the R—N—N=N group with respect to it.

Δ^2-1,2,3-Triazolines, *e.g.* (**181a**), are usually formed by 1,3-dipolar cycloaddition of azides to alkenes ⟨cf. 63AG604, 742, 63AG(E)565, 633, 68AG329, 68AG(E)321⟩. The addition always occurs *cis* and is usually regiospecific. Norbornene condenses with azides to form only the *exo*

(181a)

(181b)

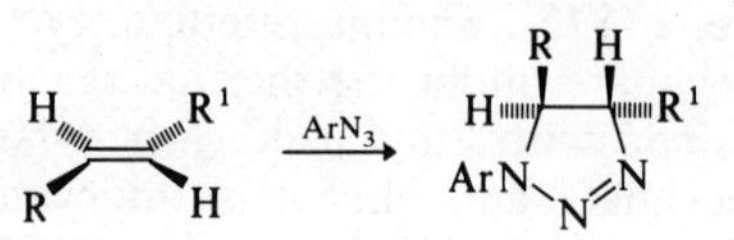

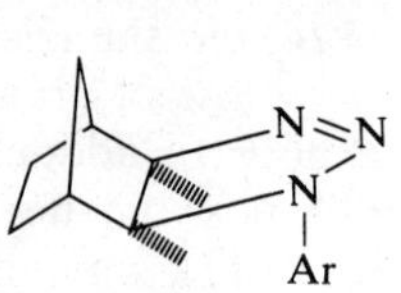

adduct (**181b**) ⟨35LA(515)185, 65CB3992, 65JA306, 66CC352⟩. Cycloaddition of *cis-* and *trans-*enol ethers produces *cis-* and *trans-*substituted Δ^2-1,2,3-triazolines, respectively, with high stereospecificity ⟨65CB1153⟩.

From examination of the ^{1}H NMR spectra of 5-alkoxy- and 5-hydroxy-Δ^2-1,2,3-triazolines, an envelope conformation of the molecule was deduced with the 5-substituent being in a pseudoaxial position at the flap and the 1-substituent being pseudoequatorial. This preferred conformation can originate a type of anomeric effect ⟨74ACS(B)425⟩. 5-Amino-1-(*p*-nitrophenyl)-Δ^2-1,2,3-triazolines have also been found to prefer the envelope conformation ⟨74ACS(B)425⟩.

4.11.4.2.2. Aromatization reactions (dehydrogenation of Δ^2-1,2,3-triazolines)

The aromatization of triazolines can take place either by rearrangement reactions, by direct oxidation or by an elimination reaction. A recent review discusses this topic ⟨78BSF(2)485⟩. From the numerous reactions some illustrative examples are described below.

Δ^2-1,2,3-Triazolines (**182**) can be smoothly converted into the corresponding 1,2,3-triazoles (**183**) with the aid of potassium permanganate ⟨72JHC717, 66CB475⟩. This oxidation to a heteroaromatic system can be successfully performed with KMnO$_4$ using a two-phase system as well as a phase-transfer catalyst ⟨78S694⟩. In some cases, suitably substituted open-chain diazo compounds can be cyclized directly to 1,2,3-triazoles by means of oxidizing agents (O$_2$, KMnO$_4$) ⟨70TL2823, 71BSF3642⟩. Spirotriazolines like (**184**) isomerize spontaneously with cleavage of the spirocyclopropane ring to afford (**185**) ⟨75JOC2042, 75JOC2045, 75DIS(B)(36)1712⟩.

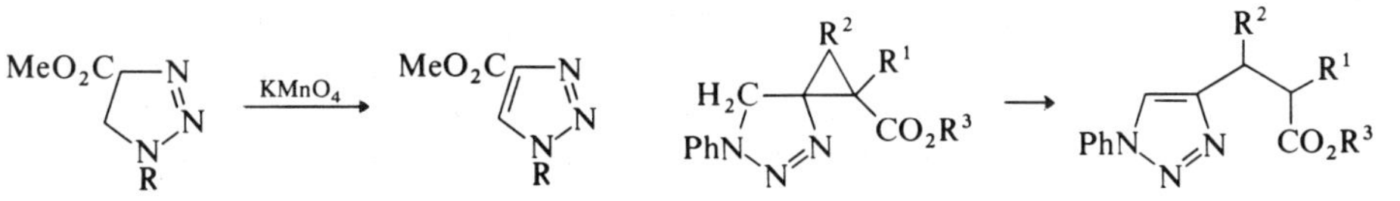

Another example of the aromatization process is represented by the rearrangement of the indolospirotriazoline (**186**) upon treatment with a basic catalyst. The triazole (**187**) is formed with cleavage of one σ-bond ⟨70LA(734)70⟩. Δ^2-1,2,3-Triazolines substituted in the 5-position with an amino, alkoxy or hydroxy group and possessing at least one 4-H as in compound (**188**) are capable of condensation reactions, eliminating RNH$_2$, ROH or H$_2$O (→ **189**), when they are heated in the presence of acid or base ⟨review: 78BSF(2)485⟩. The configurations at C-4 and C-5 play a decisive role in the facility of the elimination. Thus, the elimination of an amine is much faster if the amino substituent at C-5 is in a *trans* position with respect to H-4. Thus, the triazole (**191**) is obtained smoothly by heating of the *trans-*substituted triazoline (**190**), while the corresponding *cis-*substituted isomer (**192**) is stable under these conditions ⟨73PC41100⟩.

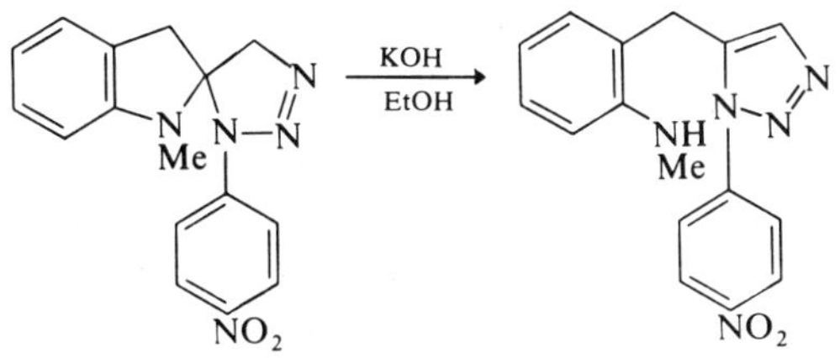

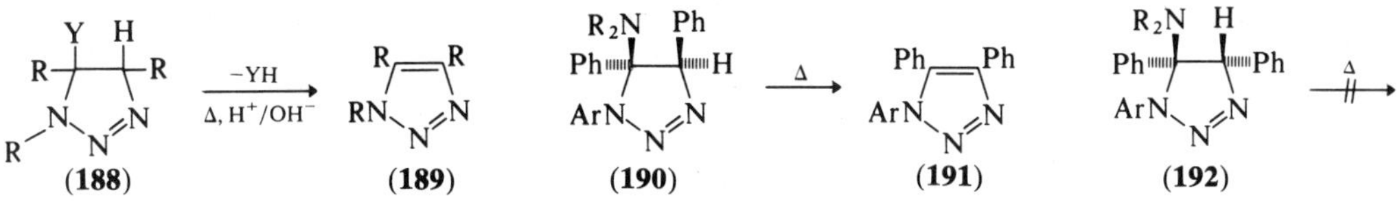

Y = OR, NHR, OH

Triazolines which are formed as intermediates after the 1,3-dipolar cycloaddition of azides to enamines cannot be isolated but directly undergo elimination to give triazoles.

Enamines with β-substituents such as an ester, ketone, cyano, nitro or sulfone group result in compounds (**193**), which, possessing an α-hydrogen atom, behave in an analogous fashion to give (**194**) ⟨65CB1138, 65CB2715, 61G849, 62G1040, 65G1220, 67G304, 68G949⟩.

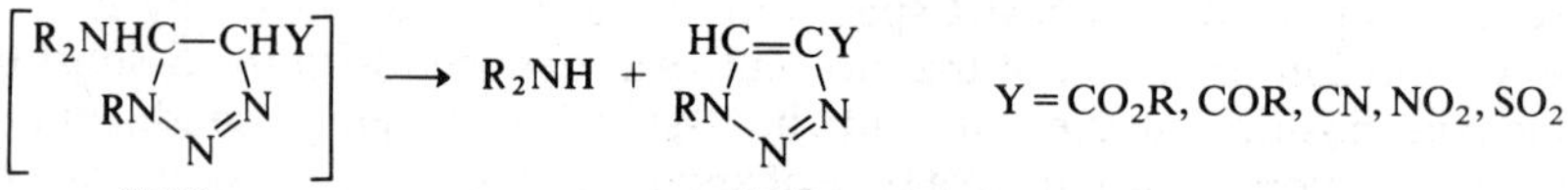

Bicyclic triazolines can be readily aromatized, *e.g.* in the presence of oxalic acid compound (**195**) gives (**196**) ⟨65MI41100⟩.

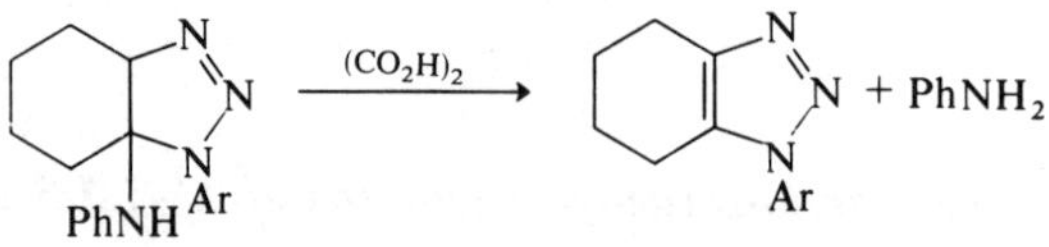

One example has been reported where, after 1,3-dipolar cycloaddition of phenyl azide to a heterocyclic β-enaminoester to give compound (**197**) ⟨cf. B-80MI41103⟩, no β-hydrogen atom is present for a regular elimination of ammonia. In this case a different aromatization reaction is observed involving an elimination–ring cleavage process resulting in the formation of the 1,2,3-triazole carbonate (**198**) ⟨71CB3510⟩. In some special cases HNO_2 can also be eliminated from an initial 1:1 cycloadduct to form, for example, the triazole (**199**) ⟨67CC918, 66JPR(305)199⟩. Furthermore, 1,2,3-triazoles are formed by the elimination of triphenylphosphine oxide from an appropriately substituted triazoline. This process is illustrated by the conversion of (**200**) into (**201**), a sequence reminiscent of the Wittig reaction. ⟨71T845, 71T5623, 69T5421, 69BSB147⟩.

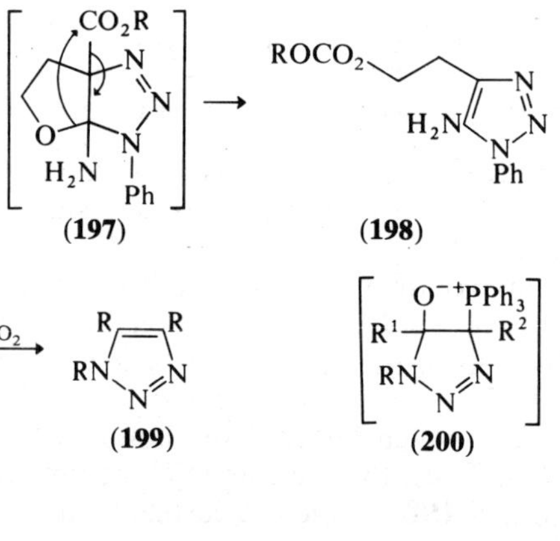

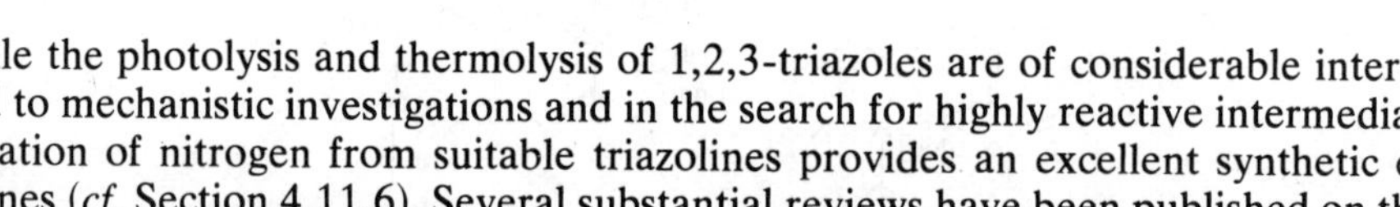

4.11.4.2.3. Ring fission of Δ^2-1,2,3-triazolines

While the photolysis and thermolysis of 1,2,3-triazoles are of considerable interest with regard to mechanistic investigations and in the search for highly reactive intermediates, the elimination of nitrogen from suitable triazolines provides an excellent synthetic entry to aziridines (*cf.* Section 4.11.6). Several substantial reviews have been published on this topic ⟨B-76PH403, B-70MI41101, B-67MI41101⟩.

Triazolines (**202**) obtained by 1,3-dipolar cycloaddition reactions lose nitrogen on pyrolysis or photolysis to give aziridines (**203**) in high yields. The isomeric imines (**204**) are also formed ⟨68T2757⟩. A higher ratio of aziridines (**203**) to imines (**204**) results from thermal reaction conditions than from the corresponding photolytic reaction ⟨68T2757⟩.

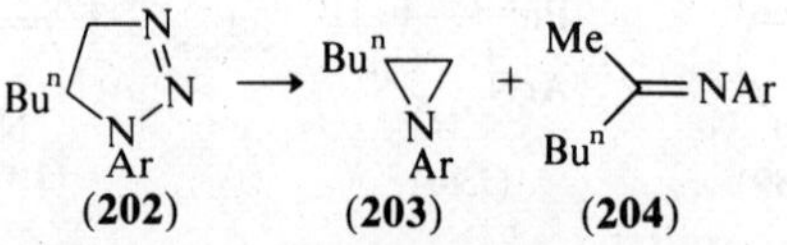

The decomposition of the diazo compounds (**207**) is 45 times slower than that of the corresponding 1,2,3-triazoline (**205**). The exclusive generation of the enamino ester (**209**)

seems to exclude an equilibrium between (205) and (207) ⟨66CB475⟩. However, the corresponding cyano compound shows in the presence of a tertiary amine an equilibrium between the triazoline and the diazo compound ⟨66CB475⟩.

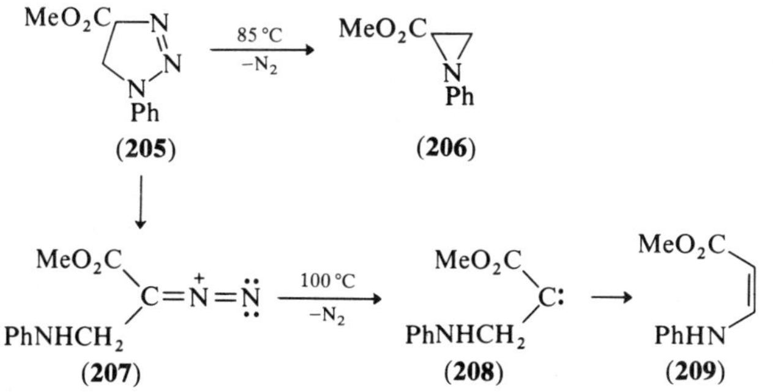

Upon pyrolysis (315 °C) bicyclic triazolines like (210) afford various mixtures of imines (211) and bicyclic aziridines (212), depending on the size of the condensed ring (see Table 19) ⟨68T2757⟩. The aziridines formed can be considered as potential azomethine ylides, and can also dimerize under the conditions of thermolysis ⟨79BSF(2)633⟩. The preparative aspect of these extrusion reactions is discussed in Section 4.11.6.

Table 19 Pyrolysis of Bicyclic 1,2,3-Triazoles ⟨68T2757⟩

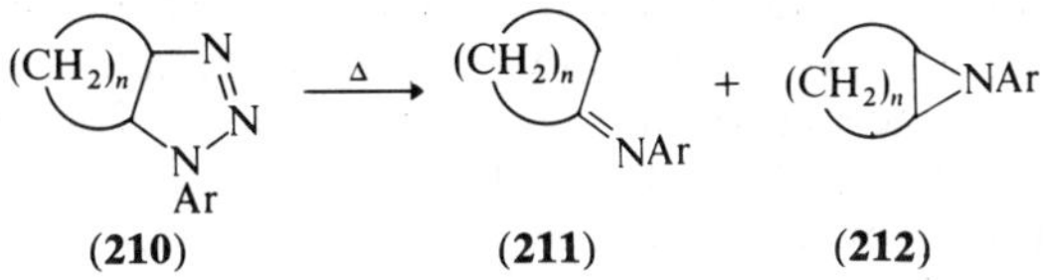

n	(211) (%)	(212) (%)
3	79	21
5	46	54
6	45	55

Treatment of 2-phenyltriazole-4-carboxylic acid (213) with barium hydroxide at elevated temperatures gives two ring-cleavage products in a concerted process ⟨66TL4387⟩.

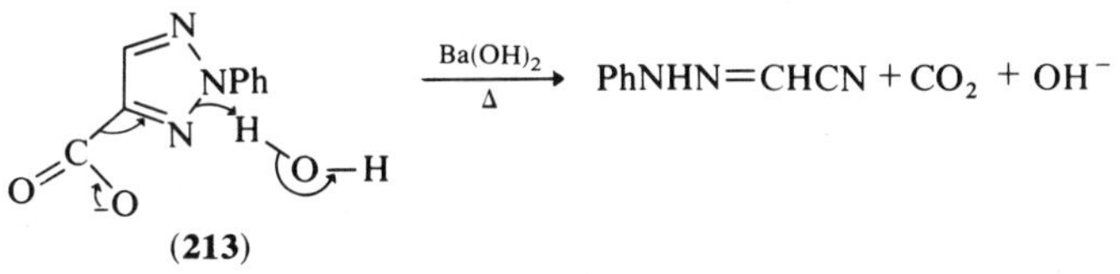

$$\text{(213)} \xrightarrow[\Delta]{Ba(OH)_2} PhNHN{=}CHCN + CO_2 + OH^-$$

4.11.5 SYNTHESIS

4.11.5.1 General Remarks

An extensive number of 1,2,3-triazoles and benzotriazoles are known today and numerous syntheses have been reported. Most of the classical methods are well documented in excellent reviews which describe the individual syntheses from different viewpoints and emphasize different concepts. These reviews may be conveniently divided according to: (a) those emphasizing different syntheses ⟨74AHC(16)33, B-61MI41100, B-71MI41104⟩ and (b) those aiming at different substitution products ⟨50CRV(46)1, 80HC(39)1⟩. As most of the major reaction pathways involve azides, several reviews are devoted especially to aspects of forming 1,2,3-triazoles in this way ⟨69CRV345, B-71MI41104⟩.

In the following discussion the methods of ring formation are divided into a number of principal synthetic routes according to Scheme 7.

The classical and well-documented pathways are described in more general terms, and more emphasis is given to recent methods.

Formation of one bond (N^1---N^2)

Formation of two bonds (C^5---N^1, C^4---N^3)

(C^4---C^5, N^1---N^2)

Scheme 7

4.11.5.2 Triazoles from Non-heterocyclic Compounds

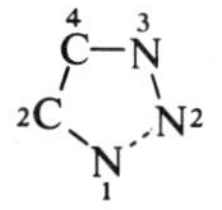

4.11.5.2.1 Formation of one bond

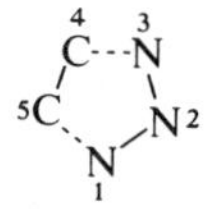

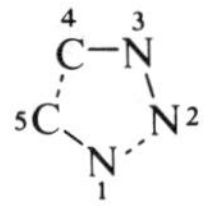

In an intramolecular cyclization of adjacent groups 1-aminotriazoles (**213b**) and their derivatives are prepared from bis-hydrazones and from substituted bis-hydrazones of α-diketones in an oxidation step involving the formation of one bond. The bis-hydrazones (**213a**) are more conveniently obtained from α-bromoketones than from α-diketones ⟨67TL3295, 71JPR882, 71ZC179⟩. The oxidation reaction is carried out by means of MnO_2 or HgO, and the bis-hydrazone (**213a**) although being unsymmetrical, gave the 1-amino-4-phenyl-1,2,3-triazole (**213b**) as the only reaction product. The mechanism has been elucidated (^{15}N labelling studies) and the structure has been confirmed by X-ray analysis ⟨71ZC179⟩. Overoxidation of the bis-hydrazones yielding alkyne compounds can be avoided by employing toluene-*p*-sulfonylhydrazones ⟨52JCS4735, 68LA(711)46, 68LA(711)65⟩.

(**213a**) (**213b**)

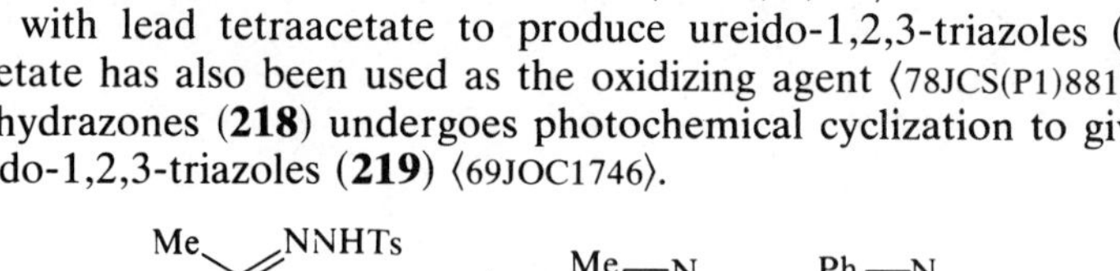

Starting with the unsymmetrical bis-toluene-*p*-sulfonylhydrazone (**214**) both possible triazoles (**215a**) and (**215b**) have been obtained ⟨73JCS(P1)555⟩. Bis-semicarbazides (**216**) react smoothly with lead tetraacetate to produce ureido-1,2,3-triazoles (**217**) ⟨76S482⟩. Mercury(II) acetate has also been used as the oxidizing agent ⟨78JCS(P1)881⟩. The dianion of α,β-bistosylhydrazones (**218**) undergoes photochemical cyclization to give in excellent yield sulfonamido-1,2,3-triazoles (**219**) ⟨69JOC1746⟩.

(**214**) (**215a**) (**215b**)

(**216**) (**217**)

(**218**) (**219**)

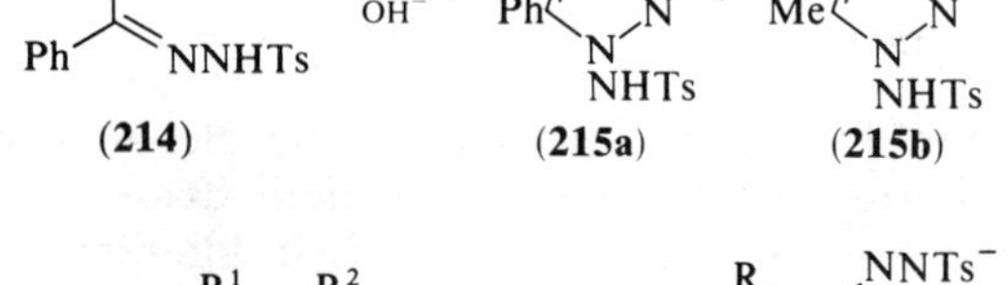

1,2,3-Triazoles and 1*H*-triazoles (**221**) can be readily obtained by ring closure of α-diazoimines (**220**), which may be prepared *in situ* from diazoketones and amines ⟨67CCC2155, 70T1393, 70LA(734)56⟩. This very general reaction may also be applied to α-diazoamides ⟨67CJC1727⟩, which on ring closure with PCl_5 afford a 5-chlorotriazole (**223**) *via* an intermediate imidoyl chloride (**222**) ⟨67ACS(B)633⟩.

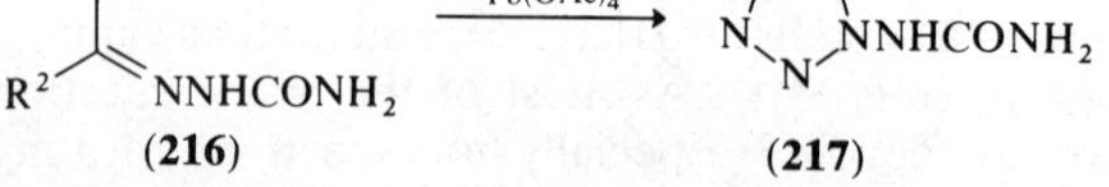

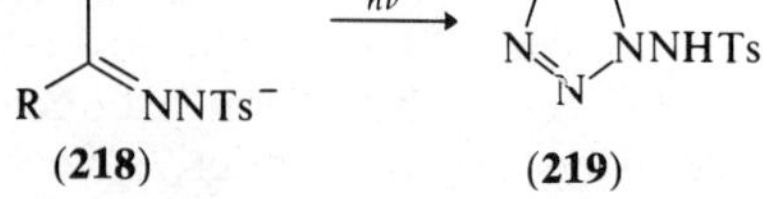

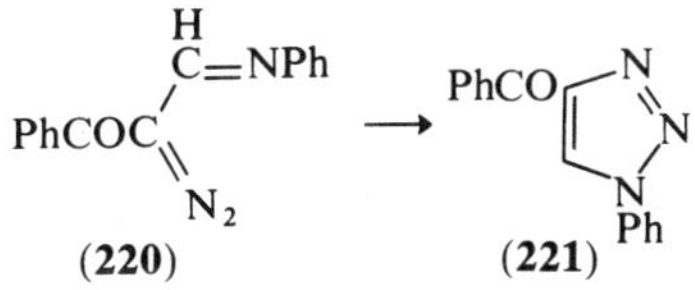

Another synthetic pathway recently used is the cyclization of linear triazenes, like (**224**), which in aprotic media with Lewis base catalysis yield 5-amino-1-aryl-1,2,3-triazoles (**225**). However, in protic media Dimroth rearrangement affords the isomeric 5-(arylamino)-1,2,3-triazoles (**226**) ⟨70JOC2215, 81JOC856⟩. The 2-tetrazene (**227**) gives methyl 1-benzyl-1,2,3-triazolecarboxylate (**228**) by photochemically induced homolysis of a single N—N bond ⟨81TL227⟩.

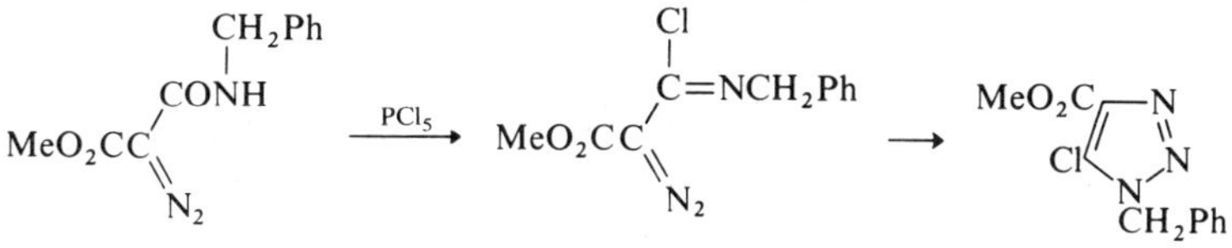

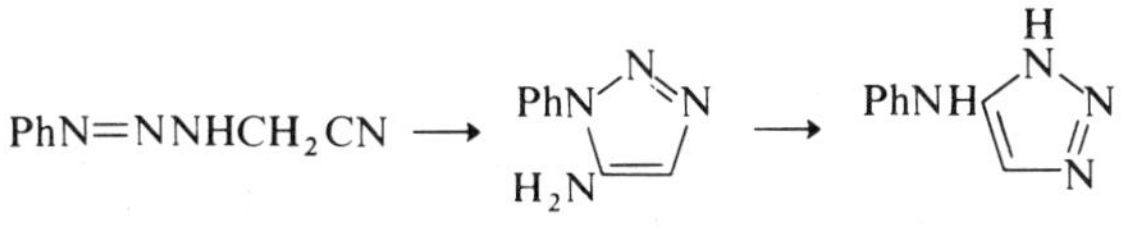

In an earlier publication the conversion of the 1,2-bis(benzeneazo)ethylene (**229**) into the triazole derivative (**230**) by treatment with mild acid was described ⟨62JOC4300⟩.

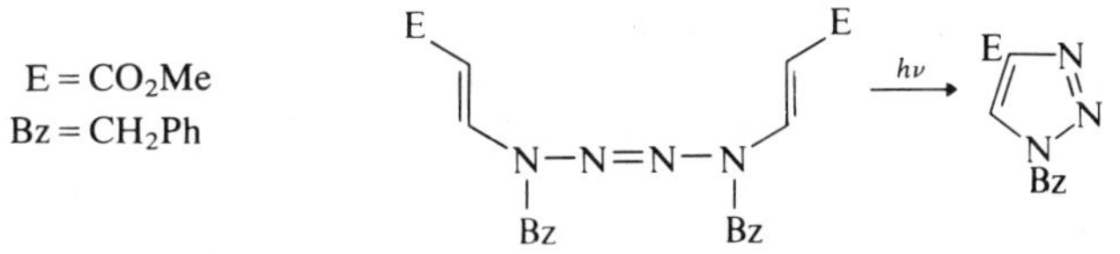

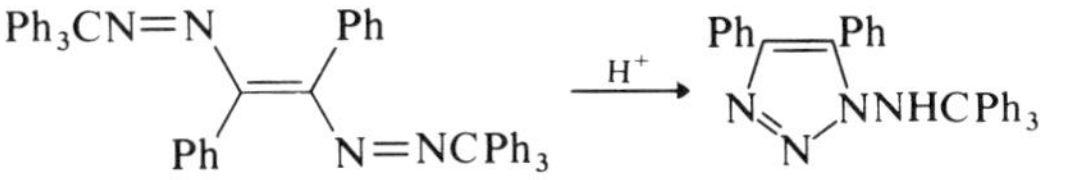

Azaallyl cations (**231**), formed from arenediazonium salts and 2-methylaminoethanol, cyclize to the triazenium salts (**232**), a new class of heterocyclic cation ⟨79CB445⟩.

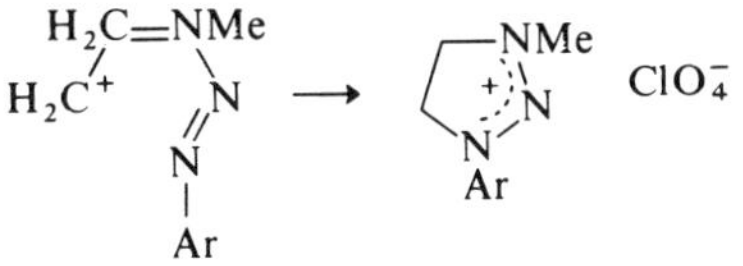

The cyclization of nitrogen derivatives of α,β-dicarbonyl compounds represents a promising approach to substituted $2H$-1,2,3-triazoles. Thus, glyoxal phenylosazone (**233**) is converted by oxidation into the bis(benzeneazo)ethylene (**234**) which is smoothly cyclized to (**235**) by the action of iron(III) chloride in dilute acid ⟨26CB1742, 48JOC815⟩. For further examples of this synthetic approach, see ⟨30G866, 46JA1799⟩. Furthermore, manganese dioxide and nickel dioxide have been found to give $1H$-triazoles in poor-to-fair yields ⟨67JOC2252, 75T1171⟩. Activated α-dicarbonyl derivatives are converted into N-oxides of 2-phenyl-4-hydroxy- (**236**) and 2,4,5-triphenyl-$2H$-1,2,3-triazoles (**237**) ⟨60JOC313, 74S198, 80AJC2447⟩. More recently, the oxidative ring closure has been effected with N-iodosuccinimide (NIS) ⟨80H(14)1279⟩.

5-Amino- and 5-hydroxy-$4H$-triazoles are formed by the reduction of α-azidonitriles and esters, e.g. (**238**) → (**239**) ⟨58M557, 562, 588, 597⟩.

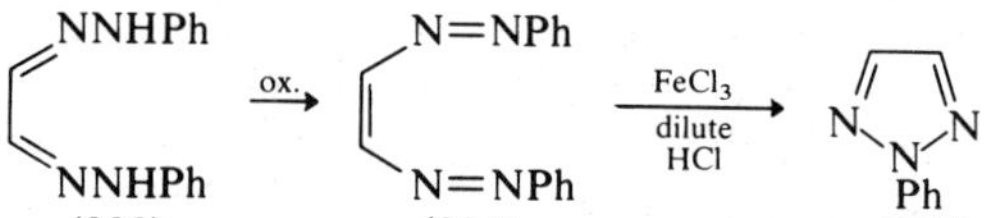

4.11.5.2.2. Formation of two bonds

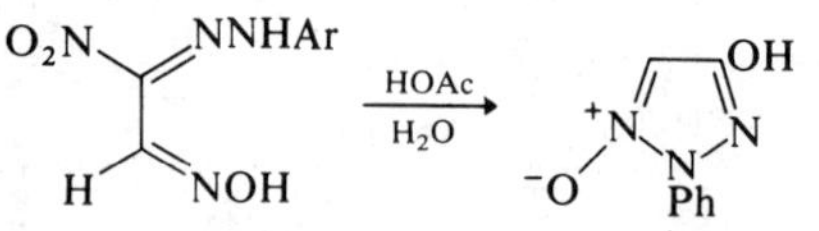

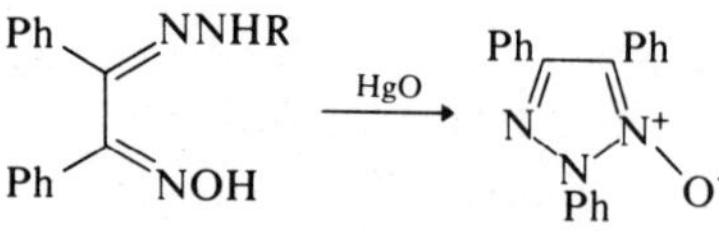

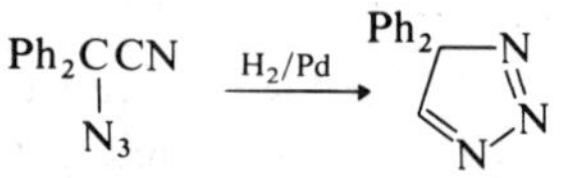

The most important syntheses in this group consist of the thermal 1,3-dipolar cycloaddition ⟨65CB4014, 67CB2494⟩ of azides to alkynes, which is, of course, one of the most versatile routes to 1H-1,2,3-triazoles. A wide range of substituents can be incorporated into the alkyne and azide components.

The parent compound, 1,2,3-triazole, has been obtained by several direct and indirect synthetic approaches: the direct addition of hydrazoic acid to alkynes with electron-withdrawing groups is the straightforward but also somewhat dangerous procedure ⟨59ACS(B)888, 64MI41100⟩; in the same way sodium azide can be used cautiously in the presence of an acid. Alternatively, the addition of trimethylsilyl azide to the carbon–carbon triple bond of the alkyne and subsequent removal of the silyl group provides a much safer procedure ⟨65AG414, 65AG(E)417⟩. Furthermore, 1H-1,2,3-triazoles (**240**) can be produced in a much safer way by cycloaddition of benzyl azide to DMAD and subsequent thermolysis and hydrogenolysis of the benzyl group ⟨56JOC190, 65LA(688)205⟩.

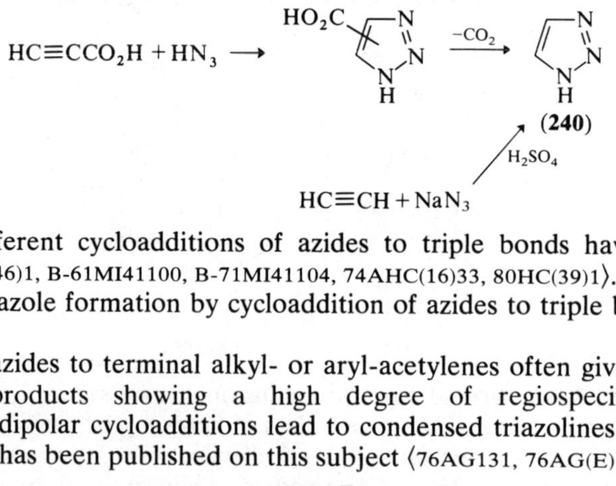

Many of the different cycloadditions of azides to triple bonds have been reviewed extensively ⟨50CRV(46)1, B-61MI41100, B-71MI41104, 74AHC(16)33, 80HC(39)1⟩. Some representative examples of triazole formation by cycloaddition of azides to triple bonds are given in Table 20.

The addition of azides to terminal alkyl- or aryl-acetylenes often gives a good yield of 1,4-disubstituted products showing a high degree of regiospecificity ⟨52HCA167⟩. Intramolecular 1,3-dipolar cycloadditions lead to condensed triazolines ⟨78ACS(B)683⟩ and an excellent review has been published on this subject ⟨76AG131, 76AG(E)123⟩.

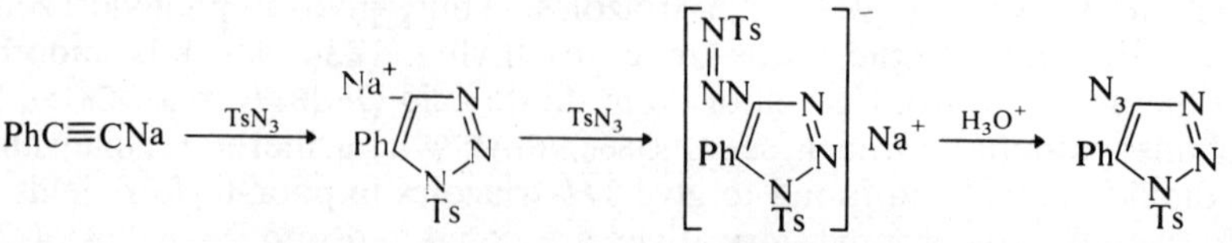

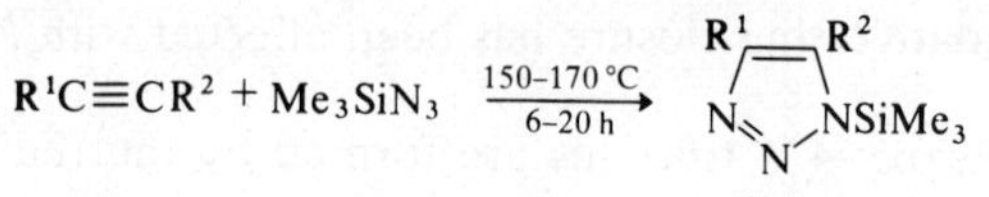

Furthermore, cycloaddition reactions have been carried out between metal acetylides and azides leading firstly to triazole triazene salts (**241**). On hydrolysis the corresponding triazole azide (**242**) is obtained ⟨58JOC1051⟩. The addition of trimethylsilyl azide to alkynes occurs at higher temperatures, but due to the considerable thermal stability of the azide, good yields of triazoles (**243**) can be achieved ⟨80MI41101⟩.

Similarly, silylated alkynes react with HN_3 to afford TMS-substituted 1,2,3-triazoles ⟨79CB2829⟩. 1,2,3-Triazoles with alkenic side chains can also be synthesized from 1,3-enyne hydrocarbons and their silane derivatives ⟨78ZOB705⟩.

Alkynes with a phosphonium substituent (**244**) cyclize with NaN_3 in DMF to (**245**), which cleaves upon saponification to give 4-phenyl-1,2,3-triazole (**246**) ⟨73JOC2708⟩. Tropylium azide can be considered in this reaction as a kind of protecting group (**247**) because of its easy removal on hydrolysis to give (**248**) ⟨65JOC638⟩. Cyanoacetylene gives 5-cyano-1,2,3-triazole and a 2:1 adduct with alumina azide ⟨69JOC1141⟩.

Table 20 Triazoles from Azides and Acetylenes

Azide	Acetylene	Product	Ref.
RN_3	$MeO_2CC{\equiv}CCO_2Me$	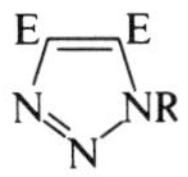 (E, E, N, N, NR triazole)	30JPR(125)498
HN_3	$HC{\equiv}CCO_2H$	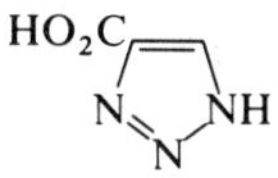 (HO_2C, N, N, NH triazole)	59ACS(B)888
TsN_3	$PhC{\equiv}CH$	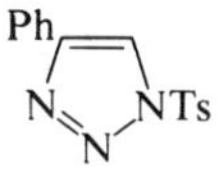 (Ph, N, N, NTs triazole)	65CB4014
NaN_3	$PhC{\equiv}CH$	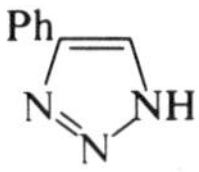 (Ph, N, N, NH triazole)	70CB1908
N_3CO_2Et	$PhC{\equiv}CH$	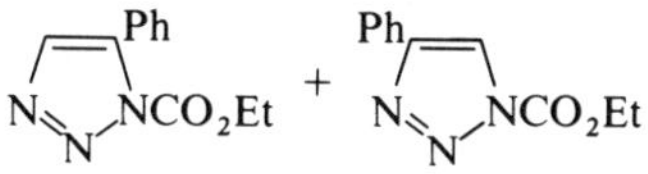 (Ph, N, N, NCO_2Et) + (Ph, N, N, NCO_2Et)	65CB2985 71T5623
$N_3C{\equiv}N$	$RC{\equiv}CR'$	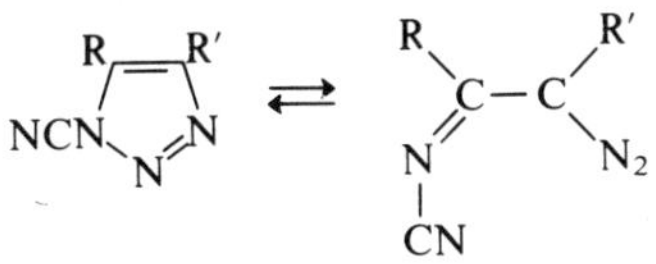 (R, R', NCN, N, N triazole) ⇌ (R, R', C=C, N, N_2, CN)	67JA4760
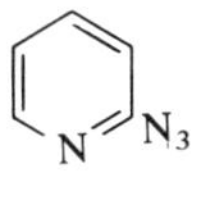 pyridyl-N_3	$PhC{\equiv}CPh$	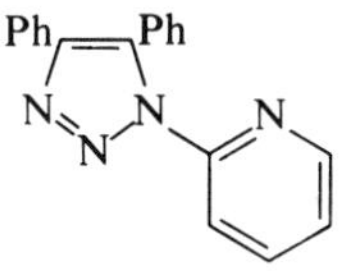 (Ph, Ph, N, N, N-pyridyl triazole)	69TL2589 71T5121 71JOC446
RN_3	$(EtO)_2CHC{\equiv}CCH(OEt)_2$	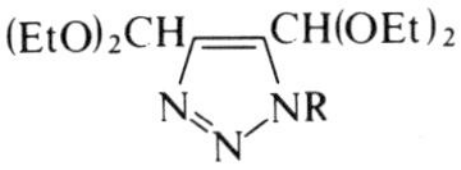 ($(EtO)_2CH$, $CH(OEt)_2$, N, N, NR)	74CB715
PhN_3	$HOCH_2C{\equiv}CCH_2OH$	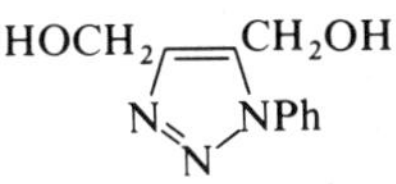 ($HOCH_2$, CH_2OH, N, N, NPh)	74CB715
PhN_3	$n\text{-}C_6F_{13}C{\equiv}CPh$	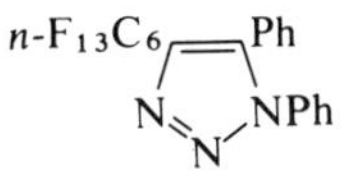 ($n\text{-}F_{13}C_6$, Ph, N, N, NPh)	75JOC810

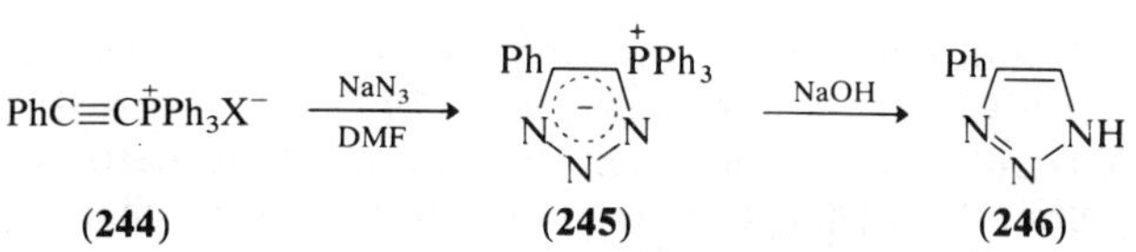

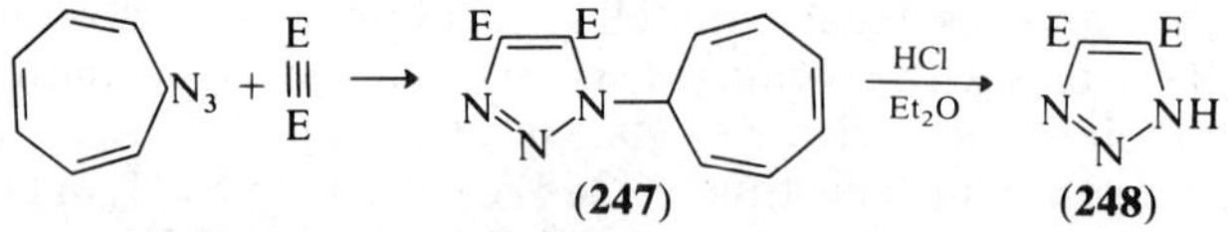

The cycloaddition of phenyl azide to the alkynic Grignard (249) provides an interesting variant of this approach ⟨67ZOR968⟩, as the intermediate can in turn be carboxylated with CO_2 to (250). Alkadiynes can serve in a successful approach to control mono- or di-addition ⟨58CB1841⟩. The products are (251) and (252).

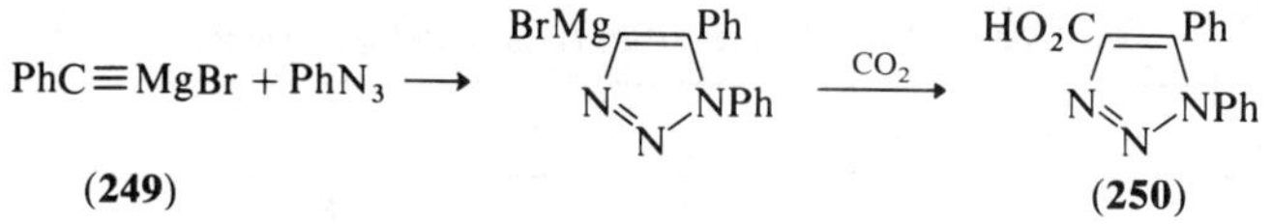

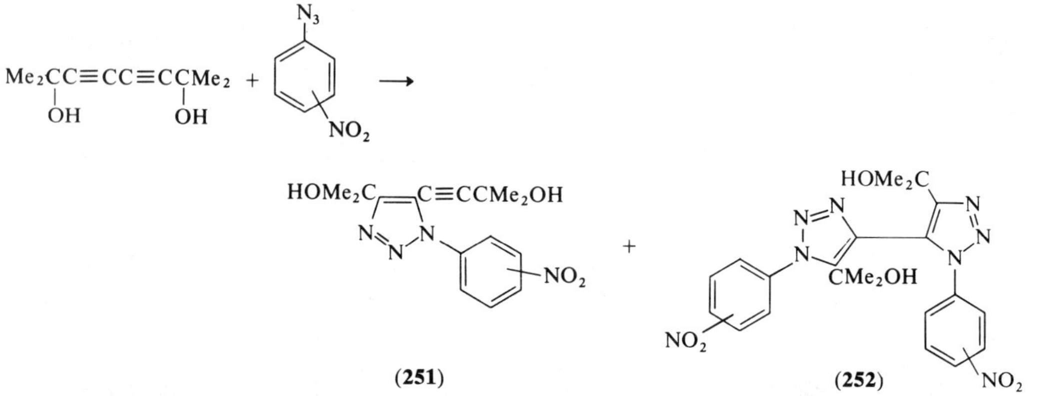

2-Trialkyl-silicon- (253) and -tin-1,2,3-triazoles (254) are obtained readily from organometallic azides and alkynes. These adducts are hydrolyzed smoothly in almost quantitative yield to give (255) ⟨66CB2512, 72JOM(44)117⟩ (cf. acylation via 2-TMS derivatives, Section 4.11.4.5).

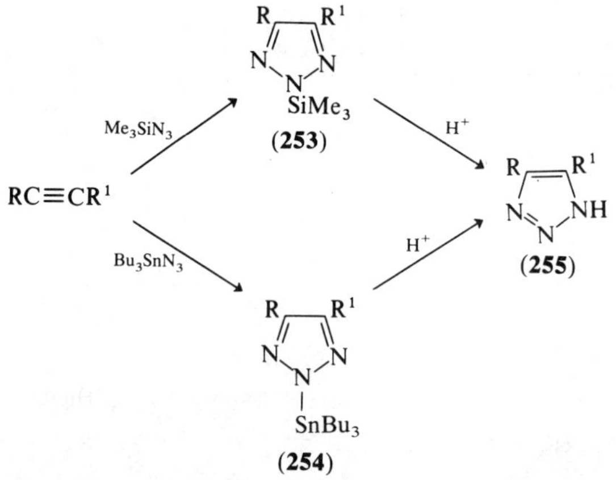

The cycloaddition of azides to alkynes is a generally applicable route to Δ^2-1,2,3-triazolines, and most triazolines have been obtained by this method. For brief reviews, see ⟨54CRV41, B-64MI41101, B-71MI141104, 78BSF(2)485⟩.

In comparison with alkynes, cycloaddition of azides to alkenes proceeds at a much slower rate. Ring strain and substitution of the double bond with electron-withdrawing groups have been found to lead to enhancement of the cycloaddition rate ⟨68T349, 2757⟩. Double bonds which are part of strained bicyclic systems are particularly reactive. Thus, the addition of phenyl azide to norbornene leads in a very fast reaction to the triazoline (256), the addition occuring at the less hindered *exo* side ⟨35LA(515)185⟩. The stereospecificity of this reaction has been established ⟨65CB1153, 66JA4759, 68JA988⟩ and mechanistic studies with norbornene and substituted phenyl azides show that the reaction occurs in a concerted manner ⟨65JA306, 65CB3992⟩. The direction of addition is controlled by electronic rather than by steric factors. For a discussion of the regioselectivity of the addition of azides to α,β-unsaturated carbonyl compounds or nitriles see ⟨66CB475⟩.

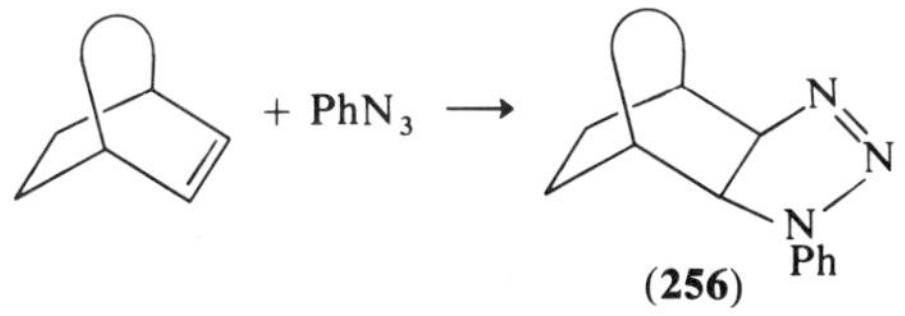

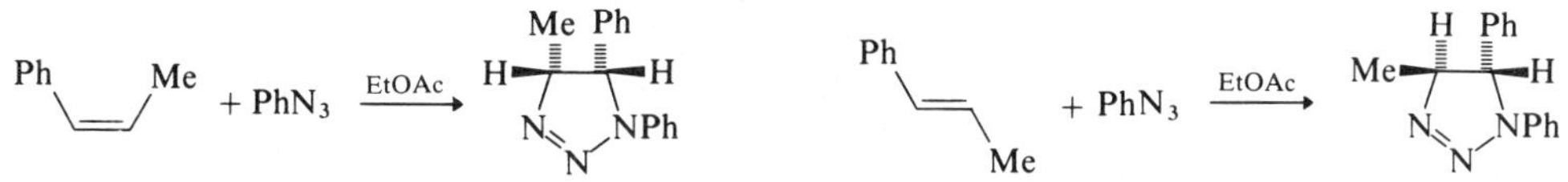

Applying the frontier orbital method to the example of phenyl azide reacting with electron-rich and conjugated dipolarophiles shows that the phenyl group is more effective in raising the energy of the HOMO than in lowering the energy of the LUMO. From this follows the orientation shown in (**257**). Phenyl azide reacting with electron-deficient alkenes consequently shows the opposite regioselectivity giving (**258**); for literature on this topic see ⟨70MI41100, 75ACR361, B-76MI41104⟩.

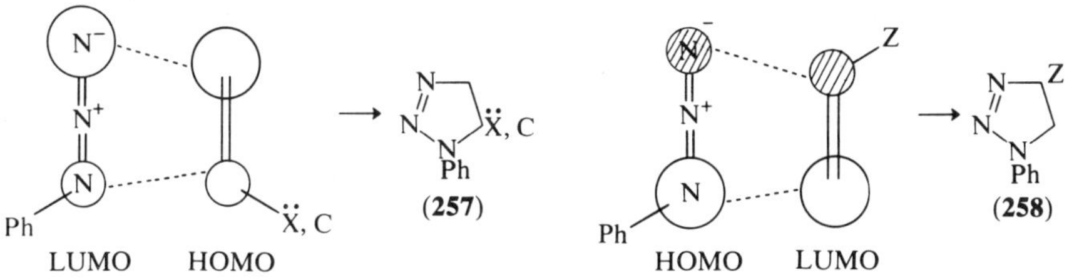

More recently, the reactivity and regioselectivity of 1-azidoadamantane with various alkenic and alkynic dipolarophiles have been described ⟨81JOC1800⟩.

Azides with electron-withdrawing substituents add to allene (**259**) to produce triazoline (**260**) ⟨68JA2131⟩. The cycloaddition of phenyl azide to cyanoallene affords 4-cyano-5-methyl-1-phenyltriazole ⟨64LA(678)95⟩.

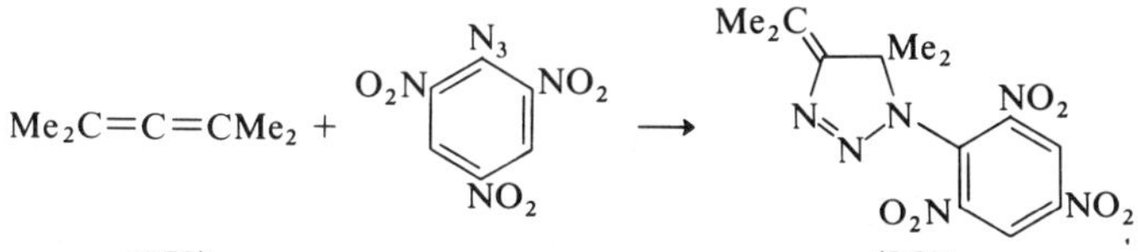

Tetrachlorocyclopropene reacts with trimethylsilyl azide to yield a bicyclic triazole (**261**) which rearranges to 4,5,6-trichloro-1,2,3-triazine (**262**) ⟨79CB1529, 81CB1546⟩. Intramolecular cycloaddition of (**263**) in the presence of potassium and Al_2O_3 leads to 4-phenyl-1,2,3-triazole ⟨70CB1908⟩. For the synthesis of 1-aryl-1,2,3-triazoles from aryl azides and sodium ethoxide see ⟨68JCS(C)1329⟩.

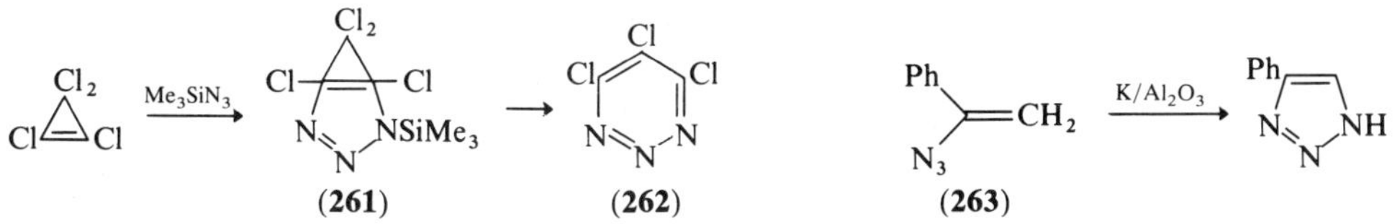

The base-catalyzed condensation of azides with activated methylene compounds is a well-established route to 1*H*-triazoles. This is a regiospecific method ideally suited for the synthesis of triazoles with a 5-amino or -hydroxy substituent, and an aryl or carbonyl function at C-4. The general scheme is shown in Scheme 8. The stepwise character of the reaction is best revealed by the anomerism of glycosyl azides during their reaction with activated methylene compounds ⟨72JA2530⟩. The proposed intermediates have been detected in solution or isolated while reacting azides with aliphatic ketones ⟨68TL3805⟩. Table 21 shows some typical examples.

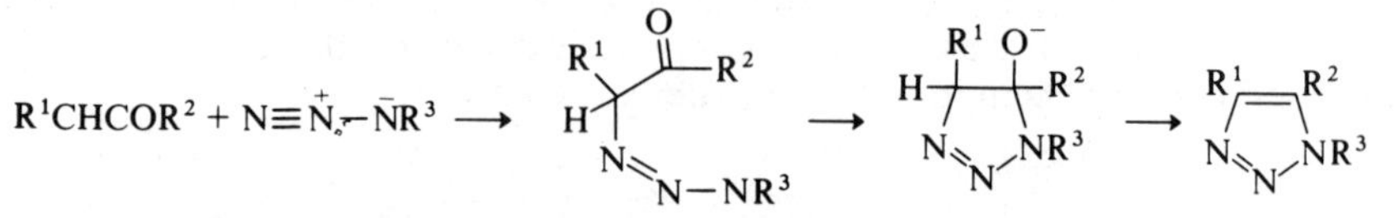

Scheme 8

With malonic esters and amides substituted at the central carbon atom, triazole formation is accompanied by decarboxylation and 4-alkyl- or 4-aryl-5-hydroxytriazoles are isolated ⟨69ACS(B)1091⟩. The 4-methoxybenzyl group has recently been found to be a versatile *N*-protecting group for the synthesis of *N*-substituted 1,2,3-triazoles ⟨82JCS(P1)627⟩.

4-Phenyltriazole (**265**) is isolated in moderate yield from the reaction of aldehydes like (**264**) and tosyl azide ⟨68CB3734, 69CB2216⟩. The reaction of azides with active methylene compounds under basic conditions also represents a very widely used approach to the versatile 1,2,3-triazole enamino esters and derivatives such as (**266**) ⟨71T5873, B-80MI41103⟩. The first example was described by Dimroth and Michaelis ⟨27LA(459)39⟩, and the reaction has been found to proceed in a highly regiospecific manner. These compounds undergo in part the Dimroth rearrangement (*cf.* Section 4.11.4.4.). Of special interest for the ring–chain tautomerism (triazole–diazoamidine equilibrium (**267a**) ⇌ (**267b**); *cf.* Section 4.11.4.1) is the cycloaddition reaction of ynamines and sulfonyl azides ⟨70TL2823, 70JOC3444⟩.

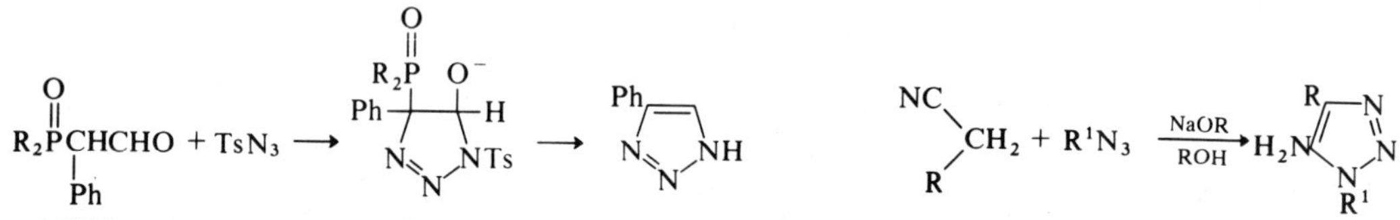

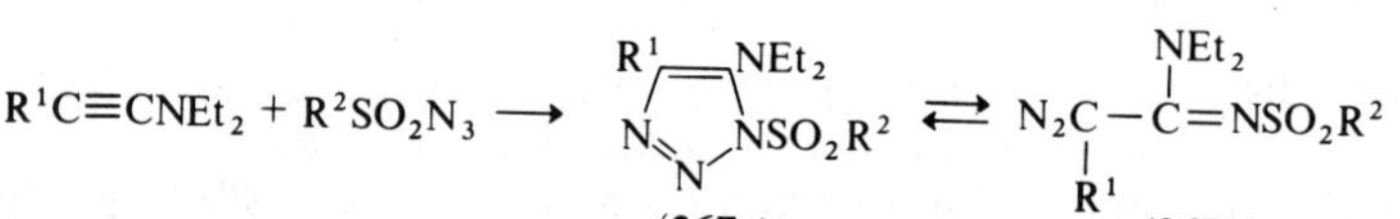

Electron-rich C=C double bonds have been observed to react very readily with azides to afford Δ^2-triazolines. With open-chain and cyclic enol ethers, triazolines are formed in regiospecific cycloadditions and nearly always in high yield ⟨63AG742, 63AG(E)633⟩. Thus, with 2-ethoxy-1-propene and 4-nitrophenyl azide the Δ^2-1,2,3-triazoline (**268**) is formed in quantitative yield; at 150 °C ethanol is eliminated to give in quantitative yield the corresponding 1,2,3-triazole (**269**) ⟨65CB1138⟩. Cyclic enol ethers, like 2,3-dihydrofuran or 3,4-dihydro-2*H*-pyran, afford similar adducts (**270**); however, upon heating nitrogen is evolved to give the iminolactones (**271**) ⟨65CB1138⟩. For a discussion of the steric course and stereoselectivity of the azide additions to enol ethers, see ⟨65CB1153⟩.

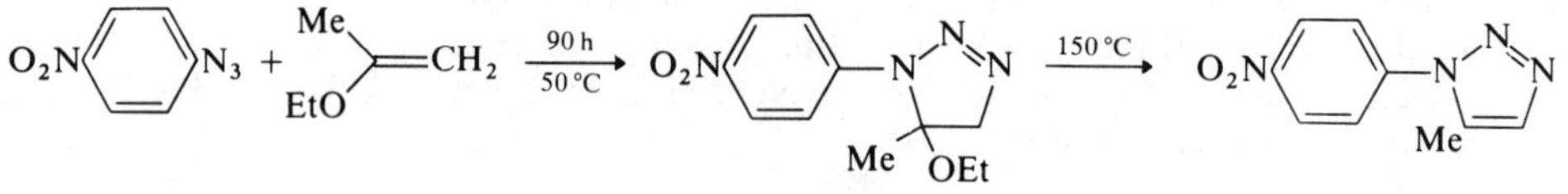

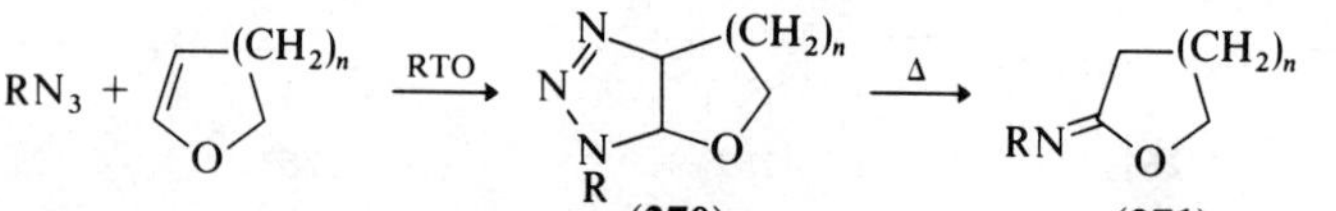

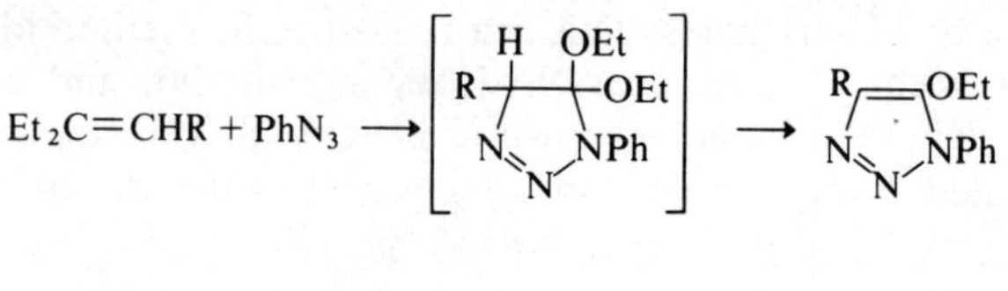

1,2,3-Triazoles and their Benzo Derivatives 713

Table 21 Some Examples of Addition of Azides to Active Methylene Compounds

Azide	Active methylene compound	Triazole	Ref.
PhCH$_2$N$_3$	PhCH$_2$CN	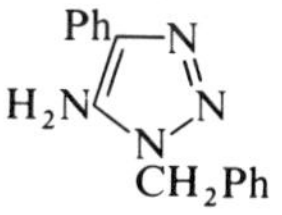 4-Ph, 5-NH$_2$, 1-CH$_2$Ph triazole	59JOC134, 71JCS(C)2156
PhN$_3$	PCH$_2$CN	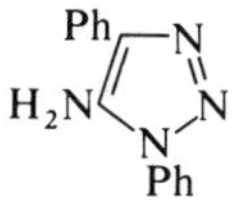 4-Ph, 5-NH$_2$, 1-Ph triazole	57JOC654, 57OS(37)26
PhN$_3$	NCCH$_2$CONH$_2$	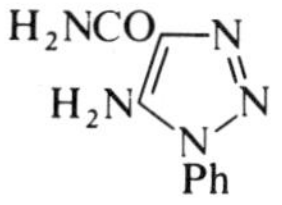 4-CONH$_2$, 5-NH$_2$, 1-Ph triazole	60CB2001
TsN$_3$	N≡CCH$_2$C≡N	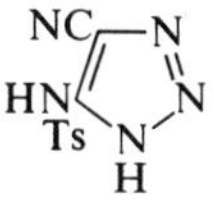 4-CN, 5-NHTs, 1-H triazole	73BSF3442
p-ClC$_6$H$_4$N$_3$	N≡CCH$_2$C≡N	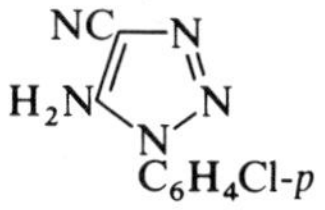 4-CN, 5-NH$_2$, 1-C$_6$H$_4$Cl-p triazole	60CB2001
PhCH$_2$N$_3$	EtO$_2$CCH$_2$CO$_2$Et	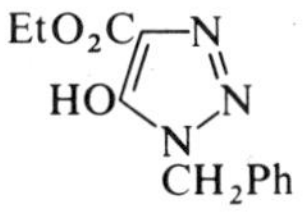 4-CO$_2$Et, 5-OH, 1-CH$_2$Ph triazole	56JA5832
ArN$_3$	MeCOCH$_2$CO$_2$Et	4-CO$_2$H, 5-Me, 1-Ar triazole	61RTC1348
PhCH$_2$N$_3$	PhCOCH$_2$CO$_2$Et	4-CO$_2$H, 5-Ph, 1-CH$_2$Ph triazole	57CB382
Bu^{t}CH=CHN$_3$	RCOCH$_2$CO$_2$Et	4-CO$_2$H, 5-R, 1-CH=CHBu^t triazole	70JHC361
 p-O$_2$N-C$_6$H$_4$-N$_3$	PhCH$_2$CN	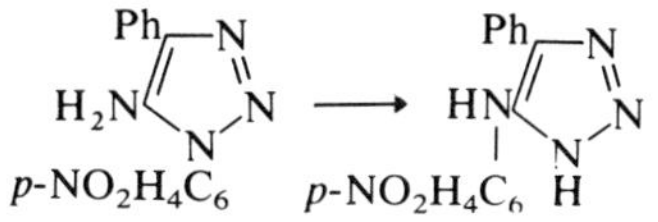 4-Ph, 5-NH$_2$, 1-C$_6$H$_4$NO$_2$-p triazole → 4-Ph, 5-NH, 2-C$_6$H$_4$NO$_2$-p triazole	27LA(459)39, 57JOC654, 59JOC134
ArN$_3$	(EtO)$_2$P(O)CH$_2$CN	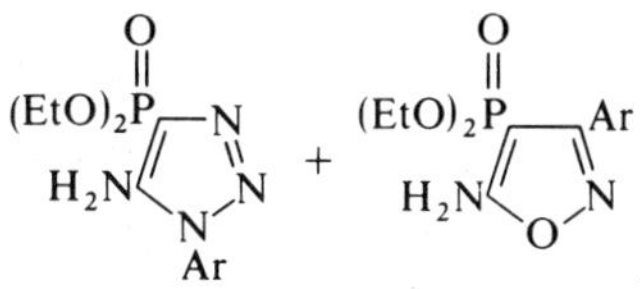 4-P(O)(OEt)$_2$, 5-NH$_2$, 1-Ar triazole + 4-P(O)(OEt)$_2$, 5-Ar, 3-NH$_2$ isoxazole	73LA578

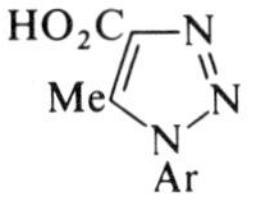

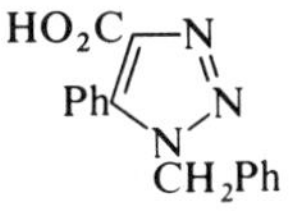

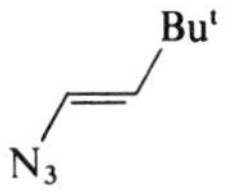

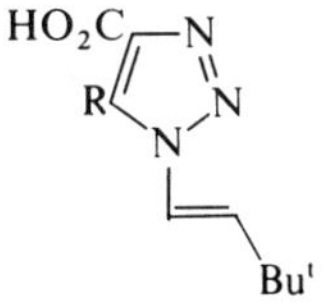

An alternative approach is the cycloaddition of bis-enol ethers (*i.e.* ketene acetals), but varying yields are obtained ⟨60G413, 63G90⟩. With temperature control and in the presence of an excess of the ketene acetal, the intermediate triazoline can be obtained ⟨68G681, 76JHC205⟩. In this context it should be mentioned that benzyl azide was found to add even to acetone with its rather small enol content to form at least 1-benzyl-5-methyl-1*H*-1,2,3-triazole (**273**) and a trisubstituted product (**274**). The primary adduct (**272**) could be isolated ⟨73ACS(B)1987⟩. For further examples, see ⟨67JOC2022, 75TL2807⟩.

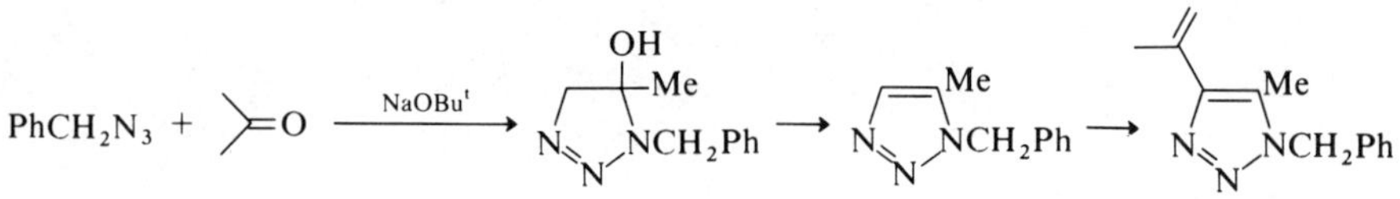

Triazolines which result from the addition of aryl azides to enamines are generally isolatable, and a considerable number of these adducts have been prepared by Fusco, Bianchetti and coworkers ⟨61G849, 933, 62G1040, 63CB802, 72JCS(P1)619, 769⟩; for reviews, see ⟨B-69MI41101, 78BSF(2)485⟩. On heating the reagents together for 1–2 hours triazolines are formed with an amino function in the 5-position. In some cases aromatization takes place *in situ*, but more usually it is effected by heating the primary adducts to give (**276**). Two characteristic reactions are shown, resulting in (**277**) and (**278**) ⟨65MI41100, 67G109⟩. The high degree of regiospecificity, *e.g.* the formation of (**279**), in these cycloadditions has been demonstrated ⟨58LA(614)1, 64JA2213⟩, and many different enamines have been employed ⟨65TL2039, 65G1220⟩.

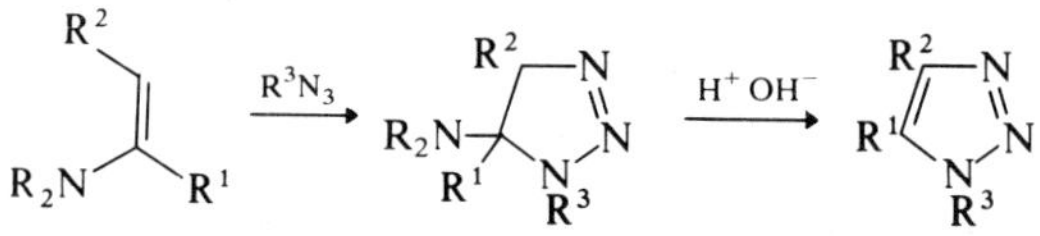

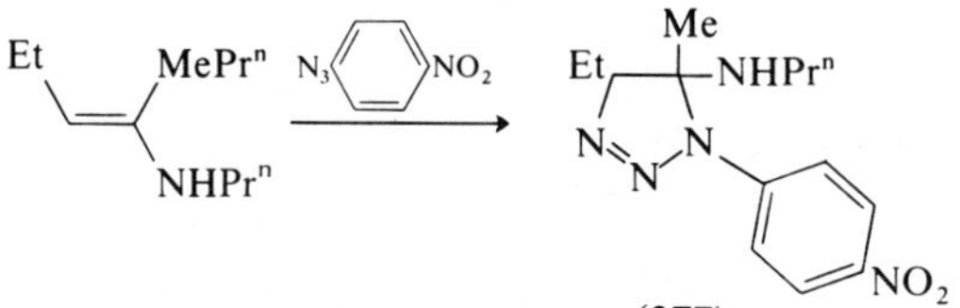

In addition, ketenimines have been used successfully leading to 1,2,3-triazolines of the type (**280**) ⟨67G289, 67G304⟩. The enamines can be generated *in situ* from ketones ⟨69G1131, 72OMR(4)247⟩ and α,β-unsaturated aldehydes ⟨72JCS(P1)619, 769⟩ leading to adducts of type (**281**). A stereochemical study has been reported ⟨72JCS(P1)997⟩ which leads to the following conclusions:

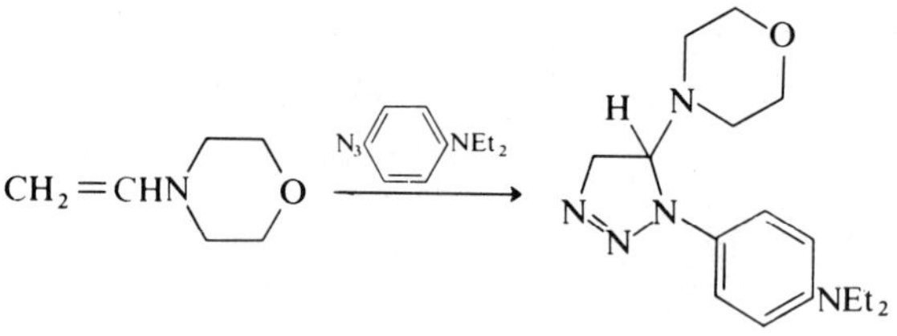

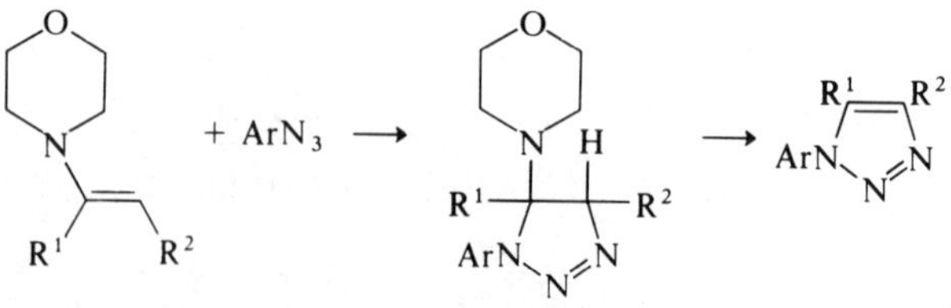

(i) kinetic control produces *trans*-triazolines,
(ii) high temperature or acid leads to a *cis–trans* products mixture,

(iii) the substituent size plays only a minor role,
(iv) a triazoline isomerization is possible.

(280)

(281)

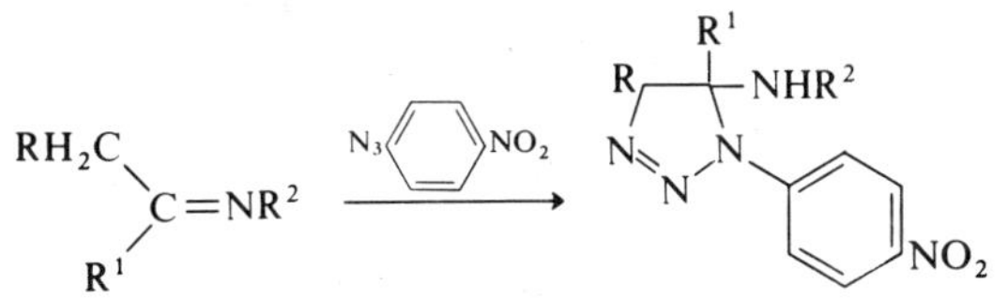

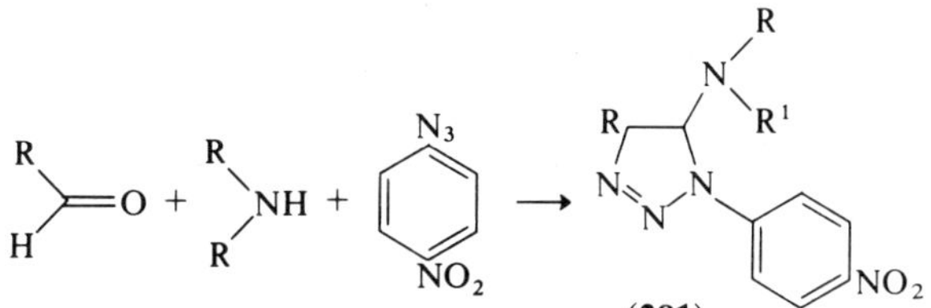

The enamine–azide 1,3-dipolar cycloaddition has been found to be thermally reversible, *e.g.* (**282**) ⇄ (**283**) ⟨75JHC505⟩. 4*H*-Triazoles (**284**), formed from sodium azide and α-chloroenamines bearing methoxycarbonyl and alkyl groups at the β-position, give on thermolysis aromatic triazoles (**285**) and/or 3-amino-2*H*-azirines (**286**). On photolysis only azirines (**286**) are obtained ⟨80CC940⟩.

(282) (283)

(284) (285) (286)

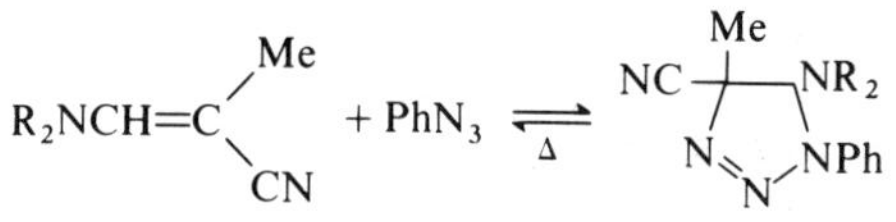

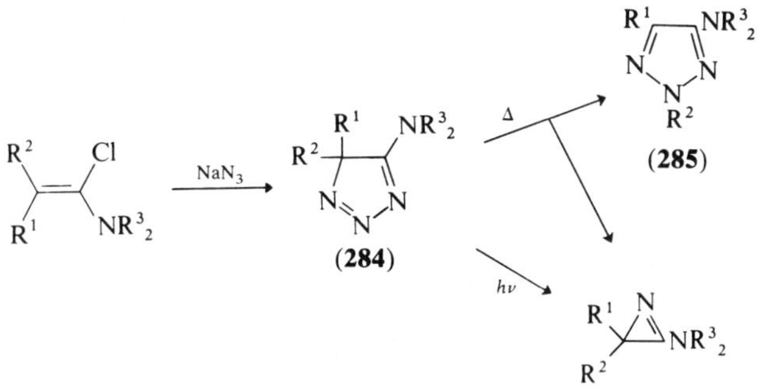

Good yields of 1-aryl-4-nitrotriazoles (**288**, **289**) are obtained on cycloaddition of 1-morpholino-2-nitroethylene (**287**) ⟨review: 81T1453⟩ with aryl azides ⟨66TL6043, 68G949⟩. The spiro cycloaddition product from (**290**) and phenyl azide isomerizes to the triazole (**291**) ⟨75JOC2045⟩.

(289) (287) (288)

(290) (291)

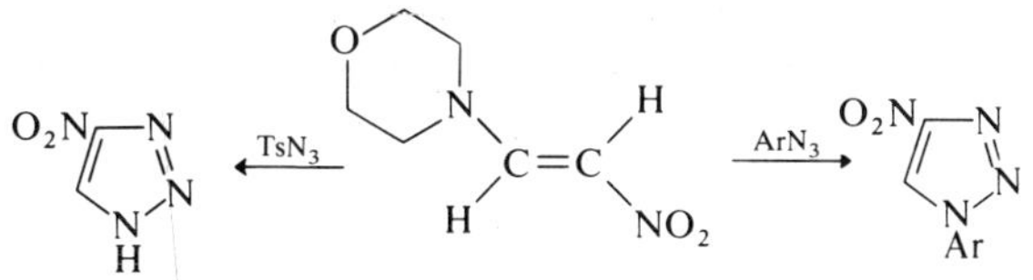

Similarly, (heterocyclic) β-enamino esters substituted with an electron donor as well as with an electron acceptor group ('push–pull') show a decreased cycloaddition tendency towards 1,3-dipoles such as azides; see ⟨71CB3510⟩ and Section 4.11.4.7.

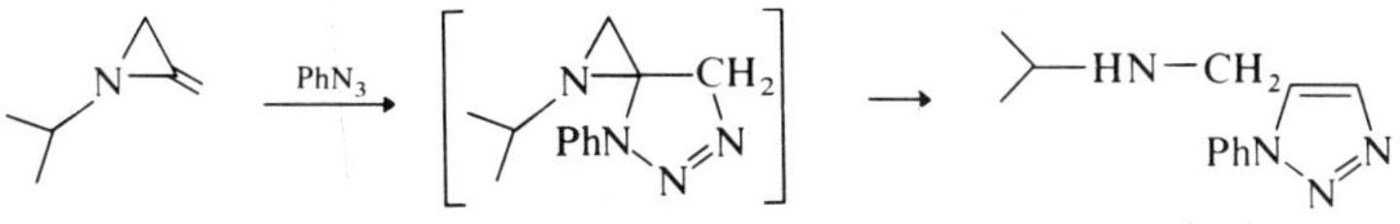

An example of an azide cycloaddition to an open-chain enamino ester (**292**) has also been reported giving (**293**) ⟨76BSF2025⟩. 4-Acyl-1,2,3-triazoles (**249**) have been obtained in fair yield by the reaction of the phenyl azide with α-keto vinyl chlorides ⟨55ZOB1366⟩.

(**292**) (**293**) (**294**)

$A = CO_2R$, COR, CN R = H, Me

α-Acylphosphorous ylides represent another versatile structural type which reacts with azides to give 1,2,3-triazoles. In a typical procedure the phosphorus ylides (**295**) and the azide are heated to 80 °C or left at room temperature. The intermediate triazolines (**296**) eliminate spontaneously the phosphine oxide to give (**297**) ⟨66JOC1587, 69T5421, 71T845, 71T5623, 72JOC3213⟩. For a more detailed discussion of this reaction see ⟨B-79MI41104⟩.

(**295**) (**296**) (**297**)

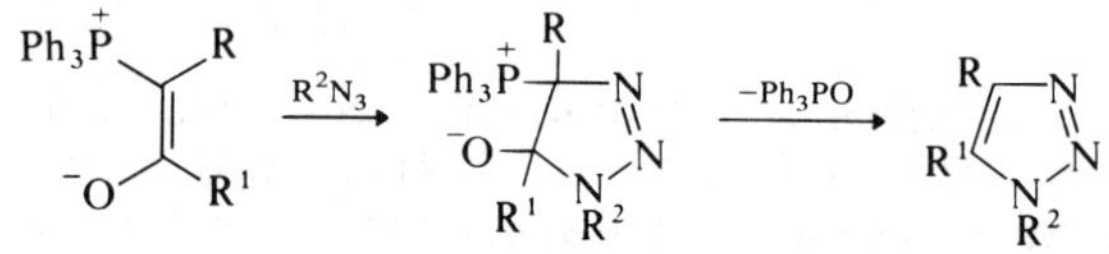

Another synthesis of 1,2,3-triazoles such as (**299**), through phosphorus-containing intermediates (**298**), has been reported ⟨73LA578⟩. Phosphorylated acetylenes (**300**) give 1,2,3-triazoles containing phosphorus substituents as in (**301**) ⟨73ZOB1683⟩. Similar results have been obtained from α-keto phosphorus ylides and ethyl azidoformate leading to a 2-ethoxycarbonyl-2*H*-1,2,3-triazole (**303**) *via* the isolable 1*H*-isomer (**302**) ⟨69TL63, 71T5623⟩. This method has been applied to the synthesis of 1,2,3-triazole nucleosides, *e.g.* (**304**), in an elegant manner ⟨80LA1455, 81CB3165⟩.

(**298**) (**299**) (**300**) (**301**)

(**302**) (**303**)

(**304**)

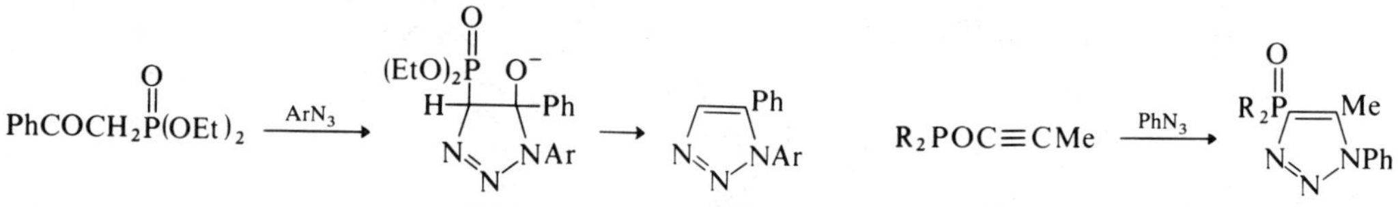

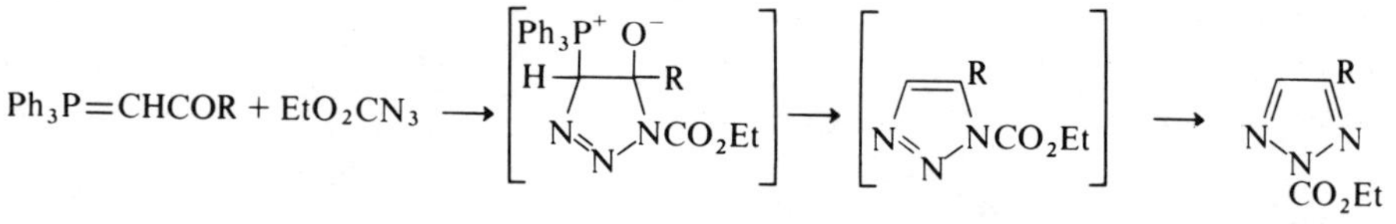

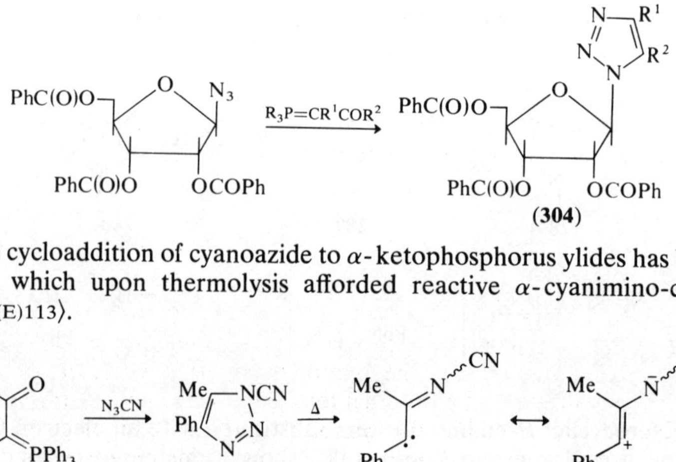

Recently, the cycloaddition of cyanoazide to α-ketophosphorus ylides has been described yielding (**305**), which upon thermolysis afforded reactive α-cyanimino-carbenes (**306**) ⟨81AG118, 81AG(E)113⟩.

(**305**) (**306**)

Another promising route consists in the cycloaddition of acetyl azide to triethylphosphoranemethylenes (**306a**) and subsequent hydrolysis to (**306b**) with yields of 60% and greater ⟨69M1438⟩. For further references see ⟨70TL5225, 73T195⟩.

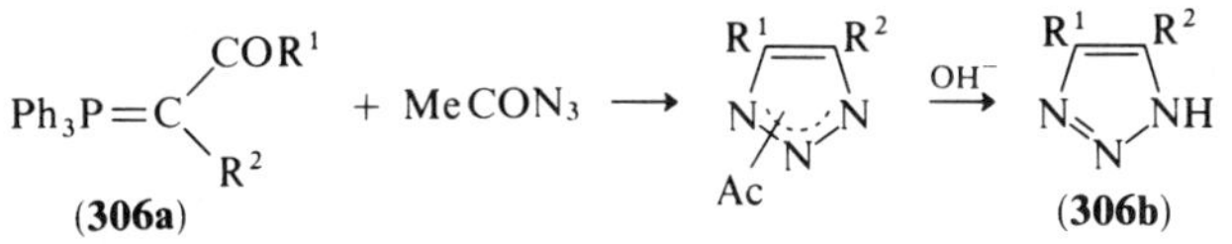

The cycloaddition of diazoalkanes to C=N double bond systems represents an additional generally applicable route to 1,2,3-triazolines like (**306c**), and this regioselective reaction can be formally regarded as a 1,3-dipolar cycloaddition. For reviews, see ⟨B-70MI41102, 78BSF(2)485⟩, and for some applications, see ⟨54JCS1850, 71MI41100, 72LA(757)9, 73S71, 76ZC102⟩.

The addition of diazomethane can be catalyzed by organoalanes ⟨67LA(707)147⟩. The cycloaddition of diazomethane to various substituted benzylidene anilines provides a facile synthesis of numerous 1,5-diaryltriazolines (**306d**) ⟨54JCS1850, 66T2453, 67JHC301, 69T3053, 72LA(757)9, 75JHC143⟩.

The cycloaddition of diazoalkanes to heterocumulenes containing nitrogen preferably takes place at the C=N bond. On adding ethyl diazoacetate to benzyl isocyanate triazoline (**307**) is obtained. In the case of ketenimines and carbodiimides, with organometallic diazoalkanes such as bis(trimethylstannyl)diazomethane, primary adducts isomerize spontaneously to give triazoles like (**308**) ⟨64CB3476, 67JPR(308)82, 71JCS(C)3910, 75CCC1199⟩.

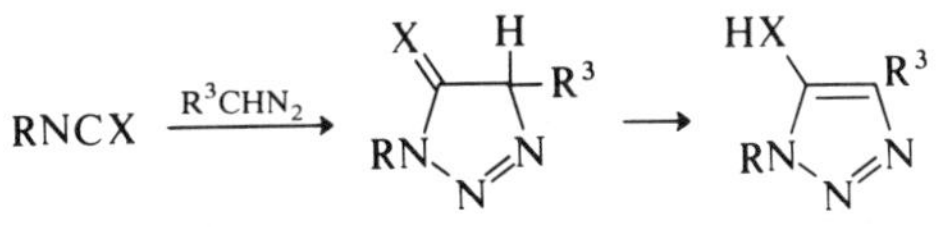

In a somewhat related reaction type ('Umpolung' ⟨cf. 69S17, 75AG1, 75AG(E)15, 79AG259, 79AG(E)239⟩), the anion of N-nitrosomethylethylamine (**309**) gives with nitriles 1,2,3-triazoles (**310**) in fair-to-good yields ⟨72AG1187, 72AG(E)1102⟩. 2,4,5-Trisubstituted 1,2,3-triazoles (**311**) are also obtained by 1,3-dipolar cycloaddition of nitrilimines to nitriles (in addition to the isomeric 1,2,4-triazoles) ⟨79CB1635⟩.

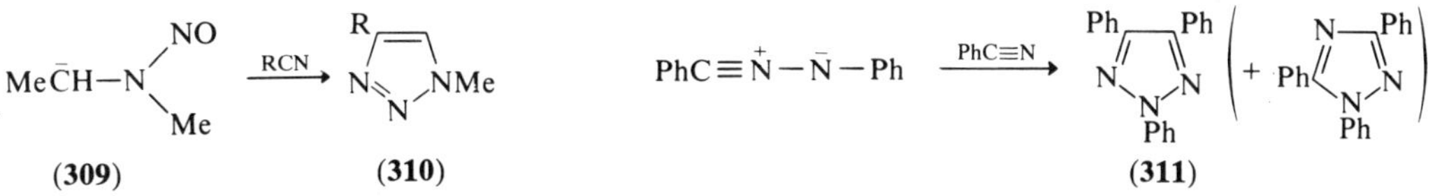

R = Ph, 4-MeC$_6$H$_4$, β-C$_{10}$H$_7$

In a special reaction mode alkenediazonium hexachloroantimonate salts (**312**) are converted into 1,2,3-triazoles (**313**) with amines ⟨81LA7, 81CB3456⟩.

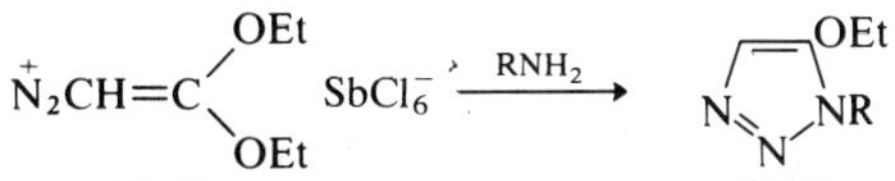

In some cases mesoionic systems have proved to be valuable precursors for 1,2,3-triazoles bearing latent functionalities. Thus, the oxatriazole derivative (**314**) decompsoes to an intermediate azide, which can be intercepted, e.g. by tolane, forming triazole (**315**) ⟨68CB536⟩. In a related photoreaction, the sydnone derivative (**316**) gives a 2H-1,2,3-triazole (**318**) by way of an intermediate nitrilimine (**317**). The reaction mechanism has been studied by ^{15}N-labelling (*) ⟨78HCA1477⟩.

(314) (315)

(316) (317) (318)

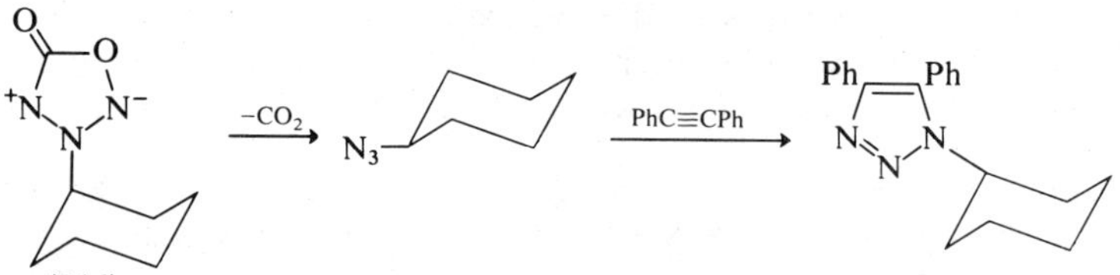

Furthermore, vinyl azides (**319**) react with dimethylsulfoxonium ylide (**320**) at the azide function to produce in high yield 1-vinyl-Δ^2-1,2,3-triazolines (**321**) ⟨67G1411, 75CI(L)278, 81TL1863⟩.

(319) (320) (321)

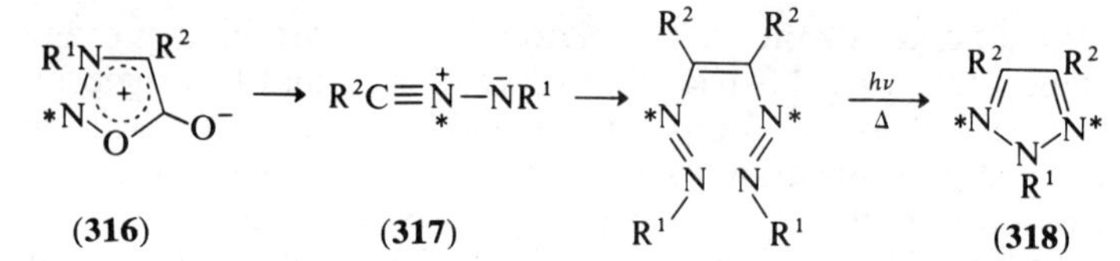

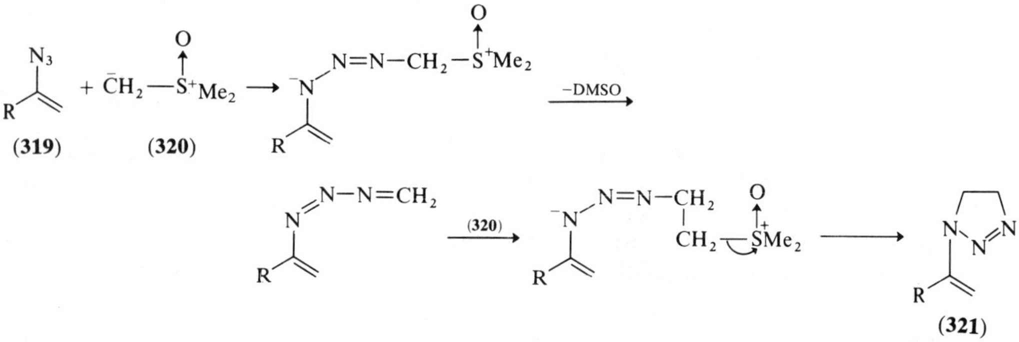

4.11.5.3. Triazoles by Transformation of Other Heterocycles

In Section 4.11.4.4 some Dimroth rearrangements and equilibrium systems, *e.g.* 1,2,3-thiadiazole (**110**) ⇄ 1,2,3-triazole (**109**), were described. Some additional examples have also been reported: 5-anilino-1,2,3-thiadiazoles (**322**) are converted under basic conditions and in high yields into 1-aryl-5-mercapto-1,2,3-triazoles (**323**) ⟨62ACS(B)1800⟩. Later compound (**322**) was suggested as an intermediate in the reaction of isothiocyanates with diazomethane ⟨74RTC317, 76CCC1551⟩. This rearrangement reaction has also been extended to 1,2,3-thiadiazole-5-carbamates (**324**) ⇄ (**325**) ⟨80GEP2852067, 79S470⟩.

(322) (323)

(324) (325)

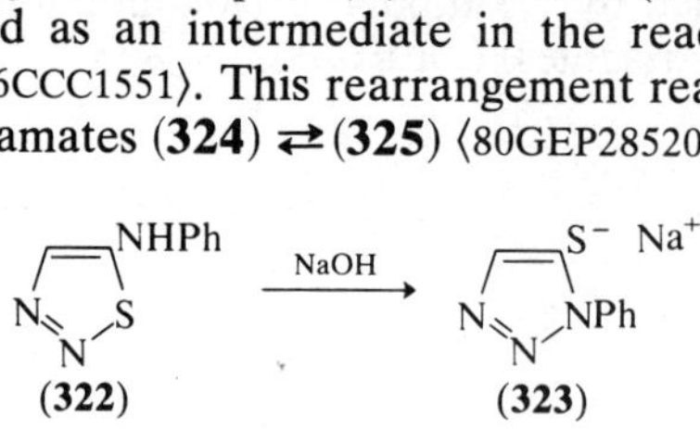

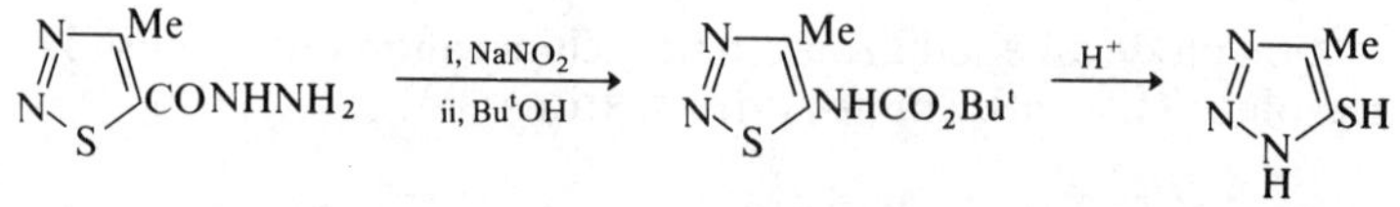

In a related, base-induced rearrangement the sydnone imines (**326**) are converted into 4-hydroxytriazoles (**327**) ⟨62HCA2441⟩. Arylazoaziridines (**328**) rearrange in the presence of sodium iodide into 1-aryl-1,2,3-triazolines (**329**) ⟨62JA993⟩.

(326) (327) (328) (329)

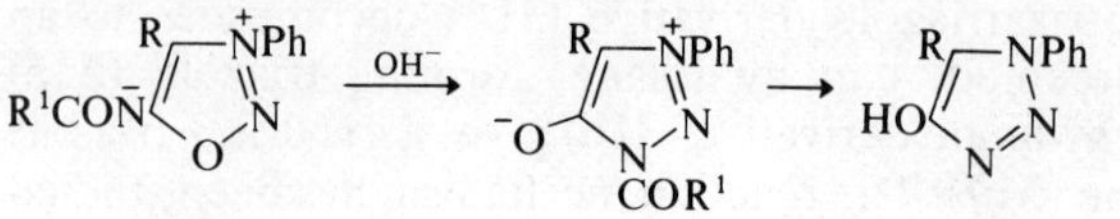

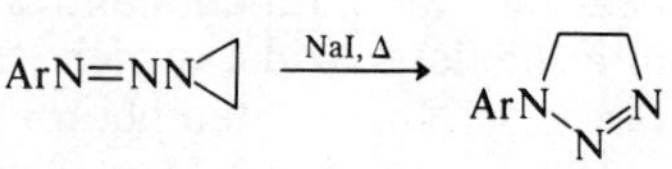

A [6 → 5] ring contraction has been reported for diphenyltriazinone (**330**). Upon treatment with chloramine, 4,5-diphenyl-1,2,3-triazole (**332**) is formed in high yield via the 2-carboxamido-2H-triazole (**331**) ⟨73JCS(P1)545, 66JOC3914⟩. Furthermore, pyridine rings undergo ring contraction under very mild conditions by diazotization of 3-aminopyridinium salts (**333**) leading to 1H-triazole-4-acroleins (**334**) ⟨65JPR(30)96, 66JPR(305)54⟩. This mechanistically interesting ring contraction reaction can be understood in terms of an intramolecular cycloaddition of a diazo group to an imine. This ring contraction process has also been successfully applied to 5-diazo-6-methoxyuracils (**335**), which upon thermolysis gave 1,2,3-triazole derivatives (**336**) and other products ⟨74JHC645, 77JHC647, 78JHC1349⟩.

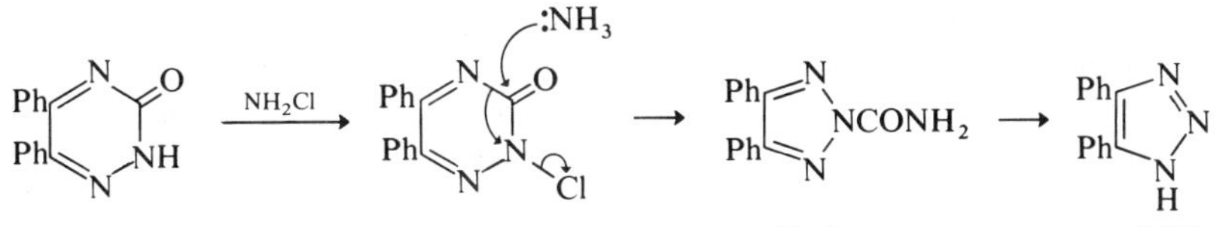

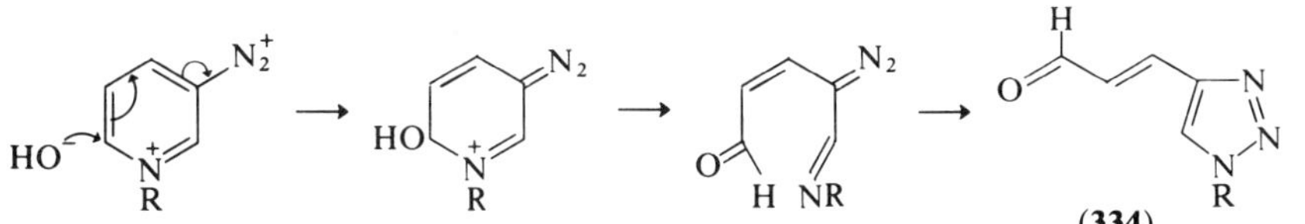

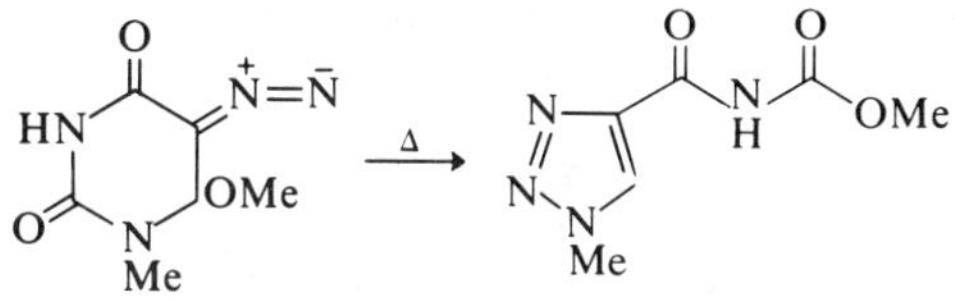

Diazotization of 4-amino-3,5-dimethylisoxazole (**337**) affords triazole (**338**) via a linear triazene ⟨62JCS2083⟩. Two German groups have independently rearranged 4-azoisoxazole derivatives into 2-substituted 2H-1,2,3-triazoles (**339**) and (**340**) ⟨79GEP2815956, 79ZC446⟩. Such 4-azoisoxazole → 2H-1,2,3-triazole rearrangements had been observed earlier by Wittig and coworkers, as well as by Freeman et al. ⟨28CB1140, 65JCS3312⟩.

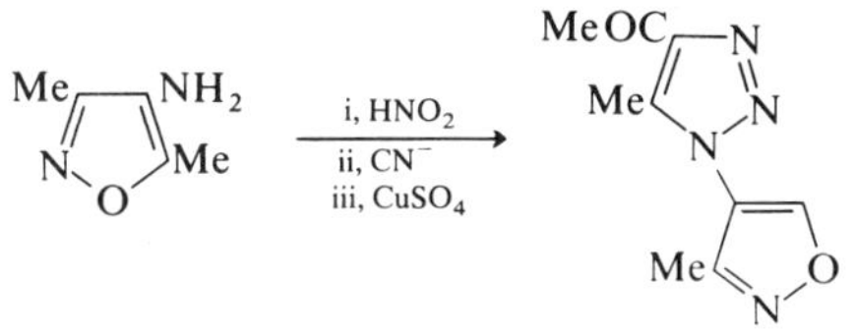

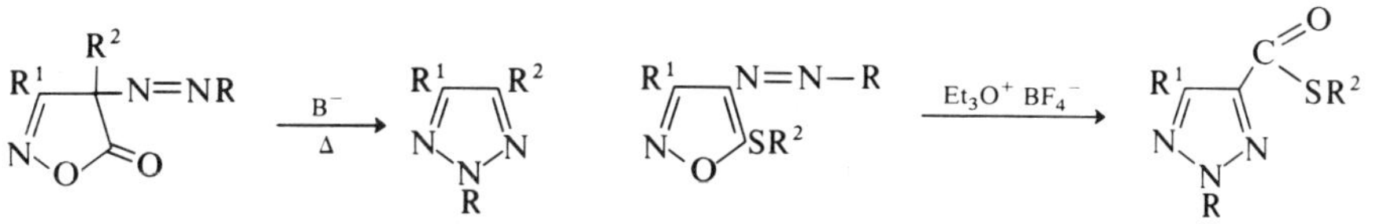

Thermal rearrangement of 4-acylisoxazolearylhydrazones (**341**) in the presence of Cu powder leads in high yields to 2H-1,2,3-triazoles (**342**) ⟨71BCJ185⟩. Another rearrangement proceeds from 5-substituted 3-aroyl-1,2,4-oxadiazoles (**342a**) with hydrazone as well as hydrazide side-chain substituents. Some of these rearrangements are carried out by heating, e.g. of (**342a**), at the melting point, or refluxing with bases or acids. In all cases 4-acylamino-2,5-diaryl-1,2,3-triazoles (**342b**) are formed ⟨review: 81AHC(29)141; 59GI654, 71BCJ185,

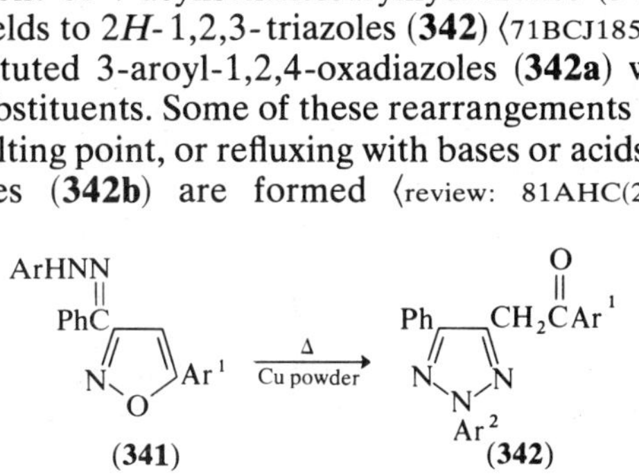

78JCS(P2)19⟩. In a similar manner, 4,4-diazido-2-pyrazolin-5-ones (343) are converted into 2H-1,2,3-triazole-4-carboxylic acids (344) ⟨80ZC437⟩.

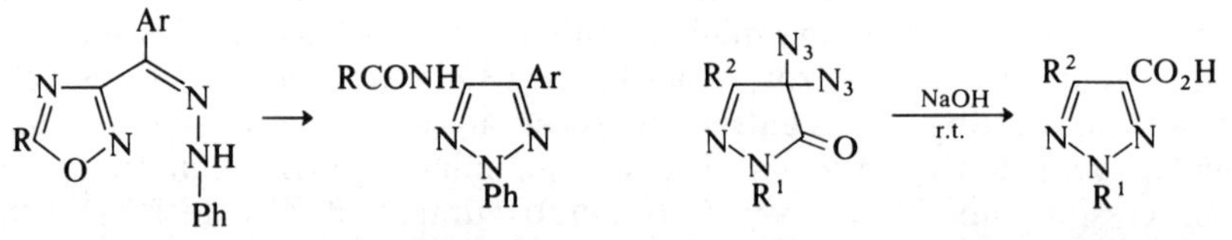

(342a)   (342b)   (343)   (344)

A photochemical method involving a heterocyclic rearrangement is represented by the tetrazole → 2H-triazole conversion (345) → (346) ⟨69JOC4118⟩. Similarly, the rearrangement of the hydrazones of 3-acyl-1,2,4-oxadiazoles provides a general route to 2-substituted 2H-1,2,3-triazoles ⟨67JCS(C)2005, 81JHC723⟩. This rearrangement has been studied with regard to its kinetics and rate ⟨76MI41100, 78JCS(P2)19⟩.

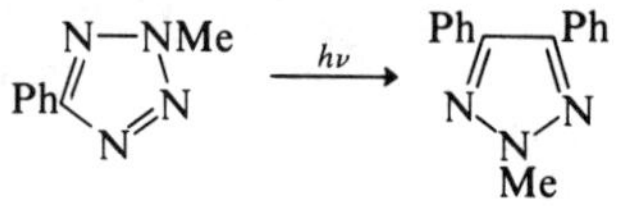

(345)   (346)

4.11.5.4 Other Methods of Synthesis of 1,2,3-Triazoles

The energy gained in producing the aromatic 1,2,3-triazole system facilitates the retro-Diels–Alder reaction of appropriate triazolines formed previously by 1,3-dipolar cycloaddition to moieties which can act as suitable diene precursors upon thermolysis. Thus, on thermolysis of norbornadiene–phenyl azide adducts (347) a smooth retrodiene reaction occurs forming the 1-phenyl-1,2,3-triazole (348) as well as a substituted cyclopentadiene. This reaction also occurs with the adduct from trimethylsilyl azide and norbornadiene ⟨50LA(566)58, 65CB3992, 72CJC3186, 76JOM(121)285⟩. Furthermore, this decomposition reaction can be extended to some heterocyclic tricyclic azide adducts, like (349). Dimethyl 1-phenyl-1,2,3-triazole-4,5-dicarboxylate (350) is produced in good-to-excellent yield on thermolysis of (349) ⟨74TL2163, 76RTC67⟩. Similarly, adduct (351) is cleaved at 70 °C into 1-phenyl-1,2,3-triazole (348) and the isoindole (352) ⟨63CB2851⟩.

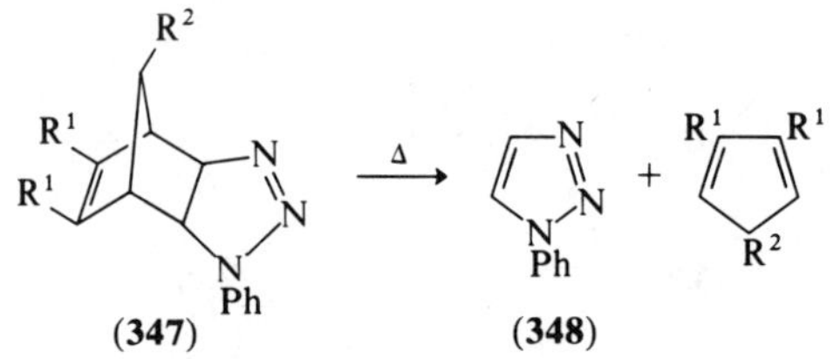

(347)   (348)

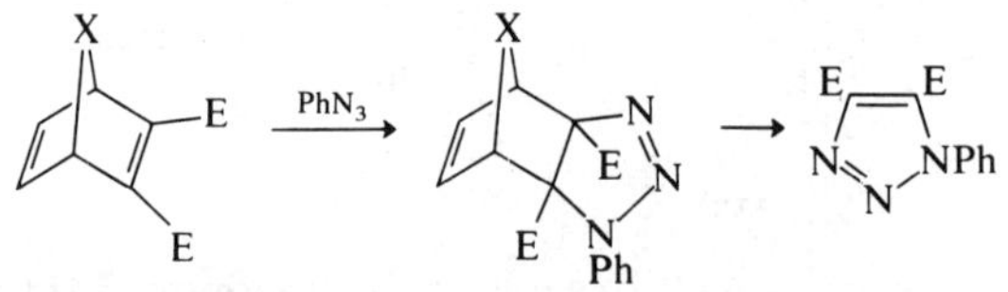

$E = CO_2Me$, $X = O$, NCO_2Et (349)   (350)

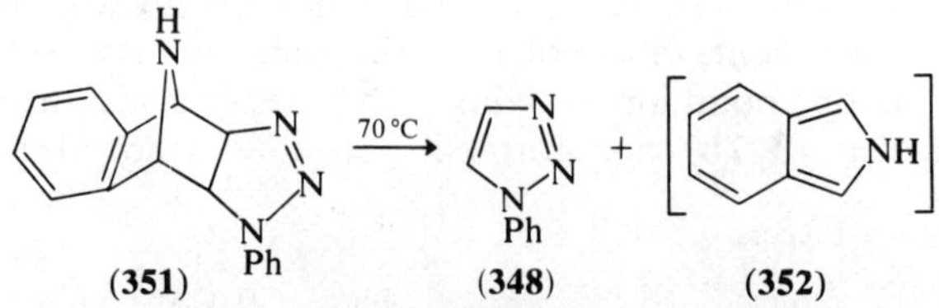

(351)   (348)   (352)

The cycloadduct of phenyl azide and tetrafluorobarrelene cannot be isolated, but instead decomposes spontaneously into tetrafluoronaphthalene and 1-phenyl-1,2,3-triazole (348) ⟨69IZV144⟩.

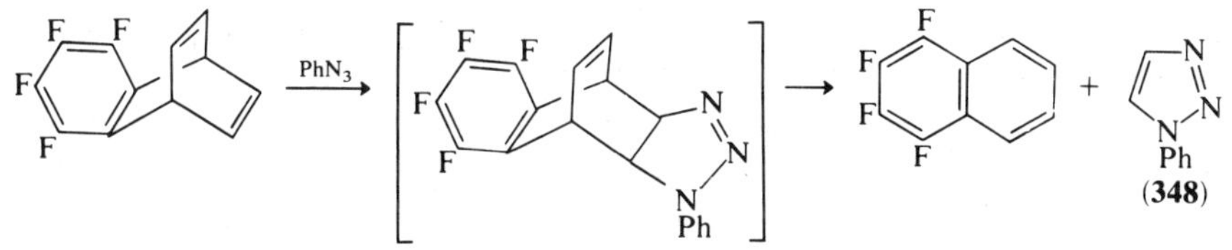

The 1,2,3-triazole moiety in benzotriazole has proved to be especially resistant to oxidation. Thus, the permanganate oxidation of several substituted benzotriazoles yields 1,2,3-triazole-4,5-dicarboxylic acids ⟨B-61MI41100⟩. In more recent work, 1- and 2-phenylbenzotriazoles have been oxidized by means of permanganate to give 1- and 2-phenyl-1,2,3-triazole-4,5-dicarboxylic acids (**353a**) and (**353b**) ⟨27LA(454)164⟩. For more examples of this oxidation see ⟨64CB3493, 67CB71⟩.

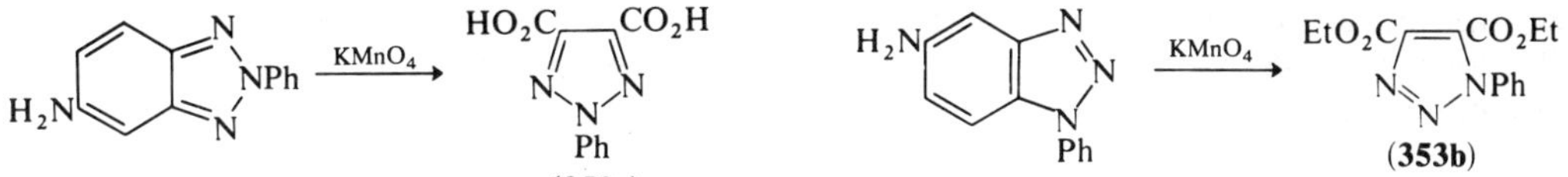

Furthermore, [1,2,3]triazolo[4,5-*d*]pyrimidines synthesized originally as purine analogs cleave to monocyclic triazoles on acidic or basic hydrolysis ⟨59JAP5910468, 63JA2967⟩. 6-Methylthio-8-azapurine is rapidly degraded in boiling 1M HCl to 4-amino-5-(methylthio)carbonyl-1,2,3-triazole (**354**) in almost quantitative yield. The 7-, 8- and 9-methyl as well as the 9-benzyl derivatives behaved similarly ⟨69JCS(C)2379, 69CC500, 69AG115, 69AG(E)132⟩. Thus, the pyrimidine ring may be used as an intermediate carrier when the 1,2,3-triazole moiety is annulated to the pyrimidine ring. *N*-Alkyl-8-azapurines can be hydrolyzed by cold aqueous 1M HCl to give 43–96% of the triazole aldehydes (**355**) ⟨77JCS(P1)1819, 79MI41101⟩. In the presence of hydrazine, bis(1,2,3-triazole-4-aldehyde)hydrazones (**356**) are obtained ⟨73JCS(P1)1625, 73JCS(P1)1634⟩.

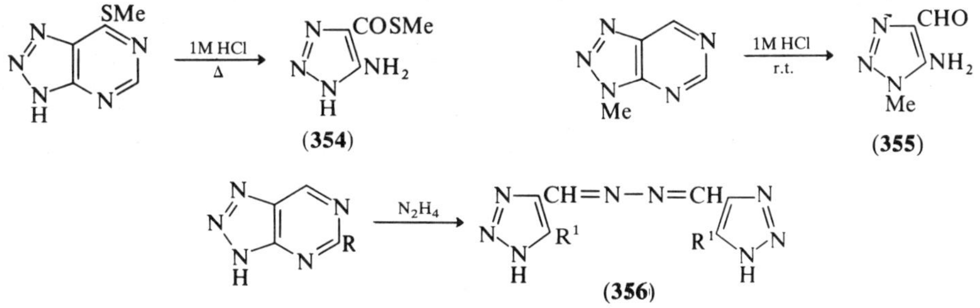

R = H, OH, SH
R¹ = NH_2, $NHCONH_2$, $NHCSNH_2$

1,2,3-Triazoles can also be obtained from more complex molecules. On heating the adduct (**357**), obtained from phenyl azide and morpholinopiperidine, in formic acid the morpholine is eliminated while the piperidine ring is cleaved reductively with CO_2 evolution to give the triazole (**358**) ⟨64CB1225⟩ (cf. also cleavage of (**197**) ⟨71CB3510⟩).

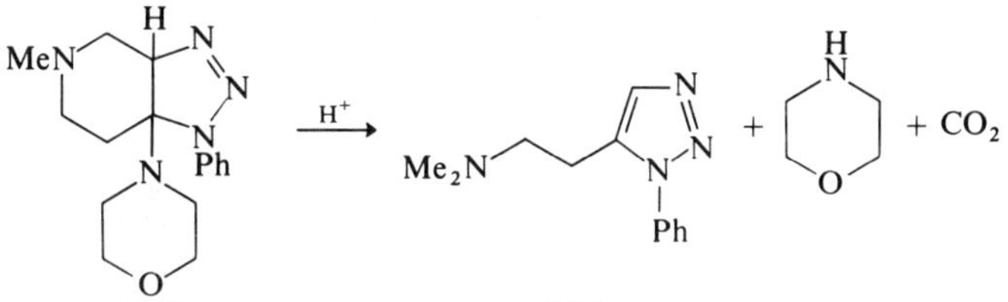

In addition condensed heterocyclic systems formed by annulation can serve as valuable sources of polyfunctional 1,2,3-triazoles. Triazolo[1,5-*a*]pyridines (**360**) are transformed ('trans-cyclization') into the 4-(pyrid-2-yl)-1,2,3-triazoles (**361**) upon heating with aniline ⟨70LA(734)56⟩.

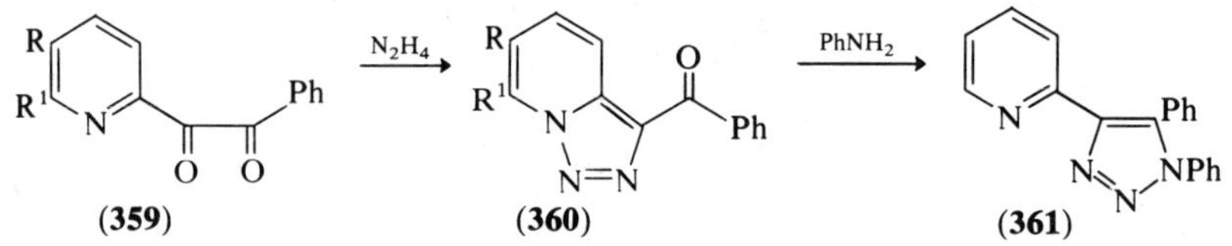

4.11.5.5 1H- and 2H-Benzotriazoles

Benzo[d][1,2,3]triazole, usually known as benzotriazole, and many of its derivatives are obtained from diazotized o-phenylenediamines or appropriately substituted derivatives. The parent compound (362) is formed on treatment of o-phenylenediamine with $NaNO_2/MeCO_2H$ ⟨40OS(20)16⟩. Numerous examples illustrate the wide variety of benzotriazoles available by this route; for reviews, see ⟨50CRV(46)1, B-61MI41100⟩. 1H-Benzotriazole (362) is a very stable material, which can be converted in boiling aqua regia into 4,5,6,7-tetrachlorobenzotriazole ⟨68JCS(C)1268⟩.

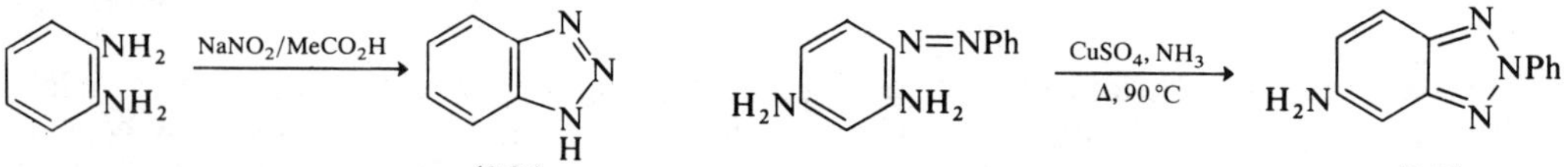

A useful preparative method of wide application for the synthesis of the 2H-isomer is the cyclization of 1,2-azoaminobenzenes to give 2-phenyl-2H-benzotriazoles (363) ⟨22GEP338926, 21CB2191, 2201⟩.

Some miscellaneous and classical cyclization methods of benzenes with 1,2-dinitrogen functions are shown in Scheme 9 ⟨1887CB1176, 21G71, 30G644⟩.

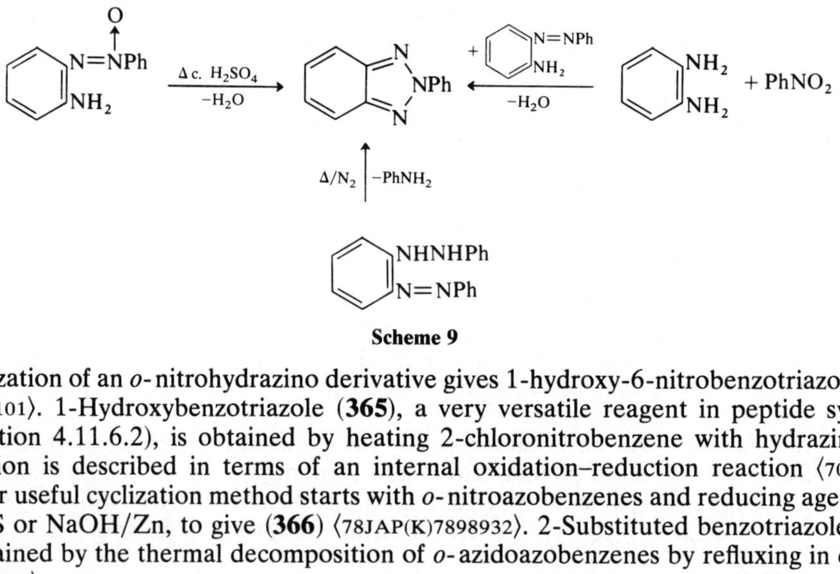

Scheme 9

Cyclization of an o-nitrohydrazino derivative gives 1-hydroxy-6-nitrobenzotriazole (364) ⟨54MI41101⟩. 1-Hydroxybenzotriazole (365), a very versatile reagent in peptide synthesis (cf. Section 4.11.6.2), is obtained by heating 2-chloronitrobenzene with hydrazine. The cyclization is described in terms of an internal oxidation–reduction reaction ⟨70CB788⟩. Another useful cyclization method starts with o-nitroazobenzenes and reducing agents, like $(NH_4)_2S$ or NaOH/Zn, to give (366) ⟨78JAP(K)7898932⟩. 2-Substituted benzotriazoles (368) are obtained by the thermal decomposition of o-azidoazobenzenes by refluxing in dioxane ⟨68JOC2954⟩.

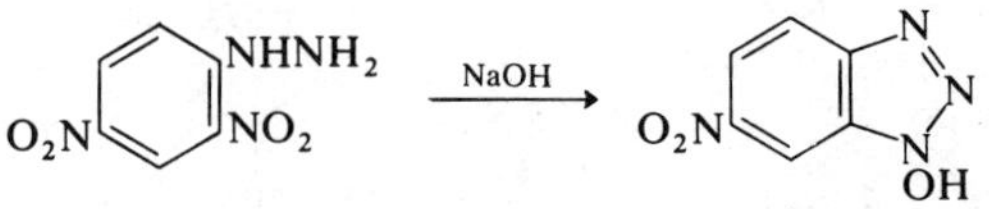

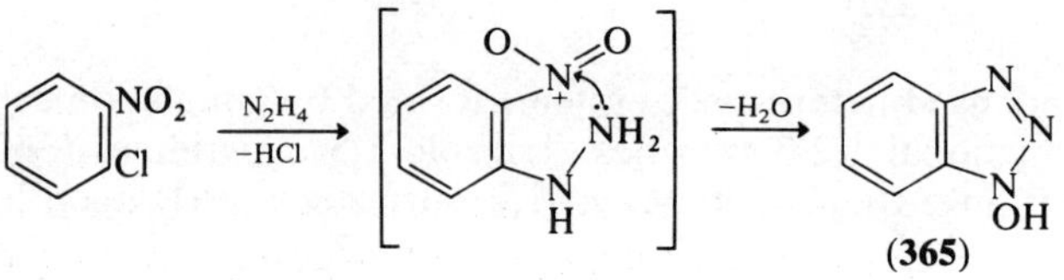

(366)

(367) (368)

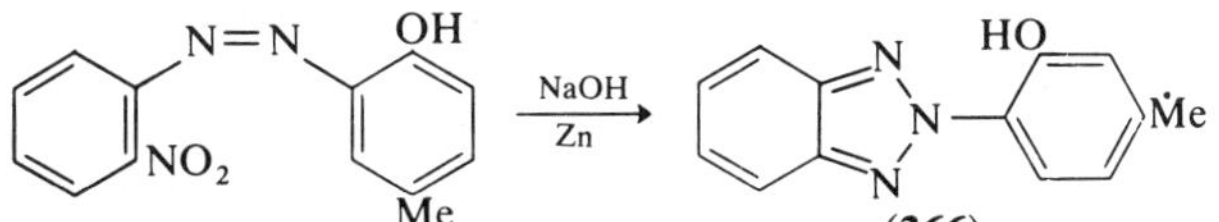

In a bimolecular cycloaddition reaction 1-phenylbenzotriazole (**369**) is obtained from dehydrobenzene (by diazotization of anthranilic acid) and phenyl azide ⟨64JOC3733, 65CB3142⟩. The synthesis of 1-glucosylbenzotriazole from glucosyl azides and dehydrobenzene has been accomplished in this way, yielding, for example, 1-(2,3,4,6-tetra-*O*-acetyl-*β*-D-glucopyranosyl)benzotriazole (**370**) ⟨68JHC699⟩.

(369)

(370)

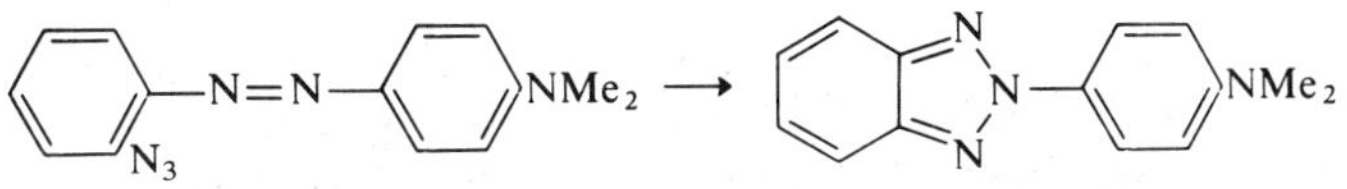

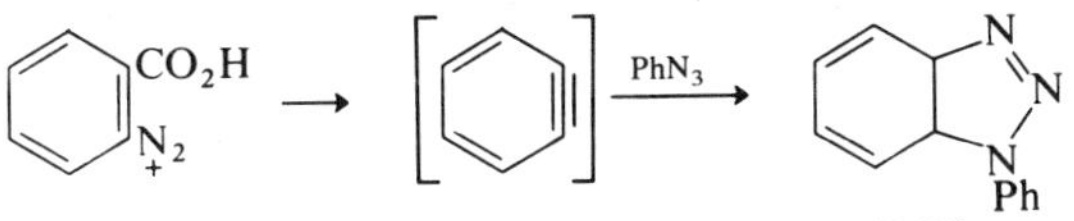

Heating 2,2′-diazidoazobenzene (**371**) at a temperature not higher than 58 °C leads to 2-(2-azidophenyl)-2*H*-benzotriazole (**372**), while above that temperature, *e.g.* at 170 °C, the 1,3a,4,6a-tetraazapentalene (**373**) is formed in 93% yield ⟨67JA2618⟩. Most probably, the second step also involves nitrene intermediates. Benzotriazoles are also accessible by ring contraction reaction of bicyclic systems, like 1,2,4-benzotriazin-3(2*H*)-one (**374**). This is converted into 1*H*-benzotriazole on treatment with chloramine (an aminating oxidant) in ether ⟨71CC532⟩.

(371) (372) (373)

(374)

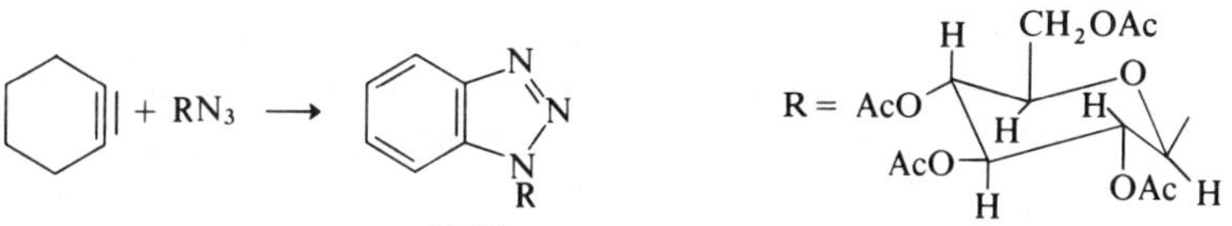

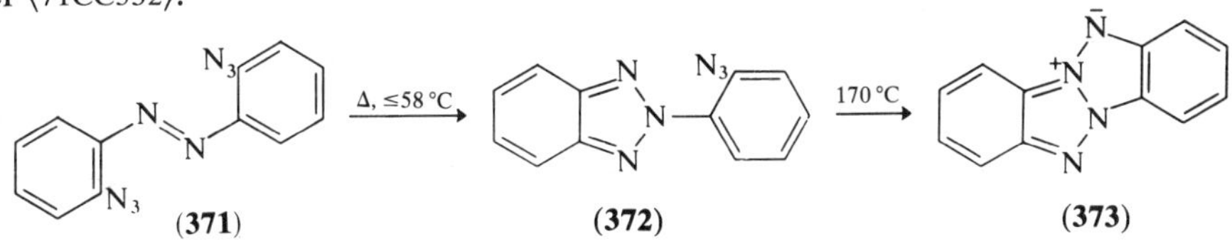

Tricyclic heterocyclic betaines containing a benzotriazole moiety as one of its component ring systems as in (**375**) are photolyzed in the presence of oxygen and Rose Bengal ($^1O_2$) to give both 1*H*- and 2*H*-benzotriazoles (**376a, b**) ⟨81CC1089⟩. The products arise formally from addition of $^1O_2$ to the two mesomeric forms of (**375**).

(375) (376a) (376b)

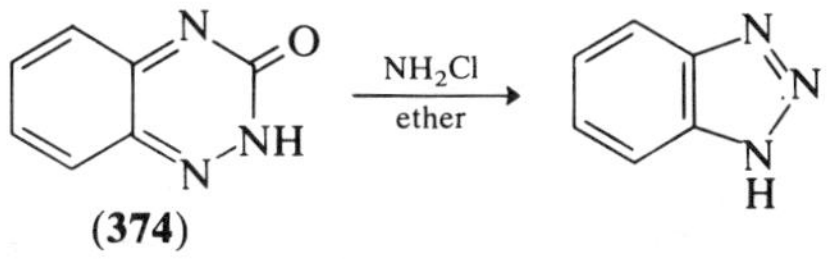

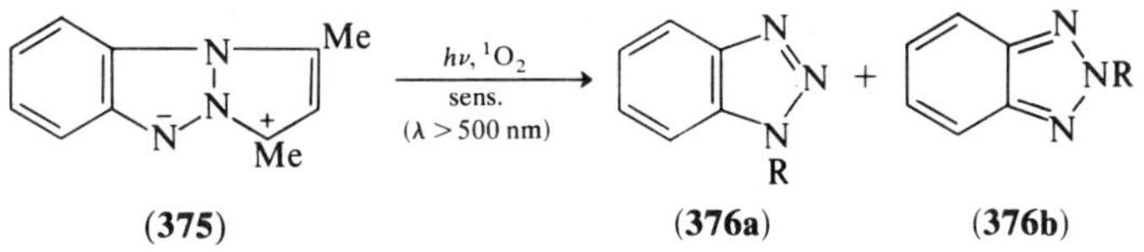

4.11.6 APPLICATIONS OF 1,2,3-TRIAZOLES AND BENZOTRIAZOLES

1,2,3-Triazoles (Δ^2-1,2,3-triazolines) and benzotriazoles have found numerous applications in organic synthesis, as well as in medicine and industry as biologicallly active systems, as dyestuffs and fluorescent compounds, as corrosion inhibitors, as photostabilizers and as agrochemicals.

4.11.6.1 Access to Compounds of Preparative Importance

As has been shown in Section 4.11.4.3, 1,2,3-triazole and Δ^2-1,2,3-triazolines can be converted either thermally or photochemically into three-membered 2H-azirines and aziridines, respectively, with evolution of nitrogen ⟨for reviews see B-67MI41101, B-70MI41101, B-76PH403⟩. In the following section some additional applications of greater preparative interest are described.

In principle, elimination of nitrogen from Δ^2-1,2,3-triazolines yields aziridines, imines or isomeric imines as shown in Scheme 10.

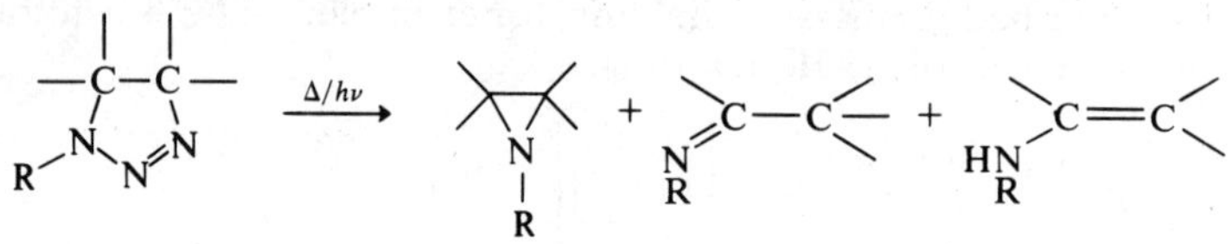

Scheme 10

Photolysis of the triazolines ($\lambda > 240$ nm) leads to better yields of aziridines ⟨66JA4759, 67JOC2022, 68JA988, 68T2757, 74JOC3739⟩. The formation of the aziridines does not occur stereospecifically, although a considerable retention of the configuration at C-4 and C-5 is observed ⟨70T4339, 66JA4759, 68JA988⟩. A typical example is presented in Table 22.

Table 22 Stereochemistry of Triazoline Photodecomposition ⟨68JA988⟩

Triazoline	Irradiation	Ph⟩⟨Me N H	(%)	Ph⟩⟨Me N H	(%)	PhCEt ‖ NPh	(%)
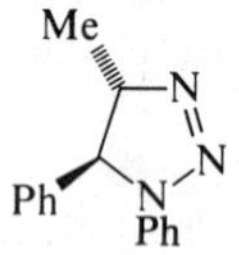 Me,,,,N Ph´`N´≈N Ph	Direct[a] Sens.[b]		66 42		22 54		12 4
Me,N Ph´`N´≈N Ph	Direct[a] Sens.[b]		17 60		65 36		18 4

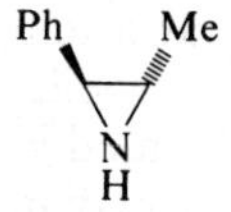

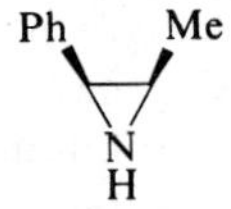

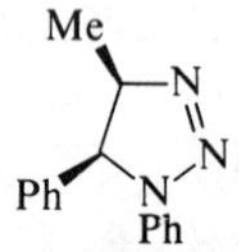

[a] $\lambda = 313$ nm in benzene. [b] $\lambda = 366$ nm with benzophenone in benzene.

On thermolysis the formation of a betaine intermediate is postulated (Scheme 11) ⟨cf. 77CR(C)(284)509, 65CC56, 65JOC4205⟩. This has alternative routes of stabilization by N_2 extrusion, either by imine/enamine formation or by cyclization to an aziridine.

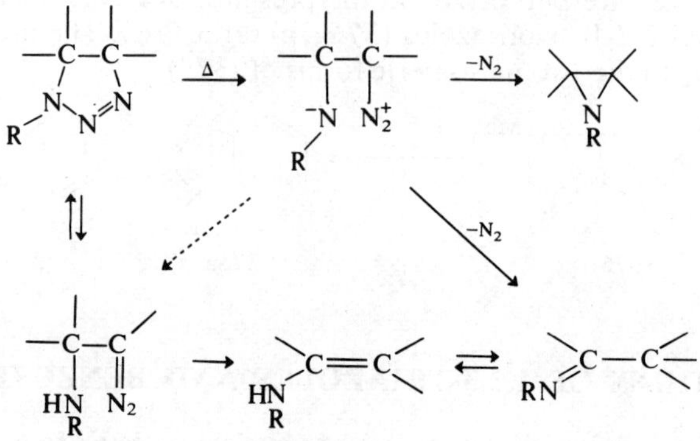

Scheme 11

Analogous to the pinacole rearrangement, bicyclic triazolines are transformed on heating into aziridines, like (**377**), and the cyclic imines (**378**) or (**379**) depending on the nature of the substituents ⟨65JA749⟩.

(**377**) (**378**) $R^1 = H$ (**379**) $R^1 = Me$

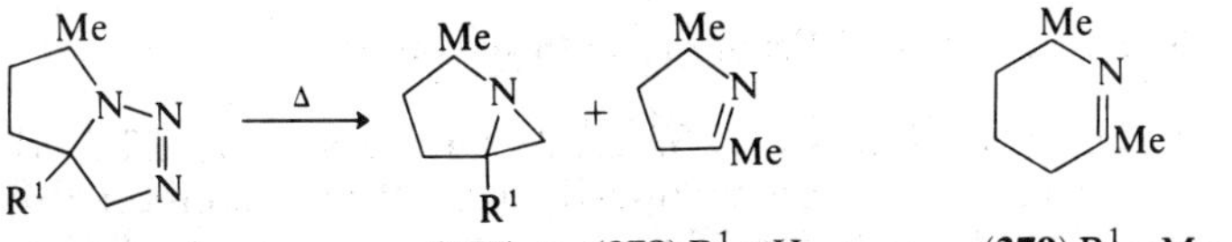

The spiro-triazoline (**380**) can be converted into both the ketenimine (**381**) and the cyclobutanimine (**382**) ⟨74JOC63⟩. Triazolines formed by cycloaddition of azides to norbornene lead to numerous interesting norbornane derivatives ⟨cf. 78BSF(2)485⟩. 5-Hydroxytriazolines (**383**) afford the ring-contracted amide (**384**) by thermal dehydration ⟨67MI41100⟩.

(**380**) (**381**) 34% (**382**) 66%

(**383**) (**384**)

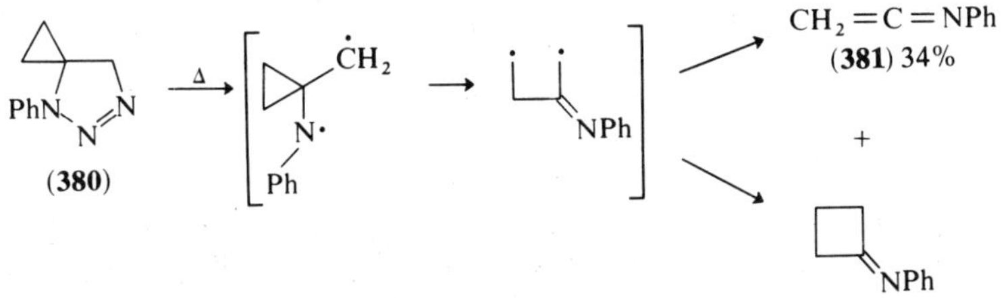

4-Cyano-1,5-diaryl-Δ^2-1,2,3-triazolines (**385**), obtained by addition of diazoacetonitrile to *N*-benzylideneanilines, have been decomposed to aziridines (**386**) and enamines (**387**) with migration of the 5-aryl substituent ⟨78T2229⟩. Thermolysis of 4,4-dimethyl-1-(1-phenylvinyl)-5-(1-pyrrolidinyl)-4,5-dihydro-1*H*-1,2,3-triazole (**388**) gives 2-methyl-*N*-(1-phenylvinyl)-1-(pyrrolidinyl)-1-propanimine (**389**) which reacts like a 2-azabutadiene, *e.g.* by [4+2] cycloaddition with electron-deficient dienophiles ⟨81BCJ2779⟩. In other solvents (**389**) is formed in much lower yields. Instead an intermediate aziridine (**390**) is obtained which undergoes a ring enlargement reaction to 2*H*-pyrroles (**391**) ⟨81BCJ2779⟩.

(**385**) (**386**) (**387**)

(**388**) (**389**)

(**390**) (**391**)

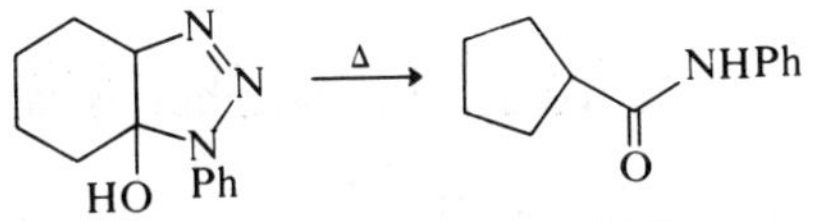

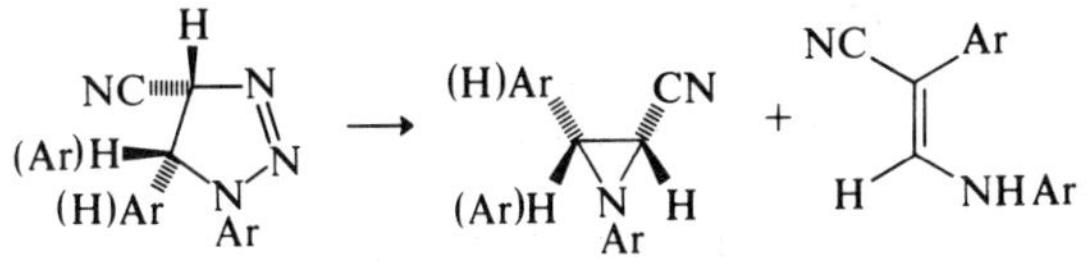

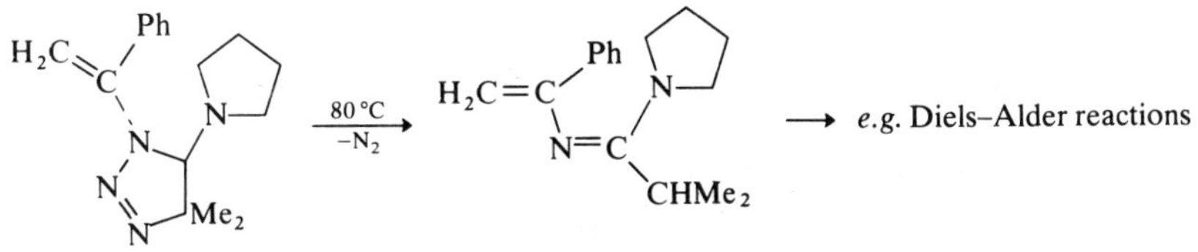

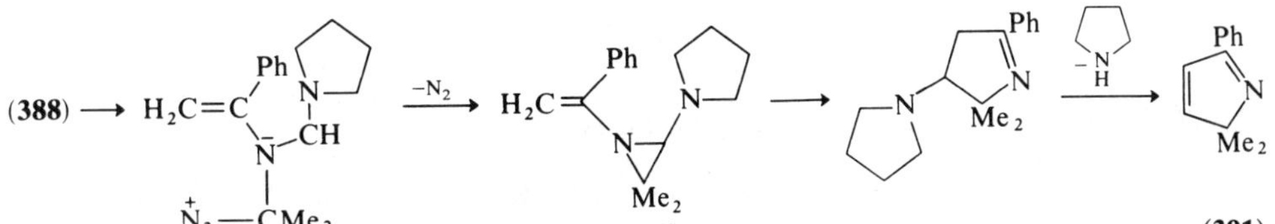

The formation of pyrrole derivatives has been reported earlier. The triazoline (**392**) was converted by heating at its melting point into the 4-oxazoline (**393**) which in turn gives the 2,5-dihydropyrrole (**394**) on reaction with ethyl acetylenedicarboxylate ⟨70CR(C)(271)958⟩.

(**392**) (**393**) (**394**)

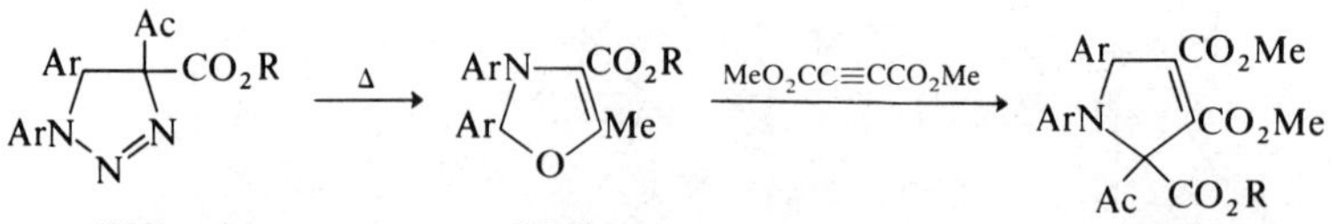

A new ring transformation reaction of 4,5-dibenzoyl-1,2,3-triazoline (**395**) has been reported, which appears to involve nitrogen expulsion followed by a C—NH proton shift. This sequence opens a novel route to enamino diketones (**396**) and tetraacylpyrrolidines (and traces of aziridine **398**) ⟨76JHC1153⟩.

(**395**) (**396**) (**397**) (**398**)

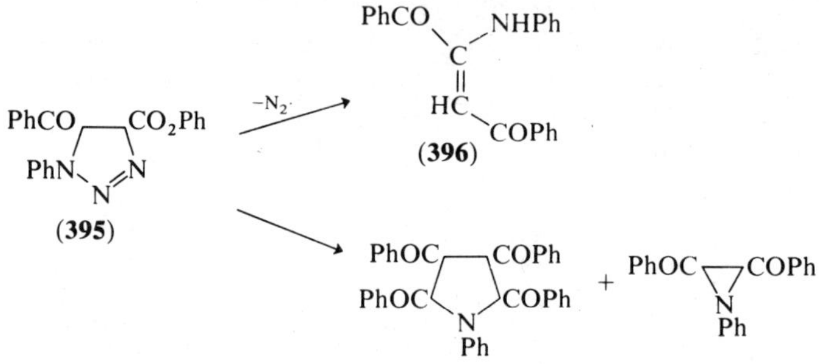

Unusual pyrrole derivatives are formed by the photolysis of 1-vinyl-4,5-dihydro-1*H*-1,2,3-triazoles ⟨81CC1519⟩.

Reaction of alkylidene cycloalkanes (**399**) with cyanogen azide (**400**) ⟨64JA4506⟩, followed by hydrolysis, affords ring-expanded cyclic ketones (**401**), a reaction applicable to a wide variety of ring sizes and to both saturated and α,β-unsaturated ketones ⟨69JA3676, 73JOC2821⟩.

Gas-phase pyrolysis of methyl- and fluoro-substituted benzotriazoles (**402**) yields cyanocyclopentadienes (**403**) ⟨67TL4379, 68CC1026⟩.

(**399**) (**400**) (**401**)

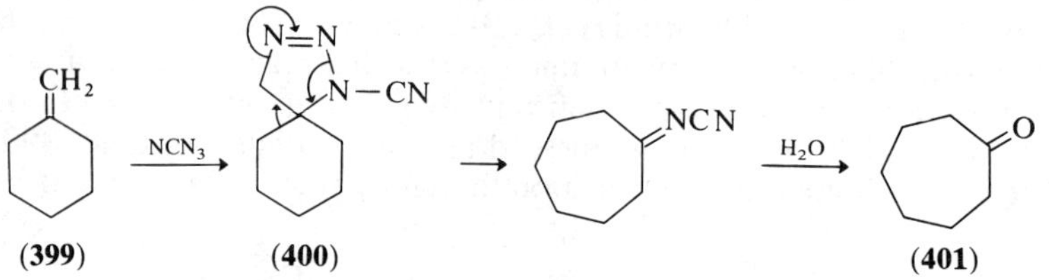

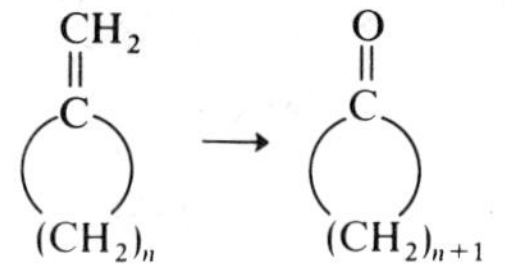

n	3	4	5	6	7	11
(%) yield	52	44	80	41	38	60

(**402**) (**403**)

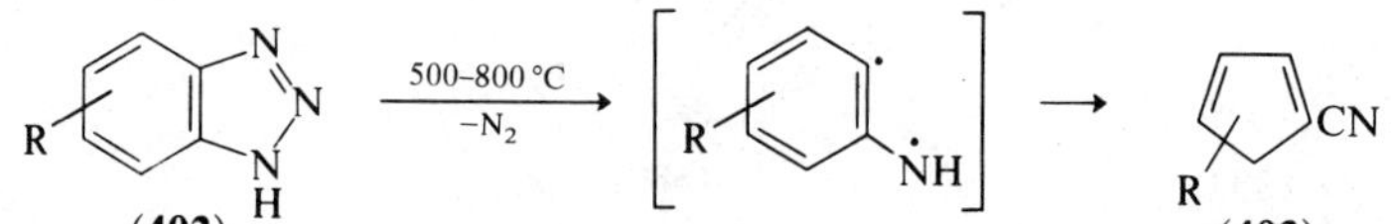

4.11.6.2 Triazoles Utilized in Organic Synthesis

A general reaction for preparing 6-amino-8-azapurines from 5-cyano-4-(dimethylaminomethylene)amino-1,2,3-triazoles has been reported ⟨B-78MI41104⟩. When the triazole

(**404**) is treated in DMF with POCl₃ at 25 °C the ring-fused system (**405**) is formed in 85% yield.

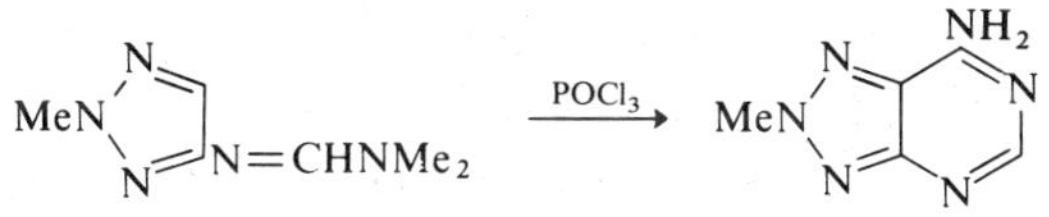

Bromo-substituted quaternary 1,2,3-triazolium salts are strong bromination reagents in basic solution ⟨69ACS(B)2733⟩.

1-Chloro-, 1-bromo- and (less) 1-iodo-benzotriazoles (easily accessible from 1-chlorobenzotriazole ⟨69JCS(C)1474⟩) have proved to be powerful oxidants ⟨69JCS(C)1478, 72S259, 77JOC592, 78JCS(P1)909⟩. Thus, cyclohexanol, 1-phenylethanol and benzyl alcohol are smoothly oxidized to give cyclohexanone, acetophenone and benzaldehyde, respectively. By treatment with 1-chlorobenzotriazole and subsequent hydrolysis of the product, ethers are cleaved oxidatively to aldehydes (Scheme 12) ⟨79AJC2787⟩.

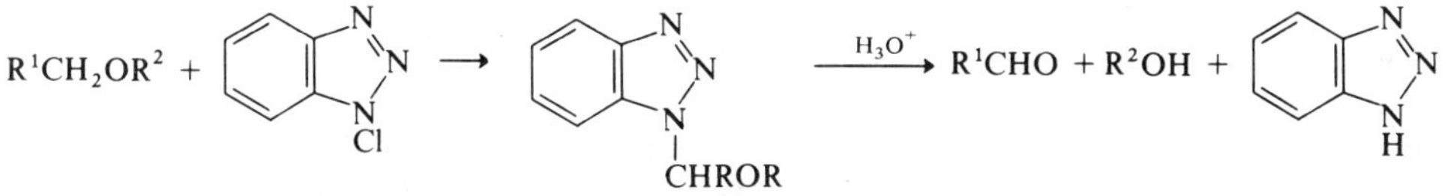

Scheme 12

5-Diazomethyl-1,4-diphenyl-1,2,3-triazole (**406**) gives products of typical carbene reactions when thermolyzed slightly above room temperature (*i.e.* 40–50 °C) in the absence of a catalyst. Carbenes generated in this way show a higher selectivity than those from other sources of methylene or ethoxycarbonylcarbene. In the case of aromatic derivatives triazolyl-substituted cycloheptatrienes (**407**) are formed ⟨68JOC1145, 74JOC1047, 81JOC679⟩.

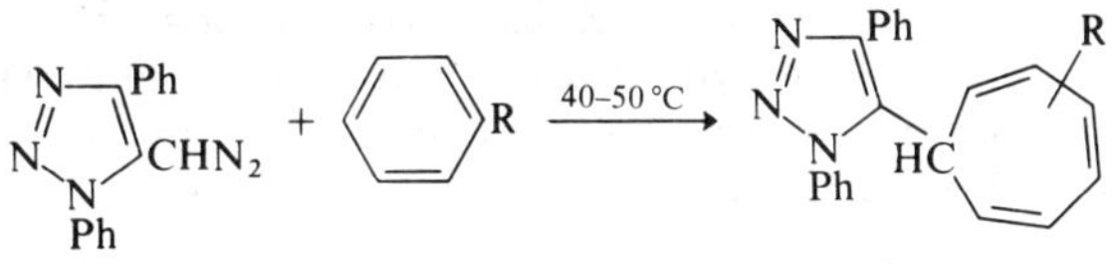

At 20 °C 4-amino-5-aminomethyl-3-benzyl-1,2,3-triazole (**408**) gives with CS₂ the dithiocarbamate (**409**), which upon heating affords the novel heterocyclic system 1,2,3-triazolo[4,5-*d*][1,3]thiazine (**410**) (and not the expected 8-azapurine) ⟨80AG319, 80AG(E)310, 80JCS(P1)2009⟩.

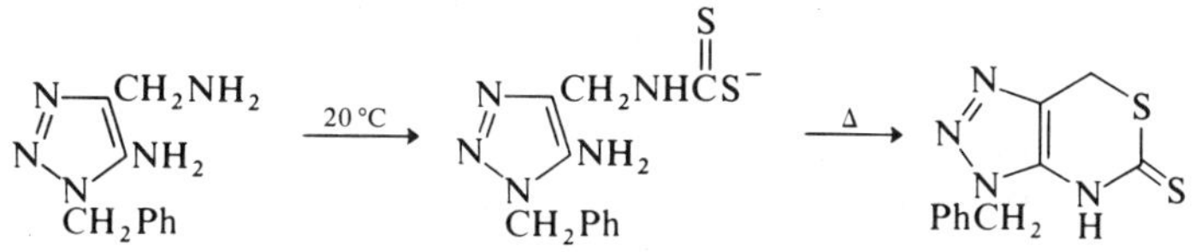

The ylidic 1-phenylimino-2,4,5-triphenyl-1,2,3-triazole (**411**) (*cf.* Section 4.11.4.6) has been shown to be an effective 1,3-dipolar system which can be reacted smoothly with CS₂, PhNCS, alkynic esters, ethyl acrylate and acrylonitrile to give [3+2] cycloadducts, *e.g.* (**412**). Upon photolysis or thermolysis the 2,4,5-triphenyl-2*H*-triazole (**413**) is formed ⟨71TL633⟩.

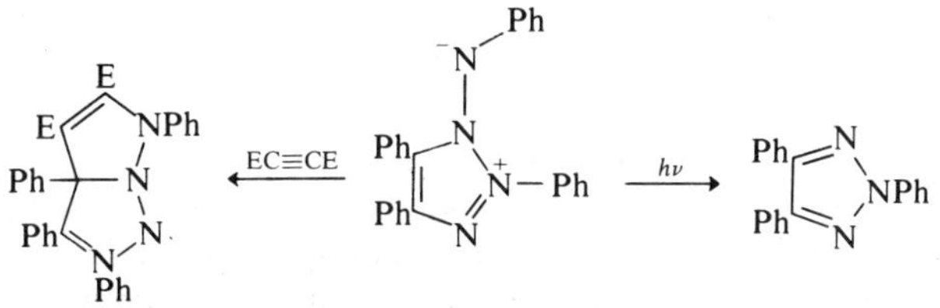

One of the most widely used benzotriazole derivatives is the 1-hydroxy compound (**414**), which has proved to be a very valuable mediator substance for activating carboxyl groups for peptide synthesis. Racemization is decreased, the formation of N-acylurea is avoided and the yields of high purity peptides are improved ⟨70CB788⟩. Some recent examples of coupling reactions in peptide, dipeptide and polypeptide synthesis, as well as protection of carboxylic acids, are given in the following references: ⟨74TL2695, 75HCA688, 77TL199, 76TL1031, 78AG63, 78AG(E)67, 79CB1057, 79TL1725, 81CB2649⟩.

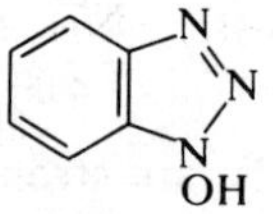

(414)

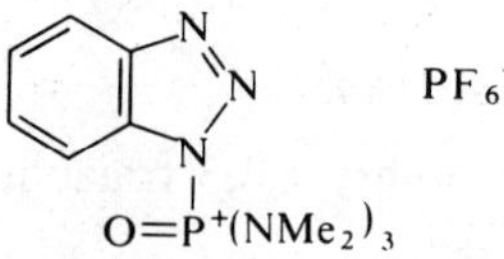

(415)

Furthermore, benzotriazolyl-N-hydroxytris(dimethylamino)phosphonium hexafluorophosphate (415) has been used for coupling reactions in peptide synthesis ⟨75TL1219⟩. Very recently, 1-ethoxycarbonylbenzotriazole has been used successfully as a C-carboxylating reagent ⟨82AG139⟩.

4.11.6.3 Triazoles in Medicinal Chemistry

Some 1,2,3-triazole derivatives have been reported to show significant biological activity, such as 4-alkyl-1,2,3-triazoles which show some slight inhibition of microorganisms ⟨54JA667⟩. Furthermore, some triazole analogs of histamine show antihistaminic and antiacetylcholine activities, while the 1,2,3-triazole ring is antagnostic to the imidazole ring in triazole analogs of adenine and guanine ⟨51JA1207, 49JA1436⟩. Ethanolamino derivatives of 1,2,3-triazoles show mydriatic action ⟨55FES235⟩.

In a more recent work sulfonyl hydrazides and their derivatives from 2-phenyl-1,2,3-triazole-4-carboxylic acid (416) have been reported to inhibit the growth of *Staphyllococcus aureus*, *Escherischia coli* and *Salmonella typhi* ⟨76JIC833⟩. In addition, 1,2,3-triazole-1,8-naphthyridine derivatives have been synthesized; compounds (417) were found to have twice the analgesic activity of phenylbutazone; both (417) and (418) showed sedative activity; and compounds (418) were fungicidal ⟨76FES797⟩. Both 5- and 6-chloro-1-aroylbenzotriazoles (419) and (420) revealed antiinflammatory activities ⟨80JHC1505⟩.

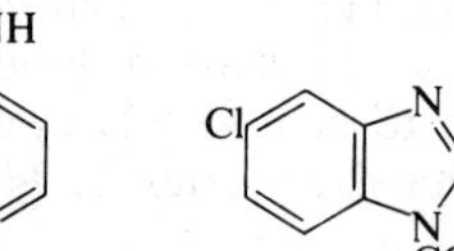

(416)
R = CONHN=CHR², R' = H
R = CO₂H, R' = SO₂NHN=CHR²

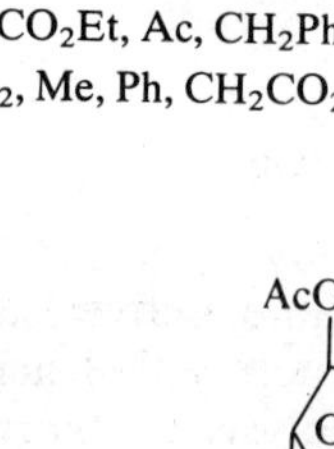

(417)
R = Ph, CO₂Et, Ac, CH₂Ph, CONHPh
R' = NH₂, Me, Ph, CH₂CO₂Et

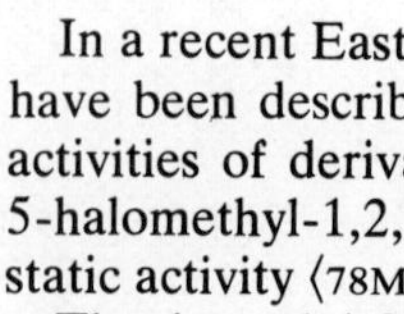

(418)

(419)
R = Ph, CO₂H

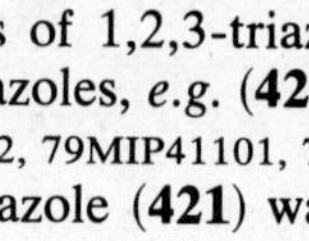

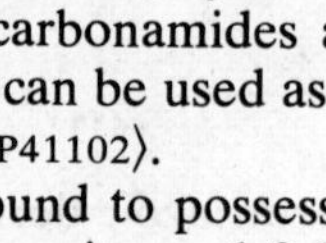

(420)

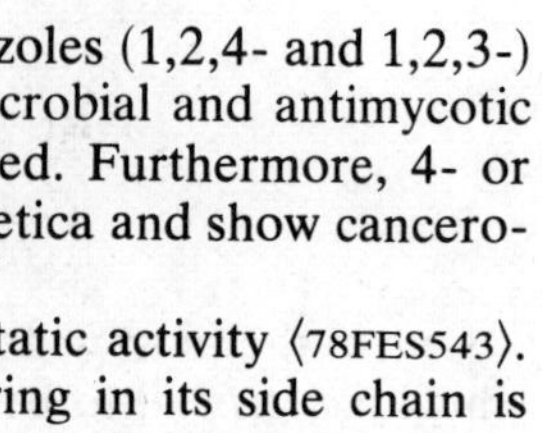

(420a) X = Br, I

In a recent East German review some biological activities of triazoles (1,2,4- and 1,2,3-) have been described ⟨81MI41100⟩. The virostatic, cytostatic, antimicrobial and antimycotic activities of derivatives of 1,2,3-triazolecarbonamides are discussed. Furthermore, 4- or 5-halomethyl-1,2,3-triazoles, *e.g.* (420a), can be used as radiomimetica and show cancerostatic activity ⟨78MI41102, 79MIP41101, 79MIP41102⟩.

The 1-tosyl-1,2,3-triazole (421) was found to possess bacteriostatic activity ⟨78FES543⟩. A typical example of an antibioticum carrying a 1,2,3-triazole ring in its side chain is cefatrizin (**422**), which belongs to the class of cephalosporins currently in use and which has activity against 342 different germs ⟨76MI41102, 78MI41103⟩. Another semisynthetic penicillin from 2-phenyl-1,2,3-triazolecarboxylic acid (**423**) has recently been reported ⟨80MI41100⟩.

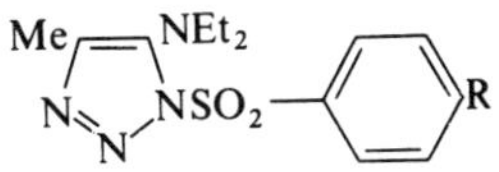

(**421**) R = H, OMe

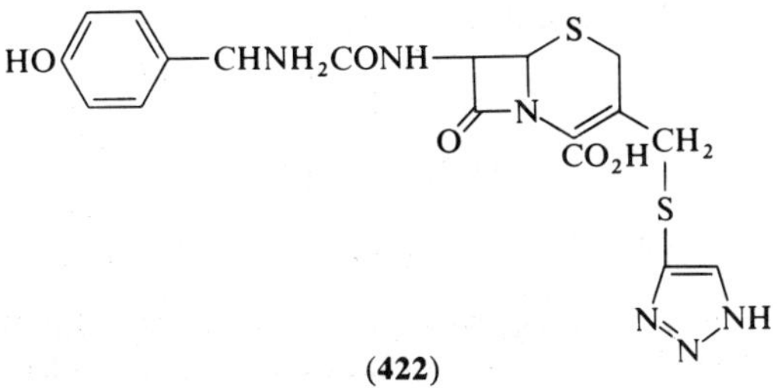

(**422**)

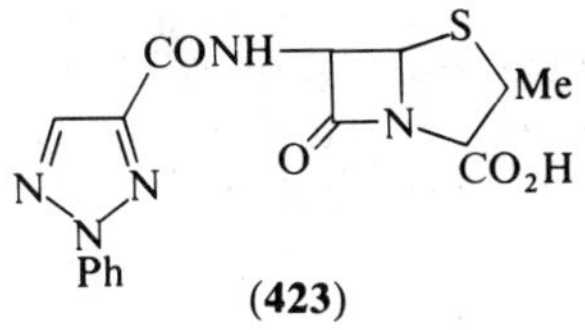

(**423**)

N-1 and N-2 derivatives of benzotriazole with glycine and different sugar residues, like β-glucopyranose, β-ribofuranose or polyacetylated furanoses, are found as cytostatica ⟨71MI41101, 71MI41102, 73MI41102⟩ and several benzo-substituted benzotriazoles (**424**) are cytostatically active against sarcomas and erythroleukomyelosis ⟨77MI41101⟩. Furthermore, 1H- and 2H-1,2,3-triazole nucleosides such as (**425**) and (**426**) have been found to possess bactericidal activity ⟨76USP3968103⟩. Other nucleosides (**427**) and (**428**) were active *in vitro* against herpes and measles virus ⟨76USP3948885, 81MI41101⟩. D-Arabinofuranosyl- and 2′-deoxy-D-ribofuranosyl-1,2,3-triazolecarboxamides have been described in ⟨77BCJ2689⟩.

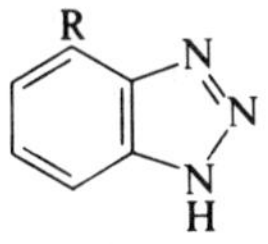

(**424**) R = H, NH_2, NHAc, $CHNH_2CO_2H$

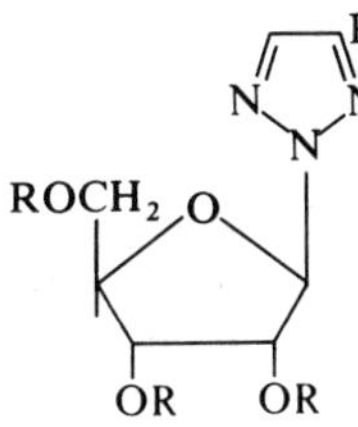

(**425**)

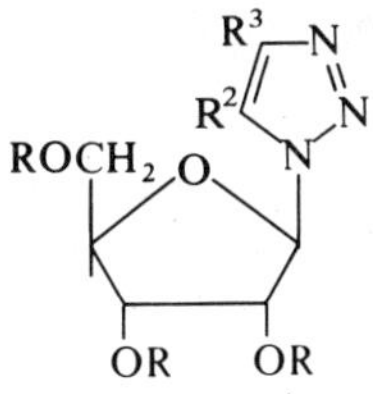

(**426**)

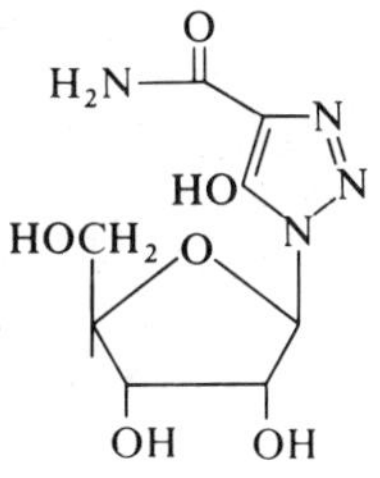

(**427**)

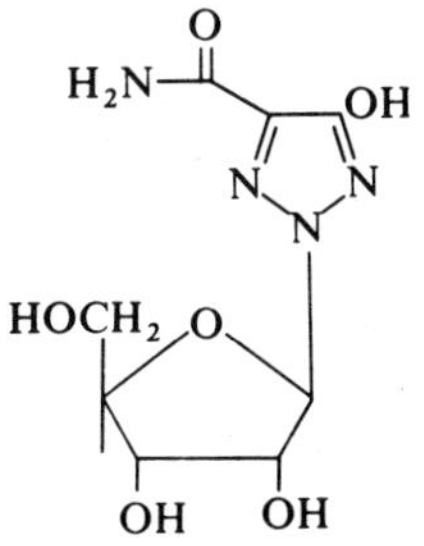

(**428**)

R = H, Ac, CH_2Ph
R^1, R^2, R^3 = H, CN, NO_2, CO_2Me, NH_2, $\underset{\underset{S}{\|}}{C}NH_2$

In two additional studies alkylated nucleosides have been investigated, and N-glycosyl(halomethyl)-1,2,3-triazoles (**429**) and N-ribosylhalomethyl-1,2,3-triazole (**430**) have been detected as a new type of alkylating agent showing cytostatic activity ⟨79JMC496⟩.

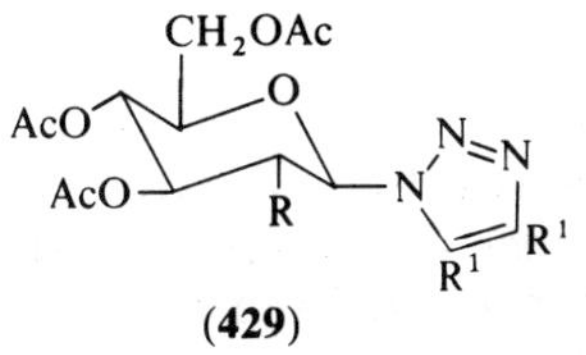

(**429**)
R = OAc, NHAc
$R^1 = CH_2X$
X = Cl, Br, I, OH

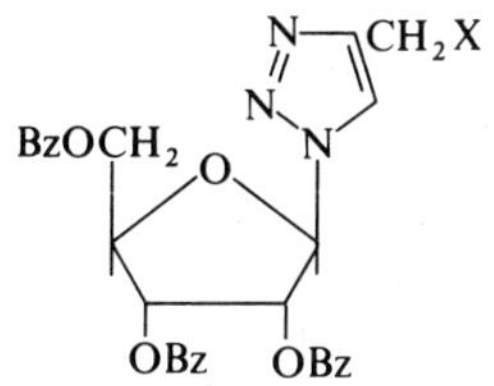

(**430**)

4.11.6.4 Industrial Applications

Besides their biological activity, some other important industrial applications of triazoles and benzotriazoles are known, including dyestuffs, fluorescent compounds, optical brighteners, corrosion inhibitors, photostabilizers and agrochemicals. Some selected applications are described in the following sections.

4.11.6.4.1 Dyestuffs, fluorescent whiteners

Numerous patents have appeared in the last few years which deal with 1,2,3-triazoles and benzotriazoles as part of optical brightening molecules. These consist in most cases of stilbene moieties with suitable heterocyclic chromophores attached ⟨B-64MI41102⟩, and as a consequence hundreds of brighteners are available today on the world market. Special applications are whitening of fibers [*e.g.* poly(ethyleneterephthalate), cellulose acetate and poly(acrylonitrile)], lacquers and plastics. Several examples of the structural types involved are shown in Scheme 13.

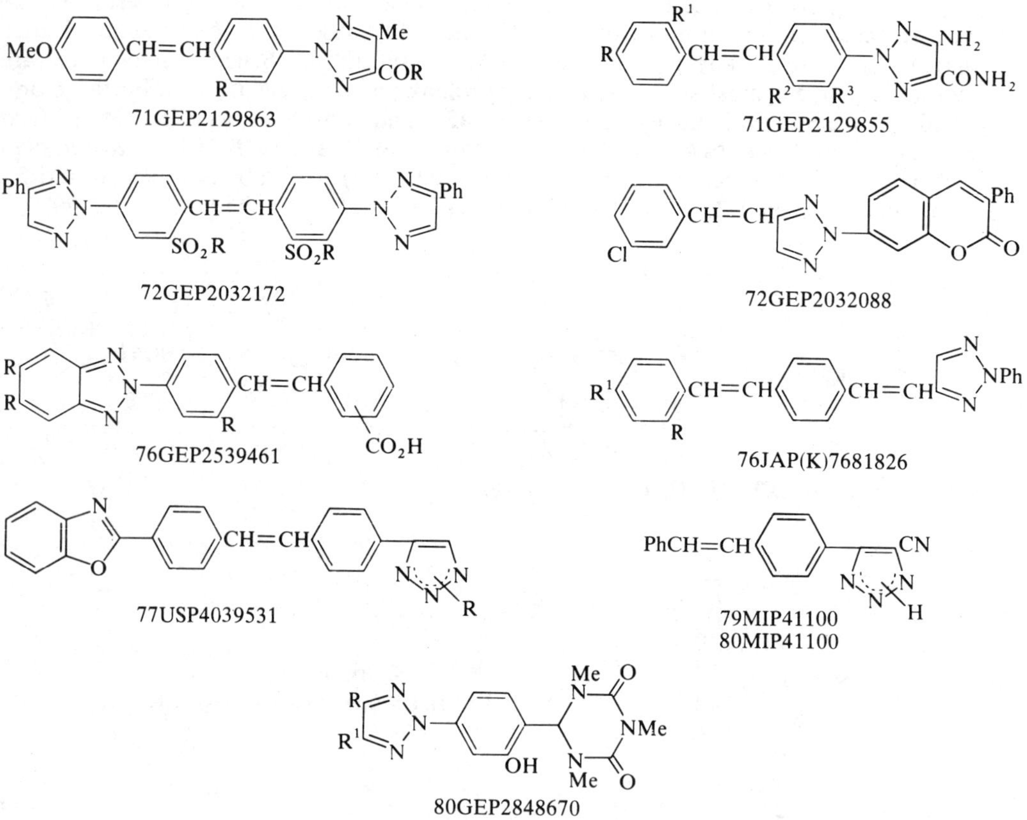

Scheme 13

A survey of the preparation of styryl and stilbenyl derivatives of 1*H*-benzotriazoles has recently been published ⟨81HCA662⟩.

4.11.6.4.2 Corrosion inhibition

Benzotriazole and its derivatives have proved to be powerful corrosion inhibitors for many metals and alloys, including copper, solder, brass, steel, cast iron and aluminum, in *e.g.* heat exchangers, heating systems or automatic radiators. For some recent literature, see ⟨79MI41102, 81JAP(K)81108882, 81JAP(K)81122884⟩.

Associated with these above properties of corrosion inhibition is the readiness of triazoles and benzotriazoles to form complexes with different metals. Benzotriazole and its 5-bromo derivatives can be used in quantitative determination of silver ion in the presence of copper, nickel, bismuth, thallium, lead, cadmium, zinc, iron, cobalt and chloride ions ⟨54MI41100⟩. Benzotriazole also forms complexes with cobalt ⟨55MI41100⟩ and palladium ⟨55JA6204, 56MI41100⟩. 4-Benzoyl- and 4-butyryltriazoles form iron complexes with iron(II) chloride or nonacarbonyliron. The latter complexes possess two triazole molecules per iron atom, which is π-bonded to the ring ⟨66DOK(170)354, 68DOK(181)1397⟩.

The synthesis and properties of chelate resins containing 1,2,3-triazole-4,5-dicarboxylic acid as an anchor group have been described. The 2-vinyltriazole (**431**) was copolymerized with divinylbenzene to yield crosslinked products and then saponified to give CO_2H-containing resins. These can be used for the complexation of heavy metals ⟨79MI41100⟩.

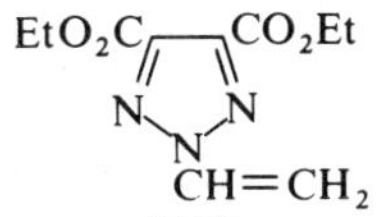

Furthermore, the extraction of some complexes of manganese(II), iron(II), iron(III) and copper(II) by a chloroform solution of 1,2,3-benzotriazole has been reported ⟨81UKZ1099⟩.

4.11.6.4.3 Photostabilizers

1,2,3-Triazole and benzotriazoles have been successfully applied as photostabilizers for fibers, plastics or dyestuffs, for example, and for the protection of human skin from harmful UV irradiation. For a brief summary, see ⟨B-71MI41103⟩.

The wavelengths of available sunlight which affect the chemical bonds usually found in polymer backbones are in the region 300–400 nm. Benzotriazoles, *e.g.* (**432**), are widely used absorbers ⟨72PAC(30)135⟩. For further literature on this topic see ⟨70JHC1113⟩. The stability of the benzotriazole series has been explained as involving a steric inhibition of rotation of the *o*-hydroxyphenyl ring with respect to the triazole ring ⟨72PAC(30)145⟩. The 2-phenyl-2*H*-1,2,3-triazoles (**433**) have proved to be UV absorbers, particularly for sunburn protection ⟨71GEP2129864⟩.

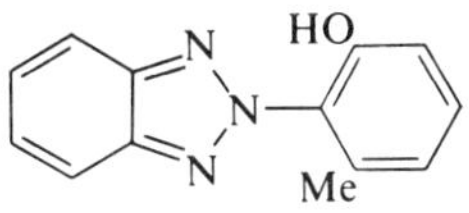

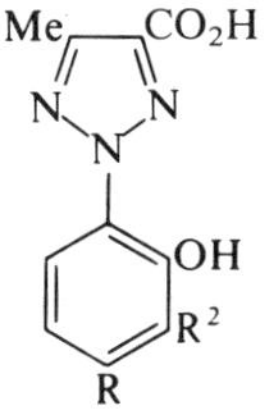

Several 2-phenylated 2*H*-1,2,3-benzotriazoles like (**434**) have been reported as light stabilizers for organic materials, especially polymers ⟨81EUP31302⟩. Tetrazoles with aryl and 2-nitrophenyl residues in the 1- and 5-positions have been converted thermally into 2-arylbenzotriazoles (**435**) which are useful as UV absorbers, especially in retarding the photodegradation of wool ⟨81AJC691⟩.

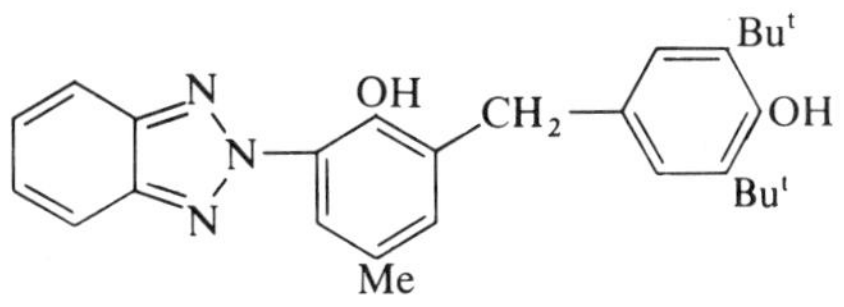

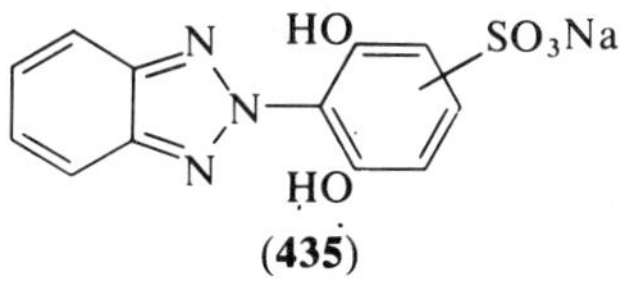

4.11.6.4.4 Agrochemicals

While 1,2,4-triazoles have found broad application as herbicides, fungicides and antibacterial agents, 1,2,3-triazoles have been detected only in the last decade to possess significant biological activity, which makes them of interest in the area of agrochemicals ⟨B-77MI41102⟩.

Thus, 5-anilino-1H-1,2,3-triazole-4-carboxylates (**436**) show antihelmintic activity ⟨70GEP2009134⟩. Büchel *et al.* have claimed triphenyl-1,2,3-triazolyl-(1′)-methanes (**437**) as fungicides ⟨75GEP2407305⟩.

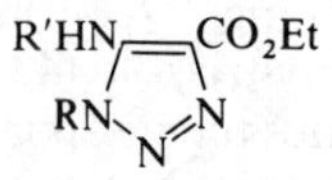

(**436**)

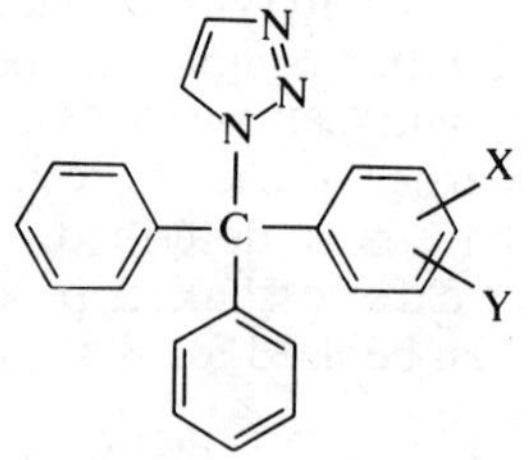

(**437**) X, Y = H, alkyl, Hal, CN

1-(3-Trifluoromethylphenyl)-5-phenyl-1,2,3-triazole and the corresponding 5-methyl and 5-ethyl derivatives (**438**) show marked to strong but non-selective herbicidal activity ⟨77MI41100⟩. 1,2,3-Triazolecarboxylic acid amides of the type (**439**) have herbicidal, defoliant, insecticidal, fungicidal and nematocidal activity ⟨80GEP2834879⟩. 1,2,3-Benzotriazoles have been shown to possess varying degrees of biological activity. Chloro substituents especially are known to increase the pesticidal activity of the molecule ⟨B75MI41101, 80JHC1115⟩.

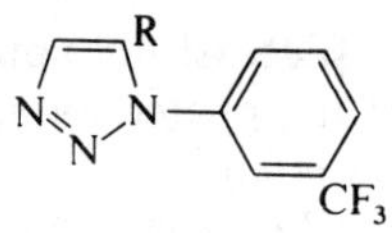

(**438**)
R = Ph, Me, Et

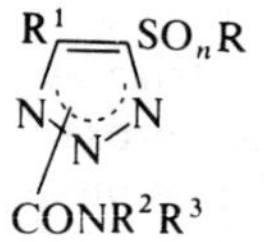

(**439**)

1,2,3-Triazoles of the type (**440**) are useful for regulating local plant growth ⟨81MIP41100⟩. Trialkyl(1,2,3-triazolyl)tin compounds (**441**) can be used as biocides ⟨81GEP2936951⟩.

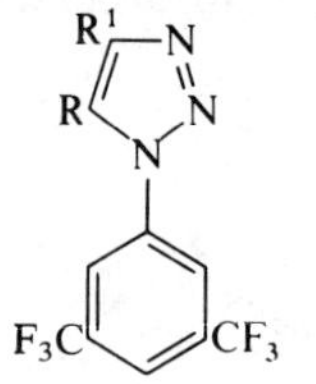

(**440**)

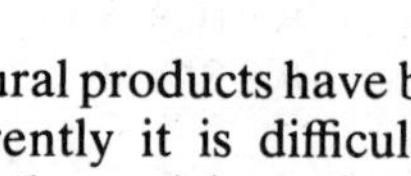

(**441**)

4.11.6.5 Important Natural Products

Up to now and to our present knowledge no natural products have been isolated containing 1,2,3-triazole or benzotriazole moieties. Apparently it is difficult for nature, and thus biochemical systems, to produce molecules with three vicinal nitrogen atoms in an open-chain or cyclic arrangement.

4.12
1,2,4-Triazoles

J. B. POLYA
University of Tasmania

4.12.1	INTRODUCTION	734
4.12.2	STRUCTURE	736
4.12.2.1	*Constitutional Problems and Methods*	736
4.12.2.2	*Molecular Geometry and Dimensions*	738
4.12.2.3	*Molecular Spectra*	739
4.12.2.3.1	*UV spectra*	739
4.12.2.3.2	*IR spectra*	739
4.12.2.3.3	*NMR spectra*	740
4.12.2.3.4	*Mass spectrometry*	741
4.12.2.4	*Theoretical Studies*	742
4.12.2.4.1	*Thermochemistry*	742
4.12.2.4.2	*Acidity*	742
4.12.2.4.3	*Dipole moments*	742
4.12.2.5	*Betaines and Mesoionic Compounds*	742
4.12.3	REACTIVITY	743
4.12.3.1	*General considerations*	743
4.12.3.2	*Reactivity at the Ring Atoms*	744
4.12.3.2.1	*Stability of the ring*	744
4.12.3.2.2	*Electrophilic attack at nitrogen*	746
4.12.3.2.3	*Electrophilic attack at carbon*	752
4.12.3.2.4	*Nucleophilic attack at carbon*	753
4.12.3.3	*Reactivity of Substituents*	755
4.12.3.3.1	*C-Linked substituents*	755
4.12.3.3.2	*N-Linked substituents*	760
4.12.4	SYNTHESIS	761
4.12.4.1	*General Considerations*	761
4.12.4.2	*Synthesis of Triazole Rings from Acyclic Compounds*	762
4.12.4.2.1	*Methods employing hydrazine derivatives*	762
4.12.4.2.2	*Nitrilimine methods*	768
4.12.4.3	*Synthesis of Triazole Rings from Other Heterocyclic Systems*	770
4.12.4.3.1	*Destruction of non-triazole rings*	771
4.12.4.3.2	*Cleavage and recyclization of non-triazole rings*	771
4.12.4.3.3	*Nitrilimines derived from non-triazole rings*	779
4.12.4.4	*Introduction and Modification of Functions on the Triazole Ring*	781
4.12.4.4.1	*Alkyl and aryl groups*	781
4.12.4.4.2	*Halo groups*	782
4.12.4.4.3	*Nitrogen functions*	782
4.12.4.4.4	*Sulfur functions*	782
4.12.4.4.5	*Carbonyl functions*	783
4.12.4.4.6	*Metal complexes*	784
4.12.4.4.7	*Triazolidinediones*	784
4.12.5	APPLICATIONS	784
4.12.5.1	*Analytical and Synthetic Applications*	785
4.12.5.1.1	*Analytical uses*	785
4.12.5.1.2	*Synthetic uses*	786
4.12.5.2	*Industrial Uses of Triazoles*	787
4.12.5.2.1	*Biological activity*	787
4.12.5.2.2	*Dyes and photographic chemicals*	788
4.12.5.2.3	*Polymers*	788
4.12.5.2.4	*Other uses*	790

4.12.1 INTRODUCTION

Simple 1,2,4-triazoles (those not forming part of a fused polynuclear system) are cyclic hydrazidines with H or some other substituent on either a hydrazide nitrogen as in (**1**) or an amide nitrogen as in (**2**). The prefixes $1H$ and $4H$ are used to distinguish (**1**) and (**2**), respectively. The name 1,3,4-triazole for (**2**) appears in some theoretical papers but it is deceptive when (**1**) and (**2**) are tautomers (with $R^1 = R^4 = H$): a catalogue of chemicals offering 1,2,4-triazole and 1,3,4-triazole, the same substance, at different prices is an example of misapplied formalism. In order to list substituents in alphabetical order or deal with rearrangements between annular nitrogens some authors have used the name '1,3,5-triazoles' thus breaking one general rule of nomenclature in order to comply with another.

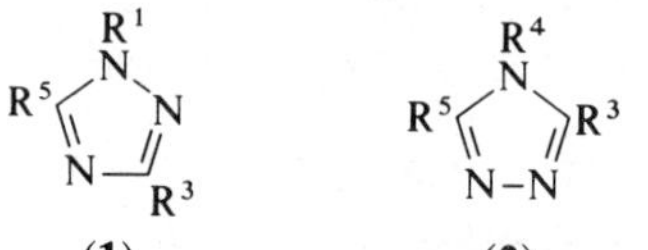

The theoretically unlikely 'isotriazole' (**3**) is best understood as the acid (**4a**) consisting of a proton balanced by the mesomeric triazolate anion more compactly represented as (**4b**) in Scheme 1 ⟨81HC(37)1⟩.

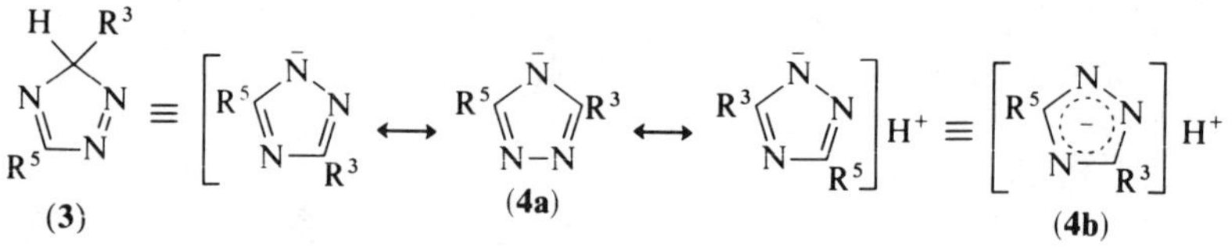

Scheme 1

In order to distinguish 1,2,3-triazoles (see Chapter 4.11) from 1,2,4-triazoles, the prefix v- is used for the former and s- for the latter. If, as in this chapter, the context clearly suggests 1,2,4-triazoles, the prefix 1,2,4- or s- may be omitted.

Partially or fully reduced triazoles (triazolines and triazolidines, respectively) with monovalent substituents are relatively unstable and of little current interest. They are not treated separately here as in other chapters. Triazolines and triazolidines without exocyclic double bonds, *i.e.* without ring-linked functions such as =O, =S, =NR'R", =CR'R", *etc.*, are not aromatic. Few representatives of these classes have been prepared ⟨81HC(37)1 p. 503, 516⟩; some of those readily available may owe their stability to substitution by, or fusion to, other heterocyclic rings. Like their aliphatic analogues, non-aromatic triazolines and triazolidines are readily hydrolyzed in acidic media; other electrophilic reagents also open the ring. Those with substituents such as =O, =S or =NR may be written as aromatic tautomers (Scheme 2) remembering, however, that graphical convenience must not be regarded as the solution of problems arising out of tautomerism or mesomerism. The same applies to the verbal or taxonomic convenience of terms such as 'hydroxytriazole' or 'mercaptotriazole' for compounds best regarded as triazolinones and triazolinethiones, respectively. With such qualifications, the term 'triazole' may be used generically, not only for 1,2,4-triazoles proper but also for triazolinones, *etc.* that are

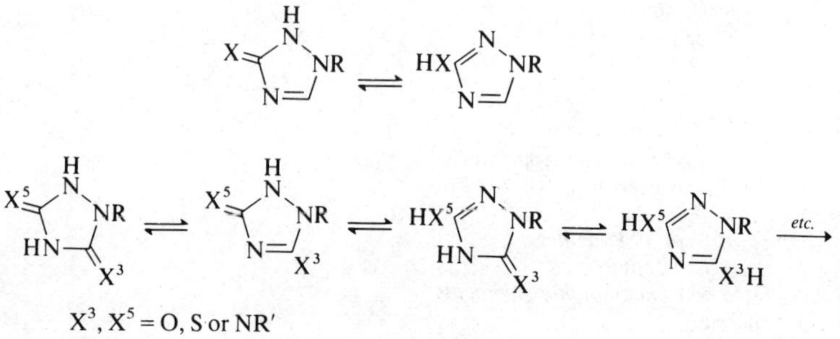

X^3, X^5 = O, S or NR'

Scheme 2

potentially aromatic, and even for unambiguously non-aromatic triazolines, triazolidines and other compounds formally derived from 1,2,4-triazoles.

A few trivial names for 1,2,4-triazoles are in common use: 3,5-dioxo-1,2,4-triazolidines (**5**) are called urazoles, and the corresponding 3,5-diamino compounds (**6**) guanazoles.

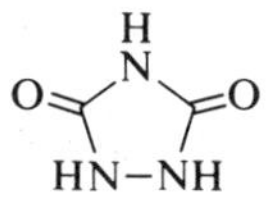

A convenient shorthand representation of 1,2,4-triazoles illustrated by formally written examples and their abbreviated equivalents (Scheme 3) may help those taking notes but will not be used in this chapter.

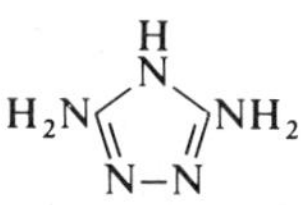

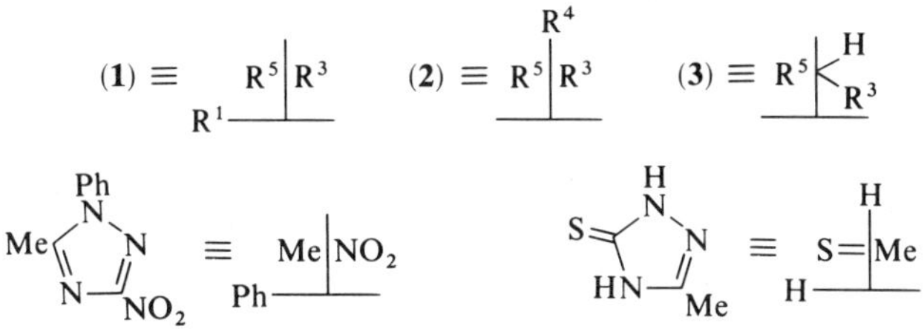

Scheme 3

The history of 1,2,4-triazoles is less than a century old and starts with the work of Bladin ⟨1885CB1544⟩ who synthesized the first representatives and coined the name for the class. The alternative term 'pyrrodiazoles' is no longer in use. Well over 20 000 'triazoles' (in the most general sense as explained above) are known but practical applications have been very few until recently. Although most triazoles are readily prepared and stored, expensive starting materials or sensitive intermediates appear to have discouraged industrial syntheses and wide application. Thanks to improved preparative techniques the field of applications has been so rapidly growing during the last two decades that its appreciation requires a review beyond the size of this monograph.

The first studies of 1,2,4-triazoles were concerned with structural isomerism. Modern instrumental and theoretical methods achieved much success in dealing with tautomeric problems, the complexity of which is one of the enduring charms of the chemistry of 1,2,4-triazoles. However, some structural and many tautomeric problems require further study; kinetic and other quantitative mechanistic studies are scarce; the stereochemistry and photochemistry of 1,2,4-triazoles are virtually unexplored.

Few books on general organic chemistry deal with 1,2,4-triazoles beyond naming the ring system. Even specialized texts of heterocyclic chemistry are not much more informative with the exception of ⟨B-67MI41200⟩ which introduces the student to such properties of triazoles that conform to the generalizable behaviour of kindred heterocycles.

Two reviews cover the first 70 years of the chemistry of 1,2,4-triazoles: ⟨61CRV87⟩ deals not only with the simple triazoles discussed in this Chapter but also with their fused polynuclear systems (see Chapter 4.15); ⟨B-61MI41200⟩ considers both s- and v-triazoles (for the latter see Chapter 4.11).

The most recent review ⟨81CHC(37)1⟩ lays stress on synthetic methods and provides an extensive tabulation of 1,2,4-triazoles up to 1976; references to reactivity and applications are numerous and interesting but subordinated to the main theme.

Another review ⟨B-76MI41200⟩ treats the structure and reactivity of azoles with references up to and including 1974, and contains valuable tables of physical properties of many triazoles in various functional classes.

This monograph aims to present a composite picture of the chemistry of 1,2,4-triazoles through examples chosen to display the variety of its problems and methods. It is also intended to draw attention to some critical matters that tend to be overlooked by those whose main concern is the ordering of a vast amount of material. Since the voluminous controversy on some points of triazole chemistry is beyond the scope of this monograph, it is hoped that readers irritated by seemingly dogmatic arbitration without full references will consult the appropriate sections of the reviews quoted above.

4.12.2 STRUCTURE

4.12.2.1 Constitutional Problems and Methods

Identification of the 1,2,4-triazole ring and location of its substituents can be decided on the evidence of synthesis and molecular formula in a few cases. However, the early literature abounds in structural errors resulting from unjustified confidence in elementary methods and from the absence of instrumental analysis that provides a variety of independent structural proofs in our days. A few examples illustrate points to be kept in mind by sceptical triazole chemists.

Addition of phenylhydrazine to cyanogen affords 'dicyanophenylhydrazine' (7) or (8); acylation and cyclization with acetic anhydride affords a triazole for which one must consider structures (9) or (10) convertible into (11) or (12) respectively (Scheme 4). The first assumption of (7), and hence (9) and (11), was proved wrong by showing the correctness of (8) ⟨1893CB2385⟩; compounds (10) and (12) were also obtained by alternative syntheses from structurally unambiguous starting materials.

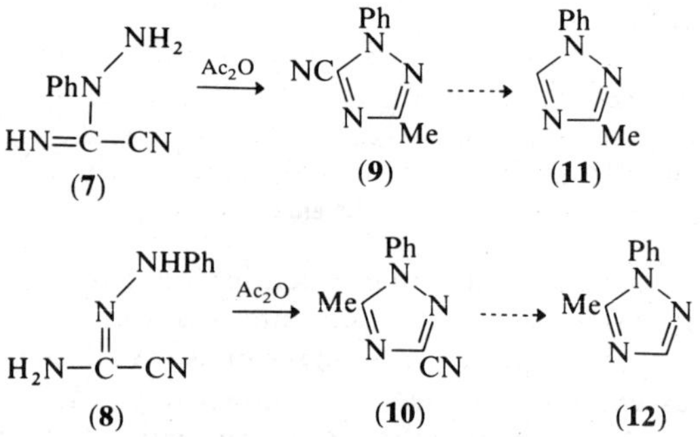

Scheme 4

Depending on substituents and solvents, the equilibrium in Scheme 5 favours the acyclic component (13) or the triazolidinone (14) ⟨78JHC806⟩. The dimerization of N-unsubstituted 1,2,4-triazolecarbaldehyde (15) ⟨71AJC393⟩ in Scheme 6 presents a similar problem of cycloisomerism (see Section 4.12.3.3). The possibility of hydrate or alcoholate formation may mask cyclization if the proof depends on elementary analyses alone; the usual methods for the determination of molecular weights are inadequate, and spectroscopic methods have to be used with care since the solvent used for analytical purposes may affect equilibria such as those presented in Schemes 5 and 6, and the same applies to the conditions implied in mass spectrometry.

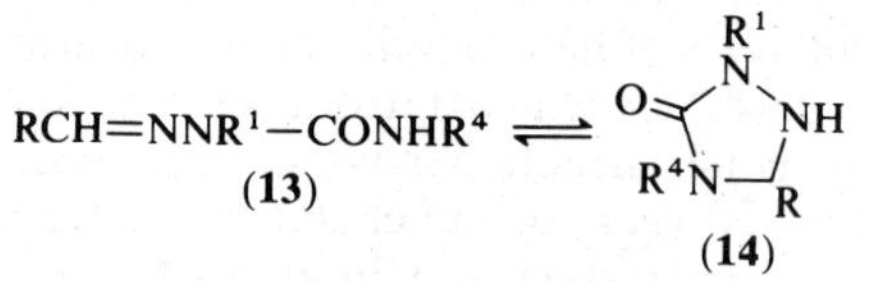

Scheme 5

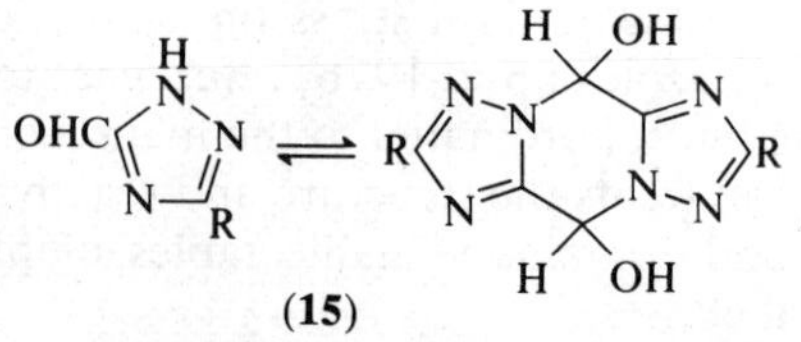

Scheme 6

Cyclizations of (16) as represented in Scheme 7 can afford triazole derivatives (17) or isomeric triazoles (18) or other heterocycles (19). The claimed isolation of two stable tautomers of (17) has been shown to be in error by demonstration of the isomerization involving (17) and (19; X = O) ⟨60LA(638)136⟩.

Compounds originally thought to have the stereochemically strained structure (20) are now understood to be mesoionic compounds ⟨57QR15⟩ (see Section 4.12.2.5).

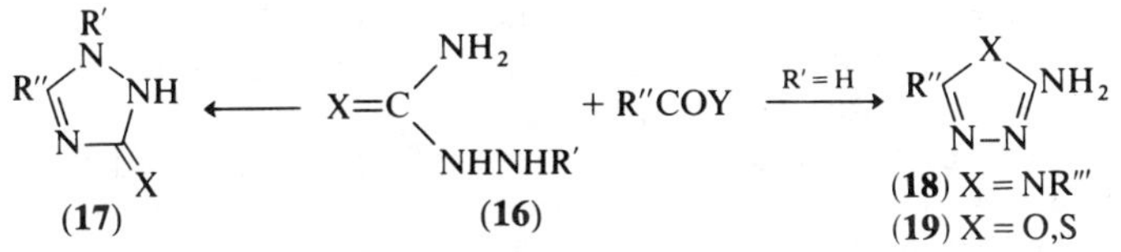

The tautomerism of 1,2,4-triazoles (see Chapter 4.01 and ⟨76AHC(S1)1⟩) may involve one or more of the following possibilities: annular prototropy, prototropy involving both the ring and substituents and tautomerism restricted to the substituent. Only the first two of these will be considered.

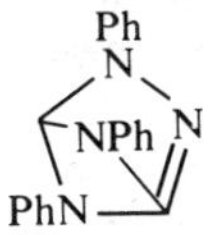

In N-unsubstituted triazoles (**1** or **2**; R^1, $R^4 = H$) prototropy between nuclear N centres may occur. When R^3 and R^5 are different, the equilibrium in Scheme 8 is not only of theoretical importance but may affect alkylation, acylation, prototropy between ring and substituents, ligand properties and so on. Most experimental and theoretical considerations favour 1H (**1**) (or 2H (**21**)) as the predominant or only tautomer; the 4H tautomer may play a comparable role. The paradox arises from the all too common oversimplification of the problem: the tautomeric status of a given triazole need not be the same in solution, melt, solid or gas phase. Variation of substituents, solvents, temperature and the often ignored variation of energy input by different analytical methods provide a field of results rather than a unanimous decision for or against a certain tautomer.

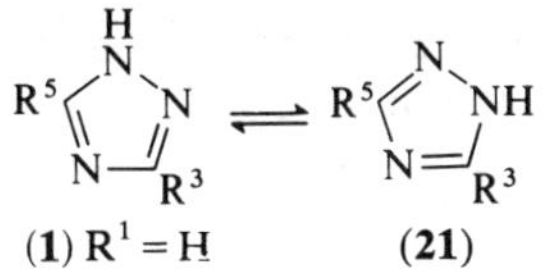

Scheme 8

Tautomerism involving 'isotriazoles' may be dismissed on considering that (**3**) lacks the aromaticity of (**4**), and the structures (**3**) and (**4**) are excluded by the molecular geometry derived from physical measurements. In addition, (**4**) exaggerates the acidity of unsubstituted triazoles.

Prototropy in triazoles is particularly complex when substituents such as OH, SH and NHR are available to donate protons to annular N. A detailed discussion of all possible sub-cases of this type is beyond the scope of this chapter, but the main aspects of this matter are reviewed in a broad critical context ⟨76AHC(S1)1, p. 388, 414, 444⟩ tabulating generalizations and their levels of reliability. The section on reactivity (Section 4.12.3) gives examples of the variability of tautomeric preference in some reactions.

Related to tautomeric problems of 1,2,4-triazoles are those concerned with hydrogen bonding. N-Unsubstituted triazoles (**1**) or (**2**) with R^1 or $R^4 = H$ may be linked by intermolecular bridges between 'pyrrole' NH and 'pyridine' —N=; if then $R^3 = 2$-pyridyl a variety of intermolecular and intramolecular hydrogen bridges are possible.

Given the aromatic nature and small size of the 1,2,4-triazole ring, the relatively few measurements on which the accepted notion of its coplanarity rests need not be questioned. If puckering could be induced through the bonding or steric repulsion of substituents, one would expect optical activity which has not been observed as yet. Triazole derivatives with asymmetric substituents (**22**) can be obtained in optically active form and as a stereospecific dienophile it has found extensive use in the resolution of homocyclic dienes ⟨80JOC5105⟩. A little information is available on the conformation of some triazoles ⟨81RCR336⟩.

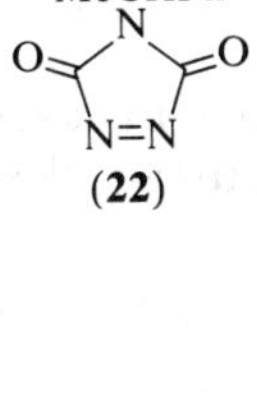

Polymorphism has been reported for a few triazoles ⟨B-61MI41201⟩. Application of modern crystallographic methods to kinetic studies of the interchange and exclusion of structural isomerization would be of interest.

Another imperfectly understood molecular multiplicity is displayed by 3-hydroxymethyl-1,5-diphenyl-1,2,4-triazole (**23**; R = Ph) ⟨62T539⟩. The labile form 'A' (m.p. 132–133 °C) changes into the stable form 'B' (m.p. 153–154 °C) in neutral solution and faster in the presence of acid. In methanol the conversion is complete in 48 hours and is accompanied by changes of the UV spectrum. The IR spectra indicate isomers of similar chemical behaviour; in fact 'A' and 'B' undergo the same reactions and yield identical derivatives. The IR spectrum indicates stronger hydrogen bonding of the alcoholic OH in 'A'. Polymerism and functional differences having been excluded, the remaining alternatives include buckling of the ring (which should have stereoisomeric consequences) and the doubtful hypothesis of valence isomers. A similar isomerism is suggested in the case of (**23**; R = 4-C_6H_4OMe).

(**23**) R = Ph, 4-C_6H_4OMe

The molecular geometry of 1,2,4-triazoles has been studied mostly in relation to annular tautomerism, especially that of simple compounds in which $1H$ and $4H$ tatuomers have different symmetries.

Although crystallographic and related methods (see Section 4.12.2.2) have been used in tautomeric studies, spectroscopic methods are most commonly used for this purpose (see Section 4.12.2.3). Theoretical methods (see Section 4.12.2.4) rely mostly on thermochemical information, data on acidity and dipole moments to be considered in the same sub-section (see Section 4.12.2.4).

Constitutional studies of mesoionic compounds make use of all the methods mentioned above, and will be treated under Section 4.12.2.5.

4.12.2.2 Molecular Geometry and Dimensions

The intuitive statistical argument from two 'hydrazinic' N centres against one 'amine' N favours the less symmetrical $1H$ rather than the symmetrical $4H$ structure for 1,2,4-triazole. On the evidence of X-ray diffraction analysis the solid parent triazole has a planar structure with hydrogen bridges between N-1 and N-4 of neighbouring rings; of the two N—H bond lengths implied, only that leading to N-1 is of the order required by covalent bonding ⟨65MI41200⟩. Confirmation from similar studies carried out at −160 °C proves the molecular dimensions as shown in Table 1 for one unit of a pleated sheet linked by hydrogen bridges ⟨69AX(B)135⟩. Slightly different values obtain at room temperature or in substituted aromatic triazoles.

Table 1 Molecular Geometry of 1,2,4-Triazole[a] at −160 °C ⟨69AX(B)135⟩

Angle (°)		Bond	Bond length (pm)
5—1—4	110.2	1—2	135.9
1—2—3	102.1	2—3	132.3
2—3—4	114.6	3—4	135.9
3—4—5	103.0	4—5	132.4
4—5—1	110.1	5—1	133.1
		N(1)—H	103.0
		C(3)—H	93.0
		C(5)—H	93.0

[a] The numbering refers to annular centres in $1H$-1,2,4-triazole.

The microwave spectrum of triazole in the vapour phase also supports the $1H$ structure ⟨71CC873⟩ for the monomer known to be the species present in the gaseous state ⟨1892CB225⟩.

Triazole is monomeric or only slightly associated in polar solvents (in which it is soluble), but solubility in non-polar solvents is slight and intermolecular bonds persist ⟨40MI41200⟩. Further evidence in support of the 1H structure appears in Section 4.12.2.4.

Although 1,2,4-triazole is formally related to pyrazole and imidazole, it resembles the latter. Its melting point (121 °C) and boiling point (260 °C) are more similar to those of imidazole than to those of pyrazole. However, in not too concentrated solutions pyrazole is a closer analogue than imidazole. For a brief review see ⟨69JCS(C)1056⟩. Substitution on N-1 lowers the melting point considerably, that on N-4 less so, but all known triazoles are solid at room temperature.

Information on structure and molecular geometry is of importance in pharmacodynamic studies: work on 2-morpholinoethyltriazoles ⟨77JHC439⟩ is a good example of the combination of crystallographic and other analytical techniques for this purpose.

Semiconductive properties of imidazole have been derived from protonic migration or proton tunnelling in an array of rotating heterocycles ⟨68MI41200⟩. The model appears to be applicable to 1,2,4-triazole which has a specific conductivity of about 10^{-10} ohm^{-1} cm^{-1}, of the same order as imidazole.

4.12.2.3 Molecular Spectra

4.12.2.3.1 UV spectra

UV spectra of representative 1,2,4-triazoles have been tabulated ⟨61CRV87, B-76MI41200, p. 267⟩. The parent triazole absorbs around 210 nm and lacks intensive $n \to \pi$ bands at higher wavelengths. Aryl substituents have a bathochromic effect that varies with the site of substitution in the order 3 (or 5) > 1 > 4. The 3- or 5-aryltriazoles have non-ionic canonical structures which are lacking for 1- and 4-isomers. The considerable bathochromic and slight hypochromic effects shown by 3-phenyl- and 3-methyl-5-phenyl-1,2,4-triazoles in the presence of alkali are similarly explained, while changes of extinction observed when N-1 is substituted by Me have been ascribed to steric effects ⟨54JCS4256⟩.

Hydroxytriazoles and tautomeric triazolinones have a common mesomeric anion, hence their spectra are affected in basic solvents. UV spectra have been used to demonstrate the predominantly dioxo structure of 1-phenyl-1,2,4-triazolidine-3,5-dione ⟨66T(S7)213⟩ and the thione structure of mercaptotriazoles ⟨66ACS57⟩ utilized also in theoretical work.

The UV spectra of C-aminotriazoles are not affected by change from neutral to acid solvents ⟨67JCS(B)641⟩. The considerable hyperchromic effects combined with hypochromic or bathochromic shifts shown by 3-amino-5-(3-pyridyl)-1,2,4-triazole in aqueous solutions containing varying concentrations of hydrochloric acid are due to the pyridine moiety ⟨54JCS4508⟩.

While N-acylation of the triazole nucleus results in a bathochromic shift, acylation of substituent amino groups has a negligible spectroscopic effect ⟨55CR(241)1049⟩.

4.12.2.3.2 IR spectra

IR spectra of N-acyltriazoles ⟨61CRV87⟩ show C—N stretching vibrations at 1208–1212 cm^{-1} and 1245–1266 cm^{-1}; the C=O stretching at 1756 cm^{-1} is in excess of that for carbonyl in Alk—COR and —CONR$_2$. A detailed comparison of IR absorptions of triazole, halotriazoles and their salts is found in ⟨70KGS248⟩.

Absorption in the region of 2400–3200 cm^{-1} characteristic for N-unsubstituted triazoles has led to the parent triazole being regarded as a salt, triazolium triazolate, but this is contradicted on all other experimental and theoretical evidence. However, a detailed study of the IR spectra of N-unsubstituted triazoles in the region 1700–3000 cm^{-1} reveals a difference between triazoles substituted on 3- (or 5- or 3,5-) by pyridyl and those with substituents such as Br, Cl, Me, CH$_2$OH, Ph and PhCH$_2$. Both groups have two sets of characteristic bands with the ratio of intensities in the range 1.6–9.6 for pyridyltriazoles and 2.0–6.6 for others. Plots of intensity ratios against band maxima show a decreasing parabolic pattern between 1700–1940 cm^{-1} and a symmetrical rise between 2473–2800 cm^{-1} for pyridyltriazoles, but such variations are not noted in the other group. Hydrogen bonding is involved in the explanation of these observations ⟨71PMH(4)272⟩. More specifically, proton tunnelling has been suggested by analogy with other systems containing strong acidic and basic centres ⟨69JCS(C)1056⟩. For a complex case of this nature (**24**) see ⟨77AJC421⟩.

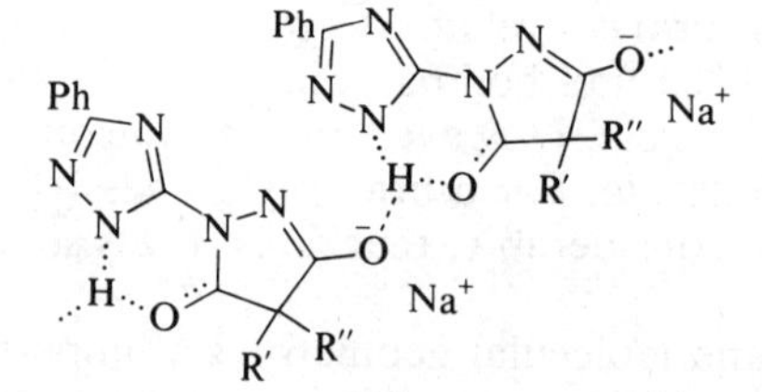

(**24**)

Characteristic examples of the use of IR spectra are structural studies on urazoles and triazolinones ⟨66T(S7)213⟩ and triazolinethiones ⟨71JCS(C)1016⟩. Tautomeric problems of oxo–hydroxy and thione–thiol systems are simplified by *N*-alkylation ⟨73CPB1342⟩. Although earlier views in favour of oxo and thio rather than hydroxy and thiol forms have been confirmed, counter-examples have been found; 3-hydroxy-2-methyl-5-phenyl-1,2,4-triazole (**25**) is preferred to the oxo tautomer (**26**) while the isomeric 4-methyl compound (**27**) follows the general rule (Scheme 9).

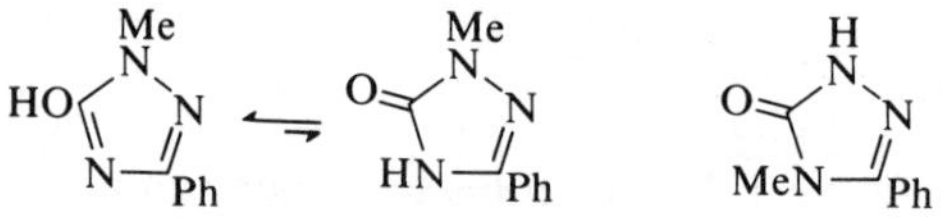

(**25**)   (**26**)   (**27**)

Scheme 9

IR data and force constant calculations suggest that the 3-nitro structure (**28**) is preferable to the 5-nitro structure (**29**) ⟨73KGS707⟩. Non-tautomeric substituents on C-3 and C-5 of *N*-unsubstituted triazoles raise questions of annular prototropy. Although the location of protons on N-1 or N-2 is favoured over the site at N-4, the choice between N-1 and N-2 is affected by solvents and substituents capable of immobilizing the proton through intermolecular bonding, *e.g.* furyl or 2-pyridyl ⟨72CPB2096⟩. The general rule ⟨75CPB955⟩ claims that location of the proton on N-1 or N-2 is determined according to whichever is nearest to the best electron-donating substituent on C-3 and C-5. Preference for the 3-nitro structure (**28**) over the 5-nitro (**29**) follows the rule but C-3 or C-5 functions capable of forming intramolecular hydrogen bonds with an annular N could promote anomalous structures such as (**30**) that would breach the general rule of preference for amino over imino groups. However, acylation studies of aminotriazoles ⟨80KGS1414⟩ argue for regular amino structures with exceptions such as the bistriazole (**31**).

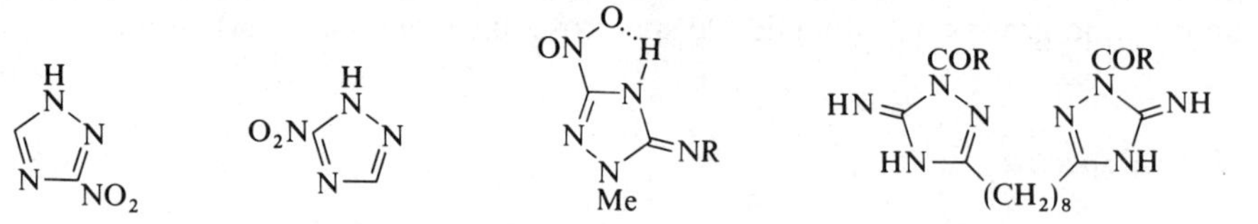

(**28**)   (**29**)   (**30**)   (**31**) R = Me, Ph

4.12.2.3.3 NMR spectra

NMR spectra of representative triazoles have been tabulated ⟨B-76MI41200, p. 277⟩. Because of rapid exchange, N—H signals of molten triazoles have been found in the range 10.88–11.40 p.p.m. downfield from DMSO ⟨69JCS(C)1056⟩, close to values for imidazole but substantially higher than those for pyrazole. On NMR evidence, association in hexadeuteroacetone increases ⟨69TL495⟩ but the bridged units of triazole do not share the bridging protons equally: the pleated sheet structure discriminates in favour of the 1*H* form.

One of the first successes of NMR analysis of triazoles was the until then much debated position of *N*-acyl groups: the protons on C-3 and C-5 are equivalent in 4-acyltriazoles but different in 1-acyl isomers; the latter alternative has been proved by two signals occurring in the ratio 1:1 ⟨62JOC2631⟩.

The parent triazole shows a broad singlet in HMPT slightly above room temperature (δ 8.03 at 37 °C) or slightly below it (δ 8.17 at 10 °C); on cooling to −34 °C two signals are obtained at δ 8.85 (H-5) and δ 7.92 (H-3). 1-Methyltriazole behaves similarly, but the single band (δ 8.20 at 37 °C) of 4-methyltriazole undergoes a slight shift to δ 8.34 on cooling to −40 °C but is not split ⟨68JOC2956⟩. This paper estimates the half-life of a proton on N-1 as 4 ns and provides thermodynamic information on the process of coalescence. For some critical remarks on this much quoted work see ⟨76AHC(S1)1, p. 22⟩.

The downfield shift of the signal for the proton on C-5 in trifluoroacetic acid has been used to establish the acidity of C-H in nitrotriazole ⟨69AJC2251⟩ and to demonstrate protonation at N-1 or N-4 according to whether N-4 or N-1 is substituted ⟨67JCS(B)516⟩. It was shown by NMR spectroscopy that the C—H nearest to the nitrogen substituted by an electron donor is most sensitive to solvent effects ⟨67BSF2630⟩ as confirmed by ⟨75CPB955⟩.

On the other hand, NMR spectroscopy is not suitable for the determination of tautomeric equilibrium constants by the interpolation method as shifts of the tautomeric equilibrium have little effect on chemical shifts ⟨67TL2109⟩.

The ratio $1H:4H$ of the parent substance in concentrated methanolic solution has been estimated as 1.5 by ^{14}N NMR ⟨72T637⟩ but this result must be qualified by solvent effects ⟨73JA324⟩. The use of ^{13}C NMR does not appear to be more suitable for the assessment of tautomeric equilibria ⟨76AHC(S1)49⟩ or in Hückel calculations ⟨68JA3543⟩. However, ^{13}C NMR has been used to establish the fused structure (32) in a complex study instead of the mechanistically more plausible (33) ⟨81KGS694⟩.

4.12.2.3.4 Mass spectrometry

Mass spectrometry of 1,2,4-triazoles has afforded many useful results, but some of its interpretations have raised controversies that are not settled as yet. An elimination pattern of the parent substance ⟨68CC727⟩ postulates a $4H$ structure decomposed *via* an 'isotriazole' and cleaved across the link between N-1 and N-2. A more substantial study of triazoles ⟨71JHC773⟩ claims elimination of nitrile leaving a diazirinium radical cation, which plausibly suggests a reversal of triazole syntheses from nitriles and nitrilium reagents (see Section 4.12.4.2.2). Assumption of autotropy, involving the identical $1H$ and $2H$ tautomers leads to an interpretation of data that assigns negligible weight to the $4H$ tautomer ⟨72OMS(6)1139⟩. A careful major study of alkyltriazoles ⟨73OMS(7)57⟩ following on a demonstration that the complex mass spectroscopic patterns of triazolinethiones require extensive labelling with ^{2}H and ^{15}N ⟨72AJC335⟩ comes to the conclusion that observed decompositions of the triazole ring are consistent with $1H$, $2H$ and $4H$ parent molecules. Mass spectroscopy of triazoles (34) may tip the balance of argument in favour of interpretations based on the $1H$ form ⟨80OMS172⟩ but the substituents of (34) may affect the tautomerism in a manner that does not apply to 1,2,4-triazole and its simple alkyl homologues.

The equilibrium between $1H$- (or $2H$-) and $4H$-forms is expected to be different in excited and unexcited molecules, hence interference with a molecule to a disruptive extent may without paradox afford results at odds with those obtained by gentler methods on quasi-static molecules.

At present the major value of mass spectrometry to structural triazole chemistry appears to be in intricate heterocyclic reactions giving rise simultaneously not only to triazoles but many other heterocycles at the same time ⟨81KGS694⟩. Identification of unstable triazolines has been aided by mass spectrometry as well as by their NMR data ⟨71BSF3296, 73BCJ1250⟩.

4.12.2.4 Theoretical Studies

4.12.2.4.1 Thermochemistry

Early data on the heat of combustion of triazoles include some for compounds selected for their exceptionally high nitrogen content ⟨57JPC261⟩. Resonance energy estimates of different authors vary in the range of 83.7–205.9 kJ mol^{-1}; of these the highest value appears most probable, and compares reasonably with that of pyrazole but less satisfactorily with that of imidazole ⟨61MI41202⟩. Calculated energy differences between azole tautomers support preference for the $1H$ over the $4H$ tautomer ⟨69JA796⟩. Similarly, the usual tautomeric preference for triazolinones over hydroxytriazoles and aminotriazoles over triazolinimines are supported on thermochemical evidence ⟨70JA2929⟩.

4.12.2.4.2 Acidity

It is remarkable that acidity constants of triazoles as determined for the first time ⟨06CB1831⟩ require only small corrections for theoretical purposes ⟨70JHC991⟩. A representative selection of acidity data with methodical notes ⟨B-76MI41200, p. 284⟩ has been drawn mainly from papers that consider the applicability of the Hammett treatment to the pK_a values of triazoles. The parent substance has pK_a values of 2.19 (as base) and 10.26 (as acid). The Hammett equation is more satisfactory for acidic than basic values of triazoles ⟨65ZC381⟩. For triazoles without acidic groups other than the annular NH, $\rho = 6.00$; for 3- (or 5-) 'hydroxytriazoles' the value is 7.62; in either case σ_m constants are most appropriate ⟨65ZC304⟩ as for nitrotriazoles ⟨70KGS558⟩.

The strong basicity of 1,2,4-triazoles on comparison with azoles other than imidazole is explained by the mesomeric stabilization of the amidinium ion formed on protonation ⟨67JCS(B)516⟩.

Acidity constants of 1,2,4-triazole and its 1-methyl and 4-methyl derivatives have been used to calculate the equilibrium constant for $1H$ and $4H$ tautomers as 10 in favour of the former, but a more general treatment and other data suggest a lower value ⟨76AHC(S1)1, p. 284⟩.

On CNDO/2 calculations with pK_a (as base) = 2.30 for triazole, the basicities of $1H$ and $4H$ tautomers are estimated to be equal, and maximum separation of protonated nitrogens (i.e. N-1 and N-4 rather than N-1 and N-2) is predicted to afford the most stable cation ⟨68TL3727⟩. Acidity constants of 1,2,4-triazoles correlate with total and π-electron densities but not with the lone pair character of the pyridine-type N in MO calculations ⟨70JCS(B)1692, 70BCJ3344⟩.

4.12.2.4.3 Dipole moments

Early calculations of electron densities in 1,2,4-triazole and its anion, beginning with ⟨51MI41201⟩ provide qualitatively correct predictions of reactivity. Most of the modern theoretical work is checked against dipole moments. Information summarized in Table 2 shows preference for the $1H$ formulation. The dipole moments of N-unsubstituted nitrotriazoles suggest 'fixation' of the proton on N-1 ⟨71KGS275⟩.

^{13}C NMR spectroscopy used in conjunction with lanthanide shift reagents supports the application of LCAO–SCF calculations to 1,2,4-triazoles ⟨74JOC357⟩.

4.12.2.5 Betaines and Mesoionic Compounds

Quaternization of the triazole amino acid (35) to (36) only superficially resembles the conversion of glycine into its betaine as in the former case the chemical system includes (37) which implies mesomeric distribution of a negative charge in the heterocyclic ring (Scheme 10). The betaine (38) ⟨69MI41200⟩ and its derivatives have attracted considerable interest from crystallographers ⟨73JCS(P2)1, 73JCS(P2)4, 73JCS(P2)9⟩.

Of wider interest are mesomerically-stabilized zwitterions of the general formula (39) with R^1 and R^4 other than H. The charge on the nucleus arises without the intervention

Table 2 Dipole Moments (10^{-30} C m) of some 1,2,4-Triazoles[a]

1,2,4-Triazole	Experimental notes	Found	Dipole moment (10^{-30} C m) Calculated		Method
			1H	4H	
Parent compound	In dioxan	10.57	—	—	—
	Dielectric constant	10.98	10.81	17.81	Vectorial
	Microwave spectrum	9.07	10.24	—	CNDO/2
	of vapour	—	7.77	—	VESCF
		—	10.24	19.12	LCAO-A
		—	10.31	18.95	LCAO[b]
		—	—	18.78	LCAO[c]
		—	12.31	17.91	LCAO–SCF
		—	10.07	—	CNDO/2[d]
1-Methyl	In dioxan	11.68	—	—	—
1-Phenyl	In benzene	9.81[e]	—	—	—
4-Phenyl	In benzene	18.78[e]	—	—	—

[a] Unless otherwise indicated, recalculated from ⟨76AHC(S1)1, p. 286⟩. [b] +σ-Electrons. [c] All electrons. [d] Fully optimized geometry. [e] ⟨43MI41200⟩.

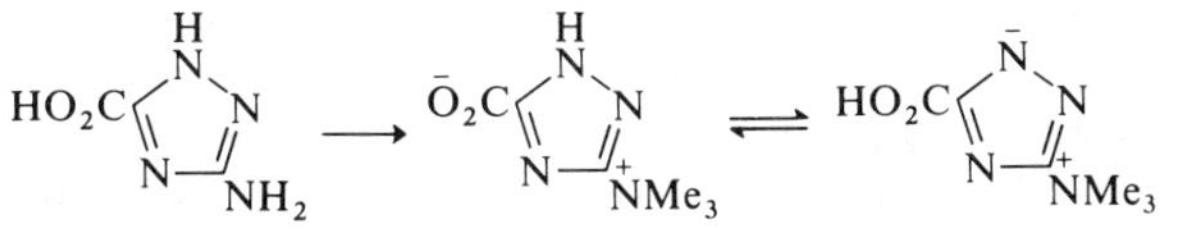

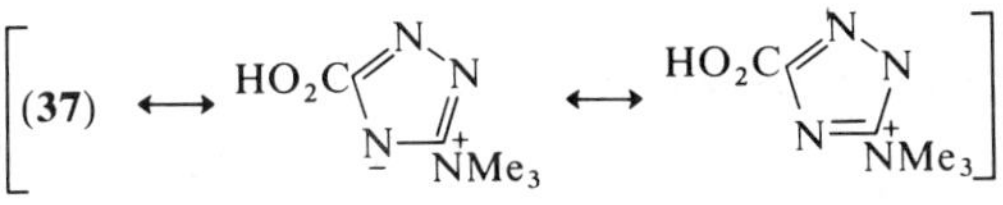

Scheme 10

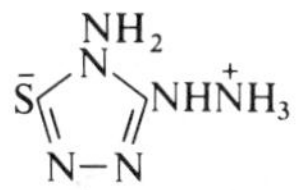

of tautomerism, and the negative charge is carried on O⁻, S⁻, RN⁻ or a carbanion. For graphical convention one may write (**40**) or (**41**), the latter implying the notional addition of water ⟨81HC(37)1, p. 599⟩.

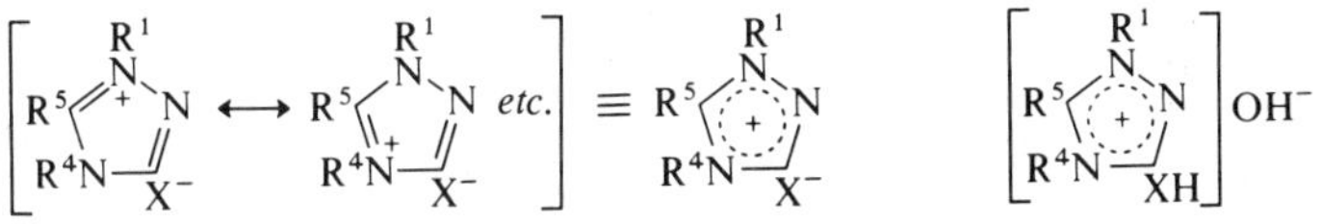

The first syntheses of mesoionic compounds included 1,2,4-triazoles and their oxa and thia analogues. The structure postulated at first as (**20**) was rejected on stereochemical grounds and on consideration of the large dipole moment, to be replaced by a charged monocyclic structure, defined as mesoionic and considerably extended to a great variety of theoretically and practically important examples ⟨57QR15, 76AHC(19)3⟩.

Establishment of structures such as (**40**) requires care as the usual methods of preparation do not exclude formation of non-triazolic isomers such as (**42**). Spectroscopic studies ⟨67AJC1779, 67JOC2245⟩, an alternative synthesis of (**40**; R^1, R^4 = Ph, R^5 = H) ⟨67TL4261⟩ and those of isomeric non-triazole heterocycles (**42**) provide conclusive proof.

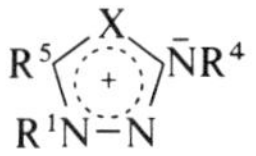

(**42**) X = O, S

4.12.3 REACTIVITY

4.12.3.1 General Considerations

On comparison of simple 1,2,4-triazoles with benzene, replacement of CH=CH in the latter by NH enhances reactivity towards electrophilic reagents but two replacements of CH by N act in the opposite direction (see Section 4.02.1.1). Comparison with other azoles is of greater heuristic value still ⟨B-76MI41200⟩.

Deactivation against electrophilic attack accounts for the difficulty or failure of nitration, sulfonation and N-oxidation of 1,2,4-triazoles proper. However, triazolate anions react readily with electrophilic reagents: alkylation and acylation have received much attention but halogenation and addition reactions less. Systematic study of the formation and reactions of salts and metallic complexes is of recent origin.

Substituents on C are exchanged in nucleophilic reactions under mild conditions and in the absence of activating groups. Reactivity of the ring towards nucleophiles is enhanced in triazolium cations and mesoionic triazole derivatives.

If in the absence of detailed mechanistic studies one cannot safely assign the site of the initial attack to C or N, it is best to consider such reactions under the heading of ring stability. The reactivities of ring atoms and those of the substituents they carry are interlinked, and it is a matter of convenience whether to regard certain reactions as chemical properties of the former or of the latter. In general, reactions involving movement of substituents to and from ring atoms will be considered under reactions of the ring atoms concerned.

Tautomerism of compounds with formal substituents such as OH, SH and NHR allows their classification not only as aromatic but also as reduced triazoles. As a matter of economy, it is best not to duplicate reference to such materials even though some of their reactions are characteristic of the aromatic tautomer while others are the outcome of the triazoline or triazolidine structure.

Protonation of triazoles as bases or deprotonation to afford triazolates or mesoionic derivatives are aspects of reactivity but usually, also here, mentioned under acidity (Section 4.12.2.4.2).

4.12.3.2 Reactivity at the Ring Atoms

4.12.3.2.1 Stability of the ring

From syntheses of 1,2,4-triazoles at raised temperatures one may infer relative thermal stability at about 300 °C. Although substituents and their location on the ring may undergo changes on heating, the ring itself is unaffected. Triazolidines are less stable as shown by the thermal cleavage of (**43**; Scheme 11) ⟨73BCJ2215⟩.

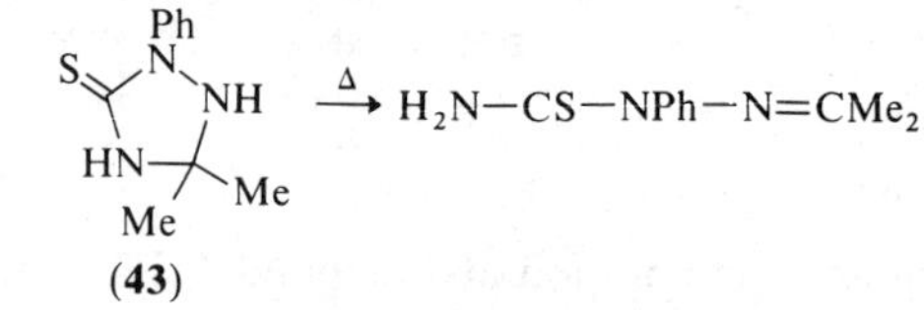

Scheme 11

Insensitivity to photolysis is suggested by the formation of 1,2,4-triazoles from some other heterocycles. However, mesoionic 4-aryl-1-phenyl-1,2,4-triazolin-3-ones (**44**) are decomposed on photolysis as in Scheme 12 ⟨72CC498⟩; triazolinethiones lose sulfur only ⟨70AJC631⟩.

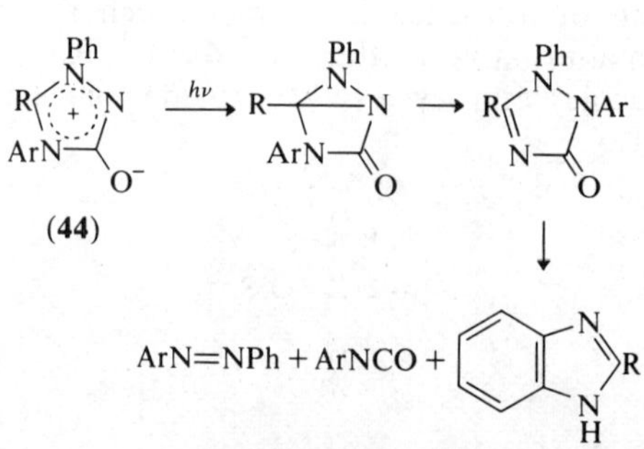

Scheme 12

The 1,2,4-triazole ring survives oxidative destruction of aromatic substituents and rings fused to it but destabilization by a nitrene intermediate (**45**) permits the oxidative cleavage of 4-aminotriazoles (**46**; Scheme 13) ⟨70TL3851⟩. Another instance of destabilization of the ring by a nitrene substituent (Scheme 14) is the thermal decomposition of azides (**47**) to

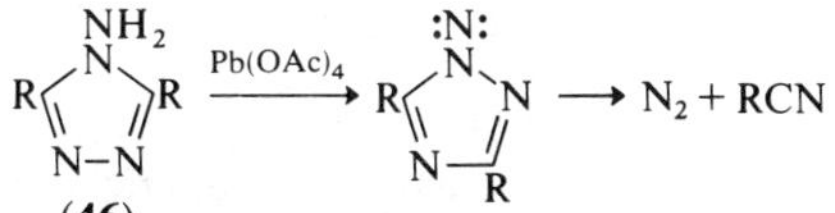

1,2,4-Triazoles

Scheme 13

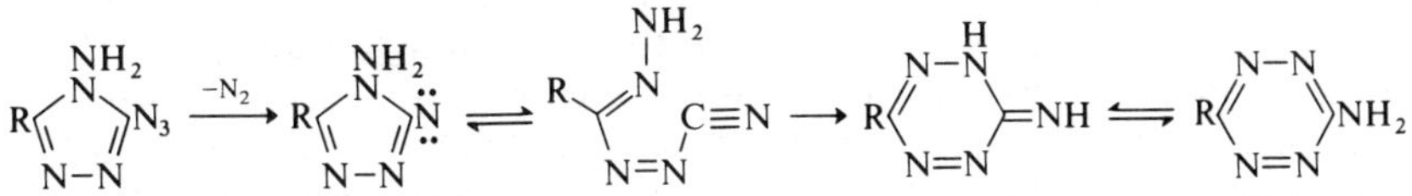

Scheme 14

tetrazines (**48**) ⟨66TL5369⟩. Triazolidines such as (**43**) are readily oxidized to acyclic compounds ⟨74M427⟩.

Triazolinones are more sensitive to reduction than aromatic triazoles. Paradoxically, reduction of the triazolinone (**49**) with lithium aluminum hydride (Scheme 15) affords the more unsaturated triazole (**50**) ⟨71BSF3296⟩.

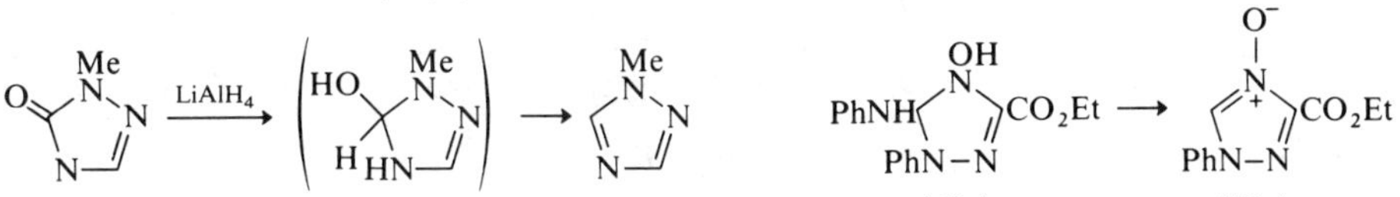

Scheme 15

The drive to aromatization accounts for the elimination of water or amine from substituted triazolines. Loss of water is illustrated in Scheme 15; aniline rather than water is lost from (**49a**; Scheme 15) to give (**50a**) ⟨65CB642⟩.

The first reported chemical cleavage of 1,2,4-triazoles subjected to the Schotten–Baumann reaction has been refuted. However, 4-benzylaminotriazolyl arenesulfonates are easily cleaved thanks to the loss of aromaticity in the triazolium salt ⟨63JOC543⟩. Another method is quaternization followed by treatment with concentrated alkali ⟨54CI(L)1458⟩ as in Scheme 16.

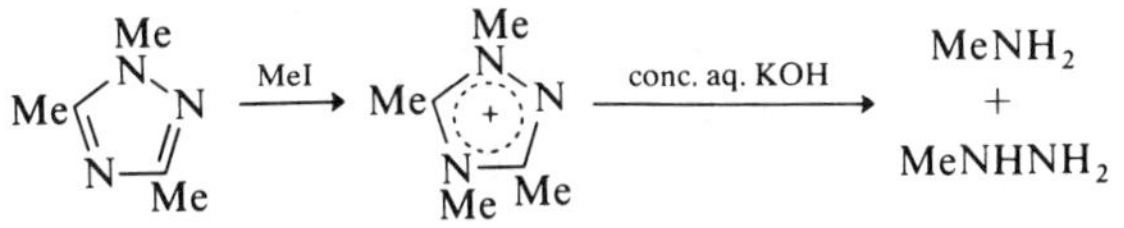

Scheme 16

It is easier to cleave triazolinones (Scheme 17) which may be regarded as cyclic amides undergoing acid hydrolysis at rates qualitatively predictable from the electronic effects of their substituents ⟨65LA(682)123⟩.

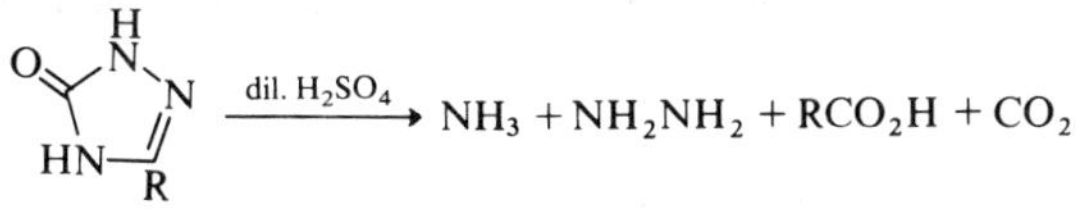

$$\text{triazolinone} \xrightarrow{\text{dil. } H_2SO_4} NH_3 + NH_2NH_2 + RCO_2H + CO_2$$

Scheme 17

Ring expansion by reaction with dichlorocarbenes noted in the pyrazole and imidazole series does not appear to occur with 1,2,4-triazoles but leads to the formation of tristriazole (**51**) ⟨69JCS(C)2251⟩ or bistriazole (**52**; Scheme 18) ⟨72BRP1269619⟩. The first of these reactions is induced by a free radical reagent but the mechanism of the second one is in doubt.

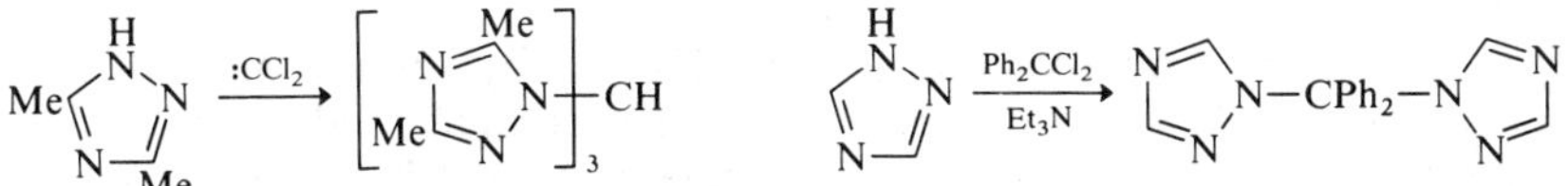

Scheme 18

4.12.3.2.2 Electrophilic attack at nitrogen

Alkylation and arylation including quaternization are well explored, and acylation has also attracted much interest. The other classes of reactions to be considered here are addition reactions involving annular NH as the addend, N-halogenation and reactions involving metallic centres.

Alkylation of triazolic NH ⟨61CRV87, B-76MI41200, p. 81⟩, either by diazomethane or sources of carbonium reactants, introduces substituents at N-1 rather than N-4. Although alkylation of simple 1,2,4-triazoles at N-4 is rare and of little practical significance, it has been observed ⟨54JCS141⟩. The observed preference for N-1 agrees with a statistical consideration which does not explain the rarity of 4-alkylation or follows from theory ⟨70JOC2246⟩. More difficult is the prediction when substitution on 3- and 5- allows a choice between N-1 and N-2 ⟨54JCS3319⟩ as in Scheme 19. Both of the expected products (**53**) and (**54**) are obtained in ratios that vary with the alkylating agent; a theoretically unexplained 'thumb-rule' suggests preference for methylation on the nitrogen adjacent to C-phenyl when diazomethane is employed. Industrial alkylating agents include epoxides ⟨68USP3394143, 74JAP74126679⟩ and ethylene carbonate ⟨74JAP7420175⟩.

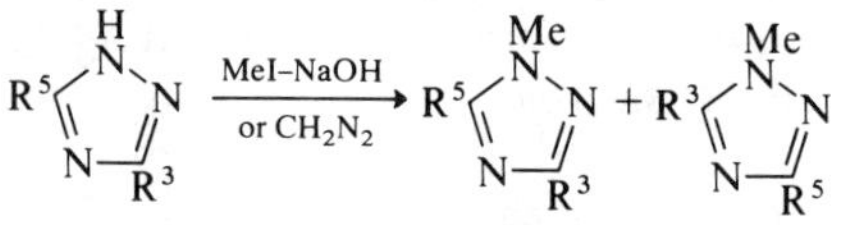

Scheme 19

N-Alkylation of 3-halotriazoles ⟨75BSF647, 75EGP111074⟩ gives N-1, N-2 and N-4 isomers, the last of which is obtained in a yield of up to 60% if methylation is carried out with dimethyl sulfate in the absence of base.

In some cases the observed alkylation on N-4 does not lessen the validity of the generalization that suggests preference for N-1. Thus 3-phenyl-5-ureido-1,2,4-triazole (**55**) is methylated or benzylated to give (**56**; Scheme 20) because the internal hydrogen bridge preempts the favoured N-1 site.

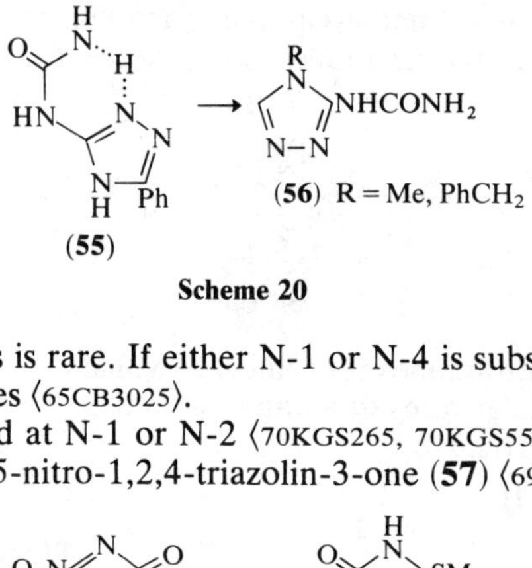

Scheme 20

Alkylation of triazolinones is rare. If either N-1 or N-4 is substituted, alkylation affords 1,4-disubstituted triazolinones ⟨65CB3025⟩.

Nitrotriazoles are alkylated at N-1 or N-2 ⟨70KGS265, 70KGS558⟩ but methylation occurs at N-1 or N-4 in the case of 5-nitro-1,2,4-triazolin-3-one (**57**) ⟨69KGS159⟩.

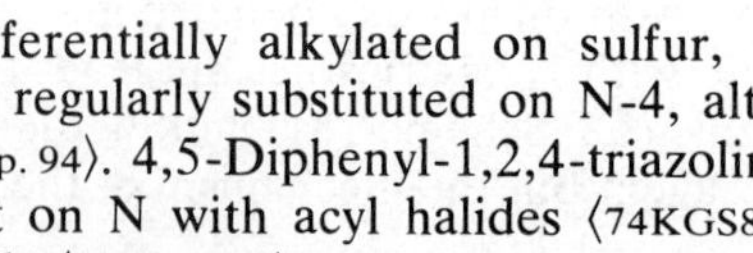

Triazolinethiones are preferentially alkylated on sulfur, but 3-methylthio-1-phenyl-1,2,4-triazolin-5-one (**58**) is regularly substituted on N-4, although some O-methylation occurs as well ⟨B-76MI41200, p. 94⟩. 4,5-Diphenyl-1,2,4-triazoline-3-thione is acylated on S in the presence of base but on N with acyl halides ⟨74KGS851⟩. The 4-phenyl-5-benzyl analogue is acylated on N only ⟨75KGS844⟩.

Generalizations based on alkylations with methyl, ethyl, benzyl and other hydrocarbon groups are not safely extended to more complex alkylating agents. Thus tetraacetylglucopyranosylation of 3-methyl-4-phenyl-1,2,4-triazoline-5-thione (**59**; Scheme 21) substitutes on S to give (**60**) which rearranges to the N-substituted isomer (**61**) on treatment with mercury(II) salt ⟨75MI41200⟩. Substitution on N rather than S takes place

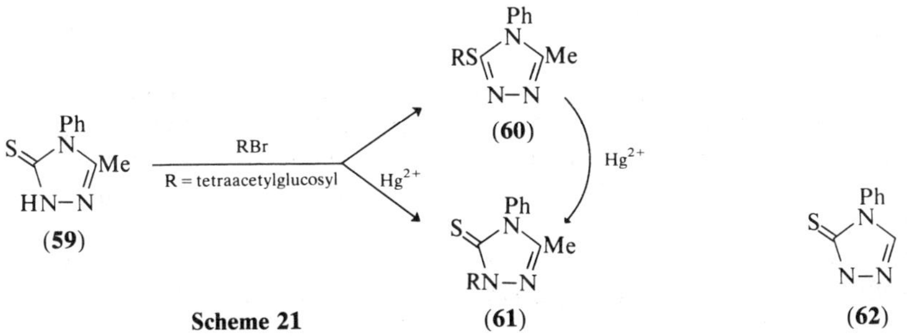

Scheme 21

when the alkylation of 4-phenyl-1,2,4-triazoline-5-thione (**62**) is being carried out with epoxide ⟨71KGS700⟩.

Methylation of 3-amino-1,2,4-triazole does not afford the methylamino derivative but all the three annular *N*-methyl derivatives are obtained in ratios that depend on the basicity of the reaction mixture ⟨73BSF1849⟩. Similar results have been obtained in the case of ribosylation and other instances ⟨81HC(37)1 p. 150⟩. Alkylation of aminotriazoles on N-4 is rare.

Activated aromatic halides and 2- or 4-pyridyl halides can be used for the *N*-arylation of 1,2,4-triazoles, best achieved by the Ullmann technique ⟨70JCS(C)85⟩. Preference for annular arylation on N-1 rather than N-4 is observed but arylations of 3-aminotriazoles with picryl chloride are anomalous: the parent triazole itself is substituted on N-1 but in 3-aminotriazole the amino group is attacked first, then the adjoining ring NH. Since 4-aminotriazole lacks annular NH groups, picrylation affects the substituent amino group only ⟨68JHC199⟩.

Triazoles with alkyl, aryl or acyl substituents on N-1 or N-4 can be quaternized. Because of the mesomeric distribution of the positive charge on triazolium compounds, representations such as (**63**) are convenient but the equivalent formula (**64**) may be used to denote the site at which quaternization has taken place (the mesomeric nature of (**64**) must be kept in mind of course). Trialkyloxonium tetrafluoroborates are powerful quaternizing reagents ⟨70JOC2246⟩.

Quaternization takes place so as to maximize the distance between substituents on annular N. Thus a triazole substituted on N-1 is quaternized on N-4 and vice versa. However, quaternization of triazolinones and triazolinethiones or diquaternization of aromatic triazoles may involve N-1 and N-2. Structural proofs of quaternized compounds are simple in cases such as that of (**63**): ready removal of the acetyl group from N-1 leaves (**65**). If then (**63**) arose from the methylation of (**66**), the quaternized site is determined unambiguously.

In early investigations, when none of the substituents on N could be removed by a selective method, interpretation of the quaternized structure has led to errors. Thus quaternization of 1,3,5-trimethyl-1,2,4-triazole (**67**; Scheme 22) could afford either or both of (**68**) and (**69**). Degradation as in Scheme 16 proves (**69**), but not quite conclusively as the reported formation of ammonia and 1,2-dimethylhydrazine is consistent with either the alternative structure of (**68**) or rearrangement between (**68**) and (**69**) under certain experimental conditions. Conclusive proof is obtained from NMR spectroscopy which distinguishes between symmetrical (**68**) and unsymmetrical (**69**) or from a chemical method ⟨59JCS3799⟩ based on the reactivity of a *C*-methyl situated between two substituted N-centres as in (**69**): cyanine dye formation from the quaternized product of (**67**) establishes the structure (**69**) for the latter.

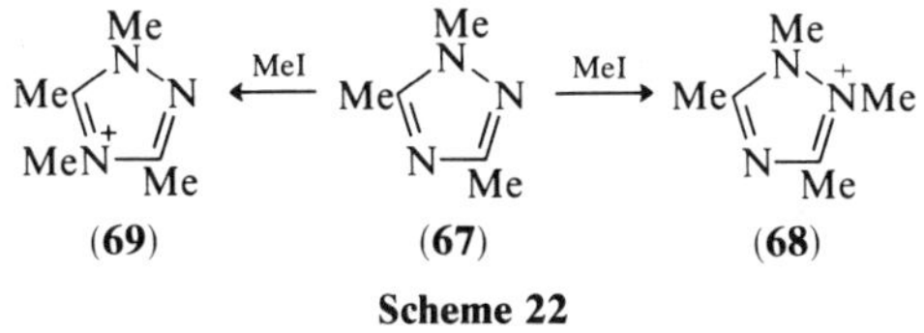

Scheme 22

Quaternization of 4-aminotriazoles at N-1 ⟨66MI41200⟩ or that of 5-amino-1-methyl-1,2,4-triazole at N-4 ⟨63MI41200⟩ follows the rule predicting attack on N-centres at maximum separation. Much the same applies to the quaternization of azo dyes derived from triazole ⟨81HC(37)1, p. 609⟩ except that 1,4- and 2,4-patterns of substitution are equivalent under this rule; even the 1,2-isomer may be formed.

Alkylation of 3-methyl-4-phenyltriazolin-5-one (**70**) with dimethyl sulfate (Scheme 23) affords not only the neutral 1-methyl derivative (**71**) but also the mesoionic compound (**72**) ⟨66AC(R)190⟩. On the other hand the isomeric 1-methyl-5-phenyltriazolin-3-one (**73**) affords the mesoionic compound (**74**) with methyl iodide but the methoxytriazole (**75**) with diazomethane ⟨76CPB1336⟩.

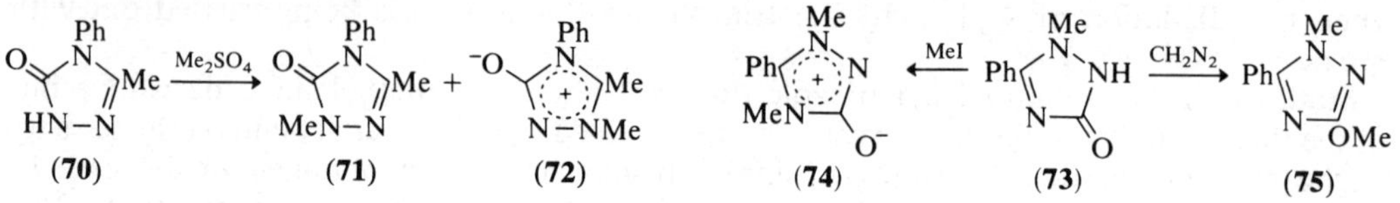

Scheme 23

In the absence of NH and potential OH, 4-phenyl-1,2,4-triazoline-3,5-dione (**76**) is converted into the unusual quaternary compound (**77**) on treatment with diphenyl-diazomethane (Scheme 24) ⟨65G33⟩. Note that (**77**) is a triazolidinium zwitterion with both charges accommodated on the nucleus, hence distinct from mesoionic compounds (**40**).

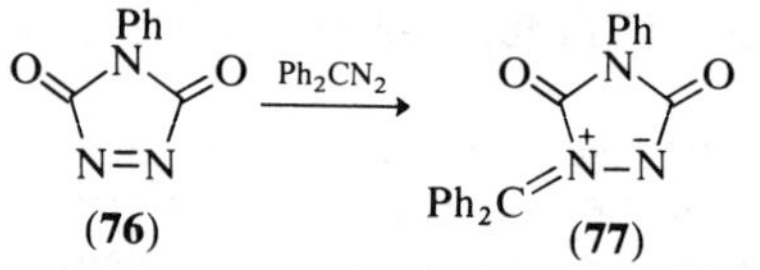

Scheme 24

Quaternization of thiones or thiols proceeds successively through alkylation to methylthio compounds to attack on annular N centres ⟨59JCS3799, 61LA(641)94, 67JOC2245⟩ to triazolium salts or mesoionic compounds. However, 3,5-bis(methylthio)-1-phenyl-1,2,4-triazole could not be quaternized ⟨54CI(L)1458⟩.

Diquaternization is possible. The structure (**78**) implies mesomeric distribution of two charges over the ring. This structure is inferred from hydrolysis as shown in Scheme 25 ⟨72JOC2259⟩.

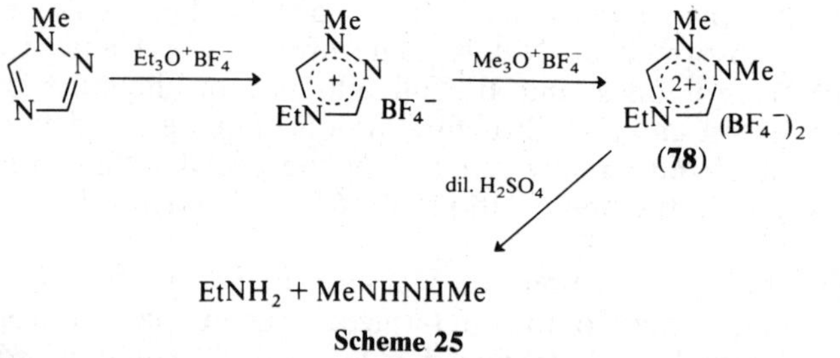

EtNH$_2$ + MeNHNHMe

Scheme 25

Although the mechanism of addition of triazolic NH or nucleophilic triazolate to polarizable π-systems is not known, one may regard such reactions as formally akin to alkylations. Vinylation of 3-aminotriazole with acetylene in the presence of base affords mainly the 1-vinyl compound in addition to some of the 4-isomer ⟨75MIP464584, 75JAP7504074⟩.

Catalytic additions to acrylic acid and arylalkenes are thought to afford N-1 derivatives ⟨55JA2572⟩. Acrylonitrile cyanoethylates 3,4-diphenyltriazoline-5-thione (**79**) on N-1 ⟨75KGS844⟩. Hydroxymethylation with formaldehyde (Scheme 26) also favours N-1 triazolinethiones such as (**80**) ⟨75KGS1000⟩, but similar reactions occur on N-4 if N-1 is substituted.

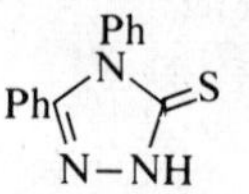

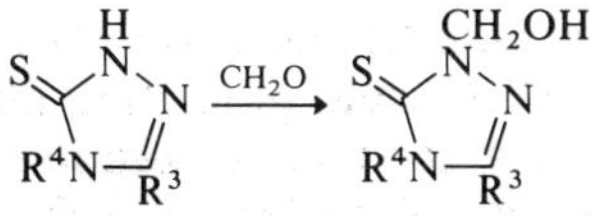

1,2,4-Triazoles

Scheme 26

An important type of *N*-alkylation involves the addition of cyclic or acyclic C=CH to reactive triazolinediones (**76**; Scheme 27) ⟨81HC(37)1, p. 530⟩. However (**76**) may behave as the π-system to which piperidine adds to yield (**81**) ⟨76JHC673⟩ or again it acts as a powerful dienophile affording (**82**) in Scheme 27 ⟨75JCS(P2)1325⟩.

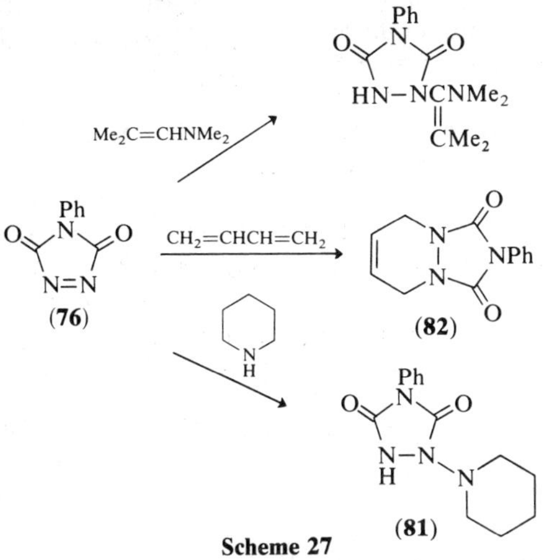

Scheme 27

Except for hydroxymethylation (Scheme 26) and self-condensation (Scheme 6), 1,2,4-triazoles do not give definable products on reactions with aldehydes but the resins formed are worth closer investigation. Reactions of aminotriazoles with β-diketo compounds are used to prepare some fused triazoles (see Chapter 4.15).

All the general references cited devote much space to acylation, mostly performed by reaction of anhydrides or acyl halides with triazolates or treatment of triazoles in inert solvents with acyl halides including thiobenzoyl chloride, ethyl chloroformate and phosgene. In the last mentioned case *N*,*N*-carbonylditriazole (**83**) is formed exothermally; acylation at N-1 has been proved by NMR spectroscopy ⟨62JOC2631⟩. Carbonyldiazoles are acylating agents: transacylation between (**83**) and benzoyl chloride affords 1-benzoyltriazole ⟨62CB1275⟩. The preparation of 1-acetyl-1,2,4-triazole from the 1-trimethylsilyl derivative and acetyl chloride ⟨60CB2804⟩ is another practical example of transacylation.

Acylation of triazolates is expected to occur at N-1 but transacylation by displacement of an electrophilic substituent from N-1 might be predicted to proceed by attack of the electrophile on N-4. The actual course of the reaction may be explained on the analogy of isotopic studies on imidazole ⟨66CB2955⟩ applied to triazole (Scheme 28): the first product of the sequence is the mixed anhydride (**84**) that reacts with the liberated triazole at N-1 to afford benzoyltriazole (**85**) and the unstable carbamic acid (**86**).

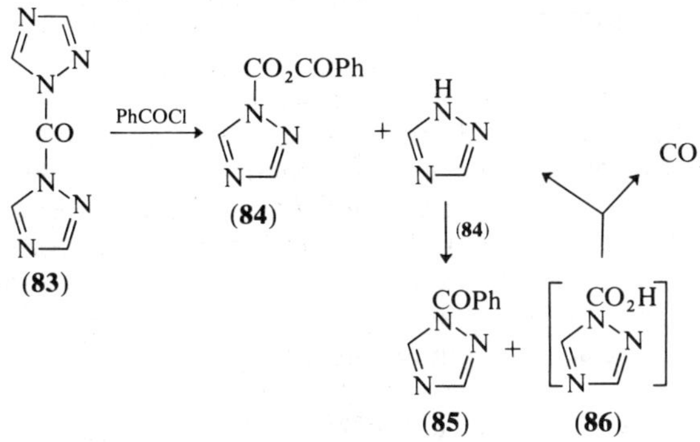

Scheme 28

Acylation with potassium cyanate, phenylisocyanate and similar reagents gives ureides ⟨81HC(37)1, p. 102⟩. According to experimental conditions, cyanogen affords the carboxamide (**87**). The anomalous reaction with *o*-anisylisocyanate results in the isourea (**88**).

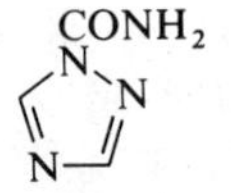

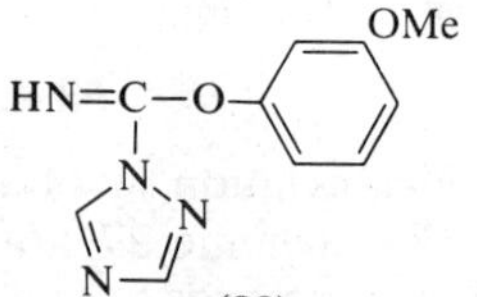

Notwithstanding controversies on *O*- against *N*-acylation of triazolinones, modern views favour the former. The substantial number of 1,2,4-triazole derivatives of phosphorus acids ⟨81HC(37)1, p. 241, 247, 316, 357⟩ are acylated on oxygen even when alternative *N*-acylation would have been possible.

This generalization does not extend to 1,2,4-triazolidine-3,5-diones: both the 1-phenyl- and 4-phenyl-urazoles (**89**) and (**90**) can be diacetylated to (**91**) and (**92**), respectively, as in Scheme 29 ⟨70JHC821, 66AP441⟩; the first of these reactions occurs with a yield of 69% while the yield in the second is almost quantitative. Phosgene does not afford carbonyl-ditriazolidines but carbamyl chlorides such as (**93**) which are useful intermediates in the synthesis of areas and urethanes ⟨74AP504⟩.

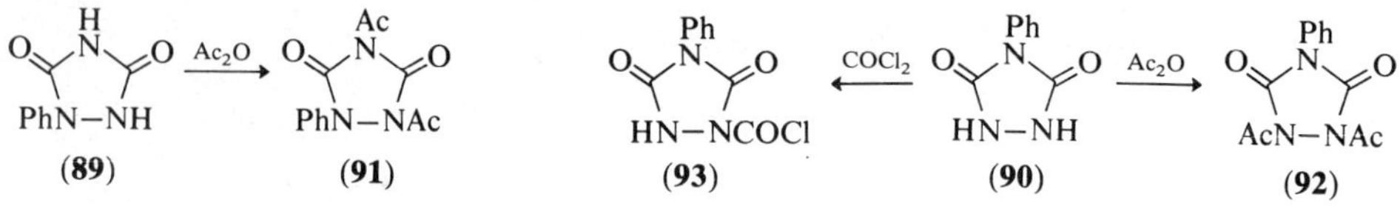

Scheme 29

Some unexpected *N*-acylations (Scheme 30) came to light in studies of 4-phenyl-1,2,4-triazoline-3,5-dione (**76**). Addition to vinyl acetate is followed by intramolecular acylation to (**94**), finally yielding (**95**) ⟨72JOC1454⟩, while reaction with methanol resulted in a 1,2-dicarboxylic acid (probably through NCH₂OH formed by reaction with formaldehyde) isolated as (**96**) ⟨76CC326⟩.

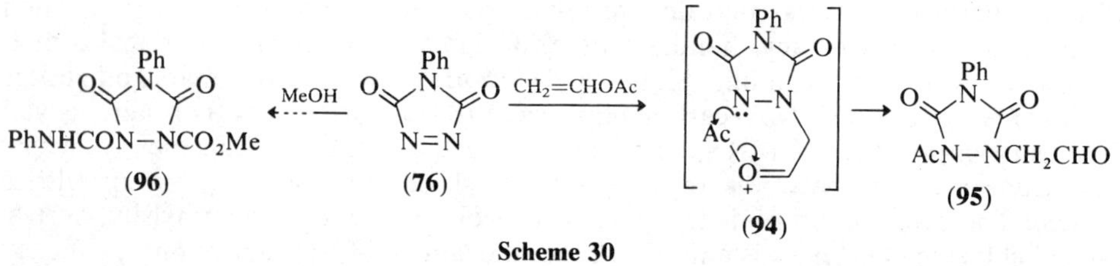

Scheme 30

Acylations of triazolinethiones by acyl halides, isocyanate or isothiocyanate under mild conditions take place on sulfur, but higher temperatures appear to favour *N*-acylation as in Scheme 31 ⟨74KGS851⟩.

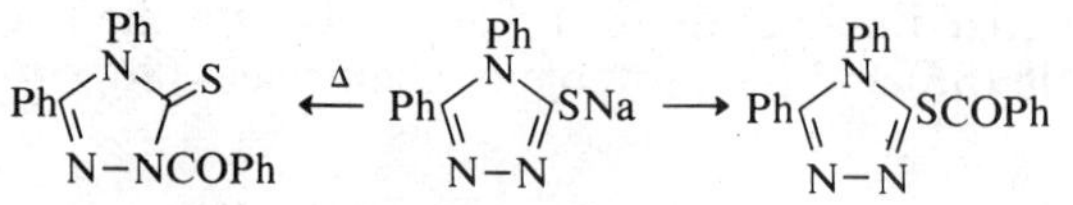

Scheme 31

The acylation of 3-aminotriazole (**97**) has received much attention and generated much controversy ⟨B-76MI41200, p. 120⟩. The general pattern for N-1 or N-2 acylation (Scheme 32) has been substantiated ⟨70JHC1149⟩ through the use of mixed acylating agents: ordinary and deuterated acetyl chloride produce (**98**)–(**104**), and action of trifluoroacetic anhydride on acetylated aminotriazoles (**98**), (**103**) and (**107**; X = Ac) affords (**105**), whence (**106**). Products of hydrolysis (**107**) are reacylated as shown. The results may be summarized in noting the rapid exchange of acyl on annular nitrogen as against the slow acylation of the substituent amino group.

Acylation of guanazole (**6**) is complex ⟨60RTC836⟩; although acetylation on N-1 or N-2 has not been rigorously established, the assignments are in agreement with later findings

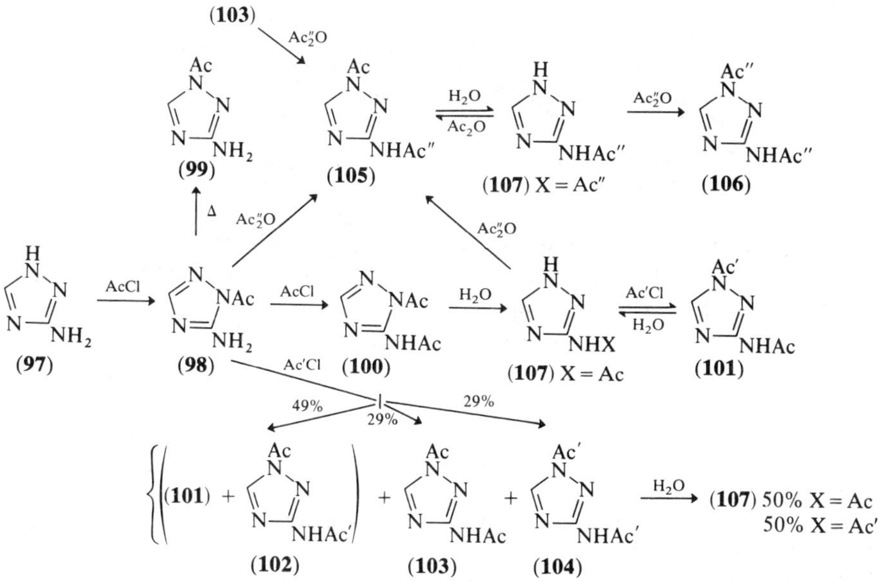

1,2,4-Triazoles

Scheme 32

Ac = MeCO; Ac' = CD$_3$CO; Ac" = CF$_3$CO

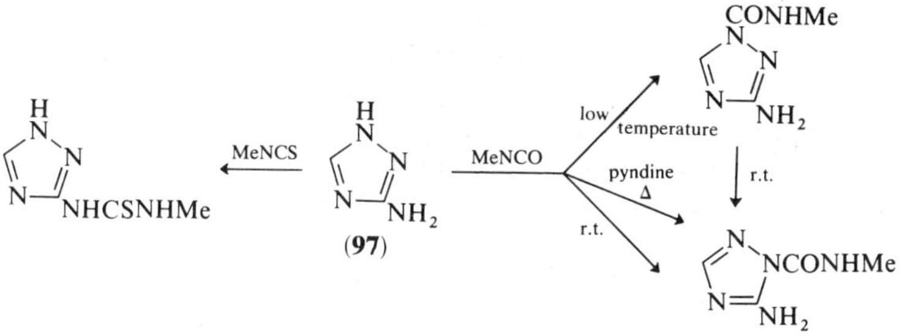

Scheme 33

such as those cited above, and partly with reactions of 3-aminotriazole with methyl isocyanate and methyl isothiocyanate summarized in Scheme 33 ⟨72JHC99, 73JCS(P1)1209⟩.

Acylation of 4-amino-1,2,4-triazole occurs on the 4-NH$_2$ group ⟨63JOC543⟩; the result is not trivial as alternatives such as triazolium salt formation are conceptually possible.

Benzoylation of hydrazinotriazoles ⟨74AJC2447, 75AJC133⟩ is in agreement with the general behaviour of aminotriazoles.

In the industrially important reaction of bis(dimethylamino)phosphoryl chloride with 3-anilinotriazole (**108**) attack at the hydrazinic sites gives a mixture of (**109**) and (**110**) as in Scheme 34 ⟨66RTC429⟩. If R = H, the main product is (**110**), but for alkyl substituents the ratio (**109**):(**110**) is 2–6, presumably for steric reasons.

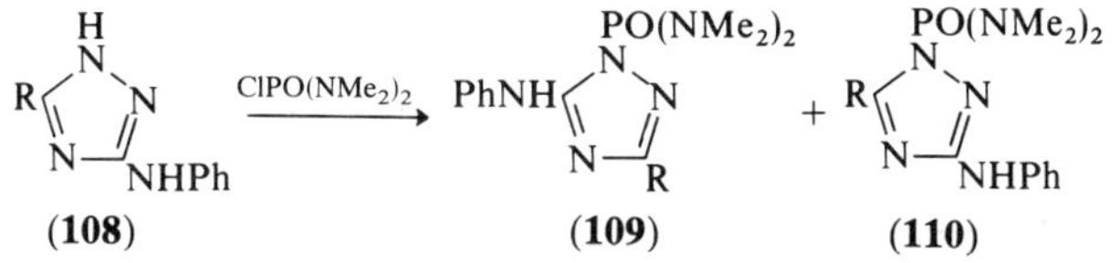

Scheme 34

A brief review ⟨B-76MI41200, p. 125⟩ does not resolve controversies on the annular arenesulfonation of 1,2,4-triazoles, only the substituent is acylated in the case of 4-aminotriazole ⟨62ZOB447⟩. For a more extensive bibliography see ⟨81HC(37)1, p. 260, 284⟩.

Although electrophilic halogenation of 1,2,4-triazoles is mostly intended as a preparative method for *C*-halo triazoles, some *N*-halo derivatives can be isolated. Chlorination in aqueous potassium hydrogen carbonate solution gives *N*-chloro-1,2,4-triazole, assumed to be the 1-compound which rearranges on heating to the *C*-halo isomer ⟨69KGS1114⟩ by a second order reaction in water to *t*-butyl alcohol ⟨72JPR923⟩. Excess of chlorine in alkaline solution allows the reaction to proceed to 1,3-dichloro- and 1,3,5-trichloro-1,2,4-triazoles. N-Chlorination occurs with 3-phenyltriazole ⟨67ZC184⟩.

Under moderate conditions, e.g. reacting potassium triazolate at 0 °C, bromine affords N-bromotriazoles. If the triazole is 3,5-substituted, N-bromotriazole may be obtained without special care; otherwise C-bromination occurs ⟨67CB2250⟩, presumably through N-bromo intermediates since 1,3-dichlorotriazole and 1,3,5-tribromo-1,2,4-triazoles may be used for the halogenation of other triazoles ⟨69ZC325⟩. Bromination of the 1,2,4-triazole-3-carboxylic acid has been shown to occur at N-1 ⟨69ZC300⟩.

In the absence of nuclear NH groups susceptible to N-bromination, as in the case of 4-phenyl-1,2,4-triazole, N-bromotriazolium bromide (**111**) is formed first. This decomposes on heating to 3-bromo-4-phenyl-1,2,4-triazole ⟨75BSF647⟩.

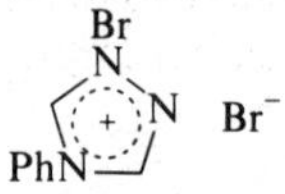

(**111**)

Iodination with iodine chloride in ethanol or aqueous alkali gives the relatively stable N-iodotriazole which does not rearrange to the C-halo isomer on heating. This acts as an iodinating agent when warmed with triazole ⟨69ZC300⟩.

For the somewhat analogous nuclear N-nitration, see Section 4.12.3.2.3.

Since nitrogen centres are responsible for the acidic and basic properties of 1,2,4-triazoles, salt formation may be considered to involve reactions at such centres. A number of inorganic salts and complexes have been noted ⟨00CB58⟩ but their exact nature and reactivity are not quite clear.

If we symbolize the parent 1,2,4-triazole as TrH (where H denotes the acidic proton), TrH·HCl may be written as $(TrH_2)^+Cl^-$ and liberates the free base on treatment with sodium hydroxide or sodium ethoxide in ethanol. This is not the case for TrH·HNO$_3$ which on such treatment affords TrH·NaNO$_3$ or less likely TrNa·HNO$_3$. Compounds such as TrH·HNO$_3$·2HgCl$_2$ and 3TrH·HNO$_3$·4AgNO$_3$ are even more surprising. Mercury(II) chloride complexes have been used as aids in the purification of triazoles, but neither their nature nor the geometry of their undoubted attachment to N centres is known, and the same applies to the copper salts noted in some of the earliest investigations of triazoles. The formation and structure of Tr$_2$Sn are known ⟨72CC544⟩.

Molten 1,2,4-triazole has been tried as a solvent in high temperature chemistry. It is merely a solvent for some simple salts such as sodium chloride but transition metal halides are complexed in an unknown manner ⟨66CI(L)600⟩. For derivatives of 1-phenyltriazole see Section 4.12.5.2.4.

Free radical formation from triazoles has been little studied. Oxidation of triazolidinediones to triazolinones may afford a free radical if dehydrogenation is blocked as in Scheme 35 ⟨74JA3335⟩. The resulting radical is stable in solution and is isolated as its dimer.

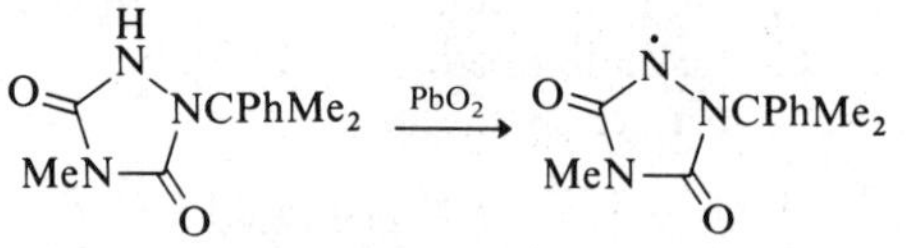

Scheme 35

4.12.3.2.3 Electrophilic attack at carbon

Nitration of triazole and its C-monoalkyl derivatives fails. Aryltriazoles are nitrated on the benzene ring but 3-p-nitrophenyltriazole (**112**) in which the benzene ring is deactivated affords the N-nitro derivative (**113**). This rearranges to the C-nitro compound (Scheme 36) through an isotriazole intermediate on kinetic evidence for a first order intramolecular rearrangement ⟨72CC37⟩.

Nitration of triazolinones (possibly reacting as 'phenolic' hydroxytriazoles) with fuming nitric acid affects the heterocyclic ring with simultaneous nitration of the benzene ring in 1- and 4-phenyl derivatives. However, 2- or 3-phenyl derivatives of 1,2,4-triazolin-5-one are nitrated in the benzene ring exclusively ⟨69CB755⟩.

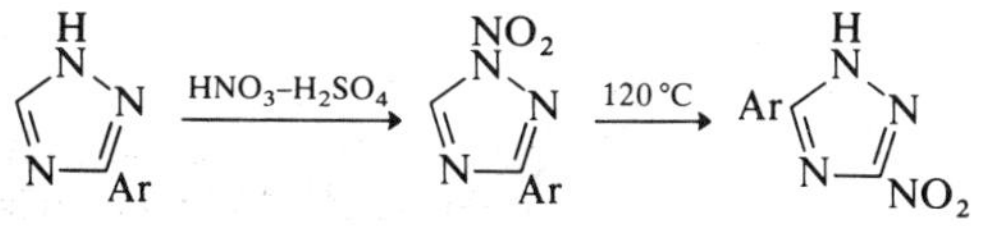

Scheme 36

(112) Ar = 4-NO$_2$C$_6$H$_4$ (113)

Halogenation proceeds through *N*-halotriazoles or triazolium derivatives that have been discussed under Section 4.12.3.2.4. Also the bromination of *N*-substituted triazoles by *N*-bromosuccinimide in chloroform ⟨75BSF647⟩ is thought to involve *N*-bromotriazolium intermediates. If so, the same applies to halogenations with *N*-halotriazoles ⟨69ZC325⟩.

Both 1,2,4-triazole and its 1-benzyl derivative undergo *C*-hydroxymethylation ⟨55JA1538⟩. The synthesis of (114) is one of the few examples of a Mannich reaction of azoles ⟨76JMC1057⟩.

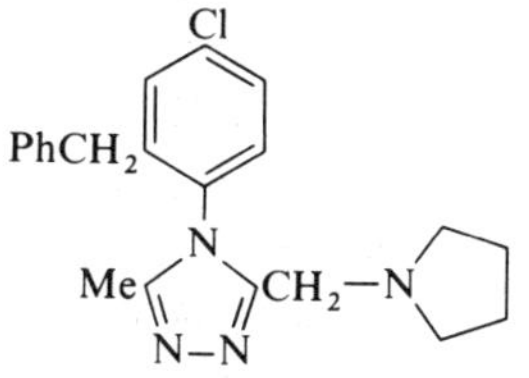

(114)

Some of the 'complexes' mentioned in Section 4.12.3.2 could include products of *C*-chloromercuration (*cf.* pyrazoles). Although *N*-unsubstituted triazoles are most likely to form triazolates on reaction with metalating agents, lithiation of 1-(2-pyridyl)-1,2,4-triazole (115) indicates the relative reactivities of the triazole and pyridine rings ⟨72AG(E)846⟩ and provides a method for the formation of bistriazoles (116; Scheme 37). Nitration of 1,4-dimethyltriazolin-5-one occurs in good yield, but that of the 1,2-dimethyl isomer fails. Similarly the *N*-benzoyltriazolium intermediate is formed in the *C*-benzoylation or dibenzoylation of 4-phenyl-1,2,4-triazole ⟨66G1084⟩. The reaction occurs on heating the triazole with benzoyl chloride and does not affect the benzene ring attached to N-4.

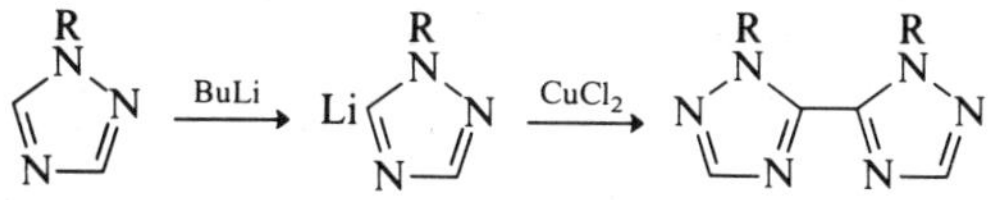

(115) R = 2-pyridyl   (116)

Scheme 37

4.12.3.2.4 Nucleophilic attack at carbon

Exchange of anionic substituents and generation of a transient carbonium ion are considered here. For some other, mechanistically less well understood reactions, see Section 4.12.3.3 which deals with the reactivity of substituents.

Many triazole derivatives are accessible by exchange of 3- or 5-halo, hydroxy, alkoxy, aryloxy, alkylthio, arylthio, amino or cyano groups ⟨67EGP59288, 69BRP1157256, 70KGS1701, 75BSF1649⟩. A kinetic comparison of haloazole, including halotriazole, reactivities is of considerable theoretical interest ⟨67JCS(B)641⟩.

The hydrolysis of 3-chloro-4,5-diphenyl-1,2,4-triazole requires the presence of mineral acid in order to increase the cationic nature of the ring ⟨62MI41200⟩. The same applies to interhalogen exchanges of halotriazolones ⟨67CB2250⟩. Thanks to a better leaving group, bromo compounds are more reactive; 5-bromo-1-methyl-1,2,4-triazole is more reactive than the isomeric 3-bromo-4-methyl compound.

The reaction of 3-iodo-1-methyl-1,2,4-triazole with 2-methylbut-3-yn-2-ol ⟨75IZV690⟩ has been used to prepare 3-ethynyl-1-methyl-1,2,4-triazole (117; Scheme 38) and the isomeric 5-ethynyl-1-methyl isomer from the corresponding iodotriazole.

An unusual feature of triazole chemistry is the ready exchange of nitro groups of nitro- and dinitro-triazoles and nitrotriazolinones ⟨67CB2250, 70KGS269, 70KGS1701⟩. The reaction of 1-methyl-3,5-dinitro-1,2,4-triazole with hydrazine ⟨70KGS997⟩ exemplifies such reactions

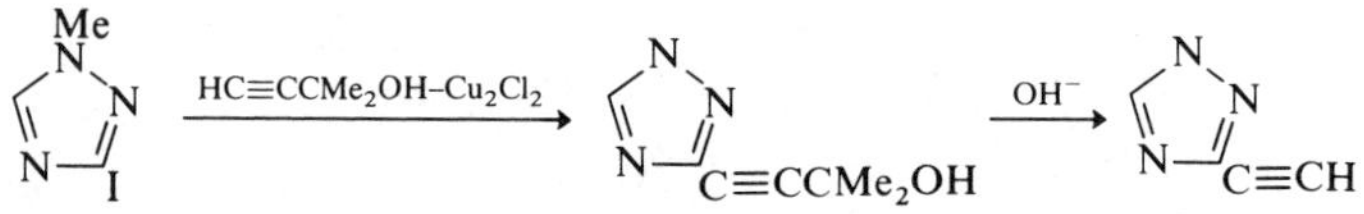

Scheme 38

in forming 5-hydrazino-1-methyl-3-nitro-1,2,4-triazole but reduction of a nitro group by the reagent also affords 5-amino-1-methyl-3-nitro-1,2,4-triazole.

Triazolinones are obtained by the alkaline hydrolysis of methyl-sulfinyl or -sulfonyl compounds (**118**; Scheme 39) ⟨73JMC312⟩. Hydrolysis of alkoxytriazoles to triazolinones ⟨68AP911⟩ is of practical importance, but the ready conversion of heteroaromatic triazoles to triazolinones may be complicated by the non-hydrolytic rearrangement of alkoxytriazoles such as (**119**) to (**120**) in Scheme 40 ⟨68JPR168⟩.

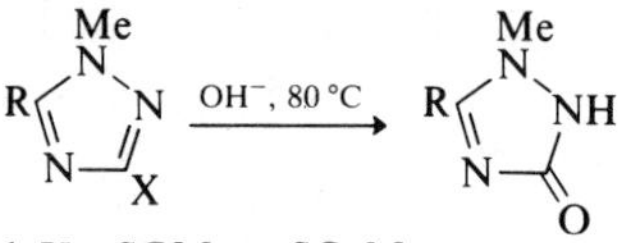

(**118**) R = 2-furyl; X = SOMe or SO$_2$Me

Scheme 39

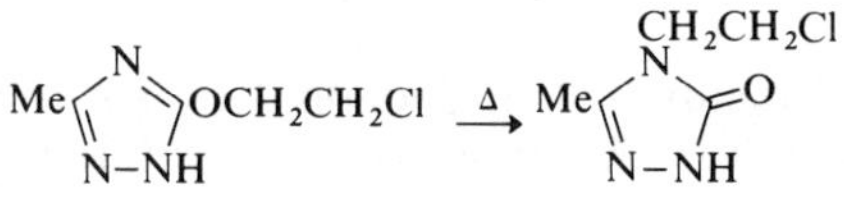

Scheme 40

The nitrosamine (**121**) and especially diazonium salts (**122**) prepared from it, as well as their alkyl and aryl derivatives, are important synthetic intermediates ⟨B-76MI41200, p. 325⟩. For conditions that decide between nitrosamine function or diazotization see ⟨73JCS(P1)1357⟩. At room temperature or in concentrated hydrochloric acid nitrosamines derived from triazoles are unstable, and their decomposition products include chlorotriazoles ⟨63LA(665)144⟩; 5-nitrosamino-3-phenyl-1,2,4-triazole affords the 5-bromo compound with hydrobromic acid at 0 °C and reacts violently with hydroiodic acid ⟨05LA(343)1⟩. Thermal arylation with triazole nitrosamines is possible ⟨73JCS(P1)1357⟩.

(**121**)   (**122**)

Halotriazoles have been prepared by treatment of (**122**) and alkyl- or aryl-substituted derivatives with halogen acids ⟨81HC(37)1, p. 228⟩. The preparation of 3-fluorotriazole required UV irradiation of (**122**; X = BF$_4$). Aminotriazolinones and 3,5-diaminotriazole are diazotized and tetrazotized, respectively, to afford 3-halo-1,2,4-triazolinones and 3,5-dihalotriazoles, respectively ⟨65LA(682)123, 67CB2250⟩.

Treatment of (**122**) with sodium nitrite affords 3-nitrotriazole ⟨69AJC2251⟩; other nitrotriazoles and dinitrotriazoles may be prepared by similar techniques ⟨70KGS259, 70KGS265⟩.

Arylation of triazole by the Gomberg–Bachmann method ⟨72MI41200⟩ has been described.

Direct oxidative thiation (Scheme 41) by heating sulfur with triazoles requires a substituent on nuclear nitrogen. The preparation of the thione (**123**) from 4-phenyl-1,2,4-triazole ⟨66MI41201⟩ is unambiguous, but 1-benzyltriazole affords the 5-thione (**124**) instead of the 3-isomer that would have been expected on steric grounds ⟨70JCS(C)2403⟩.

(**123**)   (**124**)

Scheme 41

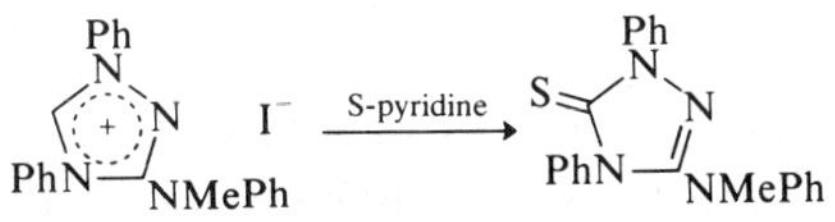

Thiation has been applied also to triazolium salts (Scheme 42) ⟨70ZN(B)421⟩. Analogous oxidation and amination reactions appear to be unknown.

4.12.3.3 Reactivity of Substituents

4.12.3.3.1 *C-Linked substituents*

The behaviour of alkyl groups situated on the triazole ring is comparable with those of benzene; thus their oxidation followed by decarboxylation has been much used in synthesis. Derived functions such as CH_2X (X = OH, hal) also behave as expected (*cf.* 55JA1538), but acetoxylation (Scheme 43) of 4-hydroxy-3-methyl-1,2,4-triazole (**125**) with acetic anhydride to give (**126**) is remarkable ⟨70JPR610⟩.

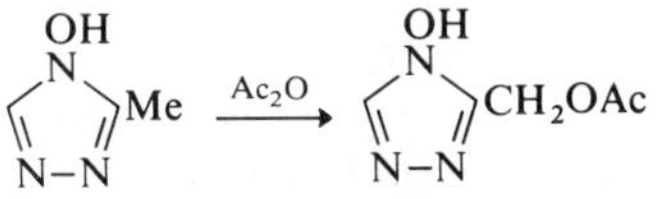

Quaternization leads to two interesting reactions of *C*-alkyl-1,2,4-triazoles. The diagnostic value of condensations undergone by a C—Me group situated between two substituted N centres has been mentioned (see Section 4.12.3.2); another example is the determination of the reactive site of (**127**) that could yield (**128**) or (**129**) on tritylation (Scheme 44). Quaternization at the N most distant from the tritylated N-1 or N-2 occurs at N-4 to afford (**130**) or (**131**). As the compound obtained does not form a Schiff base, its correct structure is (**130**) from which (**128**) is inferred as structure of the trityltriazole ⟨62JCS575⟩.

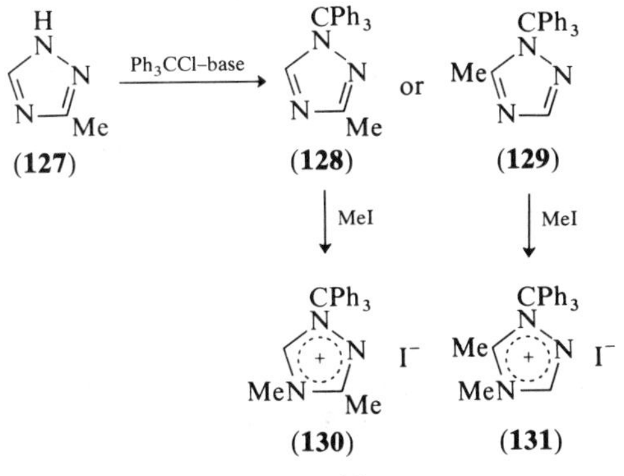

Attempts at dequaternization result in ring cleavage (Scheme 16). Alternatively ⟨71JCS(B)1648⟩ mesoionic compounds (**132**) or anhydro bases (**133**) are formed. These revert to the triazolium compound on treatment with perchloric acid (Scheme 45).

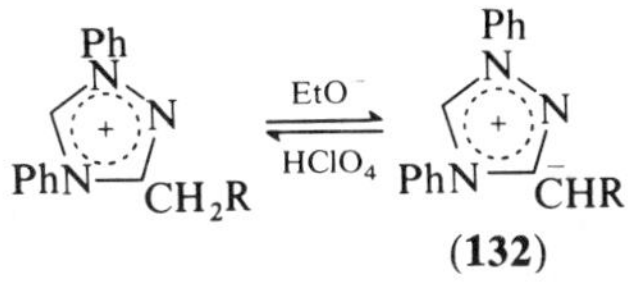

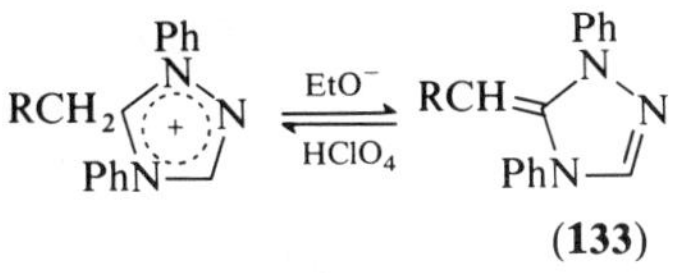

Both *C*- and *N*-aryltriazoles may have the aryl group removed by oxidation, preferably after nitration and reduction to aminoaryl. However, nitration of triazolinone rings may occur in competition with that of aryl substituents.

N-Nitroaryl groups are relatively easy to cleave by basic reagents ⟨71ZC153⟩ but the formally similar 2-pyridyl compounds resist basic and acidic hydrolysis ⟨72AG(E)846⟩.

Alkenyl- and alkynyl-triazoles have received little attention. By analogy with the behaviour of other azoles they are expected to polymerize but to be less reactive in addition reactions than alkenes or alkynes. Although the most promising polymers derived from triazoles are obtained by different methods (see Section 4.12.5.2.3), some information is available on potentially polymerizable vinyltriazole ⟨63MI41201⟩. The styryltriazole (**134**) could be oxidized to 3-methyl-1-phenyl-1,2,4-triazole, *i.e.* without affecting either the triazole or *N*-phenyl ring, but hydroxylation of the alkene chain failed ⟨54JCS4256⟩.

In some respects the behaviour of aldehydes and ketones derived from 1,2,4-triazole is comparable with that of analogous benzene derivatives. One of the major differences is the dimerization of *N*-unsubstituted 1,2,4-triazolealdehydes (Scheme 6) ⟨71AJC393⟩. Although both *N*-substituted and unsubstituted *C*-acyltriazoles give dinitrophenyl-hydrazones, only the former can be characterized as oximes; the carbaldehyde oximes resist hydrolysis. Thiosemicarbazone formation from 3-benzoyl-5-phenyl-1,2,4-triazole is exceptional. Ready hydrolysis of 3-aroyl-, or 3,5-diaroyl-triazole has been noted ⟨66G1084⟩. Only one of the two oximes of 3-benzoyl-1-phenyl-1,2,4-triazole undergoes the Beckmann rearrangement, but both oximes afford the same amide (**135**; Scheme 46) ⟨68JCS(C)824⟩. Both (**136**) and (**137**) could be considered for the constitution of the oxime reactive in the Beckmann rearrangement but all evidence favours the former. Failure of (**138**) to react has been explained by the low migratory aptitude of the triazolyl group in comparison with that of the phenyl group.

Preference for dimerization accounts for the failure of *N*-unsubstituted 1,2,4-triazolecarbaldehydes to undergo Cannizzaro or benzoin rearrangements ⟨71AJC393⟩ or Döbner and Perkin reactions ⟨62JCS575⟩.

The dimeric hemiaminals are the preferred structure in solution and aldehydes are favoured in the solid state. In either case one may assume equilibria between major and minor constituents except for the lack of reducing properties expected in the presence of free aldehyde in the equilibrium mixture.

The carboxylic acids of the 1,2,4-triazole series are readily decarboxylated on heating, but otherwise stable. Esters, amides and hydrazides react normally. Reduction of esters by lithium aluminum hydride affords triazolylmethanols.

Cyanotriazoles are known ⟨B-76MI41200, p. 329⟩. Scheme 47 illustrates some typical uses.

Halogeno-1,2,4-triazoles are reductively dehalogenated with red phosphorus and hydriodic acid ⟨05LA(343)1⟩ but, unlike *N*-halo compounds, not by bisulfite ⟨67ZC184⟩.

Although the halides undergo nucleophilic replacement with greater ease than analogues in the benzene series, and may be used for the preparation of amines, the kinetics of these reactions ⟨67JCS(B)641, 73ZOR2535⟩ are of greater interest than their synthetic value. Halogenated triazolones react more rapidly still and permit the exchange of Br for Cl ⟨67CB2250⟩

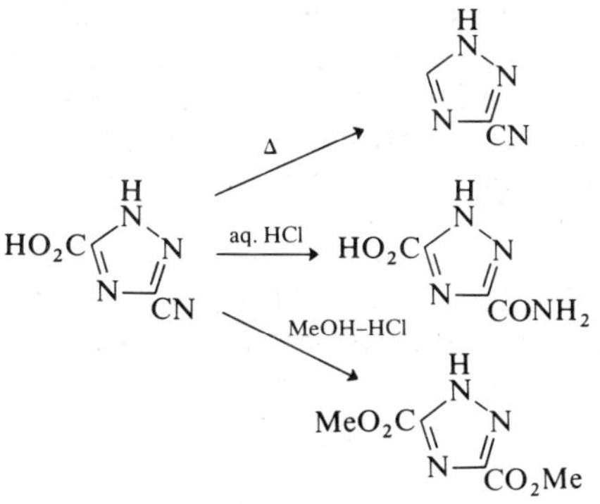

on a ring activated by protonation. Formation of triazolone by reaction with sodium hydroxide or alkoxytriazole with alkoxides is also possible ⟨B-76MI41200, p. 143⟩.

Reduction of triazolinones with phosphorus sulfide has been one of the early routes to triazoles ⟨05JCS625⟩. Milder reactions may differentiate between conjugation of the ring double bond with C=O (**139**) or the lack of it (**140**) as in Scheme 48 ⟨71BSF3296⟩. The formation of triazole in the reduction of the triazolinone (**140**) with lithium aluminum hydride ⟨71BSF3296⟩ is explicable through the formation of a hydroxytriazoline intermediate.

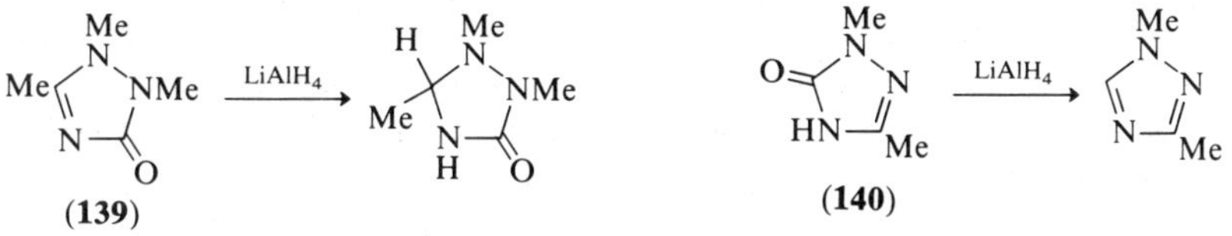

Scheme 48

Hydrolysis of alkoxytriazoles affords triazolinones ⟨67CB2250⟩. Reference has been made to the sensitivity of triazolinones to hydrolysis and photolysis (see Section 4.12.3.2.1).

Aminotriazoles are comparable to aromatic amines of the benzene series. Their alkylation and acylation in competition against nuclear nitrogen and functions containing O or S has been mentioned and some reactions of diazo compounds and related nitrosamines have been considered as examples of nucleophilic attack on a nuclear C atom.

Rearrangement of the mesoionic imines (**141**) to (**142**; Scheme 49) is an intramolecular nucleophilic reaction ⟨74JCS(P1)627⟩. Both (**141**) and (**142**) are accessible through unambiguous syntheses.

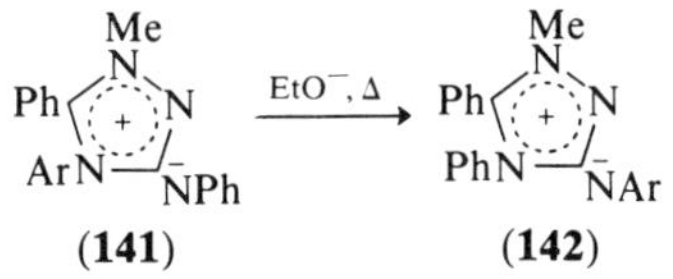

Scheme 49

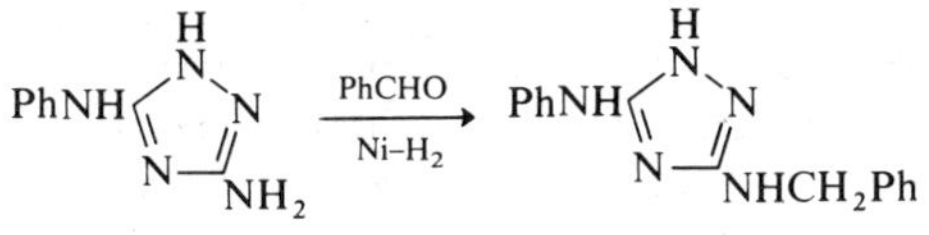

Scheme 50

Resistance of the triazole ring to reduction permits reductive condensations as in Scheme 50 ⟨72JMC694⟩.

The hydrazino group may be removed by silver acetate ⟨70KGS997⟩ or by oxidation in the presence of alkali ⟨75JCS(P1)975⟩.

Some generalized reactions linking triazoles with nitrogen functions of 3- or 5-substituted triazoles are shown in Scheme 51. The choice between (**143**) and (**144**) is governed by the reaction conditions; temperature and high concentration of acid favours the former, room temperature and more dilute acid the latter. Coupling of the diazonium compound (or nitrosamine in acid solution) to azo dyes requires acid conditions; in basic solution the reaction is inhibited. The use of nitrous acid in these reactions is difficult to generalize: in addition to reactions leading to (**143**) or (**144**), it can also act as oxidant, affording (**145**).

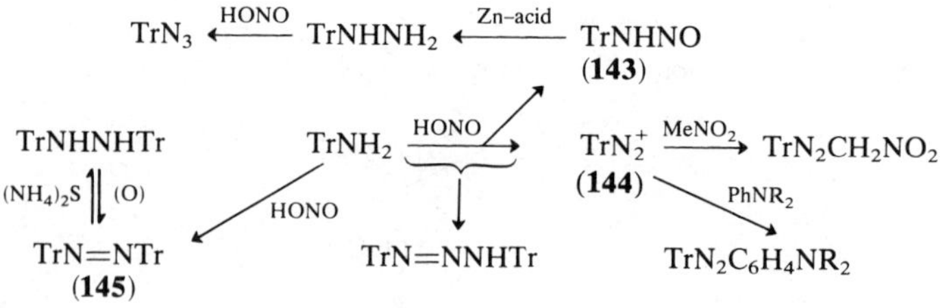

Scheme 51

Reactions leading to triazene (**146**) and polydiazoamino-1,2,4-triazole formation are shown in Scheme 52 ⟨64JOC3449⟩. Both the explosive azides and the safer nitro compounds may be used to provide leaving groups, *e.g.* for the preparation of fluorotriazole from nitrotriazole in liquid hydrofluoric acid at 150 °C ⟨73JOC4353⟩. Similar reactions are available for nitrotriazolinones. Substituted amines and hydrazines are available for nucleophilic displacement of nitro groups or by reduction. Although hydrazine can be a useful reducing agent ⟨70JOC2635⟩ the extent of reduction, or nucleophilic displacement instead of reduction, depends on conditions (Scheme 53) ⟨70KGS997⟩.

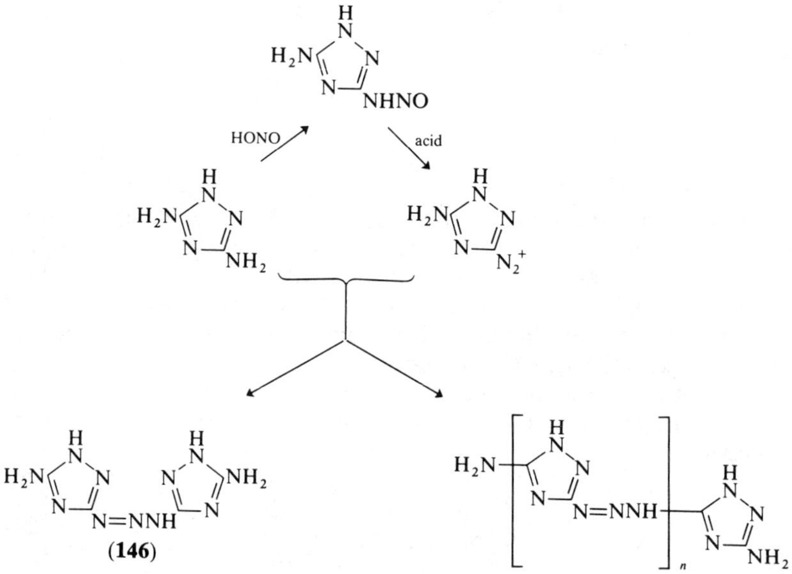

Scheme 52

The simplest *S*-containing derivatives of triazoles are thiones by structure ⟨71JCS(C)1016, 72CPB2096⟩ but behave as thiols in their most important reactions. Alkylation in competition against nuclear nitrogen and substituent oxygen has been considered earlier (Section 4.12.3.2.2). Early workers practised the oxidative desulfurization of triazolinethiones to triazoles. Modern variants include treatment with nitric acid, Raney nickel or hydrogen peroxide ⟨75LA1264⟩ or photochemical desulfurization ⟨70AJC631⟩. Careful use of the last mentioned reagent permits the preparation of sulfoxides ⟨65JCS3912⟩; peroxyphthalic acid affords sulfoxides and sulfones ⟨70JCS(C)2403⟩.

Oxidation of disulfides with halogens may lead to sulfonyl chlorides or the much less stable bromides which afford sulfonamides. However, replacement of the sulfur function by halogen or complete desulfurization may also occur. Treatment of the sulfonyl chloride does not yield the expected esters but is the best way to prepare the sulfonic acids. Electrolytic oxidation of thiones affords a mixture of sulfonic acids and desulfurized compounds. The action of ozone on disulfides is a feasible but not very convenient reaction leading to sulfonic acids. Scheme 54 summarizes some of these relationships and the unusual nitration of the sulfonamide (**147**) ⟨62BEP619423⟩; for convenience the generalized triazole Tr is shown in

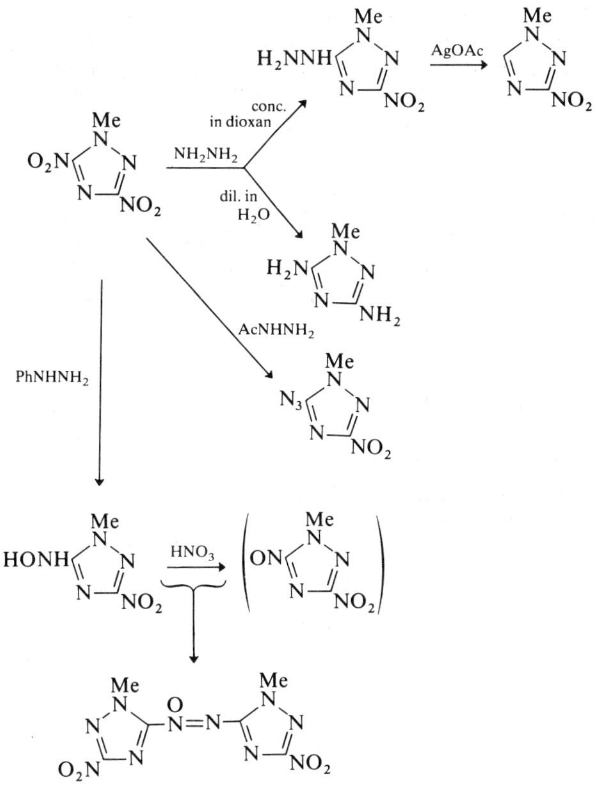

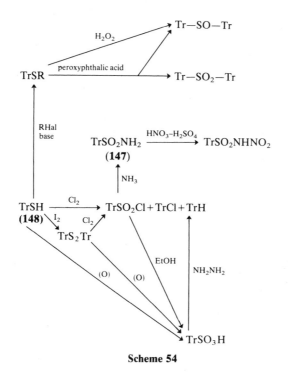

Scheme 53

Scheme 54

(**148**) as the thiol. Permanganate oxidation of thiones also affords sulfonic acids. Unlike their benzene analogues, the triazolesulfonic acids are resistant to boiling hydrobromic acid or heating with alkali. Treatment with hydrazine, however, eliminates the sulfonic acid function.

4.12.3.3.2 N-Linked substituents

Rearrangement of acyl and halo groups are the most common reactions of triazoles with functions on nitrogen. The chemistry of '4-hydroxytriazoles', in effect N-oxides, has been less extensively studied than that of triazolinones. The N—O bond is cleaved by phosphorus halides and acidic anhydride but not by oxidation with permanganate or by alkaline bromination ⟨70JPR610⟩.

The dual nature of 4-aminotriazoles as aromatic amines and unsymmetric hydrazines gives rise to many interesting reactions ⟨81HC(37)1, p. 161, 483, 607, 611⟩ some of which are shown in Schemes 55–57. Although nitrous acid normally reacts with both amino and hydrazino groups, examples of selective reactivity are shown in Scheme 58. The nitrosamine from 4-benzylamino-1,2,4-triazole is cleaved by sodium hydroxide to triazole.

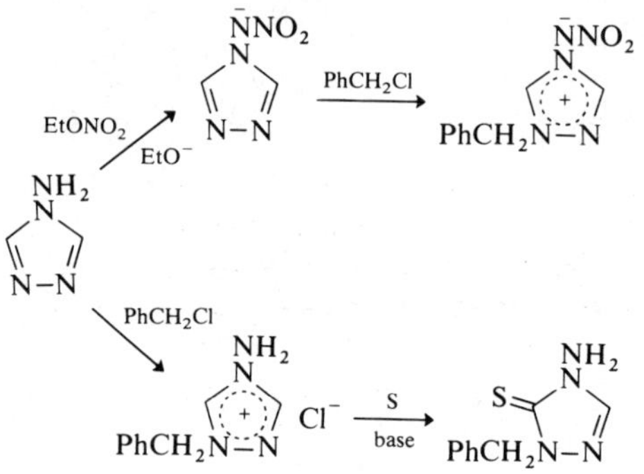

Scheme 55

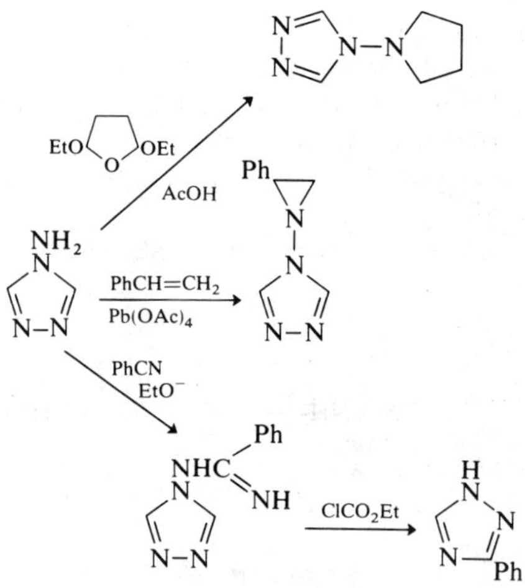

Scheme 56

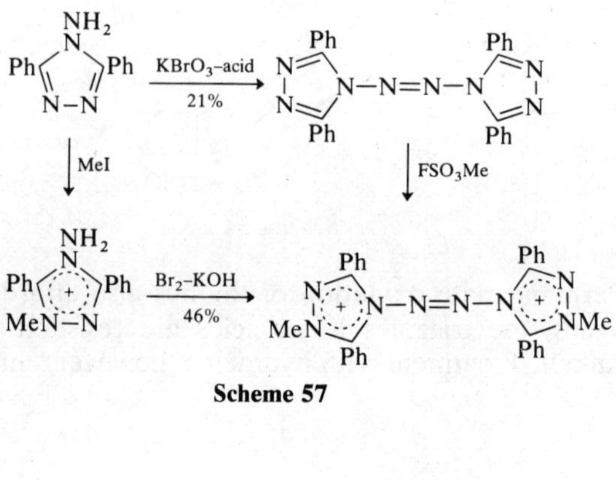

Scheme 57

Scheme 58

Trimethylsilyl-, triethoxysilyl- and trigermyl-triazoles are noted for the intermolecular exchange of such functions ⟨74JOM(70)347⟩. Reactions of 1-trimethylsilyl-1,2,4-triazole with acetyl chloride, ethyl chloroformate or phosgene affords acyltriazoles ⟨60CB2804⟩. Trimethylstannyl compounds ⟨62RTC202⟩ are usually less reactive than the corresponding silicon derivatives.

4.12.4 SYNTHESIS

4.12.4.1 General Considerations

Difficulties besetting a systematic survey of methods suitable for the preparation of 1,2,4-triazole and derivatives are not merely philosophical but are worthy of notice by practical organic chemists. In the absence of natural products derived from 1,2,4-triazole (that is, ignoring products formed through the metabolism of synthetic triazoles in biological systems), this ring is obtained either by synthesis from acyclic compounds or by transformation of other cyclic systems. In the former case, analysis of the number and nature of bonds formed is a convenient, systematic way to approach triazole synthesis but it may be misleading to the extent of obstructing progress. Synthesis of a 5-membered ring system symbolized as ABCDE from starting materials A—B and C—D—E suggests the formation of 2 bonds, say E—A and B—C, from building units containing 2 and 3 eventual ring centres, respectively. Until the mechanism of the reaction is fully understood, it may be suspected that the first step is the formation of E—A and B—C—D of a comparable classification status; or that of A and B—C—D—E, the further reaction of which is to be classified as one of 1-centre and 4-centre fragments. Other possibilities remain, of course. Since only few syntheses of 1,2,4-triazole have had their mechanisms studied and most of such studies have been carried out without the benefit of powerful modern methods, convenience of classification may not be the best guide to the choice of conditions and most reactive starting materials.

A similar problem may arise when considering the conversion of other cyclic systems into 1,2,4-triazoles. More serious is the ambivalence that arises when both 1,2,4-triazole and an isomer with a different ring system are formed from acyclic starting materials. The equivalence of ring transformation to preparation from an acyclic intermediate is evident in such cases. Most interconversions of heterocyclic ring systems occur with ring cleavage and recyclization, hence they are arbitrarily classified as distinct from preparations employing acyclic materials. Transformations of azoles without distinct bond rupture are improbable except by mechanisms with the intermediate formation of strained short-lived bicyclic systems.

Another problem of classification already encountered in other contexts is the hazy borderline between 'aromatic' and 'non-aromatic' 1,2,4-triazoles. If the preparations of triazoles and triazolines are to be discussed in separate sections, preparatively similar reactions from reactants with the skeleton type N—N—CX—N— will be duplicated with X = O or S and occasionally with X = NHR. The same problem arises in the classification of reactions consisting of a functional change on the preformed triazole nucleus.

Some of the most useful syntheses of 1,2,4-triazoles have variants which are better considered together rather than in separate sections of the possible subsets of 5 centres. Examples to illustrate such reactions are drawn mostly from cases with relatively unreactive functions that leave the main burden of synthetic strategy on ring synthesis.

In other examples with interest centred on specific functions, principles of ring synthesis need not be rehearsed and attention is directed to the choice of functions to be carried by the new-born ring.

In a third group of examples we shall consider the substitution of the preformed ring; some of this material has been treated in Section 4.12.3 but specific examples of practical synthetic interest will be given.

Still other examples of syntheses for industrial products are reserved for Section 4.12.5. In the absence of references ⟨81HC(37)1⟩ and ⟨61CRV87⟩ should be consulted.

4.12.4.2 Synthesis of Triazole Rings from Acyclic Compounds

4.12.4.2.1 Methods employing hydrazine derivatives

The ease of forming C—N and C=N bonds as compared with the difficulty of N—N formation practically prescribes the use of hydrazines in the synthesis of 1,2,4-triazoles. Scheme 59 illustrates synthetic schemes that use one of the following: (a) hydrazine, (b) an acylhydrazine, (c) amidrazone or (d) acylamidrazone. The dotted lines leave the presence or absence of a bond open. Thus in (a) possible reactants are (R^5COX, R^3COX and NH_3) or (R^5CONH_2 and R^3COX) or (R^5COX and H_2NCOR^3) or ($R^5CONHCOR^3$); X stands for a suitable leaving group, usually OH, H_2O.

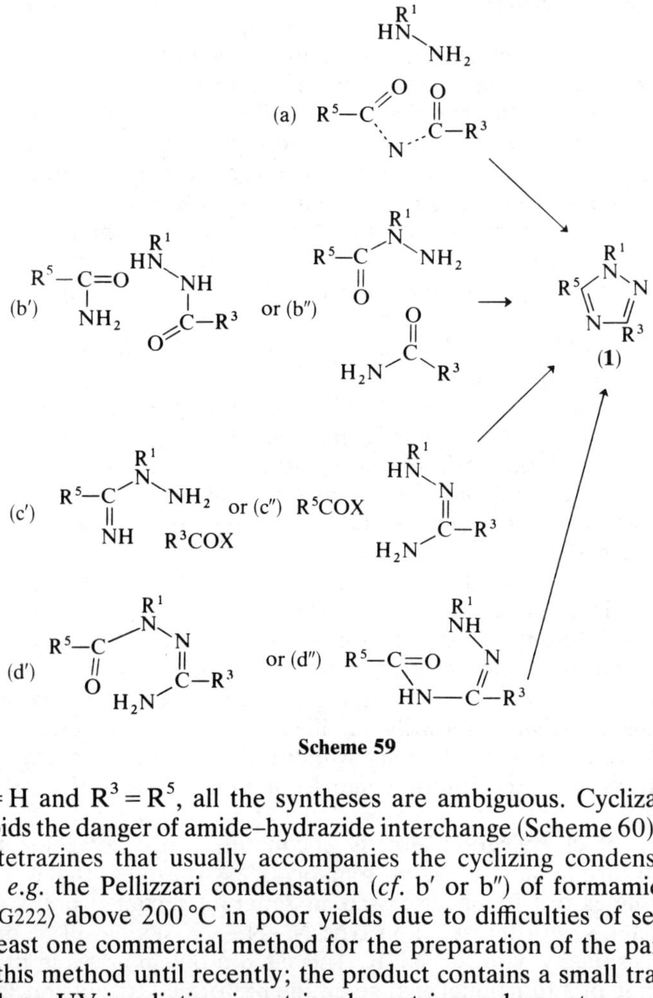

Scheme 59

Unless $R^1 = H$ and $R^3 = R^5$, all the syntheses are ambiguous. Cyclization under gentle conditions avoids the danger of amide–hydrazide interchange (Scheme 60) and the formation of polycyclic tetrazines that usually accompanies the cyclizing condensations favored by early authors, *e.g.* the Pellizzari condensation (*cf.* b′ or b″) of formamide with formylhydrazines ⟨1894G222⟩ above 200 °C in poor yields due to difficulties of separation ⟨05G373⟩. However, at least one commercial method for the preparation of the parent 1,2,4-triazole has relied on this method until recently; the product contains a small trace of non-triazole impurity which on UV irradiation in a triazole matrix produces strong green phosphorescence lasting about 20 s.

$$R'''CONH_2 + R''CONHNH_2 \rightleftharpoons R'''CONHNH_2 + R''CONH_2$$

Scheme 60

Although 1,2,4-triazole can be prepared by heating hydrazine with two (preferably more) equivalents of formamide, the reaction presumably occurs in steps, *e.g.* as in Scheme 61. Syntheses corresponding to step 3 are well known, but occur in poor yields or not at all.

Scheme 62 illustrates extensions of these methods proceeding from or through hydrazidines, amidrazones or acylamidrazones to syntheses of 1,2,4-triazoles with sub-

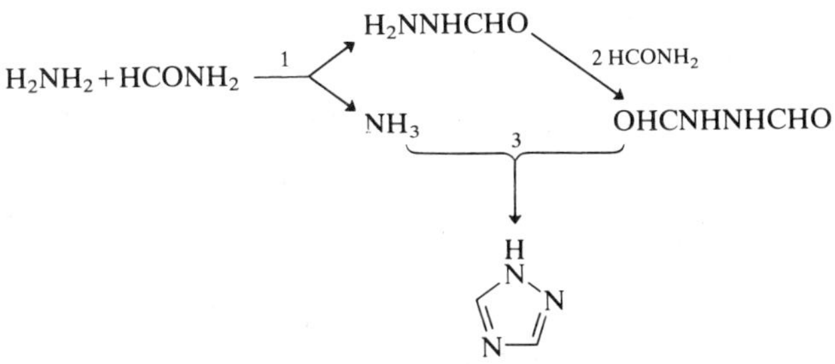

1,2,4-Triazoles

$H_2NH_2 + HCONH_2$

Scheme 61

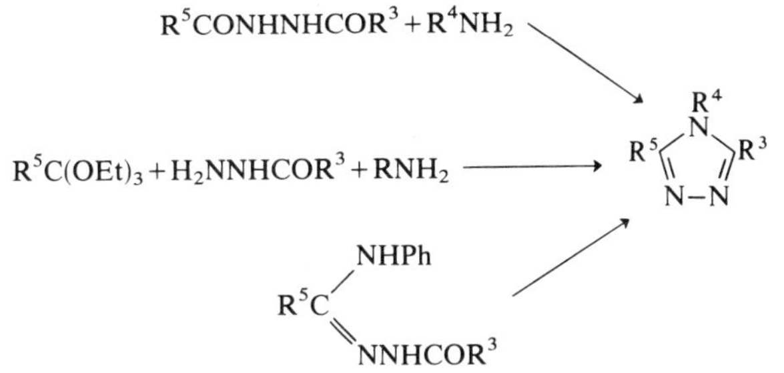

Scheme 62

stituents on N-4. Again it is assumed that all such reactions proceed through amidrazone or acylamidrazone intermediates, also that the formation of these is the rate determining step in the reactions shown in Schemes 61 and 62. Although this generalization has not been established, it is supported by the popularity of preparations starting from amidrazones or acylamidrazones, the success of other syntheses being limited by their suitability for the formation of these reactive intermediates. Another line of evidence pointing in the same direction is the elaboration of synthetic methods using starting materials other than amidrazones or acylamidrazones for reactions that are best explained by intermediate amidrazone formation. Examples of frequently used methods follow.

Synthesis starting from amidrazones (Scheme 59, c′–d″) is the most versatile preparation of the 1,2,4-triazole ring. Application to specific cases is governed by three considerations. One is the formation of the required amidrazone from a hydrazine with or without substituents and amide derivatives $RC(=NH)X$ or $RC(=NCOR')X$ where X = halide, OH, OAlk, SAlk, NH_2, etc. With two reactive hydrazinic NH groups, isomeric amidrazones can be formed but the relative basicities usually predict the correct product. Unsymmetrical hydrazines $YZNNH_2$ also form amidrazones but these cannot be cyclized unless Y or Z is reactive.

Unsubstituted hydrazine can react with an amide derivative (**149**) to yield the amidrazone (**150**) which then either reacts with more (**149**) or undergoes self-condensation to (**151**) which then cyclizes to the triazole (**152**; Scheme 63) on heating ⟨65JOC3729⟩. Self-condensation is ambiguous: the formation of (**152**) may compete against that of dihydrotetrazine (**153**) ⟨62CB2546⟩ but basic, rather than the usual neutral or acidic conditions, are required. Conversion of dihydrotetrazines into triazoles will be considered under Section 4.12.4.3.2. Tetrazine and triazole formation occur to comparable extents in the reaction of formimino ether with methylhydrazine ⟨72JOC3504⟩.

A third problem is the ambiguity of cyclizations leading to oxadiazoles or thiadiazoles as in Scheme 64 where oxadiazole is the minor product ⟨58JOC1912⟩. Conversion of the amidrazone (**154**) into the triazole (**155**; Scheme 65) takes place in basic solution, while the oxadiazole (**156**) is obtained in the presence of acid ⟨69JHC965⟩. However, dehydration of the amidrazone to triazole may take place in polyphosphoric acid.

The formation of alternative heterocycles can be the main reaction with semicarbazides or thiosemicarbazides which are used extensively in the preparations of triazolinones and triazolinethiones. Aminoguanidines can give rise to isomeric triazoles. The preferred formation of thiadiazole (**157**) from the 1-acylthiosemicarbazide (Scheme 66) is explained by the protonation of N-4 in strong acid with accompanying loss of nucleophilicity, while in the presence of base its nucleophilic character is enhanced ⟨73JOC3947⟩ and (**158**) is obtained.

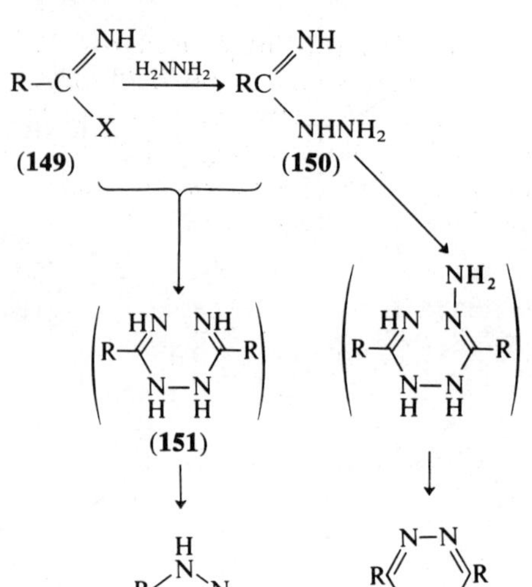

Scheme 63

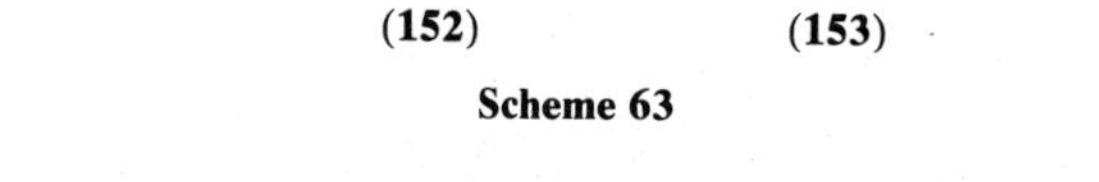

Scheme 64

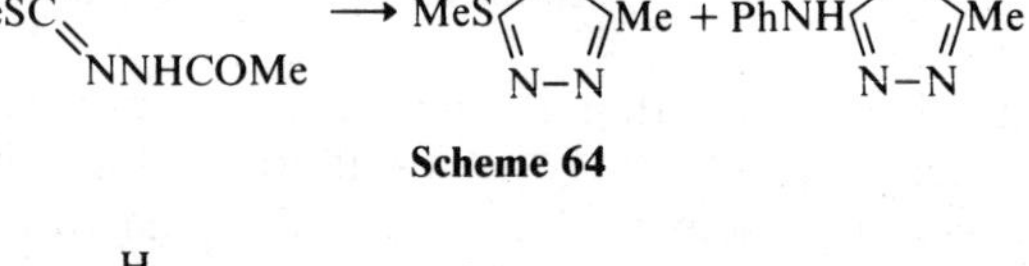

Scheme 65

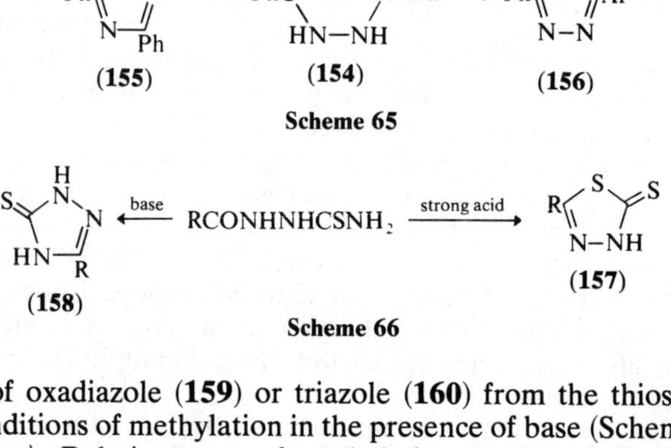

Scheme 66

The formation of oxadiazole (**159**) or triazole (**160**) from the thiosemicarbazide (**161**) under different conditions of methylation in the presence of base (Scheme 67) is less readily explained ⟨71MI41200⟩. Relative rates of methylation on S and O are involved but the different temperatures make a direct comparison impossible. The effect of acidity cannot be neglected as shown in Scheme 68 ⟨63CB1059⟩.

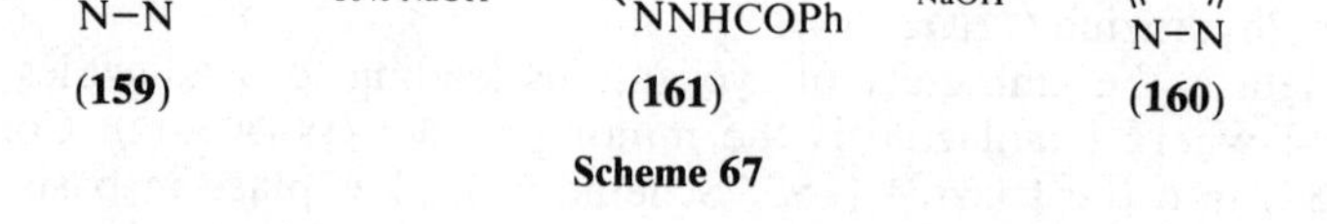

Scheme 67

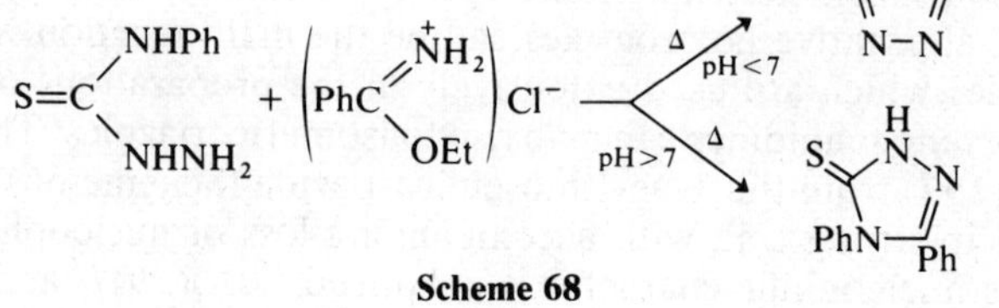

Scheme 68

The choice of acylating or cycloacylating agents chosen from RCO_2H, $RCOCl$, RCO_2Alk, $RC(OAlk)_3$, $RC(=NH)OEt$, etc., usually affects the yield only. The different reactions of the 4-arylthiosemicarbazide with diethyl carbonate in the presence of methoxide ion ⟨59CR(248)1677⟩ and with phosgene in chloroform or chlorobenzene ⟨70LA(735)158⟩ occur in different solvents and at different acidities (Scheme 69). An unusual cyclization (Scheme 70) utilizes the CCl_3 group as the acid analogue ⟨67JCS33⟩.

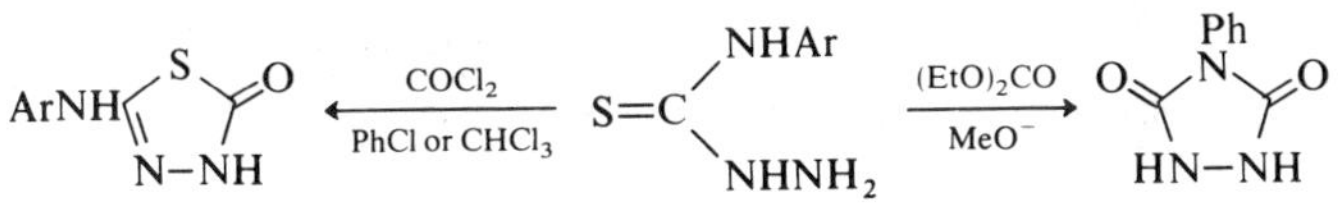

Scheme 69

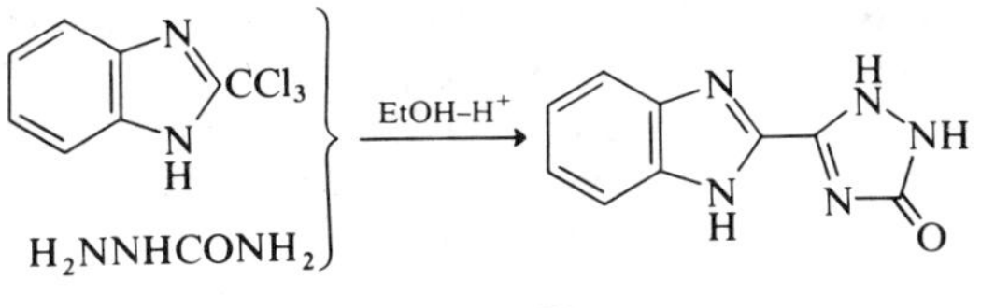

Scheme 70

Preparation of 3-, 1,5- or 1,3,5-substituted triazoles by reaction of hydrazines with diacylamines (Einhorn–Brunner reaction) is one of the earliest synthetic methods of triazole chemistry ⟨05LA(343)207, 14CB2671⟩ but appreciation of its mechanism as involving an amidrazone intermediate is of more recent origin ⟨52JCS3418, 54JCS3319⟩. However, early workers recognized that the mechanism could not be acylation of the hydrazine by diacylamines (which are much more reactive acylating agents than primary amides) since the reaction inferred does not occur under the mild conditions of the Einhorn–Brunner reaction but only at high temperatures (Pellizzari reaction). The observation that unsymmetrical diacylamines R'CONHCOR" react first at the more electrophilic carbonyl group also supports the intermediate amidrazone hypothesis. Thus formylamides RCONHCHO (R = alkyl, Ph) afford 1,5-substituted triazoles. If, however, one uses N-acylthioamides instead of diacylamines, acylamidrazone formation will occur on the thione group ⟨60CB663⟩. The use of semicarbazide in the Einhorn–Brunner reaction results in 3- or 3,5-substituted triazoles through hydrolysis of the intermediate 1-carbamoyl derivative in agreement with the lability of N-acyl-1,2,4-triazoles. The reaction may be extended to N-acylurethanes for the preparation of triazolin-5-ones. Einhorn–Brunner reactions are illustrated in Scheme 71.

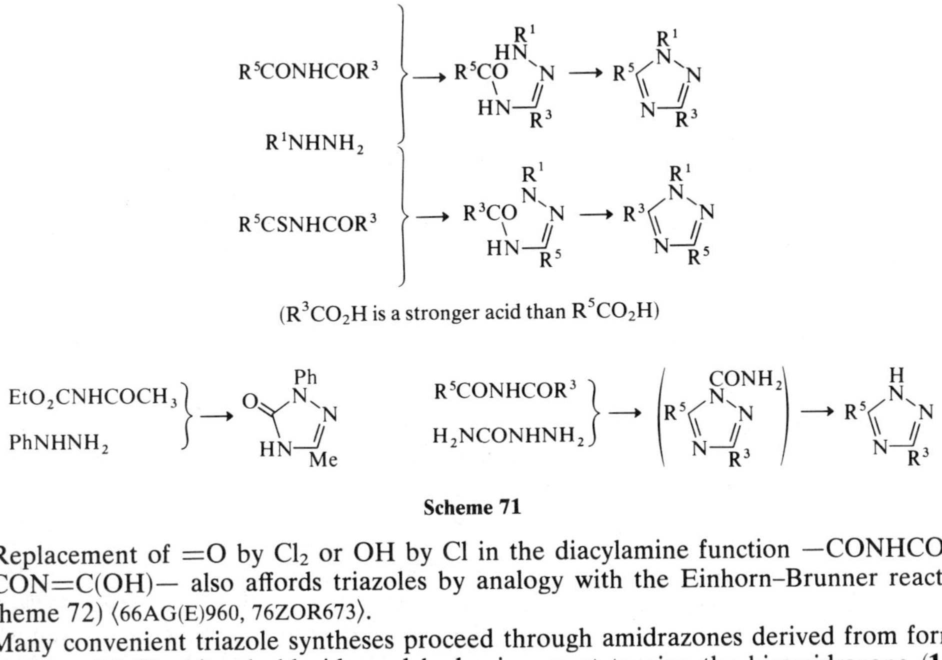

(R^3CO_2H is a stronger acid than R^5CO_2H)

Scheme 71

Replacement of =O by Cl_2 or OH by Cl in the diacylamine function —CONHCO or —CON=C(OH)— also affords triazoles by analogy with the Einhorn–Brunner reaction (Scheme 72) ⟨66AG(E)960, 76ZOR673⟩.

Many convenient triazole syntheses proceed through amidrazones derived from formic acid. Thus DMF, thionyl chloride and hydrazine react to give the bisamidrazone (**162**) which is then cyclized by transamination ⟨75JCS(P1)12⟩ (Scheme 73).

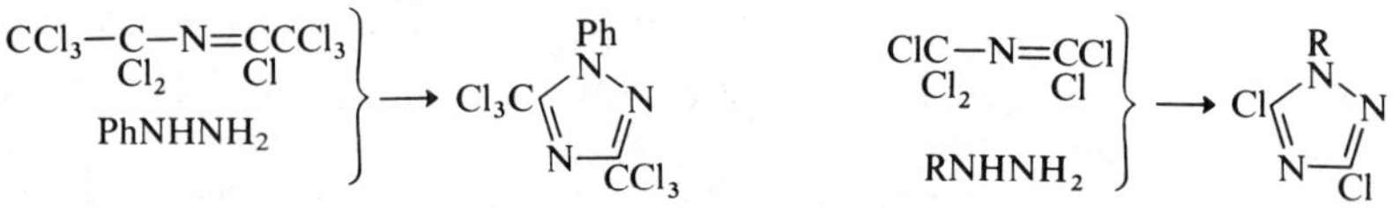

Scheme 72

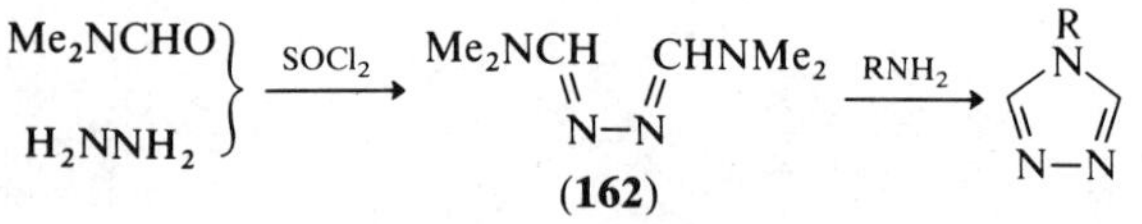

Scheme 73

One can use s-triazine (**163**) as a source of formamidrazones that react with more triazine ⟨56JOC1037⟩ or undergo self-condensation in the presence of acid ⟨62CB2546, 66CB3734⟩. In the absence of acid the acylic dihydrazone (**164**) is formed and can be cyclized to 4-aminotriazole. Scheme 74 illustrates some applications of the triazine method ⟨57JA2839⟩ and Scheme 75 the use of trichloro-s-triazine ⟨60AG956⟩.

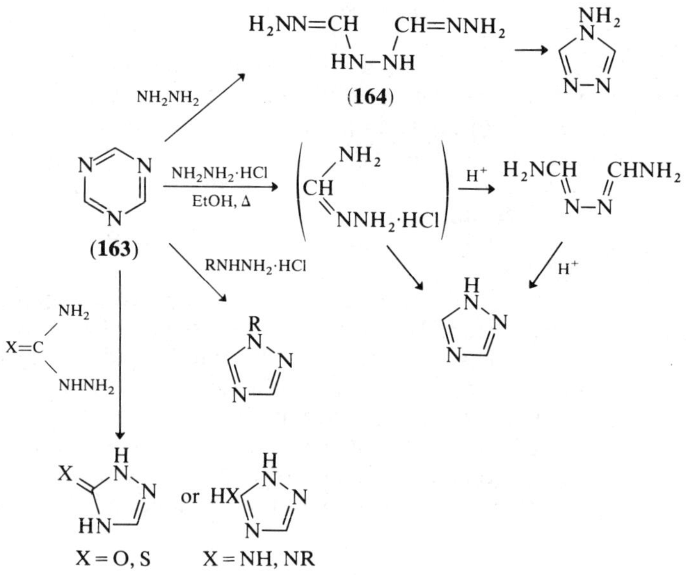

Scheme 74

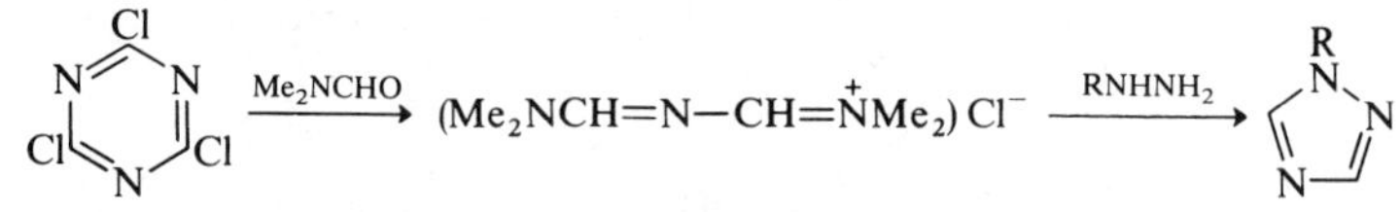

Scheme 75

Some ambiguities implied in the acylation and self-condensation of amidrazones are obviated by forming carbonyl derivatives which are then cyclized under oxidative conditions (Scheme 76). Dehydrogenation with palladium ⟨66JMC405⟩ is thought to proceed with the formation of a triazoline (**165**) ⟨70JHC1001⟩. An early method, oxidative cyclization of semicarbazones ⟨00JCS224⟩, is still in use but analogous treatment of thiosemicarbazones results in thiadiazoles. In some cases a specific oxidant is not required owing to the spontaneous oxidation of triazolines ⟨72JOC343⟩. In general, mercury(II) oxide or silver oxide in boiling xylene is recommended ⟨63CB2996⟩.

The sensitivity of triazolines to oxidation has been used for the unambiguous synthesis of 1,2,4-triazoles (see Section 4.12.4.1) but too few triazolines of sufficient keeping quality are known to permit generalization of structural features required for stability. Thus (**165a**; R = pyrazolyl, pyridazolyl, pyrimidyl, R' = Ph, 2-pyridyl) have been prepared ⟨71JHC173, 73JHC353⟩ while the more heavily arylated (**165b**; R = Ph) could not be isolated even when atmospheric oxidation was carefully excluded ⟨69CB3159⟩; compound (**165b**; R = Me) is unstable at room temperature (Scheme 76) ⟨73BCJ1250⟩.

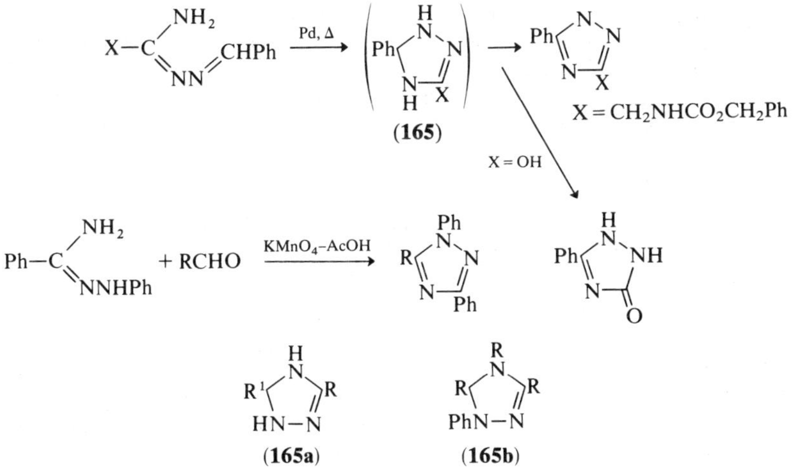

Scheme 76

The preparation of 3,5-diaryltriazoles without isolating the intermediate amidrazones can be effected by heating an arenenitrile with the arenesulfonate of an arylhydrazine ⟨54JCS3461⟩. This method, known as the Potts procedure, may also be extended to the reactions of arenenitriles with alkyl hydrazide benzenesulfonates, but not to those of alkenenitriles with arylhydrazine arenesulfonates ⟨61JCS518⟩. The yield of mixed alkyl-aryltriazoles obtained by this method (Scheme 77) is depressed by the formation of diaryltriazole and the amide corresponding to the aryl moiety. Both diaryl- and alkylaryl-triazoles are obtainable by a modification of this method that reacts free acid hydrazide and nitrile under pressure and at a somewhat higher temperature; even dialkyltriazoles are prepared this way although in low yields ⟨63CB1059⟩. On the other hand, 3,5-dialkyl-1,2,4-triazoles are obtainable in good yield by the reaction of an alkylnitrile with the zinc chloride complex of hydrazine at 140 °C ⟨76JAP7636455⟩.

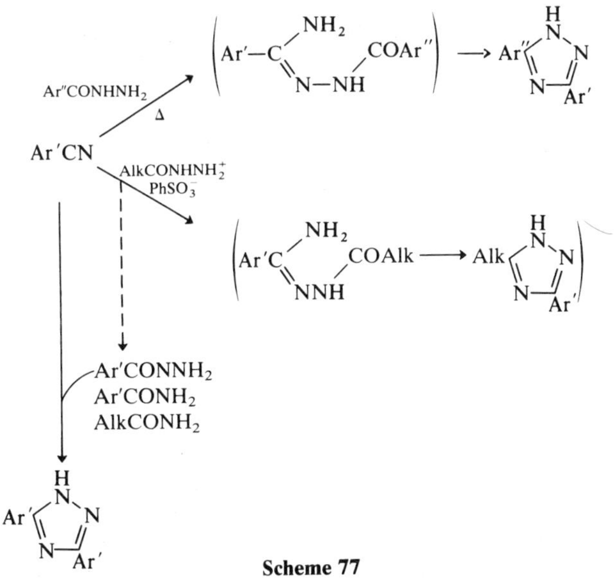

Scheme 77

Of some interesting although rarely used techniques to be mentioned, the one represented in Scheme 78 involves an amidrazone intermediate ⟨72JPR101⟩. The product is formulated as (167) rather than the more plausible 4-OH compound (166) ⟨70JPR610⟩.

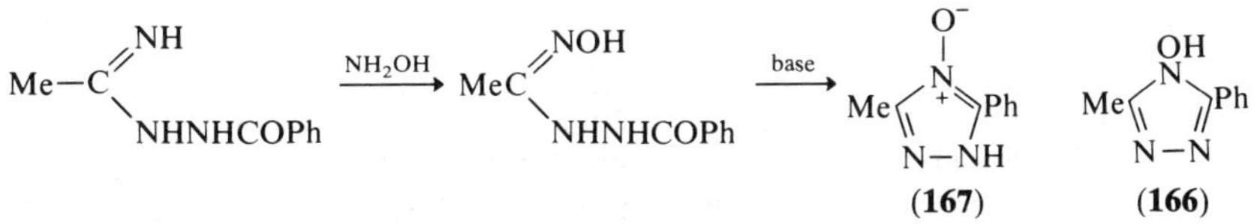

Scheme 78

The Chichibabin amidation of azines (**168**) can be used for the preparation of 3,5-diaryl triazoles (**169**) ⟨65CB928⟩ and could be interpreted to proceed through a bisamidrazone intermediate as shown in Scheme 79. However, the conversion of (**168**) to (**169**) by *t*-butoxide ⟨76JCS(P1)207⟩ occurs through cleavage of the former. Oxidation of sodium benzamidinate (**170**) with copper(II) chloride probably gives the sodium derivative, then (**169**) ⟨67AG(E)633⟩. Although not widely used, this reaction is one of the few examples of N—N bond formation as a preparative strategy for azoles.

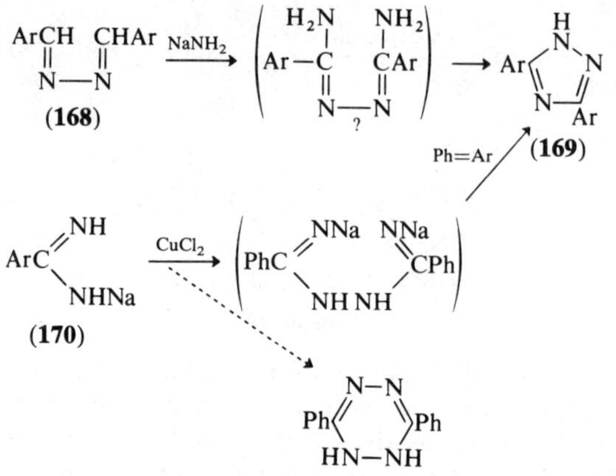

Scheme 79

Preparations of amidrazones required for triazole syntheses by the Japp–Klingemann reaction and similar methods involving malonic acid or other 1,3-diketo starting materials are too numerous for listing. A curious reaction making use of the chemistry of malonic esters for the synthesis of a triazole is shown in Scheme 80 ⟨70T3069⟩.

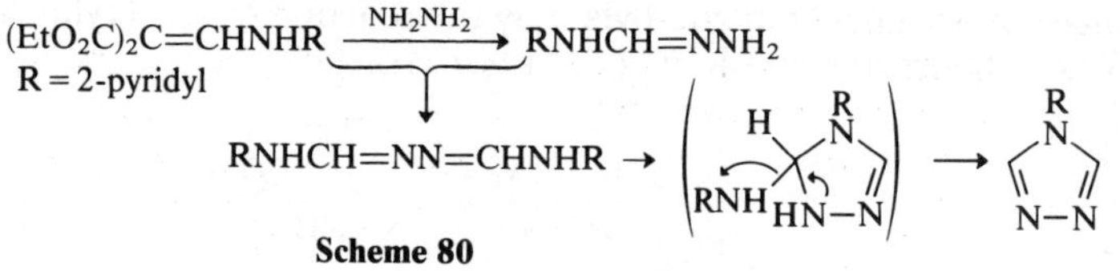

Scheme 80

Some modern triazole syntheses involve amidrazone intermediates formed through rearrangements of isocyanides ⟨75S609⟩, cyanates ⟨73AG447⟩ and isocyanates generated in the Curtius rearrangement ⟨69LA(722)29, 69LA(722)38⟩ are shown in Schemes 81, 82 and 83, respectively.

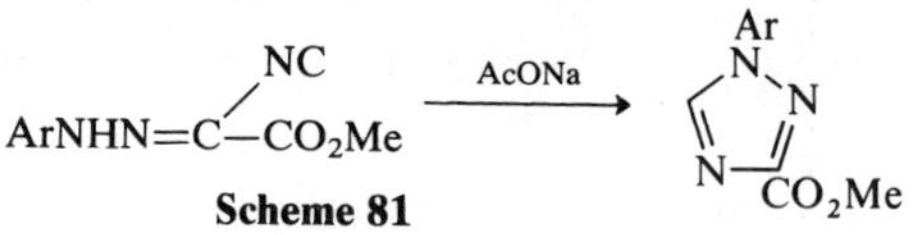

Scheme 81

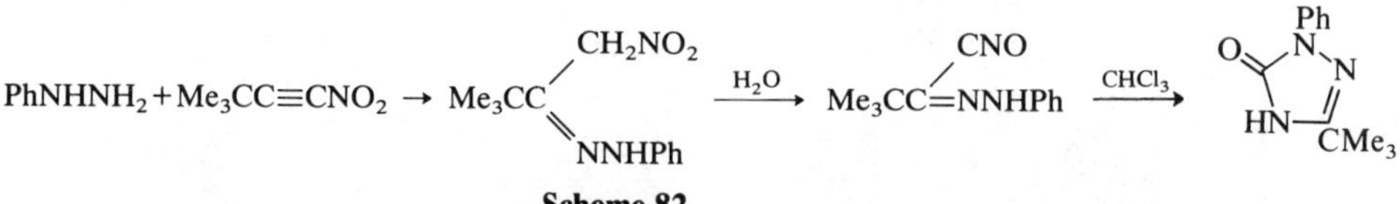

Scheme 82

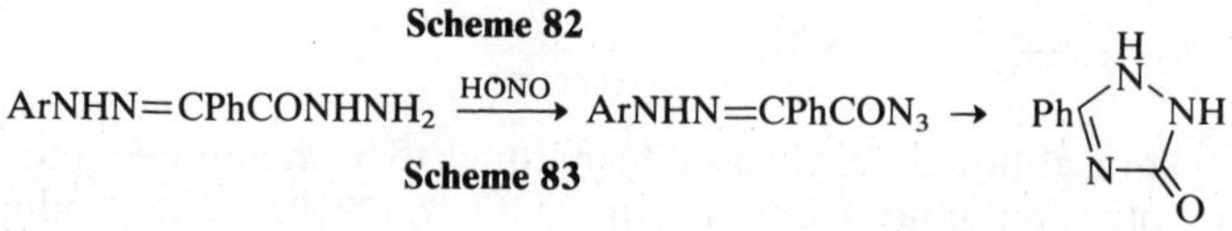

Scheme 83

4.12.4.2.2 Nitrilimine methods

Huisgen's studies of 1,3-dipolar cycloadditions leading to a great variety of heterocyclic systems ⟨61MI41203, 80RCR880⟩ are applicable to the synthesis of triazoles and derivatives. Nitrilimines (**171**) formed by dehydrohalogenation of *C*-halobenzylidenephenylhydrazones (**172**) react with C≡N, C=N (as in CNO) to afford triazoles and triazolines in yields of

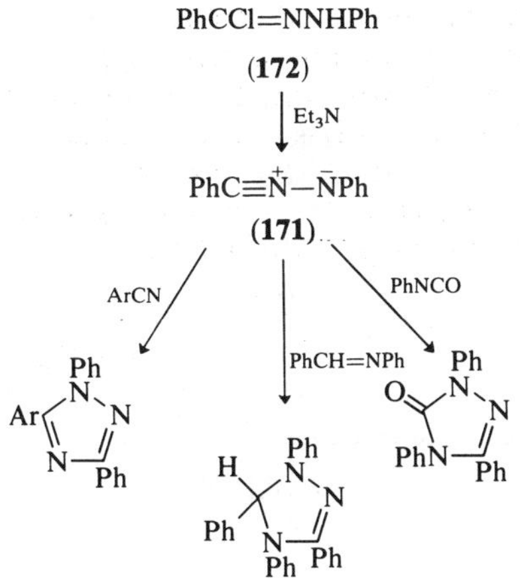

1,2,4-Triazoles

Scheme 84

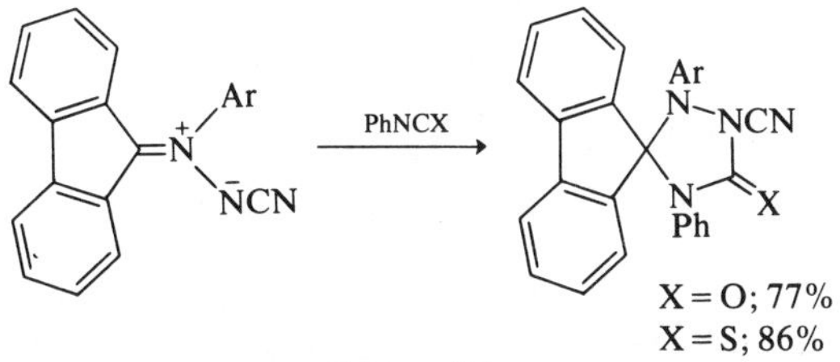

Scheme 85

50–75% (Scheme 84). Triazolidine derivatives are obtainable through similar reactions from azomethine imines, *e.g.* as shown in Scheme 85.

The versatility of triazole syntheses from nitrilimines is further illustrated in Scheme 86 ⟨65CB642⟩. Note the simultaneous formation of triazole (**173**) and triazoline (**174**) from the hypothetical bistriazoline intermediate (**175**) contrasted with the reaction of the aldoxime (**176**) which permits the stepwise isolation of triazoline (**177**) or triazole (**178**).

Scheme 86

Examples adduced in Scheme 87 indicate the caution required when a nitrilimine intermediate favours the formation of a heterocycle other than triazole. The bromo derivative (**179**) of the semicarbazone (**180**) is formally analogous to (**172**); the nitrilimine (**181**) is formed even in the presence of such weak bases as sodium acetate or water and undergoes ring closure to the 1,3,4-oxadiazole (**182**). In anhydrous acetic acid the triazolinone (**183**; R = H) is formed ⟨72JCS(P1)1918⟩; a similar reaction occurs on treating the *N*-methyl homologue of (**180**) with bromine in acetic acid to obtain (**183**; R = Me).

Another example of ambiguity (Scheme 88) arises out of the addition of phenylisocyanate in the presence of aluminum oxide to the nitrilimine (**171**) that affords a triazolinone (**184**) with some oxadiazoline (**185**) as byproduct ⟨64CB1085⟩, while the addition of phenylcyanate

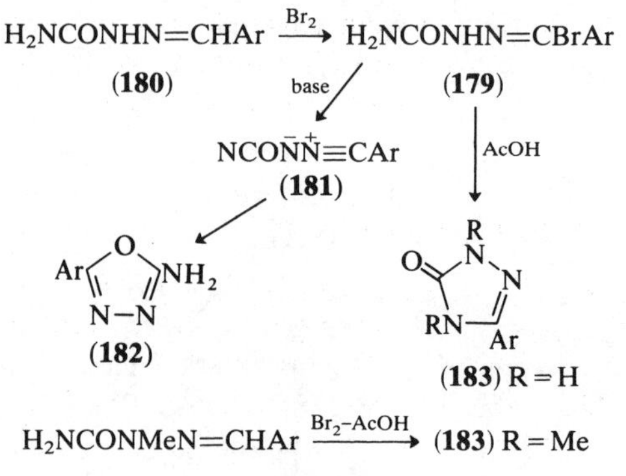

Scheme 87

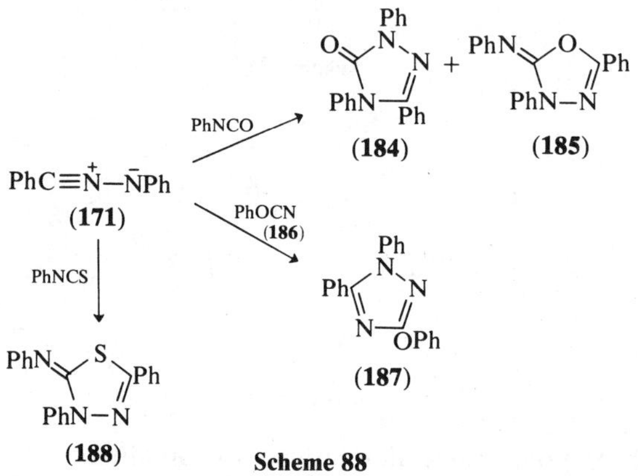

Scheme 88

(186) affords the expected triazole (187). Phenylisothiocyanate reacts with (171) like phenylisocyanate but in this case thiadiazole (188) is the major product ⟨62LA(658)169⟩.

Further examples of triazole preparations by 1,3-cycloadditions will be found under ... which deals with the formation of such reactive species from heterocyclic ... triazole.

...s to the ready availability of nitriles, their reaction with nitrilimines is a useful ...od for the preparation of bistriazoles, e.g. (189) and (190) in Scheme 89.

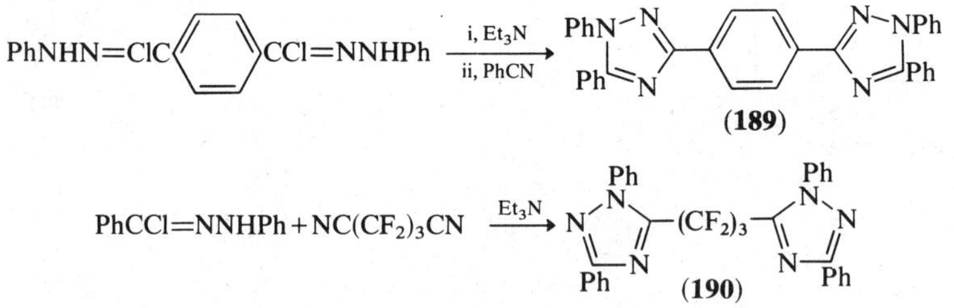

Scheme 89

A combination of amidrazone and nitrilium methods (Scheme 90) affords aminotriazole derivatives (191) in yields of 34–68% or better when hydrochlorides rather than free bases are isolated ⟨81CI(L)648⟩.

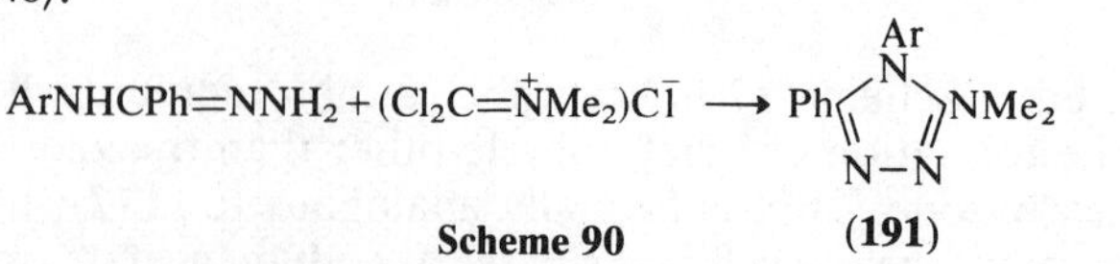

Scheme 90

4.12.4.3 Synthesis of Triazole Rings from Other Heterocyclic Systems

Transformation of other heterocycles into triazoles implies one or more of the following operations: (i) destruction of rings leaving an intact triazole core behind; (ii) cleavage of the ring to an acyclic intermediate which then recyclizes with or without rearrangement;

(iii) destruction of the original ring to form unstable species captured by an added reagent to form triazole. Conversely, reactions of this nature serve to convert simple triazoles into those with fused ring systems or into other heterocyclic rings.

4.12.4.3.1 *Destruction of non-triazole rings*

Methods of this type are best considered as reactions of the fused ring-systems in question (see Chapter 4.15). The example illustrated in Scheme 91 is of potential interest in pharmacology ⟨75T1363, *cf.* 74CPB1938⟩. The conversion of the 1-aminoadenosine (**192**) into the imidazolyltriazole (**193**) amounts to triazole formation from an amidrazone intermediate and a formyl group derived from the pyrimidine moiety.

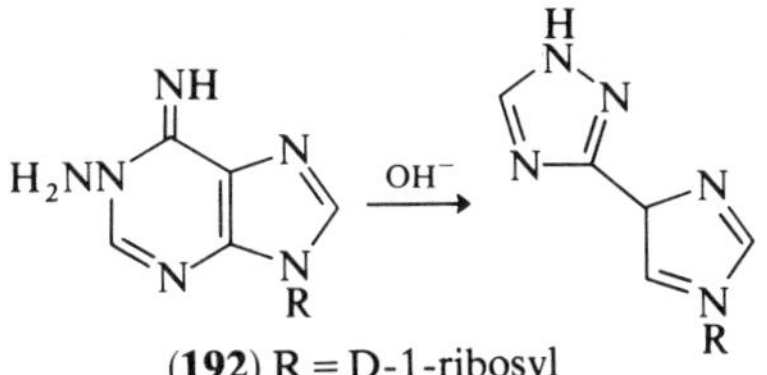

(**192**) R = D-1-ribosyl

(**193**)

Scheme 91

Other triazole syntheses by hydrazinolysis, hydrolysis or oxidation of fused ring systems are briefly treated in ⟨81HC(37)1⟩.

The ring system to be destroyed may be constructed as a measure of synthetic strategy as illustrated (Scheme 92) by the preparation of a triazole (**194**) used to synthesize fused rings of pharmacological interest. The dihydroquinazoline (**195**) provides the amidrazone intermediate and the aromatic substituent required ⟨75JHC717⟩.

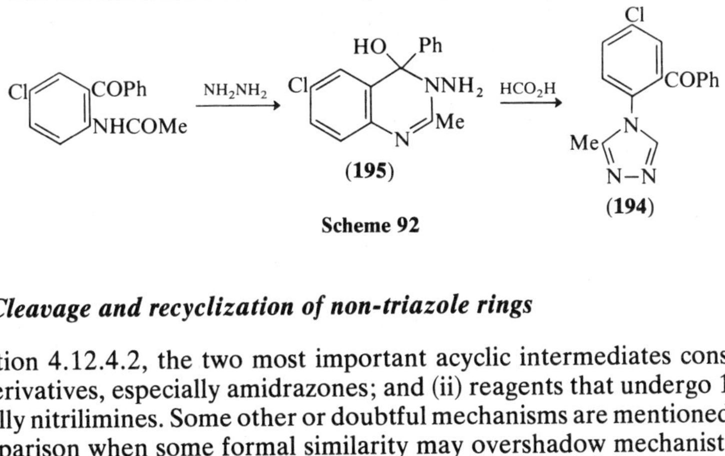

(**195**)

(**194**)

Scheme 92

4.12.4.3.2 *Cleavage and recyclization of non-triazole rings*

As in Section 4.12.4.2, the two most important acyclic intermediates considered are (i) hydrazine derivatives, especially amidrazones; and (ii) reagents that undergo 1,3-cycloaddition, especially nitrilimines. Some other or doubtful mechanisms are mentioned for reference and for comparison when some formal similarity may overshadow mechanistic differences.

Scheme 93 summarizes the overall reaction much used for the conversion of 1,3,4-oxadiazoles (**196**; X = O) and thiadiazoles (**196**; X = S) to triazoles (**197**) and (**198**). Intermediates such as (**199**) and (**200**) have been isolated in some cases and have been reasonably assumed in others, especially when such amidrazones could be converted either into (**197**) or (**198**) according to the reaction conditions ⟨see also 70KGS991⟩.

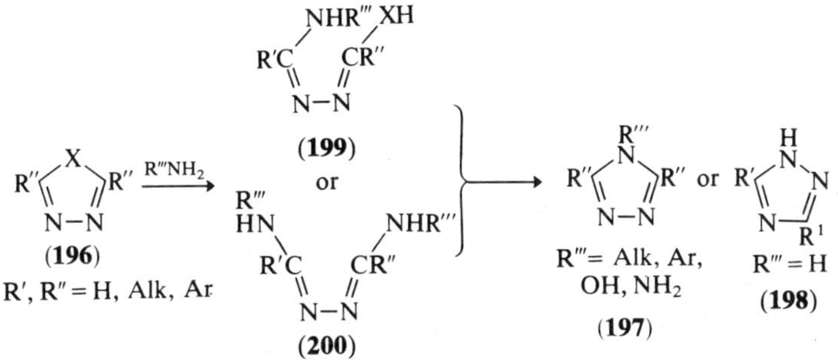

(**196**)

X = O, S; R′, R″ = H, Alk, Ar

(**199**) or (**200**)

R‴ = Alk, Ar, OH, NH_2

(**197**)

R‴ = H

(**198**)

Scheme 93

Examples have been selected partly to illustrate the care needed in the critical assessment of the plausible Scheme 93 in new situations, and partly to illustrate the general nature of the reaction for the preparation of triazolines and triazolidines.

Although both ammonia and methylamine may be used to convert the oxadiazole (**201**) into the triazoles (**202**) and (**203**) in good or excellent yields, respectively (Scheme 94), the isolated intermediates (**204**) and (**205**) of (Scheme 93) correspond to (**199**) in the former case and to (**200**) in the latter ⟨62JOC3240⟩. The conversion of (**196**; X = O, R', R" = Ar) into (**198**; R', R", R''' = Ar) occurs with considerably inferior yields without isolation of the amidrazone intermediate ⟨81HC(37)1, p. 72⟩. The dipentafluoroethyl analogue of (**201**) reacts with hydrazine *via* an intermediate related to (**204**) or (**205**) to give the expected 4-aminotriazole ⟨66JOC781⟩.

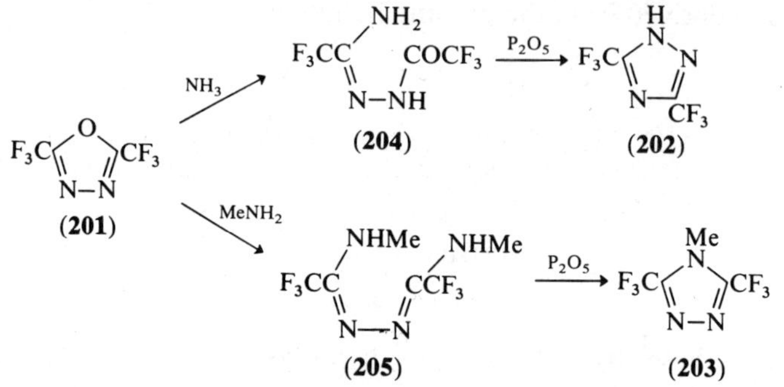

Scheme 94

In general, the action of alkali on aminooxadiazoles (**196**; X = O, R' = H, alkyl, aryl; R" = NH$_2$) affords the acylsemicarbazone intermediate (**206**) that cyclizes to the triazolinone (**207**). Ethanolic alkali reacts by ethanolysis to give (**208**) which cyclizes to ethoxytriazole (**209**; Scheme 95). Recyclization of aminoanilinooxadiazole could afford both of the triazolinones (**210a**) and (**210b**; Scheme 96) but only the latter is obtained ⟨65LA(683)149, 65LA(685)176⟩.

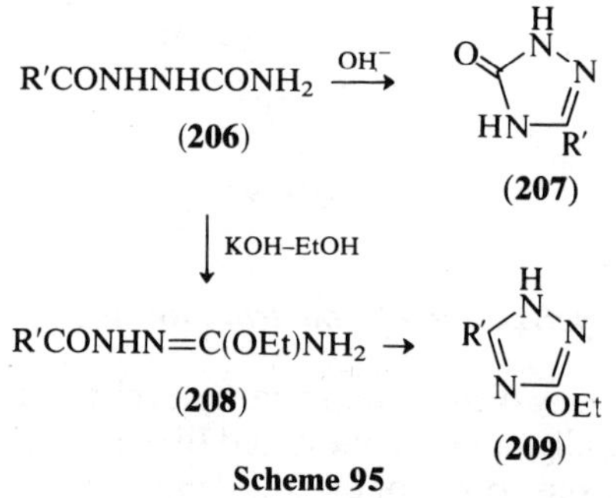

Scheme 95

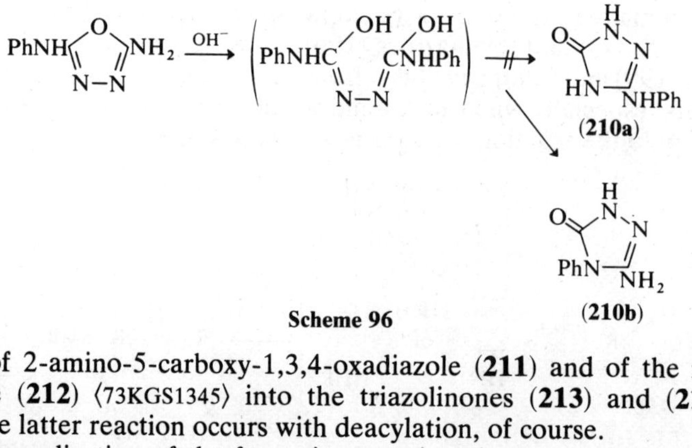

Scheme 96

Conversions of 2-amino-5-carboxy-1,3,4-oxadiazole (**211**) and of the mesoionic acylaminooxadiazole (**212**) ⟨73KGS1345⟩ into the triazolinones (**213**) and (**214**) are normal (Scheme 97). The latter reaction occurs with deacylation, of course.

An interesting application of the foregoing reactions is the preparation of azotriazoles (**215**; Scheme 98) that involves oxidation of the aminoguanidine intermediate (**216**) before cyclization ⟨69JPR213⟩.

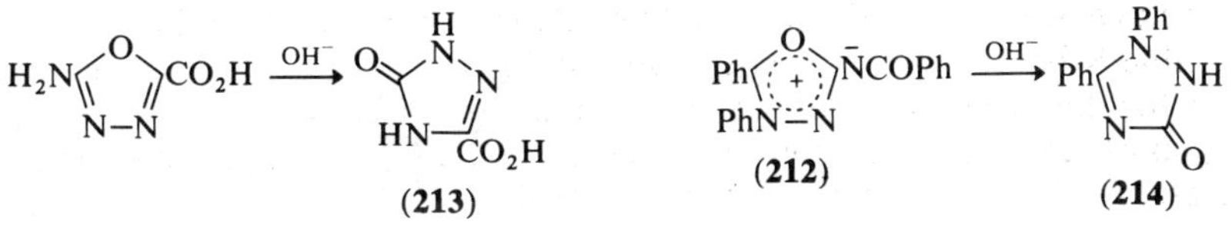

Schemes 97, 98, 99, 100, 101 (structural diagrams)

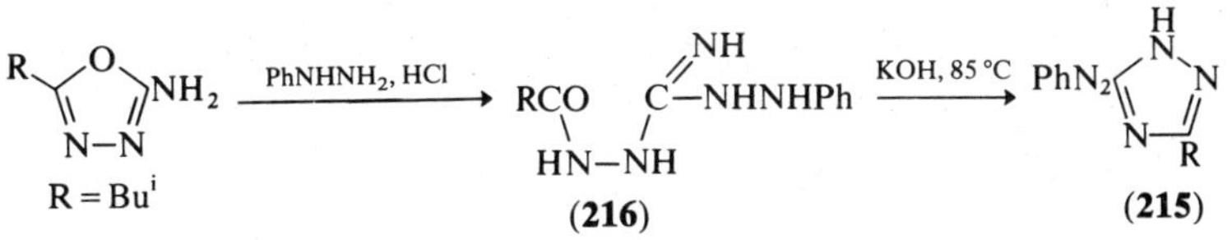

The transformation of oxazolidines (**217**) to triazolidines (**218**) (Scheme 99) requires acidic conditions to cleave the cyclic hemiaminal ⟨65RTC408⟩ and obviously does not proceed through an amidrazone intermediate.

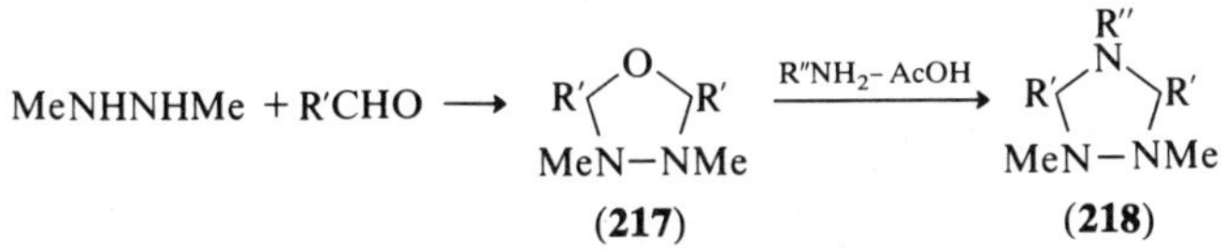

The action of diamines or aminotriazoles on 1,3,4-oxadiazoles (Scheme 100) is a synthetic route to bistriazoles such as (**219**) and (**220**) ⟨68JPR(309)168⟩. Bistriazole (**221**) formation combined with alkylation of the resulting triazolinone (*cf.* Scheme 95) can make use of glycols instead of monohydric alcohols ⟨68KGS372⟩ as in Scheme 101. Related polytriazoles with aromatic links have also been prepared ⟨81HC(37)1, p. 576⟩.

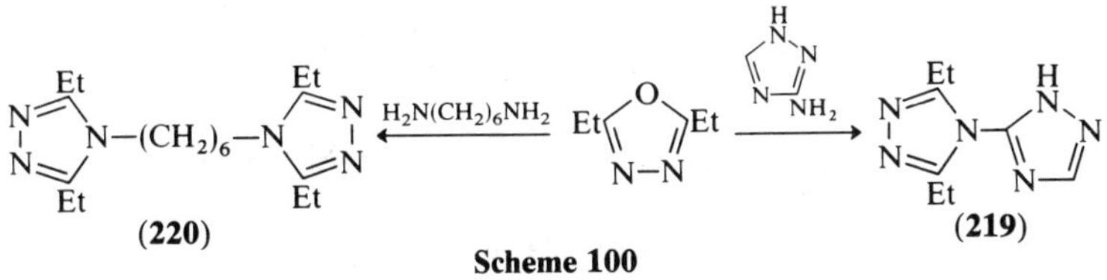

The rearrangement of 1,3,4-thiadiazoles to triazoles is similar to that of the oxa analogues, but the former are less reliable, partly because they require higher temperatures below which by-products are favoured. Thus the rearrangement of 2-methylamino-1,3,4-thiadiazole (**222**) to the triazolinethione (**223**; Scheme 102) occurs in good yield at 160 °C ⟨57CB202⟩, while a similar reaction of (**224**) to (**225**) occurs only to the extent of 30% in boiling methanol with the formation of an equal amount of (**226**) ⟨61YZ1333⟩. However, the reaction of (**224**) with hydrazine in methanol gives the 4-amino-5-thione (**227**) in good yield in a reaction proceeding through nucleophilic displacement of the halogen and ring opening to formyl thiocarbohydrazide ⟨62YZ683⟩. The same kind of ambiguity besets the reaction of (**228**) with aniline (Scheme 103) that leads to a mixture of triazolethione (**229**) and aminoanilinothiadiazole (**230**) ⟨70IJC391⟩; the latter rearranges to (**229**) in ethanolic potassium hydroxide ⟨68IJC287⟩.

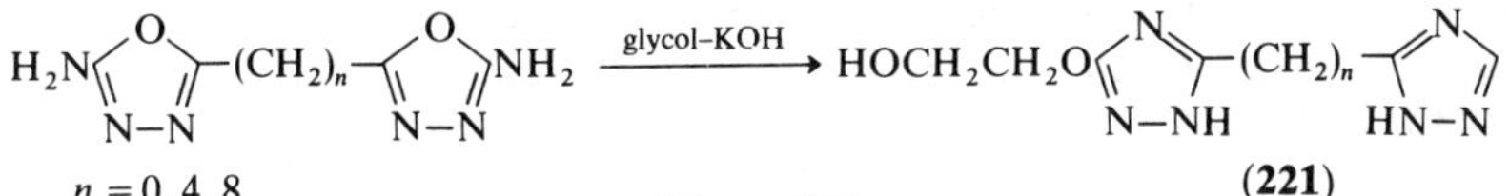

The effects of unusual leaving groups have to be kept in mind, such as that of NO_2 shown in Scheme 104 ⟨61YZ1333⟩.

A brief comparison of rearrangements of oxadiazolium and thiadiazolium salts deserves mention. The sequence of reactions represented in Scheme 105 has been well studied. The oxodiazolium salt (**231**) reacts with aniline to afford the acylamidrazone (**232**) ⟨71JCS(C)409⟩

(222) (223)

(227) (224) (225) (226)

Scheme 102

(228) (229) (230)

Scheme 103

Scheme 104

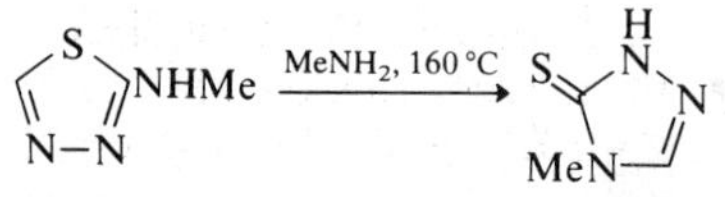

which on heating with acetic acid cyclizes to the triazolium salt (**233**) which is also available by the reaction of (**231**) with aniline in acetic acid ⟨71JCS(B)1648⟩. A related reaction (Scheme 106) leading to the triazolium salt (**234**) can be used for the reversible preparation of the triazoline (**235**) ⟨70JCS(C)807⟩.

(231) (232) (233)

Scheme 105

(234) (235)

Scheme 106

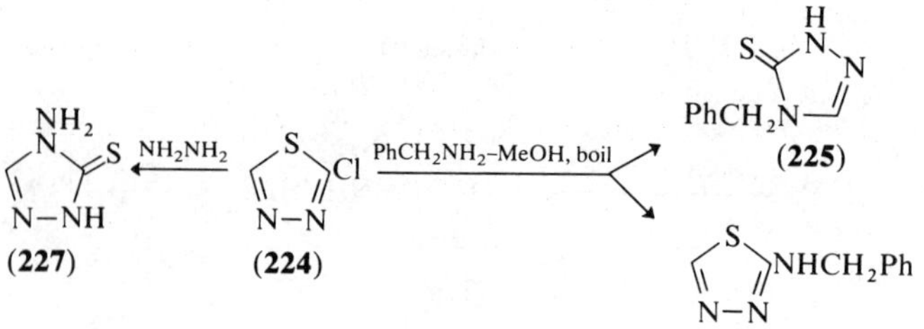

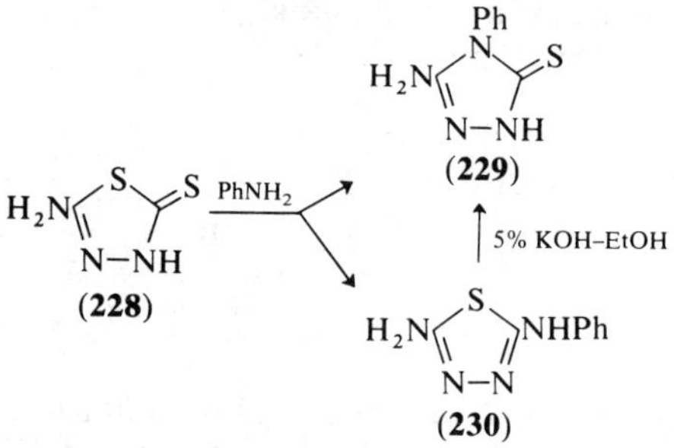

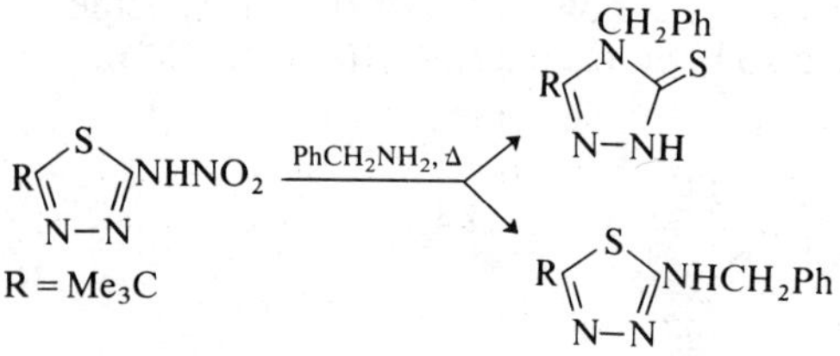

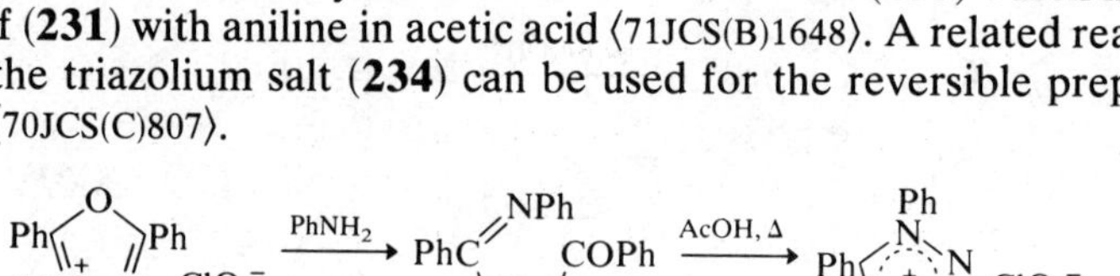

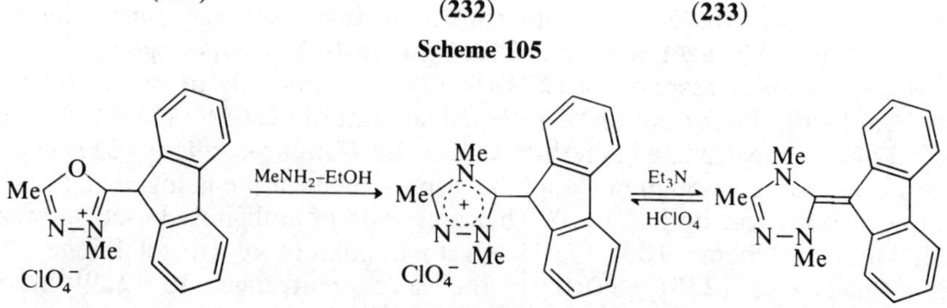

These reactions may be compared with those of thiazolium compounds (Scheme 107) which proceed analogously except when the presence of a good leaving group such as SMe permits alternatives. The formation of the thioacylamidrazone (**236**) from the thiazolium salt (**237**) reacting with an amine to give the triazolium salt (**238**) is reasonably well established ⟨74ZOR377⟩. Rearrangement of thiadiazolium compounds of the types (**239**) to triazolium compounds by bases depends on the basic strength: aliphatic amines and weaker aromatic bases may fail to react beyond acyclic intermediates related to (**236**) ⟨81HC(37)1, p. 622⟩. Methylthiothiazolium salts (**240**) also react differently with bases of different strength ⟨73IJC753⟩, e.g. as in Scheme 108. The reaction with RNH_2 occurs readily when R = Bu, but only in the presence of excess base for R = Ph. Oxadiazolium compounds give low yields of triazolium compounds irrespective of whether aliphatic or aromatic bases are used.

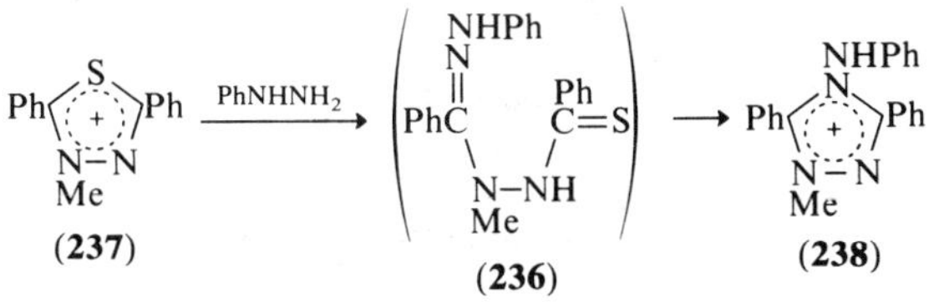

Scheme 107

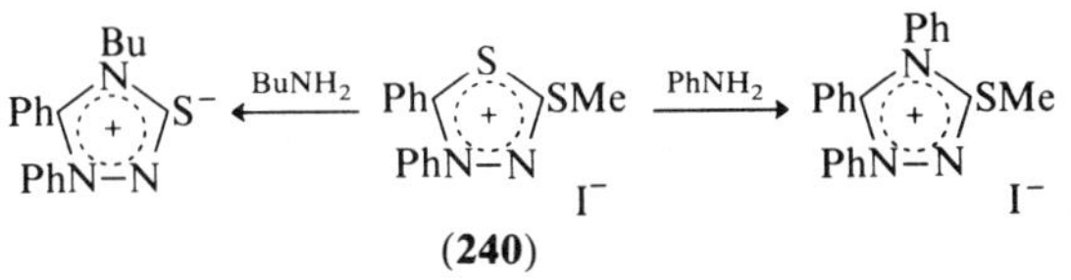

Scheme 108

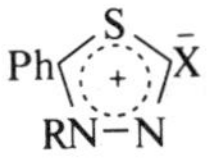

(**239**) R = Me, Ph; X = S^-, N^-Ph

Transformations of 1,3,4-oxadiazolinones into triazole ⟨81HC(37)1, p. 329⟩ derivatives are less common than those of 2-aminooxadiazoles. An unusual example (Scheme 109) is the rearrangement of the iminooxazolidinone (**241**) into the triazolidinedione (**242**) ⟨75JOC3112⟩.

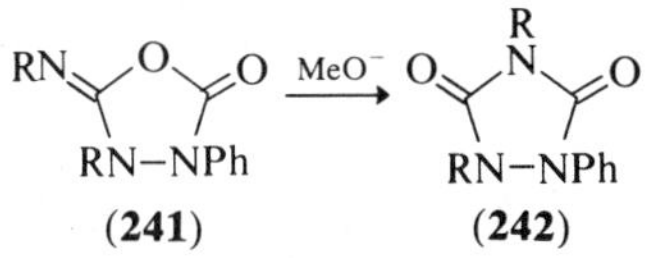

Scheme 109

An acylamidrazone (**243**) is thought to be the intermediate in the rearrangement of the arylazo-1,2,4-oxadiazole (**244**) to the triazole (**245**) ⟨74JCS(P1)81⟩ notable for the dual role of catalytic hydrogenation in Scheme 110. The acidic rearrangement of the 1,2,4-thiadiazole (**246**) to (**247**) ⟨63JCS4558⟩ may proceed through a similar intermediate. On the other hand, the thermal or base-induced rearrangement of the 1,2,4-oxadiazole (**248**) to (**249**) ⟨71JHC137⟩ necessitates N—N bond formation and is unlikely to proceed through an amidrazone intermediate (Scheme 111). The same applies to N—N bond formation (Scheme 112) in the reaarrangement of the 1,2,4-oxadiazolinone (**250**) to the triazolinone (**251**) ⟨74MI41200⟩ while the methyl ester (**252**) is formed from the aminoimidazolidinetrione (**253**) through an acylsemicarbazide intermediate (**254**) ⟨75JHC771⟩.

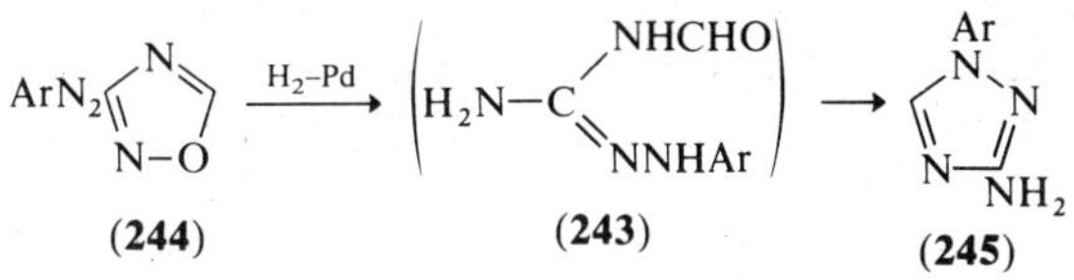

Scheme 110

Yet another example of N—N bond formation is presented by an intramolecular reaction of the oxime and imine functions of (**255**) in the preparation of the triazole (**256**; Scheme 113) ⟨74ZC270⟩.

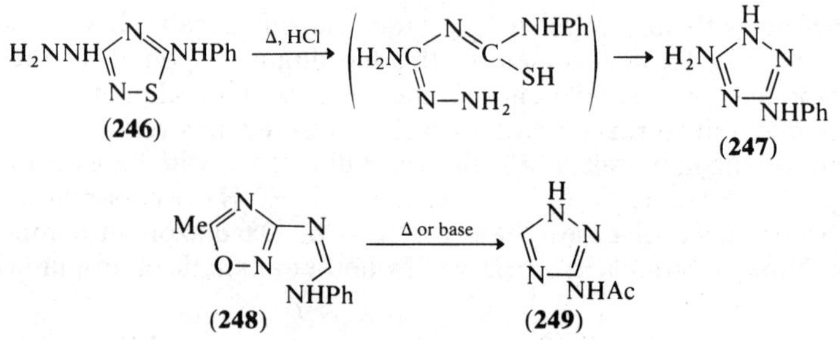

776   *1,2,4-Triazoles*

Scheme 111

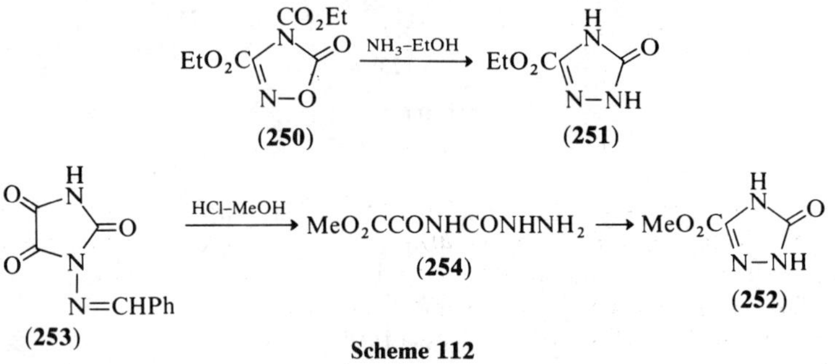

Scheme 112

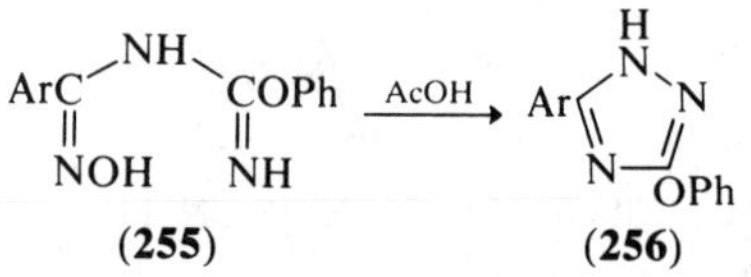

Scheme 113

Hydrazinolysis of 1,3-oxazolin-4-ones (**257**) to afford triazoles with an alcoholic side-chain (**258**) ⟨70JCS(C)1515⟩ and extension to iminooxazolidinones (**259**) to provide (**260**), an amino analogue of (**258**) ⟨74ZC267⟩, is known to proceed through acylamidrazone analogues (Scheme 114).

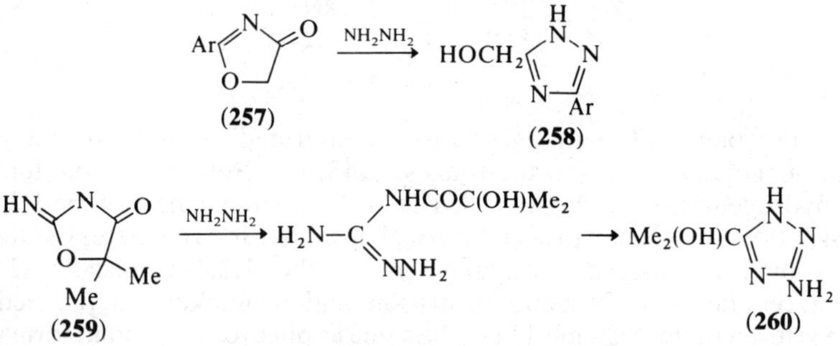

Scheme 114

Kindred methods for the preparation of mercaptomethyltriazole (**261**; Scheme 115) from thiazolidinethiones (**262**) and (**263**) are complicated by the reactivity of the thione in the case of the former ⟨71KGS39⟩ but not in that of the latter ⟨81HC(37)1, p. 409⟩. Another example illustrating the reactivity of the thione and formation of triazolinethione through an acylamidrazone intermediate ⟨73DOK(211)1369⟩ results in thioacyltriazolinone (**264**) or the hydrazone of the phenylacetyl analogue (**265**; Scheme 116).

A different synthetic strategy for the preparation of triazoles with a carboxylic acid or related functional sidechain starts from oxazolinones or thiazolinones, *i.e.* cyclic amides, to provide the carboxylic function. An ArN_2^+ or ArNHN= functional group is attached to the reactive position to obtain amidrazones with endocyclic amide and exocyclic hydrazide moieties. Note that in the examples, which are not intended to exhaust variations of this approach, 4-ones as well as 5-ones may be used.

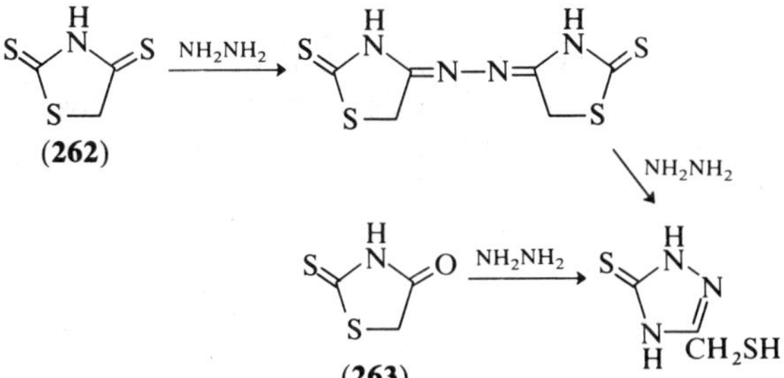

1,2,4-Triazoles

Scheme 115

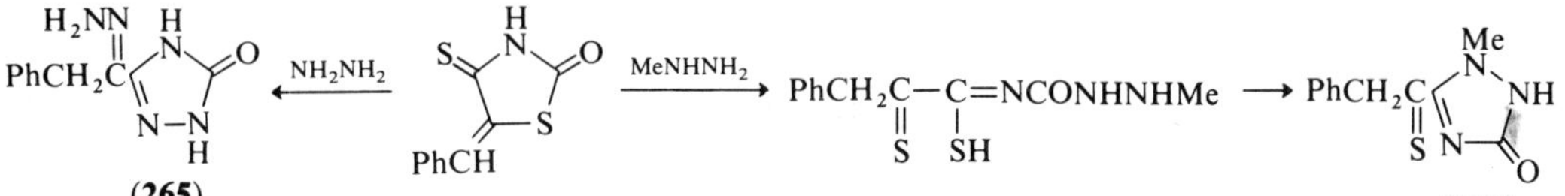

(262) → (263) → (261)

Scheme 116

(265) ← (PhCH thiazolinone) → PhCH₂C(=S)–C(SH)=NCONHNHMe → (264)

The conversion of 2-phenyl-4-phenylazooxazolin-5-one (**266**; Ar = Ph) into the ester (**267**; Ar = Ph, X = MeO) with methanolic alkali ⟨57JA1955⟩ appears to have been observed before by another worker who, on insufficient analytical data, mistook (**267**) for the acyclic intermediate (**268**), some of which was isolated and identified and then cyclized to (**267**) ⟨62JCS575⟩ in an extension of this useful reaction (Scheme 117) ⟨81HC(37)1, p. 96⟩. The reaction has been extended to rearrangements of 4-arylazo-2-alkoxy-2-thiazolin-5-ones ⟨81HC(37)1, p. 303⟩ which may occur with or without retention of the sulfur.

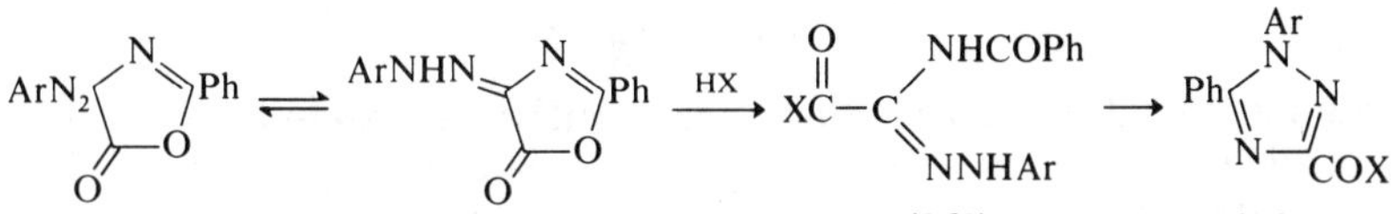

(266) ⇌ (ArNHN= intermediate) →HX→ (268) → (267)

X = O⁻, OAlk, SPh, NH₂, NHAr, NHNH₂

Scheme 117

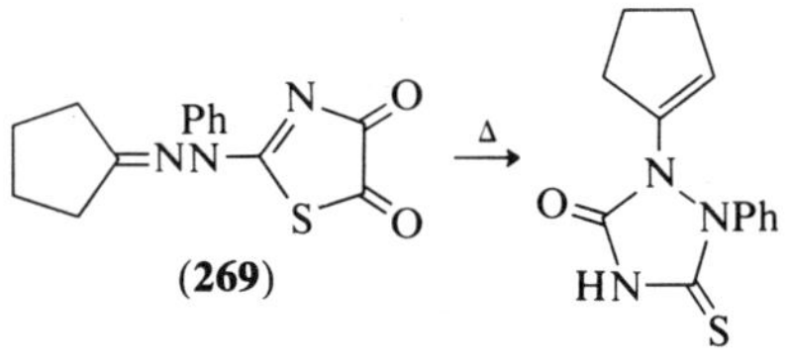

(269) →Δ→ (270)

Scheme 118

The thiazolidinedione (**269**) may be regarded as an acylamidrazone; its pyrolysis (Scheme 118) affords a triazolidinonethione (**270**) ⟨74CZ512⟩.

Although these transformations or oxazolinones and thiazolinones can be extended to imidazolones such as (**271**; Scheme 119), which also proceed through the acylamidrazone intermediate (**272**), the alkaline reagent (0.2% sodium hydroxide) may cause partial hydrolysis of the amide with the formation of acid (**273**; 2%) or the amide (**274**) ⟨69AC(R)434⟩.

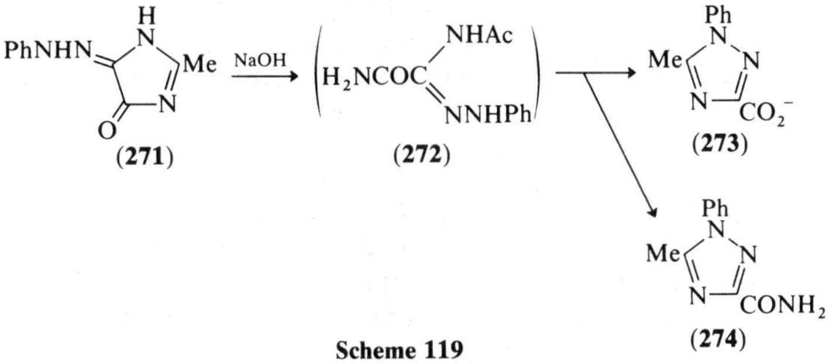

(271) →NaOH→ (272) → (273)

→ (274)

Scheme 119

The use of 6-membered ring systems as sources of amidrazones to be converted into triazoles is relatively rare; the case of 1-aminoadenine (Scheme 91) has already been mentioned. Transformations of 4,6-dihydrazinopyrimidine, hydrazines formed from haloaminopyridazine ⟨81HC(37)1, p. 8⟩ and 1,2,4-triazines ⟨81HC(37)1, pp. 409, 424, 488⟩ all involve amidrazone intermediates cyclized to triazoles with the acyl function derived from the original ring. For instructive examples of preparative links between tetrahydrotetrazine, formazans and 1,2,4-triazoles, see Scheme 120 ⟨81KGS694⟩.

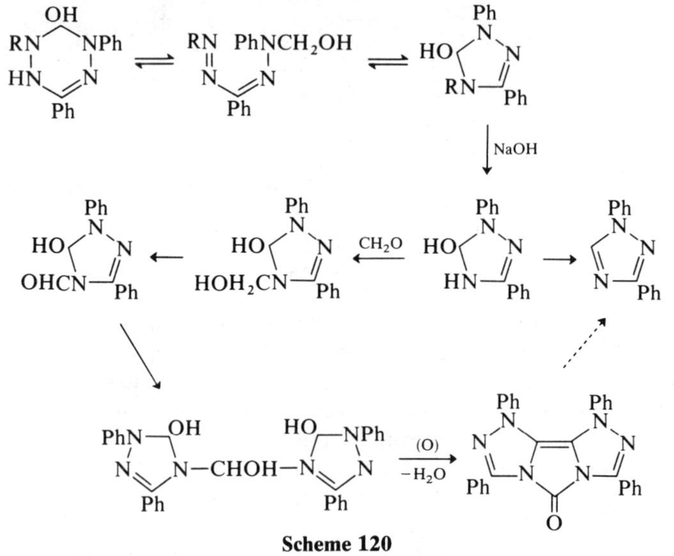

Scheme 120

The formation of dihydro or tetrahydrotetrazine intermediates in Pellizzari type reactions carried out at high temperatures has long been suspected. The conversion of diphenyltetrazine (**275**) into 3,5-diphenyltriazole (**276**) by reduction ⟨62LA(654)146⟩ occurs in two stages (Scheme 121). Formation of the aminotriazole (**277**) may be interpreted as an intramolecular reacylation of the amidrazone analogue derived from (**275**) followed by reductive cleavage of N(4)—NH$_2$. The conversion of (**275**) into (**276**) with ethanolic alkali also is consistent with the known lability of N(4)—NH$_2$.

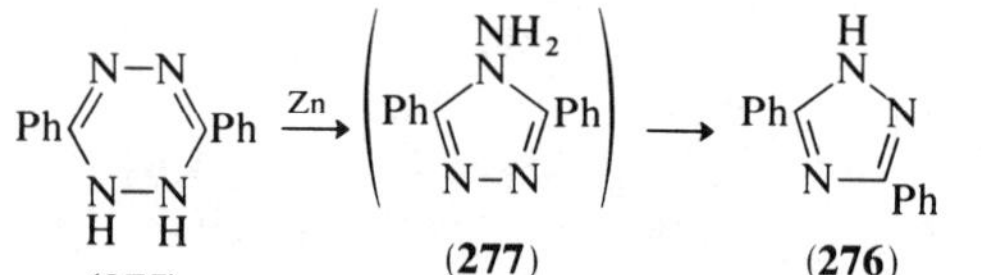

Scheme 121

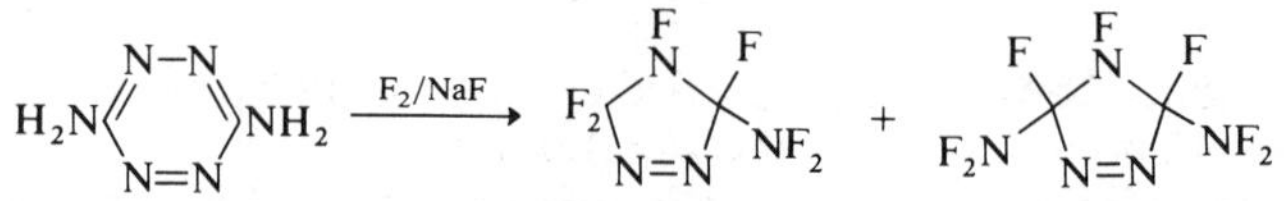

Scheme 122

Diaminotetrazine is converted by fluorination into polyfluorotriazolines (Scheme 122) ⟨67USP3326889⟩.

An unusual preparation of triazoles proceeding through an amidrazone intermediate (**278**; Scheme 123) is the transformation of 4,6-diphenyl-1,2,3,5-oxathiadiazine 2,2-dioxide (**279**) with hydrazines to 3,5- or 1,3,5-triaryltriazoles (**280**) ⟨63CB2070⟩ in yields of around 90%.

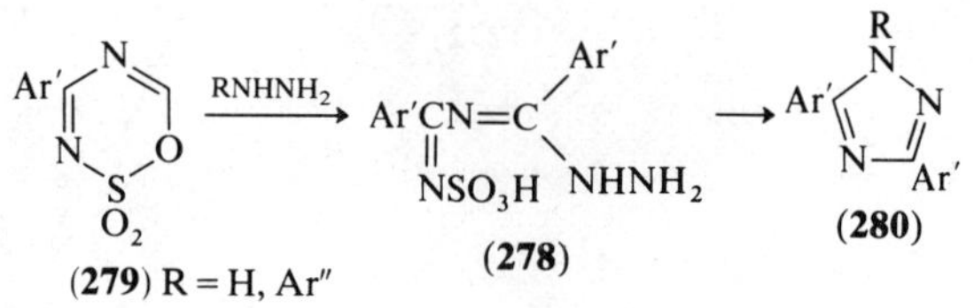

Scheme 123

Reference has been made to triazole syntheses from the use of cyclic amidrazones such as aminoquinazolines (Scheme 92). Reactions of benzoxazines with hydrazines give related intermediates (Scheme 124) leading to 4-aryltriazoles ⟨69LA(729)124⟩. Related techniques include the use of benzoxazinonethiones (**281**) ⟨72ZC175⟩ and the acidic rearrangement of 2,3-benzoxazines (**282**) ⟨72JHC581⟩ that afford 5-aryltriazoles (Scheme 125). Reaction of symmetrical hydrazines with benzoxazinone perchlorate (**283**) affords 2-aryltriazolium salts (**284**) or analogues with an aliphatic side chain (**285**) when the oxazolinium perchlorate (**286**) is used as a starting material ⟨81KGS1423⟩ (Scheme 126).

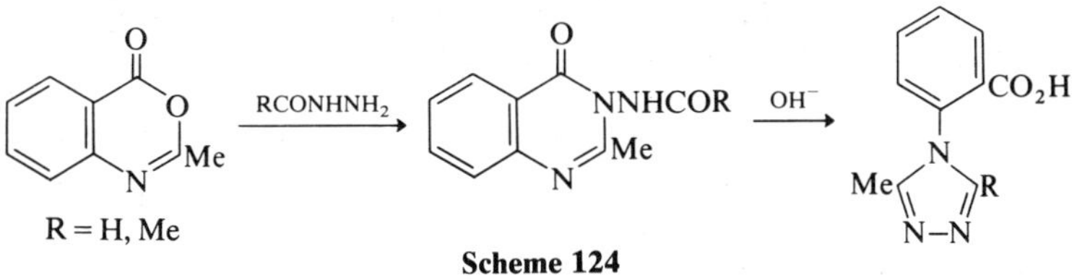

Scheme 124

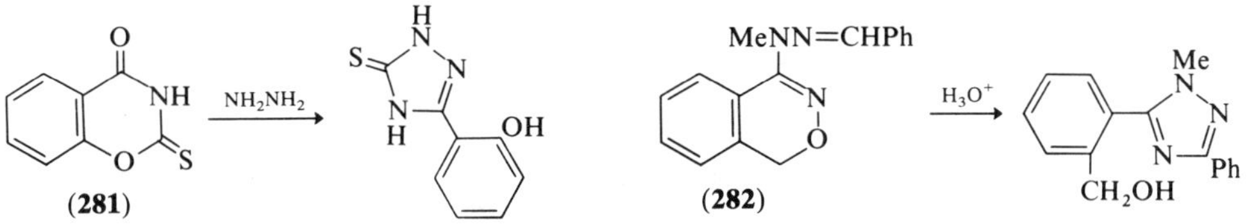

Scheme 125

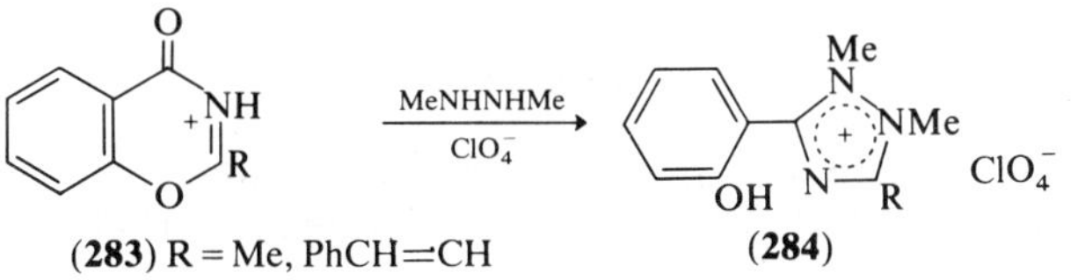

Scheme 126

The triazepinethiones (**287**) ⟨59JPR(8)298⟩ and (**288**) ⟨67JCS(C)952⟩ both give the same 1-methyl-3-phenyl-1,2,4-triazoline-5-thione (**289**), the former on treatment with base, the latter with acid (Scheme 127).

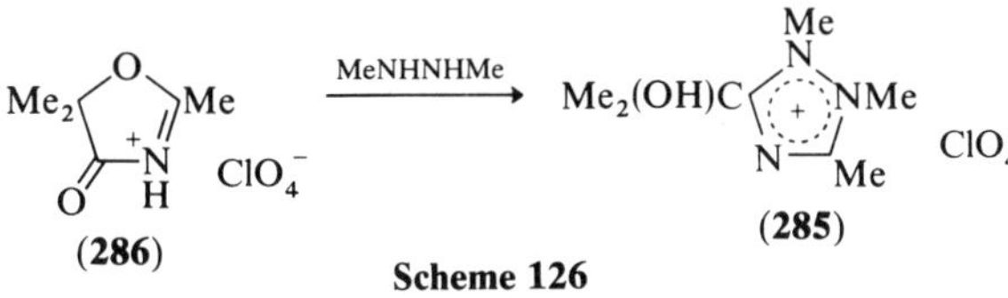

Scheme 127

4.12.4.3.3 Nitrilimines derived from non-triazole rings

Nitrilimines for the preparation of triazoles are often generated from tetrazoles ⟨61MI41203, 62LA(653)105⟩ or from 1,3,4-oxadiazoline derivatives, *e.g.* (**290**; Scheme 128) ⟨74TL1673⟩, which themselves are obtainable by thermolysis of tetrazoles. As illustrated in Scheme 129 ⟨68CB743, 70JOM(21)427⟩ self condensation of nitrilimines leads to ambiguity. The rearrangement of (**291**) to (**292**) is controversial.

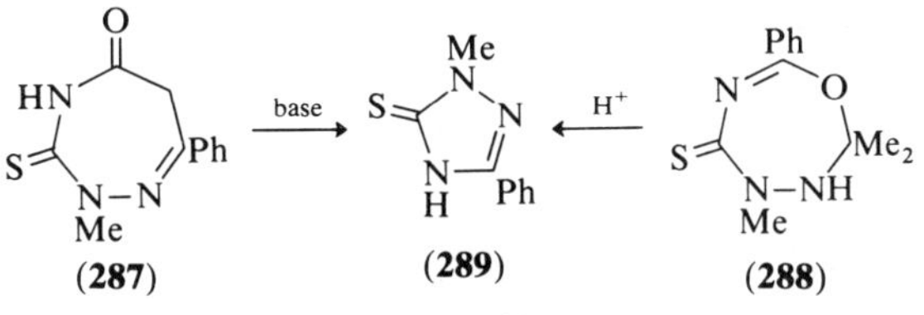

Scheme 128

An interesting application of self-condensation of the intermediate nitrilimine already incorporating the nitrile reagent ⟨67JOC1871⟩ is useful for the preparation of symmetrical 3,5-disubstituted 1,2,4-triazoles such as (**293**; Scheme 130) but the method illustrated in Scheme 129 is better for the preparation of unsymmetrical analogues (**294**).

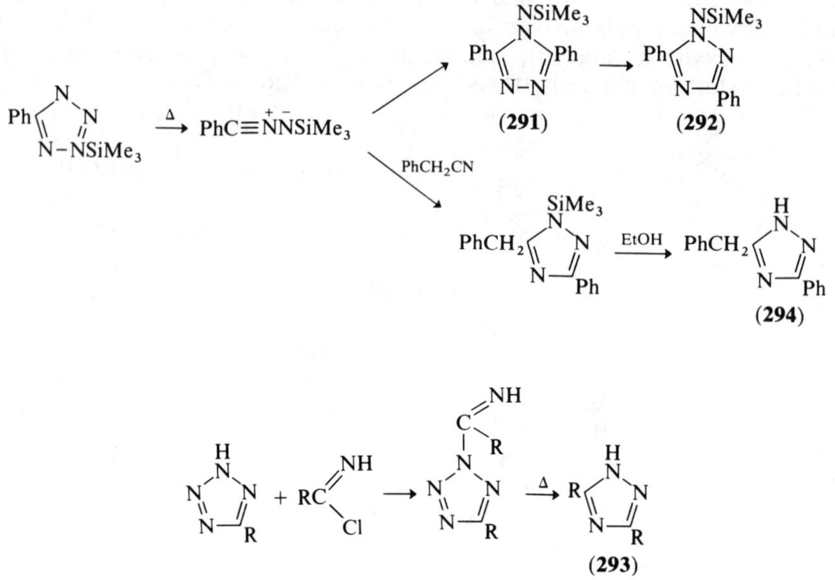

Scheme 130

Formation of a nitrilimine intermediate may be assumed in the synthesis of the triazolidinedione (**295**) from the diaziridinone (**296**; Scheme 131) ⟨76JOC2813⟩.

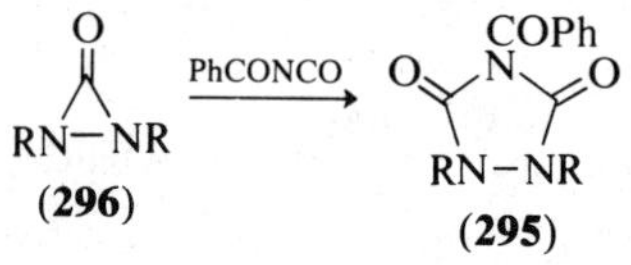

Scheme 131

Photolysis of azirine (**297**), or thermolysis of the oxazaphosphole (**298**), affords reactive azomethines (**299**) and (**300**), respectively. These react with azodicarboxylates to give triazolines (**301**) and (**302**) as in Scheme 132 ⟨74HCA1382, 73CB3421⟩. The use of diaziridine (**303**) to afford the triazolidinone (**304**) in Scheme 133 ⟨74JOC3198⟩ may be compared with the reaction shown in Scheme 131. Thermal decomposition of the aziridine (**305**) proceeds *via* an ylide (**306**) which reacts with azodicarboxylate to afford the triazolidine (**307**) in very good yield (Scheme 134) ⟨66JOC3924⟩. For similar uses of ylides see ⟨81HC(37)1, p. 516⟩.

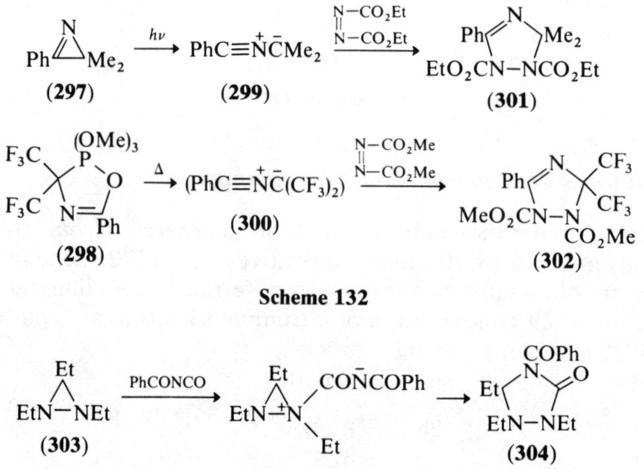

Scheme 132

Scheme 133

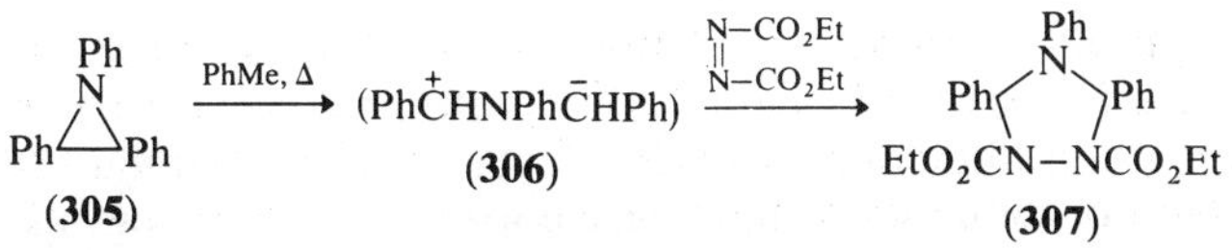

1,2,4-Triazoles

Scheme 134

Formation of the phosphatetrazepinone (**308**) intermediate is postulated in accounting for the mechanism of preparing triazoles (**309**) and (**310**) from the 'quasi dipole' (**311**; Scheme 135) ⟨69AG(E)513⟩.

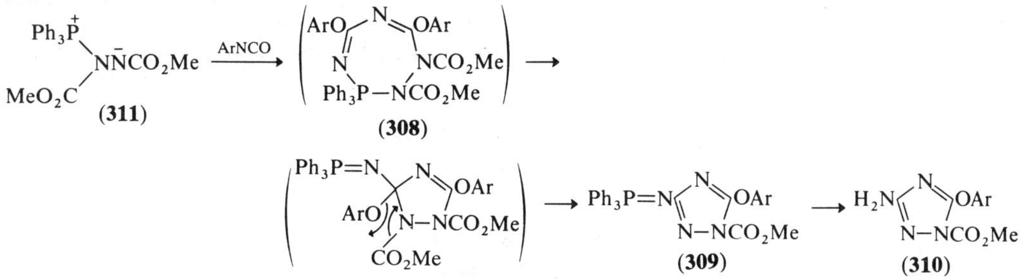

Scheme 135

4.12.4.4 Introduction and Modification of Functions on the Triazole Ring

Specific functions in specified positions of the triazole ring are available in three ways: (i) constituent portions of the ring to be formed carry the functions; (ii) the preformed triazole ring is substituted; and (iii) existing functions are modified. These synthetic approaches are not mutually exclusive. On the contrary, (i) may establish functions on the ring so as to protect a certain location so that (ii) may result in unambiguous substitution whereupon (iii) is used to remove the protective function. However, remember that (iii) may involve rearrangement, say, from N to C.

Examples of all such procedures have been given in Sections 4.12.3, 4.12.4.2 and 4.12.4.3; this section summarizes such information, much of it documented in appropriate chapters of ⟨81HC(37)1⟩.

4.12.4.4.1 Alkyl and aryl groups

Of about 300 publications dealing with the synthesis of mono-, di- and tri-alkyl- or aryl-triazoles, 76% establish the functions by method (i), 18% by (ii) and 6% by (iii); (i) is broken down further into 63% for amidrazone methods, 3% for 1,3-cycloadditions and 16% for other procedures, mainly the cleavage of fused ring systems. This rough analysis will be modified if extended to include current and future references but it may serve as a heuristic guide for those intending to devise triazole syntheses.

Alkyl- and aryl-triazolium compounds are formed mostly by quaternization of preformed triazoles, more rarely by rearrangement of oxadiazoles proceeding through acylamidrazone intermediates. An unusual method (Scheme 136) is the addition of the alkoxydiazenium fluoroborate (**312**) to the Schiff base (**313**) to form the triazolium salt (**314**) through a mechanism that may involve a triazolidine intermediate (**315**) ⟨69CB3176⟩.

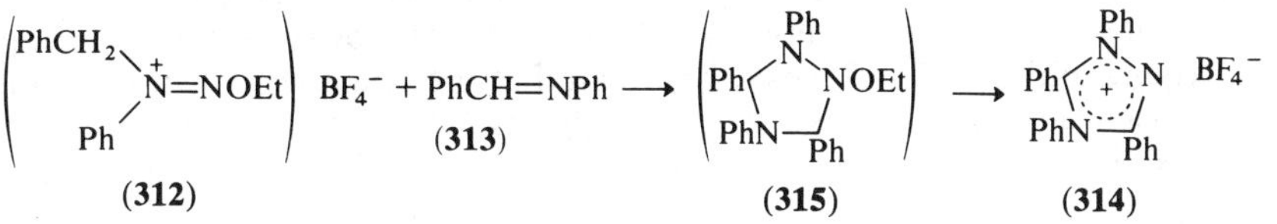

Scheme 136

Triazolinones and triazolinethiones with alkyl or aryl substituents have been prepared by the cyclization of appropriate semicarbazide or thiosemicarbazide derivatives in 84% of 500 papers; in 14% of such preparations, preformed triazoles were alkylated or arylated, but the starting materials were prepared mostly by variations of the amidrazone method. Also most of the alkylations have been carried out on thiones rather than with their oxo analogues. Similarly, of 200 modern syntheses of alkyl- or aryl-aminotriazoles, 90% made use of the cyclization of aminoguanidines or comparable amidrazones and 10% relied on 1,3-cycloadditions.

Substituted triazolidines have been obtained by Scheme 99, many by 1,3-cycloaddition and most of the triazolidinones, thiones and imines by the amidrazone method. Although chemical or electrolytic reduction of triazoles leads to ring cleavage rather than the formation of triazolidines, the reduction of triazolium salts (316) to triazolidinones or thiones (317; Scheme 137) is possible ⟨74JCS(P1)633, 74JCS(P1)642⟩. However, the reduction of triazolinones (318) and (319) with lithium aluminum hydride (Scheme 138) affords the expected triazolidinone (320) in the former case, but paradoxically results in 'oxidation' to the triazole (321) in the latter ⟨71BSF3296⟩.

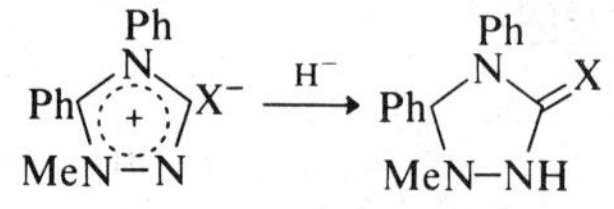

Scheme 137

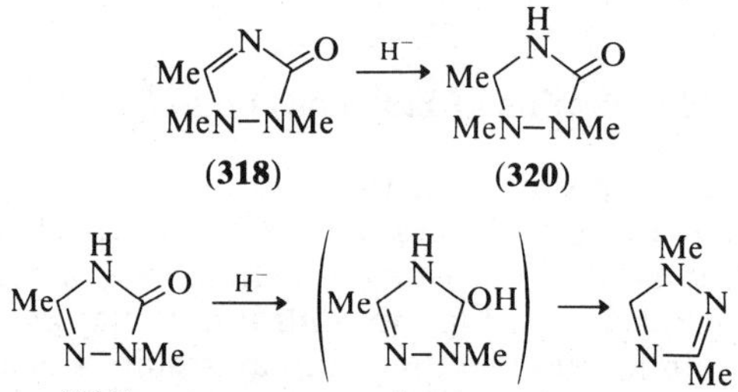

Scheme 138

4.12.4.4.2 Halo groups

The application of the amidrazone method is limited. Most of the other published preparations are equally divided between electrophilic halogenation, nucleophilic displacement of halogen or nitro groups, and displacement of diazo or nitrosamino groups. The conversion of triazolinones into chlorotriazoles with phosphorus chlorides may be regarded as a nucleophilic displacement reaction of the hydroxytriazole tautomer, while the preparation of chlorotriazoles by the action of chlorine on thiones is regarded as oxidative chlorination.

4.12.4.4.3 Nitrogen functions

Various functions have been reviewed ⟨81HC(37)1, p. 255⟩: most of their preparations rely on the reactions of amines, hydrazines, *etc.*, not specific to the chemistry of triazoles, however a few syntheses proceeding through amidrazone intermediates incorporating the phenylazo or nitramino groups have been reported.

When the *N*-function is part of another heterocyclic system further substitution, *e.g.* alkylation, implies a competition between heterocycles. The preparation and properties of some triazolylpyrazolidine-3,5-diones (322) related to the much used drug butazolidine are instructive ⟨74AJC2447, 75AJC133, 75AJC421, 76AJC2491⟩. Most of the preparations linking a triazole ring to further triazoles or other heterocycles make use of the cyclization of amidrazones.

4.12.4.4.4 Sulfur functions

For synthetic possibilities of sulfur functions see Scheme 54 and Section 4.12.3.3.1.

4.12.4.4.5 Carbonyl functions

The two major techniques for the preparation of triazolecarboxylic acids are the amidrazone method and transformations or replacements of existing functions. In the former case one can introduce the carboxylic acid through starting materials such as (**323**). Alternatively, cyclization of other amidrazones can be accomplished with derivatives of oxalic acid, *e.g.* by converting the acylamidrazone (**324**) into the amide (**325**) from which the acid may be liberated by hydrolysis. A particularly convenient method for the preparation of carboxylic acids, esters and amides is the rearrangement arylazooxazolinones (Scheme 117).

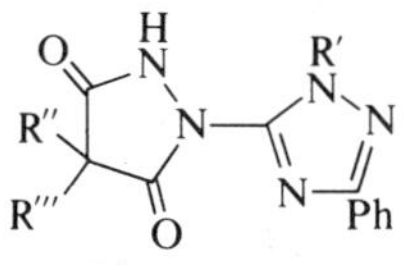

(**322**) R', R'', R''' = H, Alk 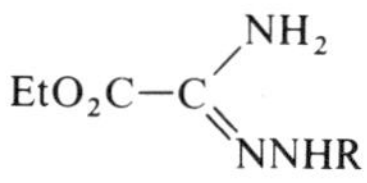   (**323**)    (**324**) 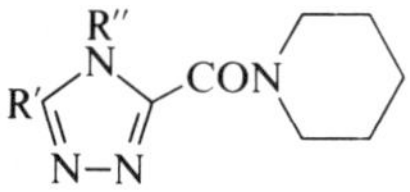   (**325**)

Conversion of existing functions include the oxidation of aliphatic or aromatic side-chains, one of the earliest preparative techniques of triazole chemistry. Hydrolysis of cyanides obtained by displacement of halo, nitro and diazo groups is of comparable importance.

Reactions of nitrilimines with ethyl cyanoformate have been little used. Given a triazolecarboxylic acid or its derivatives, further representatives of this class may be obtained by alkylation, acylation and other nuclear substitutions.

Carbonation of triazole (Scheme 139) under pressure and catalyzed by cadmium fluoride ⟨59BRP816531⟩ gives moderate yields of the mono- or the di-carboxylic acids (**326**) and (**327**) according to reaction temperature.

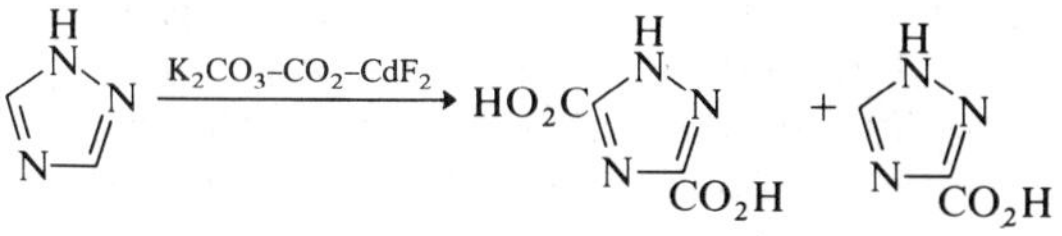

(**326**)   (**327**)

Scheme 139

Acyl derivatives other than carboxylic acids and their derivatives may be considered in two categories. One of these, *N*-acylation, has been discussed earlier (Section 4.12.3.2.2). If in view of the frequently occurring ambiguity of such procedures, an unambiguous synthesis is required, one may proceed as in Scheme 140.

Scheme 140

Triazolyl ketones have been synthesized from diacylamidrazones such as (**328**), and through oxidation of benzyltriazole ⟨68JCS(C)824⟩. The rearrangement of *N*-acyl- to *C*-acyltriazoles has been mentioned earlier as well as the reacylation of 1,1-carbonylbis-1,2,4-triazole.

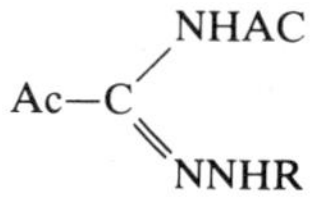

(**328**)

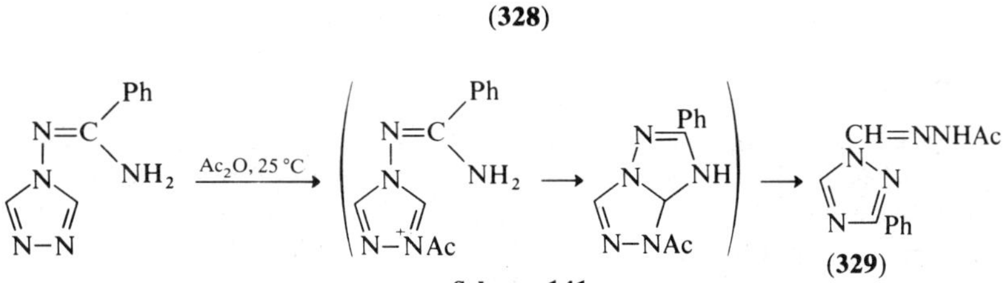

(**329**)

Scheme 141

Although 4-formyl triazoles are unstable, a curious method of preparation (Scheme 141) affords the acylhydrazone (**329**) ⟨69JPR646⟩.

The preparation of free 3-carbaldehydes (which must be *N*-substituted in order to avoid hemiaminal formation) is best carried out by oxidation of the triazolylmethanols obtainable by reduction of the corresponding esters with lithium aluminum hydride. Potassium ferricyanide in ammonia or lead tetraacetate are suitable oxidants but thermal decomposition of benzenesulfonylhydrazides (**330**) appears best ⟨62JCS575⟩.

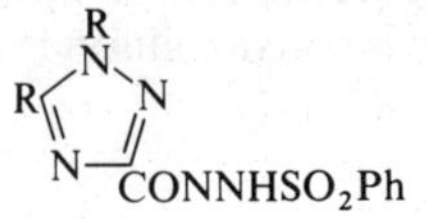

(**330**)    (**331**)

N-Unsubstituted aldehydes are available as derivatives. Acetals are prepared by the amidrazone methods; other derivatives are obtainable from the hemiaminal, *e.g.* from (**331**) ⟨70TL943⟩.

4.12.4.4.6 Metal complexes

For a brief review of metal–triazole complexes see ⟨81HC(37)1, p. 659⟩.

4.12.4.4.7 Triazolidinediones

Triazolidinediones are accessible through the general methods illustrated before and Scheme 142 presents several procedures ⟨see also 80RCR880⟩. Triazolinediones are prepared by oxidation of the triazolidinediones with lead tetraacetate, *t*-butyl hypochlorite, *N*-bromosuccinimide and the less common oxidant, activated isocyanate ⟨74JOC3506, 73TL2101, 75JCS(P2)1325, 76JHC481⟩.

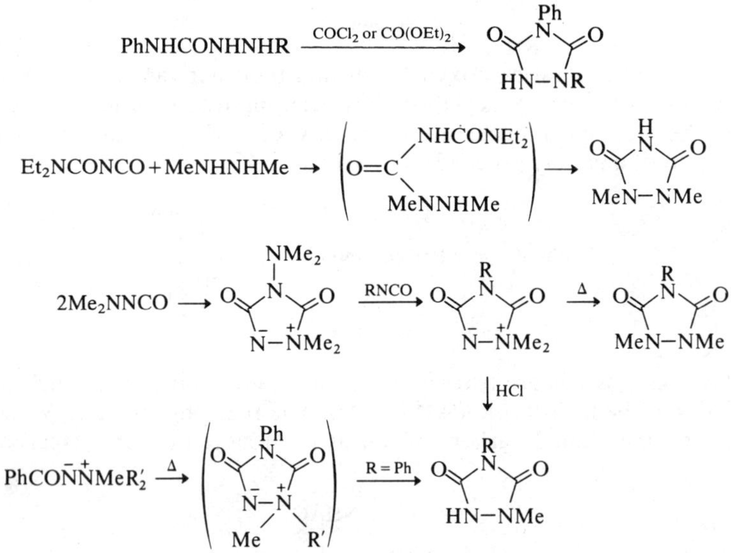

Scheme 142

4.12.5 APPLICATIONS

For the greater part of the history of 1,2,4-triazoles only two derivatives were used extensively: one in small quantities as an analytical reagent and the other on a large scale for agricultural purposes. The number of triazole derivatives in either of these categories has increased greatly but only a small fraction of approximately one thousand patents covering the preparation and use of triazoles can be treated here.

In many applications the use of 1,2,4-triazoles is not specific: the action claimed may be common to other azoles, or the compound in question may contain not only a triazole ring but so many other structural features as to make the identification of 'the' active molecular fragment difficult. In many important instances triazole derivatives are used to control weeds, fungi, insects and so on. The effectiveness of a given compound varies with the species, and such variation may be reported in hundreds of papers written by biologists rather than chemists.

The aim of this short section is to draw attention to such chemical features of 1,2,4-triazoles that have been recognized to be of practical value and discussed further in Section 4.12.5.1 which deals with analytical and synthetic applications. Uses on a larger scale are treated in Section 4.12.5.2 and are to be considered merely as examples demonstrating the wide applicability of 1,2,4-triazoles and to encourage readers with practical interests to search the vast literature in their special field of technology.

4.12.5.1 Analytical and Synthetic Applications

4.12.5.1.1 Analytical uses

The first 1,2,4-triazole to find practical application was nitron (**332**; Scheme 143) ⟨05CB856, 05CB861, 05CB4049⟩. Only 0.01% of nitron nitrate dissolves in dilute hydrochloric acid; the solubility of the perchlorate is rather less. Nitron has also been used for the determination of boron, rhenium and tungsten ⟨81HC(37)1, p. 603⟩ but the formation of a mesoionic non-stoichiometric hydrochloride of nitron ⟨80AJC2237⟩ calls for caution in its use for the development of new gravimetric procedures. In an elegant modern application ⟨79MI41200⟩, nitron is linked to polymerized *p*-chloromethylstyrene for the removal of nitrate from water. The triazolium chloride (**333**) is used for the spectrophotometric assay of cobalt ⟨68MI41201⟩, and the simple thione (**334**) for that of rhodium and platinum ⟨72MI41201⟩.

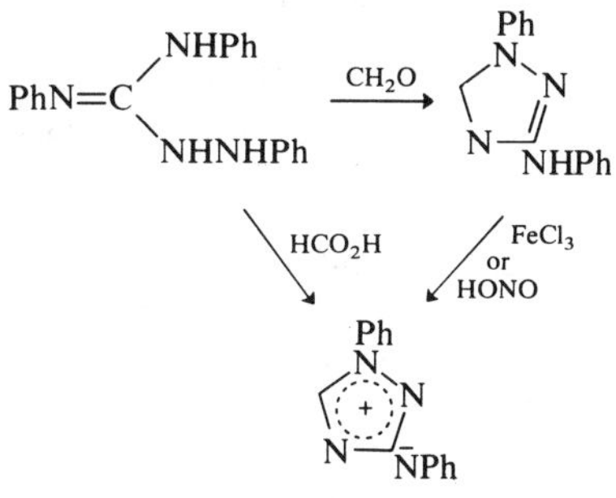

Scheme 143

(**332**)

(**333**)   (**334**)

Azo compounds derived from triazole are coloured ligands for antimony, bismuth and copper ⟨81HC(37)1, p. 609⟩. Brilliant lakes are formed with hydroxyphenylazotriazoles ⟨73AJC1585⟩; the intensity of the colour formed with various metallic ions varies with the position of the hydroxy group in the phenyl substituent.

Coupling of 3-diazo-1,2,4-triazole to proteins at pH 2.8–6.8 has been shown to affect tryptophan only, while coupling with histidine and phenols requires the usual basic conditions. This permits a specific determination of tryptophan, sharpened by preparing the reagent from 3-amino-1,2,4-(5-^{14}C)-triazole that permits assay by a scintillation method ⟨79MI41201⟩.

The reaction of 4-amino-5-hydrazino-1,2,4-triazole-3-thione (**38**) with aldehydes ⟨74OPP156⟩ has found clinical applications but the wide applicability of the reaction limits its specific use.

4.12.5.1.2 Synthetic uses

Without considering their use as synthons in preparations of fused ring systems and compounds with mixed heterocyclic portions, 1,2,4-triazoles find synthetic uses in four kinds of reaction.

Reference has been made to 1,3,5-tribromotriazole and 1,3-dichlorotriazole which act as halogenating agents (Section 4.12.3.2.2). Transfer of *N*-acyl functions to 1,2,4-triazole has been illustrated in Scheme 28 but *N*-acyltriazole and especially carbonylditriazole (**83**) may be used similarly to the analogous imidazole compounds ⟨79LA1756⟩ except that the latter are more economical.

Related to such transacylations is the ability of triazole to accept and transfer acyl groups so as to make it a catalyst for the formation of amides and esters from their constituents. In particular triazole has been used in the synthesis of peptides. Imidazole and 1,2,3-triazole which also possess both weakly acidic and weakly basic groups have similar catalytic action which, however, is accompanied by racemization not observed when 1,2,4-triazole was used ⟨65RTC213⟩. It has been found best to catalyze coupling in peptide syntheses such as Scheme 144 ⟨66LA(691)212⟩. For the use of quinolinesulfonyl-3-nitrotriazolide in nucleotide synthesis see ⟨80TL4339⟩.

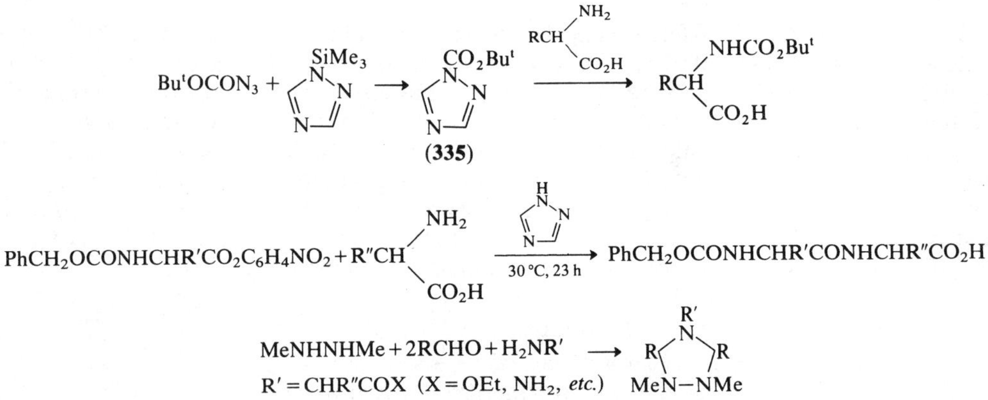

Scheme 144

Characterization or N-protection of amino acids by the *t*-butyloxycarbonyl group can be carried out either by acyl transfer from the triazole (**335**) prepared as in Scheme 144 or with *t*-butyl phenylcarbonate in the presence of triazole ⟨73TL469⟩ in lower yields. Although 1,2,4-triazole acts as a catalyst in the second of these reactions, it has to be present in an equimolar amount.

The self-condensation of glycine to triglycine at pH 6.7–8.9 is increased 10–20 fold in the presence of catalysts of which 1,2,4-triazole is one of the best. The reaction is not noted in basic solutions but the effect is maximal at the pH thought to be that of primitive oceans. The linking of triazole with prebiotic peptide syntheses is speculative and far fetched but not irrational ⟨81C59⟩.

It has been proposed that triazolidines formed from amino acids (Scheme 144) would be suitable for the characterization of amino acids and as N-protected starting material for the synthesis of peptides ⟨64RTC387⟩.

Scheme 145 represents another catalytic property of 1,2,4-triazole. Addition of the annular NH to the conjugated ketone (**336**) gives (**337**), the unstable oxime of which provides the isoxazole (**338**) not available by adding hydroxylamine to (**336**) in the absence of triazole ⟨81AG(E)885⟩.

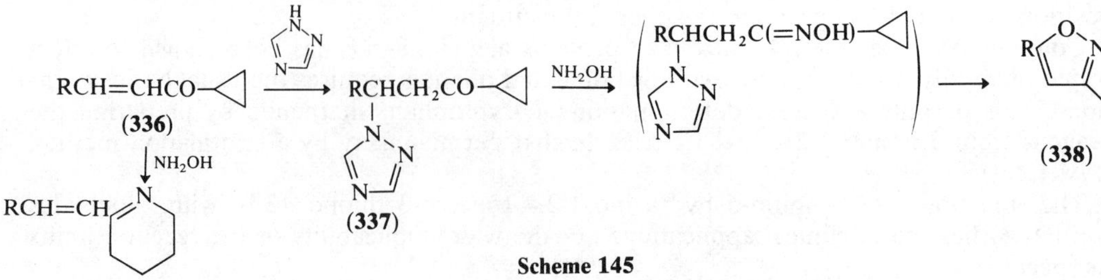

Scheme 145

The cleavage of 4-phenyltriazole by butyllithium in hexane and THF at 0 °C to afford
N-phenylcyanamide ⟨79TL3129⟩ may have advantages in laboratory scale operations but
appears too expensive for large scale purposes.

Although 1,2,4-triazoline-3,5-dione is unstable even at 0 °C, it can be reacted with dienes
in situ when formed by the oxidation of urazole (5) with dinitrogen tetroxide ⟨73TL2101⟩.
N-Substituted derivatives are more convenient to use.

4.12.5.2 Industrial Uses of Triazoles

4.12.5.2.1 Biological activity

3-Amino-1,2,4-triazole (Amitrole, Amizol, ATA, Cytrol, ENT 25445, Weedazol) was
the first 1,2,4-triazole to be manufactured on a large scale from aminoguanidine formate
⟨46OS(26)11⟩ for use as a neutral herbicide and defoliant of cotton. Its phototoxic effect and
interference with carbohydrate metabolism ⟨61CRV87, p. 121⟩ are not inconsistent with
biochemical disturbances caused through replacement of histidine with an analogue derived
from triazole rather than imidazole. Inhibition of liver and kidney catalase and impaired
thyroid function in animals and suspected carcinogenetic properties have occasionally
threatened the use of aminotriazole for the protection of edible plants (notably cranberry)
but the ill-effects observed in animals have not been found in humans. On the other hand
aminotriazole has been shown to give some protection to mice against irradiation with
X-rays.

Although some derivatives, particularly the furylidene compound, are more effective for
some purposes, none of the more complex triazole derivatives patented as herbicides has
been able to displace aminotriazole.

While aminotriazole in low concentrations may promote growth, the bistriazole (339),
prepared from triazole and dichlorodiphenylmethane ⟨72BRP1269619⟩ has the opposite effect.

(339)

4-Butyltriazole is a fungicide but injures seedlings; such damage is avoided if used as
the complex (340) ⟨71SAP7004373⟩. Other fungicides derived from triazole include *O*-
phosphorylated triazolinones. Complex triazolylglycol ethers ⟨81EUP21076, 81GEP2926280⟩
are claimed to be plant growth regulators and fungicides. Newer herbicides include
triazolidinonethiones with aromatic and lower aliphatic substituents ⟨81GEP2952685⟩. Compounds of a similar nature are insecticides, acaricides and nematicides ⟨75GEP2407304,
76GEP2554866⟩. Many alkyl and aryl derivatives of mercaptotriazole are active against fungi
and other bacteria ⟨81HC(37)1, p. 251⟩. When the substituent is attached to a cephalosporin ⟨75MI41201⟩ the antibacterial effect is best ascribed to the non-triazole heterocycle
modified by other pharmacologically active groups including the triazole portion. The same
applies to the action of (341) against *Mycobacterium tb.* ⟨73FES624⟩ or of carbamates
(342) ⟨80EUP11604⟩ against fungi and animal parasites.

(340)   (341)

(342)   (343)

Relatively simple 1,2,4-triazoles display biological activity such as inhibition of
cholinesterase ⟨55MI41200⟩, interference with mitosis ⟨51MI41200⟩ and reversible denaturation
of serum proteins ⟨52MI41200⟩, but the pharmacological significance of 1,2,4-triazoles is
slight in comparison with some other azoles ⟨61CRV87⟩. The simplest thione (334) affords

some protection of mice against radiation and (**343**) has anti-inflammatory properties comparable with the much more economically produced phenylbutazone ⟨64HCA2068⟩. The greatest promise of pharmaceuticals derived from triazoles lies in the field of polycyclic compounds outside the scope of this chapter; for virucidal amides see ⟨75GEP2511829⟩.

4.12.5.2.2 Dyes and photographic chemicals

Metalated bisazotriazoles dye fast shades ⟨73JAP7389932⟩; many dyes and whitening agents have been successfully modified by incorporating triazolium salts into their structure ⟨81HC(37)1, p. 599⟩. Cyanine dyes ⟨59JCS3799⟩ are of importance in colour photography. A combination of 1,2,4-triazole and dimethyl sulfide gold(I) chloride can be used to gild glass ⟨73GEP2245447⟩.

Like the more widely used 1,2,3-triazoles, many 1,2,4-triazole derivatives find application in photography thanks to the adsorption of thiones on silver halides, reactivity with silver compounds or the formation of disulfides as potential free radicals. Also *S*-propargylthiol (**344**) stabilizes photographic emulsions ⟨74GEP2304321⟩.

Selenium derivatives prepared by the amidrazone method from the selenium analogue of thiosemicarbazide are also used for various photographic purposes ⟨81HC(37)1, p. 510⟩.

4.12.5.2.3 Polymers

The preparation of polymers derived from 1,2,4-triazoles is currently the most important practical application of this heterocyclic system. Three recent reviews cover the substantial beginnings of a rapidly expanding field of research ⟨B-72MI41202, 77RCR76, 81HC(37)1, p. 575⟩ but only the last of them is devoted entirely to 1,2,4-triazoles. As each of the salient points summarized below is covered by several publications their documentation should be supplied by reference to these reviews.

Polymers derived from triazoles can be formed by three fundamental methods. In the simplest way 3,5-disubstituted triazoles are treated with acetylene to obtain 1-vinyl derivatives which are then polymerized with free radical catalysts. The resulting polymers do not appear to be used on their own but mixed with other polymerized alkenes they facilitate the dyeing of synthetic fabrics.

A second method is the formation of polyamides by condensation of dicarboxylic acids or diisocyanates with diamino compounds such as (**345**), the reaction of which with 2,6-naphthalenedicarboxylic acid gives heat resistant films and fibers.

The most-studied method generates the triazole ring in the course of polymerization by reactions surveyed under Section 4.12.4. Most of the techniques in use convert hydrazides or amidrazones into triazoles. Polyhydrazides are thermally cyclized with amines (Scheme 146). Of particular interest is the reaction of aliphatic or aromatic dicarboxylic acids with hydrazine that generates polyhydrazides which then react with further hydrazine to polymers with repeating units of 4-aminotriazole (Scheme 147). Polyaminotriazoles have been used to improve antistatic, dyeing and laundering properties of polyalkene fibres, also as toners in electrophotography and to protect metals against corrosion. In addition the free amino group permits the attachment of cross-linking groups and other functions, *e.g.* the carboxyl terminal of amino acids leaving the amino terminal free for anchored peptide syntheses.

Acylation of bisamidrazones followed by thermal cyclization of the linear polymer (Scheme 148) has provided valuable polymers of high thermal stability; benzene and pyridine

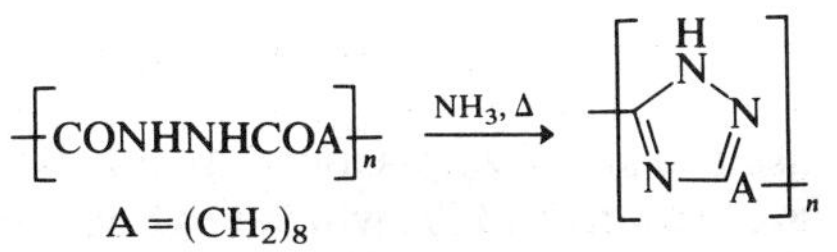

Scheme 146

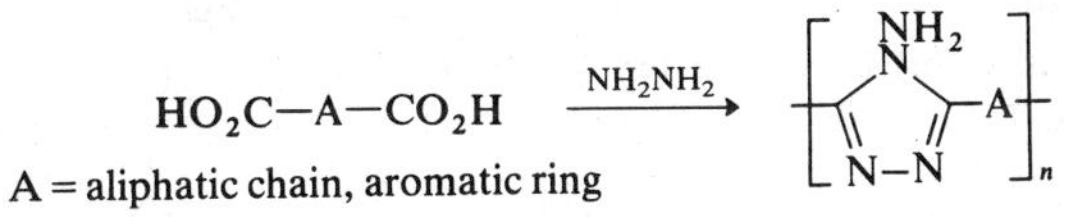

$\{CONHNHCOA\}_n \xrightarrow{NH_3, \Delta}$ [1,2,4-triazole with A bridge]$_n$

A = (CH$_2$)$_8$

Scheme 147

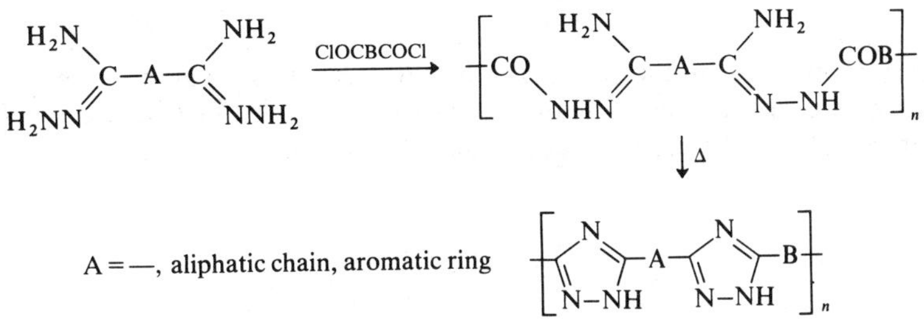

$HO_2C-A-CO_2H \xrightarrow{NH_2NH_2}$ [4-amino-1,2,4-triazole with A]$_n$

A = aliphatic chain, aromatic ring

Scheme 148

Bisamidrazone + ClOCBCOCl → [open-chain bis-acylamidrazone intermediate]$_n$ $\xrightarrow{\Delta}$ [bis-1,2,4-triazole with A and B bridges]$_n$

A = —, aliphatic chain, aromatic ring

bridges have been used. Polytriazolines formed by using dialdehydes instead of dicarboxylic acids are unstable.

Of particular interest are the 'ladder' polytriazoles formed from bisamidrazones with di- and tetra-carboxylic acids or bisbenzoxazinones, *e.g.* Scheme 149. Such polymers derived from napththoic acids have a high average molecular weight, high thermal resistance and good resistance to wear.

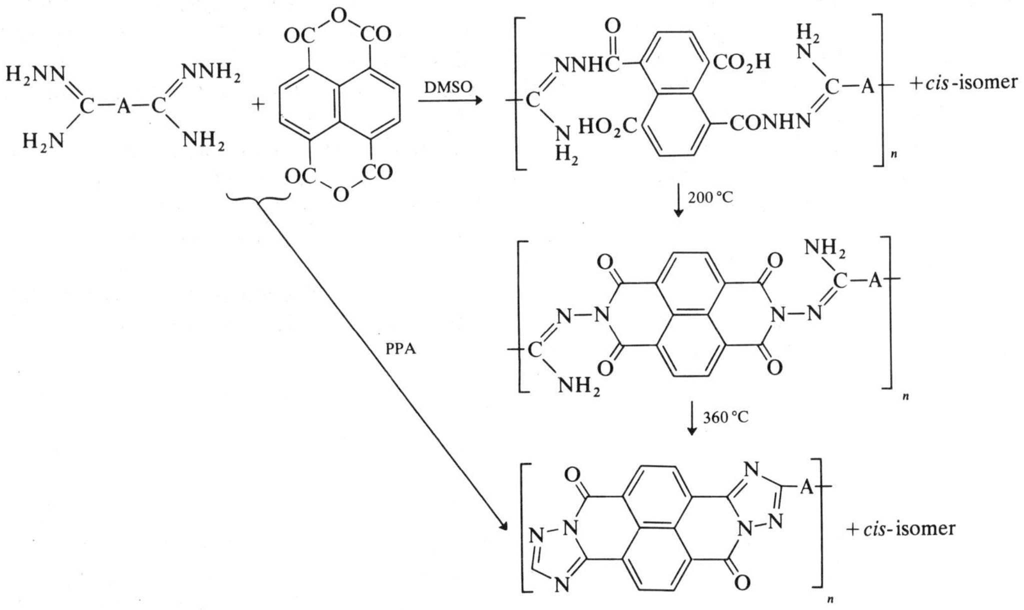

Scheme 149

Bisamidrazone + naphthalenetetracarboxylic dianhydride $\xrightarrow{DMSO}$ [open-chain intermediate]$_n$ + *cis*-isomer $\xrightarrow{200\,°C}$ [bis-imide intermediate]$_n$ $\xrightarrow{360\,°C}$ [ladder polytriazole]$_n$ + *cis*-isomer (also via PPA directly)

The transformation of polyoxazoles and polytetrazines into polytriazoles are analogous to transformations treated under Section 4.12.4.3.

Finally, mention should be made of polytriazoles formed by 1,3-addition reactions of dicyanides with bisnitrilimines conveniently generated from bistetrazoles. The polymers prepared from 1,3-dicyanohexafluoropropane were thermally stable but had a low degree of polymerization.

Detailed tables of 43 polytriazoles will be found in ⟨B-72MI41202⟩.

4.12.5.2.4 Other uses

The triazole (**346**) can be used to stabilize plastics ⟨73JAP7308667⟩. The reaction of tributyl lead hydroxide and aminotriazole affords (**347**) which has been recommended as an antiwear component of lubricating oils ⟨68FRP1525268⟩. Diazonium compounds detonate, although they are safer to handle than comparable derivatives of tetrazole. Polyfluorotriazolines ⟨67USP3326889⟩ are oxidants; their suitability as bleaching agents, for rocket fuels and constituents of pyrotechnic compositions has been claimed.

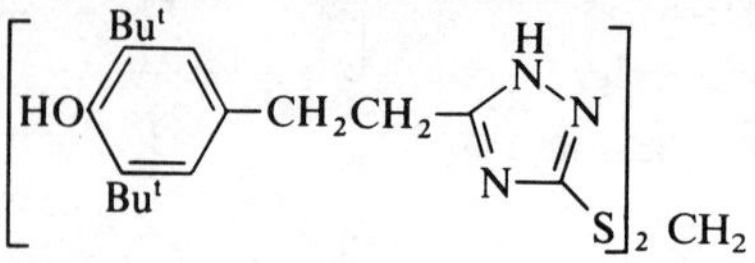

(**346**) 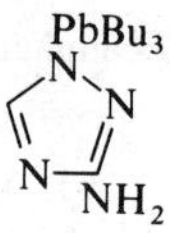   (**347**)

Polymeric complexes of 1-phenyltriazole with transition metal cyanates or thiocyanates are structurally and stereochemically understood ⟨80MI41200⟩ and an antiferromagnetic complex of triazole is of potential technical interest ⟨80ZN(A)1387⟩.

4.13
Tetrazoles

R. N. BUTLER
University College, Galway

4.13.1 INTRODUCTION	792
4.13.2 STRUCTURE	792
4.13.2.1 Possible Structures	792
4.13.2.1.1 Aromatic systems not requiring exocyclic conjugation	792
4.13.2.1.2 Aromatic systems involving exocyclic conjugation	793
4.13.2.1.3 Non-aromatic systems	794
4.13.2.2 Theoretical Methods	795
4.13.2.2.1 Dipole moments	795
4.13.2.2.2 Electronic structures and stability	795
4.13.2.3 Structural Methods	796
4.13.2.3.1 X-ray diffraction	796
4.13.2.3.2 IR and UV spectroscopy	797
4.13.2.3.3 NMR spectroscopy	798
4.13.2.3.4 ESR spectroscopy	800
4.13.2.3.5 Mass spectrometry	801
4.13.2.4 Thermodynamic Aspects	802
4.13.2.4.1 Intermolecular forces	802
4.13.2.4.2 Stability and stabilization	803
4.13.2.5 Tautomerism	804
4.13.2.5.1 Annular	804
4.13.2.5.2 Ring-substituent	805
4.13.3 REACTIVITY AT RING ATOMS: AROMATIC AND NON-AROMATIC TETRAZOLES	806
4.13.3.1 General Survey of Reactivity	806
4.13.3.1.1 Reactions with nucleophiles	806
4.13.3.1.2 Reactions with electrophiles and electrophilic ring fragmentation	807
4.13.3.2 Thermal and Photochemical Reactions Involving No Other Species	808
4.13.3.2.1 Fragmentations involving nitrilimine intermediates	808
4.13.3.2.2 Fragmentations to azacyclopropanes	810
4.13.3.2.3 Fragmentations to carbenes and nitrenes	810
4.13.3.2.4 Rearrangements	812
4.13.3.3 Electrophilic Substitution	813
4.13.3.3.1 Basicity	813
4.13.3.3.2 Electrophilic attack at nitrogen	814
4.13.3.3.3 Electrophilic attack at carbon and α-exocyclic positions	816
4.13.3.4 Nucleophilic Substitution	816
4.13.3.4.1 Acidity	816
4.13.3.4.2 Nucleophilic substitution at C-5	817
4.13.3.5 Reactions of Tetrazolate Anions: Alkylation Reactions	817
4.13.3.5.1 Ring alkylation	817
4.13.3.5.2 Exocyclic α-alkylation and polyalkylation	818
4.13.4 REACTIONS OF SUBSTITUENTS	819
4.13.4.1 General Survey	819
4.13.4.2 Aldehyde, Carbonyl and Nitrile Substituents	820
4.13.4.3 Alkoxy and Hydroxy Substituents	821
4.13.4.4 Amino Substituents	821
4.13.4.5 Sulfur Substituents	823
4.13.4.6 Halogen Substituents	823
4.13.4.7 Alkyl Substituents	824
4.13.5 SYNTHESIS	825
4.13.5.1 Ring Synthesis from Non-heterocycles	825
4.13.5.1.1 Formation of one bond	825

4.13.5.1.2 Formation of two bonds	828
4.13.5.2 Synthesis from Ring Interconversions	832
4.13.5.3 Best Practical Methods	833
4.13.6 APPLICATIONS	834
4.13.6.1 Biological and Medicinal Applications	834
4.13.6.1.1 Analeptic activity (central nervous system)	834
4.13.6.1.2 Anti-inflammatory activity	835
4.13.6.1.3 Antimicrobial activity	835
4.13.6.1.4 Antilipemic activity	835
4.13.6.1.5 Antiallergic activity	836
4.13.6.1.6 Other medicinal properties	836
4.13.6.1.7 Agricultural uses	836
4.13.6.2 Non-biological Applications	837
4.13.6.2.1 Explosives and rocket propellants	837
4.13.6.2.2 Photography	837
4.13.6.2.3 Miscellaneous applications	837

4.13.1 INTRODUCTION

The first tetrazole derivative (see compound **198**) was reported almost a century ago ⟨1885CB1544⟩. Like other higher azoles this class of compounds received little attention and 66 years later only 300 tetrazole derivatives were known ⟨47CRV(41)1⟩. By the late forties the potential of the higher azoles in the fields of explosives, photography and agriculture had been realized and this led to renewed interest. The subsequent discovery of the pharmacological and biochemical properties of tetrazole derivatives has resulted in enormous development over the past 20 years. Much interesting new chemistry has resulted from this effort. However, in contrast to earlier European textbooks, many modern textbooks of heterocyclic chemistry and organic chemistry appear to be unaware of this class of compounds and the word 'tetrazole' will rarely be found in the index. This is surprising since as a class the compounds are not only the highest azoles but also the tetrazolic acids are the formal nitrogen analogs of the carboxylic acids.

There are three major reviews covering all aspects of the chemistry of tetrazoles as it stood at the time of their writing ⟨47CRV(41)1, B-67MI41300, 77AHC(21)323⟩. Useful reviews of the chemistry of 2-substituted tetrazoles ⟨76ZC90⟩ and of the molecular and metal complexes formed by tetrazoles ⟨69CCR(4)463⟩ have been published. Shorter reviews of specific aspects of tetrazole chemistry include synthesis of tetrazoles from aminoguanidines ⟨63AG(E)468⟩, azidoazomethine–tetrazole isomerism ⟨73CI(L)371, 73S123⟩, solvent effects on 1,3-cycloadditions of azides as a route to tetrazoles ⟨73S71⟩ and tetrazolinyl radicals ⟨73AG(E)455⟩.

4.13.2 STRUCTURE

4.13.2.1 Possible Structures

4.13.2.1.1 Aromatic systems not requiring exocyclic conjugation

The tetrazole ring is a 6π-azapyrrole type system with two tautomeric forms (**1**) and (**2**). The numbering system of form (**1**) is used to number ring substituents (as against tautomeric protons) throughout this review. The tetrazole ring is different from other azole systems insofar as it represents the functional group for a full series of carbazolic acids RCN_4H, *i.e.* full nitrogen analogues of carboxylic acids RCO_2H. Replacement of one of the ring protons leads to three possible types of monosubstituted tetrazoles with substituents at the 1-, 2- or 5-positions, respectively. Replacement of the two ring protons leads to two possible types of disubstituted tetrazoles, namely the 1,5-disubstituted series and the isomeric 2,5-disubstituted derivatives.

A wide range of stable metallic and molecular complexes of tetrazole systems, *e.g.* (**3**), has been described ⟨77AHC(21)323⟩ and many crystal structures have been determined for these complexes. The tetrazole ring is invariably planar with bond lengths characteristic of an aromatic system. Tetrazolate anion salts (**4**) are stable aromatic 6π-systems and they have been used widely as nucleophiles in synthetic reactions.

Substituents bearing *p*-electrons on the tetrazole 5-carbon undergo conjugation with the tetrazole ring, *e.g.* 5-aryltetrazoles (**5**). The interannular conjugation in 5-aryltetrazole systems involves a weak resonance electron donation by the tetrazole ring, *e.g.* (**5a**) ⟨78JCS(P2)1088⟩. Introduction of *ortho* substituents into the aryl ring or N(1) substituents on the tetrazole ring of compounds (**5**) results in a conformational change causing loss of coplanarity between the rings and reduction or loss of conjugation ⟨81JCS(P1)390⟩. Conjugation effects are also present when the aryl ring is bonded to a nitrogen atom, *e.g.* compound (**6**) ⟨77JPR408⟩. The resonance electron flow is into the tetrazole ring for derivatives containing strong activating groups such as NH_2 (*cf.* structure **29**). The tetrazole ring itself possesses a strong electron-withdrawing inductive effect $(-I)$ which surpasses the weak mesomeric effect $(+M)$ and the ring is therefore a deactivating group ⟨79JCS(P2)1670, 79OMR(12)631⟩, (*cf.* Section 4.13.3.1).

(**5**) λ_{max} 241 nm   (**5a**)   (**6**) λ_{max} 236 nm

4.13.2.1.2 *Aromatic systems involving exocyclic conjugation*

Mesoionic tetrazole systems are well known. In the compound (**7**), 1,3-dimethyl-5-iminotetrazole, the exocyclic conjugation is to a nitrogen atom ⟨54JA2894⟩. Conjugation to oxygen and sulfur atoms is observed in diaryltetrazolium 5-oxides (**8a**) and (**9a**) ⟨71ACS625, 74ZOR1725⟩, in the corresponding sulfides (**8b**) and in 2,3-diphenyltetrazolium-5-thiolate (**9b**), also known as dehydrodithizone. The tetrazolyl ylide system (**10**) was obtained, along with other isomers, from the benzylation of 5-dialkylaminotetrazoles ⟨70JOC2074⟩. Tetrazolium ylides (**11**) are readily obtained by oxidation of tetrazolylformazans. Oxidation of substituted formazans readily gives 2,3,5-trisubstituted tetrazolium salts (**12**) when the *N*-substituents are not good leaving groups ⟨53CB858⟩. Alkylation of 1,5-disubstituted tetrazoles may give 1,3,5- and 1,4,5-trisubstituted tetrazolium systems, *e.g.* (**13**) and (**14**) ⟨72BCJ1471, 71JOC3807⟩. Fused tetrazolylium systems such as (**15**) ⟨73TL4295⟩ and (**16**) ⟨73JHC5, 73JHC569⟩ are also known and have been found useful in synthesis.

(**7**)   (**8a**) X = O   (**9a**) X = O   (**10**)
         (**8b**) X = S   (**9b**) X = S

(**11**)   (**12**)   (**13**)   (**14**)

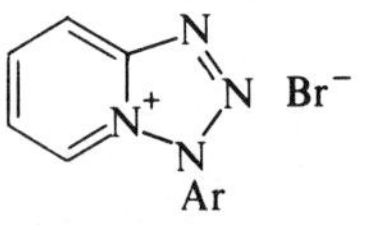

4.13.2.1.3 Non-aromatic systems

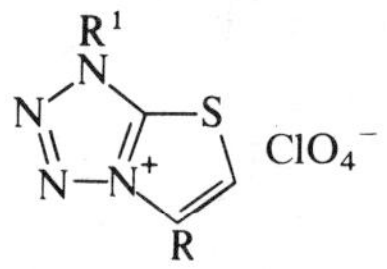

Non-aromatic (Chapter 4.01) tetrazoles appear to be relatively rare. The 1,2-disubstituted tetrazoline (**18**) has been reported as an etherate from the reaction of diazomethane with hexafluoroazomethane (**17**) ⟨62DOK(142)354⟩. Reduction of 1,4,5-trisubstituted tetrazolium iodides with NaBH$_4$ gave the Δ^2-tetrazolines (**19**) ⟨73BCJ1250⟩. These compounds were stable but upon heating at over 120 °C compound (**19**; R^1 = Me) cleaved to methyl azide and N-benzylidenemethylamine. Fused tetrazoline systems of type (**20**) have been obtained from cycloaddition reactions of dehydrodithizone (**9b**) with substituted alkynes ⟨71CC490, 76JCS(P1)1673⟩. The relative instability of the non-aromatic tetrazole system is also, to some extent, illustrated by the preferred existence of the iminoazimines (**21**) in the acylic nitrogen form (**21**) rather than the fused tetrazoline form (**22**) ⟨74CC659, 73CB1589, 74CC657⟩.

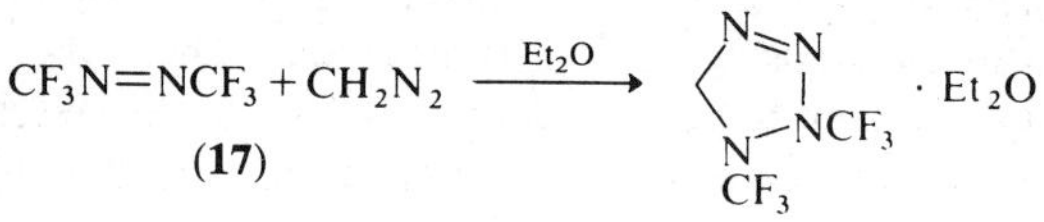

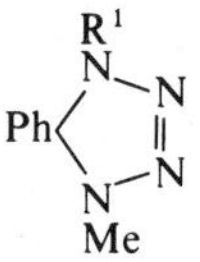

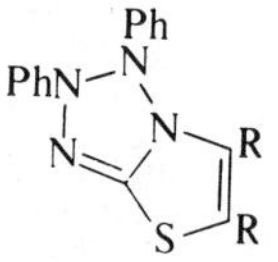

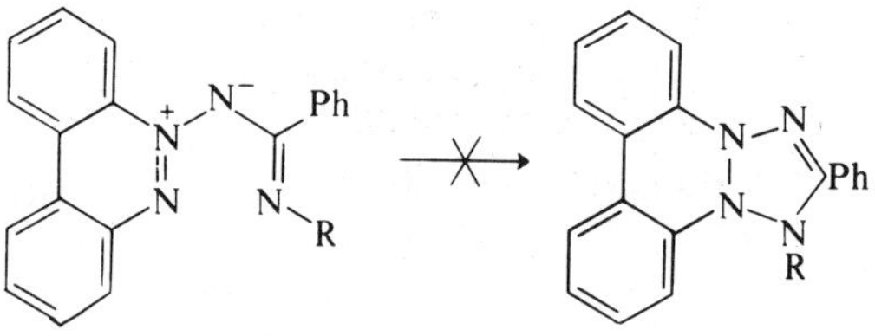

Tetrazolines with an exocyclic double bond at the C-5 atom are well known. Systems of type (**23**) are readily obtained from the further alkylation of 1-alkyl-5-aminotetrazoles ⟨54JOC439, 57JOC925⟩. The tetrazolin-5-one structures (**24**; R^1 = H) and (**25**; R^1 = H) are the normal tautomeric structures for 5-hydroxytetrazoles and tetrazole-5-thiols ⟨58CJC801, 67JOC3580, 59JA3076⟩. Alkylation of 1-substituted tetrazolin-5-ones (**24**; R^1 = H) gives 1,4-disubstituted tetrazolinones (**24**) ⟨56JA411⟩ and such tetrazolines are also obtained by addition of organic azides R^{1}N$_3$ to isocyanates RN=C=O, the addition occurring at the N=C bond ⟨73JOC675⟩. Although reactions of 1-substituted tetrazoline-5-thiones usually occur on sulfur, hydroxymethylation occurred at N-4 giving compounds of type (**25**; R^1 = CH$_2$OH) ⟨64ZOB2517⟩. Compounds (**25**; R = Ph, R^1 = R$_3$Sn) have also been obtained by treating phenyl isothiocyanates with triorganotin azides ⟨71AJC645⟩. Addition of α,β-unsaturated carbonyl compounds to 1-aryltetrazoline-5-thiones in the presence of base gave compounds (**25**; R^1 = CH$_2$CH$_2$COX) ⟨76JOC1875⟩. In the absence of base this latter reaction occurred on sulfur ⟨75ZC146⟩. Low yields of the interesting tetrazolines (**26**) were obtained on treatment of 1-alkyltetrazoline-5-thiones with acetyl chloride ⟨74ZC16⟩. The first homoaromatic tetrazole compound (**26a**), (6-amino-3-phenyl-1,2,4,5-tetrazine) has recently been reported from the addition of ammonia to 3-phenyl-1,2,4,5-tetrazine. This interesting compound shows a ring-current which causes a strong shielding of the endo-H (δ 1.51 p.p.m.) at C-6 since it extends into the shielding region of the homotetrazole plane ⟨81JOC2138⟩.

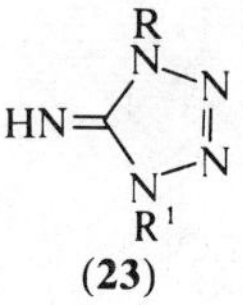

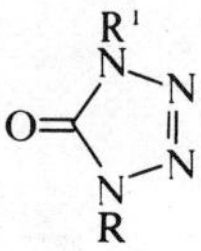

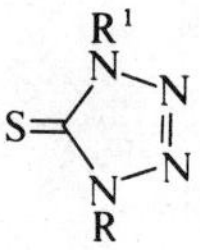

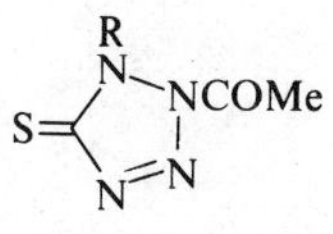

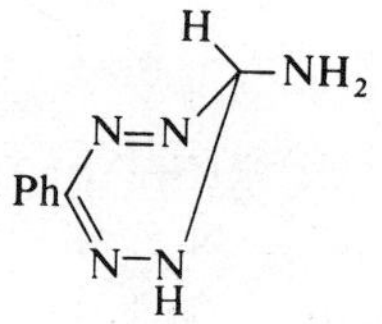

4.13.2.2 Theoretical Methods

4.13.2.2.1 Dipole moments

In the calculations which have been carried out on the tetrazole system there has generally been good agreement between calculated dipole moments and measured values. The first such calculation for tetrazole using an SCF–MO π-electron method gave μ_D 5.22 for 1,2,3,4-tetrazole (**27**) and μ_D 1.63 for 1,2,3,5-tetrazole (**28**), assuming a regular pentagon geometry for the ring with sides of 1.32 Å ⟨63JPC721⟩. A large part of the difference in the moments was due to the vectorial sum of the σ-bond moments μ_{CH} and μ_{NH}. These were additive in structure (**27**) but opposing in structure (**28**). The experimental dipole moment of tetrazole was 5.11 D in dioxane ⟨43MI41300⟩ and this suggested that the 1-NH form (**27**) dominated the annular tautomerism. CNDO–SCF calculations gave a μ_D for tetrazole (**27**) of 5.23 in good agreement, but the breakdown of the contributing components was quite different to those of the SCF–MO approach ⟨67JA6835⟩. The total energy calculated for form (**27**) by CNDO–SCF methods was 28.9 kJ mol^{-1} higher than that for form (**28**) suggesting either that the theoretical method was unreliable or that solvation favoured the 1-NH form (**27**) if this is the dominant species. In general, similar energies for both forms have resulted from many calculations and a slightly lower energy for form (**28**) has been suggested by a number of workers ⟨69BSF1097⟩. It seems likely that solvation effects are indeed the dominating factors which favour form (**27**) in solution. Microwave spectra for tetrazole in the gas phase gave μ_D 2.19 suggesting that form (**28**) predominated without a solvent ⟨74JSP(49)423⟩. The dipole moment for N-deuteriotetrazole in the gas phase was similar, 2.14 D, while that of C-deuteriotetrazole was 5.30 D suggesting that the 1-NH form of this latter compound was the dominant species.

(27)   ⇌   (28)

Measured dipole moments for a series of substituted tetrazoles gave interesting comparisons ⟨56JA4197⟩. The moments for 1-ethyltetrazole (5.46 D) and 2-ethyltetrazole (2.65 D) were at either side of the mesoionic compound (**7**) (4.02 D) and the dipole moments could not be used to identify mesoionic structures. The dipole moments for the 1,5-disubstituted tetrazole structure were consistently high, >5.3 D, while those of the 2,5-disubstituted structure were low, <2.65 D; theoretical calculations support this ⟨61T237⟩. Recently, dipole moment measurements were made on 5-(p-tolyl)tetrazole in dioxane (μ_D 4.99), its 1-methyl isomer (μ_D 6.03) and its 2-methyl isomer (μ_D 2.41) ⟨80JHC1374⟩. These results, when applied to the annular tautomerism of 5-(p-tolyl)tetrazole, suggested 60±10% of the 1-NH form in this medium.

4.13.2.2.2 Electronic structures and stability

The use of *ab initio* wave functions and full operators gave a calculated value of 5.17 D for the dipole moment of 1,2,3,4-tetrazole (**27**) and 2.54 D for 1,2,3,5-tetrazole (**28**) ⟨74JCS(P2)420⟩. The same approach gave total population figures for the azole series which indicated that introduction of a nitrogen atom adjacent to a particular ring atom C, N, or O leads to a consistent drop in the total population by *ca.* 0.10 electrons. For example, for N_β azoles total populations varied as follows: C—N_β—C, 7.272; N—N_β—C, 7.178 and N—N_β—N, 7.061. This is the theoretical basis for a useful NMR chemical shift correlation for N-alkylazoles discussed below, where isomeric structures containing the units C—N(Me)—C, N—N(Me)—C and N—N(Me)—N can be easily distinguished by the chemical shift of the Me group ⟨73CJC2315⟩.

Application of semiempirical SCF–MO methods involving the Huckel σ/π approximation, in which the total energy of the electrons is approximated by assuming additivity of bond energies and the π-energy is calculated by a Pople-type MO–LCAO–SCF treatment, to the tetrazole system gave calculated resonance energies of 134.6 kJ mol^{-1} and 108.8 kJ mol^{-1} for 1,2,3,4-tetrazole (**27**) depending on the methods used to estimate the various integrals ⟨66JCP(44)759⟩. The resonance energy for the 2H-form (**28**) was estimated at 95.66 kJ mol^{-1} or 122.3 kJ mol^{-1}. Calculated heats of combustion for compounds (**27**) and (**28**) were 2673 kJ mol^{-1} and 2698 kJ mol^{-1}, respectively. The measured value is 2710 kJ mol^{-1}. Bond lengths calculated by these SCF–MO methods ⟨66JCP(44)759⟩ are compared with some measured values in Table 1.

Table 1 Calculated and Measured Bond Lengths (Å) of Tetrazoles

Bond	Calculated	Measured[b]	Measured[c]	Calculated	Measured[d]
	(**27**)[a]			(**28**)[a]	
1—2	1.397 or 1.372	1.34	1.381	1.324 or 1.312	1.34
2—3	1.252 or 1.257	1.29	1.255	1.284 or 1.295	1.29
3—4	1.359 or 1.356	1.33	1.373	1.351 or 1.341	1.32
4—5	1.287 or 1.291	1.30	1.321	1.310 or 1.320	1.35
1—5	1.408 or 1.396	1.30	1.329	1.344 or 1.337	1.32

[a] ⟨66JCP(44)759⟩.
[b] For 1-methyltetrazole in dichlorobis(1-methyltetrazole)zinc(II) ⟨71IC661⟩.
[c] For 5-aminotetrazole monohydrate ⟨67AX308⟩.
[d] For 5-amino-2-methyltetrazole ⟨56AX(9)874⟩.

CNDO/2 calculations of electronic distributions in substituted tetrazoles generally suggest structures compatible with classical resonance concepts. For example CNDO/2 calculations on compound (**29**) showed high negative charges at the 2- and 4-positions as expected for a resonance hybrid with significant contributions from forms (**29a**) and (**29b**) ⟨73T3463⟩. A linear correlation of Hammett σ-substituent constants with CNDO-calculated charge distributions in a wide range of 5-substituted tetrazoles has been reported ⟨79ZOR844⟩.

(**29**)   (**29a**)   (**29b**)

4.13.2.3 Structural Methods

4.13.2.3.1 X-ray diffraction

X-ray crystal structures have been reported for a wide range of tetrazoles and these show the ring to be a planar resonance hybrid. Small differences in bond lengths of the order of 0.02 Å which appear to be present are probably not significant since the standard deviations are *ca.* 0.02 Å. The dimensions for 5-bromotetrazole are shown in Figure 1(a) and the full dimensions and literature for 15 other important tetrazole derivatives including those for the mesoionic compound (**9b**) and for the tetrazole ring in a number of metallic complex environments may be found in ⟨73JCS(P2)2036⟩. In the case of 5-bromotetrazole the ring proton was not located. The difference in bond lengths at C(5)—N(1) and C(5)—N(4) is of interest when compared with the similarity in the bond lengths at N(1)—N(2) and N(3)—N(4). In the tetrazolate anion (Figure 1c), all of the bond lengths are similar as expected ⟨63AX(16)596⟩. It is of interest to compare the dimensions of the tetrazolate anion with those of the tetrazolate ring in the ylide 5-(3-chlorobenzyldimethyl-ammonium)tetrazolate (Figure 1b) ⟨75JCS(P2)1200⟩.

The molecular dimensions of the hydrochloride of mesoionic 5-imino-1,3-dimethyltetrazole (**7**) are shown in Figure 1(d). These suggest a resonance hybrid with major contributions from forms such as (**30**) and (**30a**) where the C(5)—N(1) and C(5)—N(4) bonds retain some single bond character and the double bond character is more concentrated in the N(2)—N(3) and N(3)—N(4) bonds. In the mesoionic molecule (**9b**) the tetrazole ring and the sulfur atom form a coplanar aromatic resonance hybrid structure and both phenyl rings are twisted at 45° to this plane with little conjugation between the phenyl and tetrazole rings ⟨70JA1965⟩. The interring C—N bond length is 1.443 Å which suggests a normal C—N single bond. The structure of the interesting molecular complex pentamethylenetetrazole iodine monochloride (**31**, PMT·ICl) is shown in Figure 2 ⟨67JA6463⟩. In this the tetrazole ring

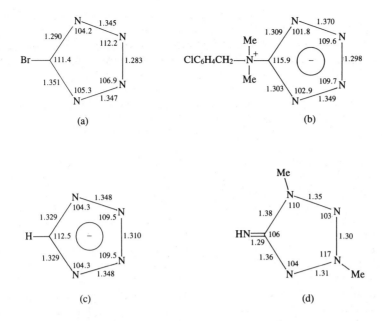

Figure 1 Bond distances (Å) and angles (°) of the tetrazole ring in: (a) 5-Bromotetrazole ⟨73JCS(P2)2036⟩. (b) 5-(3-Chlorobenzyldimethylammonium)tetrazolate ⟨75JCS(P2)1200⟩. (c) Sodium tetrazolate monohydrate ⟨63AX596⟩. (d) Hydrochloride of 1,3-dimethyl-5-iminotetrazole ⟨55AX211⟩.

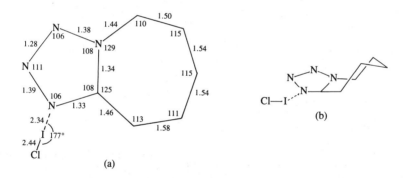

Figure 2 (a) Average bond distances (Å) and angles (°); and (b) side-view of PMT·ICl

acts as a unidentate ligand with the iodine of ICl bound to N-4 of the tetrazole ring. The N···ICl group is linear and coplanar with the tetrazole ring which is itself planar and the seven-membered ring of PMT is in a chair conformation (Figure 2b). Examples of the dimensions of other tetrazole derivatives are in Table 1. Two reviews, which discuss the structures of complexes of tetrazole derivatives, may be consulted for further detailed information ⟨77AHC(21)323, 69CCR(4)463⟩.

4.13.2.3.2 IR and UV spectroscopy

The tetrazole ring generally shows IR absorption bands in the region 1000–1100 cm^{-1} and an absorption band between 1095–1120 cm^{-1} has been reported in many tetrazole derivatives. Other absorptions which have been ascribed to the ring appear at 1000–1030 cm^{-1} and 1045–1085 cm^{-1} ⟨65CJC1154⟩. Absorptions at 1265–1298 cm^{-1}, which have been assigned to the —N—N=N— moiety in higher azole systems such as 1,2,3-triazoles and thiatriazoles, are also found in many tetrazole derivatives ⟨B-67MI41300⟩. Functional groups at the 5-position generally show their normal absorptions but 5-SH and 5-OH groups exhibit thione and ketone bands since they exist in the 1-tetrazolin-5-one form. The exocyclic C=N group of 5-iminotetrazolines, e.g. (23), absorbs at 1640–1670 cm^{-1} and the cyclic C=N of the tetrazole ring generally absorbs at 1590–1600 cm^{-1}.

In the UV region the tetrazole ring exhibits only weak end absorption at 200–220 nm. Most UV spectra of tetrazole derivatives arise from the conjugation effects of auxochromic groups. Tetrazole derivatives containing phenyl and amino substituents show absorptions in the range 235–255 nm ⟨57MI41300⟩. The UV spectra of some tetrazole derivatives illustrate interesting conjugation effects which are, however, often better probed by NMR spectroscopy. For example, in the series of phenyltetrazoles (5), (32) and (33), there is extensive interannular conjugation in compounds (5) and (33) which is reflected in their similar UV absorptions. Conjugation is much reduced in the 1-methyl isomer (32) due to a steric conformational change resulting in loss of coplanarity between the rings and the λ_{max} is shifted to shorter wavelength. Conjugation effects are also present in N-phenyltetrazoles; 2-phenyltetrazole absorbs at 250 nm while 1-phenyltetrazole (6), with λ_{max} 236 nm, shows reduced conjugation due to ortho proton interactions with the tetrazole 5-CH. The UV spectra of 1-substituted 5-aminotetrazoles show only end absorption at 218–232 nm for primary, secondary and tertiary amino groups while 2-substituted 5-aminotetrazoles show absorptions at ca. 254 nm. Mesoionic tetrazoles such as compound (7) absorb at 255 nm and the similarity of this absorption to that of 2-alkyl-5-aminotetrazoles led to some suggestions of a tautomeric mesoionic structure for these latter compounds. However, this was not the case and 5-aminotetrazoles exist predominantly in the amino form in solution ⟨77AHC(21)323⟩.

(32) λ_{max} 232 nm

(33) λ_{max} 240 nm

4.13.2.3.3 NMR spectroscopy

Proton and carbon NMR spectra have been used extensively to distinguish between isomeric tetrazole systems. A number of different approaches has been developed and the application of these to tetrazoles represents but one example of their wider use in azole chemistry. For isomeric tetrazole systems (34) and (35), containing a 5-CH moiety, long range coupling was observed over four bonds between the 5-CH and the 1-N-Me protons (J 0.45 Hz) for the structure (34) but not for the structure (35) where there was a 5-bond separation between the coupled protons ⟨80M775⟩. This effect was observed in a range of azoles. In the N-aryltetrazole series the 5-CH proton shift of 1-phenyltetrazole (6) (5-CH, δ 9.00 p.p.m., CDCl$_3$, 0.1 mol l^{-1}) was considerably more sensitive to solvent effects than that of the corresponding 2-phenyltetrazole isomer (36) (5-CH, δ 8.63 p.p.m., CDCl$_3$, 0.1 mol l^{-1}). This was characteristic of substituted N-aryltetrazoles in general and the greater solvent-induced shifts at the 5-CH arising from concentration variations or changes of solvent distinguished the 1-aryltetrazoles from their 2-aryltetrazole isomers ⟨77JPR408⟩.

(34) R = H   (35) R = H   (36)

Tetrazole isomers with methyl or alkyl substituents at the 1- and 2-positions can be readily distinguished by the ^{1}H and ^{13}C chemical shifts of the N-alkyl group. Alkyl groups bonded to N-1, e.g. (34), are more shielded by ca. 0.15–0.35 p.p.m. in the ^{1}H spectra and by ca. 2–6 p.p.m. in the ^{13}C spectra relative to their corresponding N-2 isomers, e.g. (35).

This is part of a larger shielding correlation for N-methyl groups which show increased shielding for the structural units (A) > (B) > (C) in azole systems. The 1-alkyl- and 2-alkyltetrazole structures represent (B) and (C) units respectively. The correlation is due to electron density effects arising from the electronegativity of the adjacent N atoms (cf. Section 4.13.2.2) ⟨77CJC1564⟩. Some examples of this correlation for N-methyltetrazole isomers are given in Table 2. Similar correlations are found in proton chemical shifts of amino groups bonded to the tetrazole ring although there are fewer examples. The shielding difference between NH protons of 1-N- and 2-N-aminotetrazoles ($\Delta\delta$ 0.8–1.0 p.p.m.) is higher than that for CH protons ⟨69CJC3677, 69TL1537⟩.

$$\begin{array}{ccc} \text{Me} & \text{Me} & \text{Me} \\ | & | & | \\ =\text{C}-\text{N}-\text{C}= & =\text{N}-\text{N}-\text{C}= & =\text{N}-\text{N}-\text{N}= \\ \textbf{(A)} & \textbf{(B)} & \textbf{(C)} \end{array}$$

Table 2 ^{1}H and ^{13}C NMR Chemical Shifts from TMS for Isomeric N-Methyltetrazoles (**34**) and (**35**)

R	(**34**) B-unit Me shift		(**35**) C-unit Me shift		Ref.
	^{1}H (δ, p.p.m.)	^{13}C (δ, p.p.m.)	^{1}H (δ, p.p.m.)	^{13}C (δ, p.p.m.)	
H	4.27	33.7	4.46	38.8	67JCS(B)516
Me	4.10	33.15	4.30	39–41[a]	65JOC(30)3472
NH$_2$	3.82	31.4	4.16	39–41[a]	77AHC(21)323
Br	4.13	—	4.40	—	67JCS(B)516
Cl	4.24	34.11	4.52	39–41[a]	b
Ph	4.16	35.0	4.25	39.4	81JCS(P1)390
4'-MeC$_6$H$_4$	4.16	34.95	4.34	39.7	81JCS(P1)390
4'-ClC$_6$H$_4$	4.20	35.1	4.4	39.7	81JCS(P1)390
2',6'-Cl$_2$C$_6$H$_3$	3.98	33.7	4.52	39.9	81JCS(P1)390
3'-C$_5$H$_4$N	4.20	35.1	4.48	39.8	b

[a] Overlap of solvent DMSO-d_6. [b] Author's unpublished data.

Interesting interannular conjugation effects are evident in the NMR spectra of 5-aryltetrazoles, e.g. (**5**) and (**5a**). This conjugation results in a deshielding shift for H-2' and a shielding shift for H-3' and H-4' of the phenyl ring so that when conjugation is present, as for example in compound (**37**), the difference in chemical shift of the ortho (2'-) and meta (3'-) aryl protons is as high as 0.6–0.76 p.p.m. In compounds containing an N(1)-substituent, (**38**), or a 2'-substituent in the 5-aryl ring the conjugation is inhibited by rotation of the rings and the shift difference is reduced to 0.1–0.3 p.p.m., this residual difference being due to tetrazole ring inductive and anisotropic effects on the ortho protons. In carbon spectra the interannular conjugation causes an upfield shift of ≥ 2.00 p.p.m. in C-2' of the aryl ring, e.g. (**37**), relative to unconjugated systems (**38**) and there is little difference in the 3' and 4' carbon shifts. Similar conjugation effects have been observed in N-aryltetrazoles where N(2)-aryl derivatives show much more conjugation than 1-aryltetrazoles in which loss of ring coplanarity arises due to a steric interaction between the tetrazole 5-H atom and the ortho protons of the aryl substituent ⟨81JCS(P1)390, 74ACS(B)61, 70BSF1345, 70BSF1346, 76T499, 77JPR408⟩.

(**37**)   (**38**)

A correlation of the aryl proton chemical shifts of a wide range of substituted 5-aryltetrazoles with a number of substituent parameters has confirmed the strong electron-withdrawing inductive effect of the tetrazole ring ⟨79OMR(12)631⟩. NMR conformational studies of acetyl groups directly bonded to tetrazole ring nitrogen atoms have also been reported ⟨76TL533, 73CR(C)(277)1163⟩. Nitrogen-14 NMR studies of tetrazole systems have been carried out ⟨77AHC(21)323⟩ and, in general, the signal of the singly bound nitrogen was sharper and higher upfield than that of the doubly bound nitrogens. The spectrum of trimethylenetetrazole showed four separate ^{14}N signals for the four ring nitrogen atoms. The ^{14}N shifts generally correlated with calculated π-charge densities and they probably can be used to distinguish (B) and (C) structural units also. For example, the compounds (34) and (35) gave ^{14}N-Me shifts from internal nitromethane of +150 p.p.m. and +103 p.p.m., respectively ⟨73MI41300⟩. The pure quadrupole resonance spectrum of ^{14}N in tetrazole and in the other parent azoles has been measured and correlated with electron densities. While three of the four expected resonances for imidazole and five of six expected resonances for 1,2,4-triazole were observed, only two resonances were observed for tetrazole which was piezoelectric and difficult to work with ⟨68MI41300⟩.

Electron densities control the proton shifts in tetrazoles and this is well illustrated by the 5-CH shifts of δ 9.5 p.p.m. for the neutral molecule in D$_2$O, δ 8.73 p.p.m. for the anion in 2M NaOD and δ 9.95 p.p.m. for the cation in trifluoroacetic acid ⟨67JCS(B)516⟩. The higher shielding observed in the proton NMR spectra of tetrazolate anions in general is due to increased electron densities rather than altered ring currents or anisotropic effects ⟨73JCS(P2)568⟩. This is not the case for the ^{13}C shift of the tetrazole 5-carbon which appears to be controlled more by orbital screening effects than by electron density. Thus the C-5 shifts of tetrazolate anions are generally deshielded by various amounts, rather than shielded relative to the parent tetrazole. For example, the C-5 shifts from TMS of 5-amino-, 5-methyl- and 5-phenyl-tetrazole were 156.6, 156.9 and 155.6 p.p.m., respectively in aqueous (CD$_3$)$_2$SO, and the shifts for the corresponding anions were 164.1, 160.2 and 160.35 p.p.m. ⟨77MI41300⟩. Protonation sites of a number of tetrazoles have been established by comparing the ^{1}H NMR spectra of tetrazolium salts with those of tetrazoles measured in acidic solutions. For 1-aryl- and 2-aryl-tetrazoles protonation occurred exclusively at the N(4)-position ⟨77T1399⟩. The applications of NMR spectra to tautomerism in tetrazoles is discussed in Section 4.13.2.5.

4.13.2.3.4 ESR spectroscopy

Electron spin resonance spectroscopy has been applied to tetrazole systems in the study of tetrazolinyl radicals (40) which are obtained by oxidation of formazans (39), by reduction of tetrazolium salts (41) or by disproportionations of mixtures of both ⟨73AG(E)455, 66JOC442⟩. The tetrazolinyl radicals are stabilized when R^1 and R^2 are strong resonance electron-withdrawing groups. The ESR spectra show a nine-line splitting characteristic of spin density on two pairs of equivalent ^{14}N atoms confirming the cyclic tetrazolinyl structure as against an alternative acyclic structure (42) which would have appreciable spin density on one pair of nitrogen atoms only. The g values for the tetrazolinyl radicals are ca. 2.0037 and the spin density at N(2)—N(3) [a(N$_{2,3}$) 6.1–8.6 G] is generally larger than that at N(1)—N(4) [a(N$_{1,4}$) 5.4–5.8 G]. The cyclic structure was confirmed with radicals of type (43) where hyperfine splitting due to the four ring nitrogens only was observed. Spin density on the exocyclic azo group in an acyclic structure would give a splitting pattern due to three pairs of equivalent ^{14}N atoms for this radical and this was not observed. Most of these radicals have been detected in solution as intermediates between compounds (39) and (41) but the radical (44) has been isolated as green–black needles which decompose at 126–127 °C. Its ESR spectrum in benzene consists of nine groups: a(N$_{1,4}$) 5.6; a(N$_{2,3}$) 6.1, a(H$_o$) 1.0 and a(H$_m$) 0.5 G. The radical was monomeric in the solid state and the magnetism of the solid suggested a free radical content of 94% ⟨69AG(E)520⟩.

PhN=N⁺(N-NPh)(N-NPh)(N-NPh)   Me₃C(N-NC₆H₄NO₂-p)(N-NC₆H₄NO₂-p)

(43)   (44)

4.13.2.3.5 Mass spectrometry

Mass spectral fragmentation patterns can be used to distinguish between the 1,5-disubstituted tetrazole structure (**45**) and the isomeric 2,5-disubstituted structure (**46**) ⟨69OMS(2)433, 68CJC2855⟩. The former compounds gave molecular ions while the latter gave unusual $(M+1)$ ions with no molecular ion. Compounds of type (**46**) fragmented by two main pathways. Process (a) involved loss of N_2 giving the fragment $[M-N_2]$ which was followed by loss of a hydrogen to give an even-electron fragment $[M-HN_2]$. The fragment $[M-N_2]$ also ejected $CH_2N\cdot$ giving a strong $[M-N_2-CH_2N]$ ion. The cleavage (b) involved loss of the elements $RCN_2\cdot$, probably in a number of steps, giving the ion $[Me-N_2^+]$.

R(N-N)(N-N)Me   R(N=N)(N-N)Me b,a

(45)   (46)

The fragmentation pattern for structure (**45**) was more complicated. The main processes are outlined in Figure 3. In this case loss of N_2 (cleavage g) made only a minor (5%) contribution. The most intense ion was $[M-HN_2]$ arising from process (c) and this fragment subsequently lost HCN. A composite of these two steps also occurred when CH_2N_3 was lost (process f). The fragmentation (e) gave a strong MeN_2^+ ion with loss of $RCN_2\cdot$. The loss of $N_3\cdot$ (process d), like the loss of N_2, was only of minor significance. However, important fragmentative loss of N_2 has been observed with a number of 1,5-disubstituted tetrazoles containing substituents such as CO_2Me and aryl groups at both the 1- and 5-positions. Hammett correlations of the intensity of the $[M^+ - N_2]$ ion with electronic substituent effects have been reported ⟨77ZOR2218, 67JA5958⟩. The nature of the substituents, as well as the ring substitution pattern, clearly plays an important role in influencing the preferred fragmentation pathways.

Figure 3 Main mass spectral fragmentations of 5-substituted 1-methyltetrazoles

Among individual tetrazole molecules which have displayed interesting mass spectra are 5-aminotetrazole and 1-phenyl-5-phenoxytetrazole. The former showed seven metastable ions and two main fragmentation pathways (i) $NH_2-CN_4H \rightarrow HN_3 \rightarrow HN_2$, and (ii) $NH_2-CN_4H \rightarrow CH_3N_3 \rightarrow HN_2, H_2N_2, CH_2N$ ⟨70JHC1223⟩. A 1,3 O → N phenyl migration was detected in 1-phenyl-5-phenoxytetrazole when this compound lost a CO molecule without loss of either of the phenyl rings ⟨69CC1110⟩. Mass spectral fragmentation patterns for 5-aryltetrazoles and 5-heteroaryltetrazoles have also been described. The main fragmentations of the tetrazole moiety were (a) loss of N_2 followed by HCN giving an arylnitrenium ion $Ar-N^+$, and (b) loss of HN_3 giving the species $Ar-CN^+$ ⟨77ZC65⟩. Complex mass spectra, arising from five different fragmentation pathways, have been reported for substituted 1-aryl-tetrazole-5-thiones ⟨73MI41301⟩. Mass spectral fragmentations of aryl tetrazolines of type (**23-25**; R = Ph) involved loss of the Ph—N=C=X moiety where X is the exocyclic atom at the 5-position ⟨75JHC1031⟩. Mass spectral fragmentations of $N(1)$- and $N(2)$-aryltetrazoles began by elimination of N_2 followed by HCN but different isotopic abundances allowed a distinction between the isomers to be made ⟨76OMS(11)167⟩.

4.13.2.4 Thermodynamic Aspects

4.13.2.4.1 Intermolecular forces

(i) Melting and boiling points

There has been relatively little study of the thermodynamic features of the tetrazole series and there is a paucity of quantitative data on details of associative characteristics. Since there are three doubly bound pyridine-type nitrogen atoms in the ring which can serve as proton acceptors, the availability of active hydrogens to provide intermolecular hydrogen bonding is an important factor influencing the melting and boiling points. Simple 5-substituted tetrazoles are generally white solids. Replacement of the ring NH hydrogen of a 5-substituted tetrazole by a small substituent usually causes a large decrease in the melting point if the 5-substituent has no NH or OH bonds (series 1–5 and 7; Table 3). The same pattern is not observed when the 5-substituent has NH or OH bonds (series 6, 9, 10; Table 3) since association by hydrogen bonding is still possible in the crystal.

Intermolecular hydrogen bond

Table 3 Melting and Boiling Points of Some Tetrazoles

Series no.	R	M.p. (°C)	R	R'	M.p./b.p. (°C)	R	R'	M.p./b.p. (°C)	Ref.
1	H	**156**	H	Me	**39**	H	Me	147 (atm)	47CRV(41)1
2	H	**156**	H	Et	164 (30 mm)	H	Et	71 (35 mm)	47CRV(41)1
3	Me	**148**	Me	Me	**73–74**	Me	Me	57 (13 mm)	65JOC3472
4	Br	**148**	Br	Me	**72–74**[b]	Br	Me	59–61 (4 mm)	56JA411
5	Ph	**213**	Ph	Me	**103**	Ph	Me	**51**	51JA4470
6	NH_2	**203**	NH_2	Me	**226**	NH_2	Me	**105**	56JA411
7	NMe_2	**239**[a]	H	NEt_2	56 (0.02 mm)[c]	—	—	—	—
8	NH_2	**203**	H	NH_2	142 (0.7 mm)	H	NH_2	89–91 (0.3 mm)	69CJC3677
9	BzNH	**181**	NH_2	Bz	**190**	NH_2	Bz	**86**	71JCS(C)703
10	HO[d]	**260**	HO[d]	Me	**124**	HO[d]	Me	**137**	56JA411

[a] ⟨66ACS2795⟩. [b] ⟨71CJC2139⟩. [c] ⟨65LA(688)93⟩. [d] Tetrazolone form.

Introduction of substituents with NH or OH bonds at C-5 generally raises the melting point (compare the parent compound for series 1, 6, 9, 10; Table 3) as might be expected for extra hydrogen bonding. However, since the introduction of an NMe_2 group at C-5 produces a higher melting point than an NH_2 group (parent compound series 7; Table 3) the rise in the melting point may not be due entirely to hydrogen bonding. Effects such as the overall polar character of the molecule and molecular packing in the crystalline state probably also exert important influences on melting and boiling points. The sharp contrast between the effect of an amino substituent at the 5-position and at the N(1)- and N(2)-positions (series 7 and 8; Table 3) suggests that the high melting points associated with the ring NH moiety may arise from an especially compact association of the rings in the solid state. In molecules where hydrogen bonding is not possible, for example the 1,5- and 2,5-disubstituted series (Table 3) the more polar 1,5-disubstituted form (high dipole moment) usually has a higher melting point or boiling point than the corresponding 2,5-disubstituted isomer (lower dipole moment). However such trends are precarious and there is at present no detailed information on the intermolecular forces between tetrazole molecules.

(ii) Solubility

There appears to be no quantitative solubility data available for the tetrazole series. However, a useful qualitative table of solubilities has been drawn up ⟨47CRV(41)1⟩. Insofar as the 5-substituted tetrazoles, RCN_4H, are the nitrogen analogues of the carboxylic acids, RCO_2H, solubilities comparable to those of the carboxylic acid series might be expected if due allowance is made for the fact that the tetrazole molecules are solids whereas the lower carboxylic acids are liquids. In general this expectation is fulfilled and the lower tetrazoles are highly soluble in water and probably completely miscible with it. For this reason simple tetrazoles such as 5-methyltetrazole, 5-chlorotetrazole, *etc.* cannot be easily crystallized from water and are best purified by recrystallizations from solvents such as ethyl acetate or benzene–pentane. Aromatic tetrazoles and tetrazoles where the ring NH proton has been replaced are much less soluble in water and generally are easily crystallized from aqueous alcohol. However small tetrazole molecules where the ring NH has been replaced by a small group which does not greatly disrupt the water structure are often quite soluble in water. The structures of the solutions of the *N*-unsubstituted tetrazole series are probably related to those of the corresponding carboxylic acid species. Studies of the concentration and temperature dependence of the NH proton resonance of 5-methyltetrazole in liquid SO_2 have suggested the presence of one or more types of dimeric hydrogen-bonded species which compare with the well-known dimers of acetic acid ⟨65JOC3472⟩. Whether or not tetrazole molecules might form larger aggregates of trimers, tetramers, *etc.*, as do some of the lower azole molecules, has not been reported.

5-Methyltetrazole dimer

(iii) Chromatography

Many tetrazole derivatives are solids which often decompose when heated above their melting points. Possibly for this reason there appears to be relatively little application of gaseous and liquid chromatographic techniques in the tetrazole field. However isomeric tetrazoles can generally be effectively separated by column chromatography using columns of alumina or silica gel. Paper and thin-layer chromatography have also been effectively applied to tetrazole analogs of some carboxylic acids and R_f values comparable to those of the carboxylic acids have been reported ⟨70MI41300⟩. Thin-layer chromatography has also been applied to tetrazolium salts and their formazan reduction products ⟨69MI41300⟩.

4.13.2.4.2 Stability and stabilization

(i) Thermochemistry and aromaticity

The measured heat of formation of tetrazole has been reported as 28.1 eV (2710 kJ mol^{-1}) which compares with a calculated value of 27.7 ± 0.2 eV ⟨66JCP(44)759⟩. Resonance energies ranging from 0.992 to 1.396 eV have been calculated for the tautomers of tetrazole. Similar resonance energy values have been quoted for pyrazole and the triazoles and somewhat lower values of 0.773–0.994 eV for imidazole. The heats of formation of pyrazole (39.4 eV), imidazole (39.8 eV) and 1,2,4-triazole (34.4 eV) are somewhat higher than for tetrazole ⟨66JCP(44)759⟩. These values suggest that the tetrazole ring is a normal aromatic system but probably slightly less stable than the lower azoles. Ionization potentials of 9.53–9.64 eV and 9.37–9.58 eV have been calculated for 1,2,3,5- and 1,2,3,4-tetrazole, respectively. These values, which represent the orbital energy, with opposite sign, of the highest occupied molecular orbital, are reasonably close to the values of 9.24 eV for benzene and 9.3–9.4 eV for triazoles and they differ somewhat from the values of 8.90 eV for pyrrole and 8.99 eV for imidazole ⟨67MI41301⟩.

4.13.2.5 Tautomerism

4.13.2.5.1 Annular

The main methods by which the annular tautomerism (1) ⇌ (2) has been studied are NMR spectroscopy, dipole moment measurements and theoretical calculations. The latter two approaches (Section 4.13.2.2) indicate similar energies for both forms and suggest that the predominance of the 1,2,3,4-tetrazole form (1), observed in solutions, may be due to a solvation effect. In the gas phase the 1,2,3,5-tetrazole form (2) appears to be the preferred form ⟨77AHC(21)323⟩.

The proton and ^{13}C NMR spectra of tetrazole, when compared with those of $N(1)$- and $N(2)$-alkyl isomers, agree with the dipole moment results and suggest that form (1) is the preferred form in solution. In the proton NMR spectrum of tetrazole measured in deuteroacetone at −90 °C, separate C(5)—H signals were observed, along with other signals attributed to slow reversible addition of acetone to the tetrazole nitrogen, and assigned to the tautomers (1) (85%) and (2) (15%) by comparison with N-methyl derivatives ⟨69TL495⟩. In the ^{13}C NMR spectrum of tetrazole in DMF the C-5 chemical shift was 143.9 p.p.m., similar to that of 1-methyltetrazole (144.9 p.p.m.) and different from that of 2-methyltetrazole (153.5 p.p.m.). Similar results were obtained in dioxane, water, DMSO and acetone, the C-5 chemical shift apparently showing little or no solvent sensitivity ⟨74JOC357⟩. However, this insensitivity to solvent may be a feature of the parent molecule. With a large substituent at the 5-position, *e.g.* compound (47), where the two rings are twisted orthogonally, ^{13}C spectra indicated 91% of the more polar 1,2,3,4-tetrazole form (47) in DMSO solution (ε DMSO 49) but only 68% of this 1-NH form in dioxane-d_6 (ε dioxane 2.2). A similar pattern of ^{13}C NMR results was obtained with 5-(2'-methylphenyl)tetrazole (48) where the form (48a) was indicated to the extent of 86% in DMSO and 73% in dioxane ⟨84UP41300⟩. Carbon-13 spectra for 5-(4'-methylphenyl)tetrazole and its $N(1)$- and $N(2)$-methyl isomers measured in DMSO–H$_2$O (83:17 v/v) gave a 1-NH:2-NH tautomeric ratio of 88:12, while dipole moment measurements in dioxane gave a value of 60:40 for the same ratio ⟨80JHC1373⟩. From calculated ionization constants for 1H- and 2H-tetrazoles it has been predicted that electron withdrawing substituents at C-5 should orient the tautomerism to the 2-NH form ⟨69JCS(B)1240⟩.

(47)   (48a)   (48b)

Recently the influence of some substituents at the 5-position on the tautomerism has been measured by using the C-5 chemical shift of the $N(1)$- and $N(2)$-methyl isomers as models for the C-5 chemical shift of the 1-H and 2-H tautomers, respectively. Using these models it is expected that the C-5 chemical shift of the parent 5-substituted tetrazole, representing an equilibrium mixture of both tautomers, should always fall between the extremity limits of each of the N-methyl isomers which represent the individual tautomers. This was found to be the case for all examples studied except for tetrazole itself containing a proton at the C-5 atom. In DMSO–water (4:1 v/v) the 1-NH:2-NH tautomer per cent ratios were found to be 93:7, 55:45, 48:52 and 84:14 for 5-amino, 5-methyl-, 5-chloro- and 5-phenyl-tetrazole, respectively ⟨77MI41300⟩. The influence of substituents in the phenyl ring of substituted 5-phenyltetrazoles is shown in Table 4 ⟨81JCS(P1)390⟩. In general, electron-withdrawing substituents oriented the tautomerism towards the 2-NH form. *Ortho*-substituents in the 5-phenyl ring, however, caused out-of-plane rotation of the rings and the 1-NH form was strongly dominant in such cases, *e.g.* (47). Quantitative linear correlations of log K_T with the Hammett σ values of the *para* substituents, and also with log K_a of the tetrazolic acids have been noted ⟨84UP41300⟩.

Comparisons of the ^{13}C–^{1}H coupling constant of the C(5)—H bond of tetrazole with those of $N(1)$- and $N(2)$-substituted tetrazoles also indicates the preference for the 1-NH tautomer of tetrazole in solution. Thus in DMSO the value of $J(^{13}$C–H) for tetrazole was 216 Hz while that of 1-phenyltetrazole was 216 Hz and that of 2-phenyltetrazole 211 Hz.

Table 4 ^{13}C Chemical Shifts and Tautomerism of Substituted 5-Phenyltetrazoles in DMSO

5-Phenyl substituent	Parent compound	^{13}C Chemical shift (p.p.m.)^a		1-NH form (%)	K_T	Ref.
		1-NMe form	2-NMe form			
H	155.6	154.2	164.25	86	6.14	78JCS(P2)1087
4'-Me	155.3	154.1	164.3	88	7.33	78JCS(P2)1087
2'-Me	155.35	153.85	164.6	86	6.14	81JCS(P1)390
4'-NO$_2$	155.8	152.7	162.5	68	2.12	78JCS(P2)1087
2'-NO$_2$	153.9	151.95	160.7	78	3.54	78JCS(P2)1087
4'-Cl	154.9	153.1	163.3	82	2.03	84UP41300
2'-Cl	153.45	152.6	162.4	91	10.1	81JCS(P1)390
2',6'-Cl$_2$	151.4	150.6	159.9	91	10.1	81JCS(P1)390

^{a} TMS as internal standard. ^{b} K_T = [1-NH]/[2-NH].

In acetone solution the $J(^{13}$C–H) for tetrazole was 217.5 Hz and that of 1-methyl- and 2-methyl-tetrazole 218.1 and 213.5 Hz, respectively ⟨76T499, 76JCS(P2)736⟩. The ^{14}N NMR shifts (from internal nitromethane) of tetrazole (106 p.p.m. in acetone) and its N(1)-methyl isomer (NMe, 150 p.p.m., CHCl$_3$) and N(2)-methyl isomer (NMe, 101 p.p.m., CCl$_4$) do not appear to agree with the other results and they have led to suggestions that tetrazole may exist in the 1,2,3,5 form (**2**) in solvents such as acetone ⟨72T637⟩. However, the factors involved in nitrogen chemical shifts are more complicated and less reliable than proton and carbon chemical shifts for studies of tautomerism.

4.13.2.5.2 Ring-substituent

Tetrazole–tetrazoline tautomerism of type (**49**) ⇌ (**50**) is readily observed when the exocyclic atom X is O or S. This was established by IR spectroscopy over 20 years ago and the tetrazoline form (**50**) was found to be the dominant form ⟨B-67MI41300⟩. More recently ^{13}C NMR spectral data, e.g. (**50**) and (**50a**), have confirmed the IR data ⟨75ZC402⟩. The

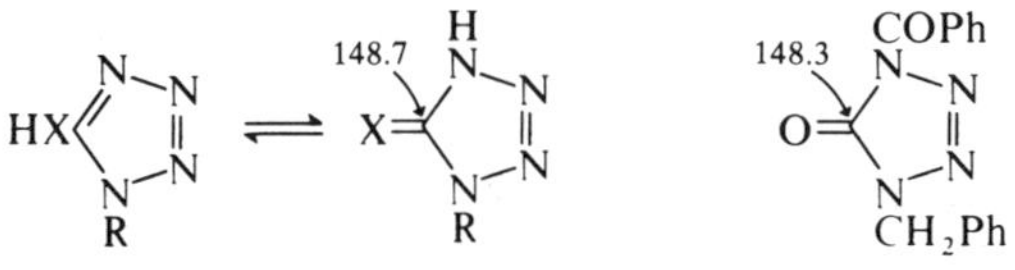

C-5 chemical shifts of compounds (**51**), (**52**) and (**53**) illustrate the thione structure of 1-substituted tetrazole-5-thiols in DMSO solution (Scheme 1) ⟨77JOC3725, 76JOC1875⟩. When the exocyclic atom X in (**49**) is an NR' moiety, this type of tautomerism does not occur to any appreciable extent in solution and the aminotetrazole form is preferred ⟨77AHC(21)323⟩. However, if the R' substituent on the amino group allows for special stabilization via an intramolecular hydrogen bond, the tetrazoline form may again dominate, e.g. structure (**54**) versus (**55**). This effect has been detected via the intramolecular hydrogen bond of structures such as (**55**). This causes a large reduction in the upfield movement of the NH signal on dilution with non-polar solvents such as CCl$_4$ which results from cleavage of intermolecular hydrogen bonds for structures (**54**). Such dilution has little effect on intramolecular hydrogen bonds, as in structure (**55**) ⟨69JCS(B)680⟩.

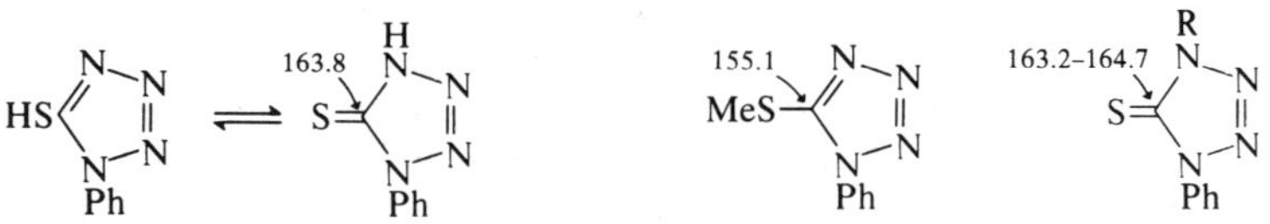

Scheme 1

The possibility of an aminotetrazole–iminotetrazoline tautomerism in the solid state has proved problematic ⟨77AHC(21)323⟩. IR spectra of some substituted 5-aminotetrazoles seem to possess an exocyclic imino absorption and laser-Raman spectra of two forms of hydrated

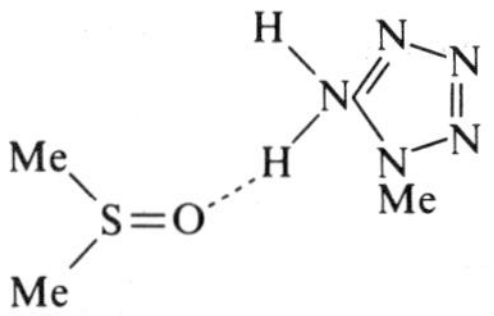

(54) 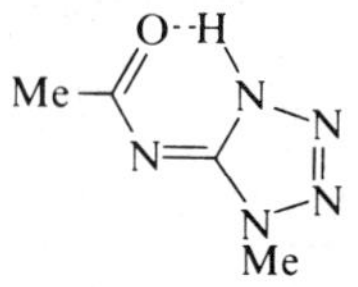   (55)

5-aminotetrazole have been interpreted in terms of both an amino form and an imino form which are interconvertible in the solid state ⟨72MI41300⟩.

4.13.3 REACTIVITY AT RING ATOMS: AROMATIC AND NON-AROMATIC TETRAZOLES

4.13.3.1 General Survey of Reactivity

4.13.3.1.1 Reactions with nucleophiles

Singly bound pyrrole-type nitrogen atoms have an electron-donating activating influence in azole rings while doubly bound pyridine-type nitrogens exert an electron-withdrawing deactivating influence. The tetrazole ring with three doubly-bound pyridine-type nitrogens is a deactivated electron-deficient system. Hammett substituent constants for the tetrazole ring are σ_m 0.64 and σ_p 0.57 which make the ring comparable with deactivating groups such as CF_3 and SO_2Me. The field (F) and resonance (R) contributions to the Hammett σ-values, in accordance with the Swain and Lupton equation $\sigma = fF + rR$, were F 1.17 and R −0.03 corresponding to a strong electron-withdrawing inductive effect comparable to that of a nitro group and a weak electron-donating resonance effect ⟨79JCS(P2)1670, 78JCS(P2)1087⟩.

Because of the deactivated π-deficient nature of the ring, nucleophilic attack can readily occur on tetrazoles. Nucleophilic replacement of halogens at the C-5 position has been widely used for the synthesis of disubstituted tetrazoles, e.g. (56). Kinetic studies of the reaction between 5-bromo-1-methyltetrazole and the 2-methyl isomer with piperidine have shown the 1-methyl compound to be considerably more reactive. Comparison of these kinetics with similar reactions for a series of azoles has suggested that two to three deactivating doubly bound nitrogens (—N=) are required to overcome the electron release from one pyrrole-type nitrogen atom in these systems ⟨67JCS(B)641⟩.

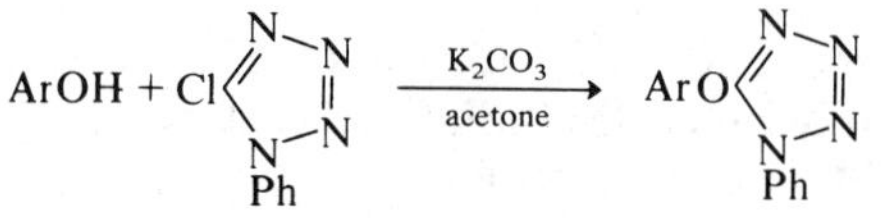

(56)

Base-induced deuterium–protium exchange at the C-5 atom of N-substituted tetrazoles occurs readily via carbanionic type intermediates, e.g. (57). The reaction for 1-substituted tetrazoles had a Hammett ρ-value of +1.3 in piperidine–MeOD–DMF ⟨69TL3377⟩. In the acidic pH range 3–5, ylide type species (58; $R^1 = H$) are also considered to be involved in the exchange arising from initial protonation at N-1 or N-2 followed by D^+ abstraction.

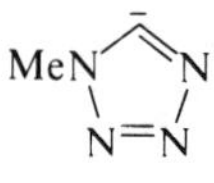

(57) 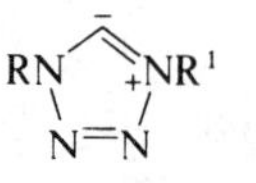   (58)

The rate of deuterium–protium exchange for 1-methyltetrazole was 10^5 times faster than that of 2-methyltetrazole in agreement with a general trend that ring protons α to singly bound nitrogens exchange much more rapidly than those located β to such nitrogens ⟨72JA5759⟩. Treatment of 1-methyltetrazole with n-butyllithium at −60 °C in THF gave the carbanionic tetrazolyl derivative (59) which was stable below −50 °C. The corresponding 1-phenyl derivative (60) was stable below −70 °C. On warming above these temperatures the tetrazolyllithiums decomposed to the corresponding substituted lithium cyanamides

(61) and nitrogen. These carbanionic tetrazoles underwent normal Grignard type reactions at the C-5 atom (Scheme 2) ⟨70CJC2139, 77AHC(21)323⟩.

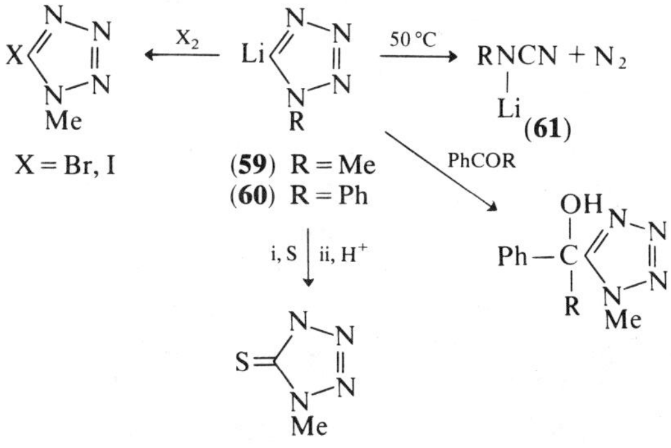

Scheme 2

The most common nucleophilic type reactions at the tetrazole nitrogens arise from the acidity of the ring N—H bond. Tetrazole, with acidic pK_a 4.89 in water, is the nitrogen analogue of formic acid (pK_a 3.75) while 5-methyltetrazole (pK_a 5.56) similarly relates to acetic acid (pK_a 4.75) ⟨B-67MI41300⟩ and in general the 5-substituted tetrazoles represent the nitrogen analogues of the carboxylic acids. These tetrazolic acids form stable anions when treated with bases and the tetrazolate anions are ambident in character and thus when used as nucleophiles may give rise to 1- or 2-substituted tetrazoles (Scheme 3). Reactions of this type are used widely for the synthesis of 1,5- and 2,5-disubstituted tetrazoles.

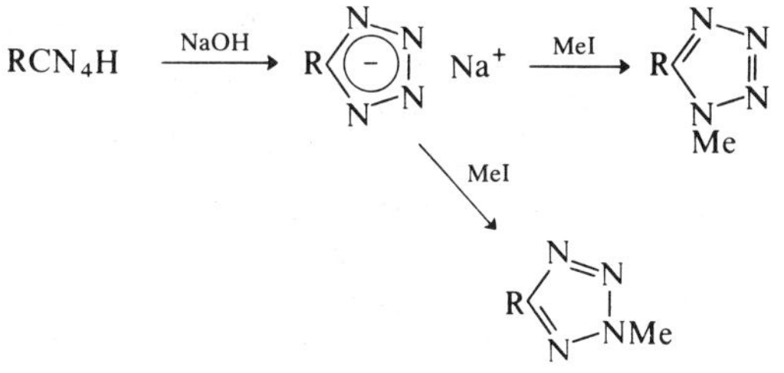

Scheme 3

4.13.3.1.2 Reactions with electrophiles and electrophilic ring fragmentations

There are few reports of electrophilic substitutions at tetrazole carbon and then only when the ring N—H bond has already been substituted. Treatment of N-phenyltetrazoles with bromine or mercury(II) acetate results in substitution at the tetrazole C-5, but nitration of N-phenyltetrazoles occurs on the phenyl ring ⟨B-67MI41300⟩. Electrophilic attack occurs readily on the tetrazole ring nitrogens. The N(2)-position is generally the preferred site for attack but substitution at the N-1 site also occurs most notably when the 5-substituent is a strongly activating group such as alkoxy or amino. Attack at the N-2 site facilitates a subsequent ready loss of N_2 with ring fragmentation (Scheme 4). This reaction, and also the thermal decomposition of 2,5-disubstituted tetrazoles (62), is now a standard method for generating nitrilimines (63; Scheme 4) ⟨77AHC(21)323⟩. A total ring fragmentation,

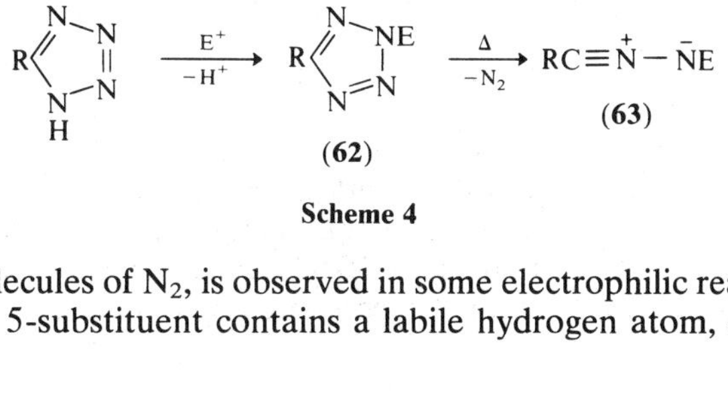

Scheme 4

involving loss of 2 molecules of N_2, is observed in some electrophilic reactions of tetrazoles, particularly when the 5-substituent contains a labile hydrogen atom, *e.g.* as in Scheme 5.

In these reactions the electrophiles are often halogens and a mechanism involving exocyclic halogenation and loss of a hydrogen halide molecule generating a tetraazafulvene intermediate (**64**) is considered to be involved (Scheme 5) ⟨77AHC(21)323, *cf.* 82TL121⟩.

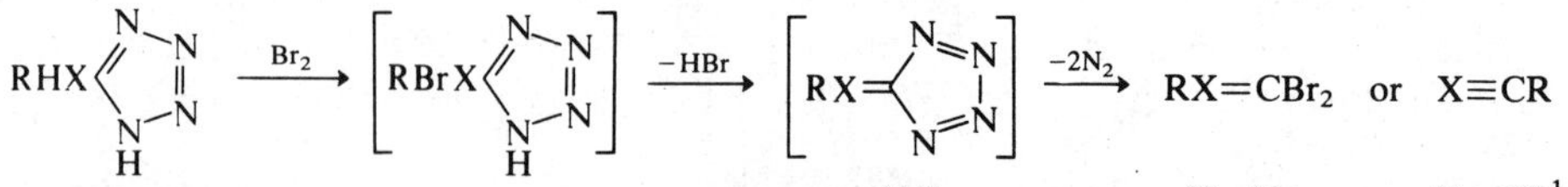

Scheme 5

4.13.3.2 Thermal and Photochemical Reactions Involving No Other Species

4.13.3.2.1 *Fragmentations involving nitrilimine intermediates*

Thermal fragmentation of 2,5-disubstituted tetrazoles results in loss of N_2 and generation of a reactive nitrilimine intermediate (**67**). The Hammett ρ-values for the *N*-phenyl and *C*-phenyl rings of diaryltetrazoles (**65**) were +1.16 and −0.23, respectively, suggesting an unsymmetrical transition state of type (**66**) ⟨68T3787⟩. Subsequent additions of the nitrilimines to unsaturated bonds such as alkenes, imines, alkynes and nitriles give an added importance to the parent tetrazoles as synthetic precursors (Scheme 6). Among the most

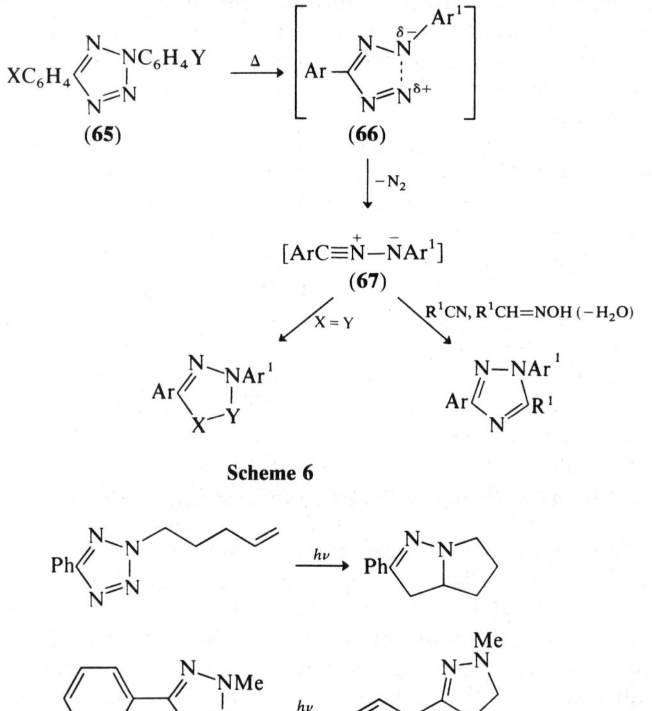

Scheme 6

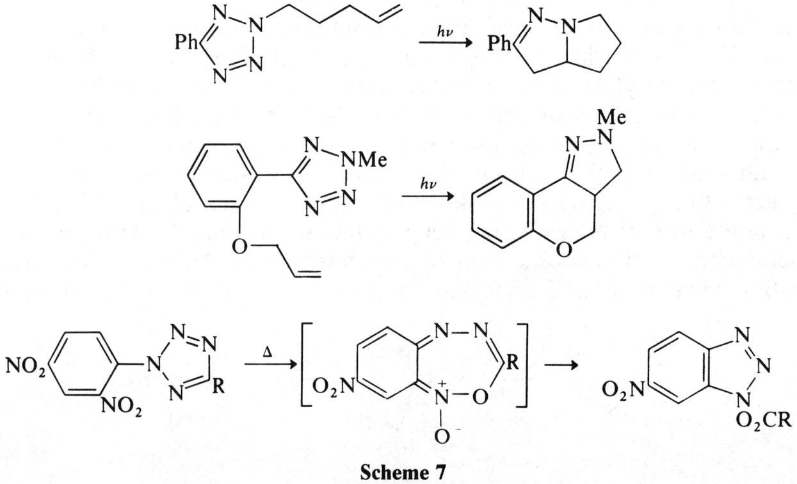

Scheme 7

interesting synthetic applications of this reaction are cases where a cyclization site exists in one of the tetrazole substituents thereby allowing an intramolecular cyclization of the nitrilimine. Some recent useful applications are shown in Scheme 7 ⟨78JOC1664, 78ZC214⟩.

The original reaction was developed using acyl groups as substituents on the tetrazole 2-position and is now widely referred to as the Huisgen reaction (Scheme 8) for synthesizing substituted oxadiazoles, thiadiazoles and 1,2,4-triazoles, e.g. (**70**) ⟨65CB2966⟩. The final step involves a heteroelectrocyclization in the nitrilimines (**69**) ⟨80AG(E)947⟩. The entire reaction may be carried out *in situ* or the compounds (**68**) may be isolated in some cases. For example, when the 5-substituent R was an aryloxy group, compounds of type (**68**; X = O) were low melting solids ⟨77AHC(21)323⟩.

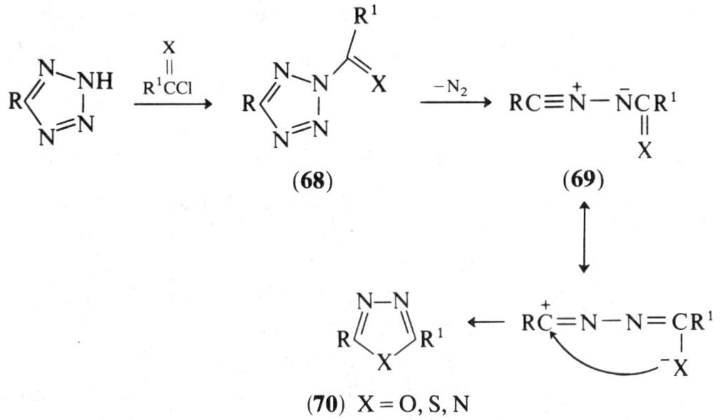

Scheme 8

Photolysis of 2,5-diphenyltetrazole in benzene solution gave 2,4,5-triphenyl-1,2,3-triazole (**72**) as the main product (28%) and benzil osazone (3%; Scheme 9) ⟨71JOC1589⟩. A nitrilimine intermediate (**71**) seems a likely precursor to these products. Recently, diphenylnitrilimine, showing IR and UV absorptions at 2228 cm^{-1} and 377 nm, respectively, has been directly detected in the photolysis of 2,5-diphenyltetrazole at 77 K in an ether–pentane–ethanol glass ⟨80JA2093⟩. On warming to 160–175 K the nitrilimine dimerized to 1,2-diphenylazo-1,2-diphenylethylene (**73**) ⟨80C(34)504⟩. Comparable triazole and dimeric products have also been reported from the photolysis of 2-methyl-5-phenyltetrazole ⟨69JOC4118⟩. Flash thermolysis of 2,5-diaryltetrazoles at 400–500 °C gave 96–100% yields of 3-arylindazoles such as (**74**) in a reaction also involving nitrilimine intermediates ⟨78JOC2037⟩. Photolysis of 5-substituted tetrazoles in THF resulted in loss of N$_2$ and the formation of 3,6-disubstituted 1,2-dihydro-1,2,4,5-tetrazines (**75**) in a reaction considered to involve dimerization of the *N*-unsubstituted nitrilimine species. Minor yields of 3,5-disubstituted 1,2,4-triazoles were also formed by further photolysis of the tetrazine products (**75**) ⟨70T2619⟩. The mechanism is supported by ^{15}N labelling experiments which indicated that ring-transposition with N-atom scrambling did not occur during photolysis and that nitrogen expulsion occurred from positions 3 and 4 ⟨72CC781⟩. Reactions involving the alternative iminonitrene species, PhC(Ṅ)=NH, did not give tetrazine-type products ⟨70JHC71⟩.

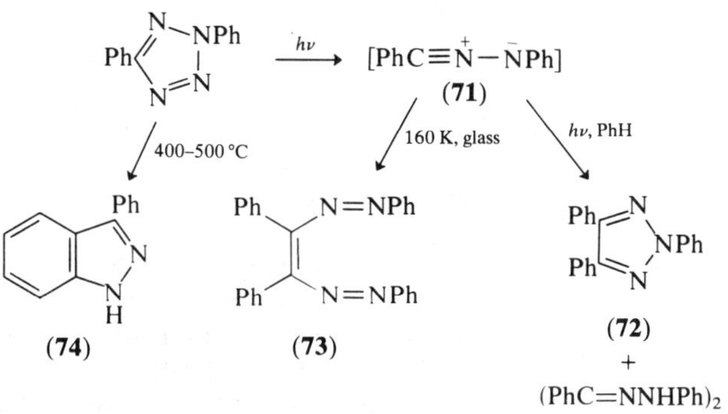

Scheme 9

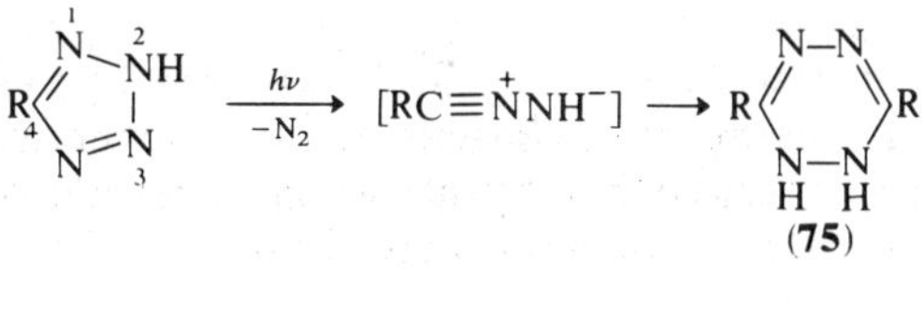

4.13.3.2.2 Fragmentations to azacyclopropanes

Elimination of N_2 from 1,4-disubstituted tetrazole derivatives (**76**) leads to substituted diaziridine compounds of type (**77**) and a number of synthetic applications of this general reaction have been developed recently. Photolysis of Δ^2-tetrazolines such as (**78**) in dichloromethane gave a ready elimination of nitrogen and 60–70% yields of 1,2,3-trisubstituted diaziridines of type (**79**) ⟨74CL185⟩. Most 1,4-disubstituted tetrazolines (**80**) contain an exocyclic double bond at C-5. Fragmentations of such systems give heterocyclic analogues of methylenecyclopropanes (Scheme 10) ⟨80AG(E)276⟩. These fragmentations were more readily achieved by photolytic rather than thermal means since the tetrazole derivatives tend to be thermally stable. With tetrazoline-5-thiones sulfur was extruded as well as N_2 giving carbodiimides and the three-membered rings were not formed. Tetrazolines containing an exocyclic alkylidene moiety gave iminoaziridines (**81**) rather than methylenediaziridines ⟨75AG(E)428⟩. Mechanistic insight into these reactions was obtained when the triplet diradical (**83**) was directly detected by its ESR spectrum on photolysis of the tetrazoline (**82**) in a butyronitrile matrix at −195 °C ⟨78JA1306⟩. The radical (**83**) is a nitrogen analogue of trimethylmethane. The synthetic applications of these and related fragmentation reactions for the synthesis of heterocyclic analogues of methylenecyclopropanes have recently been thoroughly reviewed ⟨80AG(E)276⟩.

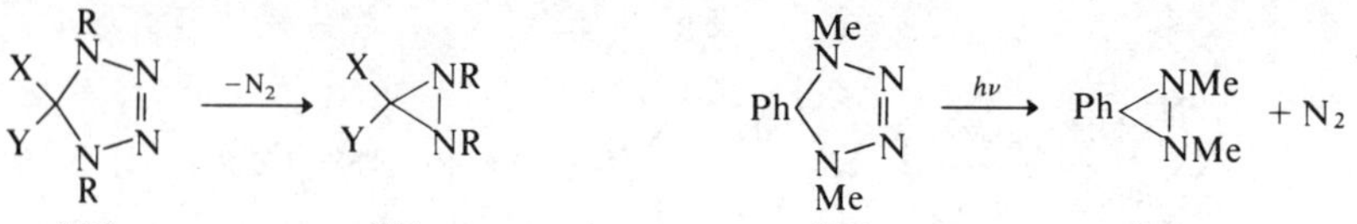

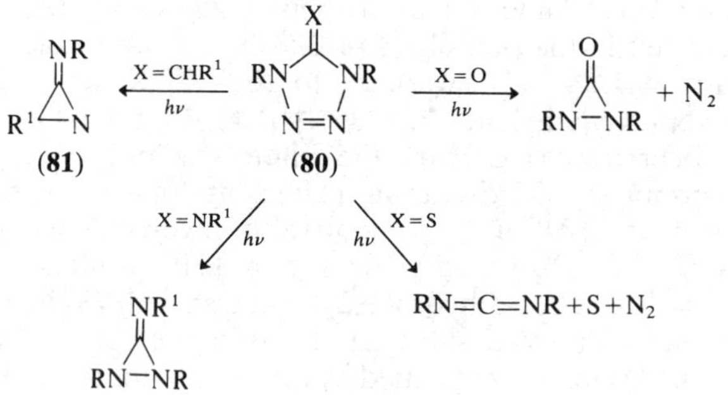

Scheme 10

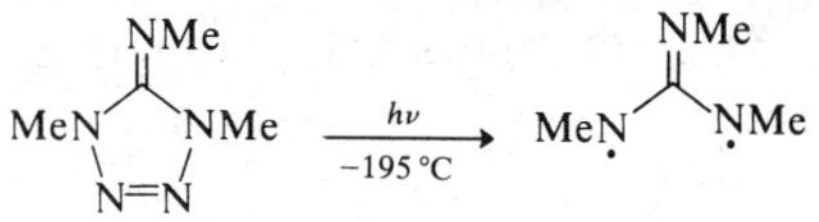

4.13.3.2.3 Fragmentations to carbenes and nitrenes

Photolysis of 5-substituted tetrazolate anions results in loss of two molecules of N_2 and generation of a carbene which gives normal insertion reactions (Scheme 11) ⟨71TL4489⟩. The reaction involved an initial excitation of the anion to a long-lived triplet state. This lost a nitrogen molecule giving a fragment which abstracted a proton from the medium generating a diazoalkane intermediate which was photolyzed to the carbene (Scheme 11). The initial triplet state could be quenched by treatment with suitable acceptors and it may be used as a photosensitizer ⟨69TL4863⟩. Rearrangements characteristic of diazoalkane species were observed with substrates such as 5-cyclopropyltetrazolate (**84**) from which the main product was the cyclobutyl methyl ether (**86**). Attempts to generate dicarbenes by photolysis of di(tetrazol-5-yl)benzenes resulted only in consecutive reactions each involving a monocarbene pathway ⟨77AHC(21)323⟩.

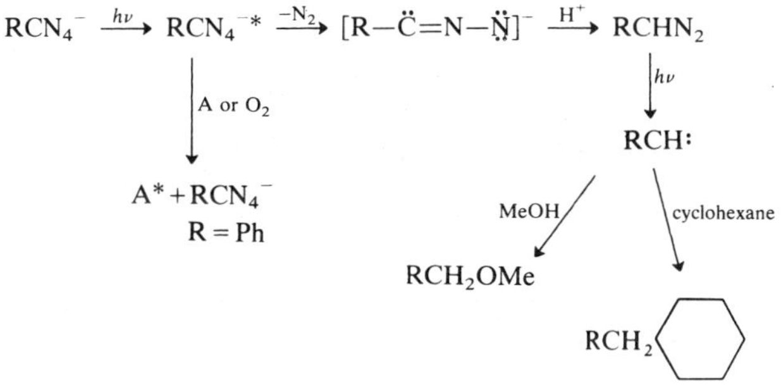

Scheme 11

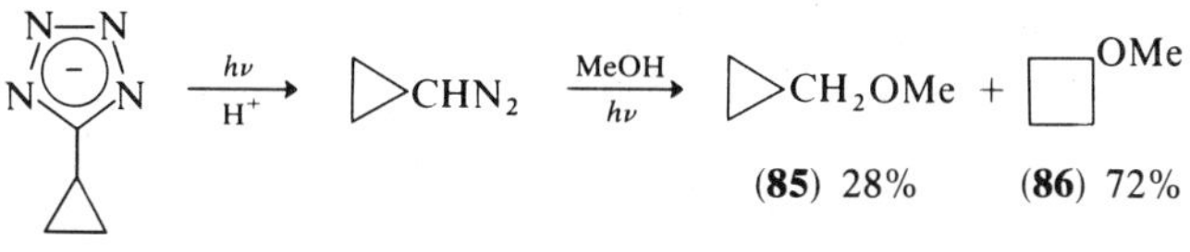

(84)

Gas-phase pyrolysis of 5-aryltetrazoles and fused tetrazoloazines has been used as a convenient route to arylcarbenes and nitrenes. The ease of synthesis of labelled tetrazole derivatives has given rise to interesting studies of the behaviour of some of these reactive intermediates, *e.g.* Scheme 12. When 2-pyridyltetrazole (**87**), carrying a ^{13}C label at the tetrazole 5-carbon, was pyrolyzed at 400 °C the carbene (**88**) was formed. This rearranged to the nitrene (**89**) which dimerized to compound (**90**) carrying the label exclusively in the *ortho*-positions. Thus when no ring contraction was observed, no scrambling of the label occurred. At higher temperatures, 670 °C, the ring-contracted product (**94**) was observed. This showed the expected label distribution for prior isomerizations to 1*H*-benzazirine (**92**) and iminocyclohexadienylidenes (**91**) and (**93**) which therefore appear to play an essential role in ring contractions of arylnitrenes ⟨76JA1258⟩.

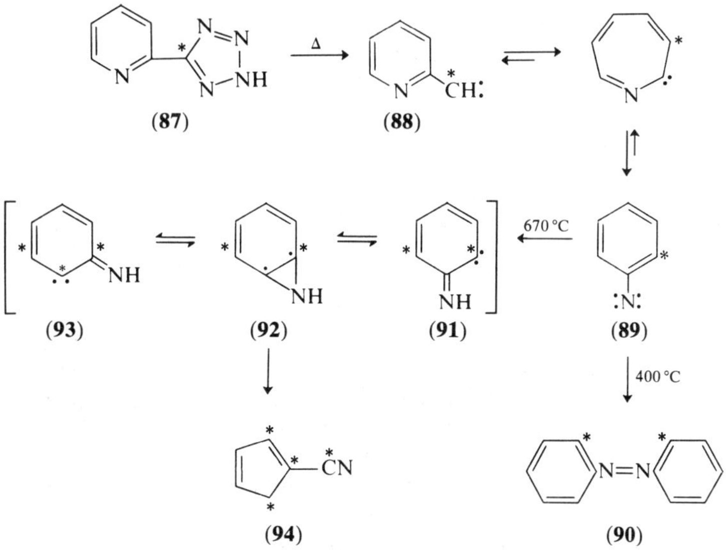

Scheme 12

UV photolysis of 1,5-disubstituted tetrazoles also results in elimination of nitrogen and gives reactions which appear to involve the iminonitrene intermediates (**95**; Scheme 13). Thus photolysis of 1,5-diphenyltetrazole and 5-phenoxy-1-phenyltetrazole gave compounds (**96**) and (**97**), respectively, as major products. Some other products also arose from further photolysis of these primary products and with 5-aryloxy-1-phenyltetrazoles a minor competitive pathway involved a homolysis of the aryl–oxygen bond and gave rise to some biphenyls when benzene was the solvent. Photolysis of dimethyl tetrazole-1,5-dicarboxylate (**98**) gave the oxadiazole (**100**) while thermolysis gave the oxadiazolone (**101**). Both reactions involved the nitrene intermediate (**99**) and it has been suggested that a singlet nitrene, favouring insertion, was involved in the thermal reaction while the photoreaction involved a longer-lived triplet nitrene ⟨71JA1534, 77AHC(21)323, 73JCS(P1)469⟩. Flash vacuum pyrolysis of 2-methyl-5-phenyltetrazole at 550 °C and 10^{-4} Torr gave Ph—CH=N—N=CH$_2$ and Ph—N=C=N—Me and this latter compound was also obtained from pyrolysis of 1-methyl-5-phenyltetrazole ⟨80CC502⟩. N-Arylbenzimidoyl nitrenes which cyclize on to the *ortho*-position have been invoked to explain the interesting series of products observed on thermolysis and photolysis of 5-phenyl-1-(2,6-dimethoxyphenyl)tetrazoles (Scheme 14) ⟨76CC414⟩. Thermolysis reactions of some 5-aminotetrazole derivatives are discussed in Section 4.13.4.4.

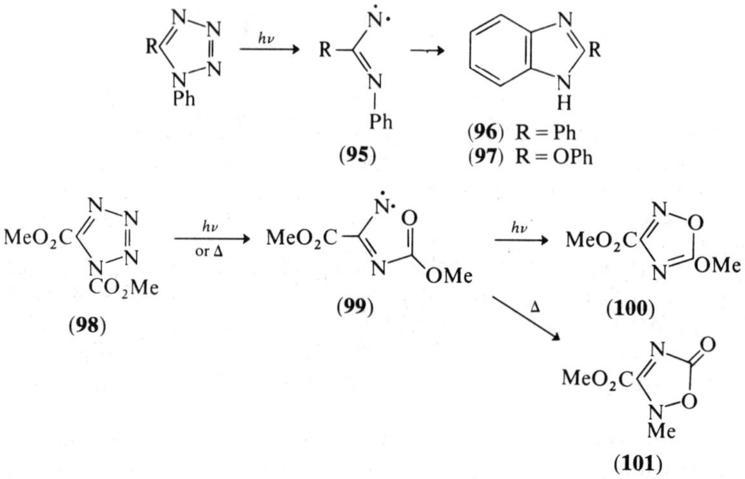

Scheme 13

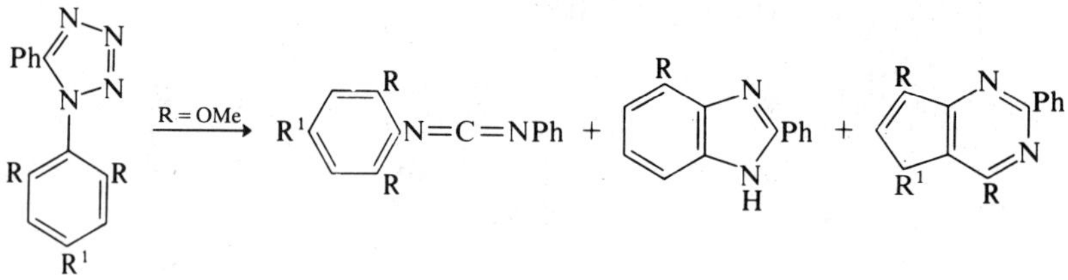

Scheme 14

4.13.3.2.4 Rearrangements

One of the best-known thermal rearrangements of tetrazoles is that of substituted 5-aminotetrazoles (**103**). In general 5-alkyl-1-arylaminotetrazoles (**103**) rearrange to 1-alkyl-5-arylaminotetrazoles (**105**) when heated (Scheme 15). The reaction is an equilibrium and both forms are usually present in the melt. The reaction is first order in tetrazole and electron-withdrawing substituents in the aryl ring favour the arylamino form (**105**). There was a linear correlation between log K and Hammett σ-values of substituents. The rearrangement involves ring opening to a reactive, bent imidoylazide (**104**) which has the required *cis*-orientation of the azido group and the imino lone pair for ring-closure but is sufficiently long-lived for rotation of the C(5)—N(4) bond and an amino–imino proton tautomeric change. In this situation competitive 6π-heteroelectrocyclization occurs preferentially on the imino nitrogen with highest electron density ⟨54JA88, 53JOC1269⟩. Thermal rearrangement of 1-aryltetrazole-5-thiones to 5-arylamino-1,2,3,4-thiatriazoles, a sulfur analogue of Scheme 15, has also been observed ⟨67JOC3580⟩. The cyclization of imidoyl azides is discussed in Section 4.13.5.

Displacement of chloride from 5-chloro-1-phenyltetrazole can be used to synthesize 5-alloxyl-1-phenyltetrazoles (**106**). These readily undergo Claisen-type 3,3-sigmatropic rearrangements when heated giving the products (**108**) with inversion. It has been suggested that ground state contributions from polar forms such as (**107**) may facilitate the rearrange-

(103) R = alkyl (105)

(104)

Scheme 15

ment ⟨67JOC2956⟩. A comparable 1,3-sigmatropic-type rearrangement of 5-alkoxy-1-aryltetrazoles (**109**) occurred when these compounds were heated in the presence of NaI ⟨74JHC491⟩. An interesting 1,4-sigmatropic-type migration has recently been reported with 1-alkoxy-5-phenyltetrazoles (**111**) on heating at 190–200°C. The rearranged 3-alkyltetrazole 1-oxides (**112**) were formed in yields of 30–53% ⟨78TL399⟩. The mesoionic 2,3-diphenyltetrazolium 5-oxide and 5-sulfide (**113**) underwent a thermal rearrangement to the corresponding 1,3-diphenyl derivatives (**114**) when heated at 150 °C as 2% w/w dispersions in acid-washed sand. The rearrangement was confirmed by X-ray crystallography of the products ⟨76CC343⟩.

(**106**) (**107**) (**108**)

(**109**) (**110**)

(**111**) (**112**)

(**113**) X = O, S (**114**) X = O, S

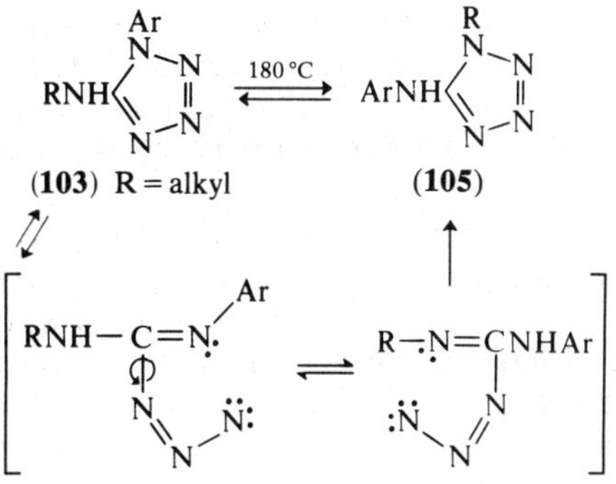

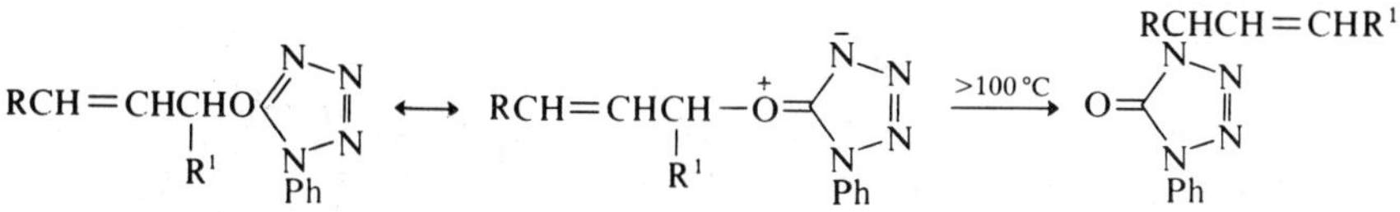

4.13.3.3 Electrophilic Substitution

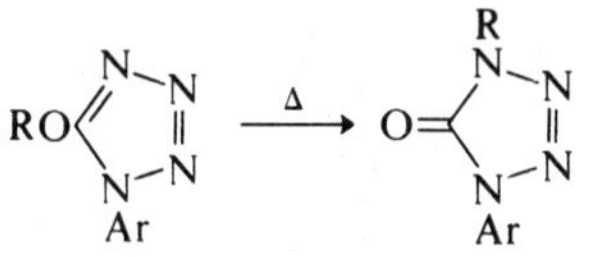

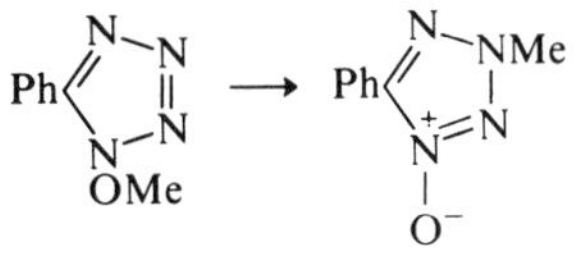

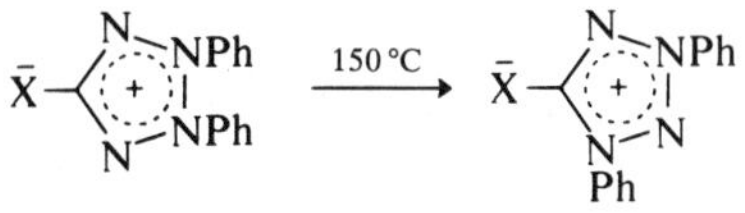

4.13.3.3.1 Basicity

The tetrazole ring is a weak base on which protonation occurs preferentially at N-4. A basic pK_a value of −3.0 has been quoted for tetrazole in aqueous H_2SO_4. Because of the weak protophilic properties of a series of 1,5-disubstituted tetrazoles it was not possible to measure basicity constants for these compounds in water. However, in formic acid pK_b values of *ca.* 2 were obtained for a range of tetrazoles and there was little difference between individual compounds because of the levelling effects of the solvent. The pK_b values in formic acid correspond to basic pK_a values of *ca.* +4.6 which are not comparable to those obtained in aqueous sulfuric acid ⟨77AHC(21)323⟩. Recently a useful linear relationship between the acidity and basicity constants of a range of 5-substituted tetrazoles has been reported: pK_a = 0.78 pK_{BH^+} + 6.37 ⟨81KGS599⟩. Tetrazoles are also weak π-donors and they form charge-transfer complexes with π-acids such as tetracyanoethylene and

1,3,5-trinitrobenzene ⟨66JPC3688⟩. The compound 1,5-pentamethylenetetrazole (PMT) was a strong electron-donor towards Lewis acids although it showed no basic properties in water. While complexes of type PMT·I$_2$ and PMT·IBr were observed in solution, the complex PMT·ICl was a stable solid. Its structure is described in Figure 2 (Section 4.13.2.3).

4.13.3.3.2 Electrophilic attack at nitrogen

Electrophiles such as acyl halides and imidoyl halides attack the 5-substituted tetrazole ring at the N(2)-position giving the Huisgen reaction outlined in Scheme 8 ⟨60CB2885, 61CB1555⟩. Further examples of similar electrophilic reactions are in Scheme 16. Cyanogen bromide reacted with 5-aryloxytetrazoles to give the 2-cyanotetrazoles (**116**). Unstable 2-tosyltetrazoles (**118**) were obtained when 5-dimethylaminotetrazole was treated with p-toluenesulfonyl chloride at 10 °C in pyridine. At normal temperatures in pyridine the products of the reaction were compounds (**120**) derived from (**118**). Mild warming of compounds (**118**) in water gave the amides (**119**) by solvent addition to the expected nitrilimine intermediate ⟨77AHC(21)323⟩. In the reaction of benzoyl chloride with 5-phenyltetrazole containing a ^{15}N label at the N(1)- and N(4)-positions to give 3,5-diphenyl-1,3,4-oxadiazole (**123**), half of the label was present in the oxadiazole product confirming N$_2$ loss from the 1,2(3,4) positions of the tetrazole ring and suggesting intermediates of type (**121**) and (**122**) ⟨61JOC2372⟩.

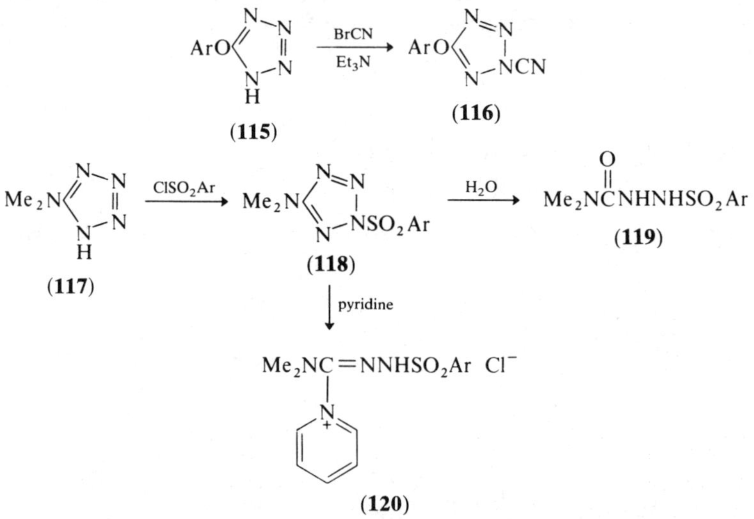

Scheme 16

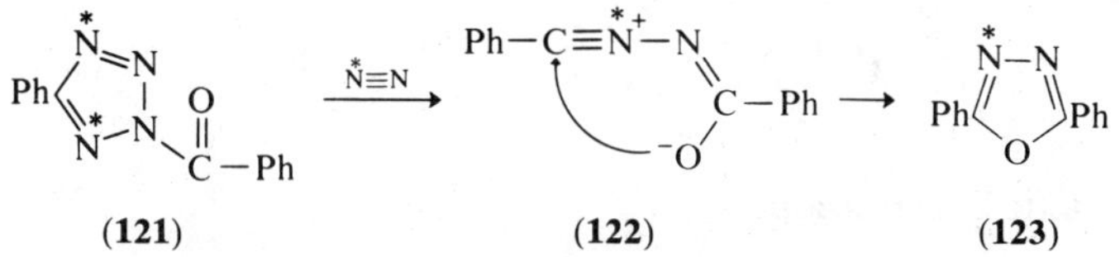

The preferred site for electrophilic attack is, however, apparently quite sensitive to the nature of the 5-substituent and strongly activating groups may orient the attack to the N-1 site. Thus attack by cyanogen bromide and methanesulfonyl chloride on 5-methoxy- and 5-ethoxy-tetrazoles occurred at the 1-position causing ring opening to imidoyl azides, e.g. ROC(N$_3$)=N—SO$_2$Me, which were obtained in high yields ⟨80JHC187⟩. Arylsulfonyltetrazoles obtained on treatment of tetrazole with arylsulfonyl chlorides, for use as condensing agents in nucleoside synthesis, have been described as 1-arylsulfonyltetrazoles, although the structural question as to why they did not open to azides as might be expected was not discussed ⟨76CJC670⟩. Other electrophiles which attack the 5-substituted tetrazole ring at the N-2 position are isocyanates and diketene. However, when the 5-substituent is an NH$_2$ group, the electrophilic substitution with these electrophiles also changes preferentially to the N(1)-position. Treatment of 5-aminotetrazole with isocyanates gave the 1-(N-alkylcarboxamido)tetrazoles (**125**) and with chlorosulfonyl isocyanate as electrophile the attack at

N-1 was accompanied by ring opening and cyclization on to the 5-amino group giving 5-azido-2H-1,2,4,6-thiatriazin-3(4H)-one 1,1-dioxide (**127**; Scheme 17) ⟨80JOC1662⟩. On treatment of 5-aryltetrazoles with diketene a normal N(2)-attack occurred giving rise to oxadiazoles by a Huisgen-type reaction. The reaction of 5-aminotetrazole with diketene, however, gave compound (**150**; Scheme 22) in quantitative yield probably via N(1)-attack followed by cyclization of the 5-amino group ⟨76CPB2549⟩. On treatment of tetrazole with vinyl ethers CH$_2$=CHOR, a 2-NH addition reaction occurred giving 2-alkoxyethyltetrazoles (**128**) ⟨80ZOR1313⟩. Treatment of 5-substituted tetrazoles with benzonitrile oxide gave a mixture of the N(1)-substituted compound (**129**) and its N(2) isomer. Thermolysis of compounds (**129**) provided a new route to 1,2,4-oxadiazoles (**130**) ⟨81BSB193⟩. An interesting electrophilic degradation of the tetrazole ring of 3-(tetrazol-5-yl)chromones (cf. structure **251**) with Vilsmeier reagents RCONR$'_2$ and POCl$_3$ to a carbonyl hydrazidine group (—CONHN=C(R)NR$'_2$) has also been reported ⟨74TL1187⟩.

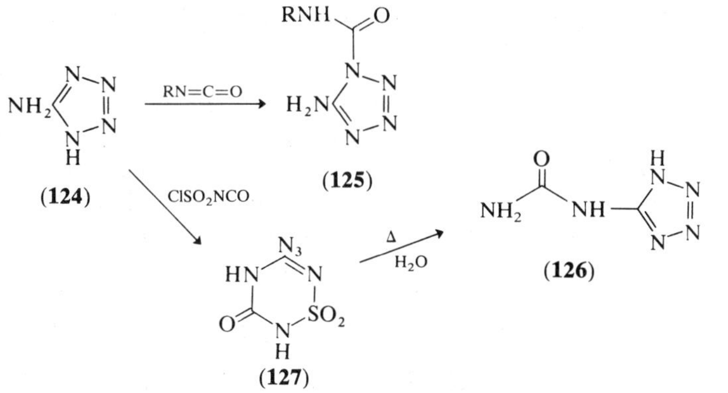

Scheme 17

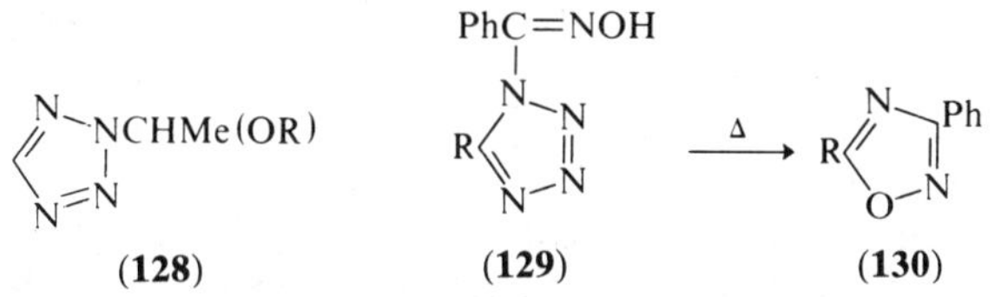

Quaternization of 5-substituted 2-aryltetrazoles occurred at the N(4)-position but further alkylation of N(1)-aryl 5-substituted derivatives occurred at both the N(2)- and N(3)-positions. An unusual conversion of 1-alkyl-5-aryltetrazoles (**131**) into the corresponding 2-alkyl isomers (**133**) occurred on heating with an alkyl iodide. The reaction has been explained in terms of a quaternization at the 3-position giving a 1,3-dialkyl-5-phenyltetrazolium iodide (**132**) which eliminated a molecule of alkyl iodide to give the thermodynamically more stable isomer ⟨77ZC261, 71JOC3807⟩. An interesting intramolecular electrophilic attack occurred at the tetrazole N(1)-position in the azacarbo cations (**133a**) which are formed by solvolysis of the corresponding hydrazonyl halides ⟨67JCS(C)239⟩. Ring cleavage to an azide occurred when the tetrazole ring was unsubstituted and a fused s-triazolotetrazole (**133b**) resulted when the tetrazole ring NH was substituted by an alkyl group.

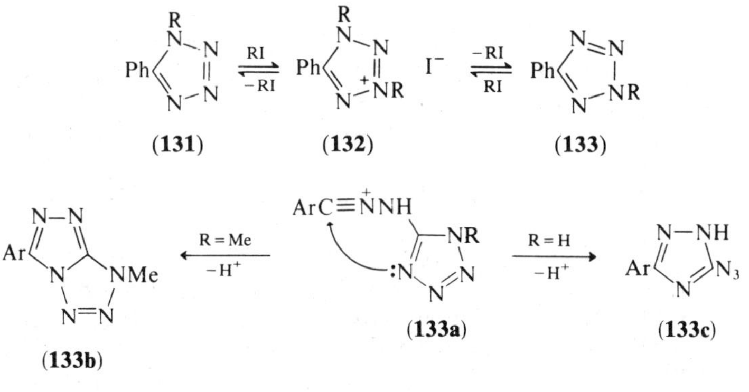

4.13.3.3.3 Electrophilic attack at carbon and α-exocyclic positions

There have been few reports of electrophilic substitutions on the tetrazole C(5)—H site of N-substituted tetrazoles, the only cases apparently being bromination and acetoxymercuration (Section 4.13.3.1.1(ii)). When the C-5 atom bears an active α-site (CH or NH), interesting ring degradations are encountered which appear to involve preferred electrophilic attack at the α-exocyclic position (Scheme 18). The presence of a labile H on the ring nitrogen atoms and a potential leaving group at the 5-exocyclic α-position, arising from the electrophilic attack, allows for loss of a small molecule and generation of a tetraazafulvene intermediate (Scheme 5). This fragments with loss of two molecules of N_2 and generates a carbene species which leads to the products (Scheme 18) ⟨77AHC(21)323, 76AG(E)113⟩.

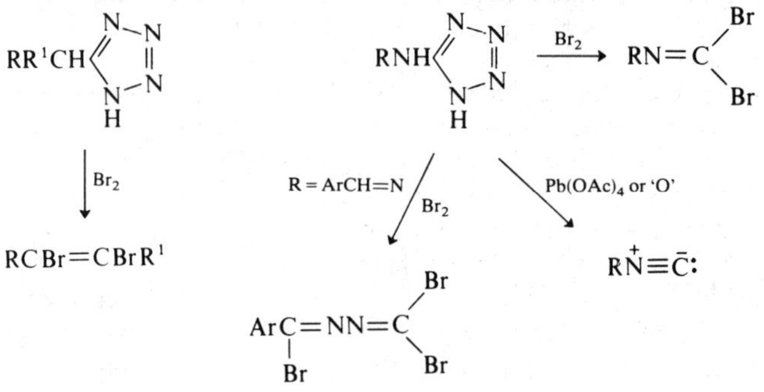

Scheme 18

4.13.3.4 Nucleophilic Substitution

4.13.3.4.1 Acidity

The tetrazole ring ($-CN_4H$, the tetrazolic acid group), represents the formal nitrogen analogue of the carboxylic acid group, $-CO_2H$ ⟨58MI41300, 56MI41300⟩. In general substituted 5-aryltetrazoles are somewhat stronger acids than the comparable benzoic acids (Table 5). This increased acidity has been ascribed to enhanced resonance stabilization of the aryltetrazolate anion and to a greater solvation of the tetrazolate anion as compared to the carboxylate anion ⟨79JCS(P2)1670⟩. Introduction of electron-withdrawing groups into the 5-phenyl ring increases the acidity of the aryltetrazolic acids and electron donating substituents have the reverse effect of decreasing acidity, in general agreement with substituent effects in benzoic acids. The well-known acid-strengthening effect of *ortho*-substituents in benzoic acids, however, does not apply in the aryltetrazolic acid series (Table 5) and *ortho*-substituted 5-phenyltetrazoles are not generally stronger acids than the corresponding *para*-substituted compounds as is the case for substituted benzoic acids. Changes of interannular conjugation effects in *ortho*-substituted 5-aryltetrazoles arising from loss of ring coplanarity are likely to contribute to these differences ⟨79JCS(P2)1670, 78JCS(P2)1087⟩.

Table 5 Comparative Acidities of 5-Aryltetrazolic Acids ($XC_6H_4CN_4H$) and Benzoic Acids ($XC_6H_4CO_2H$) in 50% Ethanol–Water ⟨79JCS(P2)1670⟩

Substituent X	Tetrazolic acid $K_a \times 10^5$	Benzoic acid $K_a \times 10^5$
H	1.05	0.178
p-Me	0.67	0.115
o-Me	0.61	—
p-MeO	0.56	0.085
m-Br	4.13	0.603
p-Br	2.06	0.447
o-Br	2.09	—
p-Cl	3.23	—
o-Cl	1.99	—
p-NO_2	20.2	2.95
o-NO_2	9.42	—

The influence of substituents directly bonded to C-5 on the acidity of the tetrazole ring follows the expected trend of electron-withdrawing groups enhancing the acidity. For the series of 5-substituents NH_2, Me, H, Cl, CF_3, the pK_a values in water were 5.93, 5.56, 4.89, 2.07 and 1.14, respectively. Many tetrazole analogues of amino acids have been synthesized and found to exhibit pK_a values in close agreement with those of the corresponding amino acids. Incorporation of these tetrazole amino acid analogues into peptides has also been achieved. These nitrogen analogues of the amino acids exist in the zwitterionic form similar to their carboxylic acid counterparts ⟨77AHC(21)323, B-67MI41300⟩ (*cf.* Section 4.13.6).

4.13.3.4.2 *Nucleophilic substitution at C-5*

Because of the deactivated nature of the tetrazole ring, leaving groups on the 5-carbon atom are readily displaced by nucleophiles (Scheme 19). The most common usage of this reaction has been with halogens as leaving groups and alcohols or phenols as nucleophiles, *e.g.* compound (**56**) ⟨77AHC(21)323⟩. The reaction may also be used to obtain substituted 5-hydrazino- and 5-amino-tetrazoles when hydrazines and amines are used as nucleophiles. Kinetic studies of this reaction with piperidine as nucleophile and base-induced protium–deuterium exchange at C-5 have been discussed in Section 4.13.3.1. Nucleophilic solvolysis type reactions at C-5 are readily observed on heating 1,4-disubstituted 5-imino- and 5-thioxo-tetrazoles in aqueous solutions when the exocyclic nitrogen and sulfur atoms are replaced by an oxygen atom giving 1,4-disubstituted tetrazolones ⟨77AHC(21)323⟩

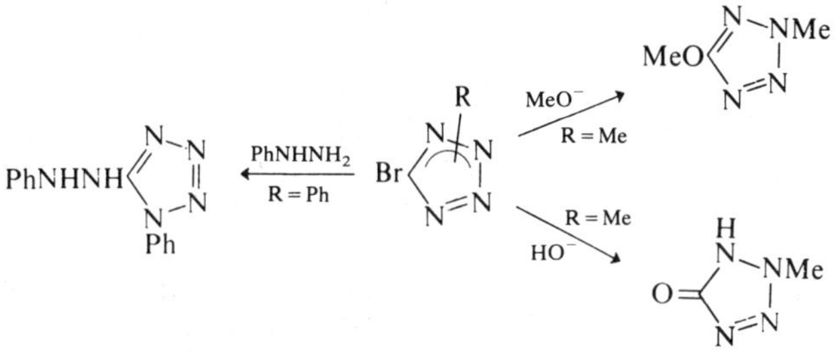

Scheme 19

4.13.3.5 Reactions of Tetrazolate Anions: Alkylation Reactions

4.13.3.5.1 *Ring alkylation*

The most common tetrazole anion is the tetrazolate species (**134a–c**). This is an ambident system in which reactions can occur at the 1- or 2-positions and also at the α-exocyclic site at C-5 if the α-atom is suitable, *e.g.* (**134c**). Anionic character at the tetrazole ring carbon has also been observed, for example in the reaction of *N*-substituted tetrazoles with alkyllithium compounds, and this is discussed in Section 4.13.3.1 (Scheme 2).

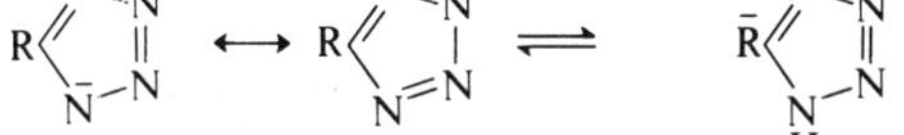

(**134c**) R = SH, NH_2, OH

The most important reactions of tetrazolate anions are alkylations in which the anion is used as a nucleophile in substitution reactions mainly with alkyl halides and sulfates. Many such reactions have been reported and a variety of conditions and reagents have been employed. The products are mixtures of $N(1)$- and $N(2)$-alkyl isomers, the relative proportions of which depend upon the conditions of the alkylation and the influence of the 5-substituent. In general, electron-donating substituents at C-5 tend to favor slightly alkylation at the $N(1)$-site while electron withdrawing substituents favor $N(2)$-alkylation. The position of alkylation was also influenced by the steric requirements of the alkylating agent and was found to be insensitive to electronic effects of *para*-substituents on benzyl halide alkylating agents. This suggests an S_N2 type process where the electronic effects of substituents on the alkyl halide often tend to cancel ⟨77AHC(21)323, B-67MI41300⟩.

In the series of substituted 5-phenyltetrazoles (**135**) electron-withdrawing substituents had a much larger effect than electron-donating groups. The isomer distribution of the products was linearly related to the Hammett σ-value of 4'-substituents but required exalted σ-values for strong resonance-electron withdrawing groups such as NO_2, and normal σ-values for strong resonance-electron donating groups such as MeO. For example, the ratio of (**137**):(**136**) for 4'-NO_2, -H and -Me groups was 6.15, 2.2 and 2.05, respectively. Substituents at the 2'- and 6'-positions caused out-of-plane rotation of the rings and, interestingly, oriented the alkylation toward the N(1)-position and not away from it as might be expected for steric reasons ⟨81JCR(S)174⟩. Linear Hammett relationships were also found for the alkylation of substituted 5-phenyltetrazoles with dimethylsulfate in acetonitrile. The product orientation was influenced as well by the nature of the cation and the temperature. For the alkylation of potassium 5-phenyltetrazolate with dimethylsulfate in acetonitrile the ratio of (**137**):(**136**) was 4.72 at 10 °C and 1.86 at 35 °C ⟨81ZOR146, 79ZOR839⟩. The net charges on the N(1)- and N(2)-sites of some tetrazolate anions have been estimated by MINDO/3 calculations. For example, the net atomic charges for the 5-methyltetrazolate anion were -0.3027 at N-1 and -0.1577 at N-2. The N(1)-position should therefore be more nucleophilic. The general and sometimes regioselective preference for reactions to occur at the N(2)-site has been ascribed to steric hindrance of the N(1)-position by the 5-substituent ⟨80JA2968⟩. The large product differences observed for alkylation of 5-(p-substituted phenyl)tetrazolate anions, where steric effects are constant, however, also indicate an important electronic influence. The observed quantitative trends suggest that the balance of ambident reactivity may reflect a harder Coulombic-controlled reaction at N-1 in competition with a softer orbital-controlled reaction at N-2 ⟨81JCR(S)174⟩.

The quantitative substituent trends which have been established for substituted 5-phenyltetrazoles are also qualitatively evident for substituents directly bonded to the 5-carbon (Table 6). However, direct comparisons under identical reaction conditions are rare and the data in Table 6 are not strictly comparable because of the variety of conditions of medium and temperature. Selective $N(1)$-alkylation has been obtained by blocking off the 2-position with a tri-n-butylstannyl group, when treatment of the 2,5-disubstituted derivatives with normal alkylating agents such as MeI and Me_2SO_4 gave up to 90% $N(1)$-alkylation and the Bu_3Sn blocking group at N-2 was removed during the reaction ⟨73BCJ2176⟩. Organic media or neutral alkylations with reagents such as diazomethane tend to favor ring-alkylation of tetrazoles at the N(2)-position. Examples for methylation with diazomethane and some recent t-butylations of 5-substituted tetrazoles are in Table 6.

4.13.3.5.2 Exocyclic α-alkylation and polyalkylation

Mono-alkylation of the exocyclic α-position at C-5 is generally not favored for 5-amino- and 5-oxo-tetrazoles and, if encountered at all, it usually occurs in ≤10%. Benzylation of 5-aminotetrazole occurred at the exocyclic 5-amino position to the extent of about 6–10% for a series of p-substituted benzyl chlorides and the main reaction was at the ring N-atoms. Benzylation of 5-dialkylaminotetrazoles was somewhat exceptional when exocyclic quaternization occurred, giving a 29% yield of the ylide (**10**) as well as the ring-alkylated isomers. Dialkylation of ring-alkylated tetrazoles may occur on an exocyclic 5-amino group and for example basic methylation of compounds (**138**) and their 2-methyl isomers occurred exclusively at the exocyclic amino group ⟨66JOC3182⟩. However, methylation of 5-amino-2-methyltetrazole (**139**) and its 1-methyl isomer occurred on N-4 giving the mesoionic compound (**7**) and 5-imino-1,4-dimethyltetrazoline, respectively. When 5-hydroxytetrazole (tetrazolinone) was treated with diazomethane, a mixture of 5-methoxy-1- and -2-methyltetrazoles and dimethylated tetrazolines was formed. Further methylation of 1-methyl-

Table 6 Alkylations of Tetrazoles

5-Substituent	Alkylating agent	Main product (%)	Ratio $N(2)$-R : $N(1)$-R	Ref.
(i) Anions				
NH_2	Me_2SO_4	55	1:3	54JA923
ArCH=NNH	MeI	60	1:6	66JOC3182
NO_2	MeI	57	3:1	54JA923
Me	MeI	—	0^a:1	65JOC3472
CF_3	MeI	—	6:1	62JOC3248
Me	EtO_2CCH_2Br	49	1:2	69CJC813
CF_3	EtO_2CCH_2Br	76	$1:0^a$	69CJC813
CF_3	n-$C_8H_{17}Br$	71	18:1	62JOC2085
$MeOCH_2$	EtO_2CCH_2Br	47	1:1.24	72ACS541
$MeOCH_2CH_2$	Me_2SO_4	47	1:1.1	61MI41300
H	EtO_2CCH_2Br	55	1:2	69CJC813
H	CH_2=$CHCH_2Br$	37	1:1.25	59JOC1565
H	Me_3SiCl	70	0^a:1	63CB2750
Cl	EtO_2CCH_2Br	63	7:1	69CJC813
EtO_2C	$PhCH_2Br$	34	1.7:1	75CB887
pyridin-3-yl	MeI	30	$1:0^a$	67JMC149
$R_2NCH_2CH_2$	EtI	43	4:3	70JMC777
(ii) Neutral molecules				
NH_2	CH_2N_2	50	1:1.06	56JA411
Me	CH_2N_2	47	5:1	65JOC3472
Br	CH_2N_2	68	5:1	56JA411
NH_2	t-BuOH, CuCl, $RNCNR^b$	58	1.4:1	76JHC391
MeS	t-BuOH, CuCl, $RNCNR^b$	59	4:1	76JHC391
Cl	t-BuOH, CuCl, $RNCNR^b$	30	1.4:1	76JHC391

^{a} Isomer either not detected or not reported. ^{b} R = cyclohexyl.

tetrazolin-5-one occurred at N-4 (80%), exocyclic oxygen (10%) and N-3 (1%) ⟨77AHC(21)323, B-67MI41300⟩.

(138), (139), (140)

Polyalkylation of N-substituted 5-thioxotetrazoles generally occurs on the exocyclic sulfur atom (*cf.* however, Section 4.13.2.3). Treatment of 1-phenyltetrazoline-5-thione with MeI or CH_2N_2 gave 5-methyl-1-phenylthiotetrazole and a similar reaction with substituted epoxides gave products of type (140) ⟨77ZC60⟩. The mesoionic compound dehydrodithizone (9b) was also alkylated preferentially on exocyclic sulfur to give 5-alkyl-2,3-diphenylthiotetrazolium salts ⟨77AHC(21)323⟩.

4.13.4 REACTIONS OF SUBSTITUENTS

4.13.4.1 General Survey

Substituents on the tetrazole ring generally behave normally and show their expected characteristic reactions. The electron-withdrawing character of the ring may influence the relative reactivity of substituent groups at the 5-carbon atom when they appear in $N(1)$- or $N(2)$-substituted tetrazole rings. The $N(1)$-substituted ring (high dipole moment) is more electron withdrawing than the $N(2)$-substituted isomer (low dipole moment) and substituents which are electron deficient in character are therefore generally more reactive towards nucleophiles in $N(1)$-substituted tetrazoles. However, these trends are not yet well established in the literature although they are clearly visible with some of the substituent groups (Figure 4).

Figure 4 Reactivity of electron deficient α-substituents to nucleophiles

4.13.4.2 Aldehyde, Carbonyl and Nitrile Substituents

Tetrazoles with an aldehyde group at the 5-position have recently been prepared as outlined in Scheme 20. The 1-substituted tetrazole-5-carbaldehydes (**141**) were more reactive than the corresponding 2-substituted isomers. They formed hydrates such as (**142**) and hemiacetal dimers (**143**). The aldehyde groups showed the range of normal aldehyde reactions for both 1- and 2-substituted tetrazole-5-carbaldehydes. On treatment of the 1-substituted aldehydes with base deformylation occurred giving the parent 1-substituted tetrazole. In particular the 1-substituted tetrazole-5-carbaldehydes were considerably more reactive towards nucleophilic reagents than the 2-substituted isomers. The parent N-unsubstituted tetrazole-5-carbaldehyde has been prepared by cycloaddition of hydrazoic acid with $(EtO)_2CHCN$ in pyridine followed by hydrolysis to the aldehyde. In the solid state this exists in the interesting dimeric structure (**141a**) ⟨75CB887, 77CZ403⟩.

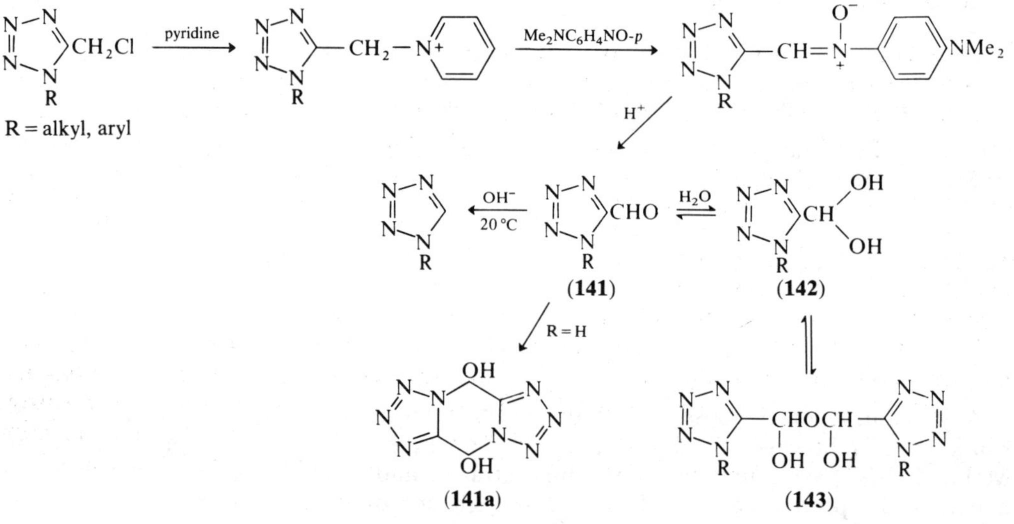

Scheme 20

Carbonyl substituents at the C(5)-position are readily introduced by oxidizing secondary alcoholic groups or methylene moieties, *e.g.* $ArCH_2$, located at the 5-position. These carbonyl groups generally behave normally, particularly when they are of the aroyl, ArCO, type. For example, oxidative degradation gives rise to 5-carboxytetrazoles, Grignard additions give the expected tertiary alcohols and halogenation reactions give the expected α-halocarbonyl compounds. However, 5-acetyltetrazoles undergo some unexpected reactions. Thus treatment of 5-acetyl-1-phenyltetrazole with NaOI gave a deacetylation to 5-iodo-1-phenyltetrazole and treatment with KCN or amines gave a degradation to 1-phenyltetrazole ⟨B-67MI41301⟩.

A nitrile group at the tetrazole 5-position also exhibits normal reactions. It is readily hydrolyzed to the amide and carboxylic acid groups and on treatment with H_2S it is converted into a thioamide function. Addition of hydroxylamine to a nitrile substituent at the tetrazole C-5 gives an amidoxime substituent, $-C(NH_2)=NOH$, at the 5-position ⟨76ZC90⟩. Treatment of 5-cyano-2-phenyltetrazole with MeMgI followed by AcCl gave 5-acetyl-2-phenyltetrazole but ring degradation to AcN(Ph)CN occurred in the comparable reaction with 5-cyano-1-phenyltetrazole ⟨81CZ194⟩.

4.13.4.3 Alkoxy and Hydroxy Substituents

Reductive cleavage of 5-aryloxy-1-phenyltetrazoles (**144**) has proved to be an effective method for removing hydroxy groups from phenols which is all the more useful because of the ease with which phenols can be converted into 5-aryloxytetrazoles. The tetrazole C(5)—O bond is apparently stronger than the phenol C—O bond hence giving rise to the elusive aryl–oxygen cleavage. The reaction has been widely used with a variety of phenol types. The strength of the tetrazole C(5)—O bond is also illustrated by the ready cleavage of 5-alkoxytetrazoles on the alkyl–oxygen bond when heated in hydrochloric acid under conditions milder than those required for normal Zeisel type cleavages of arylalkyl ethers ⟨66JA4271, 77AHC(21)323, B-67MI41300⟩.

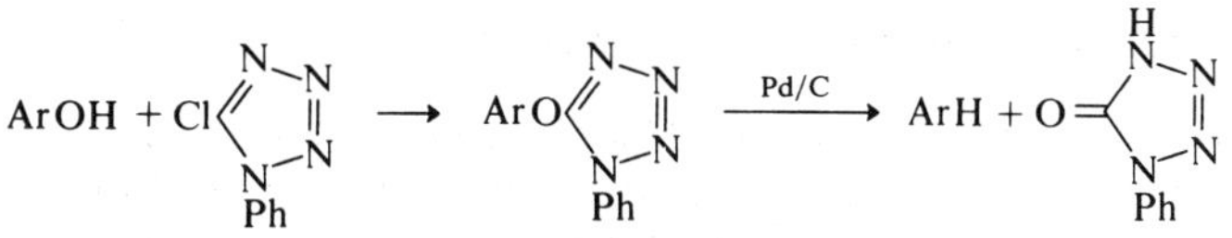

$$\text{ArOH} + \text{Cl}\underset{\underset{\text{Ph}}{|}}{\overset{\text{N}=\text{N}}{\diagup\!\!\diagdown}} \longrightarrow \text{ArO}\underset{\underset{\text{Ph}}{|}}{\overset{\text{N}=\text{N}}{\diagup\!\!\diagdown}} \xrightarrow{\text{Pd/C}} \text{ArH} + \text{O}=\underset{\underset{\text{Ph}}{|}}{\overset{\text{HN}-\text{N}}{\diagup\!\!\diagdown}}$$

(**144**)

Acylation reactions of 1-aryltetrazol-5-ones generally occur on the N(4)-position, but acylation of the exocyclic oxygen atom occurred in up to 50% yield when the acylating agent was 2-methylpropanoyl chloride ⟨73ZC429⟩. Fragmentation and rearrangement reactions of 5-alkoxytetrazoles are discussed in Section 4.13.3.2 and alkylations of these systems in Section 4.13.3.5.

4.13.4.4 Amino Substituents

Amino substituents on the tetrazole ring, whether located on the ring nitrogens or at the 5-carbon atom, generally behave normally giving the reactions expected for primary amines ⟨77AHC(21)323⟩. Thus 1-aminotetrazole (**145**) gives normal nucleophilic addition–elimination type reactions with electrophilic reagents such as aldehydes, sulfonyl halides and isocyanates (Scheme 21). The reaction with acyl halides involves acetylation of the ring N(4)-position as well as the amino group and gives rise to the substituted 1,3,4-oxadiazoles (**146**; Scheme 21) ⟨66CB850⟩. Heating of 5-aminotetrazole (**147**) with acid chlorides and anhydrides also results in ring fragmentation to oxadiazoles. When this reaction was terminated early to avoid oxadiazole formation, acylation of the 5-amino group occurred giving 5-acylamidotetrazoles as products. The ring fragmentation in this reaction may involve a second acylation at the N(2)-position of the ring as in the Huisgen reaction but the possibility of a thermal fragmentation of the initial 5-acylamidotetrazole to an imino nitrene has also been postulated ⟨76CPB2549⟩. Treatment of 5-aminotetrazole with ethoxycarbonyl or benzoyl isothiocyanate in pyridine gave the 5-amidotetrazoles (**148**; Scheme 21). The reaction involved an addition at the 5-amino group followed by elimination of HSCN ⟨73CB2925⟩.

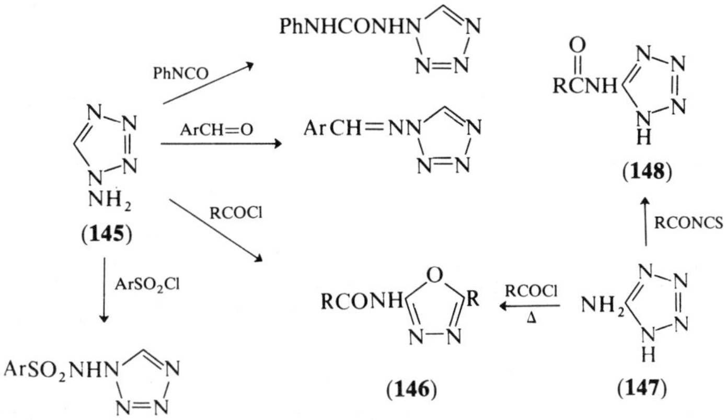

Scheme 21

5-Aminotetrazole undergoes a number of useful condensation reactions with some difunctional electrophilic reagents (Scheme 22). Cyclization with 2-alkyl-3-ethoxyacroleins gave tetrazolo[1,5-a]pyrimidines (**149**) in which the azido–tetrazole isomerism has been directly detected ⟨80AGE924⟩. The reaction with diketene gave 7-methyltetrazolo[1,5-a]-pyrimidin-5-one (**150**) ⟨76CPB2549⟩. Treatment of 5-aminotetrazole with methyl acrylate gave compound (**151**; Scheme 22) arising from addition at the 1-NH site. Heating of compound (**151**) resulted in a secondary cyclization to the 5-amino group giving the tetrahydrotetrazolopyrimidinone (**152**). Similar cyclization reactions were observed on treatment of 5-aminotetrazole with acetylenedicarboxylic esters ⟨72CB103⟩.

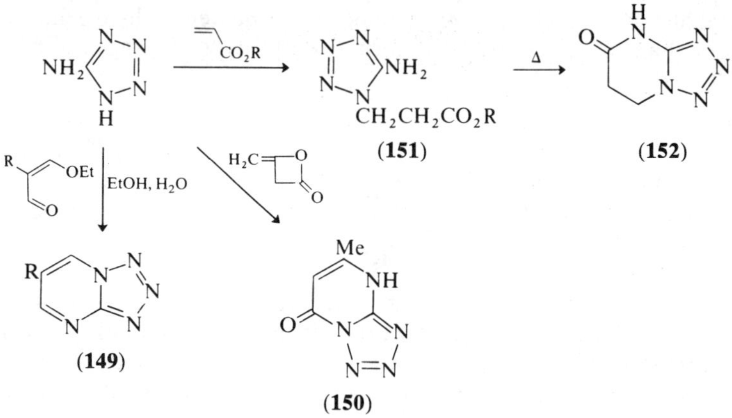

Scheme 22

Diazotization of 5-aminotetrazole with sodium nitrite in aqueous hydrochloric acid gives the explosive diazonium salt (**102**). This diazonium compound is readily reduced to tetrazol-5-ylhydrazine and it also undergoes dediazoniation reactions. Diazotization of ring N-substituted 5-aminotetrazoles with sodium nitrite in 10% hydrochloric acid gave stable primary nitrosamines of type (**153**). These nitrosamines were readily reduced to hydrazines and, when heated in aromatic hydrocarbon solvents, they displayed Gomberg–Bachmann type homolysis reactions giving 5-aryltetrazoles. When 5-aminotetrazoles are diazotized in a very dilute acidic medium, for example pentyl nitrite in acetic acid, the products are often triazenes of type (**154**) ⟨75CRV241⟩. Nitration of ring-substituted 5-aminotetrazoles occurs on the exocyclic amino group giving 5-nitroaminotetrazoles and this also occurs when the 5-amino group is a secondary amine. On treatment with oxidizing agents such as $KMnO_4$ and $Ca(OCl)_2$, 5-aminotetrazoles undergo an oxidative dimerization to azotetrazoles, e.g. (**155**) ⟨B-67MI41300⟩.

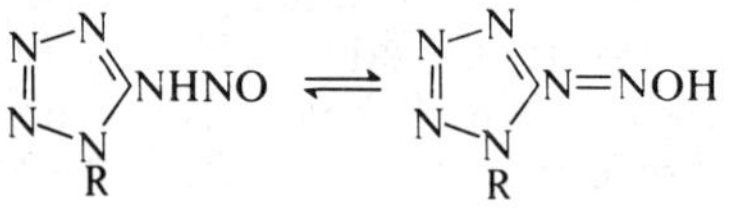

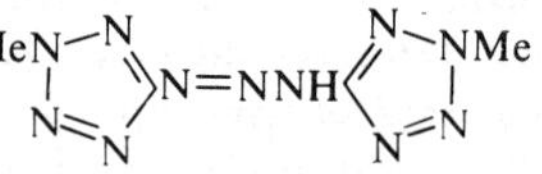

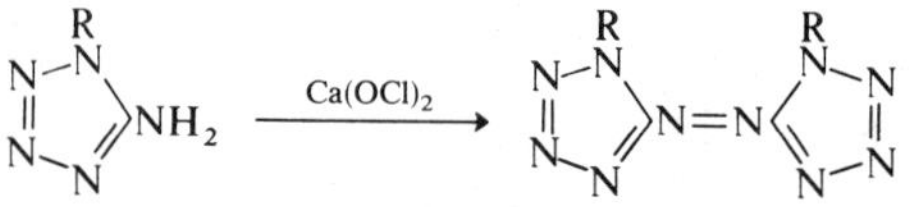

Thermolysis of 5-aminotetrazole above 200 °C gave hydrazoic acid (HN_3), hydrazonium azide (H_5N_5), cyanamide and a polymer $(CH_3N_3)_n$ and the reaction was sensitive to the medium used ⟨72CI(L)294⟩. Thermal decomposition of explosive tetrazole-5-diazonium salts (**102**) has been used to generate carbon in the atomic state ⟨74JA7830⟩.

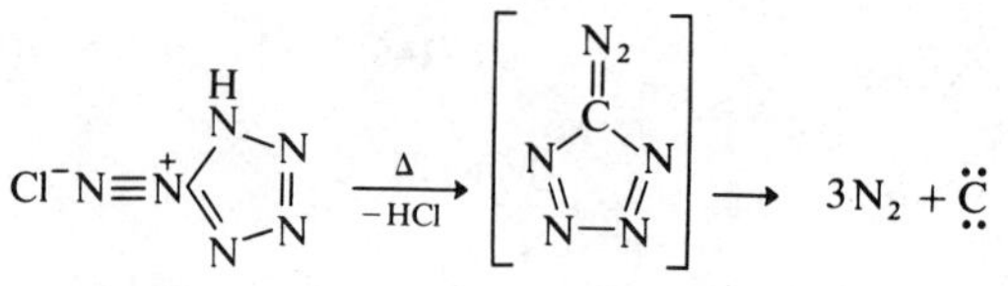

4.13.4.5 Sulfur Substituents

Alkylation reactions of 1-substituted tetrazoline-5-thiones occur on the exocyclic sulfur atom but acylation reactions may occur on the sulfur and also the ring nitrogens, usually at N-4. However, acylation of 1-substituted tetrazoline-5-thiones also occurs to some extent at N-2 and low yields of rare 1,2-disubstituted tetrazolines of type (**26**) have been reported. *S*- versus *N*-alkylation of the tetrazoline-5-thione system is commented on in Sections 4.13.2.1 and 4.13.3.5.

Disulfides (**156**) and (**157**) may be obtained by mild oxidation of tetrazoline-5-thiones or by treatment with aryl halosulfides. Oxidation of the tetrazoline-5-thione group with strong oxidizing agents such as $KMnO_4$ converts it into a sulfonic acid group, SO_3H ⟨77AHC(21)323, B-67MI41300⟩. Treatment of di(tetrazol-5-yl) disulfides (**156**) with diazoalkanes gives the interesting rearranged compounds (**158**) as the major products along with minor yields of the *S*-alkylated compounds (**159**) ⟨78CB566⟩. Oxidative desulfurization of 1-substituted tetrazole-5-thiones may be used to obtain 1-aryltetrazoles. Protection of the 5-thione function has also been achieved by treatment with hydroxybenzyl halides which alkylate the sulfur atom and can subsequently be readily removed ⟨77AHC(21)323, 78JOC1197⟩.

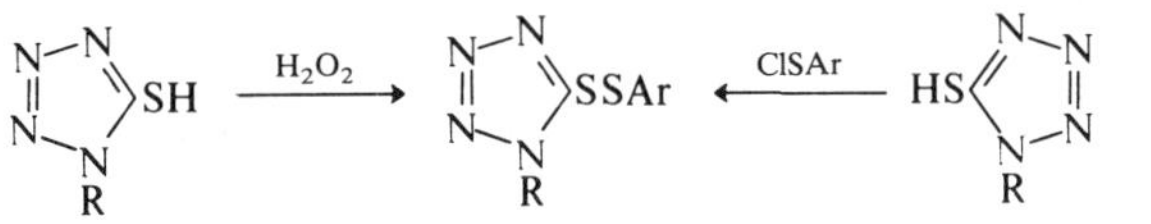

(**156**) Ar = 1-substituted tetrazol-5-yl
(**157**)

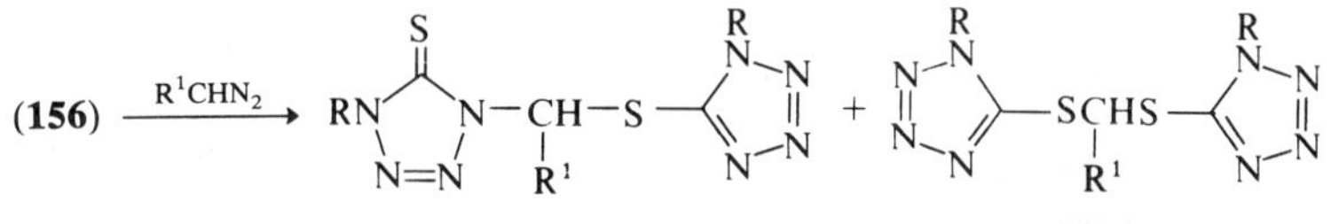

The 5-thiotetrazole system undergoes a number of interesting induced rearrangements in which the tetrazole moiety migrates away from the sulfur atom. These include the acid-promoted Smiles rearrangement (**160**→**162**; Scheme 23) which involves a migration of the tetrazole ring from sulfur to nitrogen ⟨76JOC3395⟩. Migration of the tetrazole ring from sulfur to oxygen was observed on alkylation of the compounds (**161**) with methyl iodide and potassium carbonate in acetone when the products were the tetrazolyl ethers (**163**; Scheme 23) ⟨B-76MI41300⟩. Recently the compound 1-phenyltetrazoline-5-thione has been found to be a useful agent for inducing acylations and macrocyclic lactonizations. The compound was first treated with *t*-butyl isocyanide which by insertion into the 5-SH bond gave a thioformimidate group, $-SCH=NCMe_3$, at the tetrazole 5-position. Treatment of this product with ω-hydroxyalkanoic acids gave a thiocarboxylate group $-SC(O)ROH$, at the tetrazole 5-position which was in equilibrium with the rearranged thione form containing the ω-hydroxyalkanoyl group HORCO at the tetrazole N-1 position. Such compounds were highly active acylating agents and the labile acyl moiety underwent lactonization with the ω-hydroxy group giving high yields of 16-, 17-, 18- and 20-membered lactones ⟨81AG(E)771⟩.

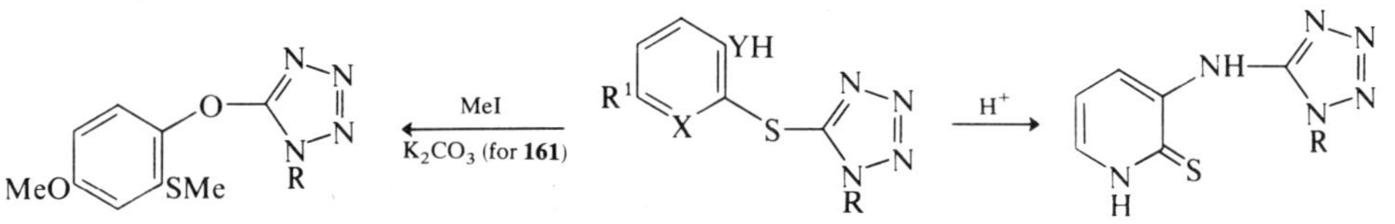

(**163**)

(**160**) X = N; Y = NH; R^1 = H
(**161**) X = CH; Y = O; R^1 = OH

(**162**)

Scheme 23

4.13.4.6 Halogen Substituents

The essential reaction of halogen substituents at the tetrazole carbon atom is that of nucleophilic displacement. This readily occurs because of the π-deficient deactivated nature of the ring. The use of the reaction in synthesis with nitrogen and oxygen nucleophiles and the results of kinetic and mechanistic studies have already been discussed and illustrated in Sections 4.13.3.1 and 4.13.3.4.2.

4.13.4.7 Alkyl Substituents

Alkyl groups on the tetrazole ring might be expected to show a significant degree of CH lability due to the electron-withdrawing effect of the ring. This is observed to a certain extent but in general the alkyl groups are not highly labile and, for example, reactions such as condensations with benzaldehyde are not observed. Treatment of *N*-aryltetrazoles containing a 5-alkyl substituent with phenyllithium or phenylsodium gives 5-(α-metallo-alkyl)tetrazole derivatives (**165**) which undergo normal substitution reactions (Scheme 24). The tetrazole 5-CH_2 moiety of 5-benzyltetrazoles may be oxidized to a carbonyl group under controlled conditions to give 5-benzoyltetrazoles. Halogen atoms on a tetrazole 5-α-alkyl substituent undergo normal nucleophilic substitution reactions (*e.g.* Schemes 20 and 24) ⟨75ZC102, 76ZC90⟩. Recently, an interesting rearrangement of the bis(1-aryltetrazol-5-yl)dichloromethanes (**166**) to the 1,2,3-triazole compounds (**167**) has been reported on heating the former with powdered copper in toluene ⟨80ZOR2623⟩. The electron-withdrawing power of the tetrazole ring allowed for the formation of stable ylides of type (**169**). When the tetrazole *N*-phenyl substituent was at the N(2)-position, the ylides were less stable due to the reduced electron-withdrawing effect of the ring. The compounds (**169**) underwent characteristic ylide substitution and addition reactions (Scheme 25) ⟨76JCS(D)909⟩.

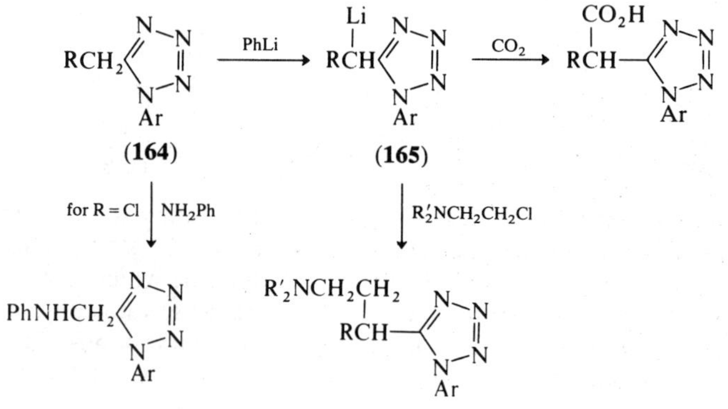

Scheme 24

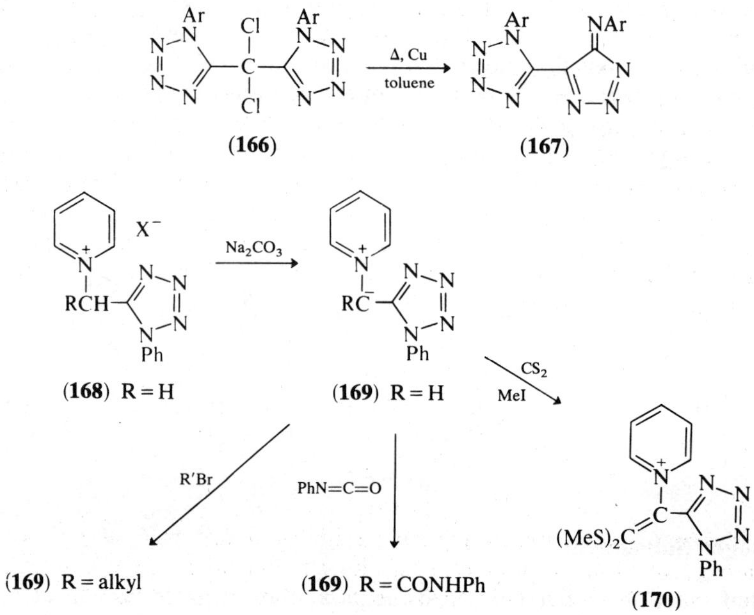

Scheme 25

4.13.5 SYNTHESIS

4.13.5.1 Ring Synthesis from Non-heterocycles

4.13.5.1.1 Formation of one bond

(i) Generation and cyclization of imidoyl azides: azidoazomethine–tetrazole isomerism

Molecules with an azide group bonded to a doubly bound carbon may exist in the acyclic or cyclic forms (**171**) and (**172**). When X is an oxygen atom, the molecule exists in the acyclic form (**171**) and when X is a sulfur atom, the cyclic form predominates. When X is an N—R' moiety, an equilibrium of both forms may be present or either form may predominate. Thus the synthesis and cyclization of imidoyl azides (**174**; Scheme 26) from imidoyl derivatives (**173**) containing a suitable leaving group is a principal route to 1-substituted and 1,5-disubstituted tetrazoles (**175**) ⟨77AHC(21)323⟩.

The cyclization of imidoyl azides (**174**) usually occurs spontaneously. In the cyclization the formation of the activated complex entails a movement of the imino lone-pair toward the terminal azido nitrogen at the same time as a lone pair is being formed on the central azido nitrogen at the expense of the azido terminal π-bond. The transition state resembles the reactant azide and the geometrical changes to the tetrazole ring structure occur late on the reaction profile ⟨76JA1685⟩. The cyclization is an example of a 1,5-heteroelectrocyclization operating in conformation with the principle of conservation of orbital symmetry and the imidoyl azide ⇌ tetrazole system is a nitrogen analogue of the hypothetical pentadienyl anion-cyclopentenyl anion system ⟨80AG(E)947⟩.

There are two main structural constraints on the cyclization. One of these is a stereoelectronic effect which arises from the required *cis* orientation of the imino lone pair and the azido group. The other arises from the electronic effects of substituents on the imino moiety. Imidoyl azides in which the azido group and the imino lone pair are in a *trans* arrangement do not cyclize without prior inversion in accordance with the requirements of the transition state, *e.g.* compounds (**176**)–(**178**). In the compounds (**178**), interestingly, the azido form is stabilized by an intramolecular hydrogen bond. These azides are readily converted into tetrazoles by treatments which bring about the required geometrical isomerizations of the imino moiety. For example, heating of the azides (**176**) with acyl chlorides gives a cyclization to 1-(acyl)hydroxy-5-substituted tetrazole derivatives and heating or changing the pH of the medium for the azides (**177**) and (**178**) gives ready cyclizations to the corresponding tetrazoles ⟨77AHC(21)323, 80JOC4302, 80JCS(P2)535⟩.

Electron-withdrawing substituents around the imino moiety favor the azido form while electron-donating substituents enhance ring closure and favor the tetrazole form. Thus compounds such as (**179**)–(**181**) exist as azides due to the deactivation of the imino group (Scheme 27) and when the CF$_3$ group of compound (**180**) was replaced by an Me group, the molecule existed as a tetrazole ⟨69LA(725)29⟩. Competitive cyclizations are possible in the addition reactions of carbodiimides with azides or hydrazoic acid. The favored tetrazole is that arising from cyclization of the intermediate imidoyl azide on to the nitrogen with the stronger electron-donating substituent, *e.g.* (**183**). When one of the imino nitrogen atoms contains a strong deactivating group such as SO_2, *e.g.* compound (**182**), cyclization does not occur to this nitrogen and the products are compounds of type (**184**; Scheme 27).

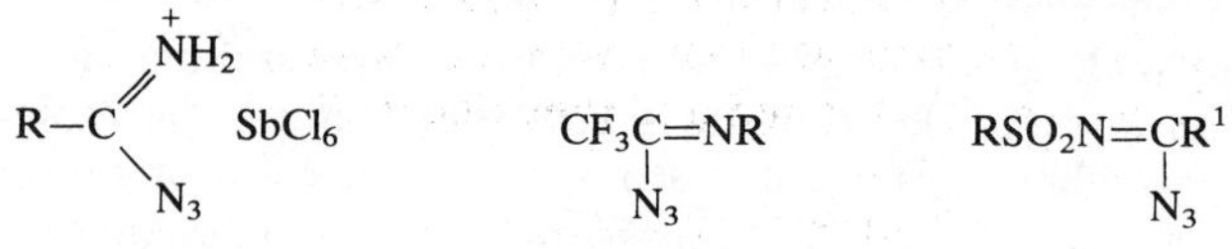

(**179**) (**180**) R = β-naphthol (**181**)

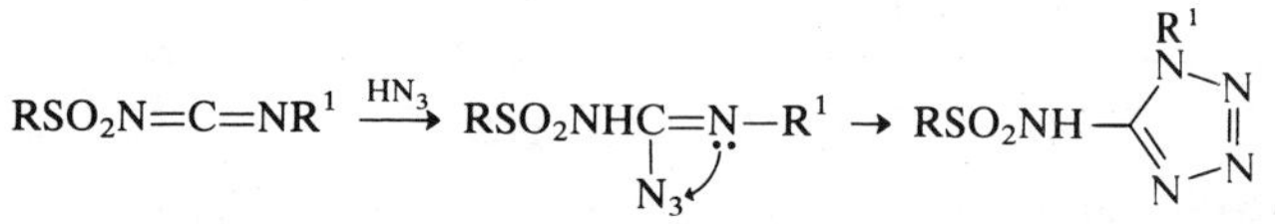

(**182**) (**183**) (**184**)

Scheme 27

Among the environmental constraints on the cyclization of imidoyl azides are temperature and the solvent system. Higher temperatures inhibit the cyclization and favor cleavage of the tetrazole ring which is an endothermic process ⟨73S123⟩. The solvent system used for the synthesis of imidoyl azides, and therefore also for their cyclization, may have a large effect on the efficiency of the process. Traditional aqueous or alcoholic media may give rise to competitive solvolysis of the imidoyl substrate (Scheme 28) and reactions in such media often involve long reaction times and tedious separations in the work-up. The use of dipolar aprotic solvents avoids many of these difficulties and these solvents also activate the azide anion through lack of hydroxylic solvation. Such media are therefore favored for tetrazole synthesis ⟨76JOC1073⟩. Acidic media in which the imidoyl azide might be protonated, *e.g.* CF_3CO_2H, also inhibit tetrazole formation while basic media, where the electron density on the imino moiety may be increased, *e.g.* through anion formation, tend to favor the cyclization process.

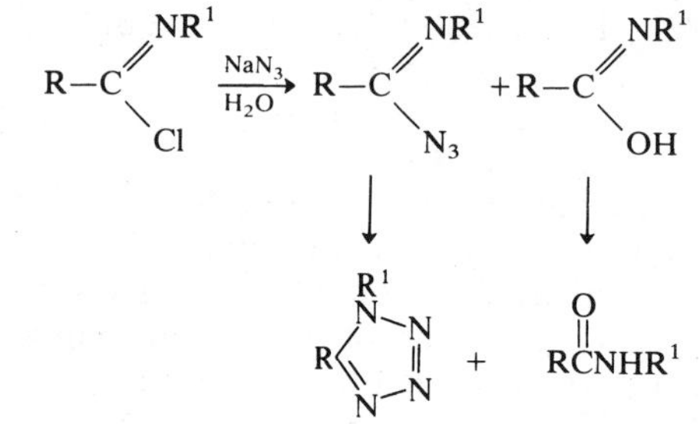

Scheme 28

Another important route to the imidoyl azide system is by diazotization of amidrazones (**185**) *via* the hydrazine form (**186**; Scheme 29). The resultant imidoyl azides (**187**) generally cyclize immediately to substituted tetrazoles (**188**) and a variety of such systems has been

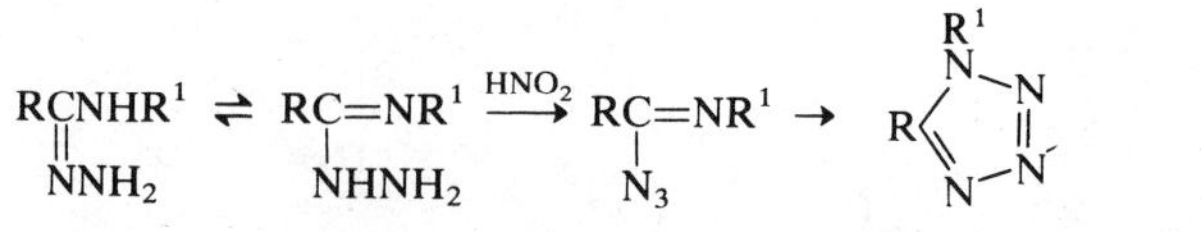

(**185**) (**186**) (**187**) (**188**) R = alkyl, aryl

Scheme 29

prepared by this reaction. If the substituent R in compounds (**187**) is an alkylamino group and R^1 a hydrogen atom, the resulting azide (**189**) may cyclize by path (a) or (b) (Scheme 30). Cyclization occurs to the imino nitrogen of highest electron density and path (a) is favored.

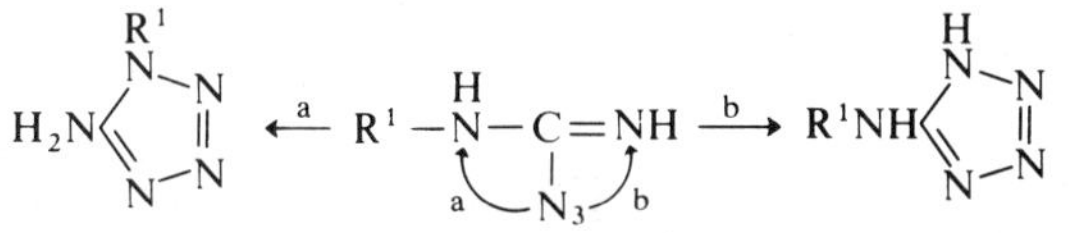

Scheme 30

R¹ = Me, 65%   (**189**)   R¹ = Me, 11%

The literature abounds in examples of 5-substituted tetrazoles, 1-substituted tetrazoles and 1,5-disubstituted tetrazoles prepared *via* imidoyl azide intermediates and the reviews mentioned in Section 4.13.1 may be consulted for further examples of these. When the imidoyl azide moiety is part of another heterocyclic ring, the cyclization gives rise to fused tetrazolo heterocycles and in many such systems both forms are observed in equilibrium ⟨77AHC(21)323⟩. Systems of this type are discussed in Chapters 4.15, 4.36 and 4.37.

(ii) Cyclization of formazan and tetrazene derivatives

Acyclic units, other than imidoyl azides which contain a potential tetrazole ring, are the formazan, —N=N—C=N—NH—, and tetrazene, —C—NH—N=N—NH—, systems. Cyclizations of derivatives of these may provide routes to substituted tetrazoles. Oxidation of formazans (**190**) readily gives tetrazolium salts (**191**) in high yields (Scheme 31). If one of the formazan N-substituents is a good leaving group, *e.g.* R¹ in compounds (**190**), the cyclization may result in a cleavage of this group thereby giving rise to 2,5-disubstituted tetrazoles (**192**; Scheme 31) ⟨76ZC90⟩. Syntheses of 2,5-disubstituted tetrazoles by this method may also be achieved without isolation of the formazan intermediates and, for example, treatment of aldehyde toluene-*p*-sulfonylhydrazones with aromatic diazonium salts in NaOH–EtOH or pyridine gives compounds (**192**) directly in up to 78% yields ⟨76BCJ1920⟩. Oxidation of tetrazolylformazans (**193**), obtained from the coupling of diazonium salts with tetrazol-5-ylhydrazones, gives interesting tetrazolium ylides of type (**194**) ⟨74KGS268, 73KGS1570⟩. Oxidation of dithizone, [S—C(N=NPh)=N—NHPh]₂ readily gives mesoionic

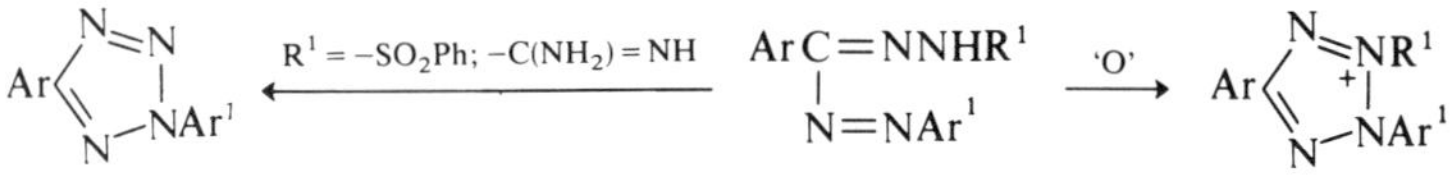

Scheme 31

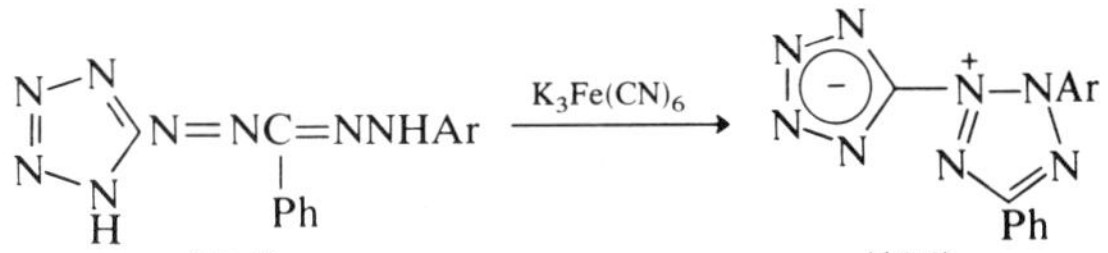

2,3-diphenyl-2*H*-tetrazolium-5-thiolate (**9b**) ⟨61JA5023, 70JA1965⟩. Treatment of 2-diazoacetophenone (**195**) with potassium *t*-butoxide in *t*-butyl alcohol gave 5-benzoyl-2-phenacyltetrazole (**196**) as well as benzoic acid and 2-phenacyltetrazole from cleavage of compound (**196**). The reaction has been explained in terms of an initial attack by base on the terminal N of the diazo moiety followed by addition of the resulting species to an

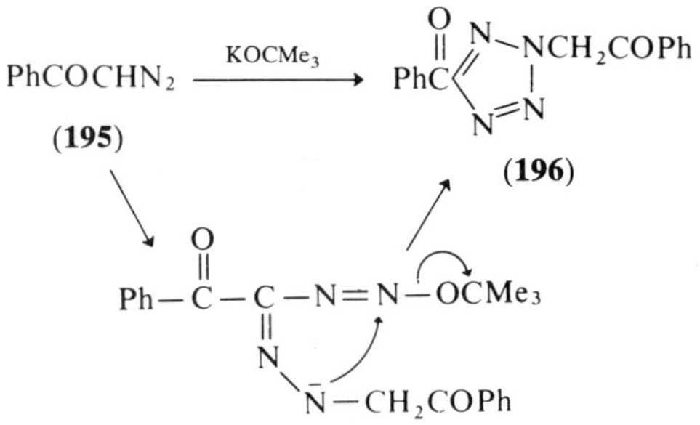

Scheme 32

unchanged molecule of starting material ultimately giving the intermediate (**197**) which cyclizes to compound (**196**; Scheme 32) ⟨69CJC3997⟩. Diazotization of substituted hydrazidines leads to good yields of 2,5-disubstituted tetrazoles (**198**) and it was this reaction which led to the first tetrazole derivative compound (**198**; R = CN, Ar = Ph) ⟨1885CB1544⟩. This was subsequently converted into tetrazole by oxidative degradations.

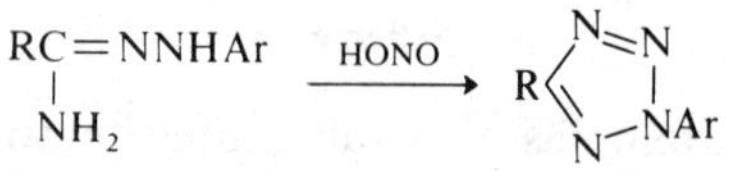

$$RC(=NNHAr)NH_2 \xrightarrow{HONO} \text{(198)}$$

Treatment of acylhydrazines with diazonium salts gives rise to tetrazenes (**199**) which can be readily dehydrated with base to give 1,5-disubstituted tetrazoles (Scheme 33). The reaction is also effective with 1,2-diacylhydrazines since one of the acyl groups is cleaved during the dehydrative cyclization, which may be conveniently carried out *in situ* without isolation of the tetrazene. This reaction can be useful for the synthesis of 1-aryltetrazoles using 1,2-diformylhydrazine as substrate ⟨B-67MI41300⟩. Mesoionic tetrazolium 5-oxides (**8a**) are obtained on treatment of bis(alkylsulfonyl)methanes with diazonium salts. Hydrazones of type (**200**) are also obtained from these reactions and the tetrazenes (**201**) derived from these are considered to be precursors to the tetrazoles (Scheme 34) ⟨74ZOR1725⟩. Treatment of arene- and heteroarene-diazonium salts with diazomethane is a related reaction which has been used to obtain 1-aryltetrazoles, *e.g.* compound (**202**). The quantity of diazomethane used in this reaction is important and if excess is available methylation of the products may occur ⟨70CB3284⟩.

$$RC(O)NHNH_2 + Cl^-N_2^+Ar \rightarrow RC(O)NHNHN=NAr \xrightarrow{NaOH} \text{tetrazole}$$

(199)

Scheme 33

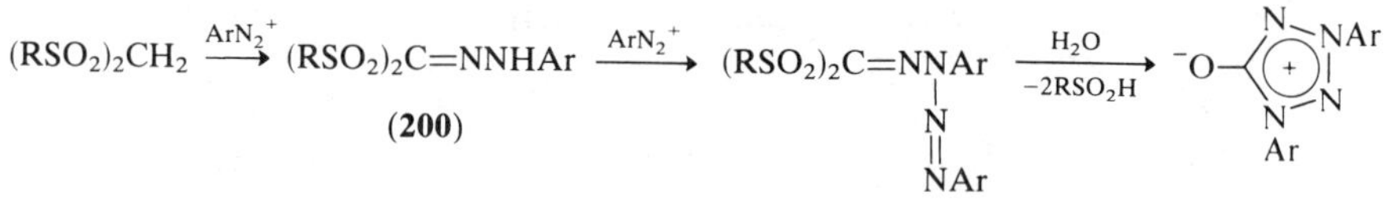

$$(RSO_2)_2CH_2 \xrightarrow{ArN_2^+} (RSO_2)_2C=NNHAr \xrightarrow{ArN_2^+} (RSO_2)_2C=NNAr \xrightarrow[-2RSO_2H]{H_2O} \text{(8a)}$$

(200)   (201)

Scheme 34

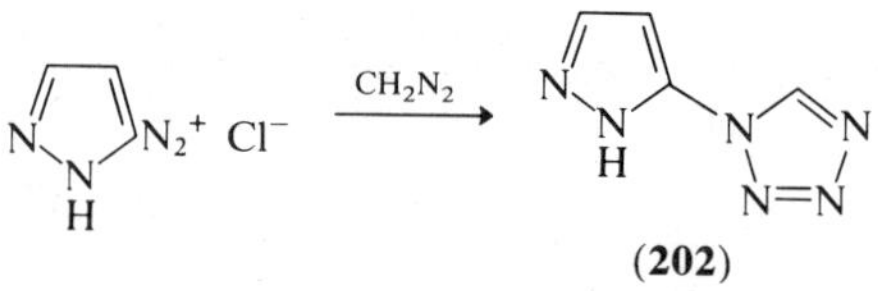

(**202**)

4.13.5.1.2 Formation of two bonds

(i) Addition of azide ion to nitriles, nitrilium ions and isonitriles

The addition of the azide anion to nitriles is the most widely used route to 5-substituted tetrazoles (**203**). This reaction can prove highly sensitive to the nature of the cation M^+ and the type of solvent used. It is also sensitive to the nature of the substituent R of the nitrile and it works best when R contains electron-withdrawing groups. In hydroxylic or alcoholic solvents the reaction usually requires a few days at reflux temperatures and the yields may be medium or poor. The best conditions tend to be in dipolar aprotic solvents such as DMF, using ammonium azide as the source of azide ion. The work-up procedures

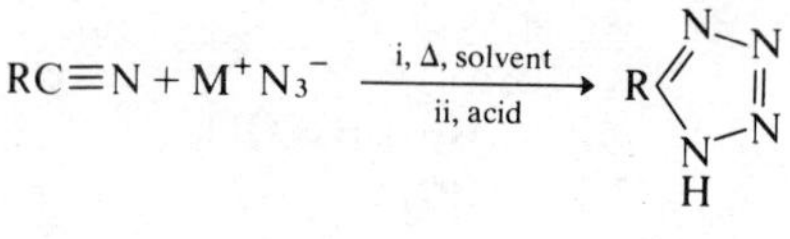

$$RC\equiv N + M^+N_3^- \xrightarrow[\text{ii, acid}]{i, \Delta, \text{solvent}} \text{(203)}$$

Tetrazoles

involve the freeing of the anionic tetrazolate salt by acid treatment. The overall reaction involves addition of hydrazoic acid (HN_3) but the use of azide ion and subsequent acid treatment is more convenient than generating solutions of hydrazoic acid ⟨77AHC(21)323, 73S71⟩.

A wide range of 5-substituted tetrazoles has been prepared using this reaction (Scheme 35). The vinyltetrazoles (**204**) were readily obtained when $Al(N_3)_3$ was the source of azide ion but the reaction failed when either NH_4N_3, LiN_3 or NaN_3 was used. Treatment of tetracyanoethylene with sodium azide at $-20\,°C$ in DMF gave the compound (**205**) and a series of substituted 5-vinyltetrazoles (**206**) has also been reported. 5-Aryltetrazoles (**207**) and heteroaryltetrazoles are readily prepared by this reaction either in dipolar aprotic solvents or in solvents such as mixtures of butanol and acetic acid. Best results are obtained when the aryl groups contain electron-withdrawing substituents. Trifluoromethyltetrazole (**209**) was obtained in 75% yield at room temperature on treatment of trifluoroacetonitrile with NaN_3 in acetonitrile as solvent, thereby illustrating the extent to which electron-withdrawing groups enhance the nucleophilic azide attack ⟨77AHC(21)323, 73S71, B-67MI41300⟩.

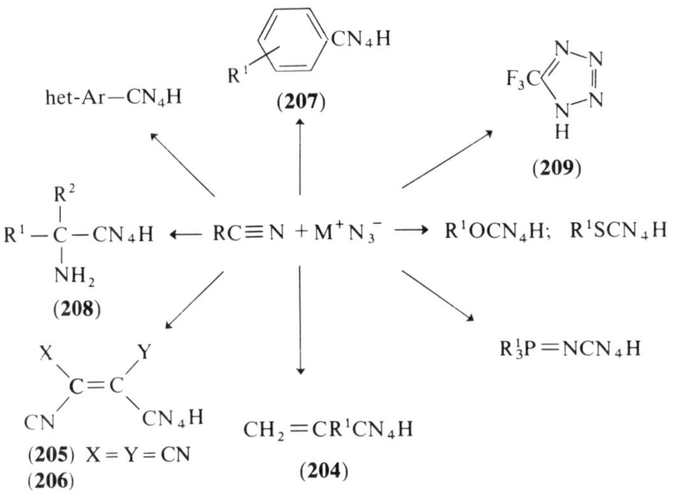

Scheme 35

Treatment of nitrilium salts (**210**) with azide ion allows for the synthesis of a range of 1,5-disubstituted tetrazoles (**211**) ⟨72JOC343⟩. The scope of this reaction has been expanded by the generation of nitrilium ions in solution from reactions of carbocations with nitriles. In the presence of azide ion (Scheme 36) such reactions give rise to 1,5-disubstituted tetrazole products. The reaction has been applied to carbocations generated by solvolysis of leaving groups (Scheme 36; $X = ArSO_3$) and by addition of electrophiles to alkenes in the presence of a nitrile solvent. For example the reaction of styrene with Br_2 and $AgClO_4$ in acetonitrile containing azide ion gives the tetrazole (**213**; Scheme 36) arising from trapping of the bromonium–styrene species by the nitrile followed by addition of azide ion ⟨71ACR10⟩.

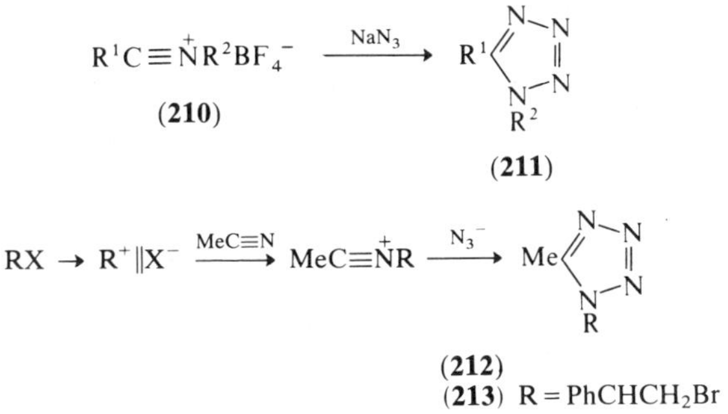

Scheme 36

Other useful tetrazole syntheses involving nitrilium ions include Beckmann reaction of oximes in the presence of azide ion ⟨83JCR(S)18⟩ and the reaction between ketones and 2 mol of hydrazoic acid in acidic media, the Schmidt reaction (Scheme 37). For example, treatment of acetophenone with 2 mol of hydrazoic acid in sulfuric acid gives mainly 5-methyl-1-phenyltetrazole (Scheme 37). When 1 mol of HN_3 is used in the Schmidt reaction, the products are amides due to attack of water on the nitrilium intermediate. The reaction has been used with steroidal ketones in recent years to synthesize compounds containing a tetrazole ring fused to a steroid system, *e.g.* compounds of type (214) ⟨76JCS(P1)1210, 77AHC(21)323⟩.

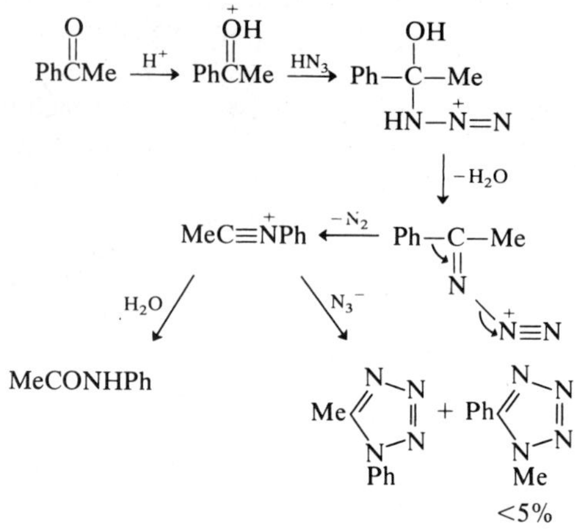

Scheme 37

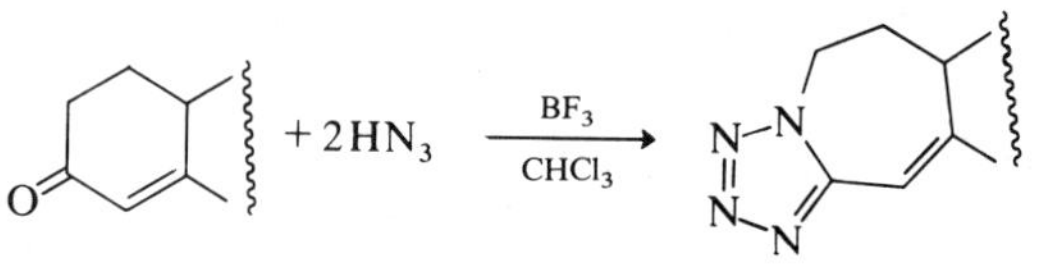

The reaction of isocyanides with hydrazoic acid has long been one of the main routes to 1-substituted tetrazoles (Scheme 38). The reaction tends to be sluggish requiring long periods of time but it is sensitive to acid catalysis and reaction times have been cut from days to hours by simply adding one drop of acid to the mixture ⟨69TL5081⟩. The acid-catalyzed reaction involves the protonated isocyanide (215) which is attacked by azide ion. When isocyanides are treated with azide ion in the presence of Mannich type reagents or enamines, 1,5-disubstituted tetrazoles are obtained with the enamine moiety bonded to the tetrazole C-5 (Scheme 39). Thus treatment of isocyanides with formaldehyde and hydrazoic acid in piperidine gives compounds of type (216). This reaction, which is known as the Ugi reaction, involves the addition of a Mannich type carbocation and azide ion to the isocyanide ⟨77AHC(21)323, B-67MI41300⟩.

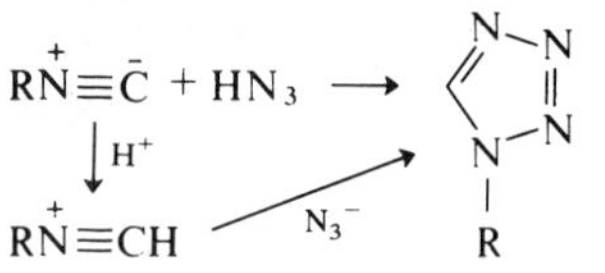

Scheme 38

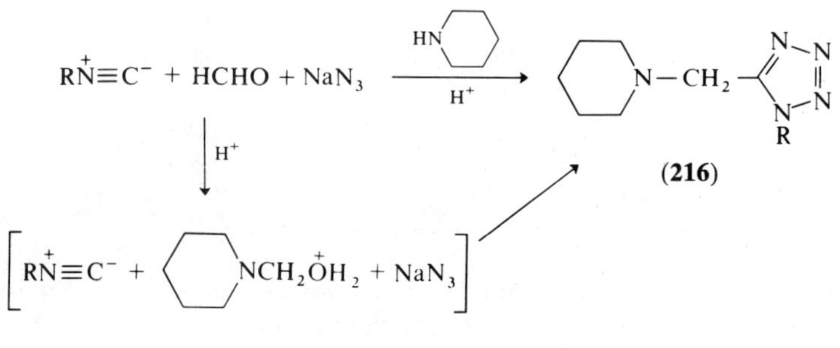

Scheme 39

(ii) Cycloaddition of azides to —C≡N and —C=N— systems

Cycloaddition of covalent azides with nitriles may lead to 1,5-disubstituted or 2,5-disubstituted tetrazoles (Scheme 40) ⟨69CRV359⟩. When the nitrile contains electron-withdrawing substituents such as perfluoroalkyl groups, the 1,5-disubstituted tetrazoles (**217**) are readily formed. This reaction was not successful unless such strong electron-withdrawing groups were present on the nitrile ⟨62JOC2085⟩. When the azide was trimethylsilyl azide, the alternative type of cycloaddition occurred with normal nitriles and gave the 2,5-disubstituted tetrazoles (**218**) ⟨68CB743, 66CB2512⟩. Recently the 5-aryloxy-1-benzyltetrazoles (**219**) have been reported to be formed in high yields from the reaction of cyanic esters, ArOCN, with benzyl azide under high pressure in nitromethane at 105 °C (Scheme 40) ⟨80IZV2668⟩. Cycloaddition of alkyl azides with nitrilium salts yields tetrazolium salts such as compounds (**220**) ⟨76TL1485⟩.

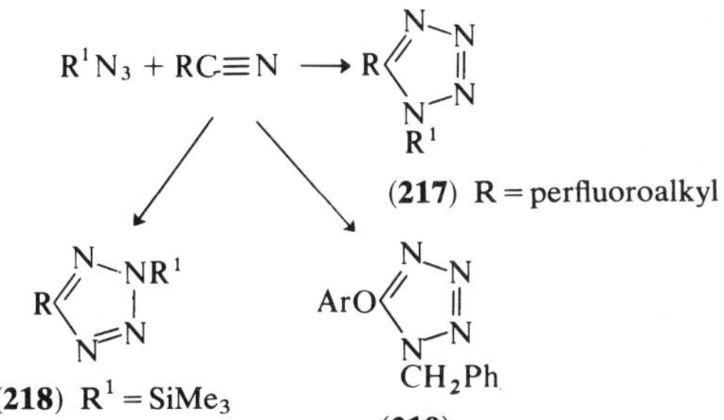

(**217**) R = perfluoroalkyl

(**218**) $R^1 = SiMe_3$

(**219**)

Scheme 40

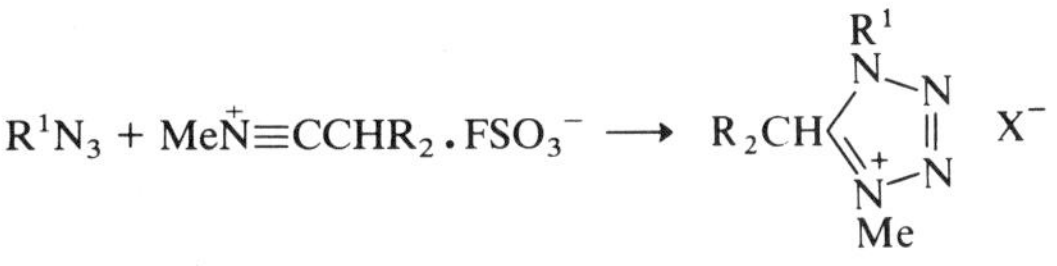

(**220**)

The reactions of covalent azides with the isocyanate system are summarized in Scheme 41. Addition of organic azides and azide ion to isocyanates occurs at the —N=C— bond and gives 1,4-disubstituted tetrazolin-5-ones (**221**; Scheme 41). The reaction of organic azides with isothiocyanates is a complicated reaction giving a mixture of five different products of which the tetrazoles (**223**) are minor constituents and the remaining compounds are sulfur heterocycles. Organometallic azides, and also azide ion, add to the —N=C—

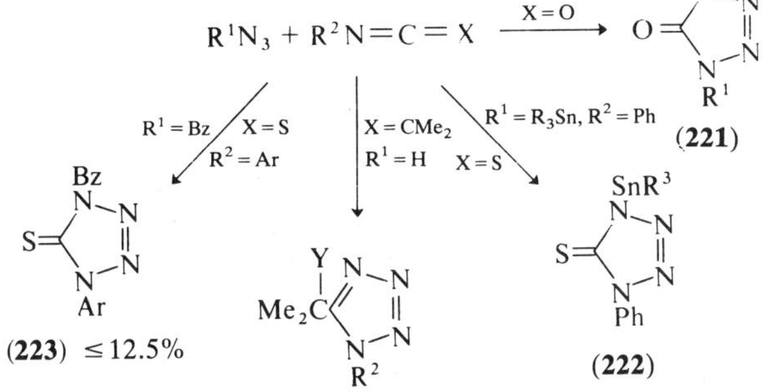

(**224**) (**225**) (**221**)

(**223**) ≤12.5%

(**226**) Y = H
(**226a**) Y = NH_2

(**222**)

Scheme 41

bond of isothiocyanates to give tetrazoline-5-thiones such as compounds (**222**; Scheme 41) ⟨73JOC675, 77JOC1159, 74JA3973, 59JA3076⟩. When the isothiocyanate contains a sulfonyl moiety bonded to the N atom, the addition of organic azides occurs on the C=S bond giving thiatriazoles (**224**; Scheme 41) (see also Chapter 4.28). Trimethylsilyl azide adds to carbodiimides to give the substituted 5-(trimethylsilylamino)tetrazoles (**225**) which could be readily desilylated. Heating or aryl isocyanates with 2 mol of trimethylsilyl azide without a solvent gave high yields of the tetrazolones (**221**; $R^1 = H$). This reaction appeared to involve initial addition of the silyl azide to the C=O bond giving an imidoyl azide, $Ar-N=C(N_3)-OSiMe_3$, which cyclized and desilylated rapidly. No tetrazoles were encountered in the reactions of trimethylsilyl azides with isothiocyanates and the products were thiatriazoles and cyanamides encountered in low yields ⟨80JOC5130⟩. Treatment of keteneimines with hydrazoic acid in ethereal solution, from which atmospheric oxygen was excluded, gave a mixture of the tetrazoles (**226**) and (**226a**). The latter compounds arose from addition of HN_3 to the C=C bond as well as the C=N bond and subsequent fragmentation of the spiro 1,2,3-triazole adduct. In the presence of air compounds (**226**; $Y = O_2H$) were also formed and for $R^2 = p$-MeC_6H_4, the yields of compounds (**226**), (**226a**) and (**226**; $Y = O_2H$) were 57, 26 and 16%, respectively ⟨78JOC3042⟩.

When aldehyde hydrazones are heated with aryl azides in basic solution, 2,5-disubstituted tetrazoles (**227**) are formed along with a primary amine derived from the azide (Scheme 42). The reaction works best when the aryl azide contains strong electron-withdrawing groups, e.g. 2,4-$(NO_2)_2C_6H_3N_3$, and a mechanism involving a [3+2] cycloaddition of the hydrazone anion with the azide followed by tautomerism and loss of $ArNH^-$ (Scheme 42) has been proposed ⟨68JHC565, 80MI41300, 68T3787⟩.

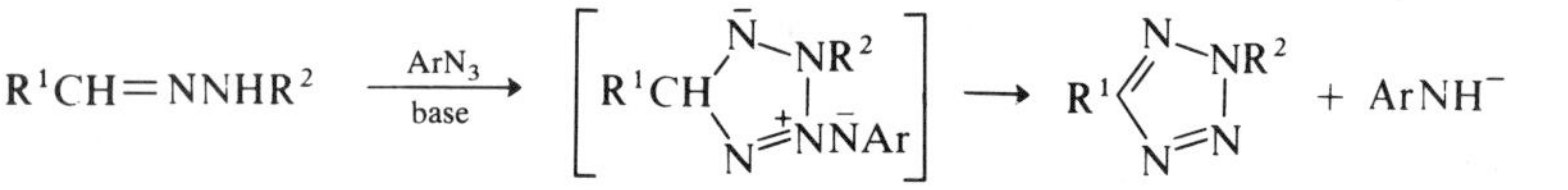

$$R^1CH=NNHR^2 \xrightarrow[\text{base}]{ArN_3} \left[\begin{array}{c} R^1CH \overset{N-NR^2}{\underset{N=NNAr}{||}} \end{array} \right] \longrightarrow R^1 \overset{N-NR^2}{\underset{N=N}{\langle|}} + ArNH^-$$

(**227**) ≈15% $R^1 = CO_2H$, Ar, fur-2-yl, indol-3-yl; $R^2 = CONH_2$, CH_2CO_2Et, Ar

Scheme 42

4.13.5.2 Synthesis from Ring Interconversions

Interconversions of lower azole ring systems into substituted tetrazole rings have been used to a limited extent to prepare tetrazole derivatives. The reactions usually involve basic conditions with substrates which often have a labile proton. Examples of the main types of ring interconversions which yield tetrazole derivatives are shown in Scheme 43. Heating 5-arylaminothiatriazoles with base gives tetrazoline-5-thiones (**229**) in a reaction involving a competitive degradation to isothiocyanate and azide ion. When the aryl group of the substrate (**228**) is replaced by an alkyl group, only the ring degradation is observed on heating with base ⟨B-67MI41300⟩. The mesoionic 1,2,3,4-oxatriazolium compounds (**230**) also rearrange to a mesoionic tetrazole system, 1,3-diaryltetrazolium 5-oxides (**8a**), on treatment with base ⟨71ACS(25)625⟩. Treatment of the 3-diazoamino-5-methylisoxazoles (**231**) with base gives tetrazole derivatives of type (**232**) in an example of a general heterocyclic rearrangement ⟨64T461, 77AHC(21)323⟩. Rearrangements of 1,2,3-triazole systems containing oxy and unsaturated nitrogen substituents at the 4- and 5-positions, e.g. compounds (**233**),

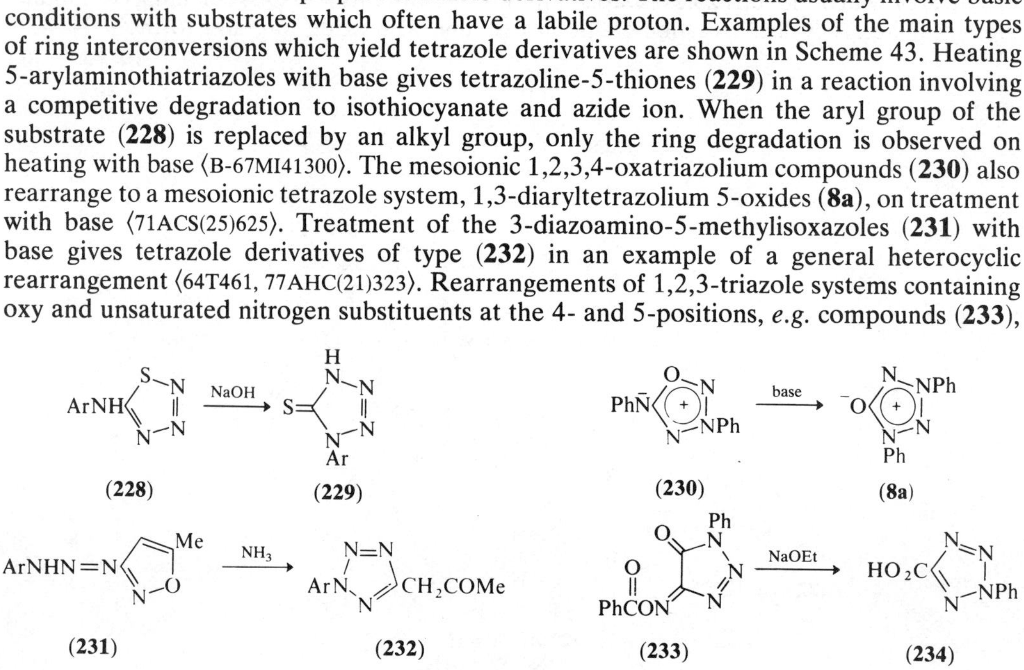

(**228**) → (**229**) NaOH

(**230**) → (**8a**) base

(**231**) → (**232**) NH_3

(**233**) → (**234**) NaOEt

Scheme 43

result in the formation of 2,5-disubstituted tetrazoles (Scheme 43) ⟨B-67MI41300⟩. Some tetrazole derivatives have also been obtained by cleavage of oxazoline and oxazolidine systems by treatment with sodium azide under acidic conditions ⟨77AHC(21)323⟩. These reactions usually involve imidoyl azide type intermediates and are not strictly ring interconversions but rather another method of generating imidoyl azides for tetrazole synthesis. Addition of HN_3 to azirines to give azidoaziridines (**235**) has recently been reported. These compounds, when heated to 80 °C, gave isomerizations to the tetrazoles (**236**) as well as decomposition to nitriles, ArCN ⟨77CB2939⟩.

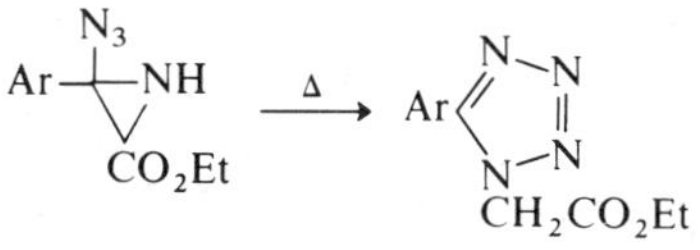

4.13.5.3 Best Practical Methods

One of the most important tetrazole derivatives for synthetic purposes is 5-aminotetrazole. This is commercially available and can be easily synthesized from the reaction of cyanamide with hydrazoic acid or by diazotization of aminoguanidine (Scheme 44) ⟨B-67MI41300⟩. Diazotization of 5-aminotetrazole gives a reactive (and explosive) diazonium salt which can be treated *in situ* to give tetrazole itself and normal dediazoniation reactions can be used to prepare a range of 5-substituted tetrazoles (Scheme 44) ⟨B-67MI41301, 77AHC(21)323⟩. Tetrazole can also be readily obtained in yields of 95% by hydrolysis of ethyl 5-tetrazolecarboxylate with KOH followed by acidification with $HClO_4$ ⟨76ZC53⟩. Treatment of nitriles with azide ion is the best route to 5-substituted tetrazoles (**237**) and optimum conditions are usually achieved by heating in DMF. The length of the heating period is determined by the reactivity of the nitrile. A number of work-up procedures may be employed depending upon the solubility of the tetrazole product ⟨73S71⟩. The most convenient route to 1,5-disubstituted tetrazoles (**238**) involves displacement of imidoyl leaving groups (usually halide) by azide ion. Imidoyl halides of type (**238**) are readily available from the dehydration of amides with PCl_5 and, without purification, they can be dissolved in DMF and the resulting solution added slowly to a DMF solution of sodium azide (Scheme 45). After appropriate stirring, the 1,5-disubstituted tetrazole can usually be precipitated by careful addition of water ⟨73S71⟩.

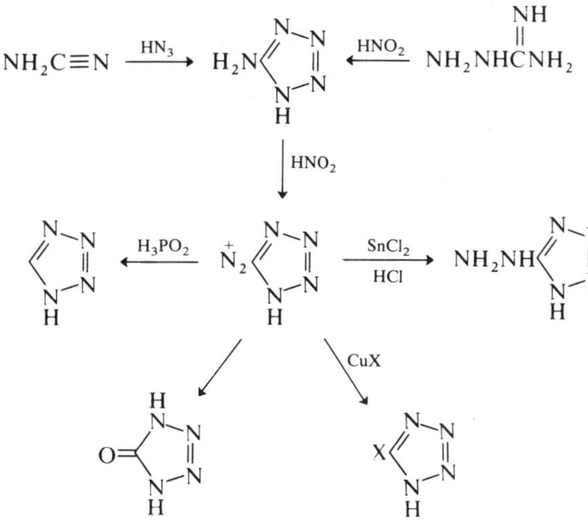

Scheme 44

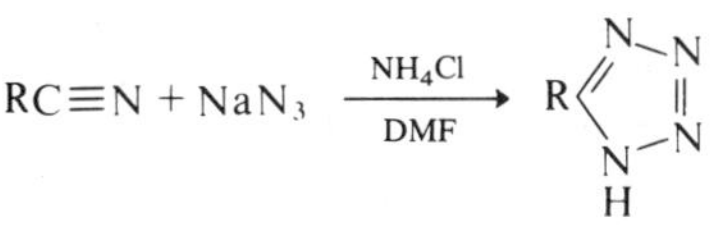

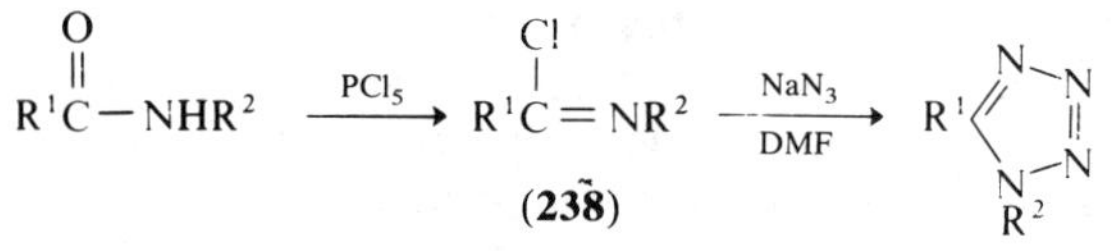

$$R^1C(=O)-NHR^2 \xrightarrow{PCl_5} R^1C(Cl)=NR^2 \xrightarrow[DMF]{NaN_3} \underset{(239)}{R^1\text{-tetrazole-}R^2}$$

$$\underset{(238)}{}$$

Scheme 45

Direct alkylation of 5-substituted tetrazoles is often the best method of obtaining simple 2,5-disubstituted tetrazoles. The 1,5-isomers, which are formed as competitive side products, are usually separated with ease by fractional crystallization or better by column chromatography on alumina or silica gel which gives excellent separations of most tetrazole isomers ⟨81JCR(S)174⟩. Another convenient synthesis of 2,5-disubstituted tetrazoles (**241**) is the formazan method (Scheme 31). By coupling aldehyde arylsulfonylhydrazones (**240**) with diazonium salts in basic solutions, compounds (**241**) may be obtained directly in yields of up to 78% ⟨76BCJ1920⟩.

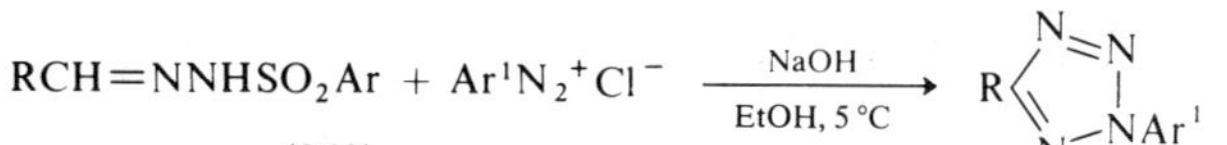

$$RCH=NNHSO_2Ar + Ar^1N_2^+Cl^- \xrightarrow[EtOH, 5\,°C]{NaOH} \underset{(241)}{R\text{-tetrazole-}NAr^1}$$

$$\underset{(240)}{}$$

For many special tetrazole derivatives there is often not a choice of routes and in general the best synthetic routes are those already outlined in Sections 4.13.5.1.1 and 4.13.5.1.2.

4.13.6 APPLICATIONS

4.13.6.1 Biological and Medicinal Applications

There is now a large number of separate tetrazole compounds reported specifically for potential biological activity. A table of biologically active tetrazole types which provides a base for a literature search has been drawn up ⟨77AHC(21)323⟩. A recent review of the medicinal applications of tetrazoles lists over 300 references ⟨80MI41301⟩. Biological activity in tetrazoles is encountered due to the special metabolism of the disubstituted tetrazole system and also because in 5-substituted tetrazole compounds the tetrazole ring is isosteric with a carboxylic acid group and of comparable acidity. Hence for all biologically active molecules possessing a carboxylic acid group (CO_2H) there is a theoretical nitrogen analogue possessing a tetrazolic acid group (CN_4H) and since the tetrazole moiety appears to be the metabolically more stable of the two a considerable exploration of these molecules is ongoing.

Tetrazole analogues of amino acids have pK_a values agreeing closely with those of the corresponding amino acids. A range of peptides with a tetrazole group incorporated has been synthesized. For example the tetrapeptide (**242**) was prepared and found to show activity for stimulating gastric acid secretion which was comparable to that of the normal peptide (**243**). IR spectra of these aminotetrazolic acids suggested that they exist in the zwitterionic form as do the carboxylic analogues ⟨59JOC1643, 69JCS(C)809, 71T2317, 71T1783, 77AHC(21)323⟩.

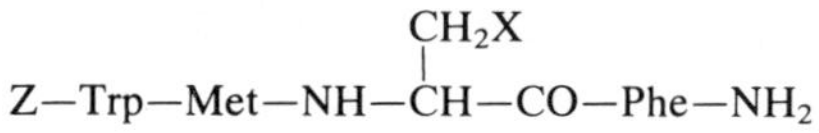

$$Z-Trp-Met-NH-\underset{|}{\overset{CH_2X}{CH}}-CO-Phe-NH_2$$

(**242**) $X = CN_4H$
(**243**) $X = CO_2H$

4.13.6.1.1 Analeptic activity (central nervous system)

In general 1,5-dialkyltetrazoles, in which the $N(1)$-substituent has four to six carbons, show activity on the central nervous system. When the 1- and 5-positions are bridged by a methylene chain, the activity is maximized. Pentamethylene tetrazole (**244**), which is probably the most important biologically active tetrazole, is a standard drug in this area and is known as leptazol (and previously as metrazol). It is used as a stimulant for the central nervous system and to counteract the effects of overdosage of barbiturates. It is more effective when administered parenterally than orally due to a slow absorption from the stomach after oral dosage. Its main metabolites are 6- and 8-hydroxypentalenetetrazoles, the main elimination pathway of the drug being through biotransformation in the liver ⟨80MI41301, 76JPS1038, 74ZPC(355)1274⟩. A number of other 1,5-disubstituted tetrazoles including 1-alkyl-5-(alkoxy- or substituted-amino) and 1-(alkoxy- or substituted-amino)-5-alkyl systems and the derivatives (245) have displayed sedative activity on the central nervous system ⟨80MI41301⟩.

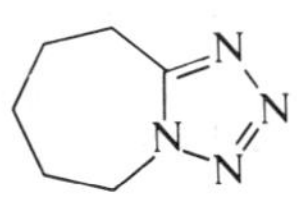

(244) 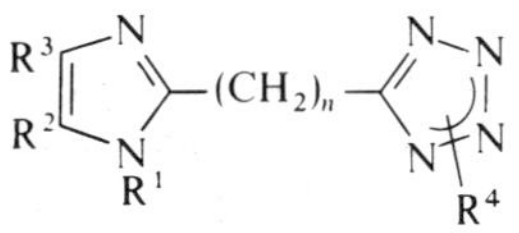   (245)

4.13.6.1.2 Anti-inflammatory activity

Anti-inflammatory activity is shown by both N-substituted and N-unsubstituted tetrazoles. The tetrazolylalkanoic acids (246) and the *ortho*-substituted 5-phenyltetrazoles (247) represent examples of both classes. Some indole systems with a 5-alkyltetrazole group at the indole 3-position *e.g.* (248), also show antiinflammatory activity ⟨80MI41301, 77AHC(21)323⟩. A series of substituted 2-(tetrazol-5-yl)benzoxazoles which showed antiinflammatory activity were used in sunburn preventing and shaving preparations ⟨73GEP2314238⟩.

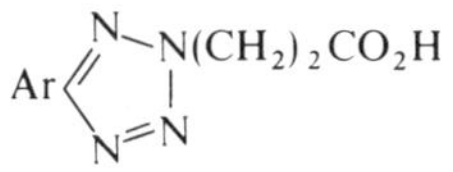

(246) 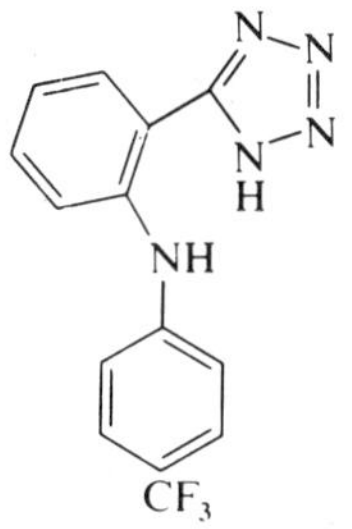   (247) 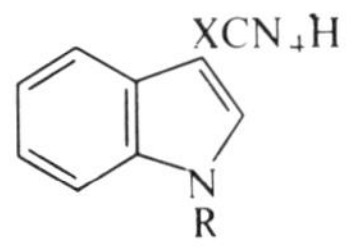   (248) X = CH_2, $S(CH_2)_n$, OCH_2

4.13.6.1.3 Antimicrobial activity

A range of cephalosporin derivatives containing N-substituted tetrazole rings has been prepared and found to display antibacterial activity ⟨80MI41301⟩. In such compounds the tetrazole moiety is a substituent group on the underlying biologically active substrate. However, some more simple tetrazole derivatives such as 1-(p-substituted phenyl)tetrazoles have also shown bactericidal properties ⟨77USP4036849⟩. Some simple 5-arylaminotetrazoles have been found to inhibit the growth of measles virus ⟨70SAP6804307⟩.

4.13.6.1.4 Antilipemic activity

Nicotinic acid exhibits hypocholesteremic effects due to its lipolysis inhibitory activity. The nitrogen analogue of nicotinic acid, compound (249), was a weaker *in vitro* lipolysis inhibitor but it depressed free fatty acid levels in dogs with a longer retention time than nicotinic acid. N-Methyl derivatives of compound (249) were inactive and in this case the tetrazole group was paralleling the carboxylic acid group but with a longer retention time due to its greater metabolic stability ⟨67JMC149⟩. The preferred tautomeric structure of compound (249) *in vitro* was the 1-NH form ⟨82JCR(S)122⟩. Compound (250) represents another example of an active antilipemic tetrazole molecule which has progressed to clinical tests ⟨68MI41301⟩. The compound (249; R = $CH_2OCONHMe$) has been claimed useful for treating thrombosis ⟨72JAP(K)7242770⟩.

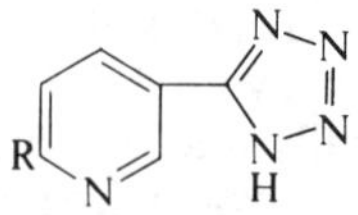

(249) R = H

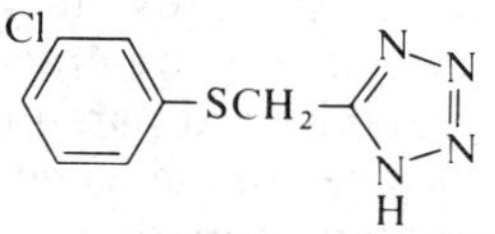

(250)

4.13.6.1.5 Antiallergic activity

Derivatives of the chromone system (251) containing carboxy groups at C-2 exhibit antiallergic properties and are used in the treatment of bronchial asthma. Such activity was also found in chromone systems bearing an *N*-unsubstituted tetrazole group at C-2, C-6 and C-8. Chromones with CO_2H groups at C-3 are inactive because of decreased acidity due to intramolecular hydrogen bonding between the carboxylic acid group and the 4-carbonyl oxygen. Such an effect was not present when the 3-CO_2H group was replaced by a CN_4H group and 3-(tetrazol-5-yl)chromones were about 2.5 times as active as when the tetrazole group was at the chromone 2-position. The acidity of the tetrazole ring was essential for the activity and *N*-methyltetrazol-5-ylchromones were inactive ⟨80MI41301⟩. A large number of substituted chromone derivatives with tetrazole groups located at C-2 and C-3 have been prepared as potential antiallergic agents ⟨77AHC(21)323⟩.

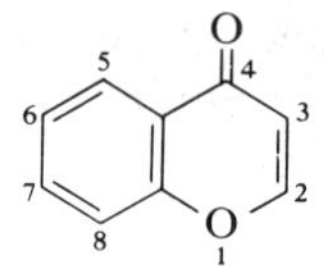

(251)

4.13.6.1.6 Other medicinal properties

Many other tetrazole systems have been explored for biological activity. Areas where significant activity has been encountered include antihypotensive activity, hormonal activity in tetrazole derivatives of potent non-steroidal oestrogens, bronchodilating effects, anticholinergic agents, diuretic agents and radioprotective agents ⟨80MI41301, 77AHC(21)323⟩.

4.13.6.1.7 Agricultural uses

Tetrazole compounds of type (246) containing aryl or heteroaryl substituents at C-5 and alkanoic acid derivatives at N-2 are effective agents for regulating plant growth. For example esters of compound (246; $Ar = p\text{-}ClC_6H_4$) at application levels of 1000 p.p.m. caused an average shoot-length decrease in apple and plum trees of 100 and 86%, respectively. A range of tetrazole molecules of this type has been found useful as inhibitors of top growth for vegetables, fruit trees, cereals and canes ⟨73GEP2310049, 75USP3865570, 77AHC(21)323⟩. Application of 5-aminotetrazole to the soil at planting time has caused temporary albinism in some plants and reduction of tetrazolium salts to red colored formazans has been used to detect the activity of reducing enzymes in seeds ⟨B-67MI41300⟩. In contrast to the growth-inhibiting tetrazoles, growth acceleration for roots of plants has been claimed for 1-(1*H*-tetrazole)acetic acid (252) ⟨71JAP(K)7135742⟩.

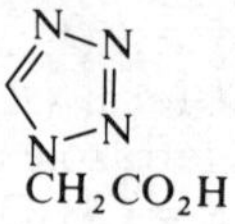

(252)

4.13.6.2 Non-biological Applications

4.13.6.2.1 *Explosives and rocket propellants*

Certain types of tetrazole derivatives are explosive and may be used as rocket propellants. Substituted 5-cyanovinyltetrazoles (**205**) and their salts and polymers have been claimed as propellants ⟨67USP3338915⟩. Metallic salts of tetrazole derivatives such as cadmium tetrazolyl azide, copper nitrotetrazole and lead azotetrazole can be used in primers. Addition of 5-aminotetrazole to nitrocellulose propellant powders rendered them flashless without loss of propellant potential ⟨B-67MI41301⟩. The silver salt of 5-azidotetrazole was hypersensitive to mechanical stimuli and highly explosive ⟨71MI41300⟩. Complexes of aluminum hydride with 1,5- and 2,5-disubstituted tetrazoles (AlH_3–tetrazole), prepared at low temperatures in ether, are also explosive ⟨68USP3396170⟩. Complexes of 5-nitramino- and 5-hydrazino-tetrazole with Ni^{2+} and Cu^{2+} were explosive on heating over copper foil but Fe^{2+} complexes did not explode ⟨79BAP21⟩. Nitrotetrazoles are potential primary explosives and studies of the deflagration to detonation transition for 1-methyl- and 2-methyl-5-nitrotetrazoles are reported and compared with PbN_3 ⟨76MI41300⟩.

4.13.6.2.2 *Photography*

Several types of tetrazole derivatives have proved useful in photography. Stabilization of silver halide emulsions against fogging is displayed by substituted 5-mercaptotetrazole derivatives either as thione forms, such as 1-phenyltetrazoline-5-thione, or as 5-*S*-substituted derivatives. Compound (**253**) is used as a component of AgBr emulsions in color films where its function is to liberate an inhibitor against over-development. Bistetrazoles, separated by an alkylene chain with 2–5 carbon atoms, *e.g.* $HN_4C-(CH_2)_n-CN_4H$, are also photographic antifoggants. In general, tetrazole derivatives have been used as additives to improve the development process and the presence of *N*-substituted tetrazolinethione compounds increases the size of the silver particles and reduces fogging when they are present in monolayer or multilayer silver halide films. Tetrazolium salts can be used to produce colored images arising from their conversion into formazans by reducing agents generated by exposure of iron(III) salts to light ⟨70FRP1596860, 72GEP2060196, 69BEP722025, 79GEP2909190, B-67MI41301⟩.

(**253**)

4.13.6.2.3 *Miscellaneous applications*

The complexing ability of a number of substituted tetrazoline-5-thiones has rendered these compounds useful for gravimetric determinations of transition metals. Ions such as Cu^{2+}, Cu^+, Ag^+, Au^+, Hg^+ and Cd^{2+} can be detected in the 0.5–50 mg range by precipitation ⟨77AHC(21)323⟩.

Ditetrazoles of type (**254**), obtained from the reaction of potassium 5-phenyltetrazolate with 1,2-dibromoethane, have proved useful for curing unsaturated elastomers in a cross-linking process that involved thermal elimination of N_2 from the ditetrazole to form dinitrilimine 1,3-dipoles ⟨72MI41301⟩.

(**254**) (**255**) (**256**)

Polymer cross-linking has also been achieved with 1-arylsulfonyl-4-butyltetrazolin-5-ones (**255**) and with bistetrazolyl-*m*-phenylene molecules of this type ⟨72GEP2226525⟩. Tetrazole polymers of type (**256**) were prepared from thermal reactions of diazides with dinitriles. They are constituted as flexible fibers and have been claimed for use as thermally stable propellant binders and as plastics ⟨68USP3386968⟩.

4.14
Pentazoles

I. UGI
Technische Universität, München

4.14.1 INTRODUCTION	839
4.14.2 STRUCTURE	839
4.14.2.1 The First ^{15}N Experiment	839
4.14.2.2 Experimentum Crucis	841
4.14.2.3 Mechanistic Alternatives	841
4.14.3 REACTIVITY	843
4.14.3.1 Structure Dependence	843
4.14.3.2 Crystalline Arylpentazoles	843
4.14.3.3 Miscellaneous	844
4.14.3.4 Perspectives	845

4.14.1 INTRODUCTION

Most of the basic types of heteroaromatic compounds were discovered in the second half of the last century, the classical age of organic chemistry ⟨B-16MI41400⟩. In the azole series, their stability towards electrophiles and oxidants was observed to increase with the number of nitrogen atoms in the heterocyclic ring, from pyrrole to tetrazole, for reasons which are nowadays quite well understood ⟨B-68MI41400⟩. Thus, it has been a challenge to prepare pentazole (**1**), the last member of the azole series. (Whether or not pentazole is a heterocycle is debatable. All its ring atoms are of the same kind, and thus pentazole may be called a homocyclic system. On the other hand, within organic molecules, nitrogen is a 'heteroatom', and pentazole is a ring of heteroatoms, which makes it a heterocycle.)

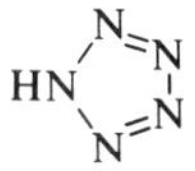

$$HN\underset{N=N}{\overset{N=N}{\diagdown\diagup}}N$$

(**1**)

Until the second half of this century none of these efforts was successful ⟨10CB2904, 15CB410, 15CB1614, 57CB414, 57JA1249, 03CB2056⟩ although some pentazole derivatives had been produced unknowingly whenever an arenediazonium ion was reacted with azide to form an aryl azide ⟨1893CB86⟩. Hantzsch ⟨03CB2056⟩ felt intuitively that pentazoles are formed by the latter reaction, and he tried to obtain arylpentazoles from arenediazonium compounds. Since the pentazoles are short-lived under ordinary conditions, there was little chance of discovering a pentazole by the experimental methods of classical organic chemistry ⟨64AHC(3)373⟩.

4.14.2 STRUCTURE

4.14.2.1 The First ^{15}N Experiments

The first experimental evidence for the existence of a pentazole derivative came from Clusius and Hürzeler ⟨54HCA789⟩ in Zürich. They studied the formation of phenyl azide and nitrogen from benzenediazonium ion and azide by ^{15}N-labelling techniques, and they found that this reaction does not follow the general pattern of the Gries–Sandmeyer type reactions ⟨53AG155⟩. In a typical experiment the aforementioned authors ⟨54HCA789⟩ diazotized unlabelled aniline (with a natural ^{15}N abundance of 0.37%) with ^{15}N-labelled nitrite (with 2.57% ^{15}N), and they reacted the resulting β-labelled benzenediazonium ion ⟨63JA122⟩ with unlabelled azide. The ^{15}N content at the various sites of the products is given (percentage in brackets) in Scheme 1.

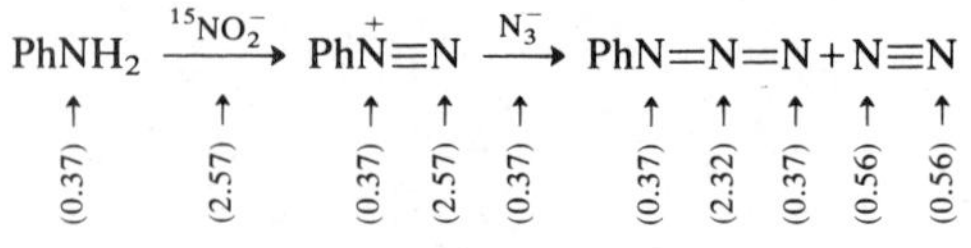

$$PhNH_2 \xrightarrow{^{15}NO_2^-} Ph\overset{+}{N}\equiv N \xrightarrow{N_3^-} PhN=N=N + N\equiv N$$

(0.37) (2.57) (0.37) (2.57) (0.37) (0.37) (2.32) (0.37) (0.56) (0.56)

Scheme 1

From the results of this and other ^{15}N-labelling experiments Clusius and Hürzeler ⟨54HCA789⟩ concluded that the considered reaction cannot proceed by a uniform reaction mechanism, and it must involve some competing, parallel reactions. They summarized their results by Scheme 2. In this scheme the outcome of the ^{15}N-labelling experiments was represented in terms of three 'components'.

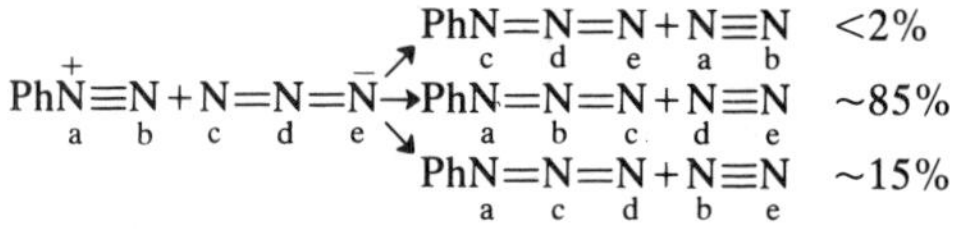

$$\overset{+}{PhN}\equiv N + N=N=\bar{N} \rightarrow \begin{array}{l} PhN=N=N + N\equiv N \quad <2\% \\ cdeab \\ PhN=N=N + N\equiv N \quad \sim 85\% \\ abcde \\ PhN=N=N + N\equiv N \quad \sim 15\% \\ acdbe \end{array}$$

Scheme 2

In retrospect, it is easy to see why this representation was unfortunate. It was bound to mislead the investigators because both its first and second components correspond in their labelling patterns to distinct mechanistic types of reactions, namely the classical Gries–Sandmeyer type and a reaction via benzenediazoazide (**3**; Scheme 3). Thus, Clusius and Hürzeler ⟨B-68MI41400⟩ were led to assume that the third pattern of labellings must also represent the ^{15}N-labelling pattern of a distinct reaction mechanism. Therefore, in their 1954 paper ⟨54HCA789⟩, these investigators excluded explicitly the intermediacy of phenylpentazole (**2**) in the reaction of the benzenediazonium ion and azide, since a reaction via (**2**) cannot be correlated with any one of the patterns in Scheme 2. The third labelling pattern would have corresponded to decomposition of (**2**) via (**2a**), while decomposition via (**2b**) would have been 'forbidden'. Since this is not compatible with the structural symmetry of (**2**), the latter interpretation cannot be valid.

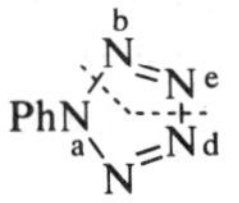

(**2a**) 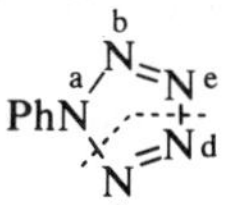   (**2b**)

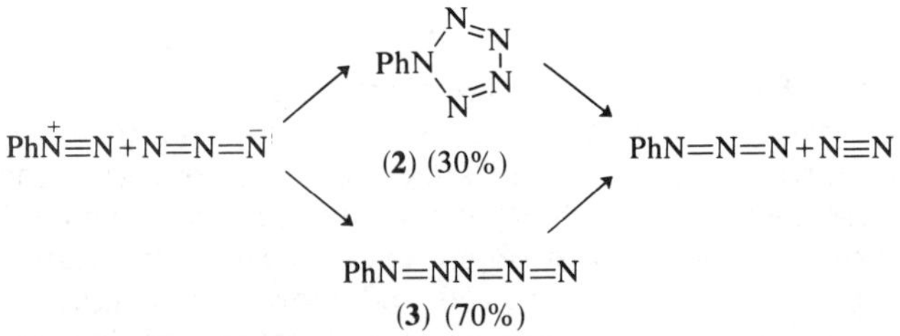

$$\overset{+}{PhN}\equiv N + N=N=\bar{N} \diagup \overset{PhN-N=N}{\underset{N=N}{\diagdown}} \diagdown PhN=N=N + N\equiv N$$

(**2**) (30%)

$$\diagdown PhN=NN=N=N \diagup$$

(**3**) (70%)

Scheme 3

At the time when the above paper ⟨54HCA789⟩ appeared Huisgen and Ugi ⟨56AG705⟩ had decided to approach the classical pentazole problem with modern methods. They studied the reaction between o-quinone diazides and aluminum azide ⟨54ZN(B)495, 54ZN(B)496⟩, in THF at low temperatures by kinetic and gas volumetric methods. There was some evidence for the formation of two distinct isomeric adducts from the above reactants. The major one decomposed with extremely rapid evolution of nitrogen, while the decomposition of the minor isomer took place in a first-order reaction that could be followed quite conveniently ⟨59TH41400⟩. It was conjectured that these isomeric adducts might be a diazoazide and a pentazole. Accordingly, we were prepared to see Scheme 3 as an alternative interpretation of the ^{15}N-experiments by Clusius and Hürzeler ⟨54HCA789⟩.

4.14.2.2 Experimentum Crucis

A kinetic investigation of the formation of phenyl azide and nitrogen from benzenediazonium chloride and lithium azide in methanol at 0 °C revealed the formation of two distinct intermediates ⟨56AG705, 57CB2914, 59TH41400⟩. The ratio of their relative amounts was approximately $Q = \bar{k}/k° = 1:2$ (Schemes 3 and 4). Both of these intermediates decomposed by first-order reactions to form phenyl azide and nitrogen. At 0 °C the major intermediate decomposes extremely rapidly, while the minor one reacts at a rate which can be conveniently followed by gas volumetric techniques. Below −25 °C the 'fast reaction' still proceeds quite rapidly, whereas, on its time scale, the 'slow reaction' seems to cease ⟨59TH41400⟩. Thus, it was apparent that under suitable conditions it would be possible to study the two mechanistic pathways separately by ^{15}N labelling methods. An *experimentum crucis* was planned accordingly, and a collaboration between Zürich and Munich was arranged for a joint execution of this experiment ⟨56AG753⟩. Lithium azide and [β-^{15}N]-benzenediazonium chloride were combined at −25 °C in methoxyethanol. The 'first' nitrogen evolved (see Scheme 4) was collected, and subsequently the reaction mixture was warmed up to 0–10 °C. Then a 'second' portion of nitrogen was produced (see Scheme 4), and taken as a separate sample $(k \gg k')$. After that, a solution of sodium hydroxide and sodium arsenite in aqueous methoxyethanol was added. The phenyl azide was reductively cleaved into aniline and nitrogen ⟨58CB2330⟩, the 'third' nitrogen sample. The ^{15}N content of the three samples was determined by Clusius and Vecchi ⟨56AG753⟩ by an analysis of the rotational fine structure of the electronic spectra. The present MS methods were not then available. A similar experiment was carried out with *p*-ethoxybenzenediazonium chloride.

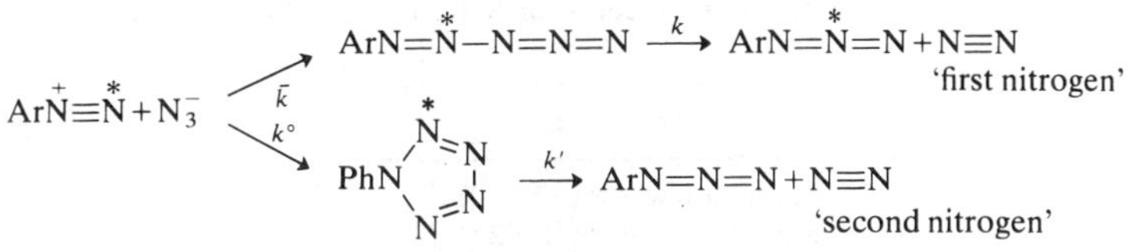

Scheme 4

The results of both experiments were fully in accord with Scheme 4. In this scheme k and k' are the rates of decomposition of (**4**) and (**5**) (see Scheme 5), and $\bar{k}$ and $k°$ are the hypothetical formation rates of (**4**) and (**5**), based on their relative amounts Q.

4.14.2.3 Mechanistic Alternatives

The existing evidence is insufficient to decide whether the formation and decomposition of the arylpentazoles (**5**) and the arenediazoazides (**4**) take place independently by some [2 + 3] cycloaddition leading to (**5**; Scheme 5), or whether the arylpentazoles (**5**) are formed from the arenediazoazides (**4**) by 1,5-electrocyclization, as has been suggested by Huisgen (Scheme 6) ⟨80AG(E)947, 80AG979⟩.

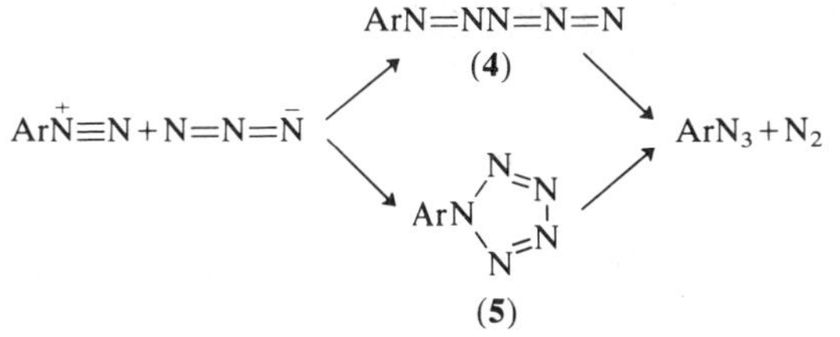

Scheme 5

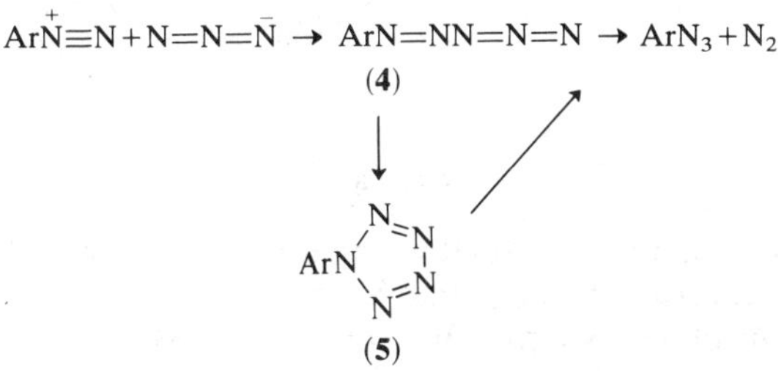

Scheme 6

Neither of the two schemes can be completely ruled out. ^{15}N-Labelling experiments, conventional kinetic measurements, or theoretical reasoning and MO calculations ⟨61CB273⟩ do not lead to a clear-cut decision. Therefore, a 'two-phase experiment' was designed in order to obtain further evidence ⟨63T1801⟩. In 80% aqueous methanol (phase I) at $-30\,°C$, benzenediazonium chloride and lithium azide react to form benzenediazoazide (3) and phenylpentazole (2) in the proportion of $Q_I = 2.03 \pm 0.05$, and benzenediazoazide decomposes at a rate of $k_I = 5.7 \times 10^{-3}\,s^{-1}$. The same reaction was carried out in the presence of a second phase (phase II: carbon tetrachloride + n-hexane, 22 + 78 wt.%; approximately of the same density as phase I, favoring intimate contact of the liquid phases), with rapid stirring. Although no measurable change in the ratio of benzenediazoazide to phenylpentazole was observed ($Q_{I+II} = 1.96 \pm 0.05$), a considerably slower rate of benzenediazoazide decomposition ($k_{I+II} = 2.3 \times 10^{-3}\,s^{-1}$) was found. From the fact that in the two-phase system the decomposition of benzenediazoazide is slower than in phase I alone, one can conclude that in the two-phase system, the benzenediazoazide formed in phase I migrates, in a significant amount, to phase II where it decomposes. A relevant difference between the two mechanisms under discussion is that, whereas in the mechanism of Scheme 5 the observed ratio Q is independent of k ($Q = \bar{k}/k°$; Scheme 4), in the alternative reaction mechanism Q is a function of k ($Q = k/k_{4 \to 5}$). As has been shown above, we have $Q_I \approx Q_{I+II}$ and $k_I > k_{I+II}$ which indicates that Q is independent of k_3. In combination with the fact that in one-phase systems the value of Q depends upon the solvent (see Table 1), the results of the 'two-phase experiment' suggest an independent formation and decomposition of benzenediazoazide and phenylpentazole according to Schemes 3–5.

Table 1 The Influence of Solvent upon the Ratio Q at $-30\,°C$

Solvent (wt.%)	Q
Methanol/water (50 + 50)	1.84
Methanol/water (75 + 25)	2.00
Methanol	2.45
Methanol/THF (50 + 50)	2.77
Methanol/n-butanol (50 + 50)	2.92
Methanol/n-butanol (10 + 90)	3.16

It could be objected that benzenediazoazide could exist in two isomeric forms (e.g. cis and trans forms), one of which would very rapidly be transformed (Scheme 7) into phenylpentazole, while the other isomer would decompose independently, and that the ratio in which these isomeric forms of benzenediazoazide are generated corresponds to the observed ratio Q. Yet such different chemical behavior of the stereoisomers of benzenediazoazide has been shown to be improbable by Roberts ⟨61CB273⟩, who calculated the 'flexibility' of the benzenediazoazide molecule. However, these MO calculations were of the simple HMO type ⟨61CB273⟩, and their results ought to be taken *cum grano salis*. Thus the outcome of the 'two-phase experiment' yields some pertinent information, but it does not really discriminate between the mechanistic alternatives.

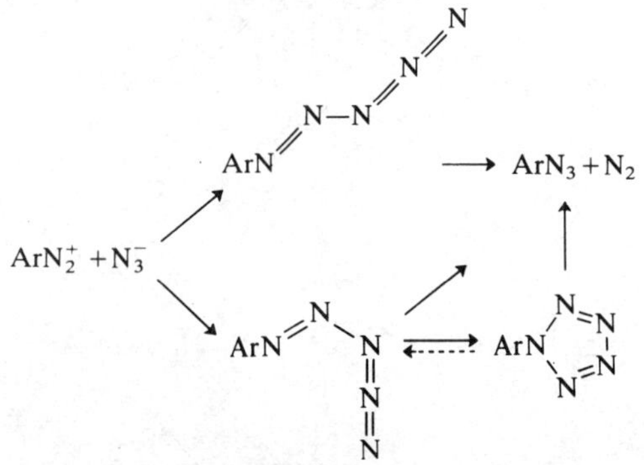

Scheme 7

If one assumes a sufficiently low relative interconversion rate for the stereoisomeric diazoazides, Scheme 7, a variation of Scheme 5, becomes an attractive mechanistic interpretation based on the 'two-phase experiment' ⟨64AHC(3)373, 63T1801⟩.

4.14.3 REACTIVITY

4.14.3.1 Structural Dependence

The relative tendency of formation, Q, and the rate of decomposition, k, of the arylpentazoles in methanol at 0 °C (see Scheme 4) show a marked structure dependence (see Table 2) ⟨58CB531⟩. The rate data for the m- and p-substituted phenylpentazoles follow a Hammett relationship ⟨B-40MI41400, 53CRV(53)191⟩ with $\rho = +1.01$. The o-substituted phenylpentazoles are not readily formed and they decompose particularly fast.

Table 2 The Relative Yields and Rates of Decomposition of the Arylpentazoles (MeOH, 0 °C).

Substituent on aryl group	Q	$k' \times 10^4$ (s^{-1})
4-NO$_2$	6.1	59
4-Me$_2$NH$^+$	—	51
3-NO$_2$	4.6	36
3-Cl	3.3	23
4-Cl	3.5	12.1
H	3.2	8.4
3-Me	2.6	7.6
3-HO	2.0	7.1
4-Me	1.9	5.6
4-HO	1.9	3.2
4-EtO	1.9	3.0
4-Me$_2$N	1.2	1.7
4-O$^-$	—	0.9
2-NO$_2$	24	92
2-Cl	8	86
2-Me	3.4	31
2,4,6-Me$_3$	13	23

4.14.3.2 Crystalline Arylpentazoles

The data in Table 2 provide some guidance as to which of the arylpentazoles have a good chance of being isolatable, namely those listed in Table 3 ⟨58CB2324⟩. The main problem in the preparation and purification of arylpentazoles is to find conditions under which they do not decompose, but can be separated from the contaminating aryl azides and inorganic compounds. The best system found for the isolation of the arylpentazoles was the two-phase combination of aqueous methanol and petroleum ether at −40 to −20 °C.

Table 3 Crystalline Arylpentazoles

Compound	Yield (%)	Decomposition point (°C)
Phenylpentazole	27	−5 to 3
p-Tolylpentazole	32	2–5
p-Chlorophenylpentazole	21	8–10
p-Methoxyphenylpentazole	38	13–15
p-Ethoxyphenylpentazole	42	26–29
p-Dimethylaminophenylpentazole	52	50–54

After the formation of an arylpentazole from an arenediazonium chloride and sodium azide, the accompanying aryl azide is soluble in the petroleum ether, and the sodium chloride stays in the aqueous methanol, whereas generally the arylpentazole is not soluble in any one of the two solvent phases ⟨56AG705, 57CB2914, 58CB2324⟩. The pentazole can be collected by filtration. Note that the crystalline arylpentazoles tend to explode violently at temperatures above their decomposition points (see Table 3); they should be handled with great care, in small amounts and at low temperatures.

With the isolation of the crystalline arylpentazoles some experiments become possible which require pure pentazole derivatives. Thus, crystalline ^{15}N-labelled p-ethoxyphenylpentazole was used to determine its molecular weight and to confirm its chemical structure according to Scheme 8 ⟨59CB1864⟩. Recently, Wallis and Dunitry ⟨83CC910⟩ have determined the structure of p-dimethylaminophenylpentazole by X-ray crystallography.

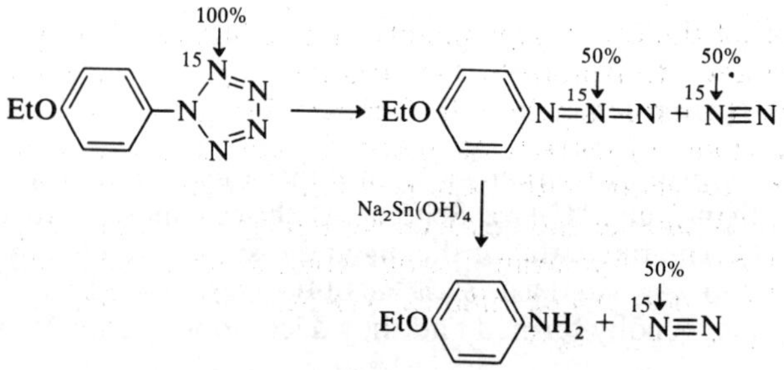

Scheme 8

The UV spectra of the (*p*-substituted phenyl)pentazoles indicate that, as a substituent of a benzenoid system, the pentazolyl group is capable of strongly interacting with electron-rich substituents on the phenyl group, in analogy to the nitro group. When this study took place, PE spectra were not available; it would be interesting to study the PE spectra of the arylpentazoles. According to kinetic studies with pure phenylpentazole, the solvent has a strong influence on its rate of decomposition ⟨58CB2324⟩; a solvent of low polarity enhances the rate of decomposition (see Table 4).

Table 4 The Decomposition of Phenylpentazole in Various Solvents

Solvent	$k' \times 10^4$ (s^{-1})
n-Hexane	45.2
Carbon tetrachloride	34.0
n-Butanol	16.9
Toluene	12.4
THF	10.4
Methanol	9.8
Chloroform	8.9
Acetone	7.7
Acetonitrile	4.1
Formic acid	2.3
Carbon tetrachloride/acetonitrile (20)	15.1
Carbon tetrachloride/acetonitrile (40)	11.2
Carbon tetrachloride/acetonitrile (60)	8.0
Carbon tetrachloride/acetonitrile (80)	7.1
Methanol/water (50)	5.7
Acetic acid/water (25)	5.8

4.14.3.3 Miscellaneous

In view of the enthalpies and activation energies of the decomposition of arylpentazoles, the activation energy of the reverse process (100–120 kJ mol^{-1}), namely the formation of arylpentazoles from aryl azides and nitrogen, should not be high enough to prohibit this reaction. In order to determine whether or not the decomposition of arylpentazoles is reversible, a [β-^{15}N]-*p*-ethoxyphenyl azide film on charcoal was exposed to unlabelled nitrogen (45–50 °C, 380 atm., 100 h), but no exchange of ^{15}N occurred ⟨59TH41400⟩. This excludes the possibility of reversible formation of an arylpentazole under the given conditions. Kinetic evidence as well as ^{15}N-labelling studies of the reaction between diazomethane and hydrazoic acid indicate that the formation of methyl azide proceeds neither *via* methanediazoazide nor *via* methylpentazole, but according to Scheme 9 ⟨58HCA1823⟩.

$$\underset{a\ \ b\ \ c\ \ d\ \ e}{CH_2=N=N + HN=N=N} \rightarrow \underset{a\ \ b\ \ c\ \ d\ \ e}{MeN=N=N + N\equiv N}$$

Scheme 9

It seems worthwhile to investigate whether or not the arylpentazoles form transition metal complexes, for example of type (**7**). These could be converted into pentazole complexes (**8**) without destruction of the pentazole system. Unsubstituted pentazole (**1**) would be expected to be a strong acid with a highly 'aromatic' anion (**6**) which could possibly form ferrocene analogs such as $M^{II}(N_5)_2$, where M^{II} represents a divalent metal ion. Unfortunately, it has not been possible to investigate the properties of unsubstituted pentazoles, since neither oxidative degradation of the carbocyclic moiety, nor reductive cleavage of the N—C bond converted arylpentazoles into pentazole. Ozonization of *p*-dimethylaminophenylpentazole in methylene chloride at approximately −60 °C caused decomposition of both the substituted benzene and the pentazole ring ⟨59TH41401, 61AG172⟩. Attempted cleavage of the N—C bond of phenylpentazole by reduction with sodium in liquid ammonia led to destruction of the pentazole ring ⟨61AG172⟩; aniline and nitrogen were formed, presumably *via* a 2,3-dihydro intermediate (see Scheme 11). A relevant analogy would seem to be the reductive degradation of the tetrazole ring ⟨40JA2002⟩. An *ab initio* LCAO–SCF–MO comparison of (**1**) and (**6**) has been carried out recently ⟨83CJC1435⟩.

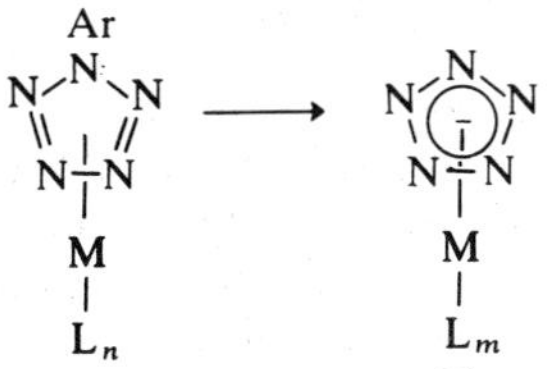

Scheme 10

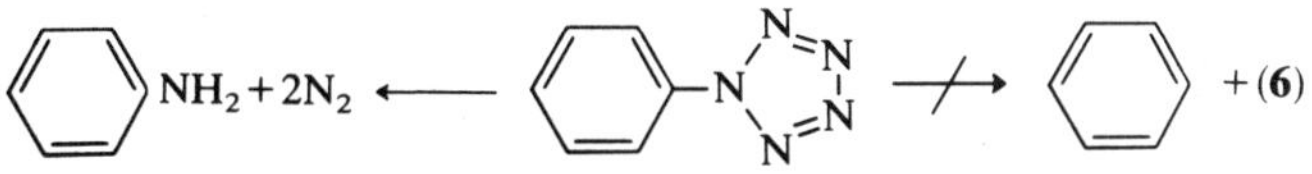

Scheme 11

4.14.3.4 Perspectives

Active experimental research in pentazole chemistry ceased about 20 years ago, not because all the problems in this field had been solved, but mainly because no new investigators have entered the field. Probably too few are cognizant of the unsolved problems of pentazole chemistry. The present author left the field because he was distracted from it by the many new challenging problems of which he became aware when he was confronted with isocyanide chemistry ⟨56AG705, 57CB2914, 62AG(E)8, 62AG9, B-71MI41400, B-75MI41400⟩. This arose from an attempt to prepare some aryl isocyanides as starting materials for aryltetrazoles. Their UV spectra were of interest for comparison with those of the corresponding pentazoles ⟨56AG705, 57CB2914⟩. The lack of preparative routes to pure isocyanides was the first step on the way to the realization that isocyanide chemistry was still *terra incognita*, with many areas waiting to be explored ⟨62AG9, 62AG(E)8, B-71MI41400, B-75MI41400, B-81MI41400⟩.

4.15

Triazoles and Tetrazoles with Fused Six-membered Rings

S. W. SCHNELLER
University of South Florida

4.15.1	INTRODUCTION	849
4.15.2	STRUCTURE	853
	4.15.2.1 *[1,2,4]Triazolopyridines*	853
	4.15.2.2 *[1,2,3]Triazolopyridines*	856
	4.15.2.3 *[1,2,4]Triazolopyridazines and [1,2,3]Triazolopyridazine*	858
	4.15.2.4 *[1,2,4]Triazolopyrimidines*	858
	4.15.2.5 *[1,2,3]Triazolopyrimidines*	858
	4.15.2.6 *[1,2,4]Triazolopyrazines and [1,2,3]Triazolopyrazines*	859
	4.15.2.7 *[1,2,4]Triazolotriazines*	859
	4.15.2.8 Tetrazolopyridines and Related Systems	859
	4.15.2.8.1 *Tetrazolo[1,5-a]pyridine*	860
	4.15.2.8.2 *Tetrazolo[1,5-b]pyridazine*	860
	4.15.2.8.3 *Tetrazolo[1,5-a]pyrimidine and tetrazolo[1,5-c]pyrimidine*	860
	4.15.2.8.4 *Tetrazolo[1,5-a]pyrazine*	860
	4.15.2.8.5 *Tetrazolo[1,5-b][1,2,4]triazine and tetrazolo[5,1-f][1,2,4]triazine*	861
4.15.3	REACTIVITY	861
	4.15.3.1 *[1,2,4]Triazolopyridines and their Fused Benzo Derivatives*	861
	4.15.3.1.1 *[1,2,4]Triazolo[4,3-a]pyridine*	861
	4.15.3.1.2 *[1,2,4]Triazolo[1,5-a]pyridine*	864
	4.15.3.1.3 *[1,2,4]Triazoloquinolines and [1,2,4]triazoloisoquinolines*	865
	4.15.3.2 *[1,2,3]Triazolopyridines and their Fused Benzo Derivatives*	866
	4.15.3.2.1 *[1,2,3]Triazolo[1,5-a]pyridine*	866
	4.15.3.2.2 *[1,2,3]Triazolo[4,5-b]pyridine and [1,2,3]triazolo[4,5-c]pyridine*	867
	4.15.3.2.3 *[1,2,3]Triazolo[1,5-a]quinoline*	868
	4.15.3.2.4 *[1,2,3]Triazolo[5,1-a]isoquinoline*	868
	4.15.3.3 *[1,2,4]Triazolopyridazines and their Fused Benzo Derivatives*	868
	4.15.3.3.1 *[1,2,4]Triazolo[4,3-b]pyridazine*	868
	4.15.3.3.2 *[1,2,4]Triazolo[3,4-a]phthalazine*	870
	4.15.3.3.3 *[1,2,4]Triazolo[1,5-b]pyridazine*	871
	4.15.3.4 *[1,2,3]Triazolo[4,5-d]pyridazine*	871
	4.15.3.5 *[1,2,4]Triazolopyrimidines and their Fused Benzo Derivatives*	872
	4.15.3.5.1 *[1,2,4]Triazolo[4,3-a]pyrimidine and [1,2,4]triazolo[4,3-c]pyrimidine*	872
	4.15.3.5.2 *[1,2,4]Triazolo[1,5-a]pyrimidine*	872
	4.15.3.5.3 *[1,2,4]Triazolo[1,5-c]pyrimidine*	873
	4.15.3.5.4 *[1,2,4]Triazolo[4,3-c]quinazoline and [1,2,4]triazolo[1,5-c]quinazoline*	873
	4.15.3.5.5 *[1,2,4]Triazolo[1,5-a]quinazoline*	873
	4.15.3.5.6 *[1,2,4]Triazolo[5,1-b]quinazoline*	874
	4.15.3.6 *[1,2,3]Triazolopyrimidines and their Fused Benzo Derivatives*	874
	4.15.3.6.1 *[1,2,3]Triazolo[1,5-a]pyrimidine*	874
	4.15.3.6.2 *[1,2,3]Triazolo[1,5-c]pyrimidine*	875
	4.15.3.6.3 *[1,2,3]Triazolo[4,5-d]pyrimidine*	875
	4.15.3.6.4 *[1,2,3]Triazolo[1,5-a]quinazoline*	876
	4.15.3.7 Triazolopyrazines and their Fused Benzo Derivatives	876
	4.15.3.7.1 *[1,2,4]Triazolo[4,3-a]pyrazine*	876
	4.15.3.7.2 *[1,2,4]Triazolo[1,5-a]pyrazine*	877
	4.15.3.7.3 *[1,2,3]Triazolo[1,5-a]pyrazine*	877
	4.15.3.7.4 *[1,2,3]Triazolo[4,5-b]pyrazine*	878
	4.15.3.7.5 *[1,2,4]Triazolo[4,3-a]quinoxaline*	878
	4.15.3.8 *[1,2,4]Triazolotriazines*	878
	4.15.3.8.1 *[1,2,4]Triazolo[1,2,4]triazines*	878
	4.15.3.8.2 *[1,2,4]Triazolo[1,3,5]triazines*	878

4.15.3.9 Tetrazolopyridines and their Fused Benzo Derivatives	879
4.15.3.9.1 Tetrazolo[1,5-a]pyridine	879
4.15.3.9.2 Tetrazolo[1,5-b]isoquinoline	879
4.15.3.9.3 Tetrazolo[1,5-a]quinoline	879
4.15.3.9.4 Tetrazolo[5,1-a]isoquinoline	879
4.15.3.10 Tetrazolo[1,5-b]pyridazine	880
4.15.3.11 Tetrazolopyrimidines and their Fused Benzo Derivatives	880
4.15.3.11.1 Tetrazolo[1,5-a]pyrimidine	880
4.15.3.11.2 Tetrazolo[1,5-c]pyrimidine	881
4.15.3.11.3 Tetrazolo[1,5-a]quinazoline	881
4.15.3.11.4 Tetrazolo[1,5-c]quinazoline	881
4.15.3.12 Tetrazolo[1,5-a]pyrazine and Tetrazolo[1,5-a]quinoxaline	881
4.15.3.12.1 Tetrazolo[1,5-a]pyrazine	881
4.15.3.12.2 Tetrazolo[1,5-a]quinoxaline	882
4.15.4 SYNTHESIS	**882**
4.15.4.1 [1,2,4]Triazolopyridines and their Fused Benzo Derivatives	882
4.15.4.1.1 [1,2,4]Triazolo[4,3-a]pyridine	882
4.15.4.1.2 [1,2,4]Triazolo[1,5-a]pyridine	884
4.15.4.1.3 [1,2,4]Triazoloquinolines and [1,2,4]triazoloisoquinolines	886
4.15.4.2 [1,2,3]Triazolopyridines and their Fused Benzo Derivatives	887
4.15.4.2.1 [1,2,3]Triazolo[1,5-a]pyridine	887
4.15.4.2.2 [1,2,3]Triazolo[4,5-b]pyridine and [1,2,3]triazolo[4,5-c]pyridine	887
4.15.4.2.3 [1,2,3]Triazolo[1,5-a]quinoline and [1,2,3]triazolo[5,1-a]isoquinoline	888
4.15.4.3 [1,2,4]Triazolopyridazines and their Fused Benzo Derivatives	888
4.15.4.3.1 [1,2,4]Triazolo[4,3-b]pyridazine	888
4.15.4.3.2 [1,2,4]Triazolo[1,5-b]pyridazine	889
4.15.4.4 [1,2,3]Triazolo[4,5-d]pyridazine	890
4.15.4.5 [1,2,4]Triazolopyrimidines and their Fused Benzo Derivatives	890
4.15.4.5.1 [1,2,4]Triazolo[4,3-a]pyrimidine	890
4.15.4.5.2 [1,2,4]Triazolo[1,5-a]pyrimidine	890
4.15.4.5.3 [1,2,4]Triazolo[4,3-c]pyrimidine	891
4.15.4.5.4 [1,2,4]Triazolo[1,5-c]pyrimidine	891
4.15.4.5.5 [1,2,4]Triazolo[4,3-c]quinazoline and [1,2,4]triazolo[1,5-c]quinazoline	892
4.15.4.5.6 [1,2,4]Triazolo[1,5-a]quinazoline	893
4.15.4.5.7 [1,2,4]Triazolo[5,1-b]quinazoline	893
4.15.4.5.8 [1,2,4]Triazolo[4,3-a]quinazoline	893
4.15.4.6 [1,2,3]Triazolopyrimidines and their Fused Benzo Derivatives	894
4.15.4.6.1 [1,2,3]Triazolo[1,5-a]pyrimidine	894
4.15.4.6.2 [1,2,3]Triazolo[1,5-c]pyrimidine	894
4.15.4.6.3 [1,2,3]Triazolo[4,5-d]pyrimidine	894
4.15.4.6.4 [1,2,3]Triazolo[1,5-a]quinazoline	895
4.15.4.6.5 [1,2,3]Triazolo[1,5-c]quinazoline	895
4.15.4.7 Triazolopyrazines and their Fused Benzo Derivatives	895
4.15.4.7.1 [1,2,4]Triazolo[4,3-a]pyrazine	895
4.15.4.7.2 [1,2,4]Triazolo[1,5-a]pyrazine	896
4.15.4.7.3 [1,2,3]Triazolo[1,5-a]pyrazine	897
4.15.4.7.4 [1,2,3]Triazolo[4,5-b]pyrazine	897
4.15.4.7.5 [1,2,4]Triazolo[4,3-a]quinoxaline	898
4.15.4.7.6 [1,2,3]Triazolo[1,5-a]quinoxaline	898
4.15.4.8 [1,2,4]Triazolotriazines	898
4.15.4.8.1 [1,2,4]Triazolo[4,3-b][1,2,4]triazine	898
4.15.4.8.2 [1,2,4]Triazolo[3,4-c][1,2,4]triazine	899
4.15.4.8.3 [1,2,4]Triazolo[5,1-c][1,2,4]triazine	899
4.15.4.8.4 [1,2,4]Triazolo[4,3-d][1,2,4]triazine	899
4.15.4.8.5 [1,2,4]Triazolo[3,4-f][1,2,4]triazine	899
4.15.4.8.6 [1,2,4]Triazolo[1,5-b][1,2,4]triazine	900
4.15.4.8.7 [1,2,4]Triazolo[1,5-d][1,2,4]triazine	900
4.15.4.8.8 [1,2,4]Triazolo[4,3-a][1,3,5]triazine	900
4.15.4.8.9 [1,2,4]Triazolo[1,5-a][1,3,5]triazine	900
4.15.4.9 Tetrazolopyridines and their Fused Benzo Derivatives	901
4.15.4.9.1 Tetrazolo[1,5-a]pyridine	901
4.15.4.9.2 Tetrazolo[1,5-b]isoquinoline	901
4.15.4.9.3 Tetrazolo[1,5-a]quinoline	902
4.15.4.9.4 Tetrazolo[5,1-a]isoquinoline	902
4.15.4.10 Tetrazolo[1,5-b]pyridazines and their Fused Benzo Derivatives	902
4.15.4.11 Tetrazolopyrimidines and their Fused Benzo Derivatives	902
4.15.4.11.1 Tetrazolo[1,5-a]pyrimidine	902
4.15.4.11.2 Tetrazolo[1,5-c]pyrimidine	902
4.15.4.11.3 Tetrazolo[1,5-a]quinazoline	902
4.15.4.11.4 Tetrazolo[1,5-c]quinazoline	903
4.15.4.12 Tetrazolo[1,5-a]pyrazine and Tetrazolo[1,5-a]quinoxaline	903
4.15.4.13 Tetrazolotriazines	903

4.15.4.13.1 Tetrazolo[1,5-b][1,2,4]triazine   903
4.15.4.13.2 Tetrazolo[5,1-c][1,2,4]triazine   903
4.15.4.13.3 Tetrazolo[1,5-d][1,2,4]triazine   903
4.15.4.13.4 Tetrazolo[5,1-f][1,2,4]triazine   903
4.15.5 APPLICATIONS   903

4.15.1 INTRODUCTION

In planning a chapter such as this whose breadth is large and whose space is restricted, a decision had to be made as to which six-membered rings are the most important upon their fusion to triazoles and tetrazoles. Based on an analysis of those ring systems attracting the greatest attention in the literature, this chapter will focus on triazoles and tetrazoles fused to pyridines (**1, 2, 8–10** and **46**), pyridazines (**13–15** and **50**), pyrimidines (**18–21, 26–28, 53** and **54**), pyrazines (**31–34** and **57**) and triazines (**37–45, 59–62**). Both 1,2,4- and 1,2,3-triazoles are considered and, in most cases, the two fused rings have one nitrogen atom in common. The only exceptions to this are generally those cases where a fused 1,2,3-triazole represents part of an 8-azapurine, a series that has no bridgehead nitrogen.

[1,2,4]Triazolopyridines

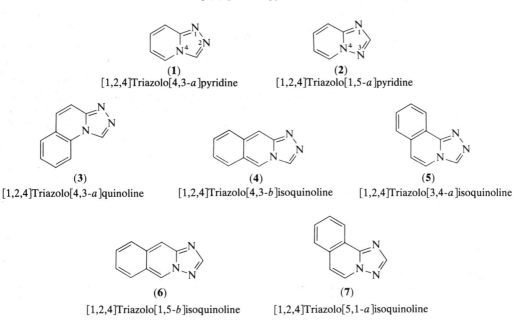

(1) [1,2,4]Triazolo[4,3-a]pyridine
(2) [1,2,4]Triazolo[1,5-a]pyridine
(3) [1,2,4]Triazolo[4,3-a]quinoline
(4) [1,2,4]Triazolo[4,3-b]isoquinoline
(5) [1,2,4]Triazolo[3,4-a]isoquinoline
(6) [1,2,4]Triazolo[1,5-b]isoquinoline
(7) [1,2,4]Triazolo[5,1-a]isoquinoline

[1,2,3]Triazolopyridines

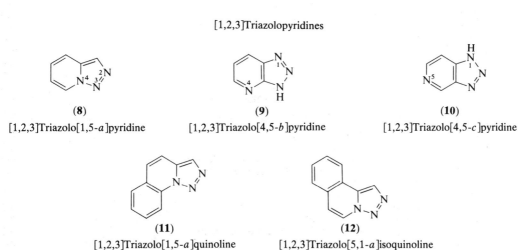

(8) [1,2,3]Triazolo[1,5-a]pyridine
(9) [1,2,3]Triazolo[4,5-b]pyridine
(10) [1,2,3]Triazolo[4,5-c]pyridine
(11) [1,2,3]Triazolo[1,5-a]quinoline
(12) [1,2,3]Triazolo[5,1-a]isoquinoline

Triazolopyridazines

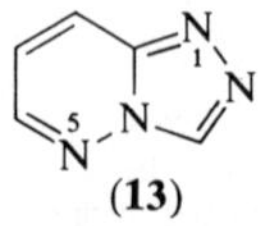

(13) [1,2,4]Triazolo[4,3-b]pyridazine

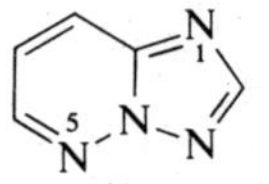

(14) [1,2,4]Triazolo[1,5-b]pyridazine

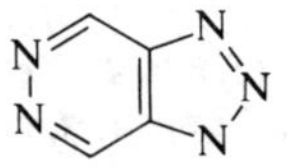

(15) [1,2,3]Triazolo[4,5-d]pyridazine

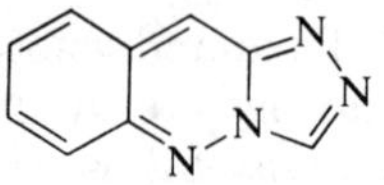

(16) [1,2,4]Triazolo[4,3-b]cinnoline

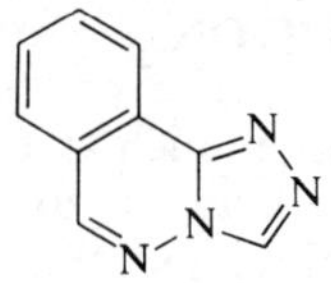

(17) [1,2,4]Triazolo[3,4-a]phthalazine

[1,2,4]Triazolopyrimidines

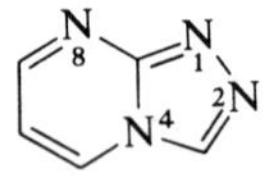

(18) [1,2,4]Triazolo[4,3-a]pyrimidine

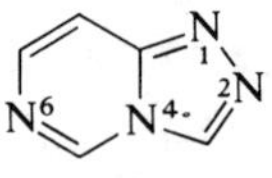

(19) [1,2,4]Triazolo[4,3-c]pyrimidine

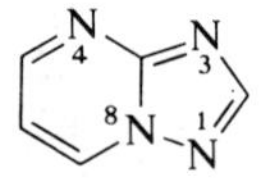

(20) [1,2,4]Triazolo[1,5-a]pyrimidine

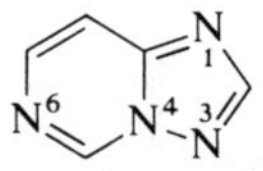

(21) [1,2,4]Triazolo[1,5-c]pyrimidine

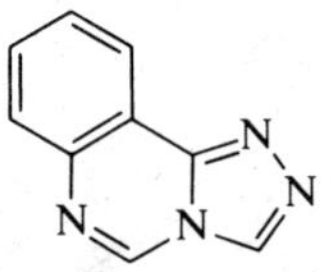

(22) [1,2,4]Triazolo[4,3-c]quinazoline

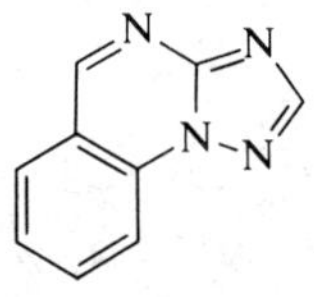

(23) [1,2,4]Triazolo[1,5-a]quinazoline

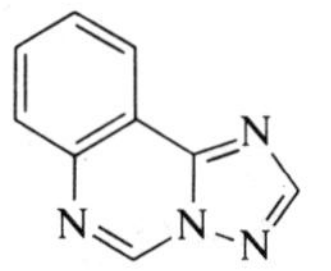

(24) [1,2,4]Triazolo[1,5-c]quinazoline

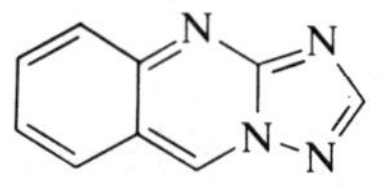

(25) [1,2,4]Triazolo[5,1-b]quinazoline

[1,2,3]Triazolopyrimidines

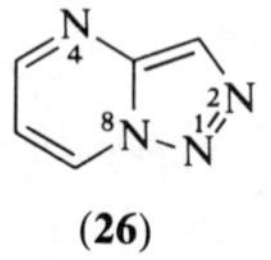

(26) [1,2,3]Triazolo[1,5-a]pyrimidine

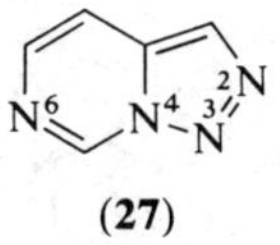

(27) [1,2,3]Triazolo[1,5-c]pyrimidine

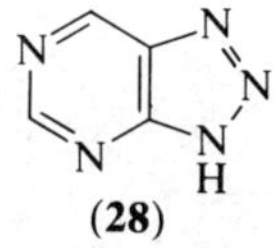

(28) [1,2,3]Triazolo[4,5-d]pyrimidine

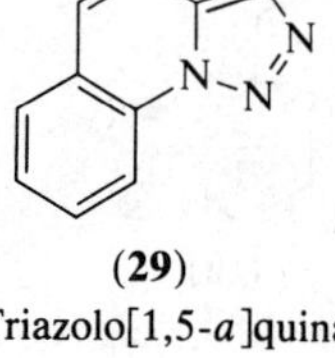

(29) [1,2,3]Triazolo[1,5-a]quinazoline

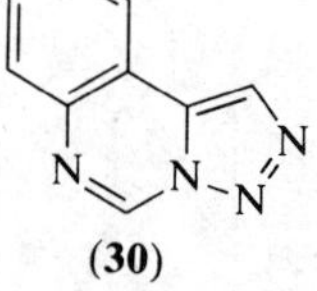

(30) [1,2,3]Triazolo[1,5-c]quinazoline

Triazolopyrazines

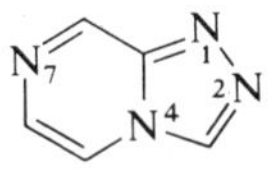

(31) [1,2,4]Triazolo[4,3-a]pyrazine

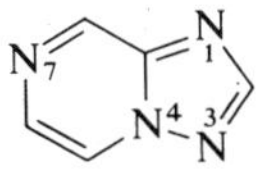

(32) [1,2,4]Triazolo[1,5-a]pyrazine

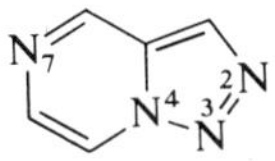

(33) [1,2,3]Triazolo[1,5-a]pyrazine

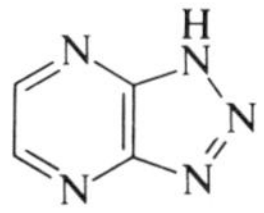

(34) [1,2,3]Triazolo[4,5-b]pyrazine

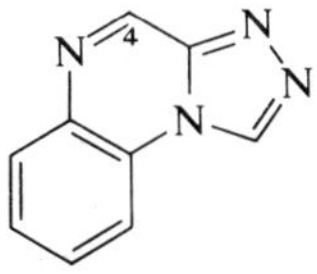

(35) [1,2,4]Triazolo[4,3-a]quinoxaline

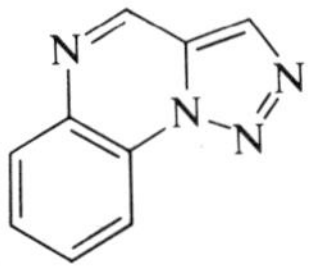

(36) [1,2,3]Triazolo[1,5-a]quinoxaline

Triazolotriazines

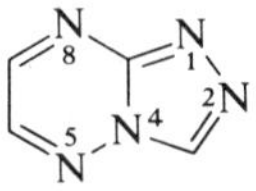

(37) [1,2,4]Triazolo[4,3-b][1,2,4]triazine

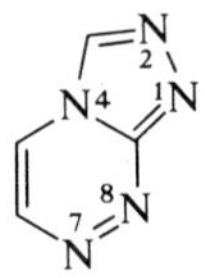

(38) [1,2,4]Triazolo[3,4-c][1,2,4]triazine

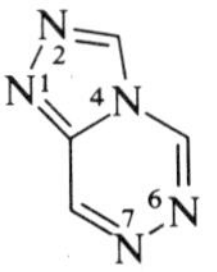

(39) [1,2,4]Triazolo[4,3-d][1,2,4]triazine

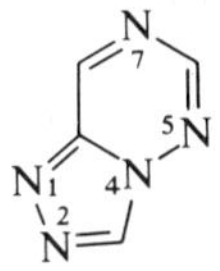

(40) [1,2,4]Triazolo[3,4-f][1,2,4]triazine

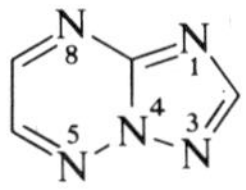

(41) [1,2,4]Triazolo[1,5-b][1,2,4]triazine

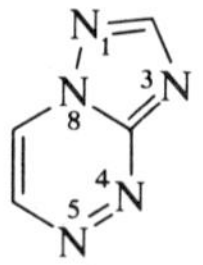

(42) [1,2,4]Triazolo[5,1-c][1,2,4]triazine

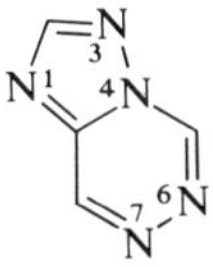

(43) [1,2,4]Triazolo[1,5-d][1,2,4]triazine

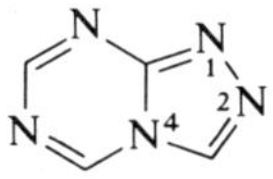

(44) [1,2,4]Triazolo[4,3-a][1,3,5]triazine

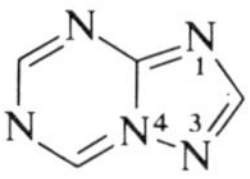

(45) [1,2,4]Triazolo[1,5-a][1,3,5]triazine

Tetrazolopyridines

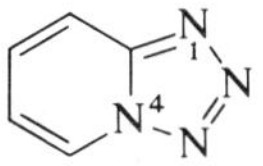

(46) Tetrazolo[1,5-a]pyridine

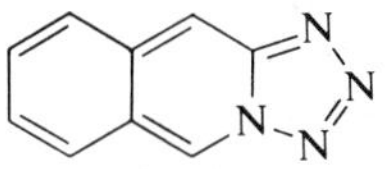

(47) Tetrazolo[1,5-b]isoquinoline

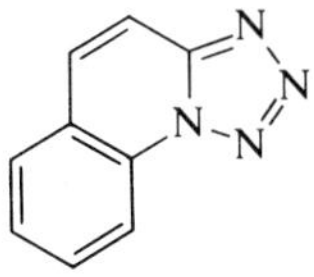

(48) Tetrazolo[1,5-a]quinoline

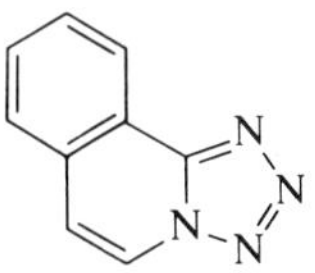

(49) Tetrazolo[5,1-a]isoquinoline

Tetrazolopyridazines

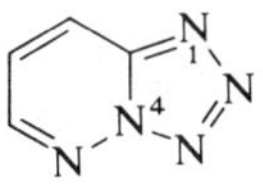

(50) Tetrazolo[1,5-b]pyridazine

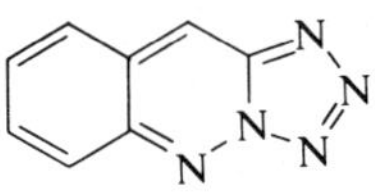

(51) Tetrazolo[1,5-b]cinnoline

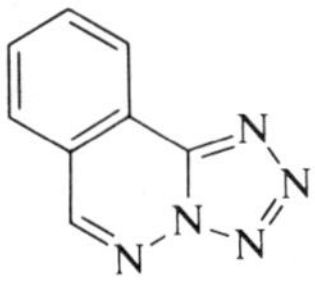

(52) Tetrazolo[5,1-a]phthalazine

Tetrazolopyrimidines

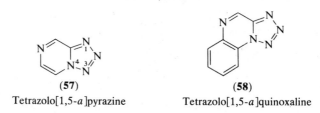

(53) Tetrazolo[1,5-a]pyrimidine   (54) Tetrazolo[1,5-c]pyrimidine

(55) Tetrazolo[1,5-a]quinazoline   (56) Tetrazolo[1,5-c]quinazoline

Tetrazolopyrazines

(57) Tetrazolo[1,5-a]pyrazine   (58) Tetrazolo[1,5-a]quinoxaline

Tetrazolo[1,2,4]triazines

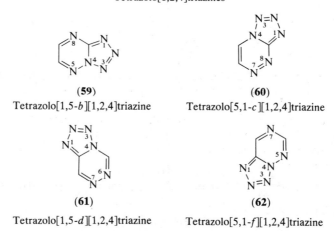

(59) Tetrazolo[1,5-b][1,2,4]triazine   (60) Tetrazolo[5,1-c][1,2,4]triazine

(61) Tetrazolo[1,5-d][1,2,4]triazine   (62) Tetrazolo[5,1-f][1,2,4]triazine

Only representative monofused benzene systems of pyridine, e.g. quinoline (**3**, **11** and **48**) and isoquinoline (**4–7**, **12**, **47** and **49**), pyridazine, e.g. cinnoline (**16** and **51**) and phthalazine (**17** and **52**), pyrimidine, e.g. quinazoline (**22–25**, **29–30** and **55–56**), and pyrazine, e.g. quinoxaline (**35**, **36** and **58**) are presented with no effort directed towards the polybenzo derivatives or any benzo derivatives of the triazines. These latter compounds seem to be of less significance than those described and have been eliminated from a detailed discussion for the sake of brevity.

Other simple bicyclic ring systems known but not covered here include [1,2,4]triazolo[1,2-a]pyridazine ⟨64ZOB390⟩, [1,2,3]triazolo[1,5-b]pyridazine (**63**) ⟨67AJC713⟩, [1,2,4]triazolo[1,2-a][1,2,4]triazine (**64**) ⟨75CJC355⟩, [1,2,3]triazolo[1,5-b][1,2,4]triazine (**65**) ⟨63CB1827⟩, [1,2,3]triazolo[5,1-c][1,2,4]triazine (**66**) ⟨72TL4719⟩, [1,2,3]triazolo[4,5-d]-[1,2,3]triazine (as 2,8-diazapurines) (**67**) ⟨66JMC733, 72JMC182⟩, [1,2,4]triazolo[4,3-b]-[1,2,4,5]tetrazine (**68**) ⟨71CHE665⟩, tetrazolo[1,5-b][1,2,4,5]tetrazine (**69**) ⟨71CHE668⟩, [1,2,3]triazolo[5,1-d][1,2,3,5]tetrazine (**70**) ⟨79TL4253⟩, [1,2,4]triazolo[3,4-c][1,4]oxazine (**71**) ⟨68CB1979⟩, [1,2,3]triazolo[5,1-c][1,4]oxazine (**72**) ⟨80ZOR730⟩, [1,2,3]triazolo[4,5-d][1,3]oxazine (**73**) ⟨70CC858⟩, tetrazolo[5,1-b][1,3]oxazine (**74**) ⟨75TL695⟩, [1,2,4]triazolo[5,1-b][1,3]thiazine (**75**) ⟨80JCS(P1)1352⟩, [1,2,4]triazolo[3,4-b][1,3]thiazine

(76) ⟨79H(12)1171⟩, [1,2,4]triazolo[3,4-c][1,2,4]thiadiazine (77) ⟨74BSF1395⟩, [1,2,4]triazolo[3,4-b][1,3,4]thiadiazine (78) ⟨78IJC(B)481⟩ and [1,2,4]triazolo[5,1-c][1,2,4,6]thiatriazine (79) ⟨77JCR(S)238⟩.

Miscellaneous Examples of Fused Triazoles and Tetrazoles

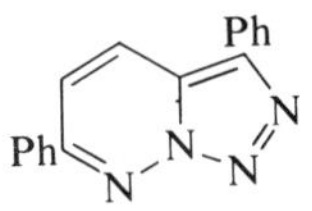

(63) 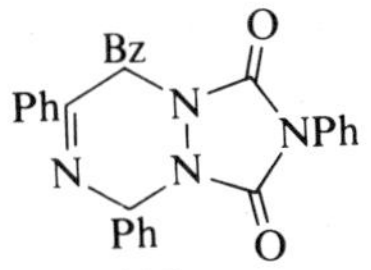 (64) 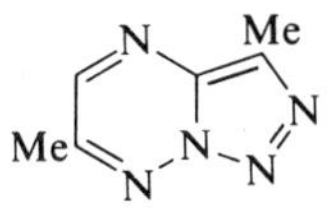 (65) 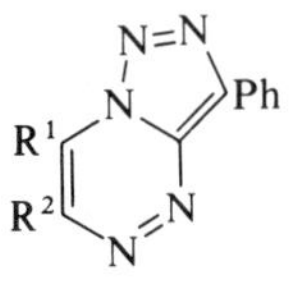 (66)
(67) 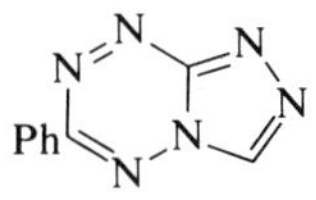 (68) 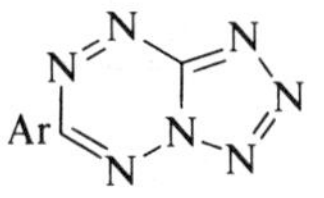 (69) 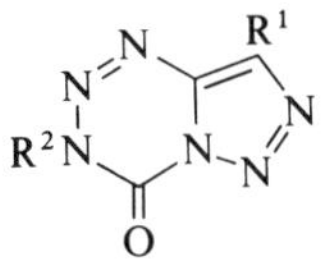 (70) 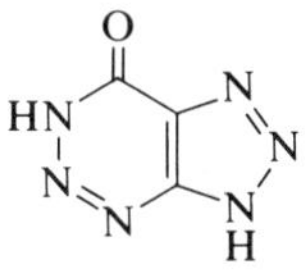
(71) 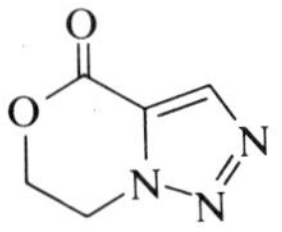 (72) 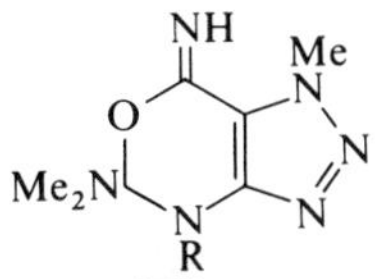 (73) 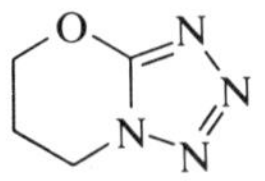 (74) 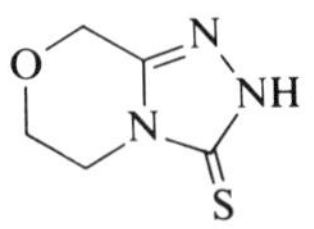
(75) 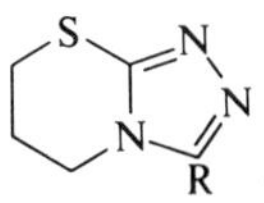 (76) 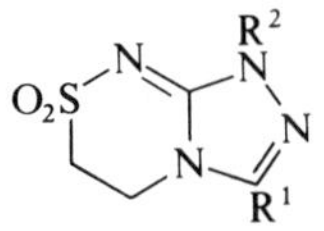 (77) 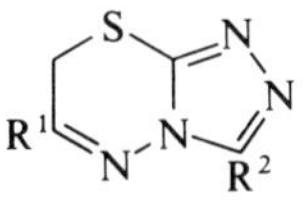 (78) 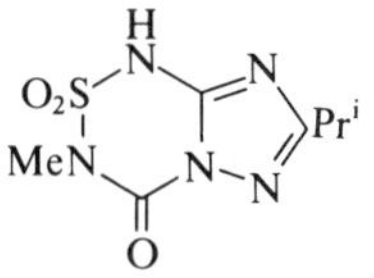 (79) 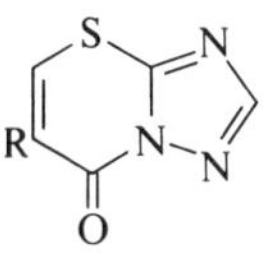

In the following presentation, structures (1)–(62) will be used, as is customary, to represent the parent heterocycle of the derivative under discussion. It should also be noted that the structures are all drawn (with several exceptions) using [1,2,4]triazolo[4,3-a]pyridine (1) as the prototype. Thus, the focus of visual orientation will be a bridgehead nitrogen in the lower center part of the ring system. This had been done not out of disrespect for the recognized molecular orientations but in an effort to simplify the relationships of these diverse systems to one another.

4.15.2 STRUCTURE

4.15.2.1 [1,2,4]Triazolopyridines

The proton NMR spectral data for (1) and (2) are shown in Table 1. It can be seen that H-3 of (1) is at a lower field relative to the H-2 of (2) and this observation is very useful in distinguishing [4,3-a] systems from their [1,5-a] isomers. As a consequence of this, only the H-3 or H-2 protons for (4)–(7) are recorded in Table 5. It is interesting to note that there is an H(3)–H(8) coupling (1 Hz) in (1) ⟨66JOC3522⟩. The ^{14}N NMR and UV spectral data for (1) and (2) are given in Tables 3 and 4, respectively. The mass spectral fragmentations of (1) and (2) ⟨71OMS(5)1⟩ indicated both systems undergo loss of hydrogen cyanide (or various nitriles for the substituted derivatives) from their molecular ions to give a three-membered ion that is common to both systems. Extensive molecular orbital calculations have produced bond orders, electron densities, etc. for systems (1) and (2) ⟨68MI41500, 66JOC265, 72JPR515, 70CB3817⟩ and these are in agreement with the chemical properties described later.

Similar spectral and molecular orbital data are available for the fused compounds (3) ⟨72JOC4410, 76AJC815, 80AJC499⟩, (5) ⟨66JHC158, 70CB3817, 70IJC894, 71CB3925, 71RTC1225, 72JOC4410⟩ and (7) ⟨71CB3925, 71CB3965, 71RTC1225⟩.

Melting points for this series of compounds are given in Table 6.

Table 1 ^{1}H NMR Spectral Data for Selected Fused Triazoles and Tetrazoles

Compound	^{1}H chemical shifts (δ, p.p.m.)								Solvent	J (Hz)	Ref.
	H-1	H-2	H-3	H-4	H-5	H-6	H-7	H-8			
(1)	—	—	8.86	—	8.21	6.88	7.28	7.79	CDCl$_3$	$J_{5,6}=6.7, J_{5,7}=1.2, J_{5,8}=1.2,$ $J_{6,7}=6.7, J_{6,8}=1.2, J_{7,8}=9.0,$ $J_{3,8}=1.0$	66JOC3522
(2)	—	8.35	—	—	8.62	7.04	7.52	7.82	CDCl$_3$	$J_{5,6}=6.6, J_{5,7}=1.3, J_{5,8}=1.2,$ $J_{6,7}=6.6, J_{6,8}=1.2, J_{7,8}=8.8$	66JOC3522
(8)	8.05	—	—	—	8.70	7.00	7.20	7.70	CDCl$_3$	$J_{1,5}=0.8, J_{7,8}=9.0, J_{6,8}=1.25,$ $J_{5,8}=1.0, J_{6,7}=6.5, J_{5,7}=1.0,$ $J_{5,6}=7.0$	78HCA1755
(14)	—	8.38	—	—	—	8.44	7.42	8.08	CDCl$_3$	$J_{6,7}=4.2, J_{7,8}=8.9, J_{6,8}=1.8$	73TL1677
(18)	—	—	9.26	—	9.00	7.13	8.73	—	(CD$_3$)$_2$SO		77AJC2515
(19)	—	—	9.40	—	9.47	7.17	7.97	7.77	(CD$_3$)$_2$SO		78AJC2505
(20)	—	8.52	—	—	8.87	—	8.95	—	CDCl$_3$	$J_{7,8}=6.0, J_{5,8}=1.0$	73TL1677
(21)	—	8.67	—	—	9.80	—	8.30	7.90	(CD$_3$)$_2$SO	$J_{5,6}=4.8, J_{5,7}=1.0, J_{6,7}=6.9$	78AJC2505
(27)	8.18	—	—	—	9.68	—	8.05	7.72	CDCl$_3$	$J_{7,8}=6, J_{5,8}=1$	78JHC1041
(28)	—	—	—	—	9.68	—	10.33	—	CF$_3$CO$_2$H	$J_{7,8}=6.7, J_{5,8}=1.6, J_{1,5}=0.6$	69CJC1129
(32)	8.30	8.50	—	—	8.58	8.19	—	9.37	CDCl$_3$	$J_{5,6}=4.5, J_{5,8}=1.5$	73TL1677
(33)	—	—	—	9.30	8.60	8.00	—	9.30	CDCl$_3$	$J_{1,5}=1.0, J_{5,8}=1.3, J_{5,6}=5.0$	78HCA1755
(34) (1-benzyl)	—	—	—	—	8.75	8.65	—	—	CDCl$_3$	$J_{5,6}=1.0$	72JOC4124
(37)	—	—	9.13	—	—	8.61	8.45	—	CDCl$_3$	$J_{6,7}=1.5$	77JOC1018
(38) (6-Me)	—	—	8.78	—	8.43	—	8.67	—	CDCl$_3$		77JOC1018
(41) (6-Me)	—	8.59	—	—	—	6.16	6.94	—	CDCl$_3$		77JOC1018
(42)	—	7.83	—	—	8.91	7.38	7.74	8.11	CDCl$_3$	$J_{6,7}=2.4$	77JOC1018
(46)	—	—	—	—	9.30	7.76	9.85	—	CDCl$_3$		71T5121
(53)	—	—	—	—	—	—	—	—	(CD$_3$)$_2$SO		65JOC826
(57)	—	—	—	—	8.83	8.36	—	9.64	CDCl$_3$	$J_{5,8}=1.6, J_{6,8}=0.5, J_{5,6}=4.7$	66JHC435

Table 2 ^{13}C NMR Spectral Data for Selected Fused Triazoles and Tetrazoles

Compound	^{13}C Chemical Shifts (p.p.m.)								Miscellaneous	Solvent	Ref.
	C-2	C-3	C-5	C-6	C-7	C-8	C-8a				
(13)	—	139.04	—	146.57	121.20	124.61	143.31	—	(CD$_3$)$_2$SO	76JHC1057
(14)	152.50	—	—	144.55	123.04	125.92	144.05	—	(CD$_3$)$_2$SO	76JHC1057
(20)-7-one (as anion)	152.50	—	154.10	97.20	160.10	—	—	158.30 (C-3a)	(CD$_3$)$_2$SO	74JOC3226
(23) (5-NH$_2$-2-Me)	161.30	—	158.70	125.60	124.70	134.10	—	114.50 (C-9) 135.30 (C-9a) 154.00 (C-3a)	(CD$_3$)$_2$SO	80HCA1
(28)-7-NH$_2$ (as anion)	—	—	155.80	—	158.90	—	—	124.00 (C-7a) 154.20 (C-3a)	(CD$_3$)$_2$SO	74JOC3226
(31)	see discussion										
(32)	see discussion										

Table 3 ^{14}N NMR Spectral Data for Selected Fused Triazoles and Tetrazoles

Compound	N-1	N-2	N-3	N-4	Solvent	Ref.
		^{14}N Chemical Shifts^a (p.p.m.)				
(1)	+60.1 ± 4.2	+118.1 ± 9.0	—	+185.5 ± 0.2	Et$_2$O	76JMR(21)185
(2)	+150.0 ± 1.0	—	+94.1 ± 2.2	+140.7 ± 0.3	Et$_2$O	76JMR(21)185
(8)	—	+40.9 ± 3.2	−30.6 ± 3.9	+119.9 ± 0.2	Et$_2$O	76JMR(21)185
(46)	+61.8 ± 1.2	+70.5 ± 11.2	+10.0 ± 2.4	+129.2 ± 0.2	Et$_2$O/Me$_2$CO	76JMR(21)185

^{a} Relative to nitromethane.

Table 4 UV Spectral Maxima (nm) for Selected Fused Triazoles and Tetrazoles

Compound	Solvent	λ (log ε or ε)	Ref.
(1)	Cyclohexane	306 (3.34), 274 (3.67), 213 (4.38)	68MI41500
(2) (2-Me)	Cyclohexane	294 (3.20), 273 (3.57), 217.5 (4.58)	68MI41500
(8)	Cyclohexane	299 (3.36), 270 (3.33), 214 (4.35)	68MI41500
(14)	EtOH	277 (3.57), 208 (4.56)	75JHC107
(18)	H$_2$O, pH 5	293 (3.35), 266 (3.24), 256 (3.23)	77AJC2515
(19)	H$_2$O	258 (3.77), 249 (3.79)	78AJC2505
(20)	H$_2$O, pH 4	272 (3.59)	77AJC2515
(21)	Dioxane	256 (3.78), 254 (3.71)	78AJC2505
(22)	MeOH	240 (4.60)	70JOC3448
(24)	MeOH	240 (4.39)	70JOC3448
(27)	Heptane	340 (2.65), 323 (3.08), 309 (3.21), 301 (3.21), 294 (3.20), 266 (3.75)	78JHC1041
(28)	H$_2$O	263 (3.87)	66JCS(B)427
(31)	EtOH	299 (3.57), 206 (4.46)	64JOC2542
(32)	EtOH	280 (3.75), 207 (4.60)	75JHC107
(33)	EtOH	283 (4935), 274 (4860)	78HCA1755
(35)	MeOH	284 (3.95)	68JHC485
(46)	EtOH	293 (3.16), 265 (3.62)	68MI41500
(57)	H$_2$O	270 (4500), 203 (12000)	66JHC435

Table 5 ^{1}H NMR Chemical Shifts for the Triazole Proton in Selected Benzo Derivatives

Compound	Chemical shift (δ, p.p.m.)	Solvent	Ref.
(4) perchlorate	8.95	(CD$_3$)$_2$SO	78JHC463
(5)	9.38	(CD$_3$)$_2$SO	71RTC1225
(6)	8.58	CDCl$_3$	77H(6)949
(7)	8.64	(CD$_3$)$_2$SO	71RTC1225
(11)	8.04	CDCl$_3$	72JOC2022
(12)	8.30	CDCl$_3$	72JOC2022
(17)	9.00	CDCl$_3$	69JOC3221
(22)	9.50	CDCl$_3$	70JOC3448
(24)	8.60	CDCl$_3$	70JOC3448

4.15.2.2 [1,2,3]Triazolopyridines

Molecules containing a fused 1,2,3-triazole ring, as in (8), (11), (12), (26), (27), (33) and (36), frequently display a diazo characteristic in their chemical reactivity that leads to homolytic and heterolytic ring fission. This is a result of an equilibrium that can exist between the closed and open forms (A) and (B) analogous to the ring–chain tautomerism exhibited by fused tetrazoles. Some studies have been undertaken ⟨71JCS(C)2156⟩ to evaluate the role that this equilibrium plays in the physical and chemical properties of fused 1,2,3-triazoles.

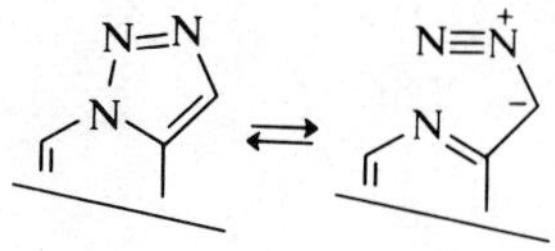

(A)   ⇌   (B)

Table 6 Melting Points (°C) for Selected Fused Triazoles and Tetrazoles

Ring system	M.p.	Ref.	Ring system	M.p.	Ref.	Ring system	M.p.	Ref.
(1) (hydrate)	36	57JCS4510	(21)	117–119	78AJC2505	(45) (5-NH$_2$)	>320 dec.	80JHC1121
(2)	102	66JOC260	(22)	213–214	70JOC3448	(46)	158–160	71T5121
(4)	227–228	78JHC463	(24)	109–110	70JOC3448	(47)	122	76JHC881
(5)	130–132	71RTC1225	(27)	130–131	78JHC1041	(48)	155–156	76JOC1073
(6)	160–161	77H(6)949	(28)	174–175	66JCS(B)427	(49)	138–140	79JHC1567
(7)	95–96.5	71RTC1225	(31)	194–195	62JOC3243	(50)	108–111	63CPB348
(8)	34	75JHC481	(32)	127	73TL1677	(51) (6,7,8,9-H$_4$)	142–143	71BSF3043
(10)	245 (dec.)	71JHC47	(33)	126–126.5	78HCA1755	(52) (6-oxo)	287	67CB1073
(11)	81.5–82	72JOC2022	(34) (1-CH$_2$Ph)	91.5	72JOC4124	(53)	121–123	57JAP570777
(12)	110–111	72JOC2022	(35)	234	60JA4044	(54)	77–79	65JOC829
(13)	156–158	76JHC639	(37)	165–166	77JOC1018	(55) (5-NHNH$_2$)	242–243	63ZOB2475
(14)	157–158	75JHC107	(38) (6-Me)	137–138	77JOC1018	(56) (5-Me)	180	80H(14)1759
(15)	308–309	70CPB1685	(41) (6-Me)	146–148	77JOC1018	(57)	90.8–91	66JHC435
(17)	184–185	69JOC3221	(42)	176–177	77JOC1018			
(18)	216–219	77AJC2515						
(19)	165–167	78AJC2505						
(20)	140–141	73TL1677						

The proton and nitrogen NMR and UV spectral data for (**8**) are given in Tables 1, 3 and 4 and a mass spectral analysis of (**9**) and (**10**) has been reported in the literature ⟨74CS(6)222⟩. The triazole proton NMR spectral peaks for (**11**) and (**12**) are given in Table 5 and the melting points for this series of compounds are presented in Table 6.

4.15.2.3 [1,2,4]Triazolopyridazines and [1,2,3]Triazolopyridazine

Little proton NMR spectral information is available on neutral compounds of this series ⟨B-73NMR288, B-73NMR302⟩. However, there are data available on the 5-oxides of (**13**) and on its N-methyl derivatives ⟨69JHC559⟩. In fact, this latter study had been quite revealing in determining the site of methylation in various [4,3-b] fused compounds. On the other hand, considerable proton NMR spectral data for the (**14**) series have been reported and the chemical shifts for (**14**) itself are given in Table 1. The ^{13}C NMR spectral data for both (**13**) and (**14**) are in Table 2 and more can be found in the literature ⟨76JHC1057⟩ for various derivatives. In the mass spectral fragmentation ⟨70JHC639⟩ of [1,2,4]triazolo[4,3-b]pyridazines, the molecular ion is the most abundant ion. Beyond this, however, the major fragments are not consistent from one derivative to the next and depend on the substituents present on the ring. Electron density calculations have been performed on (**13**) and some of its derivatives and these results confirm those found chemically upon methylation (see reactivity section) ⟨69JHC559⟩. The crystal structures of (**13**) and (**14**) have been determined and indicate that these molecules are planar with the N(5)—C(6) and C(7)—C(8) bonds of (**13**) being considerably alkenic and those same two bonds of (**14**) being aromatic ⟨78AX1136⟩. This interpretation is supported by the reactivity found for these compounds. The UV and melting point data for the ring systems of this series are given in Tables 4 and 6 with the triazolo proton NMR absorption for (**17**) given in Table 5. The proton NMR spectral data for derivatives of (**15**) have also been described ⟨B-73NMR288, B-73NMR302⟩.

4.15.2.4 [1,2,4]Triazolopyrimidines

The proton NMR spectral data for (**18**)–(**21**) are given in Table 1 and extensive analysis of this data for (**20**) has been used to evaluate its reactivities and electronic structure ⟨64CPB204, 72JPR515⟩ which correlate well with its charge distributions calculated by the simple MO methods. Another report has appeared which correlates the calculated electronic spectra with the electronic structure of derivatives of (**18**) and (**20**) ⟨72T5779⟩. Vysotskii and Zemskii ⟨80KGS984⟩ have used a quantum chemical interpretation to analyze the Dimroth-like rearrangements in (**18**)–(**21**) and Guerret, Jacquier and Maury ⟨71JHC643⟩ have compared the electron densities of a number of polyazaindolizines as a means of rationalizing the mechanisms of these rearrangements in different systems.

The ^{13}C NMR spectral data for representatives of the (**20**) and (**23**) series are given in Table 2 and the UV absorption maxima for (**18**)–(**22**) and (**24**) are presented in Table 4. The UV data have been very useful in distinguishing [1,2,4]triazolo[4,3-a]pyrimidines from their [1,5-a] isomers ⟨68T2839⟩. An X-ray structural analysis has been carried out on 2-amino-7-methyl-5-(1-propyl)-[1,2,4]triazolo[1,5-c]pyrimidine ⟨63JCS5642⟩ and this showed the two heterocyclic rings to be almost planar with the two planes inclined at an angle of 6° to each other. The bond lengths were reported to be considerably less than normal single bond lengths and they appeared more as conjugated double bonds and single bonds.

The melting point data for the ring systems of this series are given in Table 6 and the proton NMR chemical shift assignments for the triazole protons in (**22**) and (**24**) are given in Table 5.

4.15.2.5 [1,2,3]Triazolopyrimidines

The diazo character of fused triazoles that was described in Section 4.15.2.2 is also of structural significance for this series of compounds ⟨74JCS(P1)2030⟩ and is discussed in Section 4.15.3.6.1. The NMR and UV spectral data and melting points for representatives of this series are given in Tables 1, 2, 4 and 6. The mass spectral fragmentations for (**27**) have been reported with the molecular ion showing only a 47% relative intensity while the ion with *m/e* 65 was the most intense ⟨78JHC1041⟩. Molecular orbital calculations have been conducted on (**28**) and found to correlate well with its observed properties ⟨69CJC1129, 69CR(D)(268)2958⟩.

4.15.2.6 [1,2,4]Triazolopyrazines and [1,2,3]Triazolopyrazines

The proton NMR and UV spectral data and melting points for representatives of this series are given in Tables 1, 4 and 6. The ^{13}C NMR spectral data for (**31**) and (**32**) are worthy of special note since they afford a useful method for distinguishing between the isomers ⟨80JCS(P1)506⟩. The chemical shift for C-2 in the ^{13}C NMR spectrum of (**32**) was greater than the C-3 for the corresponding (**31**). This is also true for derivatives of these two ring systems. The ^{1}H NMR data can also be used for structural diagnosis since the relative positions of the signals due to H-2 and H-3 in (**32**) and (**31**), respectively, were reported ⟨80JCS(P1)506⟩ to be the reverse of those observed for the corresponding carbon atoms in the ^{13}C spectra. This was assumed ⟨80JCS(P1)506⟩ to be due to shielding by lone pairs of electrons on the nearby basic nitrogen atoms which would be the greatest for (**32**).

The mass spectral fragmentations for representatives of the (**31**) ⟨71OMS(5)663⟩, (**32**) ⟨71OMS(5)663⟩, (**33**) ⟨78HCA1755⟩ and (**36**) ⟨78HCA1755⟩ ring systems have been described. Compound (**31**) showed decomposition from its molecular ion by loss of nitrogen while its 3-alkyl derivatives lost the corresponding alkyl cyanide from the triazole nucleus. The 3-amino compounds expelled cyanamide from the molecular ion whereas a phenyl substituent produced a 1,2-phenyl migration. In the case of the parent (**32**), hydrogen cyanide was lost from the molecular ion and alkyl cyanides were eliminated from the molecular ions of the 2-alkyl derivatives.

Quantum chemical calculations have been performed on (**31**) and (**32**) to enlighten their propensity to undergo Dimroth-like rearrangements ⟨80KGS984⟩. These calculations show satisfactory correlations with what actually occurs.

4.15.2.7 [1,2,4]Triazolotriazines

Most of the structural data for this series of compounds is on substituted, often polysubstituted, derivatives and is of little value to a chapter such as this. Some NMR spectral data and melting point values for some simple representatives of the triazolotriazines are shown in Tables 1 and 6. Further NMR spectral data in the 1,2,4-triazine class are reported for the [4,3-*b*] ⟨77JOC1018⟩, [3,4-*c*] ⟨77JOC1018⟩, [4,3-*d*] ⟨69BSF3670, 81T4353⟩, [3,4-*f*] ⟨79JHC555⟩, [1,5-*b*] ⟨77JOC1018⟩, [5,1-*c*] ⟨76JCS(P1)1492⟩ and [1,5-*d*] ⟨69BSF3670⟩ systems and in the 1,3,5-triazine class they are available for the [4,3-*a*] ⟨70T3357⟩ and [1,5-*a*] ⟨74JOC2143⟩ derivatives. Many of these same references contain mass spectral and UV data on various substituted compounds of this series and there is a report ⟨70T3357⟩ on the calculated electron densities for the [1,2,4]triazolo[4,3-*a*][1,3,5]triazines.

4.15.2.8 Tetrazolopyridines and Related Systems

Tetrazoles fused to six-membered nitrogen heterocycles wherein the two rings share a nitrogen are known to exist in equilibrium with the azido tautomer (**C**) ⇌ (**D**) ⟨73S123⟩. The nucleophilicity of the common nitrogen that is part of the six-membered ring in the equilibrium is responsible for determining whether the tetrazolo or azido forms predominate in the equilibrium. If the atom is poorly nucleophilic, then the azido form is preferred, for example 1,3,5-triazines exist entirely in the azido tautomer ⟨73S123⟩; if it is highly nucleophilic (as in pyridine *versus* the diazines), then the fused tetrazole predominates. From this, it is not surprising that electron-withdrawing groups on the azine favor the azido form while electron-donating groups encourage tetrazolo fusion. Protonation of the heterocycle leads to the azido tautomer. Other factors which influence the isomerization are solvent polarity (polar solvents shift the equilibrium toward the tetrazole and less polar solvents shift it toward the azido form) and temperature (raising the temperature encourages azido formation at the expense of tetrazole) ⟨73S123⟩. This latter parameter supports the fact that the tetrazolo to azido shift is endothermic.

4.15.2.8.1 Tetrazolo[1,5-a]pyridine

Tetrazolo[1,5-a]pyridines generally exist in the tetrazolo form. As exceptions, the 6-nitro derivative exists as an equilibrium mixture of both forms whereas the 5-chloro compound prefers the azido tautomer ⟨73S123, 71T5121, 70T4969⟩. The ^{1}H and ^{14}N NMR and UV spectral data and melting point for (**46**) are given in Tables 1, 3, 4 and 6, and molecular orbital generated data ⟨68MI41500⟩ and photoelectron spectral data ⟨80JHC689⟩ on (**46**) are also available. Spectral measurements on (**48**) and (**49**) are described elsewhere ⟨76HCA259, 76AJC799, 80AJC499⟩ while their melting points and that of (**47**) are reported in Table 6.

4.15.2.8.2 Tetrazolo[1,5-b]pyridazine

Compound (**50**) and its derivatives generally exist only in the tetrazolo form even in trifluoroacetic acid ⟨73S123⟩. Electron-withdrawing groups destabilize the tetrazolo tautomer as evidenced by the preference for 3-azidopyridazine 1-oxide ⟨63CPB348⟩ and 3-azido-6-nitropyridazine 1-oxide ⟨70JOC2478, 70T4969⟩.

The mass spectral fragmentation pattern for (**50**) ⟨70JHC639⟩ and its X-ray crystal structure ⟨78AX1136⟩ have been reported. The most abundant species in the mass spectrum of (**50**) is the $[M-56]$ ion which corresponds to the loss of four nitrogens. The X-ray picture of (**50**) indicates it is a planar molecule with a great deal of alkenic character associated with the double bonds of the pyridazine ring.

4.15.2.8.3 Tetrazolo[1,5-a]pyrimidine and tetrazolo[1,5-c]pyrimidine

The parent [1,5-a] system has been reported to prefer the tetrazolo tautomer in the solid state and an equilibrium mixture of azido and tetrazolo in solution, with the azido form being exclusively present in chloroform. In comparing (**53**) and (**54**) the tetrazolo tautomer seems more stable in the former case ⟨73S123⟩. The [1,5-a]pyrimidines with electron-donating groups at C-5 (and C-7) showed no tendency to tautomerize to azides whereas those with electron-withdrawing groups at C-5 (and C-7) favored the azido form ⟨70T4969, 73S123⟩.

Tetrazolo[1,5-c]pyrimidines can also exist in equilibrium with 4-azidopyrimidines although, in the solid state, the parent compound (**54**) is predominantly in the tetrazolo form ⟨73S123⟩. The effect of substituents on this equilibrium can be correlated with that reported in the [1,5-a] system, that is, the azido tautomer is preferred when electron-withdrawing groups are in positions 5 and 7 of the tetrazolo[1,5-c]pyrimidine and electron-donating groups at C-8 favor the tetrazole product ⟨70T4969, 73S123⟩.

Tetrazolo[1,5-a]quinazolines exist almost entirely as tetrazoles whereas the [1,5-c]quinazolines are in equilibrium with the azido tautomer ⟨73S123⟩.

Other structural data for compounds of this class are in Tables 1 and 6 and can be found throughout the literature cited in Sections 4.15.3.11 and 4.15.4.11.

4.15.2.8.4 Tetrazolo[1,5-a]pyrazine

In the solid state (**57**) and its derivatives prefer the tetrazolo form whereas in solution an equilibrium exists ⟨73S123⟩. As in the other systems described here, the substituent on the pyrazine ring can greatly influence the position of the tetrazolo–azido equilibrium ⟨81MI41500⟩.

Tables 1, 4 and 6 have additional data for (**57**).

4.15.2.8.5 *Tetrazolo[1,5-b][1,2,4]triazine and tetrazolo[5,1-f][1,2,4]triazine*

Compounds derived from (59) exist exclusively in the tetrazolo tautomers and are planar and aromatic upon NMR spectral analysis ⟨76JOC2860⟩. In the [5,1-*f*] series the azido form predominates in the solid state and will remain in this tautomer for two months. However, once placed in solution, the azido tautomer shifts exclusively over to the tetrazolo tautomer. This process, which has an activation energy of 85.7 kJ mol^{-1}, is encouraged by the presence of a 6-amino group ⟨79JHC555⟩.

4.15.3 REACTIVITY

4.15.3.1 [1,2,4]Triazolopyridines and their Fused Benzo Derivatives

4.15.3.1.1 *[1,2,4]Triazolo[4,3-a]pyridine*

In order to determine the relative stability of [1,2,4]triazolo[4,3-*a*]pyridines in contrast to other bridgehead nitrogen heterocycles (namely, indolizines) oxidation of (1) with potassium permanganate at room temperature yielded 1,2,4-triazole-3-carboxylic acid (80) ⟨66JOC265⟩. Since pyrazolo[1,5-*a*]pyridine was similarly oxidized to pyrazole-3-carboxylic acid, it has been suggested ⟨66JOC265⟩ that the additional nitrogen atoms in the five-membered rings (relative to indolizine) confer greater stability on that portion of the bicyclic system.

The greater stability of the triazole rings was further evidenced by the rearrangement of (1) to the isomeric [1,2,4]triazolo[1,5-*a*]pyridine (2) in warm sodium hydroxide solution ⟨66JOC265⟩. Similarly, 3-methyl- and 3-amino-[1,2,4]triazolo[4,3-*a*]pyridine underwent rearrangement to 2-methyl- and 2-amino-[1,2,4]triazolo[1,5-*a*]pyridine (81 and 82) as shown in Scheme 1. Deamination of (82) *via* its diazonium salt and hypophosphorus acid produced (2). Treatment of the diazonium salt with cuprous chloride yielded 2-chloro-[1,2,4]triazolo[1,5-*a*]pyridine, the same product obtained from the basic treatment of 3-chloro-[1,2,4]triazolo[4,3-*a*]pyridine. The 3-methylthio-[4,3-*a*] system was rearranged to the 2-methylthio-[1,5-*a*] product but the 3-mercapto- and 3-hydroxy-[4,3-*a*] compounds decomposed, without rearrangement, under the appropriate basic conditions. It is interesting to note that 3-amino-[1,2,4]triazolo[4,3-*a*]pyridine underwent rearrangement and not hydrolysis in this process, contrary to what might have been predicted if amino–imino tautomerism had been operative. The rearrangement of these derivatives could not be effected with sodium ethoxide or dilute acid ⟨66JOC265⟩.

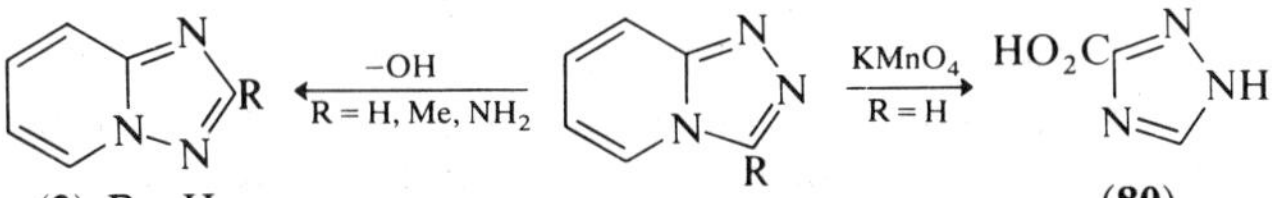

(2) R = H
(81) R = Me
(82) R = NH$_2$

(80)

Scheme 1

The rearrangement involves ⟨66JOC265, 71JHC643⟩ attack of the hydroxide ion at the C-5 position of the [1,2,4]triazolo[4,3-*a*]pyridine nucleus to give intermediate (83) which, subsequently, undergoes ring closure at the triazole N-1 which is more nucleophilic than N-4 in the 1,2,4-triazole anion (Scheme 2). This rearrangement, and a variety of others like it (*vide infra*), is related to the Dimroth rearrangement ⟨70JHC1019⟩ and was expected to go more readily with the appearance of an electron-withdrawing group on the pyridine and to be less likely to occur with an electron-donating group on this ring. To test this

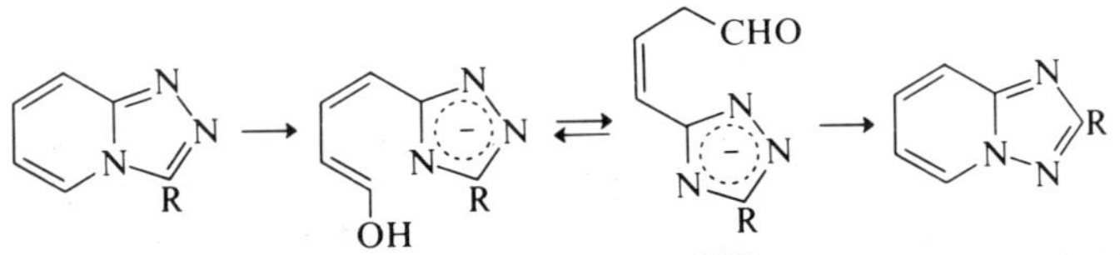

(83)

Scheme 2

hypothesis, 8-nitro-[1,2,4]triazolo[4,3-a]pyridine (**84**) was prepared and reduced catalytically to the 8-amino derivative (**85**) ⟨70JHC1019⟩. Compound (**84**), upon treatment with 10% sodium hydroxide solution for 30 minutes, or fusion or gently warming in formic acid for 1 hour or at room temperature for a longer period of time, became 8-nitro-[1,2,4]triazolo[1,5-a]pyridine (**86**). On the other hand, (**85**) required treatment with 10% sodium hydroxide solution for 48 hours to rearrange to (**87**) while formic acid produced only the 8-formamido derivative. No isomerization of (**85**) occurred upon heating.

(**84**) R = NO_2
(**85**) R = NH_2

(**86**) R = NO_2
(**87**) R = NH_2

As a result of this observation, reacting 2-hydrazino-3-nitropyridine with aliphatic acids (a typical route to [1,2,4]triazolo[4,3-a]pyridines, see later) led directly to various 8-nitro-[1,2,4]triazolo[1,5-a]pyridines. As an aside, this provides a very useful way into the 8-substituted [1,5-a] series since the 8-nitro group of (**86**) is readily reduced to the 8-amino derivative which, in turn, is susceptible to the usual diazonium conversions ⟨70JHC1019⟩. It can be assumed that this could also be applied to other 8-nitro derivatives.

The above mechanistic interpretation was supported by investigating the preparation of 6-nitro-[1,2,4]triazolo[4,3-a]pyridines (**88a**) wherein the electron-withdrawing group is adjacent to the C-5 site of the nucleophilic attack that initiates rearrangement. In this case only the [1,5-a] product (**88b**) was detectable upon ring closure of 2-hydrazino-5-nitropyridine with carboxylic acids or ortho esters. Further proof of this facile rearrangement process was the production of 2-methyl-8-nitro[1,2,4]triazolo[1,5-a]pyridine upon nitration of 3-methyl-[1,2,4]triazolo[4,3-a]pyridine using potassium nitrate–concentrated sulfuric acid at 120 °C for 10 hours ⟨70JHC1019⟩.

(**88a**)   (**88b**)

The rearrangement also occurs when 2-hydrazino-5-nitropyridine is treated with cyanogen bromide and carbon disulfide to result in 2-amino-6-nitro-[1,2,4]triazolo[1,5-a]pyridine and 6-nitro-[1,2,4]triazolo[1,5-a]pyridine-2-thiol, respectively ⟨70JHC1019⟩. The rearrangement has also been used ⟨76JOC3124⟩ for the synthesis of nucleosides of the [1,5-a] system.

This isomerization has also been reported to occur in other fused 1,2,4-triazoles ⟨71JHC643⟩ and the results from these studies support the ring opening/ring closing Dimroth type process for this occurrence (see Table 7). The extreme ease of rearrangement in the pyrimidine and 1,2,4-triazine systems can be attributed to the increase in electron deficiency at the C-5 center as a result of the additional nitrogen or two in the pyridine ring. This rearrangement is driven by the greater stability of the [1,5-a] system, a fact that has been substantiated by CNDO calculations ⟨71JHC643⟩. Further application of this process will be made as the reactivities of the other ring systems are described.

Bromination of [1,2,4]triazolo[4,3-a]pyridine in methanol gave the 3-bromo derivative (**89**), the same product obtained from the Sandmeyer reaction on 3-amino-[1,2,4]triazolo[4,3-a]pyridine ⟨66JOC265⟩. Blockage of the 3-position of (**1**) gave no nuclear

Table 7 General Survey of Dimroth-type Rearrangements in the Fused Triazoles

Starting compound	Product	Ref.
(**1**)	(**2**)	70JHC1019, 66JOC265
(**18**)	(**20**)	59JOC787, 60JCS1829, 66CB2237
(**19**)	(**21**)	63JCS5642, 65JCS3357
(**31**)	(**32**)	68JHC485, 66JCS2038, 70CC1524
(**38**)	(**42**)	69BSF2492
(**39**)	(**43**)	69BSF3670

bromination. The bromine in (**89**) would not form a Grignard reagent and could not be displaced by nucleophiles.

(**89**)   (**90**)

The ring system (**1**) did not undergo any other electrophilic substitution reactions (nitration, sulfonation, Friedel–Crafts, Vilsmeier formylation), was inert to sodamide, could not be metalated with butyllithium and could not be readily oxidized with hydrogen peroxide to an *N*-oxide ⟨66JOC265⟩. Compound (**1**) could not be reduced at atmospheric pressure by hydrogen but at 3 atm using a palladium–charcoal catalyst the 5,6,7,8-tetrahydro derivative (**90**) was obtained ⟨75G1291⟩.

As with the [1,5-*a*] analog mentioned above, 3-amino-[1,2,4]triazolo[4,3-*a*]pyridine underwent typical diazonium reactions. It could be acylated and transformed into Schiff bases. The 3-amino group had little influence as an electron-donating group for stimulating electrophilic substitution of the fused triazole nucleus ⟨66JOC265⟩.

3-Hydroxy-[1,2,4]triazolo[4,3-*a*]pyridine (**91**) was reluctant to react with the common hydroxylic chlorinating agents, giving at best, the 3-chloro compound in only 15% yield with phosphorus oxychloride–dimethylaniline. Methylation of (**91**) produced 2-methyl-[1,2,4]triazolo[4,3-*a*]pyrid-3-one (**92**) whereas methylation of the 3-thiol (**93**) resulted in methyl [1,2,4]triazolo[4,3-*a*]pyrid-3-yl sulfide (**94**), a molecule also resistant to nucleophilic substitution ⟨66JOC265⟩. Potassium ferricyanide oxidation of (**93**) resulted in the corresponding disulfide.

(**91**)   (**92**)   (**93**) R = H
                      (**94**) R = Me

Except for 3-amino-[1,2,4]triazolo[4,3-*a*]pyridine (**95**), protonation and quaternization of (**1**) and its derivatives occurred primarily at N-1 (**96**) ⟨68JHC29⟩. In some cases the *N*(2)-methyl isomer was obtained and it underwent thermal rearrangement to the N-1 analog ⟨68JHC29⟩. Molecular orbital calculations confirm this by showing N-1 to be the more basic site. In the case of (**95**), N-2 is the most basic site as a consequence of the amidine system present and reaction occurs there ⟨66JOC265, 65JCS2778⟩. 1-Methyl-[1,2,4]triazolo[4,3-*a*]pyridinium iodide (**97**) can be reduced with sodium borohydride to the unstable (**98**) ⟨66JOC265⟩. *N*-Amination of (**1**) with *O-p*-tolylsulfonylhydroxylamine gives the N-1 ammonium salt (**99**) ⟨76JCS(P1)367⟩.

(**95**)   (**96**)   (**97**)

(**98**)   (**99**)   (**100**) R = CO₂H
                       (**101**) R = CONH₂

The 3-methyl group of 3-methyl-[1,2,4]triazolo[4,3-*a*]pyridine was inert to metalation and could not be condensed with benzaldehyde ⟨66JOC265⟩. Attempts to make the 3-carboxylic acid (**100**) by hydrolysis of the 3-carboxamide (**101**) led instead to (**1**) by an apparent decarboxylation of the intermediate (**100**) ⟨66JOC251⟩.

In addition to those 6- and 8-substituted [4,3-*a*] systems mentioned previously in the Dimroth rearrangement discussion, a variety of 5-substituted derivatives has been described ⟨78JHC439⟩ from ring closure of 6-chloro-2-hydrazinopyridine (see synthesis section). A useful material obtained from this investigation was 5-chloro-[1,2,4]triazolo[4,3-*a*]pyridine (**102**) which was susceptible to nucleophilic substitution by strong nucleophiles (sodium ethoxide **103** and potassium thioethoxide **104**) but not weak ones (hydroxide, water, ammonia). On the other hand, oxidation of (**104**) with *m*-chloroperbenzoic acid to (**105**) produced a more labile C-5 substituent which upon reaction with potassium hydroxide solution produced (**106**) (shown in its predominant keto form) which was ribosylated to (**107**) ⟨78UP41500⟩. Compound (**106**) could not be prepared from ring closure of 2-hydrazinopyridin-6-ol, and (**105**) was found to be inert towards ammonia. There was little steric interference to substitution at C-5 of (**102**) by a *peri* C-3 substituent since (**108**) underwent reaction with thioethoxide to produce (**109**). Compound (**109**), however, could not be oxidized to the sulfone analogous to (**105**), apparently due to steric incompatability for the new oxygen atoms at C-5.

(**102**) R = Cl
(**103**) R = OEt
(**104**) R = SEt
(**105**) R = SO_2Et

(**106**)

(**107**)

(**108**) R = Cl
(**109**) R = SEt

Reaction of [1,2,4]triazolo[4,3-*a*]pyridine (**1**) with tetracyanoethylene in methanol has been reported to give (**110**) ⟨81IJC(B)10⟩. With dimethyl acetylenedicarboxylate in boiling benzene or toluene, the ring-fused system (**1**) gave three products, the cyanopyridopyrimidinone (**111**; 20%), the pyridylpyrazole derivative (**111a**; 20%), and the 1:2 adduct (**112**; 1%) of the initial bicyclic system. In methanol solution (**111**) was formed exclusively in 70% yield ⟨82CC1280, 81IJC(B)10⟩. The formation of these three products may be rationalized in terms of an initial Michael addition to DMAD of the triazolopyridine (**1**) *via* its two nucleophilic centers C-3 and N-1 and a subsequent series of rearrangements.

(**110**)

(**111**)

(**111a**)

(**112**)

(**113**)

Definitive evidence for the alkenic nature of the pyridine dienic unit arose when several [1,2,4]triazolo[4,3-*a*]pyridines were found to undergo photodimerization ⟨77T1247⟩. The products from this process resulted from dimerization between the 5,6-double bond in one molecule with the 7,8-double bond in another to give (**113**).

4.15.3.1.2 [1,2,4]Triazolo[1,5-a]pyridine

Beyond those reactions mentioned briefly in the previous section, for example reduction of (**86**) to (**87**), little is known about the reactivity of [1,2,4]triazolo[1,5-*a*]pyridine (**2**). It cannot be brominated or nitrated and is stable to acid, base and reducing agents. It shows similar triazole ring stability to (**1**) by its oxidation with potassium permanganate to (**80**) ⟨66JOC260⟩. Most of the reactivity data on this ring system deals with 2-methyl-[1,2,4]triazolo[1,5-*a*]pyridine (**114**). For example, nitration of (**114**) with potassium nitrate in concentrated sulfuric acid gave (**115**) and (**116**). Catalytic reduction of these two compounds produced (**117**) and (**118**) both of which were susceptible to the Sandmeyer procedure resulting in the bromo derivatives (**119**) and (**120**) ⟨66CPB523⟩. Amination of the

[1,5-*a*] system with *O-p*-tolylsulfonylhydroxylamine gave the N-1 amino derivative ⟨76JCS(P1)367⟩.

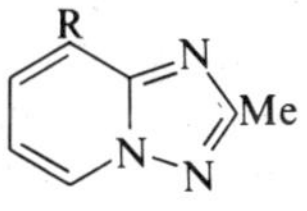

(114) R = H
(115) R = NO₂
(117) R = NH₂
(119) R = Br

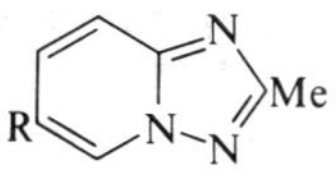

(116) R = NO₂
(118) R = NH₂
(120) R = Br

4.15.3.1.3 [1,2,4]Triazoloquinolines and [1,2,4]triazoloisoquinolines

In this series of compounds (**5**) and its derivatives have been the most extensively studied. The parent compound is stable towards mineral acid but in the presence of 4.5N potassium hydroxide solution [1,2,4]triazolo[5,1-*a*]isoquinoline (**7**) was obtained. The mechanism for this conversion is similar to that for isomerization of [1,2,4]triazolo[4,3-*a*]pyridines described earlier and is given in Scheme 3 ⟨71RTC1225⟩. Oxidation of both (**5**) and (**7**) with potassium permanganate give 3-(2-carboxyphenyl)-[1,2,4]triazole (**121**) ⟨71CB3925⟩.

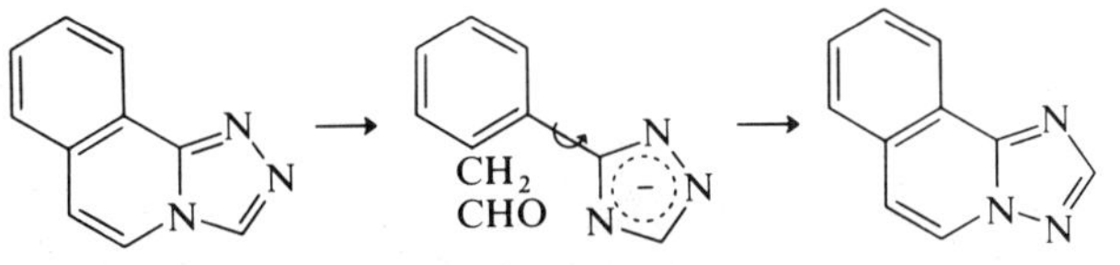

Scheme 3

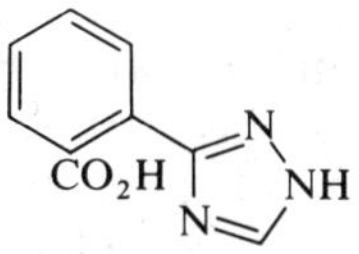

(121)

Treatment of (**5**) with bromine in methanol or acetic acid or with *N*-bromosuccinimide gave the 3-bromo derivative (**122**) which with red phosphorus/hydriodic acid was reductively dehalogenated back to (**5**). Chlorination of (**5**) with sodium hypochlorite yielded (**123**). Both (**122**) and (**123**) could also be obtained from (**124**) and phosphorus tribromide and phosphorus oxychloride, respectively. In contrast to the [4,3-*a*]pyridine series, the 3-bromo substituent of (**122**) could be displaced by nitrogen nucleophiles ⟨66JHC158⟩. Subjecting (**5**) to nitration (nitric acid/concentrated sulfuric acid) resulted in 7-nitro-[1,2,4]triazolo[3,4-*a*]isoquinoline (**125**) ⟨70IJC894, 75CB3762⟩. On the other hand, resembling the [4,3-*a*]pyridines, (**5**) was inert to Vilsmeier formylation, *N*-oxide formation, and sodamide addition ⟨70IJC894⟩.

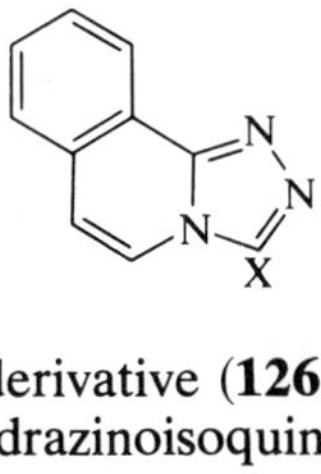

(122) X = Br
(123) X = Cl
(124) X = OH
(126) X = NH₂
(127) X = OPh
(133) X = OMe
(134) X = SH
(135) X = SMe

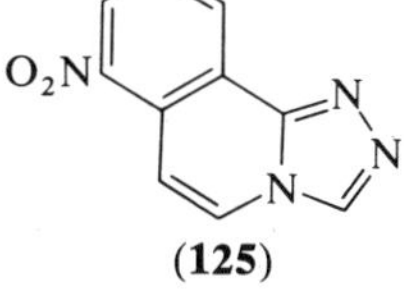

(125)

The 3-amino derivative (**126**) could not be easily synthesized by the customary method of reacting 1-hydrazinoisoquinoline with cyanogen bromide but reaction of (**127**) with ammonium acetate in a sealed tube did form (**126**) as did ring closure of 1-hydrazinoisoquinoline with *S*-methylisothiourea ⟨71CB3947⟩. It was possible to perform the standard diazonium reactions with (**126**) ⟨70IJC894⟩.

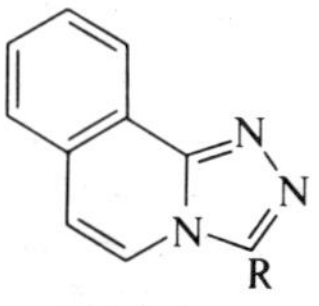

(128) R = Li
(129) R = I
(130) R = CO₂Li
(131) R = CH(OH)Me
(132) R = CO₂Me

Additional C-3 derivatives of (**5**) have arisen from the C-3 lithio derivative (**128**) (obtainable from (**122**) and butyllithium). For example, (**128**) reacts with iodine to give

(129), with carbon dioxide to produce (130) and with acetaldehyde to result in the hydroxyethyl compound (131) ⟨71CB3940⟩. Reaction of compound (130) with diazomethane provides a route to the methyl ester (132) ⟨71CB3940⟩. Also, the C-3 proton of (5) can be exchanged for deuterium in acid and basic solution ⟨70CB3817⟩.

Methylation of (5) produced ⟨75CB3762⟩ the N-1 and N-2 isomeric methyl derivatives while methylation of (124), known to exist in the lactam tautomer ⟨66JHC158⟩ under a variety of conditions, yielded the *N*(2)-methyl product ⟨71CB3955⟩. The 3-*O*-methyl isomer of this latter product (133) was available *via* the reaction of (123) with sodium methoxide. Refluxing (124) with phosphorus pentasulfide in xylene resulted in (134) which could not be dethiated with Raney nickel. However, dilute nitric acid followed by aqueous sodium hydroxide solution did transform (134) into (5). Methylation of (134) yielded (135) ⟨66JHC158⟩ whereas methylation of (126) produced the N-2 product (136) ⟨71CB3947⟩.

(136)

The fact that the C(5)—C(6) bond of (5) possesses a great deal of alkenic character has been confirmed by HMO calculations and the ease with which halogenations, hydrogenations and oxidations occur at this center ⟨71CB3925⟩. In addition, (5) undergoes photodimerization *via* a [2+2] cycloaddition similar to that described earlier for obtaining (113) ⟨77T1263⟩. Compound (3) also demonstrated a dimerization ability ⟨77T1263⟩.

[1,2,4]Triazolo[4,3-*b*]isoquinoline (4) is very unstable and, consequently, very little has been done with this ring system. On the other hand, the isomeric *o*-quinoid system [1,2,4]triazolo[1,5-*b*]isoquinoline (6) seems slightly more stable due to the contributing resonance structures (137) but its chemical reactivity remains to be explored ⟨77H(6)949⟩. Its 2-methyl derivative does undergo a Diels–Alder reaction with maleimide derivatives ⟨77H(6)949⟩ that involve the C-5/C-10 positions (see 137 for numbering).

(137)

An interesting result occurred when [1,2,4]triazolo[5,1-*a*]isoquinoline (7) was treated with butyllithium. Instead of obtaining the 2-lithio derivative as would have been expected by comparison with (5), the 5-lithio product (138) was isolated ⟨71CB3965⟩. Bromination of (138) gave (139) and reaction with acetaldehyde formed (140) and carbon dioxide yielded (141). Permitting (138) to warm from −70 to 20 °C resulted in 1-cyanaminoisoquinoline, *via* a sequence of steps requiring initial isomerization of (138) to the 2-lithio compound. As with (5), the C(5)—C(6) bond of (7) displayed typical alkenic properties ⟨71CB3965⟩.

(138) R = Li
(139) R = Br
(140) R = CH(OH)Me
(141) R = CO_2H

4.15.3.2 [1,2,3]Triazolopyridines and their Fused Benzo Derivatives

4.15.3.2.1 *[1,2,3]Triazolo[1,5-a]pyridine*

Compound (8), under Vilsmeier formylation and nitration (nitric acid–acetic anhydride) conditions, proceeded as might have been expected for electrophilic substitution reactions to give the 1-substituted products (142) and (143). Reduction of (143) gave predominantly the imidazopyridine (144) with a small amount of the tetrahydro derivative (145) ⟨81JCS(P1)78⟩. Heating (8) in D_2O led to the 1-deutero derivative ⟨78HCA1755⟩ *via* deuteration–deprotonation or *via* the tautomeric 2-(diazomethyl)pyridine tautomer of (8).

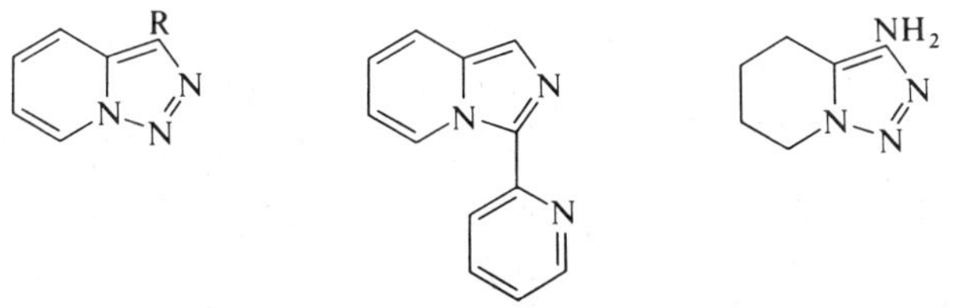

(142) R = CHO
(143) R = NO₂
(144)
(145)

On the other hand, reaction of (8) with bromine, chlorine or mercury(II) acetate led to the ring-opened products (146)–(148). This result was rationalized ⟨81JCS(P1)78⟩ as involving loss of nitrogen from either (149) or (150; Scheme 4) ⟨81JCS(P1)78⟩. If the electrophile (E) is an electron-withdrawing group (CHO, NO₂), (149) will be longer-lived and deprotonation competes successfully with the loss of nitrogen. If E is only weakly stabilizing to the diazonium intermediate (150), nucleophilic attack with loss of nitrogen is preferred. A similar explanation has been put forward ⟨81JCS(P1)78⟩ for the formation of (144) upon reduction of (143).

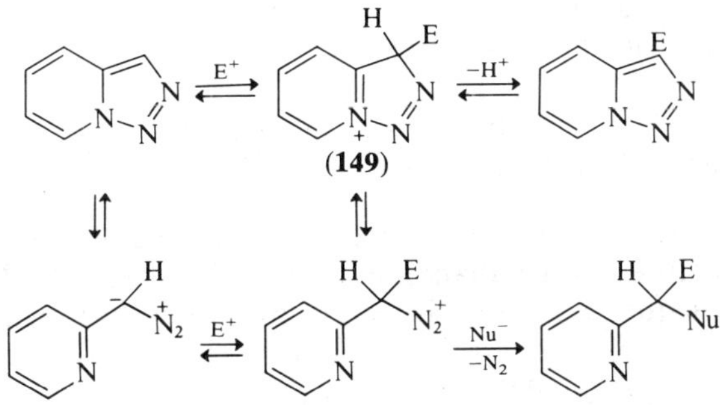

(146) E = Nu = Br
(147) E = Nu = Cl
(148) E = HgOAc, Nu = OAc

Scheme 4

Irradiation ($\lambda > 200$ nm) of (8) matrix-isolated in argon gives 1-aza-1,2,4,6-cycloheptatetraene (151) ⟨78JA282, 79RTC334⟩ via, first, 2-diazomethylpyridine and, then, triplet 2-pyridylmethylene (152). Continued irradiation leads to triplet phenylnitrene ⟨78JA6245⟩, the same product obtained from the irradiation of phenyl azide. Thermolysis of (8) has given the same results ⟨72TL2239, 74T1301, 75JA7467⟩. The existence of (152) in the photolysis of (8) has been substantiated by the reported isolation of (153)–(155) on irradiation in methanol ⟨69JHC503⟩.

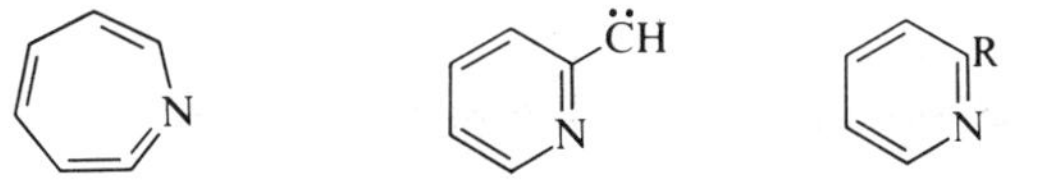

(153) R = Me
(154) R = CH₂OMe
(155) R = CH₂CH₂OH

4.15.3.2.2 [1,2,3]Triazolo[4,5-b]pyridine and [1,2,3]triazolo[4,5-c]pyridine

There is very little that can be said about the reactivity of these ring systems since most of the desired derivatives are available by diazotization of an appropriate diaminopyridine (see synthesis section) and reactivity of the bicycle is not necessary to make the various compounds. In any event, alkylation of both (9) and (10) yields all possible N-1, N-2 and N-3 isomeric products ⟨71RTC1181, 65IJC84, 76JOC1449⟩ and a chloro substituent in either system is readily displaced by nucleophiles ⟨73JOC1095, 76JOC1449⟩. A rearrangement has also been reported to occur in this series when 4,6-diamino-1H-[1,2,3]triazolo[4,5-c]pyridine (156) was treated with 12% ethanolic ammonia in a bomb at 150 °C and found to give 85% conversion to 5,7-diamino-3H-[1,2,3]triazolo[4,5-b]pyridine (157) after 120 hours. This rearrangement was believed to involve diazopyridine intermediates such as (158) and (159) and indicated that (157) was thermodynamically more stable than (156) ⟨73JOC1095⟩. No conversion of (157) into (156) was observed up to 175 °C.

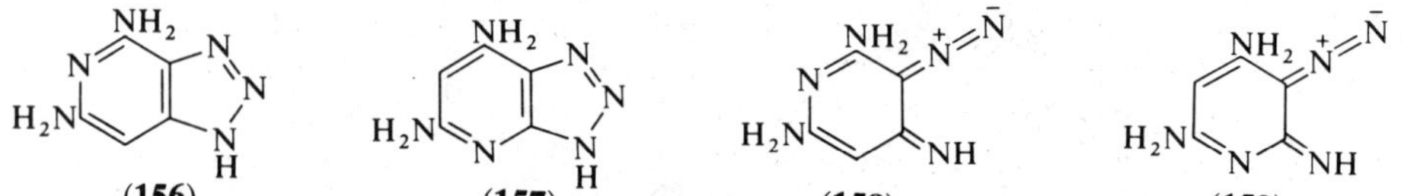

(156) (157) (158) (159)

Another interesting rearrangement in this series occurred ⟨76JOC3784⟩ when (160) was treated with sodium hydrogen sulfide in butanol and found to give the expected (161) and the [1,2,3]thiadiazolo[4,5-c]pyridine (162). Furthermore, heating (161) gave (162) whereas heating (162) in the presence of potassium carbonate gave (161).

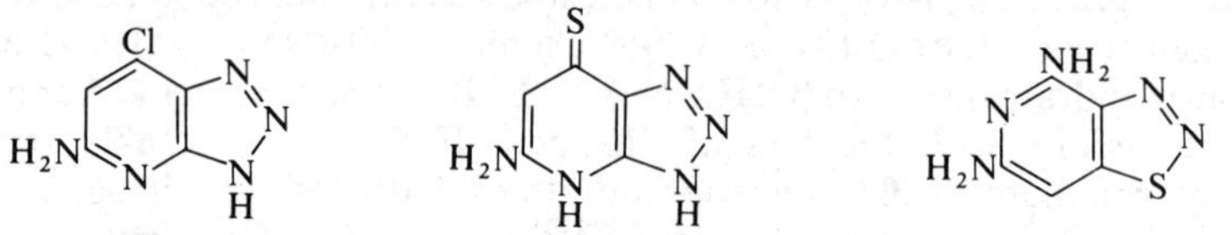

(160) (161) (162)

Pyrolysis of (9) and (10) at 500 °C gave 2-cyanopyrrole and 3-cyanopyrrole, respectively, which were interconvertible at 800 °C ⟨70T3965⟩.

4.15.3.2.3 [1,2,3]Triazolo[1,5-a]quinoline

Compared to many other fused 1,2,3-triazoles described herein, (11) is remarkably stable as evidenced by its inertness to irradiation and to heat ⟨72JOC2022⟩. This stabilization was rationalized as the consequence of extensive π-electron delocalization in the tricyclic system.

4.15.3.2.4 [1,2,3]Triazolo[5,1-a]isoquinoline

In contrast to (8), compound (12) is reasonably stable ⟨75CB3794⟩, undergoing hydrogenation and oxidation at the C(5)—C(6) double bond to give (163) and (164), respectively. Also, (12) undergoes base-promoted deuterium exchange at C-5 ⟨75CB3794⟩.

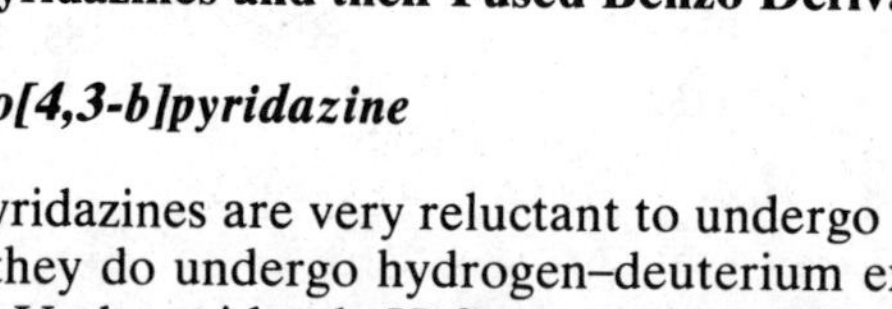

(163) (164)

4.15.3.3 [1,2,4]Triazolopyridazines and their Fused Benzo Derivatives

4.15.3.3.1 [1,2,4]Triazolo[4,3-b]pyridazine

[1,2,4]Triazolo[4,3-b]pyridazines are very reluctant to undergo electrophilic substitution reactions ⟨74MI41501⟩ but they do undergo hydrogen–deuterium exchange under acid- and base-catalyzed conditions. Under acid only H-3 was exchanged but with base the exchange reaction followed the order of H-3 > H-8 > H-7 > H-6. The mechanism for the acid-catalyzed exchange involves (165) as the initial product; (165) then rearranges to (166) via an H-3 deprotonation step ⟨74MI41501⟩. In the base process, H-3 is removed to form the C-3 anion which is then deuterated ⟨74MI41501⟩.

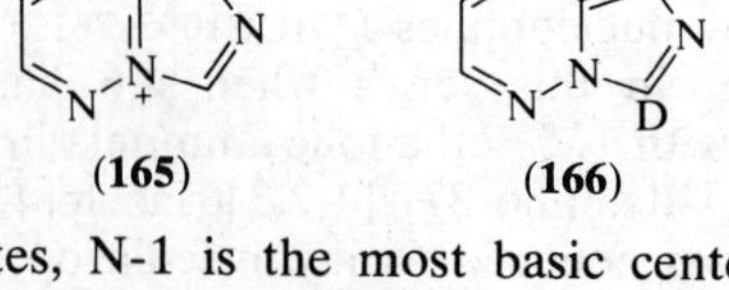

(165) (166)

As structure (165) indicates, N-1 is the most basic center in (13). This was further supported ⟨69JHC559⟩ upon methylation of (13) to give exclusively (167). However, it is interesting to note that methylation of 3-methyl- (**168**), 8-methyl- (**169**), and 3,8-dimethyl-[1,2,4]triazolo[4,3-b]pyridazines (**170**) gave mixtures of N-1 and N-2 quaternized products (**171**) and (**172**) in the ratios of 2:1, 1:2 and 1:4, respectively. From these results and MO calculations, it was concluded ⟨69JHC559⟩ that the 8-methyl group exerts a steric effect at N-1 and the 3-methyl group demonstrates an electronic effect by increasing the electron density at N-2. In the 3,8-dimethyl case, where both the steric and electronic effects favor methylation at N-2, a 4:1 ratio in favor of N-2 methylation was observed.

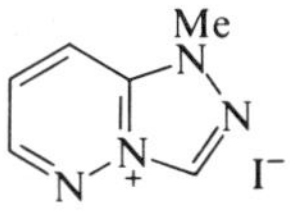

(**167**) 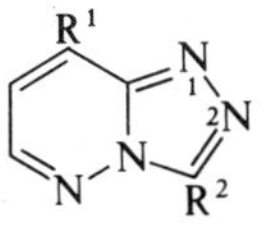   (**168**) $R^1 = H, R^2 = Me$
(**169**) $R^1 = Me, R^2 = H$
(**170**) $R^1 = R^2 = Me$
(**171**) 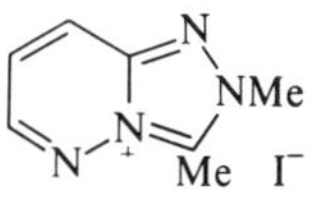   (**172**) 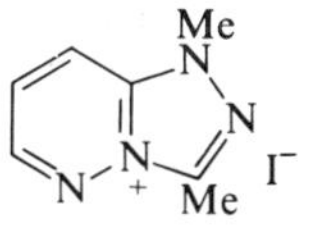

Under the reaction conditions employed, these quaternizations were irreversible and the products were stable except to heat upon which they lost the *N*-methyl group ⟨69JHC559⟩. However, upon heating these derivatives in a sealed tube, rearrangements occurred ⟨69JHC559⟩. Heating (**173**) at 255 °C for 45 minutes give a mixture (65:35) of (**173**) and (**174**). On the other hand, heating (**174**) at 215 °C for 25 minutes afforded the same products in a 33:67 ratio of (**173**) to (**174**). This thermal equilibrium is shown in Scheme 5.

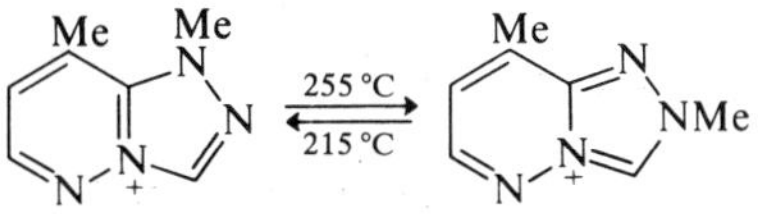

(**173**) ⇌ (**174**)
255 °C / 215 °C

Scheme 5

Free radical chlorination of the [1,2,4]triazolo[4,3-b]pyridazine ring system follows a stepwise introduction of the chlorine atoms in the following order of positions: 3 > 8 > 7 > 6. This eventually leads to the perchloro derivatives ⟨74MI41501⟩.

Oxidation of [1,2,4]triazolo[4,3-b]pyridazines with selenium dioxide has produced the 3-oxo derivatives (**175**) ⟨76JHC639⟩, compounds also available by ring closure from 3-chloropyridazines and semicarbazide ⟨71JHC415⟩. Catalytic hydrogenation ⟨75G1291⟩ of (**13**) produces the 7,8-dihydro derivative (**176**) whereas a similar hydrogenation of the 6-chloro derivative of (**13**) gave (**13**) and (**176**). On the other hand, reduction of (**13**) with sodium borohydride ⟨76JHC835⟩ gives the 5,6,7,8-tetrahydro derivative (**177**). This reduction proceeded through (**176**).

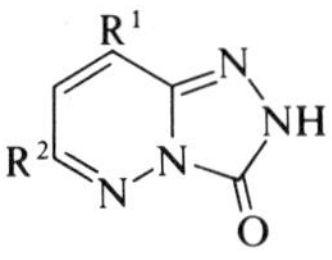

(**175**) 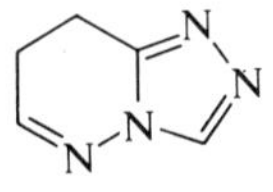   (**176**) 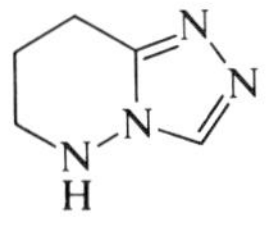   (**177**)

Hydrazinolysis of (**13**) produced a 3-(2'-pyrazolin-3'-yl)-[1,2,4]triazole (**178**) ⟨72JHC1171⟩. The mechanism for this process is given in Scheme 6 which indicates that the addition of hydrazine occurs at the N(5)—C(6) bond of (**13**) and this is followed by bond cleavage. Hydrazinolysis of (**167**) resulted in the tricyclic product (**179**) via an intramolecular cyclization ⟨74MI41501⟩. Ring opening of another quaternized derivative of (**13**) occurred when

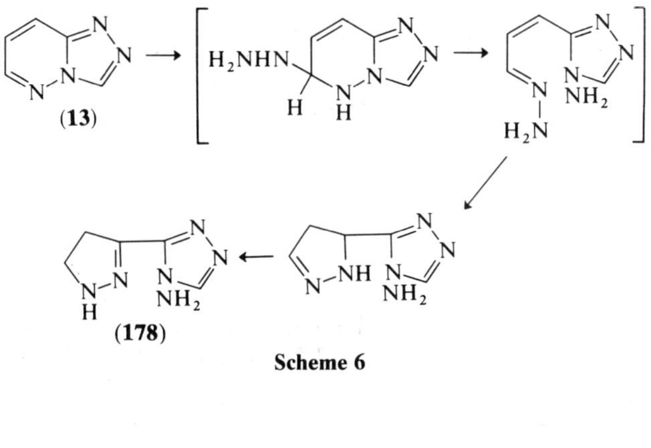

(**13**) → [H₂NHN... → ...NH₂] → (**178**)

Scheme 6

(**174**) was found to produce (**180**) and (**181**) when reacted with base ⟨74MI41501⟩. This ring opening involved breakage of the C(3)—N(4) linkage of (**174**).

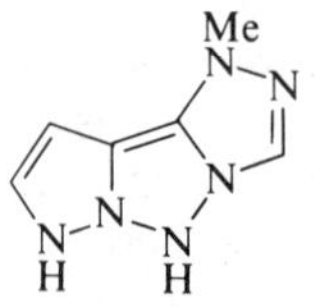

(**179**) 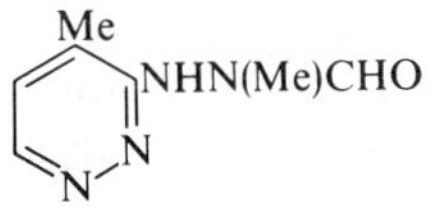 (**180**) 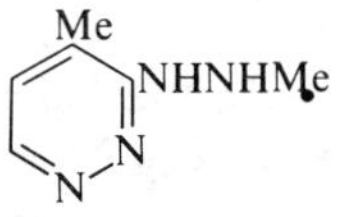 (**181**)

Oxidation of the [4,3-*b*]pyridazine ring system with potassium permanganate occurs by cleavage of the C(7)—C(8) bond. Thus, 3-phenyl-[1,2,4]triazolo[4,3-*b*]pyridazine was oxidized to 3-phenyl-1,2,4-triazole-5-carboxylic acid ⟨74MI41501⟩.

[1,2,4]Triazolo[4,3-*b*]pyridazine undergoes a number of [3 + 2] photocycloadditions with cyclic and open chain alkenes; the products result from the addition of the alkene to positions 1 and 8 of (**13**) with concurrent opening of the pyridazine ring ⟨77JHC411⟩ at N-4/N-5 as illustrated in Scheme 7. The mechanism for this reaction involves the radical addition of the alkene to the excited triazolopyridazine at its C-8 position. This is believed to be followed by a sequence of steps that include homolytic cleavage of the N(4)—N(5) bond, formation of a nitrene at the former N-5 center with concomitant ring closure by the side-chain alkyl radical (from the alkene addition) onto the former N-1 atom, and either retention (for molecules like **183**) or loss (for molecules like **182**) of hydrogen cyanide ⟨73JHC801⟩.

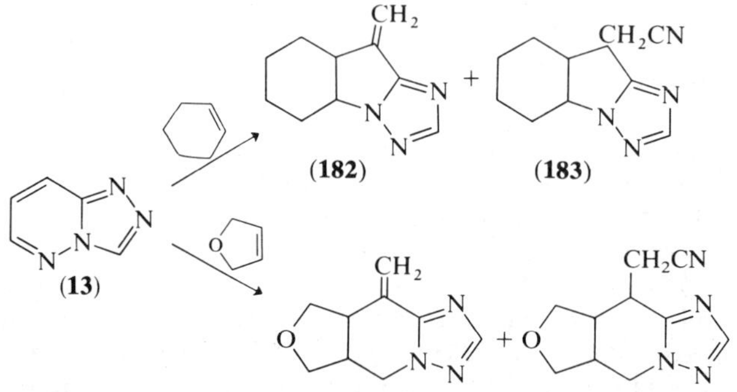

Scheme 7

Compound (**13**) has also been shown to undergo photoalkylation at its C-7 and C-8 positions when irradiated in the presence of an alcohol ⟨74JOC793, 80JOC2320⟩. These reactions proceed through the alkylated 7,8-dihydro derivatives (**184**) and (**185**) which, upon heating, undergo dehydrogenation–hydrogenolysis to (**186**) and (**187**). This is illustrated in Scheme 8. In some cases ⟨74JOC793⟩ the unsubstituted 7,8-dihydro derivative (**188**) is also obtained. The other product in this case was the aldehyde corresponding to the alcoholic side chain arising from some type of retro-aldol process involving (**184**) and (**185**).

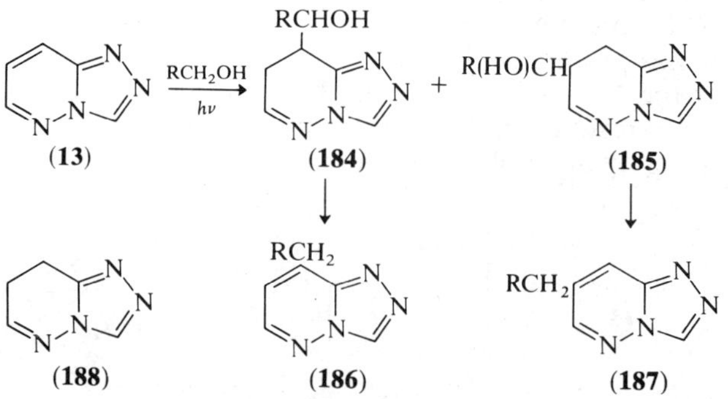

Scheme 8

4.15.3.3.2 *[1,2,4]Triazolo[3,4-a]phthalazine*

Compound (**17**) can be brominated at its 3-position with *N*-bromosuccinimide or bromine and the resulting bromo compound (**189**) is susceptible to nucleophilic substitution

⟨69JOC3221⟩. Nitration of 6-methyl-[1,2,4]triazolo[3,4-a]phthalazine in fuming nitric acid–concentrated sulfuric acid produced the 8-nitro compound (190). On the other hand, attempted nitration of (17) with copper nitrate–acetic anhydride gave only the 3-acetyl compound. The Vilsmeier formylation failed on (17) ⟨69JOC3221⟩; however, deuterium exchange occurred at C-3. The 3-carbaldehyde (see synthesis section) underwent the benzoin condensation and was oxidized to the 3-carboxylic acid.

(189) (190) (191) (192)

Methylation of (17) with methyl iodide in boiling methanol gave N-2 methylation whereas room temperature methylation gave the N-1 isomer. Reduction of (17) with lithium aluminum hydride or sodium borohydride gave the 5,6-dihydro derivative (191) ⟨69JOC3221⟩ while catalytic hydrogenation of the 6-chloro derivative of (17) produced (192) ⟨75G1291⟩, the same product isolated from treating (17) with alcoholic potassium hydroxide ⟨69JOC3221⟩. Compound (17) was stable towards hot mineral acids ⟨69JOC3221⟩.

4.15.3.3.3 [1,2,4]Triazolo[1,5-b]pyridazine

As with (13), compound (14) is very reluctant to undergo electrophilic substitution reactions ⟨74MI41501⟩ and, besides its [3 + 2] photocycloaddition with cyclohexene ⟨73JHC801⟩ and hydrazinolysis ⟨74JOC2143⟩ as with (13) (see Schemes 7 and 6, respectively), very little has been done with the unsubstituted ring system (14) ⟨74MI41501⟩. On the other hand, a great deal is known about 6-chloro-[1,2,4]triazolo[1,5-b]pyridazine (193). Its chlorine has been replaced by a large number of nucleophiles to give (194). The 6-azido compound has been prepared by nitrosation of the 6-hydrazino system and from the tetrazolo[1,5-b]pyridazine (195) by formation of the triazole ring and tetrazolo–azido isomerization ⟨74JOC2143⟩. Methylation of the 6-oxo compound (194c) afforded the 6-methoxy (194e) and N(5)-methyl compounds. Chlorination of (193) gave 6,7,8-trichloro[1,2,4]triazolo[1,5-b]pyridazine ⟨74JOC2143⟩. Hydrogenation of (193) over palladium formed the tetrahydro material (196) ⟨74JOC2143⟩. A similar set of reactions has been carried out on 6-chloro-[1,2,4]triazolo[1,5-b]pyridazine 3-oxide ⟨76JHC487⟩.

(193) X = Cl
(194a) X = NH$_2$
(194b) X = NHNH$_2$
(194c) X = OH
(194d) X = SH
(194e) X = OMe

(195) (196)

Oxidation of 2-phenyl-[1,2,4]triazolo[1,5-b]pyridazine with potassium permanganate yielded 3-phenyl-1,2,4-triazole-5-carboxylic acid ⟨74MI41501⟩.

4.15.3.4 [1,2,3]Triazolo[4,5-d]pyridazine

As mentioned in the synthesis section, most derivatives of this ring system can be prepared by the diazotization of a variety of 4,5-diaminopyridazines and reactions on (15) and its derivatives for additional representatives are not necessary. In any case, 4-methylthio-[1,2,3]triazolo[4,5-d]pyridazine (197) has proven to be a versatile intermediate due to its reactions with nucleophiles (amines, hydrazine, ammonia) ⟨69JHC93⟩ that replace the 4-methylthio group. Since the synthesis of (197) involves reactions of substituted [1,2,3]triazolo[4,5-d]pyridazines, it is described here in Scheme 9 rather than in the synthesis section. There is another report that demonstrates that the 4-methoxyl group may be more versatile than the methylthio group in the nucleophilic substitution reactions in this series ⟨70CPB1685⟩. Similar substitution reactions have been documented for the isomeric [1,2,3]triazolo[4,5-c]pyridazines ⟨64JHC247, 67JHC555⟩.

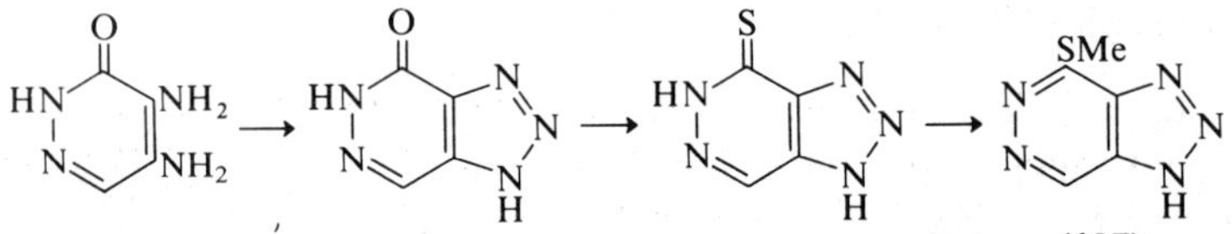

4.15.3.5 [1,2,4]Triazolopyrimidines and their Fused Benzo Derivatives

4.15.3.5.1 *[1,2,4]Triazolo[4,3-a]pyrimidine and [1,2,4]triazolo[4,3-c]pyrimidine*

The most significant reactions of ring systems (18) and (19) are the Dimroth-like rearrangements into their [1,5-*a*] and [1,5-*c*] isomers, respectively. These transformations are discussed in Sections 4.15.4.5.1 and 4.15.4.5.3.

4.15.3.5.2 *[1,2,4]Triazolo[1,5-a]pyrimidine*

Electrophilic substitution occurs in (20) at C-6 ⟨61CPB808⟩ which is also the site of acid-catalyzed deuterium exchange ⟨68JOC1087⟩; in contrast base-catalyzed deuterium exchange occurs at C-7 ⟨68JOC1087⟩. Methylation and protonation of (20) take place at N-3 to give (198a) and (198b); hydrolysis of (198a) formed 3-amino-4-methyl-1,2,4-triazole (199) ⟨68JHC691⟩. Treatment of (20) with phenyllithium gave the 7-phenyl- (200) and 5-phenyl-[1,2,4]triazolo[1,5-*a*]pyrimidines (201). This was believed to be the result of a greater delocalization of the electrons into the 5-membered ring due to resonance contributors (202) and (203) to the ground state which rendered the pyridine ring electron deficient at C-5 and C-7 ⟨72JHC1157⟩.

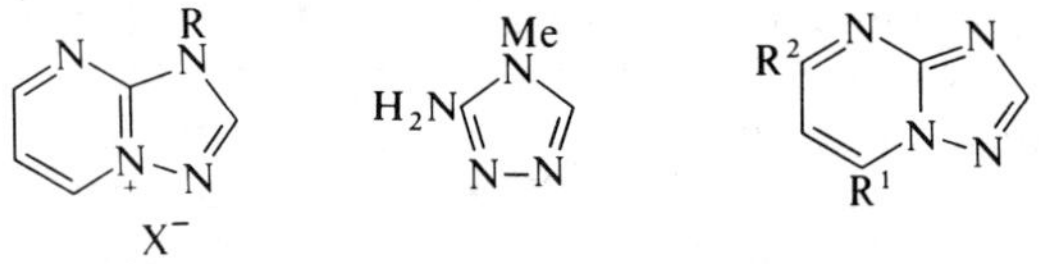

(198a) R = Me
(198b) R = H

(199)

(200) R^1 = Ph, R^2 = H
(201) R^1 = H, R^2 = Ph

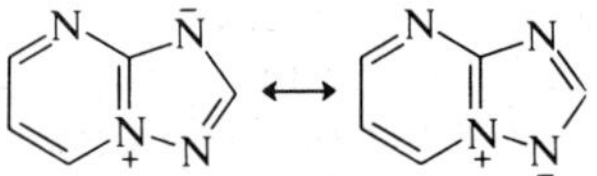

(202)   (203)

A number of 5- and 7-monosubstituted and 5,7-disubstituted [1,2,4]triazolo[1,5-*a*]pyrimidines have resulted from reactions on the 5,7-dihydroxy, -aminohydroxy and -hydroxymercapto derivatives. The hydroxy derivatives were amenable to chlorination which was followed by a variety of nucleophilic substitution reactions or hydrogenolyses. The mercapto groups were susceptible to alkylation and subsequent nucleophilic substitution ⟨61CPB801, 63CPB845⟩.

Substituted [1,2,4]triazolo[1,5-*a*]pyrimidine 1-oxides appear to be vulnerable to heat-promoted valence tautomerism into isomeric *N*-oxides *via* a nitrosoimine as the interconversion shown in Scheme 10 illustrates ⟨74CC486⟩.

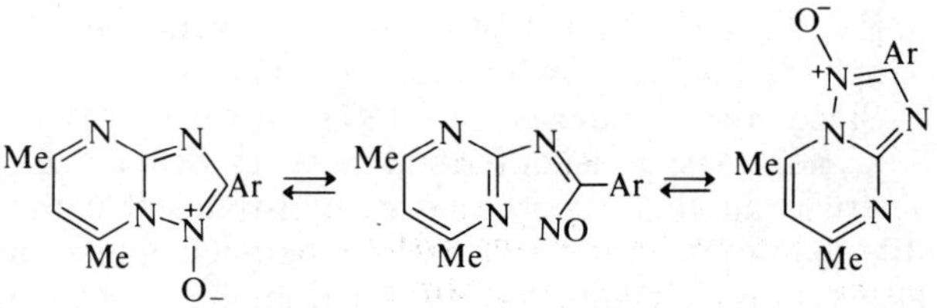

Scheme 10

4.15.3.5.3 [1,2,4]Triazolo[1,5-c]pyrimidine

Under certain hydrolytic conditions [1,2,4]triazolo[1,5-c]pyrimidines undergo ring fission to produce the same amides (**204**) as shown later in the ring opening–ring closure isomerization of the [4,3-c] series into the [1,5-c] series (see Scheme 11). There is no indication of (**204**) closing to form the thermodynamically less stable [4,3-c] system and, in some cases, it ring closes back to the initial [1,5-c] system ⟨78AJC2505⟩. There was, however, an exception to this when 5-methylthio-[1,2,4]triazolo[1,5-c]pyrimidine (**205**) was treated with hydrazine; the rearranged product (**206**) was obtained rather than the expected [1,5-c] isomer (**207**) ⟨79KGS262⟩.

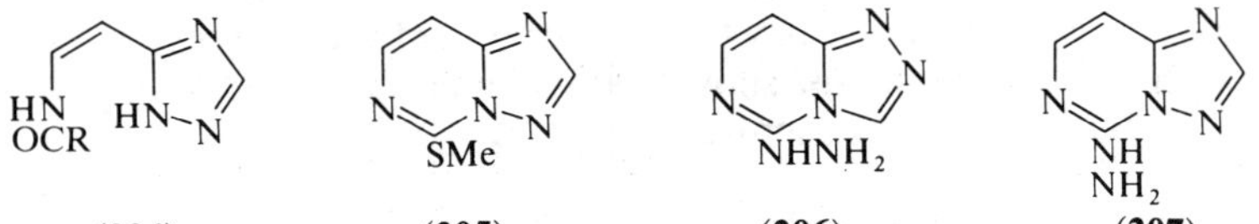

(**204**) (**205**) (**206**) (**207**)

A variety of 2-substituted [1,2,4]triazolo[1,5-c]pyrimidines were obtained when the 2-bromo derivative (Scheme 11) was treated with nucleophiles ⟨65JCS3357⟩.

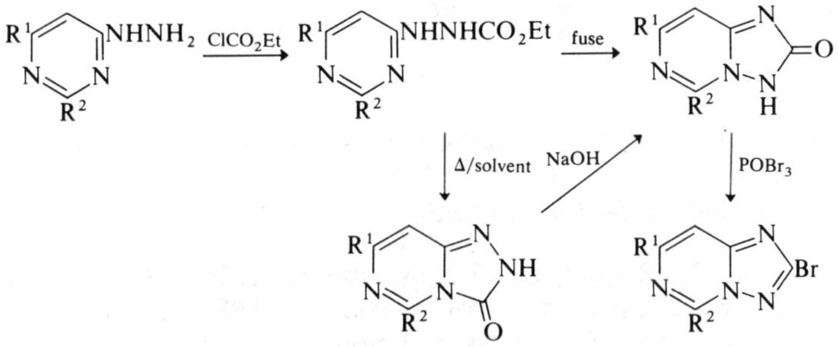

Scheme 11

4.15.3.5.4 [1,2,4]Triazolo[4,3-c]quinazoline and [1,2,4]triazolo[1,5-c]quinazoline

The most significant reactions of (**22**) and its derivatives are the Dimroth-like rearrangements to the [1,5-c] isomers discussed in the synthesis section. The [1,5-c] compounds do not rearrange into the [4,3-c] derivatives.

Treatment of (**22**) with hot aqueous sodium hydroxide gave 3-(2-aminophenyl)-1,2,4-triazole (**208**). Evidence that this latter reaction involved an initial attack of hydroxide ion at the 5-position arose from the isolation of 3-(2-benzamidophenyl)-1,2,4-triazole (**209**) upon the action of hot base on 5-phenyl-[1,2,4]triazolo[4,3-c]quinazoline (**210**). Apparently, under these rearrangement conditions, hydrolysis of the intermediate amide, for example (**209**), is faster than ring closure to the [1,5-c] isomer. Similar ring opening reactions were observed with the [1,5-c] series, a system which also underwent typical alkenic addition reactions at the C(5)—N(6) double bond ⟨70JOC3448⟩.

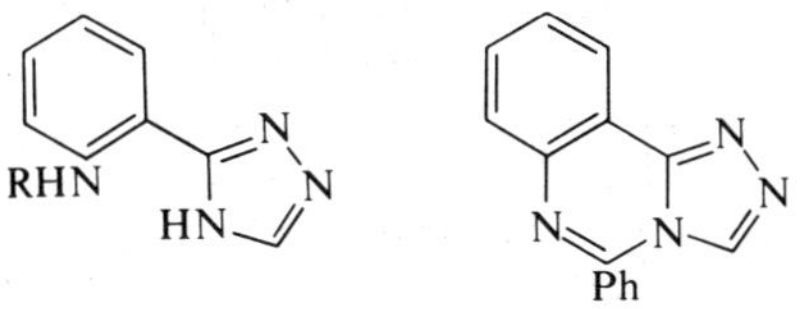

(**208**) R = H
(**209**) R = COPh
(**210**)

4.15.3.5.5 [1,2,4]Triazolo[1,5-a]quinazoline

An example of the work done on this ring system is in the ease with which the 5-chloro derivative (**211**) underwent nucleophilic substitution ⟨80HCA1⟩. Compound (**211**) was prepared from the corresponding 5-oxo substance and oxalyl chloride ⟨80HCA1⟩.

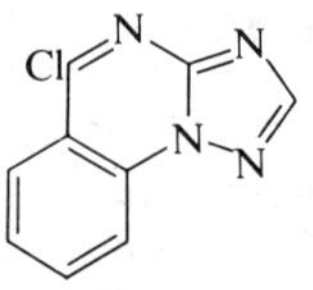

(211)

4.15.3.5.6 [1,2,4]Triazolo[5,1-b]quinazoline

Apparently, the o-quinoid arrangement in (25) has precluded it from a systematic reactivity study. This is substantiated by the preparation of the 9-oxo derivative (212) which prefers to exist in the keto tautomer shown rather than the enolic one (213) which would involve the less stable o-quinoid structure ⟨79JCS(P2)1708⟩.

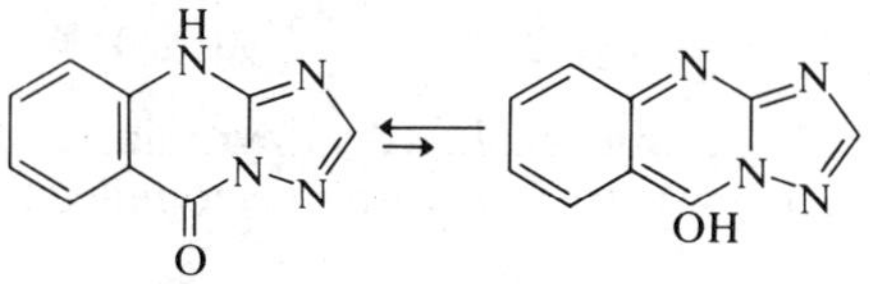

(212)   (213)

4.15.3.6 [1,2,3]Triazolopyrimidines and their Fused Benzo Derivatives

4.15.3.6.1 [1,2,3]Triazolo[1,5-a]pyrimidine

Using 5,7-dimethyl-[1,2,3]triazolo[1,5-a]pyrimidine (214), N-bromosuccinimide in chloroform yielded not the expected 3-bromo product (215), but 2-(α,α-dibromomethyl)-4,6-dimethylpyrimidine (216) and, in lesser amounts, 4,6-dimethylpyrimidine-2-carbaldehyde diethyl acetal (217) ⟨76JOC385⟩. The latter product was due to ethanolysis of (216) by the trace amounts of ethanol in the solvent since, following its removal, (216) was the only product. Use of N-chlorosuccinimide and iodine monochloride proceeded in the same way giving rise to (218) and (219), respectively. The mechanism for these transformations is illustrated in Scheme 12 and is postulated to involve a 'diazonium cation' ⟨76JOC385⟩.

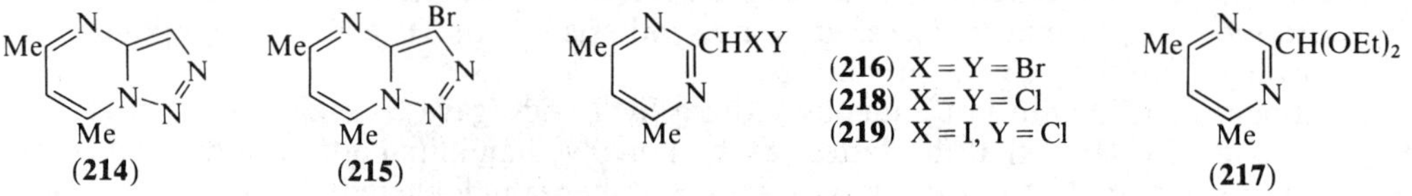

(214)   (215)   (216) X = Y = Br
                  (218) X = Y = Cl
                  (219) X = I, Y = Cl   (217)

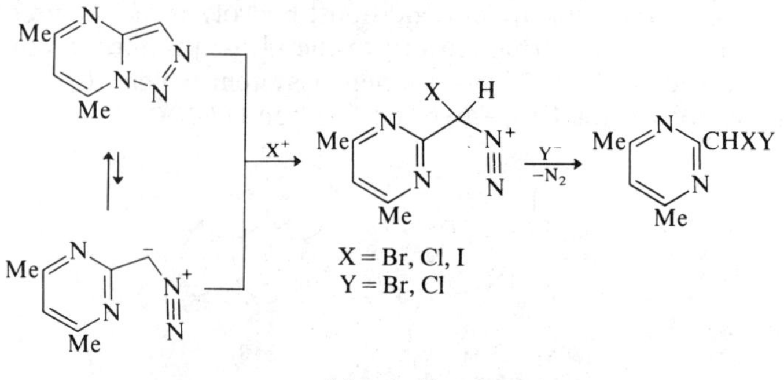

X = Br, Cl, I
Y = Br, Cl

Scheme 12

A similar mechanism has been proposed for the ring opening of the 3-phenyl derivative of (214) in acetic acid to give the acetoxybenzylpyrimidine compound (220; Scheme 13). In this reaction, the proton of the acetic acid adds to the diazonium intermediate and this is followed by acetate displacement of the nitrogen ⟨71JCS(C)2156⟩. It is interesting to note that the 3-amide (221) did not ring open in acetic acid apparently due to stabilization of the diazo zwitterionic species in equilibrium with (221) (see Scheme 12).

(220) (221) (222) (223) (224)

Scheme 13

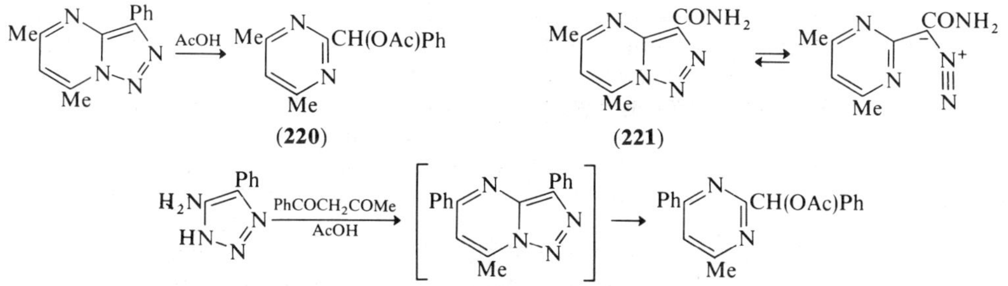

Condensation of benzoylacetone with (**222**) in acetic acid did not form (**223**) but rather (**224**). Compound (**223**) is most certainly formed in this process since a customary method for preparing the [1,2,3]triazolo[1,5-*a*]pyrimidine molecule was used; however, upon formation, (**223**) rearranged to (**224**) under the acetic acid conditions (Scheme 13) ⟨71JCS(C)2156⟩.

4.15.3.6.2 [1,2,3]Triazolo[1,5-c]pyrimidine

Since trifluoroacetic acid was known to favor the open chain tautomer in the azidoazine/tetrazoloazine isomerization (see the tetrazole portion of this section), it was felt that traces of trifluoroacetic acid would shift the equilibrium in favor of (**225**) over (**27**). However, addition of trifluoroacetic acid to a wet DMSO or chloroform solution of (**27**) gave the covalent hydrate (**226**) which was suggested to be in equilibrium with the open form (**227**; Scheme 14) ⟨78JHC1041⟩.

(27) (225) (226) (227)

Scheme 14

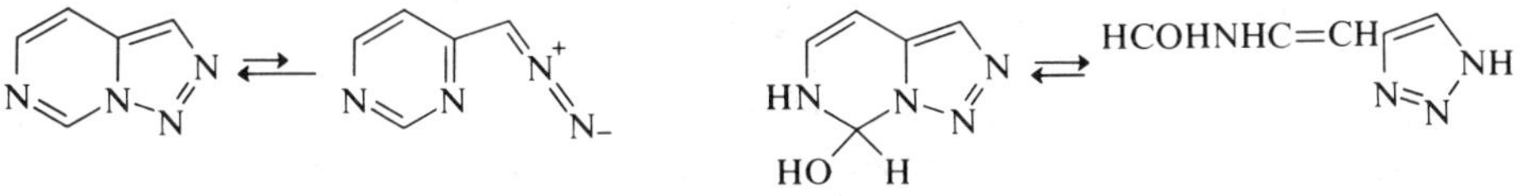

4.15.3.6.3 [1,2,3]Triazolo[4,5-d]pyrimidine

Methylation of the parent system (**28**) with dimethyl sulfate gave the 1-, 2- and 3-methyl isomers in a ratio of 1.5 : 1 : 1.1 ⟨69JCS(C)1084⟩. 3-Substituted derivatives of (**28**) underwent reaction with Grignard reagents to form the 7-alkyl-6,7-dihydro derivatives (**228**) which were transformed into the unsaturated analogs (**229**) upon oxidation with potassium ferricyanide ⟨79CPB3176⟩. The most acidic proton in 3-substituted (**28**) is at C-7 ⟨79CPB2431⟩. These systems also undergo covalent hydration ⟨66JCS(B)427, 66JCS(B)438⟩ to give the 6,7-dihydro-7-hydroxy product. At room temperature in dilute acid or base [1,2,3]triazolo[4,5-*d*]pyrimidines are hydrolyzed to 4-amino-1,2,3-triazole-5-carbaldehydes (**230**) ⟨77JCS(P1)1819⟩. The mechanism advanced ⟨77JCS(P1)1819⟩ for this is given in Scheme 15 showing that hydration occurred at C-5.

(228) (229)

(230)

Scheme 15

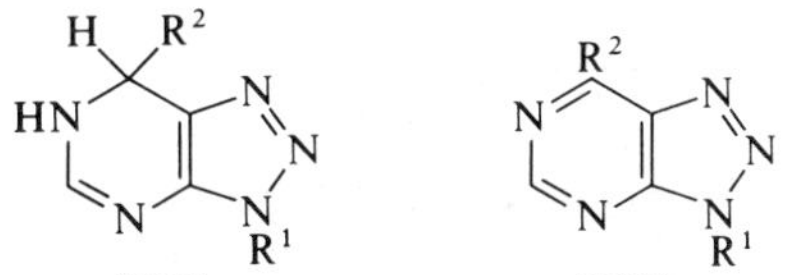

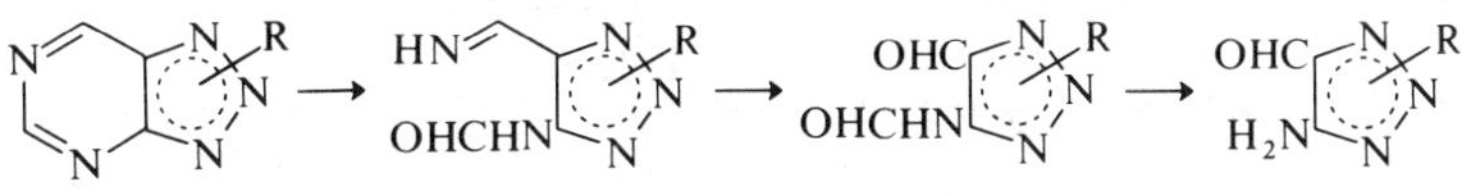

[1,2,3]Triazolo[4,5-*d*]pyrimidines are also susceptible to ring opening by active methylene compounds wherein the initial reaction takes place at C-7 ⟨79CPB2861⟩, analogous to the Grignard process described previously. For example, reaction of the parent compound with malononitrile gave (**231**) which, upon treatment with alkali, ring closed to the triazolopyridine (**232**; Scheme 16). Similar triazolopyridines (**233**)–(**235**) were obtained from (**28**) and cyanoacetamide, pentane-2,4-dione and ethyl acetoacetate ⟨73JCS(P1)1620⟩. Aliphatic and cyclic ketones also have been used in this reaction with (**28**) ⟨79CPB2861⟩.

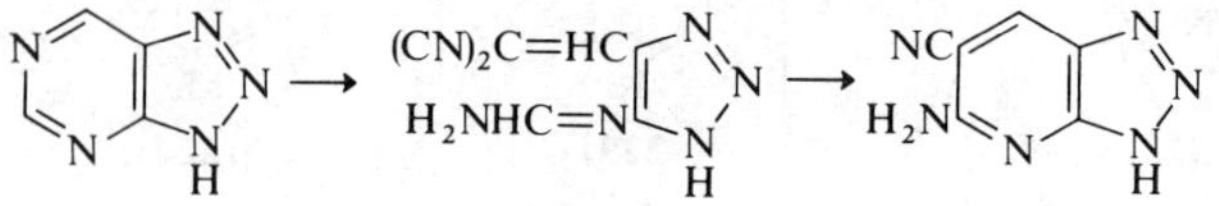

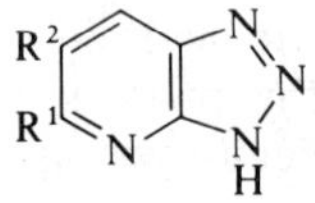

(**233**) R¹ = NH₂, R² = CONH₂
(**234**) R¹ = Me, R² = COMe
(**235**) R¹ = Me, R² = CO₂Et

Scheme 16

Pyrolysis of (**28**) did not give a cyanoimidazole as would have been expected based on the pyrolysis of [1,2,3]triazolo[4,5-*b*]pyridine which gave 2-cyanopyrrole ⟨70T3965⟩. Instead, (**28**) produced hydrogen cyanide and ammonia.

Further reactivity studies can be found in the many pyrimido-substituted [1,2,3]triazolo[4,5-*d*]pyrimidines (as 8-azapurines) which undergo various transformations (for example, nucleophilic substitution of chlorine substitutents, diazotization of amino groups, *etc.*) to produce potentially biologically useful molecules ⟨61JOC4433, 72JMC879⟩. In addition hydrogenation occurs to form the 6,7-dihydro derivatives ⟨66JCS(B)427⟩.

4.15.3.6.4 [1,2,3]Triazolo[1,5-a]quinazoline

The investigations ⟨66JCS(C)2290⟩ on the reactivity of this ring system have centered around 5-oxo-3-phenyl-4,5-dihydro-[1,2,3]triazolo[1,5-*a*]quinazoline (see Scheme 17) whose preparation is given in Section 4.15.4.6.4. The latter two reactions in Scheme 17 resemble those discussed previously for [1,2,3]triazolo[1,5-*a*]pyrimidines and depicted in Scheme 12.

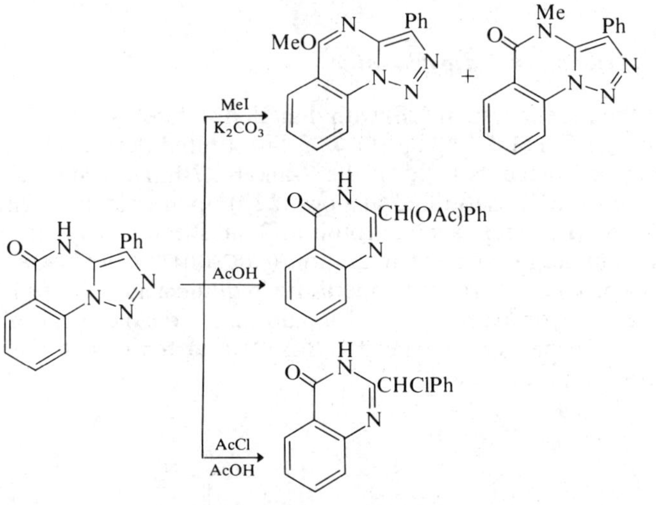

Scheme 17

4.15.3.7 Triazolopyrazines and Their Fused Benzo Derivatives

4.15.3.7.1 [1,2,4]Triazolo[4,3-a]pyrazine

As will be described in the synthesis section, [1,2,4]triazolo[4,3-*a*]pyrazines can be induced to rearrange to their [1,5-*a*] isomers in basic or acidic solution ⟨71JHC643⟩. On the other hand, the nucleus is very stable to oxidation by hot potassium permanganate ⟨68JHC485⟩ but gives the 7-oxide (236) upon treatment with peracetic acid ⟨80JCS(P1)506⟩. Reaction of (236) with acetic anhydride gave (237) and with phenyl isocyanate it gave (238).

(236)

(237)

(238) X = NHPh
(241) X = OMe
(243) X = NH$_2$ (3-Ph)

(239) X = Br
(240) X = OMe
(242) X = SPh

(244)

Bromination of (31) yielded the 5-bromo derivative (239) ⟨77JOC4197⟩. Reaction of (239) with sodium methoxide yielded the 5- and 8-methoxy compounds (240) and (241) ⟨77JOC4197⟩. The latter product arose by the *tele* substitution pathway involving attack by the methoxide ion at C-8. Inexplicably, sodium thiophenoxide produced only the C-5 product (242). If a large group is present at C-3, the C-8 substitution product is the only one obtained with no indication of the C-5 isomer ⟨77JOC4197⟩. Treatment of 5-chloro-3-phenyl-[1,2,4]triazolo[4,3-*a*]pyrazine with ammonia resulted in (243) which could also be obtained from the 8-methoxy-3-phenyl compound (244). Attack by the nucleophile at C-8 in all of these cases is promoted by the electron-withdrawing capabilities of N-7 ⟨77JOC4197⟩. With this in mind, it is not surprising that 8-chloro-[1,2,4]triazolo[4,3-*a*]pyrazines are susceptible to C-8 substitution by many nucleophiles ⟨78JHC987⟩.

The 3-amino substituent of various [1,2,4]triazolo[4,3-*a*]pyrazines can be acylated ⟨69JCS(C)1593⟩ but it is inert towards normal diazotization procedures ⟨68JHC485⟩. The 3-hydroxy compound could not be chlorinated with phosphorus oxychloride and in the IR spectrum it appeared to exist in both the keto and enol forms ⟨68JHC485⟩.

Both the [4,3-*a*] and [1,5-*a*] compounds are susceptible to hydrazinolysis and form triazole products. This ring cleavage proceeds *via* attack of the hydrazine at the electron deficient C-5 or C-8 positions of the bicyclic starting compound ⟨76H(4)943⟩.

4.15.3.7.2 [1,2,4]Triazolo[1,5-a]pyrazine

Bromination of the parent system (32) resulted in the 5-bromo derivative which with a variety of nucleophiles at elevated temperatures could be converted into other 5-substituted [1,2,4]triazolo[1,5-*a*]pyrazines (245). At room temperature, *tele* substitution occurred and the C-8 products (247) were obtained ⟨74TL4539⟩. Base-catalyzed deuterium exchange occurs at C-5 of (32) and treatment with hot, 50% aqueous sodium hydroxide formed 3-aminomethyl-1,2,4-triazole (246) and glycolic acid as a result of base attack at C-5.

(245) X = SH, SPh, OEt, NHNH$_2$

(246)

(247) X = NHNH$_2$, NHOH, NH$_2$

Oxidation of (32) with peracetic acid or *m*-chloroperbenzoic acid gave the 7-oxide (248). Compound (248) was converted into (249) with phosphorus oxychloride and the chlorine atom of (249) replaced by hydrazine to give (250) ⟨80JCS(P1)506⟩.

(248)

(249) X = Cl
(250) X = NHNH$_2$

(251)

4.15.3.7.3 [1,2,3]Triazolo[1,5-a]pyrazine

Heating [1,2,3]triazolo[1,5-*a*]pyrazine (33) in deuterium oxide caused three protons to be exchanged and for (251) to result. NMR evidence indicated that H-5 was the first one exchanged followed by H-1 and then H-8 ⟨78HCA1755⟩.

4.15.3.7.4 [1,2,3]Triazolo[4,5-b]pyrazine

1-Benzyl-5,6-dimethyl[1,2,3]triazolo[4,5-b]pyrazine was inert towards aqueous hydrochloric acid, aqueous base or hydrazine hydrate ⟨72JOC4124⟩ and the benzyl group resisted removal when the ring system was treated with sodium in liquid ammonia or hydrogen. These examples indicate that the ring system represented by (**34**) is very stable ⟨72JOC4124⟩.

4.15.3.7.5 [1,2,4]Triazolo[4,3-a]quinoxaline

As mentioned in Section 4.15.4.7.5, this ring system does not rearrange to the [1,5-a] isomers but is oxidized to 4-ones with potassium permanganate ⟨70KGS851⟩ even if a methyl group is present at the 4-position prior to oxidation.

4.15.3.8 [1,2,4]Triazolotriazines

4.15.3.8.1 [1,2,4]Triazolo[1,2,4]triazines

In addition to the rearrangements discussed later in the synthesis section (for example [3,4-c]→[5,1-c] and [4,3-d]→[1,5-d]) there are several other reactivity features of these triazolotriazines that are worthy of note. For example, [1,2,4]triazolo[4,3-b][1,2,4]triazines are stable to heat, boiling acetic acid and pyridine and upon nitration (if the 3-position is unsubstituted) give the 3-keto products as shown by (**252**) ⟨76JCS(P1)1492⟩. The [4,3-b] compounds react with Grignard reagents to produce 7-substituted derivatives ⟨77JOC1018⟩ and if the [4,3-b] and [5,1-c] systems possess a triazole amino group, they are susceptible to diazonium reactions which are often used as a route to the unsubstituted derivative ⟨76JCS(P1)1492⟩.

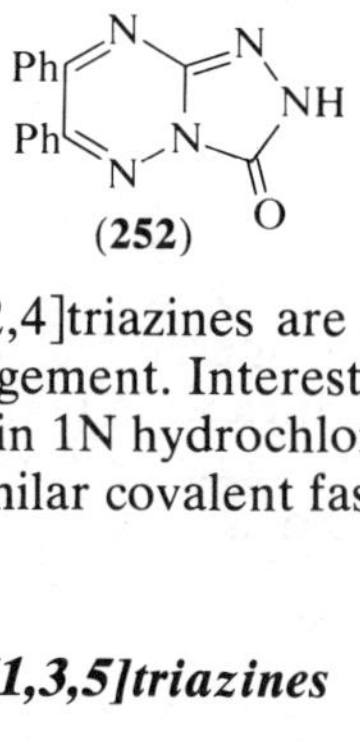

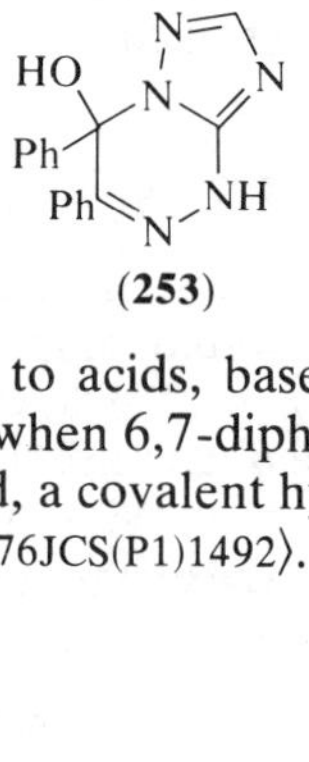

[1,2,4]Triazolo[5,1-c][1,2,4]triazines are stable to acids, bases and heat ⟨76JCS(P1)1492⟩ and do not undergo rearrangement. Interestingly, when 6,7-diphenyl[1,2,4]triazolo[5,1-c]-[1,2,4]triazine was warmed in 1N hydrochloric acid, a covalent hydrate (**253**) was obtained. Alcohols also added in a similar covalent fashion ⟨76JCS(P1)1492⟩.

4.15.3.8.2 [1,2,4]Triazolo[1,3,5]triazines

The hydrolytic reactions of the methoxy derivatives (**254**) and (**255**) were used in a structure proof ⟨70T3357⟩ and show the interesting reactivity in this series. This is shown in Scheme 18.

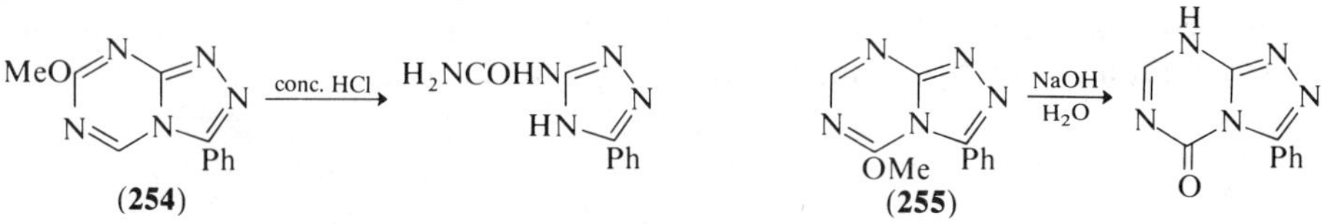

Scheme 18

In the [1,5-a][1,3,5] system it should be mentioned that 5-azaadenine could not be diazotized to 5-azahypoxanthine ⟨65JA1980⟩. Methylation of this latter compound produced the 6-methyl derivative ⟨65JA1980⟩ shown in Scheme 19.

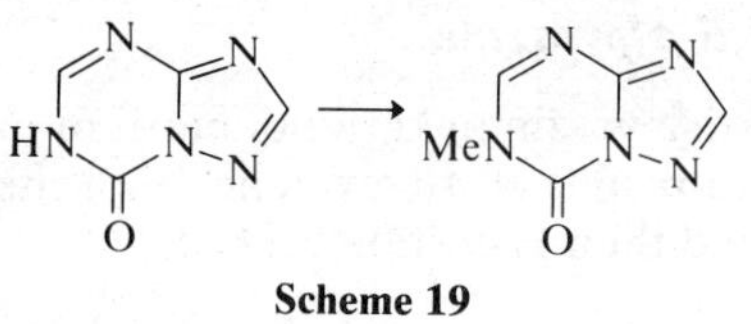

Scheme 19

4.15.3.9 Tetrazolopyridines and their Fused Benzo Derivatives

4.15.3.9.1 Tetrazolo[1,5-a]pyridine

Nitration of (46) gave (256) and bromination yielded (257) ⟨71T5121⟩. The presence of the azido tautomeric form in this series has been determined, for example, by reaction of (46) with dimethyl acetylenedicarboxylate to give (258) ⟨71T5121⟩, by cycloadditions with enamines ⟨65TL1965⟩ and by thermolysis to form, primarily, 2-cyanopyrrole ⟨76JA1259⟩. The last product was proposed to have come from 2-pyridylnitrene which, in turn, arose from loss of nitrogen from the azido tautomer ⟨70T4969, 80JA6159⟩. Methylation of (46) produced (259) ⟨75JCS(P1)1232⟩ while reaction of it and its derivatives with triphenylphosphine formed the iminophosphoranes (260) ⟨71T5359⟩.

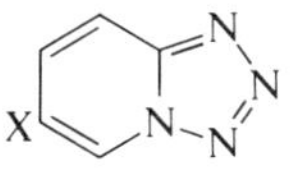

(256) X = NO$_2$
(257) X = Br
(258) R = CO$_2$Me 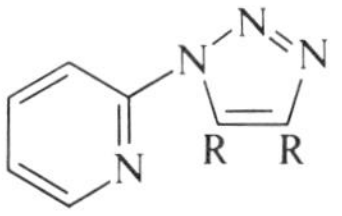
(259)
(260) 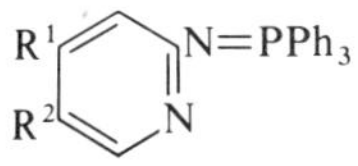

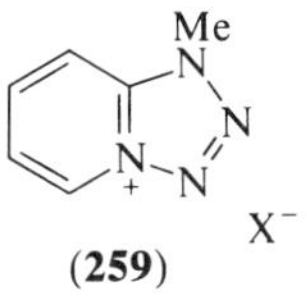

4.15.3.9.2 Tetrazolo[1,5-b]isoquinoline

Bromination of (47) led to 3-azido-4-bromoisoquinoline (261). On the other hand, trifluoroacetic acid caused only the tetrazolo tautomer (262) to exist in the equilibrium of (47) with its azido partner. This is due to elimination of the unfavored o-quinoid structure (47) from the tautomeric equilibrium and formation of the more stable naphthalenic structure (262) which, consequently, is preferred over the azido form ⟨81JOC843⟩. To extend this equilibrium study further, NMR investigations on substituted derivatives of (47) indicated that electron-releasing groups on the isoquinoline favor the tetrazolo tautomer whereas electron-attracting substituents favor the azido tautomer. Furthermore, the tetrazole was found to be preferred in DMSO solution relative to chloroform solution ⟨81JOC843⟩.

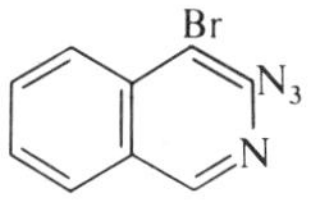

(261)
(262) 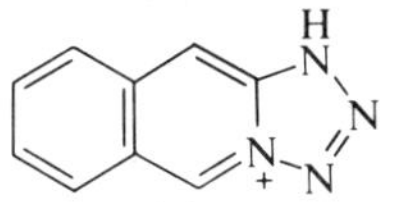

4.15.3.9.3 Tetrazolo[1,5-a]quinoline

Pyrolysis of (48) gave (263) and (264), the same products observed from pyrolysis of (49) ⟨80JA6159⟩. Both of these processes were believed to involve the common intermediate (265) which arose from the corresponding 2-quinolylnitrene or 1-isoquinolylnitrene. These latter materials resulted from the loss of nitrogen from the azido tautomers of (48) and (49) ⟨80JA6159⟩.

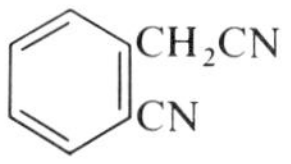

(263)
(264) 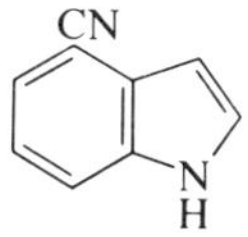
(265) 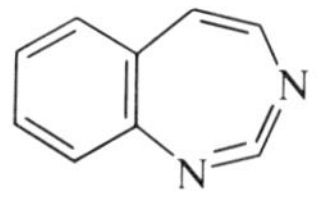

4.15.3.9.4 Tetrazolo[5,1-a]isoquinoline

The C(5)—C(6) center of (49) shows typical alkenic character by undergoing hydrogenation and oxidation at that site. Deuteration and metal exchange occur at C-5 and the metal introduced in this way can be replaced by a number of electrophilic groups (for example CO$_2$H). Chlorine atoms at C-5 or C-6 can be replaced by nucleophiles and a C-6 chlorine atom is removable by an elimination–addition sequence. Nitration of (49) takes place at C-7 ⟨75CB3780⟩.

4.15.3.10 Tetrazolo[1,5-b]pyridazine

Tetrazolo[1,5-b]pyridazine (**50**) does not undergo electrophilic substitution even under the most drastic conditions ⟨74MI41501⟩ whereas methylation, protonation and oxidation lead to monocyclic products (Scheme 20) due to a shift of the tetrazolo–azido equilibrium upon creating an electron-deficient pyridazine ring (see Section 4.15.2.8) ⟨74MI41501⟩. Unlike [1,2,4]triazolo[4,3-b]pyridazines, (**50**) did not undergo photolytically-promoted cycloaddition with alkenes (in particular, cyclohexene) ⟨73JHC801⟩; however, irradiation of (**50**) and its alkyl derivatives did give 3-cyanocyclopropenes and, in trace amounts, 3-cyanopyrazoles ⟨72CC1059, 73CPB2517⟩. Also, in contrast to [1,2,4]triazolo[4,3-b]pyridazines, (**50**) was not susceptible to photoalkylation with methanol ⟨74JOC793⟩.

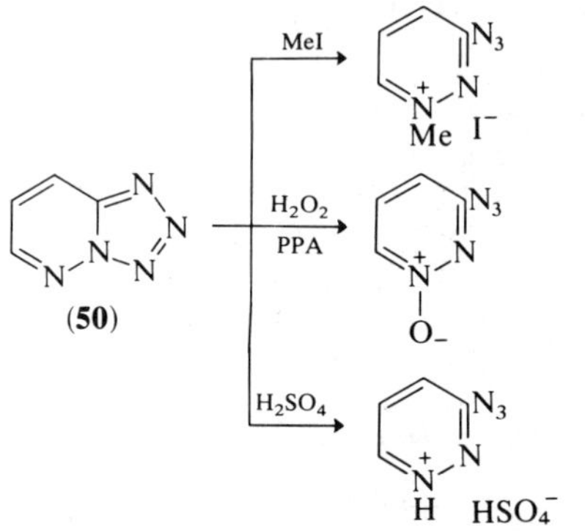

Scheme 20

Derivatives of (**50**) are reluctant to undergo hydrogen–deuterium exchange under neutral or acid conditions but with the 6-chloro derivative, base-catalyzed exchange occurs with concurrent substitution by DO$^-$ ⟨71M837⟩. Catalytic hydrogenation of (**50**) produces the 5,6,7,8-tetrahydro product ⟨75G1291⟩, the same product obtained from sodium borohydride reduction ⟨76JHC835⟩.

A number of reactions of 6-azidotetrazolo[1,5-b]pyridazine ⟨63CPB348⟩ have been reported and include (i) reduction to (**266**), (ii) substitution by alkoxide and hydroxide to (**267**) and (**268**), (iii) reaction with secondary amines ⟨76JOC3152⟩ to form (**269**) and (**270**) in a thermal process involving imines and (iv) interaction with ethyl acrylate to give the 1:2 adduct (**271**) ⟨74MI41501⟩. Since 6-azidotetrazolo[1,5-b]pyridazine can also, potentially, form a second fused tetrazole, the equilibrium between 7-methyl- and 8-methyl-6-azidotetrazolo[1,5-b]pyridazine (Scheme 21) has been studied ⟨74MI41501⟩.

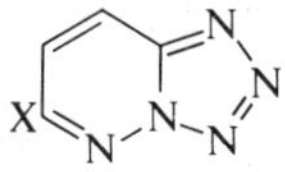

(**266**) X = NH$_2$
(**267**) X = OR
(**268**) X = OH
(**269**) X = NHR
(**270**) X = N=C(R^2)NRR^1

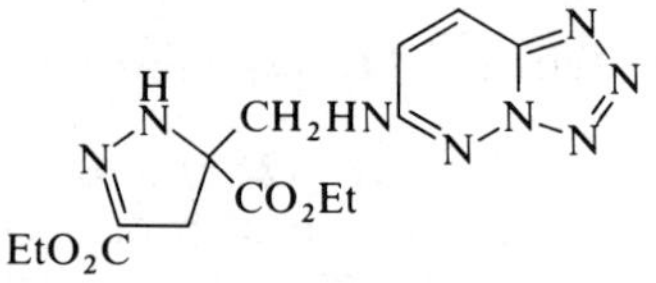

(**271**)

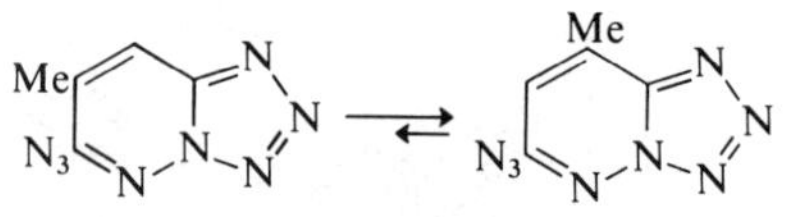

Scheme 21

No reactions analogous to 2-pyridylnitrene (see Section 4.15.3.9.1) could be observed for (**50**) decomposing to 2-pyridazinylnitrene ⟨74MI41501⟩.

4.15.3.11 Tetrazolopyrimidines and their Fused Benzo Derivatives

4.15.3.11.1 Tetrazolo[1,5-a]pyrimidine

Treatment of (**53**) with 1N sodium hydroxide gives the ring opened sodium salt of (**272**) which, upon acidification, regenerates tetrazolo[1,5-a]pyrimidine ⟨65JOC826⟩. The gas-phase thermolysis of (53) and its derivatives has been reported to yield aminopyrimidines by hydrogen abstraction and 1-cyanopyrazoles *via* ring contraction ⟨70T4915, 71T361⟩. These reactions proceeded *via* the nitrene (273) which arose from the loss of nitrogen from the azido tautomer of the corresponding tetrazolo[1,5-*a*]pyrimidine. Derivatives of (53) have also displayed cycloaddition reactions typical of azides ⟨73S123⟩.

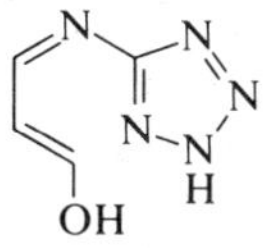

(272) 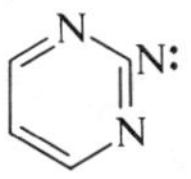 (273)

4.15.3.11.2 Tetrazolo[1,5-c]pyrimidine

In contrast to the [1,5-*a*] system, this series readily combined with water to form the covalent hydrate (274) ⟨65JOC829⟩ which arose as a result of the positive character of C-5 caused by the electron-attracting tetrazole ring. None of the azido tautomer (275) could be detected. Catalytic hydrogenation of (274) produced (276). Thermolysis of (54) and its derivatives produced 1-cyanoimidazoles by a ring contraction similar to that described in Section 4.15.3.11.1 ⟨70T4915, 72HCA2633⟩.

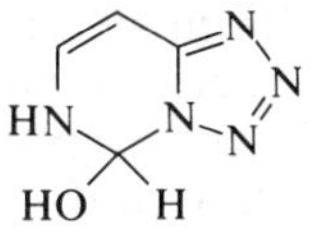

(274) 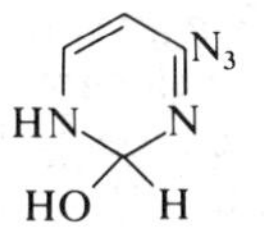 (275) 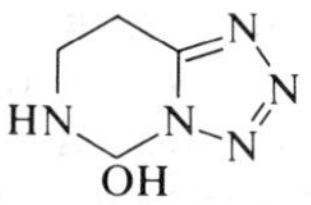 (276)

4.15.3.11.3 Tetrazolo[1,5-a]quinazoline

Refluxing 5-azidotetrazolo[1,5-*a*]quinazoline in 6N hydrochloric acid produced quinazoline-2,4-dione ⟨63ZOB2475⟩ *via* nucleophilic displacement of the '2'- and 4-azido groups by water followed by keto–enol tautomerism.

4.15.3.11.4 Tetrazolo[1,5-c]quinazoline

Similar to tetrazolo[1,5-*c*]pyrimidine, (56) underwent rapid hydration to form (277). Upon heating, this hydrate underwent ring opening to 5-(2-aminophenyl)tetrazole (278) ⟨67KGS1096⟩.

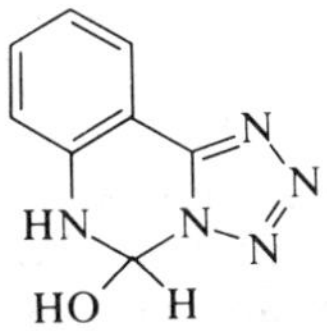

(277) 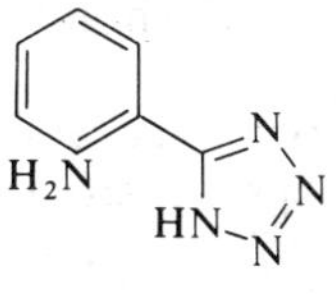 (278)

4.15.3.12 Tetrazolo[1,5-a]pyrazine and Tetrazolo[1,5-a]quinoxaline

4.15.3.12.1 Tetrazolo[1,5-a]pyrazine

The parent ring system (57) undergoes deuterium exchange at C-5 ⟨78HCA1755⟩ and thermolysis in organic solvents to produce 1-cyanoimidazoles (see Section 4.15.3.11.2) *via* ring contraction of a pyrazinylnitrene, the intermediate arising from loss of nitrogen from the azido tautomer of (57) ⟨70T4915, 72HCA565, 72HCA2633⟩. 1-Cyanobenzimidazoles arise from tetrazolo[1,5-*a*]quinoxalines ⟨72HCA2633⟩. Further evidence of the azido form came from refluxing 5,6-diphenyltetrazolo[1,5-*a*]pyrazine with acetic anhydride to produce an acetylated imidazole, and from treating the diphenyl compound with dimethyl acetylenedicarboxylate to form pyrazinylpyrazole ⟨71JOC446⟩. It is interesting to note that treatment of 5-aminotetrazolo[1,5-a]pyrazine with acetic anhydride led to 2-acetamido-6-azidopyrazine (*i.e.* retention of the azido group) being isolated as the product ⟨80JHC11⟩. This indicated that the 2-acetamido substituent stabilized the azido functionality and prevented it from losing nitrogen for further transformations or reverting to the tetrazolo tautomer.

4.15.3.12.2 *Tetrazolo[1,5-a]quinoxaline*

In addition to the thermolysis mentioned in the previous section, compounds of this series are readily oxidized to the 4-ones which, in turn, can be alkylated at N-5 ⟨70KGS851⟩.

4.15.4 SYNTHESIS

4.15.4.1 [1,2,4]Triazolopyridines and their Fused Benzo Derivatives

4.15.4.1.1 *[1,2,4]Triazolo[4,3-a]pyridine*

The preparation of [1,2,4]triazolo[4,3-a]pyridines, as with many of the compounds of this chapter, can occur by two routes: ring closure of an appropriate heterocyclic hydrazine (here, a hydrazinopyridine) with an acidic type cyclodehydrating agent which also supplies the carbon of the triazole moiety (Scheme 22), and construction of the pyridine portion of the fused ring system onto a preformed 1,2,4-triazole nucleus. The latter method introduces the possibility of isomer formation depending on the particular nitrogen of the 1,2,4-triazole nucleus to which cyclization occurs and is of little synthetic utility. On the other hand, ring closure of 2-hydrazinopyridines (**279**) has been used extensively ⟨66JOC251, 66CB2593⟩ for preparing [1,2,4]triazolo[4,3-a]pyridines.

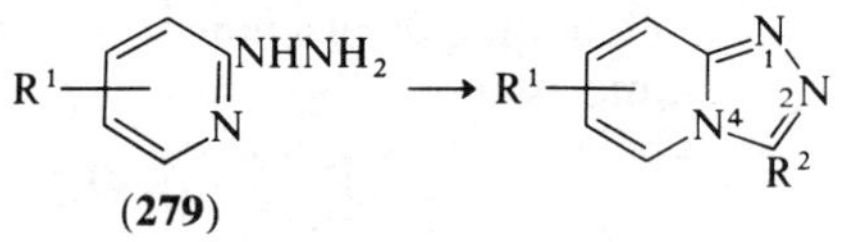

Scheme 22

A wide variety of substituted 2-hydrazinopyridines can be used in this procedure to produce a number of pyrido-substituted [1,2,4]triazolo[4,3-a]pyridines. The required 2-hydrazinopyridines are available ⟨66JOC251⟩ from the reaction of hydrazine with the appropriate 2-bromopyridines which are in turn obtainable by diazotization of the corresponding 2-aminopyridines.

[1,2,4]Triazolo[4,3-a]pyridine (**1**), and its 3-alkyl derivatives, are available from 2-hydrazinopyridines and the appropriate carboxylic acids, anhydrides or esters ⟨66JOC251⟩. A selection of the [1,2,4]triazolo[4,3-a]pyridines prepared in this way is given in Table 8. In some cases ⟨78JHC439⟩ it was best to isolate the intermediate hydrazide and carry out a subsequent cyclodehydration with, for example, phosphorus oxychloride, to realize the desired product. There seems to be some steric influence on the ring closures as reflected in the considerably depressed yields obtained for the 3,5-disubstituted derivatives ⟨66JOC251, 78JHC439⟩.

The use of ortho esters as the ring closing agent with (**279**) is less efficient, giving yields reduced by 20%, which is in contrast to the synthesis of [1,2,4]triazolo[4,3-a]pyrazines (*vide infra*) where the ortho ester is the best cyclization agent. This facile ring closure with 2-hydrazinopyrazines is suggested ⟨66JOC251⟩ to be due to differences in the basic strengths of the heterocyclic nitrogen atoms (pyridine, $pK_a = 5.2$; pyrazine, $pK_a = ca.$ 0.6) since the ring closure probably involves nucleophilic attack of the heterocyclic ring nitrogen on an intermediate carbocation of the hydrazide side-chain.

The 3-aryl-[1,2,4]triazolo[4,3-a]pyridines are best prepared by heating the hydrazinopyridine with an equimolar amount of the appropriate acid or ester or, in the

Table 8 Examples of 3-Alkyl- and 3-Aryl-[1,2,4]triazolo[4,3-a]pyridines Prepared from 2-Hydrazinopyridines and Carboxylic Acids, Anhydrides, Esters or Ortho Esters ⟨66JOC251, 78JHC439, 70JHC703, 70JHC1019⟩

R^1	R^2	R^1	R^2	R^1	R^2
H	H	7-Me	Et	5-Cl	H
5-Me	H	8-Me	Et	5-Cl	Me
6-Me	H	5,7-Me$_2$	Et	5-Cl	Et
7-Me	H	H	2-Styryl	5-Cl	Bu^t
8-Me	H	H	2-Thienyl	5-Cl	CH$_2$Cl
5,7-Me$_2$	H	H	2-Furyl	5-Ph, 7-CF$_3$	H
5-Me	Me	H	2-Pyridyl	5-Me, 7-CF$_3$	H
6-Me	Me	H	2-Thienylvinyl	8-NO$_2$	H
7-Me	Me	5-Me	Ph	8-NO$_2$	Me
5,7-Me$_2$	Me	6-Me	Ph		
H	Pr	7-Me	Ph		
H	Et	8-Me	Ph		
5-Me	Et	5,7-Me$_2$	Ph		
6-Me	Et				

case of solid aromatic acids (*e.g.* benzoic), fusing the two reagents together. Some of the derivatives made in this way are illustrated in Table 8.

The use of dicarboxylic acids or their derivatives in this procedure gave di(3-[1,2,4]triazolo[4,3-a]pyridyl)alkanes (**280**) either directly or following thermal or phosphorus oxychloride promoted ring closure of the intermediate hydrazides. Oxalic acid gave only the [1,2,4]triazolo[4,3-a]pyridine-3-carboxylic acid (**100**) whereas ethyl oxamate gave the corresponding 3-carboxamide (**101**) rather than the alternative six-membered ring product (**281**). Haloacetic acids failed when utilized as ring closing agents for preparing the 3-halomethyl compounds ⟨66JOC251⟩.

(**280**)   (**281**)   (**100**) R = OH
             (**101**) R = NH$_2$

Fusion of (**279**) with urea or ethyl allophanate or reaction with excess ethyl chloroformate yielded the 3-hydroxy derivatives (**91**). Urea seems to be the reagent of choice ⟨66JOC251⟩ and involves a semicarbazide intermediate. The spectral data for these compounds indicate that they exist predominantly as the keto tautomer (**282**).

(**91**) X = O   (**282**) X = O
(**93**) X = S   (**283**) X = S
(**95**) X = NH

Carbon disulfide, thiophosgene or potassium trithiocarbonate effectively produced the 3-thiols (**93**) when reacted with (**279**), with carbon disulfide being the most efficient reagent. With this latter reagent, an intermediate dithiocarbamic acid was formed initially which subsequently underwent ring closure with loss of hydrogen sulfide. Thiourea, in contrast to urea, failed to produce a thiol, resulting instead in 2-aminopyridine thiocyanate which arose from the decomposition of (**279**; R^1 = H) to 2-aminopyridine on prolonged heating. The 3-thiols exist predominantly as the 3-thione tautomer (**283**).

The 3-amino-[1,2,4]triazolo[4,3-a]pyridines (**95**) have been prepared by the reaction of (**279**) with cyanogen bromide ⟨66JOC251⟩. It was necessary to neutralize the product from this reaction with sodium acetate in order to obtain the free base. This synthesis proceeds *via* an intermediate cyanohydrazine. Cyanogen chloride has been used in preparing 3-amino-[1,2,4]triazolo[4,3-c]pyrimidines or 2-amino-[1,2,4]triazolo[1,5-c]pyrimidines (*vide infra*), depending on the reaction conditions ⟨63JCS5642⟩.

[1,2,4]Triazolo[4,3-a]pyridines possessing pyrido substitution other than alkyl have been only briefly studied ⟨78JHC439, 70JHC703⟩. This is due to the inaccessibility of the appropriate 2-hydrazinopyridines. A number of 5-chloro-[1,2,4]triazolo[4,3-a]pyridines have been prepared from ring closure of 6-chloro-2-hydrazinopyridine and these are given in Table 8.

A variety of 3-substituted 8-nitro-[1,2,4]triazolo[4,3-a]pyridines ⟨70JHC1019⟩ have been prepared via a number of the aforementioned ring closure methods with 2-hydrazino-3-nitropyridine. These products readily undergo rearrangement to the isomeric [1,5-a] series as described elsewhere in this chapter. Reduction of the 8-nitro substituent of several of these derivatives resulted in the corresponding 8-amino products that offer the potential starting point for other 3,8-disubstituted systems of this type.

Other less widely used methods of ring closing 2-hydrazinopyridines to [1,2,4]triazolo[4,3-a]pyridines use thioimino ethers ⟨76JOC3124, 79MI41500⟩ to produce 3-alkyl derivatives (Scheme 23) and 1,3,4-oxadiazoles (Scheme 23) ⟨69JPR71⟩ to form 3-hydrazino derivatives. An oxidative ring closure of pyrid-2-yl arylhydrazones is reported ⟨65IJC162⟩ to give the 3-aryl compounds. The 2-alkyl- or -aryl-1,2,4-triazolium salts in the [4,3-a]pyridine, [3,4-a]isoquinoline and [4,3-a]quinoline systems have been reported to arise from the reaction of pyridine, isoquinoline and quinoline, respectively, with diazenium fluoroborates as shown in Scheme 24 ⟨69CB3159⟩.

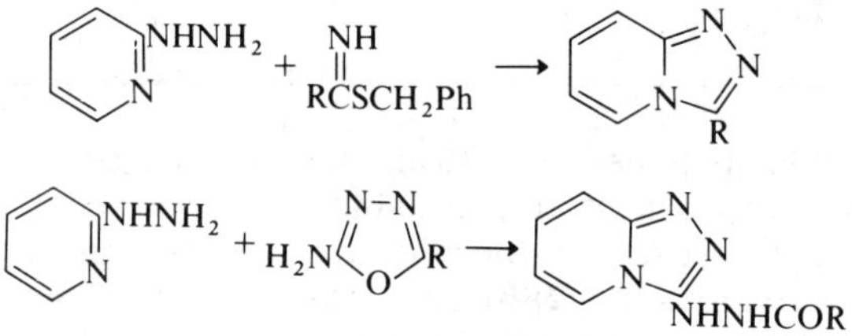

Scheme 23

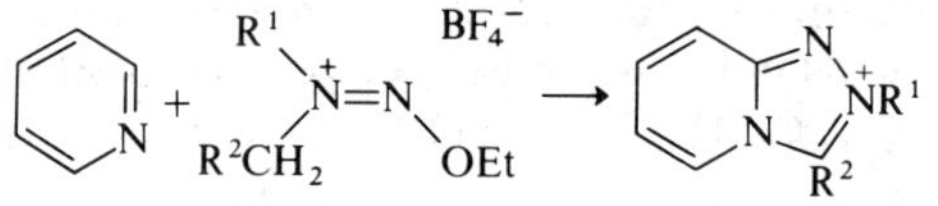

Scheme 24

4.15.4.1.2 [1,2,4]Triazolo[1,5-a]pyridine

As with the [4,3-a] series, any synthesis involving ring closure onto a preformed 1,2,4-triazole to fuse the pyridine ring has not been practical but a number of methods have been developed beginning with substituted pyridines. For example, oxidative cyclization of N-(2-pyridyl)alkyl- or -aryl-amidines (**284**) ⟨66JOC260⟩ has been accomplished using lead tetraacetate in refluxing benzene to give the compounds listed in Scheme 25 ⟨66JOC260⟩. However, since this method was limited in its ability to produce 2-unsubstituted derivatives, an alternative route was developed ⟨66JOC260, 66CPB506⟩ by treating 2-aminopyridines with hydroxylamine-O-sulfonic acid to produce the 1,2-diaminopyridinium salts (**285**) which could be converted into their free bases (**286**) by ion exchange on an Amberlite IRA-400 resin. Ring closure of either of these two products with carboxylic acids or their chlorides produced the derivatives presented in Scheme 26 ⟨66JOC260, 66CPB506⟩. Some of these same [1,2,4]triazolo[1,5-a]pyridines could also be produced by the 1,3-dipolar cycloaddition reactions of nitriles to the N-iminopyridine (**287**; Scheme 27) ⟨66CPB506⟩. This method is of limited utility since no reports of substituted N-iminopyridines undergoing this reaction

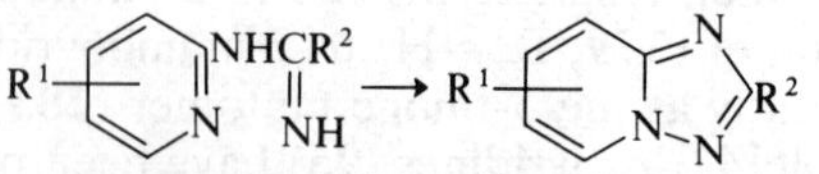

R^1	H	H	4-Me	4-Me	4-Me	5-Me	5-Me	6-Me	6-Me	7-Me	7-Me	5,7-Me$_2$	5,7-Me$_2$
R^2	Et	Ph	Me	Ph	Et	Me	Ph	Me	Ph	Me	Ph	Me	Ph

Scheme 25

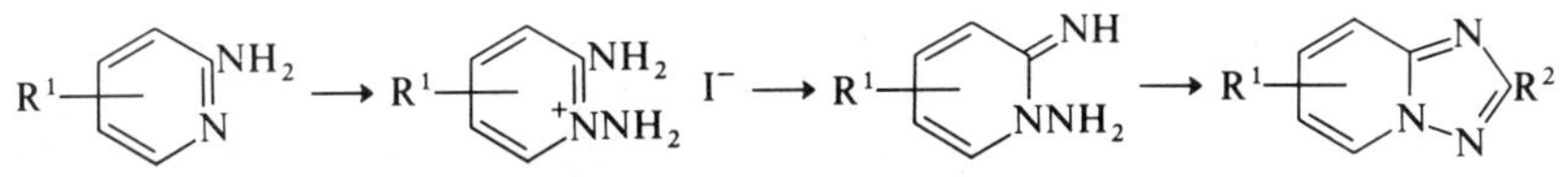

Scheme 26

R¹	H	4-Me	5-Me	6-Me	7-Me	H	H	H
R²	H	H	H	H	H	Me	Ph	4-Pyridyl

Scheme 27

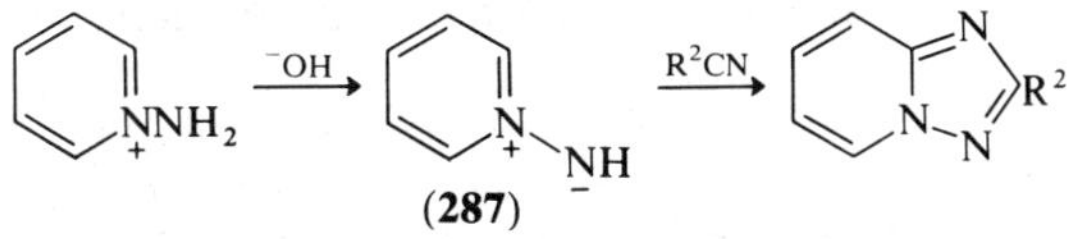

have appeared and it can be assumed that such groups would have a drastically variable effect on the dipolar cycloaddition.

[1,2,4]Triazolo[1,5-*a*]pyridines have also been produced by the isomerization of [1,2,4]triazolo[4,3-*a*]pyridines. This procedure has been described previously in the reactivity section for the [4,3-*a*] series.

Rather than forming the N(3)—N(4) bond by oxidative means (Scheme 25) ⟨66JOC260⟩, a recent report ⟨74JOC2143⟩ described a cyclodehydration process to form the same bond. In this regard, a 2-aminopyridine was reacted with *N,N*-dimethylformamide dimethyl acetal to give the corresponding *N,N*-dimethylaminomethylene derivative (**288**). Displacement of the dimethylamino group with hydroxylamine gave the hydroxyiminomethyleneamino derivative (**290**) or (**289**). Cyclodehydration of (**290**) with polyphosphoric acid resulted in the 2-unsubstituted triazolopyridine (**291**). The scope of this method remains to be determined. Rather than this stepwise procedure it has been found ⟨81JOC3123⟩ more convenient to treat the *N,N*-dimethylformamidines (**292**) with hydroxylamine-*O*-sulfonic acid in methanol/pyridine to produce [1,2,4]triazolo[1,5-*a*]pyridines. This procedure could also be used on the formimidates (**293**). The bicyclic compounds prepared by this route are given in Scheme 29.

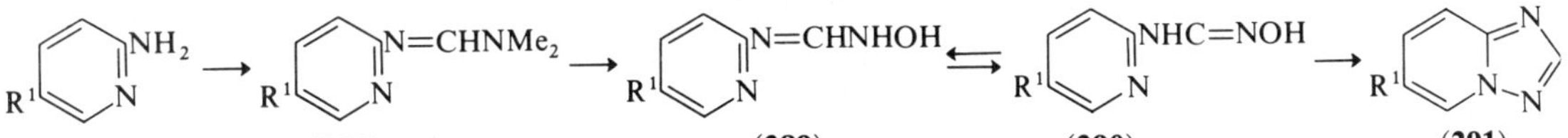

Scheme 28

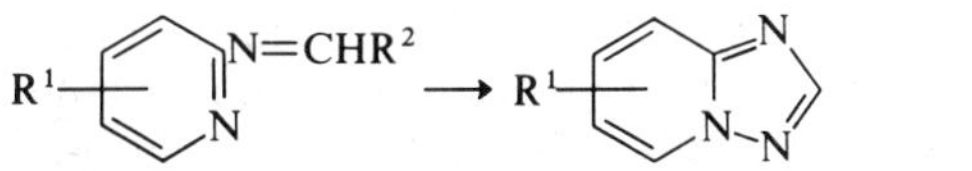

(**292**) R² = NMe₂   R¹ = H, 6-Me, 7-Me, 8-Me
(**293**) R² = OEt

Scheme 29

[1,2,4]Triazolo[1,5-*a*]pyridine 1-oxides (**294**) ⟨79H(12)1157⟩ have been synthesized by treating 1-amino-2-chloropyridinium mesylate (**295**) with hydroxylamine to form (**296**). Reaction of (**296**) with carboxylic acids gave the *N*-oxides (**294**; Scheme 30). The unsubstituted compound (**294**; R = H) was most conveniently obtained by treating (**296**) with *N,N*-dimethylformamide dimethyl acetal.

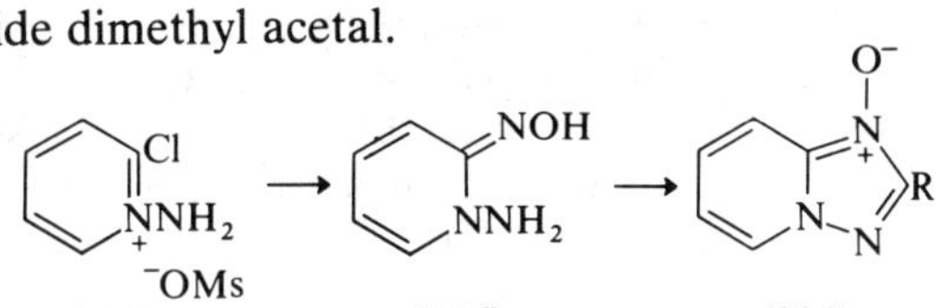

Scheme 30

A number of substituted [1,2,4]triazolo[1,5-*a*]pyridine 3-oxides (**297**) have been obtained from the reaction between 2-pyridylsulfimides (**289**) and nitrile oxides (Scheme 31) ⟨76JCS(P1)2166⟩.

(298) (297)

Scheme 31

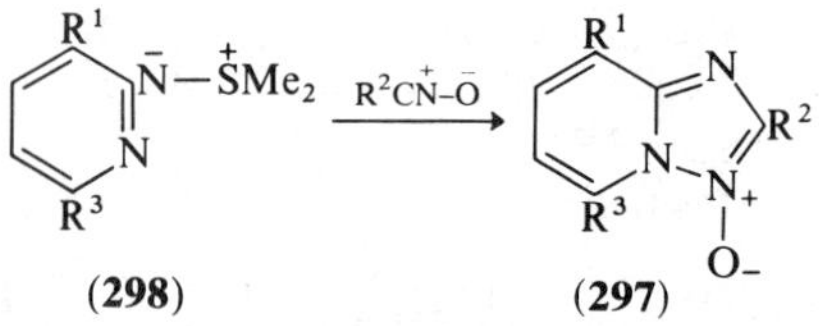

4.15.4.1.3 [1,2,4]Triazoloquinolines and [1,2,4]triazoloisoquinolines

[1,2,4]Triazolo[4,3-*a*]quinolines can be prepared by the ring closure of 2-hydrazinoquinolines with carboxylic acids ⟨74MI41502⟩ or by oxidative cyclization of quinol-2-ylhydrazones of aromatic and heterocyclic aldehydes ⟨65IJC162⟩. [1,2,4]Triazolo[3,4-*a*]isoquinolines can also be prepared in similar ways from 1-hydrazinoisoquinoline ⟨66JHC158, 71RTC1225, 65IJC162, 67IJC403⟩.

To obtain the [1,2,4]triazolo[4,3-*b*]isoquinoline series it was not possible to displace the halogen atom of 3-chloroisoquinoline with hydrazine. However, 3-hydrazinoisoquinoline 2-oxide could be prepared from the corresponding 3-chloro 2-oxide, and it could be selectively reduced by titanium trichloride to 3-hydrazinoisoquinoline, a relatively unstable compound ⟨78JHC463⟩. This hydrazino derivative was unresponsive to the usual carboxylic acid type ring closures generally used to make fused [4,3-*x*] systems. The desired products (**299**) could be obtained by cyclodehydration of the monoacylhydrazino compounds (**300**) with polyphosphoric acid. These derivatives (**299**) were, however, unstable and had to be characterized as their perchlorate salts (**301**; Scheme 32) ⟨78JHC463⟩ to eliminate the *o*-quinoid structure.

(300) (299) R = H, Me

(301)

Scheme 32

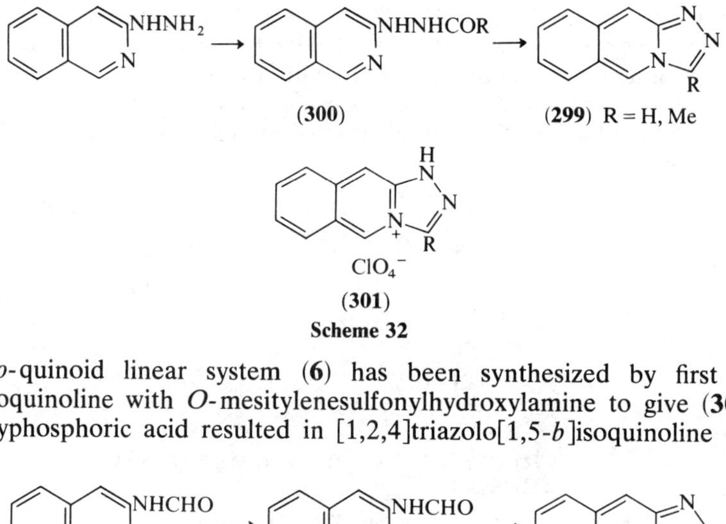

Another *o*-quinoid linear system (**6**) has been synthesized by first reacting 3-formamidoisoquinoline with *O*-mesitylenesulfonylhydroxylamine to give (**302**). Heating (**302**) in polyphosphoric acid resulted in [1,2,4]triazolo[1,5-*b*]isoquinoline (Scheme 33) ⟨77H(6)949⟩.

(302) (6)

Scheme 33

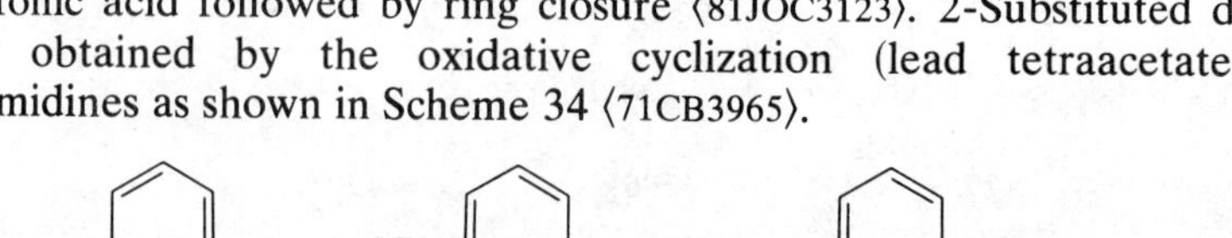

[1,2,4]Triazolo[5,1-*a*]isoquinoline (**7**) is available by the previously described method of reacting the *N'*,*N'*-dimethylformamidine derivative of 1-aminoisoquinoline with hydroxylamine-*O*-sulfonic acid followed by ring closure ⟨81JOC3123⟩. 2-Substituted derivatives of (**7**) can be obtained by the oxidative cyclization (lead tetraacetate) of *N*-(1-isoquinolyl)amidines as shown in Scheme 34 ⟨71CB3965⟩.

Scheme 34

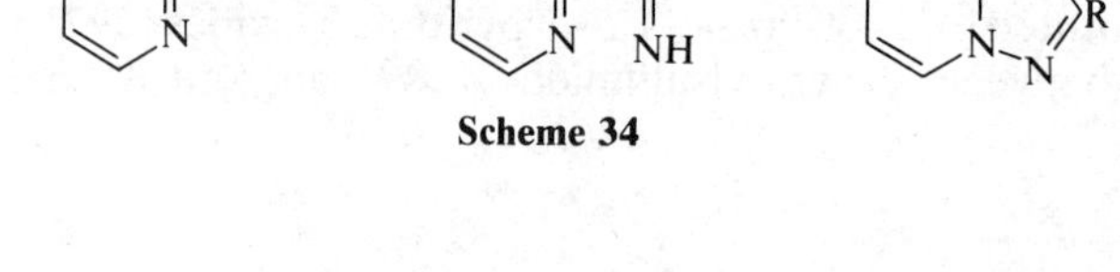

4.15.4.2 [1,2,3]Triazolopyridines and their Fused Benzo Derivatives

4.15.4.2.1 [1,2,3]Triazolo[1,5-a]pyridine

The older method of preparing [1,2,3]triazolo[1,5-*a*]pyridines by oxidation of the hydrazones of 2-pyridyl aldehydes and ketones ⟨57JA678, 57JCS4506⟩ with silver oxide, potassium ferricyanide, *etc.* has given way to a cleaner oxidant, namely nickel peroxide, which gives the 1,2,3-ring system in greater than 90% yield (Scheme 35) ⟨76SC69⟩. Other synthetic approaches include the treatment of picoline-2-carbaldehyde *p*-toluenesulfonylhydrazone with base ⟨64CB3493⟩, the treatment of pyrid-2-yl ketones with *p*-toluenesulfonyl azide (Scheme 36) ⟨66CB2918⟩ and cyclodehydration of the amine salts of picoline-2-carbaldehyde and pyrid-2-yl ketone oximes with polyphosphoric acid ⟨75JHC481⟩. The starting materials for this last procedure (**303**; Scheme 37) are available from the corresponding oximes (**304**) and *O*-mesitylenesulfonylhydroxylamine, a typical reagent for synthesizing heterocyclic *N*-amines. [1,2,3]Triazolo[1,5-*a*]quinolines (**11**) can also be made in this way from quinol-2-yl oximes ⟨75JHC481⟩.

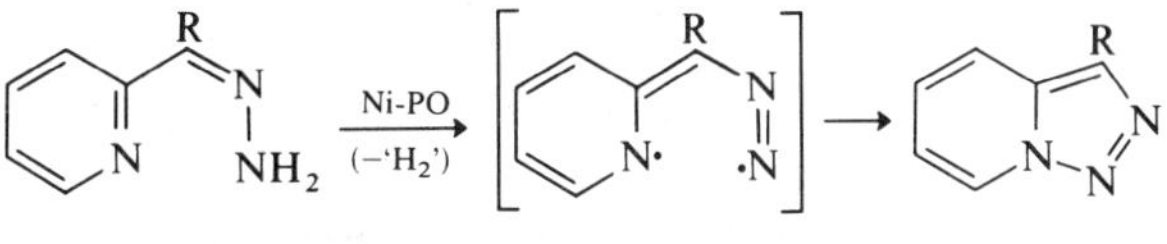

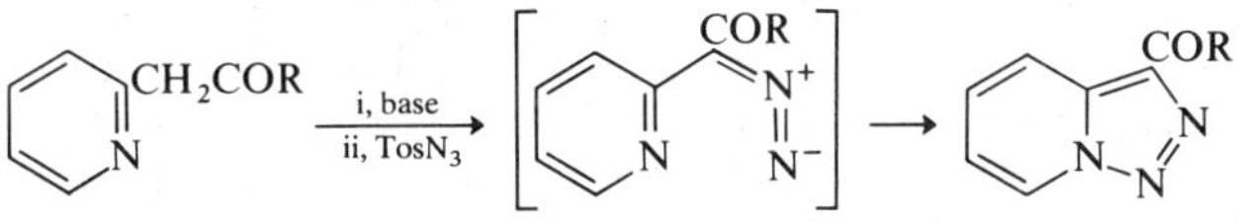

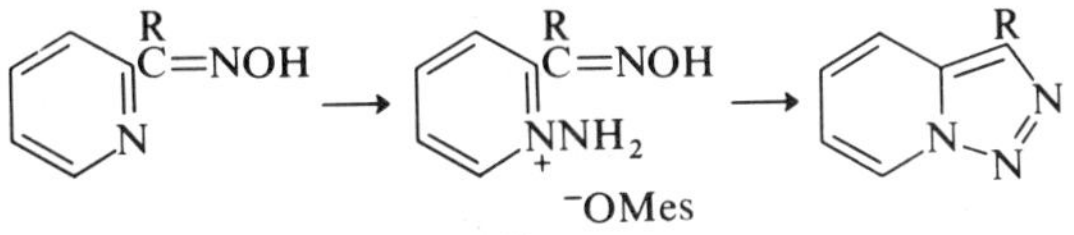

In a much poorer yield, compound (**8**) was the product from the thermal decomposition of 5-(2-pyridyl)tetrazole (**305**) at $400\,°C/10^{-3}$ Torr ⟨78HCA1755⟩. Compound (**305**) was prepared by treating pyridine-2-carbonitrile with hydrazoic acid. [1,2,3]Triazolo[1,5-*a*]pyridines are also the products from the treatment of 1-aminoquinolizinium salts (**306a**) with nitrous acid in a process proceeding *via* (**306b**) ⟨69TL1549⟩.

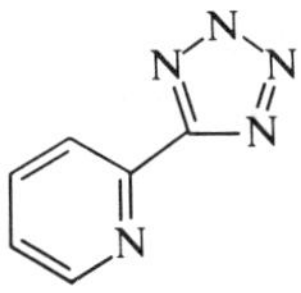

4.15.4.2.2 [1,2,3]Triazolo[4,5-b]pyridine and [1,2,3]triazolo[4,5-c]pyridine

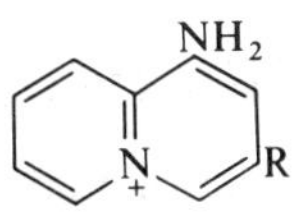

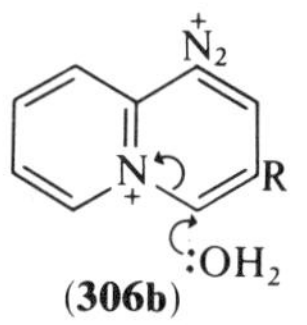

Compounds (**9**) and (**10**) have been most extensively studied as 1-deaza-8-azapurines and 3-deaza-8-azapurines, respectively, and have been almost exclusively prepared by diazotization of 2,3-diaminopyridines (for **9** and its derivatives) ⟨67CB1646, 71RTC1181, 73JOC1095⟩ and 3,4-diaminopyridines (for **10** and its derivatives) (Scheme 38) ⟨65JMC296, 73JOC1095⟩. A rather interesting synthetic entry into the [4,5-*b*] series was reported ⟨70JOC1131⟩ to involve the reaction of sodium azide with 5-nitropyridin-2(1*H*)-one to give [1,2,3]triazolo[4,5-*b*]pyridin-5(4*H*)-one after the loss of nitrous acid (Scheme 39). Pyrido[4,3-*e*][1,2,4]triazines have been shown to undergo basic rearrangement to [1,2,3]triazolo[4,5-*c*]pyridines as illustrated in Scheme 40 ⟨71JHC47⟩.

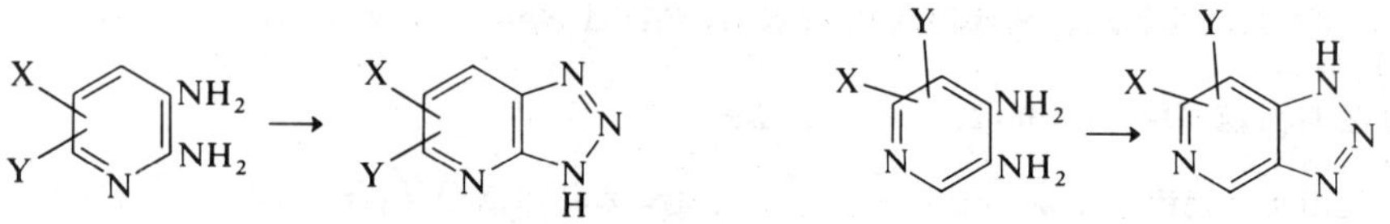

Scheme 38

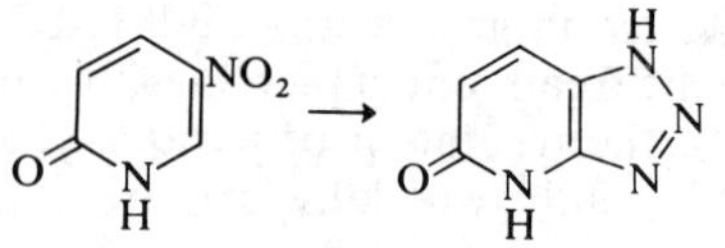

Scheme 39

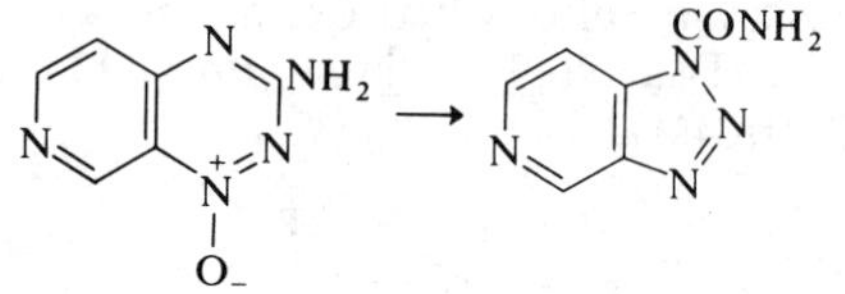

Scheme 40

[1,2,3]Triazolo[4,5-*d*]pyrimidines (8-azapurines) possessing 5-amino, 5-oxo or 5-thioxo groups undergo ring opening with a variety of active methylene compounds to give 1,2,3-triazoles of the type (**307**) and (**308**). Hydrolytic ring closure of these materials provides entry into [1,2,3]triazolo[4,5-*b*]pyridines with substituents at positions 5 and 6 (**309**) ⟨73JCS(P1)1620⟩. 8-Azapurine itself reacted with dimedone, ethyl acetoacetate and pentane-2,4-dione to give (**310**), (**311**) and (**312**), respectively, without any indication of the triazole intermediate resembling (**307**) and (**308**) ⟨73JCS(P1)1620⟩.

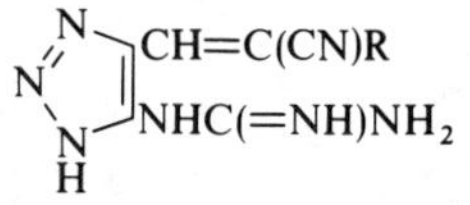

(**307**) R = CN, CO₂Et, CONH₂   (**308**) X = S, O; R = CO₂Et, CN 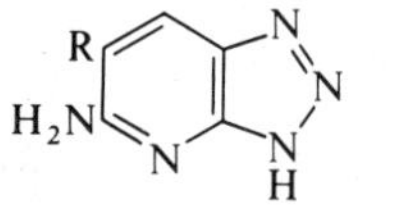   (**309**) R = CN, CONH₂

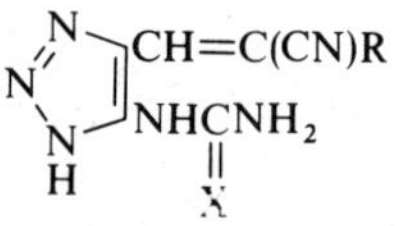

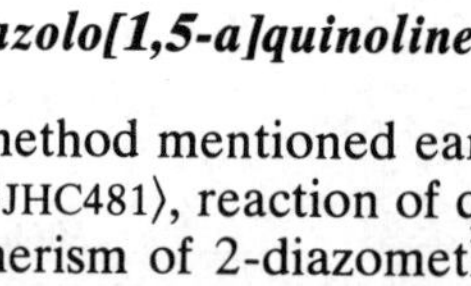

(**310**)

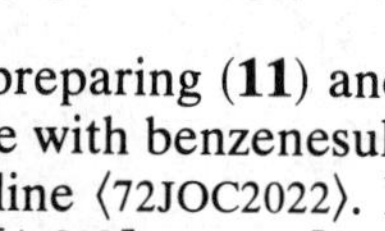

(**311**) R = OEt
(**312**) R = Me

4.15.4.2.3 [1,2,3]Triazolo[1,5-a]quinoline and [1,2,3]triazolo[5,1-a]isoquinoline

In addition to the method mentioned earlier for preparing (**11**) and its derivatives from quinol-2-yl oximes ⟨75JHC481⟩, reaction of quinaldine with benzenesulfonyl azide gave (**11**) via ring–chain tautomerism of 2-diazomethylquinoline ⟨72JOC2022⟩. Performing the same reaction on 1-methylisoquinoline gave the isomeric [1,2,3]triazolo[5,1-*a*]isoquinoline (**12**). This latter ring system and some of its derivatives have also been prepared by the oxidation of isoquinolin-1-yl ketone hydrazones or by the reaction of isoquinolin-1-yl ketones with tosylhydrazine ⟨75CB3794⟩.

4.15.4.3 [1,2,4]Triazolopyridazines and their Fused Benzo Derivatives

4.15.4.3.1 [1,2,4][Triazolo[4,3-b]pyridazine

The most fruitful and general route to (**13**) and its derivatives seems to be via oxidative ring closure (using bromine or lead tetraacetate) of the alkylidene or arylidene derivatives of the corresponding 3-hydrazinopyridazine (Scheme 41) ⟨66T2073⟩. Alternatively, thermal ring closure of the hydrazides obtained from 3-hydrazinopyridazines and acid chlorides may be utilized ⟨66T2073⟩. The ortho ester and carboxylic acid and anhydride ring closures widely employed to prepare other fused triazoles do not seem to be as useful ⟨69M671, 81JHC1523⟩ in the pyridazine series probably owing to the decreased nucleophilicity of N-2 of the six-membered ring. Thioimino ethers have also been used in this series as triazole ring-closing reagents ⟨81JHC893⟩. 4-Amino-1,2,4-triazole upon reaction with β-keto acetals provides a route to this series starting from the triazole moiety, a rarely used approach in preparing fused triazoles (Scheme 42) ⟨79JCS(P1)3085⟩.

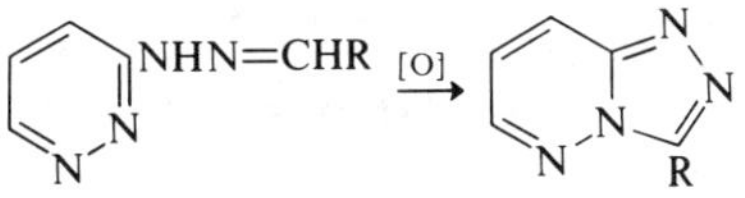

Scheme 41

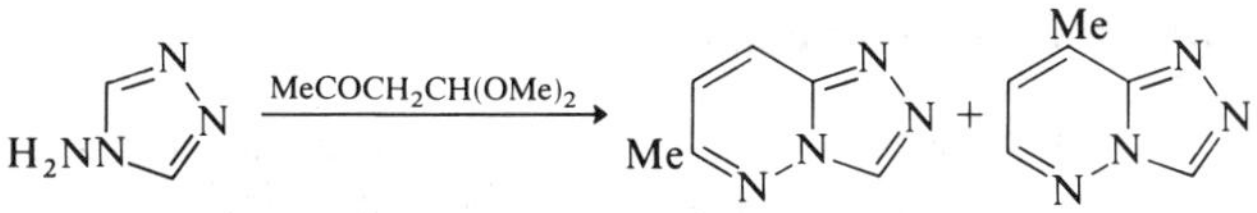

Scheme 42

The versatile 5-oxides of this series can be synthesized *via* ring closure of an appropriate 3-hydrazinopyridazine 1-oxide ⟨68JHC513⟩ or, in much lower yield, by direct oxidation of the bicyclic heterocycle ⟨70JOC2478⟩.

[1,2,4]Triazolo[3,4-*a*]phthalazine (**17**) and its derivatives are best synthesized from an appropriate 1-hydrazinophthalazine and ortho esters (3-alkyl), carbon disulfide (3-mercapto), urea (3-hydroxy) and cyanogen bromide (3-amino) ⟨69JOC3221⟩. The 3-chloro derivative was prepared by oxidative chlorination of the 3-mercapto compound or by treatment of (**17**) with sodium hypochlorite solution. Acid hydrolysis of the bis secondary amine derived from the 3-dichloromethyl derivative of (**17**) and morpholine yielded the 3-carbaldehyde ⟨69JOC3221⟩.

6,7,8,9-Tetrahydro-[1,2,4]triazolo[4,3-*b*]cinnolines have been prepared from 3-hydrazino-5,6,7,8-tetrahydrocinnolines and the reagents just described for (**17**) ⟨71BSF3043⟩, and also by condensing 2-hydroxymethylenecyclohexanone with 4-amino-1,2,4-triazole ⟨79JCS(P1)3085⟩.

4.15.4.3.2 *[1,2,4]Triazolo[1,5-b]pyridazine*

Of the routes developed to (**14**) and its derivatives two have emerged as the most useful. In the first case ⟨74JOC2143⟩, by analogy to a preparation of (**2**), various 3-aminopyridazines are first converted into their *N,N*-dimethylaminomethylene derivatives (**313**) with *N,N*-dimethylformamide dimethyl acetal. These latter derivatives with hydroxylamine become the hydroxyiminomethyleneamino compounds (**314**). Cyclization of (**314**) with polyphosphoric acid then gives the desired [1,2,4]triazolo[1,5-*b*]pyridazines. Included among these products are several possessing a 6-chloro substituent that permits easy access to a number of other derivatives by nucleophilic substitution (see reactivity section). Oxidative cyclization ⟨76JHC487⟩ of (**314b**; R = Cl) gave the versatile 3-oxide (**315**) which could be converted into (**14**).

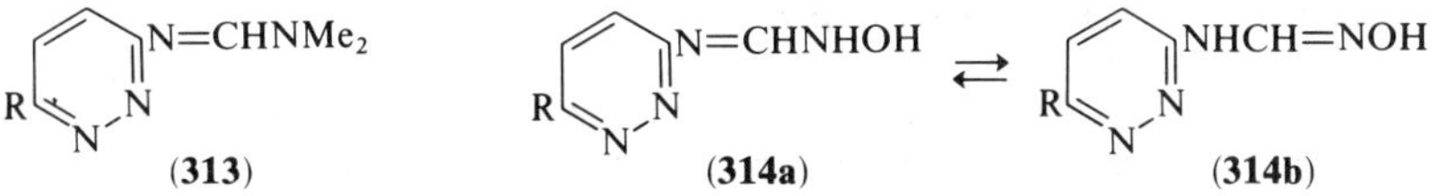

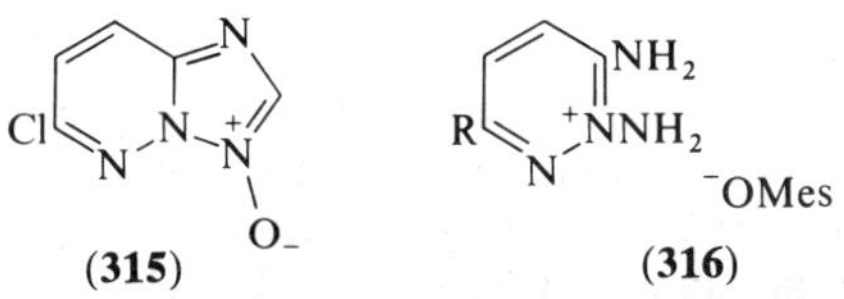

The other method to (**14**) and a large number of its derivatives ⟨75JHC107⟩ involves *N*-amination of 3-aminopyridazines with *O*-mesitylenesulfonylhydroxylamine to form the *N*-aminodiazinium salts (**316**). These salts can then be cyclized to the [1,5-*b*] system with formic acid, acetic anhydride or benzoyl chloride depending on the C-2 substituent (H, Me or Ph) desired.

The 2-phenyl derivatives have been prepared ⟨72TL4179⟩ by oxidative ring closure of the amidines (**317**) which are, in turn, available from 3-aminopyridazines and benzonitrile (Scheme 43).

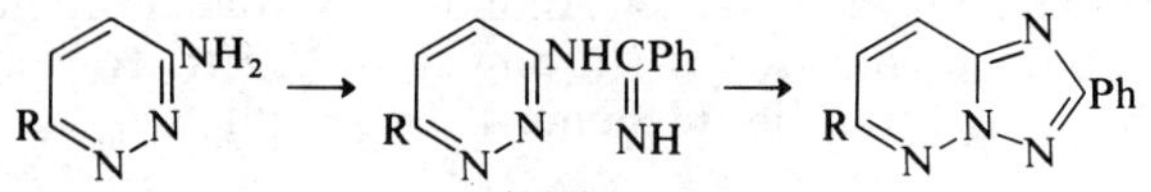

Scheme 43

4.15.4.4 [1,2,3]Triazolo[4,5-d]pyridazine

This ring system, often viewed as 2,8-diaza-3-deazapurine, and some of its derivatives have been prepared by diazotization of the appropriate 4,5-diaminopyridazine ⟨69JHC93, 70CPB1685⟩ as Scheme 44 represents for the formation of (**15**) itself. There are no other practical syntheses for this series of compounds. A similar approach has been used for preparing the isomeric [1,2,3]triazolo[4,5-c]pyridazines with the 7-chloro derivative being a useful compound ⟨67JHC555⟩.

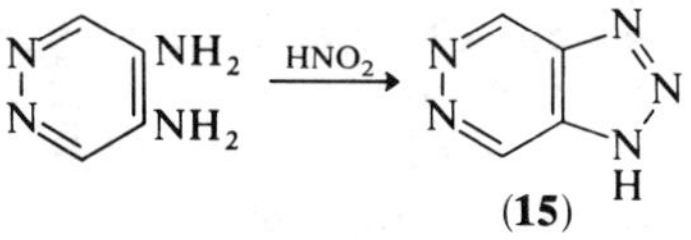

Scheme 44

4.15.4.5 [1,2,4]Triazolopyrimidines and their Fused Benzo Derivatives

4.15.4.5.1 [1,2,4]Triazolo[4,3-a]pyrimidine

Due to the rapid rearrangement of the [4,3-a] series into the [1,5-a] series (see Section 4.15.4.5.2) uncontaminated [1,2,4]triazolo[4,3-a]pyrimidines are difficult to prepare. The only useful methods available (Scheme 45) seem to be treatment of an appropriate 2-hydrazinopyrimidine with a low boiling ortho ester (for 3-H, 3-Me, 3-Et and 3-Pr) or oxidative ring closure of an arylidene derivative of 2-hydrazinopyrimidine to form the 3-aryl compounds ⟨77AJC2515⟩.

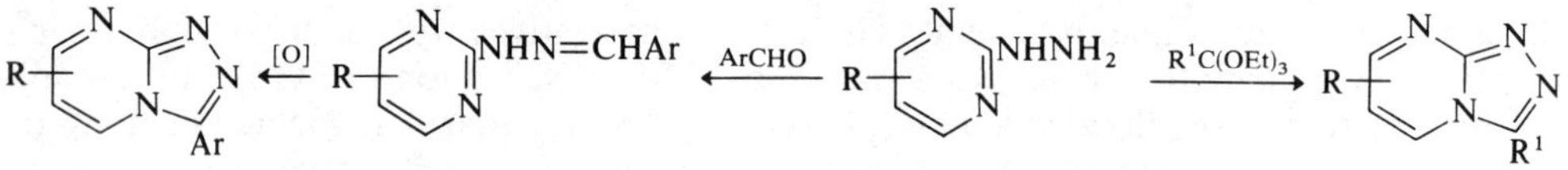

Scheme 45

4.15.4.5.2 [1,2,4]Triazolo[1,5-a]pyrimidine

The Dimroth-like rearrangement of [1,2,4]triazolo[4,3-a]pyrimidines in acid or alkali has been extensively used as a synthetic route to the isomeric [1,5-a] series (see Table 7). Thus, ring closure of a 2-hydrazinopyrimidine will initially give a [4,3-a] product which quickly rearranges to the [1,5-a] isomer, the material isolated from the reaction. The rearrangement does not occur under neutral conditions and proceeds with ring opening at N(4)—C(5) to give (**318**) in a manner similar to that described for [1,2,4]triazolo[4,3-a]pyridine (Scheme 2). The rates for isomerization are pH dependent, increasing at pH 10–12.5 for alkali rearrangement or when the pH nears the pK_a of the [4,3-a] material (pH 1.5–2.5) for acidic rearrangement ⟨77AJC2515⟩. The rearrangement was slowed by a 3-alkyl group due to electron donation retarding nucleophilic attack at C-5 and raising the pK_a. As suspected, electron-withdrawing groups at C-3 increased the rate of rearrangement

⟨77AJC2515⟩. A 5-methyl substituent retarded the rearrangement greatly, a result due to steric hindrance by the methyl group to nucleophilic attack at C-5 prior to N(4)—C(5) bond fission ⟨77AJC2515⟩.

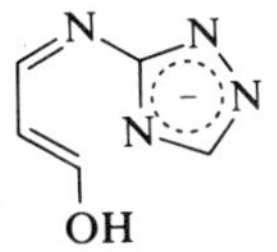

(318) (intermediate in alkali isomerization)

Several other methods have been put forward for preparing [1,2,4]triazolo[1,5-a]pyrimidines: the reaction of 1,3-dicarbonyl compounds with 3-amino-1,2,4-triazoles ⟨79JCS(P1)3085, 67JCS(C)503, 61CPB801⟩ (see Scheme 46 which relates this method with the rearrangement method) ⟨66CB2237⟩, dehydrative cyclization of oximes with polyphosphoric acid (Scheme 47) ⟨74JOC2143⟩ and ring closure of 1,2-diaminopyrimidinium mesitylenesulfonates with carboxylic acids, anhydrides or acid chlorides (Scheme 47) ⟨75JHC107⟩. Of these, the last route is preferable since the other two require somewhat more synthetically inaccessible starting materials. They do, however, make available many significant pyrimido-substituted derivatives ⟨61CPB801⟩.

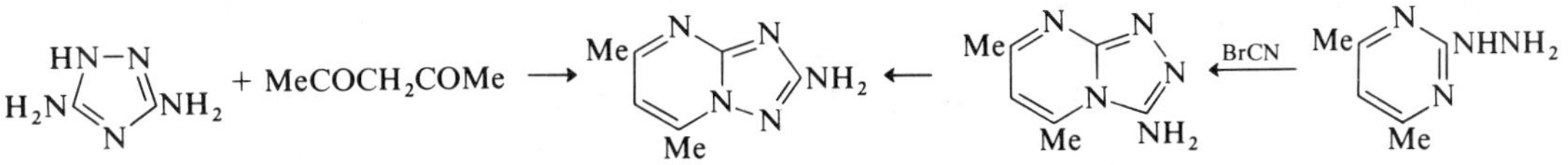

Scheme 46

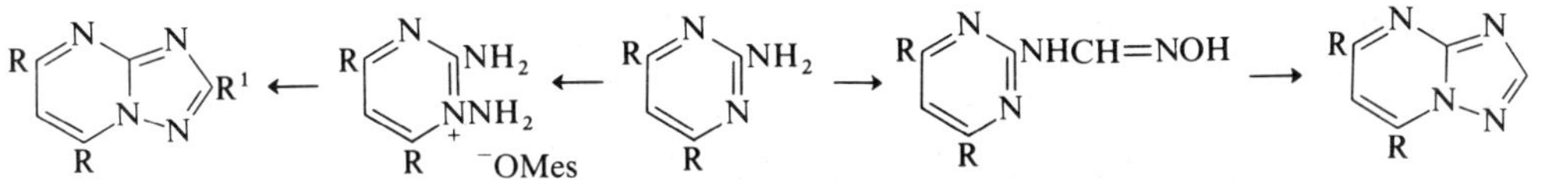

Scheme 47

Reaction of the pyrimidin-2-ylsulfimides (319) with aromatic nitrile oxides is reported to give the 3-oxides of the [1,5-a] system (320) ⟨76JCS(P1)2166⟩.

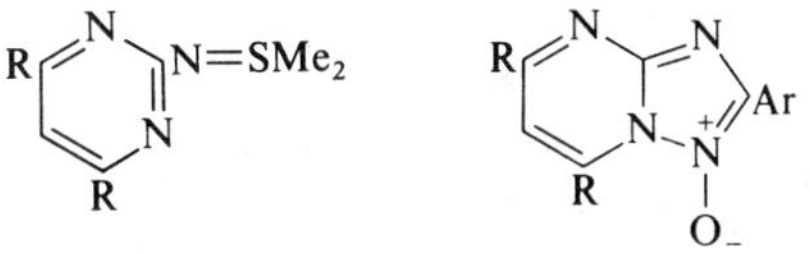

(319)   (320)

4.15.4.5.3 [1,2,4]Triazolo[4,3-c]pyrimidine

[1,2,4]Triazolo[4,3-c]pyrimidines, like their isomeric [4,3-a] compounds, are difficult to prepare without contamination by the corresponding [1,5-c] analogs ⟨76S833, 63JCS5642⟩. As with (18), ortho ester cyclization of 4-hydrazinopyrimidines ⟨78AJC2505⟩ was the most satisfactory route since the resulting [4,3-c], [1,5-c] and ethoxyalkylidene (321) derivatives could be separated without much difficulty. Heating of the alkylidene (321) in a vacuum also provided a means to (19) and its derivatives ⟨78AJC2505⟩.

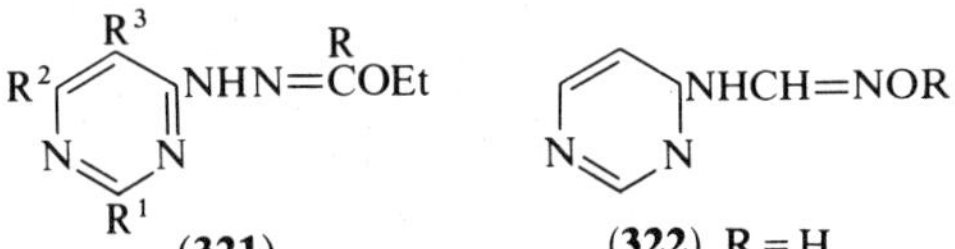

(321)

(322) R = H
(323) R = Ac

4.15.4.5.4 [1,2,4]Triazolo[1,5-c]pyrimidine

Two routes are available for preparing this ring system: (i) treatment of 4-aminopyrimidines with N,N-dimethylformamide dimethyl acetal and hydroxylamine to give

(322) followed by acetylation to (323) and warming ⟨76S833⟩ and (ii) the Dimroth-type rearrangement of the [4,3-c] series ⟨78AJC2505⟩.

In contrast to the smooth Dimroth-type transformation in the [4,3-a] → [1,5-a] systems, the [4,3-c] → [1,5-c] conversions were complicated by the stabilities of the intermediates (324), which are often isolable. These intermediates were always the terminating point in any attempted rearrangement of the [4,3-c] derivatives in aqueous buffer (pH 1–13). The fact that complete rearrangement of the [4,3-a] series to the [1,5-a] series occurred in aqueous buffer, whereas the [4,3-c] did not completely rearrange, has been explained by the nature of the respective fission products: the intermediate from the [4,3-a] compounds (325) is a reactive hydroxymethylene compound whereas the intermediate from initial ring opening of the [4,3-c] derivatives (324) is a relatively unreactive amide ⟨78AJC2505⟩.

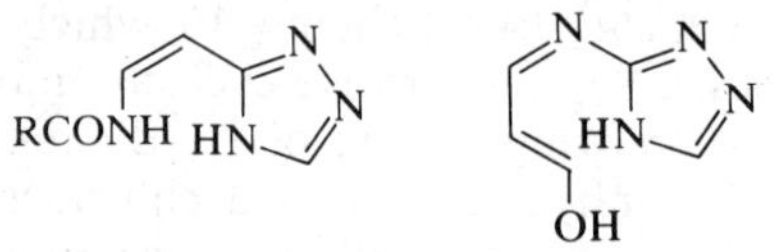

(324) (compare with 204)   (325)

A 5-methyl group on [1,2,4]triazolo[4,3-c]pyrimidines retards the rearrangement as a result of reducing the electron density at C-5 and sterically crowding this site towards nucleophilic approach. A $C(8)$-methyl group had a similar retarding effect due only to a mesomeric effect. On the other hand, a $C(7)$-methyl substituent had no effect on the rate. These results, together with the isolation of the amide intermediate (324), indicate ring opening occurs at the N(4)—C(5) center in the [4,3-c] series.

In contrast to aqueous buffer, formic acid and glacial acetic acid caused the triazolo[4,3-c]pyrimidines to rearrange completely to their [1,5-c] counterparts. This merely seems to be a catalytic effect operative on intermediate (324).

4.15.4.5.5 [1,2,4]Triazolo[4,3-c]quinazoline and [1,2,4]triazolo[1,5-c]quinazoline

Ring system (22) and its derivatives are best synthesized by treating 4-hydrazinoquinazolines with ortho esters in the presence of potassium carbonate ⟨70JOC3448⟩. Omission of the carbonate always resulted in a mixture of the [4,3-c] (22) and [1,5-c] (24) ring systems, apparently due to rearrangement of the [4,3-c] into the [1,5-c] by the traces of the appropriate acid in the ortho ester. This assumption was verified by treating 4-hydrazinoquinazolines with carboxylic acids and by warming the [4,3-c] systems with a carboxylic acid to give, in both cases, [1,2,4]triazolo[1,5-c]quinazolines (Scheme 48) ⟨70JOC3448⟩. This rearrangement, which is also promoted by heat, is the preferred route to the [1,5-c] series (24).

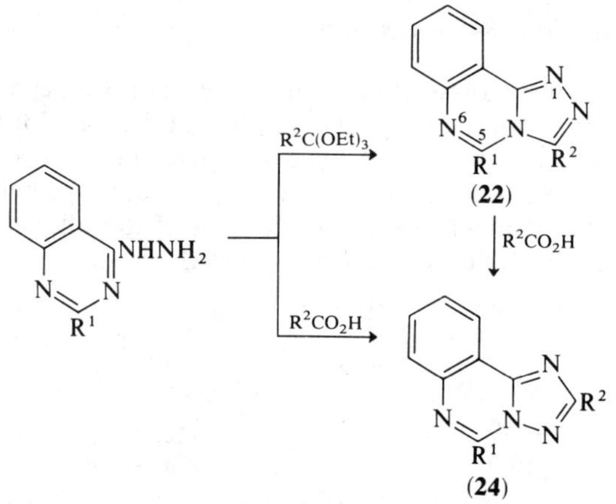

Scheme 48

The [4,3-c] → [1,5-c] isomerization is not unlike that described previously in this chapter for the [4,3-a]pyridines and [4,3-a]- and [4,3-c]-pyrimidines in that covalent hydration occurs at the C(5)—N(6) double bond of (22) with initial nucleophilic attack at C-5. This is followed by ring opening to an amide, resembling (324) from the [4,3-c]pyrimidine system, which subsequently undergoes ring closure at N-1 of the 1,2,4-triazole nucleus to give (24). The thermal isomerization may follow a different pathway involving a zwitterion ⟨70JOC3448⟩.

An alternative, more limited, route to (24) and its derivatives involves the 2-amino compound (326) which results from the reaction of aminoguanidine and the [3,1]benzoxazine (327; Scheme 49) ⟨68CB2106⟩. Diazotization of (326) and then hypophosphorus acid produces 2-substituted derivatives of (24) ⟨70JOC3448⟩.

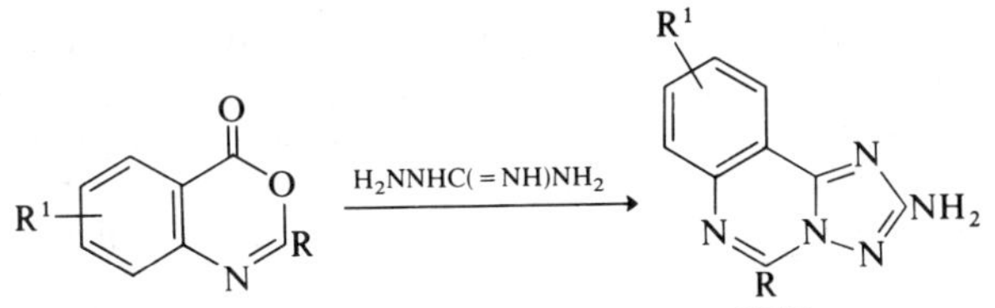

Scheme 49

4.15.4.5.6 [1,2,4]Triazolo[1,5-a]quinazoline

The synthetic studies of this ring system have been rather limited and the recent reaction of 2-hydrazinobenzoic acid with *N*-cyanoimidates (Scheme 50) seems to offer the best route to (23) and its derivatives. The 5-oxo products are susceptible to many useful transformations as described in Section 4.15.3.5.5 ⟨80HCA1⟩.

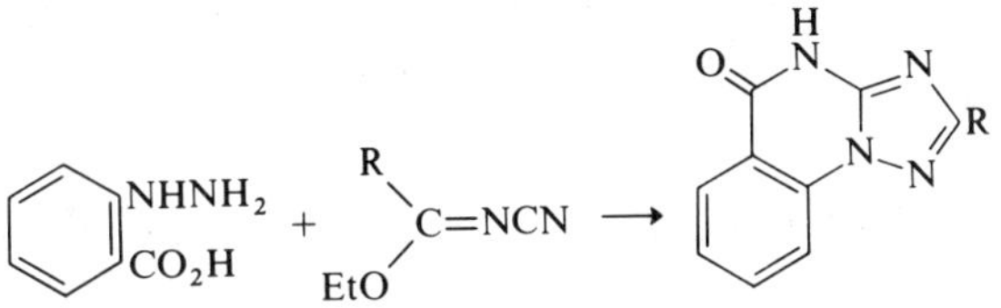

Scheme 50

4.15.4.5.7 [1,2,4]Triazolo[5,1-b]quinazoline

This ring system, in its 6,7,8,9-tetrahydro form, has been prepared ⟨79JCS(P1)3085⟩ by the reaction of 2-hydroxymethylenecyclohexanone with 3-amino-1,2,4-triazoles (Scheme 51). In the fully aromatic form, compound (25) and its [3,4-*b*] isomer are apparently not particularly stable due to the *o*-quinoid nature of the structure.

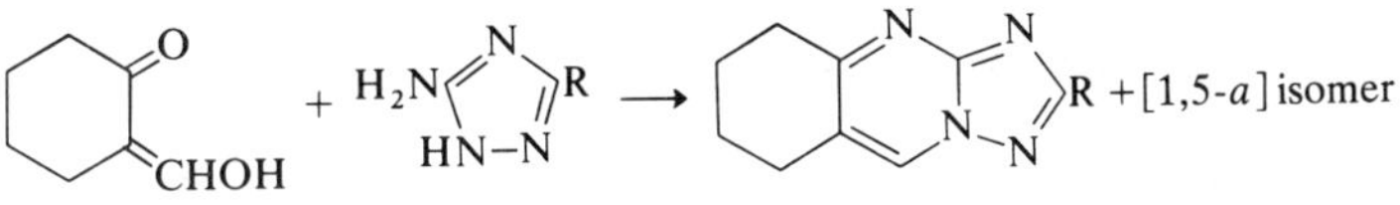

Scheme 51

4.15.4.5.8 [1,2,4]Triazolo[4,3-a]quinazoline

Very little is available on [1,2,4]triazolo[4,3-*a*]quinazolines but some representatives of this ring system have been prepared (for example 329) ⟨64ZOB1745⟩ and the ring system has pharmaceutical usefulness ⟨79MI41501⟩.

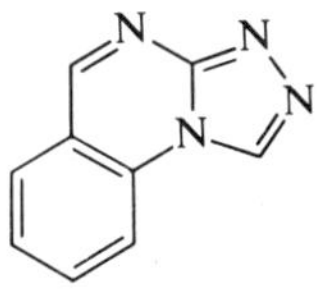

(328) 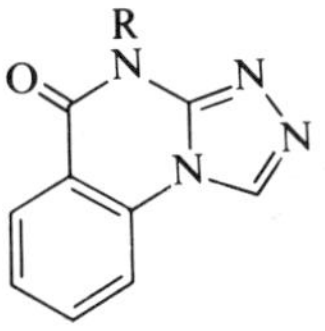   (329)

4.15.4.6 [1,2,3]Triazolopyrimidines and their Fused Benzo Derivatives

4.15.4.6.1 [1,2,3]Triazolo[1,5-a]pyrimidine

Condensation of amino-1,2,3-triazoles with β-dicarbonyl compounds is an efficient route to substituted [1,2,3]triazolo[1,5-a]pyrimidines ⟨71JCS(C)2156⟩. Scheme 52 illustrates this with pentane-2-4-dione. Care must be exercised in this condensation as the products readily undergo ring opening with loss of nitrogen to form pyrimidines (see **330**) as described in more detail in the reactivity section. Benzoylacetone and dibenzoylmethane required acetic acid to promote their reaction with the aminotriazole and, as a consequence, only the pyrimidines analogous to (**330**) were obtained ⟨71JCS(C)2156⟩ with no triazolopyrimidine detected. Reaction of the aminotriazole with diethyl malonate or ethyl benzoylacetate did not lead to the bicyclic heterocycle. Condensation of 5-amino-4-phenyl-1H-1,2,3-triazole with ethyl acetoacetate gave (**331**) ⟨73JCS(P1)943⟩.

Scheme 52

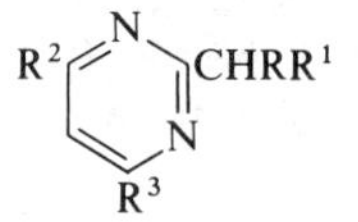

(**330**) 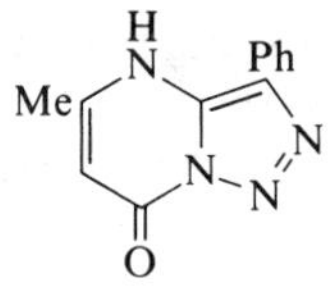   (**331**)

4.15.4.6.2 [1,2,3]Triazolo[1,5-c]pyrimidine

Little is known about this series of compounds but the parent structure (**27**) has been synthesized via the oxidative ring closure of pyrimidine-4-carbaldehyde hydrazone (Scheme 53) ⟨78JHC1041⟩.

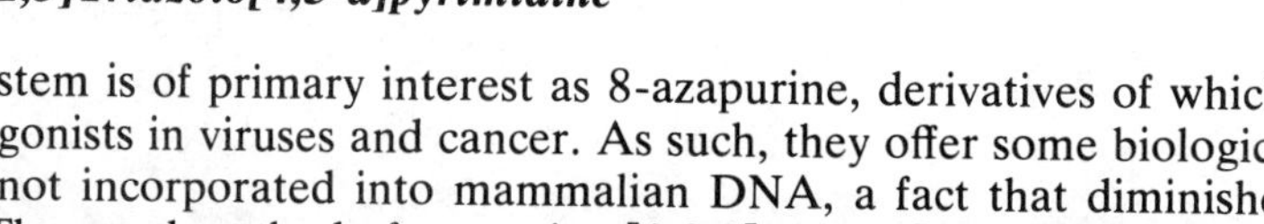

Scheme 53

4.15.4.6.3 [1,2,3]Triazolo[4,5-d]pyrimidine

This ring system is of primary interest as 8-azapurine, derivatives of which could serve as purine antagonists in viruses and cancer. As such, they offer some biological promise in that they are not incorporated into mammalian DNA, a fact that diminishes the risk of mutagenesis. The usual method of preparing [1,2,3]triazolo[4,5-d]pyrimidines ⟨61JOC4433⟩ is to treat a 4,5-diaminopyrimidine with nitrous acid (Scheme 54). This is very versatile for synthesizing pyrimido derivatives of the (**28**) ring system since 2- and 6-monosubstituted or 2,6-disubstituted 4,5-diaminopyrimidines will give the corresponding bicyclic ring system substituted at positions 5, 7 or both. Compounds of this series can also be synthesized by ring closure of a 4,5-disubstituted 1,2,3-triazole (Scheme 55) ⟨81JCS(P1)2344⟩.

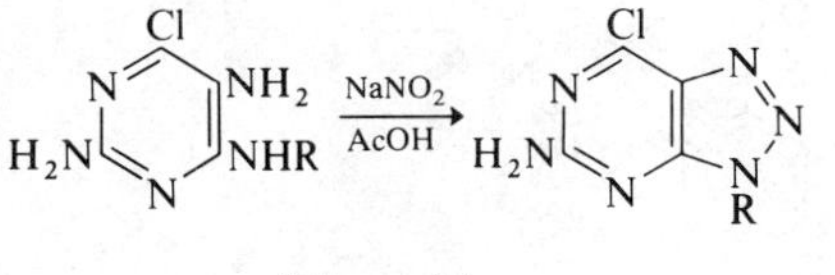

Scheme 54 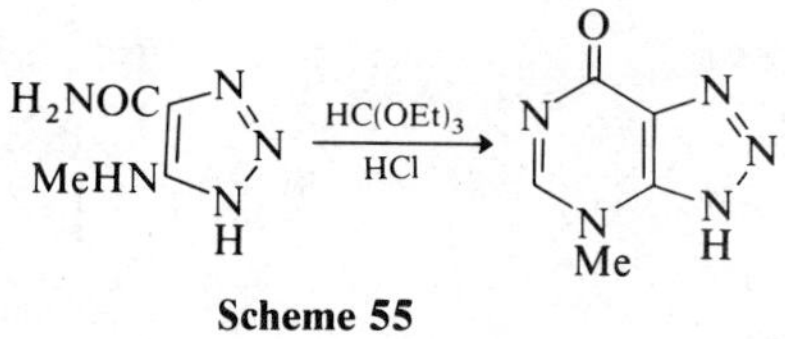   **Scheme 55**

Many 8-azapurine analogs of the naturally occurring purines have been prepared by these methods and some examples include 8-azaadenosine (**332**) ⟨70JHC215⟩, 8-azaguanine (**333**) ⟨61MI41500⟩ and 8-azatheophylline (**334**) ⟨74MI41500⟩.

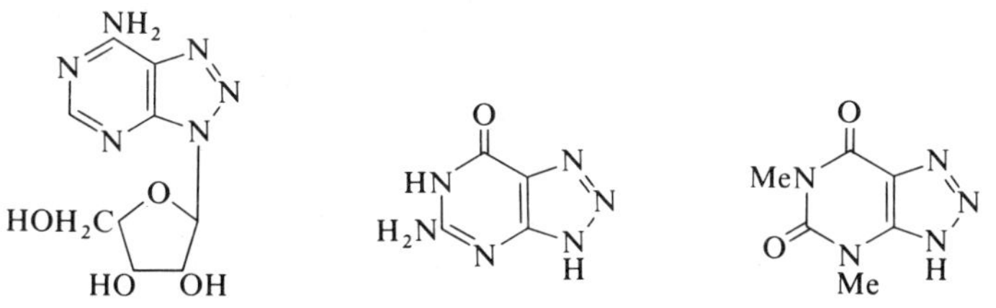

A rather interesting approach to this ring system involves, for example, treatment of 4,6-diaminopyrimidine with 2-azido-3-ethylbenzothiazolium tetrafluoroborate to give 8-azaadenine. This is illustrated in Scheme 56 ⟨78HCA108⟩ and may have general application.

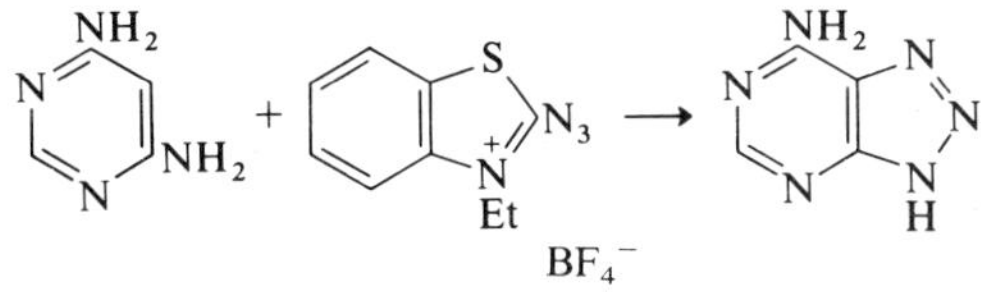

Scheme 56

4.15.4.6.4 [1,2,3]Triazolo[1,5-a]quinazoline

A representative of this ring system (namely, 5-oxo-3-phenyl-4,5-dihydro[1,2,3]triazolo[1,5-a]quinazoline (**335**)) has been prepared by the reaction between o-azidobenzoic acid and phenylacetonitrile in the presence of sodium methoxide ⟨66JCS(C)2290⟩. This is illustrated in Scheme 57.

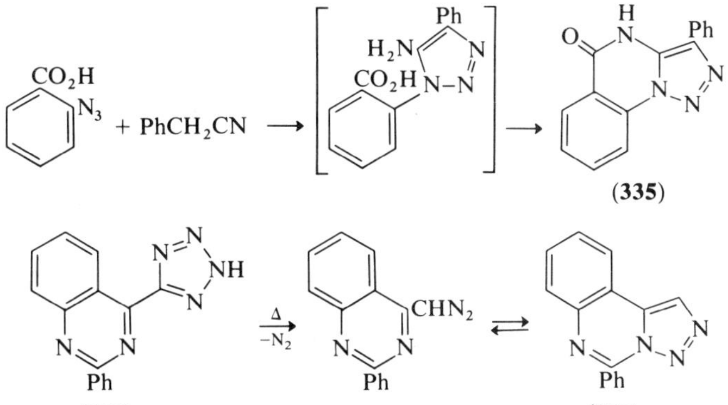

Scheme 57

4.15.4.6.5 [1,2,3]Triazolo[1,5-c]quinazoline

5-Phenyl [1,2,3]triazolo[1,5-c]quinazoline (**336**) has been prepared by heating 5-(2-phenyl-4-quinazolinyl)tetrazole (**337**) in mesitylene at 160 °C as shown in Scheme 57 ⟨78HCA1755⟩.

4.15.4.7 Triazolopyrazines and their Fused Benzo Derivatives

4.15.4.7.1 [1,2,4]Triazolo[4,3-a]pyrazine

The most efficient method of preparation of (**31**) and its alkyl derivatives was reported ⟨62JOC3243⟩ to be *via* ring closure of 2-hydrazinopyrazines with ortho esters (Scheme 58).

Now that the means for fusing triazoles to pyridine and the three diazines have been presented, it can be concluded that the ease of ring closure of the corresponding hydrazines is dependent on the pK_a of the nitrogen of the six-membered ring onto which closure takes place. Thus, the order of reactivity with ortho esters as cyclizing agents is pyrazine > pyrimidine > pyridazine > pyridine ⟨62JOC3243⟩. The procedures employed in the other triazole ring fusions (such as the carboxylic acids or anhydrides or phosphorus oxychloride cyclodehydrations on the hydrazides) failed in the pyrazine series.

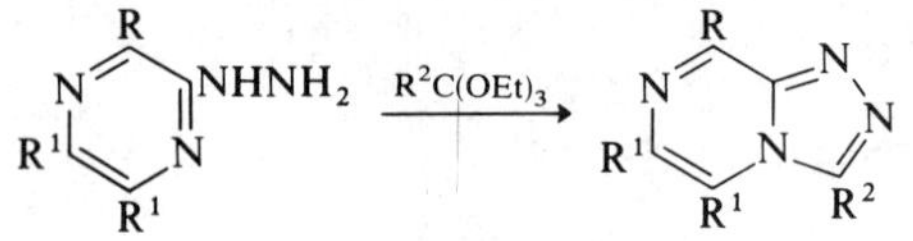

R^1 = H, Me, Ph in various combinations; R^2 = H, Me, Et

Scheme 58

Treatment of 2-hydrazinopyrazines with carbon disulfide in chloroform or cyanogen chloride produced the 3-thiol and 3-amino derivatives (**338**) and (**339**), respectively, while fusion with urea or reaction with phosgene gave the 3-hydroxy derivatives (**340**) ⟨68JHC485, 66JCS(C)2038, 69JCS(C)1593⟩. The carbon disulfide ring closure has not always been successful in the pyrazine series ⟨62JOC3243, 66JCS(C)2038⟩ so the pyrazino substituents may play an electronic role in this conversion. A number of 3,8-disubstituted [1,2,4]triazolo[4,3-a]pyrazines have been prepared by ring closure (using ortho esters or thioimino ethers) of 3-chloro-2-hydrazinopyrazines ⟨78JHC987, 79JOC1028, 79MI41500⟩. Included among these derivatives is the nucleoside (**341**).

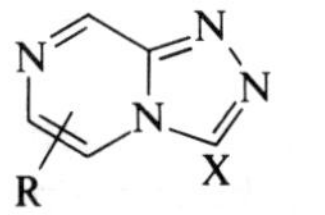

(**338**) X = SH
(**339**) X = NH$_2$
(**340**) X = OH

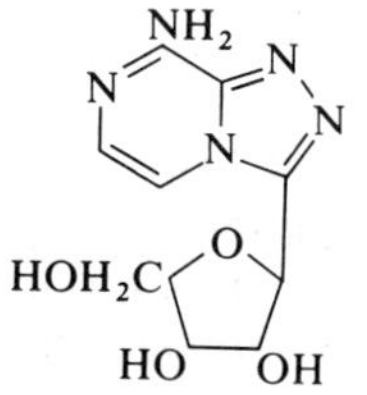

(**341**)

4.15.4.7.2 [1,2,4]Triazolo[1,5-a]pyrazine

The two most useful routes to (**32**) and some of its derivatives are (i) the ring closure of the N-aminodiazinium salt (**342**) with carboxylic acids, anhydrides or acid chlorides as previously mentioned for the pyridazines and pyrimidines ⟨75JHC107⟩ and (ii) the cyclodehydration of the hydroxyiminomethyleneamino compound (**343**) with polyphosphoric acid as reported herein for the pyridine, pyridazine and pyrimidine analogs ⟨74JOC2143⟩. These two methods are shown in Scheme 59.

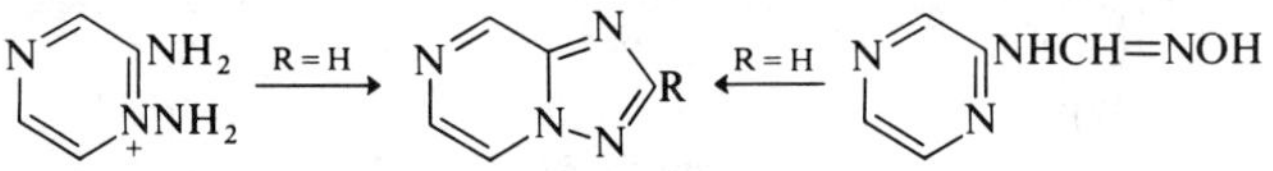

(**342**)   (**343**)

Scheme 59

2-Phenyl-[1,2,4]triazolo[1,5-a]pyrazines have been prepared by the oxidative ring closure of N(2)-pyrazinylbenzamidines (**344**) as shown in Scheme 60. The benzamidines were synthesized from 2-aminopyrazines and benzonitrile, a reaction that failed with aliphatic nitriles ⟨64JOC2542⟩.

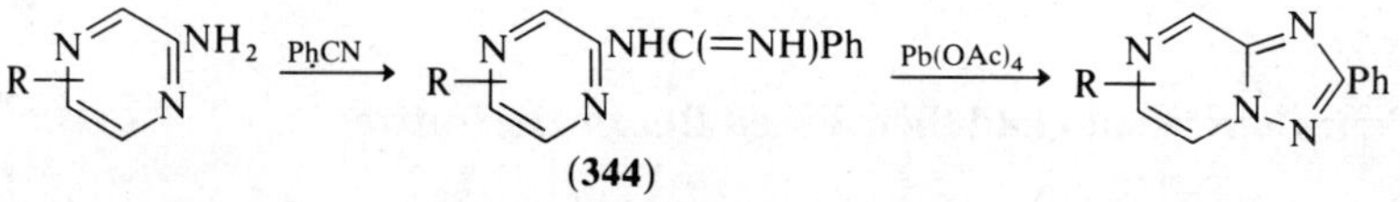

(**344**)

Scheme 60

As in the other systems discussed up to this point, the [4,3-a]pyrazine series is susceptible to the Dimroth-like rearrangement into the [1,5-a] compounds ⟨68JHC485, 66JCS(C)2038⟩.

For example, when 5,8-dimethyl-[1,2,4]triazolo[4,3-a]pyrazine was refluxed with 10% sodium hydroxide solution for 60 hours, a 20% yield of the isomeric 5,8-dimethyl-[1,2,4]triazolo[1,5-a]pyrazine was obtained. No rearrangement occurred in acid or under the influence of heat ⟨68JHC485⟩. The poor yield obtained in this isomerization may be due to two factors: (i) electronic deactivation and steric crowding caused by the 5-methyl group and/or (ii) the great π-electron density associated with C-5 (as determined by HMO calculations) for [1,2,4]triazolo[4,3-a]pyrazines in general ⟨68JHC485⟩.

On the other hand, the 3-amino derivatives rearrange under the influence of heat or hot aqueous hydrochloric acid, sodium carbonate or sodium hydroxide. For example, reaction of (345) with acid gave the rearranged product wherein the amino group had been hydrolyzed to the hydroxy compound (346; Scheme 61). That this hydrolysis took place after isomerization was proven by the inability of the corresponding 3-hydroxytriazolo[4,3-a]pyrazine to undergo this rearrangement ⟨66JCS(C)2038⟩. The 3-amino derivative unsubstituted in the pyrazine ring was too susceptible to hydrolytic breakdown to produce any product recognizable as having arisen *via* a rearrangement route.

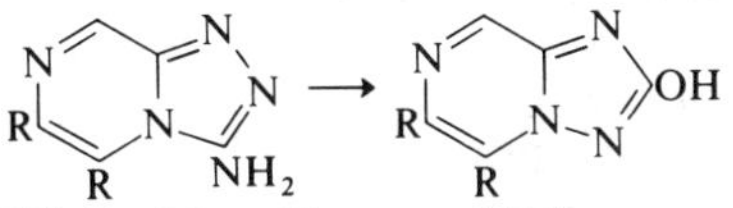

Scheme 61

6,8-Dibromo[1,2,4]triazolo[1,5-a]pyrazine 3-oxide (347), as a representative *N*-oxide in the [1,5-a] series, was obtained by the oxidative cyclization of (348) ⟨76JHC487⟩.

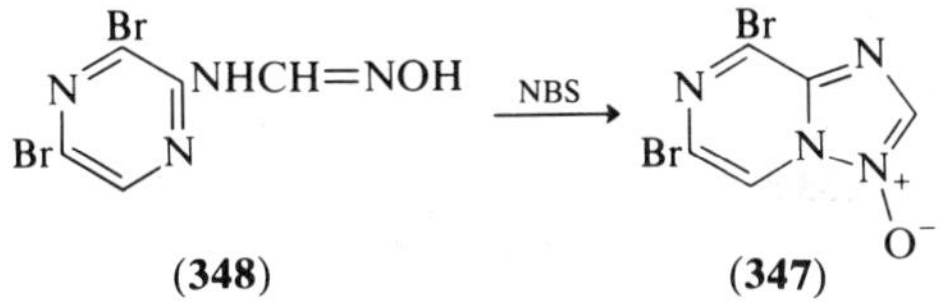

Scheme 62

A [1,2,4]triazolo[1,5-a]pyrazine analog of guanine has been prepared by cyclization of a 1,5-disubstituted triazole ⟨77JHC697⟩.

4.15.4.7.3 [1,2,3]Triazolo[1,5-a]pyrazine

Compound (33) has been prepared in 20% yield by the thermolysis of 5-(2-pyrazinyl)tetrazole as shown in Scheme 63 ⟨78HCA1755⟩.

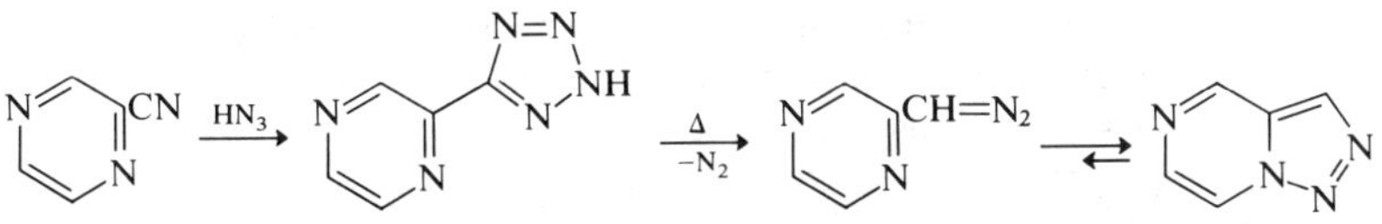

Scheme 63

4.15.4.7.4 [1,2,3]Triazolo[4,5-b]pyrazine

A variety of 1-benzyl-[1,2,3]triazolo[4,5-b]pyrazines has been prepared by the condensation between α-dicarbonyl compounds and 4,5-diamino-1-benzyl-1,2,3-triazole (Scheme 64) ⟨72JOC4124⟩. 2-Phenyl derivatives have been synthesized in a similar way using 4,5-diamino-2-phenyl-1,2,3-triazole (Scheme 64) ⟨78JOC341⟩.

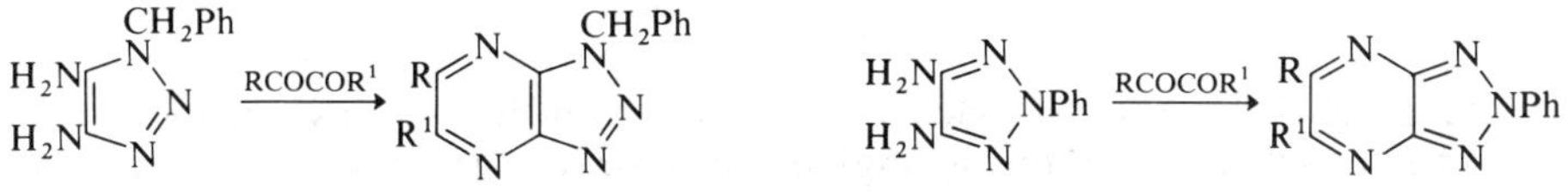

Scheme 64

4.15.4.7.5 [1,2,4]Triazolo[4,3-a]quinoxaline

Cyclization of 2-hydrazinoquinoxalines occurs very readily with carboxylic acids and esters and ortho esters. This is in contrast to the pyrazines which could only be cyclized with ortho esters ⟨60JA4044, 68JHC485⟩. Pyrolysis of the hydrazones derived from 2-hydrazinoquinoxaline have led to various compounds of this series ⟨60JA4044⟩.

The l-hydroxy, -mercapto and -amino derivatives were prepared, as in the pyrazine series, by ring closure of the hydrazine with urea, carbon disulfide and cyanogen bromide, respectively ⟨68JHC485⟩. The l-aryl compounds arose *via* phosphorus oxychloride cyclodehydration of the corresponding ring opened aroylhydrazines ⟨68JHC485⟩. As might be expected, the [4,3-*a*]quinoxalines did not rearrange to the [1,5-*a*] isomers. The 5-oxides in the [4,3-*a*]quinoxalines were prepared by ring closure of the appropriate 2-hydrazinoquinoxaline 4-oxide ⟨78JOC4125⟩.

4.15.4.7.6 [1,2,3]Triazolo[1,5-a]quinoxaline

There are two syntheses of this ring reported ⟨67JA2633, 78JOC4125⟩. The first involved the reductive cyclization of the 1-(2-nitrophenyl)-1*H*-1,2,3-triazole (**349**) to (**350**), whereas the other route gave a variety of 5-oxides (**351**) by the sodium methoxide ring closure of the tosylhydrazones (**352**). These two routes are shown in Scheme 65.

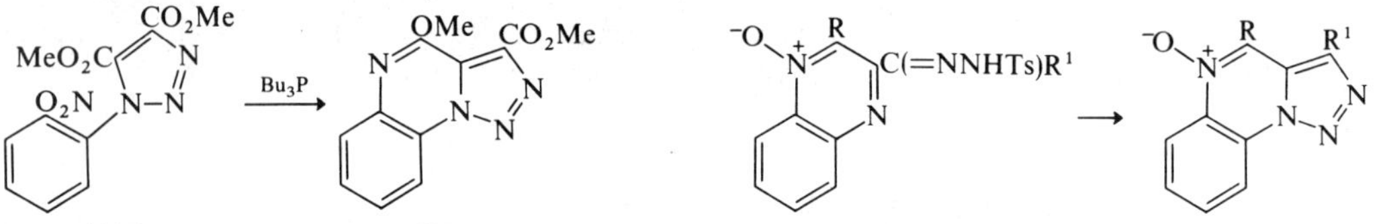

Scheme 65

4.15.4.8 [1,2,4]Triazolotriazines

4.15.4.8.1 [1,2,4]Triazolo[4,3-b][1,2,4]triazine

The [1,2,4]triazolo[4,3-*b*][1,2,4]triazines have been prepared, primarily (i) from the known ⟨68CB2747⟩ 7-oxo-[1,2,4]triazolo[4,3-*b*][1,2,4]triazines by the sequence of steps shown in Scheme 66, (ii) *via* the condensation of 3,4-diamino-1,2,4-triazoles with α-dicarbonyl reagents (Scheme 66) ⟨64CB2179⟩ and (iii) by the ring closure of 3-hydrazino-1,2,4-triazines with carboxylic acids and ortho esters (Scheme 66) ⟨77JOC1018⟩. The latter cyclization to the [4,3-*b*] product rather than the [3,4-*c*] isomer was explained by the enhanced nucleophilicity of N-2 as compared to N-4 ⟨77JOC1018, 81JHC1353⟩. Ring closure of 3-hydrazino-5,6-diphenyl-1,2,4-triazine with cyanogen bromide produced 3-amino-6,7-diphenyl-[1,2,4]triazolo[4,3-*b*][1,2,4]triazine with no evidence for the [3,4-*c*] isomer ⟨76JCS(P1)1492⟩, in agreement with the increased nucleophilicity at N-2. If N-2 was blocked, ring closure occurred at N-4 to give the [3,4-*c*] products ⟨64CB2185⟩.

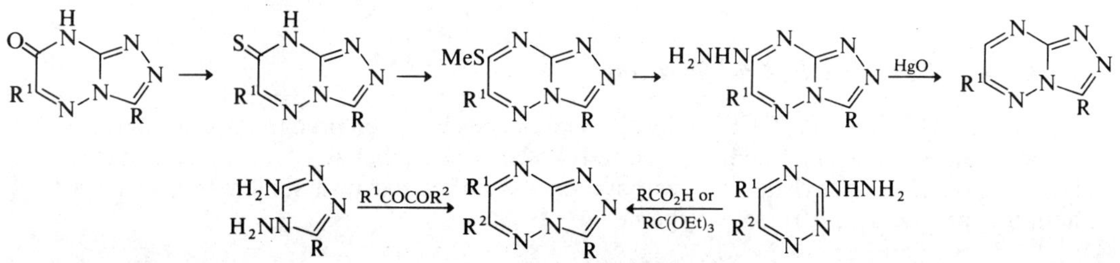

Scheme 66

In studying the N-2 *versus* N-4 closure competition further, an investigation ⟨75BSF857⟩ of the effect of the substituent R on the ring closure of 3-hydrazino-5-hydroxy-1,2,4-triazines (**353**) showed that when R was electron-donating and carboxylic acids were used as the cyclizing agents, [1,2,4]triazolo[4,3-b][1,2,4]triazines were the products. However, if R were electron-attracting, acids and ortho esters gave [1,2,4]triazolo[3,4-c][1,2,4]triazines (**38**). Beginning with 3-hydrazino-5,5,6-trimethyl[1,2,4]triazine, dihydro derivatives of both the [4,3-b] and [3,4-c] series have been obtained ⟨79JHC427⟩.

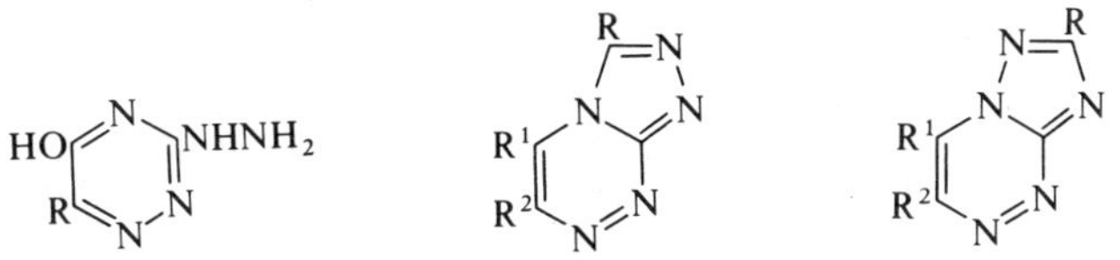

(**353**)   (**354**)   (**355**)

4.15.4.8.2 [1,2,4]Triazolo[3,4-c][1,2,4]triazine

Of the two methods just described for preparing the [3,4-c] system the one which employed the electron-attracting groups on (**353**) was the only remotely useful one but it was limited since it also produced some of the [1,2,4]triazolo[5,1-c][1,2,4]triazines (**42**) by a Dimroth related process ⟨75BSF857⟩. In an effort to circumvent this rearrangement problem, 3-hydrazino-1,2,4-triazole was condensed with α-dicarbonyl compounds under controlled conditions and found to give the derivatives represented by (**354**). However, if the temperature or reacting time were not carefully controlled, only the rearranged [5,1-c] derivative (**355**) was obtained ⟨77JOC1018⟩. The conversion of the [3,4-c] isomer (**354**) into the [5,1-c] isomer (**355**) could also be accomplished with aqueous sodium hydroxide.

4.15.4.8.3 [1,2,4]Triazolo[5,1-c][1,2,4]triazine

In addition to the Dimroth rearrangement of [1,2,4]triazolo[3,4-c][1,2,4]triazines discussed in the last section, derivatives of (**42**) would probably be accessible by the general application of a special case which involved the reaction of 1-acetonyl-5-chloro-1,2,4-triazole (**356**) with hydrazine to produce the dihydro compound (**357**). Lead tetraacetate oxidation of (**357**) formed (**358**) ⟨77JOC1018⟩.

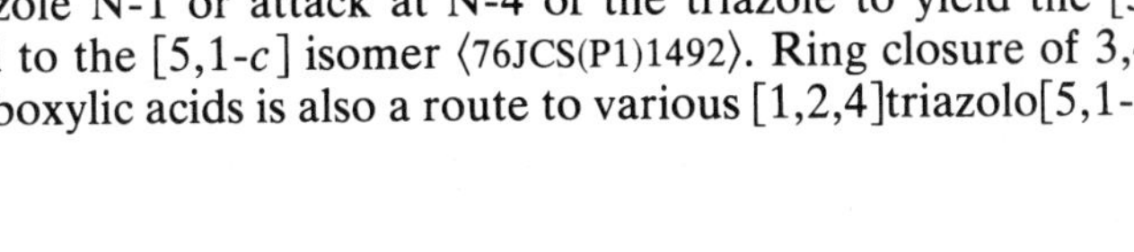

(**356**)   (**357**)   (**358**)

Treatment of 3-amino-5-hydrazino-1,2,4-triazole with benzil produced 2-amino-6,7-diphenyl-[1,2,4]triazolo[5,1-c][1,2,4]triazine *via* either direct formation involving cyclization on the triazole N-1 or attack at N-4 of the triazole to yield the [3,4-c] compound which rearranged to the [5,1-c] isomer ⟨76JCS(P1)1492⟩. Ring closure of 3,4-diamino-1,2,4-triazines with carboxylic acids is also a route to various [1,2,4]triazolo[5,1-c][1,2,4]triazines ⟨64CB2173⟩.

4.15.4.8.4 [1,2,4]Triazolo[4,3-d][1,2,4]triazine

This series of compounds is prepared by ring closure of the appropriate 4-hydrazino-1,2,4-triazines with carboxylic acids, carbon disulfide, cyanogen bromide, *etc.* to give the 3-unsubstituted or 3-substituted (alkyl, thiol, amino) derivatives ⟨67CB3467, 81JHC1353⟩.

4.15.4.8.5 [1,2,4]Triazolo[3,4-f][1,2,4]triazine

As with the [4,3-d] series of the previous section, derivatives of (**40**) are available by ring closure of various 6-hydrazino-1,2,4-triazines ⟨74JPR667, 79JHC555⟩. There is also a report whereby these compounds have been prepared from 1,2,4-triazoles ⟨70JPR669⟩.

4.15.4.8.6 [1,2,4]Triazolo[1,5-b][1,2,4]triazine

These compounds have been prepared in two ways from 3-amino-1,2,4-triazines. In the first case, the aminotriazine is converted into the amidine (**359**) with *N,N*-dimethylformamide or -acetamide acetal. Reaction of the amidine with hydroxylamine produces the amidoxime (**360**) and cyclization with phosphorus oxychloride forms the desired bicyclic product (**361**) ⟨77JOC1018⟩. Also, the desired ring system can be synthesized by treating 3-amino-1,2,4-triazines with acetonitrile in the presence of aluminum chloride to afford the acetamidines (**362**) which, in one case ($R^1 = R^2 = Me$), can be transformed into the [1,5-*b*] product with lead tetraacetate ⟨77JOC1018⟩. These two routes are shown in Scheme 67.

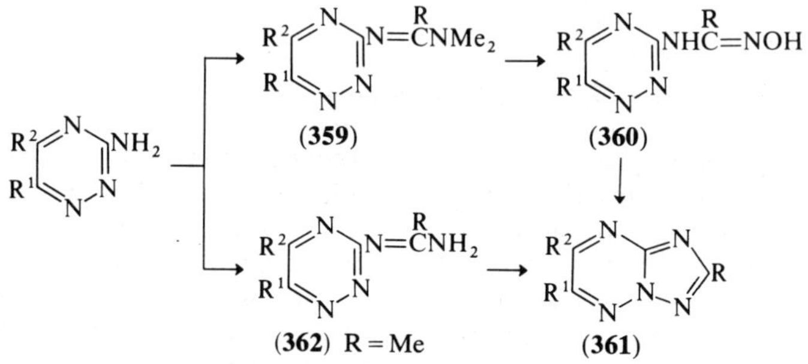

Scheme 67

4.15.4.8.7 [1,2,4]Triazolo[1,5-d][1,2,4]triazine

Ring closure of 5(3)-alkyl-3(5)-carboxyhydrazino-1,2,4-triazole with triethyl orthoformate has produced the [1,5-*d*] product along with the [4,3-*d*] (Scheme 68) ⟨81T4353⟩. There has also been a reported rearrangement of the [4,3-*d*] system to the [1,5-*d*] isomer with warm formic acid ⟨69BSF3670⟩.

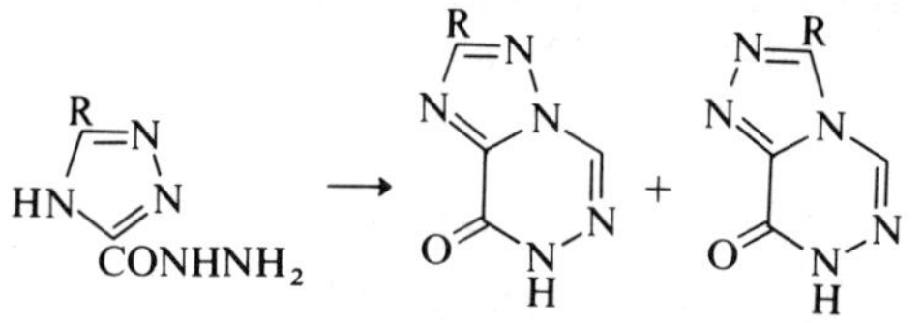

Scheme 68

4.15.4.8.8 [1,2,4]Triazolo[4,3-a][1,3,5]triazine

Ring closure of symmetrically 4,6-disubstituted (or, of course, unsubstituted) 2-hydrazino-1,3,5-triazines *via* oxidation of the alkylidene or arylidene derivatives has been shown to give the ring system represented by (**44**) with no structural ambiguity ⟨70T3357⟩. With monosubstituted and diversely substituted 2-hydrazino-1,3,5-triazines cyclization of the alkylidene or arylidene derivatives gives both possible products with electronic and steric factors controlling which product will predominate ⟨70T3357⟩. This ring system can also be prepared from 1,2,4-triazoles ⟨53JOC1610⟩.

4.15.4.8.9 [1,2,4]Triazolo[1,5-a][1,3,5]triazine

In addition to the rearrangement of the [4,3-*a*] systems into the [1,5-*a*] analogs in a Dimroth process ⟨70T3357⟩, several methods have been employed for this series of compounds, often viewed as 5-azapurines. In one case, the 5,7-dimorpholino derivative (**363**) was synthesized by treating the oxime derivative (**364**) with polyphosphoric acid ⟨74JOC2143⟩. This method has been used extensively for fused triazoles and is described in many sections of this chapter for other ring systems.

Two purine analogs (namely, 5-azaadenine **365** and 5-azahypoxanthine **366**) of this series have been prepared ⟨65JA1980⟩. The routes to these two compounds are shown in Scheme 69. The most efficient route to 5-azaadenine is *via* the reaction of 3-amino-1,2,4-triazole with ethyl *N*-cyanoformimidate ⟨80JHC1121⟩.

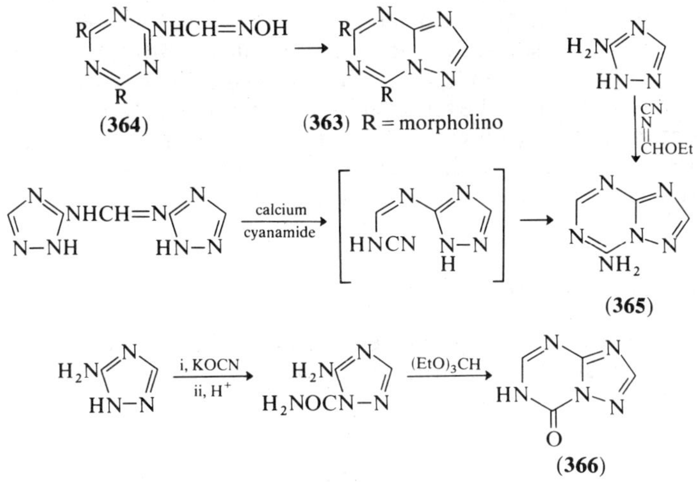

Scheme 69

4.15.4.9 Tetrazolopyridines and their Fused Benzo Derivatives

4.15.4.9.1 *Tetrazolo[1,5-a]pyridine*

Tetrazolo[1,5-*a*]pyridines have been prepared from either the corresponding halopyridine and sodium azide in the presence of acid with spontaneous electrocyclic ring closure, or by nitrosation of the appropriate 2-hydrazinopyridines ⟨71T5121, 76JOC1073⟩ as shown in Scheme 70. The reaction of 2-nitroaminopyridines with hydrazine under the influence of Raney nickel ⟨72JHC461⟩ and 2-aminopyridines with *p*-toluenesulfonyl azide ⟨79JHC1567⟩ are two other useful routes to tetrazolo[1,5-*a*]pyridines.

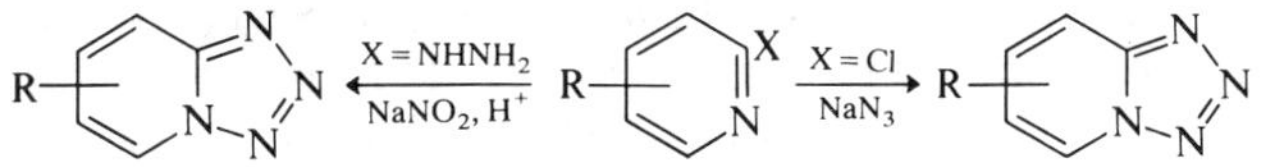

Scheme 70

4.15.4.9.2 *Tetrazolo[1,5-b]isoquinoline*

Attempts at preparing 3-azidoisoquinoline from the 3-chloro precursor were unsuccessful due to the inertness of the chloro substituent. However, if 3-chloroisoquinoline was first converted into its *N*-oxide, its susceptibility to nucleophilic substitution was enhanced. Thus, reaction of the *N*-oxide with hydrazine gave the 3-hydrazine *N*-oxide which, upon nitrosation and deoxygenation, gave the desired 3-azidoisoquinoline (**367**). This material was rather stable, with electrocyclization to (**47**) taking place only upon crystallization (Scheme 71) ⟨76JHC881, 81JOC843⟩.

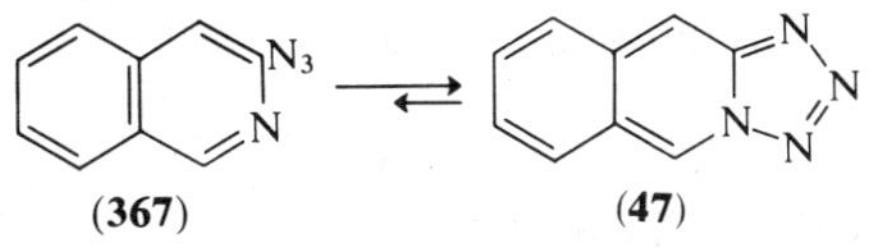

Scheme 71

Diazotization of 2,3-diaminoisoquinolinium tosylate is another route to derivatives of (**47**) ⟨81JOC843⟩.

4.15.4.9.3 Tetrazolo[1,5-a]quinoline

The ring system can be prepared in the usual way by treating a 2-chloroquinoline with ammonium azide ⟨76JOC1073⟩ to give the 2-azido derivative which cyclizes to the fused tetrazole. Thermolysis of *o*-azidocinnamonitriles has also given tetrazolo[1,5-*a*]quinolines ⟨80JOC4767⟩ *via* a cycloaddition of the azido group to the neighboring nitrile.

Comparison of the NMR spectra of the isomeric systems (**48**) and (**49**) with that of (**47**) indicates that (**47**) exists in an equilibrium with its open azido form whereas (**48**) and (**49**) are almost exclusively present as the tetrazolo valence tautomer ⟨76JHC881⟩. This equilibrium for (**47**) is solvent dependent ⟨76JHC881⟩, it is influenced by the substituents present on the molecule ⟨81JOC843⟩, and it may be an attempt by the molecule to counteract the instability introduced by ring closure to the *o*-quinoid structure (**47**).

4.15.4.9.4 Tetrazolo[5,1-a]isoquinoline

Diazotization of 1-hydrazinoisoquinolines ⟨67IJC403⟩ and reaction of *p*-toluenesulfonyl azide with 1-aminoisoquinoline ⟨79JHC1579⟩ are two methods that have been described for preparing tetrazolo[5,1-*a*]isoquinolines.

4.15.4.10 Tetrazolo[1,5-b]pyridazines and their Fused Benzo Derivatives

A large variety of derivatives of ring systems (**50**)–(**52**) has been prepared by the standard methods of reacting 3-chloropyridazines with azide anion ⟨63CPB348, 71JOC446⟩ or 3-hydrazinopyridazines ⟨63CPB348, 65JHC67, 71JOC446⟩, 3-hydrazinocinnolines ⟨71BSF3043⟩ and 1-hydrazinophthalazines ⟨67CB1073⟩ with nitrous acid. It is interesting to note that treatment of 3-chlorocinnolines with sodium azide did not give the corresponding tetrazole ⟨71BSF3043⟩.

4.15.4.11 Tetrazolopyrimidines and their Fused Benzo Derivatives

4.15.4.11.1 Tetrazolo[1,5-a]pyrimidine

The parent ring system (**53**) was prepared by diazotization of 2-hydrazinopyrimidine ⟨57JAP570777⟩. Alkyl derivatives of (**53**) have been synthesized by the condensation of β-dicarbonyl compounds with 5-aminotetrazole ⟨79JCS(P1)3085⟩. For example, treatment of 5-aminotetrazole with pentane-2,4-dione gave 5,7-dimethyltetrazolo[1,5-*a*]pyrimidine (Scheme 72) ⟨65JOC826⟩.

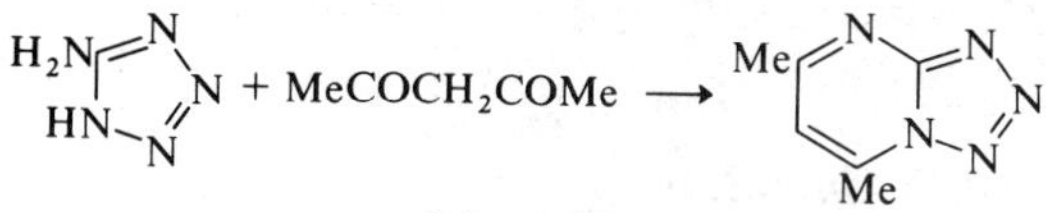

Scheme 72

4.15.4.11.2 Tetrazolo[1,5-c]pyrimidine

Many representatives of this ring system have been prepared by treating the appropriate 4-chloropyrimidine with sodium azide ⟨65JOC829⟩. It is interesting to note that attempts to synthesize 8-amino derivatives of (**54**) by diazotization of 5-amino-4-hydrazinopyrimidines yielded [1,2,3]triazolo[4,5-*d*]pyrimidines instead ⟨65JOC829⟩.

4.15.4.11.3 Tetrazolo[1,5-a]quinazoline

Reaction of 2-chloroquinazoline with sodium azide gave derivatives of (**55**) ⟨63ZOB2475⟩ whereas the condensation between 2-hydroxymethylenecyclohexanone and 5-aminotetrazole formed 5,6,7,8-tetrahydrotetrazolo[1,5-*a*]quinazoline ⟨79JCS(P1)3085⟩.

4.15.4.11.4 Tetrazolo[1,5-c]quinazoline

Nitrosation of 4-hydrazinoquinazolines has been reported as the route to the [1,5-c] system ⟨80H(14)1759⟩.

4.15.4.12 Tetrazolo[1,5-a]pyrazine and Tetrazolo[1,5-a]quinoxaline

As in the other fused tetrazoles, the pyrazines have been prepared from diazotization of a 2-hydrazinopyrazine ⟨71JOC446⟩ or *via* azide substitution on a 2-halopyrazine ⟨66JHC435, 80JHC11⟩. Likewise, the quinoxalines can be synthesized by the diazotization of 2-hydrazinoquinoxalines or the azide–halide displacement ⟨78ZC92⟩.

4.15.4.13 Tetrazolotriazines

4.15.4.13.1 Tetrazolo[1,5-b][1,2,4]triazine

Representatives of this ring system have been prepared by treating various 3-hydrazino-1,2,4-triazines with nitrous acid ⟨76JOC2860⟩. X-ray analysis has verified these products to be the [1,5-b][1,2,4] system rather than the isomeric [5,1-c][1,2,4] system which could have also resulted from the 3-hydrazino starting material ⟨76JOC2860⟩.

4.15.4.13.2 Tetrazolo[5,1-c][1,2,4]triazine

In view of the discussion in the last section which implied that 3-azido-1,2,4-triazines are very short-lived and indicated that they cyclize exclusively at N-2 to give the [1,5-b] isomer, the only means into the [5,1-c] series is by blocking the N-2 position and treating the resulting 3-hydrazino-1,2,4-triazine with nitrous acid. This has been done ⟨64CB2185⟩ and, by design, led only to derivatives with an N-8 substituent (Scheme 73).

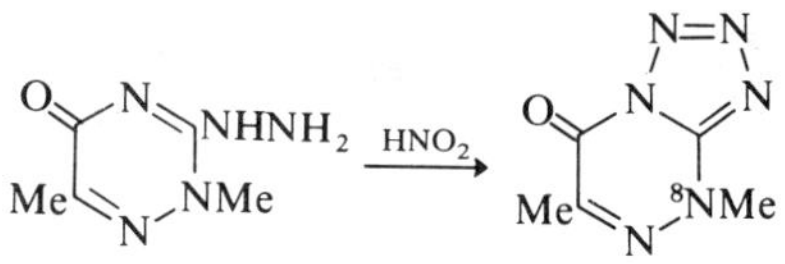

Scheme 73

4.15.4.13.3 Tetrazolo[1,5-d][1,2,4]triazine

Substituted derivatives of this ring system have been synthesized by treating an appropriate 4-hydrazino-1,2,4-triazine with nitrous acid ⟨67CB3467⟩.

4.15.4.13.4 Tetrazolo[5,1-f][1,2,4]triazine

Reaction of 3-amino-6-hydrazino-1,2,4-triazin-5(2H)-one (**368**) with nitrous acid has produced (**369**) as a representative of this class of compounds ⟨79JHC555⟩.

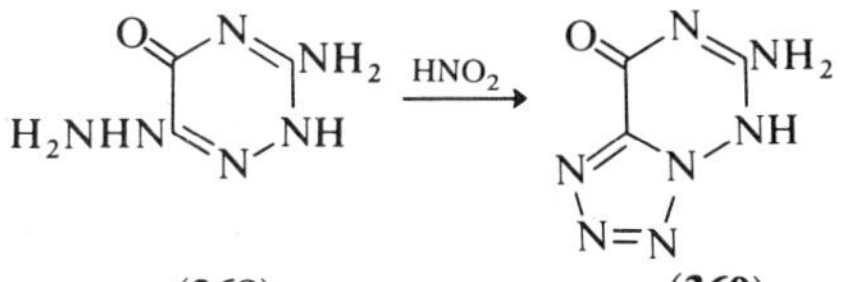

Scheme 74

4.15.5 APPLICATIONS

These are summarized in Table 9.

Table 9 Miscellaneous Applications of Various Fused Triazoles

Parent compound	Applications	Ref.
[1,2,4]Triazolo[4,3-a]pyridine	Photography	71MI41500
[1,2,4]Triazolo[3,4-a]isoquinoline	Cardiovascular	67NEP6614979
	Antiinflammatory	73FRP2135297
[1,2,3]Triazolo[1,5-a]pyridine	Organic synthon	80TL4529
[1,2,3]Triazolo[4,5-b]pyridine	Pesticide	75HCA1521
[1,2,3]Triazolo[4,5-b]quinoline	Antiallergic	80JCR(S)308
[1,2,4]Triazolo[4,3-b]pyridazine	Vasodilating, hypotensive	61FRP1248409
	Cardiovascular, photography	
	Anxiolytic	81JMC592
[1,2,4]Triazolo[3,4-a]phthalazine	C-Terminal peptide sequencing	78DIS(B)(39)233
	Hydralazine metabolism	74JMC381, 75JMC1031, 79CPB2820
	Budralazine metabolism	77CPB830
[1,2,4]Triazolo[1,5-a]pyrimidine	Photography	80MI41500
	Herbicidal	80MIP41500
	Emetic	78GEP2805483
	Seedling growth	77MI41500
[1,2,3]Triazolo[4,5-d]pyrimidine	Enzyme probe	69B2412
[1,2,4]Triazolo[4,3-a]pyrazine	Bronchodilator	69JCS(C)1593
[1,2,4]Triazolo[4,3-a]quinoxaline	Rice growth control	77USP4008322

References

EXPLANATION OF THE REFERENCE SYSTEM

Throughout this work, references are designated by a number–letter coding of which the first two numbers denote tens and units of the year of publication, the next one to three letters denote the journal, and the final numbers denote the page. This code appears in the text each time a reference is quoted; the advantages of this system are outlined in the Introduction (Chapter 1.01). The system is based on that previously used in the following two monographs: (a) A. R. Katritzky and J. M. Lagowski, 'Chemistry of the Heterocyclic N-Oxides', Academic Press, New York, 1971; (b) J. Elguero, C. Marzin, A. R. Katritzky and P. Linda, 'The Tautomerism of Heterocycles', in 'Advances in Heterocyclic Chemistry', Supplement 1, Academic Press, New York, 1976.

The following additional notes apply:

1. A list of journals which have been assigned codes is given (in alphabetical order) together with their codes immediately following these notes. Journal names are abbreviated throughout by the CASSI (Chemical Abstracts Service Source Index) system.

2. A list of journal codes in alphabetical order, together with the journals to which they refer, is given on the end papers of each volume.

3. Each volume contains all the references cited *in that volume*; no separate lists are given for individual chapters.

4. The list of references is arranged in order of (a) year, (b) journal in alphabetical order of journal code, (c) part letter or number if relevant, (d) volume number if relevant, (e) page number.

5. In the reference list the code is followed by (a) the complete literature citation in the conventional manner and (b) the number(s) of the page(s) on which the reference appears, whether in the text or in tables, schemes, *etc.*

6. For non-twentieth century references the year is given in full in the code.

7. For journals which are published in separate parts, the part letter or number is given (when necessary) in parentheses immediately after the journal code letters.

8. Journal volume numbers are *not* included in the code numbers unless more than one volume was published in the year in question, in which case the volume number is included in parentheses immediately after the journal code letters.

9. Patents are assigned appropriate three letter codes.

10. Frequently cited books are assigned codes, but the whole code is now prefixed by the letter 'B-'.

11. Less common journals and books are given the code 'MI' for miscellaneous.

12. Where journals have changed names, the same code is used throughout, *e.g.* CB refers both to *Chem. Ber.* and to *Ber. Dtsch. Chem. Ges.*

Journals

Acc. Chem. Res.	ACR
Acta Chem. Scand., Ser. B	ACS(B)
Acta Chim. Acad. Sci. Hung.	ACH
Acta Crystallogr., Part B	AX(B)
Adv. Phys. Org. Chem.	APO
Agric. Biol. Chem.	ABC

Angew. Chem.	AG
Angew. Chem., Int. Ed. Engl.	AG(E)
Ann. Chim. (Rome)	AC(R)
Ann. N.Y. Acad. Sci.	ANY
Arch. Pharm. (Weinheim, Ger.)	AP
Ark. Kemi	AK
Arzneim.-Forsch.	AF
Aust. J. Chem.	AJC
Biochem. Biophys. Res. Commun.	BBR
Biochemistry	B
Biochem. J.	BJ
Biochim. Biophys. Acta	BBA
Br. J. Pharmacol.	BJP
Bull. Acad. Pol. Sci., Ser. Sci. Chim.	BAP
Bull. Acad. Sci. USSR, Div. Chem. Sci.	BAU
Bull. Chem. Soc. Jpn.	BCJ
Bull. Soc. Chim. Belg.	BSB
Bull. Soc. Chim. Fr., Part 2	BSF(2)
Can. J. Chem.	CJC
Chem. Abstr.	CA
Chem. Ber.	CB
Chem. Heterocycl. Compd. (Engl. Transl.)	CHE
Chem. Ind. (London)	CI(L)
Chem. Lett.	CL
Chem. Pharm. Bull.	CPB
Chem. Rev.	CRV
Chem. Scr.	CS
Chem. Soc. Rev.	CSR
Chem.-Ztg.	CZ
Chimia	C
Collect. Czech. Chem. Commun.	CCC
Coord. Chem. Rev.	CCR
C.R. Hebd. Seances Acad. Sci., Ser. C	CR(C)
Cryst. Struct. Commun.	CSC
Diss. Abstr. Int. B	DIS(B)
Dokl. Akad. Nauk SSSR	DOK
Experientia	E
Farmaco Ed. Sci.	FES
Fortschr. Chem. Org. Naturst.	FOR
Gazz. Chim. Ital.	G
Helv. Chim. Acta	HCA
Heterocycles	H
Hoppe-Seyler's Z. Physiol. Chem.	ZPC
Indian J. Chem., Sect. B	IJC(B)
Inorg. Chem.	IC
Int. J. Sulfur Chem., Part B	IJS(B)
Izv. Akad. Nauk SSSR, Ser. Khim.	IZV
J. Am. Chem. Soc.	JA
J. Biol. Chem.	JBC
J. Chem. Phys.	JCP
J. Chem. Res. (S)	JCR(S)
J. Chem. Soc. (C)	JCS(C)
J. Chem. Soc., Chem. Commun.	CC
J. Chem. Soc., Dalton Trans.	JCS(D)
J. Chem. Soc., Faraday Trans. 1	JCS(F1)
J. Chem. Soc., Perkin Trans. 1	JCS(P1)
J. Gen. Chem. USSR (Engl. Transl.)	JGU
J. Heterocycl. Chem.	JHC
J. Indian Chem. Soc.	JIC
J. Magn. Reson.	JMR

J. Med. Chem.	JMC
J. Mol. Spectrosc.	JSP
J. Mol. Struct.	JST
J. Organomet. Chem.	JOM
J. Org. Chem.	JOC
J. Org. Chem. USSR (Engl. Transl.)	JOU
J. Pharm. Sci.	JPS
J. Phys. Chem.	JPC
J. Prakt. Chem.	JPR
Khim. Geterotsikl. Soedin.	KGS
Kristallografiya	K
Liebigs Ann. Chem.	LA
Monatsh. Chem.	M
Naturwissenschaften	N
Nippon Kagaku Kaishi	NKK
Nouv. J. Chim.	NJC
Org. Magn. Reson.	OMR
Org. Mass Spectrom.	OMS
Org. Prep. Proced. Int.	OPP
Org. React.	OR
Org. Synth.	OS
Org. Synth., Coll. Vol.	OSC
Phosphorus Sulfur	PS
Phytochemistry	P
Proc. Indian Acad. Sci., Sect. A	PIA(A)
Proc. Natl. Acad. Sci. USA	PNA
Pure Appl. Chem.	PAC
Q. Rev., Chem. Soc	QR
Recl. Trav. Chim. Pays-Bas	RTC
Rev. Roum. Chim.	RRC
Russ. Chem. Rev. (Engl. Transl.)	RCR
Spectrochim. Acta, Part A	SA(A)
Synth. Commun.	SC
Synthesis	S
Tetrahedron	T
Tetrahedron Lett.	TL
Ukr. Khim. Zh. (Russ. Ed.)	UKZ
Yakugaku Zasshi	YZ
Z. Chem.	ZC
Zh. Obshch. Khim.	ZOB
Zh. Org. Khim.	ZOR
Z. Naturforsch., Teil B	ZN(B)

Book Series

'Advances in Heterocyclic Chemistry'	AHC
'Chemistry of Heterocyclic Compounds' [Weissberger–Taylor series]	HC
'Methoden der Organischen Chemie (Houben-Weyl)'	HOU
'Organic Compounds of Sulphur, Selenium, and Tellurium' [R. Soc. Chem. series]	SST
'Physical Methods in Heterocyclic Chemistry'	PMH

Specific Books

Q. N. Porter and J. Baldas, 'Mass Spectrometry of Heterocyclic Compounds', Wiley, New York, 1971	MS
T. J. Batterham, 'NMR Spectra of Simple Heterocycles', Wiley, New York, 1973	NMR

'Photochemistry of Heterocyclic Compounds', ed. O. Buchardt, Wiley, New York, 1976   PH

W. L. F. Armarego, 'Stereochemistry of Heterocyclic Compounds', Wiley, New York, 1977, parts 1 and 2   SH

Patents

Belg. Pat.	BEP
Br. Pat.	BRP
Eur. Pat.	EUP
Fr. Pat.	FRP
Ger. (East) Pat.	EGP
Ger. Pat.	GEP
Neth. Pat.	NEP
Jpn. Pat.	JAP
Jpn. Kokai	JAP(K)
S. Afr. Pat.	SAP
U.S. Pat.	USP

Other Publications

All Other Books and Journals ('Miscellaneous')	MI
All Other Patents	MIP
Personal Communications	PC
Theses	TH
Unpublished Results	UP

VOLUME 5 REFERENCES

1861LA(118)151	A. Strecker; *Liebigs Ann. Chem.*, 1861, **118**, 151.	552, 596
1875LA(175)243	L. Medicus; *Liebigs Ann. Chem.*, 1875, **175**, 243.	502
1882CB253	E. Fischer; *Ber.*, 1882, **15**, 253.	558
1882LA(215)265	E. Fischer; *Liebigs Ann. Chem.*, 1882, **215**, 265.	593
1882LA(215)316	E. Fischer; *Liebigs Ann. Chem.*, 1882, **215**, 316.	534, 540
1882M(3)796	J. Horbaczewski; *Monatsh. Chem.*, 1882, **3**, 796.	568
1883LA(221)336	E. Fischer and L. Reese; *Liebigs Ann. Chem.*, 1883, **221**, 336.	540, 552
1885CB1544	J. A. Bladin; *Ber.*, 1885, **18**, 1544.	735, 792, 828
1886ZPC(10)258	J. Kossell; *Hoppe-Seyler's Z. Physiol. Chem.*, 1886, **10**, 258.	552
1887CB1176	T. Zincke and A. T. Lawson; *Ber.*, 1887, **20**, 1176.	722
1887M(8)201	J. Horbaczewski; *Monatsh. Chem.*, 1887, **8**, 201.	568
1888CB1149	L. Claisen and O. Lowman; *Ber.*, 1888, **21**, 1149.	121
1888CB2751	H. von Pechmann; *Ber.*, 1888, **21**, 2751.	670
1889CB2929	E. Buchner; *Ber.*, 1889, **22**, 2929.	142
1889LA(250)294	G. Hofmann; *Liebigs Ann. Chem.*, 1889, **250**, 294.	118
1889LA(251)235	R. Behrend and O. Roosen; *Liebigs Ann. Chem.*, 1889, **251**, 235.	582
1890CB225	G. Bruhns; *Ber.*, 1890, **23**, 225.	540
1892CB225	A. Andreocci; *Ber*, 1892, **25**, 225.	738
1892ZPC(16)329	M. Kruger; *Hoppe-Seyler's Z. Physiol. Chem.*, 1892, **16**, 329.	552
1893CB86	E. Noelting and O. Michel; *Ber.*, 1893, **26**, 86.	839

1893CB1719	R. von Rothenburg; *Ber.*, 1893, **26**, 1719.	142
1893CB2385	E. Bamberger and P. de Gruyter; *Ber.*, 1893, **26**, 2385.	736
1894G222	G. Pellizzari; *Gazz. Chim. Ital.*, 1894, **24**, 222.	762
1895CB2480	E. Fischer; *Ber.*, 1895, **28**, 2480.	557
1895MI40900	M. Gomberg; *Am. J. Chem.*, 1895, **17**, 403.	550, 551
1897AP(235)469	H. van der Slooten; *Arch. Pharm. (Weinheim, Ger.)*, 1897, **235**, 469.	529
1897CB559	E. Fischer; *Ber.*, 1897, **30**, 559.	534
1897CB1846	E. Fischer; *Ber.*, 1897, **30**, 1846.	534
1897CB2220	E. Fischer; *Ber.*, 1897, **30**, 2220.	562
1897CB2226	E. Fischer; *Ber.*, 1897, **30**, 2226.	562
1897CB2231	E. Fischer; *Ber.*, 1897, **30**, 2231.	532
1897CB2400	E. Fischer; *Ber.*, 1897, **30**, 2400.	557
1897CB2584	H. Brunner and H. Leins; *Ber.*, 1897, **30**, 2584.	551
1898CB2550	E. Fischer; *Ber.*, 1898, **31**, 2550.	529
1898CB2619	E. Fischer; *Ber.*, 1898, **31**, 2619.	557
1898CB3270	E. Fischer; *Ber.*, 1898, **31**, 3270.	601
1898CB3272	E. Fischer; *Ber.*, 1898, **31**, 3272.	562, 601
1898ZPC(24)364	M. Kruger and G. Salamon; *Hoppe-Seyler's Z. Physiol. Chem.*, 1898, **24**, 364.	598
1899CB68	T. B. Baillie and J. Tafel; *Ber.*, 1899, **32**, 68.	541
1899CB435	E. Fischer; *Ber.*, 1899, **32**, 435.	502, 534
1899CB2721	E. Fischer and F. Ach; *Ber.*, 1899, **32**, 2721.	534
00CB58	A. Hantzsch and O. Silberrad; *Ber*, 1900, **33**, 58.	752
00CB644	H. von Pechmann and W. Bauer; *Ber.*, 1900, **33**, 644.	701
00CB1371	W. Traube; *Ber.*, 1900, **33**, 1371.	571
00CB3035	W. Traube; *Ber.*, 1900, **33**, 3035.	571
00JCS(77)224	G. Young and E. Witham; *J. Chem. Soc*, 1900, **77**, 224.	766
01CB258	J. Tafel: *Ber.*, 1901, **34**, 258.	541
01CB637	F. Kunckell; *Ber.*, 1901, **34**, 637.	119
01CB1165	J. Tafel and B. Ach; *Ber.*, 1901, **34**, 1165.	541
01CB1170	J. Tafel and B. Ach; *Ber.*, 1901, **34**, 1170.	541
02CB2563	E. Fischer and H. Tullner; *Ber.*, 1902, **35**, 2563.	582
02GEP142468	Boehringer; *Ger. Pat.* 142 468 (1902).	577
03CB2056	A. Hantzsch; *Ber.*, 1903, **36**, 2056.	839
04JPR(69)509	R. Stolle and L. Gutmann; *J. Prakt. Chem.*, 1904, **69**, 509.	116
05CB856	M. Busch; *Ber.*, 1905, **38**, 856	785
05CB861	M. Busch; *Ber.*, 1905, **38**, 861.	785
05CB4049	M. Busch and G. Mehrtens; *Ber.*, 1905, **38**, 4049.	785
05G373	G. Pellizzari and A. Soldi; *Gazz. Chim. Ital.*, 1905, **35**, 373	762
05JCS(87)625	G. Young; *J. Chem. Soc.*, 1905, **87**, 625.	757
05LA(343)1	W. Manchot and R. Noll; *Liebigs Ann. Chem.*, 1905, **343**, 1.	754, 756
05LA(343)207	A. Einhorn, E. Bischkopf, C. Ladisch, T. Mauermayer, G. Schupps, E. Sprängerts and B. Szelinski; *Liebigs Ann. Chem.*, **343**, 207.	765
06AP(244)11	W. Traube; *Arch. Pharm. (Weinheim, Ger.)*, 1906, **244**, 11.	558, 559
06CB227	W. Traube and W. Nithack; *Ber.*, 1906, **39**, 227.	576
06CB1831	G. Dedichen; *Ber.*, 1906, **39**, 1831.	742
07AP(245)312	E. Schmidt and W. Schwabe; *Arch. Pharm. (Weinheim, Ger.)*, 1907, **245**, 312.	529
07CB3752	J. Tafel and J. Dodt; *Ber.*, 1907, **40**, 3752.	541
07ZPC(51)425	R. Burian; *Hoppe-Seyler's Z. Physiol. Chem.*, 1907, **51**, 425.	538
08CB176	R. Majima; *Ber.*, 1908, **41**, 176.	647
09CB659	H. von Pechmann and W. Bauer; *Ber.*, 1909, **42**, 659.	701
09LA(364)183	O. Dimroth; *Liebigs Ann. Chem.*, 1909, **364**, 183.	696
10CB2904	O. Dimroth and G. DeMontmollin; *Ber.*, 1910, **43**, 2904.	839
10CB3553	H. Biltz; *Ber.*, 1910, **43**, 3553.	540
12JBC(11)67	C. O. Johns; *J. Biol. Chem.*, 1912, **11**, 67.	596
13JBC(14)299	C. O. Johns and A. G. Hogan; *J. Biol. Chem.*, 1913, **14**, 299.	578

13JBC(14)381	C. O. Johns and E. J. Baumann; *J. Biol. Chem.*, 1913, **14**, 381.	560, 578
13JBC(15)515	C. O. Johns and E. J. Baumann; *J. Biol. Chem.*, 1913, **15**, 515.	579
13JCS1391	G. T. Morgan and F. M. G. Micklethwait; *J. Chem. Soc.*, 1913, **103**, 1391.	699
13MI41000	J. Wellisch; *Biochem. Z.*, 1913, **49**, 173.	623, 641
14CB2671	K. Brunner; *Ber.*, 1914, **47**, 2671.	765
14CB3163	M. Bachstez; *Ber.*, 1914, **47**, 3163.	113
14JBC(17)1	C. O. Johns; *J. Biol. Chem.*, 1914, **17**, 1.	534
14LA(406)22	H. Biltz and F. Damm; *Liebigs Ann. Chem.*, 1914, **406**, 22.	540
15CB410	J. Lifschitz; *Ber.*, 1915, **48**, 410.	839
15CB1614	T. Curtius, A. Darapsky and E. Müller; *Ber.*, 1915, **48**, 1614.	839
16G308	E. Oliveri-Mandala; *Gazz. Chim. Ital.*, 1916, **46**, 308.	690
16JBC(25)607	P. A. Levene and J. K. Senior; *J. Biol. Chem.*, 1916, **25**, 607.	577
B-16MI41400	E. Hjelt; 'Geschichte der Organischen Chemie', Vieweg u. Sohn, Braunschweig, 1916.	839
17LA(413)155	H, Biltz and K. Strufe; *Liebigs Ann. Chem.*, 1917, **413**, 155.	540
20CB2327	H. Biltz and F. Max; *Ber.*, 1920, **53**, 2327.	535
20MI41000	S. Frankel and K. Zeimer; *Biochem. Z.*, 1920, **110**, 234.	641
21CB1676	H. Biltz and L. Herrmann; *Ber.*, 1921, **54**, 1676.	534
21CB2191	M. P. Schmidt and A. Hagenböcker; *Ber.*, 1921, **54B**, 2191.	722
21CB2201	M. P. Schmidt and A. Hagenböcker; *Ber.*, 1921, **54B**, 2201.	722
21G71	G. G. Cusmano; *Gazz. Chim. Ital.*, 1921, **51**, 71.	722
21LA(423)200	H. Biltz and K. Strufe; *Liebigs Ann. Chem.*, 1921, **423**, 200.	533, 558
21LA(423)242	H. Biltz and K. Strufe; *Liebigs Ann. Chem.*, 1921, **423**, 242.	582
21ZPC(117)23	J. Herzig; *Hoppe-Seyler's Z. Physiol Chem.*, 1921, **117**, 23.	535
22GEP338926	Kalle & Co., *Ger. Pat.* 338 926 (1922) (*Chem. Abstr.*, 1922, **16**, 2869).	722
22LA(426)306	H. Biltz and R. Bulow; *Liebigs Ann. Chem.*, 1922, **426**, 306.	535
23LA(432)266	W. Traube, F. Schottlander, C. Goslich, R. Peter, F. A. Meyer, H. Schluter, W. Steinbach and K. Bredow; *Liebigs Ann. Chem.*, 1923, **432**, 266.	550, 574, 577
23MI41000	E. Sucharda; *Rocz. Chem.*, 1923, **3**, 236 (*Chem. Abstr.*, 1925, **19**, 72).	613
24CB1381	F. Reindel; *Ber.*, 1924, **57**, 1381.	612
24CB2092	A. E. Tschitschibabin; *Ber.*, 1924, **57**, 2092.	612, 613
24HCA713	J. Sarasin and E. Wegmann; *Helv. Chim. Acta.*, 1924, **7**, 713.	587
25CB393	F. Reindel and H. Rauch; *Ber.*, 1925, **58**, 393.	612
25CB1704	A. E. Tschitschibabin; *Ber.*, 1925, **58**, 1704.	631
25LA(441)203	A. Prüsse; *Liebigs Ann. Chem.*, 1925, **441**, 203.	582
25LA(441)215	R. Behrend; *Liebigs Ann. Chem.*, 1925, **441**, 215.	568
26CB1360	L. Schmid and B. Bangler; *Ber.*, 1926, **59**, 1360.	632
26CB1742	R. Stollé; *Ber.*, 1926, **59**, 1742.	707
26CB2048	A. E. Tschitschibabin; *Ber.*, 1926, **59**, 2048.	610, 611, 631
26CB2921	F. Reindel and H. Rauch; *Ber.*, 1926, **59**, 2921.	613
26CB2926	F. Reindel and A. von Putzer-Reybegg; *Ber.*, 1926, **59**, 2926.	612
27CB119	A. Vilsmeier and A. Haack; *Ber.*, 1927, **60**, 119.	573
27CB766	A. E. Tschitschibabin and A. V. Kirsanow; *Ber.*, 1927, **60B**, 766.	635
27LA(454)164	K. Fries; *Liebigs Ann. Chem.*, 1927, **454**, 164.	721
27LA(459)39	O. Dimroth and W. Michaelis; *Liebigs Ann. Chem.*, 1927, **459**, 39.	712, 713
28CB1140	G. Wittig, F. Bangert and H. Kleiner; *Ber.*, 1928, **61B**, 1140.	719
28CB1409	H. Biltz and H. Rakett; *Ber.*, 1928, **61**, 1409.	540, 588, 589
28JPR(118)198	H. Biltz and A. Beck; *J. Prakt. Chem.*, 1928, **118**, 198.	533
30CB2876	H. Biltz and H. Pardon; *Ber.*, 1930, **63**, 2876.	558
30G644	G. B. Crippa, G. Bellani and A. Merubini; *Gazz. Chim. Ital.*, 1930, **60**, 644.	722
30G866	M. Galotti, G. Barro and L. Salto; *Gazz. Chim. Ital.*, 1930, **60**, 866.	707
30JPR(125)498	T. Curtius and W. Klavehn; *J. Prakt. Chem.*, 1930, **125**, 498.	709
31CB752	H. Biltz and J. Sauer; *Ber.*, 1931, **64**, 752.	560
31MI40900	P. A. W. Self and W. R. Rankin; *Quart. J. Pharmacol.*, 1931, **4**, 346.	533
32JPR(134)310	H. Biltz and H. Pardon; *J. Prakt. Chem.*, 1932, **134**, 310.	535
32LA(498)1	O. Diels, K. Alder, H. Winckler and E. Petersen; *Liebigs Ann. Chem.*, 1932, **498**, 1.	633

33JCS662	J. M. Gulland and T. F. MacCrae; *J. Chem. Soc.*, 1933, 662.	533
34JA700	S. M. E. Englert and S. M. McElvain; *J. Am. Chem. Soc.*, 1934, **56**, 700.	341
34JCS1639	J.M. Gulland, E. R. Holiday and T. F. MacCrae; *J. Chem. Soc.*, 1934, 1639.	536
34JIC867	P. C. Mitter and N. Chatterjee; *J. Indian Chem. Soc.*, 1934, **11**, 867 (*Chem. Abstr.*, 1935, **29**, 2953).	664
34LA(511)168	O. Diels and J. Reese; *Liebigs Ann. Chem.*, 1934, **511**, 168.	142
35G152	M. Milone; *Gazz. Chim. Ital.*, 1935, **65**, 152.	12
35JCS955	E. Bergmann and H. Heimhold; *J. Chem. Soc.*, 1935, 955.	558
35LA(515)185	K. Alder and G. Stein; *Liebigs Ann. Chem.*, 1935, **515**, 185.	703, 711
35MI40900	G. Bargioni; *Boll. Chim-Farm.*, 1935, **64**, 869 (*Chem. Abstr.*, 1936, **30**, 2320).	564
36MI40900	G. Hunter; *Nature* (*London*), 1936, **37**, 405.	552
36MI40901	G. Hunter; *Biochem. Z.*, 1936, **30**, 1183.	552
37CB2309	R. Weidenhagen and H. Wegner; *Ber.*, 1937, **70**, 2309.	92
38CB718	H. Bredereck and G. Richter; *Ber.*, 1938, **71**, 718.	592
38CB2347	R. Weidenhagen and U. Weeden; *Ber.*, 1938, **71B**, 2347.	639
39JA351	J. R. Spies and T. H. Harris, Jr.: *J. Am. Chem. Soc.*, 1939, **61**, 351.	552
39YZ97	E. Ochiai and M. Karii; *Yakugaku Zasshi*, 1939, **59**, 97 (*Chem. Abstr.*, 1939, **33**, 3791).	647
39YZ185	E. Ochiai and M. Sibata; *Yakugaku Zasshi*, 1939, **59**, 185 (*Chem. Abstr.*, 1939, **33**, 4988).	649
40CB162	K. W. F. Kohlrausch and R. Seka; *Ber.*, 1940, **73**, 162.	686
40JA2002	R. O. Roblin, Jr., J. H. Williams, P. S. Winnek and J. P. English; *J. Am. Chem. Soc.*, 1940, **62**, 2002.	845
40MI41200	W. Hückel, N. Datow and E. Simmersbach; *Z. Phys. Chem.*, 1940, **A186**, 129.	739
B-40MI41400	L. P. Hammett; 'Physical Organic Chemistry', McGraw-Hill, New York, 1940.	843
40OS(20)16	R. E. Damschroder and W. D. Peterson; *Org. Synth.*, 1940, **20**, 16.	422
40OS(20)73	H. D. Porter and W. D. Peterson; *Org. Synth.*, 1940, **20**, 73.	288
41RTC453	H. J. Backer and G. L. Wiggerink; *Recl. Trav. Chim. Pays-Bas*, 1941, **60**, 453.	126
42CB1338	H. Specker and H. Gawrosch; *Ber.*, 1942, **75**, 1338.	684
42CB1936	R. Weidenhagen, G. Train, H. Wegner and L. Nordstrom; *Ber.*, 1942, **75**, 1936.	639
42JCS232	W. E. Allesbrook, J. M. Gulland and L. F. Story; *J. Chem. Soc.*, 1942, 232.	589
42MI41100	E. G. Taylor; *Can. J. Res.*, 1942, **20B**, 161 (*Chem. Abstr.*, 1942, **36**, 6398).	690
43JCS383	J. Baddiley, B. Lythgoe, D. McNeill and A. R. Todd; *J. Chem. Soc.*, 1943, 383.	572
43MI41100	K. A. Jensen and A. Friediger; *Kgl. Danske Videnskab. Selskab, Mat.-fys. Medd.*, 1943, **20**, 1 (*Chem. Abstr.*, 1945, **39**, 2068).	681
43MI41200	K. A. Jensen and A. Friediger; *Kgl. Danske Videnskab. Selskab Mat.-fys. Medd.*, 1943, **20**, 1 (*Chem. Abstr.*, 1945, **39**, 2068).	743
43MI41300	K. A. Jensen and A. Friediger; *Kgl. Danske Videnskab. Selskab. Mat.-fys. Medd.*, 1943, **20**, 1 (*Chem. Abstr.*, 1944, **38**, 3629).	795
43RTC580	H. J. Backer and H. Bos; *Recl. Trav. Chim. Pays-Bas*, 1943, **62**, 580.	120
44BJ(38)309	A. Neuberger; *Biochem. J.*, 1944, **38**, 309.	623, 641
44JA1805	C. F. H. Allen, J. Van Allan and C. V. Wilson; *J. Am. Chem. Soc.*, 1944, **66**, 1805.	631
44JBC(153)203	A. A. Plentl and R. Schoenheimer; *J. Biol. Chem.*, 1944, **153**, 203.	596
45HCA850	P. Ruggli and F. Buchmeier; *Helv. Chim. Acta*, 1945, **28**, 850.	86
45JCS556	G. A. Howard, B. Lythgoe and A. R. Todd; *J. Chem. Soc.*, 1945, 556.	573
45JCS751	F. G. Mann and J. W. G. Porter; *J. Chem. Soc.*, 1945, 751.	548, 552, 588, 598
46JA1799	R. F. Coles and C. F. Hamilton; *J. Am. Chem. Soc.*, 1946, **68**, 1779.	707
46JCS852	G. W. Kenner and A. R. Todd; *J. Chem. Soc.*, 1946, 852.	573
46OS(26)11	C. F. H. Allen and A. Bell; *Org. Synth.*, 1946, **26**, 11.	787
46USP2509084	W. E. Decker; *U.S. Pat.* 2 509 084 (1946) (*Chem. Abstr.*, 1950, **44**, 8366).	592
47CRV(41)1	F. R. Benson; *Chem. Rev.*, 1947, **41**, 1.	792, 802, 803
47JCS943	F. E. King and T. J. King; *J. Chem. Soc.*, 1947, 943.	578
48JA1240	L. F. Cavalieri, V. E. Blair and G. B. Brown; *J. Am. Chem. Soc.*, 1948, **70**, 1240.	577, 582, 596
48JA3109	A. Bendich, J. F. Tinker and G. B. Brown; *J. Am. Chem. Soc.*, 1948, **70**, 3109.	552, 560
48JA3429	D. Heyl, S. A. Harris and K. Folkers; *J. Am. Chem. Soc.*, 1948, **70**, 3429.	641
48JA3669	D. Heyl, E. Luz, S. A. Harris and K. Folkers; *J. Am. Chem. Soc.*, 1948, **70**, 3669.	641
48JCS967	J. Davoll, B. Lythgoe and A. R. Todd; *J. Chem. Soc.*, 1948, 967.	536

48JCS1157	W. Wilson; *J. Chem. Soc.*, 1948, 1157.	571
48JCS1389	V. Petrow and J. Saper; *J. Chem. Soc.*, 1948, 1389.	635, 636
48JCS1685	J. Davoll, B. Lythgoe and A. R. Todd; *J. Chem. Soc.*, 1948, 1685	536
48JCS1960	J. W. Cornforth and H. T. Huang; *J. Chem. Soc.*, 1948, 1960.	119, 156
48JOC815	J. L. Riebsomer; *J. Org. Chem.*, 1948, **13**, 815.	699, 707
48OS(28)87	H. D. Porter and A. Weissberger; *Org. Synth.*, 1948, **28**, 87.	288
48RTC29	F. Kogl, G. M. van der Want and C. A. Salemink; *Recl. Trav. Chim. Pays-Bas*, 1948, **67**, 29.	618, 619, 635, 640
48YZ191	J. Haginwa; *Yakugaku Zasshi*, 1948, **68**, 191 (*Chem. Abstr.*, 1953, **47**, 8074).	118
48ZOB2116	A. M. Khaletski and M. S. Eshmann; *Zh. Obshch. Khim.*, 1948, **18**, 2116.	557
48ZOB2129	E. S. Golovchinskaya; *Zh. Obshch. Khim.*, 1948, **18**, 2129.	549, 550, 593
49JA1436	J. C. Sheehan and C. A. Robinson; *J. Am. Chem. Soc.*, 1949, **71**, 1436.	728
49JA1885	J. R. Vaughan, Jr., J. Krapcho and J. P. English; *J. Am. Chem. Soc.*, 1949, **71**, 1885.	617, 618, 635, 636
49JA2297	R. A. Henry and W. M. Dehn; *J. Am. Chem. Soc.*, 1949, **71**, 2297.	699
49JA3973	L. F. Cavalieri, J. E. Tinker and G. B. Brown; *J. Am. Chem. Soc.*, 1949, **71**, 3973.	552, 589
49JA3994	R. G. Jones; *J. Am. Chem. Soc.*, 1949, **71**, 3994.	290
49JAP49177355	T. Shioi; *Jpn. Pat.* 49 177 355 (1949) (*Chem. Abstr.*, 1951, **45**, 7590).	564
49JCS786	W. Batty and B. C. L. Weedon; *J. Chem. Soc.*, 1949, 786.	140
49JCS1071	A. H. Cook, A. C. Davis, I. M. Heilbron and G. H. Thomas; *J. Chem. Soc.*, 1949, 1071.	589
49JCS2329	A. H. Cook and E. Smith; *J. Chem. Soc.*, 1949, 2329.	588, 590
49JCS2490	K. J. M. Andrews, N. Anand, A. R. Todd and A. Topham; *J. Chem. Soc.*, 1949, 2490.	558, 560
49JCS3001	A. H. Cook and E. Smith; *J. Chem. Soc.*, 1949, 3001.	577, 590
49OS(29)54	E. F. M. Stephenson; *Org. Synth.*, 1949, **29**, 54.	288
49RTC1013	C. A. Salemink and G. M. van der Want; *Recl. Trav. Chim. Pays-Bas*, 1949, **68**, 1013.	618, 619, 621, 635, 640
49USP2481953	G. Schwarz; *U.S. Pat.* 2 481 953 (1949) (*Chem. Abstr.*, 1950, **44**, 5737).	612
49YZ496	T. Takahashi and J. Shibasaki; *Yakugaku Zasshi*, 1949, **69**, 496 (*Chem. Abstr.*, 1950, **44**, 4474).	631
50BSB573	A. DeCat and A. van Dormael; *Bull. Soc. Chim. Belg.*, 1950, **59**, 573.	647
50CB201	H. Bredereck, H.-G. von Schuh and A. Martini; *Chem. Ber.*, 1950, **83**, 201.	533
50CRV(46)1	F. R. Benson and W. L. Savell; *Chem. Rev.*, 1950, **46**, 1.	670, 705, 708, 722
50HCA1183	H. Stenyl, A. Staub, C. Simon and W. Baumann; *Helv. Chim. Acta*, 1950, **33**, 1183.	322
50JA2587	L. F. Cavalieri and A. Bendich; *J. Am. Chem. Soc.*, 1950, **72**, 2587.	576
50JBC(185)423	A. Bendich, S. S. Furst and G. B. Brown; *J. Biol. Chem.*, 1950, **185**, 423.	593
50JBC(185)435	A. Bendich, W. D. Geren and G. B. Brown; *J. Biol. Chem.*, 1950, **185**, 435.	593
50JBC(185)439	E. Shaw; *J. Biol. Chem.*, 1950, **185**, 439.	583, 568, 586, 588
50JCS1884	A. H. Cook and G. H. Thomas; *J. Chem. Soc.*, 1950, 1884.	588, 589
50JCS1888	A. H. Cook and G. H. Thomas; *J. Chem. Soc.*, 1950, 1888.	590
50LA(566)58	K. Alder and W. Trimborn; *Liebigs Ann. Chem.*, 1950, **566**, 58.	720
51AG511	T. Wieland and L. Bauer; *Angew Chem.*, 1951, **63**, 511.	600
51AX81	W. Cochran; *Acta Crystallogr.*, 1951, **4**, 81.	508
51JA1207	J. C. Sheehan and C. A. Robinson; *J. Am. Chem. Soc.*, 1951, **73**, 1207.	728
51JA1650	J. Davoll and B. A. Lowy; *J. Am. Chem. Soc.*, 1951, **73**, 1650.	552
51JA1864	L. C. King and R. J. Hlavacek; *J. Am. Chem. Soc.*, 1951, **73**, 1864.	120
51JA4360	J. E. Fagel, Jr. and G. W. Ewing; *J. Am. Chem. Soc.*, 1951, **73**, 4360.	690
51JA4470	R. A. Henry; *J. Am. Chem. Soc.*, 1951, **73**, 4470.	802
51JA4609	R. Abrams and L. Clark; *J. Am. Chem. Soc.*, 1951, **73**, 4609.	572
51JA5235	G. B. Elion, E. Burgi and G. H. Hitchings; *J. Am. Chem. Soc.*, 1951, **73**, 5235.	572, 575
51JBC(189)401	D. W. Woolley and E. Shaw; *J. Biol. Chem.*, 1951, **189**, 401.	652
51JBC(193)425	R. A. Alberty, R. M. Smith and R. M. Bock; *J. Biol. Chem.*, 1951, **193**, 425.	525
51JCS2411	N. Campbell and E. B. McCall; *J. Chem. Soc.*, 1951, 2411.	631
51MI40100	C. C. J. Roothaan; *Rev. Mod. Phys.*, 1951, **23**, 69.	7
51MI40101	G. G. Hall; *Proc. R. Soc. London, Ser. A*, 1951, **205**, 541.	7
51MI40400	L. E. Orgel, T. L. Cottrell, W. Dick and L. E. Sutton; *Trans. Faraday Soc.*, 1951, **47**, 113.	172
51MI41200	W. D. Jackson and J. B. Polya; *Aust. J. Sci.*, 1951, **13**, 149.	787
51MI41201	L. E. Orgel, T. L. Cottrell, W. Dick and L. E. Sutton; *Trans. Faraday Soc.*, 1951, **47**, 113.	742
51OS(31)43	R. H. Wiley and P. E. Hexner; *Org. Synth.*, 1951, **31**, 43.	288
51YZ760	T. Matsukawa and S. Ban; *Yakugaku Zasshi*, 1951, **71**, 760.	647
52BJ(51)133	E. A. Johnson; *Biochem. J.*, 1952, **51**, 133.	524
52CB390	F. Bohlmann; *Chem. Ber.*, 1952, **85**, 390.	68
52CB1012	F. Korte; *Chem. Ber.*, 1952, **85**, 1012.	618, 635
52DOK(87)783	M. A. Prokof'ev, E. G. Antonovich and Y. P. Shvachkin; *Dokl. Akad. Nauk SSSR*, 1952, **87**, 783 (*Chem. Abstr.*, 1954, **48**, 169).	647

52HCA167	F. Moulin; *Helv. Chim. Acta*, 1952, **35**, 167.	708
52JA350	E. Schipper and A. R. Day; *J. Am. Chem. Soc.*, 1952, **74**, 350.	645
52JA411	G. B. Elion, E. Burgi and G. H. Hitchings; *J. Am. Chem. Soc.*, 1952, **74**, 411.	572
52JA3242	H. R. Snyder, F. Verbanac and D. B. Bright; *J. Am. Chem. Soc.*, 1952, **74**, 3242.	69
52JA4897	E. A. Falco, G. B. Elion, E. Burgi and G. H. Hitchings; *J. Am. Chem. Soc.*, 1952, **74**, 4897.	575, 591
52JA4917	F. R. Benson, L. W. Hartzel and W. L. Savell; *J. Am. Chem. Soc.*, 1952, **74**, 4917.	699
52JA4960	W. A. Reckhow and D. S. Tarbell; *J. Am. Chem. Soc.*, 1952, **74**, 4960.	635
52JCS3418	M. R. Atkinson and J. B. Polya; *J. Chem. Soc.*, 1952, 3418.	765
52JCS3448	F. L. Rose; *J. Chem. Soc.*, 1952, 3448.	338
52JCS4410	G. R. Ramage and G. Trappe; *J. Chem. Soc.*, 1952, 4410.	648
52JCS4735	W. R. Bamford and T. S. Stevens, *J. Chem. Soc.*, 1952, 4735.	706
B-52MI40900	H. C. Urey; 'The Planets', Yale University Press, 1952.	569
52MI41200	P. Dunn, E. A. Parkes and J. B. Polya; *Aust. J. Exp. Biol. Med. Sci.*, 1952, **30**, 437.	787
52ZOB2220	E. S. Golovchinskaya and E. S. Chaman; *Zh. Obshch. Khim.*, 1952, **22**, 2220.	549
52ZOB2225	E. S. Golovchinskaya and E. S. Chaman; *Zh. Obshch. Khim.*, 1952, **22**, 2225.	550
53AG155	E. Pfeil; *Angew. Chem.*, 1953, **65**, 155.	840
53AX369	P. J. Wheatley; *Acta Crystallogr.*, 1953, **6**, 369.	354
53BJ(54)646	A. Albert; *Biochem. J.*, 1953, **54**, 646.	525
53CB88	H. Bredereck and G. Theilig; *Chem. Ber.*, 1953, **86**, 88.	156
53CB96	H. Bredereck and G. Theilig; *Chem. Ber.*, 1953, **86**, 96.	156
53CB333	H. Bredereck, I. Hennig, W. Pfleiderer and G. Weber; *Chem. Ber.*, 1953, **86**, 333.	574
53CB858	R. Kuhn and W. Münzing; *Chem. Ber.*, 1953, **86**, 858.	793
53CR(236)2519	M. Polonovski, M. Pesson and R. Zelnik; *C.R. Hebd. Seances Acad. Sci.*, 1953, **236**, 2519.	567
53CRV(53)191	H. H. Jaffe; *Chem. Rev.*, 1953, **53**, 191.	843
53HC(6-1)	K. Hofmann; *Chem. Heterocycl. Compd.*, 1953, **6-1**.	385
53HCA1324	K. Clusius and M. Vecchi; *Helv. Chim. Acta*, 1953, **36**, 1324.	577
53JA114	J. H. Speer and A. L. Raymond; *J. Am. Chem. Soc.*, 1953, **75**, 114.	573
53JA263	R. K. Robins, K. J. Dille, C. H. Willits and B. E. Christensen; *J. Am. Chem. Soc.*, 1953, **75**, 263.	593
53JA746	I. A. Kaye, C. L. Parris and W. J. Burlant; *J. Am. Chem. Soc.*, 1953, **75**, 746.	631
53JA5753	A. R. P. Paterson and S. H. Zbarsky; *J. Am. Chem. Soc.*, 1953, **75**, 5753.	586
53JBC(204)847	R. O. Hurst, A. M. Marko and G. C. Butler; *J. Biol. Chem.*, 1953, **204**, 847.	525
53JBC(204)1019	G. B. Brown and V. S. Weliky; *J. Biol. Chem.*, 1953, **204**, 1019.	529, 603
53JOC1269	W. L. Garbrecht and R. M. Herbst; *J. Org. Chem.*, 1953, **18**, 1269.	812
53JOC1610	D. W. Kaiser, G. A. Peters and V. P. Wystrach; *J. Org. Chem.*, 1953, **18**, 1610.	900
B-53MI40900	A. I. Oparin; 'Origin of Life', Dover Publications, New York, 1953.	569
53MI40901	E. S. Golovchinskaya and I. Sbornik; *Statei Obshch. Khim. Akad. Nauk SSSR*, 1953, **1**, 702.	593
53MI40902	S. L. Miller; *Science*, 1953, **117**, 528.	569
54CB57	J. Goerdeler; *Chem. Ber.*, 1954, **87**, 57.	23
54CI(L)1458	G. F. Duffin, J. D. Kendall and H. R. J. Waddington; *Chem. Ind. (London)*, 1954, 1458.	745, 748
54CRV41	J. H. Boyer and F. C. Canter; *Chem. Rev.*, 1954, **54**, 41.	710
54HCA789	K. Clusius and H. Hürzeler; *Helv. Chim. Acta*, 1954, **37**, 789.	839, 840
54JA88	R. A. Henry, W. G. Finnegan and E. Lieber; *J. Am. Chem. Soc.*, 1954, **76**, 88.	812
54JA667	L. W. Hartzel and F. R. Benson; *J. Am. Chem. Soc.*, 1954, **76**, 667.	684, 685, 728
54JA923	R. A. Henry and W. G. Finnegan; *J. Am. Chem. Soc.*, 1954, **76**, 923.	819
54JA2894	R. A. Henry, W. G. Finnegan and E. Lieber; *J. Am. Chem. Soc.*, 1954, **76**, 2894.	793
54JA4470	C. Djerassi and G. R. Pettit; *J. Am. Chem. Soc.*, 1954, **76**, 4470.	631
54JA4933	R. H. Wiley, N. R. Smith, D. M. Johnson and J. Moffat; *J. Am. Chem. Soc.*, 1954, **76**, 4933.	685, 700
54JA5087	K. L. Dille and B. E. Christensen; *J. Am. Chem. Soc.*, 1954, **76**, 5087.	596
54JA5633	A. G. Beaman; *J. Am. Chem. Soc.*, 1954, **76**, 5633.	557, 558, 559, 562, 563, 572, 597
54JA6073	A. Bendich, P. J. Russell, Jr. and J. J. Fox; *J. Am. Chem. Soc.*, 1954, **76**, 6073.	524, 592
54JCS141	M. R. Atkinson and J. B. Polya; *J. Chem. Soc.*, 1954, 141.	746
54JCS1850	G. D. Buckley; *J. Chem. Soc.*, 1954, 1850.	717
54JCS2060	A. Albert and D. J. Brown; *J. Chem. Soc.*, 1954, 2060.	523, 524, 525, 571, 572, 575, 579, 593, 596
54JCS2701	I. M. Bassett and R. D. Brown; *J. Chem. Soc.*, 1954, 2701.	347
54JCS3319	M. R. Atkinson and J. B. Polya; *J. Chem. Soc.*, 1954, 3319.	746, 765
54JCS3461	K. T. Potts; *J. Chem. Soc.*, 1954, 3461.	767
54JCS4256	M. R. Atkinson, E. A. Parkes and J. B. Polya; *J. Chem. Soc.*, 1954, 4256.	739, 756
54JCS4508	M. R. Atkinson, A. A. Komzak, E. A. Parkes and J. B. Polya; *J. Chem. Soc.*, 1954, 4508.	739
54JGU1045	M. A. Prokof'ev and Y. P. Shvachkin; *J. Gen. Chem. USSR (Engl. Transl.)*, 1954, 1045 (*Chem. Abstr.*, 1955, **49**, 9661).	647
54JOC439	R. M. Herbst and D. F. Percival; *J. Org. Chem.*, 1954, **19**, 439.	794
54LA(590)232	D. Jerchel, M. Kracht and K. Krucher; *Liebigs Ann. Chem.*, 1954, **590**, 232.	576
54MI40900	D. J. Brown; *J. Appl. Chem.*, 1954, 358.	571

54MI40901	N. Urabe and K. Yasukochi; *J. Electrochem. Soc. Jpn.*, 1954, **22**, 469	541
54MI40902	N. Urabe and K. Yasukochi; *J. Electrochem. Soc. Jpn.*, 1954, **22**, 525.	541
54MI40903	M. Ishidate, M. Sekiya and I. Kurita; *J. Pharm. Soc. Jpn.*, 1954, **74**, 420.	580
54MI40904	I. Lasnitzki, R. E. F. Matthews and J. D. Smith; *Nature (London)*, 1954, **173**, 346.	604
54MI41100	K. L. Cheng; *Anal. Chem.*, 1954, **26**, 1038.	731
54MI41101	G. H. N. Towers and D. C. Mortimer; *Nature (London)*, 1954, **174**, 1189.	722
54ZN(B)495	E. Wiberg and H. Michaud; *Z. Naturforsch., Teil B*, 1954, **9**, 495.	840
54ZN(B)496	E. Wiberg and H. Michaud; *Z. Naturforsch., Teil B*, 1954, **9**, 496.	840
55AJC100	R. D. Brown; *Aust. J. Chem.*, 1955, **8**, 100.	346, 347, 348
55AX211	J. H. Bryden; *Acta Crystallogr.*, 1955, **8**, 211.	797
55CB1093	K. Schilling, F. Krohnke and B. Kickhofen; *Chem. Ber.*, 1955, **88**, 1093.	610, 611, 632
55CB1117	F. Krohnke, B. Kickhofen and C. Thoma; *Chem. Ber.*, 1955, **88**, 1117.	631, 632
55CB1306	H. Bredereck and A. Edenhofer; *Chem. Ber.*, 1955, **88**, 1306.	580
55CB1932	H. Priewe and A. Poljak; *Chem. Ber.*, 1955, **88**, 1932.	554
55CR(241)1049	P. Grammaticakis; *C. R. Hebd. Seances Acad. Sci.*, 1955, **241**, 1049.	739
55FES235	M. L. Stein and L. D'Antoni; *Farmaco Ed. Sci.*, 1955, **10**, 235.	728
55FES616	G. B. Crippa and A. Crippa; *Farmaco Ed. Sci.*, 1955, **10**, 616 (*Chem. Abstr.*, 1956, **50**, 1840).	645
55FES733	G. Serchi, L. Sancio and G. Buchi; *Farmaco Ed. Sci.*, 1955, **10**, 733.	537
55JA1538	R. G. Jones and C. Ainsworth; *J. Am. Chem. Soc.*, 1955, **77**, 1538.	753, 755
55JA1703	R. H. Wiley and J. Moffat; *J. Am. Chem. Soc.*, 1955, **77**, 1703.	698
55JA2351	S. L. Miller; *J. Am. Chem. Soc.*, 1955, **77**, 2351.	569
55JA2569	C. H. Willits, J. C. Decius, K. L. Dille and B. E. Christensen; *J. Am. Chem. Soc.*, 1955, **77**, 2569.	518
55JA2572	R. H. Wiley, N. R. Smith, D. M. Johnson and J. Moffat; *J. Am. Chem. Soc.*, 1955, **77**, 2572.	748
55JA5105	R. H. Wiley, K. H. Hussung and J. Moffat; *J. Am. Chem. Soc.*, 1955, **77**, 5105.	690, 700
55JA5922	F. Schaefer; *J. Am. Chem. Soc.*, 1955, **77**, 5922.	657
55JA6204	R. F. Wilson and L. E. Wilson; *J. Am. Chem. Soc.*, 1955, **77**, 6204.	731
55JA6532	C. S. Rondestvedt, Jr. and P. K. Chang; *J. Am. Chem. Soc.*, 1955, **77**, 6532.	685
55JCS896	P. R. Brook and G. R. Ramage; *J. Chem. Soc.*, 1955, 896.	648
55JCS2834	J. D. Bower and G. R. Ramage; *J. Chem. Soc.*, 1955, 2834.	614
55JGU939	M. A. Prokof'ev and Y. P. Shvachkin; *J. Gen. Chem. USSR (Engl. Transl.)*, 1955, **25**, 939 (*Chem. Abstr.*, 1956, **50**, 3458).	647
55LA(593)207	R. Hüttel and G. Welzel; *Liebigs Ann. Chem.*, 1955, **593**, 207.	698, 699
55MI40900	B. G. Bolyrev and R. G. Matrika; *J. Appl. Chem. USSR*, 1955 **28**, 399.	598
55MI41000	R. Fusco and S. Rossi; *Rend. Ist. Lomb. Sci. Lett., Cl. Sci. Mat. Nat.*, 1955, **88**, 194 (*Chem. Abstr.*, 1956, **50**, 10 743).	653
55MI41100	L. Cambi, L. Canonica and C. Sironi; *Atti Accad. Nazl. Lincei, Rend., Cl. Sci. Fis. Mat. Nat.*, 1955, **18**, 583 (*Chem. Abstr.*, 1956, **50**, 3940).	731
55MI41200	J. B. Polya; *Nature (London)*, 1955, **176**, 1175.	787
55ZOB1366	N. K. Kochetkov; *Zh. Obshch. Khim.*, 1955, **25**, 1366.	716
56AC(R)275	N. Gagnoli and A. Ricci; *Ann. Chim. (Rome)*, 1956, **46**, 275.	138
56AG705	R. Huisgen and I. Ugi; *Angew. Chem.*, 1956, **68**, 705.	840, 841, 843, 845
56AG753	I. Ugi, R. Huisgen, K. Clusius and M. Vecchi; *Angew. Chem.*, 1956, **68**, 753.	841
56AP453	G. Ehrart and I. Hennig; *Arch. Pharm. (Weinheim, Ger.)*, 1956, **289**, 453	549, 550, 551, 574, 593
56AX874	J. H. Bryden; *Acta Crystallogr.*, 1956, **9**, 874.	796
56BSF888	R. Zelnick, M. Pesson and M. Polonovski; *Bull. Soc. Chim. Fr.*, 1956, 888.	567
56CB1940	W. Otting; *Chem. Ber.*, 1956, **89**, 1940.	27
56CR(243)380	A. Pullman, B. Pullman and G. Berthier; *C.R. Hebd. Seances Acad. Sci.*, 1956, **243**, 380.	505
56CR(243)2090	N. P. Buu-Hoi and N. D. Xuong; *C. R. Hebd. Seances Acad. Sci.*, 1956, **243**, 2090.	647
56G797	D. Dal Monte, A. Mangini and R. Passerini; *Gazz. Chim. Ital.*, 1956, **86**, 797.	197
56HCA986	P. Schmidt and J. Druey; *Helv. Chim. Acta*, 1956, **39**, 986.	329
56JA159	R. G. Jones; *J. Am. Chem. Soc.*, 1956, **78**, 159.	652
56JA217	G. B. Elion, W. H. Lange and G. H. Hitchings; *J. Am. Chem. Soc.*, 1956, **78**, 217.	560
56JA242	F. L. Muehlmann and A. R. Day; *J. Am. Chem. Soc.*, 1956, **78**, 242.	645
56JA411	K. Hattori, E. Lieber and J. P. Horwitz; *J. Am. Chem. Soc.*, 1956, **78**, 411.	794, 802, 819
56JA784	R. K. Robins; *J. Am. Chem. Soc.*, 1956, **78**, 784.	323, 329
56JA1375	C. O. Miller, F. Skoog, F. S. Okumura, M. H. von Saltza and F. M. Strong; *J. Am. Chem. Soc.*, 1956, **78**, 1375.	602
56JA1618	A. F. McKay and W. G. Hatton; *J. Am. Chem. Soc.*, 1956, **78**, 1618.	647
56JA1928	J. A. Montgomery: *J. Am. Chem. Soc.*, 1956, **78**, 1928.	573, 597
56JA2418	R. K. Robins, F. W. Furcht, A. D. Grauer and J. W. Jones; *J. Am. Chem. Soc.*, 1956, **78**, 2418.	323, 324
56JA2858	G. B. Elion, W. H. Lange and G. H. Hitchings; *J. Am. Chem. Soc.*, 1956, **78**, 2858.	572
56JA3508	G. B. Elion and G. H. Hitchings; *J. Am. Chem. Soc.*, 1956, **78**, 3508	557, 563, 597, 598
56JA3511	L. B. Mackay and G. H. Hitchings; *J. Am. Chem. Soc.*, 1956, **78**, 3511.	550, 551, 593
56JA4130	D. G. Markees and G. W. Kidder; *J. Am. Chem. Soc.*, 1956, **78**, 4130.	619, 635

56JA4197	M. H. Kaufman, F. M. Ernsberger and W. S. McEwan; *J. Am. Chem. Soc.*, 1956, **78**, 4197.	795
56JA5832	J. R. E. Hoover and A. R. Day; *J. Am. Chem. Soc.*, 1956, **78**, 5832.	713
56JA5857	F. F. Blicke and R. L. Schaaf; *J. Am. Chem. Soc.*, 1956, **78**, 5857.	578
56JCS1045	V. Oakes, R. Pascoe and H. N. Rydon; *J. Chem. Soc.*, 1956, 1045.	637
56JCS3975	D. B. Ishay; *J. Chem. Soc.*, 1956, 3975.	529
56JCS4288	R. D. Brown and M. L. Heffernan; *J. Chem. Soc.*, 1956, 4288.	349
56JOC190	R. H. Wiley, K. F. Hussung and J. Moffat; *J. Org. Chem.*, 1956, **21**, 190.	708
56JOC599	L. Goldman, J. W. Marsico and A. L. Gazzola; *J. Org. Chem.*, 1956, **21**, 599.	573
56JOC1037	C. Grundmann and R. Rätz; *J. Org. Chem.*, 1956, **21**, 1037.	766
56JOC1276	M. M. Baizer, J. R. Clark, M. Dub. and A. Loter; *J. Org. Chem.*, 1956, **21**, 1276.	552
56LA(598)186	R. Huttel, O. Schafer and G. Welzel; *Liebigs Ann. Chem.*, 1956, **598**, 186.	89
56M679	K. Schlogel and H. Woidich; *Monatsh. Chem.*, 1956, **87**, 679.	649
56MI40900	R. M. Bock, N.-S. Ling, S. A. Morell and S. H. Lipton; *Arch. Biochem. Biophys.*, 1956, **62**, 253.	525
56MI41100	R. F. Wilson and L. E. Wilson; *Anal. Chem.*, 1956, **28**, 93.	731
B-56MI41300	R. M. Herbst; 'Essays in Biochemistry', ed. S. Groff; Wiley, New York, 1956, p. 141.	816
57AC(R)362	F. Cacace and R. Masironi; *Ann. Chim (Rome)*, 1957, **47**, 362.	551
57AC(R)366	F. Cacace and R. Masironi; *Ann. Chim. (Rome)*, 1957, **47**, 366.	551
57AG732	J. Jennen; *Angew. Chem.*, 1957, **69**, 732.	232
57AP20	A. Dornow and O. Hahmann; *Arch. Pharm. (Weinheim, Ger.)*, 1957, **290**, 20.	618, 636
57BSB136	R. H. Martin and Z. Tarasiejska; *Bull. Soc. Chim. Belg.*, 1957, **66**, 136.	642, 645
57CB202	J. Goerdeler and J. Galinke; *Chem. Ber.*, 1957, **90**, 202.	773
57CB382	R. Gompper; *Chem. Ber.*, 1957, **90**, 382.	713
57CB414	R. Criegee and A. Rimmelin; *Chem. Ber.*, 1957, **90**, 414.	839
57CB698	G. Huber; *Chem. Ber.*, 1957, **80**, 698.	596
57CB2914	R. Huisgen and I. Ugi; *Chem. Ber.*, 1957, **90**, 2914.	841, 843, 845
57CPB240	M. Ishidate and H. Yuki; *Chem. Pharm. Bull.*, 1957, **5**, 240.	579
57G597	S. Checchi and M. Ridi; *Gazz. Chim. Ital.*, 1957, **87**, 597.	334
57JA490	R. K. Robins and H. H. Lin; *J. Am. Chem. Soc.*, 1957, **79**, 490.	564
57JA670	B. S. Gorton and W. Shive; *J. Am. Chem. Soc.*, 1957, **79**, 670.	635
57JA678	J. H. Boyer, R. Borgers and L. T. Wolford; *J. Am. Chem. Soc.*, 1957, **79**, 678.	887
57JA1249	J. P. Horwitz and V. A. Grakauskas; *J. Am. Chem. Soc.*, 1957, **79**, 1249.	839
57JA1518	O. Vogl and E. C. Taylor; *J. Am. Chem. Soc.*, 1957, **79**, 1518.	569, 580
57JA1955	G. W. Sawdey; *J. Am. Chem. Soc.*, 1957, **79**, 1955.	777
57JA2185	J. A. Montgomery and L. B. Holum; *J. Am. Chem. Soc.*, 1957, **79**, 2185.	564, 593
57JA2839	C. Grundmann and A. Kreutzberger; *J. Am. Chem. Soc.*, 1957, **79**, 2839.	766
57JA5276	A. F. McKay, G. Y. Paris and M. E. Kreling; *J. Am. Chem. Soc.*, 1957, **79**, 5276.	139
57JA6401	R. N. Prasad and R. K. Robins; *J. Am. Chem. Soc.*, 1957, **79**, 6401.	564, 584, 590
57JA6407	R. K. Robins; *J. Am. Chem. Soc.*, 1957, **79**, 6407.	313
57JA6421	H. Graboyes and A. R. Day; *J. Am. Chem. Soc.*, 1957, **79**, 6421.	618, 619, 635, 636
57JAP570777	K. Shirakawa; *Jpn. Pat.* 57 0777 (1957) (*Chem. Abstr.*, 1958, **52**, 4699).	857, 902
57JCS442	J. W. Clark-Lewis and M. J. Thompson; *J. Chem. Soc.*, 1957, 442.	636
57JCS682	D. J. Brown and S. F. Mason; *J. Chem. Soc.*, 1957, 682.	524, 525
57JCS1556	A. Lawson and C. E. Searle; *J. Chem. Soc.*, 1957, 1556.	113
57JCS2356	E. G. Brain and I. L. Finar; *J. Chem. Soc.*, 1957, 2356.	93
57JCS2521	M. J. S. Dewar and P. M. Maitlis; *J. Chem. Soc.*, 1957, 2521	7
57JCS4506	J. D. Bower and G. R. Ramage; *J. Chem. Soc.*, 1957, 4506.	887
57JCS4510	J. D. Bower; *J. Chem. Soc.*, 1957, 4510.	307, 334, 857
57JOC568	M. W. Bullock, J. J. Hand and E. L. R. Stokstad; *J. Org. Chem.*, 1957, **22**, 568.	552
57JOC654	E. Lieber, T. S. Chao and C. N. R. Rao; *J. Org. Chem.*, 1957, **22**, 654.	713
57JOC925	D. F. Percival and R. M. Herbst; *J. Org. Chem.*, 1957, **22**, 925.	794
57JOC954	B. R. Baker, K. Hewson, H. J. Thomas and J. A. Johnson, Jr.; *J. Org. Chem.*, 1957, **22**, 954.	629
57JOC959	B. R. Baker and K. Hewson; *J. Org. Chem.*, 1957, **22**, 959.	552
57JPC261	M. M. Williams, W. S. McEwan and R. A. Henry; *J. Phys. Chem.*, 1957, **61**, 261.	742
57MI40100	H. A. Staab, W. Otting and A. Ueberle; *Z. Elektrochem.*, 1957, **61**, 1000.	27, 29
57MI40101	D. F. Othmer, P. W. Maurer, C. J. Molinary and R. C. Kowalski; *Ind. Eng. Chem.*, 1957, **49**, 125.	29
B-57MI40400	T. L. Jacobs; in 'Heterocyclic Compounds', ed. R. C. Elderfield; Wiley, New York, 1957, vol. 5, p. 57.	168
B-57MI40700	E. S. Schipper and A. R. Day; in 'Heterocyclic Compounds', ed. R. C. Elderfield; Wiley, New York, 1957, vol. 5, p. 194.	424
B-57MI40800	E. S. Schipper and A. R. Day; in 'Heterocyclic Compounds', ed. R. C. Elderfield; Wiley, New York, 1957, vol. 5, p. 194.	469, 476, 490
B-57MI40900	'Ciba Foundation Symposium on the Chemistry and Biology of Purines', ed. G. E. W. Wolstenholme and C. M. O'Connor; Churchill, London, 1957.	501, 511, 524, 556
B-57MI41100	E. H. Hoggarth; in 'Chemistry of Carbon Compounds', ed. E. H. Rodd; Elsevier, Amsterdam, 1957, p. 439.	670
57MI41300	E. Lieber, C. N. R. Rao and C. N. Pillai; *Curr. Sci.*, 1957, **26**, 167.	798

57OS(37)26	E. Lieber, T. S. Chao and C. N. R. Rao; *Org. Synth.*, 1957, **37**, 26.	713
57QR15	W. Baker and W. D. Ollis; *Q. Rev., Chem. Soc.*, 1957, **11**, 15.	736, 743
57T(1)75	R. H. Martin and J. Mathieu; *Tetrahedron*, 1957, **1**, 75.	648
57ZOB1276	T. V. Protopopova and A. P. Skoldinov; *Zh. Obshch. Khim.*, 1957, **27**, 1276.	290
57ZOB1712	G. A. Garkusha; *Zh. Obshch. Khim.*, 1957, **27**, 1712.	563
58AG571	F. Cramer and K. Randerath; *Angew. Chem.*, 1958, **70**, 571.	539
58AP368	A. Dornow, O. Hahmann and E. H. Rohe; *Arch. Pharm. (Weinheim, Ger.)*, 1958, **291**, 368.	618, 636
58AX83	D. J. Sutor; *Acta Crystallogr.*, 1958, **11**, 83.	508
58AX453	D. J. Sutor; *Acta Crystallogr.*, 1958, **11**, 453.	508
58CB531	I. Ugi and R. Huisgen; *Chem. Ber.*, 1958, **91**, 531.	843
58CB1841	A. Dornow and K. Rombusch; *Chem. Ber.*, 1958, **91**, 1841.	710
58CB2324	I. Ugi, H. Perlinger and L. Behringer; *Chem. Ber.*, 1958, **91**, 2324.	843, 844
58CB2330	I. Ugi, H. Perlinger and L. Behringer; *Chem. Ber.*, 1958, **91**, 2330.	841
58CJC801	E. Lieber, C. N. R. Rao, C. N. Pillai, J. Ramachandran and R. D. Hites; *Can. J. Chem.*, 1958, **36**, 801.	794
58CJC1441	E. Lieber, C. N. R. Rao, T. S. Chao and H. Rubinstein; *Can. J. Chem.*, 1958, **36**, 1441.	685
58CPB443	T. Takahashi, F. Yoneda and S. Yoshimura; *Chem. Pharm. Bull.*, 1958, **6**, 443.	635
58G463	S. Cusmano and M. Ruccia; *Gazz. Chim. Ital.*, 1958, **88**, 463.	404
58G977	D. Dal Monk, A. Mangini, R. Pawerini and C. Zanli; *Gazz. Chim. Ital.*, 1958, **88**, 977.	684, 685
58G1035	D. Dal Monte, A. Mangini and F. Montanardi; *Gazz. Chim. Ital.*, 1958, **88**, 1035.	684
58HCA1052	P. Schmidt, K. Eichenberger and J. Druey; *Helv. Chim. Acta*, 1958, **41**, 1052.	310
58HCA1823	K. Clusius and F. Endtinger; *Helv. Chim. Acta*, 1958, **41**, 1823.	844
58JA404	J. A. Montgomery and L. B. Holum; *J. Am. Chem. Soc.*, 1958, **80**, 404.	597
58JA421	E. C. Taylor, J. W. Barton and T. S. Osdene; *J. Am. Chem. Soc.*, 1958, **80**, 421.	314, 329
58JA599	L. A. Carpino; *J. Am. Chem. Soc.*, 1958, **80**, 599.	78
58JA1669	J. J. Fox, I. Wempen, A. Hampton and I. L. Doerr; *J. Am. Chem. Soc.*, 1958, **80**, 1669.	557
58JA2751	H. C. Koppel and R. K. Robins; *J. Am. Chem. Soc.*, 1958, **80**, 2751.	598
58JA2755	M. A. Stevens, D. I. Magrath, H. W. Smith and G. B. Brown; *J. Am. Chem. Soc.*, 1958, **80**, 2755.	539, 554, 555
58JA2759	M. A. Stevens and G. B. Brown; *J. Am. Chem. Soc.*, 1958, **80**, 2759.	595
58JA3738	H. J. Schaeffer and H. J. Thomas; *J. Am. Chem. Soc.*, 1958, **80**, 3738.	558
58JA3899	E. Shaw, *J. Am. Chem. Soc.*, 1958, **80**, 3899.	583, 584
58JA3932	A. Giner-Sorolla and A. Bendich; *J. Am. Chem. Soc.*, 1958, **80**, 3932.	524, 550, 551, 553, 556, 564
58JA4979	H. C. Brown and Y. Okamoto; *J. Am. Chem. Soc.*, 1958, **80**, 4979.	512
58JA5744	A. Giner-Sorolla and A. Bendich; *J. Am. Chem. Soc.*, 1958, **80**, 5744.	524, 574, 587
58JA6083	J. A. Carbon; *J. Am. Chem. Soc.*, 1958, **80**, 6083.	629, 652
58JA6265	T. P. Johnston, L. B. Holum and J. A. Montgomery; *J. Am. Chem. Soc.*, 1958, **80**, 6265.	560
58JA6671	R. K. Robins; *J. Am. Chem. Soc.*, 1958, **80**, 6671.	562, 563, 593
58JBC(230)717	M. Adler, B. Weissmann and A. B. Gutman; *J. Biol. Chem.*, 1958, **230**, 717.	600
58JCS2821	J. Clark and G. R. Ramage; *J. Chem. Soc.*, 1958, 2821.	648
58JCS2973	B. M. Lynch, R. K. Robins and C. C. Cheng; *J. Chem. Soc.*, 1958, 2973.	306, 309
58JCS3259	I. L. Finar and R. J. Hurlock; *J. Chem. Soc.*, 1958, 3259.	331
58JCS4069	R. Hull; *J. Chem. Soc.*, 1958, 4069.	549
58JCS4107	C. L. Leese and G. M. Timmis; *J. Chem. Soc.*, 1958, 4107.	572
58JOC125	G. B. Brown and V. S. Weliky; *J. Org. Chem.*, 1958, **23**, 125.	562
58JOC191	C. C. Cheng and R. K. Robins; *J. Org. Chem.*, 1958, **23**, 191.	319
58JOC1051	J. H. Boyer, C. H. Mack, N. Goebel and L. R. Morgan, Jr.; *J. Org. Chem.*, 1958, **23**, 1051.	709
58JOC1534	R. N. Castle and W. S. Seese; *J. Org. Chem.*, 1958, **23**, 1534.	652
58JOC1912	R. M. Herbst and J. E. Klingbeil; *J. Org. Chem.*, 1958, **23**, 1912.	763
58JOC1916	E. Leiber, C. N. R. Rao, T. S. Chao and J. Ramachandran; *J. Org. Chem.*, 1958, **23**, 1916.	690
58JOC2010	D. S. Acker and J. E. Castle; *J. Org. Chem.*, 1958, **23**, 2010.	572
58LA(614)1	W. Kirmse and L. Horner; *Liebigs Ann. Chem.*, 1958, **614**, 1.	714
58LA(614)4	W. Kirmse and L. Horner; *Liebigs Ann. Chem.*, 1958, **614**, 4.	159
58LA(615)42	W. Pfleiderer and K.-H. Schündehütte; *Liebigs Ann. Chem.*, 1958, **615**, 42.	338
58M557	K. Hohenlohe-Oehringen; *Monatsh. Chem.*, 1958, **89**, 557.	707
58M562	K. Hohenlohe-Oehringen; *Monatsh. Chem.*, 1958, **89**, 562.	707
58M588	K. Hohenlohe-Oehringen; *Monatsh. Chem.*, 1958, **89**, 588.	707
58M597	K. Hohenlohe-Oehringen; *Monatsh. Chem.*, 1958, **89**, 597.	707
58MI40200	A. K. Majumdar and M. M. Chakrabartty; *Anal. Chim. Acta*, 1958, **19**, 372.	51
58MI40900	E. S. Golovchinskaya; *Zh. Prikl. Khim.*, 1958, **31**, 918.	548
58MI40901	E. S. Golovchinskaya, V. M. Fedosova and A. A. Cherkasova; *Zh. Prikl. Khim.*, 1958, **31**, 1241.	548
58MI40902	J. W. Littlefield and D. B. Dunn; *Nature (London)*, 1958, **181**, 254.	600

58MI41300	E. Lieber and E. Oftedahl; *Trans. Illinois State Acad. Sci.*, 1958, **51**, 41 (*Chem. Abstr.*, 1962, **56**, 4151).	816
58N492	K. Winterfeld and H. Schuler; *Naturwissenschaften*, 1958, **45**, 492.	635
59ACS888	C. Pedersen; *Acta Chem. Scand.*, 1959, **13**, 888.	708, 709
59AG524	H. Bredereck, H. Herlinger and I. Graudums; *Angew. Chem.*, 1959, **71**, 524.	563
59AG753	H. Bredereck, R. Gompper, H. G. V. Schüh and W. Theilig; *Angew. Chem.*, 1959, **71**, 753.	485
59AJC543	R. D. Brown and M. L. Heffernan; *Aust. J. Chem.*, 1959, **12**, 543.	347
59BBA(36)343	B. Pullman and A. Pullman; *Biochim. Biophys. Acta*, 1959, **36**, 343.	505
59BRP816531	Henkel and Cie G.m.b.H., *Br. Pat.* 816 531 (1959) (*Chem. Abstr.*, 1960, **54**, 1552).	783
59CB566	H. Bredereck, G. Kupsch and H. Wieland; *Chem. Ber.*, 1959, **92**, 566.	574
59CB1500	J. Donat and E. Carstens; *Chem. Ber.*, 1959, **92**, 1500.	537
59CB1864	I. Ugi, H. Perlinger and L. Behringer; *Chem. Ber.*, 1959, **92**, 1864.	843
59CB2902	W. Ried and E. Torinus; *Chem. Ber.*, 1959, **92**, 2902.	576
59CPB602	T. Takahashi, F. Yoneda and R. Oishi; *Chem. Pharm. Bull.*, 1959, **7**, 602.	635
59CR(248)1677	M. Pesson, G. Polmanss and S. Dupin; *C. R. Hebd. Seances Acad. Sci.*, 1959, **248**, 1677.	765
59CR(248)1832	N. P. Buu-Hoi and N. D. Xuong; *C. R. Hebd. Seances Acad. Sci.*, 1959, **248**, 1832.	647
59G1654	M. Ruccia and D. Spinelli; *Gazz. Chim. Ital.*, 1959, **89**, 1654.	719
59HC(13)432	E. M. Smolin and L. Rapoport; *Chem. Heterocycl. Compd.*, 1959, **13**, 432.	608
59HCA349	P. Schmidt, K. Eichenberger, M. Wilhelm and J. Druey; *Helv. Chim. Acta*, 1959, **42**, 349.	329
59JA193	R. N. Prasad, C. W. Noell and R. K. Robins; *J. Am. Chem. Soc.*, 1959, **81**, 193.	574, 593
59JA1898	G. B. Elion, I. Goodman, W. Lange and G. H. Hitchings; *J. Am. Chem. Soc.*, 1959, **81**, 1898	564
59JA2442	E. C. Taylor, O. Vogl and C. Cheng; *J. Am. Chem. Soc.*, 1959, **81**, 2442.	560, 569
59JA2452	E. C. Taylor and K. S. Hartke; *J. Am. Chem. Soc.*, 1959, **81**, 2452.	318, 322
59JA2515	A. Giner-Sorolla, I. Zimmerman and A. Bendich; *J. Am. Chem. Soc.*, 1959, **81**, 2515.	524, 548, 549, 550, 593
59JA3039	T. L. Loo, M. E. Michael, A. J. Garceau and J. C. Reid; *J. Am. Chem. Soc.*, 1959, **81**, 3039.	557
59JA3076	J. P. Horwitz, B. E. Fisher and A. J. Tomasewski; *J. Am. Chem. Soc.*, 1959, **81**, 3076.	794, 832
59JA3789	S. R. Breshears, S. S. Wang, S. G. Bechtolt and B. E. Christensen; *J. Am. Chem. Soc.*, 1959, **81**, 3789.	541
59JA5997	C. W. Noell and R. K. Robins; *J. Am. Chem. Soc.*, 1959, **81**, 5997.	557
59JA6329	S. S. Chissick, M. J. S. Dewar and P. M. Maitlis; *J. Am. Chem. Soc.*, 1959, **81**, 6329.	668
59JAP5910468	S. Yamada and T. Mizoguchi; *Jpn. Pat.* 59 10 468 (1959) (*Chem. Abstr.*, 1960, **54**, 18 555).	721
59JCS2893	J. Baddiley, J. G. Buchanan, F. E. Hardy and J. Stewart; *J. Chem. Soc.*, 1959, 2893.	583
59JCS3061	A. Adams and R. Slack; *J. Chem. Soc.*, 1959, 3061.	86
59JCS3157	D. Harrison and A. C. B. Smith; *J. Chem. Soc.*, 1959, 3157.	617, 636
59JCS3799	G. F. Duffin, J. D. Kendall and H. R. J. Waddington; *J. Chem. Soc.*, 1959, 3799.	747, 748, 788
59JCS4040	G. Shaw and D. N. Butler; *J. Chem. Soc.*, 1959, 4040.	584, 585
59JOC134	E. Lieber, C. N. R. Rao and T. U. Rajkumar; *J. Org. Chem.*, 1959, **24**, 134.	713
59JOC320	C. W. Noell and R. K. Robins; *J. Org. Chem.*, 1959, **24**, 320.	535, 563
59JOC562	J. H. Burckhalter and D. R. Dill; *J. Org. Chem.*, 1959, **24**, 562.	529
59JOC787	C. F. H. Allen, H. R. Beilfuss, D. M. Burness, G. A. Reynolds, J. F. Tinker and J. A. Van Allan; *J. Org. Chem.*, 1959, **24**, 787.	862
59JOC1455	M. Israel and A. R. Day; *J. Org. Chem.*, 1959, **24**, 1455.	617, 618, 619, 635, 636
59JOC1565	W. G. Finnegan and R. A. Henry; *J. Org. Chem.*, 1959, **24**, 1565.	819
59JOC1643	J. M. McManus and R. M. Herbst; *J. Org. Chem.*, 1959, **24**, 1643.	834
59JOC1813	H. Zimmer and J. B. Mettalia; *J. Org. Chem.*, 1959, **24**, 1813.	552
59JOC2019	E. C. Taylor, C. C. Cheng and O. Vogl; *J. Org. Chem.*, 1959, **24**, 2019.	584
59JPR(8)298	G. Losse and W. Farr; *J. Prakt. Chem.*, 1959, **8**, 298.	779
59M402	F. Asinger, M. Thiel and R. Sowada; *Monatsh. Chem.*, 1959, **90**, 402.	425
59MI40100	T. Balaban; *Stud. Cercet. Chim.*, 1959, **7**, 257.	30, 33
59MI40200	R. Schindler, H. Will and L. Holleck; *Z. Elektrochem.*, 1959, **63**, 596.	73
B-59MI40400	A. Albert; 'Heterocyclic Chemistry', Athlone Press, London, 1959.	168
59MI40900	J. M. Buchanan and S. C. Hartman; *Adv. Enzymol.*, 1959, **21**, 199.	504
59MI40901	S. C. Hartman and J. M. Buchanan; *Annu. Rev. Biochem.*, 1959, **28**, 365.	504
59OS(39)27	C. Ainsworth; *Org. Synth.*, 1959, **39**, 27.	288
59TH41400	I. Ugi; Habilitation Thesis, University of Munich, 1959.	840, 841, 844
59TH41401	H. Perlinger; Doctoral Thesis, University of Munich, 1959.	845
59TL9	E. C. Taylor and C. Cheng; *Tetrahedron Lett.*, 1959, 9.	568
60AC(R)125	M. Melandri, G. Vittorina and A. Buttini; *Ann. Chim. (Rome)*, 1960, **50**, 125 (*Chem. Abstr.*, 1960, **55**, 27 306).	636
60AG708	H. Bredereck, O. Christmann, I. Graudums and W. Koser; *Angew. Chem.*, 1960, **72**, 708.	562
60AG956	H. Gold; *Angew. Chem.*, 1960, **72**, 956.	766
60AJC49	R. D. Brown and M. L. Heffernan; *Aust. J. Chem.*, 1960, **13**, 49.	172, 173

60AP203	K. Winterfeld and H. Schuller; *Arch. Pharm. (Weinheim, Ger.)*, 1960, **293**, 203.	635
60AX946	H. W. W. Ehrlich; *Acta Crystallogr.*, 1960, **13**, 946.	177, 179
60BBA(45)49	F. Bergmann, H. Ungar and A. Kalmus; *Biochem. Biophys Acta*, 1960, **45**, 49.	577
60BBR(2)207	J. Oro; *Biochem. Biophys. Res. Commun.*, 1960, **2**, 207.	569
60BBR(3)596	L. Slechta; *Biochem. Biophys. Res. Commun.*, 1960, **3**, 596.	603
60BJ(76)621	A. Albert and E. P. Serjeant; *Biochem. J.*, 1960, **76**, 621.	523
60CB663	J. Goerdeler and H. Horstmann; *Chem. Ber.*, 1960, **93**, 663.	765
60CB1106	A. Dornow and M. Siebrecht; *Chem. Ber.*, 1960, **93**, 1106.	332
60CB1206	H. Bredereck, O. Christmann and W. Koser; *Chem. Ber.*, 1960, **93**, 1206.	531
60CB2001	A. Dornow and J. Helberg; *Chem. Ber.*, 1960, **93**, 2001.	696, 713
60CB2353	W. Kirmse; *Chem. Ber.*, 1960, **93**, 2353.	45
60CB2804	L. Birkofer, P. Richter and T. Ritter; *Chem. Ber.*, 1960, **93**, 2804.	749, 761
60CB2810	L. Birkhofer, H. P. Kuhlthau and A. Ritter; *Chem. Ber.*, 1960, **93**, 2810.	576, 577
60CB2885	R. Huisgen, J. Sauer and M. Seidel; *Chem. Ber.*, 1960, **93**, 2885.	814
60CPB539	T. Takahashi, K. Kanematsu, R. Oh-ishi and T. Mizutani; *Chem. Pharm. Bull.*, 1960, **8**, 539.	616
60CPB999	T. Itai and S. Suzuki; *Chem. Pharm. Bull.*, 1960, **8**, 999.	652
60G413	P. Grünanger, D. Vita Finzi and E. Fabbri; *Gazz. Chim. Ital.*, 1960, **90**, 413.	714
60G831	M. Ruccia, G. Natale and S. Cusmano; *Gazz. Chim. Ital.*, 1960, **90**, 831.	441
60JA222	C. D. Jardetzky and O. Jardetzky; *J. Am. Chem. Soc.*, 1960, **82**, 222.	523
60JA463	J. A. Montgomery and K. Hewson; *J. Am. Chem. Soc.*, 1960, **82**, 463.	552, 553
60JA1148	M. A. Stevens, H. W. Smith and G. B. Brown; *J. Am. Chem. Soc.*, 1960, **82**, 1148.	555
60JA1469	S. C. Bell and W. T. Caldwell; *J. Am. Chem. Soc.*, 1960, **82**, 1469.	647
60JA2587	L. F. Cavalieri and A. Bendich; *J. Am. Chem. Soc.*, 1960, **72**, 2587.	539
60JA2633	G. D. Daves, Jr., C. W. Noell, R. K. Robins, H. C. Koppel and A. G. Beaman; *J. Am. Chem. Soc.*, 1960, **82**, 2633.	557, 572, 574
60JA2641	R. M. Izatt, J. H. Rytting, L. D. Hansen and J. J. Christensen; *J. Am. Chem. Soc.*, 1960, **88**, 2641.	525
60JA2654	R. K. Robins; *J. Am. Chem. Soc.*, 1960, **82**, 2654.	560
60JA3144	E. Richter, J. E. Loeffler and E. C. Taylor; *J. Am. Chem. Soc.*, 1960, **82**, 3144.	568, 584, 587
60JA3147	E. C. Taylor and P. K. Loeffler; *J. Am. Chem. Soc.*, 1960, **82**, 3147.	323, 585
60JA3189	M. A. Stevens, H. W. Smith and G. B. Brown; *J. Am. Chem. Soc.*, 1960, **82**, 3189.	629, 652
60JA3773	J. W. Jones and R. K. Robins; *J. Am. Chem. Soc.*, 1960, **82**, 3773.	538, 551, 553
60JA4044	D.-I. Shiho and S. Tagami; *J. Am. Chem. Soc.*, 1960, **82**, 4044.	898
60JA4592	J. A. Montgomery and C. Temple, Jr.; *J. Am. Chem. Soc.*, 1960, **82**, 4592.	595
60JA5007	D. W. Moore and A. G. Whittaker; *J. Am. Chem. Soc.*, 1960, **82**, 5007.	678
60JA5148	N. J. Leonard and W. K. Musker; *J. Am. Chem. Soc.*, 1960, **82**, 5148.	137
60JCS1113	J. H. Lister and G. M. Timmis; *J. Chem. Soc.*, 1960, 1113.	580
60JCS1352	A. Grimison, J. H. Ridd and B. V. Smith; *J. Chem. Soc.*, 1960, 1352.	389
60JCS1363	J. H. Ridd and B. V. Smith; *J. Chem. Soc.*, 1960, 1363.	389
60JCS1829	L. A. Williams; *J. Chem. Soc.*, 1960, 1829.	862
60JCS2536	M. A. McGee, G. T. Newbold, J. Redpath and F. S. Spring; *J. Chem. Soc.*, 1960, 2536.	318
60JCS4347	G. M. Blackburn and A. W. Johnson; *J. Chem. Soc.*, 1960, 4347.	578
60JCS4705	A. Albert; *J. Chem. Soc.*, 1960, 4705.	550
60JOC148	E. C. Taylor and C. C. Cheng; *J. Org. Chem.*, 1960, **25**, 148.	596
60JOC313	H. Rapoport and H. N. Chen; *J. Org. Chem.*, 1960, **25**, 313.	707
60JOC395	J. A. Montgomery and C. Temple, Jr.; *J. Org. Chem.*, 1960, **25**, 395.	573, 574, 592
60JOC461	H. W. Heine and H. S. Bender; *J. Org. Chem.*, 1960, **25**, 461.	155
60JOC1573	R. W. Balsiger and J. A. Montgomery; *J. Org. Chem.*, 1960, **25**, 1573.	562
60JOC2108	M. E. Freed and A. R. Day; *J. Org. Chem.*, 1960, **25**, 2108.	635
60LA(638)136	H. Gehlen and G. Blankenstein; *Liebigs Ann. Chem.*, 1960, **638**, 136.	736
60MI40900	L. Slechta; *Biochem. Pharmacol.*, 1960, **5**, 96.	603
60MI41000	S. K. Chatterjee, M. M. Dhar, N. Anand and M. L. Dhar; *J. Sci. Ind. Res., Sect. C*, 1960, **19**, 35.	615, 635
60RTC836	B. G. Van den Bos; *Recl. Trav. Chim. Pays-Bas*, 1960, **79**, 836.	750
60ZOB327	G. Ya. Urestskaya, E. I. Rybkina and G. P. Men'shikov; *Zh. Obshch. Khim.*, 1960, **30**, 327.	557
60ZOB1878	E. S. Chaman, A. A. Cherkasova and E. S. Golovchinskaya; *Zh. Obshch. Khim.*, 1960, **30**, 1878.	549
60ZOB2427	B. P. Lugovkin; *Zh. Obshch. Khim.*, 1960, **30**, 2427.	549
60ZOB3332	E. S. Golovchinskaya, I. M. Ovcharova and A. A. Cherkasova; *Zhur. Obshch. Khim.*, 1960, **30**, 3332.	548
60ZOB3339	I. M. Ovcharova and E. S. Golovchinskaya; *Zh. Obshch. Khim.*, 1960, **30**, 3339.	540
60ZOB3628	E. S. Golovchinskaya and E. S. Chaman; *Zh. Obshch. Khim.*, 1960, **30**, 3628.	593
61AC(R)1409	A. M. Bellini; *Ann. Chim. (Rome)*, 1961, **51**, 1409 (*Chem. Abstr.*, 1962, **56**, 15 509).	625, 626, 645
61AG15	P. Schmidt, K. Eichenberger and M. Wilhelm; *Angew. Chem.*, 1961, **73**, 15.	329
61AG63	H. Bredereck, F. Effenberger and G. Rainer; *Angew. Chem.*, 1961, **73**, 63.	568
61AG172	I. Ugi; *Angew. Chem.*, 1961, **73**, 172.	845
61AP404	K. Winterfeld and G. B. Singh; *Arch. Pharm. (Weinheim, Ger.)*, 1961, **294**, 404.	614

61CB273	J. D. Roberts; *Chem. Ber.*, 1961, **94**, 273.	842
61CB1555	R. Huisgen, H. J. Sturm and M. Seidel; *Chem. Ber.*, 1961, **94**, 1555.	814
61CB2503	R. Huisgen, J. Sauer and M. Seidel; *Chem. Ber.*, 1961, **94**, 2503.	146
61CI(L)292	F. Eloy, R. Lenaers and C. Moussebois; *Chem. Ind. (London)*, 1961, 292.	62, 71
61CJC2516	R. A. Abramovitch and K. A. H. Adams; *Can. J. Chem.*, 1961, **39**, 2516.	335
61CPB801	Y. Makisumi; *Chem. Pharm. Bull.*, 1961, **9**, 801.	872, 891
61CPB808	Y. Makisumi; *Chem. Pharm. Bull.*, 1961, **9**, 808.	872
61CR(253)687	C. Nofre, A. Lefier and A. Cier, *C.R. Hebd. Seances Acad. Sci.*, 1961, **253**, 687.	543
61CRV87	K. T. Potts; *Chem. Rev.*, 1961, **61**, 87.	735, 739, 746, 762, 787
61FRP1248409	Laboratoire Roger Bellon, *Fr. Pat.* 1 248 409 (1961) (*Chem. Abstr.*, 1962, **56**,10 160).	904
61G849	R. Fusco, G. Bianchetti and D. Pocar; *Gazz. Chim. Ital.*, 1961, **91**, 849.	704, 714
61G933	R. Fusco, G. Bianchetti and D. Pocar; *Gazz. Chim. Ital.*, 1961, **91**, 933.	714
61HC(15-1)460	W. L. Mosby; *Chem. Heterocycl. Compd.*, 1961, **15** (1), 460.	608, 614, 631
61HC(15-1)481	W. L. Mosby; *Chem. Heterocycl. Compd.*, 1961, **15** (1), 481.	612
61HC(15-1)643	W. L. Mosby; *Chem. Heterocycl. Compd.*, 1961, **15** (1), 643.	608
61HC(15-2)764	W. L. Mosby; *Chem. Heterocycl. Compd.*, 1961, **15** (2), 764.	608
61HC(15-2)774	W. L. Mosby; *Chem. Heterocycl. Compd.*, 1961, **15** (2), 774.	608
61HC(15-2)801	W. L. Mosby; *Chem. Heterocycl. Compd.*, 1961, **15** (2), 801.	608
61HC(15-2)802	W. L. Mosby; *Chem. Heterocycl. Compd.*, 1961, **15** (2), 802.	608
61HC(15-2)835	W. L. Mosby; *Chem. Heterocycl. Compd.*, 1961, **15** (2), 835.	608
61HC(15-2)903	W. L. Mosby; *Chem. Heterocycl. Compd.*, 1961, **15** (2), 903.	608
61HC(15-2)906	W. L. Mosby; *Chem. Heterocycl. Compd.*, 1961, **15** (2), 906.	608
61HC(15-2)916	W. L. Mosby; *Chem. Heterocycl. Compd.*, 1961, **15** (2), 916.	608, 659
61JA248	E. C. Taylor and J. A. Zoltewicz; *J. Am. Chem. Soc.*, 1961, **83**, 248.	323
61JA630	J. A. Montgomery and C. Temple, Jr.; *J. Am. Chem. Soc.*, 1961, **83**, 630.	529
61JA1113	G. A. Usbeck, J. W. Jones and R. K. Robins; *J. Am. Chem. Soc.*, 1961, **83**, 1113.	553
61JA2022	R. Mechoulam, F. Sondheimer, A. Melera and F. A. Kincl; *J. Am. Chem. Soc.*, 1961, **83**, 2022.	650
61JA2574	R. K. Robins, E. F. Godefroi, E. C. Taylor, L. R. Lewis and A. Jackson; *J. Am. Chem. Soc.*, 1961, **83**, 2574.	530
61JA2708	S. S. Chissick, M. J. S. Dewar and P. M. Maitlis; *J. Am. Chem. Soc.*, 1961, **83**, 2708.	668
61JA2934	E. Klingsberg; *J. Am. Chem. Soc.*, 1961, **83**, 2934.	91
61JA4038	A. G. Beaman and R. K. Robins; *J. Am. Chem. Soc.*, 1961, **83**, 4038.	557, 596
61JA5023	J. Ogilvie and A. Corwin; *J. Am. Chem. Soc.*, 1961, **83**, 5023.	827
61JCS504	C. L. Angell; *J. Chem. Soc.*, 1961, 504.	518
61JCS518	D. R. Liljegren and K. T. Potts; *J. Chem. Soc.*, 1961, 518.	767
61JCS2589	J. A. Davoll and K. A. Kerridge; *J. Chem. Soc.*, 1961, 2589.	329
61JCS4468	F. Bergmann and M. Tamari; *J. Chem. Soc.*, 1961, 4468.	575, 576
61JCS4845	R. N. Naylor, G. Shaw, D. V. Wilson and D. N. Butler; *J. Chem. Soc.*, 1961, 4845.	585, 595, 652
61JCS5048	J. Clark and J. H. Lister; *J. Chem. Soc.*, 1961, 5048.	573
61JOC451	M. Hauser, E. Peters and H. Tieckelmann; *J. Org. Chem.*, 1961, **26**, 451.	313
61JOC526	J. J. Fox and D. van Praag; *J. Org. Chem.*, 1961, **26**, 526.	573
61JOC1660	F. Bergmann and A. Kalmus; *J. Org. Chem.*, 1961, **26**, 1660.	557
61JOC1861	F. F. Blicke and C.-M. Lee; *J. Org. Chem.*, 1961, **26**, 1861.	666
61JOC1914	H. G. Mautner; *J. Org. Chem.*, 1961, **26**, 1914.	574
61JOC2372	R. M. Herbst; *J. Org. Chem.*, 1961, **26**, 2372.	814
61JOC2396	Y. F. Shealy, R. F. Struck, L. B. Holum and J. A. Montgomery; *J. Org. Chem.*, 1961, **26**, 2396.	652
61JOC2715	E. C. Taylor, G. A. Berchtold, N. A. Geockner and F. G. Stroehmann; *J. Org. Chem.*, 1961, **26**, 2715.	130
61JOC3386	R. W. Balsiger, A. L. Filkes, T. P. Johnston and J. A. Montgomery; *J. Org. Chem.*, 1961, **26**, 3386.	577, 578
61JOC3401	I. L. Doerr, I. Wempen, D. A. Clarke and J. J. Fox; *J. Org. Chem.*, 1961, **26**, 3401.	560, 561, 564
61JOC3446	R. W. Balsiger, A. L. Filkes, T. P. Johnston and J. A. Montgomery; *J. Org. Chem.*, 1961, **26**, 3446.	564, 566
61JOC3714	S. S. Joshi and I. R. Gambhir; *J. Org. Chem.*, 1961, **26**, 3714.	65
61JOC3837	L. R. Lewis, F. H. Schneider and R. K. Robins; *J. Org. Chem.*, 1961, **26**, 3837.	530
61JOC4433	Y. F. Shealy, R. F. Struck, J. D. Clayton and J. A. Montgomery; *J. Org. Chem.*, 1961, **26**, 4433.	876, 894
61JOC4480	E. Schipper and E. Chinery; *J. Org. Chem.*, 1961, **26**, 4480.	422
61JOC4961	E. C. Taylor, J. W. Barton and W. W. Paudler; *J. Org. Chem.*, 1961, **26**, 4961.	576
61JOC4967	E. C. Taylor and A. L. Borror; *J. Org. Chem.*, 1961, **26**, 4967.	328
61LA(641)94	S. Hünig and K. H. Oette; *Liebigs Ann. Chem.*, 1961, **641**, 94.	748
61LA(647)116	W. Ried and E.-U. Köcher; *Liebigs Ann. Chem.*, 1961, **647**, 116.	316, 332
61LA(647)155	W. Pfleiderer and G. Nubel; *Liebigs Ann. Chem.*, 1961, **647**, 155.	524
61LA(647)161	W. Pfleiderer; *Liebigs Ann. Chem.*, 1961, **647**, 161.	524, 533
61LA(648)72	H. A. Staab and W. Benz; *Liebigs Ann. Chem.*, 1961, **648**, 72.	698
61LA(649)114	H. Ballweg; *Liebigs Ann. Chem.*, 1961, **649**, 114.	550, 563
61LA(649)131	H. Lettre and C. Woenckhaus; *Liebigs Ann. Chem.*, 1961, **649**, 131.	548, 551
61MI40400	H. Zimmermann and H. Geisenfelder; *Z. Elektrochem.*, 1961, **65**, 368.	208
61MI40900	J. Oró and A. P. Kimball; *Arch. Biochem. Biophys.*, 1961, **94**, 217.	569

B-61MI41100	J. H. Boyer; in 'Heterocyclic Compounds', ed. R. E. Elderfield; Wiley, New York, 1961, vol. 7, p. 384.	670, 695, 698, 705, 708, 721, 722
B-61MI41101	A. Streitwieser; 'Molecular Orbital Theory for Organic Chemists', Wiley, New York, 1961, p. 33.	672
B-61MI41200	J. H. Boyer; in 'Heterocyclic Compounds', ed. R. C. Elderfield; Wiley, New York, 1961, vol. 7, p. 425.	735
B-61MI41201	J. H. Boyer; in 'Heterocyclic Compounds', ed. R. C. Elderfield; Wiley, New York, 1961, vol. 7, p. 460.	738
61MI41202	H. Zimmermann; *J. Elektrochem.*, 1961, **65**, 368.	742
61MI41203	R. Huisgen; *Proc. Chem. Soc.*, 1961, 357.	768, 779
61MI41300	W. G. Finnegan and S. R. Smith; *J. Chromatogr.*, 1961, **5**, 461.	819
61MI41500	K. Anzai and S. Suzuki; *J. Antibiot., Ser. A*, 1961, **14**, 253.	895
61RTC545	C. A. Salemink; *Recl. Trav. Chim. Pays-Bas*, 1961, **80**, 545.	618, 619, 635
61RTC1348	H. N. Bojarska-Dahlig; *Recl. Trav. Chim. Pays-Bas*, 1961, **80**, 1348.	713
61T(12)51	A. J. Boulton and A. R. Katritzky; *Tetrahedron*, 1961, **12**, 51.	30
61T(14)237	A. J. Owen; *Tetrahedron*, 1961, **14**, 237.	795
61YZ1333	H. Saikachi and M. Kanaoka; *Yakugaku Zasshi*, 1961, **81**, 1333 (*Chem. Abstr.*, 1962, **56**, 7304).	773
61ZOB2645	E. S. Chaman and E. S. Golovchinskaya; *Zh. Obshch. Khim.*, 1961, **31**, 2645.	550
61ZOB3406	B. P. Lugovkin; *Zh. Obshch. Khim.*, 1961, **31**, 3406.	549
62ACS155	S. Gronowitz and P. Moses; *Acta Chem. Scand.*, 1962, **16**, 155.	138
62ACS1800	T. Kindt-Larsen and C. Pedersen; *Acta Chem. Scand.*, 1962, **16**, 1800.	695, 718
62AG9	I. Ugi; *Angew. Chem.*, 1962, **74**, 9.	845
62AG(E)8	I. Ugi; *Angew. Chem., Int. Ed. Engl.*, 1962, **1**, 8.	845
62AX1179	R. F. Bryan and K.-I. Tomita; *Acta Crystallogr.*, 1962, **15**, 1179.	508
62B558	B. C. Pal; *Biochemistry*, 1962, **1**, 558.	530, 552
62BEP619423	C. Holstead; *Belg. Pat.* 619 423 (1962) (*Chem. Abstr.*, 1963, **59**, 12 826).	758
62CB403	H. Bredereck, E. Siegel and B. Fohlisch; *Chem. Ber.*, 1962, **95**, 403.	549
62CB414	H. Bredereck and B. Fohlisch; *Chem. Ber.*, 1962, **95**, 414.	549, 550
62CB937	S. Hünig and K. Hübner; *Chem. Ber.*, 1962, **95**, 937.	157
62CB956	H. Bredereck, F. Effenberger and E. H. Schweizer; *Chem. Ber.*, 1962, **95**, 956.	582
62CB1275	H. A. Staab, M. Lueking and F. H. Duerr; *Chem. Ber.*, 1962, **95**, 1275.	749
62CB1812	H. Bredereck, O. Christmann, W. Koser, P. Schellenberg and R. Nast; *Chem. Ber.*, 1962, **95**, 1812.	533
62CB2546	H. Behringer and H. J. Fischer; *Chem. Ber.*, 1962, **95**, 2546.	763, 766
62CPB620	Y. Makisumi; *Chem. Pharm. Bull.*, 1962, **10**, 620.	307, 331
62CPB1075	M. Ikehara, A. Yamazaki and T. Fujieda; *Chem. Pharm. Bull.*, 1962, **10**, 1075.	561
62DOK(142)354	V. A. Ginsburg, A. Ya. Yakubovich, A. S. Filatov, G. E. Zelenin, S. P. Makarov, V. A. Shpanskii, G. P. Kotel'nikova, L. F. Sergienko and L. L. Martynova; *Dokl. Akad. Nauk. SSSR*, 1962, **142**, 354 (*Chem. Abstr.*, 1962, **57**, 4518).	794
62G1040	R. Fusco, G. Bianchetti, D. Pocar and R. Ugo; *Gazz. Chim. Ital.*, 1962, **92**, 1040.	704, 714
62HC(16)1	D. J. Brown; *Chem. Heterocycl. Compd.*, 1962, **16**, 1.	570
62HCA37	J. Büchi, W. Vetsch and P. Fabiani; *Helv. Chim. Acta*, 1962, **45**, 37.	328
62HCA1620	P. Schmidt, K. Eichenberger and M. Wilhelm; *Helv. Chim. Acta*, 1962, **45**, 1620.	328
62HCA2383	E. Jucker and E. Rissi; *Helv. Chim. Acta*, 1962, **45**, 2383.	645
62HCA2441	H. U. Daeniker and J. Druey; *Helv. Chim. Acta*, 1962, **45**, 2441.	718
62JA62	C. D. Jardetsky; *J. Am. Chem. Soc.*, 1962, **84**, 62.	512
62JA993	H. W. Heine and D. A. Tomalia; *J. Am. Chem. Soc.*, 1962, **84**, 993.	718
62JA1412	D. L. Smith and P. J. Elving; *J. Am. Chem. Soc.*, 1962, **84**, 1412.	540
62JA1914	J. W. Jones and R. K. Robins; *J. Am. Chem. Soc.*, 1962, **84**, 1914.	531, 532, 533, 535, 552
62JA2005	K. Biemann and J. A. McCloskey; *J. Am. Chem. Soc.*, 1962, **84**, 2005.	519
62JA2148	N. J. Leonard and J. A. Deyrup; *J. Am. Chem. Soc.*, 1962, **84**, 2148.	530, 552
62JA2453	R. A. Carboni and J. E. Castle; *J. Am. Chem. Soc.*, 1962, **84**, 2453.	163
62JA3008	L. B. Townsend and R. K. Robins; *J. Am. Chem. Soc.*, 1962, **84**, 3008.	560
62JCS575	E. J. Browne and J. B. Polya; *J. Chem. Soc.*, 1962, 575.	755, 756, 777, 784
62JCS1863	K. R. H. Wooldridge and R. Slack; *J. Chem. Soc.*, 1962, 1863.	535, 557
62JCS2083	A. J. Boulton and A. R. Katritzky; *J. Chem. Soc.*, 1962, 2083.	719
62JCS2927	A. F. Bedford, P. B. Edmonson and C. T. Mortimer; *J. Chem. Soc.*, 1962, 2927.	208
62JCS2937	G. Shaw and D. V. Wilson; *J. Chem. Soc.*, 1962, 2937.	583
62JGU1876	I. I. Grandberg and G. V. Klyuchko; *J. Gen. Chem. USSR* (*Engl. Transl.*), 1962, **32**, 1876.	272
62JMC15	J. A. Montgomery, K. Hewson and C. Temple, Jr.; *J. Med. Chem.*, 1962, **5**, 15.	529
62JMC1067	A. G. Beaman and R. K. Robins; *J. Med. Chem.*, 1962, **5**, 1067.	562, 573
62JOC195	J. A. Montgomery, R. W. Balsiger, A. L. Filkes and T. P. Johnston; *J. Org. Chem.*, 1962, **27**, 195.	566
62JOC567	M. A. Stevens, A. Giner-Sorolla, H. W. Smith and G. B. Brown; *J. Org. Chem.*, 1962, **27**, 567.	539, 548, 555, 595
62JOC973	T. P. Johnston, A. L. Filkes and J. A. Montgomery; *J. Org. Chem.*, 1962, **27**, 973.	531, 564
62JOC986	A. J. Beaman, J. F. Gerster and R. K. Robins; *J. Org. Chem.*, 1962, **27**, 986.	559, 578, 598

62JOC990	L. B. Townsend and R. K. Robins; *J. Org. Chem.*, 1962, **27**, 990.	559, 594
62JOC1309	J. P. Freeman; *J. Org. Chem.*, 1962, **27**, 1309.	188
62JOC1671	C. Temple, Jr., R. L. McKee and J. A. Montgomery; *J. Org. Chem.*, 1962, **27**, 1671.	591
62JOC2085	W. R. Carpenter; *J. Org. Chem.*, 1962, **27**, 2085.	819, 831
62JOC2150	Y. F. Shealy, C. A. Krauth and J. A. Montgomery; *J. Org. Chem.*, 1962, **27**, 2150.	652
62JOC2163	R. C. De Selms; *J. Org. Chem.*, 1962, **27**, 2163.	635
62JOC2209	D. G. Farnum and P. Yates; *J. Org. Chem.*, 1962, **27**, 2209.	686
62JOC2478	G. B. Elion; *J. Org. Chem.*, 1962, **27**, 2478.	529, 532, 560, 572
62JOC2500	P. H. Laursen and B. E. Christensen; *J. Org. Chem.*, 1962, **27**, 2500.	652
62JOC2631	K. T. Potts and T. H. Crawford; *J. Org. Chem.*, 1962, **27**, 2631.	740, 749
62JOC2943	H. W. Heine and A. C. Brooker; *J. Org. Chem.*, 1962, **27**, 2943.	155
62JOC3240	H. C. Brown and M. T. Cheng; *J. Org. Chem.*, 1962, **27**, 3240.	772
62JOC3243	P. J. Nelson and K. T. Potts; *J. Org. Chem.*, 1962, **27**, 3243.	895, 896
62JOC3248	W. P. Norris; *J. Org. Chem.*, 1962, **27**, 3248.	819
62JOC3545	S. Cohen, E. Thom and A. Bendich; *J. Org. Chem.*, 1962, **27**, 3545.	548
62JOC4080	J. A. Montgomery, A. L. Fikes and T. P. Johnston; *J. Org. Chem.*, 1962, **27**, 4080.	648
62JOC4300	D. Y. Curtin, R. J. Crawford and D. K. Wedegaertner; *J. Org. Chem.*, 1962, **27**, 4300.	707
62JOC4633	C. W. Bills, S. E. Gebura, J. S. Meek and O. J. Sweeting; *J. Org. Chem.*, 1962, **27**, 4633.	589
62JPC359	R. M. Izatt and J. J. Christensen; *J. Phys. Chem.*, 1962, **66**, 359.	525
62JPR(288)199	W. Knobloch and H. Kuehne; *J. Prakt. Chem.*, 1962, **288**, 199.	622
62LA(653)105	R. Huisgen, R. Grashey, M. Seidel, G. Wallbillich, H. Knupfer and R. Schmidt; *Liebigs Ann. Chem.*, 1962, **653**, 105.	779
62LA(654)146	R. Huisgen, J. Sauer and M. Seidel; *Liebigs Ann. Chem.*, 1962, **654**, 146.	778
62LA(657)156	U. Schmidt; *Liebigs Ann. Chem.*, 1962, **657**, 156.	336
62LA(658)169	R. Huisgen, R. Grashey, M. Seidel, H. Knupfer and R. Schmidt; *Liebigs Ann. Chem.*, 1962, **658**, 169.	770
62M632	G. Spiteller and M. Spiteller-Friedmann; *Monatsh. Chem.*, 1962, **93**, 632.	519
B-62MI40900	K. Biemann; 'Mass Spectrometry', McGraw Hill, New York, 1962.	519
62MI40901	J. Altman and D. Ben-Ishai; *Bull. Res. Council Isr.*, 1962, **11A**, 4.	576
62MI40902	M. Ya. Fel'dman; *Biochem. USSR*, 1962, **27**, 378.	551
62MI40903	C. E. Liau, K. Yamashita and M. Matsui; *Agric. Biol. Chem. Jpn.*, 1962, **26**, 624.	580
62MI40904	A. G. Beaman and R. K. Robins; *J. Appl. Chem.*, 1962, **12**, 432.	597
62MI40905	J. Oró and A. P. Kimball; *Arch. Biochem. Biophys.*, 1962, **96**, 293.	569
62MI41200	S. Naqui and V. R. Srinivasan; *J. Sci. Ind. Res.*, 1962, **21B**, 195 (*Chem. Abstr.*, 1962, **57**, 12 473).	753
62OS(42)69	R. Huisgen and K. Bast; *Org. Synth.*, 1962, **42**, 69.	288
62PNA(48)1300	R. M. Kliss and C. N. Matthews; *Proc. Natl. Acad. Sci. USA*, 1962, **48**, 1300.	569
62RTC202	J. G. A. Luijten, M. J. Janssen and G. J. M. Van der Kerk; *Recl. Trav. Chim. Pays-Bas*, 1962, **81**, 202.	761
62T539	E. J. Browne and J. B. Polya; *Tetrahedron*, 1962, **18**, 539.	738
62T985	H. Hamano and H. F. Hameka; *Tetrahedron*, 1962, **18**, 985.	172, 346, 347, 348
62TL387	R. Huisgen, R. Grashey and R. Krischke; *Tetrahedron Lett.*, 1962, 387.	339
62YZ683	H. Saikachi and M. Kanaoka; *Yakugaku Zasshi*, 1962, **82**, 683 (*Chem. Abstr.*, 1963, **58**, 4543).	773
62ZC153	H. Beyer and A. Hetzheim; *Z. Chem.*, 1962, **2**, 153.	662
62ZOB447	G. I. Chipen and V. Ya. Grinshtein; *Zh. Obshch. Khim.*, 1962, **32**, 447.	751
62ZOB452	B. P. Lugovkin; *Zh. Obshch. Khim.*, 1962, **32**, 452.	549
62ZOB1655	V. M. Berezovskii and A. M. Yurkevich; *Zh. Obshch. Khim.*, 1962, **32**, 1655.	572
62ZOB2015	E. S. Chaman and E. S. Golovchinskaya; *Zh. Obshch. Khim.*, 1962, **32**, 2015.	548, 549, 593
63ACS2575	A. Hordvik; *Acta Chem. Scand.*, 1963, **17**, 2575.	9
63AF878	D. L. Aldous and R. N. Castle; *Arzneim.-Forsch.*, 1963, **13**, 878.	652
63AG604	R. Huisgen; *Angew. Chem.*, 1963, **75**, 604.	702
63AG742	R. Huisgen; *Angew. Chem.*, 1963, **75**, 742.	702, 712
63AG1101	K. Winterfeld and H. Franzke; *Angew. Chem.*, 1963, **75**, 1101.	635
63AG(E)468	F. Kurzer; *Angew. Chem., Int. Ed. Engl.*, 1963, **2**, 468.	792
63AG(E)565	R. Huisgen; *Angew. Chem., Int. Ed. Engl.*, 1963, **2**, 565.	143, 702
63AG(E)633	R. Huisgen; *Angew. Chem., Int. Ed. Engl.*, 1963, **2**, 633.	143, 702, 712
63AHC(2)245	R. Gompper; *Adv. Heterocycl. Chem.*, 1963, **2**, 245.	100
63AHC(2)365	N. K. Kochetkov and S. D. Sokolov; *Adv. Heterocycl. Chem.*, 1963, **2**, 365.	58, 59, 88, 90, 104, 106
63AX596	G. J. Palenik; *Acta Crystallogr.*, 1963, **16**, 596.	796, 797
63AX907	K. Hoogsteen; *Acta Crystallogr.*, 1963, **16**, 907.	509
63AX1157	R. H. Stanford, Jr.; *Acta Crystallogr.*, 1963, **16**, 1157.	9
63BBA(72)62	A. J. Guarino, M. L. Ibershof and R. Swain; *Biochim. Biophys. Acta*, 1963, **72**, 62.	603
63BSF1000	M. Maziere, N. P. Buu-Hoi and N. D. Xuong; *Bull. Soc. Chim. Fr.*, 1963, 1000.	647
63BSF2742	J. Renault, C. Fauran and F. Pellerin; *Bull. Soc. Chim. Fr.*, 1963, **39**, 2742.	334
63CB802	R. Fusco, G. Bianchetti, D. Pocar and R. Ugo; *Chem. Ber.*, 1963, **96**, 802.	714
63CB1059	H. Weidinger and J. Kranz; *Chem. Ber.*, 1963, **96**, 1059.	764, 767
63CB1827	R. Pfleger, E. Garthe and K. Rauer; *Chem. Ber.*, 1963, **96**, 1827.	852

63CB2070	H. Weidinger and J. Kranz; *Chem. Ber.*, 1963, **96**, 2070.	778
63CB2750	L. Birkofer, A. Ritter and P. Richter; *Chem. Ber.*, 1963, **96**, 2750.	819
63CB2851	G. Wittig and B. Reichel; *Chem. Ber.*, 1963, **96**, 2851.	720
63CB2996	A. Spassov, E. Golovinsky and G. Russev; *Chem. Ber.*, 1963, **96**, 2996.	766
63CI(L)953	J. B. Lloyd; *Chem. Ind. (London)*, 1963, 953.	563
63CJC274	B. M. Lynch and T. L. Chan; *Can. J. Chem.*, 1963, **41**, 274.	699
63CJC1807	A. F. Lewis, A. G. Beaman and R. K. Robins; *Can. J. Chem.*, 1963, **41**, 1807.	562, 563, 578, 579
63CJC2380	B. M. Lynch; *Can. J. Chem.*, 1963, **41**, 2380.	699
63CPB265	Y. Mizuno, M. Ikehara, T. Itoh and K. Saito; *Chem. Pharm. Bull.*, 1963, **11**, 265.	615
63CPB348	T. Itai and S. Kamiya; *Chem. Pharm. Bull.*, 1963, **11**, 348	857, 860, 880, 902
63CPB845	Y. Makisumi; *Chem. Pharm. Bull.*, 1963, **11**, 845.	872
63G90	R. Scarpati, D. Sica and A. Lionetti; *Gazz. Chim. Ital.*, 1963, **93**, 90.	714
63G1196	G. Palazzo and G. Corsi; *Gazz. Chim. Ital.*, 1963, **93**, 1196.	91
63IJC30	P. C. Jain, S. K. Chatterjee and N. Anand; *Indian J. Chem.*, 1963, **1**, 30.	615
63JA122	J. M. Insole and E. S. Lewis; *J. Am. Chem. Soc.*, 1963, **85**, 122.	840
63JA2026	N. J. Leonard and R. A. Laursen; *J. Am. Chem. Soc.*, 1963, **85**, 2026.	512
63JA2967	P. Yates and D. G. Farnum; *J. Am. Chem. Soc.*, 1963, **85**, 2967.	721
63JA3719	N. J. Leonard and T. Fujii; *J. Am. Chem. Soc.*, 1963, **85**, 3719.	525, 530
63JCS42	P. J. Bunyan and J. I. G. Cadogan; *J. Chem. Soc.*, 1963, 42.	335
63JCS1276	D. J. Brown and J. S. Harper; *J. Chem. Soc.*, 1963, 1276.	578, 695
63JCS1666	O. Meth-Cohn, R. K. Smalley and H. Suschitzky; *J. Chem. Soc.*, 1963, 1666.	641
63JCS2032	D. Buttimore, D. H. Jones, R. Slack and K. R. H. Wooldridge; *J. Chem. Soc.*, 1963, 2032.	105
63JCS4558	L. E. A. Godfrey and F. Kurzer; *J. Chem. Soc.*, 1963, 4558.	775
63JCS5556	N. K. Roberts; *J. Chem. Soc.*, 1963, 5556.	680
63JCS5642	G. W. Miller and F. L. Rose; *J. Chem. Soc.*, 1963, 5642	858, 862, 883, 891
63JGU525	A. N. Kost, G. K. Faizova and I. I. Grandberg; *J. Gen. Chem. USSR (Engl. Transl.)*, 1963, **33**, 525	207
63JGU2519	I. I. Grandberg, S. V. Tabak, G. K. Faizova and A. N. Kost; *J. Gen. Chem. USSR (Engl. Transl.)*, 1963, **33**, 2519.	207
63JMC289	E. Dyer, J. M. Reitz and R. E. Farris, Jr.; *J. Med. Chem.*, 1963, **6**, 289.	550, 551
63JMC471	W. A. Bowles, F. H. Schneider, L. R. Lewis and R. K. Robins; *J. Med. Chem.*, 1963, **6**, 471.	530, 536, 562
63JOC543	K. T. Potts; *J. Org. Chem.*, 1963, **28**, 543.	745, 751
63JOC945	J. F. Gerster, J. W. Jones and R. K. Robins; *J. Org. Chem.*, 1963, **28**, 945.	562, 598
63JOC981	J. R. Piper and T. P. Johnson; *J. Org. Chem.*, 1963, **28**, 981.	139
63JOC991	N. Shachat and J. J. Bagnell, Jr.; *J. Org. Chem.*, 1963, **28**, 991.	140
63JOC1305	T. P. Johnston and A. Gallagher; *J. Org. Chem.*, 1963, **28**, 1305.	517
63JOC1336	G. DeStevens, A. Halamandaris, M. Bernier and H. M. Blatter; *J. Org. Chem.*, 1963, **28**, 1336.	324
63JOC1379	S. Cohen, E. Thom and A. Bendich; *J. Org. Chem.*, 1962, **28**, 1379.	549
63JOC1662	E. Y. Sutcliffe and R. K. Robins; *J. Org. Chem.*, 1963, **28**, 1662.	529, 530, 538, 561, 595
63JOC1837	Y. Mizuno, M. Ikehara, T. Itoh and K. Saito; *J. Org. Chem.*, 1963, **28**, 1837.	615, 635
63JOC2087	T. C. Myers and L. Zeleznick; *J. Org. Chem.*, 1963, **25**, 2087.	531, 552
63JOC2310	A. J. Beaman and R. K. Robins; *J. Org. Chem.*, 1963, **28**, 2310.	529, 540, 563
63JOC2560	R. M. Cresswell and G. B. Brown; *J. Org. Chem.*, 1963, **28**, 2560.	561, 590
63JOC2677	M. H. Krackov and B. E. Christensen; *J. Org. Chem.*, 1963, **28**, 2677.	577
63JOC3041	R. K. Robins, J. K. Horner, C. V. Greco, C. W. Noell and C. G. Beames, Jr.; *J. Org. Chem.*, 1963, **28**, 3041.	619, 641
63JOC3329	Y. Mizuno, M. Ikehara, K. A. Watanabe, S. Suzaki and T. Itoh; *J. Org. Chem.*, 1963, **28**, 3329.	616
63JPC721	J. B. Lounsbury; *J. Phys. Chem.*, 1963, **67**, 721.	795
63LA(663)108	T. Pyl, S. Melde and H. Beyer; *Liebigs Ann. Chem.*, 1963, **663**, 108.	647
63LA(663)113	T. Pyl, F. Waschk and H. Beyer; *Liebigs Ann. Chem.*, 1963, **663**, 113.	116
63LA(665)144	H. Gehlen and J. Dost; *Liebigs Ann. Chem.*, 1963, **665**, 144.	754
B-63MI40400	A. L. McClellan; 'Tables of Experimental Dipole Moments', Freeman, San Francisco, 1963.	176, 282
B-63MI40600	B. Pullman and A. Pullman; 'Quantum Biochemistry', Interscience, New York, 1963.	347
63MI40900	D. S. Letham; *Life Sciences*, 1963, **2**, 569.	601
63MI40901	A. Veillard and B. Pullman: *J. Theor. Biol.*, 1963, **4**, 37.	505
B-63MI40902	J. G. Flaks and L. N. Lukens; in 'Methods in Enzymology', ed. S. P. Colowick and N. O. Kaplan; Academic, New York, 1963, vol. 6, p. 52.	504
63MI40903	R. K. Robins; *Biochem. Prep.*, 1963, **10**, 145.	598
63MI40904	F. Bergmann and M. Kleiner; *Isr. J. Chem.*, 1963, **1**, 477.	535
63MI40905	C. U. Lowe, M. W. Rees and R. Markham; *Nature (London)* 1963, **199**, 219.	569
63MI41000	B. Stanovnik and M. Tisler; *Croat. Chem. Acta*, 1963, **35**, 167.	635
63MI41001	T. Kurihara and T. Chiba; *Tohoku Yakka Daigaku Kenkyu Nenpo*, 1963, **10**, 65.	616, 635
B-63MI41100	A. L. McClellan; 'Tables of Experimental Dipole Moments', Freeman, San Francisco, 1963.	677

63MI41200	G. I. Chipen and V. Ya. Grinshtein; *Latvija PSR Zinatnu Akad. Vestis Kim. Ser.*, 1962, **255** (*Chem. Abstr.*, 1963, **59**, 12 789).	748
63MI41201	H. Hopff and M. Lippay; *J. Makromol. Chem.*, 1963, **66**, 157.	756
63N224	H. Lettre, H. Ballweg, H. Maurer and D. Rehberger; *Naturwissenschaften*, 1963, **50**, 224.	563, 564, 593
63PMH(1)1	A. Albert; *Phys. Methods Heterocycl. Chem.*, 1963, **1**, 1.	50
63PMH(1)98	A. Albert; *Phys. Methods Heterocycl. Chem.*, 1963, **1**, 98.	690
63PMH(1)177	W. Pfleiderer; *Phys. Methods Heterocycl. Chem.*, 1963, **1**, 177.	31, 32, 207
63PMH(1)189	S. Walker; *Phys. Methods Heterocycl. Chem.*, 1963, **1**, 189.	176
63PMH(2)1	S. F. Mason; *Phys. Methods Heterocycl. Chem.*, **2**, 1.	197
63PMH(2)161	A. R. Katritzky and A. P. Ambler; *Phys. Methods Heterocyl. Chem.*, 1963, **2**, 161.	24, 26, 27, 199, 200
63PNA(49)737	C. Ponnamperuma, R. M. Lemmon, R. Mariner and M. Calvin; *Proc. Natl. Acad. Sci. USA.*, 1963, **49**, 737.	569
63T1801	I. Ugi; *Tetrahedron*, 1963, **19**, 1801.	842
63TL1499	S. Matsuura and T. Goto; *Tetrahedron Lett.*, 1963, 1499.	511, 513
63YZ373	A. Takamizawa and S. Hayashi; *Yakugaku Zasshi*, 1963, **83**, 373.	216
63ZOB1650	E. S. Golovchinskaya, D. A. Kolganova, L. A. Nikolaeva and E. S. Chaman; *Zh. Obshch. Khim.*, 1963, **33**, 1650.	550, 588
63ZOB2475	I. N. Goncharova and I. Y. Postovskii; *Zh. Obshch. Khim.*, 1963, **33**, 2475.	857, 881, 902
63ZOB2942	B. P. Lugovkin; *Zh. Obshch. Khim.*, 1963, **33**, 2942.	549
63ZOB3205	B. P. Lugovkin; *Zh. Obshch. Khim.*, 1963, **33**, 3205.	549
63ZOB3342	E. S. Chaman and E. S. Golovchinskaya; *Zh. Obshch. Khim.*, 1963, **33**, 3342.	548, 550, 564
64AC(R)735	U. Croatto and A. Fava; *Ann. Chim. (Rome)*, 1964, **54**, 735 (*Chem. Abstr.*, 1965, **62**, 11 197).	685
64AG144	F. G. Willey; *Angew. Chem.*, 1964, **76**, 144.	697
64AG(E)138	F. G. Willey; *Angew. Chem., Int. Ed. Engl.*, 1964, **3**, 138.	697
64AG(E)693	A. J. Boulton, P. B. Ghosh and A. R. Katritzky; *Angew. Chem., Int. Ed. Engl.*, 1964, **3**, 693.	86
64AHC(3)1	G. F. Duffin; *Adv. Heterocycl. Chem.*, 1964, **3**, 1.	52, 77, 80
64AHC(3)263	K. A. Jensen and C. Pedersen; *Adv. Heterocycl. Chem.*, 1964, **3**, 263.	62, 102
64AHC(3)373	I. Ugi; *Adv. Heterocycl. Chem.*, 1964, **3**, 373.	839, 842
64AJC1128	P. J. Black, M. L. Heffernan, L. M. Jackman, Q. N. Porter and G. R. Underwood; *Aust. J. Chem.*, 1964, **17**, 1128.	307
64AX126	H. M. Sobell and K.-I. Tomita; *Acta Crystallogr.*, 1964, **17**, 126.	510
64B494	A. D. Broom, L. B. Townsend, J. W. Jones and R. K. Robins; *Biochemistry*, 1964, **3**, 494.	530, 552
64B880	G. B. Brown, G. Levin and S. Murphy; *Biochemistry*, 1964, **3**, 880.	554, 555
64CB934	L. Birkofer, A. Ritter and H.-P. Kuhlthau; *Chem. Ber.*, 1964, **97**, 934.	533, 535
64CB1085	R. Huisgen, R. Grashey, H. Knupfer, R. Kunz and M. Seidel; *Chem. Ber.*, 1964, **97**, 1085.	769
64CB1225	D. Pocar, G. Bianchetti and P. D. Croce; *Chem. Ber.*, 1964, **97**, 1225.	721
64CB2173	A. Dornow, H. Mensel and P. Marx; *Chem. Ber.*, 1964, **97**, 2173.	899
64CB2179	A. Dornow, W. Abele and H. Menzel; *Chem. Ber.*, 1964, **97**, 2179.	800
64CB2185	A. Dornow, H. Menzel and P. Marx; *Chem. Ber.*, 1964, **97**, 2185.	898
64CB3349	A. Dornow and W. Abele; *Chem. Ber.*, 1964, **97**, 3349.	340
64CB3476	R. Neidlein; *Chem. Ber.*, 1964, **97**, 3476.	717
64CB3493	H. Reimlinger; *Chem. Ber.*, 1964, **97**, 3493.	721, 887
64CC230	D. S. Letham, J. S. Shannon and I. R. C. McDonald; *Chem. Commun.*, 1964, 230.	519, 601
64CC231	G. Shaw and D. V. Wilson; *Chem. Commun.*, 1964, 231.	601
64CJC43	C. N. R. Rao and R. Venkataraghavan; *Can. J. Chem.*, 1964, **42**, 43.	685
64CPB204	Y. Makisumi, H. Watanabe and K. Tori; *Chem. Pharm. Bull.*, 1964, **12**, 204.	858
64CPB639	M. Ikehara, H. Uno and F. Ishikawa; *Chem. Pharm. Bull.*, 1964, **40**, 639.	556
64CPB866	Y. Mizuno, T. Itoh and K. Saito; *Chem. Pharm. Bull.*, 1964, **12**, 866.	620
64CPB1351	F. Yoneda, T. Ohtaka and Y. Nitta; *Chem. Pharm. Bull.*, 1964, **12**, 1351.	628, 650
64CRV129	F. H. C. Stewart; *Chem. Rev.*, 1964, **64**, 129.	701
64DIS2690	J. M. Gill; Ph.D. Dissertation, Indiana University, 1963 (*Diss. Abstr.*, 1964, **24**, 2690).	14
64HC(20)1	R. H. Wiley and P. Wiley; *Chem. Heterocycl. Compd.*, 1964, **20**, 1.	206, 228, 232, 239, 252, 273, 277
64HCA838	C. Moussebois and F. Eloy; *Helv. Chim. Acta*, 1964, **47**, 838.	59
64HCA942	C. Moussebois and J. F. M. Oth; *Helv. Chim. Acta*, 1964, **47**, 942.	13, 23
64HCA2068	Ph. Gold-Aubert, D. Melkonian, and L. Toribio; *Helv. Chim. Acta*, 1964, **47**, 2068.	788
64JA1242	R. E. Holmes and R. K. Robins; *J. Am. Chem. Soc.*, 1964, **86**, 1242.	540
64JA2025	P. A. S. Smith, L. O. Krbechek and W. Resemann; *J. Am. Chem. Soc.*, 1964, **86**, 2025.	697
64JA2213	M. E. Munk and Y. K. Kim; *J. Am. Chem. Soc.*, 1964, **86**, 2213.	714
64JA2948	R. Shapiro; *J. Am. Chem. Soc.*, 1964, **86**, 2948.	551

64JA4506	F. D. Marsh and M. E. Hermes; *J. Am. Chem. Soc.*, 1964, **86**, 4506.	726
64JA4721	E. C. Taylor and E. E. Garcia; *J. Am. Chem. Soc.*, 1964, **86**, 4721.	512, 554, 555
64JA5320	L. B. Townsend, R. K. Robins, R. N. Loeppky and N. J. Leonard; *J. Am. Chem. Soc.*, 1946, **86**, 5320.	552
64JCS446	M. P. L. Caton, D. H. Jones, R. Slack and K. R. H. Wooldridge; *J. Chem. Soc.*, 1964, 446.	23
64JCS792	S. Lewin; *J. Chem. Soc.*, 1964, 792.	551
64JCS4226	W. L. F. Armarego; *J. Chem. Soc.*, 1964, 4226.	307, 309, 610, 614
64JHC34	C. W. Noell and R. K. Robins; *J. Heterocycl. Chem.*, 1964, **1**, 34.	648
64JHC42	T. Kuraishi and R. N. Castle; *J. Heterocycl. Chem.*, 1964, **1**, 42.	651
64JHC115	J. A. Montgomery and H. J. Thomas; *J. Heterocycl. Chem.*, 1964, **1**, 115.	530, 552
64JHC163	E. Campaigne and F. Haaf; *J. Heterocycl. Chem.*, 1964, **1**, 163.	119
64JHC182	M. Malm and R. N. Castle; *J. Heterocycl. Chem.*, 1964, **1**, 182.	629, 652
64JHC247	G. A. Gerhardt and R. N. Castle; *J. Heterocycl. Chem.*, 1964, **1**, 247.	871
64JMC10	E. Dyer and H. S. Bender; *J. Med. Chem.*, 1964, **7**, 10.	559
64JOC942	W. J. Fanshawe, V. J. Bauer and S. R. Safir; *J. Org. Chem.*, 1964, **29**, 942.	281
64JOC1715	E. Campaigne and M. C. Wani; *J. Org. Chem.*, 1964, **29**, 1715.	663
64JOC1720	C. G. Overberger and H. A. Friedman; *J. Org. Chem.*, 1964, **29**, 1720.	663
64JOC1988	J. Bullock and O. Jardetzky; *J. Org. Chem.*, 1964, **29**, 1988.	511
64JOC2542	G. M. Badger, P. J. Nelson and K. T. Potts; *J. Org. Chem.*, 1964, **29**, 2542.	856, 896
64JOC2611	Y. Mizuno, T. Itoh and K. Saito; *J. Org. Chem.*, 1964, **29**, 2611.	619, 620
64JOC3209	A. Giner-Sorolla, E. Thom and A. Bendich; *J. Org. Chem.*, 1964, **29**, 3209.	539, 549, 556, 559, 563
64JOC3403	D. L. Garmaise and J. Komlossy; *J. Org. Chem.*, 1964, **29**, 3403.	635
64JOC3449	M. Hauser; *J. Org. Chem.*, 1964, **29**, 3449.	758
64JOC3707	E. F. Godefroi, C. A. Van der Eycken and C. Van der Westeringh; *J. Org. Chem.*, 1964, **29**, 3707.	661
64JOC3733	G. A. Reynolds; *J. Org. Chem.*, 1964, **29**, 3733.	723
64JPS962	T. G. Alexander and M. Maienthal; *J. Pharm. Sci.*, 1964, **53**, 962.	512
64LA(673)78	W. Pfleiderer and F. Sagi; *Liebigs Ann. Chem.*, 1964, **673**, 78.	533
64LA(678)95	W. Ried and H. Mengler; *Liebigs Ann. Chem.*, 1964, **678**, 95.	711
64M242	G. R. Revankar and S. Siddappa; *Monatsh. Chem.*, 1964, **95**, 242.	635
64MI40100	G. Pouzard, L. Pujol, J. Roggero and E. J. Vincent; *J. Chim. Phys. Phys. Chim. Biol.*, 1964, 613.	17
B-64MI40400	R. Huisgen, R. Grashey and J. Sauer; in 'The Chemistry of Alkenes', ed. S. Patai; Wiley-Interscience, 1964, part 1, p. 739.	169
64MI40600	G. Geisler, J. Fruwert and A. Kuennecke; *Z. Phys. Chem.*, 1964, **41**, 49 (*Chem. Abstr.*, 1964, **61**, 6892).	362
B-64MI40800	H. Bredereck, R. Gompper, H. G. V. Schüh and G. Theilig; in 'Newer Methods of Preparative Organic Chemistry', ed. W. Foerst; Academic, New York, 1964, vol. 3, p. 241.	485
64MI40900	S. Cohen and A. Vincze; *Isr. J. Chem.*, 1964, **2**, 1.	548
64MI40901	N. B. Smirnova and I. Ya. Postovskii; *Zh. Vses. Khim. Obshch. D. I. Mendeleeva*, 1964, **9**, 711 (*Chem. Abstr.*, 1964, **62**, 9130).	564
64MI41000	B. Mariani and R. Sgarbi; *Chim. Ind.* (*Milan*), 1964, **46**, 630 (*Chem. Abstr.*, 1964, **61**, 12 117).	630
64MI41001	Z. Talik and B. Brekiesz; *Rocz. Chem.*, 1964, **38**, 887 (*Chem. Abstr.*, 1965, **62**, 5271).	621
64MI41100	J. H. Boyer, R. Moriarty, B. de B. Darwent and P. A. S. Smith; *Chem. Eng. News*, 1964, **42**, 6.	708
B-64MI41101	R. Huisgen, R. Grashey and J. Sauer; in 'The Chemistry of Alkenes', ed. S. Patai, Interscience, New York, 1964, pp. 835 ff.	710
B-64MI41102	'Kirk-Othmer Encyclopedia of Chemical Technology', Wiley–Interscience, New York, 2nd edn., 1964, p. 737.	730
64N137	H. Goldner, G. Dietz and E. Carstens; *Naturwissenschaften*, 1964, **51**, 137.	580
64RTC387	J. Strating, W. E. Weening and B. Zwanenburg; *Recl. Trav. Chim. Pays-Bas*, 1964, **83**, 387.	786
64T461	H. Kano and E. Yamazaki; *Tetrahedron*, 1964, **20**, 461.	832
64TL1477	W. D. Crow and N. J. Leonard; *Tetrahedron Lett.*, 1964, **23**, 1477.	27
64TL2999	H. Tiefenthaler, W. Dörscheln, H. Göth and H. Schmid; *Tetrahedron Lett.*, 1964, 2999.	161
64USP3128274	G. H. Hitchings, G. B. Elion and L. E. Mackay; *U.S. Pat.* 3 128 274 (1964) (*Chem. Abstr.*, 1964, **60**, 14 523).	550
64ZC454	H. Goldner, G. Dietz and E. Carstens; *Z. Chem.*, 1964, **4**, 454.	580, 597
64ZOB390	Y. S. Shabarov, A. P. Smirnova and R. Y. Levina; *Zh. Obshch. Khim.*, 1964, **34**, 390 (*Chem. Abstr.*, 1964, **60**, 12 005).	852
64ZOB1137	L. A. Nikolaeva and E. S. Golovchinskaya; *Zh. Obshch. Khim.*, 1964, **34**, 1137.	540, 550
64ZOB1745	N. N. Vereshchagina and I. Y. Postovskii; *Zh. Obshch. Khim.*, 1964, **34**, 1745 (*Chem. Abstr.*, 1964, **61**, 8307).	893
64ZOB2472	I. M. Ovcharova and E. S. Golovchinskaya; *Zh. Obshch. Khim.*, 1964, **34**, 2472.	557
64ZOB2517	I. Ya. Postovskii and V. L. Nirenburg; *Zh. Obshch. Khim.*, 1964, **34**, 2517 (*Chem. Abstr.*, 1964, **61**, 14 664).	794
64ZOB3254	I. M. Ovcharova and E. S. Golovchinskaya; *Zh. Obshch. Khim.*, 1964, **34**, 3254.	564

65ACS2434	B. Bak, C. H. Christensen, T. S. Hansen, E. J. Pedersen and J. T. Nielsen; *Acta Chem. Scand.*, 1965, **19**, 2434.	58
65AF10	K. W. Merz and P. H. Stahl; *Arzneim.-Forsch.*, 1965, **15**, 10.	578
65AG414	L. Birkofer and A. Ritter; *Angew. Chem.*, 1965, **77**, 414.	698, 708
65AG(E)417	L. Birkofer and A. Ritter; *Angew. Chem., Int. Ed. Engl.*, 1965, **4**, 417.	698, 708
65AHC(4)75	R. Filler; *Adv. Heterocycl. Chem.*, 1965, **4**, 75.	66, 69, 89, 90
65AHC(4)107	R. Slack and K. R. H. Wooldridge; *Adv. Heterocycl. Chem.*, 1965, **4**, 107.	104, 135
65AHC(5)119	F. Kurzer; *Adv. Heterocycl. Chem.*, 1965, **5**, 119.	95, 96, 97, 105, 130, 135, 140
65AJC379	G. M. Badger and R. P. Rao; *Aust. J. Chem.*, 1965, **18**, 379.	336
65AX111	D. G. Watson, D. J. Sutor and P. Tollin; *Acta Crystallogr.*, 1965, **19**, 111.	509
65AX573	D. G. Watson, R. M. Sweet and R. E. Marsh; *Acta Crystallogr.*, 1965, **19**, 573.	502, 505, 507, 508, 509
65B650	I. J. Borowitz, S. M. Bloom, J. Rothschild and D. B. Sprinson; *Biochemistry*, 1965, **4**, 650.	582
65BSF769	J. Elguero and R. Jacquier; *Bull. Soc. Chim. Fr.*, 1965, 769.	254
65BSF2617	J. Filippi and J. Givern; *Bull. Soc. Chim. Fr.*, 1965, 2617.	552
65CB334	R. R. Schmidt; *Chem. Ber.*, 1965, **98**, 334.	157
65CB642	R. Huisgen, R. Grashey, E. Aufderhaar and R. Kunz; *Chem. Ber.*, 1965, **98**, 642.	745, 769
65CB928	A. Spassov and S. Robev; *Chem. Ber.*, 1965, **98**, 928.	768
65CB1138	R. Huisgen, L. Möbius and G. Szeimies; *Chem. Ber.*, 1965, **98**, 1138.	704, 712
65CB1153	R. Huisgen and G. Szeimies; *Chem. Ber.*, 1965, **98**, 1153.	17, 679, 703, 711, 712
65CB2715	G. Bianchetti, D. Pocar, P. Dalla Croce and A. Vigevani; *Chem. Ber.*, 1965, **98**, 2715.	704
65CB2966	R. Huisgen, C. Axen and H. Seidl; *Chem. Ber.*, 1965, **98**, 2966.	152, 809
65CB2985	R. Huisgen and H. Blaschke; *Chem. Ber.*, 1965, **98**, 2985.	709
65CB3025	C. F. Kröger, L. Hummel, M. Mutscher and H. Beyer; *Chem. Ber.*, 1965, **98**, 3025.	746
65CB3142	W. Ried and M. Schön; *Chem. Ber.*, 1965, **98**, 3142.	723
65CB3992	R. Huisgen, L. Möbius, G. Müller, H. Stangl, G. Szeimies and J. M. Vernon; *Chem. Ber.*, 1965, **98**, 3992.	703, 711, 720
65CB4014	R. Huisgen, R. Knorr, L. Möbius and G. Szeimies; *Chem. Ber.*, 1965, **98**, 4014.	685, 708, 709
65CC596	A. C. Oehlschlager, R. Tillman and L. H. Zalkow; *Chem. Commun.*, 1965, 596.	724
65CHE165	Z. I. Miroshnichenko and M. A. Alperovich; *Chem. Heterocycl. Compd. (Engl. Transl.)*, 1965, **1**, 165.	116
65CHE375	B. V. Ioffe and D. D. Tsitovich; *Chem. Heterocycl. Compd. (Engl. Transl.)*, 1965, **1**, 375.	216
65CHE531	B. K. Martsokha and A. M. Simonov; *Chem. Heterocycl. Compd. (Engl. Transl.)*, 1965, **1**, 531.	246
65CJC1154	M. Tremblay; *Can. J. Chem.*, 1965, **43**, 1154.	798
65CPB142	A. Takamizawa and Y. Hamashima; *Chem. Pharm. Bull.*, 1965, **13**, 142.	310
65CPB1207	A. Takamizawa and Y. Hamashima; *Chem. Pharm. Bull.*, 1965, **13**, 1207.	313
65CR(261)1872	A. LeBerre, M. Dormoy and J. Godin; *C.R. Hebd. Seances Acad. Sci.*, 1965, **261**, 1872.	341
65FES634	T. Vitali, F. Mossini and G. Bertaccini; *Farmaco Ed. Sci.*, 1965, **20**, 634 (*Chem. Abstr.*, 1966, **64**, 9710).	641
65G33	G. F. Bettinetti and L. Capretti; *Gazz. Chim. Ital*, 1965, **85**, 33.	748
65G1220	D. Pocar, G. Bianchetti and P. Dalla Croce; *Gazz. Chim. Ital.*, 1965, **95**, 1220.	704, 714
65IJC84	P. C. Jain, S. K. Chatterjee and N. Anand; *Indian J. Chem.*, 1965, **3**, 84.	867
65IJC138	S. K. Chatterjee, P. C. Jain and N. Anand; *Indian J. Chem.*, 1965, **3**, 138.	635, 640
65IJC162	S. Naqui and V. R. Srinivasan; *Indian J. Chem.*, 1965, **3**, 162.	884, 886
65JA306	P. Scheiner, J. H. Schomaker, S. Deming, W. J. Libbey and G. P. Nowack; *J. Am. Chem. Soc.*, 1965, **87**, 306.	703, 711
65JA749	A. L. Logothetis; *J. Am. Chem. Soc.*, 1965, **87**, 749.	725
65JA1772	R. E. Holmes and R. K. Robins; *J. Am. Chem. Soc.*, 1965, **87**, 1772.	564
65JA1976	E. C. Taylor and R. W. Morrison, Jr.; *J. Am. Chem. Soc.*, 1965, **87**, 1976.	580
65JA1980	E. C. Taylor and R. W. Hendess; *J. Am. Chem. Soc.*, 1965, **87**, 1980.	878, 901
65JA2225	R. J. Pugmire, D. M. Grant, R. K. Robins and G. W. Rhodes; *J. Am. Chem. Soc.*, 1965, **87**, 2225.	510, 514
65JA3023	R. J. Crawford, R. J. Dummel and A. Mishra; *J. Am. Chem. Soc.*, 1965, **87**, 3023.	188, 256
65JA3752	J. F. Gerster and R. K. Robins; *J. Am. Chem. Soc.*, 1965, **87**, 3752.	558
65JA3768	R. J. Crawford and A. Mishra; *J. Am. Chem. Soc.*, 1965, **87**, 3768.	254
65JA4976	J. P. Ferris and L. E. Orgel; *J. Am. Chem. Soc.*, 1965, **87**, 4976.	584
65JAP6522265	*Jpn. Pat.* 65 22 265 (1965) (*Chem. Abstr.*, 1966, **64**, 3566).	628
65JCS623	S. Matsuura and T. Goto; *J. Chem. Soc.*, 1965, 623.	511, 593
65JCS2778	W. L. F. Armarego; *J. Chem. Soc.*, 1965, 2778.	863, 614, 627, 642, 648, 650
65JCS3017	G. B. Barlin and N. B. Chapman; *J. Chem. Soc.*, 1965, 3017.	524, 561
65JCS3312	L. A. Summers, P. F. H. Freeman and D. J. Shields; *J. Chem. Soc.*, 1965, 3312.	719
65JCS3357	G. W. Miller and F. L. Rose; *J. Chem. Soc.*, 1965, 3357.	862, 873
65JCS3456	L. A. Cort; *J. Chem. Soc.*, 1965, 3456.	663
65JCS3770	D. J. Brown and N. W. Jacobsen; *J. Chem. Soc.*, 1965, 3770.	523
65JCS3912	F. Kurzer and K. Douraghi-Zadeh; *J. Chem. Soc.*, 1965, 3912.	758
65JCS5590	D. D. Perrin; *J. Chem. Soc.*, 1965, 5590.	385

65JCS5958	A. J. Boulton, A. C. Gripper Gray and A. R. Katritzky; *J. Chem. Soc.*, 1965, 5958.	86
65JHC53	J. P. Paolini and R. K. Robins; *J. Heterocycl. Chem.*, 1965, **2**, 53.	613, 631
65JHC67	W. D. Guither, D. G. Clark and R. N. Castle; *J. Heterocycl. Chem.*, 1965, **2**, 67.	902
65JHC196	R. J. Rousseau and R. K. Robins; *J. Heterocycl. Chem.*, 1965, **2**, 196.	621, 640
65JHC247	J. A. Gerhardt, D. L. Aldous and R. N. Castle; *J. Heterocycl. Chem.*, 1965, **2**, 247.	629, 652
65JHC253	T. Ichikawa, T. Kato and T. Takenishi; *J. Heterocycl. Chem.*, 1965, **2**, 253.	583
65JHC287	L. M. Werbel and M. L. Zamora; *J. Heterocycl. Chem.*, 1965, **2**, 287.	628, 643, 650, 653
65JHC291	N. J. Leonard, K. L. Carraway and J. P. Helgeson; *J. Heterocycl. Chem.*, 1965, **2**, 291.	530
65JHC331	C. K. Bradsher, E. F. Litzinger, Jr. and M. F. Zinn; *J. Heterocycl. Chem.*, 1965, **2**, 331.	610, 631
65JHC410	W. W. Paudler and D. E. Dunham; *J. Heterocycl. Chem.*, 1965, **2**, 410.	306
65JHC453	H. F. Ridley, R. G. W. Spickett and G. M. Timmis; *J. Heterocycl. Chem.*, 1965, **2**, 453.	575
65JMC296	M. M. Vohra, S. N. Pradhan, P. C. Jain, S. K. Chatterjee and N. Anand; *J. Med. Chem.*, 1965, **8**, 296.	635, 636, 640, 887
65JMC667	A. Giner-Sorolla and A. Bendich; *J. Med. Chem.*, 1965, **8**, 667.	563
65JMC708	J. A. Montgomery and K. Hewson; *J. Med. Chem.*, 1965, **8**, 708.	618, 619, 621, 622, 635, 636, 640
65JOC199	V. Papesch and R. M. Dodson; *J. Org. Chem.*, 1965, **30**, 199.	310, 324, 338, 342
65JOC408	R. M. Cresswell, H. K. Maurer, T. Strauss and G. B. Brown; *J. Org. Chem.*, 1965, **30**, 408.	554
65JOC638	J. J. Looker; *J. Org. Chem.*, 1965, **30**, 638.	709
65JOC732	E. Campaigne and F. Haaf; *J. Org. Chem.*, 1965, **30**, 732.	119
65JOC826	C. Temple, Jr. and J. A. Montgomery; *J. Org. Chem.*, 1965, **30**, 826.	854, 880, 902
65JOC829	C. Temple, Jr., R. L. McKee and J. A. Montgomery; *J. Org. Chem.*, 1965, **30**, 829.	857, 881, 902
65JOC1110	W. C. Coburn, M. C. Thorpe, J. A. Montgomery and K. Hewson; *J. Org. Chem.*, 1965, **30**, 1110.	511, 512
65JOC1114	W. C. Coburn, Jr., M. C. Thorpe, J. A. Montgomery and K. Hewson; *J. Org. Chem.*, 1965, **30**, 1114.	512
65JOC1916	S.-C. J. Fu, E. Chinoporos and H. Terzian; *J. Org. Chem.*, 1965, **30**, 1916.	117
65JOC2403	J. G. Lombardino; *J. Org. Chem.*, 1965, **30**, 2403.	612
65JOC2857	B. R. Baker and P. M. Tanna; *J. Org. Chem.*, 1965, **30**, 2857.	529
65JOC3235	J. A. Montgomery and H. J. Thomas; *J. Org. Chem.*, 1965, **30**, 3235.	530, 552, 594
65JOC3346	M. Charton; *J. Org. Chem.*, 1965, **30**, 3346.	364
65JOC3472	J. H. Markgraf, W. T. Bachmann and D. P. Hollis; *J. Org. Chem.*, 1965, **30**, 3472.	799, 802, 803, 819
65JOC3597	J. W. Marsico and L. Goldman; *J. Org. Chem.*, 1965, **30**, 3597.	531
65JOC3601	C. Temple, Jr., C. L. Kussner and J. A. Montgomery; *J. Org. Chem.*, 1965, **30**, 3601.	564
65JOC3729	H. C. Brown and C. R. Wetzel; *J. Org. Chem.*, 1965, **30**, 3729.	763
65JOC4066	Y. Mizuno, N. Ikekawa, T. Itoh and K. Saito; *J. Org. Chem.*, 1965, **30**, 4066.	615, 619
65JOC4081	W. W. Paudler and H. L. Blewitt; *J. Org. Chem.*, 1965, **30**, 4081.	611
65JOC4085	J. P. Paolini and R. K. Robins; *J. Org. Chem.*, 1965, **30**, 4085.	611, 613
65JOC4205	A. C. Oehlschlager and L. H. Zalkow; *J. Org. Chem.*, 1965, **30**, 4205.	724
65JOU95	V. F. Lavrushin, V. D. Bezuglyi, G. G. Belous and V. G. Tishchenko; *J. Org. Chem. USSR (Engl. Transl.)*, 1965, **1**, 95.	254
65JPR(30)96	W. König, M. Coenen, W. Lorenz, F. Bahr and A. Bassl; *J. Prakt. Chem.*, 1965, **30**, 96.	719
65LA(682)123	H. Gehlen and J. Schmidt; *Liebigs Ann. Chem.*, 1965, **682**, 123.	745, 754
65LA(683)149	H. Gehlen and W. Schade; *Liebigs Ann. Chem.*, 1965, **683**, 149.	772
65LA(685)176	H. Gehlen and K. Moeckel; *Liebigs Ann. Chem.*, 1965, **685**, 176.	772
65LA(686)134	F. Lingens and H. Schneider-Bernlöhr; *Liebigs Ann. Chem.*, 1965, **686**, 134.	142
65LA(688)93	H. Bredereck, B. Föhlisch and K. Walz; *Liebigs Ann. Chem.*, 1965, **688**, 93.	802
65LA(688)205	H. Gold; *Liebigs Ann. Chem.*, 1965, **688**, 205.	678, 698, 708
65M741	F. Asinger, W. Schafer and A. V. Grenacher; *Monatsh. Chem.*, 1965, **96**, 741.	649
65MI40100	F. Eloy; *Fortschr. Chem. Forsch.*, 1965, **4**, 807.	31
65MI40800	M. R. Grimmett; *Rev. Pure Appl. Chem.*, 1965, **15**, 101.	485
65MI40900	G. P. Hager, J. C. Krantz, J. C. Harman and R. M. Burgison; *J. Am. Pharm. Assoc.*, 1965, **43**, 152.	575
B-65MI40901	C. Ponnamperuma; 'The Origins of Prebiological Systems and their Molecular Matrices', Academic, London, 1965, p. 221.	569
B-65MI40902	K. W. Merz and P. H. Stahle; 'Beitrage Zur Biochemie und Physiologie von Naturstoffen', Fischer-Verlag, Jena, 1965, p. 285.	559
65MI40903	J. Iball and H. R. Wilson; *Proc. R. Soc. London, Ser. A*, 1965, **288**, 418.	508, 520
65MI40904	G. B. Brown, K. Suguira and R. M. Cresswell; *Cancer Res.*, 1965, **25**, 986.	555
65MI40905	H. Berthod and A. Pullman; *J. Chim. Phys., Phys. Chim. Biol.*, 1965, **62**, 942.	505
65MI40906	Z. Neiman and F. Bergman; *Isr. J. Chem.*, 1965, **3**, 161.	559, 560
65MI40907	Y. Fujimoto and N. Ono; *Pharm. Soc. Jpn.*, 1965, **81**, 3671.	580
65MI40908	G. Shaw; *New Scientist*, 1965, **25**, 788.	601
65MI41100	G. Bianchetti, P. Dalla Croce, D. Pocar and G. Gallo; *Rend. Inst. Lomb. Sci. Lett.*, 1965, **99A**, 296.	704, 714

65MI41200	H. Deuschl; *Ber. Bunsenges. Phys. Chem.*, 1965, **69**, 550.	738
65P933	P. M. Dunnill and L. Fowden; *Phytochemistry*, 1965, **4**, 933.	303
65PC40100	J. A. Elvidge; personal communication to R. Slack and K. R. H. Wooldridge; *Adv. Heterocycl. Chem.*, 1965, **4**, 114.	33
65PNA(54)1052	C. O. Miller; *Proc. Nat. Acad. Sci. USA*, 1965, **54**, 1052.	601
65RCR237	A. I. Busev, V. K. Akimov and S. I. Gusev; *Russ. Chem. Rev. (Engl. Transl.)*, 1965, **34**, 237.	169, 300
65RTC213	H. C. Beyerman, W. Maassen van den Brink, F. Weygand, A. Prox, W. König, L. Schmidhammer and E. Nintz; *Recl. Trav. Chim. Pays-Bas*, 1965, **84**, 213.	786
65RTC408	B. Zwanenburg, W. Weening and J. Strating; *Recl. Trav. Chim. Pays-Bas*, 1965, **84**, 408.	773
65T3209	D. S. Jones, G. W. Kenner, J. Preston and R. C. Sheppard; *Tetrahedron*, 1965, **21**, 3209.	643
65T3435	R. Ottinger, G. Boulvin, J. Reisse and G. Chiurdoglu; *Tetrahedron*, 1965, **21**, 3435.	512
65TL1965	R. Fusco, S. Rossi and S. Miaorana; *Tetrahedron Lett.*, 1965, 1965.	879
65TL2039	G. Bianchetti, P. Dalla Croce and D. Pocar; *Tetrahedron lett.*, 1965, 2039.	714
65TL2059	B. Shimizu and M. Miyaki; *Tetrahedron Lett.*, 1965, 2059.	530, 552
65YZ158	A. Takamizawa, S. Hayashi and H. Sato; *Yakugaku Zasshi*, 1965, **85**, 158.	216
65ZC304	J. Schmidt and H. Gehlen; *Z. Chem.*, 1965, **5**, 304.	742
65ZC381	C. F. Kröger and W. Freiberg; *Z. Chem.*, 1965, **5**, 381.	742
66AC(R)190	G. Palazzo and L. Baiocchi; *Ann. Chem., (Rome)*, 1966, **56**, 190.	748
66ACS57	J. Sandström and I. Wennerbeck; *Acta Chem. Scand.*, 1966, **20**, 57.	739
66ACS754	A. Hordvik; *Acta Chem. Scand.*, 1966, **20**, 754.	9
66ACS1555	M. Begtrup and C. Pedersen; *Acta Chem. Scand.*, 1966, **20**, 1555.	701
66ACS1907	A. Hordvik and J. Sletten; *Acta Chem. Scand.*, 1966, **20**, 1907.	9
66ACS2795	K. A. Jensen, A. Holm and S. Rachlin; *Acta Chem. Scand.*, 1966, **20**, 2795.	802
66AF541	A. Zimmer and R. W. Atchley; *Arzneim.-Forsch.*, 1966, **16**, 541.	548
66AG392	H. G. Zachau, D. Dutting and H. Feldman; *Angew. Chem.*, 1966, **78**, 392.	601
66AG(E)308	E. C. Taylor, S. Vromen, R. V. Ravindranathan and A. McKillop; *Angew. Chem., Int. Ed. Engl.*, 1966, **5**, 308.	319, 329
66AG(E)699	A. Wagner, C. W. Schellhammer and S. Petersen; *Angew. Chem., Int. Ed. Engl.*, 1966, **5**, 699.	169
66AG(E)960	H. G. Schmelzer, E. Degener and H. Holtschmidt; *Angew. Chem., Int. Ed. Engl.*, 1966, **5**, 960.	765
66AHC(6)1	J. H. Lister; *Adv. Heterocycl. Chem.*, 1966, **6**, 1.	501
66AHC(6)45	R. E. Lyle and P. S. Anderson; *Adv. Heterocycl. Chem.*, 1966, **6**, 45.	415
66AHC(6)347	A. N. Kost and I. I. Grandberg; *Adv. Heterocycl. Chem.*, 1966, **6**, 347.	47, 54, 58, 60, 65, 69, 74, 79, 88, 90, 100, 105, 106, 168, 171, 207, 222, 238, 241, 254, 260, 266, 268, 272, 273, 276, 277
66AHC(7)39	H. Prinzbach and E. Futterer; *Adv. Heterocycl. Chem.*, 1966, **7**, 39.	64, 67, 89, 97, 103, 114
66AHC(7)183	A. Hetzheim and K. Mockel; *Adv. Heterocycl. Chem.*, 1966, **7**, 183.	60, 80, 95
66AHC(7)301	G. Spiteller; *Adv. Heterocycl. Chem.*, 1966, **7**, 301.	30
66AP441	G. Zinner, R. Moll and B. Boehlke; *Arch. Pharm.*, 1966, **299**, 441.	750
66AX397	H. Ringertz; *Acta Crystallogr.*, 1966, **20**, 397.	508
66B756	R. J. Rousseau, L. B. Townsend and R. K. Robins; *Biochemistry*, 1966, **5**, 756.	619, 621, 622
66B2082	G. B. Chheda and R. H. Hall; *Biochemistry*, 1966, **5**, 2082.	565
66B3007	M. D. Litwack and B. Weissman; *Biochemistry*, 1966, **5**, 3007.	531, 552
66B3057	A. Giner-Sorolla, S. O'Bryant, J. H. Burchenal and A. Bendich; *Biochemistry*, 1966, **5**, 3057.	556
66BSF610	J. Elguero and R. Jacquier; *Bull. Soc. Chim. Fr.*, 1966, 610.	199, 255
66BSF619	J. Elguero, G. Guiraud and R. Jacquier; *Bull. Soc. Chim. Fr.*, 1966, 619.	231
66BSF775	J. Elguero, G. Guiraud, R. Jacquier and G. Tarrago; *Bull. Soc. Chim. Fr.*, 1966, 775.	231
66BSF2075	J. Elguero, A. Fruchier and R. Jacquier; *Bull. Soc. Chim. Fr.*, 1966, 2075.	186
66BSF2529	L. Petit and P. Touratier; *Bull. Soc. Chim. Fr.*, 1966, 2529.	647
66BSF2832	J. Elguero and R. Jacquier; *Bull. Soc. Chim. Fr.*, 1966, 2832.	231
66BSF2990	J. Elguero, R. Jacquier and G. Tarrago; *Bull. Soc. Chim. Fr.*, 1966, 2990.	264
66BSF3041	J. Elguero, A. Fruchier and R. Jacquier; *Bull. Soc. Chim. Fr.*, 1966, 3041.	212
66BSF3524	E. J. Vincent, R. Phan-Tan-Luu, J. Metzger and J. M. Surzur; *Bull. Soc. Chim. Fr.*, 1966, 3524 (*Chem. Abstr.*, 1967, **66**, 64 912).	15, 17
66BSF3727	J. Elguero, R. Jacquier and H. C. N. Tieu Duc; *Bull. Soc. Chim. Fr.*, 1966, 3727.	231
66BSF3744	J. Elguero, R. Jacquier and H. C. N. Tien Duc; *Bull. Soc. Chim. Fr.*, 1966, 3744.	198
66BSF3934	J. Seyden-Penne, L. T. Minh and P. Chabrier; *Bull. Soc. Chim. Fr.*, 1966, 3934.	559
66CB317	D. Martin and A. Weise; *Chem. Ber.*, 1966, **99**, 317.	686
66CB475	R. Huisgen, G. Szeimies and L. Möbius; *Chem. Ber.*, 1966, **99**, 475.	703, 705, 711
66CB618	H. Gnichtel and H. J. Schönherr; *Chem. Ber.*, 1966, **99**, 618.	274
66CB850	I. Hagedorn and H.-D. Winkelmann; *Chem. Ber.*, 1966, **99**, 850.	821
66CB944	H. Bredereck, P. Schellenberg, R. Nast, H. Heise and O. Christmann; *Chem. Ber.*, 1966, **99**, 944.	532
66CB1618	J. Goerdeler and G. Gnad; *Chem. Ber.*, 1966, **99**, 1618.	695

66CB2237	A. Kreutzberger; *Chem. Ber.*, 1966, **99**, 2237.	862, 891
66CB2512	L. Birkofer and P. Wegner; *Chem. Ber.*, 1966, **99**, 2512.	698, 710, 831
66CB2593	T. Kauffmann, K. Vogt, S. Barck and J. Schulz; *Chem. Ber.*, 1966, **99**, 2593.	882
66CB2918	M. Regitz; *Chem. Ber.*, 1966, **99**, 2918.	887
66CB2955	H. A. Staab and G. Malek; *Chem. Ber.*, 1966, **99**, 2955.	749
66CB3734	A. Spassov, E. Golovinsky and C. Demirov; *Chem. Ber.*, 1966, **99**, 3734.	766
66CC352	J. E. Baldwin, J. A. Kapecki, M. G. Newton and I. C. Paul; *Chem. Commun.*, 1966, 352.	703
66CHE134	R. G. Glushkov and O. Y. Magidson; *Chem. Heterocycl. Compd. (Engl. Transl.)*, 1966, **2**, 134 (*Chem. Abstr.*, 1966, **65**, 5460).	661
66CHE413	V. I. Minkin, S. F. Pozharskii and Yu. A. Ostroumov; *Chem. Heterocycl. Compd. (Engl. Transl.)*, 1966, **2**, 413.	172, 173, 177
66CI(L)600	A. J. Easteal and M. G. Ruthven; *Chem. Ind. (London)*, 1966, 600.	752
66CPB506	T. Okamoto, M. Hirobe, Y. Tamai and E. Yabe; *Chem. Pharm. Bull.*, 1966, **14**, 506.	884
66CPB523	T. Okamoto, M. Hirobe and E. Yabe; *Chem. Pharm. Bull.*, 1966, **14**, 523.	864
66DOK(170)354	R. A. Stukan, V. I. Gol'danskii, E. F. Makarov, B. V. Borshagovskii, N. S. Kochetkova, M. I. Rybinskaya and A. N. Nesmeyanov; *Dokl. Akad. Nauk SSSR*, 1966, **170**, 354.	690, 731
66FES811	G. Palamidessi and F. Luini; *Farmaco Ed. Sci.*, 1966, **21**, 811 (*Chem. Abstr.*, 1967, **66**, 37 886).	645
66G1084	R. Fusco, F. D'Alo and A. Masserini; *Gazz. Chim. Ital.*, 1966, **96**, 1084.	753, 756
66GEP1208303	F. Huebenett and H. Heinze; *Ger. Pat.* 1 208 303 (1966) (*Chem. Abstr.*, 1966, **64**, 8190).	57
66HC(21-1)67	D. S. Breslow and H. Skolnik; *Chem. Heterocycl. Compd.*, 1966, **21-1**, 67.	127
66HCA272	J. Büchi, P. Fabiani, H. U. Frey, A. Hofstetter and A. Schorno; *Helv. Chim. Acta*, 1966, **49**, 272.	295
66IJC146	B. D. Mehrotra, P. C. Jain and N. Anand; *Indian J. Chem.*, 1966, **4**, 146.	617, 619
66IJC403	P. C. Jain, S. K. Chatterjee and N. Anand; *Indian J. Chem.*, 1966, **4**, 403.	615, 617, 635
66JA1074	J. P. Ferris and L. E. Orgel; *J. Am. Chem. Soc.*, 1966, **88**, 1074.	570
66JA2077	J. A. Happe and M. Morales; *J. Am. Chem. Soc.*, 1966, **88**, 2077.	516
66JA2614	R. H. Hall, M. J. Robins, L. Stasiuk and R. Thedford; *J. Am. Chem. Soc.*, 1966, **88**, 2614.	601
66JA3829	J. P. Ferris and L. E. Orgel; *J. Am. Chem. Soc.*, 1966, **88**, 3829.	569, 584
66JA3963	R. J. Crawford and A. Mishra; *J. Am. Chem. Soc.*, 1966, **88**, 3963.	255
66JA4271	W. J. Musliner and J. W. Gates, Jr.; *J. Am. Chem. Soc.*, 1966, **88**, 4271.	821
66JA4759	P. Scheiner; *J. Am. Chem. Soc.*, 1966, **88**, 4759.	711, 724
66JAP6610114	T. Saito, I. Kumashiro and T. Takenishi; *Jpn. Pat.* 66 10 144 (1966) (*Chem. Abstr.*, 1966, **65**, 13 737).	569
66JBC(241)1091	J. R. Florini, H. H. Bird and P. H. Bell; *J. Biol. Chem.*, 1966, **241**, 1091.	603
66JCP(44)759	M. J. S. Dewar and G. J. Gleicher; *J. Chem. Phys.*, 1966, **44**, 759.	671, 796, 803
66JCS(B)285	G. B. Barlin; *J. Chem. Soc. (B)*, 1966, 285.	620, 621, 622, 640
66JCS(B)427	A. Albert; *J. Chem. Soc. (B)*, 1966, 427.	856, 875, 876
66JCS(B)433	J. W. Bunting and D. D. Perrin; *J. Chem. Soc. (B)*, 1966, 433.	513
66JCS(B)438	A. Albert; *J. Chem. Soc. (B)*, 1966, 438.	875
66JCS(C)10	F. Bergmann, Z. Neiman and M. Kleiner; *J. Chem. Soc. (C)*, 1966, 10.	558, 576, 577
66JCS(C)80	R. K. Smalley; *J. Chem. Soc. (C)*, 1966, 80.	641
66JCS(C)921	G. Shaw, B. M. Smallwood and D. V. Wilson; *J. Chem. Soc. (C)*, 1966, 921.	595, 601
66JCS(C)1127	M. W. Partridge and M. F. G. Stevens; *J. Chem. Soc. (C)*, 1966, 1127.	307, 330
66JCS(C)2038	S. E. Mallett and F. L. Rose; *J. Chem. Soc. (C)*, 1966, 2038.	896, 897
66JCS(C)2290	G. Tennant; *J. Chem. Soc. (C)*, 1966, 2290.	876, 895
66JHC27	C. K. Bradsher and D. F. Lohr, Jr.; *J. Heterocycl. Chem.*, 1966, **3**, 27.	121
66JHC158	G. S. Sidhu, S. Naqui and D. S. Iyengar; *J. Heterocycl. Chem.*, 1966, **3**, 158.	853, 865, 866, 886
66JHC241	L. B. Townsend and R. K. Robins; *J. Heterocycl. Chem.*, 1966, **3**, 241.	511
66JHC435	H. Rutner and P. E. Spoerri; *J. Heterocycl. Chem.*, 1966, **3**, 435.	854, 856, 857, 903
66JMC29	L. Almirante, L. Polo, A. Mugnaini, E. Provinciali, P. Rugarli, A. Gamba, A. Olivi and W. Murmann; *J. Med. Chem.*, 1966, **9**, 29.	647
66JMC38	G. Palazzo, G. Corsi, L. Baiocchi and B. Silvestrini; *J. Med. Chem.*, 1966, **9**, 38.	294
66JMC105	J. A. Montgomery and K. Hewson; *J. Med. Chem.*, 1966, **9**, 105.	619, 621, 622, 640
66JMC160	A. J. Dietz, Jr. and R. M. Burgison; *J. Med. Chem.*, 1966, **9**, 160.	577
66JMC354	J. A. Montgomery and K. Hewson; *J. Med. Chem.*, 1966, **9**, 354.	615, 618, 636
66JMC373	A. G. Beaman, W. Tautz, R. Duschinsky and E. Grunberg; *J. Med. Chem.*, 1966, **9**, 373.	551
66JMC405	H. A. Burch and W. O. Smith; *J. Med. Chem.*, 1966, **9**, 405.	766
66JMC493	M. J. Kornet; *J. Med. Chem.*, 1966, **9**, 493.	297
66JMC545	A. H. M. Raeymaekers, F. T. N. Allewijn, J. Vandenberk, P. J. A. Demoen, T. T. T. Van Offenwert and P. A. J. Janssen; *J. Med. Chem.*, 1966, **9**, 545.	139
66JMC733	Y. F. Shealy and C. A. O'Dell; *J. Med. Chem.*, 1966, **9**, 733.	701, 852
66JMC868	M. P. Mertes and N. R. Patel; *J. Med. Chem.*, 1966, **9**, 868.	645
66JOC59	P. E. Fanta and E. N. Walsh; *J. Org. Chem.*, 1966, **31**, 59.	155
66JOC251	K. T. Potts and H. R. Burton; *J. Org. Chem.*, 1966, **31**, 251.	863, 882, 883
66JOC260	K. T. Potts, H. R. Burton and J. Bhattacharyya; *J. Org. Chem.*, 1966, **31**, 260.	134, 864, 884, 885

66JOC265	K. T. Potts, H. R. Burton and S. K. Roy; *J. Org. Chem.*, 1966, **31**, 265.	853, 861, 862, 863
66JOC342	E. C. Taylor and A. Abul-Husn; *J. Org. Chem.*, 1966, **31**, 342.	322
66JOC442	O. W. Maender and G. A. Russell; *J. Org. Chem.*, 1966, **31**, 442.	800
66JOC781	H. C. Brown, H. J. Gisler, Jr. and M. T. Cheng; *J. Org. Chem.*, 1966, **31**, 781.	772
66JOC806	E. F. Godefroi, C. A. M. Van der Eycken and P. A. J. Janssen; *J. Org. Chem.*, 1966, **31**, 806.	661
66JOC809	W. W. Paudler and J. E. Kuder; *J. Org. Chem.*, 1966, **31**, 809.	627
66JOC935	C. Temple, Jr., W. C. Coburn, M. C. Thorpe and J. A. Montgomery; *J. Org. Chem.*, 1966, **31**, 935.	511, 565
66JOC966	J. C. Parham, J. Fissekis and G. B. Brown; *J. Org. Chem.*, 1966, **31**, 966.	595
66JOC1295	W. W. Paudler and H. L. Blewitt; *J. Org. Chem.*, 1966, **31**, 1295.	610
66JOC1413	H. J. Thomas and J. A. Montgomery; *J. Org. Chem.*, 1966, **31**, 1413.	552
66JOC1417	C. Temple, Jr. and J. A. Montgomery; *J. Org. Chem.*, 1966, **31**, 1417.	511, 566, 579
66JOC1587	G. R. Harvey; *J. Org. Chem.*, 1966, **31**, 1587.	716
66JOC1890	R. D. Elliott, C. Temple, Jr. and J. A. Montgomery; *J. Org. Chem.*, 1966, **31**, 1890.	636
66JOC2202	J. A. Montgomery, K. Hewson, S. J. Clayton and H. J. Thomas; *J. Org. Chem.*, 1966, **31**, 2202.	511, 531, 532
66JOC2210	C. Temple, Jr., C. L. Kussner and J. A. Montgomery; *J. Org. Chem.*, 1966, **31**, 2210.	511, 564
66JOC2380	F. B. Stocker, M. W. Fordice, J. K. Larson and J. H. Thorstenson; *J. Org. Chem.*, 1966, **31**, 2380.	641
66JOC2685	N. Nagasawa, I. Kumashiro and T. Takenishi; *J. Org. Chem.*, 1966, **31**, 2685.	532, 534
66JOC2867	L. A. Carpino, P. H. Terry and S. D. Thatte; *J. Org. Chem.*, 1966, **31**, 2867.	250, 327
66JOC3182	R. N. Butler and F. L. Scott; *J. Org. Chem.*, 1966, **31**, 3182.	818, 819
66JOC3258	J. F. Gerster and R. K. Robins; *J. Org. Chem.*, 1966, **31**, 3258.	552, 562
66JOC3522	K. T. Potts, H. R. Burton, T. H. Crawford and S. W. Thomas; *J. Org. Chem.*, 1966, **31**, 3522.	853, 854
66JOC3914	T. Sasaki and K. Minamoto; *J. Org. Chem.*, 1966, **31**, 3914.	719
66JOC3924	H. W. Heine, R. Peavy and A. J. Durbetaki; *J. Org. Chem.*, 1966, **31**, 3924.	780
66JOC4239	A. Giner-Sorolla and A. Bendich; *J. Org. Chem.*, 1966, **31**, 4239.	549
66JPC3688	T. C. Wehman and A. I. Popov; *J. Phys. Chem.*, 1966, **70**, 3688.	814
66JPR(305)54	W. König, M. Coenen, F. Bahr, B. May and A. Bassl; *J. Prakt. Chem.*, 1966, **305**, 54.	719
66JPR(305)199	G. Rembarz, B. Kirchhoff and G. Dangowski; *J. Prakt. Chem.*, 1966, **305**, 199.	704
66KGS143	L. M. Sitkina and A. M. Simonov; *Khim. Geterotsikl. Soedin.*, 1966, 143 (*Chem. Abstr.*, 1966, **65**, 88 955).	390
66LA(691)142	H. Goldner, G. Dietz and E. Carstens; *Liebigs Ann. Chem.*, 1966, **691**, 142.	555, 580, 581, 596
66LA(691)212	Th. Weiland and W. Kahle; *Liebigs Ann. Chem.*, 1966, **691**, 212.	786
66LA(692)134	H. Goldner, G. Dietz and E. Carstens; *Liebigs Ann. Chem.*, 1966, **692**, 134.	591
66LA(693)233	H. Goldner, G. Dietz and E. Carstens; *Liebigs Ann. Chem.*, 1966, **693**, 233.	554, 555, 596
66LA(695)77	W. Walter, J. Voss and J. Curts, *Liebigs Ann. Chem.*, 1966, **695**, 77.	560
66LA(698)145	H. Goldner, G. Dietz and E. Carstens; *Liebigs Ann. Chem.*, 1966, **698**, 145.	580, 596
66LA(698)174	R. Feinauer and W. Seeliger; *Liebigs Ann. Chem.*, 1966, **698**, 174.	155
66LA(699)112	T. Pyl and W. Baufeld; *Liebigs Ann. Chem.*, 1966, **699**, 112.	627
66LA(699)127	T. Pyl and W. Baufeld; *Liebigs Ann. Chem.*, 1966, **699**, 127.	627
66LA(699)145	H. Goldner, G. Dietz and E. Carstens; *Liebigs Ann. Chem.*, 1966, **699**, 145.	596
B-66MI40900	J. A. Stock; in 'Experimental Chemotherapy', Academic, London, 1966, vol. 4.	604
66MI40901	A. Giner-Sorolla; *Galenica Acta*, 1966, **19**, 97.	556, 564
66MI40902	J. J. Fox, K. A. Watanabe and A. Bloch; *Prog. Nucleic Acid Res. Mol. Biol.*, 1966, **5**, 251.	603
66MI40903	Y. Fujimoto and N. Ono; *J. Pharm. Soc. Jpn.*, 1966, **86**, 364.	593
66MI40904	J. S. Shannon and D. S. Letham; *N.Z. J. Sci.*, 1966, **9**, 833.	519
66MI40905	A. Vincze and S. Cohen; *Isr. J. Chem.*, 1966, **4**, 23.	529
66MI41200	H. G. O. Becker, H. Boettcher, T. Roethling and H. J. Timpe; *Wiss. Z. Tech. Hochsch. Chem. Leuna-Merseburg*, 1966, **8**, 22 (*Chem. Abstr.*, 1966, **64**, 9596).	748
66MI41201	I. Ya. Postovskii and I. L. Shegal; *Khim. Geterotsikl. Soedin. Akad. Nauk Latv. SSR*, 1966, 443 (*Chem. Abstr.*, 1966, **65**, 16 960).	754
66NKK71	A. Tatematsu, T. Goto and S. Matsuura; *Nippon Kagaku Zasshi*, 1966, **87**, 71.	519
66PNA(56)52	D. Klambt, G. Thies and F. Skoog; *Proc. Nat. Acad. Sci. USA*, 1966, **56**, 52.	601
66PNA(56)60	J. P. Helgeson and N. J. Leonard; *Proc. Nat. Acad. Sci. USA*, 1966, **56**, 60.	601
66RCR9	Yu. A. Naumov and I. I. Grandberg; *Russ. Chem. Rev. (Engl. Transl.)*, 1966, **35**, 9.	169, 255
66RCR122	A. F. Pozharskii, A. D. Garnovskii and A. M. Simonov; *Russ. Chem. Rev. (Engl. Transl.)*, 1966, **35**, 122.	346, 351, 356, 367, 389, 392, 405, 410, 413
66RCR122	A. F. Pozharskii, A. D. Garnovskii and A. M. Simonov; *Russ. Chem. Rev. (Engl. Transl.)*, 1966, **35**, 122.	467, 469, 470, 472, 480, 481, 490, 491
66RTC429	B. G. Van den Bos, A. Schipperheyn and F. W. Deursen; *Recl. Trav. Chim. Pays-Bas*, 1966, **85**, 429.	751
66RTC1195	C. L. Habraken, E. C. Westra, G. H. Bomhoff and R. A. Heytink; *Recl. Trav. Chim. Pays-Bas*, 1966, **85**, 1195.	260
66RTC1254	G. E. Trout and P. R. Levy; *Recl. Trav. Chim. Pays-Bas*, 1966, **85**, 1254.	588
66T835	W. Adam and A. Grimison; *Tetrahedron*, 1966, **22**, 835.	347

66T2073	A. Pollak and M. Tisler; *Tetrahedron*, 1966, **22**, 2073.	888, 889
66T2119	R. A. Olofson, J. M. Landesburg, R. O. Berry, D. Leaver, W. A. H. Robertson and D. M. McKinnon; *Tetrahedron*, 1966, **22**, 2119.	23
66T2453	P. K. Kadaba; *Tetrahedron*, 1966, **22**, 2453.	717
66T2461	P. Bouchet, J. Elguero and R. Jacquier; *Tetrahedron*, 1966, **22**, 2461.	185, 207
66T(S7)213	A. A. Gordon, A. R. Katritzky and F. D. Popp; *Tetrahedron*, 1966, **22**, Suppl. 7, 213.	739, 740
66TL2627	B. M. Lynch and H. J. M. Dou; *Tetrahedron Lett.*, 1966, 2627.	183
66TL3445	Y. Kishi, T. Goto, S. Inoue, S. Sugiura and H. Kishimoto; *Tetrahedron Lett.*, 1966, 3445.	643
66TL4043	C. H. Krauch, J. Kuhls and H.-J. Piek; *Tetrahedron Lett.*, 1966, 4043.	160
66TL4387	R. M. Carman, D. J. Brecknell and H. C. Deeth; *Tetrahedron Lett.*, 1966, 4387.	697, 705
66TL5369	H. H. Takimoto and G. C. Denault; *Tetrahedron Lett.*, 1966, 5369.	744
66TL6043	S. Maiorana, D. Pocar and P. D. Croce; *Tetrahedron Lett.*, 1966, 6043.	715
66ZOB816	E. Ebed, E. S. Chaman and E. S. Golovchinskaya; *Zh. Obshch. Khim.*, 1966, **36**, 816.	550
66ZPC(51)297	H. Schmid and P. Krenmayr; *Hoppe-Seyler's Z. Physiol. Chem.*, 1966, **51**, 297.	511
67ACS633	M. M. Begtrup and C. Pedersen; *Acta Chem. Scand.*, 1967, **21**, 633.	706
67AG786	M. Regitz; *Angew. Chem.*, 1967, **79**, 786.	670
67AG981	H. Dorn and A. Zubek; *Angew. Chem.*, 1967, **6**, 981.	272
67AG(E)258	W. Pfleiderer and F. E. Kempter; *Angew. Chem., Int. Ed. Engl.*, 1967, **6**, 258.	580
67AG(E)261	A. Messmer and A. Gelléri; *Angew. Chem., Int. Ed. Engl.*, 1967, **6**, 261.	134
67AG(E)633	Th. Kauffmann, J. Albrecht, D. Berger and J. Legler; *Angew. Chem., Int. Ed. Engl.*, 1967, **6**, 633.	768
67AHC(8)1	J. M. Tedder; *Adv. Heterocycl. Chem.*, 1967, **8**, 1.	96, 441
67AHC(8)143	B. S. Thyagarajan; *Adv. Heterocycl. Chem.*, 1967, **8**, 143.	102
67AHC(8)277	K.-H. Wunsch and A. J. Boulton; *Adv. Heterocycl. Chem.*, 1967, **8**, 277.	64, 71, 74, 85, 91, 93, 108
67AJC713	N. A. Evans, R. B. Johns and K. R. Markham; *Aust. J. Chem.*, 1967, **20**, 713.	852
67AJC1779	G. W. Evans and B. Milligan; *Aust. J. Chem.*, 1967, **20**, 1779.	743
67AJC1991	F. H. C. Stewart; *Aust. J. Chem.*, 1967, **20**, 1991.	137
67AP1000	R. Brandes and H. J. Roth; *Arch Pharm. (Weinheim, Ger.)*, 1967, **300**, 1000.	552
67AX135	C. W. Reimann, A. D. Mighell and F. A. Mauer; *Acta Crystallogr.*, 1967, **23**, 135.	227
67AX308	K. C. Britts and I. L. Karle; *Acta Crystallogr.*, 1967, **22**, 308.	796
67BSF289	L. Bardou, J. Elguero and R. Jacquier; *Bull. Soc. Chim. Fr.*, 1967, 289.	231
67BSF1833	H. Lumbroso, C. Pigenet, R. Nasielski-Hinskens and R. Promel; *Bull. Soc. Chim. Fr.*, 1967, 1833.	522
67BSF2619	J. Elguero, A. Fruchier and R. Jacquier; *Bull. Soc. Chim. Fr.*, 1967, 2619.	199, 224, 232
67BSF2630	R. Jacquier, M. L. Roumestant and P. Viallefont; *Bull. Soc. Chim. Fr.*, 1967, 2630.	741
67BSF2998	J. Elguero, E. Gonzales and R. Jacquier; *Bull. Soc. Chim. Fr.*, 1967, 2998.	678, 679
67BSF3502	P. Bouchet, J. Elguero and R. Jacquier; *Bull. Soc. Chim. Fr.*, 1967, 3502.	185, 199, 200, 207
67BSF3516	J. L. Aubagnac, J. Elguero and R. Jacquier; *Bull. Soc. Chim. Fr.*, 1967, 3516.	188, 199
67BSF4716	P. Bouchet, J. Elguero and R. Jacquier; *Bull. Soc. Chim. Fr.*, 1967, 4716.	255
67C226	K. Biemann and P. V. Fennessey; *Chimia*, 1967, **21**, 226.	519
67CB71	R. Huisgen and V. Weberndörfer; *Chem. Ber.*, 1967, **100**, 71.	721
67CB1073	W. Köhler, M. Bubner and G. Ulbricht; *Chem. Ber.*, 1967, **100**, 1073.	857, 902
67CB1646	H. Sieper; *Chem. Ber.*, 1967, **100**, 1646.	887
67CB2064	J. Goerdeler and R. Sappelt; *Chem. Ber.*, 1967, **100**, 2064.	667
67CB2250	C. F. Kröger and R. Miethchen; *Chem. Ber.*, **100**, 2250.	752, 753, 754, 756, 757
67CB2280	H. Bredereck, F. Effenberger and H. G. Osterlin; *Chem. Ber.*, 1967, **100**, 2280.	575, 576
67CB2494	R. Huisgen, G. Szeimies and L. Möbius; *Chem. Ber.*, 1967, **100**, 2494.	708
67CB2577	A. Dornow and K. Dehmer; *Chem. Ber.*, 1967, **100**, 2577.	272
67CB3418	A. Hetzheim, O. Peters and H. Beyer; *Chem. Ber.*, 1967, **100**, 3418	492
67CB3467	T. Sasaki and K. Minamoto; *Chem. Ber.*, 1967, **100**, 3467.	899, 903
67CB3485	L. Birkofer and P. Wegner; *Chem. Ber.*, 1967, **100**, 3485.	679, 698, 699
67CB3736	D. Martin and A. Weise; *Chem. Ber.*, 1967, **100**, 3736.	157
67CC488	S. N. Ege; *Chem. Commun.*, 1967, 488.	220, 221
67CC918	P. D. Callaghan and M. S. Gibson; *Chem. Commun.*, 1967, 918.	704
67CC1011	F. McCapra and Y. C. Chang; *Chem. Commun.*, 1967, 1011.	642
67CCC2155	F. M. Stojanovic and Z. Arnold; *Collect. Czech. Chem. Commun.*, 1967, **32**, 2155.	706
67CJC697	E. B. Dennler and A. R. Frasca; *Can. J. Chem.*, 1967, **45**, 697.	276
67CJC953	J. W. Lown and J. C. N. Ma; *Can. J. Chem.*, 1967, **45**, 953.	663
67CJC1727	J. H. Looker and J. W. Carpenter; *Can. J. Chem.*, 1967, **45**, 1727.	706
67CPB909	M. Saneyoshi and G. Chihara; *Chem. Pharm. Bull.*, 1967, **15**, 909.	559
67CPB1066	B. Shimizu and M. Miyaki; *Chem. Pharm. Bull.*, 1967, **15**, 1066.	552
67CR(C)(265)1507	J. Elguero and R. Wolf; *C.R. Hebd. Seances Acad. Sci., Ser. C*, 1967, **265**, 1507.	236
67EGP59288	H. G. O. Becker, V. Eisenschmidt and K. Wehner; *Ger. (East) Pat.* 59 288 (1967) (*Chem. Abstr.*, 1969, **70**, 28 922).	753
67FES821	T. Vitali, F. Mossini and G. Bertaccini; *Farmaco Ed. Sci.*, 1967, **22**, 821 (*Chem. Abstr.*, 1968, **68**, 87 234).	641

67G109	P. Ferruti, D. Pocar and G. Bianchetti; *Gazz. Chim. Ital.*, 1967, **97**, 109.	700, 714
67G289	G. Bianchetti, P. D. Croce, D. Pocar and A. Vigevani; *Gazz. Chim. Ital.*, 1967, **97**, 289.	714
67G304	G. Bianchetti, D. Pocar, P. D. Croce and R. Stradi; *Gazz. Chim. Ital.*, 1967, **97**, 304.	704, 714
67G1411	G. Gaudiano, C. Ticozzi, A. Umani-Ronchi and P. Bravo; *Gazz. Chim. Ital.*, 1967, **97**, 1411.	718
67HC(22)1	L. C. Behr, R. Fusco and C. H. Jarboe; *Chem. Heterocycl. Compd.*, 1967, **22**, 1.	168, 171, 199, 206, 222, 229, 230, 233, 238, 239, 242, 253, 254, 255, 256, 258, 259, 260, 261, 262, 265, 266, 268, 273, 274, 275, 277, 281, 282, 286
67HCA2244	H. Tiefenthaler, W. Dörscheln, H. Göth and H. Schmid; *Helv. Chim. Acta*, 1967, **50**, 2244.	46, 161, 219, 220, 221
67IJC403	M. D. Nair and S. R. Mehta; *Indian J. Chem.*, 1967, **5**, 403.	888, 902
67JA81	J. E. Anderson and J. M. Lehn; *J. Am. Chem. Soc.*, 1967, **89**, 81.	189
67JA2618	R. A. Carboni, J. C. Kauer, J. E. Castle and H. E. Simmons; *J. Am. Chem. Soc.*, 1967, **89**, 2618.	163, 723
67JA2633	J. C. Kauer and R. A. Carboni; *J. Am. Chem. Soc.*, 1967, **89**, 2633.	898
67JA2719	J. M. Rice and G. O. Dudek; *J. Am. Chem. Soc.*, 1967, **89**, 2719.	519
67JA4760	M. E. Hermes and F. D. Marsh; *J. Am. Chem. Soc.*, 1967, **89**, 4760.	692, 709
67JA5958	R. M. Moriarty, J. M. Kleigman and C. Shovlin; *J. Am. Chem. Soc.*, 1967, **89**, 5958.	159, 801
67JA5959	R. M. Moriarty, J. M. Kleigman and C. Shovlin; *J. Am. Chem. Soc.*, 1967, **89**, 5959.	159
67JA5977	W. E. Thiessen and H. Hope; *J. Am. Chem. Soc.*, 1967, **89**, 5977.	12
67JA6463	N. C. Baenziger, A. D. Nelson, A. Tulinsky, J. H. Bloor and A. I. Popov; *J. Am. Chem. Soc.*, 1967, **89**, 6463.	796
67JA6835	J. E. Bloor and D. L. Breen; *J. Am. Chem. Soc.*, 1967, **89**, 6835.	19, 795
67JCS(B)410	T. C. Chou, H. H. Lin and R. Varma; *J. Chem. Soc. (B)*, 1967, 410.	522
67JCS(B)516	G. B. Barlin and T. J. Batterham; *J. Chem. Soc. (B)*, 1967, 516.	678, 741, 742, 799, 800
67JCS(B)583	D. J. Blythin and E. S. Waight; *J. Chem. Soc. (B)*, 1967, 583.	204
67JCS(B)641	G. B. Barlin; *J. Chem. Soc., (B)*, 1967, 641.	739, 753, 756, 806
67JCS(B)954	G. B. Barlin; *J. Chem. Soc. (B)*, 1967, 954.	524, 561
67JCS(C)33	B. C. Ennis, G. Holan and E. L. Samuel; *J. Chem. Soc. (C)*, 1967, 33.	765
67JCS(C)239	R. N. Butler and F. L. Scott; *J. Chem. Soc. (C)*, 1967, 239.	815
67JCS(C)503	R. G. W. Spickett and S. H. B. Wright; *J. Chem. Soc. (C)*, 1967, 503.	891
67JCS(C)952	G. J. Durant; *J. Chem. Soc. (C)*, 1967, 952.	779
67JCS(C)1254	F. Bergman, M. Rashi, M. Kleiner and R. Knafo; *J. Chem. Soc. (C)*, 1967, 1254.	576
67JCS(C)2005	A. J. Boulton, A. R. Katritzky and A. Majid Hamid; *J. Chem. Soc. (C)*, 1967, 2005.	720
67JHC66	C. K. Bradsher and M. F. Zinn; *J. Heterocycl. Chem.*, 1967, **4**, 66.	117
67JHC301	P. K. Kadaba and N. F. Fannin; *J. Heterocycl. Chem.*, 1967, **4**, 301.	717
67JHC555	H. Murakami and R. N. Castle; *J. Heterocycl. Chem.*, 1967, **4**, 555.	628, 651, 871, 890
67JMC104	C. P. Bryant and R. E. Harmon; *J. Med. Chem.*, 1967, **10**, 104.	536
67JMC109	S. C. J. Fu, B. J. Hargis, E. Chinoporos and S. Malkiel; *J. Med. Chem.*, 1967, **10**, 109.	575
67JMC130	E. J. Reist, D. F. Calkins and L. Goodman; *J. Med. Chem.*, 1967, **10**, 130.	539
67JMC149	G. F. Holland and J. N. Pereira; *J. Med. Chem.*, 1967, **10**, 149.	819, 835
67JOC1151	J. C. Parham, J. Fissekis and G. B. Brown; *J. Org. Chem.*, 1967, **32**, 1151.	595
67JOC1825	A. Yamazaki, I. Kumashiro and T. Takenishi; *J. Org. Chem.*, 1967, **32**, 1825.	590
67JOC1871	H. C. Brown and R. J. Kassal; *J. Org. Chem.*, 1967, **32**, 1871.	780
67JOC2022	P. Scheiner; *J. Org. Chem.*, 1967, **32**, 2022.	714, 724
67JOC2245	K. T. Potts, S. K. Roy and D. P. Jones; *J. Org. Chem.*, 1967, **32**, 2245.	743, 748
67JOC2252	I. Bhatnagar and M. V. George; *J. Org. Chem.*, 1967, **32**, 2252.	707
67JOC2430	W. W. Paudler and J. E. Kuder; *J. Org. Chem.*, 1967, **32**, 2430.	614
67JOC2823	L. M. Weinstock, P. Davis, B. Handelsman and R. Tull; *J. Org. Chem.*, 1967, **32**, 2823.	127
67JOC2956	J. K. Elwood and J. W. Gates, Jr.; *J. Org. Chem.*, 1967, **32**, 2956.	813
67JOC3032	A. Yamazaki, I. Kumashiro and T. Takenishi; *J. Org. Chem.*, 1967, **32**, 3032.	590
67JOC3258	A. Yamazaki, I. Kumashiro and T. Takenishi; *J. Org. Chem.*, 1967, **32**, 3258.	558, 584, 587, 588
67JOC3580	J. C. Kauer and W. A. Sheppard; *J. Org. Chem.*, 1967, **32**, 3580.	794, 812
67JOC3856	Y. Rahamim, J. Sharvit, A. Mandelbaum and M. Sprecher; *J. Org. Chem.*, 1967, **32**, 3856.	519
67JPC2375	A. D. Mighell and C. W. Reimann; *J. Phys. Chem.*, 1967, **71**, 2375.	179
67JPR(308)82	G. Hauptmann and K. Hirschberg; *J. Prakt. Chem.*, 1967, **308**, 82.	717
67KGS1096	N. N. Vereshchagina, I. Y. Postovskii and S. L. Mertsalov; *Khim. Geterotsikl. Soedin.*, 1967, 1096.	881
67LA(707)147	H. Hoberg; *Liebigs Ann. Chem.*, 1967, **707**, 147.	698, 717
67LA(710)118	M. Regitz and A. Liedhegener; *Liebigs Ann. Chem.*, 1967, **710**, 118.	695
67MI40400	R. D. Brown and B. A. W. Coller; *Theor. Chim. Acta*, 1967, **7**, 259.	172, 176, 177
67MI40401	W. Adam and A. Grimison; *Theor. Chim. Acta*, 1967, **7**, 342.	172, 177, 183
67MI40402	J. L. Aubagnac, P. Bouchet, J. Elguero, R. Jacquier and C. Marzin; *J. Chim. Phys. Phys. Chim. Biol.*, 1967, **64**, 1649.	188
B-67MI40900	R. K. Robins; in 'Heterocyclic Compounds', ed. R. C. Elderfield; Wiley, New York, 1967.	501
67MI40901	R. A. Sanchez, J. P. Ferris and L. E. Orgel; *J. Mol. Biol.*, 1967, **30**, 223.	569

67MI40902	R. H. Hall, L. Csonka, H. David and B. McLennan; *Science*, 1967, **156**, 69.	601
67MI41000	V. G. Granik and R. G. Glushkov; *Khim.-Farm. Zh.*, 1967, **1**, 16 (*Chem. Abstr.*, 1968, **68**, 68 955).	624, 643
67MI41001	S. Takase, T. Demura and K. Tabata; *Kogyo Kagaku Zasshi*, 1967, **70**, 1826 (*Chem. Abstr.*, 1968, **68**, 114 498).	635
67MI41100	H. Krieger and M. Sodervall; *Suomen Kem.*, 1967, **40B**, 294 (*Chem. Abstr.*, 1968, **68**, 77841).	725
B-67MI41101	B. P. Stark and A. J. Duke; in 'Extrusion Reactions', Pergamon Press, Oxford, 1967, pp. 135, 327.	670, 704, 724
B-67MI41200	A. R. Katritzky and J. M. Lagowski; 'The Principles of Heterocyclic Chemistry', Academic, New York, 1968.	735
B-67MI41300	F. R. Benson; 'Heterocyclic Compounds', ed. R. C. Elderfield; Wiley, New York, 1967, vol. 8, p. 1.	792, 798, 805, 807, 817, 818, 819, 820, 821, 822, 823, 828, 829, 830, 832, 833, 836, 837
67MI41301	W. Woznicki and B. Zurawski; *Acta Phys. Polonica*, 1967, **31**, 95.	803
67MIP41000	V. G. Granik and R. G. Glushkov; *USSR Pat.* 196 876 (1967) (*Chem. Abstr.*, 1968, **68**, 95 859).	624, 643
67NEP6614979	J. R. Geigy A. G., *Neth. Pat.* 66 14 979 (1967) (*Chem. Abstr.*, 1968, **68**, 69 003).	904
67PNA(57)751	U. L. RajBhandary, S. H. Chang, A. Stuart, R. D. Faulkner, R. M. Hoskinson and H. G. Khorana; *Proc. Nat. Acad. Sci. USA*, 1967, **57**, 751.	605
67PNA(58)1279	C. Alexander, Jr. and W. Gordy; *Proc. Natl. Acad. Sci. USA*, 1967, **58**, 1279.	520
67RTC1249	P. Cohen-Fernandes and C. L. Habraken; *Recl. Trav. Chim. Pays-Bas*, 1967, **86**, 1249.	199
67T239	R. Mechoulam and A. Hirshfeld; *Tetrahedron*, 1967, **23**, 239.	649, 650
67T387	B. Stanovnik and M. Tisler; *Tetrahedron*, 1967, **23**, 387.	628, 650
67T479	D. S. Letham, J. S. Shannon and I. R. C. McDonald; *Tetrahedron*, 1967, **23**, 479.	601
67T1297	K. L. Nagpal and M. M. Dhar; *Tetrahedron*, 1967, **23**, 1297.	648
67T2513	W. Adam, A. Grimison and G. Rodríguez; *Tetrahedron*, 1967, **23**, 2513.	172, 183, 190, 347
67T2739	B. Stanovnik and M. Tisler; *Tetrahedron*, 1967, **23**, 2739.	628, 650
67T2855	G. Yagil; *Tetrahedron*, 1967, **23**, 2855.	225
67TL1317	K. Koshimizu, T. Kusaki, T. Mitsui and S. Matsubara; *Tetrahedron Lett.*, 1967, 1317.	601
67TL2029	M. E. C. Biffin, D. J. Brown and Q. N. Porter; *Tetrahedron Lett.*, 1967, 2029.	272
67TL2109	W. Freiberg, C. F. Kröger and R. Radeglia; *Tetrahedron Lett.*, 1967, 2109.	741
67TL3295	S. Hauptmann, H. Wilde and K. Moser; *Tetrahedron Lett.*, 1967, 3295.	134, 706
67TL4261	H. Kato, S. Sato and M. Ohta; *Tetrahedron Lett.*, 1967, 4261.	47, 743
67TL4379	W. D. Crow and C. Wentrup; *Tetrahedron Lett.*, 1967, 4379.	687, 726
67TL5315	P. Beak, J. L. Miesel and W. R. Messer; *Tetrahedron Lett.*, 1967, 5315.	220
67USP3326889	H. A. Brown; *U.S. Pat.* 3 326 889 (1967) (*Chem. Abstr.*, 1967, **67**, 64 407).	778, 790
67USP3338915	M. Brown; *U.S. Pat.* 3 338 915 (1967) (*Chem. Abstr.*, 1968, **68**, 87 299).	837
67USP3341549	*U.S. Pat.* 3 341 549 (1967) (*Chem. Abstr.*, 1968, **68**, 105 195).	661
67ZC184	R. Miethchen and C. F. Kröger; *Z. Chem.*, 1967, **7**, 184.	752, 756
67ZOR968	G. S. Akimova, V. N. Chistokletov and A. A. Petrov; *Zh. Org. Khim.*, 1967, **3**, 968 (*Chem. Abstr.*, 1967, **67**, 100 071).	710
67ZOR1540	R. A. Khmel'nitskii, A. P. Krasnoshchek, A. A. Polyakova and I. I. Grandberg; *Zh. Org. Khim.*, 1967, **3**, 1540.	202, 204
68ACS1655	L. A. Carlsson and J. Sandström; *Acta Chem. Scand.*, 1968, **22**, 1655.	177
68AG329	R. Huisgen; *Angew. Chem.*, 1968, **80**, 329.	702
68AG(E)321	R. Huisgen; *Angew. Chem., Int. Ed. Engl.*, 1968, **7**, 321.	702
68AHC(9)107	L. M. Weinstock and P. I. Pollak; *Adv. Heterocycl. Chem.*, 1968, **9**, 107.	44, 58, 61, 75, 86, 92, 95, 97, 105, 127
68AHC(9)165	J. Sandstrom; *Adv. Heterocycl. Chem.*, 1968, **9**, 165.	59, 92, 93, 94, 95, 97, 105
68AP911	H. Gehlen and K. H. Uteg; *Arch. Pharm. (Weinheim, Ger.)*, 1968, **301**, 911.	754
68B3721	F. E. Hruska, C. L. Bell, T. A. Victor and S. S. Danyluk; *Biochemistry*, 1968, **7**, 3721.	511
68BBR(33)436	C. E. Bugg, U. T. Thewalt and R. E. Marsh; *Biochem. Biophys. Res. Commun.*, 1968, **33**, 436.	509
68BRP1112128	*Br. Pat.* 1 112 128 (1968) (*Chem. Abstr.*, 1968, **69**, 77 258).	120
68BSF707	J. Elguero, E. Gonzalez and R. Jacquier; *Bull. Soc. Chim. Fr.*, 1968, 707.	198, 223, 224, 231
68BSF3866	J. Elguero, R. Jacquier and D. Tizané; *Bull. Soc. Chim. Fr.*, 1968, 3866.	188, 250
68BSF4222	M. Dormoy, J. Godin and A. LeBerre; *Bull. Soc. Chim. Fr.*, 1968, **44**, 4222.	341
68BSF4403	J. Elguero, A. Fruchier and R. Gil; *Bull. Soc. Chim. Fr.*, 1968, 4403.	189
68BSF5009	J. Elguero, E. Gonzalez and R. Jacquier; *Bull. Soc. Chim. Fr.*, 1968, 5009.	195, 223
68BSF5019	J. Elguero, G. Guiraud, R. Jacquier and G. Tarrago; *Bull. Soc. Chim. Fr.*, 1968, 5019.	231, 232
68CB536	R. Huisgen, H. Gotthardt and R. Grashey; *Chem. Ber.*, 1968, **101**, 536.	717
68CB611	A. Giner-Sorolla; *Chem. Ber.*, 1968, **101**, 611.	548, 549, 593
68CB743	E. Ettenhuber and K. Rühlman; *Chem. Ber.*, 1968, **101**, 743.	779, 831
68CB829	R. Huisgen, R. Grashey and H. Gotthardt; *Chem. Ber.*, 1968, **101**, 829.	260
68CB1979	J. Körösi and P. Berencsi; *Chem. Ber.*, 1968, **101**, 1979.	852
68CB2106	W. Ried and J. Valentin; *Chem. Ber.*, 1968, **101**, 2106.	893

68CB2747	T. Sasaki, K. Minamoto and S. Fukuda; *Chem. Ber.*, 1968, **101**, 2747.	898
68CB3265	H L. Finar; *J. Chem. Soc. (B)*, 1968, 725.	
		172, 173, 174, 175, 198, 199
68CB3734	M. Regitz, W. Anschütz and A. Liedhegener; *Chem. Ber.*, 1968, **101**, 3734.	712
68CC200	Z. Neiman; *Chem. Commun.*, 1968, 200.	540
68CC727	P. R. Briggs, W. L. Parker and T. W. Shannon; *Chem. Commun.*, 1968, 727.	741
68CC977	T. S. Cantrell and W. S. Haller; *Chem. Commun.*, 1968, 977.	694
68CC1002	Z. Neiman and F. Bergmann; *Chem. Commun.*, 1968, 1002.	512
68CC1026	D. W. Crow and C. Wentrup; *Chem. Commun.*, 1968, 1026.	687, 726
68CC1337	B. M. Lynch; *Chem. Commun.*, 1968, 1337.	672
68CHE322	R. G. Glushkov and A. R. Todd; *Chem. Heterocycl. Compd. (Engl. Transl.)*, 1968, **4**, 322 (*Chem. Abstr.*, 1968, **69**, 106 668).	665
68CJC2855	R. R. Fraser and K. E. Haque; *Can. J. Chem.*, 1968, **46**, 2855.	801
68CPB2011	Y. Mizuno, S. Tazawa and K. Kageura; *Chem. Pharm. Bull.*, 1968, **16**, 2011.	619, 621
68CPB2172	A. Yamazaki, I. Kumashiro, T. Takenishi and M. Ikehara; *Chem. Pharm. Bull.*, 1968, **16**, 2172.	584, 588
68CR(C)(267)1461	B. Pullman, H. Berthod, E. D. Bergmann, F. Bergmann, Z. Neiman and H. Weiler-Feilchenfeld; *C.R. Hebd. Seances Acad. Sci., Ser. C*, 1968, **267**, 1461.	521
68DOK(181)1397	A. N. Nesmeyanov, M. I. Rybinskaya, N. S. Kochetkova, V. N. Babin and G. B. Shul'pin; *Dokl. Akad. Nauk SSSR*, 1968, **181**, 1397 (*Chem. Abstr.*, 1968, **69**, 113 021).	691, 731
68FRP1525268	W. L. Perilstein and H. A. Beatty; *Fr. Pat.* 1 525 268 (1968) (*Chem. Abstr.*, 1969, **71**, 30 588).	790
68G681	R. Scarpati, M. L. Grazioni and R. A. Nicolaus; *Gazz. Chim. Ital.*, 1968, **98**, 681.	714
68G949	D. Pocar, S. Maiorana and P. D. Croce; *Gazz. Chim. Ital.*, 1968, **98**, 949.	704, 715
68IJC287	Y. R. Rao; *Indian J. Chem.*, 1968, **6**, 287.	773
68IJC616	P. C. Jain and N. Anand; *Indian J. Chem.*, 1968, **6**, 616.	615, 619
68JA988	P. Scheiner; *J. Am. Chem. Soc.*, 1968, **90**, 988.	711, 724
68JA1884	G. A. Olah and A. M. White; *J. Am. Chem. Soc.*, 1968, **90**, 1884.	33
68JA1923	E. M. Burgess, R. Carithers and L. McCullagh; *J. Am. Chem. Soc.*, 1968, **90**, 1923.	692, 694
68JA2131	R. F. Bleiholder and H. Shechter; *J. Am. Chem. Soc.*, 1968, **90**, 2131.	711
68JA2661	R. J. Roussea, R. K. Robins and L. B. Townsend; *J. Am. Chem. Soc.*, 1968, **90**, 2661.	587
68JA2970	F. P. Boer, J. Flynn, E. T. Kaiser, O. R. Zaborsky, D. A. Tomalia, A. E. Young and Y. C. Tong, *J. Am. Chem. Soc.*, 1968, **90**, 2970.	9
68JA2979	H. Linschitz and J. S. Connolly; *J. Am. Chem. Soc.*, 1968, **90**, 2979.	545
68JA3543	F. J. Weigert and J. D. Roberts; *J. Am. Chem. Soc.*, 1968, **90**, 3543.	672, 680, 741
68JA4232	R. J. Pugmire and D. M. Grant; *J. Am. Chem. Soc.*, 1968, **90**, 4232.	14
68JA6588	L. D. Hansen, B. D. West, E. J. Baca and C. L. Blank; *J. Am. Chem. Soc.*, 1968, **90**, 6588.	690
68JCS(C)344	A. Albert and K. Tratt; *J. Chem. Soc. (C)*, 1968, 344.	696
68JCS(C)824	E. J. Browne and J. B. Polya; *J. Chem. Soc. (C)*, 824.	756, 783
68JCS(C)1268	D. E. Burton, A. J. Lambie, D. W. J. Lane, G. T. Newbold and A. Percival; *J. Chem. Soc. (C)*, 1968, 1268.	722
68JCS(C)1329	H. El Khadem, H. A. R. Mansour and M. H. Meshreki; *J. Chem. Soc. (C)*, 1968, 1329.	685, 711
68JCS(C)1516	G. Shaw, B. M. Smallwood and D. V. Wilson; *J. Chem. Soc. (C)*, 1968, 1516.	595
68JCS(C)1731	D. M. G. Martin and C. B. Reese; *J. Chem. Soc. (C)*, 1968, 1731.	566
68JCS(C)1937	E. B. Mullock and H. Suschitzky; *J. Chem. Soc. (C)*, 1968, 1937.	259
68JCS(C)2510	G. Dallas, J. W. Lown and J. C. N. Ma; *J. Chem. Soc. (C)*, 1968, 2510.	663
68JHC13	N. R. Patel, W. M. Rich and R. N. Castle; *J. Heterocycl. Chem.*, 1968, **5**, 13.	629, 652
68JHC29	W. W. Paudler and R. J. Brumbaugh; *J. Heterocycl. Chem.*, 1968, **5**, 29.	863
68JHC35	J. G. Lombardino; *J. Heterocycl. Chem.*, 1968, **5**, 35.	628
68JHC199	M. D. Coburn and T. E. Jackson; *J. Heterocycl. Chem.*, 1968, **5**, 199.	747
68JHC351	A. Kovacic, B. Stanovnik and M. Tisler; *J. Heterocycl. Chem.*, 1968, **5**, 351.	628
68JHC407	C. Wijnberger and C. L. Habraken; *J. Heterocycl. Chem.*, 1968, **5**, 407.	242
68JHC485	K. T. Potts and S. W. Schneller; *J. Heterocycl. Chem.*, 1968, **5**, 485.	856, 862, 877, 896, 897, 898
68JHC513	A. Pollak, B. Stanovnik and M. Tišler; *J. Heterocycl. Chem.*, 1968, **5**, 513.	889
68JHC565	J. E. Baldwin and S. Y. Hong; *J. Heterocycl. Chem.*, 1968, **5**, 565.	832
68JHC679	J. Altman and D. Ben-Ishai; *J. Heterocycl. Chem.*, 1968, **5**, 679.	577
68JHC691	W. W. Paudler and L. S. Helmick; *J. Heterocycl. Chem.*, 1968, **5**, 691.	872
68JHC699	G. García-Muñoz, J. Iglesias, M. Lora-Tamayo and R. Mandroñero; *J. Heterocycl. Chem.*, 1968, **5**, 699.	723
68JHC863	E. P. Lira; *J. Heterocycl. Chem.*, 1968, **5**, 803.	531
68JMC52	A. Giner-Sorolla, C. Nanos, J. H. Burchenal, M. Dollinger and A. Bendich; *J. Med. Chem.*, 1968, **11**, 52.	556, 564
68JMC981	V. J. Bauer, H. P. Dalalian, W. J. Fanshawe, S. R. Safir, E. C. Tocus and C. R. Boshart; *J. Med. Chem.*, 1968, **11**, 981.	291, 292
68JMC1164	W. B. Wright, Jr. and H. J. Brabander; *J. Med. Chem.*, 1968, **11**, 1164.	128
68JOC530	C. Temple, Jr., B. H. Smith, Jr. and J. A. Montgomery; *J. Org. Chem.*, 1968, **33**, 530.	577
68JOC642	Y. Yamada, I. Kumashiro and T. Takenishi; *J. Org. Chem.*, 1968, **33**, 642.	570

68JOC1087	W. W. Paudler and L. S. Helmick; *J. Org. Chem.*, 1968, **33**, 1087.	611, 872
68JOC1097	H. W. Heine, A. B. Smith, III and J. D. Bower; *J. Org. Chem.*, 1968, **33**, 1097.	154
68JOC1145	P. A. S. Smith and J. G. Wirth; *J. Org. Chem.*, 1968, **33**, 1145.	727
68JOC1350	L. D. Spicer, M. W. Bullock, M. Garber, W. Groth, J. J. Hand, D. W. Long, J. L. Sawyer and R. S. Wayne; *J. Org. Chem.*, 1968, **33**, 1350.	139
68JOC1638	W. W. Paudler and H. G. Shin; *J. Org. Chem.*, 1968, **33**, 1638.	612
68JOC2062	V. Boekelheide and N. A. Fedoruk; *J. Org. Chem.*, 1968, **33**, 2062.	339
68JOC2291	R. Huisgen; *J. Org. Chem.*, 1968, **33**, 2291.	143
68JOC2543	F. M. Hershenson, L. Bauer and K. F. King; *J. Org. Chem.*, 1968, **33**, 2543.	645
68JOC2606	J. A. Settepani and J. B. Stokes; *J. Org. Chem.*, 1968, **33**, 2606.	281
68JOC2828	R. J. Rousseau and L. B. Townsend; *J. Org. Chem.*, 1968, **33**, 2828.	590
68JOC2954	J. H. Hall; *J. Org. Chem.*, 1968, **33**, 2954.	722
68JOC2956	L. T. Creagh and P. Truitt; *J. Org. Chem.*, 1968, **33**, 2956.	741
68JOC3766	K. T. Potts, U. P. Singh and J. Bhattacharyya; *J. Org. Chem.*, 1968, **33**, 3766.	307, 311, 338
68JOC3941	W. J. Houlihan and W. J. Theuer; *J. Org. Chem.*, 1968, **33**, 3941.	139
68JOU692	A. V. El'tsov and N. M. Omar; *J. Org. Chem. USSR (Engl. Transl.)*, 1968, **4**, 692.	254
68JPR(309)168	H. Gehlen and J. Stein; *J. Prakt. Chem.*, 1968, **309**, 168.	754, 773
68JSP(26)139	R. E. Rondeau, H. M. Rosenberg and D. J. Dunbar; *J. Mol. Spectrosc.*, 1968, **26**, 139.	680
68KGS372	M. S. Skorobogatova, N. P. Zolotareva and Y. A. Levin; *Khim. Geterotsikl. Soedin.*, 1968, 372.	773
68KGS953	R. M. Bystrova and Y. M. Yutilov; *Khim. Geterotsikl. Soedin.*, 1968, 953 (*Chem. Abstr.*, 1969, **70**, 96 718).	617
68KGS954	Y. M. Yutilov and R. M. Bystrova; *Khim. Geterotsikl. Soedin.*, 1968, 954 (*Chem. Abstr.*, 1969, **70**, 96 717).	616, 618
68LA(711)46	G. Wittig and H. L. Dorsch; *Liebigs Ann. Chem.*, 1968, **711**, 46.	697, 706
68LA(711)65	G. Wittig and J. Meske-Schüller; *Liebigs Ann. Chem.*, 1968, **711**, 65.	697, 706
68LA(713)149	W. Ried and R. Giesse; *Liebigs Ann. Chem.*, 1968, **713**, 149.	272
68LA(716)11	E. Müller and H. Meier; *Liebigs Ann. Chem.*, 1968, **716**, 11.	691
68M2157	T. Kappe, E. Lender and E. Ziegler; *Monatsh. Chem.*, 1968, **99**, 2157.	253, 272
B-68MI40900	A. Albert; 'Heterocyclic Chemistry', Athlone Press, University of London, 1968.	521
B-68MI40901	'Synthetic Procedures in Nucleic Acid Chemistry', ed. W. W. Zorbach and R. P. Tipson; Wiley-Interscience, New York, 1968, vol. 1.	501, 504, 536, 562, 592, 593, 594, 595, 597
68MI40902	R. Shapiro; *Prog. Nucleic Acid. Res. Mol. Biol.*, 1968, **8**, 73.	501
68MI40903	G. B. Brown; *Prog. Nucleic Acid. Res. Mol. Biol.*, 1968, **8**, 209.	501, 590
68MI40905	H. Weiler-Feilchenfeld and E. D. Bergmann; *Isr. J. Chem.*, 1968, **6**, 823.	522
68MI40906	R. H. Hall and B. I. S. Srivastava; *Life Sciences*, 1968, **7**, 7.	601
68MI40907	A. Pullman, E. Kochanski, M. Gilbert and A. Denis; *Theor. Chim. Acta*, 1968, **10**, 231.	505
68MI40908	F. Jordan and B. Pullman; *Theor. Chim. Acta*, 1968, **9**, 242.	505
68MI41100	H. Lund; *Discuss. Faraday Soc.*, 1968, **45**, 193.	689
B-68MI41101	D. J. Brown; in 'Mechanisms of Molecular Migrations', ed. B. S. Thyagarajan; Wiley–Interscience, New York, 1968, vol. 1, p. 209.	695
68MI41200	J. T. Daycock, G. P. Jones, J. R. N. Evans and J. M. Thomas; *Nature (London)*, 1968, **218**, 672.	739
68MI41201	C. Calzolari and L. Favretto; *Analyst*, 1968, **93**, 494.	785
68MI41300	L. Guibé and E. A. C. Lucken; *Mol. Phys.*, 1968, **14**, 73.	800
68MI41301	D. T. Nash, L. Gross, W. Haw and K. Agre; *J. Clin. Pharmacol.*, 1968, **8**, 377.	835
B-68MI41400	L. A. Paquette; 'Principles of Modern Heterocyclic Chemistry', Benjamin, New York, 1968.	839, 840
68MI41500	V. Galasso, G. De Alti and A. Bigotto; *Theor. Chim. Acta*, 1968, **9**, 222.	853, 856, 860
68N299	C. Franzke, H. Griehl, S. Grunert and E. Hollstein; *Naturwissenschaften*, 1968, **55**, 299.	600
68OS(48)8	H. Dorn and A. Zubek; *Org. Synth.*, 1968, **48**, 8.	288
68OS(48)113	J. I. G. Cadogan and R. K. Mackie; *Org. Synth.*, 1968, **48**, 113.	288
68PNA(60)450	J. J. Lichter and W. Gordy; *Proc. Natl. Acad. Sci. USA*, 1968, **60**, 450.	520
68RTC11	H. C. Beyerman, L. Maat, A. Sinnema and A. Van Veen; *Recl. Trav. Chim. Pays-Bas*, 1968, **87**, 11.	635
68T239	J. Kobe, B. Stanovnik and M. Tisler; *Tetrahedron*, 1968, **24**, 239.	628
68T349	P. Scheiner; *Tetrahedron*, 1968, **24**, 349.	710
68T1875	S. O. Lawesson, E. Schroll, J. H. Bowie and R. G. Cooks; *Tetrahedron*, 1968, **24**, 1875.	359
68T2623	A. Pollak, B. Stanovnik and M. Tisler; *Tetrahedron*, 1968, **24**, 2623.	628
68T2757	P. Scheiner; *Tetrahedron*, 1968, **24**, 2757.	704, 705, 710, 724
68T2839	A. H. Beckett, R. G. W. Spickett and S. H. B. Wright; *Tetrahedron*, 1968, **24**, 2839.	858
68T3787	S.-Y. Hong and J. E. Baldwin; *Tetrahedron*, 1968, **24**, 3787.	808, 832
68T4217	R. Buyle and H. G. Viehe; *Tetrahedron*, 1968, **24**, 4217.	132
68TL231	N. E. Alexandrou and E. D. Micromastoras; *Tetrahedron Lett.*, 1968, 231.	686
68TL325	J. Sauer and K. K. Mayer; *Tetrahedron Lett.*, 1968, 325.	160
68TL789	B. Loev and M. M. Goodman; *Tetrahedron Lett.*, 1968, 789.	653
68TL3277	H. H. Wassermann, K. Stiller and M. B. Floyd; *Tetrahedron Lett.*, 1968, 3277.	376

68TL3727	J. D. Vaughan and M. O'Donnell; *Tetrahedron Lett.*, 1968, 3727.	671, 742
68TL3805	C. E. Olsen and C. Pedersen; *Tetrahedron Lett.*, 1968, 3805.	712
68TL3873	T. Goto, S. Inoue and S. Sugiura; *Tetrahedron Lett.*, 1968, 3873.	624, 642, 643
68TL4811	M. Israel, L. C. Jones and E. J. Modest; *Tetrahedron Lett.*, 1968, 4811.	637
68TL5707	A. Morimoto, K. Noda, T. Watanabe and H. Takasugi; *Tetrahedron Lett.*, 1968, 5707.	302
68USP3386968	W. R. Carpenter; *U.S. Pat.* 3 386 968 (1968) (*Chem. Abstr.*, 1968, **69**, 28 106).	838
68USP3394143	M. Wolf; *U.S. Pat.* 3 394 143 (1968) (*Chem. Abstr.*, 1968, **69**, 77 265).	746
68USP3396170	N. R. Fetter and B. K. Bartocha; *U.S. Pat.* 3 396 170 (1968) (*Chem. Abstr.*, 1968, **69**, 87 170).	837
68ZC458	S. Hoffmann and E. Mühle; *Z. Chem.*, 1968, **8**, 458.	233
68ZOR689	A. P. Krasnoshchek, R. A. Khmel'nitskii, A. A. Polyakova and I. I. Grandberg; *Zh. Org. Khim.*, 1968, **4**, 689.	204
69AC(R)434	M. Ruccia and N. Vivona; *Ann. Chim. (Rome)*, 1969, **59**, 434.	777
69ACS1091	M. Begtrup and C. Pedersen; *Acta Chem. Scand.*, 1969, **23**, 1091.	712
69ACS2733	M. Begtrup and P. A. Kristensen; *Acta Chem. Scand.*, 1969, **23**, 2733.	700, 701, 727
69AG115	A. Albert; *Angew. Chem.*, 1969, **81**, 115.	721
69AG(E)132	A. Albert; *Angew. Chem., Int. Ed. Engl.*, 1969, **8**, 132.	721
69AG(E)513	E. Brunn and R. Huisgen; *Angew. Chem., Int. Ed. Engl.*, 1969, **8**, 513.	781
69AG(E)520	F. A. Neugebauer; *Angew. Chem., Int. Ed. Engl.*, 1969, **8**, 520.	800
69AHC(10)1	A. J. Boulton and P. B. Ghosh; *Adv. Heterocycl. Chem.*, 1969, **10**, 1.	75, 76, 85, 86, 108
69AJC2251	E. J. Browne; *Aust. J. Chem.*, 1969, **22**, 2251.	741, 754
69AX(B)135	P. Goldstein, J. Ladell and G. Abowitz; *Acta Crystallogr., Part B*, 1969, **25**, 135.	738
69AX(B)595	A. D. Mighell, C. W. Reimann and A. Santoro; *Acta Crystallogr., Part B*, 1969, **25**, 595.	227
69AX(B)1330	E. Sletten, J. Sletten and L. H. Jensen; *Acta. Crystallogr., Part B*, 1969, **25**, 1330.	509
69AX(B)1338	G. M. Brown; *Acta Crystallogr., Part B*, 1969, **25**, 1338.	509
69AX(B)1608	J. Sletten and L. H. Jensen; *Acta Crystallogr., Part B*, 1969, **25**, 1608.	508, 509
69AX(B)2355	D. L. Smith and E. K. Barrett; *Acta Crystallogr., Part B*, 1969, **25**, 2355.	209
69B2412	R. Wolfenden, J. Kaufman and J. B. Macon; *Biochemistry*, 1969, **8**, 2412.	904
69BCJ750	H. Kawashima and I. Kumashiro; *Bull. Chem. Soc. Jpn.*, 1969, **42**, 750.	539, 554, 555, 596
69BCJ3099	H. Mizuno, T. Fujiwara and K. Tomita; *Bull. Chem. Soc. Jpn.*, 1969, **42**, 3099.	508, 509
69BEP722025	J. F. Willems and F. C. Heugebaert; *Belg. Pat.* 722 025 (1969) (*Chem. Abstr.*, 1970, **72**, 66 949).	837
69BRP1157256	H. Becker and K. Wehner; *Br. Pat.* 1 157 256 (1969) (*Chem. Abstr.*, 1969, **71**, 81 375).	753
69BSB147	G. L'abbé, P. Ykman and G. Smets; *Bull. Soc. Chim. Belg.*, 1969, **78**, 147.	704
69BSF1097	M. Roche and L. Pujol; *Bull. Soc. Chim. Fr.*, 1969, 1097.	171, 172, 173, 176, 179, 187, 795, 671, 672
69BSF1687	J. Elguero, R. Jacquier and D. Tizané; *Bull. Soc. Chim. Fr.*, 1969, 1687.	187
69BSF2061	G. Coispeau, J. Elguero and R. Jacquier; *Bull. Soc. Chim. Fr.*, 1969, 2061.	272
69BSF2064	J. Elguero, A. Fruchier and R. Jacquier; *Bull. Soc. Chim. Fr.*, 1969, 2064.	230, 233
69BSF2492	J. Daunis, R. Jacquier and P. Viallefont; *Bull. Soc. Chim. Fr.*, 1969, 2492.	862
69BSF3292	J. L. Aubagnac, J. Elguero and R. Jacquier; *Bull. Soc. Chim. Fr.*, 1969, 3292.	189
69BSF3300	J. L. Aubagnac, J. Elguero and R. Jacquier; *Bull. Soc. Chim. Fr.*, 1969, 3300.	284
69BSF3302	J. L. Aubagnac, J. Elguero and R. Jacquier; *Bull. Soc. Chim. Fr.*, 1969, 3302.	254
69BSF3306	J. L. Aubagnac, J. Elguero and R. Jacquier; *Bull. Soc. Chim. Fr.*, 1969, 3306.	186, 207
69BSF3316	J. L. Aubagnac, J. Elguero and R. Jacquier; *Bull. Soc. Chim. Fr.*, 1969, 3316.	188, 189, 199, 202
69BSF3670	J. Daunis, R. Jacquier and P. Vaillefont; *Bull. Soc. Chim. Fr.*, 1969, 3670.	859, 862, 900
69BSF4075	J. Elguero, E. Gonzalez, J. L. Imbach and R. Jacquier; *Bull. Soc. Chim. Fr.*, 1969, 4075.	425
69BSF4466	C. Dittli, J. Elguero and R. Jacquier; *Bull. Soc. Chim. Fr.*, 1969, 4466.	243
69BSF4469	C. Dittli, J. Elguero and R. Jacquier; *Bull. Soc. Chim. Fr.*, 1969, 4469.	243
69BSF4474	C. Dittli, J. Elguero and R. Jacquier; *Bull. Soc. Chim. Fr.*, 1969, 4474.	243
69C109	J. P. Dubois and H. Labhart; *Chimia*, 1969, **23**, 109.	160, 221
69C303	A. Hetzheim and H. Pusch; *Chimia*, 1969, **23**, 303 (*Chem. Abstr.*, 1969, **71**, 101 827).	630, 655
69CB269	R. R. Schmidt and D. Schwille; *Chem. Ber.*, 1969, **102**, 269.	157
69CB417	M. Regitz and H. Scherer; *Chem. Ber.*, 1969, **102**, 417.	695
69CB755	C. F. Kröger, R. Miethchen, H. Frank, M. Siemer and S. Pilz; *Chem. Ber.*, 1969, **102**, 755.	752
69CB2216	M. Regitz and W. Anschütz; *Chem. Ber.*, 1969, **102**, 2216.	712
69CB3000	R. Walentowski and H.-W. Wanzlick; *Chem. Ber.*, 1969, **102**, 3000.	540, 664
69CB3159	T. Eicher, S. Hünig, H. Hansen and P. Nikolaus; *Chem. Ber.*, 1969, **102**, 3159.	766, 884
69CB3176	T. Eicher, S. Hünig and P. Nikolaus; *Chem. Ber.*, 1969, **102**, 3176.	781
69CC381	J. Kiburis and J. H. Lister; *Chem. Commun.*, 1969, 381.	563, 598
69CC458	J. A. Montgomery and H. J. Thomas; *Chem. Commun.* 1969, 458.	652
69CC500	A. Albert; *Chem. Commun.*, 1969, 500.	721
69CC818	G. C. Barrett, A. R. Khokhar and J. R. Chapman; *Chem. Commun.*, 1969, 818.	113, 138
69CC905	D. Elad and I. Rosenthal; *Chem. Commun.*, 1969, 905.	545
69CC1110	F. L. Bach, J. Karliner and G. E. Van Lear; *Chem. Commun.*, 1969, 1110.	801

69CCR(4)463	A. I. Popov; *Coord. Chem. Rev.*, 1969, **4**, 463.	792, 797
69CHE805	B. V. Ioffe, V. S. Stopskii and N. B. Burmanova; *Chem. Heterocycl. Compd. (Engl. Transl.)*, 1969, **5**, 805.	285
69CJC813	R. Raap and J. Howard; *Can. J. Chem.*, 1969, **47**, 813.	819
69CJC1129	B. M. Lynch, A. J. Robertson and J. G. K. Webb; *Can. J. Chem.*, 1969, **47**, 1139.	206, 854, 859
69CJC3677	R. Raap; *Can. J. Chem.*, 1969, **47**, 3677.	799, 802
69CJC3997	P. Yates, R. G. F. Giles and D. G. Farnum; *Can. J. Chem.*, 1969, **47**, 3997.	828
69CPB1128	A. Yamazaki, I. Kumashiro and T. Takenishi; *Chem. Pharm. Bull.*, 1969, **17**, 1128.	584
69CPB1924	A. Takamizawa and K. Hirai; *Chem. Pharm. Bull.*, 1969, **17**, 1924.	114
69CPB2209	I. Adachi; *Chem. Pharm. Bull.*, 1969, **17**, 2209.	75
69CR(C)(268)1798	Kha Vang Thang and J.-L. Olivier; *C.R. Hebd. Seances Acad. Sci., Ser. C*, 1969, **268**, 1798.	308
69CR(C)(269)570	P. Bouchet, J. Elguero, R. Jacquier and F. Forissier; *C.R. Hebd. Seances Acad. Sci., Ser. C*. 1969, **269**, 570	240
69CR(D)(268)2958	B. Pullman and H. Berthod; *C.R. Hebd. Seances Acad. Sci., Ser. D*, 1969, **268**, 2958.	859
69CRV345	G. L'abbé; *Chem. Rev.* 1969, **69**, 345.	670, 705
69CRV359	G. L'abbé; *Chem. Rev.*, 1969, **69**, 359.	831
69FRP1578366	E. J. Cragoe, Jr. and J. H. Jones; *Fr. Pat.* 1 578 366 (1969) (*Chem. Abstr.*, 1970, **72**, 111 514).	626, 646
69G1131	R. Stradi and D. Pocar; *Gazz. Chim. Ital.*, 1969, **99**, 1131.	714
69IZV144	I. N. Vorazhtsov and K. A. Barkhash; *Izv. Akad. Nauk SSSR, Ser. Khim.*, 1969, 144.	720
69IZV655	E. A. Arutyunyan, V. I. Gunar, E. P. Gracheva and S. I. Zav'yalov; *Izv. Akad. Nauk SSSR, Ser. Khim.*, 1969, 655.	557
69JA711	J. L. Brewbaker and H. Hart; *J. Am. Chem. Soc.*, 1969, **91**, 711.	275
69JA796	M. J. S. Dewar and T. Morita; *J. Am. Chem. Soc.*, 1969, **91**, 796.	208, 742
69JA3676	J. E. McMurry; *J. Am. Chem. Soc.*, 1969, **91**, 3676.	726
69JA5625	P. O. P. Ts'O, N. S. Kondo, R. K. Robins and A. D. Broom; *J. Am. Chem. Soc.*, 1969, **91**, 5625.	526
69JBC(244)2498	G. Stohrer and G. B. Brown; *J. Biol. Chem.*, 1969, **244**, 2498.	554
69JBC(244)4072	A. Myles and G. B. Brown; *J. Biol. Chem.*, 1969, **244**, 4072.	539
69JCP(51)1862	H. H. Chen and L. B. Clark; *J. Chem. Phys.*, 1969, **51**, 1862.	517
69JCS(B)270	D. J. Brown and P. B. Ghosh; *J. Chem. Soc. (B)*, 1969, 270.	33
69JCS(B)307	E. Borello, A. Zecchina and E. Guglielminotti; *J. Chem. Soc. (B)*, 1969, 307.	685
69JCS(B)680	R. N. Butler; *J. Chem. Soc. (B)*, 1969, 680.	805
69JCS(B)1240	M. Charton; *J. Chem. Soc. (B)*, 1969, 1240.	364, 804
69JCS(C)152	A. Albert; *J. Chem. Soc. (C)*, 1969, 152.	695, 696
69JCS(C)809	J. S. Morley; *J. Chem. Soc. (C)*, 1969, 809.	834
69JCS(C)1056	E. J. Browne and J. B. Polya; *J. Chem. Soc. (C)*, 1969, 1056.	739, 740
69JCS(C)1065	A. C. Day and R. N. Inwood; *J. Chem. Soc. (C)*, 1969, 1065.	188
69JCS(C)1084	A. Albert, W. Pfleiderer and D. Thacker; *J. Chem. Soc. (C)*, 1969, 1084.	875
69JCS(C)1117	G. C. Barrett and A. R. Khokhar; *J. Chem. Soc. (C)*, 1969, 1117.	138
69JCS(C)1474	C. W. Rees and R. C. Storr; *J. Chem. Soc. (C)*, 1969, 1474.	727
69JCS(C)1478	C. W. Rees and R. C. Storr; *J. Chem. Soc. (C)*, 1969, 1478.	727
69JCS(C)1593	J. Maguire, D. Paton and F. L. Rose; *J. Chem. Soc. (C)*, 1969, 1593.	877, 896, 904
69JCS(C)2251	R. L. Jones and C. W. Rees; *J. Chem. Soc. (C)*, 1969, 2251.	745
69JCS(C)2379	A. Albert; *J. Chem. Soc. (C)*, 1969, 2379.	690, 721
69JCS(C)2497	I. L. Finar and B. J. Millard; *J. Chem. Soc. (C)*, 1969, 2497.	204
69JCS(C)2624	S. N. Ege; *J. Chem. Soc. (C)*, 1969, 2624.	218, 255
69JHC93	S. F. Martin and R. N. Castle; *J. Heterocycl. Chem.*, 1969, **6**, 93.	629, 871, 890
69JHC199	J. A. White and R. C. Anderson; *J. Heterocycl. Chem.*, 1969, **6**, 199.	70
69JHC503	J. H. Boyer and R. Selvarajan; *J. Heterocycl. Chem.*, 1969, **6**, 503.	867
69JHC545	C. Wijnberger and C. L. Habraken; *J. Heterocycl. Chem.*, 1969, **6**, 545.	201
69JHC559	M. Japelj, B. Stanovnik and M. Tišler; *J. Heterocycl. Chem.*, 1969, **6**, 559.	858, 868, 869
69JHC947	J. B. Wright; *J. Heterocycl. Chem.*, 1969, **6**, 947.	201, 317
69JHC965	P. M. Hergenrother; *J. Heterocycl. Chem.*, 1969, **6**, 965.	763
69JMC43	J. Nematollahi and J. R. Nulu; *J. Med. Chem.*, 1969, **12**, 43.	652
69JMC717	A. Giner-Sorolla; *J. Med. Chem.*, 1969, **12**, 717.	554, 556
69JMC945	V. J. Bauer, W. J. Fanshawe, G. E. Wiegand and S. R. Safir; *J. Med. Chem.*, 1969, **12**, 945.	291
69JMC1124	R. P. Williams, V. J. Bauer and S. R. Safir; *J. Med. Chem.*, 1969, **12**, 1124.	291
69JOC36	L. Field and R. B. Barber; *J. Org. Chem.*, 1969, **34**, 36.	137
69JOC747	C. N. Eaton and G. H. Denny, Jr.; *J. Org. Chem.*, 1969, **34**, 747.	597
69JOC973	E. Dyer, R. E. Farris, Jr., C. E. Minnier and M. Tokizawa; *J. Org. Chem.*, 1969, **34**, 973.	521, 559
69JOC978	U. Wölcke and G. B. Brown; *J. Org. Chem.*, 1969, **34**, 978.	539, 554, 555, 595
69JOC981	U. Wölcke, W. Pfleiderer, T. Delia and G. B. Brown; *J. Org. Chem.*, 1969, **34**, 981.	555
69JOC1025	A. D. Broom and R. K. Robins; *J. Org. Chem.*, 1969, **34**, 1025.	522, 595
69JOC1141	C. Arnold, Jr. and D. N. Thatcher; *J. Org. Chem.*, 1969, **34**, 1141.	709
69JOC1170	E. C. Taylor, Y. Maki and A. McKillop; *J. Org. Chem.*, 1969, **34**, 1170.	529
69JOC1396	J. A. Montgomery and K. Hewson; *J. Org. Chem.*, 1969, **34**, 1396.	552
69JOC1746	P. K. Freeman and R. C. Johnson; *J. Org. Chem.*, 1969, **34**, 1746.	706

69JOC2129	C. K. Bradsher, R. D. Brandau, J. E. Boliek and T. L. Hough; *J. Org. Chem.*, 1969, **34**, 2129.	632
69JOC2153	I. Scheinfeld, J. C. Parham, S. Murphy and G. B. Brown; *J. Org. Chem.*, 1969, **34**, 2153.	596
69JOC2157	A. Giner-Sorolla, C. Gryte, A. Bendich and G. B. Brown; *J. Org. Chem.*, 1969, **34**, 2157.	554, 555, 595, 596
69JOC2720	P. Aeberli and W. J. Houlihan; *J. Org. Chem.*, 1969, **34**, 2720.	315
69JOC3221	K. T. Potts and C. Lovelette; *J. Org. Chem.*, 1969, **34**, 3221.	856, 871, 889
69JOC4118	R. R. Fraser, G. Gurudata and K. E. Haque; *J. Org. Chem.*, 1969, **34**, 4118.	720, 809
69JOU1480	N. M. Omar and A. V. El'tsov; *J. Org. Chem. USSR (Engl. Transl.)*, 1969, **5**, 1480.	254, 256
69JOU1790	F. I. Luknitskii, D. O. Taube and B. A. Vovsi; *J. Org. Chem. USSR (Engl. Transl.)*, 1969, **5**, 1790 (*Chem. Abstr.*, 1970, **72**, 21 671).	660
69JPR71	H. Gehlen and R. N. Neumann; *J. Prakt. Chem.*, 1969, **311**, 71.	884
69JPR213	H. Gehlen and R. Neumann; *J. Prakt. Chem.*, 1969, **311**, 213	772
69JPR646	H. G. O. Becker, J. Witthauer, N. Sauder and G. West; *J. Prakt. Chem.*, 1969, **311**, 646.	784
69JPR1058	W. Hampel; *J. Prakt. Chem.*, 1969, **311**, 1058.	272
69KGS159	G. I. Chipen and R. P. Bokaldere; *Khim. Geterotsikl. Soedin.*, 1969, 159.	746
69KGS1114	V. Ya. Grinshtein and V. Ya. Strazdin; *Khim. Geterotsikl. Soedin.*, 1969, 1114.	751
69LA(722)29	H. Neunhoeffer, M. Neunhoeffer and W. Litzius; *Liebigs Ann. Chem.*, 1969, **722**, 29.	768
69LA(722)38	H. Neunhoeffer; *Liebigs Ann. Chem.*, 1969, **722**, 38.	768
69LA(725)29	E. Zbiral and J. Ströh; *Liebigs Ann. Chem.*, 1969, **725**, 29.	826
69LA(729)73	I. Betula; *Liebigs Ann. Chem.*, 1969, **729**, 73.	540, 541
69LA(729)124	W. Ried and B. Peters; *Liebigs Ann. Chem.*, 1969, **729**, 124.	779
69M671	M. Japelj, B. Stanovnik and M. Tišler; *Monatsh. Chem.*, 1969, **100**, 671.	889
69M1250	H. Junek and I. Wrtilek; *Monatsh. Chem.*, 1969, **100**, 1250.	333
69M1438	E. Zbiral and J. Stroh; *Monatsh. Chem.*, 1969, **100**, 1438.	717
69MI40300	K. M. Pazdro; *Rocz. Chem.*, 1969, **43**, 1089.	126
69MI40301	G. C. Barrett and A. R. Khokhar; *J. Chromatogr.*, 1969, **39**, 47.	138
69MI40400	M. N. Aboul-Enein, Y. M. Abou-Zeid and S. M. El-Difrawy; *Pharm. Acta Helv.*, 1969, **44**, 570.	295
B-69MI40401	M. J. S. Dewar; 'The Molecular Orbital Theory of Organic Chemistry', McGraw-Hill, New York, 1969, pp. 378–382.	173
B-69MI40900	F. Yoneda, K. Ogiwara, M. Kanahori and S. Nikigashi; 'Proceedings of the 4th International Symposium on the Chemistry and Biology of Pteridines', 1969, 145.	592
69MI40901	J. M. Andre, M. C. Andre, D. Hahn, D. Klint and E. Clementi; *Hung. Phys. Acta*, 1969, **27**, 493.	505, 506
69MI40902	M. Sundaralingam; *Biopolymers*, 1969, **7**, 82.	523
69MI40903	P. Tollin and A. R. I. Munns; *Nature (London)*, 1969, **222**, 1170.	509
69MI40904	J. P. Ferris, J. E. Kuder and A. W. Catalano; *Science*, 1969, **166**, 765.	570
69MI40906	B. Mely and A. Pullman; *Theor. Chim. Acta*, 1969, **13**, 278.	505, 506
69MI40907	C. Giessner-Prettre and A. Pullman; *Theor. Chim. Acta*, 1969, **13**, 265.	505
69MI41000	G. Hornyak, K. Lempert and K. Zauer; *Acta Chim. (Budapest)*, 1969, **61**, 181 (*Chem. Abstr.*, 1969, **71**, 70 570).	628, 653, 655
69MI41001	S. Ostroversnik, B. Stanovnik and M. Tisler; *Croat. Chem. Acta*, 1969, **41**, 135 (*Chem. Abstr.*, 1971, **72**, 12 684).	650
69MI41002	V. I. Schvedov, L. B. Altukhova, L. A. Chernyshkova and A. N. Grinev; *Khim.-Farm. Zh.*, 1969, **3**, 15 (*Chem. Abstr.*, 1970, **72**, 66 899).	643
69MI41003	A. Nawojski; *Rocz. Chem.*, 1969, **43**, 573 (*Chem. Abstr.*, 1969, **70**, 115 142).	617
69MI41100	G. Berthier, L. Praud and J. Serre; *Quantum Aspects Heterocycl. Compd. Chem. Biochem., Proc. Int. Symp. 1969*, 1970, **2**, 40.	672
B-69MI41101	A. G. Cook; in 'Enamines, Structure, and Reactions', ed. A. G. Cook; Dekker, New York, 1969, p. 211.	714
69MI41200	R. B. Dickinson and N. W. Jacobsen; *Anal. Chem.*, 1969, **41**, 1324.	742
69MI41300	J. H. Tyrer, M. J. Eadie and W. H. Hooper; *J. Chromatogr.*, 1969, **39**, 312.	803
69OMR(1)249	J. Elguero, C. Marzin and D. Tizané; *Org. Magn. Reson.*, 1969, **1**, 249.	189, 190
69OMR(1)431	T. J. Batterham and C. Bigum; *Org. Magn. Reson.*, 1969, **1**, 431.	290
69OMS(2)433	D. M. Forkey and W. R. Carpenter; *Org. Mass Spectrom.*, 1969, **2**, 433.	801
69OMS(2)729	S. W. Tam; *Org. Mass Spectrom.*, 1969, **2**, 729.	204
69OMS(2)739	B. K. Simons, R. K. M. R. Kallury and J. H. Bowie; *Org. Mass Spectrom.*, 1969, **2**, 739.	204
69RRC763	C. Chiriac, L. Stoicescu-Crivetz and I. Zugravescu; *Rev. Roum. Chim.*, 1969, **14**, 763.	260
69RRC1263	C. Chiriac, L. Stoicescu-Crivetz and I. Zugravescu; *Rev. Roum. Chim.*, 1969, **14**, 1263.	231, 300
69RTC1263	K. B. De Roos and C. A. Salemink; *Recl. Trav. Chim. Pays-Bas*, 1969, **88**, 1263.	635, 636
69S17	D. Seebach; *Synthesis*, 1969, 17.	717
69T3053	P. K. Kadaba; *Tetrahedron*, 1969, **25**, 3053.	717
69T3287	P. Beak and W. Messer; *Tetrahedron*, 1969, **25**, 3287.	160, 377

69T3287	P. Beak and W. Messer; *Tetrahedron*, 1969, **25**, 3287.	221
69T3453	R. Buyle and H. G. Viehe; *Tetrahedron*, 1969, **25**, 3453.	132
69T4265	J. Nyitrai and K. Lempert; *Tetrahedron*, 1969, **25**, 4265.	422
69T5421	G. L'abbé, P. Ykman and G. Smets; *Tetrahedron*, 1969, **25**, 5421.	704, 716
69TL63	G. L'abbé and H. J. Bestmann; *Tetrahedron Lett.*, 1969, 63.	716
69TL271	J. Reisch and A. Fitzek; *Tetrahedron Lett.*, 1969, 271.	218
69TL289	M. Sprinzl, J. Farkaš and F. Šorm; *Tetrahedron Lett.*, 1969, 289.	272
69TL495	M. L. Roumestant, P. Viallefont, J. Elguero and R. Jacquier; *Tetrahedron Lett.*, 1969, 495.	740, 804
69TL495	M. L. Roumestant, P. Viallefont, J. Elguero and R. Jacquier; *Tetrahedron Lett.*, 1969, 495.	679
69TL785	U. Wölcke, N. J. M. Birdsall and G. B. Brown; *Tetrahedron Lett.*, 1969, 785.	555
69TL1537	B. M. Lynch; *Tetrahedron Lett.*, 1969, 1537.	799
69TL1549	L S. Davies and G. Jones; *Tetrahedron Lett.*, 1969, 1549.	887
69TL2589	R. Huisgen, K. von Fraunberg and H. J. Sturm; *Tetrahedron Lett.*, 1969, 2589.	709
69TL2595	R. Huisgen and K. V. Fraunberg; *Tetrahedron Lett.*, 1969, 2595.	46
69TL3377	A. C. Rochat and R. A. Olofson; *Tetrahedron Lett.*, 1969, 3377.	806
69TL4863	P. Scheiner; *Tetrahedron Lett.*, 1969, 4863.	810
69TL5081	D. H. Zimmerman and R. A. Olofson; *Tetrahedron Lett.*, 1969, 5081.	830
69USP3461123	J. H. Jones and E. J. Cragoe, Jr.; *U.S. Pat.* 3 461 123 (1969) (*Chem. Abstr.*, 1969, **71**, 101 883).	646
69YZ1657	Y. Kishi, S. Sugiura, S. Inoue and T. Goto; *Yakugaku Zasshi*, 1969, **89**, 1657 (*Chem. Abstr.*, 1970, **72**, 90 406).	643
69ZC27	G. Jarnecke, V. Buchholtz and K. Teubner; *Z. Chem.*, 1969, **9**, 27.	664
69ZC241	M. Wahren; *Z. Chem.*, 1969, **9**, 241.	695
69ZC300	R. Miethchen, H.-U. Seipt and C. F. Kröger; *Z. Chem.*, 1969, **9**, 300.	752
69ZC325	H. G. O. Becker, V. Eisenschmidt, M. Buhig, K. Jahnisch, N. Klein, W. Kowalski, R. Misselwitz, R. Müller, P. Reimann, C. Roth, W.-D. Sauter, W. Schössler and B. Thorein; *Z. Chem.*, 1969, **9**, 325.	752, 753
69ZOR153	B. K. Strelets and L. S. Efros; *Zh. Org. Khim.*, 1969, **5**, 153 (*Chem. Abstr.*, 1969, **70**, 86 889).	21
70ACS2137	P. Groth; *Acta Chem. Scand.*, 1970, **24**, 2137.	9
70ACS3248	F. K. Larsen, M. S. Lehmann, I. Søtofte and S. E. Rasmussen; *Acta Chem. Scand.*, 1970, **24**, 3248.	173, 179
70AG81	S. Petersen and H. Heitzer; *Angew. Chem.*, 1970, **82**, 81.	701
70AG(E)54	G. Zumach and E. Kuhle; *Angew. Chem., Int. Ed. Engl.*, 1970, **9**, 54.	130
70AG(E)67	S. Petersen and H. Heitzer; *Angew. Chem., Int. Ed. Engl.*, 1970, **9**, 67.	701
70AHC(11)1	S. T. Reid; *Adv. Heterocycl. Chem.*, 1970, **11**, 1.	45, 46, 78, 79, 81, 108
70AHC(12)103	M. R. Grimmett; *Adv. Heterocycl. Chem.*, 1970, **12**, 103.	51, 54, 59, 60, 72, 78, 86, 92, 104, 106, 126, 346, 351, 353, 355, 358, 362, 364, 384, 385, 388, 391, 400, 403, 404, 412, 419, 420, 425, 430, 443, 462, 464, 467, 468, 469, 473, 476, 480, 481, 483, 484, 485, 487, 491, 493
70AHC(12)213	H. Lund; *Adv. Heterocycl. Chem.*, 1970, **12**, 213.	73
70AJC631	A. J. -Blackman; *Aust. J. Chem.*, 1970, **23**, 631.	744, 758
70AX(B)380	J. L. Galigné and J. Falgueirettes; *Acta Crystallogr., Part B*, 1970, **26**, 380.	178
70AX(B)521	C. W. Reimann, A. Santoro and A. D. Mighell; *Acta Crystallogr., Part B*, 1970, **26**, 521.	227
70AX(B)1880	J. Berthou, J. Elguero and C. Rérat; *Acta Crystallogr., Part B*, 1970, **26**, 1880.	178, 179
70B3701	D. F. Babcock and R. O. Morris; *Biochemistry*, 1970, **9**, 3701.	601
70BCJ849	I. Hori, K. Saito and H. Midorikawa; *Bull. Chem. Soc. Jpn.*, 1970, **43**, 849.	272, 316, 320, 331
70BCJ2535	Y. Iwakura and K. Kurita; *Bull. Chem. Soc. Jpn.*, 1970, **43**, 2535.	135
70BCJ3344	M. Kamiya; *Bull. Chem. Soc. Jpn.*, 1970, **43**, 3344.	175, 198, 346, 347, 348, 349
70BCJ3344	M. Kamiya; *Bull. Chem. Soc. Jpn.*, 1970, **43**, 3344.	742
70BCJ3909	Y. Ohtsuka; *Bull. Chem. Soc. Jpn.*, 1970, **43**, 3909.	591
70BSF231	J. Elguero, R. Gélin, S. Gélin and G. Tarrago; *Bull. Soc. Chim. Fr.*, 1970, 231.	231
70BSF273	M. Roche and L. Pujol; *Bull. Soc. Chim. Fr.*, 1970, 273.	172, 177, 672
70BSF1121	J. Elguero, R. Jacquier and D. Tizané; *Bull. Soc. Chim. Fr.*, 1970, 1121.	250
70BSF1129	J. Elguero, R. Jacquier and D. Tizané; *Bull. Soc. Chim. Fr.*, 1970, 1129.	188, 199, 207
70BSF1345	J. Elguero, R. Jacquier and G. Tarrago; *Bull. Soc. Chim. Fr.*, 1970, 1345.	183, 799
70BSF1346	J. Elguero, R. Jacquier and S. Mondon; *Bull. Soc. Chim. Fr.*, 1970, 1346.	183, 231, 799
70BSF1571	J. L. Barascut, J. Elguero and R. Jacquier; *Bull. Soc. Chim. Fr.*, 1970, 1571.	216
70BSF1576	J. Elguero, R. Jacquier and S. Mondon; *Bull. Soc. Chim. Fr.*, 1970, 1576.	216
70BSF1936	J. Elguero, R. Jacquier and D. Tizané; *Bull. Soc. Chim. Fr.*, 1970, 1936.	254
70BSF1974	J. Elguero, R. Jacquier, V. Pellegrin and V. Tabacik; *Bull. Soc. Chim. Fr.*, 1970, 1974.	253, 267, 289
70BSF2717	G. Coispeau and J. Elguero; *Bull. Soc. Chim. Fr.*, 1970, 2717.	169, 277, 278
70BSF3147	J. P. Chapelle, J. Elguero, R. Jacquier and G. Tarrago; *Bull. Soc. Chim. Fr.*, 1970, 3147.	188, 216, 285
70BSF3466.	J. Elguero and C. Marzin; *Bull. Soc. Chim. Fr.*, 1970, 3466	188
70BSF4119	J. Elguero, R. Jacquier and C. Marzin; *Bull. Soc. Chim. Fr.*, 1970, 4119.	276
70C134	A. R. Katritzky; *Chimia*, 1970, **24**, 134.	36, 37, 38

Ref	Authors	Pages
70CB272	A. Hetzheim and H. Beyer; *Chem. Ber.*, 1970, **103**, 272.	662
70CB788	H. König and R. Geiger; *Chem. Ber.* 1970, **103**, 788.	722, 728
70CB1908	F. P. Woerner and H. Reimlinger; *Chem. Ber.*, 1970, **103**, 1908.	709, 711
70CB1918	H. Reimlinger, J. J. M. Vandewalle, G. S. D. King, W. R. F. Lingier and R. Merenyi; *Chem. Ber.*, 1970, **103**, 1918.	152
70CB1949	H. Reimlinger, A. Noels and J. Jadot; *Chem. Ber.*, 1970, **103**, 1949.	234, 240
70CB1954	H. Reimlinger, A. Noels, J. Jadot and A. van Overstraeten; *Chem. Ber.*, 1970, **103**, 1954.	233, 269
70CB3252	H. Reimlinger, M. A. Peiren and R. A. Mereyi; *Chem. Ber.*, 1970, **103**, 3252.	272, 331
70CB3284	H. Reimlinger and R. Merényi; *Chem. Ber.*, 1970, **103**, 3284.	828
70CB3289	G. Häfelinger; *Chem. Ber.*, 1970, **103**, 3289.	173, 179
70CB3533	A. Hetzheim, H. Pusch and H. Beyer; *Chem. Ber.*, 1970, **103**, 3533.	653
70CB3811	A. J. Hubert and H. Reimlinger; *Chem. Ber.*, 1970, **103**, 3811.	692
70CB3817	H. Reimlinger, W. R. F. Lingier and R. Merényi; *Chem. Ber.*, 1970, **103**, 3817.	853, 866
70CC56	G. B. Ansell, D. M. Forkey and D. W. Moore; *Chem. Commun.*, 1970, 56.	367
70CC286	A. D. Baker, D. Betteridge, N. R. Kemp and R. E. Kirby; *Chem. Commun.*, 1970, 286.	204
70CC289	W. R. Dolbier; *Chem. Commun.*, 1970, 289.	252
70CC858	A. Albert; *Chem. Commun.*, 1970, 858.	852
70CC1068	F. Yoneda, K. Ogiwara, M. Kanahori and S. Nishigaki; *Chem. Commun.*, 1970, 1068.	581
70CC1089	M. Ohashi, K. Tsujimoto and T. Yonezawa; *Chem. Commun.*, 1970, 1089.	694
70CC1524	F. L. Rose, G. J. Stacey, P. J. Taylor and T. W. Thompson; *Chem. Commun.*, 1970, 1524.	862
70CHE194	A. F. Pozharskii, L. M. Sitkina, A. M. Simonov and T. N. Chegolya; *Chem. Heterocycl. Compd. (Engl. Transl.)*, 1970, **6**, 194.	384, 448
70CHE628	Yu. P. Andreichikov and A. M. Simonov; *Chem. Heterocycl. Compd. (Engl. Transl.)*, 1970, **6**, 628.	430
70CHE744	B. L. Moldaver and M. E. Aronzon; *Chem. Heterocycl. Compd. (Engl. Transl.)*, 1970, 744.	199
70CHE1183	A. F. Pozharskii, É. A. Zvezdina, Yu. P. Andreichikov, A. M. Simonov, V. A. Anismova, and S. F. Popova; *Chem. Heterocycl. Compd. (Engl. Transl.)*, 1970, **6**, 1183.	440
70CHE1321	B. V. Ioffe and N. L. Zelenina; *Chem. Heterocycl. Compd. (Engl. Transl.)*, 170, 1321.	281
70CHE1339	B. A. Tertov and P. P. Onishchenko; *Chem. Heterocycl. Compd. (Engl. Transl.)*, 1970, **6**, 1339.	245
70CHE1515	P. I. Abramenko and V. G. Zhiryakov; *Chem. Heterocycl. Compd. (Engl. Transl.)*, 1970, **6**, 1515.	116
70CHE1568	B. E. Zaitsev, N. A. Andronova, R. B. Zhurin and L. B. Preobrazhenskaya; *Chem. Heterocycl. Compd. (Engl. Transl.)*, 1970, 1568.	200
70CJC1371	R. G. Micetich and C. G. Chin; *Can. J. Chem.*, 1970, **48**, 1371.	69
70CJC1716	S. Jerumanis and A. Martel; *Can. J. Chem.*, 1970, **48**, 1716.	545
70CJC2006	R. G. Micetich; *Can. J. Chem.*, 1970, **48**, 2006.	66
70CJC2139	R. Raap; *Can. J. Chem.*, 1970, **49**, 2139.	807
70CPB964	M. Ogata, H. Matsumoto, S. Takahashi and H. Kano; *Chem. Pharm. Bull.*, 1970, **18**, 964.	380
70CPB1526	A. Nakamura and S. Kamiya; *Chem. Pharm. Bull.*, 1970, **18**, 1526.	341
70CPB1685	M. Yanai, T. Kinoshita, S. Takeda, M. Mori, H. Sadaki and H. Watanabe; *Chem. Pharm. Bull.*, 1970, **18**, 1685.	629, 652, 871, 890
70CR(C)(270)1688	E. J. Vincent, R. Phan-Tan-Luu, J. Roggero and J. Metzger; *C. R. Hebd. Seances Acad. Sci, Ser. C*, 1970, **270**, 1688.	20
70CR(C)(271)958	F. Texier and R. Carrié; *C. R. Hebd. Seances Acad. Sci., Ser. C*, 1970, **271**, 958.	726
70FRP1596860	Agfa-Gevaert A.-G., *Fr. Pat.* 1 596 860 (1970) (*Chem. Abstr.*, 1971, **75**, 5907).	837
70G110	L. Pentimalli and V. Passalacqua; *Gazz. Chim. Ital.*, 1970, **100**, 110 (*Chem. Abstr.*, 1970, **72**, 121 492).	627
70G1106	L. Pentimalli and G. Milani; *Gazz. Chim. Ital.*, 1970, **100**, 1106 (*Chem. Abstr.*, 1971, **74**, 125 609).	627
70GEP1957711	E. J. J. Grabowski, E. W. Tristram and R. J. Tull; *Ger. Pat.* 1 957 711 (1970) (*Chem. Abstr.*, 1970, **73**, 66 621).	646
70GEP2009134	I. Pintér, A. Messmer, F. Ordogh and L. Pallos; *Ger. Pat.* 2 009 134 (1970) (*Chem. Abstr.*, 1970, **73**, 98 955).	732
70HCA648	J. M. J. Tronchet and F. Perret; *Helv. Chim. Acta*, 1970, **53**, 648.	303
70IC1597	R. M. Churchill, K. Gold and C. E. Maw; *Inorg. Chem.*, 1970, **9**, 1597.	227
70IJC391	M. V. Konher; *Indian J. Chem.*, 1970, **8**, 391.	773
70IJC894	D. S. Iyengar, S. Husain and G. S. Sidhu; *Indian J. Chem.*, 1970, **8**, 894.	853, 865
70IZV420	E. P. Girtcheui, Z. S. Volkova, V. I. Gunar, E. Arutyuuyan and S. I. Zavyalov; *Izv. Akad. Nauk SSSR, Ser. Khim.*, 1970, 420.	560
70JA1965	Y. Kushi and Q. Fernando; *J. Am. Chem. Soc.*, 1970, **92**, 1965.	796, 827
70JA2929	N. Bodor, M. J. S. Dewar and A. J. Harget; *J. Am. Chem. Soc.*, 1970, **92**, 2929.	742
70JA4079	A. J. Jones, D. M. Grant, M. W. Winkley and R. K. Robins; *J. Am. Chem. Soc.*, 1970, **92**, 4079.	514
70JA4088	F. E. Hruska, A. A. Grey and I. C. P. Smith; *J. Am. Chem. Soc.*, 1970, **92**, 4088.	523

70JA4751	B. Evans and R. Wolfenden; *J. Am. Chem. Soc.*, 1970, **92**, 4751.	545, 601
70JA5118	S. Trofimenko; *J. Am. Chem. Soc.*, 1970, **92**, 5118.	230
70JA6218	A. B. Evnin, D. R. Arnold, L. A. Karnischky and E. Strom; *J. Am. Chem. Soc.*, 1970, **92**, 6218.	188
70JA7441	C. E. Bugg and U. Thewalt; *J. Am. Chem. Soc.*, 1970, **92**, 7441.	509
70JA7617	K. Nakanishi, N. Furutachi, M. Funamizu, D. Grunberger and I. B. Weinstein; *J. Am. Chem. Soc.*, 1970, **92**, 7617.	605
70JAP7039709	K. Shirakawa, O. Aki, Y. Nakagawa and K. Otsu; *Jpn. Pat.* 70 39 709 (1970) (*Chem. Abstr.*, 1971, **75**, 5954).	560
70JCS(B)596	H. Weiler-Feilchenfeld and Z. Neiman; *J. Chem. Soc. (B)*, 1970, 596.	522
70JCS(B)1692	R. E. Burton and I. L. Finar; *J. Chem. Soc. (B)*, 1970, 1692.	171, 172, 173, 174, 175, 176, 177, 742
70JCS(C)63	J. K. Landquist; *J. Chem. Soc. (C)*, 1970, 63.	134
70JCS(C)85	M. A. Khan and J. B. Polya; *J. Chem. Soc. (C)*, 1970, 85.	747
70JCS(C)230	A. Albert; *J. Chem. Soc. (C)*, 1970, 230.	695
70JCS(C)807	G. V. Boyd and M. D. Harms; *J. Chem. Soc. (C)*, 1970, 807.	774
70JCS(C)1313	J. H. Lister, D. S. Manners and G. M. Timmis; *J. Chem. Soc. (C)*, 1970, 1313.	272
70JCS(C)1397	G. V. Boyd and S. R. Dando; *J. Chem. Soc. (C)*, 1970, 1397.	115
70JCS(C)1515	E. J. Browne, E. E. Nunn and J. B. Polya; *J. Chem. Soc. (C)*, 1970, 1515.	776
70JCS(C)1540	A. Albert and K. Ohta; *J. Chem. Soc. (C)*, 1970, 1540.	626, 644
70JCS(C)1842	J. Feeney, G. A. Newman and P. J. S. Pauwels; *J. Chem. Soc. (C)*, 1970, 1842.	194
70JCS(C)2206	G. Shaw and B. M. Smallwood; *J. Chem. Soc. (C)*, 1970, 2206.	531, 564
70JCS(C)2298	S. M. Mackenzie and M. F. G. Stevens; *J. Chem. Soc. (C)*, 1970, 2298.	336
70JCS(C)2403	A. J. Blackman and J. B. Polya; *J. Chem. Soc. (C)*, 2403.	754, 758
70JHC25	M. A. Kira, M. N. Aboul-Enein and M. I. Korkor; *J. Heterocycl. Chem.*, 1970, **7**, 25.	126, 277
70JHC71	J. H. Boyer and P. J. A. Frints; *J. Heterocycl. Chem.*, 1970, **7**, 71.	809
70JHC75	A. Giner-Sorolla; *J. Heterocycl. Chem.*, 1970, **7**, 75.	556
70JHC117	G. R. Revankar and L. B. Townsend; *J. Heterocycl. Chem.*, 1970, **7**, 117.	289
70JHC211	V. Sunjic, T. Fajdiga and M. Japelj; *J. Heterocycl. Chem.*, 1970, **7**, 211.	749
70JHC215	J. A. Montgomery, H. J. Thomas and S. J. Clayton; *J. Heterocycl. Chem.*, 1970, **7**, 215.	895
70JHC227	N. Blažević, F. Kajfež and V. Šunjić; *J. Heterocycl. Chem.*, 1970, **7**, 227.	364
70JHC247	M. A. Khan and B. M. Lynch; *J. Heterocycl. Chem.*, 1970, **7**, 247.	272, 331
70JHC355	H. J. Beim and A. R. Day; *J. Heterocycl. Chem.*, 1970, **7**, 355.	635
70JHC361	G. L'abbé and A. Hassner; *J. Heterocycl. Chem.*, 1970, **7**, 361.	685, 713
70JHC439	R. E. Harmon, V. L. Rizzo and S. K. Gupta; *J. Heterocycl. Chem.*, 1970, **7**, 439.	358, 445
70JHC639	V. Pirc, B. Stanovnik, M. Tišler, J. Marsel and W. W. Paudler; *J. Heterocycl. Chem.*, 1970, **7**, 639.	858, 860
70JHC703	S. Portnoy; *J. Heterocycl. Chem.*, 1970, **7**, 703.	883, 884
70JHC715	C. L. Schmidt and L. B. Townsend; *J. Heterocycl. Chem.*, 1970, **7**, 715.	627
70JHC791	I. Perillo and S. Lamden; *J. Heterocycl. Chem.*, 1970, **7**, 791.	463
70JHC821	T. Kametani, K. Sota and M. Shio; *J. Heterocycl. Chem.*, 1970, **7**, 821.	750
70JHC863	R. A. Long, J. F. Gerster and L. B. Townsend; *J. Heterocycl. Chem.*, 1970, **7**, 863.	323, 324
70JHC895	W. H. De Camp and J. M. Stewart; *J. Heterocycl. Chem.*, 1970, **7**, 895.	179, 181
70JHC923	K. R. Henery-Logan and E. A. Keiter; *J. Heterocycl. Chem.*, 1970, **7**, 923.	281
70JHC961	J. E. Oliver and J. B. Stokes; *J. Heterocycl. Chem.*, 1970, **7**, 961.	443, 700
70JHC991	L. D. Hansen, E. J. Baca and P. Scheiner; *J. Heterocycl. Chem.*, 1970, **7**, 991.	742
70JHC1001	F. H. Case; *J. Heterocycl. Chem.*, 1970, **7**, 1001.	766
70JHC1019	K. T. Potts and C. R. Surapaneni; *J. Heterocycl. Chem.*, 1970, **7**, 1019.	861, 862, 883, 884
70JHC1029	M. Israel, S. K. Tinter, D. H. Trites and E. J. Modest; *J. Heterocycl. Chem.*, 1970, **7**, 1029.	592
70JHC1113	R. Pater; *J. Heterocycl. Chem.*, 1970, **7**, 1113.	731
70JHC1149	M. D. Coburn, E. D. Loughran and L. C. Smith; *J. Heterocycl. Chem.*, 1970, **7**, 1149.	750
70JHC1223	L. E. Brady; *J. Heterocycl. Chem.* 1970, **7**, 1223.	801
70JHC1237	M. A. Khan and B. M. Lynch; *J. Heterocycl. Chem.*, 1970, **7**, 1237.	230, 231
70JHC1269	G. Alonso, M. T. García-López, G. García-Muñoz, R. Madroñero and M. Rico; *J. Heterocycl. Chem.*, 1970, **7**, 1269.	700
70JHC1329	G. R. Revankar and L. B. Towshend; *J. Heterocycl. Chem.*, 1970, **7**, 1329.	289
70JHC1391	M. D. Coburn and P. N. Neuman; *J. Heterocycl. Chem.*, 1970, **7**, 1391.	368
70JHC1435	G. Alonso, G. García-Muñoz and R. Madroñero; *J. Heterocycl. Chem.*, 1970, **7**, 1435.	289
70JMC77	M. P. Mertes and A. J. Lin; *J. Med. Chem.*, 1970, **13**, 77.	645
70JMC164	H. R. Snyder, Jr. and L. E. Benjamin; *J. Med. Chem.*, 1970, **13**, 164.	666
70JMC777	L. Huff and R. A. Henry; *J. Med. Chem.*, 1970, **13**, 777.	819
70JOC1131	H. U. Blank, I. Wempen and J. J. Fox; *J. Org. Chem.*, 1970, **35**, 1131.	887
70JOC1138	B. Stanovnik, M. Tisler, M. Ceglar and V. Bah; *J. Org. Chem.*, 1970, **35**, 1138.	650
70JOC1146	E. C. Wu and J. D. Vaughan; *J. Org. Chem.*, 1970, **35**, 1146.	290
70JOC1965	K. T. Potts and R. Armbruster; *J. Org. Chem.*, 1970, **35**, 1965.	131
70JOC2074	L. Huff, D. M. Forkey, D. W. Moore and R. A. Henry; *J. Org. Chem.*, 1970, **35**, 2074.	793

70JOC2215	P. A. S. Smith, G. J. W. Breen, M. K. Hajek and D. V. C. Awang; *J. Org. Chem.*, 1970, **35**, 2215.	98, 697, 707
70JOC2246	R. A. Olofson and R. V. Kendall; *J. Org. Chem.*, 1970, **35**, 2246.	232, 746, 747
70JOC2478	A. Pollak, B. Stanovnik and M. Tišler; *J. Org. Chem.*, 1970, **35**, 2478.	628, 860, 889
70JOC2635	J. T. Witkowski and R. K. Robins; *J. Org. Chem.*, 1970, **35**, 2635.	758
70JOC3097	A. B. Evnin, A. Lam and J. Blyskal; *J. Org. Chem.*, 1970, **35**, 3097.	426
70JOC3444	R. E. Harmon, F. Stanley, Jr., S. K. Gupta and J. Johnson; *J. Org. Chem.*, 1970, **35**, 3444.	692, 712
70JOC3448	K. T. Potts and E. G. Brugel; *J. Org. Chem.*, 1970, **35**, 3448.	857, 873, 892, 893
70JOC3451	K. T. Potts and S. Husain; *J. Org. Chem.*, 1970, **35**, 3451.	701
70JOC3768	S. P. McManus, J. T. Carroll and C. U. Pittman, Jr.; *J. Org. Chem.*, 1970, **35**, 3768.	141
70JOC3981	T. Kuneida and B. Witkop; *J. Org. Chem.*, 1970, **35**, 3981.	530
70JOM(21)427	S. S. Washburne and W. R. Peterson, Jr.; *J. Organomet. Chem.*, 1970, **21**, 427.	779
70JPR610	H. G. O. Becker, G. Goermar and H. J. Timpe; *J. Prakt. Chem.*, 1970, **312**, 610.	755, 760, 767
70JPR669	H. G. O. Becker, D. Beyer, G. Israel, R. Muller, W. Riediger and H. J. Timpe; *J. Prakt. Chem.*, 1970, **312**, 669.	899
70KGS248	V. Ya. Grinshtein, A. A. Strazdin' and A. K. Grinvalde; *Khim. Geterotsikl. Soedin.*, 1970, 248.	739
70KGS259	L. I. Bagal, M. S. Pevzner, A. N. Frolov and N. I. Sheludyakova; *Khim. Geterotsikl. Soedin.*, 1970, 259.	754
70KGS265	L. I. Bagal, M. S. Pevzner, N. I. Sheludyakova and V. M. Kerusov; *Khim. Geterotsikl. Soedin.*, 1970, 265.	746, 754
70KGS269	L. I. Bagal, M. S. Pevzner and V. Ya. Samarenko; *Khim. Geterotsikl. Soedin.*, 1970, 269.	753
70KGS558	L. I. Bagal and M. S. Pevzner; *Khim. Geterotsikl. Soedin.*, 1970, 558.	742, 746
70KGS851	N. G. Koshel', E. G. Kovalev and I. Y. Postovskii; *Khim. Geterotsikl. Soedin.*, 1970, **6**, 851.	878, 882
70KGS991	O. P. Shvaika, V. V. Artemov and S. N. Baranov; *Khim. Geterotsikl. Soedin.*, 1970, 991.	771
70KGS997	L. I. Bagal, M. S. Pevzner, A. P. Egorov and V. Ya. Samarenko; *Khim. Geterotsikl. Soedin*, 1970, 997.	753, 757, 758
70KGS1146	Y. M. Yutilov and R. M. Bystrova; *Khim. Geterotsikl. Soedin.*, 1970, 1146 (*Chem. Abstr.*, 1971, **74**, 53 655).	618
70KGS1621	A. D. Grabenko, L. N. Kulaeva and P. S. Pelkis; *Khim. Geterotsikl. Soedin.*, 1970, 1621 (*Chem. Abstr*, 1971, **74**, 76 359).	118
70KGS1701	L. I. Bagal, M. S. Pevzner, V. Ya. Samarenko and A. P. Egorov; *Khim. Geterorsikl. Soedin.*, 1970, 1701.	753
70LA(734)56	B. Eistert and E. Endres; *Liebigs Ann. Chem.*, 1970, **734**, 56.	706, 721
70LA(734)70	M. Regitz and G. Himbert; *Liebigs Ann. Chem.*, 1970, **734**, 70.	703
70LA(734)173	F. Boberg and R. Schardt, *Liebigs Ann. Chem.*, 1970, **734**, 173.	281
70LA(735)158	K. Sasse; *Liebigs Ann. Chem.*, 1970, **735**, 158.	765
70LA(742)85	A. Schoeberl and K. H. Magosch; *Liebigs Ann. Chem.*, 1970, **742**, 85.	664
70M724	J. Kobe, B. Stanovnik and M. Tisler; *Monatsh. Chem.*, 1970, **101**, 724.	631
B-70MI40100	J. A. Pople and D. L. Beveridge; 'Approximate Molecular Orbital Theory', McGraw-Hill, New York, 1970, pp. 57–84.	7
70MI40400	M. A. Khan; *Rec. Chem. Prog.*, 1970, **31**, 43.	231
B-70MI40401	G. Berthier, L. Praud and J. Serre; in 'Quantum Aspects of Heterocyclic Compounds in Chemistry and Biochemistry', ed. E. D. Bergmann and B. Pullman; Israel Academy of Science, Jerusalem, 1970, p. 40.	172, 177, 183, 190
B-70MI40402	P. A. S. Smith; in 'Nitrenes', ed. W. Lwowski; Wiley-Interscience, New York, 1970, p. 144.	263
B-70MI40403	'The Chemistry of Synthetic Dyes and Pigments', ed. H. A. Lubs; American Chemical Society, Washington, 1970.	298, 290
B-70MI40500	R. J. Suhadolnik; 'Nucleoside Antibiotics', Wiley-Interscience, New York, 1970, p. 354.	343
B-70MI40900	R. J. Suhadolnik; 'Nucleoside Antiobiotics', Wiley-Interscience, New York, 1970.	504, 601, 602, 603
70MI40901	S. Arnott; *Prog. Biophys. Mol. Biol.*, 1970, **21**, 265.	522
B-70MI40902	H. Weiler-Feilchenfeld and Z. Neiman; in 'Quantum Aspects of Heterocyclic Compounds in Chemistry and Biochemistry', ed. E. D. Bergman and B. Pullman; Academic, New York, 1970, p. 308.	521
70MI40903	B. M. Pyatin, V. G. Granik and R. G. Glushov; *Khim.-Farm. Zh.*, 1970, **4**, 22 (*Chem. Abstr.*, 1971, **74**, 112 016).	566
70MI40904	J. A. Miller; *Cancer Res.*, 1970, **30**, 559.	526
70MI40905	G. Stohrer and G. B. Brown; *Science*, 1970, **167**, 1622.	539
70MI41000	H. O. Hankovszky and K. Hideg; *Acta Chim. (Budapest)*, 1970, **63**, 447 (*Chem. Abstr.*, 1970, **72**, 100 597).	663
70MI41001	L. Del Corona, G. Massaroli and G. Signorelli; *Boll. Chim. Farm.*, 1970, **109**, 665 (*Chem. Abstr.*, 1971, **75**, 20 297).	636
70MI41100	K. Fukui; *Fortschr. Chem. Forsch.*, 1970, **15**, 1.	711
B-70MI41101	P. Scheiner; in 'Selective Organic Transformations', ed. B. S. Thyagarajan; Wiley–Interscience, New York, 1970, vol. 1, p. 327.	670, 704, 724

B-70MI41102	J. P. Anselme; in 'The Chemistry of the Carbon–Nitrogen Double Bond', ed. S. Patai; Wiley–Interscience, New York, 1970, p. 299.	717
70MI41300	Z. Grzonka; *J. Chromatogr.*, 1970, **51**, 310.	803
70OMS(3)1549	J. van Thuijl, K. J. Klebe and J. J. van Houte; *Org. Mass Spectrom.*, 1970, **3**, 1549.	202
70OMS(4)203	M. Ohashi, K. Tsujimoto, A. Yoshino and T. Yonezawa; *Org. Mass Spectrom.*, 1970, **4**, 203.	687, 688
70PAC(24)495	H. Labhart, W. Heinzelmann and J. P. Dubois; *Pure Appl. Chem.*, 1970, **24**, 495.	160
70PNA(65)27	A. J. Jones, M. W. Winkley, D. M. Grant and R. K. Robins; *Proc. Natl. Acad. Sci. USA*, 1970, **65**, 27.	515
70SAP6804307	W. B. Scanlon and W. L. Garbrecht; *S. Afr. Pat.* 68 04 307 (1970) (*Chem. Abstr.*, 1970, **73**, 56 100).	835
70T1393	G. Desimoni and G. Minoli; *Tetrahedron*, 1970, **26**, 1393.	706
70T2497	A. W. K. Chan, W. D. Crow and I. Gosney; *Tetrahedron*, 1970, **26**, 2497.	23
70T2619	P. Scheiner and J. F. Dinda, Jr.; *Tetrahedron*, 1970, **26**, 2619.	809
70T3069	C. M. Gupta, A. P. Bhaduri and N. M. Khanna; *Tetrahedron*, 1970, **26**, 3069.	768
70T3357	J. Kobe, B. Stanovnik and M. Tišler; *Tetrahedron*, 1970, **26**, 3357.	859, 878, 900
70T3965	C. Wentrup and W. D. Crow; *Tetrahedron*, 1970, **26**, 3965.	868, 876
70T4339	T. Aratani, Y. Nakanisi and H. Nozaki; *Tetrahedron*, 1970, **26**, 4339.	724
70T4915	C. Wentrup and W. D. Crow; *Tetrahedron*, 1970, **26**, 4915.	881
70T4969	C. Wentrup; *Tetrahedron*, 1970, **26**, 4969.	860, 879
70TH40400	M. Roche; Ph.D. Thesis, University of Marseille, 1970.	198
70TL943	E. J. Browne; *Tetrahedron Lett.*, 1970, 943.	93, 784
70TL1013	R. P. Panzica and L. B. Townsend; *Tetrahedron Lett.*, 1970, 1013.	646
70TL1411	J. Reiter, P. Sohar and L. Toldy; *Tetrahedron Lett.*, 1970, 1411.	647
70TL1879	C. Belzecki and J. Trojnar; *Tetrahedron Lett.*, 1970, 1879.	662
70TL2823	M. Regitz and G. Himbert; *Tetrahedron Lett.*, 1970, 2823.	692, 703, 712
70TL3099	J. Elguero, R. Jacquier and C. Marzin; *Tetrahedron Lett.*, 1970, 3099.	256
70TL3781	J. Goerdeler and K. Brüning; *Tetrahedron Lett.*, 1970, 3781.	129
70TL3851	K. Sakai and J. P. Anselme; *Tetrahedron Lett.*, 1970, 3851.	744
70TL4215	M. A. Kira, Z. M. Nofal and K. Z. Gadalla; *Tetrahedron Lett.*, 1970, 4215.	126, 277
70TL4611	M. Bobek, J. Farkaš and F. Šorm; *Tetrahedron Lett.*, 1970, 4611.	272
70TL5083	R. Grashey, M. Baumann and R. Hamprecht; *Tetrahedron Lett.*, 1970, 5083.	127
70TL5225	P. Ykman, G. L'abbé and G. Smets; *Tetrahedron Lett.*, 1970, 5225.	679, 717
70UP40400	J. Elguero, A. Fruchier and R. Gil; unpublished results, 1970.	188, 189
70USP3541093	R. J. Tull and P. I. Pollak; *U.S. Pat.* 3 541 093 (1970) (*Chem. Abstr.*, 1971, **74**, 141 878).	645
70USP3549633	E. J. J. Grabowski, E. W. Tristram and R. J. Tull; *U.S. Pat.* 3 549 633 (1970) (*Chem. Abstr.*, 1971, **74**, 53 847).	646
70YZ423	S. Sugiura, S. Inoue and T. Goto; *Yakugaku Zasshi*, 1970, **90**, 423 (*Chem. Abstr.*, 1970, **73**, 45 459).	642
70YZ441	S. Sugiura, H. Kakoi, S. Inoue and T. Goto; *Yakugaku Zasshi*, 1970, **90**, 441 (*Chem. Abstr.*, 1970, **73**, 45 462).	643
70YZ707	S. Sugiura, S. Inoue and T. Goto; *Yakugaku Zasshi*, 1970, **90**, 707 (*Chem. Abstr.*, 1970, **73**, 98 904).	643
70ZC220	R. Miethchen, A. H. Albrecht and E. Rachow; *Z. Chem.*, 1970, **10**, 220.	699
70ZC224	G. Strickmann and G. Barnikow; *Z. Chem.*, 1970, **10**, 224.	253
70ZC338	H. Lettau; *Z. Chem.*, 1970, **10**, 338.	367
70ZN(B)421	R. Walentowski and H. W. Wanzlich; *Z. Naturforsch., Teil B*, 1970, **25**, 421.	755
70ZOR908	B. V. Ioffe, Yu. P. Artsibasheva and L. M. Levina; *Zh. Org. Khim.*, 1970, **6**, 908.	281
71ACR10	A. Hassner; *Acc. Chem. Res.*, 1971, **4**, 10.	829
71ACR17	S. Trofimenko; *Acc. Chem. Res.* 1971, **4**, 17.	169
71ACS249	M. Begtrup; *Acta Chem. Scand.*, 1971, **25**, 249.	672, 700
71ACS625	C. Christophersen and S. Treppendahl; *Acta Chem. Scand.*, 1971, **25**, 625.	793, 832
71ACS2087	M. Begtrup and K. V. Poulsen; *Acta Chem. Scand.*, 1971, **25**, 2087.	52, 672, 698
71AF1532	T. Wagner-Jauregg; *Arzneim.-Forsch.*, 1969, **19**, 1532.	343
71AG(E)743	T. Kauffmann; *Angew. Chem., Int. Ed. Engl.*, 1971, **10**, 743.	168, 258, 268
71AG(E)810	K.-H. Magosch and R. Feinauer; *Angew Chem., Int. Ed. Engl.*, 1971, **10**, 810.	148
71AHC(13)77	B. Pullman and A. Pullman; *Adv. Heterocycl. Chem.*, 1971, **13**, 77.	501, 506, 521
71AJC393	E. J. Browne; *Aust. J. Chem.*, 1971, **24**, 393.	736, 756
71AJC645	P. Dunn and D. Oldfield; *Aust. J. Chem.*, 1971, **24**, 645.	794
71AJC1229	S. Beveridge and R. L. N. Harris; *Aust. J. Chem.*, 1971, **24**, 1229.	135
71AJC2405	R. K. Buckley, M. Davis and K. S. L. Srivastava; *Aust. J. Chem.*, 1971, **24**, 2405.	95
71AP121	F. Eiden and G. Evers; *Arch. Pharm. (Weinheim, Ger.)*, 1971, **304**, 121 (*Chem. Abstr.*, 1971, **75**, 5833).	316
71AP763	R. Weidlein and H. Krull; *Arch. Pharm. (Weinheim, Ger.)* 1971, **304**, 763.	127
71AX(B)573	E. Thom and A. T. Christensen; *Acta Crystallogr., Part B*, 1971, **27**, 573.	179
71AX(B)986	M. N. Sabesan and K. Venkatesan; *Acta Crystallogr., Part B*, 1971, **27**, 986.	179
71AX(B)1227	W. H. De Camp and J. M. Stewart; *Acta Crystallogr., Part B*, 1971, **27**, 1227.	178
71AX(B)1859	C. A. Kosky, P. Ganis and G. Avitabile; *Acta Crystallogr., Part B*, 1971, **27**, 1859.	227
71BCJ185	T. Sasaki, T. Yoshioka and Y. Suzuki; *Bull. Chem. Soc. Jpn.*, 1971, **44**, 185.	719
71BCJ858	T. Sasaki, K. Kanematsu and M. Uchide; *Bull. Chem. Soc. Jpn.*, 1971, **44**, 858.	340
71BJ(125)841	P. F. Swann and P. M. Magee; *Biochem. J.*, 1971, **125**, 841.	531

71BRP1232758	R. J. Tull and P. I. Pollak; *Br. Pat.* 1 232 758 (1971) (*Chem. Abstr.*, 1971, **75**, 49 134).	646
71BRP1238105	R. J. Tull and P. I. Pollak; *Br. Pat.* 1 238 105 (1971) (*Chem. Abstr.*, 1971, **75**, 88 639).	626, 646
71BRP1248146	R. J. Tull and P. I. Pollak; *Br. Pat.* 1 248 146 (1971) (*Chem. Abstr.*, 1971, **75**, 140 887).	626, 646
71BSF679	A. Lablache-Combier and M. A. Remy; *Bull. Soc. Chim. Fr.*, 1971, 679.	169
71BSF1031	P. Guerret, J.-L. Imbach, R. Jacquier, P. Martin and G. Maury; *Bull. Soc. Chim. Fr.*, 1971, 1031.	649
71BSF1040	R. Jacquier, J.-M. Lacombe and G. Maury; *Bull. Soc. Chim. Fr.*, 1971, 1040.	354
71BSF1925	J. Elguero; *Bull. Soc. Chim. Fr.*, 1971, 1925.	220, 255, 276
71BSF3043	J. Daunis, M. Guerret-Rigail and R. Jacquier; *Bull. Soc. Chim. Fr.*, 1971, 3043.	857, 889, 902
71BSF3296	J. Daunis, Y. Guindo, R. Jacquier and P. Viallefont; *Bull. Soc. Chim. Fr.*, 1971, 3296.	741, 744, 757, 782
71BSF3642	F. Texier and R. Carrié; *Bull. Soc. Chim. Fr.*, 1971, 3642.	703
71BSF4021	J. Roggero and M. Audibert; *Bull. Soc. Chim. Fr.*, 1971, 4021.	130
71CB1562	E. Brunn, E. Funke, H. Gotthardt and R. Huisgen; *Chem. Ber.*, 1971, **104**, 1562.	473, 489
71CB2053	W. Broser and U. Bollert; *Chem. Ber.*, 1971, **104**, 2053.	281
71CB3062	L. Birkofer and M. Franz; *Chem. Ber.*, 1971, **104**, 3062.	184
71CB3146	H. R. Kricheldorf; *Chem. Ber.*, 1971, **104**, 3146.	138
71CB3510	H. Wamhoff and P. Sohár; *Chem. Ber.*, 1971, **104**, 3510.	704, 715, 721
71CB3925	H. Reimlinger, J.-M. Gilles, G. Anthoine, J. J. M. Vandewalle, W. R. F. Lingier, E. de Ruiter, R. Merényi and A. Hubert; *Chem. Ber.*, 1971, **104**, 3925.	853, 865, 866
71CB3940	H. Reimlinger, W. R. F. Lingier and J. J. M. Vandewalle; *Chem. Ber.*, 1971, **104**, 3940.	866
71CB3947	H. Reimlinger, W. R. F. Lingier, J. J. M. Vandewalle and R. Merényi; *Chem. Ber.*, 1971, **104**, 3947.	865, 866
71CB3955	H. Reimlinger, J. J. M. Vandewalle, W. R. F. Lingier and F. Billiau; *Chem. Ber.*, 1971, **104**, 3955.	866
71CB3965	H. Reimlinger, W. R. F. Lingier, J. J. M. Vandewalle and R. Merényi; *Chem. Ber.*, 1971, **104**, 3965.	853, 866, 886
71CC394	J. A. Elvidge, J. R. Jones and C. O'Brien; *Chem. Commun.*, 1971, 394.	526
71CC490	P. Rajagopalan and P. Penev; *Chem. Commun.*, 1971, 490.	794
71CC532	C. W. Rees and A. A. Sale; *Chem. Commun.*, 1971, 532.	723
71CC537	P. J. Fagan and M. J. Nye; *Chem. Commun.*, 1971, 537.	327
71CC873	K. Bolton, R. D. Brown, F. R. Burden and A. Mishra; *Chem. Commun.*, 1971, 873.	738
71CC889	W. F. Cooper, N. C. Kenny, J. W. Edmonds, A. Nagel, F. Wudl and P. Coppens; *Chem. Commun.*, 1971, 889.	9
71CC1022	G. Kan, M. T. Thomas and V. Snieckus; *Chem. Commun.*, 1971, 1022.	161
71CC1519	T. L. Gilchrist, G. E. Gymer and C. W. Rees; *Chem. Commun.*, 1971, 1519.	692
71CHE132	A. M. Simonov and I. I. Popov; *Chem. Heterocycl. Compd. (Engl. Transl.)*, 1971, **7**, 132.	438
71CHE665	I. Y. Postovskii and V. A. Ershov; *Chem. Heterocycl. Compd. (Eng. Transl.)*, 1971, **7**, 665.	852
71CHE668	V. A. Ershov and I. Y. Postovskii; *Chem. Heterocycl. Compd. (Engl. Transl.)*, 1971, **7**, 668.	852
71CHE746	Yu. Ya. Usaevich, I. Kh. Fel'dman and E. I. Boksiner; *Chem. Heterocycl. Compd. (Engl. Transl.)*, 1971, **7**, 749.	424, 443
71CHE752	A. F. Pozharskii, I. S. Kashparov, Yu. P. Andreichikov, A. I. Buryak, A. A. Konstantinchenko and A. M. Simonov; *Chem. Heterocycl. Compd. (Engl. Transl.)*, 1971, **7**, 752.	368
71CHE809	A. D. Garnovskii, Yu. V. Kolodyazhnyi, O. A. Osipov, V. I. Minkin, S. A. Giller, I. B. Mazheika and I. I. Grandberg; *Chem. Heterocycl. Compd. (Engl. Transl.)*, 1971, **7**, 809.	351
71CHE1036	A. F. Pozharskii, V. V. Kuz'menko and A. M. Simonov; *Chem. Heterocycl. Compd. (Engl. Transl.)*, 1971, **7**, 1036.	410
71CHE1156	A. F. Pozharskii, É. A. Zvezdina, I. S. Kashparov, Yu. P. Andreichikov, V. M. Mar'yanovskii and A. M. Simonov; *Chem. Heterocycl. Compd. (Engl. Transl.)*, 1971, **7**, 1156.	442
71CHE1163	B. A. Tertov, A. V. Koblik and N. I. Avdyunina; *Chem. Heterocycl. Compd. (Engl. Transl.)*, 1971, **7**, 1163.	448
71CHE1534	A. P. Sineokov and V. S. Kutyreva; *Chem. Heterocycl. Compd. (Engl. Transl.)*, 1971, **7**, 1534.	139
71CJC1792	R. Raap; *Can. J. Chem.*, 1971, **49**, 1792.	69, 698
71CJC2139	R. Raap; *Can. J. Chem.*, 1971, **49**, 2139.	802
71CJC3119	J. W. Lown and K. Matsumoto; *Can. J. Chem.*, 1971, **49**, 3119.	341
71CPB1611	T. Fujii and T. Itaya; *Chem. Pharm. Bull.*, 1971, **19**, 1611.	539
71CPB1731	T. Fujii, T. Sato and T. Itaya; *Chem. Pharm. Bull.*, 1971, **19**, 1731.	554
71G569	A. C. Veronese, C. DiBello, F. Filira and F. D'Angeli; *Gazz. Chim. Ital.*, 1971, **101**, 569 (*Chem. Abstr.*, 1972, **76**, 3810).	658
71GEP1950076	E. Scheiffele; *Ger. Pat.* 1 950 076 (1971) (*Chem. Abstr.*, 1971, **75**, 36 102).	340
71GEP2129855	GAF Corporation, *Ger. Pat.* 2 129 855 (1971) (*Chem. Abstr.*, 1972, **76**, 128 833).	730

71GEP2129863	GAF Corporation, *Ger. Pat.* 2 129 863 (1971) (*Chem. Abstr.*, 1972, **76**, 128 832).	730
71GEP2129864	GAF Corporation, *Ger. Pat.* 2 129 864 (1971) (*Chem. Abstr.*, 1972, **76**, 99 674).	731
71HC(24-2)1	J. H. Lister; *Chem. Heterocycl. Compd.*, 1971, **24(2)**, 1.	501, 517, 518, 533, 538, 540, 548, 551, 558
71IC661	N. C. -Baenziger and R. J. Schultz; *Inorg. Chem.*, 1971, **10**, 661.	796
71IC2594	J. Reedijk, B. A. Stork-Blaise and G. C. Verschoor; *Inorg. Chem.*, 1971, **10**, 2594.	227
71JA1534	R. M. Moriarty and P. Serridge; *J. Am. Chem. Soc.*, 1971, **93**, 1534.	812
71JA1880	R. J. Pugmire and D. M. Grant; *J. Am. Chem. Soc.*, 1971, **93**, 1880.	514
71JA1887	R. J. Pugmire, M. J. Robins, D. M. Grant and R. K. Robins; *J. Am. Chem. Soc.*, 1971, **93**, 1887.	306
71JA3056	N. J. Leonard, A. J. Playtis, F. Skoog and R. Y. Schmitz; *J. Am. Chem. Soc.*, 1971, **93**, 3056.	601
71JA6129	E. E. van Tamelen and T. H. Whitesides; *J. Am. Chem. Soc.*, 1971, **93**, 6129.	220
71JA7045	P. Haake, L. P. Bausher and J. P. McNeal; *J. Am. Chem. Soc.*, 1971, **93**, 7045.	92
71JAP7133956	S. Kawano, S. Zoga, H. Watanabe and T. Sato; *Jpn. Pat.* 71 33 956 (1971) (*Chem. Abstr.*, 1976, **76**, 3902).	645
71JAP7133957	S. Kawano, S. Zoga, H. Watanabe and T. Sato; *Jpn. Pat.* 71 33 957 (1971) (*Chem. Abstr.*, 1971, **75**, 151 835).	645
71JAP(K)7135742	T. Kamitani, S. Tsubouchi and N. Yamamoto; *Jpn. Kokai* 71 35 742 (1971) (*Chem. Abstr.*, 1972, **76**, 14 550).	836
71JCS(B)415	G. Casalone and A. Mugnoli; *J. Chem. Soc. (B)*, 1971, 415.	9
71JCS(B)1648	G. V. Boyd and A. J. H. Summers; *J. Chem. Soc. (B)*, 1971, 1648.	755, 774
71JCS(B)2299	P. Dembeck, A. Ricci, G. Seconi and P. Vivarelli; *J. Chem. Soc. (B)*, 1971, 2299.	443
71JCS(B)2350	S. O. Chua, M. J. Cook and A. R. Katritzky; *J. Chem. Soc. (B)*, 1971, 2350.	369
71JCS(B)2365	A. G. Burton, P. P. Forsythe, C. D. Johnson and A. R. Katritzky; *J. Chem. Soc. (B)*, 1971, 2365.	223, 238
71JCS(C)409	G. V. Boyd and A. J. H. Summers; *J. Chem. Soc. (C)*, 1971, 409.	773
71JCS(C)675	J. Clark and T. Ramsden; *J. Chem. Soc. (C)*, 1971, 675.	648
71JCS(C)679	J. Clark and T. Ramsden; *J. Chem. Soc. (C)*, 1971, 679.	647, 648
71JCS(C)703	F. L. Scott and J. C. Tobin; *J. Chem. Soc. (C)*, 1971, 703.	802
71JCS(C)706	D. R. Sutherland and G. Tennant; *J. Chem. Soc. (C)*, 1971, 706.	679, 696
71JCS(C)1016	A. J. Blackman and J. B. Polya; *J. Chem. Soc. (C)*, 1971, 1016.	740, 758
71JCS(C)1193	A. J. Boulton, I. J. Fletcher and A. R. Katritzky; *J. Chem. Soc. (C)*, 1971, 1193.	87
71JCS(C)1482	G. Barker and G. P. Ellis; *J. Chem. Soc. (C)*, 1971, 1482.	136
71JCS(C)1501	N. J. Cusack, G. Shaw and G. J. Litchfield; *J. Chem. Soc. (C)*, 1971, 1501.	59
71JCS(C)1610	B. G. Hildick and G. Shaw; *J. Chem. Soc. (C)*, 1971, 1610.	316, 343
71JCS(C)1615	D. J. Dunwell and D. Evans; *J. Chem. Soc. (C)*, 1971, 1615.	655
71JCS(C)2147	T. Sasaki and K. Kanematsu; *J. Chem. Soc. (C)*, 1971, 2147.	184, 282
71JCS(C)2156	D. R. Sutherland and G. Tennant; *J. Chem. Soc. (C)*, 1971, 2156.	856, 874, 875, 894, 713
71JCS(C)2748	J. Adamson and E. E. Glover; *J. Chem. Soc. (C)*, 1971, 2748.	643
71JCS(C)3910	M. F. Lappert and J. S. Poland; *J. Chem. Soc. (C)*, 1971, 3910.	717
71JHC25	D. G. Farnum, A. T. Au and K. Rasheed; *J. Heterocycl. Chem.*, 1971, **8**, 25.	327
71JHC37	W. W. Paudler, D. J. Pokorny and J. J. Good; *J. Heterocycl. Chem.*, 1971, **8**, 37.	611, 613
71JHC47	A. Lewis and R. G. Shepherd; *J. Heterocycl. Chem.*, 1971, **8**, 47.	887
71JHC137	hM. Ruccia, N. Vivona and G. Cusmano; *J. Heterocycl. Chem.*, 1971, **8**, 137.	775
71JHC173	F. H. Case; *J. Heterocycl. Chem.*, 1971, **8**, 173.	766
71JHC311	J. van Thuijl, K. J. Klebe and J. J. van Houte; *J. Heterocycl. Chem.*, 1971, **8**, 311.	202, 290
71JHC379	G. H. Milne and L. B. Townsend; *J. Heterocycl. Chem.*, 1971, **8**, 379.	563
71JHC415	P. Francavilla and F. Lauria; *J. Heterocycl. Chem.*, 1971, **8**, 415.	869
71JHC557	J. P. Chupp; *J. Heterocycl. Chem.*, 1971, **8**, 557.	466
71JHC571	S. N. Lewis, G. A. Miller, M. Hausman and E. C. Szamborski; *J. Heterocycl. Chem.*, 1971, **8**, 571.	15
71JHC643	P. Guerret, R. Jacquier and G. Maury; *J. Heterocycl. Chem.*, 1971, **8**, 643.	627, 643, 648, 858, 861, 862, 876
71JHC657	G. A. Miller and M. Hausman; *J. Heterocycl. Chem.*, 1971, **8**, 657.	23
71JHC707	O. Tsuge and H. Samura; *J. Heterocycl. Chem.*, 1971, **8**, 707.	246
71JHC773	K. T. Potts, R. Armbruster and E. Houghton; *J. Heterocycl. Chem.*, 1971, **8**, 773.	741
71JHC785	B. Stanovnik and M. Tisler; *J. Heterocycl. Chem.*, 1971, **8**, 785.	342
71JHC835	I. Lalezari and A. Shafiee; *J. Heterocycl. Chem.*, 1971, **8**, 835.	116
71JHC1035	E. Ajello; *J. Heterocycl. Chem.*, 1971, **8**, 1035.	273
71JMC997	H. V. Secor and J. F. De Bardeleben; *J. Med. Chem.*, 1971, **14**, 997.	291, 292
71JOC10	K. T. Potts and S. Husain; *J. Org. Chem.*, 1971, **36**, 10.	128
71JOC222	R. W. Franck and S. J. M. Gilligan; *J. Org. Chem.*, 1971, **36**, 222.	117
71JOC446	T. Sasaki, K. Kanematsu and M. Murata; *J. Org. Chem.*, 1971, **36**, 446.	709, 882, 902, 903
71JOC1589	C. S. Angadiyavar and M. V. George; *J. Org. Chem.*, 1971, **36**, 1589.	160, 809
71JOC1846	K. T. Potts and R. Armbruster; *J. Org. Chem.*, 1971, **36**, 1846.	131
71JOC2635	N. J. M. Birdsall, T. C. Lee, T. J. Delia and J. C. Parshaw; *J. Org. Chem.*, 1971, **36**, 2635.	539
71JOC3084	P. Cohen-Fernandes and C. L. Habraken; *J. Org. Chem.*, 1971, **36**, 3084.	235, 259, 270
71JOC3087	M. Fraser; *J. Org. Chem.*, 1971, **36**, 3087.	309
71JOC3211	E. C. Taylor, G. P. Beardsley and Y. Maki; *J. Org. Chem.*, 1971, **36**, 3211.	591

71JOC3368	K. T. Potts and S. Hussain; *J. Org. Chem.*, 1971, **36**, 3368.	372, 402, 438
71JOC3594	H. Steinmaus, I. Rosenthal and D. Elad; *J. Org. Chem.*, 1971, **36**, 3594.	545
71JOC3807	T. Isida, S. Kozima, K. Nabika and K. Sisido; *J. Org. Chem.*, 1971, **36**, 3807.	793, 815
71JOM(27)185	D. H. O'Brien and C. P. Hrung; *J. Organomet. Chem.*, 1971, **27**, 185.	213
71JPR722	K. Fabian and H. Hartmann; *J. Prakt. Chem.*, 1971, **313**, 722.	119
71JPR882	S. Hauptmann, H. Wilde and K. Moser; *J. Prakt. Chem.*, 1971, **313**, 882.	134, 706
71JPR1148	G. Barnikow and J. Boedeker; *J. Prakt. Chem.*, 1971, **313**, 1148.	136
71JPS74	J. P. LaRocca, C. A. Gibson and B. B. Thompson; *J. Pharm. Sci.*, 1971, **60**, 74.	647
71JST(9)321	O. Martensson; *J. Mol. Struct.*, 1971, **9**, 321.	12
71JST(9)465	G. L. Blackman, R. D. Brown, F. R. Burden and A. Mishra; *J. Mol. Struct.*, 1971, **9**, 465.	290
71KGS39	O. P. Shvaika, V. N. Artemov and S. N. Baranov; *Khim. Geterotsikl. Soedin.*, 1971, 39.	776
71KGS275	M. S. Pevzner, E. Ya. Fedorova, I. N. Shokhor and L. I. Bagal; *Khim. Geterotsikl. Soedin.*, 1971, 275.	742
71KGS279	A. V. Kazymov, L. P. Schelkina and N. G. Kabirova; *Khim. Geterotsikl. Soedin.*, 1971, 279 (*Chem. Abstr.*, 1971, **75**, 35 878).	617
71KGS428	Y. M. Yutilov and I. A. Svertilova; *Khim. Geterotsikl. Soedin.*, 1971, 428 (*Chem. Abstr.*, 1972, **76**, 3755).	618, 621
71KGS471	M. O. Lozinskii, A. F. Shivanyuk and P. S. Pelkis; *Khim. Geterotsikl. Soedin.*, 1971, 471 (*Chem. Abstr.*, 1972, **76**, 25 184).	129, 131, 138
71KGS700	L. A. Vlasova and I. Ya. Postovskii; *Khim. Geterotsikl. Soedin.*, 1971, 700.	747
71KGS867	A. D. Garnovski, Y. V. Kolodyahnyi, O. A. Osipov, V. I. Minkin, S. A. Hiller, I. B. Mazheika and I. I. Grandberg; *Khim. Geterotsikl. Soedin.*, 1971, 867.	176, 177
71LA(743)50	K. D. Hesse; *Liebigs Ann. Chem.*, 1971, **743**, 50.	284
71LA(750)39	H. Dorn and D. Arndt; *Liebigs Ann. Chem.*, 1971, **750**, 39.	236
71M837	V. Pirc, B. Stanovnik and M. Tišler; *Monatsh. Chem.*, 1971, **102**, 837.	880
B-71MI40200	A. R. Katritzky and J. M. Lagowski; 'Chemistry of the Heterocyclic N-Oxides', Academic, New York, 1971, p. 51.	55
71MI40400	D. N. Kravtsov, L. A. Fedorov, A. S. Peregudov and A. A. Nesmeyanov; *Proc. Acad. Sci. USSR (Engl. Transl.)*, 1971, **196**, 18.	236
71MI40401	E. Kochanski, J. M. Lehn and B. Levy; *Theor. Chim. Acta*, 1971, **22**, 111.	195
71MI40500	H. Dahn and J. P. Fumeaux; *Bull. Soc. Vaudoise Sci. Natur.*, 1970, **70**, 313 (*Chem. Abstr.*, 1971, **75**, 140 791).	337
B-71MI40900	F. C. Steward and A. D. Krikorian; 'Plants, Chemicals and Growth', 1971, Academic, New York.	601
B-71MI40901	R. H. Hall; 'The Modified Nucleosides in Nucleic Acids', 1971, Columbia University Press, New York.	560, 561, 562, 604
B-71MI40902	A. R. Katritzky and J. M. Lagowski; 'Chemistry of the Heterocyclic N-Oxides', Academic, London, 1971.	501, 554, 562
71MI40903	Z. Neiman; *Isr. J. Chem.*, 1971, **9**, 119.	541, 543
71MI40904	L. A. Gutorov, L. A. Nikolaeva and E. S. Golovchinskaya, *Khim.-Farm. Zh.*, 1971, **5**, 17 (*Chem. Abstr.*, 1972, **76**, 126 935).	563
71MI40905	J. Filip and L. Bohacek; *Radioisotopy*, 1971, **12**, 949.	563
71MI40906	K. Nakanishi, S. H. Blobstein, M. Funamizu, N. Furutachi, G. van Lear, D. Grunberger, K. Lanks and I. B. Weinstein; *Nature (London)*, 1971, **234**, 107.	605
71MI40907	N. C. Yang, L. S. Gorelic and B. Kim; *Photochem. Photobiol.*, 1971, **13**, 275.	545
71MI40908	J. Schmidt and D. C. Borg; *Radiation Res.*, 1971, **46**, 36.	520
71MI41100	J. M. Stewart, R. L. Clarke and P. E. Pike; *J. Chem. Eng. Data*, 1971, **16**, 98 (*Chem. Abstr.*, 1971, **74**, 76 377).	717
71MI41101	L. Y. Ezerskya and N. A. Petrusha; *Onkologiya*, 1971, **2**, 45.	729
71MI41102	N. A. Petrusha; *Onkologiya*, 1971, **2**, 10.	729
B-71MI41103	D. Philips; in 'Photochemistry', Chemical Society, London, 1971, vol. 2, p. 795.	731
B-71MI41104	T. Sheradsky; in 'The Chemistry of the Azido Group', Wiley, New York, 1971, p. 331.	670, 705, 708, 710
71MI41200	I. Simiti and D. Ghiran; *Farmacia (Bucharest)*, 1971, **19**, 199 (*Chem. Abstr.*, 1971, **75**, 76 692).	764
71MI41300	A. J. Barratt, L. R. Bates, J. M. Jenkins and J. R. White; *U.S. NTIS. AD. Rep.*, 1971, No. 752 370 (*Chem. Abstr.*, 1973, **78**, 124 508).	837
B-71MI41400	I. Ugi; 'Isonitrile Chemistry', Academic, New York, 1971.	845
71MI41500	G. Fischer; *Wiss. Z. Tech. Hochsch. Chem. 'Carl Schlorlemmer' Leuna-Merseburg*, 1971, **13**, 71 (*Chem. Abstr.*, 1971, **75**, 48 939).	904
B-71MS	Q. N. Porter and J. Baldas; 'Mass Spectrometry of Heterocyclic Compounds', Wiley, New York, 1971.	30, 202
71OMR(3)595	R. M. Claramunt, J. Elguero and R. Jacquier; *Org. Magn. Reson.*, 1971, **3**, 595.	182
71OMS(5)1	K. T. Potts, E. Brugel and U. P. Singh; *Org. Mass Spectrom.*, 1971, **5**, 1.	853
71OMS(5)427	F. Compernolle and M. Dekeirel; *Org. Mass Spectrom.*, 1971, **5**, 427.	686
71OMS(5)663	K. T. Potts and E. Brugel; *Org. Mass Spectrom.*, 1971, **5**, 663.	859
71OMS(5)1101	J. van Thuijl, K. J. Klebe and J. J. van Houte; *Org. Mass Spectrom.*, 1971, **5**, 1101.	202
71PMH(3)1	A. Albert; *Phys. Methods Heterocycl. Chem.*, 1971, **3**, 1.	50, 51
71PMH(3)67	W. L. F. Armarego; *Phys. Methods Heterocycl. Chem.*, 1971, **3**, 67.	21, 23, 197, 198, 199
71PMH(3)77	W. L. F. Armarego; *Phys. Methods Heterocycl. Chem.*, 1971, **3**, 77.	684
71PMH(3)103	W. L. F. Armarego; *Phys. Methods Heterocycl. Chem.*, 1971, **3**, 103.	684

71PMH(3)297	Ya. L. Gol'dfarb, V. I. Yakerson, V. A. Ferapontov, S. Z. Taits and F. M. Stoyanovich; *Phys. Methods Heterocycl. Chem.*, 1971, **3**, 297.	32, 207
71PMH(4)21	E. A. C. Lucken; *Phys. Methods Heterocycl. Chem.*, 1971, **4**, 21.	195
71PMH(4)55	J. H. Ridd; *Phys. Methods Heterocycl. Chem.*, 1971, **4**, 55.	70, 241
71PMH(4)121	R. F. M. White and H. Williams; *Phys. Methods Heterocycl. Chem.*, 1971, **4**, 121.	13, 14, 15, 18, 678
71PMH(4)265	A. R. Katritzky and P. J. Taylor; *Phys. Methods Heterocycl. Chem.*, 1971, **4**, 265.	24, 25, 199, 201, 202, 739
71RTC594	D. A. de Bie, H. C. van der Plas and G. Guertsen; *Recl. Trav. Chim. Pays-Bas*, 1971, **90**, 594.	411
71RTC654	K. B. De Roos and C. A. Salemink; *Recl. Trav. Chim. Pays-Bas*, 1971, **90**, 654.	615, 618
71RTC1166	K. B. De Roos and C. A. Salemink; *Recl. Trav. Chim. Pays-Bas*, 1971, **90**, 1166.	618
71RTC1181	K. B. De Roos and C. A. Salemink; *Recl. Trav. Chim. Pays-Bas*, 1971, **90**, 1181.	867, 887
71RTC1225	C. Hoogzand; *Recl. Trav. Chim. Pays-Bas*, 1971, **90**, 1225.	853, 856, 865, 886
71SAP7004373	H. O. Bayer, R. S. Cook and W. C. Von Meyer; *S. Afr. Pat.* 70 04 373 (1971) (*Chem. Abstr.*, 1972, **76**, 113 224).	787
71T123	J. Elguero, R. Jacquier and D. Tizané; *Tetrahedron*, 1971, **27**, 123.	189, 211
71T133	J. Elguero, R. Jacquier and D. Tizane; *Tetrahedron*, 1971, **27**, 133.	18
71T361	C. Wentrup and W. D. Crow; *Tetrahedron*, 1971, **27**, 361.	881
71T845	G. L'abbé, P. Ykman and G. Smets; *Tetrahedron*, 1971, **27**, 845.	704, 716
71T1783	Z. Grzonka and B. Liberek; *Tetrahedron*, 1971, **27**, 1783.	834
71T2317	Z. Grzonka, E. Rekowska and B. Liberek; *Tetrahedron*, 1971, **27**, 2317 (see also other papers of the series).	834
71T2415	T. Fujii, T. Itaya, C. C. Wu and F. Tanaka; *Tetrahedron*, 1971, **27**, 2415.	554
71T2453	J. P. Albrand, A. Cogne, D. Gagnaire and J. B. Robert; *Tetrahedron*, 1971, **27**, 2453.	354, 355
71T4117	R. Neidlein and H. Reuter; *Tetrahedron*, 1971, **27**, 4117.	130
71T5121	T. Sasaki, K. Kanematsu and M. Murata; *Tetrahedron*, 1971, **27**, 5121.	709, 854, 857, 860, 879, 901
71T5359	T. Sasaki, K. Kanematsu and M. Murata; *Tetrahedron*, 1971, **27**, 5359.	879
71T5623	P. Ykman, G. L'abbé and G. Smets; *Tetrahedron*, 1971, **27**, 5623.	679, 704, 709, 716
71T5873	H. Wamhoff, H. W. Dürbeck and P. Sohár; *Tetrahedron*, 1971, **27**, 5873.	712
71TH40400	S. Mignonac-Mondon; Ph. D. Thesis, University of Montpellier, 1971.	216
71TL633	C. S. Angadiyavar, K. B. Sukumaran and M. V. George; *Tetrahedron Lett.*, 1971, 633.	727
71TL1441	E. Abushanab; *Tetrahedron Lett.*, 1971, 1441.	644
71TL1591	J. Farkas and Z. Flegelova; *Tetrahedron Lett.*, 1971, 1591.	276
71TL1993	N. K. Kochetkov, V. W. Shibaev and A. A. Kost; *Tetrahedron Lett.*, 1971, 1993.	648
71TL2725	H. Kasai, M. Goto, S. Takemura, T. Goto and S. Matsuura; *Tetrahedron Lett.*, 1971, 2725.	605
71TL4243	D. Kobelt, E. Paulus and G. Lohaus; *Tetrahedron Lett.*, 1971, **45**, 4243.	9
71TL4489	P. Scheiner; *Tetrahedron Lett.*, 1971, 4489.	810
71USP3565892	*U.S. Pat.* 3 565 892 (1971) (*Chem. Abstr.*, 1971, **75**, 36 041).	661
71ZC153	G. Henning and F. Wolff; *Z. Chem.*, 1971, **11**, 153.	756
71ZC179	J. Sieler, H. Wilde and S. Hauptmann; *Z. Chem.*, 1971, **11**, 179.	134, 706
71ZC421	H. Hartmann; *Z. Chem.*, 1971, **11**, 421.	140
72AC(R)11	G. Werber, F. Buccheri and M. L. Marino; *Ann. Chim.* (*Rome*), 1972, **62**, 11 (*Chem. Abstr.*, 1972, **77**, 48 004).	136
72ACR10	J. L. Kurz; *Acc. Chem. Res.*, 1972, **5**, 10.	601
72ACS21	L. O. Carlsson; *Acta Chem. Scand.*, 1972, **26**, 21.	209
72ACS541	A. K. Sorensen and N. A. Klitgaard; *Acta Chem. Scand.*, 1972, **26**, 541.	819
72ACS1243	M. Begtrup; *Acta Chem. Scand.*, 1972, **26**, 1243.	695
72AG1187	D. Seebach and D. Enders; *Angew. Chem.*, 1972, **84**, 1187.	717
72AG(E)846	T. Kauffman, J. Legler, E. Ludorff and H. Fischer; *Angew. Chem., Int. Ed. Engl.*, 1972, **11**, 846.	753, 756
72AG(E)1102	D. Seebach and D. Enders; *Angew. Chem., Int. Ed. Engl.*, 1972, **11**, 1102.	717
72AHC(14)1	K. R. H. Wooldridge; *Adv. Heterocycl. Chem.*, 1972, **14**, 1.	47, 58, 61, 63, 65, 69, 75, 88, 91, 93, 97, 98, 100, 104, 106, 108, 135
72AHC(14)43	M. Davis; *Adv. Heterocycl. Chem.*, 1972, **14**, 43.	47, 85, 94, 97
72AJC335	A. J. Blackman and J. H. Bowie; *Aust. J. Chem.*, 1972, **25**, 335.	741
72AX(B)791	J. Lapasset and J. Falgueriettes; *Acta Crystallogr., Part B*, 1972, **28**, 791.	178
72AX(B)1982	T. F. Lai and R. E. Marsh; *Acta Crystallogr., Part B*, 1972, **28**, 1982.	509
72AX(B)3316	J. Lapasset, A. Escande and J. Falgueirettes; *Acta Crystallogr., Part B*, 1972, **28**, 3316.	178
72B1235	M. Tomasz, J. Olson and C. M. Mercado; *Biochemistry*, 1972, **11**, 1235.	527
72BBR(46)125	H. Berthod and B. Pullman; *Biochem. Biophys. Res. Commun.*, 1972, **46**, 125.	522
72BBR(46)597	J. R. Barrio, J. A. Secrist, III and N. J. Leonard; *Biochem. Biophys. Res. Commun.*, 1972, **46**, 597.	610, 648
72BCJ1471	T. Isida, S. Kozima, S.-I. Fujimori and K. Sisido; *Bull. Chem. Soc. Jpn.*, 1972, **45**, 1471.	793

72BRP1269619	E. Regel, K. H. Buechel, R. R. Schmidt and L. Eue; *Br. Pat.* 1 269 619 (1972) (*Chem. Abstr.*, 1972, **77**, 5477).	745, 787
72BSB295	C. Draguet and M. Renson; *Bull. Soc. Chim. Belg.*, 1972, **81**, 295.	154
72BSF2481	P. Guerret, R. Jacquier and G. Maury; *Bull. Soc. Chim. Fr.*, 1972, 2481.	628
72BSF2807	J. Elguero, R. Jacquier and S. Mignonac-Mondon; *Bull. Soc. Chim. Fr.*, 1972, 2807.	12, 261, 262
72BSF3503	P. Guerret, R. Jacquier and G. Maury; *Bull. Soc. Chim. Fr.*, 1972, 3503.	647
72CB103	H. Reimlinger, M. A. Peiren and R. Merényi; *Chem. Ber.*, 1972, **105**, 103.	822
72CB188	H. Gotthardt; *Chem. Ber.*, 1972, **105**, 188.	150
72CB196	H. Gotthardt; *Chem. Ber.*, 1972, **105**, 196.	150
72CB388	E. Kranz, J. Kurz and W. Donner; *Chem. Ber.*, 1972, **105**, 388.	306, 312, 342
72CB794	H. Reimlinger, M. A. Peiren and R. Merenyi; *Chem. Ber.*, 1972, **105**, 794.	142
72CB1497	W. Pfleiderer, M. Shanshal and K. Eistetter; *Chem. Ber.*, 1972, **105**, 1497.	574
72CB1759	L. Birkofer and M. Franz; *Chem. Ber.*, 1972, **105**, 1759.	184, 267
72CB2963	G. Himbert and M. Regitz; *Chem. Ber.*, 1972, **105**, 2963.	692
72CB2975	G. Himbert and M. Regitz; *Chem. Ber.*, 1972, **105**, 2975.	692
72CC37	C. L. Habraken and P. Cohen-Fernandes; *J. Chem. Soc., Chem. Commun.*, 1972, 37.	752
72CC126	J. P. Ferris and F. R. Antonucci; *J. Chem. Soc., Chem. Commun.*, 1972, 126.	274
72CC199	F. Texier, R. Carrie and J. Jaz; *J. Chem. Soc., Chem. Commun.*, 1972, 199.	154
72CC544	P. G. Harrison; *J. Chem. Soc., Chem. Commun.*, 1972, 544.	752
72CC709	J. Castells, M. A. Merino and M. Moreno-Mañas; *J. Chem. Soc., Chem. Commun.*, 1972, 709.	212
72CC781	P. Scheiner and W. M. Litchman; *J. Chem. Soc., Chem. Commun.*, 1972, 781.	809
72CC827	A. A. Reid, J. T. Sharp and S. J. Murray; *J. Chem. Soc., Chem. Commun.*, 1972, 827.	268
72CC1059	T. Tsuchiya, H. Arai and H. Igeta; *J. Chem. Soc., Chem. Commun.*, 1972, 1059.	880
72CHE57	Yu. N. Portnov, G. A. Golubeva and A. N. Kost; *Chem. Heterocycl. Compd.* (*Engl. Transl.*), 1972, **8**, 57.	257
72CHE445	M. A. Lapitskaya, A. D. Garnovskii, G. A. Golubeva, Yu. V. Kolodyazhnyi, S. A. Alieva and A. N. Kost; *Chem. Heterocycl. Compd.* (*Engl. Transl.*), 1972, **8**, 445.	178
72CHE742	P. M. Kochergin, M. A. Klykov and I. S. Mikhailova; *Chem. Heterocycl. Compd.* (*Engl. Transl.*), 1972, **8**, 742.	447
72CHE1131	A. F. Pozharskii, V. V. Kuz'menko, Yu. V. Kolodyazhnyi and A. M. Simonov; *Chem. Heterocycl. Compd.* (*Engl. Transl.*), 1972, **8**, 1131.	346, 410
72CHE1142	O. N. Poplin and V. G. Tishchenko; *Chem. Heterocycl. Compd.* (*Engl. Transl.*), 1972, **8**, 1142.	468
72CHE1280	M. M. Medvedeva, A. F. Pozharskii and A. M. Simonov; *Chem. Heterocycl. Compd.* (*Engl. Transl.*), 1972, **8**, 1280.	407
72CHE1533	S. N. Kolodyazhnaya, A. M. Simonov and I. G. Uryukina; *Chem. Heterocycl. Compd.* (*Engl. Transl.*), 1972, **8**, 1533.	353, 429, 439
72CI(L)294	H. Reimlinger; *Chem. Ind.* (*London*), 1972, 294.	822
72CJC3186	D. M. Findlay, M. L. Roy and S. McLean; *Can. J. Chem.*, 1972, **50**, 3186.	720
72CJC3472	R. S. Glass, J. S. Blount, D. Butler, A. Perrotta and E. P. Oliveto; *Can. J. Chem.*, 1972, **50**, 3472.	351
72CPB69	A. Nakamura and S. Kamiya; *Chem. Pharm. Bull.*, 1972, **20**, 69.	314
72CPB605	Y. Maki and E. C. Taylor; *Chem. Pharm. Bull.*, 1972, **20**, 605.	134
72CPB623	H. Yamada and T. O. Kamato; *Chem. Pharm. Bull.*, 1972, **20**, 623.	568
72CPB958	T. Fujii, T. Itaya and S. Moro; *Chem. Pharm. Bull.*, 1972, **20**, 958.	554
72CPB2096	S. Kubota and M. Uda; *Chem. Pharm. Bull.*, 1972, **20**, 2096.	740, 758
72CR(C)(274)1192	J. L. Aubagnac, J. Elguero, B. Rérat, C. Rérat and Y. Uesu; *C.R. Hebd. Seances Acad. Sci., Ser. C*, 1972, **274**, 1192.	179
72CR(C)(274)2192	G. Barnathan, T. H. Dinh, A. Kolb and J. Igolen; *C.R. Hebd. Seances Acad. Sci., Ser. C*, 1972, **274**, 2192.	585
72CRV497	S. Trofimenko; *Chem. Rev.*, 1972, **72**, 497.	169, 225, 226, 235, 236
72CSC253	B. Bovio and S. Locchi; *Cryst. Struct. Commun.*, 1972, **1**, 253.	178
72E1198	G. Magnusson, J. A. Nyberg, N. O. Bodin and E. Hansson; *Experientia*, 1972, **28**, 1198.	302
72G23	G. D. Andreetti, G. Bocelli, L. Cavalca and P. Sgarabotto; *Gazz. Chim. Ital.*, 1972, **102**, 23.	9
72G491	G. Desimoni, A. Gamba, P. P. Righetti and G. Tacconi; *Gazz. Chim. Ital.*, 1972, **102**, 491.	208
72GEP2032088	CIBA-Geigy AG, *Ger. Pat.* 2 032 088 (1972) (*Chem. Abstr.*, 1972, **76**, 128 835).	730
72GEP2032172	Bayer AG, *Ger. Pat.* 2 032 172 (1972) (*Chem. Abstr.*, 1972, **76**, 128 834).	730
72GEP2060196	H. Maeder, R. Otto, P. Marx and W. Pueschel; *Ger. Pat.* 2 060 196 (1972) (*Chem. Abstr.*, 1973, **78**, 22 517).	837
72GEP2226525	G. J. Smets and J. M. Vandensavel; *Ger. Pat.* 2 226 525 (1972) (*Chem. Abstr.*, 1973, **78**, 58 984).	838
72HCA565	C. Wentrup; *Helv. Chim. Acta*, 1972, **55**, 565.	881
72HCA919	B. Jackson, M. Märky, H.-J. Hansen and H. Schmid; *Helv. Chim. Acta*, 1972, **55**, 919.	154
72HCA2633	C. Wentrup, C. Thétaz and R. Gleiter; *Helv. Chim. Acta*, 1972, **55**, 2633.	881

72JA1199	B. Singh, A. Zweig and J. B. Gallivan; *J. Am. Chem. Soc.*, 1972, **94**, 1199.	161
72JA2460	J. D. Vaughan and W. A. Smith; *J. Am. Chem. Soc.*, 1972, **94**, 2460.	241
72JA2530	R. L. Tolman, C. W. Smith and R. K. Robins; *J. Am. Chem. Soc.*, 1972, **94**, 2530.	712
72JA5759	H. Kohn, S. J. Benkovic and R. A. Olofson; *J. Am. Chem. Soc.*, 1972, **94**, 5759.	806
72JA6337	C. W. Gillies and R. L. Kuczkowski; *J. Am. Chem. Soc.*, 1972, **94**, 6337.	12
72JAP(K)7225190	T. Irikura, K. Kasuga, T. Hashizume, M. Ohashi, M. Yuge and M. Yamada; (Kyorin Pharmaceutical Co. Ltd.), *Jpn. Kokai* 72 25 190 (1972) (*Chem. Abstr.*, 1973, **78**, 4251).	635
72JAP(K)7242770	K. Sugiura, R. Ushijima and K. Shimizu; *Jpn. Kokai* 72 42 770 (1972) (*Chem. Abstr.*, 1973, **78**, 72 158).	835
72JCS(P1)461	A. Albert; *J. Chem. Soc., Perkin Trans. 1*, 1972, 461.	695
72JCS(P1)619	D. Pocar, R. Stradi and L. M. Rossi; *J. Chem. Soc., Perkin Trans. 1*, 1972, 619.	714
72JCS(P1)769	D. Pocar, R. Stradi and L. M. Rossi; *J. Chem. Soc., Perkin Trans. 1*, 1972, 769.	714
72JCS(P1)783	T. Sasaki, K. Kanematsu and K. Hayakawa; *J. Chem. Soc., Perkin Trans. 1*, 1972, 783.	328
72JCS(P1)909	G. V. Boyd and P. H. Wright; *J. Chem. Soc., Perkin Trans. 1*, 1972, 909.	113
72JCS(P1)914	G. V. Boyd and P. H. Wright; *J. Chem. Soc., Perkin Trans. 1*, 1972, 914.	113
72JCS(P1)997	G. Bianchetti, R. Stradi and D. Pocar; *J. Chem. Soc., Perkin Trans. 1*, 1972, 997.	714
72JCS(P1)1519	R. N. Butler, P. O'Sullivan and F. L. Scott; *J. Chem. Soc., Perkin Trans. 1*, 1972, 1519 and earlier references listed therein.	134
72JCS(P1)1918	F. L. Scott, T. M. Lambe and R. N. Butler; *J. Chem. Soc., Perkin Trans. 1*, 1972, 1918.	769
72JCS(P1)2567	V. Caló, F. Ciminale, L. Lopez, F. Naso and P. E. Todesco; *J. Chem. Soc., Perkin Trans. 1*, 1972, 2567.	399
72JCS(P2)68	S. Bradbury, C. W. Rees and R. C. Storr; *J. Chem. Soc., Perkin Trans. 2*, 1972, 68.	208
72JCS(P2)585	Z. Neiman; *J. Chem. Soc., Perkin Trans. 2*, 1972, 585.	521
72JCS(P2)632	M. W. Austin; *J. Chem. Soc., Perkin Trans. 2*, 1972, 632.	259
72JCS(P2)1077	P. J. Taylor; *J. Chem. Soc., Perkin Trans. 2*, 1972, 1077.	92
72JCS(P2)1420	F. Pietra and F. Del Cima; *J. Chem. Soc., Perkin Trans. 2*, 1972, 1420.	232
72JCS(P2)1654	M. R. Grimmett, S. R. Hartshorn, K. Schofield and J. B. Weston; *J. Chem. Soc., Perkin Trans. 2*, 1972, 1654.	238, 268
72JCS(P2)2218	G. G. Aloisi, E. Bordignon and A. Signor; *J. Chem. Soc., Perkin Trans. 2*, 1972, 2218.	636
72JHC41	B. T. Gillis and R. A. Izydore; *J. Heterocycl. Chem.*, 1972, **9**, 41.	328
72JHC87	R. Selvarajan and J. H. Boyer; *J. Heterocycl. Chem.*, 1972, **9**, 87.	693
72JHC99	T. Hirata, L.-M. Twanmoh, H. B. Wood, Jr., A. Goldin and J. S. Driscoll; *J. Heterocycl. Chem.* 1972, **9**, 99.	751
72JHC235	H. Höhn, T. Denzel and W. Janssen; *J. Heterocycl. Chem.*, 1972, **9**, 235.	272, 322
72JHC363	J. T. Edward and I. Lantos; *J. Heterocycl. Chem.*, 1972, **9**, 363.	367
72JHC461	J. P. Paolini; *J. Heterocycl. Chem.*, 1972, **9**, 461.	901
72JHC465	T. Itoh, S. Kitano and Y. Mizuno; *J. Heterocycl. Chem.*, 1972, **9**, 465.	615
72JHC581	G. Pifferi and P. Consonni; *J. Heterocycl. Chem.*, 1972, **9**, 581.	779
72JHC623	R. P. Panzica and L. B. Townsend; *J. Heterocycl. Chem.*, 1972, **9**, 623.	652
72JHC717	M. T. García-Lopez, G. García-Muñoz and R. Madroñero; *J. Heterocycl. Chem.*, 1972, **9**, 717.	703
72JHC939	C. L. Habraken, C. I. M. Beenakker and J. Brussee; *J. Heterocycl. Chem.*, 1972, **9**, 939.	199, 224, 225
72JHC951	V. Sprio and S. Plescia; *J. Heterocycl. Chem.*, 1972, **9**, 951.	320, 332
72JHC1157	W. W. Paudler, C. I. P. Chao and L. S. Helmick; *J. Heterocycl. Chem.*, 1972, **9**, 1157.	615, 630, 653, 872
72JHC1171	A. Bezeg, B. Stanovnik, B. Šket and M. Tišler; *J. Heterocycl. Chem.*, 1972, **9**, 1171.	869
72JHC1195	F. A. Lehmkuhl, J. T. Witkowski and R. K. Robins; *J. Heterocycl. Chem.*, 1972, **9**, 1195.	700
72JHC1373	P. B. M. W. M. Timmermans, A. P. Uijttewaal and C. L. Habraken; *J. Heterocycl. Chem.*, 1972, **9**, 1373.	182, 260, 261
72JHC1385	D. L. Trepanier and P. E. Krieger; *J. Heterocycl. Chem.*, 1972, **9**, 1385.	666
72JMC182	J. A. Montgomery and H. J. Thomas; *J. Med. Chem.*, 1972, **15**, 182.	652, 852
72JMC694	B. Blank, D. M. Nichols and P. D. Vaidya; *J. Med. Chem.*, 1972, **15**, 694.	757
72JMC841	M. Kawana, G. A. Ivanovics, R. J. Rousseau and R. K. Robins; *J. Med. Chem.*, 1972, **15**, 841.	652
72JMC879	W. Hutzenlaub, R. L. Tolman and R. K. Robins; *J. Med. Chem.*, 1972, **15**, 879.	876
72JOC343	L. A. Lee, R. Evans and J. W. Wheeler; *J. Org. Chem.*, 1972, **37**, 343.	766, 829
72JOC1454	K. B. Wagener, S. R. Turner and G. B. Butler; *J. Org. Chem.*, 1972, **37**, 1454.	750
72JOC1464	H. Alper and E. C. H. Keung; *J. Org. Chem.*, 1972, **37**, 1464.	663
72JOC1867	A. A. Watson and G. B. Brown; *J. Org. Chem.*, 1972, **37**, 1867.	590, 596
72JOC1871	G. Zvilichovsky and G. B. Brown; *J. Org. Chem.*, 1972, **37**, 1871.	596
72JOC2022	R. A. Abramovitch and T. Takaya; *J. Org. Chem.*, 1972, **37**, 2022.	856, 868, 888
72JOC2049	K. T. Potts and S. Husain; *J. Org. Chem.*, 1972, **37**, 2049.	701
72JOC2158	J. N. Wells, O. R. Tarwater and P. E. Manni; *J. Org. Chem.*, 1972, **37**, 2158.	425, 426, 467
72JOC2259	T. J. Curphey and K. J. Prasad; *J. Org. Chem.*, 1972, **37**, 2259.	748
72JOC2351	K. Sakai and J. P. Anselme; *J. Org. Chem.*, 1972, **37**, 2351.	274
72JOC3106	T. Sasaki, K. Kanematsu and A. Kakehi; *J. Org. Chem.*, 1972, **37**, 3106.	153, 339

72JOC3213	P. Ykman, G. Mathys, G. L'abbé and G. Smets; *J. Org. Chem.*, 1972, **37**, 3213.	716
72JOC3504	H. Kohn and R. A. Olofson; *J. Org. Chem.*, 1972, **37**, 3504.	763
72JOC3965	S. Hammerum and P. Volkoff; *J. Org. Chem.*, 1972, **37**, 3965.	204
72JOC3985	C. G. Tindall, Jr., R. K. Robins and R. L. Tolman; *J. Org. Chem.*, 1972, **37**, 3985.	553
72JOC4045	J. Buter, S. Wassenaar and R. M. Kellogg; *J. Org. Chem.*, 1972, **37**, 4045.	147
72JOC4124	C. A. Lovelette and L. Long, Jr.; *J. Org. Chem.*, 1972, **37**, 4124.	854, 878, 897
72JOC4136	R. W. Begland and D. R. Hartter; *J. Org. Chem.*, 1972, **37**, 4136.	127, 472
72JOC4401	R. A. Wohl and D. F. Headley; *J. Org. Chem.*, 1972, **37**, 4401.	156
72JOC4410	K. T. Potts, J. Bhattacharyya, S. L. Smith, A. M. Ihrig and C. A. Girard; *J. Org. Chem.*, 1972, **37**, 4410.	853
72JOC4464	E. C. Taylor and F. Yoneda; *J. Org. Chem.*, 1972, **37**, 4464.	581
72JOM(39)179	S. S. Eaton, G. R. Eaton and R. H. Holm; *J. Organomet. Chem.*, 1972, **39**, 179.	213
72JOM(44)117	S. Kozima, T. Itano, N. Mihara, K. Sisido and T. Isida; *J. Organomet. Chem.*, 1972, **44**, 117.	682, 710
72JPC3362	J. Hennessy and A. C. Testa; *J. Phys. Chem.*, 1972, **76**, 3362.	433
72JPR101	H. G. O. Becker, H. Goermar, H. Haufe and H. J. Timpe; *J. Prakt. Chem.*, 1972, **314**, 101.	767
72JPR515	E. Kleinpeter, R. Borsdorf, G. Fischer and H.-J. Hofmann; *J. Prakt. Chem.*, 1972, **314**, 515.	853, 858
72JPR923	H. G. O. Becker and R. Ebisch; *J. Prakt. Chem.*, 1972, **314**, 923.	751
72KGS691	N. O. Saldabol and J. Popelis; *Khim. Geterotsikl. Soedin.*, 1972, 691 (*Chem. Abstr.*, 1972, **77**, 126 503).	611
72LA(757)9	K. Burger, J. Fehn and A. Gieren; *Liebigs Ann. Chem.*, 1972, **757**, 9.	717
72LA(759)107	M. Haake, H. Fode and B. Eichenauer; *Liebigs Ann. Chem.*, 1972, **759**, 107.	131
72LA(764)112	M. Brugger and F. Korte; *Liebigs Ann. Chem.*, 1972, **764**, 112.	654
72MI40300	I. Z. Siemion, W. Steglich and L. Wilschowitz; *Rocz. Chem.*, 1972, **46**, 21.	113
72MI40400	F. A. Cotton and D. J. Ciappenelli; *Synth. React. Inorg. Metal-Org. Chem.*, 1972, **2**, 197.	213
72MI40600	C. G. Begg and M. R. Grimmett; *J. Chromatogr.*, 1972, **73**, 238.	362
B-72MI40900	G. C. Levy and G. L. Nelson; 'Carbon-13 NMR for Organic Chemists', Wiley-Interscience, New York, 1972.	510
B-72MI40901	J. B. Stothers; 'Carbon-13 NMR Spectroscopy', in 'Organic Chemistry', ed. A. T. Blomquist and H. Wasserman; Academic, New York, 1972.	510, 554, 555
B-72MI40902	A. Pullman and B. Pullman; in 'Jerusalem Symposium on Quantum Chemistry and Biochemistry', Israel Academy of Sciences and Humanities, 1972, vol. 4, p. 1.	501, 505, 507, 508, 510, 511, 512, 513, 519, 522, 523, 555, 556
72MI40903	J. Igolen, K. H. Dinh, A. Kolb and C. Perreur; *J. Med. Chim. Ther.*, 1972, **7**, 207.	585
72MI41000	B. Bobranski and J. Stankiewicz; *Diss. Pharm. Pharmacol.*, 1972, **24**, 301.	637
72MI41200	M. A. Khan and J. B. Polya; *Rev. Latinoamer. Quim.*, 1972, **3**, 119 (*Chem. Abstr.*, 1973, **78**, 72 019).	754
72MI41201	A. V. Radushev and E. N. Prokhorenko; *Zh. Anal. Khim.*, 1972, **27**, 2209 (*Chem. Abstr.*, 1973, **78**, 66 590).	785
72MI41202	R. J. Cotter and M. Matzner; 'Ring-forming Polymerizations Part B.1, Heterocyclic Rings', Academic, New York, 1972.	788, 789
72MI41300	J. H. Nelson and F. G. Baglin; *Spectrosc. Lett.*, 1972, **5**, 101 (*Chem. Abstr.*, 1972, **77**, 47 707).	806
72MI41301	N. V. Schwartz; *J. Appl. Polym. Sci.*, 1972, **16**, 2715 (*Chem. Abstr.*, 1973, **78**, 17 288).	837
72MIP41000	Y. M. Yutilov and L. I. Kovaleva; *USSR Pat.* 332 089 (1972) (*Chem. Abstr.*, 1972, **77**, 114 408).	622
72OMR(4)247	R. Stradi, D. Pocar and G. Bianchetti; *Org. Magn. Reson.*, 1972, **4**, 247.	714
72OMS(6)1139	A. Maquestiau, Y. Van Haverbeke and R. Flammang; *Org. Magn. Spectrom.*, 1972, **6**, 1139.	741
72PAC(30)135	J. E. Guillet; *Pure Appl. Chem.*, 1972, **30**, 135.	731
72PAC(30)145	H. J. Heller and H. R. Blattmann; *Pure Appl. Chem.*, 1972, **30**, 145.	731
72PMH(5)1	P. J. Wheatley; *Phys. Methods Heterocycl. Chem.*, 1972, **5**, 1.	9, 178, 179, 501, 507, 517
72RTC650	J. E. Schelling and C. A. Salemink; *Recl. Trav. Chim. Pays-Bas*, 1972, **91**, 650.	618, 619, 635
72RTC850	P. J. Lont and H. C. van der Plas; *Recl. Trav. Chim. Pays-Bas*, 1972, **91**, 850.	496
72S259	M. Cinquini and S. Colonna; *Synthesis*, 1972, 259.	727
72S618	J. Corse and J. Kuhnle; *Synthesis*, 1972, 618.	601
72T21	Y. Tamura, N. Tsujimoto, Y. Sumida and M. Ikeda; *Tetrahedron*, 1972, **28**, 21.	337
72T637	M. Witanowski, L. Stefaniek, H. Januszewski, Z. Grabowski and G. A. Webb; *Tetrahedron*, 1972, **28**, 637.	672, 684, 741, 805
72T3845	P. Bravo, G. Gaudiano, P. P. Ponti and C. Ticozzi; *Tetrahedron*, 1972, **28**, 3845.	164
72T3987	K. B. Sukumaran, C. S. Angadiyavar and M. V. George; *Tetrahedron*, 1972, **28**, 3987.	702
72T5207	F. Khuong-Huu, J. P. Le Forestier and R. Goutarel; *Tetrahedron*, 1972, **28**, 5207.	647
72T5779	C. Glier, F. Dietz, M. Scholz and G. Fischer; *Tetrahedron*, 1972, **28**, 5779.	858
72TL439	E. Zbiral and E. Hugl; *Tetrahedron Lett.*, 1972, 439.	648
72TL2235	W. D. Crow, A. R. Lea and M. N. Paddon-Row; *Tetrahedron Lett.*, 1972, 2235.	222
72TL2239	W. D. Crow, M. N. Paddon-Row and D. S. Sutherland; *Tetrahedron Lett.*, 1972, 2239.	867

72TL3637	J. L. Huppatz; *Tetrahedron Lett.*, 1972, 3637.	268
72TL4179	M. Zupan, B. Stanovnik and M. Tišler; *Tetrahedron Lett.*, 1972, 4179.	890
72TL4719	H. Mackie and G. Tennant; *Tetrahedron Lett.*, 1972, 4719.	852
72ZC130	G. Barnikov and H. Ebeling; *Z. Chem.*, 1972, **12**, 130.	135
72ZC175	G. Wagner and S. Leistner; *Z. Chem.*, 1972, **12**, 175.	779
73ACS661	I. Søtofte; *Acta Chem. Scand.*, 1973, **27**, 661.	178
73ACS1509	M. Y. Ali, J. Dale and K. Titlestad; *Acta Chem. Scand.*, 1973, **27**, 1509.	643
73ACS1845	T. La Cour and S. E. Rasmussen; *Acta Chem. Scand.*, 1973, **27**, 1845.	178, 179
73ACS1987	C. E. Olsen; *Acta Chem. Scand.*, 1973, **27**, 1987.	714
73ACS2051	M. Begtrup; *Acta Chem. Scand.*, 1973, **27**, 2051.	266
73ACS2757	H. Rasmussen and E. Sletten; *Acta Chem. Scand.*, 1973, **27**, 2757.	509
73ACS3101	M. Begtrup; *Acta Chem. Scand.*, 1973, **27**, 3101.	190, 191, 192, 354
73AG447	C. Derycke, V. Jaeger, M. van Meersche and H. G. Viehe; *Angew. Chem.*, 1973, **85**, 447.	768, 785
73AG(E)240	M. Franck-Neumann and C. Buchecker; *Angew. Chem., Int. Ed. Engl.*, 1973, **12**, 240.	318
73AG(E)405	F. Hervens and H. G. Viehe; *Angew. Chem., Int. Ed. Engl.*, 1973, **12**, 405.	277
73AG(E)455	F. A. Neugebauer; *Angew. Chem., Int. Ed. Engl.*, 1973, **12**, 455.	792, 800
73AG(E)591	W. Saenger; *Angew. Chem., Int. Ed. Engl.*, 1973, **12**, 591.	523
73AG(E)806	H. G. Viehe and Z. Janousek; *Angew. Chem., Int. Ed. Engl.*, 1973, **12**, 806.	128
73AJC1585	J. B. Polya and M. Woodruff; *Aust. J. Chem.*, 1973, **26**, 1585.	768
73AJC2435	C. G. Begg, M. R. Grimmett and P. D. Wethey; *Aust. J. Chem.*, 1973, **26**, 2435.	377, 378
73AJC2683	D. J. Gale and J. F. K. Wilshire; *Aust. J. Chem.*, 1973, **26**, 2683.	246
73AP155	J. Reisch and W. F. Ossenkop; *Arch. Pharm. (Weinheim, Ger.)*, 1973, **306**, 155.	218
73AP209	H. Moehrle and S. Mayer; *Arch. Pharm. (Weinheim, Ger.)*, 1973, **306**, 209.	633
73AP561	A. Kreutzberger and R. Schucker; *Arch. Pharm. (Weinheim, Ger.)*, 1973, **306**, 561.	656
73AP697	A. Kreutzberger and R. Schucker; *Arch. Pharm. (Weinheim, Ger.)*, 1973, **306**, 697.	657
73AP730	A. Kreutzberger and R. Schucker; *Arch. Pharm. (Weinheim, Ger.)*, 1973, **306**, 730.	657
73AP801	A. Kreutzberger and R. Schucker; *Arch. Pharm. (Weinheim, Ger.)*, 1973, **306**, 801.	656
73AX(B)714	T. P. Singh and M. Vijayan; *Acta Crystallogr., Part B*, 1973, **29**, 714.	178
73AX(B)1974	T. J. Kistenmacher; *Acta Crystallogr., Part B*, 1973, **29**, 1974.	509
73B188	S. H. Blobstein, D. Grunberger, I. B. Weinstein and K. Nakanishi; *Biochemistry*, 1973, **12**, 188.	605
73BCJ506	Y. Ohtsuka; *Bull. Chem. Soc. Jpn.*, 1973, **46**, 506.	591
73BCJ667	Y. Hayasi, H. Nakamura and H. Nozaki; *Bull. Chem. Soc. Jpn.*, 1973, **46**, 667.	154
73BCJ1250	T. Isida, T. Akiyama, N. Mihara, S. Kozima and K. Sisido; *Bull. Chem. Soc. Jpn.*, 1973, **46**, 1250.	741, 766, 794
73BCJ1836	F. Yoneda, M. Higuchi, T. Matsumura and K. Senga; *Bull. Chem. Soc. Jpn.*, 1973, **46**, 1836.	580
73BCJ2176	T. Isida, T. Akiyama, K. Nabika, K. Sisido and S. Kozima; *Bull. Chem. Soc. Jpn.*, 1973, **46**, 2176.	818
73BCJ2215	I. Arai; *Bull. Chem. Soc. Jpn.*, 1973, **46**, 2215.	744
73BCJ2600	T. Nagai, Y. Fukushima, T. Kuroda, H. Shimizu, S. Sekiguchi and K. Matsui; *Bull. Chem. Soc. Jpn.*, 1973, **46**, 2600.	443
73BSF288	J. L. Aubagnac, J. Elguero and J. L. Gilles; *Bull. Soc. Chim. Fr.*, 1973, 288.	243
73BSF1849	J. L. Barascut, R. M. Claramunt and J. Elguero; *Bull. Soc. Chim. Fr.* 1973, 1849.	747
73BSF2676	C. Chapuis, A. Gauvreau, A. Klaebe, J. J. Périé and J. Roussel; *Bull. Soc. Chim. Fr.*, 1973, 2676.	354
73BSF2996	C. Malavaud, M.-T. Boisden and J. Barrans; *Bull. Soc. Chim. Fr.*, 1973, **11**, 2996.	17
73BSF3159	P. Bouchet and C. Coquelet; *Bull. Soc. Chim. Fr.*, 1973, 3159.	286
73BSF3401	J. Elguero and C. Marzin; *Bull. Soc. Chim. Fr.*, 1973, 3401.	278
73BSF3442	R. Mertz, D. Van Assche, J. P. Fleury and M. Regitz; *Bull. Soc. Chim. Fr.*, 1973, 3442.	713
73CB288	A. Gieren; *Chem. Ber.*, 1973, **106**, 288.	179
73CB745	W. Sucrow, M. Slopianka and C. Mentzel; *Chem. Ber.*, 1973, **106**, 745.	281
73CB1589	F. A. Neugebauer and H. Fischer; *Chem. Ber.*, 1973, **106**, 1589.	794
73CB2070	J. Reisch and W. F. Ossenkop; *Chem. Ber.*, 1973, **106**, 2070.	219
73CB2415	P. Schneiders, J. Heinze and H. Baumgartel; *Chem. Ber.*, 1973, **106**, 2415.	113
73CB2925	H. J. Schrepfer, L. Capuano and H.-L. Schmidt; *Chem. Ber.*, 1973, **106**, 2925.	821
73CB3092	A. Engelmann and W. Kirmse; *Chem. Ber.*, 1973, **106**, 3092.	281
73CB3203	W. Hutzenlaub, H. Yamamato, G. B. Barlin and W. Pfleiderer; *Chem. Ber.*, 1973, **106**, 3203.	592
73CB3391	A. Haas and V. Plass; *Chem. Ber.*, 1973, **106**, 3391.	130
73CB3421	K. Burger and K. Einhellig; *Chem. Ber.*, 1973, **106**, 3421.	780
73CC7	E. E. Schweizer, C. S. Zim, C. L. Labaw and W. P. Murray; *J. Chem. Soc., Chem. Commun.*, 1973, 7.	277
73CC402	M. J. Nye, M. J. O'Hare and W. P. Tang; *J. Chem. Soc., Chem. Commun.*, 1973, 402.	201
73CC917	T. Fujii, T. Itaya, K. Mohri and T. Saito; *J. Chem. Soc., Chem. Commun.*, 1973, 917.	586
73CHE88	A. D. Garnovskii, A. M. Simonov and V. I. Minkin; *Chem. Heterocycl. Compd. (Engl. Transl.)*, 1973, **9**, 88.	347, 349

73CHE465	S. S. Novikov, L. I. Khmel'nitskii, O. V. Lebedev, V. V. Sevastyanova and L. V. Epishina; *Chem. Heterocycl. Compd. (Engl. Transl.)*, 1973, **9**, 465.	395
73CHE1074	I. N. Somin, N. I. Shapranova and S. G. Kuznetsov; *Chem. Heterocycl. Compd. (Engl. Transl.)*, 1973, **9**, 1074.	436
73CHE1173	V. I. Ofitserov, Z. V. Pushkareva, V. S. Mokrushin and K. V. Aglitskaya; *Chem. Heterocycl. Compd. (Engl. Transl.)*, 1973, **9**, 1173.	652, 667
73CHE1175	L. B. Volodarskii; *Chem. Heterocycl. Compd. (Engl. Transl.)*, 1973, **9**, 1175.	455, 474
73CI(L)371	R. N. Butler; *Chem. Ind. (London)*, 1973, 371.	792
73CJC1741	A. Taurins and V. T. Khouw; *Can. J. Chem.*, 1973, **51**, 1741.	136
73CJC2315	R. N. Butler; *Can. J. Chem.*, 1973, **51**, 2315.	182, 183, 795
73CJC3765	J. Rokach, Y. Girard and J. G. Atkinson; *Can. J. Chem.*, 1973, **51**, 3765.	450
73CPB692	A. Yamazaki, T. Furukawa, M. Akiyama, M. Okutsu, I. Kumashiro, and M. Ikehara; *Chem. Pharm. Bull.*, 1973, **21**, 692.	588
73CPB1342	S. Kubota and M. Uda; *Chem. Pharm. Bull.*, 1973, **21**, 1342.	740
73CPB1470	M. Kamiya and Y. Akahori; *Chem. Pharm. Bull.*, 1973, **21**, 1470.	521
73CPB1676	T. Fujii, C. C. Wu, T. Itaya, S. Moro and T. Saito; *Chem. Pharm. Bull.*, 1973, **21**, 1676.	554, 555
73CPB2026	H. Ogura, K. Kubo, Y. Watanabe and T. Itoh; *Chem. Pharm. Bull.*, 1973, **21**, 2026.	248
73CPB2146	S. Suzue, M. Hirobe and T. Okamoto; *Chem. Pharm. Bull.*, 1973, **21**, 2146.	338
73CPB2517	T. Tsuchiya, H. Arai and H. Igeta; *Chem. Pharm. Bull.*, 1973, **21**, 2517.	880
73CR(C)(277)1153	M. Ferrey, A. Robert and A. Foucaud; *C.R. Hebd. Seances Acad. Sci., Ser. C*, 1973, **277**, 1153.	129
73CR(C)(277)1163	M. M. L. Pappalardo, J. Elguero and C. Marzin; *C.R. Hebd. Seances Acad. Sci., Ser. C*, 1973, **277**, 1163 (*Chem. Abstr.*, 1974, **80**, 70 186).	799
73CRV93	R. N. Butler, F. L. Scott and T. A. F. O'Mahony; *Chem. Rev.*, 1973, **73**, 93.	253, 254
73CSC469	F. Bechtel, J. Gaultier and C. Hauw; *Cryst. Struct. Commun.*, 1973, **3**, 469.	178, 181, 200
73CSC473	F. Bechtel, J. Gaultier and C. Hauw; *Cryst. Struct. Commun.*, 1973, **3**, 473.	178
73DOK(211)1369	V. N. Artemov, S. N. Baranov, N. A. Kovach ahd O. P. Shvaika; *Dokl. Akad. Nauk SSSR*, 1973, **211**, 1369.	776
73FES624	A. Gasco, V. Mortarini and E. Reynaud; *Farmaco Ed. Sci.*, 1973, **28**, 624.	787
73FRP2135297	C. J. Cavallito and A. P. Gray; *Fr. Pat.* 2 135 297 (1973) (*Chem. Abstr.*, 1973, **79**, 96 989).	904
73GEP2133998	*Ger. Pat.* 2 133 998 (1973) (*Chem. Abstr.*, 1973, **78**, 97 690).	647
73GEP2210633	H. Meyer, F. Bossert, W. Vater and K. Stoepel; *Ger. Pat.* 2 210 633 (1973) (*Chem. Abstr.*, 1973, **79**, 146 519).	634
73GEP2245447	W. J. Chambers; *Ger. Pat.* 2 245 447 (1973) (*Chem. Abstr.*, 1973, **79**, 5460).	788
73GEP2310049	E. F. George and W. D. Riddell; *Ger. Pat.* 2 310 049 (1973) (*Chem. Abstr.*, 1974, **80**, 23 539).	836
73GEP2314238	H. Moeller and C. Gloxhuber; *Ger. Pat.* 2 314 238 (1973) (*Chem. Abstr.*, 1975, **82**, 4263).	835
73HC(27)801	M. Tisler and B. Stanovnik; *Chem. Heterocycl. Compd.*, 1973, **27**, 801.	608, 628
73HC(27)807	M. Tisler and B. Stanovnik; *Chem. Heterocycl. Compd.*, 1973, **27**, 807.	608
73HC(27)818	M. Tisler and B. Stanovnik; *Chem. Heterocycl. Compd.*, 1973, **27**, 818.	608
73IC508	L. J. Guggenberger, C. T. Prewitt, P. Meakin, S. Trofimenko and J. P. Jesson; *Inorg. Chem.*, 1973, **12**, 508.	227
73IJC747	J. Mohan, V. K. Chadha and H. K. Pujari; *Indian J. Chem.*, 1973, **11**, 747.	663
73IJC753	P. B. Talukdar, S. K. Sengupta and A. K. Datta; *Indian J. Chem.*, 1973, **11**, 753.	775
73JA27	P. H. Kasai and D. McLeod; *J. Am. Chem. Soc.*, 1973, **95**, 27.	206
73JA324	H. Saitô, Y. Tanaka and S. Nagata; *J. Am. Chem. Soc.*, 1973, **95**, 324.	741
73JA4452	G. A. Rogers and T. C. Bruice; *J. Am. Chem. Soc.*, 1973, **95**, 4452.	660
73JA7174	C. R. Frihart and N. J. Leonard; *J. Am. Chem. Soc.*, 1973, **95**, 7174.	537
73JA8452	W. E. McEwen, P. E. Stott and C. M. Zepp; *J. Am. Chem. Soc.*, 1973, **95**, 8452.	76
73JAP7308667	M. Minagawa and M. Akutsu; *Jpn. Pat.* 73 08 667 (1973) (*Chem. Abstr.*, 1973, **79**, 19 662).	790
73JAP7389932	T. Nakajo, S. Maeda and T. Kurahashi; *Jpn. Pat.* 73 89 932 (1973) (*Chem. Abstr.*, 1974, **81**, 65 182).	788
73JCS(P1)164	F. T. Boyle and R. A. Y. Jones; *J. Chem. Soc., Perkin Trans. 1*, 1973, 164.	269
73JCS(P1)170	F. T. Boyle and R. A. Y. Jones; *J. Chem. Soc., Perkin Trans. 1*, 1973, 170.	269
73JCS(P1)221	C. W. Rees and M. Yelland; *J. Chem. Soc., Perkin Trans. 1*, 1973, 221.	253, 326
73JCS(P1)319	H. E. Foster and J. Hurst; *J. Chem. Soc., Perkin Trans. 1*, 1973, 319.	332
73JCS(P1)469	P. D. Dobbs and P. D. Magnus; *J. Chem. Soc., Perkin Trans. 1*, 1973, 469.	812
73JCS(P1)545	C. W. Rees and A. A. Sale; *J. Chem. Soc., Perkin Trans. 1*, 1973, 545.	719
73JCS(P1)555	T. L. Gilchrist, G. E. Gymer and C. W. Rees; *J. Chem. Soc., Perkin Trans. 1*, 973, 5525.	692, 706
73JCS(P1)705	R. Fielden, O. Meth-Cohn and H. Suschitzky; *J. Chem. Soc., Perkin Trans. 1*, 1973, 705.	380, 409
73JCS(P1)943	D. R. Sutherland, G. Tennant and R. J. S. Vevers; *J. Chem. Soc., Perkin Trans. 1*, 1973, 943.	894
73JCS(P1)1209	T. Hirata, H. B. Wood, Jr. and J. S. Driscoll; *J. Chem. Soc., Perkin Trans. 1*, 1973, 1209.	751
73JCS(P1)1357	R. N. Butler, T. M. Lambe, J. C. Tobin and F. L. Scott; *J. Chem. Soc., Perkin Trans. 1*, 1973, 1357.	754

73JCS(P1)1588	D. W. Dunwell and D. Evans; *J. Chem. Soc., Perkin Trans. 1*, 1973, 1588.	439
73JCS(P1)1620	A. Albert and W. Pendergast; *J. Chem. Soc., Perkin Trans. 1*, 1973, 1620.	876, 888
73JCS(P1)1625	A. Albert and W. Pendergast; *J. Chem. Soc., Perkin Trans. 1*, 1973, 1625.	721
73JCS(P1)1634	A. Albert; *J. Chem. Soc., Perkin Trans. 1*, 1973, 1634.	721
73JCS(P1)1794	A. Albert and W. Pendergast; *J. Chem. Soc., Perkin Trans. 1*, 1973, 1794.	639
73JCS(P1)1903	S. M. Hecht and D. Werner; *J. Chem. Soc., Perkin Trans. 1*, 1973, 1903.	272, 329
73JCS(P1)2089	T. Sasaki, K. Kanematsu, A. Kakehi and G. I. Ito; *J. Chem. Soc., Perkin Trans. 1*, 1973, 2089.	153
73JCS(P1)2371	G. A. Jaffari and A. J. Nunn; *J. Chem. Soc., Perkin Trans. 1*, 1973, 2371.	230
73JCS(P1)2506	K. Yamauchi and M. Kinoshita; *J. Chem. Soc., Perkin Trans. 1*, 1973, 2506.	229
73JCS(P1)2580	Y. Tamura, Y. Miki, Y. Sumida and M. Ikeda; *J. Chem. Soc., Perkin Trans. 1*, 1973, 2580.	337
73JCS(P1)2595	E. E. Glover, K. D. Vaughan and D. C. Bishop; *J. Chem. Soc., Perkin Trans. 1*, 1973, 2595.	635
73JCS(P1)2644	R. R. Atkins, S. E. J. Glue and I. T. Kay; *J. Chem. Soc., Perkin Trans. 1*, 1973, 2644.	132
73JCS(P1)2901	H. E. Foster and J. Hurst; *J. Chem. Soc., Perkin Trans. 1*, 1973, 2901.	310, 311, 315, 339
73JCS(P2)1	R. C. Seccombe and C. H. L. Kennard; *J. Chem. Soc., Perkin Trans. 2*, 1973, 1.	742
73JCS(P2)4	R. C. Seccombe and C. H. L. Kennard; *J. Chem. Soc., Perkin Trans. 2*, 1973, 4.	742
73JCS(P2)9	R. C. Seccombe and C. H. L. Kennard; *J. Chem. Soc., Perkin Trans. 2*, 1973, 9.	742
73JCS(P2)568	S. Bradamante, G. Pagani and A. Marchesini; *J. Chem. Soc., Perkin Trans. 2*, 1973, 568.	800
73JCS(P2)936	H. Ferres, M. S. Hamdam and W. R. Jackson; *J. Chem. Soc., Perkin Trans. 2*, 1973, 936.	278
73JCS(P2)1371	B. Fernandez, I. Perillo and S. Lamden; *J. Chem. Soc., Perkin Trans. 2*, 1973, 1371.	425
73JCS(P2)1675	S. Clementi, P. P. Forsythe, C. D. Johnson and A. R. Katritzky; *J. Chem. Soc., Perkin Trans. 2*, 1973, 1675.	239
73JCS(P2)1889	J. A. Elvidge, J. R. Jones, C. O'Brien, E. A. Evans and H. C. Sheppard; *J. Chem. Soc., Perkin Trans. 2*, 1973, 1889.	526, 527
73JCS(P2)2036	G. B. Ansell; *J. Chem. Soc., Perkin Trans. 2*, 1973, 2036.	796, 797
73JCS(P2)2138	J. A. Elvidge, J. R. Jones, C. O'Brien, E. A. Evans and H. C. Sheppard; *J. Chem. Soc., Perkin Trans. 2*, 1973, 2138.	526
73JHC5	H. Alper and R. W. Stout; *J. Heterocycl. Chem.*, 1973, **10**, 5.	793
73JHC201	M. Israel and L. C. Jones; *J. Heterocycl. Chem.*, 1973, **10**, 201.	616, 637
73JHC267	R. Weber and M. Renson; *J. Heterocycl. Chem.*, 1973, **10**, 267.	21
73JHC353	F. H. Case; *J. Heterocycl. Chem.*, 1973, **10**, 353.	766
73JHC363	L. M. Werbel, A. Curry, E. F. Elslager and C. Hess; *J. Heterocycl. Chem.*, 1973, **10**, 363.	637
73JHC417	B. Rayner, C. Tapiero and J.-L. Imbach; *J. Heterocycl. Chem.*, 1973, **10**, 417.	652, 513
73JHC419	L. B. Townsend, D. W. Miles, S. J. Manning and H. Eyring; *J. Heterocycl. Chem.*, 1973, **10**, 419.	523
73JHC431	M.-T. Chenon, R. J. Pugmire, D. M. Grant, R. P. Panzica and L. B. Townsend; *J. Heterocycl. Chem.*, 1973, **10**, 431.	308
73JHC569	H. Alper and R. W. Stout; *J. Heterocycl. Chem.*, 1973, **10**, 569.	793
73JHC755	R. Jacquier, H. Lopez and G. Maury; *J. Heterocycl. Chem.*, 1973, **10**, 755.	613
73JHC801	J. S. Bradshaw, B. Stanovnik and M. Tišler; *J. Heterocycl. Chem.*, 1973, **10**, 801.	870, 871, 880
73JHC821	K. T. Potts, R. Dugas and C. R. Surapareni; *J. Heterocycl. Chem.*, 1973, **10**, 821.	339
73JHC849	R. Roe, Jr., J. S. Paul and P. O'B. Montgomery, Jr.; *J. Heterocycl. Chem.*, 1973, **10**, 849.	526
73JHC885	A. H. Albert, R. K. Robins and D. E. O'Brien; *J. Heterocycl. Chem.*, 1973, **10**, 885.	272
73JHC1021	R. P. Rao, R. K. Robins and D. E. O'Brien; *J. Heterocycl. Chem.*, 1973, **10**, 1021.	647
73JHC1055	J. W. A. M. Janssen, C. G. Kruse, H. J. Koeners and C. L. Habraken; *J. Heterocycl. Chem.*, 1973, **10**, 1055.	198, 224, 225
73JMC292	C. Temple, Jr., B. H. Smith, R. D. Elliott and J. A. Montgomery; *J. Med. Chem.*, 1973, **16**, 292.	636
73JMC312	E. B. Åkerblom and D. E. S. Campbell; *J. Med. Chem.*, 1973, **16**, 312.	754
73JMC537	J. H. Jones, W. J. Holtz and E. J. Cragoe, Jr.; *J. Med. Chem.*, 1973, **16**, 537.	646
73JMC643	R. A. Sharma, M. Bobek, F. E. Cole and A. Bloch; *J. Med. Chem.*, 1973, **16**, 643.	626, 645
73JMC1272	G. J. Durant, J. M. Loynes and S. H. B. Wright; *J. Med. Chem.*, 1973, **16**, 1272.	614
73JMC1296	H. Berner, H. Reinshagen and M. A. Koch; *J. Med. Chem.*, 1973, **16**, 1296.	622, 640
73JMR(10)130	J. P. Jacobsen, O. Snerling, E. J. Pedersen, J. T. Nielsen and K. Schaumburg; *J. Magn. Reson.*, 1973, **10**, 130.	196, 197, 290
73JMR(12)225	M. C. Thorpe and W. C. Coburn, Jr.; *J. Magn. Reson.*, 1973, **12**, 225.	354
73JOC613	C. Temple, Jr., B. H. Smith, Jr. and J. A. Montgomery; *J. Org. Chem.*, 1973, **38**, 613.	636
73JOC675	J.-M. Vandensavel, G. Smets and G. L'abbé; *J. Org. Chem.*, 1973, **38**, 675.	794, 832
73JOC806	E. C. Taylor and R. C. Portnoy; *J. Org. Chem.*, 1973, **38**, 806.	118
73JOC825	A. S. Katner; *J. Org. Chem.*, 1973, **38**, 825.	318
73JOC844	R. M. Kellogg; *J. Org. Chem.*, 1973, **38**, 844.	147
73JOC1095	C. Temple, Jr., B. H. Smith, Jr. and J. A. Montgomery; *J. Org. Chem.*, 1973, **38**, 1095.	867, 887

73JOC1437	J. E. Oliver and P. E. Sonnet; *J. Org. Chem.*, 1973, **38**, 1437.	474
73JOC1769	K. T. Potts and A. J. Elliott; *J. Org. Chem.*, 1973, **38**, 1769.	116
73JOC2049	E. Abushanab, A. P. Bindra, L. Goodman and H. Peterson, Jr.; *J. Org. Chem.*, 1973, **38**, 2049.	625, 644
73JOC2708	J. Tanaka and S. I. Miller; *J. Org. Chem.*, 1973, **38**, 2708.	700, 709
73JOC2821	J. E. McMurry and A. P. Coppolino; *J. Org. Chem.*, 1973, **38**, 2821.	726
73JOC2887	M. Kawazu, T. Kanno, S. Yamamura, T. Mizoguchi and S. Saito; *J. Org. Chem.*, 1973, **38**, 2887.	604
73JOC2891	N. Takamura, N. Taga, T. Kanno and M. Kawazu; *J. Org. Chem.*, 1973, **38**, 2891.	604
73JOC3087	K. T. Potts and J. Kane; *J. Org. Chem.*, 1973, **38**, 3087.	131
73JOC3420	J. Salomon and D. Elad; *J. Org. Chem.*, 1973, **38**, 3420.	545
73JOC3495	H. J. J. Loozen and E. F. Godefroi; *J. Org. Chem.*, 1973, **38**, 3495.	437, 489
73JOC3762	D. S. Noyce and G. T. Stowe; *J. Org. Chem.*, 1973, **38**, 3762.	395
73JOC3947	R. A. Coburn, B. Bhooshan and R. A. Glennon; *J. Org. Chem.*, 1973, **38**, 3947.	763
73JOC3995	R. Y. Ning, P. B. Madan and L. H. Sternbach; *J. Org. Chem.*, 1973, **38**, 3995.	335
73JOC4324	Y. Tamura, K. Sumoto, H. Matsushima, H. Taniguchi and M. Ikeda; *J. Org. Chem.*, 1973, **38**, 4324.	166
73JOC4353	S. R. Naik, J. T. Witkowski and R. K. Robins; *J. Org. Chem.*, 1973, **38**, 4353.	758
73JOC4379	J. V. Hay, D. E. Portlock and J. F. Wolfe; *J. Org. Chem.*, 1973, **38**, 4379.	430, 448
73JPC1629	A. Samuni and P. Neta; *J. Phys. Chem.*, 1973, **77**, 1629.	206
73JPR497	H. Hartmann, H. Schäfer and K. Gewald; *J. Prakt. Chem.*, 1973, **315**, 497.	140
73JPR539	K. Gewald, M. Hentschel and R. Heikel; *J. Prakt. Chem.*, 1973, **315**, 539.	136
73JPS839	A. Shafiee, I. Lalezari, S. Yazdany and R. Pournorouz; *J. Pharm. Sci.*, 1973, **62**, 839.	116
73JPS1011	J. R. Nulu and J. Nematollahi; *J. Pharm. Sci.*, 1973, **62**, 1011.	652
73KGS64	P. B. Terent'ev, S. M. Vinogradova, A. N. Kost and A. G. Strukovski; *Khim. Geterotsikl. Soedin.*, 1973, 64.	204, 205
73KGS138	B. A. Tertov and P. P. Onishchenko; *Khim. Geterotsikl. Soedin.*, 1973, 138.	243
73KGS138	Y. M. Yutilov and I. A. Svertilova; *Khim. Geterotsikl. Soedin.*, 1973, 138 (*Chem. Abstr.*, 1973, **78**, 97 547).	620
73KGS570	R. M. Bystrova and Y. M. Yutilov; *Khim. Geterotsikl. Soedin.*, 1973, 570 (*Chem. Abstr.*, 1973, **79**, 31 981).	616
73KGS707	V. V. Mel'nikov, V. V. Stolpakova, M. S. Pevzner and B. V. Gidaspov; *Khim. Geterotsikl. Soedin.*, 1973, 707.	740
73KGS1345	A. Ya. Lazaris, S. M. Shmuilovich and A. N. Egorochkin; *Khim. Geterotsikl. Soedin.*, 1973, 1345.	772
73KGS1570	V. P. Shchipanov, K. L. Krashina and A. A. Skachilova; *Khim. Geterotsikl. Soedin.*, 1973, 1570 (*Chem. Abstr.*, 1974, **80**, 59 898).	827
73KGS1686	T. N. Pliev, R. M. Bystrova and Y. M. Yutilov; *Khim. Geterotsikl. Soedin.*, 1973, 1686 (*Chem. Abstr.*, 1974, **80**, 82 803).	616
73LA103	H. J. Willenbrock, H. Wamhoff and F. Korte; *Liebigs Ann. Chem.*, 1973, 103.	647
73LA278	E. Hugl, G. Schulz and E. Zbiral; *Liebigs Ann. Chem.*, 1973, 278.	648
73LA380	K. Dimroth and K. Severin; *Liebigs Ann. Chem.*, 1973, 380.	353
73LA578	U. Heep; *Liebigs Ann. Chem.*, 1973, 578.	716
73LA636	W. Walter and M. Radke; *Liebigs Ann. Chem.*, 1973, 636.	698, 699
B-73MI40100	M. Witanowski, L. Stefaniak and H. Januszewski; in 'Nitrogen NMR', ed. M. Witanowski and G. A. Webb; Plenum, New York, 1973, p. 163.	20
73MI40400	F. Török, A. Hegedüs and P. Pulay; *Theor. Chim. Acta*, 1973, **32**, 145.	173, 176, 177, 181
B-73MI40401	M. Witanowski and G. A. Webb; 'Nitrogen NMR', Plenum Press, New York, 1973, 163.	195
B-73MI40402	H. C. van der Plas; 'Ring Transformations of Heterocycles', Academic Press, New York, 1973, vols. I and II.	286
B-73MI40403	A. J. Boulton: in 'Aromatic and Heteroaromatic Chemistry', The Chemical Society, London, 1973, vol. 1, p. 156.	286
73MI40700	I. Butula; *Croat. Chem. Acta*, 1973, **45**, 297 (*Chem. Abstr.*, 1973, **79**, 92 099).	419
B-73MI40701	F. G. Riddell; in 'Aliphatic, Alicyclic, and Saturated Heterocyclic Chemistry', The Chemical Society, London, 1973, vol. 1, part 3, p. 31.	425, 426, 444
B-73MI40800	M. R. Grimmett; in 'MTP International Review of Science, Organic Chemistry, Series 1', ed. K. Schofield; Butterworths, London, 1973, vol. 4, p. 55.	458, 459, 472
B-73MI40801	H. C. van der Plas; in 'Ring Transformations of Heterocycles', Academic, New York, 1973, vol. 2.	493
B-73MI40900	J. Brooks and G. Shaw; 'Origin and Development of Living Systems', Academic, London, 1973.	519, 567
B-73MI40901	'Physicochemical Properties of Nucleic Acids', ed. J. Duchesne; Academic, London, 1973, vol. 1.	501, 506, 520, 523
B-73MI40902	'Synthetic Procedures in Nucleic Acid Chemistry', ed. W. W. Zorbach and R. S. Tipson; Wiley-Interscience, 1973, vol. 2.	501, 504, 510, 511, 512, 514, 516, 517, 518, 524
73MI41100	Yu. L. Frolov, V. B. Mantisivoda, V. B. Modonov, S. N. Elovskii, E. S. Domnina and G. G. Skvortsova; *Teor. Eksp. Khim.*, 1973, **9**, 238 (*Chem. Abstr.*, 1973, **79**, 41 757).	672
73MI41101	P. Mauret, J. P. Fayet, M. Fabre, J. Elguero and M. del C. Pardo; *J. Chim. Phys. Phys. Chim. Biol.*, 1973, **70**, 1483 (*Chem. Abstr.*, 1974, **80**, 59 218).	677

73MI41102	V. P. Chernetskii, N. A. Petrusha and J. V. Alekseeva; *Fiziol. Aktiv. Veshchestra*, 1973, **5**, 121.	729
B-73MI41103	M. Witanowski and G. A. Webbs; 'Nitrogen NMR', Plenum Press, London, 1973.	684
73MI41300	M. Witanowski, L. Stefaniak, H. Januszewski and J. Elguero; *J. Chim. Phys. Phys. Chim. Biol.*, 1973, **70**, 697.	800
73MI41301	E. Lipmann, H. Löster and A. Antonowa; *J. Signal AM2*, 1973, **2**, 209 (*Chem. Abstr.*, 1975, **82**, 30 496).	802
73NKK996	K. Takahashi, K. Mitsuhashi and T. Zaima; *Nippon Kagaku Kaishi*, 1973, 996 (*Chem. Abstr.*, 1973, **79**, 31 986).	435
B-73NMR	T. J. Batterham; 'NMR Spectra of Simple Heterocycles', Wiley, New York, 1973.	13, 15, 182, 183, 184, 185, 186, 187, 188, 189, 190, 510, 678, 858
73OMR(5)453	R. Danion-Bougot and R. Carrié; *Org. Magn. Reson.*, 1973, **5**, 453.	185
73OMS(7)57	A. J. Blackman and J. H. Bowie; *Org. Mass Spectrom.*, 1973, **7**, 57.	741
73OMS(7)271	A. Maquestiau, Y. Van Haverbeke, R. Flammang, and J. Elguero; *Org. Mass Spectrom.*, 1973, **7**, 271.	686
73OMS(7)1165	J. van Thuijl, K. J. Klebe and J. J. van Houte; *Org. Mass Spectrom.*, 1973, **7**, 1165.	202
73OMS(7)1267	A. Maquestiau, Y. Van Haverbeke, R. Flammang, M. C. Pardo and J. Elguero; *Org. Mass Spectrom.*, 1973, **7**, 1267.	688
73OPP105	T. Caronna and L. Filini; *Org. Prep. Proced. Int.*, 1973, 105.	247
73P2445	D. S. Letham; *Phytochemistry*, 1973, **12**, 2445.	601
73PC41100	K. M. Kem, S. M. Masson, Y. K. Kim and M. E. Munk; personal communication, 1973.	703
73S71	P. K. Kadaba; *Synthesis*, 1973, 71.	717, 792, 829, 833
73S123	M. Tišler; *Synthesis*, 1973, 123.	792, 826, 859, 860, 881
73S300	F. Yoneda and T. Nagamatsu; *Synthesis*, 1973, 300.	337
73T101	M. K. Saxena, M. N. Gudi and M. V. George; *Tetrahedron*, 1973, **29**, 101.	248
73T195	P. Ykman, G. L'abbé and G. Smets; *Tetrahedron*, 1973, **29**, 195.	717
73T435	J. P. Marquet, J. D. Bourzat, J. Andre-Louisfert and E. Bisagni; *Tetrahedron*, 1973, **29**, 435.	309, 326
73T441	J. D. Bourzat, J. P. Marquet, A. Civier and E. Bisagni; *Tetrahedron*, 1973, **29**, 441.	308, 319
73T1135	J. P. Deleux, G. Leroy and J. Weiler; *Tetrahedron*, 1973, **29**, 1135.	185
73T1413	A. Kreutzberger and R. Schucker; *Tetrahedron*, 1973, **29**, 1413.	656
73T1833	L. Schrader; *Tetrahedron*, 1973, **29**, 1833.	252
73T1983	O. Tsuge, K. Sakai and M. Tashiro; *Tetrahedron*, 1973, **29**, 1983.	164
73T2173	S. Cradock, R. H. Findlay and M. H. Palmer; *Tetrahedron*, 1973, **29**, 2173.	205, 361, 688, 689
73T3271	Y. Tanaka, S. R. Velen and S. I. Miller; *Tetrahedron*, 1973, **29**, 3271.	699
73T3285	Y. Tanaka and S. I. Miller; *Tetrahedron*, 1973, **29**, 3285.	700
73T3463	M. A. Schroeder, R. C. Makino and W. M. Tolles; *Tetrahedron*, 1973, **29**, 3463.	672, 796
73T3469	M. A. Schroeder and R. C. Makino; *Tetrahedron*, 1973, **29**, 3469.	172, 175, 672
73T4045	M. K. Eberle, G. G. Kahle and S. M. Talati; *Tetrahedron*, 1973, **29**, 4045.	256
73T4049	M. K. Eberle and G. G. Kahle; *Tetrahedron*, 1973, **29**, 4049.	257
73TH40400	E. Henry; Ph.D. Thesis, University of Montpellier, 1973.	201
73TL469	G. Bram; *Tetrahedron Lett.*, 1973, 469.	786
73TL891	P. Bouchet, C. Coquelet, J. Elguero and R. Jacquier; *Tetrahedron Lett.*, 1973, 891.	219
73TL1137	H. Taguchi; *Tetrahedron Lett.*, 1973, 1137.	692
73TL1417	T. Yamazaki and H. Schechter; *Tetrahedron Lett.*, 1973, 1417.	318
73TL1677	S. Polanc, B. Verček, B. Stanovnik and M. Tišler; *Tetrahedron Lett.*, 1973, 1677.	854
73TL2101	J. E. Herweh and R. M. Fantazier; *Tetrahedron Lett.*, 1973, 2101.	784, 787
73TL3781	R. Baumes, J. Elguero, R. Jacquier and G. Tarrago; *Tetrahedron Lett.*, 1973, 3781.	249
73TL4295	A. Gelléri and A. Messner; *Tetrahedron Lett.*, 1973, 4295.	793
73TL4619	I. A. Korbukh, F. F. Blanco and M. N. Preobrazhenskaja; *Tetrahedron Lett.*, 1973, 4619.	289
73TL5075	K. Maeda, T. Hosokawa, S. Murahashi and I. Moritani; *Tetrahedron Lett.*, 1973, 5075.	136
73UKZ277	N. S. Miroshnichenko, O. I. Shkrebtii and A. V. Stetsenko; *Ukr. Khim. Zh.* (*Russ. Ed.*), 1973, **39**, 277 (*Chem. Abstr.*, 1973, **79**, 19 020).	636
73UKZ350	N. S. Miroshnichenko, I. G. Ryabokon and A. V. Stetsenko; *Ukr. Khim. Zh.* (*Russ. Ed.*), 1973, **39**, 350 (*Chem. Abstr.*, 1973, **79**, 32 233).	640
73UKZ703	A. V. Stetsenko and N. S. Miroshnichenko; *Ukr. Khim. Zh.* (*Russ. Ed.*), 1973, **39**, 703 (*Chem. Abstr.*, 1973, **79**, 105 515).	640
73ZC176	H. Kunzek and G. Barnckow; *Z. Chem.*, 1973, **13**, 176.	135
73ZC429	E. Lipmann, R. Widera and E. Kleinpeter; *Z. Chem.*, 1973, **13**, 429 (*Chem. Abstr.*, 1974, **80**, 82 829).	821
73ZOB1683	A. N. Pudovik, N. G. Khusainova and Z. A. Nasyballina; *Zh. Obshch. Khim.*, 1973, **43**, 1683.	716
73ZOR2006	I. I. Afonina and F. Y. Perveev; *Zh. Org. Khim.*, 1973, **9**, 2006 (*Chem. Abstr.*, 1974, **80**, 27 171).	634
73ZOR2535	N. N. Mel'nikova, M. S. Pevzner, N. M. Malysheva and L. I. Bagal; *Zh. Org. Khim.*, 1973, **9**, 2535.	756
74ACS(B)61	M. Begtrup; *Acta Chem. Scand., Ser. B*, 1974, **28**, 61.	194, 355, 674, 799
74ACS(B)425	C. E. Olsen; *Acta Chem. Scand., Ser. B*, 1974, **28**, 425.	703

Ref	Citation	Pages
74AG(E)47	J. B. Hendrickson; *Angew. Chem., Int. Ed. Engl.*, 1974, **13**, 47.	282
74AG(E)79	T. van Vyve and H. G. Viehe; *Angew. Chem., Int. Ed. Engl.*, 1974, **13**, 79.	126, 277
74AG(E)627	T. Kauffmann; *Angew. Chem., Int. Ed. Engl.*, 1974, **13**, 627.	282
74AHC(16)1	J. A. Elvidge, J. R. Jones, C. O'Brien, E. A. Evans and H. C. Sheppard; *Adv. Heterocycl. Chem.*, 1974, **16**, 1.	69, 70, 245
74AHC(16)33	T. L. Gilchrist and G. E. Gymer; *Adv. Heterocycl. Chem.*, 1974, **16**, 33.	46, 53, 54, 55, 58, 69, 92, 94, 97, 105, 106, 108, 109, 670, 695, 705, 708
74AHC(16)123	F. Minisci and O. Porta; *Adv. Heterocycl. Chem.*, 1974, **16**, 123.	73, 419
74AHC(17)99	R. Lakhan and B. Ternai; *Adv. Heterocycl. Chem.*, 1974, **17**, 99.	55, 57, 58, 59, 60, 62, 63, 65, 66, 69, 70, 74, 76, 88, 91, 92, 93, 97, 106, 107
74AHC(17)213	H.-J. Timpe; *Adv. Heterocycl. Chem.*, 1974, **17**, 213.	72, 109
74AHC(17)255	M. J. Cook, A. R. Katritzky and P. Linda; *Adv. Heterocycl. Chem.* 1974, **17**, 255.	32, 33, 208
74AJC2041	J. F. K. Wilshire; *Aust. J. Chem.*, 1974, **27**, 2041.	281
74AJC2267	N. A. Evans, D. E. Rivett and J. F. K. Wilshire; *Aust. J. Chem.*, 1974, **27**, 2267.	254
74AJC2331	B. E. Boulton and B. A. W. Coller; *Aust. J. Chem.*, 1974, **27**, 2331.	400
74AJC2343	B. E. Boulton and B. A. W. Coller; *Aust. J. Chem.*, 1974, **27**, 2343.	241
74AJC2447	M. Woodruff and J. B. Polya; *Aust. J. Chem.*, 1974, **27**, 2447.	751, 782
74AP177	T. Denzel; *Arch. Pharm. (Weinheim, Ger.)*, 1974, **307**, 177.	331
74AP211	J. Reisch and A. Fitzek; *Arch. Pharm. (Weinheim, Ger.)*, 1974, **307**, 211.	219
74AP444	H. H Otto; *Arch. Pharm. (Weinheim, Ger.)*, 1974, **307**, 444.	330
74AP504	K. Sunderdiek and G. Zinner; *Arch. Pharm. (Weinheim, Ger.)*, 1974, **307**, 504.	750
74AX(B)166	T. J. Kistenmacher and T. Shigematsu; *Acta Crystallogr., Part B*, 1974, **30**, 166.	508
74AX(B)273	L. Fanfani, A. Nunzi, P. F. Zanazzi and A. R. Zanzari; *Acta Crystallogr., Part B*, 1974, **30**, 273.	179
74AX(B)557	T. P. Singh and M. Vijayan; *Acta Crystallogr., Part B*, 1974, **30**, 557.	178
74AX(B)590	J. Toussaint, O. Dideberg and L. Dupont; *Acta Crystallogr., Part B*, 1974, **30**, 590.	179
74AX(B)1490	A. Escande, J. L. Galigné and J. LaPasset; *Acta Crystallogr., Part B*, 1974, **30**, 1490.	675
74AX(B)1528	T. J. Kistenmacher and T. Shigematsu; *Acta Crystallogr., Part B*, 1974, **30**, 1528.	508
74AX(B)2009	A. Escande and J. Lapasset; *Acta Crystallogr., Part B*, 1974, **30**, 2009.	179
74AX(B)2246	C. Brassy, A. Renard, J. Delettré and J. P. Mornon; *Acta Crystallogr., Part B*, 1974, **30**, 2246.	227
74AX(B)2500	C. Brassy, J. P. Mornon, J. Delettré and J. Lepicard; *Acta Crystallogr., Part B*, 1974, **30**, 2500.	227
74AX(B)2505	R. L. Harlow and S. H. Simonsen; *Acta Crystallogr., Part B*, 1974, **30**, 2505.	178
74BBA(361)231	S. D. Rose; *Biochim. Biophys. Acta*, 1974, **361**, 231.	627
74BCJ785	T. Kitazume and N. Ishikawa; *Bull. Chem. Soc. Jpn.*, 1974, **47**, 785.	471
74BSF673	A. Darchen and D. Peltier; *Bull. Soc. Chim. Fr.*, 1974, 673.	472
74BSF768	P. Corbon, G. Barbey, A. Dupre and C. Caullet; *Bull. Soc. Chim. Fr.*, 1974, 768.	254
74BSF1395	A. Étienne, A. Le Berre and J.-P. Giorgetti; *Bull. Soc. Chim. Fr.*, 1974, 1395.	853
74BSF1453	P. Guerret, R. Jacquier, H. Lopez and G. Maury; *Bull. Soc. Chim. Fr.*, 1974, 1453.	630, 653
74BSF2547	R. Baumes, R. Jacquier and G. Tarrago; *Bull. Soc. Chim. Fr.*, 1974, 2547.	248
74CB715	W. Winter and E. Müller; *Chem. Ber.*, 1974, **107**, 715.	709
74CB1318	W. Sucrow, C. Mentzel and M. Slopianka; *Chem. Ber.*, 1974, **107**, 1318.	189
74CB2127	W. Winter and E. Muller; *Chem. Ber.*, 1974, **107**, 2127.	229
74CB2284	H. R. Rackwitz and K. H. Scheit; *Chem. Ber.*, 1974, **107**, 2284.	597
74CB3454	C. Reichard and E. U. Würthwein; *Chem. Ber.*, 1974, **107**, 3454.	281
74CC339	G. V. Boyd, T. Norris and P. F. Lindley; *J. Chem. Soc., Chem. Commun.*, 1974, 339.	178
74CC486	T. L. Gilchrist, C. J. Harris, C. J. Moody and C. W. Rees; *J. Chem. Soc., Chem. Commun.*, 1974, 486.	872
74CC551	F. Yoneda, S. Matsumoto and M. Higuchi; *J. Chem. Soc., Chem. Commun.*, 1974, 551.	581
74CC657	J. J. Barr, R. C. Storr and J. Rimmer; *J. Chem. Soc., Chem. Commun.*, 1974, 657.	794
74CC659	M. J. Rance, C. W. Rees, P. Spagnolo and R. C. Storr; *J. Chem. Soc., Chem. Commun.*, 1974, 659.	794
74CC702	M. Begtrup; *J. Chem. Soc., Chem. Commun.*, 1974, 702.	197, 680, 681
74CC941	C. W. Rees, R. W. Stephenson and R. C. Storr; *J. Chem. Soc., Chem. Commun.*, 1974, 941.	307, 339
74CC957	S. Mansy and R. Tobias; *J. Chem. Soc., Chem. Commun.*, 1974, 957.	527
74CHE358	I. I. Popov, A. M. Simonov, V. I. Mikhailov and N. A. Sil'vanovich; *Chem. Heterocycl. Compd. (Engl. Transl.)*, 1974, **10**, 358.	437
74CHE376	V. I. Ofitserov, Z. V. Pushkareva, V. S. Mokrushin and T. V. Rapakova; *Chem. Heterocycl. Compd. (Engl. Transl.)*, 1974, **10**, 376.	629
74CHE471	V. S. Troitskaya, N. D. Konevskaya, V. G. Vinokurov and V. I. Tyulin; *Chem. Heterocycl. Compd. (Engl. Transl.)*, 1974, **10**, 471.	199, 290
74CHE813	Z. I. Moskalenko and G. P. Shumelyak; *Chem. Heterocycl. Compd. (Engl. Transl.)*, 1974, **10**, 813.	116
74CHE992	B. E. Mandrichenko, I. A. Mazur and P. M. Kochergin; *Chem. Heterocycl. Compd. (Engl. Transl.)*, 1974, **10**, 992.	647
74CHE1215	V. S. Troitskaya, Yu. D. Timoshenkova, Yu. A. Pentin and V. I. Tyulin; *Chem. Heterocycl. Compd. (Engl. Transl.)*, 1974, **10**, 1215.	199

74CHE1217	G. G. Skvortsova, N. D. Abramova and B. V. Trzhtsinskaya; *Chem. Heterocycl. Compd.* (*Engl. Transl.*), 1974, **10**, 1217.	444, 446, 450
74CHE1225	B. I. Khristich, G. M. Sudorova and A. M. Simonov; *Chem. Heterocycl. Compd.* (*Engl. Transl.*), 1974, **10**, 1225.	438
74CHE1491	I. I. Popov and A. M. Simonov; *Chem. Heterocycl. Compd.* (*Engl. Transl.*), 1974, **10**, 1491.	437
74CHE1492	M. V. Povstyanoi, P. M. Kochergin, E. V. Logachev, E. A. Yakubovskii, A. V. Akimov and V. V. Androsov; *Chem. Heterocycl. Compd.* (*Engl. Transl.*), 1974, **10**, 1492.	654
74CI(L)38	K. Akagane and G. G. Allan; *Chem. Ind.* (*London*), 1974, 38.	483
74CJC833	R. E. Wasylishen, J. B. Rowbotham and T. Schaefer; *Can. J. Chem.*, 1974, **52**, 833.	13
74CJC2296	J. W. ApSimon, J. Elguero, A. Fruchier, D. Mathieu and R. Phan Tan Luu; *Can. J. Chem.*, 1974, **52**, 2296.	189
74CJC3474	A. R. Bassindale and A. G. Brook; *Can. J. Chem.*, 1974, **52**, 3474.	254
74CJC3981	S. Fischer, L. K. Peterson and J. F. Nixon; *Can. J. Chem.*, 1974, **52**, 3981.	455
74CL185	T. Akiyama, T. Kitamura, T. Isida and M. Kawanisi; *Chem. Lett.*, 1974, 185.	810
74CL409	H. Iwasaki; *Chem. Lett.*, 1974, 409.	508
74CL951	J. H. Lee, A. Matsumoto, M. Yoshida and O. Simamura; *Chem. Lett.*, 1974, 951.	163
74CPB70	I. Adachi, R. Miyazaki and H. Kano; *Chem. Pharm. Bull.*, 1974, **22**, 70.	68
74CPB207	K. Arakawa, T. Miyasaka and H. Ochi; *Chem. Pharm. Bull.*, 1974, **22**, 207.	232, 264
74CPB214	K. Arakawa, T. Miyasaka and H. Ochi; *Chem. Pharm. Bull.*, 1974, **22**, 214.	232, 264
74CPB1269	Y. Maki, M. Suzuki, K. Izuta and S. Iwai; *Chem. Pharm. Bull.*, 1974, **22**, 1269.	338
74CPB1938	G. T. Huang, T. Okamoto, M. Maeda and Y. Kawazoe; *Chem. Pharm. Bull.* 1974, **22**, 1938.	771
74CPB2142	A. Nakemura and S. Kamiya; *Chem. Pharm. Bull.*, 1974, **22**, 2142.	314
74CPB2466	T. Fujii, S. Kawakatsu and T. Itaya; *Chem. Pharm. Bull.*, 1974, **22**, 2466.	554
74CR(C)(278)359	F. Abjean; *C.R. Hebd. Seances Acad. Sci., Ser. C*, 1974, **278**, 359.	248
74CR(C)(279)717	R. Faure, J.-R. Llinas, E. J. Vincent and J.-L. Larice; *C. R. Hebd. Seances Acad. Sci., Ser. C*, 1974, **279**, 717.	20
74CRV279	P. N. Preston; *Chem. Rev.*, 1974, **74**, 279. 390, 408, 409, 412, 420, 429, 432, 445, 353, 356, 358, 359, 361, 458, 460, 469, 470, 472, 490, 493, 495	
74CS(6)222	O. Thorstad and K. Undheim; *Chem. Scr.*, 1974, **6**, 222.	858
74CSC713	P. Domiano and A. Musati; *Cryst. Struct. Commun.*, 1974, **3**, 713.	178
74CZ512	R. Neidlein and H. G. Hege; *Chem.-Ztg.*, 1974, **98**, 512.	777
74GEP2304321	K. Lohmer, A. von Koenig and A. Mueller; *Ger. Pat.* 2 304 321 (1974) (*Chem. Abstr.*, 1974, **81**, 144 229).	788
74GEP2314488	H. Staehle, H. Koeppe, W. Kummer and W. Hoefke; *Ger. Pat.* 2 314 488 (1974) (*Chem. Abstr.*, 1974, **82**, 4317).	658
74GEP2513864	*Ger. Pat.* 2 513 864 (1974) (*Chem. Abstr.*, 1974, **84**, 5005).	647
74H(2)473	T. Nishiwaki; *Heterocycles*, 1974, **2**, 473.	202
74HCA750	F. Troxler, H. P. Weber, A. Jaunin and H. R. Loosli; *Helv. Chim. Acta*, 1974, **57**, 750.	633
74HCA873	C. B. Chapleo and A. S. Dreiding; *Helv. Chim. Acta*, 1974, **57**, 873.	308
74HCA1259	C. B. Chapleo and A. S. Dreiding; *Helv. Chim. Acta*, 1974, **57**, 1259.	318
74HCA1382	P. Gilgen, H. Heimgartner and H. Schmid; *Helv. Chim. Acta*, 1974, **57**, 1382.	780
74JA1465	G. E. Wilson, Jr. and T. J. Bazzone; *J. Am. Chem. Soc.*, 1974, **96**, 1465.	18
74JA1957	Y. Akasaki and A. Ohno; *J. Am. Chem. Soc.*, 1974, **96**, 1957.	446
74JA2010	J. P. Ferris and F. R. Antonucci; *J. Am. Chem. Soc.*, 1974, **96**, 2010.	221, 436
74JA2014	J. P. Ferris and F. R. Antonucci; *J. Am. Chem. Soc.*, 1974, **96**, 2014.	221
74JA2481	G. A. Rogers and T. C. Bruice; *J. Am. Chem. Soc.*, 1974, **96**, 2481.	660
74JA3335	P. L. Gravel and W. H. Pirkle; *J. Am. Chem. Soc.*, 1974, **96**, 3335.	752
74JA3708	G. E. Palmer, J. R. Bolton and D. R. Arnold; *J. Am. Chem. Soc.*, 1974, **96**, 3708.	251
74JA3973	G. L'abbé, E. Van Loock, R. Albert, S. Toppet, G. Verhelst and G. Smets; *J. Am. Chem. Soc.*, 1974, **96**, 3973.	832
74JA4276	K. T. Potts and D. McKeough; *J. Am. Chem. Soc.*, 1974, **96**, 4276.	116
74JA4962	R. B. Meyer, Jr., D. A. Shuman and R. K. Robins; *J. Am. Chem. Soc.*, 1974, **96**, 4962.	587, 589, 590
74JA5894	N. J. Leonard and C. R. Frihart; *J. Am. Chem. Soc.*, 1974, **96**, 5894.	537
74JA7797	A. M. Feinberg, K. Nakanishi, J. Barciszewski, A. J. Rafalski, H. Augustyniak and M. Wiewiórowski; *J. Am. Chem. Soc.*, 1974, **96**, 7797.	605
74JA7830	S. Kammula and P. B. Shevlin; *J. Am. Chem. Soc.*, 1974, **96**, 7830.	822
74JAP7420175	A. Kotone, J. Fujita and M. Hoda; *Jpn. Pat.* 74 20 175 (1974) (*Chem. Abstr.*, 1974, **81**, 105 517).	746
74JAP74126679	A. Kotone, M. Hoda and T. Fujita; *Jpn. Pat.* 74 126 679 (1974) (*Chem. Abstr.*, 1975, **82**, 170 962).	746
74JCP(71)415	A. Lautié and A. Novak; *J. Chem. Phys.*, 1974, **71**, 415.	518
74JCS(P1)81	J. A. Maddison, P. W. Seale, E. P. Tiley and W. K. Warburton; *J. Chem. Soc., Perkin Trans. 1*, 1974, 81.	775
74JCS(P1)297	D. J. Le Count and A. T. Greer; *J. Chem. Soc., Perkin Trans. 1*, 1974, 297.	281, 654
74JCS(P1)349	D. J. Brown and R. K. Lynn; *J. Chem. Soc., Perkin Trans. 1*, 1974, 349.	577, 596, 636
74JCS(P1)627	A. R. McCarthy, W. D. Ollis and C. A. Ramsden; *J. Chem. Soc., Perkin Trans. 1*, 1974, 627.	757
74JCS(P1)633	W. D. Ollis and C. A. Ramsden; *J. Chem. Soc., Perkin Trans. 1*, 1974, 633.	782

74JCS(P1)642	W. D. Ollis and C. A. Ramsden; *J. Chem. Soc., Perkin Trans. 1*, 1974, 642.	782
74JCS(P1)903	G. P. Ellis and R. T. Jones; *J. Chem. Soc., Perkin Trans. 1*, 1974, 903.	353, 470
74JCS(P1)1028	G. V. Boyd and T. Norris; *J. Chem. Soc., Perkin Trans. 1*, 1974, 1028.	262
74JCS(P1)1284	J. L. Wong and D. S. Fuchs; *J. Chem. Soc., Perkin Trans. 1*, 1974, 1284.	558
74JCS(P1)1422	R. W. Baldock, P. Hudson, A. R. Katritzky and F. Soti; *J. Chem. Soc., Perkin Trans. 1*, 1974, 1422.	74
74JCS(P1)1871	T. Nishiwaki, F. Fujiyama and E. Minamisono; *J. Chem. Soc., Perkin Trans. 1*, 1974, 1871.	221
74JCS(P1)1970	R. G. R. Bacon and S. D. Hamilton; *J. Chem. Soc., Perkin Trans. 1*, 1974, 1970.	431
74JCS(P1)2030	A. Albert; *J. Chem. Soc., Perkin Trans. 1*, 1974, 2030.	858
74JCS(P1)2229	M. Rabat, F. Bergmann and I. Tamir; *J. Chem. Soc., Perkin Trans. 1*, 1974, 2229.	534
74JCS(P2)174	J. A. Elvidge, J. R. Jones, C. O'Brien, E. A. Evans and H. C. Sheppard; *J. Chem. Soc., Perkin Trans. 2*, 1974, 174.	526
74JCS(P2)382	A. G. Burton, M. Dereli, A. R. Katritzky and H. O. Tarhan; *J. Chem. Soc., Perkin Trans. 2*, 1974, 382.	238
74JCS(P2)389	A. G. Burton, A. R. Katritzky, M. Konya and H. O. Tarhan; *J. Chem. Soc., Perkin Trans. 2*, 1974, 389.	238
74JCS(P2)420	M. H. Palmer, R. H. Findlay and A. J. Gaskell; *J. Chem. Soc., Perkin Trans. 2*, 1974, 420.	176, 177, 346
74JCS(P2)420	M. H. Palmer, R. H. Findlay and A. J. Gaskell; *J. Chem. Soc., Perkin Trans. 2*, 1974, 420.	795
74JCS(P2)420	M. H. Palmer, R. H. Findlay and A. J. Gaskell; *J. Chem. Soc., Perkin Trans. 2*, 1974, 420.	672, 673
74JCS(P2)449	P. Bouchet, C. Coquelet and J. Elguero; *J. Chem. Soc., Perkin Trans. 2*, 1974, 449.	201, 268
74JCS(P2)1298	E. N. Maslen, J. R. Cannon, A. H. White and A. C. Willis; *J. Chem. Soc., Perkin Trans. 2*, 1974, 1298.	178, 212
74JHC39	D. M. Halverson and R. N. Castle; *J. Heterocycl. Chem.*, 1974, **11**, 39.	628
74JHC77	W. A. Nasutavicus and J. Love; *J. Heterocycl. Chem.*, 1974, **11**, 77.	559
74JHC135	J. J. Bergman and B. M. Lynch; *J. Heterocycl. Chem.*, 1974, **11**, 135.	194
74JHC199	J. Kobe, R. K. Robins and D. E. O'Brien; *J. Heterocycl. Chem.*, 1974, **11**, 199.	312, 320
74JHC223	B. M. Lynch and B. P.-L. Lem; *J. Heterocycl. Chem.*, 1974, **11**, 223.	306, 311
74JHC233	R. J. Rousseau, J. A. May, Jr., R. K. Robins and L. B. Townsend; *J. Heterocycl. Chem.*, 1974, **11**, 233.	621, 641
74JHC327	I. Lalezari and Y. Levy; *J. Heterocycl. Chem.*, 1974, **11**, 327.	653
74JHC423	E. Alcalde, J. de Mendoza, J. M. Garcia-Marquina, C. Almera and J. Elguero; *J. Heterocycl. Chem.*, 1974, **11**, 423.	272, 333
74JHC491	A. Vollmar and A. Hassner; *J. Heterocycl. Chem.*, 1974, **11**, 491.	813
74JHC615	K. Hayes; *J. Heterocycl. Chem.*, 1974, **11**, 615.	456
74JHC623	S. Plescia, S. Petruso and V. Sprio; *J. Heterocycl. Chem.*, 1974, **11**, 623.	320
74JHC641	P. W. K. Woo, H. W. Dion, S. M. Lange, L. F. Dahl and L. J. Durham; *J. Heterocycl. Chem.*, 1974, **11**, 641.	601
74JHC645	T. C. Thurber, R. J. Pugmire and L. B. Townsend; *J. Heterocycl. Chem.*, 1974, **11**, 645.	682, 683, 719
74JHC691	T. Novinson, K. Senga, J. Kobe, R. K. Robins, D. E. O'Brien and A. A. Albert; *J. Heterocycl. Chem.*, 1974, **11**, 691.	325
74JHC873	T. Novinson, D. E. O'Brien and R. K. Robins; *J. Heterocycl. Chem.*, 1974, **11**, 873.	649
74JHC885	A. Walser, T. Flynn and R. I. Fryer; *J. Heterocycl. Chem.*, 1974, **11**, 885.	288
74JHC921	E. Alcalde, J. de Mendoza and J. Elguero; *J. Heterocycl. Chem.*, 1974, **11**, 921.	184, 202
74JHC991	J. Kobe, D. E. O'Brien, R. K. Robins and T. Novinson; *J. Heterocycl. Chem.*, 1974, **11**, 991.	325
74JHC1011	M. Davis, L. W. Deady and E. Homfeld; *J. Heterocycl. Chem.*, 1974, **11**, 1011.	185
74JHC1013	J. Arriau, O. Chalvet, A. Dargelos and G. Maury; *J. Heterocycl. Chem.*, 1974, **11**, 1013.	627
74JHC1093	D. L. Nagel and N. H. Cromwell; *J. Heterocycl. Chem.*, 1974, **11**, 1093.	286
74JMC381	S. B. Zak, T. G. Gilleran, J. Karliner and G. Lukas; *J. Med. Chem.*, 1974, **17**, 381.	904
74JMC645	T. Novinson, R. Hanson, M. K. Dimmitt, L. N. Simon, R. K. Robins and D. E. O'Brien; *J. Med. Chem.*, 1974, **17**, 645.	312
74JMC846	K. Okumura, K. Matsumoto, M. Fukamizu, H. Yasuo, Y. Taguchi, Y. Sugihara, I. Inou, M. Seto, Y. Sato, N. Takamura, T. Kanno, M. Kawazu, T. Mizoguchi, S. Saito, K. Takashima and S. Takeyama; *J. Med. Chem.*, 1974, **17**, 846.	604
74JMR(15)98	M. C. Thorpe, W. C. Coburn, Jr. and J. A. Montgomery; *J. Magn. Reson.*, 1974, **15**, 98.	510, 514
74JOC63	J. K. Crandall and W. W. Conover; *J. Org. Chem.*, 1974, **39**, 63.	725
74JOC95	A. Bhattacharya and A. G. Hortmann; *J. Org. Chem.*, 1974, **39**, 95.	114
74JOC357	J. Elguero, C. Marzin and J. D. Roberts; *J. Org. Chem.*, 1974, **39**, 357.	18, 191, 192, 211, 354, 355, 680, 742, 804
74JOC793	J. S. Bradshaw, M. Tišler and B. Stanovnik; *J. Org. Chem.*, 1974, **39**, 793.	870, 880
74JOC957	M. Komatsu, Y. Ohshiro, K. Yasuda, S. Ichijima and T. Agawa; *J. Org. Chem.*, 1974, **39**, 957.	18
74JOC962	R. K. Howe and J. E. Franz; *J. Org. Chem.*, 1974, **39**, 962.	14
74JOC1007	R. C. Gearhart, R. H. Wood, R. C. Thorstenson and J. A. Moore; *J. Org. Chem.*, 1974, **39**, 1007.	179
74JOC1047	P. A. S. Smith and E. M. Bruckmann; *J. Org. Chem.*, 1974, **39**, 1047.	727

74JOC1233	G. J. de Voghel, T. L. Eggerichs, Z. Janousek and H. G. Viehe; *J. Org. Chem.*, 1974, **39**, 1233.	281
74JOC1802	R. A. Abramovitch and R. B. Rogers; *J. Org. Chem.*, 1974, **39**, 1802.	636
74JOC1833	D. Fortuna, B. Stanovnik and M. Tišler; *J. Org. Chem.*, 1974, **39**, 1833.	272
74JOC1909	J. J. Wilcznyski and H. W. Johnson; *J. Org. Chem.*, 1974, **39**, 1909.	269
74JOC2023	L. B. Townsend, R. A. Long, J. P. McGraw, D. W. Miles, R. K. Robins and H. Eyring; *J. Org. Chem.*, 1974, **39**, 2023.	328
74JOC2143	S. Polanc, B. Verček, B. Šek, B. Stanovnik and M. Tišler; *J. Org. Chem.*, 1974, **39**, 2143.	859, 871, 885, 889, 891, 896, 900
74JOC2301	D. S. Noyce, G. T. Stowe and W. Wong; *J. Org. Chem.*, 1974, **39**, 2301.	430, 448
74JOC2663	J. P. Freeman and E. Janiga; *J. Org. Chem.*, 1974, **39**, 2663.	269
74JOC3198	M. Komatsu, N. Nishikaze, M. Sakamoto, Y. Ohshiro and T. Agawa; *J. Org. Chem.*, 1974, **39**, 3198.	780
74JOC3226	P. Dea, G. R. Revankar, R. L. Tolman, R. K. Robins and M. P. Schweizer; *J. Org. Chem.*, 1974, **39**, 3226.	855
74JOC3438	D. F. Wiener and N. J. Leonard; *J. Org. Chem.*, 1974, **39**, 3438.	595
74JOC3506	J. H. Wikel and C. J. Paget, *J. Org. Chem.*, 1974, **39**, 3506.	784
74JOC3608	F. Wudl, M. L. Kaplan, E. J. Hufnagel and E. W. Southwick, Jr.; *J. Org. Chem.*, 1974, **39**, 3608.	15
74JOC3651	G. A. Ivanovics, R. J. Rousseau, M. Kawana, P. C. Srivastava and R. K. Robins; *J. Org. Chem.*, 1974, **39**, 3651.	629, 652
74JOC3739	G. J. Siuta, R. W. Franck and R. J. Kempton; *J. Org. Chem.*, 1974, **39**, 3739.	724
74JOC3763	V. Nair and K. H. Kim; *J. Org. Chem.*, 1974, **39**, 3763.	153
74JOC3788	M. P. Servé; *J. Org. Chem.*, 1974, **39**, 3788.	694
74JOM(65)C51	N. F. Borkett and M. I. Bruce; *J. Organomet. Chem.*, 1974, **65**, C51.	213
74JOM(70)347	V. N. Torocheshnikov, N. M. Sergeyev, N. A. Viktorov, G. S. Goldin, V. G. Poddubny and A. N. Koltsova; *J. Organomet. Chem.*, 1974, **70**, 347.	213, 761
74JOU1542	N. I. Kudryashova, L. B. Piotrovskii, V. A. Naumev, N. B. Brovtsyna and N. B. Khromov-Borisov; *J. Org. Chem. USSR (Engl. Transl.)*, 1974, **10**, 1542.	458
74JPR147	M. I. Ali, M. A. Abou-State and A. F. Ibrahim; *J. Prakt. Chem.*, 1974, **316**, 147.	663
74JPR667	L. Heinisch; *J. Prakt. Chem.*, 1974, **316**, 667.	899
74JPR705	F. Ritschl and H. Dorn; *J. Prakt. Chem.*, 1974, **316**, 705.	182
74JSP(49)423	W. D. Krugh and L. P. Gold; *J. Mol. Spectrosc.*, 1974, **49**, 423.	12, 795
74JST(22)401	L. Nygaard, D. Christen, J. T. Nielsen, E. J. Pedersen, O. Snerling, E. Vestergaard and G. O. Sørensen; *J. Mol. Struct.*, 1974, **22**, 401.	173, 181, 290
74KGS268	V. P. Shchipanov and G. F. Grigor'eva; *Khim. Geterotsikl. Soedin.*, 1974, 268 (*Chem. Abstr.*, 1974, **80**, 133 352).	827
74KGS851	M. M. Tsitsika, S. M. Khripak and I. V. Smolanka; *Khim. Geterotsikl. Soedin.*, 1974, 851.	746, 750
74KGS854	Y. M. Yutilov and I. A. Svertilova; *Khim. Geterotsikl. Soedin.*, 1974, 854 (*Chem. Abstr.*, 1974, **81**, 120 536).	620, 622
74LA1132	H. Dürr, W. Schmidt and R. Sergio; *Liebigs Ann. Chem.*, 1974, 1132.	251
74LA1399	H. Ertel and G. Heubach; *Liebigs Ann. Chem.*, 1974, 1399.	662
74M38	F. Asinger, W. Leuchtenberger and V. Gerber; *Monatsh. Chem.*, 1974, **105**, 38.	656
74M427	J. Schantl; *Monatsh. Chem.*, 1974, **105**, 427.	745
74M834	A. Furlan, M. Furlan, B. Stanovnik and M. Tisler; *Monatsh. Chem.*, 1974, **105**, 834.	628
74MI40100	W. A. Lathan, L. A. Curtiss, W. J. Hehre, J. B. Lisle and J. A. Pople; *Prog. Phys. Org. Chem.*, 1974, **11**, 175.	7
74MI40400	J. L. Barascut, C. Tamby and J. L. Imbach; *J. Carbohydr., Nucleosides, Nucleotides*, 1974, **1**, 77.	288, 289
74MI40401	J. Elguero and A. Fruchier; *An. Quím.*, 1974, **70**, 141.	188, 189
74MI40402	B. L. Kam and J. L. Imbach; *J. Carbohydr., Nucleosides, Nucleotides*, 1974, **1**, 287.	289
74MI40403	Y. Ferré, E. J. Vincent, H. Larivé and J. Metzger; *J. Chim. Phys. Phys. Chim. Biol.*, 1974, **71**, 329.	199
74MI40404	I. A. Korbukh, M. N. Preobrazhenskaya and O. N. Judina; *J. Carbohydr., Nucleosides, Nucleotides*, 1974, **1**, 363.	289
74MI40405	V. S. Troitskaya, Yu. D. Timoshenkova, V. G. Vinokurov and V. I. Tyulin; *Vestn. Mosk. Univ., Khim.*, 1974, 677.	199, 200
74MI40406	E. Alcalde, J. M. Garía-Marquina and J. de Mendoza; *An. Quím.*, 1974, **70**, 959.	263, 266
B-74MI40407	Yu. P. Kitaev and B. I. Buzykin; 'Hydrazones', Academy of Sciences of the USSR, Moscow, 1974.	168
B-74MI40408	A. I. Myers; 'Heterocycles in Organic Synthesis', Wiley-Interscience, New York, 1974.	250
B-74MI40700	F. G. Riddell; in 'Saturated Heterocyclic Chemistry', The Chemical Society, London, 1974, vol. 2, p. 237.	427
B-74MI40900	A. Bloch; in 'Drug Design', ed. E. J. Ariens; Academic, New York, 1974, vol. 4, p. 286.	603
B-74MI40901	F. Bergmann, D. Lichtenberg, U. Reichman and Z. Neiman; 'Jerusalem Symposium of Quantum Chemistry and Biochemistry,' Israel Academy of Science, 1974, vol. 6, p. 397.	501, 560, 561
74MI40902	M. Ikehara and Y. Ogiso; *J. Carbohydr. Nucleosides, Nucleotides*, 1974, **1**, 401.	560
74MI40903	N. V. Zheltovskii and V. I. Danilov; *Biofizika*, 1974, **19**, 784.	517
74MI40904	J. Salomon and D. Elad; *Photochem. Photobiol.*, 1974, **19**, 21.	545

Ref	Citation	Page
74MI41100	P. Mauret, J. P. Fayet, M. Fabre, J. Elguero and J. De Mendoza; *J. Chim. Phys. Phys. Chim. Biol.*, 1974, **71**, 115 (*Chem. Abstr.*, 1974, **80**, 132 537).	677
B-74MI41101	C. W. Rees and R. C. Storr; *Lect. Heterocycl. Chem.*, 1974, **2**, 71.	670
74MI41200	R. Un; *Chim. Acta Turc.*, 1974, **2**, 1 (*Chem. Abstr.*, 1975, **82**, 57 623).	775
74MI41500	C. J. Coulson, R. E. Ford, E. Lunt, S. Marshall, D. L. Pain, J. H. Rogers and K. R. H. Woodridge; *Eur. J. Med. Chem.*, 1974, **9**, 313.	895
74MI41501	B. Stanovnik; *Lectures Heterocycl. Chem.*, 1974, **2**, S-27.	868, 869, 870, 871, 880
74MI41502	T. Bany, A. Maliszewska-Guz and B. Modzelewska-Banachiewicz; *Rocz. Chem.*, 1974, **48**, 1445.	886
74OMR(6)272	J. Elguero, A. Fruchier and M. C. Pardo; *Org. Magn. Reson.*, 1974, **6**, 272.	187
74OMS(8)347	L. Birkofer, M. Franz and G. Schmidtberg; *Org. Mass Spectrom.*, 1974, **8**, 347.	204
74OPP156	R. G. Dickinson and N. W. Jacobsen; *Org. Prep. Proced. Int.*, 1974, **6**, 156.	785
74PMH(6)1	E. Heilbronner, J. P. Maier and E. Haselbach; *Phys. Methods Heterocycl. Chem.*, 1974, **6**, 1.	30
74PMH(6)53	J. Sheridan; *Phys. Methods Heterocycl. Chem.*, 1974, **6**, 53.	8, 171, 181, 676
74PMH(6)95	B. C. Gilbert and M. Trenwith; *Phys. Methods Heterocycl. Chem.*, 1974, **6**, 95.	177, 206
74PMH(6)199	K. Pihlaja and E. Taskinen; *Phys. Methods Heterocycl. Chem.*, 1974, **6**, 199.	32
74RCR1089	E. S. Golovchinskaya; *Russ. Chem. Rev. (Engl. Transl.)*, 1974, **43**, 1089.	556
74RTC3	R. De Bode and C. A. Salemink; *Recl. Trav. Chim. Pays-Bas*, 1974, **93**, 3.	642
74RTC160	J. E. Schelling and C. A. Salemink; *Recl. Trav. Chim. Pays-Bas*, 1974, **93**, 160.	619, 636
74RTC317	S. Hoff and A. P. Blok; *Recl. Trav. Chim. Pays-Bas*, 1974, **93**, 317.	718
74S30	W. Schäfer, H. W. Moore and A. Aguado; *Synthesis*, 1974, 30.	137
74S32	J. R. Williams and G. M. Sarkisian; *Synthesis*, 1974, 32.	126
74S41	A. M. M. E. Omar; *Synthesis*, 1974, 41.	458
74S198	H. Lind and H. Kristinsson; *Synthesis*, 1974, 198.	707
74S815	E. Duranti and C. Balsamini; *Synthesis*, 1974, 815.	454
74T445	K. B. Sukumaran, S. Satish and M. V. George; *Tetrahedron*, 1974, **30**, 445.	134
74T1301	C. Wentrup; *Tetrahedron*, 1974, **30**, 1301.	867
74T2677	M. Maeda, K. Nushi and Y. Kawazoe; *Tetrahedron*, 1974, **30**, 2677.	543, 546
74T2903	A. Escande, J. Lapasset, R. Faure, E. J. Vincent and J. Elguero; *Tetrahedron*, 1974, **30**, 2903; erratum 1975, **31**, 2.	175, 176, 177, 179, 180, 198
74T3941	E. Breitmaier and W. Voelter; *Tetrahedron*, 1974, **30**, 3941.	515
74TH40100	D. Norbury; Ph.D. Thesis, University of Wales, 1974.	12
74TL129	J.-L. Imbach, J. L. Barascut, B. L. Kam and C. Tapiero; *Tetrahedron Lett.*, 1974, 129.	513
74TL253	M. H. Palmer and R. H. Findlay; *Tetrahedron Lett.*, 1974, 253.	677
74TL269	R. Fusco, L. Garanti and G. Zecchi; *Tetrahedron Lett.*, 1974, 269.	317
74TL1187	A. Nohara; *Tetrahedron Lett.*, 1974, 1187.	815
74TL1609	J. Vilarrasa, E. Meléndez and J. Elguero; *Tetrahedron Lett.*, 1974, 1609.	225
74TL1673	G. Scherowsky and H. Franke; *Tetrahedron Lett.*, 1974, 1673.	779
74TL1987	A. Gieren, F. Pertlik and S. Sommer; *Tetrahedron Lett.*, 1974, 1987.	278
74TL2163	D. N. Reinhoudt and C. G. Kouwenhoven; *Tetrahedron Lett.*, 1974, 2163.	720
74TL2643	K. K. Balasubramanaian and B. Venugopalan; *Tetrahedron Lett.*, 1974, 2643.	444, 446
74TL2695	D. S. Kemp, M. Trangle and K. Trangle; *Tetrahedron Lett.*, 1974, 2695.	728
74TL3893	S. Minami, Y. Kimura, T. Miyamoto and J. Matsumoto; *Tetrahedron Lett.*, 1974, 3893.	342
74TL4421	T. Yamazaki, G. Baum and H. Shechter; *Tetrahedron Lett.*, 1974, 4421.	212
74TL4539	B. Verček, B. Stanovnik and M. Tišler; *Tetrahedron Lett.*, 1974, 4539.	877
74UKZ99	M. V. Povstyanoi and P. M. Kochergin; *Ukr. Khim. Zh. (Russ. Ed.)*, 1974, **40**, 99 (*Chem. Abstr.*, 1974, **80**, 82 902).	654
74ZC16	E. Lipmann, D. Reifegerste and E. Kleinpeter; *Z. Chem.*, 1974, **14**, 16 (*Chem. Abstr.*, 1974, **81**, 25 615).	794
74ZC267	S. Leistner, G. Wagner and H. Richter; *Z. Chem.*, 1974, **14**, 267.	776
74ZC270	R. Schmidt, G. Westphal and B. Froehlich; *Z. Chem.*, 1974, **14**, 270.	775
74ZOR377	O. P. Shvaika and V. I. Fomenko; *Zh. Org. Khim.*, 1974, **10**, 377.	775
74ZOR1725	V. M. Neplynev and P. S. Pel'kis; *Zh. Org. Khim.*, 1974, **10**, 1725 (*Chem. Abstr.*, 1974, **81**, 136 059).	793, 828
74ZOR1787	I. A. Korbukh, O. N. Judina and M. N. Preobrazhenskaya; *Zh. Org. Khim.*, 1974, **10**, 1787.	289
74ZOR2467	A. V. El'tsov, V. P. Martynova, E. R. Zakhs and L. P. Shustova; *Zh. Org. Khim.*, 1974, **10**, 2467 (*Chem. Abstr.*, 1975, **82**, 57 613).	611, 612
74ZPC(355)1274	H. W. Vohland, P. E. Schulze, W. Koransky, G. Schulz and B. Acksteiner; *Hoppe-Seyler's Z. Physiol. Chem.*, 1974, **355**, 1274 (*Chem. Abstr.*, 1975, **82**, 92 876).	835
75ACR361	K. N. Houk; *Acc. Chem. Res.*, 1975, **8**, 361.	711
75ACS(A)647	K. Nielsen; *Acta Chem. Scand., Ser. A*, 1975, **29**, 647.	674
75ACS(B)1089	J. M. Bakke and K. A. Skjervold; *Acta Chem. Scand., Ser. B*, 1975, **29**, 1089.	274
75AG1	D. Seebach and D. Enders; *Angew. Chem.*, 1975, **87**, 1.	717
75AG(E)15	D. Seebach and D. Enders; *Angew. Chem., Int. Ed. Engl.*, 1975, **14**, 15.	717
75AG(E)86	C. Reichardt and K. Halbritter; *Angew. Chem., Int. Ed. Engl.*, 1975, **14**, 86.	281
75AG(E)428	H. Quast and L. Bieber; *Angew. Chem., Int. Ed. Engl.*, 1975, **14**, 428.	810

Ref	Citation	Pages
75AG(E)560	W. Sucrow, M. Slopianka and V. Bardakos; *Angew. Chem., Int. Ed. Engl.*, 1975, **14**, 560.	281
75AG(E)646	H. Dürr and B. Weiss; *Angew. Chem., Int. Ed. Engl.*, 1975, **14**, 646.	251
75AG(E)647	H. Dürr and H. Schmitz; *Angew. Chem., Int. Ed. Engl.*, 1975, **14**, 647.	251
75AG(E)665	A. Dorlars, C. W. Schellhammer and J. Schroeder; *Angew. Chem., Int. Ed. Engl.*, 1975, **14**, 665.	299
75AG(E)775	G. L'abbe; *Angew. Chem., Int. Ed. Engl.*, 1975, **14**, 775.	163
75AJC133	M. Woodruff and J. B. Polya; *Aust. J. Chem.*, 1975, **28**, 133.	281, 751, 782
75AJC421	M. Woodruff and J. B. Polya; *Aust. J. Chem.*, 1975, **28**, 421.	782
75AJC1583	M. Woodruff and J. B. Polya; *Aust. J. Chem.*, 1975, **28**, 1583.	225
75AJC1861	L. W. Deady, R. G. McLoughlin and M. R. Grimmett; *Aust. J. Chem.*, 1975, **28**, 1861.	210
75ANY(255)177	J.-L. Imbach; *Ann. N.Y. Acad. Sci.*, 1975, **255**, 177.	513
75ANY(255)292	J. A. Montgomery, R. D. Elliott and H. J. Thomas; *Ann. N. Y. Acad. Sci.*, 1975, **255**, 292.	608, 652
75APO(11)267	M. Liler; *Adv. Phys. Org. Chem.*, 1975, **11**, 267.	202, 255, 256
75AX(B)211	T. J. Kistenmacher and T. Shigematsu; *Acta. Crystallogr., Part B*, 1975, **31**, 211.	509, 513
75AX(B)548	L. Dupont, J. Toussaint, O. Dideberg, J. M. Braham and A. F. Noels; *Acta Crystallogr., Part B*, 1975, **31**, 548.	179
75AX(B)2119	J. Delettré, R. Bally and J. P. Mornon; *Acta Crystallogr., Part B*, 1975, **31**, 2119.	178
75AX(B)2862	G. S. Mandel and R. E. Marsh; *Acta. Crystallogr., Part B*, 1975, **31**, 2862.	510
75B1490	L. F. Christensen, R. B. Meyer, Jr., J. P. Miller, L. N. Simon and R. K. Robins; *Biochemistry*, 1975, **14**, 1490.	546
75BCJ1484	F. Yoneda and T. Nagamatsu; *Bull. Chem. Soc. Jpn.*, 1975, **48**, 1484.	161, 337
75BSB499	J. Elguero, C. Marzin and M. E. Peek; *Bull. Soc. Chim. Belg.*, 1975, **84**, 499.	184
75BSB1189	J. Elguero, R. Faure, J. P. Galy and E. J. Vincent; *Bull. Soc. Chim. Belg.*, 1975, **84**, 1189.	214
75BSF647	A. Bernardini, P. Viallefont, J. Daunis, M. L. Roumestant and A. B. Soulami; *Bull. Soc. Chim. Fr.*, 1975, 647.	746, 752, 753
75BSF857	J. Daunis and M. Follett; *Bull. Soc. Chim. Fr.*, 1975, 857.	898, 899
75BSF1649	J. L. Barascut, P. Viallefont and J. Daunis; *Bull. Soc. Chim. Fr.*, 1975, 1649.	753
75BSF1675	P. Mauret, J. P. Fayet and M. Fabre; *Bull. Soc. Chim. Fr.*, 1975, 1675.	176, 177, 677
75CB372	C. Metzger and J. Kurz; *Chem. Ber.*, 1975, **108**, 372.	661
75CB887	D. Moderhack; *Chem. Ber.*, 1975, **108**, 887.	819, 820
75CB3762	H. Reimlinger, J. J. M. Vandewalle, R. Merényi and W. R. F. Lingier; *Chem. Ber.*, 1975, **108**, 3762.	865, 866
75CB3780	H. Reimlinger, W. R. F. Lingier and J. J. M. Vandewalle; *Chem. Ber.*, 1975, **108**, 3780.	879
75CB3794	H. Reimlinger, W. R. F. Lingier and R. Merényi; *Chem. Ber.*, 1975, **108**, 3794.	868, 888
75CC125	J. L. Wong and J. H. Keck, Jr.; *J. Chem. Soc., Chem. Commun.*, 1975, 125.	527
75CC128	J. Dingwall and J. T. Sharp; *J. Chem. Soc., Chem. Commun.*, 1975, 128.	250
75CC146	F. Yoneda, S. Matsumoto and M. Higuchi; *J. Chem. Soc., Chem. Commun.*, 1975, 146.	581
75CC319	C.-Y. Shiue and S. H. Chu; *J. Chem. Soc., Chem. Commun.*, 1975, 319.	581
75CC964	C. C. Duke, A. J. Liepa, J. K. MacLeod, D. S. Letham and C. W. Parker; *J. Chem. Soc., Chem. Commun.*, 1975, 964.	585
75CCC1199	A. Martvoň, S. Stankovský and J. Světlík; *Collect. Czech. Chem. Commun.*, 1975, **40**, 1199.	717
75CHE81	G. K. Rogul'chenko, I. A. Mazur and P. M. Kochergin; *Chem. Heterocycl. Compd. (Engl. Transl.)*, 1975, **11**, 81.	648
75CHE123	I. I. Popov, A. M. Simonov and A. A. Zubenko; *Chem. Heterocycl. Compd. (Engl. Transl.)*, 1975, **11**, 123.	438
75CHE343	B. A. Tertov and A. S. Morkovnik; *Chem. Heterocycl. Compd. (Engl. Transl.)*, 1975, **11**, 343.	447
75CHE371	M. V. Povstyanoi, P. M. Kochergin, E. V. Logachev and E. A. Yakubovskii; *Chem. Heterocycl. Compd. (Engl. Transl.)*, 1975, **11**, 371.	654
75CHE461	I. I. Popov, P. V. Tkachenko and A. M. Simonov; *Chem. Heterocycl. Compd. (Engl. Transl.)*, 1975, **11**, 461.	450
75CHE718	P. V. Schatnev, M. S. Shvartsberg and I. Ya. Bernshtein; *Chem. Heterocycl. Compd. (Engl. Transl.)*, 1975, **11**, 718.	438
75CHE751	M. V. Povstyanoi, M. A. Klykov, N. M. Gorban' and P. M. Kochergin; *Chem. Heterocycl. Compd. (Engl. Transl.)*, 1975, **11**, 751.	656
75CHE995	Z. V. Pushkareva, V. I. Ofitserov, V. S. Mokrushin and K. V. Aglitskaya; *Chem. Heterocycl. Compd. (Engl. Transl.)*, 1975, **11**, 995.	652, 667
75CHE1416	N. D. Abramova, G. G. Skvortsova, B. V. Trzhtsinskaya and M. V. Sigalov; *Chem. Heterocycl. Compd. (Engl. Transl.)*, 1975, **11**, 1416.	450
75CI(L)278	G. L'abbe, G. Mathys and S. Toppet; *Chem. Ind. (London)*, 1975, 278.	718
75CJC119	B. M. Lynch, M. A. Khan, S. C. Sharma and H. C. Teo; *Can. J. Chem.*, 1975, **53**, 119.	306, 307, 311
75CJC355	D. H. Hunter and R. P. Steiner; *Can. J. Chem.*, 1975, **53**, 355.	852
75CJC2944	D. F. Rendle, A. Storr and J. Trotter; *Can. J. Chem.*, 1975, **53**, 2944.	237
75CL499	M. Hamaguchi and T. Ibata; *Chem. Lett.*, 1975, 499.	162
75CPB452	T. Kato and S. Masuda; *Chem. Pharm. Bull.*, 1975, **23**, 452.	337

75CPB759	R. Marumoto, Y. Yoshioka, O. Miyashita, S. Shima, K.-I. Imai, K. Kawazoe and M. Honjo; *Chem. Pharm. Bull.*, 1975, **23**, 759.	590, 664
75CPB955	S. Kubota and M. Uda; *Chem. Pharm. Bull.*, 1975, **23**, 955.	740, 741
75CPB2401	M. Seikiya and J. Suzuki; *Chem. Pharm. Bull.*, 1975, **23**, 2401.	569
75CPB2643	T. Itaya, T. Saito, K. Kawakatsu and T. Fujii; *Chem. Pharm. Bull.*, 1975, **23**, 2643.	554
75CPB2678	A. Nagel and H. C. van der Plas; *Chem. Pharm. Bull.*, 1975, **23**, 2678.	592
75CR(C)685	C. Simionescu, T. Liandru, C. Gorea and V. Gordoza; *C.R. Hebd. Seances Acad. Sci., Ser. C*, 1975, 685.	569
75CRV241	R. N. Butler; *Chem. Rev.*, 1975, **75**, 241.	822
75DIS(B)(36)1712	J. C. B. Komin; *Diss. Abstr. Int. B*, 1975, **36**, 1712.	703
75DOK(224)847	S. B. Bulgarevich, O. A. Osipov, V. S. Bolotnikov, T. V. Lifintseva, V. N. Sheinker and A. D. Garnovskii; *Dokl. Akad. Nauk SSSR*, 1975, **224**, 847 (*Chem. Abstr.*, 1976, **84**, 43 053).	672
75E996	Z. Neiman; *Experientia*, 1975, **31**, 996.	521
75EGP111074	S. Hauptmann, R. Ludwig and E. Tenor; *Ger. (East) Pat.* 111 074 (1975) (*Chem. Abstr.*, 1976, **84**, 31 078).	746
75FES197	G. Palazzo and G. Picconi; *Farmaco Ed. Sci.*, 1975, **30**, 197.	614, 635
75FES650	A. Omodei-Sale and E. Toja; *Farmaco Ed. Sci.*, 1975, **30**, 650.	644
75G1291	M. Kac, F. Kovac, B. Stanovnik and M. Tišler; *Gazz. Chim. Ital.*, 1975, **105**, 1291.	863, 869, 871, 880
75GEP2407304	H. Hoffmann, I. Hammann, B. Homeyer and W. Stendel; *Ger. Pat.* 2 407 304 (1975) (*Chem. Abstr.*, 1976, **84**, 4965).	787
75GEP2407305	Bayer AG, *Ger. Pat.* 2 407 305 (1975) (*Chem. Abstr.*, 1975, **83**, 206 290).	732
75GEP2511829	J. T. Witkowski and R. K. Robins; *Ger. Pat.* 2 511 829 (1975) (*Chem. Abstr.*, 1976, **84**, 4966).	788
75H(3)217	K. Hirai and T. Ishiba; *Heterocycles*, 1975, **3**, 217.	115
75HCA688	A. J. Bates, I. J. Galpin, A. Hallett, D. Hudson, G. W. Kenner, R. Ramage and R. C. Sheppard; *Helv. Chim. Acta*, 1975, **58**, 688.	728
75HCA1521	K. Rüfenacht; *Helv. Chim. Acta*, 1975, **58**, 1521.	904
75HCA2192	A. Edenhofer; *Helv. Chim. Acta.*, 1975, **58**, 2192.	140, 466
75IJC468	S. Sharma, R. N. Iyer and N. Anand; *Indian J. Chem.*, 1975, **13**, 468.	645
75IJC668	S. P. Gupta and P. Singh; *Indian J. Chem.*, 1975, **13**, 668.	505
75IZV690	S. F. Vasilevskii, A. N. Sinyakov, M. S. Shvartsberg and I. L. Kotlyareveskii; *Izv. Akad. Nauk SSSR, Ser. Khim.*, 1975, 690.	753
75IZV2766	N. K. Kochetkova, A. N. Kost, V. W. Shibaev, R. S. Sagitullin, A. A. Kost and Y. V. Zavy'alov; *Izv. Akad. Nauk SSSR, Ser. Khim.*, 1975, 2766 (*Chem. Abstr.*, 1976, **84**, 105 526).	648
75JA1238	E. J. Schantz, V. E. Ghazarossian, H. K. Schnoes, F. M. Strong, J. P. Springer, J. O. Pezzanite and J. Clardy; *J. Am. Chem. Soc.*, 1975, **97**, 1238.	604
75JA1285	R. C. Bingham, M. J. S. Dewar and D. H. Lo; *J. Am. Chem. Soc.*, 1975, **97**, 1285, 1294, 1302, 1307.	7
75JA2369	M. Dreyfus, G. Dodin, O. Bensaude and J. E. Dubois; *J. Am. Chem. Soc.*, 1975, **97**, 2369.	514
75JA3351	L. G. Marzilli, L. A. Epps, T. Sorrell and T. J. Kistenmacher; *J. Am. Chem. Soc.*, 1975, **97**, 3351.	531
75JA4627	M.-T. Chenon, R. J. Pugmire, D. M. Grant, R. P. Panzica and L. B. Townsend; *J. Am. Chem. Soc.*, 1975, **97**, 4627.	514
75JA4636	M. T. Chenon, R. J. Pugmire, D. M. Grant, R. P. Panzica and L. B. Townsend; *J. Am. Chem. Soc.*, 1975, **97**, 4636.	514
75JA4990	N. J. Leonard and T. R. Henderson; *J. Am. Chem. Soc.*, 1975, **97**, 4990.	570, 515
75JA6008	J. Bordner, W. E. Thiessen, H. A. Bates and H. Rapoport; *J. Am. Chem. Soc.*, 1975, **97**, 6008.	604
75JA6197	A. Holm, N. Harrit and N. H. Toubro; *J. Am. Chem. Soc.*, 1975, **97**, 6197.	150
75JA7467	C. Mayor and C. Wentrup; *J. Am. Chem. Soc.*, 1975, **97**, 7467.	867
75JAP7504074	A. Kotone, M. Hoda and T. Fujita; *Jpn. Pat.* 75 04 074 (1975) (*Chem. Abstr.*, 1975, **83**, 114 414).	748
75JCS(D)176	D. F. Rendle, A. Storr and J. Trotter; *J. Chem. Soc., Dalton Trans.*, 1975, 176.	227
75JCS(D)718	D. J. Patmore, D. F. Rendle, A. Storr and J. Trotter; *J. Chem. Soc., Dalton Trans.*, 1975, 718.	227
75JCS(D)749	K. R. Breakell, D. J. Patmore and A. Storr; *J. Chem. Soc., Dalton Trans.*, 1975, 749.	236
75JCS(P1)1	T. L. Gilchrist, G. E. Gymer and C. W. Rees; *J. Chem. Soc., Perkin Trans. 1*, 1975, 1.	686, 692, 693, 698
75JCS(P1)12	T. L. Gilchrist, C. W. Rees and C. Thomas; *J. Chem. Soc., Perkin Trans. 1*, 1975, 12.	765
75JCS(P1)41	B. M. Adger, M. Keating, C. W. Rees and R. C. Storr; *J. Chem. Soc., Perkin Trans. 1*, 1975, 41.	269
75JCS(P1)125	J. A. May, Jr. and L. B. Townsend; *J. Chem. Soc., Perkin Trans. 1*, 1975, 125.	619, 621
75JCS(P1)275	I. J. Ferguson and K. Schofield; *J. Chem. Soc., Perkin Trans. 1*, 1975, 275.	455, 456
75JCS(P1)386	N. W. Alcock, B. T. Golding, D. R. Hall, U. Horn and W. P. Watson; *J. Chem. Soc., Perkin Trans. 1*, 1975, 386.	633
75JCS(P1)406	Y. Tamura, Y. Sumida, Y. Miki and M. Ikeda; *J. Chem. Soc., Perkin Trans. 1*, 1975, 406.	339

75JCS(P1)798	G. Cooper and W. J. Irwin; *J. Chem. Soc., Perkin Trans. 1*, 1975, 798.	662
75JCS(P1)975	R. G. Dickinson and N. W. Jacobsen; *J. Chem. Soc., Perkin Trans. 1*, 1975, 975.	757
75JCS(P1)1232	S. Anderson, E. E. Glover and K. D. Vaughan; *J. Chem. Soc., Perkin Trans. 1*, 1975, 1232.	879
75JCS(P1)1695	M. H. Palmer, R. H. Findlay, S. M. F. Kennedy and P. S. McIntyre; *J. Chem. Soc., Perkin Trans. 1*, 1975, 1695.	175, 186, 230
75JCS(P1)2240	W. Pendergast; *J. Chem. Soc., Perkin Trans. 1*, 1975, 2240.	541
75JCS(P2)928	H. J. C. Yeh, K. L. Kirk, L. A. Cohen and J. S. Cohen; *J. Chem. Soc., Perkin Trans. 2*, 1975, 928.	447
75JCS(P2)1200	G. B. Ansell; *J. Chem. Soc., Perkin Trans. 2*, 1975, 1200.	796, 797
75JCS(P2)1325	M. E. Burrage, R. C. Cookson, S. S. Gupte and I. D. R. Stevens; *J. Chem. Soc., Perkin Trans. 2*, 1975, 1325.	749, 784
75JCS(P2)1600	A. R. Katritzky, B. Terem, E. V. Scriven, S. Clementi and H. O. Tarhan; *J. Chem. Soc., Perkin Trans. 2*, 1975, 1600.	238
75JCS(P2)1609	M. Dereli, A. R. Katritzky and H. O. Tarhan; *J. Chem. Soc., Perkin Trans. 2*, 1975, 1609.	238
75JCS(P2)1624	A. R. Katritzky, S. Clementi and H. O. Tarhan; *J. Chem. Soc., Perkin Trans. 2*, 1975, 1624.	238, 239
75JCS(P2)1632	A. R. Katritzky, H. O. Tarhan and B. Terem; *J. Chem. Soc., Perkin Trans. 2*, 1975, 1632.	238
75JCS(P2)1695	M. H. Palmer, R. H. Findlay, S. M. F. Kennedy and P. S. McIntyre; *J. Chem. Soc., Pekin Trans. 2*, 1975, 1695.	673
75JCS(P2)1791	D. E. McGreer, I. M. E. Masters and M. T. H. Liu; *J. Chem. Soc., Perkin Trans. 2*, 1975, 1791.	255
75JHC107	Y. Tamura, J.-H. Kim and M. Ikeda; *J. Heterocycl. Chem.*, 1975, **12**, 107.	856, 889, 891, 896
75JHC119	Y. Tamura, Y. Miki and M. Ikeda; *J. Heterocycl. Chem.*, 1975, **12**, 119.	339
75JHC143	P. K. Kadaba; *J. Heterocycl. Chem.*, 1975, **12**, 143.	717
75JHC171	A. D. Broom and G. H. Milne; *J. Heterocycl. Chem.*, 1975, **12**, 171.	526
75JHC207	E. Abushanab, A. P. Bindra and L. Goodman; *J. Heterocycl. Chem.*, 1975, **12**, 207.	625, 644
75JHC211	E. Abushanab, A. P. Bindra, D.-Y. Lee and L. Goodman; *J. Heterocycl. Chem.*, 1975, **12**, 211.	644
75JHC225	Y. Tamura, H. Hayashi, Y. Nishimura and M. Ikeda; *J. Heterocycl. Chem.*, 1975, **12**, 225.	283
75JHC279	E. Gonzalez, R. Sarlin and J. Elguero; *J. Heterocycl. Chem.*, 1975, **12**, 279.	272
75JHC379	O. Fuentes and W. W. Paudler; *J. Heterocycl. Chem.*, 1975, **12**, 379.	614, 615
75JHC471	T. Nishimura, K. Nakano, S. Shibamoto and K. Kitajima; *J. Heterocycl. Chem.*, 1975, **12**, 471.	358, 422
75JHC481	Y. Tamura, J.-H. Kim, Y. Miki, H. Hayashi and M. Ikeda; *J. Heterocycl. Chem.*, 1975, **12**, 481.	887, 888
75JHC493	C. Y. Shiue and S. H. Chu; *J. Heterocycl. Chem.*, 1975, **12**, 493.	597
75JHC505	F. Texier and J. Bourgois; *J. Heterocycl. Chem.*, 1975, **12**, 505.	715
75JHC517	J. D. Ratajczyk and L. R. Swett; *J. Heterocycl. Chem.*, 1975, **12**, 517.	272, 313, 326, 332
75JHC523	M. Winn; *J. Heterocycl. Chem.*, 1975, **12**, 523.	332
75JHC597	D. M. Mulvey and H. Jones; *J. Heterocycl. Chem.*, 1975, **12**, 597.	446
75JHC675	A. Shafiee and I. Lalezari; *J. Heterocycl. Chem.*, 1975, **12**, 675.	142
75JHC717	A. Walser, T. Flynn and R. I. Fryer; *J. Heterocycl. Chem.*, 1975, **12**, 717.	771
75JHC771	T. J. Schwan and R. L. White, Jr.; *J. Heterocycl. Chem.*, 1975, **12**, 771.	775
75JHC861	M. F. DePompei and W. W. Paudler; *J. Heterocycl. Chem.*, 1975, **12**, 861.	624, 643
75JHC883	R. C. Anderson and Y. Y. Hsiao; *J. Heterocycl. Chem.*, 1975, **12**, 883.	135
75JHC1031	N. W. Rokke, J. J. Worman and W. S. Wadsworth; *J. Heterocycl. Chem.*, 1975, **12**, 1031.	802
75JHC1039	H. J. J. Loozen, B. J. van der Beek, E. F. Godefroi and H. M. Buck; *J. Heterocycl. Chem.*, 1975, **12**, 1039.	641
75JHC1045	T. P. Seden and R. W. Turner; *J. Heterocycl. Chem.*, 1975, **12**, 1045.	530
75JHC1127	Y. C. Tong; *J. Heterocycl. Chem.*, 1975, **12**, 1127.	645
75JHC1137	L. R. Swett, J. D. Ratajczyk, C. W. Nordeen and G. H. Aynilian; *J. Heterocycl. Chem.*, 1975, **12**, 1137.	343
75JHC1191	N. P. Peet and S. Sunder; *J. Heterocycl. Chem.*, 1975, **12**, 1191.	136
75JHC1199	P. L. Southwick and B. Dhawan; *J. Heterocycl. Chem.*, 1975, **12**, 1199.	272
75JHC1233	A. T. Nielsen; *J. Heterocycl. Chem.*, 1975, **12**, 1233.	281
75JMC460	T. Novinson, J. P. Miller, M. Scholten, R. K. Robins, L. N. Simon, D. E. O'Brien and R. B. Meyer; *J. Med. Chem.*, 1975, **18**, 460.	313, 331
75JMC1031	H. Zimmer, R. Glaser, J. Kokosa, D. A. Garteiz, E. V. Hess and A. Litwin; *J. ed. Chem.*, 1975, **18**, 1031.	904
75JOC252	A. A. Macco, E. F. Godefroi and J. J. M. Drouen; *J. Org. Chem.*, 1975, **40**, 252.	431, 660
75JOC356	N. J. Leonard, A. G. Morrice and M. A. Sprecker; *J. Org. Chem.*, 1975, **40**, 356.	567
75JOC363	A. G. Morrice, M. A. Sprecker and N. J. Leonard; *J. Org. Chem.*, 1975, **40**, 363.	567
75JOC385	K. Yamauchi, M. Hayashi and M. Kinoshita; *J. Org. Chem.*, 1975, **40**, 385.	530
75JOC431	A. Holm, K. Schaumburg, N. Dahlberg, C. Christophersen and J. P. Snyder; *J. Org. Chem.*, 1975, **40**, 431.	53

75JOC746	K. Bechgaard, D. O. Cowan, A. N. Bloch and L. Henriksen; *J. Org. Chem.*, 1975, **40**, 746.	115
75JOC810	R. J. De Pasquale, C. D. Padgett and R. W. Rosser; *J. Org. Chem.*, 1975, **40**, 810.	709
75JOC915	J. W. A. M. Janssen, P. Cohen-Fernandes and R. Louw; *J. Org. Chem.*, 1975, **40**, 915.	206
75JOC1348	V. Nair and K. H. Kim; *J. Org. Chem.*, 1975, **40**, 1348.	153
75JOC1815	S. M. Hecht, D. Werner, D. D. Traficante, M. Sundaralingam, P. Prusiner, T. Ito and T. Sakurai; *J. Org. Chem.*, 1975, **40**, 1815.	192, 211, 281
75JOC2042	J. K. Crandall, W. W. Conover and J. B. Komin; *J. Org. Chem.*, 1975, **40**, 2042.	703
75JOC2045	J. K. Crandall, L. C. Crawley and J. B. Komin; *J. Org. Chem.*, 1975, **40**, 2045.	703, 715
75JOC2562	J. H. M. Hill, T. R. Fogg and H. Guttmann; *J. Org. Chem.*, 1975, **40**, 2562.	634
75JOC2600	K. T. Potts and J. Kane; *J. Org. Chem.*, 1975, **40**, 2600.	131
75JOC2604	M. H. Elnagdi, M. H. R. Elmoghayar, E. A. A. Hafez and H. H. Alnima; *J. Org. Chem.*, 1975, **40**, 2604.	140
75JOC2916	E. S. Hand and W. W. Paudler; *J. Org. Chem.*, 1975, **40**, 2916.	611
75JOC2920	P. C. Srivastava, G. A. Ivanovics, R. J. Rousseau and R. K. Robins; *J. Org. Chem.*, 1975, **40**, 2920.	660
75JOC3112	C. J. Wilkerson and F. D. Greene; *J. Org. Chem.*, 1975, **40**, 3112.	775
75JOC3296	J. Pitha; *J. Org. Chem.*, 1975, **40**, 3296.	505, 528
75JOC3373	E. Abushanab, A. P. Bindra, D.-Y. Lee and L. Goodman; *J. Org. Chem.*, 1975, **40**, 3373.	625
75JOC3376	E. Abushanab, D.-Y. Lee and L. Goodman; *J. Org. Chem.*, 1975, **40**, 3376.	625
75JOC3379	E. Abushanab, A. P. Bindra and L. Goodman; *J. Org. Chem.*, 1975, **40**, 3379.	625, 626, 644
75JOC3708	D. G. Bartholomew, P. Dea, R. K. Robins and G. R. Revankar; *J. Org. Chem.*, 1975, **40**, 3708.	627, 648
75JOM(88)115	P. G. Moniotte, A. J. Hubert and P. Teyssie; *J. Organomet. Chem.*, 1975, **88**, 115.	162
75JOU902	V. I. Ofitserov, V. S. Mokrushin and I. A. Korbukh; *J. Org. Chem. USSR (Engl. Transl.)*, 1975, **11**, 902.	629
75JOU1130	I. I. Popov, A. M. Simonov and A. A. Zubenko; *J. Org. Chem. USSR (Engl. Transl.)*, 1975, **11**, 1130.	437
75KGS690	A. S. Elina and I. S. Musatova; *Khim. Geterotsikl. Soedin.*, 1975, 690 (*Chem. Abstr.*, 1975, **83**, 114 336).	626
75KGS844	S. M. Khripak, M. M. Tsitsika and I. V. Smolanka; *Khim. Geterotsikl. Soedin.*, 1975, 844.	746, 748
75KGS1000	S. M. Khripak, M. M. Tsitsika and I. V. Smolanka; *Khim. Geterotsikl. Soedin.*, 1975, 1000.	748
75KGS1389	Y. M. Yutilov and L. I. Kovaleva; *Khim. Geterotsikl. Soedin.*, 1975, 1389 (*Chem. Abstr.*, 1976, **84**, 43 936).	619
75LA449	K. Tortschanoff, H. Kisch and O. E. Polansky; *Liebigs Ann. Chem.*, 1975, 449.	255
75LA470	C. Reichard and K. Halbritter; *Liebigs Ann. Chem.*, 1975, 470.	184, 281
75LA1264	H. Behringer and R. Ramert; *Liebigs Ann. Chem.*, 1975, 1264.	758
75LA1465	C. Rufer, K. Schwarz and E. Winterfeldt; *Liebigs Ann. Chem.*, 1975, 1465.	437
75LA1934	C. Ivancsics and E. Zbiral; *Liebigs Ann. Chem.*, 1975, 1934.	647, 648
75M417	C. Ivancsics and E. Zbiral; *Monatsh. Chem.*, 1975, **106**, 417.	648
75M1059	H. Berner and H. Reinshagen; *Monatsh. Chem.*, 1975, **106**, 1059.	619, 635
75M1461	F. Asinger, A. Saus, E. Fichtner and W. Leuchtenberger; *Monatsh. Chem.*, 1975, **106**, 1461.	422
75MI40400	R. Paredes and W. R. Dolbier; *Rev. Latinoamer. Quím*, 1975, **6**, 29.	252
75MI40401	F. D. Höppner, H. J. Hofmann and C. Weiss; *J. Signalaufzeichnungsmaterilien*, 1975, **1**, 75.	210
75MI40402	A. Garcia-Tapia, V. Gómez-Parra, R. Madroñero, L. Nebrada and S. Vega; *An. Real. Acad. Ciencias Exact. Fis. Nat.*, 1975, 421.	269, 272
75MI40403	J. Vilarrasa, C. Galvez and M. Calafell; *An. Quím.* 1975, **71**, 631.	184
B-75MI40404	S. Sternhell; in 'Dynamic Nuclear Magnetic Resonance Spectroscopy', ed. L. M. Jackman and F. A. Cotton; Academic Press, New York, 1975, p. 185.	209
B-75MI40800	A. E. A. Porter; in 'Saturated Heterocyclic Chemistry', The Chemical Society, London, 1975, vol. 3, p. 197.	458, 459, 464, 466
B-75MI40900	G. A. Le Page; in 'Antineoplastic and Immunosuppressive Agents, Part II', ed. A. C. Sartorelli and D. G. Johns; Springer-Verlag, Berlin, 1975, p. 426.	504
75MI40901	F. Yoneda; *Kagaku No Kyoiki*, 1975, **29**, 482.	501
75MI40902	L. L. Bennett, Jr. and D. L. Hill; *Mol. Pharmacol.*, 1975, **11**, 803.	603
75MI40903	T. Urushibara, Y. Furuichi, C. Nishimura and K. Miura; *FEBS Lett.*, 1975, **49**, 385.	600
75MI41000	J. A. May, Jr. and L. B. Townsend; *J. Carbohydr., Nucleosides, Nucleotides*, 1975, **2**, 271.	619
75MI41001	S. Kitano, A. Nomura, Y. Mizuno, T. Okamoto and Y. Isogai; *J. Carbohydr., Nucleosides, Nucleotides*, 1975, **2**, 299.	618, 621
75MI41002	A. S. Elina, I. S. Musatova, M. A. Muratov and M. D. Mashkovskii; *Khim.-Farm. Zh.*, 1975, **9**, 10 (*Chem. Abstr.*, 1975, **82**, 170 839).	626
75MI41100	K. Tsujimoto, M. Ohashi and T. Yonezawa; *Hukusokan Kagaku Toronkai Koen Yoshishu*, 1975, **8**, 4 (*Chem. Abstr.*, 1976, **84**, 163 670).	694

B-75MI41101	J. L. Neumeyer; in 'Principles of Medicinal Chemistry', ed. W. O. Foye; Lea and Febiger, Philadelphia, 1975, p. 774.	732
75MI41200	G. Wagner, B. Dietzsch and U. Krake; *Pharmazie*, 1975, **30**, 694 (*Chem. Abstr.*, 1976, **84**, 90 504).	746
75MI41201	R. M. De Marinis, J. R. E. Hoover, G. L. Dunn, P. Actor, J. V. Uri and J. A. Welsbach; *J. Antibiot.*, 1975, **28**, 463.	787
B-75MI41400	'Peptides 1974', ed. Y. Wolman; Wiley, New York, 1975, p. 74.	845
75MIP464584	L. P. Makhno, T. G. Ermakova, E. S. Damnina, L. A. Tatarova, G. G. Skvortsova and V. A. Lopyrev; *USSR Pat.* 464 584 (1975) (*Chem. Abstr.*, 1975, **83**, 114 416).	748
75OMR(7)455	L. Marchetti, L. Pentimalli, P. Lazzerette, L. Schenetti and F. Taddei; *Org. Magn. Reson.*, 1975, **7**, 455.	627
75OMS558	A. Maquestiau, Y. Van Haverbeke, R. Flammang, M. C. Pardo and J. Elguero; *Org. Mass Spectrom.*, 1975, **10**, 558.	203
75P2512	T. A. LaRue and J. J. Child; *Phytochemistry*, 1975, **14**, 2512.	303
75P2539	T. Sugiyama, E. Kitamura, S. Kubokawa, S. Kobayashi, T. Hashizume and S. Matsubara; *Phytochemistry*, 1975, **14**, 2539.	619
75PNA318	C. M. Wei and B. Moss; *Proc. Natl. Acad. Sci. USA*, 1975, **72**, 318.	600
75RTC153	J. E. Schelling and C. A. Salemink; *Recl. Trav. Chim. Pays-Bas*, 1975, **94**, 153.	615, 619
75S20	T. Nishiwaki; *Synthesis*, 1975, 20.	491
75S162	S. Kasina and J. Nematollahi; *Synthesis*, 1975, 162.	435
75S609	K. Matsumoto, M. Suzuki, M. Tomie, N. Yoneda and M. Miyoshi; *Synthesis*, 1975, 609.	768
75S645	K. Grohe; *Synthesis*, 1975, 645.	334
75S675	R. Lantzsch and D. Arlt; *Synthesis*, 1975, 675.	667
75S703	F. Seng and K. Ley; *Synthesis*, 1975, 703.	435, 492
75S798	S. M. S. Chauhan and H. Junjappa; *Synthesis*, 1975, 798.	281
75T1171	K. S. Balachandran, I. Hiryakkanavar and M. V. George; *Tetrahedron*, 1975, **31**, 1171.	707
75T1363	G. F. Huang, M. Maeda, T. Okamoto and Y. Kawazoe; *Tetrahedron*, 1975, **31**, 1363.	771
75T1461	A. N. Nesmeyanov, E. B. Zavelovich, V. N. Babin, N. S. Kochetkova and E. I. Fedin; *Tetrahedron*, 1975, **31**, 1461.	213
75T1463	A. N. Nesmeyanov, E. B. Zavelovich, V. N. Babin, N. S. Kochetkova and E. I. Fedin; *Tetrahedron*, 1975, **31**, 1463.	192
75T1783	A. Holm, L. Carlsen, S.-O. Lawesson and H. Kolind-Andersen; *Tetrahedron*, 1975, **31**, 1783.	55
75T2914	A. Kolb, C. Gouyette, T. H. Dinh and J. Igolen; *Tetrahedron*, 1975, **31**, 2914.	585
75TL33	R. R. Schmid and H. Huth; *Tetrahedron Lett.*, 1975, 33.	287
75TL695	A. Krantz and B. Hoppe; *Tetrahedron Lett.*, 1975, 695.	852
75TL1219	B. Castro, J. R. Dormoy, B. Evin and C. Selve; *Tetrahedron Lett.*, 1975, 1219.	728
75TL2807	B. Green and D.-W. Lin; *Tetrahedron Lett.*, 1975, 2807.	714
75TL4085	J. Elguero, G. Llouquet and C. Marzin; *Tetrahedron Lett.*, 1975, 4085.	365
75USP3865570	E. F. George and W. D. Riddell; *U.S. Pat.* 3 865 570 (1975) (*Chem. Abstr.*, 1975, **83**, 54 601).	836
75ZC102	E. Lippmann, A. Könnecke and G. Beyer; *Z. Chem.*, 1975, **15**, 102 (*Chem. Abstr.*, 1975, **83**, 28 159).	824
75ZC146	E. Lippmann and D. Reifegerste; *Z. Chem.*, 1975, **15**, 146.	794
75ZC402	A. Könnecke, E. Lippmann and E. Kleinpeter; *Z. Chem.*, 1975, **15**, 402 (*Chem. Abstr.*, 1976, **84**, 73 399).	805
75ZC482	E. Bulka and W. D. Pfeiffer; *Z. Chem.*, 1975, **15**, 482.	666
75ZN(B)622	K. Burger, H. Schickaneder and A. Meffert; *Z. Naturforsch., Teil B*, 1975, **30**, 622.	275
75ZOB1821	S. B. Bulgarevich, V. S. Bolotnikov, V. N. Sheinker, O. A. Osipov and A. D. Garnovskii; *Zh. Obshch. Khim.*, 1975, **45**, 1821 (*Chem. Abstr.*, 1975, **83**, 178 013).	672
76ACH(89)265	G. Szilágyi and H. Wamhoff; *Acta Chim. Acad. Sci. Hung.*, 1976, **89**, 265.	697
76AG131	A. Padwa; *Angew. Chem.*, 1976, **88**, 131.	708
76AG724	H. Vorbrugger and K. Krolikiewicz; *Angew. Chem.*, 1976, **88**, 724.	559
76AG(E)113	G. Höfle and B. Lange; *Angew. Chem., Int. Ed. Engl.*, 1976, **15**, 113.	816
76AG(E)123	A. Padwa; *Angew. Chem., Int. Ed. Engl.*, 1976, **15**, 123.	148, 708
76AG(E)378	U. Felcht and M. Regitz; *Angew. Chem., Int. Ed. Engl.*, 1976, **15**, 378.	271
76AHC(19)1	W. D. Ollis and C. A. Ramsden; *Adv. Heterocycl. Chem.*, 1976, **19**, 1.	2, 45, 47, 76, 171, 201, 286
76AHC(19)3	W. D. Ollis and C. A. Ramsden; *Adv. Heterocycl. Chem.*, 1976, **19**, 3.	743
76AHC(20)65	L. B. Clapp; *Adv. Heterocycl. Chem.*, 1976, **20**, 65.	13, 68, 88, 89, 105
76AHC(20)145	A. Holm; *Adv. Heterocycl. Chem.*, 1976, **20**, 145.	44, 84
76AHC(S1)1	J. Elguero, C. Marzin, A. R. Katritzky and P. Linda; *Adv. Heterocycl. Chem., Suppl. 1*, 1976, 1.	15, 35, 36, 37, 38, 169, 170, 177, 182, 185, 199, 200, 201, 210, 213, 215, 222, 223, 224, 265, 737, 741, 742, 743
76AHC(S1)32	J. Elguero, C. Marzin, A. R. Katritzky and P. Linda; *Adv. Heterocycl. Chem., Suppl. 1*, 1976, 32.	677
76AHC(S1)266	J. Elguero, C. Marzin, A. R. Katritzky and P. Linda; *Adv. Heterocycl. Chem., Suppl. 1*, 1976, 266.	366

76AHC(S1)280	J. Elguero, C. Marzin, A. R. Katritzky and P. Linda; *Adv. Heterocycl. Chem.*, *Suppl. 1*, 1976, 280.	364, 365, 366, 367
76AHC(S1)281	J. Elguero, C. Marzin, A. R. Katritzky and P. Linda; *Adv. Heterocycl. Chem.*, *Suppl. 1*, 1976, 281.	677, 691
76AHC(S1)295	J. Elguero, C. Marzin, A. R. Katritzky and P. Linda; *Adv. Heterocycl. Chem.*, *Suppl. 1*, 1976, 295.	691
76AHC(S1)296	J. Elguero, C. Marzin, A. R. Katritzky and P. Linda; *Adv. Heterocycl. Chem.*, *Suppl. 1*, 1976, 296.	35, 36, 691
76AHC(S1)502	J. Elguero, C. Marzin, A. R. Katritzky and P. Linda; *Adv. Heterocycl. Chem.*, *Suppl. 1*, 1976, 502.	501, 502, 509, 520
76AJC799	M. L. Heffernan and G. M. Irvine; *Aust. J. Chem.*, 1976, **29**, 799.	860
76AJC815	M. Heffernan and G. M. Irvine; *Aust. J. Chem.*, 1976, **29**, 815.	853
76AJC891	K. Bhushan and J. H. Lister; *Aust. J. Chem.*, 1976, **29**, 891.	537, 543
76AJC1039	A. L. Cossey, R. L. N. Harris, J. L. Huppatz and J. N. Phillips; *Aust. J. Chem.*, 1976, **29**, 1039.	632
76AJC2491	M. Woodruff and J. B. Polya; *Aust. J. Chem.*, 1976, **29**, 2491.	782
76AX(B)714	J. Meunier-Piret, P. Piret, J. P. Putzeys and M. van Meerssche; *Acta Crytallogr.*, *Part B*, 1976, **32**, 714.	675
76AX(B)853	P. Prusiner, M. Sundaralingam, T. Ito and T. Sakurai; *Acta Crystallogr., Part B*, 1976, **32**, 853.	178
76AX(B)2216	L. Dupont, O. Dideberg, J. M. Braham and A. F. Noels; *Acta Crystallogr., Part B*, 1976, **32**, 2216.	179
76AX(B)2314	R. L. Harlow and S. H. Simonsen; *Acta Crystallogr., Part B*, 1976, **32**, 2314.	179
76B898	H. Kasai, M. Goto, K. Ikeda, M. Zama, Y. Mizuno, S. Takemura, S. Matsuura, T. Sugimoto and T. Goto; *Biochemistry*, 1976, **15**, 898.	605
76BCJ143	K. Fukushima, A. Kobayashi, T. Miyamoto and Y. Sasaki; *Bull. Chem. Soc. Jpn.*, 1976, **49**, 143.	227
76BCJ1920	S. Ito, Y. Tanaka, A. Kakehi and K. Kondo; *Bull. Chem. Soc. Jpn.*, 1976, **49**, 1920.	827, 834
76BCJ1980	H. Ochi, T. Miyasaka, K. Kanada and K. Arakawa; *Bull. Chem. Soc. Jpn.*, 1976, **49**, 1980.	309, 311, 338
76BCJ3607	T. Fuchigami and K. Odo; *Bull. Chem. Soc. Jpn.*, 1976, **49**, 3607.	137
76BSB545	A. A. Ouyahia, G. Leroy, J. Weiler and R. Touillaux; *Bull. Soc. Chim. Belg.*, 1976, **85**, 545.	185, 189
76BSB697	A. Maquestiau, Y. Van Haverbeke, J. C. Vanovervelt and R. Postiaux; *Bull. Soc. Chim. Belg.*, 1976, **85**, 697.	209
76BSB795	A. Maquestiau, Y. Van Haverbeke, M. Williame and M. Begtrup; *Bull. Soc. Chim. Belg.*, 1976, **85**, 795.	687
76BSF184	P. Bouchet, C. Coquelet, J. Elguero and R. Jacquier; *Bull. Soc. Chim. Fr.*, 1976, 184.	220
76BSF192	P. Bouchet, C. Coquelet, J. Elguero and R. Jacquier; *Bull. Soc. Chim. Fr.*, 1976, 192.	380
76BSF195	P. Bouchet and C. Coquelet; *Bull. Soc. Chim. Fr.*, 1976, 195.	220
76BSF476	A. Le Berre and C. Porte; *Bull. Soc. Chim. Fr.*, 1976, 476.	281
76BSF869	J. L. Aubagnac, D. Bourgeon and R. Jacquier; *Bull. Soc. Chim. Fr.*, 1976, 869.	204
76BSF1861	H. J. M. Dou and J. Metzger; *Bull. Soc. Chim. Fr.*, 1976, 1861.	387
76BSF1983	J. L. Barascut, B. Kam and J. L. Imbach; *Bull. Soc. Chim. Fr.*, 1976, 1983.	289
76BSF2025	J. Bourgois, F. Tonnard and F. Texier; *Bull. Soc. Chim. Fr.*, 1976, 2025.	716
76CB1625	K. H. Buechel and H. Erdmann; *Chem. Ber.*, 1976, **109**, 1625.	398
76CB1638	K. H. Buechel and H. Erdmann; *Chem. Ber.*, 1976, **109**, 1638.	423
76CB1898	V. Bardakos and W. Sucrow; *Chem. Ber.*, 1976, **109**, 1898.	261
76CB2596	P. Luger, C. Tuchscherer, M. Grosse and D. Rewicki; *Chem. Ber.*, 1976, **109**, 2596.	252
76CC23	M. Baudy and A. Robert; *J. Chem. Soc., Chem. Commun.*, 1976, 23.	129
76CC48	J. C. Cass, A. R. Katritzky, R. L. Harlow and S. H. Simonsen; *J. Chem. Soc., Chem. Commun.*, 1976, 48.	391
76CC155	K. Senga, H. Kanazawa and S. Nishigaki; *J. Chem. Soc., Chem. Commun.*, 1976, 155.	581
76CC250	T. Tsuchiya, J. Kurita and K. Ogawa; *J. Chem. Soc., Chem. Commun.*, 1976, 250.	287
76CC326	L. H. Dao and D. Mackay; *J. Chem. Soc., Chem. Commun.*, 1976, 326.	750
76CC343	P. N. Preston, K. K. Tiwari, K. Turnbull and T. J. King; *J. Chem. Soc., Chem. Commun.*, 1976, 343.	813
76CC414	T. L. Gilchrist, C. J. Moody and C. W. Rees; *J. Chem. Soc., Chem. Commun.*, 1976, 414.	812
76CC658	F. Yoneda, T. Nagamura and M. Kawamura; *J. Chem. Soc., Chem. Commun.*, 1976, 658.	656
76CC685	K. Tsutsumi, I. Takagishi, H. Shizuka and K. Matsui; *J. Chem. Soc., Chem. Commun.*, 1976, 685.	253, 254
76CCC1551	M. Uher, A. Rybár, A. Martvoń and J. Leško; *Collect. Czech. Chem. Commun.*, 1976, **41**, 1551.	718
76CHE304	A. F. Pozharskii, V. V. Kuz'menko, I. S. Kashparov, Z. I. Sokolov and M. M. Medvedeva; *Chem. Heterocycl. Compd. (Engl. Transl.)*, 1976, **12**, 304.	412
76CHE433	B. V. Trzhtsinskaya, L. F. Teterina, V. K. Voronov and G. G. Skvortsova; *Chem. Heterocycl. Compd. (Engl. Transl.)*, 1976, **12**, 433.	444

Ref	Citation	Pages
76CHE437	I. I. Popov, Yu. G. Bogachev, P. V. Tkachenko, A. M. Simonov and B. A. Tertov, *Chem. Heterocycl. Compd.* (*Engl. Transl.*), 1976, **12**, 437.	407, 410
76CHE465	V. S. Mokrushin, V. I. Ofitserov, T. V. Rapakova, A. G. Tsaur and Z. V. Pushkareva; *Chem. Heterocycl. Compd.* (*Engl. Transl.*), 1976, **12**, 465.	629
76CHE696	A. F. Vlasenko, B. E. Mandrichenko, G. K. Rogul'chenko, R. S. Sinyak, I. A. Mazur and P. M. Kochergin; *Chem. Heterocycl. Compd.* (*Engl. Transl.*), 1976, **12**, 696.	647, 648
76CHE917	V. P. Piskov, V. P. Kasperovich and L. M. Yakoleva; *Chem. Heterocycl. Compd.* (*Engl. Transl.*), 1976, **12**, 917.	354, 358
76CHE924	V. I. Ofitserov, Z. S. Mokrushin, Z. V. Pushkareva and N. V. Nikiforova; *Chem. Heterocycl. Compd.* (*Engl. Transl.*), 1976, **12**, 924.	445, 666
76CHE1278	G. G. Skvortsova, B. V. Trzhtsinskaya, L. F. Teterina and V. K. Voronov; *Chem. Heterocycl. Compd.* (*Engl. Transl.*), 1976, **12**, 1278.	446
76CHE1280	V. S. Kobrin and L. B. Volodarskii; *Chem. Heterocycl. Compd.* (*Engl. Transl.*), 1976, **12**, 1280.	369, 415, 423, 455
76CJC343	R. T. Baker, S. J. Rettig, A. Storr and J. Trotter; *Can. J. Chem.*, 1976, **54**, 343.	227
76CJC617	R. E. Wasylishen and M. R. Graham; *Can. J. Chem.*, 1976, **54**, 617.	354
76CJC670	J. Stawinski, T. Hozumi and S. A. Narang; *Can. J. Chem.*, 1976, **54**, 670.	814
76CJC1329	J. Elguero, A. Fruchier and M. C. Pardo; *Can. J. Chem.*, 1976, **54**, 1329.	194, 260, 290
76CJC1752	C. Sabaté-Alduy and J. Bastide; *Can. J. Chem.*, 1976, **54**, 1752.	215
76CPB804	Y. Tanaka and M. Sano; *Chem. Pharm. Bull.*, 1976, **24**, 804.	301
76CPB1331	M. Sekiya, J. Suzuki and Y. Terao; *Chem. Pharm. Bull.*, 1976, **24**, 1331.	568, 569
76CPB1336	S. Kubota and M. Uda; *Chem. Pharm. Bull.*, 1976, **24**, 1336.	748
76CPB1561	N. Yamaji, Y. Yuasa and M. Kato; *Chem. Pharm. Bull.*, 1976, **24**, 1561.	589
76CPB1870	T. Kametani, Y. Kigawa, T. Takahashi, H. Nemoto and K. Fukimoto; *Chem. Pharm. Bull.*, 1976, **24**, 1870.	340
76CPB2267	H. Koga, M. Hirobe and T. Okamoto; *Chem. Pharm. Bull.*, 1976, **24**, 2267.	234
76CPB2274	K. Kaji and M. Kawase; *Chem. Pharm. Bull.*, 1976, **24**, 2274.	655
76CPB2549	T. Kato, T. Chiba and M. Daneshtalab; *Chem. Pharm. Bull.*, 1976, **24**, 2549.	815, 821, 822
76CRV187	C. W. Spangler; *Chem. Rev.*, 1976, **76**, 187.	249
76FES209	E. Abignente, F. Arena, P. De Caprariis and L. Parente; *Farmaco Ed. Sci.*, 1976, **31**, 209 (*Chem. Abstr.*, 1976, **84**, 164 684).	614
76FES797	O. Liri, P. L. Ferrarini and I. Tonetti; *Farmaco Ed. Sci.*, 1976, **31**, 797.	728
76GEP2429195	H. Eilingsfeld and R. Niess; *Ger. Pat.* 2 429 195 (1976) (*Chem. Abstr.*, 1976, **84**, 135 649).	135
76GEP2539461	CIBA-Geigy AG, *Ger. Pat.* 2 539 461 (1976) (*Chem. Abstr.*, 1976, **85**, 110 108).	730
76GEP2554866	T. Shigematsu, M. Tomida, T. Shibahara, M. Nakazawa and T. Munakata; *Ger. Pat.* 2 554 866, 1976 (*Chem. Abstr.*, 1976, **85**, 117 969).	787
76H(4)749	F. Yoneda and T. Nagamatsu; *Heterocycles*, 1976, **4**, 749.	592
76H(4)943	B. Verček, B. Stanovnik and M. Tišler; *Heterocycles*, 1976, **4**, 943.	624, 643, 877
76H(4)1115	M. Tišler and B. Stanovnik; *Heterocycles*, 1976, **4**, 1115.	272
76H(4)1503	F. Yoneda, M. Kawamura, T. Nagamatsu, K. Kuretani, A. Hoshi and M. Iigo; *Heterocycles*, 1976, **4**, 1503.	656
76H(4)1655	M. Ruccia, N. Vivona and G. Cusmano; *Heterocycles*, 1976, **4**, 1655.	249
76H(5)127	G. G. Habermehl and W. Ecsy; *Heterocycles*, 1976, **5**, 127.	641
76H(5)285	T. Itoh and Y. Mizuno; *Heterocycles*, 1976, **5**, 285.	615
76HCA259	C. Thétaz, F. W. Wehrli and C. Wentrup; *Helv. Chim. Acta*, 1976, **59**, 259.	860
76HCA1203	H. P. Haerter, H. Lichti, U. Stauss and O. Schindler; *Helv. Chim. Acta*, 1976, **59**, 1203.	647
76HCA1512	W. Heinzelmann, M. Märky and P. Gilgen; *Helv. Chim. Acta*, 1976, **59**, 1512.	489
76HCA2362	W. Heinzelmann, M. Märky and P. Gilgen; *Helv. Chim. Acta*, 1976, **59**, 2362.	272
76IC2788	L. D. Brown and J. A. Ibers; *Inorg. Chem.*, 1976, **15**, 2788.	675
76IC2794	L. D. Brown and J. A. Ibers; *Inorg. Chem.*, 1976, **15**, 2794.	675
76IJC(B)551	A. Sharma and G. B. Behera; *Indian J. Chem., Sect. B*, 1976, **14**, 551.	631
76JA711	C. Meali, C. S. Arcus, J. L. Wilkinson, T. J. Marks and J. A. Ibers; *J. Am. Chem. Soc.*, 1976, **98**, 711.	227
76JA1258	C. Thétaz and C. Wentrup; *J. Am. Chem. Soc.*, 1976, **98**, 1258 (see also p. 1260 and earlier papers of the series).	811
76JA1259	R. Harder and C. Wentrup; *J. Am. Chem. Soc.*, 1976, **98**, 1259.	879
76JA1492	P. D. Cook, R. J. Rousseau, A. M. Mian, P. Dea, R. B. Meyer, Jr. and R. K. Robins; *J. Am. Chem. Soc.*, 1976, **98**, 1492.	642
76JA1685	L. A. Burke, J. Elguero, G. Leroy and M. Sana; *J. Am. Chem. Soc.*, 1976, **98**, 1685.	825
76JA3737	K.-C. Chang and E. Grunwald; *J. Am. Chem. Soc.*, 1976, **98**, 3737.	363
76JA3987	N. J. Leonard, M. A. Sprecker and A. G. Morrice; *J. Am. Chem. Soc.*, 1976, **98**, 3987.	567
76JA5414	Y. Shimizu, L. J. Buckley, M. Alam, Y. Oshima, W. E. Fallon, H. Kasai, I. Miura, V. P. Gullo and K. Nakanishi; *J. Am. Chem. Soc.*, 1976, **98**, 5414.	604
76JA7401	A. H.-J. Wang, J. R. Barrio and I. C. Paul; *J. Am. Chem. Soc.*, 1976, **98**, 7401.	610
76JA7408	J. R. Barrio, P. D. Sattsangi, B. A. Gruber, L. G. Dammann and N. J. Leonard; *J. Am. Chem. Soc.*, 1976, **98**, 7408.	610, 627
76JAP7636455	T. Misawa, Y. Shimizu and Y. Nakacho; *Jpn. Pat.* 76 36 455 (1976) (*Chem. Abstr.*, 1976, **85**, 123 928).	767

76JAP(K)7681826	Mitsui Toatsu Chemicals, Inc., *Jpn. Kokai* 76 81 826 (1976) (*Chem. Abstr.*, 1976, 85, 110 109).	730
76JCS(D)455	L. Menabue and G. C. Pellacani; *J. Chem. Soc., Dalton Trans.*, 1976, 455.	137
76JCS(D)909	A. R. Katritzky and D. Moderhack; *J. Chem. Soc., Dalton Trans.*, 1976, 909.	824
76JCS(P1)75	G. Cooper and W. J. Irwin; *J. Chem. Soc., Perkin Trans. 1*, 1976, 75.	161, 450
76JCS(P1)207	J. T. A. Boyle, M. F. Grundon and M. D. Scott; *J. Chem. Soc., Perkin Trans. 1*, 1976, 207.	768
76JCS(P1)367	E. E. Glover and K. T. Rowbottom; *J. Chem. Soc., Perkin Trans. 1*, 1976, 367.	863, 865
76JCS(P1)545	G. Cooper and W. J. Irwin; *J. Chem. Soc., Perkin Trans. 1*, 1976, 545.	54, 450
76JCS(P1)1210	H. Singh, K. K. Bhutani and L. R. Gupta; *J. Chem. Soc., Perkin Trans. 1*, 1976, 1210.	830
76JCS(P1)1492	E. J. Gray and M. F. G. Stevens; *J. Chem. Soc., Perkin Trans. 1*, 1976, 1492.	859, 878, 898, 899
76JCS(P1)1547	F. Yoneda and T. Nagamatsu; *J. Chem. Soc., Perkin Trans. 1*, 1976, 1547.	581
76JCS(P1)1673	G. V. Boyd, T. Norris and P. F. Lindley; *J. Chem. Soc., Perkin Trans. 1*, 1976, 1673.	794
76JCS(P1)1722	S. Anderson, E. E. Glover and K. D. Vaughan; *J. Chem. Soc., Perkin Trans. 1*, 1976, 1722.	615
76JCS(P1)1991	D. G. Doughty, E. E. Glover and K. D. Vaughan; *J. Chem. Soc., Perkin Trans. 1*, 1976, 1991.	648
76JCS(P1)2166	T. L. Gilchrist, C. J. Harris, D. G. Hawkins, C. J. Moody and C. W. Rees; *J. Chem. Soc. Perkin Trans. 1*, 1976, 2166.	166, 885, 891
76JCS(P2)315	A. S. Afridi, A. R. Katritzky and C. A. Ramsden; *J. Chem. Soc., Perkin Trans. 2*, 1976, 315.	286
76JCS(P2)736	M. Begtrup; *J. Chem. Soc., Perkin Trans. 2*, 1976, 736.	191, 192, 681, 805
76JCS(P2)1089	R. B. Moodie, K. Schofield and J. B. Weston; *J. Chem. Soc., Perkin Trans. 2*, 1976, 1089.	396
76JCS(P2)1176	R. J. Badger and G. B. Barlin; *J. Chem. Soc., Perkin Trans. 2*, 1976, 1176.	561
76JHC155	R. R. Crenshaw, G. M. Luke and D. F. Whitehead; *J. Heterocycl. Chem.*, 1976, 13, 155.	322
76JHC205	M. L. Graziano and R. Scarpati; *J. Heterocycl. Chem.*, 1976, 13, 205.	714
76JHC257	F. G. Baddar, F. H. Al-Hajjar and N. R. El-Rayyes; *J. Heterocycl. Chem.*, 1976, 13, 257.	281
76JHC301	A. Shafiee; *J. Heterocycl. Chem.*, 1976, 13, 301.	117
76JHC391	R. A. Henry; *J. Heterocycl. Chem.*, 1976, 13, 391.	819
76JHC395	S. Plescia, E. Aiello, G. Daidone and V. Sprio; *J. Heterocycl. Chem.*, 1976, 13, 395.	272
76JHC455	H. Al-Jallo, M. Shandala, F. Al-Hajjar and N. Al-Jabour; *J. Heterocycl. Chem.*, 1976, 13, 455.	281
76JHC481	E. E. Knaus, F. M. Pasutto, C. S. Giam and E. A. Swinyard; *J. Heterocycl. Chem.*, 1976, 13, 481.	784
76JHC487	K. Babič, S. Molan, S. Polanc, B. Stanovnik, J. Stres-Bratoš, M. Tišler and B. Verček; *J. Heterocycl. Chem.*, 1976, 13, 487.	871, 889, 897
76JHC509	M. P. Servé, P. G. Seybold, W. A. Feld and M. A. Chao; *J. Heterocycl. Chem.*, 1976, 13, 509.	672
76JHC545	R. W. Hamilton; *J. Heterocycl. Chem.*, 1976, 13, 545.	294
76JHC601	G. C. Wright; *J. Heterocycl. Chem.*, 1976, 13, 601.	622, 640
76JHC639	M. Iwao and T. Kuraishi; *J. Heterocycl. Chem.*, 1976, 13, 639.	869
76JHC673	V. V. Kane, H. Werblood and S. D. Levine; *J. Heterocycl. Chem.*, 1976, 13, 673.	749
76JHC793	G. Garcia-Munoz, R. Madronero, C. Ochoa, M. Stud and W. Pfleiderer; *J. Heterocycl. Chem.*, 1976, 13, 793.	665, 666
76JHC807	B. T. Keen, D. K. Krass and W. W. Paudler; *J. Heterocycl. Chem.*, 1976, 13, 807.	653
76JHC835	P. K. Kadaba, B. Stanovnik and M. Tišler; *J. Heterocycl. Chem.*, 1976, 13, 835.	869, 880
76JHC881	G. Hajós and A. Messmer; *J. Heterocycl. Chem.*, 1976, 13, 881.	857, 901, 902
76JHC899	J. Prime, B. Stanovnik and M. Tišler; *J. Heterocycl. Chem.*, 1976, 13, 899.	273
76JHC1057	R. J. Pugmire, J. C. Smith, D. M. Grant, B. Stanovnik and M. Tišler; *J. Heterocycl. Chem.*, 1976, 13, 1057.	855, 858
76JHC1153	P. K. Kadaba; *J. Heterocycl. Chem.*, 1976, 13, 1153.	726
76JHC1305	S. Y. K. Tam, J. S. Hwang, F. G. De las Heras, R. S. Klein and J. J. Fox; *J. Heterocycl. Chem.*, 1976, 13, 1305.	272
76JIC833	V. V. Nadkarny, R. S. Rao and D. S. Fernandes; *J. Indian Chem. Soc.*, 1976, 53, 833.	728
76JMC229	M. M. Hashem, K. D. Berlin, R. W. Chesnut and N. N. Durham; *J. Med. Chem.*, 1976, 19, 229.	294
76JMC262	R. R. Crenshaw, G. M. Luke and P. Siminoff; *J. Med. Chem.*, 1976, 19, 262.	322
76JMC512	T. Novinson, B. Bhooshan, T. Okabe, G. R. Revankar, R. K. Robins, K. Senga and H. R. Wilson; *J. Med. Chem.*, 1976, 19, 512.	647
76JMC517	T. Novinson, T. Okabe, R. K. Robins and T. R. Matthews; *J. Med. Chem.*, 1976, 19, 517.	655, 656
76JMC778	G. Corsi, G. Palazzo, C. Germani, P. S. Barcellona and B. Silvestrini; *J. Med. Chem.*, 1976, 19, 778.	293
76JMC814	D. G. Bartholomew, J. H. Huffman, T. R. Matthews, R. K. Robins and G. R. Revankar; *J. Med. Chem.*, 1976, 19, 814.	627
76JMC839	W. D. Kingsbury, R. J. Gyurik, V. J. Theodorides, R. C. Parish and G. Gallagher; *J. Med. Chem.*, 1976, 19, 839.	232, 233

76JMC1057	M. Gall, J. B. Hester, Jr., A. D. Rudzik and R. A. Lahti; *J. Med. Chem.*, 1976, **19**, 1057.	753
76JMC1352	P. K. Adhikary, S. K. Das and B. A. Hess, Jr.; *J. Med. Chem.*, 1976, **19**, 1352.	612, 614
76JMR(21)185	M. Witanowski, L. Stefaniak, S. Szymański, Z. Grabowski and G. A. Webb; *J. Magn. Reson.*, 1976, **21**, 185.	856
76JOC187	K. T. Potts, D. R. Choudhury and T. R. Westby; *J. Org. Chem.*, 1976, **41**, 187.	149
76JOC385	T. Novinson, P. Dea and T. Okabe; *J. Org. Chem.*, 1976, **41**, 385.	874
76JOC568	B. N. Holmes and N. J. Leonard; *J. Org. Chem.*, 1976, **41**, 568.	537
76JOC730	H. K. Spencer, M. P. Cava, F. G. Yamagishi and A. F. Garito; *J. Org. Chem.*, 1976, **41**, 730.	162
76JOC1073	P. K. Kadaba; *J. Org. Chem.*, 1976, **41**, 1073.	826, 857, 901, 902
76JOC1244	S. Barcza, M. K. Eberle, N. Engstrom and G. G. Kahle; *J. Org. Chem.*, 1976, **41**, 1244.	257
76JOC1449	J. A. May, Jr. and L. B. Townsend; *J. Org. Chem.*, 1976, **41**, 1449.	867
76JOC1758	J. W. A. M. Janssen, C. L. Habraken and R. Louw; *J. Org. Chem.*, 1976, **41**, 1758.	270
76JOC1875	G. L'abbé, G. Vermeulen, J. Flémal and S. Toppet; *J. Org. Chem.*, 1976, **41**, 1875.	794, 805
76JOC1889	D. W. Wiley, O. W. Webster and E. P. Blanchard; *J. Org. Chem.*, 1976, **41**, 1889.	539
76JOC2120	G. Baum and H. Shechter; *J. Org. Chem.*, 1976, **41**, 2120.	252
76JOC2303	S. M. Hecht, B. L. Adams and J. W. Kozarich; *J. Org. Chem.*, 1976, **41**, 2303.	541
76JOC2813	C. A. Renner and F. D. Greene; *J. Org. Chem.*, 1976, **41**, 2813.	780
76JOC2860	M. M. Goodman, J. L. Atwood, R. Carlin, W. Hunter and W. W. Paudler; *J. Org. Chem.*, 1976, **41**, 2860.	
76JOC2874	J. E. Baldwin, O. W. Lever and N. R. Tzodikov; *J. Org. Chem.*, 1976, **41**, 2874.	188, 250
76JOC3124	T. Huynh-Dinh, J. Igolen, J.-P. Marquet, E. Bisagni and J.-M. Lhoste; *J. Org. Chem.*, 1976, **41**, 3124.	861, 903
76JOC3152	S. Polanc, B. Stanovnik and M. Tišler; *J. Org. Chem.*, 1976, **41**, 3152.	880
76JOC3395	H. W. Altland; *J. Org. Chem.*, 1976, **41**, 3395.	823
76JOC3549	E. S. Hand and W. W. Paudler; *J. Org. Chem.*, 1976, **41**, 3549.	614
76JOC3775	M. K. Eberle and L. Brzechffa; *J. Org. Chem.*, 1976, **41**, 3775.	257
76JOC3781	M. H. Elnagdi, M. R. H. El-Moghayar, D. H. Fleita, E. A. A. Hafez and S. M. Fahmy; *J. Org. Chem.*, 1976, **41**, 3781.	272
76JOC3784	C. Temple, Jr., B. H. Smith, C. L. Kussner and J. A. Montgomery; *J. Org. Chem.*, 1976, **41**, 3784.	868
76JOC4033	M. Clagett, A. Gooch, P. Graham, N. Holy, B. Mains and J. Strunk; *J. Org. Chem.*, 1976, **41**, 4033.	164
76JOM(121)285	W. R. Peterson, Jr., B. Arkles and S. S. Washburne; *J. Organomet. Chem.*, 1976, **121**, 285.	720
76JOU929	M. O. Lozinskii and V. S. Dmitrukha; *J. Org. Chem. USSR (Engl. Transl.)*, 1976, **12**, 929 (*Chem. Abstr.*, 1976, **85**, 21 284).	647
76JPS1038	H. W. Jun; *J. Pharm. Sci.*, 1976, **67**, 1038.	835
76KGS1252	Y. M. Yutilov and I. A. Svertilova; *Khim. Geterotsikl. Soedin.*, 1976, 1252 (*Chem. Abstr.*, 1977, **86**, 55 343).	618
76KGS1277	Y. M. Yutilov and I. A. Svertilova; *Khim. Geterotsikl. Soedin.*, 1976, 1277 (*Chem. Abstr.*, 1978, **88**, 22 739).	636
76KGS1396	N. O. Saldabol and S. A. Hillers; *Khim. Geterotsikl. Soedin.*, 1976, 1396 (*Chem. Abstr.*, 1977, **86**, 106 468).	631
76LA745	H. Vorbruggen and K. Krolikiewicz; *Liebigs Ann. Chem.*, 1976, 745.	557
76M171	G. Ziguener, W. B. Lintschinger, A. Fuchsgruber and K. Kollmann; *Monatsh. Chem.*, 1976, **107**, 171.	649
76M1307	H. Schindlbauer and W. Kwiecinski; *Monatsh. Chem.*, 1976, **107**, 1307.	429
76M1413	K. Gewald and G. Heinhold; *Monatsh. Chem.*, 1976, **107**, 1413.	466
76MI40100	O. L. Stiefvater; *Chem. Phys.*, 1976, **13**, 73.	12
B-76MI40200	K. Schofield, M. R. Grimmett and B. R. T. Keene; 'The Azoles', Cambridge University Press, London, 1976, p. 60.	50
B-76MI40201	K. Schofield, M. R. Grimmett and B. R. T. Keene; 'The Azoles', Cambridge University Press, London, 1976, p. 281.	50
76MI40300	G. Werber, F. Buccheri, N. Vivona and M. Gentile; *Chim. Ind. (Milan)*, 1976, **58**, 382.	136
76MI40400	K. Tsurumi, A. Abe, H. Fujimura, H. Asai, M. Nagasaka and H. Mikaye; *Folia Pharmacol. Jpn.*, 1976, **72**, 41.	294
76MI40401	G. Alonso, C. Díez, G. García-Muñoz, F. G. de las Heras and P. Navarro; *J. Carbohydr., Nucleosides, Nucleotides*, 1976, **3**, 157.	289
B-76MI40402	K. Schofield, M. R. Grimmett and B. T. R. Keene; 'Heteroaromatic Nitrogen Compounds: The Azoles', Cambridge University Press, Cambridge, 1976.	168, 171, 182, 184, 186, 197, 199, 206, 222, 224, 225, 228, 229, 230, 232, 233, 234, 237, 238, 239, 240, 241, 242, 243, 245, 246, 247, 260, 261, 262, 263, 264, 265, 266, 267, 268, 269, 270, 278
76MI40403	J. Elguero and C. Marzin; *Adv. Org. Chem.*, 1976, **9** (1), 533.	169, 255, 256, 299
B-76MI40404	'The Merck Index', ed. M. Windholz; Merck & Co., Rahway, New Jersey, 9th edn., 1976.	234, 291, 292, 293, 294, 295, 296, 297, 298, 299, 300, 301
B-76MI40500	J. Elguero, C. Marzin, A. R. Katritzky and P. Linda; 'The Tautomerism of Heterocycles', Academic, New York, 1976, p. 531.	309

B-76MI40600	K. Schofield, M. R. Grimmett and B. R. T. Keene; 'Heteroaromatic Nitrogen Compounds: The Azoles', Cambridge University Press, Cambridge, 1976.	346, 348, 350, 353, 356, 360, 367, 372
76MI40700	C. A. Matuszak and A. J. Matuszak; *J. Chem. Educ.*, 1976, **53**, 280.	384, 392
B-76MI40701	K. Schofield, M. R. Grimmett and B. R. T. Keene; 'Heteroaromatic Nitrogen Compounds: The Azoles', Cambridge University Press, Cambridge, 1976.	382, 384, 389, 93, 396, 397, 398, 400, 401, 403, 404, 405, 406, 411, 413, 415, 417, 433, 435, 436, 437, 438, 441, 443, 444, 445, 447, 448, 450, 453, 454
B-76MI40800	K. Schofield, M. R. Grimmett and B. R. T. Keene; 'Heteroaromatic Nitrogen Compounds: The Azoles', Cambridge University Press, Cambridge, 1976.	489
76MI40900	R. J. H. Davies; *Annu. Rep.*, 1976, **73B**, 375.	504
B-76MI40901	'Photochemistry and Photobiology of the Nucleic Acids', ed. S. Y. Wang; Academic, London, 1976, vols. 1 and 2.	520
76MI40902	G. Barnathan, T. H. Dinh, A. Kolb and J. Igolen; *J. Med. Chim. Ther.*, 1976, **11**, 67.	558, 585
76MI40903	F. M. Scabel, Jr., M. W. Trader and W. R. Laster, Jr.; *Proc. Am. Cancer Res.*, 1976, **17**, Abstr. 181.	603
76MI40904	M. Okutsu and A. Yamazaki; *Nucleic Acids Res.*, 1976, **3**, 231.	585
76MI40905	P. M. Schwartz, C. Shipman, Jr. and J. C. Drach; *Antimicrob. Agents, Chemother.*, 1976, **10**, 64.	603
76MI40906	G. A. LePage, L. S. Worth and A. P. Kimbal; *Cancer Res.*, 1976, **36**, 1481.	603
76MI40907	C. E. Case and T. H. A-Yeung; *Cancer Res.*, 1976, **36**, 1486.	603
76MI41000	C. Tapiero, J.-L. Imbach, R. P. Panzica and L. B. Townsend; *J. Carbohydr., Nucleosides, Nucleotides*, 1976, **3**, 191.	652
76MI41001	A. S. Elina, I. S. Musatova, N. A. Novitskaya, M. A. Muratov and M. D. Mashkovskii; *Khim.-Farm. Zh.*, 1976, **10**, 35 (*Chem. Abstr.*, 1976, **86**, 106 524).	626
76MI41002	N. K. Kochetkov, V. N. Shibaev, A. A. Kost, A. P. Razjivin and A. Y. Borisov; *Nucleic Acids Res.*, 1976, **3**, 1341.	626, 648
76MI41100	D. Spinelli, A. Corrao, V. Frenna, M. Ruccia, N. Vinova and G. Cusmano; *Chem. Ind. (Milan)*, 1976, **58**, 383.	720
76MI41101	S. Toppet, G. Wouters and G. Smets; *J. Polym. Sci. Polym. Lett.*, 1976, **14**, 389.	681
76MI41102	Y. Ueda; *Chemotherapy (Tokyo)*, 1976, **24**, 1661.	729
B-76MI41103	K. Schofield, M. R. Grimmett and B. R. T. Keene; 'Heteroaromatic Nitrogen Compounds: The Azoles', Cambridge University Press, Cambridge, 1976, p. 159.	670, 692
B-76MI41104	I. Fleming; 'Frontier Orbitals and Organic Chemical Reactions', Wiley–Interscience, London, 1976, p. 154.	711
B-76MI41200	K. Schofield, M. R. Grimmett and B. R. T. Keene; 'Heteroaromatic Nitrogen Compounds: The Azoles', Cambridge University Press, Cambridge, 1976.	735, 739, 740, 742, 743, 746, 750, 751, 754, 756, 757
76MI41300	C. M. Tarver, T. C. Goodale, R. Shaw and M. Cowperthwaite; *Off. Nav. Res. (Tech. Rep.) ACR (US)*, 1976, ACR-221, Proc. Symp. Int. Detonation 6th, 231 (*Chem. Abstr.*, 1980, **92**, 8480).	823, 837
76NKK782	N. Matsumura, Y. Otsuji and E. Imoto; *Nippon Kagaku Kaishi*, 1976, 782 (*Chem. Abstr.*, 1976, **85**, 32 929).	136
76OMS167	A. Könnecke and E. Lippmann; *Org. Mass Spectrom.*, 1976, **11**, 167.	802
B-76PH123	A. Lablache-Combier; in 'Photochemistry of Heterocyclic Compounds', ed. O. Buchardt; Wiley-Interscience, New York, 1976, p. 123.	218
B-76PH403	O. Buchardt; 'Photochemistry of Heterocyclic Compounds', Wiley–Interscience, New York, 1976, p. 403.	704, 724
76RTC67	D. N. Reinhoudt and C. G. Kouwenhoven; *Recl. Trav. Chim. Pays-Bas*, 1976, **95**, 67.	720
76RTC74	A. Rykowski and H. C. van der Plas; *Recl. Trav. Chim. Pays-Bas*, 1976, **95**, 74.	653
76RTC127	C. Kroon, A. M. van den Brink, E. J. Vlietstra and C. A. Salemink; *Recl. Trav. Chim. Pays-Bas*, 1976, **95**, 127.	636
76S52	G. Ege and P. Arnold; *Synthesis*, 1976, 52.	281
76S189	K. K. Balasubramanian and R. Nagarajan; *Synthesis*, 1976, 189.	141
76S261	M. Ferrey, A. Robert and A. Foucaud; *Synthesis*, 1976, 261.	129
76S268	N. Goetz and B. Zeeh; *Synthesis*, 1976, 268.	137
76S349	T. Wagner-Jauregg; *Synthesis*, 1976, 349.	282, 283
76S482	N. E. Alexandrou and S. Adamopoulos; *Synthesis*, 1976, 482.	706
76S489	M. Narita and C. U. Pittman, Jr.; *Synthesis*, 1976, 489.	114, 126
76S798	F. De Angelis, A. Gambacorta and R. Nicoletti; *Synthesis*, 1976, 798.	246
76S833	B. Jenko, B. Stanovnik and M. Tišler; *Synthesis*, 1976, 833.	891, 892
76SC5	C. F. Beam, R. M. Sandifer, R. S. Foote and C. R. Hauser; *Synth. Commun.*, 1976, **6**, 5.	125, 277
76SC69	S. Mineo, S. Kawamura and K. Nakagawa; *Synth. Commun.*, 1976, **6**, 69.	887
76T337	M. Araki, M. Maeda and Y. Kawazoe; *Tetrahedron*, 1976, **32**, 337.	543
76T493	B. Koren, F. Kovac, A. Petric, B. Stanovnik and M. Tišler; *Tetrahedron*, 1976, **32**, 493.	272
76T499	A. Könnecke, E. Lippmann and E. Kleinpeter; *Tetrahedron*, 1976, **32**, 499.	799, 805

76T619	M. Schneider and H. Strohäcker; *Tetrahedron*, 1976, **32**, 619.	255
76T1555	K. van der Meer and J. J. C. Mulder; *Tetrahedron*, 1976, **32**, 1555.	206
76T1909	G. De Luca, C. Panattoni and G. Renzi; *Tetrahedron*, 1976, **32**, 1909.	271
76T2165	R. M. Kellogg; *Tetrahedron*, 1976, **32**, 2165.	146
76T2421	G. Domány, J. Nyitrai and K. Lempert; *Tetrahedron*, 1976, **32**, 2421.	376
76TL45	A. Dhainaut, J. L. Montero, B. Rayner, C. Tapiero and J. L. Imbach; *Tetrahedron Lett.*, 1976, 45.	585
76TL79	M. M. Htay and O. Meth-Cohn; *Tetrahedron Lett.*, 1976, 79.	390
76TL199	S. A. Khan and K. M. Sivanandaiah; *Tetrahedron Lett.*, 1976, 199.	728
76TL571	G. M. Devasia; *Tetrahedron Lett.*, 1976, 571.	376, 479
76TL1199	Y. Maki, M. Suzuki and K. Ozeki; *Tetrahedron Lett.*, 1976, 1199.	541
76TL1485	H. Quast and L. Bieber; *Tetrahedron Lett.*, 1976, 1485.	831
76TL3313	G. Assef, J. Kister, J. Metzger, R. Faure and E. J. Vincent; *Tetrahedron Lett.*, 1976, 3313.	20, 355
76TL3483	J. E. Abola, D. J. Abraham and L. B. Townsend; *Tetrahedron Lett.*, 1976, 3483.	509
76TL3695	K. Ishikawa, Kin-Ya Akiba and N. Inamoto; *Tetrahedron Lett.*, 1976, 3695.	69
76TL3807	H. Abe, M. Uchiyama, Y. Tanaka and H. Saito; *Tetrahedron Lett.*, 1976, 3807.	605
76TL4175	M. Rull and J. Vilarrasa; *Tetrahedron Lett.*, 1976, 4175.	371
76TL4333	A. R. Katritzky, M. Michalska, R. L. Harlow and S. H. Simonsen; *Tetrahedron Lett.*, 1976, 4333.	474
76USP3939161	J. D. Ratajczyk, R. G. Stein and L. R. Swett; *U.S. Pat.* 3 939 161 (1976) (*Chem. Abstr.*, 1976, **84**, 164 835).	323
76USP3948885	ICN Pharmaceuticals, Inc., *U.S. Pat.* 3 948 885 (1976) (*Chem. Abstr.*, 1976, **85**, 6010).	729
76USP3968103	ICN Pharmaceuticals, Inc., *U.S. Pat.* 3 968 103 (1976) (*Chem. Abstr.*, 1976, **85**, 94 660).	729
76ZC16	K. Peseke; *Z. Chem.*, 1976, **16**, 16.	132
76ZC53	A. Könnecke and E. Lippmann; *Z. Chem.*, 1976, **16**, 53 (*Chem. Abstr.*, 1976, **85**, 21 225).	833
76ZC90	E. Lippmann and A. Könnecke; *Z. Chem.*, 1976, **16**, 90.	792, 820, 824, 827
76ZC102	H. Dehne and M. Süsse; *Z. Chem.*, 1976, **16**, 102.	717
76ZC280	S. Hauptmann and H. Wilde; *Z. Chem.*, 1976, **16**, 280.	689
76ZC337	M. Pulst and M. Weissenfels; *Z. Chem.*, 1976, **16**, 337.	133
76ZC398	M. Augustin and W. Dölling, *Z. Chem.*, 1976, **16**, 398.	132
76ZOR673	B. S. Drach, V. A. Kovalev and A. V. Kirsanov; *Zh. Org. Khim.*, 1976, **12**, 673.	765
77ACR403	J. A. McCloskey and S. Nishimura; *Acc. Chem. Res.*, 1977, **10**, 403.	504
77AF82	E. Winkelmann, W. Raether, H. Hartung and W. H. Wagner; *Arzneim.-Forsch.*, 1977, **27**, 82.	613
77AG(E)323	M. Franck-Neumann and J. J. Lohmann; *Angew. Chem., Int. Ed. Engl.*, 1977, **16**, 323.	251
77AHC(21)2	R. N. Butler; *Adv. Heterocycl. Chem.*, 1977, **22**, 72.	263
77AHC(21)175	R. Filler and Y. S. Rao; *Adv. Heterocycl. Chem.*, 1977, **21**, 175.	64, 78
77AHC(21)207	Y. Takeuchi and F. Furusaki; *Adv. Heterocycl. Chem.*, 1977, **21**, 207.	80
77AHC(21)323	R. N. Butler; *Adv. Heterocycl. Chem.*, 1977, **21**, 323.	46, 51, 53, 54, 69, 70, 73, 92, 94, 95, 102, 105, 106, 109, 792, 793, 797, 798, 799, 804, 805, 807, 808, 809, 810, 812, 813, 814, 816, 817, 818, 819, 821, 823, 825, 827, 829, 830, 832, 833, 834, 835, 836, 837
77AJC421	M. Woodruff and J. B. Polya; *Aust. J. Chem.*, **30**, 421.	740
77AJC2005	B. K. M. Chan, N. H. Chang and M. R. Grimmett; *Aust. J. Chem.*, 1977, **30**, 2005.	449
77AJC2515	D. J. Brown and T. Nagamatsu; *Aust. J. Chem.*, 1977, **30**, 2515.	854, 856, 890, 891
77AX(B)253	T. J. Kistenmacher and M. Rossi; *Acta Crystallogr., Part B*, 1977, **33**, 253.	509
77AX(B)413	J. P. Declercq, G. Germain, M. Van Meerssche, A. Bettencourt, Z. Janousek and H. G. Viehe; *Acta Crystallogr., Part B*, 1977, **33**, 413.	178
77BBA(491)253	K. W. Ehler, E. Girard and L. E. Orgel; *Biochim. Biophys. Acta*, 1977, **491**, 253.	649
77BCJ953	M. Takahashi, H. Tan, K. Fukushima and H. Yamazaki; *Bull. Chem. Soc. Jpn.*, 1977, **50**, 953.	133
77BCJ2375	M. Kinoshita and Y. Suzuki; *Bull. Chem. Soc. Jpn.*, 1977, **50**, 2375.	640
77BCJ2505	Y. Ogata, K. Tagaki and E. Hayashi; *Bull. Chem. Soc. Jpn.*, 1977, **50**, 2505.	697
77BCJ2689	O. Makabe, H. Suzuki and S. Umezawa; *Bull. Chem. Soc. Jpn.*, 1977, **50**, 2689.	729
77BSB281	A. Maquestiau, Y. Van Haverbeke, R. Flammang and C. de Meyer; *Bull. Soc. Chim. Belg.*, 1977, **86**, 281.	203
77BSB949	A. Maquestiau, Y. Van Haverbeke and J. C. Vanovervelt; *Bull. Soc. Chim. Belg.*, 1977, **86**, 949.	215
77BSB961	A. Maquestiau, Y. Van Haverbeke and J. C. Vanovervelt; *Bull. Soc. Chim. Belg.*, 1977, **86**, 961.	265
77BSF171	P. Bouchet, C. Coquelet and J. Elguero; *Bull. Soc. Chim. Fr.*, 1977, 171.	169, 182, 263, 267
77BSF1163	R. Lazaro, D. Mathiew, R. Fhan Tan Luu and J. Elguero; *Bull. Soc. Chim. Fr.*, 1977, 1163.	121, 278
77CB373	G. Grenner and H.-L. Schmidt; *Chem. Ber.*, 1977, **110**, 373.	510
77CB2939	G. Szeimies and K. Mannhardt; *Chem. Ber.*, 1977, **110**, 2939.	833

77CC505	P. Shu, A. N. Bloch, T. F. Carruthers and D. O. Cowan; *J. Chem. Soc., Chem. Commun.*, 1977, 505.	115
77CC556	S. Senda, K. Hirota, T. Asao and Y. Yamada; *J. Chem. Soc., Chem. Commun.*, 1977, 556.	335
77CC843	H. Dürr, S. Fröhlich, B. Schley and H. Weisgerber; *J. Chem. Soc., Chem. Commun.*, 1977, 843.	251
77CHE210	V. V. Dovlatyan, K. A. Eliazyan and L. G. Agadzhanyan; *Chem. Heterocycl. Compd. (Engl. Transl.)*, 1977, **13**, 210.	657
77CHE1110	M. A. Iradyan, A. G. Torosyan, R. G. Mirzoyan and A. A. Aroyan; *Chem. Heterocycl. Compd. (Engl. Transl.)*, 1977, **13**, 1110.	433
77CHE1160	M. I. Knyazhanskii, P. V. Gilyanovskii and O. A. Osipov; *Chem. Heterocycl. Compd. (Engl. Transl.)*, 1977, **13**, 1160.	357, 381, 382
77CJC1564	R. N. Butler, T. M. McEvoy, F. L. Scott and J. C. Tobin; *Can. J. Chem.*, 1977, **55**, 1564.	799
77CPB830	R. Moroi, K. Ono, T. Saito, T. Akimoto and M. Sano; *Chem. Pharm. Bull.*, 1977, **25**, 830.	904
77CPB1147	Y. Yamazaki, R. Moroi and M. Sano; *Chem. Pharm. Bull.*, 1977, **25**, 1147.	179
77CPB1923	T. Koyama, T. Hirota, C. Bashou, T. Nanba, S. Ohmori and M. Yamoto; *Chem. Pharm. Bull.*, 1977, **25**, 1923.	568
77CPB3087	Y. Okamoto and T. Ueda; *Chem. Pharm. Bull.*, 1977, **25**, 3087.	431
77CR(C)(284)509	J. Bourgois and F. Texier; *C. R. Hebd. Seances Acad. Sci., Ser. C*, 1977, **284**, 509.	724
77CZ403	D. Moderhack; *Chem.-Ztg.*, 1977, **101**, 403 (*Chem. Abstr.*, 1978, **88**, 37 706).	820
77DOK(234)343	N. K. Kochetkov, V. N. Shibaev, A. A. Kost, A. P. Razzhivin and S. V. Ermolin; *Dokl. Akad. Nauk SSSR*, 1977, **234**, 343 (*Chem. Abstr.*, 1977, **87**, 102 261).	648
77G91	G. Desimoni, P. Righetti, G. Tacconi and A. Vigliani; *Gazz. Chim. Ital.*, 1977, **107**, 91.	250
77GEP2700012	G. Arcari, L. Bernardi, G. Falconi, F. Luini, G. Palamidessi and U. Scarponi; *Ger. Pat.* 2 700 012 (1977) (*Chem. Abstr.*, 1977, **87**, 201 535).	623
77H(6)383	J. H. Lister; *Heterocycles*, 1977, **6**, 383.	534
77H(6)727	M. A. Khan and A. E. Guarconi; *Heterocycles*, 1977, **6**, 727.	340
77H(6)929	D. T. Hurst and J. A. Saldanha; *Heterocycles*, 1977, **6**, 929.	627
77H(6)945	K. Senga, J. Sato and S. Nishigaki; *Heterocycles*, 1977, **6**, 945.	337
77H(6)949	Y. Tamura, M. Iwaisaki, Y. Miki and M. Ikeda; *Heterocycles*, 1977, **6**, 949.	856, 866, 886
77H(6)1627	T. Fujii, T. Itaya and T. Saito; *Heterocycles*, 1977, **6**, 1627.	501
77H(8)109	Y. Tomimatsu, K. Satoh and M. Sakamoto; *Heterocycles*, 1977, **8**, 109.	166
77H(8)433	T. Itoh, J. Inaba and Y. Mizuno; *Heterocycles*, 1977, **8**, 433.	615
77HC(30-1)1	J. A. Paolini; *Chem. Heterocycl. Compd.*, 1977, **30-1**, 1.	120, 267
77HC(30-2)117	H. L. Blewitt; *Chem. Heterocycl. Compd.*, 1977, **30-2**, 117.	267, 608, 612, 613, 614
77HC(30)179	G. Maury; *Chem. Heterocycl. Compd.*, 1977, **30**, 179.	267
77HCA521	A. Edenhofer and W. Meister; *Helv. Chim. Acta*, 1977, **60**, 521.	666
77HCA3035	H. Meier and H. Heimgartner; *Helv. Chim. Acta*, 1977, **60**, 3035.	318
77IJC(B)629	K. Nagarajan, V. P. Arya, S. J. Shenoy, R. K. Shah, A. N. Goud and G. A. Bhat; *Indian J. Chem., Sect. B*, 1977, **15**, 629.	641
77JA633	W. L. Magee and H. Shechter; *J. Am. Chem. Soc.*, 1977, **99**, 633.	251
77JA714	V. Markowski, G. R. Sullivan and J. D. Roberts; *J. Am. Chem. Soc.*, 1977, **99**, 714.	510, 516
77JA2740	T. C. Clarke, L. A. Wendling and R. G. Bergman; *J. Am. Chem. Soc.*, 1977, **99**, 2740.	255, 256
77JA2818	H. Tanino, T. Nakata, T. Kaneko and Y. Kishi; *J. Am. Chem. Soc.*, 1977, **99**, 2818.	604
77JA3934	L. B. Clark; *J. Am. Chem. Soc.*, 1977, **99**, 3934.	517
77JA5096	M. F. Zady and J. L. Wong; *J. Am. Chem. Soc.*, 1977, **99**, 5096.	543, 544
77JA7257	G. Dodin, M. Dreyfuss, O. Bensaude and J.-E. Dubois; *J. Am. Chem. Soc.*, 1977, **99**, 7257.	310
77JA7806	D. Poppinger, L. Radom and J. A. Pople; *J. Am. Chem. Soc.*, 1977, **99**, 7806.	7
77JCR(M)2813	D. Bartholomew and I. T. Kay; *J. Chem. Res. (M)*, 1977, 2813.	132
77JCR(S)140	D. Clérin, B. Meyer and J. P. Fleury; *J. Chem. Res. (S)*, 1977, 140.	286
77JCR(S)238	D. Bartholomew and I. T. Kay; *J. Chem. Res. (S)*, 1977, 238.	132, 853
77JCS(P1)78	D. G. Doughty, E. E. Glover and K. D. Vaughan; *J. Chem. Soc., Perkin Trans. 1*, 1977, 78.	627
77JCS(P1)470	D. W. S. Latham, O. Meth-Cohn, H. Suschitzky and J. A. L. Herbert; *J. Chem. Soc., Perkin Trans. 1*, 1977, 470.	492
77JCS(P1)672	I. J. Ferguson, K. Schofield, J. W. Barnett and M. R. Grimmett; *J. Chem. Soc., Perkin Trans. 1*, 1977, 672.	234, 456
77JCS(P1)765	F. Yoneda, T. Nagamatsu, T. Nagamura and K. Senga; *J. Chem. Soc., Perkin Trans. 1*, 1977, 765.	336
77JCS(P1)939	M. Bonadeo, C. De Micheli and R. Gandolfi; *J. Chem. Soc., Perkin Trans. 1*, 1977, 939.	268
77JCS(P1)1819	A. Albert and C. J. Lin; *J. Chem. Soc., Perkin Trans. 1*, 1977, 1819.	721, 875
77JCS(P1)2096	J. Grimshaw and D. Mannus; *J. Chem. Soc., Perkin Trans. 1*, 1977, 2096.	220
77JCS(P2)724	M. J. S. Dewar and I. J. Turchi; *J. Chem. Soc., Perkin Trans. 2*, 1977, 724.	7
77JCS(P2)1024	G. E. Hawkes, E. W. Randall, J. Elguero and C. Marzin; *J. Chem. Soc., Perkin Trans. 2*, 1977, 1024.	194, 195, 196, 290
77JCS(P2)1779	N. A. Rodios and N. E. Alexandrou; *J. Chem. Soc., Perkin Trans. 2*, 1977, 1779.	678

77JHC37	P. Bravo and G. Gaviraghi; *J. Heterocycl. Chem.*, 1977, **14**, 37.	165
77JHC59	M. K. Eberle and P. Schirm; *J. Heterocycl. Chem.*, 1977, **14**, 59.	654
77JHC65	S. A. Lang, E. M. Lovell and E. Cohen; *J. Heterocycl. Chem.*, 1977, **14**, 65.	178, 180
77JHC155	M. H. Elnagdi, E. M. Kandeel, E. M. Zayed and Z. E. Kandil; *J. Heterocycl. Chem.*, 1977, **14**, 155.	272
77JHC195	J. A. Montgomery, A. T. Shortnacy and S. D. Clayton; *J. Heterocycl. Chem.*, 1977, **14**, 195.	619, 620, 621
77JHC227	M. H. Elnagdi, M. R. H. Elmoghayar, E. M. Kandeel and M. K. A. Ibrahim; *J. Heterocycl. Chem.*, 1977, **14**, 227.	330
77JHC375	A. S. Shawali; *J. Heterocycl. Chem.*, 1977, **14**, 375.	272, 326
77JHC387	S. E. Ealick, D. van der Helm, K. Ramalingam, G. X. Thyvelikakath and K. D. Barlin; *J. Heterocycl. Chem.*, 1977, **14**, 387.	178
77JHC411	J. S. Bradshaw, J. E. Tueller, S. L. Baxter, G. E. Maas and J. T. Carlock; *J. Heterocycl. Chem.*, 1977, **14**, 411.	870
77JHC439	J. P. Hénichart, R. Houssin, B. Lablanche, F. Baert and L. Devos; *J. Heterocycl. Chem.*, 1977, **14**, 439.	739
77JHC511	K. Okada, J. A. Kelley and J. S. Driscoll; *J. Heterocycl. Chem.*, 1977, **14**, 511.	139
77JHC517	R. J. Sundberg; *J. Heterocycl. Chem.*, 1977, **14**, 517.	448
77JHC607	D. C. H. Bigg and S. R. Purvis; *J. Heterocycl. Chem.*, 1977, **14**, 607.	358
77JHC647	T. C. Thurber and L. B. Townsend; *J. Heterocycl. Chem.*, 1977, **14**, 647.	719
77JHC697	M. V. Pickering, M. T. Campbell, J. T. Witkowski and R. K. Robins; *J. Heterocycl. Chem.*, 1977, **14**, 697.	897
77JHC813	T. Denzel and H. Hoehn; *J. Heterocycl. Chem.*, 1977, **14**, 813.	618, 635, 636
77JHC889	J. J. Baldwin, P. K. Lumma, G. S. Ponticello, F. C. Novello and J. M. Sprague; *J. Heterocycl. Chem.*, 1977, **14**, 889.	446, 447
77JHC993	M. Iwao and T. Kuraishi; *J. Heterocycl. Chem.*, 1977, **14**, 993.	635
77JHC1013	J. P. Affane-Nguema, J. P. Lavergne and P. Viallefont; *J. Heterocycl. Chem.*, 1977, **14**, 1013.	272
77JHC1163	N. W. Gilman, B. C. Holland and R. I. Fryer; *J. Heterocycl. Chem.*, 1977, **14**, 1163.	273
77JHC1171	N. W. Gilman and R. I. Fryer; *J. Heterocycl. Chem.*, 1977, **14**, 1171.	273
77JHC1175	T. Kametani, K. Kigasawa, M. Hiiragi, K. Wakisaka, O. Kusama, H. Sugi and K. Kawasaki; *J. Heterocycl. Chem.*, 1977, **14**, 1175.	273
77JHC1321	V. I. Cohen and S. Pourabass; *J. Heterocycl. Chem.*, 1977, **14**, 1321.	353, 359
77JMC386	W. E. Kirkpatrick, T. Okabe, I. W. Hillyard, R. K. Robins, A. T. Dren and T. Novinson; *J. Med. Chem.*, 1977, **20**, 386.	316
77JMC664	K. Ramalingam, L. F. Wong, K. D. Berlin, R. A. Brown, R. Fischer, J. Blunk and N. N. Durham; *J. Med. Chem.*, 1977, **20**, 664.	294
77JMC847	K. Ramalingam, G. X. Thyvelikakath, K. D. Berlin, R. W. Chesnut, R. A. Brown, N. N. Durham, S. E. Ealick and D. van der Helm; *J. Med. Chem.*, 1977, **20**, 847.	294
77JMC1189	J. J. Baldwin, P. K. Lumma, F. C. Novello, G. S. Ponticello, J. M. Sprague and D. E. Duggan; *J. Med. Chem.*, 1977, **20**, 1189.	635
77JMR(28)217	M. Witanowski, L. Stefaniak, S. Szymański and H. Januszewski; *J. Magn. Reson.*, 1977, **28**, 217.	195
77JMR(28)227	P. R. Srinivasan and R. L. Lichter; *J. Magn. Reson.*, 1977, **28**, 227.	195
77JOC542	A. V. Zeiger and M. M. Joullié; *J. Org. Chem.*, 1977, **42**, 542.	454
77JOC592	A. D. Dawson and D. Swern; *J. Org. Chem.*, 1977, **42**, 592.	727
77JOC659	M. T. Chenon, C. Coupry, D. M. Grant and R. J. Pugmire; *J. Org. Chem.*, 1977, **42**, 659.	191, 211, 213
77JOC847	N. Murai, M. Komatsu, T. Yagii, H. Nishihara, Y. Ohshiro and T. Agawa; *J. Org. Chem.*, 1977, **42**, 847.	466
77JOC941	H. Kohn, M. J. Cravey, J. H. Arceneaux, R. L. Cravey and M. R. Willcott, III; *J. Org. Chem.*, 1977, **42**, 941.	443
77JOC1018	J. Daunis, H. Lopez and G. Maury; *J. Org. Chem.*, 1977, **42**, 1018.	854, 859, 878, 898, 899, 900
77JOC1159	G. L'abbé, G. Verhelst and S. Toppet; *J. Org. Chem.*, 1977, **42**, 1159.	832
77JOC1555	R. D. Srivastava and I. M. Brinn; *J. Org. Chem.*, 1977, **42**, 1555.	17
77JOC1633	K. T. Potts, S. J. Chen, J. Kane and J. L. Marshall; *J. Org. Chem.*, 1977, **42**, 1633.	129
77JOC1639	K. T. Potts and S. J. Chen; *J. Org. Chem.*, 1977, **42**, 1639.	129
77JOC2893	C. L. Habraken and E. K. Poels; *J. Org. Chem.*, 1977, **42**, 2893.	270
77JOC3132	B. Coxon, A. J. Fatiadi, L. T. Sniegoski, H. S. Hertz and R. Schaffer; *J. Org. Chem.*, 1977, **42**, 3132.	661
77JOC3377	E. S. Hand, W. W. Paudler and S. Zachow; *J. Org. Chem.*, 1977, **42**, 3377.	612
77JOC3725	J. Bartels-Keith, M. Burgess and J. Stevenson; *J. Org. Chem.*, 1977, **42**, 3725.	19, 805
77JOC4197	J. Bradač, Z. Furek, D. Janežič, S. Molan, I. Smerkolj, B. Stanovnik, M. Tišler and B. Verček; *J. Org. Chem.*, 1977, **42**, 4197.	624, 643, 877
77JOM(127)273	T. Hitomi and S. Kozima; *J. Organomet. Chem.*, 1977, **127**, 273.	682, 683
77JOM(132)69	R. Gassend, J. C. Maire and J. C. Pommier; *J. Organomet. Chem.*, 1977, **132**, 69.	213
77JOU1817	M. V. Gorelik and T. Kh. Gladysheva; *J. Org. Chem. USSR (Engl. Transl.)*, 1977, **13**, 1817.	423
77JOU1872	V. N. Sheinker, L. G. Tishchenko, A. D. Garnovskii and A. M. Simonov; *J. Org. Chem. USSR (Engl. Transl.)*, 1977, **13**, 1872.	393
77JPR65	J. Faust; *J. Prakt. Chem.*, 1977, **319**, 65.	136

77JPR408	A. Könnecke, S. Behrendt and E. Lippmann; *J. Prakt. Chem.*, 1977, **319**, 408 (see also other papers in the series).	793, 798, 799
77JPR895	W. Freyer and G. Tomaschewski; *J. Prakt. Chem.*, 1977, **319**, 895.	209
77JPR911	W. Freyer; *J. Prakt. Chem.*, 1977, **319**, 911.	209
77JSP(65)313	G. L. Blackman, R. D. Brown, F. R. Burden and W. Garland; *J. Mol. Spectrosc.*, 1977, **65**, 313.	677
77KGS411	G. A. Mokrushina, I. Y. Postovskii and S. K. Kotovskaya; *Khim. Geterotsikl. Soedin.*, 1977, 411 (*Chem. Abstr.*, 1978, **87**, 53 170).	617, 636
77KGS556	N. O. Saldabol and J. Popelis; *Khim. Geterotsikl. Soedin.*, 1977, 556 (*Chem. Abstr.*, 1977, **87**, 68 240).	612
77KGS933	Y. M. Yutilov and A. G. Ignatenko; *Khim. Geterotsikl. Soedin.*, 1977, 993 (*Chem. Abstr.*, 1977, **87**, 167 943).	621
77LA106	S. W. Park and W. Ried; *Liebigs Ann. Chem.*, 1977, 106.	445
77LA956	W. Kliegel and G. H. Frankenstein; *Liebigs Ann. Chem.*, 1977, 956.	427
77LA1183	R. Meyer, U. Schöllkopf and P. Böhme; *Liebigs Ann. Chem.*, 1977, 1183.	358, 426, 477
B-77MI40100	K. F. Freed; in 'Modern Theoretical Chemistry. Semiempirical Methods of Electronic Structure Calculation. Part A: Techniques', ed. G. A. Segal; Plenum, New York, 1977, pp. 201–253.	7
77MI40400	K. L. Clay, W. D. Watkings and R. C. Murphy; *Drug Metab. Dispos.*, 1977, **5**, 149.	301
77MI40401	F. H. Deis, G. W. Lin and D. Lester; *Alcohol Aldehyde Metab. Syst.*, 1977, **3**, 399.	301
B-77MI40402	W. M. Jones and U. H. Brinker; in 'Pericyclic Reactions', ed. A. P. Marchand and R. E. Lehr; Academic Press, New York, 1977, vol. 1, p. 110.	251, 265
B-77MI40600	A. E. A. Porter; in 'Saturated Heterocyclic Chemistry', The Chemical Society, London, 1977, vol. 4, p. 188.	368
B-77MI40700	D. A. Wilson; in 'Saturated Heterocyclic Chemistry', The Chemical Society, London, 1977, vol. 5, p. 186.	427
B-77MI40800	D. A. Wilson; in 'Saturated Heterocyclic Chemistry', The Chemical Society, London, 1977, vol. 5, p. 186.	468
B-77MI40900	'Cyclic 3'5'-Nucleotides: Mechanism of Action', ed. H. Cramer and J. Schultz; Wiley-Interscience, London, 1977.	504
B-77MI40901	T. J. Kistenmacher and L. G. Marzilli; 'Jerusalem Symposium of Quantum Chemistry and Biochemistry', Israel Academy of Science, 1977, vol. 9, p. 7.	501
77MI40902	P. S. Song; *Annu. Rep. Prog. Chem.*, 1977, **74B**, 35.	517
77MI40903	R. P. Agarval, T. Spector and R. E. Parks, Jr.; *Biochem. Pharmacol.*, 1977, **26**, 359.	601, 603
77MI40904	R. Boone and B. Moss; *Virology*, 1977, **79**, 67.	601
77MI41000	N. O. Saldabol, J. Popelis, L. N. Alekseeva, A. K. Yalynskaya and N. D. Moskaleva; *Khim.-Farm. Zh.*, 1977, **11**, 64 (*Chem. Abstr.*, 1977, **87**, 84 898).	611
77MI41001	D. Y. Mukhametova, N. A. Akmanova, A. Y. Svetkin and Y. V. Svetkin; *Izv. Vyssh. Uchebn. Zaved., Khim. Khim. Tekhnol.*, 1977, **20**, 187 (*Chem. Abstr.*, 1977, **87**, 23 159).	632
77MI41100	V. Messori, L. Baldi and G. Bianchetti; *Chim. Ind.* (Milan), 1977, **59**, 438.	732
77MI41101	N. A. Petrusha, V. P. Chernetskii and Z. O. Rengevich; *Onkologiya*, 1977, **8**, 9.	729
B-77MI41102	K. H. Büchel; 'Pflanzenschutz und Schädlingsbekämpfung', Thieme, Stuttgart, 1977.	731
77MI41300	R. N. Butler and T. M. McEvoy; *Proc. R. Irish Acad.* (*RIC Centenary Issue*), 1977, **77B**, 359.	800, 804
77MI41500	K. Maekawa and R. Takeya; *J. Fac. Agric., Kyushu Univ.*, 1977, **21**, 99 (*Chem. Abstr.*, 1977, **87**, 17 148).	904
77OMR(9)1	U. Berg, S. Karlsson and J. Sandström; *Org. Magn. Reson.*, 1977, **9**, 1.	209
77OMR(9)235	A. Fruchier, E. Alcalde and J. Elguero; *Org. Magn. Reson.*, 1977, **9**, 235.	186, 194
77OMR(9)716	P. Bouchet, A. Fruchier, G. Joncheray and J. Elguero; *Org. Magn. Reson.*, 1977, **9**, 716.	190, 194
77RCR76	A. L. Rusanov, S. N. Leont'eva and Ts. G. Iremashvili; *Russ. Chem. Rev.* (*Engl. Transl.*) 1977, **46**, 76.	788
77RCR374	A. N. Volkov and A. N. Nikol'skaya; *Russ. Chem. Rev.* (*Engl. Transl.*), 1977, **46**, 374.	169
77RRC471	J. P. Fayet, M. C. Vertut, P. Mauret, R. M. Claramunt and J. Elguero; *Rev. Roum. Chim.*, 1977, **22**, 471.	177
77S804	K. H. Mayer, D. Lauerer and H. Heitzer; *Synthesis*, 1977, 804.	269
B-77SH(1)	W. L. F. Armarego; 'Stereochemistry of Heterocyclic Compounds', Wiley, New York, 1977, part 1.	353, 354, 702
77T45	B. Chantegrel, D. Hartmann and S. Gelin; *Tetrahedron*, 1977, **33**, 45.	272
77T751	C. Dietrich-Buchecker and M. Franck-Neumann; *Tetrahedron*, 1977, **33**, 751.	251
77T803	M. Jelińska and J. Sobkowski; *Tetrahedron*, 1977, **33**, 803.	527
77T1247	K. T. Potts, E. G. Brugel and W. C. Dunlap; *Tetrahedron*, 1977, **33**, 1247.	162, 864
77T1253	K. T. Potts, W. C. Dunlap and E. G. Brugel; *Tetrahedron*, 1977, **33**, 1253.	162
77T1263	K. T. Potts, W. C. Dunlap nd F. S. Apple; *Tetrahedron*, 1977, **33**, 1263.	866
77T1399	A. Könnecke, E. Lippmann and E. Kleinpeter; *Tetrahedron*, 1977, **33**, 1399.	800
77T2069	P. Freche, A. Gorgues and E. Levas; *Tetrahedron*, 1977, **33**, 2069.	276
77T2677	E. Lüddecke, H. Rau, H. Dürr and H. Schmitz; *Tetrahedron*, 1977, **33**, 2677.	251
77TL1031	D. S. Kemp and J. Reczek; *Tetrahedron Lett.*, 1977, 1031.	728

77TL2685	S. Inoue, H. Kakoi, M. Murata, T. Goto and O. Shimomura; *Tetrahedron Lett.*, 1977, 2685.	624
77TL3453	H.-M. Wolff and K. Hartke; *Tetrahedron Lett.*, 1977, 3453.	466
77TL4627	L. Åsbrink, C. Fridh and E. Lindholm; *Tetrahedron Lett.*, 1977, **52**, 4627.	517
77UP41100	H. Wamhoff; unpublished results.	696
77USP4008322	B. A. Dreikorn and T. D. Tibault; *U.S. Pat.* 4 008 322 (1977) (*Chem. Abstr.*, 1977, **86**, 166 387).	904
77USP4036849	W. V. Curran, A. S. Tomcufcik and A. S. Ross; *U.S. Pat.* 4 036 849 (1977) (*Chem. Abstr.*, 1977, **87**, 117 876).	835
77USP4039531	Hoechst Aktiengesellschaft, *U.S. Pat.* 4 039 531 (1977) (*Chem. Abstr.*, 1977, **87**, 186 088).	730
77ZC60	E. Lippmann, D. Reifegerste and E. Kleinpeter; *Z. Chem.*, 1977, **17**, 60 (*Chem. Abstr.*, 1977, **87**, 5875).	819
77ZC65	A. Antonowa, R. Herzschuh and S. Hauptmann; *Z. Chem.*, 1977, **17**, 65.	801
77ZC261	A. Könnecke and E. Lippmann; *Z. Chem.*, 1977, **17**, 261.	815
77ZC262	G. Kempter, W. Ehrlichmann and R. Thomann; *Z. Chem.*, 1977, **17**, 262.	616
77ZN(B)307	M. H. Elnagdi, E. M. Kandeel and M. R. H. Elmoghayar; *Z. Naturforsch., Teil B*, 1977, **32**, 307.	320
77ZN(B)430	M. H. Elnagdi, E. M. Zayed, E. M. Kandeel and S. M. Fahmy; *Z. Naturforsch., Teil B*, 1977, **32**, 430.	321
77ZOR2218	N. A. Klyuev, E. N. Istratov, R. A. Khmel'nitskii, V. A. Zyranov, V. L. Rusinov and I. Ya. Postovskii; *Zh. Org. Khim.*, 1977, **13**, 2218 (*Chem. Abstr.*, 1978, **88**, 50 091).	801
77ZOR2626	N. O. Saldabol and O. E. Lando; *Zh. Org. Khim.*, 1977, **13**, 2626 (*Chem. Abstr.*, 1978, **88**, 136 573).	611
78ACS(B)683	P. Kolsaker, P. O. Ellingsen and G. Woeien; *Acta Chem. Scand., Ser. B*, 1978, **32**, 683.	708
78AG63	H. Kunz; *Angew. Chem.*, 1978, **90**, 63.	728
78AG(E)67	H. Kunz; *Angew. Chem., Int. Ed. Engl.*, 1978, **17**, 67.	728
78AG(E)455	E. Schaumann, J. Ehlers and U. Behrens; *Angew. Chem., Int. Ed. Engl.*, 1978, **17**, 455.	9
78AHC(22)71	J. A. Zoltewicz and L. W. Deady; *Adv. Heterocycl. Chem.*, 1978, **22**, 71.	49, 51, 52, 386
78AHC(22)72	J. A. Zoltewicz and L. W. Deady; *Adv. Heterocycl. Chem.*, 1978, **22**, 72.	229
78AHC(22)184	J. Elguero, R. M. Claramunt and A. J. H. Summers; *Adv. Heterocycl. Chem.*, 1978, **22**, 184.	246, 263, 267
78AHC(23)171	I. J. Fletcher and A. E. Siegrist; *Adv. Heterocycl. Chem.*, 1978, **23**, 171.	92
78AHC(23)263	R. M. Acheson and N. F. Elmore; *Adv. Heterocycl. Chem.*, 1978, **23**, 263.	151, 233, 248
78AHC(23)265	R. M. Acheson and N. F. Elmore; *Adv. Heterocycyl. Chem.*, 1978, **23**, 265.	419, 420
78AJC179	C. G. Freeman, J. V. Turner and A. D. Ward; *Aust. J. Chem.*, 1978, **31**, 179.	647
78AJC2505	D. J. Brown and T. Nagamatsu; *Aust. J. Chem.*, 1978, **31**, 2505.	854, 856, 873, 891, 892
78AJC2675	E. R. Cole, G. Crank and E. Lye; *Aust. J. Chem.*, 1978, **31**, 2675.	381
78AX(B)293	I. Leban, B. Stanovnik and M. Tišler; *Acta Crystallogr., Part B*, 1978, **34**, 293.	179, 181
78AX(B)1136	L. Golič, I. Leban, B. Stanovnik and M. Tišler; *Acta Crystallogr., Part B*, 1978, **34**, 1136.	858, 860
78AX(B)2229	V. Langer, K. Huml and L. Lessinger; *Acta Crystallogr., Part B*, 1978, **34**, 2229.	509
78BAP291	L. Stefaniak; *Bull. Acad. Pol. Sci., Ser. Sci. Chim.*, 1978, **26**, 291.	14, 19
78BCJ251	A. Kakehi, S. Ito, Y. Konno and T. Maeda; *Bull. Chem. Soc. Jpn.*, 1978, **51**, 251.	337
78BCJ563	S. Shiraishi, T. Shigemoto and S. Ogawa; *Bull. Chem. Soc. Jpn.*, 1978, **51**, 563.	166
78BRP1528801	B. Parton and F. L. Rose; (ICI Ltd.), *Br. Pat.* 1 528 801 (1978) (*Chem. Abstr.*, 1979, **91**, 124 859).	614
78BRP1531752	B. Parton and F. L. Rose; (ICI Ltd.), *Br. Pat.* 1 531 752 (1978) (*Chem. Abstr.*, 1979, **90**, 205 761).	614
78BSB189	J. P. Fayet, M. C. Vertut, P. Mauret, R. M. Claramunt, J. Elguero and E. Alcalde; *Bull. Soc. Chim. Belg.*, 1978, **87**, 189.	177, 178
78BSF(2)485	J. Bourgois, M. Bourgois and F. Texier; *Bull. Soc. Chim. Fr., Part 2*, 1978, 485.	703, 710, 714, 717, 725
78BSF(2)602	A. Le Berre and C. Porte; *Bull. Soc. Chim. Fr., Part 2*, 1978, 602.	281
78CB566	W. Tochtermann, H. Gustmann and C. Wolff; *Chem. Ber.*, 1978, **111**, 566.	823
78CB2258	H. Dürr and H. Schmitz; *Chem. Ber.*, 1978, **111**, 2258.	318
78CB2813	H. Wamhoff and L. Lichtenthäler; *Chem. Ber.*, 1978, **111**, 2813.	632
78CC447	A. Robert and A. Le Marechal; *J. Chem. Soc., Chem. Commun.*, 1978, 447.	155
78CC453	V. M. Colburn, B. Iddon, H. Suschitzky and P. T. Gallagher; *J. Chem. Soc., Chem. Commun.*, 1978, 453.	164
78CC485	K. T. Suzuki, H. Yamada and M. Hirobe; *J. Chem. Soc., Chem. Commun.*, 1978, 485.	516
78CC652	A. R. Butler, C. Glidewell and D. C. Liles; *J. Chem. Soc., Chem. Commun.*, 1978, 652.	9
78CC857	G. J. Gainsford and A. D. Woolhouse; *J. Chem. Soc., Chem., Commun.*, 1978, 857.	376
78CCC3103	A. Holý; *Collect. Czech. Chem. Commun.*, 1978, **43**, 3103.	604
78CHE73	P. V. Tkachenko, A. M. Simonov and I. I. Popov; *Chem. Heterocycl. Compd. (Engl. Transl.)*, 1978, **14**, 73.	438

78CHE1123	N. A. Klyuev, N. M. Przhevalskii, I. I. Grandberg and A. A. Perov; *Chem. Heterocycl. Compd. (Engl. Transl.)*, 1978, **14**, 1123.	204
78CHE1366	Yu. S. Tsizin and S. A. Chernyak; *Chem. Heterocycl. Compd. (Engl. Transl.)*, 1978, **14**, 1366.	430
78CL1093	K. Matsumoto and T. Uchida; *Chem. Lett.*, 1978, 1093.	285
78CPB1929	T. Fujii, T. Itaya, T. Saito and M. Kawanishi; *Chem. Pharm. Bull.*, 1978, **26**, 1929.	586
78CPB2905	F. Yoneda, M. Higuchi, K. Mori, K. Senga, Y. Kanamori, K. Shimizu and S. Nishigaki; *Chem. Pharm. Bull.*, 1979, **26**, 2905.	576
78CPB2924	K. Takeda, K. Shudo, T. Okamoto and T. Kosuge; *Chem. Pharm. Bull.*, 1978, **26**, 2924.	614
78CPB3154	F. Yoneda, M. Noguchi, M. Noda and Y. Nitta; *Chem. Pharm. Bull.*, 1978, **26**, 3154.	656
78CPB3240	K. Senga, Y. Kanamori and S. Nishigaki; *Chem. Pharm. Bull.*, 1978, **26**, 3240.	576
78CR(C)(287)439	J. Elguero and M. Espada; *C.R. Hebd. Seances Acad. Sci., Ser. C*, 1978, **287**, 439.	269
78CS(13)157	S. Gronowitz and S. Liljefors; *Chem. Scr.*, 1978, **13**, 157.	268
78CZ361	C. Bak, K. Praefike and L. Henriksen; *Chem.-Ztg.*, 1978, **102**, 361.	119
78DIS(B)(39)233	C. S. Alleyne; *Diss. Abstr. Int. B*, 1978, **39**, 233.	904
78FES14	G. Auzzi, L. Cecchi, A. Costanzo, L. P. Vettori and F. Bruni; *Farmaco Ed. Sci.*, 1978, **33**, 14.	310
78FES543	P. L. Ferrarini and O. Livi; *Farmaco Ed. Sci.*, 1978, **33**, 543.	728
78FES799	G. Auzzi, L. Cecchi, A. Costanzo, L. P. Vettori and F. Bruni; *Farmaco Ed. Sci.*, 1978, **33**, 799.	334
78GEP2804909	R. W. Clarke and A. W. Oxford; *Ger. Pat.* 2 804 909 (1978) (*Chem. Abstr.*, 1978, **91**, 57 044).	651
78GEP2805483	G. E. Davies and D. M. Foulkes; *Ger. Pat.* 2 805 483 (1978) (*Chem. Abstr.*, 1979, **90**, 67 478).	904
78GEP2806199	*Ger. Pat.* 2 806 199 (1978) (*Chem. Abstr.*, 1978, **89**, 180 052).	647
78H(9)757	C. B. Kanner and U. K. Pandit; *Heterocycles*, 1978, **9**, 757.	633
78H(9)1207	K. Isomura, Y. Hirose, H. Shuyama, S. Abe, G. Ayabe and H. Taniguchi; *Heterocycles*, 1978, **9**, 1207.	163
78H(9)1437	K. Senga, Y. Kanamori and S. Nishigaki; *Heterocycles*, 1978, **9**, 1437.	576
78H(10)241	I. Hayakawa, K. Yamazaki, R. Dohmori and N. Koga; *Heterocycles*, 1978, **10**, 241.	618, 639
78H(11)121	K. L. Williamson and J. D. Roberts; *Heterocycles*, 1978, **11**, 121.	20
78H(11)293	T. Miyashi, Y. Nishizawa and T. Mukai; *Heterocycles*, 1978, **11**, 293.	267
78HC(33)118	H. Neunhoeffer; *Chem. Heterocycl. Compd.*, 1978, **33**, 118.	608
78HC(33)882	H. Neunhoeffer; *Chem. Heterocycl. Compd.*, 1978, **33**, 882.	608
78HC(33)885	H. Neunhoeffer; *Chem. Heterocycl. Compd.*, 1978, **33**, 885.	608
78HCA108	H. Balli and L. Felder; *Helv. Chim. Acta*, 1978, **61**, 108.	241, 895
78HCA129	J. P. Pauchard and A. E. Siegrist; *Helv. Chim. Acta*, 1978, **61**, 129.	632
78HCA234	W. Heinzelmann; *Helv. Chim. Acta*, 1978, **61**, 234.	221
78HCA618	W. Heinzelmann; *Helv. Chim. Acta*, 1978, **61**, 618.	221
78HCA1477	M. Maerky, H. Meier, A. Wunderli, H. Heimgartner, H. Schmid and H. J. Hansen; *Helv. Chim. Acta*, 1978, **61**, 1477.	717
78HCA1755	C. Wentrup; *Helv. Chim. Acta*, 1978, **61**, 1755.	854, 856, 859, 866, 877, 881, 887, 895, 897
78IC1990	J. Reedijk, J. C. Jensen, H. van Koningsveld and C. G. Kralingen; *Inorg. Chem.*, 1978, **17**, 1990.	227
78IC3026	L. D. Brown, J. A. Ibers and A. R. Siedle; *Inorg. Chem.*, 1978, **17**, 3026.	676
78IJC(B)481	S. Bala, R. P. Gupta, M. L. Sachdeva, A. Singh and H. K. Pujari; *Indian J. Chem., Sect. B*, 1978, **16**, 481.	853
78IJC(B)531	P. K. Dubey and C. V. Ratnam; *Indian J. Chem., Sect. B*, 1978, **16**, 531.	635
78IJC(B)1015	Y. Khandelwal and P. C. Jain; *Indian J. Chem., Sect. B*, 1978, **16**, 1015.	645
78IJC(B)1030	B. K. Pattanayak, D. N. Rout and G. N. Mahapatra; *Indian J. Chem., Sect. B*, 1978, **16**, 1030.	119
78JA282	O. L. Chapman and J.-P. LeRoux; *J. Am. Chem. Soc.*, 1978, **100**, 282.	867
78JA1275	C. Guimon, G. Pfister-Guillouzo and M. Begtrup; *J. Am. Chem. Soc.*, 1978, **100**, 1275.	205, 689
78JA1306	H. Quast, L. Bieber and W. C. Danen; *J. Am. Chem. Soc.*, 1978, **100**, 1306.	810
78JA2767	G. M. Brown, J. E. Sutton and H. Taube; *J. Am. Chem. Soc.*, 1978, **100**, 2767.	386
78JA3674	D. Poppinger and L. Radom; *J. Am. Chem. Soc.*, 1978, **100**, 3674.	7
78JA4617	H. Yamada, M. Hirobe, K. Higashiyama, H. Takahashi and K. T. Suzuki; *J. Am. Chem. Soc.*, 1978, **100**, 4617.	515
78JA5122	M. J. Mirbach, K. C. Liu, M. F. Mirbach, W. R. Cherry, N. J. Turro and P. S. Engel; *J. Am. Chem. Soc.*, 1978, **100**, 5122.	199, 255
78JA5701	M. D. Gordon, P. V. Alston and A. R. Rossi; *J. Am. Chem. Soc.*, 1978, **100**, 5701.	282
78JA6245	O. L. Chapman, R. S. Sheridan and J.-P. LeRoux; *J. Am. Chem. Soc.*, 1978, **100**, 6245.	867
78JA6538	E. M. Burgess and M. C. Pulcrano; *J. Am. Chem. Soc.*, 1978, **100**, 6538.	444
78JAP(K)7898932	Hitachi Chemical Co., *Jpn. Kokai Tokkyo Koho* 78 98 932 (1978) (*Chem. Abstr.*, 1979, **90**, 38 930).	722
78JBC(253)1710	S. Muthukhishnan, B. Moss, J. A. Cooper and E. S. Maxwell; *J. Biol. Chem.*, 1978, **253**, 1710.	601

78JCR(S)298	M. Winnewisser, J. Vogt and H. Albrecht; *J. Chem. Res. (S)*, 1978, 298.	694
78JCR(S)366	D. Johnston, J. Machin and D. M. Smith; *J. Chem. Res. (S)*, 1978, 366.	360
78JCR(S)407	H. Singh and C. S. Gandi; *J. Chem. Res. (S)*, 1978, 407.	138
78JCS(P1)881	R. N. Butler, A. B. Hanahoe and W. B. King; *J. Chem. Soc., Perkin Trans. 1*, 1978, 881.	706
78JCS(P1)909	M. J. Sasse and R. C. Storr; *J. Chem. Soc., Perkin Trans. 1*, 1978, 909.	727
78JCS(P1)1381	G. Mackenzie and G. Shaw; *J. Chem. Soc., Perkin Trans. 1*, 1978, 1381.	439
78JCS(P2)19	D. Spinelli, V. Frenna, A. Corrao and N. Vivona; *J. Chem. Soc., Perkin Trans. 2*, 1978, 19.	720
78JCS(P2)99	M. Begtrup, R. M. Claramunt and J. Elguero; *J. Chem. Soc., Perkin Trans. 2*, 1978, 99.	191, 194, 212, 681
78JCS(P2)865	D. J. Evans, H. F. Thimm and B. A. W. Coller; *J. Chem. Soc., Perkin Trans. 2*, 1978, 865.	346, 347, 348, 349, 400, 428
78JCS(P2)1087	R. N. Butler and T. M. McEvoy; *J. Chem. Soc., Perkin Trans. 2*, 1978, 1087.	793, 805, 806, 816
78JHC1	P. D. Cook, P. Dea and R. K. Robins; *J. Heterocycl. Chem.*, 1978, **15**, 1.	652
78JHC64	K. Senga, M. Ichiba, H. Kanazawa, S. Nishigaki, M. Higuchi and F. Yoneda; *J. Heterocycl. Chem.*, 1978, **15**, 64.	576
78JHC81	L. Grehn; *J. Heterocycl. Chem.*, 1978, **15**, 81.	135
78JHC119	A. F. Kluge; *J. Heterocycl. Chem.*, 1978, **15**, 119.	648
78JHC127	M. P. Servé, P. G. Seybold and H. P. Covault; *J. Heterocycl. Chem.*, 1978, **15**, 127.	673
78JHC185	J. Delettré, R. Bally, J. P. Mornon, E. Alcalde, J. de Mendoza, R. Faure, E. J. Vincent and J. Elguero; *J. Heterocycl. Chem.*, 1978, **15**, 185.	178, 272, 281
78JHC221	C. Ochoa and M. Stud; *J. Heterocycl. Chem.*, 1978, **15**, 221.	666
78JHC319	S. W. Schneller and R. D. Moore; *J. Heterocycl. Chem.*, 1978, **15**, 319.	272
78JHC353	A. Yamazaki and M. Okutsu; *J. Heterocycl. Chem.*, 1978, **15**, 353.	504
78JHC439	S. W. Schneller and D. G. Bartholomew; *J. Heterocycl. Chem.*, 1978, **15**, 439.	863, 882, 883, 884
78JHC463	G. Hajós and A. Messmer; *J. Heterocycl. Chem.*, 1978, **15**, 463.	856, 886
78JHC625	P. Bouchet, G. Joncheray, R. Jacquier and J. Elguero; *J. Heterocycl. Chem.*, 1978, **15**, 625.	380
78JHC807	M. Uda and S. Kubota; *J. Heterocycl. Chem.*, 1978, **15**, 807.	736
78JHC813	S. Gelin and D. Hartmann; *J. Heterocycl. Chem.*, 1978, **15**, 813.	272
78JHC839	B. L. Cline, R. P. Panzica and L. B. Townsend; *J. Heterocycl. Chem.*, 1978, **15**, 839.	615, 618, 619
78JHC937	A. Bernardini, P. Viallefont and R. Zniber; *J. Heterocycl. Chem.*, 1978, **15**, 937.	617, 618, 639
78JHC987	S. W. Schneller and J. L. May; *J. Heterocycl. Chem.*, 1978, **15**, 987.	877, 896
78JHC1021	C. V. Magatti and F. J. Villani; *J. Heterocycl. Chem.*, 1978, **15**, 1021.	634
78JHC1041	G. Maury, J.-P. Paugam and R. Paugam; *J. Heterocycl. Chem.*, 1978, **15**, 1041.	854, 856, 859, 875, 894
78JHC1043	A. W. McGee, M. P. Servé and P. G. Seybold; *J. Heterocycl. Chem.*, 1978, **15**, 1043.	672, 673
78JHC1149	D. E. Kuhla and H. A. Watson, Jr.; *J. Heterocycl. Chem.*, 1978, **15**, 1149.	631
78JHC1159	M. Robba, J. C. Lancelot, D. Maume and A. Rabaron; *J. Heterocycl. Chem.*, 1978, **15**, 1159.	273
78JHC1175	M. Kocevar, B. Stanovnik and M. Tisler; *J. Heterocycl. Chem.*, 1978, **15**, 1175.	315, 320
78JHC1227	F. Fabra, C. Gálvez, A. González, P. Viladoms and J. Vilarrasa; *J. Heterocycl. Chem.*, 1978, **15**, 1227.	364
78JHC1287	S. Plescia, G. Daidone, V. Sprio, E. Aiello, G. Dattolo and G. Cirrincione; *J. Heterocycl. Chem.*, 1978, **15**, 1287.	273
78JHC1349	S. Romani and W. Kloetzer; *J. Heterocycl. Chem.*, 1978, **15**, 1349.	719
78JHC1447	F. Fabra, J. Vilarrasa and J. Coll; *J. Heterocycl. Chem.*, 1978, **15**, 1447.	263
78JHC1451	T. Suzuki, N. Katou and K. Mitsuhashi; *J. Heterocycl. Chem.*, 1978, **15**, 1451.	652
78JHC1543	P. Bourgeois, J. Lucrece and J. Donoguès; *J. Heterocycl. Chem.*, 1978, **15**, 1543.	233
78JIC264	B. K. Pattanayak, D. N. Rout and G. N. Mahapatra; *J. Indian Chem. Soc.*, 1978, **45**, 264.	119
78JMC112	R. D. Elliott and J. A. Montgomery; *J. Med. Chem.*, 1978, **21**, 112.	617, 618
78JMC883	S.-H. Kim, D. G. Bartholomew, L. B. Allen, R. K. Robins, G. R. Revankar and P. Dea; *J. Med. Chem.*, 1978, **21**, 883.	631, 658
78JMC965	R. L. Clark, A. A. Pessolano, T.-Y. Shen, D. P. Jacobus, H. Jones, V. J. Lotti and L. M. Flataker; *J. Med. Chem.*, 1978, **21**, 965.	616, 636
78JMC1073	W. S. Di Menna, C. Piantadosi and R. G. Lamb; *J. Med. Chem.*, 1978, **21**, 1073.	604
78JMC1212	P. D. Cook, L. B. Allen, D. G. Streeter, J. H. Huffman, R. W. Sidwell and R. K. Robins; *J. Med. Chem.*, 1978, **21**, 1212.	642
78JMR(29)45	P. Büchner, W. Maurer and H. Rüterjans; *J. Magn. Reson.*, 1978, **29**, 45.	510, 516
78JOC289	P. D. Cook and R. K. Robins; *J. Org. Chem.*, 1978, **43**, 289.	642
78JOC341	N. Sato and J. Adachi; *J. Org. Chem.*, 1978, **43**, 341.	897
78JOC658	E. S. Hand and W. W. Paudler; *J. Org. Chem.*, 1978, **43**, 658.	611, 632
78JOC808	W. H. Pirkle and P. L. Gravel; *J. Org. Chem.*, 1978, **43**, 808.	206
78JOC1197	L. D. Taylor, J. M. Grasshoff and M. Pluhar; *J. Org. Chem.*, 1978, **43**, 1197.	823
78JOC1233	N. F. Haley; *J. Org. Chem.*, 1978, **43**, 1233.	260
78JOC1664	A. Padwa, S. Nahun and E. Sato; *J. Org. Chem.*, 1978, **43**, 1664.	317, 808
78JOC2037	C. Wentrup, A. Damerius and W. Reichen; *J. Org. Chem.*, 1978, **43**, 2037.	147, 809

78JOC2289	A. G. Hortmann, J.-Y. Koo and C.-C. Yu; *J. Org. Chem.*, 1978, **43**, 2289.	423
78JOC2487	J. Font, M. Torres, H. E. Gunning and O. P. Strausz; *J. Org. Chem.*, 1978, **43**, 2487.	14
78JOC2587	J. H. Keck, Jr., R. A. Simpson and J. L. Wong; *J. Org. Chem.*, 1978, **43**, 2587.	511
78JOC2665	S. Gelin, R. Gelin and D. Hartmann; *J. Org. Chem.*, 1978, **43**, 2665.	192
78JOC2900	E. S. Hand and W. W. Paudler; *J. Org. Chem.*, 1978, **43**, 2900.	613, 631
78JOC3042	G. L'abbé, J.-P. Dekerk, A. Verbruggen, S. Toppet, J. P. Declercq, G. Germain and M. Van Meerssche; *J. Org. Chem.*, 1978, **43**, 3042.	832
78JOC3370	Y. Tamaru, T. Harada and Z. Yoshida; *J. Org. Chem.*, 1978, **43**, 3370.	276
78JOC3565	T. Takeuchi, H. J. C. Yeh, K. L. Kirk and L. A. Cohen; *J. Org. Chem.*, 1978, **43**, 3565.	352
78JOC3570	Y. Takeuchi, K. L. Kirk and L. A. Cohen; *J. Org. Chem.*, 1978, **43**, 3570.	407
78JOC3732	M. Baudy, A. Robert and A. Foucaud; *J. Org. Chem.*, 1978, **43**, 3732.	129
78JOC3736	R. K. Howe, T. A. Gruner, L. G. Carter, L. L. Black and J. E. Franz; *J. Org. Chem.*, 1978, **43**, 3736.	147
78JOC4125	B. W. Cue, Jr., L. J. Czuba and J. P. Dirlam; *J. Org. Chem.*, 1978, **43**, 4125.	898
78JOC4381	K. L. Kirk; *J. Org. Chem.*, 1978, **43**, 4381.	398, 401, 414, 448
78JOC4774	E. J. Prisbe, J. P. H. Verheyden and J. G. Moffatt; *J. Org. Chem.*, 1978, **43**, 4774.	658
78JOC4784	E. J. Prisbe, J. P. H. Verheyden and J. G. Moffatt; *J. Org. Chem.*, 1978, **43**, 4784.	658
78JOC4853	N. Ranganathan and W. S. Brinigar; *J. Org. Chem.*, 1978, **43**, 4853.	455
78JPR508	W. Freyer; *J. Prakt. Chem.*, 1978, **320**, 508.	209
78JPR521	W. Freyer; *J. Prakt. Chem.*, 1978, **320**, 521.	199
78JST(43)33	M. H. Palmer and S. M. F. Kennedy; *J. Mol. Struct.*, 1978, **43**, 33.	7, 205
78JST(43)203	M. H. Palmer and S. M. F. Kennedy; *J. Mol. Struct.*, 1978, **43**, 203.	176, 177, 205
78KGS258	N. O. Saldabol and O. E. Lando; *Khim. Geterotsikl. Soedin.*, 1978, 258 (*Chem. Abstr.*, 1978, **88**, 190 733).	631
78KGS723	K. Galankiewicz; *Khim. Geterotsikl. Soedin.*, 1978, 723.	504
78KGS1525	I. N. Azerbaev, I. A. Poplavskaya, R. G. Kurmangalieva and S. F. Khalilova; *Khim. Geterotsikl. Soedin.*, 1978, 1525 (*Chem. Abstr.*, 1979, **90**, 87 319).	631
78LA1476	H. Meyer, F. Bossert and H. Horstmann; *Liebigs Ann. Chem.*, 1978, 1476.	634
78LA1491	H. Meyer and J. Kurz; *Liebigs Ann. Chem.*, 1978, 1491.	634
78LA1916	N. Engel and W. Steglich; *Liebigs Ann. Chem.*, 1978, 1916.	114, 462
78M337	F. Wille and W. Schwab; *Monatsh. Chem.*, 1978, **109**, 337.	201, 281
78M379	G. Ciurdaru, Z. Moldovan and I. Oprean; *Monatsh. Chem.*, 1978, **109**, 379.	359
78MI40400	V. Tabacik and S. Sportouch; *J. Raman Spectrosc.*, 1978, **7**, 61.	199, 200
78MI40401	U. Wrzeciono and W. Nieweglowska; *Pharmazie*, 1978, **33**, 377.	266
78MI40402	J. Elguero, E. Gonzalez and R. Sarlin; *An. Quím.*, 1978, **74**, 527.	258
78MI40403	H. J. M. Dou, J. Elguero, M. Espada and P. Hassanaly; *An. Quím.*, 1978, **74**, 1137.	183, 230
B-78MI40404	J. Bastide and O. Henri-Rousseau; in 'The Chemistry of the Carbon–Carbon Triple Bond', ed. S. Patai; Wiley-Interscience, 1978, part 2, p. 447.	282
B-78MI40405	D. S. Wulfman, G. Linstrumelle and C. F. Cooper; in 'The Chemistry of Diazonium and Diazo Groups', ed. S. Patai; Wiley-Interscience, 1978, part 2, p. 821.	282
B-78MI40500	J. G. T. Moffatt, H. P. Albrecht, G. H. Jones, D. B. Repke, G. Frummlitz, H. Ohrui and C. M. Gupta; in 'Chemistry and Biology of Nucleosides and Nucleotides', ed. R. Harmon; Academic, New York, 1978, p. 359.	334
78MI40600	C. Guimon, G. Pfister-Guillouzo, G. Salmona and E. J. Vincent; *J. Chim. Phys. Phys. Chim. Biol.*, 1978, **75**, 859.	361
B-78MI40900	G. H. Hitchings; 'Handbook of Experimental Pharmacology', Springer, Berlin, 1978.	501
B-78MI40901	'Nucleic Acid Chemistry', ed. L. B. Townsend and R. S. Tipson; Wiley-Interscience, New York, 1978, part 1.	504, 553, 592, 595, 597
B-78MI40902	P. Buchner, F. Blomberg and H. Ruterjans; in 'NMR Spectroscopy in Molecular Biology', 'Jerusalem Symposium of Quantum Chemistry and Biochemistry', ed. B. Pullman; Reidl, Amsterdam, 1978.	510, 513
B-78MI40903	'Nucleic Acid Chemistry', ed. L. B. Townsend and R. S. Tipson; Wiley-Interscience, New York, 1978, part 2.	504, 573, 592, 594, 595, 597, 598
B-78MI40905	'Advances in Cyclic Nucleotide Research', ed. W. J. George and L. J. Ignarro; Raven Press, New York, 1978, vol. 9.	504
78MI40906	S. A. Benezra and P. R. B. Foss; *Anal. Profiles Drug Subst.*, 1978, **7**, 343.	501, 516
78MI40907	T. Sato and R. Noyori; *Kagaku (Kyoto)*, 1978, **33**, 494.	504
B-78MI40908	'Chemistry and Biology of Nucleosides and Nucleotides', ed. R. E. Harmon, R. K. Robins and L. B. Townsend; Academic, New York, 1978.	504
78MI40909	H. J. Schaeffer, L. Beauchamp, P. de Miranda, G. B. Elion, D. J. Bauer and P. Collins; *Nature (London)*, 1978, **272**, 583.	603
78MI40910	K. Kimura, T. Sugryama and T. Hashizume; *Nucleic Acids Res.*, 1978, Spec. Pub. No. 5, S339.	601
78MI41000	H. Foks and M. Janowiec; *Acta Pol. Pharm.*, 1978, **35**, 281.	636
78MI41001	P. A. Bartlett, J. T. Hunt, J. L. Adams and J. C. E. Gehret; *Bioorg. Chem.*, 1978, **7**, 421.	667
78MI41002	H. G. Lenvartz, M. Hepp and W. Schunack; *Eur. J. Med. Chem. — Chim. Ther.*, 1978, **13**, 229.	649
78MI41003	J.-C. Teulade, G. Grassy, J.-P. Girard, J.-P. Chapat and M. Simeon de Buochberg; *Eur. J. Med. Chem. — Chim. Ther.*, 1978, **13**, 271.	612

78MI41004	Y. M. Yutilov and A. G. Ignatenko; *Deposited Doc.*, 1978, VINITI, 3830-78 (*Chem. Abstr.*, 1980, **92**, 198 320).	618
78MI41005	Y. M. Yutilov and I. A. Svertilova; *Deposited Doc.*, 1978, VINITI, 1193-78 (*Chem. Abstr.*, 1979, **91**, 175 261).	620, 622
78MI41006	Y. M. Yutilov and A. G. Ignatenko; *Deposited Doc.*, 1978, VINITI, 3831-78 (*Chem. Abstr.*, 1980, **92**, 198 321).	621
78MI41007	B. L. Cline, R. P. Panzica and L. B. Townsend; *Nucleic Acid Chem.*, 1978, **1**, 129.	617, 636
78MI41008	J. Biernat, J. Ciesiolka, P. Gornicki, R. W. Adamiak, W. J. Krzyzosiak and M. Wiewiorowski; *Nucleic Acids Res.*, 1978, **5**, 789.	648
78MI41009	P. Fernandez-Resa, P. Goya and M. Stud; *Nucleic Acids Res., Spec. Publ.*, 1978, **4**, 61.	666
78MI41010	A. A. Kost and S. V. Ermolin; *Nucleic Acids Res., Spec. Publ.*, 1978, **4**, 197.	648
78MI41011	J. Biernat, J. Ciesiolka, P. Gornicki, W. J. Krzyzosiak and M. Wiewiorowski; *Nucleic Acids Res., Spec. Publ.*, 1978, **4**, 203.	648
78MI41100	O. V. Voischcheva and G. V. Shatalov; *Izv. Vyssh. Uchebn. Zaved., Khim. Khim. Tekhnol.*, 1978, **21**, 1437 (*Chem. Abstr.*, 1979, **90**, 120 782).	677
78MI41101	R. Boehm; *Pharmazie*, 1978, **33**, 83.	698
78MI41102	A. Contreras, R. M. Sanchez-Perez and G. Alonso; *Cancer Chemother. Pharmacol.*, 1978, **1**, 243.	728
78MI41103	R. J. Fass and R. B. Prior; *Curr. Therap. Res.*, 1978, **24**, 352.	729
B-78MI41104	A. Albert; in 'Nucleic Acid Chemistry', ed. L. B. Townsend and R. S. Tipson; Wiley, New York, 1978, p. 97.	726
78MI41105	N. Matsuo and H. Chikako; *Transition Met. Chem.*, 1978, **3**, 366.	691
78OMR(11)27	K. A. K. Ebraheem, G. A. Webb and M. Witanowski; *Org. Magn. Reson.*, 1978, **11**, 27.	195
78OMR(11)234	J. R. Fayet, M. C. Vertut, A. Fruchier, E. M. Tjiou and J. Elguero; *Org. Magn. Reson.*, 1978, **11**, 234.	190, 192, 194
78OMR(11)385	L. Stefaniak; *Org. Magn. Reson.*, 1978, **11**, 385.	195
78OMR(11)578	S. Toppet, G. Wonters and G. Smets; *Org. Magn. Reson.*, 1978, **11**, 578.	682
78OMR(11)617	R. Faure, J. Elguero, E. J. Vincent and R. Lazaro; *Org. Magn. Reson.*, 1978, **11**, 617.	190
78OMS319	B. Gioia, L. Citerio and R. Stradi; *Org. Mass Spectrom.*, 1978, **13**, 319.	360
78OMS518	A. Maquestiau, Y. Van Haverbeke, R. Flammang, A. Menu and C. Wentrup; *Org. Mass Spectrom.*, 1978, **13**, 518.	203
78OMS571	J. L. Aubagnac, P. Campion and P. Guenot; *Org. Mass Spectrom.*, 1978, **13**, 571.	687
78OMS575	A. Atmani and J.-L. Aubagnac; *Org. Mass Spectrom.*, 1978, **13**, 575.	202, 359
78P2075	T. W. Baumann, E. Dupont Looser and H. Wanner; *Phytochemistry*, 1978, **17**, 2075.	600
78RTC35	C. Oldenhof and J. Cornelisse; *Recl. Trav. Chim. Pays-Bas*, 1978, **97**, 35.	243, 247
78RTC293	A. Noordam, L. Maat and H. C. Beyerman; *Recl. Trav. Chim. Pays-Bas*, 1978, **97**, 293.	649
78S633	L. Baiocchi, G. Corsi and G. Palazzo; *Synthesis*, 1978, 633.	169, 275, 282
78S675	J. A. M. Bastiaansen and E. F. Godefroi; *Synthesis*, 1978, 675	402
78S694	P. K. Kadaba; *Synthesis*, 1978, 694.	703
78T1139	T. Avignon, L. Bouscasse and J. Elguero; *Tetrahedron*, 1978, **34**, 1139.	190, 199, 210
78T1175	E. Gonzalez, R. Sarlin and J. Elguero; *Tetrahedron*, 1978, **34**, 1175.	326
78T2229	F. Roelants and A. Bruylants; *Tetrahedron*, 1978, **34**, 2229.	725
78T2259	O. Bensaude, M. Chevrier and J. E. Dubois; *Tetrahedron*, 1978, **34**, 2259.	211, 212, 217
78TH40400	R. Sarlin; Ph. D. Thesis, University of Perpignan, 1978.	272
78TL399	J. Plenkiewicz; *Tetrahedron Lett.*, 1978, 399.	813
78TL1009	G. Lucente, F. Pinnen and G. Zanotti; *Tetrahedron Lett.*, 1978, 1009.	643
78TL1447	K. Ienaga and W. Pfleiderer; *Tetrahedron Lett.*, 1978, 1447.	590
78TL2579	S. Nakatsuka, T. Ohgi and T. Goto; *Tetrahedron Lett.*, 1978, 2579.	565, 602
78TL4039	H. Yamada, M. Hirobe, K. Hiyashiyama, H. Takahashi and K. T. Suzuki; *Tetrahedron Lett.*, 1978, 4039.	568, 605
78TL4503	G. Le Fèvre and J. Hamelin; *Tetrahedron Lett.*, 1978, 4503.	256
78TL4581	M. J. Haddadin, A. A. Hawi and M. Z. Nazer; *Tetrahedron Lett.*, 1978, 4581.	376
78TL5007	T. Fujii, T. Saito and M. Kawahishi; *Tetrahedron Lett.*, 1978, 5007.	493, 586
78UP41500	S. W. Schneller and D. G. Bartholomew; unpublished result.	864
78YZ631	K. Kurata, H. Awaya, Y. Tominaga, Y. Matsuda and G. Kobayashi; *Yakugaku Zasshi*, 1978, **98**, 631 (*Chem. Abstr.*, 1979, **90**, 22 892).	632
78ZC92	A. Koennecke and E. Lippmann; *Z. Chem.*, 1978, **18**, 92.	903
78ZC214	A. Könnecke, P. Lepom and E. Lippmann; *Z. Chem.*, 1978, **18**, 214.	808
78ZN(B)1033	M. Konieczny and G. Sosnowsky; *Z. Naturforsch., Teil B*, 1978, **33**, 1033 (*Chem. Abstr.*, 1979, **90**, 103 895).	392, 454
78ZOB705	E. V. Komissarova, N. N. Belyaev and M. D. Stadnichuk; *Zh. Obshch. Khim.*, 1978, **48**, 705 (*Chem. Abstr.*, 1978, **88**, 190 702).	709
79ACS(A)687	I. Søtofte and K. Nielsen; *Acta Chem. Scand., Ser. A*, 1979, **33**, 687.	674
79ACS(A)693	K. Nielsen, L. Schepper and I. Søtofte; *Acta Chem. Scand., Ser. A*, 1979, **33**, 693.	674
79ACS(A)697	K. Nielsen and I. Søtofte; *Acta Chem. Scand., Ser. A*, 1979, **33**, 697.	674
79ACS(B)483	B. R. Tolf, J. Piechaczek, R. Dahlbom, H. Theorell, A. Akeson and G. Lundquist; *Acta Chem. Scand., Ser. B*, 1979, **33**, 483.	291

79AF511	Y. Ykeda, N. Takano, H. Matsushita, Y. Shiraki, T. Koide, R. Nagashima, Y. Fujimura, M. Shindo, S. Suziki and T. Iwasaki; *Arzneim.-Forsch.*, 1979, **29**, 511.	293
79AG259	D. Seebach; *Angew. Chem.*, 1979, **91**, 259.	717
79AG(E)239	D. Seebach; *Angew. Chem., Int. Ed. Engl.*, 1979, **18**, 239.	717
79AG(E)721	G. Bianchi, C. De Micheli and R. Gandolfi; *Angew. Chem., Int. Ed. Engl.*, 1979, **18**, 721.	276, 282, 283
79AHC(24)215	J. H. Lister; *Adv. Heterocycl. Chem.*, 1979, **24**, 215.	501
79AHC(25)1	J. W. Bunting; *Adv. Heterocycl. Chem.*, 1979, **25**, 1.	62, 63
79AHC(25)83	G. R. Newkome and A. Nayak; *Adv. Heterocycl. Chem.*, 1979, **25**, 83.	58, 60, 61
79AHC(25)147	B. J. Wakefield and D. J. Wright; *Adv. Heterocycl. Chem.*, 1979, **25**, 147.	46, 51, 58, 69, 71, 74, 84, 91, 93
79AHC(25)205	P. Hanson; *Adv. Heterocycl. Chem.*, 1979, **25**, 205.	60, 110, 419
79AJC387	J. H. Lister; *Aust. J. Chem.*, 1979, **32**, 387.	534
79AJC1727	K. C. Chang, M. R. Grimmett, D. D. Ward and R. T. Weavers; *Aust. J. Chem.*, 1979, **32**, 1727.	237, 266
79AJC2203	M. R. Grimmett, K. H. R. Lim and R. T. Weavers; *Aust. J. Chem.*, 1979, **32**, 2203.	389
79AJC2713	D. J. Brown, G. W. Grigg, Y. Iwai, K. N. McAndrew, T. Nagamatsu and R. Van Heeswyck; *Aust. J. Chem.*, 1979, **32**, 2713.	636
79AJC2771	J. H. Lister, *Aust. J. Chem.*, 1979, **32**, 2771.	537
79AJC2787	P. M. Pojer; *Aust. J. Chem.*, 1979, **32**, 2787.	727
79AP107	H. J. Sattler, H. G. Lennartz and W. Schunack; *Arch. Pharm. (Weinheim, Ger.)*, 1979, **312**, 107.	486
79AP610	A. Kreutzberger and K. Burgwitz; *Arch. Pharm. (Weinheim, Ger.)*, 1979, **312**, 610.	272
79AX(B)177	R. H. P. Francisco, Y. P. Mascarenhas and J. R. Lechat; *Acta Crystallogr., Part B*, 1979, **35**, 177.	227
79AX(B)1468	R. H. P. Francisco, J. R. Lechat and Y. P. Mascarenhas; *Acta Crystallogr., Part B*, 1979, **35**, 1468.	227
79AX(B)2256	L. Golic, I. Leban, B. Stanovnik and M. Tisler; *Acta Crystallogr., Part B*, 1979, **35**, 2256.	12
79BAP21	N. S. V. Subba Rao, M. C. Ganorkar, B. K. M. Murali and C. P. Ramaswamy; *Bull. Acad. Pol. Sci., Ser. Sci. Chim.*, 1979, **27**, 21 (*Chem. Abstr.*, 1980, **92**, 25 160).	837
79BAP249	D. Maciejewska and L. Skulski; *Bull. Acad. Pol. Sci., Ser. Sci. Chim.*, 1979, **27**, 249.	371
79BEP871985	G. Arcari, L. Bernardi, G. Falconi and U. Scarponi; *Belg. Pat.* 871 985 (1979) (*Chem. Abstr.*, 1979, **91**, 39 486).	623
79BSB289	A. Sayarh, M. Gelize-Duvigneau, J. Arriau and A. Maquestiau; *Bull. Soc. Chim. Belg.*, 1979, **88**, 289.	367
79BSF(2)529	J.-C. Teulade, R. Escale, G. Grassy, J.-P. Girard and J.-P. Chapat; *Bull. Soc. Chim. Fr., Part 2*, 1979, 529.	612
79BSF(2)633	M. S. Ouali, M. Vaultier and R. Carrié; *Bull. Soc. Chim. Fr., Part 2*, 1979, 633.	705
79CB445	H. Hansen, S. Hünig and K. Kishi; *Chem. Ber.*, 1979, **112**, 445.	707
79CB1057	R. Appel and L. Willms; *Chem. Ber.*, 1979, **112**, 1057.	728
79CB1529	R. Gompper and K. Schönafinger; *Chem. Ber.*, 1979, **112**, 1529.	711
79CB1635	H. Gotthardt and F. Reiter; *Chem. Ber.*, 1979, **112**, 1635.	717
79CB1712	W. Sucrow, F. Lübbe and A. Fehlauer; *Chem. Ber.*, 1979, **112**, 1712.	275
79CB1719	W. Sucrow, D. Rau, A. Fehlauer and J. Pickardt; *Chem. Ber.*, 1979, **112**, 1719.	179
79CB2829	L. Birkofer and K. Richtzenhain; *Chem. Ber.*, 1979, **112**, 2829.	709
79CC135	T. Saito and T. Fujii; *J. Chem. Soc., Chem. Commun.*, 1979, 135.	586
79CC523	J. P. Dickens, R. L. Dyer, B. J. Hamill and T. A. Harrow; *J. Chem. Soc., Chem. Commun.*, 1979, 523.	396
79CC568	R. Huisgen and H. U. Reissig; *J. Chem. Soc., Chem. Commun.*, 1979, 568.	250
79CC627	T. L. Gilchrist, C. W. Rees and J. A. R. Rodrigues; *J. Chem. Soc., Chem. Commun.*, 1979, 627.	273
79CHE115	A. S. Morkovnik and O. Yu. Okhlobystin; *Chem. Heterocycl. Compd. (Engl. Transl.)*, 1979, **15**, 115.	254
79CHE218	V. N. Doron'kin, A. F. Pozharskii and I. S. Kashparov; *Chem. Heterocycl. Compd. (Engl. Transl.)*, 1979, **15**, 218.	411
79CHE467	V. P. Semenov, A. N. Studenikov and A. A. Potekhin; *Chem. Heterocycl. Compd. (Engl. Transl.)*, 1979, **15**, 467.	460, 463
79CHE544	I. Ya. Kvitko, R. V. Khozeeva and A. V. El'tsov; *Chem. Heterocycl. Compd. (Engl. Transl.)*, 1979, **15**, 544.	372
79CHE685	V. A. Bakulev, V. S. Mokrushin, V. I. Ofitserov, Z. V. Pushkareva and A. N. Grishakov; *Chem. Heterocycl. Compd. (Engl. Transl.)*, 1979, **15**, 685.	629
79CHE939	A. F. Pozharskii; *Chem. Heterocycl. Compd. (Engl. Transl.)*, 1979, **15**, 939.	347, 349
79CHE1237	M. A. Klykov, M. V. Povstyanoi and P. M. Kochergin; *Chem. Heterocycl. Compd. (Engl. Transl.)*, 1979, **15**, 1237.	656
79CHE1255	V. P. Kruglenko and M. V. Povstyanoi; *Chem. Heterocycl. Compd. (Engl. Transl.)*, 1979, **15**, 1255.	653
79CJC139	K. R. Breakell, S. J. Rettig, A. Storr and J. Trotter; *Can. J. Chem.*, 1979, **57**, 139.	227
79CJC360	M. Mukai, T. Miura, M. Nanbu, T. Yoneda and Y. Shindo; *Can. J. Chem.*, 1979, **57**, 360.	256
79CJC813	J. Kister, G. Assef, G. Mille and J. Metzger; *Can. J. Chem.*, 1979, **57**, 813.	446
79CJC904	P. J. Fagan, E. E. Neidert, M. J. Nye, M. J. O'Hare and W. P. Tang; *Can. J. Chem.*, 1979, **57**, 904.	265

79CJC1186	W. J. Leigh and D. R. Arnold; *Can. J. Chem.*, 1979, **57**, 1186.	182, 192, 249
79CJC3034	S. C. Sharma and B. M. Lynch; *Can. J. Chem.*, 1979, **57**, 3034.	272
79CPB183	A. Matsuda, Y. Nomoto and T. Ueda; *Chem. Pharm. Bull.*, 1979, **27**, 183.	561
79CPB1153	H. Takahashi, N. Nimura, N. Obata, H. Sakai and H. Ogura; *Chem. Pharm. Bull.*, 1979, **27**, 1153.	637
79CPB1207	K. Kubo, N. Ito, Y. Isomura, I. Sozu, H. Homma and M. Murakami; *Chem. Pharm. Bull.*, 1979, **27**, 1207.	632
79CPB2116	M. Sakamoto, K. Miyazawa, K. Kuwabara and Y. Tomimatsu; *Chem. Pharm. Bull.*, 1979, **27**, 2116.	632
79CPB2431	T. Higashino, T. Katori and E. Hayashi; *Chem. Pharm. Bull.*, 1979, **27**, 2431.	875
79CPB2518	M. Saneyoshi and E. Satoh; *Chem. Pharm. Bull.*, 1979, **27**, 2518.	560
79CPB2820	A. Noda, K. Matsuyama, S.-H. Yen, K. Sogabe, Y. Aso, S. Iguchi and H. Noda; *Chem. Pharm. Bull.*, 1979, **27**, 2820.	904
79CPB2861	T. Higashino, T. Katori and E. Hayashi; *Chem. Pharm. Bull.*, 1979, **27**, 2861.	876
79CPB3176	T. Higashino, T. Katori, S. Yoshida and E. Hayashi; *Chem. Pharm. Bull.*, 1979, **27**, 3176.	875
79CRV447	M. V. George and V. Bhat; *Chem. Rev.*, 1979, **79**, 447.	254
79FES417	E. Abignente, F. Arena, M. Carola, P. De Caprariis, A. P. Caputi, F. Rossi, L. Giordano, C. Vacca, E. Lampa and E. Marmo; *Farmaco Ed. Sci.*, 1979, **34**, 417 (*Chem. Abstr.*, 1979, **91**, 39 390).	631
79FES478	G. Auzzi, L. Cecchi, A. Costanzo, L. Pecori Vettori, F. Bruni, R. Pirisino and G. B. Gottoli; *Farmaco Ed. Sci.*, 1979, **34**, 478.	331
79FES751	G. Auzzi, L. Cecchi, A. Costanzo, L. Pecori Vettori and F. Bruni; *Farmaco Ed. Sci.*, 1979, **34**, 751.	313
79FES898	G. Auzzi, L. Cecchi, A. Costanzo, L. Pecorivettori and F. Bruni; *Farmaco Ed. Sci.*, 1979, **34**, 898.	272
79GEP2815956	Bayer AG, *Ger. Pat.* 2 815 956 (1979) (*Chem. Abstr.*, 1980, **92**, 94 406).	719
79GEP2909190	K. Okauchi, T. Koitabashi and N. Fujimori; *Ger. Pat.* 2 909 190 (1979) (*Chem. Abstr.*, 1980, **92**, 172 437).	837
79H(12)186	K. Takahashi and K. Mitsuhashi; *Heterocycles*, 1979, **12**, 186.	389, 435
79H(12)657	C. Kashima, S. Shirai and Y. Yamamoto; *Heterocycles*, 1979, **12**, 657.	281
79H(12)1157	A. Tomažič, M. Tišler and B. Stanovnik; *Heterocycles*, 1979, **12**, 1157.	885
79H(12)1171	L. G. Payne, M. T. Wu and A. A. Patchett; *Heterocycles*, 1979, **12**, 1171.	853
79HC(34-1)5	J. V. Metzger, E.-J. Vincent, J. Chouteau and Gi. Mille; *Chem. Heterocycl. Compd.*, 1979, **34-1**, 5.	58, 89
79HC(34-1)67	J. V. Metzger; *Chem. Heterocycl. Compd.*, **34-1**, 67.	14
79HC(34-1)73	J. V. Metzger; *Chem. Heterocycl. Compd.*, 1979, **34-1**, 73.	14
79HC(34-1)76	J. V. Metzger; *Chem. Heterocycl. Compd.*, 1979, **34-1**, 76.	19
79HC(34-1)230	G. Vernin; *Chem. Heterocycl. Compd.*, 1979, **34-1**, 230.	120
79HC(34-1)388	J. V. Metzger; *Chem. Heterocycl. Compd.*, 1979, **34-1**, 388.	19
79HC(34-2)9	R. Barone, M. Chanon and R. Gallo; *Chem. Heterocycl. Compd.*, 1979, **34-2**, 9.	97
79HC(34-2)26	J. V. Metzger; *Chem. Heterocycl. Comp.* 1979, **34-2**, 26.	15
79HC(34-2)385	J. V. Metzger; *Chem. Heterocycl. Compd.*, 1979, **34-2**, 385.	15
79HC(34-2)421	J. V. Metzger; *Chem. Heterocycl. Compd.*, 1979, **34-2**, 421.	27
79HC(34-2)430	J. V. Metzger; *Chem. Heterocycl. Compd.*, 1979, **34-2**, 430.	27
79HC(35)365	G. W. H. Cheeseman and R. F. Cookson; *Chem. Heterocycl. Compd.*, 1979, **35**, 365.	608
79HC(35)370	G. W. H. Cheeseman and R. F. Cookson; *Chem. Heterocycl. Compd.*, 1979, **35**, 370.	608
79HC(35)376	G. W. H. Cheeseman and R. F. Cookson; *Chem. Heterocycl. Compd.*, 1979, **35**, 376.	608
79HCA497	E.-P. Krebs and E. Bondi; *Helv. Chim. Acta*, 1979, **62**, 497.	468
79IC658	G. Minghetti, G. Banditelli and F. Bonati; *Inorg. Chem.*, 1979, **18**, 658.	226
79IJC(B)428	P. K. Dubey and C. V. Ratnam; *Indian J. Chem., Sect. B*, 1979, **18**, 428.	619, 635
79IZV1618	O. A. Luk'yanov, A. A. Onishchenko, V. S. Kosygin and V. A. Tartakovskii; *Izv. Akad. Nauk SSSR, Ser. Khim.*, 1979, 1618 (*Chem. Abstr.*, 1979, **91**, 175 263).	646
79JA545	W. M. Litchman; *J. Am. Chem. Soc.*, 1979, **101**, 545.	190, 191, 192, 213
79JA1303	S. F. Dyer and P. B. Shevlin; *J. Am. Chem. Soc.*, 1979, **101**, 1303.	96
79JA1564	J. R. Barrio, F.-T. Liu, G. E. Keyser, P. vanDerLijn and N. J. Leonard; *J. Am. Chem. Soc.*, 1979, **101**, 1564.	567
79JA3647	R. Huisgen, H. U. Reissig and H. Huber; *J. Am. Chem. Soc.*, 1979, **101**, 3647.	283
79JA3976	M. Murai, M. Torres and O. P. Strausz; *J. Am. Chem. Soc.*, 1979, **101**, 3976.	44
79JA3982	S. M. Hecht, K. M. Rupprecht and P. M. Jacobs; *J. Am. Chem. oc.*, 1979, **101**, 3982.	485
79JA4899	M. J. S. Dewar and W. Thiel; *J. Am. Chem. Soc.*, 1977, **99**, 4899, 4907.	7
79JA5558	M. J. S. Dewar and G. P. Ford; *J. Am. Chem. Soc.*, 1979, **101**, 5558.	7
79JCR(M)2921	J. L. Avril, J. Mayrargue and M. Miocque; *J. Chem. Res. (M)*, 1979, 2921.	281
79JCR(S)374	P. Dapporto and F. Mani; *J. Chem. Res. (S)*, 1979, 374.	227
79JCS(D)1646	A. Murphy, B. J. Hathaway and T. J. King; *J. Chem. Soc., Dalton Trans.*, 1979, 1646.	227
79JCS(D)1867	F. Valach, J. Kohout, M. Dunaj-Jurčo, M. Hvastijová and J. Gažo; *J. Chem. Soc., Dalton Trans.*, 1979, 1867.	227
79JCS(P1)15	S. I. Miller, R.-R. Lii and Y. Tanaka; *J. Chem. Soc., Perkin Trans. 1*, 1979, 15.	687
79JCS(P1)333	D. G. Neilson, K. M. Watson and T. J. R. Weakley; *J. Chem. Soc., Perkin Trans. 1*, 1979, 333.	659
79JCS(P1)1120	R. W. Clarke, S. C. Garside, L. H. C. Lunts, D. Hartley, R. Hornby and A. W. Oxford; *J. Chem. Soc., Perkin Trans. 1*, 1979, 1120.	630, 657

79JCS(P1)1239	A. Davidson, I. E. P. Murray, P. N. Preston and T. J. King; *J. Chem. Soc., Perkin Trans. 1*, 1979, 1239.	419
79JCS(P1)1833	E. E. Glover, L. W. Peck and D. G. Doughty; *J. Chem. Soc., Perkin Trans. 1*, 1979, 1833.	615
79JCS(P1)2289	S. R. Landor, P. D. Landor, Z. T. Fomum and G. W. B. Mpango; *J. Chem. Soc., Perkin Trans. 1*, 1979, 2289.	354, 471
79JCS(P1)2786	R. E. Busby, J. Parrick, S. M. H. Rizvi and J. G. Shaw; *J. Chem. Soc., Perkin Trans. 1*, 1979, 2786.	246
79JCS(P1)3085	J. S. Bajwa and P. J. Sykes; *J. Chem. Soc., Perkin Trans. 1*, 1979, 3085.	889, 891, 893, 902
79JCS(P1)3107	T. Brown, K. Kadir, G. Mackenzie and G. Shaw; *J. Chem. Soc., Perkin Trans. 1*, 1979, 3107.	438, 584, 590
79JCS(P2)653	H. Fujiwara, A. K. Bose, M. S. Manhas and J. M. van der Veen; *J. Chem. Soc., Perkin Trans. 2*, 1979, 653.	351
79JCS(P2)1327	J. P. Riley, F. Heatley, I. H. Hillier, P. Murray-Rust and J. Murray-Rust; *J. Chem. Soc., Perkin Trans. 2*, 1979, 1327.	630
79JCS(P2)1670	J. Kaczmarek, H. Smagowski and Z. Grzonka; *J. Chem. Soc., Perkin Trans. 2*, 1979, 1670.	793, 806, 816
79JCS(P2)1708	R. A. Bowie, P. N. Edwards, S. Nicholson, P. J. Taylor and D. A. Thomson; *J. Chem. Soc., Perkin Trans. 2*, 1979, 1708.	874
79JGU1251	L. N. Nikolenko, N. S. Tolmacheva and L. D. Shustov; *J. Gen. Chem. USSR (Engl. Transl.)*, 1979, **49**, 1251.	408, 413
79JGU1877	Yu. M. Kargin, V. Z. Latypova, R. Kh. Fassakhov, A. I. Arkhipov, T. A. Eneikina and G. P. Sharnin; *J. Gen. Chem. USSR (Engl. Transl.)*, 1979, **49**, 1877.	419
79JHC53	A. Rabaron, J. C. Lancelot, D. Maume and M. Robba; *J. Heterocycl. Chem.*, 1979, **16**, 53.	273
79JHC157	J. A. Montgomery, S. D. Clayton and A. T. Shortnacy; *J. Heterocycl. Chem.*, 1979, **16**, 157.	553, 603
79JHC187	L. W. Deady and M. S. Stanborough; *J. Heterocycl. Chem.*, 1979, **16**, 187.	631
79JHC277	R. Paul and J. Menschik; *J. Heterocycl. Chem.*, 1979, **16**, 277.	630, 654
79JHC305	A. Nagel, H. C. van der Plas, G. Geurtsen and A. van der Kuilen; *J. Heterocycl. Chem.*, 1979, **16**, 305.	637
79JHC415	A. F. Janzen, G. N. Lypka and R. E. Wasylishen; *J. Heterocycl. Chem.*, 1979, **16**, 415.	454
79JHC427	J. Daunis, L. Djouai-Hifdi and H. Lopez; *J. Heterocycl. Chem.*, 1979, **16**, 427.	899
79JHC555	C. A. Lovelette; *J. Heterocycl. Chem.*, 1979, **16**, 555.	859, 861, 899, 903
79JHC571	N. A. Rodios and N. E. Alexandrou; *J. Heterocycl. Chem.*, 1979, **16**, 571.	678
79JHC585	H. T. Cory, K. Yamaizumi, D. L. Smith, D. R. Knowles, A. D. Broom and J. A. McCloskey; *J. Heterocycl. Chem.*, 1979, **16**, 585.	519
79JHC607	P. C. Joshi, S. S. Parmar and V. K. Rastogi; *J. Heterocycl. Chem.*, 1979, **16**, 607.	466
79JHC685	S. Olivella and J. Vilarrasa; *J. Heterocycl. Chem.*, 1979, **16**, 685.	371
79JHC689	M. Iwao and T. Kuraishi; *J. Heterocycl. Chem.*, 1979, **16**, 689.	635
79JHC773	Y. Van Haverbeke, A. Maquestiau and J. J. V. Eynde; *J. Heterocycl. Chem.*, 1979, **16**, 773.	272, 332
79JHC871	R. K. Sehgal and K. C. Agrawal; *J. Heterocycl. Chem.*, 1979, **16**, 871.	404, 437, 661
79JHC983	G. Holzmann, B. Krieg, H. Lautenschläger and P. Konieczny; *J. Heterocycl. Chem.*, 1979, **16**, 983.	361, 466
79JHC1005	R. Cerri, A. Boido and F. Sparatore; *J. Heterocycl. Chem.*, 1979, **16**, 1005.	459
79JHC1063	P. C. Srivastava and R. K. Robins; *J. Heterocycl. Chem.*, 1979, **16**, 1063.	642
79JHC1105	O. Migliara, S. Petruso and V. Sprio; *J. Heterocycl. Chem.*, 1979, **16**, 1105.	651
79JHC1153	J. S. Amato, V. J. Grenda, T. M. H. Liu and E. J. Grabowski; *J. Heterocycl. Chem.*, 1979, **16**, 1153.	396
79JHC1499	R. K. Sehgal and K. C. Agrawal; *J. Heterocycl. Chem.*, 1979, **16**, 1499.	441
79JHC1567	B. Stanovnik, M. Tišler, D. Gabrijelčič, M. Kunaver and J. Žmitek; *J. Heterocycl. Chem.*, 1979, **16**, 1567.	857, 901
79JHC1579	E. Barni and P. Savarino; *J. Heterocycl. Chem.*, 1979, **16**, 1579.	902
79JMC237	J. Rokach, P. Hamel, N. R. Hunter, G. Reader, C. S. Rooney, P. S. Anderson, E. J. Cragoe, Jr. and L. R. Mandel; *J. Med. Chem.*, 1979, **22**, 237.	663, 665
79JMC356	R. W. Fries, D. P. Bohlken and B. V. Plapp; *J. Med. Chem.*, 1979, **22**, 356.	291
79JMC496	F. G. de las Heras, R. Alonso and G. Alonso; *J. Med. Chem.*, 1979, **22**, 496.	729
79JMC807	M. T. García-López, R. Herranz and G. Alonso; *J. Med. Chem.*, 1979, **22**, 807.	289
79JMC944	R. B. Meyer, Jr. and E. B. Skibo; *J. Med. Chem.*, 1979, **22**, 944.	666
79JMC958	M. S. Poonian, W. W. McComas and M. J. Kramer; *J. Med. Chem.*, 1979, **22**, 958.	642
79JMC1030	C. Li, M. H. Lee and A. C. Sartorelli; *J. Med. Chem.*, 1979, **22**, 1030.	139
79JOC218	B. B. Snider, R. S. E. Conn and S. Sealfon; *J. Org. Chem.*, 1979, **41**, 218.	284
79JOC622	K. T. Potts, S. K. Datta and J. L. Marshall; *J. Org. Chem.*, 1979, **44**, 622.	164, 335
79JOC1028	T. Huynh-Dinh, R. S. Sarfati, C. Gouyette, J. Igolen, E. Bisagni, J.-M. Lhoste and A. Civier; *J. Org. Chem.*, 1979, **44**, 1028.	896
79JOC1450	M. F. Zady and J. L. Wong; *J. Org. Chem.*, 1979, **44**, 1450.	513, 543, 544
79JOC1717	W. A. Sheppard, G. W. Gukel, O. W. Webster, K. Betterton and J. W. Timberlake; *J. Org. Chem.*, 1979, **44**, 1717.	436
79JOC1740	B. Golankiewicz, J. B. Holtwick, B. N. Holmes, E. N. Duesler and N. J. Leonard; *J. Org. Chem.*, 1979, **44**, 1740.	659

79JOC1765	I. I. Schuster, C. Dyllick-Brenzinger and J. D. Roberts; *J. Org. Chem.*, 1979, **44**, 1765.	195, 196
79JOC2513	M. S. Puar, G. C. Rovnyak, A. I. Cohen, B. Toeplitz and J. Z. Gougoutas; *J. Org. Chem.*, 1979, **44**, 2513.	189
79JOC2902	H. Kimoto and L. A. Cohen; *J. Org. Chem.*, 1979, **44**, 2902.	432
79JOC2957	V. Bhat, V. M. Dixit, B. G. Ugarker, A. M. Trozzolo and M. V. George; *J. Org. Chem.*, 1979, **44**, 2957.	283
79JOC3140	N. J. Kos, H. C. van den Plas and B. van Veldhuizen; *J. Org. Chem.*, 1979, **44**, 3140.	513, 515, 542
79JOC3803	K. T. Potts and S. Kanemasa; *J. Org. Chem.*, 1979, **44**, 3803.	138, 150, 633
79JOC3835	J. B. Holtwick, B. Golankiewicz, B. N. Holmes and N. J. Leonard; *J. Org. Chem.*, 1979, **44**, 3835.	659
79JOC3858	E. P. Papadopoulos; *J. Org. Chem.*, 1979, **44**, 3858.	428, 445
79JOC4046	B.-S. Huang and J. C. Parham, *J. Org. Chem.*, 1979, **44**, 4046.	666
79JOC4156	P. Cohen-Fernandes, C. Erkelens, C. G. M. van Eendenburg, J. J. Verhoeven and C. L. Habraken; *J. Org. Chem.*, 1979, **44**, 4156.	270
79JOC4240	Y. Takeuchi, K. L. Kirk and L. A. Cohen; *J. Org. Chem.*, 1979, **44**, 4240.	397
79JOC4532	R. F. Shuman, W. E. Shearin and R. J. Tull; *J. Org. Chem.*, 1979, **44**, 4532.	461
79JOC4871	Y. Ohtsuka, E. Tohma, S. Kojima and N. Tomita; *J. Org. Chem.*, 1979, **44**, 4871.	644
79JPR881	H. Dorn and R. Ozegowski; *J. Prakt. Chem.*, 1979, **321**, 881.	272
79JPS377	M. J. Kornet and R. J. Garret; *J. Pharm. Sci.*, 1979, **68**, 377.	297
79KGS262	N. V. Volkova, V. N. Konyukhov, T. G. Koksharova, L. N. Dianova and Z. V. Pushkareva; *Khim. Geterotsikl. Soedin.*, 1979, 262 (*Chem. Abstr.*, 1979, **90**, 204 017).	873
79KGS643	Y. L. Gol'dfarb, F. M. Stoyanovich, M. A. Marakatkina and G. I. Gorushkina; *Khim. Geterotsikl. Soedin.*, 1979, 634 (*Chem. Abstr.*, 1979, **91**, 91 557).	631
79KGS805	V. M. Dziomko and B. K. Berestevich; *Khim. Geterotsikl. Soedin.*, 1979, 805.	263
79LA133	K. Burger and F. Hein; *Liebigs Ann. Chem.*, 1979, 133.	283
79LA1444	U. Schöllkopf, P. H. Porsch and H. H. Lau; *Liebigs Ann. Chem.*, 1979, 1444.	462
79LA1456	E. V. Dehmlow and K. Frake; *Liebigs Ann. Chem.*, 1979, 1456.	246
79LA1602	U. Schöllkopf and K. Hantke; *Liebigs Ann. Chem.*, 1979, 1602.	462
79LA1756	W. Walter and M. Radke; *Liebigs Ann. Chem.*, 1979, 1756.	391, 452, 786
79LA1872	K. Iluaga and W. Pfleiderer; *Liebigs Ann. Chem.*, 1979, **11**, 1872.	566
79MI40100	M. Baudet and M. Gelbcke; *Anal. Lett.*, 1979, **12**, 641.	20
B-79MI40300	C. A. Ramsden; in 'Comprehensive Organic Chemistry', ed. D. H. R. Barton and W. D. Ollis; Pergamon Press, Oxford, 1979, vol. 4, p. 1207.	127
79MI40400	J. C. Berger and L. C. Iorio; *Annu. Rep. Med. Chem.*, 1979, **14**, 27.	292, 293
79MI40401	R. D. Mackenzie; *Annu. Rep. Med. Chem.* 1979, **14**, 74.	294
79MI40402	C. Chavis, F. Grodenic and J. L. Imbach; *Eur. J. Med. Chem.*, 1979, **14**, 123.	192
79MI40403	I. A. Krol, V. M. Agre, V. K. Trunov, N. V. Rannev, O. V. Ivanov and V. M. Dziomko; *Koord. Khim.*, 1979, **5**, 126.	227
79MI40404	J. Skladzinski and M. Beskid; *Mater. Med. Pol. (Engl. Ed.)*, 1979, **11**, 355.	302
79MI40405	A. C. Grosscurt, R. van Hes and K. Wellinga; *J. Agric. Food Chem.*, 1979, **27**, 406.	298
79MI40406	W. Pett, H. Geissler, G. Dube and D. Zeigan; *J. Signalaufzeichnungsmaterialien*, 1979, **6**, 447.	299
79MI40407	R. M. Claramunt, J. Elguero, C. Marzin and J. Seita; *An. Quím.*, 1979, **75**, 701.	182, 184, 190, 191
79MI40408	J. Elguero, M. Espada, D. Mathieu and R. Phan Tan Luu; *An. Quím.*, 1979, **75**, 729.	230
79MI40409	R. Barone, P. Camps and J. Elguero; *An. Quím.*, 1979, **75**, 736.	276
79MI40410	J. Elguero and M. Espada; *An. Quím.*, 1979, **75**, 771.	191
79MI40411	A. Fruchier and J. Elguero; *Spectrosc. Lett.*, 1979, 809.	194
79MI40412	V. Agre, I. A. Krol, V. K. Trunov, V. M. Dziomko and O. V. Ivanov; *Koord. Khim.*, 1979, **5**, 1406.	227
79MI40413	I. A. Krol, V. M. Agre, V. K. Trunov and O. I. Ivanov; *Koord. Khim.*, 1979, **5**, 1569.	227
B-79MI40414	F. H. Deis and D. Lester; in 'Biochemical Pharmacology of Ethanol', ed. E. Majchrowicz and E. P. Noble; Plenum, New York, 1979, vol. 2, p. 203.	291
B-79MI40415	M. R. Grimmett; in 'Comprehensive Organic Chemistry', ed. D. H. R. Barton and W. D. Ollis; Pergamon Press, Oxford, 1979, vol. 4, p. 357.	168
B-79MI40416	C. Hansch and A. Leo; 'Substituent Constants for Correlation Analysis in Chemistry and Biology', Wiley-Interscience, New York, 1979.	207, 268
B-79MI40900	G. C. Levy and R. L. Lichter; 'NMR Spectra', Wiley-Interscience, New York, 1979.	510
B-79MI40901	'Nucleoside Analogs', ed. R. T. Walker, E. De Clercq and F. Eckstein; NATO Advanced Study Series, Plenum, New York, 1979.	504
B-79MI40902	'The Chemistry of Nucleosides and Nucleotides', ed. R. K. Robins and L. B. Townsend; Plenum, New York, 1979, vol. 1.	504
79MI40903	G. Shaw; *Annu. Rep. Prog. Chem. (B)*, 1979, 448.	504
79MI40904	M. Kaneko and B. Shimizu; *Yuki Gosei Kagaku Kyokaishi*, 1979, **37**, 40.	504
B-79MI40905	E. Lunt; in 'Comprehensive Organic Chemistry', ed. D. H. R. Barton and W. D. Ollis; Pergamon, Oxford, 1979, vol. 4, p. 493.	501

79MI40906	F. A. Savin, Ya. V. Morozov, A. V. Borodavkin, V. O. Chekhov, E. I. Budowsky and N. A. Simukova; *Int. J. Quantum Chem.*, 1979, **16**, 825.	506
B-79MI40907	'Advances in Cyclic Nucleotide Research', ed. G. Brooker, P. Greengard and G. A. Robison; Raven Press, New York, 1979, vol. 10.	504
79MI40908	M. Shimizaki, S. Kondo, K. Maeda, M. Ohno and N. Umezawa; *J. Antiobiotics*, 1979, **32**, 537.	601
79MI41000	A. Czarnocka-Janowicz, J. Sawlewicz, J. Jakubowski and M. Janowiec; *Acta Pol. Pharm.*, 1979, **36**, 529 (*Chem. Abstr.*, 1981, **94**, 65 548).	620, 622
79MI41001	Y. M. Yutilov, O. G. Eilazyan and A. G. Ignatenko; *Deposited Doc.*, VINITI, 1979, 129-79 (*Chem. Abstr.*, 1980, **92**, 215 340).	620
79MI41100	G. Manecke and C. S. Ruehl; *Makromol. Chem.*, 1979, **180**, 103.	731
79MI41101	T. Higashino, E. Hayashi and T. Katori; *Fukosokan, Kagaku Toronkai Koen Yoshishu, 12th,* 1979, 171 (*Chem. Abstr.*, 1980, **93**, 95 227).	721
79MI41102	P. G. Fox, G. Lewis and P. J. Boden; *Corrosion Sci.*, 1979, **19**, 457 (*Chem. Abstr.*, 1980, **92**, 30 906).	730
B-79MI41103	G. C. Levy and R. L. Lichter; 'Nitrogen-15 NMR Spectroscopy', Wiley–Interscience, New York, 1979.	684
B-79MI41104	E. Zbiral; in 'Organophosphorus Reagents in Organic Synthesis', ed. J. I. G. Cadogan; Academic Press, London, 1979, p. 256.	716
B-79MI41105	M. R. Grimmett; in 'Comprehensive Organic Chemistry', ed. D. H. R. Barton and W. D. Ollis; Pergamon Press, Oxford, 1979, vol. 4, p. 357.	670, 698
79MI41200	Sh. J. Chiou; *Dept. Chem. Kansas State Univ. Report,* 1979, W80-05462 (*Chem. Abstr.*, 1981, **94**, 48 103).	785
79MI41201	M. C. De Traglia, J. S. Brand and A. M. Tometsko; *Anal. Biochem.*, 1979, **99**, 464.	785
79MI41500	G. Doukhan, T. Hunyh-Dinh, E. Bisagni, J.-C. Chermann and J. Igolen; *Eur. J. Med. Chem.*, 1979, **14**, 375.	884, 896
79MI41501	A. Hagen, B. Froemmel, H. Kuechmstedt, I. Wunderlich, K. Kottke and E. Goeres; *Pharmazie,* 1979, **34**, 330.	893
79MIP41100	Hoechst Aktiengesellschaft, *Can. Pat.* 1 057 760 (1979) (*Chem. Abstr.*, 1979, **91**, 159 046).	730
79MIP41101	Consejo Superior de Investigaciones Cientificas, *Span. Pat.* 460 433 (1979) (*Chem. Abstr.*, 1979, **91**, 193 585).	728
79MIP41102	Consejo Superior de Investigaciones Cientificas, *Span. Pat.* 475 496 (1979) (*Chem. Abstr.*, 1979, **91**, 107 984).	728
79OMR(12)95	E. Gründemann, D. Martin and A. Wenzel; *Org. Magn. Reson.*, 1979, **12**, 95.	354
79OMR(12)205	P. Sohár, O. Fehér and E. Tihanyi; *Org. Magn. Reson.*, 1979, **12**, 205.	192
79OMR(12)362	F. A. L. Anet and I. Yavari; *Org. Magn. Reson.*, 1979, **12**, 362.	355
79OMR(12)476	T. Axenrod, C. M. Watnick and M. J. Wielder; *Org. Magn. Reson.*, 1979, **12**, 476.	197, 290
79OMR(12)579	R. Faure, J. Llinares, E. J. Vincent and J. Elguero; *Org. Magn. Reson.*, 1979, **12**, 579.	195
79OMR(12)587	E. Gonzalez, R. Faure, E. J. Vincent, M. Espada and J. Elguero; *Org. Magn. Reson.*, 1979, **12**, 587.	191, 192, 211
79OMR(12)631	J. Ciarkowski, J. Kaczmarek and Z. Grzonka; *Org. Magn. Reson.*, 1979, **12**, 631.	793, 799
79OMS114	A. Maquestiau, Y. Van Haverbeke, N. Vanovervelt, R. Flammang and J. Elguero; *Org. Mass Spectrom.*, 1979, **14**, 114.	203
79OMS117	A. Maquestiau, Y. Van Haverbeke, R. Flammang, H. Mispreuve and J. Elguero; *Org. Mass Spectrom.*, 1979, **14**, 117.	203
79OMS577	W. C. M. M. Luitjen and J. van Thuijl; *Org. Mass Spectrom.*, 1979, **14**, 577.	202
79P819	C. C. Duke, D. S. Letham, C. W. Parker, J. K. MacLeod and R. E. Simmons; *Phytochemistry,* 1979, **18**, 819.	601
79RCR289	S. D. Sokolov; *Russ. Chem. Rev.* (*Engl. Transl.*), 1979, **48**, 289.	169, 171, 222, 245, 248
79RTC258	A. M. van Leusen, F. J. Schaart and D. M. van Leusen; *Recl. Trav. Chim. Pays-Bas,* 1979, **98**, 258.	468
79RTC334	O. L. Chapman, R. S. Sheridan and J.-P. LeRoux; *Recl. Trav. Chim. Pays-Bas,* 1979, **98**, 334.	867
79S66	V. I. Cohen; *Synthesis,* 1979, 66.	118
79S194	B. Stanovnik, M. Tišler, M. Kočevar, B. Koren, M. Bešter and V. Kermavner; *Synthesis,* 1979, 194.	263
79S308	M. A. Ardakani, R. K. Smalley and R. H. Smith; *Synthesis,* 1979, 308.	274
79S470	K. Masuda, Y. Arai and M. Itoh; *Synthesis,* 1979, 470.	718
79S687	S. Mataka, K. Takahashi and M. Tashiro; *Synthesis,* 1979, 687.	117
79S979	V. Bertini, F. Lucchesini and A. DeMunno; *Synthesis,* 1979, 979.	127
79SA(A)1055	V. Tabacik, V. Pellegrin and H. H. Günthard; *Spectrochim. Acta, Part A,* 1979, **35**, 1055.	199, 200, 290
79T389	K. Burger, H. Schickaneder, F. Hein and J. Elguero; *Tetrahedron,* 1979, **35**, 389.	188, 189
79T511	F. A. Devillanova and G. Verani; *Tetrahedron,* 1979, **35**, 511.	357
79T663	M. Jelinska, J. Szydiowski and J. Sobkowski; *Tetrahedron,* 1979, **35**, 663.	527
79T1331	P. Bouchet, G. Joncheray, R. Jacquier and J. Elguero; *Tetrahedron,* 1979, **35**, 1331.	243, 266, 419
79TL53	K. Nagarajan, M. D. Nair and J. A. Desai; *Tetrahedron Lett.*, 1979, 53.	444

79TL279	D. H. R. Barton, C. J. R. Hedgecock, E. Lederer and B. Motherwell; *Tetrahedron Lett.*, 1979, 279.	537
79TL1281	J. Rokach, P. Hamel, Y. Girard and G. Reader; *Tetrahedron Lett.*, 1979, 1281.	664
79TL1567	G. Ege and K. Gilbert; *Tetrahedron Lett.*, 1979, 1567.	134, 166, 263
79TL1725	D. Cooper and S. Trippett; *Tetrahedron Lett.*, 1979, 1725.	728
79TL2385	N. Công-Danh, J.-P. Beaucourt and L. Pichat; *Tetrahedron Lett.*, 1979, 2385.	537, 593
79TL2621	W. Bihlmaier, R. Huisgen, H. U. Reissig and S. Voss; *Tetrahedron Lett.*, 1979, 2621.	283
79TL2991	F. Pochat; *Tetrahedron Lett.*, 1979, 2991.	192, 281
79TL3129	R. A. Olofson and J. P. Pepe; *Tetrahedron Lett.*, 1979, 3129.	787
79TL3159	N. Công-Danh, J.-P. Beaucourt and L. Pichat; *Tetrahedron Lett.*, 1979, 3159.	538
79TL3179	F. Fabra, E. Fos and J. Vilarrasa; *Tetrahedron Lett.*, 1979, 3179.	182, 263
79TL4253	G. Ege and K. Gilbert; *Tetrahedron Lett.*, 1979, 4253.	852
79TL4709	L. J. Mathias and D. Burkett; *Tetrahedron Lett.*, 1979, 4709.	387
79USP4141899	G. Arcari, L. Bernardi, G. Falconi, F. Luini, G. Palamidessi and U. Scarponi; *U.S. Pat.* 4 141 899 (1979).	623
79YZ699	S. Nagai, N. Oda and I. Ito; *Yakugaku Zasshi*, 1979, **99**, 699.	234, 270
79ZC446	J. Wrubel and R. Mayer; *Z. Chem.*, 1979, **19**, 446.	719
79ZN(B)1473	W. H. Guendel; *Z. Naturforsch., Teil B*, 1979, **34**, 1473.	636
79ZOR839	I. Yu. Shirobokov, V. A. Ostrovskii and G. I. Koldobskii; *Zh. Org. Khim.*, 1979, **15**, 839 (*Chem. Abstr.*, 1979, **91**, 19 384).	818
79ZOR844	V. A. Ostrovskii, N. S. Panina, G. I. Koldobskii, B. V. Gidaspov and I. Yu. Shirobokov; *Zh. Org. Khim.*, 1979, **15**, 844 (*Chem. Abstr.*, 1979, **91**, 19 679).	796
79ZOR2534	N. O. Saldabol, E. Liepins, J. Popelis, R. Gavars, L. Baumane and I. Birgele; *Zh. Org. Khim.*, 1979, **15**, 2534 (*Chem. Abstr.*, 1980, **93**, 26 339).	612
80ACH(105)127	A. Simay, K. Takács, K. Horváth and P. Dvortsák; *Acta Chim. Acad. Sci. Hung.*, 1980, **105**, 127.	241
80AG319	A. Albert and A. Dunand; *Angew. Chem.*, 1980, **92**, 319.	727
80AG979	R. Huisgen; *Angew. Chem.*, 1980, **92**, 979.	841
80AG(E)130	D. Günther and D. Bosse; *Angew. Chem., Int. Ed. Engl.*, 1980, **19**, 130.	473
80AG(E)276	G. L'abbé; *Angew. Chem., Int. Ed. Engl.*, 1980, **19**, 276.	810
80AG(E)310	A. Albert and A. Dunand; *Angew. Chem., Int. Ed. Engl.*, 1980, **19**, 310.	727
80AG(E)924	W. E. Hull, M. Künstlinger and E. Breitmaier; *Angew. Chem., Int. Ed. Engl.*, 1980, **19**, 924.	822
80AG(E)947	R. Huisgen; *Angew. Chem., Int. Ed. Engl.*, 1980, **19**, 947.	275, 809, 825, 841
80AHC(27)31	P. Hanson; *Adv. Heterocycl. Chem.*, 1980, **27**, 31.	73, 103
80AHC(27)151	N. Lozach and M. Stavaux; *Adv. Heterocycl. Chem.*, 1980, **27**, 151.	63, 64, 65, 66, 67, 68, 70, 71, 76, 90, 103, 105, 114
80AHC(27)241	M. R. Grimmett; *Adv. Heterocycl. Chem.*, 1980, **27**, 241.	53, 54, 62, 70, 71, 73, 77, 89, 96, 98, 108, 109, 114, 346, 348, 350, 351, 355, 358, 359 360, 361, 365, 367, 371, 378, 379, 384, 386, 387, 390, 391, 392 393, 396, 397, 400, 401, 402, 404, 407, 412, 413, 414, 417, 419 421, 422, 425, 431, 432, 436, 439, 440, 445, 450, 453, 454, 455 446, 461, 462, 464, 465, 466, 467, 468, 473, 474, 475, 477, 478 480, 481, 485, 486, 487, 488, 489, 490, 491, 493, 494, 496, 497
80AJC499	A. J. Jones, P. Hanisch, M. L. Heffernan and G. M. Irvine; *Aust. J. Chem.*, 1980, **33**, 499.	853, 860
80AJC949	N. F. Borkett, M. I. Bruce and J. D. Walsh; *Aust. J. Chem.*, 1980, **33**, 949.	235
80AJC1365	L. K. Dalton, S. Demerac and B. C. Elmes; *Aust. J. Chem.*, 1980, **33**, 1365.	281
80AJC1763	C. W. Fong; *Aust. J. Chem.*, 1980, **33**, 1763.	210, 268
80AJC2237	J. R. Cannon, C. L. Raston and A. H. White; *Aust. J. Chem.*, 1980, **33**, 2337.	785
80AJC2447	G. J. Gainsford and A. D. Woolhouse; *Aust. J. Chem.*, 1980, **33**, 2447.	707
80AX(B)159	A. M. G. Dias Rodrigues, Y. P. Mascarenhas and M. M. Rodrigues; *Acta Crystallogr., Part B*, 1980, **36**, 159.	227
80AX(B)1466	F. Iwasaki; *Acta Crystallogr., Part B*, 1980, **36**, 1466.	9
80B3773	A. B. Kremer, W. A. Gibby, C. J. Gubler, R. F. Helfand, A. F. Kluge and H. Z. Sable; *Biochemistry*, 1980, **19**, 3773.	648
80BCJ961	M. Onishi, K. Hiraki, M. Shironita, Y. Yamaguchi and S. Nakagawa; *Bull. Chem. Soc. Jpn.*, 1980, **53**, 961.	235
80BCJ1638	T. Keumi, H. Saga and H. Kitajima; *Bull. Chem. Soc. Jpn.*, 1980, **53**, 1638.	452
80BSB51	A. Maquestiau, Y. Van Haverbeke and J. J. Vanden Eynde; *Bull. Soc. Chim. Belg.*, 1980, **89**, 51.	272
80C504	H. Meier, W. Heinzelmann and H. Heimgartner; *Chimia*, 1980, **34**, 504 (*Chem. Abstr.*, 1981, **94**, 155 988).	809
80CB1884	W. Hartmann, K.-H. Scholz and H.-G. Heine; *Chem. Ber.*, 1980, **113**, 1884.	77
80CB2043	F. Sella and D. Hasselmann; *Chem. Ber.*, 1980, **113**, 2043.	605
80CB2823	A. Schönberg, E. Singer and P. Eckert; *Chem. Ber.*, 1980, **113**, 2823.	471
80CB2852	E. Öhler and E. Zbiral; *Chem. Ber.*, 1980, **113**, 2852.	165
80CB3910	A. Steigel and R. Fey; *Chem. Ber.*, 1980, **113**, 3910.	194
80CC82	K. Nakanishi, H. Komura, I. Miura and H. Kasai; *J. Chem. Soc., Chem. Commun.*, 1980, 82.	526, 538

80CC237	J. G. Buchanan, A. R. Edgar, R. J. Hutchison, A. Stobie and R. H. Wightman; *J. Chem. Soc., Chem. Commun.*, 1980, 237.	234
80CC339	R. S. Norton, R. P. Gregson and R. J. Quinn; *J. Chem. Soc., Chem. Commun.*, 1980, 339.	515
80CC444	T. Tsuchiya, J. Kurita and H. Kojima; *J. Chem. Soc., Chem. Commun.*, 1980, 444.	268
80CC502	S. Fischer and C. Wentrup; *J. Chem. Soc., Chem. Commun.*, 1980, 502.	812
80CC866	F. Wudl and D. Nalewajek; *J. Chem. Soc., Chem. Commun.*, 1980, 866.	115
80CC866	L.-Y. Chiang, T. O. Poehler, A. N. Bloch and D. O. Cowan; *J. Chem. Soc., Chem. Commun.*, 1980, 866.	115
80CC940	C. Bernard and L. Ghosez; *J. Chem. Soc., Chem. Commun.*, 1980, 940.	694, 715
80CC1263	S. N. Ege, E. J. Gess, A. Thomas, P. Umrigar, G. W. Griffin, P. K. Das, A. M. Trozzolo and T. M. Leslie; *J. Chem. Soc., Chem. Commun.*, 1980, 1263.	252
80CCC2417	V. Hanus, M. Hrazdira and O. Exner; *Collect. Czech. Chem. Commun.*, 1980, **45**, 2417.	285
80CHE1	H. Dorn; *Chem. Heterocycl. Compd. (Engl. Transl.)*, 1980, **16**, 1. 169, 241, 253, 256, 261, 262, 264, 277, 281	
80CHE66	G. P. Tolmakov, Yu. M. Udachin, N. S. Paralakha, L. K. Denisov, A. M. Lantsov and I. I. Grandberg; *Chem. Heterocycl. Compd. (Engl. Transl.)*, 1980, **16**, 66.	199
80CHE169	N. Yu. Deeva and A. N. Kost; *Chem. Heterocycl. Compd. (Engl. Transl.)*, 1980, **16**, 1969.	257
80CHE180	K. V. Mityurina, V. G. Kharchenko and L. V. Cherkesova; *Chem. Heterocycl. Compd. (Engl. Transl.)*, 1980, **16**, 180.	253
80CHE418	R. Paegle, I. Lulle, V. Krisane, I. Mazeika, E. Liepins and M. Lidaks; *Chem. Heterocycl. Compd. (Engl. Transl.)*, 1980, **16**, 418.	649
80CHE443	M. Tišler and B. Stanovnik; *Chem. Heterocycl. Compd. (Engl. Transl.)*, 1980, **16**, 443.	169, 263
80CHE628	V. A. Samsonov and L. B. Volodarskii; *Chem. Heterocycl. Compd. (Engl. Transl.)*, 1980, **16**, 628.	423, 474
80CHE936	L. A. Sviridova, G. A. Golubeva, A. V. Dovgilevich and A. N. Kost; *Chem. Heterocycl. Compd. (Engl. Transl.)*, 1980, **16**, 936.	254
80CHE1160	V. M. Dziomko, B. K. Berestevich, A. V. Kessenikh, R. S. Kuzanyan and L. V. Shmelev; *Chem. Heterocycl. Compd. (Engl. Transl.)*, 1980, **16**, 1160.	241
80CHE1190	V. V. Dovlatayan, V. A. Pivazyan, K. A. Eliazyan and R. G. Mirzoyan; *Chem. Heterocycl. Compd. (Engl. Transl.)*, 1980, **16**, 1190.	657
80CJC1080	K. S. Chong, S. J. Rettig, A. Storr and J. Trotter; *Can. J. Chem.*, 1980, **58**, 1080.	227
80CJC1091	K. S. Chong, S. J. Rettig, A. Storr and J. Trotter; *Can. J. Chem.*, 1980, **58**, 1091.	227
80CJC1880	J. Grimshaw and A. P. DeSilva; *Can. J. Chem.*, 1980, **58**, 1880.	268, 322
80CJC2624	J. G. Buchanan, A. Stobie and R. H. Wightman; *Can. J. Chem.*, 1980, **58**, 2624.	262, 270, 328
80CL299	S. Inoue, K. Okada, H. Tanino and H. Kakoi; *Chem. Lett.*, 1980, 299.	643
80CPB150	A. Yamane, A. Matsuda and T. Ueda; *Chem. Pharm. Bull.*, 1980, **28**, 150.	561
80CPB157	A. Yamane, H. Inoue and T. Ueda; *Chem. Pharm. Bull.*, 1980, **28**, 157.	559
80CPB487	S. Sugai and K. Tomita; *Chem. Pharm. Bull.*, 1980, **28**, 487.	15
80CPB932	K. Kayasuga-Mikado, T. Hashimoto, T. Negishi, K. Negishi and H. Hayatsu; *Chem. Pharm. Bull.*, 1980, **28**, 932.	648
80CS(15)102	S. Liljefors and S. Gronowitz; *Chem. Scr.*, 1980, **15**, 102.	238, 260
80CS(15)193	R. Faure, G. M. Assef, E. J. Vincent, N. de Kimpe, R. Verhe, L. de Bruyck and N. Schamp; *Chem. Scr.*, 1980, **15**, 193.	19, 355, 368
80CSC1121	M. Sikirica and I. Vicković; *Cryst. Struct. Commun.*, 1980, **9**, 1121.	178, 214
80CSC1127	M. Sikirica and I. Vicković; *Cryst. Struct. Commun.*, 1980, **9**, 1127.	178
80EUP11604	L. Gsell, P. Ackermann and R. Wehrli; *Eur. Pat.* 11 604 (1980) (*Chem. Abstr.*, 1981, **94**, 65 691).	787
80EUP13914	W. C. Lumma; *Eur. Pat. Appl.* 13 914 (1980) (*Chem. Abstr.*, 1981, **94**, 84 167).	643
80GEP2834879	Schering AG, *Ger. Pat.* 2 834 879 (1980) (*Chem. Abstr.*, 1980, **93**, 71 758).	732
80GEP2848670	Bayer AG, *Ger. Pat.* 2 848 670 (1980) (*Chem. Abstr.*, 1980, **93**, 96 810).	730
80GEP2852067	Schering AG, *Ger. Pat.* 2 852 067 (1980) (*Chem. Abstr.*, 1980, **93**, 204 664).	718
80GEP2922591	W. K. Aders, D. Mangold, J. Wahl and G. Rotermund; *Ger. Pat.* 2 922 591 (*Chem. Abstr.*, 1981, **94**, 139 804).	288
80GEP2952685	S. D. Ziman; *Ger. Pat.* 2 952 685 (1980) (*Chem. Abstr.*, 1981, **94**, 84 129).	787
80H(14)97	M. Ogata, H. Matsumoto, S. Kida and S. Shimizu; *Heterocycles*, 1980, **14**, 97.	452
80H(14)1279	S. K. Talapatra, P. Chaudhuri and B. Talapatra; *Heterocycles*, 1980, **14**, 1279.	707
80H(14)1313	S. Witek, A. Bielawska and J. Bielawski; *Heterocycles*, 1980, **14**, 1313.	427
80H(14)1725	R. Kreher and U. Bergmann; *Heterocycles*, 1980, **14**, 1725.	473
80H(14)1759	R. Chaurasia and S. K. Sharma; *Heterocycles*, 1980, **14**, 1759.	857, 903
80H(14)1963	A. Ohta, T. Watanabe, J. Nishiyama, K. Uehara and R. Hirate; *Heterocycles*, 1980, **14**, 1963.	393, 453, 495
80HC(39)1	K. T. Finley; *Chem. Heterocycl. Compd.*, 1980, **39**, 1.	670, 705, 708
80HCA1	R. Heckendorn and T. Winkler; *Helv. Chim. Acta*, 1980, **63**, 1.	855, 873, 893
80IC170	J. C. Jansen, H. van Konigsveld, J. A. C. van Ooijen and J. Reedijk; *Inorg. Chem.*, 1980, **19**, 170.	227
80IZV2668	M. M. Krayushkin, A. M. Beskopyl'nyi, S. G. Zlotin, O. A. Luk'yanov and V. M. Zhulin; *Izv. Akad. Nauk SSSR, Ser. Khim.*, 1980, 2668 (*Chem. Abstr.*, 1981, **94**, 103 260).	831

80JA525	D. M. Cheng, L. S. Kan, P. O. P. Ts'o, C. Giessner-Prettre and B. Pullman; *J. Am. Chem. Soc.*, 1980, **102**, 525.	510, 515
80JA770	P. D. Sattsangi, J. R. Barrio and N. J. Leonard; *J. Am. Chem. Soc.*, 1980, **102**, 770.	564
80JA1513	N. Niccolai, H. K. Schnoes and W. A. Gibbons; *J. Am. Chem. Soc.*, 1980, **102**, 1513.	604
80JA2093	N. H. Toubro and A. Holm; *J. Am. Chem. Soc.*, 1980, **102**, 2093.	809
80JA2968	N. E. Takach, E. M. Holt, N. W. Alcock, R. A. Henry and J. H. Nelson; *J. Am. Chem. Soc.*, 1980, **102**, 2968.	818
80JA3971	K. T. Potts, S. Kanemasa and G. Zvilichovsky; *J. Am. Chem. Soc.*, 1980, **102**, 3971.	133, 230
80JA4627	J. Lin, C. Yu, S. Peng, I. Akiyama, K. Li, L. K. Lee and P. R. LeBreton; *J. Am. Chem. Soc.*, 1980, **102**, 4627.	517
80JA6159	C. Wentrup and H.-W. Winter; *J. Am. Chem. Soc.*, 1980, **102**, 6159.	879
80JA7335	R. S. Rogers and H. Rapoport; *J. Am. Chem. Soc.*, 1980, **102**, 7335.	604
80JCR(M)0514	J. de Mendoza, M. L. Castellanos, J. P. Fayet, M. C. Vertut and J. Elguero; *J. Chem. Res. (M)*, 1980, 0514.	191, 234, 269
80JCR(S)308	D. R. Buckle; *J. Chem. Res. (S)*, 1980, 308.	904
80JCS(P1)244	E. Cawgill and N. G. Clark; *J. Chem. Soc., Perkin Trans. 1*, 1980, 244.	462
80JCS(P1)354	A. R. Katritzky, M. Michalska, R. L. Harlow and S. H. Simonsen; *J. Chem. Soc., Perkin Trans. 1*, 1980, 354.	632
80JCS(P1)481	J. S. Bajwa and P. J. Sykes; *J. Chem. Soc., Perkin Trans. 1*, 1980, 481.	272
80JCS(P1)506	C. R. Hardy and J. Parrick; *J. Chem. Soc., Perkin Trans. 1*, 1980, 506.	859, 877
80JCS(P1)574	J. Motoyoshiya, M. Nishijima, I. Yamamoto, H. Gotoh, Y. Katsube, Y. Oshiro and T. Agawa; *J. Chem. Soc., Perkin Trans. 1*, 1980, 574.	20
80JCS(P1)938	R. J. J. Dorgan, J. Parrick and C. R. Hardy; *J. Chem. Soc., Perkin Trans. 1*, 1980, 938.	272
80JCS(P1)959	E. E. Glover and L. W. Peck; *J. Chem. Soc., Perkin Trans. 1*, 1980, 959.	614
80JCS(P1)1139	I. Charles, D. W. S. Latham, D. Hartley, A. W. Oxford and D. I. C. Scopes; *J. Chem. Soc., Perkin Trans. 1*, 1980, 1139.	630, 657
80JCS(P1)1352	J. P. Clayton, P. J. O'Hanlon and T. J. King; *J. Chem. Soc., Perkin Trans. 1*, 1980, 1352.	852
80JCS(P1)1427	R. E. Busby, M. A. Khan, J. Parrick, C. J. G. Shaw and M. Iqbal; *J. Chem. Soc., Perkin Trans. 1*, 1980, 1427.	418
80JCS(P1)2009	A. Albert; *J. Chem. Soc., Perkin Trans. 1*, 1980, 2009.	727
80JCS(P1)2310	T. Brown, G. Shaw and G. J. Durant; *J. Chem. Soc., Perkin Trans. 1*, 1980, 2310.	414, 467
80JCS(P1)2608	D. Griffiths, R. Hull and T. P. Seden; *J. Chem. Soc., Perkin Trans. 1*, 1980, 2608.	462
80JCS(P1)2728	K. Kadir, G. Shaw and D. Wright; *J. Chem. Soc., Perkin Trans. 1*, 1980, 2728.	552, 585, 586
80JCS(P1)2755	Y. Kobayashi, K. Yamamoto, T. Asai, M. Nakano and I. Kumadaki; *J. Chem. Soc., Perkin Trans. 1*, 1980, 2755.	538
80JCS(P2)535	A. F. Hegarty, K. Brady and M. Mullane; *J. Chem. Soc., Perkin Trans. 2*, 1980, 535.	825
80JCS(P2)553	D. B. Bigley, C. L. Fetter and M. J. Clarke; *J. Chem. Soc., Perkin Trans. 2*, 1980, 553.	31
80JCS(P2)1350	U. Berg, R. Gallo, G. Klatte and J. Metzger; *J. Chem. Soc., Perkin Trans. 2*, 1980, 1350.	229
80JHC11	J. T. Shaw, C. E. Brotherton, R. W. Moon, M. D. Winland, M. D. Anderson and K. S. Kyler; *J. Heterocycl. Chem.*, 1980, **17**, 11.	882, 903
80JHC97	L. Citerio, E. Rivera, M. L. Saccarello, R. Stradi and B. Gioia; *J. Heterocycl. Chem.*, 1980, **17**, 97.	481
80JHC113	M. T. García-López, R. Herranz and G. Alonso; *J. Heterocycl. Chem.*, 1980, **17**, 113.	288
80JHC137	G. Tarrago, A. Ramdani, J. Elguero and M. Espada; *J. Heterocycl. Chem.*, 1980, **17**, 137.	53, 183, 229
80JHC187	O. Subba Rao and W. Lwowski; *J. Heterocycl. Chem.*, 1980, **17**, 187.	814
80JHC337	R. A. Glennon, M. E. Rogers and M. K. El-Said; *J. Heterocycl. Chem.*, 1980, **17**, 337.	647
80JHC381	Y. C. Tong; *J. Heterocycl. Chem.*, 1980, **17**, 381.	627
80JHC393	H. J. M. Dou, M. Ludwikow, P. Hassanaly, J. Kister and J. Metzger; *J. Heterocycl. Chem.*, 1980, **17**, 393.	663
80JHC461	D. L. Kern, P. D. Cook and J. C. French; *J. Heterocycl. Chem.*, 1980, **17**, 461.	602
80JHC555	J. Bourdais and A. M. M. E. Omar; *J. Heterocycl. Chem.*, 1980, **17**, 555.	635
80JHC607	T. J. Kress and S. M. Constantino; *J. Heterocycl. Chem.*, 1980, **17**, 607.	115
80JHC689	B. Kovač, L. Klasinc, B. Stanovnik and M. Tišler; *J. Heterocycl. Chem.*, 1980, **17**, 689.	860
80JHC825	W. A. Feld and M. P. Servé; *J. Heterocycl. Chem.*, 1980, **17**, 825.	694
80JHC833	A. S. Shawali and C. Párkányi; *J. Heterocycl. Chem.*, 1980, **17**, 833.	282, 284
80JHC877	A. S. Shawali, M. Sami, S. M. Sherif and C. Parkanyi; *J. Heterocycl. Chem.*, 1980, **17**, 877.	631
80JHC905	B. Cross, R. L. Arotin and C. F. Ruopp; *J. Heterocycl. Chem.*, 1980, **17**, 905.	269
80JHC1115	W. A. Feld, P. G. Seybold and M. P. Servé; *J. Heterocycl. Chem.*, 1980, **17**, 1115.	732
80JHC1121	I. Lalezari and S. Nabahi; *J. Heterocycl. Chem.*, 1980, **17**, 1121.	857, 901
80JHC1127	C. Tsoleridis, J. Stephanidou-Stephanatou and N. E. Alexandrou; *J. Heterocycl. Chem.*, 1980, **17**, 1127.	682, 683
80JHC1309	W. A. Feld, R. Paessun and M. P. Servé; *J. Heterocycl. Chem.*, 1980, **17**, 1309.	694

80JHC1373	H. Lumbroso, J. Curé and R. N. Butler; *J. Heterocycl. Chem.*, 1980, **17**, 1373.	795, 804
80JHC1505	A. Kreutzberger and J. Stratmann; *J. Heterocycl. Chem.*, 1980, **17**, 1505.	728
80JHC1527	S. Sunder and N. P. Peet; *J. Heterocycl. Chem.*, 1980, **17**, 1527.	281
80JHC1723	A. C. Veronese, G. Cavicchioni, G. Servadio and G. Vecchiati; *J. Heterocycl. Chem.*, 1980, **17**, 1723.	475
80JHC1757	R. W. Middleton and D. G. Wibberley; *J. Heterocycl. Chem.*, 1980, **17**, 1757.	617, 620, 635, 636
80JHC1777	T. Nakano, W. Rodriguez, S. Z. de Roche, J. M. Larrauri, C. Rivas and C. Pérez; *J. Heterocycl. Chem.*, 1980, **17**, 1777.	421
80JHC1789	P. Reynaud, J. D. Brion, C. Davrinche and P.-C. Dao; *J. Heterocycl. Chem.*, 1980, **17**, 1789.	471
80JMC357	J. R. Piper, A. G. Laseter and J. A. Montgomery; *J. Med. Chem.*, 1980, **23**, 357.	531
80JMC657	M. T. García-López, M. J. Domínguez, R. Herranz, R. M. Sánchez, A. Contreras and G. Alonso; *J. Med. Chem.*, 1980, **23**, 657.	289
80JMC1188	D. L. Temple, Jr., J. P. Yevich, J. D. Catt, D. Owens, C. Hanning, R. R. Covington, R. J. Seidehamel and K. W. Dungan; *J. Med. Chem.*, 1980, **23**, 1188.	647
80JOC76	J. F. Hansen, Y. I. Kim, L. J. Griswold, G. W. Hoelle, D. L. Taylor and D. E. Vietti; *J. Org. Chem.*, 1980, **45**, 76.	270
80JOC1653	A. J. Boulton, T. Kan-Woon, S. N. Balasubrahmayam, I. M. Malick and A. S. Radhakrishna; *J. Org. Chem.*, 1980, **45**, 1653.	288
80JOC1662	G. H. Denny, E. J. Cragoe, Jr, C. S. Rooney, J. P. Springer, J. M. Hirshfield and J. A. McCauley; *J. Org. Chem.*, 1980, **45**, 1662.	815
80JOC2320	D. H. Bown and J. S. Bradshaw; *J. Org. Chem.*, 1980, **45**, 2320.	870
80JOC2373	M. F. Zady and J. L. Wong; *J. Org. Chem.*, 1980, **45**, 2373.	545
80JOC2474	K. T. Potts, R. Ehlinger and S. Kanemasa; *J. Org. Chem.*, 1980, **45**, 2474.	133, 151, 663
80JOC3172	W. C. Guida and D. J. Mathre; *J. Org. Chem.*, 1980, **45**, 3172.	230, 387
80JOC3738	E. S. Hand and W. W. Paudler; *J. Org. Chem.*, 1980, **45**, 3738.	611
80JOC3750	Y. Lin, S. A. Lang, Jr. and S. R. Petty; *J. Org. Chem.*, 1980, **45**, 3750.	14
80JOC3969	V. Nair and S. G. Richardson; *J. Org. Chem.*, 1980, **45**, 3969.	511, 514, 553, 593, 594, 598
80JOC4038	N. J. Curtis and R. S. Brown; *J. Org. Chem.*, 1980, **45**, 4038.	388, 416
80JOC4302	E. P. Ahern, K. J. Dignam and A. F. Hegarty; *J. Org. Chem.*, 1980, **45**, 4302.	825
80JOC4767	L. Garanti and G. Zecchi; *J. Org. Chem.*, 1980, **45**, 4767.	902
80JOC5105	L. A. Paquette and R. F. Doehner; *J. Org. Chem.*, 1980, **45**, 5105.	737
80JOC5130	O. Tsuge, S. Urano and K. Oe; *J. Org. Chem.*, 1980, **45**, 5130.	832
80JOM(188)141	J. P. Gasparini, R. Gassend, J. C. Maire and J. Elguero; *J. Organomet. Chem.*, 1980, **188**, 141.	232
80JPR711	G. Tacconi, A. Gamba Invernizzi, P. P. Righetti and G. Desimoni; *J. Prakt. Chem.*, 1980, **322**, 711.	330
80KGS121	Y. M. Yutilov, A. G. Ignatenko and O. G. Eilazyan; *Khim. Geterotsikl. Soedin.*, 1980, 121 (*Chem. Abstr.*, 1980, **92**, 215 357).	623
80KGS528	V. A. Anisimova, N. I. Avdyunina, A. F. Pozharskii, A. M. Simonov and L. N. Talanova; *Khim. Geterotsikl. Soedin.*, 1980, 528 (*Chem. Abstr.*, 1980, **93**, 149 290).	612
80KGS984	Y. B. Vysotskii and B. P. Zemskii; *Khim. Geterotsikl. Soedin.*, 1980, 984.	858, 859
80KGS1414	Zh. N. Fidler, E. F. Shibanova, P. V. Makerov, I. D. Kalikhman, A. M. Shulunova, G. I. Sarapulova, L. V. Klyba, V. Yu. Vitkovskii, N. N. Chipanina, V. A. Lopyrev and M. G. Voronkov; *Khim. Geterotsikl. Soedin.*, 1980, 1414.	740
80LA542	S. Linke, J. Kurz, D. Lipinski and W. Gau; *Liebigs Ann. Chem.*, 1980, 542.	633
80LA1455	W. Schörkhuber and E. Zbiral; *Liebigs Ann. Chem.*, 1980, 1455.	716
80M775	J. Elguero, A. Fruchier and A. Könnecke; *Monatsh. Chem.*, 1980, **111**, 775.	186, 798
80MI40100	P. Ruoff, J. Almlö and S. Saeboe; *Chem. Phys. Lett.*, 1980, **72**, 489.	9
80MI40101	J. B. Collins and A. J. Streitwieser, Jr.; *Comput. Chem.*, 1980, **1**, 81.	7
80MI40400	J. C. Berger and L. C. Iorio; *Annu. Rep. Med. Chem.*, 1980, **15**, 26.	292, 293
80MI40401	R. H. P. Francisco, J. R. Lechat, A. C. Massabni, C. B. Melios and M. Molina; *J. Coord. Chem.*, 1980, **10**, 149.	179, 227
80MI40402	F. Bonati; *Chim. Ind. (Roma)*, 1980, **62**, 323.	169, 225, 236
80MI40403	J. Terheijden, W. L. Driessen and W. L. Groeneveld; *Transition Met. Chem.*, 1980, **5**, 346.	226
80MI40404	U. Wrzeciono and E. Linkowska, *Pharmazie*, 1980, **35**, 593.	270
80MI40405	T. Inaba, M. Lucassen and W. Kalov; *Life Sci.*, 1980, **26**, 1977.	301
B-80MI40406	'Burger's Medicinal Chemistry', Wiley, New York, 4th edn., 1980.	291, 292, 293, 295, 296, 301, 302, 303
B-80MI40407	L. S. Goodman and A. Gilman, 'The Pharmacological Basis of Therapeutics', Macmillan, New York, 1980.	295
B-80MI40408	P. E. Cassidy; 'Thermally Stable Polymers', Dekker, New York, 1980.	300, 301
80MI40900	J. B. Lambert, J. F. Marwood, L. P. Davies and K. M. Taylor; *Life Sci.*, 1980, **26**, 1069.	601
B-80MI40901	G. Shaw; in 'Rodd's Chemistry of Carbon Compounds', ed. S. Coffey; Elsevier, 1980, vol. IV(L), p. 1.	501, 504
80MI40902	G. Shaw; *Annu. Rep. Prog. Chem. (B)*, 1980, 299.	504
B-80MI40903	G. P. Ellis and R. K. Smalley; in 'Heterocyclic Chemistry', Royal Society of Chemistry, London, 1980, p. 308.	501

80MI40904	S. Yananaka and T. Utagawa; *Hakko to Kogyo*, 1980, **38**, 920.	504
80MI40905	F. A. Fuhrman, G. J. Fuhrman, Y. H. Kim, L. A. Pavelka and H. J. Mosher; *Science*, 1980, **207**, 193.	601, 602
80MI40906	P. G. Mezey, J. J. Ladik and M. Barry; *Theor. Chim. Acta*, 1980, **54**, 251.	512
80MI40907	M. Rufalska and G. Wenska; *Wiad. Chem.*, 1980, **34**, 9.	504
80MI41000	M. V. Povstyanoi, V. P. Kruglenko and M. A. Klykov; *Fiziol. Akt. Veshchestva*, 1980, **12**, 56 (*Chem. Abstr.*, 1980, **95**, 43 047).	653
80MI41100	U. Saha, A. Das, S. Chakraborty, M. Gosh and D. K. Roy; *J. Inst. Chem. (India)*, 1980, **52**, 196 (*Chem. Abstr.*, 1981, **94**, 139 681).	709, 729
80MI41101	L. Birkofer and O. Stuhl; *Top. Curr. Chem.*, 1980, **88**, 33.	699
B-80MI41102	H. Suschitzky and O. Meth-Cohn; 'Heterocyclic Chemistry', The Royal Society of Chemistry, London, 1980, vol. 1, p. 220.	670
B-80MI41103	H. Wamhoff; in 'Lectures in Heterocyclic Chemistry', ed. R. N. Castle and S. W. Schneller; Heterocorporation, Orem/Utah, 1980, vol. V, p. S-61.	704, 712
80MI41200	C. B. Donker, J. G. Haasnoot and W. L. Groeneveld; *Transition Met. Chem.*, 1980, **5**, 368.	790
80MI41300	Nguyen Dinh Trieu, Ha Thi Diep, Luong Thu Huong and Le Thi Thanh Vinh; *Tap Chi Hoa Hoc*, 1980, **18**, 22 (*Chem. Abstr.*, 1981, **94**, 121 415).	832
B-80MI41301	H. Singh, A. S. Chawla, V. K. Kapoor, D. Paul and R. K. Malhotra; 'Progress in Medicinal Chemistry', ed. G. P. Ellis and G. B. West; Elsevier/North-Holland, Biomedical Press, 1980, vol. 17, p. 151.	834, 835, 836
80MI41500	D. J. Cash and A. N. Ferguson; *J. Photographr. Sci.*, 1980, **28**, 121.	904
80MIP41100	Hoechst Aktiengesellschaft, *Swiss Pat.* 615 164 (1980) (*Chem. Abstr.*, 1980, **93**, 73 786).	730
80MIP41500	ICI Ltd., *Isr. Pat.* 51 570 (1980) (*Chem. Abstr.*, 1981, **94**, 11 549).	904
80OMR(13)197	T. Axenrod, P. Mangiaracina, C. M. Watnick, M. J. Wieder and S. Bulusu; *Org. Magn. Reson.*, 1980, **13**, 197.	196, 197, 290
80OMR(13)274	L. Stefanik, M. Witanowski and G. A. Webb; *Org. Magn. Reson.*, 1980, **13**, 274.	20
80OMR(14)129	G. C. Levy, T. Pekh and P. R. Srinivasan; *Org. Magn. Reson.*, 1980, **14**, 129.	211
80OMS144	A. Maquestiau, Y. Van Haverbeke, M. Mispreuve, R. Flammang, J. A. Harris, I. Howe and J. H. Beynon; *Org. Mass Spectrom.*, 1980, **15**, 144.	203
80OMS172	I. Simiti, H. Demian, A. M. N. Palibroda and N. Palibroda; *Org. Mass Spectrom.*, 1980, **15**, 172.	741
80OMS533	A. Atmani, J. L. Aubagnac and V. Pellegrin; *Org. Mass Spectrom.*, 1980, **15**, 533.	203
80OPP234	J. Lee, A. Guthrie and M. M. Joullie; *Org. Prep. Proced. Int.*, 1980, **12**, 234.	620
80P2239	S. Matsubara; *Phytochemistry*, 1980, **19**, 2239.	602
80RCR28	V. G. Yashunskii and L. E. Kholodov; *Russ. Chem. Rev. (Engl. Transl.)*, 1980, **49**, 28.	19
80RCR880	V. A. Galishev, U. N. Chistoklevtov and A. A. Petrov; *Russ. Chem. Rev. (Engl. Transl.)*, 1980, **49**, 880.	768, 784
80RTC20	M. J. Wanner, E. M. van Wijk, G. J. Koomen and U. K. Pandit; *Recl. Trav. Chim. Pay-Bas*, 1980, **99**, 20.	566
80RTC267	N. J. Kos, H. C. van der Plas and A. van Veldhuizen; *Recl. Trav. Chim. Pays-Bas*, 1980, **99**, 267.	542
80S842	S. Mataka, K. Takahashi, M. Tashiro and Y. Tsuda; *Synthesis*, 1980, 842.	117
80T1079	K. Kiec-Kononowicz, A. Zejc, M. Mikolajczyk, A. Zatorski, J. Karolak-Wojciechowska and M. W. Wieczorek; *Tetrahedron*, 1980, **36**, 1079.	663
80T2359	P. Caluwe; *Tetrahedron*, 1980, **36**, 2359.	272
80T2505	J. Bergman, L. Renström and B. Sjöberg; *Tetrahedron*, 1980, **36**, 2505.	414
80T3523	P. Bouchet, R. Lazaro, M. Benchidmi and J. Elguero; *Tetrahedron*, 1980, **36**, 3523.	245, 266
80TH40100	J. Llinares; Ph.D. Thesis, University of Marseilles, 1980.	20
80TH40400	J. Llinares; Ph. D. Thesis, University of Marseille, 1980.	195
80TL567	R. J. Quinn, R. P. Gregson, A. F. Cook and R. T. Bartlett; *Tetrahedron Lett.*, 1980, **21**, 567.	602
80TL1417	P. Schiess and H. Stalder; *Tetrahedron Lett.*, 1980, 1417.	249
80TL2195	P. Blatcher and D. Middlemiss; *Tetrahedron Lett.*, 1974, **21**, 2195.	615
80TL3203	T. Ogita, N. Otake, Y. Miyazaki, H. Yonehara, R. D. MacFarlane and C. J. McNeal; *Tetrahedron Lett.*, 1980, **21**, 3203.	602
80TL3723	S. P. J. M. van Nispen, C. Mensink and A. M. van Leusen; *Tetrahedron Lett.*, 1980, 3723.	477
80TL4339	J. Engels; *Tetrahedron Lett.*, 1980, **21**, 4339.	786
80TL4387	H. J. Chaves das Neves and M. S. S. Pais; *Tetrahedron Lett.*, 1980, **21**, 4387.	602
80TL4529	G. Jones and D. R. Sliskovic; *Tetrahedron Lett.*, 1980, 4529.	904
80TL4731	K. K. Balasubramanian, G. V. Bindumadhavan, M. R. Udupa and B. Krebs; *Tetrahedron Lett.*, 1980, **21**, 4731.	657
80ZC167	B. V. Ioffe; *Z. Chem.*, 1980, **20**, 167.	255
80ZC263	H. Matschiner and H. Tanneberg; *Z. Chem.*, 1980, **20**, 263.	419
80ZC413	R. Evers, E. Fischer and M. Pulkenat; *Z. Chem.*, 1980, **20**, 413.	126
80ZC437	G. Weber, G. Mann, H. Wilde and S. Hauptmann; *Z. Chem.*, 1980, **20**, 437.	720
80ZN(A)1387	D. W. Engelfriet, W. L. Groeneveld and G. M. Nap; *Z. Naturforsch., Teil A*, 1980, **35**, 1387.	790
80ZOB875	V. D. Sheludyakov, S. V. Sheludyakova, M. G. Kutnetsova, N. N. Silkina and V. F. Mironov; *Zh. Obschch. Khim.*, 1980, **50**, 875.	236, 271

80ZOR730	L. I. Vereshchagin, L. G. Tikhonova, A. V. Maksihova, E. S. Serebryakova, A. G. Proidakov and T. M. Filippova; *Zh. Org. Khim.*, 1980, **16**, 730.	852
80ZOR1313	A. M. Belousov, G. A. Gareev, Yu. M. Belousov, N. A. Cherkashina and I. G. Kaufman; *Zh. Org. Khim.*, 1980, **16**, 1313 (*Chem. Abstr.*, 1980, **93**, 239 321).	815
80ZOR2185	Yu. A. Sharanin; *Zh. Org. Khim.*, 1980, **16**, 2185.	119
80ZOR2623	A. F. Shivanyuk, M. O. Lozinskii; *Zh. Org. Khim.*, 1980, **16**, 2623 (*Chem. Abstr.*, 1981, **94**, 156 837).	824
81ACS(A)733	I. Søtofte and K. Nielsen; *Acta Chem. Scand., Ser. A*, 1981, **35**, 733.	675
81ACS(A)739	I. Søtofte and K. Nielsen; *Acta Chem. Scand., Ser. A*, 1981, **35**, 739.	676, 677
81ACS(A)747	I. Søtofte and K. Nielsen; *Acta Chem. Scand., Ser. A.*, 1981, **35**, 747.	676
81ACS(A)767	U. Anthony, G. Borch, P. Klaeboe, K. Lerstrup and P. H. Nielsen; *Acta Chem. Scand., Ser. A.*, 1981, **35**, 767.	201
81AF2096	K. Credner, M. Tauscher, L. Jozic and G. Brenner; *Arzneim.-Forsch.*, 1981, **31**, 2096.	291
81AG118	D. Danion, B. Arnold and M. Regitz; *Angew. Chem.*, 1981, **93**, 118.	716
81AG(E)113	D. Danion, B. Arnold and M. Regitz; *Angew. Chem., Int. Ed. Engl.*, 1981, **20**, 113.	165, 716
81AG(E)296	R. Gompper and U. Heinemann; *Angew. Chem., Int. Ed. Engl.*, 1981, **20**, 296.	468
81AG(E)771	U. Schmidt and M. Dietsche; *Angew. Chem., Int. Ed. Engl.*, 1981, **20**, 771.	823
81AG(E)885	C. N. Rentzea; *Angew. Chem., Int. Ed. Engl.*, 1981, **20**, 885.	786
81AHC(28)1	R. D. Chambers and C. R. Sargent; *Adv. Heterocycl. Chem.*, 1981, **28**, 1.	282
81AHC(28)231	C. Wentrup; *Adv. Heterocycl. Chem.*, 1981, **28**, 231.	44, 45, 87, 91, 96, 98, 107, 108, 109, 110
81AHC(28)232	C. Wentrup; *Adv. Heterocycl. Chem.*, 1981, **28**, 232.	246, 263, 287
81AHC(28)309	C. Wentrup; *Adv. Heterocycl. Chem.*, 1981, **28**, 309.	163
81AHC(29)71	Y. Tamura and M. Ikeda; *Adv. Heterocycl. Chem.*, 1981, **29**, 71.	55
81AHC(29)141	M. Ruccia, N. Vivona and D. Spinelli; *Adv. Heterocycl. Chem.*, 1981, **29**, 141.	158, 719
81AJC691	N. A. Evans; *Aust. J. Chem.*, 1981, **34**, 691.	731
81AJC1361	G. B. Barlin, *Aust. J. Chem.*, 1981, **34**, 1361.	651
81AP532	K. A. Kovar, W. Rohlfes and H. Auterhof; *Arch. Pharm. (Weinheim, Ger.)*, 1981, **314**, 532.	194, 237
81AX(A)C-63	M. R. Taylor and J. A. Westphalen; *Acta Crystallogr., Part A*, 1981, **37**, C-63.	509
81AX(B)1584	B. M. Craven and P. Benci; *Acta Crystallogr., Part B*, 1981, **37**, 1584.	508
81BCJ217	T. Shimizu, Y. Hayashi, K. Yamada, T. Nishio and K. Teramura; *Bull. Chem. Soc. Jpn*, 1981, **54**, 217.	342
81BCJ1579	T. Keumi, T. Yamamoto, H. Saga and H. Kitajima; *Bull. Chem. Soc. Jpn.*, 1981, **54**, 1579.	451
81BCJ2779	Y. Nomura, Y. Takeuchi, S. Tomoda and M. M. Ito; *Bull. Chem. Soc. Jpn.*, 1981, **54**, 2779.	725
81BCJ3221	S. R. F. Kagaruki, T. Kitazume and N. Ishikawa; *Bull. Chem. Soc. Jpn.*, 1981, **54**, 3221.	214
81BSB193	J. Plenkiewicz and T. Zdrojewski; *Bull. Soc. Chim. Belg.*, 1981, **90**, 193 (*Chem. Abstr.*, 1981, **95**, 62 089).	815
81BSB645	A. J. Boulton; *Bull. Soc. Chim. Belg.*, 1981, **90**, 645.	269
81C59	J. Rabinowitz and A. Hampaï; *Chimia*, 1981, **35**, 59.	303, 786
81CB1546	E. V. Dehmlow and Naser-ud-Din; *Chem. Ber.*, 1981, **114**, 1546.	711
81CB1624	E. Cuny, F. W. Lichtenthaler and U. Jahn; *Chem. Ber.*, 1980, **114**, 1624.	273
81CB2450	H. Gotthardt and F. Reiter; *Chem. Ber.*, 1981, **114**, 2450.	286
81CB2649	R. Appel and E. Hiester; *Chem. Ber.*, 1981, **114**, 2649.	728
81CB3165	W. Schörkhuber and E. Zbiral; *Chem. Ber.*, 1981, **114**, 3165.	716
81CB3456	R. W. Saalfrank and E. Ackermann; *Chem. Ber.*, 1981, **114**, 3456.	717
81CC604	J. A. Barltrop, A. C. Day, A. G. Mack, A. Shahrisa and S. Wakamatsu; *J. Chem. Soc., Chem. Commun.*, 1981, 604.	489
81CC1089	A. Albini, G. F. Bettinetti, G. Minoli and G. Vasconi; *J. Chem. Soc., Chem. Commun.*, 1981, 1089.	723
81CC1095	B. Iddon and B. L. Lim; *J. Chem. Soc., Chem. Commun.*, 1981, 1095.	416
81CC1207	J. Elguero, A. Fruchier and V. Pellegrin; *J. Chem. Soc., Chem. Commun.*, 1981, 1207.	181, 191
81CHE375	R. E. Valter, E. A. Baumanis, L. K. Stradynya and E. E. Liepin'sh; *Chem. Heterocycl. Compd. (Engl. Transl.)*, 1981, **17**, 375.	215
81CHE622	M. V. Povstyanoi, M. A. Klykov and N. A. Klyuev; *Chem. Heterocycl. Compd. (Engl. Transl.)*, 1981, **17**, 622.	656
81CHE838	V. V. Dovlatyan, V. A. Pivazyan, K. A. Eliazyan, R. G. Mirzoyan and S. M. Saakyan; *Chem. Heterocycl. Compd. (Engl. Transl.)*, 1981, **17**, 838.	657
81CI(L)648	J. S. Davidson and K. Karunaratne; *Chem. Ind. (London)*, 1981, 648.	770
81CJC2556	M. O. Chang and R. J. Crawford; *Can. J. Chem.*, 1981, **59**, 2556.	255
81CL331	A. Albini, G. F. Bettinetti and G. Minoli; *Chem. Lett.*, 1981, 331.	246
81CL1519	M. M. Ito, Y. Nomura, Y. Takeuchi and S. Tomoda; *Chem. Lett.*, 1981, 1519.	726
81CPB426	M. Sugiura, K. Kato, T. Adachi, Y. Ito, K. Hirano and S. Sawaki; *Chem. Pharm. Bull.*, 1981, **29**, 426.	598
81CPB597	K. Fukukawa, T. Ueda and T. Hirano; *Chem. Pharm. Bull.*, 1981, **29**, 597.	602, 603
81CPB1870	K. Omura, R. Marumoto and Y. Furukawa; *Chem. Pharm. Bull.*, 1981, **29**, 1870.	590, 664

81CPB2403	M. Ishino, T. Sakaguchi, I. Morimoto and T. Okitsu; *Chem. Pharm. Bull.*, 1981, **29**, 2403.	514
81CPB3214	T. Kurihara, T. Tani, K. Nasu, M. Inoue and T. Ishida; *Chem. Pharm. Bull.*, 1981, **29**, 3214.	178
81CPB3760	K. Hirota, Y. Kitade, K. Shimada and S. Senda; *Chem. Pharm. Bull.*, 1981, **29**, 3760.	286
81CZ194	D. Moderhack; *Chem.-Ztg.*, 1981, **105**, 194 (*Chem. Abstr.*, 1981, **95**, 132 767).	820
81EUP21076	B. Zeeh, E. Buschmann and J. Jung; *Eur. Pat.* 21 076 (1981) (*Chem. Abstr.*, 1981, **94**, 151 896).	787
81EUP31302	CIBA-Geigy AG, *Eur. Pat. Appl.* 31 302 (1981) (*Chem. Abstr.*, 1981, **95**, 187 267).	731
81FES315	G. A. Bistochi, G. De Meo, M. Pedini, A. Ricci, H. Brouilhet, S. Bucherie, M. Rabaud and P. Jacquignon; *Farmaco Ed. Sci.*, 1981, **36**, 315.	293
81FES1019	P. Giori, D. Mazzotta, G. Vertuani, M. Guarneri, D. Pancaldi and A. Brunelli; *Farmaco Ed. Sci.*, 1981, **36**, 1019.	242
81FES1037	G. Picciola, F. Ravenna, G. Carenini, P. Gentili and M. Riva; *Farmaco Ed. Sci.*, 1981, **36**, 1037.	293
81GEP2926280	B. Zeeh, N. Goetz, E. Ammermann and E. H. Pommer; *Ger. Pat.* 2 926 280, (1981) (*Chem. Abstr.*, 1981, **94**, 192 348).	787
81GEP2936951	Schering AG, *Ger. Pat.* 2 936 951 (1981) (*Chem. Abstr.*, 1982, **96**, 52 509).	732
81H(15)265	T. Kurihara, T. Tani and K. Nasu; *Heterocycles*, 1981, **15**, 265.	285
81H(15)943	I. Saji, K. Tamoto, H. Yamazaki and H. Agui; *Heterocycles*, 1981, **15**, 943.	377
81H(16)909	T. Fujii, T. Saito and I. Inoue; *Heterocycles*, 1981, **16**, 909.	586
81HC(37)1	C. Temple, Jr.; *Chem. Heterocycl. Compd.*, 1981, **37**, 1.	734, 735, 743, 747, 748, 749, 750, 751, 754, 760, 762, 771, 772, 773, 775, 776, 777, 778, 780, 781, 782, 784, 785, 787, 788
81HC(37)289	C. Temple, Jr.; *Chem. Heterocycl. Compd.*, 1981, **37**, 289.	58
81HCA662	A. E. Siegrist; *Helv. Chim. Acta*, 1981, **64**, 662.	730
81HCA769	G. Rihs, H. Fuhrer and A. Marxer; *Helv. Chim. Acta*, 1981, **64**, 769.	179
81IC1519	W. C. Deese, D. A. Johnson and A. W. Cordes; *Inorg. Chem.*, 1981, **20**, 1519.	227
81IC1545	G. E. Bushnell, K. R. Dixon, D. T. Eadie and S. R. Stobart; *Inorg. Chem.*, 1981, **20**, 1545.	227
81IC1553	C. W. Eigenbrot and K. N. Raymond; *Inorg. Chem.*, 1981, **20**, 1553.	227
81IJC(B)10	P. S. Ram, P. S. N. Reddy and V. R. Srinivasan; *Indian J. Chem., Sect. B*, 1980, **20**, 10.	864
81JA6977	C. F. Wichmann, W. P. Niemczura, H. K. Schoes, S. Hall, P. B. Reichardt and S. D. Darling; *J. Am. Chem. Soc.*, 1981, **103**, 6977.	604
81JA7660	E. C. Taylor, H. M. L. Davies, R. J. Clemens, H. Yanagisawa and N. F. Haley; *J. Am. Chem. Soc.*, 1981, **103**, 7660.	281
81JA7743	E. C. Taylor, N. F. Haley and R. J. Clemens; *J. Am. Chem. Soc.*, 1981, **103**, 7743.	148
81JAP8161384	*Jpn. Pat.* 81 61 384 (1981) (*Chem. Abstr.*, 1981, **95**, 115 576).	665
81JAP(K)81108882	Nippon Kasei Kogyo K. K., *Jpn. Kokai Tokkyo Koko* 81 108 882 (1981) (*Chem. Abstr.*, 1982, **96**, 56 298).	730
81JAP(K)81122884	Tatsuta Electric Wire and Cable Co., Ltd., *Jpn. Kokai Tokkyo Koko* 81 122 884 (1981) (*Chem. Abstr.*, 1982, **96**, 56 656).	730
81JCR(S)174	R. N. Butler, V. C. Garvin and T. M. McEvoy; *J. Chem. Res. (S)*, 1981, 174.	818, 834
81JCR(S)364	J. Elguero, C. Estopá and D. Ilavský; *J. Chem. Res. (S)*, 1981, 364.	107, 268
81JCS(P1)78	G. Jones, D. R. Sliskovic, B. Foster, J. Rogers, A. K. Smith, M. Y. Wong and A. C. Yarham; *J. Chem. Soc., Perkin Trans. 1*, 1981, 78.	866, 867
81JCS(P1)360	S. H. Askari, S. F. Moss and D. R. Taylor; *J. Chem. Soc., Perkin Trans. 1*, 1981, 360.	17
81JCS(P1)390	R. N. Butler and V. C. Garvin; *J. Chem. Soc., Perkin Trans. 1*, 1981, 390.	793, 799, 804, 805
81JCS(P1)403	J. de Mendoza, C. Millan and P. Rull; *J. Chem. Soc., Perkin Trans. 1*, 1981, 403.	459
81JCS(P1)503	M. Begtrup and J. Holm; *J. Chem. Soc., Perkin Trans. 1*, 1981, 503.	701
81JCS(P1)697	M. T. Ayoub, M. Y. Shandala, G. M. G. Bashi and A. Pelter; *J. Chem. Soc., Perkin Trans. 1*, 1981, 697.	194
81JCS(P1)1424	G. U. Baig and M. F. G. Stevens; *J. Chem. Soc., Perkin Trans. 1*, 1981, 1424.	656
81JCS(P1)1891	J. Barluenga, J. F. López-Ortiz, M. Tomás and V. Gotor; *J. Chem. Soc., Perkin Trans. 1*, 1981, 1891.	274
81JCS(P1)2344	A. Albert; *J. Chem. Soc., Perkin Trans. 1*, 1981, 2344.	894
81JCS(P1)2374	J. G. Buchanan, A. Stobie and R. H. Wightman; *J. Chem. Soc., Perkin Trans. 1*, 1981, 2374.	303, 325
81JCS(P1)2387	G. A. Bhat and L. B. Townsend; *J. Chem. Soc., Perkin Trans. 1*, 1981, 2387.	310, 313, 315, 325, 343
81JCS(P1)2499	M. Yokoyama, K. Motozawa, E. Kawamura and T. Imamoto; *J. Chem. Soc., Perkin Trans. 1*, 1981, 2499.	428
81JCS(P1)2991	A. M. Damas, R. O. Gould, M. M. Harding, R. M. Paton, J. F. Ross and J. Crosby; *J. Chem. Soc., Perkin Trans. 1*, 1981, 2991.	9
81JCS(P1)2997	Z. T. Fomum, S. R. Landor, P. D. Landor and G. W. P. Mpango; *J. Chem. Soc., Perkin Trans. 1*, 1981, 2997.	281
81JCS(P2)310	A. R. Butler and I. Hussain; *J. Chem. Soc., Perkin Trans. 2*, 1981, 310.	355, 482
81JCS(P2)317	A. R. Butler and I. Hussain; *J. Chem. Soc., Perkin Trans. 2*, 1981, 317.	354, 355, 471
81JCS(P2)403	B. Eliasson and U. Edlund; *J. Chem. Soc., Perkin Trans. 2*, 1981, 403.	190

81JCS(P2)414	Y. Kondo, T. Yamada and S. Kusabayashi; *J. Chem. Soc., Perkin Trans. 2*, 1981, 414.	444
81JHC9	M. A. Khan and A. A. A. Pinto; *J. Heterocycl. Chem.*, 1981, **18**, 9.	183, 228, 238
81JHC117	J. C. Sircar, T. Capiris and S. J. Kesten; *J. Heterocycl. Chem.*, 1981, **18**, 117.	342
81JHC163	T. Kurihara, K. Nasu, F. Ishimori and T. Tani; *J. Heterocycl. Chem.*, 1981, **18**, 163.	272, 320
81JHC271	O. Migliara and V. Sprio; *J. Heterocycl. Chem.*, 1981, **18**, 271.	321
81JHC333	S. Plescia, G. Daidone, J. Fabra and V. Sprio; *J. Heterocycl. Chem.*, 1981, **18**, 333.	340
81JHC559	R. P. M. Berbee and C. L. Habraken; *J. Heterocycl. Chem.*, 1981, **18**, 559.	270
81JHC723	V. Frenna, N. Vivona, D. Spinelli and G. Consiglio; *J. Heterocycl. Chem.*, 1981, **18**, 723.	720
81JHC893	M. Legraverend, E. Bisagni and J.-M. Lhoste; *J. Heterocycl. Chem.*, 1981, **18**, 893.	889
81JHC921	M. L. Malnati, R. Stradi and E. Rivera; *J. Heterocycl. Chem.*, 1981, **18**, 921.	354, 361
81JHC957	G. Adembri, A. Camparini, F. Ponticelli and P. Tedeschi; *J. Heterocycl. Chem.*, 1981, **18**, 957.	269
81JHC1073	S. Mataka, K. Takahashi and M. Tashiro; *J. Heterocycl. Chem.*, 1981, **18**, 1073.	320
81JHC1149	P. L. Anderson, J. P. Hasak, A. D. Kahle, N. A. Paolella and M. J. Shapiro; *J. Heterocycl. Chem.*, 1981, **18**, 1149.	308
81JHC1189	S. Olivella and J. Vilarrasa; *J. Heterocycl. Chem.*, 1981, **18**, 1189.	171, 172, 173, 176, 177, 208
81JHC1319	J. C. Lancelot, D. Maume and M. Robba; *J. Heterocycl. Chem.*, 1981, **18**, 1319.	273
81JHC1353	T. Sasaki and E. Ito; *J. Heterocycl. Chem.*, 1981, **18**, 1353.	898, 899
81JHC1523	R. D. Thompson and R. N. Castle; *J. Heterocycl. Chem.*, 1981, **18**, 1523.	889
81JMC592	J. D. Albright, D. B. Moran, W. B. Wright, Jr., J. B. Collins, B. Beer, A. S. Lippa and E. N. Greenblatt; *J. Med. Chem.*, 1981, **24**, 592.	904
81JMC610	K. Senga, T. Novinson, H. R. Wilson and R. K. Robins; *J. Med. Chem.*, 1981, **24**, 610.	331
81JMC735	J. C. Sircar, T. Capiris, S. J. Kesten and D. J. Herzig; *J. Med. Chem.*, 1981, **24**, 735.	342
81JMC830	J. G. Lombardino and I. G. Otterness; *J. Med. Chem.*, 1981, **24**, 830.	342
81JMC947	R. T. Bartlett, A. F. Cook, M. J. Holman, W. W. McComas, E. F. Nowaswait, M. S. Poonian, J. A. Baird-Lambert, B. A. Baldo and J. F. Marwood; *J. Med. Chem.*, 1981, **24**, 947.	664
81JMC1521	J. R. Shroff, V. Bandurco, R. Desai, S. Kobrin and P. Cervoni; *J. Med. Chem.*, 1981, **24**, 1521.	233
81JOC608	C. Gönczi, Z. Swistun and H. C. van der Plas; *J. Org. Chem.*, 1981, **46**, 608.	476
81JOC679	C. D. Bedford, E. M. Bruckmann and P. A. S. Smith; *J. Org. Chem.*, 1981, **46**, 679.	727
81JOC843	A. Messmer and G. Hajós; *J. Org. Chem.*, 1981, **46**, 843.	879, 900, 902
81JOC856	K. M. Baines, T. W. Rourke and K. Vaughan; *J. Org. Chem.*, 1981, **46**, 856.	696, 707
81JOC1781	J. P. Dickens, R. L. Dyer, B. J. Hamill, T. A. Harrow, R. H. Bible, Jr., P. M. Finnegan, K. Henrick and P. G. Owston; *J. Org. Chem.*, 1981, **46**, 1781.	396, 401
81JOC1800	T. Sasaki, S. Eguchi, M. Yamaguchi and T. Esaki; *J. Org. Chem.*, 1981, **46**, 1800.	711
81JOC2138	A. Counotte-Potman, H. C. van der Plas and B. van Veldhuizen; *J. Org. Chem.*, 1981, **46**, 2138.	794
81JOC2203	A. Chin, M. H. Hung and L. M. Stock; *J. Org. Chem.*, 1981, **46**, 2203.	546
81JOC2467	S.-F. Chen and R. P. Panzica; *J. Org. Chem.*, 1981, **46**, 2467.	52
81JOC2490	S. Knapp, B. H. Toby, M. Sebastian, K. Krogh-Jespersen and J. A. Potenza; *J. Org. Chem.*, 1981, **46**, 2490.	257
81JOC2706	G. A. Olah, S. C. Narang and A. P. Fung; *J. Org. Chem.*, 1981, **46**, 2706.	270
81JOC2819	E. K. Ryu and M. MacCoss; *J. Org. Chem.*, 1981, **46**, 2819.	540
81JOC2824	G. Neef, U. Eder and G. Sauer; *J. Org. Chem.*, 1981, **46**, 2824.	471
81JOC2872	D. Roux; *J. Org. Chem.*, 1981, **46**, 2872.	463
81JOC3026	A. Nudelman, T. E. Haran and Z. Shakked; *J. Org. Chem.*, 1981, **46**, 3026.	665
81JOC3123	Y.-i. Lin and S. A. Lang, Jr.; *J. Org. Chem.*, 1981, **46**, 3123.	885, 886
81JOC3681	J. B. Holtwick and N. J. Leonard; *J. Org. Chem.*, 1981, **46**, 3681.	59
81JOC4065	K. T. Potts, R. D. Cody and R. J. Dennis; *J. Org. Chem.*, 1981, **46**, 4065.	128
81JOC4717	C. Rav-Acha and L. A. Cohen; *J. Org. Chem.*, 1981, **46**, 4717.	431, 438
81JOM(205)147	W. Weber and K. Niedenzu; *J. Organomet. Chem.*, 1981, **205**, 147.	235
81JOM(205)247	R. Usón, L. A. Oro, M. A. Ciriano, M. T. Pinillos, A. Tiripicchio and M. Tiripicchio-Camellini; *J. Organomet. Chem.*, 1981, **205**, 247.	227
81JOM(208)309	J. P. Gasparini, R. Gassend, J. C. Maire and J. Elguero; *J. Organomet. Chem.*, 1981, **208**, 309.	271
81JOM(215)131	H. C. Clark and M. A. Mesubi; *J. Organomet. Chem.*, 1981, **215**, 131.	235
81JOM(215)157	D. Boyer, R. Gassend, J. C. Maire and J. Elguero; *J. Organomet. Chem.*, 1981, **215**, 157.	213, 236
81JOM(215)349	D. Boyer, R. Gassend, J. C. Maire and J. Elguero; *J. Organomet. Chem.*, 1981, **215**, 349.	232, 271
81JPR188	D. Zeigan, E. Kleinpeter, H. Wilde and G. Mann; *J. Prakt. Chem.*, 1981, **323**, 188.	214
81JPR279	A. Preiss, W. Walek and F. Dietzel; *J. Prakt. Chem.*, 1981, **323**, 279.	19
81JPS425	Z. Neiman and F. R. Quinn; *J. Pharm. Sci.*, 1981, **70**, 425.	524
81KGS403	K. V. Mityurina, V. G. Kharchenko and L. V. Cherkesova; *Khim. Geterotsikl. Soedin.*, 1981, 403 (*Chem. Abstr.*, 1981, **95**, 24 909).	321
81KGS599	V. A. Ostrovskii, G. I. Koldobskii, N. P. Shirokova and V. S. Poplavaskii; *Khim. Geterotsikl. Soedin.*, 1981, 599 (*Chem. Abstr.*, 1981, **95**, 42 119).	873

81KGS694	V. P. Shchipanov and N. A. Klyuev; *Khim. Geterotsikl. Soedin.*, 1981, 694.	741, 778
81KGS1423	Yu. I. Ribukhim, L. N. Faleeva and G. N. Dorofeenko; *Khim. Geterotsikl. Soedin.*, 1981, 1423.	779
81KGS1554	V. L. Rusinov, I. Y. Postovskii, A. Y. Petrov, E. O. Sidorov and Y. A. Azev; *Khim. Geterotsikl. Soedin.*, 1981, 1554.	272
81LA7	R. W. Saalfrank and E. Ackermann; *Liebigs Ann. Chem.*, 1981, 7.	717
81LA2309	R. R. Schmidt, W. Guilliard and D. Heermann; *Liebigs Ann. Chem.*, 1981, 2309.	289
81M105	W. Freyer and R. Radeglia; *Monatsh. Chem.*, 1981, **112**, 105.	228
81M245	M. H. Elnagdi, E. M. Zayed, M. A. Khalifa and S. A. Ghozlan; *Monatsh. Chem.*, 1981, **112**, 245.	320
81M675	M. A. Khan and A. C. C. Freitas; *Monatsh. Chem.*, 1981, **112**, 675.	266
81M875	O. S. Wolfbeis; *Monatsh. Chem.*, 1981, **112**, 875.	281
81MI40100	G. A. Webb, M. Witanowski and L. Stefaniak; *Annu. Rep. NMR Spectrosc.*, 1981, 118.	20
81MI40400	Y. Imai, T. Nakajima and M. Ueda; *J. Polym. Sci., Polym. Chem. Ed.*, 1981, **19**, 2161.	291, 300
81MI40401	K. Simon, K. Horváth, D. Korbonits and L. Párkányí, *J. Cryst. Mol. Struct.*, 1981, **11**, 33.	179, 227
81MI40402	J. Arriau and J. Elguero; *An. Quím., Ser. C*, 1981, **77**, 105.	215, 282
81MI40403	C. L. Chatterjee and A. Saran; *Int. J. Quantum. Chem.*, 1981, **8** (Suppl.), 129.	210
81MI40404	D. M. Bailey and L. W. Chakrin; *Annu. Rep. Med. Chem.* 1981, **16**, 213.	297
81MI40405	V. Pellegrin, A. Fruchier and J. Elguero; *J. Labelled Compd. Radiopharm.*, 1981, **18**, 999.	291
81MI40406	R. Aruga; *J. Inorg. Nucl. Chem.*, 1981, **43**, 2459.	226
B-81MI40407	M. T. García-López and F. G. de las Heras; in 'Medicinal Chemistry Advances', ed. F. G. de las Heras and S. Vega; Pergamon Press, Oxford, 1981, p. 69.	288
81MI40900	J. R. Jones and S. E. Taylor; *Chem. Soc. Rev.*, 1981, **10**, 329.	501, 527
81MI41100	R. Böhm and C. Karow; *Pharmazie*, 1981, **36**, 243.	728
81MI41101	G. R. Revankar, V. C. Solan, R. K. Robins and J. T. Witkowski; *Nucleic Acids Symp. Ser.*, 1981, **9**, 65 (*Chem. Abstr.*, 1982, **96**, 35 682).	729
B-81MI41400	I. Ugi, D. Marquarding and R. Urban; in 'Chemistry and Biochemistry of Amino Acids, Peptides and Proteins', ed. B. Weinstein; Dekker, New York, 1981, vol. 6, p. 245.	845
81MI41500	B. Stanovnik, M. Tišler, N. Trcek and B. Verček; *Vestn. Slov. Kem. Drus.*, 1981, **28**, 45 (*Chem. Abstr.*, 1981, **95**, 7220).	860
81MIP41100	Sandoz AG, *Braz. Pat.* 81 01 239 (1981) (*Chem. Abstr.*, 1982, **96**, 69 006).	732
81OMR(15)219	V. V. Lopyrev, L. I. Larina, T. I. Vakul'skaya, M. F. Larin, O. B. Nefedova, E. F. Shibanova and M. G. Voronkov; *Org. Magn. Reson.*, 1981, **15**, 219.	353
81OMS377	U. C. Pande, H. Egsgaard, E. Larsen and M. Begtrup; *Org. Mass Spectrom.*, 1981, **16**, 377.	204
81OPP371	M. Bin Mohamed and J. Parrick; *Org. Prep. Proced. Int.*, 1981, **13**, 371.	290
81RCR336	V. N. Sheinker, A. D., Garnovskii and O. A. Osipov; *Russ. Chem. Rev. (Engl. Transl.)*, 1981, **50**, 632.	209, 737
81S563	J. Barluenga, N. Gómez, F. Palacios and V. Gotor; *Synthesis*, 1981, 563.	476
81S727	P. G. Baraldi, M. Guarneri, F. Moroder and D. Simoni; *Synthesis*, 1981, 727.	323
81S925	F. Freeman; *Synthesis*, 1981, 925.	316
81T987	A. Ramdani and G. Tarrago; *Tetrahedron*, 1981, **37**, 987.	260, 261
81T991	A. Ramdani and G. Tarrago, *Tetrahedron*, 1981, **37**, 991.	228
81T1453	S. Rajappa; *Tetrahedron*, 1981, **37**, 1453.	715
81T3423	A. N. Kost, S. P. Gromov and R. S. Sagitullin; *Tetrahedron*, 1981, **37**, 3423.	312
81T4353	M. Legraverend, J. M. Lhoste and E. Bisagni; *Tetrahedron*, 1981, **37**, 4353.	859, 901
81T(S9)191	H. H. Wasserman, M. S. Wolff, K. Stiller, I. Saito and J. E. Pickett; *Tetrahedron*, 1981, **37**, Suppl. 9, 191.	406
81TL227	F. Lübbe, K.-P. Grosz, W. Hillebrand and W. Sucrow; *Tetrahedron Lett.*, 1981, 227.	707
81TL1199	A. Padwa and T. Kumagai; *Tetrahedron Lett.*, 1981, 1199.	263
81TL1863	A. Hassner, B. A. Belinka, Jr., M. Haber and P. Munger; *Tetrahedron Lett.*, 1981, **22**, 1863.	718
81TL2063	R. A. Whitney; *Tetrahedron Lett.*, 1981, 2063.	426, 443
81TL2431	R. F. Pratt and K. K. Kraus; *Tetrahedron Lett.*, 1981, 2431.	409
81TL2973	R. Gompper, M. Junius and H. U. Wagner; *Tetrahedron Lett.*, 1981, 2973.	415
81UKZ1099	V. I. Simonenko, I. V. Pyatnitskii and G. K. Nizkaya; *Ukr. Khim. Zh. (Russ. Ed.)*, 1980, **46**, 1099 (*Chem. Abstr.*, 1981, **94**, 53 670).	731
81UP40400	P. Camps, J. Elguero, C. Estopá, R. Faure, A. Fruchier, D. Ilavaský, C. Marzin and J. de Mendoza; unpublished results.	185, 190, 191, 192
81UP40600	W. Robinson and M. R. Grimmett; unpublished work in preparation.	351
81UP40700	W. Robinson and M. R. Grimmett; unpublished work in preparation.	393
81ZC187	D. Zeigan, G. Engelhardt and E. Uhlemann; *Z. Chem.*, 1981, **21**, 187.	194
81ZN(A)34	M. H. Palmer, I. Simpson and R. H. Findley; *Z. Naturforsch., Teil A*, 1981, **36**, 34.	195
81ZN(A)1246	M. H. Palmer, I. Simpson and J. R. Wheeler; *Z. Naturforsch., Teil A*, 1981, **36**, 1246.	671, 688, 689

Ref	Citation	Pages
81ZOR146	V. A. Ostrovskii, I. Yu. Shirobokov and G. I. Koldobskii; *Zh. Org. Khim.*, 1981, **17**, 146 (*Chem. Abstr.*, 1981, **95**, 23 988).	818
82ACS(B)101	B. R. Tolf, R. Dahlbom, H. Theorell and A. Akeson; *Acta Chem. Scand., Ser. B*, 1982, **36**, 101.	291
82AG139	M. V. Prostenik and I. Butula; *Angew. Chem.*, 1982, **94**, 139.	728
82AHC(30)239	S. T. Reid; *Adv. Heterocycl. Chem.*, 1982, **30**, 239.	46
82CC86	H. Nakanishi, A. Yabe and K. Honda; *J. Chem. Soc., Chem. Commun.*, 1982, 86.	208
82CC450	D. H. R. Barton, G. Lukacs and D. Wagle; *J. Chem. Soc., Chem. Commun.*, 1982, 450.	276
82CC1280	D. G. Heath and C. W. Rees; *J. Chem. Soc., Chem. Commun.*, 1982, 1280.	864
82CJC97	J. Zukerman-Schpector, E. E. Castellano, A. C. Massabni and A. D. Pinto; *Can. J. Chem.*, 1982, **60**, 97.	178, 210
82H(19)179	K. Tanaka, S. Maeno and K. Mitsuhashi; *Heterocycles*, 1982, **19**, 179.	284
82H(19)1223	S. Takano, Y. Imamura and K. Ogasawara; *Heterocycles*, 1982, **19**, 1223.	284, 302
82JA235	R. S. Hosmane, V. Bakthavachalam and N. J. Leonard; *J. Am. Chem. Soc.*, 1982, **104**, 235.	659
82JA1698	P. S. Engel, L. R. Soltero, S. A. Baughman, C. J. Nalepa, P. A. Cahill and R. B. Weisman; *J. Am. Chem. Soc.*, 1982, **104**, 1698.	255
82JA3209	R. Taylor and O. Kennard; *J. Am. Chem. Soc.*, 1982, **104**, 3209.	501, 527
82JCR(S)6	G. Ferguson, B. Kaitner, F. J. Lalor and G. Roberts; *J. Chem. Res. (S)*, 1982, 6.	227
82JCR(S)122	R. N. Butler and V. C. Garvin; *J. Chem. Res. (S)*, 1982, 122.	835
82JCS(P1)165	N. Vivona, G. Macaluso and V. Frenna; *J. Chem. Soc., Perkin Trans. 1*, 1982, 165.	288
82JCS(P1)473	M. Yogo, K. Hirota and S. Senda; *J. Chem. Soc., Perkin Trans. 1*, 1982, 473.	286
82JCS(P1)627	D. R. Buckle and C. J. M. Rockell; *J. Chem. Soc., Perkin Trans. 1*, 1982, 627.	712
82JCS(P1)759	D. Korbonits, I. Kanzel-Szvoboda and K. Horváth; *J. Chem. Soc., Perkin Trans. 1*, 1982, 759.	288
82JHC1141	S. Juliá, P. Sala, J. M. del Mazo, M. Sancho, C. Ochoa, J. Elguero, J. P. Fayet and M. C. Vertut; *J. Heterocycl. Chem.*, 1982, **19**, 1141.	230
82JHC1215	J. W. Ellis, E. A. Keiter, R. L. Keiter, T. L. Pi and R. A. Uptmor; *J. Heterocycl. Chem.*, 1982, **19**, 1215.	281
82JMC96	J. A. Montgomery, S. J. Clayton and P. K. Chiang; *J. Med. Chem.*, 1982, **25**, 96.	620
82JMC626	J. A. Montgomery, S. J. Clayton, H. J. Thomas, W. M. Shannon, G. Arnett, A. J. Bodner, I.-K. Kim, G. L. Cantoni and P. K. Chiang; *J. Med. Chem.*, 1982, **25**, 626.	640
82JMC1168	J. C. Emmett, G. J. Durant, C. R. Ganellin, A. M. Roe and J. L. Turner; *J. Med. Chem.*, 1982, **25**, 1168.	623
82JOC81	A. Citterio, F. Minisci and E. Vismara; *J. Org. Chem.*, 1982, **47**, 81.	278
82JOC167	N. Katagiri, A. Koshihara, S. Atsuumi and T. Kato; *J. Org. Chem.*, 1982, **47**, 167.	639
82JOC214	E. M. Kosower, D. Faust, M. Ben-Shoshan and I. Goldberg; *J. Org. Chem.*, 1982, **47**, 214.	250
82JOC295	G. Zvilichovski and M. David; *J. Org. Chem.*, 1982, **47**, 295.	215, 230, 281
82JOC536	J. Elguero, C. Ochoa, M. Stud, C. Esteban-Calderon, M. Martínez-Ripoll, J. P. Fayet and M. C. Vertut; *J. Org. Chem.*, 1982, **47**, 536.	229
82JOC2221	J. D. Pérez and G. I. Yranzo; *J. Org. Chem.*, 1982, **47**, 2221.	219
82JOC4256	E. Sato, Y. Kanoaka and A. Padwa; *J. Org. Chem.*, 1982, **47**, 4256.	148
82JOC4323	K. Takada, T. Kan-Woon and A. J. Boulton; *J. Org. Chem.*, 1982, **47**, 4323.	216
82JOC5132	D. S. Wofford, D. M. Forkey and J. G. Russell; *J. Org. Chem.*, 1982, **47**, 5132.	193
82JST(78)1	R. Taylor and O. Kennard; *J. Mol. Struct.*, 1982, **78**, 1.	501, 527
82JST(89)147	M. Sana, G. Leroy, G. Dive and M. T. Nguyen; *J. Mol. Struct.*, 1982, **89**, 147.	283
82KGS705	Y. M. Yutilov and I. A. Svertilova; *Khim. Geterotsikl. Soedin.*, 1982, 705 (*Chem. Abstr.*, 1982, **97**, 109 950).	623
82MI40400	H. Biere, E. Schröder, H. Ahrens, J. F. Kapp and I. Bottcher; *Eur. J. Med. Chem.*, 1982, **17**, 27.	291
82MI40401	K. Burger, F. Hein, O. Dengler and J. Elguero; *J. Fluorine Chem.*, 1982, **19**, 437.	283
82MI40402	F. Massa, M. Hanack and L. R. Subramanian; *J. Fluorine Chem.*, 1982, **19**, 601.	282
82MI41000	J. A. Montgomery; *Med. Res. Rev.*, 1982, **2**, 271.	608, 652
82MI41001	T. Itoh, T. Sugawara and Y. Mizuno; *Nucleosides, Nucleotides*, 1982, **1**, 179.	615, 616, 617, 618
82OMR(18)0	A. Fruchier, V. Pellegrin, R. Schimpf and J. Elguero; *Org. Magn. Reson.*, 1982, **18**, 10.	196, 197
82OMR(18)159	G. Aranda, M. Dessolin, M. Golfier and M.-G. Guillerez; *Org. Magn. Reson.*, 1982, **18**, 159.	19
82OMR(19)27	L. D. Colebrook and M. A. Khadim; *Org. Magn. Reson.*, 1982, **19**, 27.	355
82OMR(19)225	P. Cohen-Fernandes, C. Erkelens and C. L. Habraken; *Org. Magn. Reson.*, 1982, **19**, 225.	234
82P863	E. G. Brown, K. A. M. Flayeh and J. R. Gallon; *Phytochemistry*, 1982, **21**, 863.	303
82PC40200	R. Taft; personal communication.	49
82PC40400	B. Abarca, J. Quilez and J. Sepúlveda; personal communication.	249
82PNA4487	G. A. Olah, S. C. Narang, J. A. Olah and K. Lammertsma; *Proc. Natl. Acad. Sci. USA*; 1982, **79**, 4487.	270
82S844	R. Ketari and A. Foucaud; *Synthesis*, 1982, 844.	270

82T2193	A. de la Hoz, E. Díez-Barra, M. C. Pardo, J. P. Declercq, G. Germain, M. Van Meerssche and J. Elguero; *Tetrahedron*, 1983, **39**, 2193.	248
82TL121	P. Choi, C. W. Rees and E. H. Smith; *Tetrahedron Lett.*, 1982, 121.	808
82TL321	M. Alam, Y. Oshima and Y. Shimizu; *Tetrahedron Lett.*, 1982, **23**, 321.	604
82UP40400	J. Elguero, A. Fruchier, A. de la Hoz, and M. C. Pardo; unpublished results.	192
82UP40401	A. Fruchier, V. Pellegrin, R. M. Claramunt and J. Elguero; *Org. Magn. Reson.*, accepted for publication.	196, 290
82UP40402	A. Fruchier, B. Lupo and G. Tarrago, *Can. J. Chem.*, accepted for publication.	240
82UP40403	A. M. Cuadro, P. Navarro and J. Elguero; unpublished results.	242
82USP4315008	Bayer A. G., *U.S. Pat.* 4 315 008	282, 297
82ZC56	M. Schulz, L. Mogel, N. X. Dung and R. Radeglia; *Z. em.*, 1982, **22**, 56.	234
83AP588	H. Biere, I. Böttcher and J. F. Kapp; *Arch. Pharm. (Wienheim, Ger.)*, 1983, **316**, 588.	291
83AP608	H. Biere, I. Böttcher and J. F. Kapp; *Arch. Pharm. (Weinheim, Ger.)*, 1983, **316**, 608.	291
83BSF(2)5	R. M. Claramunt, H. Hernández, J. Elguero and S. Juliá; *Bull. Soc. Chim. Fr., Part 2*, 1983, 5.	230
83CB1520	W. Sucrow and G. Bredthaver; *Chem. Ber.*, 1983, **116**, 1520.	275
83CC910	J. D. Wallis and J. D. Dunitz; *J. Chem. Soc., Chem. Commun.*, 1983, 910.	843
83CC1144	X. Coqueret, F. Bourelle-Wargnier, J. Chuche and L. Toupet; *J. Chem. Soc., Chem. Commun.*, 1983, 1144.	281
83CJC1435	Minh-Tho Nguyen, M. Sana, G. Leroy and J. Elguero; *Can. J. Chem.*, 1983, **61**, 1435.	845
83H(20)1713	R. Faure, E. J. Vincent and J. Elguero; *Heterocycles*, 1983, **20**, 1713.	192, 211
83H(20)1717	J. Catalán, P. Pérez and J. Elguero; *Heterocycles*, 1983, **20**, 1717.	175
83HCA1537	B. C. Chen, W. von Philipsborn and K. Nagarajan; *Helv. Chim. Acta*, 1983, **66**, 1537.	193
83JA309	R. H. Nobes, W. J. Bouma and L. Radom; *J. Am. Chem. Soc.*, 1983, **105**, 309.	7
83JCR(S)18	R. N. Butler and D. A. O'Donoghue; *J. Chem. Res. (S)*, 1982, 18.	829
83JCS(P1)2273	J. Barluenga, M. Tomás, J. F. López-Ortiz and V. Gotor; *J. Chem. Soc., Perkin Trans. 1*, 1983, 2273.	274
83JCS(P2)1111	S. N. Ege, W. M. Butler, A. Bergers, B. S. Biesman, J. E. Boerma, V. I. Corondan, K. D. Locke, S. Meshinchi, S. H. Ponas and T. D. Spitzer; *J. Chem. Soc., Perkin Trans. 2*, 1983, 1111.	255
83JOC3	R. Balicki, R. S. Hosmane and N. J. Leonard; *J. Org. Chem.*, 1983, **48**, 3.	659
83MI40300	K. T. Potts; in '1,3-Dipolar Cycloadditions', ed. A. Padwa; Wiley-Interscience, New York, 1983, Chap. 8.	140, 150
83MI40400	A. Katrib, N. R. El-Rayyes and F. M. Al-Kharafi; *J. Electron Spectrosc. Related Phenom.*, 1983, **31**, 317.	205
83OMS52	E. Larsen, H. Egsgaard, U. C. Pande and M. Begtrup; *Org. Mass Spectrom.*, 1983, **18**, 52.	204
83PC40100	G. Morton; personal communication.	15
83UP40100	M. J. Sammes and A. R. Katritzky; unpublished results, 1983.	16, 19
83UP40200	M. J. Sammes and A. R. Katritzky; *Adv. Heterocycl. Chem.*, 1983, in press.	77, 78
83UP40400	J. Catalán and J. Elguero; *J. Chem. Soc., Perkin Trans. 2*, accepted for publication.	175, 224
84MI41000	N. J. Leonard; *CRC Crit. Rev. Biochem.*, 1984, **15**, 125.	608, 648
84UP41300	R. N. Butler, V. C. Garvin, H. Lumbroso and Ch. Liégois; *J. Chem. Soc., Perkin Trans. 2*, 1984, in press.	804, 805

JOURNAL CODES FOR REFERENCES
For explanation of the reference system, see p. 905.

Code	Journal	Code	Journal
ABC	Agric. Biol. Chem.	CS	Chem. Scr.
ACH	Acta Chim. Acad. Sci. Hung.	CSC	Cryst. Struct. Commun.
ACR	Acc. Chem. Res.	CSR	Chem. Soc. Rev.
AC(R)	Ann. Chim. (Rome)	CZ	Chem.-Ztg.
ACS	Acta Chem. Scand.	DIS	Diss. Abstr.
ACS(B)	Acta Chem. Scand., Ser. B	DIS(B)	Diss. Abstr. Int. B
AF	Arzneim.-Forsch.	DOK	Dokl. Akad. Nauk SSSR
AG	Angew. Chem.	E	Experientia
AG(E)	Angew. Chem., Int. Ed. Engl.	EGP	Ger. (East) Pat.
AHC	Adv. Heterocycl. Chem.	EUP	Eur. Pat.
AJC	Aust. J. Chem.	FES	Farmaco Ed. Sci.
AK	Ark. Kemi	FOR	Fortschr. Chem. Org. Naturst.
ANY	Ann. N.Y. Acad. Sci.	FRP	Fr. Pat.
AP	Arch. Pharm. (Weinheim, Ger.)	G	Gazz. Chim. Ital.
APO	Adv. Phys. Org. Chem.	GEP	Ger. Pat.
AX	Acta Crystallogr.	H	Heterocycles
AX(B)	Acta Crystallogr., Part B	HC	Chem. Heterocycl. Compd. [Weissberger–Taylor series]
B	Biochemistry		
BAP	Bull. Acad. Pol. Sci., Ser. Sci. Chim.	HCA	Helv. Chim. Acta
		HOU	Methoden Org. Chem. (Houben-Weyl)
BAU	Bull. Acad. Sci. USSR, Div. Chem. Sci.	IC	Inorg. Chem.
BBA	Biochim. Biophys. Acta	IJC	Indian J. Chem.
BBR	Biochem. Biophys. Res. Commun.	IJC(B)	Indian J. Chem., Sect. B
BCJ	Bull. Chem. Soc. Jpn.	IJS	Int. J. Sulfur Chem.
BEP	Belg. Pat.	IJS(B)	Int. J. Sulfur Chem., Part B
BJ	Biochem. J.	IZV	Izv. Akad. Nauk SSSR, Ser. Khim.
BJP	Br. J. Pharmacol.	JA	J. Am. Chem. Soc.
BRP	Br. Pat.	JAP	Jpn. Pat.
BSB	Bull. Soc. Chim. Belg.	JAP(K)	Jpn. Kokai
BSF	Bull. Soc. Chim. Fr.	JBC	J. Biol. Chem.
BSF(2)	Bull. Soc. Chim. Fr., Part 2	JCP	J. Chem. Phys.
C	Chimia	JCR(S)	J. Chem. Res. (S)
CA	Chem. Abstr.	JCS	J. Chem. Soc.
CB	Chem. Ber.	JCS(C)	J. Chem. Soc. (C)
CC	J. Chem. Soc., Chem. Commun.	JCS(D)	J. Chem. Soc., Dalton Trans.
CCC	Collect. Czech. Chem. Commun.	JCS(F1)	J. Chem. Soc., Faraday Trans. 1
CCR	Coord. Chem. Rev.	JCS(P1)	J. Chem. Soc., Perkin Trans. 1
CHE	Chem. Heterocycl. Compd. (Engl. Transl.)	JGU	J. Gen. Chem. USSR (Engl. Transl.)
CI(L)	Chem. Ind. (London)	JHC	J. Heterocycl. Chem.
CJC	Can. J. Chem.	JIC	J. Indian Chem. Soc.
CL	Chem. Lett.	JMC	J. Med. Chem.
CPB	Chem. Pharm. Bull.	JMR	J. Magn. Reson.
CR	C.R. Hebd. Seances Acad. Sci.	JOC	J. Org. Chem.
CR(C)	C.R. Hebd. Seances Acad. Sci., Ser. C	JOM	J. Organomet. Chem.
		JOU	J. Org. Chem. USSR (Engl. Transl.)
CRV	Chem. Rev.		